TEXTBOOK OF
MEDICAL PHYSIOLOGY

SIXTH EDITION

ARTHUR C. GUYTON, M.D.

Chairman and Professor of the Department of Physiology
and Biophysics, University of Mississippi, School of Medicine

W. B. SAUNDERS COMPANY   Philadelphia   London   Toronto

W. B. Saunders Company: West Washington Square
Philadelphia, PA   19105

1 St. Anne's Road
Eastbourne, East Sussex BN21 3UN, England

1 Goldthorne Avenue
Toronto, Ontario M8Z 5T9, Canada

Library of Congress Cataloging in Publication Data

Guyton, Arthur C

Textbook of medical physiology.

1. Human physiology.   I. Title.   II. Title: Medical
physiology. [DNLM: 1. Physiology. QT104 G992t]

QP34.5.G9 1980   612   80–50298

ISBN 0–7216–4394–9

Listed here are the latest translated editions of this book together
with the languages of the translation and the publishers.

Japanese (*4th Edition*)—Kirokawa Publishing Company, Tokyo, Japan

Serbo-Croatian (*4th Edition*)—Medicinska Knjiga, Belgrade, Yugoslavia

Italian (*4th Edition*)—Piccin Editore, Padova, Italy

Spanish (*5th Edition*)—Nueva Editorial Interamericana S.A. de C.V.,
                Mexico City, Mexico

Persian (*5th Edition*)—Sherkat Sahami "TCHEHR", Tehran, Iran

Portuguese (*5th Edition*)—Editora Interamericana Ltda.,
                Rio de Janeiro, Brazil

ISBN   0-7216-4394-9

Textbook of Medical Physiology

Last digit is the print number: 9   8   7   6   5   4   3   2

Dedicated to

MY FATHER
for the uncompromising principles that guided his life

MY MOTHER
for leading her children into intellectual pursuits

MY WIFE
for her magnificent devotion to her family

MY CHILDREN
for making everything worthwhile

PREFACE

When I wrote the first edition of this textbook, I had the naive belief that once the book was completed, subsequent revisions would require only simple changes. However, with each new edition I find that progress in the field of physiology is so rapid that large portions of the text must be completely recast and rewritten, and even the emphasis of the subject matter must often be changed as our knowledge becomes more penetrating. Therefore, the reader will find hardly a chapter in this edition that has not been significantly altered. Also, most of the figures have been either changed in at least some feature or replaced, and new ones have been added. Especially important, color has been added throughout the book for emphasis in both the figures and the text. Indeed, the revisions have been very extensive.

As has been true of most of the past revisions, the major change in this edition reflects a striking trend toward more fundamental physiology. The reason for this is mainly the greater success of the research physiologist in probing deeper into basic mechanisms of function than was true a few years ago. Yet I have attempted still to present physiology as an integrated study of the body's functional systems, utilizing the new fundamental knowledge to build a better understanding of the mechanisms upon which life depends.

Also a serious attempt has been made to devise and achieve techniques of expression that will help bring the physical and chemical principles of the body's complexities into the medical student's realm of understanding. In pursuit of this goal, I have kept records of those types of material with which the student has difficulty; I have quizzed students in detail to determine their levels of comprehension; and I have attempted to note inconsistencies in logic that might appear in student discussions. All these data have been used to help choose the material and methods of presentation. Thus, it has been my desire to make this book a "teaching" text as well as one that covers essentially all the basic physiology required of a student of medicine.

A special emphasis of the book is a more detailed attention than given in most other textbooks of physiology to the body's many control systems. These are the basis of what physiologists call *homeostasis*. The reason for this special emphasis is that most disease conditions of the body result from abnormal function of one or more of the control systems. Therefore, the student's comprehension of "medical" physiology depends perhaps more on a knowledge of these systems than on any other facet of physiology.

Another goal has been to make the text as accurate as possible. To help attain this, suggestions and critiques from many physiologists, students, and clinicians throughout the United States and other parts of the world have been received and utilized in checking the factual accuracy of the text. Yet, even so, because of the likelihood of error in sorting through many thousands of bits of information, I wish to issue an invitation—in fact, much more than merely an invitation, actually a request—to all readers to send along notices of error or inaccuracy. Indeed, physiologists understand how important feedback is to proper function of the human body; so, too, is feedback equally important for progressive development of a textbook of medical physiology. I hope also that those many persons who have helped already will accept my sincerest appreciation for their efforts.

A word of explanation is needed about two features of the text—first, the references, and, second, the two print sizes. The references have been chosen primarily for their up-to-dateness and for the quality of their own bibliographies. Use of these references as well as of cross-references from them can give the student almost complete coverage of the entire field of physiology.

The print is set in two sizes. The material presented in small print is of several different kinds: first, anatomical, chemical, and other information that is needed for the immediate discussion but that most students will learn in more detail in other courses; second, information that is of special importance to certain clinical fields of medicine but that is not necessary to fundamental understanding of the body's basic physiologic mechanisms; and, third, information that will be of value to those students who wish to pursue a subject more deeply than the average medical student. On the other hand, the material in large print constitutes the major bulk of physiological information that students will require in their medical studies and that they will not obtain in other courses. Those teachers who would like to present a limited course of physiology can direct students' study primarily to the large type.

Again, I wish to express my deepest appreciation to many others who have helped in the preparation of this book. I am particularly grateful to Mrs. Billie Howard, Mrs. Jane Strickland, Ms. Gwendolyn Robbins, and Mrs. Laveda Morgan for their excellent secretarial services, to Miss Tomiko Mita for her superb work on the new and colored illustrations, and to the staff of the W. B. Saunders Company for its continued editorial and production excellence.

ARTHUR C. GUYTON

Jackson, Mississippi

CONTENTS

PART I
INTRODUCTION TO PHYSIOLOGY:
THE CELL AND GENERAL PHYSIOLOGY

CHAPTER 4

**TRANSPORT THROUGH THE CELL
MEMBRANE** **41**

PART II
BLOOD CELLS, IMMUNITY, AND BLOOD CLOTTING

CHAPTER 5

**RED BLOOD CELLS, ANEMIA,
AND POLYCYTHEMIA** **56**

CHAPTER 6

**RESISTANCE OF THE BODY TO
INFECTION — THE LEUKOCYTES,
THE TISSUE MACROPHAGE SYSTEM,
AND INFLAMMATION** **65**

CHAPTER 7

IMMUNITY AND ALLERGY **74**

PART III
NERVE AND MUSCLE

PART IV
THE HEART

PART V
THE CIRCULATION

PART VI
THE BODY FLUIDS AND KIDNEYS

PART VII
RESPIRATION

PART VIII
AVIATION, SPACE, AND DEEP SEA DIVING PHYSIOLOGY

PART IX
THE NERVOUS SYSTEM

CONTENTS xxiii

PART X
THE SPECIAL SENSES

PART XI
THE GASTROINTESTINAL TRACT

PART XII
METABOLISM AND TEMPERATURE REGULATION

PART XIII
ENDOCRINOLOGY AND REPRODUCTION

Part I

INTRODUCTION TO PHYSIOLOGY: THE CELL AND GENERAL PHYSIOLOGY

1

Functional Organization of the Human Body and Control of the "Internal Environment"

Physiology is the study of function in living matter, attempting to explain the physical and chemical factors that are responsible for the origin, development, and progression of life. Each type of life, from the monomolecular virus up to the largest tree or to the complicated human being, has its own functional characteristics. Therefore, the vast field of physiology can be divided into *viral physiology, bacterial physiology, cellular physiology, plant physiology, human physiology,* and many more subdivisions.

Human Physiology. In human physiology we attempt to explain the specific characteristics and mechanisms of the human body that make it a living being. The very fact that we are alive is almost beyond our own control, for hunger makes us seek food and fear makes us seek refuge. Sensations of cold make us provide warmth, and other forces cause us to seek fellowship and to reproduce. Thus, the human being is actually an automaton, and the fact that we are sensing, feeling, and knowledgeable beings is part of this automatic sequence of life; these special attributes allow us to exist under widely varying conditions that otherwise would make life impossible.

CELLS AS THE LIVING UNITS OF THE BODY

The basic living unit of the body is the cell, and each organ is actually an aggregate of many different cells held together by intercellular supporting structures. Each type of cell is specially adapted to perform one particular function. For instance, the red blood cells, 25 trillion in all, transport oxygen from the lungs to the tissues. Though this type of cell is perhaps the most abundant, there are approximately another 50 trillion cells. The entire body, then, contains about 75 trillion cells.

However much the many cells of the body differ from each other, all of them have certain basic characteristics that are alike. For instance, each cell requires nutrition for maintenance of life, and all cells utilize almost identically the same types of nutrients. All cells use oxygen as one of the major substances from which energy is derived; the oxygen combines with carbohydrate, fat, or protein to release the energy required for cell function. The general mechanisms for changing nutrients into energy are basically the same in all cells, and all cells also deliver end-products of their chemical reactions into the surrounding fluids.

Almost all cells also have the ability to reproduce, and whenever cells of a particular type are destroyed from one cause or another, the remaining cells of this type usually divide again and again until the appropriate number is replenished.

THE EXTRACELLULAR FLUID — THE INTERNAL ENVIRONMENT

About 56 per cent of the adult human body is fluid. Though most of this fluid is inside the cells and is called *intracellular fluid,* about one-third of it is in the spaces outside the cells and is called *extracellular fluid.* This extracellular fluid is in constant motion throughout the body. It is rapidly mixed by the blood circulation and by diffusion between the blood and the tissue fluids, and in the extracellular fluid are the ions and nutrients needed by the cells for maintenance of cellular function. Therefore, all cells live in essentially the same environment, the extracellular fluid, for which reason the extracellular fluid is often called the *internal environment* of the body, or the *milieu intérieur,* a term introduced a hundred years ago by the great 19th century French physiologist, Claude Bernard.

Cells are capable of living, growing, and providing their special functions so long as the proper concentrations of oxygen, glucose, the different ions, amino acids, and fatty substances are available in this internal environment.

Differences Between Extracellular and Intracellular Fluids. The extracellular fluid contains large amounts of sodium, chloride, and bicarbonate ions, plus nutrients for the cells, such as oxygen, glucose, fatty acids, and amino acids. It also contains carbon dioxide, which is being transported from the cells to the lungs to be excreted, and other cellular products, which are being transported to the kidneys for excretion.

The intracellular fluid differs significantly from the extracellular fluid; particularly, it contains large amounts of potassium, magnesium, and phosphate ions instead of the sodium and chloride ions found in the extracellular fluid. Special mechanisms for transporting ions through the cell membranes maintain these differences. These mechanisms will be discussed in detail in Chapter 4.

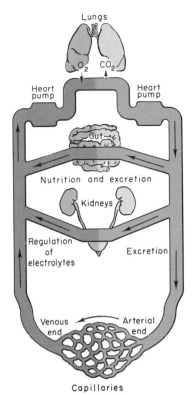

Figure 1–1. General organization of the circulatory system.

"HOMEOSTATIC" MECHANISMS OF THE MAJOR FUNCTIONAL SYSTEMS

HOMEOSTASIS

The term *homeostasis* is used by physiologists to mean *maintenance of static,* or *constant, conditions in the internal environment.* Essentially all the organs and tissues of the body perform functions that help to maintain these constant conditions. For instance, the lungs provide oxygen to the extracellular fluid to replenish continually the oxygen that is being used by the cells, the kidneys maintain constant ion concentrations, and the gut provides nutrients. A large segment of this text is concerned with the manner in which each organ or tissue contributes to homeostasis. To begin this discussion, the different functional systems of the body and their homeostatic mechanisms will be outlined briefly; then the basic theory of the control systems that cause the functional systems to operate in harmony with each other will be discussed.

THE EXTRACELLULAR FLUID TRANSPORT SYSTEM

Extracellular fluid is transported to all parts of the body in two different stages. The first stage

entails movement of blood around and around the circulatory system, and the second, movement of fluid between the blood capillaries and the cells. Figure 1–1 illustrates the overall circulation of blood, showing that the heart is actually two separate pumps, one of which propels blood through the lungs and the other through the systemic circulation. All the blood in the circulation traverses the entire circuit of the circulation an average of once each minute at rest and as many as six times each minute when a person becomes extremely active.

As blood passes through the capillaries, continual fluid exchange occurs between the plasma portion of the blood and the interstitial fluid in the intercellular spaces surrounding the capillaries. This process is illustrated in Figure 1–2. Note that the capillaries are porous so that large amounts of fluid and its dissolved constituents can *diffuse* back and forth between the blood and the tissue spaces, as illustrated by the arrows. This process of diffusion is caused by kinetic motion of the molecules in both the plasma and the interstitial fluid. That is, fluid and dissolved molecules are continually moving and bouncing in all directions, through the pores, through the tissue spaces, and so forth. Almost no cell is located more than 25 to 50 microns from a capillary, which insures diffusion of almost any substance from the capillary to the cell within a few seconds. Thus, the extra-

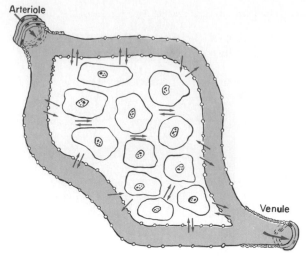

Figure 1–2. Diffusion of fluids through the capillary walls and through the interstitial spaces.

cellular fluid throughout the body, both that of the plasma and that in the interstitial spaces, is continually mixed and thereby maintains almost complete homogeneity.

ORIGIN OF NUTRIENTS IN THE EXTRACELLULAR FLUID

The Respiratory System. Figure 1–1 shows that each time the blood passes through the body it also flows through the lungs. The blood picks up oxygen in the alveoli, thus acquiring the oxygen needed by the cells. The membrane between the alveoli and the lumen of the pulmonary capillaries is only 0.4 to 2.0 microns in thickness, and oxygen diffuses through this membrane into the blood in exactly the same manner that water, nutrients, and excreta diffuse through the tissue capillaries.

The Gastrointestinal Tract. Figure 1–1 also shows that a large portion of the blood pumped by the heart passes through the walls of the gastrointestinal organs. Here, different dissolved nutrients, including carbohydrates, fatty acids, amino acids, and others, are absorbed into the extracellular fluid.

The Liver and Other Organs that Perform Primarily Metabolic Functions. Not all substances absorbed from the gastrointestinal tract can be used in their absorbed form by the cells. The liver changes the chemical compositions of many of these to more usable forms, and other tissues of the body — the fat cells, the gastrointestinal mucosa, the kidneys, and the endocrine glands — help to modify the absorbed substances or store them until they are needed at a later time.

Musculoskeletal System. Sometimes the question is asked: How does the musculoskeletal system fit into the homeostatic functions of the body?

The answer to this is obvious and simple: Were it not for this system, the body could not move to the appropriate place at the appropriate time to obtain the foods required for nutrition. The musculoskeletal system also provides motility for protection against adverse surroundings, without which the entire body, and along with it all the homeostatic mechanisms, could be destroyed instantaneously.

REMOVAL OF METABOLIC END-PRODUCTS

Removal of Carbon Dioxide by the Lungs. At the same time that blood picks up oxygen in the lungs, carbon dioxide is released from the blood into the alveoli, and the respiratory movement of air into and out of the alveoli carries the carbon dioxide to the atmosphere. Carbon dioxide is the most abundant of all the end-products of metabolism.

The Kidneys. Passage of the blood through the kidneys removes most substances from the plasma that are not needed by the cells. These substances include especially different end-products of cellular metabolism and excesses of ions and water that might have accumulated in the extracellular fluid. The kidneys perform their function by, first, filtering large quantities of plasma through the glomeruli into the tubules and then reabsorbing into the blood those substances needed by the body, such as glucose, amino acids, appropriate amounts of water, and many of the ions. However, substances not needed by the body, especially the metabolic end-products such as urea, generally are not reabsorbed but, instead, pass on through the renal tubules into the urine.

REGULATION OF BODY FUNCTIONS

The Nervous System. The nervous system is composed of three major parts: the *sensory portion,* the *central nervous system* (or *integrative portion*), and the *motor portion.* Sensory receptors detect the state of the body or the state of the surroundings. For instance, receptors present everywhere in the skin apprise one every time an object touches him at any point. The eyes are sensory organs that give one a visual image of the surrounding area. The ears also are sensory organs. The central nervous system is comprised of the brain and spinal cord. The brain can store information, generate thoughts, create ambition, and determine reactions that the body should perform in response to the sensations. Appropriate signals are then transmitted through the motor portion of the nervous system to carry out the person's desires.

A large segment of the nervous system is called the *autonomic system*. It operates at a subconscious level and controls many functions of the internal organs, including the action of the heart, the movements of the gastrointestinal tract, and the secretion by different glands.

The Hormonal System of Regulation. Located in the body are eight major endocrine glands that secrete chemical substances, the *hormones*. Hormones are transported in the extracellular fluid to all parts of the body to help regulate function. For instance, thyroid hormone increases the rates of most chemical reactions in all cells. In this way thyroid hormone helps to set the tempo of bodily activity. Likewise, insulin controls glucose metabolism; adrenocortical hormones control ion and protein metabolism; and parathormone controls bone metabolism. Thus, the hormones are a system of regulation that complements the nervous system. The nervous system, in general, regulates muscular and secretory activities of the body, while the hormonal system regulates mainly the metabolic functions.

REPRODUCTION

Reproduction sometimes is not considered to be a homeostatic function. But it does help to maintain static conditions by generating new beings to take the place of ones that are dying. This perhaps sounds like a farfetched usage of the term homeostasis, but it does illustrate that, in the final analysis, essentially all structures of the body are so organized that they help to maintain continuity of life.

THE CONTROL SYSTEMS OF THE BODY

The human body has literally thousands of control systems in it. The most intricate of all these are the genetic control systems that operate within all cells to control intracellular function, and also to control all life processes, a subject that will be discussed in detail in Chapter 3. But many other control systems operate within the organs to control functions of the individual parts of the organs, while others operate throughout the entire body to control the interrelationships between the different organs. For instance, the respiratory system, operating in association with the nervous system, regulates the concentration of carbon dioxide in the extracellular fluid. The liver and the pancreas regulate the concentration of glucose in the extracellular fluid. And the kidneys regulate the concentrations of hydrogen, sodium, potassium, phosphate, and other ions in the extracellular fluid.

EXAMPLES OF CONTROL MECHANISMS

Regulation of Oxygen and Carbon Dioxide Concentrations in the Extracellular Fluid. Since oxygen is one of the major substances required for chemical reactions in the cells, the rates of the chemical reactions, to a great extent, depend on the concentration of oxygen in the extracellular fluid. For this reason a special control mechanism maintains an almost exact and constant oxygen concentration in the extracellular fluid. This mechanism depends principally on the chemical characteristics of *hemoglobin*, which is present in all the red blood cells. Hemoglobin combines with oxygen as the blood passes through the lungs. In the tissue capillaries the hemoglobin will not release oxygen into the tissue fluid if too much oxygen is already there, but if the oxygen concentration is too little, sufficient oxygen will be released to re-establish an adequate tissue oxygen concentration. Thus, the regulation of oxygen concentration in the tissues is vested principally in the chemical characteristics of hemoglobin itself. This regulation is called the *oxygen-buffering function of hemoglobin*.

Carbon dioxide concentration in the extracellular fluid is regulated in quite a different way. Carbon dioxide is one of the major end-products of the oxidative reactions in cells. If all the carbon dioxide formed in the cells should continue to accumulate in the tissue fluids, the mass action of the carbon dioxide itself would soon halt all the energy-giving reactions of the cells. Fortunately, a nervous mechanism controls the expiration of carbon dioxide through the lungs and in this maintains a constant and reasonable concentration of carbon dioxide in the extracellular fluid. That is, a high carbon dioxide concentration *excites the respiratory center,* causing the person to breathe rapidly and deeply. This increases the rate of expiration of carbon dioxide and therefore increases its removal from the blood and extracellular fluid, and the process continues until the concentration returns to normal.

Regulation of Arterial Pressure. Several different systems contribute to the regulation of arterial pressure. One of these, the *baroreceptor system,* is very simple and an excellent example of a control mechanism. In the walls of most of the great arteries of the upper body, especially the bifurcation region of the carotids and the arch of the aorta, are many nerve receptors, called *baroreceptors,* which are stimulated by stretch of the arterial wall. When the arterial pressure becomes great, these baroreceptors are stimulated excessively, and impulses are transmitted to the medulla of the brain. Here the impulses inhibit the *vasomotor center,* which in turn decreases the number of impulses transmitted through the sympathetic nervous system to the heart and blood vessels. Lack of these impulses causes diminished pumping activity by the heart and increased ease of blood flow through the peripheral vessels, both of which lower the arterial pressure back toward normal. Conversely, a fall in arterial pressure relaxes the stretch receptors, allowing the vasomotor center to become more active than usual and thereby causing the arterial pressure to rise back toward normal.

CHARACTERISTICS OF CONTROL SYSTEMS

The above examples of homeostatic control mechanisms are only a few of the many hundreds in the body, all of which have certain characteristics in common. These are explained in the following pages.

Negative Feedback Nature of Control Systems. Most control systems of the body act by a process of *negative feedback,* which can be explained best by reviewing some of the homeostatic control systems mentioned above. In the regulation of carbon dioxide concentration, a high level of carbon dioxide in the extracellular fluid causes increased pulmonary ventilation, and this in turn causes decreased carbon dioxide concentration. In other words, the response is *negative* to the initiating stimulus. Conversely, if the carbon dioxide concentration falls too low, this causes feedback through the control system to raise the carbon dioxide concentration. This response also is negative to the initiating stimulus.

In the arterial pressure-regulating mechanisms, a high pressure causes a series of reactions that promote a lowered pressure, or a low pressure causes a series of reactions that promote an elevated pressure. In both instances these effects are negative with respect to the initiating stimulus.

Therefore, in general, if some factor becomes excessive or too little, a control system initiates *negative feedback,* which consists of a series of changes that returns the factor toward a certain mean value, thus maintaining homeostasis.

Amplification, or Gain, of a Control System. The degree of effectiveness with which a control system maintains constant conditions is called the *amplification,* or *gain,* of the system. For instance, let us assume that a person has been in a room for 24 hours, the temperature of which has been 60° F. Then suddenly the room temperature is increased to 110° F., a total increase of 50°. However, the person's body temperature rises only from 98.0° to 99.0° F. because the body automatically controls its own temperature within very narrow limits rather than following the change in atmospheric temperature. In this case, the change in atmospheric temperature is 49° more than the change in body temperature, which changes only

1°. Therefore, we say that the *gain* of the control system is −49. In other words, for each degree change in body temperature that does occur, there would be 49 times that much additional change were it not for the control system.

Thus the gain of a negative feedback control system can be calculated by using the following formula:

$$\text{Feedback gain} = \frac{\substack{\text{Amount of correction} \\ \text{of abnormality}}}{\substack{\text{Amount of abnormality} \\ \text{that still persists}}}$$

The gain of the baroreceptor system for control of arterial pressure, as measured in dogs, is approximately −2. That is, an extraneous factor that tends to increase or decrease the arterial pressure does so only one-third as much as would occur if this control system were not present. Therefore, one can see that the temperature control system is much more effective than the baroreceptor system.

Positive Feedback as a Cause of Death — Vicious Cycles. One might ask the question: why do essentially all control systems of the body operate by negative feedback rather than positive feedback? However, if you will consider the nature of positive feedback, you will immediately see that positive feedback does not lead to stability but to instability and often to death.

Figure 1–3 illustrates an instance in which death can ensue from positive feedback. This figure depicts the pumping effectiveness of the heart, showing that the heart of the normal human being pumps about 5 liters of blood per minute. However, if the person is suddenly bled 2 liters, the amount of blood in the body is decreased to such a

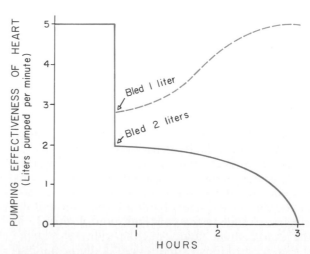

Figure 1–3. Death caused by positive feedback when 2 liters of blood are removed from the circulation.

low level that not enough is available for the heart to pump effectively. As a result, the arterial pressure falls, and the flow of blood to the heart muscle through the coronary vessels diminishes. This results in weakening of the heart, further diminished pumping, further decrease in coronary blood flow, and still more weakness of the heart; the cycle repeats itself again and again until death. Note that each cycle in the feedback results in further weakening of the heart. In other words, the initiating stimulus causes more of the same, which is *positive feedback.*

Positive feedback is better known as a "vicious cycle," but actually a mild degree of positive feedback can be overcome by the negative feedback control mechanisms of the body, and a vicious cycle will fail to develop. For instance, if the person in the above example were bled only 1 liter instead of 2 liters, the normal negative feedback mechanisms for controlling cardiac output and arterial pressure would overbalance the positive feedback, and the person would recover, as shown by the dashed curve of Figure 1–3.

AUTOMATICITY OF THE BODY

The purpose of this chapter has been to point out, first, the overall organization of the body and, second, the means by which the different parts of the body operate in harmony. To summarize, the body is actually a *social order of about 75 trillion cells* organized into different functional structures, some of which are called *organs.* Each functional structure provides its share in the maintenance of homeostatic conditions in the extracellular fluid, which is often called the *internal environment.* As long as normal conditions are maintained in the internal environment, the cells of the body will continue to live and function properly. Thus, each cell benefits from homeostasis, and in turn each cell contributes its share toward the maintenance of homeostasis. This reciprocal interplay provides continuous automaticity of the body until one or more functional systems lose their ability to contribute their share of function. When this happens, all the cells of the body suffer. Extreme dysfunction leads to death, while moderate dysfunction leads to sickness.

APPENDIX

BASIC PHYSICAL PRINCIPLES OF CONTROL SYSTEMS

Until recently the principles of control systems as applied to the human body have been taught only qualitatively rather than quantitatively. For instance, in the case of baroreceptor regulation of arterial pressure, physiologists have simply stated that an increase in

pressure causes a reflex decrease in pressure toward normal. However, this means very little unless we know *how much* effect occurs. But within the past 20 years, application of much more quantitative physical principles of control systems has begun to make physiology a much more exact science than it has been in the past.

Basic Symbols Used in Control System Analysis. Figure 1–4 illustrates common basic symbols used in control system analysis. These are the following:

The Addition-Subtraction Symbol. This symbol is shown in Figure 1–4A. For instance, let x represent the rate of intake of salt in the solid food eaten each day, y the intake of salt in liquids drunk each day, and z the rate of loss of salt in the urine each day. Then the net rate of change of salt in the body will be x + y − z, as indicated by the arrow.

The Multiplication Symbol. Figure 1–4B illustrates multiplication of three quantities, x, y, and z, to give xyz. For instance, let us assume that three separate factors are affecting arterial pressure and that these factors multiply each other. Thus, the baroreceptor system might be causing x effect to elevate arterial pressure; a hormone secreted by one of the endocrine glands might be causing y effect to elevate arterial pressure; and hemorrhage might be causing z effect to decrease arterial pressure. The net effect would be xyz.

Multiplication by a Constant Factor. Figure 1–4C illustrates multiplication by a constant factor. Let us assume that x is the *concentration* of sodium in the extracellular fluids and K is the volume of extracellular fluid; the total *quantity* of sodium in extracellular fluid would then be Kx.

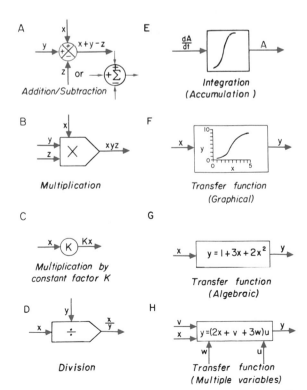

Figure 1–4. Standard symbols used in control system diagrams.

The Division Symbol. Figure 1–4D shows the value x divided by the value y to give x/y. For example, if x is the total *quantity* of sodium in the extracellular fluid and y is the extracellular fluid *volume,* then the output of this block, x/y, is the concentration of sodium in the extracellular fluid.

The Integration (or Accumulation) Symbol. Many functions of the body depend upon slow accumulation of some factor. Thus, in Figure 1–4E, if the *rate of change* of aldosterone in the body is represented by the differential term, dA/dt, the output of the block is the *quantity* of aldosterone that has accumulated in the body at any given time, t. The symbol in this block is called the *integration symbol,* or, in other words, the rate dA/dt is integrated to give A. If dA/dt is positive, the quantity of aldosterone in the body will be increasing, whereas if dA/dt is negative, the quantity of aldosterone in the body will be decreasing.

Transfer Functions. Figures 1–4F, G, and H illustrate transfer functions that show a quantity x entering a block and a quantity y leaving the block. Each block means that y is related to x in accordance with the function that is expressed inside the box. In Figure 1–4F this function is represented graphically. In Figure 1–4G it is represented by an algebraic equation. Figure 1–4H illustrates four inputs and one output, showing that three of the inputs add to each other and the other multiplies the first three.

As an example, Figure 1–4F might represent the relationship between glucose concentration, x, in the extracellular fluids and the rate of insulin secretion, y, by the pancreas. This transfer function shows that at low glucose concentrations essentially no insulin is secreted, but at high concentrations very large quantities of insulin are secreted.

General Analysis of a Control System. Figure 1–5 illustrates a general analysis that can be applied to almost any negative feedback control system in the body. The goal of most control systems is to keep some *controlled variable* at an almost constant value. In Figure 1–5 the controlled variable is the body temperature (BT), and this is represented by the quantitative values on the arrow projecting to the right. The three mathematical components of the control system are the following:

Block I illustrates the summation of the three different factors that determine the value of the controlled variable. The three inputs to this block are (1) the *normal value for the controlled variable,* that is, the mean level at which the control system attempts to maintain the variable; (2) any *disturbance* that is acting on the body to cause the variable to deviate from its normal value; and (3) any *compensation* that is caused by the control system to counteract the disturbance.

Block II calculates the difference between the actual value of the controlled variable and the normal value. The output of this block is called the *error.*

Block III is the feedback portion of the control system that responds to the error and determines the degree of compensation that will occur to overcome the disturbance.

Application of the General Analysis to the Control of Body Temperature. Now let us apply this general analysis of a control system to an actual example, the control of body temperature. At each point in the system two different sets of values are shown. The first value in each instance represents the normal when there is no disturbance and no compensation by the control system. Thus, the normal values are as follows:

Normal value for the controlled variable........ 98°(F.)
Disturbance.. 0°
Actual value of the controlled variable .... 98°
Error... 0°
Compensation .... 0°

Next, let us add a disturbance that attempts to change the body temperature, In this instance, we will assume that the air temperature increases by 50°. If the body temperature control system were not functioning, this would increase the body temperature by 50°. However, as the body temperature begins to rise, the actual temperature becomes different from the normal value, an error develops, and the control system provides sufficient compensation to overcome most of the disturbing effects of the abnormal air temperature. This compensation results from increased blood flow to the skin, decreased rate of heat production in the body, increased evaporation of sweat, and several other effects, all of which tend to reduce the body temperature.

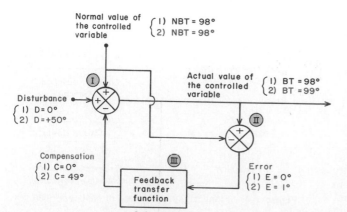

Figure 1–5. Static analysis of a control system. This figure, presenting an analysis of the control of body temperature, shows values for different variables in the system under two different conditions: (1) normal, and (2) when there is an increase in air temperature of 50° F, thereby creating a disturbance that attempts to raise the body temperature. (NBT, normal body temperature; BT, actual body temperature; E, error; C, compensation; D disturbance.)

Therefore, the second set of values shown in Figure 1–5 is the following:

Normal value of the controlled variable....... 98°(F.)
Disturbance... +50°
Actual value of the controlled variable .... 99°
Error.. 1°
Compensation .... −49°

Thus, one sees that after the control system has become effective, the actual value of the controlled variable becomes equal to the sum of three different values: (1) the normal value for the controlled variable (98°), (2) the disturbance (+50°), and (3) the compensation (−49°), giving an actual value of the controlled variable of 99°. That is, the abnormal disturbance of +50° causes a rise in body temperature of only 1°.

One can then calculate the gain of this system by dividing the compensation by the error (−49° divided by 1°), yielding a feedback gain of −49.

Figure 1–6 illustrates the effect of this 50° increase in air temperature under two conditions: (1) when the temperature control system is not functional, and (2) when it is functional. When the system is functioning, the compensation increases at the same time that the body temperature rises. Therefore, by the time the body temperature has risen from 98° to 99° the control system will have already initiated −49° of compensation. The body temperature, therefore, will rise only to 99° rather than to 148°, an increase that would occur if the control system did not exist.

More Complex Analysis of a Control System — The Glucose Control System. Figure 1–7 illustrates an analysis of the insulin control system for regulating glucose concentration in the extracellular fluid. The basic system is the following: When a person eats increased quantities of glucose, the rising glucose concentration in the extracellular fluid causes the pancreas to secrete increased quantities of insulin. The insulin, in turn, causes increased transport of glucose through the cell membranes to the interior of the cells in most parts of the body. The glucose is then used for energy. This, obviously, returns the extracellular glucose concentration back toward normal. A quantitative mathematical

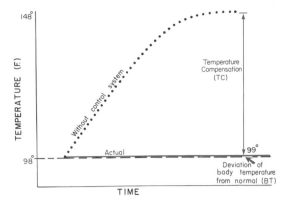

Figure 1–6. Effects on body temperature of suddenly increasing the air temperature 50° F, showing the hypothetical effect without a control system and the actual effect with a normal control system.

analysis is illustrated by the nine blocks of Figure 1–7; these blocks can be explained as follows:

Block 1 calculates the rate of change of glucose (dG/dt) in the extracellular fluids by subtracting rate of glucose transport into the cells from the rate of intake of glucose.

Block 2 integrates the rate of change of glucose to give the total extracellular glucose at any given time.

Block 3 calculates the extracellular glucose concentration by dividing the total extracellular glucose by the extracellular fluid volume.

Block 4 illustrates the effect of extracellular glucose concentration on rate of insulin secretion.

Block 5 sums the rate of insulin secretion and the rate of insulin destruction to give the rate of insulin change (dI/dt).

Block 6 integrates the rate of insulin change to give the total insulin in the body at any given time.

Block 7 calculates rate of insulin destruction by multiplying the total insulin by the constant K.

Block 8 calculates the extracellular insulin concentration by dividing total insulin by extracellular fluid volume.

Block 9 illustrates the effect of extracellular insulin

Figure 1–7. Analysis of the insulin control system for maintaining a constant glucose concentration in the extracellular fluid. By this analysis it is possible to predict transient as well as steady-state changes in variables of the system, such as the readjustments of the system after sudden changes in the rate of glucose intake.

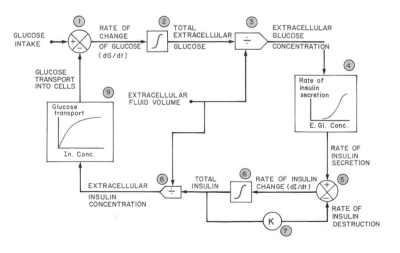

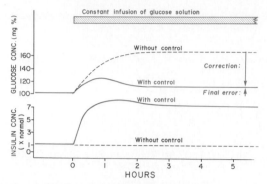

Figure 1–8. Transient changes in extracellular glucose and insulin concentrations following sudden changes in glucose intake, as predicted from the analysis of Figure 1–7 under two different conditions: (1) when the control system is not functioning, and (2) when it is functioning.

concentration on the rate of glucose transport into the body cells.

Function of the Glucose Control System When Glucose Is Infused into a Person. Figure 1–8 illustrates the effect on extracellular glucose concentration and also on insulin concentration when an infusion of glucose solution is suddenly started and maintained at a constant rate for many hours. The effect is shown for two different conditions: without function of the control system (dashed lines) and with function of the control system (solid line). Note that without the control system the glucose concentration rises to approximately 170 mg. per cent and stabilizes at this level. Note, too, that there is also no increase in secretion of insulin.

On the other hand, when the control system is operative the glucose concentration begins to rise just as rapidly as before, but it soon stops rising and returns to a value not far from the normal glucose concentration level. This is caused by the effect of the glucose on the pancreas to cause insulin secretion, followed by buildup of insulin in the body fluids, until glucose transport through the cell membranes rises to equal the rate of glucose infusion into the blood. Note that the insulin concentration remains elevated as long as glucose continues to be infused, a condition that is necessary if the glucose concentration in the body fluids is to remain near normal. Note also that at first the glucose concentration slightly overshoots the final level at which it stabilizes. This is caused by the initial buildup of glucose in the extracellular fluid before the insulin secretion and insulin function have had time to become fully active.

To the right in Figure 1–8 the degree of correction caused by the glucose control system is illustrated by the downward-directed arrow. The final error is illustrated by the upward-directed arrow. The ratio of the lengths of these two arrows represents the gain of the system. Since the correction arrow is approximately six times as long as the error arrow, the gain is approximately −6.

Steady-State Versus Transient Analyses of Control Systems. The analysis of the system for body temperature regulation shown in Figure 1–5 was a steady-state analysis. That is, it simply showed the initial conditions and the final conditions but did not characterize the transient events occurring in the control system during its activation. On the other hand, the analysis of glucose control in Figures 1–7 and 1–8 gives the transients through which the different elements of the system pass in developing the final steady-state condition. In other words, the control system does not act instantly, but instead requires a certain amount of time to develop its compensation. Furthermore, the controlled variable often overshoots the final steady-state value before it stabilizes.

OSCILLATION OF CONTROL SYSTEMS

Unfortunately, even feedback control systems sometimes become unstable and oscillate. Figure 1–9 illustrates blood pressure waves caused by oscillation of the baroreceptor system described earlier in the chapter. The cause of the oscillation is the following: Some extraneous factor causes the arterial pressure to become too high, and this activates the baroreceptor reflex. The pressure begins to fall and eventually falls even below normal. Then the very low arterial pressure activates the baroreceptor reflex in the opposite direction, causing the pressure to increase to a level above normal once again. This initiates a new cycle of oscillation, which can continue indefinitely.

Damping of Control Systems. Fortunately, oscillation of the arterial pressure, as shown in Figure 1–9, does not occur very often but only under certain peculiar conditions such as decreased blood volume or compression of the brain. The reason for lack of oscillation is that this control system, as is also true of almost all other control systems of the body, is highly *damped,* which results from the basic organization of the system itself. For instance, the glucose control system of Figures 1–7 and 1–8 shows only slight overshoot in glucose concentration and insulin concentration, which represents a very highly damped oscillation that persists for only a single overshoot cycle. The major reason for this very high degree of damping is the long period of time required for the glucose to build up its concentration in the body fluid. In the meantime, the other parts of the control system can adjust almost completely to the

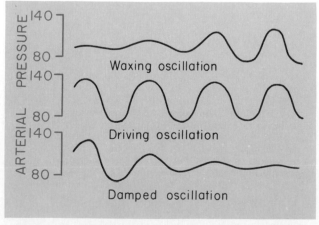

Figure 1–9. Arterial pressure waves caused by oscillation of the baroreceptor system.

changing glucose concentration, thereby preventing excessive over-response of the control system.

Figure 1–9 shows *waxing oscillation, driving oscillation,* and *damped oscillation.* When the control system is very unstable — that is, has almost no damping — even the slightest disturbance can cause mild oscillation at first; this then grows to greater and greater oscillation. If the damping is moderate, the oscillations may continue as driving oscillation indefinitely, never increasing nor decreasing. On the other hand, if the damping is very great, the oscillation fades away.

A few control systems continue to oscillate indefinitely. One of these is the system that causes respiration, which results from oscillation, also called "reverberation," of nerve signals within the respiratory center of the brain. Another continuous oscillation is responsible for the monthly sexual cycle of the female, in which hormones secreted by the anterior pituitary gland stimulate the ovaries to produce ovarian hormones, which in turn act on the pituitary to inhibit its secretion. Lack of the pituitary hormones then causes the ovaries to stop secreting their hormones. This in turn allows the pituitary to begin secreting once again. As a result, the cycle repeats itself again and again from puberty to the menopause.

REFERENCES

Adolph E. F.: Origins of Physiological Regulations. New York, Academic Press, 1968.

Adolph, E. F.: Physiological adaptations: Hypertrophies and superfunctions. *Am. Sci., 60*:608, 1972.

Bernard, C.: Lectures on the Phenomena of Life Common to Animals and Plants. Springfield, Ill., Charles C Thomas, 1974.

Borow, M.: Fundamentals of Homeostasis: A Clinical Approach to Fluid Electrolyte Acid-Base Energy Metabolism in Health and Disease. Flushing, N.Y., Medical Examination Publishing Co., 1977.

Brown, J. H. U. (ed.): Engineering Principles in Physiology. Vols. 1 and 2. New York, Academic Press, 1973.

Bruni, C., *et al.* (eds.): Systems Theory in Immunology. New York, Springer-Verlag, 1979.

Cannon, W. B.: The Wisdom of the Body. New York, W. W. Norton & Co., 1932.

Dawes, I. W., and Sutherland, I. W.: Microbial Physiology. Oxford. Blackwell Scientific Publishers, 1976.

Frisancho, A. R.: Human Adaptation. St. Louis, C. V. Mosby Co., 1979.

Guyton, A. C.: Arterial Pressure and Hypertension. Philadelphia, W. B. Saunders Co., 1980.

Guyton, A. C., and Coleman, T. G.: Quantitative analysis of the pathophysiology of hypertension. *Circ. Res., 14*:I-1, 1969.

Guyton, A. C., *et al.*: Cardiac Output and Its Regulation. Philadelphia, W. B. Saunders Co., 1973.

Guyton, A. C., *et al.*: Dynamics and Control of the Body Fluids. Philadelphia, W. B. Saunders Co., 1975.

Hemker, H. C., and Hess, B.: Analysis and Simulation of Biochemical Systems. New York, American Elsevier Publishing Co., 1972.

Huffaker, C. B. (ed.): Biological Control. New York, Plenum Press, 1974.

Iberall, A. S., and Guyton, A. C. (eds.): *Proc. Int. Symp. on Dynamics and Controls in Physiological Systems.* Regulation and Control in Physiological Systems. ISA, Pittsburgh, 1973.

Jones, R. W.: Principles of Biological Regulation: An Introduction to Feedback Systems. New York, Academic Press, 1973.

Krieger, D. T., and Aschoff, J.: Endocrine and other biological rhythms. *In* DeGroot, L. H., *et al.* (eds.): Endocrinology. Vol. 3. New York, Grune & Stratton, 1979, p. 2079.

Marmarelis, P. Z., and Marmarelis, V. Z.: Analysis of Physiological Systems: The White-Noise Approach. New York, Plenum Press, 1978.

McIntosh, J. E. A., and McIntosh, R. P.: Mathematical Modelling and Computers in Endocrinology. New York, Springer-Verlag, 1980.

Milhorn, H. T.: The Application of Control Theory to Physiological Systems. Philadelphia, W. B. Saunders Co., 1966.

Miller, S. L., and Orgel, L. E.: The Origins of Life on the Earth. Englewood Cliffs, N.J., Prentice-Hall, 1974.

Piva, F., *et al.*: Regulation of hypothalamic and pituitary function: Long, short, and ultrashort feedback loops. *In* DeGroot, L. J., *et al.* (eds.): Endocrinology. Vol. 1. New York, Grune & Stratton, 1979, p. 21.

Randall, J. E.: Microcomputers and Physiological Simulation. Reading, Mass., Addison-Wesley Publishing Co., 1980.

Reeve, E. B., and Guyton, A. C.: Physical Bases of Circulatory Transport: Regulation and Exchange. Philadelphia, W. B. Saunders Co., 1967.

Rusak, B., and Zucker, I.: Neural regulation of circadian rhythms. *Physiol. Rev., 59*:449, 1979.

Söderberg, U.: Neurophysiological aspects of homeostasis. *Annu. Rev. Physiol., 26*:271, 1964.

Sweetser, W.: Human Life (Aging and Old Age.) New York, Arno Press, 1979.

Toates, F. M.: Control Theory in Biology and Experimental Psychology, London, Hutchinson Education Ltd., 1975.

Weston, L.: Body Rhythm: The Circadian Rhythms Within You. New York, Harcourt Brace Jovanovich, 1979.

2

The Cell and Its Function

Each of the 75 trillion cells in the human being is a living structure that can survive indefinitely and, in most instances, can even reproduce itself, provided its surrounding fluids contain appropriate nutrients. To understand the function of organs and other structures of the body, it is essential that we first understand the basic organization of the cell and the functions of its component parts.

ORGANIZATION OF THE CELL

A typical cell, as seen by the light microscope, is illustrated in Figure 2–1. Its two major parts are the *nucleus* and the *cytoplasm*. The nucleus is separated from the cytoplasm by a *nuclear membrane,* and the cytoplasm is separated from the surrounding fluids by a *cell membrane*.

The different substances that make up the cell are collectively called *protoplasm*. Protoplasm is composed mainly of five basic substances: water, electrolytes, proteins, lipids, and carbohydrates.

Water. The principal fluid medium of the cell is water, which is present in a concentration of between 70 and 85 per cent. Many cellular chemicals are dissolved in the water, while others are suspended in small particulate form. Chemical reactions take place among the dissolved chemicals or at the surface boundaries between the suspended particles and the water. The fluid na-

ture of water allows both the dissolved and suspended substances to diffuse or flow to different parts of the cell, thereby providing transport of the substances from one part of the cell to another.

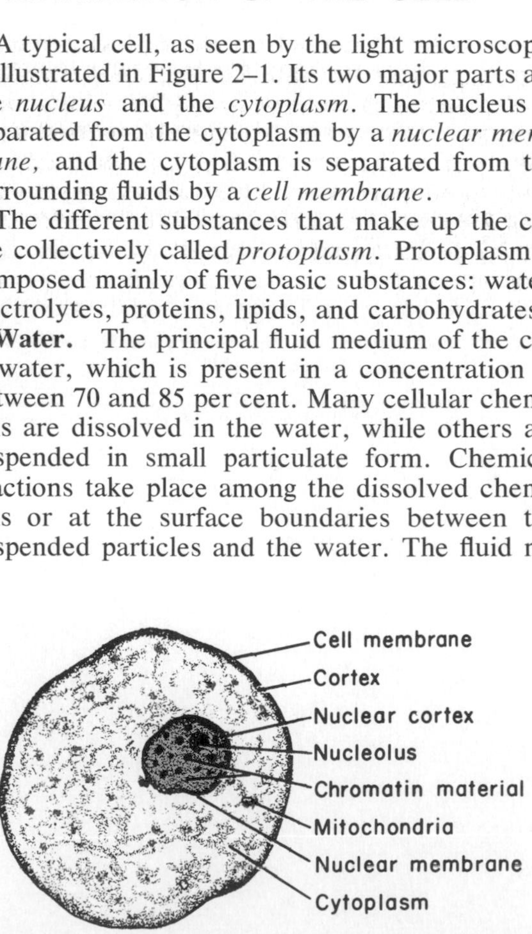

Electrolytes. The most important electrolytes in the cell are *potassium, magnesium, phosphate, sulfate, bicarbonate,* and small quantities of *sodium, chloride,* and *calcium*. These will be discussed in much greater detail in Chapter 4, which will consider the interrelationships between the intracellular and extracellular fluids.

The electrolytes are dissolved in the cell water, and they provide inorganic chemicals for cellular reactions. Also, they are necessary for operation of some of the cellular control mechanisms. For instance, electrolytes acting at the cell membrane allow transmission of electrochemical impulses in nerve and muscle fibers, and the intracellular electrolytes determine the activity of different enzymatically catalyzed reactions that are necessary for cellular metabolism.

Proteins. Next to water, the most abundant substance in most cells is proteins, which normally constitute 10 to 20 per cent of the cell mass. These can be divided into two different types, *structural proteins* and *globular proteins* that are mainly *enzymes*.

To get an idea of what is meant by *structural proteins,* one needs only to note that leather is composed principally of structural proteins, and that hair is almost entirely a structural protein. Proteins of this type are present in the cell in the form of long thin filaments that themselves are polymers of many protein molecules. The most prominent use of such intracellular filaments is to provide the contractile mechanism of all muscles. However, filaments are also organized into microtubules that provide the structures of such organelles as cilia and the mitotic spindles of mitosing cells. And, extracellularly, fibrillar proteins are found especially in the collagen and elastin fibers of connective tissue, blood vessels, tendons, ligaments, and so forth.

The *globular proteins,* on the other hand, are an entirely different type of protein, composed usual-

Cell membrane
Cortex
Nuclear cortex
Nucleolus
Chromatin material
Mitochondria
Nuclear membrane
Cytoplasm

Figure 2–1. Structure of the cell as seen with the light microscope.

ly of individual protein molecules or at most aggregates of a few molecules in a globular form rather than a fibrillar form. These proteins are mainly the enzymes of the cell and, in contrast to the fibrillar proteins, are often soluble in the fluid of the cell or are absorbed in or adherent to the surfaces of membranous structures inside the cell. The enzymes come into direct contact with other substances inside the cell and catalyze chemical reactions. For instance, the chemical reactions that split glucose into its component parts and then combine these with oxygen to form carbon dioxide and water while at the same time providing energy for cellular function are catalyzed by a series of protein enzymes. Thus, enzyme proteins control the metabolic functions of the cell.

Special types of proteins are present in different parts of the cell. Of particular importance are the *nucleoproteins*, present both in the nucleus and in the cytoplasm. The nucleoproteins of the nucleus contain *deoxyribonucleic acid (DNA)*, which constitutes the *genes*, and these control the overall function of the cell as well as the transmission of hereditary characteristics from cell to cell. These substances are so important that they will be considered in detail in Chapter 3. In addition, the chemical nature of proteins will be considered in Chapter 69, and the different structural and enzymatic functions of proteins will be subjects of discussion at numerous points throughout this text.

Lipids. Lipids are several different types of substances that are grouped together because of their common property of being soluble in fat solvents. The most important lipids in most cells are *phospholipids* and *cholesterol* which constitute about 2 per cent of the total cell mass. These are major constituents of the different membranes such as the cell membrane, the nuclear membrane, and the membranes lining intracytoplasmic organelles, e.g., the endoplasmic reticulum and the mitochondria. The special importance of phospholipid and cholesterol in the cell is that they are either insoluble or only partially soluble in water.

In addition to phospholipids and cholesterol, some cells contain large quantities of *triglyceride*, also called *neutral fat*. In the so-called "fat cells," triglycerides can account for as much as 95 per cent of the cell mass. And this fat stored in these cells represents the body's main storehouse of energy-giving nutrient that can later be dissoluted and used for energy wherever in the body it is needed.

The chemical natures of the different types of lipids and their functions in the body will be discussed in Chapter 68.

Carbohydrates. In general, carbohydrates have very little structural function in the cell except as part of glycoprotein molecules, but they play a major role in nutrition of the cell. Most human cells do not maintain large stores of carbohydrates, usually averaging about 1 per cent of their total mass. However, carbohydrate, in the form of glucose, is always present in the surrounding extracellular fluid so that it is readily available to the cell. A small amount of carbohydrate is usually stored in the cells in the form of *glycogen*, which is an insoluble polymer of glucose and can be used rapidly to supply the cells' energy needs.

PHYSICAL STRUCTURE OF THE CELL

The cell is not merely a bag of fluid, enzymes, and chemicals; it also contains highly organized physical structures called *organelles*, which are equally as important to the function of the cell as the cell's chemical constituents. For instance, without one of the organelles, the *mitochondria*, more than 95 per cent of the energy supply of the cell would cease immediately. Some principal organelles of the cell are illustrated in Figure 2–2, including the *cell membrane, nuclear membrane, endoplasmic reticulum, mitochondria,* and *lysosomes*. Others not shown in the figure are the *Golgi complex, centrioles, cilia,* and *microtubules*.

THE MEMBRANOUS STRUCTURES OF THE CELL

Essentially all physical structures of the cell are lined by membranes composed primarily of lipids

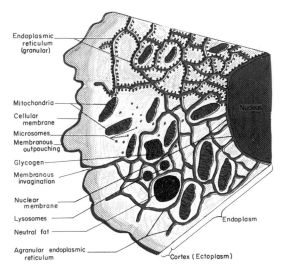

Figure 2–2. Organization of the cytoplasmic compartment of the cell.

and proteins. The lipids provide a barrier that prevents free movement of water and water-soluble substances from one cell compartment to the other. The protein molecules dissolved in the membrane, on the other hand, often interrupt the continuity of the lipid barrier and therefore provide pathways for passage of various substances through the membrane. Also, many of the membrane proteins provide enzymatic functions for a multitude of purposes that will become clear in the further discussions of this and future chapters. The different membranes include the *cell membrane,* the *nuclear membrane,* the *membrane of the endoplasmic reticulum,* and the *membranes of the mitochondria, lysosomes, Golgi complex,* and so forth.

The Cell Membrane. The cell membrane, which completely envelops the cell, is a very thin, elastic structure only 7.5 to 10 nanometers thick. It is composed almost entirely of proteins and lipids; the approximate composition is: proteins, 55 per cent; phospholipids, 25 per cent; cholesterol, 13 per cent; other lipids, 4 per cent; and carbohydrates, 3 per cent.

The Lipid Barrier of the Cell Membrane. Figure 2–3 illustrates that the basic structure of the cell membrane is a *lipid bilayer,* which is a thin film of lipids only 2 molecules thick that is continuous over the entire cell surface. Interspersed in this lipid film are large globular protein molecules.

The lipid bilayer is composed almost entirely of phospholipids and cholesterol. One part of the phospholipid and the cholesterol molecules is soluble in water, that is, *hydrophilic,* while the other part is soluble only in fats, that is, *hydrophobic.*

The phosphate radical of the phospholipid is hydrophilic and the fatty acid radicals are fat-soluble. The cholesterol has a hydroxyl radical that is water-soluble and a steroid nucleus that is fat-soluble. Because the hydrophobic portions of both these molecules are repelled by water but are mutually attracted to each other, they have a natural tendency to line up, as illustrated in Figure 2–3, with the fatty portions in the center of the membrane and the hydrophilic portions projecting to the two surfaces.

The membrane lipid bilayer is almost entirely impermeable to water and to the usual water-soluble substances such as ions, glucose, urea, and others. On the other hand, fat-soluble substances such as oxygen, carbon dioxide, and alcohols can penetrate this portion of the membrane.

A special feature of the lipid bilayer is that it is a lipid *fluid* and not a solid. Therefore, portions of the membrane can literally flow from one point to another in the membrane. Proteins or other substances dissolved in or floating in the lipid bilayer tend to diffuse to all areas of the cell membrane.

The Cell Membrane Proteins. Figure 2–3 illustrates globular masses floating in the lipid bilayer. These are membrane proteins, most of which are *glycoproteins.* Two types of proteins occur: the *integral proteins* that protrude all the way through the cell and the *peripheral proteins* that are attached only to the surface of the membrane and do not penetrate. The integral proteins, like the lipids, have fat-soluble portions and water-soluble portions; however, these proteins usually have water-soluble portions on both ends and fat-

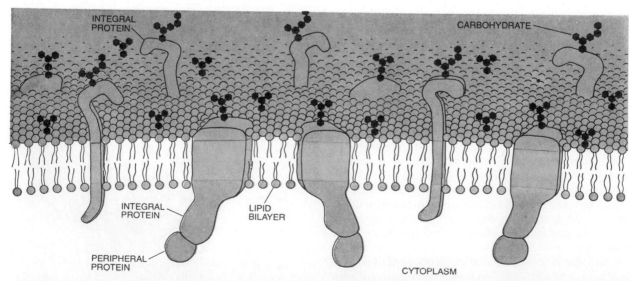

Figure 2–3. Structure of the cell membrane, showing that it is composed mainly from a lipid bilayer but with large numbers of protein molecules protruding through the layer. Also, carbohydrate moieties are attached to the protein molecules on the outside of the membrane and additional protein molecules on the inside. (From Lodish and Rothman: *Sci. Am., 240:*48, 1979. ©1979 by Scientific American, Inc.)

soluble ones in the middle. Thus, it is the fat-soluble portion of the protein that dissolves in the lipid bilayer, while the water-soluble portions protrude to the outside and inside of the membrane. The integral proteins provide structural pathways through which water and water-soluble substances, especially the ions, can diffuse between the extracellular and intracellular fluid. However, these proteins have selective properties that cause preferential diffusion of some substances more than others. Some of them can also act as enzymes.

The peripheral proteins occur either entirely or almost entirely on the inside of the membrane, and they are normally attached to one of the integral proteins. These peripheral proteins function almost entirely as enzymes.

The Membrane Carbohydrates. The membrane carbohydrates occur almost invariably on the outside of the membrane; they are the "glyco-" portion of protruding glycoprotein molecules. These carbohydrate moieties are the portion of the cell membrane that enters into immune reactions, as we shall discuss in Chapter 7, and they often act as receptor substances for binding hormones, such as insulin, that stimulate specific types of activity in the cells.

The Nuclear Membrane. The nuclear membrane is actually two membranes, one surrounding the other with a wide space in between. Each membrane is almost identical to the cell membrane, having a basic lipid bilayer structure with globular proteins floating in the lipid fluid. Very large so-called "pores," 50 to 80 nanometers in diameter, are seen in the nuclear membrane (as illustrated in Figure 2–9). These "pores" are not holes through the membrane; instead, they are areas where the two nuclear membranes fuse with each other. However, the fused membrane is so permeable in these pore areas that almost all dissolved or suspended substances, including even very large, newly formed ribosomes, can move with ease between the fluids of the nucleus and the cytoplasm.

The Endoplasmic Reticulum. Figure 2–2 illustrates in the cytoplasm a continuous network of tubular and flat vesicular stuctures, constructed of lipid bilayer-protein membranes, called the *endoplasmic reticulum.* The total surface area of this structure in some cells — the liver cells, for instance — can be as much as 30 to 40 times as great as the cell membrane area. The detailed structure of this organelle is illustrated in Figure 2–4. The space inside the tubules and vesicles is filled with *endoplasmic matrix,* a fluid medium that is different from the fluid outside the endoplasmic reticulum. Electron micrographs show that the space inside the endoplasmic reticulum is connected with the space between the two membranes of the double nuclear membrane.

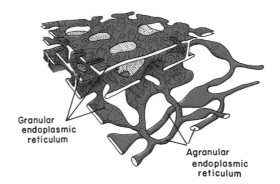

Figure 2–4. Structure of the endoplasmic reticulum. (Modified from De Robertis, Saez, and De Robertis: Cell Biology. 6th ed. Philadelphia, W. B. Saunders Company, 1975.)

Substances formed in different parts of the cell enter the spaces of the endoplasmic reticulum and are then conducted to other parts of the cell. Also, the vast surface area of the reticulum, as well as its many enzyme systems, provides the machinery for a major share of the metabolic functions of the cell.

Ribosomes and the Granular Endoplasmic Reticulum. Attached to the outer surfaces of many parts of the endoplasmic reticulum are large numbers of small granular particles called *ribosomes.* Where these are present, the reticulum is frequently called the *granular endoplasmic reticulum.* The ribosomes are composed mainly of ribonucleic acid, which functions in the synthesis of protein in the cells, as discussed later in this and in the following chapters.

The Agranular Endoplasmic Reticulum. Part of the endoplasmic reticulum has no attached ribosomes. This part is called the *agranular,* or *smooth, endoplasmic reticulum.* The agranular reticulum functions in the synthesis of lipid substances and in many other enzymatic processes of the cell.

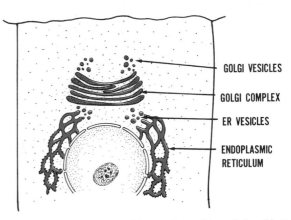

Figure 2–5. A typical Golgi complex and its relationship to the endoplasmic reticulum and the nucleus.

Golgi Complex.   The Golgi complex, illustrated in Figure 2–5, is closely related to the endoplasmic reticulum. It has membranes similar to those of the agranular endoplasmic reticulum. It is usually composed of four or more stacked layers of thin, flat vesicles lying near the nucleus. This complex is very prominent in secretory cells; in these it is located on the side of the cell from which the secretory substances will be extruded.

The Golgi complex functions mainly in association with the endoplasmic reticulum. As illustrated in Figure 2–5, small "transport vesicles" continually pinch off from the endoplasmic reticulum and shortly thereafter fuse with the Golgi complex. In this way substances are transported from the endoplasmic reticulum to the Golgi complex. The transported substances are then processed in the Golgi complex to form secretory vesicles, lysosomes, or other cytoplasmic components to be discussed later in the chapter.

THE CYTOPLASM AND ITS ORGANELLES

The cytoplasm is filled with both minute and large dispersed particles and organelles ranging in size from a few nanometers to 3 microns. The clear fluid portion of the cytoplasm in which the particles are dispersed is called *hyaloplasm;* this contains mainly dissolved proteins, electrolytes, glucose, and small quantities of phospholipids, cholesterol, and esterified fatty acids.

The portion of the cytoplasm immediately beneath the cell membrane frequently contains large numbers of microfilaments composed of fibrillar proteins. These provide a semi-solid support for the cell membrane. This zone of the cytoplasm is called the *cortex,* or *ectoplasm*. The cytoplasm between the cortex and the nuclear membrane is liquefied and is called the *endoplasm*.

Among the large dispersed particles in the cytoplasm are neutral fat globules, glycogen granules, ribosomes, secretory granules, and two especially important organelles — the *mitochondria* and *lysosomes* — which are discussed below.

Colloidal Nature of the Cytoplasm.   All the particles dispersed in the cytoplasm, whether the large lysosomes and mitochondria or the small granules, are hydrophilic — that is, attracted to water — because of ionic negative electrical charges on their surfaces. The particles remain dispersed in the cytoplasm mainly because of mutual repulsion of the charges on the different particles. Thus, the cytoplasm is actually a colloidal solution.

The Mitochondria

The mitochondria are called the "power-houses" of the cell. Without them the cells are unable to extract significant amounts of energy from the nutrients and oxygen, and as a consequence essentially all cellular functions cease. As illustrated in Figure 2–2, these organelles are present in essentially all portions of the cytoplasm, but the number per cell varies from less than a hundred to many thousand, depending upon the amount of energy required by each cell. Furthermore, the mitochondria are concentrated in those portions of the cell that are responsible for the major share of its energy metabolism. Mitochondria are also very variable in size and shape; some are only a few hundred millimicrons in diameter and globular in shape while others are as large as 1 micron in diameter, as long as 7 microns, and filamentous in shape.

The basic structure of the mitochondrion is illustrated in Figure 2–6, which shows it to be composed mainly of two lipid bilayer-protein membranes: an *outer membrane* and an *inner membrane*. Many infoldings of the inner membrane form *shelves* onto which the oxidative enzymes of the cell are attached. In addition, the inner cavity of the mitochondrion is filled with a gel *matrix* containing large quantities of dissolved enzymes that are necessary for extracting energy from nutrients. These enzymes operate in association with the oxidative enzymes on the shelves to cause oxidation of the nutrients, thereby forming carbon dioxide and water. The liberated energy is used to synthesize a high-energy substance called *adenosine triphosphate (ATP)*. ATP is then transported out of the mitochondrion, and it diffuses throughout the cell to release its energy wherever it is needed for performing cellular functions. The function of ATP is so important to the cell that it is discussed in detail later in the chapter.

Mitochondria are self-replicative, which means that one mitochondrion can form a second one, a third one, and so on, whenever there is need in the cell for increased amounts of ATP. Indeed, the mitochondria contain a special type of deoxyribo-

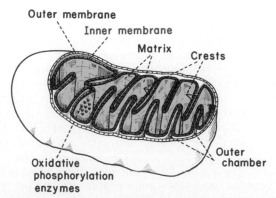

Figure 2–6.   Structure of a mitochondrion. (Modified from De Robertis, Saez, and De Robertis: Cell Biology. 6th ed. Philadelphia, W. B. Saunders Company, 1975.)

nucleic acid similar to that found in the nucleus. In the following chapter we shall see that deoxyribonucleic acid is the basic substance that controls replication of the entire cell; it has now been demonstrated that this substance plays a similar role in the mitochondrion.

The Lysosomes

Another organelle that is essential to the function of the cell, and that is dispersed throughout the cytoplasm, is the lysosome. The lysosomes provide an intracellular digestive system that allows the cell to digest and thereby remove unwanted substances and structures, especially damaged or foreign structures, such as bacteria. The lysosome, illustrated in Figure 2–2, is 250 to 750 nanometers in diameter and is surrounded by a typical lipid bilayer membrane. It is filled with large numbers of small granules 5 to 8 nanometers in diameter, which are protein aggregates of hydrolytic (digestive) enzymes. A hydrolytic enzyme is capable of splitting an organic compound into two or more parts by combining hydrogen from a water molecule with part of the compound and by combining the hydroxyl portion of the water molecule with the other part of the compound. For instance, protein is hydrolyzed to form amino acids, and glycogen is hydrolyzed to form glucose. More than 40 different *acid hydrolases* have been found in lysosomes, and the principal substances that they digest are proteins, nucleic acids, mucopolysaccharides, lipids, and glycogen.

Ordinarily, the membrane surrounding the lysosome prevents the enclosed hydrolytic enzymes from coming in contact with other substances in the cell. However, many different conditions of the cell will break the membranes of some of the lysosomes, allowing release of the enzymes. These enzymes then split the organic substances with which they come in contact into small, highly diffusible substances, such as amino acids and glucose. Some of the more specific functions of lysosomes are discussed later in the chapter.

Other Cytoplasmic Structures and Organelles

Secretory Vesicles.   One of the important functions of many cells is secretion of special substances. Almost all such secretory substances are formed by the endoplasmic reticulum–Golgi complex system and are then released from the Golgi complex inside storage vesicles, called *secretory vesicles* or sometimes *secretory granules.* Figure 2–7 illustrates typical secretory vesicles inside pancreatic acinus cells that have formed and stored protein enzymes in them. The enzymes will be secreted later through the outer cell membrane into the pancreatic duct.

Microfilaments and Microtubular Structures in the Cell.   The fibrillar proteins of the cell cytoplasm are

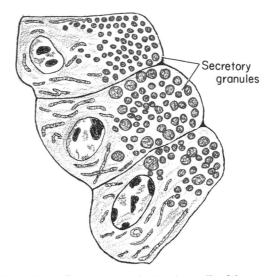

Figure 2–7.   Secretory granules in acinar cells of the pancreas.

usually organized into microfilaments or microtubules. These originate as precursor protein molecules synthesized by the ribosomes. At first they are present in dissolved form in the cytoplasm. Then, most of these polymerize to form microfilaments. We have already pointed out that large numbers of microfilaments frequently occur in the outer zone of the cytoplasm, the zone called the *ectoplasm,* to form an elastic support for the cell membrane. Also, in muscle cells microfilaments are organized into a special contractile machine that is the basis of muscle contraction throughout the body, as will be discussed in detail in Chapter 11.

Microfilaments are also frequently organized into tubular structures, the *microtubules.* Almost invariably these contain 13 microfilaments lying parallel and in a circle to form a long hollow cylinder 25 nanometers in diameter and 1 to many microns in length. These are often arranged in bundles which gives them, *en masse,* considerable structural strength. However, microtubules are stiff structures that break if bent too severely. Figure 2–8 illustrates typical microtubules that were teased from the flagellum of a sperm. Another example of microtubules is the tubular mechanical structure of cilia that gives them structural strength, radiating upward from the cell cytoplasm to the tip of the cilium. Also, the centrioles and the mitotic spindle of the mitosing cell are both composed of stiff microtubules.

Thus, a primary function of microtubules is to act as a *cytoskeleton,* providing rigid physical structures for certain parts of cells. Also, it has been noted that the cytoplasm often *streams* in the vicinity of microtubules, which might result from movement of arms that project outward from the microtubules.

THE NUCLEUS

The nucleus is the control center of the cell. It controls both the chemical reactions that occur in the cell and reproduction of the cell. Briefly, the nucleus contains large quantities of *deoxyribonucleic acid,* which we have called *genes* for many years. The genes determine the characteristics of

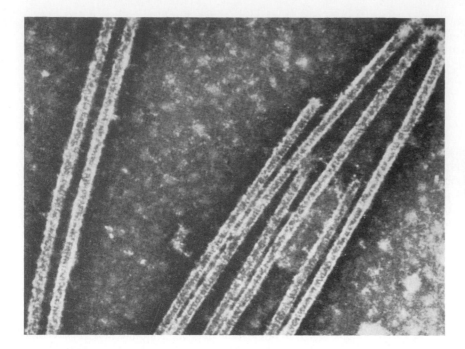

Figure 2–8. Microtubules teased from the flagellum of a sperm. (From Porter: Ciba Foundation Symposium: Principles of Biomolecular Organization. Little, Brown and Company, 1966.)

the protein enzymes of the cytoplasm, and in this way control cytoplasmic activities. To control reproduction, the genes first reproduce themselves, and after this is accomplished the cell splits by a special process called *mitosis* to form two daughter cells, each of which receives one of the two sets of genes. These activities of the nucleus are considered in detail in the following chapter.

The appearance of the nucleus under the microscope does not give much of a clue to the mechanisms by which it performs its control activities. Figure 2–9 illustrates the light microscopic appearance of the interphase nucleus (period between mitoses), showing darkly staining *chromatin material* throughout the *nuclear sap*. During mitosis, the chromatin material becomes readily identifiable as part of the highly structured *chromosomes,* which can be seen easily with the light microscope. Even during the interphase of cellular

activity the chromatin material is still organized into fibrillar chromosomal structures, but this is impossible to see except in a few types of cells.

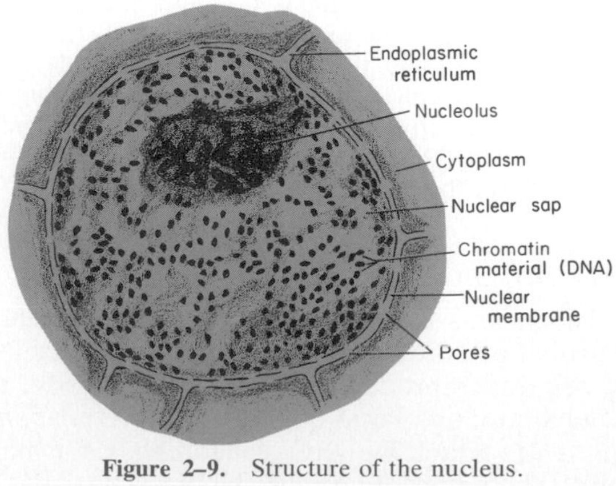

Nucleoli.   The nuclei of many cells contain one or more lightly staining structures called nucleoli. The nucleolus, unlike most of the organelles that we have discussed, does not have a limiting membrane. Instead, it is simply a protein structure that contains a large amount of *ribonucleic acid* of the type found in ribosomes. The nucleolus becomes considerably enlarged when a cell is actively synthesizing proteins. The genes of a particular chromosome pair synthesize the ribonucleic acid and then store it in the nucleolus, beginning with a loose fibrillar RNA that later condenses to form the granular ribosomes. These in turn migrate through the nuclear membrane "pores" into the cytoplasm where most of them become attached to the endoplasmic reticulum and there play an essential role for the formation of proteins, as we shall discuss in the following chapter.

COMPARISON OF THE ANIMAL CELL WITH PRECELLULAR FORMS OF LIFE

Many of us think of the cell as the lowest level of animal life. However, the cell is a very complicated organism, which probably required several billion years to develop after the earliest form of life, an organism similar to the present-day *virus,* first appeared on Earth. Figure 2–10 illustrates the relative sizes of the smallest known virus, a large virus, a *rickettsia,* a *bacterium,* and a nucleated cell, showing that the cell has a diameter about 1000 times that of the smallest virus, and, therefore, a volume about 1 billion times that of the smallest

Endoplasmic reticulum

Nucleolus

Cytoplasm

Nuclear sap

Chromatin material (DNA)

Nuclear membrane

Pores

Figure 2–9.   Structure of the nucleus.

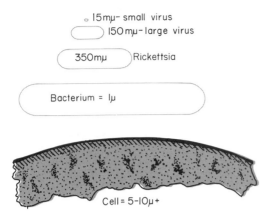

Figure 2–10. Comparison of sizes of precellular organisms with that of the average cell in the human body.

virus. Correspondingly, the functions and anatomical organization of the cell are also far more complex than those of the virus.

The essential life-giving constituent of the very small virus is a *nucleic acid* embedded in a coat of protein. This nucleic acid is similar to that of the cell, and it is capable of reproducing itself if appropriate surrounding conditions are available. Thus, the virus is capable of propagating its lineage from generation to generation, and, therefore, is a living structure in the same way that the cell and the human being are living structures.

As life evolved, other chemicals besides nucleic acid and simple proteins became integral parts of the organism, and specialized functions began to develop in different parts of the virus. A membrane formed around the virus, and inside the membrane a fluid matrix appeared. Specialized chemicals developed inside the matrix to perform special functions; many protein enzymes appeared which were capable of catalyzing chemical reactions, and, therefore, of controlling the organism's activities.

In still later stages, particularly in the rickettsial and bacterial stages, *organelles* developed inside the organism, representing physical structures of chemical aggregates that perform functions in a more efficient manner than can be achieved by dispersed chemicals throughout the fluid matrix. And, finally, in the nucleated cell, still more complex organelles developed, the most important of which is the *nucleus* itself. The nucleus distinguishes this type of cell from all lower forms of life; this structure provides a control center for all cellular activities, and it also provides for very exact reproduction of new cells generation after generation, each new cell having essentially the same structure as its progenitor.

FUNCTIONAL SYSTEMS OF THE CELL

In the remainder of this chapter we will begin discussing the functional systems of the cell that make it a living organism. However, this is such an important subject that it will continue throughout the remainder of this text. Two cellular func-

tions are so important that they will be the subjects of the following two chapters: (1) control of protein synthesis and of other cellular functions by the genes in the nucleus, and (2) transport of substances through the cell membrane. But in the present chapter, let us begin by discussing the means by which the cell ingests substances, then follow this with the energy systems of the cell and the synthesis of new substances in the cell.

INGESTION BY THE CELL

If a cell is to live and grow, it must obtain nutrients and other substances from the surrounding fluids. Substances can pass through a cell membrane in three separate ways: (1) by *diffusion* through the pores in the membrane or through the membrane matrix itself; (2) by *active transport* through the membrane, a mechanism in which enzyme systems and special carrier substances "carry" the substances through the membrane; and (3) by *endocytosis,* a mechanism by which the membrane actually engulfs particulate matter or extracellular fluid and its contents. The important subject of transport of substances by diffusion and active transport, the means by which most nutrients and other substances enter the cell, will be considered in detail in Chapter 4. Endocytosis is a specialized cellular function that merits mention here.

Endocytosis — Phagocytosis and Pinocytosis. Phagocytosis means the ingestion of large particulate matter by a cell, such as the ingestion of (a) a bacterium, (b) some other cell, or (c) particles of degenerating tissue. Pinocytosis, on the other hand, means ingestion of minute quantities of extracellular fluid and dissolved substances in the form of minute vesicles. The pinocytic vesicles are so small that they were not discovered until the advent of the electron microscope. However, phagocytosis has been known to occur from the earliest studies using the light microscope.

Thus, phagocytosis and pinocytosis are both types of *endocytosis,* and their mechanisms are essentially identical except for the sizes and natures of the ingested vesicles.

Phagocytosis occurs when certain objects contact the cell membrane. In general, those objects that have an electronegative charge are rejected while those that have an electropositive charge are especially susceptible to phagocytosis. This difference presumably results from the fact that the phagocytic cells themselves normally are electronegatively charged and therefore repel other electronegative objects. Most *normal* particulate objects in the extracellular fluid are also negatively charged; on the other hand, damaged tissues and also foreign invaders that have been especially prepared for phagocytosis by attachment to antibodies (a process called *opsonization*, that will be discussed in Chapters 6 and 7) usually acquire positive charges and therefore will be phagocytized.

Pinocytosis also occurs in response to certain types of substances that contact the cell membrane. The two most important are proteins and strong electrolyte solutions. It is especially significant that proteins cause pinocytosis, because pinocytosis is the only means by which proteins can pass through the cell membrane.

Figure 2–11 illustrates the successive steps of pinocytosis, showing first three molecules of protein attaching to the membrane by the simple process of adsorption. The presence of these proteins then causes the surface properties of the membrane to change in such a way that it invaginates and then rapidly closes over the proteins. Immediately thereafter, the invaginated portion of the membrane breaks away from the surface of the cell, forming a *pinocytic vesicle. Phagocytic vesicles* are formed in a similar manner.

What causes the cell membrane to go through the necessary contortions for forming the pinocytic and phagocytic vesicles remains a mystery. However, it is known that this process requires energy from within the cell; this is supplied by adenosine triphosphate, a high-energy substance that will be discussed later in this chapter. Also, endocytosis requires the presence of calcium ions in the extracellular fluid and probably a contractile function by the microfilaments immediately beneath the cell membrane. It is presumed that initial contact of the cell membrane with the material to be engulfed changes the surface tension of the membrane causing it to invaginate and draw the material inward. Then the microfilaments, perhaps under the stimulation of calcium ions, provide the force for pinching the vesicles away from the cell membrane.

DIGESTION OF FOREIGN SUBSTANCES IN THE CELL – FUNCTION OF THE LYSOSOMES

Almost immediately after a pinocytic or phagocytic vesicle appears inside a cell, one or more lysosomes become attached to the vesicle and empty their hydrolases into the vesicle, as illustrated in Figure 2–12. Thus, a *digestive vesicle* is formed in which the hydrolases begin hydrolyzing the proteins, glycogen, lipids, nucleic acids, mucopolysaccharides, and other substances in the vesicle. The products of digestion are small molecules of amino acids, glucose, fatty acids, phosphates, and so forth that can then diffuse through the membrane of the vesicle into the cytoplasm. What is left of the digestive vesicle, called the

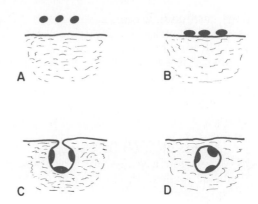

Figure 2–11. Mechanism of pinocytosis.

residual body, represents the undigestible substances. In most instances this is finally excreted through the cell membrane by a process called *exocytosis,* which is essentially the opposite of endocytosis.

Thus, the lysosomes may be called the *digestive organs* of the cells.

Regression of Tissues and Autolysis of Cells. Often, tissues of the body regress to a much smaller size than previously. For instance, this occurs in the uterus following pregnancy, in muscles during long periods of inactivity, and in mammary glands at the end of the period of lactation. Lysosomes are responsible for most of this regression. However, the mechanism by which lack of activity in a tissue causes the lysosomes to increase their activity is still unknown.

Another very special role of the lysosomes is the removal of damaged cells or damaged portions of cells from tissues — cells damaged by heat, cold, trauma, chemicals, or any other factor. Damage to the cell causes lysosomes to rupture, and the released hydrolases begin immediately to

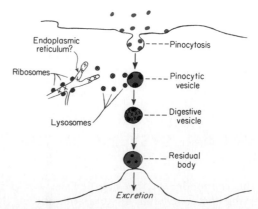

Figure 2–12. Digestion of substances in pinocytic vesicles by enzymes derived from lysosomes.

digest the surrounding organic substances. If the damage is slight, only a portion of the cell will be removed, followed by repair of the cell. However, if the damage is severe the entire cell will be digested, a process called *autolysis*. In this way, the cell is completely removed and a new cell of the same type ordinarily is formed by mitotic reproduction of an adjacent cell to take the place of the old one.

The lysosomes also contain bactericidal agents that can kill phagocytized bacteria before they can cause cellular damage. These agents include *lysozyme* that dissolves the bacterial cell membrane, *lysoferrin* that binds iron and other metals that are essential for bacterial growth, *acid* that has a pH less than 4.0, and *hydrogen peroxide* that poisons some of the bacterial metabolic systems.

Also in lysosomes are stored enzymes that, upon release into the cytoplasm, can dissolute lipid droplets and glycogen granules, making the lipid and glycogen available for use elsewhere in the cell or elsewhere in the body. In the absence of these enzymes, which results from occasional genetic disorders, extreme quantities of lipids or of glycogen often accumulate in the cells of many organs, especially the liver, and lead to early death.

EXTRACTION OF ENERGY FROM NUTRIENTS – FUNCTION OF THE MITOCHONDRIA

The principal nutrients from which cells extract energy are oxygen and one or more of the foodstuffs — carbohydrates, fats, and proteins. In the

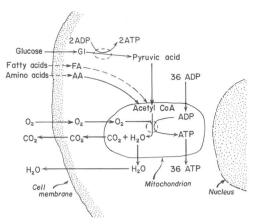

Figure 2–13. Formation of adenosine triphosphate in the cell, showing that most of the ATP is formed in the mitochondria.

human body essentially all carbohydrates are converted into glucose before they reach the cell, the proteins are converted into amino acids, and the fats are converted into fatty acids. Figure 2–13 shows oxygen and the foodstuffs — glucose, fatty acids, and amino acids — all entering the cell. Inside the cell, the foodstuffs react chemically with the oxygen under the influence of various enzymes that control their rates of reactions and channel the energy that is released in the proper direction.

Formation of Adenosine Triphosphate (ATP). The energy released from the nutrients is used to form adenosine triphosphate, generally called ATP, the formula for which is:

Adenine

Phosphate

Ribose

Note that ATP is a nucleotide composed of the nitrogenous base *adenine,* the pentose sugar *ribose,* and three *phosphate radicals.* The last two phosphate radicals are connected with the remainder of the molecule by so-called *high-energy phosphate bonds,* which are represented by the symbol ~. Each of these bonds contains about 8000 calories of energy per mole of ATP under the physical conditions of the body (7000 calories under standard conditions), which is much greater than the energy stored in the average chemical bond of other organic compounds, thus giving rise to the term ''high-energy'' bond. Furthermore, the high-energy phosphate bond is very labile so that it can be split instantly on demand whenever energy is required to promote other cellular reactions.

When ATP releases its energy, a phosphoric acid radical is split away, and *adenosine diphosphate (ADP)* is formed. Then, energy derived from the cellular nutrients causes the ADP and phosphoric acid to recombine to form new ATP, the entire process continuing over and over again. For these reasons, ATP has been called the *energy currency* of the cell, for it can be spent and remade again and again.

Chemical Processes in the Formation of ATP – Role of the Mitochondria.   On entry into the cells, glucose is subjected to enzymes in the cytoplasm that convert it into *pyruvic acid* (a process called *glycolysis*). A small amount of ADP is changed into ATP by energy released during this conversion, but this amount accounts for less than 5 per cent of the overall energy metabolism of the cell.

By far the major portion of the ATP formed in the cell is formed in the mitochondria. The pyruvic and fatty acids and most of the amino acids are all converted into the compound *citric acetyl co-A* in the matrix of the mitochondrion. This substance, in turn, is acted upon by a series of enzymes and undergoes dissolution in a sequence of chemical reactions called the *acid cycle,* or *Krebs cycle.* These chemical reactions will be explained in detail in Chapter 67.

In the citric acid cycle, acetyl co-A is split into its component parts, hydrogen atoms and carbon dioxide. The carbon dioxide, in turn, diffuses out of the mitochondria and eventually out of the cell. The hydrogen atoms combine with carrier substances and are carried to the surfaces of the shelves that protrude into the mitochondria, as shown in Figure 2–6. Attached to these shelves are the so-called *oxidative enzymes* and also protruding globules of *ATPase,* the enzyme that catalyzes the conversion of ADP to ATP or vice versa. The oxidative enzymes, by a series of sequential reactions, cause the hydrogen atoms to combine with oxygen. The enzymes are arranged on the surfaces of the shelves in such a way that the products of one chemical reaction are immediately relayed to the next enzyme, then to the next, and so on until the complete sequence of reactions has taken place. During the course of these reactions, the energy released from the combination of hydrogen with oxygen is used to activate the ATPase and drive the reaction to manufacture tremendous quantities of ATP from ADP. The ATP is then transported out of the mitochondrion into all parts of the cytoplasm and nucleoplasm where its energy is used to energize the functions of the cell.

The formation of ATP is so important to the function of the cell that many more details of this subject will be presented in Chapters 67 through 71.

Uses of ATP for Cellular Function.   ATP is used to promote three major categories of cellular functions: (1) *membrane transport,* (2) *synthesis of chemical compounds* throughout the cell, and (3) *mechanical work.* These three different uses of ATP are illustrated in Figure 2–14: (a) to supply energy for the transport of sodium through the cell membrane, (b) to promote protein synthesis by the ribosomes, and (c) to supply the energy needed during muscle contraction.

In addition to membrane transport of sodium, energy from ATP is required for transport of potassium ions and, in certain cells, calcium ions, phosphate ions, chloride ions, urate ions, hydrogen ions, and still many other special substances. Membrane transport is so important to cellular function that some cells, the renal tubular cells for instance, utilize as much as 80 per cent of the ATP formed in the cells for this purpose alone.

In addition to synthesizing proteins, cells also synthesize phospholipids, cholesterol, purines, pyrimidines, and a great host of other substances. Synthesis of almost any chemical compound requires energy. For instance, a single protein mole-

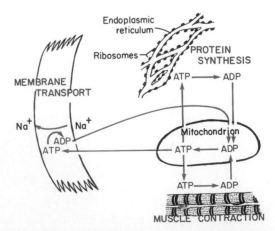

Figure 2–14.   Use of adenosine triphosphate to provide energy for three major cellular functions: membrane transport, protein synthesis, and muscle contraction.

cule might be composed of as many as several thousand amino acids attached to each other by peptide linkages; the formation of each of these linkages requires the breakdown of three high-energy bonds; thus many thousand ATP molecules must release their energy as each protein molecule is formed. Indeed, cells often utilize as much as 75 per cent of all the ATP formed in the cell simply to synthesize new chemical compounds; this is particularly true during the growth phase of cells.

The final major use of ATP is to supply energy for special cells to perform mechanical work. We shall see in Chapter 11 that each contraction of a muscle fibril requires expenditure of tremendous quantities of ATP. Other cells perform mechanical work in two additional ways, by *ciliary* or *ameboid motion,* both of which will be described later in this chapter. The source of energy for all these types of mechanical work is ATP.

In summary, therefore, ATP is always available to release its energy rapidly and almost explosively wherever in the cell it is needed. To replace the ATP used by the cell, other much slower chemical reactions break down carbohydrates, fats, and proteins and use the energy derived from these to form new ATP.

SYNTHESIS AND FORMATION OF CELLULAR STRUCTURES BY THE ENDOPLASMIC RETICULUM AND THE GOLGI COMPLEX

The extensiveness of the endoplasmic reticulum and the Golgi complex, especially in secretory cells, has already been emphasized. These two structures are made primarily of lipid bilayer membranes, and their walls are literally loaded with protein enzymes that catalyze the synthesis of most of the substances required by the cell.

In general, most synthesis begins in the endoplasmic reticulum, but the products formed in the endoplasmic reticulum are then passed on to the Golgi complex where they are further processed prior to release into the cell. But, first, let us note the specific products that are synthesized in the specific portions of the endoplasmic reticulum and the Golgi complex.

Formation of Proteins by the Granular Endoplasmic Reticulum. The granular endoplasmic reticulum is characterized by the presence of large numbers of ribosomes attached to the outer surfaces of the reticulum membrane. As we shall discuss in the following chapter, protein molecules are synthesized within the structure of the ribosomes. Furthermore, the ribosomes extrude many of the synthesized protein molecules not into the hyaloplasm but instead through the endoplasmic reticular wall into the endoplasmic matrix.

Within the endoplasmic matrix, the protein molecules are further processed during the next few minutes. In the presence of the enzymes in the endoplasmic reticular wall, the simple protein molecules are often folded, and also modified in other ways. In addition, most of them are rapidly conjugated with carbohydrate moieties to form glycoproteins.

Thus, the function of the granular reticulum is to form, in association with the ribosomes, protein products that are to be secreted by the endoplasmic reticulum and Golgi complex. It should be noted, however, that some ribosomes float in the cytoplasm, not attached to the endoplasmic reticulum, and these extrude their protein molecules directly into the cytoplasm.

Synthesis of Lipids, and Other Functions of the Smooth Endoplasmic Reticulum. The smooth endoplasmic reticulum also synthesizes substances, mainly lipids rather than proteins, including phospholipids and cholesterol. The phospholipids and cholesterol are rapidly incorporated into the lipid bilayer of the endoplasmic reticulum itself, thus causing the smooth portion of the endoplasmic reticulum to grow continually. However, small vesicles continually break away from the smooth endoplasmic reticulum; we shall see later that these vesicles mainly migrate rapidly to the Golgi apparatus.

Other significant functions of the smooth endoplasmic reticulum are:

(1) It contains the enzymes that control glycogen breakdown when glycogen is to be used for energy.

(2) It contains a vast number of enzymes that are capable of detoxifying substances that are damaging to the cell, such as drugs. It achieves this by coagulation, oxidation, hydrolysis, conjugation with glycuronic acid, and in other ways.

(3) It can synthesize a few carbohydrate moieties that are usually conjugated with protein molecules to form glycoproteins.

Synthetic Functions of the Golgi Complex. Though the major function of the Golgi complex is to process substances already formed in the endoplasmic reticulum, it also has the capability of synthesizing certain carbohydrates that cannot be formed in the endoplasmic reticulum. This is especially true of sialic acid, fructose, and galactose. In addition, it can cause the formation of saccharide polymers, the most important of which are hyaluronic acid and chondroitin sulfate. A few of the many functions of hyaluronic acid and chondroitin sulfate in the body are: (1) They are the major components of proteoglycans secreted in mucus and other glandular secretions. (2) They are the major components of the ground substance in the interstitial spaces, acting as a filler between collagen fibers and cells.

(3) They are principal components of the organic matrix in both cartilage and bone.

Processing of Endoplasmic Secretions by the Golgi Complex — Formation of Intracellular Vesicles. Figure 2–15 summarizes the major functions of the endoplasmic reticulum and Golgi complex, and also shows the formation of secretory vesicles by the Golgi complex. As substances are formed in the endoplasmic reticulum, especially the proteins, they are transported through the tubules toward the portions of the smooth endoplasmic reticulum that lie nearest the Golgi complex. At this point small "transport" vesicles of smooth endoplasmic reticulum continually break away and diffuse to the *proximal layers* of the Golgi complex, carrying inside the vesicles the synthesized proteins and other products. These vesicles instantly fuse with the Golgi complex, and their contained substances enter the vesicular spaces of the Golgi complex. Here, a few additional carbohydrate moieties are usually added to the secretions, but usually the function of the Golgi complex is mainly to compact the endoplasmic reticular secretions into highly concentrated packets. As the secretions pass toward the distal layers of the Golgi complex the compaction and processing proceed, and finally at the distal layer both small and large vesicles continually break away from the Golgi complex, carrying with them the compacted secretory substances, and they then diffuse throughout the cell.

To give one an idea of the timing of these processes, when a glandular cell is bathed in radioactive amino acids, newly formed radioactive protein molecules can be detected in the granular endoplasmic reticulum within 3 to 5 minutes. Within 20 minutes the newly formed proteins are present in the Golgi complex, and within 1 to 2 hours radioactive proteins are secreted from the surface of the cell.

Types of Vesicles Formed by the Golgi Complex — Secretory Vesicles, Lysosomes, Peroxisomes. In a highly secretory cell, the vesicles that are formed by the Golgi complex are mainly *secretory vesicles,* containing especially the protein substances that are to be secreted through the surface of the cell. These vesicles diffuse to the surface, fuse with the cell membrane, and empty their substances to the exterior by a mechanism called *exocytosis,* which is essentially the opposite of endocytosis. Exocytosis, in many cases, is stimulated by entry of calcium ions into the cell.

On the other hand, some of the vesicles are destined for intracellular use. Specialized portions of the Golgi complex form the *lysosomes* that have already been discussed, and other portions form so-called *peroxisomes,* which can both form hydrogen peroxide and also destroy hydrogen peroxide by the action of large quantities of the enzyme *catalase*. The destruction of hydrogen peroxide is essential to the life of the cell because many metabolic processes form hydrogen peroxide, and yet its continued presence in the cell is highly toxic to many of the other enzyme systems.

Use of Intracellular Vesicles to Replenish Cellular Membranes. Many of the vesicles finally fuse with the cell membrane or with the membranes of other intracellular structures such as the mitochondria and even the endoplasmic reticulum itself. This obviously increases the expanse of these membranes and thereby replenishes these membranes as they themselves are destroyed. For instance, the cell membrane loses much of its substance every time it forms a phagocytic or pinocytic vesicle, and it is vesicles from the Golgi complex that continually replenish the cell membrane.

Thus, in summary, the membranous system of the endoplasmic reticulum and the Golgi complex represents a highly metabolic organ capable of forming both new cellular structures and secretory substances to be extruded from the cell.

CELL MOVEMENT

By far the most important type of cell movement that occurs in the body is that of the specialized muscle cells in skeletal, cardiac, and smooth muscle, which comprise almost 50 per cent of the entire body mass. The specialized functions of these cells will be discussed in Chapters 11 through 13. However, two other types of movement occur in other cells, *ameboid movement* and *ciliary movement.*

Ameboid Motion. Ameboid motion means movement of an entire cell in relation to its surroundings, such as the movement of white blood cells through tissues. Typically ameboid motion begins with protrusion of a *pseudopodium* from one end of the cell. The pseudopodium projects far out away from the cell body, and then the remainder of the cell moves toward

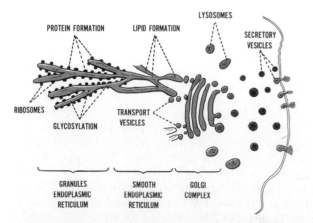

Figure 2–15. Formation of proteins, lipids, and cellular vesicles by the endoplasmic reticulum and Golgi complex.

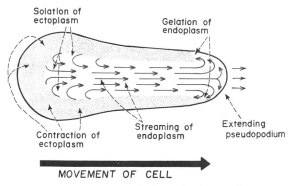

Figure 2–16. Ameboid motion by a cell.

the pseudopodium. Formerly, it was believed that the protruding pseudopodium attached itself far away from the cell and then pulled the remainder of the cell toward it. However, recent studies have changed this idea to a "streaming" concept, as follows: Figure 2–16 diagrams an elongated cell, the right-hand end of which is a protruding pseudopodium. The membrane of this end of the cell is continually moving forward, while the membrane at the left-hand end of the cell is continually following along as the cell moves.

It is believed that ameboid movement is caused in the following way: The outer portion of the cytoplasm is in a *gel* state and is called the *ectoplasm*, whereas the central portion of the cytoplasm is in a *sol* state and is called *endoplasm*. In the gel are numerous microfilaments composed of *actomyosin*, which is a highly contractile protein essentially the same as that found in muscle. These filaments contract in the presence of ATP and calcium ions. Therefore, normally there is a continual tendency for the ectoplasm to contract. However, in response to a chemical or physical stimulus the ectoplasm at one end of the cell becomes thin, causing a pseudopodium to bulge outward in the direction of the chemotaxic source. Thus, the pseudopodium moves progressively forward. However, this also causes contraction of the ectoplasm at the opposite end of the cell, which is accompanied by continuous solation — dissolution of the microfilaments — of the inner portions of the ectoplasm; then the dissolved ectoplasm "streams" forward through the endoplasm toward the pseudopodium. Thus, the combination of contraction of the ectoplasm, as well as solation of its inner layers, forces a stream of endoplasm forward toward the pseudopodium, pushing the membrane of the pseudopodium progressively forward. On reaching the pseudopodial end of the cell, the endoplasm turns toward the sides of the cell and the microfilaments reform from their constituent molecules, thus forming new ectoplasm. Therefore, at the tail end of the cell, ectoplasm is continually being solated while new ectoplasm is being formed at the pseudopodial end. The continuous repetition of this process makes the cell move in the direction in which the pseudopodium projects. One can readily see that this streaming movement inside the cell is analogous to the revolving movement of the track of a bulldozer.

Types of Cells That Exhibit Ameboid Motion. The most common cells to exhibit ameboid motion in the human body are the *white blood cells* moving out of the blood into the tissues in the form of *tissue macrophages* or *microphages*. However, many other types of cells can move by ameboid motion under certain circumstances. For instance, fibroblasts will move into any damaged area to help repair the damage, and even some of the germinal cells of the skin, though ordinarily completely sessile cells, will move by ameboid motion toward a cut area to repair the rent. Finally, ameboid motion is especially important in the development of the fetus, for embryonic cells often migrate long distances from the primordial sites of origin to new areas during the development of special structures.

Control of Ameboid Motion — "Chemotaxis." The most important factor that usually initiates ameboid motion, presumably by causing potential changes in the cell membrane or by the entry of ions into the cell — calcium ions as a possibility — is the appearance of certain chemical substances in the tissues. This phenomenon is called *chemotaxis,* and the chemical substance causing it to occur is called a *chemotaxic substance.* Most ameboid cells move toward the source of the chemotaxic substance — that is, from an area of lower concentration toward an area of higher concentration — which is called *positive chemotaxis.* However, some cells move away from the source, which is called *negative chemotaxis.*

Movement of Cilia. A second type of cellular motion, *ciliary movement,* is the movement of cilia on the surface of cells in the respiratory tract and in the fallopian tubes of the reproductive tract. As illustrated in Figure 2–17, a cilium looks like a minute, sharp-pointed hair that projects 3 to 4 microns from the surface of the cell. Many cilia can project from a single cell.

The cilium is covered by an outcropping of the cell membrane, and it is supported by 11 microtubules, 9 double tubules located around the periphery of the cilium and 2 single tubules down the center, as shown in the cross-section illustrated in Figure 2–17. Each cilium is an outgrowth of a structure that lies immediately beneath the cell membrane, called the *basal body* of the cilium.

In the inset of Figure 2–17 movement of the cilium is illustrated. The cilium moves forward with a sudden rapid stroke, 10 to 60 times per second, bending sharply where it projects from the surface of the cell. Then it moves backward very slowly in a whiplike manner. The rapid forward movement pushes the fluid lying adjacent to the cell in the direction that the cilium moves, then the slow whiplike movement in the other direction has almost no effect on the fluid. As a result, fluid is continually propelled in the direction of the forward stroke. Since most ciliated cells have large members of cilia on their surfaces, and since all the cilia are oriented in the same direction, this is a very effective means for moving fluids from one part of the surface to another; for instance, for moving mucus out of the lungs or for moving the ovum along the fallopian tube.

Mechanism of Ciliary Movement. Though all aspects of ciliary movement are not yet clear, we do know the

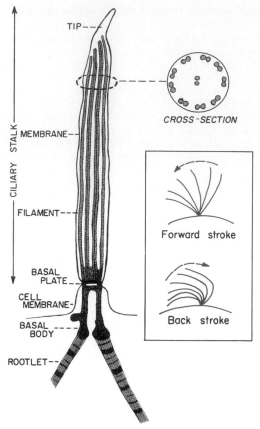

Figure 2–17. Structure and function of the cilium. (Modified from Satir: *Sci. Am., 204*:108, 1961. ©1961 by Scientific American, Inc. All rights reserved.)

Since many cilia on a cell surface contract simultaneously in a wavelike manner, it is presumed that some synchronizing signal — perhaps an electrochemical signal over the cell surface — is transmitted from cilium to cilium. This is not necessary, however, to cause the actual beating of the cilia because this will occur even in the absence of the membrane. The ATP required for the ciliary movements is provided by mitochondria near the bases of the cilia from which the ATP diffuses into the cilia.

Reproduction of Cilia. Cilia have the peculiar ability to reproduce themselves. This is achieved by the *basal body,* which is almost identical to the centriole, an important structure in the reproduction of whole cells, as we shall see in the next chapter. The basal body, as does the centriole, has the ability to reproduce itself by means not yet understood. After it reproduces itself, the new basal body then grows an additional cilium from the surface of the cell.

REFERENCES

Albrecht-Buehler, G.: The tracks of moving cells. *Sci. Am., 238*(4):68, 1978.

Allen, R. D., and Allen, N. S.: Cytoplasmic streaming in amoeboid movement. *Annu. Rev. Biophys. Bioeng., 7*:469, 1978.

Allison, A. C., and Davies, P.: Mechanisms of endocytosis and exocytosis. *Symp. Soc. Exp. Biol. (28)*:419, 1974.

Andresen, C. C.: Endocytosis in freshwater amebas. *Physiol. Rev., 57*:371, 1977.

Blake, J. R., and Sleigh, M. A.: Mechanics of ciliary locomotion. *Biol. Rev., 49*:85, 1974.

Bulger, R. E., and Strum, J. M.: The Functioning Cytoplasm. New York, Plenum Press, 1974.

Capaldi, R. A.: A dynamic model of cell membranes. *Sci. Am., 230*(3): 26, 1974.

Capaldi, R. A. (ed.): Membrane Proteins in Energy Transduction. New York, Marcel Dekker, 1979.

Chance, B., *et al.*: Hydroperoxide metabolism in mammalian organs. *Physiol. Rev., 59*:527, 1979.

Cherkin, A., *et al.* (eds.): Physiology and Cell Biology of Aging. New York, Raven Press, 1979.

de Brabander, M., *et al.* (eds.): Cell Movement and Neoplasia. New York, Pergamon Press, 1980.

De Duve, C., and Wattiaux, R.: Functions of lysosomes. *Annu. Rev. Physiol., 28*:435, 1966.

De Robertis, E. D. P., *et al.*: Cell Biology, 6th Ed. Philadelphia, W. B. Saunders Co., 1975.

Dingle, J. T.: Lysosomes in Biology and Pathology. New York, American Elsevier Publishing Co., 1973.

Ebe, T., and Kobayashi, S.: Fine Structure of Human Cells and Tissues. New York, John Wiley & Sons, 1973.

Fawcett, D. W.: The Cell. Philadelphia, W. B. Saunders Co., 1966.

Flickinger, C. J., *et al.*: Medical Cell Biology. Philadelphia, W. B. Saunders Co., 1979.

Fowler, S., and Wolinsky, H.: Lysosomes in vascular smooth muscle cells. *In* Bohr, D. F., *et al.* (eds.): *Handbook of Physiology.* Sec. 2, Vol. II. Baltimore, Williams & Wilkins Co., 1980, p. 133.

Gersh, I. (ed.): Submicroscopic Cytochemistry. New York, Academic Press, 1973.

Giese, A. C.: Cell Physiology, 5th Ed. Philadelphia, W. B. Saunders Co., 1979.

Goldman, R. D., *et al.*: Cytoplasmic fibers in mammalian cells: Cytoskeletal and contractile elements. *Annu. Rev. Physiol., 41*:703, 1979.

Hagiwara, S., and Jaffe, L. A.: Electrical properties of egg cell membranes. *Annu. Rev. Biophys. Bioeng., 8*:385, 1979.

Hammersen, F.: Histology: A Color Atlas of Cytology, Histology, and Microscopic Anatomy. Baltimore, Urban & Schwarzenberg, 1980.

Harris, A. K.: Cell surface movements related to cell locomotion. *Ciba Found. Symp., 14*:3, 1973.

Harris, H.: Nucleus and Cytoplasm, 3rd Ed. New York, Oxford University Press, 1974.

following: First, the nine double tubules and the two single tubules are all linked to each other by a complex of protein cross-linkages; this total complex of tubules and cross-linkages is called the *axoneme.* Second, even after removal of the membrane and destruction of other elements of the cilium besides the axoneme, the cilium can still beat under appropriate conditions. Third, there are two necessary conditions for continued beating of the axoneme after removal of the other structures of the cilium: (1) the presence of ATP, and (2) appropriate ionic conditions, including, especially, appropriate concentrations of magnesium and calcium. Fourth, the cilium will continue to beat even after it has been removed from the cell body. Fifth, the tubules on the front edge of the bending cilium slide outward toward the tip of the cilium while the tubules on the back edge of the bending cilium remain in place. Sixth, three protein arms with ATPase activity project from each set of peripheral tubules toward the next set.

Given the above basic information, it has been postulated that the release of energy from ATP in contact with the ATPase arms causes the arms to "crawl" along the surfaces of the adjacent pair of tubules. If the front tubules crawl outward while the back tubules remain stationary, this obviously will cause bending.

Hayflick, L.: The cell biology of human aging. *Sci. Am., 242*(1):58. 1980.

Hinkle, P. C., and McCarty, R. E.: How cells make ATP. *Sci. Am., 238*(3):104, 1978.

Inoue, S., and Stephens, R. E. (eds.): Molecules and Cell Movement. New York, Raven Press, 1975.

Jakoby, W. B., and Pastan, I. H. (eds.): Cell Culture. New York, Academic Press, 1979.

Kaplan, D. M., and Criddle, R. S.: Membrane structural proteins. *Physiol. Rev., 51*:249, 1971.

Koshland, D. E., Jr.: Bacterial chemotaxis in relation to neurobiology. *Annu. Rev. Neurosci., 3*:43, 1980.

Lenard, J.: Virus envelopes and plasma membranes. *Annu. Rev. Biophys. Bioeng., 7*:139, 1978.

Lodish, H. F., and Rothman, J. E.: The assembly of cell membranes. *Sci. Am., 240*(1):48, 1979.

Magnusson, S., *et al.* (eds.): Regulatory Proteolytic Enzymes and Their Inhibitors. New York, Pergamon Press, 1978.

Marchesi, V. T., *et al.* (eds.): Cell Surface Carbohydrates and Biological Recognition. New York, A. R. Liss, 1978.

Markham, R., *et al.* (eds.): The Generation of Subcellular Structures. New York, American Elsevier Publishing Co., 1973.

Masters, C., and Holmes, R.: Peroxisomes: New aspects of cell physiology and biochemistry. *Physiol. Rev., 57*:816, 1977.

Metcalfe, J. C. (ed.): Biochemistry of Cell Walls and Membranes II. Baltimore, University Park Press, 1978.

Nicholls, P. (ed.): Membrane Proteins. New York, Pergamon Press, 1978.

Reid, E. (ed.): Plant Organelles. New York, Halsted Press, 1979.

Satir, B.: The final steps in secretion. *Sci. Am., 233*(4):28, 1975.

Satir, P.: How cilia move. *Sci. Am., 231*(4):44, 1974.

Singer, S. J.: The molecular organization of membranes. *Annu. Rev. Physiol., 43*:805, 1974.

Singer, S. J., and Nicolson, G. L.: The fluid mosaic model of the structure of cell membranes. *Science, 175*:720, 1972.

Sloane, B. F.: Isolated membranes and organelles from vascular smooth muscle. *In* Bohr, D. F., *et al.* (eds.): Handbook of Physiology, Sec. 2, Vol. II. Baltimore, Williams & Wilkins Co., 1980, p. 121.

Staehelin, L. A., and Hull, B. E.: Junctions between living cells. *Sci. Am., 238*(5):140, 1978.

Stephens, R. E., and Edds, K. T.: Microtubules: Structure, chemistry, and function. *Physiol. Rev., 56*:709, 1976.

Toporek, M.: Basic Chemistry of Life, 2nd. Ed. New York, Appleton-Century-Crofts, 1975.

Tseng, H.: Atlas of Ultrastructure. New York, Appleton-Century-Crofts, 1980.

Wallach, D. F. H.: The Plasma Membrane: Dynamic Perspectives, Genetics, and Pathology. New York, Springer-Verlag, 1975.

Wallach, D. F. H.: Plasma Membranes and Disease. New York, Academic Press, 1979.

Wilkinson, P. C.: Chemotaxis and Inflammation. New York, Churchill Livingstone, 1973.

Williamson, J. R.: Mitochondrial function in the heart. *Annu. Rev. Physiol., 41*:485, 1979.

3

Genetic Control of Cell Function — Protein Synthesis and Cell Reproduction

Almost everyone knows that the genes control heredity from parents to children, but most persons do not realize that the same genes control the reproduction of and the day-by-day function of all cells. The genes control function of the cell by determining what substances will be synthesized within the cell — what structures, what enzymes, what chemicals.

Figure 3–1 illustrates the general schema by which the genes control cellular function. Each gene, which is a nucleic acid called *deoxyribonucleic acid (DNA),* automatically controls the formation of another nucleic acid, *ribonucleic acid (RNA),* which spreads throughout the cell and controls the formation of a specific protein. Some proteins are *structural proteins* which, in association with various lipids, form the structures of the various organelles that were discussed in the preceding chapter. But by far the majority of the proteins are *enzymes* that catalyze the different chemical reactions in the cells. For instance, enzymes promote all the oxidative reactions that supply energy to the cell, and they promote the synthesis of various chemicals such as lipids, glycogen, adenosine triphosphate, etc.

For the formation of most cellular proteins there is only one gene pair — that is, one gene on each chromosome of a chromosomal pair — in the cell. If all the DNA of the cell functioned as genes there could theoretically be more than 2 million of these. However, since most of the DNA subserves other functions besides constituting genes, it has been estimated that there is probably a maximum of about 100,000 functional genes in the human cell, which means that there could be as many as 100,000 types of protein formed in different cells, though not all of these necessarily in the same cell for reasons that we will discuss later in the chapter.

THE GENES

The genes are contained, large numbers of them attached end on end, in long, double-stranded, helical molecules of *deoxyribonucleic acid (DNA)* having molecular weights usually measured in the millions or even billions. A very short segment of such a molecule is illustrated in Figure 3–2. This molecule is composed of several simple chemical compounds arranged in a regular pattern explained in the following few paragraphs.

The Basic Building Blocks of DNA. Figure 3–3 illustrates the basic chemical compounds involved in the formation of DNA. These include (1) *phosphoric acid,*

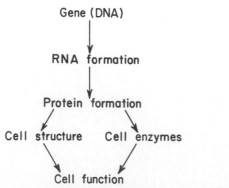

Gene (DNA)

RNA formation

Protein formation

Cell structure   Cell enzymes

Cell function

Figure 3–1. General schema by which the genes control cell function.

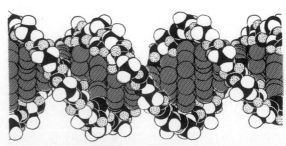

Figure 3–2. The helical, double-stranded structure of the gene. The outside strands are composed of phosphoric acid and the sugar deoxyribose. The internal molecules connecting the two strands of the helix are purine and pyrimidine bases; these determine the ''code'' of the gene.

PHOSPHORIC ACID:

DEOXYRIBOSE:

BASES:

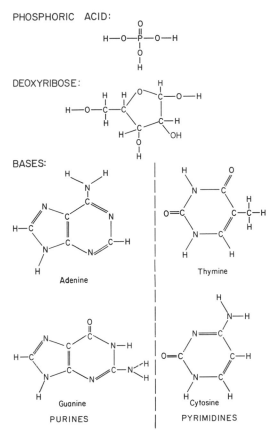

Adenine

Thymine

Guanine

Cytosine

PURINES

PYRIMIDINES

Figure 3–3. The basic building blocks of DNA.

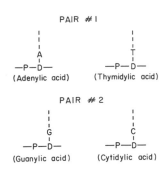

PAIR #1

A
—P—D—
(Adenylic acid)

T
—P—D—
(Thymidylic acid)

PAIR #2

G
—P—D—
(Guanylic acid)

C
—P—D—
(Cytidylic acid)

Figure 3–5. Combinations of the basic building blocks of DNA to form nucleotides. (P, phosphoric acid; D, deoxyribose. The four nucleotide bases are A, adenine; T, thymine; G, guanine; and C, cytosine.) Note that four different types of nucleotides make up DNA.

3 4 illustrates the chemical structure of adenylic acid, and Figure 3–5 illustrates simple symbols for all the four basic nucleotides that form DNA.

Note also in Figure 3–5 that the nucleotides are separated into two *complementary pairs:* (1) Adenylic acid and thymidylic acid form one pair. (2) Guanylic acid and cytidylic acid form the other pair. The *bases* of each pair can attach loosely (by hydrogen bonding) to each other, thus providing the means by which the two strands of the DNA helix are bound together: one nucleotide of a pair is on one strand of DNA, the other nucleotide of that pair is in a corresponding position on the other strand, and these are bound together by loose and reversible hydrogen bonds between the nucleotide bases.

Organization of the Nucleotides to Form DNA. Figure 3–6 illustrates the manner in which multiple numbers of nucleotides are bound together to form DNA. Note that these are combined in such a way that phosphoric acid and deoxyribose alternate with each other in the two separate strands, and these strands are held together by the respective complementary pairs of bases. Thus, in Figure 3–6 the sequence of complementary pairs of bases is CG, CG, GC, TA, CG, TA, GC, AT, and AT. However, the bases are bound together by very loose hydrogen bonding, represented in the figure by dashed lines. Because of the looseness of these bonds, the two strands can pull apart with ease, and they do so many times during the course of their function in the cell.

Now, to put the DNA of Figure 3–6 into its proper physical perspective, one needs merely to pick up the two ends and twist them into a helix. Ten pairs of

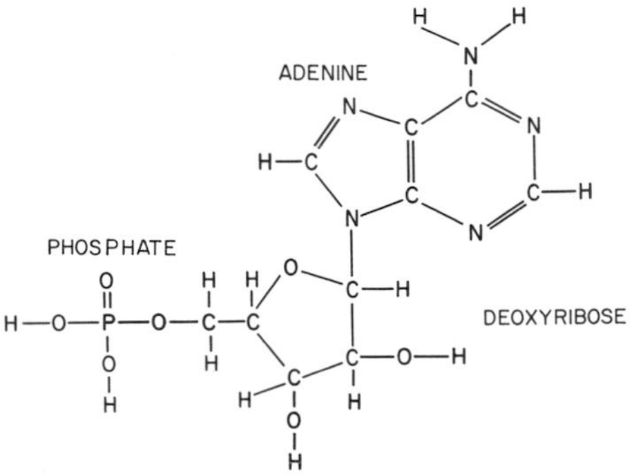

(2) a sugar called *deoxyribose,* and (3) four nitrogenous bases (two purines, *adenine* and *guanine,* and two pyrimidines, *thymine* and *cytosine*). The phosphoric acid and deoxyribose form the two helical strands of DNA, and the bases lie between the strands and connect them together.

The Nucleotides. The first stage in the formation of DNA is the combination of one molecule of phosphoric acid, one molecule of deoxyribose, and one of the four bases to form a nucleotide. Four separate nucleotides are thus formed, one for each of the four bases: *adenylic, thymidylic, guanylic,* and *cytidylic acids.* Figure

ADENINE

PHOSPHATE

DEOXYRIBOSE

Figure 3–4. Adenylic acid, one of the nucleotides that make up DNA.

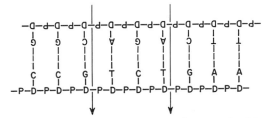

Figure 3–6. Arrangement of deoxyribose nucleotides in DNA.

nucleotides are present in each full turn of the helix in the DNA molecule, as illustrated in Figure 3–2.

THE GENETIC CODE

The importance of DNA lies in its ability to control the formation of other substances in the cell. It does this by means of the so-called genetic code. When the two strands of a DNA molecule are split apart, this exposes the purine and pyrimidine bases projecting to the side of each strand. It is these projecting bases that form the code.

Research studies in the past few years have demonstrated that the so-called *code words* consist of "triplets" of bases — that is, each three successive bases are a code word. And the successive code words control the sequence of amino acids in a protein molecule during its synthesis in the cell. Note in Figure 3–6 that each of the two strands of the DNA molecule carries its own genetic code. For instance, the top strand, reading from left to right, has the genetic code GGC, AGA, CTT, the code words being separated from each other by the arrows. As we follow this genetic code through Figures 3–7 and 3–8, we shall see that these three code words are responsible for placement of the three amino acids, *proline, serine,* and *glutamic acid,* in a molecule of protein. Furthermore, these three amino acids will be lined up in the protein molecule in exactly the same way that the genetic code is lined up in this strand of DNA.

RIBONUCLEIC ACID (RNA) – THE PROCESS OF TRANSCRIPTION

Since almost all DNA is located in the nucleus of the cell and yet most of the functions of the cell are carried out in the cytoplasm, some means must be available for the genes of the nucleus to control the chemical reactions of the cytoplasm. This is achieved through the intermediary of another type of nucleic acid, ribonucleic acid (RNA), the formation of which is controlled by the DNA of the nucleus, which is the process called *transcription*. The RNA is then transported into the cytoplasmic cavity where it controls protein synthesis.

Three separate types of RNA are important to protein synthesis: *messenger RNA, transfer RNA,* and *ribosomal RNA.* Before we describe the functions of these different RNAs in the synthesis of proteins, let us see how DNA controls the formation of RNA.

Synthesis of RNA. One strand of the DNA molecule, which contains the genes, acts as a template for synthesis of RNA molecules. (The other strand of the DNA has no genetic function but does function for replication of the gene itself, which will be discussed later in the chapter.) The code words in DNA cause the formation of *complementary* code words (or *codons*) in RNA. The stages of RNA synthesis are as follows:

The Basic Building Blocks of RNA. The basic building blocks of RNA are almost the same as those of DNA except for two differences. First, the sugar deoxyribose is not used in the formation of RNA. In its place is another sugar of very slightly different composition, *ribose*. Second, thymine is replaced by another pyrimidine, *uracil*.

Formation of RNA Nucleotides. The basic building blocks of RNA first form nucleotides exactly as described above for the synthesis of DNA. Here again, four separate nucleotides are used in the formation of RNA. These nucleotides contain the bases *adenine, guanine, cytosine,* and *uracil,* respectively, the uracil replacing the thymine found in the four nucleotides that make up DNA.

Activation of the Nucleotides. The next step in the synthesis of RNA is activation of the nucleotides. This occurs by addition to each nucleotide of two phosphate radicals to form triphosphates. These last two phosphates are combined with the nucleotide by *high-energy phosphate bonds* derived from the ATP of the cell.

The result of this activation process is that large quantities of energy are made available to each of the nucleotides, and this energy is used in promoting the subsequent chemical reactions that eventuate in the formation of the RNA chain.

Assembly of the RNA Molecule from Activated Nucleotides Using the DNA Strand as a Template — the Process of Transcription. The next stage in the formation of RNA is separation of the two strands of the DNA molecule. Then, one of these strands is used as a template on which the RNA molecule is assembled. It is this strand that contains the genes while the other strand remains genetically inactive. Assembly of the RNA molecule is accomplished in the manner illustrated in

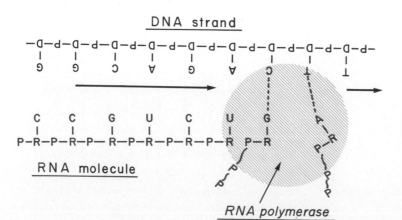

Figure 3–7. Combination of ribose nucleotides with a strand of DNA to form a molecule of ribonucleic acid (RNA) that carries the DNA code from the gene to the cytoplasm.

DNA strand

RNA molecule

RNA polymerase

Figure 3–7 under the influence of the enzyme *RNA polymerase*. The steps of this procedure are the following:

(1) First, an activated RNA nucleotide is temporarily bonded with the starting base in the DNA strand, then another nucleotide to the next base, and so forth.

(2) Each successive nucleotide is pulled into place so that it lies adjacent to the previous nucleotide.

(3) The pyrophosphate radical of each new nucleotide breaks away and in doing so liberates enough energy to cause an ester linkage between the remaining phosphate on the nucleotide and the ribose on the end of the growing RNA molecule.

(4) Immediately after bonding of the ribose and phosphate, the base of the previous RNA nucleotide breaks its bond with the DNA strand.

(5) Then the next RNA-activated nucleotide is added to the RNA molecule in the same manner.

It should be remembered that there are four different types of DNA bases and also four different types of RNA nucleotide bases. Furthermore, these always combine with each other in specific combinations. Therefore, the code that is present in the DNA strand is transmitted in *complementary* form to the RNA molecule. The ribose nucleotide bases always combine with the deoxyribose bases in the following combinations:

DNA base	RNA base
guanine	cytosine
cytosine	guanine
adenine	uracil
thymine	adenine

Once the RNA molecules are formed, they diffuse out of the nucleus and into all parts of the cytoplasm where they perform further functions. The type of RNA called *messenger RNA* carries the genetic code to the cytoplasm for formation of proteins. The *ribosomal RNA* is used for the formation of ribosomes, which is the physical and chemical structure on which protein molecules are actually assembled. *Transfer RNA* is utilized to carry activated amino acids to the ribosomes where the protein molecules are assembled.

MESSENGER RNA

Messenger RNA molecules are long straight strands that are suspended in the cytoplasm. These molecules are usually composed of several hundred to several thousand nucleotides in unpaired strands, and they contain *codons* that are exactly complementary to the code words of the genes. Figure 3–8 illustrates a small segment of a molecule of messenger RNA. Its codons are CCG, UCU, and GAA. These are the codons for

TABLE 3–1 RNA CODONS FOR THE DIFFERENT AMINO ACIDS AND FOR START AND STOP

Amino Acid	RNA Codons					
Alanine	GCU	GCC	GCA	GCG		
Arginine	CGU	CGC	CGA	CGG	AGA	AGG
Asparagine	AAU	AAC				
Aspartic acid	GAU	GAC				
Cysteine	UGU	UGC				
Glutamic acid	GAA	GAG				
Glutamine	CAA	CAG				
Glycine	GGU	GGC	GGA	GGG		
Histidine	CAU	CAC				
Isoleucine	AUU	AUC	AUA			
Leucine	CUU	CUC	CUA	CUG	UUA	UUG
Lysine	AAA	AAG				
Methionine	AUG					
Phenylalanine	UUU	UUC				
Proline	CCU	CCC	CCA	CCG		
Serine	UCU	UCC	UCA	UCG		
Threonine	ACU	ACC	ACA	ACG		
Tryptophan	UGG					
Tyrosine	UAU	UAC				
Valine	GUU	GUC	GUA	GUG		
Start (CI)	AUG	GUG				
Stop (CT)	UAA	UAG	UGA			

proline, serine, and glutamic acid. The transcription of these codons from the DNA molecule was demonstrated in Figure 3–7.

RNA Codons. Table 3–1 gives the RNA codons for the 20 common amino acids found in protein molecules. Note that several of the amino acids are represented by more than one codon; some codons represent such signals as "start manufacturing a protein molecule" or "stop manufacturing a protein molecule." In Table 3–1, these two codons are designated CI for "chain-initiating" and CT for "chain-terminating."

TRANSFER RNA

Another type of RNA that plays a prominent role in protein synthesis is called *transfer RNA* because it transfers amino acid molecules to protein molecules as the protein is synthesized. There are many different types of transfer RNA, but each type will combine specifically with only one of the 20 amino acids that are incorporated into proteins. The transfer RNA then acts as a *carrier* to transport its specific type of amino acid to the ribosomes where protein molecules are formed. In the ribosomes, each specific type of transfer RNA recognizes a particular codon on the messenger RNA, as is described below, and thereby delivers the appropriate amino acid to the appropriate place in the chain of the newly forming protein molecule.

Transfer RNA, containing only about 80 nucleotides, is a relatively small molecule in comparison with messenger RNA. It is a folded chain of nucleotides with a cloverleaf appearance similar to that illustrated in Figure 3–9. At one end of the molecule is always an adenylic acid; it is to this that the transported amino acid attaches to a hydroxyl group of the ribose in the adenylic acid. A specific enzyme causes this attachment

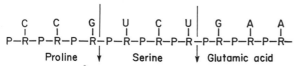

Figure 3–8. Portion of a ribonucleic acid molecule, showing three "code" words, CCG, UCU, and GAA, which represent the three amino acids *proline, serine,* and *glutamic acid.*

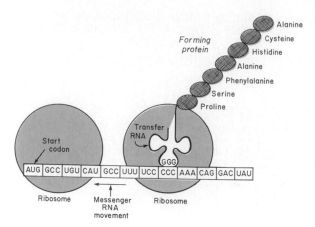

Figure 3–9. Postulated mechanism by which a protein molecule is formed in ribosomes in association with messenger RNA and soluble RNA.

for each specific type of transfer RNA; this enzyme also determines the type of amino acid that will attach to the respective type of transfer RNA.

Since the function of transfer RNA is to cause attachment of a specific amino acid to a forming protein chain, it is essential that each type of transfer RNA also have specificity for a particular codon in the messenger RNA. The specific prosthetic group in the transfer RNA that allows it to recognize a specific codon is called an *anticodon*, and this is located approximately in the middle of the transfer RNA molecule (at the bottom of the cloverleaf configuration illustrated in Figure 3–9). During formation of a protein molecule, the anticodon bases combine loosely by hydrogen bonding with the codon bases of the messenger RNA. In this way the respective amino acids are lined up one after another along the messenger RNA chain, thus establishing the appropriate sequence of amino acids in the protein molecule.

RIBOSOMAL RNA

The third type of RNA in the cell is ribosomal RNA; it constitutes about 60 per cent of the ribosome. The remainder of the ribosome is protein, containing as many as 50 different types of proteins which are both structural proteins and enzymes needed in the manufacture of protein molecules.

The ribosome is the physical and chemical structure in the cytoplasm on which protein molecules are actually synthesized. However, it always functions in association with both the other types of RNA as well: transfer RNA transports amino acids to the ribosomes for incorporation into the developing protein molecules, while messenger RNA provides the information necessary for sequencing the amino acids in proper order for each specific type of protein to be manufactured.

The ribosomes are composed of two physical subunits, called the 40S and 60S particles because of their respective rates of sedimentation in the ultra-centrifuge. Though we have only partial knowledge of the mechanism of protein manufacture in the ribosome, it is known that messenger RNA and transfer RNA first complex

with the 40S particle. Then it is believed that the 60S particle provides the enzymes that promote peptide linkages between the successive amino acids. Thus, the ribosome acts as a manufacturing plant in which the protein molecules are formed.

Formation of Ribosomes in the Nucleolus. The DNA molecules for formation of ribosomal RNA are all located in a single chromosomal pair of the nucleus. However, this chromosomal pair contains many duplicates of these ribosomal genes because of the large amount of ribosomal RNA required for cellular function.

As the ribosomal RNA forms, it collects in the *nucleolus,* a specialized structure lying adjacent to the chromosome. When large amounts of ribosomal RNA are being synthesized, as occurs in cells that manufacture large amounts of protein, the nucleolus is a very large structure, while in cells that synthesize very little protein the nucleolus may not be seen at all. The ribosomal RNA is specially processed in the nucleolus and combined with "ribosomal proteins" to form granular condensation products that are primordial forms of the ribosomes. These ribosomes are then released from the nucleolus, and they migrate through the large "pores" of the nuclear membrane to almost all parts of the cytoplasm.

FORMATION OF PROTEINS IN THE RIBOSOMES — THE PROCESS OF TRANSLATION

When a molecule of messenger RNA comes in contact with a ribosome, it travels through the ribosome, beginning at a predetermined end specified by an appropriate sequence of RNA bases. However, the protein molecule does not begin to form until a "start" (or "chain-initiating") codon enters the ribosome. Then, as illustrated in Figure 3–9, while the messenger RNA travels through the ribosome, a protein molecule is formed — a process called *translation*. Thus, the ribosome reads the code of the messenger RNA in much the same way that a tape is "read" as it passes through the playback head of a tape recorder. Then, when a "stop" (or "chain-terminating") codon slips past the ribosome, the end of a protein molecule is signaled, and the entire molecule is freed into the cytoplasm.

Polyribosomes. A single messenger RNA molecule can form protein molecules in several different ribosomes at the same time, the molecule passing through a successive ribosome as it leaves the first, as shown in Figure 3–9. The protein molecules obviously are in different stages of development in each ribosome. As a result, clusters of ribosomes frequently occur, three to eight ribosomes being attached together by a single messenger RNA at the same time. These clusters are called *polyribosomes.*

It is especially important to note that a messenger RNA can cause the formation of a protein molecule in any ribosome, and that there is no specificity of ribosomes for given types of protein. The ribosome seems to be simply the structure in which or on which the chemical reactions take place.

Attachment of Ribosomes to the Endoplasmic Reticulum. In the previous chapter it was noted that most of the ribosomes become attached to the endoplasmic

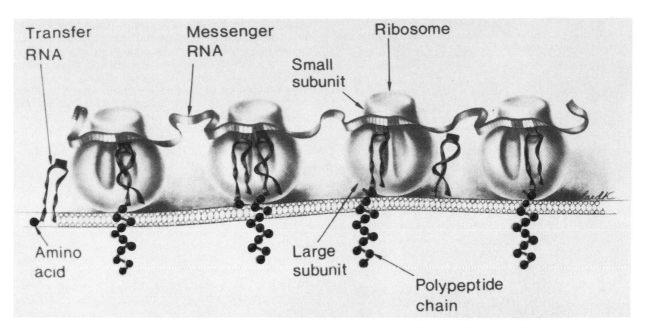

Figure 3–10. An artist's concept of the physical structure of the ribosomes as well as their functional relationship to messenger RNA, transfer RNA, and the endoplasmic reticulum during the formation of protein molecules. (From Bloom and Fawcett: A Textbook of Histology. 10th ed. Philadelphia, W. B. Saunders Company, 1975.)

reticulum. This does not occur until the ribosomes begin to form protein molecules. However, many of the protein molecules are immediately attracted to specific sites on the endoplasmic reticulum where they are transported into the endoplasmic reticular matrix. Indeed, this occurs while the protein molecule is still being formed by the ribosome. Therefore, most of the ribosomes in the cytoplasm become attached by the forming protein to the endoplasmic reticulum. This is particularly true in glandular cells that form large amounts of protein secretory vesicles, because essentially all these proteins are processed through the endoplasmic reticulum. However, some of the ribosomes do float freely in the cytoplasm and deliver their manufactured protein molecules in solution form into the cytoplasm itself.

Figure 3–10 shows the functional relationship of messenger RNA to the ribosomes and also the manner in which the ribosomes attach to the membrane of the endoplasmic reticulum. Note the process of translation occurring in several ribosomes at the same time in response to the same strand of messenger RNA. And note also the newly forming polypeptide chains passing through the endoplasmic reticulum membrane into the endoplasmic matrix, thus generating the protein molecules.

Chemical Steps in Protein Synthesis. Some of the chemical events that occur in synthesis of a protein molecule are illustrated in Figure 3–11. This figure shows representative reactions for three separate amino acids, AA_1, AA_2, and AA_{20}. The stages of the reactions are the following: (1) Each amino acid is *activated* by a chemical process in which ATP combines with the amino acid to form an *adenosine monophosphate complex with the amino acid*. (2) The activated amino acid, having an excess of energy, then *combines with its specific transfer RNA to form an amino acid–tRNA*

complex and, at the same time, releases the adenosine monophosphate. (3) The transfer RNA carrying the amino acid complex then comes in contact with the messenger RNA molecule in the ribosome where the anticodon of the transfer RNA attaches temporarily to its specific codon of the messenger RNA, thus lining up the amino acids in appropriate sequence to form a protein molecule. Then, energy from *guanosine triphosphate,* another high-energy phosphate substance almost identical to ATP, is utilized to form the peptide bonds between the successive amino acids, thus creating the protein.

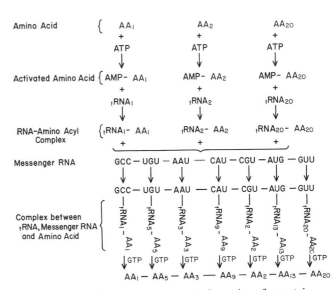

Figure 3–11. Chemical events in the formation of a protein molecule.

Peptide Linkage. The successive amino acids in the protein chain combine with each other according to the following typical reaction:

$$R-\overset{\overset{\displaystyle NH_2}{|}}{C}-\overset{\overset{\displaystyle O}{\|}}{C}-OH + H-\overset{\overset{\displaystyle H}{|}}{N}-\overset{\overset{\displaystyle R}{|}}{C}-COOH \rightarrow$$

$$R-\overset{\overset{\displaystyle NH_2}{|}}{C}-\overset{\overset{\displaystyle O}{\|}}{C}-\overset{\overset{\displaystyle H}{|}}{N}-\overset{\overset{\displaystyle R}{|}}{C}-COOH + H_2O$$

In this chemical reaction, a hydroxyl radical is removed from the COOH portion of one amino acid while a hydrogen of the NH_2 portion of the other amino acid is removed. These combine to form water, and the two reactive sites left on the two successive amino acids combine, resulting in a single molecule. This process is called *peptide linkage*.

SYNTHESIS OF OTHER SUBSTANCES IN THE CELL

Many thousand protein enzymes formed in the manner just described control essentially all the other chemical reactions that take place in cells. These enzymes promote synthesis of lipids, glycogen, purines, pyrimidines, and hundreds of other substances. We will discuss many of these synthetic processes in relation to carbohydrate, lipid, and protein metabolism in Chapters 67 through 69. It is by means of all these different substances that the many functions of the cells are performed.

CONTROL OF GENETIC FUNCTION AND BIOCHEMICAL ACTIVITY IN CELLS

There are basically two different methods by which the biochemical activities in the cell are controlled. One of these is called *genetic regulation*, in which the activities of the genes themselves are controlled, and the other *enzyme regulation*, in which the activity rates of the enzymes within the cell are controlled.

Strange as it may seem, every nucleated cell of the body contains the same set of genes as those originally present in the fertilized ovum from which the body was formed. Therefore, the vast differences in the different types of cells result not from different types of genes in the cells but instead from repression of different genes in different cells. This effect is called *differentiation* of the cells, a subject that we will discuss in more detail later in the chapter. The respective genes are repressed by the special intracellular genetic control mechanisms that allow expression of different functional characteristics in separate types of "differentiated" cells, some providing muscle activity, others glandular secretion, and still others the many other bodily functions.

Another important function of the intracellular control mechanisms is to control the rates of synthesis of the different intracellular chemicals. It would be unwise for a cell to continue forming biochemical products that are not needed in the cell. Or under other conditions, a cell may need an excess of one particular product but very little of others. For these reasons, a host of intracellular control systems is present in each cell for maintaining appropriate functional balance among the many biochemical synthetic processes.

GENETIC REGULATION

Gene function is controlled in several different ways. Some genes are normally dormant but can be activated by *inducer substances*. Other genes are naturally active but can be inhibited by *repressor substances*. As an illustration, let us describe one of the mechanisms for genetic control.

The Operon and Its Control of Biochemical Synthesis. The synthesis of a cellular biochemical product usually requires a series of reactions, and each of these reactions is catalyzed by a specific enzyme. Formation of all the enzymes needed for the synthetic process is in turn controlled by a sequence of genes all located in series one after the other on the same chromosomal DNA strand. This area of the DNA strand is called an *operon,* and the genes responsible for forming the respective enzymes are called the *structural genes*. In Figure 3–12, three respective structural genes are illustrated in an operon, and it is shown that they control the formation of three respective enzymes utilized in a particular biochemical synthetic process.

The rate at which the operon functions to transcribe RNA, and therefore to set into motion the enzymatic system for the biochemical process, is determined by the presence of two other small segments on the DNA strand called, respectively, the *promoter* and the *operator;* these are also shown in Figure 3–12. Each of these are specific sequences of DNA nucleotides, but they do not themselves serve as templates to cause the formation of RNA. Instead, they merely function as control units of the operon.

The promoter first binds with RNA polymerase, which is the first step for this polymerase to begin moving down along the operon to cause transcription of the appropriate messenger RNAs. However, lying between the promoter and the structural genes is the operator; this is a control gate that can be opened or closed. If the gate is open, the RNA polymerase will travel along the operon and initiate the transcription process; but if the gate is closed, the RNA polymerase becomes blocked at the promoter level and the operon remains dormant.

In Figure 3–12, it is shown that the presence of a critical amount of synthesized product in the cellular cytoplasm will cause negative feedback to the operator to inhibit it — that is, to close the gate. Therefore, whenever there is already enough of the required product, the operon becomes dormant. On the other hand, as the synthesized product becomes degraded in the cell and its concentration falls, the operator gate opens and the operon once again becomes active. In this way, the concentration of the synthesized product is automatically controlled.

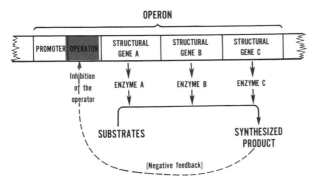

Figure 3–12. Function of the operon to control biosynthesis. Note that the synthesized product exerts a negative feedback to inhibit function of the operon, in this way automatically controlling the concentration of the product itself.

Other Mechanisms for Control of Transcription by the Operon. Variations in the basic mechanism for control of the operon have been discovered with rapidity in the last few years. Without going into detail for all these, let us merely list some of the mechanisms of control:

(1) Instead of the operator being inhibited by negative feedback from the synthesized product, it can be activated by an *inducer* derived from elsewhere within the cell or even from outside the cell. For instance, some of the steroid hormones perform their hormonal function by activating operons in this manner.

(2) Operons are frequently controlled by *regulatory genes* located elsewhere in the genetic complex of the nucleus, either on the same DNA strand as the operon or on one of the other strands. Most often the regulatory gene causes the formation of a small protein that in turn acts as a *repressor substance* that closes the operator gate and thereby turns off the operon. However, other substances can either inhibit or activate this repressor substance, thus opening or closing the gate. For instance, the operon that controls the metabolism of lactose is controlled in this manner. In the absence of lactose, a regulatory gene operating through a repressor substance inhibits the operator and thereby closes the gate. But, when lactose becomes available to the cell, it inhibits the repressor, thus opening the gate and activating the operon. And within an hour or so the cell has produced all the enzymes needed to metabolize the lactose. Thus, the lactose operates as an *inducer*.

(3) Other operons are controlled by increasing or decreasing the affinity of the promoter for RNA polymerase.

(4) Occasionally, many different operons are controlled all at the same time by the same inducer or by the same inhibitor. When this occurs, all of the operons that function together are called a *regulon*.

(5) Some synthetic processes are controlled not at the DNA level but instead at the RNA level to control the translation process for formation of proteins by messenger RNA. An example of this is the effect of excess quantities of the amino acid tryptophan in the cytoplasm to prevent translation of one or more of the protein enzymes required for tryptophan synthesis.

Because there are as many as 100,000 different genes in each cell, it is not surprising that there are a large number of different ways in which genetic activity itself can be controlled. It is through these control mechanisms, especially through negative feedback controls, that the cell is able to maintain appropriate concentrations of its necessary functional biochemicals. The genetic control systems are especially important for controlling the intracellular concentrations of amino acids, amino acid derivatives, and many if not most of the intermediary substrates of carbohydrate, lipid, and protein metabolism.

CONTROL OF ENZYME ACTIVITY

In the same way that inhibitors and activators can affect the genetic regulatory system, so also can the enzymes themselves be directly controlled by other inhibitors or activators. This, then, represents a second category of mechanisms by which cellular biochemical functions can be controlled.

Enzyme Inhibition. A great many of the chemical substances formed in the cell have a direct feedback effect to inhibit respective enzyme systems that synthesize them. Almost always the synthesized product acts on the first enzyme in a sequence, rather than on the subsequent enzymes. One can readily recognize the importance of inactivating this first enzyme: it prevents buildup of intermediary products that will not be utilized.

This process of enzyme inhibition is another example of negative feedback control; it is responsible for controlling the intracellular concentrations of some of the amino acids that are not controlled by the genetic mechanism as well as the concentrations of the purines, the pyrimidines, vitamins, and other substances.

Enzyme Activation. Enzymes that are either normally inactive or that have been inactivated by some inhibitor substance can often be activated. An example of this is the action of cyclic adenosine monophosphate (AMP) in causing glycogen to split so that energy derived from the released glucose molecules can be used to form high-energy ATP, as discussed in the previous chapter. When most of the adenosine triphosphate has been depleted in a cell, a considerable amount of cyclic AMP begins to be formed as a breakdown product of the ATP; the presence of this cyclic AMP indicates that the cellular reserves of ATP have approached a low ebb. However, the cyclic AMP immediately activates the glycogen-splitting enzyme phosphorylase, liberating glucose molecules that are rapidly used for replenishment of the ATP stores. Thus, in this case the cyclic AMP acts as an enzyme activator and thereby helps to control intracellular ATP concentration.

Another interesting instance of both enzyme inhibition and enzyme activation occurs in the formation of the purines and pyrimidines. These substances are needed by the cell in approximately equal quantities for formation of DNA and RNA. When purines are formed, they *inhibit* the enzymes that are required for formation of additional purines. However, they also *activate* the enzymes for formation of the pyrimidines. Conversely, the pyrimidines inhibit their own enzymes but activate the purine enzymes. In this way there is continual cross-feed between the synthesizing systems for these two substances, resulting in almost exactly equal amounts of the two substances in the cells at all times.

In summary, there are two different methods by which the cells control proper proportions and proper quantities of different cellular constituents: (1) the mechanism of genetic regulation, and (2) the mechanism of enzyme regulation. The genes can be either activated or inhibited, and, likewise, the enzymes can be either activated or inhibited. Most often, these regulatory mechanisms function as feedback control systems that continually monitor the cell's biochemical composition and make corrections as needed. But, on occasion, substances from without the cell (especially some of the hormones that will be discussed later in this text) also control the intracellular biochemical reactions by activating or inhibiting one or more of the intracellular control systems.

CELL REPRODUCTION

Cell reproduction is another example of the pervading, ubiquitous role that the DNA-genetic system plays in all life processes. It is the genes and their internal regulatory mechanisms that determine the growth characteristics of the cells, also when or whether these cells will divide to form new cells. In this way, this all-important genetic system controls each stage in the development of the human being from the single-cell fertilized ovum to the whole functioning body. Thus, if there is any central theme to life it is the DNA-genetic system.

As is true of almost all other events in the cell, reproduction also begins in the nucleus itself. The first step is *replication (duplication) of all DNA in the chromosomes.* The next step is *mitosis,* which consists, first, of division of the two sets of DNA between two separate nuclei and, second, splitting of the cell itself to form two new daughter cells.

The complete life cycle of a cell that is not inhibited in some way is about 10 to 30 hours from reproduction to reproduction, and the period of mitosis lasts for approximately half an hour. The period between mitoses is called *interphase.* However, in the body there are almost always inhibitory controls that slow or stop the uninhibited life cycle of the cell and give cells life cycle periods that vary from as little as 10 hours for stimulated bone marrow cells to an entire lifetime of the human body for nerve and striated muscle cells.

REPLICATION OF THE DNA

The DNA begins to be reproduced about 5 hours before mitosis takes place, and the duration of DNA replication is about four hours. The DNA normally is duplicated only once. The net result is two exact *replicates* of all DNA, which respectively become the DNA in the two new daughter cells that will be formed at mitosis. Following replication of the DNA, the nucleus continues to function normally for about one hour before mitosis begins abruptly.

Chemical and Physical Events. The DNA is duplicat-

ed in almost exactly the same way that RNA is formed from DNA. First, the two strands of the DNA helix of the gene pull apart. Second, each of these strands combines with deoxyribose nucleotides of the four types described early in the chapter as the basic building blocks of DNA. Each base on each strand of DNA in the chain attracts a nucleotide containing the appropriate *complementary* base. In this way, the appropriate nucleotides are lined up side by side. Third, enzyme mechanisms, employing *DNA polymerases* as well as several other enzymes, provide energy and cause linkage of the nucleotides to form a new DNA strand. However, a major difference between this formation of the new strand of DNA and the formation of an RNA strand is that the new strand of DNA remains attached to the old strand that has formed it, thus forming a new double-stranded DNA helix. Another major difference is that not only the DNA strand containing the genes but also the complementary DNA strand forms a new strand. Therefore, the original DNA double helix becomes two double helixes that are exact duplicates of each other, unless an error in replication occurs, which is called a *mutation.*

THE CHROMOSOMES AND THEIR REPLICATION

The chromosomes consist of two major parts: the DNA and protein. The protein, in turn, is composed mainly of small molecules. A large share of this is *histones* which probably serve to fold or otherwise compact the DNA strands into manageable sizes. On the other hand, the *nonhistone chromosomal proteins* are major components of the genetic regulatory system, acting as activators, inhibitors, and enzymes.

Recent experiments indicate that all the DNA of a particular chromosome is arranged in one long double helix and that the genes are attached end-on-end with each other. Such a molecule in the human being has a molecular weight of about 60 billion, and if spread out linearly, would be approximately 7.5 cm. long or several thousand times as long as the diameter of the nucleus itself; but the experiments also indicate that this long double helix is folded or coiled like a spring and is held in this position by its linkages to the histone molecules.

Replication of the chromosomes follows as a natural result of replication of the DNA strand. When the new double helix separates from the original double helix, it presumably carries some of the old protein with it or combines with new protein, the DNA acting as the backbone of the newly replicated chromosome.

Number of Chromosomes in the Human Cell. Each human cell contains 46 chromosomes arranged in 23 pairs. In general, the genes in the two chromosomes of each pair are identical or almost identical with each other, so that it is usually stated that the different genes exist in pairs, though occasionally this is not the case.

MITOSIS

The actual process by which the cell splits into two new cells is called mitosis. Once the genes have been duplicated and each chromosome has split to form two new chromosomes, each of which is now called a *chromatid*, mitosis follows automatically, almost without fail, within about an hour.

The Mitotic Apparatus. One of the first events of mitosis takes place in the cytoplasm, occurring during the latter part of interphase and early part of anaphase, in or around the small structures called *centrioles*. As illustrated in Figure 3–13, two pairs of centrioles lie close to each other near one pole of the nucleus. Each centriole is a small cylindrical body about 0.4 micron long and about 0.15 micron in diameter, consisting mainly of nine parallel, tubule-like structures arranged around the inner wall of the cylinder. The two centrioles of each pair lie at right angles to each other.

So far as is known, the two pairs of centrioles remain dormant during interphase until shortly before mitosis is to take place. At that time, the two pairs begin to move apart from each other. This is caused by protein microtubules growing between the respective pairs and actually pushing them apart. At the same time, microtubules grow radially away from each of the pairs, forming a spiny star, called the *aster*, in each end of the cell. Some of the spines penetrate the nucleus and will play a role in separating the two sets of DNA helixes, the chromatids, during mitosis. The set of microtubules connecting the two centriole pairs is called the *spindle*, and the entire set of microtubules plus the two pairs of centrioles is called the *mitotic apparatus*.

Prophase. The first stage of mitosis, called *prophase*, is shown in Figure 3–13A, B, and C. While the spindle is forming, the *chromatin material* of the nucleus (the DNA), which in interphase consists of long loosely coiled strands, becomes shortened into well-defined chromosomes.

Prometaphase. During this stage (Figure 3–13D) the nuclear envelope dissolutes, and microtubules from the forming mitotic apparatus become attached to the chromosomes. This attachment always occurs at the same point on each chromosome, at a small condensed portion called the *kinetochore*, located near the *centromere* where the two chromosomes of each pair are attached to each other.

Metaphase. During metaphase (Fig. 3–13E) the centriole pairs are pushed far apart by the growing spindle, and the chromosomes are thereby pulled tightly by the attached microtubules to the very center of the cell, lining up in the equatorial plane of the mitotic spindle.

Anaphase. With still further growth of the spindle, the chromatids in each pair of chromosomes are now broken apart, a stage of mitosis called anaphase (Fig. 3–13F). A microtubule connecting with one end of the mitotic spindle pulls one pair of chromatids, and a microtubule connecting with the other centriole pair pulls the opposite pair of chromatids. Thus, all 46 pairs of chromatids are separated, forming pairs of daughter chromosomes that are pulled toward one mitotic spindle and other pairs of duplicate chromosomes that are pulled toward the other mitotic spindle.

Telophase. In telophase (Fig. 3–13G and H) the mitotic spindle grows still longer, pulling the two sets of daughter chromosomes completely apart. Then the mitotic apparatus dissolutes and a new nuclear membrane develops around each set of chromosomes, this membrane being formed from portions of the endoplasmic reticulum that are already present in the cytoplasm. Shortly thereafter, the cell pinches in two midway between the two nuclei, for reasons mainly unexplained at present, except: (1) It is the two asters that initiate the cell division, each of the two daughter cells being composed of the portion of the cell occupied by a dissoluting aster. And (2) a contractile ring of *microfilaments* composed of *actin* and probably *myosin*, the two contractile proteins of muscle, develops at the juncture of the newly developing cells and pinches them off from each other.

Note, also, that each of the two pairs of centrioles are replicated during telophase. These new pairs of centrioles remain dormant through the next interphase until a mitotic apparatus is required for the next cell division.

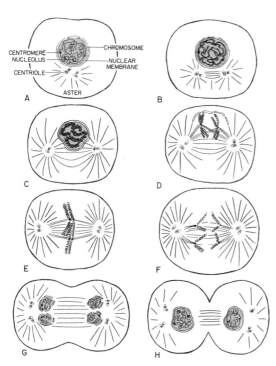

Figure 3–13. Stages in the reproduction of the cell. A, B, and C, prophase; D, prometaphase; E, metaphase; F, anaphase; G and H, telophase. (Redrawn from Mazia: *Sci. Am.*, *205*:102, 1961. © by Scientific American, Inc. All rights reserved.)

CONTROL OF CELL GROWTH AND REPRODUCTION

Cell growth and reproduction usually go together; growth normally leads to replication of the DNA of the nucleus, followed a few hours later by mitosis.

In the normal human body, regulation of cell growth and reproduction is mainly a mystery. We know that certain cells grow and reproduce all the time, such as the blood-forming cells of the bone marrow, the germinal layers of the skin, and the epithelium of the gut. However, many other cells, such as smooth muscle cells, do not reproduce for many years. And a few cells, such as the neurons and striated muscle cells, do not reproduce during the entire life of the person.

If there is an insufficiency of some types of cells in the body, these will grow and reproduce very rapidly until appropriate numbers of them are again available. For instance, seven-eighths of the liver can be removed surgically, and the cells of the remaining one-eighth will grow and divide until the liver mass returns almost to normal. The same effect occurs for almost all glandular cells, for cells of the bone marrow, the subcutaneous tissue, the intestinal epithelium, and almost any other tissue except highly differentiated cells, such as nerve and muscle cells.

We know very little about the mechanisms that maintain proper numbers of the different types of cells in the body. However, experimental studies have shown that control substances called *chalones* are secreted by the different cells that cause feedback effects to stop or slow their growth and mitosis when too many of them have been formed. We know that cells of any type removed from the body and grown in tissue culture can grow and reproduce rapidly and indefinitely if the medium in which they grow is continually replenished. Yet they will stop growing when even small amounts of their own secretions are allowed to collect in the medium, which supports the idea that control substances limit cellular growth.

Regulation of Cell Size. Cell size is determined almost entirely by the amount of DNA in the nucleus. If replication of the DNA does not occur, the cell grows to a certain size and thereafter remains at that size. On the other hand, it is possible, by use of the chemical *colchicine,* to prevent formation of the mitotic spindle and therefore to prevent mitosis also, even though replication of the DNA continues. In this event, the nucleus then contains far greater quantities of DNA than normally, and the cell grows proportionately larger. It is assumed that this results simply from increased production of RNA and cell proteins, which in turn cause the cell to grow larger.

CELL DIFFERENTIATION

A special characteristic of cell growth and cell division is *cell differentiation,* which means changes in physical and functional properties of cells as they proliferate in the embryo to form the different bodily structures. However, our problem here is not to describe the stages of differentiation but simply to discuss the theories concerning the means by which the cells change their characteristics to form all the different organs and tissues of the body.

The earliest and simplest theory for explaining differentiation was that the genetic composition of the nucleus undergoes changes during successive generations of cells in such a way that one daughter cell inherits a different set of genes from that of the other daughter cell.

However, this theory has now been almost completely disproved by the following simple experiment. The nucleus from an intestinal mucosal cell of a frog, when surgically implanted into a frog ovum from which the original nucleus has been removed, will often cause the formation of a completely normal frog. This demonstrates that even the intestinal mucosal cell, which is a reasonably well-differentiated cell, still carries all the

necessary genetic information for development of all structures required in the frog's body.

Therefore, the present idea is that instead of loss of genes during the process of differentiation, there occurs selective repression of different genetic operons. This presumably results from the buildup of different repressor substances in the cytoplasm, the repressor substances in one cell acting to repress one group of genetic characteristics and the repressor substances in another cell acting on a different group of genetic characteristics.

Embryological experiments show also that certain cells in an embryo control the differentiation of adjacent cells. For instance, the *primordial chordamesoderm* is called the *primary organizer* of the embryo because it forms a focus around which the rest of the embryo develops. It differentiates into a *mesodermal axis* containing segmentally arranged *somites* and, as a result of *inductions* in the surrounding tissues, causes formation of essentially all the organs of the body.

Another instance of induction occurs when the developing eye vesicles come in contact with the ectoderm of the head and cause it to thicken into a lens plate that folds inward to form the lens of the eye. It is possible that the entire embryo develops as a result of such inductions, one part of the body affecting another part, and this part affecting still other parts.

Thus, our understanding of cell differentiation is still hazy. We know many different control mechanisms by which differentiations *could* occur. Yet the overall basic controlling factors in cell differentiation are yet to be discovered.

CANCER

Cancer is a disease that attacks the basic life process of the cell, in almost all instances altering the cell's *genome* (the total genetic complement of the cell) and leading to wild and spreading growth of the cancerous cells. The cause of the altered genome is a *mutation* (alteration) of one or more genes; or mutation of a large segment of a DNA strand containing many genes; or, in some instances, addition or loss of large segments of chromosomes.

Only a minute fraction of the cells that mutate in the body ever lead to cancer. There are several reasons for this: First, most mutated cells have less survival capability than normal cells and therefore simply die. Second, only a few of the mutated cells that do survive lose the normal feedback controls that prevent excessive growth. And, third, those cells that are potentially cancerous are usually destroyed by the body's immune system before they grow into a cancer. This occurs in the following way: Most mutated cells form abnormal proteins within their cell bodies because of their altered genes, and these proteins then stimulate the body's immune system, causing it to form antibodies against the cancerous cells, in this way destroying them. Indeed, it is believed that all of us are continually forming cells that are potentially cancerous but that our immune system acts as a scavenger that nips these abnormal cells in the bud before they can become established. In support of this is the fact that in persons whose immune systems have been suppressed, such as those who are taking immunosuppressant drugs following transplantation of a kidney or a heart, the probability of developing a cancer is multiplied several-fold.

But what is it that causes the mutations themselves? When one realizes that many trillions of new cells are formed each year in the human being, this question should probably better be asked in the following form: Why is it that we do not develop literally millions or billions of mutant cancerous cells? The answer is the incredible precision with which DNA chromosomal strands are replicated in each cell before mitosis takes place. Indeed, even after each new strand is formed, the veracity of the replication process is "proofread" several different times. If any mistakes have been made, the new strand is cut and repaired before the mitotic process is allowed to proceed. Yet, despite all these precautions, probably one newly formed cell in every few hundred thousand to every few million still has significant mutant characteristics. We know this because it has been ascertained that each gene in a human offspring has the probability of 1 in 100,000 of being a mutant when compared with the genes of the parents.

Thus, chance alone is all that is required for mutations to take place, so we may suppose that a very large number of cancers are merely the result of an unlucky occurrence.

Yet, the probability of mutations can be increased many fold when a person is exposed to certain chemical, physical, or biological factors. Some of these are the following:

(1) It is well known that *ionizing radiation* such as x-rays, gamma rays, and particle radiations from radioactive substances, and even ultraviolet light, can predispose to cancer. Ions formed in tissue cells under the influence of such radiation are highly reactive and can rupture DNA strands, thus causing many mutations.

(2) *Chemical substances* of certain types also have a high propensity for causing mutations. Historically, it was long ago discovered that various aniline dye derivatives are very likely to cause cancer, so that workers in chemical plants producing such substances, if unprotected, have a special predisposition to cancer. Chemical substances that can cause mutation are called *carcinogens*. The carcinogens that cause by far the greatest number of deaths in our present-day society are those in cigarette smoke. These cause about one quarter of all cancer deaths.

(3) *Physical irritants* can also lead to cancer, such as continued abrasion of the linings of the intestinal tract by some types of food. The damage to the tissues leads to rapid mitotic replacement of the cells. The more rapid the mitosis, the greater the chance for mutation.

(4) In many families there is a strong *hereditary tendency to cancer*. This probably results from the fact that most cancers require not one mutation but two or more mutations before cancer will occur. In those families that are particularly predisposed to cancer, it is presumed that one or more of the genes are already mutated in the inherited genome. Therefore, far fewer additional mutations must take place in such a person before a cancer will begin to grow.

(5) In experimental animals certain types of viruses can cause some kinds of cancer, including leukemia. This occasionally results by either of two ways: First, in the case of DNA viruses, the DNA strand of the virus can insert itself directly into one of the chromosomes and thereby cause the mutation that leads to cancer. In the case of RNA viruses, some of these carry with them an enzyme called *reverse transcriptase* that will cause DNA to be transcribed from the RNA. Then the transcribed DNA inserts itself into the animal cell chromosome, thus leading to cancer. Yet, despite the demonstration that viral cancer sometimes occurs in animals, it has not yet been proved that cancer spreads this way in human beings, nor that cancer is contagious from one person to another.

Invasive Characteristic of the Cancer Cell. The two major differences between the cancer cell and the normal cell are: (1) The cancer cell does not respect usual cellular growth limits; the reason for this is that they presumably do not secrete the appropriate *chalones* that are responsible for stopping excess growth of normal cells. (2) Cancer cells are far less adhesive to each other than are normal cells. Therefore, they have a tendency to wander through the tissues, to enter the blood stream, and to be transported all through the body where they form nidi for numerous new cancerous growths.

Why Do Cancer Cells Kill? The answer to this is very simple: Cancer tissue competes with normal tissues for nutrients. Because cancer cells continue to proliferate indefinitely, their number multiplying day by day, one can readily understand that the cancer cells will soon demand essentially all the nutrition available to the body. As a result the normal tissues gradually suffer nutritive death.

REFERENCES

Atkinson, D. E., and Fox, C. F. (eds.): Modulation of Protein Function. New York, Academic Press, 1979.

Bauer, W. R.: Structure and reactions of closed duplex DNA. *Annu. Rev. Biophys. Bioeng.*, 7:287, 1978.

Baum, H., and Gergely, J. (eds.): Molecular Aspects of Medicine. New York, Pergamon Press, 1978.

Bender, D. A.: Amino Acid Metabolism. New York, John Wiley & Sons, 1978.

Butler, J. G., and Klug, A.: The assembly of a virus. *Sci. Am.*, 239(5): 62, 1978.

Cavalli-Sforza, L. L.: The genetics of human populations. *Sci. Am.*, 231(3):80, 1974.

Clark, B. F. C., et al. (eds.): Gene Expression: Protein Synthesis and Control, RNA Synthesis and Control, Chromatin Structure and Function. New York, Pergamon Press, 1978.

Cohen, S. N.: The manipulation of genes. *Sci. Am.*, 233(1):24, 1975.

Cummings, D. J. et al. (eds.): Extrachromosomal DNA. New York, Academic Press, 1979.

Dickerson, R. E.: Chemical evolution and the origin of life. *Sci. Am.*, 239(3):70, 1978.

Emmelot, P., and Kriek, E. (eds.): Environmental Carcinogenesis: Occurrence, Risk Evaluation, and Mechanisms. New York, Elsevier/North-Holland, 1979.

Fiddes, J. C.: The nucleotide sequence of a viral DNA. *Sci. Am.*, 237(6):54, 1977.

Foster, R. L.: The Nature of Enzymology. New York, John Wiley & Sons, 1979.

Frankel, E.: DNA, The Ladder of Life. New York, McGraw-Hill, 1978.

Friedman, D. L.: Role of cyclic nucleotides in cell growth and differentiation. *Physiol. Rev.*, 56:652, 1976.

Friedmann, H. C. (ed.): Enzymes. Stroudsburg, Pa., Dowden, Hutchinson & Ross, 1980.

Gallop, P. M., and Paz, M. A.: Posttranslational protein modifications, with special attention to collagen and elastin. *Physiol. Rev.*, 55:418, 1975.

Hiatt, H. H., et al. (eds.): Origins of Human Cancer. Cold Spring Harbor, N.Y., Cold Spring Harbor Laboratory, 1977.

Horowitz, M. I., and Pigman, W. (eds.): Mammalian Glycoproteins and Glycolipids. New York, Academic Press, 1977.

Horton, J. D. (ed.): Development and Differentiation of Vertebrates. New York, Elsevier/North-Holland, 1980.

Kaplan, J. G.: Membrane cation transport and the control of proliferation of mammalian cells. *Annu. Rev. Physiol., 40*:19, 1978.

Kastrup, K. W., and Neilsen, J. H. (eds.): Growth Factors: Cellular Growth Processes, Growth Factors, Hormonal Control of Growth. New York, Pergamon Press, 1978.

Knobf, M. K., *et al.*: Cancer Chemotherapy, Treatment and Care. New Haven, Conn., Com. Office, Yale Comprehensive Cancer Center, 1979.

Kornberg, A.: DNA Replication. San Francisco, W. H. Freeman, 1980.

Kouri, R. E. (ed.): Genetic Differences in Chemical Carcinogenesis. Boca Raton, Fla., CRC Press, 1980.

LaFond, R. E. (ed.): Cancer, The Outlaw Cell. Washington, American Chemical Society Publication, 1978.

Maniatis, T., and Ptashne, M.: A DNA operator-repressor system. *Sci. Am., 234*(1):64, 1976.

Molineaux, I., and Kohiyama, M. (eds.): DNA Synthesis: Present and Future. New York, Plenum Press, 1978.

Nicolson, G. L.: Cancer Metastasis. *Sci. Am., 240*(3):66, 1979.

Rapado, A., *et al.* (eds.): Purine Metabolism in Man. New York, Plenum Press, 1979.

Rich, A., and Kim, S. H.: The three-dimensional structure of transfer RNA. *Sci. Amer., 238*(1):52, 1978.

Russell, T. R., *et al.* (eds.): From Gene to Protein: Information Transfer in Normal and Abnormal Cells. New York, Academic Press, 1979.

Sandberg, A. A.: The Chromosomes in Human Cancer and Leukemia. New York, Elsevier/North-Holland, 1979.

Sarma, R. H. (ed.): Stereodynamics of Molecular Systems. New York, Pergamon Press, 1979.

Schimke, R. N.: Genetics and Cancer in Man. New York, Longman Inc., 1978.

Schopf, J. W.: The evolution of the earliest cells. *Sci. Am., 239*(3):110, 1978.

Sobell, H. M.: Symmetry in nucleic acid structure and its role in protein-nucleic acid interactions. *Annu. Rev. Biophys. Bioeng., 5*:307, 1976.

Söll, D., *et al.* (eds.): Transfer RNA: Biological Aspects. Cold Spring Harbor, N.Y., Cold Spring Harbor Laboratory, 1979.

Spreafico, F., and Arnon, R. (eds.): Tumor-Associated Antigens and Their Specific Immune Response. New York, Academic Press, 1979.

Stein, G. S., *et al.*: Chromosomal proteins and gene regulation. *Sci. Am., 232*(2):46, 1975.

Terry, W. D., and Yamamura, Y. (eds.): Immunobiology and Immunotherapy of Cancer. New York, Elsevier/North-Holland, 1979.

Walker, R. T., *et al.* (eds.): Nucleoside Analogues: Chemical, Biology, and Medical Applications. New York, Plenum Press, 1979.

Waters, H. (ed.): Immune Status in Cancer Treatment and Prognosis. New York, Garland STPM Press, 1978.

Weissman, S. M.: Gene structure and function. *In* Bondy, R. K., and Rosenberg, L. E. (eds.): Metabolic Control and Disease, 8th Ed. Philadelphia, W. B. Saunders Co., 1980, p. 1.

Wetmur, J. G.: Hybridization and renaturation kinetics of nucleic acids. *Annu. Rev. Biophys. Bioeng., 5*:337, 1976.

Wolpert, L.: Pattern formation in biological development. *Sci. Am., 239*(4):154, 1978.

Wu, R. (ed.): Recombinant DNA. New York, Academic Press, 1979.

Wynn, C. H.: The Structure and Function of Enzymes, 2nd Ed. Baltimore, University Park Press, 1979.

4

Transport Through the Cell Membrane

The fluid inside the cells of the body, called *intracellular fluid,* is very different from that outside the cells, called *extracellular fluid*. The extracellular fluid includes both the *interstitial fluid* that circulates in the spaces between the cells and also fluid of the *blood plasma* that mixes freely with the interstitial fluid through the capillary walls. It is the extracellular fluid that supplies the cells with nutrients and other substances needed for cellular function. But before the cell can utilize these substances, they must also be transported through the cell membrane.

Figure 4–1 gives the compositions of both the extracellular and intracellular fluids. Note that the extracellular fluid contains large quantities of *sodium* but only small quantities of *potassium*. Exactly the opposite is true of the intracellular fluid. Also, the extracellular fluid contains large quanti-

ties of chloride, while the intracellular fluid contains very little. But the concentrations of phosphates, essentially all of which are organic metabolic intermediates, and proteins in the intracellular fluid are considerably greater than in the extracellular fluid. These differences between the components of the intracellular and extracellular fluids are extremely important to the life of the cell. It is the purpose of this chapter to explain how these differences are brought about by the transport mechanisms in the cell membrane.

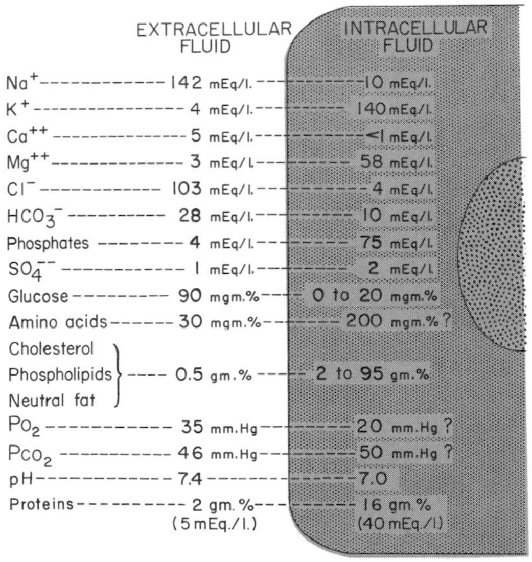

Substances are transported through the cell membrane by two major processes, *diffusion* and *active transport*. Though there are many different variations of these two basic mechanisms, as we shall see later in this chapter, basically, diffusion means free movement of substances in a random fashion caused by the normal kinetic motion of matter, whereas active transport means movement of substances in chemical combination with *carrier substances* in the membrane and also against an energy gradient, such as from a low concentration state to a high concentration state, a process that requires chemical energy to cause the movement.

DIFFUSION

All molecules and ions in the body fluids, including both water molecules and dissolved substances, are in constant motion, each particle moving its own separate way. Motion of these particles is what physicists call heat — the greater the motion, the higher is the temperature — and motion never ceases under any conditions except absolute zero temperature. When a moving molecule, A, approaches a stationary molecule, B, the electrostatic and internuclear forces of molecule A repel molecule B, adding some of the energy of motion to molecule B. Consequently, molecule B

Figure 4–1. Chemical compositions of extracellular and intracellular fluids.

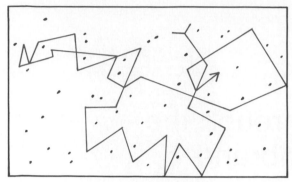

Figure 4–2. Diffusion of a fluid molecule during a fraction of a second.

gains kinetic energy of motion while molecule A slows down, losing some of its kinetic energy. Thus, as shown in Figure 4–2, a single molecule in solution bounces among the other molecules first in one direction, then another, then another, and so forth, bouncing randomly hundreds to millions of times each second. At times it travels a far distance before striking the next molecule, but at other times only a short distance.

This continual movement of molecules among each other in liquids, or in gases, is called *diffusion*. Ions diffuse in exactly the same manner as whole molecules, and even suspended colloid particles diffuse in a similar manner, except that because of their very large sizes they diffuse far less rapidly than molecular substances.

KINETICS OF DIFFUSION — THE CONCENTRATION DIFFERENCE

When a large amount of dissolved substance is placed in a solvent at one end of a chamber, it immediately begins to diffuse toward the opposite end of the chamber. If the same amount of substance is placed in the opposite end of the chamber, it begins to diffuse toward the first end, the same amount diffusing in each direction. As a result, the *net rate of diffusion* from one end to the other is zero. If, however, the concentration of the substance is greater at one end of the chamber than at the other end, the net rate of diffusion from the area of high concentration to low concentration is directly proportional to the larger concentration minus the lower concentration. The total concentration change along the axis of the chamber is called a *concentration difference,* and the concentration difference divided by the dis-

tance is called the *concentration* or *diffusion gradient*.

The rapidity with which a molecule diffuses from one point to another is less the greater the molecular size, because large particles are not impelled so intensely by collisions with other molecules. The rate of diffusion is approximately inversely proportional to the square root of the molecular weight but it is affected by the shape of the molecule as well.

If we consider all the different factors that affect the rate of diffusion of a substance from one area to another, they are the following:

(1) The greater the concentration difference, the greater is the rate of diffusion.

(2) The less the square root of the molecular weight, the greater is the rate of diffusion.

(3) The shorter the distance, the greater is the rate.

(4) The greater the cross-section of the chamber in which diffusion is taking place, the greater is the rate of diffusion.

(5) The greater the temperature, the greater is the molecular motion and also the greater is the rate of diffusion. All these can be expressed in an approximate formula, found at the bottom of this page, for diffusion in solutions.

DIFFUSION THROUGH THE CELL MEMBRANE

The cell membrane is essentially a sheet of lipid material, called the *lipid matrix,* with interspersed islands of globular protein molecules in the matrix. Some of these protein molecules penetrate all the way through the membrane. The detailed structure of the membrane was discussed in Chapter 2 and shown in the diagram of Figure 2–3. Water molecules and small dissolved molecules on each side of the membrane are believed to penetrate the protein portions of the membrane with ease, but the lipid portion of the membrane is an entirely different type of medium, acting as a limiting boundary between the extracellular and intracellular fluids.

Therefore, two different methods by which substances can diffuse through the membrane are (1) by becoming dissolved in the lipid matrix and diffusing through it in the same way that diffusion occurs in water, or (b) by diffusing through minute pores that pass directly through the membrane at wide intervals over its surface; these "pores" are probably intramolecular spaces within the structures of the protein molecules that penetrate all the way through the membrane.

$$\text{Diffusion rate} \propto \frac{\text{Concentration difference} \times \text{Cross-sectional area} \times \text{Temperature}}{\sqrt{\text{Molecular weight}} \times \text{Distance}}$$

Diffusion in the Dissolved State Through the Lipid Portion of the Membrane

A few substances are soluble in the lipid of the cell membrane as well as in water. These include oxygen, carbon dioxide, alcohol, fatty acids, and other less important substances. When one of these comes in contact with the membrane it immediately becomes dissolved in the lipid and continues to diffuse in exactly the same manner that it diffuses within the watery medium on either side of the membrane. In other words, the molecule continues its random motion within the substance of the membrane in exactly the same way that it undergoes random motion in the surrounding fluids. However, the viscosity of the lipid matrix is about 300 times that of water, and this slows the diffusion by this amount despite the lipid solubility.

Effect of Lipid Solubility on Diffusion Through the Lipid Matrix. The primary factor that determines how rapidly a substance can diffuse through the lipid matrix of a cell membrane is its solubility in lipids. If it is very soluble, it becomes dissolved in the membrane very easily and therefore passes on through. Indeed, if it is more than 300 times as soluble in lipid than in water, it actually diffuses even more rapidly through the membrane than in the water of the surrounding fluids. On the other hand, almost no substances that dissolve very poorly in lipids, such as water, glucose, and the electrolytes, pass through the lipid matrix.

Facilitated Diffusion Through the Lipid Matrix

Some substances are very insoluble in lipids and yet can still pass through the lipid matrix by a process called *carrier-mediated* or *facilitated diffusion.* This is the means by which some sugar and amino acids in particular cross the membrane. The most important of the sugars is glucose, the membrane transport of which is illustrated in Figure 4–3. This shows that glucose (G1) combines with a *carrier* substance (C) at point 1 to form the compound CG1. This combination is soluble in the lipid so that it can diffuse (or simply move by rotation of the large carrier molecule) to the other side of the membrane, where the glucose breaks away from the carrier (point 2) and passes to the inside of the cell, while the carrier moves back to the outside surface of the membrane to pick up still more glucose and transport it also to the inside. Thus, the effect of the carrier is to make the glucose soluble in the membrane; without it, glucose cannot pass through the membrane.

The rate at which a substance passes through a membrane by facilitated diffusion depends on the difference in concentration of the substance on the two sides of the membrane, the amount of carrier

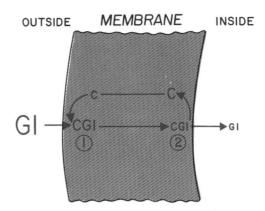

Figure 4–3. Facilitated diffusion of glucose through the cell membrane.

available, and the rapidity with which the chemical (or physical) reactions can take place. In the case of glucose transport and also transport of some of the amino acids, the overall rate is greatly increased by insulin, which is the primary hormone secreted by the pancreas. Large quantities of insulin can increase the rate of glucose transport about seven- to ten-fold, though it is not known whether this is caused by an effect of insulin to increase the quantity of carrier in the membrane or to increase the rate at which the chemical reactions take place between glucose and the carrier.

Very little is known about the carrier substance for transport of glucose through the membrane. It binds specifically with certain types of monosaccharides — glucose, mannose, galactose, xylose, and arabinose — but other monosaccharides are not transported at all or only slightly. A protein having a molecular weight of approximately 45,000 and also having the property of combining specifically and reversibly with the same types of monosaccharides that are transported has been discovered in the cell membrane. It is likely that this protein is the carrier substance for transport of glucose.

Many of the dynamics of facilitated diffusion are similar to those of active transport, which is discussed in detail later in the chapter. The primary difference between the two is that active transport can move substances through a membrane from a low concentration on one side to a high concentration on the other side, whereas in facilitated diffusion, the substance can move only from a high concentration toward a low concentration.

An important difference between facilitated diffusion and simple diffusion of free particles through a membrane is that the net diffusion rate of free particles is almost exactly proportional to the difference between the concentrations of the particles on the two sides of the membrane. However, for a carrier-transported substance, only when the concentrations of the substance are very small will this relationship hold. When the concentrations become great, the system becomes *saturated,* which means that then the rate at which facilitated diffusion will occur is determined by the quantity of available carrier or by the rates at which the chemical reactions between the mediated substance and the carrier can take place.

Diffusion Through the Membrane Pores

Some substances, such as water and many of the dissolved ions, seem to go through holes in the cell membrane called *membrane pores*. The nature of these pores is unknown. However, they are believed to be large protein molecules penetrating all the way through the cell membrane and providing pathways for movement of water-soluble substances within the intermolecular spaces of the protein. At any rate, these so-called "pores" in the cell membrane behave as if they were minute round holes approximately 0.8 nanometer (8 Angstroms) in diameter and as if the total area of the pores should equal approximately 1/5000 of the total surface area of the cell. Despite this very minute total area of the pores, water molecules diffuse so rapidly that the entire volume of fluid in some types of cells — the red blood cell for instance — can easily pass through the pores within one hundredth of a second.

Figure 4–4 illustrates a schematized structure of a pore, indicating that its surface is probably lined with negative charges, probably carboxyl or phosphate groups within the structure of the protein pore molecule. This figure shows several small particles passing through the pore and also shows that the maximum diameter of the particle that can pass through is approximately equal to the diameter of the pore itself, about 0.8 nanometer.

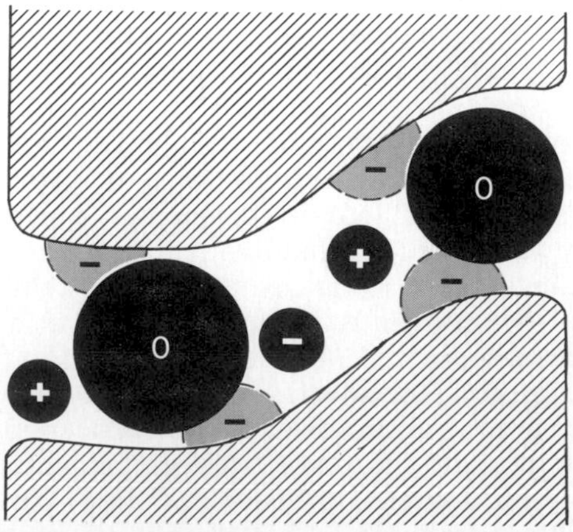

Effect of Pore Size on Diffusion Through the Pore — Permeability.   Table 4–1 gives the effective diameters of various substances in comparison with the diameter of the pore, and it also gives the relative permeability of the pores for the different substances. *The permeability can be defined as the rate of transport through the membrane for a given concentration difference.* Note that some substances, such as the water molecule, urea molecule, and chloride ion, are considerably smaller in size than the pore. All these pass through most pores, particularly those of red cells, with great ease. For instance, the rate per second of diffusion of water in each direction through the pores of a ced cell is about 100 times as great as the volume of the cell itself. It is fortunate that an identical amount of water normally diffuses in each direction, which keeps the cell from either swelling or shrinking despite the rapid rate of diffusion. The rates of diffusion of urea and chloride ions through the membrane are somewhat less than that of water, which is in keeping with the fact that their effective diameters are slightly greater than that of water.

Table 4–1 also shows that most of the sugars, including glucose, have effective diameters that are slightly greater than that of the pores. Obviously, not even a single molecule as large as these could go through a pore that is smaller than its size. For this reason essentially none of the

Figure 4–4.   Postulated structure of the pore in the mammalian red cell membrane, showing the sphere of influence exerted by charges along the surface of the pore. (Modified from Solomon: *Sci. Am. 203*:146, 1960. ©1960 by Scientific American, Inc.)

TABLE 4–1   RELATIONSHIP OF EFFECTIVE DIAMETERS OF DIFFERENT SUBSTANCES TO PORE DIAMETER AND RELATIVE PERMEABILITIES

Substance	Diameter Å	Ratio to Pore Diameter	Approximate Relative Permeability
Water molecule	3	0.38	50,000,000
Urea molecule	3.6	0.45	1,500,000
Hydrated chloride ion			
(red cell)	3.86	0.48	500,000
(nerve membrane)	—	—	0.04
Hydrated potassium ion			
(red cell)	3.96	0.49	1.1
(nerve membrane)	—	—	0.02
Hydrated sodium ion			
(red cell)	5.12	0.64	1
(nerve membrane)	—	—	0.0003
Lactate ion	5.2	0.65	?
Glycerol molecule	6.2	0.77	?
Ribose molecule	7.4	0.93	?
Pore size	8 (Ave.)	1.00	—
Galactose	8.4	1.03	?
Glucose	8.6	1.04	0.4
Mannitol	8.6	1.04	?
Sucrase	10.4	1.30	?
Lactose	10.8	1.35	?

*These data have been gathered from different sources but relate primarily to the red cell membrane. Other cell membranes have different characteristics.

sugars can pass through the pores; instead, those that do enter the cell pass through the lipid matrix by the process of facilitated diffusion.

Effect of the Electrical Charge of Ions on Their Ability to Diffuse Through Membrane Pores. The electrical charges of ions often markedly affect their ability to diffuse through pores, sometimes impeding and sometimes enhancing their diffusion. Three major types of electrical forces affect ion diffusivity: (1) the intensity of electrical charges lining the pores, (2) the hydration energy of the ions, and (3) the potential difference across the membrane. Let us explain these.

First, there is reason to believe that different membrane pores are lined with either negative or positive charges. For instance, one set of pores, called *sodium channels,* seem especially to allow easy diffusion of sodium ions. It is believed that these pores are lined with strongly *negative* charges that attract sodium ions into the pores and then shuttle the sodium ions from one negative charge to the next, thus allowing rapid movement through the pores. On the other hand, larger-sized positive ions, such as potassium ions, have considerable difficulty passing through these channels, mainly because of their diameter.

Another instance in which the electrical charge of the pore probably affects ion diffusion is the following: It is believed that one end of the sodium channel sometimes becomes blocked by the presence of a *positive* charge appearing at the opening of the pore. This charge is called a *gate* because it repels the sodium ions and impedes their passage through the channel. We will discuss this gating theory for control of sodium transport through sodium channels in Chapter 10, in relation to transmission of the nerve impulse. Thus, ionic attraction in the pore can cause enhanced ion diffusion, while ionic repulsion can diminish ion diffusion.

The second factor that affects ion diffusion through pores, the hydration energy of the ion, mainly retards movement through pores. Most ions are loosely bound with water, and most or all of the water molecules bound with the ions must be removed before the particle size is small enough to allow passage through the pores. On the other hand, the smaller the ion, the greater is the hydration energy. For instance, the hydration energy for sodium is much greater than for potassium. Consequently, this factor impedes sodium transport through some pores far more than it does potassium transport because the water molecules can be broken away from potassium much more easily than from sodium. If the pores are very strongly negatively charged, as occurs in the sodium channels, this hydration energy can be overcome easily. However, in most cell membranes there seems to be a large population of either uncharged pores or poorly charged pores that are relatively impermeable to sodium ions because of the high hydration energy of these ions but are relatively permeable to potassium ions because of their lower hydration energy. These pores are called *potassium channels.*

Third, many membranes have a different electrical potential level in the fluid on one side of the membrane than on the other side; that is, the potential is more positive on one side and more negative on the other. Negative ions are attracted through the pores toward the positive fluid and positive ions toward the negative fluid. This is such an important effect that it will be discussed again later in this chapter and especially in Chapter 10.

Thus, several different factors determine the diffusivity of ions through cell membranes. In Chapter 10, we shall see that the permeability of the nerve membrane to sodium and potassium changes extremely rapidly, as much as 50- to 5000-fold, during the course of nerve impulse transmission. These changes presumably result from rapid alterations in the electrical charges lining the pores or guarding their entrances.

Effect of Different Factors on Pore Permeability. Pore permeability does not always remain exactly the same under different conditions. For instance, excess *calcium* in the extracellular fluid causes the permeability to decrease, and diminished calcium causes considerably increased permeability. This is extremely important in the function of nerves, for the enhanced permeability in extracellular fluid calcium deficiency causes excessive diffusion of ions, which results in spurious nerve discharge throughout the body.

Another factor that has an important effect on pore permeability of some cells is *antidiuretic hormone,* which is formed in the hypothalamus and secreted through the posterior pituitary gland. This hormone has an especially important effect on the membranes of the cells lining the collecting ducts of the kidney. Increased quantities of the hormone seem to increase the pore diameter, which allows water and other substances to diffuse from the tubules back into the blood with ease.

Net Diffusion Through the Cell Membrane and Factors That Affect It

From the preceding discussion, it is evident that many different substances can diffuse either through the lipid matrix of the cell membrane or through the pores. It should be noted, however, that substances that diffuse in one direction can also diffuse in the opposite direction. Usually it is not the total quantity of substances diffusing in both directions that is important to the cell but instead the *net quantity* diffusing either into or out of the cell.

In addition to the permeability of the membrane, which has already been discussed, three other factors determine the rate of net diffusion of a substance: the concentration difference of the substance across the membrane, the electrical potential difference across the membrane, and the pressure difference across the membrane.

Effect of a Concentration Difference.   Figure 4–5A illustrates a membrane with a substance in high concentration on the outside and low concentration on the inside. The rate at which the substance diffuses *inward* is proportional to the concentration of molecules on the outside, for this concentration determines how many of the molecules strike the outside of the pore each second. On the other hand, the rate at which the molecules diffuse *outward* is proportional to their concentration inside the membrane. Obviously, therefore, the rate of net diffusion into the cell is proportional to the concentration on the outside *minus* the concentration on the inside, or

$$\text{Net diffusion} \propto P(C_o - C_i)$$

A

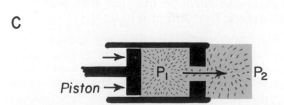

in which C_o is the concentration on the outside, C_i is the concentration on the inside, and P is the permeability of the membrane for the substance.

Effect of an Electrical Potential Difference.   If an electrical potential is applied across the membrane as shown in Figure 4–5B, ions, because of their electrical charges, will move through the membrane even though no concentration difference exists to cause their movement. Thus, to the left in Figure 4–5B, the concentrations of negative ions are exactly the same on both sides of the membrane, but a positive charge has been applied to the right side and a negative charge to the left, creating an electrical gradient across the membrane. The positive charge attracts the negative ions while the negative charge repels them. Therefore, net diffusion occurs from left to right. After much time large quantities of negative ions will have moved to the right (if we neglect, for the time being, the disturbing effects of the positive ions of the solution), creating the condition illustrated on the right in Figure 4–5B, in which a concentration difference of the same ions has developed in the direction opposite to the electrical potential difference. Obviously, the concentration difference is tending to move the ions to the left, while the electrical difference is tending to move them to the right. When the concentration difference rises high enough, the two effects exactly balance each other. At normal body temperature (38° C.), the electrical difference that will exactly balance a given concentration difference of univalent ions can be determined from the following formula called the *Nernst equation*:

$$\text{EMF (in millivolts)} = \pm 61 \ \log \frac{C_1}{C_2}$$

in which EMF is the electromotive force (voltage) between side 1 and side 2 of the membrane, C_1 is the concentration on side 1, and C_2 is the concentration on side 2. The polarity of the voltage on side 1 in the above equation is + for negative ions and − for positive ions. This relationship is extremely important in understanding the transmission of nerve impulses, for which reason it is discussed in even greater detail in Chapter 10.

Effect of a Pressure Difference.   At times considerable pressure difference develops between the two sides of a membrane. This occurs, for instance, at the capillary membrane, which has a pressure approximately 23 mm. Hg greater inside the capillary than outside. Pressure actually means the sum of all the forces of the different molecules striking a unit surface area at a given instant. Therefore, when the pressure is increased on one side of a membrane, this means that the sum of all the forces of the molecules striking the pores on that side of the membrane is greater than

B

C

Figure 4–5.   Effect of (A) concentration difference, (B) electrical difference, and (C) pressure difference on diffusion of molecules and ions through a cell membrane.

on the other side. This can result either from greater numbers of molecules striking the membrane per second or from greater kinetic energy of the average molecule striking the membrane. In either event, increased amounts of energy are available to cause net movement of molecules from the high pressure side toward the low pressure side. This effect is illustrated in Figure 4–5C, which shows a piston developing high pressure on one side of a cell membrane, thereby causing net diffusion through the membrane to the other side.

For the usual red cell membrane, 1 mm. Hg pressure difference causes approximately 10^{-4} cubic micron of net diffusion of water through each square micron of membrane each second. This appears to be only a very minute rate of fluid movement through the membrane, but in relation to the normal cell size and the very large diffusion pressures that can develop at the cell membrane because of osmotic forces, as discussed later in the chapter, this rate can represent tremendous transport of fluid in only a few seconds.

Net Movement of Water Across Cell Membranes — Osmosis Across Semipermeable Membranes

By far the most abundant substance to diffuse through the cell membrane is water. It should be recalled again that enough water ordinarily diffuses in each direction through the red cell membrane per second to equal about *100 times the volume of the cell itself.* Yet, *normally,* the amount that diffuses in the two directions is so precisely balanced that not even the slightest *net* movement of water occurs. Therefore, the volume of the cell remains constant. However, under certain conditions, a *concentration difference for water* can develop across a membrane, just as concentration differences for other substances can also occur. When this happens, net movement of water does occur across the cell membrane, causing the cell either to swell or to shrink, depending on the direction of the net movement. This process of net movement of water caused by a concentration difference is called *osmosis.*

To give an example of osmosis, let us assume that we have the conditions shown in Figure 4–6, with pure water on one side of the cell membrane and a solution of sodium chloride on the other side. Referring back to Table 4–1, we see that water molecules pass through the cell membrane with extreme ease while sodium ions pass through only with extreme difficulty. And chloride ions cannot pass through the membrane because the positive charge of the sodium ions holds the negatively charged chloride ions back to maintain a balance between the negative and positive

charges in the solution, an effect called the *principle of electroneutrality.* Therefore, sodium chloride solution is actually a mixture of diffusible water molecules and non-diffusible sodium and chloride ions, and the membrane is said to be *semipermeable,* i.e., permeable to water but not to sodium and chloride ions. Yet, the presence of the sodium and chloride has reduced the concentration of water molecules to less than that of pure water. As a result, in the example of Figure 4–6, more water molecules strike the pores on the left side where there is pure water than on the right side where the water concentration has been reduced. Thus, net movement of water occurs from left to right — that is, osmosis occurs from left to right.

Osmotic Pressure

If in Figure 4–6 pressure were applied to the sodium chloride solution, osmosis of water into this solution could be slowed or even stopped. The amount of pressure required to stop osmosis completely is called the osmotic pressure of the sodium chloride solution.

The principle of a pressure difference opposing osmosis is illustrated in Figure 4–7, which shows a semipermeable membrane separating two separate columns of fluid, one containing water and the other containing a solution of water and some solute that will not penetrate the membrane. Osmosis of water from chamber B into chamber A causes the levels of the fluid columns to become farther and farther apart, until eventually a pressure difference is developed that is great enough to oppose the osmotic effect. The pressure difference

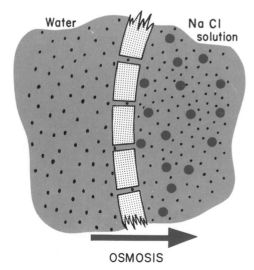

Figure 4–6. Osmosis at a cell membrane when a sodium chloride solution is placed on one side of the membrane and water on the other side.

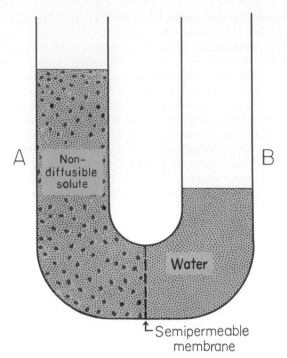

Figure 4–7. Demonstration of osmotic pressure on the two sides of a semipermeable membrane.

across the membrane at this time is the osmotic pressure of the solution containing the nondiffusible solute.

Kinetics of Osmotic Pressure. Figure 4–8 gives an analysis that can be very useful in understanding the basic cause of osmosis and osmotic pressure (even though the analysis is not strictly correct because of nonlinearities of the effects when one deals with solutions more concentrated than those of the body fluids). The actual pressure at the surface of a solution, such as a solution of sodium chloride, will be equal to atmospheric pressure. Thus, at sea level the pressure at the surface of a solution is 760 mm. Hg. Yet, strangely enough, pressure differences of many thousand mm. Hg frequently occur between the two sides of

a semipermeable membrane when the solutions on the two sides of the membrane are different from each other. This ability of the solutions to cause such pressure can be expressed in terms of *potential pressures* of the molecules and ions in the respective solutions. The potential pressure of each molecule or ion is proportional to its *chemical potential* in the solution, and the amount of chemical potential of each type of molecule or ion is directly proportional to its fractional molar concentration in the solution.

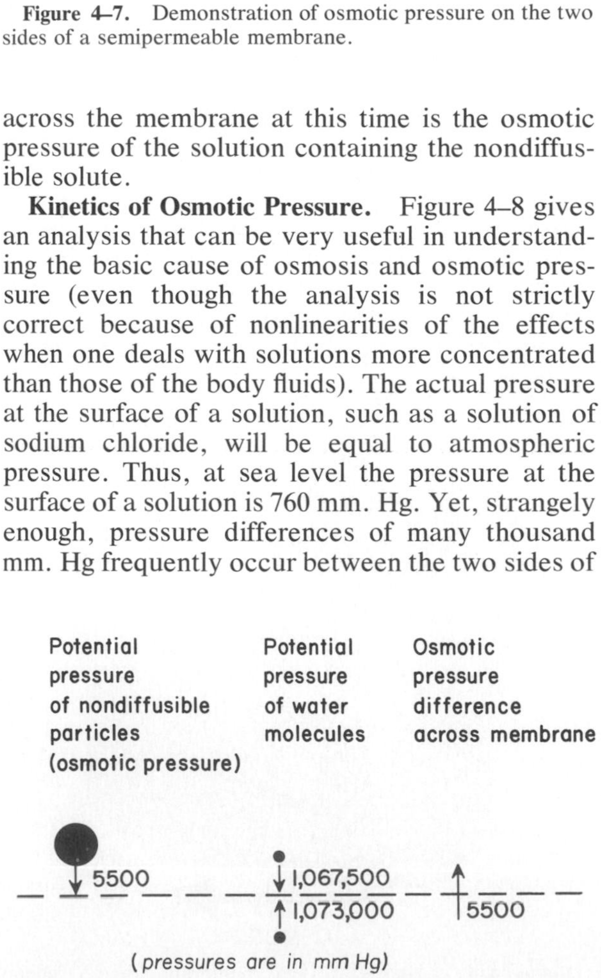

Figure 4–8 illustrates some representative potential "pressures" as they occur at cell membranes. The first potential pressure, 5500 mm. Hg, represents the total potential pressure of the nondiffusible particles in extracellular fluid — such particles as sodium, potassium, magnesium, calcium, chloride, bicarbonate, and protein ions, as well as nonionic substances such as glucose.

In the middle of the figure are shown the potential pressures of water on both sides of the membrane. On the lower side there exists pure water, and 1,073,000 mm. Hg is the theoretical total potential pressure of pure water if its entire chemical potential were converted into actual pressure (and if the principles of ideal solutions applied at high concentrations). That is, if pure water were on one side of a semipermeable membrane and pure electrolyte were on the opposite side, water theoretically could move by osmosis through the semipermeable membrane with sufficient pressure to create an actual pressure across the membrane of 1,073,000 mm. Hg at body temperature. On the top of the membrane, the potential pressure of water molecules is shown to be 1,067,500 mm. Hg, or 5500 mm. Hg less than that on the lower side. The reason for this is that the water on top has been diluted by the nondiffusible particles, making the molecular concentration of water above the membrane slightly less than that below.

Because of the greater potential pressure of water molecules below the membrane in Figure 4–8 than above, more water will diffuse from below upward than in the opposite direction, causing osmosis; and the amount of hydrostatic pressure that would be required on top of the membrane to prevent osmosis is 5500 mm. Hg. Note that this value is equal to the potential pressure of the nondiffusible particles in the upper solution.

Thus, the potential pressure of nondiffusible particles (caused by the chemical potential of these particles) is what is called the osmotic pressure of the solution. It causes osmosis of water in the direction opposite to the osmotic pressure. Therefore, the term "osmotic pressure" is actually a misnomer because it is not an actual pressure but only a potential pressure.

Lack of Effect of Molecular and Ionic Mass on

Potential pressure of nondiffusible particles (osmotic pressure)	Potential pressure of water molecules	Osmotic pressure difference across membrane
↓ 5500	↓ 1,067,500	↑
	↑ 1,073,000	↑ 5500

(pressures are in mm Hg)

Figure 4–8. Kinetics of osmosis. (See text for discussion.)

Osmotic Pressure — Importance of Numbers of Osmotic Particles (or of Molar Concentration). The osmotic pressure exerted by nondiffusible particles in a solution, whether they be molecules or ions, is determined by the *numbers* of particles per unit volume of fluid and not the mass of the particles. The reason for this is that each particle in a solution, regardless of its mass, exerts, on the average, the same amount of pressure against the membrane. That is, all particles are bouncing among each other with, on the average, equal energy. If some particles have greater kinetic energy of movement than others, their impact with the low energy particles will impart part of their energy to these, thus decreasing the energy level of the high energy particles while increasing the energy level of the others. The large particles, which have greater mass (m) than the small particles, move at lower velocities (v), while the small particles move at higher velocities in such a way that their average kinetic energies (k), determined by the equation

$$k = \frac{mv^2}{2},$$

are equal to each other. Therefore, on the average, the energy that a particle has when it strikes a membrane is approximately equal for all size particles, so that the factor that determines the osmotic pressure of a solution is the concentration of the solution in terms of numbers of particles (which is the same as its molar concentration if it is a nondissociated particle) and not in terms of mass of the solute.

Effect of Interaction Between Dissolved Particles on Osmotic Pressure. If particles interact with each other by either chemical or physical bonds, the osmotic effect becomes changed. Thus, sodium and chloride ions in solution have tremendous attractive power for each other so that neither of these ions can move totally unrestrained; therefore, their chemical activities are normally only about nine-tenths of what they would be were it not for this attraction. Because of interionic attraction (and other less powerful types of attraction) the amount of osmotic pressure in the body fluids averages only 0.93 times the maximum amount that is calculated on the basis of individual numbers of particles. However, since all the fluids of the body are affected approximately equally in this way so that the effects at the membranes cancel each other, for practical purposes this interaction between particles is usually ignored in studying osmotic effects in the body.

Osmolality. Since the amount of osmotic pressure exerted by a solute is proportional to the concentration of the solute in numbers of molecules or ions, expressing the solute concentration in terms of mass is of no value in determining osmotic pressure. To express the concentration in terms of numbers of particles, the unit called the *osmole* is used in place of grams.

One osmole is the number of particles (molecules) in 1 gram molecular weight of undissociated solute. Thus, 180 grams of glucose is equal to 1 osmole of glucose, because glucose does not dissociate. On the other hand, if the solute dissociates into two ions, 1 gram molecular weight of the solute equals 2 osmoles, because the number of osmotically active particles is now twice as great as is the case in the undissociated solute. Therefore, 1 gram molecular weight of sodium chloride, 58.5 gm., is equal to 2 osmoles.

A solution that has 1 osmole of solute dissolved in each kilogram of water is said to have an osmolality of 1 osmole per kilogram, and a solution that has 1/1000 osmole dissolved per kilogram has an osmolality of 1 milliosmol per kilogram. The normal osmolality of the extracellular and intracellular fluids is about 300 milliosmoles per kilogram.

Relationship of Osmolality to Osmotic Pressure. At normal body temperature, 38°C, a concentration of 1 osmole per liter will cause *19,300 mm. Hg* osmotic pressure in the solution. Likewise, 1 milliosmole per liter concentration is equivalent to *19.3 mm. Hg* osmotic pressure.

Osmolarity. Because of the difficulty of measuring kilograms of water in a solution, when speaking of the osmotic characteristics of body fluids, one usually uses another term, "osmolarity," which is the osmolar concentration expressed as *osmoles per liter of solution* rather than osmoles per kilogram of water. Though, strictly speaking, it is osmoles per kilogram of water (osmolality) that determines the rate of osmosis, nevertheless, for dilute solutions such as occur in the body, the quantitative differences between osmolarity and osmolality are less than 1 per cent. Since it is far more practical to use the term osmolarity than the term osmolality, this is the usual practice in almost all physiological studies.

Bulk Flow of Water Through Pores in Response to Hydrostatic and Osmotic Pressure Gradients

When either a hydrostatic or an osmotic pressure gradient is created across a membrane, the rate of movement of water molecules down the gradient is often many times as great as the rate that can be accounted for by net diffusion down the gradient. This is caused by a phenomenon called "bulk flow," which means simply that since large numbers of molecules are moving in the same direction, they tend to "stream" through the pores in unison rather than to move randomly as occurs in pure diffusion. Because of the mutual forward drag of the adjacent molecules on each other, this results in flow of far more water down either the hydrostatic or the osmotic gradient than one would predict from simple diffusion kinetics. This is often important because it accounts for far more movement of water through some membranes than would be expected on the basis of movement by diffusion alone. For instance, the osmotic movement of water through the red cell membrane is about 4 times as great and through the capillary membrane as much as 10 times as great as can be accounted for by simple diffusion. This phenomenon will be discussed in greater detail in Chapter 30.

ACTIVE TRANSPORT

Often only a minute concentration of a substance is present in the extracellular fluid, and yet a large concentration of the substance is required in the intracellular fluid. For instance, this is true of potassium ions. Conversely, other substances frequently enter cells and must be removed even though their concentrations inside are far less than outside. This is true of sodium ions.

From the discussion thus far it is evident that *no substances can diffuse against a concentration gradient,* or, as is often said, "uphill." To cause movement of substances uphill, energy must be imparted to the substance. This is analogous to the compression of air by a pump. Compression causes the concentration of the air molecules to increase but to create this greater concentration, energy must be imparted to the air molecules by the piston of the pump as they are compressed. Likewise, as molecules are transported through a cell membrane from a dilute solution to a concentrated solution, energy must be imparted to the molecules. When a cell membrane moves molecules uphill against a concentration gradient (or uphill against an electrical or pressure gradient) the process is called *active transport.*

Among the different substances that are actively transported through cell membranes are sodium ions, potassium ions, calcium ions, iron ions, hydrogen ions, chloride ions, iodide ions, urate ions, several different sugars, and the amino acids.

BASIC MECHANISM OF ACTIVE TRANSPORT

The mechanism of active transport is believed to be similar for most substances and to depend on transport by *carriers.* Figure 4–9 illustrates the basic mechanism, showing a substance S entering the outside surface of the membrane where it combines with carrier C. At the inside surface of the membrane, S separates from the carrier and is released to the inside of the cell. C then moves back to the outside to pick up more S.

One will immediately recognize the similarity between this mechanism of active transport and that of facilitated diffusion discussed earlier in the chapter and illustrated in Figure 4–3. The difference, however, is that *energy is imparted to the system* in the course of active transport, so that transport can occur *against a concentration gradient* (or against an electrical or pressure gradient).

Though the mechanism by which energy is utilized to cause active transport is not entirely known, we do know some features of this process:

First, the energy is delivered to the inside surface of the membrane from high energy substances, principally ATP, inside the cytoplasm of the cell.

Second, active transport obeys the usual laws for chemical combination of one substance (the substance to be transported) with another substance (the carrier).

Third, a specific "carrier" molecule (or combination of molecules) is required to transport each type of substance or each class of similar substances.

Fourth, a specific enzyme (or enzymes) is required to promote active transport.

From this information, we can construct a theory for active transport, as follows: Referring again to Figure 4–9, we might suppose that the carrier has a natural affinity for the substance to be transported so that at the outer surface of the membrane the carrier and the substance readily combine. Then the combination of the two *diffuses* through the membrane to the inner surface. Here an enzyme-catalyzed reaction occurs, utilizing energy from ATP to split the substance away from the carrier. In other words, the enzyme-catalyzed reaction makes the affinity of the carrier for the transported substance very low and thereby displaces it from the combination. But the released substance, being insoluble in the membrane, cannot diffuse backward through the lipid matrix of the membrane. Therefore, it is released to the inside of the membrane while the carrier diffuses back to the outside surface to transport still another molecule of substance in the inward direction.

A special characteristic of active transport is that the mechanism *saturates* when the concentration of the substance to be transported is very high. This saturation results from limitation either of quantity of carrier available to transport the substance or of enzymes to promote the chemical reactions that release the substance from the carrier. Thus, the principle of saturation of active

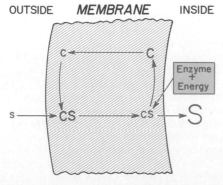

Figure 4–9.   Basic mechanism of active transport.

transport is almost identical with that of saturation of carrier-mediated facilitated diffusion, which was discussed earlier in the chapter.

Chemical Nature of Carrier Substances. Carrier substances are believed all to be proteins, conjugated proteins, or loose physical combinations of more than one protein molecule.

It has been suggested that the carrier might transport substances by a simple process of thermal motion, such as by rotating to expose its carrier site first to the outside surface of the cell membrane and then to the inside. But whether a "shuttling" of a "fixed" type of carrier exists, essentially the same principles of transport through the membrane would still apply.

Energetics of Active Transport. The amount of energy required to transport a substance actively through a membrane (aside from energy lost as heat in the chemical reactions) is determined by the degree that the substance is concentrated during transport. Compared to the energy required to concentrate a substance 10-fold, to concentrate it 100-fold requires twice as much energy, and to concentrate it 1000 fold requires three times as much. In other words, the energy required is proportional to the logarithm of the degree that the substance is concentrated as expressed by the following formula:

$$\text{Energy (in calories per osmole)} = 1400 \log \frac{C_1}{C_2}$$

Thus, in terms of calories, the amount of energy required to concentrate 1 osmole of substance 10-fold is about 1400 calories. One can see that the energy expenditure for concentrating substances in cells or for removing substances from cells against a concentration gradient can be tremendous. Some cells, such as those lining the renal tubules, expend almost 100 per cent of their energy for this purpose alone.

TRANSPORT THROUGH
INTRACELLULAR MEMBRANES

Though much less is known about the transport processes of the membranes within the cell than those of the cell membrane, we do know that at least two of these membranes, that of the mitochondrion and that of the endoplasmic reticulum, obey the same laws of both diffusion and active transport as does the cell membrane. However, their permeabilities and specificities for transport are often different. Indeed, the mitochondrion has a double membrane, with an outer membrane that is much more permeable than the inner membrane. Undoubtedly, transport through these membranes plays important roles in function of the intracellular organelles, though most of these roles are still a mystery.

ACTIVE TRANSPORT THROUGH
CELLULAR SHEETS

In many places in the body substances must be transported through an entire *cellular layer* instead of simply through the cell membrane itself. Transport of this type occurs through the intestinal epithelium, the

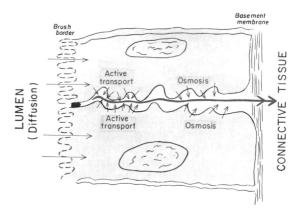

Figure 4-10. Basic mechanism of active transport through a layer of cells.

epithelium of the renal tubules, the epithelium of all exocrine glands, the membrane of the choroid plexus of the brain, and many other membranes.

One mechanism of active transport through a sheet of cells is illustrated in Figure 4-10. This figure shows two adjacent cells in a typical epithelial membrane as found in the intestine, in the gallbladder, and in certain other areas of the body. On the luminal surface of the cells is a brush border that is highly permeable to water and some solutes, allowing both of these to diffuse from the lumen of the intestine to the interior of the cell. Once inside the cell, some of the solutes are actively transported into the lateral spaces between the cells. This space is closed at the brush border of the epithelium but is wide open at the base of the cells where they rest on the basement membrane. Furthermore, the basement membrane is extremely permeable. Therefore, substances transported into this channel between the cells flow toward the connective tissue.

One of the most important substances actively transported in this manner is sodium ions; when sodium ions are transported into the space between the cells, their positive electrostatic charges pull negatively charged chloride ions through the membrane as well, illustrating once again the principle of electroneutrality. Then, when the concentration of sodium and chloride ions increases in the intercellular space, this in turn causes osmosis of water out of the cell and into the intercellular space. In consequence, the sodium and chloride flow along with the water into the connective tissue behind the basement membrane where the water and ions diffuse into the blood capillaries. These same principles of transport apply generally wherever such transport occurs through cellular sheets, whether in the intestinal epithelium, the gallbladder epithelium, the renal epithelium, or others. Such transport will be described for several of these epithelia at later points in the text.

SECONDARY ACTIVE TRANSPORT:
SODIUM "CO-TRANSPORT" OF
GLUCOSE AND AMINO ACIDS

Glucose and some amino acids are transported through the epithelial cells of the intestinal mucosa and renal tubules by a mechanism that is neither pure diffusion nor active transport but is a curious mixture of the two. This mechanism, called *secondary active transport* or *sodium co-transport,* functions for glucose transport in the following way:

First, let us remember that the epithelial cell has two functionally distinct sides, a brush border that lines the lumen of the gut or renal tubule and a base that lies adjacent to the absorptive capillaries. The basal part of the cell transports sodium out of the cell cytoplasm and into the sublying capillaries. This occurs by the usual process of active transport for sodium. The result is marked depletion of sodium ions in the interior of the epithelial cell. This in turn creates a diffusion gradient for sodium ions from the lumen of the intestine toward the interior of the epithelial cell. Consequently, sodium ions attempt to diffuse down this gradient into the cell. However, the brush border is reasonably impermeable to sodium except when the sodium combines with a carrier molecule, one type of which is the so-called *sodium-glucose carrier.* This carrier is peculiar in that it will not transport the sodium by itself but must also transport a glucose molecule at the same time. That is, when bound with both sodium and glucose, the carrier will then and only then transport both to the interior of the cell. One can readily see that it is the sodium gradient across the brush border membrane that provides the energy to promote this transport of both the sodium and the glucose. And, since energy is provided from this source, the glucose can be transported "up hill," that is, against a glucose concentration gradient. Thus, even when glucose is present in the lumen of the intestine (or of the renal tubule) in very low concentration, it can still be transported to the interior of the epithelial cell.

Once the glucose has entered the interior of the epithelial cell, it crosses the basal side of the cell by the usual process of facilitated diffusion, in the same manner that glucose crosses essentially all other membranes of the body.

There are several different types of sodium co-transport carriers. One of these, the sodium-glucose carrier discussed above, transports glucose or galactose, and four others transport different types of amino acids, as we shall note subsequently.

ACTIVE TRANSPORT OF SPECIFIC SUBSTANCES

ACTIVE TRANSPORT OF SODIUM AND POTASSIUM

Referring back to Figure 4–1, one sees that the sodium concentration outside the cell is very high in comparison with its concentration inside, and the converse is true of potassium. Also, Table 4–1 shows that minute quantities of sodium and potassium can diffuse through the pores of the cell. If such diffusion should take place over a long period of time, the concentrations of the two ions would eventually become equal inside and outside the cell unless there were some means to remove the sodium from the inside and to transport potassium back in.

Fortunately, a mechanism for active transport of sodium and potassium ions is present in all cell membranes of the body. It is called the *sodium-potassium pump.* The basic principles of this pump are illustrated in Figure 4–11. The carrier for this mechanism transports sodium from inside the cell to the outside and potassium from the outside to the inside. And, because this carrier also has the capability of splitting ATP molecules and utilizing the energy from this source to promote the sodium and potassium transport, the carrier is called *sodium-potassium ATPase*. This ATPase is composed of two protein molecules, one a globulin with a molecular weight of 95,000 and the other a glycoprotein with a molecular weight of 55,000. The larger molecule actually binds with both the sodium and potassium ions and also with the ATP, but the smaller molecule is also necessary to provide some facilitating function not yet understood.

Note in Figure 4–11 that *magnesium ATP* binds with the sodium-potassium ATPase at the inside surface of the membrane. The energy released from the ATP causes potassium ions to split away from the sodium-potassium ATPase carrier molecule and simultaneously causes sodium ions to bind. Then, at the outside surface of the mem-

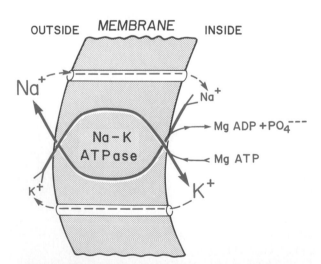

Figure 4–11. Postulated mechanism for active transport of sodium and potassium through the cell membrane, showing coupling of the two transport mechanisms and delivery of energy to the system at the inner surface of the membrane.

brane, the sodium ions split away from the carrier while potassium ions bind. But note that no energy source is required at the outside surface because the sodium release and potassium binding is an exothermic process that occurs without the need for additional energy. Thus, it is the intracellular ATP that provides the energy for transporting both the sodium and the potassium.

It should be noted especially that the sodium-potassium pump is powerful enough to transport sodium ions against concentration gradients as great as 20 to 1 and potassium gradients as great as 30 to 1.

One of the peculiarities of this sodium-potassium transport system is that it normally transports 3 sodium ions to the inside of the membrane for each 2 potassium ions transported to the outside. This gives a net transfer of 50 per cent more positive charges outward through the membrane than inward. Therefore, the sodium-potassium pump is called an *electrogenic pump* because it creates an electrical potential across the membrane whenever it pumps, creating negativity inside and positivity outside.

Another important feature of the pump is that it is strongly activated by an increase in sodium ion concentration inside the cell, the activity increasing in proportion to (sodium concentration)^3. This effect is extremely important because it allows even a slight excess buildup of sodium ions inside the cell to activate the pump very strongly and thereby return the intracellular sodium concentration back to its normal low level.

The sodium-potassium pump is so important to many different functioning systems of the body — such as to nerve and muscle fibers for transmission of impulses, various glands for the secretion of different substances, and all cells of the body to prevent cellular swelling — that we will discuss it at many different places in this text. In most instances it is the pumping of sodium rather then potassium that is especially important. Therefore, the sodium-potassium pump is called simply the *sodium pump*.

Importance of the Sodium Pump in Controlling Cell Size. One of the most important functions of the sodium pump is to prevent swelling of the cells. This may be explained as follows: All cells form many intracellular substances that are nondiffusible through the cell membrane, such as protein molecules, phosphocreatine, and adenosine triphosphate. Furthermore, almost all of these substances are electronegative so that they draw large numbers of positive ions into the cell as well. These substances and their associated positive ions tend to cause osmosis of water to the interior of the cell all the time. If some factor should not overcome the continual tendency for water to enter the cell, the cell would eventually swell until it should burst. However, the sodium transport mechanism opposes this tendency of the cell to swell by continually transporting sodium to the exterior, which initiates an opposite osmotic tenden-

cy to move water out of the cell. Whenever the metabolism of the cell ceases so that energy from adenosine triphosphate is not available to keep the sodium pump operating, the cell begins to swell immediately.

ACTIVE TRANSPORT OF OTHER IONS

Calcium and magnesium are probably transported by all cell membranes in much the same manner that sodium and potassium are transported, and certain cells of the body have the ability to transport still other ions. For instance, the glandular cell membranes of the thyroid gland can transport large quantities of iodide ion; the epithelial cells of the intestine can transport sodium, chloride, calcium, iron, hydrogen, and probably many other ions; and the epithelial cells of the renal tubules can transport calcium, magnesium, chloride, sodium, potassium, and a number of other ions.

TRANSPORT OF SUGARS

Facilitated diffusion of glucose and certain other sugars occurs in essentially all cells of the body, but secondary active transport (sodium co-transport) of sugars occurs in only a few places in the body. For instance, in the intestine and renal tubules, glucose and several other monosaccharides are continually transported into the blood even though their concentrations be minute in the lumens. Therefore, essentially none of these sugars is lost either in the intestinal excreta or the urine.

Though not all sugars are transported, almost all monosaccharides that are important to the body *are* transported, including *glucose, galactose, fructose, mannose, xylose, arabinose,* and *sorbose*. On the other hand, the disaccharides such as sucrose, lactose, and maltose are not transported at all.

The precise carrier system and chemical reactions responsible for secondary active transport of the monosaccharides are yet unknown. The one common denominator in transport of one group of sugars, including especially glucose and galactose, is the necessity for an intact –OH group on the C_2 carbon of the monosaccharide molecule. It is presumed that the monosaccharide attaches to the carrier at this point. Fructose, another monosaccharide important to the body, is transported by a separate carrier system.

TRANSPORT OF AMINO ACIDS

Most, if not all, amino acids, like glucose, are transported to the inside of essentially all cells of the body by facilitated diffusion mechanisms. Furthermore, insulin enhances this transport of amino acids in the same manner that it enhances facilitated diffusion of glucose.

Secondary active transport of amino acids (sodium co-transport) also occurs through a few membranes of the body: the epithelia of the intestines, renal tubules, and some exocrine glands. This involves at least four different carrier systems for transporting, respectively, the following different groups of amino acids: (1) neutral amino acids, (2) dibasic amino acids, (3) imino acids, and (4) dicarboxylic acids. These transport systems will be discussed further in relation to intestinal absorption in Chapter 65. It should be noted again that in the sodium co-transport mechanism it is the concentration gradient of sodium between the intestinal lumen and the interior of the cell that provides the energy for transport of the amino acid molecules, as explained earlier in the chapter.

One of the few known features of the amino acid carrier system is that transport of at least some amino acids depends on pyridoxine (vitamin B_6). Therefore, deficiency of this vitamin causes protein deficiency.

Hormonal Regulation of Amino Acid Transport. At least four different hormones are important in controlling amino acid transport: (1) *Growth hormone*, secreted by the adenohypophysis, increases amino acid transport into essentially all cells. (2) *Insulin* and (3) *glucocorticoids* increase amino acid transport at least into liver cells and possibly into other cells as well, though much less is known about these. (4) *Estradiol*, the most important of the female sex hormones, causes transport of amino acids into the musculature of the uterus, thereby promoting development of this organ.

Thus, several of the hormones exert much, if not most, of their effects in the body by controlling active transport of amino acids into all or certain cells.

PINOCYTOSIS AND PHAGOCYTOSIS

It was pointed out in Chapter 2 that the cell membrane has the ability to imbibe small amounts of substances from the extracellular fluid by the process called *pinocytosis,* and *phagocytosis* occurs by essentially the same mechanism as pinocytosis, but phagocytosis means ingestion of large particulate matter, such as bacteria or cell fragments, that is free in the extracellular fluid.

The real importance of pinocytosis to the body is that this is the only known means by which very large molecules, such as those of protein, can be transported to the interior of cells. The importance of phagocytosis is that it is used by special cells, such as the white blood cells, to rid the body of bacteria and unwanted debris in the tissues.

REFERENCES

Andreoli, T. E., and Schaefer, J. A.: Mass transport across cell membranes: The effects of antidiuretic hormone on water and solute flows in epithelia. *Annu. Rev. Physiol.,* 38:451, 1976.

Andreoli, T. E., *et al.* (eds.): Membrane Physiology. New York, Plenum Press, 1980.

Avery, J. (ed.): Membrane Structure and Mechanisms of Biological Energy Transduction. New York, Plenum Press, 1974.

Bean, C. P.: The physics of porous membranes—neutral pores. *Membranes, 1*:1, 1972.

Ellory, C., and Lew, V. L. (eds.): Membrane Transport in Red Cells. New York, Academic Press, 1977.

Fettiplace, R., and Haydon, D. A.: Water permeability of lipid membranes. *Physiol. Rev., 60*:510, 1980.

Finn, A. L.: Changing concepts of transepithelial sodium transport. *Physiol. Rev., 56*:453, 1976.

Frizzell, R. A., and Heintze, K.: Transport functions of the gallbladder. *In* Javitt, N. B. (ed.): International Review of Physiology: Liver and Biliary Tract Physiology I. Vol. 21. Baltimore, University Park Press, 1980, p. 221.

Frizzell, R. A., and Schultz, S. G.: Models of electrolyte absorption and secretion by gastrointestinal epithelia. *In* Crane, R. K. (ed.): International Review of Physiology: Gastrointestinal Physiology III. Vol. 19. Baltimore, University Park Press, 1979, p. 205.

Gilles, R. (ed.): Mechanisms of Osmoregulation: Maintenance of Cell Volume. New York, John Wiley & Sons, 1979.

Glynn, I. M., and Karlish, S. J. D.: The sodium pump. *Annu. Rev. Physiol., 37*:13, 1975.

Goresky, C. A.: Uptake in the liver: the nature of the process. *In* Javitt, N. B. (ed.): International Review of Physiology: Liver and Biliary Tract Physiology I. Vol. 21. Baltimore, University Park Press, 1980, p. 65.

Gregor, H. P., and Gregor, C. D.: Synthetic-membrane technology. *Sci. Am., 239*(1):112, 1978.

Gupta, B. L., *et al.* (eds.): Transport of Ions and Water in Animals. New York, Academic Press, 1977.

Guyton, A. C., *et al.*: Dynamics and Control of the Body Fluids. Philadelphia, W. B. Saunders Co., 1975.

Hagiwara, S., and Jaffe, L. A.: Electrical properties of egg cell membranes. *Annu. Rev. Biophys. Bioeng., 8*:385, 1979.

Hemmings, W. A. (ed.): Protein Transmission Through Living Membranes. New York, Elsevier/North-Holland, 1979.

Jacob, H. S. (ed.): Blood Cell Membranes. New York, Grune & Stratton, 1979.

Keynes, R. D.: Ion channels in the nerve-cell membrane. *Sci. Am., 240*(3):126, 1979.

Korenbrot, J. I.: Ion transport in membranes: Incorporation of biological ion-translocating proteins in model membrane systems. *Annu. Rev. Physiol., 39*:19, 1977.

Korn, E. D. (ed.): Methods in Membrane Biology. Vols. 1 and 2. New York, Plenum Press, 1974.

Korn, E. D.: Transport. New York, Plenum Publishing Corp., 1975.

Kotyk, A.: Mechanisms of nonelectrolyte transport. *Biochem. Biophys. Acta, 300*:183, 1973.

Lehninger, A. L.: The molecular organization of mitochondrial membranes. *Adv. Cytopharmacol., 1*:199, 1971.

Lodish, H. F., and Rothman, J. E.: The assembly of cell membranes. *Sci. Am., 240*(1):48, 1979.

Lux, S. E., *et al.* (eds.): Normal and Abnormal Red Cell Membranes. New York, A. R. Liss, 1979.

Macknight, A. D. C., and Leaf, A.: Regulation of cellular volume. *Physiol. Rev., 57*:510, 1977.

Metcalfe, J. C. (ed.): Biochemistry of Cell Walls and Membranes II. Baltimore, University Park Press, 1978.

Packer, L.: Biomembranes: Architecture, Biogenesis, Bioenergetics, and Differentiation. New York, Academic Press, 1975.

Parsons, D. S. (ed.): Biological Membranes. New York, Oxford University Press, 1975.

Schultz, S. G.: Principles of electrophysiology and their application to epithelial tissues. *Int. Rev. Physiol., 4*:69, 1974.

Schultz, S. G., and Curran, P. F.: Role of sodium in non-electrolyte transport across animal cell membranes. *Physiologist, 12*:437, 1969.

Schultz, S. G., and Curran, P. F.: Coupled transport of sodium and organic solutes. *Physiol. Rev., 50*:637, 1970.

Sen, A. K., and Post, R. L.: Stoichiometry and localization of adenosine triphosphate-dependent sodium and potassium transport in the erythrocyte. *J. Biol. Chem., 239*:345, 1964.

Singer, S. J.: The molecular organization of membranes. *Annu. Rev. Physiol., 43*:805, 1974.

Solomon, A. K.: Properties of water in red cell and synthetic membranes. *Biomembranes, 3*:299, 1972.

Ullrich, K. L.: Sugar, amino acid, and Na^+ cotransport in the proximal tubule. *Annu. Rev. Physiol., 41*:181, 1979.

Van Winkle, W. B., and Schwartz, A.: Ions and inotrophy. *Annu. Rev. Physiol., 38*:247, 1976.

Wallick, E. T., *et al.*: Biochemical mechanism of the sodium pump. *Annu. Rev. Physiol., 41*:397, 1979.

Wright, E. M., and Diamond, J. M.: Anion selectivity in biological systems. *Physiol. Rev., 57*:109, 1977.

Part II

BLOOD CELLS, IMMUNITY, AND BLOOD CLOTTING

5

Red Blood Cells, Anemia, and Polycythemia

With this chapter we begin a discussion of the blood cells and of other cells closely related to those of the blood: the cells of the reticuloendothelial system and of the lymphatic system. We will first present the functions of red blood cells, which are the most abundant of all the cells of the body and are necessary for delivery of oxygen to the tissues.

THE RED BLOOD CELLS

The major function of red blood cells is to transport hemoglobin, which in turn carries oxygen from the lungs to the tissues. In some lower animals hemoglobin circulates as free protein in the plasma, not enclosed in red blood cells. However, when it is free in the plasma of the human being, approximately 3 per cent of it leaks through the capillary membrane into the tissue spaces or through the glomerular membrane of the kidney into Bowman's capsule each time the blood passes through the capillaries. Therefore, for hemoglobin to remain in the blood stream, it must exist inside red blood cells.

The red blood cells have other functions besides simply transport of hemoglobin. For instance, they contain a large quantity of carbonic anhydrase, which catalyzes the reaction between carbon dioxide and

water, increasing the rate of this reaction about 250 times. The rapidity of this reaction makes it possible for blood to react with large quantities of carbon dioxide and thereby transport it from the tissues to the lungs. Also, the hemoglobin in the cells is an excellent acid-base buffer (as is true of most proteins), so that the red blood cells are responsible for approximately 70 per cent of all the buffering power of whole blood. These specific functions are discussed elsewhere in the text and, consequently, are merely mentioned at the present time.

The Shape and Size of Red Blood Cells. Normal red blood cells are biconcave disks having a mean diameter of approximately 8 microns and a thickness at the thickest point of 2 microns and in the center of 1 micron or less. The average volume of the red blood cell is 83 cubic microns.

The shapes of red blood cells can change remarkably as the cells pass through capillaries. Actually, the red blood cell is a "bag" that can be deformed into almost any shape. Furthermore, because the normal cell has a great excess of cell membrane for the quantity of material inside, deformation does not stretch the membrane, and consequently does not rupture the cell as would be the case with many other cells.

Concentration of Red Blood Cells in the Blood. In normal men the average number of red blood cells per cubic millimeter is 5,200,000 (±300,000) and in normal women 4,700,000 (±300,000). The number of red blood cells varies in the two sexes and at different ages, as shown in Figure 5–1. Also, the altitude at which the person lives affects the number of red blood cells; this is discussed later.

Quantity of Hemoglobin in the Cells. Red blood cells have the ability to concentrate hemoglobin in the cell fluid up to approximately 34 grams per 100 ml. of cells. The concentration never rises above this value, for this is a metabolic limit of the cell's hemoglobin-forming mechanism. Furthermore, in normal persons the percentage of hemoglobin is almost always near the maximum in each cell. However, when hemoglobin formation is deficient in the bone marrow, the percentage of hemoglobin in the cells may fall considerably below this value and the volume of the red cell may decrease as well because of diminished hemoglobin to fill the cell.

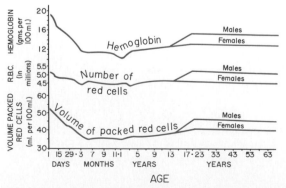

Figure 5–1. Relationship of age and sex to the hemoglobin content, red blood cell count, and hematocrit of the blood.

When the hematocrit (the percentage of the blood that is cells — normally 40 to 45 per cent) and the quantity of hemoglobin in each respective cell are normal, the whole blood of men contains an average of 16 grams of hemoglobin per 100 ml. and of women an average of 14 grams per 100 ml. As will be discussed in connection with the transport of oxygen in Chapter 41, each gram of pure hemoglobin is capable of combining with approximately 1.39 ml. of oxygen. Therefore, in normal man, over 21 ml. of oxygen can be carried in combination with hemoglobin in each 100 ml. of blood, and in normal woman 19 ml. of oxygen can be carried.

PRODUCTION OF RED BLOOD CELLS

Areas of the Body That Produce Red Blood Cells. In the early few weeks of embryonic life, primitive red blood cells are produced in the yolk sac. During the middle trimester of gestation the liver is the main organ for production of red blood cells, and at the same time a reasonable quantity of red blood cells is also produced by the spleen and lymph nodes. Then, during the latter part of gestation and after birth, red blood cells are produced exclusively by the bone marrow.

As illustrated by Figure 5–2, the bone marrow of essentially all bones produces red blood cells until age 5, but the marrow of the long bones, except for the proximal portions of the humeri and tibiae, becomes quite fatty and produces no more red blood cells after approximately the age of 20. Beyond this age most red cells are produced in the marrow of the membranous

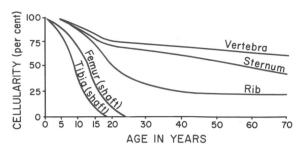

Figure 5–2. Relative rates of red blood cell production in the different bones at different ages.

bones, such as the vertebrae, the sternum, the ribs, and the pelvis. Even in these bones the marrow becomes less productive as age increases.

Sometimes when various factors stimulate the bone marrow to produce tremendous quantities of red blood cells, much of the marrow that has already stopped producing red blood cells can once again become productive, and marrow that is still producing red blood cells becomes greatly hyperplastic and produces far greater than normal quantities.

Genesis of the Red Blood Cell. The blood cells are derived from a cell known as the *hemocytoblast,* which is illustrated in Figure 5–3, and new hemocytoblasts are continually being formed from primordial *stem* cells located throughout the bone marrow. The hemocytoblast is also frequently called a *committed stem cell* or sometimes a *unipotential stem cell.* It, like the stem cell,

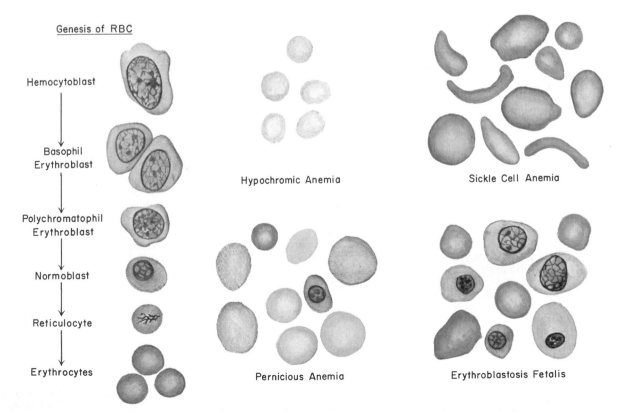

Figure 5–3. Genesis of red blood cells, and red blood cells in different types of anemias.

can reproduce itself again and again. However, it is different from the stem cell in that it can lead only to the subsequent formation of red blood cells and not to the many other types of body cells.

As illustrated in Figure 5–3, the hemocytoblast first forms the *basophil erythroblast,* which begins the synthesis of hemoglobin. The erythroblast then becomes a *polychromatophil erythroblast,* so-called because of a mixture of basophilic material and the red hemoglobin. Following this, the nucleus of the cell shrinks while still greater quantities of hemoglobin are formed, and the cell becomes a *normoblast.* During the earlier stages the different cells continue to divide so that greater and greater numbers of cells are formed. Finally, after the cytoplasm of the normoblast has become filled with hemoglobin to a concentration of approximately 34 per cent, the nucleus becomes extremely small and is extruded. At the same time, the endoplasmic reticulum is being reabsorbed. The cell at this stage of development is called a *reticulocyte* because it still contains a small amount of basophilic endoplasmic reticulum interspersed among the hemoglobin in the cytoplasm. While the cells are in the reticulocyte stage, they pass into the blood capillaries by diapedesis (squeezing through the pores of the membrane).

The remaining endoplasmic reticulum in the reticulocyte continues to produce a small amount of hemoglobin for one to two days, but by the end of that time the reticulum completely disappears. In normal blood, the total proportion of reticulocytes among all the cells is slightly less than 1 per cent. Once the reticulum has been completely resorbed, the cell is then the mature *erythrocyte.*

Regulation of Red Blood Cell Production

The total mass of red blood cells in the circulatory system is regulated within very narrow limits so that an adequate number of red cells is always available to provide sufficient tissue oxygenation and, yet, so that the cells do not become so concentrated that they impede blood flow. The little we know about this control mechanism is the following:

Tissue Oxygenation as the Basic Regulator of Red Blood Cell Production.   Any condition that causes the quantity of oxygen transported to the tissues to decrease ordinarily increases the rate of red blood cell production. Thus, when a person becomes extremely *anemic* as a result of hemorrhage or any other condition, the bone marrow immediately begins to produce large quantities of red blood cells. Also, destruction of major portions of the bone marrow by any means, especially x-ray therapy, causes hyperplasia of the remaining bone marrow, thereby attempting to supply the demand for red blood cells in the body.

At very *high altitudes,* where the quantity of oxygen in the air is greatly decreased, insufficient oxygen is transported to the tissues, and red cells are produced so rapidly that their number in the blood is considerably increased.

Therefore, it is obvious that it is not the concentration of red blood cells in the blood that controls the rate of red cell production, but instead it is the functional ability of the cells to transport oxygen to the tissues in relation to the tissue demand for oxygen.

Various diseases of the circulation that cause decreased blood flow through the peripheral vessels, and particularly those that cause failure of oxygen absorption by the blood as it passes through the lungs, also increase the rate of red cell production. This is especially apparent in prolonged *cardiac failure* and in many *lung diseases,* for the tissue hypoxia resulting from these conditions increases the rate of red cell production, with resultant increase in the hematocrit and usually some increase in the total blood volume.

Erythropoietin, Its Formation in Response to Hypoxia, and Its Function in Regulating Red Blood Cell Production.   Erythropoietin, also called *erythropoietic stimulating factor* or *hemopoietin,* is a glycoprotein that has a molecular weight between 39,000 and 70,000 and that appears in the the blood in response to hypoxia. Erythropoietin in turn acts on the bone marrow to increase the rate of red blood cell production.

There is no direct response of the bone marrow to hypoxia. Instead, hypoxia stimulates red blood cell production only through the mechanism of erythropoietin.

Role of the Kidneys in the Formation of Erythropoietin — Renal Erythropoietic Factor.   The precise mechanism for formation of erythropoietin is not clearly understood, but it is known that only minute quantities of erythropoietin are formed in animals or persons whose kidneys have been removed. Yet, strangely enough, erythropoietin cannot be isolated from the kidneys. Therefore, at present it is believed that the relationship of the kidneys to erythropoietin formation is the following:

When the kidneys become hypoxic, it is believed that they release an enzyme called *renal erythropoietic factor.* This is secreted into the blood where it acts within a few minutes on one of the plasma proteins, a globulin, to split away the glycoprotein erythropoietin molecule. Erythropoietin, in turn, circulates in the blood for about one day and during this time acts on the bone marrow to cause erythropoiesis.

In the complete absence of the kidneys, minute amounts of erythropoietin are still formed, and the quantity that is formed is slightly increased in the presence of hypoxia. Therefore, it is clear that other tissues, probably the liver in particular, can form very slight amounts of erythropoietic factor that can lead to the formation of erythropoietin. Even so, in the absence of the kidneys a person usually becomes very anemic because of the extremely low levels of circulating erythropoietin.

Effect of Erythropoietin on Erythrogenesis.   Though erythropoietin begins to be formed almost immediately upon placing an animal or person in an atmosphere of low oxygen, almost no new red blood cells appear in the circulating blood within the first two days; and it is only after five or more days that the maximum rate of new red cell production is reached. Thereafter, cells continue to be produced as long as the person remains in the low oxygen state or until he has produced enough red blood cells to carry adequate amounts of oxygen to his tissues despite the low oxygen.

Upon removal of a person from a state of low oxygen, his rate of oxygen transport to the tissues rises above normal, which causes his rate of erythropoietin formation to decrease to zero almost instantaneously and his rate of red blood cell production to fall essentially to

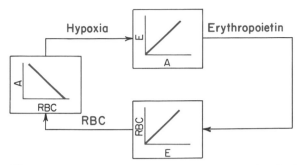

Figure 5–4. Control diagram, showing the negative feedback mechanism by which hypoxia regulates red blood cell concentration.

zero within several days. Red cell production remains at this extremely low level until enough cells have lived out their life spans and degenerated, so that the tissues again receive only their normal complement of oxygen.

The exact site of action of erythropoietin to increase the rate of red cell production is not known. However, even the earliest stages of erythroblastosis are greatly accelerated. Therefore, it is presumed that erythropoietin has an especially strong effect on increasing the rate of division of hemocytoblasts (also called the committed stem cells). It is possible that it increases the rate of conversion of stem cells into hemocytoblasts as well, though this is not yet certain.

In the complete absence of erythropoietin, few red blood cells are formed by the bone marrow. At the other extreme, when extreme quantities of erythropoietin are formed, the rate of red blood cell production can rise to as high as 6 to 8 times normal. Therefore, the erythropoietin control mechanism for red blood cell production is an extremely powerful one.

System Diagram for Control of Red Cell Numbers. Figure 5–4 illustrates a control diagram for the feedback mechanism that regulates red blood cell concentration, drawn using the denotations described in Chapter 1. This shows that hypoxia causes production of erythropoietin. Erythropoietin then causes red blood cell production, and red blood cells finally cause a negative effect on the hypoxia — that is, they transport oxygen to the tissues, and this alleviates the hypoxia. Because of the extremely powerful control of erythropoietin over red cell production, this negative feedback mechanism can adjust the red blood cell concentration to almost precisely that amount required for adequate oxygenation of the tissues.

Effect of Rate of Red Blood Cell Formation on the Type of Cell Released into the Blood

When the bone marrow produces red blood cells at a very rapid rate, many of the cells are released into the blood before they are mature erythrocytes. Thus, very rapid red cell production can cause the percentage of reticulocytes in the circulating blood to rise from less than 1 per cent to as high as 30 to 50 per cent of the total number of red blood cells. If the rate of production is even greater, a large number of normoblasts may appear

in the circulating blood. In some severe anemias, such as erythroblastosis fetalis, which is discussed later in the chapter and in greater detail in Chapter 8, the number of normoblasts may occasionally rise to as high as 5 to 20 per cent of all the circulating red blood cells, and even a small number of erythroblasts may appear in the circulating blood.

Vitamins Needed for Formation of Red Blood Cells

The Maturation Factor — Vitamin B_{12} (Cyanocobalamin). Vitamin B_{12} is an essential nutrient for all cells of the body, and growth of tissues in general is greatly depressed when this vitamin is lacking. This results from the fact that vitamin B_{12} is required for synthesis of DNA. Therefore, lack of this vitamin causes failure of nuclear maturation and division. Since tissues that produce red blood cells are among the most rapidly growing and proliferating of all the body's tissues, lack of vitamin B_{12} inhibits the rate of red blood cell production especially. Furthermore, the erythroblastic cells of the bone marrow, in addition to failing to proliferate rapidly, become larger than normal, developing into so-called *megaloblasts,* and the adult erythrocyte, called a *macrocyte,* has a flimsy membrane and is often irregular, large, and oval instead of the usual biconcave disk. These poorly formed macrocytes, after entering the circulating blood, are quite capable of carrying oxygen, but their fragility causes them to have a short life, measured in weeks instead of months as is true of normal cells. Therefore, it is said that vitamin B_{12} deficiency causes *maturation failure* in the process of erythropoiesis.

The occurrence of maturation failure does not prevent normal formation of hemoglobin. Indeed, the hemoglobin *concentration* in the cell is approximately 34 grams per cent, the same as normal, and, moreover, the average *quantity* of hemoglobin in the macrocyte is considerably greater than normal because the average volume of each cell is greater than normal.

Maturation Failure Caused by Poor Absorption of Vitamin B_{12} — Pernicious Anemia. The most common cause of maturation failure is not lack of vitamin B_{12} in the diet but instead failure to absorb vitamin B_{12} from the gastrointestinal tract. This often occurs in the disease called *pernicious anemia,* in which the basic abnormality is an *atrophic gastric mucosa* that fails to secrete normal gastric secretions. The parietal cells of the gastric glands secrete a substance (molecular weight about 50,000) called *intrinsic factor* which combines with vitamin B_{12} of the food and makes the B_{12} available for absorption by the gut. It does this in the following way: First, the intrinsic factor binds tightly with the vitamin B_{12}. In this bound state the B_{12} is protected from digestion by the gastrointestinal enzymes. Second, still in the bound state, the B_{12} and intrinsic factor become adsorbed to specific receptor sites on the brush border membranes of the mucosal cells in the ileum. Third, the combination is then transported into the cells in pinocytotic vesicles, and about four hours later the vitamin B_{12} is released into the blood.

Lack of intrinsic factor, therefore, causes loss of much of the vitamin because of enzyme action in the gut, and also results in failure of absorption.

Once vitamin B_{12} has been absorbed from the gastrointestinal tract it is stored in large quantities in the liver and then released slowly as needed to the bone marrow and other tissues of the body. The total amount of vitamin B_{12} required each day to maintain normal red cell maturation is less than 1 microgram, and the normal store in the liver is about 1000 times this amount. Therefore, many months of defective B_{12} absorption are required to cause maturation failure anemia.

Effect of Folic Acid (Pteroylglutamic Acid) on Red Cell Maturation. Occasionally a patient with maturation failure anemia responds equally as well to folic acid as to vitamin B_{12}, so that it is apparent that this vitamin is also concerned with the maturation of red blood cells. Folic acid, like B_{12}, is required for formation of DNA but in a different way. It promotes the methylation of deoxyuridylate to form deoxythymidylate, one of the nucleotides required for DNA synthesis. In other ways, folic acid is also required for RNA synthesis.

FORMATION OF HEMOGLOBIN

Synthesis of hemoglobin begins in the erythroblasts and continues through the normoblast and reticulocyte stages. Even when the reticulocytes leave the bone marrow and pass into the blood stream, they continue to form minute quantities of hemoglobin for another day or so.

Figure 5–5 shows the basic chemical steps in the formation of hemoglobin. From tracer studies with isotopes it is known that the heme portion of hemoglobin is synthesized mainly from acetic acid and glycine and that most of this synthesis occurs in the mitochondria. It is believed that the acetic acid is changed in the Krebs cycle, which will be explained in Chapter 67, into α-ketoglutaric acid, and then two molecules of this combine with one molecule of glycine to form a pyrrole compound. In turn, four pyrrole compounds combine to form a protoporphyrin compound. One of the protoporphyrin compounds, known as protoporphyrin III, then combines with iron to form the heme molecule. Finally, four heme molecules combine with one molecule of globin, a globulin that is synthesized in the ribosomes of the endoplasmic reticulum, to form hemoglobin, the formula for which is shown in Figure 5–6. Hemoglobin has a molecular weight of 64,458.

The globin portion of the hemoglobin molecule is composed of four large polypeptide chains. The nature of these chains determines the binding affinity of the

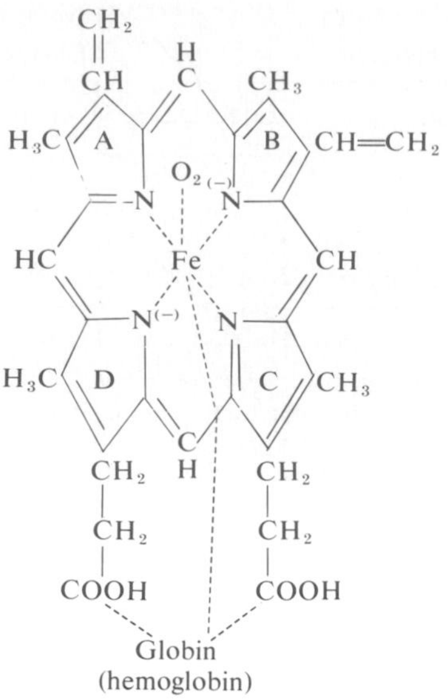

Figure 5–6. Basic structure of the hemoglobin molecule, showing one of the four heme complexes bound with the central globin core of the hemoglobin molecule.

hemoglobin for oxygen. Abnormalities of the chains can alter the physical characteristics of the hemoglobin molecule as well. For instance, in sickle cell anemia the amino acid valine is substituted for glutamic acid at one point in two of the four chains. When this type of hemoglobin is exposed to low oxygen, it forms elongated crystals inside the red blood cells that are sometimes 15 microns in length. These make it almost impossible for the cells to pass through the small capillaries, and the spiked ends of the crystals are very likely to rupture the cell membranes, thus leading to sickle cell anemia.

Combination of Hemoglobin with Oxygen. The most important feature of the hemoglobin molecule is its ability to combine loosely and reversibly with oxygen. This ability will be discussed in detail in Chapter 41 in relation to respiration, for the primary function of hemoglobin in the body depends upon its ability to combine with oxygen in the lungs and then to release this oxygen readily in the tissue capillaries where the gaseous tension of oxygen is much lower than in the lungs.

Oxygen *does not* combine with the two positive valences of the ferrous iron in the hemoglobin molecule. Instead, it binds loosely with one of the six "coordination" valences of the iron atom. This is an extremely loose bond so that the combination is easily reversible. Furthermore, the oxygen does not become ionic oxygen but is carried as molecular oxygen to the tissues where, because of the loose, readily reversible combination, it is released into the tissue fluids in the form of dissolved molecular oxygen rather than ionic oxygen.

It should be recalled, also, that each molecule of hemoglobin contains four molecules of heme. Therefore, one molecule of hemoglobin contains four iron atoms and can carry four molecules of oxygen.

Accessory Substances Needed for Formation of Hemoglobin. In addition to amino acids and iron, which are needed directly for formation of the hemoglobin molecule, a number of other substances act as catalysts or

I. 2 α-ketoglutaric acid + glycine $\longrightarrow$

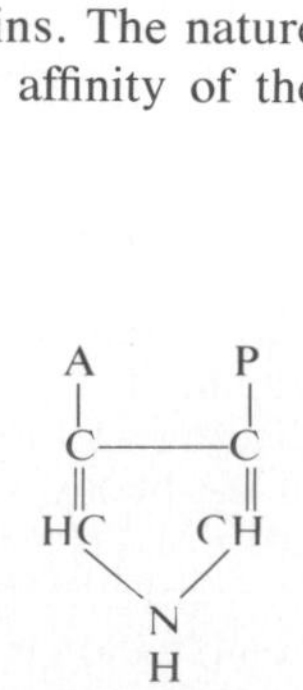

II. 4 pyrrole $\longrightarrow$ protoporphyrin III

III. protoporphyrin III + Fe $\longrightarrow$ heme

IV. 4 heme + globin $\longrightarrow$ hemoglobin

Figure 5–5. Formation of hemoglobin.

enzymes during different stages of hemoglobin formation. For instance, an average human adult requires approximately 2 mg. of *copper* each day in his diet if normal hemoglobin formation is to take place, and addition of small quantities of copper to the diet of patients who have hypochromic anemia occasionally accelerates the rate of hemoglobin formation. This occurs even though copper is not one of the substances needed as a building stone in the formation of hemoglobin. Fortunately, the quantity of copper in the normal daily diet is sufficient so that copper deficiency anemia is almost unknown in the human being.

Lack of *pyridoxine* in the diet of some animals not only decreases the rate of red blood cell formation but also depresses the rate of hemoglobin formation to an even greater extent. Also, *cobalt* deficiency can greatly depress the formation of hemoglobin in some animals, while a great excess of cobalt can cause formation of greater than normal numbers of red blood cells that contain normal quantities of hemoglobin. Finally, *nickel* has been found to take the place of cobalt to a moderate extent in aiding the synthesis of hemoglobin in the bone marrow.

Though the function of these different substances in the formation of hemoglobin is not known, the above listing of them serves mainly to emphasize the fact that hemoglobin formation results from a series of synthesis reactions, each of which depends upon appropriate building materials and also upon appropriate controlling catalysts and enzymes.

IRON METABOLISM

Because iron is important for formation of hemoglobin, myoglobin, and other substances such as the cytochromes, cytochrome oxidase, peroxidase, and catalase, it is essential to understand the means by which iron is utilized in the body.

The total quantity of iron in the body averages about 4 grams, approximately 65 per cent of which is present in the form of hemoglobin. About 4 per cent is present in the form of myoglobin, 1 per cent in the form of the various heme compounds that control intracellular oxidation, 0.1 per cent combined with the protein transferrin in the blood plasma, and 15 to 30 per cent stored mainly in the liver in the forms of ferritin and hemosiderin.

Transport and Storage of Iron. Transport, storage. and metabolism of iron in the body are illustrated in Figure 5–7, which may be explained as follows: When iron is absorbed from the small intestine, it immediately combines with a beta globulin, *transferrin,* with which it is transported in the blood plasma. The iron is very loosely combined with the globulin molecule and, consequently, can be released to any of the tissue cells at any point in the body. Excess iron in the blood is deposited in all cells of the body *but especially in the liver cells,* where about 60 per cent of the excess is stored. There it combines mainly with a protein, *apoferritin,* to form *ferritin.* Apoferritin has a molecular weight of approximately 460,000, and varying quantities of iron can combine in clusters of iron radicals with this large molecule; therefore, ferritin may contain only a small amount of iron or a relatively large amount. This iron stored in ferritin is called *storage iron.*

Smaller quantities of the iron in the storage pool are stored in an extremely insoluble form called *hemosiderin*. This is especially true when the total quantity of iron in the body is more than the apoferritin storage pool can accommodate. Hemosiderin forms large clusters in the cells and consequently can be stained and observed as large particles in tissue slices by usual histologic techniques. Ferritin can also be stained, but the ferritin particles are so small and dispersed that they usually can be seen only with the electron microscope.

When the quantity of iron in the plasma falls very low, iron is removed from ferritin quite easily but less easily from hemosiderin. The iron is then transported to the portions of the body where it is needed.

When red blood cells have lived their life span and are destroyed, the hemoglobin released from the cells is ingested by the reticuloendothelial cells. There free iron is liberated, and it can then either be stored in the ferritin pool or be reused for formation of hemoglobin.

Daily Loss of Iron. About 0.6 mg. of iron is excreted each day by the male, mainly into the feces. Additional quantities of iron are lost whenever bleeding occurs. Thus, in the female, the menstrual loss of blood brings the average iron loss to a value of approximately 1.3 mg. per day.

Obviously, the average quantity of iron derived from the diet each day must at least equal that lost from the body.

Absorption of Iron from the Gastrointestinal Tract. Iron is absorbed almost entirely in the upper part of the small intestine, mainly in the duodenum. It is absorbed by an active absorptive process, though the precise mechanism of this active absorption is unknown.

Iron is absorbed mainly in the ferrous rather than ferric form, for, in general, the greater the degree of positive charge of any ion the more difficult it is to be absorbed from the intestinal tract. This has practical significance because it means that ferrous iron compounds are more effective in treating iron deficiency than are ferric compounds.

The rate of iron absorption is extremely slow, with a maximum rate of only a few milligrams per day. This means that when tremendous quantities of iron are present in the food, only small proportions of this will be absorbed. On the other hand, if only minute quantities are present, far greater proportions will be absorbed.

Regulation of Total Body Iron by Alteration of Rate of Absorption. When the body has become saturated with iron so that essentially all of the apoferritin in the iron storage areas is already combined with iron, the rate of absorption of iron from the intestinal tract becomes greatly decreased. On the other hand, when the iron stores have been depleted of iron, the rate of absorption becomes greatly accelerated, to as much as five or more times as great as when the iron stores are saturated. Thus, the total body iron is regulated to a great extent by altering the rate of absorption.

Feedback Mechanisms by Which Iron Absorption Is Regulated. When essentially all the apoferritin in the body has become saturated with iron, it becomes difficult for transferrin to release iron to the tissues. As a consequence, the transferrin, which is normally only one-third saturated with iron, now becomes almost fully bound with iron so that the transferrin accepts almost no

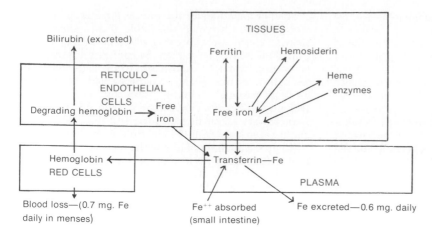

Figure 5–7. Iron transport and metabolism.

new iron from the mucosal cells. Then, as a final stage of this process, the buildup of excess iron in the mucosal cells themselves depresses active absorption of iron from the intestinal lumen and at the same time slightly enhances the rate of excretion of iron from the mucosa.

Unfortunately, the details of the method by which excess iron depresses active absorption of iron by the mucosa are yet unknown, though many different theories have been proposed. One that held sway for a number of years, the *mucosal block theory,* suggested that accumulation of excess iron in the mucosal epithelial cells blocked the active process for iron absorption, in this way regulating iron absorption. However, recent research suggests that the mucosal block theory is incorrect. Instead, a very likely mechanism for control of iron absorption is based on the ability of transferrin to accept iron from the mucosal epithelial cells. When the transferrin is already saturated with iron (which occurs when the iron storage pools are full), then no additional iron can be accepted, so that iron absorption becomes greatly diminished or ceases.

DESTRUCTION OF RED BLOOD CELLS

When red blood cells are delivered from the bone marrow into the circulatory system they normally circulate an average of 120 days before being destroyed. Even though mature red cells do not have a nucleus and also have neither mitochondria nor endoplasmic reticulum, nevertheless they do still have cytoplasmic enzymes that are capable of metabolizing glucose by the glycolytic process, thus forming small amounts of ATP. Also, additional small amounts of ATP can be formed utilizing a pathway called the hexosemonophosphate shunt (see Chapter 67) followed by oxidative formation of ATP. The ATP in turn serves the red cell in several important ways: (1) to maintain pliability of the cell membrane, (2) to maintain membrane transport of ions, (3) to keep the iron of the cell's hemoglobin in the ferrous form rather than the ferric form (which causes the formation of methemoglobin that will not carry oxygen), and (4) to prevent oxidation of the proteins in the red cell. However, these metabolic systems of the

red cell become progressively less active with time. As the cells become older they become progressively more fragile, presumably because their life processes simply wear out.

Once the red cell membrane becomes very fragile, the cell may rupture during passage through some tight spot of the circulation. Many of the red cells fragment in the spleen where the cells squeeze through the red pulp of the spleen. Here the spaces between the structural trabeculae of the pulp are only 3 microns wide in comparison with the 8 micron diameter of the red cell. When the spleen is removed, the number of abnormal cells and old cells circulating in the blood increases considerably.

Destruction of Hemoglobin. The hemoglobin released from the cells when they burst is phagocytized almost immediately by reticuloendothelial cells. During the next few days, they release the iron from the hemoglobin back into the blood to be carried by transferrin either to the bone marrow for production of new red blood cells or to the liver and other tissues for storage in the form of ferritin. The heme portion of the hemoglobin molecule is converted by the reticuloendotheial cell, through a series of stages, into the bile pigment *bilirubin,* which is released into the blood and later secreted by the liver into the bile; this will be discussed in relation to liver function in Chapter 70.

THE ANEMIAS

Anemia means a deficiency of red blood cells, which can be caused either by too rapid loss or too slow production of red blood cells. Some types of anemia and their physiological causes are the following:

Blood Loss Anemia. After rapid hemorrhage the body replaces the plasma within one to three days, but this leaves a low concentration of red blood cells. If a second hemorrhage does not occur, the red blood cell concentration returns to normal within three to four weeks.

In chronic blood loss, the person frequently cannot absorb enough iron from the intestines to form hemoglobin as rapidly as it is lost. Therefore, red cells are then produced with too little hemoglobin inside them, giving rise to *microcytic hypochromic anemia.*

Aplastic Anemia. *Bone marrow aplasia* means lack of a functioning bone marrow. For instance, the person exposed to gamma ray radiation from a nuclear bomb blast is likely to sustain complete destruction of bone marrow, followed in a few weeks by lethal anemia. Likewise, excessive x-ray treatment, certain industrial chemicals, and even drugs to which the person might be sensitive can cause the same effect.

Maturation Failure Anemia. From the earlier discussion in this chapter of vitamin B_{12}, folic acid, and intrinsic factor from the stomach mucosa, one can readily understand that loss of any one of these factors can lead to failure of maturation of the red blood cells. Thus, atrophy of the stomach mucosa, as occurs in *pernicious anemia,* or loss of the entire stomach as the result of total gastrectomy can lead to maturation failure. Also, patients who have intestinal sprue, in which folic acid, B_{12}, and other vitamin B compounds are poorly absorbed, often develop maturation failure. Because the bone marrow cannot proliferate rapidly enough to form normal numbers of red blood cells in maturation failure, the cells that are formed are oversized, of bizarre shapes, and have fragile membranes. Therefore, these cells rupture easily, leaving the person in dire need of an adequate number of red cells.

Hemolytic Anemia. Many different abnormalities of the red blood cells, many of which are hereditarily acquired, make the cells very fragile, so that they rupture easily as they go through the capillaries, especially through the spleen. Therefore, even though the number of red blood cells formed is completely normal, the red cell life span is so short that serious anemia results. Some of these types of anemia are the following:

In *hereditary spherocytosis* the red cells are very small in size, and they are *spherical* in shape rather than being biconcave discs. These cells cannot be compressed because they do not have the normal loose bag-like cell membrane structure of the biconcave discs. Therefore, on passing through the splenic pulp they are easily ruptured by even slight compression.

In *sickle cell anemia,* which is present in 0.3 to 1.0 per cent of West African and American Blacks, the cells contain an abnormal type of hemoglobin called *hemoglobin S,* caused by abnormal composition of the globin portion of the hemoglobin, as explained earlier in the chapter. When this hemoglobin is exposed to low concentrations of oxygen, it precipitates into long crystals inside the red blood cell. These crystals elongate the cell and give it the appearance of being a sickle rather than a biconcave disc. The precipitated hemoglobin also damages the cell membrane so that the cells become highly fragile, leading to serious anemia. Such patients frequently go into a vicious cycle called a sickle cell disease "crisis" in which low oxygen tension in the tissues causes sickling, which causes impediment of blood flow through the tissues, causing still further decrease in oxygen tension. Thus, once the process starts, it progresses rapidly, leading to a serious decrease in red blood cell mass within a few hours and, often, to death.

Thalassemia, which is also known as *Cooley's anemia* or *Mediterranean anemia,* is another hereditary type of anemia in which the cells are unable to synthesize adequate amounts of the polypeptide chains required to form the globin portion of hemoglobin. Therefore, hemoglobin synthesis is greatly depressed. The hemopoietic system responds by producing tremendous numbers of new red blood cells but with little hemoglobin and also cells that are small and have fragile membranes. Here again, the cells are easily ruptured upon passing through the tissues.

In *erythroblastosis fetalis,* Rh positive red blood cells in the fetus are attacked by antibodies from an Rh negative mother. These antibodies make the cells fragile and cause the child to be born with serious anemia. This will be discussed in detail in Chapter 8 in relation to the Rh factor of blood.

Hemolysis also occasionally results from transfusion reactions, from malaria, from reactions to certain drugs, and as an autoimmune process.

EFFECTS OF ANEMIA ON THE CIRCULATORY SYSTEM

The viscosity of the blood, which will be discussed in detail in Chapter 18, depends almost entirely on the concentration of red blood cells. In severe anemia the blood viscosity may fall to as low as 1.5 times that of water rather than the normal value of approximately 3 times the viscosity of water. The greatly decreased viscosity decreases the resistance to blood flow in the peripheral vessels so that far greater than normal quantities of blood return to the heart. As a consequence, the cardiac output increases to as much as 2 or more times normal as a result of decreased viscosity.

Moreover, hypoxia due to diminished transport of oxygen by the blood causes the tissue vessels to dilate, allowing further increase in the return of blood to the heart, increasing the cardiac output to a still higher level. Thus, one of the major effects of anemia is greatly *increased work load on the heart.*

The increased cardiac output in anemia partially offsets many of the effects of anemia, for, even though each unit quantity of blood carries only small quantities of oxygen, the rate of blood flow may be increased to such an extent that almost normal quantities of oxygen are delivered to the tissues. However, when a person begins to exercise, the heart is not capable of pumping much greater quantities of blood than it is already pumping. Consequently, during exercise, which greatly increases the tissue demand for oxygen, extreme tissue hypoxia results, and acute cardiac failure often ensues.

POLYCYTHEMIA

Secondary Polycythemia. Whenever the tissues become hypoxic because of too little oxygen in the atmosphere, such as at high altitudes, or because of failure of delivery of oxygen to the tissues, as occurs in cardiac failure, the blood-forming organs automatically produce large quantities of red blood cells. This condition is called *secondary polycythemia,* and the red cell count commonly rises to as high as 6 to 8 million per cubic millimeter.

A very common type of secondary polycythemia, called *physiologic polycythemia,* occurs in natives who live at altitudes of 14,000 to 17,000 feet. The blood count is generally 6 to 8 million per cubic millimeter; this is associated with an ability of these persons to perform high levels of continuous work even in the rarefied atmosphere.

Polycythemia Vera (Erythremia). In addition to those persons who have physiologic polycythemia, others have a condition known as *polycythemia vera* in which the red blood cell count may be as high as 8 to 9 million and the hematocrit as high as 70 to 80 per cent. Polycythemia vera is a tumorous condition of the organs that produce blood cells. It causes excess production of red blood cells in the same manner that a tumor of a breast causes excess production of a specific type of breast cell. It usually also causes excess production of white blood cells and platelets.

In polycythemia vera not only does the hematocrit increase but the total blood volume occasionally also increases to as much as twice normal. As a result, the entire vascular system becomes intensely engorged. In addition, many of the capillaries become plugged by the very viscous blood, for the viscosity of the blood in polycythemia vera sometimes increases from the normal of 3 times the viscosity of water to 15 times that of water.

EFFECT OF POLYCYTHEMIA ON THE CIRCULATORY SYSTEM

Because of the greatly increased viscosity of the blood in polycythemia, the flow of blood through the vessels is extremely sluggish. In accordance with the factors that regulate the return of blood to the heart as discussed in Chapter 23, it is obvious that increasing the viscosity tends to decrease the rate of venous return to the heart. On the other hand, the blood volume is greatly increased in polycythemia, which tends to increase the venous return. Actually, the cardiac output in polycythemia is not far from normal because these two factors more or less neutralize each other.

Because the total blood volume in polycythemia is sometimes twice normal, the circulation time through the body may also be increased to twice normal. In other words, the mean circulation time occasionally is as great as 120 seconds instead of the normal of approximately 60 seconds. Thus, the velocity of blood flow in any given vessel is considerably decreased in polycythemia.

The arterial pressure is normal in most persons with polycythemia, though in approximately one third of them the pressure is elevated. This means that the blood pressure–regulating mechanisms can usually offset the tendency for increased blood viscosity to increase peripheral resistance and thereby to increase arterial pressure. Yet, beyond certain limits, these regulations fail.

The color of the skin is dependent to a great extent on the quantity of blood in the subpapillary venous plexus. In polycythemia vera the quantity of blood in this plexus is greatly increased. Furthermore, because the blood passes sluggishly through the skin capillaries before entering the venous plexus, a larger than normal proportion of the hemoglobin is deoxygenated before the blood enters the plexus. The blue color of this deoxygenated hemoglobin masks the red color of the oxygenated hemoglobin. Therefore, a person with polycythemia vera ordinarily has a ruddy complexion but may at times have a bluish (cyanotic) tint to the skin. In secondary polycythemia, cyanosis is almost always evident because hypoxia is the usual cause of this type of polycythemia.

REFERENCES

Baum, S. J., and Ledney, G. D. (eds.): Experimental Hematology Today, 1978. New York, Springer-Verlag, 1978.

Bessis, M., *et al.* (eds.): Red Cell Rheology. New York, Springer-Verlag, 1978.

Beutler, E.: Hemolytic Anemia in Disorders of Red Cell Metabolism. New York, Plenum Press, 1978.

Botez, M. I., and Reynolds, E. H. (eds.): Folic Acid in Neurology, Psychiatry, and Internal Medicine. New York, Raven Press, 1979.

Bottomley, S. S., and Whitcomb, W. H.: Erythropoiesis. *In* Frohlich, E. D. (ed.): Pathophysiology, 2nd Ed. Philadelphia, J. B. Lippincott Co., 1976, p. 567.

Cokelet, G. R., *et al.* (eds.): Erythrocyte Mechanics and Blood Flow. New York, A. R. Liss, 1979.

Doss, M. (ed.): Diagnosis and Therapy of Porphyrias and Lead Toxication. New York, Springer-Verlag, 1978.

Dunn, C. D. R.: The Differentiation of Haemopoietic Stem Cells. Baltimore, Williams & Wilkins, 1971.

Ellory, C., and Lew, V. L. (eds.): Membrane Transport in Red Cells. New York, Academic Press, 1977.

Erslev, A. J., and Gabuzda, T. G.: Pathophysiology of Blood. Philadelphia, W. B. Saunders Co., 1979.

Fevery, J., and Heirwegh, K. P. M.: Bilirubin metabolism. *In* Javitt, N. B. (ed.): International Review of Physiology: Liver and Biliary Tract Physiology I. Vol. 21. Baltimore, University Park Press, 1980, p. 171.

Fung, Y. C.: Red blood cells and their deformability. *In* Kaley, G., and Altura, B. M. (eds.): Microcirculation. Vol. 1. Baltimore, University Park Press, 1977, p. 255.

Heimpel, H. *et al.* (eds.): International Symposium on Aplastic Anemia. New York, Springer-Verlag, 1979.

Hoffbrand, A. V., *et al.* (eds.): Recent Advances in Haematology. New York, Longman, Inc., 1977.

Kass, L.: Bone Marrow Interpretation. Philadelphia, J. B. Lippincott, Co., 1979.

Kelemen, E., *et al.*: Atlas of Human Hemopoietic Development. New York, Springer-Verlag, 1979.

Konigsberg, W.: Protein structure and molecular dysfunction: Hemoglobin. *In* Bondy, P. K., and Rosenberg, L. E. (eds.): Metabolic Control and Disease, 8th Ed. Philadelphia, W. B. Saunders Co., 1980, p. 27.

Kruckeberg, W. C., *et al.* (eds.): Erythrocyte Membranes. New York, A. R. Liss, 1978.

Lux, S. E., *et al.* (eds.): Normal and Abnormal Red Cell Membranes. New York, A. R. Liss, 1979.

Maclean, N.: Haemoglobin. London, Edward Arnold, 1978.

Maslow, W. C., *et al.*: Hematologic Disease. Boston, Houghton Mifflin, 1979.

Munro, H. N., and Linder, M. C.: Ferritin: Structure, biosynthesis, and role in iron metabolism. *Physiol. Rev., 58*:317, 1978.

Peschle, C.: Erythropoiesis. *Annu. Rev. Med., 31*:303, 1980.

Petz, L. D., and Garratty, G.: Acquired Immune Hemolytic Anemias. New York, Churchill Livingstone, 1979.

Platt, W. R.: Color Atlas and Textbook of Hematology. Philadelphia, J. B. Lippincott Co., 1978.

Spivak, J. L. (ed.): Fundamentals of Clinical Hematology, Hagerstown, Md., Harper & Row, 1980.

Stamatoyannopoulos, G., and Nienhuis, A. W. (eds.): Cellular and Molecular Regulation of Hemoglobin Switching. New York, Grune & Stratton, 1979.

Swisher, S. N., (ed.): Immune Hemolytic Anemias. New York, Grune & Stratton, 1976.

Tschudy, D. P., and Lamon, J. M.: Porphyrin metabolism and the porphyrias. *In* Bondy, P. K., and Rosenberg, L. E. (eds.): Metabolic Control and Disease, 8th Ed. Philadelphia, W. B. Saunders Co., 1980, p. 939.

Zagalak, B., and Friedrich, W. (eds.): Vitamin B12. New York, Walter De Gruyter, 1979.

6

Resistance of the Body to Infection — The Leukocytes, the Tissue Macrophage System, and Inflammation

Our bodies normally are exposed to bacteria, viruses, fungi, and parasites, which occur especially in the skin, the mouth, the respiratory passageways, the colon, the lining membranes of the eyes, and even the urinary tract. Many of these agents are capable of causing serious disease if they invade the deeper tissues. In addition, we are exposed intermittently to other highly infectious bacteria and viruses besides those that are normally present in our bodies, and these cause lethal diseases such as pneumonia, streptococcal infection, and typhoid fever.

Fortunately, our bodies have a special system for combatting the different infectious and toxic agents. This is composed of the *leukocytes* (also called white blood cells), the *tissue macrophage system* (frequently but incorrectly called the reticuloendothelial system), and the *lymphoid tissue*. These tissues function in two different ways to prevent disease: (1) by actually destroying invading agents by the process of phagocytosis and (2) by forming antibodies and sensitized lymphocytes, one or both of which may destroy the invader. The present chapter is concerned with the first of these methods, while the following chapter is concerned with the second.

THE LEUKOCYTES (WHITE BLOOD CELLS)

The leukocytes are the *mobile units* of the body's protective system. They are formed partially in the bone marrow (the *granulocytes* and *monocytes,* and a few *lymphocytes*) and partially in the lymph tissue (*lymphocytes* and *plasma cells*), but after formation they are transported in the blood to the different parts of the body where they are to be used. The real value of the white blood cells is that most of them are specifically transported to areas of serious inflammation, thereby providing a rapid and potent defense against any infectious agent that might be present. As we shall see later, the granulocytes and monocytes have a special capability to "seek out and destroy" any foreign invader.

GENERAL CHARACTERISTICS OF LEUKOCYTES

The Types of White Blood Cells. Six different types of white blood cells are normally found in the blood. These are *polymorphonuclear neutrophils, polymorphonuclear eosinophils, polymorphonuclear basophils, monocytes, lymphocytes* and *plasma cells*. In addition, there are large numbers of *platelets*, which are fragments of a seventh type of white cell found in the bone marrow, the *megakaryocyte*. The three types of polymorphonuclear cells have a granular appearance, as illustrated in Figure 6–1, for which reason they are called *granulocytes,* or in clinical terminology they are often called simply "polys."

The granulocytes and the monocytes protect the body against invading organisms by ingesting them — that is, by the process of *phagocytosis*. One of the functions of lymphocytes also is to attach to specific invading organisms and to destroy them; this is part of the immune system and will be discussed in the following chapter. Finally, the function of platelets is to activate the blood clotting mechanism, which will be discussed in Chapter 9.

Concentrations of the Different White Blood Cells in the Blood. The adult human being has approximately 7000 white blood cells per cubic millimeter of blood. The normal percentages of the different types of white blood cells are approximately the following:

Polymorphonuclear neutrophils	62.0%
Polymorphonuclear eosinophils	2.3%
Polymorphonuclear basophils	0.4%
Monocytes	5.3%
Lymphocytes	30.0%

The number of platelets, which are only cell fragments, in each cubic millimeter of blood is normally about 300,000.

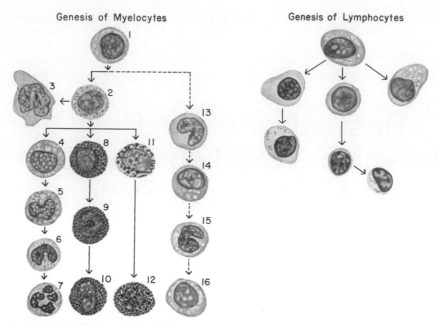

Figure 6–1.   Genesis of the white blood cells. The different cells of the myelogenous series are: 1, myeloblast; 2, promyelocyte; 3, megakaryocyte; 4, neutrophil myelocyte; 5, young neutrophil metamyelocyte; 6, "band" neutrophil metamyelocyte; 7, polymorphonuclear neutrophil; 8, eosinophil myelocyte; 9, eosinophil metamyelocyte; 10, polymorphonuclear eosinophil; 11, basophil myelocyte; 12, polymorphonuclear basophil; 13–16, stages of monocyte formation.

GENESIS OF THE LEUKOCYTES

Figure 6–1 illustrates the stages in the development of white blood cells. The polymorphonuclear cells and monocytes are normally formed only in the bone marrow. On the other hand, lymphocytes and plasma cells are produced in the various lymphogenous organs, including the lymph glands, the spleen, the thymus, the tonsils, and various lymphoid rests in the gut and elsewhere.

The white blood cells formed in the bone marrow, especially the granulocytes, are stored within the marrow until they are needed in the circulatory system. Then when the need arises, various factors that are discussed later cause them to be released. Normally, about three times as many granulocytes are stored in the marrow as circulate in the entire blood. This represents about a six day supply of granulocytes.

As illustrated in Figure 6–1, megakaryocytes are also formed in the bone marrow and are part of the myelogenous group of bone marrow cells. These megakaryocytes fragment in the bone marrow, the small fragments known as *platelets* or *thrombocytes* passing then into the blood.

Materials Needed for Formation of White Blood Cells.   In general, the white blood cells need essentially the same vitamins and amino acids as most of the other cells of the body for their formation. Especially does lack of folic acid, a compound of the vitamin B complex, block the formation of white blood cells as well as prevent maturation of red blood cells, which was pointed out in Chapter 5. Also, in extreme debilitation, production of white blood cells may be greatly reduced, despite the fact that these cells are needed more during such a state than usually.

LIFE SPAN OF THE WHITE BLOOD CELLS

The main reason white blood cells are present in the blood is simply to be transported from the bone marrow or lymphoid tissue to the areas of the body where they are needed. Therefore, it is to be expected that the transit time of the white blood cells in the blood would be short.

The subsequent life of the granulocytes once released from the bone marrow is normally six to eight hours circulating in the blood and another two to three days in the tissues. In times of serious tissue infection, this total life span is often shortened to only a few hours because the granulocytes then proceed rapidly to the infected area, ingest the invading organisms, and in the process are themselves destroyed.

The monocytes also have a short transit time in the blood before wandering through the capillary membranes into the tissues. However, once in the tissues they swell to much larger sizes to become tissue macrophages and in this form can live for months or even years unless destroyed by performing phagocytic function. These tissue macrophages form the basis of the tissue macrophage system that provides a first line of defense in the tissues against infection, as we shall discuss later in the chapter.

Lymphocytes enter the circulatory system continually along with the drainage of lymph from the lymph nodes. The total number entering the blood from the thoracic duct in each 24 hours is usually several times the total number of lymphocytes present in the blood stream at any given time. Therefore, the span of time that the lymphocytes remain in the blood must be only a few hours. However, studies using radioactive lymphocytes have shown that almost all of these pass by diapedesis back into the tissues, then re-enter the lymph and return to the blood again and again; thus, there is continual circulation of the lymphocytes through the tissues. And many of these cells have life spans of 100 to 300 days, or in some instances perhaps even years, but this also depends on the tissue's need for these cells.

The platelets in the blood are totally replaced approximately once every 10 days; in other words, about 30,000 platelets are formed each day for each cubic millimeter of blood.

PROPERTIES OF NEUTROPHILS, MONOCYTES, AND MACROPHAGES

It is mainly the neutrophils and the monocytes that attack and destroy invading bacteria, viruses, and other injurious agents. The neutrophils are mature cells that can attack and destroy bacteria and viruses even in the circulating blood. On the other hand, the blood monocytes are immature cells that have very little ability to fight infectious agents. However, once they enter the tissues they begin to swell, often increasing their diameters as much as five-fold, to as great as 80 microns, a size that can be seen with the naked eye. Also, extremely large numbers of lysosomes and mitochondria develop in the cytoplasm, giving the cytoplasm the appearance of a bag filled with granules. These cells are now called *macrophages*, and they have extreme capability of combatting disease agents.

Diapedesis. Neutrophils and monocytes can squeeze through the pores of the blood vessels by the process of diapedesis. That is, even though a pore is much smaller than the size of the cell, a small portion of the cell slides through the pore at a time, the portion sliding through being momentarily constricted to the size of the pore, as illustrated in Figure 6–2.

Ameboid Motion. Both neutrophils and macrophages move through the tissues by ameboid motion, which was described in Chapter 2. Some of the cells can move through the tissues at rates as great as 40 microns per minute — that is, they can move at least three times their own length each minute.

Chemotaxis. A number of different chemical substances in the tissues cause both neutrophils and macrophages to move either toward or away from the source of the chemical. This phenomenon is known as *chemotaxis.* When a tissue becomes inflamed, a number of different products can cause chemotaxis of both neutrophils and macrophages, causing them to move toward the inflamed area. These include (a) some of the bacterial toxins, (b) degenerative products of the inflamed tissues themselves, (c) several reaction products of the "complement complex" (discussed in the following chapter), (d) several reaction products caused by plasma clotting in the inflamed area, and (e) still other substances.

As illustrated in Figure 6–2, chemotaxis depends on a concentration gradient of the chemotactic substance. The concentration is greatest near the source, which causes directional movement of the leukocytes — including the neutrophils, monocytes, and other white blood cells as well — toward the inflamed area. Chemotaxis is very effective up to 100 microns away from an inflamed tissue; since almost no tissue area is more than 30 to 50 microns away from a capillary, the chemotactic

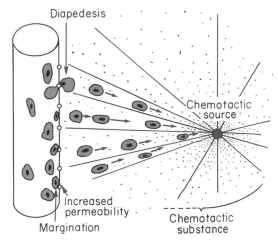

Figure 6–2. Movement of neutrophils by the process of *chemotaxis* toward an area of tissue damage.

signal can easily move vast hordes of leukocytes from the capillaries into the inflamed area.

Margination and Diapedesis of Leukocytes Through the Capillary Membrane. Some of these same products that cause chemotaxis, as well as enzymes and necrotic products released from the inflamed tissues, have a direct effect on the local capillaries to cause margination and diapedesis of white blood cells. Margination means sticking of the white blood cells to the wall. Even normally, granulocytes and monocytes, partly because of their large sizes and partly because of a natural stickiness, adhere to the capillary walls. In fact, about three fifths of the granulocytes and three fourths of the monocytes in the blood are sequestered in the capillaries in this manner. However, after the inflammatory products have acted on the capillary walls, the endothelial cells become especially sticky to the white blood cells, causing extreme amounts of margination of both the granulocytes and monocytes as illustrated in Figure 6–2. At the same time, the pores of the capillaries open more widely than normally to allow rapid diapedesis of the cells into the tissues, and thence chemotaxis toward the inflamed site.

Phagocytosis. The most important function of the neutrophils and macrophages is phagocytosis.

Obviously, the phagocytes must be selective of the material that is phagocytized, or otherwise some of the normal cells and structures of the body would be ingested. Whether or not phagocytosis will occur depends especially upon three selective procedures. First, if the surface of a particle is rough, the likelihood of phagocytosis is increased. Second, most natural substances of the body have electronegative surface charges that repel the phagocytes, which also carry electronegative surface charges. On the other hand, dead tissues and foreign particles are frequently electropositive and are therefore subject to phagocytosis. Third, the body has a specific means for recognizing certain foreign materials. This is the function of the immune system that will be described in the following chapter. The immune system develops antibodies against infectious agents like bacteria. These antibodies then adhere to the bacterial membranes and thereby make the bacteria especially susceptible to phagocytosis. In this case, the antibody is called an *opsonin*. After adhering to the bacteria, the opsonins then combine with *complement,* which is another part of the immune system that will be discussed in the following chapter. Some of the elements of this complement complex then activate the neutrophils and macrophages to initiate the phagocytic process.

Phagocytosis by Neutrophils.   The neutrophils entering the tissues are already mature cells that can immediately begin phagocytosis. On approaching a particle to be phagocytized, the neutrophil projects pseudopodia in all directions around the particle, and the pseudopodia meet each other on the opposite side and fuse. This creates an enclosed chamber containing the phagocytized particle. Then the chamber invaginates to the inside of the cytoplasmic cavity, and the portion of the cell membrane that surrounds the phagocytized particle breaks away from the outer cell membrane to form a free-floating *phagocytic vesicle* (also called a *phagosome*) inside the cytoplasm.

A neutrophil can usually phagocytize 5 to 20 bacteria before the neutrophil itself becomes inactivated and dies.

Phagocytosis by Macrophages.   Macrophages are much more powerful phagocytes than the neutrophils, often capable of phagocytizing as many as 100 bacteria. They have the ability to engulf much larger particles and often 5 or more times as many particles as the neutrophils. And they can even phagocytize whole red blood cells or malarial parasites, whereas neutrophils are not capable of phagocytizing particles much larger than bacteria. Also, macrophages have the ability to phagocytize necrotic tissue and even dead neutrophils, which is a very important function performed by these cells in chronic infection.

Enzymatic Digestion of the Phagocytized Particles.   Once a foreign particle has been phagocytized, lysosomes immediately come in contact with the phagocytic vesicle, and their membranes fuse with those of the vesicle, thereby dumping many digestive enzymes of the lysosomes into the vesicle. Thus, the phagocytic vesicle now becomes a *digestive vesicle,* and digestion of the phagocytized particle begins immediately.

Neutrophils and macrophages both have an abundance of lysosomes filled with *proteolytic enzymes* especially geared for digesting bacteria and other foreign protein matter. The lysosomes of macrophages also contain large amounts of *lipases,* which digest the thick lipid membranes possessed by tubercle bacteria, leprosy bacteria, and others.

In addition to the lysosomal enzymes that actually digest the ingested particles, the phagocytic cells, especially the neutrophils, also contain bactericidal agents that kill bacteria before they can multiply. The neutrophil, for instance, contains special vesicles capable of synthesizing hydrogen peroxide which, upon entering the digestive vesicle, exerts an especially potent bactericidal effect based on its ability to oxidize the organic substances of the bacteria. Indeed, in a rare hereditary disease in which these peroxide vesicles are missing from the neutrophils, removal of some types of bacteria from the tissues is so incomplete that fulminating infection often leads to death.

It is especially significant that the macrophages digest bacteria and viruses into larger units than the neutrophils. The importance of this is that many of the digestive end-products released from the macrophages still retain their capability to activate the immune process. Therefore, the macrophages play a special role in initiating immunity against specific organisms, a function that we shall discuss in the following chapter.

Death of the Neutrophils as a Result of Phagocytosis.   Neutrophils continue to ingest and digest foreign particles until toxic substances from these, as well as enzymes released inside the neutrophils from the lysosomes, kill the neutrophils themselves. This usually occurs after a neutrophil has phagocytized from 5 to 25 bacteria. Then the macrophages set about to phagocytize and digest the dead neutrophils.

The macrophages often also are killed in their process of phagocytizing infectious agents. However, this is not always the case because the macrophage, in contradistinction to the neutrophil, is capable of extruding the residual breakdown products of the phagocytized agent and therefore continuing to perform its function sometimes for weeks, months, or even years.

THE TISSUE MACROPHAGE SYSTEM (THE RETICULOENDOTHELIAL SYSTEM)

In the above paragraphs we have described the macrophages mainly as mobile cells that are capable of wandering through the tissues. However, a vast majority of the monocytes, on entering the tissues and after becoming macrophages, become attached to the tissues and remain attached for months or even years unless they are called upon to perform specific protective functions. They have the same capabilities as the mobile macrophages to phagocytize large quantities of bacteria, viruses, necrotic tissue, or other foreign particles in the tissue. And, when appropriately stimulated, they can break away from their attachments and become mobile macrophages that respond to chemotaxis and all the other stimuli related to the inflammation process.

The combination of mobile macrophages and fixed tissue macrophages is collectively called the *reticuloendothelial system.* The reason for this name is that it was formerly believed that a major share of the blood vessel endothelial cells could perform phagocytic functions similar to those performed by the macrophage system. However, recent studies have disproved this. Therefore, the reticuloendothelial system is actually a misnomer. Yet, because the term is so widely used, it should be remembered that it is almost synonymous with the tissue macrophage system.

The tissue macrophages in various tissues differ in appearances because of environmental differences, and they are known by different names: *Kupffer cells* in the liver; *reticulum cells* in lymph nodes, spleen, and bone marrow; *alveolar macrophages* in the alveoli of the lungs; *tissue histiocytes, clasmatocytes,* or *fixed macrophages* in the subcutaneous tissues; and *microglia* in the brain. Let us describe briefly the function of the tissue macrophages in some areas of the body that are especially exposed to infectious agents.

Tissue Macrophages in the Skin and Subcutaneous Tissues (Histiocytes).   Though the skin is normally impregnable to infectious agents, this no longer holds true when the skin is broken. When infection does begin in

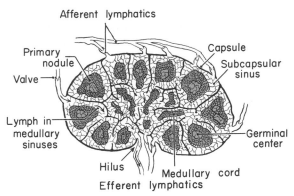

Figure 6–3. Functional diagram of a lymph node. (Redrawn from Ham: Histology. Philadelphia, J. B. Lippincott Company, 1971.)

the subcutaneous tissues and local inflammation ensues, the tissue macrophages can divide *in situ* and form more macrophages. Then they perform the usual functions of attacking and devouring the infectious agents, as described earlier.

Macrophages of the Lymph Nodes (Reticulum Cells). Essentially no particulate matter that enters the tissues can be absorbed directly through the capillary membranes into the blood. Instead, if the particles are not destroyed locally in the tissues, they enter the lymph and flow through the lymphatic vessels to the lymph nodes located intermittently along the course of the lymphatics. The foreign particles are trapped there in a meshwork of sinuses lined by tissue macrophages called *reticulum cells.*

Figure 6–3 illustrates the general organization of the lymph node, showing lymph entering by way of the *afferent lymphatics,* flowing through the *medullary si-*

nuses, and finally passing out of the *hilus* into the *efferent lymphatics.* Large numbers of reticulum cells line the sinuses, and, if any particles enter the sinuses, these cells phagocytize them and prevent general dissemination throughout the body.

Alveolar Macrophages. Another route by which invading organisms frequently enter the body is through the respiratory system. Fortunately, large numbers of tissue macrophages are present as integral components of the alveolar walls. These can phagocytize particles that become entrapped in the alveoli. If the particles are digestible, the macrophages can also digest them and release the digestive products into the lymph. If the particle is not digestible the macrophages often form a "giant cell" capsule around the particle until such time, if ever, that it can be slowly dissoluted. Such capsules are frequently formed around tubercle bacilli, silica dust particles, and even carbon particles.

Tissue Macrophages (Kupffer Cells) in the Liver Sinuses. Still another favorite route by which bacteria invade the body is through the gastrointestinal tract. Large numbers of bacteria constantly pass through the gastrointestinal mucosa into the portal blood. However, before this blood enters the general circulation, it must pass through the sinuses of the liver; these sinuses are lined with tissue macrophages called *Kupffer cells,* illustrated in Figure 6–4. These cells form such an effective particulate filtration system that almost none of the bacteria from the gastrointestinal tract succeeds in passing from the portal blood into the general systemic circulation. Indeed, motion pictures of phagocytosis by Kupffer cells have demonstrated phagocytosis of single bacteria in less than 1/100 second.

Macrophages (Reticulum Cells) of the Spleen and Bone Marrow. If an invading organism does succeed in entering the general circulation, there still remain other lines of defense by the tissue macrophage system, especially by *reticulum cells* of the spleen and bone marrow. In both of these tissues, macrophages have become entrapped by the reticular meshwork of the two organs, and when foreign particles come in contact with the reticulum cells in this meshwork the particles are phagocytized.

The spleen is similar to the lymph nodes, except that blood, instead of lymph, flows through the substance of the spleen. Figure 6–5 illustrates the spleen's general structure, showing a small peripheral segment. Note that a small artery penetrates from the splenic capsule into the *splenic pulp,* and terminates in small capillaries. The capillaries are highly porous, allowing large numbers of whole blood cells to pass out of the capillaries into the *cords of the red pulp.* These cells then gradually *squeeze* through the tissue substance of the cords and eventually return to the circulation through the endothelial walls of the *venous sinuses.* The cords of the red pulp are loaded with macrophages (reticulum cells), and in addition the venous sinuses are also lined with macrophages. This peculiar passage of blood through the cords of the red pulp provides an exceptional means for phagocytosis of unwanted debris in the blood, especially old and abnormal red blood cells. The spleen is also an important organ for phagocytic removal of abnormal platelets, blood parasites, and any bacteria that might succeed in entering the general circulating blood.

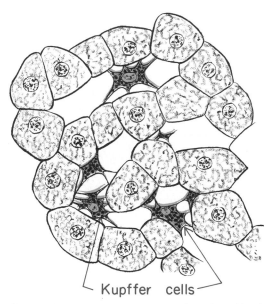

Figure 6–4. Kupffer cells lining the liver sinusoids, showing phagocytosis of India ink particles. (Redrawn from Copenhaver et al.: Bailey's Textbook of Histology. Baltimore, Williams and Wilkins Company, 1969.)

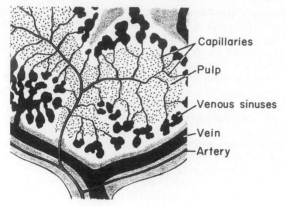

Figure 6–5. Functional structures of the spleen. (Modified from Bloom and Fawcett: Textbook of Histology. Philadelphia, W. B. Saunders Company, 1975.)

In a similar way, reticulum cells of the bone marrow also help to remove unwanted debris and pathologic agents from the blood.

INFLAMMATION AND FUNCTION OF NEUTROPHILS AND MACROPHAGES

THE PROCESS OF INFLAMMATION

Inflammation is a complex of sequential changes in the tissues in response to injury. When tissue injury occurs, whether it be caused by bacteria, trauma, chemicals, heat, or any other phenomenon, large quantities of *histamine, bradykinin, serotonin,* and other substances are liberated by the damaged tissue into the surrounding fluids. These, especially the histamine, increase the local blood flow and also increase the permeability of the venous capillaries and venules, allowing large quantities of fluid and protein, including fibrinogen, to leak into the tissues. Local extracellular edema results, and the extracellular fluid and lymphatic fluid both clot because of the coagulating effect of tissue exudates on the leaking fibrinogen. Thus, *brawny edema* develops in the spaces surrounding the injured cells.

The "Walling Off" Effect of Inflammation. It is clear that one of the first results of inflammation is to "wall off" the area of injury from the remaining tissues. The tissue spaces and the lymphatics in the inflamed area are blocked by fibrinogen clots so that fluid barely flows through the spaces. Therefore, walling off the area of injury delays the spread of bacteria or toxic products.

The intensity of the inflammatory process is usually proportional to the degree of tissue injury. For instance, staphylococci invading the tissues liberate extremely lethal cellular toxins. As a result, the process of inflammation develops rapidly — indeed, much more rapidly than the staphylococci themselves can multiply and spread. Therefore, staphylococcal infection is characteristically walled off rapidly. On the other hand, streptococci do not cause such intense local tissue destruction. Therefore the walling off process develops slowly while the streptococci reproduce and migrate. As a result, streptococci have far greater tendency to spread

through the body and cause death than do staphylococci, even though staphylococci are far more destructive to the tissues.

The Macrophage and Neutrophil Response to Inflammation

Soon after inflammation begins, the inflamed area becomes invaded by both neutrophils and macrophages, and these set about to perform their scavenger functions to rid the tissue of infectious or toxic agents. However, the macrophage and neutrophil responses occur in several different stages.

The Tissue Macrophages as the First Line of Defense. The macrophages that are already present in the tissues, whether they be the histiocytes in the subcutaneous tissues, the alveolar macrophages in the lungs, the microglia in the brain, and so forth, immediately begin their phagocytic actions. Therefore, they are the first line of defense against infection during the first hour or so. However, their numbers often are not very great.

Neutrophilia and Neutrophil Invasion of the Inflamed Area — the Second Line of Defense. The term *neutrophilia* means an increase in a number of neutrophils in the blood. The term "leukocytosis" is also often used to mean the same as neutrophilia, though this term actually means excess number of all white cells, whatever their types.

Within a few hours after the onset of acute inflammation, the number of neutrophils in the blood sometimes increases as much as four- to five-fold — to as high as 15,000 to 25,000 per cubic millimeter. This results from a combination of chemical substances that are released from the inflamed tissues, collectively called *leukocytosis-inducing factor.* This factor diffuses from the inflamed tissue into the blood and is carried to the bone marrow. There it is believed to dilate the venous sinusoids of the marrow, thus causing release of many leukocytes, especially neutrophils, that are already stored in these venous sinusoids. In this way large numbers of neutrophils are almost immediately transferred from the bone marrow storage pool into the circulating blood.

Movement of Neutrophils to the Area of Inflammation. Products from the inflamed tissues also cause neutrophils to move from the circulation into the inflamed area. They do this in three ways:

First, they damage the capillary walls and thereby cause neutrophils to stick, which is the process called *margination* that was illustrated in Figure 6–2.

Second, they greatly increase the permeability of the capillaries and small venules, and this allows the neutrophils to pass by *diapedesis* into the tissue spaces.

Third, there is also the phenomenon of *chemotaxis* that causes the neutrophils to migrate toward the injured tissues. This is caused by bacterial or cellular products, clotting reaction factors, antigen-antibody products, and some of the "complement" components that will be described in the following chapter.

Thus, within several hours after tissue damage begins, the area becomes well supplied with neutrophils. Since neutrophils are already mature cells, they are ready to begin immediately their scavenger functions for removal of foreign matter from the inflamed tissues.

Increased Production of Neutrophils by the Bone Marrow — Colony-Stimulating Factor. Inflammation not only causes release of leukocytes from the bone marrow's storage pool but causes an increased rate of production of leukocytes as well. A number of different factors seem to cause this; one that has been isolated and characterized is called *colony-stimulating factor* because it causes colonies of leukocytes to proliferate in tissue culture as well as causing increased leukocyte production in the bone marrow. Colony-stimulating factor is a glycoprotein with a molecular weight of about 45,000. It is mainly produced by the macrophages in the inflamed tissues, probably in response to bacterial toxins, breakdown products of neutrophils, and other products of the inflamed tissues.

Once released from the macrophages, colony-stimulating factor acts on the entire mitotic pool of the bone marrow white blood cells, causing rapid division of the *granulopoietic stem cells, myeloblasts, promyelocytes,* and *myelocytes*. Though the time required for producing white blood cells, beginning with committed granulopoietic stem cells to their final release into the blood, is normally about 14 days, during inflammation this process can be speeded up to less than half the time. Thus, stimulation of increased production of granulocytes is important for maintaining a high level of neutrophilia for many days, weeks, or even months — but not for the early stages of neutrophilia.

Macrophage Proliferation and the Monocyte Response — the Third Line of Defense. Though the first and second lines of defense in tissue infection are the already present tissue macrophages and the rapid production and movement of neutrophils into the inflamed area, still a third line of defense is a slow but long-continuing increase in the number of macrophages as well. This results partly from reproduction of the already present tissue macrophages but also from migration of large numbers of monocytes into the inflamed area. Though the monocytes are still immature cells and do not have the capability of phagocytosis when they first enter the tissues, over a period of 8 to 12 hours they swell markedly, form greatly increased quantities of cytoplasmic lysosomes, exhibit increased ameboid motion, and move chemotactically toward the damaged tissues.

Next, the rate of production of monocytes by the bone marrow also increases. This possibly results partly from stimulation by colony-stimulating factor, but there seem to be other, yet undefined stimulating factors as well because in long-term chronic infection there is progressively increasing production of monocytes, thereby increasing the ratio of macrophages to neutrophils in the tissues. Therefore, the long-term chronic defense against infection is mainly a macrophage response rather than a neutrophil response.

As has already been pointed out, the macrophages can phagocytize far more bacteria and far larger particles, including even neutrophils and large quantities of necrotic tissue. Also, the macrophages play an important role in initiating the development of antibodies, as we shall discuss in the following chapter.

Formation of Pus

When the neutrophils and macrophages engulf large numbers of bacteria and necrotic tissue, essentially all of the neutrophils and many if not most of the macrophages themselves eventually die. After several days, a cavity is often excavated in the inflamed tissues containing varying portions of necrotic tissue, dead neutrophils, and dead macrophages. Such a mixture is commonly known as *pus*.

Ordinarily, pus formation continues until all infection is suppressed. Sometimes the pus cavity eats its way to the surface of the body or into an internal cavity and in this way empties itself. At other times the pus cavity remains closed even after tissue destruction has ceased. When this happens the dead cells and necrotic tissue in the pus gradually autolyze over a period of days, and the end-products of autolysis are usually absorbed into the surrounding tissues until most of the evidence of tissue damage is gone.

NEUTROPHILIA CAUSED BY CONDITIONS OTHER THAN INFLAMMATION

Almost any factor that causes some degree of tissue destruction will cause neutrophilia. For instance, persons debilitated by cancer exhibit an increase in neutrophils from the normal of 4500 per cubic millimeter sometimes up to 15,000 or more. Even extreme fatigue can cause neutrophilia. Acute hemorrhage, poisoning, operative procedures, very slight hemorrhage into the peritoneal cavity, and injection of foreign protein into the body all cause considerable increase in the number of neutrophils in the circulatory system. In summary, neutrophilia results from almost any tissue-damaging process in the body, whether or not this process is associated with inflammation.

For instance, neutrophilia is one of the diagnostic features of coronary thrombosis. Presumably, when the coronary vessel becomes blocked the ischemic musculature of the heart begins to necrose, and degenerative substances liberated into the blood promote the release of neutrophils from the bone marrow.

Physiological Neutrophilia. The number of neutrophils in the circulatory system can increase to as much as 2 to 3 times normal after a single minute of extremely hard exercise or after injection of norepinephrine. This can be explained as follows: When blood flow is sluggish through the tissues, large numbers of white blood cells, especially neutrophils, adhere to the walls of the capillaries — a process called *margination* — and, therefore, are sequestered from the usual circulation. Hard exercise or stimulation of the circulation by norepinephrine, with rapid flow of blood through essentially all capillaries, can mobilize the leukocytes.

Approximately one hour after physiological neutrophilia has resulted from exercise or any other stimulus, the number of leukocytes in the blood is usually back to normal because most of the leukocytes will again be sequestered in the capillaries.

THE EOSINOPHILS

The eosinophils normally comprise 1 to 3 per cent of all the blood leukocytes. Eosinophils are weak phagocytes, and they exhibit chemotaxis, but in comparison with the neutrophils, it is doubtful that the eosinophils are of significant importance in protection against usual types of infection.

Eosinophils enter the blood in large numbers after foreign protein injection. Furthermore, many eosinophils are present in the mucosa of the intestinal tract and in the tissues of the lungs, where foreign proteins normally enter the body. It has been suggested that the function of these is to detoxify the proteins before they can cause damage to the body.

Eosinophils also migrate into blood clots where they probably release the substance *profibrinolysin*. This substance then becomes activated to form *fibrinolysin*, which is an enzyme that digests fibrin, a subject that will be discussed in Chapter 9. Therefore, eosinophils are possibly important for dissolution of old clots.

Eosinophils have a special propensity to collect at sites of antigen-antibody reactions in the tissues. They also have a special capability to phagocytize and digest the combined antigen-antibody complex after the immune process has performed its function. This is particularly true when the complex precipitates. Also, the total number of eosinophils increases greatly in the circulating blood during allergic reactions, presumably because the tissue reactions release products that specifically increase bone marrow production of eosinophils.

Large numbers of eosinophils also frequently enter inflamed areas when the inflammation is in its last stages of resolution. It is believed that these eosinophils clean up much of the remaining debris, especially that they remove histamine, bradykinin, serotonin, and other products of inflamed tissues. In fact, these same substances are especially powerful in causing chemotaxis of eosinophils to the inflamed site.

Probably the most common cause of extremely large numbers of eosinophils in the blood is infection with parasites. In the condition known as *trichinosis,* which results from invasion of the muscles by the *Trichinella* parasite ("pork worm") after eating uncooked pork, the percentage of eosinophils in the circulating blood may rise to as high as 25 to 50 per cent of all the leukocytes. The means by which parasitic infections cause an increase in eosinophils also is not known, though if the function of eosinophils is to detoxify proteins, it is presumed that the parasite proteins are a constant source of eosinophil stimulation.

THE BASOPHILS

The basophils in the circulating blood are very similar to though maybe not identical with the large *mast* cells located immediately outside many of the capillaries in the body. These cells liberate *heparin* into the blood, a substance that can prevent blood coagulation and that can also speed the removal of fat particles from the blood after a fatty meal. Therefore, it is probable that the basophils in the circulating blood perform similar functions within the blood stream, or it is even possible that the blood simply transports basophils to tissues where they then become mast cells and perform the function of heparin liberation.

The mast cells and basophils also release histamine, as well as smaller quantities of bradykinin and serotonin. Indeed, it is mainly the mast cells in inflamed tissues that release these substances during inflammation.

The mast cells and basophils play an exceedingly important role in some types of allergic reactions because the type of antibody that causes allergic reactions, the IgE type (see Chapter 7), has a special propensity to become attached to mast cells and basophils. Then, when the specific antigen subsequently reacts with the antibody, the resulting attachment of the antigen to the antibody causes the mast cell or basophil to rupture and release exceedingly large quantities of histamine, bradykinin, serotonin, and a number of lysosomal enzymes. These in turn cause local vascular and tissue reactions that cause the allergic manifestations. These effects will be discussed in greater detail in the following chapter.

AGRANULOCYTOSIS

A clinical condition known as "agranulocytosis" occasionally occurs, in which the bone marrow stops producing white blood cells, leaving the body unprotected against bacteria and other agents that might invade the tissues.

Normally, the human body lives in symbiosis with many bacteria, for all the mucous membranes of the body are constantly exposed to large numbers of bacteria. The mouth almost always contains various spirochetal, fusiform, pneumococcal, and streptococcal bacteria, and these same bacteria are present to a lesser extent in the entire respiratory tract. The gastrointestinal tract is especially loaded with colon bacilli. Furthermore, one can almost always find bacteria in the eyes, the urethra, and the vagina. Therefore, any decrease in the number of neutrophils immediately allows invasion of the tissues by the bacteria that are already present in the body. Within two days after the bone marrow stops producing white blood cells, ulcers may appear in the mouth and colon, or the person develops some form of severe respiratory infection. Bacteria from the ulcers then rapidly invade the surrounding tissues and the blood. Without treatment, death often ensues three to six days after acute agranulocytosis begins.

Irradiation of the body by gamma rays caused by a nuclear explosion, or exposure to drugs and chemicals containing benzene or anthracene nuclei is quite likely to cause aplasia of the bone marrow. Indeed, some of the common drugs, such as the sulfonamides, chloramphenicol, thiouracil (used to treat thyrotoxicosis), and even the various barbiturate hypnotics, on occasion cause agranulocytosis (or *bone marrow aplasia* in which no cells of any type — red cells included — are produced in the bone marrow), thus setting off the entire infective sequence of this malady.

After irradiation injury to the bone marrow, a large number of stem cells, myeloblasts, and hemocytoblasts usually remain undestroyed and are capable of regenerating the bone marrow, provided sufficient time is available. Therefore, the patient properly treated with antibiotics and other drugs to ward off infection will usually develop enough new bone marrow within several weeks to several months that his blood cell concentrations can return to normal.

THE LEUKEMIAS

Uncontrolled production of white blood cells is caused by cancerous mutation of a myelogenous or a lymphogenous cell. This causes leukemia, which is usually characterized by greatly increased numbers of abnormal white blood cells in the circulating blood.

Types of Leukemia. Ordinarily, leukemias are divided into two general types: the *lymphogenous leukemias* and the *myelogenous leukemias*. The lymphogenous leukemias are caused by cancerous production of lymphoid cells, beginning first in a lymph node or other lymphogenous tissue and then spreading to other areas of the body. The second type of leukemia, myelogenous leukemia, begins by cancerous production of young myelogenous cells in the bone marrow and then spreads throughout the body so that white blood cells are produced in many extramedullary organs.

In myelogenous leukemia, the cancerous process occasionally produces reasonably differentiated cells, resulting in *neutrophilic leukemia, eosinophilic leukemia, basophilic leukemia,* or *monocytic leukemia*. More frequently, however, the leukemia cells are bizarre and undifferentiated, and not identical with any of the normal white blood cells.

Leukemic cells, especially the very undifferentiated cells, are often nonfunctional, so that they cannot provide the usual protection associated with white blood cells.

EFFECTS OF LEUKEMIA ON THE BODY

The first effect of leukemia is metastatic growth of leukemic cells in abnormal areas of the body. The leukemic cells of the bone marrow may reproduce so greatly that they invade the surrounding bone, causing pain and eventually a tendency to easy fracture. Almost all leukemias spread to the spleen, the lymph nodes, the liver, and other especially vascular regions, regardless of whether the origin of the leukemia is in the bone marrow or in the lymph nodes. In each of these areas the rapidly growing cells invade the surrounding tissues, utilizing the metabolic elements of these tissues and consequently causing tissue destruction.

Very common effects in leukemia are the development of infections, severe anemia, and bleeding tendency caused by thrombocytopenia (lack of platelets). These effects result mainly from displacement of the normal bone marrow by the leukemic cells.

Finally, perhaps the most important effect of leukemia on the body is the excessive use of metabolic substrates by the growing cancerous cells. The leukemic tissues reproduce new cells so rapidly that tremendous demands are made on the body fluids for foodstuffs, especially the amino acids and vitamins. Consequently, the energy of the patient is greatly depleted, and the excessive utilization of amino acids causes rapid deterioration of the normal protein tissues of the body. Thus, while the leukemic tissues grow, the other tissues are debilitated. Obviously, after metabolic starvation has continued long enough, this alone is sufficient to cause death.

REFERENCES

Allison, A. C., et al.: Inflammation. New York, Springer-Verlag, 1978.
Carr, I.: Lymphoreticular Disease: An Introduction for the Pathologist and Oncologist. Philadelphia, J. B. Lippincott Co., 1977.
Crowther, D. G. (ed.): Leukemia and Non-Hodgkin Lymphoma. New York, Pergamon Press, 1979.
Dick, G. (ed.): Immunological Aspects of Infectious Disease. Baltimore, University Park Press, 1978.
Escobar, M. R., and Friedman, H. (eds.): Macrophages and Lymphocytes; Nature, Functions and Interaction. New York, Plenum Press, 1979.
Friedman, H., et al. (eds.): The Reticuloendothelial System. New York, Plenum Press, 1979.
Gadebusch, H. J. (ed.): Phagocytosis and Cellular Immunity. West Palm Beach, Fla., CRC Press, 1979.
Gowans, J. L. (in honour of): Blood Cells and Vessel Walls: Functional Interactions. Princeton, N.J., Excerpta Medica, 1980.
Güttler, F., et al. (eds.): Inborn Errors of Immunity and Phagocytosis. Baltimore, University Park Press, 1979.
Ham, A. W., et al.: Blood Cell Formation and the Cellular Basis of Immune Responses. Philadelphia, J. B. Lippincott Co., 1979.
Hershey, S. G.: The reticulo-endothelial system: Relationship to shock and host defense. In Kaley, G., and Altura, B. M. (eds.): Microcirculation. Vol. III. Baltimore, University Park Press, 1977.
Houck, J. C. (ed.): Chemical Messengers of the Inflammatory Process. New York, Elsevier/North-Holland, 1979.
Janoff, A.: Neutrophil chemotaxis and mediation of tissue damage. In Kaley, G., and Altura, B. M. (eds.): Microcirculation. Vol. III. Baltimore, University Park Press, 1977.
Jeljaszewicz, J., and Wadström, T. (eds.): Bacterial Toxins and Cell Membranes. New York, Academic Press, 1978.
Kaley, G.: Mechanisms of inflammation. In Kaley, G., and Altura, B. M. (eds.): Microcirculation. Vol. III. Baltimore, University Park Press, 1977.
Kass, L.: Bone Marrow Interpretation. Philadelphia, J. B. Lippincott Co., 1979.
Keleman, E., et al.: Atlas of Human Hemopoietic Development. New York, Springer-Verlag, 1979.
Kinney, J. M., and Felig, P.: The metabolic response to injury and infection. In DeGroot, L. J., et al. (eds.): Endocrinology. Vol. 3. New York, Grune & Stratton, 1979, p. 1963.
Klebanoff, S. J., and Clark, R. A.: The Neutrophil: Function and Clinical Disorders. New York, Elsevier/North-Holland, 1978.
Kokubun, Y., and Kobayashi, N. (eds.): Phagocytosis, Its Physiology and Pathology. Baltimore, University Park Press, 1979.
Lichtman, M. A. (ed.): Hematology and Oncology. New York, Grune & Stratton, 1980.
Lisiewicz, J.: Human Neutrophils. Bowie, Md., Charles Press Publishers, 1979.
Logue, G. L., and Shimm, D. S.: Autoimmune granulocytopenia. Annu. Rev. Med., 31:191, 1980.
Movat, H. Z. (ed.): Inflammatory Reaction. New York, Springer-Verlag, 1979.
Neth, R., et al. (eds.): Modern Trends in Human Leukemia III. New York, Springer-Verlag, 1979.
Quastel, M. R. (ed.): Cell Biology and Immunology of Leukocyte Function. New York, Academic Press, 1979.
Rebuck, J. W., et al. (eds.): The Reticuloendothelial System. Huntington, N.Y., R. E. Krieger Publishing Co., 1980.
Rossof, A. H., and Robinson, W. A. (eds.): Lithium Effects on Granulopoiesis and Immune Function. New York, Plenum Press, 1980.
Ryan, G. B., and Majno, G.: Inflammation. Kalamazoo, Mich., Upjohn Co., 1977.
Simone, J. V. (ed.): Acute Leukemia. Philadelphia, W. B. Saunders, Co., 1978.
Spivak, J. L. (ed.): Fundamentals of Clinical Hematology, Hagerstown, Md., Harper & Row, 1980.
Styckmans, P. A.: Treatment of chronic myeloid leukemia. Annu. Rev. Med., 31:159, 1980.
Vane, J. R., and Ferreira, S. H. (eds.): Anti-Inflammatory Drugs. New York, Springer-Verlag, 1978.
Van Furth, R. (ed.): Mononuclear Phagocytes: Functional Aspects of Mononuclear Phagocytes. Boston, M. Nijhoff, 1979.
Wilkinson, A. W., and Cuthbertson, D. (eds.): Metabolism and the Response to Injury. Tunbridge Wells, England, Pitman Medical Publishers, 1976.

7

Immunity and Allergy

INNATE IMMUNITY

The human body has the ability to resist almost all types of organisms or toxins that tend to damage the tissues and organs. This capacity is called *immunity*. Much of the immunity is caused by a special immune system that forms antibodies and sensitized lymphocytes that attack and destroy the specific organisms or toxins. This type of immunity is *acquired immunity*. However, an additional portion of the immunity results from general processes rather than from processes directed at specific disease organisms. This is called *innate immunity*. It includes the following:

1. Phagocytosis of bacteria and other invaders by white blood cells and cells of the tissue macrophage system, as described in the previous chapter.

2. Destruction by the acid secretions of the stomach and by the digestive enzymes of organisms swallowed into the stomach.

3. Resistance of the skin to invasion by organisms.

4. Presence in the blood of certain chemical compounds that attach to foreign organisms or toxins and destroy them. Some of these are (a) *lysozyme,* a mucolytic polysaccharide that attacks bacteria and causes them to dissolute; (b) *basic polypeptides,* which react with and inactivate certain types of gram-positive bacteria; (c) *properdin,* a very large protein that can, in association with complement, react directly with gram-negative bacteria and destroy them; and (d) very small amounts of naturally occurring antibodies in the blood that have the specific ability to destroy certain bacteria, viruses, or toxins (these antibodies are similar to those that will be described later in the chapter as part of the acquired immunity system, but they occur without the necessity of previous exposure to the invading agent).

This innate immunity makes the human body resistant to such diseases as dysentery, some paralytic virus diseases of animals, hog cholera, cattle plague, and distemper, a viral disease that kills a large percentage of dogs that become afflicted with it. On the other hand, animals are resistant to many human diseases, such as poliomyelitis, mumps, human cholera, measles, and syphilis, which are very destructive or even lethal to the human being.

ACQUIRED IMMUNITY (OR ADAPTIVE IMMUNITY)

In addition to its innate immunity, the human body also has the ability to develop extremely powerful specific immunity against individual invading agents such as lethal bacteria, viruses, toxins, and even foreign tissues from other animals. This is called *acquired immunity* or *adaptive immunity*. It is with this immune mechanism and with very closely allied sequelae of the immune mechanism — the allergies — that most of this chapter will be concerned.

Acquired immunity can often bestow extreme protection. For instance, certain toxins such as the paralytic toxin of botulinum or the tetanizing toxin of tetanus can be protected against in doses as high as 100,000 times the amount that would be lethal without immunity. This is the reason the process known as "vaccination" is so extremely important in protecting human beings against disease and against toxins, as will be explained in the course of this chapter.

TWO BASIC TYPES OF ACQUIRED IMMUNITY

Two basic, but closely allied, types of acquired immunity occur in the body. In one of these the body develops circulating *antibodies,* which are globulin molecules that are capable of attacking the invading agent. This type of immunity is called *humoral immunity*. The second type of acquired immunity is achieved through the formation of large numbers of highly specialized lymphocytes that are specifically sensitized against the foreign agent. These *sensitized lymphocytes* have the special ability to attach to the foreign agent and to destroy it. This type of immunity is called *cellular immunity* or, sometimes, *lymphocytic immunity*.

We shall see shortly that both the antibodies and the sensitized lymphocytes are formed in the lymphoid tissue of the body. First, let us discuss the initiation of the immune process by *antigens*.

ANTIGENS

Since acquired immunity does not occur until after first invasion by a foreign organism or toxin, it is clear that the body must have some mechanism for recognizing the initial invasion. Each toxin or each type of organism contains one or more specific chemical compounds in its makeup that are different from all other compounds. In general, these are proteins, large polysaccharides, or large lipoprotein complexes, and it is they that cause the acquired immunity. These substances are called *antigens*.

For a substance to be antigenic it usually must have a high molecular weight, 8000 or greater. Furthermore, the process of antigenicity depends upon regularly recurring prosthetic radicals on the surface of the large molecule, which perhaps explains why proteins and polysaccharides are almost always antigenic, for they both have this type of stereochemical characteristic.

Haptens. Though substances with molecular weights less than 8000 only rarely act as antigens, immunity can nevertheless be developed against substances of low molecular weight in a very special way, as follows: If the low molecular weight compound, which is called a *hapten*, first combines with a substance that *is* antigenic, such as a protein, then the combination will elicit an immune response. The antibodies or sensitized lymphocytes that develop against the combination can then react either against the protein or against the hapten. Therefore, on second exposure to the hapten, the antibodies or lymphocytes react with it before it can spread through the body and cause damage.

The haptens that elicit immune responses of this type are usually drugs, chemical constituents in dust, breakdown products of dandruff from animals, degenerative products of scaling skin, various industrial chemicals, the toxin of poison ivy, and so forth.

ROLE OF LYMPHOID TISSUE IN ACQUIRED IMMUNITY

Acquired immunity is the product of the body's lymphoid tissue. In persons who have a genetic lack of lymphoid tissue or whose lymphoid tissue has been destroyed by radiation or by chemicals, no acquired immunity whatsoever can develop. And almost immediately after birth such a person dies of fulminating infection unless treated by heroic measures. Therefore, it is clear that the lymphoid tissue is essential to survival of the human being.

The lymphoid tissue is located most extensively in the lymph nodes, but it is also found in special lymphoid tissues such as that of the spleen, submucosal areas of the gastrointestinal tract, and, to a slight extent, the bone marrow. The lymphoid tissue is distributed very advantageously in the body to intercept the invading organisms or toxins before they can spread too widely. For instance, the lymphoid tissue of the gastrointestinal tract is exposed immediately to antigens invading through the gut. The lymphoid tissue of the throat and pharynx (the tonsils and adenoids) is extremely well located to intercept antigens that enter by way of the upper respiratory tract. The lymphoid tissue in the lymph nodes is exposed to antigens that invade the peripheral tissues of the body. And, finally, the lymphoid tissue of the spleen and bone marrow plays the specific role of intercepting antigenic agents that have succeeded in reaching the circulating blood.

Two Types of Lymphocytes That Promote, Respectively, Cellular Immunity and Humoral Immunity — the "T" and the "B" Lymphocytes. Though most of the lymphocytes in normal lymphoid tissue look alike when studied under the microscope, these cells are distinctly divided into two major populations. One of the popula-

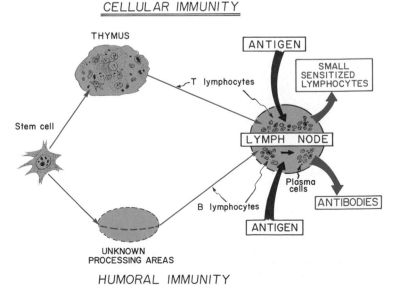

CELLULAR IMMUNITY

Figure 7–1. Formation of antibodies and sensitized lymphocytes by a lymph node in response to antigens. This figure also shows the origin of *thymic* ("T") and *bursal* ("B") lymphocytes that are responsible for the cellular and humoral immune processes of the lymph nodes.

HUMORAL IMMUNITY

tions is responsible for forming the sensitized lymphocytes that provide cellular immunity and the other for forming the antibodies that provide humoral immunity.

Both of these types of lymphocytes are derived originally in the embryo from *stem cells* that differentiate and become committed to form lymphocytes. The lymphocytes that are formed eventually end up in the lymphoid tissue, but before doing so they are further differentiated or "preprocessed" in the following ways:

Those lymphocytes that are eventually destined to form sensitized lymphocytes first migrate to and are preprocessed in the *thymus* gland, for which reason they are called "T" lymphocytes. These are responsible for cellular immunity.

The other population of lymphocytes — those that are destined to form antibodies — are processed in some unknown area of the body, possibly the fetal liver, the bone marrow, or gastrointestinal mucosa. However, this population of cells was first discovered in birds in which the preprocessing occurs in the *bursa of Fabricius,* a structure not found in mammals. For this reason these lymphocytes are called "B" lymphocytes, and they are responsible for humoral immunity.

Figure 7–1 illustrates the two separate lymphocyte systems for the formation, respectively, of the sensitized lymphocytes and the antibodies.

PREPROCESSING OF THE T AND B LYMPHOCYTES

Though all the lymphocytes of the body originate from the lymphocytic stem cells of the embryo, these stem cells are themselves incapable of forming either sensitized lymphocytes or antibodies. Before they can do so, they must be further differentiated in appropriate processing areas in the thymus or in the B cell processing area.

Role of the Thymus Gland for Preprocessing the T Lymphocytes. Most of the preprocessing of the T lymphocytes of the thymus gland occurs shortly before the birth of the baby and for a few months after birth. Therefore, beyond this period of time, removal of the thymus gland usually will not seriously impair the T lymphocytic immune system, the system necessary for cellular immunity. However, removal of the thymus several months before birth can completely prevent development of all cellular immunity. Since it is this cellular type of immunity that is mainly responsible for rejection of transplanted organs such as hearts and kidneys, one can transplant organs with little likelihood of rejection if the thymus is removed from an animal a reasonable period of time before birth.

Thymic Hormone. In addition to preprocessing the T lymphocytes, the thymus probably also secretes a hormone, called *thymopoietin,* that circulates through the body fluids and increases the activity of the T lymphocytes that have already left the thymus gland and have migrated to the lymphoid tissue. This hormone is believed to cause further proliferation and increased activity of these lymphocytes. Otherwise, little is known about either the nature or the function of this hormone.

Role of the Bursa of Fabricius for Preprocessing B Lymphocytes in Birds. It is during the latter part of fetal life that the bursa of Fabricius preprocesses the B lymphocytes and prepares them to manufacture antibodies. Here again, this process continues for a while after birth. In mammals, recent experiments indicate that it is lymphoid tissue in the fetal liver, and perhaps to a slight extent lymphoid tissue in the bone marrow and gastrointestinal mucosa, that performs this same function.

Spread of Processed Lymphocytes to the Lymphoid Tissue. After formation of processed lymphocytes in both the thymus and the bursa, these first circulate freely in the blood for a few hours but then become entrapped in the lymphoid tissue. Thus, the lymphocytes do not originate primordially in the lymphoid tissue, but, instead, are transported to this tissue by way of the preprocessing areas of the thymus and probably fetal liver.

MECHANISMS FOR DETERMINING SPECIFICITY OF SENSITIZED LYMPHOCYTES AND ANTIBODIES—LYMPHOCYTE CLONES

Earlier in the chapter it was pointed out that the lymphocytes of the lymphoid tissue can form sensitized lymphocytes and antibodies that can react highly specifically against particular types of invading agents. This effect is believed to occur in the following way:

Specificity of the Sensitized Lymphocytes or Antibodies Formed by a Single Type of Lymphocyte — Lymphocyte Clones. When a lymphocyte in the lymphoid tissue is stimulated to form either sensitized lymphocytes or antibodies, it always forms a sensitized lymphocyte or an antibody having specificity for a specific antigen. If more than one type of sensitized lymphocyte or antibody is to be formed, then a separate population of lymphocytes must be stimulated for each type. Because it is known that the lymphocytes of the lymphoid tissue can form 10,000 to 100,000 different types of sensitized lymphocytes or antibodies, each specific for a different antigen, it is also almost certain that equally as many different types of precursor lymphocytes pre-exist in the lymph nodes for formation of the many specific types of sensitized lymphocytes or antibodies.

All the lymphocytes of one specific type in the lymphoid tissue — those that form one specific type of sensitized lymphocyte or one specific type of antibody — are called a *clone of lymphocytes.* That is, all of the lymphocytes in each clone are alike and are probably derived originally from one or a few early lymphocytes of the specific type.

Origin of the Many Clones of Lymphocytes. The way in which the many different clones of lymphocytes are originally formed is not known, but there are two main theories for the origin. The first theory suggests that each clone is genetically determined, that is, there is a separate gene for each clone. This theory proposes that in the thymus in which the T lymphocytes are processed and in the processing area for the B lymphocytes the respective genes for the different lymphocytic clones are brought to expression, causing differentiation of the lymphocytic committed stem cells into the multiple clones preordained to form single types of sensitized lymphocyte or single types of antibody.

The second theory assumes that the lymphocytes in the thymus or in the B cell processing area simply differentiate wildly into a whole host of random clones of lymphocytes.

Excitation of a Clone of Lymphocytes. Each clone of lymphocytes is responsive to only a single type of antigen (or to a group of antigens that have almost exactly the same stereochemical characteristics). When excited by the clone's specific antigen, all the cells of the clone proliferate madly, forming tremendous numbers of progeny, and these in turn lead to the formation of large quantities of antibodies if the clone is B lymphocytes, or to the formation of sensitized lymphocytes if the clone is T lymphocytes. During this process, the total number of lymphocytes in the lymphoid tissue increases markedly.

Role of Macrophages in Stimulating the Clones of Lymphocytes. Aside from the lymphocytes that are entrapped in the lymphoid tissue, many monocytes also become entrapped, and these swell to become tissue macrophages that are called *reticulum cells,* as was discussed in the previous chapter. These line the sinusoids of the lymph nodes, spleen, and other lymphoid tissue, and they lie in apposition to the lymph node lymphocytes. Most invading organisms are first phagocytized by the macrophages, and the antigenic products are then liberated from the invader. It is believed that these antigens then pass directly from the macrophages to the lymphocytes to stimulate the specific lymphocytic clones.

It is also believed that initial excitation of T lymphocytes can lead to secondary excitation of B lymphocytes so that both sensitized lymphocytes and antibodies can be formed against the same invading agent. In other words, there appears to be an element of cooperation between the two systems of acquired immunity.

TOLERANCE OF THE ACQUIRED IMMUNITY SYSTEM TO ONE'S OWN TISSUES – ROLE OF THE THYMUS AND THE BURSA

Obviously, if a person should become immune to his or her own tissues, the process of acquired immunity would destroy the individual's own body. Fortunately, the immune mechanism normally "recognizes" a person's own tissues as being completely distinctive from those of invaders, and his immunity system forms very few antibodies or sensitized lymphocytes against his own antigens. This phenomenon is known as *tolerance* to the body's own tissues.

Mechanism of Tolerance. It is believed that tolerance develops during the processing of the lymphocytes in the thymus and in the B lymphocyte processing area. The reason for this belief is that injecting a strong antigen into a fetus at the time that the lymphocytes are being processed in these two areas will prevent the

development of clones of lymphocytes in the lymphoid tissue that are specific for the injected antigen. Also, experiments have shown that specific immature lymphocytes in the thymus, when exposed to a strong antigen, become lymphoblastic, proliferate considerably, and then combine with the stimulating antigen — an effect that is believed to cause the cells themselves to be destroyed before they can migrate to and colonize the lymphoid tissue.

Therefore, it is believed that during the processing of lymphocytes in the thymus and in the B lymphocyte processing area, all those clones of lymphocytes that are specific for the body's own tissues are self-destroyed because of their continual exposure to the body's antigens.

Failure of the Tolerance Mechanism — Autoimmune Diseases. Unfortunately, people frequently lose some of their immune tolerance to their own tissues. This occurs to a greater extent the older a person becomes. It usually results from destruction of some of the body's tissues, which releases considerable quantities of antigens that circulate in the body and cause acquired immunity in the form of either sensitized lymphocytes or antibodies. Some of these antigens perhaps combine with other proteins, such as proteins from bacteria or viruses, to form a new type of antigen that can then cause immunity. Then the resulting sensitized lymphocytes and antibodies attack the body's own tissues. Also, it is believed that some of the proteins of the body are normally sequestered from the immune system during embryonic development of tolerance so that tolerance to these proteins never forms in the first place. For instance, the proteins of the cornea do not seem to circulate in the fluids of the fetus; this is also true of the thyroglobulin molecule of the thyroid; therefore, tolerance to these never develops. When damage occurs to either of these two tissues, these protein molecules can then elicit immunity, and the immunity in turn can attack the cornea in the first instance or the thyroid gland in the second instance to cause corneal opacity or thyroiditis.

Other diseases that result from autoimmunity include: *rheumatic fever,* in which the body becomes immunized against tissues in the heart and joints following exposure to a specific type of streptococcal toxin; one type of *glomerulonephritis* in which the person becomes immunized against the basement membranes of his glomeruli; *myasthenia gravis,* in which immunity develops against the muscle portion of the neuromuscular junction, causing paralysis; and *lupus erythematosus,* in which the person becomes immunized against many different body tissues at the same time, a disease that causes extensive damage, often causing rapid death.

SPECIFIC ATTRIBUTES OF THE B LYMPHOCYTE SYSTEM – HUMORAL IMMUNITY AND THE ANTIBODIES

Formation of Antibodies by the Plasma Cells. Prior to exposure to a specific antigen, the clones of B lymphocytes remain dormant in the lymphoid tissue. However, upon entry of a foreign antigen, the lymphoid tissue macrophages phagocytize the antigen and then present

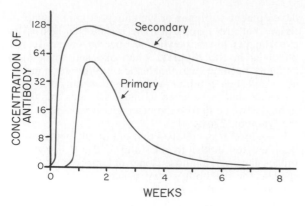

Figure 7–2. The time course of the antibody response to a *primary* injection of antigen and to a *secondary* injection several weeks later.

it to the adjacent lymphocytes. Those lymphocytes specific for that antigen immediately enlarge and take on the appearance of *lymphoblasts*. Some of these then further differentiate to form *plasmablasts,* which are the precursors of *plasma cells*. In these cells the cytoplasm expands, and the endoplasmic reticulum proliferates. They then begin to divide at a rate of approximately once every ten hours for about nine divisions, giving in four days a total population of about 500 cells for each original plasmablast. The mature plasma cell then produces gamma globulin antibodies at an extremely rapid rate — about 2000 molecules per second for each plasma cell. The antibodies are secreted into the lymph and are carried to the circulating blood. This process continues for several days or weeks until death of the plasma cells.

Formation of "Memory" Cells — Difference Between the Primary Response and the Secondary Response. Some of the lymphoblasts formed by activation of a clone of B lymphocytes do not go on to form plasma cells but, instead, form moderate numbers of new B lymphocytes similar to those of the original clone. In other words, the population of the specifically activated clone becomes greatly enhanced. And the new B lymphocytes are added to the original lymphocytes of the clone. These then remain dormant in the lymphoid tissue until activated once again by a new quantity of the same antigen. They are called *memory cells*. Obviously, subsequent exposure to the same antigen will then cause a much more rapid and much more potent antibody response.

Figure 7–2 illustrates the differences between the *primary response* that occurs on first exposure to a specific antigen and the *secondary response* that occurs following a second exposure to the same antigen. Note the delay in the appearance of the primary response, its weak potency, and its short life. The secondary response, by contrast, begins rapidly after exposure to the antigen, is far more potent, and forms antibodies for many months rather than for only a few weeks.

The increased potency and duration of the secondary response explains why *vaccination* is usually accomplished by injecting antigen in multiple doses with periods of several weeks or several months between injections.

The Nature of the Antibodies

The antibodies are gamma globulins called *immunoglobulins,* and they have molecular weights between approximately 150,000 and 900,000.

All of the immunoglobulins are composed of combinations of *light* and *heavy polypeptide chains,* most of which are a combination of two light and two heavy chains, as illustrated in Figure 7–3. Some of the immunoglobulins, though, have combinations of greater than two heavy and two light chains, which gives rise to the much larger molecular weight immunoglobulins. Yet, in all of these, each heavy chain is paralleled by a light chain at one of its ends, thus forming a heavy-light pair, and there are always at least two such pairs in each immunoglobulin molecule.

Figure 7–3 shows a designated end of each of the light and each of the heavy chains called the "variable portion," and the remainder of each chain is called the "constant portion." The variable portion is different for each specificity of antibody, and it is this portion that allows the antibody to attach specifically to a particular type of antigen. The constant portion of the antibody determines the gross physical and chemical properties of the antibody, establishing such factors as mobility of the antibody in the tissues, adherence of the antibody to specific structures within the tissues, attachment to complement, the ease with which the antibodies pass through membranes, and other biological properties of the antibody.

Specificity of Antibodies. Each antibody that is specific for a particular antigen has a different organization of amino acid residues in the variable portions of both the light and heavy chains. These have a specific steric shape for each antigen specificity so that when an antigen comes in contact with it, the prosthetic radicals of the antigen fit as a mirror image with those of the antibody, thus allowing a rapid and tight chemical or physical bond between the antibody and the antigen.

The constant portions of the antibody, on the other hand, provide means for attachment of the antibody to cells or other tissues and also provide means by which the antibody can combine with other chemical substances, most particularly the *complement complex,* which will be discussed subsequently.

Note, especially, in Figure 7–3 that there are two variable sites on the antibody for attachment of an-

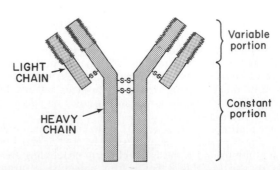

Figure 7–3. Structure of the typical IgG antibody, showing it to be composed of two heavy polypeptide chains and two light polypeptide chains. The antigen binds at two different sites on the variable portions of the chains.

tigens. Thus, most antibodies are *bivalent*. However, a small proportion of the antibodies, which have high molecular weight combinations of light and heavy chains, have more than two reactive sites.

Classes of Antibodies. There are five general classes of antibodies, respectively named *IgM, IgG, IgA, IgD, and IgE.* Ig stands for immunoglobulin, and the other five respective letters simply designate the respective classes of immunoglobulins.

For the purpose of our present limited discussion, two of these classes of antibodies are of particular importance: IgG, which comprises about 75 per cent of the antibodies of the normal person; and IgE, which constitutes only a small per cent of the antibodies but which is especially involved in allergy.

Mechanisms of Action of Antibodies

Antibodies can act in three different ways to protect the body against invading agents: (1) by direct attack on the invader, (2) by activation of the complement system that then destroys the invader, or (3) by activation of the anaphylactic system that changes the local environment around the invading antigen and in this way probably prevents its virulence.

Direct Action of Antibodies on Invading Agents. Figure 7–4 illustrates antibodies (designated by the bars) reacting with antigens (designated by the darkened dumbbells). Because of the bivalent nature of the antibodies and the multiple antigen sites on most invading agents, the antibodies can inactivate the invading agent in one of several ways, as follows:

1. *Agglutination,* in which multiple antigenic agents are bound together into a clump.

2. *Precipitation,* in which the complex of soluble antigen (such as tetanus toxin) and antibody becomes insoluble and precipitates.

3. *Neutralization,* in which the antibodies cover the toxic sites of the antigenic agent.

4. *Lysis,* in which some very potent antibodies are rarely capable of directly attacking membranes of cellular agents and thereby causing rupture of the cell.

However, the direct actions of antibodies attacking the antigenic invaders probably, under normal conditions, are not strong enough to play a major role in protecting the body against the invader. Most of the protection comes through the *amplifying* effects of the

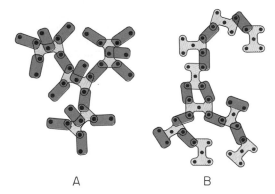

Figure 7–4. Reactions between antigens and antibodies. The antibodies are represented by bar structures, and the antigens are represented by stippled dumbbell structures. A, Reaction when excess antibody is present. B, Reaction when excess antigen is present.

complement and anaphylactic effector systems described below.

The Complement System for Antibody Action. Complement is a system of nine different enzyme precursors (designated C-1 through C-9) plus several other associated substances which are found normally in the plasma and other body fluids, but the enzymes are normally inactive. However, when an antibody combines with an antigen, a reactive site on the "constant" portion of the antibody becomes uncovered or activated, and this in turn sets into motion a "cascade" of sequential reactions in the complement system, illustrated in Figure 7–5. Only a few antigen-antibody combinations are required to activate enzyme precursor molecules in the first stage of the complement system, and the enzymes thus formed then activate successively increasing quantities of the enzymes in the later stages of the system. The activated enzymes then attack the invading agent in several different ways as well as initiate local tissue reactions that also provide protection against damage by the invader. Among the more important effects that occur are the following:

1. *Lysis.* The proteolytic enzymes of the complement system digest portions of the cell membrane, thus causing rupture of cellular agents such as bacteria or other types of invading cells.

Figure 7–5. Cascade of reactions during activation of the classical pathway of complement. (Reprinted from Alexander and Good: Fundamentals of Clinical Immunology. Philadelphia, W. B. Saunders Company, 1977.)

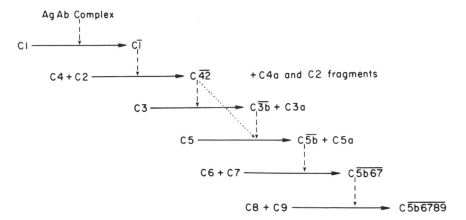

2. *Opsonization and phagocytosis.* The complement enzymes attack the surfaces of bacteria and other antigens, making these highly susceptible to phagocytosis by neutrophils and tissue macrophages. This process is called *opsonization*. It often enhances the number of bacteria that can be destroyed many hundred-fold.

3. *Chemotaxis.* One or more of the complement products cause chemotaxis of neutrophils and macrophages, thus greatly enhancing the number of these phagocytes in the local region of the antigenic agent.

4. *Agglutination.* The complement enzymes also change the surfaces of some of the antigenic agents so that they adhere to each other, thus causing agglutination.

5. *Neutralization of viruses.* The complement enzymes frequently attack the molecular structures of viruses and thereby render them nonvirulent.

6. *Inflammatory effects.* The complement products elicit a local inflammatory reaction, leading to hyperemia, coagulation of proteins in the tissues, and other aspects of the inflammation process, thus preventing movement of the invading agent through the tissues.

Activation of the Anaphylactic System by Antibodies. Some of the antibodies, particularly the IgE antibodies, attach to the membranes of cells in the tissues and blood. Among the most important cells are the *mast cells* in tissues surrounding the blood vessels and the *basophils* circulating in the blood. When an antigen reacts with one of the antibody molecules attached to the cell, there is an immediate swelling and then rupture of the cell, with the release of a large number of factors that affect the local environment. These factors include:

1. *Histamine,* which causes local vasodilatation and increased permeability of the capillaries.

2. *Slow-reacting substance of anaphylaxis,* which causes prolonged contraction of certain types of smooth muscle such as the bronchi.

3. *Chemotactic factor,* which causes chemotaxis of neutrophils and macrophages into the area of the antigen-antibody reaction. The chemotactic factor, especially, causes chemotaxis of large numbers of eosinophils into the area. It has been suggested that eosinophils play a special role in phagocytizing the products of the antibody-antigen reactions.

4. *Lysosomal enzymes,* which elicit a local inflammatory reaction.

These anaphylactic reactions can frequently be very harmful to the body, often causing the harmful reactions of allergy, as will be discussed subsequently. However, it is also known that in persons who are genetically unable to respond with the anaphylactic reaction, many types of infection spread much more rapidly through the body than occurs when the reaction can take place. Therefore, this reaction presumably helps to immobilize the antigenic invader.

SPECIAL ATTRIBUTES OF THE T LYMPHOCYTE SYSTEM – CELLULAR IMMUNITY AND SENSITIZED LYMPHOCYTES

Release of Sensitized Lymphocytes from Lymphoid Tissue and Formation of Memory Cells. Upon exposure to the proper antigens, sensitized lymphocytes are released from lymphoid tissue in ways that parallel antibody release. The only real difference is that instead of releasing antibodies, whole sensitized lymphocytes are formed and released into the lymph. These then pass into the circulation where they remain a few minutes to a few hours, at most; instead, they filter out of the circulation into all the tissues of the body.

Also, lymphocyte *memory cells* are formed in the same way that memory cells are formed in the humoral antibody system. Thus, when T lymphocytes are activated by an antigen, a large number of newly formed lymphocytes become additional T lymphocytes of that specific clone and remain in the lymphoid tissue, thus greatly increasing the population of this type of T lymphocyte. Therefore, upon subsequent exposure to the same antigen, the release of sensitized lymphocytes occurs much more rapidly and much more powerfully than in the first response.

Mechanism of Sensitization of T Lymphocytes. It is believed that T lymphocytes become sensitized against specific antigens by forming on their surfaces a type of "antibody." This antibody is composed of a *variable unit* similar to the variable portion of the humoral antibody, but it has no constant portion. Instead, multiple variable units are attached directly to the cell membrane of the T lymphocyte.

Persistence of Cellular Immunity. An important difference between cellular immunity and humoral immunity is its persistence. Humoral antibodies rarely persist more than a few months, or at most, a few years. On the other hand, sensitized lymphocytes probably have an indefinite life span and seem to persist until they eventually come in contact with their specific antigen. There is reason to believe that such sensitized lymphocytes might persist as long as ten years in some instances, a fact which makes cellular immunity far more persistent than humoral immunity.

Types of Organisms Resisted by Sensitized Lymphocytes. Although the humoral antibody mechanism for immunity is especially efficacious against more acute bacterial diseases, the cellular immunity system is activated much more potently by the more slowly developing bacterial diseases such as tuberculosis, brucellosis, and so forth. Also, this system is active against cancer cells, cells of transplanted organs, and fungus organisms, all of which are far larger than bacteria. And, finally, the system is very active against some viruses.

Therefore, cellular immunity is especially important in protecting the body against some virus diseases, in destroying many early cancerous cells before they can cause cancer, and unfortunately in causing rejection of transplanted tissues from one person to another.

Mechanism of Action of Sensitized Lymphocytes

The sensitized lymphocyte, on coming in contact with its specific antigen, combines with the antigen. This combination in turn leads to a sequence of reactions whereby the sensitized lymphocyte destroys the invader. As is also true of the humoral immunity system, the sensitized lymphocyte destroys the invader either directly or indirectly.

Direct Destruction of the Invader. Figure 7–6 illustrates sensitized lymphocytes that have bound with

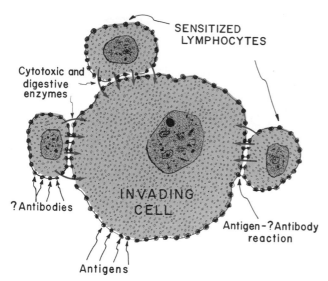

Figure 7–6. Direct destruction of an invading cell by sensitized lymphocytes.

antigens in the membrane of an invading cell such as a cancer cell, a heart transplant cell, or a parasitic cell of another type. The immediate effect of this attachment is swelling of the sensitized lymphocyte and release of cytotoxic substances from the lymphocyte to attack the invading cell. The cytotoxic substances are probably mainly lysosomal enzymes manufactured in the lymphocytes. However, these direct effects of the sensitized lymphocyte in destroying the invading cell are relatively weak in comparison with the indirect effects, as follows:

The Indirect "Amplifying" Mechanisms of Cellular Immunity. When the sensitized lymphocytes combine with their specific antigens, they release a number of different substances into the surrounding tissues that lead to a sequence of reactions. These reactions in turn are much more potent than the original attack on the invader. Some of these reactions are the following:

Release of Transfer Factor. The sensitized lymphocytes release a polypeptide substance with a molecular weight of 2000 to 8000, called *transfer factor*. This then reacts with other small lymphocytes in the tissues that are of the nonsensitized variety. On entering these lymphocytes, transfer factor causes them to take on characteristics similar to those of the original sensitized cells and to attack the invader along with the original cells. Furthermore, there is much reason to believe that these newly sensitized lymphocytes might also be specific for the same antigen as were the original sensitized lymphocytes. Thus, this mechanism multiplies the effect of the sensitized lymphocytes.

Attraction and Activation of Macrophages. A second product of the activated sensitized lymphocyte is a *macrophage chemotactic factor* that causes as many as 1000 macrophages to enter the vicinity of the activated sensitized lymphocyte. A third factor, called *migration inhibition factor,* then stops the migration of the macrophages once they come into the vicinity of the activated lymphocyte. Thus, a single lymphocyte can collect as many as 1000 macrophages around it. Finally, a fourth substance increases the phagocytic activity of

the macrophages. Therefore, the macrophages play a major role in removing the foreign antigenic invader.

Thus, it is by a combination of a weak direct effect of the sensitized lymphocytes on the antigen invader and much more powerful indirect reactions that the cellular immunity system destroys the invader.

VACCINATION

The process of vaccination has been used for many years to cause acquired immunity against specific diseases. A person can be vaccinated by injecting dead organisms that are no longer capable of causing disease but which still have their chemical antigens. This type of vaccination is used to protect against typhoid fever, whooping cough, diphtheria, and many other types of bacterial diseases. Also, immunity can be achieved against toxins that have been treated with chemicals so that their toxic nature has been destroyed even though their antigens for causing immunity are still intact. This procedure is used in vaccinating against tetanus, botulism, and other similar toxic diseases. And, finally, a person can be vaccinated by infecting him with live organisms that have been "attenuated." That is, these organisms either have been grown in special culture mediums or have been passed through a series of animals until they have mutated enough that they will not cause disease but do still carry the specific antigens. This procedure is used to protect against poliomyelitis, yellow fever, measles, smallpox, and many other viral diseases.

PASSIVE IMMUNITY

Thus far, all the acquired immunity that we have discussed has been *active immunity.* That is, the person's body develops either antibodies or sensitized lymphocytes in response to invasion of the body by a foreign antigen. However, temporary immunity can be achieved in a person without injecting any antigen whatsoever. This is done by infusing antibodies, sensitized lymphocytes, or both from someone else or from some other animal that has been actively immunized against the antigen. The antibodies will last for two to three weeks, and during that time the person is protected against the invading disease. Sensitized lymphocytes will last for a few weeks if transfused from another person, and for a few hours to a few days if transfused from an animal. The transfusion of antibodies or lymphocytes to confer immunity is called *passive immunity.*

INTERFERON — ANOTHER TYPE OF ACQUIRED IMMUNITY

Another type of acquired immunity has been discovered recently. Unfortunately, though, we still do not know how important it might be for protection against disease. This system is the following:

When cells in the body are attacked by viruses, many of them form protein substances called *interferons* that inactivate the virus that is attacking. These substances prevent the ribosomes from replicating either the RNA or the DNA virus particles and, therefore, inhibit its further damaging qualities. In addition, the interferons are released from the infected cells and carried through the body fluids to cells elsewhere in the body where they also prevent translation of the viral message. Therefore, infection of some cells by viral particles tends to protect other cells in the body from the same virus.

Though it is not known yet whether this is an important mechanism of acquired immunity, it is known that some virus diseases such as influenza begin to recede at the very time the concentration of interferon reaches its peak.

ALLERGY

One of the important side effects of immunity is the development, under some conditions, of allergy. There are at least three different types of allergy, two of which can occur in any person, and a third that occurs only in persons who have a specific allergic tendency.

ALLERGIES THAT OCCUR IN NORMAL PEOPLE

Delayed-Reaction Allergy. This type of allergy frequently causes skin eruptions in response to certain drugs or chemicals, particularly some cosmetics and household chemicals, to which one's skin is often exposed. Another example of such an allergy is the skin eruption caused by exposure to poison ivy.

Delayed-reaction allergy is caused by sensitized lymphocytes and not by antibodies. In the case of poison ivy, the toxin of poison ivy in itself does not cause much harm to the tissues. However, upon repeated exposure it does cause the formation of sensitized lymphocytes. Then, following subsequent exposure to the poison ivy toxin, within a day or so the sensitized lymphocytes diffuse in sufficient numbers into the skin to combine with the poison ivy toxin and elicit a cellular immunity type of reaction. Remembering that cellular immunity can cause release of many toxic substances from the sensitized lymphocytes, as well as extensive invasion of the tissues by macrophages and their subsequent effects, one can well understand that the eventual result of some delayed-reaction allergies can be serious tissue damage.

Allergies Caused by Reaction Between IgG Antibodies and Antigens. When a person becomes strongly immunized against an antigen and has developed a very high titer of IgG antibodies (the most usual type of antibody), subsequent sudden exposure of that person to a high concentration of the same antigen can cause a serious tissue reaction. The antigen-antibody complex that is formed precipitates, and some of it deposits as granules in the walls of the small blood vessels. These granules also activate the complement system, setting off extensive release of proteolytic enzymes. The result of these two effects is severe inflammation and destruction of the small blood vessels.

This type of allergy is especially manifest in the reaction called the *Arthus reaction:* this occurs when a large amount of antigen is injected into the tissues of a person who is strongly immunized to this same antigen.

The reaction between the IgG antibodies and the antigen elicits potent local vascular and other effects that cause almost total destruction of the local tissue — the damage beginning within a few minutes and leading to death and dissolution of the local tissue area within a few days.

Another manifestation of this type of reaction is *serum sickness.* Serum injected into a person can cause subsequent formation of IgG antibodies. When these begin to appear, they react with the protein of the injected serum and elicit a widespread antigen-antibody reaction throughout the body. Fortunately, this reaction occurs slowly over a period of days as the antibodies are formed, and usually it is not lethal. However, it can be lethal on occasion, and on other occasions it can cause widespread inflammation and edema throughout the body with development of a circulatory shocklike syndrome.

ALLERGIES IN THE "ALLERGIC" PERSON

Some persons have an "allergic" tendency. This phenomenon is genetically passed on from parent to child, and it is characterized by the presence of large quantities of *IgE antibodies*. These antibodies are called *reagins* or *sensitizing antibodies* to distinguish them from the more common IgG antibodies. When an *allergen* (defined as an antigen that reacts specifically with a specific type of IgE reagin antibody) enters the body, an allergen-reagin reaction takes place, and a subsequent allergic reaction takes place.

As was pointed out earlier in the discussion of immunity, the IgE antibodies (the reagins) attach to cells throughout the body, especially to mast cells and basophils; therefore, the allergen-reagin reaction damages the cells. The result is *anaphylactoid types of immune reactions*. These result primarily from the rupture of the mast cells and basophils with consequent release of *histamine, slow-reacting substance of anaphylaxis, eosinophil chemotactic substance, lysosomal enzymes,* and other less important substances.

Among the different types of allergic reaction of this type are:

Anaphylaxis. When a specific allergen is injected directly into the circulation it can react in widespread areas of the body with the basophils of the blood and the mast cells located immediately outside the small blood vessels. Therefore, the anaphylactic type of reaction occurs everywhere. The histamine released into the circulation causes widespread peripheral vasodilatation as well as increased permeability of the capillaries and marked loss of plasma from the circulation. Often, persons experiencing this reaction die of circulatory shock within a few minutes unless treated with norepinephrine to oppose the effects of the histamine. But also released from the cells is the substance called slow-reacting substance of anaphylaxis, which sometimes causes spasm of the smooth muscle of the bronchioles, eliciting an asthma-like attack and sometimes causing death by suffocation.

Urticaria. Urticaria results from antigen entering specific skin areas and causing localized anaphylactoid reactions. *Histamine* released locally causes (a) vasodilatation that induces an immediate *red flare* and (b) increased permeability of the capillaries that leads to swelling of the skin in another few minutes. The swellings are commonly called "hives." Administration of antihistamine drugs to a person prior to exposure will prevent the hives.

Hay Fever. In hay fever, the allergen-reagin reaction occurs in the nose. *Histamine* released in response to this causes local vascular dilatation with resultant increased capillary pressure, and it also causes increased capillary permeability. Both of these effects cause rapid fluid leakage into the tissues of the nose, and the nasal linings become swollen and secretory. Here again, use of antihistamine drugs can prevent this swelling reaction. However, other products of the allergen-reagin reaction still cause irritation of the nose, still eliciting the typical sneezing syndrome despite drug therapy.

Asthma. In asthma, the allergen-reagin reaction occurs in the bronchioles of the lungs. Here, the most important product released from the mast cells seems to be the *slow-reacting substance of anaphylaxis,* which causes spasm of the bronchiolar smooth muscle. Consequently, the person has difficulty breathing until the reactive products of the allergic reaction have been removed. Unfortunately, administration of antihistaminics has little effect on the course of asthma, because histamine does not appear to be the major factor eliciting the asthmatic reaction.

REFERENCES

Alexander, J. W.: Autoimmune mechanisms and transplantation. *In* Frohlich, E. D. (ed.): Pathophysiology, 2nd Ed. Philadelphia, J. B. Lippincott Co., 1976, p. 821.

Amos, D. B., *et al.* (eds.): Immune Mechanisms and Disease. New York, Academic Press, 1979.

Baram, P., *et al.* (eds.): Immunologic Tolerance and Macrophage Function. New York, Elsevier/North-Holland, 1979.

Bellanti, J. A.: Immunology II. Philadelphia, W. B. Saunders Co., 1978.

Benacerraf, B., and Unanue, E. R.: Textbook of Immunology. Baltimore, Williams & Wilkins, 1979.

Bruni, C., *et al.* (eds.): Systems Theory in Immunology. New York, Springer-Verlag, 1979.

Burke, D. C.: The status of interferon. *Sci. Am., 236*(4):42, 1977.

Capra, J. D., and Edmundson, A. B.: The antibody combining site. *Sci. Am., 236*(1):50, 1977.

Cochran, A. J.: Man, Cancer, and Immunity. New York, Academic Press, 1978.

Cohen, A. S. (ed.): Rheumatology and Immunology. New York, Grune & Stratton, 1979.

Dick, G. (ed.): Immunological Aspects of Infectious Diseases. Baltimore, University Park Press, 1978.

Fudenberg, H. H., and Smith, C. L. (eds): The Lymphocyte in Health And Disease. New York, Grune & Stratton, 1979.

Gerard, J. W.: Food Allergy. Springfield, Ill., Charles C Thomas, 1980.

Good, R. A., and Finstad, J.: Adaptive immunity. *In* Frohlich, E. D. (ed.): Pathophysiology, 2nd Ed. Philadelphia, J. B. Lippincott Co., 1976, p. 777.

Güttler, F., *et al.* (eds.): Inborn Errors of Immunity and Phagocytosis. Baltimore, University Park Press, 1979.

Ham, A. W., *et al.*: Blood Cell Formation and the Cellular Basis of Immune Responses. Philadelphia, J. B. Lippincott Co., 1979.

Hemmings, W. A. (ed.): Antigen Absorption by the Gut. Baltimore, University Park Press, 1978.

Holborow, E. J., and Reeves, W. G.: Immunology in Medicine: A Comprehensive Guide to Clinical Immunology. New York, Grune & Stratton, 1977.

Hood, L. E., *et al.*: Immunology. Menlo Park, Cal., Benjamin/Cummings Publishing Co., 1978.

Jirsch, D. W. (ed.): Immunological Engineering. Baltimore, University Park Press, 1978.

Johnson, F. (ed.): Allergy, Including IGE in Diagnosis and Treatment. Chicago, Year Book Medical Publishers, 1979.

Kaplan, J. G. (ed.): The Molecular Basis of Immune Cell Function. New York, Elsevier/North-Holland, 1979.

Koffler, D.: The Immunology of Rheumatoid Diseases. Summit, N.J., CIBA Pharmaceutical Company, 1979.

Lance, E. M., *et al.* (eds.): An Introduction to Immunology. London, Wildwood House, 1977.

Litman, G. W., and Good, R. A. (eds.): Immunoglobulins. New York, Plenum Press, 1977.

Maini, R. N., *et al.*: Immunology of the Rheumatic Diseases. London, Edward Arnold, 1977.

Marchesi, V. T., *et al.* (eds.): Cell Surface Carbohydrates and Biological Recognition. New York, A. R. Liss, 1978.

McGiven, A. R. (ed.): Immunological Investigation of Renal Disease. New York, Churchill Livingstone, 1980.

Metzger, H.: Early molecular events in antigen-antibody cell activation. *Annu. Rev. Pharmacol. Toxicol., 19*:427, 1979.

Mygind, N.: Nasal Allergy. Oxford, Blackwell Scientific Publications, 1978.

Nahmias, A. J., and O'Reilly, R. (eds.): Immunology of Human Infection. New York, Plenum Press, 1979.

Pernis, B., and Vogel, H. J. (eds.): Cells of Immunoglobulin Synthesis. New York, Academic Press, 1979.

Pinchera, A., and Fenzi, G.: Endocrine autoimmune diseases. *In* DeGroot, L. J., *et al.* (eds.): Endocrinology. Vol. 3. New York, Grune & Stratton, 1979, p. 2063.

Pollara, B., *et al.* (eds.): Inborn Errors of Specific Immunity. New York, Academic Press, 1979.

Proffitt, M. R. (ed.): Virus-Lymphocyte Interactions: Implications of Disease. New York, Elsevier/North-Holland, 1979.

Raff, M. C.: Cell-surface immunology. *Sci. Am., 234*(5):30, 1976.

Roitt, I. M.: Essential Immunology. Oxford, Blackwell Scientific Publications, 1977.

Salton, M. R. J. (ed.): Immunochemistry of Enzymes and Their Antibodies. Huntington, N.Y., R. E. Krieger Publishing Co., 1980.

Samter, M. (ed.): Immunological Diseases. Boston, Little, Brown, 1978.

Schiff, G. M.: Active immunization for adults. *Annu. Rev. Med., 31*: 441, 1980.

Schwartz, L. M.: Compendium of Immunology. New York, Van Nostrand Reinhold, 1979.

Sercarz, E. E., and Cunningham, A. J. (eds.): Strategies of Immune Regulation. New York, Academic Press, 1980.

Singhal, S. K., and Sinclair, N. R. St. C. (eds.): Suppressor Cells in Immunity. London, Ont., University of Western Ontario, 1975.

Sterzl, J., and Riha, I. (eds.): Developmental Aspects of Antibody Formation and Structure. New York, Academic Press, 1970.

Stuart, F. P., and Fitch, F. W., (eds.): Immunological Tolerance and Enhancement. Baltimore, University Park Press, 1979.

Terry, W. D., and Yamamura, Y. (eds.): Immunobiology and Immunotherapy of Cancer. New York, Elsevier/North-Holland, 1979.

Voller, A., and Friedman, H. (eds.): New Trends and Developments in Vaccines. Baltimore, University Park Press, 1978.

Weksler, M. E. (ed.): Immune Effector Mechanisms in Disease. New York, Grune & Stratton, 1977.

Wells, J. H., and Cain, W. A.: Allergic mechanisms. *In* Frohlich, E. D. (ed.): Pathophysiology, 2nd Ed. Philadelphia, J. B. Lippincott Co., 1976, p. 803.

8

Blood Groups; Transfusion;
Tissue and Organ
Transplantation

ANTIGENICITY AND IMMUNE
REACTIONS OF BLOOD

When blood transfusions from one person to another were first attempted the transfusions were successful in some instances, but, in many more, immediate or delayed agglutination and hemolysis of the red blood cells occurred. Soon it was discovered that the bloods of different persons usually have different antigenic and immune properties so that antibodies in the plasma of one blood react with antigens on the red cells of another. Furthermore, the antigens and the antibodies are almost never precisely the same in one person as in another. For this reason, it is easy for blood from a donor to be mismatched with that of a recipient. Fortunately, if proper precautions are taken, one can determine ahead of time whether or not appropriate antibodies and antigens are present in the donor and recipient bloods to cause a reaction, but, on the other hand, lack of proper precautions often results in varying degrees of red cell agglutination and hemolysis, resulting in a typical transfusion reaction that can lead to death.

Multiplicity of Antigens in the Blood Cells. At least 30 commonly occurring antigens, each of which can at times cause antigen-antibody reactions, have been found in human blood cells, especially on the surfaces of the cell membranes. In addition to these, more than 300 others of less potency or that occur in individual families rather than having widespread occurrence are known to exist. Among the 30 or more common antigens, certain ones are highly antigenic and regularly cause transfusion reactions if proper precautions are not taken, whereas others are of importance principally for studying the inheritance of genes and therefore for establishing parentage, race, and so forth. Essentially all of these antigens are either glycolipids or mucopolysaccharides.

Two particular groups of antigens are more likely than the others to cause blood transfusion reactions. These are the so-called *O-A-B* system of antigens and the *Rh* system. Bloods are divided into different *groups* and *types* in accordance with the types of antigens present in the cells.

O-A-B BLOOD GROUPS

THE A AND B ANTIGENS — CALLED "AGGLUTINOGENS"

Two different but related antigens — type A and type B — occur on the surfaces of the red blood cells in different persons. Because of the way these antigens are inherited, people may have neither of them in their cells, they may have one, or they may have both simultaneously.

As will be discussed below, some bloods also contain strong antibodies that react specifically with either the type A or type B antigens in the cells, causing agglutination and hemolysis. Because the type A and type B antigens in the cells make the cells susceptible to agglutination, these antigens are called *agglutinogens*. It is on the basis of the presence or absence of these agglutinogens in the red blood cells that blood is grouped for the purpose of transfusion.

The Four Major O-A-B Blood Groups. In transfusing blood from one person to another, the bloods of donors and recipients are normally classified into four major O-A-B groups, as illustrated in Table 8–1, depending on the presence or absence of the two agglutinogens. When neither A nor B agglutinogen is present, the blood group is *group O*. When only type A agglutinogen is present, the blood is *group A*. When only type B agglutinogen is present, the blood is *group B*. And when both A and B agglutinogens are present, the blood is *group AB*.

Relative Frequencies of the Different Blood Types. The prevalence of the different blood types among Caucasoids is approximately as follows:

Type	Per cent
O	47
A	41
B	9
AB	3

It is obvious from these percentages that the O and A genes occur frequently but the B gene is infrequent.

84

TABLE 8–1 THE BLOOD GROUPS WITH THEIR GENOTYPES AND THEIR CONSTITUENT AGGLUTINOGENS AND AGGLUTININS

Genotypes	Blood Groups	Agglutinogens	Agglutinins
OO	O	—	Anti-A and Anti-B
OA or AA	A	A	Anti-B
OB or BB	B	B	Anti-A
AB	AB	A and B	—

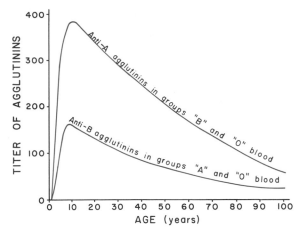

Figure 8–1. Average titers of anti-A and anti-B agglutinins in the blood of persons in group B and group A at different ages.

Genetic Determination of the Agglutinogens. Genes on two adjacent chromosomes, one gene on each chromosome, determine the O-A-B blood groups. These are allelomorphic genes that can be any one of three different types, but only one type on each chromosome: type O, type A, or type B. There is no dominance among the three different allelomorphs. However, the type O gene is either functionless or almost functionless, so that it causes either no type O agglutinogen in the cells or such a weak agglutinogen that it is normally insignificant. On the other hand, the type A and type B genes do cause strong agglutinogens in the cells. Therefore, if either of the genes on the two respective chromosomes is type A, the red blood cells will contain type A agglutinogen, and likewise, if either of the two genes is type B, the red blood cells will contain type B agglutinogen. Or, if the gene on one chromosome is type A and on the other is type B, the red blood cells will contain both A and B agglutinogens.

The six possible combinations of genes, as shown in Table 8–1, are OO, OA, OB, AA, BB, and AB. These different combinations of genes are known as the *genotypes*, and each person is one of the six different genotypes.

One can observe from the table that a person with genotype OO produces no agglutinogens at all, and, therefore, the blood group is O. A person with either genotype OA or AA produces type A agglutinogens and, therefore, has blood group A. Genotypes OB and BB give group B blood, and genotype AB gives group AB blood.

THE AGGLUTININS

When type A agglutinogen *is not present* in a person's red blood cells, antibodies known as "anti-A" agglutinins develop in his plasma. Also, when type B agglutinogen *is not present* in the red blood cells, antibodies known as "anti-B" agglutinins develop in the plasma.

Thus referring once again to Table 8–1, it will be observed that group O blood, though containing no agglutinogens, does contain both *anti-A* and *anti-B agglutinins*, while group A blood contains type A agglutinogens and *anti-B agglutinins*, and group B blood contains type B agglutinogens and *anti-A agglutinins*. Finally, group AB blood contains both A and B agglutinogens but no agglutinins at all.

Titer of the Agglutinins at Different Ages. Immediately after birth the quantity of agglutinins in the plasma is almost zero. Two to eight months after birth, the infant begins to produce agglutinins — anti-A agglutinins when type A agglutinogens are not present in the cells and anti-B agglutinins when type B agglutinogens are not in the cells. Figure 8–1 illustrates the changing titer of alpha and beta agglutinins at different ages. A maximum titer is usually reached at 8 to 10 years of age, and this gradually declines throughout the remaining years of life.

Origin of the Agglutinins in the Plasma. The agglutinins are gamma globulins, as are other antibodies, and are produced by the same cells that produce antibodies to any other antigens. Most of them are IgM and IgG immunoglobulin molecules.

It is difficult to understand how agglutinins are produced in individuals who do not have the respective antigenic substances in their red blood cells. However, small amounts of group A and B antigens enter the body in the food, in bacteria, and in other ways, and these substances presumably initiate the development of the anti-A or anti-B agglutinins. One of the reasons for believing this is that injection of group A or group B antigen into a recipient having another blood type causes a typical immune response with formation of greater quantities of agglutinins than ever. Also, the newborn baby has few if any agglutinins, showing that agglutinin formation occurs almost entirely after birth.

THE AGGLUTINATION PROCESS IN TRANSFUSION REACTIONS

When bloods are mismatched so that anti-A or anti-B agglutinins are mixed with red blood cells containing A or B agglutinogens respectively, the red cells agglutinate by the following process: The agglutinins attach themselves to the red blood cells. Because the agglutinins are bivalent (IgG type) or polyvalent (IgM type), a single agglutinin can attach to two different red blood cells at the same time, thereby causing the cells to adhere to each other. This causes the cells to clump. Then these clumps plug small blood vessels throughout the circulatory system. During the ensuing few hours to few days, the phagocytic white blood cells and the reticuloendothelial system destroy the agglutinated cells, releasing hemoglobin into the plasma.

Hemolysis in Transfusion Reactions. Sometimes, when recipient and donor bloods are mismatched, immediate hemolysis of red cells occurs in the circulating blood. In this case the antibodies cause lysis of the red blood cells by activating the complement system. This in turn releases proteolytic enzymes that rupture the cell membranes, as was described in Chapter 7.

However, immediate intravascular hemolysis is far less common than agglutination, because not only does there have to be a very high titer of antibodies for this to occur but also a different type of antibody seems to be required; these antibodies are called *hemolysins*. However, even agglutination eventually leads to hemolysis of the agglutinated red cells because the phagocytic white blood cells and the reticuloendothelial cells rupture the agglutinated cells within a few hours after agglutination has taken place.

BLOOD TYPING

Prior to giving a transfusion, it is necessary to determine the blood group of the recipient and the group of the donor blood so that the bloods will be appropriately matched. This "typing" of blood is performed as follows:

A usual method of blood typing is the slide technique. In using this technique a drop or more of blood is removed from the person to be typed. This is then diluted approximately 50 times with saline so that clotting will not occur. This leaves essentially a suspension of red blood cells in saline. Two separate drops of this suspension are placed on a microscope slide, and a drop of anti-A agglutinin serum is mixed with one of the drops of cell suspension while a drop of anti-B agglutinin serum is mixed with the second drop of cell suspension. After allowing several minutes for the agglutination process to take place, the slide is observed under a microscope to determine whether or not the cells have clumped. If they have clumped, one knows that an immune reaction has resulted between the serum and the cells.

Table 8–2 illustrates the reactions that occur with each of the four different types of blood. Group O red blood cells have no agglutinogens and, therefore, do not react with either the anti-A or the anti-B serum. Group A blood has A agglutinogens and therefore agglutinates with anti-A agglutinins. Group B blood has B agglutinogens and agglutinates with the anti-B serum. Group AB

TABLE 8–2 BLOOD TYPING—SHOWING AGGLUTINATION OF CELLS OF THE DIFFERENT BLOOD GROUPS WITH ANTI-A AND ANTI-B AGGLUTININS

| Red Blood Cells | Sera | |
	Anti-A	Anti-B
O	−	−
A	+	−
B	−	+
AB	+	+

blood has both A and B agglutinogens and agglutinates with both types of serum.

CROSS-MATCHING

An additional procedure to determine compatibility of donor and recipient bloods is to test the bloods against each other to determine whether or not agglutination will occur. To do this, one prepares, first, a suspension of red cells from the donor, and, second, a small quantity of defibrinated serum from the recipient. Then the serum from the recipient is mixed with the cells from the donor to determine whether or not agglutination occurs. In a second test, the cells of the recipient are "cross-matched" against the serum of the donor. If no agglutination of either the donor's or the recipient's cells occurs, it can be assumed that the two bloods are probably compatible enough to proceed with a transfusion even though the actual blood types might be unknown and even though the bloods might contain other red cell antigens in addition to the O-A-B antigens.

THE Rh BLOOD TYPES

In addition to the O-A-B blood group system, several other systems are sometimes important in the transfusion of blood. The most important of these is the Rh system. The one major difference between the O-A-B system and the Rh system is the following: In the O-A-B system, the agglutinins responsible for causing transfusion reactions develop spontaneously, while in the Rh system spontaneous agglutinins almost never occur. Instead, the person must first be massively exposed to an Rh antigen, usually by transfusion of blood, before enough agglutinins to cause a significant transfusion reaction are developed.

The Rh Antigens — "Rh Positive" and "Rh Negative" Persons. There are six common types of Rh antigens, each of which is called an Rh factor. These types are designated C, D, E, c, d, and e. A person who has a C antigen will not have the c antigen, but the person missing the C antigen will always have the c antigen. The same is true for the D-d and E-e antigens. Also, because of the manner of inheritance of these factors, each person will have one of each of the three pairs of antigens.

Only the C, D, and E antigens are usually antigenic enough to cause significant development of anti-Rh antibodies which are capable of causing transfusion reactions. Therefore, anyone who has any one of these three antigens, or any combination of them, is said to be *Rh positive*. A person who has no C, D, or E antigens but instead has only c, d, and e antigens is said to be Rh negative.

Inheritance of Rh Factors. There are three separate loci on one pair of chromosomes, respectively, for the three different pairs of Rh factors, C-c, D-d, and E-e. In each case, the Rh positive factors — C, D, and E — are mendelian dominant, so that if either chromosome contains a gene for an Rh positive factor, this factor will be present in the blood. To be Rh negative, a person must have no genes that code for Rh positive factors.

Approximately 85 per cent of all Caucasoids are Rh positive and 15 per cent, Rh negative. In American Blacks, the percentage of Rh positives is about 95.

Typing Bloods for Rh Factors. Typing for Rh factors is performed in a manner similar to that used in typing the A-B-O agglutinogens. This is usually achieved using from 4 to 6 different anti-Rh sera. However, the anti-Rh antibodies are far less potent in their capability of causing red cell agglutination than are the anti-A and the anti-B antibodies. Therefore, to cause agglutination in the presence of the Rh antibodies, a small amount of protein must be added to the reactant mixture; this protein provides cross-linkages between the antibodies after they attach to the red blood cells.

THE Rh IMMUNE RESPONSE

Formation of Anti-Rh Agglutinins. When red blood cells containing one or more Rh positive factors (C, D, or E), or even protein breakdown products of such cells, are injected into an Rh negative person, anti-Rh agglutinins develop very slowly, the maximum concentration of agglutinins occurring approximately two to four months later. This immune response occurs to a much greater extent in some people than in others. On multiple exposure to the Rh factor, the Rh negative person eventually becomes strongly "sensitized" to the Rh factor — that is, he or she develops a very high titer of anti-Rh agglutinins.

Characteristics of Rh Transfusion Reactions. If an Rh negative person has never before been exposed to Rh positive blood, transfusion of Rh positive blood into him causes no immediate reaction at all. However, in some of these persons anti-Rh antibodies develop in sufficient quantities during the next two to four weeks to cause agglutination of the transfused cells that are still in the blood. These cells are then hemolyzed by the reticuloendothelial system. Thus, a delayed transfusion reaction occurs, though it is usually mild. Yet, on subsequent transfusion of Rh positive blood into the same person, who is now immunized against the Rh factor, the transfusion reaction is greatly enhanced and can be as severe as the reactions that occur with types A and B bloods.

Erythroblastosis Fetalis. Erythroblastosis fetalis is a disease of the fetus and newborn infant characterized by progressive agglutination and subsequent phagocytosis of the red blood cells. In most instances of erythroblastosis fetalis the mother is Rh negative, and the father is Rh positive; the baby has inherited the Rh positive characteristic from the father, and the mother has developed anti-Rh agglutinins that have diffused through the placenta into the fetus to cause red blood cell agglutination.

Prevalence of the Disease. An Rh negative mother having her first Rh positive child usually does not develop sufficient anti-Rh agglutinins to cause any harm. However, an Rh negative mother having her second Rh positive child often will have become "sensitized" by the first child and therefore will often develop anti-Rh agglutinins rapidly upon becoming pregnant with the second child. Approximately 3 per cent of these second babies exhibit some signs of erythroblastosis fetalis; approximately 10 per cent of the third babies exhibit the disease;

and the incidence rises progressively with subsequent pregnancies.

The Rh negative mother develops anti-Rh agglutinins only when the fetus is Rh positive. Many of the Rh positive fathers are heterozygous (about 55 per cent), causing about one fourth of the offspring to be Rh negative. Therefore, after an erythroblastotic child has been born, it is not certain that future children will also be erythroblastotic.

Prevention of Erythroblastosis in Babies with Rh Negative Mothers. The Rh negative mother usually becomes sensitized to the Rh positive factor in her child during the first few days following birth of the baby and not while she is carrying the child. At this time large quantities of degenerating products of the placenta release their antigens into the blood of the mother, giving her a healthy dose of the baby's Rh positive antigen. If this Rh positive antigen can be destroyed at this time before it can initiate the antibody response, the mother will not become presensitized against the Rh factor for subsequent pregnancies. This result can be achieved by passively immunizing the mother against the Rh positive factor. This is done by injecting into the mother serum from another Rh negative person who has already formed anti-Rh agglutinins. These agglutinins circulate in the mother's blood for three to eight weeks and destroy all the Rh positive factor from the placenta, thus preventing development of the mother's anti-Rh agglutinins. With this treatment it is possible to eliminate most future occurrences of erythroblastosis fetalis.

Effect of the Mother's Antibodies on the Fetus. After anti-Rh antibodies have formed in the mother, they diffuse very slowly through the placental membrane into the fetus' blood. There they cause slow agglutination of the fetus' blood. The agglutinated red blood cells gradually hemolyze, releasing hemoglobin into the blood. The reticuloendothelial cells then convert the hemoglobin into bilirubin, which causes yellowness (jaundice) of the skin. The antibodies probably also attack and damage many of the other cells of the body.

Clinical Picture of Erythroblastosis. The newborn, jaundiced, erythroblastotic baby is usually anemic at birth, and the anti-Rh agglutinins from the mother usually circulate in the baby's blood for one to two months after birth, destroying more and more red blood cells. Therefore, the hemoglobin level of the untreated erythroblastotic baby often falls for approximately the first 45 days after birth, and, if the level falls below 6 to 8 grams per cent, the baby usually dies.

The hemopoietic tissues of the baby attempt to replace the hemolyzing red blood cells. The liver and the spleen become greatly enlarged and produce red blood cells in the same manner that they normally do during the middle of gestation. Because of the very rapid production of cells, many early forms, including many nucleated blastic forms, are emptied into the circulatory system, and it is because of the presence of these in the blood that the disease has been called "erythroblastosis fetalis."

Though the severe anemia of erythroblastosis fetalis is usually the cause of death, many children who barely survive from the anemia exhibit permanent mental impairment or damage to motor areas of the brain because of precipitation of bilirubin in the neuronal cells, causing their destruction, a condition called *kernicterus.*

Treatment of the Erythroblastotic Baby. The usual treatment for erythroblastosis fetalis is to replace the newborn infant's blood with Rh negative blood. Approximately 400 ml. of Rh negative blood is infused over a period of 1.5 or more hours while the baby's own Rh positive blood is being removed. This procedure may be repeated several times during the first few weeks of life, mainly to keep the bilirubin level low and thereby to prevent kernicterus. By the time the Rh negative cells are replaced with the baby's own Rh positive cells, a process that requires six or more weeks, the anti-Rh agglutinins that had come from the mother will have been destroyed.

Erythroblastosis Fetalis Occurring in Babies of Rh Positive Mothers. Approximately 7 per cent of the babies who have erythroblastosis fetalis are born of Rh positive mothers rather than Rh negative mothers. A very few cases are caused by anti-A or anti-B agglutinins that diffuse from the mother into the fetus and agglutinate the fetus' red blood cells. Some of the other instances result from the fact that the so-called "Rh positive" mother is sometimes Rh positive for only one or two Rh factors and yet Rh negative for the other Rh factors. Finally, similar reactions have also been noted on rare occasions as a result of the c or e Rh antigens or other less well-known antigenic factors in the fetus' blood, some of which are discussed in the following section.

OTHER BLOOD FACTORS

Many antigenic proteins besides the O, A, B, and Rh factors are present in red blood cells of different persons, but these other factors only rarely cause transfusion reactions and, therefore, are mainly of academic and legal importance. Some of these different blood factors are the M, N, S, s, P, Kell, Lewis, Duffy, Kidd, Diego, and Lutheran factors. Occasionally, multiple transfusions of red cells containing one of these factors cause specific agglutinins to develop in the recipient, and transfusion reactions can then occur when the same type of blood is infused again.

Method for Studying Obscure Blood Factors. One of the means by which different blood factors, including the Rh factor, have been studied in the human being has been to immunize lower animals, such as rabbits, with human blood cells. Specific antibodies develop in the animal against the antigens of the human cells, and specific immune sera prepared from the plasma of these animals can then be used for determining the presence or absence of the same antigens in red blood cells of other persons.

Blood Typing in Legal Medicine. In the past three decades the use of blood typing has become an important legal procedure in cases of disputed parentage. Including all the blood types, as many as 50 common blood group genes can be determined for each person by blood-typing procedures. After the mother's and child's genes have been determined, many of the father's genes, and his corresponding blood factors, are then known immediately, because any gene present in a child but not present in the mother must be present in the father. If a suspected man is missing any one of the necessary blood factors, he is not the father. A falsely charged man can be cleared in about 75 per cent of the cases using the usually available antisera for blood typing, and in essentially all cases using all possible types of antisera.

TRANSFUSION

Indications for Transfusion. The most common reason for transfusion is decreased blood volume, which will be discussed in detail in Chapter 28 in relation to shock. Also, transfusions are often used for treating anemia or to supply the recipient with some other constituent of whole blood besides red blood cells, such as to supply a thrombocytopenic patient with new platelets. Also, hemophilic patients can be rendered temporarily nonhemophilic by plasma transfusion, and, occasionally, the quantity of "complement" in a recipient must be supplemented by fresh plasma infusions before certain antigen-antibody reactions can take place.

TRANSFUSION REACTIONS RESULTING FROM MISMATCHED BLOOD GROUPS

If blood of one blood group is transfused to a recipient of another blood group, a transfusion reaction is likely to occur in which the red blood cells *of the donor blood* are agglutinated. It is very rare that the transfused blood ever causes agglutination *of the recipient's cells* for the following reason: The plasma portion of the donor blood immediately becomes diluted by all the plasma of the recipient, thereby decreasing the titer of the infused agglutinins to a level too low to cause agglutination. On the other hand, the infused blood does not dilute the agglutinins in the recipient's plasma to a major extent. Therefore, the recipient's agglutinins can still agglutinate the donor cells.

Hemolysis of Red Cells Following Transfusion Reactions. All transfusion reactions resulting from mismatched blood groups eventually cause hemolysis of the red blood cells. Occasionally, the antibodies are potent enough and are composed of the appropriate class of immunoglobulins to cause immediate hemolysis, but more frequently the cells agglutinate first and then are mainly entrapped in the peripheral vessels. Over a period of hours to days the entrapped cells are phagocytized, thereby liberating hemoglobin into the circulatory system.

When the rate of hemolysis is rapid, the concentration of hemoglobin in the plasma can rise to extremely high values. A small quantity of hemoglobin can become attached to one of the plasma proteins, *haptoglobin,* and continue to circulate in the blood without causing any harm. However, above the threshold value of 100 mg. of hemoglobin per 100 ml. of plasma, the excess hemoglobin remains in the free form and diffuses out of the circulation into the tissue spaces or through the renal glomeruli into the kidney tubules, as is discussed below. The hemoglobin remaining in the circulation or passing into the tissue spaces is gradually ingested by phagocytic cells and converted into bilirubin, which will be discussed in Chapter 70. The concentration of bilirubin in the body fluids sometimes rises high enough to cause *jaundice* — that is, the person's tissues become tinged with yellow pigment. But, if liver function is normal, jaundice usually does not appear unless more than 300 to 500 ml. of blood is hemolyzed in less than a day.

Acute Kidney Shutdown Following Transfusion Reactions. One of the most lethal effects of transfusion reactions is acute kidney shutdown, which can begin within a few minutes to a few hours and continue until the person dies of renal failure.

The kidney shutdown seems to result from three different causes: First, the antigen-antibody reaction of the transfusion reaction releases toxic substances from the hemolyzing blood that cause powerful renal vasoconstriction. Second, the loss of circulating red cells along with production of toxic substances from the cells and from the immune reaction often causes circulatory shock; the arterial blood pressure falls very low and the renal blood flow and urinary output decrease. Third, if the total amount of free hemoglobin in the circulating blood is greater than that quantity which can bind with haptoglobin, much of the excess leaks through the glomerular membranes into the kidney tubules. If this amount is still slight, it can be reabsorbed through the tubular epithelium into the blood and will cause no harm, but, if it is great, then only a small percentage is reabsorbed. Yet water continues to be reabsorbed, causing the tubular hemoglobin concentration to rise so high that it precipitates and blocks many of the tubules; this is especially true if the urine is acidic. Thus, renal vasoconstriction, circulatory shock, and tubular blockage all add together to cause acute renal shutdown. If the shutdown is complete, the patient dies within a week to 12 days, as explained in Chapter 38, unless treated with the artificial kidney.

Physiological Principles of Treatment Following Mismatched Blood Transfusion Reactions. The immediate treatment of a transfusion reaction is directed toward (1) preventing circulatory shock and (2) preventing hemoglobin precipitation in the kidney tubules. Both results can often be achieved by rapid infusion of dilute intravenous fluids to expand the blood volume and at the same time to cause diuresis. Sometimes mannitol, an osmotic substance that prevents reabsorption of water by the tubules, is also given to prevent excessive concentration of the hemoglobin in the tubules and thereby to prevent its precipitation.

However, in many instances it is not possible to prevent either complete or partial renal shutdown. Even so, if the patient is treated for a week or more using the artificial kidney to replace his or her own renal function, the kidneys will usually return to full function with no permanent consequences.

OTHER TYPES OF TRANSFUSION REACTIONS

Pyrogenic Reactions. Transfusion reactions resulting from mismatched blood types are very rare using present-day transfusion procedures. Much more common are pyrogenic reactions that cause chills and fever in the recipient. These result mainly from the presence in the donor plasma of proteins to which the recipient is allergic.

Anaphylactic Reactions. Occasionally the recipient has an anaphylactic reaction caused by the transfused blood. This sometimes results from allergy to proteins in the donor plasma, but it also occurs on occasion because antibodies in the recipient react with white blood cells in the transfused blood to release toxic substances, includ-ing histamine from the basophils. If the anaphylaxis is severe, the blood pressure might fall drastically, and norepinephrine injections are then necessary to save the person's life.

Serum Hepatitis. One of the most troublesome sequelae of transfusion is the development of hepatitis. This occurs because many persons are "carriers" of serum hepatitis virus and do not themselves show symptoms of the disease. The virus often causes severe acute hepatitis in the recipient, and this sometimes leads to permanent liver damage. Fortunately, only a very small percentage of transfusion bloods contains hepatitis virus.

Transfusion Reactions Resulting from Anticoagulants. The usual anticoagulant used for transfusion is a citrate salt. As discussed in the following chapter, citrate operates as an anticoagulant by combining with the calcium ions of the plasma so that these become nonionizable. Without the presence of ionizable calcium the coagulation process cannot take place. Normal nerve and muscle function also cannot occur in the absence of calcium ions. Therefore, if large quantities of blood containing citrate anticoagulant are administered rapidly, the recipient may experience tetany due to low calcium, which is discussed in detail in Chapter 79. Such tetany can kill the patient within a few minutes because of respiratory muscle spasm.

Ordinarily the liver can remove citrate from the blood within a few minutes and convert it into glycogen or utilize it directly for energy. Therefore, when blood transfusions are given at slow rates (less than 1 liter per hour), the person is usually completely safe from the citrate type of reaction. On the other hand, if liver damage is present, the rate of transfusion must be decreased more than usual to prevent a low calcium ion reaction.

TRANSPLANTATION OF TISSUES AND ORGANS

Relation of Genotypes to Transplantation. In this modern age of surgery, many attempts are being made to transplant tissues and organs from one person to another, or, occasionally, from lower animals to the human being. Many of the different antigenic proteins of red blood cells that cause transfusion reactions plus still many more are present in the other cells of the body as well. Consequently, any foreign cells transplanted into a recipient can cause immune responses and immune reactions. In other words, most recipients are just as able to resist invasion by foreign cells as to resist invasion by foreign bacteria.

Isografts, Allografts, and Xenografts. A transplant of a tissue or whole organ from one identical twin to another is called an *isograft*. A transplant from one human being to another or from any animal to another animal of the same species is called an *allograft*. Finally, a transplant from a lower animal to a human being or from an animal of one species to one of another species is called a *xenograft*.

Transplantation of Cellular Tissues. In the case of isografts, cells in the transplant will almost always live indefinitely if an adequate blood supply is provided, but, in the case of allografts and xenografts, immune reactions almost always occur, causing death of all the cells in the graft 3 to 10 weeks after transplantation unless some specific therapy is used to prevent the immune reaction. The cells of allografts usually persist longer than those of xenografts because the antigenic structure of the allograft is more nearly the same as that of the recipient's tissues than is true of the xenograft. And when the tissues are properly "typed" and are very similar prior to transplant, completely successful allografts occasionally result. The greater the difference in antigenic structure, the more rapid and the more severe are the immune reactions to the graft.

Some of the different cellular tissues and organs that have been transplanted either experimentally or for temporary benefit from one person to another are skin, kidney, heart, liver, glandular tissue, bone marrow, and lung. Many kidney allografts have been successful for as long as five to ten years, a rare liver and heart transplant for one to five years, and lung transplants for one month.

Transplantation of Noncellular Tissues. Certain tissues that have no cells or in which the cells are unimportant to the purpose of the graft, such as the cornea, tendon, fascia, and bone, can usually be grafted from one person to another with considerable success. In these instances the grafts act merely as a supporting latticework into which or around which the surrounding living tissues of the recipient grow. Indeed, some such grafts — bone and fascia — are occasionally successful even when they come from a lower animal rather than from another human being.

ATTEMPTS TO OVERCOME THE ANTIGEN-ANTIBODY REACTIONS IN TRANSPLANTED TISSUE

Because of the extreme potential importance of transplanting certain tissues and organs, such as skin, kidneys, and lungs, serious attempts have been made to prevent the antigen-antibody reactions associated with transplants. The following specific procedures have met with certain degrees of clinical or experimental success.

Tissue Typing. In the same way that red blood cells can be typed to prevent reactions between recipient and donor, so also is it possible to "type" tissues to help prevent graft rejection, though thus far this procedure has met with far less success than has been achieved in red blood cell typing.

The most important antigens that cause graft rejection are a group of antigens called the HLA antigens. These are a group of 50 or more different antigens in the tissue cell membranes that are determined by four separate genes at the so-called *HLA genetic locus.* The genes are allelomorphic and, therefore, can code for only four of the HLA antigens in any one person. That is, only four of the antigens can occur in any one individual, but even this represents a multitude of possible different combinations.

The same HLA antigens occur in the white blood cells. Therefore, the means for tissue typing is to type for these antigens in the membranes of lymphocytes separated from the person's blood. The lymphocytes are mixed with appropriate antisera and complement, and after appropriate incubation the cells are tested for membrane damage, usually by testing the rate of uptake by the lymphocytic cells of a supravital dye.

Fortunately, some of the HLA antigens are not severely antigenic, for which reason precise match of some of the antigens between donor and recipient is not absolutely essential to allow allograft acceptance. Nevertheless, by obtaining the best possible match between donor and recipient, the grafting procedure has become far less hazardous. The best success has been tissue-type matches between members of the same family. Of course, the match in identical twins is exact so that transplants between twins almost never reject because of immune reactions.

Glucocorticoid Therapy (Cortisone, Hydrocortisone, and ACTH). The glucocorticoid hormones from the adrenal gland greatly suppress the formation of both antibodies and immunologically competent lymphocytes. Therefore, administration of large quantities of these, or of ACTH which causes the adrenal gland to produce glucocorticoids, helps tremendously in preventing transplant rejection and is a mainstay of many treatment programs.

Suppression of Antibody Formation. Occasionally, a person has naturally suppressed antibody formation resulting from (a) congenital agammaglobulinemia, in which case gamma globulins are not produced, and (b) destructive diseases of the lymphoid system. Transplants of allografts into such individuals are occasionally successful, or at least their destruction is delayed. Also, irradiative destruction of most of the lymphoid tissue by either x-rays or gamma rays renders a person much more receptive than usual to an allograft. And treatment with certain drugs, such as azathioprine (Imuran), which suppresses antibody formation, also increases the likelihood of success; indeed, this, along with glucocorticoids, is the basis for most immunosuppressive therapy. Unfortunately, all of these procedures also leave the person unprotected from disease.

Use of Antilymphocyte Serum. It was pointed out in the previous chapter that grafted tissues are usually destroyed by lymphocytes that become sensitized against the graft. These lymphocytes invade the graft and then cause the cells of the graft to swell, their membranes to become very permeable, and finally their cell membranes to rupture. Simultaneously, macrophages move in to clean up the debris. Within a few days to a few weeks after this process begins, the tissue often is completely destroyed even though the graft had been completely viable and functioning normally only a short time earlier.

Therefore, an effective procedure for preventing rejection of grafted tissues has been to inoculate the recipient with antilymphocyte serum. This serum is made in animals by injecting human lymphocytes into them; the antibodies that develop in these animals will then attack human lymphocytes. When this serum is injected into the transplanted recipient, the number of circulating small lymphocytes can be decreased to as little as 5 to 10 per cent of normal, and there is a resulting decrease in the intensity of the graft rejection reaction. Unfortunately, the procedure does not continue to work well after the first few injections of the antiserum because the recipient soon begins to build up antibodies against the animal antiserum itself.

Removal of the Thymus Gland to Prevent Antibody Formation. It was pointed out in Chapter 7 that the thymus gland during fetal and early postnatal life forms thymic lymphocytes that migrate throughout the body and become the precursors of the sensitized lymphocytes. Also, for the first few weeks or months after birth, the thymus secretes a hormone that enhances the proliferation of lymphoid tissue throughout the body. Therefore, if the thymus gland is removed during fetal life or immediately after birth, the immune system is greatly impaired for the remainder of the animal's life. Allografts are frequently successful in such animals, but the loss of immunity to disease makes it difficult for the animal to live.

To summarize, transplantation of living tissues in human beings up to the present has been mainly an experiment except in the case of kidney transplants, which are now successful in more than 50 per cent of the cases, but only when massive immunosuppressive treatment is used simultaneously. But when someone succeeds in blocking the immune response of the recipient to a donor organ without at the same time destroying the recipient's specific immunity for disease, this story will change overnight.

REFERENCES

Amos, D. B., and Ward, F. E.: Immunogenetics of the HL-A system. *Physiol. Rev., 55*:206, 1975.

Ballantyne, D. L., and Converse, J. M.: Experimental Skin Grafts and Transplantation Immunity: A Recapitulation. New York, Springer-Verlag, 1979.

Barrow, E. M., and Graham, J. B.: Blood coagulation factor VIII (antihemophilic factor): With comments on von Willebrand's disease and Christmas disease. *Physiol. Rev., 54*:23, 1974.

Carpenter, C. B., and Miller, W. V. (eds.): Clinical Histocompatibility Testing. New York, Grune & Stratton, 1977.

Cerottini, J. C., and Brunner, K. T.: Cell-mediated cytotoxicity, allograft rejection, and tumor immunity. *Adv. Immunol., 18*:67, 1974.

Chatterjee, S. N. (ed.): Symposium on Organ Transplantation. Philadelphia, W. B. Saunders Co., 1978.

Chatterjee, S. N. (ed.): Renal Transplantation: A Multidisciplinary Approach. New York, Raven Press, 1980.

Clarke, C. A., and McConnell, R. B.: Prevention of Rh-Hemolytic Disease. Springfield, Ill., Charles C Thomas, 1972.

Cunningham, B. A.: The structure and function of histocompatibility antigens. *Sci. Am., 234*(4):96, 1977.

Dick, H. F., and Kissmeyer-Nielsen, F. (eds.): Histocompatibility Techniques. New York, Elsevier/North-Holland, 1979.

Ferrone, S., (ed.): HLA Antigens in Clinical Medicine and Biology. New York, Garland STPM Press, 1978.

Festenstein, H., and Démant, P.: HLA and H-2: Basic Immonogenetics, Biology, and Clinical Relevance. London, Edward Arnold, 1978.

Gamma globulins [Series of papers] *N. Engl. J. Med., 275*:480, 536, 591, 652, 709, 769, 826; 1966.

Hubbell, R. C. (ed.): Advances in Blood Transfusion. Arlington, Va., American Blood Commission, 1979.

Issitt, P. D.: Serology and Genetics of the Rhesus Blood Group System. Cincinnati, Montgomery Scientific Publications, 1979.

Mayer, V., and Mitrová, E.: Viral Infections and Their Mitigation in Experimental Immunosuppression. Bratislava:Veda, Publishing House of Slovak Academy of Sciences, 1975.

Mohn, J. F., et al. (eds.): Human Blood Groups, New York, S. Karger, 1977.

Mollison, P. L.: Blood Transfusion in Clinical Medicine, 5th Ed. Philadelphia, J. B. Lippincott Co., 1972.

Mourant, A. E.: Blood Groups and Diseases: A Study of Associations of Diseases with Blood Groups and Other Polymorphisms. New York, Oxford University Press, 1977.

Nickander, R., et al.: Nonsteroidal anti-inflammatory agents. *Annu. Rev. Pharmacol. Toxicol., 19*:469, 1979.

Oleinick, S. R.: Mechanisms of tumor immunology. *In* Frohlich, E. D. (ed.): Pathophysiology, 2nd Ed. Philadelphia, J. B. Lippincott Co., 1976, p. 839.

Roberts, J. A. F., and Pembrey, M. E.: An Introduction to Medical Genetics. New York, Oxford University Press, 1978.

Rosenfield, R. E., et al.: Genetic model for the Rh blood-group system. *Proc. Natl. Acad. Sci. USA, 70*:1303, 1973.

Rossman, P., and Jirka, J.: Rejection Nephropathy. New York, Elsevier/North-Holland, 1978.

Selwood, N., and Hedges, A.: Transplantation Antigens: A Study in Serological Data Analysis. New York, John Wiley & Sons, 1978.

Stiller, C. R., et al. (eds.): Immunologic Monitoring of the Transplant Patient. New York, Grune & Stratton, 1978.

Touraine, J. L., et al. (eds.): Transplantation and Clinical Immunology. New York, Elsevier/North-Holland, 1980.

Unanue, E. R.: Cellular events following binding of antigen to lymphocytes. *Am. J. Pathol., 77*:2, 1974.

Wiener, A. S., and Socha, W. W.: Macro- and microdifferences in blood group antigens and antibodies. *Int. Arch. Allergy Appl. Immunol., 47*:547, 1974.

9

Hemostasis and Blood Coagulation

EVENTS IN HEMOSTASIS

The term hemostasis means prevention of blood loss. Whenever a vessel is severed or ruptured, hemostasis is achieved by several different mechanisms including (1) vascular spasm, (2) formation of a platelet plug, (3) blood coagulation, and (4) growth of fibrous tissue into the blood clot to close the hole in the vessel permanently.

VASCULAR SPASM

Immediately after a blood vessel is cut or ruptured, the wall of the vessel contracts; this instantaneously reduces the flow of blood from the vessel rupture. The contraction results from both nervous reflexes and local myogenic spasm. The nervous reflexes presumably are initiated by pain impulses originating from the traumatized vessel or from nearby tissues. However, most of the spasm probably results from local myogenic contraction of the blood vessels initiated by direct damage to the vascular wall, which presumably causes transmission of action potentials along the vessel wall for several centimeters and results in constriction of the vessel. The more of the vessel that is traumatized, the greater is the degree of spasm; this means that a sharply cut blood vessel usually bleeds much more than does a vessel ruptured by crushing. This local vascular spasm lasts for as long as 20 to 30 minutes, during which time the ensuing processes of platelet plugging and blood coagulation can take place.

The value of vascular spasm as a mean of hemostasis is illustrated by the fact that persons whose legs have been severed by crushing types of trauma sometimes have such intense spasm in vessels as large as the anterior tibial artery that there is not lethal loss of blood.

FORMATION OF THE PLATELET PLUG

The second event in hemostasis is an attempt by the platelets to plug the rent in the vessel. To understand this it is important that we first understand the nature of platelets themselves.

Platelets are minute round or oval discs about 2 microns in diameter. They are fragments of *megakaryocytes,* which are extremely large cells of the hemopoietic series formed in the bone marrow. The megakaryocytes disintegrate into platelets while they are still in the bone marrow and release the platelets into the blood. The normal concentration of platelets in the blood is between 200,000 and 400,000 per cubic millimeter.

Mechanism of the Platelet Plug. Platelet repair of vascular openings is based on several important functions of the platelet itself: When platelets come in contact with a damaged vascular surface, such as the collagen fibers in the vascular wall or even damaged endothelial cells, they immediately change their characteristics drastically. They begin to swell; they assume irregular forms with numerous irradiating processes protruding from their surfaces; they become sticky so that they stick to the collagen fibers; and they secrete large quantities of *ADP* and enzymes that cause formation of *thromboxane A* in the plasma. The ADP and thromboxane A, in turn, act on nearby platelets to activate them as well, and the stickiness of these additional platelets causes them to adhere to the originally activated platelets. Therefore, at the site of any rent in a vessel, the damaged vascular wall or extravascular tissues elicit a vicious cycle of activation of successively increasing numbers of platelets; these accumulate to form a *platelet plug*. This is a fairly loose plug, but it is usually successful in blocking the blood loss if the vascular opening is small. Then, during the subsequent process of blood coagulation, to be described in subsequent paragraphs, *fibrin threads* form that attach to the platelets, thus forming a tight and unyielding plug.

Importance of the Platelet Method for Closing Vascular Holes. If the rent in a vessel is small, the platelet plug by itself can stop blood loss completely, but if there is a large hole, a blood clot in addition to the platelet plug is required to stop the bleeding.

The platelet plugging mechanism is extremely important to close the minute ruptures in very small blood vessels that occur hundreds of times daily, including those through the endothelial cells themselves. A person who has very few platelets develops literally hundreds of small hemorrhagic areas under his skin and throughout his internal tissues, but this does not occur in the normal person. The platelet plugging mechanism usually does not occlude the vessel itself but merely plugs the hole, so that the vessel continues to function normally. This often is not true when a blood clot occurs.

BLOOD COAGULATION IN THE RUPTURED VESSEL

The third mechanism for hemostasis is formation of the blood clot. The clot begins to develop in 15 to 20 seconds if the trauma of the vascular wall has been severe and in one to two minutes if the trauma has been minor. Activator substances both from the traumatized vascular wall and from platelets and blood proteins adhering to the traumatized vascular wall initiate the clotting process. The physical events of this process are illustrated in Figure 9–1, and the chemical events will be discussed in detail later in the chapter.

Within three to six minutes after rupture of a vessel, the entire cut or broken end of the vessel is filled with clot. After 30 minutes to an hour, the clot retracts; this closes the vessel still further. Platelets play an important role in this clot retraction, as will also be discussed later in the chapter.

FIBROUS ORGANIZATION OR DISSOLUTION OF THE BLOOD CLOT

Once a blood clot has formed, it can follow two separate courses: it can become invaded by fibroblasts, which subsequently form connective tissue all through the clot; or it can dissolute. The usual course for a clot that forms in a small hole of a vessel wall is invasion by fibroblasts, beginning within a few hours after the clot is formed and continuing to complete organization of the clot into fibrous tissue within approximately 7 to 10 days. On the other hand, when a large amount of blood

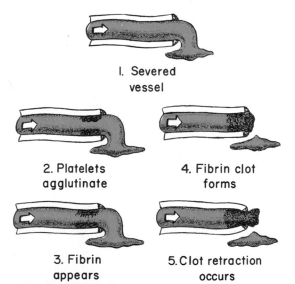

1. Severed vessel

2. Platelets agglutinate

3. Fibrin appears

4. Fibrin clot forms

5. Clot retraction occurs

Figure 9–1. The clotting process in the traumatized blood vessel.

coagulates to form one large blood clot, such as blood that has leaked into tissues, special substances within the clot itself become activated, and these then function as enzymes to dissolute the clot itself, as will be discussed later in the chapter.

MECHANISM OF BLOOD COAGULATION

Basic Theory. Over 30 different substances that affect blood coagulation have been found in the blood and tissues, some promoting coagulation, called *procoagulants,* and others inhibiting coagulation, called *anticoagulants.* Whether or not the blood will coagulate depends on the degree of balance between these two groups of substances. Normally the anticoagulants predominate and the blood does not coagulate, but when a vessel is ruptured the activity of the procoagulants in the area of damage becomes much greater than that of the anticoagulants, and then a clot does develop.

General Mechanism. Almost all research workers in the field of blood coagulation agree that clotting takes place in three essential steps:

First, a substance or complex of substances called *prothrombin activator* **is formed in response to rupture of the vessel or damage to the blood itself.**

Second, the prothrombin activator catalyzes the conversion of prothrombin into *thrombin.*

Third, the thrombin acts as an enzyme to convert fibrinogen into *fibrin threads* **that enmesh platelets, blood cells, and plasma to form the clot itself.**

Unfortunately, the detailed mechanisms by which prothrombin activator is formed are still incompletely understood. On the other hand, the mechanisms by which prothrombin is converted to thrombin and by which thrombin then acts to cause the formation of fibrin threads are much better established. Therefore, let us first discuss the basic mechanism by which the blood clot is formed, beginning with the conversion of prothrombin to thrombin; then we will come back to the initiating stages in the clotting process by which prothrombin activator is formed.

CONVERSION OF PROTHROMBIN TO THROMBIN

After prothrombin activator has been formed as a result of rupture of the blood vessel or as a result of damage to special activator substances in the blood itself, the prothrombin activator can then cause conversion of prothrombin to thrombin, which in turn causes polymerization of fibrinogen molecules into fibrin threads within another 10 to 15 seconds. Thus the rate-limiting factor in causing blood coagulation is usually the formation of prothrombin activator and not the subsequent reactions beyond that point.

Prothrombin and Thrombin. Prothrombin is a plasma protein, an alpha$_2$-globulin, having a molecular weight of 68,700. It is present in normal plasma in a concentration of about 15 mg. per 100 ml. It is an unstable protein that can split easily into smaller com-

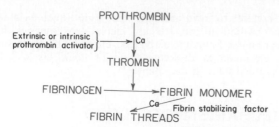

Figure 9–2. Schema for conversion of prothrombin to thrombin, and polymerization of fibrinogen to form fibrin threads.

pounds, one of which is *thrombin,* which has a molecular weight of 33,700, almost exactly half that of prothrombin.

Prothrombin is formed continually by the liver, and it is continually being used throughout the body for blood clotting. If the liver fails to produce prothrombin, its concentration in the plasma falls too low within 24 hours to provide normal blood coagulation. Vitamin K is required by the liver for normal formation of prothrombin; therefore, either lack of vitamin K or the presence of liver disease that prevents normal prothrombin formation can often decrease the prothrombin level so low that a bleeding tendency results.

Effect of Prothrombin Activator to Form Thrombin from Prothrombin. Figure 9–2 illustrates the conversion of prothrombin to thrombin under the influence of prothrombin activator and calcium ions. The rate of formation of thrombin from prothrombin is almost directly proportional to the quantity of prothrombin activator available, which in turn is approximately proportional to the degree of trauma to the vessel wall or to the blood. In turn, the rapidity of the clotting process is proportional to the quantity of thrombin formed.

CONVERSION OF FIBRINOGEN TO FIBRIN — FORMATION OF THE CLOT

Fibrinogen. Fibrinogen is a high molecular weight protein (340,000) occurring in the plasma in quantities of 100 to 700 mg. per 100 ml. Fibrinogen is formed in the liver, and liver disease occasionally decreases the concentration of circulating fibrinogen, as it does the concentration of prothrombin which was pointed out previously.

Because of its large molecular size, very little fibrinogen normally leaks into the interstitial fluids, and, since it is one of the essential factors in the coagulation process, interstitial fluids ordinarily coagulate poorly if at all. Yet, when the permeability of the capillaries becomes pathologically increased, fibrinogen does then appear in the tissue fluids in sufficient quantities to allow clotting in much the same way that plasma and whole blood clot.

Action of Thrombin on Fibrinogen to Form Fibrin. Thrombin is a protein *enzyme* with proteolytic capabilities. It acts on fibrinogen to remove two low molecular weight peptides from each molecule of fibrinogen, forming a molecule of *fibrin monomer* which has the automatic capability of polymerizing with other fibrin monomer molecules. Therefore, many fibrin monomer

molecules polymerize within seconds into *long fibrin threads* that form the *reticulum* of the clot.

In the early stages of this polymerization, the fibrin threads are not cross-linked with each other, and the resultant clot is weak and can be broken apart with ease. However, still another process occurs during the following few minutes that greatly strengthens the fibrin reticulum. This involves a substance called *fibrin-stabilizing factor* that is normally present in small amounts in the plasma globulins but that is also released from platelets entrapped in the clot. Before fibrin-stabilizing factor can have an effect on the fibrin threads it must itself be activated. Fortunately, the same thrombin that causes fibrin formation also activates the fibrin-stabilizing factor. Then this activated substance operates as an enzyme to cause covalent cross-linking bonds between the adjacent fibrin threads, thus adding tremendously to the three-dimensional strength of the fibrin meshwork.

The Blood Clot. The clot is composed of a meshwork of fibrin threads running in all directions and entrapping blood cells, platelets, and plasma. The fibrin threads adhere to damaged surfaces of blood vessels; therefore, the blood clot becomes adherent to any vascular opening and thereby prevents blood loss.

Clot Retraction — Serum. Within a few minutes after a clot is formed, it begins to contract and usually expresses most of the fluid from the clot within 30 to 60 minutes. The fluid expressed is called *serum,* because all of its fibrinogen and most of the other clotting factors have been removed; in this way, serum differs from plasma. Serum obviously cannot clot because of lack of these factors.

Platelets are necessary for clot retraction to occur. Therefore, failure of clot retraction is an indication that the number of platelets in the circulating blood is low. Electron micrographs of platelets in blood clots show that they become attached to the fibrin threads in such a way that they actually bond different threads together. Furthermore, platelets entrapped in the clot continue to release procoagulant substances, one of which is fibrin stabilizing factor that causes more and more cross-linking bondage between the adjacent fibrin threads. It is possible that other coagulation factors from the platelets also cause stronger bonding within the fibrin threads themselves, thereby causing them to contract.

As the clot retracts, the edges of the broken blood vessel are pulled together, thus possibly or probably contributing to the ultimate state of hemostasis.

THE VICIOUS CYCLE OF CLOT FORMATION

Once a blood clot has started to develop, it normally extends within minutes into the surrounding blood. That is, the clot itself initiates a vicious cycle to promote more clotting. One of the most important causes of this is the fact that the proteolytic action of thrombin allows it to act on many of the other blood clotting factors in addition to fibrinogen. For instance, thrombin has a direct proteolytic effect on prothrombin itself, tending to split this into still more thrombin, and it acts on some of the blood clotting factors responsible for the formation of prothrombin activator. (These effects, to be

TABLE 9–1 CLOTTING FACTORS IN THE BLOOD
AND THEIR SYNONYMS

Clotting Factor	Synonym
Fibrinogen	Factor I
Prothrombin	Factor II
Tissue thromboplastin	Factor III
Calcium	Factor IV
Factor V	Proaccelerin; labile factor; Ac-globulin; Ac-G
Factor VII	Serum prothrombin conversion accelerator; SPCA; convertin; stable factor
Factor VIII	Antihemophilic factor; AHF; antihemophilic globulin; AHG; antihemophilic factor A
Factor IX	Plasma thromboplastin component; PTC; Christmas factor; antihemophilic factor B
Factor X	Stuart factor; Stuart-Prower factor; antihemophilic factor C
Factor XI	Plasma thromboplastin antecedent; PTA; antihemophilic factor C
Factor XII	Hageman factor; antihemophilic factor D
Factor XIII	Fibrin stabilizing factor
Prothrombin activator	Thrombokinase; complete thromboplastin

discussed in subsequent paragraphs, include (a) acceleration of the actions of Factors VIII, IX, X, XI, and XII, and (b) aggregation of platelets.) Once a critical amount of thrombin is formed, a vicious cycle develops that causes still more blood clotting and more thrombin to be formed; thus, the blood clot continues to grow until something stops its growth.

BLOCK OF CLOT GROWTH BY BLOOD FLOW

Fortunately, when a clot develops, the vicious cycle of continued clot formation occurs only where the blood is not moving because flowing blood carries the thrombin and the other procoagulants released during the clotting process away so rapidly that their concentrations cannot rise high enough to promote further clotting. Thus, extension of the clot almost always stops where it comes in contact with blood that is flowing faster than a certain velocity.

In addition, the clotting process fortunately is not self-propagating until the concentrations of the procoagulants rise above critical concentrations. At lower concentrations, many inhibitor substances in the blood, some of which will be discussed later in the chapter, continually block the actions of the procoagulants or destroy them. In addition, the reticuloendothelial system, particularly that of the liver and of the bone marrow, removes most of the circulating procoagulants within a few minutes.

INITIATION OF COAGULATION: FORMATION OF PROTHROMBIN ACTIVATOR

Now that we have discussed the clotting process initiated by the formation of thrombin from prothrombin, we must turn to the more complex mechanisms that activate the prothrombin. These mechanisms can be set

into play by trauma to the tissues, trauma to the blood, or contact of the blood with damaged endothelial cells or with special substances such as collagen outside the blood vessel endothelium. In each instance, they lead to the formation of *prothrombin activator*, which then causes prothrombin conversion to thrombin.

There are two basic ways in which prothrombin activator can be formed: (1) by the *extrinsic pathway* that begins with trauma to the vascular wall or to the tissues outside the blood vessels, or (2) by the *intrinsic pathway* that begins in the blood itself.

In both the extrinsic and intrinsic pathways a series of different plasma proteins, especially beta-globulins, play major roles. These, along with the other factors already discussed that enter into the clotting process, are called *blood clotting factors* and for the most part they are inactive forms of proteolytic enzymes. When converted to the active forms, their enzymatic actions cause the successive reactions of the clotting process.

Most of the clotting factors are designated by Roman numerals, as listed in Table 9–1. In the sections dealing with intrinsic and extrinsic pathways we will specifically discuss blood clotting factor V and factors VII through XII.

The Extrinsic Mechanism for Initiating Clotting

The extrinsic mechanism for initiating the formation of prothrombin activator begins with blood coming in contact with traumatized vascular wall or extravascular tissues and occurs according to the following three basic steps, as illustrated in Figure 9–3.

(1) *Release of tissue factor and tissue phospholipids.* The traumatized tissue releases two factors that set the clotting process into motion. These are (a) *tissue factor*, which is a proteolytic enzyme, and (b) *tissue phospholipids*, which are mainly phospholipids of the tissue cell membranes.

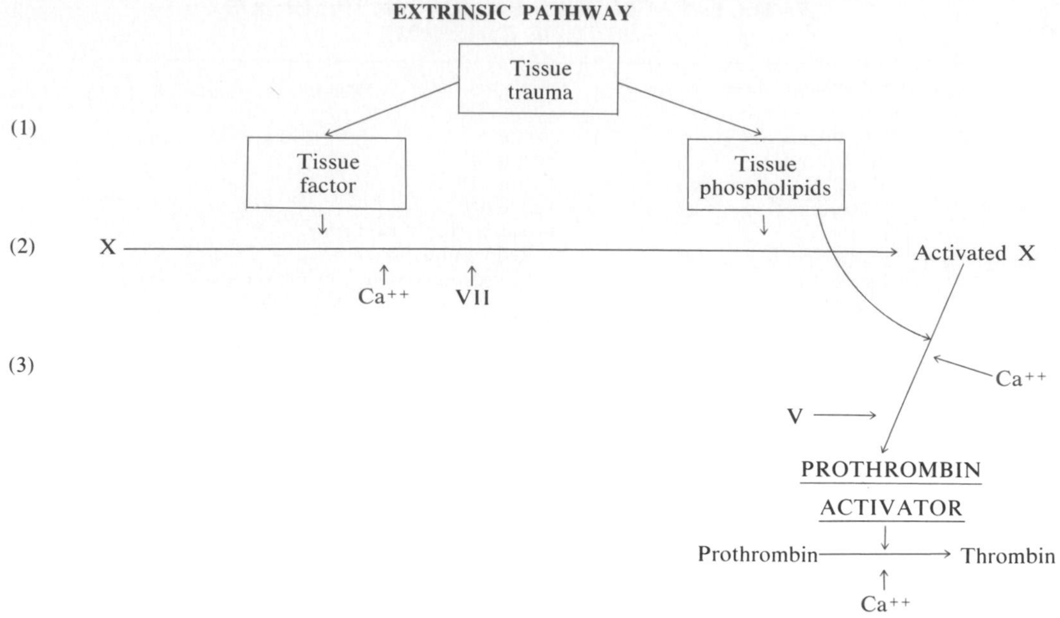

EXTRINSIC PATHWAY

Figure 9–3. The extrinsic pathway for initiating blood clotting.

(2) *Activation of factor X to form activated factor X — role of factor VII and tissue factor.* The tissue factor complexes with blood coagulation factor VII, and this complex, in the presence also of tissue phospholipids, acts enzymatically on factor X to form *activated factor X.*

(3) *Effect of activated factor X to form prothrombin activator — role of factor V.* The activated factor X complexes immediately with the tissue phospholipids released from the traumatized tissue and also with factor V to form the complex called *prothrombin activator.* Within a few seconds this splits prothrombin to form thrombin, and the clotting process proceeds as has already been explained.

The Intrinsic Mechanism for Initiating Clotting

The second mechanism for initiating the formation of prothrombin activator, and therefore for initiating clotting, begins with trauma to the blood itself and continues through the following series of cascading reactions, as illustrated in Figure 9–4.

(1) *Activation of factor XII and release of platelet phospholipids by blood trauma.* Trauma to the blood alters two important clotting factors in the blood — factor XII and the platelets. When factor XII is disturbed, such as by coming into contact with collagen or with a wettable surface such as glass, it takes on a new configuration that converts it into a proteolytic enzyme called "activated factor XII."

Simultaneously, the blood trauma also damages the platelets, either because of adherence to collagen or to a wettable surface (or by damage in other ways), and this releases platelet phospholipid, frequently called *platelet factor III,* which also plays a role in subsequent clotting reactions.

(2) *Activation of factor XI.* The activated factor XII acts enzymatically on factor XI to activate this as well, which is the second step in the intrinsic pathway.

(3) *Activation of factor IX by activated factor XI.* The activated factor XI then acts enzymatically on factor IX to activate this factor, also.

(4) *Activation of factor X — role of factor VIII.* The activated factor IX, acting in concert with factor VIII and with the platelet phospholipids from the traumatized platelets, activates factor X. It is clear that when either factor VIII or platelets are in short supply, this step is deficient. Factor VIII is the factor that is missing in the person who has classical *hemophilia,* for which reason it is called *antihemophilic factor.* Platelets are the clotting factor that is lacking in the bleeding disease called *thrombocytopenia.*

(5) *Action of activated factor X to form prothrombin activator — role of factor V.* This step in the intrinsic pathway is essentially the same as the last step in the extrinsic pathway. That is, activated factor X combines with factor V and platelet phospholipids to form the complex called *prothrombin activator.* The only difference is that the phospholipids in this instance come from the traumatized platelets rather than from traumatized tissues. The prothrombin activator in turn initiates within seconds the cleavage of prothrombin to form thrombin, thereby setting into motion the final clotting process, as described earlier.

INTRINSIC PATHWAY

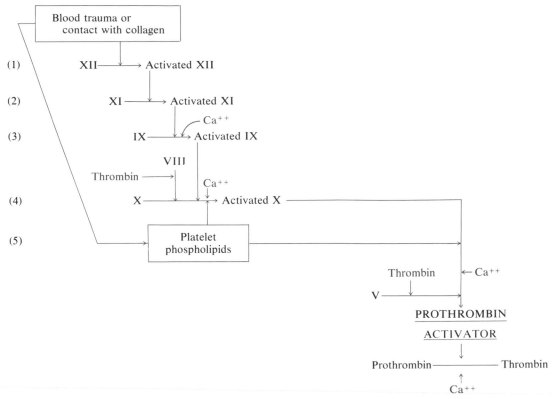

Figure 9–4. The intrinsic pathway for initiating blood clotting.

Role of Calcium Ions in the Intrinsic and Extrinsic Pathways

Except for the first two steps in the intrinsic pathway, calcium ions are required for promotion of all of the reactions. Therefore, in the absence of calcium ions, blood clotting will not occur.

Fortunately, in the living body the calcium ion concentration never falls low enough to affect significantly the kinetics of blood clotting. The reason for this is that long before calcium ion concentration can fall this low, the diminished level of calcium ions will kill the person by causing muscle tetany throughout the body, especially of the respiratory muscles.

On the other hand, when blood is removed from a person, it can be prevented from clotting by reducing the calcium ion concentration below the threshold level for clotting, either by deionizing the calcium by reacting it with substances such as *citrate ion* or by precipitating the calcium with substances such as the *oxalate ion*.

Summary of Blood Clotting Initiation

It is clear from the above schemas of the intrinsic and extrinsic systems for initiating blood clotting that clotting is initiated after rupture of blood vessels by both of the pathways. The tissue factor and tissue phospholipids initiate the extrinsic pathway, while contact of factor XII and the platelets with collagen in the vascular wall initiates the intrinsic pathway.

In contrast, when blood is removed from the body and held in a test tube, it is the intrinsic pathway alone that must elicit the clotting. This usually results from contact of factor XII and platelets with the wall of the vessel, which activates both of these and initiates the intrinsic mechanism. If the surface of the container is very "nonwettable," such as a siliconized surface, blood clotting can sometimes be prevented for an hour or more.

Intravascular clotting sometimes results also from other factors that activate the intrinsic pathway. For instance, antigen-antibody reactions sometimes initiate the clotting process, and the same is true of some drugs or circulating debris that might happen to enter the circulation.

An especially important difference between the extrinsic and intrinsic pathways is that the extrinsic pathway is explosive in nature; once initiated, its speed of occurrence is limited only by the amount of tissue factor and tissue phospholipids released from the traumatized tissues, and by the quantities of factors X, VII, and V in the blood. With severe tissue trauma, clotting can occur in as little as 15 seconds. On the other hand, the intrinsic pathway is much slower to proceed, usually requiring 2 to 6 minutes to cause clotting. Also, various inhibitors in the blood set up roadblocks along the way for the intrinsic pathway. At times, these inhibitors can block the intrinsic pathway entirely; some of them will be discussed in more detail later in the chapter.

PREVENTION OF BLOOD CLOTTING IN THE NORMAL VASCULAR SYSTEM — THE INTRAVASCULAR ANTICOAGULANTS

Endothelial Surface Factors. Probably the two most important factors for preventing clotting in the normal vascular system are, first, the smoothness of the endothelium which prevents contact activation of the intrinsic clotting system and, second, a monomolecular layer of negatively charged protein adsorbed to the inner surface of the endothelium that repels the clotting factors and platelets, thereby preventing activation of clotting. When the endothelial wall is damaged, its smoothness and its negative electrical charge are both lost, which is believed to help activate factor XII and to set off the intrinsic pathway of clotting. And, if the factor XII comes in contact with the subendothelial collagen, the specific effect of this interaction is a powerful initiator of the clotting process.

The Antithrombin Action of Fibrin and of Antithrombin-Heparin Cofactor. Among the most important anticoagulants in the blood itself are those that remove thrombin from the blood. The two most powerful of these are the *fibrin threads* that are formed during the process of clotting and an alpha-globulin called *antithrombin III* or also *antithrombin-heparin cofactor*.

While a clot is forming, approximately 85 to 90 per cent of the thrombin formed from the prothrombin becomes adsorbed to the fibrin threads as they develop. This obviously helps to prevent the spread of thrombin into the remaining blood and therefore prevents excessive spread of the clot.

The thrombin that does not adsorb to the fibrin threads soon combines with antithrombin-heparin cofactor, which by a binding process blocks the effect of the thrombin on the fibrinogen and then inactivates the bound thrombin during the next 12 to 20 minutes.

Heparin. Small amounts of heparin, a powerful anticoagulant, are normally present in the blood. Heparin is a conjugated polysaccharide found in the cytoplasm of many types of cells, including even the cytoplasm of unicellular animals. Therefore, heparin is produced by many different cells of the human body, though especially large quantities are formed by the basophilic *mast cells* located in the pericapillary connective tissue throughout the body. These cells continually secrete small quantities of heparin, and the heparin then diffuses into the circulatory system. The *basophil cells* of the blood, which seem to be functionally almost identical with the mast cells, also release small quantities of heparin into the plasma.

Mast cells are extremely abundant in the tissue surrounding the capillaries of the lungs and to a lesser extent the capillaries of the liver. It is easy to understand why large quantities of heparin might be needed in these areas, for the capillaries of the lungs and liver receive many embolic clots formed in the slowly flowing venous blood; sufficient formation of heparin might prevent further growth of the clots.

The concentration of heparin in normal blood has been estimated to be as much as 0.01 mg. per 100 ml. of blood. Though this concentration is 10 to 100 times less than that often used clinically to prevent blood clotting, it is probably sufficient to aid in preventing blood coagulation in the normal circulatory system, for only minute quantities of procoagulants are normally formed, and only minute amounts of heparin are therefore needed to prevent clotting.

Mechanism of Heparin Action. Heparin prevents blood coagulation almost entirely by combining with antithrombin-heparin cofactor, which in turn makes this factor combine with thrombin 1000 times as rapidly as normally. Therefore, in the presence of an excess of heparin, the removal of thrombin from the circulating blood is almost instantaneous.

This complex of heparin and antithrombin-heparin cofactor also reacts in a similar way with several of the other activated coagulation factors of both the intrinsic and extrinsic pathways, thus inactivating their proteolytic (and blood clotting) functions. The ones specifically inactivated in this way include the activated forms of factors XII, XI, IX, and X, which in each case further decrease the rate of blood coagulation.

Alpha$_2$-Macroglobulin. *Alpha$_2$-macroglobulin* is a very large globulin molecule having a molecular weight of 360,000. It is similar to antithrombin-heparin cofactor in that it combines with the proteolytic coagulation factors. However, its activity is not accelerated by heparin. Its function is mainly to act as a binding agent for the coagulation factors until they can be destroyed. In this way it probably plays an important role even normally in preventing blood clotting.

LYSIS OF BLOOD CLOTS — PLASMIN

The plasma proteins contain a euglobulin called *plasminogen* or *profibrinolysin* which, when activated, becomes a substance called *plasmin* or *fibrinolysin*. Plasmin is a proteolytic enzyme that resembles trypsin, the most important digestive enzyme of pancreatic secretion. It digests the fibrin threads and also digests other substances in the surrounding blood, such as fibrinogen, factor V, factor VIII, prothrombin, and factor XII. Therefore, whenever plasmin is formed in a blood clot, it can cause lysis of the clot and also destruction of many of the clotting factors, thereby causing hypocoagulability of the blood.

Formation of Plasmin and Lysis of Clots. When a clot is formed a large amount of plasminogen is incorporated in the clot along with other plasma proteins. However, this will not become plasmin and will not cause lysis of the clot until it is activated. Fortunately, the tissues and blood contain substances that can activate plasminogen to plasmin, including (1) thrombin, (2) activated factor XII, (3) lysosomal enzymes from damaged tissues, and (4) factors from the vascular endothelium. Within a day or two after blood has leaked into a tissue and clotted, these activators cause the formation of enough plasmin that it in turn dissolves the clot.

Clots that occur inside blood vessels can also be dissolved, though this occurs more readily in the small vessels than in the large ones. This illustrates that activator systems also occur within the blood itself.

An activator called *urokinase* is found in the urine and is believed to be important in the lysis of clots that develop in the renal tubules. It is possible that urokinase also plays a role as an intravascular activator before it is excreted into the urine by the kidneys.

Certain bacteria also release activator enzymes; for instance, streptococci release a substance called *streptokinase*, which acts on plasminogen to form plasmin. When streptococcal infection occurs in the tissues, the plasmin generated in response to the streptokinase dissolves clotted lymph and clotted tissue fluids and allows the streptococci to spread extensively through the tissues instead of being blocked by the body's "walling off" process described in Chapter 6.

Significance of the Fibrinolysin System. The lysis of blood clots allows slow clearing (over a period of several days) even of extraneous blood from the tissues and sometimes allows reopening of clotted vessels. Unfortunately, reopening of large vessels occurs only rarely. But an important function of the fibrinolysin system is to remove very minute clots from the millions of tiny peripheral vessels that eventually would all become occluded were there no way to cleanse them.

CONDITIONS THAT CAUSE EXCESSIVE BLEEDING IN HUMAN BEINGS

Excessive bleeding can result from deficiency of any one of the many different blood clotting factors. Three particular types of bleeding tendencies that have been studied to the greatest extent will be discussed: (1) bleeding caused by vitamin K deficiency, (2) hemophilia, and (3) thrombocytopenia (platelet deficiency).

DECREASED PROTHROMBIN, FACTOR VII, FACTOR IX, AND FACTOR X CAUSED BY VITAMIN K DEFICIENCY

Hepatitis, cirrhosis, acute yellow atrophy, and other diseases of the liver can all depress the formation of prothrombin and factors VII, IX, and X so greatly that the patient develops a severe tendency to bleed.

Another cause of depressed levels of these substances is vitamin K deficiency. Vitamin K is necessary for some of the intermediate stages in the formation of all of them. Fortunately, vitamin K is continually synthesized in the gastrointestinal tract by bacteria so that vitamin K deficiency rarely if ever occurs simply because of its absence from the diet. However, vitamin K deficiency does often occur as a result of poor absorption of fats from the gastrointestinal tract, because vitamin K is fat soluble and ordinarily is absorbed into the blood along with the fats.

One of the most prevalent causes of vitamin K deficiency is failure of the liver to secrete bile into the gastrointestinal tract (which occurs either as a result of obstruction of the bile ducts or as a result of liver disease), for lack of bile prevents adequate fat digestion and absorption. Therefore, liver disease often causes decreased production of prothrombin and the other factors both because of poor vitamin K absorption and

because of dysfunctional liver cells. Because of this, vitamin K is injected into all patients with liver disease or obstructed bile ducts prior to performing any surgical procedure. Ordinarily, if vitamin K is given to a deficient patient four to eight hours prior to operation and the liver parenchymal cells are at least one-half normal in function, sufficient clotting factors will be produced to prevent excessive bleeding during the operation.

HEMOPHILIA

The term hemophilia is loosely applied to several different hereditary deficiencies of coagulation, all of which cause bleeding tendencies hardly distinguishable from one another. The three most common causes of hemophilic syndrome are deficiency of (1) factor VIII (classical hemophilia) — about 83 per cent of the total, (2) factor IX — about 15 per cent, and (3) factor XI — about 2 per cent.

Many persons with hemophilia die in early life, though many others with less severe bleeding have a normal life span. Very commonly, the person's joints become severely damaged because of repeated joint hemorrhage following exercise or trauma.

Regardless of the precise type of hemophilia, transfusion of normal fresh plasma or of the appropriate purified protein clotting factor — factor VIII for classical hemophilia — into the hemophilic person usually relieves his bleeding tendency for a few days.

THROMBOCYTOPENIA

Thrombocytopenia means the presence of a very low quantity of platelets in the circulatory system. Persons with thrombocytopenia have a tendency to bleed as do hemophiliacs, except that the bleeding is usually from many small capillaries rather than from larger vessels, as in hemophilia. As a result, small punctate hemorrhages occur throughout all the body tissues. The skin of such a person displays many small, purplish blotches, giving the disease the name *thrombocytopenic purpura*. It will be remembered that platelets are especially important for repair of minute breaks in capillaries and other small vessels. Indeed, platelets can aggregate to fill such ruptures without actually causing clots.

Ordinarily, bleeding does not occur until the number of platelets in the blood falls below a value of approximately 50,000 per cubic millimeter rather than the normal of 200,000 to 400,000. Levels as low as 10,000 per cubic millimeter are frequently lethal.

Even without making specific platelet counts on the blood, one can sometimes suggest the existence of thrombocytopenia by simply noting whether or not a clot of the person's blood retracts, for, as pointed out earlier, clot retraction is normally dependent upon the presence of large numbers of platelets entrapped in the fibrin mesh of the clot.

Most persons with thrombocytopenia have the disease known as *idiopathic thrombocytopenia*, which means simply "thrombocytopenia of unknown cause." However, in the past few years it has been discovered that in most of these persons specific antibodies are destroying the platelets. Occasionally these have developed because of transfusions from other persons, but usually they result from development of autoimmunity to the person's own platelets, the cause of which, however, is not known.

In addition to idiopathic thrombocytopenia, the number of thrombocytes (platelets) in the blood may be greatly depressed by any abnormality that causes aplasia of the bone marrow. For instance, *irradiation injury* to the bone marrow, aplasia of the bone marrow resulting from *drug sensitivity,* and even *pernicious anemia* can cause sufficient decrease in the total number of platelets that thrombocytopenic bleeding results.

Relief from bleeding for one to four days can often be effected in the thrombocytopenic patient by giving *fresh whole blood transfusions*. To do this, the blood is best removed from the donor into a siliconized chamber and then rapidly placed in the recipient so that the platelets are damaged as little as possible. *Cortisone,* which suppresses immune reactions, is often beneficial in the idiopathic type of thrombocytopenia. And, *splenectomy* often helps because the spleen removes large numbers of platelets, particularly damaged ones, from the blood.

THROMBOEMBOLIC CONDITIONS IN THE HUMAN BEING

Thrombi and Emboli. An abnormal clot that develops in a blood vessel is called a *thrombus*. Once a clot has developed, continued flow of blood past the clot is likely to break it away from its attachment, and such freely flowing clots are known as *emboli*. Emboli generally do not stop flowing until they come to a narrow point in the circulatory sytem. Thus, emboli originating in large arteries or in the left side of the heart eventually plug either smaller systemic arteries or arterioles. On the other hand, emboli originating in the venous system and in the right side of the heart flow into the vessels of the lung to cause pulmonary arterial embolism.

Causes of Thromboembolic Conditions. The causes of thromboembolic conditions in the human being are usually two-fold: First, any *roughened endothelial surface of a vessel* — as may be caused by arteriosclerosis, infection, or trauma — is likely to initiate the clotting process. Second, blood often clots *when it flows very slowly* through blood vessels, for small quantities of thrombin and other procoagulants are always being formed. These are generally removed from the blood by the reticuloendothelial cells, mainly the Kupffer cells of the liver. If the blood is flowing too slowly, the concentrations of the procoagulants in local areas often rise high enough to initiate clotting, but when the blood flows rapidly these are rapidly mixed with large quantities of blood and are removed during passage through the liver.

FEMORAL THROMBOSIS AND MASSIVE PULMONARY EMBOLISM

Because clotting almost always occurs when blood flow is blocked for many hours in any vessel of the body, the immobility of bed patients plus the practice of propping the knees up with underlying pillows often causes intravascular clotting because of blood stasis in one or more of the leg veins for hours at a time. Then the clot grows, especially in the direction of the slowly moving blood, sometimes growing the entire length of the leg veins and occasionally even up into the common iliac vein and inferior vena cava. Then, about 1 time out of every 10, a large part of the clot disengages from its attachments to the vessel wall and flows freely with the venous blood into the right side of the heart and thence into the pulmonary arteries to cause *massive pulmonary embolism*. If the clot is large enough to occlude both the pulmonary arteries, immediate death ensues. If only one pulmonary artery or a smaller branch is blocked, death may not occur, or the embolism may lead to death a few hours to several days later because of further growth of the clot within the pulmonary vessels.

DISSEMINATED INTRAVASCULAR CLOTTING

Occasionally, the clotting mechanism becomes activated in widespread areas of the circulation, giving rise to the condition called *disseminated intravascular clotting*. Frequently, the clots are small but numerous, and they plug a large share of the small peripheral blood vessels. This effect occurs especially in septicemic shock, in which either circulating bacteria or bacterial toxins — especially *endotoxins* — activate the clotting mechanisms. The plugging of the small peripheral vessels greatly diminishes the delivery of oxygen and other nutrients to the tissues — a situation which exacerbates the shock picture. It is partly for this reason that full-blown septicemic shock is lethal in 85 per cent or more of the patients.

A peculiar effect of disseminated intravascular clotting is that the patient frequently begins to bleed. The reason for this is that so many of the clotting factors are removed by the widespread clotting that too few procoagulants remain to allow normal hemostasis of the remaining blood.

ANTICOAGULANTS FOR CLINICAL USE

In some thromboembolic conditions it is desirable to delay the coagulation process to a certain degree. Therefore, various anticoagulants have been developed for treatment of these conditions. The ones most useful clinically are heparin and the coumarins.

HEPARIN AS AN INTRAVENOUS ANTICOAGULANT

Commercial heparin is extracted from animal tissues from several organs and is prepared in almost pure form. Injection of relatively small quantities, approximately 0.5 to 1 mg. per kilogram of body weight, causes the blood clotting time to increase from a normal of approximately 6 minutes to 30 or more minutes. Furthermore, this change in clotting time occurs instantaneously, thereby immediately preventing further development of the thromboembolic condition.

The action of heparin lasts approximately three to four hours. It is believed that the injected heparin is destroyed by an enzyme in the blood known as *heparinase*. Also, much of the injected heparin is entrapped in the reticuloendothelial cells or diffuses into the interstitial fluids and therefore becomes unavailable as a blood anticoagulant.

In the treatment of a patient with heparin, too much heparin is sometimes given, and *serious* bleeding crises occur. In these instances, *protamine* acts specifically as an antiheparin, and the clotting mechanism can be reverted to normal by administering this substance. This substance combines with heparin and inactivates it because it carries positive electrical charges, whereas heparin carries negative charges.

COUMARINS AS ANTICOAGULANTS

When a coumarin, such as *warfarin,* is given to a patient, the plasma levels of prothrombin and factors VII, IX, and X, all formed by the liver, begin to fall, indicating that warfarin has a potent depressant effect on liver formation of all these compounds. Warfarin possibly causes this effect by competing with vitamin K for reactive sites in the intermediate processes for formation of prothrombin and the other three clotting factors, thereby blocking the action of vitamin K.

After administration of an effective dose of warfarin, the coagulant activity of the blood decreases to approximately 50 per cent of normal by the end of 12 hours and to approximately 20 per cent of normal by the end of 24 hours. In other words, the coagulation process is not blocked immediately, but must await the consumption of the prothrombin and other factors already present in the plasma. Normal coagulation returns one to three days after discontinuing therapy.

PREVENTION OF BLOOD COAGULATION OUTSIDE THE BODY

Though blood removed from the body and held in a glass test tube normally clots in about six minutes, blood collected in *siliconized containers* often does not clot for as long as an hour or more. The reason for this delay is that preparing the surfaces of the containers with silicone prevents contact activation of factor XII, which initiates the intrinsic clotting mechanism. On the other hand, untreated glass containers allow contact activation and rapid development of clots.

Heparin can be used for preventing coagulation of blood outside the body as well as in the body, and heparin is occasionally used as an anticoagulant when blood is removed from a donor to be transfused later into a recipient. Also, heparin is used in all surgical procedures in which the blood is passed through a heart-lung machine and then back into the person.

Various substances that *decrease the concentration of calcium ions* in the blood can be used for preventing blood coagulation outside the body. For instance, soluble *oxalate* compounds mixed in very small quantity with a sample of blood cause precipitation of calcium oxalate from the plasma and thereby decrease the ionic calcium levels so much that blood coagulation is blocked.

A second calcium deionizing agent used for preventing coagulation is *sodium, ammonium,* or *potassium citrate*. The citrate ion combines with calcium in the blood to cause an un-ionized calcium compound, and the lack of ionic calcium prevents coagulation. Citrate anticoagulants have a very important advantage over the oxalate anticoagulants, for oxalate is toxic to the body, whereas moderate quantities of citrate can be injected intravenously. After injection, the citrate ion is removed from the body within a few minutes by the liver and is polymerized into glucose, and then metabolized in the usual manner. Consequently, 500 ml. of blood that has been rendered incoagulable by citrate can ordinarily be injected into a recipient within a few minutes without any dire consequences. If the liver is damaged or if large quantities of citrated blood or plasma are given too rapidly, the citrate ion may not be removed quickly enough, and the citrate can then greatly depress the level of calcium ion in the blood, which results in tetany and convulsive death.

BLOOD COAGULATION TESTS

BLEEDING TIME

When a sharp knife is used to pierce the tip of the finger or lobe of the ear, bleeding ordinarily lasts three to six minutes. However, the time depends largely on the depth of the wound and on the degree of hyperemia in the finger at the time of the test. Lack of several of the clotting factors can prolong the bleeding time, but this is especially prolonged by lack of platelets.

CLOTTING TIME

Many methods have been devised for determining clotting times. The one most widely used is to collect blood in a chemically clean glass test tube and then to tip the tube back and forth approximately every 30 seconds until the blood has clotted. By this method, the normal clotting time ranges between five and eight minutes.

Procedures using multiple test tubes have been devised for determining clotting time more accurately. However, clotting times are also very dependent on the condition of the glass itself and even on the size of the tube, which makes a high degree of standardization necessary to obtain accurate results.

PROTHROMBIN TIME

The prothrombin time gives an indication of the total quantity of prothrombin in the blood. Figure 9–5 shows the relationship of prothrombin concentration to prothrombin time. The means for determining prothrombin time is the following:

Blood removed from the patient is immediately oxalated so that none of the prothrombin can change into thrombin. At any time later, a large excess of calcium ion and tissue extract is suddenly mixed with the oxalated blood. The calcium nullifies the effect of the oxalate, and the tissue extract activates the prothrombin-to-thrombin reaction by means of the ex-

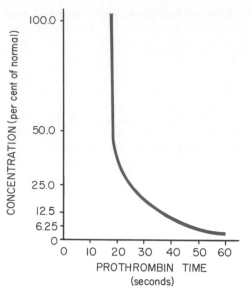

Figure 9–5. Relationship of prothrombin and concentration in the blood to the prothrombin time.

trinsic clotting pathway. The time required for coagulation to take place is known as the "prothrombin time." The normal prothrombin time is approximately 12 seconds, though this depends to a certain extent on the exact procedure employed. In each laboratory a curve relating prothrombin concentration to prothrombin time, such as that shown in Figure 9–5, is ordinarily drawn for the method used so that the significance of the prothrombin time can be evaluated.

Tests similar to that for prothrombin time have been devised to determine the relative quantities of other clotting factors in the body. In performing the tests excesses of all the factors besides the one being tested are added to oxalated blood all at once, and then the time of coagulation is determined in the same manner as the usual prothrombin time. If the factor is deficient, the time will be considerably prolonged. In determining the usual prothrombin time, a deficiency of some of these other factors can also prolong the measured time. Therefore, an increased prothrombin time as performed in the usual hospital laboratory does not always mean decreased quantity of prothrombin but may mean decreased quantity of some other factor, such as fibrinogen.

REFERENCES

Biggs, R. (ed.): Human Blood Coagulation, Hemostasis and Thrombosis. Philadelphia, J. B. Lippincott Co., 1972.
Biggs, R. (ed.): The Treatment of Haemophilia A and B and Von Willebrand's Disease. Philadelphia, J. B. Lippincott Co., 1978.
Born, G. V. R.: Arterial thrombosis and its prevention. In Hayase, S., and Murao, S. (eds.): Cardiology; Proceedings of the VIII World Congress of Cardiology, Tokyo, 1978. New York, Elsevier/North-Holland, 1979, p. 81.
Collen, D., et al. (eds.): The Physiological Inhibitors of Blood Coagulation and Fibrinolysis. New York, Elsevier/North-Holland, 1979.
Cooper, H. A., et al.: The platelet: Membrane and surface reactions. Annu. Rev. Physiol. 38:501, 1976.
Davison, J. F., et al.: Progress in Chemical Fibrinolysis and Thrombolysis. New York, Raven Press, 1975.
Donaldson, V. H.: Complement, coagulation and fibrinolysis: Mechanisms of the fluid interphase. In Frohlich, E. D. (ed.): Pathology, 2nd Ed., Philadelphia, J. B. Lippincott Co., 1976, p. 639.
Engelbert, H.: Heparin. New York, S. Karger, 1978.
Gaffney, P. J. (ed.): Fibrinolysis; Current Fundamental and Clinical Concepts. New York, Academic Press, 1978.
Goldsmith, H. L.: Blood flow and thrombosis. Thromb. Diath. Haemorrh., 32:35, 1974.
Gowans, J. L. (in honour of): Blood Cells and Vessel Walls: Functional Interactions. Princeton, N.J., Excerpta Medica, 1980.
Hampton, J. W., and Brinkhouse, K. M.: Thrombopoiesis. In Frohlich, E. D. (ed.): Pathophysiology, 2nd Ed. Philadelphia, J. B. Lippincott Co., 1976, p. 609.
Hirsh, J., et al.: Concepts in Hemostasis and Thrombosis. New York, Churchill Livingstone, 1979.
Jacques, L. B.: Heparin. Sem. Thromb. Hemostasis, 4:275, 1978.
Joist, J. H., and Sherman, L. A. (eds.): Venous and Arterial Thrombosis: Pathogenesis, Diagnosis, Prevention and Therapy. New York, Grune & Stratton, 1979.
Kline, D. L., and Reddy, K. N. N. (eds.): Fibrinolysis. Boca Raton, Fla., CRC Press, 1980.
Lewis, J. H., et al.: Bleeding Disorders. Garden City, N.Y., Medical Examination Publishing Co., 1979.
Malinovsky, N. N., and Kozlov, V. A.: Anticoagulant and Thrombolytic Therapy in Surgery. St. Louis, C. V. Mosby, 1979.
Markwardt, F. (ed.): Fibrinolytics and Antifibrinolytics. New York, Springer-Verlag, 1978.
Markwardt, F., and Caen, J. P.: Current status of platelet inhibitors in the prevention of thromboembolism. In Hayase, S., and Murao, S. (eds.): Cardiology: Proceedings of the VIII World Congress of Cardiology, Tokyo, 1978. New York, Elsevier/North-Holland, 1979, p. 249.
McDuffie, N. M. (ed.): Heparin; Structure, Cellular Functions, and Clinical Application. New York, Academic Press, 1979.
Mielke, C. H., Jr., and Rodvien, R. (eds.): Mechanisms of Hemostasis and Thrombosis. Miami, Fla., Symposia Specialists, 1978.
Minna, J. D., et al.: Disseminated Intravascular Coagulation in Man. Springfield, Ill., Charles C Thomas, 1974.
Mitchell, J. R. A., and Domenet, J. G. (eds.): Thromboembolism: A New Approach to Therapy. New York, Academic Press, 1977.
Murano, G., and Bick, R. L. (eds.): Basic Concepts of Hemostasis and Thrombosis: Clinical Laboratory Evaluation of Thrombohemorrhagic Phenomena. Boca Raton, Fla., CRC Press, 1980.
Mustard, J. F., et al.: Prostaglandins and platelets. Annu. Rev. Med. 31:89, 1980.
Quick, A. J.: The Hemorrhagic Diseases and the Pathology of Hemostasis. Springfield, Ill., Charles C Thomas, 1974.
Ratnoff, O. D.: Some recent advances in the study of hemostasis. Circ. Res., 35:1, 1974.
Ratnoff, O. D.: Hemostasis, coagulation and fibrinolysis: Mechanisms of the fluid interphase. In Frohlich, E. D. (ed.): Pathophysiology, 2nd Ed., Philadelphia, J. B. Lippincott Co., 1976, p. 621.
Seegers, W. H.: Blood clotting mechanisms: Three basic reactions. Annu. Rev. Physiol., 31:269, 1969.
Suttie, J. W. (ed.): Vitamin K Metabolism and Vitamin K–Dependent Proteins. Baltimore, University Park Press, 1979.
Suttie, J. W., and Jackson, C. M.: Prothrombin structure, activation, and biosynthesis. Physiol. Rev., 57:1, 1977.
Thomson, J. M. (ed.): Blood Coagulation and Haemostasis: A Practical Guide. London, Churchill Livingstone, 1979.
Thrombosis Detection. New York, Grune & Stratton, 1977.
Tullis, J. L.: Clot. Springfield, Ill., Charles C Thomas, 1975.
Wall, R. T., and Harker, L. A.: The endothelium and thrombosis. Annu. Rev. Med., 31:361, 1980.

Part III
NERVE AND MUSCLE

10

Membrane Potentials, Action Potentials, Excitation, and Rhythmicity

Electrical potentials exist across the membranes of essentially all cells of the body, and some cells, such as nerve and muscle cells, are "excitable" — that is, capable of self-generation of electrochemical impulses at their membranes and, in some instances, utilization of these impulses to transmit signals along the membranes. In still other types of cells, such as glandular cells, macrophages, and ciliated cells, changes in membrane potentials probably play significant roles in controlling many of the cell's functions. However, the present discussion is concerned with membrane potentials generated both at rest and during action by nerve and muscle cells.

BASIC PHYSICS OF MEMBRANE POTENTIALS

Before beginning this discussion, let us first recall that the fluids both inside and outside the cells are electrolytic solutions containing 150 to 160 mEq. per liter of positive ions and the same concentration of negative ions. Generally, a very minute excess of negative ions (anions) accumulates immediately inside the cell membrane along its inner surface, and an equal number of positive ions (cations) accumulates immediately outside the membrane. The effect of this is the establishment of a *membrane potential* between the inside and outside of the cell.

The two basic means by which membrane potentials can develop are: (1) active transport of ions through the membrane, thus creating an imbalance of negative and positive charges on the two sides of the membrane, and (2) diffusion of ions through the membrane as a result of ion concentration differences between the two sides

of the membrane, thus also creating an imbalance of charges.

MEMBRANE POTENTIALS CAUSED BY ACTIVE TRANSPORT — THE "ELECTROGENIC PUMP"

Figure 10–1A illustrates how the process of active transport can create a membrane potential. In this figure equal concentrations of anions, which are *negatively charged,* are present both inside and outside the nerve fiber. However, the sodium "pump," which was discussed in Chapter 4, has transported some of the *positively charged* sodium ions to the exterior of the fiber. Thus, more negatively charged anions than positively charged sodium ions remain inside the nerve fiber, causing negativity on the inside. On the other hand, outside the fiber there are more positively charged sodium ions than negatively charged anions, thus causing positivity outside the fiber. A pump such as this, which causes the development of a membrane potential, is called an *electrogenic pump*.

In Chapter 4, it was pointed out that the sodium pump is also a potassium pump. That is, the same ATPase that acts as a carrier to transport sodium out of the cell also transports potassium inward at the same time. However, this pump normally transports three sodium ions outward for every two potassium ions inward. Thus, there is always more transfer of positively charged ions outward than inward. Because of this imbalance, the pump is still electrogenic. Therefore, as illustrated in Figure 10–1B, operation of this pump can still create electronegativity inside the nerve fiber membrane. We shall see in later sections of this chapter that this electrogenic pump is important in

ELECTROGENIC MEMBRANE POTENTIALS

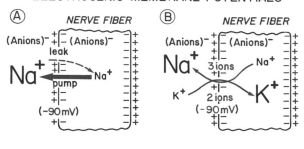

DIFFUSION MEMBRANE POTENTIALS

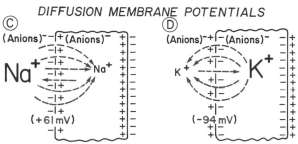

Figure 10–1. A, Establishment of a membrane potential as a result of active transport of sodium ions out of the nerve fiber. B, Establishment of a membrane potential as a result of sodium and potassium pumping through the nerve membrane by the sodium-potassium electrogenic pump, three sodium ions passing out of the membrane for each two potassium ions entering the membrane. C, Establishment of a diffusion membrane potential caused by permeability of the membrane only to sodium ions while the membrane is impermeable to all other ions. D, Establishment of a diffusion membrane potential because of permeability of the membrane only to potassium ions while the membrane is impermeable to all other ions. Note that the internal membrane potential is positive in the case of sodium ion permeability and negative in the case of potassium ion permeability because of opposite concentration gradients.

establishing the membrane potential of probably all cells in the body.

MEMBRANE POTENTIALS CAUSED BY DIFFUSION

Figure 10–1C and D illustrates the nerve fiber when there is no active transport of either sodium or potassium. In Figure 10–1C the sodium concentration is very great outside the membrane while that inside is very low. Furthermore, the membrane is very permeable to the sodium ions but not to the anions. Because of the large sodium concentration gradient from the outside toward the inside, there is a strong tendency for sodium ions to diffuse inward. As they do so, they carry positive charges to the inside, thus creating a state of electropositivity inside the membrane, and electronegativity on the outside because of the negative anions that remain behind, that do not diffuse inward along with the sodium. This potential difference across the membrane tends to repel the sodium ions in the backward direction from inside

toward the outside. And within a millisecond or so the potential becomes great enough to block further diffusion of sodium ions to the interior. The potential at this point is called the Nernst potential, as we shall discuss below.

Figure 10–1D illustrates the same effect as that in Figure 10–1C but with a high concentration of potassium ions inside the membrane and a low potassium concentration outside. These ions are also positively charged. Also, the membrane is highly permeable to the potassium ions but impermeable to the anions. Diffusion of the potassium ions to the outside creates a membrane potential now of opposite polarity, with negativity inside and positivity outside, and, again, the membrane potential rises high enough within milliseconds to block further net diffusion of the potassium ions to the outside.

Thus, in both Figure 10–1C and D, we see that a concentration difference of ions across a semipermeable membrane can, under appropriate conditions, cause the creation of a membrane potential. In later sections of this chapter, we shall also see that many of the membrane potential changes observed during the course of nerve and muscle impulse transmission result from the occurrence of rapidly changing diffusion membrane potentials.

Relationship of the Diffusion Potential to the Concentration Difference — The Nernst Equation. When a concentration difference of a single type of ions across a membrane causes diffusion of ions through the membrane, thus creating a membrane potential, the magnitude of the potential *inside* the membrane versus the outside is determined by the ratio of the tendency for the ions to diffuse in one direction versus the other direction, which is determined for positive ions by the following formula (*at body temperature, 38° C.*):

$$\text{EMF (millivolts)} = -61 \log \frac{\text{Conc. inside}}{\text{Conc. outside}}$$

Thus, when the concentration of positive ions on the inside of a membrane is 10 times that on the outside, the log of 10 is 1, and the potential difference calculates to be −61 millivolts. This equation is called the *Nernst equation*.

However, two conditions are necessary for this *Nernst potential* to develop as a result of diffusion: (1) The membrane must be semipermeable, allowing ions of one charge to diffuse through the pores while ions of the opposite charge do not diffuse. (2) The concentration of the diffusible ions must be greater on one side of the membrane than on the other side.

Using the above formula, let us now calculate the Nernst potential across the nerve membrane, first, when the membrane is permeable only to sodium ions, and, second, when the membrane is permeable only to potassium ions:

The normal concentration of sodium ions inside the nerve membrane is approximately 14 mEq. and outside approximately 142 mEq. Thus, the ratio of these two is 0.10, and the logarithm of 0.10 is −1.00. Multiplying this by −61 millivolts gives a Nernst potential for sodium of +61 millivolts inside the nerve fiber membrane.

The normal concentration of potassium ions inside the nerve fiber is approximately 140 mEq. per liter and 4 mEq. per liter on the outside. The ratio of these two is 35. The logarithm of 35 is 1.54; this times −61 millivolts equals a Nernst potential for potassium of −94 millivolts inside the membrane.

Therefore, if there were no pumping of ions through the nerve membrane and if the membrane were permeable only to sodium but not at all to other ions, the potential inside the nerve fiber would be +61 millivolts. Conversely, if the membrane were permeable to potassium but not permeable to any other ions, the membrane potential would be −94 millivolts. We shall see later in this discussion that under resting conditions the membrane potential averages about −90 millivolts, which is very near to the −94 millivolts potassium Nernst potential. This is true because in the resting state the membrane is very permeable to potassium and only slightly permeable to sodium. On the other hand, when a nerve impulse is transmitted, the membrane, for a minute fraction of a second, becomes much more permeable to sodium than to potassium. Therefore, during this split second, the membrane potential rises to approximately +45 millivolts, which is much nearer the sodium Nernst potential than the potassium Nernst potential.

CALCULATION OF THE MEMBRANE POTENTIAL WHEN THE MEMBRANE IS PERMEABLE TO SEVERAL DIFFERENT IONS

When a membrane is permeable to several different ions, the diffusion potential that will develop depends on three factors: (1) the polarity of the electrical charge of each ion, (2) the permeability of the membrane *(P)* to each ion, and (3) the concentration of the respective ions on the two sides of the membrane. Thus, the following formula, called the *constant field equation* or the *Goldman equation,* gives the calculated membrane potential on the *inside* of the membrane when two univalent positive ions (cations, C) and two univalent negative ions (anions, A) are involved.

$$\text{EMF (millivolts)} = -61 \log \frac{C_{1_i}^+ P_1 + C_{2_i}^+ P_2 + A_{3_o}^- P_3 + A_{4_o}^- P_4}{C_{1_o}^+ P_1 + C_{2_o}^+ P_2 + A_{3_i}^- P_3 + A_{4_i}^- P_4} \quad (2)$$

Note that a positive ion gradient from the inside *(i)* to the outside *(o)* of a membrane causes electronegativity inside the membrane, while a negative ion gradient in *exactly the opposite direction* also causes electronegativity on the inside.

The Cell Membrane as a Capacitor. Note in Figure 10–1A that the negative and positive ionic charges are shown lined up against the membrane. This occurs because the negative charges inside the membrane and the positive charges on the outside pull each other together at the membrane barrier, thus creating an abrupt change in electrical potential across the cell membrane.

This alignment of electrical charges on the two sides of the membrane is exactly the same process that takes place when an electrical capacitor becomes charged with electricity. In the cell membrane, the lipid matrix of the membrane is the *dielectric,* much as mica, paper, and mylar are frequently used as dielectrics in electrical capacitors. It will be recalled that a capacitor's capacity for holding electrical charges is inversely proportional to the thickness of the membrane. Because of the extreme thinness of the cell membrane (70 to 100 Angstroms), the capacitance of the cell membrane is tremendous for its area — about 1 microfarad per square centimeter (though only 1/1000 this much when insulated with a myelin sheath in the large nerve fibers).

Though the figures of this chapter will show large numbers of negative or positive ions (or charges) lined up against the nerve membrane, in actuality very few positive or negative ions need to line up on the two sides of the membrane to create the usual membrane potential of nerves. In the case of the usual large nerve fiber, only about 1/50,000 to 1/500,000 of the positive charges inside the nerve fiber need to be transferred to the outside of the fiber to create the normal nerve potential of −90 millivolts inside the fiber.

ORIGIN OF THE NERVE CELL MEMBRANE POTENTIAL

Distinction Between Original Development of a Membrane Potential and the Instantaneous Reestablishment of the Membrane Potential after Nerve Impulse Transmission. Let us assume that, in the beginning, the concentrations of all ions are the same inside and outside the nerve fiber. Under these conditions there will be no membrane potential. However, the normal cell will automatically develop a membrane potential. How does this come about? The answer to this lies in the function of the sodium-potassium pump and in the electrogenicity of this pump. That is, the continual pumping of more positive charges to the outside of the membrane than to the inside (three sodium ions pumped outward for every two potassium ions inward) eventually leads to the negative membrane potential inside the cell. Thus, the original development of the membrane potential results from this electrogenicity of the sodium-potassium pump.

But now let us write another scenario: First, assume that the sodium-potassium pump is nonfunctional but that there is already a high potassium concentration inside the cell and low potassium outside. Second, assume that the cell membrane is highly permeable to potassium while poorly permeable to sodium. Therefore, large numbers of potassium ions will diffuse to the outside while only a few sodium ions will diffuse to the inside. This results in far more transfer of positive ions to the outside than in the other direction. Consequently, once again a negative membrane potential develops inside the nerve membrane. It is in this manner that the negative membrane potential is re-established within less than a millisecond immediately following transmission of a nerve impulse along the nerve membrane.

Therefore, from the outset it is very important for the student to recognize the difference between the original development of the membrane potential, which is an electrogenic phenomenon, and the instantaneous re-establishment of the membrane potential following each nerve impulse (that is, following each action potential) which is a diffusion phenomenon. Now, let us discuss in detail the original development of the membrane potential.

Role of the Sodium-Potassium Pump and of Nondiffusible Anions in the Original Development of the Cell Membrane Potential. Before attempting to explain the original development of the cell membrane potential, several basic facts need to be understood:

1. The nerve membrane is endowed with a sodium-potassium pump, three sodium ions being pumped to the exterior for each two potassium ions pumped to the interior. This pump was discussed earlier in this chapter, and in even greater detail in Chapter 4.

2. The resting nerve membrane is normally 50 to 100 times as permeable to potassium as to sodium. Therefore, potassium diffuses with relative ease through the resting membrane, whereas sodium diffuses only with difficulty.

3. Inside the nerve fiber are large numbers of anions (negatively charged) that cannot diffuse through the nerve membrane at all or that diffuse

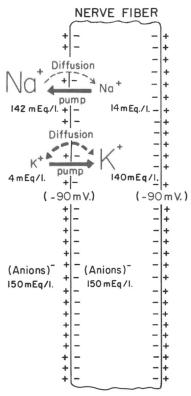

Figure 10–2. Establishment of a membrane potential of −90 millivolts in the normal resting nerve fiber and development of concentration differences of sodium and potassium ions between the two sides of the membrane. The dashed arrows represent diffusion and the solid arrows represent active transport (pumps).

very poorly. These anions include especially organic phosphate ions, sulfate ions, and protein ions.

Putting the above facts together, we can see how the resting nerve membrane potential comes about. First, sodium is pumped to the outside of the fiber, while potassium is pumped to the inside. However, because three sodium ions are pumped out of the fiber for every two potassium ions pumped in, more positive ions are continually being pumped out of the fiber than into it. Since most of the negatively charged anions inside the nerve fiber are nondiffusible, the negative charges remain inside the nerve fiber so that the inside of the fiber becomes electronegative, while the outside becomes electropositive, as illustrated in Figure 10–2.

Equilibration of Sodium Ion Transfer in the Two Directions Across the Membrane — Maximum Potential That Can Be Achieved by the Electrogenic Sodium-Potassium Pump. As progressively more sodium ions are pumped out of the nerve fiber, they begin to diffuse back into the nerve fiber for two reasons: (1) a sodium concentration gradient develops from the outside of the fiber toward the inside, and (2) a negative membrane potential develops inside the fiber and attracts the positively charged sodium ions inward. Eventually, there comes a point at which the inward diffusion equals the outward pumping by the sodium pump. When this occurs, the pump has reached its maximum capability to cause net transfer of sodium ions to the outside. This occurs when the sodium ion concentration inside the nerve fiber falls to about 14 mEq. per liter (in contrast to 142 mEq. per liter in the extracellular fluids) and when the membrane potential inside the fiber falls to -70 to -100 millivolts, averaging about -90 millivolts. Therefore, this -90 millivolts becomes the *resting potential* of the nerve membrane.

Equilibration of Potassium Ion Transfer Across the Membrane.   At the same time that the sodium-potassium pump pumps sodium ions to the exterior, it pumps about two thirds as many potassium ions to the interior. However, the resting membrane is 50 to 100 times as permeable to potassium ions as to sodium ions, which means that each time potassium ions are pumped in they tend to diffuse back out of the fiber almost immediately. Therefore, the potassium pump by itself can build up only a slight excess of potassium ions inside the nerve fiber.

Yet, it is known that potassium ions do build up to a high concentration inside the nerve fiber. How could this be, if it is not caused by the potassium pump? The answer is the high degree of negativity created inside the fiber by the electrogenic sodium-potassium pump. The -90 millivolts inside the fiber attract the positive potassium ions from the exterior to the inside, and this attraction is essential for most of the buildup of potassium ions inside the fiber.

Thus, from a quantitative point of view, once the membrane has reached a steady state condition, three factors affect the equilibration of potassium ions on the two sides of the membrane: (1) the potassium pump pumping potassium ions *inward,* (2) the electrical gradient causing *inward* diffusion of potassium ions, and (3) the concentration gradient caused by the buildup of potassium ions inside the fiber causing diffusion of potassium ions *outward.* In the steady state condition, these three factors all come to exact quantitative equilibrium; that is, the two factors causing inward movement of potassium become exactly equal to the single factor causing outward movement.

Role of Chloride and Other Ions.   Chloride ions are not pumped through the nerve membrane in either direction. However, they readily diffuse through the membrane. Because there is no pump to force buildup of a concentration difference of chloride ions across the membrane, the distribution of chloride ions on the two sides of the membrane is determined entirely by the degree and polarity of the membrane potential. That is,

the negativity that develops inside the membrane repels chloride ions from inside the fiber and causes the chloride concentration to fall to a very low value, about 4 mEq./liter in comparison with an extracellular fluid concentration of about 103 mEq./liter. Because there is no chloride pump, the Nernst equation can be used to calculate the exact concentration ratio of chloride ions on the two sides of the membrane when the membrane potential is known and when the membrane is in an equilibrium state. Thus, chloride ions play only a passive role. Later in the chapter we will discuss instantaneous changes in membrane potential caused by the so-called "action potential." Under these conditions, chloride ions move rapidly through the membrane and affect the duration and the magnitude of the action potential, but again their role is a passive one.

Essentially the same principles as those for sodium, potassium, and chloride distribution across the membrane also apply to other ions. Magnesium ions are affected in much the same way as potassium, and calcium ions in much the same manner as sodium ions. However, the concentrations and permeabilities of magnesium and calcium ions are small enough that the total number of electrical charges actually involved is also small. Yet both of these ions, especially calcium ions, affect membrane potentials in another way, by altering the permeability of the membrane to other ions, a subject that is discussed at greater length later in the chapter.

Membrane Potentials Measured in Nerve and Muscle Fibers.   Figure 10–3 illustrates a method that has been used for measuring the resting membrane potential. A micropipet is made from a minute capillary glass tube so that the tip of the pipet has a diameter of only 0.25 to 2 microns. Inside the pipet is a solution of very concentrated potassium chloride, which acts as an electrical conductor. The fiber whose membrane is to be measured is pierced with the pipet and electrical connections are made from the pipet to an appropriate meter, as illustrated in the figure. The resting membrane potential measured in many different nerve and skeletal muscle fibers of mammals has usually ranged between -70 and -100 millivolts, with -90 millivolts as a reasonable average of the different measurements.

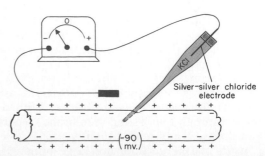

Figure 10–3.   Measurement of the membrane potential of the nerve fiber using a microelectrode.

THE ACTION POTENTIAL

In the above section we described the original development of the *resting membrane potential* and pointed out that it is normally about −90 mv. Now we will describe how membrane potentials are employed by nerve fibers to transmit nerve impulses, which is the means by which informational signals are transmitted from one part of the nervous system to another. This is achieved by means of *action potentials,* which are abrupt pulse-like changes in the membrane potential, lasting a few ten thousandths to a few thousandths of a second as illustrated in Figures 10–5, 10–6, 10–10, 10–11, and 10–12. And, for reasons that will be discussed below, the action potential moves along the length of the nerve, thus giving rise to the nerve signals.

An action potential can be elicited in a nerve fiber by almost any factor that suddenly increases the permeability of the membrane to sodium ions — such factors as electrical stimulation, mechanical compression of the fiber, application of chemicals to the membrane, or almost any other event that disturbs the normal resting state of the membrane. We shall see in Chapter 48 that the various nervous sensory receptors are also stimulated in one of these ways to initiate signals in the nervous system.

Role of Sodium and Potassium Membrane Diffusion Potentials in the Generation of the Action Potential. In the above discussion of the original development of the resting membrane potential, it became clear that the sodium-potassium pump is basically responsible for generating this resting potential. Furthermore, the sodium-potassium pump creates large sodium and potassium differences across the membranes, with very high sodium concentration outside the membrane and very high potassium concentration inside. But it is important to emphasize that once the sodium and potassium concentration differences have been created across the cell membrane, the occurrence of action potentials does not depend upon operation of the sodium-potassium pump. Indeed, many thousands of impulses can be transmitted even after the pump has been poisoned before the sodium and potassium concentration differences "run down." The action potential results from rapid changes in membrane permeability to sodium and potassium ions — the sodium permeability increasing about 5000-fold at the onset of the action potential, followed instantaneously by return of the sodium permeability to normal, and then the potassium permeability increases greatly. As a result, the membrane potential changes rapidly from its normal negative value to an instantaneous positive valve, then equally rapidly back to

its negative value. These changes will be discussed in the following sections.

Membrane Depolarization and Membrane Repolarization — the Two Stages of the Action Potential. The action potential occurs in two separate stages called *membrane depolarization* and *membrane repolarization,* each of which may be explained by referring to Figure 10–4. Figure 10–4A illustrates the resting state of the membrane, with negativity inside and positivity outside. When the permeability of the membrane to sodium ions suddenly increases, many of the sodium ions that are present in very high concentration outside the fiber rush to the inside, carrying enough positive charges to the inside to cause complete disappearance of the normal negative resting potential, and usually enough charges actually to develop a positive state inside the fiber. This sudden loss of normal negative potential inside the fiber is called *depolarization.* The positive potential that develops momentarily inside the fiber is called the *reversal potential.*

Almost immediately after depolarization takes place, the pores of the membrane again become almost impermeable to sodium ions, but at the same time considerably more permeable than normally to potassium ions. Therefore, sodium ions stop moving to the inside of the fiber, and, instead, potassium ions move to the outside because of the high potassium concentration on the inside. Thus, because the potassium ions are positively charged, the excess positive charges inside the fiber are transferred back out of the fiber, and the normal negative resting membrane potential returns. This effect, called *repolarization,* is illustrated in Fig-

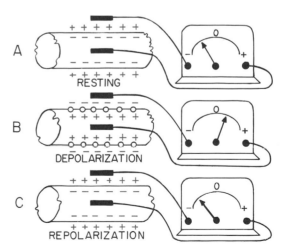

Figure 10–4. Sequential events during the action potential, showing: A, the normal resting potential, B, development of a reversal potential during depolarization, and C, reestablishment of the normal resting potential during repolarization.

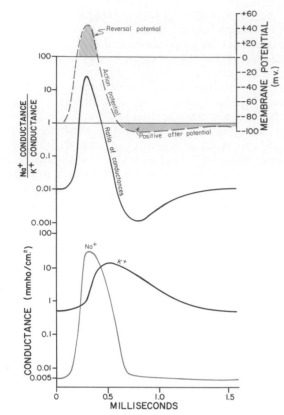

Figure 10–5. Changes in sodium and potassium conductances during the course of the action potential. Note that sodium conductance increases several thousand-fold during the early stages of the action potential, while potassium conductance increases only about 30-fold during the latter stages of the action potential and for a short period thereafter. (Curves constructed from data in Hodgkin and Huxley papers but transposed from squid axon to apply to the membrane potentials of large mammalian nerve fibers.)

ure 10–4C. The curve of Figure 10–5 shows the sequence of the recorded action potential inside the fiber membrane. Now let us explain in more detail *how* this sequence of events occurs.

Sodium "Channels" and Potassium "Channels." Most physiologists believe that the sodium and potassium ions diffuse mainly through separate types of pores. The pores through which the sodium ions diffuse are called *sodium channels* and those through which the potassium ions diffuse are called *potassium channels*. The actual anatomical structures of these channels are not known, but they have physiological properties *as if* the sodium channels were oval-shaped pores with dimensions of about 0.3 by 0.5 nanometer, and the potassium channels were round pores with dimensions of about 0.3 by 0.3 nanometer.

Each channel is believed to be guarded by an electrically charged "gate" that can open and close the channel. Under resting conditions, the

gates of the sodium channels are almost completely closed, while those of the potassium channels are only partially closed; therefore, the potassium channels are some 50 to 100 times as permeable as the sodium channels. This difference allows far more potassium diffusion than sodium diffusion in the resting state, as has been explained. Yet, when the gates are widely opened, the permeability of the sodium channels can increase as much as 5000-fold, and the permeability of the potassium channels, which are already partially opened, can increase an additional 50-fold. It is this opening and closing of the gates to the sodium and potassium channels that is responsible for the depolarization and repolarization processes.

A Positive Feedback Regenerative Process that Opens the Sodium Gates to Cause Depolarization. The initial event in the causation of an action potential is a slight increase in the permeability of the membrane to sodium ions — that is, a slight opening of the sodium gates. When these gates open even a slight amount, a positive feedback effect immediately begins and initiates a regenerative cycle to open the gates still wider and wider until the sodium channel permeability increases within a few ten thousandths of a second about 5000-fold. This effect is illustrated by the abrupt rise in the sodium conductance curve (conductance = permeability × ion concentration) in Figure 10–5. At this instant the membrane is approximately 20 to 30 times as permeable to sodium as to potassium, because only the sodium channels open widely during the depolarization process. The result is that far more sodium ions now diffuse to the interior of the fiber than potassium ions to the exterior, carrying positive charges to the inside and leaving negative anions on the outside.

Mechanism of the Positive Feedback – The Gating Potential. Unfortunately, we do not know the exact means by which the gates of the sodium and potassium channels open and close. However, it is believed that the so-called gates are positive electrical charges which in turn are controlled by *gating potentials* that occur in the lipid matrix surrounding the sodium and potassium channels. When the gating potential is strong, the electrical charges are believed to repel the sodium or potassium ions and thereby close the gates. When weak, the channels open.

At the onset of the action potential, the first partial opening of the sodium channels allows enough sodium ions to flow to the interior of the fiber to cause partial depolarization. That is, the membrane potential changes from −90 mv. to approximately −60 mv. This change of potential inside the fiber is *in the positive direction* and it causes a *capacitative* change in the electrical field at the lining surfaces of the sodium channels to reduce the gating potentials of these channels. Consequently, the gates open more widely, still more sodium ions flow to the interior, the potential inside the fiber becomes still more positive, the gating potential becomes still weaker, the pores open even more widely, and still more sodium ions diffuse to the interior. Thus, the positive feedback regenerative cycle continues over and over again until the channels become fully opened. And the electrical potential inside the fiber becomes positive, about +45 mv., which is the *reversal potential,* as illustrated at the top of the action potential in Figure 10–5 (the upper curve).

Mechanism of Repolarization — Closure of the Sodium Channels and Opening of the Potassium Channels. Once the sodium channels become widely opened and the positive reversal potential appears inside the fiber, the sodium and potassium channel permeabilities make an about-face. The exact cause of this is not known, but it is known that the positive interior potential now causes electrical repulsion of the incoming sodium ions until their incoming movement slows or stops. This slowing of the sodium ion movement, for reasons not yet understood but it is possibly also an electrical field effect, causes the gating potential of the sodium channels to become strong once again. Consequently, at this point the sodium channels begin to close, but, just as rapidly, the potassium channels begin to open. This presumably results from the fact that the positive potential inside the membrane is now pushing potassium ions to the exterior at an extremely rapid rate. Loss of the potassium ions to the exterior carries positive charges to the exterior, and the potential on the interior of the fiber now returns toward the normal resting negative state. As the membrane becomes more negative, capacitative radiation of this negative field through the lipid matrix to the potassium channel gates opens the potassium channels.

Thus, during the repolarization process, a positive feedback regenerative cycle develops for opening the potassium channels, as illustrated by the increasing potassium conductance in Figure 10–5, in exactly the same way that a similar cycle had occurred during depolarization to open the sodium channels. More and more potassium ions pass to the exterior, and the electrical potential inside the fiber returns to the normal resting level of approximately −90 mv.

Once the membrane potential returns back to its negative resting level, it then remains at this level until the membrane is disturbed once again. The reason is that the potassium channels never close completely but only partially. Therefore, potassium ions continue to diffuse through the membrane with relative ease during the resting state, even though the permeability of the membrane to sodi-

um ions is greatly restricted. Thus, in the resting state, positive charges (the potassium ions) tend to diffuse out of the nerve fiber to the exterior, creating and maintaining negativity on the inside until the next action potential.

Role of Calcium Ions in Closing the Sodium Gates. When calcium ions are deficient in the extracellular fluid, the sodium gates will not close fully between action potentials, and the membrane remains very leaky to sodium ions — sometimes so much so that the membrane remains depolarized continuously or fires repetitively. Though the cause of this effect is not fully understood, a possible answer is the following: Since calcium ions tend to bind relatively strongly with proteins, it is possible that binding with the protein linings of the sodium channels creates an electrical field in or near the channels that blocks entry of sodium ions.

Quantity of Ions Lost from the Nerve Fiber During the Action Potential. The actual quantity of ions that must pass through the nerve membrane to cause the action potential — that is, to cause a 135 millivolt increase in the membrane potential and then to return this potential back to its normal resting level — is extremely slight. For large myelinated nerve fibers only about 1/100,000 to 1/500,000 of the ions normally inside the fiber are exchanged during this process. Nevertheless, this does cause a very slight increase of sodium ions inside the fiber and a corresponding very slight decrease in potassium ions. As we shall see later in the chapter, the active transport processes restore these ions within a few milliseconds or seconds.

RELATIONSHIP OF THE ACTION POTENTIAL TO THE POTASSIUM AND SODIUM NERNST POTENTIALS

Now that we have explained the events that transpire during the course of the action potential, let us review for a moment the relationship of this action potential to the potassium and sodium Nernst potentials. It was pointed out earlier in the chapter that under resting conditions the nerve membrane is highly permeable to potassium but only slightly permeable to sodium. Also, a calculation of the potassium Nernst potential earlier in the chapter showed this potential to be only a few millivolts more negative than the resting membrane potential. This is illustrated again in Figure 10–6, showing the potassium Nernst potential to be −94 mv. and a resting potential, −90 mv. Then, during the course of the action potential the membrane potential reverses and becomes approximately +45 mv, thus approaching the sodium Nernst potential of +61 mv., as also illustrated in the figure. However, the reversal potential is not as near to the sodium Nernst potential as the

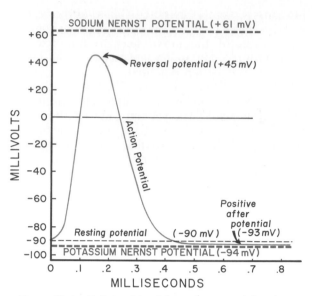

Figure 10–6. Relationship of the action potential to the potassium and sodium Nernst potentials.

resting potential is to the potassium Nernst potential. The reason for this is that the ratio of sodium permeability to potassium permeability even at the peak of the action potential, about 20 to 1, never rises as high as the ratio of potassium permeability to sodium permeability during the resting state of the nerve fiber, about 75 to 1.

The reversal potential lasts for only a few fractions of a millisecond in a large myelinated nerve fiber, because the state of high permeability to sodium ions is a short-lived event. And when the permeability of the membrane to sodium decreases back to the normal level, the permeability to potassium ions now becomes highly dominant once again, which causes the resting potential to return to a level very near the potassium Nernst potential.

Thus, it can be seen that the potassium and sodium Nernst potentials represent the lower and upper limits of the membrane potential and that the action potential is displayed between these limits.

SOME EXPERIMENTAL METHODS THAT HAVE BEEN USED TO STUDY THE ACTION POTENTIAL

The events of the action potential, as discussed in the preceding few pages, have been studied most widely in the *squid axon,* a nerve fiber that is sometimes as large as 1 mm. in diameter. This fiber is large enough that potentials are easily measured on its inside, and it is even possible to remove the axoplasm from inside the fiber and to replace this with artificial solution. Also, electrodes can be placed inside the fiber to excite it or to "clamp" the voltage across the membrane — that is, to fix the voltage across the membrane at a constant level

by electronically controlling the current flow through the electrodes. Use of this clamping process has made it possible to study the effect of voltage changes on sodium and potassium permeabilities.

Two drugs have proved to be invaluable in the study of the sodium and potassium channels. One drug, *tetrodotoxin,* will block the sodium channels. However, it will block these channels only when applied to the outside of the nerve fiber. Nevertheless, when the sodium channels are blocked it is then possible to study the permeability of the potassium channels independently of the sodium channels. In a similar manner, application of *tetraethylammonium ion* to the inside of the fiber membrane blocks the potassium channels. In this state, the sodium channels can be studied independently of the potassium channels. Thus, by a painstaking process of elimination, the separate effects on the two types of channels are gradually being elucidated.

PROPAGATION OF THE ACTION POTENTIAL

In the preceding paragraphs we have discussed the action potential as it occurs at one spot on the membrane. However, an action potential elicited at any one point on an excitable membrane usually excites adjacent portions of the membrane, resulting in propagation of the action potential. The mechanism of this is illustrated in Figure 10–7. Figure 10–7A shows a normal resting nerve fiber, and Figure 10–7B shows a nerve fiber that has been excited in its midportion — that is, the midportion has suddenly developed increased permeability to sodium. The arrows illustrate a local circuit of current flow between the depolarized and the resting membrane areas; positive current flows inward through the depolarized membrane and outward through the resting membrane, thus completing a circuit. In some way not understood, *the outward current flow through the resting mem-*

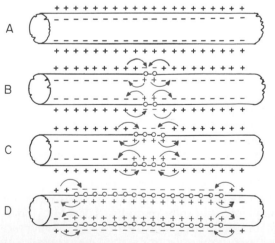

Figure 10–7. Propagation of action potentials in both directions along a conductive fiber.

brane now increases the membrane's permeability to sodium, which immediately allows sodium ions to diffuse inward through the membrane, thus setting up the vicious cycle of increasing sodium permeability discussed earlier in the chapter. As a result, depolarization occurs at this area of the membrane as well. Therefore, as illustrated in Figure 10–7C and D, successive portions of the membrane become depolarized. And these newly depolarized areas cause local circuits of current flow still farther along the membrane, causing progressively more and more depolarization. Thus, the depolarization process travels in both directions along the entire extent of the fiber. The transmission of the depolarization process along a nerve or muscle fiber is called a *nerve* or *muscle impulse.*

Direction of Propagation. It is now obvious that an excitable membrane has no single direction of propagation, but that the impulse can travel in both directions away from the stimulus — and even along all branches of a nerve fiber — until the entire membrane has become depolarized.

The All-or-Nothing Principle. It is equally obvious that, once an action potential has been elicited at any point on the membrane of a normal fiber, the depolarization process will travel over the entire membrane. This is called the all-or-nothing principle, and it applies to all normal excitable tissues. Occasionally, though, when the fiber is in an abnormal state the impulse will reach a point on the membrane at which the action potential does not generate sufficient voltage to stimulate the adjacent area of the membrane. When this occurs the spread of depolarization will stop. Therefore, for normal propagation of an impulse to occur, the ratio of action potential to threshold for excitation, called the *safety factor,* must at all times be greater than unity.

Propagation of Repolarization. The action potential normally lasts almost the same length of time at each point along a fiber. Therefore, repolarization normally occurs first at the point of original stimulus and then spreads progressively along the membrane, moving in the same direction that depolarization had previously spread. Figure 10–8 illustrates the same nerve fiber as that in Figure 10–7, showing that the polarization process

is propagated in the same direction as the depolarization process but a few ten thousandths of a second later.

"RECHARGING" THE FIBER MEMBRANE – IMPORTANCE OF ENERGY METABOLISM

Transmission of each impulse along the nerve fiber reduces the concentration differences of sodium and potassium between the inside and outside of the membrane. For a single action potential, this effect is so minute that it cannot even be measured. Indeed, 100,000 to 500,000 impulses can be transmitted by a large nerve fiber before the concentration differences have run down to the point that action potential conduction ceases. Yet, even so, with time it becomes necessary to reestablish these sodium and potassium membrane concentration differences. This is achieved by the action of the sodium and potassium pump in exactly the same way as that described in the first part of the chapter for establishment of the original resting potential. That is, the sodium ions that, during the action potentials, have diffused to the interior of the cell and the potassium ions that have diffused to the exterior are returned to their original state by the sodium and potassium pump. Since this pump requires energy for operation, this process of "recharging" the nerve fiber is an active metabolic one, utilizing energy derived from the adenosine triphosphate energy "currency" system of the cell.

A special feature of the sodium-potassium ATP-ase membrane pumping system is that its degree of activity is very strongly stimulated by excess sodium ions inside the cell membrane. In fact, the pumping activity increases approximately in proportion to the third power of the sodium concentration. That is, if the internal sodium concentration rises from 10 to 20 mEq. per liter, the activity of the pump does not merely double but instead

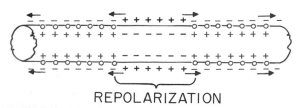

REPOLARIZATION

Figure 10–8. Propagation of repolarization in both directions along a conductive fiber.

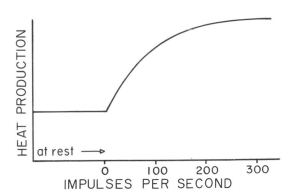

Figure 10–9. Heat production in a nerve fiber at rest and at progressively increasing rates of stimulation.

increases approximately 8-fold. Therefore, it can easily be understood how the recharging process of the nerve fiber can rapidly be set into motion whenever the concentration differences of sodium and potassium across the membrane begin to "run down."

Heat Production by the Nerve Fiber. Figure 10–9 illustrates the relationship of heat production in a nerve fiber to the number of impulses transmitted by the fiber each second. The rate of heat production is a measure of the rate of metabolism in the nerve, because heat is always liberated as a product of the chemical reactions of energy metabolism. Note that the heat production increases markedly as the number of impulses per second increases. It is this increased energy that causes the "recharging" process.

THE SPIKE POTENTIAL AND THE AFTER-POTENTIALS

Figure 10–10 illustrates an action potential recorded with a much slower time scale than that illustrated in Figure 10–5; many milliseconds of recording are shown in comparison with only the first 1.5 milliseconds of the action potential in Figure 10–5.

The Spike Potential. The initial, very large change in membrane potential shown in Figure 10–10 is called the spike potential. In large, type A myelinated nerve fibers it lasts for about 0.4 millisecond. The spike potential is the same as the first part of the action potential that has been discussed in the preceding paragraphs and is also called the *nerve impulse*.

The Negative After-Potential. At the termination of the spike potential, the membrane potential sometimes fails to return all the way to its resting level for another few milliseconds, as shown in Figure 10–10. This is particularly likely to occur after a series of rapidly repeated action potentials. This is called the *negative after-potential*. It is believed to result from a buildup of potassium ions immediately outside the membrane; this causes the concentration ratio of potassium across the

membrane to be temporarily less than normal and therefore prevents full return of the normal resting membrane potential for a few additional milliseconds.

The Positive After-Potential. Once the membrane potential has returned to its resting value, it then becomes a little more negative than its normal resting value; this excess negativity is called the *positive after-potential*. It is a fraction of a millivolt to a few millivolts more negative than the normal resting membrane potential, but it lasts from 50 milliseconds to as long as many seconds.

The first part of this positive after-potential is caused by the excess permeability of the nerve membrane to potassium ions at the end of the spike potential, which drives the membrane potential nearer to the potassium Nernst potential, as was illustrated in Figure 10–6. However, the prolonged continuance of this potential is caused principally by the electrogenic pumping of excess sodium outward through the nerve fiber membrane, which is the recharging process that was discussed previously. If the active transport processes are poisoned, most of the positive after-potential is lost, though both the action potential and the negative after-potential continue to occur.

(The student might wonder why greater negativity in the resting membrane potential is called a positive rather than a negative after-potential, and, likewise, why the so-called negative after-potential is not named positive. The reason is that these potentials were first measured *outside* the nerve fibers rather than inside, and all potential changes on the outside are of exactly opposite polarity, whereas modern terminology expresses membrane potentials in terms of the inside potential rather than the outside potential.)

PLATEAU IN THE ACTION POTENTIAL

In some instances the excitable membrane does not repolarize immediately after depolarization, but, instead, the potential remains on a plateau near the peak of the spike sometimes for many milliseconds before repolarization begins. Such a plateau is illustrated in Figure 10–11, from which one can readily see that the plateau greatly prolongs the period of depolarization. It is this type of action potential that occurs in the heart, where the plateau lasts for as long as two- to three-tenths second and causes contraction of the heart muscle during this entire period of time.

The cause of the action potential plateau is probably a combination of several different factors. First, there is delay in closure of the sodium channels, which allows extra sodium ions to continue flowing into the fiber. Second, there is a small amount of calcium current flowing into the fiber at the same time, and these two currents together maintain the positive state inside the membrane that causes the plateau. However, third, probably equally as important is the fact that the permeability of the potassium channels decreases about five-fold at the onset of the action potential in exitable membranes that exhibit pla-

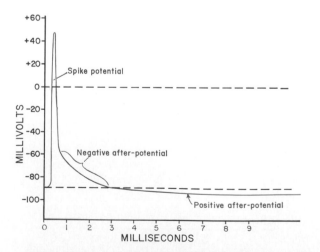

Figure 10–10. An idealized action potential, showing: the initial spike followed by a negative after-potential and a positive after-potential.

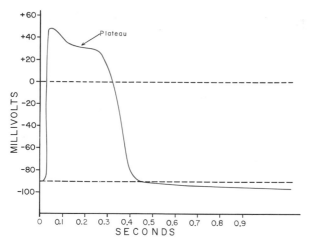

Figure 10–11. An action potential from a Purkinje fiber of the heart, showing a "plateau."

teaus, and this prevents rapid outflow of potassium ions to the outside of the fiber and, therefore, delays the repolarization process.

Yet, when closure of the sodium channels does begin, the process proceeds unabated; simultaneously, the potassium permeability of the membrane increases a hundred-fold or more. Therefore, sodium and calcium ions stop diffusing to the interior of the fiber, while potassium ions diffuse outward extremely rapidly. Consequently, the membrane potential returns quickly to its normal negative level, as illustrated in Figure 10–11, by the rapid decline of the potential at the end of the plateau.

RHYTHMICITY OF CERTAIN EXCITABLE TISSUES – REPETITIVE DISCHARGE

All excitable tissues can discharge repetitively if the threshold for stimulation is reduced low enough. For instance, even nerve fibers and skeletal muscle fibers, which normally are highly stable, discharge repetitively when they are placed in a solution containing the drug veratrine or when the calcium ion concentration falls below a critical value. Repetitive discharges, or rhythmicity, occur normally in the heart, in most smooth muscle, and also in many of the neurons of the central nervous system. It is these rhythmical discharges that cause the heart beat, that cause peristalsis, and that cause such neronal events as the rhythmical control of breathing.

The Re-Excitation Process Necessary for Rhythmicity. For rhythmicity to occur, the membrane, even in its natural state, must be already permeable enough to sodium ions to allow automatic membrane depolarization. Thus, Figure 10–12 shows that the "resting" membrane potential is only −60 to −70 millivolts. This is not enough

negative voltage to keep the sodium channels closed. That is, (a) sodium ions flow inward, (b) this further increases the membrane permeability, (c) still more sodium ions flow inward, (d) the permeability increases more, and so forth, thus eliciting the regenerative process of sodium channel opening until an action potential is generated. Then, at the end of the action potential the membrane repolarizes. But shortly thereafter, the depolarization process begins again and a new action potential occurs spontaneously — this cycle continuing again and again and causing self-induced rhythmical excitation of the excitable tissue.

Yet why does the membrane not remain depolarized all the time instead of repolarizing, only to become depolarized again shortly thereafter? The answer to this can be found by referring back to Figure 10–5, which shows that toward the end of all action potentials, and continuing for a short period thereafter, the membrane becomes excessively permeable to potassium. The excessive outflow of potassium ions carries tremendous numbers of positive charges to the outside of the membrane, creating inside the fiber considerably more negativity than would otherwise occur for a short period after the preceding action potential is over. That is, the excessive potassium permeability counterbalances the naturally high sodium permeability, thus drawing the membrane potential nearer to the potassium Nernst potential. This is a state called *hyperpolarization*, which is illustrated in Figure 10–12. As long as this state exists, re-excitation will not occur; but gradually the excess potassium conductance (and the state of hyperpolarization) disappears, thereby allowing the onset of a new action potential.

Figure 10–12 illustrates this relationship between repetitive action potentials and potassium conductance. The state of hyperpolarization is established immediately after each preceding action potential; but it gradually recedes, and the membrane potential correspondingly increases until it reaches the *threshold* for excitation; then

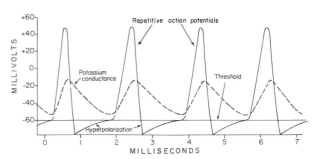

Figure 10–12. Rhythmic action potentials, and their relationship to potassium conductance and to the state of hyperpolarization.

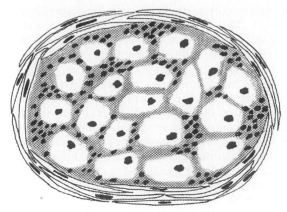

Figure 10–13. Cross-section of a small nerve trunk containing myelinated and unmyelinated fibers.

suddenly a new action potential results, the process occurring again and again.

SPECIAL ASPECTS OF IMPULSE TRANSMISSION IN NERVES

Myelinated and Unmyelinated Nerve Fibers. Figure 10–13 illustrates a cross-section of a typical small nerve trunk, showing a few very large nerve fibers that compose most of the cross-sectional area and many more small fibers lying between the large ones. The large fibers are *myelinated* and the small ones are *unmyelinated*. The average nerve trunk contains about twice as many unmyelinated fibers as myelinated fibers.

Figure 10–14 illustrates a typical myelinated fiber. The central core of the fiber is the *axon,* and the membrane of the axon is the actual *conductive membrane.* The axon is filled in its center with *axoplasm,* which is a viscid intracellular fluid. Surrounding the axon is a *myelin sheath* that is approximately as thick as the axon itself, and about once every millimeter along the extent of the axon the myelin sheath is interrupted by a *node of Ranvier.*

The myelin sheath is deposited around the axon by Schwann cells in the following manner: The membrane of a Schwann cell first envelops the axon. Then the cell rotates around the axon several times, laying down multiple layers of cellular membrane containing the lipid substance *sphingomyelin.* This substance is an excellent insulator that prevents almost all flow of ions. In fact, it increases the resistance to ion flow through the membrane approximately 5000-fold and also decreases the membrane capacitance as much as 1000-fold. However, at the juncture between each two successive Schwann cells along the axon, a small uninsulated area remains where ions can flow with ease between the extracellular fluid and the axon. This area is the node of Ranvier.

Saltatory Conduction in Myelinated Fibers from Node to Node. Even though ions cannot flow to a significant extent through the thick myelin sheaths of myelinated nerves, they can flow with considerable ease through the nodes of Ranvier. Indeed, the membrane at this point is 500 times as permeable as the membranes of some unmyelinated fibers.

Action potentials are conducted from node to node by the myelinated nerve rather than continuously along the entire fiber as occurs in the unmyelinated fiber. This process, illustrated in Figure 10–15, is called *saltatory conduction.* That is, electrical current flows through the surrounding extracellular fluids and also through the axoplasm from node to node, exciting successive nodes one after another. Thus, the impulse jumps down the fiber, which is the origin of the term "saltatory."

I mm. length

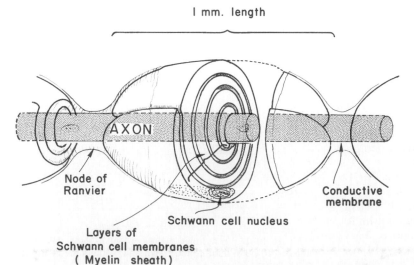

Figure 10–14. The myelin sheath and its formation by Schwann cells. (Modified from Elias and Pauley: Human Microanatomy. Philadelphia, F. A. Davis Company, 1966.)

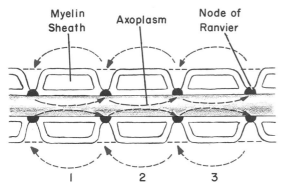

Figure 10–15. Saltatory conduction along a myelinated axon.

Saltatory conduction is of value for two reasons: First, by causing the depolarization process to jump long intervals along the axis of the nerve fiber, this mechanism greatly increases the velocity of nerve transmission in myelinated fibers. Second, saltatory conduction conserves energy for the axon, for only the nodes depolarize, allowing several hundred times less loss of ions than would otherwise be necessary and therefore requiring little extra metabolism for re-establishing the sodium and potassium concentration differences across the membrane after a series of nerve impulses.

VELOCITY OF CONDUCTION IN NERVE FIBERS

The velocity of conduction in nerve fibers varies from as little as 0.5 meter per second in very small unmyelinated fibers up to as high as 130 meters per second (the length of a football field) in very large myelinated fibers. The velocity increases approximately with the fiber diameter in myelinated nerve fibers and approximately with the square root of fiber diameter in unmyelinated fibers.

EXCITATION — THE PROCESS OF ELICITING THE ACTION POTENTIAL

Chemical Stimulation. Basically, any factor that causes sodium ions to begin to diffuse inward through the membrane in sufficient numbers will set off the automatic, regenerative opening of the sodium channels, as noted earlier in the chapter, that eventuates in the action potential. Thus, certain chemicals can stimulate a nerve fiber by increasing the membrane permeability. Such chemicals include acids, bases, almost any salt solution of very strong concentration, and, most importantly, the substance *acetylcholine*. Many nerve fibers, when stimulated, secrete acetylcholine at their endings where they synapse with other neurons or where they end on muscle fibers. The acetylcholine in turn stimulates the successive neuron or muscle fiber by opening pores in the membrane with diameters of 0.6 to 0.7 nanometer, large enough for sodium (as well as most other ions) to go through with ease. This is discussed in much greater detail in Chapter 12, and it is one of the most important means by which nerve and muscle fibers are stimulated. Likewise, *norepinephrine* secreted by sympathetic nerve endings can stimulate cardiac muscle fibers and some smooth muscle fibers, and still other hormonal substances can stimulate successive neurons in the central nervous system.

Mechanical Stimulation. Crushing, pinching, or pricking a nerve fiber can cause a sudden surge of sodium influx and, for obvious reasons, can elicit an action potential. Even slight pressure on some specialized nerve endings can stimulate these; this will be discussed in Chapter 48 in relation to sensory perception.

Electrical Stimulation. Electrical stimulation also can initiate an action potential. An electrical charge artificially induced across the membrane causes excess flow of ions through the membrane; this in turn can initiate an action potential. However, not all methods of applying electrical stimuli result in excitation, and, since this is the usual means by which nerve fibers are excited when they are studied in the laboratory, the process of electrical excitation deserves more detailed comment.

Cathodal Versus Anodal Currents. Figure 10–16 illustrates a battery connected to two electrodes on the surface of a nerve fiber. At the cathode, or negative electrode, the potential outside the membrane is negative with respect to that on the inside, and the positive current that flows outward through the membrane at this point is called *cathodal current*. At the anode, the electrode is positive with respect to the potential immediately inside the membrane, and the inward positive current flow at this point is called *anodal current*.

A cathodal current excites the fiber whereas an anodal current actually makes the fiber more resistant to excitation than normal. Though the cause of this difference between the two types of current cannot be explained completely, it is known that the normal impermeability of the membrane to sodium results partially from the high resting membrane potential across the membrane, and any condition that lessens this potential causes the membrane to become progressively more permeable to sodium. Obviously, at the cathode the applied voltage is opposite to the resting potential of the membrane, and this reduces the net potential. As a

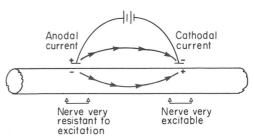

Figure 10–16. Effects of anodal and cathodal currents on excitability of the nerve membrane.

result, the membrane becomes far more permeable than usual to sodium followed by subsequent development of an action potential.

On the other hand, at the anode, the applied potential actually enhances the membrane potential. This makes the membrane less permeable to sodium than ever, resulting in increased resistance of the membrane to stimulation by other means.

Threshold for Excitation and "Acute Subthreshold Potential." A very weak cathodal potential cannot excite the fiber. But, when this potential is progressively increased, there comes a point at which excitation takes place. Figure 10–17 illustrates the effects of successively applied cathodal stimuli of progressing strength. A very weak stimulus at point A causes the membrane potential to change from −90 to −85 millivolts, but this is not a sufficient change for the automatic regenerative processes of the action potential to develop. At point B the stimulus is greater, but, here again, the intensity still is not enough to set off the automatic action potential. Nevertheless, the membrane voltage is disturbed for as long as a millisecond or more after both of the weak stimuli; the potential changes during these short intervals of time are called *acute subthreshold potentials,* as illustrated in the figure.

At point C in Figure 10–17 the stimulus elicits an acute membrane potential that is not subthreshold but slightly more than the threshold value, and, after a short "latent period," it initiates an action potential. At point D the stimulus is still stronger, and the acute membrane potential initiates the action potential even sooner. Thus, this figure shows that even a very weak stimulus always causes a local potential change at the membrane, but that the intensity of the *local potential* must rise to a *threshold value* before the automatic action potential will be set off.

"Accommodation" to Stimuli. When a cathodal potential applied to a nerve fiber is made to increase very slowly, rather than rapidly, the threshold voltage required to cause firing is considerably increased. This phenomenon is called *accommodation;* in other words, the excitable membrane is said to "accommodate" itself to slowly increasing potentials rather than firing. It is probable that a slowly increasing stimulatory current allows time for ions to build up (or become depleted) in the areas immediately adjacent to the fibers, these

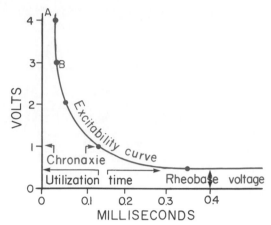

Figure 10–18. Excitability of a large myelinated nerve fiber.

changes in ion concentrations partially opposing the stimulus.

Excitability Curve of Nerve Fibers. A so-called "excitability curve" of a nerve fiber is shown in Figure 10–18. To obtain this curve a high voltage stimulus (4 volts, in this instance) is first applied to the fiber, and the minimum duration of stimulus required to excite the fiber is found. The voltage and minimal time are plotted as point A. Then a stimulus voltage of 3 volts is applied, and the minimal time required is again determined; the results are plotted as point B. The same is repeated at 2 volts, 1 volt, 0.5 volt, and so forth, until the least voltage possible at which the membrane is stimulated has been reached. On connection of these points, the excitability curve is generated.

The excitability curve of Figure 10–18 is that of a large myelinated nerve fiber. The least possible voltage at which it will fire is called the *rheobase,* and the time required for this least voltage to stimulate the fiber is called the *utilization time.* Then, if the voltage is increased to twice the rheobase voltage, the time required to stimulate the fiber is called the *chronaxie;* the chronaxie is often used as a means of expressing relative excitabilities of different excitable tissues. For instance, the chronaxie of a large type A fiber is about 0.0001 to 0.0002 second; of smaller myelinated nerve fibers, approximately 0.0003 second; of unmyelinated fibers, 0.0005 second; of skeletal muscle fibers, 0.00025 to 0.001 second; and of heart muscle, 0.001 to 0.003 second.

The Refractory Period. A second action potential cannot occur in an excitable fiber as long as the membrane is still depolarized from the preceding action potential. Therefore, even an electrical stimulus of maximum strength applied before the first spike potential is almost over will not elicit a second one. This interval of inexcitability is called the *absolute refractory period.* The absolute refractory period of large myelinated nerve fibers is about 1/2500 second. Therefore, one can readily calculate that such a fiber can carry a maximum of about 2500 impulses per second. Following the absolute refractory period is a *relative refractory period* lasting about one quarter as long. During this period, stronger than normal stimuli are required to

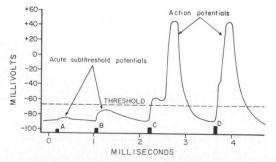

Figure 10–17. Effect of stimuli on the potential of the excitable membrane, showing the development of "acute subthreshold potentials" when the stimuli are below the threshold value required for eliciting an action potential.

excite the fiber. In some types of fibers, a short period of supernormal excitability follows the relative refractory period.

FACTORS THAT INCREASE MEMBRANE EXCITABILITY

Any condition that increases the natural permeability of the membrane usually causes it to become more excitable than usual. For instance, the drug *veratrine* has a direct action on the membrane to increase its permeability to sodium, and, as a consequence, the strength of stimulus needed to elicit an impulse is greatly reduced; or, on occasion, the fiber becomes so excitable that spontaneous impulses are generated without any extraneous excitation.

Low Calcium Tetany. An extremely important potentiator of excitability is low concentration of calcium ions in the extracellular fluids. Calcium ions normally decrease the permeability of the membrane to sodium. If sufficient calcium ions are not available, however, the permeability becomes increased, and, as a result, the membrane excitability greatly increases — sometimes so greatly that many spontaneous impulses result and cause muscular spasm. This condition is known as low calcium tetany. It often occurs in patients who have lost their parathyroid glands and who therefore cannot maintain normal calcium ion concentrations. This condition will be discussed in Chapter 79.

INHIBITION OF EXCITABILITY – "STABILIZERS" AND LOCAL ANESTHETICS

Contrary to the factors that increase excitability, still others called *membrane stabilizing factors* can decrease excitability. For instance, a *high calcium ion concentration* decreases the membrane permeability and simultaneously reduces its excitability. Therefore, calcium ions are said to be a "stabilizer." Also, *low potassium ion concentration* in the extracellular fluids, because it increases the negativity of the resting membrane potential (a process called "hyperpolarization"), likewise acts as a stabilizer and reduces membrane excitability. Indeed, in a hereditary disease known as *familial periodic paralysis,* the extracellular potassium ion concentration is often so greatly reduced that the person actually becomes paralyzed but reverts to normal instantly after intravenous administration of potassium.

Local Anesthetics and the "Safety Factor." Among the most important stabilizers are the many substances used clinically as local anesthetics, including *cocaine, procaine, tetracaine,* and many other drugs. These act directly on the membrane, decreasing its permeability to sodium and, therefore, also reducing membrane excitability. When the excitability has been reduced so low that the ratio of *action potential strength to excitability threshold* (called the "safety factor") is reduced below unity, a nerve impulse fails to pass through the anesthetized area.

RECORDING MEMBRANE POTENTIALS AND ACTION POTENTIALS

The Cathode Ray Oscilloscope. Earlier in this chapter we noted that the membrane potential changes very rapidly throughout the course of an action potential. Indeed, most of the action potential complex of large nerve fibers takes place in less than $1/1000$ second. In some figures of this chapter a meter has been shown recording these potential changes. However, it must be understood that any meter capable of recording them must be capable of responding extremely rapidly. For practical purposes the only type of meter that is capable of responding accurately to the very rapid membrane potential changes of most excitable fibers is the cathode ray oscilloscope.

Figure 10–19 illustrates the basic components of a cathode ray oscilloscope. The cathode ray tube itself is composed basically of an *electron gun* and a *fluorescent surface* against which electrons are fired. Where the electrons hit the surface, the fluorescent material glows.

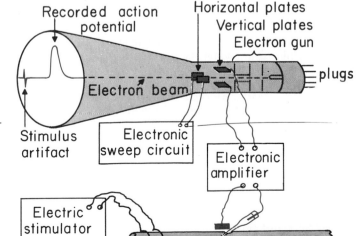

Figure 10–19. The cathode ray oscilloscope for recording transient action potentials.

If the electron beam is moved across the surface, the spot of glowing light also moves and draws a fluorescent line on the screen.

In addition to the electron gun and fluorescent surface, the cathode ray tube is provided with two sets of plates: one set, called the *horizontal deflection plates,* positioned on either side of the electron beam, and the other set, called the *vertical deflection plates,* positioned above and below the beam. If a negative charge is applied to the left-hand plate and a positive charge to the right-hand plate, the electron beam will be repelled away from the left plate and attracted toward the right plate, thus bending the beam toward the right, and this will cause the spot of light on the fluorescent surface of the cathode ray screen to move to the right. Likewise, positive and negative charges can be applied to the vertical deflection plates to move the beam up or down.

Since electrons travel at extremely rapid velocity and since the plates of the cathode ray tube can be alternately charged positively or negatively within less than a millionth of a second, it is obvious that the spot of light on the face of the tube can also be moved to almost any position in less than a millionth of a second. For this reason, the cathode ray tube oscilloscope can be considered to be an inertialess meter capable of recording with extreme fidelity almost any change in membrane potential.

To use the cathode ray tube for recording action potentials, two electrical circuits must be employed. These are (1) an *electronic sweep circuit* that controls the voltages on the horizontal deflection plates and (2) an *electronic amplifier* that controls the voltages on the vertical deflection plates. The sweep circuit automatically causes the spot of light to begin at the left-hand side and move slowly toward the right. When the spot reaches the right side it jumps back immediately to the left-hand side and starts a new trace.

The electronic amplifier amplifies signals that come from the nerve. If a change in membrane potential occurs while the spot of light is moving across the screen, this change in potential will be amplified and will cause the spot to rise above or fall below the mean level of the trace, as illustrated in the figure. In other words, the sweep circuit provides the lateral movement of the electron beam while the amplifier provides the vertical movement in direct proportion to the changes in membrane potentials picked up by appropriate electrodes.

Figure 10–19 also shows an electric stimulator used to stimulate the nerve. When the nerve is stimulated, a small *stimulus artifact* usually appears on the oscilloscope screen prior to the action potential.

Recording the Monophasic Action Potential. Throughout this chapter "monophasic" action potentials have been shown in the different diagrams. To record these, an electrode such as that illustrated earlier in the chapter in Figure 10–3 must be inserted into the interior of the fiber. Then, as the action potential spreads down the fiber, the changes in the potential inside the fiber are recorded as illustrated earlier in the chapter in Figures 10–5, 10–6, and 10–11.

Recording a Biphasic Action Potential. When one wishes to record impulses from a whole nerve trunk, it is not feasible to place electrodes inside the nerve fibers. Therefore, the usual method of recording is to place two electrodes on the outside of fibers. However, the record that is obtained is then biphasic for the following reasons: When an action potential moving down the nerve fiber reaches the first electrode, it becomes charged negatively while the second electrode is still unaffected. This causes the oscilloscope to record in the negative direction. Then as the action potential is proceeding down the nerve, there comes a point when the membrane beneath the first electrode becomes repolarized while the second electrode is still negative, and the oscilloscope records in the opposite direction. When these changes are recorded by the oscilloscope, a graphic record such as that illustrated in Figure 10–20 is recorded, showing a potential change first in one direction and then in the opposite direction.

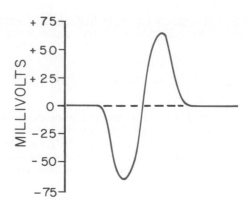

Figure 10–20.   Recording of a biphasic action potential.

REFERENCES

Adrian, R. H.: Charge movement in the membrane of striated muscle. *Annu. Rev. Biophys. Bioeng.,* 7:85, 1978.

Agin, D.: Negative conductance and electrodiffusion in excitable membrane systems. *Membranes,* 1:249, 1972.

Aguayo, A. J., and Karpati, G. (eds.): Current Topics in Nerve and Muscle Research. New York, Elsevier/North-Holland, 1979.

Almers, W.: Gating currents and charge movements in excitable membranes. *In* Adrian, R. H., *et al.* (eds.): Reviews of Physiology, Biochemistry, and Pharmacology. New York, Springer-Verlag, 1978, p. 96.

Asbury, A. K., and Johnson, P. C.: Pathology of Peripheral Nerve. Philadelphia, W. B. Saunders Co., 1978.

Baker, P. F., and Reuter, H.: Calcium Movement in Excitable Cells. New York, Pergamon Press, 1975.

Baker, P. F., *et al.*: Replacement of the axoplasm of giant nerve fibers with artificial solutions. *J. Physiol. (Lond.),* 164:330, 1962.

Baker, P. F., *et al.*: The effects of changes in internal ionic concentrations on the electrical properties of perfused giant axons. *J. Physiol. (Lond.),* 164:355, 1962.

Brinley, F. J., Jr.: Calcium buffering in squid axons. *Annu. Rev. Biophys. Bioeng.,* 7:363, 1978.

Caldwell, P. C.: Factors governing movement and distribution of inorganic ions in nerve and muscle. *Physiol. Rev.,* 48:1, 1968.

Carmeliet, E., and Vereecke, J.: Electrogenesis of the action potential and automaticity. *In* Berne, R. M., *et al.* (eds.): Handbook of Physiology. Sec. 2, Vol. 1. Baltimore, Williams & Wilkins, 1979, p. 269.

Ceccarelli, B., and Clementi, F. (eds.): Neurotoxins, Tools in Neurobiology. New York, Raven Press, 1979.

Cohen, L. B.: Changes in neuron structure during action potential propagation and synaptic transmission. *Physiol. Rev.,* 53:373, 1973.

Cohen, L. B., and De Weer, P.: Structural and metabolic processes directly related to action potential propagation. *In* Brookhart, J. M., and Mountcastle, V. B. (eds.): Handbook of Physiology. Sec. 1, Vol. 1, Baltimore, Williams & Wilkins, 1977, p. 137.

Cole, K. S.: Electrodiffusion models for the membrane of squid giant axon. *Physiol. Rev., 45*:340, 1965.

Cuénod, M., *et al.* (eds.): Development and Chemical Specificity of Neurons. New York, Elsevier/North-Holland, 1979.

De Weer, P.: Aspects of the recovery process in nerve. *In* MTP International Review of Science: Physiology. Baltimore, University Park Press, 1975, Vol. 3, p. 231.

Ehrenstein, G., and Lecar, H.: The mechanism of signal transmission in nerve axons. *Annu. Rev. Biophys. Bioeng., 1*:347, 1972.

Finkelstein, A., and Mauro, A.: Physical principles and formalisms of electrical excitability. *In* Brookhart, J. M., and Mountcastle, V. B. (eds.): Handbook of Physiology. Sec. 1, Vol. 1. Baltimore, Williams & Wilkins, 1977, p. 161.

Fozzard, H. A.: Conduction of the action potential. *In* Berne, R. M., *et al.* (eds.): Handbook of Physiology. Sec. 2, Vol. 1., Baltimore, Williams & Wilkins, 1979, p. 335.

Glynn, I. M., and Karlish, S. J. D.: The sodium pump. *Annu. Rev. Physiol., 37*:13, 1975.

Greene, L. A., and Shooter, E. M.: The nerve growth factor: Biochemistry, synthesis, and mechanism of action. *Annu. Rev. Neurosci., 3*:353, 1980.

Guroff, G.: Molecular Neurobiology. New York, Marcel Dekker, 1979.

Hille, B.: Gating in sodium channels of nerve. *Annu. Rev. Physiol., 38*:139, 1976.

Hille, B.: Ionic basis of resting and action potentials. *In* Brookhart, J. M., and Mountcastle, V. B. (eds.): Handbook of Physiology. Sec. 1, Vol. 1. Baltimore, Williams & Wilkins, 1977, p. 99.

Hodgkin, A. L.: The Conduction of the Nervous Impulse. Springfield, Ill., Charles C Thomas, 1963.

Hodgkin, A. L., and Horowicz, P.: The effect of sudden changes in ionic concentrations on the membrane potential of single muscle fibres. *J. Physiol. (Lond.), 153*:370, 1960.

Hodgkin, A. L., and Huxley, A. F.: Movement of sodium and potassium ions during nervous activity. *Cold Spr. Harb. Symp. Quant. Biol., 17*:43, 1952.

Hodgkin, A. L., and Huxley, A. F.: Quantitative description of membrane current and its application to conduction and excitation in nerve. *J. Physiol. (Lond.), 117*:500, 1952.

Jack, J. J. B., *et al.*: Electric Current Flow in Excitable Cells. New York, Oxford University Press, 1975.

Jewett, D. L., and McCarroll, H. R., (eds.): Nerve Repair and Regeneration: Its Clinical and Experimental Basis. St. Louis, C. V. Mosby, 1979.

Jones, A. W.: Content and fluxes of electrolytes. *In* Bohr, D. F., *et al.* (eds.): Handbook of Physiology. Sec. 2, Vol. II. Baltimore, Williams and Wilkins, 1980, p. 253.

Katz, B.: Nerve, Muscle, and Synapse. New York, McGraw-Hill, 1968.

Keynes, R. D.: Ion channels in the nerve-cell membrane. *Sci. Am. 240*(3):126, 1979.

Kristensson, K.: Retrograde transport of macromolecules in axons. *Annu. Rev. Pharmacol. Toxicol., 18*:97, 1978.

Landowne, D., *et al.*: Structure-function relationships in excitable membranes. *Annu. Rev. Physiol., 37*:485, 1975.

Menaker, M., *et al.*: The physiology of circadian pacemakers. *Annu. Rev. Physiol., 40*:501, 1978.

Montal, M.: Experimental membranes and mechanisms of bioenergy transductions. *Annu. Rev. Biophys. Bioeng., 5*:119, 1976.

Mystrom, R. A.: Membrane Physiology. Englewood Cliffs, N.J., Prentice-Hall, 1973.

Narahashi, T.: Chemicals as tools in the study of excitable membranes. *Physiol. Rev., 54*:813, 1974.

Neher, E., and Stevens, C. F.: Conductance fluctuations and ionic pores in membranes. *Annu. Rev. Biophys. Bioeng., 6*:345, 1977.

Nelson, P. G.: Nerve and muscle cells in culture. *Physiol. Rev., 55*:1, 1975.

Noble, D.: Applications of Hodgkin-Huxley equations to excitable tissues. *Physiol. Rev., 46*:1, 1966.

Palay, S. L., and Chan-Palay, V.: General morphology of neurons and neuroglia. *In* Brookhart, J. M., and Mountcastle, V. B. (eds.): Handbook of Physiology. Sec. 1, Vol. 1. Baltimore, Williams & Wilkins, 1977, p. 5.

Patrick, J., *et al.*: Biology of cultured nerve and muscle. *Annu. Rev. Neurosci., 1*:417, 1978.

Pfenninger, K. H.: Organization of neuronal membrane. *Annu. Rev. Neurosci., 1*:445, 1978.

Rall, W.: Core conductor theory and cable properties of neurons. *In* Brookhart, J. M., and Mountcastle, V. B. (eds.): Handbook of Physiology. Sec. 1, Vol. 1. Baltimore, Williams & Wilkins, 1977, p. 39.

Ritchie, J. M.: A pharmacological approach to the structure of sodium channels in myelinated axons. *Annu. Rev. Neurosci., 2*:341, 1979.

Schultz, S. G.: Principles of electrophysiology and their application to epithelial tissues. *In* MTP International Review of Science: Physiology. Baltimore, University Park Press, 1974, Vol. 4, p. 69.

Schwartz, J. H.: Axonal transport: components, mechanisms, and specificity. *Annu. Rev. Neurosci., 2*:476, 1979.

Shanes, A. M.: Electrochemical aspects of physiological and pharmacological action in excitable cells. *Pharmacol. Rev., 10*:59, 165, 1958.

Somjem, G. G.: Electrogenesis of sustained potentials. *Prog. Neurobiol., 1*:201, 1973.

Spitzer, N. C.: Ion channels in development. *Annu. Rev. Neurosci., 2*:363, 1979.

Stevens, C. F.: The neuron. *Sci. Am., 241*(3):54, 1979.

Tasaki, I., and Hallett, M.: Bioenergetics of nerve excitation. *J. Bioenerg., 3*:65, 1972.

Trautwein, W.: Membrane currents in cardiac muscle fibers. *Physiol. Rev., 53*:793, 1973.

Ulbricht, W.: Ionic channels and gating currents in excitable membranes. *Annu. Rev. Biophys. Bioeng., 6*:7, 1977.

Varon, S. S., and Bunge, R. P.: Trophic mechanisms in the peripheral nervous system. *Annu. Rev. Neurosci., 1*:327, 1978.

Waggoner, A. S.: Dye indicators of membrane potential. *Annu. Rev. Biophys. Bioeng., 8*:47, 1979.

Walker, J. L., and Brown, H. M.: Intracellular ionic activity measurements in nerve and muscle. *Physiol. Rev., 57*:729, 1977.

Wallick, E. T., *et al.*: Biochemical mechanism of the sodium pump. *Annu. Rev. Physiol., 41*:397, 1979.

Waxman, S. G., (ed.): Physiology and Pathobiology of Axons. New York, Raven Press, 1978.

Zachar, J.: Electrogenesis and Contractility in Skeletal Muscle Cells. Baltimore, University Park Press, 1973.

11

Contraction of Skeletal Muscle

Approximately 40 per cent of the body is skeletal muscle and another 5 to 10 per cent is smooth and cardiac muscle. Many of the same principles of contraction apply to all these different types of muscle, but in the present chapter the function of skeletal muscle is considered mainly, while the specialized functions of smooth muscle will be discussed in the following chapter, and cardiac muscle in Chapter 13.

PHYSIOLOGIC ANATOMY OF SKELETAL MUSCLE

THE SKELETAL MUSCLE FIBER

Figure 11–1 illustrates the organization of skeletal muscle, showing that all skeletal muscles are made of numerous fibers ranging between 10 and 80 microns in diameter. Each of these fibers in

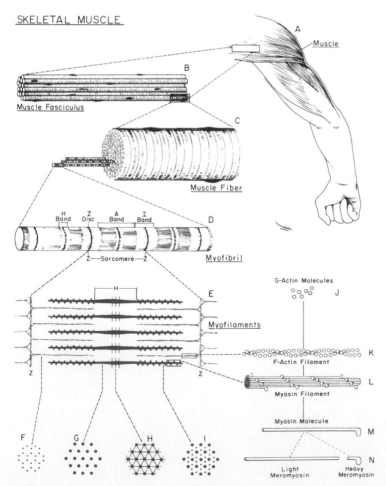

Figure 11–1. Organization of skeletal muscle, from the gross to the molecular level. F, G, H, and I are cross-sections at the levels indicated. (Drawing by Sylvia Colard Keene. From Bloom and Fawcett: A Textbook of Histology. Philadelphia, W. B. Saunders Company, 1975.)

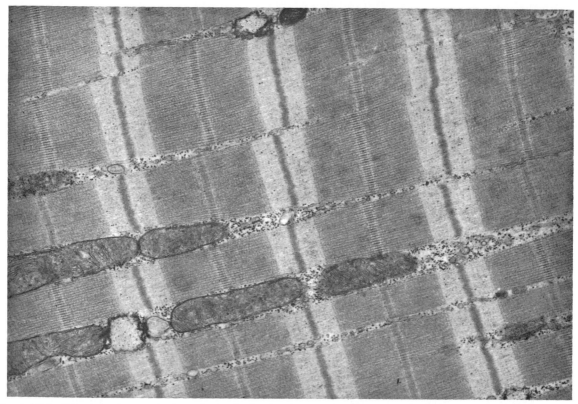

Figure 11–2. Electron micrograph of muscle myofibrils, showing the detailed organization of actin and myosin filaments. Note the mitochondria lying between the myofibrils. (From Fawcett: The Cell. Philadelphia, W. B. Saunders Company, 1966).

turn is made up of successively smaller subunits, also illustrated in Figure 11–1, that will be described in subsequent paragraphs.

In most muscles the fibers extend the entire length of the muscle, and, except for about 2 per cent of the fibers, each is innervated by only one nerve ending, located near the middle of the fiber.

The Sarcolemma. The sarcolemma is the cell membrane of the muscle fiber. However, the sarcolemma consists of a true cell membrane, called the *plasma membrane,* and a thin layer of polysaccharide material similar to that of the basement membrane surrounding blood capillaries; thin collagen fibrillae are also present in the outer layer of the sarcolemma. At the ends of the muscle fibers, these surface layers of the sarcolemma fuse with tendon fibers, which in turn collect into bundles to form the muscle tendons, and thence insert into the bones.

Myofibrils; Actin and Myosin Filaments. Each muscle fiber contains several hundred to several thousand *myofibrils,* which are illustrated by the many small open dots in the cross-sectional view of Figure 11–1C. Each myofibril (Figure 11–1D) in turn has, lying side-by-side, about 1500 *myosin filaments* and 3000 *actin filaments,* which are large polymerized protein molecules that are respons-

ible for muscle contraction. These can be seen in longitudinal view in the electron micrograph of Figure 11–2, and are represented diagrammatically in Figure 11–1E. The thick filaments are *myosin* and the thin filaments are *actin.* Note that the myosin and actin filaments partially interdigitate and thus cause the myofibrils to have alternate light and dark bands. The light bands, which contain only actin filaments, are called *I bands* because they are mainly *isotropic* to polarized light. The dark bands, which contain the myosin filaments as well as the ends of the actin filaments where they overlap the myosin, are called *A bands* because they are *anisotropic* to polarized light. Note also the small projections from the sides of the myosin filaments. These are called *cross-bridges.* They protrude from the surfaces of the myosin filaments along the entire extent of the filament, except in the very center. It is interaction between these cross-bridges and the actin filaments that causes contraction.

Figure 11–1E also shows that the actin filaments are attached to the so-called *Z membrane* or *Z disc,* and the filaments extend on either side of the Z membrane to interdigitate with the myosin filaments. The Z membrane also passes from myofibril to myofibril, attaching the myofibrils to each other all the way across the muscle fiber. Therefore, the entire muscle fiber has light and dark bands, as is also true of the individual myofibrils. It is these bands that give skeletal and cardiac muscle their striated appearance.

The portion of a myofibril (or of the whole muscle fiber) that lies between two successive Z membranes is called a *sarcomere*. When the muscle fiber is at its normal fully stretched resting length, the length of the sarcomere is about 2.0 microns. At this length, the actin filaments completely overlap the myosin filaments and are just beginning to overlap each other. We shall see later that it is at this length that the sarcomere also is capable of generating its greatest force of contraction.

When a muscle fiber is stretched beyond its resting length, as it is in Figure 11–1, the ends of the actin filaments pull apart, leaving a light area in the center of the A band. This light area, called the *H zone,* is illustrated in Figure 11–2. Such an H zone rarely occurs in the normally functioning muscle because normal sarcomere contraction occurs when the length of the sarcomere is between 2.0 microns and 1.6 microns. In this range the ends of the actin filaments not only overlap the myosin filaments but also overlap each other.

The Sarcoplasm.   The myofibrils are suspended inside the muscle fiber in a matrix called *sarcoplasm,* which is composed of usual intracellular constituents. The fluid of the sarcoplasm contains large quantities of potassium, magnesium, phosphate, and protein enzymes. Also present are tremendous numbers of *mitochondria* that lie between and parallel to the myofibrils, a condition which is indicative of the great need of the contracting myofibrils for large amounts of ATP formed by the mitochondria.

The Sarcoplasmic Reticulum.   Also in the sarcoplasm is an extensive endoplasmic reticulum, which in the muscle fiber is called the *sarcoplasmic reticulum*. This reticulum has a special organization that is extremely important in the control of muscle contraction, which will be discussed later in the chapter. The electron micrograph of Figure 11–3 illustrates the arrangement of this sarcoplasmic reticulum and shows how extensive it can be. The more rapidly contracting types of muscle have especially extensive sarcoplasmic reticulums, indicating that this structure is important in causing rapid muscle contraction, as will also be discussed later.

MOLECULAR MECHANISM OF MUSCLE CONTRACTION

Sliding Mechanism of Contraction.   Figure 11–4 illustrates the basic mechanism of muscle contraction. It shows the relaxed state of a sarcomere (above) and the contracted state (below). In the relaxed state, the ends of the actin filaments derived from two successive Z membranes barely overlap each other while at the same time completely overlapping the myosin filaments. On the other hand, in the contracted state these actin filaments have been pulled inward among the myosin filaments so that they now overlap each other to a major extent. Also, the Z membranes have been pulled by the actin filaments up to the ends of the myosin filaments. Indeed, the actin filaments can be pulled together so tightly that the ends of the myosin filaments actually buckle during very intense contraction. Thus, muscle contraction occurs by a *sliding filament mechanism*.

But what causes the actin filaments to slide inward among the myosin filaments? Unfortunate-

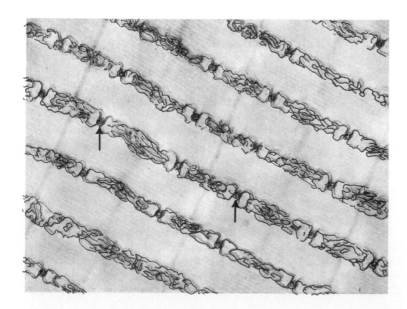

Figure 11–3. Sarcoplasmic reticulum surrounding the myofibril, showing the longitudinal system paralleling the myofibrils. Also shown in cross-section are the T tubules that lead to the exterior of the fiber membrane and that contain extracellular fluid (arrows). (From Fawcett: The Cell. Philadelphia, W. B. Saunders Company, 1966.)

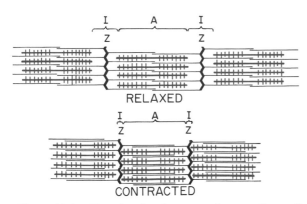

Figure 11–4. The relaxed and contracted states of a myofibril, showing sliding of the actin filaments into the channels between the myosin filaments.

ly, we do not completely know the answer to this question. Yet we do know that it is caused by attractive forces that develop between the actin and myosin filaments. Almost certainly, these attractive forces are the result of mechanical, chemical, or electrostatic forces generated by the interaction of the cross-bridges of the myosin filaments with the actin filaments.

Under resting conditions, the attractive forces between the actin and myosin filaments are inhibited, but when an action potential travels over the muscle fiber membrane, this causes the release of large quantities of calcium ions into the sarcoplasm surrounding the myofibrils. These calcium ions activate the attractive forces between the filaments and contraction begins. But energy is also needed for the contractile process to proceed. This energy is derived from the high energy bonds of adenosine triphosphate (ATP), which is degraded to adenosine diphosphate (ADP) to give the energy required.

In the next few sections we will describe what is known about the details of the molecular processes of contraction. To begin this discussion, however, we must first characterize in detail the myosin and actin filaments.

MOLECULAR CHARACTERISTICS OF THE CONTRACTILE FILAMENTS

The Myosin Filament. The myosin filament is composed of approximately 200 myosin molecules, each having a molecular weight of 490,000. Figure 11–5, section A illustrates an individual molecule; section B illustrates the organization of the molecules to form a myosin filament as well as its interaction with the ends of two actin filaments.

The myosin molecule is composed of two parts; *light meromyosin* and *heavy meromyosin*. The light meromyosin consists of two peptide strands wound around each other in a helix. The heavy meromyosin in turn consists of two parts: first, a double helix similar to that of the light meromyosin; second, a *head* attached to the end of the double helix. The head itself is a composite of two globular protein masses.

It is believed that the myosin molecule is especially flexible at two points — the juncture between the light meromyosin and the heavy meromyosin and between the body of the heavy meromyosin and the head. These two areas are called *hinges*.

In section B of Figure 11–5 the central portion of a myosin filament is illustrated. The body of this filament is composed of parallel strands of light meromyosin from multiple myosin molecules. In fact, whenever myosin molecules are precipitated from solution, it is found that the light meromyosin portions of the myosin molecules have a natural tendency to aggregate together to form filaments almost precisely like those of the myosin filaments found in muscle. On the other hand, the heavy meromyosin portions of myosin molecules protrude from all sides of the myosin filament, as illustrated in the figure. These protrusions constitute the *cross-bridges*. The heads of the cross-bridges lie in apposition to the actin filaments, whereas the helix portions of the cross-bridges act as hinged arms that allow the heads to extend either far outward from the body of the myosin filament or to lie close to the body.

Note also that the arms of the cross-bridges extend in both directions away from the center-

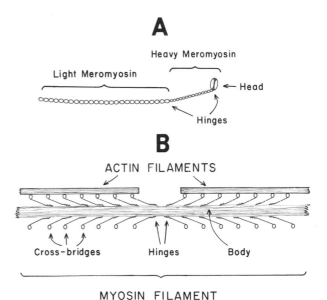

Figure 11–5. A, The myosin molecule. B, Combination of many myosin molecules to form a myosin filament. Also shown are the cross-bridges and the interaction between the heads of the cross-bridges and adjacent actin filaments.

most part of the filament. Therefore, in the very center of the myosin filament, for a length of about 0.2 micron, there are no cross-bridge heads.

The total length of the myosin filament is 1.6 microns, and the 200 myosin molecules allow the formation of 100 pairs of cross-bridges — 50 pairs on each end of the myosin filament.

Now, to complete the picture, the myosin filament is twisted so that it makes one complete revolution for each three pairs of cross-bridges. Each twist has a length of 42.9 nanometers with each pair of cross-bridges axially displaced from the previous pair by 120 degrees.

The Actin Filament. The actin filament is also complex. It is composed of three different components; *actin, tropomyosin,* and *troponin.*

The backbone of the actin filament is a double-stranded F-actin protein molecule, illustrated in Figure 11–6. The two strands are wound in a helix in the same manner as the myosin molecule, but with a complete revolution every 70 nanometers.

Each strand of the double F-actin helix is composed of polymerized G-actin molecules, each having a molecular weight of 47,000. There are approximately 13 of these molecules in each revolution of each strand of helix. Attached to each one of the G-actin molecules is one molecule of ADP. It is believed that these ADP molecules are the active sites on the actin filaments with which the cross-bridges of the myosin filaments interact to cause muscle contraction. The active sites on the two F-actin strands of the double helix are staggered, giving one active site on the overall actin filament approximately every 2.7 nanometers.

The Tropomyosin Strands. The actin filament also contains two additional protein strands that are polymers of *tropomyosin* molecules, each molecule having a molecular weight of 70,000 and extending a length of 40 nanometers. It is believed that each tropomyosin strand is loosely attached to an F-actin strand and that in the resting state it physically covers the active sites of the actin strands so that interaction cannot occur between the actin and myosin to cause contraction.

Troponin and Its Role in Muscle Contraction. Attached approximately two-thirds the distance along each tropomyosin molecule is a com-

plex of three globular protein molecules called *troponin.* One of the globular proteins has a strong affinity for actin, another for tropomyosin, and a third for calcium ions. This complex is believed to attach the tropomyosin to the actin. The strong affinity of the troponin for calcium ions is believed to initiate the contraction process, as will be explained in the following section.

Interaction of Myosin and Actin Filaments to Cause Contraction

Inhibition of the Actin Filament by the Troponin-Tropomyosin Complex; Activation by Calcium Ions. A pure actin filament without the presence of the troponin-tropomyosin complex binds strongly with myosin molecules in the presence of magnesium ions and ATP, both of which are normally abundant in the myofibril. But, if the troponin-tropomyosin complex is added to the actin filament, this binding does not take place. Therefore, it is believed that the normal active sites on the normal actin filament of the relaxed muscle are inhibited (or perhaps physically covered) by the troponin-tropomyosin complex. Consequently, they cannot interact with the myosin filaments to cause contraction. Before contraction can take place the inhibitory effect of the troponin-tropomyosin complex must itself be inhibited.

Now, let us discuss the role of the calcium ions. In the presence of large amounts of calcium ions the inhibitory effect of the troponin-tropomyosin on the actin filaments is itself inhibited. The mechanism of this is not known, but one suggestion is the following: When calcium ions combine with the calcium binding subunit of troponin, which has an extremely high affinity for calcium ions even when they are present in minute quantities, the troponin complex supposedly undergoes a conformational change that in some way tugs on the tropomyosin protein strand. At the same time, the bonds between the troponin and the actin become loosened. This combination of effects supposedly moves the tropomyosin strand deeper into the groove between the two actin strands and thereby "uncovers" the active sites of the actin, thus allowing contraction to proceed. Though this is a hypothetical mechanism, nevertheless it does emphasize that the normal relationship between the tropomyosin-troponin complex and actin is altered by calcium ions — a condition which leads to contraction.

Interaction Between the "Activated" Actin Filament and the Myosin Molecule — The Ratchet Theory of Contraction. As soon as the actin filament becomes activated by the calcium ions, it is believed that the heads of the cross-bridges from the myosin filaments immediately become attracted to the active sites of the actin filament, and this in

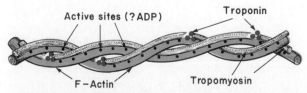

Figure 11–6. The actin filament, composed of two helical strands of F-actin and two tropomyosin strands that lie in the grooves between the actin strands. Attaching the tropomyosin to the actin are several troponin complexes.

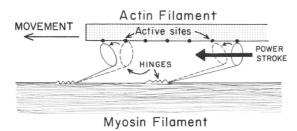

Figure 11–7. The ratchet mechanism for contraction of the muscle.

some way causes contraction to occur. Though the precise manner by which this interaction between the cross-bridges and the actin causes contraction is still unknown, a suggested hypothesis for which considerable circumstantial evidence exists is the so-called *ratchet theory of contraction.*

Figure 11–7 illustrates the postulated ratchet mechanism for contraction. This figure shows the heads of two cross-bridges attaching to and disengaging from the active sites of an actin filament. It is postulated that when the head attaches to an active site this attachment simultaneously causes profound changes in the intramolecular forces in the head and arm of the cross-bridge. The new alignment of forces causes the head to tilt toward the arm, and to drag the actin filament along with it. This tilt of the head of the cross-bridge is called the *power stroke.* Then, immediately after tilting, the head automatically splits away from the active site and returns to its normal perpendicular direction. In this position it combines with an active site further down along the actin filament; then, a similar tilt takes place again to cause a new power stroke, and the actin filament moves another step. Thus, the heads of the cross-bridges bend back and forth and step by step pull the actin filament toward the center of the myosin filament. Thus, the movements of the cross-bridges use the active sites of the actin filaments as cogs of a *ratchet.*

Each one of the cross-bridges is believed to operate independently of all others, each attaching and pulling in a continuous, alternating ratchet cycle. Therefore, the greater the number of cross-bridges in contact with the actin filament at any given time, the greater, theoretically, is the force of contraction.

ATP as the Source of Energy for Contraction — Chemical Events in the Ratchet Cycle. When a muscle contracts against a load, work is performed, and energy is required. It is found that large amounts of ATP are cleaved to form ADP during the contraction process. Furthermore, the greater the amount of work performed by the muscle, the greater is the amount of ATP that is cleaved, which is called the *Fenn effect.* However, unfortunately, it is still not known exactly how

ATP is used to provide the energy for contraction. Yet, the following is a sequence of events that has been suggested as the means by which this occurs:

1. When the inhibitory effect of the troponin-tropomyosin complex has itself been inhibited by calcium ions, the heads of the cross-bridges bind with uncovered sites on the actin filament, as illustrated in Figure 11–7.

2. It is assumed that the bond between the head of the cross-bridge and the active site of the actin filament causes a conformational change in the head, thus causing the head to tilt and to provide the power stroke for pulling the actin filament. This power stroke is believed to result from energy that has already been stored in the heavy meromyosin molecule and not from energy derived from ATP.

3. Once the head of the cross-bridge is tilted, the conformational change in the head exposes a site in the head where ATP can bind. Therefore, one molecule of ATP binds with the head, and this binding in turn causes detachment of the head from the active site.

4. Once the head bound with ATP splits away from the active site, the ATP is itself cleaved by a very potent ATPase activity of the heavy meromyosin. The energy released supposedly tilts the head back to its normal perpendicular condition and theoretically "cocks" the head in this position.

5. Then, when the "cocked" head, with its stored energy derived from the cleaved ATP, binds with a new active site on the actin filament, it becomes uncocked and once again provides the power stroke.

6. Thus, the process proceeds again and again until the actin filament pulls the Z membrane up against the ends of the myosin filaments or until the load on the muscle becomes too great for further pulling to occur.

RELATIONSHIP BETWEEN ACTIN AND MYOSIN FILAMENT OVERLAP AND TENSION DEVELOPED BY THE CONTRACTING MUSCLE

Figure 11–8 illustrates the relationship between the length of sarcomere and the tension developed by a single, contracting, *isolated* muscle fiber. To the right are illustrated different degrees of overlap of the myosin and actin filaments at different sarcomere lengths. At point D on the diagram, the actin filament has pulled all the way out to the end of the myosin filament with no overlap at all. At this point, the tension developed by the activated muscle is zero. Then, as the sarcomere shortens and the actin filament overlaps the myosin filament progressively more and more, the tension

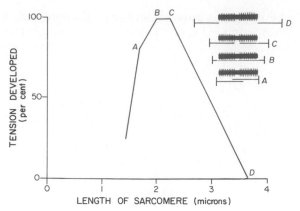

Figure 11–8. Length-tension diagram for a single sarcomere, illustrating maximum strength of contraction when the sarcomere is 2.0 to 2.2 microns in length. At the upper right are shown the relative positions of the actin and myosin filaments at different sarcomere lengths from point A to point D. (Modified from Gordon, Huxley, and Julian: *J. Physiol., 171*:28P, 1964.)

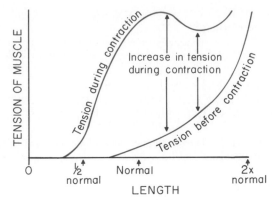

Figure 11–9. Relation of muscle length to force of contraction.

increases progressively until the sarcomere length decreases to about 2.2 microns. At this point the actin filament has already overlapped all the cross-bridges of the myosin filament but has not yet reached the center of the myosin filament. Upon further shortening, the sarcomere maintains full tension until point B at a sarcomere length of approximately 2.0 microns. It is at this point that the ends of the two actin filaments begin to overlap. As the sarcomere length falls from 2 microns down to about 1.65 microns, at point A, the strength of contraction decreases. It is at this point that the two Z membranes of the sarcomere abut the ends of the myosin filaments. Then, as contraction proceeds to still shorter sarcomere lengths, the ends of the myosin filaments are actually crumpled, and as illustrated in Figure 11–8, the strength of contraction also decreases precipitously.

This diagram illustrates that maximum contraction occurs when there is maximum overlap between the actin filaments and the cross-bridges of the myosin filaments, and it supports the idea that the greater the number of cross-bridges pulling the actin filaments, the greater is the strength of contraction.

Relation of Force of Contraction of the Intact Muscle to Muscle Length. The upper curve of Figure 11–9 is similar to that in Figure 11–8, but this illustrates the intact whole muscle rather than the isolated muscle fiber. The whole muscle has a large amount of connective tissue in it; also, the sarcomeres in different parts of the muscle do not necessarily contract exactly in unison. Therefore, the curve has somewhat different dimensions from those illustrated for the individual muscle fiber, but it nevertheless exhibits the same form.

Note in Figure 11–9 that when the muscle is at its normal resting length and is then activated, it contracts with maximum force of contraction. If the muscle is stretched to much greater than normal length prior to contraction, a large amount of *resting tension* develops in the muscle even before contraction takes place; this tension results from the elastic forces of the connective tissue, of the sarcolemma, the blood vessels, the nerves, and so forth. However, the *increase* in tension during contraction, called *active tension,* decreases as the muscle is stretched beyond its normal length.

Note also in Figure 11–9 that when the resting muscle is shortened to less than its normal length, the maximum tension of contraction decreases progressively and reaches zero when the muscle has shortened to approximately half its normal resting length.

RELATION OF VELOCITY OF CONTRACTION TO LOAD

A muscle contracts extremely rapidly when it contracts against no load — to a state of full

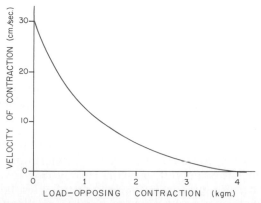

Figure 11–10. Relation of load to velocity of contraction in a skeletal muscle 8 cm. long.

contraction in approximately $1/20$ second for the average muscle. However, when loads are applied, the velocity of contraction becomes progressively less as the load increases, as illustrated in Figure 11–10. When the load increases to equal the maximum force that the muscle can exert, then the velocity of contraction becomes zero, and no contraction at all results, despite activation of the muscle fiber.

This decreasing velocity with load seems to be caused mainly by the fact that a load on a contracting muscle is a reverse force that opposes the contractile force caused by muscle contraction. Therefore, the net force that is available to cause velocity of shortening is correspondingly reduced.

INITIATION OF MUSCLE CONTRACTION: EXCITATION-CONTRACTION COUPLING

THE MUSCLE ACTION POTENTIAL

Initiation of contraction in skeletal muscle begins with action potentials in the muscle fibers. These elicit electrical currents that spread to the interior of the fiber where they cause release of calcium ions from the sarcoplasmic reticulum. It is the calcium ions that in turn initiate the chemical events of the contractile process.

Almost everything discussed in Chapter 10 regarding initiation and conduction of action potentials in nerve fibers applies equally well to skeletal muscle fibers, except for quantitative differences. Some of the quantitative aspects of muscle potentials are the following:

1. Resting membrane potential: Approximately −90 millivolts in skeletal fibers — the same as in large myelinated nerve fibers.

2. Duration of action potential: 1 to 5 milliseconds in skeletal muscle — about five times as long as large myelinated nerves.

3. Velocity of conduction: 3 to 5 meters per second — about $1/18$ the velocity of conduction in the large myelinated nerve fibers that excite skeletal muscle.

Excitation of Skeletal Muscle Fibers by Nerves.   In normal function of the body, skeletal muscle fibers are excited by large myelinated nerve fibers. These attach to the skeletal muscle fibers at the neuromuscular junction, which will be discussed in detail in the following chapter. Except for 2 per cent of the muscle fibers, there is only one neuromuscular junction to each muscle fiber; this junction is located near the middle of the fiber. Therefore, the action potential spreads from the middle of the fiber towards its two ends. This

spreading is important because it allows nearly coincident contraction of all sarcomeres of the muscles so that they can all contract together rather than separately.

SPREAD OF THE ACTION POTENTIAL TO THE INTERIOR OF THE MUSCLE FIBER BY WAY OF THE TRANSVERSE TUBULE SYSTEM

The skeletal muscle fiber is so large that action potentials spreading along the membrane cause almost no current flow deep within the fiber. Yet, to cause contraction, these electrical currents must penetrate to the vicinity of all the separate myofibrils. This is achieved by transmission of the action potentials along *transverse tubules* (T tubules) that penetrate all the way through the muscle fiber from one side to the other. The T tubule action potentials in turn cause the sarcoplasmic reticulum to release calcium ions in the immediate vicinity of all the myofibrils, and it is these calcium ions that in turn cause contraction. Now, let us describe this system in much greater detail.

The Transverse Tubule–Sarcoplasmic Reticulum System.   Figure 11–11 illustrates a group of myofibrils surrounded by the transverse tubule–sarcoplasmic reticulum system. The transverse tubules penetrate all the way from one side of the muscle fiber to the opposite side. Not shown in the figure is the fact that these tubules branch among themselves so that they form entire *planes* of T tubules interlacing among all the separate myofibrils. Also, it should be noted that where the T tubules originate from the cell membrane they are open to the exterior. Therefore, they communicate with the fluid surrounding the muscle fiber and contain extracellular fluid in their lumens. In other words, the T tubules are internal extensions of the cell membrane. Therefore, when an action potential spreads over a muscle fiber membrane, it spreads along the T tubules to the deep interior of the muscle fiber as well. The action potential currents surrounding these transverse tubules then elicit the muscle contraction.

Figure 11–11 shows the extensiveness of the *sarcoplasmic reticulum* as well. This is composed of two major parts: (1) long *longitudinal tubules* that terminate in (2) large chambers called *terminal cisternae*. The cisternae in turn abut the transverse tubules. When the muscle fiber is sectioned longitudinally and electron micrographs are made, one sees this abutting of the cisternae against the transverse tubule, which gives the appearance of a *triad* with a small central tubule and a large cisterna on either side. This is illustrated in Figure 11–11 and is also seen in the electron micrograph of Figure 11–3.

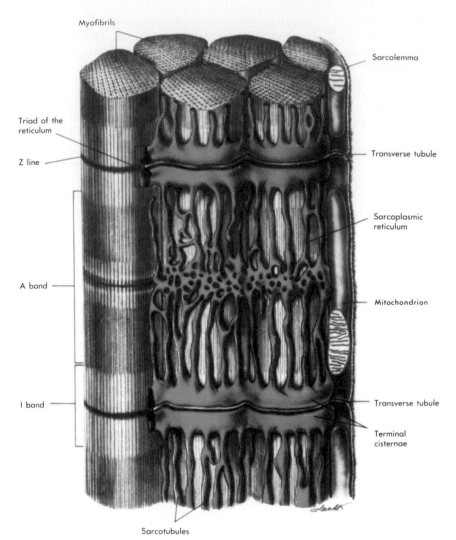

Myofibrils

Triad of the reticulum

Z line

A band

I band

Sarcolemma

Transverse tubule

Sarcoplasmic reticulum

Mitochondrion

Transverse tubule

Terminal cisternae

Sarcotubules

Figure 11–11. The transverse tubule–sarcoplasmic reticulum system. Note the *longitudinal tubules* that terminate in large *cisternae*. The cisternae in turn abut the transverse tubules. Note also that the transverse tubules communicate with the outside of the cell membrane. This illustration was drawn from frog muscle, which has one transverse tubule per sarcomere, located at the Z line. A similar arrangement is found in mammalian heart muscle, but mammalian skeletal muscle has two transverse tubules per sarcomere, located at the A-I junctions. (From Bloom and Fawcett: A Textbook of Histology. Philadelphia, W. B. Saunders Company, 1975. Modified after Peachey: *J. Cell Biol. 25*:209, 1965. Drawn by Sylvia Colard Keene.)

In the muscle of lower animals such as the frog, there is a single T tubule network for each sarcomere, located at the level of the Z membrane as illustrated in Figure 11–11. Cardiac muscle also has this type of T tubule system. However, in mammalian skeletal muscle there are two T tubule networks for each sarcomere located near the two ends of the myosin filaments, which are the points where the actual mechanical forces of muscle contraction are created. Thus, mammalian skeletal muscle is optimally organized for rapid excitation of muscle contraction.

RELEASE OF CALCIUM IONS BY THE CISTERNAE OF THE SARCOPLASMIC RETICULUM

One of the special features of the sarcoplasmic reticulum is that it contains calcium ions in very high concentration, and many of these ions are released when the adjacent T tubule is excited.

Figure 11–12 shows that the action potential of the T tubule causes current flow through the cisternae where they abut the T tubule. The cisternae project *junctional feet* that surround the T tubule, presumably facilitating passage of electrical current from the T tubule into the cisternae. It is possible that the current entering the cisternae elicits action potentials along the membranes of the sarcoplasmic reticulum in the same manner that action potentials spread along the T tubules. Indeed, such action potentials have been demonstrated under special conditions, but it is not known whether they occur normally. At any rate, current flow into the cisternae from the T tubules does cause rapid release of calcium ions from the cisternae. Presumably this results from the opening of calcium pores similar to the opening of sodium pores at the onset of the action potential, though the actual mechanism is still unknown.

The calcium ions that are thus released from the cisternae diffuse to the adjacent myofibrils where

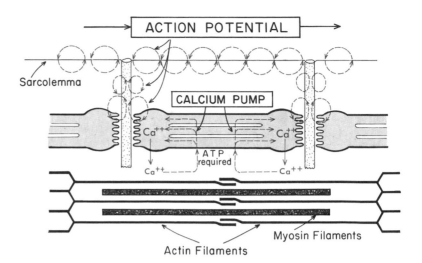

Figure 11–12. Excitation-contraction coupling in the muscle, showing an action potential that causes release of calcium ions from the sarcoplasmic reticulum and then re-uptake of the calcium ions by a calcium pump.

they bind strongly with troponin, as discussed in an earlier section, and this in turn elicits the muscle contraction, as has also been discussed. However, the calcium ions also bind less strongly with the myosin filaments, which could theoretically be another factor in initiating contraction.

The Calcium Pump for Removing Calcium Ions from the Sarcoplasmic Fluid. Once the calcium ions have been released from the cisternae and have diffused to the myofibrils, muscle contraction will then continue as long as the calcium ions are still present in high concentration in the sarcoplasmic fluid. However, a continually active calcium pump located in the walls of the sarcoplasmic reticulum pumps calcium ions out of the sarcoplasmic fluid back into the vesicular cavities of the reticulum. This pump can concentrate the calcium ions about 2000-fold inside the reticulum, a condition which allows massive buildup of calcium in the sarcoplasmic reticulum and also causes almost total depletion of calcium ions in the fluid of the myofibrils. Therefore, except immediately after an action potential, the calcium ion concentration in the myofibrils is kept at an extremely low level.

The Excitatory "Pulse" of Calcium Ions. The normal concentration (about 10^{-7} molar) of calcium ions in the sarcoplasm that bathes the myofibrils is too little to elicit contraction. Therefore, in the resting state, the troponin-tropomyosin complex keeps the actin filaments inhibited and maintains a relaxed state of the muscle.

On the other hand, full excitation of the T tubule–sarcoplasmic reticulum system causes enough release of calcium ions to increase the concentration in the myofibrillar fluid to as high as 2×10^{-4} molar concentration, which causes maximum muscle contraction. Immediately thereafter, the calcium pump depletes the calcium ions again. The total duration of this calcium "pulse" in the

usual skeletal muscle fiber lasts about $1/30$ of a second, though it may last several times as long as this in some skeletal muscle fibers and be several times shorter in others (in heart muscle the pulse lasts for as long as 0.3 second). It is during this calcium pulse that muscle contraction occurs. If the contraction is to continue without interruption for longer intervals, a series of such pulses must be initiated by a continuous series of repetitive action potentials, as will be discussed in more detail later in the chapter.

THE SOURCE OF ENERGY FOR MUSCLE CONTRACTION

We have already seen that muscle contraction depends upon energy supplied by ATP. Most of this energy is required to actuate the ratchet mechanism by which the cross-bridges pull the actin filaments, but small amounts are required for (1) pumping calcium from the sarcoplasm into the sarcoplasmic reticulum, and (2) pumping sodium and potassium ions through the muscle fiber membrane to maintain an appropriate ionic environment for the propagation of action potentials.

However, the amount of ATP that is present in the muscle fiber is sufficient to maintain full contraction for less than 1 second. Fortunately, after the ATP is broken into ADP, as was described in Chapter 2, the ATP is rephosphorylated to form new ATP within a fraction of a second. There are several sources of the energy required for this rephosphorylation:

The first source of energy that is used to reconstitute the ATP is the substance *creatine phosphate,* which carries a high-energy phosphate bond similar to those of ATP. The high-energy phosphate bond of the creatine phosphate is cleaved and the released energy causes bonding of a new phosphate ion to ADP to reconstitute the ATP. However, the total amount of creatine phosphate is also very little — only about five times as great as the ATP. Therefore, the combined energy of both the stored ATP and the creatine phosphate in the muscle is still capable of causing maximal muscle contraction for no longer than a few seconds.

The next source of energy used to reconstitute both the creatine phosphate and the ATP is energy released from the foodstuffs — from carbohydrates, fats, and proteins. Most of this energy is released in the course of oxidation of these foodstuffs. This oxidative release of energy takes place almost entirely in the mitochondria, which utilize the released energy to form new ATP. Thus, the ultimate source of energy for muscle contraction is the basic food substances and oxygen. The detailed mechanisms of these energetic processes are discussed in Chapters 67 through 71.

Efficiency of Muscle Contraction. The "efficiency" of an engine or a motor is calculated as the percentage of energy input that is converted into work instead of heat. The percentage of the input energy to a muscle (the chemical energy in the nutrients) that can be converted into work is less than 20 to 25 per cent, the remainder becoming heat. Maximum efficiency can be realized only when the muscle contracts at a moderate velocity. If the muscle contracts very slowly, large amounts of *maintenance heat* are released during the process of contraction, thereby decreasing the efficiency. On the other hand, if contraction is too rapid, large proportions of the energy are used to overcome the viscous friction within the muscle itself, and this, too, reduces the efficiency of contraction. Ordinarily, maximum efficiency is developed when the velocity of contraction is about 30 per cent of maximum.

CHARACTERISTICS OF A SINGLE MUSCLE TWITCH

Many features of muscle contraction can be especially well demonstrated by eliciting single *muscle twitches*. This can be accomplished by instantaneously exciting the nerve to a muscle or by passing a short electrical stimulus through the muscle itself, giving rise to a single, sudden contraction lasting for a fraction of a second.

Isometric Versus Isotonic Contraction. Muscle contraction is said to be *isometric* when the muscle does not shorten during contraction and *isotonic* when it shortens but the tension on the muscle remains constant. Systems for recording the two types of muscle contraction are illustrated in Figure 11–13.

To the right is the isometric system in which the muscle is suspended between a solid rod and a lever of an electronic force transducer. This transducer records force with almost zero movement of the lever; therefore, in effect, the muscle is bound between fixed points so that it cannot contract significantly. To the left is shown an isotonic recording system: the muscle simply

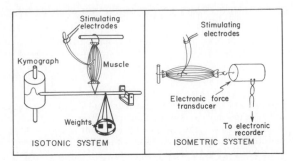

Figure 11–13. Isotonic and isometric recording systems.

lifts a pan of weights so that the force against which the muscle contracts remains constant, though the length of the muscle changes considerably.

There are several basic differences between isometric and isotonic contractions. First, isometric contraction does not require much sliding of myofibrils among each other. Second, in isotonic contraction a load is moved, which involves the phenomena of inertia. That is, the weight or other type of object being moved must first be accelerated, and once a velocity has been attained the load has momentum that causes it to continue moving even after the contraction is over. Therefore an isotonic contraction is likely to last considerably longer than an isometric contraction of the same muscle. Third, isotonic contraction entails the performance of external work. Therefore, in accordance with the Fenn effect discussed previously, a greater amount of energy is used by the muscle.

In comparing the rapidity of contraction of different types of muscle, isometric recordings such as those illustrated in Figure 11–14 are usually used instead of isotonic recordings because the duration of an isotonic recording is almost as dependent on the inertia of the recording system as on the contraction itself, and this makes it difficult to compare time relationships of contractions from one muscle to another.

Muscles can contract both isometrically and isotonically in the body, but most contractions are actually a mixture of the two. When standing, a person tenses the quadriceps muscles to tighten the knee joints and to

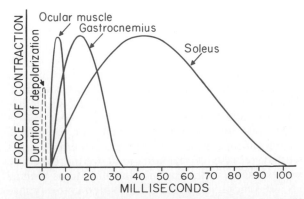

Figure 11–14. Duration of isometric contractions of different types of mammalian muscles, showing also a latent period between the action potential and muscle contraction.

keep the legs stiff. This is isometric contraction. On the other hand, when a person lifts a weight using the biceps, this is mainly an isotonic contraction. Finally, contractions of leg muscles during running are a mixture of isometric and isotonic contractions —isometric mainly to keep the limbs stiff when the legs hit the ground and isotonic mainly to move the limbs.

The Series Elastic Component of Muscle Contraction.   When muscle fibers contract against a load, those portions of the muscle that do not contract — the tendons, the sarcolemmal ends of the muscle fibers where they attach to the tendons, and perhaps even the hinged arms of the cross-bridges — will stretch slightly as the tension increases. Consequently, the muscle must shorten an extra 3 to 5 per cent to make up for the stretch of these elements. The elements of the muscle that stretch during contraction are called the *series elastic component* of the muscle.

Characteristics of Isometric Twitches Recorded from Different Muscles.   The body has many different sizes of skeletal muscles — from the very small stapedius muscle of only a few millimeters length and a millimeter or so in diameter up to the very large quadriceps muscle. Furthermore, the fibers may be as small as 10 microns in diameter or as large as 80 microns. And, finally, the energetics of muscle contraction vary considerably from one muscle to another. These different physical and chemical characteristics often manifest themselves in the form of different characteristics of contraction, some muscles contracting rapidly while others contract slowly.

Fast Versus Slow Muscle.   Figure 11–14 illustrates isometric contractions of three different types of skeletal muscles: an ocular muscle, which has a duration of contraction of less than $1/100$ second; the gastrocnemius muscle, which has a duration of contraction of about $1/30$ second; and the soleus muscle, which has a duration of contraction of about $1/10$ second. It is interesting that these durations of contractions are adapted to the function of each of the respective muscles, for ocular movements must be extremely rapid to maintain fixation of the eyes upon specific objects, the gastrocnemius muscle must contract moderately rapidly to provide sufficient velocity of limb movement for running and jumping, while the soleus muscle is concerned principally with slow reactions for continual support of the body against gravity.

Thus, there is a wide range of gradations from *fast* to *slow* muscles. In general, the slow muscles are used more for prolonged performance of work. Their muscle fibers generally are smaller, are surrounded by more blood capillaries, and have far more mitochondria than the fast muscles. They also have a large amount of *myoglobin* in the sarcoplasm (myoglobin is a substance similar to the hemoglobin in red blood cells and can combine with oxygen and store this oxygen inside the muscle cell until it is needed by the mitochondria). On the other hand, the fast muscles generally have a much more extensive sarcoplasmic reticulum, which allows very rapid release of calcium ions and then rapid re-uptake of the calcium ions so that the contraction can be rapid. The slow muscle is frequently called *red muscle* because of the reddish tint caused by the myoglobin as well as by the large amount of red blood cells in the capillaries; the fast muscle is called *white muscle*

because of a whitish appearance caused by a paucity of these elements.

MECHANICS OF SKELETAL MUSCLE CONTRACTION

THE MOTOR UNIT

Each motor neuron that leaves the spinal cord usually innervates many different muscle fibers, the number depending on the type of muscle. All the muscle fibers innervated by a single motor nerve fiber are called a *motor unit.* In general, small muscles that react rapidly and whose control is exact have few muscle fibers (as few as 2 to 3 in some of the laryngeal muscles) in each motor unit and have a large number of nerve fibers going to each muscle. On the other hand, the large muscles which do not require very fine degree of control, such as the gastrocnemius muscle, may have several hundred muscle fibers in a motor unit. An average figure for all the muscles of the body can be considered to be about 150 muscle fibers to the motor unit.

Usually muscle fibers of adjacent motor units overlap, with small bundles of 10 to 15 fibers from one motor unit lying among similar bundles of the second motor unit. This interdigitation allows the separate motor units to contract in support of each other rather than entirely as individual segments.

Macromotor Units.   Loss of some of the nerve fibers to a muscle causes the remaining nerve fibers to sprout forth and innervate many of the paralyzed muscle fibers. When this occurs, such as following poliomyelitis, one occasionally develops *macromotor units,* which can contain as many as 5 times the normal number of muscle fibers. This obviously decreases the degree of control that one has over the muscles, but, nevertheless, allows the muscles to regain function.

SUMMATION OF MUSCLE CONTRACTION

Summation means the adding together of individual muscle twitches to make strong and concerted muscle movements. In general, summation occurs in two different ways: (1) by increasing the number of motor units contracting simultaneously and (2) by increasing the rapidity of contraction of individual motor units. These are called, respectively, *multiple motor unit summation* and *wave summation* (or spatial summation and temporal summation).

Multiple Motor Unit Summation.   Even within a single muscle, the numbers of muscle fibers and their sizes in the different motor units vary tremendously, so that one motor unit may be as much as 50 times as strong as another. The smaller motor units are far more easily excited than are the larger ones because they are innervated by smaller nerve fibers whose cell bodies in the spinal cord have a naturally high level of excitability. This effect causes the gradations of muscle strength during weak muscle contraction to occur in very small steps, while the steps become progressively greater as the intensity of contraction increases because the larger motor units then begin to contract.

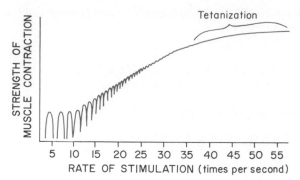

Figure 11-15. Wave summation and tetanization.

Wave Summation. Figure 11-15 illustrates the principles of wave summation, showing in the lower left-hand corner several single muscle twitches followed by successive muscle twitches at various frequencies. When the frequency of twitches rises above 10 per second, the first muscle twitch is not completely over by the time the second one begins. Therefore, since the muscle is already in a partially contracted state when the second twitch begins, the degree of muscle shortening this time is slightly greater than that which occurs with the single muscle twitch. At more rapid rates of contraction, the degree of summation of successive contractions becomes greater and greater, because the successive contractions appear at earlier times following the preceding contraction.

Tetanization. When a muscle is stimulated at progressively greater frequencies, a frequency is finally reached at which the successive contractions fuse together and cannot be distinguished one from the other. This state is called *tetanization,* and the lowest frequency at which it occurs is called the *critical frequency.*

Tetanization results partly from the viscous properties of the muscle and partly from the nature of the contractile process itself. The muscle fibers are filled with sarcoplasm, which is a viscous fluid, and the fibers are encased in fasciae and muscle sheaths that have a viscous resistance to change in length. Therefore, undoubtedly these viscous factors play a role in causing the successive contractions to fuse with each other.

But in addition to the viscous property of muscle, the activation process itself lasts for a definite period of time, and successive pulsatile states of activation of the muscle fiber can occur so rapidly that they fuse into a long continual state of activation; that is, free calcium ions persist continuously in the myofibrils and provide an uninterrupted stimulus for maintenance of contraction. Once the critical frequency for tetanization is reached, further increase in rate of stimulation increases the force of contraction only a few more per cent, as shown in Figure 11-15.

Asynchronous Summation of Motor Units. Actually it is rare for either multiple motor unit summation or wave summation to occur separately from each other in normal muscle function. Instead, special neurogenic mechanisms in the spinal cord normally increase both the impulse rate and the number of motor units firing at the same time. If a motor unit fires at all, it usually fires at least 5 times per second, but this can increase to as high as 50 per second for most muscles or much more than this for the very fast muscles — to frequencies sufficient to cause complete tetanization.

Yet, even when tetanization of individual motor units of a muscle is not occurring, the tension exerted by the whole muscle is still continuous and nonjerky because *the different motor units fire asynchronously;* that is, while one is contracting another is relaxing; then another fires, followed by still another, and so forth. Consequently, even when motor units fire as infrequently as 5 times per second, the muscle contraction, though weak, is nevertheless very smooth.

Maximum Strength of Contraction. The maximum strength of tetanic contraction of a muscle operating at a normal muscle length is about 3.5 kilograms per square centimeter of muscle, or 50 pounds per square inch. Since a quadriceps muscle can at times have as much as 16 square inches of muscle belly, as much as 800 pounds of tension may at times be applied to the patellar tendon. One can readily understand, therefore, how it is possible for muscles sometimes to pull their tendons out of the insertions in bones. This often occurs where the patellar tendon inserts in the tibia, and it occurs even more frequently where the Achilles tendon of the gastrocnemius muscle inserts at the heel.

Changes in Muscle Strength at the Onset of Contraction — The Staircase Effect (Treppe). When a muscle begins to contract after a long period of rest, its initial strength of contraction may be as little as one-half its strength 30 to 50 muscle twitches later. That is, the strength of contraction increases to a plateau, a phenomenon called the *staircase effect* or *treppe.* This phenomenon has interested physiologists greatly because it gives possible insights into the mechanism of muscle contraction.

Though all the possible causes of the staircase effect are not yet known, it is believed to be caused primarily by electrolyte changes that occur when a series of contractions begins. For instance, there is a net increase in calcium ions inside the muscle fiber because of movement of calcium ions inward through the membrane with each action potential. There is probably also further increase of calcium ions in the sarcoplasm because of release of these ions from the sarcoplasmic reticulum and failure to recapture the ions immediately. In addition, there is decreased potassium inside the cell as well as increased sodium; it has been suggested that the changes in these two ions increase the rate of liberation of calcium ions from the sarcoplasmic reticulum. Recalling the earlier discussion of the relationship of calcium ions to the contractile process, one can readily understand that progressive increase in calcium ion concentration in the sarcoplasm, caused either directly or as a consequence of sodium and potassium movement, could progressively increase the strength of muscle contraction, giving rise to the staircase effect.

SKELETAL MUSCLE TONE

Even when muscles are at rest, a certain amount of tautness usually remains. This residual degree of contraction in skeletal muscle is called *muscle tone.* Since skeletal muscle fibers do not contract without an actual action potential to stimulate the fibers except in certain pathological conditions, it is believed that skeletal muscle tone results entirely from nerve impulses coming from the spinal cord. These in turn are controlled partly by impulses transmitted from the brain to the appropriate anterior motor neurons and partly by impulses that originate in *muscle spindles* located in the muscle itself.

Muscle spindles are sensory receptors that exist throughout essentially all skeletal muscles to detect the degree of muscle contraction. These will be discussed in detail in Chapter 51, but, briefly, they transmit impulses almost continually through the posterior roots into the spinal cord, where they excite the anterior motor neurons, which in turn provide the necessary nerve stimuli for muscle tone. Simply cutting the posterior roots, thereby blocking the muscle spindle impulses, usually reduces muscle tone to such a low level that the muscle becomes almost completely flaccid.

Many other neurogenic factors, originating especially in the brain, enter into the control of muscle tone. These will be discussed in detail in relation to muscle spindle and spinal cord function in Chapter 51.

MUSCLE FATIGUE

Prolonged and strong contraction of a muscle leads to the well-known state of muscle fatigue. This results simply from inability of the contractile and metabolic processes of the muscle fibers to continue supplying the same work output. The nerve continues to function properly, the nerve impulses pass normally through the neuromuscular junction into the muscle fiber, and even normal action potentials spread over the muscle fibers, but the contraction becomes weaker and weaker because of reduction of ATP formation in the muscle fibers themselves.

Interruption of blood flow through a contracting muscle leads to almost complete muscle fatigue in a minute or more because of the obvious loss of nutrient supply.

THE LEVER SYSTEMS OF THE BODY

Muscles obviously operate by applying tension to their points of insertion into bones, and the bones in turn form various types of lever systems. Figure 11–16

Figure 11–16. The lever system activated by the biceps muscle.

illustrates the lever system activated by the biceps muscle to lift the forearm. If we assume that a large biceps muscle has a cross-sectional area of 6 square inches, then the maximum force of contraction would be about 300 pounds. When the forearm is exactly at right angles with the upper arm, the tendon attachment of the biceps is about 2 inches anterior to the fulcrum at the elbow, and the total length of the forearm lever is about 14 inches. Therefore, the amount of lifting power that the biceps would have at the hand would be only one seventh of the 300 pounds force, or about 43 pounds. When the arm is in the fully extended position the attachment of the biceps is much less than 2 inches anterior to the fulcrum, and the force with which the forearm can be brought forward is much less than 43 pounds.

In short, an analysis of the lever systems of the body depends on (a) a discrete knowledge of the point of muscle insertion and (b) its distance from the fulcrum of the lever, as well as (c) the length of the lever arm and (d) the position of the lever. Obviously, many different types of movement are required in the body, some of which need great strength and others large distances of movement. For this reason there are all varieties of muscles; some are long and contract a long distance and some are short but have large cross-sectional areas and therefore can provide extreme strengths of contraction over short distances. The study of different types of muscles, lever systems, and their movements is called *kinesiology* and is a very important phase of human physioanatomy.

Accommodation of Muscle Length to the Length of the Lever System. If a bone is broken and then heals in a shortened state, the force of contraction of the muscles lying along this broken bone would obviously become decreased because of the shortened lengths of muscles. However, muscles shortened in this manner undergo *physical shortening* during the next few weeks. That is, the muscle fibers actually shorten and re-establish new muscle lengths approximately equal to the maximum length of the lever system itself, thus re-establishing optimum force of contraction by the muscles.

The same shortening process also occurs in muscles of limbs immobilized for several weeks in casts if the muscles during this time are in a shortened position. When the cast is removed, the muscles must often be restretched over a period of weeks before full mobility is restored.

SPECIAL FEATURES AND ABNORMALITIES OF SKELETAL MUSCLE FUNCTION

MUSCLE HYPERTROPHY

Forceful muscular activity causes the muscle size to increase, a phenomenon called hypertrophy. The diameters of the individual muscle fibers increase, the sarcoplasm increases, and the fibers gain in various nutrient and intermediary metabolic substances, such as adenosine triphosphate, creatine phosphate, glycogen, intracellular lipids, and even many additional mitochondria. It is likely that the myofibrils also increase in size and perhaps in numbers as well, but this has not been proved. Briefly, muscular hypertrophy increases both the motive power of the muscle and the nutrient mechanisms for maintaining increased motive power.

Weak muscular activity, even when sustained over long periods of time, does not result in significant hypertrophy. Instead, hypertrophy results mainly from *very* forceful muscle activity, though the activity might occur for only a few minutes each day. For this reason, strength can be developed in muscles much more rapidly when "resistive" or "isometric" exercise is used rather than simply prolonged mild exercise. Indeed, essentially no enlargement of the muscle fibers occurs unless the muscle contracts to at least 75 per cent of its maximum tension.

On the other hand, prolonged muscle activity does increase muscle endurance, causing increases in the oxidative enzymes, myoglobin, and even blood capillaries — all of which are essential to increased muscle metabolism.

MUSCLE ATROPHY

Muscle atrophy is the reverse of muscle hypertrophy; it results any time a muscle is not used or even when a muscle is used only for very weak contractions. Atrophy is particularly likely to occur when limbs are placed in casts, thereby preventing muscular contraction. As little as one month of disuse can sometimes decrease the muscle size to one-half normal.

Atrophy Caused by Muscle Denervation. When a muscle is denervated it immediately begins to atrophy, and the muscle continues to decrease in size for several years. If the muscle becomes reinnervated during the first three to four months, full function of the muscle usually returns, but after four months of denervation some of the muscle fibers usually will have degenerated. Re-innervation after two years rarely results in return of any function at all. Pathological studies show that the muscle fibers have by that time been replaced by fat and fibrous tissue.

Prevention of Muscle Atrophy by Electrical Stimulation. Strong electrical stimulation of denervated muscles, particularly when the resulting contractions occur against loads, will delay and in some instances prevent muscle atrophy despite denervation. This procedure is used to keep muscles alive until reinnervation can take place.

Physical Contracture of Muscle Following Denervation. When a muscle is denervated, its fibers tend to shorten if the muscle is kept in a shortened position, and even the associated nerves and fasciae shorten. All this is a natural characteristic of protein fibers called "creep." That is, unless continual movement keeps stretching the muscle and other structures, they will creep toward a shortened length. This is one of the most difficult problems in the treatment of patients with denervated muscles, such as occur in poliomyelitis or nerve trauma. Unless passive stretching is applied daily to the muscles, they may become so shortened that even when re-innervated they will be of little value. But, more important, the shortening can often result in extremely contorted positions of different parts of the body.

RIGOR MORTIS

Several hours after death all the muscles of the body go into a state of *contracture* called rigor mortis; that is, the muscle contracts and becomes rigid even without action potentials. It is believed that this rigidity is caused by loss of all the ATP, which is required to cause separation of the cross-bridges from the actin filaments during the relaxation process. The muscles remain in rigor until the muscle proteins are destroyed, which usually results from autolysis caused by enzymes released from the lysosomes some 15 to 25 hours later.

FAMILIAL PERIODIC PARALYSIS

Occasionally, a hereditary disease called familial periodic paralysis occurs. In persons so afflicted, the extracellular fluid potassium concentration periodically falls to very low levels, causing various degrees of paralysis. The paralysis is caused in the following manner: A great decrease in extracellular fluid potassium increases the muscle fiber membrane potential to a very high value. This results in strong *hyperpolarization* of the membrane (a membrane potential more negative than the normal −90 millivolts) making the fiber almost totally inexcitable; that is, the membrane potential is so high that the normal stimulus at the neuromuscular junction is incapable of exciting the fiber.

THE ELECTROMYOGRAM

Each time an action potential passes along a muscle fiber a small portion of the electrical current spreads away from the muscle as far as the skin. If many muscle fibers contract simultaneously, the summated electrical potentials at the skin may be very great. By placing two electrodes on the skin or inserting needle electrodes into the muscle, an electrical recording called the electromyogram can be made when the muscle is stimulated. Figure 11–17 illustrates a typical electromyographic recording from the gastrocnemius muscle during a moderate contraction. Electromyograms are frequently used clinically to discern abnormalities of muscle excitation. Two such abnormalities are muscle *fasciculation* and *fibrillation*.

Muscle Fasciculation. When an abnormal impulse occurs in a motor nerve fiber its whole motor unit contracts. This often causes sufficient contraction in the muscle that one can see a slight ripple in the skin over the muscle. This process is called fasciculation.

Fasciculation occurs especially following destruction of anterior motor neurons in poliomyelitis or following

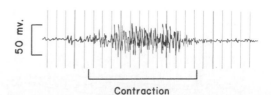

Figure 11–17. Electromyogram recorded during contraction of the gastrocnemius muscle.

traumatic interruption of a nerve. As the peripheral nerve fibers die, spontaneous impulses are generated during the first few days, and fasciculatory muscle movements result in the muscle. Typical electromyographic records of weak periodic potentials can be obtained from the skin overlying the muscle.

Muscle Fibrillation. After all nerves to a muscle have been destroyed and the nerve fibers themselves have become nonfunctional, which requires three to five days, spontaneous impulses begin to appear in the denervated muscle fibers. At first, these occur at a rate of once every few seconds, but, after a few more days or a few weeks, the impulses become as rapid as 3 to 10 times per second. Thus, skeletal muscle fibers, when released from innervation, develop an intrinsic rhythmicity. After several more weeks the muscle fibers atrophy to such an extent that the fibrillatory impulses finally cease. To record an electromyogram of fibrillation, minute bipolar needle electrodes must be inserted into the muscle belly itself because adjacent muscle fibers do not fire simultaneously and, therefore, do not summate. As a result, the potentials are not strong enough to record from the surface of the skin.

REFERENCES

Aguayo, A. J., and Karpati, G. (eds.): Current Topics in Nerve and Muscle Research. New York, Elsevier/North-Holland, 1979.

Baker, P. F., and Reuter, H.: Calcium Movement in Excitable Cells. New York, Pergamon Press, 1975.

Basmajian, J. V.: Muscles Alive; Their Functions Revealed by Electromyography. Baltimore, Williams & Wilkins, 1978.

Becker, P. E.: Myotonia Congenita and Syndromes Associated with Myotonia: Clinical-Genetic Studies of the Nondystrophic Myotonias. Stuttgart, Thieme, 1977.

Bourne, G. H. (ed.): The Structure and Function of Muscle, 2nd Ed. New York, Academic Press, 1973.

Buchthal, F., and Schmalbruch, H.: Motor unit of mammalian muscle. Physiol. Rev., 60:90, 1980.

Buller, A. J.: The physiology of skeletal muscle. In MTP International Review of Science: Physiology. Baltimore, University Park Press, 1975, Vol. 3, p. 279.

Caputo, C.: Excitation and contraction processes in muscle. Annu. Rev. Biophys. Bioeng., 7:63, 1978.

Clausen, J. P.: Effect of physical training on cardiovascular adjustments to exercise in man. Physiol. Rev., 57:779, 1977.

Cohen, C.: The protein switch of muscle contraction. Sci. Am., 233(5):36, 1975.

Cosgrove, M.: Your Muscles and Ways to Exercise Them. New York, Dodd, Mead, 1980.

Costantin, L. L.: Activation in striated muscle. In Brookhart, J. M., and Mountcastle, V. B. (eds.): Handbook of Physiology. Sec. 1, Vol. 1. Baltimore, Williams & Wilkins, 1977, p. 215.

Curtin, N. A., and Woledge, R. C.: Energy changes in muscular contraction. Physiol. Rev., 58:690, 1978.

Ebashi, S.: Regulatory mechanism of muscle contraction with special reference to the Ca-troponin-tropomyosin system. Essays Biochem., 10:1, 1974.

Ebashi, S.: Excitation-contraction coupling. Annu. Rev. Physiol., 38:293, 1976.

Endo, M.: Calcium release from the sarcoplasmic reticulum. Physiol. Rev., 57:71, 1977.

Fabiato, A., and Fabiato, F.: Calcium and cardiac excitation-contraction coupling. Annu. Rev. Physiol., 41:473, 1979.

Feldman, S. A.: Muscle Relaxants, 2nd Ed. Philadelphia, W. B. Saunders Co., 1979.

Fozzard, H. A.: Heart: Excitation-contraction coupling. Annu. Rev. Physiol., 39:201, 1977.

Fuchs, F.: Striated muscle. Annu. Rev. Physiol., 36:461, 1974.

Grimby, G., and Saltin, B.: Physiological effects of physical training. Scand. J. Rehabil. Med., 3:6, 1971.

Harper, P. S.: Myotonic Dystrophy. Philadelphia, W. B. Saunders Co., 1979.

Hess, A.: Vertebrate slow muscle fibers. Physiol. Rev., 50:40, 1970.

Hill, A. L., et al.: Physiology of voluntary muscle. Br. Med. Bull., 12(Sept.): 1956.

Holloszy, J. O., and Booth, F. W.: Biochemical adaptations to endurance exercise in muscle. Annu. Rev. Physiol., 38:273, 1976.

Holman, M. E., and Hirst, G. D. S.: Junctional transmission in smooth muscle and the autonomic nervous system. In Brookhart, J. M., and Mountcastle, V. B. (eds.): Handbook of Physiology. Sec. 1, Vol. 1. Baltimore, Williams & Wilkins, 1977, p. 417.

Homsher, E., and Kean, C. J.: Skeletal muscle energetics and metabolism. Annu. Rev. Physiol., 40:93, 1978.

Hughes, J. T.: Pathology of Muscle. Philadelphia, W. B. Saunders Co., 1975.

Huxley, A. F., and Gordon, A. M.: Striation patterns in active and passive shortening of muscle. Nature (Lond.), 193:280, 1962.

Huxley, H. E.: Muscular contraction and cell motility. Nature, 243:445, 1973.

Ingels, N. B., Jr., (ed.): The Molecular Basis of Force Development in Muscle. Palo Alto, Cal., Palo Alto Medical Research Foundation, 1979.

Johnson, E. W., (ed.): Practical Electromyography. Baltimore, Williams & Wilkins, 1979.

Kaldor, G., and DiBattista, W. J. (eds.): Aging in Muscle. New York, Raven Press, 1978.

Lyman, R. W.: Kinetic analysis of myosin and actinomysin ATPase. Annu. Rev. Biophys. Bioeng., 8:145, 1979.

Mauro, A., and Bischoff, R., (eds.): Muscle Regeneration. New York, Raven Press, 1979.

Nakano, K. K.: Neurology of Musculoskeletal and Pneumatic Disorders. Boston, Houghton Mifflin, 1978.

Nelson, P. G.: Nerve and muscle cells in culture. Physiol. Rev., 55:1, 1975.

Northrip, J. W., et al.: Introduction to Biomechanic Analysis of Sport. Dubuque, Iowa, W. C. Brown Co., 1979.

Rasch, P. J., and Burke, R. K.: Kinesiology and Applied Anatomy: The Science of Human Movement. Philadelphia, Lea & Febiger, 1978.

Sahlin, K.: Intracellular pH and Energy Metabolism in Skeletal Muscle of Man: with Special Reference to Exercise. Stockholm, Acta Physiologica Scandinavica, 1978.

Schneck, A. G., and Vandecasserie, C., (eds.): Myoglobin. Brussels, l'Université de Bruxelles, Faculté des Sciences, 1977.

Sugi, H., and Pollack, G. H., (eds.): Cross-Bridge Mechanism in Muscle Contraction. Baltimore, University Park Press, 1979.

Tada, M., et al.: Molecular mechanism of active calcium transport by sarcoplasmic reticulum. Physiol. Rev., 58:1, 1978.

Taylor, C. R.: Exercise and environmental heat loads: different mechanisms for solving different problems: Int. Rev. Physiol., 15:119, 1977.

Toida, N., et al.: Obliquely striated muscle. Physiol. Rev., 55:700, 1975.

Tregear, R. T., and Marston, S. B.: The crossbridge theory. Annu. Rev. Physiol., 41:723, 1979.

Walton, J. N. (ed.): Disorders of Voluntary Muscle. New York, Churchill Livingstone, 1974.

Weber, A., and Murray, J. M.: Molecular control mechanisms in muscle contraction. Physiol. Rev., 53:612, 1973.

Winter, D. A.: Biomechanics of Human Movement. New York, John Wiley & Sons, 1979.

Zak, R., and Rabinowitz, M.: Molecular aspects ot cardiac hypertrophy. Annu. Rev. Physiol., 41:539, 1979.

12

Neuromuscular Transmission; Function of Smooth Muscle

TRANSMISSION OF IMPULSES FROM NERVES TO SKELETAL MUSCLE FIBERS: THE NEUROMUSCULAR JUNCTION

The skeletal muscles are innervated by large myelinated nerve fibers that originate in the large motoneurons of the anterior horns of the spinal cord. It was pointed out in the previous chapter that each nerve fiber normally branches many times and stimulates from 3 to several hundred skeletal muscle fibers. The nerve ending makes a junction, called the *neuromuscular junction* or the *myoneural junction,* with the muscle fiber approximately at the fiber's midpoint so that the action potential in the fiber travels in both directions. With the exception of about 2 per cent of the muscle fibers there is only one such junction per muscle fiber.

Physiologic Anatomy of the Neuromuscular Junction. Figure 12–1, Parts A and B, illustrates the neuromuscular junction between a large myelinated nerve fiber and a skeletal muscle fiber. The nerve fiber branches at its end to form a complex of branching nerve *terminals* called the *end-plate,* which invaginates into the muscle fiber but lies entirely outside the muscle fiber plasma membrane. The entire structure is covered by one or more Schwann cells that insulate the end-plate from the surrounding fluids.

Figure 12–1C shows an electronmicrographic sketch of the juncture between a single-branch axon terminal and the muscle fiber membrane. The invagination of the membrane is called the *synaptic gutter* or *synaptic trough,* and the space

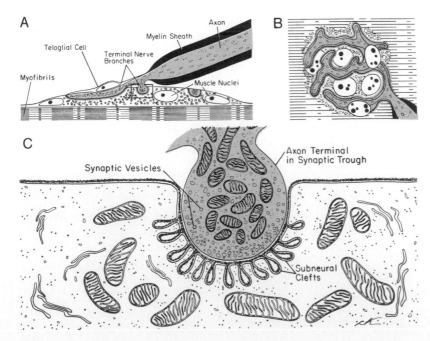

Figure 12–1. Different views of the motor end-plate. A, Longitudinal section through the end-plate. B, Surface view of the end-plate. C, Electronmicrographic appearance of the contact point between one of the axon terminals and the muscle fiber membrane, representing the rectangular area shown in A. (From Bloom and Fawcett, as modified from R. Couteaux: A Textbook of Histology. Philadelphia, W. B. Saunders Company, 1975.)

between the terminal and the fiber membrane is called the *synaptic cleft*. The synaptic cleft is 20 to 30 nanometers wide and is filled with a gelatinous "ground" substance through which diffuses extracellular fluid. At the bottom of the gutter are numerous *folds* of the muscle membrane which form *subneural clefts* that greatly increase the surface area at which the synaptic transmitter can act. In the axon terminal are many mitochondria that supply energy mainly for synthesis of the excitatory transmitter *acetylcholine* that, in turn, excites the muscle fiber. The acetylcholine is synthesized in the cytoplasm of the terminal but is rapidly absorbed into many small synaptic vesicles, approximately 300,000 of which are normally in all the terminals of a single end-plate. In the matrix of the subneural clefts are large quantities of the enzyme *cholinesterase,* which is capable of destroying acetylcholine, as is explained in further detail below.

Secretion of Acetylcholine by the Axon Terminals. When a nerve impulse reaches the neuromuscular junction, about 300 vesicles of acetylcholine are released by the terminals into the synaptic clefts between the terminals and the muscle fiber membrane. This results from movement of calcium ions from the extracellular fluid into the membranes of the terminals when the action potential depolarizes their membranes. The calcium ions cause the vesicles of acetylcholine to rupture through the membrane. In the absence of calcium or in the presence of excess magnesium, the release of acetylcholine is greatly depressed.

Destruction of the Released Acetylcholine by Cholinesterase. Within approximately 1 millisecond after acetylcholine is released by the axon terminal, a small portion of it diffuses out of the synaptic gutter and no longer acts on the muscle fiber membrane, but the greater bulk of it is destroyed by the cholinesterase in the subneural clefts. The very short period of time that the acetylcholine remains in contact with the muscle fiber membrane — about 1 millisecond — is almost always sufficient to excite the muscle fiber, and yet the rapid removal of the acetylcholine prevents re-excitation after the muscle fiber has recovered from the first action potential.

The "End-Plate Potential" and Excitation of the Skeletal Muscle Fiber. Even though the acetylcholine released into the space between the end-plate and the muscle membrane lasts for only a minute fraction of a second, during this period of time it can affect the muscle membrane sufficiently to open large pores with diameters of 0.6 to 0.7 nanometers (6 to 7 Angstroms) in the membrane. This makes the membrane very permeable to sodium ions (as well as to most other ions), allowing rapid influx of sodium into the muscle fiber. As a result, the membrane potential rises *in the local area of the end-plate* as much as 50 to 75 millivolts, creating a *local potential* called the *end-plate potential.*

The mechanism by which acetylcholine increases the permeability of the muscle membrane is probably the following: It is believed that the muscle membrane contains special protein molecules called *acetylcholine receptors* to which the acetylcholine binds. Under the influence of the acetylcholine this receptor supposedly undergoes a conformational change that increases the permeability of the membrane to ions. Rapid influx of sodium ions ensues, and this elicits the end-plate potential that is responsible for initiation of the action potential at the muscle fiber membrane.

An end-plate potential is illustrated at point A in Figure 12–2. This potential was recorded after the muscle had been poisoned with curare, which made the end-plate potential too weak to excite an action potential in the muscle fiber. However, in the normal muscle fiber, the effect shown at point B occurs: The end-plate potential begins, but before it can complete its course the intense local current flow created by this potential initiates an action potential; this in turn spreads in both directions along the muscle fiber. The threshold voltage at which the skeletal muscle fiber is stimulated is approximately −50 millivolts. Thus, if the resting membrane potential of the muscle is at the normal level of about −90 millivolts, then an end-plate potential of +40 millivolts is required to elicit an action potential in the muscle fiber.

At point C, another situation is illustrated in which only a small quantity of acetylcholine is released at the end-plate. This also causes an end-plate potential that is less than the threshold value required to cause an action potential. Note that the potential dies in a few milliseconds. Thus,

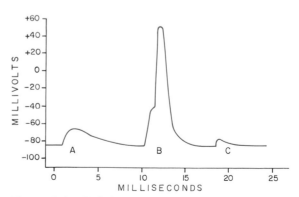

Figure 12–2. End-plate potentials: A, weakened end-plate potential recorded in a curarized muscle, too weak to elicit an action potential; B, normal end-plate potential eliciting a muscle action potential; and C, weakened end-plate potential caused by botulinum toxin that decreases end-plate release of acetylcholine, again too weak to elicit a muscle action potential.

failure to transmit a signal to the muscle fiber can result from either diminished receptivity of the muscle membrane, as occurs when the muscle has been poisoned with curare, or decreased release of acetylcholine, as occurs when the end-plates have been poisoned with botulinum toxin.

"Safety Factor" for Transmission at the Neuromuscular Junction; Fatigue of the Junction. Ordinarily, each impulse that arrives at the neuromuscular junction creates an end-plate current flow about three to four times that required to stimulate the muscle fiber. Therefore, the normal neuromuscular junction is said to have a very high *safety factor*. However, artificial stimulation of the nerve fiber at rates greater than 100 times per second for several minutes often diminishes the number of vesicles of acetylcholine released with each impulse so much that impulses then often fail to pass into the muscle fiber. This is called *fatigue* of the neuromuscular junction, and it is analogous to fatigue of the synapse in the central nervous system. However, under normal functioning conditions, fatigue of the neuromuscular junction rarely occurs, because almost never do the spinal nerves stimulate even the most active neuromuscular junctions more than 100 times per second.

Drugs that Affect Transmission at the Neuromuscular Junction. *Drugs that Stimulate the Muscle Fiber by Acetylcholine-Like Action.* Many different compounds, including *methacholine, carbachol,* and *nicotine,* have the same effect on the muscle fiber as does acetylcholine. The difference between these drugs and acetylcholine is that they are not destroyed by cholinesterase or are destroyed very slowly, so that when once applied to the muscle fiber the action persists for many minutes to several hours. Moderate quantities of the above three drugs applied to a muscle fiber cause localized areas of depolarization, and every time the muscle fiber becomes repolarized elsewhere, these depolarized areas, by virtue of their leaking ions, cause new action potentials, thereby causing a state of spasm. On the other hand, when extreme doses of these drugs are used, so much of the membranes becomes depolarized that the fibers can no longer pass impulses at all, and a state of flaccid paralysis exists instead of the spasm that occurs with moderate dosages.

Drugs that Block Transmission at the Neuromuscular Junction. A group of drugs, known as the *curariform drugs,* can prevent passage of impulses from the end-plate into the muscle. Thus, D-tubocurarine affects the membrane, probably by competing with acetylcholine for the receptor sites of the membrane, so that the acetylcholine cannot increase the permeability of the membrane sufficiently to initiate a depolarization wave.

Drugs that Stimulate the Neuromuscular Junction by Inactivating Cholinesterase. Three particularly well-known drugs, *neostigmine, physostigmine,* and *diisopropyl fluorophosphate,* inactivate cholinesterase so that the cholinesterase normally in the synapses will not hydrolyze acetylcholine released at the end-plate. As a result, acetylcholine increases in quantity with successive nerve impulses so that extreme amounts of ace-

tylcholine can accumulate and repetitively stimulate the muscle fiber. This causes *muscular spasm* when even a few nerve impulses reach the muscle; this can cause death due to laryngeal spasm, which smothers the person.

Neostigmine and physostigmine combine with cholinesterase to inactivate it for several hours, after which they are displaced from the cholinesterase so that it once again becomes active. On the other hand, diisopropyl fluorophosphate, which has military potential as a very powerful "nerve" gas, actually inactivates cholinesterase for several weeks, which makes this a particularly lethal drug.

MYASTHENIA GRAVIS

The disease *myasthenia gravis,* which occurs in rare instances in human beings, causes the person to become paralyzed because of inability of the neuromuscular junctions to transmit signals from the nerve fibers to the muscle fibers. Pathologically, the number of subneural clefts in the synaptic gutter is reduced and the synaptic cleft itself is widened as much as 50 per cent. Also, antibodies that attack the muscle fibers have been demonstrated in the bloods of many of these patients. Therefore, it has been postulated that myasthenia gravis is an autoimmune disease in which patients have developed antibodies against their own muscles. And the antibodies in turn have damaged the muscle fibers — one of the effects being partial destruction of the receptor membrane of the neuromuscular junction.

Though it has not been possible to make precise neurophysiological measurements at the neuromuscular junctions in these patients, circumstantial evidence suggests that the end-plate potentials in these patients are greatly reduced, perhaps or probably because of the damaged receptor membrane of the junction.

Regardless of the cause, the end-plate potentials developed in the muscle fibers are too weak to stimulate the muscle fibers adequately. If the disease is intense enough, the patient dies of paralysis — in particular, of paralysis of the respiratory muscles. However, the disease can usually be ameliorated with several different drugs, as follows:

Treatment with Drugs. When a patient with myasthenia gravis is treated with a drug, such as neostigmine, that is capable of inactivating or destroying cholinesterase, the acetylcholine secreted by the end-plate is not destroyed immediately. If a sequence of nerve impulses arrives at the end-plate, the quantity of acetylcholine present at the membrane increases progressively until finally the end-plate potential caused by the acetylcholine rises above threshold value for stimulating the muscle fiber. Thus, it is possible by diminishing the quantity of cholinesterase in the muscles of a patient with myasthenia gravis to allow even the inadequate quantities of acetylcholine secreted at the end-plates to effect almost normal muscular activity.

CONTRACTION OF SMOOTH MUSCLE

In the previous chapter and thus far in the present chapter, the discussion has been concerned with skeletal muscle. We now turn to smooth muscle, which is composed of far smaller fibers — usually 2 to 5 microns in diameter and only 50 to 200 microns in length — in contrast to the skeletal muscle fibers that are as much as 20 times as large (in diameter) and thousands of times as long. Nevertheless, many of the same principles of contraction apply to both smooth muscle and skeletal muscle. Most important, essentially the same chemical substances cause contraction in smooth muscle as in skeletal muscle, but the physical arrangement of smooth muscle fibers is entirely different, as we shall see.

TYPES OF SMOOTH MUSCLE

The smooth muscle of each organ is distinctive from that of most other organs in several different ways: physical dimensions, organization into bundles or sheets, response to different types of stimuli, characteristics of its innervation, and function. Yet, for the sake of simplicity, smooth muscle can generally be divided into two major types, which are illustrated in Figure 12–3: *multiunit smooth muscle* and *visceral smooth muscle*.

Multiunit Smooth Muscle. This type of smooth muscle is composed of discrete smooth muscle fibers. Each fiber operates entirely independently of the others and is often innervated by a single nerve ending, as occurs for skeletal muscle fibers. Furthermore, the outer surfaces of these fibers, like those of skeletal muscle fibers, are covered by a thin layer of "basement membrane-like" substance, a glycoprotein that helps to insulate the separate fibers from each other.

The most important characteristic of multiunit smooth muscle fibers is that their control is exerted almost entirely by nerve signals and very little by other stimuli such as local tissue factors. This is in contrast to a major share of the control of visceral smooth muscle by non-nervous stimuli. An additional characteristic is that they rarely exhibit spontaneous contractions.

Some examples of multiunit smooth muscle found in the body are the smooth muscle fibers of the ciliary muscle of the eye, the iris of the eye, the nictitating membrane that covers the eyes in some lower animals, the piloerector muscles that cause erection of the hairs when stimulated by the sympathetic nervous system, and the smooth muscle of many of the larger blood vessels.

Visceral Smooth Muscle. Visceral smooth muscle fibers are similar to multiunit fibers except that they are usually arranged in sheets or bundles and the cell membranes contact each other at multiple points to form many *gap junctions,* or *nexi.* Thus the fibers form a *functional syncytium* that usually contracts in large areas at once. For this reason, this type of smooth muscle is also known as *single-unit* or *unitary smooth muscle.* This type of muscle is found in most of the organs of the body, especially in the walls of the gut, the bile ducts, the ureters, the uterus, and so forth.

When one portion of a visceral muscle tissue is stimulated, the action potential is conducted to the surrounding fibers by *ephaptic conduction.* This means that the action potential generated in one area of the muscle electrically excites the adjacent fibers without secretion of any excitatory substance. Visceral smooth muscle fibers can transmit action potentials one to another in this manner because the gap junctions between adjacent fibers exhibit greatly enhanced permeability so that the electrical resistance between the inside of one fiber and the next is only a fraction of the normal membrane resistance. This obviously allows easy flow of current from the interior of one cell to the next and therefore allows ease of transmission of action potentials over the surface membranes of the muscle mass.

THE CONTRACTILE PROCESS IN SMOOTH MUSCLE

The Chemical Basis for Contraction. Smooth muscle contains both *actin* and *myosin filaments,* having chemical characteristics similar to but not exactly the same as those of the actin and myosin filaments in skeletal muscle. Smooth muscle also contains *tropomyosin,* but it is doubtful whether troponin or a troponin-like substance exists in smooth muscle. This raises a question about the mechanism for control of smooth muscle contraction, which will be discussed in more detail in a subsequent section of this chapter.

Chemical studies have shown that actin and myosin derived from smooth muscle interact with each other in the same way that this occurs for actin and myosin derived from skeletal muscle. Furthermore, the contractile process is activated by calcium ions, and ATP is degraded to ADP to provide the energy for contraction.

On the other hand, there are major differences

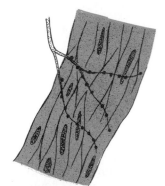

VISCERAL   MULTIUNIT

Figure 12–3. Visceral and multiunit smooth muscle fibers.

in the physical organization of smooth muscle and skeletal muscle, as well as differences in other aspects of smooth muscle function such as excitation-contraction coupling, control of the contractile process by calcium ions, duration of contraction, and amount of energy required for the contractile process.

The Physical Basis for Smooth Muscle Contraction. Smooth muscle does not have the same striated arrangement of the actin and myosin filaments as that found in skeletal muscle. And, for a long time, it was impossible to discern even in electronmicrographs any specific organization in the smooth muscle cell that could account for contraction. However, recent special staining techniques suggest the physical organization illustrated in Figure 12–4. This shows large numbers of actin filaments attached to so-called *dense bodies*. Some of these bodies in turn are attached to the cell membrane while others are dispersed throughout the sarcoplasm. There also appear to be enough cross-attachments from one dense body to another to hold these in relatively fixed positions within the cell. Interspersed among the actin filaments are a few thick filaments about 2.5 times the diameter of the thin actin filaments. These are assumed to be myosin filaments. However, there are only one twelfth to one fifteenth as many of these ''myosin filaments'' as actin filaments.

Despite the relative paucity of myosin filaments, it is assumed that they have sufficient

cross-bridges to attract the many actin filaments and cause contraction by the sliding filament mechanism in essentially the same way that this occurs in skeletal muscle. And it is especially interesting to note that the maximum strength of contraction of smooth muscle is approximately equal to that of skeletal muscle, about 2 to 3 kg. per square centimeter of cross-sectional area of the muscle.

Slowness of Contraction and Relaxation of Smooth Muscle. Though each smooth muscle tissue in the body has its own characteristics quite distinctive from the others, a typical smooth muscle tissue will begin to contract 50 to 100 milliseconds after it is excited, and will reach full contraction about half a second later. Then the contraction declines in another 1 to 2 seconds, giving a total contraction time of 1 to 3 seconds, which is about 30 times as long as the single-twitch contraction of skeletal muscle. However, smooth muscle contractions as short as 0.2 second and as long as 30 seconds have been recorded.

A major share of the prolonged contractile state of smooth muscle seems to be caused by slowness of the chemical reactions that cause the contraction. For instance, assuming that smooth muscle contraction occurs by the same ratchet mechanism as that proposed for skeletal muscle, it has been calculated that the frequency of power strokes by the heads of the cross-bridges is only one fiftieth to one hundredth as rapid as in skeletal muscle. This apparently results from the fact that the cross-bridge heads have far less ATPase activity than that exhibited by the cross-bridge heads in skeletal muscle.

Energy Required to Sustain Smooth Muscle Contraction. Measurements have shown that as little as one five-hundredth as much energy is required to sustain the same tension of contraction in smooth muscle as in skeletal muscle. This presumably results from the very slow activity of the myosin ATPase and also from the fact that there are far fewer myosin filaments in smooth muscle than in skeletal muscle.

This economy of energy utilization by smooth muscle is exceedingly important to overall function of the body, because organs such as the intestines, the urinary bladder, the gallbladder, and other viscera must maintain moderate degrees of muscle contractile tone day in and day out.

MEMBRANE POTENTIALS AND ACTION POTENTIALS IN SMOOTH MUSCLE

Smooth muscle exhibits membrane potentials and action potentials similar to those that occur in skeletal muscle fibers. Furthermore, smooth muscle contraction can be elicited by depolarization of the membrane in the same way that contraction is initiated by depolarization of skeletal muscle fibers. However, there are both quantitative and qualitative differences in the membrane potentials and action potentials of smooth muscle that require special attention.

Membrane Potentials in Smooth Muscle. The

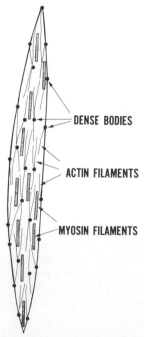

Figure 12–4. Arrangement of actin and myosin filaments in the smooth muscle cell. Note the attachment of the actin filaments to ''dense bodies,'' some of which are themselves attached to the cell membrane.

DENSE BODIES

ACTIN FILAMENTS

MYOSIN FILAMENTS

quantitative value of the membrane potential of smooth muscle is variable from one type of smooth muscle to another, and it also depends on the momentary condition of the muscle. However, in the normal resting state, the membrane potential is usually about −50 to −60 millivolts, or about 30 millivolts less negative than in skeletal muscle.

Action Potentials in Visceral Smooth Muscle. Action potentials occur in visceral smooth muscle in the same way that they occur in skeletal muscle. However, action potentials probably do not normally occur in multiunit types of smooth muscle, as will be discussed in a subsequent section.

The action potentials of visceral smooth muscle occur in two different forms: (1) spike potentials and (2) action potentials with plateaus.

Spike Potentials. Typical spike action potentials, such as those seen in skeletal muscle, occur in most types of visceral smooth muscle. The duration of this type of action potential is 10 to 50 milliseconds, as illustrated in Figure 12–5A. Such action potentials can be elicited in many ways, such as by electrical stimulation, by the action of hormones on the smooth muscle, by the action of transmitter substances from nerve fibers, or as a result of spontaneous generation in the muscle fiber itself, as discussed below.

Action Potentials with Plateaus. Figure 12–6 illustrates an action potential with a plateau. The onset of this action potential is similar to that of the typical spike potential. However, instead of rapid repolarization of the muscle fiber membrane, the repolarization is delayed for several hundred

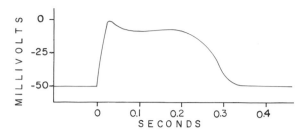

Figure 12–6. Monophasic action potential from a smooth muscle fiber of the rat uterus.

to several thousand milliseconds. Plateaus as long as 30 seconds have been recorded. The importance of the plateau is that it can account for the prolonged periods of contraction that occur in at least some types of smooth muscle. This type of action potential often occurs in the ureter, in the uterus under some conditions, and in some types of vascular smooth muscle. (Also, this is the type of action potential seen in cardiac muscle fibers that have a prolonged period of contraction, as we shall discuss in the next two chapters.)

Slow Wave Potentials in Visceral Smooth Muscle and Spontaneous Generation of Action Potentials. Some smooth muscle is self-excitatory. That is, action potentials arise within the smooth muscle itself without an extrinsic stimulus. This is usually associated with a basic *slow wave rhythm* of the membrane potential. A typical slow wave of this type is illustrated in Figure 12–5B. The slow wave itself is not an action potential. It is not a self-regenerative process that spreads progressively over the membranes of the muscle fibers. Instead, it is a local property of the smooth muscle fibers that make up the muscle mass.

The cause of the slow wave rhythm is as yet unknown; one suggestion, for which there is circumstantial evidence, is that the slow waves are caused by waxing and waning of the pumping of sodium outward through the muscle fiber membrane; the membrane potential becomes more negative when sodium is pumped rapidly and less negative when the sodium pump becomes less active.

The importance of the slow waves lies in the fact that they can initiate action potentials. The slow waves themselves cannot cause muscle contraction, but when the potential of the slow wave rises above the level of approximately −35 millivolts (the approximate threshold for eliciting action potentials in most visceral smooth muscle), an action potential develops and spreads over the visceral smooth muscle mass, and then contraction does occur. Figure 12–5B illustrates this effect, showing that at each peak of the slow wave, one or more action potentials occur. This effect can obviously promote a series of rhythmical contractions of the smooth muscle mass. There-

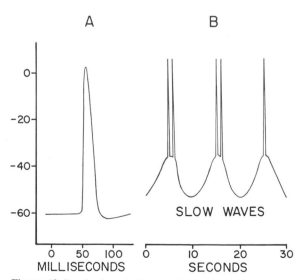

Figure 12–5. A, A typical smooth muscle action potential (spike potential) elicited by an external stimulus. B, A series of spike action potentials elicited by rhythmical slow electrical waves occurring spontaneously in the smooth muscle wall of the intestine.

fore, the slow waves are frequently called *pacemaker waves*. This type of activity is especially prominent in tubular types of smooth muscle masses, such as in the gut, the ureter, and so forth. In Chapter 63 we shall see that this type of activity controls the rhythmical contractions of the gut.

Spread of Action Potentials Through Visceral Smooth Muscle. Because of the *gap junctions* that allow easy spread of electrical current from one smooth muscle fiber to the next in visceral smooth muscle, once an action potential begins, it spreads slowly through the muscle mass. For instance, if it begins at the upper end of the gastrointestinal tract, it can spread variable distances downward along the intestinal wall, creating a constrictive ring that moves forward. The constrictive ring propels the intestinal contents forward. This process is called *peristalsis,* an important end result of smooth muscle function.

Excitation of Visceral Smooth Muscle by Stretch. When visceral smooth muscle is stretched sufficiently, spontaneous action potentials are usually generated. These result from a combination of the normal slow wave potentials plus a decrease in the membrane potential caused by the stretch itself. This response to stretch is an especially important function of visceral smooth muscle because it allows a hollow organ that is excessively stretched to contract automatically and therefore to resist the stretch. For instance, when the gut is overstretched by intestinal contents, a local automatic contraction sets up a peristaltic wave that moves the contents away from the super-stretched intestine.

Depolarization of Multiunit Smooth Muscle Without Action Potentials. The smooth muscle fibers of multiunit smooth muscle normally contract only in response to nerve stimuli. The nerve endings secrete acetylcholine in the case of some multiunit smooth muscles and norepinephrine in the case of others. In both instances, these transmitter substances cause depolarization of the smooth muscle membrane, and this response in turn elicits the contraction. However, action potentials most often do not develop. The presumed reason for this is that the fibers are too small to generate an action potential. In fact, when action potentials are elicited in visceral smooth muscle, as many as 30 to 40 smooth muscle fibers must depolarize simultaneously before a self-propagating action potential ensues. Yet, even without an action potential in the multiunit smooth muscle fibers, the local depolarization caused by the nerve transmitter substance itself spreads "electrotonically" over the entire fiber and is all that is needed to cause the muscle contraction.

Role of Calcium Ions in Causing Smooth Muscle Action Potentials. It will be recalled from the discussion in Chapter 10 that the depolarization process during the action potential of nerve fibers (and also of skeletal muscle fibers) is caused almost entirely by rapid influx of sodium ions to the interior of the cell membrane. In the action potential of smooth muscle fibers, the rapid influx of ions includes not only sodium ions but also a large quantity of calcium ions. Indeed, it is believed that for many types of smooth muscle, the onset of depolarization is caused mainly by influx of calcium ions rather than of sodium ions. This is particularly important in view of the fact that influx of calcium ions seems also to be the major means by which muscle contraction itself is elicited, as will be discussed in the following section.

EXCITATION-CONTRACTION COUPLING – ROLE OF CALCIUM IONS

In the previous chapter it was pointed out that the actual contractile process in skeletal muscle is activated by calcium ions. This is also true in smooth muscle. However, the source of the calcium ions differs in smooth muscle because the sarcoplasmic reticulum of smooth muscle is poorly developed, in contrast to the sarcoplasmic reticulum of skeletal muscle which is very extensive and is the source of very nearly 100 per cent of the contraction-inducing calcium ions.

In some types of smooth muscle, most of the calcium ions that cause contraction enter the muscle fiber from the extracellular fluid at the time of the action potential. There is a reasonably high concentration of calcium ions in the extracellular fluid, and as was pointed out in the previous section, the action potential is caused at least partly by influx of calcium ions into the muscle fiber. Because the smooth muscle fibers are extremely small (in contrast to the sizes of the skeletal muscle fibers), these calcium ions can diffuse to all parts of the smooth muscle and elicit the contractile process. The time required for this diffusion to occur is usually 200 to 300 milliseconds and is called the *latent period* before the contraction begins; this latent period is some 50 times as great as that for skeletal muscle contraction.

Yet, in some smooth muscle there is a moderately developed sarcoplasmic reticulum. However, there are no T tubules. Instead, the cisternae of the reticulum abut the cell membrane. Therefore, it is believed that the membrane action potentials in these smooth muscle fibers cause release of calcium ions from these cisternae, thereby providing a greater degree of contraction than would occur on the basis of calcium ions entering through the cell membrane alone.

The Calcium Pump. To cause relaxation of the smooth muscle contractile elements, it is necessary to remove the calcium ions. This removal is achieved by a calcium pump that pumps the calcium ions out of the smooth muscle fiber and back into the extracellular fluid, or pumps the calcium ions into the sarcoplasmic reticulum. However, this pump is very slow-acting in comparison with the fast-acting sarcoplasmic reticulum pump in skeletal muscle. Therefore, the duration of smooth muscle contraction is often in the order of seconds rather than in tens of milliseconds, as occurs for skeletal muscle.

Mechanism by Which Calcium Ions Excite Contraction in Smooth Muscle. In skeletal muscle, calcium ions activate contraction by binding with troponin. This in turn causes conformational changes in the tropomyosin, then activation of the actin filament, and finally the contractile process itself. However, troponin has not been proved to be present in smooth muscle cells. Therefore, it is doubtful that calcium ions initiate smooth muscle contraction in the same way that they initiate skeletal muscle contraction. Instead, in smooth muscle, calcium ions strongly excite the ATPase activity of the heads of the myosin cross-bridges. In skeletal muscle this activation by calcium ions is weak, but in smooth muscle it occurs at very low calcium ion concentrations. In contrast to skeletal muscle, therefore, this activation of the myosin ATPase system is believed to initiate smooth muscle contraction. That is, the ATPase begins to split ATP; in turn, the released energy sets into motion the contractile process. It is possible that the troponin-tropomyosin system is not involved.

NEUROMUSCULAR JUNCTIONS OF SMOOTH MUSCLE

Physiologic Anatomy of Smooth Muscle Neuromuscular Junctions. Neuromuscular junctions of the type found on skeletal muscle fibers do not occur in smooth muscle. Instead, the nerve fibers generally branch diffusely on top of a sheet of muscle fibers, as was illustrated in Figure 12–3. In most instances these fibers do not make direct contact with the smooth muscle fibers at all but instead form so-called *diffuse junctions* that secrete their transmitter substance into the interstitial fluid a few microns away from the muscle cells; the transmitter substance then diffuses to the cells. Furthermore, where there are many layers of muscle cells, the nerve fibers often innervate only the outer layer, and the muscle excitation then travels from this outer layer to the inner layers by direct action potential conduction or by subsequent diffusion of the transmitter substance. But, less often, terminal branches of the axons do penetrate into the muscle mass.

The axons innervating smooth muscle fibers also do not have typical end-feet as observed in the end-plate on skeletal muscle fibers. Instead, the fine terminal axons have multiple varicosities spread along their axes. At these points the Schwann cells are interrupted so that transmitter substance can be secreted through the walls of the

varicosities. In the varicosities are vesicles similar to those present in the skeletal muscle end-plate containing transmitter substance. However, in contrast to the vesicles of skeletal muscle junctions that contain only acetylcholine, the vesicles of the autonomic nerve fiber varicosities contain acetylcholine in some fibers and norepinephrine in others.

In a few instances, particularly in the multiunit type of smooth muscle, the varicosities lie directly on the muscle fiber membrane with a separation from this membrane of only 20 nanometers — the same width as the synaptic cleft that occurs in the skeletal muscle junction. These *contact junctions* function in much the same way as the skeletal muscle neuromuscular junction, and the latent period of contraction of these smooth muscle fibers is considerably shorter than of fibers stimulated by the diffuse junctions.

Excitatory and Inhibitory Transmitter Substances at the Smooth Muscle Neuromuscular Junction. Two different transmitter substances known to be secreted by the autonomic nerves innervating smooth muscle are *acetylcholine* and *norepinephrine*. Acetylcholine is an excitatory substance for smooth muscle fibers in some organs but an inhibitory substance for smooth muscle in other organs. And when acetylcholine excites a muscle fiber, norepinephrine ordinarily inhibits it, or when acetylcholine inhibits a fiber norepinephrine excites it.

It is believed that *receptor substances* in the membranes of the different smooth muscle fibers determine which will excite them, acetylcholine or norepinephrine. Thus, there are *excitatory receptors* and *inhibitory receptors*. These receptor substances will be discussed in more detail in Chapter 57 in relation to function of the autonomic nervous system.

Excitation of Action Potentials in Smooth Muscle Fibers — The Junctional Potential. Transmission of impulses from terminal nerve fibers to smooth muscle fibers occurs in very much the same manner as transmission at the neuromuscular junction of skeletal muscle fibers except for temporal differences. When an action potential reaches the terminal of an excitatory nerve fibril, there is a typical latent period of 50 milliseconds before any change in the membrane potential of the smooth muscle fiber can be detected. Then the potential rises to a maximal level in approximately 100 milliseconds. If an action potential does not occur, this potential gradually disappears at a rate of approximately one-half every 200 to 500 milliseconds. This complete sequence of potential changes is called the *junctional potential;* it is analogous to the end-plate potential of the skeletal muscle fibers except that its duration is 20 to 100 times as long.

If the junctional potential rises to the threshold

level for discharge of the smooth muscle membrane, an action potential will occur in the smooth muscle fiber in exactly the same way that an action potential occurs in a skeletal muscle fiber. A typical smooth muscle fiber has a normal resting membrane potential of −50 to −60 millivolts, and the threshold potential at which the action potential occurs is about −30 to −35 millivolts.

Inhibition at the Smooth Muscle Neuromuscular Junction. When a transmitter substance at the nerve ending interacts with an inhibitory receptor instead of an excitatory receptor, the membrane potential of the muscle fiber becomes more negative than ever, for instance −70 millivolts; that is, it becomes *hyperpolarized* and therefore becomes much more difficult to excite than is usually the case.

SMOOTH MUSCLE CONTRACTION WITHOUT ACTION POTENTIALS — EFFECT OF LOCAL TISSUE FACTORS AND HORMONES

Though we have thus far discussed smooth muscle contraction elicited only by nervous signals and smooth muscle membrane action potentials, we must quickly disavow the fact that all smooth muscle contraction occurs in this way. In fact, probably half or more of all smooth muscle contraction is initiated not by action potentials but by stimulatory factors acting directly on the smooth muscle contractile machinery. The two types of non-nervous and nonaction potential stimulating factors most often involved are (1) local tissue factors, and (2) various hormones.

Smooth Muscle Contraction in Response to Local Tissue Factors. In Chapter 20 we shall discuss the control of contraction of the arterioles, meta-arterioles, and precapillary sphincters. The smaller of these vessels have little or no nervous supply. Yet, the smooth muscle is highly contractile, responding rapidly to changes in local conditions in the surrounding interstitial fluid. In this way a powerful local feedback control system exists to maintain appropriate local fluid environmental conditions. Some of the specific control factors are:

(1) Lack of oxygen in the local tissues causes smooth muscle relaxation and therefore vasodilatation.
(2) Excess carbon dioxide causes vasodilatation.
(3) Increased hydrogen ion concentration also causes increased vasodilatation.

And such factors as lactic acid, increased potassium ions, diminished calcium ion concentration, and decreased body temperature will also cause local vasodilatation.

Effects of Hormones on Smooth Muscle Contrac-tion. Most of the circulating hormones in the body affect smooth muscle contraction at least to some degree, and some often have very profound effects. Some of the more important hormones are norepinephrine, epinephrine, acetylcholine, angiotensin, vasopressin, oxytocin, serotonin, and histamine.

A hormone will cause contraction of smooth muscle when the smooth muscle cells contain an *excitatory receptor* for the respective hormone. However, the hormone will cause inhibition instead of contraction if the cells contain an *inhibitory receptor* rather than an excitatory receptor. Thus, most of the hormones will cause excitation of some smooth muscle but inhibition of other muscle.

Some of the hormones — especially norepinephrine, vasopressin, and angiotensin — have such a powerful excitatory effect that they can cause smooth muscle spasm for hours.

Mechanism of Muscle Excitation by Local Tissue Factors and Hormones. It is believed that the local tissue factors and hormones that cause smooth muscle contraction do so by activating the calcium mechanism for control of the contractile process. Some of these factors change the membrane potential a moderate amount but without necessarily causing an action potential, and this increases the flow of calcium ions to the interior of the cell. However, most of them can activate contraction even when the membrane potential is not altered and even when calcium ions are not available to enter the cell. In these circumstances calcium ions probably are released from the sarcoplasmic reticulum. In some instances, possibly the affinity of the contractile mechanism for calcium ions is increased by the local factor or the hormone.

MECHANICAL CHARACTERISTICS OF SMOOTH MUSCLE CONTRACTION

From the foregoing discussion of the many different types of smooth muscle and the different ways in which contraction can be elicited, one can readily understand why smooth muscle in different parts of the body has many different characteristics of contraction. For instance, the multiunit smooth muscle of the large blood vessels contracts mainly in response to nerve impulses, whereas in many types of visceral smooth muscle — the smaller blood vessels, the ureter, the bile ducts, and other glandular ducts — a self-excitatory process controlled mainly by local factors and by hormones causes continuous rhythmic contraction.

Tone of Smooth Muscle. Smooth muscle can maintain a state of long-term, steady contraction that has been called either *tonus* contraction of smooth muscle or simply *smooth muscle tone*. This is an important feature of smooth muscle contraction because it allows prolonged or even indefinite continuance of the smooth muscle function. For instance, the arterioles are maintained in a state of tonic contraction almost throughout the entire life of the person. Likewise, tonic contraction in the gut wall maintains steady pressure on the contents of the gut, and tonic contraction of the urinary bladder wall maintains a moderate amount of pressure on the urine in the bladder.

Tonic contractions of smooth muscle can be caused in either of two ways:

(1) They are sometimes caused by *summation of individual contractile pulses;* each contractile pulse is initiated by a separate action potential in the same way that tetanic contractions are produced in skeletal muscle.

(2) However, most smooth muscle tonic contractions probably result from *prolonged direct smooth muscle excitation* without action potentials, usually caused by local tissue factors or circulating hormones. For instance, prolonged tonic contractions of the blood vessels without the mediation of action potentials are regularly caused by angiotensin, vasopressin, or norepinephrine, and these play an important role in the long-term regulation of arterial pressure, as will be discussed in Chapters 21 and 22.

A typical example of a tonic contraction (also frequently called simply *tonus*) is illustrated in Figure 12–7. This figure also shows that rhythmic contractions can be superimposed onto the tonic contraction because of simultaneous (a) rhythmical nerve discharges or (b) rhythmical pacemaker slow waves that periodically excite the smooth muscle.

Degree of Shortening of Smooth Muscle During Contraction. A special characteristic of smooth muscle — one that is also different from skeletal muscle — is its ability to shorten a far greater percentage of its length than can skeletal muscle. Skeletal muscle has a useful distance of contraction equal to only 25 to 35 per cent of its length, while smooth muscle can often contract quite effectively from a length two times its normal length to as short as one-half its normal length, giving as much as a four-fold distance of contraction. This allows smooth muscle to perform important functions in the hollow viscera — for instance, allowing the gut, the bladder, blood vessels, and other internal structures of the body to change their lumen diameters from almost zero up to very large values.

Stress-Relaxation of Smooth Muscle. A very important characteristic of smooth muscle is its ability to change length greatly without marked changes in tension. This results from a phenomenon called *stress-relaxation*, which may be explained as follows:

If a segment of smooth muscle 1 inch long is suddenly stretched to 2 inches, the tension between the two ends increases tremendously at first, but the extra tension also begins to disappear immediately, and within a few minutes it has returned almost to its level prior to the stretch, even though the muscle is now twice as long. This possibly results from the loose arrangement of the actin and myosin filaments in smooth muscle. Over a period of time, the filaments of the stretched muscle presumably rearrange their bonds and gradually allow the sliding process to take place, thus allowing the tension to return almost to its original amount.

Exactly the converse effect occurs when smooth muscle is shortened. Thus, if the 2 inch segment of smooth muscle is shortened back to 1 inch, essentially all tension will be lost from the muscle immediately. Gradually, over a period of 1 minute or more, much of the tension returns, this again presumably resulting from slow sliding of the filaments. This is called *reverse stress-relaxation*.

REFERENCES

Baker, P. F., and Reuter, H.: Calcium Movement in Excitable Cells. New York, Pergamon Press, 1975.

Bolton, T. B.: Mechanisms of action of transmitters and other substances on smooth muscle. *Physiol. Rev.,* 59:606, 1979.

Bulbring, E., et al. (eds.): Physiology of Smooth Muscles; Twenty-sixth International Congress of Physiological Sciences. New York, Raven Press, 1975.

Ceccarelli, B., and Hurlbut, W. P.: Vesicle hypothesis of the release of quanta of acetylcholine. *Physiol. Rev.,* 60:396, 1980.

Chamley-Campbell, J., et al.: Smooth muscle cell in culture. *Physiol. Rev.,* 59:1, 1979.

Daniel, E. E., and Sarna, S.: The generation and conduction of activity in smooth muscle. *Annu. Rev. Pharmacol. Toxicol.,* 18:145, 1978.

Dau, P. C. (ed.): Plasmapheresis and the Immobiology of Myasthenia Gravis. Boston, Houghton Mifflin, 1979.

De Robertis, E., and Schacht, J. (eds.): Neurochemistry of Cholinergic Receptors. New York, American Elsevier Publishing Co., 1974.

Devine, C. E.: Morphology and ultrastructure. *In* Kaley, G., and Altura, B. M. (eds.): Microcirculation. Vol. II. Baltimore, University Park Press, 1977.

Fambrough, D. M.: Control of acetylcholine receptors in skeletal muscle. *Physiol. Rev.,* 59:165, 1979.

Gage, P. W.: Generation of end-plate potentials. *Physiol. Rev.,* 56:177, 1976.

Guyton, A. C., and MacDonald, M. A.: Physiology of botulinus toxin. *Arch. Neurol. Psychiat.,* 57:578, 1947.

Guyton, A. C., and Reeder, R. C.: The dynamics of curarization. *J. Pharmacol. Exp. Ther.,* 97:322, 1949.

Hall, Z. W., et al.: Chemistry of Synaptic Transmission. Newton, Mass., Chiron Press, 1974.

Hartshorne, D. J., and Gorecka, A.: Biochemistry of the contractile proteins of smooth muscle. *In* Bohr, D. F., et al. (eds.): Handbook of Physiology. Sec. 2, Vol. II. Baltimore, Williams & Wilkins, 1980, p. 93.

Heuser, J. E., and Reese, T. S.: Structure of the synapse. *In* Brookhart, J. M., and Mountcastle, V. B. (eds.): Handbook of Physiology. Sec. 1, Vol. 1. Baltimore, Williams & Wilkins, 1977, p. 261.

Johansson, B.: Vascular smooth muscle biophysics. *In* Kaley, G., and Altura, B. M. (eds.): Microcirculation. Vol. II. Baltimore, University Park Press, 1977.

Johansson, B., and Somlyo, A. P.: Electrophysiology and excitation-contraction coupling. *In* Bohr, D. F., et al. (eds.): Handbook of Physiology. Sec. 2, Vol. II. Baltimore, Williams & Wilkins, 1980, p. 301.

Landmesser, L. T.: The generation of neuromuscular specificity. *Annu. Rev. Neurosci., 3*:279, 1980.

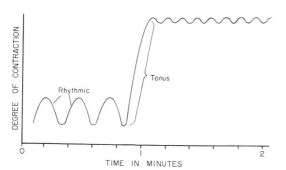

Figure 12–7. Record of rhythmic and tonic smooth muscle contraction.

Lester, H. A.: The response to acetylcholine. *Sci. Am., 236*(2):106, 1977.

Lunt, G. G., and Marchbanks, R. M.: The Biochemistry of Myasthenia Gravis and Muscular Dystrophy. New York, Academic Press, 1978.

Mark, R. F.: Synaptic repression at neuromuscular junctions. *Physiol. Rev., 60*:355, 1980.

Martin, A. R.: Junctional transmission. II. Presynaptic mechanisms. *In* Brookhart, J. M., and Mountcastle, V. B. (eds.): Handbook of Physiology. Sec. 1, Vol. I. Baltimore, Williams & Wilkins, 1977, p. 329.

McGeachie, J. K.: Smooth Muscle Regeneration; A Review and Experimental Study. Basel, S. Karger, 1975.

Minton, S. A., Jr.: Venom Diseases. Springfield, Ill., Charles C Thomas, 1974.

Murphy, R. A.: Filament organization and contractile function in vertebrate smooth muscle. *Annu. Rev. Physiol., 41*:737, 1979.

Paul, R. J.: Chemical energetics of vascular smooth muscle. *In* Bohr, D. F., *et al.* (eds.): Handbook of Physiology. Sec. 2, Vol. II. Baltimore, Williams & Wilkins, 1980, p. 201.

Paul, R. J., and Rüegg, J. C.: Biochemistry of vascular smooth muscle: Energy metabolism and proteins. *In* Kaley, G., and Altura, B. M.

(eds.): Microcirculation. Vol. II. Baltimore, University Park Press, 1977.

Pepeu, G., *et al.* (eds.): Receptors for Neurotransmitters and Peptide Hormones. New York, Raven Press, 1980.

Prosser, C. L.: Smooth muscle. *Annu. Rev. Physiol., 36*:503, 1974.

Prosser, C. L.: Evolution and diversity of nonstriated muscles. *In* Bohr, D. F., *et al.* (eds.): Handbook of Physiology. Sec. 2, Vol. II. Baltimore, Williams & Wilkins, 1980, p. 635.

Rémond, A., and Izard, C. (eds.): Electrophysiological Effects of Nicotine. New York, Elsevier/North-Holland, 1979.

Somlyo, A. P.: Excitation-contraction coupling in vertebrate smooth muscle: Correlation of ultrastructure with function. *Physiologist, 15*: 338, 1972.

Somlyo, A. P.: Ultrastructure of vascular smooth muscle. *In* Bohr, D. F., *et al.* (eds.): Handbook of Physiology. Sec. 2, Vol. II. Baltimore, Williams & Wilkins, 1980, p. 33.

Takeuchi, A.: Junctional transmission. I. Postsynaptic mechanisms. *In* Brookhart, J. M., and Mountcastle, V. B. (eds.): Handbook of Physiology. Sec. 1, Vol. I. Baltimore, Williams & Wilkins, 1977, p. 295.

Part IV
THE HEART

13

Heart Muscle; The Heart as a Pump

With this chapter we begin discussion of the heart and circulatory system. The heart is a pulsatile, four-chamber pump composed of two atria and two ventricles. The atria function principally as entryways to the ventricles, but they also pump weakly to help move the blood into the ventricles. The ventricles supply the main force that propels blood through the lungs and through the peripheral circulatory system.

Special mechanisms in the heart maintain cardiac rhythmicity and transmit action potentials throughout the cardiac musculature to initiate its contraction. These mechanisms will be explained in detail in the following chapter. In the present chapter we will explain how the heart operates as a pump: that is, explain the function of the muscle, of the valves, and of the various chambers of the heart. Therefore, we will discuss first the basic physiology of cardiac muscle itself, especially how it differs from skeletal muscle, which was discussed in Chapter 11.

PHYSIOLOGY OF CARDIAC MUSCLE

The heart is composed of three major types of cardiac muscle: atrial muscle, ventricular muscle, and specialized excitatory and conductive muscle fibers. The atrial and ventricular types of muscle contract in much the same way as the skeletal muscle fibers. On the other hand, the specialized excitatory and conductive fibers contract only feebly because they contain few contractile fibrils; instead, because of their rhythmical properties and their rapidity of conduction, they provide an excitatory system for the heart and a transmission system for rapid conduction of impulses throughout the heart.

PHYSIOLOGIC ANATOMY OF CARDIAC MUSCLE

Figure 13–1 illustrates a typical histologic picture of cardiac muscle, showing the cardiac muscle fibers arranged in a latticework, the fibers dividing, then recombining, and then spreading again. One notes immediately from this figure that cardiac muscle is *striated* in the same manner as typical skeletal muscle. Furthermore, cardiac muscle has typical myofibrils that contain *actin* and *myosin filaments* almost identical to those found in skeletal muscle, and these filaments interdigitate and slide along each other during the process of contraction in the same manner as occurs in skeletal muscle. (See Chapter 11.)

Cardiac Muscle as Syncytium. The angulated dark areas crossing the cardiac muscle fibers in Figure 13–1 are called *intercalated discs*; however, they are actually cell membranes that separate individual cardiac muscle cells from each other. That is, cardiac muscle fibers are made up of many

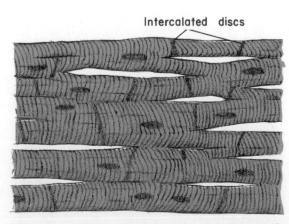

Figure 13–1. The "syncytial" nature of cardiac muscle.

cardiac muscle cells connected in series with each other. Yet electrical resistance through the intercalated disc is only one four-hundredth the resistance through the outside membrane of the cardiac muscle fiber, because the cell membranes fuse with each other to form "tight junctions" that allow almost completely free diffusion of ions. Therefore, from a functional point of view, ions move with ease along the axes of the cardiac muscle fibers so that action potentials travel from one cardiac muscle cell to another, past the intercalated discs, without significant hindrance. Therefore, cardiac muscle is a *syncytium*, in which the cardiac muscle cells are so tightly bound that when one of these cells becomes excited, the action potential spreads to all of them, spreading from cell to cell and spreading throughout the latticework interconnections.

The heart is composed of two separate syncytiums, the *atrial syncytium* and the *ventricular syncytium*. These are separated from each other by the fibrous tissue surrounding the valvular rings, but an action potential can be conducted from the atrial syncytium into the ventricular syncytium by way of a specialized conductive system, the *A-V bundle*, which will be discussed in detail in the following chapter.

All-or-Nothing Principle as Applied to the Heart.   Because of the syncytial nature of cardiac muscle, stimulation of any single atrial muscle fiber causes the action potential to travel over the entire atrial muscle mass, and, similarly, stimulation of any single ventricular fiber causes excitation of the entire ventricular muscle mass. If the A-V bundle is intact, the action potential passes also from the atria to the ventricles. This is called the all-or-nothing principle; and it is precisely the same as that discussed in Chapter 10 for nerve fibers. However, because the cardiac muscle fibers interconnect with each other, the all-or-nothing principle applies to the entire functional syncytium of the heart rather than to single muscle fibers as in the case of skeletal muscle fibers.

ACTION POTENTIALS IN CARDIAC MUSCLE

The *resting membrane potential* of normal cardiac muscle is approximately −85 to −95 millivolt (mv.) and approximately −90 to −100 mv. in the specialized conductive fibers, the Purkinje fibers, which are discussed in the following chapter.

The *action potential* recorded in ventricular muscle, shown at the bottom of Figure 13–2, is 105 mv., which means that the membrane potential rises from its normally very negative value to a slightly positive value of about +20 mv. Because of this change of potential from negative to positive, the positive portion is called the *reversal*

potential. The basic principles of the genesis of resting, action, and reversal potentials were discussed in Chapter 10.

Cardiac muscle has a peculiar type of action potential. After the initial *spike* the membrane remains depolarized for 0.15 for atrial muscle to 0.3 second for ventricular muscle, exhibiting a *plateau* as illustrated in Figure 13–2, followed at the end of the plateau by abrupt repolarization. The presence of this plateau in the action potential causes the action potential to last 20 to 50 times as long in cardiac muscle as in skeletal muscle and causes a correspondingly increased period of contraction.

At this point we must ask the question: Why does the action potential of cardiac muscle have a plateau while that of skeletal muscle does not? There are at least two major differences between the membrane properties of these two types of muscle that presumably account for the plateau in cardiac muscle. First, a moderate quantity of calcium ions diffuses to the inside of the cardiac muscle fiber during the action potential, while only a very small quantity diffuses into skeletal muscle. Furthermore, the calcium ion influx does not occur only at the onset of the action potential, as is true for sodium ions, but instead continues for 0.2 to 0.3 second. The plateau occurs during this prolonged influx of calcium ions.

The second major functional difference between cardiac muscle and skeletal muscle that helps to account for the plateau is this: immediately after the onset of the action potential the permeability of the cardiac muscle membrane for potassium *decreases* about five-fold, an effect that does not occur in skeletal muscle. It is believed that this decreased potassium permeability is caused by the excess calcium influx noted above. The decreased potassium permeability greatly decreases the out-

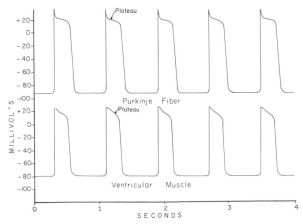

Figure 13–2. Rhythmic action potentials from a Purkinje fiber and from a ventricular muscle fiber, recorded by means of microelectrodes.

flux of potassium ions for the next 0.2 to 0.3 second, which prevents rapid repolarization of the membrane and thus gives rise to the plateau. Then, at the end of this time, the membrane permeability for potassium begins to increase so that more and more potassium leaves the fiber. This becomes a self-regenerative process; that is, the greater the rate of potassium outflux, the greater becomes the permeability of the membrane for potassium so that the potassium outflux becomes still greater. This regenerative cycle causes an almost explosive increase in potassium permeability, and the rapid loss of potassium from the fiber returns the membrane potential to its resting level, thus ending the action potential.

Velocity of Conduction in Cardiac Muscle. The velocity of conduction of the action potential in both atrial and ventricular muscle fibers is about 0.3 to 0.5 meter per second, or about 1/250 the velocity in very large nerve fibers and about one tenth the velocity in skeletal muscle fibers. The velocity of conduction in the specialized conductive system varies from 0.02 to 4 meters per second in different parts of the system, as is explained in the following chapter.

Refractory Period of Cardiac Muscle. Cardiac muscle, like an excitable tissue, is refractory to restimulation during the action potential. An extremely strong electrical stimulus can sometimes initiate a new spike at the very end of the action potential plateau, but this is a very abnormal situation and the spike is not propagated along the muscle. Therefore, the refractory period of the heart is usually stated to be the interval of time, as shown to the left in Figure 13–3, during which a normal cardiac impulse cannot re-excite an already excited area of cardiac muscle. The normal refractory period of the ventricle is 0.25 to 0.3 second, which is approximately the duration of the action potential. There is an additional *relative refractory period* of about 0.05 second during

which the muscle is more difficult than normal to excite but nevertheless can be excited, as illustrated by the early premature contraction in Figure 13–3.

The refractory period of atrial muscle is much shorter than that for the ventricles (about 0.15 second), and the relative refractory period is another 0.03 second. Therefore, the rhythmical rate of contraction of the atria can be much faster than that of the ventricles.

CONTRACTION OF CARDIAC MUSCLE

Excitation-Contraction Coupling — Function of Calcium Ions and of the T Tubules. The term excitation-contraction coupling means the mechanism by which the action potential causes the myofibrils of muscle to contract. This was discussed in detail for skeletal muscle in Chapter 11. However, once again there are differences in this mechanism in cardiac muscle that have important effects on the characteristics of cardiac muscle contraction.

As is true for skeletal muscle, when an action potential passes over the cardiac muscle membrane, the action potential also spreads to the interior of the cardiac muscle fiber along the membranes of the T tubules. The T tubule action potentials in turn cause instantaneous release of calcium ions into the muscle sarcoplasm from the cisternae of the sarcoplasmic reticulum. Then the calcium ions diffuse in another few thousandths of a second into the myofibrils where they catalyze the chemical reactions that promote sliding of the actin and myosin filaments along each other; this in turn produces the muscle contraction. Thus far, this mechanism of excitation-contraction coupling is the same as that for skeletal muscle, but at this point a major difference begins to appear. In addition to the calcium ions that are released into the sarcoplasm from the cisternae of the sarcoplasmic reticulum, large quantities of calcium ions also diffuse during the action potential from the T tubules into the sarcoplasm. Indeed, without this extra calcium from the T tubules, it is probable that the strength of cardiac muscle contraction would be considerably reduced, because the cisternae of cardiac muscle are less well developed than those of skeletal muscle and do not store enough calcium. On the other hand, the T tubules of cardiac muscle have a diameter 5 times as great as that of the skeletal muscle tubules; inside the T tubules is a large quantity of mucopolysaccharides that are electronegatively charged and bind an abundant storehouse of calcium ions, keeping this always available for diffusion to the interior of the cardiac muscle fiber when the T tubule action potential occurs. This extra supply of calcium from the T tubules is at least one of the factors that

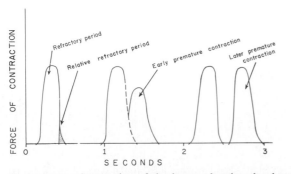

Figure 13–3. Contraction of the heart, showing the durations of the refractory period and the relative refractory period, the effect of an early premature contraction, and the effect of a later premature contraction. Note that the premature contractions do not cause wave summation as occurs in skeletal muscle.

prolongs the cardiac muscle action potential and maintains cardiac muscle contraction for as long as a third of a second rather than one-tenth that time as occurs in skeletal muscle.

At the end of the plateau of the action potential the supply of new calcium ions to the interior of the muscle fiber is suddenly cut off, and the calcium ions in the sarcoplasm are rapidly pumped back into the sarcoplasmic reticulum and T tubules — thus, the end of contraction until a new action potential occurs.

It is especially interesting that the strength of contraction of cardiac muscle depends partly upon the concentration of calcium ions in the extracellular fluids, which normally is not the case for skeletal muscle. The probable reason for this is that the quantity of calcium ions in the T tubules is directly proportional to the extracellular fluid calcium concentration, because the T tubules themselves are actually filled with extracellular fluid. Consequently, the availability of calcium ions to cause cardiac muscle contraction directly depends upon the extracellular fluid calcium.

Another difference between the tubules of cardiac muscle and those of skeletal muscle is that the T tubules in the skeletal muscle are located adjacent to actin and myosin filaments where they overlap each other so that there are two T tubule systems to each sarcomere. On the other hand, in cardiac muscle there is only one T tubule system per sarcomere located at the Z line but composed of much larger T tubules. This difference in structure is compatible with the fact that cardiac muscle contracts much more slowly than skeletal muscle so that there is adequate time for calcium ions to diffuse from the Z line to the middle of the sarcomere where the contractile process occurs.

Duration of Contraction. Cardiac muscle begins to contract a few milliseconds after the action potential begins and continues to contract for a few milliseconds after the action potential ends. Therefore, the duration of contraction of cardiac muscle is mainly a function of the duration of the action potential —about 0.15 second in atrial muscle and 0.3 second in ventricular muscle.

Effect of Heart Rate on Duration of Contraction. When the heart rate increases, the duration of each total cycle of the heart, including both the contraction phase and the relaxation phase, obviously decreases. The duration of the action potential and the period of contraction (systole) also decrease but not as great a percentage as does the relaxation phase (diastole). At a normal heart rate of 72 beats per minute, the period of contraction is about 0.4 of the entire cycle. At 3 times normal heart rate, this period is about 0.65 of the entire cycle, which means that the heart under some conditions does not remain relaxed long enough to allow complete filling of the cardiac chambers prior to the next contraction.

THE CARDIAC CYCLE

Figure 13–4 illustrates the physical structure of the heart and also the pattern of blood flow

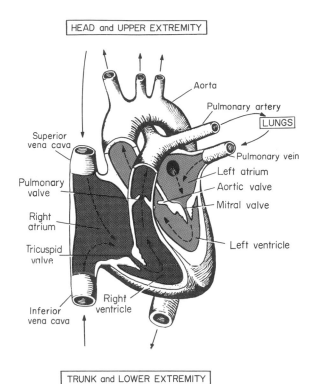

Figure 13–4. Structure of the heart, and course of blood flow through the heart chambers.

through the heart. It is also clear that the heart is in reality four separate pumps: two *primer pumps,* the *atria,* and two *power pumps,* the *ventricles.*

The period from the end of one heart contraction to the end of the next is called the *cardiac cycle.* Each cycle is initiated by spontaneous generation of an action potential in the S-A node, as will be explained in detail in the following chapter. This node is located in the posterior wall of the right atrium near the opening of the superior vena cava, and the action potential travels rapidly through both atria and thence through the A-V bundle into the ventricles. However, because of a special arrangement of the conducting system from the atria into the ventricles, there is a delay of more than $1/10$ second between passage of the cardiac impulse from the atria into the ventricles. This allows the atria to contract ahead of the ventricles, thereby pumping blood into the ventricles prior to the very strong ventricular contraction. Thus, the atria act as primer pumps for the ventricles, and the ventricles then provide the major source of power for moving blood through the vascular system.

SYSTOLE AND DIASTOLE

The cardiac cycle consists of a period of relaxation called *diastole* followed by a period of contraction called *systole.*

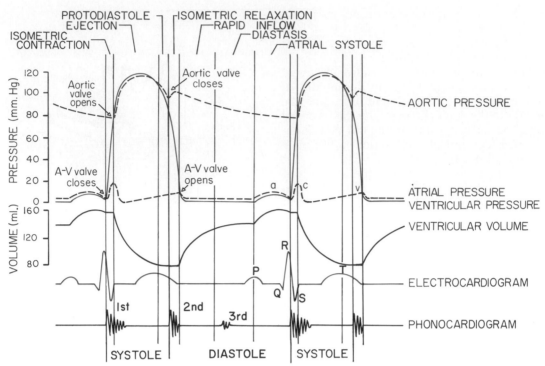

Figure 13–5. The events of the cardiac cycle, showing changes in left atrial pressure, left ventricular pressure, aortic pressure, ventricular volume, the electrocardiogram, and the phonocardiogram.

Figure 13–5 illustrates the different events during the cardiac cycle. The top three curves show the pressure changes in the aorta, the left ventricle, and the left atrium, respectively. The fourth curve depicts the changes in ventricular volume, the fifth the electrocardiogram, and the sixth a phonocardiogram, which is a recording of the sounds produced by the heart as it pumps. It is especially important that the student study in detail the diagram of this figure and understand the causes of all the events illustrated. These are explained as follows:

RELATIONSHIP OF THE ELECTROCARDIOGRAM TO THE CARDIAC CYCLE

The electrocardiogram in Figure 13–5 shows the *P, Q, R, S,* and *T* waves, which will be discussed in Chapters 15 through 17. These are electrical voltages generated by the heart and recorded by the electrocardiograph from the surface of the body. The *P wave* is caused by the *spread of depolarization* through the atria, and this is followed by atrial contraction, which causes a slight rise in the atrial pressure curve immediately after the P wave. Approximately 0.16 second after the onset of the P wave, the *QRS waves* appear as a result of depolarization of the ventricles, which initiates contraction of the ventricles and causes the ventricular pressure to begin rising, as illustrated in the figure. Therefore, the QRS complex begins slightly before the onset of ventricular systole.

Finally, one observes the *ventricular T wave* in the electrocardiogram. This represents the stage of repolarization of the ventricles at which time the ventricular muscle fibers begin to relax. Therefore, the T wave occurs slightly prior to the end of ventricular contraction.

FUNCTION OF THE ATRIA AS PUMPS

Blood normally flows continually from the great veins into the atria; approximately 70 per cent of this flows directly through the atria into the ventricles even before the atria contract. Then, atrial contraction causes an additional 30 per cent filling of the ventricles. Therefore, the atria simply function as primer pumps that increase the ventricular pumping effectiveness by approximately 30 per cent. Yet, the heart can continue to operate quite satisfactorily under normal resting conditions even without this extra 30 per cent effectiveness because it normally has the capability of pumping 300 to 500 per cent more blood than is required by the body anyway. Therefore, the difference is unlikely to be noticed unless a person exercises; then acute signs of heart failure occasionally develop, especially shortness of breath.

Pressure Changes in the Atria — The a, c, and v Waves. In the atrial pressure curve of Figure 13–5, three major pressure elevations called the *a, c,* and *v atrial pressure waves can be noted.*

The *a wave* is caused by atrial contraction. Ordinarily, the *right* atrial pressure rises 4 to 6 mm. Hg during atrial contraction, while the *left* atrial pressure rises about 7 to 8 mm. Hg.

The *c wave* occurs when the ventricles begin to contract, and it is caused mainly by two factors: (1) bulging of the A-V valves backward toward the atria because of increasing pressure in the ventricles, and (2) pulling on the atrial muscle by the contracting ventricles.

The *v wave* occurs toward the end of ventricular contraction; it results from slow buildup of blood in the atria while the A-V valves are closed during ventricular contraction. Then, when ventricular contraction is over, the A-V valves open, allowing blood to flow rapidly into the ventricles and causing the v wave to disappear.

FUNCTION OF THE VENTRICLES AS PUMPS

Filling of the Ventricles. During ventricular systole large amounts of blood accumulate in the atria because of the closed A-V valves, and the atrial pressures become considerably elevated. Therefore, just as soon as systole is over and the ventricular pressures fall again to their low diastolic values, the high pressures in the atria immediately push the A-V valves open and allow blood to flow rapidly into the ventricles, as shown by the *ventricular volume curve* in Figure 13–5. This is called the *period of rapid filling of the ventricles.* The atrial pressures fall to within a fraction of a millimeter of the ventricular pressures because the normal A-V valve openings are so large that they offer almost no resistance to blood flow.

The period of rapid filling lasts approximately the first third of diastole. During the middle third of diastole only a small amount of blood normally flows into the ventricles; this is blood that continues to empty into the atria from the veins and passes on through the atria directly into the ventricles. This middle third of diastole, when the inflow of blood into the ventricles is almost at a standstill, is called *diastasis.*

During the latter third of diastole, the atria contract and give an additional thrust to the inflow of blood into the ventricles; this accounts for approximately 30 per cent of the filling of the ventricles during each heart cycle.

Emptying of the Ventricles During Systole. *Period of Isometric (Isovolumic) Contraction.* Immediately after ventricular contraction begins, the ventricular pressure abruptly rises, as shown in Figure 13–5, causing the A-V valves to close. Then an additional 0.02 to 0.03 second is required for the ventricle to build up sufficient pressure to push the semilunar (aortic and pulmonary) valves open against the pressures in the aorta and pulmonary artery. Therefore, during this period of time, contraction is occurring in the ventricles, but there is no emptying. This period is called the period of isometric or isovolumic contraction, meaning by these terms that tension is increasing in the muscle but no shortening of the muscle fibers is occurring. (This is not strictly true because there is apex-to-base shortening and circumferential elongation.)

Period of Ejection. When the left ventricular pressure rises slightly above 80 mm. Hg and the right ventricular pressure slightly above 8 mm. Hg, the ventricular pressures now push the semilunar valves open. Immediately, blood begins to pour out of the ventricles, with about 60 per cent of the emptying occurring during the first quarter of systole and usually most of the remaining 40 per cent during the next two quarters. These three-quarters of systole are called the period of ejection.

Protodiastole. During the last one-fourth of ventricular systole, little blood flows from the ventricles into the large arteries; yet the ventricular musculature remains contracted. This period is called protodiastole. The arterial pressure falls during this period, because large quantities of blood are flowing from the arteries through the peripheral vessels.

The ventricular pressure actually falls to a value *below* that in the aorta during protodiastole despite the fact that some blood is still leaving the ventricles. The reason is that the blood that is flowing out of the ventricles has built up momentum. As this momentum decreases during the latter part of systole, the kinetic energy of the momentum is converted into pressure in the large arteries, which makes the arterial pressure slightly greater than the pressure inside the ventricles.

Period of Isometric (Isovolumic) Relaxation. At the end of systole, ventricular relaxation begins suddenly, allowing the intraventricular pressures to fall rapidly. The elevated pressures in the large arteries immediately push blood back toward the ventricles, which snaps the aortic and pulmonary valves closed. For another 0.03 to 0.06 second, the ventricular muscle continues to relax, and the intraventricular pressures fall rapidly back to their very low diastolic levels. Then the A-V valves open to begin a new cycle of ventricular pumping.

End-Diastolic Volume, End-Systolic Volume, and Stroke Volume Output. During diastole, filling of the ventricles normally increases the volume of each ventricle to about 120 to 130 ml. This volume is known as the *end-diastolic volume.* Then, as the ventricles empty during systole, the volume decreases about 70 ml., which is called the *stroke volume output*. The remaining volume in each ventricle, about 50 to 60 ml., is called the *end-systolic volume*.

When the heart contracts strongly, the end-systolic volume can fall to as little as 10 to 30 ml. On the other hand, when large amounts of blood flow into the ventricles during diastole, their end-diastolic volumes can become as great as 200 to 250 ml. in the normal heart. And by both increasing the end-diastolic volume and decreasing the end-systolic volume, the stroke volume output can be increased to more than double normal.

FUNCTION OF THE VALVES

The Atrioventricular Valves. The *A-V valves* (the *tricuspid* and the *mitral* valves) prevent backflow of blood from the ventricles to the atria during systole, and the *semilunar valves* (the *aortic* and *pulmonary* valves) prevent backflow from the aorta and pulmonary arteries into the ventricles during diastole. All these valves, which are illustrated in Figure 13–6, close and open *passively*. That is, they close when a backward pressure gradient pushes blood backward, and they open when a forward pressure gradient forces blood in the forward direction. For obvious anatomical reasons, the thin, filmy A-V valves require almost no backflow to cause closure while the much heavier semilunar valves require rather strong backflow for a few milliseconds.

Function of the Papillary Muscles. Figure 13–6 also illustrates the papillary muscles that attach to

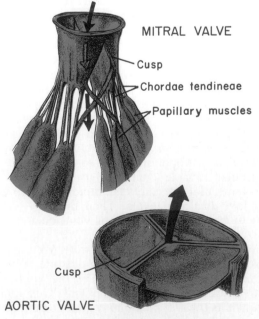

Figure 13–6. Mitral and aortic valves.

MITRAL VALVE

Cusp

Chordae tendineae

Papillary muscles

Cusp

AORTIC VALVE

the vanes of the A-V valves by the *chordae tendineae*. The papillary muscles contract when the ventricular walls contract, but, contrary to what might be expected, they *do not* help the valves to close. Instead, they pull the vanes of the valves inward toward the ventricles to prevent their bulging too far backward toward the atria during ventricular contraction. If a chorda tendinea becomes ruptured or if one of the papillary muscles becomes paralyzed, the valve bulges far backward, sometimes so far that it leaks severely and results in severe or even lethal cardiac incapacity.

The Aortic and Pulmonary Valves. There are differences between the operation of the aortic and pulmonary valves and that of the A-V valves. First, the high pressures in the arteries at the end of systole cause the semilunar valves to snap to the closed position in comparison with a much softer closure of the A-V valves. Second, the velocity of blood ejection through the aortic and pulmonary valves is far greater than that through the much larger A-V valves. Because of the rapid closure and rapid ejection, the edges of the semilunar valves are subjected to much greater mechanical abrasion than are the A-V valves, which also are supported by the chordae tendineae. It is obvious from the anatomy of the aortic and pulmonary valves, as illustrated in Figure 13–6, that they are well adapted to withstand this extra physical trauma.

THE AORTIC PRESSURE CURVE

The pressure that develops in the aorta depends on many other factors in addition to contraction of the heart. The aortic pressure curve will be discussed in detail in Chapter 19, but it is desirable now simply to point out those features of the curve that relate especially to the cardiac cycle.

When the left ventricle contracts, the ventricular pressure rises rapidly until the aortic valve opens. Then the pressure in the ventricle rises much less thereafter, as illustrated in Figure 13–5, because blood immediately flows out of the ventricle into the aorta.

The entry of blood into the arteries causes the walls of these arteries to stretch and the pressure to rise. Then, at the end of systole, after the left ventricle stops ejecting blood and the aortic valve closes, the elastic recoil of the arteries maintains a moderately high pressure in the arteries even during diastole.

A so-called *incisura* occurs in the aortic pressure curve when the aortic valve closes. This is caused by a short period of backward flow of blood immediately prior to closure of the valve.

After the aortic valve has closed, pressure in the aorta falls slowly throughout diastole because blood stored in the distended elastic arteries flows continually through the peripheral vessels back to the veins. By the time the ventricle contracts again, the aortic pressure usually has fallen to approximately 80 mm. Hg (diastolic pressure), which is two-thirds the maximal pressure of 120 mm. Hg (systolic pressure) occurring in the aorta during ventricular contraction.

The pressure curve in the pulmonary artery is similar to that in the aorta, except that the pressures are much less, as will be discussed in Chapter 24.

RELATIONSHIP OF THE HEART SOUNDS TO HEART PUMPING

When listening to the heart with a stethoscope, one does not hear the opening of the valves, for this is a relatively slowly developing process that makes no noise. However, when the valves close, the vanes of the valves and the surrounding fluids vibrate under the influence of the sudden pressure differentials that develop, giving off sound that travels in all directions through the chest. When the ventricles first contract, one hears a sound that is caused by closure of the A-V valves. The vibration is low in pitch and relatively long continued and is known as the *first heart sound*. When the aortic and pulmonary valves close, one hears a relatively rapid snap, for these valves close extremely rapidly, and the surroundings vibrate for only a short period of time. This sound is known as the *second heart sound*. The precise causes of these sounds will be discussed in Chapter 27 in relation to auscultation.

Occasionally, one can hear an *atrial sound* when the atria beat, presumably because of vibrations associated with the flow of blood into the ventricles. Also, a *third heart sound* sometimes occurs at the end of the first third of diastole or in the middle of diastole. This is said to be caused by blood flowing with a rumbling motion into the almost-filled ventricles. The atrial sound and the third heart sound can usually be recorded with special recording instruments but can be heard with the stethoscope only with great difficulty.

The Terms "Systole" and Diastole" in Clinical Usage. Strictly speaking, systole means "contraction." Therefore, physiologically it is probably best that systole be considered to begin approximately with the closure of the A-V valves and to end approximately with the opening of the A-V valves.

Clinically, it is not possible to determine when the A-V valves open. Yet, because only a short interval elapses between the closure of the aortic and pulmonary valves and the opening of the A-V valves, the clinician measures systole as the time interval between the first heart sound and the second heart sound or, in other words, the time interval between closure of the A-V valves and closure of the semilunar valves. Diastole is considered to be the interval between closure of the aortic and pulmonary valves and closure of the A-V valves.

WORK OUTPUT OF THE HEART

The work output of the heart is the amount of energy that the heart converts to work while pumping blood into the arteries. This is in two forms: First, by far the major proportion is used to move the blood from the low pressure veins to the high pressure arteries. This is *potential energy of pressure*. Second, a minor proportion of the energy is used to accelerate the blood to its velocity of ejection through the aortic and pulmonary valves. This is *kinetic energy of blood flow*.

Stroke Work Output. The work performed by the left ventricle to raise the pressure of the blood during each heartbeat is equal to *stroke volume output × (left ventricular mean ejection pressure* minus *left atrial pressure)*. Likewise, the work performed by the right ventricle to raise the pressure of the blood is called the stroke work output. This is equal to *stroke volume output × (right ventricular mean ejection pressure* minus *right atrial pressure)*. When pressure is expressed in *dynes per square centimeter* and stroke volume output in *milliliters*, the work output is in *ergs*. Right ventricular work output is usually about one-seventh the work output of the left ventricle because of the difference in systolic pressure against which the two ventricles must pump.

The Kinetic Energy of Blood Flow. The work output of each ventricle required to create kinetic energy of blood flow is proportional to the mass of blood ejected times the square of velocity of ejection. That is,

$$\text{Kinetic energy} = \frac{mv^2}{2}$$

When the mass is expressed in *grams* of blood ejected and the velocity in *centimeters per second*, the work output is in *ergs*.

Ordinarily, the work output of the left ventricle required to create kinetic energy of blood flow is about 1 per cent of the total work output of the ventricle. Most of this energy is required to cause the rapid acceleration of blood during the first quarter of systole. In certain abnormal conditions, such as aortic stenosis, in which the blood flows with great velocity through the stenosed valve, as much as 50 per cent of the total work output may be required to create kinetic energy of blood flow.

ENERGY FOR CARDIAC CONTRACTION

Heart muscle, like skeletal muscle, utilizes chemical energy to provide the work of contraction. This energy is derived mainly from metabolism of fatty acids and to a lesser extent from metabolism of other nutrients, especially lactate and glucose. The different reactions that liberate this energy will be discussed in detail in Chapters 67 and 68.

The amount of energy expended by the heart is related to its work load in the following manner: The energy expended is *approximately proportional to the peak tension generated by the heart musculature during contraction*. In addition, the amount of time that the tension is maintained also plays a lesser role in determining the energy expenditure.

Efficiency of Cardiac Contraction. During muscular contraction most of the chemical energy is converted into heat and a small portion into work output. The ratio of work output to chemical energy expenditure is called the efficiency of cardiac contraction, or simply *efficien-*

cy of the heart. The efficiency of the normal heart is between 20 and 25 per cent.

REGULATION OF CARDIAC FUNCTION

When a person is at rest, the heart pumps only 4 to 6 liters of blood each minute. However, during severe exercise it may be required to pump as much as 4 to 7 times this amount. The present section discusses the means by which the heart can adapt itself to such extreme increases in cardiac output.

The two basic means by which the volume pumped by the heart is regulated are (1) intrinsic autoregulation of pumping in response to changes in volume of blood flowing into the heart and (2) reflex control of the heart by the autonomic nervous system.

INTRINSIC AUTOREGULATION OF CARDIAC PUMPING — THE FRANK-STARLING LAW OF THE HEART

In Chapter 23 we shall see that one of the major factors determining the amount of blood pumped by the heart each minute is the rate of blood flow into the heart from the veins, which is called *venous return.* That is, each peripheral tissue of the body controls its own blood flow, and whatever amount of blood flows through all the peripheral tissues returns by way of the veins to the right atrium. The heart in turn automatically pumps this incoming blood on into the systemic arteries so that it can flow around the circuit again. Thus, the heart must adapt itself from moment to moment or even second to second to widely varying inputs of blood, sometimes falling as low as 2 to 3 liters per minute and at other times rising to as high as 25 or more liters per minute.

This intrinsic ability of the heart to adapt itself to changing loads of inflowing blood is called the *Frank-Starling law of the heart,* in honor of Frank and Starling, two great physiologists over half a century ago. Basically, the Frank-Starling law states that the greater the heart is filled during diastole, the greater will be the quantity of blood pumped into the aorta. Or another way to express this law is: *Within physiological limits, the heart pumps all the blood that comes to it without allowing excessive damming of blood in the veins.* In other words, the heart can pump either a small amount of blood or a large amount, depending on the amount that flows into it from the veins; and it automatically adapts to whatever this load might be as long as the total quantity does not rise above the physiologic limit that the heart can pump.

Mechanism of the Frank-Starling Law. The pri-mary mechanism by which the heart adapts to changing inflow of blood is the following: When the cardiac muscle becomes stretched an extra amount, as it does when extra amounts of blood enter the heart chambers, the stretched muscle contracts with a greatly increased force, thereby automatically pumping the extra blood into the arteries. This ability of stretched muscle to contract with increased force is characteristic of all striated muscle and not simply of cardiac muscle. Referring back to Chapter 11, one will see that stretching a skeletal muscle, within its physiological limit, also increases its force of contraction. As was pointed out in Chapter 11, the increased force of contraction is probably caused by the fact that the actin and myosin filaments are brought to a more nearly optimal degree of interdigitation for achieving contraction. This ability of the heart to contract with increased force as its chambers are stretched is sometimes called *heterometric autoregulation of the heart.*

Effect of Heart Rate and "Homeometric" Autoregulation. In addition to the important effect of stretching the heart muscle, at least two other, less important factors increase heart pumping effectiveness when its volume is increased. First, stretch of the right atrial walls increases the heart rate by as much as 10 to 30 per cent; this in itself can increase the amount of blood pumped each minute. Second, changes in heart metabolism that occur when the heart is stretched cause an additional increase in contractile strength. It takes approximately 30 seconds for this effect to develop fully, an effect called *homeometric autoregulation* because the increased contractile strength returns the muscle fiber lengths nearly to their original lengths, thus giving a considerable increase in output with very little change in heart muscle length.

Failure of Arterial Pressure Load to Alter Cardiac Output. One of the most important features of the Frank-Starling law of the heart is that, within reasonable limits. changes in arterial pressure load against which the heart pumps have almost no effect on the rate at which blood is pumped by the heart each minute (the cardiac output). This effect is illustrated in Figure 13–7, which is a curve extrapolated to the human being from data in dogs in which the arterial pressure was progressively changed by constricting the arteries while the cardiac output was measured simultaneously. The significance of this effect is the following: Regardless of the arterial pressure, the most important factor determining the amount of blood pumped by the heart is still the right atrial pressure generated by the entry of blood into the heart.

Figure 13–7 shows that when the arterial pressure rises above approximately 170 mm. Hg, the arterial pressure load then causes the heart to

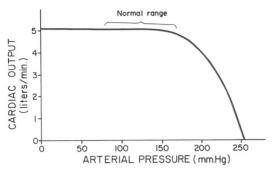

Figure 13–7. Constancy of cardiac output even in the face of wide changes in arterial pressure. Only when the arterial pressure rises above the normal operating pressure range does the pressure load cause the heart to begin to fail.

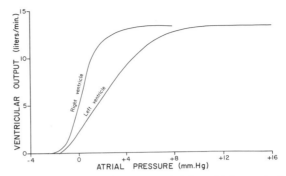

Figure 13–9. Approximate normal right and left ventricular output curves for the human heart as extrapolated from data obtained in dogs.

begin to fail. However, this figure also shows that the normal daily range of arterial pressures is between approximately 80 and 170 mm. Hg, again emphasizing that within the normal operating range, the output of the heart is independent of changes in arterial pressure.

The independence of the cardiac output of changes in arterial pressure load is partly caused by the fact that the heart is a two-stage pump, for the following reasons: Even if the left ventricle were to begin to fail moderately, this would raise the left atrial pressure a few millimeters Hg. This elevation of left atrial pressure then would raise the pulmonary arterial pressure only a fraction of a millimeter because left atrial pressure is not transmitted backward through the lungs to a great extent until this pressure becomes abnormally elevated. Therefore, the right ventricle experiences almost no change in load and continues to pump normal quantities of blood into the lungs, forcing the left ventricle (even though it might be beginning to fail) to continue to pump a normal cardiac output.

Ventricular Function Curves. One of the best ways to express the functional ability of the ventri-

cles to pump blood is by ventricular function curves, as shown in Figures 13–8 and 13–9. Figure 13–8 illustrates a type of ventricular function curve called the *stroke work output curve*. Note that as the atrial pressures increase, the stroke work output also increases until it reaches the limit of the heart's ability.

Figure 13–9 illustrates another type of ventricular function curve called the *minute ventricular output curve*. These two curves represent function of the two ventricles of the human heart based on data extrapolated from lower animals. As each atrial pressure rises, the respective ventricular volume output per minute also increases.

Thus, ventricular function curves are another way of expressing the Frank-Starling law of the heart. That is, as the ventricles fill to higher atrial pressures, the strength of cardiac contraction increases, causing the heart to pump increased quantities of blood into the arteries. In later chapters we shall see that ventricular function curves are exceedingly important in analyzing overall function of the circulation, for it is by such means that one can express the quantitative capabilities of the heart as a pump.

CONTROL OF THE HEART BY NERVES

The heart is well supplied with both sympathetic and parasympathetic (vagal) nerves, as illustrated in Figure 13–10. These nerves affect cardiac pumping in two ways: (1) by changing the heart rate, and (2) by changing the strength of contraction of the heart. The effect of nerve stimulation on heart rate and rhythm will be discussed in detail in the following chapter. For the present, suffice it to say that parasympathetic stimulation decreases heart rate, and sympathetic stimulation increases heart rate. The range of control is from as little as 20 to 30 heart beats per minute with maximum vagal stimulation up to as high as 250 or, rarely, 300 heart beats per minute with maximum sympathetic stimulation.

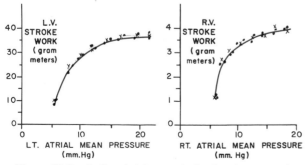

Figure 13–8. Left and right ventricular function curves in a dog, depicting ventricular stroke work output as a function of left and right mean atrial pressures. (Curves reconstructed from data in Sarnoff: *Physiol. Rev., 35*:107, 1955.)

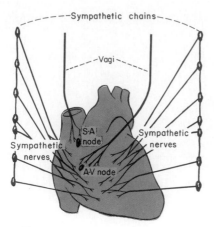

Figure 13–10.   The cardiac nerves.

Effect of Heart Rate on Function of the Heart as a Pump.

In general, the more times the heart beats per minute, the more blood it can pump, but there are important limitations to this effect. For instance, once the heart rate rises above a critical level, the heart strength itself decreases, presumably because of overutilization of metabolic substrates in the cardiac muscle. In addition, the period of diastole between the contractions becomes so reduced that blood does not have time to flow adequately from the atria into the ventricles. For these reasons, when the heart rate is increased artificially by electrical stimulation, the heart has its peak ability to pump large quantities of blood at a heart rate between 100 and 150 beats per minute. On the other hand, when its rate is increased by sympathetic stimulation, it reaches its peak ability to pump blood at a heart rate between 170 and 250 beats per minute. The reason for this difference is that sympathetic stimulation not only increases the heart rate but also increases heart strength as well.

When the atrial pressures are high, the higher heart rates are especially effective in increasing the output of the heart.

Nervous Regulation of Contractile Strength of the Heart.

The two atria are especially well supplied with large numbers of both sympathetic and parasympathetic nerves, but the ventricles are supplied mainly by sympathetic nerves and far fewer parasympathetic fibers. In general, sympathetic stimulation increases the strength of heart muscle contraction, whereas parasympathetic stimulation decreases the strength of contraction.

Under normal conditions the sympathetic nerve fibers to the heart continually discharge at a slow rate that maintains a strength of ventricular contraction about 20 per cent above its strength with no sympathetic stimulation at all. Therefore, one method by which the nervous system can decrease the strength of ventricular contraction is simply to

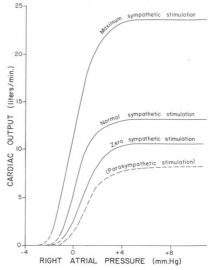

Figure 13–11.   Effect on the cardiac output curve of different degrees of sympathetic and parasympathetic stimulation.

slow or stop the transmission of sympathetic impulses to the heart. On the other hand, maximal sympathetic stimulation can increase the strength of ventricular contraction to approximately 100 per cent greater than normal.

Maximal parasympathetic stimulation of the heart decreases ventricular contractile strength about 30 per cent. Thus, the parasympathetic effect is, by contrast with the sympathetic effect, relatively small.

Effect of Autonomic Control of Cardiac Function As Depicted by Cardiac Function Curves.

Figure 13–11 illustrates four separate cardiac function curves. These are much the same as the ventricular function curves of Figures 13–8 and 13–9, except that they represent function of the entire heart rather than of a single ventricle; they show the relationship between the right atrial pressure at the input of the heart and cardiac output from the left ventricle.

The curves of Figure 13–11 demonstrate that at any given right atrial pressure, the cardiac output increases with increasing sympathetic stimulation and decreases with increasing parasympathetic stimulation. Note in particular that the changes in the function curves of Figure 13–11 are brought about by both *changes in heart rate* and *changes in contractile strength of the heart,* for both of these affect cardiac output.

EFFECT OF HEART DEBILITY ON CARDIAC FUNCTION — THE HYPOEFFECTIVE HEART

Any factor that damages the heart, whether it be damage to the myocardium, to the valves, to the conducting system, or otherwise, is likely to make the heart a poorer pump, and the heart under these conditions is called a hypoeffective heart. Figure 13–12 illustrates by the very dark curve the normal cardiac function curve and by the three curves below this the effect of different degrees of hypoeffectiveness on cardiac function. Obviously, the more serious the damage, the less will be the cardiac output at any given right atrial pressure.

Different factors that can cause a hypoeffective heart include:

Myocardial infarction
Valvular heart disease
Vagal stimulation of the heart
Inhibition of the sympathetics to the heart
Congenital heart disease
Myocarditis
Cardiac anoxia
Diphtheritic or other types of myocardial damage

EFFECT OF EXERCISE ON THE HEART — THE HYPEREFFECTIVE HEART

Chronic heavy exercise over a period of many weeks or months leads to hypertrophy of the cardiac muscle and also to enlargement of the ventricular chambers. As a result, the overall strength of the heart becomes greatly enhanced, and the effectiveness of the heart as a pump increases. The upper three function curves of Figure 13–12 illustrate the effect of different degrees of cardiac hypereffectiveness on the cardiac function curve; maximal degrees of hypereffectiveness can increase pumping by the heart more than 100 per cent.

Other factors besides hypertrophy that can cause a hypereffective heart are:

Sympathetic stimulation (50 to 100 per cent increase)
Parasympathetic inhibition (10 to 20 per cent increase)

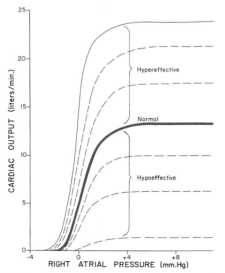

Figure 13–12. Cardiac output curves for various degrees of hypo- and hypereffective hearts. (From Guyton, Jones, and Coleman: Circulatory Physiology: Cardiac Output and Its Regulation. Philadelphia, W. B. Saunders Company, 1973.)

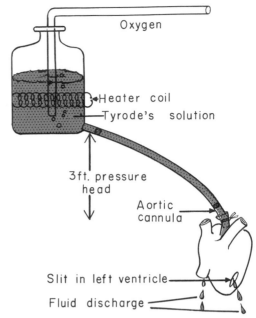

Figure 13–13. System for perfusing the heart.

EFFECT OF VARIOUS IONS ON HEART FUNCTION

In the discussion of membrane potentials in Chapter 10, it was pointed out that three particular cations —potassium, sodium, and calcium — all have marked effects on action potential transmission, and in Chapter 11 it was noted that calcium ions play an especially important role in initiating the muscle contractile process. Therefore, it is to be expected that the concentrations of these three ions in the extracellular fluids will also have important effects on cardiac function.

The action of individual ions on the heart can be studied by perfusing the isolated heart, as illustrated in Figure 13–13, using the so-called *Langendorf preparation*. In making this preparation, the heart is removed rapidly from the animal's body, and the aorta is cannulated immediately. The cannula is connected to a perfusion bottle located approximately three feet higher than the heart itself so that the fluid in the aorta will be under pressure. The pressure causes the aortic valve to close and the fluid to flow through the coronary vessels. When the perfusion fluid contains appropriate nutrients (glucose and oxygen) and ions, and the heart is kept appropriately warmed, the perfused heart will beat for many hours.

Effect of Potassium Ions. Excess potassium in the extracellular fluids causes the heart to become extremely dilated and flaccid and slows the heart rate. Very large quantities can also block conduction of the cardiac impulse from the atria to the ventricles through the A-V bundle. Elevation of potassium concentration to only 8 to 12 mEq./liter — two to three times the normal value — will usually cause such weakness of the heart and abnormal rhythm that it will cause death.

All these effects of potassium excess are almost certainly caused by decreased negativity of the resting membrane potential which results from the high potassium concentration in the extracellular fluids. As the membrane potential decreases, the intensity of the action potential also decreases, which makes the contraction of the heart progressively weaker, for the strength of the action potential determines to a great extent the strength of contraction.

Effect of Calcium Ions. An excess of calcium ions causes effects almost exactly opposite to those of potassium ions, causing the heart to go into spastic contraction. This is probably caused by the direct effect of calcium ions to excite the cardiac contractile process, as explained earlier in the chapter. Conversely, a deficiency of calcium ions causes cardiac flaccidity, similar to the effect of potassium.

However, the calcium ion concentration rarely changes sufficiently during life to alter cardiac function greatly, for greatly diminished calcium ion concentration will usually kill a person because of tetany before it will significantly affect the heart, and elevation of the calcium ion concentration to a level that will significantly affect the heart almost never occurs because calcium ions are precipitated in bone or occasionally elsewhere in the body's tissues as insoluble calcium salts before such a level can be reached.

Effect of Sodium Ions. An excess of sodium ions depresses cardiac function, an effect similar to that of potassium ions but for an entirely different reason. Sodium ions compete with calcium ions at some yet unexplained point in the contractile process of muscle in such a way that the greater the sodium ion concentration in the extracellular fluids the less the effectiveness of the calcium ions in causing contraction when an action potential occurs.

However, from a practical point of view, the sodium ion concentration in the extracellular fluids probably never becomes abnormal enough even in serious pathological conditions to cause significant change in cardiac strength. However, very low sodium concentration, as occurs in water intoxication, often causes death because of cardiac fibrillation, a phenomenon explained in the following chapter.

EFFECT OF TEMPERATURE ON THE HEART

The effect of temperature on the heart can also be studied using the perfusion system of Figure 13–13. Increased temperature causes greatly increased heart rate, and decreased temperature causes greatly decreased rate. These effects presumably result from the heat causing increased permeability of the muscle membrane to the ions, resulting in acceleration of the self-excitation process.

Contractile strength of the heart is often enhanced temporarily by a moderate increase in temperature, but prolonged elevation of the temperature exhausts the heart and causes weakness.

THE HEART-LUNG PREPARATION

A method frequently used to demonstrate the capabilities of the heart as a pump is the heart-lung preparation illustrated in Figure 13–14. In this preparation, blood leaving the heart by way of the aorta is diverted into an external system of tubes and then back again into the

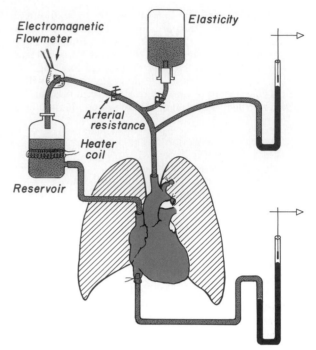

Figure 13–14. The heart-lung preparation.

right atrium. The only parts of the animal's body that remain alive are the heart and lungs. The blood is oxygenated by artificial respiration of the lungs, and glucose is added periodically to the blood to supply nutrition for cardiac contraction.

As blood flows through the external circuit, the right atrial pressure, aortic pressure, and blood flow are measured. Also, an adjustable venous reservoir allows the pressure in the right atrium to be increased or decreased. A screw clamp is provided on the main outflow tube from the heart so that the arterial resistance can be increased or decreased. Finally, a special reservoir containing air in its upper part and blood in its lower part is used as an "elastic buffer" *(windkessel)* so that the blood pressure will not rise and fall excessively with each contraction of the heart. Some of the principles of cardiac function that can be demonstrated with the heart-lung preparation are the following:

First, the *effect of end-diastolic volume or pressure on stroke volume output* can be demonstrated by raising and lowering the venous reservoir. When the reservoir is raised, blood flows into the heart rapidly, distending the cardiac chambers and causing increased stroke volume output, thus demonstrating the Frank-Starling law of the heart.

Second, one can show that *increasing the arterial resistance does not significantly decrease the stroke volume output as long as the right atrial pressure remains constant.* This illustrates that the aortic pressure, within physiological limits, has rather little effect on the output, though very high pressures can overload the heart so much that the stroke volume output does then become reduced.

Third, the *effect of different factors on heart rate* can be studied, such as a slight increase in heart rate that occurs when the load of inflowing blood increases, an increase in heart rate with increasing temperature, an increase in heart rate when epinephrine is injected into blood perfusing the heart, a decrease in heart rate when acetylcholine is injected, and changes in heart rate when the nerves to the heart are stimulated.

ASSESSMENT OF CONTRACTILITY

Though it is very easy to determine the heart rate by simply timing the pulse, it has always been difficult to determine the strength of contraction of the heart, commonly called *cardiac contractility*. Very often the change in contractility is exactly opposite to the change in heart rate. Indeed, this effect occurs almost invariably in heart-debilitating diseases.

One of the ways in which cardiac contractility can be determined with great precision is to record one or more of the cardiac function curves. However, this can be done only in experimental animals, and even then only with considerable difficulty. Therefore, many physiologists and clinicians have searched for methods to assess the cardiac contractility in a simple way. One of these methods is to determine the so-called dP/dt.

dP/dt as a Measure of Cardiac Contractility. dP/dt means the *rate of change of the ventricular pressure* with respect to time. The dP/dt record is generated by an electronic computer that *differentiates* the ventricular pressure wave, thus giving a record of the rate of change of the ventricular pressure. Figure 13–15 illustrates two separate recordings of the ventricular pressure wave as well as simultaneous recordings (in color) of the dP/dt. In the upper part of the figure the heart was beating normally, and in the lower part the heart had been stimulated by isoproterenol, a drug that has essentially the same effect on the heart as sympathetic stimulation, as will be discussed in more detail in Chapter 57.

Note in the upper record that at the same time that the ventricular pressure is increasing at its most rapid rate, the recording of the dP/dt record also reaches its greatest height. On the other hand, at the time that the ventricular pressure is falling most rapidly, the dP/dt record reaches its lowest level. When the ventricular pressure is neither rising nor falling, the dP/dt record is at zero value.

Experimental studies have shown that the rate of rise of ventricular pressure, the dP/dt, in general correlates very well with the strength of contraction of the ventricle. This effect is illustrated by a comparison of the dP/dt record in the upper part of Figure 13–15 with that in the lower part. Note that the peak value for dP/dt in the upper record is only 1800 mm. Hg per second, whereas in the lower record it rises to approximately 2600 mm. Hg per second, illustrating the stimulatory effect of isoproterenol on the contractility of the ventricle. Thus, the *peak* dP/dt is often used as a means for comparing the contractilities of hearts in different functional states.

Unfortunately, the quantitative value for peak dP/dt is also affected by other factors that are not related to cardiac contractility. For instance, the value is increased by both increased input pressure to the left ventricle (the end-diastolic ventricular pressure) and the pressure in the aorta into which the heart is pumping the

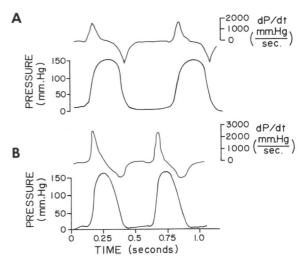

Figure 13–15. Simultaneous recordings of ventricular pressure and dP/dt. A shows results from a normal heart and B, from a heart stimulated by isoproterenol. (Modified from Mason et al., in Sodeman and Sodeman (eds.): Pathologic Physiology. 5th ed. Philadelphia, W. B. Saunders Company, 1974.)

blood. Therefore, it is often difficult to use dP/dt as a measure of contractility in comparing hearts from one person to another because one of these factors may differ. For this reason other quantitative measures have also been used in attempts to assess cardiac contractility. One of these has been to use dP/dt divided by the instantaneous pressure in the ventricle, or (dP/dt)/P.

Use of V_{max} as a Measure of Contractility. Experimental studies have also shown that the rate of shortening of the cardiac muscle is often a good measure of the contractile strength. However, it is very difficult to measure this rate of shortening directly. On the other hand, several indirect methods for estimating the rate of shortening have been achieved. It will not be possible to explain these procedures here, but the peak estimated velocity of ventricular muscle shortening at zero intraventricular pressure (called V_{max}) is often used, especially in experimental situations, to assess cardiac contractility. However, here again, a number of factors besides cardiac contractility enter into determining the level of V_{max}.

Thus, both the dP/dt and the V_{max} determinations are useful in comparative studies of ventricular contractility under many conditions. Since both of them also display serious inaccuracies in some cardiac states, however, they are at present considered to have considerable empirical experimental value but not necessarily to represent absolute measures of cardiac contractility.

REFERENCES

Alpert, N. R., *et al.*: Heart muscle mechanics. *Annu. Rev. Physiol.*, 41:521, 1979.
Atsumi, K., and Nose, Y. (chairpersons): Mechanical circulatory devices. *In* Hayase, S., and Murao, S. (eds.): Cardiology: Proceedings of the VIII World Congress of Cardiology, Tokyo, 1978. New York, Elsevier/North-Holland, 1979, p. 999.
Baan, J., *et al.* (eds.): Cardiac Dynamics. Boston, M. Nijhoff, 1979.
Bishop, V. S., *et al.*: Factors influencing cardiac performance. *Int. Rev. Physiol.*, 9:239, 1976.

Boom, H. B. K.: Elasticity of the Heart: Instantaneous Pressure Volume Relations of the Left Ventricle Throughout the Cardiac Cycle. Rotterdam, Bronder-Offset, 1971.

Brady, A. J.: Mechanical properties of cardiac fibers. *In* Berne, R. M., *et al.* (eds.): Handbook of Physiology. Sec. 2, Vol. I. Baltimore, Williams & Wilkins, 1979, p. 461.

Braunwald, E., and Ross, J., Jr.: Control of cardiac performance. *In* Berne, R. M., *et al.* (eds.): Handbook of Physiology. Sec. 2, Vol. I. Baltimore, Williams & Wilkins, 1979, p. 533.

Bruce, T. A., and Douglas, J. E.: Dynamic cardiac performance. *In* Frohlich, E. D. (ed.): Pathophysiology, 2nd Ed. Philadelphia, J. B. Lippincott Co., 1976, p. 5.

Brutsaert, D. L., and Sonnenblick, E. H.: Cardiac muscle mechanics in the evaluation of myocardial contractility and pump function: Problems, concepts, and directions. *Prog. Cardiovasc. Dis., 16*:337, 1973.

Cowley, A. W., Jr., and Guyton, A. C.: Heart rate as a determinant of cardiac output in dogs with arteriovenous fistula. *Am. J. Cardiol., 28*:321, 1971.

Ebashi, S.: Modern concepts of myocardial contraction. *In* Hayase, S., and Murao, S. (eds.): Cardiology: Proceedings of the VIII World Congress of Cardiology, Tokyo, 1978. New York, Elsevier/North-Holland, 1979, p. 92.

Fabiato, A., and Fabiato, F.: Calcium and cardiac excitation-contraction coupling. *Annu. Rev. Physiol., 41*:473, 1979.

Guyton, A. C.: Determination of cardiac output by equating venous return curves with cardiac response curves. *Physiol. Rev., 35*:123, 1955.

Guyton, A. C., *et al.*: Circulatory Physiology: Cardiac Output and Its Regulation, 2nd Ed. Philadelphia, W. B. Saunders Co., 1973.

Jewell, B. R.: The physiology of cardiac muscle contraction. *In* Dickinson, C. J., and Marks, J. (eds.): Developments in Cardiovascular Medicine. Lancaster, England, MTP Press, 1978, p. 129.

Johnson, E. A.: Force-interval relationship of cardiac muscle. *In* Berne, R. M., *et al.* (eds.): Handbook of Physiology. Sec. 2, Vol. 1. Baltimore, Williams & Wilkins, 1979, p. 475.

Johnson, E. A., and Lieberman, M.: Heart: Excitation and contraction. *Annu. Rev. Physiol., 33*:479, 1971.

Krebs, R., and Kersting, F.: The Effect of Barbiturates on the Myocardium and Its Reversibility. New York, Fischer, 1979.

Langer, G. A.: Heart: Excitation-contraction coupling. *Annu. Rev. Physiol., 35*:55, 1973.

Langer, G. A., and Brady, A. J.: The Mammalian Myocardium. New York, John Wiley & Sons, 1974.

Langer, G. A., *et al.*: The myocardium. *Int. Rev. Physiol., 9*:191, 1976.

Levy, M. N., and Martin, P. J.: Cardiac excitation and contraction. *In* MTP International Review of Science: Physiology. Baltimore. University Park Press, 1974, Vol. 1, p. 49.

Lieberman, M., and Sano, T. (eds.): Developmental and Physiological Correlates of Cardiac Muscle. New York, Raven Press, 1975.

Lüllmann, H., and Peters, T.: Action of Cardiac Glycosides on the Excitation-Contraction Coupling in Heart Muscle. New York, Fischer, 1979.

Mirsky, I.: Elastic properties of the myocardium: A quantitative approach with physiological and clinical applications. *In* Berne, R. M., *et al.* (eds.): Handbook of Physiology. Sec. 2, Vol. I. Baltimore, Williams & Wilkins, 1979, p. 497.

Navaratnam, V.: The structure of cardiac muscle. *In* Dickinson, C. J., and Marks, J. (eds.): Developments in Cardiovascular Medicine. Lancaster, England, MTP Press, 1978, p. 119.

Parmley, W. W., and Talbot, W.: Heart as a pump. *In* Berne, R. M., *et al.* (eds.): Handbook of Physiology. Sec. 2, Vol. I. Baltimore, Williams & Wilkins, 1979, p. 429.

Sarnoff, S. J.: Myocardial contractility as described by ventricular function curves. *Physiol. Rev., 35*:107, 1955.

Sommer, J. R., and Johnson, E. A.: Ultrastructure of cardiac muscle. *In* Berne, R. M., *et al.* (eds.): Handbook of Physiology. Sec. 2, Vol. I. Baltimore, Williams & Wilkins, 1979, p. 113.

Starling, E. H.: The Linacre Lecture on the Law of the Heart. London, Longmans Green & Co., 1918.

Stone, H. L., *et al.*: Cardiac function after embolization of coronaries with microspheres. *Am. J. Physiol., 204*:16, 1963.

Sugimoto, T., *et al.*: Quantitative effect of low coronary pressure on left ventricular performance. *Jap. Heart J., 9*:46, 1968.

Sugimoto, T., *et al.*: Effect of maximal work load on cardiac function. *Jap. Heart J., 14*:146, 1973.

Trautwein, W.: Membrane currents in cardiac muscle fibers. *Physiol. Rev., 53*:793, 1973.

Van der Werf, T.: Proceedings of the 2nd Workshop on Contractile Behavior of the Heart, Utrecht. 1973. New York, American Elsevier Publishing Co., 1974.

Weisfeldt, M. L. (ed.): The Heart in Old Age: Its Function and Response to Stress. New York, Raven Press, 1980.

Winegrad, S.: Electromechanical coupling in heart muscle. *In* Berne, R. M., *et al.* (eds.): Handbook of Physiology. Sec. 2, Vol. I. Baltimore, Williams & Wilkins, 1979, p. 393.

14

Rhythmic Excitation of the Heart

The heart is endowed with a special system (a) for generating rhythmical impulses to cause rhythmical contraction of the heart muscle and (b) for conducting these impulses rapidly throughout the heart. Many of the ills of the heart, especially the cardiac arrhythmias, are based on abnormalities of this special excitatory and conductive system.

THE SPECIAL EXCITATORY AND CONDUCTIVE SYSTEM OF THE HEART

The adult human heart normally contracts at a rhythmic rate of about 72 beats per minute. Figure 14–1 illustrates the special excitatory and conductive system of the heart that controls these cardiac contractions. The figure shows: (A) the *S-A node* in which the normal rhythmic self-excitatory impulse is generated, (B) the *internodal pathways*

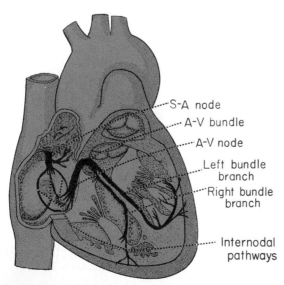

Figure 14–1. The S-A node and the Purkinje system of the heart.

that conduct the impulse from the S-A node to the A-V node, (C) the *A-V node* in which the impulse from the atria is delayed before passing into the ventricles, (D) the *A-V bundle,* which conducts the impulse from the atria into the ventricles, and (E) the *left* and *right bundles of Purkinje fibers,* which conduct the cardiac impulse to all parts of the ventricles.

THE SINO-ATRIAL NODE

The sino-atrial (S-A) node is a small, crescent strip of specialized muscle approximately 3 mm. wide and 1 cm. long; it is located in the posterior wall of the right atrium immediately beneath and medial to the opening of the superior vena cava. The fibers of this node are each 3 to 5 microns in diameter, in contrast to a diameter of 15 to 20 microns for the surrounding atrial muscle fibers. However, the S-A fibers are continuous with the atrial fibers so that any action potential that begins in the S-A node spreads immediately into the atria.

Automatic Rhythmicity of the Sino-Atrial Fibers. Most cardiac fibers have the capability of *self-excitation,* a process which can cause automatic rhythmical contraction. This is especially true of the fibers of the heart's specialized conducting system; the portion of this system that displays self-excitation to the greatest extent is the fibers of the S-A node. For this reason, the sino-atrial (S-A) node ordinarily controls the rate of beat of the entire heart, as will be discussed in detail later in this chapter. First, however, let us describe this automatic rhythmicity.

The sino-atrial fibers are somewhat different from most other cardiac muscle fibers, having a resting membrane potential of only −55 to −60 millivolts in comparison with −85 to −95 millivolts in most of the other fibers. This low resting potential is caused by a natural leakiness of the membranes to sodium ions. And it is also this leakage of sodium that causes the self-excitation

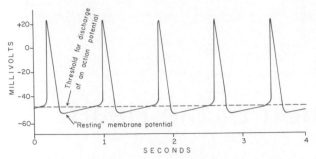

Figure 14–2. Rhythmic discharge of an S-A nodal fiber.

of the S-A fibers. Figure 14–2 illustrates the rhythmical repetitiveness of the action potentials in the S-A node, which can be explained as follows:

Immediately after each action potential is over, the membrane is even more permeable to potassium ions than normally, as was discussed in Chapter 10, and this contrasts with a very low permeability at this instant for sodium ions. Recalling the Goldman equation from Chapter 10, one may note that when the ratio of potassium ion to sodium ion permeability is very high, the potential inside the membrane becomes very highly negative because potassium ions diffusing out of the cell carry positive charges away from the interior of the cell. Therefore, immediately after the action potential is over, the membrane potential reaches its greatest degree of negativity, a state called *hyperpolarization*. At this time the membrane potential is much too negative for the fiber to fire again. However, this does not last long because the high degree of negativity inside the membrane causes the membrane to become progressively less permeable to potassium during the next few tenths of a second, and the natural leakiness of the membrane to sodium returns and causes the membrane potential to "drift" slowly back toward a less negative value. This is illustrated after each one of the action potentials in Figure 14–2.

After a few tenths of a second, the so-called "resting" potential will have drifted enough for it finally to reach the threshold level for excitation of the fiber. When this occurs, the flux of sodium ions to the interior of the fiber accelerates, causing the fiber to become even more conductive to sodium. Thus a self-regeneration process begins that leads to extremely high sodium permeability and rapid total depolarization of the membrane, followed by overshoot of the membrane potential to a positive potential of approximately +20 millivolts, called a *reversal potential*, as pointed out in Chapter 10. For slightly over 0.10 second, the membrane remains depolarized; then the permeability of the membrane for potassium increases while that for sodium decreases to a low value.

After approximately 0.12 second the permeabili-

ty for potassium is now great enough that a self-regenerative process begins in the opposite direction, with accelerating flux of potassium ions to the exterior; this recreates the negative potential inside the fiber. After the action potential is over, an especially high potassium permeability persists again for a few tenths of a second, thus maintaining a new state of hyperpolarization. Then the process begins again, with gradual decrease in permeability of the membrane to potassium, more and more leakage of sodium, and then a new regenerative cycle that causes depolarization. This process continues over and over throughout the life of the person, thereby providing rhythmical excitation of the S-A nodal fibers at a normal resting rate of about 72 times per minute.

INTERNODAL PATHWAYS AND TRANSMISSION OF THE CARDIAC IMPULSE THROUGH THE ATRIA

The ends of the S-A nodal fibers fuse with the surrounding atrial muscle fibers, and action potentials originating in the S-A node travel outward into these fibers. In this way, the action potential spreads through the entire atrial muscle mass and eventually also to the A-V node. The velocity of conduction in the atrial muscle is approximately 0.3 meter per second. However, conduction is somewhat more rapid in several small bundles of atrial muscle fibers, some of which pass directly from the S-A node to the A-V node and conduct the cardiac impulse at a velocity of 0.45 to 0.6 meter per second. Some physiologists claim that these bundles contain specialized conduction fibers similar to the Purkinje fibers of the ventricles, which will be discussed subsequently. However, other physiologists believe that this more rapid conduction is caused simply by the greater mass of muscle and the more direct orientation of the muscle fibers in the special bundles. Nevertheless, these pathways do conduct the impulse from the S-A node to the A-V node more rapidly than the conduction in the general mass of atrial muscle; these *internodal pathways* are illustrated in Figure 14–1.

THE ATRIOVENTRICULAR (A-V) NODE AND THE PURKINJE SYSTEM

Delay in Transmission at the A-V Node. Fortunately, the conductive system is organized so that the cardiac impulse will not travel from the atria into the ventricles too rapidly; this allows time for the atria to empty their contents into the ventricles before ventricular contraction begins. It is primarily the A-V node and its associated conductive fibers that delays this transmission of the cardiac impulse from the atria into the ventricles.

Figure 14–3 shows diagrammatically the different parts of the A-V node and its connections with the atrial internodal pathway fibers and the A-V bundle. The figure also shows the approximate intervals of time in fractions of a second between the genesis of the cardiac impulse in the S-A node and its appearance at different points in the A-V nodal system. Note that the impulse, after traveling through the internodal pathway, reaches the A-V node approximately 0.04 second after its origin in the S-A node. However, between this time and the time that the impulse emerges in the A-V bundle, another 0.11 second elapses. About one half of this time lapse occurs in the *junctional fibers,* which are very small fibers that connect the normal atrial fibers with the fibers of the node itself (illustrated in Figures 14–3 and 14–4). The velocity of conduction in these fibers is about 0.02 meter per second (about one twenty-fifth that in normal cardiac muscle), which greatly delays entrance of the impulse into the A-V node. After entering the node proper, the velocity of conduction in the *nodal fibers* is still quite low, only 0.1 meter per second, about one-fourth the conduction velocity in normal cardiac muscle. Therefore, a further delay in transmission occurs as the impulse travels through the A-V node into the *transitional fibers* and finally into the *A-V bundle* (also called the *bundle of His*).

Figure 14–4 illustrates a functional diagram of the A-V nodal region showing the atrial fibers leading into the minute junctional fibers, then progressive enlargement of the fibers again as they spread through the node, through the transitional region, and into the A-V bundle. The characteristics of the action potential in the different parts of the node are illustrated at the bottom.

The cause of the extremely slow conduction in

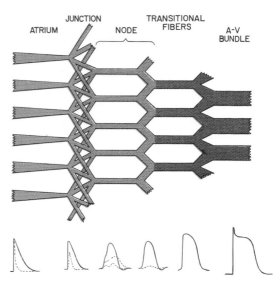

Figure 14–4. Functional diagram of the anatomical relationships in the region of the atrioventricular node. The action potentials at the bottom were recorded under normal conditions (solid lines) and under the influence of acetylcholine (dashed lines), showing blockage of conduction by this substance. (From Hoffman and Cranefield: Electrophysiology of the Heart. New York, McGraw-Hill Book Company, 1960.)

the junctional and other A-V nodal fibers is probably three-fold: (1) The fibers are very small, a fact which in itself makes the conduction rate very slow. (2) The number of tight junctions at the intercalated discs from one cardiac cell to the next in these fibers is much less than in the usual cardiac muscle fibers — that is, the number of *nexi* (points of tight fusion between the cell membranes) is greatly reduced between these cardiac muscle cells, thereby reducing the rapidity of ionic transport along the axis of the fiber. (3) These fibers are made up of a much more embryonic type of cell with much less differentiation than in the usual cardiac muscle cell, this also presumably reducing the capability of the cell to transmit the cardiac impulse.

TRANSMISSION IN THE PURKINJE SYSTEM

The *Purkinje fibers* that lead from the A-V node through the A-V bundle and into the ventricles have functional characteristics quite the opposite of those of the A-V nodal fibers; they are very large fibers, even larger than the normal ventricular muscle fibers, and they transmit impulses at a velocity of 1.5 to 4.0 meters per second, a velocity about 6 times that in the usual cardiac muscle and 150 times that in the junctional fibers. This allows almost immediate transmission of the cardiac impulse throughout the entire ventricular system.

The very rapid transmission of action potentials

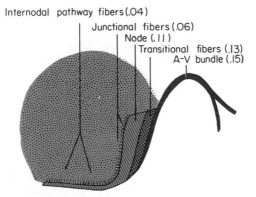

Figure 14–3. Organization of the A-V node. The numbers represent the interval of time from the origin of the impulse in the S-A node. The values have been extrapolated to the human being. (This figure is based on studies in lower animals discussed and illustrated in Hoffman and Cranefield: Electrophysiology of the Heart. New York, McGraw-Hill Book Company, 1960.)

by Purkinje fibers is probably caused by increased numbers of nexi between the successive cardiac cells that make up the Purkinje fibers. At these nexi, ions are transmitted easily from one cell to the next, thus enhancing the velocity of transmission. The Purkinje fibers also have very few myofibrils, which means that they barely contract during the course of impulse transmission.

Distribution of the Purkinje Fibers in the Ventricles. The Purkinje fibers, after originating in the A-V node, form the A-V bundle, which then threads through the fibrous tissue between the valves of the heart and thence into the ventricular system, as shown in Figure 14–1. The A-V bundle divides almost immediately into the *left* and *right bundle branches* that lie beneath the endocardium of the respective sides of the septum. Each of these branches spreads downward toward the apex of the respective ventricle, but then divides into small branches and spreads around each ventricular chamber and finally back toward the base of the heart along the lateral wall. The terminal Purkinje fibers form swirls underneath the endocardium and penetrate about one-third of the way into the muscle mass to terminate on the muscle fibers.

From the time that the cardiac impulse first enters the A-V bundle until it reaches the terminations of the Purkinje fibers, the total time that lapses is only 0.03 second; therefore, once a cardiac impulse enters the Purkinje system, it spreads almost immediately to the entire endocardial surface of the ventricular muscle.

TRANSMISSION OF THE CARDIAC IMPULSE IN THE VENTRICULAR MUSCLE

Once the cardiac impulse has reached the ends of the Purkinje fibers, it is then transmitted through the ventricular muscle mass by the ventricular muscle fibers themselves. The velocity of transmission is now only 0.4 to 0.5 meter per second, one-sixth that in the Purkinje fibers.

The cardiac muscle is coiled around the heart in a double spiral with fibrous septa between the spiralling layers; therefore, the cardiac impulse does not necessarily travel directly outward toward the surface of the heart but instead angulates toward the surface along the directions of the spirals. Because of this, transmission from the endocardial surface to the epicardial surface of the ventricle requires as much as another 0.03 second, approximately equal to the time required for transmission through the entire Purkinje system. Thus, the total time for transmission of the cardiac impulse from the origin of the Purkinje system to the last of the ventricular muscle fibers in the normal heart is about 0.06 second.

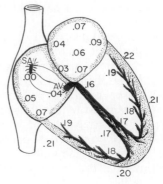

Figure 14–5. Transmission of the cardiac impulse through the heart, showing the time of appearance (in fractions of a second) of the impulse in different parts of the heart.

SUMMARY OF THE SPREAD OF THE CARDIAC IMPULSE THROUGH THE HEART

Figure 14–5 illustrates in summary form the transmission of the cardiac impulse through the human heart. The numbers on the figure represent the intervals of time in fractions of a second that lapse between the origin of the cardiac impulse in the S-A node and its appearance at each respective point in the heart. Note that the impulse spreads at moderate velocity through the atria but is delayed more than 0.1 second in the A-V nodal region before appearing in the A-V bundle. Once it has entered the bundle, it spreads rapidly through the Purkinje fibers to the entire endocardial surfaces of the ventricles. Then the impulse spreads slowly through the ventricular muscle to the epicardial surfaces.

It is extremely important that the student learn in detail the course of the cardiac impulse through the heart and the times of its appearance in each separate part of the heart, for a quantitative knowledge of this process is essential to the understanding of electrocardiography, which is discussed in the following three chapters.

CONTROL OF EXCITATION AND CONDUCTION IN THE HEART

THE S-A NODE AS THE PACEMAKER OF THE HEART

In the above discussion of the genesis and transmission of the cardiac impulse through the heart, it was stated that the impulse normally arises in the S-A node. However, this need not be the case under abnormal conditions, for other parts of the heart can exhibit rhythmic contraction in the same way that the fibers of the S-A node can; this is particularly true of the A-V nodal and Purkinje fibers.

The A-V nodal fibers, when not stimulated from some outside source, discharge at an intrinsic rhythmic rate of 40 to 60 times per minute, and the Purkinje fibers discharge at a rate of somewhere between 15 and 40 times per minute. These rates are in contrast to the normal rate of the S-A node of 70 to 80 times per minute.

Therefore, the question that we must ask is: Why does the S-A node control the heart's rhythmicity rather than the A-V node or the Purkinje fibers? The answer to this is simply that the rate of the S-A node is considerably greater than that of either the A-V node or the Purkinje fibers. Each time the S-A node discharges, its impulse is conducted into both the A-V node and the Purkinje fibers, discharging their excitable membranes. Then these tissues, as well as the S-A node, recover from the action potential and become hyperpolarized. But the S-A node loses this hyperpolarization much more rapidly than does either of the other two and emits a new impulse before either one of them can reach its own threshold for self-excitation. The new impulse again discharges both the A-V node and Purkinje fibers. This process continues on and on, the S-A node always exciting these other potentially self-excitatory tissues before self-excitation can actually occur.

Thus, the S-A node controls the beat of the heart, because its rate of rhythmic discharge is greater than that of any other part of the heart. Therefore, it is said that the S-A node is the normal *pacemaker* of the heart.

Abnormal Pacemakers — The Ectopic Pacemaker. Occasionally some other part of the heart develops a rhythmic discharge rate that is more rapid than that of the S-A node. For instance, this frequently occurs in the A-V node or in the Purkinje fibers. In either of these cases, the pacemaker of the heart shifts from the S-A node to the A-V node or to the excitable Purkinje fibers. Under rare conditions a point in the atrial or ventricular muscle develops excessive excitability and becomes the pacemaker.

A pacemaker elsewhere than the S-A node is called an *ectopic pacemaker*. Obviously, an ectopic pacemaker causes an abnormal sequence of contraction of the different parts of the heart.

ROLE OF THE PURKINJE SYSTEM IN CAUSING SYNCHRONOUS CONTRACTION OF THE VENTRICULAR MUSCLE

It is clear from the above description of the Purkinje system that the cardiac impulse arrives at almost all portions of the ventricles within a very narrow span of time, exciting the first ventricular muscle fiber only 0.06 second ahead of excitation of the last ventricular muscle fiber. Since the ventricular muscle fibers remain contracted for a total period of 0.30 second, one can see that this rapid spread of excitation throughout the entire ventricular muscle mass causes all portions of the ventricular muscle in both ventricles to contract at almost exactly the same time. Effective pumping by the two ventricular chambers requires this synchronous type of contraction. If the cardiac impulse traveled through the ventricular muscle very slowly, then much of the ventricular mass would contract prior to contraction of the remainder, in which case the overall pumping effect would be greatly depressed. Indeed, in some types of cardiac debilities, some of which will be discussed in Chapters 16 and 17, such slow transmission does indeed occur, and the pumping effectiveness of the ventricles is decreased perhaps as much as 20 to 30 per cent.

FUNCTION OF THE PURKINJE SYSTEM IN PREVENTING ARRHYTHMIAS

Another extremely important function of the Purkinje system is to prevent the development of stray impulses in the heart that can cause ventricular fibrillation and other cardiac arrhythmias, as will be discussed later in this chapter and in relation to electrocardiography in Chapter 17. This function may be explained as follows:

The refractory period of the cardiac muscle is approximately 0.3 second, which is also the duration of the muscle contraction, as was discussed in the previous chapter. The ratio of this refractory period to the rapidity of transmission of the cardiac impulse throughout the ventricles plays an important role in preventing serious cardiac arrhythmias. That is, the normal amount of time required for the cardiac impulse to travel from the A-V bundle all the way through the entire ventricular mass in the normal heart is only 0.06 second. Therefore, the first ventricular muscle that is stimulated is still very much in the refractory state when the last portion of the ventricular muscle becomes stimulated. Consequently, the action potential in this last muscle now finds no remaining ventricular muscle to excite, and all the action potentials in all portions of the ventricular muscle simply die. We shall see later in this chapter that any serious delay in transmission of the impulse through the ventricle can make it possible for the impulse from the last excited ventricular muscle to re-enter the first muscle. This, in turn, sets up a re-entrance cycle in the ventricles that can continue on and on, creating the condition of ventricular fibrillation that causes the ventricles to remain in a semicontracted state indefinitely and, therefore, not to provide cardiac pumping, as will be discussed in detail later in the chapter.

Refractory Period of the Purkinje Fibers and Its Functional Significance. The action potential and refractory period of the Purkinje fibers lasts about 25 per cent longer than the action potential and refractory period of the usual ventricular cardiac muscle. This has very important functional importance for maintaining the normal heart rhythm, as follows: By the time the Purkinje fibers are no longer refractory, the ventricular muscle fibers have already long been out of the refractory state. As a result, all portions of the ventricular muscle are already available to accept a new action potential and, therefore, are ready to contract. Therefore, this long refractory period of the Purkinje fibers in relation to the somewhat shorter refractory period of the ventricular muscle protects the ventricular muscle from being excited at too early a time in the cardiac cycle.

CONTROL OF HEART RHYTHMICITY AND CONDUCTION BY THE AUTONOMIC NERVES

The heart is supplied with both sympathetic and parasympathetic nerves, as illustrated in Figure 13–10 of the previous chapter. The parasympathetic nerves are distributed mainly to the S-A and A-V nodes, to a lesser extent to the muscle of the two atria, and even less to the ventricular muscle. The sympathetic nerves are distributed to these same areas but with a strong representation to the ventricular muscle as well as to the other parts of the heart.

Effect of Parasympathetic (Vagal) Stimulation on Cardiac Rhythm and Conduction — Ventricular Escape. Stimulation of the parasympathetic nerves to the heart (the vagi) causes the hormone acetylcholine to be released at the vagal endings. This hormone has two major effects on the heart. First, it decreases the rate of rhythm of the S-A node, and, second, it decreases the excitability of the A-V junctional fibers between the atrial musculature and the A-V node, thereby slowing transmission of the cardiac impulse into the ventricles. Very strong stimulation of the vagi can completely stop the rhythmic contraction of the S-A node or completely block transmission of the cardiac impulse through the A-V junction. In either case, rhythmic impulses are no longer transmitted into the ventricles. The ventricles stop beating for 4 to 10 seconds, but then some point in the Purkinje fibers, usually in the A-V bundle, develops a rhythm of its own and causes ventricular contraction at a rate of 15 to 40 beats per minute. This phenomenon is called *ventricular escape*.

Mechanism of the Vagal Effects. The acetylcholine released at the vagal nerve endings greatly increases the permeability of the fiber membranes to potassium, which allows rapid leakage of potassium to the exterior. This causes increased negativity inside the fibers, an effect called *hyperpolarization,* which makes excitable tissue much less excitable, as was explained in Chapter 10.

In the A-V node, the state of hyperpolarization makes it difficult for the minute junctional fibers, which can generate only small quantities of current during the action potential, to excite the nodal fibers. Therefore, the *safety factor* for transmission of the cardiac impulse through the junctional fibers and into the nodal fibers decreases. A moderate decrease in this simply delays conduction of the impulse, but a decrease in safety factor below unity (which means so low that the action potential of one fiber cannot cause an action potential in the successive fiber) completely blocks conduction.

Effect of Sympathetic Stimulation on Cardiac Rhythm and Conduction. Sympathetic stimulation causes essentially the opposite effects on the heart to those caused by vagal stimulation as follows: First, it increases the rate of S-A nodal discharge. Second, it increases the rate of conduction and the excitability in all portions of the heart. Third, it increases greatly the force of contraction of all the cardiac musculature, both atrial and ventricular, as discussed in the previous chapter.

In short, sympathetic stimulation increases the overall activity of the heart. Maximal stimulation can almost triple the rate of heartbeat and can increase the strength of heart contraction as much as two-fold.

Mechanism of the Sympathetic Effect. Stimulation of the sympathetic nerves releases the hormone norepinephrine at the sympathetic nerve endings. The precise mechanism by which this hormone acts on cardiac muscle fibers is still somewhat doubtful, but the present belief is that it increases the permeability of the fiber membrane to sodium and calcium. In the S-A node, an increase of sodium permeability would cause increased tendency for the resting membrane potential to decay to the threshold level for self-excitation, which obviously would accelerate the onset of self-excitation after each successive heartbeat and therefore increase the heart rate.

In the A-V node, increased sodium permeability would make it easier for each fiber to excite the succeeding fiber, thereby decreasing the conduction time from the atria to the ventricles.

The increase in permeability to calcium ions is probably at least partially responsible for the increase in contractile strength of the cardiac muscle under the influence of sympathetic stimulation, because calcium ions play a powerful role in exciting the contractile process of the myofibrils.

ABNORMAL RHYTHMS OF THE HEART

Abnormal cardiac rhythms can be caused by (1) abnormal rhythmicity of the pacemaker itself, (2) shift of the pacemaker from the S-A node to other parts of the heart, (3) blocks at different points in the transmission of the impulse through the heart, (4) abnormal pathways of impulse transmission through the heart, and (5) spontaneous generation of abnormal impulses in almost any part of the heart. Some of these will be discussed in Chapter 17 in relation to electrocardiographic analysis of cardiac arrhythmias, but the major disturbances and their causes are presented here to illustrate some aberrations that can occur in the function of the rhythmicity and conducting systems of the heart.

PREMATURE CONTRACTIONS — ECTOPIC FOCI

Often, a small area of the heart becomes much more excitable than normal and causes an occasional abnormal impulse to be generated in between the normal impulses. A depolarization wave spreads outward from the irritable area and initiates a *premature contraction* of the heart. The focus at which the abnormal impulse is generated is called an *ectopic focus*.

The usual cause of an ectopic focus is an irritable area of cardiac muscle resulting from a local area of ischemia (too little coronary blood flow to the muscle), over-use of stimulants such as caffeine or nicotine, lack of sleep, anxiety, or other debilitating state.

Shift of the Pacemaker to an Ectopic Focus.   Sometimes an ectopic focus becomes so irritable that it establishes a rhythmic contraction of its own at a more rapid rate than that of the S-A node. When this occurs the ectopic focus becomes the pacemaker of the heart. The most common point for development of an ectopic pacemaker is the A-V node or the A-V bundle, as will be discussed in more detail in Chapter 17.

HEART BLOCK

Occasionally, transmission of the impulse through the heart is blocked at a critical point in the conductive system. One of the most common of these points is between the atria and the ventricles; this condition is called *atrioventricular block*. Another common point is in one of the *bundle branches* of the Purkinje system. Rarely a block also develops between the S-A node and the atrial musculature.

Atrioventricular Block.   In the human being, a block between the atria and the ventricles can result from localized damage or depression of the *A-V junctional* fibers or of the *A-V bundle*. The causes include different types of infectious processes, excessive stimulation by the vagus nerves (which depresses conductivity in the junctional fibers), localized destruction of the A-V bundle as a result of a coronary infarct, pressure on the A-V bundle by arteriosclerotic plaques, or depression caused by various drugs.

Figure 14–6 illustrates a typical record of atrial and ventricular contraction while the A-V bundle was being progressively compressed to cause successive stages of block. During the first three contractions of the record, ventricular contraction followed in orderly sequence approximately 0.16 second after atrial contraction. Then, A-V bundle compression was begun, and the interval of time between the beginning of atrial contraction and the beginning of ventricular contraction increased steadily during the next five heartbeats from 0.16 to 0.32 second. Beyond this point further compression completely blocked impulse transmission. Thereafter, the atria continued to beat at their normal rate of rhythm, while the ventricles failed to contract at all for approximately 7 seconds. Then, "ventricular escape" occurred, and a rhythmic focus in the ventricles suddenly began to act as a ventricular pacemaker, causing ventricular contractions at a rate of approximately 30 per minute; these were completely dissociated from the atrial contractions.

FLUTTER AND FIBRILLATION

Frequently, either the atria or the ventricles begin to contract extremely rapidly and often incoordinately. The low frequency, more coordinate contractions up to 200 to 300 beats per minute are generally called *flutter* and the very high frequency, incoordinate contractions, *fibrillation*.

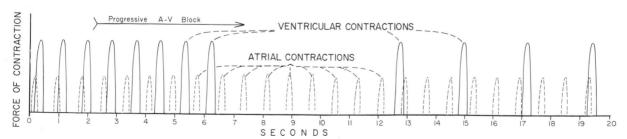

Figure 14–6.   Contraction of the atria and ventricles of a heart, showing the effect of progressive A-V block. Note the progressive increase in the interval between the onset of atrial contraction and the onset of ventricular contraction. Note also the final complete asynchronism between atrial and ventricular contraction.

Two basic theories of flutter and fibrillation have been proposed; these are (1) a *single* or *multiple ectopic foci* emitting many impulses one after another in rapid succession and (2) a *circus movement,* in which the impulse travels around and around through the heart muscle, never stopping.

Ectopic Foci. The ectopic focus theory is easy to understand. An area of the heart simply becomes so irritable that is keeps sending rapid impulses in all directions, resulting in rapid rates of contraction. Though most instances of flutter and fibrillation are probably caused by circus movements, nevertheless, there is reason to believe that at least some instances of atrial flutter may result from this cause.

The Circus Movement (Re-entry of the Impulse). There is also reason to believe that many instances of atrial flutter result from circus movements around the atria, and since only the circus movement theory has explained adequately the course of events in most instances of atrial and ventricular fibrillation, we will discuss it in detail.

Figure 14–7 illustrates several small cardiac muscle strips cut in the form of circles. If such a strip is stimulated at the 12 o'clock position *so that the impulse travels in only one direction,* the impulse spreads progressively around the circle until it returns to the 12 o'clock position. If the originally stimulated muscle fibers are still in a refractory state, the impulse then dies out, for refractory muscle cannot transmit a second impulse. However, there are three different conditions that can cause this impulse to continue to travel around the circle, that is, to cause "re-entry" of the impulse into muscle that has already been excited.

First, if the *length of the pathway around the circle is long,* by the time the impulse returns to the 12 o'clock position the originally stimulated muscle will no longer be refractory, and the impulse will continue around the circle again and again.

Second, if the length of the pathway remains constant but the *velocity of conduction becomes decreased* enough, an increased interval of time will elapse before the impulse returns to the 12 o'clock position. By this time the originally stimulated muscle might be out of the

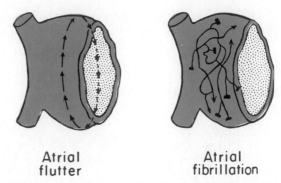

Figure 14–8. Pathways of impulses in atrial flutter and atrial fibrillation.

refractory state, and the impulse can continue around the circle again and again.

Third, *the refractory period of the muscle might become greatly shortened.* In this case, the impulse could also continue around and around the circle.

All three of these conditions occur in different pathological states of the human heart as follows: (1) A long pathway frequently occurs in dilated hearts. (2) Decreased rate of conduction frequently results from blockage of the Purkinje system, ischemia of the muscle, high blood potassium, and many other factors. (3) A shortened refractory period frequently occurs in response to various drugs, such as epinephrine, or following repetitive electrical stimulation. Thus, in many different cardiac disturbances circus movements can cause abnormal cardiac rhythmicity that completely ignores the pacesetting effects of the S-A node.

Atrial Flutter Resulting from a Circus Pathway

The left-hand panel of Figure 14–8 illustrates a circus pathway around and around the atria from top to bottom, the pathway lying to the left of the superior and inferior venae cavae. Such circus pathways have been initiated experimentally in the atria of dogs' hearts, and electrocardiographic records, which will be discussed in Chapter 17, indicate that this type of circus pathway also develops in the human heart when the atria become greatly dilated as a result of valvular heart disease. The rate of flutter is usually 200 to 350 times per minute.

Partial Block at the A-V Node During Atrial Flutter. The functional refractory period of the Purkinje fibers and ventricular muscle is approximately 1/200 minute so that not over 200 impulses per minute can be transmitted into the ventricles. Therefore, when the atrium contracts as rapidly as 300 times per minute, only one of every two impulses passes into the ventricles, thus causing the atria to beat at a rate 2 times that of the ventricles. Occasionally, a 3:1 or, rarely, a 4:1 rhythm of the heart develops in the same manner.

Fibrillation — The "Chain Reaction" Mechanism

Fibrillation, whether it occurs in the atria or in the ventricles, is a very different condition from flutter. One

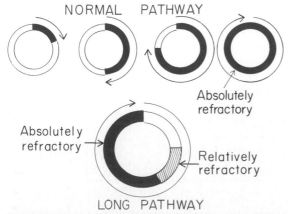

Figure 14–7. The circus movement, showing annihilation of the impulse in the short pathway and continued propagation of the impulse in the long pathway.

can see many separate contractile waves spreading in different directions over the cardiac muscle at the same time in either atrial or ventricular fibrillation. Obviously, then, the circus movement in fibrillation is entirely different from that in flutter. One of the best ways to explain the mechanism of fibrillation is to describe the initiation of fibrillation by stimulation with 60 cycle alternating electrical current.

Fibrillation Caused by 60 Cycle Alternating Current. At a central point in the ventricles of heart A in Figure 14–9, a 60 cycle electrical stimulus is applied through a stimulating electrode. The first cycle of the electrical stimulus causes a depolarization wave to spread in all directions, leaving all the muscle beneath the electrode in a refractory state. After about 0.25 second, this muscle begins to come out of the refractory state, some portions of the muscle coming out of refractoriness prior to other portions. This state of events is depicted in heart A by many light patches, which represent excitable cardiac muscle, and dark patches, which represent still refractory muscle. New stimuli from the electrode can now cause impulses to travel in certain directions through the heart but not in all directions. It will be observed in heart A that certain impulses travel for short distances until they reach refractory areas of the heart and then are blocked. Other impulses, however, pass between the refractory areas and continue to travel in the excitable patches of muscle. Now, several events transpire in rapid succession, all occurring simultaneously and eventuating in a state of fibrillation. These are:

First, block of the impulses in some directions but successful transmission in other directions creates one of the necessary conditions for a circus movement to develop — that is, *transmission of at least some of the depolarization waves around the heart in only one direction.* As a result, these waves do not run into waves traveling in the opposite direction and therefore do not annihilate themselves on the opposite side of the heart but can continue around and around the ventricles.

Second, the rapid stimulation of the heart causes two changes in the cardiac muscle itself, both of which predispose to circus movement: (1) The *velocity of conduction through the heart becomes decreased,* which allows a longer time interval for the impulses to travel around the heart. (2) The *refractory period of the muscle becomes shortened,* allowing re-entry of the impulse into previously excited heart muscle within a much shorter period of time than normally.

Third, one of the most important features of fibrillation is the *division of impulses,* as illustrated in heart A. When a depolarization wave reaches a refractory area in the heart, it travels to both sides around the area. Thus, a single impulse becomes two impulses. Then when each of these reaches another refractory area it, too, divides to form still two more impulses. In this way many different impulses are continually being formed in the heart by a progressive *chain reaction* until, finally, there are many small depolarization waves traveling in many different directions at the same time. Furthermore, this irregular pattern of impulse travel causes a *circuitous route for the impulses to travel, greatly lengthening the conductive pathway, which is one of the conditions leading to fibrillation.* It also results in a continual irregular pattern of patchy refractory areas in the heart. One can readily see that a vicious cycle has been initiated: more and more impulses are formed; these cause more and more patches of refractory muscle; and the refractory patches cause more and more division of the impulses. Therefore, any time a single area of cardiac muscle comes out of refractoriness, an impulse is always close at hand to re-enter the area.

Heart B in Figure 14–9 illustrates the final state that develops in fibrillation. Here one can see many impulses traveling in all directions, some dividing and increasing the number of impulses while others are blocked entirely by refractory areas. In the final state of fibrillation, the number of new impulses being formed exactly equals the number of impulses that are being blocked by refractory areas. Thus, a steady state has developed with a certain average number of impulses traveling all the time in all directions through the cardiac syncytium.

As is the case with many other cardiac arrhythmias, fibrillation is usually confined to either the atria or the ventricles alone and not to both syncytial masses of muscle at the same time, because these two masses of muscles are electrically insulated from each other by the rings of fibrous tissue around the heart valves.

Demonstration of the Chain Reaction Mechanism of Fibrillation in the "Iron Heart." In the human or animal heart the chain reaction mechanism is difficult to demonstrate for two reasons: (1) It is impossible to follow exactly the wave fronts of the electrical impulses traveling through the heart muscle, and (2) so many minute waves of contraction are spreading at the same time that the eyes cannot follow these continually. An *iron heart model* has been developed in our laboratory that does show easily the chain reaction mechanism. This is a large iron bob suspended in nitric acid. Under appropriate conditions an oxide film develops on the surface of the iron. Then, a single electrical stimulus to the surface will cause a single excitation wave to travel over the entire surface, comparable to normal stimulation of the heart. But multiple stimuli either all at once or in rapid succession will cause the typical chain reaction just described, with resultant fibrillation. One can easily

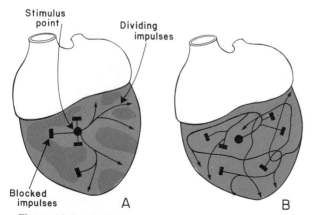

Figure 14–9. A, Initiation of fibrillation in a heart when patches of refractory musculature are present. B, Continued propagation of *fibrillatory impulses* in the fibrillating ventricle.

see the chain reaction on the surface of the iron bob, and it can be recorded electrically or photographically. Also, typical fibrillatory bipolar and monopolar electrocardiographic patterns can be recorded.

Atrial Fibrillation

Atrial fibrillation is completely different from atrial flutter because the circus movement does not travel in a regular pathway. Instead, many different excitation waves can be seen to travel over the surface of the atria at the same time. Atrial fibrillation occurs frequently when the atria become greatly overdilated — in fact, many times as frequently as flutter. When flutter does occur, it usually becomes fibrillation after a few days or weeks. To the right in Figure 14–8 are illustrated the pathways of fibrillatory impulses traveling through the atria.

Obviously, atrial fibrillation results in complete incoordination of atrial contraction so that atrial pumping ceases entirely.

Effect of Atrial Fibrillation on the Overall Pumping Effectiveness of the Heart. The normal function of the atria is to help fill the ventricles. However, the atria are probably responsible for not more than 25 to 30 per cent of the normal ventricular filling, which was explained in Chapter 13. Therefore, even when the atria fail to act as primer pumps because of atrial fibrillation, the ventricles can still fill enough that the effectiveness of the heart as a pump is reduced only 25 to 30 per cent, which is well within the "cardiac reserve" of all but severely weakened hearts. For this reason, atrial fibrillation can continue for many years without serious cardiac debility.

Irregularity of Ventricular Rate During Atrial Fibrillation. When the atria are fibrillating, impulses arrive at the A-V node rapidly but also irregularly. Since the A-V bundle will not pass a second impulse for approximately 0.3 second after a previous one, at least 0.3 second must elapse between one ventricular contraction and the next, and an additional but variable interval of 0 to 0.6 second occurs before one of the irregular fibrillatory impulses happens to arrive at the A-V node. Thus, the interval between successive ventricular contractions varies from a minimum of about 0.3 second to a maximum of about 0.9 second, causing a very irregular heart beat. In fact, this irregularity is one of the clinical findings used to diagnose the condition.

Ventricular Fibrillation

Ventricular fibrillation is extremely important because at least one quarter of all persons die in ventricular fibrillation. For instance, the hearts of most patients with coronary infarcts fibrillate shortly before death. In only a few instances on record have fibrillating human ventricles been known to return of their own accord to a rhythmic beat.

The likelihood of circus movements in the ventricles and, consequently, of ventricular fibrillation is greatly increased when the ventricles are dilated or when the rapidly conducting *Purkinje system is blocked* so that impulses cannot be transmitted rapidly. Also, *electric shock,* particularly with 60 cycle electric current, as discussed above, or *ectopic foci,* which are discussed below, are common initiating causes of ventricular fibrillation.

Inability of the Heart to Pump Blood During Ventricular Fibrillation. When the ventricles begin to fibrillate, the different parts of the ventricles no longer contract simultaneously. For the first few seconds the ventricular muscle undergoes rather coarse contractions which may pump a few milliliters of blood with each contraction. However, the impulses in the ventricles rapidly become divided into many much smaller impulses, and the contractions become so fine and asynchronous, rather than coarse, that they pump no blood whatsoever. The ventricles dilate because of failure to pump the blood that is flowing into them, and within 60 to 90 seconds the ventricular muscle becomes too weak, because of lack of coronary blood supply, to contract strongly even if coordinate contraction should return. Therefore, death is immediate when ventricular fibrillation begins.

Irritable Foci as the Usual Cause of Ventricular Fibrillation. The cause of ventricular fibrillation is most often an irritable focus in the ventricular muscle caused by coronary insufficiency or by compression from an arteriosclerotic plaque. Most impulses originating in an irritable focus travel around the heart in all directions, meet on the opposite side of the heart, and annihilate themselves. However, when impulses from a focus become frequent the refractory period of the stimulated muscle becomes shortened, impulse conduction becomes slowed, and patchy areas of refractoriness develop in the ventricles. These are the essential conditions for initiating the chain reaction mechanism, and fibrillation begins in the same manner as that described above for stimulation with a 60 cycle electrical current. Thus, any ectopic focus that emits rapid impulses is likely to initiate fibrillation.

Electrical Defibrillation of the Ventricles. Though a weak alternating current almost invariably throws the ventricles into fibrillation, a very strong electrical current passed through the ventricles for a short interval of time can stop fibrillation by throwing all the ventricular muscle into refractoriness simultaneously. This is accomplished by passing intense current through electrodes placed on two sides of the heart. The current penetrates most of the fibers of the ventricles, thus stimulating essentially all parts of the ventricles simultaneously and causing them to become refractory. All impulses stop, and the heart then remains quiescent for 3 to 5 seconds, after which it begins to beat again, with the S-A node or, often, some other part of the heart becoming the pacemaker. Occasionally, however, the same irritable focus that had originally thrown the ventricles into fibrillation is still present, and fibrillation begins again immediately.

When electrodes are applied directly to the two sides of the heart, fibrillation can usually be stopped with 70 to 100 volts of 60 cycle alternating current applied for 0.10 second or 1000 volts direct current applied for a few thousandths of a second. When applied through the chest wall, as illustrated in Figure 14–10, the usual procedure is to charge a large electrical capacitor up to several thousand volts and then cause the capacitor to discharge in a few thousandths of a second through the electrodes and the heart. In our laboratory the heart of a single anesthetized dog was defibrillated 130 times through the chest wall, and the animal remained in perfectly normal condition.

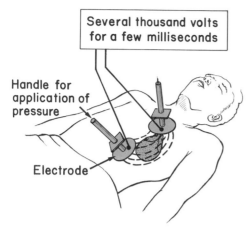

Several thousand volts for a few milliseconds

Handle for application of pressure

Electrode

Figure 14–10. Application of electrical current to the chest to stop ventricular fibrillation.

Hand Pumping of the Heart ("Cardiac Massage") as an Aid to Defibrillation. Unless defibrillated within one minute after fibrillation begins, the heart is usually too weak to be revived. However, it is still possible to revive the heart by preliminarily pumping it by hand and then defibrillating it later. In this way small quantities of blood are delivered into the aorta, and a renewed coronary blood supply develops. After a few minutes, electrical defibrillation often becomes possible. Indeed, fibrillating hearts have been pumped by hand as long as 90 minutes before defibrillation. In recent years, a technique of pumping the heart without opening the chest has been developed; this technique consists of intermittent, very powerful thrusts of pressure on the chest wall.

Lack of blood flow to the brain for more than 5 to 10 minutes usually results in permanent mental impairment or even total destruction of the brain. Even though the heart should be revived, the person might die from the effects of brain damage or live with permanent mental impairment.

CARDIAC ARREST

When cardiac metabolism becomes greatly disturbed as a result of any one of many possible conditions, the rhythmic contractions of the heart occasionally stop. One of the most common causes of cardiac arrest is hypoxia of the heart, for severe hypoxia prevents the muscle fibers from maintaining normal ionic differentials across their membranes. Therefore, polarization of the membranes becomes reduced, and the excitability may be so affected that the automatic rhythmicity disappears.

Occasionally, patients with severe myocardial disease develop cardiac arrest, which obviously can lead to death. In many cases, however, rhythmic electrical impulses from an implanted electronic cardiac "pacemaker" have been used successfully to keep patients alive for many years.

REFERENCES

Brown, H., et al.: The initiation of the heartbeat and its control by autonomic transmitters. In Dickinson, C. J., and Marks, J. (eds.): Developments in Cardiovascular Medicine. Lancaster, England, MTP Press, 1978, p. 31.

Brutsaert, D. L., and Paulus, W. J.: Contraction and relaxation of the heart as muscle and pump. In Guyton, A. C., and Young, D. B. (eds.): International Review of Physiology: Cardiovascular Physiology III. Vol. 18. Baltimore, University Park Press, 1979, p. 1.

Chung, E. K. (ed.): Artificial Cardiac Pacing: Practical Approach. Baltimore, Williams & Wilkins, 1978.

Cobb, L. A., et al.: Community cardiopulmonary resuscitation. Annu. Rev. Med., 31:453, 1980.

Cranefield, P. F.: The Conduction of the Cardiac Impulse. New York, Futura Publishing Co., 1975.

Del Negro, A. A., and Fletcher, R. D.: Indications for and use of Artificial Cardiac Pacemakers. Chicago, Year Book Medical Publishers, 1978.

Durrer, D., et al.: Human cardiac electrophysiology. In Dickinson, C. J., and Marks, J. (eds.): Developments in Cardiovascular Medicine. Lancaster, England, MTP Press, 1978, p. 53.

Erslev, A. J.: Renal biogenesis of erythropoietin. Am. J. Med., 58:25, 1975.

Ewy, G. A.: Cardiac Arrest and Resuscitation: Defibrilators and Defibrillation. Chicago, Year Book Medical Publishers, 1978.

Fozzard, H. A.: Heart: Excitation-contraction coupling. Annu. Rev. Physiol., 39:201, 1977.

Guyton, A. C., and Satterfield, J.: Factors concerned in electrical defibrillation of the heart, particularly through the unopened chest. Am. J. Physiol., 167:81, 1951.

Irisawa, H.: Comparative physiology of the cardiac pacemaker mechanism. Physiol. Rev., 58:461, 1978.

James, T. N., and Oliver, M. F.: Sudden cardiac death. In Hayase, S., and Murao, S. (eds.): Cardiology: Proceedings of the VIII World Congress of Cardiology, Tokyo, 1978. New York, Elsevier/North-Holland, 1979, p. 480.

Jones, P.: Cardiac Pacing. New York, Appleton-Century-Crofts, 1980.

Josephson, M. E., and Seides, S. F.: Clinical Cardiac Electrophysiology Techniques and Interpretations. Philadelphia, Lea & Febiger, 1979.

Kulbertus, H. E., and Wellens, H. J. J. (eds.): Sudden Death. Hingham, Mass., Kluwer Boston, 1980.

Levy, M. N., and Martin, P. J.: Cardiac excitation and contraction. In MTP International Review of Science: Physiology. Baltimore, University Park Press, 1974, Vol. 1, p. 49.

Levy, M. N., and Martin, P. J.: Neural control of the heart. In Berne, R. M., et al. (eds.): Handbook of Physiology. Sec. 2, Vol. I. Baltimore, Williams & Wilkins, 1979, p. 581.

Narula, O. S. (ed.): Cardiac Arrhythmias: Electrophysiology, Diagnosis, and Management. Baltimore, Williams & Wilkins, 1979.

Nobel, D.: The Initiation of the Heartbeat. New York, Oxford University Press, 1979.

Pick, A., and Langendorf, R.: Interpretation of Complex Arrhythmias. Philadelphia, Lea & Febiger, 1980.

Reuter, H.: Properties of two inward membrane currents in the heart. Annu. Rev. Physiol., 41:413, 1979.

Reuter, H., and Winegrad, S.: Excitation-contraction coupling of the heart: Neurohumoral effects. In Hayase, S., and Murao, S. (eds.): Cardiology: Proceedings of the VIII World Congress of Cardiology, Tokyo, 1978. New York, Elsevier/North-Holland, 1979, p. 800.

Rusy, B. F.: Pharmacology of antiarrhythmic drugs. Med. Clin. North Am., 58:987, 1974.

Samet, P., and El-Sherif, N. (eds.): Cardiac Pacing, 2nd ed. New York, Grune & Stratton, 1979.

Sperelakis, N.: Origin of the cardiac resting potential. In Berne, R. M., et al. (eds.): Handbook of Physiology. Sec. 2, Vol. I. Baltimore, Williams & Wilkins, 1979, p. 187.

Sperelakis, N.: Propagation mechanisms in heart. Annu. Rev. Physiol., 41:441, 1979.

Stent, G. S., et al.: Neural control of heartbeat in the leech and in some other invertebrates. Physiol. Rev., 59:101, 1979.

Stull, J. T., and Mayer, S. E.: Biochemical mechanisms of adrenergic and cholinergic regulation of myocardial contractility. In Berne, R. M., et al. (eds.): Handbook of Physiology. Sec. 2, Vol. I. Baltimore, Williams & Wilkins, 1979, p. 741.

Thalen, H. J. T., and Hori, M.: New trends in cardiac pacing. In Hayase, S., and Murao, S. (eds.): Cardiology: Proceedings of the VIII World Congress of Cardiology, Tokyo, 1978. New York, Elsevier/North-Holland, 1979, p. 972.

Thalen, H. J., and Meere, C. C. (eds.): Fundamentals of Cardiac Pacing. Hingham, Mass., Kluwer Boston, 1979.

Varraile, P., and Naclerio, E. A.: Cardiac Pacing: A Concise Guide to Clinical Practice. Philadelphia, Lea & Febiger, 1979.

Vasselle, M.: Electrogenesis of the plateau and pacemaker potential. Annu. Rev. Physiol., 41:425, 1979.

Yamada, K., and Rosenbaum, M. B.: Conduction disturbances. In Hayase, S., and Murao, S. (eds.): Cardiology: Proceedings of the VIII World Congress of Cardiology, Tokyo, 1978. New York, Elsevier/North-Holland, 1979, p. 913.

15

The Normal Electrocardiogram

Transmission of the depolarization wave, also commonly called the *cardiac impulse*, through the heart has been discussed in detail in Chapter 14. As the wave passes through the heart, electrical currents spread into the tissues surrounding the heart, and a small proportion of these spreads all the way to the surface of the body. If electrodes are placed on the skin on opposite sides of the heart, electrical potentials generated by the heart can be recorded; the recording is known as an *electrocardiogram*. A normal electrocardiogram for two beats of the heart is illustrated in Figure 15–1.

CHARACTERISTICS OF THE NORMAL ELECTROCARDIOGRAM

The normal electrocardiogram is composed of a P wave, a "QRS complex," and a T wave. The QRS complex is actually three separate waves, the Q wave, the R wave, and the S wave.

The P wave is caused by electrical currents generated as the atria depolarize prior to contraction, and the QRS complex is caused by currents generated when the ventricles depolarize prior to contraction, that is, as the depolarization wave spreads through the ventricles. Therefore, both the P wave and the components of the QRS complex are *depolarization waves*. The T wave is caused by currents generated as the ventricles recover from the state of depolarization. This process occurs in ventricular muscle 0.25 to 0.30 second after depolarization, and this wave is known as a *repolarization wave*.

Thus, the electrocardiogram is composed of both depolarization and repolarization waves. The principles of depolarization and repolarization were discussed in Chapter 10. However, the distinction between depolarization waves and repolarization waves is so important in electrocardiography that further clarification is needed, as follows:

DEPOLARIZATION WAVES VERSUS REPOLARIZATION WAVES

Figure 15–2 illustrates a muscle fiber in four different stages of depolarization and repolarization. During the process of "depolarization" the normal negative potential inside the fiber is lost and the membrane potential actually reverses; that is, it becomes slightly positive inside and negative outside.

In Figure 15–2A the process of depolarization, illustrated by positivity inside and negativity outside, is

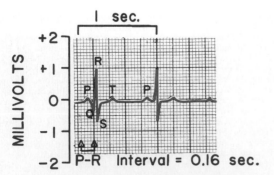

Figure 15–1. The normal electrocardiogram.

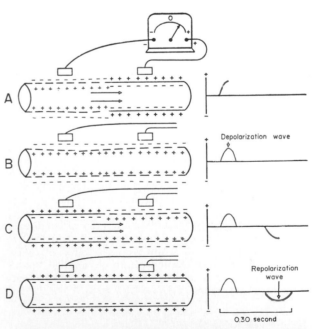

Figure 15–2. Recording the *depolorization wave* and the *repolarization wave* from a cardiac muscle fiber.

traveling from left to right, and the first half of the fiber has already depolarized while the remaining half is still polarized. Therefore, the left electrode on the fiber is in an area of negativity where it touches the outside of the fiber, while the right electrode is in an area of positivity; this causes the meter to record positively. To the right of the muscle fiber is illustrated a record of the potential between the electrodes as recorded by a high-speed recording meter at this particular stage of depolarization. Note that when depolarization has reached this halfway mark, the record has risen to a maximum positive value.

In Figure 15–2B depolarization has extended over the entire muscle fiber, and the recording to the right has returned to the zero base line because both electrodes are now in areas of equal negativity. The completed wave is a *depolarization wave* because it results from spread of depolarization along the entire extent of the muscle fiber.

Figure 15–2C illustrates the repolarization process in the muscle fiber, which has proceeded halfway along the fiber from left to right. At this point, the left electrode is in an area of positivity, while the right electrode is in an area of negativity. This is opposite to the polarity in Figure 15–2A. Consequently, the recording, as illustrated to the right, becomes negative.

Finally, in Figure 15–2D the muscle fiber has completely repolarized, and both electrodes are in areas of positivity so that no potential is recorded between them. Thus, in the recording to the right, the potential returns once more to the zero level. This completed negative wave is a *repolarization wave* because it results from spread of the repolarization process over the muscle fiber.

Relationship of the Monophasic Action Potential of Cardiac Muscle to the QRS and T Waves. The monophasic action potential of ventricular muscle, which was discussed in the preceding chapter, normally lasts between 0.25 and 0.30 second. The top part of Figure 15–3 illustrates a monophasic action potential recorded from a microelectrode inserted to the inside of a single ventricular muscle fiber. The upsweep of this action potential is caused by *depolarization*, and the return of the potential to the base line is caused by *repolarization*.

Note below the simultaneous recording of the electrocardiogram from this same ventricle, which shows the QRS wave appearing at the beginning of the monophasic action potential and the T wave appearing at the end. Note especially that *no potential at all is recorded in the electrocardiogram when the ventricular muscle is either completely polarized or completely depolarized*. It is only when the muscle is partly polarized and partly depolarized that current flows from one part of the ventricles to another part and therefore also flows to the surface of the body to cause the electrocardiogram.

RELATIONSHIP OF ATRIAL AND VENTRICULAR CONTRACTION TO THE WAVES OF THE ELECTROCARDIOGRAM

Before contraction of muscle can occur, a depolarization wave must spread through the muscle to initiate the chemical processes of contraction. The P wave results from spread of the depolarization wave through the atria, and the QRS wave from spread of the depolarization wave through the ventricles. Therefore, the P wave occurs immediately before the *beginning of contraction of the atria*, and the QRS wave occurs immediately before the *beginning of contraction of the ventricles*. The ventricles remain contracted until a few milliseconds after repolarization has occurred, that is, until after the end of the T wave.

The atria repolarize approximately 0.10 to 0.20 second after the depolarization wave. However, this is just at the moment that the QRS wave is being recorded in the electrocardiogram. Therefore, the atrial repolarization wave, known as the *atrial T wave*, is usually totally obscured by the much larger QRS wave. For this reason, an atrial T wave is rarely observed in the electrocardiogram.

On the other hand, the ventricular repolarization wave is the T wave of the normal electrocardiogram. Ordinarily, ventricular muscle begins to repolarize in some fibers approximately 0.15 second after the beginning of the depolarization wave, but in many other fibers repolarization does not occur until as long as 0.30 second after onset of depolarization. Thus, the process of repolarization extends over a fairly long period of time, about 0.15 second. For this reason the T wave in the normal electrocardiogram is a fairly prolonged wave, but the voltage of the T wave is considerably less than the voltage of the QRS complex, partly because of its prolonged length.

VOLTAGE AND TIME CALIBRATION OF THE ELECTROCARDIOGRAM

All recordings of electrocardiograms are made with appropriate calibration lines on the recording paper. Either these calibration lines are already ruled on the paper, as is the case when a pen recorder is used, or they are recorded on the paper at the same time that the electrocardiogram is recorded, which is the case with the photographic types of electrocardiographs.

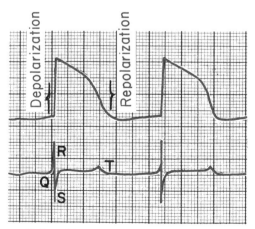

Figure 15–3. *Above:* Monophasic action potential from a ventricular muscle fiber during normal cardiac function, showing rapid depolarization and then repolarization occurring slowly during the plateau stage but very rapidly toward the end. *Below:* Electrocardiogram recorded simultaneously.

As illustrated in Figure 15–1, the calibration lines are arranged so that 10 small divisions in the vertical direction in the standard electrocardiogram represent 1 millivolt, with positivity in the upward direction and negativity in the downward direction.

The vertical lines on the electrocardiogram are time calibration lines. Each inch in the horizontal direction is 1 second, and each inch in turn is usually broken into five segments by dark vertical lines; the intervals between these lines represent 0.20 second. These intervals are then broken into five smaller intervals by thin lines, and each of these represents 0.04 second.

Normal Voltages in the Electrocardiogram. The voltages of the waves in the normal electrocardiogram depend on the manner in which the electrodes are applied to the surface of the body. When one electrode is placed directly over the heart and the second electrode is placed elsewhere on the body, the voltage of the QRS complex may be as great as 3 to 4 millivolts (mv.). Even this voltage is very small in comparison with the monophasic action potential of 120 mv. recorded directly at the heart muscle membrane. When electrocardiograms are recorded from electrodes on the two arms or on one arm and one leg, the voltage of the QRS complex usually is approximately 1 mv. from the top of the R wave to the bottom of the S wave, the voltage of the P wave between 0.1 and 0.3 mv., and that of the T wave between 0.2 and 0.3 mv.

The P-Q or P-R Interval. The duration of time between the beginning of the P wave and the beginning of the QRS wave is the interval between the beginning of contraction of the atrium and the beginning of contraction of the ventricle. This period of time is called the P-Q interval. The normal P-Q interval is approximately 0.16 second. This interval is sometimes also called the P-R interval because the Q wave is frequently absent.

The Q-T Interval. Contraction of the ventricle lasts essentially between the beginning of the Q wave and the end of the T wave. This interval of time is called the Q-T interval and ordinarily is approximately 0.30 second.

The Rate of the Heart as Determined from Electrocardiograms. The rate of heartbeat can be determined easily from electrocardiograms, because the time interval between two successive beats is the reciprocal of the heart rate. If the interval between two beats as determined from the time calibration lines is 1 second, the heart rate is 60 beats per minute. The normal interval between two successive QRS complexes is approximately 0.83 second. This is a heart rate of 60/0.83 times per minute, or 72 beats per minute.

METHODS FOR RECORDING ELECTROCARDIOGRAMS

The electrical currents generated by the cardiac muscle during each beat of the heart sometimes change potentials and polarity in less than 0.01 second. Therefore, it is essential that any apparatus for recording electrocardiograms be capable of responding rapidly to these changes in electrical potentials. In general, two different types of recording apparatuses are used for this purpose, as follows:

THE PEN RECORDER

Most older types of electrocardiographic apparatus used optical and photographic methods for recording the electrocardiogram. In many of these a beam of light was focused on a mirror mounted on a galvanometer coil located in a powerful magnetic field. When current flowed through the coil, the coil rotated in the field, and the mirror swept the reflected beam of light back and forth across a moving photographic paper. However, because the records required photographic development before the recording could be viewed, direct pen writing recorders have become the vogue in recent years.

The pen writing recorders write the electrocardiogram with a pen directly on a moving sheet of paper. The pen is often a thin tube connected at one end to an inkwell, and its recording end is connected to a powerful electromagnet system that is capable of moving the pen back and forth at high speed. As the paper moves forward, the pen records the electrocardiogram. The movement of the pen in turn is controlled by means of appropriate amplifiers connected to electrocardiographic electrodes on the patient.

Other pen recording systems use special paper that does not require ink in the recording stylus. One such paper turns black when it is exposed to heat; the stylus itself is made very hot by electrical current flowing through its tip. Another type turns black when electrical current flows from the tip of the stylus through the paper to an electrode at its back. This leaves a black line at every point on the paper that the stylus touches.

RECORDING ELECTROCARDIOGRAMS WITH THE OSCILLOSCOPE

Electrocardiograms can also be viewed on the screen of an oscilloscope by the method discussed for nerve potentials in Chapter 10, or they can be photographed from the oscilloscopic screen. However, because of the cost of the oscilloscope, and because extremely high frequency electrical potentials do not need to be recorded, the less complicated and less expensive pen recorder just described is ordinarily used in clinical electrocardiography.

FLOW OF CURRENT AROUND THE HEART DURING THE CARDIAC CYCLE

RECORDING ELECTRICAL POTENTIALS FROM A PARTIALLY DEPOLARIZED MASS OF SYNCYTIAL CARDIAC MUSCLE

Figure 15–4 illustrates a syncytial mass of cardiac muscle that has been stimulated at its centralmost point. Prior to stimulation, all the exteriors of the muscle cells had been positive and the interiors, negative. However, for reasons presented in Chapter 10 in the discussion of membrane potentials, as soon as an area of the cardiac syncytium becomes depolarized, negative charges leak

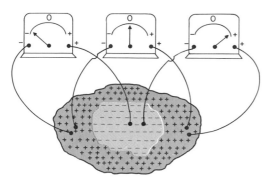

Figure 15–4. Instantaneous potentials developed on the surface of a cardiac muscle mass that has been depolarized in its center.

to the outsides of the depolarized muscle fibers, making this surface area electronegative, as represented by the negative signs in the figure, with respect to the remaining surface of the heart which is still polarized in the normal manner, as represented by the positive signs. Therefore, a meter connected with its negative terminal on the area of depolarization and its positive terminal on one of the still-polarized areas, as illustrated to the right in the figure, will record positively.

Two other possible electrode placements and meter readings are also illustrated in Figure 15–4. *These should be studied carefully and the student should explain to himself or herself the causes of the respective meter readings.* Obviously, since the process of depolarization spreads in all directions through the heart, the potential differences shown in the figure last for only a few milliseconds, and the actual voltage measurements can be accomplished only with a high-speed recording apparatus.

FLOW OF ELECTRICAL CURRENTS AROUND THE HEART IN THE CHEST

Figure 15–5 illustrates the ventricular muscle mass lying within the chest. Even the lungs, though filled with air, conduct electricity to a surprising extent, and fluids of the other tissues surrounding the heart conduct electricity even more easily. Therefore, the heart is actually suspended in a conductive medium. When one portion of the ventricles becomes electronegative with respect to the remainder, electrical current flows from the depolarized area to the polarized area in large circuitous routes, as noted in the figure.

It will be recalled from the discussion of the Purkinje system in Chapter 14 that the cardiac impulse first arrives in the ventricles in the septum and shortly thereafter on the endocardial surfaces of the remainder of the ventricles, as shown by the shaded areas and the negative signs in Figure 15–5. This provides electronegativity on the insides of the ventricles and electropositivity on the outer walls of the ventricles, and current flows through the fluids surrounding the ventricles along elliptical paths, as illustrated in the figure. If one algebraically averages all the lines of current flow (the elliptical lines), one finds that the average current flow is from the base of the heart toward the apex. During most

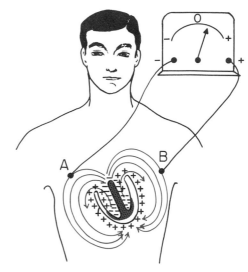

Figure 15–5. Flow of current in the chest around a partially depolarized heart.

of the remainder of the depolarization process, the current continues to flow in this direction as the depolarization wave spreads from the endocardial surface outward through the ventricular muscle. However, immediately before the depolarization wave has completed its course through the ventricles, the direction of current flow reverses for about 1/100 second, flowing then from the apex toward the base because the very last part of the heart to become depolarized is the outer walls of the ventricles near the base of the heart.

Thus, in the normal heart it may be considered that current flows primarily in the direction from the base toward the apex during almost the entire cycle of depolarization except at the very end. Therefore, if a meter is connected to the surface of the body as shown in Figure 15–5, the electrode nearer the base will be negative with respect to the electrode nearer the apex, and the recording meter will show a slight positive potential between the two electrodes. In making electrocardiographic recordings, various standard positions for placement of electrodes are used, and whether the polarity of the recording during each cardiac cycle is positive or negative is determined by the orientation of electrodes with respect to the current flow in the heart. Some of the conventional electrode systems, commonly called *electrocardiographic leads*, are discussed below.

ELECTROCARDIOGRAPHIC LEADS

THE THREE STANDARD LIMB LEADS

Figure 15–6 illustrates electrical connections between the limbs and the electrocardiograph for recording electrocardiograms from the so-called "standard" limb leads. The electrocardiograph in each instance is illustrated by mechanical meters in the diagram, though the

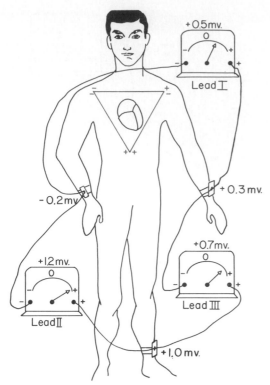

Figure 15–6. Conventional arrangement of electrodes for recording the standard electrocardiographic leads. Einthoven's triangle is superimposed on the chest.

actual electrocardiograph is a high-speed recording meter.

Lead I. In recording limb lead I, the *negative terminal of the electrocardiograph is connected to the right arm* and the *positive terminal to the left arm*. Therefore, when the point on the chest where the right arm connects to the chest is electronegative with respect to the point where the left arm connects, the electrocardiograph records positively — that is, above the zero voltage line in the electrocardiogram. When the opposite is true, the electrocardiograph records below the line.

Lead II. In recording limb lead II, the *negative terminal of the electrocardiograph is connected to the right arm* and the *positive terminal to the left leg*. Therefore, when the right arm is negative with respect to the left leg, the electrocardiograph records positively.

Lead III. In recording limb lead III, the *negative terminal of the electrocardiograph is connected to the left arm* and the *positive terminal to the left leg*. This means that the electrocardiograph records positively when the left arm is negative with respect to the left leg.

Einthoven's Triangle. In Figure 15–6, a triangle, called *Einthoven's triangle,* is drawn around the area of the heart. This is a diagrammatic means of illustrating that the two arms and the left leg form apices of a triangle surrounding the heart. The two apices at the upper part of the triangle represent the points at which the two arms connect electrically with the fluids around the heart, and the lower apex is the point at which the left leg connects with the fluids.

Einthoven's Law. Einthoven's law states simply that if the electrical potentials of any two of the three standard electrocardiographic leads are known at any given instant, the third one can be determined mathematically from the first two by simply summing the first two (but note that the positive and negative signs of the different leads must be observed when making this summation). For instance, let us assume that momentarily, as noted in Figure 15–6, the right arm is 0.2 mv. negative with respect to the average potential in the body, the left arm is 0.3 mv. positive, and the left leg is 1.0 mv. positive. Observing the meters in the figure, it will be seen that lead I records a positive potential of 0.5 mv., lead III records a positive potential of 0.7 mv., and lead II records a positive potential of 1.2 mv., because these are the instantaneous potential differences between the respective pairs of limbs.

Note that *the sum of the voltages in leads I and III equals the voltage in lead II*. That is, 0.5 plus 0.7 equals 1.2. Mathematically, this principle, called Einthoven's law, holds true at any given instant while the electrocardiogram is being recorded.

Normal Electrocardiograms Recorded by the Three Standard Leads. Figure 15–7 illustrates simultaneous recordings of the electrocardiogram in leads I, II, and III. It is obvious from this figure that the electrocardiograms in these three standard leads are very similar to each other, for they all record positive *P* waves and positive *T* waves, and the major portion of the QRS complex is also positive in each electrocardiogram.

On analysis of the three electrocardiograms, it can be shown with careful measurements that at any given instant the sum of the potentials in leads I and III equals the potential in lead II, thus illustrating the validity of Einthoven's law.

Because the recordings from all the standard limb leads are similar to each other, it does not matter greatly which lead is recorded when one wishes to diagnose the different cardiac arrhythmias, for diagnosis of arrhythmias depends mainly on the time relationships between the different waves of the cardiac cycle. On the other

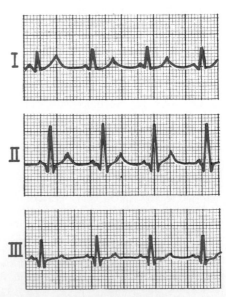

Figure 15–7. Normal electrocardiograms recorded from the three standard electrocardiographic leads.

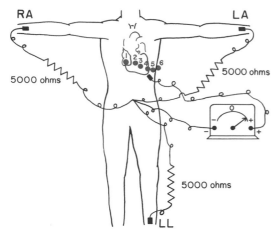

Figure 15–8. Connections of the body with the electrocardiograph for recording chest leads.

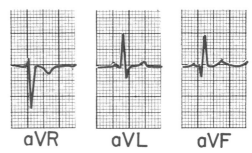

Figure 15–10. Normal electrocardiograms recorded from the three augmented unipolar limb leads.

hand, when one wishes to diagnose damage in the ventricular or atrial muscle or conducting system, it does matter greatly which leads are recorded, for abnormalities of the cardiac muscle change the patterns of the electrocardiograms markedly in some leads and yet may not affect other leads.

Electrocardiographic interpretation of these two types of conditions — cardiac myopathies and cardiac arrhythmias — are discussed separately in the following two chapters.

CHEST LEADS (PRECORDIAL LEADS)

Often electrocardiograms are recorded with one electrode placed on the anterior surface of the chest over the heart, as illustrated by the six separate points in Figure 15–8. This electrode is connected to the positive terminal of the electrocardiograph, and the negative electrode, called the *indifferent electrode*, is normally connected simultaneously through electrical resistances to the right arm, left arm, and left leg, as also shown in the figure. Usually six different standard chest leads are recorded from the anterior chest wall, the chest electrode being placed respectively at the six points illustrated in the diagram. The different leads recorded by the method illustrated in Figure 15–8 are known as leads V_1, V_2, V_3, V_4, V_5, and V_6.

Figure 15–9 illustrates the electrocardiograms of the normal heart as recorded from the six standard chest leads. Because the heart surfaces are close to the chest wall, each chest lead records mainly the electrical potential of the cardiac musculature immediately beneath the electrode. Therefore, relatively minute abnormalities in the ventricles, particularly in the anterior ventricular wall, frequently cause marked changes in the electrocardiograms recorded from chest leads.

In leads V_1 and V_2, the QRS recordings of the normal heart are mainly negative because, as illustrated in Figure 15–8, the chest electrode in these leads is nearer the base of the heart than the apex, which is the direction of electronegativity during most of the ventricular depolarization process. On the other hand, the QRS complexes in leads V_4, V_5, and V_6 are mainly positive because the chest electrode in these leads is nearer the apex, which is the direction of electropositivity during depolarization.

AUGMENTED UNIPOLAR LIMB LEADS

Another system of leads in wide use is the "augmented unipolar limb lead." In this type of recording, two of the limbs are connected through electrical resistances to the negative terminal of the electrocardiograph while the third limb is connected to the positive terminal. When the positive terminal is on the right arm, the lead is known as the aV_R lead; when on the left arm, the aV_L lead; and when on the left leg, the aV_F lead.

Normal recordings of the augmented unipolar limb leads are shown in Figure 15–10. These are all similar to the standard limb lead recordings except that the recording from the aV_R lead is inverted. The reason for this inversion is that the polarity of the electrocardiograph in this instance is connected backward to the major direction of current flow in the heart during the cardiac cycle.

Each augmented unipolar limb lead records the potential of the heart on the side nearest to the respective limb. Thus, when the recording in the aV_R lead is negative, this means that the side of the heart nearest to the right arm is negative in relation to the remainder of the heart; when the recording in the aV_F lead is positive, this means that the apex of the heart, which is the part of the heart nearest the foot, is positive with respect to the remainder of the heart.

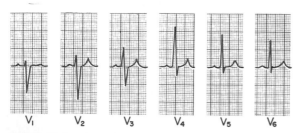

Figure 15–9. Normal electrocardiograms recorded from the six standard chest leads.

REFERENCES

See bibliography at end of Chapter 17.

16

Electrocardiographic Interpretation in Cardiac Myopathies — Vectorial Analysis

From the discussion in Chapter 14 of impulse transmission through the heart, it is obvious that any change in the pattern of this transmission can cause abnormal electrical currents around the heart and, consequently, can alter the shapes of the waves in the electrocardiogram. For this reason, almost all serious abnormalities of the heart muscle can be detected by analyzing the contours of the different waves in the different electrocardiographic leads. The purpose of the present chapter is to discuss the various alterations in the electrocardiograms when either the muscle of the heart or the conduction system, especially of the ventricles, functions abnormally.

PRINCIPLES OF VECTORIAL ANALYSIS OF ELECTROCARDIOGRAMS

USE OF VECTORS TO REPRESENT ELECTRICAL POTENTIALS

Before it is possible to understand how cardiac abnormalities affect the contours of waves in the electrocardiogram, one must first become familiar with the concept of vectors and vectorial analysis as applied to electrical currents flowing in and around the heart.

Several times in the preceding chapter it was pointed out that heart currents flow in a particular direction at a given instant in the cardiac cycle. A vector is an arrow that points in the direction of current flow *with the arrowpoint in the positive direction.* Also, by convention, the length of the arrow is drawn *proportional to the voltage generated by the current flow.*

In Figure 16–1, several different vectors are shown between areas of negativity and positivity in a syncytial mass of cardiac muscle. Note, first, that each vector points from negative to positive and, second, that the

length of each vector is proportional to the amount of charges causing the current flow.

The Summated Vector in the Heart at Any Given Instant. Figure 16–2 shows, via the shaded area and the negative signs, depolarization of the ventricular septum and parts of the lateral endocardial walls of the two ventricles. Electrical currents flow between these depolarized areas inside the heart and the nondepolarized areas on the outside of the heart as indicated by the elliptical arrows. Currents also flow inside the heart chambers directly from the depolarized areas toward the polarized areas. Even though a small amount of current flows upward inside the heart, a considerably greater quantity flows downward on the outside of the ventricles toward the apex. Therefore, the summated vector of currents at this particular instant is drawn through the center of the ventricles in a direction from the base of the heart toward the apex. Furthermore, because these currents are considerable in quantity, the vector is relatively long.

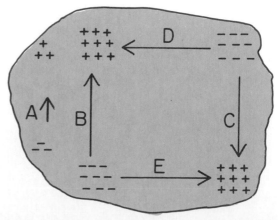

Figure 16–1. Vectors depicting current flow within a mass of cardiac muscle. See text for discussion.

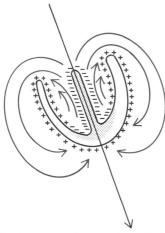

Figure 16–2.   A summated vector through the partially depolarized heart. Two additional vectors are shown, one to each side of the heart. These have the same significance as the vector through the center of the heart. See text for discussion.

DENOTING THE DIRECTION OF A VECTOR IN TERMS OF DEGREES

When a vector is horizontal and directed toward the subject's left side, it is said that the vector extends in the direction of 0 degrees, as is illustrated in Figure 16–3. From this zero reference point, the scale of vectors rotates clockwise; when the vector extends from above downward, it has a direction of 90 degrees; when it extends from the subject's left to the right, it has a direction of 180 degrees; and, when it extends upward, it has a direction of −90 or +270 degrees.

In a normal heart the average direction of the vector of the heart during spread of the depolarization wave is approximately 59 degrees, which is illustrated by vector A drawn through the center of Figure 16–3 in the 59 degree direction. This means that during most of the depolarization wave, the apex of the heart remains

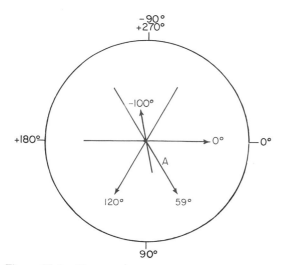

Figure 16–3.   Vectors drawn to represent direction of current flow and potentials for several different hearts.

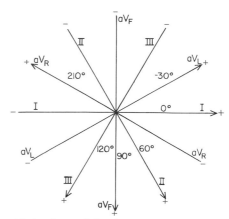

Figure 16–4.   Axes of the three standard leads and of the three unipolar leads.

positive with respect to the base of the heart, as is discussed later in the chapter.

"AXIS" OF EACH OF THE STANDARD AND UNIPOLAR LEADS

In the preceding chapter the three standard leads and the three unipolar leads were described. Each lead is actually a pair of electrodes connected to the body on opposite sides of the heart, and the direction from the negative to the positive electrode is called the *axis of the lead*. Lead I is recorded from two electrodes placed respectively on the two arms. Since the electrodes lie in the horizontal direction with the positive electrode to the left, the axis of lead I is 0 degrees.

In recording lead II, electrodes are placed on the right arm and left leg. The right arm connects to the torso in the upper right-hand corner and the left leg to the lower left-hand corner. Therefore, the direction of this lead is approximately 60 degrees.

By a similar analysis it can be seen that lead III has an axis of approximately 120 degrees, lead aV_R 210 degrees, aV_F 90 degrees, and aV_L −30 degrees. The directions of the axes of all these different leads are shown in Figure 16–4. Also, the polarities of the electrodes are illustrated by the plus and minus signs. *The student must learn these axes and their polarities, particularly for the standard leads I, II, and III, in order to understand the remainder of this chapter.*

VECTORIAL ANALYSIS OF POTENTIALS RECORDED IN DIFFERENT LEADS

Now that we have discussed, first, the conventions for representing current flow and potentials across the heart by means of vectors and, second, the axes of the leads, it becomes possible to put these two together to determine the potential that will be recorded in each lead for a given vector in the heart.

Figure 16–5 illustrates a partially depolarized heart; vector A represents the direction of current flow in the heart and its potential. In this instance the direction of current flow is 55 degrees, and the potential will be assumed to be 2 mv. Through the base of vector A is

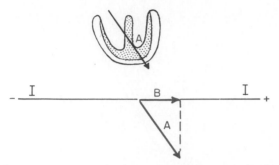

Figure 16–5. Determination of a resultant vector B along the axis of lead I when vector A represents the current flow in the ventricles.

drawn the axis of lead I in the 0 degree direction. From the tip of vector *A* a perpendicular is dropped to the lead I axis, and a so-called *resultant vector (B)* is drawn along the axis. The head of this vector is in the positive direction, which means that the record momentarily being recorded in the electrocardiogram of lead I will be positive. The voltage recorded will be equal to the length of *B* divided by the length of *A* times 2 mv., or approximately 1 mv.

Figure 16–6 illustrates another example of vectorial analysis. In this example, vector *A* represents the current flow at a given instant during ventricular depolarization in another heart, in which the left side of the heart becomes depolarized somewhat more rapidly than the right. In this instance the vector has a direction of 100 degrees, and the voltage is again 2 mv. To determine the potential in lead I, we drop a perpendicular to the lead I axis and find the resultant vector *B*. Vector *B* is very short and this time in the negative direction, indicating that at this particular instant the recording in lead I will be negative (below the zero line), and the voltage recorded will be slight. This figure illustrates that *when the vector in the heart is in a direction almost perpendicular to the axis of the lead, the voltage recorded in the electrocardiogram of this lead is very low.* On the other hand, *when the heart vector has almost exactly the same axis as the lead, essentially the entire voltage of the vector will be recorded.*

Vectorial Analysis of Potentials in the Three Standard

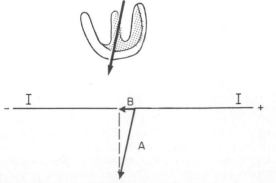

Figure 16–6. Determination of a resultant vector B along the axis of lead I when vector A represents the current flow in the ventricles.

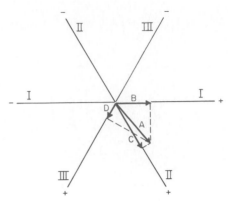

Figure 16–7. Determination of resultant vectors in leads I, II, and III when vector A represents the current flow in the ventricles.

Leads. In Figure 16–7, vector *A* depicts the instantaneous direction and potential of current in a partially depolarized ventricle. To determine the potential recorded at this instant in the electrocardiogram of each one of the three standard leads, perpendiculars are dropped to all the lines representing the different leads as illustrated in the figure. The resultant vector *B* depicts the potential recorded at that instant in lead I; vector *C* depicts the potential in lead II; and vector *D,* in lead III. In each of these the record in the electrocardiogram is positive — that is, above the zero line — because the resultant vectors lie in the positive directions along the axes of the leads. The potential in lead I is approximately one-half that of the vector through the heart; in lead II it is almost exactly equal to that in the heart; and in lead III it is about one-third that in the heart.

An identical analysis can be used to determine potentials recorded in augmented limb leads, except that the respective axes of these leads (see Figure 16–4) are used in place of the standard lead axes used in Figure 16–7.

VECTORIAL ANALYSIS OF THE NORMAL ELECTROCARDIOGRAM

VECTORS OCCURRING DURING DEPOLARIZATION OF THE VENTRICLES — THE QRS COMPLEX

When the cardiac impulse enters the ventricles through the A-V bundle, the first part of the ventricles to become depolarized is the left endocardial surface of the septum. This depolarization spreads rapidly to involve both endocardial surfaces of the septum, as illustrated by the shaded portion of the ventricle in Figure 16–8A. Then the depolarization spreads along the endocardial surfaces of the two ventricles, as shown in Figures 16–8B and C. Finally, it spreads through the ventricular muscle to the outside of the heart, as shown progressively in Figures 16–8C, D, and E.

At each stage of depolarization of the ventricles in Figures 16–8A to E, the instantaneous, summated current flow is represented by a vector superimposed on

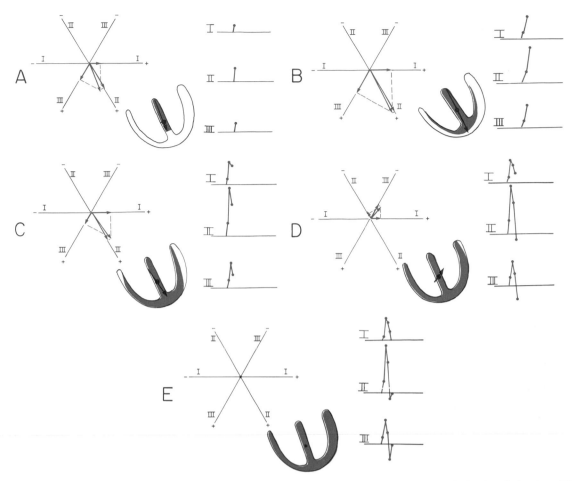

Figure 16–8. (A) The ventricular vectors and the QRS complexes 0.01 second after onset of ventricular depolarization. (B) 0.02 second after onset of depolarization. (C) 0.035 second after onset of depolarization. (D) 0.05 second after onset of depolarization, and (E) after depolarization of the ventricles is complete, 0.06 second after onset.

the ventricle in each figure. Each of these vectors is analyzed by the method described in the preceding section to determine the voltages that will be recorded at each instant in each of the three standard electrocardiographic leads. To the right is shown the progressive development of the QRS complex. *Keep in mind that a positive vector in a lead will cause the recording in the electrocardiogram to be above the zero line, while a negative vector will cause the recording to be below the line.*

Before proceeding with any further consideration of vectorial analysis, it is essential that this analysis of the successive normal vectors presented in Figure 16–8 be understood. Each of these analyses should be studied in detail by the procedure given above. A short summary of this sequence is the following:

In Figure 16–8A the ventricular muscle has just begun to be depolarized, representing an instant about 0.01 second after the onset of depolarization. At this time, the vector is short because only a small portion of the ventricles — the septum — is depolarized. Therefore, all electrocardiographic voltages are low, as recorded to the right of the ventricular muscle for each of the leads. The voltage in lead II is greater than the voltages in

leads I and III because the heart vector extends mainly in the same direction as the axis of lead II.

In Figure 16–8B, which represents approximately 0.02 second after onset of depolarization, the heart vector is long because much of the ventricles has now become depolarized. Therefore, the voltages in all electrocardiographic leads have increased.

In Figure 16–8C — about 0.035 second after onset of depolarization — the heart vector is becoming shorter and the recorded EKG voltages are less because the outside of the apex of the heart is now electronegative, neutralizing much of the negativity on the endocardial surfaces of the heart. Also, the axis of the vector is shifting toward the left side of the chest because the left ventricle is slightly slower to depolarize than the right. The ratio of the voltage in lead I to that in lead III is increasing.

In Figure 16–8D — about 0.05 second after onset of depolarization — the heart vector points toward the base of the left ventricle, and it is short because only a minute portion of the ventricular muscle is still polarized. Because of the direction of the vector at this time, the voltages recorded in leads II and III are both negative — that is, below the line.

In Figure 16–8E — about 0.06 second after onset of depolarization — the entire ventricular muscle mass is depolarized so that no current flows around the heart at all. The vector becomes zero, and the voltages in all leads become zero.

Thus, the QRS complexes are completed in the three standard electrocardiographic leads. The QRS complex sometimes has a slight negative depression at its beginning, which is not shown in Figure 16–8; this is the Q wave. When it occurs, it is caused by initial depolarization of the septum nearer to the apex of the septum than to the base and/or depolarization of the left side of the septum before the right side, which occasionally creates a weak vector in the apex-to-base direction or from left to right for a minute fraction of a second before the usual base-to-apex vector occurs. The major positive deflection shown in Figure 16–8 is the R wave, and the final negative deflection is the S wave.

THE ELECTROCARDIOGRAM DURING REPOLARIZATION — THE T WAVE

Once the ventricular muscle has become depolarized, approximately 0.15 second elapses before sufficient repolarization begins for it to be observed in the electrocardiogram; then repolarization proceeds throughout the ventricular muscle until it is complete at about 0.30 second. It is this process of repolarization that causes the T wave in the electrocardiogram.

Because the septum and other endocardial areas of the ventricular muscle depolarize first, it seems logical that these areas should repolarize first, but this is not the usual case because the septum and the other endocardial areas have a longer period of contraction and depolarization than do other areas of the heart. Actually, many sections of the ventricles begin to repolarize almost simultaneously. Yet *the greatest portion of ventricular muscle to repolarize first is that located over the entire outer surface of the ventricles and especially near the apex of the heart.* And the endocardial areas, on the average, repolarize last. The reason for this abnormal sequence of repolarization is believed to be that high pressure in the ventricles during contraction greatly reduces coronary blood flow to the endocardium, thereby slowing the repolarization process in the endocardial areas.

Therefore, because the outer, apical surfaces of the ventricles repolarize before the inner and basal surfaces, the positive end of the heart vector during repolarization is toward the apex of the heart. Thus, *the predominant direction of the vector through the heart during* repolarization *of the ventricles is from base to apex, which is also the predominant direction of the vector during* depolarization. *As a result, the T wave in the normal electrocardiogram is positive, which is also the polarity of most of the normal QRS complex.*

In Figure 16–9 five stages of repolarization of the ventricles are denoted by the progressive increase of the white areas — the repolarized areas. At each stage, the vector extends from the base toward the apex. At first the vector is relatively small because the area of repolarization is small. Later the vector becomes stronger and stronger because of greater degrees of repolarization. Finally, the vector becomes weaker again because

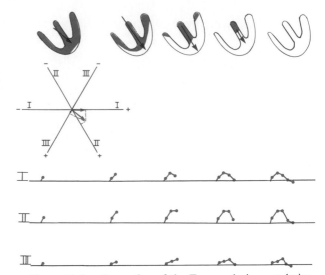

Figure 16–9. Generation of the T wave during repolarization of the ventricles, showing a vectorial analysis of the first stage. The total time from the beginning of the T wave to the end is about 0.15 second.

the areas of depolarization still persisting become so slight that the total quantity of current flow begins to decrease. These changes illustrate that the vector is greatest when approximately half the heart is in the polarized state and approximately half is depolarized.

The changes in the electrocardiograms of the three standard leads during the process of repolarization are noted under each of the ventricles, depicting the progressive stages of repolarization. Over approximately 0.15 second, the period required for the whole process to take place, the T wave of the electrocardiogram is generated.

DEPOLARIZATION OF THE ATRIA — THE P WAVE

Depolarization of the atria begins in the S-A node and spreads in all directions over the atria. Therefore, the point of original electronegativity in the atria is approximately at the base of the superior vena cava where the S-A node lies, and the direction of current flow in the atrium at the beginning of depolarization is in the direction noted in Figure 16–10. Furthermore, the vector remains generally in this direction throughout the process of atrial depolarization.

Thus, the vector of current flow during depolarization in the atria points in almost the same direction as that in the ventricles. And, because this direction is in the directions of the axes of the standard leads I, II, and III, the electrocardiograms recorded from the atria during the process of depolarization usually are positive in all three standard leads, as illustrated in Figure 16–10. The record of atrial depolarization is known as the P wave.

Repolarization of the Atria — The Atrial T Wave. Spread of the depolarization wave through the atrial muscle is *much slower than in the ventricles.* Therefore, the musculature around the S-A node becomes depolarized a long time before the musculature in

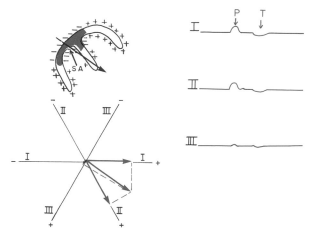

Figure 16–10. Depolarization of the atria and generation of the P wave, showing the vector through the atria and the resultant vectors in the three standard leads. At right are the atrial P and T waves.

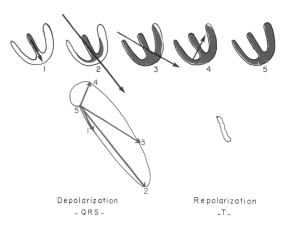

Figure 16–11. The QRS and T vectorcardiograms.

distal parts of the atria. Because of this, *the area in the atria that becomes repolarized first is the S-A nodal region, the area that had originally become depolarized first,* which is an entirely different situation from that in the ventricles. Thus, when repolarization begins, the region around the S-A node becomes positive with respect to the rest of the atria. Therefore, the atrial repolarization vector is *backward to the vector of depolarization.* Note again that this is opposite to the effect that occurs in the ventricles. Therefore, as noted to the right in Figure 16–10, the so-called atrial T wave follows about 0.15 second after the atrial P wave, but this T wave is on the opposite side of the zero reference line from the P wave — that is, it is normally negative rather than positive. In the normal electrocardiogram this T wave appears at approximately the same time that the QRS complex of the ventricle appears. Therefore, it is almost always totally obscured by the larger QRS complex, though in some abnormal states it does play a role in the recorded electrocardiogram.

THE VECTORCARDIOGRAM

It was noted in the preceding discussions that the vector of current flow through the heart changes rapidly as the impulse spreads through the myocardium. It changes in two aspects: First, the vector increases and decreases in length because of the increasing and decreasing potential of the vector. Second, the vector changes its direction because of changes in the average direction of current flow around the heart. The so-called *vectorcardiogram* depicts these changes in the vectors at the different times during the cardiac cycle, as illustrated in Figure 16–11.

In the vectorcardiogram of Figure 16–11, point 5 is the zero reference point, and this point is the negative end of all the vectors. While the heart is quiescent, the positive end of the vector also remains at the zero point because there is no current flow. However, just as soon as current begins to flow through the heart, the positive end of the vector leaves the zero reference point.

When the septum first becomes depolarized, the

vector extends downward toward the apex of the heart, but it is relatively weak, thus generating the first portion of the vectorcardiogram, as illustrated by the positive end of vector 1. As more of the heart becomes depolarized, the vector becomes stronger and stronger, usually swinging slightly to one side. Thus vector 2 of Figure 16–11 represents the state of depolarization of the heart about 0.02 second after vector 1. After another 0.02 second, vector 3 represents the current flow in the heart, and vector 4 occurs in still another 0.01 second. Finally, the heart becomes totally depolarized, and the vector becomes zero once again, as shown at point 5.

The elliptical figure generated by the positive ends of the vectors is called the *QRS vectorcardiogram.*

Vectorcardiograms can be recorded instantaneously on an oscilloscope by connecting electrodes from above and below the heart to the vertical plates of the oscilloscope and connecting electrodes from each side of the heart to the horizontal plates. When the vector changes, the spot of light on the oscilloscope follows the course of the positive end of the changing vector, thus inscribing the vectorcardiogram on the oscilloscopic screen.

The "T" Vectorcardiogram. Changing vectors in the heart occur not only during the depolarization process, for vectors depicting current flow around the ventricles also reappear during repolarization. Therefore, a second, smaller vectorcardiogram — the *T vectorcardiogram* — is inscribed during repolarization of the muscular mass; this is depicted to the right in Figure 16–10. Also, a still smaller *P vectorcardiogram* is inscribed during atrial depolarization.

THE MEAN ELECTRICAL AXIS OF THE VENTRICLE

The vectorcardiogram of the ventricular depolarization wave (the QRS vectorcardiogram) shown in Figure 16–11 is that of a normal heart. Note from this vectorcardiogram that the preponderant direction of the vectors of the ventricles is normally toward the apex of the heart — that is, during most of the cycle of ventricular depolarization, current flows from the base of the ventricles toward the apex. This preponderant direction of current flow during depolarization is called the *mean electrical axis of the ventricles.* The mean electrical axis of the normal ventricles is 59 degrees. However, in certain pathological conditions of the heart this direction of current flow is changed markedly — sometimes even to opposite poles of the heart.

DETERMINING THE ELECTRICAL AXIS FROM STANDARD LEAD ELECTROCARDIOGRAMS

Clinically, the electrical axis of the heart is usually determined from the standard lead electrocardiograms rather than from the vectorcardiogram. Figure 16–12 illustrates a method for doing this. After recording the various standard leads, one determines the maximum potential and polarity of the recording in two of the leads. In lead I of the figure the recording is positive and in lead III the recording is mainly positive, but negative during part of the cycle. If any part of the recording is negative, *this negative potential is subtracted from the positive potential* to determine the net vector for that lead, as illustrated by the arrows to the right of the QRS complexes of leads I and III. After subtracting the negative portion of the QRS wave in lead III from the positive portion, each net vector is plotted on the axes of the respective leads, with the base of the vector at the point of intersection of the axes, as illustrated in Figure 16–12.

If the vector of lead I is positive, it is plotted in a positive direction along the line depicting lead I. On the other hand, if this vector is negative, it is plotted in a negative direction. Also, for lead III, the vector is placed with its base at the point of intersection, and, if positive, it is plotted in the positive direction along the line depicting lead III. If it is negative it is plotted in the negative direction.

In order to determine the actual vector of current flow in the ventricles, one draws perpendicular lines from the apices of the two vectors of leads I and III, respectively. The point of intersection of these two perpendicular lines represents, by vectorial analysis, the apex of the actual vector in the ventricles, and the point of intersection of the two lead axes represents the negative end of the actual vector. Therefore, the *ventricular vector* is drawn between these two points. The average potential generated by the ventricles during depolarization is represented by the length of the vector, and the mean

electrical axis is represented by the direction of the vector. Thus, the vector of the mean electrical axis of the normal ventricles, as determined in Figure 16–12, is 59 degrees.

The vector of current flow through the ventricles determined by this method does not determine the vector at a given instant in the depolarization process of the ventricles but, instead, determines approximately how much and in what direction the *average* net current flows during the entire depolarization period.

ABNORMAL VENTRICULAR CONDITIONS THAT CAUSE AXIS DEVIATION

Though the mean electrical axis of the ventricles averages approximately 59 degrees, this average can swing to the left even in the normal heart to approximately 20 degrees or to the right to approximately 100 degrees. The causes of the normal variations are simply anatomical differences in the Purkinje fiber distribution or in the musculature itself of different hearts. Yet, a number of conditions can cause axis deviation even beyond these normal limits, as follows:

Change in the Position of the Heart. Obviously, if the heart itself is angulated to the left, the mean electrical axis of the heart will also shift to the left. Such shift occurs (a) during expiration, (b) when a person lies down, because the abdominal contents press upward against the diaphragm, and (c) quite frequently in stocky, fat persons whose diaphragms normally press upward against the heart all the time.

Likewise, angulation of the heart to the right causes the mean electrical axis of the ventricles to shift to the right. This condition occurs (a) during inspiration, (b) when a person stands up, and (c) normally in tall, lanky persons whose hearts hang downward.

Hypertrophy of One Ventricle. When one ventricle greatly hypertrophies, the axis of the heart shifts toward the hypertrophied ventricle for two reasons: First, far greater quantity of muscle exists on the hypertrophied side of the heart than on the other side, and this allows excess generation of electrical currents on that side. Second, more time is required for the depolarization wave to travel through the hypertrophied ventricle than through the normal ventricle. Consequently, the normal ventricle becomes depolarized considerably in advance of the hypertrophied ventricle, and this causes a strong vector from the normal side of the heart toward the hypertrophied side. Thus the axis deviates toward the hypertrophied ventricle.

Vectorial Analysis of Left Axis Deviation Resulting from Hypertrophy of the Left Ventricle. Figure 16–13 illustrates the three standard leads of an electrocardiogram in which an analysis of the axis direction shows left axis deviation with the mean electrical axis pointing in the −15 degree direction. This is a typical electrocardiogram resulting from increased muscular mass of the left ventricle. In this instance the axis deviation was caused by *hypertension* (high blood pressure), which caused the left ventricle to hypertrophy in order to pump blood against the elevated systemic arterial pressure. However, a similar picture of left axis deviation occurs when the left ventricle hypertrophies as a result of *aortic*

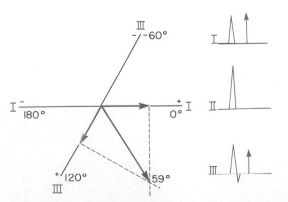

Figure 16–12. Plotting the mean electrical axis of the heart from two electrocardiographic leads.

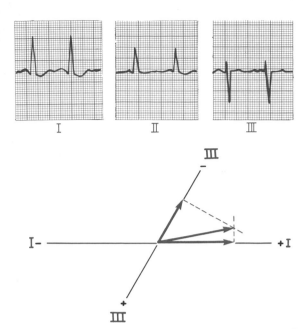

Figure 16–13. Left axis deviation in hypertensive heart disease. Note the slightly prolonged QRS complex.

valvular stenosis, aortic valvular regurgitation, or any of a number of *congenital heart conditions* in which the left ventricle enlarges while the right side of the heart remains relatively normal in size.

Vectorial Analysis of Right Axis Deviation Resulting from Hypertrophy of the Right Ventricle. The electrocardiogram of Figure 16–14 illustrates intense right axis deviation with an electrical axis of approximately 170 degrees, which is 111 degrees to the right of the normal mean electrical axis of the ventricles. The right axis deviation illustrated in this figure was caused by hypertrophy of the right ventricle as a result of *pulmonary stenosis.* However, right axis deviation may also occur in other congenital heart conditions, such as *tetralogy of Fallot* or *interventricular septal defect.* Also, hypertrophy of the right ventricle as a result of *increased pulmonary vascular resistance* can cause right axis deviation.

Bundle Branch Block. Ordinarily, the two lateral walls of the ventricles depolarize at almost the same time, because both the left and right bundle branches transmit the cardiac impulse to the endocardial surfaces of the two ventricular walls at almost the same instant. As a result, the currents flowing from the walls of the two ventricles almost neutralize each other. However, if one of the major bundle branches is blocked, depolarization of the two ventricles does not occur even nearly simultaneously, and the depolarization currents do not neutralize each other. As a result, axis deviation occurs as follows:

Vectorial Analysis of Left Bundle Branch Block. When the left bundle branch is blocked, cardiac depolarization spreads through the right ventricle approximately 3 times as rapidly as through the left ventricle. Consequently, much of the left ventricle remains polarized for a long time after the right ventricle has become totally depolarized. Thus, the right ventricle becomes

electronegative with respect to the left ventricle during most of the depolarization process, and a very strong vector projects from the right ventricle toward the left ventricle. In other words, there is intense left axis deviation because the positive end of the vector points toward the left ventricle. This is illustrated in Figure 16–15, which shows typical left axis deviation resulting from left bundle branch block. Note that the axis is approximately −50 degrees.

Because of slowness of impulse conduction when the Purkinje system is blocked, axis deviation resulting from bundle branch block also greatly increases the duration of the QRS complex, which one can see by observing the excessive widths of the QRS waves in Figure 16–15. This will be discussed in greater detail later in the chapter. This prolonged QRS complex differentiates this condition from axis deviation caused by hypertrophy.

Vectorial Analysis of Right Bundle Branch Block. When the right bundle branch is blocked, the left ventricle depolarizes far more rapidly than the right ventricle so that the left becomes electronegative while the right remains electropositive. A very strong vector develops with its negative end toward the left ventricle and its positive end toward the right ventricle. In other words, intense right axis deviation occurs.

Right axis deviation caused by right bundle branch block is illustrated and its vector analyzed in Figure 16–16, which shows an axis of approximately 105 degrees and a prolonged QRS complex because of blocked

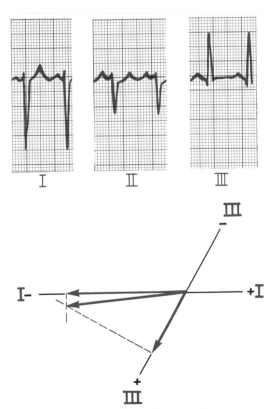

Figure 16–14. High-voltage electrocardiogram in pulmonary stenosis with right ventricular hypertrophy. Intense right axis deviation is also seen.

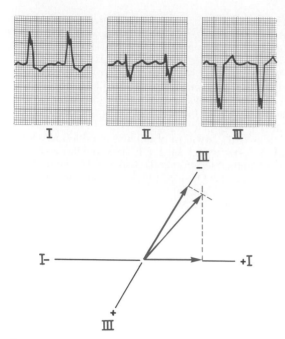

Figure 16–15. Left axis deviation due to left bundle branch block. Note the greatly prolonged QRS complex.

conduction. Note the great prolongation of the QRS complex.

Axis Deviation Caused by Muscular Destruction. Abnormal deviations of the axis of the ventricles can also occur following heart attacks that result in destruction of part of the cardiac muscle and replacement of this muscle with fibrous tissue. The major causes of the axis deviation are two-fold: First, part of the muscular mass itself is destroyed and is replaced by scar tissue so that a smaller quantity of muscle may then be available on one side of the heart than on the other side to generate electrical currents, thus creating an imbalance of current flow towards one side or the other. Second, and probably more important, is blockage of the conduction of depolarization at one or more local points in the smaller branches of the Purkinje system. When blockage occurs, the impulse must then be conducted by the

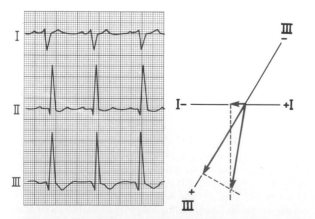

Figure 16–16. Right axis deviation due to right bundle branch block. Note the greatly prolonged QRS complex.

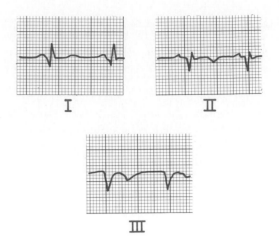

Figure 16–17. Low-voltage electrocardiogram, with evidences of local damage throughout the ventricles caused by old myocardial infarction.

muscle itself. This conduction occurs at a slow velocity, and it often has to detour around the scar tissue, allowing the muscular tissue beyond the block to remain electropositive for long periods of time after other portions of the heart have already become totally depolarized. Thus, even local blocks in the ventricles can cause considerable shift of the axis of the heart.

Muscular destruction causes the axis to shift in the direction opposite to the side of destruction, but conduction block causes the axis to shift toward the side of destruction. Thus, the two effects may or may not neutralize each other. Consequently, old infarctions in the heart do not produce consistent changes in axis deviation. Therefore, when other signs of old infarction are present in the electrocardiogram, such as low voltage and a prominent Q wave (as shown in Figure 16–17), increased duration of the QRS complex, and bizarre spiking patterns of the QRS complex, the diagnosis is usually definite, and it is generally unimportant to determine the axis of the heart.

CONDITIONS THAT CAUSE ABNORMAL VOLTAGES OF THE QRS COMPLEX

INCREASED VOLTAGE IN THE STANDARD LEADS

Normally, the voltages in the three standard electrocardiographic leads, as measured from the peak of the R wave to the bottom of the S wave, vary between 0.5 and 2.0 mv., with lead III usually recording the lowest voltage and lead II the highest. However, these relationships are not invariably true even in the normal heart. In general, when the sum of the voltages of all the QRS complexes of the three standard leads is greater than 4 mv., one considers that the patient has a high-voltage electrocardiogram.

The cause of high-voltage QRS complexes is most often increased muscular mass of the heart, which ordinarily results from *hypertrophy of the muscle* in response to excessive load on one part of the heart or the other. For instance, the right ventricle hypertrophies when it must pump blood through a stenotic pulmonary valve, and the left ventricle hypertrophies when a person has high blood pressure. The increased quantity of muscle allows generation of increased quantities of current around the heart. As a result, the electrical potentials recorded in the electrocardiographic leads are considerably greater than normal, as shown in Figures 16–13 and 16–14.

DECREASED VOLTAGE IN THE STANDARD LEADS

There are three major causes of decreased voltage in the electrocardiograms of the three standard electrocardiographic leads. These are, first, abnormalities of the cardiac muscle itself that prevent generation of large quantities of currents; second, abnormal conditions around the heart so that currents cannot be conducted from the heart to the surface of the body with ease; and, third, rotation of the apex of the heart to point toward the anterior chest wall so that the electrical currents of the heart flow mainly anteroposteriorly in the chest rather than in the frontal plane of the body.

Decreased Voltage Caused by Cardiac Myopathies. One of the most usual causes of decreased voltage of the QRS complex is a series of *old myocardial infarctions* associated with *diminished muscle mass*. This causes the depolarization wave to move through the ventricles slowly and prevents major portions of the heart from becoming massively depolarized all at once. Consequently, this condition causes moderate prolongation of the QRS complex along with decreased voltage. Figure 16–17 illustrates the typical low voltage electrocardiogram with prolongation of the QRS complex, which one often finds after multiple small infarctions of the heart that have resulted in local blocks and loss of muscle mass throughout the ventricles.

Decreased Voltage Caused by Conditions Surrounding the Heart. One of the most important causes of decreased voltage in the electrocardiographic leads is *fluid in the pericardium*. Because extracellular fluid conducts electrical currents with great ease, a large portion of currents flowing out of the heart is conducted from one part of the heart to another through the pericardial effusion. Thus this effusion effectively "short-circuits" the current that is generated by the heart so that the proportion of this current reaching the surface of the body is greatly diminished. *Pleural effusion* to a lesser extent can also "short" the currents around the heart so that the voltages at the surface of the body and in the electrocardiograms are decreased.

Pulmonary emphysema can decrease the electrocardiographic potentials but by a different method from that of pericardial effusion. In pulmonary emphysema, conduction of electrical current through the lungs is considerably depressed because of the excessive quantity of air in the lungs. Also, the chest cavity enlarges, and the lungs tend to envelop the heart to a greater extent than normally. Therefore, the lungs act as an insulator to prevent spread of currents from the heart to the surface of the body, and this in general results in decreased electrocardiographic potentials in the various leads.

PROLONGED AND BIZARRE PATTERNS OF THE QRS COMPLEX

PROLONGED QRS COMPLEX AS A RESULT OF CARDIAC HYPERTROPHY OR DILATATION

The QRS complex lasts as long as the process of depolarization continues to spread through the ventricles — that is, as long as part of the ventricles is depolarized and part is still polarized. Therefore, the cause of a prolonged QRS complex is always *prolonged conduction* of the impulse through the ventricles. Such prolongation often occurs when one or both ventricles are hypertrophied or dilated, owing to the longer pathway that the impulse must then travel. The normal QRS complex lasts about 0.06 second, whereas in hypertrophy or dilatation of the left or right ventricle the QRS complex may be prolonged to 0.09 second or occasionally 0.10 second.

PROLONGED QRS COMPLEX RESULTING FROM PURKINJE SYSTEM BLOCKS

Block of the Purkinje fibers requires that the cardiac impulse be conducted by the ventricular muscle instead of through the specialized conduction system, thereby decreasing the velocity of impulse conduction to approximately one-third to one-fourth normal. Therefore, if complete block of one of the bundle branches occurs, the duration of the QRS complex is increased to 0.14 second or greater.

In general, a QRS complex is considered to be abnormally long when it lasts more than 0.08 second, and when it lasts more than 0.12 second the prolongation is almost certain to be caused by pathological block of the conduction system somewhere in the ventricles, as illustrated by the electrocardiograms for bundle branch block in Figures 16–15 and 16–16.

CONDITIONS CAUSING BIZARRE QRS COMPLEXES

Bizarre patterns of the QRS complex are most frequently caused by two conditions: first, destruction of cardiac muscle in various areas throughout the ventricular system with replacement of this muscle by scar tissue, and, second, local blocks in the conduction of impulses by the Purkinje system.

Sometimes local blocks occur in both the right and left ventricles. If the blocks are such that the impulse reaches the blocked area in the right ventricle much later than it reaches the blocked area in the left ventricle, but the total quantity of blocked muscle in the right ventricle is greater than the total quantity in the left ventricle, then a situation might occur in which the axis of the heart first shifts to the left and then, when the left ventricle has become totally depolarized but a portion of the right ventricle still remains polarized, rapidly shifts diametrically oppositely toward the right ventricle. This effect causes double or even triple peaks in some of the electrocardiographic leads, such as those illustrated in Figure 16–15.

CURRENT OF INJURY

Many different cardiac abnormalities, especially those that damage the heart muscle itself, often cause part of the heart to remain partially or totally *depolarized all the time*. When this occurs, current flows between the pathologically depolarized and the normally polarized areas. This is called the *current of injury*. Note especially that *the injured part of the heart is negative because it is this part that is depolarized while the remainder is positive*.

Some of the abnormalities that can cause current of injury are: (1) *mechanical trauma,* which makes the membranes remain so permeable that full repolarization cannot take place, (2) *infectious processes* that damage the muscle membranes, and (3) *ischemia of local areas of muscle caused by coronary occlusion,* which is by far the most common cause of current of injury in the heart; during ischemia enough energy simply is not available to maintain normal function of the muscle.

EFFECT OF CURRENT OF INJURY ON THE QRS COMPLEX

In Figure 16–18 a shaded area in the base of the left ventricle is newly infarcted. Therefore, during the T-P interval — that is, when the normal ventricular muscle is polarized — current flows from the base of the left ventricle toward the rest of the ventricles. The vector of this "current of injury" is in a direction of approximately 125 degrees, with the base of the vector, the *negative end,* toward the injured muscle. As illustrated in the lower portions of the figure, even before the QRS complex begins, *this current flow causes an initial record in lead I below the zero potential line* because the resultant vector of the current of injury points

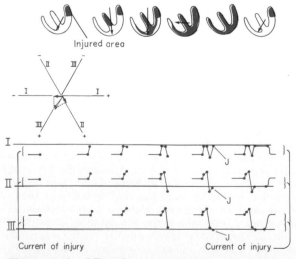

Figure 16–18. Effect of a current of injury on the electrocardiogram.

toward the negative end of the lead I axis. In lead II the record is above the line because the resultant vector points toward the positive terminal of lead II. In lead III the vector of the current flow is also in the same direction as the polarity of lead III so that the record is positive. Furthermore, because the vector of the current of injury lies almost exactly along the axis of lead III, the potential of the current of injury in lead III is much higher than in either of the other two records.

As the heart then proceeds through its normal process of depolarization, the septum first becomes depolarized, and the depolarization spreads down to the apex and back toward the bases of the ventricles. The last portion of the ventricles to become totally depolarized is the base of the right ventricle, because the base of the left ventricle is already totally and permanently depolarized. By vectorial analysis, as illustrated in the figure, the electrocardiogram generated by the depolarization wave traveling through the ventricles may be constructed graphically, as shown in Figure 16–18.

When the heart becomes totally depolarized at the end of the depolarization process, as noted by the next to last stage in Figure 16–18, all of the ventricular muscle is in a negative state. Therefore, at this instant in the electrocardiogram, absolutely no current flows around the musculature of the ventricles, because now the injured heart muscle and the contracting muscle are both completely depolarized.

As repolarization then takes place, all the heart finally repolarizes except the area of permanent depolarization in the injured base of the left ventricle. Thus, the repolarization causes a return of the current of injury in each lead, as noted at the far right in Figure 16–18.

THE J POINT — THE ZERO REFERENCE POTENTIAL OF THE ELECTROCARDIOGRAM

One would think that the electrocardiograph machines for recording electrocardiograms could determine when no current is flowing around the heart. However, many stray currents exist in the body, such as currents resulting from "skin potentials" and from differences in ionic concentrations in different parts of the body. Therefore, when two electrodes are connected between the arms or between an arm and a leg, these stray currents make it impossible for one to predetermine the exact zero reference level in the electrocardiogram. For these reasons, the following procedure must be used to determine the zero potential level: First, one notes *the exact point at which the wave of depolarization just completes its passage through the heart,* which occurs at the very end of the QRS complex. At exactly this point, all parts of the ventricles are depolarized so that no currents are flowing around the heart. Even the current of injury disappears at this point. Therefore, the potential of the electrocardiogram at this instant is at exactly zero voltage. This point is known as the "J point" in the electrocardiogram, as illustrated in Figures 16–18 and 16–19.

For analysis of the electrical axis of the current of injury, a horizontal line is drawn through the electrocardiogram at the level of the J point, and this horizontal line is the zero potential line in the electrocardiogram from which all potentials caused by currents of injury must be measured.

Use of the J Point in Plotting the Axis of a Current of Injury. Figure 16–19 illustrates electrocardiograms recorded from leads I and III, both of which show currents of injury. In other words, the J point and the S-T segment of each of these two electrocardiograms are not on the same line as the T-P segment. A horizontal line has been drawn through the J point to represent the zero potential level in each of the two recordings. The potential of the current of injury in each lead is the difference between the level of the T-P segment of the electrocardiogram (which is recorded between heart beats when there is a current of injury) and the zero potential line, as illustrated by the arrows. In lead I the recorded current of injury is above the zero potential line and is, therefore, positive. On the other hand, in lead III the T-P segment is below the zero potential line; therefore, the current of injury in lead III is negative.

At the bottom in Figure 16–19, the potentials of the current of injury in leads I and III are plotted on the coordinates of these leads, and the vector of the current of injury is determined by the method already described. In this instance, the vector of the current of injury extends from the right side of the heart toward the left and slightly upward, with an axis of approximately −30 degrees.

If one places the vector of the current of injury directly over the ventricles, *the negative end of the vector points toward the permanently depolarized, "injured" area of the ventricles.* In the instance illustrated in Figure 16–19, the injured area would be in the lateral wall of the right ventricle.

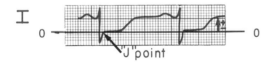

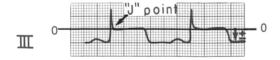

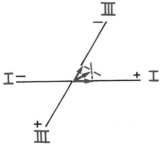

Figure 16–19. The "J" point as the zero reference voltage of the electrocardiogram. Also, method for plotting the axis of a current of injury.

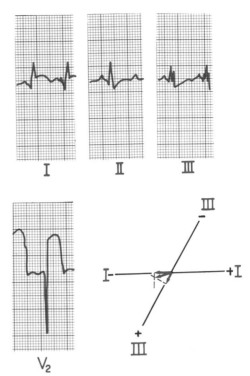

Figure 16–20. Current of injury in acute anterior wall infarction. Note the intense current of injury in lead V_2.

CORONARY ISCHEMIA AS A CAUSE OF CURRENT OF INJURY

Insufficient blood flow to the cardiac muscle depresses metabolism of the muscle for three different reasons: oxygen lack, excess carbon dioxide, and lack of sufficient nutrients. Consequently, polarization of the membranes cannot occur in areas of severe myocardial ischemia. Often, the heart muscle does not die because the blood flow is sufficient to maintain life of the muscle even though it is not sufficient to cause repolarization of the membranes. As long as this state exists, a current of injury continues to flow during diastole.

Extreme ischemia of the cardiac muscle occurs following coronary occlusion, and strong currents of injury flow from the infarcted area of the ventricles during the period between heartbeats, as is illustrated in Figures 16–20 and 16–21. Therefore, one of the most important diagnostic features of electrocardiograms recorded following acute coronary thrombosis is the current of injury.

Acute Anterior Wall Infarction. Figure 16–20 illustrates the electrocardiogram in the three standard leads and in one chest lead recorded from a patient with acute anterior wall cardiac infarction. The most important diagnostic feature of this electrocardiogram is the intense current of injury in the chest lead. If one draws a zero potential line through the J point of this electrocardiogram, he finds a strong *negative* current of injury during diastole. In other words, the negative end of the current of injury vector is against the chest wall. This means that the current of injury is emanating from the anterior wall of the ventricles, which is the main reason

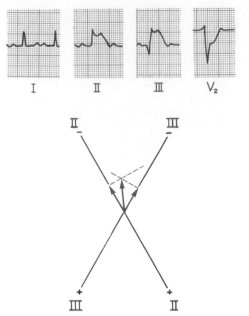

Figure 16–21. Current of injury in acute posterior wall, apical infarction.

for diagnosing this condition as anterior wall infarction.

If one analyzes the currents of injury in leads I and III, he finds a negative vector for the current of injury in lead I and a positive vector for the current of injury in lead III. This means that the resultant vector of the current of injury in the heart is approximately +150 degrees, with the negative end of the vector pointing toward the left ventricle and the positive end pointing toward the right ventricle. Thus, in this particular electrocardiogram the current of injury appears to be coming mainly from the left ventricle and from the anterior wall of the heart. Therefore, one would suspect that this anterior wall infarction is probably caused by thrombosis of the anterior descending limb of the left coronary artery.

Posterior Wall Infarction. Figure 16–21 illustrates the three standard leads and one chest lead from a patient with posterior wall infarction. The major diagnostic feature of this electrocardiogram is also in the chest lead. If a zero potential reference line is drawn through the J point of this lead, it is readily apparent that during the T-P interval the vector of the current of injury is positive. This means that the positive end of the vector is at the chest wall and the negative end (injured end) is away from the chest wall. In other words, the current of injury is coming from the side of the heart opposite that portion adjacent to the chest wall, which is the reason why this type of electrocardiogram is the basis for diagnosing posterior wall infarction.

If one analyzes the currents of injury in leads II and III of Figure 16–21, it is readily apparent that the current of injury is negative in both leads. By vectorial analysis as shown in the figure, one finds that the vector of the current of injury is approximately –95 degrees, with the negative end of the vector pointing downward and the positive end upward. Thus, because the infarct, as

indicated by the chest lead, is on the posterior wall of the heart and, as indicated by the currents of injury in leads II and III, is in the apical portion of the heart, one would suspect that this infarct is close to the apex on the posterior wall of the left ventricle.

Infarction in Other Parts of the Heart. By the same procedures as those illustrated in the preceding two discussions of anterior and posterior wall infarctions, it is possible to determine the locus of an infarcted area emitting a current of injury regardless of which part of the heart is involved. In making such vectorial analyses, it must always be remembered that *the positive end of the vector points toward the normal cardiac muscle and the negative end points toward the abnormal portion of the heart that is emitting the current of injury.*

Recovery from Coronary Thrombosis. Figure 16–22 illustrates a V_3 chest lead from a patient with posterior infarction, showing the change in the electrocardiogram of this lead from the day of the attack to one week later, then three weeks later, and finally one year later. From this electrocardiogram it can be seen that the current of injury is strong immediately after the acute attack (T-P segment displaced positively from the J point and the S-T segment), but after approximately one week the current of injury has diminished considerably and after three weeks it is completely gone. After that, the electrocardiogram does not change greatly during the following year. This is the usual recovery pattern following cardiac infarction of moderate degree when the collateral coronary blood flow is sufficient to reestablish appropriate nutrition to most of the infarcted area.

On the other hand, when all the coronary vessels throughout the heart are fairly well sclerosed, it may not be possible for the adjacent coronary vessels to supply sufficient blood to the infarcted area for recovery. Therefore, in some patients with coronary infarction, the infarcted area never redevelops an adequate coronary blood supply, some of the heart muscle dies, and relative coronary insufficiency persists in this area of the heart indefinitely. If the muscle does not die and become replaced by scar tissue, it continually emits a current of injury as long as the relative ischemia exists, particularly during bouts of exercise when the heart is overloaded.

Old Recovered Myocardial Infarction. Figure 16–23 illustrates leads I and III following anterior infarction

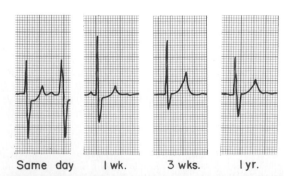

Figure 16–22. Recovery of the myocardium following moderate posterior wall infarction, illustrating disappearance of the current of injury (lead V_3).

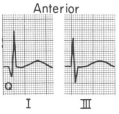
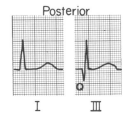

Figure 16-23. Electrocardiograms of old anterior and posterior wall infarctions, illustrating the Q wave in lead I in old anterior wall infarction and the Q wave in lead III in old posterior wall infarction.

and posterior infarction as these leads appear approximately a year after the acute episode. These are what might be called the "ideal" configurations of the QRS complex in these types of recovered myocardial infarction. It is illustrated in these figures that usually a Q wave develops at the beginning of the QRS complex in lead I in anterior infarction because of loss of muscle mass in the anterior left wall of the left ventricle; whereas in posterior infarction a Q wave develops at the beginning of the QRS complex in lead III because of loss of muscle in the posterior apical part of the ventricle.

These configurations are certainly not those found in all cases of old anterior and posterior cardiac infarction. Local loss of muscle and local areas of conduction block can cause the following abnormalities of the QRS complex: bizarre patterns (the prominent Q waves, for instance), decreased voltage, and prolongation.

Current of Injury in Angina Pectoris. "Angina pectoris" means simply pain in the pectoral regions of the upper chest, this pain usually radiating into the neck and down the left arm. The pain is caused by *relative* ischemia of the heart. No pain is usually felt as long as the person is perfectly quiet, but, just as soon as the patient overworks the heart, the pain appears.

A current of injury often occurs during an attack of severe angina pectoris, for the relative coronary insufficiency occasionally becomes great enough then to prevent adequate repolarization of the membranes in some areas of the heart during diastole.

ABNORMALITIES IN THE T WAVE

Earlier in the chapter it was pointed out that the T wave is normally positive in all the standard leads and that this is caused by repolarization of the apex of the heart ahead of the endocardial surfaces of the ventricles. This direction that repolarization spreads over the heart is backward to the direction in which depolarization takes place. (If the basic principles of the upright T wave in the standard leads have not been understood by now, the student should become familiar with the earlier, more detailed discussion of this before proceeding to the following few sections.)

The T wave becomes abnormal when the normal sequence of repolarization does not occur. Several factors can change this sequence of repolarization, as follows:

EFFECT OF SLOW CONDUCTION OF THE DEPOLARIZATION WAVE ON THE T WAVE

Referring back to Figure 16-15, note that the QRS complex is considerably prolonged. The reason for this prolongation is delayed conduction in the left ventricle as a result of left bundle branch block. The left ventricle becomes depolarized approximately 0.10 second after depolarization of the right ventricle. The refractory periods of the right and left ventricular muscle masses are not greatly different from each other. Therefore, the right ventricle begins to repolarize long before the left ventricle; this causes positivity in the right ventricle and negativity in the left ventricle. In other words, the mean axis of the T wave is from left to right, which is opposite to the mean electrical axis of the QRS complex in this same electrocardiogram. Thus, when conduction of the impulse through the ventricles is greatly delayed, the T wave is almost always of opposite polarity to that of the QRS complex.

In Figure 16-16 and in several figures in the following chapter, conduction also does not occur through the Purkinje system. As a result, the rate of conduction is greatly slowed, and in each instance the T wave is of opposite polarity to that of the QRS complex, whether the condition causing this delayed conduction happens to be left bundle branch block, right bundle branch block, ventricular extrasystole, or otherwise.

PROLONGED DEPOLARIZATION IN PORTIONS OF THE VENTRICULAR MUSCLE AS A CAUSE OF ABNORMALITIES IN THE T WAVE

If the apex of the ventricles should have an abnormally long period of depolarization, that is, a prolonged action potential, then repolarization of the ventricles would not begin at the apex as it normally does. Instead, the base of the ventricles would repolarize ahead of the apex, and the vector of repolarization would point from the apex toward the base of the heart. This is opposite to the usual vector of repolarization, and, consequently, the T wave in all three standard leads would be negative rather than the usual positive. Thus, the simple fact that the apical muscle of the heart has a prolonged period of depolarization is sufficient to cause marked changes in the T wave, even to the extent of changing the entire polarity, as is illustrated in Figure 16-24.

Ischemia is by far the most common cause of increased period of depolarization of cardiac muscle, and, when the ischemia occurs in only one area of the heart, the depolarization period of this area increases out of

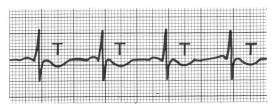

Figure 16-24. Inverted T wave resulting from mild ischemia of the apex of the ventricles.

proportion to that in other portions. As a result, definite changes in the T wave may take place. The ischemia may result from chronic, progressive coronary occlusion, acute coronary occlusion, or relative coronary insufficiency occurring during exercise.

One means for detecting mild coronary insufficiency is to have the patient exercise and then to record the electrocardiogram immediately thereafter, noting whether or not changes occur in the T waves. The changes in the T waves need not be specific, for any change in the T wave in any lead — inversion, for instance, or a biphasic wave — is often evidence enough that some portion of the ventricular muscle has increased its period of depolarization out of proportion to the rest of the heart, and this is probably caused by relative coronary insufficiency.

All the other conditions that can cause currents of injury, including pericarditis, myocarditis, and mechanical trauma of the heart, can also cause changes in the T wave. A current of injury occurs when the period of depolarization of some muscle is so long that the muscle fails to repolarize completely before the next cardiac cycle begins. Therefore, a current of injury is actually an exacerbated form of abnormal T wave, for both of these result from an increased depolarization period of one or more portions of cardiac muscle, and the difference is only a matter of degree.

Effect of Digitalis on the T Wave. As discussed in Chapter 26, digitalis is a drug that can be used during relative coronary insufficiency to increase the strength of cardiac muscular contraction. However, digitalis also increases the period of depolarization of cardiac muscle. It usually increases this period by approximately the same proportion in all or most of the ventricular muscle, but when overdosages of digitalis are given, the depo-

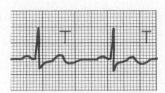

Figure 16–25. Biphasic T wave caused by digitalis toxicity.

larization period of one part of the heart may be increased out of proportion to that of other parts. As a result, nonspecific changes, such as T wave inversion or biphasic T waves, may occur in one or more of the electrocardiographic leads. A biphasic T wave caused by excessive administration of digitalis is illustrated in Figure 16–25. There is a slight amount of current of injury, too. This probably results from continuous depolarization of part of the ventricular muscle.

Changes in the T wave during digitalis administration are the earliest signs of digitalis toxicity. If still more digitalis is given to the patient, strong currents of injury may develop, and, as noted in the following chapter, digitalis can also block conduction of the cardiac impulse to various portions of the heart, so that various arrhythmias can result. It is desirable clinically to keep the effects of digitalis from going beyond the stage of mild T wave abnormalities. Therefore, the electrocardiograph is used routinely in following digitalized patients.

REFERENCES

See bibliography at end of Chapter 17.

17

Electrocardiographic Interpretation of Cardiac Arrhythmias

The rhythmicity of the heart and some abnormalities of rhythmicity were discussed in Chapter 14. The major purpose of the present chapter is to discuss still other abnormalities of rhythm and to describe the electrocardiograms recorded in the conditions known clinically as "cardiac arrhythmias."

ABNORMAL SINUS RHYTHMS

TACHYCARDIA

The term "tachycardia" means fast heart rate, usually defined as faster than 100 beats per minute. An electrocardiogram recorded from a patient with tachycardia is illustrated in Figure 17–1. This electrocardiogram is normal except that the rate of heartbeat, as determined from the time intervals between QRS complexes, is approximately 150 per minute instead of the normal 72 per minute.

The three general causes of tachycardia are *increased body temperature, stimulation of the heart by the autonomic nerves,* and *toxic conditions of the heart.*

The rate of the heart increases approximately 10 beats per minute for each degree Fahrenheit (18 beats per degree Celsius) increase in body temperature up to a body temperature of about 105°; beyond this the heart rate may actually decrease, owing to progressive weakening of the heart muscle as a result of the fever. Fever causes tachycardia because increased temperature increases the rate of metabolism of the S-A node, which in turn directly increases its excitability and rate of rhythm.

The various factors that can cause the autonomic nervous system to excite the heart will be discussed in Chapter 22 in relation to the reflexes that control the circulatory system. For instance, when a patient loses blood and passes into a state of shock or semishock, reflex stimulation of the heart increases its rate to as high as 150 to 180 beats per minute. Also, simple weakening of the myocardium usually increases the heart rate because the weakened heart does not pump blood into the arterial tree to a normal extent, and this elicits reflexes to increase the rate of the heart.

BRADYCARDIA

The term "bradycardia" means, simply, a slowed heart rate, usually defined as less than 60 beats per minute. Bradycardia is illustrated by the electrocardiogram in Figure 17–2.

Bradycardia in Athletes. The athlete's heart is considerably stronger than that of a normal person, a fact that allows the athlete's heart to pump a greater stroke volume output per beat. The excessive quantities of blood pumped into the arterial tree with each beat presumably initiate circulatory reflexes to cause the bradycardia.

Vagal Stimulation as a Cause of Bradycardia. Obviously, any circulatory reflex that stimulates the vagus nerve can cause the heart rate to decrease considerably. Perhaps the most striking example of this occurs in patients with the *carotid sinus syndrome.* In these patients an arteriosclerotic process in the carotid sinus region of the carotid artery causes excessive sensitivity of the pressure receptors (baroreceptors) located in the arterial wall; as a result, mild pressure on the neck elicits a strong baroreceptor reflex, causing intense

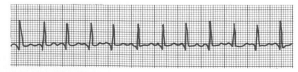

Figure 17–1. Sinus tachycardia (lead I).

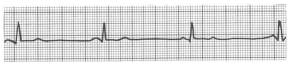

Figure 17–2. Sinus bradycardia (lead III).

Figure 17–3. Sinus arrhythmia as detected by a cardiotachometer. To the left is the recording taken when the subject was breathing normally; to the right, when breathing deeply.

vagal stimulation of the heart and extreme bradycardia. Indeed, such pressure may actually stop the heart.

SINUS ARRHYTHMIA

Figure 17–3 illustrates a *cardiotachometer* recording of the heart rate during normal and deep respiration. A cardiotachometer is an instrument that records by the height of successive spikes the duration of the interval between each two QRS complexes in the electrocardiogram. Note from this record that the heart rate increases and decreases approximately 5 per cent during the various phases of the quiet respiratory cycle. However, during deep respiration, as shown to the right in Figure 17–3, the heart rate even normally increases and decreases by as much as 30 per cent.

Sinus arrhythmia is presumably the result of several circulatory reflexes that will be described in detail in Chapter 22. First, when the blood pressure rises and falls during each cycle of respiration, the *baroreceptors are alternately stimulated and depressed*, causing reflex slowing and speeding up of the heart. Second, during each respiratory cycle, the negative intrapleural pressure increases and decreases, which increases and decreases the effective pressure in the veins of the chest. This elicits a *waxing and waning Bainbridge reflex* that also increases and decreases the cardiac rate. Third, when the respiratory center of the medulla is excited during each respiratory cycle, some of the *impulses "spill over" from the respiratory center into the vasomotor center*, causing alternate increase and decrease in the number of impulses transmitted to the heart through the sympathetics and vagus nerves.

ABNORMAL RHYTHMS RESULTING FROM IMPULSE CONDUCTION BLOCK

SINO-ATRIAL BLOCK

In rare instances the impulse from the S-A node is blocked before it enters the atrial muscle. This phenomenon is illustrated in Figure 17–4, which shows the

sudden cessation of P waves with resultant standstill of the atrium. However, the ventricle picks up a new rhythm, the impulse usually originating in the A-V node or in the A-V bundle so that the ventricular QRS-T complex is not altered.

ATRIOVENTRICULAR BLOCK

The only means by which impulses can ordinarily pass from the atria into the ventricles is through the *A-V bundle*, which is also known as the *bundle of His*. The different conditions that can either decrease the rate of conduction of the impulse through this bundle or can totally block the impulse are:

1. *Ischemia of the A-V junctional fibers* often delays or blocks conduction from the atria to the ventricles. Coronary insufficiency can cause ischemia of the A-V junction in the same manner that it can cause ischemia of the myocardium.

2. *Compression of the A-V bundle* by scar tissue or by calcified portions of the heart can depress or block conduction from the atria to the ventricles.

3. *Inflammation of the A-V bundle or fibers of the A-V junction* can depress conductivity between the atria and the ventricles. Inflammation results frequently from different types of myocarditis such as occur in diphtheria and rheumatic fever.

4. *Extreme stimulation of the heart by the vagus nerves* in rare instances blocks impulse conduction through the A-V junctional fibers. Such vagal excitation occasionally results from strong stimulation of the baroreceptors in persons with the *carotid sinus syndrome*, which was just discussed in relation to bradycardia.

Incomplete Heart Block. *Prolonged P-R (or P-Q) Interval — "First Degree Block."* The normal lapse of time between the *beginning* of the P wave and the *beginning* of the QRS complex is approximately 0.16 second when the heart is beating at a normal rate. This P-R interval decreases in length with faster heartbeat and increases with slower heartbeat. In general when the P-R interval increases above a value of approximately 0.20 second in the heart beating at normal rate, the P-R interval is said to be prolonged, and the patient is said to have *first degree incomplete heart block*. Figure 17–5 illustrates an electrocardiogram with a prolonged P-R interval, the interval in this instance being approximately 0.30 second.

The P-R interval rarely increases above 0.35 to 0.45 second, for by the time the rate of conduction through the A-V bundle is depressed to such an extent, conduction stops entirely. Thus, when a patient's P-R interval is approaching these limits, additional slight increase in the severity of the condition will completely block impulse conduction rather than simply delay conduction further.

Obviously, one of the means for determining the

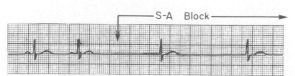

Figure 17–4. S-A nodal block with A-V nodal rhythm (lead III).

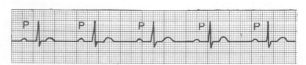

Figure 17–5. Prolonged P-R interval (lead II).

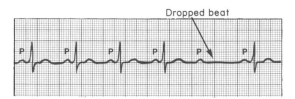

Figure 17–6.   Partial atrioventricular block (lead V₃).

severity of some heart diseases — rheumatic fever, for instance — is to measure the P-R interval.

Second Degree Block.   When conduction through the A-V junction is slowed until the P-R interval is 0.25 to 0.45 second, the action potentials traveling through the A-V junctional fibers sometimes are strong enough to pass on into the A-V node and at other times are not strong enough. Often the impulse passes into the ventricles following one atrial contraction and fails to pass the next one or two atrial contractions, thus alternating between conduction and nonconduction. In this instance, the atria beat at a considerably faster rate than the ventricles, and it is said that there are "dropped beats" of the ventricles. This condition is called *second degree incomplete heart block.*

Figure 17–6 illustrates P-R intervals of 0.30 second, and it also illustrates one dropped beat as a result of failure of conduction from the atria to the ventricles.

At times every other beat of the ventricles is dropped so that a "2:1 rhythm" develops in the heart, with the atria beating twice for every single beat of the ventricles. Sometimes other rhythms, such as 3:2 or 3:1, also develop.

Complete Atrioventricular Block (Third Degree Block).   When the condition causing poor conduction in the A-V bundle becomes extremely severe, complete block of the impulse from the atria into the ventricles occurs. In this instance the P waves become completely dissociated from the QRS-T complexes, as illustrated in Figure 17–7. Note that the rate of rhythm of the atria in this electrocardiogram is approximately 100 beats per minute, while the rate of ventricular beat is less than 40 per minute. Furthermore, there is no relationship whatsoever between the rhythm of the atria and that of the ventricles, for the ventricles have "escaped" from control by the atria, and they are beating at their own natural rate.

Stokes-Adams Syndrome — Ventricular Escape.   In some patients with atrioventricular block, the total block comes and goes — that is, impulses are conducted from the atria into the ventricles for a period of time, and then suddenly no impulses at all are transmitted. The duration of total block may be a few seconds, a few minutes, a few hours, or it may be weeks or even longer before conduction returns. In particular, this condition occurs in hearts with borderline ischemia.

Immediately after A-V conduction is first blocked, the ventricles stop contracting entirely for about 5 to 10 seconds. Then some part of the Purkinje system beyond the block, usually in the A-V bundle itself, begins discharging rhythmically at a rate of 15 to 40 times per minute and acting as the pacemaker of the ventricles. This is called *ventricular escape*. Because the brain cannot remain active for more than 3 to 5 seconds without blood supply, patients usually faint when complete block occurs because the heart then does not pump any blood for 5 to 10 seconds until the ventricles "escape." However, even the slowly beating ventricles usually pump enough blood to allow rapid recovery from the faint and then to sustain the person. These periodic fainting spells are known as the Stokes-Adams syndrome.

Occasionally the interval of ventricular standstill at the onset of complete block is so long that it either becomes detrimental to the health of the patient or causes death. Consequently, many of these patients would be better off with total heart block than constantly shifting back and forth between total heart block and atrial control of the ventricles. Such patients are frequently treated with drugs that depress conduction in the A-V bundle so that the total heart block persists all the time. Commonly, digitalis is the drug used for this purpose, but quinidine or barium ion can also convert a Stokes-Adams syndrome into total heart block.

However, most of these patients are now provided with an *artificial pacemaker*, which is a small battery-operated electrical stimulator planted beneath the skin, the electrodes from which are connected to the heart. This pacemaker provides continued rhythmic impulses that take over control of the heart. The batteries are replaced about once every five years (more than 10 years for a new atomic-powered battery).

INCOMPLETE INTRAVENTRICULAR BLOCK — ELECTRICAL ALTERNANS

Most of the same factors that can cause A-V block can also block impulse conduction in peripheral portions of the ventricular Purkinje system. At times, *incomplete* block occurs so that the impulse is sometimes transmitted and sometimes not transmitted, causing block of the impulse during some heart cycles and not during others. The QRS complex may be considerably abnormal during those cycles in which the impulse is blocked. Figure 17–8 illustrates the condition known as *electrical alternans,* which results from intraventricular block every other heartbeat. This electrocardiogram also illustrates tachycardia, which is probably the reason the block has occurred, for when the rate of the

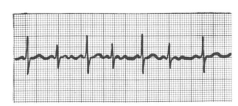

Figure 17–8.   Partial intraventricular block — electrical alternans (lead III).

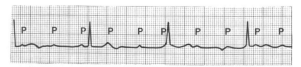

Figure 17–7.   Complete atrioventricular block (lead II).

heart is very rapid, it may be impossible for portions of the Purkinje system to recover from the refractory period rapidly enough to respond during each succeeding heartbeat. Also, many conditions that depress the heart, such as ischemia, myocarditis, and digitalis toxicity, can cause incomplete intraventricular block and electrical alternans.

PREMATURE CONTRACTIONS

A premature contraction is a contraction of the heart prior to the time that normal contraction would have been expected. This condition is also frequently called *extrasystole, premature beat*, or *ectopic contraction*.

Causes of Premature Contractions. Most premature contractions result from *ectopic foci* in the heart, which emit abnormal impulses at odd times during the cardiac rhythm. The possible causes of ectopic foci are (1) local areas of ischemia, (2) small calcified plaques at different points in the heart, which press against the adjacent cardiac muscle so that some of the fibers are irritated, and (3) toxic irritation of the A-V node, Purkinje system, or myocardium caused by drugs, nicotine, or caffeine. Mechanical initiation of ectopic contractions is also frequent during cardiac catheterization, large numbers of premature contractions often occurring when the catheter enters the right ventricle and presses against the endocardium.

PREMATURE ATRIAL CONTRACTIONS

Figure 17–9 is an electrocardiogram showing a single premature atrial contraction. The P wave of this beat is relatively normal and the QRS complex is also normal, but the P-R interval and the interval between the preceding contraction and the premature contraction is shortened. Also, the interval between the premature contraction and the next succeeding contraction is slightly prolonged. The reason for this is that the premature contraction originated in the atrium some distance from the S-A node, and the impulse of it had to travel through a considerable amount of atrial muscle before it discharged the S-A node. Consequently, the S-A node discharged very late in the premature cycle, and this made the succeeding heartbeat also late in appearing.

Premature atrial contractions occur frequently in healthy persons and, indeed, are often found in athletes or others whose hearts are certain to be in healthy condition. Yet mild toxic conditions resulting from such factors as excess smoking, lack of sleep, too much coffee, alcoholism, and use of various drugs can also initiate such contractions.

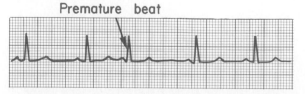

Figure 17–9. Atrial premature contraction (lead I).

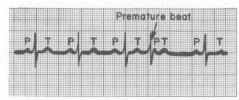

Figure 17–10. A-V nodal premature contraction (lead III).

Pulse Deficit. When the heart contracts ahead of schedule, the ventricles will not have filled with blood normally, and the stroke volume output during that contraction is depressed or sometimes almost absent. Therefore, the pulse wave passing to the periphery following a premature contraction may be so weak that the pulse cannot be felt at all in the radial artery, whereas the succeeding pulse may be extra strong because of compensatory overfilling. Thus, a deficit in the number of pulses felt in the radial pulse occurs in relation to the number of contractions of the heart.

Bigeminal Pulse. At times every other beat of the heart may be a premature contraction. This causes the patient to have a bigeminal pulse — that is, two pulses close together, then a longer diastolic interval, then two again, and so on.

A-V NODAL OR A-V BUNDLE PREMATURE CONTRACTIONS

Figure 17–10 illustrates a premature contraction originating either in the A-V node or in the A-V bundle. The P wave is missing from the record of the premature contraction. Instead, the P wave is superimposed on the QRS-T complex of the premature contraction because the cardiac impulse travels backward into the atria at the same time that it travels forward into the ventricles; this P wave distorts the complex, but the P wave itself cannot be discerned as such.

In general, A-V nodal premature contractions have the same significance and causes as atrial premature contractions.

PREMATURE VENTRICULAR CONTRACTIONS (PVCs)

The electrocardiogram of Figure 17–11 illustrates a series of premature ventricular contractions alternating with normal contractions. Premature ventricular contractions result from an ectopic focus in the ventricles and cause several effects in the electrocardiogram:

First, the QRS complex is usually considerably prolonged. The reason is that the impulse is conducted mainly through the slowly conducting muscle of the ventricle rather than through the Purkinje system.

Second, the QRS complex has a very high voltage for the following reason: When the normal impulse passes through the heart, it passes through both ventricles approximately simultaneously; consequently, the depolarization waves of the two sides of the heart partially neutralize each other. However, when a ventricular premature contraction occurs, the impulse travels in only one direction so that there is no such neutralization effect, and one entire side of the heart is depolarized

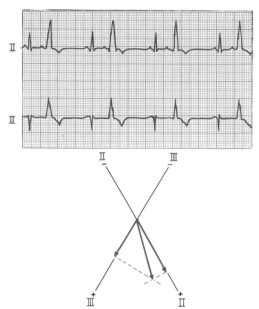

Figure 17–11. Premature ventricular contractions (PVCs) illustrated by the large abnormal QRS-T complexes (leads II and III). Axis of the premature contractions is plotted in accord with the principles of vectorial analysis explained in Chapter 16.

while the other entire side is still polarized; this causes intense electrical potentials.

Third, following almost all premature ventricular contractions the T wave has a potential opposite to that of the QRS complex, because the *slow conduction of the impulse* through the cardiac muscle causes the area first depolarized also to repolarize first. As a result, the direction of current flow in the heart during repolarization is opposite to that during depolarization, and the potential of the T wave is reversed to that of the QRS complex. This is not true of the normal T wave, as was explained in Chapter 16.

Some premature ventricular contractions are relatively benign in their origin and result from simple factors such as cigarettes, coffee, lack of sleep, various mild toxic states, and even emotional irritability. On the other hand, a large share of premature ventricular contractions results from actual pathology of the heart. For instance, many premature ventricular contractions occur following coronary thrombosis because of stray impulses originating around the borders of the infarcted area of the heart. Therefore, the presence of such PVCs is not to be taken lightly. Statistics show that persons with significant numbers of such PVCs have a much higher than normal chance of developing spontaneous lethal ventricular fibrillation, presumably initiated by one of the PVCs itself.

Vector Analysis of the Origin of an Ectopic Premature Ventricular Contraction. In Chapter 16 the principles of vectorial analysis were explained. Applying these principles, one can determine from the electrocardiogram in Figure 17–11 the point of origin of the premature ventricular contraction as follows: Note that the vectors of the premature contractions in leads II and III are both strongly positive. Plotting these vectors on the axes of

leads II and III and solving by vectorial analysis for the actual vector in the heart, one finds that the vector of the premature contraction has its negative end (origin) at the base of the heart and its positive end toward the apex. Thus, the first portion of the heart to become depolarized during the premature contractions lies near the base of the heart, which therefore is the locus of the ectopic focus.

PAROXYSMAL TACHYCARDIA

Abnormalities in any portion of the heart, including the atria, the Purkinje system, and the ventricles, can sometimes cause rapid rhythmic discharge of impulses which spread in all directions throughout the heart. Because of the rapid rhythm in the irritable focus, this focus becomes the pacemaker of the heart.

The term "paroxysmal" means that the heart rate usually becomes very rapid in paroxysms, with the paroxysms beginning suddenly and lasting for a few seconds, a few minutes, a few hours, or sometimes much longer. Then the paroxysms usually end as suddenly as they had begun, the pacemaker of the heart then shifting back to the S-A node.

Paroxysmal tachycardia can often be stopped by eliciting a vagal reflex. A strange type of vagal reflex sometimes elicited for this purpose is one that occurs when painful pressure is applied to the eyes. Also, pressure on the carotid sinuses can sometimes elicit enough of a vagal reflex to stop the tachycardia. Various drugs may also be used. Two frequently used are quinidine and lidocaine, both of which depress the normal increase in sodium permeability of the cardiac muscle membrane during the generation of the action potential, thereby often blocking the rhythmic discharge of the irritable focus that is causing the paroxysmal attack.

ATRIAL PAROXYSMAL TACHYCARDIA

Figure 17–12 illustrates in the middle of the record a sudden increase in rate of heartbeat from approximately 95 to approximately 150 beats per minute. On close observation of the electrocardiogram it will be seen that an inverted P wave occurs before each of the QRS-T complexes during the paroxysm of rapid heartbeat, and this P wave is partially superimposed on the normal T wave of the preceding beat. This indicates that the origin of this paroxysmal tachycardia is in the atrium, but, because the P wave is abnormal, the origin is not near the S-A node.

A-V Nodal Paroxysmal Tachycardia. Paroxysmal tachycardia very often results from an irritable focus in either the A-V node or the A-V bundle. This initiates normal impulses in the ventricles, thereby causing nor-

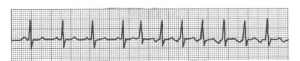

Figure 17–12. Atrial paroxysmal tachycardia — onset in middle of record (lead I).

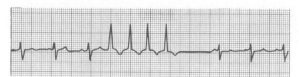

Figure 17–13. Ventricular paroxysmal tachycardia (lead III).

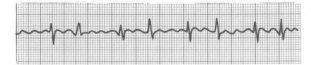

Figure 17–14. Atrial flutter — 2:1 and 3:1 rhythm (lead II).

mal QRS-T complexes. However, P waves cannot be seen because the atrial impulse travels backward from the A-V node through the atria at the same time that the ventricular impulse travels through the ventricles. Therefore, the P wave is obscured by the QRS complex.

Atrial or A-V nodal paroxysmal tachycardia usually occurs in young, otherwise healthy persons. In general, atrial paroxysmal tachycardia frightens the individual tremendously and may cause weakness during the paroxysms, but only rarely does permanent harm come from the attacks.

VENTRICULAR PAROXYSMAL TACHYCARDIA

Figure 17–13 illustrates a typical short paroxysm of ventricular tachycardia. The electrocardiogram of ventricular paroxysmal tachycardia has the appearance of a series of ventricular premature beats occurring one after another without any normal beats interspersed.

Ventricular paroxysmal tachycardia is usually a serious condition for two reasons. First, this type of tachycardia usually does not occur unless considerable damage is present in the ventricles. Second, ventricular tachycardia very frequently initiates ventricular fibrillation, which is almost invariably fatal. The reasons why ventricular tachycardia predisposes to ventricular fibrillation are: First, rapid stimulation decreases the refractory period of the heart and slows conduction, both of which predispose to circus movements. Second, one of the ventricular impulses frequently occurs at an instant when part of the ventricular muscle is still in a refractory state and part in a nonrefractory state; this causes the impulse to take a devious route through the heart, avoiding the refractory areas and initiating a long circuitous pathway, as was discussed in Chapter 14. The long pathway allows the impulse to continue around and around the ventricles, resulting in ventricular fibrillation.

Digitalis intoxication sometimes causes ventricular tachycardia. On the other hand, quinidine or lidocaine, which increases the refractory period of cardiac muscle and also increases its threshold for excitation, may be used to block irritable foci causing ventricular tachycardia.

ABNORMAL RHYTHMS RESULTING FROM CIRCUS MOVEMENTS

The circus movement phenomenon was discussed in detail in Chapter 14, and it was pointed out that these

movements can cause atrial flutter, atrial fibrillation, and ventricular fibrillation.

ATRIAL FLUTTER

Figure 17–14 illustrates lead II of the electrocardiogram in atrial flutter. The rate of atrial contraction (P waves) is approximately 300 times per minute, while the rate of ventricular contraction (QRS-T waves) is only 125 times per minute. From the record it will be seen that sometimes either a 2:1 or a 3:1 rhythm occurs. In other words, the atria contract two or three times for every one impulse that is conducted through the A-V bundle into the ventricles.

An important distinguishing feature of atrial flutter is the relatively high voltage, 0.2 to 0.3 mv., of the P waves in lead II and sometimes in lead III. Leads II and III record higher voltage than lead I because the circus movement pathway of the impulse causing atrial flutter passes around and around the atria from top to bottom. This means that the current flow in the atria extends up and down the body more or less parallel to the axes of leads II and III, which causes strong voltages in these leads but weak voltages in lead I.

ATRIAL FIBRILLATION

Figure 17–15 illustrates the electrocardiogram during atrial fibrillation. As discussed in Chapter 14, numerous small depolarization waves spread in all directions through the atria during atrial fibrillation. Because the waves are weak and because many of them are of opposite polarity at any given time, they usually almost completely neutralize each other. Therefore, in the electrocardiogram one can see either no P waves from the atria or a fine, high frequency, very low voltage wavy record. On the other hand the QRS-T complexes are completely normal unless there is some pathology of the ventricles, but their timing is very irregular for the following reasons:

Since the fibrillatory depolarization waves in the atria are completely irregular, the intervals between impulses arriving at the A-V node are extremely variable. Therefore, an impulse may arrive at the A-V node immediately after the node itself is out of its refractory period from its previous discharge, or it may not arrive there for

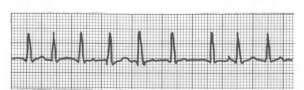

Figure 17–15. Atrial fibrillation (lead I).

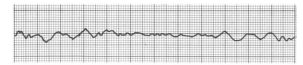

Figure 17–16. Ventricular fibrillation (lead II).

several additional tenths of a second. Consequently, the rhythm of the ventricles is very irregular, many of the ventricular beats falling quite close together and many far apart, as illustrated in Figure 17–15. The overall ventricular rate is 125 to 150 beats per minute in most instances.

The pumping effectiveness of the heart in atrial fibrillation is considerably depressed because the ventricles often do not have sufficient time to fill between beats. Therefore, one of the principles of treating patients with atrial fibrillation is to slow the rate of the ventricles. This is ordinarily done by administering digitalis, which causes a prolonged refractory period of the Purkinje system, probably by stimulating the vagal nerves to the A-V node and A-V bundle, thereby decreasing the number of impulses conducted from the atria into the ventricles and preventing any two ventricular beats from occurring close together. Heart function may be greatly improved when the rate is thus depressed to less than 100 beats per minute.

Pulse Deficit in Atrial Fibrillation.   In the same manner that premature contractions can cause a pulse deficit in the radial pulse, so can ventricular beats occurring close together in atrial fibrillation also cause a pulse deficit. Therefore, before treatment of atrial fibrillation one frequently finds a pulse deficit, whereas, after slowing the ventricular rate, the pulse deficit often disappears.

VENTRICULAR FIBRILLATION

In ventricular fibrillation the electrocardiogram is extremely bizarre, as seen in Figure 17–16, and ordinarily shows no tendency toward a rhythm of any type. In the early phases of ventricular fibrillation relatively large masses of muscle contract simultaneously, and this causes strong though irregular waves in the electrocardiogram. However, after only a few seconds the coarse contractions of the ventricles disappear, and the electrocardiogram changes into a new pattern of low voltage, very irregular waves. Thus, no repetitive electrocardiographic pattern can be ascribed to ventricular fibrillation except that the electrical potentials constantly and spasmodically change, because the currents in the heart flow first in one direction, then in another, and rarely repeat any specific cycle.

The voltages of the waves in the electrocardiogram in ventricular fibrillation are usually about 0.5 mv. when ventricular fibrillation first begins, but these decay rapidly so that after 20 to 30 seconds they are usually only 0.2 to 0.3 mv. Minute voltages of 0.1 mv. or less may be recorded for 10 minutes or longer after ventricular fibrillation begins. As already pointed out, ventricular fibrillation is lethal unless it is stopped by some heroic therapy such as immediate electroshock through the heart, as explained in Chapter 14.

REFERENCES

Braun, H. A., and Diettert, G. A.: ECG Arrhythmia Interpretation: A Programmed Text for Health Care Personnel. Reston, Va., Reston Publishing Co., 1979.

Burch, G. E., and Winsor, T.: A Primer of Electrocardiography, 6th Ed. Philadelphia, Lea & Febiger, 1972.

Chou, T.: Electrocardiography in Clinical Practice. New York, Grune & Stratton, 1979.

Chung, E. K.: Electrocardiography. Hagerstown, Md., Harper & Row, 1974.

Chung, E. K.: Ambulatory Electrocardiography. New York, Springer-Verlag, 1979.

Chung, E. K.: Electrocardiography: Practical Applications With Vectorial Principles, 2nd Ed. Hagerstown, Md., Harper & Row, 1980.

Coumel, P., and Marquez-Montes, J.: Management of the pre-excitation syndrome. In Hayase, S., and Murao, S. (eds.): Cardiology: Proceedings of the VIII World Congress of Cardiology, Tokyo, 1978. New York, Elsevier/North-Holland, 1979, p. 943.

Cranefield, P. F., and Wit, A. L.: Cardiac arrhythmias. Annu. Rev. Physiol., 41:459, 1979.

Durrer, D., and Lie, K. I.: Electrophysiology of the heart. In Hayase, S., and Murao, S. (eds.): Cardiology: Proceedings of the VIII World Congress of Cardiology, Tokyo, 1978. New York, Elsevier/North-Holland, 1979, p. 33.

Fletcher, G. F.: Dynamic Electrocardiographic Recording. Mt. Kisco, N.Y., Futura Publishing Co., 1979.

Goldberger, A. L.: Myocardial Infarction: Electrocardiographic Differential Diagnosis, 2nd Ed. St. Louis, C. V. Mosby, 1979.

Guyton, A. C., and Crowell, J. W.: A stereovectorcardiograph. J. Lab. Clin. Med., 40:726, 1952.

Halhuber, M. J., et al.: ECG, An Introductory Course: A Practical Introduction to Clinical Electrocardiography. New York, Springer-Verlag, 1979.

Hamer, J.: An Introduction to Electrocardiography: A Primer for Students, Graduates, Practitioners and Nurses Concerned With Coronary Care and Other Forms of Intensive Care, 2nd Ed. Boston, G. K. Hall, 1979.

Han, J. (ed.): Cardiac Arrhythmias. Springfield, Ill., Charles C Thomas, 1972.

Harris, C. C.: A Primer of Cardiac Arrhythmias: A Self-Instructional Program. St. Louis, C. V. Mosby, 1979.

Josephson, M. E., and Seides, S. F.: Clinical Cardiac Electrophysiology Techniques and Interpretations. Philadelphia, Lea & Febiger, 1979.

Krikler, D. M.: The application of electrocardiography to clinical medicine. In Dickinson, C. J., and Marks, J. (eds.): Developments in Cardiovascular Medicine. Lancaster, England, MTP Press, 1978, p. 105.

Mangiola, S., and Ritota, M. C.: Cardiac Arrhythmias. Philadelphia, J. B. Lippincott Co., 1974.

Moe, G. K.: Mechanisms of cardiac dysrhythmias. In Frohlich, E. D. (ed.): Pathophysiology, 2nd Ed. Philadelphia, J. B. Lippincott Co., 1976, p. 83.

Moss, A. J.: Prediction and prevention of sudden cardiac death. Annu. Rev. Med. 31:1, 1980.

Narula, O. S. (ed.): Cardiac Arrhythmias: Electrophysiology, Diagnosis, and Management. Baltimore, Williams & Wilkins, 1979.

Olsson, S. B., and Dreifus, L. S.: Value and limitations of ECG and VCG. In Hayase, S., and Murao, S. (eds.): Cardiology: Proceedings of the VIII World Congress of Cardiology, Tokyo, 1978. New York, Elsevier/North-Holland, 1979, p. 1037.

Phibbs, B.: The Cardiac Arrhythmias. St. Louis, C. V. Mosby, 1978.

Pick, A., and Langendorf, R.: Interpretation of Complex Arrhythmias. Philadelphia, Lea & Febiger, 1980.

Pordy, L.: Computer Electrocardiography: Present Status and Criteria. Mt. Kisco, N.Y., Futura Publishing Co., 1977.

Puech, P.: Human intercardiac electrocardiography. In Dickinson, C. J., and Marks, J. (eds.): Developments in Cardiovascular Medicine. Lancaster, England, MTP Press, 1978, p. 77.

Sano, T., and Puech, P.: Electrophysiological basis of arrhythmias. In Hayase, S., and Murao, S. (eds.): Cardiology: Proceedings of the VIII World Congress of Cardiology, Tokyo, 1978. New York, Elsevier/North-Holland, 1979, p. 894.

Scher, A. M., and Spach, M. S.: Cardiac depolarization and repolarization and the electrocardiogram. In Berne, R. M., et al. (eds.): Handbook of Physiology. Sec. 2, Vol. 1. Baltimore, Williams & Wilkins, 1979, p. 357.

Simpson, W. T., and Caldwell, A. D. S. (eds.): Amiodarone in Cardiac Arrhythmias. New York, Grune & Stratton, 1979.

Watanabe, Y., and Sloman, J. G.: Antiarrhythmic agents. In Hayase, S., and Murao, S. (eds.): Cardiology: Proceedings of the VIII World Congress of Cardiology, Tokyo, 1978. New York, Elsevier/North-Holland, 1979, p. 920.

Part V
THE CIRCULATION

18

Physics of Blood, Blood Flow, and Pressure: Hemodynamics

THE CIRCULATORY SYSTEM AS A "CIRCUIT"

The most important feature of the circulation that must always be kept in mind is that it is a continuous circuit. That is, if a given amount of blood is pumped by the heart, this same amount must also flow through each subdivision of the circulation.

Figure 18–1 illustrates the general plan of the circulation, showing the two major subdivisions, the *systemic circulation* and the *pulmonary circulation*. In the figure the arteries of each subdivision are represented by a single distensible

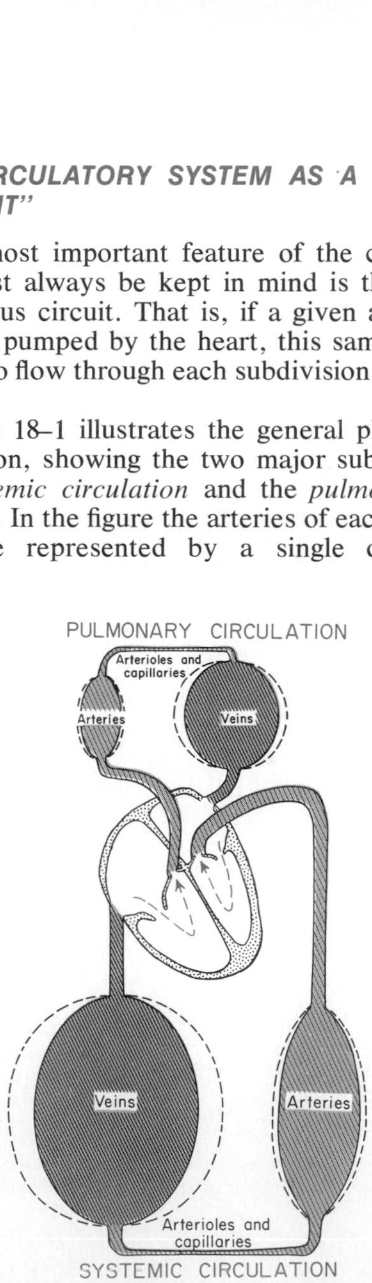

Figure 18–1. Representation of the circulation, showing both the distensible and the resistive portions of the systemic and pulmonary circulations.

chamber and all the veins by another, even larger distensible chamber, and the arterioles and capillaries represent very small connections between the arteries and veins. Blood flows with almost no resistance in all the larger vessels of the circulation, but this is not the case in the arterioles and capillaries, where considerable resistance does occur. To cause blood to flow through these small "resistance" vessels, the heart pumps blood into the arteries under high pressure — normally at a systolic pressure of about 120 mm. Hg in the systemic system and 22 mm. Hg in the pulmonary system.

As a first step toward explaining the overall function of the circulation, this chapter will discuss the physical characteristics of blood itself and then the physical principles of blood flow through the vessels, including especially the interrelationships among pressure, flow, and resistance. The study of these interrelationships and other basic physical principles of blood circulation is called *hemodynamics*.

THE PHYSICAL CHARACTERISTICS OF BLOOD

Blood is a viscous fluid composed of *cells* and *plasma*. More than 99 per cent of the cells are red blood cells; this means that for practical purposes the white blood cells play almost no role in determining the physical characteristics of the blood.

THE HEMATOCRIT

The per cent of the blood that is cells is called the hematocrit. Thus, if a person has a hematocrit of 40, 40 per cent of the blood volume is cells and the remainder is plasma. The hematocrit of normal man averages about 42, while that of normal woman averages about 38. These values vary tremendously, depending upon whether or not the person has anemia, the degree of bodily activity, and the altitude at which the person resides. These effects were discussed in relation to the red blood cells and their function in Chapter 5.

Blood hematocrit is determined by centrifuging blood in a calibrated tube such as that shown in Figure 18–2. The calibration allows direct reading of the per cent of cells.

Effect of Hematocrit on Blood Viscosity. Blood is several times as viscous as water, and this viscosity increases the difficulty with which the blood flows through the small vessels. The greater the percentage of cells in the blood — that is, the greater the hematocrit — the more friction there is between successive layers of blood, and this friction determines viscosity. Therefore, the viscosity of blood increases drastically as the hematocrit increases, as illustrated in Figure 18–3. If we arbitrarily consider the viscosity of water to be 1, then the viscosity of whole blood at normal hematocrit is about 3; this means that 3 times as much pressure is required to force whole blood through a given tube than to force water through the same tube. Note that when the hematocrit rises to 60 or 70, which it often does in polycythemia as discussed in Chapter 5, the blood viscosity can become as great as 10 times that of water, and its flow through blood vessels is greatly retarded.

Another factor that affects blood viscosity is the concentration and types of proteins in the plasma, but these effects are so much less important than the effect of hematocrit that they are not significant considerations in most hemodynamic studies.

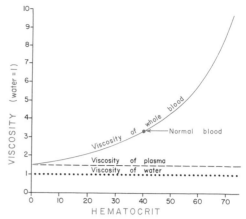

Figure 18–3. Effect of hematocrit on viscosity.

The viscosity of blood plasma is about 1.5 times that of water.

Since most resistance in the circulatory system occurs in the very small blood vessels, it is especially important to know how blood viscosity affects blood flow in these minute vessels. At least three additional factors besides hematocrit and plasma proteins affect blood viscosity in these vessels:

1. Blood flow in very minute vessels exhibits far less viscous effect than it does in large vessels. This effect, called the *Fahraeus-Lindqvist effect*, begins to appear when the vessel diameter falls below approximately 1.5 mm. In vessels as small as capillaries, this effect is so prominent that the viscosity of whole blood is as little as one-half that in large vessels. The Fahraeus-Lindqvist effect is probably caused by alignment of the red cells as they pass through the vessels. That is, the red cells, instead of moving randomly, line up and move through the vessels as a single plug, thus eliminating the viscous resistance that occurs internally in the blood itself. The Fahraeus-Lindqvist effect, however, is probably more than offset by the following two effects under most conditions.

2. The viscosity of blood increases drastically as its velocity of flow decreases. Since the velocity of blood flow in the small vessels is extremely minute, often less than 1 mm. per second, blood viscosity can increase as much as 10-fold from this factor alone. This effect is presumably caused by adherence of the red cells to each other (formation of rouleaux and larger aggregates) and to the vessel walls.

3. Cells also often become stuck at constrictions in small blood vessels; this happens especially in capillaries where the nuclei of endothelial cells protrude into the capillary lumen. When this occurs, blood flow can become totally blocked for a fraction of a second, for several seconds, or for much longer periods of time, thus giving an apparent effect of greatly increased viscosity.

Because of these special effects that occur in the minute vessels of the circulatory system, it has been impossible to determine the exact manner in which hematocrit affects viscosity in the minute vessels, which is the place in the circulatory system where viscosity almost certainly plays its most important role. Never-

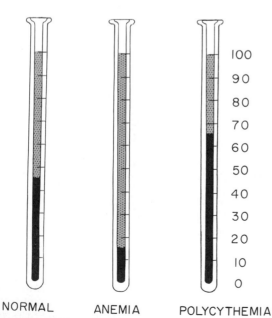

Figure 18–2. Hematocrits in the normal person and in patients with anemia and polycythemia.

theless, because some of these effects tend to decrease viscosity and others tend to increase viscosity, it is perhaps best at present simply to assume that the overall viscous effects in the small vessels are approximately equivalent to those that occur in the larger vessels.

PLASMA

Plasma is part of the extracellular fluid of the body. It is almost identical to the interstitial fluid found between the tissue cells except for one major difference: plasma contains about 7 per cent protein, while interstitial fluid contains an average of only 2 per cent protein. The reason for this difference is that plasma protein leaks only slightly through the capillary pores into the interstitial spaces. As a result, most of the plasma protein is held in the circulatory system, and that which does leak is eventually returned to the circulation by the lymph vessels. Therefore, the plasma protein concentration is about 3.5 times that of the fluid outside the capillaries.

The Types of Protein in Plasma. The plasma protein is divided into three major types, as follows:

	Grams Per Cent
Albumin	4.5
Globulins	2.5
Fibrinogen	0.3

The primary function of the *albumin* (and of the other types of protein to a lesser extent) is to cause osmotic pressure at the capillary membrane. This pressure, called *colloid osmotic pressure*, prevents the fluid of the plasma from leaking out of the capillaries into the interstitial spaces. This function is so important that it will be discussed in detail in Chapter 30.

The globulins are divided into three major types: alpha, beta, and gamma globulins. The *alpha* and *beta globulins* perform diverse functions in the circulation, such as transporting other substances by combining with them, acting as substrates for formation of other substances, and transporting protein itself from one part of the body to another. The *gamma globulins,* and to a lesser extent the *beta globulins*, play a special role in protecting the body against infection, for it is these globulins that are mainly the *antibodies* that resist infection and toxicity, thus providing the body with what we call *immunity*. The function of immunity was discussed in detail in Chapter 7.

The *fibrinogen* of plasma is of basic importance in blood clotting and was discussed in Chapter 9.

INTERRELATIONSHIPS AMONG PRESSURE, FLOW, AND RESISTANCE

Flow through a blood vessel is determined entirely by two factors: (1) the *pressure difference*

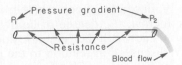

Figure 18–4. Relationships among pressure, resistance, and blood flow.

between the two ends of the vessel, which is the force that pushes the blood through the vessel, and (2) the impediment to blood flow through the vessel, which is called vascular *resistance*. Figure 18–4 illustrates these relationships, showing a blood vessel segment located anywhere in the circulatory system.

P_1 represents the pressure at the origin of the vessel; at the other end the pressure is P_2. The flow through the vessel can be calculated as follows:

$$Q = \frac{\Delta P}{R} \qquad (1)$$

in which Q is blood flow, ΔP is the pressure difference $(p_1 - p_2)$ between the two ends of the vessel, and R is the resistance.

It should be noted especially that it is the *difference* in pressure between the two ends of the vessel that determines the rate of flow and not the absolute pressure in the vessel. For instance, if the pressure at both ends of the segment were 100 mm. Hg and yet no difference existed between the two ends, there would be no flow despite the presence of the 100 mm. Hg pressure.

The above formula expresses the most most important of all the relationships that the student needs to understand to comprehend the hemodynamics of the circulation. Because of the extreme importance of this formula the student should also become familiar with its other two algebraic forms:

$$\Delta P = Q \times R \qquad (2)$$

$$R = \frac{\Delta P}{Q} \qquad (3)$$

BLOOD FLOW

Blood flow means simply the quantity of blood that passes a given point in the circulation in a given period of time. Ordinarily, blood flow is expressed in *milliliters* or *liters per minute*, but it can be expressed in milliliters per second or in any other unit of flow.

The overall blood flow in the circulation of an adult person at rest is about 5000 ml. per minute. This is called the *cardiac output* because it is the amount of blood pumped by the heart in a unit period of time.

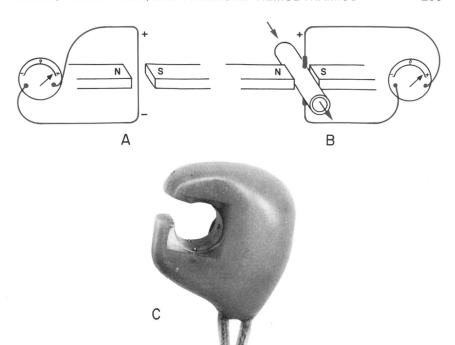

Figure 18–5. A flowmeter of the electromagnetic type, showing (A) generation of an electromotive force in a wire as it passes through an electromagnetic field, (B) generation of an electromotive force in electrodes on a blood vessel when the vessel is placed in a strong magnetic field and blood flows through the vessel, and (C) a modern electromagnetic flowmeter "probe" for chronic implantation around blood vessels.

Methods for Measuring Blood Flow. Many different mechanical or mechanoelectrical devices can be inserted in series with a blood vessel, or in some instances applied to the outside of the vessel, to measure flow. These are called simply *flowmeters*.

The Electromagnetic Flowmeter. In recent years several new devices have been developed that can be used to measure blood flow in a vessel without opening it. One of the most important of these is the electromagnetic flowmeter, the principles of which are illustrated in Figure 18–5. Figure 18–5A shows generation of electromotive force in a wire that is moved rapidly through a magnetic field. This is the well-known principle for production of electricity by the electric generator. Figure 18–5B shows that exactly the same principle applies for generation of electromotive force in blood when it moves through a magnetic field. In this case, a blood vessel is placed between the poles of a strong magnet, and electrodes are placed on the two sides of the vessel perpendicular to the magnetic lines of force. When blood flows through the vessel, electrical voltage proportional to the rate of flow is generated between the two electrodes, and this is recorded using an appropriate meter or electronic apparatus. Figure 18–5C illustrates an actual "probe" that is placed on a large blood vessel to record its blood flow. This probe contains both the strong magnet and the electrodes.

An additional advantage of the electromagnetic flowmeter is that it can record changes in flow that occur in less than 0.01 second, allowing accurate recording of pulsatile changes in flow as well as steady flow.

The Ultrasonic Doppler Flowmeter. Another type of flowmeter that can be applied to the outside of the vessel and that has many of the same advantages as the electromagnetic flowmeter is the ultrasonic Doppler flowmeter illustrated in Figure 18–6. A minute piezoelectric crystal is mounted in the wall of the device. This crystal, when energized with an appropriate electronic apparatus, transmits sound at a frequency of several million cycles per second downstream along the flowing blood. A portion of the sound is reflected by the flowing red blood cells so that reflected sound waves travel backward from the blood toward the crystal. However, these reflected waves have a lower frequency than the transmitted wave because the red cells are moving away from the transmitter crystal. This is called the Doppler effect. It is the same effect that one experiences when a train approaches a listener and passes by while blowing the whistle. Once the whistle has passed by the person, the pitch of the sound from the whistle suddenly becomes much lower than when the train is approaching. The transmitted wave is intermittently cut off and the reflected wave is received back onto the crystal, then amplified greatly by the electronic apparatus. Another portion of the apparatus determines the frequency difference between the transmitted wave and the reflected wave, thus also determining the velocity of blood flow.

Like the electromagnetic flowmeter, the ultrasonic Doppler flowmeter is capable of recording very rapid, pulsatile changes in flow as well as steady flow.

Laminar Flow of Blood in Vessels. When blood flows at a steady rate through a long, smooth vessel, it flows in *streamlines*, with each layer of

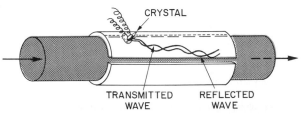

CRYSTAL

TRANSMITTED WAVE   REFLECTED WAVE

Figure 18–6. An ultrasonic Doppler flowmeter.

blood remaining the same distance from the wall. Also, the central portion of the blood stays in the center of the vessel. This type of flow is called *laminar flow* or *streamline flow*, and it is opposite to *turbulent flow*, which is blood flowing in all directions in the vessel and continually mixing within the vessel, as will be discussed below.

Parabolic Velocity Profile During Laminar Flow. When laminar flow occurs, the velocity of flow in the center of the vessel is far greater than that toward the outer edges. This is illustrated by the experiment shown in Figure 18–7. In vessel A are two different fluids, the one to the left colored by a dye and the one to the right a clear fluid, but there is no flow in the vessel. Then the fluids are made to flow; and a parabolic interface develops between the two fluids, as shown 1 second later in vessel B, illustrating that the portion of fluid adjacent to the vessel wall has hardly moved at all, the portion slightly away from the wall has moved a small distance, and the portion in the center of the vessel has moved a long distance. This effect is called a parabolic profile for the velocity of blood flow.

The cause of the parabolic profile is the following: The fluid molecules touching the wall hardly move because of adherence to the vessel wall. The next layer of molecules slips over these, the third layer over the second, the fourth layer over the third, and so forth. Therefore, the fluid in the middle of the vessel can move rapidly because many layers of the molecules exist between the middle of the vessel and the vessel wall, all of these capable of slipping over each .other, while those portions of fluid near the wall do not have this advantage.

Turbulent Flow of Blood Under Some Conditions. When the rate of blood flow becomes too great, when it passes by an obstruction in a vessel, when it makes a sharp turn, or when it passes over a rough surface, the flow may then become *turbulent* rather than streamline. Turbulent flow means that the blood flows crosswise in the vessel as well as along the vessel, usually forming whorls in the blood called *eddy currents* These are similar to the whirlpools that one frequently sees in a rapidly flowing river at a point of obstruction.

When eddy currents are present, blood flows with much greater resistance than when the flow is streamline because the eddies add tremendously to the overall friction of flow in the vessel.

The tendency for turbulent flow increases in direct proportion of the velocity of blood flow, in direct proportion to the diameter of the blood vessel, and inversely proportional to the viscosity of the blood divided by its density in accordance with the following equation:

$$\mathrm{Re} = \frac{v \cdot d}{\dfrac{\eta}{\rho}} \qquad (4)$$

in which Re is *Reynolds' number* and is the measure of the tendency for turbulence to occur, v is the velocity of blood flow (in centimeters), η is the viscosity (in poises) and ρ is density. The viscosity of blood is normally about 1/30 poise, and the density is only slightly greater than 1. When Reynolds' number rises above 200 to 400, turbulent flow will occur at some branches of vessels but will die out along the smooth portions of the vessels. However, when Reynolds' number rises above approximately 2000, turbulence will usually occur even in a straight, smooth vessel. Reynolds' number for flow in the vascular system even normally rises to 200 to 400 in large arteries; as a result there is almost always some turbulence of flow at the branches of these vessels. In the proximal portions of the aorta and pulmonary artery, Reynolds' number can rise to several thousand during the rapid phase of ejection by the ventricles; this causes considerable turbulence in the proximal aorta and pulmonary artery where many conditions are appropriate for turbulence: (1) high velocity of blood flow, (2) pulsatile nature of the flow, (3) sudden change in vessel diameter, and (4) large vessel diameter.

However, in small vessels, Reynolds' number is almost never high enough to cause turbulence.

BLOOD PRESSURE

The Standard Units of Pressure. Blood pressure is almost always measured in *millimeters of mercury (mm. Hg)* because the mercury manometer (shown in Figure 18–8) has been used as the standard reference for measuring blood pressure throughout the history of physiology. Actually, blood pressure means the *force exerted by the blood against any unit area of the vessel wall.* When one says that the pressure in a vessel is 50 mm. Hg, this means that the force exerted is sufficient to push a column of mercury up to a level 50 mm. high. If the pressure is 100 mm. Hg, it will push the column of mercury up to 100 mm.

Occasionally, pressure is measured in *centimeters of water.* A pressure of 10 cm. of water means a pressure sufficient to raise a column of water to a height of 10 cm. *One millimeter of mercury equals 1.36 cm. of water* because the specific gravity of mercury is 13.6 times that of water, and 1 cm. is 10 times as great as 1 mm. Dividing 13.6 by 10, we derive the factor 1.36.

Measurement of Blood Pressure Using the Mercury Manometer. Figure 18–8 illustrates a stand-

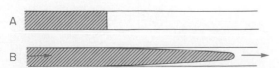

Figure 18–7. An experiment illustrating laminar blood flow. A, Two separate fluids before flow begins; B, the same fluids 1 second after flow begins.

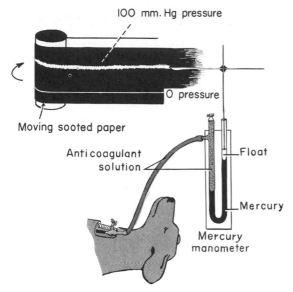

Figure 18–8. Recording arterial pressure with a mercury manometer, a method that has been used in the manner shown above for recording pressure throughout the history of physiology.

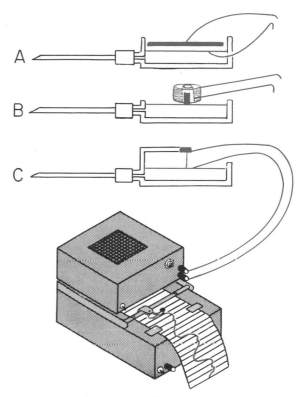

Figure 18–9. Principles of three different types of electronic transducers for recording rapidly changing blood pressures.

ard mercury manometer for measuring blood pressure. A cannula or catheter is inserted into an artery, a vein, or even the heart, and the pressure from the cannula or catheter is transmitted to the left-hand side of the manometer where it pushes the mercury down while raising the right-hand mercury column. The difference between the two levels of mercury is approximately equal to the pressure in the circulation in terms of millimeters of mercury. (To be more exact, it is equal to 104 per cent of the true pressure because of the weight of the water on the left-hand column of mercury.)

High-Fidelity Methods for Measuring Blood Pressure. Unfortunately, the mercury in the mercury manometer has so much *inertia* that it cannot rise and fall rapidly. For this reason the mercury manometer, though excellent for recording steady pressures, cannot respond to pressure changes that occur more rapidly than approximately one cycle every 2 to 3 seconds. Whenever it is desired to record rapidly changing pressures, some other type of pressure recorder is needed. Figure 18–9 demonstrates the basic principles of three electronic pressure *transducers* commonly used for converting pressure into electrical signals and then recording the pressure on a high-speed electrical recorder. Each of these transducers employs a very thin and highly stretched metal membrane which forms one wall of the fluid chamber. The fluid chamber in turn is connected through a needle or catheter with the vessel in which the pressure is to be measured. Pressure variations in the vessel cause changes of pressure in the chamber beneath the membrane. When the pressure is high the membrane bulges outward slightly, and when low it returns toward its resting position.

In Figure 18–9A a simple metal plate is placed a few

thousandths of an inch above the membrane. When the membrane bulges outward, the *capacitance* between the plate and membrane increases, and this change in capacitance can be recorded by an appropriate electronic system.

In Figure 18–9B a small iron slug rests on the membrane, and this can be displaced upward into a coil. Movement of the iron changes the *inductance* of the coil, and this, too, can be recorded electronically.

Finally, in Figure 18–9C a very thin, stretched resistance wire is connected to the membrane. When this wire is greatly stretched its resistance increases, and when less stretched the resistance decreases. These changes also can be recorded by means of an electronic system.

With some of these high-fidelity types of recording systems, pressure cycles up to 500 cycles per second have been recorded accurately, and in common use are recorders capable of registering pressure changes occurring as rapidly as 20 to 100 cycles per second.

RESISTANCE OF BLOOD FLOW

Units of Resistance. Resistance is the impediment to blood flow in a vessel, but it cannot be measured by any direct means. Instead, resistance must be calculated from measurements of blood flow and pressure difference in the vessel. If the pressure difference between two points in a vessel is 1 mm. Hg and the flow is 1 ml./sec., then the resistance is said to be 1 *peripheral resistance unit*, usually abbreviated *PRU*.

Expression of Resistance in CGS Units.   Occasionally, a basic physical unit called the CGS (centimeters, grams, seconds) unit is used to express resistance. This unit is *dyne seconds/centimeters⁵*. Resistance in these units can be calculated by the following formula:

$$R \left(in \ \frac{\text{dyne sec.}}{\text{cm.}^5} \right) = \frac{1333 \times \text{mm. Hg}}{\text{ml./sec.}} \qquad (5)$$

Total Peripheral Resistance and Total Pulmonary Resistance.   The rate of blood flow through the circulatory system when a person is at rest is close to 100 ml./sec., and the pressure difference from the systemic arteries to the systemic veins is about 100 mm. Hg. Therefore, in round figures the resistance of the entire systemic circulation, called the *total peripheral resistance,* is approximately 100/100 or 1 PRU. In some conditions in which the blood vessels throughout the body become strongly constricted, the total peripheral resistance rises to as high as 4 PRU, and when the vessels become greatly dilated it can fall to as little as 0.2 PRU.

In the pulmonary system the mean arterial pressure averages 16 mm. Hg and the mean left atrial pressure averages 4 mm. Hg, giving a net pressure difference of 12 mm. Therefore, in round figures the *total pulmonary resistance* at rest calculates to be about 0.12 PRU. This can increase in disease conditions to as high as 1 PRU and can fall in certain physiological states, such as exercise, to as low as 0.03 PRU.

"Conductance" of Blood in a Vessel and Its Relationship to Resistance.   Conductance is a measure of the blood flow through a vessel for a given pressure difference. This is generally expressed in terms of ml./sec./mm. Hg pressure, but it can also be expressed in terms of liters/sec./mm. Hg or in any other units of blood flow and pressure.

It is immediately evident that conductance is the reciprocal of resistance in accord with the following equation:

$$\text{Conductance} = \frac{1}{\text{Resistance}} \qquad (6)$$

Effect of Vascular Diameter on Conductance.   Slight changes in the diameter of a vessel cause tremendous changes in its ability to conduct blood when the blood flow is streamline. This is illustrated forcefully by the experiment in Figure 18–10A, which shows three separate vessels with relative diameters of 1, 2, and 4 but with the same pressure difference of 100 mm. Hg between the two ends of the vessels. Though the diameters of these vessels increase only four-fold, the respective flows are 1, 16, and 256 ml./mm., which is a 256-fold

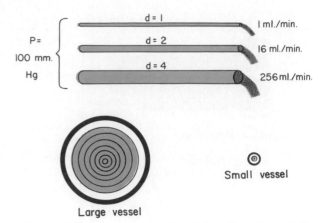

Figure 18–10.   A, Demonstration of the effect of vessel diameter on blood flow. B, Concentric rings of blood flowing at different velocities; the farther away from the vessel wall, the faster the flow.

increase in flow. Thus, the conductance of the vessel increases in proportion to the *fourth power of the diameter,* in accord with the following formula:

$$\text{Conductance} \propto \text{Diameter}^4 \qquad (7)$$

The cause of this great increase in conductance with an increase in diameter can be explained by referring to Figure 18–10B. This illustrates cross-sections of a large and a small vessel. The concentric rings inside these vessels indicate that the velocity of flow in each ring is different from that in the other rings because of *laminar* flow, which was discussed earlier in the chapter. That is, the blood in the ring touching the wall of the vessel is flowing hardly at all because of its adherence to the vascular endothelium. The next ring of blood slips past the first ring and, therefore, flows at a more rapid velocity. The third, fourth, fifth, and sixth rings likewise flow at progressively increasing velocities. Thus, the blood that is very near the wall of the vessel flows extremely slowly, while that in the middle of the vessel flows extremely rapidly.

In the small vessel essentially all of the blood is very near the wall so that the extremely rapidly flowing central stream of blood simply does not exist.

By integrating the velocities of all the concentric rings of flowing blood one can derive the following formula relating *mean velocity* of blood flow to vascular radius:

$$v = \frac{\Delta P \cdot r^2}{8 \ \eta l} \qquad (8)$$

in which v is velocity (in centimeters per second), ΔP is the pressure gradient (in dynes per square centimeter), r is the radius of the vessel (in centimeters), η is the viscosity (in poises), and 1 is the length of the vessel (in centimeters).

Poiseuille's Law. The quantity of blood that will flow through a vessel in a given period of time is equal to the velocity of flow times the cross-sectional area according to the following equation:

$$Q = v \pi r^2 \qquad (9)$$

in which Q is the rate of blood flow (in milliliters per second) and πr^2 is the cross-sectional area (in square centimeters)

Now let us substitute the value for velocity of blood flow from Equation 8 into Equation 9. This gives the following equation, which is known as Poiseuille's law:

$$Q = \frac{\pi \Delta P r^4}{8 \, \eta l} \qquad (10)$$

Note particularly in this equation that the rate of blood flow is directly proportional to the *fourth power of the radius* of the vessel, which illustrates once again that the diameter of a blood vessel plays by far the greatest role of all factors in determining the rate of blood flow through the vessel.

Summary of the Different Factors That Affect Conductance and Resistance. In the equation representing Poiseuille's law, Q represents flow, and ΔP represents the pressure difference. The remainder of the equation represents the conductance in accordance with the following equation:

$$C = \frac{\pi r^4}{8 \, \eta l} \qquad (11)$$

And, since conductance is the reciprocal of resistance, the following equation shows the factors that affect resistance:

$$R = \frac{8 \, \eta l}{\pi r^4} \qquad (12)$$

Thus, note that the resistance of a vessel is directly proportional to the *blood viscosity* (η) and *length* (l) of the vessel but inversely proportional to the *fourth power of the radius* (r^4).

Resistance to Blood Flow Through Series Vessels. In Figure 18–11A are two vessels connected in series, having resistances of R_1 and R_2. It is immediately evident that the total resistance is equal to the sum of the two, or

$$R_{(total)} = R_1 + R_2 \qquad (13)$$

Furthermore, it is equally evident that any number of resistances in series with each other must be added together. For instance, the total peripheral resistance is equal to the resistance of the arteries plus that of the arterioles plus that of the capillaries plus that of the veins.

Resistance of Vessels in Parallel. Shown in Figure 18–11B are four vessels, *connected in parallel*, with respective resistances of R_1, R_2, R_3, and R_4. However, the diameters of the vessels are not exactly the same. It is obvious that for a given pressure difference, far greater amounts of blood will flow through this system than through any one of the vessels alone. Therefore, the total resistance is far less than the resistance of any single vessel.

To calculate the total resistance in Figure 18–11B one first determines the *conductance* of each of the vessels, which is equal to the reciprocal of the resistance, or $\frac{1}{R_1}$, $\frac{1}{R_2}$, $\frac{1}{R_3}$, and $\frac{1}{R_4}$. The total conductance of [C (total)] or all the vessels is equal to the sum of the individual conductances:

$$C_{(total)} = \frac{1}{R_1} + \frac{1}{R_2} + \frac{1}{R_3} + \frac{1}{R_4} \qquad (14)$$

And resistance through the parallel circuit is the reciprocal of the total conductance, or

$$R_{(total)} = \frac{1}{\dfrac{1}{R_1} + \dfrac{1}{R_2} + \dfrac{1}{R_3} + \dfrac{1}{R_4}} \qquad (15)$$

Effect of Pressure on Vascular Resistance — Critical Closing Pressure

Since all blood vessels are distensible, increasing the pressure inside the vessels causes the vascular diameters also to increase. This in turn reduces the resistance of the vessel. Conversely, reduction in vascular pressures increases the resistance.

The middle curve of Figure 18–12 illustrates the normal effect on blood flow through a small tissue vascular bed caused by changing the arterial pressure. As the arterial pressure falls from 130 mm. Hg, the flow decreases rapidly at first because of two factors: (1) the decreasing pressure difference between the artery and the vein of the tissue, and (2) the decreasing diameters of the vessels. At 20 mm. Hg blood flow ceases entirely. This point at which the blood stops flowing is called the *critical closing pressure, because at this point the small*

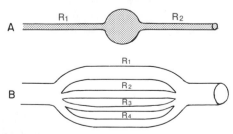

Figure 18–11. Vascular resistances: A, in series and B, in parallel.

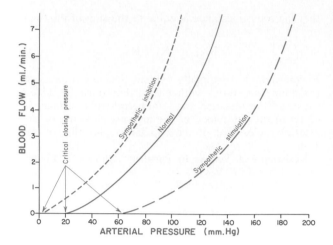

Figure 18–12. Effect of arterial pressure on blood flow through a blood vessel at different degrees of vascular tone, showing also the effect of vascular tone on critical closing pressure.

vessels, the arterioles in particular, close so completely that all flow ceases. The mechanism of this can be explained as follows:

Mechanism of Critical Closing Pressure. The vasomotor tone of the arterioles is always attempting to constrict these vessels to smaller diameters, while on the other hand the pressure inside the arterioles is tending to dilate them. As the pressure falls progressively lower and lower, it finally reaches a point at which the pressure inside the vessel is no longer capable of keeping the vessel open; this pressure is called the critical closing pressure.

A physical law, the *law of Laplace,* helps to explain the closure of vessels at an exact critical pressure level. This law states that the circumferential force tending to stretch the muscle fibers in the vascular wall is proportional to the *diameter of the vessel* times the *pressure,* or $F \propto D \cdot P$. Therefore, as the pressure in the vessel falls, thus allowing the vascular diameter also to decrease, the force tending to keep the vascular wall stretched decreases extremely rapidly. There comes a pressure below which the elastic tension in the wall that is tending to close the vessel becomes greater than the stretching force caused by the pressure; then the vessel can no longer remain open.

Another factor that contributes to the cessation of blood flow when the arterial pressure falls low is the size of the blood cells themselves, for, when the arteriolar diameter falls below a certain critical diameter, the red cells cannot pass through, and they will actually block the flow of plasma as well. The average critical closing pressure is about 20 mm. Hg when whole blood is flowing through the vessels and 5 to 10 mm. Hg when only plasma is flowing through.

Effect of Sympathetic Inhibition and Stimulation on Vascular Flow. Essentially all blood vessels of the body are normally stimulated by sympathetic impulses even under resting conditions. Different circulatory reflexes, which will be discussed in Chapter 21, can cause these impulses either to disappear, called *sympathetic inhibition,* or to increase to many times their normal rate,

called *sympathetic stimulation.* Sympathetic impulses in most parts of the body cause an increase in tone of the vascular smooth muscle. Therefore, as illustrated by the dashed curves of Figure 18–12, sympathetic inhibition allows far more blood to flow through a tissue for a given pressure than is normally true, whereas sympathetic stimulation vastly decreases the rate of blood flow.

Note also that sympathetic inhibition decreases the critical closing pressure, sometimes to as low as 5 mm. Hg. On the other hand, sympathetic stimulation can increase the critical closing pressure to as high as 100 or more mm. Hg, which means that strong sympathetic stimulation can actually stop blood flow through some tissues even when the arterial pressure is at its normal level of 100 mm. Hg.

VASCULAR DISTENSIBILITY — PRESSURE-VOLUME CURVES

The diameter of blood vessels, unlike that of metal pipes and glass tubes, increases as the internal pressure increases, because blood vessels are *distensible.* However, the vascular distensibilities differ greatly in different segments of the circulation, and, as we shall see, this affects significantly the operation of the circulatory system under many changing physiological conditions.

Units of Vascular Distensibility. Vascular distensibility is normally expressed as the fractional increase in volume for each millimeter mercury rise in pressure in accordance with the following formula:

$$\text{Vascular distensibility} = \frac{\text{Increase in volume}}{\text{Increase in pressure} \times \text{Original volume}} \quad (16)$$

That is, if 1 mm. Hg causes a vessel originally containing 10 ml. of blood to increase its volume by 1 ml., then the distensibility would be 0.1 per mm. Hg or 10 per cent per mm. Hg.

Difference in Distensibility of the Arteries and the Veins. Anatomically, the walls of arteries are far stronger than those of veins. Consequently, the veins, on the average, are about 6 to 10 times as distensible as the arteries. That is, a given rise in pressure will cause about 6 to 10 times as much extra blood to fill a vein as an artery of comparable size.

In the pulmonary circulation the veins are very similar to those of the systemic veins. However, the pulmonary arteries, which normally operate under pressures about one-seventh those in the systemic arterial system, have distensibilities only about one-half those of veins, rather than one-eighth, as is true of the systemic arteries.

VASCULAR COMPLIANCE (OR CAPACITANCE)

Usually in hemodynamic studies it is much more important to know the *total quantity of blood* that can be stored in a given portion of the circulation for each mm. Hg pressure rise than to know the distensibility of the individual vessels. This value is sometimes called the *overall distensibility* or *total distensibility,* or it can be expressed still more precisely by either of the terms *compliance* or *capacitance,* which are physical terms meaning the increase in volume that causes a given increase in pressure as follows:

$$\text{Vascular compliance} = \frac{\text{Increase in volume}}{\text{Increase in pressure}} \quad (17)$$

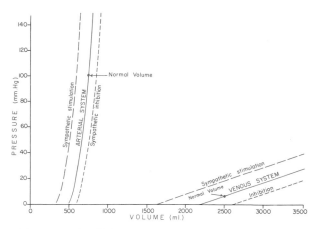

Figure 18–13. Volume-pressure curves of the systemic arterial and venous systems, showing also the effects of sympathetic stimulation and sympathetic inhibition.

Compliance and distensibility are quite different. A highly distensible vessel that has a very slight volume may have far less compliance than a much less distensible vessel that has a very large volume, for *compliance is equal to distensibility × volume.*

The compliance of a vein is about 24 times that of its corresponding artery because it is about 8 times as distensible and it has a volume about 3 times as great (8 × 3 = 24).

VOLUME-PRESSURE CURVES OF THE ARTERIAL AND VENOUS CIRCULATIONS

A convenient method for expressing the relationship of pressure to volume in a vessel or in a large portion of the circulation is the so-called *volume-pressure curve* (also frequently called the *pressure-volume curve*). The two solid curves of Figure 18–13 represent respectively the volume-pressure curves of the normal arterial and venous systems, showing that when the arterial system, including the larger arteries, small arteries, and arterioles, is filled with approximately 750 ml. of blood the mean arterial pressure is 100 mm. Hg, but when filled with only 500 ml. the pressure falls to zero.

The volume of blood normally in the entire venous tree is about 2500 ml. and tremendous changes in this volume are required to change the venous pressure only a few millimeters of mercury.

Difference in Compliance of the Arterial and Venous Systems.   Referring once again to Figure 18–13, one can see that a change of 1 mm. Hg requires a very large change in venous volume but much less change in arterial volume. That is, the *compliance of the venous system is far greater than the compliance of the arteries — about 24 times as great.*

This difference in compliance is particularly important because it means that tremendous amounts of blood can be stored in the veins with only slight changes in pressure. Therefore, the veins are frequently called the *storage areas* of the circulation.

Effect of Sympathetic Stimulation or Sympathetic Inhibition on the Volume-Pressure Relationships of the Arterial and Venous Systems.   Also shown in Figure 18–13 are the volume-pressure curves of the arterial and venous systems during moderate sympathetic stimulation and during sympathetic inhibition. It is evident that sympathetic stimulation, with its concomitant increase in smooth muscle tone in the vascular walls, increases the pressure at each volume of the arteries or veins while, on the other hand, sympathetic inhibition decreases the pressure at each respective volume. Obviously, control of the vessels in this manner by the sympathetics can be valuable for diminishing the dimensions of one segment of the circulation, thus transferring blood to other segments. For instance, an increase in vascular tone throughout the systemic circulation often causes large volumes of blood to shift into the heart, which is a major way in which pumping by the heart is increased.

Sympathetic control of vascular capacity is also especially important during hemorrhage. Enhancement of the sympathetic tone of the vessels, especially of the veins, reduces the dimensions of the circulatory system, and the circulation continues to operate almost normally even when as much as 25 per cent of the total blood volume has been lost.

"MEAN CIRCULATORY FILLING PRESSURE" AND VOLUME-PRESSURE CURVES OF THE ENTIRE CIRCULATORY SYSTEM

THE MEAN CIRCULATORY FILLING PRESSURE

The mean circulatory filling pressure (also called the "mean circulatory pressure" or the "static pressure") is a measure of the degree of filling of the circulatory system. That is, it is the pressure that would be measured in the circulation if one could instantaneously stop all blood flow and bring all the pressures in the circulation immediately to equilibrium. The mean circulatory filling pressure has been measured reasonably accurately in dogs within 2 to 5 seconds after the heart has been stopped. To do this the heart is thrown into fibrillation by an electrical stimulus, and blood is pumped rapidly from the systemic arteries to the veins to cause equilibrium between the two major chambers of the circulation.

The mean circulatory filling pressure measured in the above manner in the pentobarbital-anesthetized dog is almost exactly 7 mm. Hg and almost never varies more than 1 mm. from this value. It is believed to be about this same value in the human being. However, many different factors can change it, including especially change in the blood volume and increased or decreased sympathetic stimulation.

The mean circulatory filling pressure is *one of the major factors that determines the rate at which blood flows from the vascular tree into the right atrium of the heart, which in turn determines the cardiac output.* This is so important that that it will be explained in detail in Chapter 23; it will also be discussed in relation to blood volume regulation in Chapter 36.

VOLUME-PRESSURE CURVES OF THE ENTIRE CIRCULATION

Figure 18–14 illustrates the changes in the mean circulatory filling pressure as the total blood volume increases (1) under normal conditions, (2) during strong sympathetic stimulation, and (3) during complete sympathetic inhibition. The point marked by the arrow is the operating point of the normal circulation: a mean circulatory filling pressure of 7 mm. Hg and a blood volume of 5000 ml. However, *if blood is lost from the circulatory system,* the mean circulatory filling pressure falls to a lower value. If increased amounts of blood are added, the mean circulatory filling pressure rises accordingly.

The *compliance* of the entire circulatory system

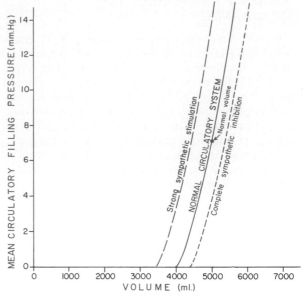

Figure 18–14. Volume-pressure curves of the entire circulation, illustrating the effect of strong sympathetic stimulation and complete sympathetic inhibition.

in the human being — that is, the increase in volume when the filling pressure increases — is approximately 100 ml. for each 1 mm. rise in mean circulatory filling pressure.

Sympathetic stimulation and *inhibition* affect the volume-pressure curves of the entire circulatory system in the same way that they affect the volume-pressure curves of the individual parts of the circulation, as illustrated by the two dashed curves of Figure 18–14. That is, for any given blood volume, the mean circulatory filling pressure rises two- to fourfold with strong sympathetic stimulation and falls markedly when the sympathetics are inhibited. This is an extremely important factor in the regulation of blood flow into the heart and thereby for regulating the cardiac output. For instance, during exercise, sympathetic activity increases the mean circulatory filling pressure several-fold, in this way helping to increase the cardiac output.

RELATIONSHIP BETWEEN MEAN CIRCULATORY FILLING PRESSURE, "MEAN SYSTEMIC FILLING PRESSURE," AND "MEAN PULMONARY FILLING PRESSURE"

As pointed out, the term *mean circulatory filling pressure* refers to the pressure that is measured in the entire circulatory system when all pressures are instantaneously brought to equilibrium. On the other hand, the term *mean systemic filling pressure* is the pressure measured in the systemic vessels when the root of the aorta and the great veins entering the heart are suddenly clamped and all pressures in the systemic system brought instantaneously to equilibrium. And, finally, the *mean pulmonary filling pressure* is the pressure measured in the pulmonary system when the pulmonary artery and larger pulmonary veins are suddenly clamped and all pulmonary pressures brought instantaneously to equilibrium.

In measurements made in dogs, the normal mean systemic filling pressure has averaged about 7 mm. Hg, which is equal to the mean circulatory filling pressure. The mean pulmonary filling pressure has measured about 10 mm. Hg. However, these values change markedly in certain pathological conditions for the following reason:

If the right heart becomes much weaker than the left heart, blood is dammed in the systemic circulation, while it is pumped out of the pulmonary system. As a consequence, the mean systemic filling pressure rises while the mean pulmonary filling pressure falls. Because of the small compliance of the pulmonary system compared with that of the systemic system (about one-seventh as great), the decrease in mean pulmonary filling pressure is about 7 times as great as the rise in mean systemic filling pressure. Conversely, damage to the left heart causes exactly the opposite effect.

DELAYED COMPLIANCE (STRESS-RELAXATION) OF VESSELS

The term "delayed compliance" means that a vessel whose pressure is increased by increased volume gradually loses much of this pressure over a period of many minutes because of progressive stretch of the vessel. Likewise, a vessel exposed to constantly increased pressure becomes progressively enlarged.

Figure 18–15 diagrams one of the effects of delayed compliance. In this figure, the pressure is being recorded in a small segment of a vein that is occluded at both ends. Then, an extra volume of blood is suddenly injected into the segment until the pressure rises from 5

to 12 mm. Hg. Even though none of the blood is removed after it is injected, the pressure nevertheless begins to fall immediately and approaches approximately 9 mm. Hg after several minutes. In other words, the volume of blood injected caused immediate *elastic* distention of the vein, but then the smooth muscle fibers of the vein began to "creep" to longer lengths, and their tensions correspondingly decreased. Therefore, the pressure rose markedly at first but then decreased with time. This effect is a characteristic of all smooth muscle tissue called *stress-relaxation*, which was explained in Chapter 12.

After the delayed increase in compliance had taken place in the experiment illustrated in Figure 18–15, the extra blood volume was suddenly removed, and the pressure immediately fell to a very low value. Subsequently, the smooth muscle fibers began to readjust their tensions back to their initial values, and after a number of minutes the normal vascular pressure of 5 mm. Hg returned.

Delayed compliance occurs only slightly in the arteries but to a much greater extent in the veins. As a result, prolonged elevation of venous pressure can often double the blood volume in the venous tree. This is a valuable mechanism by which the circulation can accommodate much extra blood when necessary, such as following too large a transfusion. Also, delayed compliance in the reverse direction is one of the ways by which the circulation automatically adjusts itself over a period of minutes or hours to diminished blood volume after serious hemorrhage.

REFERENCES

Bergel, D. H. (ed.): Cardiovascular Fluid Dynamics. New York, Academic Press, 1972.

Bergel, D. H., and Schultz, D. L.: Arterial elasticity and fluid dynamics. *Prog. Biophys. Mol. Biol.*, 22:1, 1971.

Bessis, M., *et al.* (eds.): Red Cell Rheology. New York, Springer-Verlag, 1978.

Braasch, D.: Red cell deformability and capillary blood flow. *Physiol. Rev.*, 51:679, 1971.

Caro, C. G., *et al.*: Mechanics of the circulation. *In* MTP International Review of Science: Physiology. Baltimore, University Park Press, 1974, Vol. 1, p. 1.

Cokelet, G. R., *et al.* (eds.): Erythrocyte Mechanics and Blood Flow. New York, A. R. Liss, 1979.

Dobrin, P. B.: Mechanical properties of arteries. *Physiol. Rev.*, 58:397, 1978.

Fitz-Gerlad, J. M.: Plasma motions in narrow capillary flow. *J. Fluid Mech.*, 51:463, 1972.

Fung, Y. C.: Introduction to biophysical aspects of microcirculation. *In* Kaley, G., and Altura, B. M. (eds.). Microcirculation. Vol. 1. Baltimore, University Park Press, 1977, p. 253.

Fung, Y. C.: Rheology of blood in microvessels. *In* Kaley, G., and Altura, B. M. (eds.): Microcirculation. Vol. 1. Baltimore, University Park Press, 1977, p. 279.

Gaehtgens, P., Meiselman, H. J., and Wayland, H.: Erythrocyte flow velocities in mesenteric microvessels of the cat. *Microvasc. Res.*, 2:151, 1970.

Ghista, D. N., *et al.* (eds.): Theoretical Foundations of Cardiovascular Processes. New York, S. Karger, 1979.

Gow, B. S.: Circulatory correlates: Vascular impedance, resistance, and capacity. *In* Bohr, D. F., *et al.* (eds.): Handbook of Physiology. Sec. 2, Vol. II. Baltimore, Williams & Wilkins, 1980, p. 353.

Green, H. D.: Circulation: Physical principals. *In* Glasser, O. (ed.): Medical Physics. Chicago, Year Book Medical Publishers, 1944.

Gross, J. F., and Popel, A. (eds.): Mathematics of Microcirculation Phenomena. New York, Raven Press, 1980.

Guyton, A. C.: Arterial Pressure and Hypertension. Philadelphia, W. B. Saunders Co., 1980.

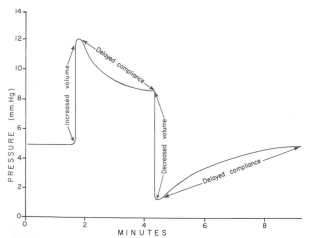

Figure 18–15. Effect on the intravascular pressure of injecting a small volume of blood into a venous segment, illustrating the principles of stress-relaxation.

Guyton, A. C., Greganti, F. P.: A physiologic reference point for measuring circulatory pressures in the dog—particularly venous pressure. *Am. J. Physiol., 185*:137, 1956.

Guyton, A. C., *et al.*: Pressure-volume curves of the entire arterial and venous systems in the living animal. *Am. J. Physiol., 184*:253, 1956.

Guyton, A. C., *et al.*: Cardiac Output and Its Regulation. Philadelphia, W. B. Saunders Co., 1973.

Harlan, J. C., *et al.*: Pressure volume curves of systemic and pulmonary circuit. *Am. J. Physiol., 213*:1499, 1967.

Huang, C., and Copley, A. L. (eds.): Biorheology. New York, American Institute of Chemical Engineers, 1978.

Hwang, M. H. C., *et al.* (eds.): Quantitative Cardiovascular Studies: Clinical Research Applications of Engineering Principles. Baltimore, University Park Press, 1978.

Intaglietta, M., *et al.*: Velocity pressure, flow, and elastic properties in microvessels of cat omentum. *Am. J. Physiol., 221*:922, 1971.

James, D. G. (ed.): Circulation of the Blood. Baltimore, University Park Press, 1978.

Lee, J.: Pressure-flow relationships of single vessels and organs. *In* Kaley, G., and Altura, B. M. (eds.): Microcirculation. Vol. 1. Baltimore, University Park Press, 1977, p. 335.

Lighthill, M. J.: Physiological fluid dynamics: a survey. *J. Fluid Mech., 52*:475, 1972.

Manning, R. D., Jr., *et al.*: Essential role of mean circulatory filling pressure in salt-induced hypertension. *Am. J. Physiol., 236*:R40, 1979.

McDonald, D. A.: Blood Flow in Arteries, 2nd Ed. Baltimore, Williams & Wilkins, 1974.

Murphy, R. A.: Mechanics of vascular smooth muscle. *In* Bohr, D. F., *et al.* (eds.): Handbook of Physiology. Sec. 2, Vol. II. Baltimore, Williams & Wilkins, 1980, p. 325.

Patel, D. J., *et al.*: Hemodynamics. *Annu. Rev. Physiol., 36*:125, 1974.

Pedley, T. J.: The Fluid Mechanics of Large Blood Vessels. New York, Cambridge University Press, 1979.

Porciuncula, C. I., *et al.*: Delayed compliance in external jugular vein of the dog. *Am. J. Physiol., 207*:728, 1964.

Roach, M. R.: Biophysical analyses of blood vessel walls and blood flow. *Annu. Rev. Physiol., 39*:51, 1977.

Rosenblum, W. I.: Viscosity: In vitro versus in vivo. *In* Kaley, G., and Altura, B. M. (eds.): Microcirculation. Vol. 1. Baltimore, University Park Press, 1977, p. 325.

Schmid-Schönbein, H.: Microrheology of erythrocytes, blood viscosity, and the distribution of blood flow in the microcirculation. *Int. Rev. Physiol., 9*:1, 1976.

Schneck, D. J., and Vawter, D. L. (eds.): Biofluid Mechanics. New York, Plenum Press, 1980.

Smith, J. J., and Kampine, J. P.: Circulatory Physiology; The Essentials. Baltimore, Williams & Wilkins, 1979.

Stehbens, W. E. (ed.): Hemodynamics and the Blood Vessel Wall. Springfield, Ill., Charles C Thomas, 1978.

Taylor, M. G.: Hemodynamics. *Annu. Rev. Physiol., 35*:87, 1973.

19

The Systemic Circulation

The circulation is divided into the *systemic circulation* and the *pulmonary circulation*. And, because the systemic circulation supplies all the tissues of the body except the lungs with blood flow, it is also frequently called the *greater circulation* or *peripheral circulation*.

Though the vascular system in each separate tissue of the body has its own special characteristics, some general principles of vascular function nevertheless apply in all parts of the systemic circulation. It is the purpose of the present chapter to discuss these general principles.

The Functional Parts of the Systemic Circulation. Before attempting to discuss the details of function in the systemic circulation, it is important to understand the overall role of each of its parts, as follows:

The function of the *arteries* is to transport blood *under high pressure* to the tissues. For this reason the arteries have strong vascular walls, and blood flows rapidly in the arteries.

The *arterioles* are the last small branches of the arterial system, and they act as *control valves* through which blood is released into the capillaries. The arteriole has a strong muscular wall that is capable of closing the arteriole completely or of allowing it to be dilated several-fold, thus having the capability of vastly altering blood flow to the capillaries.

The function of the *capillaries* is to exchange fluid, nutrients, electrolytes, hormones, and other substances between the blood and the interstitial spaces. For this role, the capillary walls are very thin and permeable to small molecular substances.

The *venules* collect blood from the capillaries; they gradually coalesce into progressively larger veins.

The *veins* function as conduits for transport of blood from the tissues back to the heart. Since the pressure in the venous system is very low, the venous walls are thin. Even so, they are muscular, and this allows them to contract or expand and

thereby to act as a reservoir for extra blood, either a small or large amount depending upon the needs of the body.

PHYSICAL CHARACTERISTICS OF THE SYSTEMIC CIRCULATION

Quantities of Blood in the Different Parts of the Circulation. By far the greater amount of the blood in the circulation is contained in the systemic veins. Figure 19–1 shows this, illustrating that approximately 84 per cent of the entire blood volume of the body is in the systemic circulation, with 64 per cent in the veins, 15 per cent in the arteries, and 5 per cent in the capillaries. The heart contains 7 per cent of the blood, and the pulmonary vessels, 9 per cent. Most surprising is the very low blood volume in the capillaries of the systemic circulation, only about 5 per cent of the total, for it is here that the most important function of the systemic circulation occurs, namely, diffusion of substances back and forth between the blood and the tissues. This function is so important that it will be discussed in detail in Chapter 30.

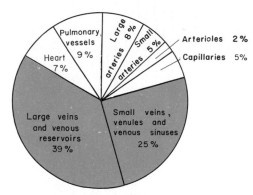

Figure 19–1. Percentage of the total blood volume in each portion of the circulatory system.

Cross-Sectional Areas and Velocities of Blood Flow.   If all vessels of each type were put side by side, their total cross-sectional areas would be:

	cm^2
Aorta	2.5
Small arteries	20
Arterioles	40
Capillaries	2500
Venules	250
Small veins	80
Venae cavae	8

Note particularly the large cross-sectional areas of the veins, averaging about 4 times those of the corresponding arteries. This explains the very large storage of blood in the venous system in comparison with that in the arterial system.

The velocity of blood flow in each segment of the circulation is inversely proportional to its cross-sectional area. Thus, under resting conditions, the velocity averages 33 cm. per second in the aorta, but 1/1000 this in the capillaries, or about 0.3 mm. per second. However, since the capillaries have a typical length of only 0.3 mm. to 1 mm., each segment of flowing blood remains in the capillaries for only 1 to 3 seconds, a very surprising fact, since all diffusion that takes place through the capillary walls must occur in this exceedingly short time.

Pressures and Resistances in the Various Portions of the Systemic Circulation.   Because the heart pumps blood continually into the aorta, the pressure in the aorta is obviously high, averaging approximately 100 mm. Hg. And, also, because the pumping by the heart is pulsatile, the arterial pressure fluctuates between a *systolic level* of 120 mm. Hg and a *diastolic level of* 80 mm. Hg, as illustrated in Figure 19–2. As the blood flows through the systemic circulation, its pressure falls progressively to approximately 0 mm. Hg by the time it reaches the right atrium.

The decrease in arterial pressure in each part of the systemic circulation is directly proportional to the vascular resistance. Thus, in the aorta the resistance is almost zero; therefore, the mean arterial pressure at the end of the aorta is still almost 100 mm. Hg. Likewise, the resistance in the large arteries is very slight so that the mean arterial pressure in arteries as small as 3 mm. in diameter is still 95 to 97 mm. Hg. Then the resistance begins to increase rapidly in the very small arteries, causing the pressure to drop to approximately 85 mm. Hg at the beginning of the arterioles.

The resistance of the *arterioles* is greatest of any part of the systemic circulation, accounting for about half the resistance in the entire systemic circulation. Thus, the pressure decreases about 55 mm. Hg in the arterioles so that the pressure of the blood as it leaves the arterioles to enter the capillaries is only about 30 mm. Hg. Arteriolar resistance is so important to the regulation of blood flow in different tissues of the body that it is discussed in detail later in the chapter and also in the following few chapters, which consider the regulation of the systemic circulation.

The pressure at the arterial ends of the *capillaries* is normally about 30 mm. Hg and at the venous ends about 10 mm. Hg. Therefore, the pressure decrease in the capillaries is only 20 mm. Hg, which illustrates that the capillary resistance is about two-fifths that of the arterioles.

The pressure at the beginning of the venous system, that is, at the *venules,* is about 10 mm.

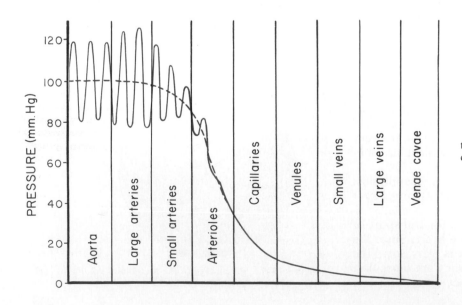

Figure 19–2. Blood pressures in the different portions of the systemic circulatory system.

Hg, and this decreases to almost exactly 0 mm. Hg at the right atrium. This large decrease in pressure in the veins indicates that the veins have far more resistance than one would expect for vessels of their large sizes. Much of this resistance is caused by compression of the veins from the outside, which keeps many of them, especially the venae cavae, collapsed a large share of the time. This effect is discussed later in the chapter.

PRESSURE PULSES IN THE ARTERIES

Since the heart is a pulsatile pump, blood enters the arteries intermittently with each heart beat, causing *pressure pulses* in the arterial system. In the normal young adult the pressure at the height of a pulse, the *systolic pressure,* is about 120 mm. Hg and at its lowest point, the *diastolic pressure,* is about 80 mm. Hg. The difference between these two pressures, about 40 mm. Hg, is called the *pulse pressure*.

Figure 19–3 illustrates a typical, idealized *pressure pulse curve* recorded in the ascending aorta of a human being, showing a very rapid rise in arterial pressure during ventricular systole, followed by a maintained high level of pressure for 0.2 to 0.3 second. This is terminated by a sharp *incisura* or *notch* at the end of systole, followed by a slow decline of pressure back to the diastolic level. The incisura occurs immediately before the aortic valve closes and is caused as follows: When the ventricle relaxes, the intraventricular pressure begins to fall rapidly, and backflow of blood from the aorta into the ventricle allows the aortic pressure also to begin falling. However, the backflow suddenly snaps the aortic valve closed. The momentum that has built up in the backflowing blood brings still more blood into the root of the aorta

even after the valve has closed, and this raises the pressure again, and thus giving the short positive pressure wave in the record immediately after the incisura.

FACTORS THAT AFFECT THE PULSE PRESSURE

Effect of Stroke Volume and Arterial Compliance. Two major factors affect the pulse pressure: (1) the *stroke volume output* of the heart and (2) the *compliance (total distensibility)* of the arterial tree. A third, less important factor is the character of ejection from the heart during systole.

In general, the greater the stroke volume output, the greater is the amount of blood that must be accommodated in the arterial tree with each heartbeat and, therefore, the greater is the pressure rise and fall during systole and diastole, thus causing a greater pulse pressure.

On the other hand, the greater the compliance of the arterial system, the less will be the rise in pressure for a given stroke volume of blood pumped into the arteries. In effect, then, the pulse pressure is determined approximately by the *ratio of stroke volume output to compliance of the arterial tree.* Therefore, any condition of the circulation that affects either of these two factors will also affect the pulse pressure.

Factors That Affect the Pulse Pressure by Changing the Stroke Volume Output. So many different circulatory conditions change the stroke volume output that only a few of these can be mentioned:

An *increase in heart rate* while the cardiac output remains constant causes the stroke volume output to decrease in inverse proportion to the increased rate, and the pulse pressure decreases accordingly.

A *decrease in total peripheral resistance* allows rapid flow of blood from the arteries to the veins. This increases the venous return to the heart and increases the stroke volume output. Therefore, the pulse pressure is also greatly increased.

An *increase in mean circulatory filling pressure*, if all other circulatory factors remain constant, increases the rate of venous return to the heart and consequently increases both the stroke volume output and the pulse pressure. For instance, this is the effect that occurs when a rapid transfusion is given to a person, increasing the blood volume and mean circulatory pressure.

The *character of ejection from the heart* affects the pulse pressure in two ways: First, if the duration of systole is long, a large portion of the stroke volume output runs off through the systemic circulation while it is being ejected into the aorta; therefore, the magnitude of the stroke volume effect on pulse pressure is decreased. Second, rapid onset of heart contraction sometimes occurs when a heart is beating vigorously, and this causes the pressure in the aorta to rise very high before the blood can run off through the peripheral circulation. Therefore, sudden ejection causes a greater pulse pressure than does more prolonged ejection.

Decreased Arterial Compliance in Old Age — Arteriosclerosis. In contrast to the many different conditions that can change the stroke volume output, only one major factor often alters the arterial compliance: this is

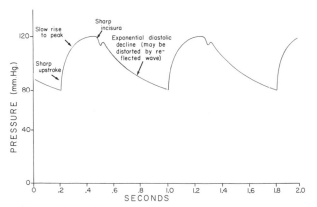

Figure 19–3. A normal pressure pulse contour recorded from the ascending aorta. (From Opdyke: *Fed. Proc., 11*:734, 1952.)

pathological changes that affect the distensibility of the arterial walls. This occurs mainly in old age when the arterial walls lose much of their elastic and muscular tissues and are replaced by fibrous tissue and sometimes even calcified plaques that cannot stretch a significant amount. This is the well-known condition called *arteriosclerosis.* It occurs at a much earlier age in some people than others, but it eventually occurs in all if they live long enough.

The vascular changes in arteriosclerosis greatly decrease the compliance of the arterial system, which in turn causes the arterial pressure to rise very high during systole and to fall greatly during diastole as blood runs off from the arteries to the veins. The middle curve of Figure 19–4 illustrates the very great pulse pressure that occurs in arteriosclerosis.

ABNORMAL PRESSURE PULSE CONTOURS

Some conditions of the circulation also cause abnormal contours to the pressure pulse wave in addition to altering the pulse pressure. Especially distinctive among these are patent ductus arteriosus and aortic regurgitation.

Patent Ductus Arteriosus. This is an abnormal condition in which the ductus arteriosus, which carries blood from the pulmonary artery to the aorta during fetal life, fails to close after birth. Instead, the blood now flows backward from the aorta through the open ductus into the pulmonary artery, allowing very rapid runoff of blood from the arterial tree after each heartbeat and a greatly decreased diastolic pressure. However, this is compensated for by a greater than normal stroke volume output so that the systolic pressure rises much higher than normal. These effects give the pressure pulse contour shown by the lowest curve in Figure 19–4. Here one finds an elevated systolic pressure, a greatly depressed diastolic pressure, a greatly increased pulse pressure, and the incisura occurring at a very low point on the downslope because an excessively large amount of blood runs off from the arteries even before systole is over.

Aortic Regurgitation. Aortic regurgitation means backward flow of blood through the aortic valve when it fails to close. This usually results from heart diseases that destroy the valve, especially rheumatic fever. In aortic regurgitation much of the blood that is pumped into the aorta during systole flows back into the left ventricle during diastole, thus giving a low diastolic pressure. However, this backflow is compensated for by a much greater than normal stroke volume output during systole, and consequently a high systolic pressure. Thus, the pressure pulse contour is very similar to that of patent ductus arteriosus, as illustrated by the lower curve of Figure 19–4, but is not always identical, for in aortic regurgitation the valve sometimes fails to close even partially. When this is true the incisura is entirely absent.

TRANSMISSION OF THE PRESSURE PULSE TO THE PERIPHERY

When the heart ejects blood into the aorta during systole, only the proximal portion of the aorta becomes distended at first, and it is only in this portion of the arterial tree that the pressure rises immediately. The cause of this is the inertia of the blood in the aorta, which prevents its sudden movement away from the central arteries into the peripheral arteries.

However, the rising pressure in the central aorta rapidly overcomes the inertia of the blood, causing the rising pressure to move farther and farther out in the arterial tree. Figure 19–5 illustrates this *transmission of the pressure pulse* down the aorta, the more distal portions of the aorta becoming distended as the pressure wave moves forward.

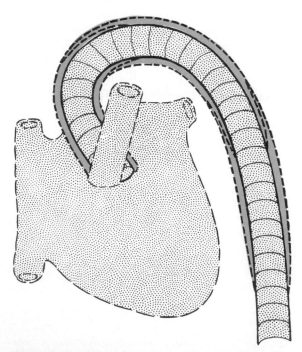

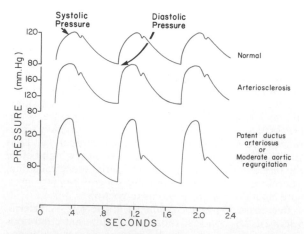

Figure 19–4. Pressure pulse contours in arteriosclerosis, patent ductus arteriosus, and moderate aortic regurgitation.

Figure 19–5. Progressive stages in the transmission of the pressure pulse along the aorta.

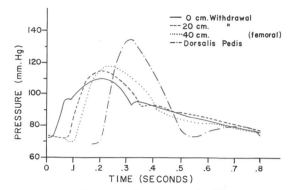

Figure 19–6. Pressure pulse contours in different segments of the arterial tree in humans, showing (1) the delay in the pressure pulse as it spreads more and more peripherally and (2) augmentation of the pulse pressure, especially in the dorsalis pedis artery.

The velocity of transmission of the pressure pulse along the normal aorta is 3 to 5 meters per second, along the large arterial branches 7 to 10 meters per second, and the smaller arteries 15 to 35 meters per second. In general, the greater the compliance of each vascular segment, the slower is the velocity of transmission, which explains the slowness of transmission in the aorta versus the rapidity of transmission in the far less compliant, small, distal arteries.

The *velocity of transmission of the pressure pulse is much greater than the velocity of blood flow*. During transmission of the pressure pulse, only a small amount of blood entering the proximal aorta pushes the more distal blood forward sufficiently to elevate the pressure in the very distal arteries. Therefore, the actual blood ejected by the heart may have traveled only a few centimeters by the time the pressure wave has already reached the distal ends of the arteries. In the aorta, the velocity of the pressure pulse is approximately 15 times that of blood flow, while in the more distal arteries the velocity of the pressure pulse may be as great as 100 times the velocity of blood flow.

Augmentation of the Pulse Pressure in the Peripheral Arteries. An interesting phenomenon that often occurs in transmission of pressure pulses to the periphery is an increase in pulse pressure. This effect is illustrated in Figure 19–6, which shows considerable enhancement of the pulse pressure more and more peripherally. This effect is caused primarily by *reflection* of the pulse wave from peripheral arteries, and it may be explained as follows:

When a pressure pulse enters the peripheral arteries and distends them, the rapidly decreasing compliance of the smaller arteries makes it difficult for the pressure pulse to travel any farther. Instead, the surge of pressure in the narrow arteries causes the pulse wave to begin traveling backward along the same vessels from which it had come, thus reflecting the wave backward toward the heart. This is analogous to a wave traveling in a bowl of water until it hits the edge. On striking the edge, the wave turns around and travels back onto the surface of the water. If the returning wave strikes an oncoming wave, the two ''summate,'' causing a much higher wave than would otherwise occur. Such is the

case in the arterial tree. The first portion of the pressure wave is reflected before the latter portion of the same wave reaches the peripheral arteries. Therefore, the first portion summates with the latter portion, causing a much higher pressure than would otherwise be recorded.

Though this peripheral augmentation of the pulse pressure is of little significance to the function of the circulation, this phenomenon must be remembered whenever pressure measurements are made in peripheral arteries, for systolic pressure is sometimes as much as 20 to 30 per cent above that in the central aorta, and diastolic pressure is often reduced as much as 10 to 15 per cent.

Damping of the Pressure Pulse in the Small Arteries and Arterioles. The pressure pulse becomes less and less intense as it passes through the small arteries and arterioles, until it becomes almost absent in the capillaries. This was illustrated in Figure 19–2 and is called *damping* of the pressure pulse.

Damping of the pressure pulse is caused mainly by a combined effect of vascular distensibility and vascular resistance. That is, for a pressure wave to travel from one area of an artery to another area, a small amount of blood must flow between the two areas. The resistance in the small arteries and arterioles is great enough that this small flow of blood, and consequently the transmission of pressure, is greatly impeded. At the same time, the distensibility of the small arteries is great enough that the small amount of blood that is caused to flow during each pressure pulse produces progressively less pressure rise and fall the more distal the pulse wave proceeds in the vessels.

Figure 19–7 illustrates the extreme effect of resistance on damping. The curve to the left is a normal pressure pulse contour recorded from the dorsalis pedis artery in a dog. To the right is the same pressure pulse contour recorded during contraction of the femoral artery. The damping effect of the increased resistance is readily apparent. It is important for the clinician to remember this effect when feeling the peripheral pulses in patients with occlusive vascular disease.

Capillary Pulsation. Pressure pulses are not completely damped by the time they reach the capillaries — the pulse pressure usually still averages 1 mm. Hg or more. And under some conditions very severe capillary pulsation results. Two major factors can cause abnormal capillary pulsation: (1) It occurs when the central pressure pulse is greatly exacerbated, as when the *heart rate is very slow* or when the stroke volume output is greatly increased as a result of *aortic regurgitation, patent ductus arteriosus,* or *even extreme increase in*

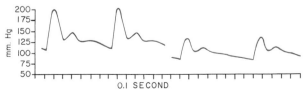

Figure 19–7. Pressure pulse contour in the dorsalis pedis artery of a dog recorded, first, under normal conditions and, second, during stimulation of the sympathetic nerves supplying the femoral artery.

venous return. (2) Another cause of capillary pulsation is *extreme dilatation of the small arteries and arterioles,* which reduces the resistance of these vessels and thereby reduces the damping.

Abnormal capillary pulsation can be readily demonstrated by bending the anterior portion of the fingernail downward so that the blood is pressed out of the anterior capillaries of the nail bed. This causes blanching distally while the proximal half of the nail remains red because of blood still in the capillaries. If significant pulsation is occurring in the capillaries, the border between the red and white areas will shift forward as more capillaries become filled during the high pressure phase and backward during the low pressure phase.

THE RADIAL PULSE

Clinically, it has been the habit for many years for a physician to feel the radial pulse of each patient. This is done to determine the rate of the heartbeat or, frequently, because of the psychic contact that it gives the doctor with the patient. Under certain circumstances, however, the character of the pulse can also be of value in the diagnosis of circulatory diseases.

Weak Pulse. A weak pulse at the radial artery usually indicates either (1) greatly decreased central pulse pressure, such as occurs when the stroke volume output is low, or (2) increased damping of the pulse wave caused by vascular spasm; the latter occurs when the sympathetic nervous system becomes overly active, for instance, following blood loss or when a person is having a chill.

Pulsus Paradoxus. Occasionally the strength of the pulse becomes strong, then weak, then strong, continuing in synchrony with the phases of respiration. This is caused by alternate increase and decrease in cardiac output with each respiration. During inspiration, all the blood vessels of the lungs increase in size because of increased negative pressure in the thorax. Therefore, blood collects in the lungs, and the stroke volume output and pulse strength decrease. During expiration, opposite effects occur. This is a normal phenomenon in all persons, but it becomes extremely distinct in some conditions, such as in very deep breathing or in cardiac tamponade (compression of the heart from the outside by fluid in the pericardial sac or by a constricted pericardium).

Pulse Deficit. The rhythm of the heart is very irregular in atrial fibrillation or in the case of ectopic beats. In these arrhythmias, which were discussed in Chapter 17, two beats of the heart often come so close together that the second beat pumps no blood or very little blood because the left ventricle has too little time to fill between the beats. In this circumstance, one can hear the second beat of the heart with a stethoscope applied directly over the heart but cannot feel a pulsation in the radial artery, an effect called a *pulse deficit.* The greater the pulse deficit each minute, the more serious, ordinarily, is the arrhythmia.

Pulsus Alternans. In a few conditions the heart beats strongly with one beat and then weakly with the next, causing alternation in the strength of the pulse as well, a condition called *pulsus alternans.* This alternation is frequently caused by enlarged ventricles, the stroke volume output alternating in amount from one heartbeat to the next.

Some instances of pulsus alternans probably result from a mechanical oscillation that occurs in the circulation as follows: A strong heartbeat forces a large amount of blood to the periphery, leaving a deficit of blood in the central vessels. Therefore, the next heartbeat is weak. But by the third heartbeat, the large surge of blood to the periphery now returns to the input of the heart and a strong heartbeat ensues once again. Or the oscillation might also be from one side of the heart to the other, one ventricle strong while the other is weak, then the two ventricles repeatedly alternating with each other. Both of these effects have been demonstrated in our laboratory in a computer analysis of circulatory function, but neither has been proved yet in a patient.

THE ARTERIOLES AND CAPILLARIES

Blood flow in each tissue is controlled almost entirely by the degree of contraction or dilatation of the arterioles, and it is in the capillaries that the important process of exchange between blood and the interstitial fluid occurs. These two segments of the circulation are so important that they will receive special discussion in chapters to follow: arteriolar regulation of blood flow in Chapters 20 and 21, and capillary exchange phenomena in Chapters 30, 31, and 32.

Upon leaving the small arteries, blood courses through the arterioles, which are only a few millimeters in length and have diameters from 8 to 50 microns. Each arteriole branches many times to supply 10 to 100 capillaries.

There are approximately 10 billion capillaries in the peripheral tissues, having more than 500 square meters of surface area. The thickness of the capillary wall is usually less than 1 micron, and there are small pores in the wall (as will be explained in Chapter 30) through which substances can diffuse.

Though the capillary walls are very thin and, therefore, very weak, their diameters are also very small. It will be recalled from the discussion of vascular wall tension and vascular pressure in the previous chapter that the wall tension is directly proportional to the *pressure times the diameter* (law of Laplace). Therefore, since the diameter is extremely minute, the tension developed in the wall is also extremely minute, which explains why the very thin-walled capillaries can withstand the pressure therein.

EXCHANGE OF FLUID THROUGH THE CAPILLARY MEMBRANE

Though the detailed dynamics of fluid exchange through the capillary membrane will be presented in Chapter 30, it is important to introduce this subject briefly here.

The capillary membrane is highly permeable to water as well as to all the substances dissolved in plasma and tissue fluids *except the plasma proteins*. This failure of the plasma proteins to go through the pores of the capillaries allows the proteins to cause osmotic pressure, called *colloid osmotic pressure* or *oncotic pressure,* at the membrane.

Therefore, two different types of pressure gradients can cause fluid movement through the capillary membrane: (1) the hydrostatic pressure gradient between the inside and outside of the membrane and (2) the colloid osmotic pressure gradient between the two sides. The greater the difference between the intracapillary hydrostatic pressure and the hydrostatic pressure in the tissue spaces surrounding the capillaries, the greater will be the tendency for fluid to move out of the capillaries into the interstitial spaces. On the other hand, the greater the difference between the plasma colloid osmotic pressure and the tissue fluid colloid osmotic pressure, the greater will be the osmotic tendency for fluid to move from the tissue spaces into the capillary. Under normal conditions, the hydrostatic and osmotic pressures are approximately in equilibrium, so that net exchange of fluid volume across the capillary membrane is very slight — only a minute normal outward flow which provides barely enough fluid for formation of the lymph that returns by way of the lymphatic vessels to the circulation.

On the other hand, any significant increase in the capillary pressure from its normal value will cause loss of fluid out of the circulation into the tissue spaces, or a decrease in capillary pressure will cause osmotic movement of fluid into the circulation from the tissue spaces. Therefore, we will often refer to loss of fluid from the circulation when the capillary pressure rises too high or gain of fluid into the circulation when the capillary pressure falls below normal.

Another important feature of capillary function is two-way diffusion through the capillary membrane of dissolved substances between the plasma and tissue fluids. Thus, sodium ions diffuse in both directions in approximately equal amounts so that the sodium concentration remains almost exactly the same in both the blood and the tissue fluids. Likewise, when the cells of the tissues deplete the oxygen in the tissue fluids, oxygen diffuses from the blood toward the cells. Conversely, when the cells form excess carbon dioxide this diffuses toward the blood. In this way, the capillaries provide nutrition to the cells and remove the end-products of metabolism from the cells. All these functions will be discussed in much more detail in Chapter 30, which is devoted specifically to capillary function.

THE VEINS AND THEIR FUNCTIONS

For years the veins have been considered to be nothing more than passageways for flow of blood into the heart, but it is rapidly becoming apparent that they perform many functions that are necessary to the operation of the circulation. They are capable of constricting and enlarging, of storing large quantities of blood and making this blood available when it is required by the remainder of the circulation, of actually propelling blood forward by means of a so-called "venous pump," and even of helping to regulate cardiac output, a function so important that it will be described in detail in Chapter 23.

RIGHT ATRIAL PRESSURE (CENTRAL VENOUS PRESSURE) AND ITS RELATION TO VENOUS PRESSURE

To understand the various functions of the veins, it is first necessary to know something about the pressures in the veins and how they are regulated. Blood from all the systemic veins flows into the right atrium; therefore, the pressure in the right atrium is frequently called the *central venous pressure.* The pressures in the peripheral veins to a great extent depend on the level of this pressure; that is, anything that affects right atrial pressure usually affects venous pressure everywhere in the body — particularly so when the atrial pressure rises above 5 to 10 mm. Hg.

Right atrial pressure is regulated by a balance between, first, *the ability of the heart to pump blood out of the right atrium* and, second, *the tendency for blood to flow from the peripheral vessels back into the right atrium.*

If the heart is pumping strongly, the right atrial pressure tends to decrease. On the other hand, weakness of the heart tends to elevate the right atrial pressure. Likewise, any effect that causes rapid inflow of blood into the right atrium from the veins tends to elevate the right atrial pressure. Some of the factors that increase this tendency for venous return (and also to increase the right atrial pressure) are (1) increased blood volume, (2) increased large vessel tone throughout the body with resultant increased peripheral venous pressures, and (3) dilatation of the arterioles, which decreases the peripheral resistance and allows rapid flow of blood from the arteries to the veins.

The same factors that regulate right atrial pressure also enter into the regulation of cardiac output, for the amount of blood pumped by the heart depends both on the ability of the heart to pump and the tendency for blood to flow into the heart from the peripheral vessels. Therefore, we will discuss the regulation of right atrial pressure in much more depth in Chapter 23 in connection with the regulation of cardiac output.

The *normal right atrial pressure* is approximately 0 mm. Hg, which is about equal to the atmospheric pressure around the body. However, it can rise to as high as 20 to 30 mm. Hg under very abnormal conditions, such as (a) serious heart failure or (b) following massive transfusion of blood, which will cause excessive quantities of blood to attempt to flow into the heart from the peripheral vessels.

The lower limit to the right atrial pressure is usually about −4 to −5 mm. Hg, which is the pressure in the pericardial and intrapleural spaces that surround the heart. The right atrial pressure approaches these very low values when the heart pumps with exceptional vigor or when the flow of blood into the heart from the peripheral vessels is greatly depressed, such as following severe hemorrhage.

VENOUS RESISTANCE AND PERIPHERAL VENOUS PRESSURE

Large veins have almost no resistance *when they are distended.* However, as illustrated in Figure 19–8, most of the large veins entering the thorax are compressed at many points by the surrounding tissues so that blood flow is impeded. For instance, the veins from the arms are compressed by their sharp angulation over the first rib. Second, the pressure in the neck veins often falls so low that the atmospheric pressure on the outside of the neck causes them to collapse. Finally, veins coursing through the abdomen are often compressed by different organs and by the intra-abdominal pressure so that usually they are at least partially collapsed to an ovoid or slitlike

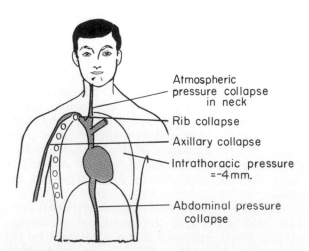

Atmospheric pressure collapse in neck

Rib collapse

Axillary collapse

Intrathoracic pressure =-4mm.

Abdominal pressure collapse

Figure 19–8. Factors tending to collapse the veins entering the thorax.

state. For these reasons the *large veins do usually offer considerable resistance to blood flow,* and because of this the pressure in the peripheral veins is usually 4 to 9 mm. Hg greater than the right atrial pressure.

Note, however, that the veins inside the thorax *are not collapsed,* because the *negative pressure* inside the chest distends these veins.

Effect of High Right Atrial Pressure on Peripheral Venous Pressure. When the right atrial pressure rises above its normal value of 0 mm. Hg, blood begins to back up in the large veins and to open them up. The pressures in the peripheral veins do not rise until all the collapsed points between the peripheral veins and the large veins have opened up. This usually occurs when the right atrial pressure rises to about 4 to 6 mm. Hg. If the right atrial pressure then rises still further, the additional increase in pressure is reflected by a corresponding rise in peripheral venous pressure. Since the heart must be greatly weakened to cause a rise in right atrial pressure to as high as 4 to 6 mm. Hg, one often finds that the peripheral venous pressure is not elevated in the early stages of cardiac failure.

Effect of Abdominal Pressure on Venous Pressures of the Leg. The normal pressure in the peritoneal cavity averages about 2 mm. Hg, but at times it can rise to as high as 15 to 20 mm. Hg as a result of pregnancy, large tumors, or excessive ascites in the peritoneal cavity as explained in Chapter 32. When this happens, the pressure in the veins of the legs must rise *above* the abdominal pressure before the abdominal veins will open and allow the blood to flow from the legs to the heart. Thus, if the intra-abdominal pressure is 20 mm. Hg, the lowest possible pressure in the femoral veins is 20 mm. Hg.

EFFECT OF HYDROSTATIC PRESSURE ON VENOUS PRESSURE

In any body of water, the pressure at the surface of the water is equal to atmospheric pressure, but the pressure rises 1 mm. Hg for each 13.6 mm. distance below the surface. This pressure results from the weight of the water and therefore is called *hydrostatic pressure.*

Hydrostatic pressure also occurs in the vascular system of the human being because of the weight of the blood in the vessels, as is illustrated in Figure 19–9. When a person is standing, the pressure in the right atrium remains approximately 0 mm. Hg because the heart pumps into the arteries any excess blood that attempts to accumulate at this point. However, in an adult *who is standing absolutely still* the pressure in the veins of the feet is approximately +90 mm. Hg simply because of the distance from the feet to the heart and the weight of the blood in the veins between the heart and the feet. The venous pressures at other levels

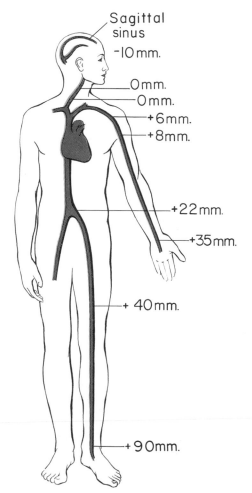

Sagittal
sinus
−10 mm.

0 mm.
0 mm.
+6 mm.
+8 mm.

+22 mm.

+35 mm.

+40 mm.

+90 mm.

Figure 19–9.   Effect of hydrostatic pressure on the venous pressures throughout the body.

fall below this level collapses the veins still more, which increases their resistance and again returns the pressure back to zero.

The veins inside the skull, however, are in a noncollapsible chamber, and they will not collapse. Consequently, *negative pressure can exist in the dural sinuses of the head;* in the standing position the venous pressure in the sagittal sinus is approximately −10 mm. Hg because of the hydrostatic "suction" between the top of the skull and the base of the skull. Therefore, if the sagittal sinus is opened during surgery, air can be sucked immediately into this vein; and the air may even pass downward to cause air embolism in the heart so that the heart valves will not function satisfactorily, and death often ensues.

Effect of the Hydrostatic Factor on Arterial and Other Pressures.   The hydrostatic factor also affects the peripheral pressures in the arteries and capillaries as well as in the veins. For instance, a standing person who has an arterial pressure of 100 mm. Hg at the level of the heart has an arterial pressure in the feet of about 190 mm. Hg. Therefore, any time one states that the arterial pressure is 100 mm. Hg, it generally means that this is the pressure at the hydrostatic level of the heart.

VENOUS VALVES AND THE "VENOUS PUMP"

Because of hydrostatic pressure, the venous pressure in the feet would always be about +90 mm. Hg in a standing adult were it not for the valves in the veins. However, every time one moves the legs one tightens the muscles and compresses the veins either in the muscles or adjacent to them, and this squeezes the blood out of the veins. Yet, the valves in the veins, as illustrated in Figure 19–10, are arranged so that the direction of blood flow can be only toward the heart. Consequently, every time a person moves the legs or even tenses the muscles, a certain amount of blood is propelled toward the heart, and the pressure in the dependent veins of the body is lowered. This pumping system is known as the "venous pump" or "muscle pump," and it is efficient enough that under ordinary circumstances the venous pressure in the feet of a walking adult remains less than 25 mm. Hg.

If the human being stands perfectly still, the venous pump does not work, and the venous pressures in the lower part of the leg can rise to the full hydrostatic value of 90 mm. Hg in about 30 seconds. Under such circumstances the pressures within the capillaries also increase greatly, and fluid leaks from the circulatory system into the tissue spaces. As a result, the legs swell, and the blood volume diminishes. Indeed, as much as 15 to 20 per cent of the blood volume is frequently

of the body lie proportionately between 0 and 90 mm. Hg.

In the arm veins, the pressure at the level of the top rib is usually about +6 mm. Hg because of compression of the subclavian vein as it passes over this rib. The hydrostatic pressure down the length of the arm is then determined by the distance below the level of this rib. Thus, if the hydrostatic difference between the level of the rib and the hand is 29 mm. Hg, this hydrostatic pressure is added to the 6 mm. Hg pressure caused by compression of the vein as it crosses the rib, making a total of 35 mm. Hg pressure in the veins of the hand.

The neck veins collapse almost completely all the way to the skull owing to atmospheric pressure on the outside of the neck. This collapse causes the pressure in these veins to remain zero along their entire extent. The reason for this is that any tendency for the pressure to rise above this level opens the veins and allows the pressure to fall back to zero, and any tendency for the pressure to

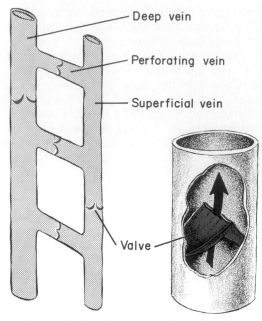

Figure 19–10.   The venous valves of the leg.

lost from the circulatory system within the first 15 minutes of standing absolutely still, as occurs when a soldier is made to stand at absolute attention.

Varicose Veins.   The valves of the venous system are frequently destroyed. This occurs particularly when the veins have been overstretched by an excess of venous pressure for a prolonged period of time, as occurs in pregnancy or when one stands on his feet most of the time. Stretching the veins, obviously, increases their cross-sectional areas, but the valves do not increase in size. Therefore, the vanes of the valves will no longer close completely and block reverse blood flow in the enlarged veins. When such a situation develops, the pressure in the veins of the legs increases still more owing to failure of the venous pump; this further increases the size of the veins and finally destroys the function of the valves entirely. Thus, the person develops "varicose veins," which are characterized by large bulbous protrusions of the veins beneath the skin of the entire leg and particularly of the lower leg. The venous and capillary pressures become very high because of the incompetent venous pump, and leakage of fluid from the capillary blood into the tissues causes constant edema in the legs of these persons whenever they stand for more than a few minutes. The edema in turn prevents adequate diffusion of nutritional materials from the capillaries to the muscle and skin cells so that the muscles become painful and weak, and the skin frequently becomes gangrenous and ulcerates. Obviously, the best treatment for such a condition is continual elevation of the legs to a level at least as high as the heart, but tight binders on the legs are also of considerable aid in preventing the edema and its sequelae.

PRESSURE REFERENCE LEVEL FOR MEASURING VENOUS AND OTHER CIRCULATORY PRESSURES

In previous discussions we have often spoken of right atrial pressure as being 0 mm. Hg and arterial pressure as being 100 mm. Hg, but we have not stated the hydrostatic level in the circulatory system to which this pressure is referred. There is one point in the circulatory system at which hydrostatic pressure factors caused by changes in body position usually do not affect the pressure measurement by more than 1 mm. Hg. This is at the level of the tricuspid valve, as shown by the crossed axes in Figure 19–11. Therefore, all pressure measurements discussed in this text are referred to the level of the tricuspid valve, which is called the *reference level for pressure measurement.*

The reason for lack of hydrostatic effects at the tricuspid valve is that the heart automatically prevents significant hydrostatic changes in pressure at this point in the following way:

If the pressure at the tricuspid valve rises slightly above normal, then the right ventricle fills to a greater extent than usual, causing the heart to pump blood more rapidly than usual and therefore to decrease the pressure at the tricuspid valve back toward the normal mean

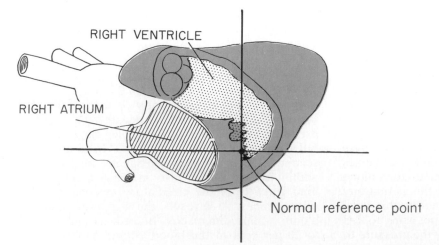

Figure 19–11.   Location of the reference point for pressure measurement at the tricuspid valve.

value. On the other hand, if the pressure at this point falls, the right ventricle fails to fill adequately, its pumping decreases, and blood dams up in the venous system until the tricuspid pressure again rises to a normal value. In other words, *the heart acts as a feedback regulator of pressure* at the tricuspid valve.

The three axes of the reference level, as extrapolated from animals to the human being, are (1) *approximately 61 per cent of the thickness of the chest in the front of the back,* (2) *almost exactly in the midline,* and (3) *approximately one quarter of the distance above the lower end of the sternum.* A person can be standing, lying on his back, lying on his stomach, lying on either side, or even in the head-down position, and his central venous pressure referred to this reference level will remain almost exactly constant regardless of the position of the body. This does not mean that the pressure at the tricuspid valve is always zero. It may be as low as −4 mm. Hg or as high as +20 mm. Hg, but, whatever its value, changing the position of the body does not alter it more than 1 mm. Hg.

In making arterial pressure measurements, the precise hydrostatic level to which pressures are referred matters much less than for venous pressure measurements because percentagewise a hydrostatic error of as much as 10 or more centimeters (equivalent to 7.4 mm. Hg error) still does not affect the arterial pressure reading to a great extent. However, in venous pressure measurements the reference level must be very exact if the measurements are to be significant, because abnormalities of venous pressure as small as 1 mm. Hg can often result in changes in cardiac output as great as 50 to 100 per cent.

PRESSURE PULSES IN THE VEINS

The pressure pulses in the arteries are almost completely damped out before they pass through the capillaries into the systemic veins. However, pulsations are sometimes transmitted backward from the heart to cause pressure pulses in the large veins that have characteristics entirely different from those in the arteries. This backward transmission of pulses usually does not occur to a significant extent in the normal circulation because most of the veins, where they lead into the thoracic cavity, are compressed by surrounding tissues; this causes sufficient resistance to damp out the pulsations before they can be transmitted backward into the peripheral veins. However, whenever the right atrial pressure is high, especially in heart failure, the veins are then well filled with blood and can easily transmit the pulsations.

It is the *atrial pulsations,* which were discussed in Chapter 13, that are transmitted backward in the distended veins. Briefly, these pulsations consist of three separate waves during each heart cycle: (1) the *a wave,* caused by contraction of the atrium, (2) the *c wave,* caused by contraction of the ventricle, and (3) the *v wave,* caused by continued inflow of blood into the atrium when the A-V valves are closed during ventricular systole.

In severe cardiac failure when the venous pressure is high, the venous pressure waves usually become prominent enough that one can see the veins of the patient's neck pulsating. In earlier days these pulsations were recorded for diagnostic purposes. The *a-c interval* in the recording is approximately equal to the P-Q interval of the electrocardiogram, and before the days of electrocardiography this was often used as a measure of the delay between atrial excitation and ventricular excitation. These waves, therefore, were then of considerable value in diagnosing different degrees of heart block.

MEASUREMENT OF VENOUS PRESSURE

Clinical Estimation of Venous Pressure. The venous pressure can be estimated by simply observing the degree of distension of the peripheral veins — especially the neck veins. For instance, in the sitting position, the neck veins are never distended in the normal person. However, when the right atrial pressure becomes increased to as much as 10 mm. Hg, the lower veins of the neck begin to protrude even when one is sitting; when the right atrial pressure rises to as high as 15 mm. Hg, essentially all the veins in the neck become distended, and the venous pulse can then usually be seen in the walls of the protruding veins.

Rough estimates of the venous pressure can also be made by raising or lowering an arm of a reclining person while observing the degree of distension of the antecubital or hand veins. As the arm is progressively raised, the veins suddenly collapse, and the level at which they collapse, when referred to the level of the heart, is a rough measure of the peripheral venous pressure.

Direct Measurement of Venous Pressure and Right Atrial Pressure. Venous pressure can be measured with ease by inserting a syringe needle connected to a pressure recorder or to a water manometer directly into a vein. The venous pressure is expressed in relation to the level of the tricuspid valve, i.e., the height in centimeters of water above the level of the tricuspid valve; this can be converted to mm. Hg by dividing by a factor of 1.36.

The only means by which *right atrial pressure* can be measured accurately is by inserting a catheter through the veins into the right atrium. This catheter can then be connected to an appropriate pressure measuring apparatus.

BLOOD RESERVOIR FUNCTION OF THE VEINS

In discussing the general characteristics of the systemic circulation earlier in the chapter, it was pointed out that over 60 per cent of all the blood in the circulatory system is in the systemic veins. For this reason it is frequently said that the systemic veins act as a *blood reservoir* for the circulation. Also, relatively large quantities of blood are present in the veins of the lungs so that these, too, are considered to be blood reservoirs.

When blood is lost from the body to the extent that the arterial pressure begins to fall, pressure reflexes are elicited from the carotid sinuses and other pressure sensitive areas of the circulation, as will be discussed in Chapter 21; these in turn cause sympathetic nerve signals that constrict the veins, thus automatically taking up much of the slack in the circulatory system caused by the lost blood. Indeed, even after as much as 20 to 25 per cent of the total blood volume has been lost, the circulatory system often functions almost normally because of this variable reservoir system of the veins.

Specific Blood Reservoirs. Certain portions of the circulatory system are so extensive that they are specifically called "blood reservoirs." These include (1) the *spleen,* which can sometimes decrease in size sufficiently to release as much as 150 ml. of blood into other areas of the circulation, (2) the *liver,* the sinuses of which can release several hundred milliliters of blood into the remainder of the circulation, (3) the *large abdominal veins,* which can contribute as much as 300 ml., and (4) the *venous plexus beneath the skin,* which can probably contribute several hundred milliliters. The *heart* itself and the *lungs,* though not parts of the systemic venous reservoir system, must also be considered to be blood reservoirs. The heart, for instance, becomes reduced in size during sympathetic stimulation and in this way can contribute about 100 ml. of blood, and the lungs can contribute another 100 to 200 ml. when the pulmonary pressures fall to low values.

ASSESSMENT OF VENOUS FUNCTION BY MEASUREMENT OF MEAN CIRCULATORY FILLING PRESSURE

It was pointed out in the previous chapter that the *mean circulatory filling pressure* can be measured by stopping the heart and pumping blood rapidly from the systemic arteries to the veins, bringing the pressures in the two major vascular reservoirs to equilibrium within a few seconds; this equilibrium pressure is the mean circulatory filling pressure. This pressure is actually a measure of how tightly the vascular system is filled with blood. Therefore, the greater the degree of filling (especially of the veins), the higher is the mean circulatory filling pressure. The normal value is approximately 7 mm. Hg, but following massive transfusion this value can rise to as high as 30 to 40 mm. Hg; and in congestive heart failure, in which the circulatory blood volume is greatly increased, it can rise to as high as 20 to 25 mm. Hg.

Measurements of mean circulatory filling pressure are also important in assessing venous contraction for the following reasons: The systemic venous system has many times as much compliance as all the remainder of the circulation put together. The mean circulatory filling pressure is about three-fourths determined by venous tone and venous filling with blood, in contrast with only about one-fourth by all the remainder of the circulation. Therefore, even slight changes in degree of venous contraction will cause marked changes in the mean circulatory filling pressure, whereas even large changes in degree of contraction of other segments of the circulation, such as the arterioles, the pulmonary vessels, and so forth, have very little effect. Thus one of the most sensitive barometers of the degree of contraction of the

venous system is the change that this causes in the mean circulatory filling pressure.

REFERENCES

Attinger, E. O.: Analysis of pulsatile blood flow. *Adv. Biomed. Eng. Med. Phys., 1*:1, 1968.

Caro, C. G., *et al.*: Mechanics of the circulation. *In* MTP International Review of Science: Physiology. Baltimore, University Park Press, 1974, Vol. 1, p. 1.

D'Agrosa, L. S.: Patterns of venous vasomotion in the bat wing. *Am. J. Physiol., 218*:530, 1970.

Folkow, B.: Role of the nervous system in the control of vascular tone. *Circulation, 21*:706, 1960.

Friedman, S. M., and Friedman, C. L.: Effects of ions on vascular smooth muscle. *In* Hamilton, W. F. (ed.): Handbook of Physiology. Sec. 2, Vol. 2. Baltimore, Williams & Wilkins, 1963, p. 1135.

Goodman, A. H., *et al.*: A television method for measuring capillary red cell velocities. *J. Appl. Physiol., 37*:126, 1974.

Guyton, A. C.: The venous system and its role in the circulation. *Mod. Conc. Cardiov. Dis., 27*:483, 1958.

Guyton, A. C.: Peripheral circulation. *Annu. Rev. Physiol., 21*:239, 1959.

Guyton, A. C.: Cardiac output and regional circulation. *In* Gordon, B. L. (ed.): Clinical Cardiopulmonary Physiology. New York, Grune & Stratton, 1969, pp. 28–38.

Guyton, A. C.: Arterial Pressure and Hypertension. Philadelphia, W. B. Saunders Co., 1980.

Guyton, A. C., and Jones, C. E.: Central venous pressure: Physiological significance and clinical implications. *Am. Heart J., 86*:431, 1973.

Guyton, A. C., *et al.*: Evidence for tissue oxygen demand as the major factor causing autoregulation. *Circ. Res., 14*:60, 1964.

Guyton, A. C., *et al.*: Cardiac Output and Its Regulation. Philadelphia, W. B. Saunders Co., 1973.

Haddy, F. J.: Local effects of sodium, calcium, and magnesium upon small and large blood vessels of the dog forelimb. *Circ. Res., 8*:57, 1960.

Herd, J. A.: Overall regulation of the circulation. *Annu. Rev. Physiol., 32*:289, 1970.

James, D. G. (ed.): Circulation of the Blood. Baltimore, University Park Press, 1978.

Korner, P. I.: Circulatory adaptations in hypoxia. *Physiol. Rev., 39*:687, 1959.

Lundgren, O., and Jodal, M.: Regional blood flow. *Annu. Rev. Physiol., 37*:395, 1975.

McDonald, D. A.: The relation of pulsatile pressure to flow in arteries. *J. Physiol., 127*:533, 1955.

Mellander, S.: Systematic circulation. *Annu. Rev. Physiol., 32*:313, 1970.

Pedley, T. J.: The Fluid Mechanics of Large Blood Vessels. New York, Cambridge University Press, 1979.

Remington, J. W., and O'Brien, L. J.: Construction of aortic flow pulse from pressure pulse. *Am. J. Physiol., 218*:437, 1970.

Rovick, A. A., and Randall, W. C.: Systemic circulation. *Annu. Rev. Physiol., 29*:225, 1967.

Saunders, J. B. deC. M.: The history of venous valves. *In* Dickinson, C. J., and Marks, J. (eds.): Developments in Cardiovascular Medicine. Lancaster, England, MTP Press, 1978, p. 335.

Schmidt-Nielsen, K., and Pennycuik, P.: Capillary density in mammals in relation to body size and oxygen consumption. *Am. J. Physiol., 200*:746, 1961.

Schneck, D. J., and Vawter, D. L. (eds.): Biofluid Mechanics. New York, Plenum Press, 1980.

Shepherd, J. T., and Vanhoutte, P. M.: The Human Cardiovascular System: Facts and Concepts. New York, Raven Press, 1979.

Sonnenschein, R. R., and White, F. N.: Systemic circulation. *Annu. Rev. Physiol., 30*:147, 1968.

Stainsby, W. N.: Autoregulation of blood flow in skeletal muscle during increased metabolic activity. *Am. J. Physiol., 202*:273, 1962.

Stainsby, W. N., and Renkin, E. M.: Autoregulation of blood flow in resting skeletal muscle. *Am. J. Physiol., 207*:117, 1961.

Stehbens, W. E. (ed.): Hemodynamics and the Blood Vessel Wall. Springfield, Ill., Charles C Thomas, 1978.

Uchida, E., and Bohr, D. F.: Myogenic tone in isolated perfused resistance vessels from rats. *Am. J. Physiol., 216*:1343, 1969.

Uchida, E., and Bohr, D. F.: Myogenic tone in isolated perfused vessels: Occurrence among vascular beds and along vascular trees. *Circ. Res., 25*:549, 1969.

Vanhoutte, P. M., and Leusen, I. (eds.): Mechanics of Vasodilatation. New York, S. Karger, 1978.

Widmer, L. K., *et al.* (eds.): New Trends in Venous Diseases. H. Huber, 1977.

Wiedeman, M. P.: Dimensions of blood vessels from distributing artery to collecting vein. *Circ. Res., 12*:375, 1963.

Wolf, S., and Werthessen, N. T. (eds.): Dynamics of Arterial Flow. New York, Plenum Press, 1979.

Wolf, S., *et al.* (eds.): Structure and Function of the Circulation. New York, Plenum Press, 1979.

Wood, E.: The Veins. Boston, Little, Brown, 1965.

Ziegler, M. G.: Postural hypotension. *Annu. Rev. Med., 31*:239, 1980.

20

Local Control of Blood Flow by the Tissues, and Nervous and Humoral Regulation

The circulatory system is provided with a complex system for control of blood flow to the different parts of the body. In general, the controls are of three major types:

1. Local control of blood flow in each individual tissue, the flow being controlled mainly in proportion to that tissue's need for blood perfusion.

2. Nervous control of blood flow, which often affects blood flow in large segments of the systemic circulation, such as shifting blood flow from the nonmuscular vascular beds to the muscles during exercise or changing the blood flow in the skin to control body temperature.

3. Humoral control, in which various substances dissolved in the blood such as hormones, ions, or other chemicals can cause either local increase or decrease in tissue flow or widespread generalized changes in flow.

LOCAL CONTROL OF BLOOD FLOW BY THE TISSUES THEMSELVES

Acute Local Regulation in Response to Tissue Need for Flow. In most tissues the blood flow is controlled in proportion to the need for nutrition, such as the need for delivery of oxygen, especially, but also glucose, amino acids, fatty acids, and other nutrients. However, in some tissues the local flow performs other functions. In the skin its purpose is to transfer heat from the body to the surrounding air. In the kidneys its purpose is to deliver substances to the kidneys for excretion. And in the brain it is to determine, to a great extent, the carbon dioxide and hydrogen ion con-

centrations of the brain fluids, which in turn play important roles in controlling the level of brain activity.

Fortunately, local blood flow can be increased in response to many different local factors in the tissues — at times to lack of oxygen, at times to excess of carbon dioxide or hydrogen ion concentration, and at other times to still other factors. It is these many different control factors that help to distribute the blood flow to the different parts of the body in proportion to the respective needs of the tissues.

Table 20–1 gives the approximate *resting* blood flows through the different organs of the body. Note the tremendous flows through the brain, liver, and kidneys despite the fact that these

TABLE 20–1 BLOOD FLOW TO DIFFERENT ORGANS AND TISSUES UNDER BASAL CONDITIONS*

	Per cent	*ml./min.*
Brain	14	700
Heart	4	200
Bronchial	2	100
Kidneys	22	1100
Liver	27	1350
Portal	(21)	(1050)
Arterial	(6)	(300)
Muscle (inactive state)	15	750
Bone	5	250
Skin (cool weather)	6	300
Thyroid gland	1	50
Adrenal glands	0.5	25
Other tissues	3.5	175
Total	100.0	5000

*Based mainly on data compiled by Dr. L. A. Sapirstein.

organs represent only a small fraction of the total body mass. Yet, even under basal conditions, the need for flow in each one of these tissues is very great; in the liver to support its high level of metabolic activity, in the brain to provide nutrition and to prevent the carbon dioxide and hydrogen ion concentrations from becoming too great, and in the kidneys to maintain adequate excretion.

The skeletal muscle of the body represents 35 to 40 per cent of the total body mass, and yet, in the inactive state, the blood flow through all the skeletal muscle is only 15 to 20 per cent of the total cardiac output. This accords with the fact that inactive muscle has a very low metabolic rate. Yet, when the muscles become active, their metabolic rate sometimes increases as much as 50-fold, and the blood flow in individual muscles can increase as much as 20-fold, illustrating a marked increase in blood flow in response to the increased need of the muscle for nutrients.

Functional Anatomy of the Systemic Microcirculation. Though the student will recall that each tissue has its own characteristic vascular system, Figure 20–1 presents a typical capillary bed; this one is in the connective tissue of the mesentery, which is very easily studied and its components easily analyzed. This figure shows that blood enters the capillary bed through a small *arteriole* and leaves by way of a small *venule*. From the arteriole the blood usually divides and flows through several *metarterioles* before entering the *capillaries*. Some of the capillaries are very large, and they course almost directly to the venule. These are called *preferential channels*. However, most of the capillaries, called the *true capillaries*, branch mainly from the metarterioles and then finally terminate in a venule.

The arterioles have a strong muscular coat, and the metarterioles are surrounded by sparse but highly active smooth muscle fibers. In addition, in many tissues, at each point at which a capillary leaves a metarteriole, a small muscular *precapillary sphincter,* consisting of a single spiraling smooth muscle fiber, surrounds the origin of the capillary. In those tissues that do not have pre-

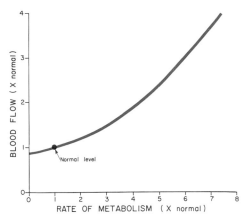

Figure 20–2. Effect of increasing rate of metabolism on tissue blood flow.

capillary sphincters, such as muscle, the metarterioles serve the same constrictor function for controlling blood flow into the capillaries.

The venules also have a smooth muscle coat, but one that is much less extensive than that of the arterioles.

As will be discussed in more detail later in the chapter, the arterioles especially, and the venules to a much lesser extent, are supplied by extensive innervation from the sympathetic nervous system, and the degree of contraction of these structures is strongly influenced by the intensity of sympathetic signals transmitted from the central nervous system to the blood vessels.

On the other hand, innervation of the metarterioles and the precapillary sphincters is usually very sparse, or even absent in most instances. Instead, the muscle fibers of these two structures are controlled almost entirely by the local factors in the tissues, that is, by the concentrations of oxygen, carbon dioxide, hydrogen ions, electrolytes, and other substances in each individual tissue area. These local factors, therefore, are major controllers of blood flow in the local tissue areas.

Effect of Tissue Metabolism on Local Blood Flow. Figure 20–2 illustrates the approximate quantitative effect on blood flow of increasing the rate of metabolism in a local tissue such as muscle. Note that an increase in metabolism up to 8 times normal increases the blood flow about four-fold. The increase in flow at first is less than the increase in metabolism. However, once the metabolism rises high enough to remove most of the nutrients from the blood, further increase in metabolism can occur only with a concomitant increase in blood flow to supply the required nutrients.

Local Blood Flow Regulation When Oxygen Availability Changes. One of the most necessary

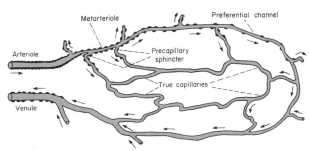

Figure 20–1. Overall structure of a capillary bed. (From Zweifach: Factors Regulating Blood Pressure. New York, Josiah Macy, Jr., Foundation, 1950.)

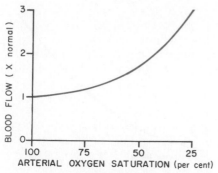

Figure 20–3. Effect of atrial oxygen saturation on blood flow through an isolated dog leg.

of the nutrients is oxygen. Whenever the availability of oxygen to the tissues decreases, such as at high altitude, in pneumonia, in carbon monoxide poisoning (which poisons the ability of hemoglobin to transport oxygen), or in cyanide poisoning (which poisons the ability of the tissues to utilize oxygen), the blood flow through the tissues increases markedly. Figure 20–3 shows that as the arterial oxygen saturation falls to about 25 per cent of normal, the blood flow through an isolated leg increases about three-fold; that is, the blood flow increases almost enough, but not quite, to make up for the decreased amount of oxygen in the blood, thus automatically maintaining an almost constant supply of oxygen to the tissues. Cyanide poisoning of local tissue areas can cause a local blood flow increase as much as seven-fold, thus illustrating the extreme effect of oxygen deficiency in increasing blood flow.

There are two basic theories for the regulation of local blood flow when either the rate of tissue metabolism changes or the availability of oxygen changes. These are (1) the *vasodilator theory* and (2) *oxygen demand theory*.

The Vasodilator Theory for Local Blood Flow Regulation.   According to this theory, the greater the rate of metabolism, or the less the blood flow, or the less the availability of oxygen and other nutrients to a tissue, the greater becomes the rate of formation of a *vasodilator substance*. The vasodilator substance then supposedly diffuses back to the precapillary sphincters, metarterioles, and arterioles to cause dilatation. Some of the different vasodilator substances that have been suggested are carbon dioxide, lactic acid, adenosine, adenosine phosphate compounds, histamine, potassium ions, and hydrogen ions.

Some of the vasodilator theories assume that the vasodilator substance is released from the tissue in response to oxygen deficiency. For instance, it has been demonstrated that decreased availability of oxygen can cause both lactic acid and adenosine to be released from the tissues; these are substances that can cause vasodilatation

and therefore could be responsible, or partially responsible, for the local blood flow regulation.

Recently, many physiologists have suggested that the substance *adenosine* is an especially important local vasodilator that might play a major role in regulating local blood flow. For instance, vasodilator quantities of adenosine are released from heart muscle cells whenever coronary blood flow becomes too little, and it is believed that this causes local vasodilatation in the heart and thereby returns the blood flow back toward normal. Also, whenever the heart becomes overly active and the heart's metabolism increases, this too causes excessive utilization of oxygen, decreased oxygen concentration in the local tissues, increased degradation of adenosine triphosphate, and, therefore, increased formation of adenosine, as explained in Chapters 2 and 67. Here again the increased concentration of adenosine is believed to cause coronary vasodilation, with consequent increase in coronary blood flow to supply the nutrient demands of the more active heart.

Though the research evidence is less clear, some physiologists have also suggested that the same adenosine mechanism might control blood flow in skeletal muscle and other tissues of the body, as well as in the heart.

The problem with the different vasodilator theories of local blood flow regulation has been the following: It has been difficult to prove that sufficient quantities of any single vasodilator substance are indeed formed in the tissues to cause all of the measured increase in blood flow in states of increased tissue metabolic demand. On the other hand, perhaps a combination of all the different vasodilators could increase the blood flow sufficiently.

The Oxygen Demand Theory for Local Blood Flow Control.   Though the vasodilator theory is accepted by most physiologists, several critical facts have made a few physiologists favor still another theory, which can be called either the oxygen demand theory or, more accurately, the *nutrient demand theory* (because probably other nutrients besides oxygen are involved). Oxygen (and other nutrients as well) is required to maintain vascular muscle contraction. Therefore, in the absence of an adequate supply of oxygen and other nutrients, it is reasonable to believe that the blood vessels would naturally dilate. Also, increased utilization of oxygen in the tissues as a result of increased metabolism would theoretically decrease the local tissue oxygen availability, and this too would cause local vasodilatation.

A mechanism by which the oxygen demand theory could operate is illustrated in Figure 20–4. This figure shows what might be called a "tissue unit," a single capillary and its surrounding tissue. At the origin of the capillary is a *precapillary*

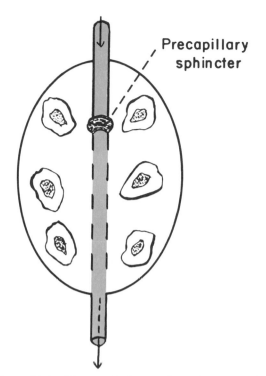

Figure 20–4. Diagram of a tissue unit area for explanation of local feedback control of blood flow.

sphincter. Observing under a microscope a thin tissue such as a bat's wing, one sees that the precapillary sphincters are normally either completely open or completely closed. The number of precapillary sphincters that are open at any given time is approximately proportional to the requirements of the tissue for nutrition. In addition, the precapillary sphincters often open and close cyclically several times per minute, with the duration of the open phases approximately proportional to the metabolic needs of the tissues. The cyclic opening and closing of the sphincter is called *vasomotion*.

Now, let us explain how oxygen concentration in the local tissue could regulate blood flow through the area. Since smooth muscle requires oxygen to remain contracted, we might assume that the strength of the contraction of the precapillary sphincter would increase with an increase in oxygen concentration. Consequently, when the oxygen concentration in the tissue rises above a certain level, the precapillary sphincter presumably closes and remains closed until the tissue cells consume the excess oxygen. When the oxygen concentration falls low enough, the sphincter opens once more to begin the cycle again.

However, in most thick tissues, the precapillary sphincters do not open and close rhythmically; instead, a certain proportion of them remain open while the others remain closed. This is the effect that one would expect when several capillaries are

supplying the same tissue area, because interference between the capillaries would cause just exactly the right number of capillary sphincters to open to supply the required oxygen, and only occasionally would one of the sphincters close or open.

The evidence against the oxygen demand theory is that many types of smooth muscle can remain contracted for long periods of time in the presence of extremely minute concentrations of oxygen — concentrations even below those normally found in the tissues. A possible answer to this is that the smooth muscle in the microvessels might be genetically more sensitive to oxygen lack than are the types of smooth muscle that have thus far been studied. Indeed, in very small, perfused arteries (with internal diameters of approximately 0.5 mm.), marked vasodilation does occur at oxygen concentrations found even normally in the tissues.

Thus, on the basis of presently available data, either a vasodilator theory or an oxygen demand theory could explain local blood flow regulation in response to the metabolic needs of the tissues. Perhaps the truth lies in a combination of the two mechanisms.

Possible Role of Other Nutrients Besides Oxygen in the Control of Local Blood Flow. Under special conditions, it has been shown that long-term lack of glucose can cause local tissue vasodilation. It is possible that this same effect occurs when other nutrients are also deficient. At least one clinical condition of nutritional deficiency leads to marked peripheral vasodilation. This is the vitamin deficiency disease *beriberi,* in which the patient usually has deficiencies of thiamine, niacin, and riboflavin. In this disease the peripheral vascular blood flow increases as much as 2- to 3-fold. Since these vitamins are all concerned with the oxidative phosphorylation mechanism for generating ATP in the local tissues, one could suspect that deficiency of these vitamins leads to diminished smooth muscle contractile ability and therefore leads to the local vasodilation.

"Autoregulation" of Blood Flow When the Arterial Pressure Changes. In dead organs or in tissues in which the local blood flow regulatory mechanisms are nonfunctional, an increase in arterial pressure always causes an increase in blood flow at least as great as the increase in pressure. However, in almost all normally functioning tissues of the body, the arterial pressure can be changed over a very wide range, and the blood flow through the tissues will remain almost normal. This effect is illustrated by the dark curve in Figure 20–5, which shows the effect on blood flow through a muscle of increasing the arterial pressure progressively from a very low value to a very high value over a period of about 20 minutes. Note that between an arterial pressure of approximately 75 mm. Hg and 175 mm. Hg the blood flow remains within ± 10 to 15 per cent of the normal value. The dashed curve of this figure illustrates

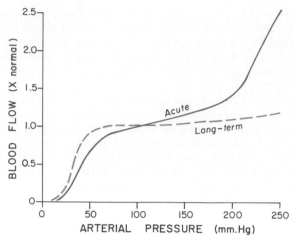

Figure 20–5. Effect on blood flow through a muscle of increasing arterial pressure. The solid curve shows the effect if the arterial presure is raised over a period of a few minutes. The dashed curve shows the effect if the arterial pressure is raised extremely slowly over a period of many weeks.

the long-term effects of arterial pressure on blood flow when the pressure changes slowly over a period of days or weeks, as will be discussed in more detail later in the chapter.

To readily understand how autoregulation of blood flow can occur when the arterial pressure changes, simply apply the basic principles of local blood flow regulation, as already discussed in earlier sections. Thus, when the arterial pressure becomes too great, the excess flow either will provide too many nutrients to the tissues or will flush out all vasodilator substances, either of which will cause the blood vessels to constrict. Therefore, the increased pressure will not increase flow because the simultaneous constriction nullifies the effect of the pressure. On the other hand, whenever the arterial pressure falls too low, the diminished flow of nutrients into the tissues or the ischemic release of vasodilator substances causes vasodilatation, and the blood flow returns almost to normal despite the reduced arterial pressure.

Reactive Hyperemia. When the blood supply to a tissue is blocked for either a short or a long period of time and then is unblocked, the flow through the tissue usually increases to about 5 times normal for a few seconds if the block has lasted a few seconds and sometimes for as long as many hours if the blood flow has been stopped for an hour or more. This phenomenon is called *reactive hyperemia*. Reactive hyperemia is almost certainly another manifestation of the local blood flow regulation mechanism; that is, lack of flow sets into motion all of those factors that cause vasodilatation. Following short periods of vascular occlusion, the extra blood flow during the reactive hyperemia phase lasts long enough to repay almost exactly the tissue oxygen deficit that has accrued during the period of occlusion. This mechanism emphasizes the close connection between local blood flow regulation and delivery of nutrients to the tissues.

Active Hyperemia. When any tissue becomes highly active, such as a muscle during exercise, a gastrointesti-

nal gland during a hypersecretory period, or even the brain during rapid mental activity, the rate of blood flow through the tissue increases. Here again, by simply applying the basic principles of local blood flow control, one can easily understand this so-called *active hyperemia*. The increase in local metabolism causes the cells to devour the tissue fluid nutrients extremely rapidly and also to release large quantities of vasodilator substances. The result obviously would be to dilate the local blood vessels, and therefore, to increase local blood flow. In this way, the active tissue will receive the additional nutrients required to sustain its new level of function.

SPECIAL TYPES OF LOCAL BLOOD FLOW REGULATION

In certain tissues, blood flow is regulated in proportion to other factors besides the requirement for nutrients. This is true particularly in the kidney and in the brain, as follows:

"Autoregulation" in the Kidneys. The rate of blood flow through the kidneys remains constant, even more constant than in most other tissues, despite marked changes in arterial pressure. Yet the availability of oxygen to the kidneys can be changed tremendously without changing renal blood flow significantly. Therefore, the "autoregulation" of renal blood flow when the arterial pressure changes is undoubtedly different from the usual type of local blood flow regulation.

The two factors that change renal blood flow to the greatest extent are sodium concentration in the blood and the concentration of end-products of protein metabolism; an increase in either of these increases renal blood flow. Recent experiments suggest that it is the concentrations of some of these substances in the renal tubules that control renal blood flow. It has been postulated that feedback control of afferent arteriolar constriction occurs at the juxtaglomerular apparatus where the distal tubule comes in contact with the afferent arteriole. The degree of constriction of the afferent arteriole of each nephron supposedly changes in proportion to the concentration of one or more substances in the distal tubule. In Chapter 35 this mechanism will be discussed separately in relation to the physiology of the kidney.

Local Regulation of Cerebral Blood Flow in Response to Tissue CO_2 and Hydrogen Ion Concentration. In the brain, a powerful mechanism for local regulation of blood flow is based on carbon dioxide and hydrogen ion concentrations. When the concentrations of these substances increase, the blood vessels dilate, allowing more rapid blood flow to remove much of the carbon dioxide from the tissues, thereby reducing the CO_2 back toward a normal mean value. This also causes the hydrogen ion concentration to return toward normal because removal of CO_2 also removes carbonic acid, thereby removing hydrogen ions. Conversely, a decrease in CO_2 and hydrogen ions causes vasoconstriction, which allows these substances to accumulate in the tissues until they rise back toward the normal mean values. This is an important regulation because changes in CO_2 and hydrogen ion concentrations drastically change the degree of activity of all neurons.

LONG-TERM LOCAL BLOOD FLOW REGULATION

The local blood flow regulation that has been discussed thus far occurs acutely, within a minute or more after local tissue conditions have changed. For instance, if the arterial pressure is suddenly increased from 100 mm. Hg to 150 mm. Hg, the blood flow increases almost instantaneously about 100 per cent, partly because the increase in pressure directly increases the flow and partly because the high pressure dilates the peripheral vessels. But within one to two minutes the blood flow decreases back to only 15 per cent above the original control value. This illustrates the rapidity of the acute type of local regulation, but the regulation is still very incomplete because there remains a 15 per cent increase in blood flow.

However, over a period of hours, days, and weeks a long-term type of local blood flow regulation develops in addition to the acute regulation, and this long-term regulation gives far more complete regulation than does the acute mechanism. For instance, in the above example, if the arterial pressure remains at 150 mm. Hg indefinitely, within a few weeks the blood flow through the tissues will gradually reapproach almost exactly the normal value. Figure 20–5 illustrates by the dashed curve the extreme effectiveness of this long-term local blood flow regulation. Note that once the long-term regulation has had time to occur changes in arterial pressure between 50 and 250 mm. Hg have very little effect on the rate of local blood flow.

Long-term regulation also occurs when the metabolic demands of a tissue change. Thus, if a tissue becomes chronically overactive and therefore requires chronically increased quantities of nutrients, the blood supply gradually increases to match the needs of the tissue.

Long-Term Local Flow Regulation in Patients with Coarctation of the Aorta. Coarctation of the aorta means occlusion of the aorta somewhere along its course. This occasionally occurs during fetal life, and the person goes through life with an occluded aorta. When such an occlusion occurs in the descending thoracic aorta, blood must flow to the lower part of the body through high-resistance collateral blood vessels in the chest wall. Therefore, the arterial pressure in the upper part of the body is very high — sometimes 1.6 times normal — while the arterial pressure in the lower body is almost exactly normal. Yet, despite this tremendous difference in pressure between the upper and lower body, the blood flow per unit mass of tissue is essentially the same above and below the coarctation. Studies of coarctation have given us most of our quantitative information on the powerful

nature of the long-term local flow regulatory mechanism. Indeed, this long-term mechanism seems to be several times as potent as is the mechanism of acute local flow regulation.

Mechanism of Long-Term Regulation — Change in Tissue Vascularity. The mechanism of long-term local blood flow regulation is almost certainly a change in the degree of vascularity of the tissues. That is, if the arterial pressure falls to 60 mm. Hg and remains at this level for many weeks, the number and sizes of vessels in the tissue increase; if the pressure then rises to a very high level, the number and sizes of vessels decrease. Likewise, if the metabolism in a given tissue becomes elevated for a prolonged period of time, vascularity increases; or if the metabolism is decreased, vascularity decreases.

Thus, there is continual day-by-day reconstruction of the tissue vasculature to meet the needs of the tissues. This reconstruction occurs very rapidly (within days) in extremely young animals. It also occurs rapidly in new growth of tissue, such as in scar tissue and in cancerous tissue; on the other hand, it occurs very slowly in old, well-established tissues. Therefore, the time required for long-term regulation to take place may be only a few days in the neonate or as long as months or even years in the elderly person. Furthermore, the final degree of response is much greater in younger tissues than in older, so that in the neonate the vascularity will adjust to match almost exactly the needs of the tissue for blood flow, whereas in older tissues, vascularity frequently lags far behind the needs of the tissues.

Role of Oxygen in Long-Term Regulation. A probable stimulus for increased or decreased vascularity in many if not most instances is need of the tissue for oxygen. The reason for believing this is that increased vascularity occurs in animals that live at high altitudes where the atmospheric oxygen is low. This effect is also demonstrated dramatically in newborn infants who are put into an oxygen tent for therapeutic purposes. The excess oxygen causes almost immediate cessation of new vascular growth in the retina of the eye and even causes degeneration of some of the capillaries that have already formed. Then when the baby is taken out of the oxygen, there is explosive overgrowth of new vessels to make up for the sudden decrease in available oxygen; indeed, there is so much overgrowth that the vessels grow into the vitreous humor and eventually cause blindness. (This condition is called *retrolental fibroplasia*.)

The mechanisms by which changes in oxygen availability also cause changes in vascularity are not yet clear. Lack of oxygen causes enlargement of the vessels already in the tissue and, under appropriate conditions, growth of new vessels as well. It is possible that the change in vascularity results secondarily to the acute local blood flow regulation mechanism in the following way: In the absence of oxygen, the blood vessels dilate. Because the tension in the vascular wall increases directly in proportion to the diameter of the vessel, the vascular wall elements then become subjected to far greater than normal stretch. This stretch theoretically could cause permanent enlargement of the vessel or outgrowth of new vessels. In support of this concept is the fact that venous occlusion causes more growth of new small vessels than does arterial occlusion; both types of occlusion reduce tissue oxygenation, but venous occlusion especially increases the intraluminal pressure.

DEVELOPMENT OF COLLATERAL CIRCULATION AS A PHENOMENON OF LONG-TERM LOCAL BLOOD FLOW REGULATION

The development of collateral circulation when blood flow to a tissue is blocked is a type of long-term local blood flow regulation. Thus, if the femoral artery becomes occluded, the small vessels that by-pass the femoral occlusion become greatly enlarged, and the leg, especially in younger persons, usually will develop an adequate blood supply. Within seconds after an occlusion, blood flow to the lower leg, which occurs by way of the collateral vessels, falls to about one-eighth the normal flow. But acute dilatation of the collateral vessels occurs within the first 1 to 2 minutes, so that blood flow to the leg returns to approximately one-half normal. Then, over a period of a week or more the blood flow returns almost all the way to normal, indicating progressive opening of the collateral vessels.

Thus, the opening of collateral vessels follows almost identically the same pattern as that observed in other types of acute and long-term local blood flow regulation. Furthermore, the acute opening of the collateral vessels seems to be metabolically initiated, because reduction of blood flow to a tissue by reducing arterial pressure, without causing vascular occlusion, also causes the collateral vessels to open completely; that is, insufficiency of blood flow to the tissues opens the collaterals, even though the main artery is not closed.

The long-term increase in blood flow through the collaterals results from marked enlargement of the collateral vessels. The cross-sectional area of all the collaterals often becomes even greater than the original occluded artery.

SIGNIFICANCE OF LONG-TERM LOCAL REGULATION — THE METABOLIC MASS TO TISSUE VASCULARITY PROPORTIONALITY

From these discussions it should already be apparent to the student that there is a built-in mechanism in most tissues to keep the degree of vascularity of the tissue almost exactly that required to supply the metabolic needs of the tissue.

Thus, one can state as a general rule that the vascularity of most tissues of the body is directly proportional to the local metabolism. If ever this proportionality constant becomes abnormal, the long-term local regulatory mechanism automatically readjusts the degree of vascularity over a period of weeks or months. In young persons this degree of readjustment is usually very exact; in old persons it is only partial.

NERVOUS REGULATION OF THE CIRCULATION

Superimposed onto the intrinsic local tissue regulation of blood flow are two additional types of regulation: (1) *nervous* and (2) *humoral*. These regulations are not necessary for most normal functions of the circulation, but they do provide greatly increased effectiveness of control under special conditions such as exercise or hemorrhage.

There are two very important features of nervous regulation of the circulation: First, nervous regulation can function extremely rapidly, some of the nervous effects beginning to occur within 1 second and reaching full development within 5 to 30 seconds. Second, the nervous system provides a means for controlling large parts of the circulation simultaneously, often in spite of the effect that this has on the blood flow to individual tissues. For instance, when it is important to raise the arterial pressure temporarily, the nervous system can arbitrarily cut off, or at least greatly decrease, blood flow to major segments of the circulation despite the fact that the local blood flow regulatory mechanisms oppose this.

THE AUTONOMIC NERVOUS SYSTEM

The autonomic nervous system will be discussed in detail in Chapter 56. However, it is so important to the regulation of the circulation that its specific anatomical and functional characteristics relating to the circulation deserve special attention here.

By far the most important part of the autonomic nervous system for regulation of the circulation is the *sympathetic nervous system*. The *parasympathetic nervous system* is important only for its regulation of heart function, as we shall see later in the chapter.

The Sympathetic Nervous System. Figure 20–6 illustrates the anatomy of sympathetic nervous control of the circulation. Sympathetic vasomotor nerve fibers leave the spinal cord through all the thoracic and the first one to two lumbar spinal nerves. These pass into the sympathetic chain and thence by two routes to the blood vessels through-

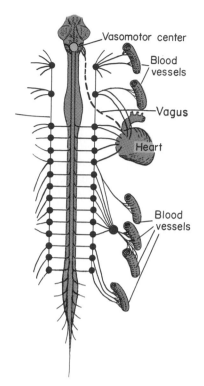

Figure 20–6. The vasomotor center and its control of the circulatory system through the sympathetic and vagus nerves.

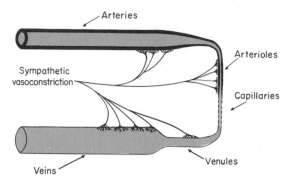

Figure 20–7. Innervation of the systemic circulation.

out the body: (1) through the *peripheral sympathetic nerves* and (2) through the *spinal nerves*. The precise pathways of these fibers in the spinal cord and in the sympathetic chains will be discussed in Chapter 56. It suffices to say here that, with rare exceptions, all vascular areas of the body are supplied with sympathetic nerve fibers.

Distribution of Sympathetic Nerve Fibers to the Peripheral Vasculature. Figure 20–7 illustrates the distribution of sympathetic nerve fibers to the peripheral blood vessels and shows that all the vessels, except the capillaries, sphincters, and most of the metarterioles, are innervated.

The innervation of the small arteries and arterioles allows sympathetic stimulation to increase the *resistance* and thereby to change the rate of blood flow through the tissues. The innervation of large vessels, particularly of the veins, makes it possible for sympathetic stimulation to change the volume of these vessels and thereby to alter the volume of the peripheral circulatory system, which plays a major role in the regulation of cardiovascular function, as we shall see later in this chapter and subsequent chapters.

Sympathetic Nerve Fibers to the Heart. In addition to sympathetic nerve fibers supplying the blood vessels, these fibers also go to the heart. This innervation was discussed in Chapter 13. It will be recalled that sympathetic stimulation mark-

edly increases the activity of the heart, increasing the heart rate and enhancing its strength of pumping.

Parasympathetic Control of Heart Function, Especially Heart Rate. Though the parasympathetic nervous system is exceedingly important for many other autonomic functions of the body, it plays only a minor role in regulation of the circulation. Its only really important effect is its control of heart rate. It also has a slight effect on control of cardiac contractility; however, this effect is far overshadowed by the sympathetic nervous system control of contractility. Parasympathetic nerves pass to the heart in the vagus nerve, as illustrated in Figure 20–6.

The effects of parasympathetic stimulation on heart function were discussed in detail in Chapter 13. Principally, parasympathetic stimulation causes a marked *decrease* in heart rate and a slight decrease in contractility.

The Sympathetic Vasoconstrictor System and Its Control by the Central Nervous System

The sympathetic nerves carry both vasoconstrictor and vasodilator fibers, but by far the most important of these are the *sympathetic vasoconstrictor* fibers. Sympathetic vasoconstrictor fibers are distributed to essentially all segments of the circulation. However, this distribution is greater in some tissues than in others. It is less potent in both skeletal and cardiac muscle and in the brain, while it is powerful in the kidneys, the gut, the spleen, and the skin.

The Vasomotor Center and Its Control of the Vasoconstrictor System — Vasomotor Tone. Located bilaterally in the reticular substance of the lower third of the pons and upper two-thirds of the medulla, as illustrated in Figure 20–8, is an area called the *vasomotor center*. This center transmits impulses downward through the cord and thence through the vasoconstrictor fibers to all the blood vessels of the body.

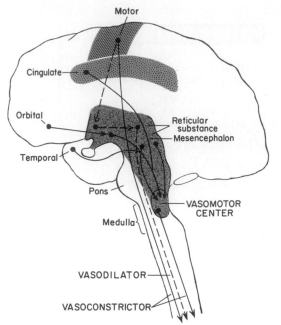

Figure 20–8. Areas of the brain that play important roles in the nervous regulation of the circulation.

the blood vessels, a state called *vasomotor tone*.

The brain stem can be severed above the lower third of the pons without significantly changing the normal activity of the vasomotor center. This center remains tonically active and continues to transmit approximately normal numbers of impulses to the sympathetic vasoconstrictor fibers throughout the body.

Figure 20–9 demonstrates the significance of vasoconstrictor tone. In the experiment of this figure, total spinal anesthesia was administered to an animal, which completely blocked all transmission of nerve impulses from the central nervous system to the periphery. As a result, the arterial pressure fell from 100 to 50 mm. Hg, illustrating the effect of loss of vasoconstrictor tone throughout the body. A few minutes later a small amount of the hormone norepinephrine was injected — norepinephrine is the substance secreted at the endings of sympathetic nerve fibers throughout the body. As this hormone was transported in the blood to all the blood vessels, the vessels once again became constricted, and the arterial pressure rose to a level even greater than normal for a minute or two until the norepinephrine was destroyed.

The Inhibitory Area of the Vasomotor Center. The medial and lower portion of the vasomotor center does not participate in excitation of the vasoconstrictor fibers. Instead, stimulation of this area transmits *inhibitory* impulses into the upper lateral parts of the vasomotor center, thereby *decreasing the degree of sympathetic vasoconstrictor tone* and consequently allowing dilatation of the blood vessels. Thus, the vasomo-

The upper and lateral portions of the vasomotor center are *tonically active*. That is, they have an inherent tendency to transmit nerve impulses all the time, thereby maintaining even normally a slow rate of firing in essentially all vasoconstrictor nerve fibers of the body at a rate of about one-half to two impulses per second. This continual firing is called *sympathetic vasoconstrictor tone*. These impulses maintain a partial state of contraction in

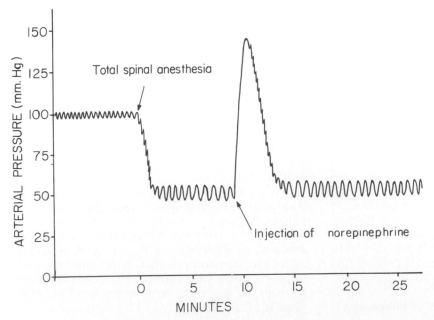

Figure 20–9. Effect of total spinal anesthesia on the arterial pressure, showing a marked fall in pressure resulting from loss of vasomotor tone.

tor center is composed of two parts, a bilateral *excitatory part* that can excite the vasoconstrictor fibers and cause vascular constriction and a medial *inhibitory part* that can inhibit vasoconstriction, thus allowing vasodilatation.

Control of Heart Activity by the Vasomotor Center. At the same time that the vasomotor center is controlling the degree of vascular constriction, it also controls heart activity. The lateral portions of the vasomotor center transmit excitatory impulses through the sympathetic nerve fibers to the heart to increase heart rate and contractility, while the medial portion of the vasomotor center, which lies in immediate apposition to the *dorsal motor nucleus of the vagus nerve,* transmits impulses through the vagus nerve to the heart to decrease heart rate. Therefore, the vasomotor center can either increase or decrease heart activity, this ordinarily increasing at the same time that vasoconstriction occurs throughout the body and ordinarily decreasing at the same time that vasoconstriction is inhibited. However, these interrelationships are not invariable, because some signals that pass down the vagus nerves to the heart can by-pass the vasomotor center.

Control of the Vasomotor Center by Higher Nervous Centers. Large numbers of areas throughout the *reticular substance* of the *pons, mesencephalon,* and *diencephalon* can either excite or inhibit the vasomotor center. This reticular substance is illustrated in Figure 20–8 by the diffuse shaded area. In general, the more lateral and superior portions of the reticular substance cause excitation, while the more medial and inferior portions cause inhibition.

The *hypothalamus* plays a special role in the control of the vasoconstrictor system, for it can exert powerful excitatory or inhibitory effects on the vasomotor center. The *posterolateral portions* of the hypothalamus cause mainly excitation, while the *anterior part* can cause mild excitation or inhibition, depending on the precise part of the anterior hypothalamus stimulated.

Many different parts of the *cerebral cortex* can also excite or inhibit the vasomotor center. Stimulation of the *motor cortex,* for instance, excites the vasomotor center because of impulses transmitted downward into the hypothalamus and thence to the vasomotor center. Also, stimulation of the *anterior temporal lobe,* the *orbital areas of the frontal cortex,* the *anterior part of the cingulate gyrus,* the *amygdala,* the *septum,* and the *hippocampus* can all either excite or inhibit the vasomotor center, depending on the precise portion of these areas that is stimulated and upon the intensity of the stimulus.

Thus, widespread areas of the brain can have profound effects on the vasomotor center and in

turn on the sympathetic vasoconstrictor system of the body, either further enhancing the degree of vasoconstriction or inhibiting the vasoconstrictor tone and thereby causing vasodilatation.

Norepinephrine — The Sympathetic Vasoconstrictor Transmitter Substance. The substance secreted at the endings of the vasoconstrictor nerves is norepinephrine. Norepinephrine acts directly on the smooth muscle of the vessels to cause vasoconstriction, as will be discussed in Chapter 56.

The Adrenal Medullae and Their Relationship to the Sympathetic Vasoconstrictor System. Sympathetic impulses are transmitted to the adrenal medullae at the same time that they are transmitted to all the blood vessels. These impulses cause the medullae to secrete both epinephrine and norepinephrine into the circulating blood, as will be described in Chapter 56. These two hormones are carried in the bloodstream to all parts of the body, where they act directly on the blood vessels usually to cause vasoconstriction, but sometimes the epinephrine causes vasodilatation, as will be discussed later in the chapter.

The Sympathetic Vasodilator System and Its Control by the Central Nervous System

The sympathetic nerves to skeletal muscles carry sympathetic vasodilator fibers as well as constrictor fibers. In lower animals, such as the cat, these fibers release *acetylcholine* at their endings, though in primates the vasodilator substance might be epinephrine. The vasodilator acts on the smooth muscle of the blood vessels to cause vasodilatation in contrast to the vasoconstrictor effect of norepinephrine.

The pathways for central nervous system control of the vasodilator system are illustrated by the dashed lines in Figure 20–8. The principal area of the brain controlling this system is the *anterior hypothalamus,* which transmits impulses to a relay station in the *subcollicular region* of the mesencephalon. From here impulses are transmitted down the cord to the *sympathetic preganglionic neurons* in the lateral horns of the cord. These then transmit vasodilator impulses to the blood vessels of the muscles.

Note also in Figure 20–8 that stimulation of the *motor cortex excites* the *sympathetic vasodilator system* by transmitting impulses downward into the hypothalamus.

Importance of the Sympathetic Vasodilator System. It is doubtful that the sympathetic vasodilator system plays a very important role in the control of the circulation, because complete block of the sympathetic nerves to the muscles hardly affects the ability of the muscles to control their own blood flow in response to their needs. Yet, it is possible, if not probable, that at the onset of exercise the sympathetic vasodilator system causes initial vasodilation in the skeletal muscles to allow an *anticipatory increase in blood flow* even before the muscles require increased nutrients.

"PATTERNS" OF CIRCULATORY RESPONSES ELICITED BY DIFFERENT CENTRAL NERVOUS SYSTEM CENTERS

Stimulation of the Vasomotor Center — The Mass Action Effect. Stimulation of the lateral portions of the vasomotor center causes widespread activation of the vasoconstrictor fibers throughout the body, while stimulation of the medial portions of the vasomotor center causes widespread *inhibition of vasoconstriction*. In many conditions the entire vasomotor center acts as a unit, stimulating all vasoconstrictors throughout the body and also the heart, as well as stimulating the adrenal medullae to secrete epinephrine and norepinephrine that circulate in the blood to excite the circulation still further. The results of this "mass action" are three-fold: First, the peripheral resistance increases in most parts of the circulation, thereby elevating the arterial pressure. Second, the capacity vessels, particularly the veins, are excited at the same time, greatly decreasing their capacity; this forces increased quantities of blood into the heart, thereby increasing the cardiac output. Third, the heart is simultaneously stimulated so that it can handle the increased cardiac output.

Thus, the overall effect of this "mass action" is to prepare the circulation for increased delivery of blood flow to the body.

Stimulation of the Hypothalamus — The "Alarm" Pattern. Under normal conditions, the hypothalamus probably does not transmit large numbers of impulses into the sympathetic vascular control system. However, on occasion the hypothalamus becomes strongly stimulated and can then activate both the vasoconstrictor and vasodilator systems or either of them separately.

Diffuse stimulation of the hypothalamus (or of closely allied areas such as the septum) activates the vasodilator system to the muscles, thereby increasing the blood flow through the muscles; at the same time it causes intense vasoconstriction throughout the remainder of the body and an intense increase in heart activity. The arterial pressure rises, the cardiac output increases, the heart rate increases, and the circulation is ready to supply nutrients to the muscles if there be need. Also, impulses are transmitted simultaneously throughout the central nervous system to cause a state of generalized excitement and attentiveness, these often increasing to such a pitch that the overall pattern of the reaction is that of *alarm*.

Thus, the "alarm" pattern contains all the ingredients of the "mass action" pattern as well as the elements of muscle vasodilation and extreme psychic excitement. This pattern seems to have the purposeful effect of preparing the animal or person to perform on a second's notice whatever activity is required.

The "Motor" Pattern of Circulatory Stimulation. When the motor cortex transmits signals to the skeletal muscle to cause motor activities, it also sends signals to the circulatory system to cause the following effects: First, it excites the alarm system just described, and this generally activates the heart and increases arterial pressure, making the circulatory system ready to supply increased blood flow to the muscles. This effect also causes vasoconstriction in most nonmuscular parts of the body, such as the kidneys, the skin, and the gut, thereby forcing a larger share of the blood flow through the active muscles. Second, impulses pass directly from the motor cortex to the sympathetic neurons of the spinal cord. These enhance the vasoconstriction in the nonmuscular parts of the body, thereby raising the arterial pressure still more and helping to increase muscle blood flow. They possibly also inhibit the vasoconstrictor nerves to the excited muscles.

The motor pattern of circulatory responses can also be elicited by stimulation of several areas of the reticular substance in the brain stem. Stimulation of one of these areas, the *fields of Forel*, elicits a response almost precisely the same as that caused by direct activation of the motor cortex. Therefore, it is possible, if not likely, that this area of the reticular substance plays a major role in control of the circulatory system in exercise.

Emotional Fainting — Vasovagal Syncope. A particularly interesting circulatory reaction occurs in persons who faint because of intense emotional experiences. In this condition, the muscle vasodilator system becomes powerfully activated so that blood flow through the muscles increases several-fold. Intense vagal stimulation of the heart also occurs, causing the heart rate to slow markedly. This overall effect is called *vasovagal syncope*. The arterial pressure falls instantly, which in turn reduces the blood flow to the brain and causes the person to lose consciousness. It is probable, therefore, that emotional fainting results from powerful stimulation of the anterior hypothalamic vasodilator center.

REFLEX REGULATION OF THE CIRCULATION

In addition to the many ways in which signals transmitted from the brain to the blood vessels and heart can alter circulatory function, a system of specific circulatory reflexes also helps to regulate the circulation. Most of these reflexes are concerned with the regulation of arterial pressure or with the regulation of blood flow to specific local areas of the body. Therefore, these will be discussed in detail in the following chapter in relation to arterial pressure regulation, as well as at other points in the text. However, let us summarize here a few of the more important reflex regulations of the circulation.

Arterial Pressure Reflexes. The most important of the arterial pressure reflexes is the *baroreceptor reflex*. An increase in arterial pressure stretches the walls of the major arteries in the chest and neck, and this, in turn, excites stretch receptors, the *baroreceptors*. Signals are transmitted to the vasomotor center of the brain stem, and reflex signals are transmitted back to the heart and blood vessels to slow the heart and to dilate the vessels, thereby reducing the arterial pressure toward normal. Thus, the baroreceptor reflex helps to stabilize the arterial pressure.

Reflexes for Control of Blood Volume. When the blood volume increases, the volume of blood in the central veins and in the right and left atria usually also increases. This stretches the walls of the atria and large veins, again exciting stretch receptors. These stretch receptors also transmit signals into the vasomotor center, and reflex signals are then transmitted to the kidneys to increase urinary output of fluid. Also, signals are transmitted to the hypothalamus to decrease the rate of secretion of antidiuretic hormone, thereby also increasing the output of fluid by the kidneys. Thus, this reflex mechanism helps to control the total amount of fluid in the body, and in this way also helps to control blood volume.

Reflexes for Control of Body Temperature. When the body temperature rises too high, special neurons in the anterior hypothalamus become excited; these in turn send signals through the sympathetic nervous system to dilate the skin blood vessels, thereby allowing transfer of major amounts of internal body heat to the skin. This heat then passes from the skin to the surroundings. As a result, the body temperature falls back toward normal. This reflex is also associated with sweating reflexes and with other aspects of body temperature control, all of which will be discussed in Chapter 72.

HUMORAL REGULATION OF THE CIRCULATION

The term humoral regulation means regulation by substances — such as hormones, ions, and so forth — in the body fluids. Among the most important of these factors are the following:

Vasoconstrictor Agents

Epinephrine and Norepinephrine. Earlier in the chapter it was pointed out that the adrenal medullae secrete both epinephrine and norepinephrine, usually more of the former than the latter. When the sympathetic nervous system throughout the body is stimulated to cause direct effects on the blood vessels, it also causes the adrenal medullae to secrete these two hormones, which then circulate everywhere in the body fluids and act on all vasculature. Norepinephrine has vasoconstrictor effects in almost all vascular beds of the body, and epinephrine has similar effects in most, but not all, beds. For instance, epinephrine often causes mild vasodilatation in both skeletal and cardiac muscle. These two hormones will be discussed in more detail in Chapter 57 in the discussion of the autonomic nervous system.

Angiotensin. Angiotensin is one of the most powerful vasoconstrictor substances known. As little as *one ten-millionth* of a gram can increase the arterial pressure of a human being as much as 10 to 20 mm. Hg under some conditions. Since this substance is very important in relation to arterial pressure regulation, it will be discussed in detail in the following two chapters.

Briefly, either a decrease in arterial pressure or a decrease in quantity of sodium in the body fluids will cause the kidneys to secrete the substance *renin*. The renin in turn acts on one of the plasma proteins, *renin substrate,* to split away the vasoactive peptide angiotensin. The angiotensin in turn has a number of important effects on the circula-

tion related to arterial pressure control: (1) it causes marked constriction of the peripheral arterioles; (2) it causes moderate constriction of the veins, thereby reducing the vascular volume and also probably decreasing vascular compliance; and (3) it causes constriction of the renal arterioles, thereby causing the kidneys to retain both water and salt, thus increasing the body fluid volume, which helps to raise the arterial pressure. Hence, an initial decrease in arterial pressure or decrease in sodium causes a compensatory build-up of body fluid and sodium as well as an increase in arterial pressure, thereby compensating for the original deficit.

Vasopressin. Vasopressin, also called *antidiuretic hormone,* is formed in the hypothalamus (see Chapter 75) but is secreted through the posterior pituitary gland. It is even more powerful than angiotensin as a vasoconstrictor, thus making it perhaps the body's most potent constrictor substance. Therefore, it is clear that vasopressin could have very intense effects on circulatory function. Yet, normally, only very minute amounts of vasopressin are secreted, so that most physiologists have thought that vasopressin plays little role in vascular control. On the other hand, recent experiments have shown that the concentration of circulating vasopressin during severe hemorrhage can rise enough to increase the arterial pressure as much as 40 to 60 mm. Hg, and in many instances this can by itself bring the arterial pressure almost back up to normal.

Also, vasopressin has an *all-important* function to control water reabsorption in the renal tubules, which will be discussed in Chapter 36, and therefore to help control body fluid volume. That is why this hormone is also called antidiuretic hormone.

Vasodilator Agents

Bradykinin. Several substances called *kinins* that can cause vasodilatation have been isolated from blood and tissue fluids. One of these substances is *bradykinin.*

The kinins are small polypeptides that are split away from alpha$_2$-globulins in the plasma or tissue fluids. Different types of proteolytic enzymes can split the kinins from the globulin. An enzyme of particular importance is *kallikrein,* which is present in the blood and tissue fluids in an inactive form. Kallikrein can be activated in several different ways, such as by maceration of the blood, dilution of the blood, contact of the blood with glass, and other similar chemical and physical effects on the blood. As kallikrein becomes activated, it acts immediately on the alpha$_2$-globulin to release bradykinin. Once formed, the bradykinin persists for only a few minutes, because it is digested by the enzyme *carboxypeptidase.* The activated kallikrein enzyme is destroyed by a kallikrein inhibitor also present in the body fluids.

Bradykinin causes very powerful *vasodilatation* and

also *increased capillary permeability*. For instance, injection of 1 *microgram* of bradykinin into the brachial artery of a man increases the blood flow through the arm as much as six-fold, and even smaller amounts injected locally into tissues can cause marked edema because of the increase in capillary pore size.

Though unfortunately we know little about the function of the kinins in the control of the circulation, these extremely powerful effects, coupled with the fact that kinins can develop anywhere in the circulatory system or tissues with ease, indicate that they must play important roles in circulatory regulation. There is reason to believe that kinins play special roles in regulating blood flow, as well as capillary release of fluids in inflamed tissues. It has also been claimed — though it is still not certain — that bradykinin plays a role in regulating skin blood flow and blood flow in gastrointestinal glands.

Serotonin. Serotonin (5-hydroxytryptamine) is present in large concentrations in the chromaffin tissue of the intestine and other abdominal structures. Also, it is present in high concentration in the platelets. Serotonin can have either a vasodilator or vasoconstrictor effect, depending on the condition or the area of the circulation. And, even though these effects can sometimes be powerful, the functions of serotonin in regulation of the circulation are almost entirely unknown. Occasionally, tumors composed of chromaffin tissue develop, called *carcinoid tumors*. These secrete tremendous quantities of serotonin and cause mottled areas of vasodilatation in the skin; but the very fact that these tremendous quantities of serotonin do not drastically disturb the circulation makes it doubtful that serotonin plays a widespread general role in regulation of circulatory function.

Histamine. Histamine is released by essentially every tissue of the body whenever it becomes damaged. Most of the histamine is probably derived from eosinophils and mast cells in the damaged tissues.

Histamine has a powerful vasodilator effect on the arterioles, and like bradykinin, also has the ability to greatly increase capillary porosity, allowing leakage of both fluid and plasma protein into the tissues. Though the role of histamine in normal regulation of the circulation is unknown, in many pathological conditions the intense arteriolar dilatation and increased capillary porosity caused by histamine cause tremendous quantities of fluid to leak out of the circulation into the tissues, inducing edema. The effects of histamine were discussed in Chapter 7 in relation to allergic reactions.

Prostaglandins. Almost every tissue of the body contains small to moderate amounts of several chemically related substances called prostaglandins. These substances are released into the local tissue fluids and into the circulating blood under both physiological and pathological conditions. Though some of the prostaglandins cause vasoconstriction, most of the more important ones seem to be mainly vasodilator agents. Thus far, no specific pattern of function of the prostaglandins in circulatory control has been found. However, their widespread prevalence in the tissues and their myriad effects on the circulation make them ideal candidates for special roles in circulatory control, especially for control in local vascular areas. For this reason these substances are presently under intensive research inves-

tigation, though the unequivocal results thus far are not enough to justify further discussion here.

Effects of Chemical Factors on Vascular Constriction

Many different chemical factors can either dilate or constrict local blood vessels. Though the roles of these substances in the overall *regulation* of the circulation generally are not known, their specific effects can be listed as follows:

An increase in *calcium ion* concentration causes vasoconstriction. This results from the general effect of calcium to stimulate smooth muscle contraction, as discussed in Chapter 12.

An increase in *potassium ion* concentration causes vasodilatation. This results from the ability of potassium ions to inhibit smooth muscle contraction.

An increase in *magnesium ion* concentration causes powerful vasodilatation, for magnesium ions inhibit smooth muscle generally.

Increased *sodium ion* concentration causes arteriolar dilatation. This results from an increase in osmolality of the fluids rather than from a specific effect of sodium ion itself. *Increased osmolality* of the blood caused by increased quantities of *glucose* or other nonvasoactive substances also causes arteriolar dilatation. Decreased osmolality causes arteriolar constriction.

The only anions to have significant effects on blood vessels are *acetate* and *citrate,* both of which cause mild degrees of vasodilatation.

An *increase in hydrogen ion* concentration (decrease in pH) causes dilatation of the arterioles. A slight *decrease in hydrogen ion* concentration causes arteriolar constriction, but an intense decrease causes dilatation, which is the same effect as that which occurs with increased hydrogen ion concentration.

An increase in carbon dioxide concentration causes moderate vasodilatation in most tissues and marked vasodilatation in the brain. However, carbon dioxide, acting on the vasomotor center, has an extremely powerful indirect vasoconstrictor effect that is transmitted through the sympathetic vasoconstrictor system.

REFERENCES

Altura, B. M.: Humoral, hormonal, and myogenic mechanisms. *In* Kaley, G., and Altura, B. M. (eds.): Microcirculation. Vol. II. Baltimore, University Park Press, 1977.

Baez, S.: Microcirculation. *Annu. Rev. Physiol., 39*:391, 1977.

Bevan, J. A., et al.: Adrenergic regulation of vascular smooth muscle. *In* Bohr, D. F., et al. (eds.): Handbook of Physiology. Sec. 2. Vol. II. Baltimore, Williams & Wilkins, 1980, p. 515.

Bevan, J. A., et al. (eds.): Vascular Neuroeffector Mechanisms. New York, Raven Press, 1980.

Bhoola, K. D., and Erdös, E. G. (eds.): Bradykinin, Kallidin, and Kallikrein. New York, Springer-Verlag, 1979.

Bohr, D. F., et al.: Mechanisms of action of vasoactive agents. *In* Kaley, G., and Altura, B. M. (eds.): Microcirculation. Vol. II. Baltimore, University Park Press, 1977.

Borchard, F.: The Adrenergic Nerves of the Normal and the Hypertrophied Heart. Stuttgart, Thieme, 1978.

Brody, M. J., and Abboud, F. M.: Tissue perfusion. *In* Frohlich, E. D. (ed.): Pathophysiology, 2nd Ed. Philadelphia, J. B. Lippincott Co., 1976, p. 29.

Burnstock, G.: Cholinergic and purinergic regulation of blood vessels. *In* Bohr, D. F., et al. (eds.): Handbook of Physiology. Sec. 2, Vol. II. Baltimore, Williams & Wilkins, 1980, p. 567.

Cowley, A. W., Jr., and Guyton, A. C.: Quantification of intermediate steps in the renin-angiotensin-vasoconstrictor feedback loop in the dog. *Circ. Res., 30*:557, 1972.

Donald, D. E., and Ferguson, D. A.: Study of the sympathetic vasoconstrictor nerves to the vessels of the dog hind limb. *Circ. Res., 26*:171, 1970.

Duling, B.: Oxygen, metabolism, and microcirculatory regulation. *In* Kaley, G., and Altura, B. M. (eds.): Microcirculation. Vol. II. Baltimore, University Park Press, 1977.

Eriksson, E., and Zarem, H. A.: Growth and differentiation of blood vessels. *In* Kaley, G., and Altura, B. M. (eds.): Microcirculation. Vol. 1. Baltimore, University Park Press, 1977, p. 393.

Folkman, J.: The vascularization of tumors. *Sci. Am., 234*(5):58, 1976.

Frölich, J. C. (ed.): Methods in Prostaglandin Research. New York, Raven Press, 1978.

Fujii, S., *et al.* (eds.): Kinins II. New York, Plenum Press, 1979.

Galli, C., *et al.* (eds.): Phospholipases and Prostaglandins. New York, Raven Press, 1978.

Granger, H. J., and Guyton, A. C.: Autoregulation of the total systemic circulation following destruction of the central nervous system in the dog. *Circ. Res., 25*:379, 1969.

Guyton, A. C.: Integrative hemodynamics. *In* Sodeman, W. A., Jr., and Sodeman, T. M. (eds.): Pathologic Physiology: Mechanisms of Disease, 6th Ed. Philadelphia, W. B. Saunders Co., 1979, p. 169.

Guyton, A. C.: Arterial Pressure and Hypertension. Philadelphia, W. B. Saunders Co., 1980.

Guyton, A. C., *et al.*: Circulation: Overall regulation. *Annu. Rev. Physiol., 34*:13, 1972.

Guyton, A. C., *et al.*: Cardiac Output and Its Regulation. Philadelphia, W. B. Saunders Co., 1973.

Haddy, F. J.: Local effects of sodium, calcium, and magnesium upon small and large blood vessels of the dog forelimb. *Circ. Res., 8*:57, 1960.

Horn, L.: Electrical properties of blood vessels. *In* Kaley, G., and Altura, B. M. (eds.): Microcirculation. Vol. II. Baltimore, University Park Press, 1977.

Johnson, P. C.: The microcirculation and local and humoral control of the circulation. *In* MTP International Review of Science: Physiology. Baltimore, University Park Press, 1974, Vol. 1, p. 163.

Johnson, P. C.: The myogenic response. *In* Bohr, D. F., *et al.* (eds.): Handbook of Physiology. Sec. 2. Vol. II. Baltimore, Williams & Wilkins, 1980, p. 409.

Kahlson, G., and Rosengren, E.: Biogenesis and Physiology of Histamine. Baltimore, Williams & Wilkins, 1971.

Keatinge, W. R., and Harman, M. C.: Local Mechanisms Controlling Blood Vessels. New York, Academic Press, 1979.

Korner, P. I.: Circulatory adaptations in hypoxia. *Physiol. Rev., 39*:687, 1959.

Kramer, G. L., and Hardman, J. G.: Cyclic nucleotides and blood vessel contraction. *In* Bohr, D. F., *et al.* (eds.): Handbook of Physiology. Sec. 2, Vol. II. Baltimore, Williams & Wilkins, 1980, p. 179.

Lands, W. E. M.: The biosynthesis and metabolism of prostaglandins. *Annu. Rev. Physiol., 41*:633, 1979.

Landsberg, L., and Young, J. B.: Catecholamines and the adrenal medulla. *In* Bondy, P. K., and Rosenberg, L. E. (eds.): Metabolic Control in Disease, 8th Ed. Philadelphia, W. B. Saunders Co., 1980, p. 1621.

Lee, J. B.: Cardiovascular-renal effects of prostaglandins: The antihypertensive, natriuretic renal "endocrine" function. *Arch. Intern. Med., 133*:56, 1974.

Levine, R. J.: Serotonin and the carcinoid syndrome; Histamine and mastocytosis. *In* Bondy, P. K., and Rosenberg, L. E. (eds.): Duncan's Diseases of Metabolism, 7th Ed. Philadelphia, W. B. Saunders Co., 1974, p. 1651.

Mellander, S.: Systemic circulation. *Annu. Rev. Physiol., 32*:313, 1970.

Nakano, J.: Cardiovascular responses to neurohypophysial hormones. *In* Greep, R. O., and Astwood, E. B. (eds.): Handbook of Physiology. Sec. 7, Vol IV. Baltimore, Williams & Wilkins, 1974, p. 395.

Needleman, P., and Isakson, P. C.: Intrinsic prostaglandin biosynthesis in blood vessels. *In* Bohr, D. F., *et al.* (eds.): Handbook of Physiology. Sec. 2, Vol. II. Baltimore, Williams & Wilkins, 1980, p. 613.

Ramwell, P. W., and Leovey, E. M. K.: Prostaglandins and humoral regulation. *In* DeGroot, L. J., *et al.* (eds.): Endocrinology. Vol. 3. New York, Grune & Stratton, 1979, p. 1711.

Renkin, E. M.: Nutritive and shunt flow. *In* Kaley, G., and Altura, B. M. (eds.): Microcirculation. Vol. II. Baltimore, University Park Press, 1977.

Rosenthal, S. L., and Guyton, A. C.: Hemodynamics of collateral vasodilatation following femoral artery occlusion in anesthetized dogs. *Circ. Res., 23*:239, 1968.

Schacter, M., and Barton, S.: Kallikreins (kininogenases) and kinins. *In* DeGroot, L. J., *et al.* (eds.): Endocrinology. Vol. 3. New York, Grune & Stratton, 1979, p. 1699.

Scott, J. B., *et al.*: Role of osmolarity, K^+, H^+, Mg^{++}, and O_2 in local blood flow regulation. *Am. J. Physiol., 218*:338, 1970.

Scriabine, A., *et al.*: Prostaglandins in Cardiovascular and Renal Function. Jamaica, N.Y., Spectrum Publications, 1980.

Share, L.: Blood pressure, blood volume, and the release of vasopressin. *In* Greep, R. O., and Astwood, E. B. (eds.): Handbook of Physiology. Sec. 7, Part I, Vol. IV. Baltimore, Williams & Wilkins, 1974, p. 243.

Shepherd, A. P., *et al.*: Local control of tissue oxygen delivery and its contribution to the regulation of cardiac output. *Am. J. Physiol., 225*:747, 1973.

Silverman, W. A.: The lesson of retrolental fibroplasia. *Sci. Am., 236*(6):100, 1977.

Sparks, H. V., Jr.: Effect of local metabolic factors on vascular smooth muscle. *In* Bohr, D. F., *et al.* (eds.): Handbook of Physiology. Sec. 2, Vol. II. Baltimore, Williams & Wilkins, 1980, p. 475.

Sparks, H. V., Jr., and Belloni, F. L.: The peripheral circulation: Local regulation. *Annu. Rev. Physiol., 40*:67, 1978.

Stainsby, W. N.: Local control of regional blood flow. *Annu. Rev. Physiol., 35*:151, 1973.

Tsakiris, A. G., *et al.*: Cardiovascular responses to hypertension and hypotension in dogs with denervated hearts. *J. Appl. Physiol., 27*:817, 1969.

Vane, J. R., and Berstrom, S. (eds.): Prostacyclin. New York, Raven Press, 1979.

Vanhoutte, P. M.: Physical factors of regulation. *In* Bohr, D. F., *et al.* (eds.): Handbook of Physiology. Sec. 2, Vol. II. Baltimore, Williams & Wilkins, 1980, p. 443.

Walker, J. R., and Guyton, A. C.: Influence of blood oxygen saturation on pressure-flow curve of dog hindleg. *Am. J. Physiol., 212*:506, 1967.

Whalen, W. J.: Intracellular PO_2 in heart and skeletal muscle. *Physiologist, 14*:69, 1971.

Wolff, J. R.: Ultrastructure of the terminal vascular bed as related to function. *In* Kaley, G., and Altura, B. M. (ed.): Microcirculation. Vol. 1. Baltimore, University Park Press, 1977, p. 95.

Wolthuis, R. A., *et al.*: Physiological effects of locally applied reduced pressure in man. *Physiol. Rev., 54*:566, 1974.

21

Short-Term Regulation of Mean Arterial Pressure: Nervous Reflex and Hormonal Mechanisms for Rapid Pressure Control

In the previous chapter we pointed out that each tissue can control its own blood flow by simply dilating or constricting its local arterioles. For this mechanism to work, it is necessary that the arterial pressure remain constant or nearly constant because with a variable arterial pressure one would never know whether dilating the blood vessels would necessarily increase the local blood flow. Fortunately, the circulation has an intricate system for regulation of the arterial pressure. It maintains the normal mean arterial pressure in the young adult within rather narrow limits between 90 mm. Hg and 110 mm. Hg. Some of the pressure regulatory mechanisms (mainly nervous and hormonal mechanisms) act very rapidly, and some (mainly mechanisms related to kidney function and blood volume regulation) act very slowly. In the present chapter we will discuss the rapid nervous and hormonal pressure control mechanisms. In the following chapter we will discuss both the long-term regulation of arterial pressure, based primarily on renal and body-fluid mechanisms, and the clinical problem of hypertension or "high blood pressure," which is caused by abnormalities of the long-term pressure regulatory mechanisms.

But first, let us discuss some of the normal arterial pressure values in a human being and the clinical method for measuring these pressures.

NORMAL ARTERIAL PRESSURES

Arterial Pressures at Different Ages. Figure 21–1 illustrates the typical diastolic, systolic, and mean arterial pressures from birth to 80 years of age. From this figure it can be seen that the systolic pressure of a

normal young adult averages about 120 mm. Hg and the diastolic pressure about 80 mm. Hg — that is, arterial pressure is said to be 120/80. The shaded areas on either side of the curves depict the normal ranges of systolic and diastolic pressures, showing considerable variation from person to person.

The increase in arterial pressure at older ages is usually associated with developing arteriosclerosis. In this disease the systolic pressure especially increases; in approximately one-tenth of all old people it eventually rises above 200 mm. Hg.

THE MEAN ARTERIAL PRESSURE

The mean arterial pressure is the average pressure throughout each cycle of the heartbeat. Offhand one might expect that it would be equal to the average of systolic and diastolic pressures but this is not true; the arterial pressure usually remains nearer to diastolic level than to systolic level during a greater portion of the

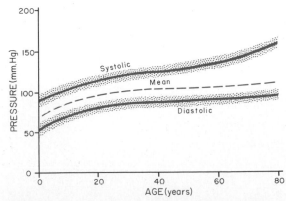

Figure 21–1. Changes in systolic, diastolic, and mean arterial pressures with age. The shaded areas show the normal range.

pulse cycle, which can be seen in all of the pictures of pressure pulses shown in Chapter 19. Therefore, the mean arterial pressure is usually slightly less than the average of systolic and diastolic pressures, as is evident in Figure 21–1.

Mean Arterial Pressure in the Human Being. The mean arterial pressure of the normal young adult averages about 96 mm. Hg, which is slightly less than the average of the systolic and diastolic pressures, 120 and 80 mm. Hg, respectively. However, for purposes of discussion, the mean arterial pressure is usually considered to be 100 mm. Hg because this value is easy to remember.

The mean arterial pressure, like the systolic and diastolic pressures, is lowest immediately after birth, measuring about 70 mm. Hg at birth and reaching an average of about 110 mm. Hg in the normal old person or as high as 130 mm. Hg in the person with arteriosclerosis. From adolescence to middle age the mean arterial pressure does not vary greatly from the normal value of 100 mm. Hg.

Significance of Mean Arterial Pressure. Mean arterial pressure is the average pressure tending to push blood through the systemic circulatory system. Therefore, *from the point of view of tissue blood flow, it is generally the mean arterial pressure that is important.*

CLINICAL METHODS FOR MEASURING SYSTOLIC AND DIASTOLIC PRESSURES

Obviously, it is impossible to use the various pressure recorders that require needle insertion into an artery as described in Chapter 18 for making routine pressure measurements in human patients, although they are used on occasion when special studies are necessary. Instead, the clinician determines systolic and diastolic pressures by indirect means, most usually by the auscultatory method.

The Auscultatory Method. Figure 21–2 illustrates the auscultatory method for determining systolic and diastolic arterial pressures. A stethoscope is placed over the

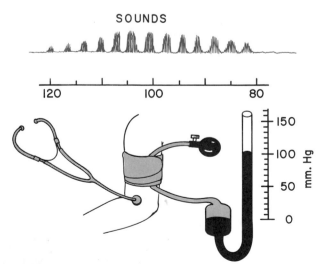

Figure 21–2. The auscultatory method for measuring systolic and diastolic pressures.

antecubital artery while a blood pressure cuff is inflated around the upper arm. As long as the cuff presses against the arm with so little pressure that the artery remains distended with blood, no sounds whatsoever are heard by the stethoscope despite the fact that the blood within the artery is pulsating. But when the cuff pressure is great enough to close the artery during part of the arterial pressure cycle, a sound is heard in the stethoscope with each pulsation. These sounds are called *Korotkoff sounds.*

The exact cause of Korotkoff sounds is still debated, but they are believed to be caused by blood jetting through the partly occluded vessel. The jet causes turbulence in the open vessel beyond the cuff, and this sets up the vibrations heard through the stethoscope.

In determining blood pressure by the auscultatory method, the pressure in the cuff is first elevated well above arterial systolic pressure. As long as this pressure is higher than systolic pressure, the brachial artery remains collapsed and no blood whatsoever flows into the lower artery during any part of the pressure cycle. Therefore, no Korotkoff sounds are heard in the lower artery. But then the cuff pressure is gradually reduced. Just as soon as the pressure in the cuff falls below systolic pressure, blood slips through the artery beneath the cuff during the peak of systolic pressure, and one begins to hear *tapping* sounds in the antecubital artery in synchrony with the heart beat. As soon as these sounds are heard, the pressure level indicated by the manometer connected to the cuff is approximately equal to the systolic pressure.

As the pressure in the cuff is lowered still more, the Korotkoff sounds change in quality, having less of the tapping quality but more of a rhythmic, harsh quality. Then, finally, when the pressure in the cuff falls to equal diastolic pressure, the artery no longer closes during diastole, which means that the basic factor causing the sounds (the jetting of blood through a squeezed artery) is no longer present. Therefore, the sounds suddenly change to a muffled quality, and they usually disappear entirely after another 5 to 10 mm. drop in cuff pressure. One notes the manometer pressure when the Korotkoff sounds change to the muffled quality, and this pressure is approximately equal to the diastolic pressure.

The auscultatory method for determining systolic and diastolic pressures is not entirely accurate, but it usually gives values within 10 per cent of those determined by direct measurement from the arteries.

Radial Pulse and Oscillometric Methods for Estimating Arterial Pressure. Arterial blood pressure can also be estimated by feeling the radial pulse or by recording the pulsation in the lower arm with an *oscillometer* while a cuff is inflated over the upper arm. An oscillometer consists of a recording apparatus that can register pulsations in a lightly inflated blood pressure cuff around the forearm. A blood pressure recorder of the aneroid type makes an excellent oscillometer for estimating arterial pressure.

In applying these methods, the pressure in the cuff on the upper arm is raised well above the systolic level and then it is progressively reduced. No pulsation will be felt in the radial artery or registered by the oscillometer until the cuff pressure falls below the systolic pressure level. But, just as soon as this point is reached, a distinct radial pulse can be felt, or a distinct oscillation will be recorded by the oscillometer. Further decrease in the upper arm cuff pressure results, at first, in progressive increase in intensity of radial or oscillometer pulsation. Then there is a slight decrease in these pulsations at approximately the diastolic level of pressure, but this change is so indistinct that estimation of diastolic pressure is likely to be in severe error.

The oscillometric method is valuable for measuring pressures when distinct sounds cannot be heard from the forearm arteries. This often results in (1) very young children or (2) adults when the arteries are in a state of spasm, such as when a patient is in *shock*. These instances are often among the most important in which it is essential to have at least some estimation of arterial pressure, thus illustrating the value of knowing this accessory method for measuring systolic arterial pressure in addition to the auscultatory method.

RELATIONSHIP OF ARTERIAL PRESSURE TO CARDIAC OUTPUT AND TOTAL PERIPHERAL RESISTANCE

Before discussing the overall regulation of arterial pressure, it is good to remember the basic relationship between arterial pressure, cardiac output, and total peripheral resistance, which was discussed in detail in Chapter 18, as follows:

Arterial Pressure = Cardiac Output
                      × Total Peripheral Resistance

It is obvious from this formula that any condition that increases either the cardiac output or total peripheral resistance (if the other factor does not change) will cause an increase in mean arterial pressure. Both of these factors are often manipulated in the control of arterial pressure, as we shall see in the remainder of this chapter.

THE OVERALL SYSTEM FOR ARTERIAL PRESSURE REGULATION

Arterial pressure is not regulated by a single pressure controlling system but instead by several interrelated systems that perform specific functions. When a person bleeds severely so that the pressure falls suddenly, two problems immediately confront the pressure control system. The first is to return the arterial pressure immediately to a high enough level that the person can live through the acute hemorrhagic episode. The second is to return the blood volume eventually to its normal level so that the circulatory system can re-establish full normality, including return of the arterial pressure all the way back to its normal

value. These two problems characterize two major types of arterial pressure control systems in the body: (1) a system of rapidly acting pressure control mechanisms, and (2) a system for long-term control of the basic arterial pressure level.

To illustrate the multiplicity of pressure control mechanisms, let us look for a moment at Figure 21–3. This figure illustrates the reaction (expressed as feedback gain) of eight different arterial pressure control systems, showing also the times required for them to begin to respond and their durations of response following an acute abnormal change in the arterial pressure. These control mechanisms can be divided into those that react very rapidly and those that subserve long-term arterial pressure regulation.

Rapidly Acting Pressure Control Mechanisms. Note in Figure 21–3 that three different pressure control mechanisms begin to react within seconds. All are nervous pressure control mechanisms: the baroreceptor feedback mechanism, the central nervous system ischemic mechanism, and the chemoreceptor mechanism. Thus, the first line of defense against abnormal pressures is subserved by the nervous mechanisms for control of arterial pressure.

Within minutes several other pressure control mechanisms also come into play. Three of these that are shown in Figure 21–3 are the renin-angiotensin-vasoconstrictor mechanism, stress relaxation changes in the vasculature, and shift of fluid through the capillaries from the tissues into or out of the circulation to readjust the blood volume as needed. These three mechanisms become fully active within 30 minutes to several

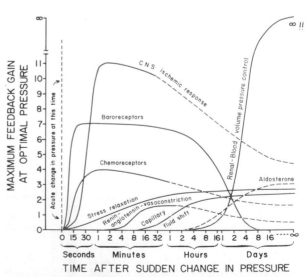

Figure 21–3. Potency of various arterial pressure control mechanisms at different time intervals after the onset of a disturbance to the arterial pressure. Note especially the infinite gain of the renal–body fluid pressure control mechanism that occurs after a few days' time.

hours, in contrast to the nervous mechanisms that usually become fully active within a minute or so.

Long-Term Mechanisms for Arterial Pressure Regulation. The nervous regulators of arterial pressure, though acting very rapidly and powerfully to correct acute abnormalities of arterial pressure, generally lose their power to control arterial pressure after a few hours to a few days because the nervous pressure receptors "adapt"; that is, they lose their responsiveness. Therefore, except under unusual circumstances, the nervous mechanisms for arterial pressure control do not play a major role in long-term regulation of arterial pressure. Long-term regulation, instead, is vested mainly in a renal–body fluid–pressure control mechanism which is illustrated by the curve farthest to the right in Figure 21–3. As we shall discuss in the following chapter, this mechanism involves control of blood volume with its consequent effects on arterial pressure, and part of this mechanism involves control of kidney function by several different hormonal systems, including especially the renin-angiotensin system and the hormone aldosterone secreted by the adrenal cortex.

Basically, the renal–body fluid–pressure control mechanism works as follows: When the arterial pressure falls, this fall in pressure itself causes the kidneys to retain fluid and the blood volume, therefore, to rise. This, in turn, eventually will raise the blood pressure back to normal. The special value of this mechanism is that it will not stop increasing the blood volume until the pressure rises all the way back to normal. This characteristic of the mechanism makes it an extremely powerful one, as will be described in the following chapter.

RAPIDLY ACTING NERVOUS MECHANISMS FOR ARTERIAL PRESSURE CONTROL

THE ARTERIAL BARORECEPTOR CONTROL SYSTEM — BARORECEPTOR REFLEXES

By far the best known of the mechanisms for arterial pressure control is the *baroreceptor reflex.* Basically, this reflex is initiated by stretch receptors, called either *baroreceptors* or *pressoreceptors,* located in the walls of the large systemic arteries. A rise in pressure stretches the baroreceptors and causes them to transmit signals into the central nervous system, and other signals are in turn sent to the circulation to reduce arterial pressure back toward the normal level.

Physiologic Anatomy of the Barorecep-

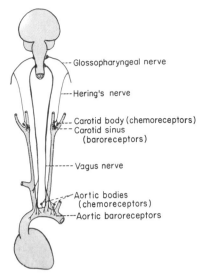

Figure 21–4. The baroreceptor system.

tors. Baroreceptors are spray-type nerve endings lying in the walls of the arteries; they are stimulated when stretched. A few baroreceptors are located in the wall of almost every large artery of the thoracic and neck regions; but, as illustrated in Figure 21–4, baroreceptors are extremely abundant, in (1) the walls of the internal carotid arteries slightly above the carotid bifurcations, areas known as the *carotid sinuses,* and (2) the walls of the aortic arch.

Figure 21–4 also shows that impulses are transmitted from each carotid sinus through the very small *Hering's nerve* to the glossopharyngeal nerve and thence to the medullary area of the brain stem. Impulses from the arch of the aorta are transmitted through the vagus nerves also to the medulla. Hering's nerve is especially important in physiologic experiments because baroreceptor impulses can be recorded from it with ease.

Response of the Baroreceptors to Pressure. Figure 21–5 illustrates the effect of different arterial pressures on the rate of impulse transmission in a Hering's nerve. Note that the carotid sinus baroreceptors are not stimulated at all by pressures between 0 and 60 mm. Hg, but above 60 mm. Hg they respond progressively more and more rapidly and reach a maximum at about 180 mm. Hg. The responses of the aortic baroreceptors are similar to those of the carotid receptors, except that they respond, in general, at pressure levels about 30 mm. Hg higher.

Note especially that the increase in number of impulses for each unit change in arterial pressure, expressed as $\Delta I/\Delta P$ in the figure, is greatest at a pressure level near the normal mean arterial pressure. This means that the increase in baroreceptor control of arterial pressure occurs just at the level where the response needs to be most marked.

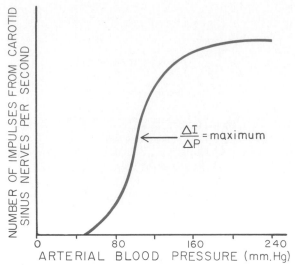

Figure 21–5. Response of the baroreceptors at different levels of arterial pressure.

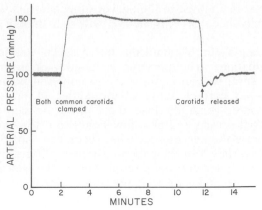

Figure 21–6. Typical carotid sinus reflex effect on arterial pressure caused by clamping both common carotids.

That is, in the normal operating range of arterial pressure, even a slight change in pressure causes strong sympathetic reflexes to readjust the arterial pressure back toward normal.

The baroreceptors respond extremely rapidly to changes in arterial pressure; the number of impulses even increases during systole and decreases again during diastole. Furthermore, the baroreceptors *respond much more to a rapidly changing pressure* than to a stationary pressure. That is, if the mean arterial pressure is 150 mm. Hg but at that moment is rising rapidly, the rate of impulse transmission may be as much as twice that when the pressure is stationary at 150 mm. Hg. On the other hand, if the pressure is falling, the rate might be as little as one-quarter that for a stationary pressure.

The Reflex Initiated by the Baroreceptors. The baroreceptor impulses *inhibit the vasoconstrictor center* of the medulla and *excite* the vagal center. The net effects are (1) *vasodilatation* throughout the peripheral circulatory system and (2) *decreased heart rate* and *strength of contraction*. Therefore, excitation of the baroreceptors by pressure in the arteries reflexly *causes the arterial pressure to decrease*. Conversely, low pressure has opposite effects, reflexly causing the pressure to rise back toward normal.

Figure 21–6 illustrates a typical reflex change in arterial pressure caused by clamping the common carotids. This procedure reduces the carotid sinus pressure; as a result, the baroreceptors become inactive and lose their inhibitory effect on the vasomotor center. The vasomotor center then becomes much more active than usual, causing the arterial pressure to rise and to remain elevated during the ten minutes that the carotids are clamped. Removal of the clamps allows the pressure to fall immediately to slightly below normal as a momentary overcompensation and then to return to normal in another minute or so.

Function of the Baroreceptors During Changes in Body Posture. The ability of the baroreceptors to maintain relatively constant arterial pressure is extremely important when a person sits or stands after having been lying down. Immediately upon

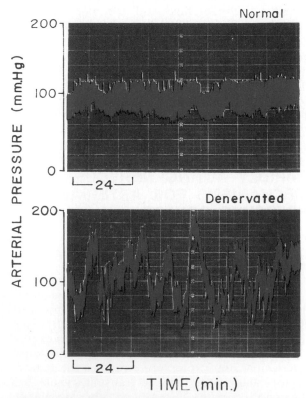

Figure 21–7. Two hour records of arterial pressure in a normal dog (above) and in the same dog (below) several weeks after the baroreceptors had been denervated. (Courtesy of Dr. Allen W. Cowley, Jr.)

standing, the arterial pressure in the head and upper part of the body obviously tends to fall, and marked reduction of this pressure can cause loss of consciousness. Fortunately, however, the falling pressure at the baroreceptors elicits an immediate reflex, resulting in strong sympathetic discharge throughout the body, and this minimizes the decrease in pressure in the head and upper body.

The "Buffer" Function of the Baroreceptor Control System. Because the baroreceptor system opposes increases and decreases in arterial pressure, it is often called a *pressure buffer system,* and the nerves from the baroreceptors are called *buffer nerves.*

Figure 21–7 illustrates the importance of this buffer function of the baroreceptors. The upper record in this figure shows an arterial pressure recording for 2 hours from a normal dog and the lower record from a dog whose baroreceptor nerves from both the carotid sinuses and the aorta had previously been removed. Note the extreme variability of pressure in the denervated dog caused by simple events of the day such as lying down, standing, excitement, eating, defecation, noises, and so forth.

Figure 21–8 illustrates the frequency distributions of the arterial pressures recorded for a full 24-hour day in both the normal dog and the denervated dog. Note that when the baroreceptors were intact, the arterial pressure remained throughout the day within the narrow range of from 85 mm. Hg to 115 mm. Hg — most of the day at almost exactly 100 mm. Hg. On the other hand, after removal of the baroreceptors the pressure range was increased 2.5-fold, frequently falling to

as low as 50 mm. Hg or rising to over 160 mm. Hg. Thus, one can see the extreme variability of pressure in the absence of the arterial baroreceptor system.

In summary, we can state that the primary purpose of the arterial baroreceptor system is to reduce the daily variation in arterial pressure to about one-half to one-third that which would occur were the baroreceptor system not present. But note, as shown in both Figures 21–7 and 21–8, that the average arterial pressure over any prolonged period of time is almost exactly the same whether the baroreceptors are present or not, illustrating the unimportance of the baroreceptor system for long-term regulation of arterial pressure even though it is extremely potent in preventing the rapid changes of arterial pressure that occur moment-by-moment or hour-by-hour.

Unimportance of the Baroreceptor System for Long-Term Regulation of Arterial Pressure — Adaptation of the Baroreceptors. The baroreceptor control system is probably of no importance whatsoever in long-term regulation of arterial pressure for a very simple reason: The baroreceptors themselves adapt in one to three days to whatever pressure level they are exposed to. That is, if the pressure rises from the normal value of 100 mm. Hg up to 200 mm. Hg, extreme numbers of baroreceptor impulses are at first transmitted. During the next few seconds, the rate of firing diminishes considerably; then it diminishes very slowly during the next one to two days, at the end of which time the rate will have returned essentially to the normal level despite the fact that the arterial pressure remains 200 mm. Hg. Conversely, when the arterial pressure falls to a very low value, the baroreceptors at first transmit no impulses at all, but gradually over a period of several days the rate of baroreceptor firing returns again to the original control level.

This adaptation of the baroreceptors obviously prevents the baroreceptors reflex from functioning as a control system to buffer arterial pressure changes that last longer than a few days at a time. Therefore, prolonged regulation of arterial pressure requires other control systems, principally the renal–body fluid–pressure control system (along with its associated hormonal mechanisms) to be discussed in the following chapter.

The Carotid Sinus Syndrome. Strong pressure on the neck over the bifurcations of the carotids in the human being can excite the baroreceptors of the carotid sinuses, causing the arterial pressure to fall as much as 20 mm. Hg in the normal person. In some older persons, particularly after calcified arteriosclerotic plaques have developed in the carotid arteries, pressure on the carotid sinuses often causes such strong responses that the heart stops completely, or at least the arterial pressure falls drastically. Even tight collars can cause the arterial pressure to fall low enough to cause fainting in these

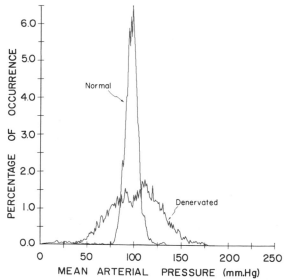

Figure 21–8. Frequency distribution curves of the arterial pressure for a 24-hour period in a normal dog and in the same dog several weeks after the baroreceptors had been denervated. (Courtesy of Dr. Allen W. Cowley, Jr.)

persons, an effect called *carotid sinus syncope.* Fortunately, when the reflex causes the heart to stop, the ventricles usually "escape" from the vagal inhibition in approximately 7 to 10 seconds and begin to beat with their own intrinsic rhythm. However, occasionally the ventricles fail to escape, and the patient dies of cardiac arrest. Treatment of the condition consists of surgically stripping the nerves from the carotid arteries above and below the bifurcations.

CONTROL OF ARTERIAL PRESSURE BY THE CAROTID AND AORTIC CHEMORECEPTORS — EFFECT OF OXYGEN LACK ON ARTERIAL PRESSURE

The pressure in the aorta and in the carotid arteries indirectly controls arterial pressure in another way besides through the baroreceptor reflexes. It achieves this by stimulating the *chemoreceptors* when the pressure falls too low.

The chemoreceptors are chemosensitive cells located in several small organs 1 to 2 mm. in size, two *carotid bodies* that lie in the bifurcations of the two common carotid arteries, and several *aortic bodies* adjacent to the aorta. The chemoreceptors excite nerve fibers that pass with the baroreceptor fibers through Hering's nerves and the vagus nerves into the vasomotor center.

Each carotid or aortic body is supplied with an abundant blood flow through a small nutrient artery so that the chemoreceptors are always in close contact with arterial blood. Whenever the arterial pressure falls below a critical level, the chemoreceptors become stimulated because of diminished availability of oxygen and also because of excess carbon dioxide and hydrogen ions that are not removed by the slow flow of blood. The signals from the chemoreceptors are transmitted into the vasomotor center to *excite* the vasomotor center, this reflexly elevating the arterial pressure. Obviously, this reflex helps to return the arterial pressure back toward the normal level whenever it falls too low. However, it is not a powerful arterial pressure controller in a normal arterial pressure range, because it does not respond strongly until the arterial pressure falls below 80 mm. Hg.

The chemoreceptor mechanism also increases the arterial pressure whenever the concentration of oxygen in the arterial blood falls below normal or the concentrations of carbon dioxide and hydrogen ions rise above normal. The increased pressure helps the circulation to deliver increased quantities of oxygen to the tissues and to remove excess carbon dioxide and hydrogen ions from the tissues.

The chemoreceptors will be discussed in much more detail in Chapter 42 in relation to respiratory control, where they play an even more important role than in pressure control.

ATRIAL AND PULMONARY ARTERY REFLEXES THAT HELP TO REGULATE ARTERIAL PRESSURE

Both the atria and the pulmonary arteries have stretch receptors, called *low pressure receptors,* in their walls similar to the baroreceptor stretch receptors of the large systemic arteries. When these low pressure receptors are intact, the arterial pressure changes far less in response to changes in blood volume than when they are not present. To give an example, if 300 ml. of blood are infused into a dog with no receptors at all — no pulmonary artery receptors, no atrial receptors, and no baroreceptors — the arterial pressure of the animal will increase approximately 120 mm. Hg. If the baroreceptors are intact, but the low pressure receptors are not, the arterial pressure will rise 50 mm. Hg. If all the receptors are intact, the pressure will rise only 15 mm. Hg.

Thus, one can see that even though the low pressure receptors in the pulmonary artery and in the atria cannot detect the systemic arterial pressure, these receptors, nevertheless, do detect the simultaneous increase in pressure in the low pressure areas of the circulation caused by the increase in volume, and they elicit reflexes parallel to the baroreceptor reflexes to make the total reflex system much more potent for control of arterial pressure.

The receptors in the pulmonary artery operate in almost identically the same way as the baroreceptors from the systemic arteries. On the other hand, the atrial receptors operate somewhat differently, as follows.

Atrial Reflexes to Decrease Arterial Pressure.   Recent experiments have demonstrated that stretching the atria causes slight reflex vasodilatation of the peripheral arterioles. This reduces the total peripheral resistance and, therefore, decreases the arterial pressure back toward the normal level. This effect also plays an important role in reducing the blood volume back toward normal in the following way: The decrease in arteriolar resistance causes rapid blood flow into the capillaries, thereby raising the capillary pressure. This increased capillary pressure causes fluid to filter out of the circulation into the tissue spaces so that some of the excess blood volume is temporarily dumped into the tissues. This mechanism will be discussed, especially in relation to blood volume regulation, in Chapter 36.

Atrial Reflexes to the Kidneys — the Volume Reflex.   Stretch of the atria also causes reflex dilatation of the afferent arterioles in the kidneys, which is the same reflex effect that occurs in other peripheral arterioles, but is usually more potent in the kidneys than elsewhere. Signals are transmitted simultaneously to the hypothalamus to decrease the secretion of antidiuretic hormone, thereby indirectly affecting kidney function. The decreased afferent arteriolar resistance causes the glomerular capillary pressure to rise, with resultant increase in filtration of fluid into the kidney tubules. The diminution of antidiuretic hormone diminishes the reabsorption of water from the tubules. The combination of these two effects therefore causes rapid loss of fluid into the urine, which also serves as a powerful means to return the blood volume back toward normal.

Obviously, all these mechanisms that tend to return the blood volume back toward normal following a volume overload act indirectly as pressure controllers as well as volume controllers because excess volume drives the heart to greater cardiac output and leads therefore to greater arterial pressure.

Atrial Reflexes for Control of Heart Rate (the Bainbridge Reflex). An increase in atrial pressure also causes an increase in heart rate, sometimes increasing the heart rate as much as 75 per cent. Part of this increase in heart rate is caused by the direct effect of the increased atrial volume to stretch the S-A node; it was pointed out in Chapter 14 that such direct stretch can increase the heart rate as much as 15 per cent. An additional 40 to 60 per cent increase in rate is caused by a reflex called the *Bainbridge reflex*. The stretch receptors of the atria that elicit the Bainbridge reflex transmit their afferent signals through the vagus nerves to the medulla of the brain. Then, efferent signals are transmitted back through both the vagal and sympathetic nerves to increase the heart's rate and strength of contraction. Thus, this reflex helps to prevent damming of blood in the veins, the atria, and the pulmonary circulation. This reflex obviously has a different purpose from that of controlling arterial pressure and is actually detrimental to pressure control for short periods of time.

CONTROL OF ARTERIAL PRESSURE BY THE VASOMOTOR CENTER IN RESPONSE TO DIMINISHED BRAIN BLOOD FLOW — THE CNS ISCHEMIC RESPONSE

Normally, most nervous control of blood pressure is achieved by reflexes originating in the baroreceptors, the chemoreceptors, and the low pressure receptors, all of which are located in the peripheral circulation outside the brain. However, when blood flow to the vasomotor center in the lower brain stem becomes decreased enough to cause nutritional deficiency, a condition called *ischemia,* the neurons in the vasomotor center itself respond directly to the ischemia and become strongly excited. When this occurs, the systemic arterial pressure often rises to a level as high as the heart can possibly pump. This effect is believed to be caused by failure of the slowly flowing blood to carry carbon dioxide away from the vasomotor center; the local concentration of carbon dioxide increases greatly and has an extremely potent effect in stimulating the sympathetic nervous system. It is possible that other factors, such as the buildup of lactic acid and other acidic substances, also contribute to the marked stimulation of the vasomotor center and to the elevation in pressure. This arterial pressure elevation in response to cerebral ischemia is known as the *central nervous system ischemic response* or simply *CNS ischemic response.*

The magnitude of the ischemic effect on vasomotor activity is tremendous; it can elevate the mean arterial pressure for as long as ten minutes sometimes to as high as 270 mm. Hg. *The degree of sympathetic vasoconstriction caused by intense cerebral ischemia is often so great that some of the peripheral vessels become totally or almost totally occluded.* The kidneys, for instance, will entirely cease their production of urine because of arteriolar constriction in response to the sympathetic discharge. Therefore, *the CNS ischemic response is one of the most powerful of all the activators of the sympathetic vasoconstrictor system.*

Importance of the CNS Ischemic Response as a Regulator of Arterial Pressure. Despite the extremely powerful nature of the CNS ischemic response, it does not become very active until the arterial pressure falls far below normal, down to *levels below 50 mm. Hg* and below, reaching its greatest degree of stimulation at a pressure of 15 to 20 mm. Hg. Therefore, it is not one of the mechanisms for regulating normal arterial pressure. Instead it operates principally as an *emergency arterial pressure control system that acts rapidly and extremely powerfully to prevent further decrease in arterial pressure whenever blood flow to the brain decreases dangerously close to the lethal level.* It is sometimes called the "last ditch stand" pressure control mechanism.

The Cushing Reaction. The so-called Cushing reaction is a special type of CNS ischemic response that results from increased pressure in the cranial vault. For instance, when the cerebrospinal fluid pressure rises to equal the arterial pressure, it compresses the arteries in the brain and cuts off the blood supply to the brain. Obviously, this initiates a CNS ischemic response, which causes the arterial pressure to rise. When the arterial pressure has risen to a level higher than the cerebrospinal fluid pressure, blood flows once again into the vessels of the brain to relieve the ischemia. Ordinarily, the blood pressure comes to a new equilibrium level slightly higher than the cerebrospinal fluid pressure, thus allowing blood to continue flowing to the brain at all times. A typical Cushing reaction is illustrated in Figure 21–9.

The Cushing reaction helps to protect the vital centers

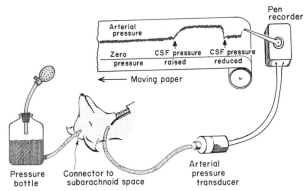

Figure 21–9. The Cushing reaction, showing a rise in arterial pressure resulting from increased cerebrospinal fluid pressure.

of the brain from loss of nutrition if ever the cerebrospinal fluid pressure rises high enough to compress the cerebral arteries.

Depressant Effect of Extreme Ischemia on the Vasomotor Center. If cerebral ischemia becomes so severe that maximum rise in mean arterial pressure still cannot relieve the ischemia, the neuronal cells begin to suffer metabolically, and within 3 to 10 minutes they become totally inactive. The arterial pressure then falls to about 40 to 50 mm. Hg, which is the level to which the pressures fall when the vasomotor center loses all of its control of the circulation so that all tonic vasoconstrictor activity is lost. Therefore, it is fortunate that the ischemic response is extremely powerful so that arterial pressure can usually rise high enough to correct brain ischemia before it causes nutritional depression and death of the neuronal cells.

PARTICIPATION OF THE VEINS IN NERVOUS REGULATION OF CARDIAC OUTPUT AND ARTERIAL PRESSURE

Thus far, we have discussed primarily the ability of the nervous system to regulate arterial pressure by altering arterial resistance. However, much of the regulatory effect of the nervous system is carried out through sympathetic vasoconstrictor fibers to the veins. Indeed, the veins constrict in response to even weaker sympathetic stimuli than do the arterioles and arteries.

Sympathetic Alterations in Venous Capacity and Cardiac Output. The veins offer relatively little resistance to blood flow in comparison with the arterioles and arteries. Therefore, sympathetic constriction of the veins does not significantly change the overall total peripheral resistance. Instead, the important effect of sympathetic stimulation of the veins is a *decrease in their capacity*. This means that the veins then hold less blood at any given venous pressure, which causes translocation of blood out of the systemic veins into the heart, lungs, and systemic arteries. The distension of the heart in turn causes the heart to pump with increasing effectiveness in accordance with the Frank-Starling law of the heart, as discussed in Chapter 13. Therefore, the net effect of sympathetic stimulation of the veins is to increase the cardiac output, which in turn elevates the arterial pressure.

The veins participate in all of the reflexes and reactions that have been discussed thus far, including the baroreceptor reflex, the CNS ischemic response, the chemoreceptor reflex, and also the atrial reflexes. To comprehend the potency of the venous reaction in some of these nervous mechanisms, consider the following: The mean circulatory filling pressure (mainly determined by the degree of contraction of the veins) rises from the normal value of 7 mm. Hg to approximately 10 mm. Hg when the baroreceptor reflex is excited fully and rises to approximately 18 mm.

Hg — enough to increase the pressure forcing blood into the heart by as much as 2.5-fold — when a maximal CNS ischemic response is elicited.

ROLE OF THE SKELETAL NERVES AND SKELETAL MUSCLES TO INCREASE CARDIAC OUTPUT AND ARTERIAL PRESSURE

Though most nervous control of the circulation is effected through the autonomic nervous system, at least two conditions in which the skeletal nerves and muscles also play major roles in circulatory responses are the following:

The Abdominal Compression Reflex. When a baroreceptor or chemoreceptor reflex is elicited, or whenever almost any other factor stimulates the sympathetic vasoconstrictor system, the vasomotor center and other areas of the reticular substance of the brain stem transmit impulses simultaneously through skeletal nerves to skeletal muscles of the body, particularly to the abdominal muscles. This obviously increases the basal tone of these muscles, and contraction of the abdomen compresses all the venous reservoirs of the abdomen, helping to translocate blood out of the abdominal vascular reservoirs toward the heart. As a result, increased quantities of blood are made available to the heart to be pumped. This overall response is called the *abdominal compression reflex*. The resulting effect on the circulation is the same as that caused by sympathetic vasoconstrictor impulses when they constrict the veins, namely, an increase in both cardiac output and arterial pressure.

The abdominal compression reflex is probably much more important than has been realized in the past, for it is well known that persons whose skeletal muscles have been paralyzed are considerably more prone to hypotensive episodes than are normal persons.

Increased Mean Circulatory Filling Pressure Caused by Skeletal Muscle Contraction During Exercise. When the skeletal muscles contract during exercise, they compress blood vessels throughout the body. And even anticipation of exercise tightens the muscles, thereby compressing the vessels. The resulting effect is to increase the *mean circulatory filling pressure* from its normal value of about 7 mm. Hg to as high as 20 to 30 mm. Hg. We shall see in Chapter 23 that this high filling pressure acts to translocate blood from the peripheral vessels into the heart and lungs and therefore to increase the cardiac output. This is probably an essential effect to help cause the 5- to 6-fold increase in cardiac output that sometimes occurs in severe exercise. It also helps to raise the arterial pressure a small-to-moderate amount during exercise.

RESPIRATORY WAVES IN THE ARTERIAL PRESSURE

With each cycle of respiration, the arterial pressure rises and falls usually 4 to 6 mm. Hg in a wavelike fashion because of several different effects, some of which are reflex in nature, as follows:

First, many impulses arising in the respiratory center of the medulla "spill over" into the vasomotor center.

Second, every time a person inspires, the pressure in the thoracic cavity becomes more negative than usual, causing the blood vessels in the chest to expand. This reduces the quantity of blood returning to the left side of the heart and thereby momentarily decreases the cardiac output and arterial pressure.

Third, the pressure changes caused in the thoracic vessels by respiration can excite baroreceptors.

Though it is difficult to analyze the exact relationship of all these factors in causing the so-called *respiratory pressure waves,* the net result during normal respiration is usually an increase in arterial pressure during the very late part of inspiration and early part of expiration and a fall in pressure during the remainder of the respiratory cycle. During deep respiration the blood pressure can rise and fall as much as 20 mm. Hg with each respiratory cycle.

ARTERIAL PRESSURE VASOMOTOR WAVES — OSCILLATION OF THE PRESSURE REFLEX CONTROL SYSTEMS

Often in recording arterial pressure in an animal, in addition to the small pressure waves caused by the respiration, some much larger waves are noted — as great as 20 to 40 mm. Hg at times — that also rise and fall much more slowly than the respiratory waves. The duration of each cycle varies from 26 seconds in the dog to as short as 7 to 10 seconds in the human being. These waves are called *vasomotor waves* or frequently "Mayer waves" or "Traube-Hering waves." Such records are illustrated in Figure 21–10, showing the cyclical rise and fall in arterial pressure.

The cause of vasomotor waves is usually oscillation of one or more nervous pressure control mechanisms, some of which are the following:

Oscillation of the Baroreceptor and Chemoreceptor Reflexes. The vasomotor waves of Figure 21–10B are the common vasomotor waves that are seen almost daily in experimental pressure recordings. They are caused mainly by oscillation of the *baroreceptor reflex.* That is, a high pressure excites the baroreceptors; this

then inhibits the sympathetic nervous system and lowers the pressure. The decreased pressure reduces the baroreceptor stimulation and allows the vasomotor center to become active once again, elevating the pressure again to a high value. This high pressure then initiates another cycle, and the oscillation continues on and on.

The *chemoreceptor reflex* can also oscillate to give the same type of waves. Usually, this reflex oscillates simultaneously with the baroreceptor reflex. In fact, it probably plays the major role in causing vasomotor waves when the arterial pressure is in the range of 40 to 80 mm. Hg, because in this range chemoreceptor control of the circulation becomes powerful while baroreceptor control becomes weak.

Oscillation of the CNS Ischemic Response. The record in Figure 21–10A resulted from oscillation of the CNS ischemic pressure control mechanism. In this experiment the cerebrospinal fluid pressure was raised to 160 mm. Hg, which compressed the cerebral vessels and initiated a CNS ischemic response. When the arterial pressure rose above 160 mm. Hg, the ischemia was relieved, and the sympathetic nervous system became inactive. As a result, the arterial pressure fell rapidly back to a lower value, causing medullary ischemia once again. The ischemia then initiated another rise in pressure. Then, again the ischemia was relieved, and the pressure fell. This repeated itself cyclically as long as the cerebrospinal fluid pressure remained elevated.

Thus, any pressure control mechanism can oscillate if the intensity of "feedback" is strong enough. The vasomotor waves are of considerable theoretical importance because they show that the nervous reflexes that control arterial pressure obey identically the same principles as those applicable to mechanical and electrical control systems. For instance, if the feedback is too great in the guiding mechanism of an automatic pilot for an airplane, the plane oscillates from side to side instead of following a straight course.

HORMONAL MECHANISMS FOR RAPID CONTROL OF ARTERIAL PRESSURE

In addition to the rapidly acting nervous mechanisms for control of arterial pressure, there are at least three hormonal mechanisms that also provide either rapid or moderately rapid control of arterial pressure. These mechanisms are:

1. The norepinephrine-epinephrine vasoconstrictor mechanism.

2. The renin-angiotensin vasoconstrictor mechanism.

3. The vasopressin vasoconstrictor mechanism.

THE NOREPINEPHRINE-EPINEPHRINE VASOCONSTRICTOR MECHANISM

In the previous chapter it was pointed out that stimulation of the sympathetic nervous system not only causes direct nervous excitation of the blood

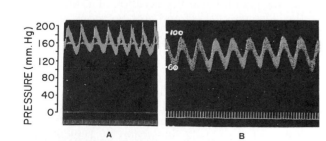

Figure 21–10. A, Vasomotor waves caused by oscillation of the CNS ischemic response. B, Vasomotor waves caused by baroreceptor reflex oscillation.

vessels and heart but also causes release by the adrenal medullae of norepinephrine and epinephrine into the circulating blood. These two hormones circulate to all parts of the body and cause essentially the same effects on the circulatory system as direct sympathetic stimulation. That is, they excite the heart, they constrict most of the blood vessels, and they constrict the veins.

Therefore, the different reflexes that regulate arterial pressure by exciting the sympathetic nervous system cause the pressure to rise in two ways: by direct circulatory stimulation and by indirect stimulation through the release of norepinephrine and epinephrine into the blood.

Norepinephrine and epinephrine circulate in the blood for 1 to 3 minutes before being destroyed, thus maintaining a slightly prolonged excitation of the circulation. Also, these hormones can reach some parts of the circulation that have no sympathetic nervous supply at all, including some of the very minute vessels such as the metarterioles. And these hormones have especially potent actions on some vascular beds, particularly the skin vasculature.

In general, the norepinephrine and epinephrine system can be considered to be a part of the total sympathetic mechanism for arterial pressure control. There are slight differences between the actions of epinephrine and norepinephrine that will be discussed in Chapter 57.

THE RENIN-ANGIOTENSIN VASOCONSTRICTOR MECHANISM FOR CONTROL OF ARTERIAL PRESSURE

The hormone *angiotensin II* is one of the most potent vasoconstrictors known. Whenever the arterial pressure falls very low, large quantities of angiotensin II appear in the circulation. This results from a special mechanism involving the kidneys and the release of the enzyme *renin* from the kidneys when the arterial pressure falls too low.

The overall schema for formation of angiotensin and the effect of angiotensin II to increase arterial pressure are illustrated in Figure 21–11. When blood flow through the kidneys is decreased, the *juxtaglomerular cells* (cells located in the walls of the afferent arterioles immediately proximal to the glomeruli) secrete renin into the blood. Renin itself is an enzyme that splits the end off one of the plasma proteins, called *renin substrate,* to release a decapeptide, *angiotensin I*. The renin persists in the blood for as long as 1 hour and continues to cause formation of angiotensin I during the entire time. Within a few seconds after formation of the angiotensin I, two additional amino acids are split from it to form the octapeptide *angiotensin II*. This conversion occurs almost entirely in the

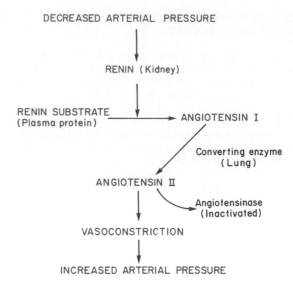

Figure 21–11.   The renin-angiotensin-vasoconstrictor mechanism for arterial pressure control.

small vessels of the lungs, catalyzed by an enzyme called *converting enzyme*. Angiotensin II persists in the blood for a minute or so but is rapidly inactivated by a number of different blood and tissue enzymes collectively called *angiotensinase*.

During its persistence in the blood, angiotensin II has several effects that can elevate arterial pressure. One of these effects occurs very rapidly — vasoconstriction especially of the arterioles and to a lesser extent of the veins at the same time. Constriction of the arterioles increases the peripheral resistance and thereby raises the arterial pressure back toward normal, as illustrated at the bottom of the schema in Figure 21–11. Also, mild constriction of the veins increases the mean circulatory filling pressure, sometimes as much as 20 per cent, and this promotes increased tendency for venous return of blood to the heart, thereby helping the heart to pump against the extra pressure load.

The other effects of angiotensin are mainly related to the body fluid volumes: (1) angiotensin has a direct effect on the kidneys to cause decreased excretion of both salt and water; and (2) angiotensin stimulates the secretion of aldosterone by the adrenal cortex, and this hormone in turn also acts on the kidneys to cause decreased excretion of both salt and water. Both these effects tend to elevate the blood volume — an important factor in long-term regulation of arterial pressure, as will be discussed in the following chapter.

Rapidity of Action and Pressure Controlling Power of the Renin-Angiotensin Vasoconstrictor System. Figure 21–12 illustrates a typical experiment showing the effect of hemorrhage on the arterial pressure under two separate conditions:

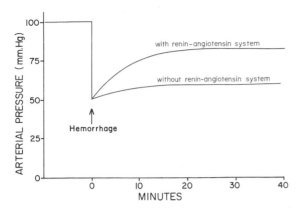

Figure 21–12. Pressure compensating effect of the renin-angiotensin-vasoconstrictor system following severe hemorrhage. (Data from experiments by Dr. Royce Brough.)

(1) with the renin-angiotensin system functioning, and (2) without the system functioning (the system was interrupted by a renin-blocking antibody). Note that following hemorrhage, which caused an acute fall of the arterial pressure to 50 mm. Hg, the arterial pressure rose back to 83 mm. Hg when the renin-angiotensin system was functional. On the other hand, it rose to only 60 mm. Hg when the renin-angiotensin system was blocked. This illustrates that the renin-angiotensin system is powerful enough to return the arterial pressure at least halfway back to normal following severe hemorrhage. Therefore, it can sometimes be of life-saving service to the body, especially in circulatory shock.

Note that the renin-angiotensin vasoconstrictor system requires approximately 20 minutes to become fully active. Therefore, it is far slower to act than are the nervous reflexes and the norepinephrine-epinephrine system; however, it also has a correspondingly longer duration of action.

There is reason to believe that the renin-angiotensin vasoconstrictor system is much more powerful under some conditions than others. For instance, some patients with diseased kidneys secrete tremendous quantities of renin, and the pressure control action of the system is then likely to be very powerful.

ROLE OF VASOPRESSIN IN RAPID CONTROL OF ARTERIAL PRESSURE

When the arterial pressure falls low, the hypothalamus secretes large quantities of vasopressin by way of the posterior pituitary gland — a mechanism that will be discussed in detail in relation to the pituitary hormones in Chapter 75. The vasopressin in turn has a direct vasoconstrictor effect on the blood vessels, thereby increasing both the total peripheral resistance and the mean circulatory filling pressure, raising the arterial pressure back toward normal.

Until recently, most physiologists believed that the amount of vasopressin secreted in low blood pressure states was not sufficient to play a significant role in compensating for the low pressure. However, recent experiments in animals in which the baroreceptor pressure-controlling mechanism has been removed have shown that the circulating amounts of vasopressin found in the blood following hemorrhage can then increase the arterial pressure as much as 35 to 50 mm. Hg. In a recent study by Cowley, the vasopressin system was shown to return the blood pressure about 75 per cent of the way back toward normal within a few minutes after acute hemorrhage had decreased the arterial pressure to as low as 50 mm. Hg. Furthermore, it has been discovered that vasopressin is an even more potent vascular constrictor than angiotensin.

Therefore, it is almost certain that vasopressin plays a very important role to re-establish normal arterial pressure when the pressure falls acutely to dangerously low levels.

Vasopressin also plays an indirect role in the long-term control of arterial pressure through its effects on the kidneys to cause decreased excretion of water. Because of this effect, vasopressin is also called *antidiuretic hormone*. When even minute amounts of vasopressin are secreted, the kidney excretion of water decreases to a minimal amount, an effect that helps to increase the blood volume any time the arterial pressure falls too low. The increased blood volume then helps to bring the pressure back to normal. Thus, vasopressin not only plays an important role in acute regulation of arterial pressure but also in long-term regulation, which will be discussed in the following chapter.

TWO INTRINSIC CIRCULATORY MECHANISMS FOR ARTERIAL PRESSURE REGULATION

In addition to the nervous and hormonal mechanisms for rapid control of arterial pressure, two intrinsic physical mechanisms of the circulation also help to control arterial pressure, usually beginning to act within a few minutes and reaching full function within a few hours. These are: (1) the capillary fluid shift mechanism, and (2) the vascular stress-relaxation mechanism.

Capillary Fluid Shift. When the arterial pressure changes, this is usually also associated with a similar change in capillary pressure. This causes fluid to begin moving across the capillary membrane between the blood and the interstitial fluid compartment. Within a few minutes to an hour a new state of equilibrium usually will be achieved, but in the meantime this shift of fluid will have played a very beneficial role in the control of arterial pressure. For instance, if the arterial pressure rises too high, loss of fluid through the capillaries into the interstitial spaces causes the blood volume to fall and thereby causes return of arterial pressure back toward normal. The gain of this system is about 3, which means that it can return the arterial pressure about three-fourths of the way back toward its normal mean value; but it takes effect much more slowly than do the nervous reflex mechanisms.

Stress-Relaxation. When the arterial pressure falls, the pressure usually also falls in most of the blood storage areas such as in the veins, the liver, the spleen, the lungs, and so forth. Conversely, a rise in arterial pressure is often associated with a rise in pressure in the same storage areas. As pointed out in Chapter 18, a pressure change causes vessels gradually to adapt to a new size, thereby accommodating the amount of blood that is available. This phenomenon is called *stress-relaxation* or *reverse stress-relaxation*. Thus, following massive transfusion, the arterial pressure rises markedly at first, but because of relaxation of the circulation during the next 10 minutes to an hour, the arterial pressure returns essentially to normal even though the blood volume may be as great as 30 per cent above normal. Conversely, following severe bleeding, the reverse stress-relaxation mechanism can cause the blood vessels gradually to tighten around the amount of blood that is left, thereby again re-establishing almost normal circulatory dynamics.

Unfortunately, the stress-relaxation mechanism has very definite limits, so that acute changes in blood volume of more than approximately plus 30 per cent or minus 15 per cent cannot be corrected by this mechanism.

REFERENCES

Brough, R. B., *et al.*: Quantitative analysis of the acute response to hemorrhage of the renin-angiotensin-vasoconstrictor feedback loop in areflexic dogs. *Cardiovasc. Res.*, 9:722–733, 1975.

Brown, A. M.: Cardiac reflexes. *In* Berne, R. M., *et al.* (eds.): Handbook of Physiology. Sec. 2, Vol. I. Baltimore, Williams & Wilkins, 1979, p. 677.

Campese, V. M., and DeQuattro, V.: Orthostatic hypotension: Causes and therapy. *In* DeGroot, L. J., *et al.* (eds.): Endocrinology. Vol. 2. New York, Grune & Stratton, 1979, p. 1289.

Coleman, T. G., *et al.*: Angiotensin and the hemodynamics of chronic salt deprivation. *Am. J. Physiol.*, 229:167–171, 1975.

Coleridge, J. C. G., and Coleridge, H. M.: Chemoreflex regulation of the heart. *In* Berne, R. M., *et al.* (eds.): Handbook of Physiology. Sec. 2, Vol. I. Baltimore, Williams & Wilkins, 1979, p. 653.

Cooper, M. D., and Lawton, A. R., III: The Physiology of the giraffe. *Sci. Am.*, 231(5):96, 1974.

Cowley, A. W., Jr., and Guyton, A. C.: Baroreceptor reflex contribution in angiotensin-II-induced hypertension. *Circulation*, 50:61, 1974.

Cowley, A. W., Jr., *et al.*: Interaction of vasopressin and the baroreceptor reflex system in the regulation of arterial pressure in the dog. *Circ. Res.*, 34:505, 1974.

Cushing, H.: Concerning a definite regulatory mechanism of the vasomotor center which controls blood pressure during cerebral compression. *Bull. Johns Hopkins Hosp.*, 12:290, 1901.

Davis, J. O., and Freeman, R. H.: Mechanisms regulating renin release. *Physiol. Rev.*, 56:1, 1976.

Downing, S. E.: Baroreceptor regulation of the heart. *In* Berne, R. M., *et al.* (eds.): Handbook of Physiology. Sec. 2, Vol. I. Baltimore, Williams & Wilkins, 1979, p. 621.

Fagard, R. H., *et al.*: Renal responses to slight elevations of renal arterial plasma angiotensin II concentration in dogs. *Clin. Exp. Pharmacol. Physiol.*, 3:19, 1976.

Goetz, K. L., *et al.*: Atrial receptors and renal function. *Physiol. Rev.*, 55:157, 1975.

Guyton, A. C.: Acute hypertension in dogs with cerebral ischemia. *Am. J. Physiol.*, 154:45, 1948.

Guyton, A. C.: Arterial Pressure and Hypertension. Philadelphia, W. B. Saunders Co., 1980.

Guyton, A. C., and Satterfield, J. H.: Vasomotor waves possibly resulting from CNS ischemic reflex oscillation. *Am. J. Physiol.*, 170:601, 1952.

Guyton, A. C., *et al.*: Method for studying competence of the body's blood pressure regulatory mechanisms and effect of pressoreceptor denervation. *Am. J. Physiol.*, 164:360, 1951.

Guyton, A. C., *et al.*: Synthesis of endocrine control in hypertension. *Clin. Sci. Molec. Med.*, 51:319, 1976.

Hainsworth, R., and Linden, R. J.: Reflex control of vascular capacitance. *In* International Review of Physiology: Cardiovascular Physiology III. Vol. 18. Baltimore, University Park Press, 1979, p. 67.

Hall, J. W., and Guyton, A. C.: Changes in renal hemodynamics and renin release caused by increased plasma oncotic pressure. *Am. J. Physiol.*, 231:1550, 1976.

Kaley, G.: Microcirculatory-endocrine interactions. *In* Kaley, G., and Altura, B. M. (eds.): Microcirculation. Vol. II. Baltimore, University Park Press, 1977.

Kirchheim, H. R.: Systemic arterial baroreceptor reflexes. *Physiol. Rev.*, 56:100, 1976.

Krieger, E. M.: Time course of baroreceptor resetting in acute hypertension. *Am. J. Physiol.*, 218:486, 1970.

Mancia, G., *et al.*: Reflex control of circulation by heart and lungs. *Int. Rev. Physiol.*, 9:111, 1976.

Oberg, B.: Overall cardiovascular regulation. *Annu. Rev. Physiol.*, 38:537, 1976.

Peach, M. J.: Renin-angiotensin system: Biochemistry and mechanisms of action. *Physiol. Rev.*, 57:313, 1977.

Reid, I. A., *et al.*: The renin-angiotensin system. *Annu. Rev. Physiol.*, 40:377, 1978.

Rosell, S.: Nervous control of microcirculation. *In* Kaley, G., and Altura, B. M. (eds.): Microcirculation. Vol. II. Baltimore, University Park Press, 1977.

Rowell, L. B.: Human cardiovascular adjustments to exercise and thermal stress. *Physiol. Rev.*, 54:75, 1974.

Sagawa, K., *et al.*: Quantitation of cerebral ischemic pressor response in dogs. *Am. J. Physiol.*, 200:1164, 1961.

Sagawa, K., *et al.*: Dynamic performance and stability of cerebral ischemic pressor response. *Am. J. Physiol.*, 201:1164, 1961.

Sagawa, K., *et al.*: Elicitation of theoretically predicted feedback oscillation in arterial pressure. *Am. J. Physiol.*, 203:141, 1962.

Sagawa, K., *et al.*: Nervous control of the circulation. *In* MTP International Review of Science: Physiology. Baltimore, University Park Press, 1974. Vol. 1, p. 197.

Smith, E. E., and Guyton, A. C.: Center of arterial pressure regulation during rotation of normal and abnormal dogs. *Am. J. Physiol.*, 204:979, 1963.

Smith, O. A.: Reflex and central mechanisms involved in the control of the heart and circulation. *Annu. Rev. Physiol.*, 36:93, 1974.

Uther, J. B., and Guyton, A. C.: Cardiovascular regulation following changes in central nervous perfusion pressure in the unanesthetized rabbit. *Aust. J. Exp. Biol. Med. Sci.*, 51:295, 1973.

Youmans, W. B., *et al.*: The Abdominal Compression Reaction. Baltimore, Williams & Wilkins, 1963.

Youmans, W. B., *et al.*: Control of involuntary activity of abdominal muscles. *Am. J. Phys. Med.*, 53:57, 1974.

Ziegler, M. G.: Postural hypotension. *Annu. Rev. Med.*, 31:239, 1980.

22

Long-Term Regulation of Mean
Arterial Pressure: The
Renal-Body Fluid Pressure
Control System; Long-Term
Functions of the
Renin-Angiotensin System; and
Mechanisms of Hypertension

SHORT-TERM VERSUS LONG-TERM PRESSURE CONTROL MECHANISMS

In the previous chapter, the important short-term mechanisms for arterial pressure control were described. We have seen that these mechanisms have the capability of reacting within seconds, minutes, or hours to correct almost completely, but never entirely, most abnormal arterial pressures that occur. In the present chapter we will discuss those mechanisms that are of special importance for control of the day-by-day, week-by-week, month-by-month, and year-by-year arterial pressure — that is, long-term control of arterial pressure.

The mechanisms for long-term arterial pressure control are considerably different from the short-term mechanisms of control. Referring back to Figure 21–1, we see that the feedback gains of many of the short-term pressure control mechanisms — the baroreceptor mechanism, for instance — diminish drastically as time proceeds. On the other hand, the feedback gain of at least one of the long-term pressure control systems is almost zero for the first few hours but then approaches infinity over a period of days and weeks. This is the renal–body fluid pressure control system. Because of this extreme long-term potency, this system plays a central role in long-term pressure control. However, it is aided in this role by a large number of accessory mechanisms, including special effects of the renin-angiotensin system, of the nervous system, of the aldosterone system, and so forth. The purpose of this chapter is to describe first the basic renal–body fluid pressure control system, and then to discuss briefly how some of the accessory control mechanisms function in association with this system. Finally, we will use some of the knowledge gained in this and the previous chapter to analyze some of the major causes of hypertension, which is also called "high blood pressure."

THE RENAL–BODY FLUID SYSTEM FOR ARTERIAL PRESSURE CONTROL

The Phenomenon of Pressure Diuresis and Pressure Natriuresis As a Basis for Arterial Pressure Control. From the very earliest studies of the kidneys it immediately became evident that increased arterial pressure increases greatly the rate at which the kidneys excrete both water and salt, effects called *pressure diuresis* and *pressure natriuresis*. Or, to state this another way, an increase in pressure causes marked loss of extracellular fluid volume from the body, because extracellular fluid is composed almost entirely of water and salt.

And this decreases the blood volume. By now, it should be clear to the student that blood volume is one of the principal components for successful pumping of blood by the heart and for the maintenance of arterial pressure. Therefore, whenever the extracellular fluid volume decreases, there is a tendency for the arterial pressure to decrease.

Now, let us put this total system in perspective as a blood pressure control mechanism. Elevated arterial pressure causes pressure diuresis and natriuresis, which decrease the arterial pressure back to normal. Conversely, when the arterial pressure falls too low, the kidneys retain fluid, the blood volume increases, and the arterial pressure returns again to normal.

Figure 22–1 illustrates this pressure diuresis and natriuresis mechanism, showing the normal *renal output curve* for an experimentally perfused isolated kidney. It shows that at a blood pressure of 50 to 60 mm. Hg, the urinary output of water and salt is essentially zero. At 100 mm. Hg, it is normal, and at 200 mm. Hg, about 6 to 8 times normal. Thus, it is already clear that this mechanism is not a weak one. Later in the chapter we will show that the renal output curve becomes far steeper still when one considers the accessory control mechanisms that help to make the kidney control of body fluid even more effective.

Some More Details of the Renal–Body Fluid System. Because of the importance of the renal–body fluid system and because there are still more complexities to it, we will now present the system in detail. Figure 22–2 illustrates the most important components, the essential steps of which are the following:

A decrease in the *arterial pressure* (Block 1) causes a decrease in renal output of salt and water (Block 2).

If the renal output of salt and water is less than the net intake of salt and water (Block 10), then the rate of

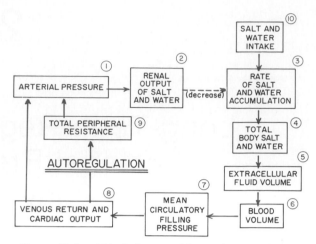

Figure 22–2. A block diagram of the renal–body fluid mechanism for long-term control of arterial pressure.

accumulation of salt and water in the body will be *positive* (Block 3), and the total body salt and water (Block 4) will increase progressively over a period of days.

The increased body salt and water increases the extracellular fluid volume (Block 5).

The increased extracellular fluid volume increases the blood volume (Block 6).

The increased blood volume increases the mean circulatory filling pressure (Block 7).

The increased mean circulatory filling pressure increases venous return and cardiac output (Block 8).

The increased cardiac output increases the arterial pressure in two ways: one is by the direct effect of the increased cardiac output on the pressure and the other by an indirect increase in total peripheral resistance (Block 9), which results from local vascular autoregulation (that is, when the cardiac output becomes too great, the blood vessels in all the local tissues of the body automatically constrict, this constriction becoming progressively greater over a period of days and weeks). Most of the early increase in arterial pressure is caused by the direct effect of increased cardiac output. However, after several weeks 80 to 90 per cent of the increase in pressure is caused by the increased total peripheral resistance and only 10 to 20 per cent by the direct effect of increased cardiac output.

The Minuteness of the Fluid Volume Changes Required to Cause Marked Changes in Pressure. One of the features of the renal–body fluid mechanism for arterial pressure control that is most often misunderstood is how small the changes in quantities of fluid in the body have to be to cause marked changes in pressure. To give an example, a 2 per cent chronic increase in blood volume can increase the mean circulatory filling pressure as much as 5 per cent, and this in turn can increase the venous return and cardiac output also as much as 5 per cent. Finally, an increase in cardiac output of 5 per cent can increase the total peripheral resistance 25 to 50 per cent; this figure multiplied by the 5 per cent increase in cardiac

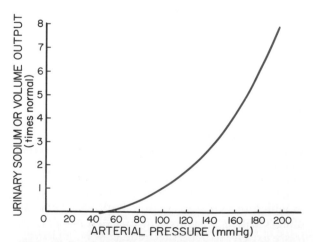

Figure 22–1. A typical renal output curve measured in a perfused isolated kidney, showing both pressure diuresis (excess output of water) and pressure natriuresis (excess output of sodium) when the arterial pressure rises above normal.

output gives an increase in the arterial pressure of as much as 30 to 57 per cent. Thus, one can understand very easily that a chronic increase of only a few hundred milliliters of extracellular fluid can lead to a hypertensive state. Indeed, in patients with hypertension, the arterial pressure can often be reduced back to normal by administering a diuretic and natriuretic (a drug that causes excess water and sodium excretion) that do nothing more than reduce the extracellular fluid volume by about 500 ml.

One of the reasons it has been difficult in the past to understand the minuteness of the fluid volumes involved is that large volumes of fluid can be infused into a person *acutely* without causing hypertension. However, it must be remembered that for the first few minutes to several hours after such an infusion the nervous reflex control mechanism and other short-term pressure control mechanisms prevent a significant pressure rise.

An Experiment Demonstrating the Renal–Body Fluid System for Arterial Pressure Control. Figure 22–3 illustrates an experiment in dogs in which all the nervous reflex mechanisms for blood pressure control were blocked and the arterial pressure was then suddenly elevated by infusing 300 ml. of blood. Note the instantaneous increase in cardiac output to approximately double its normal level and the increase in arterial pressure to 115

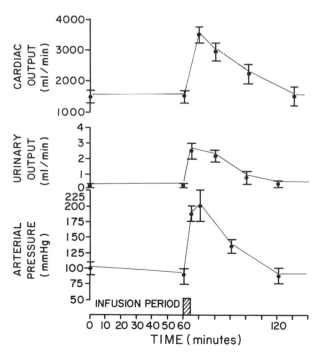

Figure 22–3. Increase in cardiac output, arterial pressure, and urinary output caused by increased blood volume in animals whose nervous pressure control mechanisms had been blocked. This figure shows the return of arterial pressure to normal after about an hour of fluid loss into the urine. (Courtesy of Dr. William Dobbs.)

mm. Hg above its resting level. Also shown, by the middle curve, is the effect of this increased arterial pressure on urinary output. The output increased 12-fold, and both the cardiac output and the arterial pressure returned back to normal during the subsequent hour. Thus, one sees the extreme capability of the kidneys to readjust the blood volume and in so doing to return the arterial pressure back to normal.

Infinite Gain of the Renal–Body Fluid Pressure Control System. A special feature of this mechanism for pressure control is its ability to return the arterial pressure *all the way* back to normal — not merely a certain proportion of the way back. This is quite different from the nervous reflex mechanisms that were discussed in the previous chapter. For instance, the baroreceptor and other nervous mechanisms for arterial pressure control, all working together, are capable of returning the arterial pressure approximately seven-eighths of the way back toward normal following a blood volume change, but never all the way. From the discussion of the effectiveness of a feedback control system in Chapter 1, remember that the "feedback gain" of a system is calculated by dividing the total amount of compensation caused by the feedback by the remaining error that fails to be compensated. In the case of the nervous pressure control system, the compensation is seven eighths, and the remaining error is one eighth. Therefore, the gain is seven. On the other hand, in the case of the renal–body fluid system, the compensation is *100 per cent of the way back to normal,* and the final error is zero. Thus, the gain of the system is 100 divided by 0, or *infinity,* which is the reason that this system is so all-important for long-term control of arterial pressure.

FACTORS THAT INCREASE THE EFFECTIVENESS OF THE RENAL–BODY FLUID SYSTEM OF PRESSURE CONTROL: (1) THE RENIN-ANGIOTENSIN SYSTEM, (2) THE ALDOSTERONE SYSTEM, AND (3) THE NERVOUS SYSTEM

When the intake of water and salt increases, several factors in addition to the increase in pressure also increase the renal output of water and salt to levels far greater than those caused by pressure diuresis and natriuresis alone. For instance, the increase in fluid intake, as well as the rise in the pressure, causes (1) *decreased* secretion of renin by the kidneys, (2) *decreased* secretion of aldosterone by the adrenal cortices, and (3) *decreased* sympathetic signals to the kidneys. All these, in turn, increase the renal output of water and salt by mechanisms that we shall discuss in detail in Chapter 35. But, for the present, let us see how these factors affect urinary output when

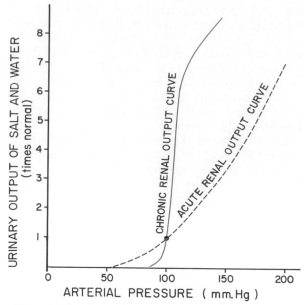

Figure 22–4. Comparison of the *chronic* renal output curve and the *acute* renal output curve, showing an extremely steep relationship between arterial pressure and urinary output in the chronic state when all the factors that affect urinary output in response to pressure have had time to become fully operative. (See explanation in the text.)

one's fluid intake is increased and the arterial pressure rises 15 mm. Hg. First, this increase in pressure by itself has a direct effect on the kidneys approximately to double the urinary output of water and salt. However, because the above-mentioned three mechanisms further enhance the output of water and salt, the 15 mm. Hg increase in arterial pressure is associated with about a ten-fold increase instead of a two-fold increase in output.

This increased output is illustrated in Figure 22–4, which compares the "acute renal output curve" and the "chronic renal output curve." The acute curve is the same as that illustrated in Figure 22–1, and it shows the direct effect of arterial pressure alone in causing increased output of water and salt by the kidneys. The chronic curve, on the other hand, shows the relationship of urinary output to arterial pressure when all the different factors that affect output are included: the direct effect of pressure on the kidneys and also the effects of the three additional mechanisms listed above.

Thus, it is clear that the chronic renal output curve depicts a very steep relationship between arterial pressure and urinary output. Therefore, in the normal animal, even a slight elevation of arterial pressure that lasts long enough for all the above mechanisms to become effective causes a marked increase in urinary fluid loss from the body, and, conversely, a slight decrease in arterial

pressure causes almost total cessation of fluid loss. One can readily see that the steepness of this relationship between arterial pressure and urinary output makes the fluid excretory mechanism an extremely potent arterial pressure control mechanism.

GRAPHICAL ANALYSIS OF THE FUNCTION OF THE KIDNEYS IN ARTERIAL PRESSURE CONTROL

Figure 22–5 illustrates a graphical method that one can use to analyze the control of arterial pressure. Note first the "intake" level, and the "renal output curve." The only point on the graph at which the normal level of intake can be in exact balance with urinary output is at point A, where the intake level crosses the output curve. This point defines the level at which the pressure stabilizes, in this case 100 mm. Hg.

Now, note that the "high intake" level crosses the output curve at point B, which is at a pressure level of 106 mm. Hg. Thus, a high intake of water and salt in a person with normal kidneys hardly affects the arterial pressure, but when the kidneys are abnormal the intake level can be a critical determinant of the pressure, as we shall see later in the chapter.

Let us assume, however, that the arterial pressure for some reason rises to 110 mm. Hg. This

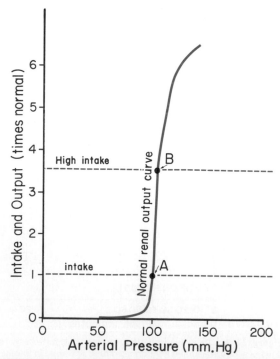

Figure 22–5. Graphical procedure for predicting the steady-state long-term arterial pressure level by equating the renal output curve with the fluid and salt intake level. (See explanation in the text.)

rise causes the urinary output to become far greater than normal. Therefore, the person now loses fluid at a rapid rate, the blood volume decreases, and the arterial pressure returns to normal. When the arterial pressure has returned to exactly 100 mm. Hg, the urinary output will have returned to Point A on the output curve, and the output will again exactly balance the intake.

Thus, the system always automatically adjusts the output of urine to equal exactly the intake, and in so doing it also returns the arterial pressure to its exact original level *if neither the intake level nor the renal output curve has changed in the meantime.* Therefore, this graphical analysis of pressure control again demonstrates the principle of infinite gain of this pressure control system.

Effect of Altering the Renal Output Curve. The kidneys of one person frequently function quite differently from those of other persons. This can result from pathological kidney conditions or from factors originating elsewhere in the body but acting on the kidneys to change their function, such as stimulation of the kidneys by excess angiotensin, aldosterone, sympathetic signals, or so forth. We will discuss some of these in detail in relation to hypertension later in the chapter.

One can see readily from Figure 22–5 that any factor that shifts the renal output curve to the right — that is, to higher arterial pressure range — will also cause the point at which the intake line crosses the renal output curve to rise to a higher arterial pressure as well. Therefore, any factor that alters the renal output curve in this way will also, in the long run, alter the long-term level at which the arterial pressure of the person is controlled.

The Two Determinants of the Long-Term Arterial Pressure Level. From the above discussion and from the principles illustrated by the graphical analysis of Figure 22–5, one can readily see that two primary factors determine the long-term level of arterial pressure. These are called the *long-term determinants of arterial pressure;* they are (1) the *pressure range of the renal output curve,* and (2) the *net rate of fluid intake.* A change in either of these can change the long-term level of arterial pressure. Furthermore, *the long-term level of arterial pressure cannot be changed in any way other than by changing one or both of these two determinants.*

RESPECTIVE ROLES OF TOTAL PERIPHERAL RESISTANCE AND CARDIAC OUTPUT IN THE LONG-TERM CONTROL OF ARTERIAL PRESSURE

It is frequently stated that the arterial pressure is controlled by (1) total peripheral resistance and (2) cardiac output, because it is well known that

arterial pressure is equal to the product of these two. In a sense, it is true that these two factors do control the arterial pressure. However, neither of them is a fundamental controller of pressure because each of them is in turn controlled by other factors. For instance, the renal–body fluid system controls both of them as follows: (1) An increase in fluid volume increases the cardiac output. (2) The increase in cardiac output then causes local vascular autoregulation because of excess blood flow through the tissues, and this increases the total peripheral resistance. Thus, both cardiac output and total peripheral resistance are "operator" pawns of the renal–body fluid system. Indeed, this renal–body fluid system does control the arterial pressure by operating through these two factors.

Likewise, the other pressure control mechanisms also work by changing either total peripheral resistance or cardiac output. For instance, activation of the renin-angiotensin system causes a marked increase in total peripheral resistance, and this increases the arterial pressure. Also, a sudden increase in the strength of the heart, which increases the cardiac output, can increase the arterial pressure, but only acutely, not chronically. In both these instances, unless one of the two determinants of the long-term arterial pressure level is changed at the same time, as discussed above, these initial effects of total peripheral resistance or cardiac output on arterial pressure will be compensated for by the secondary effects of the renal–body fluid system. Therefore, it is never safe to assume that a simple change in total peripheral resistance or cardiac output will necessarily alter the long-term level of arterial pressure. We will give a more detailed example in relation to total peripheral resistance in the following section.

Inability of a Primary Change in Total Peripheral Resistance (If Renal Resistance Does Not Change) to Affect the Long-Term Arterial Pressure Level. Let us at this point make a categorical statement and then see if we can support it. It is this: A primary increase in total peripheral resistance, if it does not affect renal function, will not cause a long-term increase in arterial pressure. This statement is quite contrary to what is frequently taught. However, in many instances in both normal physiology and in the clinical practice of medicine the total peripheral resistance changes markedly without affecting the long-term level of arterial pressure. Here are a few:

(1) Opening and closing a large shunt from the aorta to the vena cava (an arteriovenous fistula) will cause only a temporary change in arterial pressure. After a few days the arterial pressure returns exactly to the original level even though the total peripheral resistance sometimes may have been changed as much as 100 per cent.

(2) In thyrotoxicosis, in which the total peripheral resistance sometimes decreases to two-thirds normal, the arterial pressure may rise a little, but certainly does not fall.

(3) In the disease beriberi, the total peripheral resistance often decreases to less than one-half normal. However, the arterial pressure does not decrease except when the patient is about to die. Furthermore, when the patient is treated with appropriate vitamins, within hours the total peripheral resistance returns to normal — often as much as a 100 per cent increase — and yet the long-term level of arterial pressure does not change.

The same effect occurs in all other conditions that cause increased or decreased total peripheral resistance, except when the resistance to blood flow through the kidney is altered.

All the above-cited instances represent primary changes in total peripheral resistance. Yet the long-term level of arterial pressure does not change, for this reason: Remember, first, that neither the fluid intake level nor the renal output curve, the two determinants of the long-term arterial level, is altered by any of these conditions. Therefore, the acute changes in arterial pressure caused by the increase or decrease in total peripheral resistance immediately set the renal–body fluid pressure controller into operation, and fluid is either lost from the body or retained in the body until the arterial pressure returns *all the way back* to its original level. This is a manifestation of the infinite gain feature of the renal–body fluid pressure control system.

But Why Is Total Peripheral Resistance Usually High in Hypertensive Patients?

On the other hand, in at least 95 per cent of patients with hypertension, the total peripheral resistance is increased by approximately the same percentage as the increase in pressure. Why? It is because the same factors that shift the renal output curve to higher pressure ranges and thereby cause hypertension almost always also increase the total peripheral resistance, either directly or indirectly. For instance, we shall see shortly that angiotensin II shifts the renal function curve to a higher pressure range, thus causing chronic hypertension; but, simultaneously, angiotensin also constricts the peripheral arterioles, thereby increasing the total peripheral resistance. And we have already seen that when hypertension results from increased blood volume, the increased volume at first increases the cardiac output, but this in turn causes local vascular autoregulation throughout the body which increases the total peripheral resistance.

Therefore, the research evidence at present seems to be quite clear that a primary change in total peripheral resistance will not affect the long-term level of arterial pressure. Yet, for other reasons, in almost all instances of hypertension the total peripheral resistance is increased.

ROLE OF THE RENIN-ANGIOTENSIN-ALDOSTERONE SYSTEM IN LONG-TERM CONTROL OF ARTERIAL PRESSURE

Many research workers have emphasized the importance of the renin-angiotensin-aldosterone system in the long-term control of arterial pressure. Most of these workers have suggested two properties of this system that are important for this control:

(1) *Vasoconstriction.* Whenever blood flow to the kidneys diminishes, as explained in the previous chapter, renin is secreted and angiotensin is formed. The angiotensin in turn causes widespread vasoconstriction throughout the body with consequent greatly increased total peripheral resistance. It has been suggested that this increased resistance is important not only for short-term control of pressure, as explained in the previous chapter, but also for long-term control. (But we have seen in the above discussion that a primary increase in total peripheral resistance without a change in renal function will not cause a long-term change in the pressure.)

(2) *Increased aldosterone secretion.* Other research workers have emphasized a second effect of the renin-angiotensin system. The formation of angiotensin II affects the adrenal cortex, causing increased secretion of aldosterone. The aldosterone in turn causes the kidneys to retain salt and water and therefore to increase extracellular fluid volume, blood volume, cardiac output, and arterial pressure. This certainly can have a long-term effect for increasing the arterial pressure, because it shifts the renal output curve to a higher level.

A third function of angiotensin: the direct effect on the kidneys to decrease renal output. In addition to the first two functions of the renin-angiotensin system, recent experiments have demonstrated a third that is even more important in causing long-term increases in the arterial pressure. This is an effect of angiotensin II acting directly on the kidneys, not mediating through aldosterone, to cause salt and water retention. This effect is illustrated in Figure 22–6 which shows (1) the normal renal output curve for sodium and (2) this curve measured when angiotensin II was infused continuously over a period of a month at a rate equal to only 3 times the normal rate of angiotensin formation in the body. Note

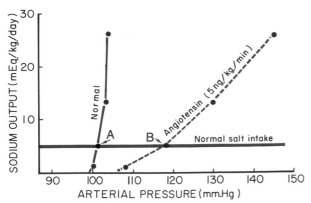

Figure 22–6. Effect on the chronic renal output curve caused by continuous infusion of angiotensin II at a low rate, showing marked shift of the curve toward a higher pressure range. (Drawn from data obtained by J. DeClue.)

that at each respective pressure level the output of salt (and also of water) is markedly decreased. Thus, the entire renal output curve is shifted to a higher pressure range. Therefore, one of the primary determinants of the long-term arterial pressure level, the renal output curve, is significantly altered by the renin-angiotensin system. Aldosterone causes a similar but lesser shift in the renal output curve toward higher pressure ranges, as pointed out above. Therefore, both the direct effect of the angiotensin and its indirect effect acting through increased aldosterone secretion will cause a definite long-term increase in the arterial pressure of a person.

Decreased Cardiac Output and Blood Volume in Types of Hypertension Producing Large Quantities of Angiotensin II (or Other Vasoconstrictors). Studies in patients with hypertension have shown that the blood volume and the cardiac output are both usually decreased below normal when large quantities of angiotensin are present in the circulating blood. On first thought, this seems to contradict the renal–body fluid mechanism for blood pressure control, but it does not. Let us explain this. First, remember that the renal–body fluid system for pressure control controls the blood volume to whatever level is required to attain the pressure needed for balance between fluid intake and output. And sometimes the volume required is less than the normal volume. For instance, in the presence of a large amount of angiotensin, potent vasoconstriction causes the vascular container to be reduced in size so that less volume is required. Also, the total peripheral resistance is very high so that less than normal cardiac output is required. Therefore, for good reason, the renal–body fluid system, in the presence of large quantities of angiotensin, will regulate both blood volume and the cardiac output at levels below normal, even though at the same time the arterial pressure is far above normal.

Yet, the student should think about the following statement for a long time until it is understood: Even in these low volume and low cardiac output states, it is still the renal–body fluid system that *determines* the level to which the arterial pressure will be regulated. And it is

the two determinants of this renal–body fluid system (the long-term determinants of the arterial pressure level discussed above) that remain the controllers of the long-term arterial pressure level: (1) the pressure range of the renal output curve and (2) the intake level of water and salt.

HYPERTENSION (HIGH BLOOD PRESSURE)

Now that we have discussed the principles of long-term control of arterial pressure, we can call on these principles to discuss the mechanisms by which a person can develop hypertension. In all of this discussion it will be good to keep the following basic dictum in mind: *Any factor that increases the pressure range of the renal output curve can also cause hypertension.* In the following pages, we will see how this occurs.

SOME CHARACTERISTIC TYPES OF HYPERTENSION

Figure 22–7 illustrates the normal renal output curve and four abnormal curves that frequently occur in hypertension. These abnormal curves are caused, respectively, by (1) reduced kidney mass, (2) excess aldosterone secretion by the adrenal glands or excess circulating angiotensin, (3) increased renal arterial resistance (Goldblatt kidneys), and (4) reduced glomerular filtration coefficient. Now, let us see how each of these conditions causes hypertension.

VOLUME-LOADING HYPERTENSION

Hypertension Caused by Excess Water and Salt Intake in Patients with Low Kidney Mass. The curve in Figure

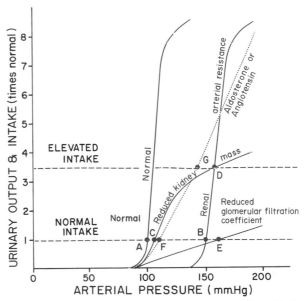

Figure 22–7. Abnormal urinary output curves and graphical analysis of the long-term level of arterial pressure in various hypertensive states caused by the abnormal renal function.

22–7 labeled "reduced kidney mass" illustrates the relationship between arterial pressure and urinary output when approximately 70 per cent of the kidney mass has been removed or destroyed but the remaining 30 per cent is still functioning normally. Point C illustrates the "equilibrium point" where this output curve crosses the normal intake curve; therefore, point C also depicts the level to which the arterial pressure will be controlled when the fluid intake is normal; this level is only 6 mm. Hg above normal. The reason for the smallness of this rise is that the remaining one third of the kidney mass is still sufficient to excrete the normal amounts of water and salt ingested each day with only a slight increase in arterial pressure.

On the other hand, when the fluid intake level is progressively increased, finally a limit comes above which the kidneys are unable to excrete the excess ingested water and salt unless the arterial pressure rises markedly. Therefore, when a person with reduced kidney mass ingests excess fluid and salt (especially excess salt), the extracellular fluid volume increases, the blood volume increases, the cardiac output increases, and the arterial pressure finally rises high enough to cause the kidneys to excrete the increased water and salt load. Thus, Figure 22–7 shows that when the fluid and salt intake is increased to 3.5 times normal, the pressure rises to point D where the elevated fluid intake line crosses the "reduced kidney mass" renal output curve. Thus, the pressure must rise to 160 mm. Hg before balance between intake and output will be achieved. Once the pressure does rise to this level, it will stabilize there and continue at this level as long as the intake remains high.

Figure 22–8 illustrates a typical experiment showing volume-loading hypertension in a group of dogs with 70 per cent of the renal mass removed. At the first circled point on the curve, the two poles of one of the kidneys were removed, and at the second circled point the entire opposite kidney was removed, leaving the animals with only 30 per cent of the normal renal mass. Note that removal of this amount of mass increased the arterial

pressure an average of only 6 mm. Hg. However, at this point the dogs were required to drink salt solution instead of normal drinking water. Because salt solution fails to quench the thirst, the dogs drank two to four times the normal amounts of volume, and within a few days their average arterial pressure had adjusted to a much higher level. After two weeks the dogs were allowed to drink tap water again instead of salt solution; the pressure returned to normal within two days. Finally, at the end of the experiment, the dogs were required to drink salt solution again, and this time the pressure rose much more rapidly and to a higher level because the dogs had already learned to tolerate the salt solution and therefore drank much more. Thus, this experiment demonstrates the principles of volume-loading hypertension.

Returning to Figure 22–7, one can see that the above experiment demonstrates the effects of changing the pressure control level successively from Point C to Point D, then back to Point C again, and finally once more to Point D, by simply changing the fluid and salt intake. Note also in Figure 22–7, however, that a similar increase in intake in an animal with normal kidneys would have increased the arterial pressure only 6 mm. Hg, which is exactly what is found when such an experiment is performed in normal animals.

Volume-Loading Hypertension in Patients Who Have No Kidneys but Are Being Maintained on an Artificial Kidney. When a patient is maintained on an artificial kidney, it is especially important to keep the body fluid volume at a normal level — that is, to remove the appropriate amount of water and salt each time that the patient is dialyzed. Figure 22–9 illustrates a study in a patient whose body fluid volume increased only 3 liters. This is shown by the lowermost curve labeled "weight," which is the means by which one assesses sudden increases in fluid volume in these patients. Note that when the weight increased, the cardiac output also increased markedly, and this in turn caused the arterial pressure to rise from an original normal value of 95 mm. Hg to a hypertensive level of 140 mm. Hg.

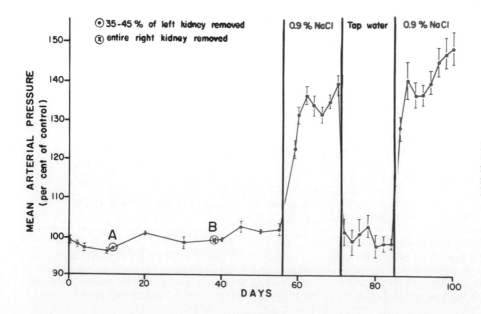

Figure 22–8. Effect on arterial pressure of drinking 0.9 per cent saline solution instead of water in four dogs with 70 per cent of their renal tissue removed. (Courtesy of Dr. Jimmy B. Langston.)

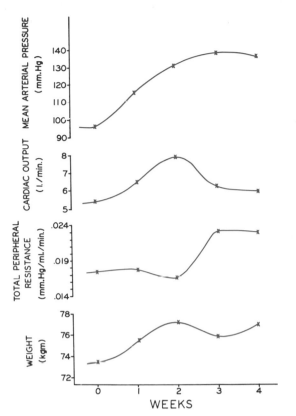

Figure 22–9. Hypertension in a patient whose two kidneys had been removed and whose body fluid volume had been increased by 3 liters. Note not only the increase in cardiac output at the onset of the hypertension, but also the rise in total peripheral resistance only *after* the hypertension had already developed. (Courtesy of Dr. Thomas Coleman.)

Role of "Autoregulation" in Volume-Loading Hypertension — Secondary Elevation of Total Peripheral Resistance. It is important now to observe very carefully the changes that occurred in total peripheral resistance in the patient studied in Figure 22–9. When the body fluid volume first increased, the cardiac output and the arterial pressure increased, but at first the total peripheral resistance did not increase. This same effect has been demonstrated many times in dogs in which this same type of condition has been created. That is, the initial rise in arterial pressure is caused entirely by elevation of the cardiac output. Then, during the subsequent week or so, the total peripheral resistance rises while the cardiac output returns back toward normal. Thus, the total peripheral resistance increases *after,* rather than before, the arterial pressure has become elevated. To state this another way, the rise in total peripheral resistance is secondary, rather than primary, to the hypertension. This effect can be explained as follows:

During the initial onset of the hypertension, blood flow through all the body's tissues is greatly increased. Because this flow is far greater than that required to supply the tissue's needs, the tissue vessels slowly constrict. Consequently, the total peripheral resistance increases while the cardiac output returns toward normal, a manifestation of the local tissue blood flow autoregulation mechanism discussed in Chapter 20. Thus, the effect of the increase in total peripheral resistance is not to cause the increased arterial pressure (because this has already risen before the resistance itself has risen) but to return the cardiac output back near to normal after the hypertension has occurred. Reduction of the cardiac output back to near-normal is very valuable to the body because it reduces the work load of the heart and therefore reduces the likelihood of heart failure.

Thus, it should always be remembered that pure *volume-loading hypertension is a high total peripheral resistance type of hypertension — not a high cardiac output type —* even though the increase in resistance occurs as a result of the hypertension and is not its primary cause.

Relative Importance of Salt Retention and of Water Retention in the Causation of Volume-Loading Hypertension. When one speaks of volume-loading hypertension, an increase in extracellular fluid volume is usually meant; therefore, the quantity of both salt and water increases. But which of these is it that causes the hypertension? In animals with their kidneys removed and which are maintained by using an artificial kidney, one can increase the salt or the water in the body independently of one another. When the salt is allowed to increase so that its concentration rises to as much as 20 per cent above normal, no hypertension occurs unless the volume rises simultaneously. On the other hand, when the volume increases along with the increase in salt, hypertension does ensue. Therefore, as would be predicted from an understanding of the fluid volume system for pressure control, it is the increase in volume and not the increase in salt *per se* that causes hypertension.

However, so long as the kidneys are even slightly functional it is usually almost impossible to increase the fluid volume of the body without a simultaneous increase in salt. The reason for this is simply because the kidneys will not retain water when there is not enough salt to go with it — a process that will be explained in Chapter 36. Furthermore, when a person with poorly excreting kidneys ingests large amounts of salt, this ingestion automatically causes him to become thirsty and to drink a commensurate amount of water. Thus, salt is the usual factor that determines the volume of extracellular fluid in the body. For this reason, most clinicians and experimenters dealing with experimental hypertension are prone to think more in terms of salt retention rather than water retention as the cause of hypertension, even though experiments demonstrate that salt retention causes hypertension only when it also increases the volume as well.

VASOCONSTRICTOR HYPERTENSION — HYPERTENSION CAUSED BY CONTINUOUS INFUSION OF ANGIOTENSIN II OR BY A RENIN-SECRETING TUMOR

A type of hypertension that shows sharp contrasts to the volume-loading type of hypertension is that caused by continuous infusion of angiotensin II or by a tumor of the renal juxtaglomerular cells that secretes renin. Fig-

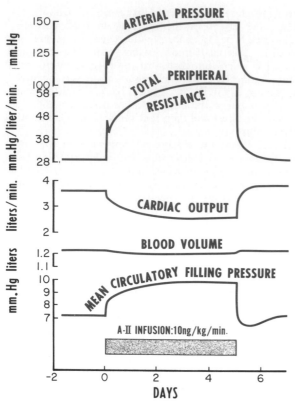

Figure 22–10. Effect of angiotensin infusion at a rate of 10 ng./kg./minute (approximately 6 times the normal rate of angiotensin formation in the body), illustrating marked increases in total peripheral resistance and arterial pressure, marked decrease in cardiac output, and a slight decrease in blood volume. (Drawn from data obtained in experiments in dogs by D. Young and A. Cowley.)

ure 22–10 illustrates some of the characteristics of this type of hypertension, showing the changes in arterial pressure, total peripheral resistance, cardiac output, and blood volume observed in a 10 day study of hypertension caused in a group of dogs by infusing 10 ng./kg. of angiotensin II per minute. Note the marked increases in both total peripheral resistance and arterial pressure but, conversely, a slight decrease in blood volume and a considerable decrease in cardiac output.

All the effects illustrated in Figure 22–10 can be explained by the basic principles of blood pressure regulation that were presented earlier in the chapter:

The increase in total peripheral resistance obviously results from the very potent effect of angiotensin II in constricting the arterioles. It has usually been stated that it is this increase in total peripheral resistance that causes the hypertension. However, we shall see in the discussion below that it is the effect of the angiotensin II on kidney excretion of salt and water that is critical in the maintenance of the long-term hypertension.

The slight decrease in blood volume is caused by a very interesting interplay between the increase in total peripheral resistance and the effect of angiotensin II on the kidneys. The increase in total peripheral resistance has an immediate effect of increasing the arterial pressure. On the other hand, the effect of angiotensin on the kidneys is to shift the renal output curve toward a higher

pressure range. If the rise in arterial pressure caused by the increase in total peripheral resistance is greater than the shift in the renal output curve, then one finds that the initial rise in arterial pressure is greater than can be sustained by the shift in the kidney output curve. Therefore, this excess pressure causes pressure diuresis and natriuresis, thus causing loss of fluid from the body and decreasing the blood volume slightly. In addition, part of the decrease in blood volume also results from a mild constrictor effect that angiotensin II exerts on the veins.

The *decrease in cardiac output* results almost entirely from the intense arteriolar constriction. In the following chapter we shall discuss the regulation of cardiac output, and we will see that an increase in the resistance to peripheral blood flow has a potent effect on decreasing venous return to the heart, which, in turn, causes the decreased cardiac output because the heart automatically adjusts the cardiac output to equal the venous return.

Thus, it is clear from Figure 22–10 that the hypertension caused by angiotension-induced vasoconstriction is, like volume-loading hypertension, a high resistance type of hypertension. However, there are subtle differences. In volume-loading hypertension, the total peripheral resistance is not increased quite as much as the increase in pressure, and the cardiac output is slightly above normal. On the other hand, in vasoconstrictor hypertension, the total peripheral resistance is increased somewhat in excess of the increase in arterial pressure, while the cardiac output is decreased to below normal.

Vasoconstrictor Hypertension Caused by a Pheochromocytoma. Another type of vasoconstrictor hypertension is that caused by a tumor called a *pheochromocytoma,* which is a tumor of the adrenal medulla that secretes large amounts of epinephrine and norepinephrine. Essentially the same type of vasoconstriction occurs as that in angiotensin hypertension, the total peripheral resistance rises very high, and the blood volume and cardiac output are likely to be low (except when the epinephrine stimulates the body metabolism so much that it sometimes elevates the cardiac output for this reason).

Role of the Renal–Body Fluid Pressure Control Mechanism in Determining Long-Term Arterial Pressure in Vasoconstrictor Hypertension. Let us for a moment suppose that angiotensin were infused into a person at a rate high enough to cause a very large increase in total peripheral resistance and therefore an acute rise in arterial pressure. Let us further suppose that the angiotensin had no effect at all on the renal output curve — that is, the angiotensin did not in any way affect the normal capability of the kidneys to excrete salt and water. In such a theoretical instance as this, the initial marked rise in arterial pressure would cause very intense pressure diuresis and pressure natriuresis. Consequently, the blood volume would decrease rapidly and the arterial pressure eventually return to normal, because it is only at a normal arterial pressure that urinary output would come to balance with the salt and water intake. Therefore, the increase in total peripheral resistance could not sustain a long-term increase in arterial pressure. Instead, to sustain the increased pressure, it is essential that the kidney output curve be shifted to a high pressure range. This effect of angiotensin on the renal output curve was illustrated in Figure 22–6. Thus, strange as it may seem, even in the vasoconstrictor type of hypertension it is the effect of the vasoconstrictor on kidney output of salt and water that determines the arterial pressure level at which the hypertension stabilizes. This concept will help the student tremendously in understanding both blood pressure regulation and the different clinical hypertensive conditions.

We shall see in subsequent discussions in this chapter that most clinical types of hypertension show various degrees of either the pure volume-loading type of hypertension or the pure vasoconstrictor type of hypertension. Therefore, it is useful to keep in mind the characteristics of these two extremes among the different types of hypertension.

GOLDBLATT HYPERTENSION

One-kidney Goldblatt Hypertension. When one kidney is removed and a constrictor is placed on the renal artery of the remaining kidney, as illustrated in Figure 22–11, the initial effect is greatly reduced renal arterial

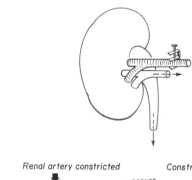

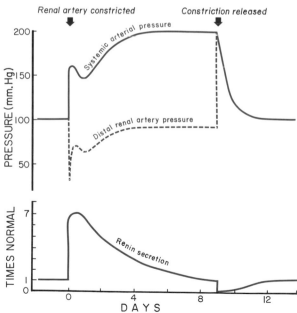

Figure 22–11. Effect of placing a constricting clamp on the renal artery of one kidney after the other kidney has been removed. Note the changes in systemic arterial pressure, renal artery pressure distal to the clamp, and rate of renin secretion. The resulting hypertension is called "one-kidney Goldblatt hypertension."

pressure, shown by the dashed curve in the lower portion of the figure. However, within a few minutes the systemic arterial pressure begins to rise and continues to rise for several days. The pressure usually rises rapidly for the first two hours, and during the next day returns to a slightly lower level, to be followed a day or so later by a second rise to a much higher pressure. When the systemic arterial pressure reaches its new stable pressure level, the renal arterial pressure will have returned almost all the way back to normal. The hypertension produced in this way is called *Goldblatt hypertension* in honor of Dr. Harry Goldblatt, who first studied the important quantitative features of hypertension caused by renal artery constriction.

The early rise in arterial pressure in Goldblatt hypertension is caused by the renin-angiotensin mechanism. Because of the poor blood flow through the kidney after acute application of the constrictor, large quantities of renin are secreted by the kidneys, as illustrated by the lowermost curve in Figure 20–11, and this causes angiotensin to be formed in the blood, as described in the previous chapter; the angiotensin in turn raises the arterial pressure acutely. However, the secretion of renin rises to a peak in a few hours but returns all the way back to normal within five to seven days because the renal arterial pressure by that time has also risen back to normal so that the kidney is no longer ischemic.

The second rise in arterial pressure is caused by fluid retention; within five to seven days the fluid volume has increased enough to raise the arterial pressure to its new sustained level. The quantitative value of this sustained pressure level is determined by the degree of constriction of the renal artery. Referring back to Figure 22–7, one notes that the renal output curve is shifted far to the right when the constrictor is applied to the renal artery (the curve labeled "renal arterial resistance"). This shift in the curve results from the fact that the aortic pressure must now rise to a much higher than normal level in order to make the renal arterial pressure distal to the constrictor rise high enough to cause normal urinary output.

Referring again to Figure 22–7, you will see that the shift in the renal output curve far to the right causes the arterial pressure to change from point A (100 mm. Hg) to point B (148 mm. Hg). That is, it is only at this high pressure level that the output of fluid and salt will be in balance with the intake.

Note especially from this analysis of the so-called "one-kidney Goldblatt" hypertension that it has two phases. The first phase is a vasoconstrictor type of hypertension caused by angiotensin, but this is only transient. The second stage is a volume-loading type of hypertension. However, it is often very difficult to tell that this second stage is true volume-loading hypertension because neither the blood volume nor the cardiac output are significantly elevated. Instead, the total peripheral resistance is increased. To understand this, recall that in pure volume-loading hypertension blood volume and cardiac output are elevated only the first few days during the onset; but after the first few days, volume-loading hypertension is a high resistance hypertension exactly as seen in the second stage of one-kidney Goldblatt hypertension. And the transient increases in blood volume and cardiac output that are normally seen in pure volume-loading hypertension are obscured by the angiotension vasoconstriction that occurs in Goldblatt hypertension during these early days.

Two-Kidney Goldblatt Hypertension. Hypertension also often results when the artery to one kidney is constricted while the artery to the other kidney is still normal. This hypertension results from the following mechanism: The renal output curve of the constricted kidney shifts far to the right because of decreased renal arterial pressure in this kidney. Also, the renal output curve of the "normal" kidney shifts to the right because of the renin produced by the ischemic kidney. This renin causes the formation of angiotension which circulates to the opposite kidney to cause it also to retain sodium and water. Thus, both kidneys, but for different reasons, become sodium and water retainers. Consequently, hypertension develops. Furthermore, the excessive secretion of renin by the ischemic kidney never ceases, because the so-called "normal" kidney is just normal enough that it will not allow the arterial pressure to rise high enough ever to return the blood flow to the ischemic kidney entirely to normal.

Hypertension Caused by Diseased Kidneys That Secrete Renin Chronically. Often patchy areas of one or both kidneys are diseased and ischemic while other areas of the kidneys are normal. When this occurs, the almost identical effects occur as in the "two-kidney" type of Goldblatt hypertension. That is, the patchy ischemic kidney tissue secretes renin, and this in turn, acting through the formation of angiotensin II, causes the remaining kidney mass also to retain salt and water. Indeed, one of the most common causes of renal hypertension is such patchy ischemic kidney disease.

HYPERTENSION IN TOXEMIA OF PREGNANCY

During pregnancy many patients develop hypertension, which is one of the manifestations of the syndrome called *toxemia of pregnancy*. The pathological abnormality that causes this hypertension is believed to be thickening of the glomerular membranes, which reduces the rate of fluid filtration from the glomeruli into the renal tubules. For obvious reasons, the pressure range of the renal output curve is elevated, and the long-term level of arterial pressure becomes correspondingly elevated. These patients are especially prone to hypertension when they eat large quantities of salt. This type of hypertension is illustrated in Figure 22–7 by the renal output curve labeled "reduced glomerular filtration coefficient." Note that this curve equates with the normal intake line at point E, which depicts the level of arterial pressure where the pressure stabilizes — in this instance at a pressure of 163 mm. Hg. And note also that the pressure rises markedly as the net fluid and salt intake increases.

NEUROGENIC HYPERTENSION

Many nervous disorders can cause temporary hypertension and sometimes even permanent hypertension. Temporary hypertension can be caused a few hours at a time by sympathetic stimulation of the blood vessels, which increases the vascular resistance throughout the body. However, to cause long-term hypertension the renal arterioles must be constricted continuously by the sympathetic stimulation for days at a time. This will increase the pressure range of the kidney output curve; therefore, the renal mechanism for pressure control will then maintain the pressure at an elevated level as long as the sympathetic stimulation continues.

Hypertension Caused by Artificial Stimulation of the Renal Sympathetic Nerves. Direct stimulation of the sympathetic nerves to the kidneys by an electrical stimulator will increase the arterial pressure. When the stimulator is first turned on, the pressure begins to rise and reaches a new steady-state elevated level in approximately three days, and it remains at this level as long as the stimulus continues. After removal of the stimulus, the pressure returns to normal in about two days. This type of experiment demonstrates once again the importance of the kidneys even in hypertension that is of neurogenic origin.

Hypertension Caused by Frustration or Pain. Prolonged sympathetic stimulation has been caused in animals by continual frustration or pain, such as by tantalizing the animal with food barely out of reach or by repeated electric shocks. Such animals develop mild to moderate hypertension as long as the abnormal conditions exist. However, within two to three weeks after the cause has been removed, the pressure usually returns to normal. Yet, it is believed by many clinicians that such abnormal sympathetic stimulation of the kidneys in patients for prolonged periods of time — perhaps for years — causes actual structural changes to occur gradually in the kidneys so that permanent pathological elevation of the kidney output curve develops. Then, even if the sympathetic stimulation is removed, the hypertension still persists. That is, neurogenic factors are believed finally to lead to hypertension that is sustained by a secondary renal abnormality.

HYPERTENSION CAUSED BY PRIMARY ALDOSTERONISM

A small tumor that secretes large quantities of aldosterone occasionally occurs in the adrenal glands. As will be discussed in detail in Chapter 35, the aldosterone increases the rate of reabsorption of salt and water in the distal tubules of the kidneys, thereby greatly reducing the rate of excretion of water and salt. Thus, the renal output curve is shifted to a higher pressure range, as illustrated in Figure 22–7. Consequently, mild to moderate hypertension occurs even at normal fluid intake, and when the intake of water and salt is greatly increased, severe hypertension results. Several weeks or months of excess aldosterone stimulation of the kidneys seems to cause pathological changes in the kidneys that cause the renal output curve to remain shifted to the right even after the tumor is removed — and therefore to cause continued hypertension.

ESSENTIAL HYPERTENSION

Approximately 90 per cent of all persons who have hypertension are said to have "essential hypertension," meaning hypertension of unknown origin. A few years ago, many of the patients who are now known to have one of the types of hypertension described earlier in this chapter would have been said to have essential hypertension. Thus, the more we learn about hypertension the more we can diagnose the cause precisely.

One group of patients formerly diagnosed to have essential hypertension are now known to have hypertension caused by glomerulosclerosis. These patients in early life had mild acute glomerulonephritis. They all supposedly recovered completely from the glomerulonephritis; then, during the ensuing 20 years they gradually developed what was called essential hypertension. Recent biopsies on the kidneys of these patients have demonstrated a sclerotic process in their glomeruli, with obviously decreased filtering capacity by the glomeruli. The effect of this process on renal function is almost the same as that which occurs in toxemia of pregnancy, which was illustrated in Figure 22–7 by the curve labeled "reduced glomerular filtration coefficient." Thus, it is very clear that this type of patient, who would have been said to have essential hypertension a few years ago, in reality has a type of hypertension caused by a kidney abnormality.

The Kidney Output Curve in Patients with Essential Hypertension.   Even though we do not know the cause of hypertension in most hypertensive patients and therefore must resort to the name "essential hypertension," we do know that the renal output curve is shifted to a higher pressure level in these patients than in the normal person. This is illustrated in Figure 22–12 by the curve labeled "essential hypertension." The curve shows that the urinary output is exactly normal (Point B) in these patients as long as the arterial pressure is at the hypertensive level of 150 mm. Hg. However, if the patient bleeds or if any other factor causes the arterial pressure to fall to 110 mm. Hg (almost to the pressure level for a normal person), the urinary output falls to zero (Point C) even though in the normal person the urinary output would be excessive at this pressure.

Though the cause of the shifted pressure level of the

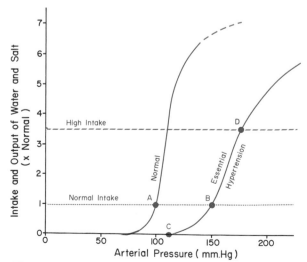

Figure 22–12.   Graphical analysis of elevated arterial pressure in essential hypertension. Note the shift of the urinary output curve to a much higher arterial pressure level in a patient with essential hypertension. (Reprinted from Guyton et al.: *Am. J. Med.*, 52:584, 1972.)

renal output curve in patients with essential hypertension is unknown, possible causes are: (1) thickening of the glomerular membranes or reduced surface areas of the glomeruli, either of which would cause a reduced glomerular filtration coefficient; or (2) increased afferent arteriolar resistance caused by vascular sclerosis. Either of these abnormalities would allow the patient with essential hypertension to have completely normal excretion of the urinary waste products, as will be explained in the following section.

Normal Excretion of Waste Products by Patients with Essential Hypertension.   Many clinicians have believed that essential hypertension does not have a renal origin because the kidneys of these patients excrete urinary waste products normally. Therefore, how could the kidneys be responsible for the hypertension, when the urinary function of the kidneys is so normal? The answer to this lies in the following observations: when the arterial pressure falls from the hypertensive level to normal, the kidneys of the patient with essential hypertension then retain the waste products in the blood, and the patient becomes uremic if the pressure is kept at the normal level. Thus, the elevated pressure is essential for the normal excretory function of the kidneys to occur, illustrating that the kidneys are not functioning normally.

Provided the lesion affects all nephrons equally, the normal excretion of waste products by the kidneys is exactly what one would expect for any kidney lesion that occurs somewhere in the kidneys proximal to the tubules. The reason for this is that once the arterial pressure has risen to its hypertensive level, glomerular filtration in each glomerulus will then be exactly normal, and the glomerular filtrate will pass through the normal tubules and be processed normally. Therefore, one would expect normal urinary output of waste products — but only as long as the arterial pressure remains high enough to provide normal glomerular filtration.

Treatment of Essential Hypertension.   Essential hypertension can generally be treated by two different types of drugs: (1) a drug that will increase glomerular filtration, or (2) a drug that will decrease tubular reabsorption of salt and water. Those drugs that will increase glomerular filtration are the various vasodilator drugs. Some of these act by inhibiting sympathetic impulses to the kidney, and others act by direct paralysis of the smooth muscle in the walls of the blood vessels. The drugs that reduce reabsorption of water and salt by the tubules include, especially, drugs that block active transport of salt through the tubular wall; the blockage in turn prevents the osmotic reabsorption of water as well. Such substances that reduce reabsorption of salt (and consequently, of water as well) are called *natriuretics* or *diuretics;* they will be discussed in greater detail in Chapter 38.

TYPES OF KIDNEY DISEASE THAT CAUSE HYPERTENSION; TYPES THAT CAUSE UREMIA

Not all types of kidney disease cause hypertension. Indeed, simple loss of kidney mass will cause hypertension only if the patient simultaneously ingests too much salt and water, which increases the body fluid volume and sets into play the typical sequence of volume-loading hypertension. However, loss of kidney mass always causes retention of the end-products of metabolism, such as urea, creatinine, uric acid, hydrogen ions, phosphates, and so forth.

From a physiological point of view, one can determine which type of kidney disease will cause uremia and which type will cause hypertension according to the following classification:

1. Any kidney disease that reduces kidney mass but allows all the remaining nephrons to remain normal always causes a uremic tendency but rarely causes hypertension.

2. Any kidney abnormality that tends to reduce glomerular filtration without altering the ability of the kidney to reabsorb the filtrate, such as diseases that decrease the glomerular filtration coefficient or that increase the renal vascular resistance, will cause hypertension but no uremic tendency if all the glomeruli are affected equally. That is, retention of water and salt will occur until the arterial pressure rises high enough to form normal quantities of glomerular filtrate. Once glomerular filtration is normal, the tubules process the urine as usual, and no retention symptoms occur except for the retention of the water and salt that causes the hypertension.

3. Any factor that causes excess reabsorption by the tubules, particularly excess reabsorption of salt and water as occurs in primary aldosteronism, will cause buildup of water and salt in the body, with resultant hypertension. If anything, these persons have lower than normal retention of the metabolic end-products. Thus, there is pure hypertension with no tendency toward uremia unless there is superimposed kidney damage as well.

4. Any kidney disease that causes pure destruction of tubular epithelium will cause decreased reabsorption of water and salt, and, therefore, will cause a tendency to *hypotension* unless the person ingests large amounts of water and salt. This is a rare condition, but it does occur.

EFFECTS OF HYPERTENSION ON THE BODY

Hypertension can be very damaging because of two primary effects: (1) increased work load on the heart and (2) damage to the arteries themselves by the excessive pressure.

Effects of Increased Work Load on the Heart. Cardiac muscle, like skeletal muscle, hypertrophies when its work load increases. In hypertension, the very high pressure against which the left ventricle must beat causes it to increase in weight as much as 2- to 3-fold. This increase is not accompanied by quite as much increase in coronary blood supply as there is increase in muscle tissue itself. Therefore, *relative ischemia* of the left ventricle develops as the hypertension becomes more and more severe. In the late stages of hypertension, this can become serious enough that the person develops angina pectoris. Also, the very high pressure in the coronary arteries causes rapid development of coronary arteriosclerosis so that hypertensive patients

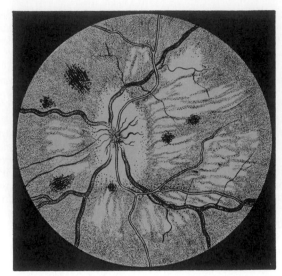

Figure 22–13. The retina of a person with malignant hypertension, illustrating edema of the optic disc, spasm of the arterioles, dark areas of retinal hemorrhages, and light areas of protein exudate.

tend to die of coronary occlusion at much earlier ages than do normal persons.

Effects of the High Pressure in the Arteries. High pressure in the arteries not only causes coronary sclerosis but also sclerosis of blood vessels throughout all the remainder of the body. The arteriosclerosis process causes blood clots to develop in the vessels and also causes the blood vessels to become weakened. Therefore these vessels frequently thrombose, or they rupture and bleed severely. In either case, marked damage can occur in organs throughout the body. The two most important types of damage which occur in hypertension are:

1. Cerebral hemorrhage, which means bleeding of a cerebral vessel with resultant destruction of local areas of brain tissue.

2. Hemorrhage of renal vessels inside the kidneys, which destroys large areas of the kidneys and therefore causes progressive deterioration of the kidneys and further exacerbation of the hypertension.

To illustrate the damaging effects of hypertension on the small vessels of the microcirculation, Figure 22–13 shows the retina of a person with very severe hypertension called *malignant hypertension.* This figure shows edema of the optic disc, spasm of the arterioles, dark areas in the retina that are the result of hemorrhages from the small arteries, and very light areas in the retina that represent protein that has exuded out of the blood vessels into the surrounding tissues.

REFERENCES

Beller, F. K., and MacGillivray, I. (eds.): Hypertensive Disorders in Pregnancy. Littleton, Mass., PSG Publishing Company, 1978.
Bianchi, G., and Bazzato, G. (eds.): The Kidney in Arterial Hypertension. Baltimore, University Park Press, 1979.
Blair-West, J. R.: Renin-angiotensin system and sodium metabolism. *Int. Rev. Physiol.,* 11:95, 1976.

Brown, J. J., et al.: Renal mechanisms in renovascular hypertension: Relevance to essential hypertension. In Hayase, S., and Murao, S. (eds.): Cardiology: Proceedings of the VIII World Congress of Cardiology, Tokyo, 1978. New York, Elsevier/North-Holland, 1979, p. 69.

Cevese, A., and Guyton, A. C.: Ischemic blood volume expansion in normal and areflexive dogs. Am. J. Physiol., 231:104–111, 1976.

Coleman, T. G., and Guyton, A. C.: Hypertension caused by salt loading in the dog. III. Onset transients of cardiac output and other circulatory variables. Circ. Res., 25:153, 1969.

Coleman, T. G., and Guyton, A. C.: The pressor role of angiotensin in salt deprivation and renal hypertension. Clin. Sci., 48:458–488, 1975.

Coleman, T. G., et al.: Regulation of arterial pressure in the anephric state. Circulation, 42:509, 1970.

Coleman, T. G., et al.: Experimental hypertension and the long-term control of arterial pressure. In MTP International Review of Science: Physiology. Baltimore, University Park Press, 1974, Vol. 1, p. 259.

Cowley, A. W., Jr., et al.: Open-loop analysis of the renin-angiotensin system in the dog. Circ. Res., 28:568, 1971.

Crawford, M. P., et al.: Renal servocontrol of arterial blood pressure. J. Appl. Physiol., 22:139, 1967.

DeClue, J. W., et al.: Subpressor angiotensin infusion, renal sodium handling, and salt-induced hypertension in the dog. Circ. Res., 43(4);503, 1978.

DeQuattro, V., and Campese, V. M.: Pheochromocytoma: Diagnosis and therapy. In DeGroot, L. J., et al. (eds.): Endocrinology. Vol. 2. New York, Grune & Stratton, 1979, p. 1279.

DeQuattro, V., and Myers, M. R.: The sympathetic nervous system and primary hypertension in man. In DeGroot, L. J., et al. (eds.): Endocrinology. Vol. 2. New York, Grune & Stratton, 1979, p. 1297.

Douglas, B. H., et al.: Hypertension caused by salt loading. II: Fluid volume and tissue pressure changes. Am. J. Physiol., 207:669, 1964.

Doyle, A. E., et al.: Pharmacological and Therapeutic Aspects of Hypertension. West Palm Beach, Fla., CRC Press, 1979.

Dustan, H. P.: Obesity and hypertension. In Lauer, R. M., and Shekelle, R. B. (eds.): Childhood Prevention of Atherosclerosis and Hypertension. New York, Raven Press, 1980.

Folkow, B., et al.: Importance of adaptive changes in vascular design for establishment of primary hypertension, studied in man and in spontaneously hypertensive rats. Circ. Res., (Suppl. 1) 32:2, 1973.

Freis, E. D.: Hemodynamics of hypertension. Physiol. Rev., 40:27, 1960.

Freis, E. D.: Salt in hypertension and the effects of diuretics. Annu. Rev. Pharmacol. Toxicol., 19:13, 1979.

Gant, N. F. (ed.): Pregnancy-Induced Hypertension. New York, Grune & Stratton, 1978.

Gordon, D. B. (ed.). Hypertension; The Renal Basis. New York, Academic Press, 1979.

Granger, H. J., and Guyton, A. C.: Autoregulation of the total systemic circulation following destruction of the central nervous system in the dog. Circ. Res., 25:379, 1969.

Guyton, A. C.: Acute hypertension in dogs with cerebral ischemia. Am. J. Physiol., 154:45, 1948.

Guyton, A. C.: Essential cardiovascular regulation—the control linkages between bodily needs and circulatory function. In Dickinson, C. J., and Marks, J. (eds.): Developments in Cardiovascular Medicine. Lancaster, England, MTP Press, 1978, p. 265.

Guyton, A. C.: Arterial Pressure and Hypertension. Philadelphia, W. B. Saunders Co., 1980.

Guyton, A. C., and Coleman, T. G.: Long-term regulation of the circulation; Interrelationships with body fluid volumes. In Physical Bases of Circulatory Transport Regulation and Exchange. Philadelphia, W. B. Saunders Co., 1967.

Guyton, A. C., and Coleman, T. G.: Quantitative analysis of the pathophysiology of hypertension. Circ. Res., 24:I-1, 1969.

Guyton, A. C., et al.: Physiological control of arterial pressure. Bull. N.Y. Acad. Med., 45:811, 1969.

Guyton, A. C., et al.: Arterial pressure regulation: Overriding dominance of the kidneys in long-term regulation and in hypertension. Am. J. Med., 52:584, 1972.

Guyton, A. C., et al.: A systems analysis approach to understanding long-range arterial blood pressure control and hypertension. Circ. Res., 35:159, 1974.

Guyton, A. C., et al.: Integration and control of circulatory function. Int. Rev. Physiol., 9:341, 1976.

Guyton, A. C., et al.: Synthesis of endocrine control in hypertension. Clin. Sci. Molec. Med., 51:319, 1976.

Guyton, A. C., et al.: The Hemodynamics of Hypertension. New York, Excerpta Medica, 1977, p. 5.

Guyton, A. C., et al.: Modern concepts of circulatory function. In Hayase, S., and Murao, S. (eds.): Cardiology: Proceedings of the VIII World Congress of Cardiology, Tokyo, 1978. New York, Elsevier/North-Holland, 1979, p. 60.

Guyton, A. C., et al.: Physiology of blood pressure regulation. In Lauer, R. M., and Shekelle, R. B. (eds.): Childhood Prevention of Atherosclerosis and Hypertension. New York, Raven Press, 1980.

Guyton, A. C., et al.: Salt balance and long-term blood pressure control. Annu. Rev. Med., 31:15–27, 1980.

Haber, E., and Zanchetti, A.: Pathogenesis of hypertension. In Hayase, S., and Murao, S. (eds.): Cardiology: Proceedings of the VIII World Congress of Cardiology, Tokyo, 1978. New York, Elsevier/North-Holland, 1979, p. 421.

Hall, J. E., et al.: Renal hemodynamics in acute and chronic angiotensin II hypertension. Am. J. Physiol., 235(3):F174, 1978.

Hall, J. E., et al.: Control of arterial pressure and renal function during glucocorticoid excess in dogs. Hypertension, 2:139, 1980.

Hart, J. T.: Hypertension. New York, Churchill Livingstone. 1980.

Hollenberg, N. K.: Pharmacologic interruption of the renin-angiotensin system. Annu. Rev. Pharmacol. Toxicol., 19:559, 1979.

Kaley, G.: Mechanisms of experimental hypertension. In Kaley, G., and Altura, B. M. (eds.): Microcirculation. Vol. III. Baltimore, University Park Press, 1977.

Langston, J. B., et al.: Effect of changes in salt intake on arterial pressure and renal function in partially nephrectomized dogs. Circ. Res., 12:508, 1963.

Laragh, J. H., and Sealey, J.: The renin-angiotensin-aldosterone hormonal system and regulation of sodium, potassium, and blood pressure homeostasis. In Orloff, F., and Berliner, R. W. (eds.): Handbook of Physiology. Sec. 8. Baltimore, Williams & Wilkins, 1973, p. 831.

Laragh, J. H., et al.: The renin axis and vasoconstriction volume analysis for understanding and treating renovascular and renal hypertension. Am. J. Med., 58:4, 1975.

Ledingham, J. M.: Blood pressure regulation in renal failure. J. Roy. Coll. Physicians Lond., 5:103, 1971.

Liard, J. F., et al.: Renin-aldosterone, body fluid volumes, and baroreceptor reflex in the development and reversal of Goldblatt hypertension in conscious dogs. Circ. Res., 34:549, 1974.

Lohmeier, T. E., et al.: Failure of chronic aldosterone infusion to increase arterial pressure in dogs with angiotensin-induced hypertension. Circ. Res., 43(3):381, 1978.

Mandal, A. K., and Wenzl, J. E.: Electron Microscopy of the Kidney in Renal Disease and Hypertension: A Clinicopathological Approach. New York, Plenum Press, 1978.

Manning, R. D., Jr., et al.: Essential role of mean circulatory filling pressure in salt-induced hypertension. Am. J. Physiol., 236:R40, 1979.

Manning, R. D., Jr., et al.: Hypertension in dogs during antidiuretic hormone and hypotonic saline infusion. Am. J. Physiol., 236:H314, 1979.

Marshall, A. J., and Barritt, D. W. (eds.): The Hypertensive Patient. Baltimore, University Park Press, 1979.

McCaa, R. E., et al.: Role of aldosterone in experimental hypertension. J. Endocrinol., 81:69, 1979.

Meyer, P.: Hypertension; Mechanisms and Clinical and Therapeutic Aspects. New York, Oxford University Press, 1980.

Meyer, P., and Schmitt, H. (eds.): Nervous System and Hypertension. New York, John Wiley & Sons, 1979.

Norman, R. A., Jr., et al.: Arterial pressure–urinary output relationship in hypertensive rats. Am. J. Physiol., 234(3):R98, 1978.

Norman, R. A., Jr., et al.: Renal function curves in normotensive and spontaneously hypertensive rats. Am. J. Physiol., 234:R98, 1978.

Onesti, G., and Klimt, C. R. (eds.): Hypertension: Determinants, Complications, and Intervention. New York, Grune & Stratton, 1978.

Peach, M. J.: Renin-angiotensin system: Biochemistry and mechanisms of action. Physiol. Rev., 57:313, 1977.

Ram, C. V. S., and Kaplan, N. M.: Alpha- and Beta-Receptor Blocking Drugs in the Treatment of Hypertension. Chicago, Year Book Medical Publishers, 1979.

Reid, I. A., et al.: The renin-angiotensin system. Annu. Rev. Physiol., 40:377, 1978.

Scriabine, A. (ed.): Pharmacology of Antihypertensive Drugs. New York, Raven Press, 1980.

Tan, S. Y., and Mulrow, P. J.: Aldosterone in hypertension and edema. In Bondy, P. K., and Rosenberg, L. E. (eds.): Metabolic Control in Disease, 8th Ed. Philadelphia, W. B. Saunders Co., 1980, p. 1501.

Vander, A. J.: Control of renin release. Physiol. Rev., 47:359, 1967.

Vaughan, E. D., Jr., and Peach, M. J. (eds.): Saralasin. New York, Springer-Verlag, 1979.

Yamori, Y., et al. (eds.): Prophylactic Approach to Hypertensive Diseases. New York, Raven Press, 1979.

Young, D. B., and Guyton, A. C.: Steady state aldosterone dose-response relationships. Circ. Res., 40(2):138, 1977.

23

Cardiac Output, Venous Return, and Their Regulation

Cardiac output is perhaps the most important single factor that we have to consider in relation to the circulation, for it is cardiac output that is responsible for transport of substances to and from the tissues.

Cardiac output is the quantity of blood pumped by the left ventricle into the aorta each minute, and *venous return* is the quantity of blood flowing from the veins into the right atrium each minute. Obviously, over any prolonged period of time, venous return must equal cardiac output. However, for a few heartbeats venous return and cardiac output need not be the same, since blood can temporarily increase or decrease in the heart and lungs.

NORMAL VALUES FOR CARDIAC OUTPUT

The normal cardiac output for the young healthy male adult averages approximately 5.6 liters per minute and for women, 10 to 20 per cent less. However, if we consider older people as well, the cardiac output for the average adult is very close to 5 liters per minute.

Cardiac Index. The cardiac output changes markedly with body size. Therefore, it has been important to find some means by which the cardiac outputs of different sized persons can be compared with each other. Experiments have shown that the cardiac output increases approximately in proportion to the surface area of the body. Therefore, the cardiac output is frequently stated in terms of the *cardiac index,* which is the *cardiac output per square meter of body surface area.* The normal human being weighing 70 kg. has a body surface area of approximately 1.7 square meters, which means that the normal average cardiac index for adults of all ages, both men and women, is approximately 3.0 liters per minute per square meter. The body surface area of persons of dif-

ferent heights and weights can be determined from the chart in Figure 71–7, Chapter 71.

Effect of Age. Figure 23–1 illustrates the change in cardiac index with age. Rising rapidly to a level greater than 4 liters per minute per square meter at 10 years of age, the cardiac index declines to about 2.4 liters per minute at the age of 80.

Effect of Posture. When a person rises from the reclining to the standing position, the cardiac output falls about 20 per cent if the person stands quietly because much of the blood "pools" in the lower part of the body. However, if the muscles become taut, as occurs when one prepares for exercise, the cardiac output rises 1 to 2 liters per minute. The causes of these changes are explained later in the chapter.

Effect of Metabolism and Exercise. The cardiac output usually remains almost proportional to the overall metabolism of the body. That is, the greater the degree of activity of the muscles and other organs, the greater

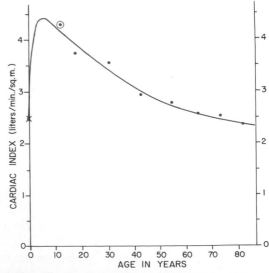

Figure 23–1. Cardiac index at different ages. (From Guyton, Jones, and Coleman: Circulatory Physiology: Cardiac Output and Its Regulation. Philadelphia, W. B. Saunders Company, 1973.)

274

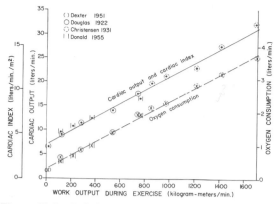

Figure 23–2. Relationship between cardiac output and work output (solid curve) and between oxygen consumption and work output (dashed curve) during different levels of exercise. (From Guyton, Jones, and Coleman: Circulatory Physiology: Cardiac Output and Its Regulation. Philadelphia, W. B. Saunders Company, 1973.)

also will be the cardiac output. This relationship is illustrated in Figure 23–2, which shows that as the work output during exercise increases, the cardiac output also increases in almost linear proportion. Note that in very intense exercise the cardiac output can rise to as high as 30 to 35 liters per minute in the young, well-trained athlete, which in about 5 to 7 times the normal control value.

Figure 23–2 also demonstrates that oxygen consumption increases in almost direct proportion to work output during exercise. We shall see later in the chapter that the increase in cardiac output probably results primarily from increased oxygen consumption.

REGULATION OF CARDIAC OUTPUT

PRIMARY ROLE OF THE PERIPHERAL CIRCULATION IN THE CONTROL OF CARDIAC OUTPUT: PERMISSIVE ROLE OF THE HEART

Control of Cardiac Output by Venous Return — the Frank-Starling Law of the Heart. It is worthwhile at this point to recall the Frank-Starling law of the heart, which was discussed in Chapter 13. This law states that, within the physiological limit of the heart, the heart will pump whatever amount of blood enters the right atrium and will do so without significant buildup of back pressure in the right atrium. In other words, the heart is an automatic pump that is capable of pumping far more than the 5 liters per minute that normally returns to it from the peripheral circulation. Consequently, the factor that normally determines how much blood will be pumped by the heart is the amount of blood that flows into the heart from the systemic circulation — not the pumping capacity

of the heart. To state this still another way, under normal circumstances the major factor determining the cardiac output is the *rate of venous return*.

Therefore, a large share of our discussion in this chapter will concern the factors in the peripheral circulation that determine the rate at which blood returns to the heart. However, there are times when the amount of blood attempting to return to the heart is greater than the amount that the heart can pump. Under these conditions the heart then becomes the limiting factor in cardiac output control, and the heart is said to *fail*.

Thus, the heart normally plays a *permissive role* in cardiac output regulation. That is, it is capable of pumping a certain amount of blood each minute and, therefore, will *permit* the cardiac output to be regulated at any value below this given permitted level. Thus, the normal human heart that is *not stimulated by the autonomic nervous system* permits a maximum heart pumping of about 13 to 15 liters per minute, but the actual cardiac output under resting conditions is only approximately 5 liters per minute because this is the normal level of venous return. Therefore, it is the peripheral circulatory system, not the heart, that sets this level of 5 liters per minute.

The concept that the heart plays a permissive role in cardiac output regulation is so important that it needs to be explained in still another way, as follows: The normal heart, beating at a normal heart rate, and having a normal strength of contraction, neither excessively stimulated by the autonomic nervous system nor suppressed, will pump whatever amount of blood flows into the right atrium, up to about 13 to 15 liters per minute. If any more than this tries to flow into the right atrium, the heart will not pump it without cardiac stimulation. Under normal resting conditions the amount of blood that normally flows into the right atrium from the peripheral circulation is about 5 liters; and since this 5 liters is within the permissive range of heart pumping, it is pumped on into the aorta. Therefore, the venous return from the peripheral circulatory system controls the cardiac output whenever the permissive level of heart pumping is greater than venous return.

Increase in Permissive Level for Heart Pumping in Hypertrophied Hearts. Heavy athletic training causes the heart to enlarge sometimes as much as 50 per cent. Coincident with this enlargement is an increase in the permissive level to which the heart can pump. Thus, even when the heart is not stimulated by the nervous system, the permissive level for a well-trained athlete might be as great as 20 liters per minute, rather than the normal value of about 13 to 15 liters per minute.

Increase in Permissive Level of Heart Pumping by Autonomic Stimulation of the Heart. There are times when the cardiac output must rise temporarily to levels greater than the normal permissive level of the heart. For instance, in heavy exercise by well-trained athletes, cardiac outputs as high as 35 liters per minute have been measured. Obviously, the resting heart would not be able to pump this amount of blood. On the other hand, *stimulation of the heart by the sympathetic nervous system increases the permissive level of heart pumping to approximately double normal.* This effect comes about by autonomic enhancement of both heart rate and strength of heart contraction. Furthermore, the increase in permissive level of heart pumping occurs within a few seconds after exercise begins, even before most of the increase in venous return occurs.

Reduction in Permissive Level for Heart Pumping in Heart Disease. Though the normal permissive level for heart pumping is usually much higher than the venous return, this is not necessarily true when the heart is diseased. Such conditions as myocardial infarction, valvular heart disease, myocarditis, and congenital heart abnormalities can reduce the pumping effectiveness of the heart. In these instances, the permissive level for heart pumping may fall below 5 liters per minute, indeed for a few hours even as low as 2 to 3 liters per minute. When this happens, the heart becomes unable to cope with the amount of blood that is attempting to flow into the right atrium from the peripheral circulation. Therefore, the heart is said to *fail,* meaning simply that it fails to pump the amount of blood that is demanded of it. Under these conditions, the heart becomes the limiting factor in cardiac output control. However, this is not the normal state. (We shall discuss cardiac failure in Chapter 26).

ROLE OF TOTAL PERIPHERAL RESISTANCE IN DETERMINING NORMAL VENOUS RETURN AND CARDIAC OUTPUT

Let us recall the formula for blood flow through blood vessels that was presented in Chapter 18:

$$\text{Blood Flow} = \frac{\text{Pressure (input)} - \text{Pressure (output)}}{\text{Resistance}}$$

Now if we apply this to venous return, the formula becomes:

$$\text{Venous Return} = \frac{\text{Arterial Pressure} - \text{Right Atrial Pressure}}{\text{Total Peripheral Resistance}}$$

Since the right atrial pressure remains very nearly zero, it is clear that the amount of blood that flows into the heart each minute (venous return) and that is pumped each minute (cardiac output) is determined by two prime factors: (1) the arterial pressure, and (2) the total resistance. When the arterial pressure remains normal, as it usually does, venous return and cardiac output are then inversely proportional to the total peripheral resistance. To state this still another way, every time a peripheral blood vessel dilates, the venous return and cardiac output increase. Furthermore, the more vessels that dilate in the peripheral circulation, the greater the cardiac output becomes.

Cardiac Output Regulation as the Sum of the Local Blood Flow Regulations Throughout the Body. As long as the arterial pressure remains normal, each local tissue in the body can control its own blood flow by simply dilating or constricting its local blood vessels. This mechanism, which was discussed in detail in Chapter 20, is the means by which each tissue protects its own nutrient supply, controlling the blood flow in response to its own needs.

Therefore, since the venous return to the heart is the sum of all the local blood flows through all the individual tissues of the body, all the local blood flow regulatory mechanisms throughout the peripheral circulation are the true controllers of cardiac output under normal conditions. This is an automatic mechanism that allows the heart to respond instantaneously to the needs of each individual tissue. If some tissues need extra blood flow and their local blood vessels dilate, the venous return increases automatically, and the cardiac output increases by an equivalent amount. If all the tissues throughout the body require increased blood flow at the same time, the venous return becomes very great, and the cardiac output increases accordingly.

Thus, the whole theory of normal cardiac output regulation is that the tissues control the output in accordance with their needs. Again, it must be stated that it is not the heart that controls the cardiac output under normal conditions; instead, the heart plays a permissive role that allows the tissues to do the controlling. The heart does this by always maintaining a permissive pumping capacity that is somewhat above the actual venous return — that is, except when the heart fails.

Effect of Local Tissue Metabolism on Cardiac Output Regulation, and Importance of Simultaneous Arterial Pressure Regulation. The most important factor that controls the local blood flows in the individual tissues is the metabolic rates of the respective tissues. Therefore, venous return and cardiac output are normally controlled in relation to the level of metabolism of the body. This effect was illustrated in Figure 23–2, which showed that the cardiac output increases directly in relation to work output during exercise and that the cardiac

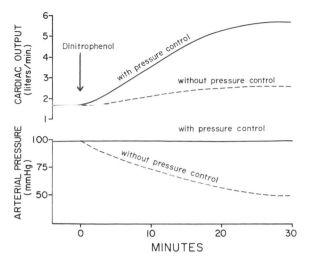

Figure 23–3. An experiment in a dog to demonstrate the importance of arterial pressure control as a prerequisite for cardiac output control. Note that with pressure control the metabolic stimulant dinitrophenol increases cardiac output; without pressure control, the arterial pressure falls and the cardiac output rises very little. (Data from experiments by Dr. M. Banet.)

output increases parallel to the increase in oxygen consumption, a measure of the rate of metabolism.

However, it is essential that the arterial pressure be maintained at a normal level if changes in metabolism are to regulate cardiac output. This is illustrated in Figure 23–3, which shows changes in cardiac output in response to an approximately five-fold increase in tissue metabolism in a dog. The increase in metabolism was caused by the toxic substance dinitrophenol, which causes the metabolic systems to increase their metabolic use of oxygen tremendously and, simultaneously, to dilate the local blood vessels supplying the tissues. Note in this figure that when the arterial pressure was normally controlled, the cardiac output increased approximately 300 per cent. On the other hand, when the nervous mechanisms for control of arterial pressure were blocked, the cardiac output increased only 100 per cent.

Therefore, once again, we can state that, under normal conditions, venous return and cardiac output are determined almost entirely by the degree of dilatation of the local blood vessels in the tissues throughout the body. However, this mechanism operates properly only in case the mechanisms for maintaining a normal arterial pressure, which were discussed in the preceding two chapters, are also functioning properly.

EFFECT OF ARTERIOVENOUS FISTULAE ON CARDIAC OUTPUT

The effect of an A-V fistula on cardiac output is discussed here because of the extremely important

lessons that can be learned from this condition. An A-V fistula is a direct opening from an artery into a vein that allows rapid flow of blood directly from the artery to the vein. In experimental animals, such a fistula can be opened and closed artificially, and one can study the changes in fistula flow, cardiac output, and arterial pressure. A typical experiment of this type in the dog is illustrated in Figure 23–4, showing an initial cardiac output of 1300 ml. per minute. After 15 seconds of recording normal conditions, a fistula from the aorta to the vena cava is opened through which 1200 ml. of blood flows per minute. Note that the cardiac output increases instantaneously about 1000 ml. per minute — about 84 per cent as much increase in cardiac output as there is fistula flow. (If the fistula is left open for 24 hours or more, the cardiac output increases 100 per cent as much as the fistula flow because of renal retention of fluid that returns the initially decreased arterial pressure to normal.) Then, when the fistula is closed, the cardiac output returns back to its original level.

Now for the lesson to be learned from this study: When the fistula is opened, the pumping ability of the heart is not changed at all; there is only a change in the peripheral circulation —opening of the A-V fistula. Nevertheless, within a few heartbeats after the fistula is opened, the cardiac output increases almost as much as there is increase in blood flow through the fistula. In other words, the fistula flow is added to the flow through the remainder of the body, and the two of these summate to determine the venous return and the cardiac output.

This experiment demonstrates the two cardinal features of normal cardiac output regulation: (1)

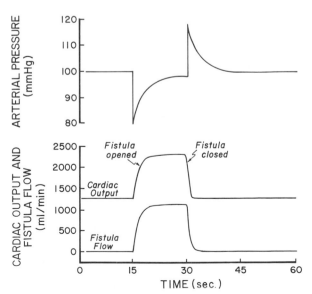

Figure 23–4. Effect of suddenly opening and suddenly closing an A-V fistula, showing changes in fistula flow, cardiac output, and arterial pressure.

almost all normal cardiac output regulation is determined by changes in total peripheral resistance, and (2) the normal heart has a permissive level of pumping that is considerably above the actual cardiac output. The heart, therefore, does not have to be stimulated for the increase in cardiac output to come about unless the venous return is excessive.

IMPORTANCE OF THE "MEAN SYSTEMIC FILLING PRESSURE" IN CARDIAC OUTPUT REGULATION

If the quantity of blood in the circulatory system is too little to fill the system adequately, blood will flow very poorly from the peripheral vessels into the heart. Therefore, the degree of filling of the circulation is one of the most important factors to determine the venous return to the heart and therefore also to determine the cardiac output.

The student may recall from Chapter 18 that the degree of filling of the circulatory system with blood can be expressed in terms of the *mean circulatory filling pressure*, and that the degree of filling of the systemic portion of the circulation can be expressed in terms of the *mean systemic filling pressure*. The mean systemic filling pressure is the pressure in all parts of the systemic circulation that is measured when blood flow in the circulatory system is suddenly stopped and the blood in the systemic vessels is redistributed so that the pressure in all the vessels is equal. Normally, the mean systemic filling pressure is 7 mm. Hg, but an increase in blood volume of 15 to 30 per cent doubles the mean systemic filling pressure, whereas a decrease in blood volume of this same amount reduces the mean systemic filling pressure to zero.

The mean systemic filling pressure is the *average* effective pressure of the blood in the peripheral circulation that tends to push the blood toward the heart. Experiments have demonstrated that, *when the peripheral resistance remains constant,* the rate of *venous return* from the systemic vessels through the veins to the heart is directly proportional to the *mean systemic filling pressure* minus *the right atrial pressure*. Therefore, whenever the mean systemic filling pressure falls to zero, the flow of blood returning to the heart likewise approaches zero.

Consequently, it is very important that the mean systemic filling pressure remain high enough at all times to supply the peripheral pressure needed to push blood from the peripheral vessels back to the heart. The quantitative relationship of the mean systemic filling pressure to cardiac output regulation will be considered in much more detail later in the chapter in connection with graphical analysis of cardiac output regulation.

REGULATION OF CARDIAC OUTPUT IN HEAVY EXERCISE, REQUIRING SIMULTANEOUS PERIPHERAL AND CARDIAC ADJUSTMENTS

Heavy exercise is one of the most stressful conditions to which the body is ever subjected. The tissues can require as much as 20 times normal amounts of oxygen and other nutrients, so that simply to transport enough oxygen from the lungs to the tissues sometimes demands a minimal cardiac output increase of five- to six-fold. This is greater than the amount of cardiac output the normal, unstimulated heart can pump. Therefore, to insure the massive increase in cardiac output that is required during heavy exercise, almost all factors that are known to increase cardiac output are called into play.

Muscle Vasodilatation Resulting from Increased Metabolism. By far the most important factor that increases the cardiac output during exercise is the vasodilatation that occurs in all exercising muscles. The cause of the vasodilatation is the large increase in muscle metabolism during exercise. This leads to tremendously increased use of oxygen and other nutrients by the muscles, as well as formation of vasodilator substances, all of which act together to cause marked local vascular dilatation and greatly increased local blood flow. This local vasodilatation requires 5 to 15 seconds to reach full development after a person begins to exercise strongly; but once it does reach full development, the very great decrease in vascular resistance allows extreme quantities of blood to flow through the muscles and thence into the veins to be returned to the heart, thereby markedly increasing venous return and cardiac output.

Role of the Heart in Strenuous Exercise. When large numbers of muscles are exercising simultaneously, the peripheral vasodilatation may be so great and the venous return to the heart so voluminous that the "resting" heart simply cannot pump this extra amount of blood. Therefore, it is essential that the *permissive level of pumping* by the heart be greatly increased from its normal level of 13 to 15 liters per minute. This is achieved by mainly sympathetic stimulation of the heart (but also partly by a decrease in parasympathetic stimulation), which increases the permissive level to as high as 20 to 25 liters per minute in a normal person or as high as 35 liters in the trained athlete.

Therefore, in exercise, the cardiac output can increase two- to three-fold without stimulation of the heart. But to go above this amount it is absolutely essential that the heart be nervously excited to increase the permissive level of heart pumping.

Special Functions of the Sympathetic Nervous System During Exercise. In light exercise, the heart and the circulation do not need to be stimulated by the sympathetic nervous system. But in heavy exercise, sympathetic stimulation is an absolute essential. Sympathetic stimulation has multiple effects on the circulation that are critical in increasing the cardiac output to the very high levels required in heavy exercise. These effects are:

1. Strong sympathetic stimulation can increase the permissive level of heart pumping from 13 to 15 liters per minute to as high as 20 to 25 liters per minute (or even higher in the trained athlete) as noted.

2. Sympathetic stimulation constricts almost all the blood vessels throughout the body, especially the veins, and this increases the mean systemic filling pressure to as much as 2.5 times normal. As discussed, this increase in mean systemic filling pressure provides an extra push on the blood in the peripheral vessels to return this blood to the heart. This is absolutely essential to achieve enough venous return, and therefore to achieve enough cardiac output, in heavy exercise.

3. It is believed by some physiologists that a special component of the sympathetic nervous system, a so-called "vasodilator" component, is stimulated during exercise to cause direct vasodilatation of the muscle blood vessels. This, theoretically, could further enhance blood flow through the muscles and thereby help to supply the required nutrients, but its effects are still in doubt.

Mechanisms for Stimulating the Sympathetic Nervous System During Exercise. It is believed that at least three different mechanisms enhance sympathetic activity during exercise.

1. Simply thinking about exercise has the psychic effect of exciting the autonomic nervous centers, which increases the heart rate, increases the strength of heart contraction, and constricts the blood vessels throughout the body to increase the mean systemic filling pressure. Together, these effects can increase the cardiac output instantaneously as much as 50 per cent even before the exercise begins.

2. The same signals from the motor cortex that excite activity in the muscles are believed also to excite the sympathetic nervous system. This causes vasoconstriction in most tissues of the body besides in the exercising muscles, thus increasing the arterial pressure. The increase in pressure in turn provides an extra push to force blood through the muscles. Also, the signals from the motor cortex are believed to stimulate *sympathetic vasodilator fibers* that go directly to the muscles and increase muscle blood flow. It seems certain that this vasodilator mechanism operates during the first few seconds of exercise to provide increased blood flow before the local metabolic vasodilatation occurs in the muscles.

3. It has also been demonstrated that contraction of the muscles themselves elicits reflex stimulation of the sympathetic nervous system. Sensory endings in the muscles are probably excited by metabolic products produced during contraction, and signals then pass to the vasomotor center to excite the sympathetic system still more. In fact, some physiologists believe that this reflex stimulation of the sympathetic nervous system is the most important of the sympathetic stimulatory mechanisms, but this is not yet certain.

Role of Abdominal Contraction for Enhancing Venous Return and Cardiac Output. At the very onset of exercise, the initial tenseness of the body that is associated with exercise causes abdominal contraction and compression of the large venous reservoirs in the abdomen. This immediately increases the mean systemic filling pressure two- to three-fold, which can increase venous return and cardiac output as much as 30 to 80 per cent within one to two beats of the heart.

Summary. It is clear that a complex set of controls comes into play during exercise to allow the heart and the circulation to increase the cardiac output to the tremendous levels required for nutrient supply to the muscles. The increase in cardiac output occurs first when the person begins to anticipate exercise, which stimulates the sympathetic nervous system. Then, tensing of the abdominal muscles gives an additional surge of venous return that also increases the cardiac output. Finally, within seconds after the exercise actually begins, intense local metabolic vasodilatation occurs in the muscles themselves. Combined with the hyperdynamic state of the heart and the increased mean systemic filling pressure, this causes the further surge in the cardiac output up to levels between 20 and 35 liters per minute.

ABNORMALLY LOW AND ABNORMALLY HIGH CARDIAC OUTPUTS

LOW OUTPUT

Figure 23–5 illustrates the cardiac outputs in different pathological conditions, showing at the far right the conditions that cause abnormally low outputs and at the left the conditions that cause abnormally high outputs.

It is easy to understand those abnormalities of the circulation that cause low cardiac output. These fall into two different categories: (1) those abnormalities that cause the pumping effectiveness of the heart to fall too low, and (2) those that cause venous return to fall too low.

Decreased Cardiac Output Caused by Cardiac Factors. Whenever the heart becomes severely damaged, from whatever cause, its permissive level of pumping may fall below that needed for adequate blood flow to the tissues. Some examples of this include *severe myocar-*

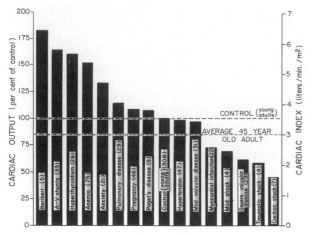

Figure 23–5. Cardiac output in different pathological conditions. The numbers in parentheses indicate number of patients studied in each condition. (From Guyton, Jones, and Coleman: Circulatory Physiology: Cardiac Output and Its Regulation. Philadelphia, W. B. Saunders Company, 1973.)

dial infarction, severe valvular heart disease, myocarditis, cardiac tamponade, and certain *cardiac metabolic derangements.* The effect of several of these is illustrated on the right in Figure 23–5, showing the low cardiac outputs that result.

When the cardiac output falls so low that the tissues throughout the body begin to suffer nutritional deficiency, the condition is called *cardiac shock.* This will be discussed fully in Chapter 26 in relation to cardiac failure.

Decrease in Cardiac Output Caused by Peripheral Factors — Decreased Venous Return. Any factor that interferes with venous return also can lead to decreased cardiac output. Some of these are:

1. *Decreased blood volume.* By far the most common peripheral factor that leads to decreased cardiac output is decreased blood volume, resulting most often from hemorrhage. It is very clear why this decreases the cardiac output: loss of blood decreases the mean systemic filling pressure to such a low level that there simply is not enough blood in the peripheral vessels to create the pressure required to push blood back to the heart. In other words, the mean systemic filling pressure falls to or near to zero, so that venous return also approaches zero.

2. *Acute venous dilatation.* On some occasions the peripheral veins become acutely vasodilated. This results most often when the sympathetic nervous system suddenly becomes inactive. For instance, fainting often results from sudden loss of sympathetic nervous system activity, which causes the peripheral capacitative vessels, especially the veins, to dilate markedly. This decreases the mean systemic filling pressure because the blood volume can no longer create adequate pressure in the now flaccid peripheral blood vessels. As a result the blood "pools" in the vessels and will not return to the heart.

3. *Obstruction of the veins.* On rare occasions, the large veins leading into the heart become obstructed, in which case the blood in the peripheral vessels cannot

flow back into the heart. Consequently the cardiac output falls markedly.

Regardless of the cause of low cardiac output, whether it be a peripheral factor or a cardiac factor, if ever the cardiac output falls below that level required for adequate nutrition of the tissues, the person is said to suffer *circulatory shock.* This can be lethal within a few minutes to a few hours. Circulatory shock is such an important clinical problem that it will be discussed in detail in Chapter 28.

HIGH CARDIAC OUTPUT — ROLE OF REDUCED TOTAL PERIPHERAL RESISTANCE IN CHRONIC HIGH CARDIAC OUTPUT CONDITIONS

The left side of Figure 23–5 identifies conditions that commonly cause cardiac outputs higher than normal. One of the distinguishing features of these conditions is that *they all result from chronically reduced total peripheral resistance.* And none of them result from excessive excitation of the heart itself, which we will explain below. For the present, let us look at some of the peripheral factors that can increase the cardiac output to above normal:

(1) *Beriberi.* This is a disease caused by insufficient quantity of the vitamin thiamine in the diet. Lack of this vitamin causes diminished ability of the tissues to utilize cellular nutrients, which in turn causes marked peripheral vasodilatation. The total peripheral resistance decreases sometimes to as little as one-half normal. Consequently, the long-term level of cardiac output also often increases to as much as 2 times normal.

(2) *Arteriovenous fistula (shunt).* Earlier in the chapter we pointed out that whenever a fistula (also called a shunt) occurs between a major artery and a major vein, tremendous amounts of blood will flow directly from the artery into the vein. This, too, greatly decreases the total peripheral resistance, and likewise increases the venous return and cardiac output.

(3) *Hyperthyroidism.* In hyperthyroidism, the metabolism of all the tissues of the body becomes greatly increased. Oxygen usage increases, and vasodilator products are released from the tissues. Therefore, the total peripheral resistance decreases markedly, and the cardiac output often increases to as much as 40 to 80 per cent above normal.

(4) *Anemia.* In anemia, two peripheral effects greatly decrease the total peripheral resistance. One of these is reduced viscosity of the blood, resulting from the decreased concentration of red blood cells. The other is diminished delivery of oxygen to the tissues because of the decreased hemoglobin, which causes local vasodilatation. As a consequence, the total peripheral resistance decreases greatly, and the cardiac output increases.

And, any other factor that decreases the total peripheral resistance chronically will also increase the cardiac output.

Failure of Increased Cardiac Pumping to Cause Prolonged Increase of the Cardiac Output. If the heart is suddenly stimulated excessively, the cardiac output often increases as much as 50 per cent. However, this is maintained no longer than a few minutes even though the heart continues to be strongly stimulated. There are two reasons for this. (1) Excess blood flow through the tissues causes automatic vasoconstriction of the blood vessels because of the autoregulation mechanism discussed in previous chapters, and this reduces the venous return and cardiac output back toward normal. (2) The slightly increased arterial pressure that results following acute cardiac stimulation raises the capillary pressure, and fluid filters out of the capillaries into the tissues thereby decreasing the blood volume and also decreasing the venous return back toward normal. Also, the increased pressure causes the kidneys to lose fluid volume as well until the arterial pressure and cardiac output return to normal.

Thus, all the known conditions that cause *chronic* elevation of the cardiac output result from decreased total peripheral resistance and not increased cardiac activity.

GRAPHICAL ANALYSIS OF CARDIAC OUTPUT REGULATION

The discussions of cardiac output regulation thus far are adequate for an understanding of the factors that control cardiac output in most simple conditions, and they also present a reasonably complete *qualitative* picture of cardiac output control. However, it is often important to obtain a much more quantitative understanding of cardiac output regulation. For instance, in cardiac failure and circulatory shock, the qualitative type of analysis is only partially satisfactory. More quantitative analyses will allow one to understand many details of these critical clinical conditions that are impossible to comprehend through qualitative thinking alone.

To perform quantitative analyses of cardiac output regulation, it is necessary to distinguish separately the two primary factors that are concerned with cardiac output regulation: (1) the pumping ability of the heart, and (2) the peripheral factors that affect flow of blood into the heart. Then, these two factors, the cardiac and the peripheral, must be put together in a quantitative way to see how they interact with each other to determine cardiac output. This process has been frequently accomplished by using electronic computers, but it can also be done more simply by using a type of graphical analysis that one can carry in his mind and that can be used daily in solving some of the quantitative problems of cardiac output regulation. In the following sections we will discuss this type of analysis. To begin, we will first characterize the pumping ability of the heart by use of *cardiac output curves,* and then we characterize the peripheral factors that affect venous return by use of *venous return curves.*

CARDIAC OUTPUT CURVES

Use of cardiac function curves to depict the ability of the heart to pump blood was discussed in Chapter 13, and the effects of several different factors on these function curves were demonstrated in Figures 13–10 and 13–11. For the purpose of understanding cardiac output regulation, the type of cardiac function curve required is that illustrated in Figures 23–6 and 23–7, which depicts the *cardiac output at different right atrial pressures* and is

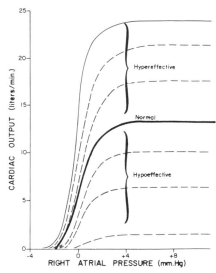

Figure 23–6. Cardiac output curves for the normal heart and for hypo- and hypereffective hearts. (From Guyton, Jones, and Coleman: Circulatory Physiology: Cardiac Output and Its Regulation. Philadelphia, W. B. Saunders Company, 1973.)

called the *cardiac output curve.* The normal cardiac output curve is labeled in each figure, while the other curves represent cardiac output in special conditions.

The basic factors that determine the precise characteristics of the cardiac output curve are (1) the *effectiveness of the heart as a pump* and (2) the *extracardiac pressure.*

Effectiveness of the Heart as a Pump. A heart that is capable of pumping greater quantities of blood than can the normal heart, such as a heart strongly stimulated by sympathetic impulses or a hypertrophied heart, is called a *hypereffective heart.* A weakened heart is called a *hypoeffective heart.* Figure 23–6 illustrates the effect of hypereffectivity or hypoeffectivity on the cardiac output curves, showing that the curve rises greatly in the case of the hypereffective heart and falls equally as much in the

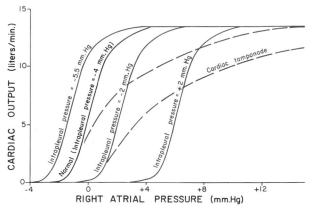

Figure 23–7. Cardiac output caused by changes in intrapleural pressure and by cardiac tamponade. (From Guyton, Jones, and Coleman: Circulatory Physiology: Cardiac Output and Its Regulation. Philadelphia, W. B. Saunders Company, 1973.)

case of the hypoeffective heart. Obviously, there are all degrees of hypereffectivity and hypoeffectivity.

Factors That Can Cause a Hypereffective Heart. Factors that can make the heart a better pump than normal include:

1. *Sympathetic stimulation,* which can increase the cardiac function curve to approximately 2 times normal.

2. *Hypertrophy of the heart,* which can increase the cardiac function curve to perhaps as much as 2 times normal.

3. *Inhibition of the parasympathetics to the heart,* which removes the parasympathetic tone, allowing the heart rate to increase, and, therefore, increases the pumping effectiveness of the heart.

Factors That Can Cause a Hypoeffective Heart. Obviously, any factor that decreases the ability of the heart to pump blood can cause hypoeffectivity. Some of the numerous factors are:

> *Myocardial infarction*
> *Valvular heart disease*
> *Increased arterial pressure*
> *Parasympathetic stimulation of the heart*
> *Inhibition of the sympathetics to the heart*
> *Congenital heart disease*
> *Myocarditis*
> *Cardiac anoxia*
> *Diphtheritic or other types of myocardial damage or toxicity*

Effect of Extracardiac Pressure on Cardiac Output Curves. Figure 23–7 illustrates the effect of changes in extracardiac pressure on the cardiac output curve. The normal extracardiac pressure, which is also equal to the normal intrapleural pressure, is −4 mm. Hg. Note in the figure that a rise in intrapleural pressure to −2 mm. Hg shifts the entire curve to the right by this same amount. This shift occurs because an extra 2 mm. Hg right atrial pressure is now required to overcome the increased pressure on the outside of the heart. Likewise, an increase in intrapleural pressure to +2 mm. Hg requires a 6 mm. Hg increase in right atrial pressure, which obviously shifts the entire curve 6 mm. Hg to the right. And, in the same manner, a more negative intrapleural pressure shifts the curve to the left.

Some of the factors that alter the intrapleural pressure and thereby shift the cardiac output curve include:

1. *Cyclic changes during normal respiration.*

2. *Breathing against a negative pressure,* which shifts the curve to a more negative atrial pressure (to the left).

3. *Positive pressure breathing,* which shifts the curve to the right.

4. *Opening the thoracic cage,* which increases the intrapleural pressure and shifts the curve to the right.

5. *Cardiac tamponade,* which means an accumulation of large quantities of fluid in the pericardial cavity around the heart with resultant increase in extracardiac pressure and shifting of the curve to the right. Note in Figure 23–7 that cardiac tamponade shifts the upper parts of the curves farther to the right than the lower parts because the extracardiac pressure rises to higher values as the chambers of the heart fill to increased volumes during high cardiac output.

Combinations of Different Patterns of Cardiac Output Curves. Figure 23–8 illustrates that the cardiac output curve can change as a result of simultaneous changes in

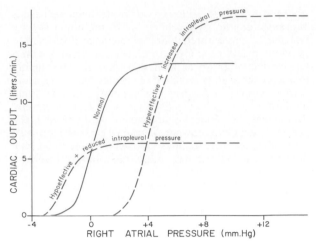

Figure 23–8. Combinations of two major patterns of cardiac output curves, showing the effect of alterations in both extracardiac pressure and effectiveness of the heart as a pump. (From Guyton, Jones, and Coleman: Circulatory Physiology: Cardiac Output and Its Regulation. Philadelphia, W. B. Saunders Company, 1973.)

extracardiac pressure and effectiveness of the heart as a pump. Thus, knowing what is happening to the extracardiac pressure and also the approximate capability of the heart as a pump, one can express the momentary ability of the heart to pump blood by a single cardiac output curve.

VENOUS RETURN CURVES

There still remains the entire systemic circulation that must be considered before a total analysis of circulatory function can be made. To analyze the function of the systemic circulation, we remove the heart and lungs from the circulation of an animal and replace these with a pump and artificial oxygenator system. Then, different factors, such as changes in blood volume and changes in right atrial pressure, are altered to determine how the systemic circulation operates in different circulatory states. In these studies, one finds two principal factors that affect the function of the systemic circulation in relation to cardiac output regulation. These are: (1) *the degree of filling of the systemic circulation,* which is measured by the *mean systemic filling pressure,* discussed in Chapter 18, and (2) the *resistance to blood flow* in the different segments of the systemic circulation.

The Normal Venous Return Curve. In the same way that the cardiac output curve relates cardiac output to right atrial pressure, the *venous return curve relates venous return to right atrial pressure.*

The very dark curve in Figure 23–9 is the normal venous return curve. This curve shows that as the right atrial pressure increases, it causes back pressure on the systemic circulation and thereby decreases venous return of blood to the heart. *If all circulatory reflexes are prevented from acting,* venous return decreases to zero when the right atrial pressure rises to equal the normal mean systemic filling pressure of about 7 mm. Hg. Such a slight rise in right atrial pressure causes a drastic decrease in venous return because the systemic circulation is a very distensible bag so that any increase in back pressure causes blood to dam up in this bag instead of

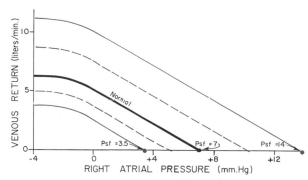

Figure 23–9. Venous return curves, showing the normal curve when the mean systemic filling pressure *(Psf)* is 7 mm. Hg, and showing the effect of altering the mean systemic filling pressure. (From Guyton, Jones, and Coleman: Circulatory Physiology: Cardiac Output and Its Regulation. Philadelphia, W. B. Saunders Company, 1973.)

returning to the heart. Lack of venous flow into the heart decreases the cardiac output, thereby decreasing the arterial pressure. Thus, at the same time that the right atrial pressure is rising, the arterial pressure is falling, and these two pressures come to equilibrium when all flow in the systemic circulation ceases at a pressure of 7 mm. Hg, which, by definition, is the mean systemic filling pressure.

Plateau in the Venous Return Curve Caused by Collapse of the Veins. When the right atrial pressure falls *below* zero — that is, below atmospheric pressure — venous return does not increase significantly. By the time the right atrial pressure has fallen to about −2 mm. Hg, the venous return will have reached a plateau; then it remains at this plateau level even though the right atrial pressure falls as low as −20 to −50 mm. Hg. This plateau is caused by collapse of the veins entering the chest. Low right atrial pressure sucks the walls of the veins together where they enter the chest, which prevents the negative pressure from sucking blood through the veins. Instead, the pressure in the veins immediately outside the chest remains almost exactly equal to atmospheric pressure (zero pressure). Therefore, for all practical purposes, the central venous pressure never falls below a value of 0 mm. Hg despite the fact that the right atrial pressure may fall to very negative values.

Effect on the Venous Return Curve of Changes in Mean Systemic Filling Pressure. Figure 23–9 also illustrates the effect of increasing or decreasing the mean systemic filling pressure. By definition, the mean systemic filling pressure is the pressure in all parts of the systemic circulation when there is no flow in the circulation. Therefore, at the point where the venous return curve reaches the zero venous return level, the right atrial pressure becomes equal to the mean systemic filling pressure. Note in Figure 23–9 that the normal mean systemic filling pressure *(Psf)* is 7 mm. Hg. For the uppermost curve in the figure, the mean systemic filling pressure is 14 mm. Hg, and for the lowermost curve it is 3.5 mm. Hg. These curves demonstrate that the greater the mean systemic filling pressure, which also means the greater the "tightness" with which the circulatory system is filled with blood, the more the venous return curve shifts upward and to the right. Conversely, the lower the mean systemic filling pressure, the more the curve shifts

downward and to the left. To express this another way, the greater the system is filled, the easier it is for blood to flow into the heart. And the less the filling, the more difficult it is for blood to flow into the heart.

Factors That Can Alter the Mean Systemic Filling Pressure. The factors that can alter the mean systemic filling pressure were discussed in Chapter 18. The three primary factors are: (1) The mean systemic filling pressure increases rapidly with an *increase in blood volume* — an acute increase in blood volume of 15 per cent or a chronic increase of about 30 per cent increasing the mean systemic filling pressure to approximately double normal. (2) Maximal *sympathetic stimulation* can increase the mean systemic filling pressure from 7 mm. Hg to about 17 mm. Hg, approximately a 2½-fold increase. (3) An *increase in contraction* of all the skeletal muscles throughout the body can increase the mean systemic filling pressure by compressing the blood vessels from the outside to about 25 mm. Hg.

Factors that decrease the mean systemic filling pressure include principally (1) *loss of sympathetic tone,* which decreases the mean systemic filling pressure from 7 mm. Hg to about 4 mm. Hg, or (2) *loss of blood volume,* which can decrease the mean systemic filling pressure to as little as zero.

"Pressure Gradient for Venous Return." When the right atrial pressure rises to equal the mean systemic filling pressure, all other pressures in the systemic circulation also approach this same pressure. Therefore, there becomes no pressure gradient for flow of blood toward the heart. However, when the right atrial pressure falls progressively lower than the mean systemic filling pressure, the flow to the heart increases proportionately, as one can see by studying any of the venous return curves in Figure 23–9. That is, *the greater the difference between the mean systemic filling pressure and the right atrial pressure the greater becomes the venous return.* Therefore, the difference between these two pressures is called the *pressure gradient for venous return.*

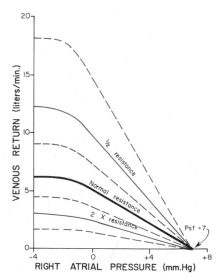

Figure 23–10. Venous return curves, depicting the effect of altering the "resistance to venous return." (From Guyton, Jones, and Coleman: Circulatory Physiology: Cardiac Output and Its Regulation. Philadelphia, W. B. Saunders Company, 1973.)

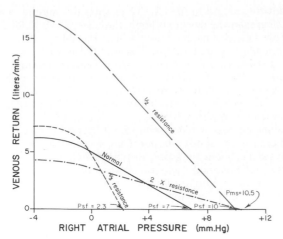

Figure 23–11. Combinations of the major patterns of venous return curves, illustrating the effects of simultaneous changes in mean systemic filling pressure *(Psf)* and in "resistance to venous return." (From Guyton, Jones, and Coleman: Circulatory Physiology: Cardiac Output and Its Regulation. Philadelphia, W. B. Saunders Company, 1973.)

Effect of Systemic Resistance on the Venous Return Curve. Figure 23–10 illustrates the effect of different systemic resistances on the venous return curve, showing that a decrease in resistance to one-half normal *rotates the curve upward*, while an increase in resistance *rotates the curve downward*. The venous return still becomes zero when the right atrial pressure rises to equal the mean systemic filling pressure, because when there is no pressure gradient to cause flow of blood, it makes no difference what the resistance is in the circulation — the flow is still zero.

Obviously, any factor that dilates the peripheral vessels decreases the resistance and rotates the function curve upward, while any factor that constricts the peripheral vessels rotates the curve downward. However, changes in venous resistance affect venous return about 8 times as much as similar resistance changes in the arteries, because far more blood is stored in the distensible blood vessels proximal to the veins than in the vessels proximal to the arteries.

Combinations of Venous Return Curve Patterns. Figure 23–11 illustrates the effects on the systemic function curve caused by simultaneous changes in mean systemic pressure and resistance to venous return, illustrating that both these factors can operate simultaneously.

ANALYSIS OF CARDIAC OUTPUT AND RIGHT ATRIAL PRESSURE USING CARDIAC OUTPUT AND VENOUS RETURN CURVES

In the complete circulation, the heart and the systemic circulation must operate together. This means: (1) The venous return from the systemic circulation must equal the cardiac output from the heart. (2) The right atrial pressure is the same for both the heart and for the systemic circulation.

Therefore, one can predict the cardiac output and right atrial pressure in the following way: (1) determine the

momentary pumping ability of the heart and depict this in the form of a cardiac output curve, (2) determine the momentary state of the systemic circulation and depict this in the form of a venous return curve, and (3) "equate" these two curves against each other as shown in Figure 23–12.

The two solid curves of Figure 23–12 depict both the *normal cardiac output curve* and the *normal venous return curve*. Obviously, there is only one point on the graph, point A, at which the venous return equals the cardiac output and at which the right atrial pressure is the same in relation to both the heart and the systemic circulation. Therefore, in the normal circulation the right atrial pressure, cardiac output, and venous return are all depicted by point A, called the *equilibrium point*.

Effect of Increased Blood Volume on Cardiac Output

A sudden increase in blood volume of about 20 per cent increases the cardiac output to about 2.5 to 3 times normal. An analysis of this effect is illustrated by the dashed curve of Figure 23–12. Immediately upon infusing the large quantity of extra blood, the increased filling of the system causes the mean systemic filling pressure to rise to 16 mm. Hg, which shifts the venous return curve upward and to the right. At the same time, the increased blood volume distends the blood vessels, thus reducing their resistance and thereby rotating the curve upward. As a result of these two effects, the venous return curve of Figure 23–12 changes from the solid curve to the dashed curve. This new curve equates with the cardiac output curve at point B, showing that the cardiac output increases 2.5 to 3 times and that the right atrial pressure rises to about +8 mm. Hg.

Compensatory Effects Initiated in Response to the Increased Blood Volume. The increased cardiac output caused by the increased blood volume lasts only a few

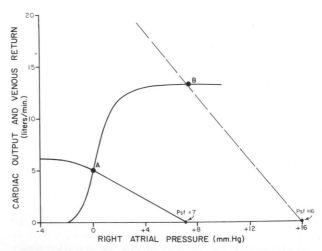

Figure 23–12. The two solid curves demonstrate an analysis of cardiac output and right atrial pressure when the cardiac output and venous return curves are known. Transfusion of blood equal to 20 per cent of the blood volume causes the venous return curve to become the dashed curve; as a result the cardiac output and right atrial pressure shift from point A to point B.

minutes, because several different compensatory effects immediately begin to occur: (1) The increased cardiac output increases the capillary pressure so that fluid begins to transude out of the capillaries into the tissues, thereby returning the blood volume toward normal. (2) The increased pressure in the veins causes the veins to distend gradually by the mechanism called *stress-relaxation,* especially causing the venous blood reservoirs such as the liver and spleen to distend. (3) The excess blood flow through the peripheral tissues causes autoregulatory increase in the peripheral resistance. These factors cause the mean systemic filling pressure to return back toward normal and also cause the resistance vessels of the systemic circulation to constrict. Therefore, gradually over a period of 20 to 40 minutes, the cardiac output returns almost to normal.

Effect of Sympathetic Stimulation on Cardiac Output

Sympathetic stimulation affects both the heart and the systemic circulation: (1) It makes the heart a stronger pump. (2) In the systemic circulation, it increases the mean systemic filling pressure because of contraction of the peripheral vessels, and it also increases the resistance to venous return. The uppermost curves of Figure 23–13 depict analyses of the effects of moderate and also maximal sympathetic stimulation on the cardiac output and right atrial pressure. The normal cardiac output and venous return curves are depicted by the very dark lines; moderate sympathetic stimulation is depicted by the long-dashed curves, and maximal sympathetic stimulation by the dot-dash colored curves.

Note that maximal sympathetic stimulation (the colored curves) increases the mean systemic filling pressure (depicted by the point at which the venous return curve reaches the zero venous return level) to 17 mm. Hg, and it also increases the pumping effectiveness of the heart by about 70 per cent. As a result, the cardiac output rises from the normal value at equilibrium point A to approximately double normal at equilibrium point D. Thus, different degrees of sympathetic stimulation can increase

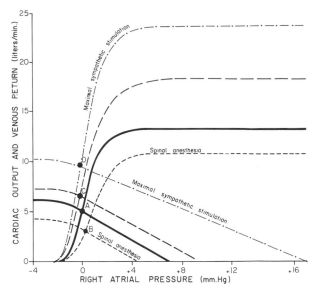

Figure 23–13. Analysis of the effect on cardiac output of (1) moderate sympathetic stimulation, (2) maximal sympathetic stimulation (colored curves), and (3) sympathetic inhibition caused by total spinal anesthesia. (From Guyton, Jones, and Coleman: Circulatory Physiology: Cardiac Output and Its Regulation. Philadelphia, W. B. Saunders Company, 1973.)

the cardiac output progressively to about 2 times normal — at least for short periods of time.

Effect of Sympathetic Inhibition

The sympathetic nervous system can be blocked completely by inducing *total spinal anesthesia* or by using some drug, such as *hexamethonium,* that blocks transmission of nerve impulses through the autonomic ganglia. The two lowermost curves in Figure 23–13, the short-dashed curves, show the effect of total sympathetic inhibition caused by spinal anesthesia, illustrating that

Figure 23–14. Graphical analysis of the changes in cardiac output and right atrial pressure with the onset of strenuous exercise.

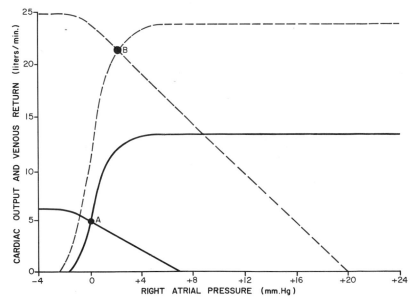

(1) the mean systemic filling pressure falls to about 4 mm. Hg, and (2) the effectiveness of the heart as a pump decreases to about 80 per cent of normal. The cardiac output falls from point A to point B, which is a decrease to about 60 per cent of normal.

Increase in Cardiac Output in Exercise

Several different factors operate in exercise to increase the cardiac output. These include: (1) intense sympathetic stimulation, which strengthens the heart and also increases the mean systemic filling pressure, (2) contraction of the muscles around the blood vessels, which further increases the mean systemic filling pressure, and (3) intense dilatation of the resistance vessels in the muscles, thereby decreasing the resistance to venous return. Figure 23–14 depicts the effects of these factors on the circulation, showing that (1) compression of the blood vessels by the muscles plus sympathetic stimulation of the vessel walls increases the mean systemic filling pressure from 7 mm. Hg to 20 mm. Hg, (2) sympathetic stimulation increases the strength of the heart as a pump by about 70 per cent, and (3) the local vasodilatation in the muscles rotates the venous return curve upward to the right. As a result, the cardiac output and venous return curves in exercise (the dashed curves) equate at point B, which is a cardiac output of 22 liters per minute, or 4½ times normal. The right atrial pressure hardly changes; indeed, when the degree of heart stimulation is great, the right atrial pressure can fall to below normal despite a very high cardiac output.

RIGHT VENTRICULAR OUTPUT VERSUS LEFT VENTRICULAR OUTPUT — BALANCE BETWEEN THE VENTRICLES

Obviously, the output of one ventricle must remain almost exactly the same as that of the other. Therefore, an intrinsic mechanism must be available for automatically adjusting the outputs of the two ventricles to each other. This operates as follows:

Let us assume that the strength of the left ventricle decreases suddenly so that left ventricular output falls below right ventricular output. This immediately causes more blood to be pumped into the lungs than is pumped into the systemic circulation. Consequently, the mean pulmonary filling pressure and the left atrial pressure rise while the mean systemic filling pressure and the right atrial pressure fall. If we apply the principles of analysis presented above to each of the ventricles separately, we will see that the increased *pulmonary filling pressure* increases left heart venous return and left ventricular output while the decrease in *systemic filling pressure* decreases right heart venous return and right ventricular output. This process continues until the output of the left ventricle rises to equal the falling output of the right ventricle. Thus, the outputs of the two ventricles become rebalanced with each other within a few beats of the heart. The same effects occur in the opposite direction when the strength of the right heart diminishes.

Problems of balance between the two ventricles are

especially important when myocardial failure occurs in one ventricle or when valvular lesions cause poor pumping by one side of the heart. These conditions will be discussed in more detail in Chapter 26 in relation to cardiac failure.

METHODS FOR MEASURING CARDIAC OUTPUT

In lower animals, the aorta, the pulmonary artery, or the great veins entering the heart can be cannulated and the cardiac output measured by any type of flowmeter. Also, an electromagnetic or ultrasonic flowmeter can be placed on the aorta or pulmonary artery to measure cardiac output. However, except in rare instances, in the human being, cardiac output is measured by indirect methods that do not require surgery. Two of the methods commonly used are the *oxygen Fick method* and the *indicator dilution method*.

PULSATILE OUTPUT OF THE HEART AS MEASURED BY AN ELECTROMAGNETIC OR ULTRASONIC FLOWMETER

Figure 23–15 illustrates a recording in the dog of blood flow in the root of the aorta made using an electromagnetic flowmeter. It demonstrates that the blood flow rises rapidly to a peak during systole and, then, at the end of systole, actually reverses for a small fraction of a second. It is this reverse flow that causes the aortic valve to close. And a minute amount of reverse flow continues throughout diastole to supply blood to the coronary vessels.

MEASUREMENT OF CARDIAC OUTPUT BY THE OXYGEN FICK METHOD

The Fick procedure is best explained by Figure 23–16, which shows the absorption of 200 ml. of oxygen from the lungs into the pulmonary blood each minute and also illustrates that the blood entering the right side of the heart has an oxygen concentration of approximately 160 ml. per liter of blood, while that leaving the left side has an oxygen concentration of approximately 200 ml. per liter of blood. From these data we see that each liter of blood passing through the lungs picks up 40 ml. of oxygen. And, since the total quantity of oxygen absorbed into the blood from the lungs each minute is 200 ml., a total of 5 1-liter portions of blood must pass through the pulmonary circulation each minute to absorb this amount

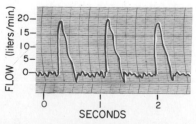

Figure 23–15. Pulsatile blood flow in the root of the aorta recorded by an electromagnetic flowmeter.

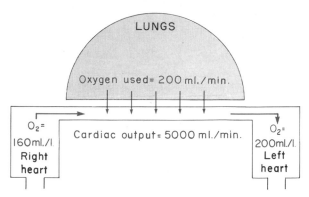

Figure 23–16. The Fick principle for determining cardiac output.

of oxygen. Therefore, the quantity of blood flowing through the lungs each minute is 5 liters, which is also a measure of the cardiac output. Thus, the cardiac output can be calculated by the following formula:

Cardiac output (liters/min.) =

$$\frac{O_2 \text{ absorbed per minute by the lungs (ml./min.)}}{\text{Arteriovenous } O_2 \text{ difference (ml./liter of blood)}}$$

In applying the Fick procedure, accurate determination of venous oxygen concentration can be achieved only by sampling blood directly from the right ventricle or preferably even the pulmonary artery, for blood in any one vein usually has a concentration of oxygen different from that in other veins, and even blood in the right atrium usually has not yet mixed satisfactorily. To obtain such a sample of mixed venous blood a catheter is usually inserted up the brachial vein of the forearm, through the subclavian vein, down to the right atrium, and finally into the right ventricle or pulmonary artery.

Blood used for determining the oxygen saturation in arterial blood can be obtained from any artery in the body, because all arterial blood is thoroughly mixed before it leaves the heart and therefore has the same oxygen concentration.

The rate of oxygen absorption by the lungs is usually measured by a "respirometer," which will be described in Chapter 71. In essence, this device is a floating chamber containing oxygen that sinks in the water as oxygen is removed from it, thus measuring the rate of oxygen usage by the rate at which it falls.

THE INDICATOR DILUTION METHOD

In measuring the cardiac output by the indicator dilution method a small amount of indicator, such as a dye, is injected into a large vein or preferably into the right side of the heart itself. This then passes rapidly through the right heart, the lungs, the left heart, and finally into the arterial system. If one records the concentration of the dye as it passes through one of the peripheral arteries, a curve such as one of the colored curves illustrated in Figure 23–17 will be obtained. In each of these instances 5 mg. of Cardio-Green dye was injected at zero time. In the top recording none of the dye passed into the arterial tree until approximately 3 sec-

onds after the injection, but then the arterial concentration of the dye rose rapidly to a maximum in approximately 6 to 7 seconds. After that, the concentration fell rapidly. However, before the concentration reached the zero point, some of the dye had already circulated all the way through some of the peripheral vessels and returned through the heart for a second time. Consequently, the dye concentration in the artery began to rise again. For the purpose of calculation, however, it is necessary to extrapolate the early downslope of the curve to the zero point, as shown by the dashed portion of the curve. In this way, the *time-concentration curve* of the dye in an artery can be measured in its first portion and estimated reasonably accurately in its latter portion.

Once the time-concentration curve has been determined, one can then calculate the mean concentration of dye in the arterial blood for the duration of the curve. In Figure 23–17, this was done by measuring the area under the entire curve, and then averaging the concentration of dye for the duration of the curve; one can see from the shaded rectangle straddling the upper curve of the figure that the average concentration of dye was approximately 0.25 mg./100 ml. blood and that the duration of the curve was 12 seconds. However, a total of 5 mg. of dye was injected at the beginning of the experiment. In order for blood carrying only 0.25 mg. of dye in each 100 ml. to carry the entire 5 mg. of dye through the heart and lungs in 12 seconds, it would be necessary for a total of 20 100-ml. portions of blood to pass through the heart during this time, which would be the same as a cardiac output of 2 liters per 12 seconds, or 10 liters per minute.

In the bottom curve of Figure 23–17, the blood flow through the heart was considerably slower, and the dye did not appear in the arterial system until approximately 6 seconds after it had been injected. It reached a maximum height in 12 to 13 seconds and was extrapolated to 0 at approximately 30 seconds. Averaging the dye concentrations over the 24-second duration of the curve, one finds again an average concentration of 0.25 mg. of dye in each 100 ml. of blood, but this time for a 24-second time interval instead of 12 seconds. To transport the total 5 mg. of dye, 20 100-ml. portions of blood would have had

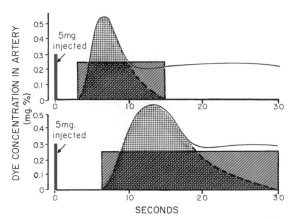

Figure 23–17. Dye concentration curves used to calculate the cardiac output by the dilution method. (The rectangular areas are the calculated average concentrations of dye in the arterial blood for the durations of the respective curves.)

to pass through the heart during the 24-second time interval. Therefore, the cardiac output was 2 liters per 24 seconds, or 5 liters per minute.

To summarize, the cardiac output can be determined from the following formula:

Cardiac output (ml./min.) =

$$\frac{\text{Milligrams of dye injected} \times 60}{\left(\begin{array}{l}\text{Average concentration of dye} \\ \text{in each milliliter of blood} \\ \text{for the duration of the curve}\end{array}\right) \times \left(\begin{array}{l}\text{Duration of} \\ \text{the curve} \\ \text{in seconds}\end{array}\right)}$$

Substances That Can Be Injected for Determining Cardiac Output by the Indicator Dilution Method. Almost any substance that can be analyzed satisfactorily in the arterial blood can be injected when making use of the indicator dilution method for determining cardiac output. However, for optimum accuracy it is necessary that the injected substance not be lost into the tissues of the lungs during its passage to the sampling site. The most widely used substance is *Cardio-Green*, a dye that combines with the plasma proteins and, therefore, is not lost from the blood.

REFERENCES

Banet, M., and Guyton, A. C.: Effect of body metabolism on cardiac output: Role of the central nervous system. *Am. J. Physiol., 220*:662, 1971.

Bishop, V. S., and Stone, H. L.: Quantitative description of ventricular output curves in conscious dogs. *Circ. Res., 20*:581, 1967.

Bishop, V. S., *et al.*: Cardiac function curves in conscious dogs. *Am. J. Physiol., 207*:677, 1964.

Brecher, G. A.: Venous Return. New York, Grune & Stratton, 1956.

Bruce, T. A., and Douglas, J. E.: Dynamic cardiac performance. *In* Frohlich, E. D. (ed.): Pathophysiology, 2nd Ed. Philadelphia, J. B. Lippincott Co., 1976, p. 5.

Coleman, T. G., *et al.*: Control of cardiac output by regional blood flow distribution. *Ann. Biomed. Eng., 2*:149, 1974.

Dobbs, W. A., Jr., *et al.*: Relative importance of nervous control of cardiac output and arterial pressure. *Am. J. Cardiol., 27*:507, 1971.

Dodge, H. T., and Kennedy, J. W.: Cardiac output, cardiac performance, hypertrophy, dilatation, valvular disease, ischemic heart disease, and pericardial disease. *In* Sodeman, W. A., Jr., and Sodeman, T. M. (eds.): Pathologic Physiology: Mechanisms of Disease, 6th Ed. Philadelphia, W. B. Saunders Co., 1979, p. 271.

Donald, D. E., and Shepherd, J. T.: Response to exercise in dogs with cardiac denervation. *Am. J. Physiol., 205*:393, 1963.

Dow, P.: Estimations of cardiac output and central blood volume by dye dilution. *Physiol. Rev., 36*:77, 1956.

Fermoso, J. D., *et al.*: Mechanism of decrease in cardiac output caused by opening the chest. *Am. J. Physiol., 207*:1112, 1964.

Green, J. F.: Determinants of systemic blood flow. *In* Guyton, A. C., and Young, D. B. (eds.): International Review of Physiology: Cardiovascular Physiology, III. Vol. 18. Baltimore, University Park Press, 1979, p. 33.

Grodins, F. S.: Integrative cardiovascular physiology: A mathematical synthesis of cardiac and blood vessel hemodynamics. *Q. Rev. Biol., 34*:93, 1959.

Guyton, A. C.: Determination of cardiac output by equating venous return curves with cardiac response curves. *Physiol. Rev., 35*:123, 1955.

Guyton, A. C.: Venous return. *In* Hamiton, W. F. (ed.): Handbook of Physiology. Sec. 2, Vol. 2. Baltimore, Williams & Wilkins, 1963, p. 1099.

Guyton, A. C.: Regulation of cardiac output. *N. Engl. J. Med., 277*:805, 1967.

Guyton, A. C.: An overall analysis of cardiovascular regulation. *Anesth. Analg., 56*:761, 1977.

Guyton, A. C.: Essential cardiovascular regulation—the control linkages between bodily needs and circulatory function. *In* Dickinson, C. J., and Marks, J. (eds.): Developments in Cardiovascular Medicine. Lancaster, England, MTP Press, 1978, p. 265.

Guyton, A. C., *et al.*: Venous return at various right atrial pressures and the normal venous return curve. *Am. J. Physiol., 189*:609, 1957.

Guyton, A. C., *et al.*: Relative importance of venous and arterial resistance in controlling venous return and cardiac output. *Am. J. Physiol., 196*:1008, 1959.

Guyton, A. C., *et al.*: Instantaneous increase in mean circulatory pressure and cardiac output at onset of muscular activity. *Circ. Res., 11*:431, 1962.

Guyton, A. C., *et al.*: Autoregulation of the total systemic circulation and its relation to control of cardiac output and arterial pressure. *Circ. Res., 28 (Suppl. 1)*:93, 1971.

Guyton, A. C., *et al.*: Circulation: Overall regulation. *Annu. Rev. Physiol., 34*:13, 1972.

Guyton, A. C., *et al.*: Systems analysis of arterial pressure regulation and hypertension. *Ann. Biomed. Eng., 1*:254, 1972.

Guyton, A. C., *et al.*: Cardiac Output and Its Regulation. Philadelphia, W. B. Saunders Co., 1973.

Ishikawa, N., *et al.*: Direct recording of cardiac output– and venous return–curves. *Jap. Heart J., 19*:775, 1978.

Jones, C. E., *et al.*: Cardiac output and physiological mechanisms in circulatory shock. *In* MTP International Review of Science: Physiology. Vol. 1. Baltimore, University Park Press, 1974, p. 233.

Keul, J.: The relationship between circulation and metabolism during exercise. *Med. Sci. Sports, 5*:209, 1973.

Knoop, A. A.: Physiological aspects of circulatory dynamics especially related to ageing as studied by displacement ballistocardiography and other cardiovascular methods. *Bibl. Cardiol., 30*:87, 1973.

Levy, M. N., and Zieske, H.: A closed circulatory system model. *Physiologist, 10*:419, 1967.

Longhurst, J. C., and Mitchell, J. H.: Reflex control of the circulation by afferents from skeletal muscle. *In* Guyton, A. C., and Young, D. B. (eds.): International Review of Physiology: Cardiovascular Physiology III. Vol. 18. Baltimore, University Park Press, 1979, p. 125.

Mitchell, J. H., and Wildenthal, K.: Static (isometric) exercise and the heart: Physiological and clinical considerations. *Annu. Rev. Med., 25*:369, 1974.

Nichols, W. W.: Continuous cardiac output derived from the aortic pressure waveform: A review of current methods. *Biomed. Eng., 8*:376, 1973.

Prather, J. W., *et al.*: Effect of blood volume, mean circulatory pressure, and stress relaxation on cardiac output. *Am. J. Physiol., 216*:467, 1969.

Sarnoff, S., and Mitchell, J. H.: The regulation of the performance of the heart. *Am. J. Med., 30*:747, 1961.

Stone, H. L., *et al.*: Ventricular function in cardiac-denervated and cardiac-sympathectomized conscious dogs. *Circ. Res., 20*:587, 1967.

Sugimoto, T., *et al.*: Effect of tachycardia on cardiac output during normal and increased venous return. *Am. J. Physiol., 211*:288, 1966.

Varat, M. A., *et al.*: Cardiovascular effects of anemia. *Am. Heart J., 83*:415, 1972.

Weisel, R. D., *et al.*: Current concepts measurement of cardiac output by thermodilution. *N. Engl. J. Med., 292*:682, 1975.

24

The Pulmonary Circulation

The quantity of blood flowing through the lungs is essentially equal to that flowing through the systemic circulation. However, there are problems related to distribution of blood flow and other hemodynamics that are special to the pulmonary circulation. Therefore, the present discussion is concerned specifically with the special features of blood flow in the pulmonary circuit and the function of the right side of the heart in maintaining this flow.

PHYSIOLOGIC ANATOMY OF THE PULMONARY CIRCULATORY SYSTEM

The Right Side of the Heart. As illustrated in Figure 24–1, the right ventricle is wrapped halfway around the

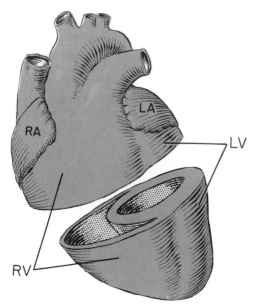

Figure 24–1. The anatomical relationship of the right ventricle to the left ventricle, showing the globular shape of the left ventricle and the half-moon shape of the right ventricle as it drapes around the left ventricle.

left ventricle. The cause of this is the difference in pressures developed by the two ventricles during systole. Because the left ventricle contracts with extreme force in comparison with the right ventricle, the left ventricle assumes a globular shape, and the septum protrudes into the right heart. Yet each side of the heart pumps essentially the same quantity of blood; therefore, the external wall of the right ventricle bulges far outward and extends around a large portion of the left ventricle, in this way accommodating about the same quantity of blood as does the left ventricle.

The muscle of the right ventricle is slightly more than one-third as thick as that of the left ventricle; this results from the difference in pressures between the two sides of the heart. Indeed, the wall of the right ventricle is only about 3 times as thick as the atrial walls, while the left ventricular muscle is about 8 times as thick.

The Pulmonary Vessels. The pulmonary artery extends only 4 centimeters beyond the apex of the right ventricle and then divides into the right and left main branches, which supply blood to the two respective lungs. The pulmonary artery is also thin, with a wall thickness approximately twice that of the venae cavae and one-third that of the aorta. The pulmonary arterial branches are all very short. However, all the pulmonary arteries, even the smaller arteries and arterioles, have much larger diameters than their counterpart systemic arteries. This, combined with the fact that the vessels are very thin and distensible, gives the pulmonary arterial tree a very large compliance, averaging 4 ml. per mm. Hg, which is almost equal to that of the entire systemic arterial tree. This large compliance allows the pulmonary arteries to accommodate the stroke volume output of the right ventricle.

The pulmonary veins, like the pulmonary arteries, are also short, but their distensibility characteristics are similar to those of the veins in the systemic circulation.

The Bronchial Vessels. Blood also flows to the lungs through several bronchial arteries, amounting to about 1 to 2 per cent of the total cardiac output. This bronchial arterial blood is *oxygenated* blood, in contrast to the partially deoxygenated blood in the pulmonary arteries. It supplies the supporting tissues of the lungs, including the connective tissue, the septa, and the large and small bronchi. After this bronchial arterial blood has passed through the supporting tissues, it empties into the pulmonary veins and *enters the left atrium* rather than passing back to the right atrium. Therefore, the left ventricular output is slightly greater than the right ventricular output.

The Lymphatics. Lymphatics extend from all the supportive tissues of the lung, beginning in the perivascular and peribronchial spaces and coursing to the hilum of the lung and thence mainly into the right lymphatic duct. Particulate matter entering the alveoli is usually removed rapidly via these channels, and protein is also removed from the lung tissues, thereby preventing edema.

PRESSURES IN THE PULMONARY SYSTEM

The Pressure Pulse Curve in the Right Ventricle. The pressure pulse curves of the right ventricle and pulmonary artery are illustrated in the lower portion of Figure 24–2. These are contrasted with the much higher aortic pressure curve shown above. The systolic pressure in the right ventricle of the normal human being averages approximately 22 mm. Hg, and the diastolic pressure averages about 0 to 1 mm. Hg, values that are only one-fifth to one-sixth those for the left ventricle.

Pressures in the Pulmonary Artery. During systole, the pressure in the pulmonary artery is essentially equal to the pressure in the right ventricle, as is also shown in Figure 24–2. However, after the pulmonary valve closes at the end of systole, the ventricular pressure falls, while the pulmonary arterial pressure remains elevated and then falls gradually as blood flows through the capillaries of the lungs.

As shown in Figure 24–3, the systolic pulmonary arterial pressure averages approximately 22 mm. Hg in the normal human being; the diastolic pulmonary arterial pressure, approximately 8 mm. Hg; and the mean pulmonary arterial pressure, 13 mm. Hg.

Pulmonary Arterial Pulse Pressure. The pulse pressure in the pulmonary arteries averages 14

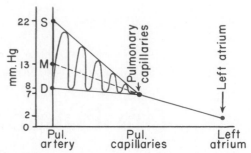

Figure 24–3. Pressures in the different vessels of the lungs.

mm. Hg, which is almost two-thirds as much as the systolic pressure. In previous discussions of pulse pressure in the systemic circulation, it has been pointed out that the less the compliance of an elastic reservoir that receives pulsatile injections of blood, the greater will be the pulse pressure. If it were not for the large compliance of the thin pulmonary arteries, the pulmonary arterial pulse pressure would be even greater than it is. However, another factor that keeps the pulse pressure from being higher is that about one-half the blood ejected by the right ventricle runs off from the pulmonary arteries into the pulmonary veins and left atrium at the same time that it is being ejected.

Pulmonary Capillary Pressure. The mean pulmonary capillary pressure, as diagrammed in Figure 24–3, has been estimated by indirect means to be approximately 7 mm. Hg. This will be discussed in detail later in the chapter in relation to fluid exchange functions of the capillary. However, we should note here that the capillary pressure of 7 mm. Hg is almost exactly halfway between the mean pulmonary arterial pressure of 13 mm. Hg and the left atrial pressure of 2 mm. Hg, indicating that the *arterial* and *venous resistances* of the lungs are approximately equal. This relationship is in marked contrast to that in the systemic circulation, in which the arterial resistance is 4 to 7 times as great as the venous resistance.

Left Atrial and Pulmonary Venous Pressure. The mean pressure in the left atrium and in the major pulmonary veins averages approximately 2 mm. Hg in the human being, varying from as low as 1 mm. Hg to as high as 4 mm. Hg.

It usually is not reasonable to measure the left atrial pressure directly in the normal human being because it is difficult to pass a catheter through the heart chambers into the left atrium. However, the left atrial pressure often can be estimated closely by measuring the so-called *pulmonary wedge pressure.* This is achieved by inserting a catheter through the right heart and pulmonary artery into one of the small branch pulmonary arteries and then pushing the catheter until it wedges tightly in

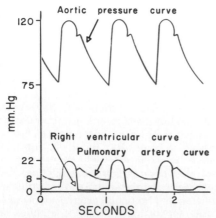

Figure 24–2. Pressure pulse contours in the right ventricle, pulmonary artery, and aorta.

the artery. The pressure then measured through the catheter, called the "wedge pressure," is about 5 mm. Hg. Since all blood flow has been stopped in the small artery and the blood vessels extending from the artery make direct connection through the pulmonary capillaries with the blood in the pulmonary veins, this wedge pressure is usually only 2 to 3 mm. Hg greater than the left atrial pressure. And when the left atrial pressure rises to high values, so also rises the pulmonary wedge pressure. This procedure is used frequently for studying left-heart dynamics in congestive heart failure.

THE BLOOD VOLUME OF THE LUNGS

The blood volume of the lungs is approximately 450 ml., about 9 per cent of the total blood volume of the circulatory system. About 70 ml. of this is in the capillaries, and the remainder is divided about equally between the arteries and veins.

The Lungs as a Blood Reservoir. Under different physiological and pathological conditions, the quantity of blood in the lungs can vary from as little as 50 per cent of normal up to as high as 200 per cent of normal. For instance, when a person blows air out so hard that high pressure is built up in the lungs — such as when blowing a trumpet — as much as 250 ml. of blood can be expelled from the pulmonary circulatory system into the systemic circulation. Also, loss of blood from the systemic circulation by hemorrhage can be partly compensated for by automatic shift of blood from the lungs into the systemic vessels.

Shift of Blood Between the Pulmonary and Systemic Circulatory Systems as a Result of Cardiac Pathology. Failure of the left side of the heart or increased resistance to blood flow through the mitral valve as a result of mitral stenosis or mitral regurgitation causes blood to dam up in the pulmonary circulation, thus sometimes increasing the pulmonary blood volume as much as 100 per cent, and also causing corresponding increases in the pulmonary vascular pressures.

On the other hand, exactly the opposite effects take place when the right side of the heart fails.

Because the volume of the systemic circulation is about 7 times that of the pulmonary system, a shift of blood from one system to the other affects the pulmonary system greatly but usually has only mild systemic effects.

BLOOD FLOW THROUGH THE LUNGS AND ITS DISTRIBUTION

The blood flow through the lungs is essentially equal to the cardiac output. Therefore, the factors that control cardiac output — mainly peripheral factors, as discussed in Chapter 23 — also control pulmonary blood flow. Under most conditions, the pulmonary vessels act as passive, distensible tubes that enlarge with increasing pressure and narrow with decreasing pressure. But, for adequate aeration it is important for the blood to be distributed to those segments of the lungs where the alveoli are best oxygenated. This is achieved by the following mechanism:

Effect of Low Alveolar Oxygen Pressure on Pulmonary Vascular Resistance — Automatic Local Control of Pulmonary Blood Flow Distribution. When alveolar oxygen concentration becomes very low, the adjacent blood vessels slowly constrict during the ensuing 3 to 10 minutes, the vascular resistance increasing to as much as triple normal. It should be noted specifically that this effect is *opposite to the effect* normally observed in systemic vessels, which dilate rather than constrict in response to low oxygen. However, this constrictor effect of low oxygen concentration does not occur in pulmonary arteries that have been isolated from all lung tissue. Therefore, it is believed that the low oxygen concentration causes some vasoconstrictor substance to be released from the lung tissue, this substance in turn promoting small arterial and arteriolar constriction. Unfortunately, research workers have not yet been able to isolate the vasoconstrictor substance.

This effect of low oxygen on pulmonary vascular resistance has an important function: to distribute blood flow where it is most effective. That is, when some of the alveoli are poorly ventilated so that the oxygen concentration in them becomes low, the local vessels constrict. This in turn causes most of the blood to flow through other areas of the lungs that are better aerated, thus providing an automatic control system for distributing blood flow through different pulmonary areas in proportion to their degrees of ventilation.

Paucity of Autonomic Nervous Control of Blood Flow in the Lungs. Though nerves innervate the lung tissues profusely, it is doubtful that these have a major function in normal control of pulmonary blood flow. Normally, stimulation of the vagal fibers to the lungs causes a very slight decrease in pulmonary vascular resistance, and stimulation of the sympathetics causes a slight-to-moderate increase in resistance. Sympathetic stimulation causes considerably more vasoconstriction in the presence of alveolar hypoxia, which suggests that the sympathetic nervous system might contribute significantly to the hypoxia mechanism discussed above to redistribute the blood flow.

Often research workers describe reflexes in the pulmonary vascular system that might be of clinical importance. For instance, it has been claimed that small emboli occluding the small pulmonary arteries cause a reflex that elicits sympathetic vasoconstriction throughout the lungs, this vasoconstriction then leading to an intense increase in pulmonary arterial pressure. However, the significance of this reflex is still not certain.

EFFECT OF HYDROSTATIC PRESSURE GRADIENTS IN THE LUNGS ON REGIONAL PULMONARY BLOOD FLOW

In Chapter 19 it was pointed out that the pressure in the foot of a standing person can be as much as 90 mm. Hg greater than the pressure at the level of the heart. This is caused by *hydrostatic pressure* — that is, by the weight of the blood itself. The same effect, but to a lesser degree, occurs in the lungs. In the normal, upright adult person, the lowest point in the lungs is about 30 cm. below the highest point. This represents a 23 mm. Hg pressure difference, about 15 mm. Hg of which are above the heart and 8 below. That is, the pulmonary arterial pressure in the uppermost portion of the lung of a standing person is about 15 mm. Hg less than the mean pulmonary arterial pressure, and the pressure in the lowest portion of the lungs is about 8 mm. Hg greater. Such pressure differences have profound effects on blood flow through the different areas of the lungs, which is illustrated by the lowest curve in Figure 24–4, which plots flow against hydrostatic level in the lung. Note that in the standing position there is very little blood flow in the top of the lung but 5 to 10 times this much flow in the lower lung. To help explain the reasons for these differences, the lung is often divided into three different zones, as illustrated in Figure 24–5, in which the patterns of blood flow are quite different. Let us explain these differences.

Zone 1. When the pulmonary arterial pressure is very low and the person is in a standing position, the pressure in the capillaries at the apices of the lungs may be less than the pulmonary alveolar pressure (which normally averages 0 mm. Hg). In such a condition, the capillaries remain collapsed all the time because the pulmonary capillary pressure is not great enough to oppose the tendency of the alveolar pressure to collapse the capillaries. Therefore, there will be zero blood flow. Thus, *zone 1* is an area of no flow.

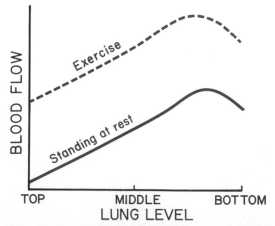

Figure 24–4. Blood flow at different levels in the lung of an upright person, both at rest and during exercise. Note that when the person is at rest, the blood flow is almost zero at the top of the lungs and most of the flow is through the lower lung.

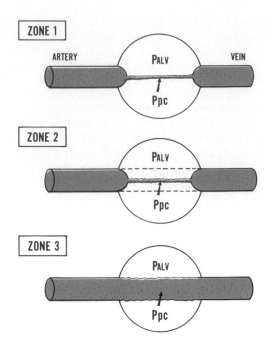

Figure 24–5. Mechanics of blood flow in the three different blood flow zones of the lung: *zone 1,* no flow, because alveolar pressure is greater than arterial pressure; *zone 2,* intermittent flow, because systolic arterial pressure rises higher than alveolar pressure but diastolic pressure falls below alveolar pressure; *zone 3,* continuous flow, because arterial pressure remains greater than alveolar pressure at all times.

Note, however, that a zone 1 area does not occur in the normal lung but instead occurs only in abnormal conditions that will be discussed in a subsequent section.

Zone 2. Recall that the normal pulmonary arterial pressure ranges between 8 mm. Hg during diastole and 22 mm. Hg during systole. Since some portions of the normal upright lung are at a hydrostatic pressure level as much as 15 mm. Hg above the mean level of the heart, during diastole the pulmonary arterial pressure cannot force blood into the upper portions of the normal lung. Yet, during systole blood does flow. Therefore, as illustrated in the middle panel of Figure 24–5, the alveolar capillaries collapse during the diastolic portion of the heart cycle but open during systole. Thus, in *zone 2,* blood flows intermittently.

In the normal lung under normal standing conditions, the lower part of zone 2 begins 7 to 10 cm. above the level of the heart and extends all the way to the top of the lungs.

Zone 3. In the lower part of the lung, the pulmonary vascular pressures are always higher than the pulmonary pressure, even during diastole. Therefore, as illustrated in the lower part of Figure 24–5, the alveolar capillaries remain distended all the time, and blood flows continuously from the arteries to the veins.

The zone 3 portion of the lung, the portion in which the pulmonary capillaries remain open all the time, normally begins 7 to 10 cm. above the level of the heart and extends all the way to the lowermost portions of the lung.

Factors That Can Change the Locations of Zones 1, 2,

and 3 in the Lungs. Though normally there is no zone 1 blood flow in the lungs, this very often does occur when the pulmonary arterial pressure falls. This is very common following hemorrhage. However, the most common cause of a zone 1 area in the lungs is breathing against a high air pressure, as occurs when a person is blowing on a musical instrument. Under these conditions, the alveolar pressure rises very high, so high that the normal pulmonary arterial systolic pressure cannot overcome the compressive effects on the capillaries of the alveolar pressure. Consequently, no blood then flows through this portion of the lungs.

Another factor that affects the zones is position of the body. In the lying position, all portions of the lung become zone 3, and blood flow in the different parts of the lung then becomes rather evenly distributed. In the upside-down position, blood flow to the apices can actually be greater than blood flow to the base of the lung.

Effect of Exercise on Blood Flow Through the Lungs. Referring again to Figure 24–5, one sees that the blood flow in all parts of the lung increases during exercise. However, the increase in flow in the top of the lung may be as much as 1000 per cent, while the increase in the lower part of the lung may be no more than 100 per cent. The reason for these effects is the considerably higher pulmonary pressures that occur during exercise, which effectively converts the entire lung into a zone 3 pattern of flow.

EFFECT OF INCREASED CARDIAC OUTPUT ON THE PULMONARY CIRCULATION DURING HEAVY EXERCISE

During heavy exercise the lungs are frequently called upon to absorb up to 20 times as much oxygen into the blood as they do normally. This absorption is achieved in two ways: (1) by increasing the number of open capillaries so that oxygen can diffuse more readily between the alveolar gas and the blood (a mechanism discussed above), and (2) by increasing cardiac output, with its concomitant increase in blood flow through the lungs — the blood thus picking up greater quantities of oxygen.

Fortunately, the cardiac output can increase to 4 to 6 times normal before pulmonary arterial pressure becomes excessively elevated; this effect is illustrated in Figure 24–6. As the blood flow into the lungs becomes increased, more and more capillaries open up; also, the pulmonary arterioles and capillaries expand. Therefore, the excess flow passes on through the capillary system without excessive increase in pulmonary arterial pressure. Indeed, as shown in the figure, the pressure rarely rises more than 2-fold despite as much as a 5- to 6-fold increase in flow.

This ability of the lungs to accommodate greatly increased blood flow during exercise with relatively little increase in pulmonary vascular pressure is important for at least two reasons: (1) it obviously conserves the energy of the right heart, and (2) it prevents a significant rise in pulmonary capillary pressure and therefore also prevents development of pulmonary edema during the increased cardiac output.

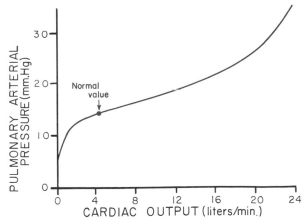

Figure 24–6. Effect on the pulmonary arterial pressure of increasing the cardiac output.

FUNCTION OF THE PULMONARY CIRCULATION WHEN THE LEFT ATRIAL PRESSURE RISES AS A RESULT OF LEFT HEART FAILURE

When the left heart fails or when mitral valvular disease blocks the outflow of blood from the left atrium, the left atrial pressure is increased and blood becomes dammed up in the pulmonary system. The initial effect is to increase the volume of blood in the pulmonary veins. Then, with further increase in left atrial pressure, the capillaries and the pulmonary arteries also become excessively filled with blood, and this damming of blood in the pulmonary arteries eventually raises the pulmonary arterial pressure. Also, an important effect of the rising capillary pressure is transudation of fluid into the pulmonary tissues and alveoli, an event called *pulmonary edema,* which will be discussed subsequently.

Quantitative Relationship Between Left Atrial Pressure and Pulmonary Arterial Pressure. Figure 24–7 shows the approximate relationship between the level of left atrial pressure and its effect on the pulmonary arterial pressure, provided that the pulmonary blood flow is normal. This figure shows that an increase in left atrial pressure from zero up to about 7 mm. Hg has relatively little effect on the pulmonary arterial pressure. The reason for this is that this moderate increase in left atrial pressure causes more and more opening up of capillaries. This opening of blood vessels progressively reduces the pulmonary vascular resistance, which mainly compensates for the back pressure effect of the rising left atrial pressure. However, once all these vessels have been opened, they do not expand much more with ease. Therefore, any further rise in left atrial pressure above the 7 mm. Hg level causes a marked rise in pulmonary arterial pressure, as shown in the figure.

Failure of Increasing Load on the Normal Left Ventricle to Affect Significantly the Pulmonary Circulation. When the work load of the left ventricle is greatly increased as a result of increased systemic arterial pressure or as a result of increased work output of the heart, the left atrial pressure rises a few mm. Hg — up to perhaps 5 to 6 mm. Hg under most maximal loads. However, as long as the left ventricle does not actually fail in its job of pumping blood, the left atrial pressure

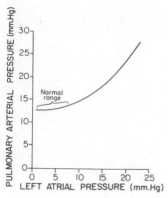

Figure 24–7. Effect of left atrial pressure on pulmonary arterial pressure.

will not rise high enough to alter pulmonary circulatory function significantly. Also, Figure 24–7 illustrates that these small increases in left atrial pressure have little back pressure effect on pulmonary arterial pressure and therefore do not measurably increase the load on the right ventricle.

PULMONARY CAPILLARY DYNAMICS

Exchange of gases between the alveolar air and the pulmonary capillary blood will be discussed in Chapter 40. It is important for us to note here that the alveolar walls are filled with capillaries that in most places lie so close to each other that they literally touch. Therefore, it has often been said that the capillary blood flows as a sheet rather than in individual vessels.

Pulmonary Capillary Pressure. Unfortunately, no direct measurements of pulmonary capillary pressure have been made. However, "isogravimetric" measurement of pulmonary capillary pressure, using a technique described in Chapter 30, has given a value of 7 mm. Hg. This is probably very nearly correct because the mean left atrial pressure is about 2 mm. Hg and the mean pulmonary arterial pressure is only 13 mm. Hg, so that the mean pulmonary capillary pressure must lie somewhere between these two values.

Length of Time Blood Stays in the Capillaries. From histologic study of the total cross-sectional area of all the pulmonary capillaries, it can be calculated that when the cardiac output is normal, blood passes through the pulmonary capillaries in about 1 second. Increasing the cardiac output shortens this time sometimes to less than 0.4 second, but the shortening would be much greater were it not for the fact that additional capillaries, which normally remain collapsed, open up to accommodate the increased blood flow. Thus, in less than 1 second, blood passing

through the capillaries becomes oxygenated and loses its excess carbon dioxide.

CAPILLARY EXCHANGE OF FLUID IN THE LUNGS

Negative Interstitial Fluid Pressure in the Lungs and Its Significance. Fluid exchange through the capillary membrane will be discussed in detail in Chapter 30, where it is pointed out that the most significant difference between pulmonary capillary dynamics and capillary dynamics elsewhere in the body is the very low capillary pressure in the lungs, about 7 mm. Hg, in comparison with a considerably higher "functional" capillary pressure elsewhere in the body, probably about 17 mm. Hg. Because of the very low pulmonary capillary pressure, the hydrostatic force tending to push fluid out the capillary pores into the interstitial spaces is also very slight. Yet, the colloid osmotic pressure of the plasma, about 28 mm. Hg, is a large force tending to pull fluid into the capillaries. Therefore, there is continual osmotic tendency to dehydrate the interstitial spaces of the lungs. Referring to Chapter 30 again, it is calculated there that the normal pulmonary interstitial fluid pressure is probably about −8 mm. Hg. That is, there is approximately 8 mm. Hg tending to pull the alveolar epithelial membrane toward the capillary membrane, thus squeezing the pulmonary interstitial space down to almost nothing. Electron micrographic studies have demonstrated this fact, the interstitial space at times being so narrow that the basement membrane of the alveolar epithelium is fused with the basement membrane of the capillary endothelium. As a result, the distance between the air in the alveoli and the blood in the capillaries is minimal, averaging about 0.4 micron in distance; this obviously allows very rapid diffusion of oxygen and carbon dioxide.

The details of fluid exchange through the pulmonary capillary membrane will be discussed in relation to overall capillary function in Chapter 30.

Mechanism by Which the Alveoli Remain Dry. Another consequence of the negative pressure in the interstitial spaces is that it pulls fluid from the alveoli through the alveolar membrane and into the interstitial spaces, thereby normally keeping the alveoli dry.

Pulmonary Edema. Pulmonary edema means excessive quantities of fluid either in the pulmonary interstitial spaces or in the alveoli. The normal pulmonary interstitial fluid volume is approximately 20 per cent of the lung mass, but this can increase perhaps 50 per cent in pulmonary edema. In addition, many times this much fluid can enter the alveoli and cause *intra-alveolar edema;* intra-alveolar fluid sometimes reaches 500 to 1000 per cent of the normal interstitial fluid.

Safety Factor Against Pulmonary Edema. The most common cause of pulmonary edema is greatly elevated pulmonary capillary pressure resulting from failure of the left heart and consequent damming of blood in the lungs. However, the pulmonary capillary pressure usually must rise to very high values before serious pulmonary edema will develop. The reason for this is the very high dehydrating force of the colloid osmotic pressure of the blood in the lungs. This effect is illustrated in Figure 24–8, which shows development of lung edema in dogs subjected to progressively increasing left atrial pressure. In this experiment no edema fluid developed in the lungs until left atrial pressure rose above 23 mm. Hg, which was approximately 3 mm. Hg greater than the colloid osmotic pressure of dog blood, about 20 mm. Hg. This experiment demonstrates that the hydrostatic pressure in the pulmonary capillary usually must rise a few mm. Hg above the colloid osmotic pressure before serious pulmonary edema can ensue. In the human being the colloid osmotic pressure is about 28 mm. Hg, so that pulmonary edema will rarely develop below 30 mm. Hg pulmonary capillary pressure. Thus, if the capillary pressure in the lungs is normally 7 mm. Hg and this pressure must usually rise above 30 mm. Hg before edema will occur, the lungs have a *safety factor against edema* of approximately 23 mm. Hg.

Safety Factor Against Edema in Chronic Elevation of Capillary Pressure. Patients with chronic elevation of pulmonary capillary pressure occasionally will not develop pulmonary edema even with pulmonary capillary pressures as high as 45 mm. Hg. The reason for this is probably extremely rapid run-off of fluid from the pulmonary interstitial spaces through the lymphatics, because when the pulmonary capillary pressure remains elevated for more than approximately two weeks, the pulmonary lymphatics enlarge as much as 6 to 10-fold, and lymph flow can increase as much as 20-fold above the normal resting level. This extra lymph flow gives a safety factor perhaps 15 or more mm. Hg above that which one normally has against edema.

Pulmonary Edema as a Result of Capillary Dam-age. Pulmonary edema can also result from local capillary damage in the lungs. This effect is often caused by bacterial infection, such as occurs in pneumonia, or by irritant gases such as chlorine, sulfur dioxide, or war gases — mustard gas, for instance. All these directly damage the alveolar epithelium and the endothelium of the capillaries, allowing rapid transudation of both fluid and protein into the alveoli and interstitial spaces.

Alveolar Fluid in Pulmonary Edema. Though mild degrees of pulmonary edema can be limited to an increase only of the pulmonary interstitial fluid, serious edema almost always causes fluid transudation into the alveoli themselves. It seems that the alveolar epithelium does not have enough strength to resist any significant degree of positive pressure in the interstitial spaces. Therefore, minute or large ruptures occur in the epithelium, and fluid flows readily out of the interstitial spaces into the alveoli.

PATHOLOGICAL CONDITIONS THAT OBSTRUCT BLOOD FLOW THROUGH THE LUNGS

Removal of Lung Tissue. When one *entire lung* is removed, the flow of blood through the remaining lung usually is well within the limits of compensation as long as the patient remains inactive. However, the patient thereafter has far less *pulmonary circulatory reserve* than does the normal person, for if the cardiac output increases more than 100 to 150 per cent above normal, the pulmonary arterial pressure begins to rise rapidly. This compares with the normal individual whose pulmonary circulatory reserve is 300 to 400 per cent above normal before the pressure rises seriously.

Massive Pulmonary Embolism. One of the most severe postoperative calamities in surgical practice is massive pulmonary embolism. Patients lying immobile in bed tend to develop extensive clots, especially in the veins of the legs, because of sluggish blood flow. Also, women, after delivery of their babies, frequently develop massive clots in the hypogastric veins. Such clots often break away from the initial sites of formation, particularly when the patient first walks after a long period of immobilization. The clots then flow to the right side of the heart and into the pulmonary artery. Such a free-moving clot is called an embolus.

Total blockage of only one of the major branches of the pulmonary artery usually is not immediately fatal because the opposite lung can accommodate all the blood flow. However, blood clots, as was discussed in Chapter 9, have a tendency to grow. Consequently, the embolus becomes larger and larger, and, as it extends into the other major branch of the pulmonary artery, the few remaining vessels that do not become plugged are taxed beyond their limit, and death ensues because of an inordinate rise in pulmonary arterial pressure and right-sided heart failure.

If the blood is rendered less coagulable by administration of anticoagulants so that the clot cannot grow, the life of the patient can often be saved; also, surgical removal of the clot is sometimes successful.

Diffuse Pulmonary Embolism. Sometimes many small blood clot emboli or small emboli of other substances besides blood clots enter the venous blood and

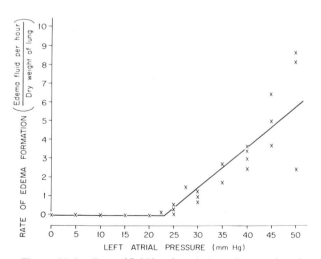

Figure 24–8. Rate of fluid loss into the lung tissues when the left atrial pressure (and also pulmonary capillary pressure) is increased. (From Guyton and Lindsey: *Circ. Res.,* 7:649, 1959.)

eventually clog the vessels of the lungs. One such instance is *fat embolism*. This occurs when a large volume of fatty tissue, such as that in the breast, becomes traumatized or infected so that the fatty material liquefies and enters the venous blood. It passes to the lungs, and because of the high surface tension of fat, the pulmonary arterial pressure often cannot force the fat globules through the small capillary vessels. Consequently, many of the small vessels of the lungs become blocked.

Another instance is *air embolism*. The air, like fat, has a large surface tension at the interface between the blood and air so that the globules of air cannot be deformed enough to be pushed through the capillaries.

The physical effects of diffuse pulmonary embolism on the pulmonary circulatory system are similar to those of massive pulmonary embolism — that is, increased pulmonary circulatory resistance with resulting increase in pulmonary arterial pressure and failure of the right side of the heart. However, with diffuse embolism, the pulmonary vessels appear to develop considerable vasospasm which adds additional resistance to flow besides that caused by the emboli themselves. It is believed that this vasospasm is caused by a sympathetic reflex, and, clinically, the stellate ganglion is often anesthetized to block the reflex.

Patients with this type of embolism exhibit a rapid respiratory rate because of local irritation by the emboli in the lungs and also because of resultant ischemia throughout the body.

Emphysema. Pulmonary emphysema means literally too much air in the lungs, and it is usually characterized by destruction of many of the alveolar walls. This causes the adjacent alveoli to become confluent, thereby forming large *emphysematous cavities* rather than the usual small alveoli. Obviously, loss of the alveolar septa greatly decreases the total alveolar surface area of the lungs and hinders gas exchange between the alveoli and the blood. This effect will be discussed in connection with the gas exchange functions of the lungs in Chapter 43.

Emphysema also has an important effect on the pulmonary vasculature, for each time an alveolar wall is destroyed, some of the small blood vessels of the pulmonary system are also destroyed, thus progressively increasing the pulmonary resistance and elevating the pulmonary arterial pressure.

In addition, a *physiologic increase in vascular resistance* also occurs in emphysema in the following way: Some of the emphysematous alveoli have poor exchange of air with the atmosphere. This poor ventilation results in alveolar hypoxia so that the adjacent blood vessels come continually under the influence of the hypoxic stimulus discussed earlier in the chapter, resulting in chronic vasoconstriction, which compounds the difficulties already caused by the pathological destruction of pulmonary vessels. Furthermore, the generalized hypoxia throughout the body causes increased cardiac output, and this, combined with the increased pulmonary resistance, sometimes results in serious pulmonary hypertension and right heart failure.

The physiologic increase in resistance can often be treated effectively by simple oxygen therapy. Therefore, emphysematous patients who also have right heart failure often experience rapid reversal of the failure along with decrease of the dyspneic symptoms after only a short period of oxygen breathing.

Unfortunately, the prevalence of emphysema is increasing rapidly because of cigarette smoking. Therefore, the physiological principles of its treatment are becoming progressively more important.

Diffuse Sclerosis of the Lungs. A number of pathological conditions cause excessive fibrosis in the supportive tissues in the lungs, and the fibrous tissue in turn contracts around the vessels. Some of these conditions are silicosis, tuberculosis, syphilis, and, to a lesser extent, anthracosis.

In early stages of diffuse sclerosis, the pulmonary arterial pressure often is normal as long as the person is not exercising, but just as soon as even mild exercise is performed, the pulmonary arterial pressure rises inordinately because the vessels do not have the ability to expand as much as normal pulmonary vessels do. In late stages of diffuse sclerosis the pulmonary arterial pressure remains elevated constantly; as a result, the right ventricle hypertrophies, and it may fail. Diffusion of gases through the alveolar membranes in most instances is impaired either because of decreased surface area or because of diminished blood flow to some alveoli.

Atelectasis. Atelectasis is the clinical term for collapse of a lung or part of a lung. This occurs often when the bronchi become plugged, because the pulmonary blood rapidly absorbs the air in the entrapped alveoli, which causes the alveoli to collapse.

Atelectasis also occurs when the chest cavity is opened to atmospheric pressure, for, when air is allowed to enter the pleural space, the elastic nature of the lungs causes them to collapse immediately.

When the elastic tissues of the lungs contract during atelectasis, they constrict not only the alveoli but also the blood vessels. This automatically decreases blood flow in the atelectatic portions of the lungs to as little as one-fourth normal, shifting the remaining three-fourths to the aerated portions. This is an important safety mechanism, for it prevents flow of major quantities of blood through collapsed, nonaerated pulmonary areas.

REFERENCES

Aviado, D. M.: The Lung Circulation. Vols. 1 and 2. New York, Pergamon Press, 1965.

Bakhle, Y. S., and Vane, J. R.: Pharmacokinetic function of the pulmonary circulation. *Physiol. Rev.,* 54:1007, 1974.

Bergofsky, E. H.: Mechanisms underlying vasomotor regulation of regional pulmonary blood flow in normal and disease states. *Am. J. Med.,* 57:378, 1974.

Cournand, A.: Some aspects of the pulmonary circulation in normal man and in chronic cardiopulmonary diseases. *Circulation,* 2:641, 1950.

Cumming, G.: The pulmonary circulation. *In* MTP International Review of Science: Physiology. Baltimore, University Park Press, Vol. 1, 1974, p. 93.

Fishman, A. P.: Dietary pulmonary hypertension. *Circ. Res.,* 35:657, 1974.

Fishman, A. P.: Hypoxia in the pulmonary circulation. *Circ. Res.,* 38:221, 1976.

Fishman, A. P., and Pietra, G.: Hemodynamic pulmonary edema. *In* Fishman, A. P., and Renkin, E. M. (eds.): Pulmonary Edema. Baltimore, Waverly Press, 1979, p. 79.

Fratantoni, J., and Wessler, S. (eds.): Prophylactic Therapy of Deep Vein Thrombosis and Pulmonary Embolism. Bethesda, Md., U.S. Dept. of Health, Education and Welfare, Public Health Service, National Institutes of Health, 1975.

Gaar, K. A., *et al.*: Pulmonary capillary pressure and filtration coefficient in the isolated perfused lung. *Am. J. Physiol., 213*:910, 1967.

Gil, J.: Influence of surface forces on pulmonary circulation. *In* Fishman, A. P., and Renkin, E. M. (eds.): Pulmonary Edema. Baltimore, Waverly Press, 1979, p. 53.

Giuntini, C.: Effects of cardiovascular disease on the lungs. *In* Hayase, S., and Murao, S. (eds.): Cardiology: Proceedings of the VIII World Congress of Cardiology, Tokyo, 1978. New York, Elsevier/North-Holland, 1979, p. 658.

Guyton, A. C.: Introduction to Part I: Pulmonary alveolar-capillary interface and interstitium. *In* Fishman, A. P., and Hecht, H. H. (eds.): The Pulmonary Circulation and Interstitial Space. Chicago, University of Chicago Press, 1969, p. 3.

Guyton, A. C., *et al.*: Dynamics of subatmospheric pressure in the pulmonary interstitial fluid. *In* Lung Liquids. Ciba Symposium. New York, Elsevier/North-Holland, 1976, p. 77.

Guyton, A. C., *et al.*: Forces governing water movement in the lung. *In* Pulmonary Edema. Washington, D.C., American Physiological Society, 1979, p. 65.

Harlan, J. C., *et al.*: Pressure-volume curves of systemic and pulmonary circuits. *Am. J. Physiol., 213*:1499, 1967.

Hebb, C.: Motor innervation of the pulmonary blood vessels of mammals. *In* Fishman, A. P., and Hecht, H. H. (eds.): The Pulmonary Circulation and Interstitial Space. Chicago, University of Chicago Press, 1969, p. 195.

Hughes, J. M. B.: Pulmonary circulatory and fluid balance. *Int. Rev. Physiol., 14*:135, 1977.

Levine, O. R., *et al.*: Extravascular lung water and distribution of pulmonary blood flow in the dog. *J. Appl. Physiol., 28*:166, 1970.

McIntyre, K. M., and Sasahara, A. A.: Hemodynamic and ventricular responses to pulmonary embolism. *Prog. Cardiovasc. Dis., 17*:175, 1974.

Meyer, B. J., *et al.*: Interstitial fluid pressure V. Negative pressure in the lungs. *Circ. Res., 22*:263, 1968.

Mobin-Uddin, K. (ed.): Pulmonary Thromboembolism. Springfield, Ill., Charles C Thomas, 1974.

Parker, J. C., *et al.*: Pulmonary interstitial and capillary pressures estimated from intra-alveolar fluid pressures. *J. Appl. Physiol., 44*(2):267, 1978.

Parker, J. C., *et al.*: Pulmonary transcapillary exchange and pulmonary edema. *In* Guyton, A. C., and Young, D. B. (eds.): International Review of Physiology: Cardiovascular Physiology III, Vol. 18. Baltimore, University Park Press, 1979, p. 261.

Permutt, S., *et al.*: Effect of lung inflation on static pressure-volume characteristics of pulmonary vessels. *J. Appl. Physiol., 16*:64, 1961.

Pietra, G. G., *et al.*: Bronchial veins and pulmonary edema. *In* Fishman, A. P., and Renkin, E. M. (eds.): Pulmonary Edema. Baltimore, Waverly Press, 1979, p. 195.

Racz, G. B.: Pulmonary blood flow in normal and abnormal states. *Surg. Clin. North Am., 54*:967, 1974.

Reed, J. H., Jr., and Wood, E. H.: Effect of body position on vertical distribution of pulmonary blood flow. *J. Appl. Physiol., 28*:303, 1970.

Reeves, J. T.: Pulmonary vascular response to high altitude residence. *Cardiovasc. Clin., 5*:81, 1973.

Staub, N. C.: Pulmonary edema. *Physiol. Rev., 54*:678, 1974.

Staub, N. C. (ed.): Lung Water and Solute Exchange. New York, Marcel Dekker, 1978.

Staub, N. C.: Pathways for fluid and solute fluxes in pulmonary edema. *In* Fishman, A. P., and Renkin, E. M. (eds.): Pulmonary Edema. Baltimore, Waverly Press, 1979, p. 113.

Stein, M., and Levy, S. E.: Reflex and humoral responses to pulmonary embolism. *Prog. Cardiovasc. Dis., 17*:167, 1974.

Taylor, A. E.: Permeability of the alveolar membrane to solutes. *Circ. Res., 16*:353, 1965.

Taylor, A. E., *et al.*: Na^{24} space, D$_2$O space, and blood volume in isolated dog lung. *Am. J. Physiol., 211*:66, 1966.

Wagenvoort, C. A., and Gurtner, H. P.: Idiopathic pulmonary hypertension. *In* Hayase, S., and Murao, S. (eds.): Cardiology: Proceedings of the VIII World Congress of Cardiology, Tokyo, 1978. New York, Elsevier/North-Holland, 1979, p. 691.

Webb, W. R., *et al.*: Microscopic studies of the pulmonary circulation in situ. *Surg. Clin. North Am., 54*:1067, 1974.

West, J. B.: Ventilation/Blood Flow and Gas Exchange. Oxford, Blackwell Scientific Publications, 1965.

West, J. B.: Blood flow to the lung and gas exchange. *Anesthesiology, 41*:124, 1974.

Widimsky, J. (ed.): Pulmonary Hypertension. New York, S. Karger, 1975.

25

The Coronary Circulation and Ischemic Heart Disease

Approximately one third of all deaths result from coronary artery disease, and almost all elderly persons have at least some impairment of the coronary artery circulation. For this reason, the normal and pathological physiology of the coronary circulation is one of the most important subjects in the entire field of medicine. The purpose of this chapter is to present this subject, emphasizing also the physiology of coronary occlusion and myocardial infarction.

NORMAL CORONARY BLOOD FLOW AND ITS VARIATIONS

PHYSIOLOGIC ANATOMY OF THE CORONARY BLOOD SUPPLY

Figure 25–1 illustrates the heart with its coronary blood supply. Note that the main coronary arteries lie on the surface of the heart, and small arteries penetrate into the cardiac muscle mass. It is almost entirely through these arteries that the heart receives its nutritive blood supply. Only the inner 75 to 100 microns of the endocardial surface

can obtain significant amounts of nutrition directly from the blood in the cardiac chambers.

The *left coronary artery* supplies mainly the anterior part of the left ventricle, while the *right coronary artery* supplies most of the right ventricle as well as the posterior part of the left ventricle in 80 to 90 per cent of all persons. In about half of all human beings, more blood flows through the right coronary artery than through the left, in about 30 per cent the arteries are about equal, and in 20 per cent the left artery predominates.

Most of the venous blood flow from the left ventricle leaves by way of the *coronary sinus —* which is about 75 per cent of the total coronary blood flow — and most of the venous blood from the right ventricle flows through the small *anterior cardiac veins,* which empty directly into the right atrium and are not connected with the coronary sinus. A small amount of coronary blood flows back into the heart through *thebesian veins,* which empty directly into all chambers of the heart.

NORMAL CORONARY BLOOD FLOW

The resting coronary blood flow in the human being averages approximately 225 ml. per minute, which is about 0.7 to 0.8 per gram of heart muscle, or 4 to 5 per cent of the total cardiac output.

In strenuous exercise the heart increases its cardiac output as much as four-to six-fold, and it pumps this blood against a higher than normal arterial pressure. Consequently, the work output of the heart under severe conditions may increase as much as six- to eight-fold. The coronary blood flow increases four- to five-fold to supply the extra nutrients needed by the heart. Obviously, this increase is not quite as much as the increase in work load, which means that the ratio of coronary blood flow to energy expenditure by the heart decreases. However, the "efficiency" of cardiac utilization of energy increases to make up for this relative deficiency of blood supply.

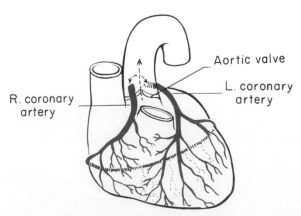

Figure 25–1. The coronary vessels.

Aortic valve

L. coronary artery

R. coronary artery

Phasic Changes in Coronary Blood Flow — Effect of Cardiac Muscle Compression. Figure 25–2 illustrates the average blood flow *through the small nutrient vessels* of the coronary system in milliliters per minute in the human heart during systole and diastole as *calculated* from experiments in lower animals. Note from this diagram that the blood flow in the left ventricle falls to a low value during *systole,* which is opposite to the flow in all other vascular beds of the body. The reason for this is the strong compression of the left ventricular muscle around the intramuscular vessels during systole.

During *diastole,* the cardiac muscle relaxes completely and no longer obstructs the blood flow through the left ventricular capillaries, so that blood now flows rapidly during all of diastole.

Blood flow through the coronary capillaries of the right ventricle undergoes phasic changes similar to those in the coronary capillaries of the left ventricle during the cardiac cycle, but, because the force of contraction of the right ventricle is far less than that of the left ventricle, these inverse phasic changes are relatively mild compared with those in the left ventricle, as is shown in Figure 25–2.

Epicardial Versus Subendocardial Blood Flow — Effect of Intramyocardial Pressure. During cardiac contraction all the cardiac muscle squeezes toward the centers of the ventricles. That is, the ventricular muscle adjacent to the ventricular chambers (the subendocardial muscle) squeezes the blood in the ventricle; the muscle in the middle layer of the ventricle squeezes the blood in the ventricle *and also the subendocardial muscle*; and the outermost muscle squeezes both the middle and the subendocardial muscle as well as the blood in the ventricle. Therefore, during systole, a gradient of *intramyocardial pressure* develops, with the pressure in the subendocardial muscle having a pressure almost as great as the pressure inside the ventricle, while the pressure in the outer layer of the heart muscle

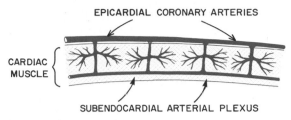

Figure 25–3. Diagram of the epicardial, intramuscular, and subendocardial coronary vasculature.

is only slightly above atmospheric pressure. The importance of this pressure gradient is that *the intramyocardial pressure compresses the subendocardial blood vessels far more than the outer vessels.*

Figure 25–3 illustrates the special arrangement of the coronary vessels at different depths in the heart, showing on the surface of the cardiac muscle the large epicardial coronary arteries that supply the heart. Smaller intramuscular arteries penetrate the muscle, supplying the needed nutrients en route to the endocardium. Then immediately outside the endocardium lies a plexus of subendocardial arteries. During systole, blood flow through the subendocardial plexus of the left ventricle, where the contractile force of the muscle is very great, falls almost to zero. To compensate for this almost total lack of flow during systole, the subendocardial arteries are much larger than the nutrient arteries in the middle and outer layers of the heart. Therefore, during diastole, blood flow in the subendocardial arteries is considerably greater than is blood flow in the outermost arteries. Later in the chapter we shall see that this peculiar difference between blood flow in the epicardial and subendocardial arteries plays an important role in certain types of coronary ischemia.

CONTROL OF CORONARY BLOOD FLOW

LOCAL METABOLISM AS THE PRIMARY CONTROLLER OF CORONARY FLOW

Blood flow through the coronary system is regulated almost entirely by vascular response to the local needs of the cardiac musculature for nutrition. This mechanism works equally well when the nerves to the heart are intact or are removed. Whenever the vigor of contraction is increased, regardless of cause, the rate of coronary blood flow simultaneously increases, and, conversely, decreased activity is accompanied by decreased coronary flow. It is immediately obvious that this local regulation of blood flow is almost identical with that which occurs in many other tissues, especially in the skeletal muscles of all the body.

Oxygen Demand as a Major Factor in Local Blood Flow Regulation. Blood flow in the coronaries is regulated almost exactly in proportion to the need of the cardiac musculature for oxygen.

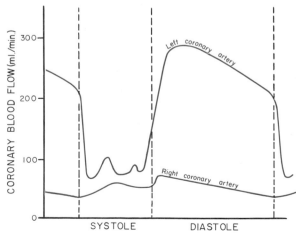

Figure 25–2. Phasic flow of blood through the coronary capillaries of the left and right ventricles.

Even in the normal resting state, 65 to 70 per cent of the oxygen in the arterial blood is removed as the blood passes through the heart; and, because not much oxygen is left, little additional oxygen can be removed from the blood unless the blood flow increases. Fortunately, the blood flow does increase, and almost directly in proportion to the metabolic consumption of oxygen by the heart.

Yet, the exact means by which increased oxygen consumption causes coronary dilatation has not been determined. It is speculated that a decrease in the oxygen concentration in the heart causes vasodilator substances to be released from the muscle cells, and that these dilate the arterioles. The substance with the greatest vasodilator propensity is *adenosine*. In the presence of very low concentrations of oxygen in the muscle cells, a large proportion of the cell's adenosine triphosphate degrades to adenosine monophosphate; then enzymes that are presumably associated with the cell membrane split small quantities of adenosine from this compound and release this into the tissue fluids of the heart muscle. After the adenosine causes vasodilatation it is destroyed within a few seconds so that it cannot go elsewhere in the circulation to cause unwarranted vasodilatation.

Adenosine is not the only vasodilator product that has been identified. Others include potassium ions, hydrogen ions, carbon dioxide, bradykinin, and possibly prostaglandins.

Yet, difficulties still exist with the vasodilator hypothesis. For one, infusion of maximal amounts of adenosine into the coronary arteries does not dilate these vessels nearly as much as does maximal increase in cardiac muscle metabolism. Second, agents that block or partially block the vasodilator effect of adenosine do not prevent the coronary vasodilatation in response to increased muscle activity.

Therefore, yet another theory to explain the coronary artery dilatation should be remembered until proved false: in the absence of adequate amounts of oxygen in the cardiac muscle, not only does the muscle itself suffer oxygen deficiency but so do the arteriolar muscle walls as well. This could easily cause local vasodilatation because of lack of the required energy to keep the coronary vessels contracted against the high arterial pressure. But this, too, has its problems, because the coronary arteries require only minute amounts of oxygen to maintain full contraction.

The Determinants of Oxygen Consumption. Since the rate of oxygen consumption is the major factor that determines coronary blood flow, it is important to know the different factors that can alter myocardial oxygen consumption.

In general, the rate of cardiac oxygen consumption is closely related to work performed by the heart — the greater the work, the greater the oxygen consumption, and consequently the greater the coronary flow. However, this is not exactly true for the following reasons: Work output of the heart is determined by the arterial pressure against which the heart is pumping × the cardiac output pumped per minute. When the pressure increases, the oxygen consumption increases almost directly in proportion to the increase in pressure. On the other hand, when the cardiac output increases without an increase in pressure, the oxygen consumption increases only a slight to moderate amount rather than in direct proportion to the increase in cardiac output.

Peak Muscle Tension as the Primary Determinant of Oxygen Consumption. Perhaps the best relationship that has yet been found between cardiac function and oxygen consumption is: *oxygen consumption is proportional to peak myocardial muscle tension*. Thus, when the arterial pressure rises, the muscle tension increases and oxygen consumption also increases. Likewise, when the heart dilates, which makes it necessary for the muscle to generate increased tension to pump against even a normal arterial pressure, oxygen consumption also increases even though the work output of the heart does not increase. This results from the law of Laplace, which states that the tension required to generate a given pressure increases in proportion to the diameter of the heart.

Other Causes of Increased Oxygen Consumption. Other factors that increase cardiac oxygen consumption are stimulation of the heart by epinephrine, norepinephrine, thyroxine, digitalis, and calcium ions and increased temperature of the heart. All these factors increase the metabolic activity of the cardiac muscle fibers themselves, which in turn increases the rate of oxygen usage even though they might not increase the work output of the heart. They also increase coronary blood flow approximately in proportion to the increase in rate of oxygen usage.

Importance of the Increase in Coronary Blood Flow in Response to Myocardial Oxygen Usage. The resting heart extracts most of the oxygen from the coronary blood as it flows through the heart muscle, and very little of the heart's oxygen need can be met by additional extraction of oxygen from the coronary blood. Therefore, the only significant way in which the heart can be supplied with additional amounts of oxygen is through an increase in blood flow. Consequently, it is essential that the coronary blood flow increase whenever the heart muscle demands additional oxygen. When the coronary blood flow fails to increase appropriately, the strength of the muscle diminishes rapidly and drastically, often causing acute heart failure. Also, this *relative* ischemia of the muscle can cause severe pain, called *anginal pain,* which will be discussed later in the chapter.

Reactive Hyperemia in the Coronary System. If the coronary flow to the heart is completely occluded for a

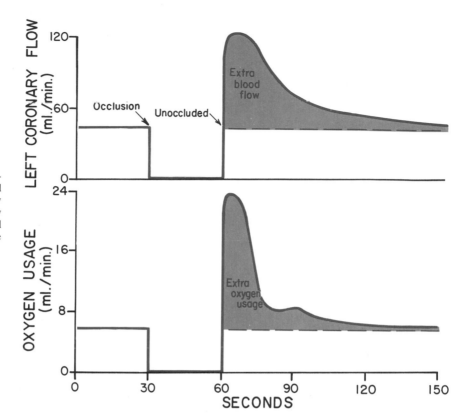

Figure 25–4. Reactive hyperemia in the coronary system caused by a 30 second period of coronary occlusion. Note the extra blood flow called "reactive hyperemia" and the extra oxygen usage after the period of occlusion was over.

few seconds to a few minutes and then suddenly unoccluded, the blood flow increases to as high as 3 to 6 times normal, as shown in Figure 25–4. It remains high for a few seconds to several minutes, depending on the period of occlusion. This extra flow of blood is called reactive hyperemia, which was explained in Chapter 20. During the period of excess flow, the heart removes a large amount of extra oxygen from the blood, as shown by the bottom curve of the figure, to make up for the deficiency of oxygen during the period of occlusion. Reactive hyperemia is simply another manifestation of the ability of the coronary system to adjust its flow to the metabolic needs of the heart.

NERVOUS CONTROL OF CORONARY BLOOD FLOW

Stimulation of the autonomic nerves to the heart can affect coronary blood flow in two ways — directly and indirectly. The direct effects result from direct action of the nervous transmitter substances, acetylcholine and norepinephrine, on the coronary vessels themselves. The indirect effects result from secondary changes in coronary blood flow caused by increased or decreased activity of the heart.

The indirect effects play by far the more important role in normal control of coronary blood flow. Thus, sympathetic stimulation increases both heart rate and heart contractility as well as its rate of metabolism. In turn, the increased activity of the heart sets off local blood flow regulatory mechanisms for dilating the coronary vessels, the blood flow increasing approximately in proportion to the metabolic needs of the heart muscle. In contrast, parasympathetic stimulation slows the heart and also has a slight depressive effect on cardiac contractility. Both these effects decrease cardiac oxygen consumption and therefore constrict the coronaries.

Direct Effects of Nervous Stimuli on the Coronary Vasculature. The distribution of parasympathetic (vagal) nerve fibers to the ventricular coronary system is so slight that parasympathetic stimulation has almost negligible direct effect on coronary blood flow, though perhaps a very slight effect to dilate the coronaries. On the other hand, there is extensive sympathetic innervation of the coronary vessels. In Chapter 56 we shall see that the sympathetic transmitter substances, norepinephrine and epinephrine, can have either dilator or constrictor effects depending on the presence or absence of specific receptors in the blood vessel walls. The constrictor receptors are called *alpha receptors,* and the dilator receptors are called *beta receptors*. Both alpha and beta receptors are known to exist in the coronary vessels. In general, the epicardial coronary vessels have a preponderance of alpha receptors, whereas the intramuscular arteries have a preponderance of beta receptors. Therefore sympathetic stimulation can cause either slight overall coronary constriction or dilatation, but perhaps usually a little more constriction. Yet, in some persons, the alpha vasoconstrictor effects seem to be disproportionately severe, and these persons can have vasospastic myocardial ischemia during periods of excess sympathetic drive, often with resultant anginal pain.

It must be pointed out again, however, that metabolic factors — especially myocardial oxygen consumption — are the major controllers of myocardial blood flow. Therefore, whenever nervous stimulation alters the coronary blood flow, the metabolic factors usually within seconds return the flow most of the way back toward normal.

THE SUBSTRATES OF CARDIAC METABOLISM

The general principles of cellular metabolism, which will be discussed in Chapters 67 through 70, apply to cardiac muscle, though there are some quantitative differences. Most importantly, under resting conditions cardiac muscle normally utilizes fats mainly for its energy, with approximately 70 per cent of the normal metabolism being derived from fatty acids. However, as is true of other tissues, under anaerobic or ischemic conditions cardiac metabolism must call upon the anaerobic glycolysis mechanism for energy. This, unfortunately, can supply very little extra energy in relation to the large energy requirements of the heart. Also, it utilizes tremendous quantities of the blood glucose and at the same time forms large amounts of lactic acid in the cardiac tissue, which is probably one of the causes of cardiac pain in cardiac ischemic conditions, as will be discussed later in the chapter.

As is true in other tissues, more than 95 per cent of the metabolic energy liberated from the foods is used to form ATP in the mitochondria. This ATP in turn transfers its energy through the mitochondrial membrane to ATP and creatinine phosphate in the hyaloplasm. Finally, these latter two supply the energy for cellular function. In coronary ischemia, the ATP degrades to ADP, AMP, and adenosine. Because the cell membrane is permeable to adenosine, much of this is rapidly lost from the hyaloplasm into the circulating blood. This released adenosine is believed to be one of the substances that causes dilatation of the coronary arterioles during coronary hypoxia. However, the loss of adenosine has a very serious cellular consequence as well. Within as little as half an hour of severe coronary ischemia, as occurs after a myocardial infarct or during cardiac arrest, very large amounts of the adenine base can be lost from the cardiac cellular hyaloplasm. Furthermore, this can be replaced by new synthesis of adenosine at a rate of only 2 per cent per hour. Therefore, once a serious bout of ischemia has persisted for half an hour or so, relief of the coronary ischemia may be too late to save the lives of the cardiac cells. This is almost certainly the major cause of cardiac cellular death following myocardial ischemia and also one of the most important causes of cardiac debility in the late stages of circulatory shock, as will be discussed in Chapter 28.

ISCHEMIC HEART DISEASE

The single most common cause of death is ischemic heart disease, which results from insufficient coronary blood flow. Approximately 35 per cent of all human beings in the United States die of this cause. Some deaths occur suddenly as a result of an acute coronary occlusion or of fibrillation of the heart, whereas others occur slowly over a period of weeks to years as a result of progressive weakening of the heart pumping process. In the present chapter we will discuss the coronary ischemia problem itself, as well as acute coronary occlusion and myocardial infarction. In the following chapter we will discuss congestive heart failure, the most frequent cause of which is progressive coronary ischemia.

Atherosclerosis as the Cause of Ischemic Heart Disease. The most frequent cause of diminished coronary blood flow is atherosclerosis. Although the atherosclerotic process will be discussed in connection with lipid metabolism in Chapter 68, briefly, this process is the following: In certain persons who have a genetic predisposition to atherosclerosis or in persons who eat excessive quantities of cholesterol and other fats, large quantities of cholesterol gradually become deposited beneath the intima at many points in the arteries. Later, these areas of deposit become invaded by fibrous tissue, and they also frequently become calcified. The net result is the development of atherosclerotic plaques that protrude into the vessels and either block or partially block blood flow.

A very common site for development of atherosclerotic plaques is the first few centimeters of the coronary arteries.

Acute Coronary Occlusion. Acute occlusion of the coronary artery frequently occurs in a person who already has serious underlying atherosclerotic coronary heart disease, but almost never in a person with a normal coronary circulation. This condition can result from any one of several different effects, two of which are the following:

1. The atherosclerotic plaque can cause a local blood clot called a *thrombus,* which in turn occludes the artery. The thrombus usually begins where the plaque has grown so much that it has broken through the intima, thus coming in contact with the flowing blood. Because the plaque presents an unsmooth surface to the blood, platelets begin to adhere to it, fibrin begins to be deposited, and blood cells become entrapped and form a clot that grows until it occludes the vessel. Or occasionally the clot breaks away from its attachment on the atherosclerotic plaque and flows to a more peripheral branch of the coronary arterial tree where it blocks the artery at that point. A thrombus that flows along the artery in this way and occludes the vessel more distally is called an *embolus.*

2. It is believed by many clinicians that local spasm of a coronary artery can also cause sudden occlusion. The spasm might result from irritation of the smooth muscle of the arterial wall by the edges of an arteriosclerotic plaque, or it might result from nervous reflexes that cause the coronary contraction. The spasm may then lead to *secondary* thrombosis of the vessel.

Collateral Circulation in the Heart. The degree of damage to the heart caused either by slowly developing atherosclerotic constriction of the coronary arteries or by sudden occlusion is determined to a great extent by the degree of collateral circulation that is already developed or that can develop within a short period of time after the occlusion.

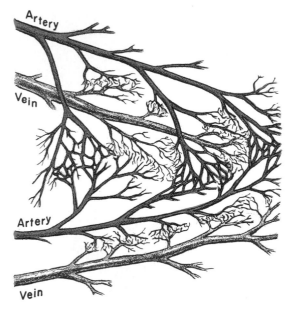

Figure 25–5. Minute anastomoses of the coronary arterial system.

Unfortunately, in a normal heart, relatively few communications exist among the larger coronary arteries. But many anastomoses do exist among the smaller arteries sized 20 to 250 microns in diameter, as shown in Figure 25–5.

When a sudden occlusion occurs in one of the larger coronary arteries, the sizes of the minute anastomoses increase to their maximum physical diameters within a few seconds. The blood flow through these minute collaterals is less than one-half that needed to keep alive the cardiac muscle that they supply; and unfortunately the diameters of the collateral vessels do not enlarge further for the next 8 to 24 hours. But then collateral flow does begin to increase, doubling by the second or third day and often reaching normal or almost normal coronary supply within about one month; it is capable of increasing even further with increased metabolic loads. It is because of these developing collateral channels that a patient recovers from the various types of coronary occlusion.

When atherosclerosis constricts the coronary arteries slowly over a period of many years rather than suddenly, collateral vessels can develop at the same time that the atherosclerosis does. Therefore, the person may never experience an acute episode of cardiac dysfunction. Eventually, however, the sclerotic process develops beyond the limits of even the collateral blood supply to provide the needed blood flow, and even the collaterals develop atherosclerosis. When this occurs, the heart muscle becomes severely limited in its work output, sometimes so much so that the heart cannot pump even the normally required amounts of blood flow. This is the most common cause of cardiac failure.

MYOCARDIAL INFARCTION

Immediately after an acute coronary occlusion, blood flow ceases in the coronary vessels beyond the occlu-

sion except for small amounts of collateral flow from surrounding vessels. The area of muscle that has either zero flow or so little flow that it cannot sustain cardiac muscle function is said to be *infarcted*. The overall process is called a *myocardial infarction*.

Soon after the onset of the infarction small amounts of collateral blood seep into the infarcted area, and this, combined with progressive dilatation of the local blood vessels, causes the area to become overfilled with stagnant blood. Simultaneously, the muscle fibers utilize the last vestiges of the oxygen in the blood, causing the hemoglobin to become totally reduced and very dark blue in color. Therefore, the infarcted area takes on a dark bluish hue, and the blood vessels of the area appear to be engorged despite the lack of blood flow. In later stages, the vessels become highly permeable and leak fluid, the tissue becomes edematous, and the cardiac muscle cells begin to swell because of diminished cellular metabolism. Finally, many of the cells die.

Cardiac muscle requires approximately 1.3 ml. of oxygen per 100 grams of muscle tissue per minute simply to remain alive. This is in comparison with approximately 8 ml. of oxygen per 100 grams delivered to the normal resting left ventricle each minute. Therefore, if there is even as much as 15 to 30 per cent of normal resting coronary blood flow, the muscle will not die. In the central portion of a large infarct, however, the blood flow is usually less than this so that this muscle does die.

Myocardial Infarction Caused by Myocardial Ischemia but WITHOUT Coronary Thrombosis. In studying the hearts of patients who have died from myocardial infarctions, pathologists have found up to 50 per cent of the patients with no evidence of coronary thrombosis. However, all these patients do have atherosclerotic narrowing of one or more major coronary arteries. It is believed that some special stress on the heart — such as excess exercise, abnormal emotional disturbance, or any event that either increases the work load of the heart or temporarily diminishes the coronary blood flow — then leads to an inadequate ratio of oxygen supply to the work load of the muscle, and part of the ischemic muscle stops contracting. As a result, a vicious cycle is presumed to develop in the following manner:

1. The small area of noncontracting muscle, even though it may still be alive, nevertheless causes overall diminution of ventricular pumping. To compensate, the remaining ventricular muscle now increases its work load and oxygen consumption. Consequently, additional portions of the heart muscle now develop an inadequate ratio of oxygen supply to work load, so this muscle too becomes noncontracting. This process continues until much of the muscle with marginal blood supply becomes noncontracting.

2. Next, because of the very low coronary blood flow in the noncontracting muscle area, the local vascular resistance increases rapidly during the next few hours. Part of this is caused by development of edema in the surrounding tissues and part, perhaps, by such factors as release of vasoconstrictor substances from the damaged muscle cells. At any rate, the coronary blood flow becomes progressively less and less, and the cycle continues until much of the muscle in the area where the blood supply is poor has become nonfunctional and infarcted.

Thus, it is possible — if not highly probable — that many myocardial infarctions occur entirely in the absence of acute coronary thrombosis. However, infarction will almost never occur unless the internal diameter of one or more of the major coronary arteries has been reduced by at least 70 per cent. This emphasizes the fact that pre-existing coronary artery stenosis is generally a prerequisite to the occurrence of myocardial infarction.

Subendocardial Myocardial Infarction. Myocardial infarction frequently occurs in the subendocardial muscle even when the epicardial portions of the heart muscle remain uninfarcted. This form of infarction occurs especially when the diastolic arterial pressure is very low or when the diastolic intraventricular pressure is very high. From the discussion earlier in the chapter of the subendocardial arterial plexus, it was pointed out that most blood flow into this area of the heart occurs during diastole. Therefore, when the diastolic arterial pressure is very low — as occurs in patients who have aortic regurgitation, patent ductus arteriosus, or to a lesser extent arteriosclerosis — one can expect a high incidence of subendocardial myocardial infarction.

CAUSES OF DEATH FOLLOWING ACUTE CORONARY OCCLUSION

The four major causes of death following acute myocardial infarction are decreased cardiac output; damming of blood in the pulmonary or systemic veins with death resulting from edema, especially pulmonary edema; fibrillation of the heart; and, occasionally, rupture of the heart.

Decreased Cardiac Output — Cardiac Shock. When some of the cardiac muscle fibers are not functioning at all and others are too weak to contract with great force, the overall pumping ability of the affected ventricle is proportionately depressed. Indeed, the overall pumping strength of the heart is often decreased more than one might expect because of the phenomenon of *systolic stretch,* which is illustrated in Figure 25–6. When the normal portions of the ventricular muscle contract, the ischemic muscle, whether this be dead or simply nonfunctional, instead of contracting is actually forced outward by the pressure that develops inside the ventricle. Therefore, much of the pumping force of the ventricle is dissipated by bulging of the area of nonfunctional cardiac muscle.

When the heart becomes incapable of contracting with sufficient force to pump enough blood into the arterial tree, cardiac failure and death of the peripheral tissues ensue as a result of peripheral ischemia. This condition is called *coronary shock, cardiogenic shock, cardiac shock,* or *low cardiac output failure.* It will be discussed in the following chapter. Cardiac shock almost always occurs when more than 40 per cent of the left ventricle is infarcted. Death occurs in more than 80 per cent of those patients who develop cardiac shock.

Damming of Blood in the Venous System. When the heart is not pumping blood forward, it must be damming blood in the venous system of the lungs or the systemic circulation. Obviously, this increases both the left and right atrial pressures and also leads to increased capillary pressures, particularly in the lungs. These effects often cause very little difficulty during the first few

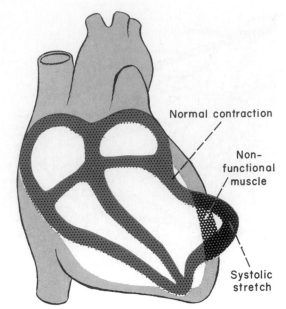

Figure 25–6. Systolic stretch in an area of ischemic cardiac muscle.

hours after the myocardial infarction. Instead, the symptoms develop a few days later for the following reason: The diminished cardiac output leads to diminished blood flow to the kidneys, and, for reasons that will be discussed in the following chapter, the kidneys retain large quantities of fluid. This adds progressively to the venous congestive symptoms. Therefore, many patients who seemingly are getting along well will suddenly develop acute pulmonary edema several days after a myocardial infarction and often will die within a few hours after appearance of the initial edema symptoms.

Rupture of the Infarcted Area. During the first day of an acute infarct there is little danger of rupture of the ischemic portion of the heart, but a few days after a large infarct occurs, the dead muscle fibers begin to degenerate, and the dead heart musculature is likely to become very thin. If this happens, the dead muscle bulges outward with each heart contraction, an effect called systolic stretch as explained above, and this stretch becomes greater and greater until finally the heart ruptures. In fact, one of the means used in assessing the progress of a severe myocardial infarction is to record by radiograph whether the degree of systolic stretch is getting worse or better.

When a ventricle does rupture, the loss of blood into the pericardial space causes rapid development of *cardiac tamponade* — that is, compression of the heart from the outside by blood collecting in the pericardial cavity. Because the heart is compressed, blood cannot flow into the right atrium with ease, and the patient dies of suddenly decreased cardiac output.

Fibrillation of the Ventricles Following Myocardial Infarction. Many persons who die of coronary occlusion die because of ventricular fibrillation. The tendency to develop fibrillation is especially great following a large infarction, but fibrillation can occur following small occlusions as well. Indeed, many patients with coronary insufficiency die acutely from fibrillation without any infarction at all.

There are two especially dangerous periods during which fibrillation is most likely to occur. The first is during the first 10 minutes after the infarction occurs. Then there is a period of relative safety, followed by a secondary period of cardiac irritability beginning three to five hours after the infarction and lasting for hours to days thereafter.

At least four different factors enter into the tendency for the heart to fibrillate:

First, acute loss of blood supply to the cardiac muscle causes rapid depletion of potassium from the ischemic musculature. This increases the potassium concentration in the extracellular fluids surrounding the cardiac muscle fibers. Experiments in which potassium has been injected into the coronary system have demonstrated that an elevated extracellular potassium concentration increases the irritability of the cardiac musculature.

Second, ischemia of the muscle causes an "injury current," which was described in Chapter 16 in relation to electrocardiograms in patients with acute myocardial infarction. The ischemic musculature cannot repolarize its membranes so that this muscle remains negative with respect to the normal polarized cardiac muscle membrane. Therefore, electrical current flows from this ischemic area of the heart to the normal area and can elicit abnormal impulses which can cause fibrillation.

Third, powerful sympathetic reflexes develop following massive infarction, principally because the heart does not pump an adequate volume of blood into the arterial tree. The sympathetic stimulation also increases the irritability of the cardiac muscle and thereby predisposes to fibrillation.

Fourth, the myocardial infarction itself often causes the ventricle to dilate excessively. This increases the pathway length for impulse conduction in the heart and also frequently causes abnormal conduction pathways around the infarcted area of the cardiac muscle. Both of these effects predispose to development of circus movements. As was discussed in Chapter 14, any prolongation of the conduction pathway in the ventricles will allow an impulse to re-enter muscle that is already recovering from refractoriness, thereby initiating a subsequent cycle of excitation and causing the process to continue on and on.

THE STAGES OF RECOVERY FROM ACUTE MYOCARDIAL INFARCTION

The upper part of Figure 25–7 illustrates the effects of acute coronary occlusion, on the left, in a patient with a small area of muscle ischemia and, on the right, in a patient with a large area of ischemia. When the area of ischemia is small, little or no death of the muscle cells may occur, but part of the muscle often does become temporarily nonfunctional because of inadequate nutrition to support muscle contraction.

When the area of ischemia is large, some of the muscle fibers in the very center of the area die rapidly, within about one hour in an area of total cessation of coronary blood supply. Immediately around the dead area is a nonfunctional area because of failure of contraction and usually also failure of impulse conduction. Then, extend-

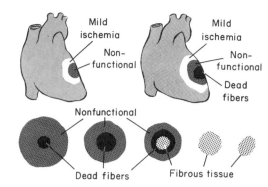

Figure 25–7. *Top:* small and large areas of coronary ischemia. *Bottom:* stages of recovery from myocardial infarction.

ing circumferentially around the nonfunctional area is an area that is still contracting but weakly so because of mild ischemia. In this mildly ischemic area the small arterial collaterals are sufficient to supply the cardiac musculature, provided the patient is kept at rest, but even this area may become nonfunctional if the coronary blood flow is diverted to normal musculature during exercise. This illustrates the necessity for rest in patients following coronary heart attacks.

Replacement of Dead Muscle by Scar Tissue. In the lower part of Figure 25–7, the various stages of recovery following a myocardial infarction are illustrated. Shortly after the occlusion, the muscle fibers die in the very center of the ischemic area. Then during the ensuing days, this area of dead fibers grows because many of the marginal fibers finally succumb to the prolonged ischemia. At the same time, owing to the enlargement of the collateral arterial channels growing into the outer rim of the infarcted area, the nonfunctional area of muscle becomes smaller and smaller. After a few days to three weeks most of the nonfunctional area of muscle becomes functional again or dead — one or the other. In the meantime, fibrous tissue begins developing among the dead fibers, for ischemia stimulates growth of fibroblasts and promotes development of greater than normal quantities of fibrous tissue. Therefore, the dead muscular tissue is gradually replaced by fibrous tissue. Then, because it is a general property of fibrous tissue to undergo progressive elastomeric contraction and dissolution, the size of the fibrous scar becomes smaller and smaller over a period of several months to a year.

During progressive recovery of the infarcted area of the heart, the development of a strong, fibrous scar which becomes smaller and smaller finally stops the original systolic stretch, and the functional musculature once again becomes capable of exerting its entire force for pumping blood rather than for stretching the dead area of the heart. Furthermore, the normal areas of the heart gradually hypertrophy to compensate at least partially for the lost cardiac musculature. By these means the heart recovers.

Value of Rest in Treating Myocardial Infarction. The degree of cellular death is determined by the *degree of ischemia × the degree of metabolism* of the heart muscle. When the metabolism of the heart muscle is greatly increased, such as during exercise, in severe emotional strain, or as a result of fatigue, the heart musculature needs increased oxygen and other nutrients for sustaining its life. Furthermore, anastomotic blood vessels that supply blood to ischemic areas of the heart must still supply the areas of the heart that they normally supply. When the heart becomes excessively active, the vessels of the normal musculature become greatly dilated, and this allows most of the blood flowing into the coronary vessels to flow through the normal muscle tissue, thus leaving little blood to flow through the small anastomotic channels into the ischemic area. This condition is called the ''coronary steal'' syndrome. Consequently, one of the most important factors in the treatment of a patient with myocardial infarction is observance of absolute rest. The greater the degree of rest during the first few weeks following sudden infarction, the less the degree of cellular death.

FUNCTION OF THE HEART FOLLOWING RECOVERY FROM MYOCARDIAL INFARCTION

Occasionally, a heart that has recovered from a large myocardial infarction returns to full functional capability, but more frequently its pumping capability is decreased below that of a normal heart. This does not mean that the person is necessarily a cardiac invalid or that the resting cardiac output is depressed below normal, because the normal person's heart is capable of pumping 400 per cent more blood per minute than the body requires — that is, a person has a ''cardiac reserve'' of 400 per cent. Even when the cardiac reserve is reduced to as little as 100 per cent, the person can still perform normal activity of a quiet, restful type, but not strenuous exercise that would overload the heart.

The effect of myocardial infarction and other myocardial debilities on cardiac output is discussed in the following chapter in relation to the regulation of cardiac output in cardiac failure.

PAIN IN CORONARY DISEASE

Normally, a person cannot ''feel'' his or her heart, but ischemic cardiac muscle does exhibit pain sensation. Exactly what causes this pain is not known, but it is believed that ischemia causes the muscle to release acidic substances such as lactic acid or other pain-promoting products such as histamine or kinins that are not removed rapidly enough by the slowly moving blood. The high concentrations of these abnormal products then stimulate the pain endings in the cardiac muscles, and pain impulses are conducted through the sympathetic afferent nerve fibers into the central nervous system.

ANGINA PECTORIS

In most persons who develop progressive constriction of their coronary arteries, cardiac pain, called *angina pectoris,* begins to appear whenever the load on the heart becomes too great in relation to the coronary blood flow. This pain is usually felt beneath the upper sternum and is often also transferred to surface areas of the body, most often to the left arm and left shoulder but also frequently to the neck and even to the side of the face or to the opposite arm and shoulder. The reason for this distribution of pain is that the heart originates during embryonic life in the neck, as do the arms. Therefore, both of these structures receive pain nerve fibers from the same spinal cord segments.

In general, most persons who have chronic angina pectoris feel the pain when they exercise, also when they experience emotions that increase metabolism of the heart or temporarily constrict the coronary vessels because of sympathetic vasoconstrictor nerve signals. Usually the pain lasts for only a few minutes. However, some patients have such severe and lasting ischemia that the pain is present all the time. The pain is frequently described as dull, pressing, and constricting; it is of such quality that it usually makes the patient stop all activity and come to a complete state of rest.

When a person has frequent attacks of angina, the likelihood of developing an acute coronary occlusion is generally very great.

Treatment with Vasodilator Drugs.   Several vasodilator drugs, when administered during an acute anginal attack, will often give immediate relief from the pain. Two commonly used drugs are nitroglycerin and amyl nitrite. The original theory for use of these drugs was that they caused coronary vasodilatation and that this relieved the pain by increasing coronary blood flow. However, in the normal person the drugs have almost no effect on dilatation of the coronary arterioles and do not cause a measurable increase in the coronary blood flow. But they do have two other effects that probably explain their efficacy in relieving anginal pain. First, they cause marked venous dilatation throughout the body, which reduces venous return to the heart, cardiac output, and arterial pressure. Therefore, the work load of the heart is immediately diminished, thereby decreasing the formation of the ischemic products that are presumed to be the cause of the pain. Second, even though these drugs do not dilate the coronary arterioles, they do dilate the large epicardial coronary arteries. In the person who has severely constricted epicardial arteries, this slight dilatation can probably be of additional benefit in relieving the anginal pain even though in the normal heart such dilatation causes no measurable increase in coronary blood flow.

Treatment with Sympathetic Blocking Drugs.   Another medical procedure for relieving anginal pain is to block the sympathetic nervous stimulation of the heart. This is most frequently accomplished by administered *propranolol,* which blocks the beta adrenergic receptors. It is stimulation of these receptors that enhances cardiac activity. Therefore, when they are blocked, the level of cardiac muscle contractility and the heart rate both decrease, and the cardiac work output also decreases; at the same time, the heart's usage of oxygen diminishes as much as 20 per cent. In this way, the ratio of oxygen need to oxygen supply is reduced enough to eliminate the pain.

SURGICAL TREATMENT OF CORONARY DISEASE

Treatment of the Pain of Angina Pectoris.   Rarely, for treating intractable angina pectoris the sympathetic chain is removed from T-2 through T-5 to block the pathway of cardiac pain fibers into the spinal cord. Such an operation occasionally is successful when performed only on the left side, but often it is necessary to remove the sympathetic chain on both sides of the vertebral column.

Treatment of Ischemia — Aortic-Coronary By-Pass Surgery. In many patients with coronary ischemia, the constricted areas of the coronary vessels are located at only a few discrete points, and the coronary vessels beyond these points are of normal or almost normal size. A surgical procedure has been developed in the past few years, called *aortic-coronary by-pass,* for anastomosing small vein grafts to the aorta and to the sides of the more peripheral coronary vessels. Usually, one to three such grafts are performed during the operation, each of which supplies a peripheral coronary artery beyond a block. However, on occasion, as many as five grafts have been inserted in a single person. The vein that is used for the graft is usually the long superficial saphenous vein removed from the leg of the recipient.

The acute results from this type of surgery have been especially good, causing this to be the most common cardiac operation performed. Anginal pain is relieved in most patients. Unfortunately, it is still too early to determine whether the operative procedure will prolong the lives of the patients because other complications, such as secondary closure of the grafts, may in the long run be more detrimental than the original disease.

MEASUREMENT OF CORONARY BLOOD FLOW IN MAN

Obviously, it would be almost impossible to insert a flowmeter into the coronary circulatory system of human beings. However, a satisfactory indirect method for measuring coronary flow has been achieved, the principles of which are the following:

The blood flow through any organ of the body can be determined by means of the so-called Fick principle, which was discussed in Chapter 23 in relation to the measurement of cardiac output. That is, if a constituent of the blood is removed as it flows through an organ, then the blood flow can be determined from two measurements: first, the amount removed from each milliliter of blood as it goes through the organ and, second, the total quantity of the substance that is removed by the organ in a given period of time. The total blood flow during the period of measurement can be calculated by dividing the amount of substance removed from each milliliter of blood into the total quantity removed.

In measuring blood flow through the heart the subject suddenly begins to breathe nitrous oxide of a given concentration. Samples of arterial blood are removed from any artery in the body and samples of venous blood are obtained from the coronary sinus through a catheter that has been passed into the sinus by way of the venous system. The concentrations of the nitrous oxide in both of these bloods are then plotted for 10 minutes, as shown in Figure 25–8. The quantity of nitrous oxide lost to the heart by each milliliter of blood as it passes through the heart at each instant of the experiment can be determined from the difference in height of the two curves. This is called the *A-V difference.* The A-V difference is determined each minute for a period of 10 minutes and then averaged.

Now, to determine the amount of nitrous oxide that is

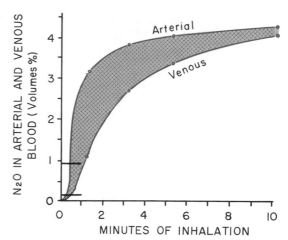

Figure 25–8. Method for measuring human coronary blood flow as explained in the text.

removed from the blood by the cardiac muscle, one proceeds as follows: It will be noted from the curves of Figure 25–8 that the concentrations of nitrous oxide begin to reach an equilibrium level toward the end of the 10 minute experiment. It is also known that the total amount of nitrous oxide absorbed by the cardiac muscle is directly proportional to its concentration in the arterial blood at the end of the 10 minute run. A *proportionality factor* has been determined in experiments on isolated cardiac muscle so that at the end of 10 minutes the total amount of nitrous oxide absorbed by each 100 grams of cardiac muscle can be determined by multiplying the arterial concentration of nitrous oxide times the proportionality factor.

Once the average *A-V difference* and the *amount of nitrous oxide absorbed* by each 100 grams of heart muscle have been determined, the coronary blood flow per 100 grams of muscle is then calculated by dividing the average A-V difference into the quantity absorbed per minute. As an example, if 8 ml. of nitrous oxide is absorbed by 100 grams of heart in 10 minutes and the average A-V difference during this period of time is 0.01 ml. for each ml. of blood, a total of 800 ml. of blood would have passed through the 100 grams of heart during the 10 minute interval. In other words, the blood flow would be 80 ml. per 100 grams of heart per minute. Assuming a heart mass of 280 grams, the blood flow would be 224 ml. per minute.

REFERENCES

Amsterdam, E. A., *et al.*: Indirect assessment of myocardial oxygen consumption in the evaluation of mechanisms and therapy of angina pectoris. *Am. J. Cardiol., 33*:737, 1974.

Baltaxe, H. A., *et al.*: Coronary Angiography. Springfield, Ill., Charles C Thomas, 1974.

Bell, J. R., and Fox, A. C.: Pathogenesis of subendocardial ischemia. *Am. J. Med. Sci. 268*:3, 1974.

Beneken, J. E. W., *et al.*: Coronary perfusion pressure and left ventricular function. *Pflugers Arch., 305*:76, 1969.

Berne, R. M., and Rubio, R.: Adenine nucleotide metabolism in the heart. *Circ. Res., 35*:109, 1974.

Berne, R. M., and Rubio, R.: Regulation of coronary blood flow. *Adv. Cardiol., 12*:303, 1974.

Berne, R. M., and Rubio, R.: Coronary circulation. *In* Berne, R. M., *et al.* (eds.): Handbook of Physiology. Sec. 2. Vol. I. Baltimore, Williams & Wilkins, 1979, p. 873.

Blackburn, H.: The epidemiology and prevention of coronary heart disease. *In* Hayase, S., and Murao, S. (eds.): Cardiology: Proceedings of the VIII World Congress of Cardiology, Tokyo, 1978. New York, Elsevier/North-Holland, 1979, p. 23.

Bloor, C. M.: Functional significance of the coronary collateral circulation. A review. *Am. J. Pathol., 76*:561, 1974.

Bodem, G., and Dengler, H. J. (eds.): Cardiac Glycosides. New York, Springer-Verlag, 1978.

Braunwald, E., and Kimura, E.: Protection of the ischemic myocardium. *In* Hayase, S., and Murao, S. (eds.): Cardiology: Proceedings of the VIII World Congress of Cardiology, Tokyo, 1978. New York, Elsevier/North-Holland, 1979, p. 278.

Braunwald, E., and Maroko, P. R.: Limitation of Infarct Size. Chicago, Year Book Medical Publishers, 1978.

Budassi, S. A. (ed.): Cardiopulmonary Resuscitation. Germantown, Md., Aspen Systems Corp., 1978.

Chandler, A. B.: Mechanisms and frequency of thrombosis in the coronary circulation. *Thromb. Res., 4* (Suppl. 1): 3, 1974.

Corday, E., and Murao, S.: Pathogenesis of angina pectoris. *In* Hayase, S., and Murao, S. (eds.): Cardiology: Proceedings of the VIII World Congress of Cardiology, Tokyo, 1978. New York, Elsevier/North-Holland, 1979, p. 207.

Cosby, R. S., *et al.*: Coronary collateral circulation. *Chest, 66*:27, 1974.

Fletcher, G. F., and Cantwell, J. D.: Exercise and Coronary Heart Disease: Role in Prevention, Diagnosis, Treatment. Springfield, Ill., Charles C Thomas, 1978.

Fox, S. M., 3rd, *et al.*: Physical activity and cardiovascular health. I. Potential for prevention of coronary heart disease and possible mechanisms. *Mod. Concepts Cardiovasc. Dis., 41*:17, 1972.

Gibbs, C. L., and Chapman, J. B.: Cardiac energetics. *In* Berne, R. M., *et al.* (eds.): Handbook of Physiology. Sec. 2, Vol. I. Baltimore, Williams & Wilkins, 1979, p. 775.

Gregg, D. E.: Coronary Circulation in Health and Disease. Philadelphia, Lea & Febiger, 1950.

Gregg, D. E.: The natural history of coronary collateral development. *Circ. Res., 35*:335, 1974.

Grover, R. F.: Mechanisms augmenting coronary arterial oxygen extraction. *Adv. Cardiol., 9*:89, 1973.

Guyton, R. A., and Daggett, W. M.: The evolution of myocardial infarction: Physiological basis for clinical intervention. *Int. Rev. Physiol., 9*:305, 1976.

Hamby, R. I.: Clinical-Anatomical Correlates in Coronary Artery Disease. Mt. Kisco, N.Y., Futura Publishing Co., 1979.

Hatch, F. T.: Interactions between nutrition and heredity in coronary heart disease. *Am. J. Clin. Nutr., 27*:80, 1974.

Hutchins, G. M.: Pathological changes in aortocoronary bypass grafts. *Annu. Rev. Med., 31*:289, 1980.

Julian, D. G.: Diagnosis and Management of Angina Pectoris. Chicago, Year Book Medical Publishers, 1977.

Kirk, E. S., *et al.*: Problems in cardiac performance: Regulation of coronary blood flow and the physiology of heart failure. *In* MTP International Review of Science: Physiology, Vol. 1. Baltimore, University Park Press, 1974, p. 299.

Klocke, F. J., and Ellis, A. K.: Control of coronary blood flow. *Annu. Rev. Med., 31*:489, 1980.

Long, C. (ed.): Prevention and Rehabilitation in Ischemic Heart Disease. Baltimore, Williams & Wilkins, 1980.

Lüllmann, H., and Peters, T.: Action of Cardiac Glycosides on the Excitation-Contraction Coupling in Heart Muscle. New York, Fischer, 1979.

Manninen, V., and Halonen, P. I. (eds.): Sudden Coronary Death. New York, S. Karger, 1978.

Maseri, A., *et al.* (eds.): Primary and Secondary Angina Pectoris. New York, Grune & Stratton, 1978.

Mason, D. T., *et al.*: (eds.): Florence International Meeting on Myocardial Infarction, May 8–12, 1979. New York, Elsevier/North-Holland, 1979.

Murray, R.: Myocardial Imaging. New York, MSS Information Corp., 1977.

Naughton, J., and Hellerstein, H. K. (eds.): Exercise Testing and Exercise Training in Coronary Heart Disease. New York, Academic Press, 1973.

Neely, J. R., and Morgan, H. E.: Relationship between carbohydrate and lipid metabolism and the energy balance of the heart. *Annu. Rev. Physiol., 36*:413, 1974.

Oliver, M. F.: Metabolism of the normal and ischaemic myocardium. *In* Dickinson, C. J., and Marks, J. (eds.): Developments in Cardiovascular Medicine. Lancaster, England, MTP Press, 1978, p. 145.

Pujadas, G., *et al.*: Coronary Angiography in the Medical and Surgical Treatment of Ischemic Heart Disease. New York, McGraw-Hill, 1980.

Purcell, J. A., *et al.*: Angina Pectoris. Atlanta, Prichett & Hull Associates, 1979.

Rackley, C. E., and Russell, R. O., Jr.: Coronary Artery Disease: Recognition and Management. Mt. Kisco, N.Y., Futura Publishing Co., 1979.

Randle, P. J., and Tubbs, P. K.: Carbohydrate and fatty acid metabolism. *In* Berne, R. M., *et al.* (eds.): Handbook of Physiology. Sec. 2, Vol. I. Baltimore, Williams & Wilkins, 1979, p. 805.

Roskamm, H., and Schumuziger, M. (eds.): Coronary Heart Surgery: A Rehabilitation Measure. New York, Springer-Verlag, 1979.

Sabiston, D. C., Jr.: The William F. Reinhoff, Jr. lecture: The coronary circulation. *Johns Hopkins Med. J., 134*:314, 1974.

Sarnoff, S. J., *et al.*: Hemodynamic determinants of oxygen consumption of the heart with special reference to the tension-time index. *Am. J. Physiol., 192*:148, 1958.

Schaper, W.: The Collateral Circulation of the Heart. New York, American Elsevier Publishing Company, 1971.

Schaper, W. (ed.): The Pathophysiology of Myocardial Perfusion. New York, Elsevier/North-Holland, 1979.

Stone, H. L., *et al.*: Cardiac function after embolization of coronaries with microspheres. *Am. J. Physiol., 204*:16, 1963.

Stone, H. L., *et al.*: Progressive changes in cardiovascular function after unilateral heart irradiation. *Am. J. Physiol., 206*:289, 1964.

Strauss, H. W. (ed.): Cardiovascular Nuclear Medicine. St. Louis, C. V. Mosby, 1979.

Sugimoto, T., *et al.*: Quantitative effect of low coronary pressure on left ventricular performance. *Jap. Heart J., 9*:46, 1968.

Winbury, M. A., and Abiko, Y. (eds.): Ischemic Myocardium and Antianginal Drugs. New York, Raven Press, 1979.

26

Cardiac Failure

Perhaps the most important ailment that must be treated by the physician is cardiac failure, which can result from any heart condition that reduces the ability of the heart to pump blood. Usually the cause is decreased contractility of the myocardium from diminished coronary blood flow, but failure to pump adequate quantities of blood can also be caused by damage to the heart valves, external pressure around the heart, vitamin deficiency, primary cardiac muscle disease, or any other abnormality that makes the heart a hypoeffective pump. In the present chapter, we will discuss primarily cardiac failure caused by ischemic heart disease. In the following chapter we will discuss valvular and congenital heart disease.

Definition of Cardiac Failure. The term cardiac failure means simply *failure of the heart, because of some malfunction of the heart itself, to pump enough blood.* Cardiac failure may be manifest in either of two ways: (1) by a decrease in cardiac output or (2) by a damming up of the blood in the veins behind the left or right heart even though the cardiac output is normal or at times even above normal, as we shall discuss later.

Unilateral Versus Bilateral Cardiac Failure. Since the left and right sides of the heart are two separate pumping systems, it is possible for one of these to fail independently of the other. For instance, unilateral failure can result from coronary thrombosis in one or the other of the ventricles. However, cardiac failure from this cause occurs approximately 30 times as often in the left ventricle as in the right ventricle, so that there is a tendency among clinicians to view failure following myocardial infarction as almost always primarily left-sided. Occasionally, however, right-sided failure does occur with no left-sided failure at all, but this happens most frequently in persons with pulmonary stenosis or some other congenital disease affecting primarily the right heart.

In the first part of this chapter we will consider the whole heart failing as a unit and then return to the specific features of unilateral left- and right-sided failure.

DYNAMICS OF THE CIRCULATION IN CARDIAC FAILURE

ACUTE EFFECTS OF MODERATE CARDIAC FAILURE

If a heart suddenly becomes severely damaged in any way, such as by myocardial infarction, the pumping ability of the heart is immediately depressed. As a result, two essential effects occur: (a) reduced cardiac output and (b) damming of blood in the veins resulting in increased systemic venous pressure. These two effects are shown graphically in Figure 26–1. This figure illus-

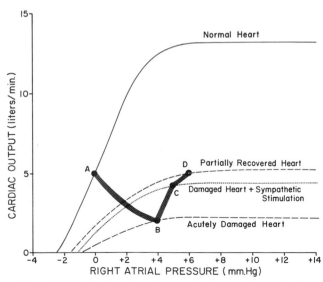

Figure 26–1. Progressive changes in the cardiac output curve following acute myocardial infarction. The cardiac output and right atrial pressure change progressively from point A to point D, as explained in the text.

309

trates, first, a normal cardiac output curve, depicting the state of the circulation prior to the cardiac damage. Point A represents the normal state of the circulation, showing that the normal cardiac output under resting conditions is 5 liters per minute and the right atrial pressure 0 mm. Hg.

Immediately after the heart becomes damaged, the cardiac output curve becomes greatly reduced, falling to the lower, long-dashed curve. Within a few seconds after the acute heart attack, a new circulatory state is established at point B rather than point A, showing that the cardiac output has fallen to 2 liters per minute, about two-fifths normal, while the right atrial pressure has risen to 4 mm. Hg because blood returning to the heart is dammed up in the right atrium. This low cardiac output is still sufficient to sustain life, but it is likely to be associated with fainting. Fortunately, this acute stage lasts for only a few seconds because sympathetic reflexes occur immediately that can compensate to a great extent for the damaged heart as follows:

Compensation for Acute Cardiac Failure by Sympathetic Reflexes. When the cardiac output falls precariously low, many of the different circulatory reflexes discussed in Chapter 21 are immediately activated. The best known of these is the baroreceptor reflex, which is activated by diminished arterial pressure. It is probable that the chemoreceptor reflex, the central nervous system ischemic response, and possibly even reflexes originating in the damaged heart itself also contribute to a lesser extent to the nervous response. But whatever all the reflexes might be, the sympathetics become strongly stimulated within a few seconds, and the parasympathetics become reciprocally inhibited at the same time.

Strong sympathetic stimulation has two major effects on the circulation: first, on the heart itself, and, second, on the peripheral vasculature. Even the damaged myocardium responds to the sympathetic stimulation with increased force of contraction. If all the musculature is diffusely damaged, sympathetic stimulation strengthens this damaged musculature. Likewise, if part of the muscle is totally nonfunctional while part of it is still normal, the normal muscle is strongly stimulated by sympathetic stimulation, in this way compensating for the nonfunctional muscle. Thus, *the heart one way or another becomes a stronger pump, often as much as 100 per cent stronger, under the influence of the sympathetic impulses*. This effect is also illustrated in Figure 26–1, which shows elevation of the cardiac output curve after sympathetic compensation (the dotted curve).

Sympathetic stimulation also increases the tendency for venous return, for it increases the tone of most of the blood vessels of the circulation, especially of the veins, *raising the mean systemic filling pressure* to 12 to 14 mm. Hg, almost 100 per cent above normal. As will be recalled from the discussion in Chapter 23, this greatly increases the tendency for blood to flow back to the heart. Therefore, the damaged heart becomes primed with more inflowing blood than usual, and the right atrial pressure rises still further, which helps the heart to pump larger quantities of blood. Thus, in Figure 26–1 the new circulatory state is depicted by Point C, showing a cardiac output of 4.2 liters per minute and a right atrial pressure of 5 mm. Hg.

The sympathetic reflexes become maximally devel-oped in about 30 seconds. Therefore, a person who has a sudden moderate heart attack might experience nothing more than cardiac pain and a few seconds of fainting. Shortly thereafter, with the aid of the sympathetic reflex compensations as described above, the cardiac output may return to a level entirely adequate to sustain the person who remains quiet, though the pain might persist.

THE CHRONIC STAGE OF FAILURE

After the first few minutes of an acute heart attack, a prolonged secondary state begins. This is characterized mainly by two events: (1) retention of fluid by the kidneys, and (2) progressive recovery of the heart itself over a period of several weeks to months, as was discussed in the previous chapter.

Renal Retention of Fluid

A low cardiac output has a profound effect on renal function, sometimes causing anuria when the cardiac output falls to as low as one-half to two-thirds normal. In general, the urinary output is reduced as long as the cardiac output is significantly less than normal, and it usually does not return to normal after an acute heart attack until the cardiac output rises either all the way back to normal or almost to normal. This relationship of renal function to cardiac output is one of the most important of all the factors affecting the dynamics of the circulation in chronic cardiac failure.

Causes of Renal Retention of Fluid in Cardiac Failure. There are three known causes of the reduced renal output during cardiac failure, all of which are perhaps equally important but in different ways.

1. Decreased Glomerular Filtration. A decrease in cardiac output has a tendency to reduce the glomerular pressure in the kidneys because of (a) *reduced arterial pressure* and (b) *intense sympathetic constriction of the different arterioles of the kidney*. As a consequence, except in the mildest degrees of heart failure, the glomerular filtration rate becomes less than normal. It will become evident from the discussion of kidney function in Chapter 35 that *even a very slight decrease in glomerular filtration often decreases urine output markedly*. When the cardiac output falls to about one-half normal, this factor alone can result in almost complete anuria.

2. Activation of the Renin-Angiotensin System and Increased Reabsorption of Water and Salt by the Kidneys. The reduced blood flow to the kidneys causes marked increase in renin output, and this in turn causes the formation of angiotensin by the mechanism described in Chapter 21. The angiotensin has a direct effect on the arterioles of the kidneys to decrease further the blood flow through the kidneys. Also, increased efferent arterial resistance decreases peritubular capillary pressure, thus promoting greatly increased reabsorption of both water and salt from the renal tubules. Therefore, the net loss of water and salt into the urine is greatly decreased, and the quantities of salt and water in the body fluids increase.

3. Increased Aldosterone Secretion. In the chronic stage of heart failure, large quantities of aldosterone are secreted by the adrenal cortex. This probably results mainly from the effect of angiotensin in stimulating aldosterone secretion. However, recent research suggests that much of the increase in aldosterone secretion might also result from increased plasma potassium, because excess potassium is the most powerful stimulus known for aldosterone secretion, and the potassium concentration does rise in response to the reduced renal function in cardiac failure. Regardless of the cause of the increased aldosterone, it further increases the reabsorption of sodium from the renal tubules, and this in turn leads to a secondary increase in water reabsorption for two reasons: First, as the sodium is reabsorbed, it reduces the osmotic pressure in the tubules while increasing the osmotic pressure in the renal interstitial fluids; these changes promote osmosis of water into the blood. Second, the absorbed sodium increases the osmotic concentration of the extracellular fluid and elicits *antidiuretic hormone* secretion by the supraoptico–posterior pituitary system, which is discussed in Chapter 36. The antidiuretic hormone then promotes increased tubular reabsorption of water.

The Beneficial Effects of Moderate Fluid Retention in Cardiac Failure. Though many cardiologists formerly considered fluid retention always to have a detrimental effect in cardiac failure, it is now known that a moderate increase in body fluid and blood volume is actually a very important factor helping to compensate for the diminished pumping ability of the heart. It does this by increasing the tendency for venous return. The increased blood volume increases the venous return in two ways: First, it increases the mean systemic filling pressure, which *increases the pressure gradient for flow of blood toward the heart.* Second, it distends the veins, which *reduces the venous resistance* and thereby allows increased ease of flow of blood to the heart.

If the heart is not too greatly damaged, this increased tendency for venous return can often fully compensate for the heart's diminished pumping ability — so much so, in fact, that if the heart's pumping ability is reduced to as little as 30 to 50 per cent of normal, still the increased venous return will cause an entirely normal cardiac output.

Detrimental Effects of Excess Fluid Retention in the Severe Stages of Cardiac Failure. In contrast to the beneficial effects of moderate fluid retention in cardiac failure, in severe failure with extreme excesses of fluid retention the fluid then begins to have very serious physiological consequences, including overstretching of the heart thus weakening the heart still more, filtration of fluid into the lungs to cause pulmonary edema and consequent deoxygenation of the blood, and, often, development of extensive edema in all of the peripheral tissues of the body as well. These very detrimental effects of excessive fluid will be discussed in subsequent sections of the chapter.

Recovery of the Myocardium Following Myocardial Infarction

After a heart becomes suddenly damaged as a result of myocardial infarction, the natural reparative processes of the body begin immediately to help restore normal cardiac function. Thus, a new collateral blood supply begins to penetrate the peripheral portions of the infarcted area, often completely restoring the muscle function. Also, the undamaged musculature hypertrophies, in this way offsetting much of the cardiac damage.

Obviously, the degree of recovery depends on the type of cardiac damage, and it varies from no recovery at all to almost complete recovery. Ordinarily, after myocardial infarction the heart recovers rapidly during the first few days and weeks and will have achieved most of its final state of recovery within four to six months.

Cardiac Output Curve After Partial Recovery. The short-dashed curve of Figure 26–1 illustrates function of the partially recovered heart a week or so after the acute myocardial infarction. By this time, considerable fluid has been retained in the body, and the tendency for venous return has increased markedly; therefore, the right atrial pressure has also risen. As a result, the state of the circulation is now changed from Point C to Point D, which represents a *normal* cardiac output of 5 liters per minute but with a right atrial pressure elevated to 6 mm. Hg.

Since the cardiac output has returned to normal, renal output also will have returned to normal and no further fluid retention will occur. Therefore, except for the high right atrial pressure represented by Point D in this figure, the person now has essentially normal cardiovascular dynamics *as long as he remains at rest.*

If the heart itself recovers to a significant extent and if adequate fluid retention occurs, the sympathetic stimulation gradually abates toward normal for the following reasons: The partial recovery of the heart can do the same thing for the cardiac curve as sympathetic stimulation, and fluid retention in the circulatory system can do the same thing for venous return as sympathetic stimulation. Thus, as these two factors develop, the fast pulse rate, cold skin, sweating, and pallor resulting from sympathetic stimulation in the acute stage of cardiac failure gradually disappear.

SUMMARY OF THE CHANGES THAT OCCUR FOLLOWING ACUTE CARDIAC FAILURE — "COMPENSATED HEART FAILURE"

To summarize the events discussed in the past few sections describing the dynamics of circulatory changes following an acute, moderate heart attack, we may divide the stages into (1) the instantaneous effect of the cardiac damage, (2) compensation by the sympathetic nervous system, and (3) chronic compensations resulting from partial cardiac recovery and renal retention of fluid. All these changes are shown graphically by the very heavy line in Figure 26–1. The progression of this line shows: the normal state of the circulation (point 'A), the state a few seconds after the heart attack but before sympathetic reflexes have occurred (point B), the rise in cardiac output toward normal caused by sympathetic stimulation (point C), and final return of the cardiac output to normal following several days to several weeks of cardiac recovery and fluid retention (point D). This final state is called compensated heart failure.

Compensated Heart Failure. Note especially in Figure 26–1 that the pumping ability of the heart, as depicted by the cardiac function curve, is still depressed to less than one-half normal. This illustrates that factors that increase the right atrial pressure (principally retention of fluid) can maintain the cardiac output at a normal level despite continued weakness of the heart itself. However, one of the results of chronic cardiac weakness is this chronic increase in right atrial pressure itself; in Figure 26–1 it is shown to be 6 mm. Hg. There are many persons, especially in old age, who have completely normal resting cardiac outputs but mildly to moderately elevated right atrial pressures because of compensated heart failure. These persons may not know that they have cardiac damage because the damage more often than not has occurred a little at a time, and the compensation has occurred concurrently with the progressive stages of damage.

Reduced Cardiac Reserve in Compensated Failure. The slight elevation of right atrial pressure has little harmful effect on the circulatory system (perhaps a slightly enlarged liver and tendency toward ankle edema). However, the person with compensated heart failure certainly does not have a normal circulatory system, for should that person try to exercise strongly or should any other stress be placed on the circulatory system, such as might occur in some disease condition, the heart would be unable to respond normally. Instead, the person becomes very weak and often develops severe dyspnea because the circulation cannot deliver the needed oxygen to the nervous system and also cannot remove the carbon dioxide. In addition, the sympathetic system becomes excessively stimulated in a futile attempt to excite the circulation to a greater level of cardiac output — the heart rate becomes very rapid, the skin pallid, and sweat beads often appear.

To state this another way, the normal person has far greater *cardiac reserve* than the person with compensated heart disease. Cardiac reserve is discussed at greater length later in the chapter.

DYNAMICS OF SEVERE CARDIAC FAILURE — DECOMPENSATED HEART FAILURE

If the heart becomes severely damaged, then no amount of compensation, either by sympathetic nervous reflexes or by fluid retention, can make this weakened heart pump a normal cardiac output. As a consequence, the cardiac output cannot rise to a high enough value to bring about return of normal renal function. Fluid continues to be retained, the person develops progressively more and more edema, and this state of events eventually leads to death. This is called *decompensated heart failure*. The main basis of decompensated heart failure is *failure of the heart to pump sufficient blood to make the kidneys function adequately*.

Analysis of Decompensated Heart Failure. Figure 26–2 illustrates a greatly depressed cardiac output curve, depicting the function of a heart that has become extremely weakened and cannot be strengthened. Point A on this curve represents the approximate state of the circulation before any compensation has occurred, and Point B the state after the first few minutes of acute compensation by sympathetic stimulation, as described above. At Point B the cardiac output has risen to 4 liters per minute and the right atrial pressure to 5 mm. Hg. The person appears to be in reasonably good condition, but this state will not remain static for the following reason: The cardiac output has not risen quite high enough to cause adequate kidney excretion of fluid; therefore, fluid retention continues unabated and can eventually be the cause of death. These events can be explained quantitatively in the following way:

Note the dashed line at a cardiac output level of 5 liters in Figure 26–2. This is the critical cardiac output level that is required for re-establishment of normal fluid balance. At any cardiac output below this level, all the fluid-retaining mechanisms discussed in the earlier section remain in play, and the body fluid volumes will increase progressively. Because of this progressive increase in fluid volume, the mean systemic filling pressure continues to rise, and this forces progressively increasing quantities of blood into the right atrium, thus increasing the right atrial pressure. After a day or so, the state of the circulation changes from Point B in Figure 26–2 to Point C — the right atrial pressure rising to 7 mm. Hg. and the cardiac output to 4.2 liters per minute. However, note again that the cardiac output is still not high enough to cause normal renal output of fluid; therefore, fluid continues to be retained, and after another day or so the right atrial pressure rises to 9 mm. Hg, and the circulatory state becomes that depicted by Point D. Still, the cardiac output is not enough to establish normal fluid balance.

After another few days the right atrial pressure has risen still further, but by now the cardiac function curve is beginning to decline toward a lower level. This decline is caused by overstretch of the heart, edema of the heart muscle, and other factors that diminish the pumping performance of the heart. It is especially evident, however, that further retention of fluid from

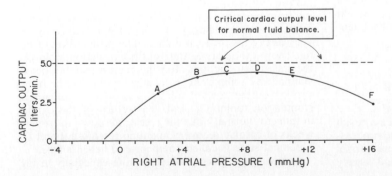

Figure 26–2. Greatly depressed cardiac output curve that indicates decompensated heart disease. Progressive fluid retention raises the right atrial pressure, and the cardiac output progresses from point A to point F, as explained in the text.

then on will be more and more detrimental to the condition. Still, the cardiac output is not high enough to bring about normal fluid balance, and fluid retention not only continues but actually accelerates because of the falling cardiac output. Consequently, within a few days the state of the circulation has reached Point F on the curve, with the cardiac output less than 2.5 liters per minute and the right atrial pressure 16 mm. Hg. This is a state that now has approached or reached incompatibility with life, and the patient dies in *decompensation.*

Thus, one can see from this analysis that failure of the cardiac output ever to rise to the critical level required for normal renal function results in (a) progressive retention of fluid, which causes (b) progressive elevation of the mean systemic filling pressure, and (c) progressive elevation of the right atrial pressure until finally the heart is so overstretched or so edematous that it becomes unable to pump even moderate quantities of blood, and, therefore, fails completely. Clinically, one detects this serious condition of decompensation principally by the progressive edema, especially edema of the lungs, which leads to bubbling rales and dyspnea (air hunger). All clinicians know that failure to institute appropriate therapy when this state of events occurs leads to rapid death.

Myocardial Deterioration in Decompensated Failure. In addition to the severe problem of progressive fluid retention in decompensated heart failure, another important complicating problem is progressive deterioration of the myocardium itself. The basic causes of this are probably such factors as: (1) diminished blood supply to the cardiac muscle, (2) overstretching of the muscle, (3) edema of the muscle, (4) growth of fibrous tissue in the myocardium, (5) overworking of the remaining functional musculature, and so on. But, regardless of the cause or causes of the deterioration, some of the important results are the following:

First, the contractility of the heart muscle itself diminishes.

Second, the calcium mechanism for causing excitation-contraction coupling is depressed. This results partly from the fact that the longitudinal tubules of the sarcoplasmic reticulum fail to accumulate normal amounts of calcium. Therefore, when the muscle is stimulated, not enough calcium is available to cause full contraction. In addition, the troponin-tropomyosin system for initiating interaction between the actin and myosin filaments seems to respond less intensely to the calcium that is available. These effects also explain why digitalis is often very effective in improving the contractile strength of the failing heart, because digitalis has a specific function to increase the buildup of calcium ions in cardiac muscle.

Third, the quantity of stored norepinephrine in the sympathetic nerve endings decreases markedly in the failing myocardium. And the rate of secretion of norepinephrine in response to sympathetic stimuli is correspondingly reduced. Therefore, the failing myocardium responds very poorly to sympathetic stimulation. This means that the heart loses one of its very important stimulating mechanisms.

One can well understand that such cardiac deterioration occurring in the late stages of cardiac failure can lead to still more failure. Once this begins, it has a tendency to perpetuate itself and even to become a vicious cycle. Therefore, intensive therapy in this stage of heart disease is essential if the person is to survive.

Treatment of Decompensation. The two ways in which the decompensation process can often be stopped are: (1) by strengthening the heart in any one of several ways, especially by administration of a cardiotonic drug, such as digitalis, so that it can pump adequate quantities of blood to make the kidneys function normally again, or (2) by administering diuretics and reducing water and salt intake, which brings about a balance between fluid intake and output despite the low cardiac output.

Both methods stop the decompensation process by re-establishing normal fluid balance so that at least as much fluid leaves the body as enters it.

Mechanism of Action of the Cardiotonic Drugs. Cardiotonic drugs, such as digitalis, have little effect on increasing the contractile strength of normal cardiac muscle. On the other hand, when administered to a person with chronic cardiac failure these same drugs can frequently increase the strength of the failing myocardium as much as two-fold. Therefore, they are the mainstay of therapy in chronic heart failure.

The way in which digitalis and the other cardiotonic glycosides strengthen heart contraction is believed to be by increasing the quantity of calcium ions in the muscle fibers. It was pointed out above that the cardiac muscle sarcoplasmic reticulum fails to accumulate normal quantities of calcium in severe heart failure and therefore cannot release enough calcium to cause full contraction of the muscle. Consequently, this accumulation of extra calcium in the muscle fibers at least partially corrects this problem.

UNILATERAL CARDIAC FAILURE

In the discussions thus far in this chapter, we have considered failure of the heart as a whole. Yet in a large number of patients, especially those with early acute failure, left-sided failure predominates over right-sided failure, and in rare instances, especially in congenital heart disease, the right side may fail without significant failure of the left side. Therefore, we now need to discuss the special features of unilateral failure.

UNILATERAL LEFT HEART FAILURE

When the left side of the heart fails without concomitant failure of the right side, blood continues to be pumped into the lungs with usual right heart vigor while it is not pumped adequately out of the lungs into the systemic circulation. As a result, the *mean pulmonary filling pressure* rises because of shift of large volumes of blood from the systemic circulation into the pulmonary circulation.

As the volume of blood in the lungs increases, the pulmonary vessels enlarge, and, if the pulmonary capillary pressure rises above 28 mm. Hg, that is, above the colloid osmotic pressure of the plasma, fluid begins to filter out of the capillaries into the interstitial spaces and alveoli, resulting in pulmonary edema.

Thus, among the most important problems of left

heart failure are *pulmonary vascular congestion* and *pulmonary edema,* which are discussed in detail in Chapters 24 and 31 in relation to the pulmonary circulation and capillary dynamics. As long as the pulmonary capillary pressure remains less than the normal colloid osmotic pressure of the blood, ranging between 23 and 33 mm. Hg, the lungs remain "dry." But even a few millimeters' rise in capillary pressure above this critical level causes progressive transudation of fluid into the interstitial spaces and alveoli, leading rapidly to death. Pulmonary edema can occur so rapidly that it can cause death after only 20 to 30 minutes of severe acute left heart failure.

Course of Events for Several Days After Acute Left Heart Failure

During the several days after the onset of left heart failure, one additional feature must be added to the acute picture. This is retention of fluid resulting from reduced renal function. In moderate acute left heart failure the pulmonary capillary pressure sometimes rises only to 15 to 20 mm. Hg, as illustrated in Figure 26–3, not enough to cause pulmonary edema. Yet following retention of fluid for the next few days, the blood volume increases, and more blood is pumped into the lungs by the right ventricle. Then, the pulmonary capillary pressure rises still more, often rising above the plasma colloid osmotic pressure, resulting in severe pulmonary edema as shown in the figure. Indeed, this is a common occurrence: the patient suddenly develops severe pulmonary edema a week or so after the acute attack and dies a respiratory death, not a death resulting from diminished cardiac output.

UNILATERAL RIGHT HEART FAILURE

In unilateral right heart failure, blood is not pumped adequately from the systemic circulation into the lungs. Therefore, blood shifts from the lungs into the systemic circulation. However, the effect of this on the systemic circulation is hardly noticeable for the following reason: The total amount of blood in the lungs is only about one-eighth that in the systemic circulation. Therefore, even in severe acute right heart failure, the systemic blood volume increases only a few per cent because of

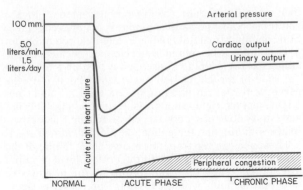

Figure 26–4. The overall effects, acute and chronic, of right-sided heart failure.

blood shift from the lungs, and this is not sufficient to cause significant systemic congestion.

Low Cardiac Output in Acute Right Heart Failure. On the other hand, acute right heart failure can cause greater depression of the cardiac output than acute left heart failure of the same degree. This again stems from the far greater compliance of the systemic circulation than of the pulmonary circulation. Not enough blood can transfer from the lungs into the systemic vessels to raise the systemic pressures to a very high level, and these pressures are not enough to make the weakened right ventricle pump adequate quantities of blood. Therefore, in those rare conditions in which the right heart does fail acutely, the cardiac output falls greatly, often leading to such low cardiac output that death ensues rapidly.

The Chronic Stage of Unilateral Right Heart Failure. The chronic stage of unilateral right heart failure is much the same as that discussed earlier for the entire heart. The depressed cardiac output results in progressively more and more retention of fluid by the kidneys until the cardiac output either rises back nearly to normal or until the person goes into decompensation and dies. Figure 26–4 illustrates the progressive changes that occur in chronic right-sided heart failure, showing a gradual return of cardiac output to normal or near normal, a return of arterial and urinary output, and, finally, progressive development of peripheral congestion and edema.

"HIGH CARDIAC OUTPUT FAILURE" — OVERLOADING OF THE HEART

A condition called "high cardiac output failure" frequently occurs in persons who have cardiac outputs much higher than normal but who have signs of cardiac failure because the left or right or both atrial pressures are very high. Sometimes there is also accumulation of edema. The true problem is often not failure of the pumping ability of the heart but instead *overloading of the heart with too much venous return.* The basic causes of the increased cardiac output are (1) decreased systemic resistance or (2) increased mean systemic filling pressure. These *may* or *may not* be associated with a

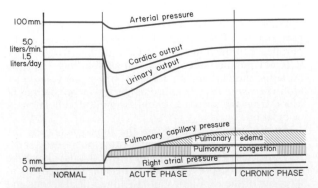

Figure 26–3. The overall effects, acute and chronic, of left-sided heart failure.

weak heart, depending on the condition causing the high output failure.

Decreased Systemic Resistance.   Diseases that cause high cardiac output failure by decreasing the systemic resistance include (a) *arteriovenous fistulae,* in which the blood is shunted directly from the arteries to the veins, (b) *beriberi heart disease,* in which avitaminosis of the B vitamins results in profound systemic vasodilatation, and (c) *thyrotoxicosis,* which, because of its stimulatory effect on metabolism, causes generalized systemic vasodilatation. Obviously, the decreased resistance results in rapid venous return, often causing more blood to flow into the heart than can be properly handled and resulting in greatly increased right and left atrial pressures but at the same time resulting in a cardiac output sometimes as much as 2 or 3 times normal.

Increased Mean Systemic Filling Pressure.   The conditions that cause high cardiac output failure by increasing the mean systemic filling pressure include (a) too rapid and too much *transfusion* and (b) fluid retention resulting from excess secretion of different *steroid hormones* — especially aldosterone. These *increase the pressure gradient for venous return,* thereby forcing large excesses of blood into the heart and overloading it.

CARDIOGENIC SHOCK

In the description of acute heart failure, it was pointed out that the cardiac output can fall very low immediately after heart damage occurs. This obviously leads to greatly diminished blood flow throughout the body and can lead to a typical picture of circulatory shock as described in Chapter 28. This means, simply, that the cardiac output falls too low to supply the body with adequate blood flow. As a result, the tissues deteriorate rapidly and death ensues. Sometimes death comes in less than an hour; at other times, it comes over a period of several days. The circulatory shock that is caused by inadequate cardiac pumping is called *cardiogenic shock* or *cardiac shock,* and it is sometimes also called the *power failure syndrome.*

Cardiogenic shock is extremely important to the clinician because approximately one-tenth of all patients who have acute myocardial infarction will have enough power failure to die of circulatory shock before the physiologic compensatory measures can come into play to save life. Once cardiac shock has become well established after myocardial infarction, all the typical events occur that also occur in the late stages of other types of circulatory shock, as described in Chapter 28, especially rapid deterioration of almost all bodily functions.

Vicious Cycle of Cardiac Deterioration in Cardiogenic Shock.   The discussion of circulatory shock in Chapter 28 will emphasize the tendency for the heart itself to become progressively damaged when its coronary blood supply is reduced during the course of shock. That is, the low arterial pressure that occurs during shock reduces the coronary supply, which makes the heart still weaker, which makes the shock still worse, the process eventually becoming a vicious cycle of cardiac deterioration. In cardiogenic shock caused by myocardial infarction, this problem is greatly compounded by the already existing coronary thrombosis. For instance, in a normal heart, the arterial pressure usually must be reduced below about 45 mm. Hg before cardiac deterioration sets in. However, in a heart that already has a major coronary vessel blocked, deterioration will set in when the arterial pressure falls as low as 80 to 90 mm. Hg. In other words, even the minutest amount of fall in arterial pressure can set off a vicious cycle of cardiac deterioration following myocardial infarction, though this cycle will occur in a normal heart only in the very late stages of shock. For this reason, in treating myocardial infarction it is extremly important to prevent even short periods of hypotension.

Physiology of Treatment.   Often a patient dies of cardiogenic shock before the various compensatory processes can return the cardiac output to a life-sustaining level. Therefore, treatment of this condition is one of the most important problems in the management of acute heart attacks. Immediate digitalization of the heart is often employed for strengthening the heart if the remaining functioning ventricular muscle shows signs of deterioration. Also, infusion of whole blood, plasma, or a blood pressure–raising drug is used to sustain the arterial pressure. If the arterial pressure can be raised high enough, the coronary blood flow can often be elevated to a high enough value to prevent the vicious cycle of deterioration until appropriate compensatory mechanisms in the body can correct the shock. Even with the best therapy, however, once the shock syndrome has begun, with the arterial pressure remaining as much as 20 mm. Hg. below normal for as long as an hour, 85 per cent of the patients die.

EDEMA IN PATIENTS WITH CARDIAC FAILURE

Much of the preceding discussion has emphasized the edema that develops in cardiac failure. On first thought one would think that it should be easy to understand the cause of edema when the heart fails. One's natural logic directs that when the heart fails, blood becomes dammed up behind the heart, thereby increasing the venous and capillary pressures until fluid transudes out of the capillaries into the tissues. However, peripheral edema will not occur until large quantities of water and salt are retained by the kidneys. Consequently, there are special quirks of logic that must be explained before one can understand some of the problems related to edema in cardiac failure.

Role of the Kidneys in Cardiac Edema — Inability of Acute Cardiac Failure to Cause Peripheral Edema.   Though acute left heart failure can cause terrific congestion of the lungs, with rapid development of pulmonary edema, acute heart failure of any type never causes immediate development of peripheral edema. This can be explained best by referring to Figure 26–5. When a previously normal heart acutely fails as a pump the arterial pressure falls and the right atrial pressure rises. The two pressures approach each other at an equilibrium value of about 13 mm. Hg. It is obvious that capillary pressure must also fall from its

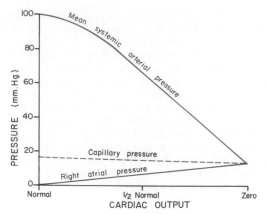

Figure 26-5. Progressive changes in mean systemic arterial pressure and right atrial pressure as the cardiac output falls from normal to zero. This figure also shows the effect of decreased cardiac output on capillary pressure.

normal value of 17 mm. Hg to the equilibrium pressure of 13 mm. Hg. Thus, *severe acute cardiac failure actually causes a fall in capillary pressure rather than a rise.* And animal experiments as well as experience in human beings show that acute cardiac failure does not cause immediate development of peripheral edema. In fact, during the first hours fluid is actually absorbed into the capillaries from the interstitial spaces.

On the other hand, edema does develop during the succeeding days *because of fluid retention by the kidneys*. The retention of fluid increases the mean systemic filling pressure, resulting in increased tendency for blood to return to the heart. This now elevates the right atrial pressure to a still higher value, and it also returns the arterial pressure to normal. Therefore, the rising right atrial pressure *now causes the capillary pressure to rise markedly,* thus causing loss of fluid into the tissues and development of severe edema.

The mechanisms of fluid retention by the kidneys were discussed earlier in the chapter, but these are so important that they need to be relisted here for our present discussion: (1) decreased glomerular pressure because of slightly decreased arterial pressure and especially because of sympathetic constriction of the afferent arterioles; (2) formation of angiotensin, which causes efferent arteriolar constriction with resultant excessive reabsorption of water and salt by the renal tubules; (3) marked increase in aldosterone secretion, which also causes greatly increased salt and water reabsorption by the renal tubules; and (4) increased ADH secretion, which markedly increases water reabsorption by the renal tubules.

Acute Pulmonary Edema in Chronic Heart Failure—A Lethal Vicious Cycle

A frequent cause of death is acute pulmonary edema occurring in patients who have had chronic heart failure for a long time. Sometimes this edema is caused by a new myocardial infarction, but in many instances no new pathology is found in the heart.

When acute pulmonary edema occurs in a person without new cardiac damage, it usually follows some temporary overload of the heart, such as might result from a bout of heavy exercise, some emotional experience, or even a severe cold. This acute pulmonary edema is believed to result from the following vicious cycle:

1. A temporarily increased load on the already weak left ventricle results from increased venous return from the peripheral circulation. Because of the limited pumping capacity of the left heart, blood begins to dam up in the lungs.

2. The increased blood in the lungs increases the pulmonary capillary pressure, and a small amount of fluid begins to transude into the lung tissues and alveoli.

3. The increased fluid in the lungs diminishes the degree of oxygenation of the blood.

4. The decreased oxygen in the blood further weakens the heart and also causes peripheral vasodilatation.

5. The peripheral vasodilatation increases venous return from the peripheral circulation still more.

6. The increased venous return further increases the damming of the blood in the lungs, leading to still more transudation of fluid, more arterial oxygen desaturation, more venous return, and so forth. Thus, a vicious cycle has been established.

Once this vicious cycle has proceeded beyond a certain critical point, it will continue until death of the patient unless heroic therapeutic measures are employed. The types of heroic therapeutic measures that can reverse the process and save the life of the patient include:

a. Putting tourniquets on all four limbs to sequester much of the blood in the veins and therefore to decrease the work load on the left heart.

b. Bleeding the patient.

c. Giving a rapidly acting diuretic such as furosemide to cause rapid loss of fluid from the body.

d. Giving the patient pure oxygen to breathe to reverse the blood desaturation, the heart deterioration, and peripheral vasodilatation.

e. Giving the patient a rapidly acting cardiotonic drug such as ouabain to strengthen the heart.

Unfortunately, this vicious cycle of acute pulmonary edema can proceed so rapidly that death can occur within 30 minutes to an hour. Therefore, any procedure that is to be successful must be instituted immediately.

PHYSIOLOGICAL CLASSIFICATION OF CARDIAC FAILURE

From the above discussions, it is apparent that the symptoms of cardiac failure fall into the following three physiological classifications:

> Low cardiac output
> Pulmonary congestion
> Systemic congestion.

Low cardiac output usually occurs immediately after a heart attack. If the attack is mainly right-sided, this may be the only symptom. If the acute heart attack is mainly left-sided, concurrent pulmonary congestion almost always occurs along with the low cardiac output, but the pulmonary congestion may be mild (without pulmonary edema) until after considerable fluid has been retained by the kidneys. Thus, low cardiac output may be the only significant clinical effect observed in many persons who have sudden heart attacks. This results in the following symptoms:

1. Generalized weakness
2. Fainting
3. Symptoms of increased sympathetic activity such as high heart rate, thready pulse, cold skin, sweating, and so forth.

Pulmonary congestion may be the only effect in patients with pure left-sided *chronic* heart failure, because in the chronic stage enough fluid will have been retained to return the cardiac output to normal despite the weak left ventricle — but this occurs at the expense of greatly elevated pulmonary vascular pressures. And, since the right heart is not failing, pulmonary congestive symptoms alone can occur with essentially no systemic congestion nor low cardiac output.

Systemic congestion alone can occur in pure right-sided *chronic* heart failure. In this condition there is no pulmonary congestion, and, if sufficient fluids have been retained in the blood to prime the heart sufficiently, the heart may pump a normal cardiac output.

Obviously, all the above classes of heart failure can occur together or in any combination.

CARDIAC RESERVE

Fortunately, the normal heart can increase its output to 4 to 5 times normal under conditions of stress in most younger persons and to 6 to 7 times normal in endurance athletes. The maximum percentage that the cardiac output can increase above normal is called the *cardiac reserve*. Thus, in the normal young adult the cardiac reserve is 300 to 400 per cent. In the athletically trained person it is occasionally as high as 500 to 600 per cent, while in the asthenic person it may be as low as 200 per cent. As an example, during severe exercise the cardiac output of the normal healthy young adult can rise to about 5 times normal; this is an increase above normal of 400 per cent — that is, a cardiac reserve of 400 per cent.

Any factor that prevents the heart from pumping blood satisfactorily decreases the cardiac reserve. This can result from ischemic heart disease, primary myocardial disease, vitamin deficiency damage to the myocardium, valvular heart disease, and many other factors, some of which are illustrated in Figure 26–6.

Diagnosis of Low Cardiac Reserve — The Exercise Test. So long as people with low cardiac reserve remain in a state of rest, they probably will not know that they have heart disease. However, a diagnosis of low cardiac reserve can usually be made easily by requiring a person to exercise either on a treadmill or by walking up and down steps. The increased load on the heart rapidly uses up the small amount of reserve that is available, and the cardiac output fails to rise high enough to sustain the body's new level of activity. The acute effects are:

1. Immediate and sometimes extreme shortness of

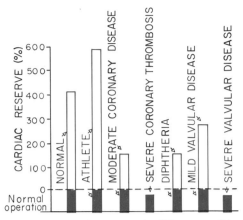

Figure 26–6. Cardiac reserve in different conditions.

breath (dyspnea) resulting from the heart's not pumping sufficient blood to the tissues, thereby causing tissue ischemia and creating a sensation of air hunger.

2. Extreme muscle fatigue resulting from muscle ischemia, thus limiting the person's ability to continue with the exercise.

Therefore, exercise tests are part of the armamentarium of the cardiologist. These tests take the place of cardiac output measurements that unfortunately cannot be made with ease in most clinical settings.

APPENDIX TO CHAPTER 26: A QUANTITATIVE GRAPHICAL METHOD FOR ANALYSIS OF CARDIAC FAILURE

Though it is possible to understand most of the general principles of cardiac failure using mainly qualitative logic, one can grasp the importance of the different factors in cardiac failure with far greater depth by using quantitative approaches. One such approach is the graphical method for analysis of cardiac output regulation that was introduced in Chapter 23. In the following few paragraphs, we will analyze several aspects of cardiac failure utilizing this graphical technique.

Graphical Analysis of Acute Heart Failure and of Chronic Compensation. Figure 26–7 illustrates cardiac output and venous return curves for different states of the heart and of the peripheral circulation. The two *solid*

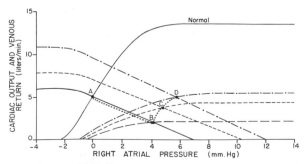

Figure 26–7. The progressive changes in cardiac output and right atrial pressure during different stages of cardiac failure.

curves are (1) the normal cardiac output curve, and (2) the normal venous return curve. As pointed out in Chapter 23, there is only one point on each of these two curves at which the circulatory system can operate so long as these two curves remain normal. This point is where the two curves cross at Point A. Therefore, the normal state of the circulation is a cardiac output and venous return of 5 liters per minute and a right atrial pressure of 0 mm. Hg.

During the first few seconds after a moderately severe heart attack, the cardiac output curve falls to the lowermost long-dashed curve. During this few seconds, the venous return curve still has not changed because the peripheral circulatory system is still operating normally. Therefore, the new state of the circulation is depicted by Point B, where the new cardiac output curve crosses the normal venous return curve. Thus, the right atrial pressure rises to 4 mm. Hg while the cardiac output falls to 2 liters per minute.

Within the next 30 seconds, the sympathetic reflexes become very active. These affect both of the curves, raising them to the short-dashed curves. Sympathetic stimulation can increase the plateau level of the cardiac output curve as much as 100 per cent. It can also increase the mean systemic filling pressure (depicted by the point where the venous return curve crosses the zero venous return axis) by several mm. Hg — in this figure from a normal value of 7 mm. Hg up to 10.5 mm. Hg. This increase in mean systemic filling pressure shifts the entire venous return curve to the right and upward. The new cardiac output and venous return curves now equilibrate at Point C, that is, at a right atrial pressure of 5 mm. Hg and at a cardiac output of 4 liters per minute.

During the ensuing week, the cardiac output and venous return curves rise to the dot-dashed curves because of (1) some recovery of the heart and (2) renal retention of salt and water, which raises the mean systemic filling pressure still further — this time up to 12 mm. Hg. The two new curves, the dot-dashed curves, now equilibrate at Point D. Thus, the cardiac output has now returned entirely to normal. The right atrial pressure, however, has risen still further to 6 mm. Hg. Because the cardiac output is now normal, renal output is also normal so that a new state of equilibrated fluid balance has been achieved. Therefore, the circulatory system will continue to function at Point D and will remain stable, with a normal cardiac output and an elevated right atrial pressure, until some additional extrinsic factor changes either the cardiac output curve or the venous return curve.

Utilizing this technique for analysis, one can see especially the importance of fluid retention and how it eventually leads to a new stable state of the circulation; one can also see the interrelationship between the mean systemic filling pressure and cardiac pumping at various degrees of cardiac failure.

Note especially that the events described in Figure 26–7 are the same as those presented in Figure 26–1 but presented in a more quantitative manner.

Graphical Analysis of "Decompensated" Cardiac Failure. The cardiac output curve in Figure 26–8 is the same as that in Figure 26–2, a very low curve that in this case has already reached a degree of recovery as great as this heart can achieve. In this figure we have added

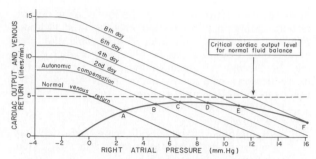

Figure 26–8. Graphical analysis of decompensated heart disease, showing progressive shift of the venous return curve to the right as a result of fluid retention.

venous return curves that occur during successive days following the acute fall of the cardiac output curve to this very low level. At Point A the curve equates with the normal venous return curve to give a cardiac output of approximately 3 liters per minute. However, stimulation of the sympathetic nervous system, caused by this low cardiac output, increases the mean systemic filling pressure within 30 seconds from 7 mm. Hg to 10.5 mm. Hg and shifts the venous return curve upward and to the right to give the curve labeled "autonomic compensation." Thus, the new venous return curve equates with the cardiac output curve at Point B. The cardiac output has been improved up to a level of 4 liters per minute but at the expense of an additional rise in right atrial pressure to 5 mm. Hg.

The cardiac output of 4 liters per minute is still too low to cause the kidneys to function normally. Therefore, fluid continues to be retained, the mean systemic filling pressure rises from 10.5 almost to 13 mm. Hg, and the venous return curve becomes that labeled "second day." This equilibrates with the cardiac output curve at Point C, and the cardiac output rises to 4.2 liters per minute while the right atrial pressure rises to 7 mm. Hg.

During the succeeding days the cardiac output never rises high enough to re-establish normal fluid balance. Therefore, fluid continues to be retained, the mean systemic filling pressure continues to rise, the venous return curve continues to shift to the right, and the equilibrium point between the venous return curve and the cardiac output curve shifts progressively to Point D, to Point E, and finally to Point F. The equilibration process is now on the downslope of the cardiac output curve so that further retention of fluid causes only a detrimental effect on cardiac output. Therefore, the condition accelerates in a downhill direction until death occurs.

Thus, the process of "decompensation" results from the fact that the cardiac output curve never rises up to the critical level of 5 liters per minute required to re-establish normal balance between fluid input and output.

Treatment of Decompensated Heart Disease with Digitalis. Let us assume that the stage of decompensation has already reached Point E in Figure 26–8, and let us proceed to the same Point E in Figure 26–9. At this point, digitalis is given to strengthen the heart and to raise the cardiac output curve up to the heavy curve in Figure 26–9. This does not immediately change the

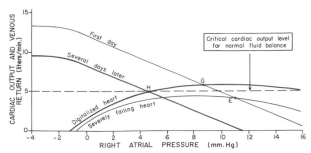

Figure 29–9. Treatment of decompensated heart disease, showing the effect of digitalis in elevating the cardiac output curve, this in turn causing progressive shift of the venous return curve to the left.

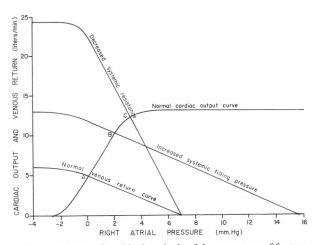

Figure 26–10. Graphical analysis of the two types of factors that can cause high cardiac output failure.

venous return curve. Therefore, the new cardiac output curve equates with the venous return curve at Point G. The cardiac output is now 5.7 liters per minute, a value greater than the critical level of 5 liters required for normal fluid balance. Therefore, the kidneys eliminate far more fluid than normally, causing diuresis, a well known effect of digitalis.

The progressive loss of fluid over a period of several days reduces the mean systemic filling pressure back down to 11.5 mm. Hg, and the new venous return curve becomes the heavy curve labeled "several days later." This curve equates with the cardiac output curve of the digitalized heart at Point H, at an output of 5 liters per minute and a right atrial pressure of 4.6 mm. Hg. This cardiac output is precisely that required for normal fluid balance. Therefore, no additional fluid will be lost, and none will be gained. Consequently, the circulatory system has now stabilized, or, in other words, the decompensation of the heart failure has been "compensated." To state this another way, the final steady-state condition of the circulation is defined by the crossing point of three different curves: the cardiac output curve, the venous return curve, and the critical level for normal fluid balance. The compensatory mechanisms automatically stabilize the circulation when all of these three curves cross at the same point.

Graphical Analysis of High Cardiac Output "Failure." Figure 26–10 shows an analysis of two types of high cardiac output "failure" caused by overloading the heart but not by diminished pumping capability of the heart. In this analysis the cardiac output curve remains normal at all times. As long as the venous return curve is also normal, the cardiac output and venous return curves equate with each other at point A, at a normal cardiac output of 5 liters per minute and a normal right atrial pressure of 0 mm. Hg.

Now, let us assume that the systemic resistance (the total peripheral resistance) becomes greatly decreased (such as might occur with the opening of a very large arteriovenous fistula). The venous return curve rotates upward to give the curve labeled "decreased systemic resistance." This venous return curve equates with the cardiac output curve at point C, with a cardiac output of 12.5 liters per minute and a right atrial pressure of 3 mm. Hg. Thus, the cardiac output has become greatly elevated, the right atrial pressure is slightly elevated, and there are mild signs of peripheral congestion. If the person attempts to exercise, he will have very little

cardiac reserve because the heart is already being used almost to maximum capacity simply to pump the extra blood flow through the arteriovenous fistula. Therefore, this condition resembles a failure condition and is called "high output failure," but in reality the heart is simply overloaded by excess venous return.

The second illustrated cause of high output failure in Figure 26–10 is excessive blood volume, which increases the mean systemic filling pressure to a level far above normal. The venous return curve, labeled "increased systemic filling pressure," is caused by overtransfusion of a person, causing a mean systemic filling pressure of 16 mm. Hg and a venous return curve shifted far to the right and elevated. This curve equilibrates with the cardiac output curve at point B, with a cardiac output of 11 liters per minute and a right atrial pressure of 2 mm. Hg. Here again, mild symptoms of congestion occur, which in some ways resemble cardiac failure, but in reality this is also a condition of overloading the heart.

In other instances of high cardiac output failure, the cardiac output curve may actually be reduced below normal, but because of enhanced venous return the cardiac output is still greater than normal. This frequently occurs in beriberi heart disease and in thyrotoxicosis. The student should draw into Figure 26–10 such a cardiac function curve to show how high cardiac output can actually occur in a person even in the presence of a weakened heart — a truly "failing" heart.

REFERENCES

Barger, A. C.: The kidney in congestive heart failure. *Circulation*, 21:124, 1960.

Bradley, R. D.: Studies in Acute Heart Failure. London, Edward Arnold, 1977.

Braunwald, E.: The determinants of myocardial oxygen consumption. *Physiologist*, 12:65, 1969.

Braunwald, E.: Mechanics and energetics of the normal and failing heart. *Trans. Assoc. Am. Physicians*, 84:63, 1971.

Braunwald, E.: Heart failure—pathophysiological considerations. *In* Dickinson, C. J., and Marks, J. (eds.): Developments in Cardiovascular Medicine. Lancaster, England, MTP Press, 1978, p. 213.

Braunwald, E., *et al.*: Mechanism of Contractility of the Normal and Failing Heart. Boston, Little, Brown, 1968.

Brender, D., *et al.*: Potentiation of adrenergic venomotor responses in dogs by cardiac glycosides. *Circ. Res.*, 25:597, 1969.

Bruce, T. A., and Douglas, J. E.: Dynamic cardiac performance. *In* Frohlich, E. D. (ed.): Pathophysiology, 2nd Ed. Philadelphia, J. B. Lippincott Co., 1976, p. 5.

Burch, G. E.: Interesting aspects of geriatric cardiology. *Am. Heart J., 89*:99, 1975.

Conradsson, T. B., and Werkö, L.: Management of heart disease in pregnancy. *Prog. Cardiovasc. Dis., 16*:407, 1974.

Corday, E., and Swan, H. J. C. (ed.): Clinical Strategies in Ischemic Heart Disease. Baltimore, Williams & Wilkins, 1979.

Dock, W.: Cardiomyopathies of the senescent and senile. *Cardiovasc. Clin., 4*:362, 1972.

Friedberg, C. K.: Diseases of the Heart, 3rd Ed. Philadelphia, W. B. Saunders Co., 1966.

Gibbs, C. L.: Cardiac energetics. *Physiol. Rev., 58*:174, 1978.

Guyton, A. C.: The systemic venous system in cardiac failure. *J. Chronic Dis., 9*:465, 1959.

Guyton, A. C., *et al.*: Cardiac Output and Its Regulation. Philadelphia, W. B. Saunders Co., 1973.

Harrison, T. R., and Reeves, T. J.: Principles and Problems of Ischemic Heart Disease. Chicago, Year Book Medical Publishers, 1968.

Hurst, W.: The Heart. New York, McGraw-Hill Book Company, 1966.

Jacob, R. (ed.): The Hypertrophied Heart: Biophysical, Biochemical and Morphological Aspects of Hypertrophy. Darnstadt, D. Steinkopff, 1977.

Kirk, E. S., *et al.*: Problems in the cardiac performance regulation of coronary blood flow and the physiology of heart failure. *In* MTP International Review of Science: Physiology. Vol. 1. Baltimore, University Park Press, 1974, p. 299.

Kones, R. J.: The catecholamines: Reappraisal of their use for acute myocardial infarction and the low cardiac output syndromes. *Crit. Care Med., 1*:203, 1973.

Kones, R. J.: Cardiogenic shock: Therapeutic implications of altered myocardial energy balance. *Angiology, 25*:317, 1974.

Kones, R. J.: Cardiogenic Shock: Mechanisms and Management. New York, Futura Publishing Co., 1975.

Koppes, G., *et al.*: Treadmill Exercise Testing. Chicago, Year Book Medical Publishers, 1977.

Marks, B. H., and Weissler, A. M. (eds.): Basic and Clinical Pharmacology of Digitalis. Springfield, Ill., Charles C Thomas, 1972.

Mashima, H., and Brutsaert, D. L.: Cardiac muscle dynamics and heart failure. *In* Hayase, S., and Murao, S. (eds.): Cardiology: Proceedings of the VIII World Congress of Cardiology, Tokyo, 1978. New York, Elsevier/North-Holland, 1979, p. 816.

Mason, D. T., and Loogen, F.: Prevention and management of heart failure. *In* Hayase, S., and Murao, S. (eds.): Cardiology: Proceedings of the VIII World Congress of Cardiology, Tokyo, 1978. New York, Elsevier/North-Holland, 1979, p. 836.

Rapaport, E. (ed.): Current Controversies in Cardiovascular Disease. Philadelphia, W. B. Saunders Co., 1980.

Resnekov, L.: Hemodynamic effects of acute myocardial infarction. *Med. Clin. North Am., 57*:243, 1973.

Stone, H. L., *et al.*: Progressive changes in cardiovascular function after unilateral heart irradiation. *Am. J. Physiol., 206*:289, 1964.

Stone, H. L., *et al.*: Ventricular function following radiation damage of the right ventricle. *Am. J. Physiol., 211*:1209, 1966.

Svennerholm, L., *et al.* (eds.): Structure and Functure of Gangliosides. New York, Plenum Press, 1979.

Swan, H. J. C.: Treatment of cardiac failure with vasodilator drugs. *In* Hayase, S., and Murao, S. (eds.): Cardiology: Proceedings of the VIII World Congress of Cardiology, Tokyo, 1978. New York, Elsevier/North-Holland, 1979, p. 868.

Swan, H. J. C., and Parmley, W. W.: Congestive heart failure. *In* Sodeman, W. A., Jr., and Sodeman, T. M. (eds.): Pathologic Physiology: Mechanisms of Disease, 6th Ed. Philadelphia, W. B. Saunders Co., 1979, p. 313.

Wenger, N. K. (ed.): Exercise and the Heart. Philadelphia, F. A. Davis Co., 1978.

Willis, J. (ed.): The Heart: Update. New York, McGraw-Hill, 1979.

Zelis, R., *et al.*: Peripheral circulatory control mechanisms in congestive heart failure. *Am. J. Cardiol., 32*:481, 1973.

27

Heart Sounds; Dynamics of Valvular and Congenital Heart Defects

THE HEART SOUNDS

The function of the heart valves was discussed in Chapter 13, and it was pointed out that closure of the valves is associated with audible sounds, though no sounds usually occur when the valves open. The purpose of the present section is to discuss the factors that cause the sounds in the heart, under both normal and abnormal conditions.

NORMAL HEART SOUNDS

Listening with a stethoscope to a normal heart, one hears a sound usually described as "lub, dub, lub, dub - - -." The "lub" is associated with closure of the A-V valves at the beginning of systole and the "dub" with closure of the semilunar valves at the end of systole. The "lub" sound is called the *first heart sound* and the "dub" the *second heart sound* because the normal cycle of the heart is considered to start with the beginning of systole.

Causes of the First and Second Heart Sounds. Closure of the valves in any pump system usually causes a certain amount of noise because the valves close solidly and suddenly over some opening, setting up vibrations in the fluid or walls of the pump. In the heart, the valves are cushioned by blood, so that it is difficult to understand why these valves create as much sound as they do.

The earliest suggestion for the cause of the heart sounds was that the slapping together of the valve leaflets themselves sets up vibrations, but this has now been shown to cause little if any of the sound because of the cushioning effect of the blood. Instead, the cause is *vibration of the taut valves immediately after closure,* as well as *vibration of the adjacent blood, walls of the heart, and major vessels around the heart.* That is, contraction of the ventricle causes sudden backflow of blood against the A-V valves, causing them to bulge toward the atria until the chordae tendineae abruptly stop the backbulging. The elastic tautness of the valves

then causes the backsurging blood to bounce forward again into each respective ventricle. This sets the blood and the ventricles as well as the valves into vibration and also causes vibrating turbulence in the blood. The vibrations then travel to the chest wall where they can be heard as sound by the stethoscope. This is the cause of a major part of the first heart sound.

The second heart sound results from vibration of the taut, closed semilunar valves and from vibration of the blood and the walls of the pulmonary artery, the aorta, and, to much less extent, the ventricles. When the semilunar valves close, they bulge backward toward the ventricles, and their elastic stretch recoils the blood back into the arteries, which causes a short period of reverberation of blood back and forth between the walls of the arteries and the valves. The vibrations set up in the arterial walls are then transmitted along the arteries at the velocity of the pulse wave. When the vibrations of the vessels come into contact with a "sounding board," such as the chest wall, they create sound that can be heard.

Durations and Frequencies of the First and Second Heart Sounds. The duration of each of the heart sounds is slightly more than 0.1 second; the first sound lasts about 0.14 second and the second about 0.11 second. Both of them are described as very low-pitched sounds, the first lower than the second.

Sound consists of vibrations of different frequencies. Figure 27-1 illustrates by the shaded area the amplitudes of the different frequencies in the heart sounds and murmurs, showing that these are composed of frequencies ranging all the way from a few cycles per second to more than 1000 cycles per second, with the maximum amplitude of vibration occurring at a frequency of about 24 cycles per second, which is actually below the audible range of the ear.

Also shown in Figure 27-1 is a curve called the "threshold of audibility," which depicts the capability of the ear to hear sounds of different amplitudes. Note that in the very low frequency range the heart vibrations have a high degree of amplitude, but the threshold of audibility is so high that ordinarily the heart vibrations

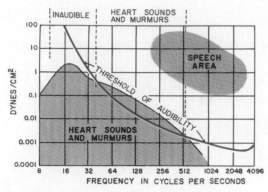

Figure 27–1. Amplitude of different frequency vibrations in the heart sounds and heart murmurs in relation to the threshold of audibility, showing that the range of sounds that can be heard is between about 40 and 500 cycles per second.

below approximately 30 to 50 cycles per second are not heard by the ears. Then above about 500 cycles per second, the heart sounds are so weak that, despite a low threshold of audibility, no frequencies in this range are heard. For practical purposes, then, we can consider that all the *audible* heart sounds lie in the range of approximately 40 to 500 cycles per second despite the fact that the maximum amplitude of vibration occurs at the very low frequency of 24 cycles per second.

Both the first and second heart sounds have a mixture of frequencies in the entire audible range of the heart sounds, though the first sound has slightly more low frequency sound than does the second heart sound.

The reason the frequency of the first heart sound is lower than that of the second sound is probably twofold: First, the *elastic modulus* of the A-V valves and of the walls of the ventricles is far less than the elastic modulus of the semilunar valves and the arterial walls. It is well known that any mechanical vibrating system having a low elastic modulus oscillates at a lower frequency than a system having a greater modulus. Second, the *mass of blood* in the ventricles is much greater than that in the great vessels, which means that the inertia of the vibrating mass is also much greater. This also would cause the first heart sound to have a lower frequency than the second.

The reason the second heart sound is shorter in duration than the first is probably that the second sound is "damped" out by the vascular walls much more rapidly than is the first heart sound by the ventricular walls.

Loudness of the First and Second Heart Sounds.   The loudness of the first and second heart sounds is almost directly proportional to the *rate of change* of the respective pressure differences across the A-V and semilunar valves. For instance, when the onset of systole is very rapid, the intraventricular pressure rises very rapidly during the isometric period of ventricular contraction; yet the pressure on the opposite side of the A-V valves remains only a few mm. Hg. Thus, the rate of change of the pressure difference between the two sides of the valves is very great, and as a result the first heart sound is loud. Also, when the heart is very active, such as during and immediately following exercise, the force of

contraction of the ventricle is greatly enhanced, so that the first heart sound is in this instance also greatly accentuated. Conversely, in a weakened heart in which the onset of contraction is sluggish, the loudness of the first sound is greatly diminished.

In the case of the second heart sound, it is the rate of decrease in ventricular pressure at the end of systole that determines the loudness. The magnitude of this rate of decrease is determined mainly by the level of ventricular systolic pressure at the time the valve closes. In a person who has hypertension, the pressure at the time the aortic valve closes may be as great as 200 mm. Hg, so that the intraventricular pressure falls twice as rapidly as normal, all the way to zero in a few hundredths of a second. Therefore, the aortic sound is markedly accentuated. Likewise, in pulmonary hypertension the pulmonic sound is greatly accentuated. On the other hand, when the arterial pressure is low, such as in shock or in the terminal stages of cardiac failure, the second heart sound is diminished to a very low intensity.

The Third Heart Sound.   Occasionally a third heart sound is heard at the beginning of the middle third of diastole. A logical, but yet unproved, explanation of this sound is oscillation of blood back and forth between the walls of the ventricles initiated by inrushing blood from the atria. This is analogous to running water from a faucet into a sack, the inrushing water reverberating back and forth between the walls of the sack to cause vibrations in the walls.

The third heart sound is an extremely weak rumble of such low frequency that it usually cannot be heard with a stethoscope, but it can be recorded frequently in the phonocardiogram. The very low frequency of this sound presumably results from the flaccid, inelastic condition of the heart during diastole. Also, the reason the third heart sound does not occur until after the first third of diastole is over is presumably that in the early part of diastole the heart is not filled sufficiently to create even the small amount of elastic tension in the ventricles necessary for reverberation. The reason the third heart sound does not continue into the latter part of diastole is presumably that little blood flows into the ventricles during the latter part of diastole, so that no initiating stimulus then exists for causing reverberation.

The Atrial Heart Sound (Fourth Heart Sound).   An atrial heart sound can be recorded in many persons in the phonocardiogram, but it can almost never be heard with a stethoscope because of its low frequency — usually 20 cycles per second or less. This sound occurs when the atria contract, and presumably it is caused by inrush of blood into the ventricles, which initiates vibrations similar to those of the third heart sound.

AREAS FOR AUSCULTATION OF NORMAL HEART SOUNDS

Listening to the sounds of the body, usually with the aid of a stethoscope, is called *auscultation.* Figure 27–2 illustrates the areas of the chest wall from which the different valvular sounds can best be distinguished. With the stethoscope placed in any one of the special valvular areas, the sounds from all the other valves can also still be heard, though the sound from the special valve is as loud, *relative to the other sounds,* as it ever

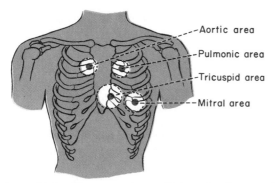

Figure 27–2. Chest areas from which each valve sound is best heard.

will be. The cardiologist distinguishes the sounds from the different valves by a process of elimination; that is, he moves the stethoscope from one area to another, noting the loudness of the sounds in different areas and gradually picking out the sound components from each valve.

The areas for listening to the different heart sounds are not directly over the valves themselves. The aortic area is upward along the aorta, the pulmonic area is upward along the pulmonary artery, the tricuspid area is over the right ventricle, and the mitral area is over the apex of the heart, which is the only portion of the left ventricle near the surface of the chest because the heart is rotated so that most of the left ventricle lies behind the right ventricle. In other words, the sounds caused by the A-V valves are transmitted to the chest wall through each respective ventricle, and the sounds from the semilunar valves are transmitted especially along the great vessels leading from the heart. This transmission of sounds is in keeping with the concept that vibrations in the ventricle or large arteries are the cause of the heart sounds.

THE PHONOCARDIOGRAM

If a microphone specially designed to detect low, frequency sound waves is placed on the chest, the heart sounds can be amplified. Recording is possible by a high-speed recording apparatus, such as an oscilloscope or a high-speed pen recorder; these were described in Chapters 10 and 15 for recording nerve potentials and electrocardiograms. The recording is called a *phonocardiogram,* and the heart sounds appear as waves, as illustrated schematically in Figure 27–3. Record A is a recording of normal heart sounds, showing the vibrations of the first, second, and third heart sounds and even the atrial sound. Note specifically that the third and atrial heart sounds are each a very low rumble. The third heart sound can be recorded in only one third to one half of all persons, and the atrial heart sound can be recorded in perhaps one fourth of all persons.

VALVULAR LESIONS

Rheumatic Valvular Lesions. By far the greatest number of valvular lesions results from rheumatic fever. Rheumatic fever is an autoimmune or allergic disease in

which the heart valves are likely to be damaged or destroyed. It is initiated by streptococcal toxin in the following manner:

The entire sequence of events almost always begins with a preliminary streptococcal infection (caused by Group A hemolytic streptococci), such as a sore throat, scarlet fever, or middle ear infection. The streptococci release several different proteins against which antibodies are formed, the most important of which seems to be a protein called the "M" antigen. The antibodies then react with many different tissues of the body, causing either immunologic or allergic damage. These reactions continue to take place as long as the antibodies persist in the blood — six months or more. As a result, rheumatic fever causes damage in many parts of the body but especially in certain very susceptible areas such as the heart valves. The degree of heart valve damage is directly correlated with the titer and the persistence of these antibodies. Principles of immunity relating to this type of reaction were discussed in Chapter 7, and it is also noted in Chapter 38 that acute glomerular nephritis has a similar basis.

In rheumatic fever, large hemorrhagic, fibrinous, bulbous lesions grow along the inflamed edges of the heart valves. Because the mitral valve receives more trauma during valvular action than do any of the other valves, this valve is the one most often seriously damaged, and the aortic valve is second most frequently damaged. The tricuspid and pulmonary valves are also often involved, but much less severely, probably because the stresses acting on these valves are slight compared with those in the left ventricle.

Scarring of the Valves. The lesions of acute rheumatic fever frequently occur on adjacent valve leaflets simultaneously so that the edges of the leaflets become stuck together. Then, weeks, months, or years later, the lesions become scar tissue, permanently fusing portions of the leaflets. Also, the free edges of the leaflets, which are normally filmy and free-flapping, become solid, scarred masses.

A valve in which the leaflets adhere to each other so extensively that blood cannot flow through satisfactorily is said to be *stenosed.* On the other hand, when the valve edges are so destroyed by scar tissue that they cannot close when the ventricles contract, *regurgita-*

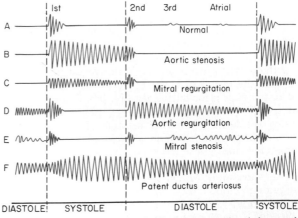

Figure 27–3. Phonocardiograms from normal and abnormal hearts.

tion, or backflow, of blood occurs when the valve should be closed. Stenosis usually does not occur without the coexistence of at least some degree of regurgitation, and vice versa. Therefore, when a person is said to have stenosis or regurgitation, it is usually meant that one predominates over the other.

Other Causes of Valvular Lesions. Stenosis or lack of one or more leaflets of a valve frequently occurs as a congenital defect. Complete lack of leaflets is rare, though stenosis is common, as is discussed later in this chapter.

ABNORMAL HEART SOUNDS CAUSED BY VALVULAR LESIONS

As illustrated by the phonocardiograms of Figure 27–3, many abnormal heart sounds, known as "murmurs," occur when there are abnormalities of the valves, as follows:

The Murmur of Aortic Stenosis. In aortic stenosis, blood is ejected from the left ventricle through only a small opening of the aortic valve. Because of the resistance to ejection, the pressure in the left ventricle rises sometimes to as high as 350 mm. Hg while the pressure in the aorta is still normal. Thus, a nozzle effect is created *during systole,* with blood jetting at tremendous velocity through the small opening of the valve. This causes *severe turbulence* of the blood in the root of the aorta. The turbulent blood impinging against the aortic walls causes intense vibration, and a loud murmur is transmitted throughout the upper aorta and even into the larger arteries of the neck. This sound is harsh and occasionally so loud that it can be heard several feet away from the patient. Also, the sound vibrations can often be felt with the hand on the upper chest and lower neck, a phenomenon known as a "thrill."

The Murmur of Aortic Regurgitation. In aortic regurgitation no sound is heard during systole, but *during diastole* blood flows backward from the aorta into the left ventricle, causing a "blowing" murmur of relatively high pitch and with a swishing quality heard maximally over the left ventricle. This murmur results from *turbulence* of blood jetting backward into the blood already in the left ventricle.

The sound of aortic regurgitation is not nearly so loud as that of aortic stenosis, mainly because the pressure differential between the aorta and left ventricle in regurgitation is not so great as it usually is in stenosis.

If the aortic valve is so badly destroyed that essentially all the return of blood from the arterial tree into the heart takes place during the first portion of diastole, the sound of aortic regurgitation may not be heard at all during the latter portion of diastole. Therefore, loud aortic regurgitation murmurs lasting throughout diastole sometimes mean less severely damaged valves than weaker murmurs heard only during the early part.

The Murmur of Mitral Regurgitation. In mitral regurgitation blood flows backward through the mitral valve *during systole*. This also causes a high frequency "blowing," swishing sound, which is transmitted most strongly into the left atrium, but the left atrium is so deep within the chest that it is difficult to hear this sound directly over the atrium. As a result, the sound of mitral regurgitation is transmitted to the chest wall mainly through the left ventricle, and it is usually heard best at the apex of the heart.

The blowing, swishing quality of the mitral regurgitation murmur, like that of aortic regurgitation, is presumably caused by the turbulence of blood ejected backward through the mitral valve into the blood already in the left atrium or against the atrial wall. The quality of mitral regurgitation murmur is almost exactly the same as that of aortic regurgitation, but it occurs during systole rather than diastole.

The Murmur of Mitral Stenosis. In mitral stenosis, blood passes with difficulty from the left atrium into the left ventricle and, because the pressure in the left atrium rarely rises above 35 mm. Hg except for short periods of time, a great pressure differential forcing blood from the left atrium into the left ventricle never develops. Consequently, the abnormal sounds heard in mitral stenosis are usually weak.

During the early part of diastole, the ventricle has so little blood in it and its walls are so flabby that blood does not reverberate back and forth between the walls of the ventricle. For this reason, even in severe mitral stenosis, no murmur at all might be heard during the first third of diastole. However, after the first third the ventricle is stretched enough for blood to reverberate, and a low rumbling murmur then often begins. This murmur is of such low pitch that it is difficult to hear, but, with the aid of a proper stethoscope 'the "bell" type), one can usually discern very low frequency sounds of 30 to 50 cycles per second. In mild stenosis the murmur lasts only during the first half of the middle third of diastole, but in severe stenosis it can begin early in diastole and persist for the whole remainder of diastole. On the other hand, one can often feel low frequency vibrations, called a "thrill," over the apex of the heart despite the fact that the sound itself may be weak. The reason for this is that the frequency of vibration in the ventricle is often so low that it cannot be heard but yet can be felt.

Often in early stages of mitral stenosis a *presystolic murmur* may be heard. This presystolic murmur is caused by the momentarily increased left atrial pressure resulting from atrial contraction.

Phonocardiograms of Valvular Murmurs. Phonocardiograms B, C, D, and E of Figure 27–3 illustrate, respectively, idealized records obtained from patients with aortic stenosis, mitral regurgitation, aortic regurgitation, and mitral stenosis. It is obvious from these phonocardiograms that the aortic stenotic lesion causes the loudest of all these murmurs, and the mitral stenotic lesion causes the weakest, a murmur of very low frequency and rumbling quality. The phonocardiograms show how the intensity of the murmurs varies during differential portions of systole and diastole, and the relative timing of each murmur is also evident. Note especially that the murmurs of aortic stenosis and mitral regurgitation occur only during systole, while the murmurs of aortic regurgitation and mitral stenosis occur only during diastole — if a student does not understand this timing, a moment's pause should be taken until it is understood.

ABNORMAL CIRCULATORY DYNAMICS IN VALVULAR HEART DISEASE

DYNAMICS OF THE CIRCULATION IN AORTIC STENOSIS AND AORTIC REGURGITATION

Aortic stenosis means a constricted aortic valve with a valvular opening too small for easy ejection of blood from the left ventricle. *Aortic regurgitation* means failure of the aortic valve to close completely and, therefore, failure to prevent backflow of blood from the aorta into the left ventricle during diastole. In aortic stenosis the left ventricle fails to empty adequately, while in aortic regurgitation blood returns to the ventricle after the ventricle has been emptied. Therefore, in either case, the *net stroke* volume output of the heart is reduced, and this in turn tends to reduce the cardiac output, resulting eventually in typical circulatory failure.

However, several important compensations take place that can ameliorate the severity of the circulatory defects. Some of these are the following:

Hypertrophy of the Left Ventricle. In both aortic stenosis and regurgitation, the left ventricular musculature hypertrophies, and in regurgitation the ventricle also enlarges to hold all the regurgitant blood. Sometimes the left ventricle muscle mass increases as much as four- to five-fold, creating a tremendously large left heart. When the aortic valve is seriously stenosed, this hypertrophied muscle allows the left ventricle to develop as much as 450 mm. Hg intraventricular pressure during occasional periods of peak activity; even at rest the pressure differential across the stenotic valve is often 150 mm. Hg. In severe aortic regurgitation, the hypertrophied muscle allows the left ventricle sometimes to pump a stroke volume output as high as 300 ml., though as much as three fourths of this blood on occasion returns to the ventricle during diastole.

Increase in Blood Volume. Another effect that helps to compensate for the diminished net pumping by the left ventricle is increased blood volume. This results from an initial slight decrease in arterial pressure plus peripheral circulatory reflexes that this decrease induces, both of which diminish renal output of urine until the blood volume increases and the pressure returns to normal. Also, red cell mass increases because of a slight degree of tissue hypoxia.

The increase in blood volume tends to increase venous return to the heart. This in turns increases the ventricular end-diastolic volume, causing the left ventricle to pump its very high pressure in aortic stenosis or its very high stroke volume output in aortic regurgitation.

Eventual Failure of the Left Ventricle, and Development of Pulmonary Edema

In the early stages of aortic stenosis or aortic regurgitation, the intrinsic ability of the left ventricle to adapt to increasing loads prevents significant abnormalities in

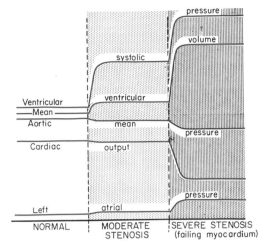

Figure 27–4. Circulatory dynamics in aortic stenosis.

circulatory function other than increased work output required of the left ventricle. Therefore, marked degrees of aortic stenosis or aortic regurgitation often occur before the person knows that he has serious heart disease, such as resting left ventricular systolic pressures as high as 200 to 300 mm. Hg in aortic stenosis, or left ventricular stroke volume outputs as high as double normal in aortic regurgitation.

However, beyond critical stages of development of these two aortic lesions, the left ventricle finally cannot keep up with the work demand, and as a consequence the left ventricle dilates and cardiac output begins to fall while blood simultaneously dams up in the left atrium and lungs behind the failing left ventricle. The left atrial pressure rises progressively, and at pressures above 30 to 40 mm. Hg mean atrial pressure edema appears in the lungs, as is discussed in detail in Chapter 31.

The progressive changes in circulatory dynamics are diagrammed in Figures 27–4 and 27–5 for aortic stenosis and regurgitation, respectively.

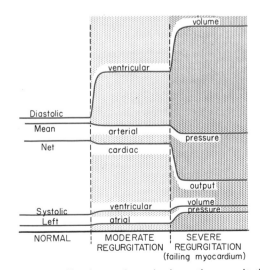

Figure 27–5. Circulatory dynamics in aortic regurgitation.

Myocardial Ischemia in Aortic Valvular Disease. Because of the very high intraventricular pressure during systole in aortic stenosis, very little blood flows through the coronaries during systole. Therefore, extra blood flow is required during diastole to make up the difference. Unfortunately, the hypertrophied muscle of the left ventricle also frequently has a relatively deficient coronary vasculature. Finally, the intraventricular pressure sometimes remains high during diastole, thereby compressing the inner layers of the heart and also diminishing coronary blood flow. For all these reasons, the patient frequently experiences severe degrees of coronary ischemia and angina.

In aortic regurgitation the problem is further compounded because the diastolic aortic pressure frequently falls very low as the aortic blood regurgitates back into the ventricle. Since most of the left ventricular coronary blood flow occurs during diastole, this low pressure can be particularly detrimental to the flow. The effect is especially serious for the subendocardial myocardium in which almost zero flow occurs during systole. Therefore, again, coronary ischemia occurs with concomitant anginal pain.

DYNAMICS OF MITRAL STENOSIS AND MITRAL REGURGITATION

In mitral stenosis blood flow from the left atrium into the left ventricle is impeded, and in mitral regurgitation much of the blood that has flowed into the left ventricle leaks back into the left atrium during systole rather than being pumped into the aorta. Therefore, the effect is reduced net movement of blood from the left atrium into the left ventricle.

Pulmonary Edema in Mitral Valvular Disease. Obviously, the buildup of blood in the left atrium causes progressive increase in left atrial pressure, and this can result eventually in the development of serious pulmonary edema. Ordinarily, lethal edema will not occur until the mean left atrial pressure rises at least above 30 mm. Hg; more often it must rise to as high as 40 mm. Hg because the lung lymphatic vasculature enlarges manyfold and can carry fluid away from the lung tissues extremely rapidly.

Enlarged Left Atrium and Atrial Fibrillation. The high left atrial pressure also causes progressive enlargement of the left atrium, which increases the distance that the cardiac impulse must travel in the atrial wall. Eventually, this pathway becomes so long that it predisposes to the development of circus movements. Therefore, in late stages of mitral valvular disease, especially stenosis, atrial fibrillation usually occurs. This state further reduces the pumping effectiveness of the heart and, therefore causes still further cardiac debility.

Compensations in Mitral Valvular Disease. As also occurs in aortic valvular disease and in many types of congenital heart disease, the blood volume increases in mitral valvular disease. This increases venous return to the heart, thereby helping to overcome the effect of the cardiac debility to reduce cardiac output. Therefore, cardiac output does not fall more than minimally until the late stages of mitral valvular disease.

As the left atrial and pulmonary capillary pressures rise, blood also begins to dam up in the pulmonary artery, and the incipient edema of the lungs causes intense

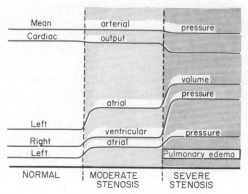

Figure 27–6. Circulatory dynamics in mitral stenosis.

pulmonary arteriolar constriction; these two effects together then increase pulmonary arterial pressure sometimes to as high as 60 mm. Hg. This, in turn, causes hypertrophy of the right heart, which partially compensates for its increased work load.

Figure 27–6 illustrates the progressive changes in circulatory dynamics in mitral stenosis. Essentially the same effects occur in mitral regurgitaion. Indeed, these two conditions more often than not occur together in varying degrees.

CIRCULATORY DYNAMICS DURING EXERCISE IN PATIENTS WITH VALVULAR LESIONS

During exercise very large quantities of venous blood are returned to the heart from the peripheral circulation. Therefore, all of the dynamic abnormalities that occur in the different types of valvular heart disease become tremendously exacerbated. Even in mild valvular heart disease, in which the symptoms may be completely unrecognizable at rest, severe symptoms often develop during heavy exercise. For instance, in patients with aortic valvular lesions, exercise can cause acute left ventricular failure followed by acute pulmonary edema. Also, in patients with mitral disease, exercise can cause so much damming of blood in the lungs that serious pulmonary edema ensues within minutes.

Even in the mildest cases of valvular disease, the patient finds that his cardiac reserve is diminished in proportion to the severity of the valvular dysfunction. That is, the cardiac output does not increase as it should during exercise. Therefore, the muscles of the body fatigue rapidly.

ABNORMAL CIRCULATORY DYNAMICS IN CONGENITAL HEART DEFECTS

Occasionally, the heart or its associated blood vessels are malformed during fetal life; the defect is called a *congenital anomaly.* Basically, there are three major types of congenital anomalies of the heart and its associated vessels: (1) *stenosis* of the channel of blood flow at some point in the heart or in a closely allied major vessel, (2) an abnormality that allows blood to flow directly from the left heart or aorta to the right heart or pulmonary artery, thus by-passing the systemic circulation — this is called a *left-to-right shunt* — and (3) an abnormality that allows blood to flow from the right heart or pulmonary artery directly into the left heart or aorta, thus by-passing the lungs — this is called a *right-to-left shunt*.

The three most common types of stenotic lesions are *coarctation of the aorta, congenital pulmonary stenosis,* and *congenital aortic stenosis.* The effect of coarctation of the aorta on the circulation in relation to the regulation of arterial pressure was discussed in Chapter 22, and congenital aortic stenosis causes almost exactly the same alterations in circulatory dynamics as those noted above for acquired aortic stenosis. The dynamics of pulmonary stenosis will be discussed later in this chapter.

In all left-to-right shunts, blood is shunted directly from the left heart back to the right heart without going through the systemic circulation. Compensatory effects then occur in the circulatory system to increase the total cardiac output, thereby making up for most of the extra blood flow through the shunt. For instance, if 50 per cent of all the blood pumped by the left heart is shunted directly back to the right heart, then the compensatory mechanisms normally increase the cardiac output to almost double normal so that the amount of blood flowing through the systemic vessels is still almost normal despite the shunt. However, this causes an extra load on the heart and therefore usually causes the heart to fail at an early age. Also, it reduces the cardiac reserve of the person throughout his life. Some of the types of left-to-right shunts are *patent ductus arteriosus, interatrial septal defect, interventricular septal defect,* and a *direct communication between the aorta and pulmonary artery at their bases.*

In all right-to-left shunts, much of the blood by-passes the lungs and therefore fails to become oxygenated. Consequently, large quantities of venous blood enter the arterial system directly, and the patient is usually cyanotic (blue) all the time. The most common type of right-to-left shunt is the *tetralogy of Fallot,* which is described later; several other right-to-left shunts are some variation of this abnormality.

Causes of Congenital Anomalies. One of the most common causes of congenital heart defects is a virus infection in the mother during the first trimester of pregnancy when the fetal heart is being formed. Defects are particularly prone to develop when the mother contracts German measles at this time — so often indeed that obstetricians advise termination of pregnancy if German measles occurs in the first trimester. However, some congenital defects of the heart are believed to be hereditary because the same defect has been known to occur in identical twins and also in succeeding generations. Children of patients surgically treated for congenital heart disease have ten times as much chance of having congenital heart disease as do other children. Congenital defects of the heart are frequently also associated with other congenital defects of the body.

PATENT DUCTUS ARTERIOSUS – A LEFT-TO-RIGHT SHUNT

During fetal life the lungs are collapsed, and the elastic factors that keep the alveoli collapsed also keep the blood vessels collapsed. Therefore, in the collapsed lung the resistance to blood flow is about five times as great as it is in the inflated lung. For this reason, the pulmonary arterial pressure is high in the fetus. Because of very low resistance through the large vessels of the placenta, the pressure in the aorta is lower than in the pulmonary artery, causing almost all the pulmonary arterial blood to flow through the ductus arteriosus into the aorta rather than through the lungs. This allows immediate recirculation of the blood through the systemic arteries of the fetus. Obviously, this lack of blood flow through the lungs is not detrimental to the fetus because the blood is oxygenated by the placenta of the mother.

Closure of the Ductus. As soon as the baby is born its lungs inflate; and not only do the alveoli fill, but also the resistance to blood flow through the pulmonary vascular tree decreases tremendously, allowing pulmonary arterial pressure to fall. Simultaneously, the aortic pressure rises because of sudden cessation of blood flow through the placenta. Thus, the pressure in the pulmonary artery falls, while that in the aorta rises. As a result, forward blood flow through the ductus ceases suddenly at birth, and blood even flows backward from the aorta to the pulmonary artery. This new state of blood flow causes the ductus arteriosus to become occluded within a few hours to a few days in most babies so that blood flow through the ductus does not persist. The ductus probably closes because the aortic blood now flowing through the ductus has about two times as high an oxygen concentration as the pulmonary blood, and the oxygen constricts the ductus muscle. This will be discussed further in Chapter 83. In many instances it takes several months for the ductus to close completely, and in about 1 out of every 5500 babies the ductus never closes, causing the condition known as *patent ductus arteriosus,* which is illustrated in Figure 27–7.

Dynamics of Persistent Patent Ductus. During the early months of an infant's life a patent ductus usually does not cause severely abnormal dynamics because the blood pressure of the aorta then is not much higher than the pressure in the pulmonary artery, and only a small amount of blood flows backward into the pulmonary system. However, in most instances, as the child grows older the differential between the pressure in the aorta and that in the pulmonary artery progressively increases, with corresponding increase in the backward flow of blood from the aorta to the pulmonary artery. Also, the diameter of the partially closed ductus often increases with time, making the condition worse.

Recirculation Through the Lungs. In the older child with a patent ductus, as much as half to two thirds of the blood flows into the pulmonary artery, then through the lungs, into the left atrium, and finally back into the left ventricle, passing through this lung circuit two or more times for every one time that it passes through the systemic circulation.

These persons do not show cyanosis until the heart fails or until the lungs become congested. Indeed, the arterial blood is often better oxygenated than normally because of the extra times of passage through the lungs. Furthermore, in the early stages of patent ductus arteriosus the quantity of blood flowing into the systemic aorta remains essentially normal because the quantity of blood returning to the heart from the peripheral circulatory system is normal. Yet, because of the tremendous

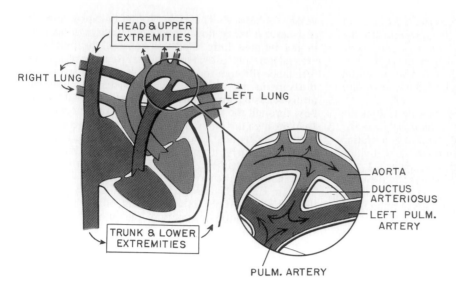

Figure 27-7. Patent ductus arteriosus, illustrating the degree of blood oxygenation in the different parts of the circulation.

accessory flow of blood around and around through the lungs and left side of the heart, the output of the left ventricle in patent ductus arteriosus is often two to three times normal.

Diminished Cardiac and Respiratory Reserve. The major effects of patent ductus arteriosus on the patient are low cardiac and respiratory reserve. The left ventricle is already pumping approximately two or more times the normal cardiac output, and the maximum that it can possibly pump is about four to six times normal. Therefore, during exercise the cardiac output can be increased much less than usual. Under basal conditions, patients usually appear normal except for possible heaving of the chest with each beat of the heart, but with even moderately strenuous exercise they are likely to become weak and occasionally even faint from momentary heart failure. Also, the high pressures in the pulmonary vessels often lead to pulmonary congestion.

The entire heart usually hypertrophies greatly in patent ductus arteriosus. The left ventricle hypertrophies because of the excessive work load that it must perform in pumping a far greater than normal cardiac output, while the right ventricle hypertrophies because of increased pulmonary arterial pressure resulting from, first, increased flow of blood through the lungs caused by the extra blood from the patent ductus and, second, increased resistance to blood flow through the lungs caused by progressive sclerosing of the vessels as they are exposed year after year to excessive pulmonary blood flow.

As a result of the increased load on the heart and especially because the pulmonary congestion and vascular sclerosis effects become progressively more severe with age, most patients with uncorrected patent ductus die between the ages of 20 and 40.

The Machinery Murmur. In the infant with patent ductus arteriosus, occasionally no abnormal heart sounds are heard because the quantity of reversed blood flow may be insufficient. As the baby grows older, reaching the age of one to three years, a harsh, blowing murmur begins to be heard in the pulmonic area of the chest. This sound is much more intense during systole

when the aortic pressure is high and much less intense during diastole when the aortic pressure falls very low, so that the murmur waxes and wanes with each beat of the heart, creating the so-called "machinery murmur." The idealized phonocardiogram of this murmur is shown in Figure 27-3F.

Surgical Treatment. Surgical treatment of patent ductus arteriosus is extremely simple, for all one needs to do is to ligate the patent ductus or to divide it and sew the two ends.

INTERVENTRICULAR SEPTAL DEFECT – A LEFT-TO-RIGHT SHUNT

Because the systolic pressure in the left ventricle is normally about six times that in the right ventricle, a large amount of blood flows from the left to the right ventricle whenever a hole occurs in the septum, called an interventricular defect. The excess flow of blood into the right ventricle in turn increases the pressure in the right ventricle. As a result, the right ventricle hypertrophies, sometimes to such an extent that its muscular wall approximately equals that of the left ventricle.

Diagnosis of an interventricular septal defect is based on (1) the presence of a systolic blowing murmur heard over the anterior projection of the heart, (2) high right ventricular systolic pressure recorded from a catheter, and (3) the presence of oxygenated blood in a blood sample removed through the catheter from the right ventricle, this blood having leaked backward from the left ventricle.

Blood flowing from the left ventricle into the right ventricle passes one or more times through the lungs and then back to the left ventricle again before finally entering the peripheral circulatory system. This condition, therefore, is analogous to patent ductus arteriosus except that the flow of blood from the systemic circulation to the pulmonary circulation occurs only during systole rather than during both systole and diastole as in patent ductus.

The septal defect can be treated surgically by placing a patch over the defect.

INTERATRIAL SEPTAL DEFECTS – A LEFT-TO-RIGHT SHUNT

Closure of the Foramen Ovale. During fetal life much of the blood entering the right atrium courses directly through the foramen ovale into the left atrium and thence out into the systemic circulation, thus by-passing the right ventricle and the lungs. This mechanism aids the ductus arteriosus in shunting blood around the lungs, thereby relieving the fetal right heart from the unnecessary load. Immediately after birth of the child, the pressure in the pulmonary artery decreases while that in the aorta increases, as discussed above. These changes also decrease and increase respectively the loads on the right and left ventricles so that the right atrial pressure decreases and the left atrial pressure increases. As a result, blood then attempts to flow from the left atrium back into the right atrium. However, the foramen ovale is covered by a small valvelike vane, which closes over the foramen. In two thirds of all persons the foramen later becomes totally occluded by fibrous tissue, but in one third the foramen never becomes totally occluded. Yet this nonoccluded opening does not cause any physiological abnormality in the heart because the left atrial pressure remains 1 to 3 mm. Hg higher than the right atrial pressure, keeping the valve permanently closed.

Dynamics of Interatrial Defects. Occasionally, the valve does not cover the foramen ovale, and a hole persists permanently between the left atrium and the right atrium. If this hole is small, only a small amount of blood passes from the left atrium back to the right atrium, but occasionally almost the entire wall between the two atria is missing. In this case a tremendous quantity of blood flows from the left atrium into the right atrium and recirculates, sometimes as much as three times, through the right ventricle and lungs before finally passing into the systemic circulation. Therefore, large atrial septal defects greatly reduce the cardiac reserve, and early death results from right ventricle failure and pulmonary congestion.

The diagnosis of an interatrial septal defect is difficult to make, for no murmurs can be heard, the patient's arterial blood is well oxygenated, and the usual radiologic studies are relatively nonspecific. The diagnosis can be made accurately by (1) angiocardiograms (x-ray films made after injecting a radiopaque dye into the heart) and (2) finding oxygenated blood in a catheter specimen from the right atrium in the absence of a detectable murmur.

As in interventricular septal defect, in interatrial septal defect a surgically applied patch over the opening can cause dramatic reversal of the course of the disease.

TETRALOGY OF FALLOT – A RIGHT-TO-LEFT SHUNT

Tetralogy of Fallot is illustrated in Figure 27–8, from which it will be noted that four different abnormalities of the heart occur simultaneously.

First, the aorta originates from the right ventricle rather than the left, or it overrides the septum as shown in the figure.

Second, the pulmonary artery is stenosed so that much less than normal amounts of blood pass from the right ventricle into the lungs; instead the blood passes into the aorta.

Third, blood from the left ventricle flows through a ventricular septal defect into the right ventricle and then into the aorta or directly into the overriding aorta.

Fourth, because the right side of the heart must pump large quantities of blood against the high pressure in the aorta, its musculature is highly developed, causing an enlarged right ventricle.

Abnormal Dynamics. It is readily apparent that the major physiological difficulty caused by tetralogy of Fallot is the shunting of blood past the lungs without its becoming oxygenated. As much as 75 per cent of the venous blood returning to the heart may pass directly from the right ventricle into the aorta without becoming oxygenated. Tetralogy of Fallot is the major cause of cyanosis in babies (''blue babies'').

A diagnosis of tetralogy of Fallot is usually based on (1) the fact that the baby is blue, (2) records of high systolic pressure in the right ventricle recorded through a catheter, (3) characteristic changes in the radiologic silhou-

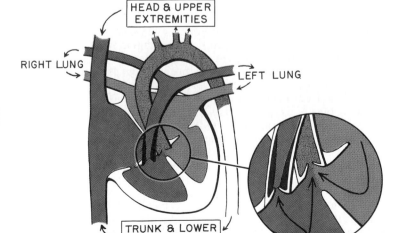

Figure 27–8. Tetralogy of Fallot, illustrating the degree of blood oxygenation in the different parts of the circulation.

ette of the heart showing an enlarged right ventricle, and (4) angiograms showing abnormal blood flow through the interventricular septal defect, the overriding aorta, and the pulmonary stenotic area.

Surgical Treatment. In recent years tetralogy of Fallot has been treated relatively successfully by surgery. The usual operation is to open the pulmonary stenosis, to close the septal defect, and to reconstruct the flow pathway into the aorta. When successful, the average life expectancy increases from 5 to 10 years up to 50 or more years.

PULMONARY STENOSIS

Often a child is born with pulmonary stenosis but without other congenital abnormalities. Such severe pulmonary stenosis occasionally occurs that the right side of the heart is likely to fail at an early age because blood flow from the right ventricle into the lungs is greatly impeded, the right ventricular systolic pressure rising to 75 to 100 mm. Hg instead of the normal 22 mm. Hg. The right side of the heart dilates, and the muscle becomes greatly hypertrophied in order to withstand the load. Also, a loud stenotic murmur is heard over the pulmonary valve area. In most cases of pulmonary stenosis, the stenotic area can be enlarged by surgical operation so that the heart resumes normal function.

USE OF EXTRACORPOREAL CIRCULATION DURING CARDIAC SURGERY

It is almost impossible to repair intracardiac defects while the heart is still pumping. Therefore, many different types of artificial *heart-lung machines* have been developed to take the place of the heart and lungs during the course of operation. Such a system is called an *extracorporeal circulation*. The system consists principally of (1) a pump and (2) an oxygenating device. Almost any type of pump that does not cause hemolysis of the blood seems to be suitable.

The different principles that have been used for oxygenating blood are (1) bubbling oxygen through the blood and then removing the bubbles from the blood before passing it back into the patient, (2) dripping the blood downward over the surfaces of large areas of plastic sheet in the presence of oxygen, (3) passing the blood over the surfaces of rotating discs, and (4) passing the blood between thin membranes or through thin tubes that are porous to oxygen and carbon dioxide.

The different oxygenators have been fraught with many difficulties, including hemolysis of the blood, development of small clots in the blood, likelihood of small bubbles of oxygen or small emboli of antifoaming agent passing into the arteries of the patient, necessity for large quantities of blood to prime the entire system, failure to exchange adequate quantities of oxygen, and the necessity to use heparin in the system to prevent blood coagulation, the heparin also preventing adequate hemostasis during the surgical procedure. Yet, despite these difficulties, in the hands of experts patients can be kept on artificial heart-lung machines for many hours while operations are performed on the inside of the heart.

HYPERTROPHY OF THE HEART IN VALVULAR AND CONGENITAL HEART DISEASE

Hypertrophy of cardiac muscle is one of the most important mechanisms by which the heart adapts to increased work loads, whether these loads be caused by increased pressure against which the heart muscle must contract or by increased volume that must be pumped. Some investigators believe that it is the increased work load itself that causes the hypertrophy; others believe the increased metabolic rate of the muscle to be the primary stimulus and the work load simply to be the cause of the increase in the metabolic rate. Regardless of which of these is correct, one can calculate approximately how much hypertrophy will occur in each chamber of the heart by multiplying ventricular output times the pressure against which the ventricle must work, with special emphasis on the pressure. Thus, hypertrophy occurs in most types of valvular and congenital disease, as follows:

In *aortic stenosis* and *aortic regurgitation,* the left ventricular musculature hypertrophies tremendously, sometimes to as much as four to five times normal, so that the weight of the heart on occasion may be as great as 1000 grams instead of the normal 300 grams.

In *mitral stenosis,* the left atrium hypertrophies and dilates, and the right ventricle hypertrophies slightly, but no left ventricular hypertrophy occurs. In *mitral regurgitation* moderate hypertrophy of the left ventricle occurs, and some hypertrophy of the right ventricle also develops owing to back pressure effects through the lungs causing elevation of the pulmonary arterial pressure.

In *patent ductus arteriosus,* the work load of both ventricles is increased. The left ventricle pumps an average of twice the quantity of blood that it normally pumps; therefore, it would be expected to hypertrophy. On the other hand, the right ventricle must pump its blood against a much higher than normal pulmonary arterial pressure because of the large quantity of blood refluxing from the aorta into the pulmonary artery. Consequently, the right ventricle also hypertrophies.

In *tetralogy of Fallot,* one of the cardinal signs is marked right ventricular hypertrophy, for the right ventricle must pump against the pressure in the aorta, and it also must pump an increased volume of blood. On the other hand, the work load of the left ventricle is actually less than normal because of the reduced blood flow from the lungs into the left heart. Therefore, the right ventricle in tetralogy of Fallot is often larger than the normal left ventricle, while the left ventricle may be relatively small.

Detrimental Effects of the Late Stages of Hypertrophy. Though physiological hypertrophy of heart muscle is usually very beneficial to cardiac function, extreme degrees of hypertrophy sometimes lead to failure. One of the reasons for this is that coronary blood flow does not always increase to the same extent as the mass of muscle. A second reason is that fibrosis often develops in the muscle, the fibrous tissue replacing degenerating muscle fibers. Because of the sometimes disproportionate increase in muscle mass relative to coronary flow, relative ischemia sometimes develops as the muscle hypertrophies, and coronary insufficiency easily ensues. Therefore, anginal pain is a frequent accompaniment of many valvular and congenital heart diseases.

REFERENCES

Alpert, N. (ed.): Cardiac Hypertrophy. New York, Academic Press, 1971.

Badeer, H. S.: Development of cardiomegaly. *Cardiology, 57*:247, 1972.

Barry, A.: Aortic and Tricuspid Valvular Disease. New York, Appleton-Century-Crofts, 1980.

Bayer, L. M., and Honzik, M. P.: Children with Intracardiac Defects. Springfield, Ill., Charles C Thomas, 1974.

Bernstein, E. F. (ed.): Noninvasive Diagnostic Techniques in Vascular Disease. St. Louis, C. V. Mosby, 1978.

Butterworth, J. S., *et al.*: Cardiac Auscultation–Including Audio-Visual Principles, 2nd Ed. New York, Grune & Stratton, 1960.

Comroe, J. H., Jr., and Dripps, R. D.: The Top Ten Clinical Advances in Cardiovascular-Pulmonary Medicine and Surgery Between 1945 and 1975; How They Came About: Final Report. Washington, U.S. Govt. Printing Office, 1978.

Cooley, D. A., and Hallman, G. L.: Surgical Treatment of Congenital Heart Disease. Philadelphia, Lea & Febiger, 1966.

Elliott, L. P.: An Angiocardiographic and Plain Film Approach to Complex Congenital Heart Disease. Chicago, Year Book Medical Publishers, 1978.

Grossman, W. (ed.): Cardiac Catheterization and Angiography, 2nd Ed. Philadelphia, Lea & Febiger, 1980.

Haft, J. I., and Horowitz, M. S.: Clinical Echocardiography. Mount Kisco, N.Y., Futura Publishing Co., 1978.

Hayase, S., and Murao, S.: Cardiology. New York, Elsevier/North-Holland, 1979.

Heymann, M. A., and Rudolph, A. M.: Control of the ductus arteriosus. *Physiol. Rev., 55*:62, 1975.

Hugenholtz, P. G., *et al.*: Current status of echocardiology. *In* Dickinson, C. J., and Marks, J. (eds.): Developments in Cardiovascular Medicine. Lancaster, England, MTP Press, 1978, p. 15.

Kidd, L., and Somerville, J.: Long-term survival after operations for congenital heart disease. *In* Hayase, S., and Murao, S. (eds.): Cardiology: Proceedings of the VIII World Congress of Cardiology, Tokyo, 1978. New York, Elsevier/North-Holland, 1979, p. 774.

Kobayashi, T., *et al.* (eds.): Cardiac Adaptation. Baltimore, University Park Press, 1978.

Kotler, M. N., and Segal, B. L. (eds.). Clinical Echocardiography. Philadelphia, F. A. Davis Co., 1978.

Kreel, L.: Computed tomography and the cardiovascular system. *In* Dickinson, C. J., and Marks, J. (eds.): Developments in Cardiovascular Medicine. Lancaster, England, MTP Press, 1978, p. 3.

Kremkau, F. W.: Diagnostic Ultrasound; Physical Principles and Exercises. New York, Grune & Stratton, 1980.

Lear, M. W.: Heartsounds. New York, Simon & Schuster, 1979.

Lindsay, J., Jr., and Hurst, J. W. (eds.): The Aorta. New York, Grune & Stratton, 1979.

Lundström, N. (ed.): Echocardiography in Congenital Heart Disease. New York, Elsevier/North-Holland, 1978.

Mair, D. D., and Ritter, D. G.: The physiology of cyanotic congenital heart disease. *Int. Rev. Physiol., 9*:275, 1976.

Mitchell, J. H., and Wildenthal, K.: Static (isometric) exercise and the heart: Physiological and clinical considerations. *Annu. Rev. Med., 25*:369, 1974.

Nadas, A. S. (ed.): Pulmonary Stenosis, Aortic Stenosis, Ventricular Septal Defect. Dallas, American Heart Association, 1977.

Nanda, N. C., and Gramiak, R.: Clinical Echocardiography. St. Louis, C. V. Mosby, 1978.

New York Heart Association Criteria Committee: Nomenclature and Criteria for Diagnosis of Diseases of the Heart and Great Vessels, 8th Ed. Boston, Little, Brown, 1979.

Nimura, Y., and Henry, W.: Newer aspects of echocardiography. *In* Hayase, S., and Murao, S.: Cardiology: Proceedings of the VIII World Congress of Cardiology, Tokyo, 1978. New York, Elsevier/North-Holland, 1979, p. 1053.

Perloff, J. K.: The Clinical Recognition of Congenital Heart Disease. Philadelphia, W. B. Saunders Co., 1978.

Pomerance, A., and Davis, M. J. (eds.): Pathology of the Heart. Philadelphia, J. B. Lippincott Co., 1975.

Rapaport, E. (ed.): Current Controversies in Cardiovascular Disease. Philadelphia, W. B. Saunders Co., 1980.

Reiser, S. J.: The medical influence of the stethoscope. *Sci. Am., 240*(2):148, 1979.

Roberts, W. C., and Spray, T. L.: Pericardial Heart Disease. Chicago, Year Book Medical Publishers, 1977.

Senning, A.: Surgical treatment of valvular disease. *In* Hayase, S., and Murao, S. (eds.): Cardiology: Proceedings of the VIII World Congress of Cardiology, Tokyo, 1978. New York, Elsevier/North-Holland, 1979, p. 751.

Shah, P. M., and Roberts, D. L.: Diagnosis and Treatment of Aortic Valve Stenosis. Chicago, Year Book Medical Publishers, 1977.

Stapleton, J. F., and Harvey, W. P.: Heart sounds, murmurs, and precordial movements. *In* Sodeman, W. A., Jr., and Sodeman, T. M. (eds.): Pathologic Physiology: Mechanisms of Disease, 6th Ed. Philadelphia, W. B. Saunders Co., 1979, p. 335.

Stollerman, G. H., and Shiokawa, Y.: Current status of rheumatic fever and rheumatic carditis. *In* Hayase, S., and Murao, S. (eds.): Cardiology: Proceedings of the VIII World Congress of Cardiology, Tokyo, 1978. New York, Elsevier/North-Holland, 1979, p. 500.

Taussig, H.: Congenital Malformations of the Heart. Vol. 1: General Considerations, 2nd Ed. Vol. 2: Specific Malformations, 2nd Ed. Cambridge, Mass., Harvard University Press, 1960.

Zak, R., and Rabinowtz, M.: Molecular aspects of cardiac hypertrophy. *Annu. Rev. Physiol., 41*:539, 1979.

28

Circulatory Shock and Physiology of Its Treatment

Circulatory shock means generalized inadequacy of blood flow throughout the body, to the extent that the tissues are damaged because of too little flow, especially too little delivery of oxygen to the tissue cells. Even the cardiovascular system itself — the heart musculature, the walls of the blood vessels, the vasomotor system, and other circulatory parts — begins to deteriorate so that the shock becomes progressively worse.

In this discussion, shock will be divided into three major stages: (1) a *nonprogressive stage* (sometimes called the *compensated stage*), (2) a *progressive stage,* and (3) an *irreversible stage*. In the nonprogressive stage, tissue perfusion is deficient but not deficient enough to cause a vicious cycle of cardiovascular deterioration. In the progressive stage, the shock has progressed to the point that the circulatory system begins to deteriorate, thus leading to a vicious cycle that eventuates in death unless treatment is instituted. In the irreversible stage, the shock has progressed to the point that all forms of therapy will be inadequate to save the life of the person even though the person is at the moment still alive.

Cardiac Output in Shock.   Usually, the cause of inadequate tissue perfusion in circulatory shock is inadequate cardiac output, meaning simply that not enough blood is pumped by the heart to supply adequate blood flow to the tissues. However, under special circumstances the cardiac output itself is normal or even greater than normal but still is inadequate to supply the tissue needs. This condition may result from too high a rate of metabolism in the tissues or from abnormal flow patterns in the peripheral vasculature that prevent adequate diffusion of nutrients and other substances between the circulatory system and the tissue cells — both these effects occur in some forms of septic shock, as we shall see.

Arterial Pressure in Shock.   There are times when a person is in severe shock and still has a normal arterial pressure because of nervous reflexes that keep the pressure from falling. At other times the arterial pressure can fall to as low as one-half normal but the person can still have normal tissue perfusion and not be in shock.

However, in many types of shock, especially that caused by severe blood loss, the arterial pressure does usually decrease at the same time that the cardiac output decreases, though usually not as much as the decrease in output. Nevertheless, measurements of arterial pressure are usually of major value in assessing the degree of shock.

PHYSIOLOGICAL CAUSES OF SHOCK

Since shock usually results from inadequate cardiac output, any factor that can reduce cardiac output can also cause shock. The different factors that can do this were discussed in Chapter 23 and were grouped into two categories:

1. Those that decrease the ability of the heart to pump blood.

2. Those that tend to decrease venous return. Thus, serious myocardial infarction or any other factor that damages the heart so severely that it cannot pump adequate quantities of blood can cause a type of shock called *cardiogenic shock,* which was discussed in Chapter 26. Also, all the factors that reduce venous return, including (a) diminished blood volume, (b) decreased vasomotor tone, or (c) greatly increased resistance to blood flow, can result in shock. The present chapter will deal primarily with these factors.

SHOCK CAUSED BY HYPOVOLEMIA — HEMORRHAGIC SHOCK

Hypovolemia means diminished blood volume, and hemorrhage is perhaps the most common cause of hypovolemic shock. Therefore, a discussion of hemorrhagic shock will serve to explain many of the basic principles of the shock problem.

Hemorrhage *decreases the mean systemic filling pressure* and as a consequence decreases venous return. As a result, the cardiac output falls below normal, and shock ensues. Obviously, all degrees of shock can result from hemorrhage, from the mildest diminishment of cardiac output to almost complete cessation of output.

RELATIONSHIP OF BLEEDING VOLUME TO CARDIAC OUTPUT AND ARTERIAL PRESSURE

Figure 28–1 illustrates the effect on both cardiac output and arterial pressure of removing blood from the circulatory system over a period of about half an hour. Approximately 10 per cent of the total blood volume can be removed with no significant effect on arterial pressure or cardiac output, but greater blood loss usually diminishes the cardiac output first and later the pressure, both of these falling to zero when about 35 to 45 per cent of the total blood volume has been removed.

Sympathetic Reflex Compensation in Shock. Fortunately, the decrease in arterial pressure caused by blood loss initiates powerful sympathetic reflexes (initiated mainly by the baroreceptors) that stimulate the sympathetic vasoconstrictor system throughout the body, resulting in three important effects: (1) The arterioles constrict in most parts of the body, thereby greatly increasing the total peripheral resistance. (2) The veins and venous reservoirs constrict, thereby helping to maintain adequate venous return despite diminished blood volume. And (3) heart activity increases markedly, sometimes increasing the heart rate from the normal value of 72 beats per minute to as much as 200 beats per minute.

Value of the Reflexes. In the absence of the sympathetic reflexes, only 15 to 20 per cent of the blood volume can be removed over a period of half an hour before a person will die; this is in contrast to 30 to 40 per cent when the reflexes are intact. Therefore, the reflexes extend the amount of blood loss that can occur without causing death to about two times that which would be possible in their absence.

Greater Effect of the Reflexes in Maintaining Arterial Pressure Than in Maintaining Output. Referring again to Figure 28–1, one sees that the arterial pressure is maintained at or near normal levels in the hemorrhaging person longer than is the cardiac output. The reason for this is that the sympathetic reflexes are geared more for maintenance of arterial pressure than for maintenance of output. They increase the arterial pressure to a great extent by increasing the total peripheral resistance, which has no beneficial effect on cardiac output; but the sympathetic constriction of the veins is important to keep venous return and cardiac output from falling too much.

Especially interesting is the second plateau in the arterial pressure curve of Figure 28–1. This results from activation of the *CNS ischemic response,* which causes extreme stimulation of the sympathetic nervous system, as discussed in Chapter 21. This effect of the CNS ischemic response can be called the "last ditch stand" of the sympathetic reflexes in their attempt to keep the arterial pressure from falling too low.

Protection of Coronary and Cerebral Blood Flow by the Reflexes. A special value of the maintenance of normal arterial pressure even in the face of decreasing cardiac output is protection of blood flow through the coronary and cerebral circulatory systems. Sympathetic stimulation does not cause significant constriction of either the cerebral or cardiac vessels. In addition, in both these vascular beds local autoregulation is excellent, which prevents moderate changes in arterial pressure from significantly affecting their blood flows. Therefore, blood flow through the heart and brain is maintained essentially at normal levels as long as the arterial pressure does not fall below about 70 mm. Hg, despite the fact that blood flow in many other areas of the body might be decreased almost to zero because of vasospasm.

NONPROGRESSIVE AND PROGRESSIVE HEMORRHAGIC SHOCK

Figure 28–2 illustrates an experiment that we performed in dogs to demonstrate the effects of different degrees of hemorrhage on the subsequent course of arterial pressure. The dogs were bled rapidly until their arterial pressures fell to different levels. Those dogs whose pressures fell immedi-

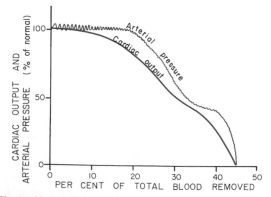

Figure 28–1. Effect of hemorrhage on cardiac output and arterial pressure.

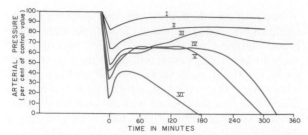

Figure 28–2. Course of arterial pressure in dogs after different degrees of acute hemorrhage. Each curve represents the average results from six dogs.

ately no lower than 45 mm. Hg (Groups I, II, and III) all eventually recovered; the recovery occurred rapidly if the pressure fell only slightly (Group I) but occurred slowly if it fell almost to the 45 mm. Hg level (Group III). On the other hand, when the arterial pressure fell below 45 mm. Hg (Groups IV, V, and VI), all the dogs died, though many of them hovered between life and death for many hours before the circulatory system began to deteriorate.

This experiment demonstrates that the circulatory system can recover as long as the degree of hemorrhage is no greater than a certain critical amount. However, crossing this critical amount by even a few milliliters of blood loss makes the eventual difference between life and death. Thus, hemorrhage beyond a certain critical level causes shock to become *progressive*. That is, *the shock itself causes still more shock,* the condition becoming a vicious cycle that leads eventually to complete deterioration of the circulation and to death.

Nonprogressive Shock – Compensated Shock

If shock is not severe enough to cause its own progression, the person eventually recovers. Therefore, shock of this lesser degree can be called *nonprogressive shock.* It is also frequently called *compensated shock,* meaning that the sympathetic reflexes and other factors have compensated enough to prevent deterioration of the circulation.

The factors that cause a person to recover from moderate degrees of shock are the negative feedback control mechanisms of the circulation that attempt to return cardiac output and arterial pressure to normal levels. These include:

1. The *baroreceptor reflexes,* which elicit powerful sympathetic stimulation of the circulation;

2. The *central nervous system ischemic response,* which elicits even more powerful sympathetic stimulation throughout the body but is not activated until the arterial pressure falls below 50 mm. Hg;

3. *Reverse stress-relaxation of the circulatory system,* which causes the blood vessels to contract down around the diminished blood volume so that the blood volume that is available will more adequately fill the circulation;

4. *Formation of angiotensin,* which constricts the peripheral arteries and causes increased conservation of water and salt by the kidneys, both of which help to prevent progression of the shock;

5. *Formation of vasopressin (antidiuretic hormone),* which constricts the peripheral arteries and veins and causes greatly increased water retention by the kidneys;

6. *Compensatory mechanisms that return the blood volume back toward normal,* including absorption of large quantities of fluid from the intestinal tract, absorption of fluid from the interstitial spaces of the body, conservation of water and salt by the kidneys, and increased thirst and increased appetite for salt which make the person drink water and eat salty foods if able.

The sympathetic reflexes provide immediate help toward bringing about recovery, for they become maximally activated within 30 seconds after hemorrhage. The reverse stress-relaxation that causes contraction of the blood vessels and venous reservoirs around the blood requires some 10 minutes to an hour to occur completely, but, nevertheless, this aids greatly in increasing the mean systemic filling pressure and thereby increasing the return of blood to the heart. Finally, the readjustment of blood volume by absorption of fluid from the interstitial spaces and from the intestinal tract, as well as the ingestion and absorption of additional quantities of fluid and salt, may require from 1 to 48 hours, but eventually recovery takes place provided the shock does not become severe enough to enter the progressive stage.

Progressive Shock – The Vicious Cycle of Cardiovascular Deterioration

Once shock has become severe enough, the structures of the circulatory system themselves begin to deteriorate, and various types of positive feedback develop that can cause a vicious cycle of progressively decreasing cardiac output. Figure 28–3 illustrates some of these different types of positive feedback that further depress the cardiac output in shock. These are the following:

Cardiac Depression. When the arterial pressure falls low enough, coronary blood flow decreases below that required for adequate nutrition of the myocardium itself. This obviously weakens the heart and thereby decreases the cardiac output still more. As a consequence, the arterial pressure falls still further, and the coronary blood flow decreases more, making the heart still weaker.

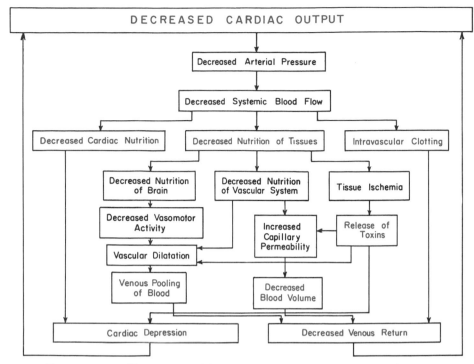

Figure 28–3. Different types of feedback that can lead to progression of shock.

Thus, a positive feedback cycle has developed whereby the shock becomes more and more severe.

Figure 28–4 illustrates cardiac output curves from experiments in dogs, showing progressive deterioration of the heart at different times following the onset of shock. A dog was bled until the arterial pressure fell to 30 mm. Hg, and the pressure was held at this level by further bleeding or retransfusion of blood as required. Note that

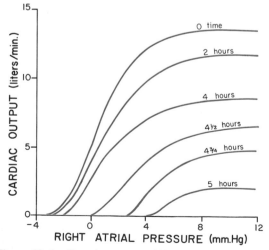

Figure 28–4. Function curves of the heart at different times after hemorrhagic shock begins. (These curves are extrapolated to the human heart from data obtained in dog experiments by Dr. J. W. Crowell.)

there was little deterioration of the heart during the first two hours, but by four hours the heart had deteriorated about 40 per cent; then, rapidly, during the last hour of the experiment the heart deteriorated almost completely.

Thus, one of the important features of progressive shock, whether it be hemorrhagic in origin or caused in any other way, is eventual progressive deterioration of the heart. In the early stages of shock, this plays very little role in the condition of the person, partly because deterioration of the heart itself is not very severe during the first hour or so of shock but mainly because the heart has tremendous reserve that makes it normally capable of pumping 300 to 400 per cent more blood than is required by the body for adequate nutrition. However, in the late stages of shock, deterioration of the heart is probably the most important factor in the further progression of the shock.

Aside from the myocardial depression caused by poor coronary blood flow in shock, the myocardium can also be depressed by toxic factors transported to the heart from other parts of the body, especially by a factor called *myocardial toxic factor* (MTF), but also by such substances as excess lactic acid, bacterial toxins from the gut, degeneration products from dying tissues, and so forth. These factors will be discussed later in this chapter.

Vasomotor Failure. In the early stages of shock various circulatory reflexes cause intense activity of the sympathetic nervous system. This, as discussed above, helps to delay depression of the cardiac output and especially helps to prevent decreased arterial pressure. However, there comes a point at which diminished blood flow to the vasomotor center itself so depresses the center that it becomes progressively less active and finally totally inactive. For instance, complete circulatory arrest to the brain causes, during the first four to eight minutes, the most intense of all sympathetic discharges, resulting in a typical CNS ischemic response, but by the end of 10 to 15 minutes the vasomotor center becomes so depressed that no evidence of sympathetic discharge at all can be demonstrated. Fortunately, though, the vasomotor center does not usually fail in the early stages of shock — only in the late stages.

Thrombosis of the Minute Vessels — Sludged Blood. Recently, several research workers have shown that thrombosis occurring in many of the minute vessels in the circulatory system can be one of the causes of shock progression. That is, blood flow through many tissues becomes extremely sluggish, but tissue metabolism continues so that large amounts of acid, either carbonic acid or lactic acid, continue to empty into the blood. This acid, plus other deterioration products from the ischemic tissues, causes blood agglutination or actual blood clots, thus leading to minute plugs in the small vessels. Even if the vessels do not become plugged, the tendency for the cells to stick to each other makes it more difficult for blood to flow through the microvasculature, giving rise to the term *sludged blood.*

Increased Capillary Permeability. After many hours of capillary hypoxia the permeability of the capillaries gradually increases and large quantities of fluid begin to transude into the tissues. This further decreases the blood volume, with resultant further decrease in cardiac output, thus making the shock still more severe.

Fortunately, capillary hypoxia does not cause increased capillary permeability until the very late stages of extremely prolonged shock. Therefore, this factor plays a significant role in few instances of shock.

Release of Toxins by Ischemic Tissues. Throughout the history of research in the field of shock, it has been suggested time and again that shock causes tissues to release toxic substances like histamine that then cause further deterioration of the circulatory system. Critical quantitative studies, however, have failed to prove the significance of histamine and serotonin in shock. On the other hand, two toxic factors are now known to play important roles in at least some types of shock. These are *myocardial toxic factor* and *endotoxin.*

Myocardial Toxic Factor (MTF). During the course of shock, the splanchnic arterioles become strongly constricted, and the arterioles of the pancreas appear to become constricted even more than those in other areas of the abdomen. The extreme ischemia that occurs in the pancreas allows activation of some of the pancreatic enzymes, including trypsin itself. This sets into play degenerative processes within the pancreatic tissues, and several toxic factors are released into the circulating blood. One of these is a peptide with a molecular weight of 500 to 1000 that is called *myocardial toxic factor,* more commonly known as *MTF.* The prominent effect of MTF is that of a direct depressant on the heart itself, frequently depressing cardiac contractility as much as 50 per cent. The toxin seems to interfere with the function of calcium ions in the excitation-contraction coupling process because increase in calcium ions in the blood or administration of ouabain, one of the cardiac glycosides that causes increased concentrations of calcium in the cardiac muscle cell, nullifies the depressant effect of MTF.

Regardless of the precise mechanism by which MTF reduces cardiac contractility, its generation during the course of shock further exacerbates the shock syndrome and, therefore, is part of the positive feedback deteriorative process that causes progression of the shock.

Endotoxin. Another toxic factor that probably plays a role in the progression of shock is *endotoxin,* which is a toxin released from the bodies of dead gram-negative bacteria in the intestines. Diminished blood flow to the intestines causes enhanced absorption of this toxic substance from the intestines, and it then causes extensive vascular dilatation and cardiac depression. Though this toxin can play a major role in some types of shock, especially septic shock discussed later in the chapter, it still is not clear how much endotoxin is released during hemorrhagic shock and whether or not it is an important factor in the progression of this type of shock.

Generalized Cellular Deterioration. As shock becomes very severe, many signs of generalized cellular deterioration occur throughout the body. One organ especially affected is the *liver,* primarily because of the lack of enough nutrients to support the normally high rate of metabolism in liver cells, but partly because of the extreme vascular exposure of the liver cells to any toxic or other abnormal metabolic factors in shock. Among the different effects that are known to occur are:

1. Active transport of sodium and potassium through the cell membrane is greatly diminished. As a result, sodium and chloride accumulate in the cells and potassium is lost from the cells. In addition the cells begin to swell.

2. Mitochondrial activity in the liver cells, as

well as in many other tissues of the body, becomes severely depressed.

3. Lysosomes begin to split in widespread tissue areas, with intracellular release of hydrolases that cause further intracellular deterioration.

4. Cellular metabolism of nutrients such as glucose becomes greatly depressed. The activities of some hormones are depressed as well, including as much as 200-fold depression in the action of insulin.

Obviously, all these effects contribute to further deterioration of many different organs of the body, particularly of the liver, and also of the heart thereby further depressing the contractility of the heart.

Tissue Necrosis in Severe Shock — Effect of Pattern of Vascular Blood Flow in Different Organs. Not all cells of the body are equally damaged by shock because some tissues have better blood supplies than others. For instance, the cells adjacent to the arterial ends of capillaries receive better nutrition than the cells adjacent to the venous ends of the same capillaries. Therefore, one would expect more nutritive deficiency around the venous ends of capillaries than elsewhere. This is precisely the effect that Crowell has found in studying tissue areas in many parts of the body. Figure 28–5 illustrates necrosis in the center of a liver lobule, the portion of the lobule that is last to be bathed by the blood as it passes through the liver sinusoids. The first sign of damage is swelling of the cells, and this swelling then compresses the sinusoids, often causing total blockage of blood flow in their central ends; the result is rapidly accelerating

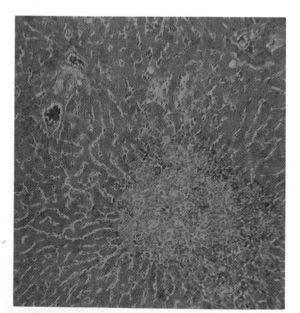

Figure 28–5. Necrosis of the central portion of a liver lobule in severe shock. (Courtesy of Dr. Jack Crowell.)

deterioration of the central portions of the liver lobules.

Similar punctate lesions occur in heart muscle, though here a definite repetitive pattern such as occurs in the liver cannot be demonstrated. Nevertheless, the cardiac lesions probably play an important role in leading to the final irreversible stage of shock. Similar deteriorative lesions occur in the kidneys, especially in the kidney tubules, leading to kidney failure and subsequent uremic death several days later.

Acidosis in Shock. Most of the metabolic derangements that occur in shocked tissue can lead to acidosis. Especially important is the poor delivery of oxygen to the tissues, which greatly diminishes oxidative metabolism of the foodstuffs. When this occurs, the cells obtain their energy by the anaerobic process of glycolysis that leads to tremendous quantities of *excess lactic acid* in the blood. In addition, the poor blood flow through the tissues prevents normal removal of carbon dioxide; the carbon dioxide reacts locally in the cells with water to form very high concentrations of intracellular carbonic acid; this in turn reacts with the various tissue buffers to form still other intracellular acidic substances.

Thus, another deteriorative effect of shock is both generalized and local tissue acidosis, leading to still further progression of the shock itself.

Positive Feedback and the Vicious Cycle of Progressive Shock. All the factors just discussed that can lead to further progression of shock are types of *positive feedback*. That is, each increase in the degree of shock causes a further increase in the shock.

However, positive feedback does not always necessarily lead to a vicious cycle. Whether or not a vicious cycle develops depends on the intensity of the positive feedback. In mild degrees of shock, the negative feedback mechanisms — the sympathetic reflexes, the plasticity of the circulation, the recovery of blood volume, and others — can easily overcome the positive feedback influences and can therefore cause recovery. In severe degrees of shock, however, the positive feedback mechanisms becomes more and more powerful, thus leading to such rapid deterioration of the circulation that the negative feedback systems cannot return the cardiac output to normal.

Thus, considering once again the principles of positive feedback as discussed in Chapter 1, one will see that a vicious cycle of deterioration occurs in shock only when the positive feedback effect decreasing cardiac output further is greater than the initial decrease in output. This explains why there is a critical cardiac output level above which a person in shock recovers and below which the person enters a vicious cycle of circulatory deterioration.

IRREVERSIBLE SHOCK

After shock has progressed to a certain stage, transfusion or any other type of therapy becomes incapable of saving the life of the person. Therefore, the person is then said to be in the *irreversible stage of shock*. Ironically, even in this irreversible stage, therapy can on occasion still return the arterial pressure and even the cardiac output to normal for short periods of time, but the circulatory system nevertheless continues to deteriorate and death ensues in another few minutes to few hours.

Figure 28–6 illustrates this effect, showing that transfusion during the irreversible stage can sometimes cause the cardiac output (as well as the arterial pressure) to return to normal. However, the cardiac output soon begins to fall again, and subsequent transfusions have less and less effect. Thus, something changes in the overall function of the circulatory system during shock that may not necessarily affect the *immediate* ability of the heart to pump blood but over a long period of time does depress this ability and results in death. Now the question remains: What factor or factors lead to the eventual total deterioration of circulatory function?

The answer to this question seems to be, simply, that beyond a certain point so much tissue damage has occurred, so many destructive enzymes have been released into the body fluids, so much acidosis has developed, and so many other destructive factors are now in progress that even a normal cardiac output cannot reverse the continuing deterioration. Therefore, in severe shock frequently a stage is reached beyond which a person is destined to die even though vigorous therapy can still return the cardiac output to normal for short periods of time.

Depletion of Cellular High Energy Phosphate Reserves in Irreversible Shock.   The high energy phosphate reserves in the tissues of the body, especially in the liver and in the heart, are greatly diminished in severe degrees of shock. Essentially all of the creatinine phosphate is degraded, and almost all of the sarcoplasmic ATP has been degraded to ADP, AMP, or adenosine. Much of the adenosine that is derived from degradation of the ATP diffuses out of the cells into the circulating blood and is converted into uric acid, a substance that cannot re-enter the cells to reconstitute the adenosine phosphate system. Unfortunately, new ATP can be synthesized at the rate of only about 2 per cent an hour, meaning that, once depleted, the high energy phosphate stores of the cells are difficult to replenish. Therefore, one of the most important of the end-results of deterioration in shock, and one that is perhaps the most significant of all in the development of the final state of irreversibility, is this cellular depletion of the high energy compounds.

Deterioration of the Heart as a Primary Cause of Irreversible Shock.   It is clear that deterioration can occur in many different organ systems in shock and that the degeneration in any of these systems could become so severe that it would eventually be incompatible with continued life. However, there is reason to believe that in most instances it is deterioration of the heart itself that makes the shock irreversible. The reason for believing this is the following: Modern therapy is very effective in producing adequate venous return. Administration of blood and other substitution fluids can almost always provide adequate inflow pressure to the heart even in the most severe degrees of shock. But, still, in the late stages of shock, the heart fails to pump this inflowing blood. Therefore, the heart is the final weak link in the system.

On the other hand, if one does not use all forms of available therapy, a person can die of shock because of peripheral abnormalities such as continued loss of fluid into the tissues, pooling of blood in greatly distended blood vessels, respiratory failure, acidosis, and so forth. Unfortunately, in such instances, the patient dies of shock that is still reversible if adequate therapy is provided. Therefore, we can state again that, *with modern therapy,* the heart is usually the final weak link that leads to the irreversibility.

Special Role of Oxygen Deficiency in Shock and in Irreversibility

Though poor blood flow leads to tissue deterioration because of deficiency of many different nutrients, deficiency of oxygen almost certainly is the most important. For instance, in a large series of dogs, Crowell measured the accumulated deficit of oxygen usage of animals (a) in mild shock, (b) in moderate shock, and (c) in very severe shock. In some animals, this deficit accumulated slowly, while in others it accumulated very rapidly. But in each group of animals, when the average accumulated oxygen deficit reached 120 milliliters of oxy-

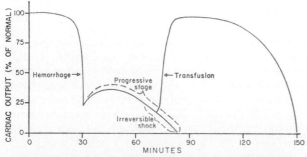

Figure 28–6.   Failure of transfusion to prevent death in irreversible shock.

gen per kilogram of body mass, 50 per cent of the animals died regardless of how long it took to accumulate this amount of oxygen deficit.

Therefore, it seems to be clear that the one most important nutrient necessary to prevent cellular deterioration and death during shock is oxygen.

HYPOVOLEMIC SHOCK CAUSED BY PLASMA LOSS

Loss of plasma from the circulatory system, even without the loss of whole blood, can sometimes be severe enough to reduce the total blood volume markedly, in this way causing typical hypovolemic shock similar in almost all details to that caused by hemorrhage. Severe plasma loss occurs in the following conditions:

1. *Intestinal obstruction* is often a cause of reduced plasma volume. The resulting distention of the intestine causes fluid to leak from the intestinal capillaries into the intestinal walls and intestinal lumen. This loss of fluid might result from elevated capillary pressure caused by increased resistance in the stretched veins over the surface of the intestine, or it might be caused by direct damage to the capillaries themselves. Nevertheless, the lost fluid has a very high content of protein, thereby reducing the total plasma protein as well as the plasma volume.

2. Often, in patients who have *severe burns* or other denuding conditions of the skin, so much plasma is lost through the exposed areas that the plasma volume becomes markedly reduced.

The hypovolemic shock that results from plasma loss has almost the same characteristics as the shock caused by hemorrhage, except for one additional complicating factor — the blood viscosity increases greatly as a result of plasma loss, and this further exacerbates the sluggishness of blood flow.

3. Loss of fluid from all fluid compartments of the body is called *dehydration;* this, too, can reduce the blood volume and cause hypovolemic shock very similar to that resulting from hemorrhage. Some of the causes of this type of shock are: (a) excessive sweating; (b) fluid loss in severe diarrhea or vomiting; (c) excess loss of fluid by nephrotic kidneys; (d) inadequate intake of fluid and electrolytes; and (e) destruction of the adrenal cortices, with consequent failure of the kidneys to reabsorb sodium, chloride, and water.

HYPOVOLEMIC SHOCK CAUSED BY TRAUMA

One of the most common causes of circulatory shock is trauma to the body. Often the shock results simply from hemorrhage caused by the trauma, but it can also occur even without hemorrhage, for contusion of the body can often damage the capillaries sufficiently to allow excessive loss of plasma into the tissues. This results in greatly reduced plasma volume with resultant hypovolemic shock. Thus, whether or not hemorrhage occurs when a person is severely traumatized, the blood volume can still be markedly reduced.

The *pain* associated with serious trauma can be an additional aggravating factor in traumatic shock, for pain sometimes strongly inhibits the vasomotor center, thereby increasing the vascular capacitance and reducing the venous return. Various attempts have also been made to implicate toxic factors released by the traumatized tissues as one of the causes of shock following trauma. However, cross-transfusion experiments have failed to show any such toxic element.

In summary, traumatic shock seems to result mainly from hypovolemia, though there may also be a moderate degree of concomitant neurogenic shock caused by the pain.

NEUROGENIC SHOCK — INCREASED VASCULAR CAPACITY

Occasionally shock results without any loss of blood volume whatsoever. Instead, the *vascular capacity* increases so much that even the normal amount of blood becomes incapable of adequately filling the circulatory system. One of the major causes of this is *loss of vasomotor tone* throughout the body, and the resulting condition is then known as *neurogenic shock.*

The relationship of vascular capacity to blood volume was discussed in Chapter 18, where it was pointed out that either an increase in vascular capacity or a decrease in blood volume *reduces the mean systemic filling pressure*, which in turn reduces the venous return to the heart. This effect is often called "venous pooling" of blood.

Causes of Neurogenic Shock. Some of the different factors that can cause loss of vasomotor tone include:

1. *Deep general anesthesia* often depresses the vasomotor center enough to cause vasomotor collapse, with resulting neurogenic shock.

2. *Spinal anesthesia*, especially when this extends all the way up the spinal cord, blocks the sympathetic outflow from the nervous system and is a common cause of neurogenic shock.

3. *Brain damage* is often a cause of vasomotor collapse. Many patients who have had brain concussion or contusion of the basal regions of the brain develop profound neurogenic shock. Also, even though short periods of medullary ischemia cause extreme vasomotor activity, prolonged ischemia can result in inactivation of the vasomotor neurons and can cause development of severe neurogenic shock.

4. *Depressed vasomotor center* frequently occurs during fever, after excessive loss of sleep, or in metabolic disturbances. This often leads to *fainting,* in which the peripheral blood vessels become greatly dilated. As a result, blood pools, and the cardiac output falls drastically. Such a person held in an upright position will go into the progressive stage of shock and can die as a result. Fortunately, on fainting, a person usually falls to a horizontal position so that almost normal cardiac output ordinarily re-ensues almost immediately.

Vasovagal Syncope — Emotional Fainting. The circulatory collapse that results from "emotional" fainting usually is not caused by vasomotor failure but instead by strong emotional excitation of the parasympathetic nerves to the heart and of the vasodilator nerves to the skeletal muscles, thereby slowing the heart and reducing the arterial pressure. Therefore, the fainting that results from an emotional disturbance is called *vasovagal syncope* to differentiate it from the other types of fainting which result from diminished sympathetic activity throughout the body or from other causes of reduced cardiac output.

ANAPHYLACTIC SHOCK

"Anaphylaxis" is an allergic condition in which the cardiac output and arterial pressure often fall drastically. This was discussed in Chapter 7. It results primarily from an antigen-antibody reaction that takes place all through the body immediately after an antigen to which the person is sensitive has entered the circulatory system. Such a reaction is detrimental to the circulatory system in several important ways. First, if the antigen-antibody reaction takes place in direct contact with the vascular walls or cardiac musculature, damage to these tissues presumably can result directly. Second, cells damaged anywhere in the body by the antigen-antibody reaction release several highly toxic substances into the blood. Among these is *histamine* or a *histamine-like substance,* released mainly from the circulating basophils and the mast cells outside the capillaries. The histamine in turn causes (1) an increase in vascular capacity because of venous dilatation, (2) dilatation of the arterioles with resultant greatly reduced arterial pressure, and (3) greatly increased capillary permeability with rapid loss of fluid into the tissue spaces. Unfortunately, all the precise relationships of the above factors in anaphylaxis have not been determined, but the sum total is a great reduction in venous return and often such serious shock that the person dies within minutes.

Intravenous injection of large amounts of histamine causes "histamine shock," which has characteristics almost identical with those of anaphylactic shock, though usually less severe.

SEPTIC SHOCK

The condition that was formerly known by the popular name of "blood poisoning" is now called *septic shock* by most clinicians. This simply means widely disseminated infection in many areas of the body, with the infection being borne through the blood from one tissue to another and causing extensive damage. Actually, there are many different varieties of septic shock because of the many different types of bacterial infection that can cause it and also because infection in one part of the body will produce different effects from those caused by infection elsewhere in the body.

Septic shock is extremely important to the clinician because it is this type of shock that, more frequently than any other kind of shock besides cardiogenic shock, causes patient death in the modern hospital. Some of the typical causes of septic shock include:

1. Peritonitis caused by spread of infection from the uterus and fallopian tubes, frequently resulting from instrumental abortion.

2. Peritonitis resulting from rupture of the gut, sometimes caused by intestinal disease and sometimes by wounds.

3. Generalized infection resulting from spread of a simple skin infection such as streptococcal or staphylococcal infection.

4. Generalized gangrenous infection resulting specifically from gas gangrene bacilli, spreading first through the tissues themselves and finally by way of the blood to the internal organs, especially to the liver.

5. Infection spreading into the blood from the kidney or urinary tract, often caused by colon bacilli.

Special Features of Septic Shock. Because of the multiple types of septic shock, it is difficult to categorize this condition. However, some features often seen in septic shock are the following:

1. High fever.

2. Marked vasodilatation throughout the body, especially in the infected tissues.

3. High cardiac output in perhaps half of the patients, caused by vasodilatation in the infected tissues and also by high metabolic rate and vasodilatation elsewhere in the body resulting from the high body temperature.

4. Sludging of the blood, presumably caused by red cell agglutination in response to degenerating tissues.

5. Development of microclots in widespread areas of the body, a condition called *disseminated intravascular coagulation.* Also, this causes the clotting factors to be used up so that hemorrhages occur into many tissues, especially into the gut wall into the intestinal tract.

In the early stages of septic shock the patient usually does not have signs of circulatory collapse but, instead, only signs of the bacterial infection itself. However, as the infection becomes more severe, the circulatory system usually becomes involved either directly or as a secondary result of toxins from the bacteria, and *there finally comes a point at which deterioration of the circulation becomes progressive in the same way that progression occurs in all other types of shock. Therefore, the end stages of septic shock are not greatly different from the end stages of hemorrhagic shock,* even though the initiating factors are markedly different in the two conditions.

Endotoxin Shock. A special type of septic shock is known as endotoxin shock. It frequently occurs when a large segment of the gut becomes strangulated and loses most of its blood supply. The gut rapidly becomes gangrenous, and the bacteria in the gut multiply rapidly.

Most of these bacteria are so-called "gram-negative" bacteria, mainly colon bacilli, that contain a toxin called *endotoxin*. Another condition that also frequently causes colon bacilli septicemia is extension of urinary tract infections into the blood.

On entering the circulation, endotoxin causes an effect very similar to that of anaphylaxis, sometimes resulting in severe shock. Indeed, the cause of this shock is possibly an anaphylactic reaction caused by the endotoxin itself. However, endotoxin has direct effects on the circulation as well, including (1) toxic depression of the heart and (2) vascular dilatation.

Severe shock also occurs during the course of diseases caused by other gram-negative bacterial infections, such as dysentery, tularemia, brucellosis, and typhoid fever. It is likely that this shock results at least partially from endotoxin released from the gram negative bacteria causing these conditions.

EFFECTS OF SHOCK ON THE BODY

Decreased Metabolism in Hypovolemic Shock. The decreased cardiac output in hypovolemic shock reduces the amount of oxygen and other nutrients available to the different tissues, and this in turn reduces the level of metabolism that can be maintained by the different cells of the body. Usually a person can continue to live for only a few hours if the cardiac output falls to as low as 40 per cent of normal.

Muscular Weakness. One of the earliest symptoms of shock is severe muscular weakness which is also associated with profound and rapid fatigue whenever patients attempt to use their muscles. This obviously results from the diminished supply of nutrients — especially oxygen — to the muscles.

Body Temperature. Because of the depressed metabolism in shock, the amount of heat liberated in the body is reduced (except in septic shock in which the infection may cause an opposite effect). As a result, the body temperature tends to decrease if the body is exposed to even the slightest cold.

Mental Function. In the early stages of shock the person is usually conscious, though signs of mental haziness may be noted. As the shock progresses, the person falls into a state of stupor, and in the last stages of shock even the subconscious mental functions, including vasomotor control and respiration, fail.

A person who recovers from shock usually exhibits no permanent impairment of mental functions. However, following complete circulatory arrest, in which the blood flow is completely cut off from the brain for many minutes, the brain does often suffer severe permanent impairment, the cause of which is discussed in a separate section on circulatory arrest later in the chapter.

Reduced Renal Function. The very low blood flow during shock greatly diminishes urine output or even causes cessation of output, because glomerular pressure falls below the critical value required for filtration of fluid into Bowman's capsule, as explained in Chapters 33 and 34. Also, the kidney has such a high rate of metabolism and requires such large amounts of nutrients that the reduced blood flow often causes *tubular necrosis*, which means death of the tubular epithelial cells, with subsequent sloughing and blockage of the tubules, causing total loss of function of the respective nephrons. This is often a serious aftereffect of shock that occurs during major surgical operations; the patient sometimes survives the shock associated with the surgical procedure and then dies a week or so later of uremia.

PHYSIOLOGY OF TREATMENT IN SHOCK

REPLACEMENT THERAPY

Blood and Plasma Transfusion. If a person is in shock caused by hemorrhage, the best possible therapy is usually transfusion of whole blood. If the shock is caused by plasma loss, the best therapy is administration of plasma; when dehydration is the cause, administration of the appropriate electrolytic solution can correct the shock.

Unfortunately, whole blood is not always available, such as under battlefield conditions. However, plasma can usually substitute adequately for whole blood because it increases the blood volume and restores normal hemodynamics. Plasma cannot restore a normal hematocrit, but the human being can usually stand a decrease in hematocrit to about one-third normal before serious consequences result if the cardiac output is adequate. Therefore, in acute conditions it is reasonable to use plasma in place of whole blood for treatment of hemorrhagic and most other types of hypovolemic shock.

Sometimes plasma also is unavailable. For these instances, various *plasma substitutes* have been developed that perform almost exactly the same hemodynamic functions as plasma. One of these is the following:

Dextran Solution as a Plasma Substitute. The principal requirement of a truly effective plasma substitute is that it remain in the circulatory system — that is, not filter through the capillary pores into the tissue spaces. But, in addition, the solution must be nontoxic and must contain appropriate electrolytes to prevent derangement of the extracellular fluid electrolytes on administration. To remain in the circulation the plasma substitute must contain some substance that has a large enough molecular size to exert colloid osmotic pressure.

One of the most satisfactory substances that has been developed thus far for this purpose is dextran, a large polysaccharide polymer of glucose. Certain bacteria secrete dextran as a by-product of their growth, and commercial dextran is manufactured by a bacterial culture procedure. By varying the growth conditions of the bacteria, the molecular weight of the dextran can be controlled to the desired value. Dextrans of appropriate molecular size do not pass through the capillary pores and, therefore, can replace plasma proteins as colloid osmotic agents.

Fortunately, few toxic reactions have been observed when using dextran to provide colloid osmotic pressure; therefore, solutions of this substance have proved to be a satisfactory substitute for plasma in much fluid replacement therapy.

TREATMENT OF SHOCK WITH SYMPATHOMIMETIC AND SYMPATHOLYTIC DRUGS OR OTHER THERAPY

A *sympathomimetic drug* mimics sympathetic stimulation. These drugs include norepinephrine, epinephrine, and a large number of long-acting drugs that have the same effects as epinephrine and norepinephrine. A *sympatholytic drug,* on the other hand, blocks the action of normal sympathetic stimuli on the circulatory system.

Both sympathomimetic and sympatholytic drugs have been used in treating different types of shock, including hemorrhagic shock. However, there is no proof that either plays a very beneficial role in this type of shock. The sympathomimetic drugs usually have little benefit because the circulatory reflexes are usually already maximally activated in severe hypovolemic shock and are already causing marked endogenous secretion of norepinephrine and epinephrine. On the other hand, the proponents of the use of sympatholytic drugs have claimed that these drugs increase blood flow through the tissues and thereby prevent nutritional damage. Unfortunately, these drugs also usually cause paralysis of the vasculature, resulting in increased pooling of blood in the venous system, which reduces the cardiac output and negates much if not all of the direct effect of the drugs in increasing local blood flow. Yet a combination of *transfusion plus sympatholytic drugs* probably is beneficial in the treatment of many patients with shock; the transfusion fills the expanded vascular bed and prevents the pooling, while the sympatholytic drug allows rapid blood flow to the tissues.

OTHER THERAPY

Treatment by the Head-Reclining Position. When the pressure falls too low in most types of shock, especially hemorrhagic and neurogenic shock, placing the patient with the head as much as a foot lower than the feet will help tremendously in promoting venous return and thereby increasing cardiac output. Therefore, this is the first essential in the treatment of many types of shock.

Oxygen Therapy. Since the major deleterious effect of most types of shock is too little delivery of oxygen to the tissues, giving the patient oxygen to breathe can be of benefit in some instances. However, this frequently is of far less value than one might expect because the problem usually is not inadequate oxygenation of the blood in the lungs, but inadequate transport of the blood after it is oxygenated.

Treatment with Glucocorticoids. Glucocorticoids are frequently given to patients in severe shock for several reasons: (1) experiments have shown empirically that glucocorticoids frequently increase the strength of the heart in the late stages of shock; (2) glucocorticoids stabilize the lysosomal membranes and prevent release of lysosomal enzymes into the cytoplasm of the cells, thus preventing deterioration from this source; (3) glucocorticoids have a specific action to oppose release of MTF from the pancreas; and (4) glucocorticoids might also aid in the metabolism of glucose by the severely damaged cells.

Treatment of Neurogenic Shock. In neurogenic shock, sympathomimetic drugs actually reverse the basic cause of the shock itself by increasing the vasomotor tone throughout the body, but even in this condition these drugs may not be required, for, as long as the patient is in the *horizontal or head-down position,* adequate venous return to the heart usually still occurs and maintains a sufficient cardiac output to prevent progressive shock.

Treatment of Anaphylactic Shock. Unfortunately, anaphylactic shock occurs so rapidly that one often cannot institute any therapy before death ensues, but if therapy can be instituted the condition can often be ameliorated or almost completely reverted by rapid administration of norepinephrine or some other sympathomimetic drug. This does not correct the basic cause of the anaphylaxis, but it does cause vasoconstriction, which opposes the vasodilatation caused by histamine in anaphylaxis. Also, if one suspects that anaphylaxis might occur in a patient, its severity can usually be reduced by preliminary administration of *cortisol,* which attenuates the allergic reaction responsible for the anaphylaxis, or preliminary administration of *antihistaminics,* which reduce the effects of the histamine released during anaphylaxis. However, in both these instances the therapy must be given *before* anaphylaxis takes place, for which reason they are usually of little value in therapy of anaphylaxis.

Pain-Relieving Drugs in Traumatic Shock. The use of pain-relieving drugs, such as morphine, has been found clinically to reduce the severity of most traumatic shock, supposedly by removing the neurogenic element of the shock.

CIRCULATORY ARREST

A condition closely allied to circulatory shock is circulatory arrest, in which all blood flow completely stops. This occurs frequently on the surgical operating table as a result of *cardiac arrest* or of *ventricular fibrillation.*

Ventricular fibrillation can usually be stopped by strong electroshock of the heart, the basic principles of which were described in Chapter 14.

Cardiac arrest usually results from too little oxygen in the anesthetic gaseous mixture or from a depressant effect of the anesthesia itself. A normal cardiac rhythm can usually be restored by removing the anesthetic and then applying cardiopulmonary resuscitation procedures for a few minutes while supplying the patient's lungs with adequate quantities of ventilatory oxygen.

EFFECT OF CIRCULATORY ARREST ON THE BRAIN

The real problem in circulatory arrest is usually not to restore cardiac function but instead to prevent detrimental effects in the brain as a result of the circulatory arrest. In general, four to five minutes of circulatory arrest causes permanent brain damage in over half the patients, and circulatory arrest for as long as 10 minutes almost universally destroys most, if not all, of the mental powers.

For many years it has been taught that these detrimental effects on the brain are caused by the cerebral hypoxia that occurs during circulatory arrest. However, recent studies by Crowell and others have shown that dogs can almost universally stand up to 30 minutes of circulatory arrest without permanent brain damage *if the blood is removed from the brain circulation prior to the arrest.* On the basis of these studies, it is postulated that the circulatory arrest causes vascular *clots* to develop throughout the brain and that these cause permanent or semipermanent ischemia of brain areas. This accords well with the results in human beings who have undergone long periods of circulatory arrest, for complete destruction of large areas in one side of the brain often occurs while corresponding areas in the opposite side of the brain, which should also be affected if hypoxia were the cause of the damage, are not affected even in the slightest.

REFERENCES

Bergentz, S. E., and Leandoer, L.: Disseminated intravascular coagulation in shock. *Ann. Chir. Gynaecol. Fenn., 60*:175, 1971.

Chien, S.: Role of the sympathetic nervous system in hemorrhage. *Physiol. Rev., 47*:214, 1967.

Crowell, J. W.: Cardiac deterioration as the cause of irreversibility in shock. *In* Mills, L. J., and Moyer, J. H. (eds.): Shock and Hypotension: Pathogenesis and Treatment. Grune & Stratton, 1965, p. 605.

Crowell, J. W., and Guyton, A. C.: Evidence favoring a cardiac mechanism in irreversible hemorrhagic shock. *Am. J. Physiol., 201*:893, 1961.

Crowell, J. W., and Read, W. L.: *In vivo* coagulation–a probable cause of irreversible shock. *Am. J. Physiol., 183*:565, 1955.

Crowell, J. W., and Smith, E. E.: Oxygen deficit and irreversible hemorrhagic shock. *Am. J. Physiol., 206*:313, 1964.

Crowell, J. W., *et al.*: The mechanism of death after resuscitation following acute circulatory failure. *Surgery, 38*:696, 1955.

Crowell, J. W., *et al.*: The effect of varying the hematocrit ratio on the susceptibility to hemorrhagic shock. *Am. J. Physiol., 192*:171, 1958.

Crowell, J. W., *et al.*: Effect of allopurinol on hemorrhagic shock. *Am. J. Physiol., 216*:744, 1969.

Franklin, J., and Doelp, A.: Shocktrauma. New York, St. Martin's Press, 1980.

Guyton, A. C., and Crowell, J. W.: Dynamics of the heart in shock. *Fed. Proc., 20*:51, 1961.

Guyton, A. C., *et al.*: Cardiac Output and Its Regulation. Philadelphia, W. B. Saunders Co., 1973.

Hershey, S. G.: The reticulo-endothelial system: Relationship to shock and host defense. *In* Kaley, G., and Altura, B. M. (eds.): Microcirculation. Vol. III. Baltimore, University Park Press, 1977.

Jamieson, G. A., and Greenwalt, T. J. (eds.): Blood Substitutes and Plasma Expanders. New York, A. R. Liss, 1978.

Jones, C. E., *et al.*: A cause-effect relationship between oxygen deficit and irreversible hemorrhagic shock. *Surgery, 127*:93, 1968.

Jones, C. E., *et al.*: Significance of increased blood uric acid following extensive hemorrhage. *Am. J. Physiol., 214*:1374, 1968.

Jones, C. E., *et al.*: Cardiac output and physiological mechanisms in circulatory shock. *In* MTP International Review of Science: Physiology. Vol. 1. Baltimore. University Park Press, 1974, p. 233.

Kovach, A. G. B., and Sandor, P.: Cerebral blood flow and brain function during hypotension. *Annu. Rev. Physiol., 38*:571, 1976.

Lefer, A. M., and Martin, J.: Mechanism of the protective effect of corticosteroids in hemorrhagic shock. *Am. J. Physiol., 216*:314, 1969.

Lefer, A. M., and Martin, J.: Relationship of plasma peptides to the myocardial depressant factor in hemorrhagic shock in cats. *Circ. Res., 26*:59, 1970.

Lefer, A. M., *et al.*: Characterization of a myocardial depressant factor present in hemorrhagic shock. *Am. J. Physiol., 213*:492, 1967.

Lewis, D. H.: Microcirculation in low-flow states. *In* Kaley, G., and Altura, B. M. (eds.): Microcirculation. Vol. III. Baltimore, University Park Press, 1977.

Moyer, C. A., and Butcher, H. R.: Burns, Shock, and Plasma Volume Regulation. St. Louis, C. V. Mosby, 1967.

Nagler, A.: Circulatory manifestations of endotoxemia. *In* Kaley, G., and Altura, B. M. (eds.): Microcirculation. Vol. III. Baltimore, University Park Press, 1977.

O'Riordan, J. P., *et al.*: The Indications for the Use of Albumin, Plasma Protein Solutions and Plasma Substitutes. Strasbourg, European Public Health Committee, 1978.

Rosenblum, W. I.: Blood viscosity and disease. *In* Kaley, G., and Altura, B. M. (eds.): Microcirculation. Vol. III. Baltimore, University Park Press, 1977.

Rothe, C. F., *et al.*: Control of total vascular resistance in hemorrhagic shock in the dog. *Circ. Res., 12*:667, 1963.

Schmid-Schönbein, H., and Teitel, P. (eds.): Basic Aspects of Blood Trauma. Hingham, Mass., Kluwer Boston, 1979.

Schumer, W., and Nyhus, L. M. (eds.): Treatment of Shock. Philadelphia, Lea & Febiger, 1974.

Sheiner, N. M.: Peripheral Vascular Surgery. Chicago, Year Book Medical Publishers, 1978.

Shoemaker, W. C.: Pattern of pulmonary hemodynamic and functional changes in shock. *Crit. Care Med., 2*:200, 1974.

Smith, E. E., and Crowell, J. W.: Effect of hemorrhagic hypotension on oxygen consumption of dogs. *Am. J. Physiol., 207*:647, 1964.

Suteu, I., *et al.*: Shock: Pathology, metabolism, shock cell treatment. Tunbridge Wells, Abacus Press, 1977.

Weil, M. H., and Shubin, H.: Shock. Baltimore, Williams & Wilkins, 1967.

Wilkinson, A. W.: Body Fluids in Surgery, 4th Ed. New York, Churchill Livingstone, Div. of Longman, 1973.

Zweifach, B. W.: Mechanisms of blood flow and fluid exchange in microvessels: hemorrhagic hypotension model. *Anesthesiology, 41*:157, 1974.

29

Muscle Blood Flow During Exercise; Cerebral, Splanchnic, and Skin Blood Flows

The blood flow in many special areas of the body, such as the lungs and the heart, has already been discussed in previous chapters, and the circulation in the kidney, in the uterus, and in the fetus will be discussed later in the text. In the present chapter the characteristics of blood flow in some of the other important tissues, such as the muscles, the brain, the splanchnic system, and the skin, are presented.

BLOOD FLOW THROUGH SKELETAL MUSCLES AND ITS REGULATION IN EXERCISE

Very strenuous exercise is the most stressful condition that the normal circulatory system faces. This is true because the blood flow in muscles can increase more than 20-fold (a greater increase than in any other tissue of the body) and also because there is such a very large mass of skeletal muscle in the body. The product of these two factors is so great that the total muscle blood flow can become great enough to increase the cardiac output in the normal young adult to as much as five times normal and in the well-trained athlete to as much as six to seven times normal.

RATE OF BLOOD FLOW THROUGH THE MUSCLES

During rest, blood flow through skeletal muscle averages 3 to 4 ml. per minute per 100 grams of muscle. However, during extreme exercise this rate can increase as much as 15- to 25-fold, rising to 50 to 80 ml. per 100 grams of muscle.

Intermittent Flow During Muscle Contraction. Figure 29–1 illustrates a study of blood flow

changes in the calf muscles of the human leg during strong rhythmic contraction. Note that the flow increases and decreases with each muscle contraction, decreasing during the contraction phase and increasing between contractions. At the end of the rhythmic contractions, the blood flow remains very high for a few seconds but then gradually fades toward normal during the next few minutes.

The cause of the decreased flow during sustained muscle contraction is compression of the blood vessels by the contracted muscle. During strong *tetanic* contraction, blood flow can be almost totally stopped.

Opening of Muscle Capillaries During Exercise. During rest, only 20 to 25 per cent of the muscle capillaries are open. But during strenuous exercise all the capillaries open up, which can be demonstrated by studying histologic specimens removed from muscles appropriately stained dur-

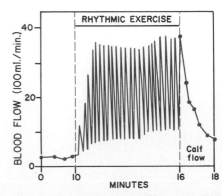

Figure 29–1. Effects of muscle exercise on blood flow in the calf of a leg during strong rhythmic contraction. The blood flow was much less during contraction than between contractions. (From Barcroft and Dornhorst: *J. Physiol., 109*:402, 1949.)

ing exercise. It is this opening up of dormant capillaries that allows most of the increased blood flow. It also diminishes the distance that oxygen and other nutrients must diffuse from the capillaries to the muscle fibers and contributes a much increased surface area through which nutrients can diffuse from the blood.

CONTROL OF BLOOD FLOW THROUGH THE SKELETAL MUSCLES

Local Regulation.   The tremendous increase in muscle blood flow that occurs during skeletal muscle activity is caused primarily by local effects in the muscles acting directly on the arterioles to cause vasodilatation.

This local increase in blood flow during muscle contraction is probably caused by several different factors all operating at the same time. One of the most important of these is reduction of dissolved oxygen in the muscle tissues. That is, during muscle activity the muscle utilizes oxygen very rapidly, thereby decreasing its concentration in the tissue fluids. This in turn causes vasodilatation either because the vessel walls cannot maintain contraction in the absence of oxygen or because oxygen deficiency causes release of vasodilator substances. The vasodilator substance that has been suggested most widely in recent years has been adenosine.

Other vasodilator substances released during muscle contraction include potassium ions, acetylcholine, adenosine triphosphate, lactic acid, and carbon dioxide. Unfortunately, we still do not know quantitatively how great a role each of these plays in increasing muscle blood flow during muscle activity; this subject was discussed in more detail in Chapter 20.

Nervous Control of Muscle Blood Flow.   In addition to the local tissue regulatory mechanism, the skeletal muscles are also provided with sympathetic vasoconstrictor nerves and, in some species of animals, sympathetic vasodilator nerves as well.

Sympathetic Vasoconstrictor Nerves.   The sympathetic vasoconstrictor nerve fibers secrete norepinephrine and when maximally stimulated can decrease blood flow through the muscles to about one-fourth normal. This represents rather poor vasoconstriction in comparison with that caused by sympathetic nerves in some other areas of the body in which blood flow can be almost completely blocked. Yet even this degree of vasoconstriction is of physiological importance in circulatory shock and during other periods of stress when it is desirable to reduce blood flow through the many muscles of the body.

In addition to the norepinephrine secreted at the sympathetic vasoconstrictor nerve endings, the adrenal medullae secrete large amounts of additional norepinephrine and epinephrine into the circulating blood during strenuous exercise. The circulating norepinephrine acts on the muscle vessels to cause a vasoconstric- tor effect similar to that caused by direct sympathetic nerve stimulation. The epinephrine, on the other hand, has a vasodilator effect. It is believed that this results from the fact that epinephrine excites the *beta* receptors of the vessels, which are vasodilator receptors, in contrast to the *alpha* receptors excited by the norepinephrine. These receptors are discussed in Chapter 57.

Sympathetic Vasodilator Fibers.   In the cat and some other lower animals there are also sympathetic *vasodilator* fibers that secrete acetylcholine — this hormone, in turn, causing the vasodilatation. However, such fibers have not been proved to occur in the human being. Instead, as noted above, circulating epinephrine from the adrenal medullae, acting at beta receptors in the muscle arterioles, causes vasodilatation, and this may subserve the same function as the vasodilator system of the lower animals.

In the cat, maximal stimulation of the sympathetic vasodilator fibers to the skeletal muscles can increase blood flow by 400 per cent. These fibers are activated by a special nervous pathway beginning in the cerebral cortex, in close association with the motor areas for control of muscular activity, and passing downward through the hypothalamus and brain stem to the spinal cord. When the motor cortex initiates muscle activity, it simultaneously excites the vasodilator fibers to the active muscles, and vasodilatation occurs immediately, several seconds before the local vasodilatation can take place. Thus, it seems that this vasodilator system has the important *function of initiating extra blood flow through the muscles at the onset of muscular activity*. However, the vasodilator fibers are probably of little or no importance in maintaining increased blood flow during prolonged muscular activity, for both in lower animals and in human beings the final degree of vasodilatation that occurs during exercise is not discernibly different when the muscle is normally innervated with sympathetics or denervated.

CIRCULATORY READJUSTMENTS DURING EXERCISE

Three major effects occur during exercise that are essential for the circulatory system to supply the tremendous blood flow required by the muscles. These effects are (1) mass discharge of the sympathetic nervous system throughout the body with consequent stimulatory effects on the circulation, (2) increase in cardiac output, and (3) increase in arterial pressure.

Mass Sympathetic Discharge.   At the onset of exercise, signals are transmitted not only from the brain to the muscle to cause muscle contraction but also from the higher levels of the brain into the vasomotor center to initiate mass sympathetic discharge. Simultaneously, the parasympathetic signals to the heart are greatly attenuated. Therefore, two major circulatory effects result. First, the heart is stimulated to greatly increased heart rate and pumping strength as a result of the sympathetic drive to the heart, as well as release of the heart from the normal parasympathetic inhibition. Second, all the blood vessels of the peripheral circulation are strongly contracted except the vessels in the active muscles, which are strongly vasodilated by the local vasodilator effects in the muscles themselves. Thus, the heart is stimulated to supply the increased blood flow required by the muscles, and blood flow through most nonmuscular areas of the body is temporarily reduced, thereby temporarily ''lending'' their blood supply to the muscles. This effect accounts for as much as 2.5 liters of extra blood flow to the muscles. It is exceedingly important when one thinks of the wild animal running for its life, because even a fractional increase in running speed may make the difference between life and death. However, two of the organ circulatory systems, the coronary and cerebral systems, are spared this vasoconstrictor effect because both of these circulatory areas have very poor vasoconstrictor innervation — fortunately so because both the heart and the brain are as essential to exercise as are the skeletal muscles themselves.

A Muscle Reflex That Stimulates the Sympathetic Nervous System. Aside from the sympathetic stimulation caused by direct signals from the brain, reflex signals from the contracting muscles are also believed to pass up the spinal cord to the vasomotor center and to excite the sympathetic nerves. These signals probably are initiated by metabolic end-products acting on small sensory nerve endings in the muscle tissue.

Increase in Cardiac Output. The increase in cardiac output that occurs during exercise results mainly from the intense local vasodilatation in active muscles. As was explained in Chapter 23 in relation to the basic theory of cardiac output regulation, local vasodilatation increases the venous return of blood back to the heart. The heart in turn pumps this extra returning blood and sends it immediately back to the muscles through the arteries. Thus, it is mainly the muscles themselves that determine the amount of increase in cardiac output — up to the limit of the heart's ability to respond.

Another factor that helps greatly to cause the large increase in venous return is the strong sympathetic stimulation of the veins. This stimulation greatly increases the mean systemic filling pressure, sometimes as high as 30 mm. Hg (four times normal), and is therefore important in increasing venous return.

Mechanisms by Which the Heart Increases Its Output. One of the principal mechanisms by which the heart increases its output during exercise is the Frank-Starling mechanism, which was discussed in Chapter 13. Via this mechanism, when increased quantities of blood flow from the veins into the heart and dilate its chambers, the heart muscle contracts with increased force, thus also pumping an increased volume of blood with each heart beat. However, in addition to this basic intrinsic cardiac mechanism, the heart is also strongly stimulated by the sympathetic nervous system, and the normal parasympathetic inhibition is reduced or eliminated. The net effects are greatly increased heart rate (occasionally to as high as 200 beats per minute) and almost doubling of the cardiac muscle strength of contraction. These two effects combine to make the heart capable of pumping at least 100 per cent more blood than would be true based on the Frank-Starling mechanism alone.

Increase in Arterial Pressure. The mass sympathetic discharge throughout the body during exercise and the resultant vasoconstriction of most of the blood vessels besides those in the active muscles almost always increase the arterial pressure during exercise. This increase can be as little as 20 mm. Hg or as great as 80 mm. Hg, depending on the conditions under which the exercise is performed. For instance, when a person performs exercise under very tense conditions but uses only a few muscles, the sympathetic response still occurs throughout the body but vasodilatation occurs in only a few muscles. Therefore, the net effect is mainly one of vasoconstriction, often increasing the mean arterial pressure to as high as 180 mm. Hg. Such a condition occurs in a person standing on a ladder and nailing with a hammer on the ceiling above. The tenseness of the situation is obvious, and yet the amount of muscle vasodilatation is relatively slight.

On the other hand, when a person performs whole-body exercise, such as running or swimming, the increase in arterial pressure is usually only 20 to 40 mm. Hg. The lack of a tremendous rise in pressure results from the extreme vasodilatation occurring in large masses of muscle.

In rare instances, persons are found in whom the sympathetic nervous system is absent, either because of congenital absence or because of surgical removal. When such a person exercises, instead of the arterial pressure rising, the pressure actually falls — sometimes to as low as one-half normal, and the cardiac output rises only about one-third as much as it does normally. Therefore, one can readily understand the major importance of increased sympathetic activity during exercise.

Importance of the Arterial Pressure Rise During Exercise. In the well-trained athlete it has been calculated the muscle blood flow can increase at least 20-fold. Though most of this increase results from vasodilatation in the active muscles, the increase in arterial pressure also plays an important role. If one remembers that an increase in pressure not only forces extra blood through the muscle because of the increased pressure itself but also dilates the blood vessels, he can see that as little as a 20 to 40 mm. Hg rise in pressure can at times actually double peripheral blood flow.

It is especially important to note that in animals or human beings who do not have a sympathetic nervous system, the fall in arterial pressure that occurs during exercise has a strong negating effect on the rise in cardiac output that normally occurs. In such instances, the cardiac output can almost never be increased more than two-fold, instead of the four- to seven-fold that can occur when the arterial pressure rises above normal.

The Cardiovascular System as the Limiting Factor in Heavy Exercise. The capability of an athlete to enhance cardiac output and consequently to deliver increased quantities of oxygen and other nutrients to his or her tissues is the major factor that determines the degree of prolonged heavy exercise that the athlete can sustain. For instance, the speed of a marathon runner is almost directly proportional to the ability to enhance cardiac output. Therefore, the ability of the circulatory system to adapt to exercise is equally as important as the muscles themselves in setting the limit for the performance of muscle work.

THE CEREBRAL CIRCULATION

NORMAL RATE OF CEREBRAL BLOOD FLOW

The normal blood flow through brain tissue averages 50 to 55 ml. per 100 grams of brain per minute. For the entire brain of the average adult, this is approximately 750 ml. per minute, or 15 per cent of the total resting cardiac output.

REGULATION OF CEREBRAL BLOOD FLOW

As in most other tissues of the body, cerebral blood flow is highly related to the metabolism of the cerebral tissue. At least three different metabolic factors have been shown to have very potent effects on cerebral blood flow. These are carbon dioxide concentration, hydrogen ion concentration, and oxygen concentration. An increase in either the carbon dioxide or the hydrogen ion concentration increases cerebral blood flow, whereas a decrease in oxygen concentration increases the flow.

Regulation of Cerebral Blood Flow in Response to Excess Carbon Dioxide or Hydrogen Ion Concentration. An increase in carbon dioxide concentration in the arterial blood perfusing the brain greatly increases cerebral blood flow. This is illustrated in Figure 29–2, which shows that doubling

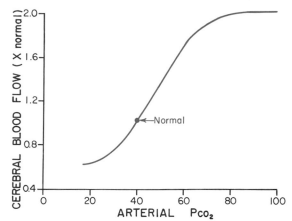

Figure 29-2. Relationship between arterial P_{CO_2} and cerebral blood flow.

the arterial P_{CO_2} by breathing carbon dioxide also approximately doubles the blood flow.

Carbon dioxide increases cerebral blood flow by combining with water in the body fluids to form carbonic acid, with subsequent dissociation to form hydrogen ions. The hydrogen ions then cause vasodilatation of the cerebral vessels — the dilatation being almost directly proportional to the increase in hydrogen ion concentration.

Any other substance that increases the acidity of the brain tissue, and therefore also increases the hydrogen ion concentration, increases blood flow as well. Such substances include lactic acid, pyruvic acid, or any other acidic material formed during the course of metabolism.

Importance of the Carbon Dioxide and Hydrogen Control of the Cerebral Blood Flow. Increased hydrogen ion concentration greatly depresses neuronal activity; conversely, diminished hydrogen ion concentration greatly increases neuronal activity. Therefore, it is fortunate that an increase in hydrogen ion concentration causes an increase in the blood flow, which in turn carries both carbon dioxide and dissolved acids away from the brain tissues. Loss of the carbon dioxide removes carbonic acid from the tissues, and this, along with the removal of other acids, reduces the hydrogen ion concentration back toward normal. Thus, this mechanism helps to maintain a very constant hydrogen ion concentration in the cerebral fluids and therefore also to maintain a normal level of neuronal activity.

Oxygen Deficiency as a Regulator of Cerebral Blood Flow. Except during periods of intense brain activity, the utilization of oxygen by the brain tissue remains within very narrow limits — within a few per cent of 3.5 ml. of oxygen per 100 grams of brain tissue per minute. If the blood flow to the brain ever becomes insufficient to supply this needed amount of oxygen, the oxygen deficiency mechanism for vasodilatation, which was discussed in Chapter 20 and which functions in essentially all tissues of the body, immediately causes vasodilatation, returning the blood flow and transport of oxygen to the cerebral tissues near to normal. Thus, this local blood flow regulatory mechanism is much the same as that existing in the coronary and skeletal muscle circulations and in many other circulatory areas of the body.

Experiments have shown that a decrease in cerebral *venous* blood P_{O_2} below approximately 30 mm. Hg (normal value is about 35 mm. Hg) will begin to increase cerebral blood flow. It is very fortunate that flow does respond at this critical level because brain function begins to become deranged at not much lower values of P_{O_2}, especially so at levels below 20 mm. Hg. Thus, the oxygen mechanism for local regulation of cerebral blood flow is a very important protective response against diminished cerebral neuronal activity and, therefore, against derangement of mental capability.

Measurement of Cerebral Blood Flow and Effect of Cerebral Activity on the Flow. A method has recently been developed to record blood flow in as many as 256 isolated segments of the human cerebral cortex simultaneously. A radioactive substance, usually radioactive xenon, is injected into the carotid artery, then the radioactivity of the cortex as the radioactive substance passes through the brain tissue is recorded. Two hundred and fifty-six small radioactive scintillation detectors are focused on the same number of separate parts of the cortex; the rate of decay of the radioactivity after it once appears in each tissue segment is a direct measure of the rate of blood flow through the segment.

Using this technique, it has become clear that the blood flow in each individual segment of the brain changes within seconds in response to changes in local neuronal activity. For instance, simply clasping the hand causes an immediate increase in blood flow in the motor cortex of the opposite side of the brain. Reading a book increases the blood flow in multiple areas of the brain, especially in the occipital cortex and in the language areas of the temporal cortex. This measuring procedure can also be used to localize the origin of epileptic attacks, for the blood flow increases acutely and markedly in the focal point of the attack at its very onset.

Illustrating the effect of local neuronal activity on cerebral blood flow, Figure 29–3 shows an increase in occipital blood flow recorded in a cat when intense light was shone into its eyes for a period of one-half minute.

Autoregulation of Cerebral Blood Flow When the Arterial Pressure Is Changed. Cerebral blood flow is autoregulated extremely well between the pressure limits

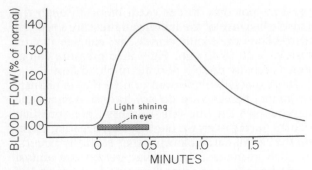

Figure 29–3. Increase in blood flow to the occipital regions of the brain when a light is flashed in the eyes of an animal.

of 60 and 140 mm. Hg. That is, the arterial pressure can be decreased acutely to as low as 60 mm. Hg or increased to as high as 140 mm. Hg without a significant change in cerebral blood flow. In persons who have hypertension, this autoregulatory range shifts to even higher pressure levels, up to as high as 180 to 200 mm. Hg. This effect is illustrated in Figure 29–4, which shows cerebral blood flows measured both in normal human beings and in hypertensive patients. Note the extreme constancy of cerebral blood flow between the limits of 60 and 180 mm. Hg mean arterial pressure. On the other hand, if the arterial pressure does fall below 60 mm. Hg, cerebral blood flow then becomes severely compromised, and if the pressure rises above the upper limit of autoregulation, the blood flow rises rapidly and can cause severe overstretch of the cerebral blood vessels, resulting in serious brain edema.

Minor Importance of Autonomic Nerves in Regulating Cerebral Blood Flow. Sympathetic nerves from the cervical sympathetic chain pass upward along the cerebral arteries to supply the superficial cerebral vessels. Also, parasympathetic fibers from the great superficial petrosal and facial nerves supply some of these vessels as well. However, transection of either the sympathetic or parasympathetic nerves causes no measurable effect on cerebral blood flow. On the other hand, stimulation does affect blood flow slightly, the sympathetics causing

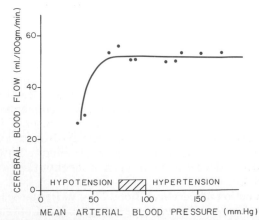

Figure 29–4. Relationship of mean arterial pressure to cerebral blood flow in normotensive, hypotensive, and hypertensive persons. (Modified from Lassen: *Physiol. Rev., 39*:183, 1959.)

mild vasoconstriction of the large vessels but not of the small vessels, and the parasympathetics causing mild vasodilatation. Though these nerves normally play little or no role in the control of cerebral blood flow, sympathetic reflexes probably do cause vasospasm in the superficial vessels in some instances of stroke, subdural hematoma, tumor, and other types of brain damage.

Brain Edema

One of the most serious complications of abnormal cerebral hemodynamics is the development of brain edema. Because the brain is encased in a solid vault, the accumulation of edema fluid compresses the blood vessels, with eventual depression of blood flow and destruction of brain tissue.

The usual cause of brain edema is either greatly increased capillary pressure or damage to the capillary endothelium. One cause of excessively high capillary pressure is a sudden increase in the cerebral blood pressure to levels too high for the autoregulatory mechanism to cope with. However, the most common cause is brain concussion, in which the brain tissues and capillaries are traumatized and capillary fluid leaks into the traumatized tissues. Once brain edema begins, it often initiates a vicious cycle because of the following positive feedback: The edema compresses the vasculature. This in turn decreases the blood flow and causes brain ischemia. The ischemia causes arteriolar dilatation with increased capillary pressure. The increased capillary pressure then causes more edema fluid, so that the edema becomes progressively worse. Once this vicious cycle has begun, heroic measures must be used to prevent total destruction of the brain. One such measure is to infuse intravenously a concentrated osmotic substance, such as 50 per cent sucrose solution. This pulls fluid by osmosis from the brain tissue and breaks up the vicious cycle. Another procedure is to remove fluid quickly from the ventricles of the brain, thereby relieving the intracerebral pressure.

THE SPLANCHNIC CIRCULATION

A large share of the cardiac output flows through the vessels of the intestines and through the spleen, finally coursing into the portal venous system and then through the liver, as illustrated in Figure 29–5. This is called the portal circulatory system, and it, plus the arterial blood flow into the liver, is called the splanchnic circulation.

BLOOD FLOW THROUGH THE LIVER

About 1100 ml. of portal blood enter the liver each minute. This flows through the *hepatic sinuses* in close contact with the cords of liver parenchymal cells. Then it enters the *central veins* of the liver and from there flows into the vena cava.

In addition to the portal blood flow, approximately 350 ml. of blood flows into the liver each

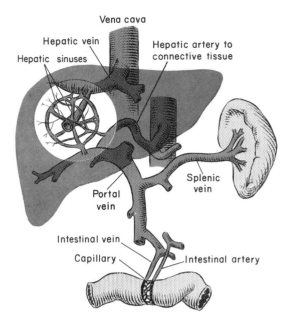

Figure 29–5. The portal and hepatic circulations.

minute through the hepatic artery, making a total hepatic flow of almost 1500 ml. per minute, or an average of 29 per cent of the total cardiac output. The hepatic arterial blood flow maintains nutrition of the connective tissue and especially of the wall of the bile ducts. Therefore, loss of hepatic arterial flow can be lethal because this often causes necrosis of the basic liver structures. The blood from the hepatic artery, after it supplies the structural elements of the liver, empties into the hepatic sinuses to mix with the portal blood.

Control of Liver Blood Flow. Three quarters of the blood flow through the liver is derived from the portal blood flow into the liver; this flow is controlled by the various factors that determine flow through the gastrointestinal tract and spleen — factors that will be discussed in subsequent sections of this chapter.

The additional one quarter of the blood flow through the liver is derived from the hepatic arteries; its flow rate is determined primarily by local metabolic factors in the liver itself. For instance, a decrease in oxygen in the hepatic blood causes an increase in hepatic arterial blood flow, indicating that the need to deliver nutrients to the liver tissues has a direct vasodilating effect.

Reservoir Function of the Liver. Because the liver is an expandable and compressible organ, large quantities of blood can be stored in its blood vessels. Its normal blood volume, including both that in the hepatic veins and hepatic sinuses, is about 500 ml., or 10 per cent of the total blood volume. However, when high pressure in the right atrium causes back pressure in the liver, the liver expands, and as much as 1 liter of extra blood occasionally is thereby stored in the hepatic veins and sinuses. This occurs especially in cardiac failure with peripheral congestion, which was discussed in Chapter 26.

Thus, in effect, the liver is a large expandable venous organ capable of acting as a valuable blood reservoir in times of excess blood volume and capable of supplying extra blood in times of diminished blood volume.

Control of Blood Storage in the Liver by the Sympathetic Nervous System. The storage portions of the liver, particularly the large veins and much less so the sinusoids, are vasoconstricted by sympathetic stimulation. Therefore, in times of circulatory stress when the sympathetic nervous system is discharging strongly, a large portion of blood stored in the liver is expelled into the general circulation within one to four minutes. In the normal human being this can amount to as much as 350 ml. of blood. Thus, the liver is probably the single most important source of blood to fill the other portions of the circulation in times of need, especially during heavy exercise or after severe hemorrhage.

Permeability of the Hepatic Sinuses. The hepatic sinuses are lined with an endothelium similar to that of the capillaries, but its permeability is extreme in comparison with that of usual capillaries — so much so that even the proteins of the blood diffuse into the extravascular spaces of the liver almost as easily as fluids. Fortunately, the pressure in the sinuses is only 6 to 8 mm. Hg so that most of the proteins that diffuse out of the sinuses also diffuse back in. Yet, the remainder passes into the liver lymphatics, so that liver lymph normally contains almost as great a protein concentration as that of the plasma itself. Also, the lymph from the liver accounts for one-half to two-thirds of the total body lymph flow.

This extreme permeability of the liver sinuses brings the fluids of the hepatic blood into extremely close contact with the liver parenchymal cells, thus facilitating rapid exchange of nutrient materials between the blood and the liver cells.

The Blood-Cleansing Function of the Liver. Blood flowing through the intestinal capillaries picks up many bacteria from the intestines. Indeed, a sample of blood from the portal system almost always grows colon bacilli when cultured, whereas growth of colon bacilli from blood in the systemic circulation is extremely rare. Special high-speed motion pictures of the action of Kupffer cells, the large phagocytic cells that line the hepatic sinuses, have demonstrated that these cells can cleanse blood extremely efficiently as it passes through the sinuses; when a bacterium comes into momentary contact with a Kupffer cell, in less than 0.01 second the bacterium passes inward through the wall of the Kupffer cell to become permanently lodged therein until it is digested. Probably not over 1 per cent of the bacteria entering the portal blood from the intestines succeeds in passing through the liver into the systemic circulation.

Measurement of Blood Flow Through the Liver in Man. Several indirect methods for measuring blood flow through the liver, similar to those used for measuring blood flow through the heart and brain, have been developed. The most important of these is the following: One of the several substances that are removed from the blood only by the liver is the dye *indocyanine green*. To measure the liver blood flow, this dye is infused into a vein at a continuous rate until its concentration becomes steady in the circulating blood. Then a sample of blood is removed from a peripheral vein while at the same time another sample of blood is removed from a catheter previously placed into the hepatic vein under fluoroscopic control. Because the liver removes much of the dye as the blood passes through it, there is a concentration difference between the systemic blood and the liver venous blood. Also, because the concentration in the systemic blood is constant, the rate of removal of the dye by the liver is equal to the rate of injection of dye. Consequently, the liver blood flow can be calculated using the *Fick equation* for flow that was explained for measurement of cardiac output in Chapter 23. That is:

$$\text{Liver blood flow} = \frac{\text{Rate of infusion of dye}}{\text{Concentration difference}}$$

BLOOD FLOW THROUGH THE INTESTINAL VESSELS

About four fifths of the portal blood flow originates in the intestines and stomach (about 850 ml. per minute), and the remaining one fifth originates in the spleen and pancreas. Over two thirds of the intestinal flow is to the mucosa to supply the energy needed for forming the intestinal secretions and for absorbing the digested food.

Control of Gastrointestinal Blood Flow. Blood flow in the gastrointestinal tract is controlled in almost exactly the same way as in most other areas of the body: by local regulatory mechanisms mainly. Furthermore, blood flow to the mucosa and submucosa, where the glands are located and where absorption occurs, is controlled separately from blood flow to the musculature. When glandular secretion increases, so does mucosal and submucosal blood flow. Likewise, when motor activity of the gut increases, blood flow in the muscle layers increases.

However, the precise mechanisms by which alterations in gastrointestinal activity alter the blood flow are not completely understood. It is known that decreased availability of oxygen to the gut increases local blood flow in the same way that this occurs elsewhere in the body, therefore local regulation of blood flow in the gut might occur entirely secondarily to changes in metabolic rate. On the other hand, it is also known that various peptide hormones are released from the mucosa of the intestinal tract during the digestive process and that these in turn cause mucosal vasodilatation. The best known of these hormones are *gastrin*, *secretin*, and *cholecystokinin*. Also, it has been claimed that some or all of the gastrointestinal glands form the substance *bradykinin* at the same time that they release their secretions. The bradykinin in turn has been postulated to cause mucosal vasodilatation. However, crucial experiments have not yet proved this mechanism.

Nervous Control of Gastrointestinal Blood Flow. Stimulation of the parasympathetic nerves (the vagi) to the *stomach* and *lower colon* increases local blood flow at the same time that it increases glandular secretion. However, this increased flow probably results from the increased glandular activity.

Sympathetic stimulation, in contrast, has a direct effect on essentially all blood vessels of the gastrointestinal tract to cause intense vasoconstriction. However, after a few minutes of this vasoconstriction, the flow returns to or almost to normal via a mechanism called "autoregulatory escape." That is, the local metabolic vasodilator mechanisms that are elicited by ischemia become prepotent over the sympathetic vasoconstriction and, therefore, redilate the arterioles, thus causing return of the necessary nutrient blood flow to the gastrointestinal glands and muscle.

A major value of sympathetic vasoconstriction in the gut is that it allows shutting off of splanchnic blood flow for short periods of time during heavy exercise when increased flow is needed by the skeletal muscle and heart.

The vasoconstriction of the intestinal and mesenteric *veins* that is caused by sympathetic stimulation does not "escape." Instead, both short and prolonged periods of sympathetic stimulation decrease the volume of these veins and thereby displace large amounts of blood into other parts of the circulation. In hemorrhagic shock or other states of low blood volume, this mechanism can provide several hundred milliliters of extra blood to sustain the general circulation.

PORTAL VENOUS PRESSURE

The liver offers a moderate amount of resistance to blood flow from the portal system to the vena cava. As a result, the pressure in the portal vein averages 8 to 10 mm. Hg, which is considerably higher than the almost zero pressure in the vena cava. Because of this high portal venous pressure, the pressures in the intestinal venules and capillaries have a much greater tendency to become abnormally high than is true elsewhere in the body.

Blockage of the Portal System. Frequently, extreme amounts of fibrous tissue develop within the liver structure, destroying many of the parenchymal cells and eventually contracting around the blood vessels, thereby greatly impeding the flow of portal blood through the liver. This disease process is known as *cirrhosis of the liver*. It results most frequently from alcoholism, but it can also follow ingestion of poisons such as carbon tetrachloride, virus diseases such as infectious hepatitis, or infectious processes in the bile ducts.

The portal system is also occasionally blocked by a large clot developing in the portal vein or in its major branches.

When the portal system is suddenly blocked, the return of blood from the intestines and spleen to the systemic circulation is tremendously impeded, the capillary pressure rising as much as 15 to 20 mm. Hg, and the patient often dies within a few hours because of excessive loss of fluid from the capillaries into the lumina and walls of the intestines.

Collateral Circulation from the Portal Veins to the Systemic Veins. When portal blood flow is blocked slowly rather than suddenly, such as occurs in slowly developing cirrhosis of the liver, large collateral vessels develop between the portal veins and the systemic veins. The most important of these are collaterals from the splenic to the esophageal veins; these collaterals frequently become so large that they protrude deeply into the lumen of the esophagus and are then called *esophageal varicosities*. The esophageal mucosa overlying these varicosities eventually becomes eroded; and in many if not most of these patients, the erosion finally penetrates all the way into the varicosity and causes the person to bleed severely, often to bleed to death.

Ascites as a Result of Portal Obstruction. Ascites is free fluid in the peritoneal cavity. It results from exudation of fluid either from the surface of the liver or from the surfaces of the gut and its mesentery. Ascites usually will develop only in case outflow of blood from the liver into the inferior vena cava is blocked. This causes extremely high pressure in the liver sinusoids, which in turn causes fluid to weep from the surfaces of the liver. The weeping fluid is almost pure plasma, containing tremendous quantities of protein. The protein, because it causes a high colloid osmotic pressure in the abdominal fluid, then pulls by osmosis additional fluid from the surfaces of the gut and mesentery.

On the other hand, obstruction of the portal vein, without directly involving the liver, rarely causes ascites. If obstruction occurs acutely, the person is likely to die of shock within hours because of fluid loss into the gut; if it occurs slowly, collateral vessels can usually develop enough to prevent either the intestinal loss of fluid or the development of ascites.

THE SPLENIC CIRCULATION

The Spleen as a Reservoir. The capsule of the spleen in many lower animals contains large amounts of smooth muscle; sympathetic stimulation causes intense contraction of the spleen. Conversely, sympathetic inhibition results in considerable splenic expansion with consequent storage of blood.

In humans, the splenic capsule is nonmuscular, but even so, dilatation of vessels within the spleen can still cause the spleen to store several hundred milliliters of blood at times, and then, under the influence of sympathetic stimulation, constriction of the vessels will express most of this blood into the general circulation. Unfortunately, these ef-

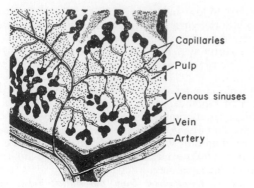

Figure 29–6. The functional structures of the spleen. (Modified from Bloom and Fawcett: Textbook of Histology. Philadelphia, W. B. Saunders Company, 1975.)

fects are poorly understood; but the spleen is so small, only 150 to 200 ml., that this reservoir function in humans is of relatively little importance.

As illustrated in Figure 29–6, two areas exist in the spleen for the storage of blood: the venous sinuses and the pulp. Small vessels flow directly into the venous sinuses, and when the spleen distends, the venous sinuses swell, thus storing blood.

In the splenic pulp, the capillaries are very permeable, so that much of the blood passes first into the pulp and then oozes through this before entering the venous sinuses. As the spleen enlarges, many cells become stored in the pulp. Therefore, the net quantity of red blood cells in the general circulation decreases slightly when the spleen enlarges. The spleen can store enough cells that splenic contraction can cause the hematocrit of the systemic blood to increase in humans as much as 1 to 2 per cent and as much as 3 to 4 per cent in some lower animals.

This ability of the spleen to store and release blood is important, at least in lower animals, in times of stress such as during strenuous exercise, for the circulatory system then needs an increased volume of blood so that rapid return of blood to the heart can be maintained. At the same time, the blood that flows through the muscles needs to carry larger quantities of oxygen than normally. Therefore, the increased hematocrit of the circulating blood resulting from splenic contraction is also an aid to the body during these periods of stress.

The Blood-Cleansing Function of the Spleen-Removal of Old Cells. Blood passing through the splenic pulp before it enters the sinuses undergoes thorough squeezing. Indeed, it is believed that the red blood cells re-enter the venous sinuses from the pulp by "diapedesis"; that is, by squeezing through pores much smaller than the size of the cells themselves. Under these circumstances it is

to be expected that fragile red blood cells would not withstand the trauma. For this reason, many of the red blood cells destroyed in the body have their final demise in the spleen. After the cells rupture, the released hemoglobin and the cell stroma are ingested by the reticuloendothelial cells of the spleen.

Reticuloendothelial Cells of the Spleen. The pulp of the spleen contains many large phagocytic reticuloendothelial cells, and the venous sinuses are lined with similar cells. These cells act as a cleansing system for the blood, similar to that in the venous sinuses of the liver. When the blood is invaded by infectious agents, the reticuloendothelial cells of the spleen rapidly remove debris, bacteria, parasites, etc. Also, in many infectious processes the spleen enlarges in the same manner that lymph glands enlarge and then performs its cleansing function even more adequately.

Much of the spleen is filled with *white pulp*, which is in reality a large quantity of lymphocytes and plasma cells. These function in exactly the same way in the spleen as in the lymph glands to cause either humoral or lymphocytic immunity against toxins, bacteria, and so forth, as described in Chapter 7.

The Spleen as a Hemopoietic Organ. During fetal life, the splenic pulp produces blood cells in exactly the same manner that the red bone marrow in the adult produces cells. As the normal fetus approaches birth, the spleen normally loses this ability to produce cells, but, in some diseases, the spleen continues to produce cells even after birth. For instance, in the disease *erythroblastosis fetalis,* which results from excessive destruction of red blood cells by abnormal antibodies in the plasma, as discussed in Chapter 5, the fetus must produce 10 or more times as many red blood cells as normally. As a result, the hemopoietic function of the spleen persists for several weeks after birth.

CIRCULATION IN THE SKIN

PHYSIOLOGIC ANATOMY OF THE CUTANEOUS CIRCULATION

Circulation through the skin subserves two major functions: first, *nutrition of the skin tissues* and, second, *conduction of heat* from the internal structures of the body to the skin so that the heat can be removed from the body. To perform these two functions the circulatory apparatus of the skin is characterized by two major types of vessels, illustrated diagrammatically in Figure 29–7: (1) the usual nutritive arteries, capillaries, and veins and (2) vascular structures concerned with heating the skin, consisting principally of (a) an extensive *subcutaneous venous plexus,* which holds large quantities of blood that can heat the surface of the skin, and (b) in some skin areas, *arteriovenous anastomoses,* which are large vascular communications directly between the arteries and the venous plexuses. The walls of these anastomoses

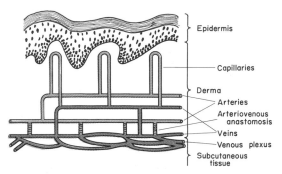

Figure 29-7. The skin circulation.

have strong muscular coats innervated by sympathetic vasoconstrictor nerve fibers that secrete norepinephrine. When constricted, they reduce the flow of blood into the venous plexuses to almost nothing; or when maximally dilated, they allow extremely rapid flow of warm blood into the plexuses. The arteriovenous anastomoses are found principally in the volar surfaces of the hands and feet, the lips, the nose, and the ears, which are areas of the body most often exposed to maximal cooling.

Rate of Blood Flow Through the Skin. The rate of blood flow through the skin is among the most variable of any part of the body, because the flow required to regulate body temperature changes markedly in response to, first, the rate of metabolic activity of the body, and, second, the temperature of the surroundings. This will be discussed in detail in Chapter 72. The blood flow required for nutrition is slight, so that this plays almost no role in controlling normal skin blood flow. At ordinary skin temperatures, the amount of blood flowing through the skin vessels to subserve heat regulation is about 10 times as much as that needed to supply the nutritive needs of the tissues. But, when the skin is exposed to extreme cold, the blood flow may become so slight that nutrition begins to suffer — even to the extent, for instance, that the fingernails grow considerably more slowly in arctic climates than in temperate climates.

Under ordinary cool conditions the blood flow to the skin is about 0.25 liter/sq. meter of body surface area, or a total of about 400 ml. per minute in the average adult. This can decrease to as little as 50 ml. per minute in severe cold, and, when the skin is heated until maximal vasodilatation has resulted, the blood flow can be as much as 2.8 liters per minute, thus illustrating both the extreme variability of skin blood flow and the great drain on cardiac output that can occur under hot conditions. Indeed, many persons with borderline cardiac failure develop severe failure in hot weather because of the extra load on the heart and then revert from failure in cool weather.

REGULATION OF BLOOD FLOW IN THE SKIN

Nervous Control of Cutaneous Blood Flow. Since most of the blood flow through the skin is to control body temperature, and since this function in turn is regulated by the nervous system, the blood flow through the skin is principally regulated by nervous mechanisms rather than by local regulation, which is opposite to the regulation in most parts of the body.

Temperature Control Center of the Hypothalamus. Located in the preoptic region of the anterior hypothalamus is a small center that is capable of controlling body temperature. Heating this area causes vasodilation of essentially all the skin vessels of the body and also causes sweating. Cooling of the center causes vasoconstriction and cessation of sweating. The detailed function of this center will be discussed in Chapter 72 in relation to body temperature. The important point in the present discussion is that the hypothalamus controls blood flow through the skin in response to changes in body temperature by two mechanisms: (1) a vasoconstrictor mechanism and (2) a vasodilator mechanism.

The Vasoconstrictor Mechanism. The skin throughout the body is supplied with sympathetic vasoconstrictor fibers that secrete norepinephrine at their endings. This constrictor system is extremely powerful in the feet, hands, lips, nose, and ears, which are the areas most frequently exposed to severe cold and which are also the areas where large numbers of arteriovenous anastomoses are found. At normal body temperature the sympathetic vasoconstrictor nerves keep these anastomoses almost totally closed, but when the body becomes overly heated the number of sympathetic impulses is greatly reduced so that the anastomoses dilate and allow large quantities of warm blood to flow into the venous plexuses, thereby promoting loss of heat from the body.

In the remainder of the body — that is, over the surfaces of the arms, legs, and body trunk — almost no arteriovenous anastomoses are present, but nevertheless vasoconstrictor control of the nutritive vessels can still effect major changes in blood flow. When the body becomes overheated, the sympathetic vasoconstrictor impulses cease, and the blood flow to the skin vessels increases about two-fold.

Extreme Sensitivity of the Skin Blood Vessels to Circulating Norepinephrine and Epinephrine. In addition to the direct sympathetic vasoconstrictor effect in the skin, the skin blood vessels are also extremely sensitive to circulating norepinephrine and epinephrine. Therefore, even in areas of the skin that might have lost their sympathetic innervation, any mass discharge of the sympathetic nervous system will still cause intense skin vaso-constriction. Sometimes the sensitivity of the skin to circulating norepinephrine and epinephrine is so great that the vasoconstriction can actually damage the skin. This response will be discussed in more detail later in the chapter in relation to *Raynaud's disease*.

The Vasodilator Mechanism. When the body temperature becomes excessive and sweating begins to occur, the blood flow through the skin of the forearms and trunk increases an additional three-fold, which occurs as a result of so-called "active" vasodilatation of the vessels. However, the basic mechanism by which this active dilation occurs is not completely known. Yet this additional increase in blood flow does not occur except in the presence of sweating, and it does not occur in lower animals that do not have sweat glands. Therefore, it has been postulated that the sympathetic fibers that secrete acetylcholine to activate the sweat glands cause a secondary vasodilatation as follows: The increased activity of the sweat glands is postulated to cause these glands to release an enzyme called *kallikrein*, which in turn splits the polypeptide *bradykinin* from globulin in the interstitial fluids. Bradykinin in turn is a powerful vasodilator substance that could account for the greatly increased blood flow when sweating begins to occur. In opposition to this theory, however, is the fact that inhibition of the bradykinin mechanism does not completely block the increased blood flow associated with sweating.

Effect of Cold on the Skin Circulation. When cold is applied directly to the skin, the skin vessels constrict more and more down to a temperature of about 15° C., at which point they reach their maximum degree of constriction. This constriction results primarily from increased sensitivity of the vessels to nerve stimulation, but it probably also results at least partly from a reflex that passes to the cord and then back to the vessels. At temperatures below 15° C., the vessels begin to dilate. This dilation is caused by a direct local effect of the cold on the vessels themselves — probably paralysis of the contractile mechanism of the vessel wall or block of the nerve impulses coming to the vessels. In any event, at temperatures approaching 0° C. the vessels frequently reach maximum vasodilation. This intense vasodilatation in severe degrees of cold plays a purposeful role in preventing freezing of the exposed portions of the body, particularly the hands and ears.

Local Regulation of Blood Flow in the Skin. Though local regulation of blood flow usually plays only a small role in skin blood flow control, it does have an effect in those few instances in which skin blood flow becomes greatly decreased. For instance, if one sits on his buttocks for 30 minutes or more and then stands, he will note intense reddening of the affected skin. That is, the blood flow to the skin area has now increased to a marked extent, which is typical *reactive hyperemia* resulting from diminished availability of nutrients to the tissues during the period of compression. Thus, the local regulatory mechanism present in essentially all other tissues of the body is also present in the skin and can be called into play when needed to prevent nutritional damage to the tissues.

Shift of Blood from the Skin to the Remainder of the Circulation in Times of Circulatory Stress. The skin venous plexuses of the entire body, including those of the hands, feet, arms, legs, and trunk, are all strongly supplied with sympathetic vasoconstrictor innervation. In times of circulatory stress, such as during exercise, following severe hemorrhage, or even in states of anxiety, sympathetic stimulation of these venous plexuses can force large quantities of blood, estimated to be as much as 5 to 10 per cent of the blood volume, into the internal vessels. Thus, the subcutaneous veins of the skin act as an important blood reservoir, often providing blood to subserve other circulatory functions when needed.

COLOR OF THE SKIN IN RELATION TO SKIN TEMPERATURE

Much of the skin color is due to the color of the blood in the skin capillaries and veins. Therefore, when the skin is hot and arterial blood is flowing rapidly through these vessels, the skin is usually red. On the other hand, when the skin is cold and the blood is flowing extremely slowly, most of the oxygen is removed for nutritive purposes before the blood can leave the capillaries. Consequently, the capillaries and veins then contain large amounts of dark, deoxygenated blood that gives the skin a bluish hue.

Another cutaneous vascular effect that often affects skin color is severe constriction of the cutaneous vessels, which expresses most of the blood out of the skin into other parts of the circulation. When this occurs the skin takes on the color of the subcutaneous connective tissue, which is composed principally of collagen fibers and has a whitish hue. Thus, in conditions in which the cutaneous vessels become greatly constricted, the skin has an ashen white pallor.

PHYSIOLOGY OF VASCULAR DISEASES OF THE LIMBS

RAYNAUD'S DISEASE

The local vasoconstrictor reflex that occurs normally during exposure to cold becomes especially marked in the limbs of some persons, particularly in many women who live in cold climates. This effect, called *Raynaud's disease,* occurs most frequently in the hands, though occasionally also in the feet. When the hands become even slightly cooled, they are likely to exhibit extreme vascular spasm — so much so that the hands do not receive sufficient flow of blood even for maintaining adequate metabolism. The tissues then develop *ischemic pain* in exactly the same manner that the heart develops pain in angina pectoris and as other tissues develop pain during ischemia. After the ischemia persists for several minutes, *hyperemia* often supervenes, causing dilatation of all the vessels of the hands so that the final stage is sometimes *intense reddening* of the hands. This effect is probably caused by the same factors that cause reactive hyperemia following temporary vascular occlusion, as discussed in Chapter 20. Strangely enough, the pain becomes even more severe at this moment, probably because of mobilization of the abnormal metabolic products that cause pain.

Though Raynaud's disease in its early stages is merely a physiologic entity that passes through the phases of blanching of the skin, pain, and reddening of the skin, in some patients it progresses over a period of years until the vessels become continuously constricted. In severe cases the blood flow can become decreased so much that the fingers become gangrenous.

Sympathectomy for Raynaud's Disease. To treat Raynaud's disease the sympathetic reflex is often blocked. This is done best by cutting the preganglionic fibers in the sympathetic chain at T-2 and T-3, which interrupts the sympathetic nerve impulses from the spinal cord to the hand. The postganglionic fibers to the hand originate mainly in the stellate ganglion, but the stellate ganglion is not removed, for removal of postganglionic sympathetic fibers allows the blood vessels to become excessively sensitized to circulating norepinephrine and epinephrine. If this happens the hands will still exhibit Raynaud's syndrome every time the adrenal glands are excited. This phenomenon of sensitization will be discussed in Chapter 57 in connection with the various reactions of the sympathetic nervous system.

BUERGER'S DISEASE

Buerger's disease, also known as *thromboangiitis obliterans,* is caused by inflammatory lesions of the small blood vessels in the triangular sheaths containing the nerve, the artery, and the vein. This condition occurs most frequently in the legs but occasionally in the arms. Even though the disease is caused by an inflammatory process that constricts the vessels, it is made much worse by the effects of nicotine incident to smoking.

The patient with Buerger's disease suffers a symptom known as *intermittent claudication* — that is, when walking, the calves become extremely painful because of ischemic products collecting in the muscles and not removed by adequate blood flow. The nature of the ischemic pain is similar to that of Raynaud's disease and also to the ischemic cardiac pain of angina pectoris but involves the muscles.

Sympathectomy for Buerger's Disease. People with Buerger's disease can be benefited greatly by sectioning the sympathetic nerves to the area afflicted. This is done for a leg by removing the L-1, L-2, and L-3 ganglia of the sympathetic chain. Even though the pathologic changes of Buerger's disease are not caused by excessive sympathetic activity, removal of the sympathetic nerves blocks normal vasomotor tone, which allows vasodilatation in the affected areas. Approximately 90 per cent of the patients who otherwise might lose toes or even legs from Buerger's disease can save these by sympathectomy.

PERIPHERAL ARTERIOSCLEROSIS

Arteriosclerosis, which causes intra-arterial plaques to develop, has already been discussed in connection with the coronary arteries. This same process can occur throughout the arterial tree and not only in the coronaries. Therefore, in old age many of the peripheral arteries will often have their intimal surfaces almost completely covered with lipid deposits or calcified plaques.

Progressive obliteration of the peripheral arterial tree by arteriosclerosis results in all the symptoms that one would expect. It causes pain in the muscles during activity and also in other tissues because of ischemia, and it is likely to cause gangrene and ulceration of the skin because of poor blood supply.

A special difference between arteriosclerotic peripheral vascular disease and either Buerger's disease or Raynaud's disease is that sympathectomy is not nearly so effective for treatment as it is in the other two diseases. This is true probably because this vascular disease is caused by obstruction of the large arteries rather than the small arteries, and the small arteries, which are affected most by the sympathetics, probably remain maximally vasodilated in severe arteriosclerosis whether the sympathetics are removed or not.

REFERENCES

Abramson, D. I.: Vascular Disorders of the Extremities, 2nd Ed. Hagerstown, Md., Harper & Row, 1974.

Abramson, D. I. (ed.): Circulatory Diseases of the Limbs: A Primer. New York, Grune & Stratton, 1978.

Appenzeller, O., and Atkinson, R. (eds.): Health Aspects of Endurance Training. New York, S. Karger, 1978.

Apple, D. F., Jr., and Cantwell, J. D.: Medicine for Sport. Chicago, Year Book Medical Publishers, 1979.

Barker, W. F.: Peripheral Arterial Disease, 2nd Ed. Philadelphia, W. B. Saunders Co., 1975.

Betz, E.: Cerebral blood flow: Its measurement and regulation. *Physiol. Rev., 52*:595, 1972.

Bevegard, B. S., and Shepherd, J. T.: Regulation of the circulation during exercise in man. *Physiol. Rev., 47*:178, 1967.

Boullin, D. J. (ed.): Cerebral Vasospasm. New York, John Wiley & Sons, 1980.

Cerebrovascular Disorders: A Clinical and Research Classification. Albany, N.Y., World Health Organization, 1978.

Chapman, C. B.: The physiology of exercise. *Sci. Am. 212*:88, 1965.

Clarke, D. H.: Exercise Physiology. Englewood Cliffs, N.J. Prentice-Hall, 1975.

Fein, J. M.: Microvascular surgery for stroke. *Sci. Am., 238*(4), 1978.

Fox, E. L.: Sports Physiology. Philadelphia, W. B. Saunders Co., 1979.

Goldstein, M., *et al.* (eds.): Cerebrovascular Disorders and Stroke. New York, Raven Press, 1979.

Grayson, J.: The gastrointestinal circulation. *In* MTP International Review of Science: Physiology. Vol. 4. Baltimore, University Park Press, 1974, p. 105.

Green, H. D., and Kepchar, J. H.: Control of peripheral resistance in major systemic vascular beds. *Physiol. Rev., 39*:617, 1959.

Greenleaf, J. E.: Hyperthermia and exercise. *In* Robertshaw, D. (ed.): International Review of Physiology: Environmental Physiology III. Vol. 20. Baltimore, University Park Press, 1979, p. 157.

Greenway, C. V., and Stark, R. D.: Hepatic vascular bed. *Physiol. Rev., 51*:23, 1971.

Guyton, A. C., *et al.*: Cardiac Output and Its Regulation. Philadelphia, W. B. Saunders Co., 1973.

Haddy, F. J., and Scott, J. B.: Active hyperemia, reactive hyperemia, and autoregulation of blood flow. *In* Kaley, G., and Altura, B. M. (eds.): Microcirculation. Vol. II. Baltimore, University Park Press, 1977.

Heistad, D. D., and Abboud, F. M.: Factors that influence blood flow in skeletal muscle and skin. *Anesthesiology, 41*:139, 1974.

Helwig, E. B., and Mostofi, F. K. (eds.): The Skin. New York, R. E. Krieger Co., 1980.

Jacobson, E. D.: The gastrointestinal circulation. *Annu. Rev. Physiol., 30*:133, 1968.

Jensen, C. R., and Fisher, G.: Scientific Basis of Athletic Conditioning. Philadelphia, Lea & Febiger, 1978.

Jonsson, B., and Naughton, J.: Exercise cardiology. *In* Hayase, S., and Murao, S. (eds.): Cardiology: Proceedings of the VIII World Congress of Cardiology, Tokyo, 1978. New York, Elsevier/North-Holland, 1979, p. 175.

Juergens, J. L., *et al.* (eds.): Allen, Barker, Hines Peripheral Vascular Disease, 5th Ed. Philadelphia, W. B. Saunders Co., 1980.

Korner, P. I.: Control of blood flow to special vascular areas: Brain, kidney, muscle, skin, liver, and intestine. *In* MTP International Review of Science: Physiology. Vol. 1. Baltimore, University Park Press, 1974, p. 123.

Kuschinsky, W., and Wahl, M.: Local chemical and neurogenic regulation of cerebral vascular resistance. *Physiol. Rev., 58*:656, 1978.

Lassen, N. A.: Control of cerebral circulation in health and disease. *Circ. Res., 34*:749, 1974.

Lassen, N. A.: Study of local cerebral blood flow. *In* Dickinson, C. J., and Marks, J. (eds.): Developments in Cardiovascular Medicine. Lancaster, England, MTP Press, 1978, p. 9.

Lassen, N. A., *et al.*: Brain function and blood flow. *Sci. Am., 239*(4):62, 1978.

Lundgren, O., and Jodal, M.: Regional blood flow. *Annu. Rev. Physiol., 37*:395, 1975.

Macpherson, A. I. S., *et al.*: The Spleen. Springfield, Ill., Charles C Thomas, 1973.

Mannick, J. A. (ed.): Symposium on Peripheral Vascular Surgery. Philadelphia, W. B. Saunders Co., 1979.

McHenry, L. C., Jr.: Cerebral Circulation and Stroke. St. Louis, W. H. Green, 1978.

Meyer, J. S., *et al.* (eds.): Research on the Cerebral Circulation. Springfield, Ill., Charles C Thomas, 1973.

Moskalenko, Y. E., *et al.*: Biophysical Aspects of Cerebral Circulation. New York, Pergamon Press, 1979.

Nicoll, P. A., and Cortese, T. A., Jr.: The physiology of skin. *Annu. Rev. Physiol., 34*:177, 1972.

Purves, M. J.: The Physiology of the Cerebral Circulation. (Physiological Society Monographs). New York, Cambridge University Press, 1972.

Rappaport, A. M.: Hepatic blood flow: Morphologic aspects and physiologic regulation. *In* Javitt, N. B. (ed.): International Review of Physiology: Liver and Biliary Tract Physiology I. Vol. 21. Baltimore, University Park Press, 1980, p. 1.

Rosell, S., and Belfrage, E.: Blood circulation in adipose tissue. *Physiol. Rev., 59*:1078, 1979.

Smith, E. E., *et al.*: Integrative mechanisms and cardiovascular responses and control in exercise in the normal human. *Prog. Cardiovasc. Dis., 18*:421, 1976.

Strauss, R. H. (ed.): Sports Medicine and Physiology. Philadelphia, W. B. Saunders Co., 1979.

Svanik, J., and Lundgren, O.: Gastrointestinal circulation. *Int. Rev. Physiol., 12*:1, 1977.

Wahren, J.: Metabolic adaptation to physical exercise in man. *In* DeGroot, L. J., *et al.* (eds.): Endocrinology. Vol. 3. New York, Grune & Stratton, 1979, p. 1911.

Wilkins, R. H. (ed.): Cerebral Vasospasm. Baltimore, Williams & Wilkins, 1980.

Part VI

THE BODY FLUIDS
AND KIDNEYS

30

Capillary Dynamics, and Exchange of Fluid Between the Blood and Interstitial Fluid

In the capillaries the most purposeful function of the circulation occurs, namely, interchange of nutrients and cellular excreta between the tissues and the circulating blood. About 10 billion capillaries, having a total surface area probably 500 to 700 square meters, provide this function. Indeed, it is rare that any single functional cell of the body is more than 20 to 30 microns away from a capillary.

The purpose of this chapter is to discuss the transfer of substances between the blood and interstitial fluid and especially to discuss the factors that affect the transfer of fluid volume itself between the circulating blood and the interstitial fluids.

STRUCTURE OF THE CAPILLARY SYSTEM

Figure 30–1 illustrates the structure of a "unit" capillary bed as seen in the mesentery, illustrating that blood enters the capillaries through an *arteriole* and leaves by way of a *venule*. Blood from the arteriole passes into a series of *metarterioles*, which have a

structure midway between that of arterioles and capillaries. After leaving the metarteriole, the blood enters the *capillaries,* some of which are large and are called *preferential channels* and others of which are small and are *true capillaries*. After passing through the capillaries the blood enters the venule and returns to the general circulation.

The arterioles are highly muscular, and their diameters can change many-fold, as was discussed in Chapter 18. The metarterioles do not have a continuous muscular coat, but smooth muscle fibers encircle the vessel at intermediate points, as illustrated in Figure 30–1 by the large black dots to the sides of the metarteriole.

At the point where the true capillaries originate from the metarterioles a smooth muscle fiber usually encircles the capillary. This is called the *precapillary sphincter*. This sphincter can open and close the entrance to the capillary.

The venules are considerably larger than the arterioles and have a much weaker muscular coat. Yet, it must be remembered that the pressure in the venules is much less than that in the arterioles so that the venules can still contract considerably.

This typical arrangement of the capillary bed is not found in all parts of the body; however, some similar arrangement is found, beginning with arterioles, passing through metarterioles and capillaries, and returning through venules.

Structure of the Capillary Wall. Figure 30–2 illustrates the ultramicroscopic structure of the capillary wall. Note that the wall is composed of a unicellular layer of endothelial cells and is surrounded by a basement membrane on the outside. The total thickness of the wall is about 0.5 micron.

The diameter of the capillary is 5 to 9 microns, barely large enough for red blood cells and other blood cells to squeeze through.

"Pores" in the Capillary Membrane. Studying Figure 30–1, one sees two minute passageways connecting the interior of the capillary with the exterior. One of these is the *intercellular cleft*

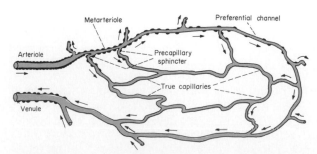

Figure 30–1. Structure of the mesenteric capillary bed. (From Zweifach: Factors Regulating Blood Pressure. New York, Josiah Macy, Jr., Foundation, 1950.)

358

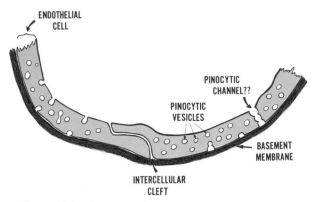

Figure 30–2. Structure of the capillary wall. Note especially the *intercellular cleft* at the junction between adjacent endothelial cells; it is believed that most water-soluble substances diffuse through the capillary membrane along this cleft.

which is a thin slit that lies between adjacent endothelial cells. The cells are held apart approximately 6 to 7 nanometers (60 to 70 Angstroms) by loose reticular fibrillae composed of proteoglycans, mainly hyaluronic acid. These intercellular clefts are located only at the edges of the endothelial cells and therefore represent no more than one-thousandth of the total surface area of the capillary. Nevertheless, it is believed that most water-soluble ions and molecules pass between the interior and exterior of the capillary through these slit-pores.

Also present in the endothelial cells are many pinocytic vesicles. These form at one surface of the cell and move to the opposite surface where they discharge their contents, often carrying large molecules and even solid particles through the capillary membrane. Note also in Figure 30–2 that some of these pinocytic vesicles occasionally coalesce with each other and form a continuous channel through the endothelial membrane as illustrated by the so-called *pinocytic channel* shown to the right in the figure. However, it is still doubted whether large amounts of substances pass through the capillary membrane via such "pores" as this.

In addition, in some capillary membranes, such as the membranes of the glomerular tufts in the kidney, many large *fenestrae,* which are small oval windows, develop through the very middle of the endothelial cells and allow passage of tremendous quantities of substances.

FLOW OF BLOOD IN THE CAPILLARIES — VASOMOTION

Blood usually does not flow at a continuous rate through the capillaries. Instead, it flows intermittently. The cause of this intermittency is the phenomenon called *vasomotion,* which means intermittent contraction of the metarterioles and precapillary sphincters. These constrict and relax in an alternating cycle usually 5 to 10 times per minute.

Regulation of Vasomotion. The most important factor found thus far to affect the degree of opening and closing of the metarterioles and precapillary sphincters is the concentration of *oxygen* in the tissues. When the oxygen concentration is very low, the intermittent periods of blood flow occur more often, and the duration of each period of flow lasts for a longer time, thereby allowing the blood to carry increased quantities of oxygen (as well as other nutrients) to the tissues. It follows also that the greater the use of oxygen by the tissue, the greater will be the amount of blood that flows. Thus, by this intermittent opening and closing of the precapillary sphincters and metarterioles, the blood flow to the tissue is *autoregulated,* as was discussed in Chapter 20.

AVERAGE FUNCTION OF THE CAPILLARY SYSTEM

Despite the fact that blood flow through each capillary is intermittent, so many capillaries are present in the tissues that their overall function becomes averaged. That is, there is an *average rate of blood flow* through each tissue capillary bed, an *average capillary pressure* within the capillaries, an *average rate of transfer of substances* between the blood of the capillaries and the surrounding interstitial fluid. In the remainder of this chapter, we will be concerned with these averages, though one must remember that the average functions are in reality the functions of literally billions of individual capillaries, each operating intermittently.

EXCHANGE OF NUTRIENTS AND OTHER SUBSTANCES BETWEEN THE BLOOD AND INTERSTITIAL FLUID

DIFFUSION THROUGH THE CAPILLARY MEMBRANE

By far the most important means by which substances are transferred between the plasma and interstitial fluids is by diffusion. Figure 30–3 illustrates this process, showing that as the blood traverses the capillary, tremendous numbers of water molecules and dissolved particles diffuse back and forth through the capillary wall, providing continual mixing between the interstitial fluids and the plasma, as was explained in Chapter 4. Diffusion results from thermal motion of the water molecules and the dissolved substances in the fluid, the different particles moving first in one direc-

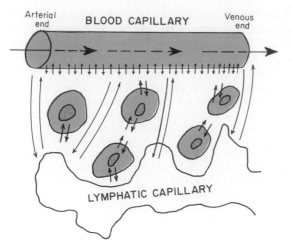

Figure 30–3. Diffusion of fluid and dissolved substances between the capillary and interstitial fluid spaces.

tion, then another, moving randomly in every direction.

Diffusion of Lipid-Soluble Substances Through the Capillary Membrane. If a substance is lipid-soluble, it can diffuse directly through the cell membranes of the capillary without having to go through the pores. Such substances include especially oxygen and carbon dioxide. Since these can permeate all areas of the capillary membrane, their rates of transport through the capillary membrane are about 2 times the rate for water and many times the rates for most lipid-insoluble substances such as sodium ions, glucose, and so forth.

Other lipid-soluble substances that are rapidly transported through the capillary membrane include various anesthetic gases and alcohol.

Diffusion of Water Molecules Through the Capillary Membrane. After the lipid-soluble molecules, water molecules are the next most rapid to diffuse through the capillary membrane. Much of the water diffusion occurs directly through the endothelial cell wall, first into the intracellular fluid of the endothelial cell and then out the membrane on the other side of the cell. However, water diffusion occurs through the "pores" in the capillary membrane as well, especially through the intercellular clefts.

Diffusion of Water-Soluble, Lipid-Insoluble Substances Through the Capillary Membrane. Many substances needed by the tissues are soluble in water but cannot pass through the lipid membranes of the endothelial cells; such substances include sodium ions, chloride ions, glucose, and so forth. These substances diffuse between the plasma and interstitial fluids only through the capillary pores, which themselves are filled with water.

Despite the fact that not over $1/1000$ of the surface area of the capillaries is represented by the inter-

cellular junctions, the velocity of thermal motion is so great that even this small area is sufficient to allow tremendous diffusion of water and water-soluble substances through these pores. To give one an idea of the extreme rapidity with which substances diffuse, *the rate at which water molecules diffuse through the capillary membrane is approximately 80 times as great as the rate at which plasma itself flows linearly along the capillary.* That is, the water of the plasma is exchanged with the water of the interstitial fluids 80 times before the plasma can go the entire distance through the capillary. Yet, despite this rapid rate of diffusion, the rates of diffusion out of the capillary and into the capillary are so nearly equal that the rate of net movement of fluid volume through the capillaries is thousands of times less than the rate of two-way diffusional exchange.

Effect of Molecular Size on Pore Permeability. The width of the capillary intercellular slit-pores, 6 to 7 nanometers, is about 20 times the diameter of the water molecule, which is the smallest molecule that normally passes through the capillary pores. On the other hand, the diameters of plasma protein molecules are slightly greater than the width of the pores. Other substances, such as sodium ions, chloride ions, glucose, and urea, have intermediate diameters. Therefore, it is obvious that the permeability of the capillary pores for different substances will vary according to their molecular diameters.

Table 30–1 gives the relative permeabilities of the capillary pores in muscle for substances commonly encountered by the capillary membrane, illustrating for instance that the permeability for glucose molecules is 0.6 times as great as that for water molecules, while the permeability for albumin molecules is less than $1/10,000$ that for water molecules. Thus, the membrane is almost impermeable to albumin, which causes a significant concentration difference to develop between the albumin of the plasma and that of the interstitial fluid, as will become evident later in the chapter.

However, a word of caution must be issued at

TABLE 30–1 RELATIVE PERMEABILITY OF MUSCLE CAPILLARY PORES TO DIFFERENT-SIZED MOLECULES

Substance	Molecular Weight	Permeability
Water	18	1.00
NaCl	58.5	0.96
Urea	60	0.8
Glucose	180	0.6
Sucrose	342	0.4
Inulin	5,000	0.2
Myoglobin	17,600	0.03
Hemoglobin	68,000	0.01
Albumin	69,000	<0.0001

Modified from Pappenheimer.

this point: The capillaries in different tissues have extreme differences in their permeabilities. For instance, the membrane of the liver capillary sinusoids is so permeable that even plasma proteins leak through the walls of these sinusoids almost as easily as water and other substances. Also, the permeability of the renal glomerular membrane for water and electrolytes is about 500 times the permeability of the muscle capillaries, but the glomerular and muscle permeabilities for protein are about the same. When we study these different organs later in this text, it will become clear why some tissues require greater degrees of capillary permeability than others — the liver, for instance, to transfer tremendous amounts of nutrients between the blood and the liver parenchymal cells, and the kidneys to allow filtration of large quantities of fluid for the formation of urine.

Effect of Concentration Difference on Net Rate of Diffusion Through the Capillary Membrane. As explained in Chapter 4, the "net" rate of diffusion of a substance through any membrane is proportional to the *concentration difference* between the two sides of the membrane. That is, the greater the difference between the concentrations of any given substance on the two sides of the capillary membrane, the greater will be the net movement of the substance through the membrane. Thus, the concentration of oxygen in the blood is normally greater than that in the interstitial fluids. Therefore, large quantities of oxygen normally move from the blood toward the tissues. Conversely, the concentration of carbon dioxide is greater in the tissues than in the blood, which obviously causes carbon dioxide to move into the blood and to be carried away from the tissues.

Fortunately, the rates of diffusion through the capillary membranes of most nutritionally important substances are so great that only slight concentration differences suffice to cause more than adequate transport between the plasma and interstitial fluid. For instance, the concentration of oxygen in the interstitial fluid immediately outside the capillary is probably no more than 1 per cent less than the concentration in the blood, and yet this 1 per cent difference causes enough oxygen to move from the blood into the interstitial spaces to provide all the oxygen required for tissue metabolism.

Transport Through the Capillary Membrane by Pinocytosis

Because the intercellular clefts (where the capillary endothelial cells join each other) have a width of only 6 to 7 nanometers, plasma proteins, which have a minimum diameter of about 8 nanometers, normally cannot pass by this route from the interior to the exterior of the capillary. Furthermore, many much larger substances in the plasma, such as lipoprotein molecules, large polysaccharide molecules like dextran, and proteoglycan molecules, certainly cannot go through these slit-pores. Yet, physiological studies have shown that small quantities of all these substances do pass through the capillary membranes. On the basis of electron microscopic studies, it is believed that a large share of this passage, if not nearly all of it, occurs by means of pinocytosis. Basically, this process means that the endothelial cells of the capillary wall ingest small amounts of plasma and form many small intracellular pinocytic vesicles; these vesicles then migrate from the inner surface of the endothelial cell to the outer surface where the plasma and its contents are released. Under some conditions such vesicles can move from one side of the endothelial membrane to the other side in as little as a few seconds. Therefore, it is likely that a significant amount of the large molecular weight substances are transported through the capillary membrane by means of pinocytosis.

Effect of Hydrostatic Pressure Difference on Movement of Substances Through the Capillary Membrane — Bulk Flow

When the hydrostatic pressure is different on the two sides of a membrane, the greater pressure on one side causes slightly increased diffusion of substances toward the opposite side. However, far more water and dissolved substances move through the capillary membrane than can be accounted for by increased diffusion alone. The reason this occurs is that the water and other substances actually *flow* in bulk through the capillary pores when a hydrostatic pressure difference occurs. This is exactly the same phenomenon that occurs in a blood vessel when a pressure difference develops between the two ends of the vessel, namely, flow of the blood along the axis of the blood vessel from the high pressure area to the low pressure area. So also do fluid and its dissolved substances flow through the capillary pores when a hydrostatic pressure difference develops across the capillary membrane. This phenomenon is called *bulk flow* of fluid through the membrane.

Effect of Osmotic Pressure Difference on Movement of Substances Through the Capillary Membrane — Bulk Flow Caused by This Means Also

When a substance has a molecular weight too great for it to move through the slit-pores of the capillary membrane, it creates osmotic pressure at the end of the slit where the movement of the molecule is impeded. This pressure then acts exactly as does a hydrostatic pressure difference across the pore to cause bulk flow through the pore. The osmotic substance that normally is most important in causing bulk flow through the capillary membrane is the plasma proteins, as will be explained later in the chapter.

THE INTERSTITIUM AND THE INTERSTITIAL FLUID

Approximately one sixth of the body tissues is spaces between cells, which collectively are called the *interstitium*. The fluid in these spaces is the interstitial fluid.

The structure of the interstitium is illustrated in Figure 30–4. It' has two major types of solid structures: (1) collagen fiber bundles and (2) proteoglycan filaments. Figure 30–4 illustrates that the collagen fiber bundles extend long distances in the interstitium. They are extremely strong and therefore provide most of the tensional strength of the tissues. The proteoglycan filaments, on the other hand, are extremely thin, long, coiled molecules composed of about 98 per cent hyaluronic acid and 2 per cent protein. These molecules are so thin that they can never be seen with a light microscope and are very difficult to demonstrate even with the electron microscope. Nevertheless, they form a mat of very fine reticular filaments aptly described as a "brush pile." This brush pile of proteoglycan filaments is ubiquitous in the interstitium. It fills all the spaces between the collagen fibers, the crannies between the cells, and almost all the other minute spaces of the tissues.

Characteristics of "Tissue Gel" in the Interstitium. The fluid in the interstitium is an ultrafiltrate of plasma derived from the capillaries. In the interstitium this fluid is mainly entrapped in the minute spaces among the proteoglycan filaments. Thus, the combination of the proteoglycan filaments and the fluid entrapped within them has characteristics of a *gel,* therefore called the *tissue gel*.

Because of the large number of proteoglycan filaments, fluid *flows* through the tissue gel only very poorly. Instead, it mainly *diffuses* through the gel; that is, it moves molecule by molecule from one place to another by the process of kinetic motion rather than by large numbers of molecules moving together. In contrast to the severe impediment of flow through the gel, diffusion through the gel occurs about 95 to 99 per cent as effectively as it does through free fluid. This diffusion allows rapid transport through the interstitium not only of the water molecules, but also electrolytes, nutrients, cellular excreta, oxygen, carbon dioxide, and so forth.

Free Fluid in the Interstitium. Though almost all the fluid in the interstitium is normally entrapped within the tissue gel, occasionally small *rivulets of free fluid* and also small *free fluid vesicles* are also present. When a dye is injected into the circulating blood, it can often be seen to flow through the interstitium in the small rivulets that usually course along the surfaces of collagen fibers or surfaces of cells. In quantity, however, the amount of free fluid present in normal tissues is very slight. On the other hand, when the tissues develop acute edema, it is the small pockets of free fluid that expand tremendously.

Relationship of the Free Fluid to the Gel Fluid. The free fluid and the gel fluid are continually interchanging with each other by diffusion. At the surface between the free fluid and the gel, there is a slight osmotic pressure difference. Though this has never been measured accurately, the measurements that are available suggest that the osmotic pressure of the gel is about 2 mm. Hg greater than that of the free fluid. This osmotic pressure is caused mainly by the Donnan equilibrium effect, which can be explained as follows: The proteoglycan molecules have a strong negative charge and therefore attract large numbers of positively charged ions, especially sodium ions, into the meshwork of filaments. These sodium ions are responsible for almost all the osmotic pressure of the gel.

Because of the uncertainty of the quantitative value of the osmotic pressure of the gel and therefore the uncertainty of the hydrostatic pressure of the fluid between the gel filaments, in most of the discussions of this chapter and the next we will speak mainly of the free fluid pressure rather than the intragel fluid pressure. There are other reasons for this as well: First, the fluid as it enters the interstitium from the capillary is in the form of free fluid. Second, as it leaves the interstitium to enter the lymphatic vessels the fluid is also in the form of free fluid. Third, when acute edema develops, the free fluid portion of the tissue mainly swells, and this is the most important clinical consideration relating to the interstitial fluids.

Nevertheless, as we quote pressures for the free fluid in the interstitium, the student should remem-

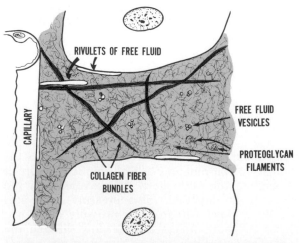

Figure 30–4. Structure of the interstitium. Proteoglycan filaments fill the spaces between the collagen fiber bundles. Free fluid vesicles are seen and small amounts of free fluid in the form of rivulets.

ber that the pressures in the minute spaces of the gel reticulum probably average about 2 mm. Hg more positive than the pressures of the free fluid.

DISTRIBUTION OF FLUID VOLUME BETWEEN THE PLASMA AND INTERSTITIAL FLUID

Despite the tremendous rates of diffusion of substances both out of the capillary into the interstitial spaces and in the opposite direction, these rates in both directions so nearly equal each other that the rate of net volume movement across the capillary membrane is normally very low. Consequently, the volumes of both the blood and the interstitial fluids normally change very little from hour to hour or even from day to day. Yet, under abnormal conditions, fluid can leak rapidly out of the circulation into the interstitial spaces, sometimes causing circulatory shock because of decreased blood volume or tissue edema because of excess fluid in the interstitial spaces.

The pressure in the capillaries continuously tends to force fluid and its dissolved substances through the capillary pores into the interstitial spaces. But, in contrast, osmotic pressure caused by the plasma proteins (called *colloid* osmotic pressure) tends to cause fluid movement by osmosis from the interstitial spaces into the blood; it is mainly this osmotic pressure that prevents continual loss of fluid volume from the blood into the interstitial spaces. Yet, the process is much more complicated than this and includes the role of the lymphatic system to return back to the circulation the small amounts of protein and fluid that do leak continuously into the interstitial spaces. In the following few paragraphs we will discuss all of the factors that play a significant role in the movement of fluid volume through the capillary membrane, and in the following chapter we will discuss the role of the lymphatic system in this overall mechanism.

The Four Primary Factors That Determine Fluid Movement Through the Capillary Membrane. Figure 30–5 illustrates the four primary factors that determine whether fluid will move out of the blood into the interstitial fluid or in the opposite direction; these are:

1. The *capillary pressure* (Pc), which tends to move fluid outward through the capillary membrane.

2. The *interstitial fluid pressure* (Pif), which tends to move fluid inward through the capillary membrane when Pif is positive but outward when Pif is negative.

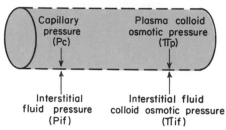

Figure 30–5. Forces operative at the capillary membrane tending to move fluid either outward or inward through the membrane.

3. The *plasma colloid osmotic pressure* (IIp), which tends to cause osmosis of fluid inward through the membrane.

4. The *interstitial fluid colloid osmotic pressure* (IIif), which tends to cause osmosis of fluid outward through the membrane.

The regulation of fluid volumes in the blood and interstitial fluid is so important that each of these factors is discussed in turn in the following sections.

CAPILLARY PRESSURE

Unfortunately, the exact capillary pressure is not known because it has been impossible to measure capillary pressure under absolutely normal conditions. Yet two different methods have been used to estimate the capillary pressure: (1) *direct cannulation of the capillaries,* which has given an average mean capillary pressure of about 25 mm. Hg, and (2) *indirect functional measurement of the capillary pressure,* which has given a capillary pressure averaging about 17 mm. Hg. These methods are the following:

Cannulation Method for Measuring Capillary Pressure. To measure pressure in a capillary by cannulation, a microscopic glass pipet is thrust directly into the capillary, and the pressure is measured by an appropriate micromanometer system. Using this method, capillary pressures have been measured in capillaries of exposed tissues of lower animals and in large capillary loops of the eponychium at the base of the fingernail in human beings. These measurements have given pressures of 30 to 40 mm. Hg in the arterial ends of the capillaries, 10 to 15 mm. Hg in the venous ends, and about 25 mm. Hg in the middle.

Isogravimetric and Isovolumetric Methods for Indirectly Measuring Mean Capillary Pressure. Figure 30–6 illustrates an *isogravimetric* method for estimating capillary pressure. This figure shows a section of gut held by one arm of a gravimetric balance. Blood is perfused through the gut. When the arterial pressure is decreased, the resulting decrease in capillary pressure allows the osmotic pressure of the plasma proteins to cause

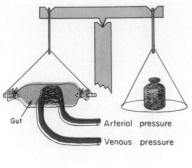

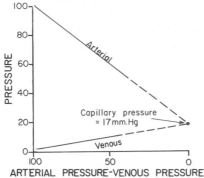

Figure 30–6. Isogravimetric method for measuring capillary pressure (explained in the text).

absorption of fluid out of the gut wall and makes the weight of the gut decrease. This immediately causes displacement of the balance arm. However, to prevent this weight decrease, the venous pressure is raised an amount sufficient to overcome the effect of decreasing the arterial pressure.

In the lower part of the figure, the changes in arterial and venous pressures that exactly nullify their respective effects on the weight of the gut are illustrated. The arterial and venous curves meet each other at a value of 17 mm. Hg. Therefore, the capillary pressure must have remained at this same level of 17 mm. Hg throughout these maneuvers. Thus, in a roundabout way, the "functional" capillary pressure is measured to be about 17 mm. Hg.

The *isovolumetric* method for measuring capillary pressure is essentially the same as the isogravimetric method, except that the *volume* of the tissue is recorded rather than the weight. The arterial and venous pressures are gradually brought toward each other and are continually adjusted so that the tissue neither gains nor loses volume. On extrapolating the arterial and venous pressures, one again finds that the capillary pressure estimated by this means is considerably lower than that measured by direct cannulation, again averaging about 17 mm. Hg.

Functional Capillary Pressure. Since the cannulation method does not give the same pressure measurement as the isogravimetric and isovolumetric methods, one must decide which of these measurements is probably the true functional capillary pressure of the tissues. The isovolumetric and isogravimetric measurements are probably much nearer to the normal values for capillary pressure than are the micropipet measurements, for several reasons:

First, the metarterioles and precapillary sphincters of the capillary system are normally closed during a greater part of the vasomotion cycle than they are open. When they are closed the pressure in the entire capillary system beyond the closures should be almost exactly equal to the pressure at the venous ends of the capillaries, about 10 mm. Hg. Therefore, if one considers a *weighted* average of the pressures in all capillaries, one would expect the *functional* mean capillary pressure to be much nearer to the pressure in the venous ends of the capillaries than to the pressure in the arterial ends.

Second, the surface area of the venous capillaries is several times as great as the surface area of the arterial capillaries. Therefore, the mean pressure in the venous capillaries, 10 mm. Hg, plays far more role in determining the *functional* capillary pressure than does the mean arterial capillary pressure, 35 mm. Hg.

Third, the venous capillaries are several times as permeable as the arterial capillaries. Therefore, for determining fluid movement through the capillary membrane, the venous capillary pressures are much more important than the arterial capillary pressures.

Thus, there are many reasons for believing *that the normal functional mean capillary pressure is about 17 mm. Hg.*

What is the *average* functional pressure in the arterial ends of the capillaries? This we do not know, but it must be considerably below the 30 to 40 mm. Hg measured in the micropipet in *open* arterial capillaries, because even the arterial ends of the capillaries are closed off most of the time because of metarteriole and precapillary sphincter closure. We might estimate this pressure to be about 25 mm. Hg, which is somewhat below the pipet measurements and somewhat above the functional mean capillary pressure.

INTERSTITIAL FLUID PRESSURE — INTRAGEL PRESSURE AND FREE FLUID PRESSURE

The interstitial fluid pressure, like capillary pressure, has been difficult to measure, primarily because the maximum width of the spaces between the reticular fibers that make up the solid structure of the interstitium is only 10 to 40 nanometers, much too small to cannulate for direct measurement of the pressure. Therefore, in general, indirect methods have been used for measuring this pressure.

Prior to 1961, it had been believed universally that the interstitial fluid pressure was always slightly greater than atmospheric pressure. However, on the basis of measurements made in our laboratory in 1961, we concluded that the interstitial fluid pressure is probably subatmospheric in most tissues of the body, usually 2 to 9 mm. Hg less than the pressure of the surrounding air. Since that time, a continuing controversy has existed concerning whether or not the interstitial fluid pressure is truly less than the atmospheric level.

Also, the question must be asked: Is it the pressure of the fluid inside the gel, the *intragel pressure,* that is important, or is it the pressure of the free fluid, the *free fluid pressure,* that is important? From the point of view of understanding tissue function, the answer is that both of these are important. However, from the point of view of the practice of medicine, it is the free fluid pressure that determines whether or not the tissues will be nonedematous or edematous. Therefore, most attempts to measure the interstitial fluid pressure have been concerned primarily with measuring the free fluid pressure, and that is what we will be considering in the present chapter rather than the intragel pressure. As discussed earlier in the chapter, measurements of the osmotic pressure of gel have indicated that the intragel fluid pressure is probably about 2 mm. Hg more positive than the free fluid pressure.

With this as background, let us now summarize some of the methods that have been used for measuring the interstitial free fluid pressure.

Measurement of Interstitial Free Fluid Pressure in Implanted Perforated Hollow Capsules. Figure 30–7 illustrates an indirect method for measuring interstitial fluid pressure which may be explained as follows: A small hollow plastic capsule perforated by several hundred small holes is implanted in the tissues, and the surgical wound is allowed to heal for approximately one month. At the end of that time, tissue will have grown inward through the holes to line the inner surface of the sphere. Furthermore, the cavity is filled with fluid

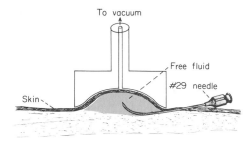

Figure 30–8. The equilibrium method for measuring interstitial fluid pressure.

that flows freely through the perforations back and forth between the fluid in the interstitial spaces and the fluid in the cavity. Therefore, the pressure in the cavity should equal the free fluid pressure in the interstitial fluid spaces. A needle is inserted through the skin and through one of the perforations to the interior of the cavity, and the pressure is measured by use of an appropriate manometer.

Interstitial free fluid pressure measured by this method in normal tissues averages about −6.3 mm. Hg. That is, the pressure is *less than atmospheric pressure* or, in other words, is a semivacuum or a suction.

Recently, instead of using capsules with large perforations as illustrated in Figure 30–6, we have made these capsules of a porous plastic material that has pore sizes too small for tissues to grow into the cavity of the capsule. Using this new procedure, the pressures that have been measured have been about −5 mm. Hg.

The significance of the negativity of interstitial free fluid pressure is that it causes *suction of fluid out of the capillaries,* as we shall see in subsequent sections of this chapter. The mechanism of its development is discussed in the following chapter in relation to interstitial fluid dynamics.

Measurement of Interstitial Free Fluid Pressure in Fluid Spaces Developed Underneath the Skin. Figure 30–8 illustrates another method that has been used to measure the interstitial fluid pressure. A small vacuum cup is placed over the skin, and the skin is sucked up into the cup. Gradually, over about 24 hours, free fluid collects beneath the mound of skin, and its pressure comes to equilibrium. A needle is then inserted into the tissue so that its tip is within this free fluid, and the pressure is recorded. Since the fluid is in free communication with the fluid in the surrounding tissues, the pressure measured here should equal the pressure in the other interstitial spaces. The pressure measured by this method is also negative, averaging again about −6 mm. Hg.

Measurement of Interstitial Free Fluid Pressure by Means of a Cotton Wick. A new method used recently is to insert into a tissue a small Teflon tube with about eight cotton fibers protruding from its end. Each cotton fiber is itself a very minute tube containing large numbers of small side openings. The fluid inside the cotton fibers, therefore, makes excellent contact with

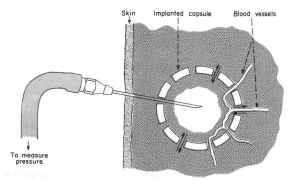

Figure 30–7. The perforated capsule method for interstitial fluid pressure.

the tissue fluids and transmits interstitial fluid pressure into the Teflon tube; the pressure can then be measured from the tube by usual manometric means. Pressures measured by this technique have also been negative, usually measuring about −3 mm. Hg.

Measurement of the Interstitial Fluid Pressure by Means of Tissue Cannulas. A method that was used for many years to "measure" interstitial fluid pressure was to insert a needle into the tissue, inject a small bolus of fluid through the tip of the needle, and then measure the pressure in the fluid bolus before it was absorbed. The result was always a pressure measurement slightly greater than atmospheric pressure in subcutaneous tissue. However, this was a false measurement of pressure because the preliminary injection of the bolus of fluid caused local tissue edema.

Recently, microscopic pipets have been placed into the tissue in attempts to measure the interstitial fluid pressure. Earlier measurements in this way suggested slightly positive pressures, but negative pressures of −1 to −3 mm. Hg have more recently been made. Since the pipets are 30 to 50 times as large as the spaces between the reticular fibrillae of the interstitium, one wonders how much the distortion of the tissues that is caused by the method affects the measurement.

Validity of the Interstitial Free Fluid Pressure Measurements

The implanted capsule method has been used to test *changes* in interstitial free fluid pressure caused by various physiological phenomena. The changes that have been recorded have been almost exactly those that would be predicted for the experimental procedures. For instance, if concentrated dextran solution is injected intravenously, one would expect absorption of fluid from the tissue spaces and therefore a quantifiable decrease in the interstitial fluid pressure. The capsule technique demonstrates such predicted pressure changes. Similarly, the predictions have been validated for other physiological experiments, such as changes in arterial and venous pressure and the degree of hydration of the tissues. This type of validation has not been shown for the other methods. For this reason, and other reasons that will be presented in the following chapter, we have chosen to use in the discussions of this chapter and the following chapters the pressure measurements made by the implanted capsule technique.

PLASMA COLLOID OSMOTIC PRESSURE

Colloid Osmotic Pressure Caused by Proteins. The proteins are the only dissolved substances of the plasma that do not diffuse readily into the interstitial fluid. Furthermore, when small quantities of protein do diffuse into the interstitial fluid, these are soon removed from the interstitial spaces by way of the lymph vessels. Therefore, the concentration of protein in the plasma averages almost 4 times as much as that in the interstitial fluid, 7.3 gm./100 ml. in the plasma versus 2 gm./100 ml. in the interstitial fluid.

In the discussion of osmotic pressure in Chapter

4, it was pointed out that only those substances that fail to pass through the pores of a semipermeable membrane exert osmotic pressure. Since the proteins are the only dissolved constituents that do not readily penetrate the pores of the capillary membrane, it is the dissolved proteins of the plasma and interstitial fluids that are responsible for the osmotic pressure at the capillary membrane. To distinguish this osmotic pressure from that which occurs at the cell membrane, it is called either *colloid osmotic pressure* or *oncotic pressure*. The term "colloid" osmotic pressure is derived from the fact that a protein solution resembles a colloidal solution despite the fact that it is actually a true solution. (The osmotic pressure that results at the cell membrane is often called *total osmotic pressure* to distinguish it from the colloid osmotic pressure because essentially all dissolved substances of the body fluids exert osmotic pressure at the cell membrane.)

The Donnan Effect on the Colloid Osmotic Pressure. The so-called *Donnan effect* causes the colloid osmotic pressure to be about 50 per cent greater than that caused by the proteins alone. This results from the fact that the proteins are negative ions, and to balance these negative ions a large number of positively charged ions (cations), mainly sodium ions, are attracted to the electronegative charges of the proteins. These extra cations, therefore, increase the number of osmotically active substances wherever the proteins occur and increase the total colloid osmotic pressure. Even more important, however (for mathematical reasons that cannot be explained here), the Donnan equilibrium effect becomes progressively more significant the higher the concentration of proteins. This means, as illustrated in Figure 30–9, that the initial few grams of protein in each 100 ml. of plasma or interstitial fluid has much less colloid osmotic effect than do the next few grams.

Normal Values for Plasma Colloid Osmotic Pressure. The colloid osmotic pressure of normal human plasma averages approximately 28 mm. Hg; 19 mm of this is caused by the dissolved protein, and 9 mm. by the cations held in the plasma by the Donnan effect of the proteins, as was just discussed.

Note particularly that the 28 mm. Hg. colloid osmotic pressure that can develop at the capillary membrane is only 1/200 the total osmotic pressure that would develop at a *cell* membrane if normal interstitial fluid were on one side of the membrane and pure water on the other side. Thus, the colloid osmotic pressure of the plasma is actually a weak osmotic force; but, even so, it plays an exceedingly important role in the maintenance of normal blood and interstitial fluid volumes, as will be evident in the remainder of this chapter and in the following chapters.

Effect of the Different Plasma Proteins on Colloid Osmotic Pressure. The plasma proteins are a mixture of proteins that contains albumin, with an average molecular weight of 69,000; globulins, 140,000; and fibrinogen, 400,000. Thus, 1 gram of globulin contains

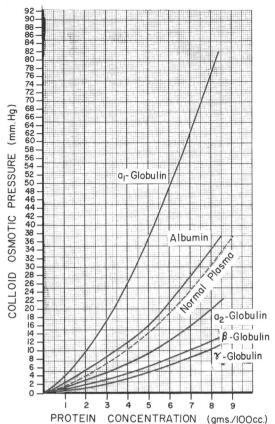

Figure 30–9. The osmotic pressure of five fractions of the plasma proteins at different concentrations. Also, the dashed line shows the osmotic pressure of normal plasma proteins, which are a mixture of the others.

only half as many molecules as 1 gram of albumin, and 1 gram of fibrinogen contains only one-sixth as many molecules as 1 gram of albumin. (It will be recalled from the discussion of osmotic pressure in Chapter 4 that the osmotic pressure is determined by the *number of molecules* dissolved in a fluid rather than by the weight of these molecules.) The average relative concentrations of the different types of proteins in the plasma and their respective colloid osmotic pressures are:

	gm. %	IIp (mm. Hg.)
Albumin	4.5	21.8
Globulins	2.5	6.0
Fibrinogen	0.3	0.2
TOTAL	7.3	28.0

Because each gram of albumin exerts twice the osmotic pressure of a gram of globulins and because there is almost twice as much albumin in the plasma as globulins, about 75 per cent of the total colloid osmotic pressure of the plasma results from the albumin fraction, and 25 per cent from the globulins, and almost none from the fibrinogen. Therefore, from the point of view of capillary dynamics, it is mainly albumin that is important.

Figure 30–8 illustrates graphically the colloid osmotic pressures exerted by different concentrations of albumin and four different fractions of globulins.

INTERSTITIAL FLUID COLLOID OSMOTIC PRESSURE

Though the size of the usual capillary pore is smaller than the molecular sizes of the plasma proteins, this is not true of all the pores. Therefore, small amounts of plasma proteins do leak through the pores into the interstitial spaces, and perhaps additional amounts are transported by pinocytosis.

The total quantity of protein in the entire 12 liters of interstitial fluid of the body is slightly greater than the total quantity of protein in the plasma itself, but since this volume is four times the volume of plasma the average protein *concentration* of the interstitial fluid is only a little more than one-fourth that in plasma, or approximately 2 grams per cent. Referring to the diagram in Figure 30–8, one finds that the average colloid osmotic pressure for this concentration of proteins in the interstitial fluids is approximately 5 mm. Hg.

EXCHANGE OF FLUID VOLUME THROUGH THE CAPILLARY MEMBRANE

Now that the different factors affecting capillary membrane dynamics have been discussed, it is possible to put all these together to see how normal capillaries function.

The capillary pressure at the arterial ends of the capillaries is 15 to 20 mm. Hg greater than at the venous ends. Because of this difference, fluid "filters" out of the capillaries at their arterial ends and then is reabsorbed into the capillaries at their venous ends. Thus, a small amount of fluid actually "flows" through the tissues from the arterial ends of the capillaries to the venous ends. The dynamics of this flow are the following:

Analysis of the Forces Causing Filtration at the Arterial End of the Capillary. The forces operative at the arterial end of the capillary to cause movement through the capillary membrane are:

	mm. Hg
Forces tending to move fluid outward:	
Capillary pressure	25.0
Negative interstitial free fluid pressure	6.3
Interstitial fluid colloid osmotic pressure	5.0
TOTAL OUTWARD FORCE	36.3
Force tending to move fluid inward:	
Plasma colloid osmotic pressure	28.0
TOTAL INWARD FORCE	28.0
Summation of forces:	
Outward	36.3
Inward	28.0
NET OUTWARD FORCE	8.3

Thus, the summation of forces at the arterial end of the capillary shows a net *filtration pressure* of 8.3 mm. Hg. tending to move fluid out of the arterial ends of the capillaries into the interstitial spaces.

This 8.3 mm. Hg filtration pressure causes an average of about 0.3 per cent of the fluid of the plasma to filter out of the arterial ends of the capillaries into the interstitial spaces. (This percentage varies tremendously from very, very low in the brain to very high in the liver.)

Analysis of Reabsorption at the Venous End of the Capillary. The low pressure at the venous end of the capillary changes the balance of forces in favor of absorption as follows:

	mm. Hg
Force tending to move fluid inward:	
Plasma colloid osmotic pressure	28.0
TOTAL INWARD FORCE	28.0
Forces tending to move fluid outward:	
Capillary pressure	10.0
Negative interstitial free fluid	
pressure	6.3
Interstitial fluid colloid osmotic	
pressure	5.0
TOTAL OUTWARD FORCE	21.3
Summation of forces:	
Inward	28.0
Outward	21.3
NET INWARD FORCE	6.7

Thus, the force that causes fluid to move into the capillary, 28 mm. Hg, is greater than that opposing reabsorption, 21.3 mm. Hg. The difference, 6.7 mm. Hg, is the *reabsorption pressure*.

The reabsorption pressure causes about nine-tenths of the fluid that has filtered out the arterial ends of the capillaries to be reabsorbed at the venous ends. The other one-tenth flows into the lymph vessels, as is discussed in the following chapter.

Flow of Fluid Through the Interstitial Spaces. The 0.3 per cent of the plasma fluid that filters out of the arterial ends of the capillaries *flows* through the tissue spaces to the venous ends of the capillaries where all but about one-tenth of it is reabsorbed. (A much higher proportion than this is reabsorbed in the muscles where very little protein leaks through the capillary membranes, and much less is reabsorbed in the liver where tremendous amounts of protein leak.)

Distinction Between Filtration and Diffusion. It is especially important to distinguish between *filtration* and *diffusion* through the capillary membrane. Diffusion occurs in both directions, while filtration is the *net* movement of fluid out of the capillaries at the arterial ends. The rate of diffu-

sion of water through all the capillary membranes of the entire body is about 240,000 ml./minute, while the normal rate of filtration at the arterial ends of all the capillaries is only 16 ml./minute, a difference of about 15,000-fold. Thus, the quantitative rate of diffusion of water and nutrients back and forth between the capillaries and the interstitial spaces is tremendous in comparison with the minute rate of "flow" of fluid through the tissues.

THE STARLING EQUILIBRIUM FOR CAPILLARY EXCHANGE

E. H. Starling pointed out almost a century ago that under normal conditions a state of near-equilibrium exists at the capillary membrane whereby the amount of fluid filtering outward through the arterial capillaries equals that quantity of fluid that is returned to the circulation by reabsorption at the venous ends of the capillaries. This near-equilibrium is caused by near-equilibration of the *mean* forces tending to move fluid through the capillary membranes. If we assume that the mean capillary pressure is 17 mm. Hg, the normal mean dynamics of the capillary are the following:

	mm. Hg
Mean forces tending to move fluid outward:	
Mean capillary pressure	17.0
Negative interstitial free fluid	
pressure	6.3
Interstitial fluid colloid osmotic	
pressure	5.0
TOTAL OUTWARD FORCE	28.3
Mean force tending to move fluid inward:	
Plasma colloid osmotic pressure	28.0
TOTAL INWARD FORCE	28.0
Summation of mean forces:	
Outward	28.3
Inward	28.0
NET OUTWARD FORCE	0.3

Thus, we find a near-equilibrium but nevertheless a slight imbalance of forces at the capillary membranes that causes slightly more filtration of fluid into the interstitial spaces than reabsorption. This slight excess of filtration is called the *net filtration,* and it is balanced by fluid return to the circulation through the lymphatics. The normal rate of net filtration in the entire body is about 1.7 to 3.5 ml./min. These figures also represent the rate of fluid flow into the lymphatics each minute.

The Filtration Coefficient. In the above example an average net imbalance of forces at the capillary membranes of 0.3 mm. Hg causes a net rate of fluid filtration in the entire body of 1.7 to 3.5 ml./min. Expressing this per mm. Hg, one finds a net filtration rate of 5.7 to 11.4 ml. fluid/min./mm. Hg for the entire body. This expression is the *filtration coefficient*.

The filtration coefficient can also be expressed for different areas of the body in terms of the rate of filtration/min./mm. Hg/100 gm. of tissue. On this basis the filtration coefficient of the average tissue is 0.008 to 0.015 ml./min./mm. Hg/100 gm. of tissue. However, because of extreme differences in permeabilities of the capillary systems in different tissues, this coefficient varies greatly from one tissue to another. It is very small in both brain and muscle, moderately great in subcutaneous tissue, large in the intestine, and extreme in the liver where the pores of the liver sinusoids are almost wide open. By the same token, the permeation of proteins through the capillary membranes varies in approximately the same way. The concentration of protein in the interstitial fluid of muscles is about 1.5 gm. per cent, in subcutaneous tissue 2 gm. per cent, in intestine 4 gm. per cent, and in liver 6 gm. per cent.

Effect of Imbalance of Forces at the Capillary Membrane. If the mean capillary pressure rises above 17 mm. Hg the net force tending to cause filtration of fluid into the tissue spaces obviously rises. Thus a 20 mm. Hg rise in mean capillary pressure causes an increase in the net filtration pressure from 0.3 mm. Hg to 20.3 mm. Hg, which results in 68 times as much net filtration of fluid into the interstitial spaces as normally occurs, and this would require also 68 times the normal flow of fluid into the lymphatic system, an amount that is usually too much for the lymphatics to carry away. As a result, fluid begins to accumulate in the interstitial spaces, and edema results.

Conversely, if the capillary pressure falls very low, net reabsorption of fluid into the capillaries occurs instead of net filtration, and the blood volume increases at the expense of the interstitial fluid volume. The effects of these imbalances at the capillary membrane are discussed in the following chapter in relation to the formation of edema.

REFERENCES

Allen, P. C., *et al.*: Plasma proteins: Analytical and preparative techniques. Philadelphia, J. B. Lippincott Co., 1977.

Aschheim, E.: Passage of substances across the walls of blood vessels: Kinetics and mechanism. *In* Kaley, G., and Altura, B. M. (eds.): Microcirculation. Vol. 1. Baltimore, University Park Press, 1977, p. 213.

Baez, S.: Microcirculation. *Annu. Rev. Physiol.*, 39:391, 1977.

Bing, D. H. (ed.): The Chemistry and Physiology of the Human Plasma Proteins. New York, Pergamon Press, 1979.

Brace, R. A., and Guyton, A. C.: Effect of hindlimb isolation procedure on isogravimetric capillary pressure and transcapillary fluid dynamics in dogs. *Circ. Res.*, 38:192, 1976.

Brace, R. A., and Guyton, A. C.: Interaction of transcapillary Starling forces in the isolated dog forelimb. *Am. J. Physiol.*, 233:H136, 1977.

Brace, R. A., *et al.*: Determinants of isogravimetric capillary pressure in the isolated dog hindlimb. *Am. J. Physiol.*, 233:H130, 1977.

Comper, W. D., and Laurent, T. C.: Physiological function of connective tissue polysaccharides. *Physiol. Rev.*, 58:255, 1978.

Crone, C., and Christensen, O.: Transcapillary transport of small solutes and water. *In* Guyton, A. C., and Young, D. B. (eds.): International Review of Physiology: Cardiovascular Physiology III. Vol. 18. Baltimore, University Park Press, 1979, p. 149.

Duling, B. R., and Berne, R. M.: Propagated vasodilation in the microcirculation of the hamster cheek pouch. *Circ. Res.*, 26:163, 1970.

Gabbiani, G., and Majno, G.: Fine structure and endothelium. *In* Kaley, G., and Altura, B. M. (eds.): Microcirculation. Vol. 1. Baltimore, University Park Press, 1977, p. 133.

Gross, J. F., and Popel, A. (eds.): Mathematics of Microcirculation Phenomena. New York, Raven Press, 1980.

Guyton, A. C.: Concept of negative interstitial pressure based on pressures in implanted perforated capsules. *Circ. Res.*, 12:399, 1963.

Guyton, A. C.: Interstitial fluid pressure: II. Pressure-volume curves of interstitial space. *Circ. Res.*, 16:452, 1965.

Guyton, A. C.: Interstitial fluid pressure-volume relationships and their regulation. *In* Wolstenholme, G. E. W., and Knight, J. (eds.): Ciba Foundation Symposium on Circulatory and Respiratory Mass Transport. London, J. & A. Churchill Ltd., 1969, p. 4.

Guyton, A. C., *et al.*: Interstitial fluid pressure: III. Its effect on resistance to tissue fluid mobility. *Circ. Res.*, 19:412, 1966.

Guyton, A. C., *et al.*: Interstitial fluid pressure: IV. Its effect on fluid movement through the capillary wall. *Circ. Res.*, 19:1022, 1966.

Guyton, A. C., *et al.*: Interstitial fluid pressure. *Physiol. Rev.*, 51:527, 1971.

Guyton, A. C., *et al.*: Circulatory Physiology II. Dynamics and Control of the Body Fluids. Philadelphia, W. B. Saunders Co., 1975.

Haddy, F. J., Scott, J. B., and Grega, G. J.: Peripheral circulation: Fluid transfer across the microvascular membrane. *Int. Rev. Physiol.*, 9:63, 1976.

Hruza, Z.: Connective tissue. *In* Kaley, G., and Altura, B. M. (eds.): Microcirculation. Vol. 1. Baltimore, University Park Press, 1977, p. 167.

Intaglietta, M.: The measurement of pressure and flow in the microcirculation. *Microvasc. Res.*, 5:357, 1973.

Intaglietta, M.: Transcapillary exchange of fluid in single microvessels. *In* Kaley, G., and Altura, B. M. (eds.): Microcirculation. Vol. 1. Baltimore, University Park Press, 1977, p. 197.

Johnson, P. C.: The microcirculation and local and humoral control of the circulation. *In* MTP International Review of Science: Physiology. Vol. 1. Baltimore, University Park Press, 1974, p. 163.

Landis, E. M.: Capillary pressure and capillary permeability. *Physiol. Rev.*, 14:404, 1934.

Landis, E. M., and Papenheimer, J. R.: Exchange of substances through the capillary walls. *In* Hamilton, W. F. (ed.): Handbook of Physiology. Sec. 2. Vol. 2. Baltimore, Williams & Wilkins, 1963, p. 961.

Leonard, E. F., and Jorgensen, S. B.: The analysis of convection and diffusion in capillary beds. *Annu. Rev. Biophys. Bioeng.*, 3:293, 1974.

Mayerson, H. S.: The physiologic importance of lymph. *In* Hamilton, W. F. (ed.): Handbook of Physiology. Sec. 2. Vol. 2. Baltimore. Williams & Wilkins, 1963, p. 1035.

Nicoll, P. A., and Taylor, A. E.: Lymph formation and flow. *Annu. Rev. Physiol.*, 39:73, 1977.

Pappenheimer, J. R.: Passage of molecules through capillary walls. *Physiol. Rev.*, 33:387, 1953.

Peeters, H., and Wright, P. H. (eds.): Plasma Protein Pathology: A Workshop on Plasma Proteins, Their Availability, Assay, and Therapeutic Uses. New York, Pergamon Press, 1979.

Prather, J. W., *et al.*: Direct continuous recording of plasma colloid osmotic pressure of whole blood. *J. Appl. Physiol.*, 24:602, 1968.

Prockop, D. J.: Collagen, elastin, and proteoglycans: matrix for fluid accumulation in the lung. *In* Fishman, A. P., and Renkin, E. M. (eds.): Pulmonary Edema. Baltimore, Waverly Press, 1979, p. 125.

Rhodin, J. A. G.: Architecture of the vessel wall. *In* Bohr, D. F., *et al.* (eds.): Handbook of Physiology. Sec. 2. Vol. II. Baltimore, Williams & Wilkins, 1980, p. 1.

Rothschild, M. A., *et al.*: Albumin synthesis. *In* Javitt, N. B. (ed.): International Review of Physiology: Liver and Biliary Tract Physiology I. Vol. 21. Baltimore, University Park Press, 1980, p. 249.

Sharon, N.: Glycoproteins. *Sci. Am.*, 230(5):78, 1974.

Simionescu, M.: Transendothelial movement of large molecules in the microvasculature. *In* Fishman, A. P., and Renkin, E. M. (eds.): Pulmonary Edema. Baltimore, Waverly Press, 1979, p. 39.

Stromme, S. B., *et al.*: Interstitial fluid pressure in terrestrial and semi-terrestrial animals. *J. Appl. Physiol.*, 27:123, 1969.

31

The Lymphatic System, Interstitial Fluid Dynamics, Edema, and Pulmonary Fluid

THE LYMPHATIC SYSTEM

The lymphatic system represents an accessory route by which fluids can flow from the interstitial spaces into the blood. And, most important of all, the lymphatics can carry proteins and large particulate matter away from the tissue spaces, neither of which can be removed by absorption directly into the blood capillary. We shall see that this removal of proteins from the interstitial spaces is an absolutely essential function, without which we would die within about 24 hours.

THE LYMPH CHANNELS OF THE BODY

All tissues of the body, with the exception of a very few, have lymphatic channels that drain excess fluid directly from the interstitial spaces. The exceptions include the superficial portions of the skin, the central nervous system, deeper portions of peripheral nerves, the endomysium of muscles, and the bones. However, even these tissues have minute interstitial channels called *prelymphatics* through which interstitial fluid can flow; eventually this fluid flows into lymphatic vessels or, in the case of the brain, flows into the cerebrospinal fluid and thence directly back into the blood.

Essentially all the lymph from the lower part of the body — even some of that from the legs — flows up the *thoracic duct* and empties into the venous system at the juncture of the *left* internal jugular vein and subclavian vein, as illustrated in Figure 31–1. However, small amounts of lymph from the lower part of the body can enter the veins in the inguinal region and perhaps also at various points in the abdomen.

Lymph from the left side of the head, the left arm, and left chest region also enters the thoracic duct before it empties into the veins. Lymph from the right side of the neck and head, from the right arm, and from parts of the right thorax enters the *right lymph duct*, which then empties into the venous system at the juncture of the *right* subclavian vein and internal jugular vein.

370

The Lymphatic Capillaries and Their Permeability. Most of the fluid filtering from the arterial capillaries flows among the cells and is finally reabsorbed back into the *venous* capillaries; but, on the average, about *one-tenth* of the fluid enters the *lymphatic* capillaries and returns to the blood through the lymphatic system rather than through the venous capillaries.

The minute quantity of fluid that returns to the circulation by way of the lymphatics is extremely important because substances of high molecular

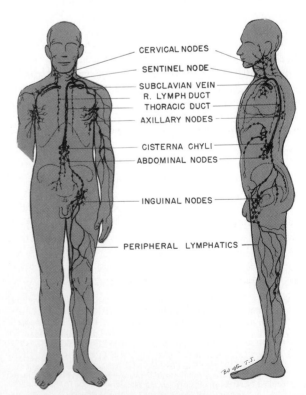

CERVICAL NODES
SENTINEL NODE
SUBCLAVIAN VEIN
R. LYMPH DUCT
THORACIC DUCT
AXILLARY NODES
CISTERNA CHYLI
ABDOMINAL NODES
INGUINAL NODES
PERIPHERAL LYMPHATICS

Figure 31–1. The lymphatic system.

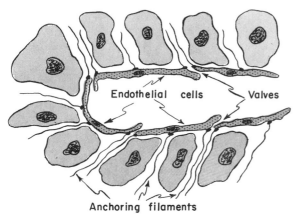

Figure 31–2. Special structure of the lymphatic capillaries that permits passage of substances of high molecular weight back into the circulation.

weight such as proteins cannot pass with ease through the pores of the venous capillaries, but they can enter the lymphatic capillaries almost completely unimpeded. The reason for this is a special structure of the lymphatic capillaries, illustrated in Figure 31–2. This figure shows the endothelial cells of the capillary attached by *anchoring filaments* to the connective tissue between the surrounding tissue cells. However, at the junctions of adjacent endothelial cells there are usually very loose connections between the cells. Instead, the edge of one endothelial cell usually overlaps the edge of the adjacent one in such a way that the overlapping edge is free to flap inward, thus forming a minute valve that opens to the interior of the capillary. Interstitial fluid, along with its suspended particles, can push the valve open and flow directly into the capillary. But this fluid cannot leave the capillary once it has entered because any backflow will close the flap valve. Thus, the lymphatics have valves at the very tips of the terminal lymphatic capillaries as well as valves along their larger vessels up to the point where they empty into the blood circulation.

FORMATION OF LYMPH

Lymph is interstitial fluid that flows into the lymphatics. Therefore, lymph has almost the same composition as the tissue fluid in the part of the body from which the lymph flows. There is, however, a growing belief among many physiologists that the lymphatic vessels might concentrate the plasma proteins in the lymph by filtering water and electrolytes outward through the lymphatic walls.

The protein concentration in the interstitial fluid averages about 2 gm. per cent, and the protein concentration of lymph flowing from most of the peripheral tissues is near this value or a little

more concentrated. On the other hand, lymph formed in the liver has a protein concentration as high as 6 gm. per cent, and lymph formed in the intestines has a protein concentration as high as 3 to 5 gm per cent. Since more than half of the lymph is derived from the liver and intestines, the thoracic lymph, which is a mixture of lymph from all areas of the body, usually has a protein concentration of 3 to 5 gm. per cent.

The lymphatic system is also one of the major routes for absorption of nutrients from the gastrointestinal tract, being responsible principally for the absorption of fats, as will be discussed in Chapter 65. Indeed, after a fatty meal, thoracic duct lymph sometimes contains as much as 1 to 2 per cent fat.

Finally, even large particles, such as bacteria, can push their way between the endothelial cells of the lymphatic capillaries and in this way enter the lymph. As the lymph passes through the lymph nodes, these particles are removed and destroyed, as was discussed in Chapter 6.

The Mechanism by Which Proteins Become Concentrated in Interstitial Fluids. Fluid filtering from the arterial ends of the capillaries in peripheral tissues such as the subcutaneous tissue usually has a protein concentration of about 0.2 per cent, though the average concentration of protein in the interstitial fluids is almost 10 times this value. The reason for this difference is that only a small proportion of the protein that leaks into the tissue spaces is reabsorbed at the venous ends of the capillaries, even though most of the filtered water and ions are reabsorbed. In this way, the proteins become concentrated in the interstitial fluid before it flows into the lymphatics.

TOTAL RATE OF LYMPH FLOW

Approximately 100 ml. of lymph flow through the thoracic duct of a resting man per hour, and perhaps another 20 ml. of lymph flow into the circulation each hour through other channels, making a total estimated lymph flow of perhaps 120 ml. per hour. This is about 1/120,000 of the calculated rate of fluid *diffusion* back and forth through the capillary membranes, and it is one-tenth the rate of fluid *filtration* from the arterial ends of the capillaries into the tissue spaces in the entire body. These facts illustrate that the flow of lymph is relatively small in comparison with the total exchange of fluid between the plasma and the interstitial fluid.

Factors That Determine the Rate of Lymph Flow

Interstitial Fluid Pressure. Elevation of interstitial *free* fluid pressure above its normal level of

−6.3 mm. Hg increases the flow of interstitial fluid into the lymphatic capillaries and consequently also increases the rate of lymph flow. The increase in flow becomes progressively greater as the interstitial fluid pressure rises until this pressure reaches a value slightly greater than zero mm. Hg; at that point the flow rate reaches a maximum, but by that time it has risen to 10 to 50 times normal. Therefore, any factor (besides obstruction of the lymphatic system itself) that tends to increase interstitial pressure increases the rate of lymph flow. Such factors include:

  Elevated capillary pressure
  Decreased plasma colloid osmotic pressure
  Increased interstitial fluid protein
  Increased permeability of the capillaries

The Lymphatic Pump. Valves exist in all lymph channels; a typical valve is illustrated in Figure 31–3 in a *collecting lymphatic* into which the lymphatic capillaries empty. In the large lymphatics, valves exist every few millimeters, and in the smaller lymphatics the valves are much closer together than this, which illustrates the widespread existence of the valves. Every time the lymph vessel is compressed by pressure from any source, lymph tends to be squeezed in both directions, but because the valves open only in the central direction, the lymph moves unidirectionally. The lymph vessels can be compressed either by contraction of the walls of the lymphatics or by pressure from surrounding structures.

Motion pictures taken of exposed lymph vessels, both in animals and human beings, have shown that any time a lymph vessel becomes stretched with fluid the smooth muscle in the wall of the vessel automatically contracts. Furthermore, each segment of the lymph vessel between successive valves functions as a separate automatic pump. That is, filling of a segment causes it to contract, and the fluid is then pumped through the next valve into the following lymphatic segment. This fills the subsequent segment so that within a few seconds it too contracts, the process continuing all along the lymphatic until the fluid is finally emptied. In a large lymph vessel this lymphatic pump can generate pressure as high as 25 to 50 mm. Hg if the outflow from the vessel becomes blocked.

In addition to the pumping caused by intrinsic contraction of the lymph vessel walls, any external factor that compresses the lymph vessel can also cause pumping. In order of their importance, such factors are:

  Contraction of muscles
  Movements of the parts of the body
  Arterial pulsations
  Compression of the tissues by objects outside
    the body

Obviously, the lymphatic pump becomes very active during exercise, often increasing lymph flow as much as 5- to 15-fold. On the other hand, during periods of rest, lymph flow is very sluggish.

The Lymphatic Capillary Pump. Many physiologists believe that the lymphatic capillary is also capable of pumping lymph, in addition to the lymphatic pump of the larger lymph vessels. As was explained earlier in the chapter, the walls of the lymphatic capillaries are tightly adherent to the surrounding tissue cells by means of their anchoring filaments. Therefore, each time excess fluid enters the tissue and the tissue swells, the anchoring filaments pull the lymphatic capillary open, and fluid flows into the capillary through the junctions between the endothelial cells. Then, when the tissue is compressed, the pressure inside the capillary increases and tends to push the fluid in two directions: first, backward through the openings between the endothelial cells, and, second, forward into the collecting lymphatic. However, since the edges of the endothelial cells normally overlap each other on the inside of the lymphatic capillary as illustrated in Figure 31–2, any backward flow through this opening closes the overlapping cell over the opening. Thus, the opening closes as if it were a one-way *valve,* and very little fluid flows backward into the tissues. On the other hand, the lymph that travels forward into the collecting lymphatic will not return to the capillary after the compression cycle is over because the many valves in the collecting lymphatic block any backward flow.

Thus, any factor that causes compression of the lymphatic capillaries probably causes them to pump fluid in the same way that compression of the larger lymphatics causes pumping.

Flow of Lymph into the Lymphatic Capillary Despite Negative Pressure in the Interstitial Spaces. It has been difficult for many students of physiology to understand how fluid can flow out of the interstitial spaces into the lymphatic capillary in the face of the negative pressure in the interstitial spaces averaging −6 mm. Hg, as was discussed in the previous chapter. The resolution to this difficulty lies in the fact that the lymphatic capillaries, during their periodic expansion process, can almost certainly create small amounts of suction. Indeed, this has even been demonstrated

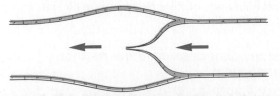

Figure 31–3.   A valve in a collecting lymphatic.

to occur in some of the large lymphatic vessels, because a manometer connected to the central end of a cut lymphatic will record a suction of several mm. Hg. Another way in which fluid could move from the tissues into the lymphatic vessels despite the negative interstitial fluid pressure is the following: Every time a tissue is compressed, interstitial fluid pressure in the local area of compression rises temporarily to a positive value. This causes minute amounts of fluid to move into the lymphatics and subsequently to be pumped away from the tissues. Then, upon removal of the compression, the recoiling of the elastic structures in the tissues, particularly of the reticular structure of the tissue gel, creates suction in the tissue spaces. Thus, except during the momentary periods of compression, a negative pressure could in this way be maintained in the tissue spaces.

Summary of Factors That Determine Lymph Flow. From the above discussion one can see that the two primary factors that determine lymph flow are the interstitial fluid pressure and the activity of the lymphatic pump. Therefore, one can state that, roughly, *the rate of lymph flow is determined by the product of interstitial fluid pressure and the activity of the lymphatic pump.*

Maximum Limit to the Rate of Lymph Flow. Figure 31–4 illustrates the relationship between interstitial free fluid pressure (P_T) and the rate of lymph flow. Note that at the normal interstitial fluid pressure of -6 to -7 mm. Hg the lymph flow is very low. However, as the interstitial fluid pressure rises to slightly greater than 0

mm. Hg, the flow increases more than 20 times, but at this point it reaches a plateau beyond which it will not rise any more even though the interstitial fluid pressure continues to rise.

There seem to be two major reasons why lymph flow reaches a maximum limit: (1) Once the tissues become edematous, the lymphatic capillaries also become tremendously dilated. This causes the flap valves between the endothelial cells of the capillaries to become separated from each other so that they are no longer reliable; therefore, the lymphatic capillary pump begins to fail. (2) The interstitial fluid pressure presses on the outside of the larger lymph channels to cause them to collapse; therefore, the input pressures at the tips of the lymphatic capillaries are opposed by a compression of the lymphatic walls of equal magnitude.

This maximum limit of lymph flow is of major significance because it demonstrates that most of the compensation to prevent edema caused by increased lymph flow occurs before the edema actually appears. That is, this mechanism prevents the development of edema *before* it occurs rather than responding after edema is present. It is only those persons whose abnormalities have gone beyond the limits of this lymphatic compensating mechanism who actually develop edema.

CONTROL OF INTERSTITIAL FLUID PROTEIN CONCENTRATION AND INTERSTITIAL FLUID PRESSURE

The fact that interstitial fluid pressure is negative (that is, subatmospheric) was discovered only a few years ago, though it has now been confirmed by a number of different independent methods described in the previous chapter. Even so, it has been difficult for many students and even professional physiologists to understand the significance of the negative pressure. To explain this, it is first necessary to discuss the regulation of interstitial fluid protein concentration because the problem of interstitial fluid pressure is inextricably bound with that of interstitial fluid protein concentration, as we shall see in the following paragraphs.

REGULATION OF INTERSTITIAL FLUID PROTEIN BY LYMPHATIC PUMPING

Since protein continually leaks from the capillaries into the interstitial fluid spaces, it must also be removed continually, or otherwise the tissue colloid osmotic pressure will become so high that normal capillary dynamics can no longer continue. Unfortunately, only a small proportion of the protein that leaks into the tissue spaces can diffuse

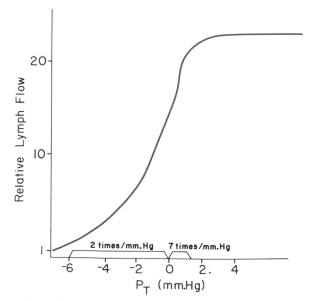

Figure 31–4. Relationship between interstitial fluid pressure and lymph flow. Note that lymph flow reaches a maximum as the interstitial pressure rises slightly above atmospheric pressure (0 mm. Hg). (Courtesy of Drs. Harry Gibson and Aubrey Taylor.)

back into the capillaries because there is almost four times as much protein in the plasma as in the interstitial fluid. Therefore, by far the most important of all the lymphatic functions is the maintenance of low protein concentration in the interstitial fluid. The mechanism of this is the following:

As fluid leaks from the arterial ends of the capillaries into the interstitial spaces, only small quantities of protein accompany it, but then, as fluid is reabsorbed at the venous ends of the capillaries, most of the protein is left behind. Therefore, *protein progressively accumulates in the interstitial fluid* and this in turn *increases the tissue colloid osmotic pressure*. The osmotic pressure decreases reabsorption of fluid by the capillaries, thereby *promoting increased tissue fluid volume* and *less negative interstitial fluid pressure*. The less negative pressure then allows the lymphatic pump to pump the interstitial fluid into the lymphatic capillaries, and this fluid carries with it the excess protein that has accumulated. This continual *washout* of the protein keeps its concentration at a low level in the interstitial fluid.

To summarize, an increase in tissue fluid protein increases the rate of lymph flow, and this washes the proteins out of the tissue spaces, automatically returning the protein concentration to its normal low level.

The importance of this function of the lymphatics cannot be stressed too strongly, for *there is no other route besides the lymphatics through which excess proteins can return to the circulatory system*. If it were not for this continual removal of proteins, the dynamics of fluid exchange at the blood capillaries would become so abnormal within only a few hours that life could no longer continue. There is certainly no other function of the lymphatics that can even approach this in importance.

MECHANISM OF NEGATIVE INTERSTITIAL FLUID PRESSURE

Until recent measurements of the interstitial fluid pressure demonstrated that the interstitial *free* fluid pressure is negative rather than positive, as explained in the preceding chapter, it had been taught that the normal interstitial fluid pressure ranges between +1 and +4 mm. Hg, and it has still been difficult to understand how negative pressure can develop in the interstitial fluid spaces. However, we can explain this negative interstitial fluid pressure by the following considerations:

First, it was pointed out above that fluid can flow into lymphatic vessels from interstitial spaces even when the interstitial fluid pressure is negative, mainly because the lymphatic pump can create slight degrees of suction. The continual

movement of interstitial fluid into the lymphatics keeps the protein concentration of the interstitial fluid at a low value and thereby keeps the colloid osmotic pressure also at a low value, usually at about 5 mm. Hg in most peripheral tissues such as the muscles.

Second, the negativity of the interstitial fluid pressure can then be explained mainly on the basis of the balance of forces at the capillary membrane. If we add all the other forces besides the interstitial fluid pressure that cause movement of fluid across the capillary membrane, we find the following:

	mm. Hg
Outward force:	
Capillary pressure	17
Interstitial fluid colloid osmotic pressure	5
TOTAL	22
Inward force:	
Colloid osmotic pressure	28
DIFFERENCE (Interstitial fluid pressure)	−6

Thus, we see that the interstitial fluid pressure required to balance the other forces across the capillary membrane is −6 mm. Hg. Thus, −6 mm. of the negative interstitial fluid pressure is caused by this imbalance of forces at the capillary membrane. Indirectly, this results from the continual pumping of *protein* into the lymphatic vessels. Another −0.3 mm. Hg is caused by the continual pumping of *fluid* into the lymphatic vessels, giving a total negativity of −6.3 mm. Hg.

Significance of Negative Interstitial Fluid Pressure as a Means for Holding the Body Tissues Together. In the past it has been assumed that the different tissues of the body are held together entirely by connective tissue fibers. However, at many places in the body, connective tissue fibers are absent. This occurs particularly at points where tissues slide over each other. Yet, even at these places, the tissues are held together by the negative interstitial fluid pressure, which is actually a partial vacuum. When the tissues lose their negative pressure, fluid accumulates in the spaces, and the condition known as *edema* occurs, which is discussed later in the chapter.

Significance of the Normally "Dry" State of the Interstitial Spaces. The normal tendency for the capillaries to absorb fluid from the interstitial spaces and thereby to create a partial vacuum causes all the minute structures of the interstitial spaces to be *compacted*. Figure 31–5 illustrates a physical model of the tissues constructed to illustrate this effect. To the left, positive pressure is

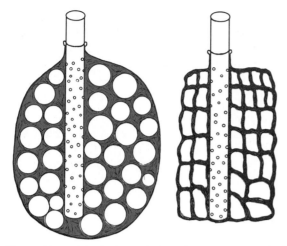

Figure 31–5. Physical model of the tissues constructed of a rubber bag, a perforated tube to simulate a capillary, balloons filled with water to simulate cells, and cotton between the balloons to simulate intercellular elements. *Left:* edematous state. *Right:* nonedematous state.

present and excessive quantities of fluid are present in the "interstitial spaces." To the right, negative pressure has been applied through the perforated tube, which simulates a capillary, and the tissue elements are pulled tightly together. This represents a "dry" state; that is, no *excess* fluid is present besides that required simply to fill the crevices between the tissue elements.

EDEMA

Edema means the presence of excess interstitial fluid in the tissues. Referring again to Figure 31–5, one sees that the left-hand portion of the figure represents the edematous state, while the right-hand portion represents the nonedematous state.

Obviously, any factor that increases the interstitial fluid pressure high enough can cause excess interstitial fluid volume and thereby cause edema. However, to explain the conditions under which edema develops, we must first characterize the *pressure-volume curve* of the interstitial fluid spaces.

PRESSURE-VOLUME CURVE OF THE INTERSTITIAL FLUID SPACES

Figure 31–6 illustrates the average relationship between pressure and volume in the interstitial fluid spaces in the human body as extrapolated from measurements in the dog. The shape of the curve was determined in the following manner: A dog's leg was removed from its body and then perfused with concentrated dextran solution having a colloid osmotic pressure about twice that of normal plasma. This high colloid osmotic force inside the capillaries caused absorption of fluid from the interstitial spaces and caused the weight of the leg to decrease. Measurement of this change in weight provided a means for measuring the

decrease in interstitial fluid volume. Simultaneously, the free fluid pressure in the interstitial fluid spaces was measured using the implanted capsule method described in the preceding chapter. Later in the experiment, fluid having no colloid osmotic pressure was perfused through the limb, and this caused tremendous quantities of fluid to leak out of the capillaries into the interstitial spaces, thereby increasing the interstitial fluid volume. The curve of Figure 31–6 represents the average results obtained in experiments of this type.

The Slight Interstitial Fluid Volume Change in the Negative Pressure Range. One of the most significant features of the curve in Figure 31–6 is that so long as the interstitial fluid pressure remains in the negative range there is little change in interstitial fluid volume despite marked change in pressure. Therefore, edema will not occur so long as the interstitial free fluid pressure remains negative. Indeed, in several hundred measurements of interstitial free fluid pressure made in experimental animals, no edema has ever been recorded in the presence of negative interstitial pressure.

Tremendous Increase in Interstitial Fluid Volume When the Interstitial Free Fluid Pressure Becomes Positive. Note in Figure 31–6 that just as soon as the interstitial free fluid pressure rises to equal atmospheric pressure (zero pressure), the slope of the pressure-volume curve suddenly changes and the volume increases precipitously. An additional increase in interstitial free fluid pressure of only 1 to 3 mm. Hg now causes the interstitial fluid volume to increase several hundred per cent. Finally, at the very top of the figure, the skin begins to be stretched so that the volume now increases much less rapidly.

Similarity of the Tissue Spaces to an Elastic Bag. If one will think for a moment, one will realize that a

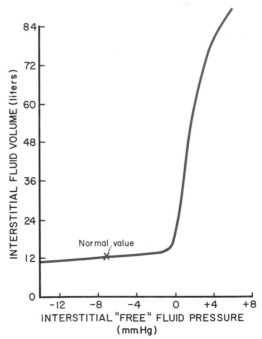

Figure 31–6. Pressure-volume curve of the interstitial spaces. (Extrapolated to the human being from data obtained in dogs.)

pressure-volume curve similar to that illustrated in Figure 31–6 can also be recorded from almost any collapsible elastic bag, such as a rubber balloon. When negative pressure is applied to the balloon, its volume remains constant at very near zero even though the pressure becomes very negative. But when the pressure is increased above atmospheric pressure, the balloon suddenly begins to expand. Almost no additional pressure is then required to fill the balloon until its walls begin to stretch. Then, further filling is impeded by the stretching wall. These are precisely the characteristics illustrated in Figure 31–6 for the interstitial fluid spaces of the body. These spaces are in effect a collapsed bag which expands greatly when the interstitial free fluid pressure rises above the surrounding atmospheric pressure.

Tissue Space Compliance in Different Pressure Ranges. Another way to express the pressure-volume characteristics of the interstitial fluid spaces is in terms of compliance, which is defined as the change in volume for a given change in pressure. In the negative pressure range, the compliance of the interstitial spaces is slight, about 400 ml./mm. Hg for the entire human body (as extrapolated from measurements in dogs). But, just as soon as interstitial free fluid pressure rises into the positive range, this compliance increases tremendously, rising to about 10,000 ml./mm. Hg. Thus, the compliance increases approximately 25-fold between the negative pressure range and the positive pressure range.

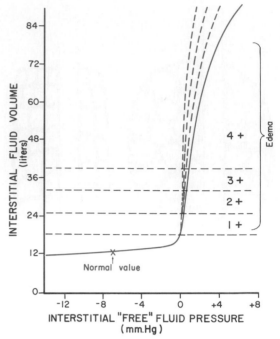

Figure 31–7. Relationship of edema to the pressure-volume curve of the interstitial spaces. The dashed curves show the effect of prolonged edema on the pressure-volume curve.

POSITIVE INTERSTITIAL FLUID PRESSURE AS THE PHYSICAL BASIS FOR EDEMA

After studying the pressure-volume curve of Figure 31–6, one can readily see that whenever the interstitial free fluid pressure rises above the surrounding atmospheric pressure, the tissue spaces begin to swell. Therefore, *the physical cause of edema is positive pressure (that is, supra-atmospheric pressure) in the interstitial fluid spaces.*

Degree of Edema in Relation to the Degree of Positive Pressure. The solid curve of Figure 31–7 is the same pressure-volume curve as that shown in Figure 31–6, but an edema scale has been added to the figure. One-plus edema means edema that is barely detectable, and 4+ edema means edema in which the limbs are swollen to diameters 1.5 to 2 times normal.

Note in Figure 31–7 that edema usually is not detectable in tissues until the interstitial fluid volume has risen to about 30 per cent above normal. And note also that the interstitial fluid volume increases to several hundred per cent above normal in seriously edematous tissues.

Stretch of the Tissue Spaces in Chronic Edema. If edema persists even for a few hours, and especially if it persists for weeks, months, or years, the tissue spaces gradually become stretched. As a result, the pressure-volume curve changes from the solid curve in Figure 31–7 to the progressively higher dashed curves. In other words, in chronic edema the tissue "bag" expands, which increases the ease with which the tissues can develop severe edema. Even a 1 to 2 mm. Hg pressure rise above atmospheric pressure can cause 4+ edema once the tissue spaces have been stretched for many

days. This phenomenon of stretch of the tissues is called *delayed compliance* or *stress-relaxation of the tissue spaces.*

The Phenomenon of "Pitting" Edema. If one presses a finger on the skin over an edematous area and then suddenly removes the finger, a small depression called a "pit" remains. Gradually, within 5 to 30 seconds, the pit disappears. The cause of pitting is that edema fluid has been translocated away from the area beneath the pressure point. The fluid simply flows through the tissue spaces to other tissue areas. Then, when the finger is removed, 5 to 30 seconds is required for the fluid to flow back into the area from which it had been displaced.

Nonpitting Edema. Occasionally, the fluid in seriously edematous tissues cannot be mobilized to other areas by pressure. The usual cause of this is coagulation of the fluid in the tissues. For instance, in an infected or traumatized area, large quantities of fluid can collect, but coagulation of the fluid prevents the fluid from being expulsed by pressure. Also, swelling of the tissue cells, which occurs when cells are traumatized, diseased, or fail to receive adequate nutrition can also cause nonpitting edema. This type of edema is frequently also called *brawny* edema.

THE CONCEPT OF A "SAFETY FACTOR" BEFORE EDEMA DEVELOPS

Safety Factor Caused by the Normal Negative Interstitial Free Fluid Pressure. One can readily see from Figures 31–6 and 31–7 that the interstitial fluid pressure must rise from the normal value of −6.3 to above zero mm. Hg before edema begins to appear. Thus, there is a safety factor of 6.3 mm. Hg caused by the normal negative interstitial free fluid pressure before edema appears.

Safety Factor Caused by Flow of Lymph from the Tissues. Another safety factor that helps to prevent edema is increased lymph flow. When the interstitial free fluid pressure rises above the normal value of −6.3 mm. Hg, lymph flow increases greatly, which removes a greater portion of the extra fluid entering the interstitial spaces. And this obviously helps to prevent the development of edema.

One can estimate that maximally increased lymph flow gives approximately a 7 mm. Hg safety factor, for the extra lymph flow can carry away from the tissues approximately the extra amount of fluid that is formed by a 7 mm. Hg excess in capillary pressure.

Safety Factor Caused by Washout of Protein From the Interstitial Spaces. In addition to removal of fluid volume from the interstitial fluid spaces, increased lymph flow also washes out most of the proteins from the interstitial fluid spaces, decreasing the colloid osmotic pressure of the interstitial fluid from the normal value of 5 mm. Hg down to about 1 mm. Hg. This provides another 4 mm. Hg safety factor.

Total Safety Factor and Its Significance. Now, let us add all the above safety factors:

	mm. Hg
Negative interstitial fluid pressure	6.3
Lymphatic flow	7.0
Lymphatic washout of proteins	4.0
TOTAL	17.3

Thus we find that a total safety factor of about 17 mm. Hg is present to prevent edema. This means that the capillary pressure must rise about 17 mm. Hg above its normal value of 17 mm. Hg — that is, above 34 mm. Hg — before edema begins to appear. Or the plasma colloid osmotic pressure must fall from the normal level of 28 mm. Hg to below 11 mm. Hg before edema begins to appear. This explains why the normal human being does not become edematous until severe abnormalities occur in the circulatory system.

EDEMA RESULTING FROM ABNORMAL CAPILLARY DYNAMICS

From the discussion of capillary and interstitial fluid dynamics in the preceding and present chapters, it is already evident that several different abnormalities in these dynamics can increase the tissue pressure and in turn cause extracellular fluid edema. The different causes of extracellular fluid edema are:

Increased Capillary Pressure as a Cause of Edema. Figure 31–8A shows the effect of increased mean capillary pressure on the dynamics of fluid exchange at the capillary membrane. When the mean capillary pressure first becomes abnormally high, more fluid flows out of the capillary than returns into the capillary and, therefore, it collects in the tissue spaces until the interstitial free fluid pressure rises high enough to balance the excessive level of pressure in the capillaries. In Figure 31–8A the mean capillary pressure is 41 mm. Hg instead of the usual normal 17 mm. Hg.

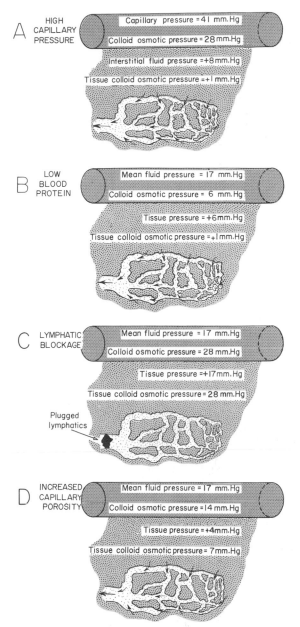

Figure 31–8. Various causes of edema.

Consequently, in this instance enough fluid flows into the tissue spaces to raise the interstitial fluid pressure to +8 mm. Hg. This is far above atmospheric pressure of 0 mm. Hg, and therefore causes progressive enlargement of the tissue spaces with tremendous expansion of the extracellular fluid volume.

Causes of Increased Capillary Pressure. Increased capillary pressure can result from any clinical condition that causes either venous obstruction or arteriolar dilatation. Large venous blood clots frequently cause local areas of venous obstruction, which block the return of blood to the heart and promote edema in the tissues normally drained by the obstructed veins.

More frequently, capillary pressure is increased by obstruction of venous return due to cardiac failure, for, when the heart no longer pumps blood out of the veins with ease, blood dams up in the venous system. The capillary pressure rises, and serious "cardiac edema" occurs. The dynamics of this type of edema are complicated, however, and were discussed in detail in Chapter 26.

When arteriolar dilatation occurs in localized areas of the body, blood flows rapidly through the locally dilated arterioles and the capillary pressure increases tremendously. Therefore, local edema results. Such local edema occurs commonly in allergic conditions and in the condition known as *angioneurotic edema*. Allergic reactions (discussed in Chapter 7) cause the release of histamine into the tissues; histamine relaxes the smooth muscle of arterioles and, when present in large amounts, constricts the venules. The localized edematous areas that result are called "hives" or *urticaria*.

Angioneurotic edema apparently is caused by localized decrease in arteriolar tone due to abnormal vascular control by the autonomic nervous system. When a person is emotionally upset, such edema frequently occurs in the larynx and causes hoarseness.

Decreased Plasma Proteins as a Cause of Edema. Figure 31–8B illustrates the abnormal dynamics that occur at the capillary membrane when the quantity of plasma protein falls to abnormally low values. The major effect is a markedly lowered colloid osmotic pressure of the plasma. Consequently, the capillary pressure far overbalances the colloid osmotic pressure, increasing the tendency for fluid to leave the capillaries and enter the tissue spaces. As a result, fluid collects in the tissue spaces, and the interstitial free fluid pressure rises. As long as the pressure remains excessively elevated, the tissue spaces continually enlarge, with the edema becoming progressively worse.

As is also the case with changes in capillary pressure, the decrease in plasma colloid osmotic pressure must be extreme before edema begins to develop. Since the normal safety factor is about 17 mm. Hg, one can calculate that edema begins to appear when the plasma colloid osmotic pressure falls below approximately 11 mm. Hg.

Conditions That Decrease the Plasma Protein Concentration. Albumin is often lost from the plasma in large quantities when the skin is extensively burned. Therefore, one of the complications of severe burns is not only severe edema in the tissues surrounding the burned area but also edema throughout the body because of lowered colloid osmotic pressure.

Often large quantities of protein, especially albumin, are lost through the kidneys into the urine in the disease known as *nephrosis*. Sometimes as much as 20 to 30 grams of albumin are lost each day, and the colloid osmotic pressure of the plasma may fall to one-half normal or even less. This results in severe edema, and the edema itself is likely to contribute to death by means that are discussed later in the chapter.

Finally, persons who do not have sufficient protein in their diets are unable to form adequate quantities of plasma protein and, therefore, are likely to develop protein deficiency edema, which is called *nutritional edema*. This occurs frequently in famine areas, especially in some tribal areas of Africa.

Lymphatic Obstruction as a Cause of Edema. A small amount of protein leaks continually from the capillaries into the tissue spaces, but this protein cannot be reabsorbed into the circulatory system through the capillary membrane. The only route by which the protein can be returned to the circulatory system is through the lymphatics. If the lymphatic drainage from any area of the body becomes blocked, more and more protein collects in the local tissue spaces until finally the concentration of this protein may approach the concentration of protein in the plasma. As shown in Figure 31–8C, the colloid osmotic pressure of the tissue fluids may theoretically rise to as high as 28 mm. Hg, and, to balance this, fluid collects in the tissues until the interstitial fluid pressure rises to a value equal to the capillary pressure, about 17 mm. Hg. Such elevated tissue pressure rapidly expands the tissue spaces, with resultant edema of the severest kind.

Causes of Lymphatic Obstruction. One of the most common causes of lymphatic obstruction is *filariasis* — that is, infection by nematodes of the superfamily Filarioidea. The disease is widespread in the tropics, where larvae (microfilariae) are transmitted to human hosts by mosquitoes. The larvae pass out of the capillaries into the interstitial fluid and then by way of the lymph into the lymph nodes. Subsequent inflammatory reactions progressively obstruct the lymphatic channels of these nodes with scar tissue. After several years, the lymphatic drainage from one of the peripheral parts of the body may become almost totally occluded. Thus, a leg can swell to such a size that it might weigh as much as all the remainder of the body. Because of this extreme degree of edema, the swollen condition is frequently called *elephantiasis*. An interesting type of elephantiasis occasionally occurs in the scrotum, which has been known to enlarge so much that the person must carry it in a wheelbarrow in order to move about.

Lymphatic obstruction also occurs following operations for removal of cancerous tissue. Because the lymph nodes draining a cancerous area of the body must be removed in order to prevent possible spread of the cancer, the return of lymph to the circulatory system from that area will be blocked. Occasionally a radical mastectomy for removal of a cancerous breast causes the corresponding arm to swell to as much as twice its normal size, but usually the swelling regresses during the following two to three months as new lymph channels develop.

Increased Permeability of the Capillaries as a Cause of Edema. Figure 31–8D illustrates a capillary whose membrane has become so permeable that even protein molecules pass from the plasma into the interstitial spaces with ease. The protein content of the plasma decreases while that of the interstitial spaces increases. In the example of the figure the tissue pressure rises to +7 mm. Hg in order to balance the changes in plasma and tissue colloid osmotic pressure occasioned by the leakage of protein. The elevated interstitial free fluid pressure in turn causes progressive edema.

Cause of Increased Capillary Permeability. Capillaries become excessively permeable when any factor destroys the integrity of the capillary endothelium. Burns are a frequent cause of increased permeability of the capillaries because overheated capillaries become friable, and their pores enlarge. Allergic reactions also frequently cause the release of histamine or various polypeptides that damage the capillary membranes and cause increased permeability.

A bacterial toxin produced by *Clostridium oedematiens* can often cause such extreme increase in capillary permeability that plasma loss into the tissues kills the patient within a few hours.

EDEMA CAUSED BY KIDNEY RETENTION OF FLUID

When the kidney fails to excrete adequate quantities of urine, and the person continues to drink normal amounts of water and ingest normal amounts of electrolytes, the total amount of extracellular fluid in the body increases progressively. This fluid is absorbed from the gut into the blood and elevates the capillary pressure. This in turn causes most of the fluid to pass into the interstitial fluid spaces, elevating the interstitial fluid pressure as well. Therefore, simple retention of fluid by the kidneys can result in extensive edema. Furthermore, if the retained fluid is mainly water, intracellular edema will also result, as will be discussed in Chapter 33.

THE PRESENCE AND IMPORTANCE OF GEL IN THE INTERSTITIAL SPACES

Up to this point we have talked about interstitial fluid as if it were entirely in a mobile, "free" state. However, in the normal interstitial spaces, the interstitial fluid is entrapped in a *gel-matrix* comprised of large *proteoglycan molecules* (also called *mucopolysaccharides*). These molecules generally have molecular weights greater than one million, are folded into odd shapes, and are entangled among each other, which is what causes the gel-like nature of the normal interstitial fluid. The widths of the spaces between the molecules are usually only 20 to 40 nanometers, which is so small that water molecules and dissolved substances in the interstitial fluids can *flow* through this gel-matrix only with considerable difficulty. Therefore, the interstitial fluid in normal tissues is relatively immobilized.

Even though the fluid in the interstitial gel cannot "flow" easily from one part of the interstitium to another, the individual molecules still do move by random kinetic motion as rapidly as ever. Furthermore, since these molecules generally have diameters 20 or more times less than the sizes of the spaces between the proteoglycan molecules, they can still move by the process of diffusion through the interstitium more than 95 per cent as effectively as in free fluid. Therefore, nutrients can diffuse from the capillaries to the cells almost equally as well through the gel as through free fluid.

There are several important advantages of having a gel-matrix in the interstitium. Some of these are:

(1) The proteoglycan molecules act as a "filler" to hold the cells apart. This creates large enough spaces for fluid and nutrients to diffuse from the capillaries to those cells that are located some distance away from the capillaries.

(2) Since the fluid in the tissue spaces is mainly immobilized by the gel, this prevents fluid from flowing through the tissue spaces from the upper part of the body into the lower body. Otherwise, all of the interstitial fluid (16 per cent of the total body weight) would flow within minutes into the tissue spaces of the legs and feet.

(3) The proteoglycan meshwork not only immobilizes fluid but also immobilizes bacteria and keeps them from spreading through the tissues.

Relation of Edema Fluid to the Gel. When excessive amounts of fluid begin to accumulate in the interstitial spaces, the gel at first entraps this additional fluid as well, and the entire gel-matrix of the interstitium swells. However, once the gel has swollen more than 30 to 50 per cent, the entanglements of the proteoglycan molecules begin to break up, and then free fluid spaces develop all through the interstitium. As still more fluid accumulates, the free fluid spaces become so large that they coalesce to form large free fluid channels in the tissues. Once this has occurred, fluid then flows freely through the tissues.

Figure 31–9 illustrates the volume relationships between free interstitial fluid, gel fluid, and total interstitial fluid in both the nonedematous and the edematous states. Note that under normal conditions, when the interstitial free fluid pressure is in its normal negative pressure range, there is almost imperceptible free fluid in the tissues. Instead, almost all of the fluid is in the gel phase and it is also highly immobile. On the other hand, as the interstitial free fluid pressure rises and as the state of edema approaches, the gel swells some 30 to 50 per cent, after which it swells no more. With still more increase in interstitial free fluid pressure, all the additional edema fluid that accumulates is free fluid that is highly mobile through the tissue spaces. It is this high degree of mobility that causes the edema to be of the pitting type, as was explained earlier in the chapter.

Relationship of Interstitial Fluid Gel to the Regulation of Interstitial Fluid Volume. Since approximately 16 per cent of the average tissue is composed of interstitial fluid and normally almost all of it is in the gel state, one can derive the following theory for the regulation of interstitial fluid volume. The mechanism previously described for creating a negative pressure in the tissue spaces is actually a "drying" mechanism that always attempts to remove any *free* fluid that appears in the tissues. Thus, normally, essentially all free fluid is removed as rapidly as it is formed, leaving in the normal tissues only the gel, which still constitutes about 16 per cent of the tissue volume. The question remains, though, why does this drying mechanism remove only a small amount of fluid from the gel? The answer to this has two components: First, the fine reticular filaments of the gel are composed of hyaluronic acid molecules that are coiled like springs and are compressed against each other. Therefore, the elastic forces of these mole-

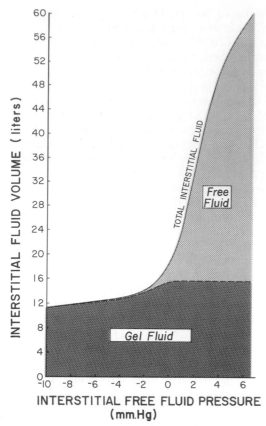

Figure 31–9. Effect of increasing interstitial fluid pressure on the volumes of total interstitial fluid, gel fluid, and free fluid. Note that significant amounts of free fluid occur only when the interstitial free fluid pressure becomes positive. (Modified from Guyton, Granger, and Taylor: *Physiol. Rev., 51*:527, 1971.)

cules prevent further compression in the same way that cotton fibers in absorbent cotton prevent compression beyond a certain point. Second, the gel has a slight amount of osmotic pressure caused by the Donnan equilibrium effect; that is, the gel reticulum has negative electrostatic charges that hold small mobile positive ions — mainly sodium ions — within the gel. These ions in turn cause osmosis of water into the gel. The quantity of mucopolysaccharides in the tissue gel is sufficient to give an osmotic absorptive pressure in the gel that is calculated to be about 2 mm. Hg. The elastic recoil of the hyaluronic acid "springs" gives approximately another 5 mm. Hg of recoil force, making a total of about 7 mm. Hg to resist dehydration caused by the −6.3 mm. Hg pressure in the free fluid of the tissue spaces.

PULMONARY INTERSTITIAL FLUID DYNAMICS

The dynamics of the pulmonary interstitial fluid are essentially the same as those for fluid in the peripheral tissues except for the following important quantitative differences:

1. The pulmonary capillary pressure is very low in comparison with the systemic capillary pressure, approximately 7 mm. Hg in comparison with 17 mm. Hg.

2. The interstitial free fluid pressure in the lung interstitium has been measured to be −8 mm. Hg in comparison with −6 mm. Hg in subcutaneous tissue.

3. The pulmonary capillaries are relatively leaky to protein molecules so that the protein concentration of lymph leaving the lungs is relatively high, averaging about 4 gm. per cent instead of 2 gm. per cent in the peripheral tissues.

4. The rate of lymph flow from the lungs is also very high, mainly because of the continuous pumping motion of the lungs.

5. The interstitial spaces of the alveolar portions of the lungs are very narrow, represented by the minute spaces between the capillary endothelium and the alveolar epithelium.

6. The alveolar epithelia are not strong enough to resist very much positive pressure. They are probably ruptured by any positive pressure in the interstitial spaces greater than atmospheric pressure (0 mm. Hg), which allows dumping of fluid from the interstitial spaces into the alveoli.

Now let us see how these quantitative differences affect pulmonary fluid dynamics.

Interrelationship Between Interstitial Fluid Pressure and Other Pressures in the Lung. Figure 31–10 illustrates a pulmonary capillary, a pulmonary alveolus, and a lymphatic capillary draining the interstitial space between the capillary and the alveolus. Note the balance of forces at the capillary membrane as follows:

PRESSURES CAUSING FLUID MOVEMENT

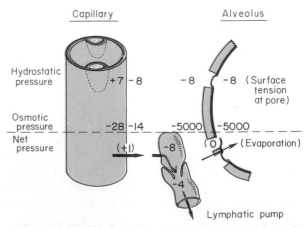

Figure 31–10. Hydrostatic and osmotic forces at the capillary (left) and alveolar membrane (right) of the lungs. Also shown is a lymphatic (center) that pumps fluid from the pulmonary interstitial spaces. (Modified from Guyton, Taylor, and Granger: Dynamics of the Body Fluids. Philadelphia, W. B. Saunders Company, 1975.)

	mm. Hg
Forces tending to cause movement of fluid outward from the capillaries and into the pulmonary interstitium:	
Capillary pressure	7
Interstitial fluid colloid osmotic pressure	14
TOTAL OUTWARD FORCE	21

	mm. Hg
Forces tending to cause absorption of fluid into the capillaries:	
Plasma colloid osmotic pressure	28
Interstitial free fluid pressure	−8
TOTAL INWARD FORCE	20

Note that the normal outward forces are slightly greater than the inward forces. The *net mean filtration pressure* at the pulmonary capillary membrane can be calculated as follows:

	mm. Hg
Total outward force	+21
Total inward force	−20
NET MEAN FILTRATION PRESSURE	+1

This net filtration pressure causes a slight continual flow of fluid from the pulmonary capillaries into the interstitial spaces, and, except for a small amount that evaporates in the alveoli, this fluid is pumped back to the circulation through the pulmonary lymphatic system.

Fluid Exchange at the Pulmonary Alveolar Membrane; the Mechanism for Keeping the Alveoli "Dry." The alveolar epithelial membrane is quite different from the pulmonary capillary membrane in the following way: The pulmonary capillaries, like other capillaries of the body, have very large slit-pores between the adjacent endothelial cells. Ions such as sodium, chloride, and potassium as well as crystalloid molecules such as glucose, urea, and so forth can pass through these large capillary pores with ease. On the other hand, the alveolar epithelial membrane contains no such large openings. Therefore, all of the above molecules can cause osmotic pressure effects at the alveolar membrane even though they have no such effect at the pulmonary capillary membrane. For instance, when water enters the alveoli, the high concentration of the different dissolved substances in the pulmonary interstitial fluid causes almost instantaneous osmosis of the water from the alveoli into the interstitial fluid, and the fluid is then absorbed into the pulmonary capillaries because of the *colloid* osmotic pressure of the plasma. Indeed, in a person who drowns in fresh water, enough fluid can be absorbed from the alveoli into the blood within two to three minutes to cause fibrillation of the heart because of dilution of the blood electrolytes.

In addition to osmosis of fluid from the alveoli, small amounts of fluid can also be moved from the alveoli into the interstitial spaces as a result of suction by the negative pressure in these spaces. Even saline solution, the ions of which prevent its osmosis into the interstitial fluid, moves slowly from the alveoli into the interstitial spaces because of the negative interstitial pressure.

But much more important than fluid absorption from the alveoli is the following question: How is it that the fluid normally present in the interstitial spaces is prevented from flooding the alveoli? The answer to this is again the negative interstitial fluid pressure of approximately −8 mm. Hg, which continually tends to pull fluid inward through the alveolar membrane and therefore also prevents fluid loss in the outward direction.

The only fluid that does go outward through the alveolar membrane is that small amount that moves by the mechanism of *capillarity* through the cellular pores of the epithelial cells and then creeps along the lining surfaces of the alveoli to keep them moist.

PULMONARY EDEMA

Pulmonary edema occurs in the same way that it occurs elsewhere in the body. Any factor that causes the pulmonary interstitial fluid pressure to rise from the negative range into the positive range will cause sudden filling of the pulmonary interstitial spaces and in more severe cases even the alveoli with large amounts of free fluid.

The usual causes of pulmonary edema are:

1. Left heart failure or mitral valvular disease with consequent great increase in pulmonary capillary pressure and flooding of the interstitial spaces.

2. Damage to the pulmonary capillary membrane caused by breathing noxious substances such as chloride gas and sulfur dioxide gas.

3. Decrease in plasma colloid osmotic pressure to a low enough level that fluids transude from the blood into the pulmonary interstitial spaces (occurs only rarely).

Pulmonary "Interstitial Fluid" Edema Versus Pulmonary "Alveolar" Edema. The interstitial fluid volume of the lungs usually cannot increase more than about 50 per cent (representing less than 100 milliliters of fluid) before the alveolar epithelial membranes rupture and fluid begins to pour from the interstitial spaces into the alveoli. The cause of this is simply the almost infinitesimal tensional strength of the pulmonary alveolar epithelium; that is, any positive pressure in the interstitial fluid spaces seems to cause immediate rupture of the alveolar epithelium.

Therefore, except in the mildest cases of pulmonary edema, edema fluid always enters the alveoli; if this edema becomes severe enough, it can cause death by suffocation, as has already been discussed in Chapters 26 and 27.

Safety Factor Against Pulmonary Edema. All the factors that tend to prevent edema in the peripheral tissues also tend to prevent edema in the lungs. That is, before positive interstitial fluid pressure can occur and cause edema all the following factors must be overcome: (1) the normal negativity of the interstitial fluid pressure of the lungs, (2) the lymphatic pumping of fluid out of the interstitial spaces, and (3) the increased osmosis of fluid into the pulmonary capillaries caused by decreased protein in the interstitial fluid when the lymph flow increases. In experiments in animals it has been found that the pulmonary capillary pressure normally must rise to a value at least equal to the plasma colloid osmotic pressure before significant pulmonary edema will occur. Thus, in the human being, who normally has a plasma colloid osmotic pressure of 28 mm. Hg, one can predict that the pulmonary capillary pressure must rise from the normal level of 7 mm. Hg to over 28 mm. Hg to cause pulmonary edema, giving a safety factor against edema of about 21 mm. Hg.

Safety Factor in Chronic Conditions. When the pulmonary capillary pressure is elevated chronically (for at least two weeks), the lungs become even more resistant to pulmonary edema because the lymph vessels expand greatly, increasing their capability for carrying fluid away from the interstitial spaces as much as 10-fold. Therefore, a patient with chronic mitral stenosis frequently has a pulmonary capillary pressure as great as 40 to 45 mm. Hg without having significant pulmonary edema.

Thus, in chronic pulmonary edema, the safety factor against edema can rise to as high as 35 to 40 mm. Hg in comparison with a normal value of 21 mm. Hg under acute conditions.

Rapidity of Death in Acute Pulmonary Edema. When the pulmonary capillary pressure does rise above the safety factor level, lethal pulmonary edema can occur within hours if it is only slightly above the safety factor, and within 20 to 30 minutes if it is as much as 25 to 30 mm. Hg above the safety factor level. Thus, in acute left heart failure, in which the pulmonary capillary pressure occasionally rises to as high as 50 mm. Hg, death frequently ensues within 30 minutes from acute pulmonary edema.

REFERENCES

Bengis, R. G., and Guyton, A. C.: Some pressure and fluid dynamic characteristics of the canine epidural space. *Am. J. Physiol., 232*:H-255, 1977.

Brace, R. A., et al.: Time course of lymph protein concentration in the dog. *Microvasc. Res., 14*:243, 1977.

Brigham, K. L.: Lung lymph composition and flow in experimental pulmonary edema. *In* Fishman, A. P., and Renkin, E. M. (eds.): Pulmonary Edema. Baltimore, Waverly Press, 1979, p. 161.

Casley-Smith, J. R.: Lymph and lymphatics. *In* Kaley, G., and Altura, B. M. (eds.): Microcirculation. Vol. 1. Baltimore, University Park Press, 1977, p. 423.

Clodius, L. (ed.): Lymphedema, Stuttgart, Thieme, 1977.

Drinker, C. K.: The Lymphatic System. Stanford, Stanford University Press, 1942.

Fishman, A. P., and Pietra, G. G.: Hemodynamic pulmonary edema. *In* Fishman, A. P., and Renkin, E. M. (eds.): Pulmonary Edema. Baltimore, Waverly Press, 1979, p. 79.

Fishman, A. P., and Renkin, E. M. (eds.): Pulmonary Edema. Baltimore, American Physiological Society, 1979.

Gaar, K. A., Jr., et al.: Effect of capillary pressure and plasma protein on development of pulmonary edema. *Am. J. Physiol., 213*:79, 1967.

Gaar, K. A., Jr., et al.: Effect of lung edema on pulmonary capillary pressure. *Am. J. Physiol., 216*:1370, 1969.

Gabbiani, G., and Majno, G.: Pathophysiology of small vessel permeability. *In* Kaley, G., and Altura, B. M. (eds.): Microcirculation. Vol. III. Baltimore, University Park Press, 1977.

Gauer, O. H., Henry, J. P., and Behn, C.: The regulation of extracellular fluid volume. *Annu. Rev. Physiol., 32*:547, 1970.

Gross, J. F., and Popel, A. (eds.): Mathematics of Microcirculation Phenomena. New York, Raven Press, 1980.

Guyton, A. C.: Interstitial fluid pressure: II. Pressure-volume curves of interstitial space. *Circ. Res., 16*:452, 1965.

Guyton, A. C.: Interstitial fluid pressure-volume relationships and their regulation. *In* Wolstenholme, G. E. W., and Knight, J. (eds.): CIBA Foundation Symposium on Circulatory and Respiratory Mass Transport. London, J. & A. Churchill, 1969, p. 4.

Guyton, A. C.: Introduction to Part I: Pulmonary alveolar-capillary interface and interstitium. *In* Fishman, A. P., and Hecht, H. H. (eds.): The Pulmonary Circulation and Interstitial Space. Chicago, University of Chicago Press, 1969, p. 3.

Guyton, A. C., and Coleman, T. G.: Regulation of interstitial fluid volume and pressure. *Ann. N.Y. Acad. Sci., 150*:537, 1968.

Guyton, A. C., and Lindsey, A. W.: Effect of elevated left atrial pressure and decreased plasma protein concentration on the development of pulmonary edema. *Circ. Res., 7*:649, 1959.

Guyton, A. C., et al.: Dynamics and Control of the Body Fluids. Philadelphia, W. B. Saunders Co., 1975.

Guyton, A. C., et al.: Interstitial fluid pressure. *Physiol. Rev., 51*:527, 1971.

Guyton, A. C., et al.: Circulatory Physiology II. Dynamics and Control of the Body Fluids. Philadelphia, W. B. Saunders Co., 1975.

Guyton, A. C., et al.: Dynamics of subatmospheric pressure in the pulmonary interstitial fluid. *In* Lung Liquids. CIBA Symposium. Amsterdam, Elsevier/North-Holland, 1976, p. 77.

Guyton, A. C., et al.: Forces governing water movement in the lung. *In* Fishman, A. P., and Renkin, E. M. (eds.): Pulmonary Edema. Baltimore, Waverly Press, 1979, p. 65.

Kennedy, J. F.: Proteoglycans: Biological and chemical aspects of human life. New York, Elsevier Scientific Publishing Co., 1979.

Leak, L. V., and Burke, J. F.: Ultrastructural studies on the lymphatic anchoring of filaments. *J. Cell Biol., 36*:129, 1968.

Lennarz, W. J. (ed.): The Biochemistry of Glycoproteins and Proteoglycans. New York, Plenum Press, 1979.

Mayerson, H. S.: The physiologic importance of lymph. *In* Hamilton, W. F. (ed.): Handbook of Physiology. Sec. 2. Vol. 2. Baltimore, Williams & Wilkins, 1963, p. 1035.

Meyer, B. J., et al.: Interstitial fluid pressure. V. Negative pressure in the lungs. *Circ. Res., 22*:263, 1968.

Nicoll, P. A., and Taylor, A. E.: Lymph formation and flow. *Annu. Rev. Physiol., 39*:73, 1977.

Parker, J. C., et al.: Pulmonary interstitial and capillary pressures estimated from intra-alveolar fluid pressures. *J. Appl. Physiol., 44*(2):267, 1978.

Parker, J. C., et al.: Pulmonary transcapillary exchange and pulmonary edema. *In* Guyton, A. C., and Young, D. B. (eds.): International Review of Physiology: Cardiovascular Physiology III. Vol. 18. Baltimore, University Park Press, 1979, p. 261.

Renkin, E. M.: Lymph as a measure of the composition of interstitial fluid. *In* Fishman, A. P., and Renkin, E. M. (eds.): Pulmonary Edema. Baltimore, Waverly Press, 1979, p. 145.

Robin, E. D.: Permeability pulmonary edema. *In* Fishman, A. P., and Renkin, E. M. (eds.): Pulmonary Edema. Baltimore, Waverly Press, 1979, p. 217.

Schneeberger, E. E.: Barrier function of intracellular junctions in adult and fetal lungs. *In* Fishman, A. P., and Renkin, E. M. (eds.): Pulmonary Edema. Baltimore, Waverly Press, 1979, p. 21.

Staub, N. C. (ed.): Lung Water and Solute Exchange. New York, Marcel Dekker, 1978.

Staub, N. C.: Pathways for fluid and solute fluxes in pulmonary edema. *In* Fishman, A. P., and Renkin, E. M. (eds.): Pulmonary Edema. Baltimore, Waverly Press, 1979, p. 113.

Teplitz, C.: Pulmonary cellular and interstitial edema. *In* Fishman, A. P., and Renkin, E. M. (eds.): Pulmonary Edema. Baltimore, Waverly Press, 1979, p. 97.

Visscher, M. B., et al.: The physiology and pharmacology of lung edema. *Pharm. Rev., 8*:389, 1956.

Weibel, E. R., and Bachofen, H.: Structural design of the alveolar septum and fluid exchange. *In* Fishman, A. P., and Renkin, E. M. (eds.): Pulmonary Edema. Baltimore, Waverly Press, 1979, p. 1.

Yoffey, J. M., and Courtice, F. C. (eds.): Lymphatics, Lymph and Lymphomyeloid Complex. New York, Academic Press, 1970.

32

The Special Fluid Systems of the Body — Cerebrospinal, Ocular, Pleural, Pericardial, Peritoneal, and Synovial

Several special fluid systems exist in the body, each performing functions peculiar to itself. For instance, the cerebrospinal fluid supports the brain in the cranial vault, the intraocular fluid maintains distension of the eyeballs so that the optical dimensions of the eye remain constant, and the potential spaces, such as the pleural and pericardial spaces, provide lubricated chambers in which the internal organs can move. All these fluid systems have characteristics that are similar to each other and that are also similar to those of the interstitial fluid system. However, they are also sufficiently different that they require special consideration.

THE CEREBROSPINAL FLUID SYSTEM

The entire cavity enclosing the brain and spinal cord has a volume of approximately 1650 ml., and about 150 ml. of this volume is occupied by cerebrospinal fluid. This fluid, as shown in Figure 32–1, is found in the ventricles of the brain, in the cisterns around the brain, and in the subarachnoid space around both the brain and the spinal cord. All these chambers are connected with each other, and the pressure of the fluid is regulated at a constant level.

CUSHIONING FUNCTION OF THE CEREBROSPINAL FLUID

A major function of the cerebrospinal fluid is to cushion the brain within its solid vault. Fortunately, the brain and the cerebrospinal fluid have approximately the same specific gravity, so that the brain simply floats in the fluid. Therefore, a blow on the head moves the entire brain simultaneously, causing no one portion of the brain to be momentarily contorted by the blow.

Contrecoup. When a blow to the head is extremely severe, it usually does not damage the brain on the side of the head where the blow is struck, but, instead, on the opposite side. This phenomenon is known as "contrecoup," and the reason for this effect is the following: When the blow is struck, the fluid on the struck side is so incompressible that, as the skull moves, the fluid pushes the brain at the same time. However, on the opposite side, the sudden movement of the skull causes it to pull away from the brain momentarily, creating for a short instant a vacuum in the cranial vault at this point. Then, when the skull is no longer being accelerated by the blow, the vacuum suddenly collapses and the brain strikes the inner surface of the skull. Because of this effect, the damage to the brain of a boxer usually does not occur in the frontal regions but, instead, in the occipital regions.

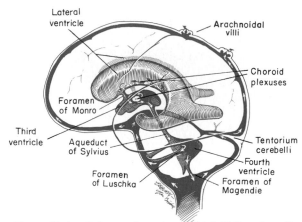

Figure 32–1. Pathway of cerebrospinal fluid flow from the choroid plexuses in the lateral ventricles to the arachnoidal villi protruding into the dural sinuses.

383

FORMATION, FLOW, AND ABSORPTION OF CEREBROSPINAL FLUID

Cerebrospinal fluid is formed at a rate of approximately 800 ml. each day, which is 5 to 6 times as much as the total volume of fluid in the entire cerebrospinal fluid cavity. Most of this fluid originates as a secretion from the choroid plexus in each of the four ventricles, though most of it by far in the two lateral ventricles. Additional amounts of fluid are secreted by all the ependymal surfaces of the ventricles.

Figure 32–1 illustrates the main channel of fluid flow from the choroid plexuses and then through the cerebrospinal fluid system. The fluid secreted in the lateral ventricles passes into the third ventricle through the *foramina of Monro*, combines with that secreted in the third ventricle, and then passes along the *aqueduct of Sylvius* into the fourth ventricle where a small amount of additional fluid is added. It then passes out of the fourth ventricle through three small openings, two lateral *foramina of Luschka* and a midline *foramen of Magendie*, entering the *cisterna magna*, a large fluid space that lies behind the medulla and beneath the cerebellum. The cisterna magna is continous with the *subarachnoid space* that surrounds the entire brain and spinal cord, and the cerebrospinal fluid flows upward through this space toward the cerebrum; but before it can reach the cerebrum it must first flow through the small *tentorial opening* around the mesencephalon where the flow is sometimes impeded. From the cerebral subarachnoid spaces, the fluid flows into arachnoidal villi that project mainly into the large sagittal venous sinus. Finally, the fluid empties into the venous blood through the surfaces of these villi.

Secretion by the Choroid Plexus.   The choroid plexus, which is illustrated in Figure 32–2, is a cauliflower-like growth of blood vessels covered by a thin coat of epithelial cells. This plexus projects into (a) the tem-

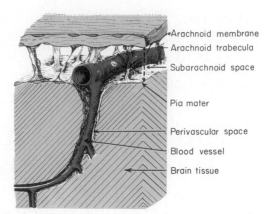

Figure 32–3.   Drainage of the perivascular spaces into the subarachnoid space. (From Ranson and Clark: Anatomy of the Nervous System. Philadelphia, W. B. Saunders Company, 1959.)

poral horn of each lateral ventricle, (b) the posterior portion of the third ventricle, and (c) the roof of the fourth ventricle.

The fluid secreted by the choroid plexus is similar to an ultrafiltrate of plasma but not identical with it. Instead, its concentration of sodium is 7 per cent greater than the sodium concentration in extracellular fluid; its glucose concentration is 30 per cent less; its potassium concentration is 40 per cent less; and its concentrations of other ions and molecular solutes are similarly different. When a metabolic poison is applied to the choroid, some of these differences disappear; it has also been demonstrated in other ways that different substances are transported through the choroid epithelium by carrier-mediated *facilitated diffusion* or *active transport*.

Absorption of Cerebrospinal Fluid Through the Arachnoidal Villi.   The *arachnoidal villi*, sometimes also called *arachnoidal granulations*, are finger-like projections of the arachnoidal membrane through the walls of the venous sinuses. The endothelial cells covering the villi have been shown by electron microscopy to form large vesicular holes directly through the bodies of the cells. These are large enough to allow relatively free flow of cerebrospinal fluid, protein molecules, and even particles as large as red blood cells into the venous blood.

The Perivascular Spaces and Cerebrospinal Fluid.   The blood vessels entering the substance of the brain pass first along the surface of the brain and then penetrate inward, carrying a layer of *pia mater* with them, as shown in Figure 32–3. The pia is only loosely adherent to the vessels, so that a space, the *perivascular space*, exists between it and each vessel. Perivascular spaces follow both the arteries and the veins into the brain as far as the arterioles and venules but not to the capillaries.

The Lymphatic Function of the Perivascular Spaces.   As is true elsewhere in the body, a small amount of protein leaks out of the parenchymal capillaries into the interstitial spaces of the brain, and since no true lymphatics are present in brain tissue, this protein leaves the tissue mainly through the perivascular spaces but partly also by direct diffusion through the pia mater into the subarachnoid spaces. On reaching the

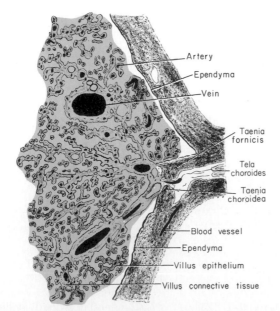

Figure 32–2.   The choroid plexus. (Modified from Clara: Das Nervensystem des Menschen. Barth.)

subarachnoid spaces, the protein flows along with the cerebrospinal fluid to be absorbed through the *arachnoidal villi* into the cerebral veins. Therefore, the perivascular spaces, in effect, are a modified lymphatic system for the brain.

In addition to transporting fluid and proteins, the perivascular spaces also transport extraneous particulate matter from the brain into the subarachnoid space. For instance, whenever infection occurs in the brain, dead white blood cells are carried away through the perivascular spaces.

CEREBROSPINAL FLUID PRESSURE

The normal pressure in the cerebrospinal fluid system when one is lying in a horizontal position averages 130 mm. water (10 mm. Hg), though this may be as low as 70 mm. water or as high as 180 mm. water even in the normal person. These values are considerably greater than the −6.3 mm. Hg pressure in the interstitial spaces elsewhere in the body.

The cerebrospinal fluid pressure is regulated by the product of, first, the *rate of fluid formation* and, second, the *resistance to absorption through the arachnoidal villi*. When either of these is increased, the pressure rises; when either is decreased, the pressure falls.

Cerebrospinal Fluid Pressure in Pathological Conditions of the Brain. Often a large *brain tumor* elevates the cerebrospinal fluid pressure by decreasing the rate of absorption of fluid. For instance, if the tumor is above the tentorium and becomes so large that it compresses the brain downward, the upward flow of fluid through the tentorial opening may become blocked and the absorption of fluid greatly curtailed. As a result, the cerebrospinal fluid pressure can rise to as high as 500 mm. water (37 mm. Hg) or more.

The pressure also rises considerably when *hemorrhage* or *infection* occurs in the cranial vault. In both of these conditions, large numbers of cells suddenly appear in the cerebrospinal fluid, and these can cause serious blockage of the small channels for absorption through the arachnoidal villi. This sometimes elevates the cerebrospinal fluid pressure to as high as 400 to 600 mm. water.

Many babies are born with very high cerebrospinal fluid pressure. Frequently this is caused by excess formation of fluid, but abnormal absorption is also an occasional cause. This is discussed later in connection with *hydrocephalus*.

Measurement of Cerebrospinal Fluid Pressure. The usual procedure for measuring cerebrospinal fluid pressure is the following: First, the subject lies exactly horizontally on his side so that the spinal fluid pressure is equal to the pressure in the cranial vault. A spinal needle is then inserted into the lumbar spinal canal below the lower end of the cord and is connected to a glass tube. The spinal fluid is allowed to rise in the tube as high as it will. If it rises to a level 100 mm. above the level of the needle, the pressure is said to be 100 mm. water pressure or, dividing this by 13.6, which is the specific gravity of mercury, about 7.5 mm. Hg pressure.

Effect of High Cerebrospinal Fluid Pressure on the Optic Disc — Papilledema. Anatomically, the dura of the brain extends as a sheath around the optic nerve and then connects with the sclera of the eye. When the pressure rises in the cerebrospinal fluid system, it also rises in the optic nerve sheath. The retinal artery and vein pierce this sheath a few millimeters behind the eye and then pass with the optic nerve into the eye itself. The high pressure in the optic sheath impedes the flow of blood in the retinal vein, thereby also increasing the retinal capillary pressure throughout the eye, which results in retinal edema. The tissues of the optic disc are much more distensible than those of the remainder of the retina, so that the disc becomes far more edematous than the remainder of the retina and swells into the cavity of the eye. The swelling of the disc, which can be observed with an ophthalmoscope, is called *papilledema*, and neurologists can estimate the cerebrospinal fluid pressure by assessing the extent to which the optic disc protrudes into the eyeball.

OBSTRUCTION TO THE FLOW OF CEREBROSPINAL FLUID

Hydrocephalus. "Hydrocephalus" means excess water in the cranial vault. This condition is frequently divided into *communicating hydrocephalus* and *noncommunicating hydrocephalus*. In communicating hydrocephalus fluid flows readily from the ventricular system into the subarachnoid space, while in noncommunicating hydrocephalus fluid flow out of one or more of the ventricles is blocked.

Usually the *noncommunicating* type of hydrocephalus is caused by a block in the aqueduct of Sylvius so that as fluid is formed by the choroid plexuses in the two lateral and the third ventricles, the volumes of these three ventricles increase greatly. This flattens the brain into a thin shell against the skull. In newborn babies the increased pressure also causes the whole head to swell because the skull bones still have not fused.

The communicating type of hydrocephalus is usually caused by blockage of fluid flow in the subarachnoid space around the basal regions of the brain or blockage of the arachnoidal villi themselves. Fluid therefore collects both inside the ventricles and on the outside of the brain, causing the head to swell tremendously and often damaging the brain severely.

The most effective therapy for hydrocephalus is surgical institution of a silicone rubber tube shunt all the way from one of the ventricles to the peritoneal cavity, where the fluid can then be absorbed through the peritoneum.

THE BLOOD – CEREBROSPINAL FLUID AND BLOOD-BRAIN BARRIERS

It has already been pointed out that the constituents of the cerebrospinal fluid are not exactly the same as those of the extracellular fluid elsewhere in the body. Furthermore, many large molecular substances hardly pass at all from the blood into the cerebrospinal fluid or into the interstitial fluids of the brain even though these same substances pass readily into the usual interstitial fluids of the body. Therefore, it is said that barriers, called the *blood–cerebrospinal fluid barrier* and the *blood-brain barrier*, exist between the blood and the cerebrospinal fluid and brain fluid, respectively. These barriers exist in the choroid plexus and in essentially all areas of the brain parenchyma *except the hypothalamus*, where substances diffuse with ease into the tissue spaces. This ease of diffusion is very important because the hypothalamus responds to many different changes in the body fluids, such as changes in osmolality, glucose concentration, and so forth; these responses provide the signals for feedback regulation of each of the factors.

In general, the blood–cerebrospinal fluid and blood-brain barriers are highly permeable to water, carbon dioxide, oxygen, and most liquid-soluble substances such as alcohol and most anesthetics; slightly permeable to the electrolytes, such as sodium, chloride and potassium; and almost totally impermeable to substances like arsenic, sulfur, and gold. Though these latter substances are not of physiological importance, they are important occasionally for certain types of drug therapy; the blood–cerebrospinal fluid and blood-brain barriers often make it impossible to achieve effective concentrations of some such drugs in the cerebrospinal fluid or parenchyma of the brain.

The cause of the low permeability of the blood–cerebrospinal fluid and blood-brain barriers is the manner in which the endothelial cells of the capillaries are joined to each other: They are joined by so-called tight junctions. That is, the membranes of the adjacent endothelial cells are almost fused with each other rather than having slit-pores between them, as is the case in most other capillaries of the body.

Interstitial Fluid of the Brain. Research studies have shown that the interstitial fluid in brain tissue is about 12 per cent of the tissue weight, in contrast to 17 per cent elsewhere in the body. The cause of this difference is lack of collagen fibers between the cells.

Diffusion Between the Cerebrospinal Fluid and the Brain Interstitial Fluid. The surfaces of the ventricles are lined with a thin cuboidal epithelium called the *ependyma*, and the cerebrospinal fluid on the outer surfaces of the brain is separated from the brain tissue by a thin membrane called the *pia mater*. Both the ependyma and the pia mater are extremely permeable so that almost all substances that enter the cerebrospinal fluid can also diffuse readily into the brain interstitial fluid. Or, likewise, substances in the interstitial fluid can diffuse in the other direction as well. Therefore, many drugs that have no effect at all on the brain when introduced into the blood stream nevertheless can have very important effects on the brain when injected into the cerebrospinal fluid.

Physiological Importance of the Blood-Brain Barrier, and Importance of the Special Composition of the Cerebrospinal Fluid

The neurons of the brain require a very exactly controlled environment, or else their function becomes abnormal and so also does the function of the entire brain. The blood-brain barrier protects the cerebral tissue from detrimental substances in the blood, and the transport processes of the choroid plexuses and the brain capillary endothelium help to provide the appropriate fluid environment for the brain.

One of the most important results of this special brain fluid control system is the low concentration of potassium ions in the brain interstitial fluid. Experiments have shown that even when the circulating blood potassium rises to values almost 2 times normal, the potassium concentration in the cerebrospinal fluid still remains at its normal low value. Thus, the barrier system, along with its carrier-mediated transport of potassium, not only maintains a low potassium concentration but also keeps this concentration very constant, allowing the neurons to generate very high electrical potentials that do not change with the vagaries of the rest of the body.

The blood-brain barrier also prevents such substances as acetylcholine, norepinephrine, dopamine, and glycine from entering the brain from the blood even though their concentrations might become quite high in the circulating blood. This is exceedingly important, because all these are very powerful synaptic transmitter substances and could have devastating effects on brain function.

THE INTRAOCULAR FLUID

The eye is filled with intraocular fluid which maintains sufficient pressure in the eyeball to keep it distended. Figure 32–4 illustrates that this fluid can be divided into two portions, the *aqueous humor*, which lies in front and to the sides of the lens, and the *vitreous humor*, which lies between the lens and the retina. The aqueous humor is a freely flowing fluid, while the vitreous humor, sometimes called the *vitreous body*, is a gelatinous mass held together by a fine fibrillar network. Substances can *diffuse* slowly in the vitreous humor, but there is little *flow* of fluid.

Aqueous humor is continually being formed and reabsorbed. The balance between formation and reabsorption of aqueous humor regulates the total volume and pressure of the intraocular fluid.

FORMATION OF AQUEOUS HUMOR BY THE CILIARY BODY

Aqueous humor is formed in the eye *at an average rate of 2 to 3 cubic millimeters each minute*. Essentially all of this is secreted by the *ciliary processes*, which are linear folds projecting from the *ciliary body* into the space behind the iris where the lens ligaments also attach to the eyeball. A cross-section of these ciliary processes is illustrated in Figure 32–5, and their relationship to the fluid chambers of the eye can be seen in Figure 32–4. Because of their folded architecture, the total surface area of the ciliary processes is approximately 6 square centimeters in each eye — a large area, considering the small size of the ciliary body. The surfaces of these processes are covered by epithelial cells, and immediately beneath these is a highly vascular area.

Aqueous humor is formed by the ciliary processes in much the same manner that cerebrospinal fluid is

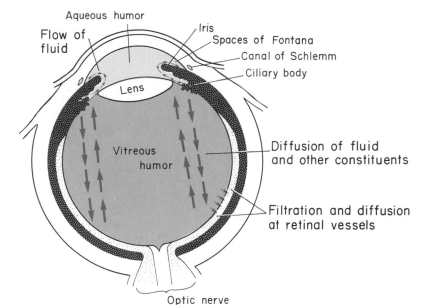

Figure 32–4.   Formation and flow of fluid in the eye.

formed by the choroid plexus. The postulated mechanism is the following: It is believed that the ciliary epithelium actively secretes sodium, chloride, and probably bicarbonate ions into the spaces between the cells. This in turn causes osmosis of water into these spaces, and the resulting solution then oozes out of the surfaces of the ciliary processes. In addition, several nutrients are transported across the epithelium by active transport or facilitated diffusion; these include amino acids, ascorbic acid, and probably also glucose.

OUTFLOW OF AQUEOUS HUMOR FROM THE EYE

After aqueous humor is formed by the ciliary processes, it flows, as shown in Figure 32–4, *between the ligaments of the lens*, then *through the pupil*, and finally *into the anterior chamber of the eye*. Here, the fluid flows into the *angle between the cornea and the iris* and thence through a meshwork of *trabeculae*, finally into the *canal of Schlemm*. Figure 32–6 illustrates the anatomical structures at the irido-corneal angle, showing that the spaces between the trabeculae extend all the way from the anterior chamber to the canal of Schlemm. The canal of Schlemm in turn is a thin-walled vein that extends circumferentially all the way around the eye. Its endothelial membrane is so porous that even large protein molecules, as well as small particulate matter, can pass from the anterior chamber into the canal of Schlemm. Even though the canal of Schlemm is actually a venous blood vessel, so much aqueous humor normally flows into it that it is filled only with aqueous humor rather than with blood. Also, the small veins that lead from the canal of Schlemm to the larger veins of the eye usually contain only aqueous humor, and these are called *aqueous veins*.

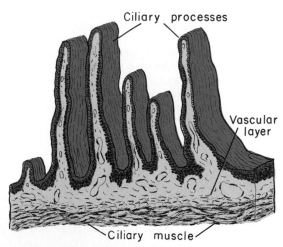

Figure 32–5.   Anatomy of the ciliary processes.

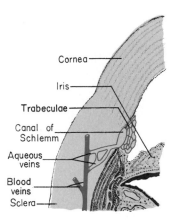

Figure 32–6.   Anatomy of the iridocorneal angle, showing the system for outflow of aqueous humor into the conjunctival veins.

INTRAOCULAR PRESSURE

The average normal intraocular pressure is approximately 16 mm. Hg, with a range from 12 to 20.

Tonometry. Because it is impractical to pass a needle into a patient's eye for measurement of intraocular pressure, this pressure is measured clinically by means of a tonometer, the principle of which is illustrated in Figure 32–7. The cornea of the eye is anesthetized with a local anesthetic, and the footplate of the tonometer is placed on the cornea. A small force is then applied to a central plunger, causing the central part of the cornea to be displaced inward. The amount of displacement is recorded on the scale of the tonometer, and this in turn is calibrated in terms of intraocular pressure.

Regulation of Intraocular Pressure. The intraocular pressure of the normal eye remains almost exactly constant throughout life, illustrating that the pressure regulating mechanism is very effective. But the precise operation of this mechanism is not clear. The pressure is regulated mainly by the outflow resistance from the anterior chamber into the canal of Schlemm, presumably in the following way: The "trabeculae" guarding the entrance of the fluid into the canal of Schlemm are shown in cross-section in Figure 32–6. When studied in three dimensions, these are actually laminar plates that lie one on top of the other, as illustrated in Figure 32–8. Each of the plates is penetrated by numerous small holes. When the plates are compressed against each other, each successive plate partially blocks the holes in the next plate. An increase in pressure above normal is believed to distend the spaces between the plates and therefore to open the holes, thus causing rapid flow into the canal of Schlemm and decrease of the pressure back to normal. On the other hand, a decrease in pressure below normal allows the plates to impinge upon each other, thus preventing fluid loss until the pressure rises again back to normal. Thus, this mechanism acts as an automatic feedback regulatory system for keeping the intraocular pressure at a nearly constant level day in and day out.

Cleansing of the Trabecular Spaces and of the Intraocular Fluid. When large amounts of debris occur in

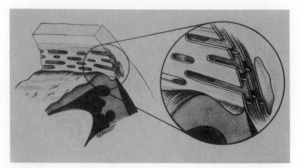

Figure 32–8. Perforated laminar plates that overlie the canal of Schlemm. (From Ashton, Brini, and Smith: *Br. J. Ophthalmol., 40*:257, 1956.)

the aqueous humor, as occurs following hemorrhage into the eye or during intraocular infection, the debris is likely to accumulate in the trabecular spaces, therefore preventing adequate reabsorption of fluid from the anterior chamber and sometimes causing glaucoma, as explained below. However, on the surfaces of the trabecular plates are large numbers of phagocytic cells. Also, immediately outside the canal of Schlemm — between the canal and the trabecular plates — is a layer of interstitial gel containing large numbers of reticuloendothelial cells that have an extremely high capacity for both engulfing debris and degrading it into small molecular substances that can then be absorbed. Thus, this phagocytic system keeps the trabecular spaces cleaned.

In addition, the surface of the iris and other surfaces of the eye behind the iris are covered with an epithelium that is capable of phagocytizing proteins and small particles from the aqueous humor, thereby helping to maintain a perfectly clear fluid. This epithelium can also transport many toxic substances out of the aqueous humor, thereby helping to maintain the chemical purity of the internal environment of the eye.

Glaucoma. Glaucoma is one of the most common causes of blindness. It is a disease of the eye in which the intraocular pressure becomes pathologically high, sometimes rising to as high as 70 mm. Hg. Pressures rising above as little as 25 to 30 mm. Hg can cause loss of vision when maintained for many years. And the extremely high pressures can cause blindness within days. As the pressure rises, the retinal artery, which enters the eyeball at the optic disc, is compressed, thus reducing the nutrition to the retina. This often results in permanent atrophy of the retina and optic nerve, with consequent blindness.

In essentially all cases of glaucoma the abnormally high pressure results from increased resistance to fluid outflow at the irido-corneal junction. In most patients, the cause of this is unknown, but in some it results from infection or trauma to the eye. As explained above, red blood cells, white blood cells, and tissue debris block the outflow of fluid, thereby greatly increasing the intraocular pressure.

Administration of *Diamox* is usually effective in reducing the pressure in glaucoma. This drug acts by reducing the rate of formation of aqueous humor, pos-

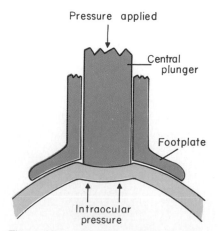

Pressure applied

Central plunger

Footplate

Intraocular pressure

Figure 32–7. Principles of the tonometer.

sibly because of an effect that reduces bicarbonate and sodium ion secretion.

Sometimes, when all other therapeutic procedures fail to reinstitute adequate outflow of aqueous humor, a *trephine* operation is performed — that is, a small hole is made through the corneoscleral junction into the anterior chamber, a portion of the iris is removed, and the conjunctiva is sewed over the hole. Fluid therafter leaks from the anterior chamber into the areolar tissue beneath the conjunctiva and is absorbed into the lymphatic channels.

FLUID CIRCULATION IN THE POTENTIAL SPACES OF THE BODY

Among the potential spaces of the body are the *pleural cavity,* the *pericardial cavity,* the *peritoneal cavity,* the cavity of the *tunica vaginalis* surrounding the testis, and the *synovial cavities* including both the joint cavities and the bursae. Normally, these are all empty except for a few milliliters of viscous fluid that lubricates movement of the surfaces against each other. But in certain abnormal conditions the spaces can swell to contain tremendous quantities of fluid, which is the reason they are called "potential" spaces.

FLUID EXCHANGE BETWEEN THE CAPILLARIES AND THE POTENTIAL SPACES

The membrane of a potential space usually does not offer significant resistance to the passage of fluid, electrolytes, or even proteins back and forth between the space and the interstitial fluid of the surrounding tissue. Consequently, fluid leaving a capillary adjacent to the potential space membrane, as shown in Figure 32–9, not only diffuses into the interstitial fluid but also into the potential space. Likewise, fluid can diffuse back out of the space into the interstitial fluid and thence into the capillary.

Lymphatic Drainage of the Potential Spaces. In the same manner that protein collects in the interstitial spaces because of leakage out of the capillaries, so does it tend to collect in the potential spaces, and it must be removed through lymphatics or some other channel. Fortunately, each potential space is either directly or indirectly connected with lymphatic drainage systems, which will be discussed in connection with each different type of potential space.

Function of the Lymphatic Pump in the Drainage of the Potential Spaces. Almost all the potential spaces are located where there is continual movement: the joints are moved, the bursae are compressed, the heart moves in the pericardial cavity, the lungs move in the pleural cavity with respiration, and the peritoneal spaces are also squeezed with each respiratory cycle or with other body movement. Each movement compresses fluid into the lymphatics surrounding the potential spaces, and, because of the valves in the lymphatics and the pumping mechanisms of the lymphatics, the fluid flows continually away from the space. Therefore, any time that excess protein collects in the potential space and osmotically pulls any excess fluid whatever into the space, this fluid automatically washes the protein into the lymphatics and from there back into the blood stream.

Altered Capillary Dynamics as a Cause of Increased Pressure and Fluid in the Potential Spaces. Any abnormal changes that can occur in capillary dynamics to cause extracellular edema of the tissue spaces, as described in the preceding chapter, can also cause increased pressure and fluid in the potential spaces. Thus, *increased capillary permeability, increased capillary pressure, decreased plasma colloid osmotic pressure,* and *blockage of the lymphatics* from a potential space can all cause swelling of the space. The fluid that collects is called a *transudate.*

One of the most common causes of swelling in a potential space is *infection.* White blood cells and other debris caused by the infection block the lymphatics, resulting in (1) buildup of protein in the space, (2) increased colloid osmotic pressure, and (3) consequent failure of fluid reabsorption.

THE PLEURAL CAVITY

Figure 32–9 shows specifically the diffusion of fluid into and out of the pleural cavity at the parietal and the visceral pleural surfaces. This occurs in precisely the same manner as in the usual tissue spaces, except that a very porous mesenchymal *serous membrane,* the *pleura*, is interspersed between the capillaries and the pleural cavity.

Large numbers of lymphatics drain from the mediastinal and lateral surfaces of the parietal pleura, and with each expiration the intrapleural pressure rises, forcing small amounts of fluid into the lymphatics; also the respiratory movements alternately compress the lymphatic vessels, promoting continuous flow along the lymphatic channels.

Maintenance of a Negative Pressure in the Pleural Cavity. The visceral pleura of the lungs continually absorbs fluid with considerable "absorptive force." This is caused by the low capillary pressure — about

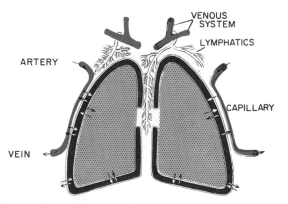

Figure 32–9. Dynamics of fluid exchange in the intrapleural spaces.

5 to 10 mm. Hg — in the pulmonary system. In contrast to this low pressure, the plasma proteins exert about 28 mm. Hg colloid osmotic pressure, causing an absorption pressure at the visceral pleura of up to 20 mm. Hg at all times. As a result, the pressure of the *fluid* in the intrapleural space remains negative at all times, averaging −8 to −10 mm. Hg. This negative pressure is much greater than the elastic force of the lungs (−4 mm. Hg) that tends to collapse the lungs and to pull the lungs away from the chest wall. Therefore, it keeps the lungs expanded.

THE PERICARDIAL CAVITY

The space around the heart operates with essentially the same dynamics as those of the pleural cavity. The pressure of the fluid in the pericardial cavity, like that in the pleural cavity, is negative. Here again, during expiration as well as during excessive filling of the heart, the pericardial pressure rises intermittently, forcing excess fluid into lymphatic channels of the mediastinum.

THE PERITONEAL CAVITY

The peritoneal cavity of the abdomen is subject to the same fluid dynamics as are the other potential spaces; fluid is filtered into the peritoneal space through the serous membrane called the *visceral peritoneum* covering the viscera and the *parietal peritoneum* lining the outer walls of the abdominal cavity. Fluid is also absorbed through the peritoneum.

The peritoneal cavity is more susceptible to the development of excess fluids than are most of the other cavities, for two reasons: (1) Any time the pressure in the liver sinusoids rises more than 5 to 10 mm. Hg above normal, fluid containing large amounts of protein begins to transude through the liver surface into the abdominal cavity, and (2) the capillary pressure in the visceral peritoneum is probably higher than elsewhere in the body; this higher pressure is caused by the resistance to portal blood flow through the liver. High venous pressure caused by heart failure or extra resistance in the liver in pathological states, such as *cirrhosis, carcinoma,* or *portal vein obstruction*, frequently results in marked transudation of fluid into the abdomen; the transudate in this case is called *ascites.*

Numerous large lymphatic channels lead from the peritoneal cavity, especially from the lower surface of the diaphragm. With each diaphragmatic movement relatively large quantities of lymph flow out of the peritoneal cavity into the thoracic duct. This can be shown effectively by injecting radioactive red blood cells into the abdomen. A significant proportion of these cells, still in the whole form, is found in the blood within 10 to 20 minutes. Occasionally, however, cancer spreads so widely throughout the abdomen that it blocks the lymphatics, thereby preventing return of protein to the blood. As a result, the colloid osmotic pressure also rises and severe chronic ascites ensues.

THE SYNOVIAL CAVITIES

The joint cavities and the bursae are known as *synovial cavities.* The synovial membrane is not a true membrane at all but only a collection of dense fibrous tissue cells that line the surface between the interstitial spaces and the cavities. For this reason these cavities might be considered to be nothing more than enlarged tissue spaces. However, the synovial cavities do contain a large amount of proteoglycans, much more than normally present in the interstitial fluids. The origin of this is not known, though presumably it is secreted by the surrounding connective tissue cells.

In the synovial cavities, as in the other potential spaces, excess proteins are likely to collect, and these must be returned to the circulatory system through the lymphatics; otherwise the space swells. Since the synovial membrane offers little or no barrier to the transfer of fluid into the surrounding tissues, the protein can flow into the lymphatics of the area.

The pressure in joint cavities usually measures about −4 to −8 mm. Hg. This negative pressure presumably results from the same factors that cause negative pressures throughout most of the interstitial spaces of the body; these factors were discussed in detail in the preceding chapter.

REFERENCES

Agostoni, E.: Mechanics of the pleural space. *Physiol. Rev.,* 52:57, 1972.

Allen, L., and Weatherford, T.: Role of fenestrated basement membrane in lymphatic absorption from peritoneal cavity. *Am. J. Physiol.,* 197:551, 1959.

Bell, W. E., and McCormick, W. F.: Increased intracranial pressure in children. *Major Probl. Clin. Pediatr.,* 8:3, 1972.

Bill, A.: Blood circulation and fluid dynamics in the eye. *Physiol. Rev.,* 55:383, 1975.

Chandler, P. A., et al.: Glaucoma, 2nd Ed. Philadelphia, Lea & Febiger, 1979.

Crick, R. P., and Caldwell, A. D. S. (eds.): Glaucoma. New York, Grune & Stratton, 1979.

Cserr, H. F.: Physiology of the choroid plexus. *Physiol. Rev., 51*:273, 1971.

Davson, H.: The Physiology of the Cerebrospinal Fluid. Boston, Little, Brown, 1967.

Davson, H.: Physiology of the Eye, 3rd Ed. New York, Churchill Livingstone, Div. of Longman, 1972.

Elliot, R. H.: A Treatise on Glaucoma. Huntington, N.Y., R. E. Krieger, 1979.

Freese, A. S.: Glaucoma: Diagnosis, Treatment, Prevention. New York, Public Affairs Committee, 1979.

Guyton, A. C., et al.: Circulatory Physiology. II. Dynamics and Control of the Body Fluids. Philadelphia, W. B. Saunders Co., 1975.

Hamerman, D., et al.: The structure and chemistry of the synovial membrane in health and disease. *In* Bittar, E. E., and Bittar, N. (eds.): The Biological Basis of Medicine. Vol. 3. New York, Academic Press, 1969, p. 269.

Katzman, R., and Pappius, H.: Brain Electrolytes and Fluid Metabolism. Baltimore, Williams & Wilkins, 1973.

Leydhecker, W., and Krieglstein, G. K. (eds.): Recent Advances in Glaucoma. New York, Springer-Verlag, 1979.

Millen, J. W., and Woollam, D. H. M.: The Anatomy of the Cerebrospinal Fluid. New York, Oxford University Press, 1962.

Oldendorf, W. H.: Blood-brain barrier permeability to drugs. *Annu. Rev. Pharmacol., 14*:239, 1974.

Pappenheimer, J. R., et al.: Perfusion of the cerebral ventricular system in unanesthetized goats. *Am. J. Physiol., 203*:763, 1962.

Shulman, K. (ed.): Intracranial Pressure IV. New York, Springer-Verlag, 1980.

33

Partition of the Body Fluids: Osmotic Equilibria Between Extracellular and Intracellular Fluids

The body fluids are so important to the basic physiology of bodily function that we discuss them at several points in this text but each time in different contexts. In Chapter 4 it was pointed out that the body fluids can be divided mainly into extracellular and intracellular fluids, and the basic differences between these were discussed, as well as the manner in which these differences come about as the result of cell membrane transport. In Chapter 22 the relationship of blood volume to the overall regulation of the circulation was presented, and in Chapters 30 through 32 the interrelationships between capillary dynamics and interstitial fluids were discussed.

In the present chapter and in the following chapters on the kidneys, we will be concerned with the body fluids from the total point of view,

including regulation of body fluid volume, regulation of the constituents in the extracellular fluid, regulation of acid-base balance, and factors that govern gross interchange of fluid between the extracellular and the intracellular compartments — especially the osmotic relationships between these compartments. The present chapter will discuss the general distribution of body fluids and their osmotic interrelationships.

TOTAL BODY WATER

The total amount of water in a man of average weight (70 kg.) is approximately 40 liters (see Fig. 33–1), averaging 57 per cent of his total body weight. In a newborn infant this may be as high as 75 per cent of the body weight, but it progressively

Figure 33–1. Diagrammatic representation of the body fluids, showing the extracellular fluid volume, intracellular fluid volume, blood volume, and total body fluids.

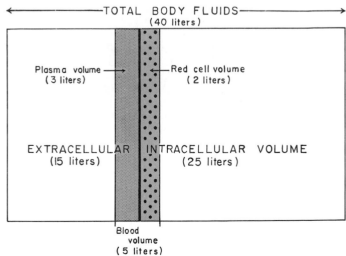

decreases from birth to old age, most of the decrease occurring in the first 10 years of life. Also, obesity decreases the percentage of water in the body, sometimes down to as low as 45 per cent.

INTAKE VERSUS OUTPUT OF WATER

Daily Intake of Water. Most of our daily intake of water enters by the oral route. Approximately two thirds is in the form of pure water or some other beverage, and the remainder is in the food that is eaten. A small amount is also synthesized in the body as the result of oxidation of hydrogen in the food; this quantity ranges between 150 and 250 ml. per day, depending on the rate of metabolism. The normal intake of fluid, including that synthesized in the body, averages about 2300 ml. per day.

Daily Loss of Body Water. Table 33–1 shows the routes by which water is lost from the body under different conditions. Normally, at an atmospheric temperature of about 68°F., approximately 1400 ml. of the 2300 ml. of water intake is lost in the *urine*, 100 ml. is lost in the *sweat*, and 100 ml. in the *feces*. The remaining 700 ml. is lost by *evaporation from the respiratory tract* or by *diffusion through the skin*.

Insensible Water Loss. Loss of water by diffusion through the skin and by evaporation from the respiratory tract is known as *insensible water loss* because we do not know that we are actually losing water at the time that it is leaving the body.

The average loss of water by diffusion through the skin is approximately 300 to 400 ml. per day; this amount is lost even in a person who is born with congenital lack of sweat glands. In other words, the water molecules themselves actually diffuse through the cells of the skin. Fortunately, the cholesterol-filled, cornified layer of the skin acts as a protector against still much greater loss of water by diffusion. When the cornified layer becomes denuded, as after extensive burns, the rate of evaporation can increase to as much as 3 to 5 liters each day.

All air that enters the respiratory tract becomes totally saturated with moisture, to a vapor pressure of approximately 47 mm. Hg, before it is expelled. Since the vapor pressure of the inspired atmospheric air is usually far less than 47 mm. Hg, the average water loss through the lungs is about 300 to 400 ml. per day. The atmospheric vapor pressure normally decreases with decreasing temperature so that the loss is greatest in very cold and least in very warm weather. This explains the dry feeling in the respiratory passages in cold weather.

Loss of Water in Hot Weather and During Exercise. In very hot weather, water loss in the sweat is occasionally increased to as much as 3.5 liters an hour, which obviously can rapidly deplete the body fluids. Sweating will be discussed in Chapter 72.

Exercise increases the loss of water in two ways: First, it increases the rate of respiration, which promotes increased water loss through the respiratory tract in proportion to the increased ventilatory rate. Second, and much more important, exercise increases the body heat and consequently is likely to result in excessive sweating.

BODY FLUID COMPARTMENTS

THE INTRACELLULAR COMPARTMENT

About 25 of the 40 liters of fluid in the body are inside the approximately 75 trillion cells of the body and are collectively called the *intracellular fluid*. The fluid of each cell contains its own individual mixture of different constituents, but the concentrations of these constituents are reasonably similar from one cell to another. For this reason the intracellular fluid of all the different cells is considered to be one large fluid compartment, though in reality it is an aggregate of trillions of minute compartments.

THE EXTRACELLULAR FLUID COMPARTMENT

All the fluids outside the cells are called *extracellular fluid*, and these fluids are constantly mixing, as was explained in Chapter 1. The total amount of fluid in the extracellular compartment averages 15 liters in a 70 kg. adult.

The extracellular fluid can be divided into *interstitial fluid, plasma, cerebrospinal fluid, intraocular fluid, fluids of the gastrointestinal tract*, and *fluids of the potential spaces*.

Interstitial Fluid. The interstitial fluid lies in the spaces between the cells. A minute portion of it is free in the form of actual flowing fluid, while probably more than 99 per cent of it is held in the gel of the interstitial spaces, as discussed in Chapter 30.

The fluids in the cerebrospinal system, the eye, the potential spaces, and the gastrointestinal tract are discussed in Chapter 32 and 64. These fluids, on the average, are similar to interstitial fluid and often are considered to be part of the interstitial compartment.

TABLE 33–1 DAILY LOSS OF WATER
(in Milliliters)

	Normal Temperature	Hot Weather	Prolonged Heavy Exercise
Insensible Loss:			
Skin	350	350	350
Respiratory tract	350	250	650
Urine	1400	1200	500
Sweat	100	1400	5000
Feces	100	100	100
Total	2300	3300	6600

Plasma. The plasma is the noncellular portion of blood. It is part of the extracellular fluid and communicates continually with the interstitial fluid through pores in the capillaries. Loss of plasma from the circulatory system through the capillary pores is minimized by the colloid osmotic pressure exerted by the plasma proteins, which was explained in Chapter 30. Yet the capillaries are porous enough for most dissolved substances and water molecules to *diffuse* through them freely, allowing constant mixing between the plasma and interstitial fluid of almost all substances except the protein.

The plasma volume averages 3 liters in the normal adult.

BLOOD VOLUME

Blood contains both extracellular fluid (the fluid of the plasma) and intracellular fluid (the fluid in the red blood cells). However, since blood is contained in a closed chamber all its own — the circulatory system — its volume and its special dynamics are exceedingly important.

The average blood volume of a normal adult is almost exactly 5000 ml. On the average, approximately 3000 ml. of this is plasma, and the remainder, 2000 ml., is red blood cells. However, these values vary greatly in different individuals; also, sex, weight, and many other factors affect the blood volume.

Effect of Weight and Sex on Blood Volume. In persons who have a minimum of adipose tissue the blood volume varies almost directly in proportion to the body weight, normally averaging about 79 ml./kg. ± 10 per cent for both *lean* males and *lean* females. However, the greater the ratio of fat to body weight, the less the blood volume per unit weight, because fat tissue has little vascular volume. Figure 33–2 illustrates this effect of fat, showing that the heavier the person becomes, the less on the average becomes the blood volume in relation to weight. Furthermore, in very heavy females, who are especially prone to have high ratios of fat tissue to lean tissue, the blood volume per kilogram decreases much more than for males.

The Hematocrit. The hematocrit is the percentage of red blood cells in the blood as determined by centrifuging blood in a "hematocrit tube" until the cells become packed tightly in the bottom of the tube. The percentage of red blood cells in the blood can be determined roughly from the level of the packed cells. Such centrifuged blood in hematocrit tubes was illustrated in Figure 18–2 of Chapter 18. Unfortunately, it is impossible for the red blood cells to be packed completely together; about 3 to 8 per cent of plasma remains entrapped among the cells. Therefore, the true hematocrit (H) averages about 96 per cent of the measured hematocrit (Hct); that is:

$$H = 0.96 \text{ Hct}$$

The true hematocrit (H) is approximately 40 for a normal man and 36 for a normal woman.

In severe anemia, the hematocrit may fall to as low as 10, but this small quantity of red blood cells is barely sufficient to sustain life. On the other hand, a few conditions cause excessive production of red blood cells, resulting in polycythemia. In these instances, the hematocrit often rises to 65, and occasionally to 80. Obviously, there is an upper limit to the level of the hematocrit in polycythemic blood, because excessive hematocrit causes the blood to become so viscous that death ensues as a result of multiple plugging of the peripheral vascular tree.

The Body Hematocrit. The hematocrit of the blood in the capillaries, arterioles, and other very small vessels of the body is considerably less than that in the large veins and arteries. The cause of this is *axial streaming* of blood cells in blood vessels, which was explained in Chapter 18. In general, red cells cannot flow near the walls of the vessels nearly so easily as can plasma. Therefore, the cells tend to migrate to the center of the vessels, while a large portion of the plasma remains near the walls. In the very large vessels the ratio of wall surface to total volume is slight, so that the accumulation of plasma near the walls does not affect the hematocrit significantly. However, in the small vessels this ratio of wall surface to volume is great, causing the ratio of plasma to cells to be far greater than in the large vessels.

If one averages the hematocrit in both the large and small vessels, he determines a value called *body hematocrit*. For normal man the body hematocrit (H_0) averages 91 per cent of the large vessel hematocrit; that is:

$$H_0 = 0.91 \text{ H} = 0.87 \text{ Hct}$$

MEASUREMENT OF BODY FLUID VOLUMES

THE DILUTION PRINCIPLE FOR MEASURING FLUID VOLUMES

The volume of any fluid compartment of the body can be measured by placing a substance in

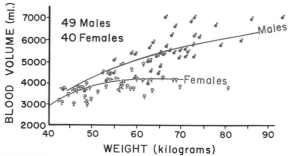

Figure 33–2. Relationship of blood volume to sex and weight.

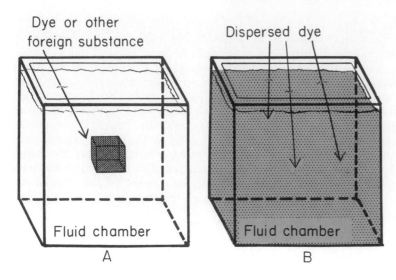

Dye or other
foreign substance

Dispersed dye

Figure 33–3. Principles of the dilution method for measuring fluid volumes (explained in the text).

Fluid chamber

A

Fluid chamber

B

the compartment, allowing it to disperse evenly throughout the fluid, and then measuring the extent to which the substance becomes diluted. Figure 33–3 illustrates this "dilution" principle for measuring the volume of any fluid compartment of the body. In this example, a small quantity of dye or other foreign substance is placed in fluid chamber A, and the substance is allowed to disperse throughout the chamber until it becomes mixed in equal concentrations in all areas, as shown in chamber B. Then a sample of the dispersed fluid is removed and the concentration of the substance is analyzed chemically, photoelectrically, or by any other means. The volume of the chamber can then be determined from the following formula:

Volume in ml.

$$= \frac{\text{Quantity of test substance instilled}}{\text{Concentration per ml. of dispersed fluid}}$$

Note that all one needs to know is (1) the *total quantity of the test substance* put into the chamber and (2) the *concentration in the fluid after dispersement.*

DETERMINATION OF BLOOD VOLUME

Substances Used in Determining Blood Volume. A substance used for measuring blood volume must be capable of dispersing throughout the blood with ease, and it must remain in the circulatory system long enough for measurements to be made. The two major groups of substances that satisfy these conditions are (1) substances that combine with the red blood cells and (2) substances that combine with the plasma proteins, for both the red blood cells and the plasma proteins remain reasonably well in the circulatory system, and any foreign substance that combines with

either of them likewise remains in the blood stream.

Examples of substances that combine with red blood cells and that are used for determining blood volume are *radioactive iron, radioactive chromium*, and *radioactive phosphate*. Examples of substances that combine with plasma proteins are the *vital dyes* and *radioactive iodine*.

Radioactive Red Blood Cells. The method most often used to make red blood cells radioactive is to tag the red blood cells with radioactive chromium (Cr^{51}). A small quantity of Cr^{51} is mixed with a few milliliters of blood removed from the person, and this is incubated at 36°C. for half an hour or more. After this time, most of the Cr^{51} will have entered the red blood cells, but to remove the extra chromium from the mixture, the red blood cells are washed in saline. Their total content of Cr^{51} is then determined with a Geiger or scintillation counter (apparatuses for measuring the total number of radioactive disintegrations occurring in the sample per minute). Then the radioactive cells are reinjected into the person. After mixing in the circulatory system has continued for approximately 10 minutes, blood is removed from the circulatory system, and the radioactivity in this blood is determined. Using the above dilution formula, the total blood volume is calculated.

To be accurate, this calculated blood volume must now be corrected to determine the true value, because the blood sample is removed from the veins where the hematocrit is not equal to the body hematocrit, as discussed earlier in the chapter. This correction is made as follows:

Actual blood volume = 1.1 × Measured blood volume

Vital Dyes for Measurement of Plasma Volume. A number of dyes, generally known as "vital dyes," have the ability to combine with proteins. When such a dye is injected into the blood, it immediately forms a tight union with the plasma proteins. Thereafter, the dye travels where the proteins travel.

The dye almost universally used for measuring plasma volume is *T-1824*, also called *Evans blue*. In making determinations of plasma volume, a known quantity of the dye is injected, and it immediately combines with the proteins and disperses throughout the circulatory system within approximately 10 minutes. A sample of the blood is then taken, and the red blood cells are removed from the plasma by centrifugation. Then, by spectrophotometric analysis of the plasma, one can determine the exact quantity of dye in the sample of plasma. From the determined quantity of dye in each milliliter of plasma and the known quantity of dye injected, the *plasma volume* is calculated using the above dilution formula.

To be even more exact in measuring the plasma volume, the rate of loss of dye from the circulatory system during the interval of mixing must also be considered. On the average, T-1824 is lost from the circulatory system at a rate of about 5 per cent per hour, part of this being excreted into the urine and part being carried into the interstitial spaces by the leakage of plasma proteins through the capillary walls. To offset this error, three different samples of blood are usually removed from the circulation at 10, 20, and 30 minutes. Then the plasma volumes are calculated from these samples and plotted on semilog graph paper, as illustrated in Figure 33–4. A straight line is drawn through the measured points and extrapolated back to zero time. Since the percentage loss is approximately the same during the first 10 minutes as during each of the two succeeding 10-minute intervals, the extrapolated point indicates approximately the true plasma volume. In Figure 33–4 this value is 2500 ml.

Note that neither T-1824 nor any other vital dye enters the red blood cells. Therefore, this method *does not measure the total blood volume*. However, the blood volume can be calculated from the plasma volume, provided the hematocrit is determined, by using the following formula:

Blood volume
$$= \text{Plasma volume} \times \frac{100}{100 - 0.87 \text{ Hematocrit}}$$

Radioactive Protein. If a sample of plasma is allowed to incubate with radioactive iodine (I^{131}) for 30 minutes or more, some of the protein combines with the iodine,

and the iodinated protein can be separated from the remaining iodine by dialysis. The radioactive protein is then injected into the subject, and plasma and blood volumes are determined in the same manner as that discussed for the vital dyes.

MEASUREMENT OF THE EXTRACELLULAR FLUID VOLUME

To use the dilution principle for measuring the volume of the extracellular fluid, one injects into the blood stream a substance that can diffuse readily throughout the entire extracellular fluid chamber, passing easily through the capillary membranes but as little as possible through the cell membranes into the cells. After half an hour or more of mixing, a sample of extracellular fluid is obtained by removing blood and separating the plasma from the cells by centrifugation. The plasma, which is actually a part of the extracellular fluid, is then analyzed for the injected substance.

Substances Used in Measuring Extracellular Fluid Volume — The Concept of "Fluid Space." Substances that have been used for measuring extracellular fluid volume are *radioactive sodium, radioactive chloride, radioactive bromide, radioiothalamate, thiosulfate ion, thiocyanate ion, inulin,* and *sucrose.* Some of these, inulin especially, do not diffuse readily into all the out-of-the-way places of the extracellular fluid compartment. Therefore, the volume of extracellular fluid measured with inulin is likely to be lower than the actual volume of the compartment. On the other hand, others of these substances — radioactive chloride, radioactive bromide, radioactive sodium, and thiocyanate ion, for instance — are likely to penetrate into the cells to a slight extent and, therefore, are likely to measure a volume somewhat in excess of the extracellular fluid volume.

Because there is no single substance that measures the exact extracellular fluid volume, one usually speaks of the *sodium space*, the *thiocyanate space*, the *inulin space*, and so forth, rather than the extracellular volume. At present it is impossible to say exactly which "space" most nearly approximates the true extracellular volume. Measurements for the normal 70 kg. adult, when different ones of the above substances have been used, have ranged from 9 liters to 22 liters; the average measurement has been about 15 liters.

Correction for Loss of the Test Substance. In measuring the extracellular fluid volume of a normal person, at least 30 minutes is required for almost complete dispersion of the test substance throughout the extracellular compartment, and, if major amounts of the substance are lost from the compartment during this time, especially in the urine, an error will be made in the measurement. For this reason, measurements employing each of the different test substances require different systems of correction but they generally employ an extrapolation back to zero time, as explained above for plasma volume determination.

MEASUREMENT OF TOTAL BODY WATER

The total body water can be measured in exactly the same way as the extracellular fluid volume except that a

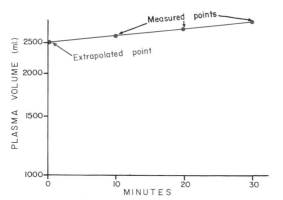

Figure 33–4. Determination of blood volume. The measured points represent plasma volumes calculated from blood samples removed 10, 20, and 30 minutes after injection of T-1824. The extrapolated point represents the calculated blood volume at the instant of dye injection.

substance must be used that will diffuse into the cells as well as throughout the extracellular compartment. The substances that give best results are either *tritiated water* (which can be analyzed with radiation-measuring instruments) or *heavy water* (which can be analyzed quantitatively either by accurate specific gravity measurements of water samples or by infrared spectrophotometry). After administration of the "tagged" water, several hours are required for complete mixing with all the water of the body, and appropriate corrections must be made for any fluid that is lost either into the urine or otherwise during this period of mixing. The concentration of the tagged water in the total body water can, at the end of the period of mixing, be determined by simply measuring the tagged water concentration in the plasma.

Measurements of total body water in the 70 kg. adult have ranged from as low as 30 liters to as high as 50 liters, with a reasonable average of approximately 40 liters, or 57 per cent of the total body mass.

CALCULATION OF INTERSTITIAL FLUID VOLUME

Since any substance that passes into the interstitial fluid also passes into almost all other portions of the extracellular fluid, there is no direct method for measuring interstitial fluid volume separately from the entire extracellular fluid volume. However, if the extracellular fluid volume and plasma volume have both been measured, the interstitial fluid volume can be approximated by *subtracting the plasma volume from the total extracellular fluid volume*. This calculation gives a normal interstitial fluid volume of 12 liters in the 70 kg. adult.

CONSTITUENTS OF EXTRACELLULAR AND INTRACELLULAR FLUIDS

Figure 33–5 diagrams the major constituents of the extracellular and intracellular fluids. (The actual values for these constituents were given in tabular form in Figure 4–1 of Chapter 4.) The quantities of the different substances are represented in Figure 33–5 in *milliequivalents* or *millimoles per liter*. However, the protein molecules and some of the nonelectrolyte molecules are extremely large compared with the more numerous small ions. Therefore, *in terms of mass*, the proteins and nonelectrolytes actually comprise about 90 per cent of the dissolved constituents in the plasma, about 60 per cent in the interstitial fluid, and about 97 per cent in the intracellular fluid.

Figure 33–6 illustrates the distribution of the nonelectrolytes in the plasma, and most of these same substances are also present in almost equal concentrations in the interstitial fluid, except for some of the fatty compounds that are present in the plasma in large suspended particles, the *lipoproteins*.

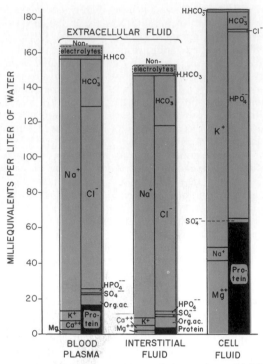

Figure 33–5. The compositions of plasma, interstitial fluid, and intracellular fluid. (Modified from Gamble: Chemical Anatomy, Physiology, and Pathology of Extracellular Fluid: A Lecture Syllabus. Harvard University Press, 1954.)

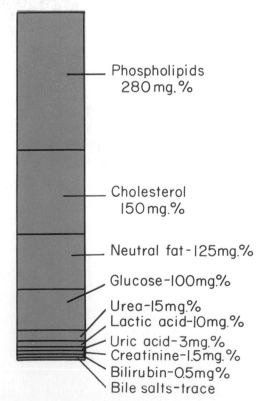

Figure 33–6. The nonelectrolytes of the extracellular fluid.

The Extracellular Fluid. Referring again to Figure 33–5, one sees that extracellular fluid, both that of the blood plasma and of the interstitial fluid, contains large quantities of *sodium* and *chloride ions*, reasonably large quantities of *bicarbonate ion*, but only small quantities of potassium, calcium, magnesium, phosphate, sulfate, and organic acid ions. In addition, plasma contains a large amount of protein while interstitial fluid contains much less. (The proteins in these fluids and their significance were discussed in detail in Chapter 30.)

In Chapter 1 it was pointed out that the extracellular fluid is called the *internal environment* of the body and that its constituents are accurately regulated so that the cells remain bathed continually in a fluid containing the proper electrolytes and nutrients for continued cellular function. The regulation of most of these constituents will be presented in Chapter 36.

The Intracellular Fluid. From Figure 33–5 it is also readily apparent that the intracellular fluid contains only small quantities of sodium and chloride ions and almost no calcium ions; but it does contain large quantities of *potassium* and *phosphate* and moderate quantities of *magnesium* and *sulfate ions*, all of which are present in only small concentrations in the extracellular fluid. In addition, cells contain large amounts of protein, approximately four times as much as the plasma.

OSMOTIC EQUILIBRIA AND FLUID SHIFTS BETWEEN THE EXTRACELLULAR AND INTRACELLULAR FLUIDS

One of the most troublesome of all problems in clinical medicine is maintenance of adequate body fluids and proper balance between the extracellular and intracellular fluid volumes in seriously ill patients. The purpose of the following discussion, therefore, is to explain the interrelationships between extracellular and intracellular fluid volumes and the osmotic factors that cause shifts of fluid between the extracellular and intracellular compartments.

BASIC PRINCIPLES OF OSMOSIS AND OSMOTIC PRESSURE

The basic principles of osmosis and osmotic pressure were presented in Chapter 4. However, these principles are so important to the following discussion that they are reviewed here briefly.

Whenever a membrane between two fluid compartments is permeable to water but not to some of the dissolved solutes — that is, the membrane is *semipermeable* — and the concentration of the solutes is greater on one side of the membrane than on the other, water passes through the membrane toward the side with the greater concentration of solutes. This phenomenon is called *osmosis*.

Osmosis results from the kinetic motion of the individual particles — both the molecules and the ions — in the solutions on the two sides of the membrane. This can be explained in the following way: When the temperature is the same in both solutions, the particles on both sides of the membrane, on the average, all have the same amount of kinetic energy of motion. However, the nondiffusible particles in the two solutions displace some of the water molecules. As a result the so-called chemical potential of the water is reduced on each side of the membrane in direct proportion to the concentration of nondiffusible particles. Therefore, on the side of the membrane where there is the lower concentration of nondiffusible particles, the concentration of water molecules is the greater; that is, the so-called *chemical potential* of the water is greater. This means that more water molecules will strike each pore of the membrane each second on the side of the membrane that has the lowest concentration of nondiffusible solutes. As a result, net diffusion of water molecules will occur from this side toward the solution on the opposite side of the membrane where the concentration of nondiffusible particles is greater. This net rate of diffusion is called the *rate of osmosis*.

Osmosis of water molecules can be opposed by applying a pressure across the semipermeable membrane in the direction opposite to that of the osmosis. The amount of pressure required exactly to oppose the osmosis is called the *osmotic pressure*.

Relationship of the Molecular Concentration of a Solution to Its Osmotic Pressure. Each nondiffusible molecule dissolved in water reduces the chemical potential of the water by a given amount. Consequently, the tendency for the water in the solution to diffuse through a membrane is reduced in direct proportion to the concentration of nondiffusible molecules. And, as a corollary, the osmotic pressure of the solution is also proportional to the concentration of nondiffusible molecules in the solution. This relationship holds true for all nondiffusible molecules almost regardless of their molecular weight. For instance, one molecule of albumin with a molecular weight of 70,000 has the same osmotic effect as a molecule of glucose with a molecular weight of 180.

Osmotic Effect of Ions. Nondiffusible ions cause osmosis and osmotic pressure in exactly the same manner as do nondiffusible molecules. Furthermore, when a molecule dissociates into two or more ions, each of the ions then exerts osmotic pressure individually. Therefore, to determine the osmotic effect, all the nondiffusible ions must be added to all the nondiffusible molecules; but note that a bivalent ion, such as calcium, exerts no more osmotic pressure than does a univalent ion, such as sodium.

Osmoles. The ability of solutes to cause osmosis and osmotic pressure is measured in terms of "osmoles"; the osmole is a measure of the total number of particles. *One gram mol of nondiffusible and nonionizable substance is equal to 1 osmole.* On the other hand, if a substance ionizes into two ions (sodium chloride into sodium and chloride ions, for instance), then 0.5 gram mole of the substance equals 1 osmole. The obvious reason for using the osmole is that osmotic pressure is determined by the number of particles instead of the mass of the solute.

In general, the osmole is too large a unit for satisfactory use in expressing osmotic activity of solutes in the body. Therefore, the term *milliosmole,* which equals ¹/₁₀₀₀ osmole, is commonly used.

Osmolality and Osmolarity. The osmolal concentration of a solution is called its *osmolality* when the concentration is expressed in osmoles per kilogram of water; it is called *osmolarity* when it is expressed as osmoles per liter of solution. The term "osmolality" is generally preferred because the osmotic pressure of a solution is considerably more closely related to its osmolality than to its osmolarity in very concentrated solutions. However, in the very dilute solutions of the normal human body, the differences are so slight that the terms are frequently used interchangeably. Furthermore, it is so much easier to express the body fluid quantities in liters than in kilograms of water that almost all calculations are based on osmolarities rather than on osmolalities. It is common practice for many physiologists to speak in terms of osmolality even though they make their calculations in terms of osmolarities.

Relationship of Osmotic Pressure to Osmolality. The osmotic pressure of a solution *at body temperature* can be determined approximately from the following formula:

Osmotic pressure (mm. Hg)
= 19.3 × Osmolality (milliosmole/liter)

OSMOLALITY OF THE BODY FLUIDS

Table 33–2 lists the osmotically active substances in plasma, interstitial fluid, and intracellular fluid. The milliosmoles of each of these per liter of water is given. Note especially that approximately four-fifths of the total osmolality of the interstitial fluid and plasma is caused by sodium and chloride ions, while approximately half of the intracellular osmolality is caused by potassium ions, the remainder being divided among the many other intracellular substances.

As noted at the bottom of Table 33–2, the total osmolality of each of the three compartments is approximately 300 milliosmoles per liter, with that

TABLE 33–2 OSMOLAR SUBSTANCES IN EXTRACELLULAR AND INTRACELLULAR FLUIDS

	Plasma (mOsmole/L. of H_2O)	Interstitial (mOsmole/L. of H_2O)	Intracellular (mOsmole/L. of H_2O)
Na^+	146	142	14
K^+	4.2	4.0	140
Ca^{++}	2.5	2.4	0
Mg^{++}	1.5	1.4	31
Cl^-	105	108	4
HCO_3^-	27	28.3	10
HPO_4^{--}, $H_2PO_4^-$	2	2	11
SO_4	0.5	0.5	1
Phosphocreatine			45
Carnosine			14
Amino acids	2	2	8
Creatine	0.2	0.2	9
Lactate	1.2	1.2	1.5
Adenosine triphosphate			5
Hexose monophosphate			3.7
Glucose	5.6	5.6	
Protein	1.2	0.2	4
Urea	4	4	4
TOTAL mOsmole/L.	302.9	301.8	302.2
Corrected osmolar activity (mOsmole/L.)	282.6	281.3	281.3
Total osmotic pressure at 37° C. (mm. Hg)	5453	5430	5430

of the plasma 1.3 milliosmoles greater than that of the interstitial and intracellular fluids. This slight difference between plasma and interstitial fluid is caused by the osmotic effect of the plasma proteins, which maintains about 23 mm. Hg greater pressure in the capillaries than in the surrounding interstitial fluid spaces, as was explained in Chapter 30.

Corrected Osmolar Activity of the Body Fluids. At the bottom of Table 33–2 is shown a corrected osmolar activity of plasma, interstitial fluid, and intracellular fluid. The reason for this correction is the following: All molecules and ions in solution exert either *intermolecular attraction* or *intermolecular repulsion*, and these two effects can cause, respectively, a decrease or an increase in the osmotic "activity" of the dissolved substance. In general, there is more intermolecular attraction than repulsion, so that the overall osmotic activity of the substances is only about 93 per cent of that which one would calculate from the number of milliosmoles present. For this reason, the actual osmotic pressure of the body fluids is proportional to the corrected osmolar activity, which amounts to approximately 280 milliosmoles/liter.

However, since osmotic effects in the body are usually determined by relative, rather than absolute, osmolar concentrations, the correction factor is often ignored.

Total Osmotic Pressure Exerted by the Body Fluids. At the bottom of Table 33–2 is shown the total osmotic pressure in mm. Hg that would be exerted by each of the different fluids if it were placed on one side of a cell membrane with pure water on the other side. Note that this total pressure averages about 5450 mm. Hg for plasma and also that the osmotic pressure of plasma is 23 mm. Hg greater than that of the interstitial fluids, this difference equaling the approximate hydrostatic pressure difference between the pressure of the blood inside the capillaries and the negative pressure in the interstitial fluid outside the capillaries.

MAINTENANCE OF OSMOTIC EQUILIBRIUM BETWEEN EXTRACELLULAR AND INTRACELLULAR FLUIDS

The tremendous osmotic pressure that can develop across the cell membrane when one side is exposed to pure water — more than 5400 mm. Hg — illustrates how much force can become available to push water molecules through the membrane when the solutions on the two sides of the membrane are not in osmotic equilibrium. For instance, in Figure 33–7A, a cell is placed in a solution that has an osmolality far less than that of

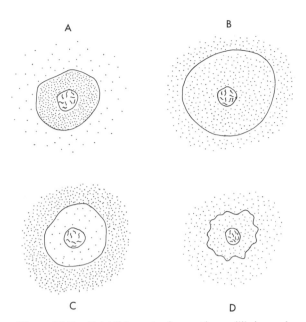

Figure 33–7. Establishment of osmotic equilibrium when cells are placed in a hypo- or hypertonic solution.

the intracellular fluid. As a result, osmosis of water begins immediately from the extracellular fluid to the intracellular fluid, causing the cell to swell and diluting the intracellular fluid while concentrating the extracellular fluid. When the fluid inside the cell becomes diluted sufficiently to equal the osmolal concentration of the fluid on the outside, further osmosis then ceases. This condition is shown in Figure 33–7B. In Figure 33–7C, a cell is placed in a solution having a much higher concentration outside the cell than inside. This time, water passes by osmosis to the exterior, diluting the extracellular fluid. In this process the cell shrinks until the two concentrations become equal, as shown in Figure 33–7D.

Rapidity of Attaining Extracellular and Intracellular Osmotic Equilibrium. The transfer of water through the cell membrane by osmosis occurs so rapidly that any lack of osmotic equilibrium between the two fluid compartments in any given tissue is usually corrected within a few seconds and at most within a minute or so. However, this rapid transfer of water does not mean that complete equilibration occurs between the extracellular and intracellular compartments throughout the whole body within this same short period of time. The reason for this is that fluid usually enters the body through the gut and must then be transported by the blood to all tissues before complete equilibration can occur. In the normal person it may take as long as 30 minutes to achieve reasonably good equilibration everywhere in the body after drinking water.

Isotonicity, Hypotonicity, and Hypertonicity. A

fluid into which normal body cells can be placed without causing either swelling or shrinkage of the cells is said to be *isotonic* with the cells. A 0.9 per cent solution of sodium chloride or a 5 per cent glucose solution is approximately isotonic.

A solution that will cause the cells to swell is said to be *hypotonic*; any solution of sodium chloride with less than 0.9 per cent concentration is hypotonic.

A solution that will cause the cells to shrink is said to be *hypertonic*; sodium chloride solutions of greater than 0.9 per cent concentration are all hypertonic.

CHANGES IN THE VOLUMES AND OSMOLALITIES OF THE EXTRACELLULAR AND INTRACELLULAR FLUID COMPARTMENTS IN ABNORMAL STATES

CALCULATION OF FLUID SHIFTS BETWEEN THE EXTRACELLULAR AND INTRACELLULAR FLUID COMPARTMENTS

Among the different factors that can cause extracellular or intracellular volumes to change markedly are ingestion of water, dehydration, intravenous infusion of different types of solutions, loss of large quantities of fluids from the gastrointestinal tract, or loss of abnormal quantities of fluid by sweating or through the kidneys.

The changes in both extracellular and intracellular fluid volumes can be calculated easily if one will keep the following two basic principles in mind:

1. *The osmolalities of the extracellular and intracellular fluids remain exactly equal to each other except for a few minutes after a change in one of the fluids occurs.*

2. *The number of osmoles of osmotically active substance in each compartment, in the extracellular fluid or in the intracellular fluid, remains constant unless one of the osmotically active substances moves through the cell membranes to the other compartment or is lost from or added to one of the two compartments in some other way.*

Using these two basic principles, we can now analyze the effects of different abnormal fluid conditions on extracellular and intracellular fluid volumes and osmolalities.

EFFECT OF ADDING WATER TO THE EXTRACELLULAR FLUID

Water can be added to the extracellular fluid by injection into the blood stream, by injection beneath the skin, or by ingesting water and, following this, by absorption from the gastrointestinal tract into the blood. The water dilutes the extracellular fluid, causing it to become hypotonic with respect to the intracellular fluids. Osmosis begins immediately at the cell membranes, with large amounts of water passing to the interiors of the cells. Within a few minutes the water becomes distributed almost evenly among all the extracellular and intracellular fluid compartments.

Calculation of the Changes in Extracellular and Intracellular Volumes and Osmolalities. Table 33–3 illustrates the progressive changes that would result from injecting 10 liters of water into the extracellular fluid. This table gives the volumes and milliosmoles per liter of the extracellular fluid, the intracellular fluid, and the total body water.

Initially the extracellular fluid volume is 15 liters, the intracellular fluid volume 25 liters, and the total body water 40 liters; the milliosmoles per liter is 300 in each of these. The solution that is added, 10 liters of water, has zero osmolality, and when first injected it is added both to the extracellular fluid and to the total body water, but not to the intracellular fluid.

The third line of the table shows the instantaneous effect on the volumes and osmolalities caused by addition of this fluid (before any osmosis occurs). The extracellular fluid volume rises to 25 liters, and its concentration becomes diluted immediately to 180 milliosmoles/liter. During this instantaneous interval of time nothing has happened to the intracellular fluids. But within a few seconds to a few minutes, the dilute extracellular fluid diffuses throughout the interstitial spaces of the body and comes in contact with the cells, allowing large portions of the water to pass by osmosis into the intracellular fluid. Osmosis will not stop until the milliosmolalities of the two fluids become equal. As a result, the milliosmolality of the extracellular fluid rises upward from 180 while that of the intracellular fluid falls downward from 300 until the two meet each other. After a few moments the conditions in the last line of the table obtain, with the extracellular fluid volume 18.75 liters, the intracellular volume 31.25 liters, and a milliosmolality in both compartments of 240.

TABLE 33–3 EFFECT OF ADMINISTERING 10 LITERS OF WATER INTRAVENOUSLY

	Extracellular			Intracellular			Total Body Water		
	Volume (Liters)	Concentration (mOsmole/L.)	Total mOsmole	Volume (Liters)	Concentration (mOsmole/L.)	Total mOsmole	Volume (Liters)	Concentration (mOsmole/L.)	Total mOsmole
Initial	15	300	4500	25	300	7500	40	300	12000
Solution added	10	0	0	0	0	0	10	0	0
Instantaneous effect	25	180	4500	25	300	7500	50	No equilibrium	12000
After osmotic equilibrium	18.75	240	4500	31.25	240	7500	50	240	12000

To calculate the changes occurring in the above example, one needs only to keep accurate accounting of the total number of milliosmoles in each fluid compartment and also in the total body water in the following manner: After the 10 liters of water are added the total body water becames 50 liters instead of 40, but the total milliosmoles in the whole body remain the same, 12,000. Dividing 50 into 12,000, the average milliosmolar concentration in each liter of the body water under the new conditions is found to be 240. One can readily understand that, after osmotic equilibrium has occurred throughout the body, this calculated value of 240 is the milliosmolality in both the extracellular and intracellular fluid compartments. Then, dividing 240 milliosmoles into the total milliosmoles in the extracellular fluid compartment, 4500, one finds that the new volume of extracellular fluid is 18.75 liters. Dividing 240 into the total intracellular milliosmoles, 7500, gives the new intracellular fluid volume of 31.25 liters.

EFFECT OF DEHYDRATION

Water can be removed from the body by evaporation from the skin, evaporation from the lungs, or excretion of a very dilute urine. In all these conditions the water leaves the extracellular fluid compartment, but on doing so some of the intracellular water passes immediately into the extracellular compartment by osmosis, thus keeping the osmolalities of the extracellular and intracellular fluids equal to each other. The overall effect is called *dehydration*.

EFFECT OF ADDING SALINE SOLUTION TO THE EXTRACELLULAR FLUID

If an *isotonic* saline solution is added to the extracellular fluid compartment, the osmolality of the extracellular fluid does not change, and no osmosis results. The only effect is an increase in extracellular fluid volume.

However, if a *hypertonic* solution is added to the extracellular fluid, the osmolality increases and causes osmosis of water out of the cells into the extracellular compartment.

Finally, if a *hypotonic* solution is added, the osmolality of the extracellular fluid decreases, and some of the extracellular fluids pass into the cells.

Table 33–4 gives the calculations of the effects which would occur if 2 liters of 4.4 per cent (five times isotonic concentration) sodium chloride solution were added to the extracellular fluid compartment. The added solution

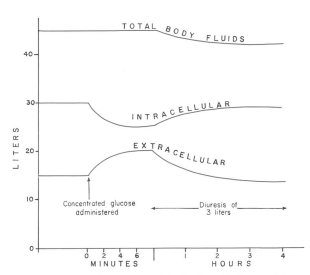

Figure 33–8. Time course of the body fluid changes following infusion of very concentrated glucose into the extracellular fluids.

contains a large total number of milliosmoles, 3000, which are added to the extracellular fluid compartment and also to the total body fluids, and the volume of the total body water rises from 40 to 42. Dividing 42 into the new total milliosmoles, 15,000, one finds that the new milliosmolar concentration is 357. Now, dividing 357 into the total milliosmoles in the extracellular fluid, 7500, one finds that the new extracellular fluid volume is 21 liters. And, dividing 357 into the total intracellular milliosmoles, 7500, the intracellular fluid is found also to be 21 liters.

The student should become completely familiar with this method of calculation, for an understanding of the mathematical aspects of osmotic equilibria between the two compartments is essential to the understanding of almost all fluid problems of the body.

EFFECT OF INFUSING HYPERTONIC GLUCOSE, MANNITOL, OR SUCROSE SOLUTIONS

Very concentrated glucose, mannitol, or sucrose solutions are often injected into patients to cause immediate decrease in intracellular fluid volume. For instance, often in severe cerebral edema the patient dies because of too much pressure in the cranial vault, which obstructs the flow of blood to the brain. The condition can

TABLE 33–4   EFFECT OF ADDING 2 LITERS OF 4.4 PER CENT SODIUM CHLORIDE SOLUTION

	Extracellular			*Intracellular*			*Total Body Water*		
	Volume (Liters)	Concentration (mOsmole/L.)	Total mOsmole	Volume (Liters)	Concentration (mOsmole/L.)	Total mOsmole	Volume (Liters)	Concentration (mOsmole/L.)	Total mOsmole
Initial	15	300	4500	25	300	7500	40	300	12000
Solution added	2	1500	3000	0	0	0	2	1500	3000
Instantaneous effect	17	441	7500	25	300	7500	42	No equilibrium	15000
After osmotic equilibrium	21	357	7500	21	357	7500	42	357	15000

be relieved, however, in a few minutes by injecting a hypertonic solution of a substance that will not enter the intracellular compartment. The dynamics of the resulting changes are shown in Figure 33–8, illustrating that the intracellular fluid volume can be decreased by several liters in a few minutes.

But mannitol and sucrose are both excreted rapidly by the kidneys, and glucose is metabolized by the cells for energy (or excreted as well if its concentration rises too high). Therefore, within two to four hours the osmotic effects of these substances are lost so that large quantities of water can then rediffuse into the intracellular compartment. Thus, this procedure is only temporarily beneficial, lasting only a few hours. Nevertheless, from the standpoint of saving the life of a patient it is often very valuable.

GLUCOSE AND OTHER SOLUTIONS ADMINISTERED FOR NUTRITIVE PURPOSES

Many different types of solutions are often administered intravenously to provide nutrition to patients who cannot otherwise take adequate amounts of food. Especially used are glucose solutions and to a lesser extent amino acid solutions. Rarely used are homogenized fat solutions. When all these are being administered, their concentrations are adjusted nearly to isotonicity, or they are given slowly enough that they do not upset the osmotic equilibria of the body fluids. However, after the glucose or other nutrient is metabolized, an excess of water often remains. Ordinarily, the kidneys excrete this in the form of a very dilute urine. Thus, the net result is only addition of the nutrient to the body.

Hyperalimentation. When nutritive solutions are given intravenously into a peripheral vein, their concentrations must be adjusted very nearly to isotonicity; otherwise, the abnormal osmolality will severely damage the intima of the blood vessel and the vessel will usually become plugged by a blood clot. On the other hand, a patient receiving only isotonic solutions cannot tolerate enough water each day to allow total nutritive support.

However, it has recently been learned that very concentrated solutions can be infused through a catheter directly into one of the major central veins where the blood flows very rapidly. The solution is diluted quickly enough so that it will not harm the vascular system. Using this procedure, enough calories can be given to a patient each 24 hours to provide full nutritive support even though only 2 to 3 liters of fluid are given to the patient each day. Therefore, it is now possible to maintain patients with no intake of food by mouth whatsoever for many months at a time.

REFERENCES

Adolph, E. F.: Physiology of Man in the Desert. New York, Interscience Publishers, 1947.

Andersson, B.: Regulation of body fluids. *Annu. Rev. Physiol., 39*:185, 1977.

Andreoli, T. E., *et al.* (eds.): Membrane Physiology. New York, Plenum Press, 1980.

Artz, C. P., *et al.* (eds.): Burns, A Team Approach. Philadelphia, W. B. Saunders Co., 1979.

Bing, D. H. (ed.): The Chemistry and Physiology of the Human Plasma Proteins. New York, Pergamon Press, 1979.

Borow, M.: Fundamentals of Homeostasis: A Clinical Approach to Fluid Electrolyte Acid-Base Energy Metabolism in Health and Disease. Flushing, N.Y., Medical Examination Publishing Co., 1977.

Borut, A., and Shkolnik, A.: Physiological adaptations to the desert environment. *Int. Rev. Physiol., 7*:185, 1974.

Bradbury, M. W.: Physiology of body fluids and electrolytes. *Br. J. Anaesthesiol., 45*:937, 1973.

Burton, R. F.: The significance of ionic concentrations in the internal media of animals. *Biol. Rev., 48*:195, 1973.

Coleman, T. G., *et al.*: Dynamics of water-isotope distribution. *Am. J. Physiol., 223*:1371, 1972.

Epstein, A. N., *et al.*: The Neuropsychology of Thirst. Washington, V. H. Winston & Sons, 1973.

Gilles, R. (ed.): Mechanisms of Osmoregulation: Maintenance of Cell Volume. New York, John Wiley & Sons, 1979.

Goldberger, E.: A Primer of Water, Electrolyte, and Acid-Base Syndromes. Philadelphia, Lea & Febiger, 1980.

Gupta, B. L., *et al.* (eds.): Transport of Ions and Water in Animals. New York, Academic Press, 1977.

Guyton, A. C., *et al.*: Circulatory Physiology II: Dynamics and Control of Body Fluids. Philadelphia, W. B. Saunders Co., 1975.

House, C. R.: Water Transport in Cells and Tissues. London, Edward Arnold, 1974.

Jamieson, G. A., and Greenwalt, T. J. (eds.): Blood Substitutes and Plasma Expanders. New York, A. R. Liss, 1978.

Kay, R. L. (ed.): The Physical Chemistry of Aqueous Systems. New York, Plenum Publishing Corp., 1974.

Lightfoot, E. N.: Transport Phenomena and Living Systems; Biomedical Aspects of Momentum and Mass Transport. New York, John Wiley & Sons, 1974.

Mason, E. E.: Fluid, Electrolyte, and Nutrient Therapy in Surgery. Philadelphia, Lea & Febiger, 1974.

Parsa, M. H., *et al.*: Safe Central Venous Nutrition. Springfield, Ill., Charles C Thomas, 1974.

Peters, T., and Sjöholm, I. (eds.): Albumin: Structure, Biosynthesis, Function. New York, Pergamon Press, 1978.

Schreiber, G., and Urban, J.: The synthesis and secretion of albumin. *In* Adrian, R. H., *et al.* (eds.): Reviews of Physiology, Biochemistry, and Pharmacology. New York, Springer-Verlag, 1978, p. 27.

Shoemaker, V. H., and Nagy, K. A.: Osmoregulation in amphibians and reptiles. *Annu. Rev. Physiol., 39*:449, 1977.

Smith, K.: Fluids and Electrolytes; A Conceptual Approach. New York, Churchill Livingstone, 1980.

Solomon, A. K.: The state of water in red cells. *Sci. Am., 2*:88, 1971.

Vaamonde, C. A., and Papper, S.: Maintenance of body tonicity. *In* Frohlich, E. D. (ed.): Pathophysiology, 2nd Ed. Philadelphia, J. B. Lippincott Co., 1976, p. 265.

Wilkinson, A. W.: Body Fluids in Surgery, 4th Ed. New York, Churchill Livingstone, Div. of Longman, 1973.

Wolf, A. V., and Crowder, N. A.: Introduction to Body Fluid Metabolism. Baltimore, Williams & Wilkins, 1964.

34

Formation of Urine by the Kidney: Glomerular Filtration, Tubular Function, and Plasma Clearance

The kidneys perform two major functions: first, they excrete most of the end-products of bodily metabolism, and, second, they control the concentrations of most of the constituents of the body fluids. The purpose of the present chapter is to discuss the principles of urine formation and especially the mechanisms by which the kidneys excrete the end-products of metabolism.

PHYSIOLOGIC ANATOMY OF THE KIDNEY

The two kidneys together contain about 2,400,000 nephrons, and each nephron is capable of forming urine by itself. Therefore, in most instances, it is not necessary to discuss the entire kidney but merely the function of the single nephron to explain the function of the kidney.

The nephron is composed basically of (1) a *glomerulus* from which fluid is filtered, and (2) a long *tubule* in which the filtered fluid is converted into urine on its way to the *pelvis* of the kidney. Figure 34–1 shows the general organizational plan of the kidney, illustrating especially the distinction between the *cortex* of the kidney and the *medulla*. And Figure 34–2A illustrates the basic anatomy of the nephron, which may be described as follows: Blood enters the glomerulus through the *afferent arteriole* and then leaves through the *efferent arteriole*. The glomerulus is a network of up to 50 parallel capillaries covered by epithelial cells and encased in *Bowman's capsule*. Pressure of the blood in the glomerulus causes fluid to filter into Bowman's capsule, from which it flows first into the *proximal tubule* that lies in the *cortex* of the kidney along with the glomerulus. From there the fluid passes into the *loop of Henle*. Those

nephrons that have glomeruli lying close to the renal *medulla* are called *juxtamedullary nephrons*, one of which is illustrated in Figure 34–2B, and they have long extended loops of Henle that dip deep into the medulla; the lower portion of the loop has a very thin wall and therefore is called the *thin segment* of the loop of Henle. From the loop of Henle the fluid flows back to the renal cortex through the *distal tubule*. Finally, it flows into the *collecting duct*, which collects fluid from several nephrons. The collecting duct passes from the

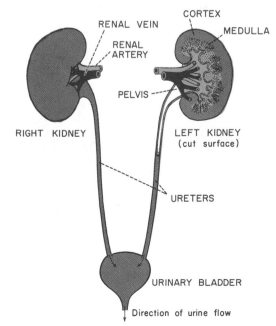

Figure 34–1. The general organizational plan of the urinary system.

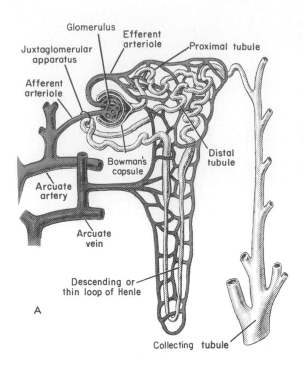

A

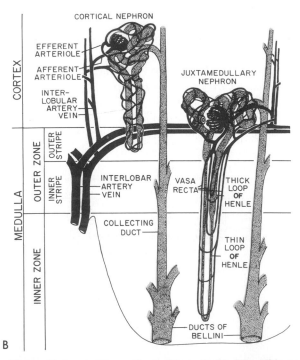

B

Figure 34–2. *A*, The nephron. (From Smith: The Kidney: Structure and Functions in Health and Disease. New York, Oxford University Press, 1951.) *B*, Differences between a cortical and a juxtamedullary nephron. (From Pitts: Physiology of the Kidney and Body Fluids. Chicago, Year Book Medical Publishers, 1974.)

cortex back downward through the medulla, paralleling the loops of Henle. Then it empties into the pelvis of the kidney.

As the glomerular filtrate flows through the tubules, most of its water and varying amounts of

its solutes are reabsorbed into the *peritubular capillaries* and small amounts of other solutes are secreted into the tubules. The remaining tubular water and solutes become urine.

Nephrons that have glomeruli lying close to the surface of the kidney are called *cortical nephrons*, as illustrated in Figure 34–2B; these have very short, thin segments in their loops of Henle, and their loops of Henle fail to penetrate all the way into the medulla. Other than this difference, these cortical nephrons are much the same as the juxtamedullary nephrons.

After blood passes into the efferent arteriole from the glomerulus, most of it flows through the *peritubular capillary network* that surrounds the cortical portions of the tubules. The remainder of the blood, most of it coming from the juxtamedullary glomeruli, flows into straight capillary loops called *vasa recta* that extend downward into the medulla to envelop the lower parts of the thin segments before looping back upward to empty into the cortical veins.

Functional Diagram of the Nephron. Figure 34–3 illustrates a simplified diagram of the "physiologic nephron." This diagram contains most of the nephron's functional structures, and it is used in the present discussion to explain many aspects of renal function.

BASIC THEORY OF NEPHRON FUNCTION

The basic function of the nephron is to clean, or "clear," the blood plasma of unwanted substances as it passes through the kidney. The

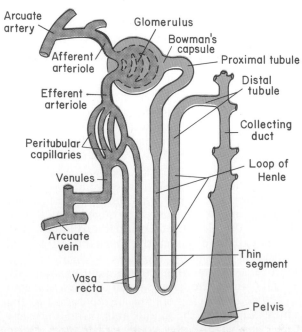

Figure 34–3. The functional nephron.

substances that must be cleared include particularly the end-products of metabolism such as urea, creatinine, uric acid, and urates. In addition, many other substances, such as sodium ions, potassium ions, chloride ions, and hydrogen ions tend to accumulate in the body in excess quantities; it is the function of the nephron also to clear the plasma of the excesses.

The principal mechanism by which the nephron clears the plasma of unwanted substances is: (1) It filters a large proportion of the plasma, usually about one-fifth of it, through the glomerular membrane into the tubules of the nephron. (2) Then, as this filtered fluid flows through the tubules, the unwanted substances fail to be reabsorbed while the wanted substances, especially the water and many of the electrolytes, are reabsorbed back into the plasma of the peritubular capillaries. In other words, the wanted portions of the tubular fluid are returned to the blood, while the unwanted portions pass into the urine.

A second mechanism by which the nephron clears the plasma of unwanted substances is by *secretion*. That is, substances are secreted from the plasma directly through the epithelial cells lining the tubules and into the tubular fluid. Thus, the urine that is eventually formed is composed mainly of *filtered* substances but also small amounts of *secreted* substances.

RENAL BLOOD FLOW AND PRESSURES

BLOOD FLOW THROUGH THE KIDNEYS

The rate of blood flow through both kidneys of a 70 kg. man is about 1200 ml./minute.

The portion of the total cardiac output that passes through the kidneys is called the *renal fraction*. Since the normal cardiac output of a 70 kg. adult male is about 5600 ml./minute, and the blood flow through both kidneys is about 1200 ml./minute, one can calculate that the normal renal fraction is about 21 per cent. This can vary from as little as 12 per cent to as high as 30 per cent in the normal resting person.

Special Aspects of Blood Flow Through the Nephron. Note in Figure 34–3 that there are two capillary beds supplying the nephron: (1) the *glomerulus* and (2) the *peritubular capillaries*. The glomerular capillary bed receives its blood from the *afferent arteriole,* and this bed is separated from the peritubular capillary bed by the *efferent arteriole,* which offers considerable resistance to blood flow. As a result, the glomerular capillary bed is a *high pressure bed* while the peritubular capillary bed is a *low pressure bed*. Because of the high pressure in the glomerulus, it functions in much the same way as the usual arterial ends of

the tissue capillaries, with fluid filtering continually out of the glomerulus into Bowman's capsule. On the other hand, the low pressure in the peritubular capillary system causes it to function in much the same way as the venous ends of the tissue capillaries, with fluid being absorbed continually into the capillaries.

The Vasa Recta. A special portion of the peritubular capillary system is the vasa recta, which are a network of capillaries that descend around the lower portions of the loops of Henle. These capillaries form loops in the medulla of the kidney and then return to the cortex before emptying into the veins. The vasa recta play a special role in the formation of concentrated urine, which will be discussed in the following chapter.

Only a small proportion of the total renal blood flow, about 1 to 2 per cent, flows through the vasa recta. In other words, blood flow through the medulla of the kidney is sluggish in contrast to the rapid blood flow in the cortex.

PRESSURES IN THE RENAL CIRCULATION

Figure 34–4 gives the approximate pressures in the different parts of the renal circulation and

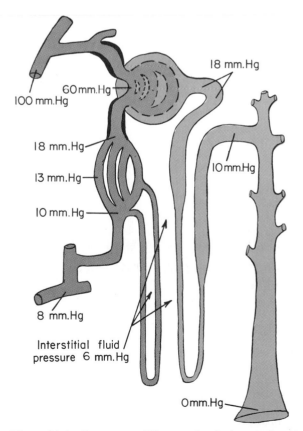

Figure 34–4. Pressures at different points in the vessels and tubules of the functional nephron and in the interstitial fluid.

tubules, showing an initial pressure of approximately 100 mm. Hg in the large arcuate arteries and about 8 mm. Hg in the veins into which the blood finally drains. The two major areas of resistance to blood flow through the nephron are (1) the *afferent arteriole* and (2) the *efferent arteriole*. In the afferent arteriole the pressure falls from 100 mm. Hg at its arterial end to an estimated mean pressure of about 60 mm. Hg in the glomerulus. (This pressure is still in serious doubt, having been calculated to be as high as 70 mm. Hg in the dog and measured to be as low as 45 mm. Hg in the rat. Therefore, 60 mm. Hg is only an average estimate.) As the blood flows through the efferent arterioles from the glomerulus to the peritubular capillary system, the pressure falls another 47 mm. Hg to a mean peritubular capillary pressure of 13 mm. Hg. Thus, the high pressure capillary bed in the glomerulus operates at a mean pressure of about 60 mm. Hg and therefore causes rapid filtration of fluid, while the low pressure capillary bed in the peritubular capillary system operates at a mean capillary pressure of about 13 mm. Hg, therefore allowing rapid absorption of fluid because of the high osmotic pressure of the plasma.

"INTRARENAL PRESSURE" AND RENAL INTERSTITIAL FLUID PRESSURE

The kidney is encased in a tight *fibrous capsule*. When a needle is inserted into the kidney and the pressure in the needle is gradually raised until fluid flows into the kidney tissue, the pressure at which flow begins is between 10 and 18 mm. Hg, averaging perhaps 12 mm. Hg. This "needle" pressure is called the *intrarenal pressure*. It was pointed out in Chapter 30 that such needle pressures do not measure the interstitial fluid pressure but instead measure the *total tissue pressure* (the pressure that tends to collapse blood vessels and tubules).

Recent attempts to measure the *interstitial fluid pressure* of the kidney utilizing implanted perforated capsules as described in Chapter 30 have given an average mean value of +6 mm. Hg, which, at present, is probably the best estimate of interstitial fluid pressure of the kidney.

FUNCTION OF THE PERITUBULAR CAPILLARIES

Tremendous quantities of fluid, about 180 liters each day, are filtered through all the glomeruli; all but 1 to 1.5 liters of this is reabsorbed from the tubules into the renal interstitial spaces and thence into the peritubular capillaries. This represents about 4 times as much fluid as that reabsorbed at the venous ends of all the other capillaries of the entire body. Therefore, one can readily see that reabsorption of fluid into the peritubular capillaries presents a special problem. However, the peritubular capillaries are extremely porous in comparison with those in other body tissues, so that extremely rapid osmosis of fluid resulting from the colloid osmotic pressure of the plasma proteins can account for the rapid absorption that is required.

GLOMERULAR FILTRATION AND THE GLOMERULAR FILTRATE

The Glomerular Membrane and Glomerular Filtrate. The fluid that filters through the glomerulus into Bowman's capsule is called *glomerular filtrate*, and the membrane of the glomerular capillaries is called the *glomerular membrane*. Though, in general, this membrane is similar to that of other capillaries throughout the body, it has several differences. First, it has three major layers: (1) the endothelial layer of the capillary itself, (2) a basement membrane, and (3) a layer of epithelial cells that is illustrated on the outer surfaces of the glomerular capillaries in Figure 34–3. Yet, despite the number of layers, the permeability of the glomerular membrane is from 100 to 1000 times as great as that of the usual capillary.

The tremendous permeability of the glomerular membrane is caused by its special structure, which is illustrated in Figure 34–5. The capillary *endothelial cells* lining the glomerulus are perforated by literally thousands of small holes called *fenestrae*. Then, outside the endothelial cells is a basement membrane composed mainly of a meshwork of proteoglycan fibrillae. A final layer of the glomerular membrane is a layer of epithelial cells that line the outside of the glomerulus. However, these cells are not continuous but instead consist

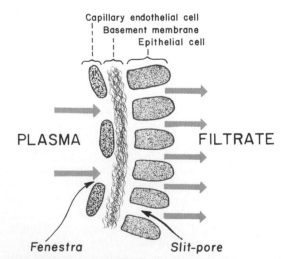

Figure 34–5. Functional structure of the glomerular membrane.

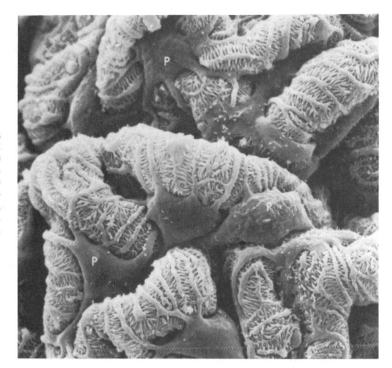

Figure 34–6. Scanning electron micrograph of a glomerulus from a normal rat kidney. The visceral epithelial cells, or podocytes (P), extend multiple processes outward from the main cell body, which wrap around the individual capillary loops. Note that immediately adjacent pedicels, or foot processes, arise from different podocytes. (Magnification, × 3300.) (From Brenner and Rector: The Kidney. Philadelphia, W. B. Saunders Company, 1976, p. 16.)

mainly of finger-like projections that cover the outer surface of the basement membrane. Figure 34–6 is a scanning electron micrograph of the outer surfaces of the glomerular capillaries, illustrating the multiple "fingers" of the epithelial cells. These fingers form slits called *slit-pores* through which the glomerular filtrate filters. Thus, the glomerular filtrate must pass through three different layers before entering Bowman's capsule.

Experiments have shown that the fenestrae of the capillary endothelial cells are small enough in diameter that they prevent the filtration of all particles with an average size greater than 16 millimicrons (160 A). The meshwork of the basement membrane prevents filtration of all particles greater in size than 11 millimicrons. And the slit-pores prevent filtration of all particles with diameters greater than 7 millimicrons. Since plasma proteins are slightly larger than the 7 millimicron diameter, it is possible for the glomerular membrane to prevent the filtration of all substances with molecular weights equal to or greater than those of the plasma proteins. Yet, the great numbers of fenestrae and slit-pores allow tremendously rapid filtration of fluid and small molecular weight substances from the plasma into Bowman's capsule.

Overall, the permeability of the glomerular membrane to substances of different molecular weights (expressed as the ratio of concentration of the dissolved substance on the filtrate side of the membrane to its concentration on the plasma side) is approximately as follows:

Molecular Weight	Permeability	Example Substance
5200	1.00	Inulin
30,000	0.5	Very small protein
69,000	0.005	Albumin

This means that at a molecular weight of 5000, the dissolved substance filters just as easily as water, but at a molecular weight of 69,000 only 0.005 per cent of the dissolved substance filters. Note that the molecular weight of the smallest plasma protein, albumin, is 69,000. Therefore, for practical purposes the glomerular membrane is almost completely impermeable to all plasma proteins but is highly permeable to essentially all other dissolved substances in normal plasma.

Composition of the Glomerular Filtrate. The glomerular filtrate has almost exactly the same composition as the fluid that filters from the arterial ends of the capillaries into the interstitial fluids. It contains no red blood cells and about 0.03 per cent protein, or about $1/_{200}$ the protein in the plasma.

The electrolyte and other solute composition of glomerular filtrate is also similar to that of the interstitial fluid. However, because of the paucity of negatively charged protein ions in the filtrate, a Donnan equilibrium effect occurs that causes the concentration of the other negative ions, including chloride and bicarbonate ions, to be about 5 per cent higher in the glomerular filtrate than in plasma; and the concentration of positive ions is about 5 per cent lower.

To summarize: for all practical purposes, glomerular filtrate is the same as plasma except that it has no significant amount of proteins.

THE GLOMERULAR FILTRATION RATE

The quantity of glomerular filtrate formed each minute in all nephrons of both kidneys is called the *glomerular filtration rate*. In the normal person this averages approximately 125 ml./min.; however, in different normal functional states of the kidneys, it can vary from a few milliliters to 200 ml./min. To express this differently, the total quantity of glomerular filtrate formed each day averages about 180 liters, or more than two times the total weight of the body. Over 99 per cent of the filtrate is usually reabsorbed in the tubules, the remainder passing into the urine, as explained later in the chapter.

The Filtration Fraction. The filtration fraction is the fraction of the renal plasma flow that becomes glomerular filtrate. Since the normal plasma flow through both kidneys is 650 ml./min. and the normal glomerular filtration rate in both kidneys is 125 ml., *the average filtration fraction is approximately* $^{125}/_{650}$, *or 19 per cent*. Here again, this value can vary tremendously, both physiologically and pathologically.

DYNAMICS OF GLOMERULAR FILTRATION

Glomerular filtration occurs in almost exactly the same manner that fluid filters out of any high pressure capillary in the body. That is, *pressure inside the glomerular capillaries* causes filtration of fluid through the capillary membrane into Bowman's capsule. On the other hand, *colloid osmotic pressure in the blood and pressure in Bowman's capsule* oppose the filtration. (Ordinarily, the amount of protein in Bowman's capsule is too slight to be of any significance, but if this ever

becomes increased to a significant amount, its colloid osmotic pressure will obviously also be active at the membrane, promoting increased filtration of fluid through the membrane.)

Glomerular Pressure. The glomerular pressure is the average pressure in the glomerular capillaries. This unfortunately has been measured directly only in one mammal, the rat, in which the average value is about 45 mm. Hg. However, from various indirect measurements it has been calculated to be from 55 to 70 mm. Hg in the dog. Because man is a large mammal, *a reasonable average value can be considered to be 60 mm. Hg*, though, as noted below, this can increase or decrease considerably under varying conditions.

Pressure in Bowman's Capsule. In lower animals, pressure measurements have actually been made in Bowman's capsule and at different points along the renal tubules by inserting micropipets into the lumen. On the basis of these studies, *capsular pressure in the human being is estimated to be 18 mm. Hg.*

Colloid Osmotic Pressure in the Glomerular Capillaries. Because approximately one fifth of the plasma in the capillaries filters into the capsule, the protein concentration increases about 20 per cent as the blood passes from the arterial to the venous ends of the glomerular capillaries. If the normal colloid osmotic pressure of blood entering the capillaries is 28 mm. Hg, it rises to approximately 36 mm. Hg by the time the blood reaches the venous ends of the capillaries, and the average colloid osmotic pressure is about 32 mm. Hg. (See Figure 30–8 in Chapter 30, which shows the relationship between protein concentration and colloid osmotic pressure.)

Filtration Pressure, Filtration Coefficient, and Glomerular Filtration Rate. The filtration pressure is the net pressure forcing fluid through the glomerular membrane, and this is *equal to the glomerular pressure minus the sum of glomerular colloid osmotic pressure and capsular pressure*. In Figure 34–7A, *the normal filtration pressure is shown to be about 10 mm. Hg*.

The filtration coefficient, called K_f, is a con-

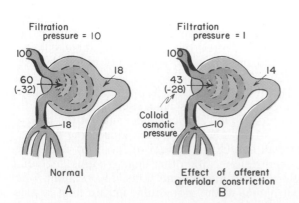

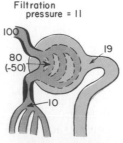

Figure 34–7. *A*, Normal pressures at different points in the nephron, and the normal filtration pressure. *B*, Effect of afferent arteriolar constriction on pressures in the nephron and on filtration pressure. *C*, Effect of efferent arteriolar constriction on pressures in the nephron and on filtration pressure.

stant; it is the glomerular filtration rate for both kidneys per mm. Hg of filtration pressure. That is, the glomerular filtration rate is equal to the filtration pressure times the filtration coefficient, or

$$GFR = \text{Filtration pressure} \cdot K_f$$

The normal filtration coefficient is 12.5 ml. per min. per mm. Hg of filtration pressure. Thus, at a normal mean filtration pressure of 10 mm. Hg, the total filtration rate of both kidneys is 125 ml. per min.

(Note: From studies in rats, several investigators have come to the conclusion that the filtration coefficient is much greater than given here and that the filtration pressure is much lower, but these values are still under discussion.)

FACTORS THAT AFFECT THE GLOMERULAR FILTRATION RATE

It is clear from the above equation that the filtration pressure and the filtration coefficient determine the glomerular filtration rate. The filtration coefficient probably does not change greatly from normal except when the kidneys become diseased.

On the other hand, the three factors that determine filtration pressure — (1) glomerular pressure, (2) plasma colloid osmotic pressure, and (3) Bowman's capsule pressure — do play very significant roles in determining glomerular filtration rate. In general, the greater the glomerular pressure, the greater will be the filtration rate; conversely, the greater the plasma colloid osmotic pressure or Bowman's capsule pressure, the less will be the glomerular filtration rate.

Effect of Renal Blood Flow on Glomerular Filtration Rate. The glomerular filtration rate is also affected to a great extent by the rate of blood flow through the nephrons. This effect can be explained as follows: It is not the plasma colloid osmotic pressure in the arterial blood that determines filtration through the glomerulus but instead the colloid osmotic pressure of the plasma in the glomerulus itself. Since a very large proportion of the plasma is filtered through the glomerular membrane, the colloid osmotic pressure in the glomerulus rises very high and opposes further filtration. Therefore, once a certain proportion of the plasma has filtered, no more will filter until new plasma flows into the glomerulus. Consequently, the greater the rate of flow of plasma into the glomerulus, the greater the glomerular filtration rate.

For mathematical reasons that cannot be explained here, the greater the filtration coefficient of the glomeruli, the greater is the effect of blood flow on glomerular filtration rate. On the other hand, the lower the filtration coefficient, the greater the effect glomerular pressure has on filtration rate.

Effect of Afferent Arteriolar Constriction on Glomerular Filtration Rate. Afferent arteriolar constriction decreases the rate of blood flow into the glomerulus and also decreases the glomerular pressure, both of these effects decreasing the filtration rate. This effect is illustrated in Figure 34–7B. Conversely, dilatation of the afferent arteriole increases the glomerular pressure, with a corresponding increase in glomerular filtration rate.

Effect of Efferent Arteriolar Constriction. Constriction of the efferent arteriole increases the resistance to outflow from the glomeruli. This obviously increases the glomerular pressure and usually increases the glomerular filtration rate, as illustrated in Figure 34–7C. However, the blood flow decreases at the same time, and if the degree of efferent arteriolar constriction is severe, the plasma will remain for a long period of time in the glomerulus, and extra large portions of plasma will filter out. This will increase the plasma colloid osmotic pressure to excessive levels and will cause glomerular filtration to fall paradoxically to a low value despite the elevated glomerular pressure.

Effect of Sympathetic Stimulation. During sympathetic stimulation of the kidneys, the afferent arterioles are constricted preferentially, thereby decreasing the glomerular filtration rate. With very strong sympathetic stimulation, glomerular blood flow and the glomerular pressure are reduced so greatly that the urinary output can fall to zero for as long as 5 to 10 minutes

Effect of Arterial Pressure. One would expect an increase in arterial pressure to cause a proportionate increase in all pressures in the nephron and therefore to increase the glomerular filtration rate to a great extent. In actual fact, this effect is greatly blunted because of a phenomenon called *autoregulation*, which is explained in the following chapter. Briefly, when the arterial pressure rises, afferent arteriolar constriction occurs automatically; this prevents a significant rise in glomerular pressure despite the rise in arterial pressure. Therefore, the glomerular filtration rate increases only a few per cent even when the mean arterial pressure rises from its normal value of 100 mm. Hg to as high as 150 mm. Hg.

Nevertheless, we shall also see in the following chapter that even a small percentage increase in glomerular filtration rate can cause a many-fold increase in urinary output. Therefore, an increase in arterial pressure can greatly increase urinary output even though it affects glomerular filtration rate only slightly.

REABSORPTION AND SECRETION IN THE TUBULES

The glomerular filtrate entering the tubules of the nephron flows (1) through the *proximal tubule*, (2) through the *loop of Henle*, (3) through the *distal tubule*, and (4) through the *collecting duct* into the pelvis of the kidney. Along this course, substances are selectively reabsorbed or secreted by the tubular epithelium, and the resultant fluid entering the pelvis is *urine*. Reabsorption plays a much greater role than does secretion in this formation of urine, but secretion is especially important in determining the amounts of potassium ions, hydrogen ions, and a few other substances in the urine, as is discussed later.

Ordinarily, more than 99 per cent of the water in the glomerular filtrate is reabsorbed as it passes through the tubules. Therefore, if some dissolved constituent of the glomerular filtrate is not reabsorbed at all along the entire course of the tubules, this reabsorption of water obviously concentrates the substance more than 99-fold. On the other hand, some constituents, such as glucose and amino acids, are reabsorbed almost entirely so that their concentrations decrease almost to zero before the fluid becomes urine. In this way the tubules separate substances that are to be conserved in the body from those that are to be eliminated in the urine.

BASIC MECHANISMS OF ABSORPTION AND SECRETION IN THE TUBULES

The basic mechanisms for transport through the tubular membrane are essentially the same as those for transport through other membranes of the body. These can be divided into *active transport* and *passive transport*. The basic essentials of these mechanisms are described here, but for additional details the reader should refer to Chapter 4.

Active Transport Through the Tubular Wall

Figure 34–8 illustrates, by way of example, the mechanism for active transport of sodium from the lumen of the proximal tubule into the peritubular capillary. Note, first, the character of the epithelial cells that line the tubule. Each epithelial cell has a "brush" border on its luminal surface. This brush is composed of literally thousands of very minute microvilli that multiply the surface area of luminal exposure of the cell about 20-fold. The base of the cell sits on the basement membrane, but it is pocked by an extensive system of *basal channels* that multiply the basal surface area manyfold as well. The epithelial cells are attached

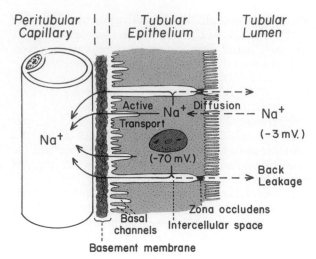

Figure 34–8. Mechanism for active transport of sodium from the tubular lumen into the peritubular capillary, illustrating active transport at the base and sides of the epithelial cell and diffusion through the luminal border of the cell.

to each other only where they touch each other adjacent to the brush border, a tight attachment area called the *zonula occludens*.

Active transport of sodium occurs from inside the epithelial cell through its side and basal membranes into the basal channels and into the spaces between the cells. This transport outward from the cell diminishes the sodium concentration inside the cell. Then, because this low concentration inside the cell establishes a sodium ion concentration gradient from the tubule to the inside of the cell, sodium ions diffuse from the tubule through the brush border into the cell. Once inside the cell, the sodium is carried by the active transport process the rest of the way into the peritubular fluid of the basal channels and the spaces between the cells.

Note in the figure that the electrical potential inside the epithelial cell, caused by the continual active transport of the sodium out of the cells, is approximately −70 millivolts. This very negative intracellular voltage is important because it, as well as the low concentration of sodium inside the epithelial cell, is an additional important factor that causes sodium diffusion from the tubular lumen into the cell. These two factors together are called the "electrochemical" gradient. The rapid diffusion of sodium into the cell is also facilitated by a very high permeability of the brush border membrane to sodium, primarily because of the extensive surface area of the thousands of microvilli.

Once the sodium has been transported into the basal channels of the epithelial cells and also into the intercellular spaces, the sodium can then move on into the peritubular capillary, or it can leak backward through the zonula occludens into the tubular lumen. The way it will go is determined by several different factors, including especially the hydrostatic pressures in both the tubule and the peritubular capillary and the colloid osmotic pressure in the peritubular capillary. The effects of these factors will be discussed in the following chapter.

Other substances besides sodium that are actively absorbed through the tubular epithelial cells include *glucose, amino acids, calcium ions, potassium ions, chloride ions, phosphate ions, urate ions*, and others. As was explained in Chapter 4, both glucose and the amino acids are transported from the tubular lumen through the brush border by a process called *secondary active transport* or *sodium co-transport*. That is, the glucose or amino acid molecule binds with a carrier molecule, and a sodium ion also binds with the same carrier. Then, as the sodium diffuses inward through the brush border membrane, it literally pulls the carrier and the glucose or amino acid along with it. Once inside the epithelial cell, the sodium and the glucose or amino acid split from the carrier. The glucose or amino acid then diffuses through the basal membrane of the cell and thence into the peritubular capillaries. For further details of this transport process, the reader is referred to a more complete discussion in Chapter 4.

In addition, some substances are actively *secreted* into all or some portions of the tubules; these include especially *hydrogen ions, potassium ions,* and *urate ions.* Active secretion occurs in the same way as active absorption except that the cell membrane transports the secreted substance in the opposite direction; for some of the actively secreted substances, it is the brush border rather than the base and sides of the cell that provides the active transport mechanism.

Passive Absorption of Water: Osmosis of Water Through the Tubular Epithelium

When the different solutes are transported out of the tubule and through the tubular epithelium, their total concentration decreases inside the tubular lumen and increases outside. This obviously creates a concentration difference that causes osmosis of water in the same direction that the solutes have been transported.

However, some portions of the tubular system are far more permeable to water than are others. In those portions that are highly permeable, such as the proximal tubules, osmosis of water occurs so rapidly that the osmolar concentration of solutes on the peritubular side of the membrane is almost never more than a few milliosmoles greater than on the intratubular side. For instance, referring again to Figure 34–8, when sodium ions are

transported from a tubule into the peritubular fluid of the intercellular spaces and basal channels, the osmotic pressure in this fluid increases while that in the tubule decreases. Since the proximal tubular membrane is highly permeable to water, osmosis occurs almost instantaneously into the peritubular fluid to re-equilibrate the osmolar concentrations on the two sides of the membrane.

On the other hand, the diluting segment of the distal tubule is an example of a tubular area that is almost completely impermeable to water, a fact that plays a very important role in the mechanism for controlling urine concentration. This will be discussed later in this chapter and in the following chapter.

Passive Absorption of Urea and Other Nonactively Transported Solutes by the Process of Diffusion

When water is reabsorbed by osmosis, about half of the urea in the tubular fluid remains behind; therefore, the concentration of urea in the tubular fluid rises, which obviously establishes a concentration difference for urea between the tubular and peritubular fluids. This in turn causes urea also to diffuse from the tubular fluid into the peritubular fluid. This same effect also occurs for other tubular solutes that are not actively reabsorbed but that are diffusible through the tubular membrane.

The rate of resorption of a nonactively reabsorbed solute is determined by (1) the amount of water that is reabsorbed, because this determines the tubular concentration of the solute, and (2) the permeability of the tubular membrane for the solute. The permeability of the membrane for urea in most parts of the tubules is far less than that for water, which means that far less urea is reabsorbed than water. Therefore, a large proportion of the urea remains in the tubules and is lost in the urine — usually about 50 per cent of all that enters the glomerular filtrate. The permeability of the tubular membrane for reabsorption of creatinine, inulin (a large polysaccharide), mannitol (a monosaccharide), and sucrose is zero, which means that once these substances have filtered into the glomerular filtrate, 100 per cent of that which enters the glomerular filtrate passes on into the urine.

Diffusion Caused by Electrical Differences Across the Tubular Membrane. The active absorption of sodium from the tubule creates negativity inside the tubule with respect to the peritubular fluid, as was illustrated in Figure 34–8. In the early proximal tubule, this electrical potential is approximately −3 millivolts. In the distal end of the distal tubule, it ranges from −10 to −70 millivolts. In some other segments of the tubules, this potential is a few millivolts positive — for instance, in the so-called "diluting segment" of the distal tubule, as will be discussed in more detail later in this chapter.

When the electrical potential is negative inside the tubule, this repels negative ions, such as chloride and phosphate ions, causing them to diffuse out of the tubules and into the peritubular fluid (if the tubular membrane is permeable to the ions). On the other hand, it will attract positive ions, such as sodium and potassium ions, from the peritubular fluid into the tubules. Conversely, when the intratubular fluid is positive, positive ions then move toward the peritubular fluid and negative ions toward the tubular lumen.

Passive Secretion by the Process of Diffusion

A few substances are secreted by diffusion in a manner similar to the reabsorption of water and other substances. For instance, ammonium ions are synthesized inside the epithelial cells and then diffuse into the tubular lumen; they help to control the degree of acidity of the tubular fluid, as will be discussed in Chapter 37. Also, potassium ions diffuse through the luminal border of the epithelial cells in the distal tubules and collecting ducts as a result of the strong negative potentials in these tubular segments and in this way potassium is "secreted." Active transport, however, plays a major role in the movement of potassium ions from the peritubular blood into the interior of the epithelial cells through their basal borders. This will be discussed further in Chapter 36.

ABSORPTIVE CAPABILITIES OF DIFFERENT TUBULE SEGMENTS

In subsequent sections of this chapter the absorption and secretion of specific substances in different segments of the tubular system will be discussed. However, it is important first to point out basic differences between the absorptive and secretory capabilities of the different types of tubules.

Proximal Tubular Epithelium. Figure 34–9 illustrates the cellular characteristics of the tubular membrane in (1) the proximal tubule, (2) the thin segment of the loop of Henle, (3) the distal tubule,

and (4) the collecting tubule. The proximal tubular cells have the appearance of being highly metabolic cells, having large numbers of mitochondria to support extremely rapid active transport processes; and, true enough, one finds that about 65 per cent of all reabsorptive and secretory processes that occur in the tubular system take place in the proximal tubules. That is, only 35 per cent of the glomerular filtrate normally passes the whole distance through the proximal tubules because 65 per cent is reabsorbed before reaching the loops of Henle. As already described in relation to sodium transport, the proximal tubular epithelial cells have an extensive brush border. They also have a far-ranging labyrinth of intercellular and basal channels which provide an extensive membrane area on the peritubular side of the epithelium, the side where abundant active transport of sodium and other ions occurs.

Thin Segment of the Loop of Henle. The epithelium of the thin segment of the loop of Henle, as the name implies, is very thin. The cells have no brush border and very few mitochondria, indicating a minimal level of metabolic activity. The descending portion of this thin segment is highly permeable to water and moderately permeable to urea, sodium, and most other ions. Therefore, it appears to be adapted primarily for simple diffusion of substances through its walls.

The ascending portion of the thin segment, on the other hand, is believed to be different in one very important characteristic; it is believed to be far less permeable to water and urea than is the descending portion. This fact is important for explaining the function of the countercurrent mechanism for concentrating urine that will be discussed in the following chapter.

The Distal Tubular Epithelium. The distal tubule begins in the ascending limb of the loop of Henle where the epithelial cells become grossly thickened; this portion of the ascending limb is also frequently called the *thick segment of the ascending limb of the loop of Henle*. This segment ascends all the way back to the same glomerulus from which the tubule originated and passes very snugly through the angle between the afferent and efferent arterioles, forming a complex with these arterioles called the *juxtaglomerular complex*. This structure will be discussed in much greater detail in the following chapter because it plays a

Proximal   Loop of Henle   Distal   Collecting
Tubule   Thin Segment   Tubule   Tubule

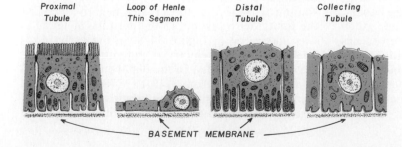

Figure 34–9. Characteristics of the epithelial cells in different tubular segments.

BASEMENT MEMBRANE

very important role in controlling nephron function. Beyond the juxtaglomerular complex, the distal tubule becomes convoluted and at the end of the convolution leads into the upper end of the collecting duct.

The distal tubule is divided into two important functional segments, the *diluting segment* and the *late distal tubule*.

The Diluting Segment. The function of the diluting segment, as its name implies, is to dilute the tubular fluid. This segment includes the entire thick portion of the ascending limb of the loop of Henle and about half of the convoluted portion of the distal tubule. The epithelial cells are similar to those in the proximal tubules except that they have only a rudimentary brush border, fewer basal channels, and much tighter zonulae occludentes where the cells attach to each other. The cells are specifically adapted for active transport of chloride ions from inside the tubular lumen into the peritubular fluid. This transport of the negative chloride ions out of the tubules creates a slight positive potential inside the tubule of about +6 millivolts; the resulting electrical gradient then causes sodium ions to diffuse also from the tubular lumen into the peritubular fluid. (It is possible, but not certain, that a small amount of sodium ions is actively transported through the tubular membrane as well.) To provide energy for pumping the chloride ions against a steep electrochemical gradient, large numbers of mitochondria lie in close proximity to the basal membrane of the epithelial cell.

This diluting segment of the distal tubule is almost impermeable to water and even more nearly impermeable to urea. Therefore, after most of the ions are transported out of the tubular fluid, the remaining fluid is very dilute except that its concentration of urea is still high.

The Late Distal Tubule. The late distal tubule has epithelial cells that are very similar in appearance to those of the diluting segment. However, functionally they are mainly adapted for active transport of sodium and some of the other positive ions and much less so for chloride ions. Yet, like the diluting segment, the late distal tubule is almost entirely impermeable to urea so that the urea passes on into the collecting duct to be excreted in the urine.

The permeability of the late distal tubule to water varies in different species; in some it is very impermeable; in others it is permeable in the presence of antidiuretic hormone but impermeable when this hormone is absent, providing a method for controlling the degree of dilution of the urine, a subject that we shall discuss in detail in the following chapter.

Collecting Duct Epithelium. The collecting duct, like the distal tubule, is composed of two important segments that are similar in physical characteristics but distinctly different in functional characteristics: the *cortical collecting tubule* and the *inner medullary collecting duct*. The cortical collecting tubule begins in the renal cortex at the termination of the convoluted distal tubule; it fuses with several other cortical collecting tubules before turning downward from the cortex toward the renal papilla. The functional characteristics of the cortical collecting tubule change markedly as it passes through the outer portion of the medulla. It is here that it becomes the inner medullary collecting duct that then passes all the way to the renal papilla to empty into the pelvis of the kidney.

The epithelial cells of both portions of the collecting duct are nearly cuboidal in shape, with smooth surfaces, and contain relatively few mitochondria and other organelles. Even so, these epithelial cells can provide pumping of sodium, potassium, hydrogen, calcium, and other ions against very large electrochemical gradients, as will be discussed later in this chapter and in the following chapters. In the collecting ducts urine becomes either highly concentrated or highly dilute, highly acidic or highly basic. The collecting duct epithelium appears to be well designed to resist these extremes of tubular fluid.

The cortical collecting tubule is almost totally impermeable to urea. On the other hand, the inner medullary collecting duct is moderately permeable to urea, especially so when antidiuretic hormone is present, so that large quantities of urea are normally reabsorbed from this portion of the collecting duct into the medullary interstitium; this is very important for concentrating the urine, as will be discussed in the following chapter.

The permeability of the epithelium to water in both portions of the collecting duct is determined mainly by the concentration of antidiuretic hormone in the blood. When large quantities of this hormone are present, the epithelium becomes very permeable to water, and most of the water will be reabsorbed from the tubules and returned to the blood; in the absence of antidiuretic hormone, very little water is reabsorbed, most instead passing on into the urine.

REABSORPTION AND SECRETION OF INDIVIDUAL SUBSTANCES IN DIFFERENT SEGMENTS OF THE TUBULES

Transport of Water and Flow of Fluid at Different Points in the Tubular System. Water transport occurs entirely by osmotic diffusion. That is, whenever some solute in the glomerular filtrate is absorbed either by active reabsorption or by diffusion caused by an electrochemical gradient, the resulting decreased concentration of solute in the tubular fluid and increased concentration in the

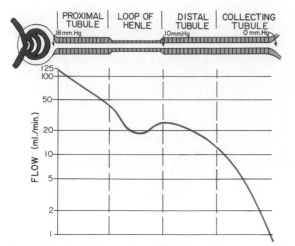

Figure 34–10. Volume flow of fluid in each segment of the tubular system per minute. Note that the flow is plotted on a semi-logarithmic scale, illustrating the tremendous difference in flow between the earlier and later segments of the tubules.

peritubular fluid causes osmosis of water out of the tubules.

Figure 34–10 depicts the rates of water flow axially along the tubular system at different points in the tubules. In both kidneys of human beings, the total fluid volumes flowing into each segment of the tubular system each minute (under normal resting conditions) are the following:

	ml./min.
Glomerular filtrate	125
Flowing into the loops of Henle	45
Flowing into the distal tubules	25
Flowing into the collecting ducts	12
Flowing into the urine	1

From this chart one can also deduce the per cent of the glomerular filtrate water that is reabsorbed in each segment of the tubules, as follows:

	Per Cent
Proximal tubules	65
Loop of Henle	15
Distal tubules	10
Collecting ducts	9.3
Passing into the urine	0.7

We shall see later in the chapter that these values vary greatly under different operational conditions of the kidney, particularly when the kidney is forming very dilute or very concentrated urine.

Reabsorption of Substances of Nutritional Value to the Body — Glucose, Proteins, Amino Acids, Acetoacetate Ions, and Vitamins. Five different substances in the glomerular filtrate of particular importance to bodily nutrition are glucose, proteins, amino acids, acetoacetate ions, and the vitamins. Normally all of these are completely or almost completely reabsorbed by active processes in the *proximal tubules* of the kidney, as shown in Figure 34–11 for glucose, protein, and amino acids. Therefore, almost none of these substances remain in the tubular fluid entering the loop of Henle.

Special Mechanism for Absorption of Protein. As much as 30 grams of protein filter into the glomerular filtrate each day. This would be a great metabolic drain on the body if the protein were not returned to the body fluids. Because the protein molecule is much too large to be transported by the usual transport processes, protein is absorbed through the brush border of the proximal tubular epithelium by pinocytosis, which means simply that the protein attaches itself to the membrane and this portion of the membrane then invaginates to the interior of the cell. Once inside the cell the protein is digested into its constituent amino acids which are then absorbed through the base and sides of the cell into the peritubular fluids. Details of the pinocytosis mechanism were discussed in Chapter 4.

Poor Reabsorption of the Metabolic End-Products: Urea, Creatinine, and Others. Figure 34–11 also illustrates the rates of flow of three major metabolic end-products in the different seg-

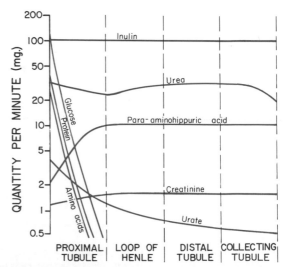

Figure 34–11. Rates of flow of important organic substances through the tubules at each point along the tubular course. (Two of these substances, inulin and para-aminohippuric acid, are not normally present but are important test substances.) This figure demonstrates reabsorption of all nutritionally important substances in the proximal tubules, and poor reabsorption of the metabolic end-products in all segments of the tubules. Note the total absence of reabsorption of inulin and the secretion of para-aminohippuric acid into the proximal tubules.

ments of the tubular system — urea, creatinine, and urate ions. Note, especially, that only moderate quantities of *urea* — about 50 per cent of the total — are reabsorbed during the entire course through the tubular system.

Creatinine is not reabsorbed in the tubules at all; indeed, small quantities of creatinine are actually secreted into the tubules by the proximal tubules so that the total quantity of creatinine increases about 20 per cent.

The *urate ion* is absorbed much more than urea — about 86 per cent reabsorption. But even so, large quantities of urate remain in the fluid that finally issues into the urine. Several other end-products, such as *sulfates, phosphates,* and *nitrates,* are transported in much the same way as urate ions. All of these are normally reabsorbed to a far less extent than is water so that their concentrations become greatly increased as they flow along the tubules. Yet, *each is actively reabsorbed to some extent,* which keeps their concentrations in the extracellular fluid from ever falling too low.

Transport of Inulin and Para-aminohippuric Acid by the Tubules. Note also in Figure 34–11 that the rate of flow of *inulin,* a large polysaccharide, remains exactly the same throughout the entire tubular system. The cause of this is simply that inulin is neither reabsorbed nor secreted in any segment of the tubules.

Finally, Figure 34–11 shows that the rate of flow of *para-aminohippuric acid* (PAH) increases about five-fold as the tubular fluid passes through the proximal tubules; then its rate of flow remains constant in the other tubules. This is because large quantities of PAH are *secreted* into the tubular fluid by the proximal tubular epithelial cells, and it is not reabsorbed in any segment of the tubular system.

These two substances play an important role in experimental studies of tubular function, as will be discussed later in the chapter.

Transport of Different Ions by the Tubular Epithelium — Sodium, Chloride, Bicarbonate, and Potassium. Figure 34–12 illustrates the rates of flow of different important ions — sodium, chloride, bicarbonate, and potassium — in different segments of the tubular system. Note that all these decrease markedly because of reabsorption as the tubular fluid progresses from glomerular filtrate to urine.

In most segments of the tubules, positive ions are generally transported through the tubular epithelium by active transport processes; negative ions are usually transported passively as a result of electrical differences developed across the membrane when the positive ions are transported. For instance, when sodium ions are transported out of the proximal tubular fluid, the resulting electronegativity that develops in the tubular fluid

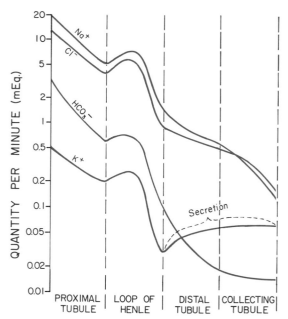

Figure 34–12. Rates of flow of the different ions at different points in the tubules each minute. Note not only the reabsorption of the ions but also the secretion of potassium into the distal and collecting tubules.

causes chloride ions to follow in the wake of the sodium ions. But, despite the general rule that negative ions are usually passively reabsorbed, exactly the opposite is true in the diluting segment of the distal tubule where chloride ions are actively absorbed and sodium and other positive ions are mainly absorbed passively.

The increased flow of all ions in the first half of the loop of Henle is caused by *diffusion* of ions into the tubules from the peritubular fluids of the medulla, which is explained in the following chapter in relation to the mechanism for concentrating urine.

Secretion of Potassium and Hydrogen Ions. Note also in Figure 34–12 that the flow of potassium ions normally increases as the tubular fluid passes through the distal tubules and collecting tubules, as a result of *secretion of potassium.* This will be discussed at greater length in Chapter 36 in relation to the regulation of potassium concentration in the extracellular fluids.

Hydrogen ions are actively secreted in the proximal tubules, distal tubules, and collecting ducts. This secretion is controlled by the hydrogen ion concentration of the extracellular fluid, which will be discussed in Chapter 37.

Special Aspects of Bicarbonate Ion Transport. Bicarbonate ion is probably mainly reabsorbed in the form of carbon dioxide rather than in the form of bicarbonate ion itself. This occurs as follows: The bicarbonate ions in the tubular fluid first combine with hydrogen ions that are secreted into the fluid by the epithelial cells. The reaction forms carbonic acid which then dissociates into water and carbon dioxide. The carbon dioxide, being highly lipid-soluble, diffuses rapidly through the tubular membrane into the peritubular capillary blood. When more bicarbonate ions are present than there are available hydrogen ions, most of the excess bicarbonate ions flow on into the urine because the tubules are only slightly permeable to them.

Transport of Other Ions. Though we know much less about the specific means of transport of other ions besides the four illustrated in Figure 34-12, in general essentially all of them can be reabsorbed either by active transport or as a result of electrical differences across the membrane. Thus, calcium, magnesium, and other positive ions are actively reabsorbed, and many of the negative ions are reabsorbed as a result of electrical differences that develop when the positive ions are reabsorbed. In addition, certain negative ions—urate, phosphates, sulfate, and nitrate—can be reabsorbed by active transport, this occurring to the greatest extent in the proximal tubules.

CONCENTRATIONS OF DIFFERENT SUBSTANCES AT DIFFERENT POINTS IN THE TUBULES

Whether or not a substance becomes concentrated in the tubular fluid as it moves along the tubules is determined by the *relative reabsorption of the substance versus the reabsorption of water.* If a greater percentage of water is reabsorbed, the substance becomes more concentrated. Conversely, if a greater percentage of the substance is reabsorbed, it becomes more dilute. Therefore, we can combine the curves of Figures 34-10, 34-11, and 34-12 to determine which substances become concentrated in the tubular system and which become diluted. Figure 34-13 illustrates the results of this combination, showing three different classes of substances as follows:

First, the nutritionally important substances—glucose, protein, and amino acids—are reabsorbed so much more rapidly than water that their concentrations fall extremely rapidly in the proximal tubules and remain essentially zero throughout the remainder of the tubular system as well as in the urine.

Second, the concentrations of the metabolic end-products as well as of the artificially injected substances inulin and para-aminohippuric acid (PAH) become progressively greater throughout the tubular system, because all these substances are reabsorbed to a far lesser extent than is water. Note that potassium ions also fall into this catego-

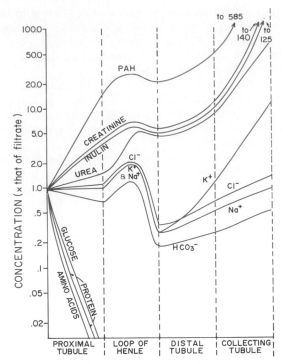

Figure 34-13. Composite figure, showing average concentrations of different substances at different points in the tubular system.

ry because a far greater proportion of potassium ions is normally removed each day from the extracellular fluid than that of water.

Third, many other ions are normally excreted into the urine in concentrations not greatly different from those in the glomerular filtrate and extracellular fluid. That is, sodium and chloride ions, on the average, are normally reabsorbed from the tubules in not too dissimilar proportions from water.

Table 34-1 summarizes the concentrating ability of the tubular system for different substances. It also gives the actual quantities of the different substances normally handled by the tubules each minute.

Variability of Concentrating Power. A word of caution must be given at this point: the quantitative values for reabsorption of different substances as given in Figures 34-10, 34-11, and 34-12 and for concentrating power of the tubules as given in Figure 34-13 and Table 34-1 are all *average* values. These values change greatly in different physiological states. For instance, when essentially no antidiuretic hormone is secreted by the hypothalamic–posterior pituitary mechanism, reabsorption of water by the collecting ducts becomes greatly reduced; the presence of the extra water in the urine dilutes essentially all the substances listed in Table 34-1 by as much as fourfold. Conversely, secretion of large quan-

TABLE 34–1 RELATIVE CONCENTRATIONS OF SUBSTANCES IN THE GLOMERULAR FILTRATE AND IN THE URINE

	Glomerular Filtrate (125 ml./min.)		Urine (1 ml./min.)		Conc. Urine/ Conc. Plasma (Plasma Clearance per Minute)
	Quantity/min.	Concentration	Quantity/min.	Concentration	
Na^+	17.7 mEq.	142 mEq./l.	0.128 mEq.	128 mEq./l.	0.9
K^+	0.63	5	0.06	60	12
Ca^{++}	0.5	4	0.0048	4.8	1.2
Mg^{++}	0.38	3	0.015	15	5.0
Cl^-	12.9	103	0.134	134	1.3
HCO_3^-	3.5	28	0.014	14	0.5
$H_2PO_4^-$ HPO_4^{--}	0.25	2	0.05	50	25
SO_4^{--}	0.09	0.7	0.033	33	47
Glucose	125 mg.	100 mg./100 ml.	0 mg.	0 mg./100 ml.	0.0
Urea	33	26	18.2	1820	70
Uric acid	3.8	3	0.42	42	14
Creatinine	1.4	1.1	1.96	196	140
Inulin	....	....	....	....	125
Diodrast	....	....	....	....	560
PAH	....	....	....	....	585

tities of antidiuretic hormone causes such excessive reabsorption of water that all the substances can be concentrated to as much as 4 to 5 times the values given in Table 34–1.

THE CONCEPT OF "PLASMA CLEARANCE"

The term "plasma clearance" is used to express the ability of the kidneys to clean, or "clear," the plasma of various substances. Thus, if the plasma passing through the kidneys contains 0.1 gram of a substance in each 100 ml., and 0.1 gram of this substance also passes into the urine each minute, then 100 ml. of the plasma are cleaned or cleared of the substance per minute.

Referring again to Table 34–1, note that the normal concentration of urea in each milliliter of plasma and glomerular filtrate is 0.26 mg., and the quantity of urea that passes into the urine each minute is approximately 18.2 mg. Therefore, the equivalent quantity of plasma that completely loses its entire content of urea each minute can be calculated by dividing the quantity of urea entering the urine each minute by the quantity of urea in each milliliter of plasma. Thus $18.2 \div 0.26 = 70$; that is, 70 ml. of plasma is *cleaned* or *cleared* of urea each minute. This amount that is cleared each minute is known as the *plasma clearance* of urea. Obviously, the plasma clearance is a measure of the effectiveness of the kidney to remove substances from the extracellular fluid.

Plasma clearance for any substance can be calculated by the formula at the bottom of the page.

The plasma clearances of the usual constituents of urine are shown in the last column of Table 34–1.

$$\text{Plasma clearance (ml./min.)} = \frac{\text{Quantity of urine (ml./min.)} \times \text{Concentration in urine}}{\text{Concentration in plasma}}$$

INULIN CLEARANCE AS A MEASURE OF GLOMERULAR FILTRATION RATE

Inulin is a polysaccharide that has the specific attributes of not being reabsorbed to a significant extent by the tubules of the nephron and yet being of small enough molecular weight (about 5200) that it passes through the glomerular membrane as freely as the crystalloids and water of the plasma. Also, inulin is not actively secreted even in the minutest amount by the tubules. Consequently, glomerular filtrate contains the same concentration of inulin as does plasma, and as the filtrate flows down the tubules all the filtered inulin continues on into the urine. Thus, *all the glomerular filtrate formed is cleared of inulin.* Therefore, the plasma clearance per minute of inulin is equal to the glomerular filtration rate.

As an example, let us assume that it is found by chemical analysis that the inulin concentration in the plasma is 0.1 gram in each 100 ml., and that 0.125 gram of inulin passes into the urine per minute. Therefore, by dividing 0.1 into 0.125, one finds that 1.25 *100-milliliter portions* of glomerular filtrate must be formed each minute in order to deliver to the urine the analyzed quantity of inulin. In other words, 125 ml. of glomerular filtrate is formed per minute, and this is also the plasma clearance of inulin.

Inulin is not the only compound that can be used for determining the quantity of glomerular filtrate formed each minute, for the plasma clearance of any other substance that is totally diffusible through the glomerular membrane but is neither absorbed nor secreted by the tubular walls is equal to the glomerular filtration rate. *Mannitol* is a monosaccharide that is frequently used instead of inulin for such measurements, and *radioactive iothalamate* is still another substance frequently used because its radioactivity allows easy quantitative analysis.

PARA-AMINOHIPPURIC ACID (PAH) CLEARANCE AS A MEASURE OF PLASMA FLOW THROUGH THE KIDNEYS

PAH, like inulin, passes through the glomerular membrane with perfect ease. However, it is different from inulin in that almost all the PAH remaining in the plasma after the glomerular filtrate is formed is secreted into the tubules by the tubular epithelium if the plasma concentration of PAH is very low. Indeed, only about one tenth of the original PAH remains in the plasma by the time the blood leaves the kidneys.

One can use the clearance of PAH for estimating the *flow of plasma* through the kidneys. As an example, let us assume that 1 mg. of PAH is present in each 100 ml. of plasma, and that 5.85 mg. of PAH passes into the urine per minute. Consequently, 585 ml. of plasma is cleared of PAH each minute. Obviously, if this much plasma is cleared of PAH, *at least this much plasma must have passed through the kidneys* in this same period of time. And, since it is known that almost all the PAH is cleared from the blood as it passes through the kidneys, 585 ml. would be a reasonable first approximation of the actual plasma flow per minute.

Yet, to be still more accurate, one can correct for the average amount of PAH that is still in the blood when it leaves the kidney. In many different experiments the PAH clearance has averaged 91 per cent of the plasma load of PAH entering the kidneys. Therefore, the 585 ml. of plasma calculated above would be only 91 per cent of the total amount of plasma flowing through the kidneys. Dividing 585 by 0.91 gives a total plasma flow per minute of approximately 650 ml.

The 91 per cent removal of PAH as it passes through the kidneys is called the *extraction ratio.*

One can calculate the *total blood flow* through the kidneys each minute from the plasma flow and the hematocrit (the percentage of red blood cells in the blood). If the hematocrit is 45 per cent and the plasma flow 650 ml./min., the total blood flow through both kidneys is $650 \times (100/55)$, or 1182 ml./min.

Clearance of *Diodrast* can also be used for estimating plasma flow or blood flow through the two kidneys. However, the extraction ratio of Diodrast averages about 0.85, slightly less than that of PAH.

CALCULATING THE FILTRATION FRACTION FROM PLASMA CLEARANCES

To calculate the filtration fraction — that is, the fraction of the plasma that filters through the glomerular membrane — one must determine (1) the plasma flow through the two kidneys (PAH clearance) and (2) the glomerular filtration rate per minute (inulin clearance). Using 650 ml. plasma flow and 125 ml. glomerular filtration rate as values, we find that the calculated filtration fraction is 125/650, or, to express this as a percentage, 19 per cent.

EFFECT OF "TUBULAR LOAD" AND "TUBULAR TRANSPORT MAXIMUM" ON URINE CONSTITUENTS

Tubular Load. The *tubular load* of a substance is the total amount of the substance that filters through the glomerular membrane into the tubules each minute. For instance, if 125 ml. of glomerular filtrate is formed each minute with a glucose concentration of 100 mg. per cent, the tubular load of glucose is 100 mg. $\times$ 1.25, or 125 mg. of glucose per minute. Similarly, the load of sodium that enters the tubules each minute is approximately 18 mEq./min., the load of chloride ion is about 13 mEq./min., and the load of urea is approximately 33 mg./min., etc.

Maximum Rate of Transport of Actively Reabsorbed or Secreted Substances — The "Tubular Transport Maximum" (Tm). Since each substance that is actively reabsorbed requires a specific transport system in the tubular epithelial cells, the maximum amount that can be reabsorbed often depends on the maximum rate at which the transport system itself can operate, and this in turn depends on the total amounts of carrier and specific enzymes available. Consequently, for almost every actively reabsorbed substance, there is a maximum rate at which it can be reab-

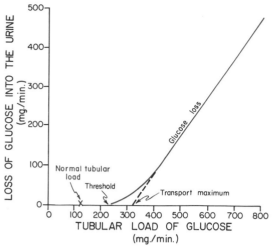

Figure 34–14. Relationship of tubular load of glucose to loss of glucose into the urine.

sorbed; this is called the *tubular transport maximum* for the substance, and the abbreviation is *Tm*. For instance, the Tm for glucose averages 320 mg./min. for the adult human being, and if the tubular load becomes greater than 320 mg./min., the excess above this amount is not reabsorbed but instead passes on into the urine. Ordinarily, though, the tubular load of glucose is only 125 mg./min., so that for practical purposes, all of it is reabsorbed.

Figure 34–14 demonstrates the relationship between tubular load of glucose, tubular transport maximum for glucose, and rate of glucose loss into the urine. Note that when the tubular load is at its normal level of 125 mg./min. there is no detectable loss of glucose into the urine. However, when the tubular load rises above about 220 mg./min., significant quantities of glucose begin to appear in the urine. And, once the load has risen above approximately 400 mg./min., the loss into the urine thereafter remains *320 mg./min. less than the tubular load.* Thus, at 400 mg./min. tubular load, the loss is 80 mg./min., and at 800 mg./min. tubular load it is 480 mg./min. In other words, 320 mg./min. of the tubular load, which represents the

tubular transport maximum for glucose, is reabsorbed, and all the remainder is lost into the urine.

Threshold for Substances That Have a Tubular Transport Maximum. Every substance that has a transfer maximum also has a "threshold" concentration in the plasma below which none of it appears in the urine and above which progressively larger quantities appear.

Thus, Figure 34–14 shows that glucose begins to spill into the urine when its tubular load exceeds 220 mg./min. The threshold concentration of glucose in plasma that gives this tubular load is 180 mg. per cent when the kidneys are operating at their normal glomerular filtration rate of 125 ml./min.

Tubular Transport Maximums of Important Substances Absorbed from the Tubules. Some of the important tubular transport maximums for substances absorbed from the tubules are the following:

Glucose	320	mg./min.
Phosphate	0.1	mM./min.
Sulfate	0.06	mM./min.
Amino acids	1.5	mM./min.
Urate	15	mg./min.
Plasma protein	30	mg./min.
Hemoglobin	1	mg./min.
Lactate	75	mg./min.
Acetoacetate	variable	(about 30 mg./min)

It is particularly significant that sodium ions *do not* exhibit a tubular transport maximum.

Tubular Transport Maximums for Secretion. Many substances secreted by the tubules also exhibit tubular transport maximums as follows:

	mg./min.
Creatinine	16
PAH	80
Diodrast	57 (of iodine)
Phenol red	56

REFERENCES

See bibliography at end of Chapter 35.

35

Renal Mechanisms for Concentrating and Diluting the Urine and for Urea, Sodium, Potassium, and Fluid Volume Excretion

In the previous chapter we discussed the mechanisms by which glomerular filtrate is formed and the way in which it is then processed in the tubules to become urine. The present chapter will be devoted to the kidney's capability to change the composition of urine from moment to moment to reflect the needs of the body for excreting different substances. For instance, the body at times has great excesses of water and, therefore, must excrete a dilute urine; at other times, it has a deficit of water and, therefore, must excrete a minimum amount of water but still excrete the end-products of metabolism as well as other substances that must be removed from the body fluids.

In this chapter we will discuss principally the ways in which the kidneys can change (1) the osmolar concentration of the urine, (2) the volume of both water and dissolved substances excreted each day, and (3) the rates of excretion of urea, sodium, and potassium. In Chapter 36 we will discuss the feedback systems that control these various excretory functions and in Chapter 37 the ability of the kidneys to alter the urine concentrations of hydrogen ions, chloride ions, and bicarbonate ions and thereby to control the acid-base balance of the extracellular fluid.

DILUTING MECHANISM OF THE KIDNEY — THE MECHANISM FOR EXCRETING EXCESS WATER

One of the most important functions of the kidney is to control the osmolality of the body fluids. It does this by excreting excessive amounts of water in the urine when the body fluids are too dilute or by excreting excessive amounts of solutes when the body fluids are too concentrated.

Shift of the kidneys from excreting excess water to excreting excess solutes is controlled by *antidiuretic hormone*. In the absence of antidiuretic hormone, the kidneys excrete excessive amounts of water, but when the blood concentration of antidiuretic hormone is high the kidneys excrete excessive amounts of solutes.

When the glomerular filtrate is formed in the glomerulus, its osmolality is almost exactly the same as that of the plasma, approximately 300 milliosmoles per liter. To excrete excess water, it is necessary to dilute the filtrate as it passes through the tubules. This is achieved by reabsorbing a higher proportion of solutes than water. Figure 35–1 illustrates this process. The colored arrows in this figure represent rapid reabsorption of tubular solute, and the thickened walls of the more distal tubular segments indicate that these portions are relatively impermeable to water when antidiuretic hormone is *not* present in the circulating body fluids.

In the diluting segment of the distal tubule (which begins with the thick portion of the ascending limb of the loop of Henle), the absorption of solutes results primarily from active absorption of chloride ions, and this causes electrogenic passive absorption of the positive ions sodium, potassium, calcium, and magnesium. On the other hand, in the late distal tubule and the collecting ducts, the major driving force is active transport of sodium ions, and this then causes electrogenic passive

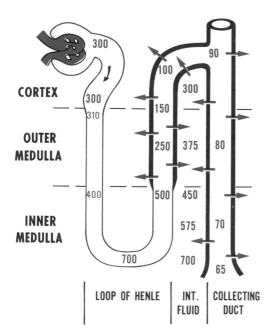

Figure 35–1. The renal mechanism for forming a dilute urine. The darkened walls of the distal portions of the tubular system indicate that these portions of the tubules are relatively impermeable to the reabsorption of water in the absence of antidiuretic hormone. The solid arrows indicate active processes for absorption of most of the solutes besides the urinary waste products. (Numerical values are in milliosmoles per liter.)

absorption of the anions, mainly chloride ions. Consequently, most of the ionic substances of the tubular fluids are reabsorbed from these distal segments of the tubular system before the fluid is emptied as urine, but the water remains and the urine is dilute. Note in Figure 35–1 that in the diluting segment of the distal tubule, the osmolality of the fluid falls rapidly from its initial value of 300 milliosmoles per liter to about 100 milliosmoles per liter in the late distal tubule, 90 mOsm./liter in the cortical collecting tubule, and as low as 65 to 70 mOsm./liter in the last portions of the inner medullary collecting duct and in the urine.

To summarize, the process for excreting a dilute urine is simply one of absorbing solutes from the distal segments of the tubules while water fails to be reabsorbed.

CONCENTRATING MECHANISM OF THE KIDNEY; EXCRETION OF EXCESS SOLUTES — THE COUNTER-CURRENT MECHANISM

The process for concentrating urine is not nearly so simple as for diluting it. Yet, at times it is exceedingly important to concentrate the urine as much as possible so that excess solutes can be eliminated with as little loss of water from the body as possible — for instance, when one is exposed to desert conditions with an inadequate supply of water. Fortunately, the kidneys have developed a special mechanism for concentrating the urine called the *counter-current mechanism*.

The counter-current mechanism depends on a special anatomical arrangement of the loops of Henle and the vasa recta. In the human being, the loops of Henle of one-third to one-fifth of the nephrons dip deep into the medulla and then return to the cortex; some dip all the way to the tips of the papillae that project into the renal pelvis. This group of nephrons with the long loops of Henle is called the *juxtamedullary nephrons*. Paralleling the long loops of Henle are loops of peritubular capillaries called the *vasa recta*; these loop down into the medulla from the cortex and then back out to the cortex again. These arrangements of the different parts of the juxtamedullary nephron and the vasa recta are diagrammed in Figure 35–2.

Hyperosmolality of the Medullary Interstitial Fluid, and Mechanisms for Achieving It

The first step in the excretion of excess solutes in the urine — that is, for excretion of a concentrated urine — is to create hyperosmolality of the medullary interstitial fluid. As we shall see later, this in turn is necessary for concentrating the urine. But, first, let us explain the mechanism for creating this hyperosmolality in the medullary interstitium.

The normal osmolality of the fluids in almost all parts of the body is about 300 mOsm./liter. However, as shown by the numbers in Figure 35–2, the osmolality of the interstitial fluid in the medulla of the kidney is much higher than this, and it becomes progressively greater the deeper one goes into the medulla, increasing from 300 mOsm./liter in the cortex to 1200 mOsm./liter (occasionally as high as 1400 mOsm./liter) in the pelvic tip of the medulla. Four different solute-concentrating mechanisms are responsible for this hyperosmolality; these are:

First, the principal cause of the greatly increased medullary osmolality is active transport of chloride ions (plus electrogenic passive absorption of sodium ions) out of the thick portion of the ascending limb of the loop of Henle. The large colored arrows shown in this tubular segment in Figure 35–2 illustate this transport of chloride and sodium ions (as well as potassium, calcium, and magnesium ions to a lesser extent) out of the thick portion of the loop of Henle and into the outer medullary interstitial fluid. All these solutes be-

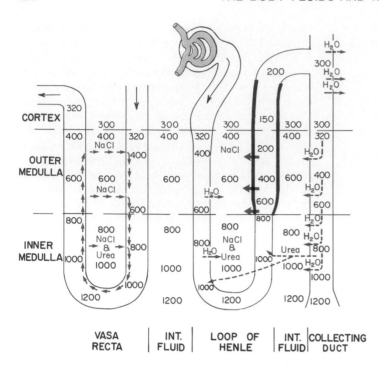

Figure 35–2. The counter-current mechanism for concentrating the urine. (Numerical values are in milliosmoles per liter.)

come concentrated in this fluid. These ions are also carried downward into the inner medulla by the flowing blood in the vasa recta, as we shall see shortly.

Second, ions are also transported into the medullary interstitial fluid from the collecting duct, mainly resulting from active transport of sodium ions and electrogenic passive absorption of chloride ions along with the sodium ions.

Third, when the concentration of antidiuretic hormone is high in the blood, large amounts of urea are also absorbed into the medullary fluid from the inner medullary collecting duct. The reason is: In the presence of antidiuretic hormone, the inner medullary portion of the collecting duct becomes moderately permeable to urea and highly permeable to water. The fact that it becomes highly permeable to water causes rapid reabsorption of water out of the collecting duct, and this greatly increases the concentration of urea in the duct. Now, because of this high urea concentration, the urea too diffuses through the collecting duct wall into the medullary interstitium. Consequently, the urea concentration in the medullary interstitial fluid rises almost to equal the concentration in the collecting duct. In the human being, during maximal antidiuretic hormone stimulation this may be to as high as 400 to 500 mOsm./liter, which obviously greatly increases the osmolality of the inner medullary interstitial fluid.

And, fourth, the final event that causes increased osmolal concentration of the medullary interstitial fluid is transport of sodium and chloride ions into the inner medullary interstitium from the

thin segment of the loop of Henle. Some physiologists believe that this results from active transport from the ascending portion of this thin segment. However, most recent experiments suggest that this is very unlikely, which also fits with the fact that the epithelial cells in the thin segment have very few mitochondria and almost certainly could not support much active transport. But, if there is not active transport, what is the force that causes sodium and chloride ions to be transported into the medullary interstitium? The answer seems to be the following: When the concentration of urea rises very high in the medullary interstitium because of urea absorption from the collecting duct, this immediately promotes osmosis of water out of the descending thin limb of the loop of Henle. Therefore, the concentration of sodium chloride inside the thin limb rises almost to twice normal. And now, because of this high concentration, both sodium and chloride ions diffuse passively out of the thin segment into the interstitium.

In summary, at least four different factors contribute to the marked increase in osmolality in the medullary interstitial fluid. These are: (1) active transport of the ions into the interstitium by the thick portion of the ascending limb of the loop of Henle, (2) active transport of ions from the collecting duct into the interstitium, (3) passive diffusion of large amounts of urea from the collecting duct into the interstitium, and (4) transport of additional sodium and chloride into the interstitium from the thin segment of the loop of Henle, a transport that is probably also passive as explained above. The net result is an increase in the osmolality of the medullary interstitial fluid, when adequate amounts of antidiuretic hormone are present, to as high as 1200 to 1400 mOsm./liter near the tips of the papillae.

The combination of all these mechanisms for creating hyperosmolality in the medullary interstitium is frequently called the counter-current multiplier mechanism of the loop of Henle because some aspects of the different mechanisms require counter-current fluid flow in the loop of Henle and collecting duct.

Counter-Current Exchange Mechanism in the Vasa Recta

We have now discussed the mechanisms by which high concentrations of solutes are achieved in the medullary interstitium. However, without a special medullary vascular system as well, the flow of blood through the interstitium would rapidly remove the excess solutes and keep the concentration from rising very high. Fortunately, the medullary blood flow has two characteristics, both exceedingly important, for maintaining the high solute concentration in the medullary interstitial fluids. These are:

First, the medullary blood flow is very slight in quantity, amounting to only 1 to 2 per cent of the total blood flow of the kidney. Because of this very sluggish blood flow, removal of solutes is minimized.

Second, the vasa recta functions as a *counter-current exchanger* that prevents washout of solutes from the medulla. This can be explained as follows: A counter-current fluid exchange mechanism is one in which fluid flows through a long U-tube, with the two arms of the U lying in close proximity to each other so that fluid and solutes can exchange readily between the two arms. This obviously also requires that each of the two arms of the U be highly permeable, which is true for the vasa recta. When the fluids and solutes in the two parallel streams of flow can exchange rapidly, tremendous concentrations of solute can be maintained at the tip of the loop with relatively negligible washout of solute.

Thus, in Figure 35–2, as blood flows down the descending limbs of the vasa recta, sodium chloride and urea diffuse into the blood from the interstitial fluid while water diffuses outward into the interstitium, and these two effects cause the osmolal concentration in the tubular fluid to rise progressively higher, to a maximum concentration of 1200 mOsm./liter at the tips of the vasa recta. Then, as the blood flows back up the ascending limbs, the extreme diffusibility of all molecules through the capillary membrane allows essentially all the extra sodium chloride and urea to diffuse back out of the blood into the interstitial fluid

while water diffuses back into the blood. Therefore, by the time the blood finally leaves the medulla, its osmolal concentration is only slightly greater than that of the blood that had initially entered the vasa recta. As a result, blood flowing through the vasa recta carries only a minute amount of the medullary interstitial solutes away from the medulla.

Mechanism for Excreting a Concentrated Urine

Now that we have explained how the kidney creates hyperosmolality in the medullary interstitium, it becomes a simple matter to explain the mechanism for excreting a concentrated urine, thus causing loss of excess solutes from the body fluids while at the same time retaining as much water as possible. When the concentration of *antidiuretic hormone* in the blood is high, the epithelium of the entire collecting duct, and in some species of animals of the late distal tubule also, becomes highly permeable to water. This is illustrated in Figure 35–2 by the thin walls of the collecting duct and late distal tubule. As the tubular fluid flows through the collecting duct, water is pulled by osmosis into the highly concentrated fluid of the medullary interstitium. Thus, the collecting duct fluid also becomes highly concentrated, and it issues from the papilla into the pelvis of the kidney at a concentration of about 1200 mOsm./liter, almost exactly equal to the osmolal concentration of the solutes in the medullary interstitium near the papilla.

Summary of the Osmolal Concentration Changes in the Different Segments of the Tubules

Figure 35–3 illustrates the changes in osmolality of tubular fluid as it passes through the different

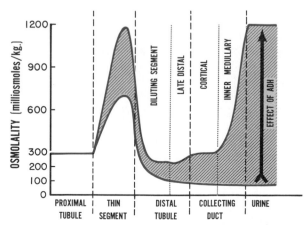

Figure 35–3. Changes in osmolality of the tubular fluid as it passes through the tubular system.

segments of the tubules. In the proximal tubules, the tubular membranes are so highly permeable to water that whenever a solute is transported across the membrane an almost exact proportionate amount of water crosses the membrane by osmosis at the same time; therefore, the osmolality of the fluid remains almost exactly equal to that of the glomerular filtrate, 300 mOsm./liter, throughout the entire extent of the proximal tubule.

Next, in the loop of Henle, the osmolality rises rapidly because of the counter-current mechanism explained in the previous few paragraphs. When a concentrated urine is being formed — that is, when the blood concentration of ADH is high — the loop of Henle osmolality rises much higher than when a dilute urine is being formed, because of the large quantity of urea that is passively reabsorbed into the medullary interstitium from the collecting ducts.

Next, in the diluting segment of the distal tubule, beginning at the thick portion of the ascending limb of the loop of Henle, the osmolality falls once again to a very low level, usually to about 100 mOsm./liter.

Finally, in the late distal tubule and collecting duct the osmolality depends entirely on the presence or absence of ADH. In the absence of ADH, very little water is reabsorbed from either the late distal tubule or the collecting duct. Therefore, the osmolality remains less than 100 mOsm./liter and even falls still another few milliosmoles because of active transport of ions through the epithelium of these tubules. Therefore, a very dilute urine is formed.

But, in the presence of excess ADH, the entire collecting duct (and in some animal species the late distal tubule as well) becomes highly permeable to water so that most of the water is reabsorbed, thus producing a very concentrated urine.

Note especially the shaded areas in Figure 35–3. These denote the ranges of concentration of the tubular fluid, and also the range of urine concentration between the extremes of 65 and 1200 mOsm./liter in human beings, depending on the blood concentration of antidiuretic hormone at any given time. The overall mechanism for control of antidiuretic hormone concentration, as well as control of osmolality of the body fluids, is discussed in the following chapter.

OSMOLAR CLEARANCE; FREE WATER CLEARANCE

One can calculate the clearance of osmolar substances in terms of the volume of plasma cleared per minute in the same way that one can calculate the clearance of any single substance. That is, osmolar clearance (C_{osm}) can be calculated by means of the following formula:

$$C_{osm} = \frac{\text{Osmoles entering urine per minute}}{\text{Plasma osmolar concentration}}$$

For instance, if the plasma osmolality is 300 milliosmoles/kg., and the milliosmoles entering the urine per minute is 1.5, then the osmolar clearance will be 1.5/300 liters/min., or 5 ml./min.

Free Water Clearance. When the kidney forms urine that is osmotically more dilute than plasma, it is obvious that a higher proportion of the water in the glomerular filtrate is excreted than of the osmolar substances. The excess water that is excreted is called *free water*, and the total plasma volume that is cleared of this excess water each minute is called the *free water clearance*. The free water clearance can be calculated by first determining the osmolar clearance and then subtracting this rate from the rate of urine flow per minute. Thus, the formula for free water clearance (C_{H_2O}) is:

$$C_{H_2O} = \text{Urine volume per minute} - C_{osm}$$

The free water clearance is important because it determines how rapidly the kidneys are changing the balance between water and osmotic substances in the body fluid — that is, the effect of renal excretion on the rate of change of body fluid osmolarity. The free water clearance can be either positive, in which case excess water is being removed; or it can be negative, in which case excess solutes are being removed.

UREA EXCRETION

The body forms an average of 25 to 30 grams of urea each day — more than this in persons who eat a very high protein diet and less in persons who are on a low protein diet. All this urea must be excreted in the urine; otherwise, it will accumulate in the body fluids. Its normal concentration in plasma is approximately 26 mg./100 ml., but it has been recorded in rare abnormal states to be as high as 800 mg./100 ml.; patients with renal insufficiency frequently have levels as high as 200 mg./100 ml.

The two major factors that determine the rate of urea excretion are (1) the concentration of urea in the plasma, and (2) the glomerular filtration rate. These factors increase urea excretion mainly because the load of urea entering the proximal tubules is equal to the product of the plasma urea concentration and the glomerular filtration rate. And, in general, the quantity of urea that passes on through the tubules into the urine averages about 50 to 60 per cent of the urea load that enters the proximal tubules. However, this holds true only when the glomerular filtration rate does not vary too greatly from normal.

Effect of Extremes of Glomerular Filtration Rate on Urea Excretion. When the glomerular filtration rate is very low, the filtrate remains in the tubules for a prolonged period of time before it finally becomes urine. Because all the tubules are at least slightly permeable to urea, the longer the tubular fluid remains in the tubules, the greater is the percentage of reabsorption of the urea into the blood; the proportion of the filtered urea that reaches the urine decreases considerably.

On the other hand, when the glomerular filtration rate is very high, the fluid passes through the tubular system so rapidly that very little urea is reabsorbed. Thus, at extremely high glomerular filtration rates, almost 100 per cent of the tubular load of urea passes into the urine.

An important lesson to be learned from these relationships is that in patients with renal insufficiency it is important for the glomerular filtration rate to be maintained at a high level. When the glomerular filtration rate falls too low, the blood urea concentration rises to a proportionately higher level.

Effect of Degree of Urine Concentration on Rate of Urea Excretion. When excreting a highly concentrated urine, one normally excretes very little urine volume because the person is actively conserving water in the body. Also, when excreting a concentrated urine the blood ADH is high, and this increases the permeability of the inner medullary collecting duct to urea as well as to water. As a result, excessive amounts of urea are reabsorbed from the collecting duct into the medullary interstitium. For all these reasons, one would expect that the rate of excretion of urea would decrease markedly when a person excretes a concentrated urine. However, the decrease in urea excretion is very slight, for reasons that are illustrated in Figure 35–4. Let us explain:

Figure 35–4 illustrates urea concentrations in milliosmoles per liter at different points in the tubular system and in the renal interstitium when a very concentrated urine is being formed. Note the high concentration in the lower collecting duct and also the rapid reabsorption of urea into the medullary interstitium, which increases the urea concentration there to high levels, as high as 400 to 500 mOsm./liter. A large share of this urea is then reabsorbed into the thin limb of the loop of Henle so that it passes upward through the distal tubule and back down through the collecting duct again. In this way, urea recirculates through these terminal portions of the tubular system several times before it is excreted. As a result, its concentration rises to the very high levels shown in the figure. However, the distal tubules and the cortical collecting tubules are almost entirely impermeable to urea. Therefore, despite this recirculating of urea several times, almost none of the urea is reabsorbed back into the blood. Instead, it eventually is excreted into the urine in a very highly concentrated form even though very little water is excreted along with it.

Thus, this urea recirculation mechanism through the loop of Henle, the distal tubule, and the collecting duct is a way to concentrate urea in the medullary interstitium and in the urine at the same time. Since urea is the most abundant of the waste products that must be excreted by the kidneys, this mechanism for concentrating urea before it is excreted is essential to the economy of the body fluid when a person must live on short rations of water.

SODIUM EXCRETION

Sodium Reabsorption from the Proximal Tubules and Loops of Henle. From the previous chapter it will be recalled that approximately 65 per cent of glomerular filtrate is reabsorbed in the proximal tubules. This reabsorption is caused primarily by the active transport of sodium through the proximal tubular epithelium. When the sodium is reabsorbed it causes diffusion of negative ions through the membrane as well, and the cumulative reabsorption of ions creates an osmotic pressure that then moves water through the membranes, too. The epithelium is so permeable to water that almost identically the same proportion of water and sodium ions is reabsorbed.

In the thin segments of the loops of Henle, very little sodium and water are absorbed. However, in the diluting segment of the distal tubules, the active transport of chloride ions causes sodium (and other positive ions) to be absorbed along the chloride ions, as has already been ex As a result, the concentrations of s

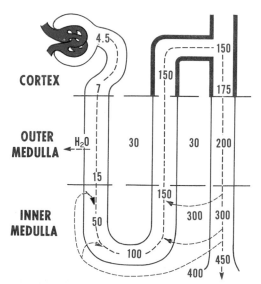

Figure 35–4. Re-circulation of the urea absorbed from the collecting duct, this passing into the loop of Henle, through the distal tubule, and finally back to the collecting duct. (Numerical values are milliosmolalities of urea during antidiuresis caused by the presence of large amounts of antidiuretic hormone.)

chloride in the tubular fluid often fall to as low as one-third to one-fifth their concentrations in the original glomerular filtrate. Therefore, on the average, less than 10 per cent of the sodium chloride in the original glomerular filtrate still remains in the tubular fluid by the time the fluid reaches the late distal tubules.

Sodium Reabsorption in the Late Distal Tubules and Collecting Ducts. Sodium reabsorption in the late distal tubules and collecting ducts is highly variable, depending mainly on the concentration of aldosterone, a hormone secreted by the adrenal cortex. In the presence of large amounts of aldosterone, almost the last vestiges of the tubular sodium are reabsorbed from the late distal tubules and collecting ducts so that essentially none of the sodium issues into the urine. On the other hand, in the absence of aldosterone, a very large proportion of the sodium is not reabsorbed and passes into the urine. Thus, the sodium excretion may be as little as one-tenth of a gram per day or as great as 30 to 40 grams. This ability of the tubular system to reabsorb almost all the sodium that filters through the glomeruli is a remarkable feat when one recognizes that an average of 600 grams of sodium filter each day. Indeed, the importance of this conservation of sodium is even more apparent when one realizes that almost 10 times as much sodium enters the glomerular filtrate each day as is present in the entire body.

Mechanism of Sodium Transport Through the Late Distal Tubule and Collecting Duct Epithelium. The mechanism for sodium transport through the distal tubular epithelium is illustrated in Figure 35–5. The movement is caused primarily by active transport of sodium from the interior of the tubular epithelial cell into the lateral intercellular spaces and into the extensive channels at the base of the cell. The transport out of the cell creates a very low concentration of sodium inside the cell and also creates a negative electrical potential of approximately −70 millivolts within the cell. The low sodium concentration and the very negative

potential create a high electrochemical gradient that causes sodium ions to diffuse from the tubular lumen into the cell, thus replacing the sodium as it is transported into the peritubular fluid.

The active transport of sodium out of the epithelial cell is coupled at least partially with active transport of potassium into the cell — potassium being exchanged for sodium. However, the exchange process is generally in favor of more sodium transport than potassium transport, which is the cause of the very negative potential inside the cell. The rates of transport of both sodium and potassium in this exchange process are determined almost entirely by the concentration of aldosterone in the body fluids. Thus, the rates of sodium and potassium transport through both the late distal tubule and collecting duct epithelium are almost entirely under the control of this hormone, a very important factor in the control of both sodium and potassium ion concentration in the extracellular fluids, as will be discussed in the following chapter.

Mechanism by Which Aldosterone Enhances Sodium and Potassium Transport. Upon entering a tubular epithelial cell, aldosterone combines with a *receptor protein;* this combination diffuses within minutes into the nucleus where it activates the DNA molecules to form one or more types of messenger RNA. The RNA is then believed to cause formation of carrier proteins or protein enzymes that are necessary for the sodium and potassium transport process.

Ordinarily, aldosterone has no effect on sodium and potassium transport for the first 45 minutes after it is administered; after this time the specific proteins important for transport begin to appear in the epithelial cells, followed by progressive increase in transport during the ensuing few hours.

POTASSIUM EXCRETION

Potassium Transport in the Proximal Tubules and Loops of Henle. Potassium is transported through the epithelium of the proximal tubules and of the diluting segment of the distal tubule in almost exactly parallel fashion to the transport of sodium. That is, both the potassium and the sodium are transported from the tubule into the blood. Thus, approximately 65 per cent of the potassium in the glomerular filtrate is absorbed in the proximal tubules (as is also true for sodium), and another 25 per cent is absorbed in the diluting segment of the distal tubules so that by the time the tubular fluid reaches the late distal tubules, the total quantity of potassium delivery to the late distal tubules each minute is less than 10 per cent of that in the original glomerular filtrate (as is also true for sodium).

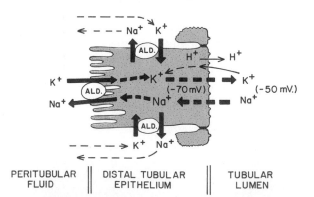

Figure 35–5. Mechanisms of Na⁺ and K⁺ transport through the distal tubular epithelium.

Active Secretion of Potassium in the Late Distal Tubules and Collecting Ducts. Normally, considerable amounts of potassium are secreted into the distal tubules, and some also into the collecting ducts. Figure 35–5 illustrates this secretion process. It shows that as sodium is transported from the cytoplasm of the epithelial cell into the peritubular fluid, potassium is simultaneously transported in the opposite direction to the interior of the cell. It is believed that this inward transport of potassium is not rigidly coupled with the outward transport of sodium. Instead, either all or most of the potassium movement into the cell is probably caused by the very negative electrical potential created inside the cell when sodium is pumped out into the peritubular fluid — this negativity attracting the positively charged potassium ions into the cell. Once the potassium has entered the epithelial cell, it then diffuses passively from the cell into the tubular lumen.

This secretory transport of potassium into the distal tubules is extremely important for the control of plasma potassium concentration for the following simple reason: The total quantity of potassium delivered from the loops of Henle into the distal tubules each day averages only about 70 mEq. Yet, the human being regularly eats this much potassium each day and on occasion eats as much as several hundred mEq. per day. Even if all of the 70 mEq. that enters the distal tubule should pass on into the urine, this still would not be enough potassium elimination. Therefore, it is essential that the excess potassium be removed by the process of secretion; otherwise, death might ensue from potassium toxicity. Indeed, cardiac arrhythmias usually appear when the potassium concentration rises from the normal value of 4 to 5 mEq./liter up to a level of 8 mEq./liter. A still higher potassium concentration than this can end in cardiac arrest.

Occasional Reabsorption of Potassium by the Distal Tubules and Collecting Ducts. In addition to the secretion of potassium by the distal tubules and collecting ducts, there is also continual active reabsorption of potassium through the luminal membrane of the epithelial cell to the inside of the cell, as illustrated by the light solid arrow in Figure 35–5. However, this reabsorption of potassium from the tubule is usually far overshadowed by potassium secretion into the tubule. Yet, in the absence of aldosterone, in which case potassium secretion falls essentially to zero, the reabsorptive process then becomes dominant. Thus, in the presence of aldosterone potassium secretion occurs, and in its absence potassium reabsorption occurs.

Importance of Aldosterone Control of Potassium Secretion. It is exceedingly important to emphasize that aldosterone plays equally as important a role in determining potassium secretion as in determining sodium reabsorption from the late distal tubules and collecting ducts even though the potassium secretion is mainly a secondary result of the sodium reabsorption. Indeed, in the following chapter when we discuss the control of the concentrations of these two ions in the extracellular fluids, we shall see that aldosterone actually plays far more of a role in controlling extracellular fluid potassium concentration than in controlling sodium concentration.

FLUID VOLUME EXCRETION

Up to this point we have considered the intrarenal mechanisms that determine the *concentrations* of various substances in the urine — water, urea, sodium, and potassium. Now it is important to consider the different factors that determine the rate of fluid *volume* excretion.

Glomerulotubular Balance and Its Relationship to Fluid Volume Excretion. Unfortunately, the term *glomerulotubular balance* means different things to different physiologists. The precise meaning of the words themselves is: exact balance between glomerular filtration rate and tubular reabsorption rate — in which case there would be zero urine output. However, by the term glomerulotubular balance most physiologists mean that whenever the glomerular filtration rate increases, all of the *additional* filtrate is reabsorbed and does not pass into the urine. This is *almost* true for the normal kidney, so that it is usually said that the kidney normally obeys the principle of glomerulotubular balance.

Glomerulotubular "Imbalance" and Its Importance to Fluid Volume Regulation. Though the concept of glomerulotubular balance is important to explain the way in which glomerular filtration rate and rate of tubular reabsorption automatically adjust to one another, very precise measurements show that 100 per cent glomerulotubular balance, even in the restricted sense employed by most physiologists, very rarely occurs. For instance, the following table gives approximate values for glomerular filtration rates, rates of fluid reabsorption, and rates of urine output for the average human adult:

Glomerular Filtration Rate	Rate of Tubular Reabsorption	Rate of Urine Output
ml.	*ml.*	*ml.*
50	49.8	0.2
75	74.7	0.3
100	99.5	0.5
125	124.0	1.0
150	145.0	5.0
175	163.0	12.0

If we examine these figures critically, we see that glomerular filtration rate and rate of tubular reabsorption actually do appear to parallel each other very closely. On the other hand, the degree of imbalance that occurs causes far greater change, proportionately, in urine output than in either glomerular filtration rate or tubular reabsorption rate. For instance, let us study the increase in glomerular filtration rate from 100 ml./min. up to 150 ml./min. The rate of reabsorption increases from 99.5 ml./min. up to 145 ml./min., representing only slight glomerulotubular imbalance. Nevertheless, this 50 per cent increase in glomerular filtration rate causes a 1000 per cent increase in rate of urine output! Thus, even a very slight degree of glomerulotubular imbalance leads to a tremendous increase in urine output when the glomerular filtration rate is increased. Also, very slight changes in rate of reabsorption of tubular fluid can cause equally as great alterations in urine output.

Therefore, the various factors that can alter either glomerular filtration rate or rate of tubular reabsorption are also the factors that play significant roles in determining the rate of fluid volume excretion. The five most important of these are: (1) the tubular osmolar clearance, (2) the plasma colloid osmotic pressure, (3) the degree of sympathetic stimulation of the kidneys, (4) the arterial pressure, and (5) the effect of antidiuretic hormone on tubular reabsorption.

1. EFFECT OF TUBULAR OSMOLAR CLEARANCE ON RATE OF FLUID VOLUME EXCRETION

Under most normal conditions, approximately proportional quantities of solutes and water are reabsorbed from the tubules. That is, whenever an osmolar substance is reabsorbed, so also will a proportionate amount of water be reabsorbed because the change in osmotic gradient across the tubular wall causes osmosis of water from the tubules into the peritubular spaces. Conversely, the greater the quantity of osmolar substances that *fails* to be reabsorbed by the tubules, the greater the quantity of water that fails to be reabsorbed. To state this another way, when the osmolar clearance is great, the volume of urine usually increases by approximately the same percentage.

Osmotic Diuresis. The effect of increased osmolar clearance in causing a proportionate amount of water to be lost in the urine at the same time is called *osmotic diuresis*. A particularly interesting type of osmotic diuresis occurs in diabetes mellitus in which the proximal tubules fail to reabsorb all the glucose, as normally occurs. Instead, the nonreabsorbed glucose passes the entire distance through the tubules and carries with it a large portion of the tubular water. Therefore, in diabetes mellitus (the word "diabetes" means diuresis) the urine output occasionally increases to as high as 4 to 5 liters per day.

Osmotic diuresis also occurs when substances that cannot be reabsorbed by the tubules are filtered in excessive quantities from the plasma into the glomerular filtrate. For instance, sucrose, mannitol, and urea, when in the circulating plasma in large quantities, are all filtered into the glomerular filtrate and cause large tubular loads of osmotic substances that either are not reabsorbed at all or are reabsorbed very poorly. Therefore, they too cause extreme osmotic diuresis.

2. EFFECT OF PLASMA COLLOID OSMOTIC PRESSURE ON RATE OF FLUID VOLUME EXCRETION

Another factor that greatly affects the rate of volume excretion is the plasma colloid osmotic pressure. A sudden increase in plasma colloid osmotic pressure instantaneously decreases the rate of fluid volume excretion. The cause of this effect is two-fold; (1) an increase in plasma colloid osmotic pressure *decreases* glomerular filtration rate, and (2) an increase in plasma colloid osmotic pressure *increases* tubular reabsorption. Both of these effects add together to decrease greatly the urine volume excretion.

Decreased Glomerular Filtration Rate. An increase in plasma colloid osmotic pressure of 1 mm. Hg decreases glomerular filtration rate as much as 5 to 10 per cent. The cause of this is simply the decrease in glomerular filtration pressure.

Increased Tubular Reabsorption. The effect of increased plasma colloid osmotic pressure to increase tubular reabsorption is less well understood, but the mechanism is believed to be that illustrated in Figure 35–6. This figure shows transport of sodium and water from the tubule into the spaces surrounding the epithelial cells. Once the sodium and water (and the other substances that move along with these two) have entered the intercellular spaces and the labyrinthine channels at the bases of the epithelial cells, the fluid can then go one of two ways: either it can leak backward into the tubule through the junctions between the epithelial cells, or it can move forward through the basement membrane into the peritubular capillary.

One of the major factors that determines whether or not the sodium and water will leak backward into the tubular lumen or will move forward into the peritubular capillary is the balance of hydrostatic and colloid osmotic pressures at the capillary membrane. At the bottom of Figure 35–6 the normal balance of pressures is

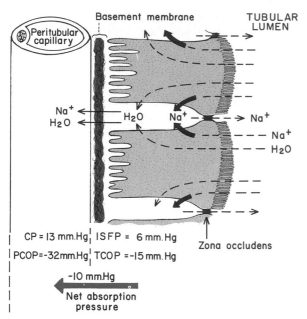

Figure 35–6. Absorption of fluid at the peritubular capillary membrane, and the effect of this on the absorption of water and sodium through the tubular epithelium.

depicted. If any one of the four factors affecting the balance at the capillary membrane changes, this obviously will also change the net absorption pressure. Thus, an increase in the plasma colloid osmotic pressure will increase the rate at which the capillary pulls fluid through the basement membrane, which also decreases the rate at which fluid leaks backward into the tubular lumen. To a very great extent, therefore, the rate of fluid absorption from the tubular lumen into the peritubular capillaries is determined by the plasma colloid osmotic pressure.

3. EFFECT OF SYMPATHETIC STIMULATION ON RATE OF FLUID VOLUME EXCRETION

Sympathetic stimulation has an especially powerful effect on constriction of the afferent arterioles. It greatly decreases the glomerular pressure and simultaneously decreases glomerular filtration rate. As has already been pointed out, a decrease in glomerular filtration rate often causes 10 times as much proportional decrease in urine output because of the slight degree of glomerulotubular imbalance that occurs even normally, as was discussed earlier.

Conversely, a decrease of sympathetic stimulation to below normal causes a mild degree of afferent arteriolar dilatation, which increases the glomerular filtration rate a slight amount. Consequently, decreased sympathetic stimulation leads to increased urine volume excretion.

4. EFFECT OF ARTERIAL PRESSURE ON RATE OF FLUID VOLUME EXCRETION

If all other factors remain constant but the renal arterial pressure is changed, the rate of urine output changes markedly. This effect is illustrated in Figure 35–7, showing that when the arterial pressure rises from 100 mm. Hg to 200 mm. Hg the increase in urine output is approximately sevenfold. Conversely, when the arterial pressure falls from 100 mm. Hg to 60 mm. Hg, the urine output falls either to zero or near to zero. We have already pointed out in Chapter 22 that this pressure effect on urine output plays an extremely important role in the feedback regulation of arterial pressure. It also plays an extremely important role in the feedback regulation of body fluid volume, as we shall discuss in the following chapter. Now, however, let us discuss the mechanism of the increased urine output caused by the increase in arterial pressure. This results from two separate effects: (1) the increase in arterial pressure increases glomerular pressure, which in turn increases glomerular filtration rate, thus leading to increased urine output; (2) the increase in arterial pressure also increases the peritubular capillary pressure, thereby decreasing tubular reabsorption. The combination of these two effects causes considerable glomerulotubular imbalance and therefore also causes marked increase in urine output, as illustrated in Figure 35–7.

5. EFFECT OF ANTIDIURETIC HORMONE ON RATE OF FLUID VOLUME EXCRETION

When excess antidiuretic hormone is secreted by the hypothalamic–posterior pituitary system, this secretion causes an acute effect to decrease

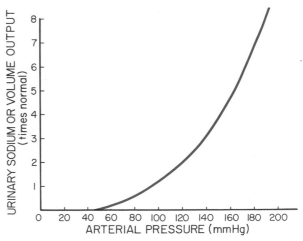

Figure 35–7. Effect of arterial pressure change on urinary output.

the urinary volume output. The reason for this is that antidiuretic hormone causes increased water reabsorption from the collecting ducts and perhaps to a slight extent from the late distal tubules as well. Therefore, less urinary volume is excreted; on the other hand, the urine that is excreted is highly concentrated.

When excess antidiuretic hormone is secreted for long periods of time, the acute effect to decrease urinary output is not sustained. The reason for this is that other factors, such as the arterial pressure, colloid osmotic pressure, and concentrations of the osmolar substances in the glomerular filtrate change — all of which lead eventually to a urinary volume output equal to the daily intake of fluid minus the losses of fluid in other ways from the body. In other words, balance between body fluid intake and body fluid output is re-established. Therefore, contrary to what is often taught, long-term secretion of excess antidiuretic hormone plays only a small role in the regulation of body fluid volume. As we shall see in discussion of the condition called "inappropriate ADH syndrome" in the following chapter, continued long-term secretion of very large quantities of antidiuretic hormone causes severe abnormalities in body fluid ionic composition even though the body fluid volume is only slightly altered.

Mechanism by Which Antidiuretic Hormone Increases Water Reabsorption. The precise mechanism by which antidiuretic hormone increases water reabsorption by the collecting tubules is not known. However, several established facts about the mechanism are the following: Stimulation of the epithelial cells of the collecting tubules by antidiuretic hormone activates *adenyl cyclase* in the epithelial cell membrane on the peritubular side of the cell, and this causes formation of cyclic adenosine monophosphate (cyclic AMP) in the cell cytoplasm. The increase in cyclic AMP is then associated — for reasons that are not yet known — with marked increase in permeability of the *luminal* membrane of the epithelial cells to water; this in turn is responsible for the increase in water reabsorption by the collecting ducts.

SUMMARY OF THE CONTROL OF FLUID VOLUME EXCRETION

From the preceding few sections, it is clear that many different factors have effects on the regulation of urine volume excretion. Some of the factors cause extreme acute changes in urinary output — for instance, the acute decrease in urine volume output caused by antidiuretic hormone or the acute increase in urine volume output caused by increased tubular osmotic loads. However, over a longer period of time, other longer acting factors, such as changes in arterial pressure and

colloid osmotic pressure, tend to nullify the extreme acute effects of these first factors.

Especially important in the long-term regulation of urinary output is the increase in output caused by elevated arterial pressure, an effect that seems to be sustained indefinitely. Indeed, present data indicate that this effect, instead of fading, actually increases with time. Therefore, the level of arterial pressure is probably the single, most important factor that determines the long-term level of urinary volume excretion, which is a subject that we shall discuss in much more detail in the following chapter in relation to the control of extracellular fluid volume.

AUTOREGULATION OF GLOMERULAR FILTRATION RATE

Even though a change in arterial pressure causes a marked change in urinary output, this pressure can change from as little as 75 mm. Hg to as high as 160 mm. Hg while causing very little change in glomerular filtration rate. This effect is illustrated in Figure 35–8 and it is called *autoregulation of glomerular filtration rate*. During the first few minutes after an arterial pressure change, renal blood flow is also autoregulated. However, over a period of hours the glomerular filtration rate remains well autoregulated while renal blood flow then becomes poorly autoregulated, as will be explained in the following sections.

Importance of Autoregulation of Glomerular Filtration Rate. When *excess* glomerular filtrate flows rapidly through the tubules, tubular reabsorption is usually much too little before the fluid passes into the urine. As a result, the urine volume is extreme, and excessive amounts of both water

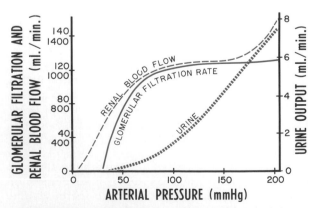

Figure 35–8. Autoregulation of glomerular filtration rate (GFR) and renal blood flow (RBF) when the arterial pressure is increased but there is lack of autoregulation of urine flow.

and solutes are lost from the body. Therefore, the kidney fails in one of its primary duties, namely, to conserve those substances needed by the body.

Conversely, when too little glomerular filtrate is formed, too much of the fluid is reabsorbed and urine excretion is greatly reduced. And, again, the kidney fails in another of its primary functions: to rid the body of the end-products of metabolism and excess electrolytes as rapidly as possible.

Thus, it is clear that the nephron requires an optimal rate of glomerular filtration if it is to perform its function, neither too great a rate nor too little. Recent analyses of tubular function have shown that even a 5 per cent too great or too little rate of glomerular filtration can have profound effects in causing either excess loss in the urine or too little excretion of the necessary waste products.

Mechanism of Autoregulation of Glomerular Filtration Rate – Tubuloglomerular Feedback

The precision with which glomerular filtration rate must be autoregulated demands that there be an effective means for controlling this filtration rate. Fortunately, each nephron is provided with not one but *two* special feedback mechanisms which add together to provide the degree of glomerular filtration autoregulation that is required. These two mechanisms are (1) an *afferent arteriolar vasodilator feedback mechanism* and

(2) an *efferent arteriolar vasoconstrictor feedback mechanism*. In each of these mechanisms, the degree to which the fluid in the distal tubule has been processed is detected, and appropriate signals are fed back to the afferent and efferent arterioles to increase or decrease the glomerular filtration rate as required. The combination of these two feedback mechanisms is called *tubuloglomerular feedback*. And the feedback process probably occurs either entirely or almost entirely at the *juxtaglomerular complex,* which has the following characteristics:

The Juxtaglomerular Complex.   Figure 35–9 illustrates the juxtaglomerular complex, showing that the distal tubule passes in the angle between the afferent and efferent arterioles, actually abutting each of these two arterioles. Furthermore, those epithelial cells of the distal tubule that come in contact with the arterioles are more dense than the other tubular cells and are collectively called the *macula densa*. The position in the distal tubule where the macula densa is located is about midway in the diluting segment of the distal tubule, at the upper end of the thick portion of the ascending limb of the loop of Henle. The macula densa cells appear to secrete some substance toward the arterioles because the Golgi apparatus, an intracellular secretory organelle, is directed toward the arterioles and not toward the lumen of the tubule, in contrast to all the other tubular epithelial cells. Note also in Figure 35–9 that the smooth muscle cells of both the afferent and efferent arterioles are

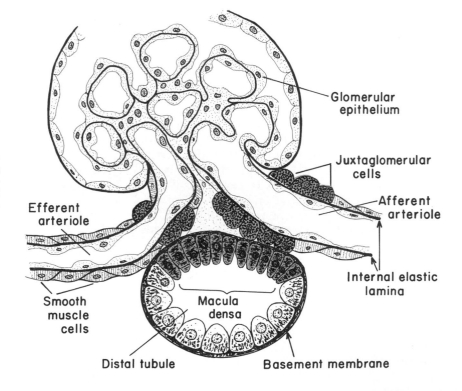

Figure 35–9.   Structure of the juxtaglomerular apparatus, illustrating its possible feedback role in the control of nephron function. (Modified from Ham: Histology. J. B. Lippincott Co.)

swollen and contain dark granules where they come in contact with the macula densa. These cells are called *juxtaglomerular cells,* and the granules are composed mainly of inactive *renin.* The whole complex of macula densa and juxtaglomerular cells is called the *juxtaglomerular complex.*

Thus, the anatomical structure of the juxtaglomerular apparatus suggests strongly that the fluid in the distal tubules in some way plays an important role in helping to control nephron function by providing feedback signals to both the afferent and efferent arterioles.

The Afferent Arteriolar Vasodilator Feedback Mechanism. A low rate of glomerular filtration causes over-reabsorption of chloride and therefore decreases the chloride ion concentration at the macula densa. This decrease in chloride ions in turn initiates a signal from the macula densa to dilate the afferent arterioles. Putting these two facts together, the following is the postulated mechanism by which the afferent arteriolar vasodilator feedback mechanism controls glomerular filtration rate:

(1) Too little flow of glomerular filtrate into the tubules causes decreased chloride ion concentration at the macula densa.

(2) The decreased chloride concentration causes afferent arteriolar dilatation.

(3) This in turn increases the rate of blood flow into the glomerulus and increases the glomerular pressure.

(4) The increased glomerular pressure increases the glomerular filtration rate back toward the required level.

Thus, this is a typical negative feedback mechanism for controlling the glomerular filtration rate at a steady rate. This mechanism also autoregulates the renal blood flow at the same time, as we shall discuss.

The Efferent Arteriolar Vasoconstrictor Feedback Mechanism. Too few chloride ions at the macula densa are believed also to cause the juxtaglomerular cells to release renin, and this in turn causes formation of angiotensin. The angiotensin then constricts the renal arterioles, but mainly the efferent arteriole because it is more sensitive to angiotensin II than is the afferent arteriole, probably also because it is downstream along the flowing blood from the juxtaglomerular complex while the afferent arteriole is upstream from it.

With these facts in mind, we can now describe the efferent arteriolar vasoconstrictor mechanism that helps to maintain a constant glomerular filtration rate:

(1) A too low glomerular filtration rate causes excess reabsorption of chloride ions from the filtrate, reducing the chloride ion concentration at the macula densa.

(2) The low concentration of chloride ions then causes the JG cells to release renin from their granules.

(3) The renin causes formation of angiotensin II.

(4) The angiotensin II constricts the efferent arterioles, which causes the pressure in the glomerulus to rise.

(5) The increased pressure then causes the glomerular filtration rate to return back toward normal.

Thus, this is still another negative feedback mechanism that helps to maintain a very constant glomerular filtration rate; it does so by constricting the efferent arterioles at the same time that the afferent vasodilator mechanism described above dilates the afferent arterioles. When both of these mechanisms function together, the glomerular filtration rate increases only a few per cent even though the arterial pressure changes between the limits of 75 mm. Hg and 160 mm. Hg.

AUTOREGULATION OF RENAL BLOOD FLOW

When the arterial pressure is changed for only a few minutes at a time, renal blood flow and the glomerular filtration rate are autoregulated at the same time. This is illustrated in Figure 35–8, which shows a relatively constant renal blood flow between the limits of 70 and 160 mm. Hg arterial pressure.

It is the afferent arteriolar vasodilator feedback mechanism described above that causes this renal blood flow autoregulation. This can be explained as follows: When the renal blood flow becomes too little, the glomerular pressure falls and the glomerular filtration rate also becomes too little. As a consequence, the feedback mechanism causes afferent arteriolar dilatation to return the glomerular filtration rate back toward normal. At the same time, the dilatation also increases the blood flow back toward normal despite the low arterial pressure. Conversely, when the blood flow becomes too great, the same feedback mechanism now operates in the opposite direction to cause afferent arteriolar vasoconstriction. And again the renal blood flow returns toward normal.

Failure of Blood Flow Autoregulation During Long-Term Changes in the Arterial Pressure. When the arterial pressure remains depressed or elevated for hours instead of minutes, the glomerular filtration rate continues to be autoregulated to almost exactly the normal level. But blood flow autoregulation is not maintained for this reason: During the first few minutes of *glomerular filtration rate autoregulation*, this autoregulation is caused almost entirely by the afferent mechanism, and it is also this mechanism that causes *blood flow autoregulation.* On the other hand, within 5 to 10 minutes, the efferent mechanism, which is a vasoconstrictor mechanism resulting from the formation of large amounts of angiotensin II, begins to become as potent as the afferent mechanism. Over a period of hours, it becomes even more potent. Yet, this efferent mechanism causes decreased rather than increased renal blood flow. Therefore, when the arterial pressure remains low for more than 10 to 20 minutes, blood flow autoregulation disappears. Instead, the increasing efferent arteriolar resistance now *decreases* the renal blood flow, which is opposite to renal blood flow autoregulation.

If one will think for a moment, the wisdom of using the efferent mechanism to maintain glomerular filtration rate when the arterial pressure remains low for long periods of time will be understood. This allows glomerular filtration to continue unabated, but the marked decrease in blood flow through the kidney helps to return the arterial pressure back toward normal.

ROLE OF THE RENIN-ANGIOTENSIN SYSTEM AND OF THE EFFERENT VASOCONSTRICTOR MECHANISM IN CONSERVING WATER AND SALT BUT ELIMINATING UREA DURING ARTERIAL HYPOTENSION

The efferent arteriolar vasoconstrictor mechanism not only helps to maintain normal glomerular filtration when the arterial pressure falls too low but also provides a means for controlling urea excretion separately from the excretion of water and salt. In arterial hypotension it is very important to conserve as much water and salt as possible. Yet, on the other hand, it is equally as important to continue excreting the body's waste products, the most abundant of which is urea. Therefore, let us explain this:

Earlier in the chapter it was pointed out that the rate of urea excretion is almost directly proportional to the rate of glomerular filtration. Therefore, as long as the efferent arteriolar vasoconstrictor mechanism can maintain a high glomerular filtration even in the presence of low arterial pressure, so also will almost normal amounts of urea be excreted in the urine. Therefore, hypotension down to arterial pressure levels as low as 65 to 75 mm. Hg do not cause significant retention of urea.

On the other hand, as angiotensin II builds up within the kidneys and also in the circulating blood during arterial hypotension, this causes marked retention by the kidney of water and of the various ions — sodium, chloride, potassium, and others. Thus, this provides a means for conserving water and ions despite the fact that urea continues to be excreted.

The cause of the conservation of the water and the ions is not yet entirely clear. However, it probably results from the sum of three different effects: (1) The increase in efferent resistance reduces renal blood flow and therefore also reduces peritubular capillary pressure. This in turn increases the rate of reabsorption of water and electrolytes from the tubular system, as explained earlier in this chapter. (2) The decreased blood flow into the glomerulus allows an extremely high proportion of the fluid in the plasma to filter through the glomerulus. Therefore, the concentration of plasma proteins becomes very high in the plasma that flows from the glomerulus into the peritubular capillaries. This increases the colloid osmotic pressure in these capillaries, which causes still additional reabsorption of water and ions, as was also explained earlier in the chapter. And (3) some research workers have suggested that angiotensin has a direct effect on the distal tubules in causing increased active reabsorption of sodium, which in turn could cause reabsorption of water and other substances in increased amounts.

Regardless of all the mechanisms involved, the function of the renin-angiotensin system in causing conservation of water and electrolytes while at the same time promoting continued high excretion of waste products is an exceedingly important life-saving event when hypotension occurs. This mechanism is also activated when animals are dehydrated, such as in animals exposed to desert conditions, because dehydration alone also activates the intrarenal renin system to conserve water and salt but still allowing the waste products to be excreted.

Possible Role of the Renin-Angiotensin System to Maintain Balance of Nephron Function Between the Two Kidneys

With 2.4 million nephrons in the human kidneys, it is important to have some mechanism that will maintain approximately equal function among all these nephrons. Angiotensin can possibly play at least part of this role. It supposedly does so by helping to provide each nephron with an individual feedback control mechanism for maintaining waste product excretion, as described above. However, in addition, the release of renin into the circulating blood by ischemic nephrons, and the subsequent formation of angiotensin II, distributes angiotensin to all the other nephrons that themselves may not be forming large amounts. This angiotensin arrives first at the afferent arterioles and causes a moderate degree of afferent arteriolar constriction and then constricts the efferent arterioles as well; both of these processes cause water and salt retention. Therefore, if one set of nephrons becomes ischemic and their function is reduced as a result of this ischemia, the circulating angiotensin can cause the other nephrons also to retain water and salt. Thus, ischemia of one kidney can cause water and salt retention not only by the ischemic kidney but also by the opposite kidney as well, an effect that has been demonstrated to occur when only one renal artery is constricted. The water and salt retention then causes increased arterial pressure, which helps to relieve the ischemia of the malfunctioning nephrons. This is sometimes of life-saving importance because the increased elimination of waste products by these marginal nephrons is often required to prevent uremia.

REFERENCES

Andersson, B.: Regulation of body fluids. *Annu. Rev. Physiol., 39*:185, 1977.

Andreoli, T. E., and Schaefer, J. A.: Mass transport across cell membranes: The effects of antidiuretic hormone on water and solute flows in epithelia. *Annu. Rev. Physiol., 38*:451, 1976.

Andreoli, T., *et al.* (eds.): Physiology of Membrane Disorders. New York, Plenum Press, 1978.

Aukland, K.: Renal blood flow. *Int. Rev. Physiol., 11*:23, 1976.

Barger, A. C., and Herd, J. A.: Renal vascular anatomy and distribution of blood flow. *In* Orloff, F., and Berliner, R. W. (eds.): Handbook of Physiology. Sec. 8. Baltimore, Williams & Wilkins, 1973, p. 249.

Bianchi, C.: Measurement of the glomerular filtration rate. *Prog. Nucl. Med., 2*:21, 1972.

Brenner, B. M.: Adaptation of glomerular forces and flows to renal injury. *Yale J. Biol. Med., 51*(3):301, 1978.

Brenner, B. M., and Berliner, R. W.: Transport of potassium. *In* Orloff, F., and Berliner, R. W. (eds.): Handbook of Physiology. Sec. 8. Baltimore, Williams & Wilkins, 1973, p. 497.

Brenner, B. M., and Deen, W. M.: The physiological basis of glomerular ultrafiltration. *In* MTP International Review of Science: Physiology. Vol. 6. Baltimore, University Park Press, 1974, p. 335.

Brenner, B. M., *et al.*: Determinations of glomerular filtration rate. *Annu. Rev. Physiol., 38*:9, 1976.

Brenner, B. M., *et al.*: Transport of molecules across renal glomerular capillaries. *Physiol. Rev., 56*:502, 1976.

Burg, M., and Stoner, L.: Renal tubular chloride transport and the mode of action of some diuretics. *Annu. Rev. Physiol., 38*:37, 1976.

Churg, J. (ed.): The Kidney. Baltimore, Williams & Wilkins, 1979.

Deetjen, P., *et al.*: Physiology of the Kidney and of Water Balance. New York, Springer-Verlag, 1975.

de Wardener, H. E.: Mechanisms influencing urinary sodium excretion. *In* Dickinson, C. J., and Marks, J. (eds.): Developments in Cardiovascular Medicine. Lancaster, England, MTP Press, 1978, p. 179.

Edmonds, C. J.: Salts and water. *Biomembranes, 4B*:711, 1974.

Giebisch, G., and Stanton, B.: Potassium transport in the nephron. *Annu. Rev. Physiol., 41*:241, 1979.

Giebisch, G., and Windhager, E.: Electrolyte transport across renal tubular membranes. *In* Orloff, F., and Berliner, R. W. (eds.): Handbook of Physiology. Sec. 8. Baltimore, Williams & Wilkins, 1973, p. 315.

Glynn, I. M., and Karlish, S. J. D.: The sodium pump. *Annu. Rev. Physiol., 37*:13, 1975.

Goldberg, M., *et al.*: Renal handling of calcium and phosphate. *Int. Rev. Physiol., 11*:211, 1976.

Gottschalk, C. W., and Lassiter, W. E.: Micropuncture methodology. *In* Orloff, F., and Berliner, R. W. (eds.): Handbook of Physiology. Sec. 8. Baltimore, Williams & Wilkins, 1973, p. 129.

Gottschalk, C. W., *et al.*: Micropuncture study of composition of loop of Henle fluid in desert rodents. *Am. J. Physiol., 204*:532, 1963.

Grantham, J. J., *et al.*: Studies of isolated renal tubules in vitro. *Annu. Rev. Physiol., 40*:249, 1978.

Greger, R., *et al.*: Renal excretion of purine metabolites, urate and allantoin by the mammalian kidney. *Int. Rev. Physiol., 11*:257, 1976.

Guyton, A. C., *et al.*: Dynamics and Control of the Body Fluids. Philadelphia, W. B. Saunders Co., 1975.

Hall, J. E., *et al.*: A single-injection method for measuring glomerular filtration rate. *Am. J. Physiol., 232*:F72, 1977.

Hall, J. E., *et al.*: Dissociation of renal blood flow and filtration rate autoregulation by renin depletion. *Am. J. Physiol., 232*:F215, 1977.

Hierholzer, K., and Lange, S.: The effects of adrenal steroids on renal function. *In* MTP International Review of Science: Physiology. Vol. 6. Baltimore, University Park Press, 1974, p. 273.

Katz, A. I., and Lindheimer, M. D.: Actions of hormones on the kidney. *Annu. Rev. Physiol., 39*:97, 1977.

Kinne, R.: Membrane-molecular aspects of tubular transport. *Int. Rev. Physiol., 11*:169, 1976.

Knox, F. G., and Davis, B. B.: Role of physical and neuroendocrine factors in proximal electrolyte reabsorption. *Metabolism, 23*:793, 1974.

Knox, F. G., and Diaz-Buxo, J. A.: The hormonal control of sodium excretion. *Int. Rev. Physiol., 16*:173, 1977.

Kramer, H. J., and Krück, F. (eds.): Natriuretic Hormone. New York, Springer-Verlag, 1978.

Lameire, N. H., *et al.*: Heterogeneity of nephron function. *Annu. Rev. Physiol., 39*:159, 1977.

Lassiter, W. E.: Kidney. *Annu. Rev. Physiol., 37*:371, 1975.

Latta, H.: Ultrastructure of the glomerulus and juxtaglomerular apparatus. *In* Orloff, F., and Berliner, R. W. (eds.): Handbook of Physiology. Sec. 8. Baltimore, Williams & Wilkins, 1973, p. 1.

Lohmeier, T. E., *et al.*: Effects of endogenous angiotensin II on renal sodium excretion and renal hemodynamics. *Am. J. Physiol., 233*:F388, 1977.

Maude, D. L.: Mechanism of tubular transport of salt and water. *In* MTP International Review of Science: Physiology. Vol. 6. Baltimore, University Park Press, 1974, p. 39.

Maunsbach, A. B.: Cellular mechanisms of tubular protein transport. *Int. Rev. Physiol., 11*:145, 1976.

Mercer, P. F., *et al.*: Current concepts of sodium chloride and water transport by the mammalian nephron. *West. J. Med., 120*:33, 1974.

Morel, F., and de Rouffignac, C.: Kidney. *Annu. Rev. Physiol., 35*:17, 1973.

Moses, A. M., and Share, L. (eds.): Neurohypophysis. New York, S. Karger, 1977.

Mudge, G. H., *et al.*: Tubular transport of urea, glucose, phosphate, uric acid, sulfate, and thiosulfate. *In* Orloff, F., and Berliner, R. W. (eds.): Handbook of Physiology. Sec. 8. Baltimore, Williams & Wilkins, 1973, p. 587.

Navar, L. G., *et al.*: Effect of alterations in plasma osmolality on renal blood flow autoregulation. *Am. J. Physiol., 211*:1387, 1966.

Pitts, R.: Physiology of the Kidney and Body Fluids. 3rd Ed. Chicago, Year Book Medical Publishers, 1974.

Pollak, V. E., and Pesce, A. J.: Maintenance of body protein homeostasis. *In* Frohlich, E. D. (ed.): Pathophysiology, 2nd Ed. Philadelphia, J. B. Lippincott Co., 1976, p. 221.

Renkin, E. M., and Gilmore, J. P.: Glomerular filtration. *In* Orloff, F., and Berliner, R. W. (eds.): Handbook of Physiology. Sec. 8. Baltimore, Williams & Wilkins, 1973, p. 185.

Renkin, E. M., and Robinson, R. R.: Glomerular filtration. *N. Engl. J. Med., 290*:785, 1974.

Schafer, J. A., and Andreoli, T. E.: Rheogenic and passive Na+ absorption by the proximal nephron. *Annu. Rev. Physiol., 41*:211, 1979.

Schrier, R. W.: Effects of adrenergic nervous system and catecholamines on systemic and renal hemodynamics, sodium and water excretion, and renin secretion. *Kidney Int., 6*:291, 1974.

Smith, H. W.: The Kidney: Structure and Function in Health and Disease. New York, Oxford University Press, 1951.

Stephenson, J. L.: Countercurrent transport in the kidney. *Annu. Rev. Biophys. Bioeng., 7*:315, 1978.

Stoff, J. S., *et al.*: Recent advances in renal tubular biochemistry. *Annu. Rev. Physiol., 38*:46, 1976.

Ullrich, K. L.: Sugar, amino acids, and Na+ cotransport in the proximal tubule. *Annu. Rev. Physiol., 41*:181, 1979.

Vander, A. J.: Renal Physiology. New York, McGraw-Hill, 1980.

Walser, M.: Divalent cations: Physicochemical state in glomerular filtrate and urine and renal excretion. *In* Orloff, F., and Berliner, R. W. (eds.): Handbook of Physiology. Sec. 8. Baltimore, Williams & Wilkins, 1973, p. 555.

Weitzman, R., and Kleeman, C. R.: Water metabolism and the neurohypophysial hormones. *In* Bondy, P. K., and Rosenberg, L. E. (eds.): Metabolic Control and Disease, 8th Ed. Philadelphia, W. B. Saunders Co., 1980, p. 1241.

Windhager, E. E.: Kidney, water, and electrolytes. *Annu. Rev. Physiol., 31*:117, 1969.

Wright, F. S.: Intrarenal regulation of glomerular filtration rate. *N. Engl. J. Med., 291*:135, 1974.

Wright, F. S.: Potassium transport by the renal tubule. *In* MTP International Review of Science: Physiology. Vol. 6. Baltimore, University Park Press, 1974, p. 79.

Wright, F. S., and Briggs, J. P.: Feedback control of glomerular blood flow, pressure and filtration rate. *Physiol. Rev., 59*:958, 1979.

36

Regulation of Blood Volume, Extracellular Fluid Volume, and Extracellular Fluid Composition by the Kidneys and by the Thirst Mechanism

The kidneys, more than any other organ, play determining roles in regulating important characteristics of the body fluids, including: (1) blood volume, (2) extracellular fluid volume, (3) osmolality of the body fluids — that is, the ratio of water to dissolved substances, (4) specific concentrations of the different ions, and (5) degree of acidity of the body fluids. The first four of these will be discussed in the present chapter, and the last — regulation of hydrogen ion concentration — will be discussed in the following chapter. In controlling some of these characteristics the thirst mechanism also plays a key role; this mechanism will also be discussed.

CONTROL OF BLOOD VOLUME

Constancy of the Blood Volume. The extreme degree of precision with which the blood volume is controlled is illustrated in Figure 36–1. This shows the effect of changing the daily fluid intake, including both water and dissolved electrolytes, from very low values to very high values. It indicates that almost no change in blood volume occurs despite tremendous changes in intake (except when the intake becomes so low that it is not sufficient to make up for fluid losses caused by evaporation from the surface of the body, evaporation from the lungs, and a small amount of "obligatory" loss of fluid by the kidneys needed for continued excretion of minimal amounts of waste products). To state this another way, even when the intake of water and salt is increased many-fold, the blood volume is hardly altered. Conversely, a decrease in fluid intake to as little as one-third normal causes hardly a change.

BASIC MECHANISM FOR BLOOD VOLUME CONTROL

The basic mechanism for blood volume control is illustrated in Figure 36–2. This is essentially the same as the basic mechanism for arterial pressure control that was presented in Chapter 22. It was pointed out in the discussion in this earlier chapter that extracellular fluid volume, blood volume, cardiac output, arterial pressure, and urine output are all mainly or partially controlled by a single common basic feedback mechanism. The basic features of this common mechanism are illustrated again in Figure 36–2, but this time to emphasize those factors that are important mainly in blood volume and extracellular fluid volume regulation. Therefore, let us explain the six major steps in this mechanism, illustrated by each of the six blocks in the figure:

Block 1 shows that an increase in blood volume increases cardiac output.

Block 2 illustrates the relationship between cardiac output and arterial pressure, showing that an increase in cardiac output increases arterial pressure.

Block 3 shows that an increase in arterial pressure increases urinary output. The dashed curve is the effect that occurs in acute experiments when the arterial pressure is raised, and the solid curve is the more chronic effect, illustrating the extreme

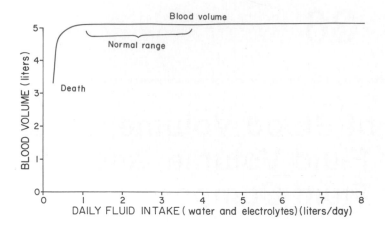

Figure 36–1. Effect on blood volume of marked changes in daily fluid intake. Note the precision of blood volume control in the normal range.

increase in urinary output when the arterial pressure rises only slightly above normal, as was explained in Chapter 22 in relation to arterial pressure control.

Block 4 gives a summation of the fluid intake minus the fluid losses from the body, losses including both urinary output and other fluid loss. The output of this block is the rate of change of extracellular fluid volume. If the intake is greater than the output, the rate of change will be positive; if the output is greater than the intake, the rate of change will be negative.

Block 5 integrates the rate of change of extracellular fluid volume — that is, it shows the accumulation of greater or lesser fluid volume with time, depending on whether the rate of change is positive or negative. The output of Block 5 is the actual extracellular fluid volume.

Block 6 gives the relationship between extracellular fluid volume and blood volume, showing that, in general, as the extracellular fluid volume increases, the blood volume also increases. We will discuss this basic relationship in greater detail later in connection with extracellular fluid volume control.

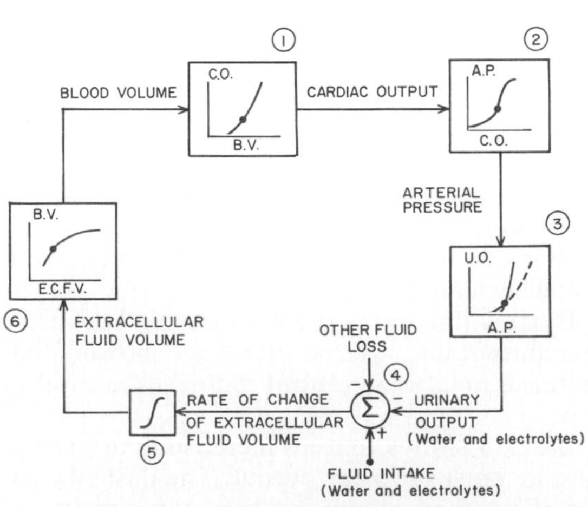

Summary of the Basic Blood Volume Control Mechanisms. To summarize the principles illustrated in Figure 36–2, we can trace what happens when the blood volume becomes abnormal. When the blood volume becomes too great, the cardiac output and arterial pressure increase. This, in turn, has a profound effect on the kidneys, causing loss of fluid from the body and returning the blood volume back to normal. Conversely, if the blood volume falls below normal, the cardiac output and arterial pressure decrease, the kidneys retain fluid, and progressive accumulation of the fluid intake builds the blood volume eventually back to normal. Obviously, parallel processes will also occur to reconstitute red cell mass, plasma proteins, and so forth if these have become abnormal at the same time. (If the red cell volume remains abnormal, however, the plasma volume will simply make up the difference, the volume becoming essentially normal despite the low red cell mass.)

Reason for the Precision of the Blood Volume Regulating Mechanism. Studying the diagram of Figure 36–2 very carefully, one can see why the blood volume remains almost exactly constant as illustrated in Figure 36–1 despite extreme changes in daily fluid intake. The reason for this is that the slopes of the curves in Blocks 1, 2, and 3 are all very steep, meaning that a slight change in blood volume causes a marked change in cardiac output, a slight change in cardiac output causes a marked change in arterial pressure, and a slight change in arterial pressure causes a marked change in urinary output. These factors all multiply together to give an extremely high gain for the feedback control of blood volume.

ROLE OF THE VOLUME RECEPTORS IN BLOOD VOLUME CONTROL

It was pointed out in Chapter 21 that "volume receptor" reflexes help to control blood volume. These

Figure 36–2. Basic feedback mechanism for control of blood volume and extracellular fluid volume (the points on each of the curves represent normal values).

modify the function of the above basic mechanism for blood volume control in a very specific way, as follows:

The basic mechanism of Figure 36–2 is slow to come to equilibrium. That is, when the blood volume increases, the effects on cardiac output, arterial pressure, and urinary output are slow to build up, often requiring many hours to develop full effect; even after they do develop full effect, additional hours are still required for the excess fluid to be eliminated from the body. However, the volume receptor reflexes can greatly accelerate this process.

The volume receptors are mainly stretch receptors located in the walls of the right and left atria. When the blood volume becomes excessive, a large share of this volume accumulates in the central veins of the thorax and causes increased pressure in the two atria. The resultant stretch of the atrial walls transmits nerve signals into the brain, and these in turn elicit responses that accelerate the return of blood volume to normal. The various responses that occur include the following:

1. The sympathetic nervous signals to the kidneys are inhibited, thus slightly to moderately increasing the rate of urinary output.

2. The secretion of antidiuretic hormone by the supraopticohypophyseal system is reduced, allowing increased water excretion by the kidneys.

3. The peripheral arterioles throughout the body are dilated because of reflex reduction of sympathetic stimulation, thus increasing capillary pressure and allowing much of the excess blood volume to filter temporarily into the tissue spaces for a few hours until the excess fluid can be excreted through the kidneys.

In most instances, these volume receptor reflex effects can cause the blood volume to return almost all the way to normal within an hour or so, but the final determination of the precise level to which the blood volume will be adjusted is still a function of the basic volume control mechanism illustrated in Figure 36–2. The reason for this is that over a period of one to three days the volume receptors adapt completely so that they no longer transmit any corrective signals. Therefore, they are of value only to help readjust the volume during the first few hours or few days after an abnormality occurs, but not for long-term monitoring of volume or for precise adjustments of the long-term level of blood volume.

Role of the Baroreceptors and Other Stretch Receptors in the Volume Receptor Response. Ordinarily, when the blood volume increases, the systemic arterial pressure and the pulmonary arterial pressure also increase at least to some extent. Therefore, the baroreceptors of the carotid and aortic regions and also of the pulmonary arteries are excited. These, too, cause essentially the same effects as the atrial volume receptors. Therefore, the baroreceptors add still more to the volume receptor reflex effect.

OTHER FACTORS THAT HELP TO CONTROL BLOOD VOLUME

Capacity of the Circulation. Any change in the basic capacity of the circulation will automatically change the level to which the blood volume is regulated. For instance, when a person develops very severe varicose veins, the total capacity of the circulatory system often increases as much as 1 liter. Thereafter, the volume controlling mechanisms will automatically control the blood volume to this higher level.

Other factors that sometimes change the capacity of the system include vasoconstrictor factors, vasodilator factors, aneurysms, and so forth. For instance, when a person is under the chronic influence of strong sympathetic vasoconstrictor stimulation or of vasoconstrictor agents such as norepinephrine, the blood volume becomes regulated to a lower level. That is, the volume control system is geared to adjust to the volume of blood to fill the capacity of the system itself.

Effect of Antidiuretic Hormone and Aldosterone on Blood Volume. The two hormones antidiuretic hormone and aldosterone both play important roles in some aspects of fluid and electrolyte economy of the body. Therefore, both of them have been extolled as very important blood volume regulators. However, measurements show that these normally affect blood volume relatively little. Therefore, let us discuss only briefly their effects on blood volume regulation.

Antidiuretic Hormone. Complete absence of antidiuretic hormone in the condition called *diabetes insipidus* usually will cause no measurable decrease in blood volume even though it may increase the output of urine by as much as 3- to 10-fold. The thirst mechanism simply causes the person to drink enough water to make up the difference. On the other hand, if the person is prevented from obtaining water to drink, then the blood volume does decrease drastically, and this causes circulatory shock; this combination of effects only rarely occurs.

When the secretion of antidiuretic hormone is tremendous, as occurs in the condition called *syndrome of inappropriate ADH secretion,* the blood volume again increases almost imperceptibly — perhaps as much as 3 to 5 per cent. The reason for only this minute increase in blood volume is that the slight volume increase that does occur simply increases the arterial pressure enough (it requires only a minute rise in pressure) to overcome the effect of the antidiuretic hormone to cause water retention.

Aldosterone. Aldosterone causes excessive salt reabsorption from the late distal tubules and collecting ducts of the kidneys, and this in turn causes osmotic reabsorption of water. Therefore, the immediate effect is to decrease urine output greatly. The extracellular fluid volume and the blood volume both increase. However, before these can increase more than a few per cent, the basic feedback mechanism for blood volume control (Figure 36–2) comes into play and overbalances the retention of fluid by the kidneys. It does this by raising the arterial pressure a few mm. Hg, increasing glomerular filtration rate a few per cent, and therefore increasing the urinary output back to equal the intake of fluid.

Consequently, even in persons who have tremendous secretion of aldosterone (patients with *primary aldosteronism*) the extracellular fluid volume and blood volume rarely rise more than 5 to 10 per cent, at most. On the other hand, in persons who have no secretion of aldosterone (patients with *Addison's disease*) the late distal tubules and collecting ducts fail to reabsorb salt and water, and the kidneys lose tremendous quantities of fluids into the urine. If the person eats enough salt and drinks a concomitant amount of water, the blood volume will still regulate in the normal range. But if salt and water intake are not sufficient, the patient can develop severe dehydration and hypovolemia with resultant circulatory shock.

CONTROL OF EXTRACELLULAR FLUID VOLUME

It is already clear from the above discussion of the basic mechanisms for blood volume control that extracellular fluid volume is controlled at the same time. That is, fluid first goes into the blood, but it rapidly becomes distributed between the interstitial spaces and the plasma. Therefore, it is impossible to control blood volume to any given level without controlling the extracellular fluid volume at the same time. Yet, the relative volumes of distribution between the interstitial spaces and the blood can vary greatly, depending on the physical characteristics of the circulatory system and of the interstitial spaces. Under normal conditions, the interstitial spaces are in a relatively "dry" state. That is, the fluid in the interstitial spaces is bound in a gel-like matrix of hyaluronic acid molecules, and there is essentially no free fluid. At other times, however, abnormal conditions can cause edema to occur. These abnormal conditions were discussed in detail in Chapter 31. The principal factors that can cause edema are: (1) increased capillary pressure, (2) decreased plasma colloid osmotic pressure, (3) increased tissue colloid osmotic pressure, and (4) increased permeability of the capillaries. Whenever any one of these conditions occurs, a very high proportion of the extracellular fluid becomes distributed to the interstitial spaces.

Normal Distribution of Fluid Volume Between the Interstitial Spaces and the Vascular System. Figure 36–3 illustrates the approximate normal relationship between extracellular fluid volume and blood volume. In the normal operating range for both the circulatory system and the interstitial fluid system, an increase in extracellular fluid volume is associated with an increase in blood volume of one-sixth to one-third as much. The remainder of the fluid is distributed to the interstitial spaces. However, when the extracellular fluid volume rises considerably above normal, there comes a point, as shown in the figure, at which very little of the additional fluid will remain in the blood — almost all of it instead going into the interstitial spaces. This occurs when the interstitial fluid pressure rises from its normal negative (subatmospheric) value to a positive value, because a positive pressure in the tissues causes them to blow up like a balloon, as was explained in

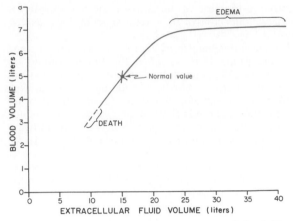

Figure 36–3. Relationship between extracellular fluid volume and blood volume, showing a nearly linear relationship in the normal range but indicating failure of the blood volume to continue rising when the extracellular fluid volume becomes excessive.

Chapter 31 in relation to the discussion of edema. The tissue spaces can then hold as much as 10 to 30 liters of fluid with very little further rise in the interstitial fluid pressure. And these spaces literally become an "overflow" reservoir for excess fluid. This obviously causes edema, but it also acts as an important overflow release valve for the circulatory system, a well-known phenomenon that is utilized daily by the clinician to allow administration of almost unlimited quantities of intravenous fluid and yet not to force the heart into cardiac failure.

To summarize, extracellular fluid volume is controlled simultaneously with the control of blood volume, but the relative ratio of the extracellular fluid volume to blood volume depends upon the physical properties of the circulation and of the interstitial spaces, including their compliances and their dynamics.

CONTROL OF EXTRACELLULAR FLUID SODIUM CONCENTRATION AND EXTRACELLULAR FLUID OSMOLALITY

Relationship of Sodium Concentration to Extracellular Fluid Osmolality. The osmolality of the extracellular fluids (and also of the intracellular fluids, since they remain in osmotic equilibrium with the extracellular fluids) is determined almost entirely by the extracellular fluid sodium concentration. The reason for this is that sodium is by far the most abundant positive ion of the extracellular fluid. Furthermore, the acid-base control mechanisms of the kidneys, which will be discussed in the following chapter, adjust the negative ion concen-

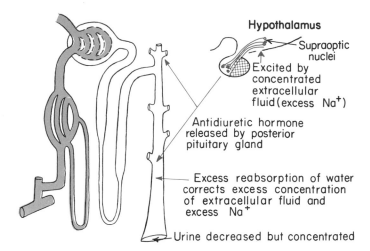

Figure 36–4. Control of extracellular fluid osmolality and sodium ion concentration by the osmo-sodium receptor–antidiuretic hormone feedback control system.

trations of the body fluids to equal those of the positive ions. Also, the glucose and urea, which are the most abundant of the non-ionic osmolar solutes in the extracellular fluids, normally represent only 3 per cent of the total osmolality, and even then the urea exerts very little effective osmotic pressure because it penetrates cells far more easily than does sodium. Therefore, in effect, the sodium ion of the extracellular fluid controls 90 to 95 per cent of the *effective* osmotic pressure of the extracelular fluid. Consequently, we can generally talk in terms of control of sodium concentration and control of osmolality at the same time.

Two separate control systems operate in close association to regulate extracellular sodium concentration and osmolality. These are: (1) the osmo-sodium receptor–antidiuretic hormone system, and (2) the thirst mechanism.

THE OSMO-SODIUM RECEPTOR–ANTIDIURETIC HORMONE FEEDBACK CONTROL SYSTEM

Figure 36–4 illustrates the osmo-sodium receptor–antidiuretic hormone system for control of extracellular fluid sodium concentration and osmolality. It is a typical feedback control system that operates by the following steps:

1. An increase in osmolality (excess sodium and the negative ions that go with it) excites *osmoreceptors* located in the supraoptic nuclei of the hypothalamus.

2. Excitation of the supraoptic nuclei causes release of antidiuretic hormone.

3. The antidiuretic hormone increases the permeability of the collecting ducts, as explained in the previous chapter, and therefore causes increased conservation of water by the kidneys.

4. The conservation of water but loss of sodium and other osmolar substances in the urine causes dilution of the sodium and other substances in the extracellular fluid, thus correcting the initial excessively concentrated extracellular fluid.

Conversely, when the extracellular fluid becomes too dilute (hypo-osmotic), less antidiuretic hormone is formed, and excess water is lost in comparison with the extracellular fluid solutes, thus concentrating the body fluids back toward normal.

The Osmoreceptors (or Osmo-Sodium Receptors). Located in the supraoptic nuclei of the anterior hypothalamus, as shown in Figure 36–5, are specialized neuronal cells called *osmoreceptors*. These respond to changes in osmolality (sodium concentration) of the extracellular fluid. When the extracellular osmolality becomes low, osmosis of water into the osmoreceptors causes them to swell. This decreases their rate of impulse discharge. Conversely, increased osmolality in the extracellular fluid pulls water out of the osmoreceptors, causing them to shrink and thereby to increase their rate of discharge.

The osmoreceptors respond to changes in extracellular fluid sodium concentration but not to changes in potassium concentration and only

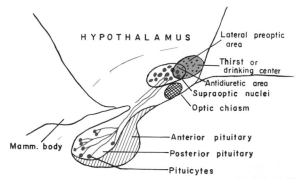

Figure 36–5. The supraoptico-pituitary antidiuretic system and its relationship to the thirst center in the hypothalamus.

slightly to changes in urea and glucose concentrations. Therefore, it must be emphasized again that, for all practical purposes, the osmoreceptors are actually sodium concentration receptors — hence the name *osmo-sodium receptors.*

The impulses from the osmoreceptors are transmitted from the supraoptic nuclei through the pituitary stalk into the posterior pituitary gland where they promote the release of antidiuretic hormone (ADH); the details of the secretion and release of ADH will be discussed in Chapter 75 in relation to the endocrinology of the pituitary gland.

Thus, ADH secretion is controlled by the osmolality (or sodium concentration) of the extracellular fluid — the greater the osmolality, the greater the rate of ADH secretion. The straight line and the open circles of Figure 36–6 illustrate the effect of different levels of extracellular fluid osmolality on antidiuretic hormone concentration in the body fluids, showing a very marked and very important effect on ADH concentration caused by only a 1 to 5 per cent change in osmolality. (The second curve of this figure illustrates the effect of hemorrhage on ADH concentration, an effect discussed earlier in the chapter as one of the mechanisms for blood volume control. Note especially that the loss of blood volume must be very marked before this causes an increase in the ADH concentration; usually a 10 per cent loss in blood volume is required before a significant effect can be observed.)

Water Diuresis.   When a person drinks a large

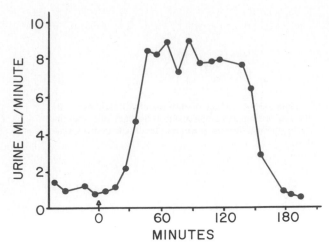

Figure 36–7.   Water diuresis in a human being following ingestion of 1000 ml. of water. (Redrawn from Smith: The Kidney: Structure and Functions in Health and Disease. Oxford University Press, 1951.)

amount of water, a phenomenon called *water diuresis* ensues, a typical record of which is shown in Figure 36–7. In this example, a man drank 1 liter of water, and approximately 30 minutes later his urine output had increased to 8 times normal. It remained at this level for two hours — that is, until the osmolality of the extracellular fluid had returned essentially to normal. The delay in onset of water diuresis is caused partly by delay in absorption of the water from the gastrointestinal tract but mainly by the time required for destruction of the antidiuretic hormone that had already been released by the pituitary gland prior to drinking the water.

Diabetes Insipidus.   Destruction of the supraoptic nuclei or high level destruction of the nerve tract (above the median eminence) from the supraoptic nuclei to the posterior pituitary gland causes antidiuretic hormone secretion to cease or at least to become greatly reduced. When this happens, the person thereafter excretes a dilute urine, and the daily urine volume is increased to 5 to 15 liters per day — a condition called *diabetes insipidus.* In diabetes insipidus, the body fluid volumes remain almost normal so long as the thirst mechanism is still functional because this ordinarily makes the person drink enough water to make up for the increased loss of water in the urine. On the other hand, any factor that prevents adequate intake of fluid, such as unconsciousness, results rapidly in a state of dehydration, tremendous hyperosmolality, and excessive concentration of sodium in the extracellular fluid.

The different lesions in the hypothalamus or pituitary stalk that can cause diabetes insipidus are discussed in Chapter 75.

Syndrome of Inappropriate ADH Secretion.   Certain types of tumors, especially bronchogenic tumors of the lungs or also tumors of the basal regions of the brain, often secrete antidiuretic or a similar hormone. This condition is called the *syndrome of inappropriate ADH secretion.* This excess ADH causes only a slight increase in extracellular fluid volume. Instead, its principal effect is *to decrease greatly the sodium concentration of the extracellular fluid.* The explanation of this

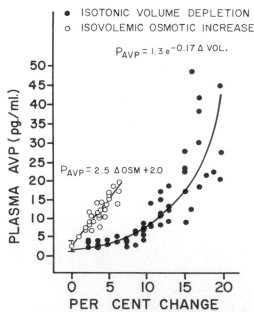

Figure 36–6.   Effect of changes in plasma osmolality or blood volume on the level of plasma ADH (arginine vasopressin — AVP). (From Dunn, et al.: *J. Clin. Invest., 52:*3212, 1973.)

effect is the following: The ADH at first causes a decrease in urine output and a simultaneous slight increase in blood volume. This in turn activates the basic mechanism for blood volume control that was illustrated in Figure 36–2. That is, a slight rise in arterial pressure occurs, and this, combined with the dilution of all the body fluids, then causes a secondary *increase* in urinary output. Furthermore, the urine that is excreted is tremendously concentrated because of the tendency of the kidneys to retain water. Consequently, the kidneys excrete extreme amounts of sodium into the urine but keep the water in the extracellular fluids. Therefore, the sodium concentration becomes seriously reduced, sometimes falling from a normal value of 142 mEq./liter down to as low as 110 to 120 mEq./liter. At values this low, patients frequently die sudden deaths because of coma and convulsions.

This disease is especially instructive because it illustrates the extreme importance of the antidiuretic hormone mechanism for control of sodium concentration, and yet its relatively mild effect on control of body fluid volume.

THIRST, AND ITS ROLE IN CONTROLLING SODIUM CONCENTRATION AND OSMOLALITY

The phenomenon of thirst is equally as important for regulating body water and sodium concentration as is the osmoreceptor-renal mechanism discussed above, because the amount of water in the body at any one time is determined by the balance between both *intake* and *output* of water each day. Thirst, the primary regulator of the intake of water, is defined as the *conscious desire for water.*

Neural Integration of Thirst — the "Thirst" Center

Referring again to Figure 36–5, one sees a small area located slightly anterior to the supraoptic nuclei in the lateral preoptic area of the hypothalamus called the thirst center. Electrical stimulation of this center by implanted electrodes causes an animal to begin drinking within seconds and to continue drinking until the electrical stimulus is stopped. Also, injection of hypertonic salt solutions into the area, which causes osmosis of water out of the neuronal cells and the cells to shrink, also causes drinking. Thus, the neuronal cells of the thirst center function in almost identically the same way as the osmoreceptors of the supraoptic nuclei.

An increase in osmotic pressure of the cerebrospinal fluid in the third ventricle has essentially the same effect — to promote thirst — as an increase of the osmotic pressure of the circulating extracellular fluid. Therefore, it has been suggested that the primary site for detecting changes in osmolality might be at the surface of the third ventricle and that appropriate signals are then transmitted to the thirst center to cause drinking. However, this is still debated.

Basic Stimulus for Exciting the Thirst Center — Intracellular Dehydration. Any factor that will cause *intracellular dehydration* will in general cause the sensation of thirst. The most common cause of this is increased osmolar concentration of the extracellular fluid, especially increased sodium concentration, which causes osmosis of fluid from the neuronal cells of the thirst center. However, another important cause is excessive loss of potassium from the body, which reduces the intracellular potassium of the thirst cells and therefore decreases their volume.

Other Stimuli That Lead to Thirst

Stimulation of Thirst by Angiotensin. Excess angiotensin II in the circulating body fluids often increases the rate of drinking by animals as much as 2- to 3-fold. Therefore, almost any circulatory condition that leads to increased production of angiotensin II will also lead to thirst and drinking. In addition, injection of angiotensin into the third ventricle especially strongly promotes drinking, but this can be prevented by destroying the anteroventral surface of the third ventricle. Therefore, it has been suggested that the major site of action of angiotensin in causing drinking is not on the thirst center itself but instead on a specialized center located immediately beneath the surface of the third ventricle, and that signals are conducted from this area to the thirst center to increase drinking.

Stimulation of Thirst by Hemorrhage and Low Cardiac Output. A small amount of hemorrhage ordinarily does not cause thirst, but loss of as much as 10 per cent of the blood volume, which also reduces the cardiac output significantly, usually does lead to thirst. Also, patients with low cardiac output resulting from cardiac failure frequently develop intense thirst. In both of these conditions excessive amounts of angiotensin II are formed, and it is possible that this is the stimulus for the thirst.

Dryness of the Mouth. Almost everyone is aware that a dry mouth is often associated with thirst. The probable explanation for this is that the same factors that cause intracellular dehydration and therefore stimulate the thirst center also cause a dry mouth. Therefore, we have come to associate a dry mouth with the thirst sensation. However, in opposition to this concept is the fact that in animal experiments in which a dry mouth has been achieved by blocking secretion by the salivary glands, the animals do not drink excessively except under one condition: when they are eating food. This seems to result from the need for lubricating the food and not from thirst. The same effect occurs in human beings whose salivary glands do not secrete saliva.

Temporary Relief of Thirst Caused by the Act of Drinking

A thirsty person receives relief from thirst immediately after drinking water even before the water has been absorbed from the gastrointestinal tract. In fact, in persons who have esophageal fistulae (a condition in which the water never goes into the gastrointestinal tract), partial relief of thirst still occurs following the act of drinking, but this relief is only temporary, and the thirst returns after 15 minutes or more. If the water does enter the stomach, distension of the stomach and other portions of the upper gastrointestinal tract provides still further temporary relief from thirst. For instance, simple inflation of a balloon in the stomach can relieve thirst for 5 to 30 minutes.

One might wonder what the value of this temporary relief from thirst could be, but there is good reason for its occurrence. After a person has drunk water, as long as one-half to one hour may be required for all of the water to be absorbed and distributed throughout the body. Were the thirst sensation not temporarily relieved after drinking of water, the person would continue to drink more and more. When all this water should finally become absorbed, the body fluids would be far more diluted than normal, and an abnormal condition opposite to that which the person was attempting to correct would have been created. It is well known that a thirsty animal almost never drinks more than the amount of water needed to relieve the state of dehydration. Indeed, it is uncanny that the animal usually drinks almost exactly the right amount.

Role of Thirst in Controlling Osmolality and Sodium Concentration of the Extracellular Fluid

Threshold for Drinking — The Tripping Mechanism. The kidneys are continually excreting fluid, and water is also lost by evaporation from the skin and lungs. Therefore, a person is continually being dehydrated, causing the volume of extracellular fluid to decrease and its concentration of sodium and other osmolar elements to rise. When the sodium concentration rises approximately 2 mEq./liter above normal (or the osmolality rises approximately 4 mOsm./liter above normal) the drinking mechanism becomes "tripped" because the person by then reaches a level of thirst that is strong enough to activate the necessary motor effort to cause drinking. The person

ordinarily drinks precisely the required amount of fluid to bring the extracellular fluids back to normal — that is, to a state of *satiety*. Then the process of dehydration and sodium begins again, and the drinking act is tripped again, the process continuing on and on indefinitely.

In this way, both the sodium concentration and the osmolality of the extracellular fluid are very precisely controlled.

COMBINED ROLES OF THE ANTIDIURETIC AND THIRST MECHANISMS FOR CONTROL OF EXTRACELLULAR FLUID SODIUM CONCENTRATION AND OSMOLALITY

When either the antidiuretic hormone mechanism or the thirst mechanism fails, the other ordinarily can still control both sodium concentration and extracellular fluid osmolality with reasonable effectiveness. On the other hand, if both of them fail simultaneously, neither sodium nor osmolality is then adequately controlled.

Figure 36–8 dramatically demonstrates the overall capability of the ADH-thirst system to control extracellular fluid sodium concentration. This figure demonstrates the ability of the same animal to control its extracellular fluid sodium concentration in two different conditions: (1) in the normal

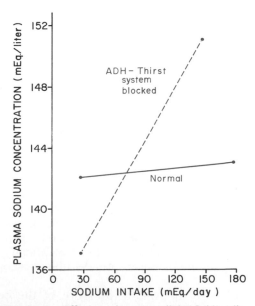

Figure 36–8. Effect on the extracellular fluid sodium concentration in dogs caused by tremendous changes in sodium intake (1) under normal conditions, and (2) after the antidiuretic hormone and thirst feedback systems had been blocked. This figure shows lack of sodium ion control in the absence of these systems. (Courtesy of Dr. David B. Young.)

state, and (2) after both the antidiuretic hormone and thirst mechanisms had been blocked. Note that in the normal animal a six-fold increase in sodium intake caused the sodium concentration to change only two-thirds of 1 per cent (from 142 mEq./liter up to 143 mEq./liter) — an excellent degree of sodium concentration control. Now note the dashed curve of the figure, which shows the change in sodium concentration when the ADH-thirst system was blocked. In this case, sodium concentration increased 10 per cent with only a five-fold increase in sodium intake (a change in sodium concentration from 137 mEq./liter up to 151 mEq./liter), which is an extreme change in sodium concentration when one realizes that the normal sodium concentration rarely rises or falls more than 1 per cent from day to day.

Therefore, the major feedback mechanism for control of sodium concentration (and also for extracellular osmolality) is the ADH-thirst mechanism. In the absence of this mechanism there is no feedback mechanism that will cause the body to increase water ingestion or water conservation by the kidneys when excess sodium enters the body. Therefore, the sodium concentration simply increases.

EFFECT OF ALDOSTERONE ON SODIUM CONCENTRATION

A second hormonal system that plays a *small* role in controlling extracellular fluid sodium concentration is the aldosterone feedback system. Figure 36–9 illustrates how slight the effect of the aldosterone system is in controlling plasma sodium concentration. This figure shows the effect on sodium concentration of more than a six-fold increase in sodium intake in the same dog (a) under normal conditions, and (b) after the aldosterone system had been blocked — that is, the adrenal glands had been removed and the animals were infused at a constant rate of aldosterone that could neither change upward nor downward. Note that in the normal state the sodium concentration changed exactly 1 per cent, while when the aldosterone system was blocked it changed exactly 2 per cent. In other words, even without a functional aldosterone feedback system (because the aldosterone could neither change upward nor downward) sodium concentration was still very well regulated.

Because of the great effect that aldosterone has on tubular sodium reabsorption, this lack of importance of aldosterone for regulation of sodium concentration seems to be a paradox, but it results from the following simple effect: When the aldosterone causes sodium reabsorption from the tubules, as was discussed in the previous chapter, this causes a simultaneous reabsorption of water and an increase in extracellular fluid volume. An increase of only a few per cent in the extracellular fluid volume eventually leads to an increase in arterial pressure, and the increase in arterial pressure to increased glomerular filtration rate, a well known effect in the presence of excess aldosterone. The rapid flow of filtrate down the tubular system then compensates for the excessive reabsorptive effect of the aldosterone and thereby almost completely nullifies the effect of aldosterone on extracellular fluid sodium concentration.

Furthermore, as was explained above, the ADH-thirst system is an extremely powerful controller of sodium concentration — much more powerful than the aldosterone feedback system — so that the ADH-thirst system greatly overshadows the aldosterone system for sodium control under normal conditions. Indeed, even in patients who have primary aldosteronism (these patients secrete tremendous quantities of aldosterone) the sodium concentration still rises only 2 to 3 mEq./liter above normal.

CONTROL OF SODIUM INTAKE — APPETITE AND CRAVING FOR SALT

Maintenance of normal extracellular sodium requires not only the control of sodium excretion but also the control of sodium intake. Unfortunately, we know very little about this except that salt-depleted persons (or persons who have lost blood) develop a desire for salt; as an example, this occurs in persons who have *Addison's disease,* a condition in which the adrenal cortices no longer secrete aldosterone so that the salt stores of the body become depleted. The salt-depleted person craves and eats naturally salty foods. Likewise, it is well known that animals living in areas far removed from the seashore actively search out "salt licks." This craving for salt is analogous to thirst, and it is also analogous to appetite for other types of foods, which is still another homeostatic mechanism that will be discussed in Chapter 73.

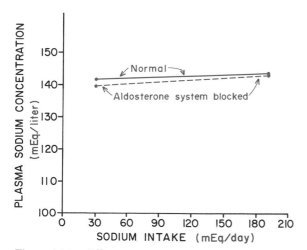

Figure 36–9. Effect on extracellular fluid sodium concentration in dogs caused by tremendous changes in sodium intake (1) under normal conditions, and (2) after the aldosterone feedback system had been blocked. Note that sodium is exceedingly well controlled with or without aldosterone feedback control. (Courtesy of Dr. David B. Young.)

CONTROL OF EXTRACELLULAR POTASSIUM CONCENTRATION — ROLE OF ALDOSTERONE

In the previous chapter, it was pointed out that aldosterone not only causes increased sodium reabsorption by the tubules, but it also causes greatly increased tubular secretion of potassium as well, and therefore increased loss of potassium in the urine. Though the antidiuretic hormone and thirst mechanisms can override aldosterone control of extracellular fluid sodium concentration, as was discussed above, this is not true for the control of potassium concentration. Therefore, aldosterone does play an exceedingly important role in the control of extracellular fluid potassium ion concentration. We will explain this control system in the following sections.

Effect of Potassium Ion Concentration on Rate of Aldosterone Secretion

In a properly functioning feedback system, the factor that is controlled almost invariably has a feedback effect to control the controller; this is precisely true for the aldosterone-potassium control system because the rate of aldosterone secretion is controlled very strongly by the extracellular fluid potassium concentration. Figure 36–10 illustrates the results from a series of dogs in which different rates of potassium infusion were maintained for several weeks while the aldosterone secretion rate was measured. Note the tremen-

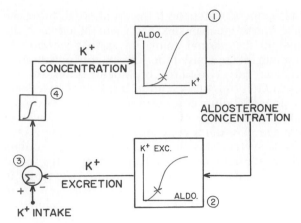

Figure 36–11. Simplified schema of the aldosterone system for control of extracellular fluid potassium concentration.

dous increases in aldosterone secretion rate caused by very minute increases in potassium ion concentration. This extreme feedback effect is the hallmark of a very potent feedback control system.

Basic Mechanism for Aldosterone Control of Potassium Concentration

Putting the effects illustrated in Figure 36–10 together with the fact that aldosterone greatly increases renal excretion of potassium, one can construct a very simple system for negative feedback control of potassium concentration, as illustrated in Figure 36–11. That is, an increase in potassium concentration causes an increase in aldosterone concentration in the circulating blood (Block 1). The increase in aldosterone concentration then causes a marked increase in potassium excretion by the kidneys (Block 2). The increased potassium excretion then decreases the extracellular fluid potassium concentration back toward normal (Blocks 3 and 4).

Importance of the Aldosterone Feedback System for Control of Potassium Concentration. Without a functioning aldosterone feedback system, an animal can easily die from either hypopotassemia or hyperpotassemia.

Figure 36–12 illustrates the potent effect of the aldosterone feedback system to control potassium concentration. In the experiment of this figure, a series of dogs was subjected to an almost seven-fold increase in potassium intake in two different states: (1) the normal state, and (2) after the aldosterone feedback system had been blocked by removing the adrenal glands and the animals given a fixed rate of aldosterone infusion.

Note that in the normal animal the seven-fold increase in potassium intake caused an increase in plasma potassium concentration of only 2.4 per

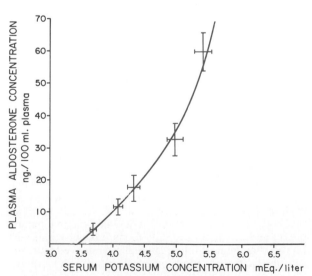

Figure 36–10. Effect on extracellular fluid aldosterone concentration of potassium ion concentration changes. Note the extreme change in aldosterone concentration for very minute changes in potassium concentration. (Courtesy of Dr. R. E. McCaa.)

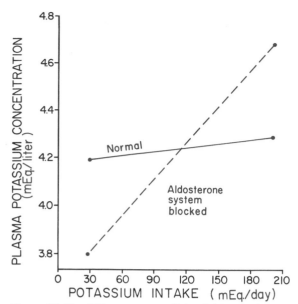

Figure 36–12. Effect on extracellular fluid potassium concentration of tremendous changes in potassium intake (1) under normal conditions, and (2) after the aldosterone feedback system had been blocked. This figure demonstrates that potassium concentration is very poorly controlled after block of the aldosterone system. (Courtesy of Dr. David B. Young.)

cent — from a concentration of 4.2 mEq./liter to 4.3 mEq./liter. Thus, when the aldosterone feedback system was functioning normally, the potassium concentration remained very precisely controlled despite the tremendous change in potassium intake.

On the other hand, the dashed curve in the figure shows the effect after the aldosterone system had been blocked. Note that the same increase in potassium intake now caused a 26 per cent increase in potassium concentration! Thus, the control of potassium concentration in the normal animals was many times as effective as in the animals without an aldosterone feedback mechanism.

Effect of Primary Aldosteronism and Addison's Disease on Extracellular Fluid Potassium Concentration. Primary aldosteronism is caused by a tumor of the zona glomerulosa of one of the adrenal glands, the tumor secreting tremendous quantities of aldosterone. One of the most important effects of this disease is a severe decrease in extracellular fluid potassium concentration, so much so that many of these patients experience paralysis caused by failure of nerve transmission resulting from hyperpolarization of the nerve membranes.

Conversely, in the patient with untreated Addison's disease, whose adrenal glands have been destroyed, the extracellular fluid potassium concentration frequently rises to as high as double

normal. This is often the cause of death in these patients, resulting in cardiac arrest.

OTHER FACTORS THAT AFFECT POTASSIUM ION CONCENTRATION

Other factors that affect the extracellular fluid potassium ion concentration include (1) changes in the hydrogen ion concentration (because hydrogen competes with potassium for secretion by the kidney tubules), and (2) the level of sodium intake (because sodium is reciprocally transported through the late distal tubule and collecting duct epithelium in exchange for potassium). However, the simple experiment illustrated in Figure 36–12, in which potassium ion concentration became almost totally uncontrolled in the absence of a functional aldosterone feedback system, illustrates that the aldosterone mechanism is the prepotent one.

CONTROL OF THE EXTRACELLULAR CONCENTRATIONS OF OTHER IONS

Regulation of Calcium Ion Concentration

The role of calcium in the body and control of its concentration in the extracellular fluid will be discussed in detail in Chapter 79 in relation to the endocrinology of parathyroid hormone, calcitonin, and bone. However, briefly, it is the following:

The day-by-day calcium ion concentration is controlled principally by the effect of parathyroid hormone on bone reabsorption. When the extracellular fluid concentration of calcium falls too low, the parathyroid glands are directly stimulated to promote increased secretion of parathyroid hormone, and this hormone in turn acts directly on the bones to increase the reabsorption of bone salts, thus releasing large amounts of calcium into the extracellular fluid and elevating the calcium level back to normal. On the other hand, when the calcium concentration becomes too great, parathyroid hormone secretion becomes depressed so that almost no bone reabsorption then occurs. Yet, the osteoblastic system to form new bone does continue to deposit calcium, thus removing calcium from the extracellular fluid, in this way reducing the calcium ion concentration back to normal.

However, the bones are not an inexhaustible supply of calcium, and eventually the bones will run out of calcium. Therefore, long-term control of calcium ion concentration results from the effect of parathyroid hormone on reabsorption of calcium from the kidney tubules and absorption of calcium from the gut through the gastrointestinal mucosa, both of which effects are markedly increased by parathyroid hormone.

The effect of parathyroid hormone on calcium handling by the kidneys almost exactly parallels the effect of aldosterone on sodium handling by the kidneys. That is, even in the absence of parathyroid hormone, most of the calcium is reabsorbed from the tubular fluid in the proximal tubules, the loop of Henle, and the diluting segment of the distal tubules, but about 10 per cent of the filtered load of calcium still remains to enter the late distal tubules. Then, if large amounts of parathyroid hormone are present in the body fluids, essentially all the remaining calcium will be reabsorbed from the late distal tubules and collecting ducts, thus conserving the calcium in the body. So the principal mechanism for long-term control of calcium ion concentration is: low calcium in the extracellular fluid leads to parathyroid hormone secretion and the parathyroid hormone then promotes intense conservation of calcium by the kidneys and also greatly increased calcium absorption from the gastrointestinal tract. These effects will be discussed in more detail in Chapter 79.

Regulation of Magnesium Ion Concentration

Much less is known about the regulation of magnesium ion concentration than of calcium ion concentration. However, magnesium ions are reabsorbed by all portions of the renal tubules, and it is the late distal tubules and collecting ducts that finally control urine excretion of magnesium. When the extracellular fluid magnesium concentration is high, excess amounts of magnesium are excreted, and, conversely, when the magnesium concentration is low, magnesium is conserved.

Unfortunately, the feedback mechanisms by which extracellular fluid magnesium ion concentration controls urinary excretion of magnesium are not yet known. It is interesting, however, that excess aldosterone causes increased loss of magnesium in the urine in the same way that it causes increased loss of potassium. It is also known that magnesium handling by most body cells is similar to the handling of potassium. Thus, there are at least several similarities between the control of potassium and the control of magnesium.

Regulation of Phosphate Concentration

Phosphate concentration is regulated primarily by an *overflow* mechanism, which can be explained as follows: The renal tubules have a normal transport maximum of 0.1 millimole of phosphate per minute. When less than this "load" of phosphate is present in the glomerular filtrate, all of it is reabsorbed. When more than this amount is present, the excess is excreted. Therefore, normally, phosphate ion spills into the urine when its concentration in the plasma is above the threshold value of approximately 0.8 millimole/liter. Any time the concentration falls below this value, all the phosphate is conserved in the plasma, and the daily ingested phosphate accumulates in the extracellular fluid until its concentration rises above the threshold. On the other hand, whenever the phosphate concentration rises above this level, the excess is excreted into the urine. Since most people ingest large quantities of phosphate day in and day out, either in milk or in meat, the concentration of phosphate is usually maintained at a level of about 1.0 millimole/liter, a level at which there is continual overflow of excess phosphate into the urine.

Role of Parathyroid Hormone in Phosphate Ion Regulation. Parathyroid hormone, which plays a major role in

regulation of calcium ion concentration as explained above, also affects phosphate ion concentration in two different ways. First, parathyroid hormone promotes bone reabsorption, thereby dumping large quantities of phosphate ions into the extracellular fluid from the bone salts. Second, this hormone decreases the transport maximum for phosphate by the renal tubules so that a greater proportion of the tubular phosphate is lost in the urine. The combination of these factors causes marked loss of phosphates in the urine. The interrelationships between phosphate, calcium, and parathyroid hormone control will be discussed in Chapter 79.

Regulation of Other Negative Ions. Other important negative ions in the body fluid include sulfates, nitrates, urates, lactates, and the amino acids. Essentially all of these, like phosphate, have definite transport maximums. When the concentration of each is below its respective threshold, it is conserved in the extracellular fluid, but when above this threshold the excess spills into the urine. Thus, the concentrations of most of these negative ions are regulated by the overflow mechanism in the same way that phosphate ion concentration is regulated.

REFERENCES

Andersson, B.: Regulation of body fluids. *Annu. Rev. Physiol., 39*:185, 1977.

Andersson, B.: Regulation of water intake. *Physiol. Rev., 58*:582, 1978.

Andreoli, T. E., and Schaefer, J. A.: Mass transport across cell membranes: The effects of antidiuretic hormone on water and solute flows in epithelia. *Annu. Rev. Physiol., 38*:451, 1976.

Brenner, B. M., and Stein, J. H. (eds.): Acid-Base and Potassium Homeostasis. New York, Churchill Livingstone, 1978.

Brenner, B. M., and Stein, J. H. (eds.): Hormonal Function and the Kidney. New York, Churchill Livingstone, 1979.

Crawford, M. P., *et al.*: Renal servocontrol of arterial blood pressure. *J. Appl. Physiol., 22*:139, 1967.

Cross, B. A., and Wakerley, J. B.: The neurohypophysis. *Int. Rev. Physiol., 16*:1, 1977.

Dennis, V. W., *et al.*: Renal handling of phosphate and calcium. *Annu. Rev. Physiol., 41*:257, 1979.

Earley, L. E., and Schrier, R. W.: Intrarenal control of sodium excretion by hemodynamic and physical factors. *In* Orloff, F., and Berliner, R. W. (eds.): Handbook of Physiology. Sec. 8. Baltimore, Williams & Wilkins, 1973, p. 721.

Ehrlich, E. N.: Adrenocortical regulation of salt and water metabolism: Physiology, pathophysiology, and clinical syndromes. *In* DeGroot, L. J., *et al.* (eds.): Endocrinology. Vol. 3. New York, Grune & Stratton, 1979, p. 1883.

Eisenbach, G. M., and Brod, J. (eds.): Non-Vasoactive Renal Hormones. New York, S. Karger, 1978.

Eisenbach, G. M., and Brod, J. (eds.): Vasoactive Renal Hormones. New York, S. Karger, 1978.

Epstein, M.: Renal effects of head-out water immersion in man: Implications for an understanding of volume homeostasis. *Physiol. Rev., 58*:529, 1978.

Fitzsimons, J. T.: Thirst. *Physiol. Rev., 52*:468, 1972.

Gauer, O. H., and Henry, J. P.: Neurohormonal control of plasma volume. *Int. Rev. Physiol., 9*:145, 1976.

Gertz, K. H., and Boylan, J. W.: Glomerular-tubular balance. *In* Orloff, F., and Berliner, R. W. (eds.): Handbook of Physiology. Sec. 8. Baltimore, Williams & Wilkins, 1973, p. 763.

Goetz, K. L., *et al.*: Atrial receptors and renal function. *Physiol. Rev., 55*:157, 1975.

Gottschalk, C. W.: Renal nerves and sodium excretion. *Annu. Rev. Physiol., 41*:229, 1979.

Gottschalk, C. W., and Mylle, M.: Micropuncture study of the mammalian urinary concentrating mechanism: Evidence for the countercurrent hypothesis. *Am. J. Physiol., 196*:927, 1959.

Grantham, J. J.: Action of antiduretic hormone in the mammalian kidney. *In* MTP International Review of Science: Physiology. Vol. 6. Baltimore, University Park Press, 1974, p. 247.

Guignard, J. P., and Filloux, B.: Studies on compensatory adaptation of renal functions. *Yale J. Biol. Med., 51*(3):247, 1978.

Guyton, A. C., *et al.*: Theory for renal autoregulation by feedback at the juxtaglomerular apparatus. *Circ. Res., 14*:187, 1964.

Guyton, A. C., *et al.*: Dynamics and Control of the Body Fluids. Philadelphia, W. B. Saunders Co., 1975.

Guyton, A. C., *et al.*: A systems analysis of volume regulation. Alfred Benzon Symposium XI. Munksgaard, 1978.

Hall, J. E., *et al.*: Control of glomerular filtration rate by renin-angiotensin system. *Am. J. Physiol., 233*:F366, 1977.

Hall, J. E., *et al.*: Dissociation of renal blood flow and filtration rate autoregulation by renin depletion. *Am. J. Physiol., 232*:F215, 1977.

Hall, J. E., *et al.*: Intrarenal control of electrolyte excretion by angiotensin II. *Am. J. Physiol., 232*:F538, 1977.

Hall, J. E., *et al.*: Intrarenal role of angiotensin II and [des-Asp^1] angiotensin II. *Am. J. Physiol., 236*:F252, 1979.

Hayslett, J. P.: Functional adaptation of reduction in renal mass. *Physiol. Rev., 59*:137, 1979.

Hayward, J. N.: Neural control of the posterior pituitary. *Annu. Rev. Physiol., 37*:191, 1975.

Hierholzer, K., and Lange, S.: The effects of adrenal steroids on renal function. *In* MTP International Review of Science: Physiology. Vol. 6. Baltimore, University Park Press, 1974, p. 273.

Jackson, T. E., *et al.*: Transient response of glomerular filtration rate and renal blood flow to step changes in arterial pressure. *Am. J. Physiol., 233*:F396, 1977.

Jamison, R. L.: Countercurrent system. *In* MTP International Review of Science: Physiology. Vol. 6. Baltimore, University Park Press, 1974, p. 199.

Katz, A. I., and Lindheimer, M. D.: Actions of hormones on the kidney. *Annu. Rev. Physiol., 39*:97, 1977.

Kaufmann, W., and Krause, D. K. (eds.): Central Nervous Control of NA$^+$ [NA] Balance: Relations to the Renin-Angiotensin System. Stuttgart, Thieme, 1976.

Kleeman, C. R., and Vorherr, H.: Water metabolism and the neurohypophysial hormones. *In* Bondy, P. K., and Rosenberg, L. E. (eds.): Duncan's Diseases of Metabolism, 7th Ed. Philadelphia, W. B. Saunders Co., 1974, p. 1459.

Knox, F. G., and Diaz-Buxo, J. A.: The hormonal control of sodium excretion. *Int. Rev. Physiol., 16*:173, 1977.

Lee, J., and deWardener, H. E.: Neurosecretion and sodium excretion. *Kidney Int., 6*:323, 1974.

Linden, R. J.: Neurocirculatory control of sodium and water excretion. *In* Dickinson, C. J., and Marks, J. (eds.): Developments in Cardiovascular Medicine. Lancaster, England, MTP Press Limited, 1978, p. 191.

Livingston, D. M., and Wacker, W. E. C.: Magnesium metabolism. *In* Greep, R. O., and Astwood, G. D. (eds.): Handbook of Physiology. Sec. 7, Vol. 7. Baltimore, Williams & Wilkins, 1976, p. 215.

Maher, J. F., and Bartter, F. C.: Maintenance of dynamic equilibrium of body fluids and electrolytes. *In* Frohlich, E. D. (ed.): Pathophysiology, 2nd Ed. Philadelphia, J. B. Lippincott Co., 1976, p. 241.

Maxwell, M. H., and Kleeman, C. R. (eds.): Clinical Disorders of Fluid and Electrolyte Metabolism, 3rd Ed. New York, McGraw-Hill, 1979.

McCaa, R. E., *et al.*: Increased plasma aldosterone concentration in response to hemodialysis in nephrectomized man. *Circ. Res., 31*:473, 1972.

Moses, A. M., and Miller, M.: Osmotic influences on the release of vasopressin. *In* Greep, R. O., and Astwood, E. B. (eds.): Handbook of Physiology. Sec. 7. Vol. 4, Part 1. Baltimore, Williams & Wilkins, 1974, p. 225.

Navar, L. G., *et al.*: Effect of alterations in plasma osmolality on renal blood flow autoregulation. *Am. J. Physiol., 211*:1387, 1966.

Ott, C. E., *et al.*: Pressures in static and dynamic states from capsules implanted in the kidney. *Am. J. Physiol., 34*:235, 1971.

Robertson, G. L.: Vasopressin in osmotic regulation in man. *Annu. Rev. Med., 25*:315, 1974.

Sawyer, W. H.: The mammalian antidiuretic response. *In* Greep, R. O., and Astwood, E. B. (eds.): Handbook of Physiology. Sec. 7, Vol. 4, Part 1. Baltimore, Williams & Wilkins, 1974, p. 443.

Schmidt-Nielsen, B., and Laws, D. F.: Invertebrate mechanisms for diluting and concentrating the urine. *Annu. Rev. Physiol., 25*:631, 1963.

Scriabine, A., *et al.* (eds.): Prostaglandins in Cardiovascular and Renal Function. Jamaica, N.Y., Spectrum Publications, 1980.

Share, L., and Claybaugh, J. R.: Regulation of body fluids. *Annu. Rev. Physiol., 34*:235, 1972.

Sharp, G. W. G., and Leaf, A.: Effects of aldosterone and its mechanism of action on sodium transport. *In* Orloff, F., and Berliner, R. W. (eds.): Handbook of Physiology. Sec. 8. Baltimore, Williams & Wilkins, 1973, p. 815.

Smith, M. J., Jr., *et al.*: Acute and chronic effects of vasopressin on blood pressure, electrolytes, and fluid volumes. *Am. J. Physiol., 237*(3):F232, 1979.

Stein, J. H., and Reineck, H. J.: Effect of alterations in extracellular fluid volume on segmental sodium transport. *Physiol. Rev., 55*:127, 1975.

Trippodo, N. C., *et al.*: Effect of prolonged angiotensin II infusion on thirst. *Am. J. Physiol., 230*:1063, 1976.

Trippodo, N. C., *et al.*: Intrarenal role of angiotensin II in controlling sodium excretion during dehydration in dogs. *Clin. Sci. Molec. Med., 52*:545, 1977.

Vaamonde, C. A., and Papper, S.: Maintenance of body tonicity. *In* Frohlich, E. D. (ed.): Pathophysiology, 2nd Ed. Philadelphia, J. B. Lippincott Co., 1976, p. 265.

Valtin, H.: Renal Dysfunction: Mechanisms Involved in Fluid and Solute Imbalance. Boston, Little, Brown, 1979.

Verney, E. B.: Absorption and excretion of water; Antidiuretic hormone. *Lancet, 2*:739, 1946.

Weitzman, R., and Kleeman, C. R.: Water metabolism and the neurohypophysial hormones. *In* Bondy, P. K., and Rosenberg, L. E. (eds.): Metabolic Control and Disease, 8th Ed. Philadelphia, W. B. Saunders Co., 1980, p. 1241.

Windhager, E. E.: Glomerulo-tubular balance of salt and water. *Physiologist, 11*:103, 1968.

Wolf, A. V.: Thirst: Physiology of the Urge to Drink and Problems of Water Lack. Springfield, Ill., Charles C Thomas, 1958.

Wolf, G., McGovern, J. F., and Dicara, L. V.: Sodium appetite: Some conceptual and methodologic aspects of a model drive system. *Behav. Biol., 10*:27, 1974.

Wright, F. S.: Potassium transport by the renal tubule. *In* MTP International Review of Physiology. Vol. 6. Baltimore, University Park Press, 1974, p. 79.

Young, D. B., *et al.*: Effectiveness of the aldosterone-sodium and -postassium feedback control system. *Am. J. Physiol., 231*:945, 1976.

Young, D. B., *et al.*: Control of extracellular sodium concentration by antidiuretic hormone–thirst feedback mechanism. *Am. J. Physiol., 232*:R145, 1977.

Zerbe, R., *et al.*: Vasopressin function in the syndrome of inappropriate antidiuresis. *Annu. Rev. Med., 31*:315, 1980.

37

Regulation of Acid-Base Balance

When one speaks of the regulation of acid-base balance, regulation of hydrogen ion concentration in the body fluids is actually meant. The hydrogen ion concentration in different solutions can vary from less than 10^{-14} equivalents per liter to higher than 10^0, which means a total variation of more than a quadrillion-fold. On a logarithmic basis, the hydrogen ion concentration in the human body is approximately midway between these two extremes.

Only slight changes in hydrogen ion concentration from the normal value can cause marked alterations in the rates of chemical reactions in the cells, some being depressed and others accelerated. For this reason the regulation of hydrogen ion concentration is one of the most important aspects of homeostasis. Later in the chapter the overall effects of high hydrogen ion concentration (acidosis) and low hydrogen ion concentration (alkalosis) are discussed. In general, when people become acidotic they are likely to die in coma, and when they become alkalotic they may die of tetany or convulsions.

Normal Hydrogen Ion Concentration and Normal pH of the Body Fluids — Acidosis and Alkalosis. The hydrogen ion concentration in the extracellular fluid is normally regulated at a constant value of approximately 4×10^{-8} Eq./liter; this value can vary from as low as 1.0×10^{-8} to as high as 1.0×10^{-7} without causing death.

From these values, it is already apparent that expressing hydrogen ion concentration in terms of actual concentrations is a cumbersome procedure. Therefore, the symbol *pH* has come into usage for expressing the concentration, and pH is related to actual hydrogen ion concentration by the following formula (when H^+ conc. is expressed in equivalents per liter):

$$pH = \log \frac{1}{H^+ \text{ conc.}} = -\log H^+ \text{ conc.} \qquad (1)$$

Note from this formula that a low pH corresponds to a high hydrogen ion concentration, which is called *acidosis;* and, conversely, a high pH corresponds to a low hydrogen ion concentration, which is called *alkalosis*.

The normal pH of arterial blood is 7.4, while the pH of venous blood and of interstitial fluids is about 7.35 because of extra quantities of carbon dioxide that form carbonic acid in these fluids.

Since the normal pH of the arterial blood is 7.4, a person is considered to have acidosis whenever the pH is below this value and to have alkalosis when it rises above 7.4. The lower limit at which a person can live more than a few hours is about 7.0, and the upper limit approximately 8.0.

Intracellular pH. On the basis of indirect measurements, it has been found that the intracellular pH usually ranges between 6.0 and 7.4 in different cells, perhaps averaging about 7.0. A *rapid rate of metabolism* in cells increases the rate of carbon dioxide formation and consequently decreases pH. Also, *poor blood flow* to any tissue causes carbon dioxide accumulation and a decrease in pH.

DEFENSE AGAINST CHANGES IN HYDROGEN ION CONCENTRATION

To prevent acidosis or alkalosis, several special control systems are available: (1) All the body fluids are supplied with acid-base *buffer systems* that immediately combine with any acid or alkali and thereby prevent excessive changes in hydrogen ion concentration. (2) If the hydrogen ion concentration does change measurably, the *respiratory center is immediately stimulated* to alter the rate of breathing. As a result, the rate of carbon dioxide removal from the body fluids is automatically changed, and, for reasons that will be presented later, this causes the hydrogen ion concentration to return toward normal. (3) When the hydrogen ion concentration changes from normal, *the kidneys excrete either an acid or alkaline urine,* thereby also helping to readjust the hydrogen ion concentration of the body fluids back toward normal.

The buffer systems can act within a fraction of a second to prevent excessive changes in hydrogen ion concentration. On the other hand, it takes 1 to 15 minutes for the respiratory system to readjust the hydrogen ion concentration after a sudden change has occurred. Finally, the kidneys, though providing the most powerful of all the acid-base regulatory systems, require several hours to several days to readjust the hydrogen ion concentration.

FUNCTION OF ACID-BASE BUFFERS

An acid-base buffer is a solution of two or more chemical compounds that prevents marked changes in hydrogen ion concentration when either an acid or a base is added to the solution. As an example, if only a few drops of concentrated hydrochloric acid are added to a beaker of pure water, the pH of the water might immediately fall to as low as 1.0. However, if a satisfactory buffer system is present, the hydrochloric acid combines instantaneously with the buffer, and the pH falls only slightly. Perhaps the best way to explain the action of an acid-base buffer is to consider an actual simple buffer system, such as the bicarbonate buffer, which is extremely important in regulation of acid-base balance in the body.

THE BICARBONATE BUFFER SYSTEM

A typical bicarbonate buffer system consists of a *mixture* of carbonic acid (H_2CO_3) and sodium bicarbonate ($NaHCO_3$) in the same solution. It must first be noted that carbonic acid is a very weak acid for two reasons: First, its degree of dissociation into hydrogen ions and bicarbonate ions is poor in comparison with that of many other acids. Second, about 999 parts out of 1000 of any carbonic acid in a solution almost immediately dissociate into carbon dioxide and water, the net result being a high concentration of dissolved carbon dioxide but only a weak concentration of acid.

When a strong acid, such as hydrochloric acid, is added to a buffer solution containing bicarbonate salt, the following reaction takes place:

$$HCl + NaHCO_3 \rightarrow H_2CO_3 + NaCl \qquad (2)$$

From this equation it can be seen that the strong hydrochloric acid is converted into the very weak carbonic acid. Therefore, the HCl lowers the pH of the solution only slightly.

On the other hand, if a strong base, such as sodium hydroxide, is added to a buffer solution

containing carbonic acid, the following reaction takes place:

$$NaOH + H_2CO_3 \rightarrow NaHCO_3 + H_2O \qquad (3)$$

This equation shows that the hydroxyl ion of the sodium hydroxide combines with the hydrogen ion from the carbonic acid to form water and that the other product formed is sodium bicarbonate. The net result is exchange of the strong base NaOH for the weak base $NaHCO_3$.

Though this bicarbonate buffer system has been illustrated in the above reactions as a *mixture* of carbonic acid and *sodium* bicarbonate, in the intracellular fluid, where little sodium bicarbonate is present, the bicarbonate ion is provided mainly as potassium and magnesium bicarbonate.

Quantitative Dynamics of Buffer Systems

Dissociation of Carbonic Acid. All acids are ionized to a certain extent, and the percentage of ionization is called the *degree of dissociation*. Equation 4 illustrates the reversible relationship between undissociated carbonic acid and the two ions that it forms, H^+ and HCO_3^-.

$$H_2CO_3 \rightleftarrows H^+ + HCO_3^- \qquad (4)$$

A physicochemical law has been found to apply to the dissociation of all acids, but as applied in this case specifically to carbonic acid, is expressed by the following formula:

$$\frac{H^+ \times HCO_3^-}{H_2CO_3} = K \qquad (5)$$

This formula states that in any given carbonic acid solution the concentration of hydrogen ions times the concentration of bicarbonate ions divided by the concentration of undissociated carbonic acid is equal to a constant, K.

However, it is almost impossible to measure the concentration of undissociated carbonic acid in a solution because it is also continually in reversible equilibrium with dissolved carbon dioxide in the solution. Ordinarily, the amount of dissolved carbon dioxide is approximately 1000 times the concentration of the undissociated acid. On the other hand, it is possible to measure the total amount of dissolved carbon dioxide, and, since the amount of undissociated carbonic acid is proportional to the amount of dissolved carbon dioxide, Formula 5 above can also be expressed as follows:

$$\frac{H^+ \times HCO_3^-}{CO_2} = K' \qquad (6)$$

The only real difference between the above two formulas is that the constant K is approximately 1000 times the constant K'.

Formula 6 can be changed into the following form:

$$H^+ = K' \cdot \frac{CO_2}{HCO_3^-} \qquad (7)$$

If we take the logarithm of each of the two sides of Formula 7, it becomes the following:

$$\log H^+ = \log K' + \log \frac{CO_2}{HCO_3^-} \qquad (8)$$

Now, the signs of the log H^+ and of log K are changed from positive to negative and the carbon dioxide and bicarbonate are inverted in the last term, which is the same as changing its sign also, giving the following formula:

$$-\log H^+ = -\log K' + \log \frac{HCO_3^-}{CO_2} \qquad (9)$$

It will be recalled from earlier in the chapter that $-\log H^+$ is equal to the pH of the solution. Likewise, $-\log K$ is called the pK of a buffer. Therefore, this formula can be changed still further to the following:

$$pH = pK' + \log \frac{HCO_3^-}{CO_2} \qquad (10)$$

The Henderson-Hasselbalch Equation. For the bicarbonate buffer system the pK is 6.1, and Formula 10 may be expressed as follows:

$$pH = 6.1 + \log \frac{HCO_3^-}{CO_2} \qquad (11)$$

This is called the Henderson-Hasselbalch equation, and by using it one can calculate the pH of a solution with reasonable accuracy if the molar concentrations of bicarbonate ion and dissolved carbon dioxide are known. If the bicarbonate concentration is equal to the dissolved carbon dioxide concentration, the second member of the right-hand portion of the equation becomes log of 1, which is equal to zero. Therefore, under these conditions the pH of the solution is equal to the pK.

From the Henderson-Hasselbalch equation one can readily see that an increase in bicarbonate ion concentration causes the pH to rise, or, in other words, shifts the acid-base balance toward the alkaline side. On the other hand, an increase in the concentration of dissolved carbon dioxide decreases the pH, or shifts the acid-base balance toward the acid side. It will be apparent later in this chapter that one can change the concentration of dissolved CO_2 in the body fluids by increasing or decreasing the rate of respiration. In this way the respiratory system can change to a certain extent the pH of the body fluids. On the other hand, the kidneys can increase or decrease the concentration of bicarbonate ion in the body fluids, in this way increasing or decreasing the pH. Thus, these two major hydrogen ion regulatory systems operate principally by altering one or the other of the two elements of the bicarbonate buffer system.

The Reaction Curve of the Bicarbonate Buffer System. Figure 37–1 shows the changes in pH of the body fluids when the ratio of bicarbonate ion to carbon dioxide changes. Note that when the concentrations of the two elements of the buffer are equal, the pH of the solution is 6.1, which is equal to the pK' of the bicarbon-

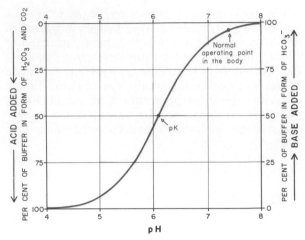

Figure 37–1. Reaction curve for the bicarbonate buffer system.

ate buffer system. When base is added to the buffer a large proportion of the dissolved carbon dioxide and carbonic acid is converted into bicarbonate ions, and the ratio is altered. As a result, the pH rises as indicated by the forward slope of the curve. On the other hand, when acid is added, a large proportion of the bicarbonate ion is converted first into carbonic acid and then into dissolved carbon dioxide so that the ratio changes in favor of the acidic side, and the pH falls, as illustrated by the downslope of the curve.

Buffering Power of the Bicarbonate Buffer System. Referring once again to Figure 37–1, note that at the central point of the curve addition of a slight amount of acid or base causes minimal change in pH. However, toward each end of the curve addition of a slight amount of acid or base causes the pH to change greatly. Thus, the so-called *buffering power* of the buffer system *is greatest when the pH is equal to the pK'*, which is in the exact center of the curve. The buffering power is still reasonably effective until the ratio of one element of the buffer system to the other reaches as much as 8:1 or 1:8, but beyond these limits the buffering power diminishes rapidly. And, when all the carbon dioxide has been converted into bicarbonate ion or when all the bicarbonate ion has been converted into carbon dioxide, the system has no more buffering power at all.

A second factor that determines the buffering power is the concentrations of the two elements in the buffer solution. Obviously, if the concentrations are slight, only a small amount of acid or base need be added to the solution to change the pH considerably. Thus, *the buffering power of a buffer is also directly proportional to the concentrations of the buffer substances.*

THE BUFFER SYSTEMS OF THE BODY FLUIDS

The three major buffer systems of the body fluids are the *bicarbonate buffer,* which has been described above, the *phosphate buffer,* and the *protein buffer.* Each of these performs major buffering functions under different conditions.

The Bicarbonate Buffer System. The bicarbonate system is not an exceedingly powerful buffer for two reasons. First, the pH in the extracellular fluids is about 7.4, while the pK' of the bicarbonate buffer system is 6.1. This means that approximately 20 times as much of the bicarbonate buffer is in the form of bicarbonate ion as in the form of dissolved carbon dioxide. For this reason, the system operates on a portion of its buffering curve where the buffering power is poor. Second, the concentrations of the two elements of the bicarbonate system, CO_2 and HCO_3^-, are not great.

Yet, despite the fact that the bicarbonate buffer system is not especially powerful, it is probably equally as important as all the others in the body because the concentrations of each of the two elements of the bicarbonate system can be regulated, carbon dioxide by the respiratory system and bicarbonate ion by the kidneys. As a result, the pH of the blood can be shifted up or down by the respiratory and renal regulatory systems.

The Phosphate Buffer System. The phosphate buffer system acts in almost identically the same manner as the bicarbonate buffer system, but it is composed of the following two elements: $H_2PO_4^-$ and HPO_4^{--}. When a strong acid, such as hydrochloric acid, is added to a mixture of these two substances, the following reaction occurs:

$$HCl + Na_2HPO_4 \rightarrow NaH_2PO_4 + NaCl \quad (12)$$

The net result of this reaction is that the hydrochloric acid is removed, and in its place an additional quantity of NaH_2PO_4 is formed. NaH_2PO_4 is only weakly acidic, so that the added strong acid is immediately traded for a very weak acid, and the pH changes relatively slightly.

Conversely, if a strong base, such as sodium hydroxide, is added to the buffer system, the following reaction takes place:

$$NaOH + NaH_2PO_4 \rightarrow Na_2HPO_4 + H_2O \quad (13)$$

Here sodium hydroxide is decomposed to form water and Na_2HPO_4. That is, a strong base is traded for the very weak base Na_2HPO_4, allowing only a slight shift in pH toward the alkaline side.

The phosphate buffer system has a pK of 6.8, which is not far from the normal pH of 7.4 in the body fluids; this allows the phosphate system to operate near its maximum buffering power. However, despite the fact that this buffer system operates in a reasonably good portion of the buffer curve, its concentration in the extracellular fluid is only one-twelfth that of the bicarbonate buffer. Therefore, its total buffering power *in the extracellular fluid* is even far less than that of the bicarbonate system.

On the other hand, the phosphate buffer is especially important in the tubular fluids of the kidneys for two reasons: First, phosphate usually becomes greatly concentrated in the tubules, thereby also greatly increasing the buffering power of the phosphate system. Second, the tubular fluid usually becomes more acidic than the extracellular fluid, bringing the operating range of the buffer closer to the pK of the system.

The phosphate buffer is also very important in the intracellular fluids because the concentration of phosphate in these fluids is many times that in the extracellular fluids, and also because the pH of the intracellular fluids is usually closer to the pK of the phosphate buffer system than is the pH of the extracellular fluid.

The Protein Buffer System The most plentiful buffer of the body is the proteins of the cells and plasma, mainly because of their very high concentrations. There is a slight amount of diffusion of hydrogen ions through the cell membrane, and even more important, carbon dioxide can diffuse readily through cell membranes and bicarbonate ions can diffuse to some extent (they require several hours to come to equilibrium in most cells other than the red blood cells). The diffusion of the two elements of the bicarbonate buffer system causes the pH in the intracellular fluids to change approximately in proportion to the changes in pH in the extracellular fluids. Thus, all the buffer systems inside the cells help to buffer the extracellular fluids as well. These include the extremely large amounts of proteins inside the cells. Indeed, experimental studies have shown that about three quarters of all the *chemical* buffering power of the body fluids is inside the cells and most of this results from the intracellular proteins. However, except for the red blood cells, the slowness of movement of hydrogen and bicarbonate ions through the cell membranes often delays the ability of the intracellular buffers to buffer extracellular acid-base abnormalities for several hours.

The method by which the protein buffer system operates is precisely the same as that of the bicarbonate buffer system. It will be recalled that a protein is composed of amino acids bound together by peptide linkages, but some of the different amino acids have free acidic radicals that can dissociate into base plus H^+. Furthermore, the pKs of a few of the protein buffering systems are not far from 7.4. This, too, helps to make the protein buffering systems the most powerful of the body.

THE ISOHYDRIC PRINCIPLE

Each of the above buffer systems has been discussed as if it could operate individually in the body fluids. However, they all actually work together, for the hydrogen is common to the chemical reactions of all the systems. Therefore, whenever any condition causes the hydrogen ion concentration to change, it causes the balance of all the buffer systems to change at the same time. This phenomenon, called the *isohydric principle,* is presented by the following formula:

$$H^+ = \frac{K_1 \times HA_1}{A_1^-} = \frac{K_2 \times HA_2}{A_2^-} = \frac{K_3 \times HA_3}{A_3^-} \quad (14)$$

in which K_1, K_2, and K_3 are the dissociation constants of three respective acids, HA_1, HA_2, and HA_3, and A_1^-, A_2^-, and A_3^- are the concentrations of the free negative ions of the acids.

The important feature of this principle is that any condition that changes the balance of any one of the buffer systems also changes the balance of all the others, for *the buffer systems actually buffer each other.*

RESPIRATORY REGULATION OF ACID-BASE BALANCE

In the discussion of the Henderson-Hasselbalch equation, it was noted that an *increase in carbon dioxide* concentration in the body fluids *decreases the pH* toward the acidic side, whereas a decrease in carbon dioxide raises the pH toward the alkaline side. It is on the basis of this effect that the respiratory system is capable of altering the pH either up or down.

Balance Between Metabolic Formation of Carbon Dioxide and Pulmonary Expiration of Carbon Dioxide. Carbon dioxide is continually being formed in the body by the different intracellular metabolic processes, the carbon in the foods being oxidized by oxygen to form carbon dioxide. This in turn diffuses into the interstitial fluids and blood, and is transported to the lungs where it diffuses into the alveoli and is transferred to the atmosphere by pulmonary ventilation. However, several minutes are required for this passage of carbon dioxide from the cells to the atmosphere so that an average of 1.2 millimoles/liter of dissolved carbon dioxide is normally in the extracellular fluids at all times.

If the rate of metabolic formation of carbon dioxide becomes increased, its concentration in the extracellular fluids is likewise increased. Conversely, decreased metabolism decreases the carbon dioxide concentration.

On the other hand, if the rate of pulmonary ventilation is increased, carbon dioxide is blown off from the lungs, and the amount of carbon dioxide in the extracellular fluids decreases.

Effect of Increasing or Decreasing the Alveolar Ventilation on pH of the Extracellular Fluids

If we assume that the rate of metabolic formation of carbon dioxide remains constant, then the

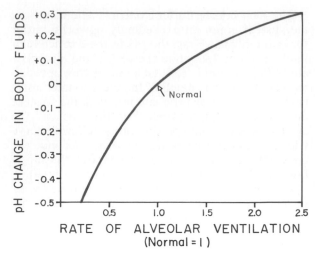

Figure 37–2. Approximate change in body fluid pH caused by increased or decreased rate of alveolar ventilation.

only factor that affects the carbon dioxide concentration in the body fluids is the rate of alveolar ventilation as expressed by the following formula:

$$CO_2 \propto \frac{1}{\text{Alveolar ventilation}} \quad (15)$$

And, since an increase in carbon dioxide decreases the pH, changes in alveolar ventilation also change the hydrogen ion concentration.

Figure 37–2 illustrates the approximate change in pH in the blood that can be effected by increasing or decreasing the rate of alveolar ventilation. Note that an increase in alveolar ventilation to 2 times normal raises the pH of the extracellular fluids by about 0.23 pH unit. This means that if the pH of the body fluids has been 7.4 with normal alveolar ventilation, doubling the ventilation raises the pH to 7.63. Conversely, a decrease in alveolar ventilation to one-quarter normal reduces the pH 0.4 pH unit. That is, if at normal alveolar ventilation the pH had been 7.4, reducing the ventilation to one-quarter reduces the pH to 7.0. Since alveolar ventilation can be reduced to zero ventilation or increased to about 15 times normal, one can readily understand how much the pH of the body fluids can be changed by alterations in the activity of the respiratory system.

Effect of Hydrogen Ion Concentration on Alveolar Ventilation

Not only does the rate of alveolar ventilation affect the hydrogen ion concentration of the body fluids, but, in turn, the hydrogen ion concentration affects the rate of alveolar ventilation. This results from a *direct action of hydrogen ions on the respiratory center in the medulla oblongata* that controls breathing, which will be discussed in detail in Chapter 42.

Figure 37–3 illustrates the changes in alveolar ventilation caused by changing the pH of arterial blood from 7.0 to 7.6. From this graph it is evident that a decrease in pH from the normal value of 7.4 to the strongly acidic range can increase the rate of alveolar ventilation to as much as four to five times normal, while an increase in pH into the alkaline range can decrease the rate of alveolar ventilation to as little as 50 to 75 per cent of normal.

Feedback Regulation of Hydrogen Ion Concentration by the Respiratory System. Because of the ability of the respiratory center to respond to hydrogen ion concentration, and because changes in alveolar ventilation in turn alter the hydrogen ion concentration in the body fluids, the respiratory system acts as a typical feedback regulatory system for controlling hydrogen ion concentration. That is, any time the hydrogen ion concentration becomes high, the respiratory system becomes more active, and alveolar ventilation increases. As a result, the carbon dioxide concentration in the extracellular fluids decreases, thus reducing the hydrogen concentration back toward a normal value. Conversely, if the hydrogen ion concentration falls too low, the respiratory center becomes depressed, alveolar ventilation also decreases, and the hydrogen ion concentration rises back toward normal.

Efficiency of Respiratory Regulation of Hydrogen Ion Concentration. Unfortunately, respiratory control cannot return the hydrogen ion concentration all the way to the normal value of 7.4 when some abnormality outside the respiratory system has altered the pH from the normal. The reason for this is that, as the pH returns toward normal, the stimulus that has been causing either increased or decreased respiration will itself begin to be lost. Ordinarily, the respiratory mechanism for regulation of hydrogen ion concentration has a control effectiveness of between 50 and 75 per cent (a feedback gain of 1 to 3). That is, if the hydrogen ion concentration should suddenly be decreased from 7.4 to 7.0 by some extraneous factor, the respiratory system, in 3 to 12 minutes, returns the pH to a value of about 7.2 to 7.3.

Buffering Power of the Respiratory System. In effect, respiratory regulation of acid-base balance is a *physiological type of buffer system* having almost identically the same importance as the chemical buffering systems of the body discussed earlier in the chapter. The overall "buffering power" of the respiratory system is 1 to 2 times as great as that of all the chemical buffers combined. That is, 1 to 2 times as much acid or base can normally be buffered by this mechanism as by the chemical buffers.

RENAL REGULATION OF HYDROGEN ION CONCENTRATION

In the earlier discussion of the Henderson-Hasselbalch equation it was pointed out that the kidneys regulate hydrogen ion concentration principally by increasing or decreasing the bicarbonate ion concentration in the body fluid. To do this, a complex series of reactions occurs in the renal tubules. The following sections describe the different tubular mechanisms that help to regulate the hydrogen ion concentration of the body fluids.

TUBULAR SECRETION OF HYDROGEN IONS

The epithelial cells of the proximal tubules, distal tubules, and collecting ducts all secrete

Figure 37–3. Effect of blood pH on the rate of alveolar ventilation. (Constructed from data obtained by Gray: Pulmonary Ventilation and Its Regulation. Charles C Thomas.)

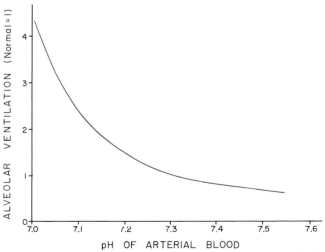

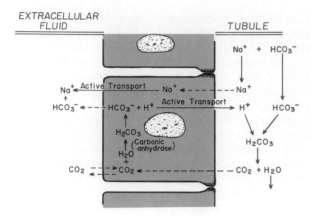

Figure 37–4. Chemical reactions for (1) hydrogen ion secretion, (2) sodium ion absorption in exchange for a hydrogen ion, and (3) combination of hydrogen ions with bicarbonate ions in the tubules.

hydrogen ions into the tubular fluid. The postulated mechanism by which this occurs is illustrated in Figure 37–4. The secretory process *begins with carbon dioxide* that either diffuses into or is formed by metabolism in the tubular epithelial cells. The carbon dioxide, under the influence of an enzyme, *carbonic anhydrase,* combines with water to form *carbonic acid.* This then dissociates into *bicarbonate ion* and *hydrogen ion,* and the hydrogen ion is secreted by active transport through the luminal border of the cell membrane into the tubule.

In the collecting ducts hydrogen ion secretion can continue until the concentration of hydrogen ions in the tubules becomes as much as 900 times that in the extracellular fluid or, in other words, until the pH of the tubular fluids falls to about 4.5. This represents a limit to the ability of the tubular epithelium to secrete hydrogen ions.

About 84 per cent of all the hydrogen ions secreted by the tubules are secreted in the proximal tubules, but the maximum concentration gradient that can be achieved here is only about three- to four-fold instead of the 900-fold that can be achieved in the collecting tubules. That is, the pH can be decreased only to about 6.9, 0.5 pH unit below 7.4, the pH of the glomerular filtrate. The distal tubules can decrease the pH to about 6.0 to 6.5 which is between that achieved by the proximal tubules and collecting tubules.

Regulation of Hydrogen Ion Secretion by the Carbon Dioxide Concentration in the Extracellular Fluid. Since the chemical reactions for secretion of hydrogen ions begin with carbon dioxide, the greater the carbon dioxide concentration in the extracellular fluid, the more rapidly the reactions proceed, and the greater becomes the rate of hydrogen ion secretion. Therefore, any factor that increases the carbon dioxide concentration in the extracellular fluids, such as decreased respiration

or increased metabolic rate, also increases the rate of hydrogen ion secretion. Conversely, any factor that decreases the carbon dioxide, such as excess pulmonary ventilation or decreased metabolic rate, decreases the rate of hydrogen ion secretion.

At normal carbon dioxide concentrations, the rate of hydrogen ion secretion is about 3.5 millimoles per minute, but this rises or falls directly in proportion to changes in extracellular carbon dioxide.

Interaction of Bicarbonate Ions with Hydrogen Ions in the Tubules — "Reabsorption" of Bicarbonate Ions

It is already clear from previous discussions that the bicarbonate ion concentration in the extracellular fluid plays an extremely important role in the acid-base buffer system and, therefore, in the control of extracellular fluid hydrogen ion concentration. Therefore, it is important that the kidney tubules help to regulate the extracellular fluid bicarbonate ion concentration. Yet, the tubules are not very permeable to the bicarbonate ion because it is a large ion and also is electrically charged. However, the bicarbonate ion can, in effect, be "reabsorbed" by a special process which is also illustrated in Figure 37–4.

The reabsorption of bicarbonate ions is initiated by a reaction in the tubules between the bicarbonate ions and the hydrogen ions secreted by the tubular cells, as illustrated in the figure. The carbonic acid then dissociates into carbon dioxide and water. The water becomes part of the tubular fluid, while the carbon dioxide, having the capability to diffuse extremely readily through all cellular membranes, instantaneously diffuses into the epithelial cells or all the way into the blood where it combines with water to form new bicarbonate ions. *If an excess of hydrogen ions is secreted by the tubules, the bicarbonate ions will be almost completely removed from the tubules* so that, for practical purposes, *none* will remain to pass into the urine.

If we now note in Figure 37–4 the chemical reactions that are responsible for formation of hydrogen ions in the epithelial cells, we will see that each time a hydrogen ion is formed a bicarbonate ion is formed inside these cells by the dissociation of H_2CO_3. This bicarbonate ion then diffuses into the peritubular fluid in combination with a sodium ion that has been absorbed from the tubule.

The net effect of all these reactions is a mechanism for reabsorption of bicarbonate ions from the tubules, though the bicarbonate ions that enter the peritubular fluid are not the same bicarbonate ions that are removed from the tubular fluid.

Normal Rates of Bicarbonate Ion Filtration and

Hydrogen Ion Secretion into the Tubules — Titration of Bicarbonate Ions Against Hydrogen Ions. Under normal conditions, the rate of hydrogen ion secretion is about 3.50 millimoles/minute, and the rate of filtration of bicarbonate ions in the glomerular filtrate is about 3.49 millimoles/minute. Thus, the quantities of the two ions entering the tubules are almost equal, and they combine with each other and actually annihilate each other, the end-products being carbon dioxide and water. Therefore, it is said that the bicarbonate ions and hydrogen ions normally "titrate" each other in the tubules.

However, note also that this titration process is not quite complete, for usually a slight excess of hydrogen ions (the acidic component) remains in the tubules to be excreted in the urine. The reason for this is that under normal conditions a person's metabolic processes continually form a small amount of excess acid that gives rise to the slight excess of hydrogen ions over bicarbonate ions in the tubules.

On rare occasions the bicarbonate ions are in excess, as we shall see in subsequent discussions. When this occurs, the titration process again is not quite complete; this time, excess bicarbonate ions (the basic component) are left in the tubules to pass into the urine.

Thus, the basic mechanism by which the kidney corrects either acidosis or alkalosis is by incomplete titration of hydrogen ions against bicarbonate ions, leaving one or the other of these to pass into the urine and therefore to be removed from the extracellular fluid.

From 80 to 99 per cent of this titration process occurs in the proximal tubules, and the carbonic acid formed by the titration reaction is then split very rapidly into its end products, carbon dioxide and water. The water passes down the tubules, and the carbon dioxide diffuses into the extracellular fluid. To promote this rapid dissociation of carbonic acid into carbon dioxide and water, the luminal brush border surface of the proximal tubules (but not of the other tubules) has a large amount of attached carbonic anhydrase that accelerates the reaction.

RENAL CORRECTION OF ALKALOSIS— DECREASE IN BICARBONATE IONS IN THE EXTRACELLULAR FLUID

Now that we have described the mechanisms by which the renal tubules secrete hydrogen ions and reabsorb bicarbonate ions, we can explain the manner in which the kidneys readjust the pH of the extracellular fluids when it becomes abnormal.

The initial step in this explanation is to understand what happens to the concentrations of carbon dioxide and bicarbonate ions in the extra-cellular fluids in alkalosis and acidosis. First, let us consider *alkalosis*. Referring again to Equation 11, the Henderson-Hasselbalch equation, we see that the *ratio* of bicarbonate ions to dissolved carbon dioxide molecules increases when the pH rises into the alkalosis range above 7.4. The effect of this on the titration process in the tubules is to increase the *ratio* of bicarbonate ions filtered into the tubules to hydrogen ions secreted. This increase occurs because the high extracellular bicarbonate ion concentration also increases its concentration in the glomerular filtrate, and the low carbon dioxide concentration decreases the secretion of hydrogen ions. Therefore, the fine balance that normally exists in the tubules between the hydrogen and bicarbonate ions no longer occurs. Instead, far greater quantities of bicarbonate ions than hydrogen ions now enter the tubules. Since no bicarbonate ions can be reabsorbed without first reacting with hydrogen ions, all the excess bicarbonate ions pass into the urine and carry with them sodium ions or other positive ions. Thus, in effect, sodium bicarbonate is removed from the extracellular fluid.

Loss of sodium bicarbonate from the extracellular fluid decreases the bicarbonate ion portion of the bicarbonate buffer system, and, in accordance with the Henderson-Hasselbalch equation, this shifts the pH of the body fluids back in the acid direction. Furthermore, because of the isohydric principle, all the other body buffers shift back in the acid direction too. Thus, the alkalosis is corrected.

RENAL CORRECTION OF ACIDOSIS— INCREASE IN BICARBONATE IONS IN THE EXTRACELLULAR FLUID

In acidosis, the *ratio* of carbon dioxide to bicarbonate ions in the extracellular fluid increases, which is exactly opposite to the effect in alkalosis. Therefore, in acidosis, the *rate of hydrogen ion secretion* rises to a level far greater than the *rate of bicarbonate ion filtration* into the tubules. As a result, a great excess of hydrogen ions is secreted into the tubules which have far too few bicarbonate ions to react with. These excess hydrogen ions combine with the buffers in the tubular fluid, as explained in the following paragraphs, and are excreted into the urine.

Figure 37–4 shows that each time a hydrogen ion is secreted into the tubules two other effects occur simultaneously: first, a bicarbonate ion is formed in the tubular epithelial cell and, second, a sodium ion is absorbed from the tubule into the epithelial cell. The sodium ion and bicarbonate ion then diffuse together from the epithelial cell into the peritubular fluid. Thus, *the net effect of secreting excess hydrogen ions into the tubules is to increase the quantity of sodium bicarbonate in the extracellular fluid.* This increases the bicarbonate portion of the bicarbonate buffer system, which, in accordance with the Henderson-Hasselbalch equation and the isohydric principle, shifts all the buffers in the alkaline direction, increasing the pH in the process, and thereby correcting the acidosis.

COMBINATION OF THE EXCESS HYDROGEN IONS WITH TUBULAR BUFFERS AND THEIR TRANSPORT INTO THE URINE

When excess hydrogen ions are secreted into the tubules, only a small portion of these can be carried in the free form by the tubular fluid into the urine. The reason for this is that the maximum hydrogen ion concentration that the tubular system can achieve is $10^{-4.5}$ molar, which corresponds to a pH of 4.5. At normal daily urine flows, this concentration represents only 1 per cent of the daily excretion of excess hydrogen ions.

Therefore, to carry the excess hydrogen ions into the urine the hydrogen ions must combine with buffers in the tubular fluid to keep the hydrogen ion concentration itself from rising too high. Otherwise, as the hydrogen ion concentration of the tubular fluid approaches the maximum limit that can be achieved (a concentration of $10^{-4.5}$ molar), the rate of secretion of hydrogen ions falls to near-zero.

The tubular fluids have two very important buffer systems for transport of the excess hydrogen ions into the urine: (1) the phosphate buffer and (2) the ammonia buffer. In addition, there are a number of weak buffer systems such as urate, citrate, and similar systems, as well as the bicarbonate buffer system.

Transport of Excess Hydrogen Ions into the Urine by the Phosphate Buffer. The phosphate buffer is composed of a mixture of HPO_4^{--} and $H_2PO_4^-$. Both of these become considerably concentrated in the tubular fluid because of their relatively poor reabsorption and because of removal of water from the tubular fluid. Therefore, even though the phosphate buffer is very weak in the blood, it is a much more powerful buffer in the tubular fluid.

The quantity of HPO_4^{--} in the glomerular filtrate is normally about 4 times as great as that of $H_2PO_4^-$. Excess hydrogen ions entering the tubules combine with the HPO_4^{--}, as illustrated in Figure 37–5, forming $H_2PO_4^-$, which passes on into the urine. Sodium ion is absorbed into the extracellular fluid in place of the hydrogen ion involved in the reaction, and at the same time a *bicarbonate ion,* formed in the process of secreting the hydrogen ion, is also released into the extracellular fluid. Thus, the net effect of this reaction is to increase the amount of sodium bicarbonate in the extracellular fluids, which is the

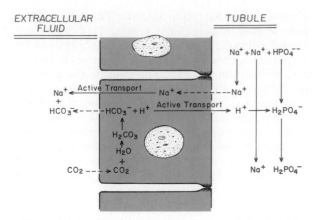

Figure 37–5. Chemical reactions in the tubules involving hydrogen ions, sodium ions, and the phosphate buffer system.

kidney's way of reducing the degree of acidosis in the body fluids.

Transport of Excess Hydrogen Ions into the Urine by the Ammonia Buffer System. Another very potent buffer system of the tubular fluid is composed of ammonia (NH_3) and the ammonium ion (NH_4^+). The epithelial cells of all the tubules besides those of the thin segment of the loop of Henle continually synthesize ammonia, and this diffuses into the tubules. The ammonia then reacts with hydrogen ions, as illustrated in Figure 37–6, to form ammonium ions. These are then excreted into the urine in combination with chloride ions and other tubular anions. Note in the figure that the net effect of these reactions is, again, *to increase the bicarbonate concentration* in the extracellular fluid.

This ammonium ion mechanism for transport of excess hydrogen ions in the tubules is especially important for two reasons: (1) Each time an ammonia molecule combines with a hydrogen ion to form an ammonium ion the concentration of ammonia in the tubular fluid becomes decreased,

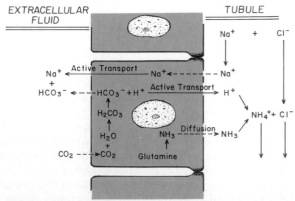

Figure 37–6. Secretion of ammonia by the tubular epithelial cells, and reaction of the ammonia with hydrogen ions in the tubules.

which causes still more ammonia to diffuse from the epithelial cells into the tubular fluid. Thus, the rate of ammonia secretion into the tubular fluid is actually controlled by the amount of excess hydrogen ions to be transported. (2) Most of the negative ions of the tubular fluid are chloride ions. Only a few hydrogen ions could be transported into the urine in direct combination with chloride, because hydrochloric acid is a very strong acid and the tubular pH would fall rapidly below the critical value of 4.5 so that further hydrogen ion secretion would cease. However, when hydrogen ions combine with ammonia and the resulting ammonium ions then combine with chloride, the pH does not fall significantly because ammonium chloride is only very weakly acidic.

Sixty per cent of the ammonia secreted by the tubular epithelium is derived from *glutamine*, and the remaining 40 per cent from different amino acids, particularly glycine and alanine.

Enhancement of the Ammonia Buffer System in Chronic Acidosis. If the tubular fluids remain highly acidic for long periods of time, the formation of ammonia steadily increases during the first two to three days, rising as much as 10-fold. For instance, immediately after acidosis begins, as little as 30 millimoles of ammonia might be secreted each day, but after several days as much as 300 to 450 millimoles can be secreted, illustrating that the ammonia-secreting mechanism can adapt readily to handle greatly increased loads of acid elimination.

RAPIDITY OF ACID-BASE REGULATION BY THE KIDNEYS

Figure 37-7 illustrates the effect of extracellular fluid pH on the rate at which bicarbonate ions are lost from or gained by the body fluids each minute. For instance, at a pH of 7.0, approximately 2.3

millimoles of bicarbonate ions is gained each minute, but as the pH returns toward the normal value of 7.4, the rate of gain falls to zero. Then, when the pH becomes significantly greater than normal, bicarbonate ions are lost by the extracellular fluids. For instance, at a pH of 7.6 about 1.5 millimoles of bicarbonate ions are lost each minute.

The total amount of buffers in the entire body (within the range of pH 7.0 to 7.8) is approximately 1000 millimoles. If all these should be suddenly shifted to the alkaline or acidic side by injecting an alkali or an acid, the kidneys would be able to return the pH of the body fluids back amost to normal in one to three days. Though this mechanism is slow to act, it continues acting until the pH returns almost exactly to normal rather than a certain percentage of the way. Therefore, the real value of the renal mechanism for regulating hydrogen ion concentration is not rapidity of action but instead its ability in the end to neutralize completely any excess acid or alkali that enters the body fluids, unless the excess continues to enter.

Ordinarily, the kidneys can remove up to about 500 millimoles of acid or alkali each day. If greater quantities than this enter the body fluids, the kidneys are unable to cope with the extra load, and severe acidosis or alkalosis ensues.

Range of Urinary pH. In the process of adjusting the hydrogen ion concentration of the extracellular fluid, the kidneys often excrete urine at pH's as low as 4.5 or as high as 8.0. When acid is being excreted the pH falls, and when alkali is being excreted the pH rises. Even when the pH of the extracellular fluids is at the normal value of 7.4, a fraction of a millimole of acid is still lost each minute. The reason for this is that about 50 to 80 millimoles more acid than alkali are formed in the body each day, and this acid must be removed continually. Because of the presence of this excess acid in the urine, the normal urine pH is about 6.0 instead of 7.4, the pH of the blood.

RENAL REGULATION OF PLASMA CHLORIDE CONCENTRATION—THE CHLORIDE TO BICARBONATE RATIO

In the above discussions we have emphasized the ability of the kidneys to conserve bicarbonate ion in the extracellular fluids whenever a state of acidosis develops, or to remove bicarbonate ions in a state of alkalosis. Thus, the bicarbonate ion is shuttled back and forth between high and low values as one of the principal means of adjusting the acid-base balance of the extracellular buffer systems and therefore also for adjusting the extracellular fluid pH.

However, in the process of juggling the extracellular fluid concentration of bicarbonate ion, it is

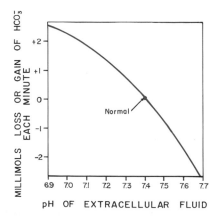

Figure 37-7. Effect of extracellular fluid pH on the rate of loss or gain of bicarbonate ions in the body fluids each minute.

essential to remove some other anion from the extracellular fluids each time the bicarbonate is increased, or to increase some other anion when the bicarbonate concentration is decreased. In general, the anion that is reciprocally juggled up or down with the bicarbonate ion is the chloride ion because this is the anion in greatest concentration in the extracellular fluid.

Function of the Ammonia Buffer System in Controlling the Bicarbonate Ion to Chloride Ion Ratio. It was pointed out above that the ammonia buffer system plays an extremely important role in removing excess hydrogen ions from the tubules. Now, let us study Figure 37–6 once again. We see that in the process of transporting excess hydrogen ions into the urine in combination with ammonia, for each hydrogen ion transported a chloride ion also passes into the urine and a bicarbonate ion simultaneously enters the extracellular fluid. Thus, this ammonia system substitutes a bicarbonate ion in the extracellular fluid for a chloride ion that is lost from the extracellular fluids. Conversely, when a person is alkalotic, the ammonia system becomes inoperative; bicarbonate ions instead of chloride ions then pass into the urine, and a concomitant excess of chloride is reabsorbed.

Thus, in the process of controlling the pH of the body fluids, the renal acid-base regulating system also regulates the ratio of chloride ions to bicarbonate ions in the extracellular fluid.

CLINICAL ABNORMALITIES OF ACID-BASE BALANCE

RESPIRATORY ACIDOSIS AND ALKALOSIS

From the discussions earlier in the chapter it is obvious that any factor that decreases the rate of pulmonary ventilation increases the concentration of dissolved carbon dioxide in the extracellular fluid, which in turn leads to increased carbonic acid and hydrogen ions, thus resulting in acidosis. Because this type of acidosis is caused by an abnormality of respiration, it is called *respiratory acidosis*.

On the other hand, excessive pulmonary ventilation reverses the process and decreases the hydrogen ion concentration, thus resulting in alkalosis; this condition is called *respiratory alkalosis*.

A person can cause respiratory acidosis in himself by simply holding his breath, which he can do until the pH of the body fluids falls to as low as perhaps 7.0. On the other hand, he can voluntarily overbreathe and cause alkalosis to a pH of about 7.9.

Respiratory acidosis frequently results from pathological conditions. For instance, damage to the respiratory center in the medulla oblongata causing reduced breathing, obstruction of the passageways in the respiratory tract, pneumonia, decreased pulmonary membrane surface area, and any other factor that interferes with the exchange of gases between the blood and alveolar air results in respiratory acidosis.

On the other hand, only rarely do pathologic conditions cause *respiratory alkalosis*. However, occasionally a psychoneurosis causes overbreathing to the extent that a person becomes alkalotic. A physiological type of respiratory alkalosis occurs when a person ascends to a *high altitude*. The low oxygen content of the air stimulates respiration, which causes excess loss of carbon dioxide and development of mild respiratory alkalosis.

METABOLIC ACIDOSIS AND ALKALOSIS

The terms metabolic acidosis and metabolic alkalosis refer to all other abnormalities of acid-base balance besides those caused by excess or insufficient carbon dioxide in the body fluids. Use of the word "metabolic" in this instance is unfortunate, because carbon dioxide is also a metabolic product. Yet, by convention, carbonic acid resulting from dissolved carbon dioxide is called a *respiratory acid* while any other acid in the body, whether it be formed by metabolism or simply ingested by the person, is called a *metabolic acid,* or a *fixed acid.*

Causes of Metabolic Acidosis. Metabolic acidosis can result from (1) failure of the kidneys to excrete the metabolic acids normally formed in the body, (2) formation of excessive quantities of metabolic acids in the body, (3) intravenous administration of metabolic acids, or (4) addition of metabolic acids by way of the gastrointestinal tract. Metabolic acidosis can result also from (5) loss of alkali from the body fluids. Some of the specific conditions that cause metabolic acidosis are the following:

Diarrhea. Severe diarrhea is one of the most frequent causes of metabolic acidosis for the following reasons: The gastrointestinal secretions normally contain large amounts of sodium bicarbonate. Therefore, excessive loss of these secretions during a bout of diarrhea is exactly the same as excretion of large amounts of sodium bicarbonate into the urine. In accordance with the Henderson-Hasselbalch equation, this results in a shift of the bicarbonate buffer system toward the acid side and results in metabolic acidosis. In fact, acidosis resulting from severe diarrhea can be so serious that it is one of the most common causes of death in young children.

Vomiting. A second cause of metabolic acidosis is vomiting. Vomiting of gastric contents alone, which occurs rarely, causes a loss of acid and leads to alkalosis, but vomiting of contents from deeper in the gastrointestinal tract, which often occurs, causes loss of alkali and results in metabolic acidosis.

Uremia. A third common type of acidosis is uremic acidosis, which occurs in severe renal disease. The cause of this is failure of the kidneys to rid the body of even the normal amounts of acids formed each day by the metabolic processes of the body.

Diabetes Mellitus. A fourth and extremely important cause of metabolic acidosis is diabetes mellitus. In this condition, lack of insulin secretion by the pancreas prevents normal use of glucose for metabolism. Instead, fat is split into acetoacetic acid, and this in turn is metabolized by the tissues for energy in place of glucose. Simultaneously, the concentration of acetoacetic acid in the extracellular fluids often rises very high, and large quantities of it are excreted in the urine, sometimes as much as 500 to 1000 millimoles per day.

Acidosis Caused by Carbonic Anhydrase Inhibitors. Administration of the common carbonic anhydrase inhibitor acetazolamide (Diamox), a drug that is frequently used to cause diuresis, also causes a mild degree of acidosis. This occurs because inhibition of the carbonic anhydrase on the luminal surface of the proximal tubular epithelium prevents adequate "reabsorption" of bicarbonate ions. Loss of these into the urine causes a decrease in bicarbonate ions in the extracellular fluid, thus leading to acidosis.

Acidosis Caused by High Extracellular Fluid Potassium Concentration. Potassium ions, like hydrogen ions, are also secreted by the distal tubules and collecting ducts, as was explained in Chapter 34. When the plasma concentration of potassium ions is high, the tubular epithelium secretes extra large quantities of these into the tubules, and these compete with hydrogen ions to combine with the available buffer anions of the tubular fluid. Therefore, an excess of potassium decreases the quantity of excess hydrogen ions that can be carried into the urine, and this in turn *decreases* the secretion of hydrogen ions. Consequently, normal removal of hydrogen ions cannot occur, thereby leading to acidosis.

Causes of Metabolic Alkalosis. Metabolic alkalosis does not occur nearly as often as metabolic acidosis. However, there are several common causes of metabolic alkalosis, as follows:

Alkalosis Caused by Administering Diuretics (Except the Carbonic Anhydrase Inhibitors). All diuretics cause increased flow of fluid along the tubules, and this increase usually causes a great excess of sodium ions to flow into the distal and collecting tubules, leading also to rapid reabsorption of sodium ions from these tubules. This rapid reabsorption is often coupled with enhanced hydrogen ion secretion, therefore excessive loss of hydrogen ions from the body and resultant extracellular fluid alkalosis.

Excessive Ingestion of Alkaline Drugs. Perhaps the second most common cause of alkalosis is excessive ingestion of alkaline drugs, such as sodium bicarbonate, for the treatment of gastritis or peptic ulcer.

Alkalosis Caused by Loss of Chloride Ions. Excessive vomiting of gastric contents without vomiting of lower gastrointestinal contents causes excessive loss of hydrochloric acid secreted by the stomach mucosa. This loss leads to reduction of chloride ions and enhancement of bicarbonate ions in the extracellular fluid. The bicarbonate ions are derived from the stomach glandular cells that secrete the hydrogen ions; these form bicarbonate ion in the same way that the tubular cells do when they secrete hydrogen ions. The net result is loss of acid from the extracellular fluids and development of metabolic alkalosis.

Alkalosis Caused by Excess Aldosterone. When excess quantities of aldosterone are secreted by the adrenal glands, the extracellular fluid becomes slightly alkalotic. This is caused in the following way: The aldosterone promotes extensive reabsorption of sodium ions from the distal segments of the tubular system, but coupled with this is increased secretion of hydrogen ions and loss from the extracellular fluids, thus promoting alkalosis.

EFFECTS OF ACIDOSIS AND ALKALOSIS ON THE BODY

Acidosis. The major effect of acidosis is depression of the *central nervous system*. When the pH of the blood falls below 7.0, the nervous system becomes so depressed that the person first becomes disoriented and, later, comatose. Therefore, patients dying of diabetic acidosis, uremic acidosis, and other types of acidosis usually die in a state of coma.

In metabolic acidosis the high hydrogen ion concentration causes increased rate and depth of respiration. Therefore, one of the diagnostic signs of *metabolic* acidosis is increased pulmonary ventilation. On the other hand, *in respiratory acidosis, respiration is usually depressed* because this is the cause of the acidosis, which is opposite to the effect in metabolic acidosis.

Alkalosis. The major effect of alkalosis on the body is *overexcitability of the nervous system*. This occurs both in the central nervous system and in the peripheral nerves, but usually the peripheral nerves are affected before the central nervous system. The nerves become so excitable that they automatically and repetitively fire even when they are not stimulated by normal stimuli. As a result, the muscles go into a state of *tetany,* which means a state of tonic spasm. This tetany usually appears first in the muscles of the forearm, then spreads rapidly to the muscles of the face, and finally all over the body. Extremely alkalotic patients may die from tetany of the respiratory muscles.

Occasionally an alkalotic person develops severe symptoms of central nervous system overexcitability. The symptoms may manifest themselves as extreme nervousness or, in susceptible persons, as convulsions. For instance, in persons who are predisposed to epileptic fits, simply overbreathing often results in an attack. Indeed, this is one of the clinical methods for assessing one's degree of epileptic predisposition.

RESPIRATORY COMPENSATION OF METABOLIC ACIDOSIS OR ALKALOSIS

It was pointed out above that the high hydrogen ion concentration of metabolic acidosis causes increased pulmonary ventilation, which in turn results in rapid removal of carbon dioxide from the body fluids and reduces the hydrogen ion concentration back toward normal. Thus, this respiratory effect helps to compensate for the metabolic acidosis. However, this compensation occurs only partway. Ordinarily, the respiratory system is capable of compensating between 50 and 75 per cent. That is, if the metabolic factor makes the pH of the blood fall to 7.0 at normal pulmonary ventilation, the rate of pulmonary ventilation normally increases sufficiently to return the pH of the blood to 7.2 to 7.3, as was pointed out earlier in the chapter.

The opposite effect occurs in metabolic alkalosis. That is, alkalosis diminishes the pulmonary ventilation, which in turn causes increased hydrogen ion concentration. Here again the compensation can take place to about 50 to 75 per cent.

RENAL COMPENSATION OF RESPIRATORY ACIDOSIS OR ALKALOSIS

If a person develops respiratory acidosis that continues for a prolonged period of time, the kidneys will secrete an excess of hydrogen ions, resulting in an increase in sodium bicarbonate in the extracellular fluids. After one to six days, the pH of the body fluids will have returned about 65 to 75 per cent of the way to normal even though the person continues to breathe poorly.

Exactly the converse effect occurs in respiratory alkalosis. Large amounts of sodium bicarbonate are lost into the urine, decreasing the extracellular bicarbonate ion and thereby decreasing the pH also almost back to normal.

PHYSIOLOGY OF TREATMENT IN ACIDOSIS OR ALKALOSIS

Obviously, the best treatment for acidosis or alkalosis is to remove the condition causing the abnormality, but, if this cannot be effected, different drugs can be used to neutralize the excess acid or alkali.

To neutralize excess acid, large amounts of sodium bicarbonate can be ingested by mouth. This is absorbed into the blood stream and increases the bicarbonate ion portion of the bicarbonate buffer, thereby shifting the pH to the alkaline side. Sodium bicarbonate is occasionally used also for intravenous therapy, but this has such strong and often dangerous physiological effects that other substances are often used instead, such as sodium lactate or sodium gluconate. The lactate and gluconate portions of the molecules are metabolized in the body, leaving the sodium in the extracellular fluids in the form of sodium bicarbonate, and thereby shifting the pH of the fluids in the alkaline direction.

For treatment of alkalosis, ammonium chloride is often administered by mouth. When this is absorbed into the blood, the ammonia portion of the ammonium chloride is converted by the liver into urea; this reaction liberates hydrochloric acid, which immediately reacts with the buffers of the body fluids to shift the hydrogen ion concentration in the acid direction. Occasionally, ammonium chloride is infused intravenously, but the ammonium ion is highly toxic and this procedure can be dangerous. Another substance occasionally used is *lysine monohydrochloride*.

CLINICAL MEASUREMENTS AND ANALYSIS OF ACID-BASE ABNORMALITIES

pH Measurements. In studying a patient with acidosis or alkalosis it is desirable to know the pH of the body fluids. This can be determined easily by measuring the pH of the plasma with a glass electrode pH meter. However, extreme care must be exercised in removing the plasma and in making the measurement, because even the slightest diffusion of carbon dioxide out of the plasma into the air shifts the bicarbonate buffer system in the alkaline direction, resulting in a much elevated pH measurement.

The Concept of "Buffer Base." The term buffer base means the anion components of the buffer system. Thus, in plasma the principal buffer base is bicarbonate ion. In whole blood it is mainly bicarbonate ion plus the hemoglobin ion in the red blood cell. In the entire body it is mainly the bicarbonate ion plus all of the proteins in the body. In addition, there are many buffer bases of lesser quantity such as the phosphate ion, the urate ion, and so forth.

The normal quantity of buffer base in whole blood is approximately 45 mEq./liter. About half of this is usually bicarbonate ion and the other half the hemogloblin buffer base. Whenever the bicarbonate portion of the buffer base is greater than normal, one suspects metabolic alkalosis, but when less than normal, metabolic acidosis.

The pH-Bicarbonate Diagram. The so-called pH-bicarbonate diagram, which is illustrated in Figure 37–8, can be used to determine the *type* of acidosis or alkalosis a person has and its *severity*. Its use may be explained as follows:

The more vertical curves of the diagram depict different carbon dioxide concentrations. The normal carbon dioxide concentration of 1.2 mM./liter is denoted by the colored line. The points along this line represent the possible combinations of bicarbonate concentration and pH that can exist in the body fluids when the carbon dioxide concentration is normal.

The more horizontal lines represent the concentrations of metabolic acids and bases in the body fluids. The colored line, denoted by the numeral zero, indicates a balance between both of these. That is, the points on this line represent the possible combinations of bicarbonate concentration and pH that can occur as long as the metabolic acids and bases in the body fluids are normal. The upper two horizontal lines indicate respectively additions of 5 to 10 mM./liter of extra metabolic base to the body fluids, and the two lower horizontal lines indicate additions of 5 to 10 extra mM./liter of metabolic acid.

To use this diagram we simply find the pH of the blood and the bicarbonate concentration, then plot the appropriate point on the diagram. For instance, if the

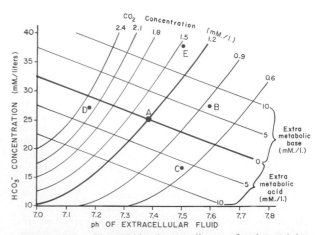

Figure 37–8. The pH-bicarbonate diagram for determining relative degrees of metabolic and respiratory acidosis or alkalosis in a patient. (Modified from Davenport: The ABC of Acid-Base Chemistry. University of Chicago Press.)

pH is found to be the normal value of 7.4 and bicarbonate concentration the normal value of 25 mM./liter, we plot point A, which represents the normal condition.

Now using data obtained from another patient, we plot a new point with a pH of 7.63 and a bicarbonate concentration of 28 mM./liter. This is point B on the diagram, which represents a carbon dioxide concentration of 0.8 mM./liter and 7 mM./liter extra metabolic base. Thus, this person has *metabolic alkalosis* because he has considerable extra metabolic base in his body fluids, but he also has *respiratory alkalosis* because he is overventilating to an extent that his carbon dioxide concentration is considerably less than normal.

Similarly, from other patients, we plot points C, D, and E. Point C represents 6 mM./liter of *metabolic acidosis* and sufficient *respiratory alkalosis* to reduce the carbon dioxide concentration to 0.7 mM./liter. A person with such a finding as this could have respiratory alkalosis that has been partially compensated by metabolic acidosis produced by the kidneys. Or, conversely, it is possible that the metabolic acidosis was primary, and the person has overcompensated with respiratory alkalosis.

Point D represents *mild metabolic acidosis, 2 mM./liter,* combined with *severe respiratory acidosis.* A person could get into this state with severe primary respiratory acidosis and mild metabolic acidosis resulting from some other cause.

Point E represents *mild respiratory acidosis and severe metabolic alkalosis.* Presumably, in this case the metabolic alkalosis was primary, and respiratory compensation caused mild respiratory acidosis in an attempt to compensate for the metabolic alkalosis.

In summary, by using the pH-bicarbonate diagram, one can determine both the degree of metabolic acidosis or alkalosis and the degree of respiratory acidosis or alkalosis in a patient at the same time.

The Astrup and Siggaard-Andersen Procedure for Calculating Abnormalities of Buffer Base — The Concept of Base Excess. Astrup and Siggaard-Andersen developed a method that is very simple to use in clinical patients for determining both the buffer base content of blood and a quantity called "base excess." The base excess is the quantity of extra base above normal (principally bicarbonate ions) after exclusion of the hemoglobin content. In other words, the deviation of the buffer base content from normal is calculated, and then this is corrected for the hemoglobin content to give the value called base excess. If the base excess is a positive number, then the person has an excess of metabolic base in the body fluids and therefore has metabolic alkalosis. If the base excess is a negative number, then the person has a deficit of metabolic acid in the blood and thus has metabolic acidosis.

The Astrup and Siggaard-Andersen procedure for determining base excess is the following: The nomogram illustrated in Figure 37–9 has been calculated and constructed to determine both the buffer base and the base excess in the blood. To use this nomogram several steps must first be followed: A small amount of blood is removed from a patient. Then the blood is equilibrated with carbon dioxide in a small chamber at

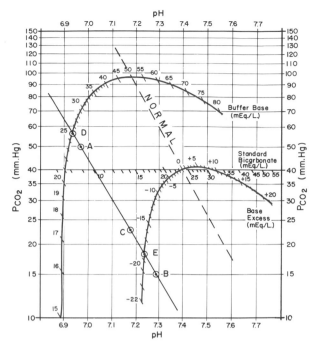

Figure 37–9. The Siggaard-Andersen nomogram, a method widely used clinically for diagnosis of acid-base abnormalities. (See text for explanation.)

a known carbon dioxide tension, and the pH of the blood is measured while equilibrated at this tension. Next, the carbon dioxide tension is changed to another value and the pH measured again. Points A and B in Figure 37–9 illustrate a pair of such measurements for the blood from a given patient. Next, a line is drawn through these two points. The line is shown to intersect with three different curves on the nomogram. One of these curves is labeled "buffer base"; the point at which the experimental line crosses this curve gives the buffer base of the blood. The point at which the experimental line crosses the "base excess" curve gives the base excess for the blood. Then, the point at which the experimental line crosses the line labeled "standard bicarbonate" gives the bicarbonate ion concentration in the blood that would be measured were the carbon dioxide tension at a normal level of 40 mm. Hg. Finally, the pH of the blood when it was first collected — before equilibrating it with CO_2 — is plotted on the experimental line to tell what the P_{CO_2} of the blood of that particular patient was at the time of measurement. Point C illustrates this.

Now let us summarize the information that one can derive from the nomogram: Points A and B are the experimental determinations when the blood is equilibrated with different known gaseous tensions of P_{CO_2}. Point C gives the blood pH and also the P_{CO_2} of the fresh blood. If this P_{CO_2} is above 40 mm. Hg, then the person has a degree of respiratory acidosis, that is, too little respiration and accumulation of carbonic acid in the blood. If the P_{CO_2} is below 40 mm. Hg, then the patient has respiratory alkalosis. In this case the P_{CO_2} is 23 mm. Hg; therefore, there is definite respiratory alkalosis. Point D illustrates the buffer base in the blood, a value of 25 mEq./liter. Point E gives the base excess in the blood. Note that this is −18.5 mEq./liter, which means that the person has a marked degree of metabolic acidosis.

Therapy Based on the Astrup and Siggaard-Andersen Nomogram. One of the principal values of this type of nomogram is that it can be used readily and quickly as a guide for treating metabolic acidosis or alkalosis in a patient because the degree of abnormality is given directly in terms of base excess or base deficiency. To calculate the *total body* base excess or base deficiency one multiplies the base excess or deficiency per liter of blood by 0.3 and by body weight in kilograms. This value can then be used as an estimate of the total amount of acid or base that will have to be given therapeutically to correct the whole body acid-base abnormality.

REFERENCES

Arruda, J. A. L., and Kurtzman, N. A.: Relationship of renal sodium and water transport to hydrogen ion secretion. *Annu. Rev. Physiol., 40*:43, 1978.

Bank, N., *et al.*: A micropuncture study of HCO₃ reabsorption by the hypertrophied proximal tubule. *Yale J. Biol. Med., 51*(3):275, 1978.

Brenner, B. M., and Stein, J. H. (eds.): Acid-Base and Potassium Homeostasis. New York, Churchill Livingstone, 1978.

Chan, J. C.: The influence of dietary intake on endogenous acid production. Theoretical and experimental background. *Nutr. Metab., 16*:1, 1974.

Christensen, H. N.: Body Fluids and the Acid-Base Balance, 2nd Ed. Philadelphia, W. B. Saunders Co., 1964.

Goldberger, E.: A Primer of Water, Electrolyte, and Acid-Base Syndromes. Philadelphia, Lea & Febiger, 1980.

Gottschalk, C. W., *et al.*: Localization of urine acidification in the mammalian kidney. *Am. J. Physiol., 198*:581, 1960.

Guder, W. G., and Schmidt, U. (eds.): Biochemical Nephrology. Bern, H. Huber, 1978.

Hills, A. G.: Acid-Base Balance. Baltimore, Williams & Wilkins, 1973.

Irvine, R. D.: Diet and drugs in renal acidosis and acid-base regulation. *Prog. Biochem. Pharmacol., 7*:146, 1972.

Jones, N. L.: Blood Gases and Acid-Base Physiology. New York, B. C. Decker, 1980.

Kintner, E. P.: Acid base, blood gas, and electrolyte balances. *Prog. Clin. Pathol., 4*:143, 1972.

Lemann, J., Jr., and Lennon, E. J.: Role of diet, gastrointestinal tract, and bone in acid-base homeostasis. *Kidney Int., 1*:275, 1972.

Lennon, E. J.: Body buffering mechanisms. *In* Frohlich, E. D. (ed.): Pathophysiology, 2nd Ed. Philadelphia, J. B. Lippincott Co., 1976, p. 287.

Levitin, H.: Acid-base balance. *In* Bondy, P. K., and Rosenberg, L. E. (eds.): Duncan's Diseases of Metabolism, 7th Ed. Philadelphia, W. B. Saunders Co., 1974, p. 1531.

Malnic, G.: Tubular handling of H. *In* MTP International Review of Science: Physiology. Vol. 6. Baltimore, University Park Press, 1974, p. 107.

Malnic, G., and Giebisch, G.: Symposium on acid-base homeostasis. Mechanism of renal hydrogen ion secretion. *Kidney Int., 1*:280, 1972.

Pitts, R. F.: Production and excretion of ammonia in relation to acid-base regulation. *In* Orloff, F., and Berliner, R. W. (eds.): Handbook of Physiology. Sec. 8. Baltimore, Williams & Wilkins, 1973, p. 455.

Quintero, J. A.: Acid-Base Balance: A Manual for Clinicians. St. Louis, W. H. Green, 1979.

Rahn, H.: Body temperature and acid-base regulation. (Review article). *Pneumonologie, 151*:87, 1974.

Rattenborg, C. C.: Acid-base monitoring. *Clin. Anesth., 9*:139, 1973.

Rector, F. C., Jr.: Acidification of the urine. *In* Orloff, F., and Berliner, R. W. (eds.): Handbook of Physiology. Sec. 8. Baltimore, Williams & Wilkins, 1973, p. 431.

Relman, A. S.: Metabolic consequences of acid-base disorders. *Kidney Int., 1*:347, 1972.

Robinson, J. R.: Fundamentals of Acid-Base Regulation, 4th Ed. Philadelphia, J. B. Lippincott Co., 1972.

Siesjö, B. K.: Symposium on acid-base homeostasis. The regulation of cerebrospinal fluid pH. *Kidney Int., 1*:360, 1972.

Steinmetz, P. R.: Cellular mechanisms of urinary acidification. *Physiol. Rev., 54*:890, 1974.

Tannen, R. L.: Control of acid excretion by the kidney. *Annu. Rev. Med., 31*:35, 1980.

Waddell, W. J., and Bates, R. G.: Intracellular pH. *Physiol. Rev., 49*:285, 1969.

Warnock, D., and Rector, F.: Protein secretion by the kidney. *Annu. Rev. Physiol., 41*:197, 1979.

Weiner, I. M.: Transport of weak acids and bases. *In* Orloff, F., and Berliner, R. W. (eds.): Handbook of Physiology. Sec. 8. Baltimore, Williams & Wilkins, 1973, p. 521.

38

Renal Disease, Diuresis, and Micturition

RENAL DISEASE

Renal disease can be classified into five different physiological categories: (1) *acute renal failure,* in which the kidneys stop working entirely or almost entirely, (2) *chronic renal failure,* in which progressively more nephrons are destroyed until the kidneys simply cannot perform all the necessary functions, (3) *hypertensive kidney disease,* in which vascular or glomerular lesions cause hypertension but not renal failure, (4) *nephrotic syndrome,* in which the glomeruli have become far more permeable than normal so that large amounts of protein are lost into the urine, and (5) *specific tubular abnormalities* that cause abnormal reabsorption or lack of reabsorption of certain substances by the tubules.

ACUTE RENAL FAILURE

Almost any condition that seriously interferes with kidney function can cause acute renal failure. Two of the most common causes are (1) acute glomerulonephritis and (2) acute damage and obstruction of the tubules.

Renal Failure Caused by Acute Glomerulonephritis. Acute glomerulonephritis is a disease caused by an abnormal immune reaction. In about 95 per cent of the patients, this occurs one to three weeks following an infection elsewhere in the body caused by certain types of Group A beta streptococci. The infection may have been a streptococcal sore throat, streptococcal tonsillitis, scarlet fever, or even streptococcal infection of the skin. It is not the infection itself that causes damage to the kidneys. Instead, as antibodies develop during the succeeding few weeks against the streptococcal antigen, the antibodies and antigen react with each other to form an insoluble immune complex that becomes entrapped in the glomerulus, especially in the basement membrane portion of the glomerulus.

Once the immune complex has deposited in the glomeruli, all the cells of the glomerulus begin to proliferate, but mainly the epithelial cells and mesangial cells that lie between the endothelium and epithelium. In addition, large numbers of white blood cells become entrapped in the glomeruli. Many of the glomeruli become blocked entirely by this inflammatory reaction, and those that are not blocked usually become excessively permeable, allowing both protein and red blood cells to leak into the glomerular filtrate. In the severest cases, either total or almost total renal shutdown occurs.

The acute inflammation of the glomeruli usually subsides in ten days to two weeks, and in most patients the kidneys return to normal function within the next few weeks to few months. Sometimes, however, many of the glomeruli are destroyed beyond repair, and in a small percentage of the patients progressive renal deterioration continues indefinitely, similar to that described for chronic glomerulonephritis in a subsequent section.

Tubular Necrosis as a Cause of Acute Renal Failure. Another common cause of acute renal shutdown is *tubular necrosis,* which means destruction of epithelial cells in the tubules, as illustrated in Figure 38–1. Common causes of tubular necrosis are (1) *various poisons* that destroy the tubular epithelial cells, and (2) *severe acute ischemia* of the kidneys.

Renal Poisons. Among the different renal poisons are *carbon tetrachloride* and the heavy metals such as the *mercuric ion.* These substances have specific nephrotoxic action on the tubular epithelial cells, causing death of many of them. As a result, the epithelial cells slough away from the basement membrane and plug the tubules. In some instances the basement membrane also is destroyed, but, if not, new tubular epithelial cells can usually grow along the surface of the membrane so that the tubule becomes repaired within 10 to 20 days.

Severe Acute Renal Ischemia. Severe ischemia of the kidney is likely to result from *severe circulatory shock.* In shock, the heart simply fails to pump sufficient amounts of blood to supply adequate nutrition to the different parts of the body, and renal blood flow is particularly likely to suffer because of strong sympathetic constriction of the renal vessels. Therefore, lack of adequate nutrition often destroys many tubular epithelial cells and thereby plugs many of the nephrons.

Transfusion Reaction as a Cause of Acute Renal Failure. A transfusion reaction normally results in hemol-

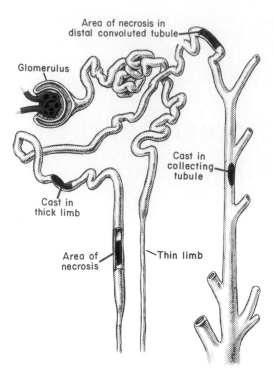

Figure 38–1. Damage to the distal tubules as a result of shock. (Modified from MacLean: Acute Renal Failure. Charles C Thomas.)

ysis of large amounts of red blood cells with release of hemoglobin into the plasma. The size of the hemoglobin molecule is slightly less than that of the pores in the glomerular membrane so that much of the hemoglobin passes through the membrane into the glomerular filtrate. Therefore, after a transfusion reaction the tubular load of hemoglobin becomes far greater than can be reabsorbed by the proximal tubules. The excess hemoglobin is likely to precipitate in the nephron and in this way cause blockage. In addition, hemolysis of red blood cells probably also releases vasoconstrictor agents into the blood stream, and it is believed that the vasoconstriction can promote poor blood supply to the tubules, this acting as a further cause of tubular damage.

Physiological Effects of Acute Renal Failure

When the degree of acute renal failure is moderate, the major physiological effect is retention of salt and water. At first the tissues become edematous but the person has few other symptoms. However, within a few days the patient also develops hypertension, usually with a blood pressure increase of 30 to 40 mm. Hg; this usually continues until the acute renal failure subsides.

In the most severe cases, uremic retention of waste products ensues and acidosis soon develops. In complete renal shutdown without treatment, the patient dies within 8 to 14 days. Other effects of renal retention will be discussed in the following section in relation to chronic renal failure.

CHRONIC RENAL FAILURE — DECREASE IN NUMBER OF FUNCTIONAL NEPHRONS

In many types of kidney disease large numbers of nephrons are destroyed or damaged so severely that the remaining nephrons simply cannot perform the normal functions of the kidney. Some of the different causes of this include *chronic glomerulonephritis, traumatic loss of kidney tissue, congenital absence of kidney tissue, congenital polycystic disease* (in which large cysts develop in the kidneys and destroy surrounding nephrons by compression), *urinary tract obstruction* resulting from renal stones, *pyelonephritis,* and diseases of the renal vasculature.

Chronic Glomerulonephritis. Chronic glomerulonephritis is caused by any one of several different diseases that damage principally the glomeruli but often the tubules as well. The basic glomerular lesion is usally very similar to that which occurs in acute glomerulonephritis. It seems to begin with accumulation of precipitated antigen-antibody complex in the glomerular membrane, though in only a few instances is this caused by streptococcal infection. The result is inflammation of the glomeruli. The glomerular membrane becomes progressively thickened and is eventually invaded by fibrous tissue. In the later stages of the disease, the glomerular filtration coefficient becomes greatly reduced because of decreased numbers of filtering capillaries in the glomerular tufts and because of thickened glomerular membranes. In the final stages of the disease many of the glomeruli are completely replaced by fibrous tissue, and the function of these nephrons is thereafter lost forever.

Pyelonephritis. Pyelonephritis is an infectious and inflammatory process that usually begins in the renal pelvis but extends progressively into the renal parenchyma. The infection can result from many different types of bacteria, but especially from the colon bacillus that originates from fecal contamination of the urinary tract. Invasion of the kidneys by these bacteria results in progressive destruction of renal tubules, glomeruli, and any other structures in the path of the invading organisms. Consequently, large portions of the functional renal tissue are lost.

A particularly interesting feature of pyelonephritis is that the invading infection usually affects the medulla of the kidney more than it affects the cortex. Since one of the primary functions of the medulla is to provide the counter-current mechanism for concentrating the urine, patients with pyelonephritis frequently have reasonably normal renal function except for inability to concentrate their urine.

Destruction of Nephrons by Renal Vascular Disease — Benign Nephrosclerosis. Though many different types of vascular lesions can lead to renal ischemia and death of the renal tissue, the most common of these are: (1) *atherosclerosis* of the larger renal arteries with progressive sclerotic constriction of the vessels, (2) *fibromuscular hyperplasia* of one or more of the large arteries, which also causes occlusion of these large vessels, and (3) *benign nephrosclerosis,* a very common condition caused by sclerotic lesions of the smaller arteries and arterioles.

Arteriosclerotic or hyperplastic lesions of the larger arteries frequently affect one kidney more than the

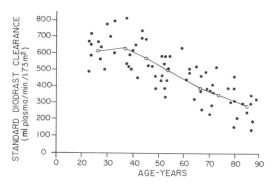

Figure 38–2. Effect of aging on the Diodrast clearance of the kidneys. (Modified from Wolstenholme et al.: Ciba Foundation Colloquia on Ageing. Little, Brown and Co.)

other and therefore cause unilaterally diminished renal function.

Benign nephrosclerosis is believed to begin with leakage of plasma through the intimal membrane of the small arteries and arterioles. This causes a fibrinoid deposit to develop in the medial layers of these vessels followed by progressive fibrous tissue invasion that eventually constricts the vessel — in many instances totally occluding it. Since there is essentially no collateral circulation among the smaller renal arteries, destruction of one of these also causes destruction of a comparable amount of renal mass. Therefore, much of the kidney tissue becomes replaced by minute areas of fibrous tissue; the kidneys become greatly reduced in size and progressively develop a nodular surface. This process occurs at least to some extent in almost all persons in old age and causes progressive decrease in both renal blood flow and renal plasma clearance. Figure 38–2 illustrates the Diodrast clearance of the kidneys (as a measure of renal blood flow) of otherwise normal persons at different ages, showing that even in the "normal" person, kidney blood flow decreases on the average to about 45 per cent of normal by the age of 80, and there is accompanying reduction of excretory function.

Abnormal Nephron Function in Chronic Renal Failure

Inability of Failing Kidneys to Excrete Excess "Loads" of Excretory Products. The most important effect of renal failure is the inability of the kidneys to cope with large "loads" of electrolytes or other substances that must be excreted. Normally, only one third of the normal number of nephrons can eliminate essentially all the normal load of waste products from the body without serious accumulation of any of these in the body fluids. However, further reduction of numbers of nephrons leads to urinary retention, and death ensues when the number of nephrons falls below 10 to 20 per cent of normal.

Function of the Remaining Nephrons in Renal Failure. In renal failure, the still-functioning nephrons usually become greatly overloaded in several different ways. First, for reasons not understood, the blood flow through the glomerulus and the amount of glomerular

filtrate formed each minute by each nephron often increase to more than double normal. Second, extra large concentrations of excretory substances, such as urea, phosphates, sulfates, uric acid, and creatinine, accumulate in the extracellular fluid. On entering the glomerular filtrate of the few remaining functioning nephrons, these constitute markedly increased tubular loads of substances that are poorly reabsorbed. These increasing loads are partially compensated for by as much as 50 per cent increase in reabsorptive power of each tubule, but even this increase is often far too little to keep pace with the loads, which may be increased as much as 1000 per cent for each nephron. Therefore, only a small fraction of the solutes are reabsorbed, and the remaining solutes act like an osmotic diuretic, resulting in rapid flushing of tubular fluid through the tubules. Consequently, the volume of urine formation by each nephron can rise to as much as 20 times normal, and a person will occasionally have *as much as 3 times normal total urine output* despite significant renal insufficiency. This paradoxical situation is caused by a greater increase in urine volume output per nephron than reduction in numbers of nephrons.

Isosthenuria.   Another effect of rapid flushing of fluid through the tubules is that the normal concentrating and diluting mechanisms of the kidney do not have time to function properly, especially the concentrating mechanism. Therefore, as progressively more nephrons are destroyed, the specific gravity of the urine approaches that of the glomerular filtrate, approximately 1.008. These effects are illustrated in Figure 38–3, which gives the approximate upper and lower limits of urine specific gravity as the number of nephrons decreases. Since the concentrating mechanism is impaired more than the diluting mechanism, an important renal function test is to determine how well the kidneys can concentrate urine when the person is dehydrated for 12 or more hours.

Effects of Renal Failure on the Body Fluids — Uremia

The effect of renal failure on the body fluids depends to a great extent on the water and food intake of the person. Assuming that the person continues to ingest

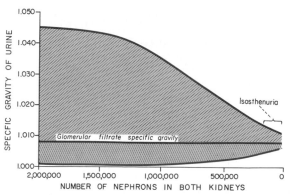

Figure 38–3. Development of *isosthenuria* in patients with decreased numbers of active nephrons.

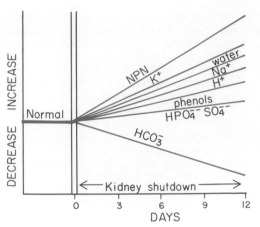

Figure 38–4. Effect of kidney shutdown on extracellular fluid constituents.

moderate amounts of water and food, the concentration changes of different substances in the extracellular fluid are approximately those shown in Figure 38–4. The most important effects are (1) *generalized edema* resulting from water and salt retention, (2) *acidosis* resulting from failure of the kidneys to rid the body of normal acidic products, (3) *high concentrations of the nonprotein nitrogens,* especially *urea,* resulting from failure of the body to excrete the metabolic end-products, and (4) *high concentration of other urinary retention products* including *creatinine, uric acid, phenols, guanidine bases, sulfates, phosphates,* and *potassium.* This condition is called *uremia* because of the high concentrations of normal urinary excretory products that collect in the body fluids.

Water Retention and Edema. If treatment is begun immediately after acute renal failure, the total body fluid content may not change at all. However, when the patient drinks water in response to normal desire, the body fluids begin increasing immediately and rapidly. If the patient takes in no electrolytes at the same time, approximately five-eighths of the water enters the cells and three-eighths remains in the extracellular fluids. Thus, a patient with edema in renal failure will usually have both *intracellular* and *extracellular edema.* On the other hand, if the patient should ingest large amounts of salt along with the water, the edema might be only extracellular because of greatly increased amounts of osmotically active substances in the extracellular compartment.

Acidosis in Renal Failure. Each day the metabolic processes of the body normally produce 50 to 100 millimoles more metabolic acid than metabolic alkali. Therefore, any time the kidneys fail to function, acid begins to accumulate in the body fluids. Normally, the buffers of the fluids can buffer up to a total of 500 to 1000 millimoles of acid without severe depression of the extracellular fluid pH, and the phosphate compounds in the bones can buffer an additional few thousand millimoles, but gradually this buffering power is used up so that the pH falls drastically. The patient becomes *comatose* at about this same time, and it is believed that this is partly caused by the acidosis as is discussed below.

Elevated Potassium Concentration in Uremia. The amount that the potassium concentration increases depends principally on the rate of protein catabolism and on potassium intake after failure has occurred. Breakdown of the cellular proteins releases potassium from combination with the proteins, and the extra potassium then passes into the extracellular fluid. Obviously, failure of the kidneys to excrete the potassium causes an elevated potassium concentration in the extracellular fluid. When the potassium concentration rises to 8 mEq./liter, it begins to have a cardiotoxic effect, resulting in dilatation of the heart, and at higher levels it leads to cardiac arrest and death.

Increase in Urea and Other Nonprotein Nitrogens (Azotemia) in Uremia. The nonprotein nitrogens include urea, uric acid, creatinine, and a few less important compounds. These, in general, are the end-products of protein metabolism and must be removed from the body continually to insure continued protein metabolism in the cells. The concentrations of these, particularly of urea, can rise to as high as 10 times normal during one to two weeks of renal failure. However, even these high levels do not seem to affect physiological function nearly so much as the high concentrations of hydrogen and potassium ions and some of the other less obvious substances such as very toxic guanidine bases, ammonium ions, and others. Yet one of the most important means for assessing the degree of renal failure is to measure the concentrations of the nonprotein nitrogens.

Uremic Coma. After a week or more of renal failure the sensorium becomes clouded, and the patient soon progresses into a state of coma. The acidosis is believed to be one of the principal factors responsible for the coma because acidosis caused by other conditions, such as severe diabetes mellitus, also causes coma. However, many other abnormalities could also be contributory — the generalized edema, the high potassium concentration, and possibly even the high nonprotein nitrogen concentration.

The respiration usually is deep and rapid in coma, which is a respiratory attempt to compensate for the metabolic acidosis. In addition to this, during the last day or so before death, the arterial pressure falls progressively, then rapidly in the last few hours. Death occurs usually when the pH of the blood falls to about 6.9.

Anemia in Renal Failure. A patient with chronic renal failure of severe degree almost always also develops severe *anemia.* The probable cause of this is the following: The kidneys normally secrete an enzyme that splits the substance *erythropoietin* from a plasma *protein.* The erythropoietin in turn stimulates the bone marrow to produce red blood cells. Obviously, if the kidneys are seriously damaged they are unable to initiate adequate production of this substance, which leads to diminished red blood cell production and consequent anemia. However, other factors, such as the high concentration of urea, hydrogen ions, and potassium, might also play important roles in causing the anemia.

Osteomalacia in Renal Failure. Prolonged renal failure also causes osteomalacia, a condition in which the bones are partially absorbed and therefore greatly weakened, as will be explained in relation to the physiology of bone in Chaper 79. The most important cause of this condition is the following: Vitamin D must be converted by a two-stage process, first in the liver and then in the kidney, into 1,25-dihydroxycholecalciferol before it is able to promote calcium absorption from the intestine. Therefore, serious damage to the kidneys greatly reduces the availability of calcium to the bones.

Dialysis of Uremic Patients with the Artificial Kidney

Artificial kidneys have now been used for almost 30 years to treat patients with severe renal failure. In certain types of acute renal failure, such as that following mercury poisoning or following circulatory shock, the artificial kidney is used simply to tide the patient over for a few weeks until the renal damage heals so that the kidneys can resume function. Also, the artificial kidney has now been developed to the point that several thousand persons with permanent renal failure or even total kidney removal are being maintained in health for years at a time, their lives depending entirely on the artificial kidney.

The basic principle of the artificial kidney is to pass blood through very minute blood channels bounded by a thin membrane. On the other side of the membrane is a *dialyzing fluid* into which unwanted substances in the blood pass by diffusion.

Figure 38–5 illustrates an artificial kidney in which blood flows continually between two thin sheets of cellophane; on the outside of the sheets is the dialyzing fluid. The cellophane is porous enough to allow all constituents of the plasma except the plasma proteins to diffuse freely in both directions — from plasma into the dialyzing fluid and from the dialyzing fluid back into the plasma. If the concentration of a substance is greater in the plasma than in the dialyzing fluid, there will be net transfer of the substance from the plasma into the dialyzing fluid. The amount of the substance that transfers depends on: (1) the difference between the concentrations on the two sides of the membrane, (2) molecular size, the smaller molecules diffusing more rapidly than larger ones, and (3) the length of time that the blood and the fluid remain in contact with the membrane.

In normal operation of the artificial kidney, blood continually flows from an artery, through the kidney, and back into a vein. The total amount of blood in the artificial kidney at any one time is usually less than 500 ml., the rate of flow may be several hundred ml. per minute, and the total diffusing surface is usually between 10,000 and 20,000 square centimeters. To prevent coagulation of blood in the artificial kidney, heparin is infused into the blood as it enters the "kidney." Then, to prevent bleeding in the patient as a result of heparin, an anti-heparin substance, protamine, is infused into the blood as it is returned to the patient.

The Dialyzing Fluid. Table 38–1 compares the constituents in a typical dialyzing fluid with those in normal

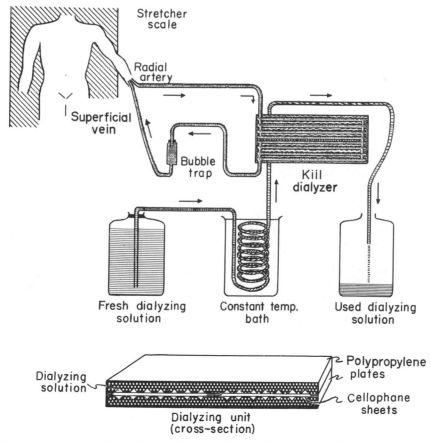

Figure 38–5. Diagram of the artificial kidney.

TABLE 38-1 COMPARISON OF DIALYZING FLUID WITH NORMAL AND UREMIC PLASMA

Constituent	Normal Plasma	Dialyzing Fluid	Uremic Plasma
Electrolytes (mEq./Liter)			
Na$^+$	142	142	142
K$^+$	5	4	7
Ca^{++}	3	3	2
Mg^{++}	1.5	1.5	1.5
Cl$^-$	107	107	107
HCO$_3^-$	27	27	14
Lactate$^-$	1.2	1.2	1.2
HPO$_4^{--}$	3	0	9
Urate$^-$	0.3	0	2
Sulfate^{--}	0.5	0	3
Nonelectrolytes (mg. %)			
Glucose	100	125	100
Urea	26	0	200
Creatinine	1	0	6

plasma and uremic plasma. Note that sodium, potassium, and chloride concentrations in the dialyzing fluid and in normal plasma are nearly identical, but in the uremic plasma the potassium concentration is considerably greater. This ion diffuses through the dialyzing membrane so rapidly that its concentration falls almost to equal that in the dialyzing fluid within only three to four hours' exposure to the dialyzing fluid.

On the other hand, there is no phosphate, urea, urate, sulfate, or creatinine in the dialyzing fluid, but they are present in high concentrations in the uremic blood. Therefore, when the uremic patient is dialyzed, these substances are lost in large quantities into the dialyzing fluid, thereby removing major proportions of them from the plasma.

Thus, the constituents of the dialyzing fluid are chosen so that those substances in excess in the extracellular fluid in uremia can be removed at rapid rates, while the normal electrolytes remain essentially normal.

Effectiveness of the Artificial Kidney. The effectiveness of an artificial kidney is expressed in terms of the amount of plasma that can be cleared of different substances each minute, which, as will be recalled from Chapter 35, is also the principal means for expressing the functional effectiveness of the kidneys themselves. Most artificial kidneys can clear 100 to 200 ml. of plasma per minute of urea, which shows that, at least in the excretion of this substance, the artificial kidney can function about 2 times as rapidly as the two normal kidneys together, whose urea clearance is only 70 ml. per minute. Yet the artificial kidney can be used for no more than 12 hours every three to four days because of danger from excess heparin, hemolysis of blood, and infection. Therefore, the overall plasma clearance is still somewhat limited when the artificial kidney replaces the normal kidneys.

HYPERTENSIVE KIDNEY DISEASE

Most of the same types of kidney disease that lead to chronic renal failure can also cause hypertension. How-

ever, this is not invariably true because damage to certain portions of the kidney is very prone to cause hypertension, whereas damage to other portions will cause uremia without hypertension. A classification of renal disease relative to its hypertensive or nonhypertensive effects is the following:

Hypertensive Renal Lesions. Essentially all renal lesions that cause *diminished vascular supply* or *diminished glomerular filtration* will cause hypertension. Both of these conditions tend to decrease the glomerular filtration rate, which in turn leads to a sequence of events (explained in Chapter 22) that begins with retention of salt and water and eventually leads to hypertension. Once the hypertension has developed, the glomerular filtration rate may return entirely to normal. If all the tubules are normal, the filtrate is then processed normally in these so that urinary excretion may be completely normal, and there may be absolutely no sign of renal failure.

Renal Diseases That Lead to Uremia but Might Not Lead to Hypertension. Loss of large numbers of whole nephrons, such as occurs from loss of one kidney and part of another kidney, will always lead to uremia if the amount of kidney tissue lost is great enough. However, if the remaining nephrons are completely normal, this condition frequently will not cause hypertension because even a slight rise in arterial pressure will increase the glomerular filtration rate sufficiently to promote rapid water and salt loss in the urine — even with few nephrons. On the other hand, a patient with this type of kidney abnormality who eats large quantities of salt will then develop very severe hypertension because the kidneys simply cannot clear adequate quantities of salt under these conditions.

Kidney Lesions That Tend to Cause Hypotension. An occasional person has a type of kidney disease that leads to tremendous salt and water loss in the urine even though there may be simultaneous retention of waste products and a state of uremia. This most frequently results from interstitial kidney disease, such as *medullary pyelonephritis*. As long as the person ingests sufficient quantities of salt, the pressure will remain normal. Placing a patient of this type on a normal salt regime, or on treatment with a diuretic, however, tends to decrease arterial pressure. At the same time, excessive amounts of renin and angiotensin are usually formed, and this keeps the pressure from falling too much but leads to further uremic symptoms. Also, further stress to the arterial pressure control system can lead to acute hypotension, often culminating in death within a few hours.

Hypertension Caused by Renal Secretion of Renin. When one part of the kidney mass is ischemic while the remainder is nonischemic (such as occurs when one renal artery is severely constricted), the ischemic renal tissue secretes large quantities of renin. This secretion leads to the formation of angiotensin II, which in turn leads to hypertension. The most probable cause of the chronic hypertension, as was discussed in Chapter 22, is (1) the ischemic kidney tissue itself excretes less than normal amounts of water and salt, and (2) the angiotensin affects the nonischemic kidney tissue, causing it also to retain water and salt. (Many physiologists have believed in the past that peripheral vasoconstriction caused by the angiotensin is the cause of this type of hypertension. However, recent experiments have demonstrated almost conclusively that this is not the case, as was discussed in Chapter 22.)

THE NEPHROTIC SYNDROME — INCREASED GLOMERULAR PERMEABILITY

Large numbers of patients with renal disease develop a so-called *nephrotic syndrome,* which is characterized especially by *loss of large quantities of plasma proteins into the urine.* In some instances this occurs without evidence of any other abnormality of renal function, but more often it is associated with some degree of renal failure.

The cause of the protein loss in the urine is increased permeability of the glomerular membrane. Therefore, any disease condition that can increase the permeability of this membrane can cause the nephrotic syndrome. Such diseases include *chronic glomerulonephritis* (in the previous discussion, it was noted that this disease primarily affects the glomeruli and often causes a greatly increased permeability of the glomerular membrane), *amyloidosis,* which results from deposition of an abnormal proteinoid substance in the walls of the blood vessels and seriously damages the basement membrane of the glomerulus, and *lipoid nephrosis,* a disease found mainly in young children.

Lipoid Nephrosis. Lipoid nephrosis is very common in children, occurring most often before the age of 4 years but occurring occasionally in adults as well. Its basic cause is unknown, but the resulting renal lesion increases the permeability of the glomerular membrane and causes loss of proteins into the urine. This lesion is the following: The epithelial cells that line the outer surface of the glomerulus are greatly swollen, and they form distorted foot processes that cover the Bowman's capsule surface of the glomerulus. (It will be recalled from the discussion of the glomerular membrane in Chapter 34 that the openings between these foot processes are very small, and it is believed to be the smallness of these openings that prevents the passage of protein through the glomerular membrane.) Therefore, in the presence of abnormally shaped foot processes, tremendous quantities of protein leak into the tubules even though larger elements of the blood, such as red cells, usually are prevented from leaking.

The name "lipoid nephrosis" is derived from the fact that large quantities of lipid droplets are found in the epithelial cells of the tubules and also from the fact that the concentration of lipid substances in the blood is usually increased (this is also true of some other types of nephrosis to a lesser extent). The lipid deposits in the tubules apparently play no role in the increased loss of proteins.

Administration of glucocorticoids such as hydrocortisone will usually cause complete remission of the disease, although glucocorticoids will not cause remission of most other types of nephrotic syndrome.

Protein Loss. In the nephrotic syndrome, as much as 30 grams of plasma proteins can be lost into the urine each day. Though the resulting low plasma protein concentration stimulates the liver to produce far more plasma proteins than usual, nevertheless the liver often cannot keep up with the loss. Therefore, in severe nephrosis the colloid osmotic pressure sometimes falls extremely low, often from the normal level of 28 mm. Hg to as low as 6 to 8 mm. Hg.

Edema. The low colloid osmotic pressure in turn allows large amounts of fluid to filter into the interstitial spaces and also into the potential spaces of the body, thus causing serious *edema.* The nephrotic person has been known on occasion to develop as much as 40 liters of excess extracellular fluid, and as much as 15 liters of this has been *ascites* in the abdomen. Also, the joints swell, and the pleural cavity and the pericardium become partially filled with fluid.

In severe nephrosis, the loss of fluid into the interstitial spaces causes the blood volume to diminish and the arterial pressure to fall 10 to 20 mm. Hg. These effects in turn lead to increased angiotensin and aldosterone production, both of which cause the kidneys to reabsorb excessive quantities of sodium chloride and water, which further compounds the seriousness of the edema.

A nephrotic person can be greatly benefited by intravenous infusion of large quantities of concentrated plasma proteins. Yet this is of only temporary benefit because enough protein can be lost into the urine in only a day or two to return the person to the original predicament.

SPECIFIC TUBULAR DISORDERS

In the discussion of active reabsorption and secretion by the tubules in Chapter 34, it was pointed out that the active transport processes are carried out by various carriers and enzymes in the tubular epithelial cells. Furthermore, there are a number of different carrier mechanisms for the different individual substances. In Chapter 3 it was also pointed out that each cellular enzyme and probably also each carrier substance is formed in response to a respective gene in the nucleus. If any required gene happens to be absent or abnormal, the tubules might be deficient in one of the appropriate enzymes or carriers. For this reason many different specific tubular disorders are known to occur for the transport of individual or special groups of substances through the tubular membrane. Essentially all these are hereditary disorders. Some of the more important ones are:

Renal Glycosuria. In this condition the blood glucose concentration may be completely normal, but the transport maximum for reabsorption of glucose each minute is greatly limited. Consequently, despite the normal blood glucose level, large quantities of glucose pass into the urine each day. Because one of the tests for diabetes mellitus (which results from lack of insulin secretion by the pancreas) is the presence of glucose in the urine, renal glycosuria (a benign condition which causes essentially no dysfunction of the body) must always be ruled out before making a diagnosis of diabetes mellitus.

Nephrogenic Diabetes Insipidus. Occasionally the renal tubules do not respond completely to antidiuretic hormone secreted by the supraopticohypophyseal system, and as a consequence large quantities of dilute urine are continually excreted. As long as the person is supplied with plenty of water, this condition rarely causes any severe difficulty. However, when adequate quantities of water are not received, the person rapidly becomes dehydrated.

Renal Tubular Acidosis. In this condition the person

is unable to secrete adequate quantities of hydrogen ions, and, as a result, large amounts of sodium bicarbonate are continually lost into the urine for reasons discussed in the preceding chapter. This causes a continual state of metabolic acidosis. However, adequate replacement therapy by continual administration of alkali can maintain normal bodily function.

Renal Hypophosphatemia. In renal hypophosphatemia the renal tubules fail to reabsorb adequate quantities of phosphate ions even when the phosphate concentration of the body fluids falls very low. This condition does not cause any serious immediate abnormalities, because the phosphate level of the extracellular fluids can vary widely without significant cellular dysfunction. However, over a long period of time the low phosphate level results in diminished calcification of the bones and causes the person to develop rickets. Furthermore, this type of rickets is refractory to vitamin D therapy in contrast to the rapid response of the usual type of rickets, which is discussed in Chapter 79.

Aminoacidurias. Some amino acids share mutual carrier systems for reabsorption, while other amino acids have distinct carrier systems of their own. Rarely, a condition called *generalized aminoaciduria* results from deficient reabsorption of all amino acids, but, more frequently, deficiencies of specific carrier systems may result in (1) *essential cystinuria,* in which large amounts of cystine fail to be reabsorbed and often crystallize in the urine to form renal stones, (2) simple *glycinuria,* in which glycine fails to be reabsorbed, or (3) *beta-aminoisobutyric aciduria,* which occurs in about 5 per cent of all people but apparently has no clinical significance.

RENAL FUNCTION TESTS

The renal function tests can be divided into three categories: (1) determination of renal clearances, (2) measurement of substances in the blood that are normally excreted by the kidneys, and (3) chemical and physical analyses of the urine.

Renal Clearance Tests. Any of the renal clearance tests, including clearance of para-aminohippuric acid, Diodrast, inulin, mannitol, or other substances, as described in Chapter 35, can be used as renal function tests. Indeed, if all these are performed one can determine the glomerular filtration rate, the effective blood flow through the kidney per minute, the filtration fraction, and many other characteristics of renal function. However, because most of these clearance tests are difficult to do, several other clearance tests in which radiopaque or radioactive substances are excreted from the blood into the renal pelvis are usually employed. Some of these are the following:

Intravenous Pyelography. Several substances containing large quantities of iodine in their molecules — Diodrast, Hippuran, and Iopax — are excreted into the urine both by glomerular filtration and by active tubular secretion. Consequently, their concentration in the urine becomes very high within a few minutes after intravenous injection. Also, the iodine in the compounds makes them relatively opaque to x-rays. Therefore, x-ray pictures can be made showing shadows of the renal pelves, of the ureters, and even of the urinary bladder. Ordinarily, a sufficient quantity is excreted within five minutes after injection to give good shadows of the kidney pelves. Failure to show a distinct shadow within this time indicates depressed renal clearance.

Radioactive Clearance Studies. If any of the above substances (or many others) is prepared with *radioactive* iodine or some other radioactive nuclide, one can measure the radioactivity from both renal pelves by placing appropriate radioactivity counters over the kidneys. Only a trace amount of the substance need be injected intravenously and the degree of radioactivity recorded during the following few minutes to determine roughly the renal clearances.

A special value of both x-ray and radioactive pyelography is that they measure function of each kidney independently of the other rather than total function of both kidneys together, as is measured by the other renal function tests.

Blood Analyses as Tests of Renal Function. One can also estimate how well the kidneys are performing their functions by measuring the concentrations of various substances in the blood. For instance, the normal concentration of *urea* in the blood is 26 mg./100 ml., but in severe cases of renal insufficiency this can rise to as high as 300 mg. per cent. The normal concentration of *creatinine* in the blood is 1.1 mg. per cent, but this, too, can increase as much as 10-fold. To determine the degree of metabolic acidosis resulting from renal dysfunction, one can measure the *base excess* in the blood, as discussed in the preceding chapter. Though these different tests are not as satisfactory as clearance tests for determining the functional capabilities of the kidneys, they are easy to perform, and they do tell the physician how seriously the internal environment has been disturbed.

Physical Measurements of the Urine as Renal Function Tests. Obviously, one of the most important urinary measurements is the *volume of urine* formed each day. In acute renal failure this can fall to zero, and in chronic renal failure it usually is diminished. On the other hand, moderate renal failure may actually increase the urinary output, as was described above, because of vast overfunction of the remaining nephrons when the majority have been destroyed.

A second factor frequently measured is the *specific gravity* of the urine. Depending on the types of substances being cleared, this can vary tremendously; its upper extreme can go as high as 1.045 and it can fall to as little as 1.002. To test the ability of the kidneys to dilute the urine, the patient drinks large quantities of water, and measurements are made of the minimal specific gravity that can be attained. Then, at another time, the patient goes without water for 12 or more hours, and the maximum concentration of the urine is measured. Referring back to Figure 38–3, note that the concentrating ability of the kidneys is especially compromised as the number of nephrons decreases.

DIURETICS AND MECHANISMS OF THEIR ACTIONS

A diuretic is a substance that increases the rate of urine output. Most diuretics act by decreasing the rate of reabsorption of fluid from the tubules.

The principal use of diuretics is to reduce the total amount of fluid in the body. They are especially important in treating edema and hypertension.

When using a diuretic it is usually important that the rate of sodium loss in the urine also be increased as well as the rate of water loss. The reason for this is the following: If water alone were removed from the body fluids, the fluids would become hypertonic and elicit an osmoreceptor response, followed by marked secretion of antidiuretic hormone. Consequently, large amounts of water would be reabsorbed immediately from the tubules, which would nullify the effect of the diuretic. However, if sodium is lost along with the water, this nullification will not result. Therefore, all valuable diuretics caused marked *natriuresis* (sodium loss) as well as diuresis.

Several significant types of diuretics are the following:

The Osmotic Diuretics. Injection into the blood stream of *urea, sucrose, mannitol,* or any other substance not easily reabsorbed by the tubules causes a great increase in the osmotically active substances in the tubules. The osmotic pressure of these prevents water reabsorption, so that large amounts of tubular fluid flush on into the urine.

The same effect occurs when the glucose concentration of the blood rises to very high levels in diabetes mellitus. Above a glucose concentration of about 250 mg. per cent, very little glucose is reabsorbed by the tubules; instead it acts as an osmotic diuretic and causes rapid loss of fluid into the urine. The name "diabetes" refers to the prolific urine flow.

Diuretics That Diminish Active Reabsorption. Any substance that inhibits carrier systems in the tubular epithelial cells and thereby diminishes active reabsorption of tubular solutes increases the tubular osmotic pressure and causes osmotic diuresis. Some of the most commonly used drugs of this type are the following:

The "Loop" Diuretics, Furosemide and Ethacrynic Acid. Furosemide and ethacrynic acid are the most powerful of all the clinically used diuretics. They are called loop diuretics because they block active reabsorption of chloride ions from the ascending limb of the loop of Henle as well as from all the remainder of the diluting segment of the distal tubule. And, when chloride ions are not reabsorbed, the positive ions normally reabsorbed along with the chloride ions — mainly sodium ions — also are not reabsorbed. This blockage of reabsorption of chloride and sodium ions causes diuresis for two reasons: (1) It allows greatly increased quantities of solutes to be delivered into the distal portions of the nephrons, and these then act as osmotic agents to prevent water reabsorption as well. (2) Failure to absorb sodium and chloride ions from the loop of Henle into the medullary interstitium decreases the concentration of the medullary interstitial fluid. Consequently, the concentrating ability of the kidney is greatly decreased so that the reabsorption of fluid is further decreased. Because of these two effects as much as 20 to 30 per cent of the glomerular filtrate may be delivered into the urine, causing under acute conditions urine outputs as great as 25 times normal for a period of a few minutes.

Chlorothiazide. Chlorothiazide and other thiazide derivatives act primarily on the convoluted portions of the distal tubules and probably also on the cortical collecting tubules to prevent active sodium reabsorption; under favorable conditions they cause as much as 8 per cent of the glomerular filtrate to pass into the urine.

Carbonic Anhydrase Inhibitors — Acetazolamide. Acetazolamide (Diamox) and other carbonic anhydrase inhibitors primarily block reabsorption of bicarbonate ions from the proximal tubules. They do this by inhibiting the carbonic anhydrase that lines the luminal border of the tubular epithelial cells and normally catalyzes the dissociation of carbonic acid into water and carbon dioxide. This prevents removal of bicarbonate ions from the tubular fluid, and these remain in the tubules to act as an osmotic diuretic. However, use of this drug also causes some degree of acidosis because of excessive loss of bicarbonate ions from the body fluids.

Competitive Aldosterone Inhibitors — Spironolactone. Spironolactone and several other similar substances compete with aldosterone for receptor sites in the distal nephron epithelial cells and thereby block the sodium reabsorption promoting effect of aldosterone. As a consequence, the sodium remains in the tubules and acts as an osmotic diuretic. These drugs also block the effect of aldosterone of promoting potassium secretion into the tubules. Therefore, in some instances, the extracellular fluid potassium concentration becomes dangerously elevated.

Diuretics that Inhibit the Secretion of Antidiuretic Hormone. The single diuretic that is most important in inhibiting the secretion of antidiuretic hormone (ADH) is *water*. This was discussed in Chapter 36. When large amounts of water are ingested, the body fluids become diluted and ADH is no longer secreted by the supraoptico-posterior pituitary system. As a result, water fails to be reabsorbed by the distal tubules and collecting ducts, and large amounts of dilute urine flow. In addition to water, various psychic factors as well as certain drugs that affect the central nervous system, such as *narcotics, hypnotics,* and *anesthetics,* can inhibit ADH secretion. Also, *alcohol* inhibits ADH secretion. For this reason, these can all cause increases in urinary output.

MICTURITION

Micturition is the process by which the urinary bladder empties when it becomes filled. Basically the bladder (1) progressively fills until the tension in its walls rises above a threshold value, at which time (2) a nervous reflex called the "micturition reflex" occurs that either causes micturition or, if it fails in this, at least causes a conscious desire to urinate.

PHYSIOLOGIC ANATOMY OF THE BLADDER AND ITS NERVOUS CONNECTIONS

The urinary bladder, which is illustrated in Figure 38-6, is a smooth muscle chamber composed of three principal parts: (1) the *body,* which is composed mainly of the *detrusor muscle,* (2) the *trigone,* a small triangular area near the neck of the bladder through which both the *ureters* and the *urethra* pass, and (3) the *bladder neck,* which is also called the *posterior urethra.*

During bladder expansion the body of the bladder stretches, and during micturition the detrusor muscle contracts to empty the bladder.

Each ureter enters the bladder through its posterolateral margin, coursing obliquely through the detrusor muscle and then passing still another 1 to 2 cm. underneath the bladder mucosa before emptying at the upper corner of the trigone.

The bladder neck (posterior urethra) is 2 to 3 cm. long, and its wall is composed of detrusor muscle interlaced among a large amount of elastic tissue. The muscle in this area is frequently called the *internal sphincter,* and its natural tone prevents emptying of the bladder until the pressure in the body of the bladder rises above a critical threshold.

Beyond the bladder neck, the urethra passes through the *urogenital diaphragm* which contains a layer of muscle called the *external sphincter* of the bladder. This muscle is a voluntary skeletal muscle, in contrast to the muscle of the bladder body and bladder neck, which is entirely smooth muscle. This external muscle is under voluntary control of the nervous system and can be used to prevent urination even when the involuntary controls are attempting to empty the bladder.

TRANSPORT OF URINE THROUGH THE URETERS

The ureters are small, smooth muscle tubes that originate in the pelves of the two kidneys and pass downward to enter the bladder. Each ureter is innervated by both sympathetic and parasympathetic nerves, and each also has an intramural plexus of neurons and nerve fibers that extends along its entire length.

As urine collects in the pelvis, the pressure in the pelvis increases and initiates a peristaltic contraction beginning in the pelvis and spreading downward along the ureter to force urine toward the bladder. A peristaltic wave, traveling at a velocity of about 3 cm./sec., occurs from once every 10 seconds to once every 2 to 3 minutes. The peristaltic wave can move urine against an obstruction with a pressure as high as 25 to 50 mm. Hg. Parasympathetic stimulation increases and sympathetic stimulation decreases the frequency. Transmission of the peristaltic wave is probably caused mainly by nerve impulses passing along the intramural plexus in the same manner that the intramural plexus functions in the gut.

At the lower end, the ureter penetrates the bladder obliquely through the *trigone,* as illustrated in Figure 38–6. The ureter courses for several centimeters under the bladder epithelium so that pressure in the bladder compresses the ureter, thereby preventing backflow of urine when pressure builds up in the bladder during micturition.

Pain Sensations from the Ureters, and the Ureterorenal Reflex. The ureters are well supplied

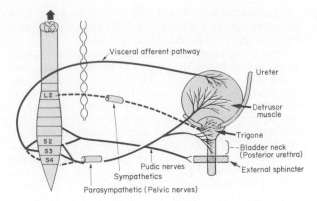

Figure 38–6. The urinary bladder and its innervation.

with pain nerve fibers. When the ureter becomes blocked, such as by a ureteral stone, intense reflex constriction associated with very severe pain occurs. In addition, the pain impulses probably cause a sympathetic reflex back to the kidney to constrict the renal arterioles, thereby decreasing urinary output from that kidney. This effect is called the *ureterorenal reflex;* it obviously is important to prevent excessive flow of fluid into the pelvis of a kidney with a blocked ureter.

TONE OF THE BLADDER WALL, AND THE CYSTOMETROGRAM DURING BLADDER FILLING

The solid curve of Figure 38–7 is called the *cystometrogram* of the bladder. It shows the changes in intravesical pressure as the bladder fills with urine. When no urine at all is in the bladder, the intravesical pressure is approximately zero, but by the time 100 ml. of urine has collected, the pressure will have risen to 5 to 10 cm. water. Additional urine up to 300 to 400 ml. can collect

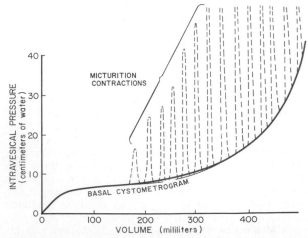

Figure 38–7. A normal cystometrogram showing also acute pressure waves (the dashed curves) caused by micturition reflexes.

with only a small amount of additional rise in pressure; this is caused by *intrinsic tone* of the bladder wall itself. Beyond 400 to 500 ml., collection of more urine causes the pressure to rise very rapidly.

Superimposed on the tonic pressure changes during filling of the bladder are periodic acute increases in pressure, which last from a few seconds to more than a minute. The pressure can rise only a few centimeters of water or it can rise to over 100 cm. water. These are *micturition waves* in the cystometrogram caused by the micturition reflex, which is discussed below.

THE MICTURITION REFLEX

Referring again to Figure 38–7, one sees that as the bladder fills, many superimposed *micturition contractions* begin to appear. These are the result of a stretch reflex initiated by stretch receptors in the bladder wall. Sensory signals are conducted to the sacral segments of the cord through the pelvic nerves and then back again to the bladder through the parasympathetic fibers in these same nerves.

Once a micturition reflex begins, it is "self-regenerative." That is, initial contraction of the bladder further activates the receptors to cause still further increase in afferent impulses from the bladder, which causes further increase in reflex contraction of the bladder, the cycle thus repeating itself again and again until the bladder has reached a strong degree of contraction. Then, after a few seconds to more than a minute, the reflex begins to fatigue, and the regenerative cycle of the micturition reflex ceases, allowing rapid reduction in bladder contraction. In other words, the micturition reflex is a single complete cycle of (a) progressive and rapid increase in pressure, (b) a period of sustained pressure, and (c) return of the pressure to the basal tonic pressure of the bladder. Once a micturition reflex has occurred and has not succeeded in emptying the bladder, the nervous elements of this reflex usually remain in an inhibited state for at least a few minutes to sometimes as long as an hour or more before another micturition reflex occurs. However, as the bladder becomes more and more filled, micturition reflexes occur more and more often and more and more powerfully.

Once the micturition reflex becomes powerful enough and the fluid pressure in the bladder great enough to force the bladder neck open despite the tonic contraction of the bladder neck muscle, stretch of the neck then causes still another reflex. This reflex passes to the sacral portion of the spinal cord and then back to the external sphincter to inhibit it. If this inhibition is more potent than the voluntary constrictor signals from the brain, then urination will occur. If not, urination still will

not occur until the bladder fills still more and the micturition reflex becomes more powerful.

Control of Micturition by the Brain. The micturition reflex is a completely automatic cord reflex, but it can be inhibited or facilitated by centers in the brain. These include (a) strong *facilitatory and inhibitory centers in the brain stem,* probably located in the pons, and (b) several *centers located in the cerebral cortex* that are mainly inhibitory but can at times become excitatory.

The micturition reflex is the basic cause of micturition, but the higher centers normally exert final control of micturition by the following means:

1. The higher centers keep the micturition reflex partially inhibited all the time except when it is desired to micturate.

2. The higher centers prevent micturition, even if a micturition reflex occurs, by continual tonic contraction of the external urinary sphincter until a convenient time presents itself.

3. When the time to urinate arrives, the cortical centers can (a) facilitate the sacral micturition centers to initiate a micturition reflex and (b) inhibit the external urinary sphincter so that urination can occur.

ABNORMALITIES OF MICTURITION

The Atonic Bladder. Destruction of the sensory nerve fibers from the bladder to the spinal cord prevents transmission of stretch signals from the bladder and, therefore, also prevents micturition reflex contractions. Therefore, the person loses all bladder control despite intact efferent fibers from the cord to the bladder and despite intact neurogenic connections with the brain. Instead of emptying periodically, the bladder fills to capacity and overflows a few drops at a time through the urethra. This is called *overflow dribbling.*

The atonic bladder was a common occurrence when syphilis was widespread, because syphilis frequently causes constrictive fibrosis around the dorsal nerve root fibers where they enter the spinal cord and, subsequently, destroys these fibers. This condition is called *tabes dorsalis,* and the resulting bladder condition is called a *tabetic bladder.* Another common cause of this condition is crushing injuries to the sacral region of the cord.

The Automatic Bladder. If the spinal cord is damaged above the sacral region but the sacral segments are still intact, typical micturition reflexes still occur. However, they are no longer controllable by the brain. During the first few days to several weeks after the damage to the cord has occurred, the micturition reflexes are completely suppressed because of sudden loss of facilitatory impulses from the brain stem and cerebrum. However, if the bladder is emptied periodically by catheterization to prevent physical bladder injury, the excitability of the micturition reflex gradually increases until typical micturition reflexes return.

It is especially interesting that scratching or otherwise

stimulating the skin in the genital region can sometimes elicit a micturition reflex in this condition, thus providing a means by which some patients can still control urination.

The Uninhibited Neurogenic Bladder.   Another common abnormality of micturition is the so-called "uninhibited neurogenic bladder," which results in frequent and relatively uncontrollable micturition. This condition derives from damage in the spinal cord or brain stem that interrupts most of the inhibitory signals. Therefore, facilitatory impulses passing continually down the cord keep the sacral centers so excitable that even a very small quantity of urine will elicit an uncontrollable micturition reflex and thereby promote urination.

REFERENCES

Bailey, G. L. (ed.): Hemodialysis: Principles and Practice. New York, Academic Press, 1973.

Bennett, W. M., et al.: Drugs and Renal Disease. New York, Churchill Livingstone, 1978.

Bissada, N. K., and Finkbeiner, A. E.: Lower Urinary Tract Function and Dysfunction. New York, Appleton-Century-Crofts, 1978.

Boyarsky, S.: Ureteral Dynamics. Baltimore, Williams & Wilkins, 1972.

Boyarsky, S., et al.: Care of the Patient with Neurogenic Bladder. Boston, Little, Brown, 1979.

Burg, M., and Stoner, L.: Renal tubular chloride transport and the mode of action of some diuretics. Annu. Rev. Physiol., 38:37, 1976.

Carriere, S.: Compensatory renal hypertrophy in dogs: Single nephron glomerular filtration rate. Yale J. Biol. Med., 51(3):307, 1978.

Chapman, A. (ed.): Acute Renal Failure. New York, Churchill Livingstone, 1980.

Coleman, T. G., et al.: Chronic hemodialysis and circulatory function. Simulation, Nov.:222, 1970.

Darmady, E. M., and MacIver, A. G.: Renal Pathology. Boston, Butterworths, 1979.

DeLuca, H. F.: The kidney as an endocrine organ involved in calcium homeostasis. Kidney Int., 4:80, 1973.

Diamond, L. H., and Balow, J. E. (eds.): Nephrology Reviews, 1980. New York, John Wiley & Sons, 1980.

Diezi, J., et al.: Studies on possible mechanisms of early functional compensatory adaptation in the remaining kidney. Yale J. Biol. Med., 51(3):265, 1978.

Dirks, J. H., and Wong, N. L. M.: Acute functional adaptation to nephron loss: Micropuncture studies. Yale J. Biol. Med., 51(3):255, 1978.

Drukker, W., et al. (eds.): Replacement of Renal Function by Dialysis. Boston, M. Nijhoff Medical Division, 1979.

Earley, L. E., and Gottschalk, C. W. (eds.): Strauss and Welt's Diseases of the Kidney, 3rd Ed. Boston, Little, Brown, 1979.

Gennari, F. J., and Kassirer, J. P.: Osmotic diuresis. N. Engl. J. Med., 291:714, 1974.

Ginzler, E. M., et al.: The natural history and response to therapy of lupus nephritis. Annu. Rev. Med., 31:463, 1980.

Goldberg, M.: The renal physiology of diuretics. In Orloff, F., and Berliner, R. W. (eds.): Handbook of Physiology. Sec. 8. Baltimore, Williams & Wilkins, 1973, p. 1003.

Grantham, J. J.: Fluid secretion in the nephron: Relation to renal failure. Physiol. Rev., 56:248, 1976.

Hepinstall, R. H.: Pathology of the Kidney, 2nd Ed. Boston, Little, Brown, 1975.

Juncos, L. I.: Physiological Basis of Diuretic Therapy: A Programmed Course. Springfield, Ill., Charles C Thomas, 1978.

Katz, A. I., et al.: The role of renal "work" in compensatory kidney growth. Yale J. Biol. Med., 51(3):331, 1978.

Kim, Y., and Michael, A. F.: Idiopathic membranoproliferative glomerulonephritis. Annu. Rev. Med. 31:273, 1980.

Kincaid-Smith, P., et al. (eds.): Glomerulonephritis: Morphology, Natural History, and Treatment. New York, John Wiley & Sons, 1973.

Kincaid-Smith, P., et al. (eds.): Progress in Glomerulonephritis. New York, John Wiley & Sons, 1979.

Kirschenbaum, M. A.: Renal Disease. Boston, Houghton Mifflin, 1978.

Knox, F. G. (ed.): Textbook of Renal Pathophysiology. Hagerstown, Md., Harper & Row, 1978.

Krane, R. J., and Siroky, M. B. (eds.): Clinical Neuro-Urology. Boston, Little, Brown, 1979.

Leaf, A., and Cotran, R. S.: Renal Pathophysiology. 2nd Ed. New York, Oxford University Press, 1980.

Leaf, A., and Giebisch, G. (eds.): Renal Pathophysiology – Recent Advances. New York, Raven Press, 1979.

Nowinski, W., and Goss, R. J.: Compensatory Renal Hypertrophy. New York, Academic Press, 1970.

Oberley, E. T., and Oberley, T. D.: Understanding Your New Life with Dialysis: A Patient Guide for Physical and Psychological Adjustment to Maintenance Dialysis. Springfield, Ill., Charles C Thomas, 1978.

O'Reilly, P. H., et al. (eds.): Nuclear Medicine in Urology and Nephrology. Boston, Butterworths, 1979.

Papper, S.: Clinical Nephrology, 2nd Ed. Boston, Little, Brown, 1978.

Pesce, A. J., and First, M. R.: Proteinuria: An Integrated Review. New York, Marcel Dekker, 1979.

Pollak, V. E., and Pesce, A. J.: Maintenance of body protein homeostasis. In Frohlich, E. D. (ed.): Pathophysiology, 2nd Ed. Philadelphia, J. B. Lippincott Co., 1976, p. 221.

Ritz, E. et al. (eds.): Pathophysiological Problems in Clinical Nephrology. New York, S. Karger, 1978.

Rubin, M. I., and Barratt, T. M. (eds.): Pediatric Nephrology. Baltimore, Williams & Wilkins, 1975.

Schrier, R. W., et al.: Role of solute excretion in prevention of norepinephrine-induced acute renal failure. Yale J. Biol. Med., 51(3):355, 1978.

Segal, S., and Thier, S. O.: Renal handling of amino acids. In Orloff, F., and Berliner, R. W. (eds.): Handbook of Physiology. Sec. 8. Baltimore, Williams & Wilkins, 1973, p. 653.

Strauss, J. (ed.): Nephrotic Syndrome. New York, Garland Press, 1979.

Striker, G. E., et al.: Use and Interpretation of Renal Biopsy. Philadelphia, W. B. Saunders Co., 1978.

Thurau, K. (ed.): Experimental Acute Renal Failure. New York, Springer-Verlag, 1976.

Wardle, E. N.: Renal Medicine. Baltimore, University Park Press, 1979.

Weller, J. M. (ed.): Fundamentals of Nephrology. Hagerstown, Md., Harper & Row, 1979.

Whelpton, D. (ed.): Renal Dialysis. Philadelphia, J. B. Lippincott Co., 1974.

Wilson, C. B. (ed.): Immunological Mechanisms of Renal Disease. New York, Churchill Livingstone, 1979.

Wong, N. L. M., et al.: Chronic reduction in renal mass. Micropuncture studies of response to volume expansion and furosemide. Yale J. Biol. Med., 51(3):289, 1978.

Yoshitoshi, Y. (ed.): Glomerulonephritis. Baltimore, University Park Press, 1979.

Part VII

RESPIRATION

39

Pulmonary Ventilation

The process of respiration can be divided into four major mechanistic events: (1) pulmonary ventilation, which means the inflow and outflow of air between the atmosphere and the lung alveoli, (2) diffusion of oxygen and carbon dioxide between the alveoli and the blood, (3) transport of oxygen and carbon dioxide in the blood and body fluids to and from the cells, and (4) regulation of ventilation and other facets of respiration. The present chapter and the three following discuss, respectively, these four major aspects of respiration. In subsequent chapters pulmonary disorders and special respiratory problems related to aviation medicine and deep sea diving physiology are discussed to illustrate some of the basic principles of respiratory physiology.

MECHANICS OF PULMONARY VENTILATION

BASIC MECHANISMS OF LUNG EXPANSION AND CONTRACTION

The lungs can be expanded and contracted in two ways, (1) by downward and upward movement of the diaphragm to lengthen or shorten the chest cavity, and (2) by elevation and depression of the ribs to increase and decrease the anteroposterior diameter of the chest cavity. Figure 39–1 illustrates these two methods.

Normal quiet breathing is accomplished almost entirely by inspiratory movement of the diaphragm. During inspiration the diaphragm pulls the lower surfaces of the lungs downward. Then, during expiration, the diaphragm simply relaxes and the elastic recoil of the lungs, chest wall, and abdominal structures compresses the lungs. During heavy breathing, however, the elastic forces are not powerful enough to cause the necessary rapid expiration, so this is achieved by contraction of the abdominal muscles, which forces the abdominal contents upward against the bottom of the diaphragm.

The second method for expanding the lungs is to raise the rib cage. This expands the lungs because, in the natural resting position, the ribs slant downward, thus allowing the sternum to fall backward toward the spinal column. But, when the rib cage is elevated, the ribs project directly forward so that the sternum now also moves forward away from the spine, making the anteroposterior thickness of the chest about 20 per cent greater during maximum inspiration than during expiration. Therefore, those muscles that elevate the chest cage can be classified as muscles of inspiration and those muscles that depress the chest cage, as muscles of expiration. The muscles that raise the rib cage include (1) the *sternocleidomastoid* muscles that lift upward on the sternum, (2) the *anterior serrati* that lift many of the ribs, (3) the *scaleni* that lift the first two ribs, and the *external intercostals*, the function of which will be explained below. The muscles that pull the rib cage downward during expiration are (1) the *abdominal recti* that have the powerful effect of pulling downward on the lower ribs at the same time that

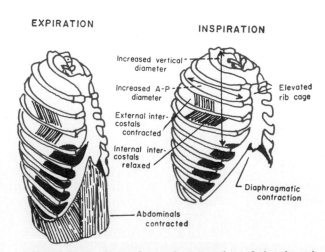

Figure 39–1. Expansion and contraction of the thoracic cage during expiration and inspiration, illustrating especially diaphragmatic contraction, elevation of the rib cage, and function of the intercostals.

they and the other abdominal muscles also compress the abdominal contents upward toward the diaphragm, and (2) the *internal intercostals*.

Figure 39–1 illustrates the mechanism by which the external and internal intercostals act to cause inspiration and expiration. To the left, the ribs during expiration are angled downward and the external intercostals are stretched in a forward and downward direction. As they contract, they pull the upper ribs forward in relation to the lower ribs, and this causes leverage on the ribs to raise them upward. Conversely, in the inspiratory position, the internal intercostals are stretched, and their contraction pulls the upper ribs backward in relation to the lower ribs. This causes leverage in the opposite direction and lowers the chest cage.

RESPIRATORY PRESSURES

Intra-alveolar Pressure. The respiratory muscles cause pulmonary ventilation by alternatively compressing and distending the lungs, which in turn causes the pressure in the alveoli to rise and fall. During inspiration the intra-alveolar pressure becomes slightly negative *with respect to atmospheric pressure,* normally slightly less than −1 mm. Hg, and this causes air to flow inward through the respiratory passageways. During normal expiration, on the other hand, the intra-alveolar pressure rises to slightly less than +1 mm. Hg, which causes air to flow outward through the respiratory passageways. Note especially, how little pressure is required to move air into and out of the normal lung, an effect that often is seriously compromised in many lung diseases.

During maximum expiratory effort with the glottis closed the intra-alveolar pressure can be increased to over 100 mm. Hg in the strong healthy male, and during maximum inspiratory effort it can be reduced to as low as −80 mm. Hg.

Recoil Tendency of the Lungs, and the Intrapleural Pressure. The lungs have a continual elastic tendency to collapse and therefore to recoil away from the chest wall. This elastic tendency is caused by two different factors. First, throughout the lungs are many *elastic fibers* that are stretched by lung inflation and therefore attempt to shorten. Second, and even more important, the *surface tension* of the fluid lining the alveoli also causes a continual elastic tendency for the alveoli to collapse. This effect is caused by intermolecular attraction between the surface molecules of the fluid that tends continually to reduce the surface areas of the individual alveoli; all these minute forces added together tend to collapse the whole lung and therefore to cause its recoil away from the chest wall.

Ordinarily, the elastic fibers in the lungs account for about one-third of the recoil tendency, and the surface tension phenomenon accounts for about two-thirds.

The total recoil tendency of the lungs can be measured by the amount of negative pressure in the intrapleural spaces required to prevent collapse of the lungs, and this pressure is called the *intrapleural pressure* or, occasionally, the *recoil pressure.* It is normally about −4 mm. Hg. That is, when the alveolar spaces are open to the atmosphere through the trachea so that their pressure is at atmospheric pressure, a pressure of −4 mm. Hg in the intrapleural space is required to keep the lungs expanded to normal size. When the lungs are stretched to very large size, such as at the end of deep inspiration, the intrapleural pressure required then to expand the lungs may be as great as −12 to −18 mm. Hg.

"Surfactant" in the Alveoli, and Its Effect on the Collapse Tendency. A lipoprotein mixture called "surfactant" is secreted by special *surfactant-secreting cells* that are component parts of the alveolar epithelium. This mixture, containing especially the phospholipid *dipalmitoyl lecithin,* decreases the surface tension of the fluids lining the alveoli. In the absence of surfactant, lung expansion is extremely difficult, often requiring intrapleural pressures as low as −20 to −30 mm. Hg to overcome the collapse tendency of the alveoli. This illustrates that surfactant is exceedingly important to minimize the effect of surface tension in causing collapse of the lungs.

A few newborn babies, especially premature babies, do not secrete adequate quantities of surfactant, which makes lung expansion difficult. Without immediate and very careful treatment, most of these die soon after birth because of inadequate ventilation. This condition is called *hyaline membrane disease* or *respiratory distress syndrome.*

Surfactant acts by forming a monomolecular layer at the interface between the fluid lining the alveoli and the air in the alveoli. This prevents the development of a water-air interface, which has 2 to 14 times as much surface tension as the surfactant-air interface.

Role of Surfactant to "Stabilize" the Alveoli. The effect of surface tension in causing collapse of an alveolus becomes very much greater as the diameter of the alveolus decreases. This effect is explained by the law of Laplace, which was discussed in Chapter 18 in relation to the blood vessels. That is, the transalveolar pressure that is required to keep the alveolus expanded is directly proportional to the tension in the alveolar wall (in this instance, the surface tension of the fluid) *divided by* the diameter. Therefore, as the diameter becomes less, the required pressure becomes proportionately greater. Now, let us see what would happen if two occluded alveoli were connected together and one of them were slightly smaller than the other. Obviously the pressure generated in the small alveolus, because of its greater collapse tendency, would be considerably greater than the pressure generated in the larger alveolus. Consequently, air would be displaced from the smaller alveolus into the larger one, and the smaller alveolus would thus become still smaller and its pressure still greater while the larger alveolus would become still larger but its pressure less. And this process would continue until the smaller alveolus would collapse entirely while displacing all of its air into the larger one. This phenomenon is called *instability of the alveoli*.

On the other hand, surfactant plays a very important role in preventing this unstable condition. Experiments with surfactant have shown that as an alveolus becomes smaller and the surfactant becomes more concentrated at the surface of the alveolar lining fluid, the surface tension becomes progressively more reduced. On the other hand, as an alveolus becomes larger and the surfactant is spread more thinly on the fluid surface, the surface tension becomes much greater. Thus, this special characteristic of surfactant helps to "stabilize" the sizes of the alveoli, causing the larger alveoli to contract more and the smaller ones to contract less. This effect obviously helps to insure that the alveoli in any one area of the lung all remain approximately the same size.

Still another factor also helps to insure that the alveoli maintain approximately equal diameter. This is the phenomenon called *interdependence*. That is, the walls of adjacent alveoli are mutually attached to each other; therefore, it is difficult physically for one alveolus to contract without the other one contracting at the same time, this also helping to assure equality of alveolar size.

Role of Surfactant in Preventing Accumulation of Edema Fluid in the Alveoli. The surface tension of the fluid in the alveoli not only tends to cause collapse of the alveoli, but also tends to pull fluid into the alveoli from the alveolar wall. In the normal lung, when there are adequate amounts of surfactant, the surface tension even normally is great enough to pull fluid away from the wall with an average pressure of −3 mm. Hg. In the absence of surfactant, the average pulling force may become as great as −10 to −20 mm. Hg, and this is now enough to cause massive filtration of fluid out of the alveolar wall capillaries into the alveoli, thus filling the alveoli with fluid. Therefore, one of the consequences of the absence of surfactant is severe pulmonary edema. For instance, in the respiratory distress syndrome of the newborn, the amount of surfactant is greatly decreased, and a major number of the alveoli are filled with fluid; this is one of the factors that causes the severe respiratory distress.

EXPANSIBILITY OF THE LUNGS AND THORAX: "COMPLIANCE"

The expansibility of the lungs and thorax is called *compliance*. This is expressed as the *volume increase in the lungs for each unit increase in intra-alveolar pressure*. The compliance of the normal lungs and thorax combined is 0.13 liter per centimeter of water pressure. That is, every time the alveolar pressure is increased by 1 cm. water, the lungs expand 130 ml.

Compliance of the Lungs Alone. The lungs alone, when removed from the chest, are almost twice as distensible as the lungs and thorax together, because the thoracic cage itself must also be stretched when the lungs are expanded *in situ*. Thus, the compliance of the normal lungs when removed from the thorax is about 0.22 liter per cm. water. This illustrates that the muscles of inspiration must expend energy not only to expand the lungs but also to expand the thoracic cage around the lungs.

Measurement of Lung Compliance. Compliance of the lungs is measured in the following way: The person's glottis must first be completely open and remain so. Then air is inspired in steps of approximately 50 to 100 ml. at a time, and pressure measurements are made from an intra-esophageal balloon (which measures almost exactly the intrapleural pressure) at the end of each step, until the total volume of air in the lungs is equal to the normal tidal volume of the person. Then the air is expired, also in steps, until the lung volume returns to the expiratory resting level. The relationship of lung volume to pressure is then plotted as illustrated in Figure 39–2. This graph shows that the plot during inspiration is a different curve from that during expiration, which is caused by the viscous properties of the lungs discussed

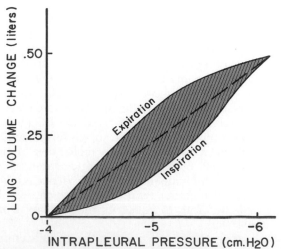

Figure 39–2. Compliance diagram in a normal person. This diagram shows the compliance of the lungs alone.

above. The average compliance is represented by the dashed line in the figure. Thus, the lung volume increases 220 ml. for a change in translung pressure (atmospheric pressure in the lung minus intra-esophageal pressure) of 1 cm. water. Therefore, the compliance in this instance is 0.22 liter per cm. water.

Slight modifications of this procedure can be used to measure the compliance of the lungs and thoracic cage together.

Factors That Cause Abnormal Compliance. Any condition that destroys lung tissue, causes it to become fibrotic or edematous, blocks the bronchioles, or in any other way impedes lung expansion and contraction causes decreased lung compliance. When considering the compliance of both the lung and thorax together, one must also include any abnormality that reduces the expansibility of the thoracic cage. Thus, deformities of the chest cage, such as kyphosis, severe scoliosis, and other restraining conditions, such as fibrotic pleurisy or paralyzed and fibrotic muscles, can all reduce the expansibility of the lungs and thereby reduce the total pulmonary compliance.

THE "WORK" OF BREATHING

We have already pointed out that during normal quiet respiration respiratory muscle contraction occurs only during inspiration, while expiration is entirely a passive process caused by elastic recoil of the lung and chest cage structures. Thus, the respiratory muscles normally perform work only to cause inspiration and not at all to cause expiration.

The work of inspiration can be divided into three different fractions: (1) that required to expand the lungs against its elastic forces, called *compliance work*, (2) that required to overcome the viscosity of the lung and chest wall structures, called *tissue resistance work*, and (3) that required to overcome airway resistance during the movement of air into the lungs, called *airway resistance work*. These three different types of work are illustrated graphically in Figure 39–3. In this diagram the curve labeled "inspiration" illustrates the progressive changes in intrapleural pressure and lung volume during inspiration, and the total shaded area of the figure represents the total work performed by the expiratory muscles during the act of inspiration. The shaded area, in turn, is divided into three different segments representing the three different types of work performed during inspiration. These can be explained as follows:

Compliance Work. The dotted area represents the compliance work that is required to expand the lungs against the elastic forces. This can be calculated by multiplying the volume of expansion times the average pressure required to cause the expansion, and this is equal to the area represented by the dots. That is,

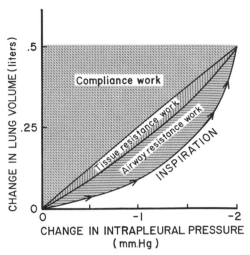

Figure 39–3. Graphical representation of the three different types of work accomplished during inspiration: (1) compliance work, (2) tissue resistance work, and (3) airway resistance work.

$$\text{Compliance work} = \frac{\Delta V \cdot \Delta P}{2}$$

where ΔV is the increase in volume, and ΔP is the increase in intrapleural pressure.

Tissue Resistance Work. The area shown by the vertical bars represents the work expenditure required to overcome the viscosity of the lungs and chest cage. This effect was also illustrated for the lungs alone in Figure 39–2.

Airway Resistance Work. Finally, the area in Figure 39–3 illustrated by the horizontal bars represents the work required to overcome the resistance to airflow through the respiratory passageways.

Comparison of the Different Types of Work. It is clear from Figure 39–3 that during normal quiet breathing most of the work performed by the respiratory muscles is used simply to expand the lungs. A small amount is used to overcome tissue resistance (tissue viscosity), and somewhat more is used to overcome airway resistance. On the other hand, during very heavy breathing, when air must flow through the respiratory passageways at very high velocity, the greater proportion of the work is used to overcome airway resistance.

In pulmonary disease, all three of the different types of work are frequently vastly increased. Compliance work and tissue resistance work are especially increased by diseases that cause fibrosis of the lungs, and airway resistance work is especially increased by diseases that obstruct the airways.

Ordinarily there is no muscle "work" performed during expiration because expiration results from elastic recoil of the lungs and chest. But in heavy breathing or when airway resistance and tissue resistance are great, then expiratory work does of course occur.

Energy Required for Respiration. During normal quiet respiration, only 2 to 3 per cent of the total energy expended by the body is required to energize the pulmonary ventilatory process. During very heavy exercise, the absolute amount of energy required for pulmonary ventilation can increase as much as 25-fold. However, this still does not represent a significant increase in *percentage* of total energy expenditure because the total energy release in the body increases at the same time as much as 15- to 20-fold. Thus, even in heavy exercise only 3 to 4 per cent of the total energy expended is used for ventilation.

On the other hand, pulmonary diseases that decrease the pulmonary compliance, that increase airway resistance, or that increase the viscosity of the lung or chest wall can at times increase the work of breathing so much that one third or more of the total energy expended by the body is for respiration alone. Such respiratory diseases can proceed to the point that this excess work load alone is the cause of death.

THE PULMONARY VOLUMES AND CAPACITIES

RECORDING CHANGES IN PULMONARY VOLUME – SPIROMETRY

A simple method for studying pulmonary ventilation is to record the volume movement of air into and out of the lungs, a process called *spirometry*. A typical spirometer is illustrated in Figure 39–4. This consists of a drum inverted over a chamber of water, with the drum counterbalanced by a weight. In the drum is a breathing mixture of gases, usually air or oxygen; a tube connects the mouth with the gas chamber. When one breathes in and out of the chamber the drum rises and falls, and an appropriate recording is made on a moving sheet of paper. With this type of spirometer, only a few breaths can be recorded at each recording session because of the buildup of carbon dioxide and loss of oxygen from the gas chamber. However, a more elaborate type of spirometer, such as that illustrated in Figure 71–4 of Chapter 71, can be used for continual recording of the spirograph. This apparatus chemically removes the carbon dioxide as it is formed, and oxygen can be injected at a rate equal to its consumption.

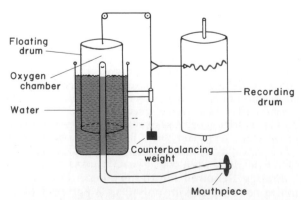

Figure 39–4. A spirometer.

Figure 39–5 illustrates a spirogram showing changes in lung volume under different conditions of breathing. For ease in describing the events of pulmonary ventilation, the air in the lungs has been subdivided at different points on this diagram into four different *volumes* and four different *capacities*, which are as follows:

THE PULMONARY "VOLUMES"

To the left in Figure 39–5 are listed four different pulmonary lung "volumes" which, when added together, equal the maximum volume to which the lungs can be expanded. The significance of each of these volumes is the following:

1. The *tidal volume* is the volume of air inspired or expired with each normal breath, and it amounts to about 500 ml.

2. The *inspiratory reserve volume* is the extra volume of air that can be inspired over and beyond the normal tidal volume, and it is usually equal to approximately 3000 ml.

3. The *expiratory reserve volume* is the amount of air that can still be expired by forceful expiration after the end of a normal tidal expiration; this normally amounts to about 1100 ml.

4. The *residual volume* is the volume of air still remaining in the lungs after the most forceful expiration. This volume averages about 1200 ml.

THE PULMONARY "CAPACITIES"

In describing events in the pulmonary cycle, it is sometimes desirable to consider two or more of the above volumes together. Such combinations are called pulmonary capacities. To the right in Figure 39–5 are listed the different pulmonary capacities, which can be described as follows:

1. The *inspiratory capacity* equals the *tidal volume* plus the *inspiratory reserve volume*. This is the amount of air (about 3500 ml.) that a person can breathe beginning at the normal expiratory level and distending his lungs to the maximum amount.

2. The *functional residual capacity* equals the *expiratory reserve volume* plus the *residual volume*. This is the amount of air remaining in the lungs at the end of normal expiration (about 2300 ml.).

3. The *vital capacity* equals the *inspiratory reserve volume* plus the *tidal volume* plus the *expiratory reserve volume*. This is the maximum amount of air that a person can expel from the lungs after first filling the lungs to their maximum extent and then expiring to the maximum extent (about 4600 ml.).

4. The *total lung capacity* is the maximum volume to which the lungs can be expanded with

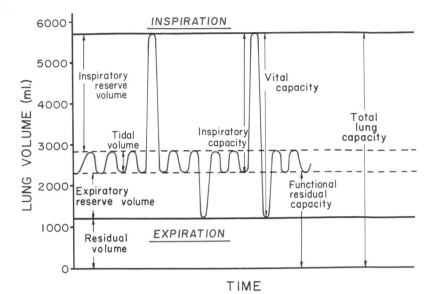

Figure 39–5. Diagram showing respiratory excursions during normal breathing and during maximal inspiration and maximal expiration.

the greatest possible inspiratory effort (about 5800 ml.).

All pulmonary volumes and capacities are about 20 to 25 per cent less in the female than in the male, and they obviously are greater in large and athletic persons than in small and asthenic persons.

Resting Expiratory Level. Normal pulmonary ventilation is accomplished almost entirely by the muscles of inspiration. On relaxation of the inspiratory muscles the elastic properties of the lungs and thorax cause the lungs to contract passively. Therefore, when all inspiratory muscles are completely relaxed the lungs return to a relaxed state called the *resting expiratory level.* The volume of air in the lungs at this level is equal to the functional residual capacity, or about 2300 ml. in the young adult.

SIGNIFICANCE OF THE PULMONARY VOLUMES AND CAPACITIES

In normal persons the volume of air in the lungs depends primarily on body size and build. Furthermore, the different "volumes" and "capacities" change with the position of the body, most of them decreasing when the person lies down and increasing on standing. This change with position is caused by two major factors: first, a tendency for the abdominal contents to press upward against the diaphragm in the lying position, and, second, an increase in the pulmonary blood volume in the lying position, which correspondingly decreases the space available for pulmonary air.

Significance of the Residual Volume. The residual volume represents the air that cannot be removed from the lungs even by forceful expiration. This is important because it provides air in the alveoli to aerate the blood even between breaths. Were it not for the residual air, the concentrations of oxygen and carbon dioxide in the blood would rise and fall markedly with each respira-

tion, which would certainly be disadvantageous to the respiratory process.

Significance of the Vital Capacity. Other than the anatomical build of a person, the major factors that affect vital capacity are (1) the position of the person during the vital capacity measurement, (2) the strength of the respiratory muscles, and (3) the distensibility of the lungs and chest cage, which is called "pulmonary compliance."

The average vital capacity in the young adult male is about 4.6 liters, and in the young adult female about 3.1 liters, though these values are much greater in some persons of the same weight than in others. A tall, thin person usually has a higher vital capacity than an obese person, and a well-developed athlete may have a vital capacity as great as 30 to 40 per cent above normal — that is, 6 to 7 liters.

Vital Capacity Following Paralysis of the Respiratory Muscles. Paralysis of the respiratory muscles, which often occurs following spinal cord injuries or poliomyelitis, can cause a great decrease in vital capacity, to as low as 500 to 1000 ml. — barely enough to maintain life. This decrease may be even lower in the case of respirator patients.

Decreased Vital Capacity Caused by Diminished Pulmonary Compliance. Obviously, any factor that reduces the ability of the lungs to expand also reduces the vital capacity. Thus, tuberculosis, chronic asthma, lung cancer, chronic bronchitis, and fibrotic pleurisy can all reduce the pulmonary compliance and thereby decrease the vital capacity. For this reason vital capacity measurements are among the most important and yet simplest of all clinical respiratory measurements for assessing the progress of different types of pulmonary fibrotic diseases.

Changes in Vital Capacity Resulting from Pulmonary Congestion. In left heart disease or any other disease that causes pulmonary vascular congestion and edema, the vital capacity becomes reduced because excess fluid in the lungs decreases lung compliance.

Vital capacity measurements made periodically in left-sided heart disease are a good means for determining whether the person's condition is progressing or getting better, for these measurements can indicate the degree of pulmonary edema.

ABBREVIATIONS AND SYMBOLS USED IN PULMONARY FUNCTION STUDIES

With this introduction of spirometry and its usefulness in studying many features of pulmonary ventilatory function, the student also has an introduction to the modern clinical pulmonary function laboratory. Spirometry is only one of many measurement procedures that the pulmonary physician uses daily. The more important of these will be presented later in this chapter and in the following four chapters. It will become clear as we proceed that the clinical practice of respiratory medicine probably rests on basic physiological principles more than almost any other type of clinical medicine. Furthermore, we shall see that many of the measurement procedures depend heavily on mathematical computations; one will be discussed in the following section describing the measurement of functional residual capacity. To simplify presenting pulmonary function data, a number of abbreviations and symbols have become standardized. Some of the more important of these are given in Table 39–1. Using these symbols, we present here a few simple algebraic exercises showing some of the interrelationships among the pulmonary volumes and capacities; the student should think through and verify these interrelationships:

$$VC = IRV + V_T + ERV$$
$$VC = IC + ERV$$
$$TLC = VC + RV$$
$$TLC = IC + FRC$$
$$FRC = ERV + RV$$

DETERMINATION OF FUNCTIONAL RESIDUAL CAPACITY — THE HELIUM DILUTION METHOD

The functional residual capacity, which is the amount of air in the lungs at the end of normal expiration, represents the amount of air that remains in the lungs between breaths. Therefore, it is an important factor in function of the lungs. Its value changes markedly in some types of pulmonary disease, for which reason it is often desirable to measure the functional residual capacity. However, the spirometer cannot measure the residual volume of the lungs (the volume remaining after maximum expiration), and this volume composes about half of the functional residual capacity. Therefore, an indirect method must be used, usually the helium dilution method.

A spirometer of known volume is filled with air mixed with helium at a known concentration. Before breathing from the spirometer, the person expires normally. At the end of this expiration the remaining volume of gases in the lungs is exactly equal to the functional residual capacity. At this point the subject immediately begins to breathe from the spirometer, and the gases of the spirometer begin to mix with the gases of the lungs. As a result the helium becomes diluted by the functional residual capacity gases, and the volume of the functional residual capacity can then be calculated from the degree of dilution of the helium, using the following formula:

$$FRC = \left(\frac{Ci_{He}}{Cf_{He}} - 1\right) Vi_{Spir}$$

in which

FRC is *functional residual capacity*
Ci_{He} is *initial concentration of helium in the spirometer*

TABLE 39–1   LIST OF ABBREVIATIONS AND SYMBOLS

V_T	tidal volume		P_B	atmospheric pressure
FRC	functional residual capacity		P_{alv}	alveolar pressure
ERV	expiratory reserve volume		P_{pl}	pleural pressure
RV	residual volume		P_{O_2}	partial pressure of oxygen
IC	inspiratory capacity		P_{CO_2}	partial pressure of carbon dioxide
IRV	inspiratory reserve volume		P_{N_2}	partial pressure of nitrogen
TLC	total lung capacity		Pa_{O_2}	partial pressure of oxygen in arterial blood
VC	vital capacity		Pa_{CO_2}	partial pressure of carbon dioxide in arterial blood
Raw	resistance of tracheobronchial tree to flow of air into the lung		PA_{O_2}	partial pressure of oxygen in alveolar gas
C	compliance		PA_{CO_2}	partial pressure of carbon dioxide in alveolar gas
V_D	volume of dead space gas			
V_A	volume of alveolar gas		PA_{H_2O}	partial pressure of water in alveolar gas
$\dot{V}$	gas volume per minute		R	respiratory exchange ratio
$\dot{V}_I$	inspired volume of ventilation per minute		$\dot{Q}$	cardiac output
$\dot{V}_E$	expired volume of ventilation per minute		$\dot{Q}s$	shunt flow
$\dot{V}_{O_2}$	rate of oxygen uptake per minute		Ca_{O_2}	concentration of oxygen in arterial blood
$\dot{V}_{CO_2}$	amount of carbon dioxide eliminated per minute		$C\bar{v}_{O_2}$	concentration of oxygen in mixed venous blood
$\dot{V}_{CO}$	rate of carbon monoxide uptake per minute		S_{O_2}	percentage saturation of hemoglobin with oxygen
$D_{L_{O_2}}$	diffusing capacity of the lung for oxygen		Sa_{O_2}	percentage saturation of hemoglobin with oxygen in arterial blood
$D_{L_{CO}}$	diffusing capacity of the lung for carbon monoxide			

Cf_{He} is *final concentration of helium in the spirometer*
Vi_{Spir} is *initial volume of the spirometer*

Once the functional residual capacity has been determined, the residual volume can then be determined by subtracting the expiratory reserve volume from the functional residual capacity. Also, the total lung capacity can be determined by adding the inspiratory capacity to the functional residual capacity. That is,

$$RV = FRC - ERV$$

and

$$TLC = FRC + IC$$

THE MINUTE RESPIRATORY VOLUME — RESPIRATORY RATE AND TIDAL VOLUME

The *minute respiratory volume* is the total amount of new air moved into the respiratory passages each minute, and this is equal to the *tidal volume* × the *respiratory rate*. The normal tidal volume is about 500 ml., and the normal respiratory rate is approximately 12 breaths per minute. Therefore, the *minute respiratory volume averages about 6 liters per minute.* A person can occasionally live for short periods of time with a minute respiratory volume as low as 1.5 liters per minute and with a respiratory rate as low as two to four breaths per minute.

The respiratory rate occasionally rises to as high as 40 to 50 per minute, and the tidal volume can become as great as the vital capacity, about 4600 ml. in the young adult male. However, at rapid breathing rates, a person usually cannot sustain a tidal volume greater than about one-half the vital capacity.

Maximum Expiratory Flow

When a person expires with progressively increasing force, the expiratory air flow reaches a maximum rate despite still further increase in expiratory force. This effect can be explained by reference to Section A of Figure 39–6. When pressure is applied to the lungs by chest cage compression, the same amount of pressure is applied to the outsides of both the alveoli and the respiratory passageways, as indicated by the arrows. Therefore, not only is the pressure increased in the alveoli to force air to the exterior, but the terminal bronchioles are collapsed at the same time, which increases the airway resistance. Beyond a certain expiratory effort these two effects have equal but opposing results on air flow, thus preventing further increase in flow.

Section B of Figure 39–6 illustrates the effect of terminal bronchiole collapse on the maximum expiratory flow. The curve recorded in Section B is the expiratory flow achieved by a normal person who first

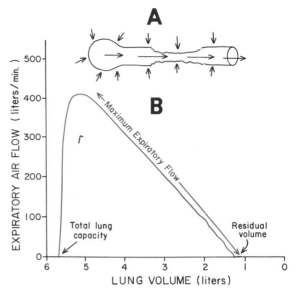

Figure 39–6. *A*, Collapse of the respiratory passageway during maximum expiratory effort, an effect that limits the expiratory flow rate. *B*, Effect of lung volume on the maximum expiratory air flow, showing decreasing maximum expiratory air flow as the lung volume becomes smaller.

inhales as much air as possible and then expires with maximum expiratory effort until he can expire no more. Note that he quickly reaches an expiratory air flow of over 400 liters/min. But it does not matter how much additional expiratory effort he exerts, this is the *maximum expiratory flow* that he can achieve.

Note also that as the lung volume becomes smaller this maximum expiratory flow also becomes less. The main reason for this is that in the enlarged lung the bronchi are held open partially via elastic pull on their outsides by lung structural elements; however, as the lung becomes smaller, these structures are relaxed so that the bronchi collapse more easily.

Abnormalities of the Maximum Expiratory Flow-Volume Curve. The maximum expiratory flow-volume curve is often recorded in the pulmonary function laboratory to determine abnormalities of pulmonary ventilation. Figure 39–7 illustrates once more the normal curve and curves recorded in two different types of lung diseases: (1) constricted lungs and (2) airway obstruction.

Note that the *constricted lungs* have both reduced total lung capacity (TLC) and reduced residual volume (RV). Furthermore, since the lung cannot expand to its normal volume, even with the greatest possible expiratory effort the maximal expiratory flow cannot rise to equal that of the normal curve. Constricted lung diseases include fibrotic diseases of the lung itself such as *tuberculosis, silicosis,* and others, and also diseases that constrict the chest cage such as *kyphosis, scoliosis,* and *fibrotic pleurisy.*

In diseases with *airway obstruction*, it is usually much more difficult to expire than to inspire because the expiratory closing tendency of the airways is greatly increased, while the negative intrapleural pressure of inspiration actually "pulls" the airways open. There-

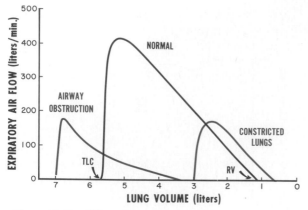

Figure 39–7. Effect of two different respiratory abnormalities — constricted lungs and airway obstruction — on the maximum expiratory flow-volume curve.

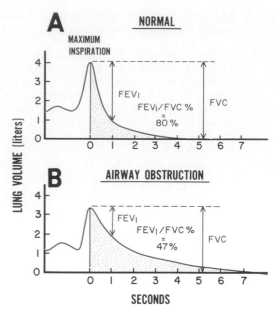

Figure 39–8. Recordings during a forced vital capacity maneuver, (A) in the normal person, and (B) in the person with airway obstruction.

fore, air tends to enter the lung easily and become trapped in the lungs. Thus, both the total lung capacity and residual volume increase markedly, as illustrated in Figure 39–7. Also, because of partial obstruction of many of the airways and also because these collapse more easily than normal airways, the maximum expiratory flow is greatly reduced. The classic disease that causes severe airway obstruction is *asthma*. However, serious airway obstruction also occurs in some stages of *emphysema*.

FORCED EXPIRATORY VITAL CAPACITY AND FORCED EXPIRATORY VOLUME

Another exceedingly useful clinical pulmonary test, and one that is also very simple, is to make a record on a spirometer of the *forced expiratory vital capacity* (FVC). Such a record is illustrated in Figure 39–8A for a person with normal lungs and in Figure 39–8B for a person with airway obstruction. In performing the forced expiratory vital capacity maneuver the person first inspires maximally to the total lung capacity, then exhales into the spirometer with maximum expiratory effort as rapidly and as completely as possible. The total excursion of the record represents the "forced vital capacity," as illustrated in the figure.

Now, let us observe the difference between the two records for normal lungs and airway obstruction. The forced vital capacities are nearly equal, indicating only moderate difference in basic lung volumes in the two persons. On the other hand, there is a major difference in the flow rate at which the persons can expire. And the most important difference in this flow rate occurs within the first second. Therefore, it is customary to record the forced expiratory volume during the first second (FEV_1) and to use this for comparison between the normal and the abnormal. In the normal person the percentage of the forced vital capacity that is expired in the first second ($FEV_1/FVC\%$) is about 80. However, note in Figure 39–8B that in an airway obstruction this value has decreased to only 47 per cent. In serious airway obstruction, as often occurs in acute asthma, this can sometimes decrease to less than 20 per cent.

VENTILATION OF THE ALVEOLI

The truly important factor of the entire pulmonary ventilatory process is the rate at which the air is renewed each minute by atmospheric air in the gas exchange area of the lungs, the alveoli, the alveolar sacs, the alveolar ducts, and the respiratory bronchioles; this is called *alveolar ventilation*. One can readily understand that alveolar ventilation per minute is not equal to the minute respiratory volume because a large portion of the inspired air goes to fill respiratory passageways of which the membranes are not capable of significant gaseous exchange with the blood.

Diffusion of Gases Between the Terminal Bronchioles and the Alveoli. During inspiration, only a small portion of the inspired air actually flows *en masse* beyond the terminal bronchioles into the alveoli. Instead, by the time the inspired air reaches these small passageways, the total airway cross-sectional area has become so great and the velocity of flow so slow that the rate of the *diffusion* of the air is now greater than the rate of flow. Diffusion is caused by the kinetic motion of molecules, each gas molecule moving at high velocity until it strikes another molecule or an airway wall. It is by means of this rapid diffusion that the gases move the remainder of the way to the alveoli. Furthermore, all the gases in the alveoli, the new as well as the old, become completely mixed with each other within a fraction of a second.

THE DEAD SPACE

Effect of Dead Space on Alveolar Ventilation. The air that goes to fill the respiratory passages with each breath is called *dead space air*.

On inspiration, much of the new air must first fill the different dead space areas — the nasal passageways, the pharynx, the trachea, and the bronchi — before any reaches the alveoli. Then, on expiration, all the air in the dead space is expired first before any of the air from the alveoli reaches the atmosphere. *The volume of air that enters the alveoli (including collectively the alveolar ducts and respiratory bronchioles as well) with each breath, therefore, is equal to the tidal volume minus the dead space volume.*

Measurement of the Dead Space Volume. A simple method for measuring dead space volume is illustrated by Figure 39–9. In making this measurement the subject first breathes normal air and then suddenly takes an inspiration of oxygen. This, obviously, fills the entire dead space with pure oxygen, and some of the oxygen also mixes with the alveolar air. Then the person expires through a rapidly recording nitrogen meter, which makes the record shown in the figure. The first portion of the expired air contains only pure oxygen, and the per cent of nitrogen is zero, but about one-quarter of the way into expiration, as the alveolar air reaches the nitrogen meter, the nitrogen concentration rises, in this instance up to 60 per cent, and then levels off. The total volume of air expired in this instance was 500 ml., and it can readily be seen that the area covered by the dots represents the air that has no nitrogen in it. Therefore, this area also represents the dead space portion of the expired air. The area covered by the hatching represents the air containing nitrogen, and therefore is the alveolar portion of the expired air. Thus, one can determine the amount of dead space air from the following equation:

$$V_D = \frac{\text{Area of dots} \times V_E}{\text{Area of hatching} + \text{Area of dots}}$$

where V_D is *dead space air,* and V_E is *total volume of expired air.*

Let us assume, for instance, that the area of the dots on the graph is 30 cm.^{2}, and the area of the hatching is 70 cm.^{2}, and the total volume expired is 500 ml. The dead space then would be

$$\frac{30}{30 + 70} \times 500, \text{ or } 150 \text{ ml.}$$

Normal Dead Space Volume. The normal dead space air in the young adult is about 150 ml. This increases slightly with age.

Anatomic versus Physiologic Dead Space. The method just described for measuring the dead space measures the volume of all the spaces of the respiratory system besides the gas exchange areas, the alveoli and terminal ducts; this is called the *anatomic dead space.* On occasion, however, some of the alveoli themselves are not functional or only partially functional because of absent or poor blood flow through adjacent pulmonary capillaries and, therefore, must be considered also to be dead space. When the alveolar dead space is included in the total measurement of dead space this is then called *physiologic dead space,* in contradistinction to the anatomic dead space. In the normal person, the anatomic and the physiologic dead spaces are nearly equal because all alveoli are functional in the normal lung, but in persons with partially functional or nonfunctional alveoli in some parts of the lungs, the physiologic dead space is sometimes as much as 10 times the anatomic dead space, or as much as 1 to 2 liters. These problems will be discussed further in Chapter 40 in relation to pulmonary gaseous exchange and in Chapter 43 in relation to certain pulmonary diseases.

RATE OF ALVEOLAR VENTILATION

Alveolar ventilation per minute is the total volume of new air entering the alveoli each minute. It is equal to the respiratory rate times the amount of new air that enters the alveoli with each breath:

$$\dot{V}_I = \text{Freq} \cdot (V_T - V_D)$$

where $\dot{V}_I$ is the *inspired volume of ventilation per minute,* Freq is the *frequency of respiration per minute,* V_T is the *tidal volume,* and V_D is the *dead space volume.*

Thus, with a normal tidal volume of 500 ml., a normal dead space of 150 ml., and a respiratory rate of 12 times per minute, alveolar ventilation equals $12 \times (500 - 150)$, or 4200 ml. per minute.

Theoretically, when the tidal volume falls to equal the dead space volume, no new air at all enters the alveoli with each breath, and the al-

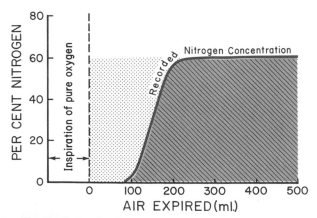

Figure 39–9. Continuous record of the changes in nitrogen concentration in the expired air following a previous inspiration of pure oxygen. This record can be used to calculate dead space as discussed in the text.

veolar ventilation per minute becomes zero however rapidly the person breathes. (This relationship is not entirely true because all the dead space air is never completely expired before some of the alveolar air begins to be expired, and the same is true for inspiration. Therefore, there can be a slight amount of alveolar ventilation even with tidal volumes as little as 60 to 75 ml.)

On the other hand, when the tidal volume is several liters, the effect of dead space volume on alveolar ventilation obviously becomes almost insignificant.

Alveolar ventilation is one of the major factors determining the concentrations of oxygen and carbon dioxide in the alveoli. Therefore, almost all discussions of gaseous exchange problems in the following chapters emphasize alveolar ventilation. *The respiratory rate, the tidal volume, and the minute respiratory volume are of importance only insofar as they affect alveolar ventilation.*

FUNCTIONS OF THE RESPIRATORY PASSAGEWAYS

FUNCTIONS OF THE NOSE

As air passes through the nose, three distinct functions are performed by the nasal cavities: First, the *air is warmed* by the extensive surfaces of the turbinates and septum, a total area of about 160 cm.², which are illustrated in Figure 39–10. Second, the *air is almost completely humidified* even before it passes beyond the nose. Third, the *air is filtered*. All these functions together are called the *air conditioning function* of the

upper respiratory passageways. Ordinarily, the air rises to within 2 to 3 per cent of body temperature and within 2 to 3 per cent of full saturation with water vapor before it reaches the trachea. When a person breathes air through a tube directly into the trachea (as through a tracheostomy), the cooling and, especially, the drying effect in the lower lung can lead to serious lung crusting and infection.

Filtration Function of the Nose. The hairs at the entrance to the nostrils are important for the removal of large particles. Much more important, though, is the removal of particles by *turbulent precipitation*. That is, the air passing through the nasal passageways hits many obstructing vanes: the turbinates, the septum, and the pharyngeal wall. Each time air hits one of these obstructions it must change its direction of movement; the particles suspended in the air, having far more mass and momentum than air, cannot change their direction of travel as rapidly as can the air. Therefore, they continue forward, striking the surfaces of the obstructions.

All the surfaces of the nose are coated with a thin layer of *mucus,* which is secreted by the *mucous membrane* covering the surfaces. Furthermore, the epithelium of the nasal passageways is ciliated, and these *cilia* constantly beat toward the pharynx. Therefore, after particles are entrapped in the mucus, this mucus is moved like a sliding sheet at a rate of about 1 cm. per minute toward the pharynx and finally is either expectorated or swallowed.

Size of Particles Entrapped in the Respiratory Passages. The nasal turbulence mechanism for removing particles from air is so effective that almost no particles larger than 4 to 6 microns in diameter enter the lungs through the nose. This size is smaller than the size of red blood cells. Consequently, by far the greater proportion of dust and other large particles in air is removed before the air finally reaches the lungs.

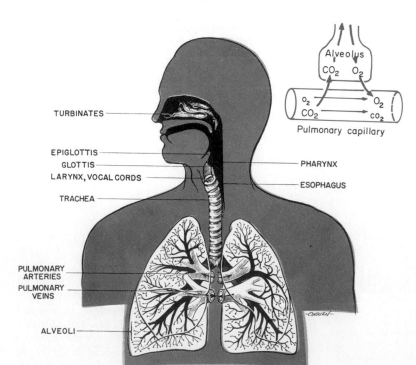

Figure 39–10. The respiratory passages.

Of the remaining particles, many with sizes between 1 and 5 microns *settle* out in the smaller bronchioles as a result of *gravitational precipitation*. For instance, terminal bronchiolar disease is very common in coal miners because of settled dust particles. Some of the still smaller particles (smaller than 1 micron in diameter) *diffuse* against the walls of the alveoli and adhere to the alveolar fluid. But many particles smaller than 0.5 micron in diameter remain suspended in the alveolar air and are later expelled by expiration. For instance, the particles of cigarette smoke have a particle size of approximately 0.3 micron. Almost none of these are precipitated in the respiratory passageways before they reach the alveoli. However, up to one-third of them do precipitate in the alveoli by the diffusion process, with the balance remaining suspended and expelled in the expired air.

Particles that become entrapped in the alveoli are slowly removed mainly by macrophages. An excess of particles causes growth of fibrous tissue in the alveolar septa, leading to permanent debility.

THE COUGH REFLEX

The cough reflex is almost essential to life, for the cough is the means by which the passageways of the lungs are maintained free of foreign matter.

The bronchi and the trachea are so sensitive that any foreign matter or other cause of irritation initiates the cough reflex. The larynx and carina (the point where the trachea divides into the bronchi) are especially sensitive, and the terminal bronchioles and alveoli are especially sensitive to corrosive chemical stimuli such as sulfur dioxide gas and chlorine. Afferent impulses pass from the respiratory passages mainly through the vagus nerve to the medulla. There, an automatic sequence of events is triggered by the neuronal circuits of the medulla, causing the following effects:

First, about 2.5 liters of air is inspired. Second, the epiglottis closes, and the vocal cords shut tightly to entrap the air within the lungs. Third, the abdominal muscles contract forcefully, pushing against the diaphragm while other expiratory muscles, such as the internal intercostals, also contract forcefully. Consequently, the pressure in the lungs rises to as high as 100 or more mm. Hg. Fourth, the vocal cords and the epiglottis suddenly open widely so that air under pressure in the lungs *explodes* outward. Indeed, this air is sometimes expelled at velocities as high as 75 to 100 miles an hour. Furthermore, and very important, the strong compression of the lungs also collapses the bronchi and trachea (the noncartilaginous part of the trachea invaginating inward) so that the exploding air actually passes through *bronchial* and *tracheal slits*. The rapidly moving air usually carries with it any foreign matter that is present in the bronchi or trachea.

THE SNEEZE REFLEX

The sneeze reflex is very much like the cough reflex except that it applies to the nasal passageways instead of to the lower respiratory passages. The initiating stimulus of the sneeze reflex is irritation in the nasal passageways, the afferent impulses passing in the fifth nerve to the medulla where the reflex is triggered. A series of reactions similar to those for the cough reflex takes place; however, the uvula is depressed so that large amounts of air pass rapidly through the nose, as well as through the mouth, thus helping to clear the nasal passages of foreign matter.

ACTION OF THE CILIA TO CLEAR RESPIRATORY PASSAGEWAYS

In addition to the cough mechanism, the respiratory passageways of the trachea and lungs, like the nose, are lined with a ciliated, mucus-coated epithelium that aids in clearing the passages, for the cilia beat toward the pharynx and move the mucus as a continually flowing sheet. Thus, small foreign particles and mucus are mobilized at a velocity of as much as a centimeter per minute along the surface of the trachea toward the pharynx, in the same manner that foreign matter in the nasal passageways is also mobilized toward the pharynx and is swallowed.

VOCALIZATION

Speech involves the respiratory system particularly, but it also involves (1) specific speech control centers in the cerebral cortex, which will be discussed in Chapter 55, (2) respiratory centers of the brain stem, and (3) the articulation and resonance structures of the mouth and nasal cavities. Basically, speech is composed of two separate mechanical functions: (1) *phonation,* which is achieved by the larynx, and (2) *articulation,* which is achieved by the structures of the mouth.

Phonation. The larynx is specially adapted to act as a vibrator. The vibrating element is the *vocal cords,* which are folds along the lateral walls of the larynx that are stretched and positioned by several specific muscles within the confines of the larynx itself.

Figure 39–11A illustrates the basic structure of the

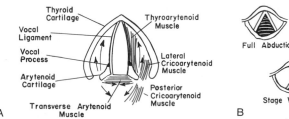
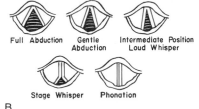

Figure 39–11. Laryngeal function in phonation. (Modified from Greene: The Voice and Its Disorders. 3rd Ed. Philadelphia, J. B. Lippincott Company, 1972.)

larynx, showing that each vocal cord is stretched between the *thyroid cartilage* and an *arytenoid cartilage*. The specific muscles within the larynx that position and control the degree of stretch of the vocal cords are also shown. Thus, one can see in the figure that contraction of the *posterior cricoarytenoid* muscles pulls the arytenoid cartilages away from the thyroid cartilage and thereby stretches the vocal cords. The *transverse arytenoid muscle* pulls the arytenoid cartilages together and, therefore, approximates the two vocal cords so that they vibrate in a stream of expired air. Conversely, contraction of the *lateral cricoarytenoid muscles* pulls the arytenoid cartilages forward and apart to allow normal respiration.

The *thyroarytenoid muscles* are made up of many small slips of muscle controlled separately by different nerve fibers. The slips of muscle adjacent to the edges of the vocal cord can contract separately from those adjacent to the wall of the larynx, and other individual portions of these muscles can also contract independently of each other. These contractions control the *shape* of the vocal cords — whether thick or thin, or sharp or blunt edges — during different types of phonation.

Vibration of the Vocal Cords.   One might suspect that the vocal cords would vibrate in the direction of the flowing air. However, this is not the case. Instead, they vibrate laterally. The cause of the vibration is the following: When the vocal cords are approximated and air is expired, pressure of the air from below first pushes the vocal cords apart, which allows rapid flow of air between their margins. The rapid flow of air then immediately creates a partial vacuum between the vocal cords, which pulls them once again toward each other. This stops the flow of air, pressure builds up behind the cords, and the cords open once more, thus continuing in a vibratory pattern.

Frequency of Vibration.   The pitch of the sound emitted by the larynx can be changed in two different ways. First, a change can be achieved by *stretching or relaxing the vocal cords*. The mechanisms involved were partly explained above in the discussion of the intrinsic laryngeal muscles; but, in addition to the effects of the intrinsic muscles, the muscles attached to the external surfaces of the larynx can also pull against the cartilages and thereby help to stretch or relax the vocal cords. For instance, the entire larynx is moved upward by the external laryngeal muscles, which helps to stretch the vocal cords when one wishes to emit a very high frequency sound, and the larynx is moved downward, with corresponding loosening of the vocal cords, when one wishes to emit a very bass sound.

The second means for changing the sound frequency is *to change the shape and mass of the vocal cord edges*. When very high frequency sounds are emitted, different slips of the thyroarytenoid muscles contract in such a way that the edges of the vocal cords are sharpened and thinned, whereas when bass frequencies are emitted, the thyroarytenoid muscles contract in a different pattern so that broad edges with a large mass are approximated. Figure 39–11B shows some of the positions and shapes of the vocal cords during different types of phonation.

Articulation and Resonance.   The three major organs of articulation are the *lips*, the *tongue*, and the *soft palate*. These need not be discussed in detail because all of us are familiar with their movements during speech and other vocalizations.

The resonators include the *mouth*, the *nose* and *associated nasal sinuses*, the *pharynx*, and even the *chest cavity* itself. Here again we are all familiar with the resonating qualities of these different structures. For instance, the function of the nasal resonators is illustrated by the change in quality of the voice when a person has a severe cold.

ARTIFICIAL RESPIRATION

MOUTH-TO-MOUTH BREATHING

A very successful method of artificial respiration is mouth-to-mouth breathing, in which the operator rapidly inspires a deep breath and then breathes into the mouth of the subject. This method has often been shunned in the past because of the belief that the expired air of the operator would not be beneficial to the subject. This is not true, because normal expired air usually still has an adequate amount of oxygen to sustain life in almost anyone. Furthermore, the carbon dioxide in the expired air is sometimes actually desirable because it helps to stimulate the respiratory center of the subject.

MECHANICAL METHODS OF ARTIFICIAL RESPIRATION

The Resuscitator.   Many types of resuscitators are available, and each has its own characteristic principles of operation. Basically, the resuscitator, illustrated in Figure 39–12A, consists of a supply of oxygen or air, a mechanism for applying intermittent positive pressure and, with some machines, negative pressure as well, and a mask that fits over the face of the patient. This apparatus forces air through the mask into the lungs of the patient during the positive pressure cycle and then either allows the air to flow out of the lungs during the remainder of the cycle or pulls the air out by negative pressure.

Earlier resuscitators often caused such severe damage to the lungs because of excessive positive pressure that their usage was at one time greatly decried. However, most resuscitators now have safety valves that prevent the positive pressure from rising usually above +14 mm. Hg and the negative pressure from falling below −9 mm. Hg. These pressure limits are adequate to cause excellent artificial respiration of *normal lungs* and yet are slight enough to prevent damage.

The Tank Respirator.   Figure 39–12B illustrates the usual tank respirator with a patient's body inside the tank and his head protruding through a flexible but airtight collar. At the end of the tank opposite to the patient's head is a motor-driven leather diaphragm that moves back and forth with sufficient excursion to raise and lower the pressure inside the tank. As the leather diaphragm moves inward, positive pressure develops around the body and causes expiration; and as the diaphragm moves outward, negative pressure causes inspiration. Check valves on the respirator control the positive and negative pressure. Ordinarily these pres-

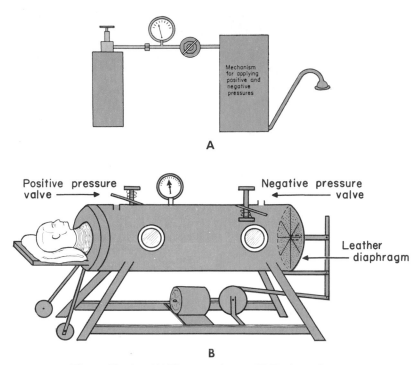

Figure 39–12. (A) The resuscitator. (B) Tank respirator.

sures are adjusted so that the negative pressure that causes inspiration falls to −10 to −20 cm. water and the positive pressure rises to 0 to +5 cm. water.

Effect of the Resuscitator and the Tank Respirator on Venous Return. When air is forced into the lungs under positive pressure, or when the pressure around the patient's body is greatly reduced, as in the case of the tank respirator, the pressure inside the chest cavity becomes greater than the pressure everywhere else in the body. Therefore, the flow of blood into the chest from the peripheral veins becomes impeded. As a result, use of excessive positive pressures with either the resuscitator or the tank respirator can reduce the cardiac output — sometimes to lethal levels. A person can usually survive as much as 20 mm. Hg continuous positive pressure in the lungs, but exposure for more than a few minutes to greater than 30 mm. Hg usually causes death.

REFERENCES

Agostoni, E.: Thickness and pressure of the pleural liquid. *In* Fishman, A. P., and Hecht, H. H. (eds.): The Pulmonary Circulation and Interstitial Space. Chicago, University of Chicago Press, 1969, p. 65.

Bradley, G. W.: Control of the breathing pattern. *Int. Rev. Physiol., 14*:185, 1977.

Bryant, C.: The Biology of Respiration. Baltimore, University Park Press, 1979.

Cherniack, R. M.: Ventilation, perfusion and gas exchange. *In* Frohlich, E. D. (ed.): Pathophysiology, 2nd Ed. Philadelphia, J. B. Lippincott Co., 1976, p. 149.

Clements, J. A., and Tierney, D. F.: Alveolar instability associated with altered surface tension. *In* Fenn, W. O., and Rahn, H. (eds.): Handbook of Physiology. Sec. 3. Vol. 2. Baltimore, Williams & Wilkins, 1965, p. 1565.

Comroe, J. H., Jr., et al.: The Lung: Clinical Physiology and Pulmonary Function Tests, 2nd Ed. Chicago, Year Book Medical Publishers, 1963.

Cumming, G.: Alveolar ventilation: Recent model analysis. *In* MTP International Review of Science: Physiology. Vol. 2. Baltimore, University Park Press, 1974, p. 139.

Davis, J. N.: Control of the muscles of breathing. *In* MTP International Review of Science: Physiology. Vol. 2. Baltimore, University Park Press, 1974, p. 221.

Ellis, P. D., and Billings, D. M.: Cardiopulmonary Resuscitation: Procedures for Basic and Advanced Life Support. St. Louis, C. V. Mosby, 1979.

Engel, L. A., and Macklem, P. T.: Gas mixing and distribution in the lung. *Int. Rev. Physiol., 14*:37, 1977.

Feldman, S. A., and Crawley, B. E. (eds.): Tracheostomy and Artificial Ventilation in the Treatment of Respiratory Failure. Baltimore, Williams & Wilkins, 1977.

Fink, B. R.: The Human Larynx: A Functional Study. New York, Raven Press, 1975.

Fishman, A. P.: Assessment of Pulmonary Function. New York, McGraw-Hill, 1980.

Forster, R. E.: Pulmonary ventilation and blood gas exchange. *In* Sodeman, W. A., Jr., and Sodeman, W. A. (eds.): Pathologic Physiology: Mechanisms of Disease, 5th Ed. Philadelphia, W. B. Saunders Co., 1974, p. 371.

Greene, M.: The Voice and Its Disorders, 2nd Ed. Philadelphia, J. B. Lippincott Co., 1965.

Guyton, A. C.: Electronic counting and size determination of particles in aerosols. *J. Indust. Hyg. Toxicol., 28*:133, 1946.

Guyton, A. C.: Measurement of the respiratory volumes of laboratory animals. *Am. J. Physiol., 150*:70, 1947.

Guyton, A. C.: Analysis of respiratory patterns in laboratory animals. *Am. J. Physiol., 150*:78, 1947.

Hawker, R. W.: Notebook of Medical Physiology: Cardiopulmonary, with Aspects of Clinical Measurement and Monitoring. New York, Longman, Inc., 1979.

Healy, G. B., and McGill, T. J. I. (eds.): Laryngo-Tracheal Problems in the Pediatric Patient. Springfield, Ill., Charles C Thomas, 1979.

Heinemann, H. O., and Fishman, A. P.: Nonrespiratory functions of mammalian lung. *Physiol. Rev., 49*:1, 1969.

Hong, S. K., et al.: Mechanics of respiration during submersion in water. *J. Appl. Physiol., 27*:535, 1969.

Hoppin, G. G., Jr., et al.: Distribution of pleural surface pressures in dogs. *J. Appl. Physiol., 27*:863, 1969.

Horsfield, K.: The regulation between structure and function in the airways of the lung. *Br. J. Dis. Chest, 68*:145, 1974.

Hyatt, R. E., and Black, L. F.: The flow-volume curve. A current perspective. *Am. Rev. Resp. Dis., 107*:191, 1973.

Jarvis, J. F.: An Introduction to the Anatomy and Physiology of Speech and Hearing. Cape Town, South Africa, Juta & Co., 1978.

Kao, F. F.: An Introduction to Respiratory Physiology. New York, American Elsevier Publishing Co., 1972.

Lapp, N. L.: Physiological approaches to detection of small airway disease. *Environ. Res., 6*:253, 1973.

Luchsinger, R., and Arnold, G.: Voice-Speech-Language: Clinical Communicology—Its Physiology and Pathology. Belmont, Cal., Wadsworth Publishing Co., 1965.

Macklem, P. T.: Airway obstruction and collateral ventilation. *Physiol. Rev., 51*:368, 1971.

Macklem, P. T.: Respiratory mechanics. *Annu. Rev. Physiol., 40*:157, 1978.

Mead, J.: Respiration: Pulmonary mechanics. *Annu. Rev. Physiol., 35*:169, 1973.

Milic-Emili, J.: Pulmonary statics. *In* MTP International Review of Science: Physiology. Vol. 2. Baltimore, University Park Press, 1974, p. 105.

Murray, J. F.: The Normal Lung. Philadelphia, W. B. Saunders Co., 1976.

Nadel, J. A., *et al.*: Control of mucus secretion and ion transport in airways. *Annu. Rev. Physiol., 41*:369, 1979.

Nagaishi, C.: Functional Anatomy and Histology of the Lung. Baltimore, University Park Press, 1973.

Phillipson, E. A.: Respiratory adaptations in sleep. *Annu. Rev. Physiol., 40*:133, 1978.

Rahn, H., *et al.*: The pressure-volume diagram of the thorax and lung. *Am. J. Physiol., 146*:161, 1946.

Said, S. I.: Metabolic functions of the lung. *In* Frohlich, E. D. (ed.): Pathophysiology, 2nd Ed. Philadelphia, J. B. Lippincott Co., 1976, p. 189.

Sawashima, M., and Cooper, F. S. (eds.): Dynamic Aspects of Speech Production: Current Results, Emerging Problems, and New Instrumentation. Tokyo, University of Tokyo Press, 1977.

Scarpelli, E., and Auld, P. A. M. (eds.): Pediatric Pulmonary Physiology. Philadelphia, Lea & Febiger, 1975.

Scheich, O. C. H., and Schreiner, C. (eds.): Hearing Mechanisms and Speech. New York, Springer-Verlag, 1979.

Shearer, W. M.: Illustrated Speech Anatomy. Springfield, Ill., Charles C Thomas, 1978.

Singh, R. P.: Anatomy of Hearing and Speech. New York, Oxford University Press, 1980.

Staub, N. C.: Respiratory. *Annu. Rev. Physiol., 31*:173, 1969.

Strang, L. B.: Fetal and newborn lung. *In* MTP International Review of Science: Physiology. Vol. 2. Baltimore, University Park Press, 1974, p. 31.

Stuart, B. O.: Deposition of inhaled aerosols. *Arch. Intern. Med., 131*:60, 1973.

Thurlbeck, W. M.: Structure of the lungs. *Int. Rev. Physiol., 14*:1, 1977.

Thurlbeck, W. M., and Wang, N. S.: The structure of the lungs. *In* MTP International Review of Science: Physiology. Vol. 2. Baltimore, University Park Press, 1974, p. 1.

Tierney, D. F.: Lung metabolism and biochemistry. *Annu. Rev. Physiol., 36*:209, 1974.

Weibel, E. R., and Backofen, H.: Structural design of the alveolar septum and fluid exchange. *In* Fishman, A. P., and Renkin, E. M. (eds.): Pulmonary Edema. Baltimore, Waverly Press, 1979, p. 1.

West, J. B.: Respiratory. *Annu. Rev. Physiol., 34*:91, 1972.

West, J. B.: Respiratory Physiology. Baltimore, Williams & Wilkins, 1974.

White, F. N.: Comparative aspects of vertebrate cardiorespiratory physiology. *Annu. Rev. Physiol., 40*:471, 1978.

Wilson, D. K.: Voice Problems of Children. Baltimore, Williams & Wilkins, 1978.

Wyke, B. D. (ed.): Ventilatory and Phonatory Control Systems. New York, Oxford University Press, 1974.

Wyman, R. J.: Neural generation of the breathing rhythm. *Annu. Rev. Physiol., 39*:417, 1977.

40

Physical Principles of Gaseous Exchange; Diffusion of Oxygen and Carbon Dioxide Through the Respiratory Membrane

After the alveoli are ventilated with fresh air, the next step in the respiratory process is *diffusion* of oxygen from the alveoli into the pulmonary blood and diffusion of carbon dioxide in the opposite direction — from the pulmonary blood into the alveoli. The process of diffusion is simple, involving merely random molecular motion of molecules, these intertwining their ways in both directions through the respiratory membrane. However, in respiratory physiology we are concerned not only with the basic mechanism by which diffusion occurs but also with the *rate* at which it occurs, and this is a much more complicated problem, requiring a rather deep understanding of the physics of diffusion and gaseous exchange. Therefore, a brief review of this subject is presented here as a prelude to the main text of this chapter.

PHYSICS OF DIFFUSION AND GAS PRESSURES

THE MOLECULAR BASIS OF GASEOUS DIFFUSION

All the gases that are of concern in respiratory physiology are simple molecules that are free to move among each other, which is the process called "diffusion." This is also true of the gases dissolved in the fluids and tissues of the body.

However, for diffusion to occur, there must be a source of energy. This is provided by the kinetic motion of the molecules themselves. That is, except at absolute zero temperature, all molecules of all matter are continually undergoing some type of motion. For free mole-cules that are not physically attached to others, this means linear movement at high velocity of the molecules until they strike other molecules. Then they bounce away in new directions and continue again until striking still other molecules. In this way the molecules move rapidly among each other.

Net Diffusion of a Gas in One Direction — Effect of a Concentration Gradient. However, if a gas chamber or a solution has a high concentration of a gas at one end of the chamber and a low concentration at the other end, net diffusion of the gas will occur from the high concentration area toward the low area. This effect is illustrated in Figure 40–1 which shows high concentration of dissolved oxygen at end A of a chamber and low concentration at end B. Random motion of the high concentration of oxygen molecules at end A will cause these molecules to move indiscriminately throughout the entire chamber. Likewise, these molecules at end B will also move indiscriminately throughout the chamber. However, since at the beginning there are more molecules at end A than at end B, the net movement of the oxygen molecules will be from A to B. Thus, the arrows in the figure illustrate that a large amount of diffusion occurs from A to B and only a small amount from B to A. And *net* diffusion is represented by the difference between the lengths of the two arrows.

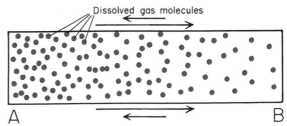

Figure 40–1. Net diffusion of oxygen from one end of a chamber to the other.

GAS PRESSURES IN A MIXTURE OF GASES — PARTIAL PRESSURES OF INDIVIDUAL GASES

The cause of the pressure that a gas exerts against a surface is constant impaction of the kinetically moving molecules against the surface. Obviously, the greater the concentration of the gas, the greater also will be the summated force of impaction of all the molecules striking the surface at any given instant. Therefore, the pressure of a gas is directly proportional to its concentration. It is also directly proportional to the average kinetic energy of the molecules, which is directly proportional to temperature. Therefore, the greater the temperature, the greater also is the pressure; but in the body, the temperature remains relatively constant at 37°C so this is usually not a factor of major consideration in respiratory problems.

Now, let us consider the pressure exerted against the surface by each one of the gases in a mixture of gases. For instance, consider air which has an approximate composition of 79 per cent nitrogen and 21 per cent oxygen. The total pressure of this mixture is 760 mm. Hg. A portion of this total is caused by nitrogen and another portion by oxygen. It is clear from the above description of the molecular basis of pressure that each gas contributes to the total pressure in direct proportion to its relative concentration. Therefore, 79 per cent of the 760 mm. Hg is caused by nitrogen (about 600 mm. Hg) and 21 per cent by oxygen (about 160 mm. Hg). Thus, the "partial pressure" of nitrogen in the mixture is 600 mm. Hg, and the "partial pressure" of oxygen is 160 mm. Hg, while the total pressure is 760 mm. Hg, which is the sum of the individual partial pressures.

The partial pressures of the individual gases in a mixture are designated by the terms, P_{O_2}, P_{CO_2}, P_{N_2}, P_{H_2O}, P_{He}, and so forth.

PARTIAL PRESSURE OF GASES IN WATER AND TISSUES

When a gas under pressure is impressed onto a water interface, instead of bouncing back from the interface some of the molecules will move on into the water and become dissolved. However, as more and more molecules become dissolved, they also begin to diffuse backward to the interface, and some escape back into the gas phase. Once the concentration of dissolved molecules reaches a certain level, the number of molecules leaving the solution to enter the gas phase becomes exactly equal to the number of molecules moving in the opposite direction from the gas into the solution. Thus, a state of *equilibrium* has occurred. In this equilibrium state the pressure of the dissolved gas is exactly equal to the pressure of the gas in the gas state, each pushing against the other at the interface with equal force. Thus, gases in solution exert pressures in exactly the same way as they do in gas phase mixtures. And the partial pressures of the separate dissolved gases are designated similarly as for the gases in the gaseous state, i.e., P_{O_2}, P_{CO_2}, P_{N_2}, P_{He}.

Factors That Determine the Concentration of a Gas Dissolved in a Fluid. The concentration of a gas in a solution is determined not only by its pressure but also by the *solubility coefficient* of the gas. That is, some types of molecules, especially carbon dioxide, are physically or chemically attracted to water molecules while others are repelled. Obviously, when molecules are attracted, far more of them can then become dissolved without building up excess pressure within the solution. On the other hand, those that are repelled will develop excessive pressures for very little degree of solubility.

Thus, since both pressure and solubility coefficient determine the volume of gas that will be dissolved in a given volume of fluid — which is the concentration of the dissolved gas — this can be expressed by the following formula, which is *Henry's law:*

Concentration of dissolved gas = pressure × solubility coefficient

When concentration is expressed in volume of gas dissolved in each volume of water at zero degrees centigrade and pressure is expressed in atmospheres, the solubility coefficients for important respiratory gases at body temperature are the following:

Oxygen	0.024
Carbon dioxide	0.57
Carbon monoxide	0.018
Nitrogen	0.012
Helium	0.008

THE VAPOR PRESSURE OF WATER

Everywhere in the body that a gas mixture in the gas phase occurs, it comes in contact with water in the surrounding tissues. And water has the propensity to evaporate into the gas mixture and to humidify it. This results from the fact that water molecules, like the different dissolved gas molecules, are continually escaping from the water surface into the gas phase. The pressure that the water molecules exert to escape through the surface is called the *vapor pressure* of water. At normal body temperature, 37°C., this vapor pressure is 47 mm. Hg. Therefore, once the gas mixture has become fully humidified — that is, in "equilibrium" with the surrounding water — the partial pressure of the water vapor in the gas mixture is also 47 mm. Hg. This partial pressure, like the other partial pressures, is designated P_{H_2O}.

The vapor pressure of water depends entirely on the temperature of the water. The greater the temperature, the greater is the kinetic activity of the molecules, and therefore the greater is the likelihood that water molecules will escape from the surface of the water into the

TABLE 40–1 VAPOR PRESSURE OF WATER

Temp. (°C.)	Vapor pressure (mm.)	Temp. (°C.)	Vapor pressure (mm.)
0	4.6	37	46.6
10	9.1	40	54.9
20	17.4	50	92.0
30	31.5	70	233.3
35	41.8	100	760.0

gas phase. Consequently, water vapor pressure also increases as the temperature increases. The water vapor pressure at various temperatures from 0° to 100°C. is given in Table 40–1. The most important value to remember is the vapor pressure at body temperature, 47 mm. Hg; this value will appear in most of our subsequent discussions.

DIFFUSION OF GASES THROUGH LIQUIDS — THE PRESSURE GRADIENT FOR DIFFUSION

Now, let us return to the problem of diffusion. From the above discussion it is already clear that when the concentration, or pressure, of a gas is greater in one area than in another area of a chamber, there will be net diffusion from the high pressure area toward the low pressure area. For instance, returning to Figure 40–1 one can readily see that the molecules in the area of high pressure, because of their greater number, have more statistical chance of moving randomly into the area of low pressure than do molecules attempting to go in the other direction. However, some molecules do bounce from the area of low pressure toward the area of high pressure. Therefore, the *net diffusion* of gas from the area of high pressure to the area of low pressure is equal to the number of molecules bouncing in this direction minus the number bouncing in the opposite direction, and this in turn is proportional to the gas pressure difference between the two areas. The pressure in area A of Figure 40–1 minus the pressure in area B divided by the distance of diffusion is known as the *pressure gradient for diffusion* or simply the *diffusion gradient*. The rate of net gas diffusion from area A to area B is directly proportional to this gradient.

The principle of diffusion from an area of high pressure to an area of low pressure holds true for diffusion of gases in a gaseous mixture, diffusion of dissolved gases in a solution, and even diffusion of gases from the gaseous phase into the dissolved state in liquids. That is, *there is always net diffusion from areas of high pressure to areas of low pressure.*

As more and more gas diffuses from area A to area B in Figure 40–1, the pressure in area A falls while that in area B rises so that the two pressures approach each other; as a result, the net rate of diffusion becomes less and less. After a reasonable length of time, the gaseous pressures in both ends of the chamber become essentially equal, and, thereafter, no net diffusion of gas occurs from one end to the other end. This does not mean that no molecules of gas diffuse, but merely that as many molecules then diffuse in one direction as the other.

Quantifying the Net Rate of Diffusion. In addition to the pressure difference, several other factors affect the rate of gas diffusion in a fluid. These are (1) the solubility of the gas in the fluid, (2) cross-sectional area of the fluid, (3) the distance through which the gas must diffuse, (4) the molecular weight of the gas, and (5) the temperature of the fluid. In the body, the temperature remains reasonably constant and usually need not be considered.

Obviously, the greater the solubility of the gas, the greater will be the number of molecules available to

diffuse for any given pressure difference. Also, the greater the cross-sectional area of the chamber, the greater will be the total number of molecules to diffuse. On the other hand, the greater the distance that the molecules must diffuse, the longer it will take the molecules to diffuse the entire distance. Finally, the greater the velocity of kinetic movement of the molecules, which at any given temperature is inversely proportional to the square root of the molecular weight, the greater is the rate of diffusion of the gas. All of these factors can be expressed in a single formula, as follows:

$$D \propto \frac{\Delta P \times A \times S}{d \times \sqrt{MW}}$$

in which D is the diffusion rate, ΔP is the pressure difference between the two ends of the chamber, A is the cross-sectional area of the chamber, S is the solubility of the gas, d is the distance of diffusion, and MW is the molecular weight of the gas.

It is obvious from this formula that the characteristics of the gas itself determine two factors of the formula: solubility and molecular weight. Therefore, the *diffusion coefficient* — that is, the rate of diffusion through a given area for a given distance and pressure difference — for any given gas is proportional to $S/\sqrt{MW}$. Considering the diffusion coefficient for oxygen to be 1, the *relative* diffusion coefficients for different gases of respiratory importance in the body fluids are:

Oxygen	1.0
Carbon dioxide	20.3
Carbon monoxide	0.81
Nitrogen	0.53
Helium	0.95

DIFFUSION OF GASES THROUGH TISSUES

The gases that are of respiratory importance are highly soluble in lipids and, consequently, are also highly soluble in cell membranes. Because of this, these gases diffuse through the cell membranes with very little impediment. Instead, the major limitation to the movement of gases in tissues is the rate at which the gases can diffuse through the tissue water instead of through the cell membranes. Therefore, diffusion of gases through the tissues, including through the respiratory membrane, is almost equal to the diffusion of gases through water, as given in the above list of diffusion rates for the important respiratory gases.

COMPOSITION OF ALVEOLAR AIR — ITS RELATION TO ATMOSPHERIC AIR

Alveolar air does not have the same concentrations of gases as atmospheric air by any means, which can readily be seen by comparing the alveolar air composition in column 3 of Table 40–2

TABLE 40–2 PARTIAL PRESSURES OF RESPIRATORY GASES AS THEY ENTER AND LEAVE THE LUNGS (AT SEA LEVEL)—PER CENT CONCENTRATIONS ARE GIVEN IN PARENTHESES

	Atmospheric Air* (mm. Hg)		Humidified Air (mm. Hg)		Alveolar Air (mm. Hg)		Expired Air (mm. Hg)	
N_2	597.0	(78.62%)	563.4	(74.09%)	569.0	(74.9%)	566.0	(74.5%)
O_2	159.0	(20.84%)	149.3	(19.67%)	104.0	(13.6%)	120.0	(15.7%)
CO_2	0.3	(0.04%)	0.3	(0.04%)	40.0	(5.3%)	27.0	(3.6%)
H_2O	3.7	(0.50%)	47.0	(6.20%)	47.0	(6.2%)	47.0	(6.2%)
TOTAL	760.0	(100.0%)	760.0	(100.0%)	760.0	(100.0%)	760.0	(100.0%)

*On an average cool, clear day.

with the composition of atmospheric air in column 1. There are several reasons for the differences. First, the alveolar air is only partially replaced by atmospheric air with each breath. Second, oxygen is constantly being absorbed from the alveolar air. Third, carbon dioxide is constantly diffusing from the pulmonary blood into the alveoli. And, fourth, dry atmospheric air that enters the respiratory passages is humidified even before it reaches the alveoli.

Humidification of the Air as It Enters the Respiratory Passages. Column 1 of Table 40–2 shows that atmospheric air is composed almost entirely of nitrogen and oxygen; it normally contains almost no carbon dioxide and little water vapor. However, as soon as the atmospheric air enters the respiratory passages, it is exposed to the fluids covering the respiratory surfaces. Even before the air enters the alveoli, it becomes totally humidified.

The partial pressure of water vapor at normal body temperature of 37°C. is 47 mm. Hg, which, therefore, is the partial pressure of water in the alveolar air. Since the total pressure in the alveoli cannot rise to more than the atmospheric pressure, this water vapor simply expands the volume of the air and thereby *dilutes* all the other gases in the inspired air. In column 2 of Table 40–2 it can be seen that humidification of the air has diluted the oxygen partial pressure at sea level from an average of 159 mm. Hg in atmospheric air to 149 mm. Hg in the humidified air, and it has diluted the nitrogen partial pressure from 597 to 563 mm. Hg.

RATE AT WHICH ALVEOLAR AIR IS RENEWED BY ATMOSPHERIC AIR

In the preceding chapter it was pointed out that the *functional residual capacity* of the lungs, which is the amount of air remaining in the lungs at the end of normal expiration, measures approximately 2300 ml. Furthermore, only 350 ml. of new air is brought into the alveoli with each normal respiration, and the same amount of old alveolar air is expired. Therefore, the amount of alveolar air replaced by new atmospheric air with each

breath is only one seventh of the total, so that many breaths are required to exchange most of the alveolar air. Figure 40–2 illustrates this slow rate of renewal of the alveolar air. In the first alveolus of the figure an excess amount of a gas has been placed momentarily in all the alveoli. The second alveolus shows slight dilution of this gas with the first breath; the next alveolus shows still further dilution with the second breath, and so forth for the third, fourth, eighth, twelfth, and sixteenth breaths. Note that even at the end of 16 breaths the excess gas still has not been completely removed from the alveoli.

Figure 40–3 illustrates graphically the rate at which an excess of a gas in the alveoli is normally removed, showing that with normal alveolar ventilation approximately half the gas is removed in 17 seconds. When a person's rate of alveolar ventilation is only half normal, half the gas is removed in 34 seconds, and, when the rate of ventilation is 2 times normal, half is removed in about 8 seconds.

Importance of the Slow Replacement of Alveolar Air. This slow replacement of alveolar air is of particular importance in preventing sudden changes in gaseous concentrations in the blood.

This makes the respiratory control mechanism much more stable than it would otherwise be and helps to prevent excessive increases and decreases in tissue oxygenation, tissue carbon dioxide concentration, and tissue pH when respiration is temporarily interrupted.

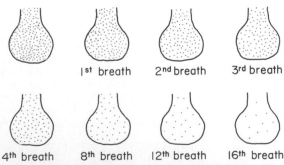

Figure 40–2. Expiration of a gaseous excess from the alveoli with successive breaths.

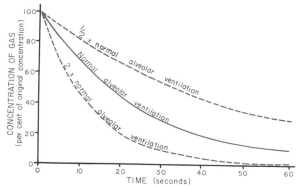

Figure 40–3. Rate of removal of excess gas from the alveoli.

OXYGEN CONCENTRATION AND PARTIAL PRESSURE IN THE ALVEOLI

Oxygen is continually being absorbed into the blood of the lungs, and new oxygen is continually entering the alveoli from the atmosphere. The more rapidly oxygen is absorbed, the lower becomes its concentration in the alveoli; on the other hand, the more rapidly new oxygen is brought into the alveoli from the atmosphere, the higher becomes its concentration. Therefore, oxygen concentration in the alveoli, and therefore its partial pressure as well, is controlled by, first, the rate of absorption of oxygen into the blood and, second, the rate of entry of new oxygen into the lungs by the ventilatory process.

Figure 40–4 illustrates the effect both of alveolar ventilation and of rate of oxygen absorption into the blood on the alveolar partial pressure of oxygen (PA_{O_2}). The solid curve represents oxygen absorption at a rate of 250 ml. per minute, and the dotted curve at 1000 ml. per minute. At a normal ventilatory rate of 4.2 liters per minute and an oxygen consumption of 250 ml. per minute, the normal operating point in Figure 40–4 is point A. During exercise, the rate of oxygen utilization is increased in proportion to the intensity of the exercise. From Figure 40–4 it can be seen that when 1000 ml. of oxygen is being absorbed each minute, the rate of alveolar ventilation must increase four-fold to maintain the alveolar P_{O_2} at the normal value of 104 mm. Hg, and at still higher rates of oxygen absorption, the rate of ventilation must rise proportionately to maintain normal alveolar P_{O_2}.

Another effect illustrated in Figure 40–4 is that an extremely marked increase in alveolar ventilation can never increase the alveolar P_{O_2} above 149 mm. Hg as long as the person is breathing normal atmospheric air, for this is the maximum content of oxygen in humidified air. However, if the person breathes gases containing concentrations of oxygen higher than 149 mm. Hg, the alveolar P_{O_2} can approach these higher concentrations as alveolar ventilation approaches maximum.

CO₂ CONCENTRATION AND PARTIAL PRESSURE IN THE ALVEOLI

Carbon dioxide is continually being formed in the body, then discharged into the alveoli; and it is continually being removed from the alveoli by the process of ventilation. Therefore, the two factors that determine alveolar concentration of carbon dioxide and also its partial pressure (PA_{CO_2}) are (1) the rate of excretion of carbon dioxide from the blood into the alveoli and (2) the rate at which carbon dioxide is removed from the alveoli by alveolar ventilation.

Figure 40–5 illustrates the effects on the alveolar P_{CO_2} of both alveolar ventilation and the rate of carbon dioxide excretion. The dark curve represents a normal rate of carbon dioxide excretion of 200 ml. per minute. At the normal rate of alveolar ventilation of 4.2 liters per minute, the operating point for alveolar P_{CO_2} is at point A in Figure 40–5 — that is, 40 mm. Hg.

Two other facts are also evident from Figure 40–5: First, *the alveolar P_{CO_2} increases directly in*

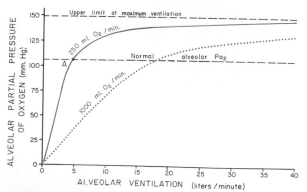

Figure 40–4. Effect of alveolar ventilation and of rate of oxygen absorption from the alveoli on the alveolar P_{O_2}.

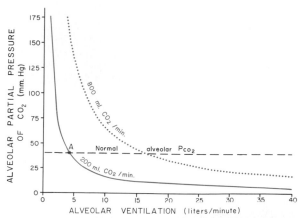

Figure 40–5. Effect on alveolar P_{CO_2} of alveolar ventilation and rate of carbon dioxide excretion from the blood.

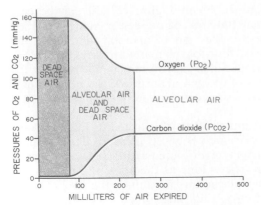

Figure 40–6. Oxygen and carbon dioxide partial pressures in the various portions of normal expired air.

proportion to the rate of carbon dioxide excretion, as represented by the dotted curve for 800 ml. CO_2 excretion per minute. Second, *the alveolar P_{CO_2} decreases in inverse proportion to alveolar ventilation.* Therefore, the concentrations and partial pressures of both oxygen and carbon dioxide in the alveoli are determined by the rates of absorp-

tion or excretion of the two gases, and also by the alveolar ventilation.

EXPIRED AIR

Expired air is a combination of dead space air and alveolar air, and its overall composition is, therefore, determined by, first, the proportion of the expired air that is dead space air and the proportion that is alveolar air. Figure 40–6 shows the progressive changes in oxygen and carbon dioxide partial pressures in the expired air during the course of expiration. The very first portion of this air, the dead space air, is typical humidified air as shown in column 2 of Table 40–2. Then, progressively more and more alveolar air becomes mixed with the dead space air until all the dead space air has finally been washed out and nothing but alveolar air remains. Thus, one of the means for collecting alveolar air for study is simply to collect a sample of the last portion of expired air.

Normal expired air, containing both dead space air and alveolar air, has gaseous concentrations (and partial pressures) approximately as shown in column 4 of Table 40–2 — that is, concentrations somewhere between those of humidified atmospheric air and alveolar air, under normal resting conditions about two-thirds alveolar air and one-third dead space air.

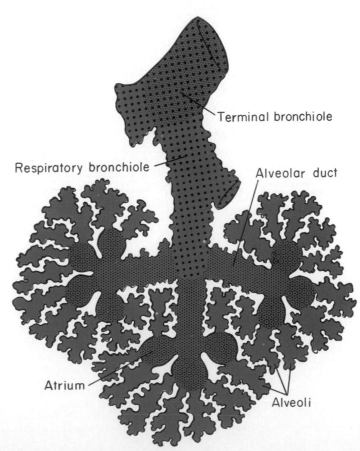

Figure 40–7. The respiratory lobule. (From Miller: The Lung. Charles C Thomas.)

DIFFUSION OF GASES THROUGH THE RESPIRATORY MEMBRANE

The Respiratory Unit.   Figure 40–7 illustrates the respiratory unit, which is comprised of a *respiratory bronchiole, alveolar ducts, atria, and* *alveoli* (of which there are about 300 million in the two lungs, each alveolus having an average diameter of about 0.25 mm.). The alveolar walls are extremely thin, and within them is an almost solid network of interconnecting capillaries, as illustrated in Figure 40–8. Indeed, the flow of blood in the alveolar wall has been described as a "sheet" of

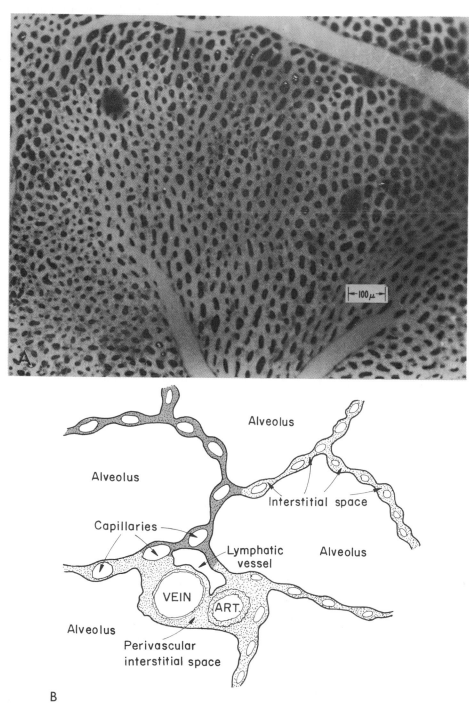

Figure 40–8.   A, Surface view of capillaries in an alveolar wall. (From Maloney and Castle: *Resp. Physiol.,* 7:150, 1969. Reproduced by permission of ASP Biological and Medical Press, North-Holland Division.) B, Cross-sectional view of alveolar walls and their vascular supply.

flowing blood. Thus, it is obvious that the alveolar gases are in close proximity to the blood of the capillaries. Consequently, gaseous exchange between the alveolar air and the pulmonary blood occurs through the membranes of all the terminal portions of the lungs. These membranes are collectively known as the *respiratory membrane,* also called the *pulmonary membrane.*

The Respiratory Membrane. Figure 40–9 illustrates the ultrastructure of the respiratory membrane. It also shows the diffusion of oxygen from the alveolus into the red blood cell and diffusion of the carbon dioxide in the opposite direction. Note the following different layers of the respiratory membrane:

1. A layer of fluid lining the alveolus and containing surfactant that reduces the surface tension of the alveolar fluid.

2. The alveolar epithelium comprised of very thin epithelial cells.

3. An epithelial basement membrane.

4. A very thin interstitial space between the alveolar epithelium and capillary membrane.

5. A capillary basement membrane that in many places fuses with the epithelial basement membrane.

6. The capillary endothelial membrane.

Despite the large number of layers, the overall

thickness of the respiratory membrane in some areas is as little as 0.2 micron and averages perhaps 0.5 micron.

From histologic studies it has been estimated that the total surface area of the respiratory membrane is approximately 70 square meters in the normal adult. This is equivalent to the floor area of a room 30 feet long by 25 feet wide. The total quantity of blood in the capillaries of the lung at any given instant is 60 to 140 ml. If this small amount of blood were spread over the entire surface of a 25 by 30 foot floor, one could readily understand how respiratory exchange of gases occurs as rapidly as it does.

The average diameter of the pulmonary capillaries is only 8 microns, which means that red blood cells must actually squeeze through them. Therefore, the red blood cell membrane usually touches the capillary wall so that oxygen and carbon dioxide need not pass through significant amounts of plasma as they diffuse between the alveolus and the red cell. Obviously, this increases the rapidity of diffusion.

FACTORS THAT AFFECT RATE OF GAS DIFFUSION THROUGH THE RESPIRATORY MEMBRANE

Referring to the above discussion of diffusion through water, one can apply the same principles and same formula to diffusion of gases through the respiratory membrane. Thus, the factors that determine how rapidly a gas will pass through the membrane are (1) the *thickness of the membrane,* (2) the *surface area of the membrane,* (3) the *diffusion coefficient* of the gas in the substance of the membrane — that is, in water, and (4) the *pressure difference* between the two sides of the membrane.

The *thickness of the respiratory membrane* occasionally increases, often as a result of edema fluid in the interstitial space of the membrane and in the alveoli, so that the respiratory gases must diffuse not only through the membrane but also through this fluid. Also, some pulmonary diseases cause fibrosis of the lungs, which can increase the thickness of some portions of the respiratory membrane. Because the rate of diffusion through the membrane is inversely proportional to the thickness of the membrane, any factor that increases the thickness to more than 2 to 3 times normal can interfere very significantly with normal respiratory exchange of gases.

The *surface area of the respiratory membrane* may be greatly decreased by many different conditions. For instance, removal of an entire lung decreases the surface area to half normal. Also, in *emphysema* many of the alveoli coalesce, with

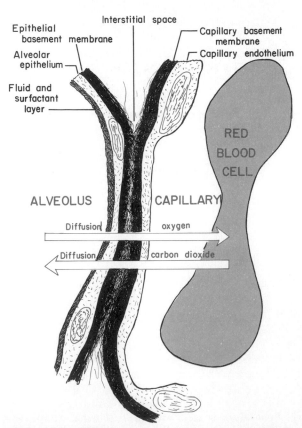

Figure 40–9. Ultrastructure of the respiratory membrane.

dissolution of many alveolar walls. Therefore, the new chambers are much larger than the original alveoli, but the total surface area of the respiratory membrane is considerably decreased because of loss of the alveolar walls. When the total surface area is decreased to approximately one-third to one-fourth normal, exchange of gases through the membrane is impeded to a significant degree *even under resting conditions*. And, during competitive sports and other strenuous exercise, even the slightest decrease in surface area of the lungs can be a serious detriment to respiratory exchange of gases.

The *diffusion coefficient* for the transfer of each gas through the respiratory membrane depends on its *solubility* in the membrane and inversely on the *square root* of its *molecular weight*. The rate of diffusion in the respiratory membrane is almost exactly the same as that in water, for reasons explained above. Therefore, for a given pressure difference, carbon dioxide diffuses through the membrane about 20 times as rapidly as oxygen. Oxygen in turn diffuses about 2 times as rapidly as nitrogen.

The *pressure difference* across the respiratory membrane is the difference between the partial pressure of the gas in the alveoli and the pressure of the gas in the blood. The partial pressure represents a measure of the total number of molecules of a particular gas striking a unit area of the alveolar surface of the membrane in unit time, and the pressure of the gas in the blood represents the number of molecules striking the same area of the membrane from the opposite side. Therefore, the difference between these two pressures is a measure of the *net tendency* for the gas to move through the membrane. Obviously, when the partial pressure of a gas in the alveoli is greater than the pressure of the gas in the blood, as is true for oxygen, net diffusion from the alveoli into the blood occurs, but, when the pressure of the gas in the blood is greater than the partial pressure in the alveoli, as is true for carbon dioxide, net diffusion from the blood into the alveoli occurs.

DIFFUSING CAPACITY OF THE RESPIRATORY MEMBRANE

The overall ability of the respiratory membrane to exchange a gas between the alveoli and the pulmonary blood can be expressed in terms of its *diffusing capacity,* which is defined as the *volume of a gas that diffuses through the membrane each minute for a pressure difference of 1 mm. Hg.*

Obviously, all the factors discussed above that affect diffusion through the respiratory membrane can affect the diffusing capacity.

The Diffusing Capacity for Oxygen. In the average young male adult the diffusing capacity for oxygen under resting conditions averages 21 ml. per minute per mm. Hg. The mean oxygen pressure difference across the respiratory membrane during normal, quiet breathing is approximately 11 mm. Hg. Multiplication of this pressure by the diffusing capacity (11×21) gives a total of about 230 ml. of oxygen normally diffusing through the respiratory membrane each minute, and this is equal to the rate at which the body uses oxygen.

Change in Oxygen-Diffusing Capacity During Exercise. During strenuous exercise, or during other conditions that greatly increase pulmonary activity, the diffusing capacity for oxygen increases in young male adults to a maximum of about 65 ml. per minute per mm. Hg, which is 3 times the diffusing capacity under resting conditions. This increase is caused by several different factors, among which are: (1) opening up of a number of previously dormant pulmonary capillaries, thereby increasing the surface area of the blood into which the oxygen can diffuse, and (2) dilatation of all the pulmonary capillaries that were already open, thereby further increasing the surface area. Therefore, during exercise, the oxygenation of the blood is increased not only by increased alveolar ventilation but also by a greater capacity of the respiratory membrane for transmitting oxygen into the blood.

Diffusing Capacity for Carbon Dioxide. The diffusing capacity for carbon dioxide has never been measured because of the following technical difficulty: Carbon dioxide diffuses through the respiratory membrane so rapidly that the average P_{CO_2} in the pulmonary blood is not far different from the P_{CO_2} in the alveoli — the average difference is less than 1 mm. Hg. — and with the available techniques, this difference is too small to be measured.

Nevertheless, measurements of diffusion of other gases have shown that the diffusing capacity varies directly with the diffusion coefficient of the particular gas. Since the diffusion coefficient of carbon dioxide is 20 times that of oxygen, one would expect a diffusing capacity for carbon dioxide under resting conditions of about 400 to 450 ml. and during exercise of about 1200 to 1300 ml. per minute per mm. Hg.

The importance of these high diffusing capacities for carbon dioxide is this: When the respiratory membrane becomes progressively damaged, its capacity for transmitting oxygen into the blood is often impaired enough to cause death of the person long before serious impairment of carbon dioxide diffusion occurs. The only time that a low diffusing capacity for carbon dioxide causes any significant difficulty·is when lung damage is far beyond that which ordinarily causes death but when the person's life is being maintained by intensive oxygen therapy that overcomes the reduction in oxygen-diffusing capacity.

Figure 40–10 compares the measured or calculated diffusing capacities of oxygen, carbon dioxide, and carbon monoxide at rest and during exercise, showing the extreme diffusing capacity of carbon dioxide and also the effect of exercise on the diffusing capacities of all the gases.

Measurement of Diffusing Capacity. The oxygen-diffusing capacity can be calculated from measurements of (1) alveolar P_{O_2}, (2) P_{O_2} in the pulmonary capillary blood, and (3) the rate of oxygen utilization. However, to measure the P_{O_2} in the pulmonary capillary blood is so difficult and so imprecise that it is not practical to measure oxygen-diffusing capacity by such a direct procedure, except on an experimental basis.

To obviate the difficulties encountered in measuring oxygen-diffusing capacity, physiologists usually measure carbon monoxide–diffusing capacity instead, and calculate the oxygen-diffusing capacity from this. The principle of the carbon monoxide method is the following: A small amount of carbon monoxide is breathed into the alveoli, and the partial pressure of the carbon monoxide in the alveoli is measured from appropriate alveolar air samples. Under most conditions it can be assumed that the carbon monoxide pressure in the blood is essentially zero because the hemoglobin combines with this so rapidly that the pressure never has time to build up. Therefore, the pressure difference of carbon monoxide across the respiratory membrane is equal to its partial pressure in the alveoli. Then, by measuring the volume of carbon monoxide absorbed in a short period of time and dividing this by the alveolar carbon monoxide partial pressure, one can determine accurately the carbon monoxide–diffusing capacity.

To convert carbon monoxide–diffusing capacity to oxygen-diffusing capacity, the value is multiplied by a factor of 1.23 because the diffusion coefficient for oxygen is 1.23 times that for carbon monoxide. Thus, the average diffusing capacity for carbon monoxide in young male adults in 17 ml. per minute per mm. Hg, and the diffusing capacity for oxygen is 1.23 times this, or 21 ml. per minute per mm. Hg.

EFFECT OF THE VENTILATION-PERFUSION RATIO ON ALVEOLAR GAS CONCENTRATION

Earlier in the chapter we discussed the importance of ventilation in determining both the P_{O_2} and P_{CO_2} in the alveoli. But we must also hasten to state that the rate of blood flow through the alveolar capillaries also affects alveolar P_{O_2} and P_{CO_2}. In fact, it is really the ratio of ventilation to pulmonary capillary blood flow, called the ventilation-perfusion ratio, or simply $\dot{V}_A/\dot{Q}$, that actually determines what the alveolar gas composition will be.

Furthermore, the ventilation-perfusion ratio is exceedingly important in determining the effectiveness of gas exchange across the respiratory membrane — especially oxygen exchange but in some instances carbon dioxide exchange as well. To begin this discussion let us consider the two extremes of $\dot{V}_A/\dot{Q}$: when this ratio is zero and when it is infinity.

Zero Gas Exchange, and Alveolar Partial Pressures When $\dot{V}_A/\dot{Q}$ Equals Zero. A $\dot{V}_A/\dot{Q}$ equal to zero means that there is zero alveolar ventilation but still blood flow through the alveolar capillaries. Obviously, without ventilation there will be zero gas exchange between the atmosphere and the blood for the simple reason that no gas can move from the alveoli to the exterior air.

However, even when $\dot{V}_A/\dot{Q}$ is equal to zero, the air in the alveolus still comes to equilibrium with the blood gases, because gases continue to diffuse between the blood and the alveolar air. Since the blood that perfuses the capillaries is venous blood returning to the lungs from the systemic circulation, it is the gases in this blood that come to equilibrium with the alveolar gases. In the next chapter we shall see that the average venous blood ($\bar{v}$) has a normal P_{O_2} of 40 mm. Hg and a P_{CO_2} of 45 mm. Hg. Therefore, these are also the usual partial pressures of these two gases in alveoli that have blood flow but no ventilation.

Zero Gas Exchange, and Alveolar Partial Pressures When $\dot{V}_A/\dot{Q}$ Equals Infinity. A $\dot{V}_A/\dot{Q}$ equal to infinity means that there is alveolar ventilation but no blood flow through the alveolar capillaries. Obviously, this condition is also incompatible with

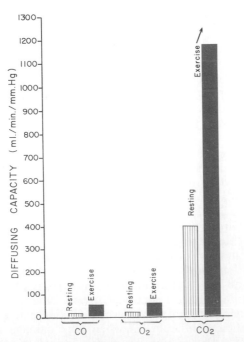

Figure 40–10. Lung diffusing capacities for carbon monoxide, oxygen, and carbon dioxide in the normal lungs.

exchange of any gas whatsoever between the atmosphere and the blood, for the simple reason that there is no blood to carry the gases away from the alveoli.

Yet, the effect on the alveolar gas partial pressures is entirely different from the effect when $\dot{V}A/\dot{Q}$ equals zero. Instead of the alveolar gases coming to equilibrium with the venous blood, the alveolar air now becomes equal to the inspired air. That is, the air that is inspired loses no oxygen to the blood and it gains no CO_2. And, since normal inspired and humidified air has a P_{O_2} of 149 mm. Hg and a P_{CO_2} of zero mm. Hg, these are also the partial pressures of these two gases in the alveoli.

Gas Exchange, and Alveolar Partial Pressures When $\dot{V}A/\dot{Q}$ Is Normal. When there is both normal ventilation and also normal pulmonary capillary blood flow, exchange of oxygen and carbon dioxide through the respiratory membrane is approximately optimal. Oxygen is absorbed out of the inspired air so that the alveolar P_{O_2} lies somewhere between that of the inspired air and that of the venous blood. Likewise, carbon dioxide is transferred from the venous blood into the alveoli, which makes the alveolar P_{CO_2} rise to some level between that of the inspired air and that of the venous blood. Thus, under normal conditions, the alveolar air P_{O_2} averages 104 mm. Hg and the P_{CO_2}, 40 mm. Hg.

The P_{O_2}-P_{CO_2}, $\dot{V}A/\dot{Q}$ Diagram. The concepts presented in the above sections can also be presented very usefully in graphical form as illustrated in Figure 40–11. The diagram is called the P_{O_2}-P_{CO_2}, $\dot{V}A/\dot{Q}$ diagram. The curve in the diagram represents all possible P_{O_2} and P_{CO_2} combinations between the limits of $\dot{V}A/\dot{Q}$ equals zero and $\dot{V}A/\dot{Q}$ equals infinity. Thus, point $\bar{v}$ is the point plot of P_{O_2} and P_{CO_2} when $\dot{V}A/\dot{Q}$ equals zero. At this point, the P_{O_2} is 40 mm. Hg and the P_{CO_2} is 45 mm. Hg, which are the values in venous blood.

At the other end of the curve, when $\dot{V}A/\dot{Q}$ equals infinity, point I represents inspired air, and the P_{O_2} is 149 mm. Hg while the P_{CO_2} is zero.

Also plotted on the curve is the point representing normal alveolar air when the $\dot{V}A/\dot{Q}$ is normal. At this point, P_{O_2} is 104 mm. Hg and P_{CO_2} is 40 mm. Hg.

However, note once again that as $\dot{V}A/\dot{Q}$ decreases below normal — that is, ventilation becomes too little in relation to blood flow — the alveolar P_{O_2} and P_{CO_2} approach those values found in venous blood. On the other hand, as $\dot{V}A/\dot{Q}$ rises above normal and approaches infinity, the alveolar air composition approaches that of inspired air.

The Concept of "Physiologic Shunt" (When $\dot{V}A/\dot{Q}$ Is Below Normal)

Whenever $\dot{V}A/\dot{Q}$ is below normal, there obviously is not ventilation enough to provide the oxygen needed to oxygenate the blood flowing through the alveolar capillaries. Therefore, a certain fraction of the venous blood passing through the pulmonary capillaries does not become oxygenated. This fraction is called *shunted blood*. Still more blood flows through the bronchial vessels rather than through the alveolar capillaries, normally about 2 per cent of the cardiac output; this too is unoxygenated, shunted blood.

The total quantitative amount of shunted blood per minute is called the *physiologic shunt*. This physiologic shunt is measured in the clinical pulmonary function laboratory by analyzing the concentration of oxygen in both mixed venous blood and arterial blood. From these values the physiologic shunt can then be calculated using the following equation:

$$\frac{\dot{Q}_{PS}}{\dot{Q}_T} = \frac{Ci_{O_2} - Ca_{O_2}}{Ci_{O_2} - C\bar{v}_{O_2}}$$

in which:

$\dot{Q}_{PS}$ is the physiologic shunt blood flow per minute

$\dot{Q}_T$ is cardiac output per minute

Ci is the concentration of oxygen in the arterial blood if there is "ideal" ventilation-perfusion ratio

Ca_{O_2} is the measured concentration of oxygen in the arterial blood

$C\bar{v}_{O_2}$ is the measured concentration of oxygen in the mixed venous blood

Obviously, the greater the physiologic shunt, the greater is the amount of blood that fails to be oxygenated as it passes through the lungs.

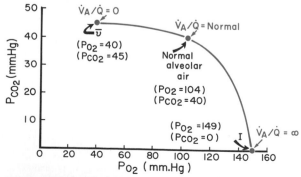

Figure 40–11. The normal P_{O_2}-P_{CO_2}, $\dot{V}A/\dot{Q}$ diagram.

The Concept of "Physiologic Dead Space" (When $\dot{V}_A/\dot{Q}$ Is Greater Than Normal)

When the ventilation is great, but blood flow is low, there is then far more available oxygen in the alveoli than can be transported away from the alveoli by the flowing blood. Thus, a large portion of the ventilation is said to be *wasted*. The ventilation of the dead space areas of the lungs is also wasted. The sum of these two types of wasted ventilation is called the *physiologic dead space*. This is measured in the clinical pulmonary function laboratory by making appropriate blood and expiratory gas measurements and using the following equation, called the Bohr equation:

$$\frac{V_{D_{phys}}}{V_T} = \frac{Pa_{CO_2} - P\bar{E}_{CO_2}}{Pa_{CO_2}}$$

in which:

$V_{D_{phys}}$ is the physiologic dead space
V_T is the tidal volume
Pa_{CO_2} is the partial pressure of carbon dioxide in the arterial blood
$P\bar{E}_{CO_2}$ is the average partial pressure of carbon dioxide in the entire expired air

Obviously, when the physiologic dead space is very great, much of the work of ventilation is wasted effort because so much of the ventilated air never reaches the blood.

Abnormalities of Ventilation to Perfusion Ratio

Abnormal $\dot{V}_A/\dot{Q}$s in the Upper and Lower Normal Lung. In a normal person in the upright position, both blood flow and alveolar ventilation are considerably less in the upper part of the lung than in the lower part; however, blood flow is decreased far more than ventilation. Therefore, at the top of the lung, $\dot{V}_A/\dot{Q}$ is as much as 3 times as great as the ideal value, which causes a moderate degree of *physiologic dead space* in this area of the lung.

At the other extreme, in the bottom of the lung there is slightly too little ventilation in relation to blood flow, with $\dot{V}_A/\dot{Q}$ as low as 0.6 times the ideal value. Therefore, in this area a small fraction of the blood fails to become normally oxygenated, and this represents a *physiologic shunt*.

Therefore, in both extremes of the lung, inequalities of ventilation and perfusion decrease the effectiveness of the lung slightly for exchange of oxygen and carbon dioxide. However, during exercise, the blood flow to the upper part of the lung increases markedly so that then far less physiologic dead space occurs and the effectiveness of gas exchange approaches optimum.

Abnormal $\dot{V}_A/\dot{Q}$ in Chronic Obstructive Lung Disease. Most persons who smoke for prolonged periods of time develop some degree of bronchial obstruction;

indeed, in a large share of them this eventually becomes so severe that they develop serious pulmonary air trapping and resultant emphysema. In the emphysematous lung many of the alveolar walls also become destroyed. Thus, in this disease there are two abnormalities that can cause abnormal $\dot{V}_A/\dot{Q}$. First, because many of the small bronchioles are obstructed, the alveoli beyond the obstructions are unventilated, causing a $\dot{V}_A/\dot{Q}$ that approaches zero. Second, in those areas of the lung where the alveolar walls have been mainly destroyed but there is still alveolar ventilation, most of the ventilation is wasted because of inadequate blood flow to transport the blood gases. Thus, in chronic obstructive lung disease, some areas of the lung exhibit very serious physiologic shunt and other areas very serious physiologic dead space. Both of these tremendously decrease the effectiveness of the lungs as gas exchange organs, sometimes reducing the effectiveness to as little as one-tenth normal. In fact, this is the most prevalent cause of pulmonary disability today.

REFERENCES

Bauer, C., *et al.* (eds.): Biophysics and Physiology of Carbon Dioxide. New York, Springer-Verlag, 1980.

Burrows, B.: Arterial oxygenation and pulmonary hemodynamics in patients with chronic airways obstruction. *Am. Rev. Resp. Dis., 110*:64, 1974.

Cherniack, R. M.: Ventilation, perfusion and gas exchange. *In* Frohlich, E. D. (ed.): Pathophysiology, 2nd ed. Philadelphia, J. B. Lippincott Co., 1976, p. 149.

Cumming, G.: Alveolar ventilation: Recent model analysis. *In* MTP International Review of Science: Physiology. Vol. 2. Baltimore, University Park Press, 1974, p. 139.

Cunningham, D. J.: Time patterns of alveolar carbon dioxide and oxygen: The effects of various patterns of oscillations on breathing in man. *Sci. Basis Med.,* 333, 1972.

Forster, R. E., and Crandall, E. D.: Pulmonary gas exchange. *Annu. Rev. Physiol., 38*:69, 1976.

Forster, R. E.: Diffusion of gases. *In* Fenn, W. O.,and Rahn, H. (eds.): Handbook of Physiology. Sec. 3, Vol. 1. Baltimore, Williams & Wilkins, 1964, p. 839.

Forster, R. E.: Interpretation of measurements of pulmonary diffusing capacity. *In* Fenn, W. O., and Rahn, H. (eds.): Handbook of Physiology. Sec. 3, Vol. 2. Baltimore, Williams & Wilkins, 1965, p. 1435.

Forster, R. E.: Pulmonary ventilation and blood gas exchange. *In* Sodeman, W. A., Jr., and Sodeman, W. A. (eds.): Pathologic Physiology: Mechanisms of Disease, 5th Ed. Philadelphia, W. B. Saunders Company, 1974, p. 371.

Guyton, A. C., *et al.*: An arteriovenous oxygen difference recorder. *J. Appl. Physiol., 10*:158, 1957.

Hughes, J. M.: Proceedings: Regional differences in gas exchange. *Proc. R. Soc. Med., 66*:974, 1973.

Johansen, K.: Comparative physiology: Gas exchange and circulation in fishes. *Annu. Rev. Physiol., 33*:569, 1971.

Jones, N. L.: Blood Gases and Acid-Base Physiology. New York, B. C. Decker, 1980.

Milhorn, H. T., Jr., and Pulley, P. E., Jr.: A theoretical study of pulmonary capillary gas exchange and venous admixture. *Biophys. J., 8*:337, 1968.

Moran, F., and Pack, A. I.: Proceedings: Measurement of ventilation-perfusion distribution. *Proc. R. Soc. Med., 66*:975, 1973.

Morrow, P. E.: Alveolar clearance of aerosols. *Arch. Intern. Med., 131*:101, 1973.

Otis, A. B.: Quantitative relationships in steady-state gas exchange. *In* Fenn, W. O., and Rahn, H. (eds.): Handbook of Physiology. Sec. 3, Vol. 1. Baltimore, Williams & Wilkins, 1964, p. 681.

Piiper, J., and Scheid, P.: Respiration: Alveolar gas exchange. *Annu. Rev. Physiol., 33*:131, 1971.

Piiper, J., and Scheid, P.: Comparative physiology of respiration: Functional analysis of gas exchange organs in vertebrates. *Int. Rev. Physiol., 14*:219, 1977.

Radford, E. P., Jr.: The physics of gases. *In* Fenn, W. O., and Rahn, H. (eds.): Handbook of Physiology. Sec. 3, Vol. 1. Baltimore, Williams & Wilkins, 1964, p. 125.

Rahn, H., and Farhi, L. E.: Ventilation, perfusion, and gas exchange — the Va/Q concept. *In* Fenn, W. O., and Rahn, H. (eds.): Handbook of Physiology. Sec. 3, Vol. 1. Baltimore, Williams & Wilkins, 1964, p. 125.

Rahn, H., and Fenn, W. O.: A Graphical Analysis of Respiratory Gas Exchange. Washington, American Physiological Society, 1955.

Rahn, H., *et al.:* How bird eggs breathe. *Sci. Am. 240*(2):46, 1979.

Riley, R. L., and Permutt, S.: The four-quadrant diagram for analyzing the distribution of gas and blood in the lung. *In* Fenn, W. O., and Rahn, H. (eds.): Handbook of Physiology. Sec. 3, Vol. 2. Baltimore, Williams & Wilkins, 1965, p. 1413.

Thurlbeck, W. M., and Wang, N. S.: The structure of the lungs. *In* MTP International Review of Science: Physiology. Vol. 2. Baltimore, University Park Press, 1974, p. 1.

Wagner, P. D.: Diffusion and chemical reaction in pulmonary gas exchange. *Physiol. Rev., 57*:257, 1977.

Weibel, E. R.: Morphological basis of alveolar capillary gas exchange. *Physiol. Rev., 53*:419, 1973.

West, J. B.: Ventilation/Blood Flow and Gas Exchange, 2nd Ed. Philadelphia, J. B. Lippincott Co., 1970.

West, J. B.: Pulmonary gas exchange. *Int. Rev. Physiol., 14*:83, 1977.

41

Transport of Oxygen and Carbon Dioxide in the Blood and Body Fluids

Once oxygen has diffused from the alveoli into the pulmonary blood, it is transported principally in combination with hemoglobin to the tissue capillaries where it is released for use by the cells. The presence of hemoglobin in the red cells of the blood allows the blood to transport 30 to 100 times as much oxygen as could be transported simply in the form of dissolved oxygen in the water of the blood.

In the tissue cells oxygen reacts with various foodstuffs to form large quantities of carbon dioxide. This in turn enters the tissue capillaries and is transported back to the lungs. Carbon dioxide, like oxygen, also combines with chemical substances in the blood that increase carbon dioxide transport 15- to 20-fold.

The purpose of the present chapter, therefore, is to present both qualitatively and quantitatively the physical and chemical principles of oxygen and carbon dioxide transport in the blood and body fluids.

PRESSURES OF OXYGEN AND CARBON DIOXIDE IN THE LUNGS, BLOOD, AND TISSUES

In the discussions of the preceding chapter it was pointed out that gases can move from one point to another by diffusion and that the cause of this movement is always a pressure difference from the first point to the other. Thus, oxygen diffuses from the alveoli into the pulmonary capillary blood because the oxygen pressure (P_{O_2}) in the alveoli is greater than the P_{O_2} in the pulmonary blood. Then in the tissues a much higher P_{O_2} in the capillary blood causes oxygen to diffuse to the cells.

Conversely, when oxygen is metabolized in the cells, the carbon dioxide pressure (P_{CO_2}) rises to a high value, which causes carbon dioxide to diffuse into the tissue capillaries. Similarly, it diffuses out of the blood into the alveoli because the P_{CO_2} in the alveoli is lower than that in the pulmonary capillary blood.

Basically, then, the transport of oxygen and carbon dioxide by the blood depends on both diffusion and the movement of blood. We now need to consider quantitatively the factors responsible for these effects as well as their significance in the overall physiology of respiration.

UPTAKE OF OXYGEN BY THE PULMONARY BLOOD

The top part of Figure 41–1 illustrates a pulmonary alveolus adjacent to a pulmonary capillary, showing diffusion of oxygen molecules between the alveolar air and the pulmonary blood. However, the P_{O_2} of the venous blood entering the capillary is only 40 mm. Hg because a large amount of oxygen has been removed from this blood as it has passed through the body. The P_{O_2} in the alveolus is 104 mm. Hg, giving an initial pressure difference for diffusion of oxygen into the pulmonary capillary of 104 − 40 or 64 mm. Hg. Therefore, far more oxygen diffuses into the pulmonary capillary than in the opposite direction. The curve below the capillary shows the progressive rise in blood P_{O_2} as the blood passes through the capillary. This curve illustrates that the P_{O_2} rises essentially to equal that of the alveolar air before passing through a third of the capillary, becoming approximately 104 mm. Hg.

The average pressure difference for oxygen diffusion through the pulmonary capillary during

504

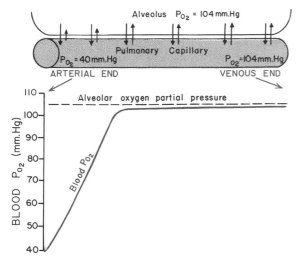

Figure 41–1. Uptake of oxygen by the pulmonary capillary blood. (The curve in this figure was constructed from data in Milhorn and Pulley: *Biophys. J., 8*:337, 1968.)

normal respiration is about 11 mm. Hg. This is a "time-integrated" average and not simply an average of the 64 mm. Hg pressure difference at the beginning of the capillary and the final zero pressure difference at the end of the capillary, because the initial pressure difference lasts for only a fraction of the transit time in the pulmonary capillary, while the low pressure difference lasts for a long time.

Uptake of Oxygen by the Pulmonary Blood During Exercise. During strenuous exercise, a person's body may require as much as 20 times the normal amount of oxygen. However, because of the increased cardiac output, the time that the blood remains in the capillary may be reduced to less than one-half normal. Therefore, oxygenation of the blood could suffer for two reasons: First, the blood remains in contact with the alveoli for short periods of time and, second, far larger quantities of oxygen are needed to oxygenate the blood. Yet, because of a great *safety factor* for diffusion of oxygen through the pulmonary membrane, the blood is still *almost completely saturated* with oxygen when it leaves the pulmonary capillaries. The reasons for this are:

First, it was pointed out in the previous chapter that the diffusing capacity for oxygen increases about three-fold during exercise; this results mainly from increased numbers of capillaries participating in the diffusion but also from dilatation of both the alveoli and capillaries, as well as a more ideal ventilation-perfusion ratio in the upper part of the lungs.

Second, note in Figure 41–1 that during normal pulmonary blood flow the blood becomes almost saturated with oxygen by the time it has passed through one-third of the pulmonary capillary, and

little additional oxygen enters the blood during the latter two-thirds of its transit. That is, the blood normally stays in the lung capillaries about 3 times as long as necessary to cause full oxygenation. Therefore, even with the shortened time of exposure in exercise, the blood still can become fully oxygenated.

TRANSPORT OF OXYGEN IN THE ARTERIAL BLOOD

Normally, the systemic arterial blood is composed 98 to 99 per cent of oxygenated blood that passes through the pulmonary capillaries and another 1 to 2 per cent of unoxygenated blood that passes through the bronchial circulation. Therefore, as illustrated in Figure 41–2, the blood leaving the pulmonary capillaries has a P_{O_2} of approximately 104 mm. Hg while the arterial blood, which contains the bronchial blood as well, has a P_{O_2} averaging only 95 mm. Hg. However, this great fall in P_{O_2} actually represents only about a 1 per cent decrease in oxygen concentration because, as we shall see later in the chapter, the combining affinity of oxygen with hemoglobin is very nonlinear so that as it approaches full saturation, the P_{O_2} changes considerably for very slight changes in the amount of oxygen bound with the hemoglobin.

DIFFUSION OF OXYGEN FROM THE CAPILLARIES TO THE INTERSTITIAL FLUID

As illustrated in Figure 41–3, the P_{O_2} in the interstitial fluid, though very variable, averages about 40 mm. Hg, while that in the arterial blood entering the capillaries is high, about 95 mm. Hg. Therefore, at the arterial end of the capillary, a pressure difference of 55 mm. Hg causes rapid diffusion of oxygen into the tissues so that the capillary P_{O_2} approaches the 40 mm. Hg oxygen

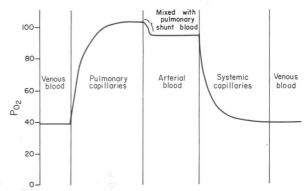

Figure 41–2. Changes in P_{O_2} in the pulmonary capillary blood, the arterial blood, and the systemic capillary blood, illustrating the effect of "venous admixture."

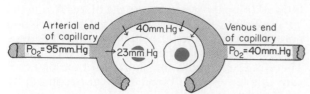

Figure 41–3. Diffusion of oxygen from a tissue capillary to the cells.

pressure in the tissue fluids. Consequently, the venous blood leaving the tissue capillaries contains oxygen at essentially the same pressure as that in the interstitial fluid immediately outside the tissue capillaries, 40 mm. Hg.

Effect of Rate of Blood Flow on Interstitial Fluid P_{O_2}. If the blood flow through a particular tissue becomes increased, greater quantities of oxygen are transported into the tissue in a given period of time, and the tissue P_{O_2} becomes correspondingly increased — an effect illustrated in Figure 41–4. Note that an increase in flow to 400 per cent of normal increases the P_{O_2} from Point A to Point B, that is, from 40 mm. Hg to 66 mm. Hg. However, the upper limit to which the P_{O_2} can rise, even with maximal blood flow, is 95 mm. Hg, which is the oxygen pressure in the arterial blood.

Effect of Rate of Tissue Metabolism on Interstitial Fluid P_{O_2}. If the cells utilize far more oxygen for metabolism than normally, this tends to reduce the interstitial fluid P_{O_2}. Figure 41–4 also illustrates this effect, showing that increased tissue oxygen consumption greatly reduces the interstitial fluid P_{O_2}, while reduced consumption greatly increases the P_{O_2}.

Effect of Hemoglobin Concentration on Interstitial Fluid P_{O_2}. Because approximately 97 per cent of the oxygen transported in the blood is carried by hemoglobin, a decrease in hemoglobin concentration has the same effect on interstitial fluid P_{O_2} as does a decrease in blood flow. Thus, reducing the hemoglobin concentration to one-quarter normal while maintaining normal blood flow reduces the interstitial fluid P_{O_2} to point C in Figure 41–4 — that is, to about 13 mm. Hg.

In summary, tissue P_{O_2} is determined by a balance between (a) the rate of oxygen transport to the tissues in

the blood and (b) the rate at which the oxygen is utilized by the tissues.

DIFFUSION OF OXYGEN FROM THE CAPILLARIES TO THE CELLS

Since oxygen is always being used by the cells, the intracellular P_{O_2} remains lower than the interstitial fluid P_{O_2}. Nevertheless, as pointed out in Chapter 4, oxygen diffuses through cell membranes extremely rapidly. Therefore, the intracellular P_{O_2} is almost equal to that of its immediately surrounding interstitial fluids. Yet, in many instances, there is considerable distance between the capillaries and the cells. Therefore, the normal intracellular P_{O_2} ranges from as low as 5 mm. Hg to as high as 60 mm. Hg, averaging (by direct measurement in lower animals) 23 mm. Hg. Since only 3 mm. Hg of oxygen pressure is normally required for full support of the metabolic processes of the cell, one can see that even this low cellular P_{O_2}, 23 mm. Hg, is adequate and actually provides a considerable safety factor.

DIFFUSION OF CARBON DIOXIDE FROM THE CELLS TO THE TISSUE CAPILLARIES

Because of the continual large quantities of carbon dioxide formed in the cells, the intracellular P_{CO_2} tends to rise. However, carbon dioxide diffuses about 20 times as easily as oxygen, diffusing from the cells extremely rapidly into the interstitial fluids and thence into the capillary blood. Thus, in Figure 41–5 the intracellular P_{CO_2} is shown to be about 46 mm. Hg, while that in the interstitial fluid is about 45 mm. Hg, a pressure differential of only 1 mm. Hg.

Arterial blood entering the tissue capillaries contain carbon dioxide at a pressure of approximately 40 mm. Hg. As the blood passes through the capillaries, the blood P_{CO_2} rises to approach the 45 mm. Hg P_{CO_2} of the interstitial fluid. Because of the very large diffusion coefficient for carbon dioxide, the P_{CO_2} of venous blood is also about 45 mm. Hg, within a fraction of a millimeter of reaching complete equilibrium with the P_{CO_2} of the interstitial fluid.

**Effect of Tissue Metabolism and Blood Flow on

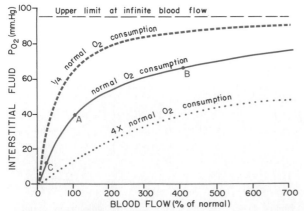

Figure 41–4. Effect of blood flow and rate of oxygen consumption on tissue P_{O_2}.

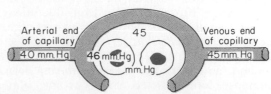

Figure 41–5. Uptake of carbon dioxide by the blood in the capillaries.

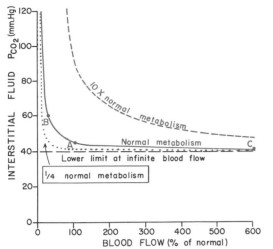

Figure 41–6. Effect of blood flow and metabolic rate on tissue P_{CO_2}.

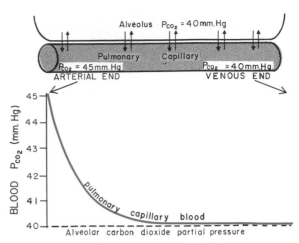

Figure 41–7. Diffusion of carbon dioxide from the pulmonary blood into the alveolus. (This curve was constructed from data in Milhorn and Pulley: *Biophys. J.*, 8:337, 1968.)

Interstitial Fluid P_{CO_2}. Blood flow and tissue metabolism affect tissue P_{CO_2} in a way exactly opposite to the way that they affect tissue P_{O_2}. Figure 41–6 shows the effects of normal metabolism, one-quarter normal metabolism, and 10 times normal metabolism on interstitial fluid P_{CO_2} at different rates of blood flow. Note that the lower limit to which the interstitial fluid P_{CO_2} can possibly fall is the P_{CO_2} of the arterial blood entering the tissue capillaries — normally about 40 mm. Hg. The elevation of tissue P_{CO_2} above this value depends on both blood flow and metabolic rate. Point A illustrates the P_{CO_2} when both the blood flow and metabolism are normal. Point B shows that a decrease in blood flow to one-quarter normal increases the tissue P_{CO_2} to 60 mm. Hg, while point C shows that an increase in blood flow to six times normal reduces the tissue P_{CO_2} to 41 mm. Hg, almost to its lower limit of 40 mm. Hg. It can also be seen that increasing the rate of metabolism, especially at low blood flow rates, can increase the tissue P_{CO_2} to extremely high levels, while reducing the rate of metabolism correspondingly reduces the tissue P_{CO_2}.

REMOVAL OF CARBON DIOXIDE FROM THE PULMONARY BLOOD

On arriving at the lungs, the P_{CO_2} of the venous blood is about 45 mm. Hg while that in the alveoli is 40 mm. Hg. Therefore, as illustrated in Figure 41–7, the initial pressure difference for diffusion is only 5 mm. Hg, which is far less than for diffusion of oxygen across the membrane. Yet, even so, because of the 20 times as great diffusion coefficient for carbon dioxide as for oxygen, the excess carbon dioxide in the blood is rapidly transferred into the alveoli. Indeed, the figure shows that the

P_{CO_2} of the pulmonary capillary blood becomes almost equal to that of the alveoli within the first third of the blood's transit through the pulmonary capillary.

TRANSPORT OF OXYGEN IN THE BLOOD

Normally, about 97 per cent of the oxygen transported from the lungs to the tissues is carried in chemical combination with hemoglobin in the red blood cells, and the remaining 3 per cent in the dissolved state in the water of the plasma and cells. Thus, *under normal conditions* oxygen is carried to the tissues almost entirely by hemoglobin.

THE REVERSIBLE COMBINATION OF OXYGEN WITH HEMOGLOBIN

The chemistry of hemoglobin was presented in Chapter 5, where it was pointed out that the oxygen molecule combines loosely and reversibly with the heme portion of the hemoglobin. When the P_{O_2} is high, as in the pulmonary capillaries, oxygen binds with the hemoglobin, but, when the P_{O_2} is low, as in the tissue capillaries, oxygen is released from the hemoglobin. This is the basis for oxygen transport from the lungs to the tissues.

The Oxygen-Hemoglobin Dissociation Curve. Figure 41–8 illustrates the oxygen-hemoglobin dissociation curve, which shows the progressive increase in the per cent of the hemoglobin that is bound with oxygen as the P_{O_2} increases. This is called the *per cent saturation of the hemoglobin.* Since the blood leaving the lungs usually has a P_{O_2} of about 100 mm. Hg, one can see from the

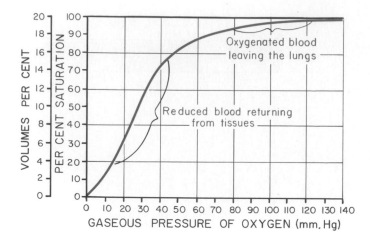

Figure 41–8. The oxygen-hemoglobin dissociation curve.

dissociation curve that the *usual oxygen saturation of arterial blood is about 97 per cent*. On the other hand, in normal venous blood returning from the tissues the P_{O_2} is about 40 mm. Hg and *the saturation of the hemoglobin is about 70 per cent*.

Maximum Amount of Oxygen That Can Combine with the Hemoglobin of the Blood. The blood of a normal person contains approximately 15 grams of hemoglobin in each 100 ml. of blood, and each gram of hemoglobin can bind with a maximum of about 1.34 ml. of oxygen (1.39 ml. when the hemoglobin is chemically pure, but this is reduced about 4 per cent by impurities such as methemoglobin). Therefore, on the average, the hemoglobin in 100 ml. of blood can combine with a total of almost exactly 20 ml. of oxygen when the hemoglobin is 100 per cent saturated. This is usually expressed as 20 *volumes per cent*. The oxygen-hemoglobin dissociation curve for the normal person, therefore, can be expressed in terms of volume per cent of oxygen, as shown in Figure 41–9, rather than per cent saturation of hemoglobin.

Amount of Oxygen Released from the Hemoglo-

bin **in the Tissues.** The total quantity of oxygen *bound with hemoglobin* in normal arterial blood, which is normally 97 per cent saturated, is approximately 19.4 ml. This is illustrated in Figure 41–9. However, on passing through the tissue capillaries, this amount is reduced, on the average, to 14.4 ml. (P_{O_2} of 40 mm. Hg, 75 per cent saturated). Thus, *under normal conditions about 5 ml. of oxygen is transported by each 100 ml. of blood during each cycle through the tissues*.

Transport of Oxygen During Strenuous Exercise. In heavy exercise the muscle cells utilize oxygen at a rapid rate, which causes the interstitial fluid P_{O_2} to fall to as low as 15 mm. Hg. At this pressure only 4.4 ml. of oxygen remains bound with the hemoglobin in each 100 ml. of blood, as shown in Figure 41–9. Thus 19.4 − 4.4, or 15 ml., is the total quantity of oxygen transported by each 100 ml. of blood during each cycle through the tissues. This, obviously, is 3 times as much as that normally transported by the same amount of blood, illustrating that simply an increase in rate of oxygen utilization by the tissues causes an automatic increase in the rate of oxygen release from the hemoglobin.

The Utilization Coefficient. The fraction of the blood that gives up its oxygen as it passes through the tissue capillaries is called the *utilization coefficient*. Normally, this is approximately 0.25, or 25 per cent, of the blood. That is, *the normal utilization coefficient is approximately one fourth*. During strenuous exercise, as much as 75 to 85 per cent of the blood can give up its oxygen; the utilization coefficient is then approximately 0.75 to 0.85. These values are about the highest utilization coefficients that can be attained in the body overall even when the tissues are in extreme need of oxygen. However, in local tissue areas where the blood flow is very slow or the metabolic rate very high, utilization coefficients approaching 100 per cent have been recorded — that is, essentially all the oxygen is removed.

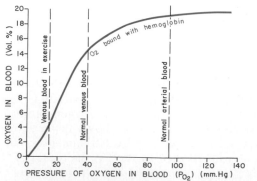

Figure 41–9. Effect of blood P_{O_2} on the quantity of oxygen bound with hemoglobin in each 100 ml. of blood.

THE OXYGEN BUFFER FUNCTION OF HEMOGLOBIN

Though hemoglobin is necessary for transport of oxygen to the tissues, it performs still another major function essential to life. This is its function as an "oxygen buffer" system. That is, the hemoglobin in the blood is mainly responsible for controlling the oxygen pressure in the tissues. This can be explained as follows:

Value of Hemoglobin for Maintaining Constant Po_2 in the Tissue Fluids.   Under basal conditions the tissues require about 5 ml. of oxygen from each 100 ml. of blood passing through the tissue capillaries. Referring back to the oxygen-hemoglobin dissociation curve in Figure 41–9, one can see that for the 5 ml. of oxygen to be released, the Po_2 must fall to about 40 mm. Hg. Therefore, the tissue capillary Po_2 cannot rise above this 40 mm. Hg level, for if such should occur, the oxygen needed by the tissues could not be released from the hemoglobin. In this way, the hemoglobin normally sets an upper limit on the gaseous pressure in the tissues at approximately 40 mm. Hg.

On the other hand, in heavy exercise extra large amounts of oxygen must be delivered from the hemoglobin to the tissues. But this can be achieved with very little further decrease in tissue Po_2 because of the steep slope of the dissociation curve — that is, a small fall in Po_2 causes extreme amounts of oxygen to be released. Therefore, the Po_2 rarely falls below 20 mm. Hg.

It can be seen, then, that hemoglobin automatically delivers oxygen to the tissues at a pressure between approximately 20 and 40 mm. Hg. This seems to be a wide range of Po_2s in the interstitial fluid, but, when one considers how much the interstitial fluid Po_2 might possibly change during exercise and other types of stress, this range of 20 to 40 mm. Hg is relatively narrow.

Value of Hemoglobin for Maintaining Constant Tissue Po_2 When Atmospheric Oxygen Concentration Changes Markedly.   The normal Po_2 in the alveoli is approximately 104 mm. Hg, but, as one ascends a mountain or goes high in an airplane, the Po_2 can easily fall to less than half this. Or, when one enters areas of compressed air, such as deep below the sea or in tunnels, the Po_2 may rise to 10 times this level. Yet, even so, the tissue Po_2 changes very little. Let us explain this.

It will be seen from the oxygen-hemoglobin dissociation curve in Figure 41–8 that, when the alveolar Po_2 is decreased to as low as 60 mm. Hg, the arterial hemoglobin is still 89 per cent saturated, only 8 per cent below the normal saturation of 97 per cent. Furthermore, the tissues still move approximately 5 ml. of oxygen from every 100 ml. of blood passing through the tissues; to remove this oxygen, the Po_2 of the venous blood falls to only slightly less than 40 mm. Hg. Thus, the tissue Po_2 hardly changes despite the marked fall in alveolar Po_2 from 104 to 60 mm. Hg.

On the other hand, when the alveolar Po_2 rises far above the normal value of 104 mm. Hg, the maximum oxygen saturation of hemoglobin can never rise above 100 per cent. Therefore, even though the oxygen in the alveoli should rise to a partial pressure of 500 mm. Hg, or even more, the increase in the saturation of hemoglobin would be only 3 per cent because, even at 104 mm. Hg Po_2, 97 per cent of the hemoglobin is already combined with oxygen; and only a small amount of additional oxygen dissolves in the fluid of the blood, as will be discussed subsequently. Then, when the blood passes through the tissue capillaries, it still loses several milliliters of oxygen to the tissues, and this loss automatically reduces the Po_2 of the capillary blood to a value only a few millimeters greater than the normal 40 mm. Hg.

Consequently, alveolar oxygen may vary greatly — from 60 to more than 500 mm. Hg Po_2 — and still the Po_2 in the tissue does not vary more than a few millimeters from normal.

FACTORS THAT CAUSE THE HEMOGLOBIN DISSOCIATION CURVE TO SHIFT

The hemoglobin dissociation curves of Figures 41–8 and 41–9 are those for normal average blood. However, a number of different factors can displace the dissociation curve in one direction or the other in the manner illustrated in Figure 41–10. This figure shows that when the blood becomes slightly acidic, with the pH decreased from the normal value of 7.4 to 7.2, the hemoglobin dissociation curve shifts on the average about 15 per cent to the right. On the other hand, an increase in the pH to 7.6 shifts the curve a similar amount to the left.

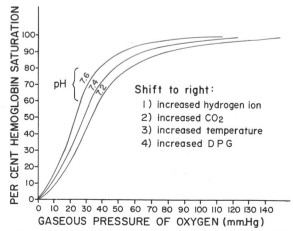

Figure 41–10.   Shift of the oxygen-hemoglobin dissociation curve to the right by increases in (1) hydrogen ions, (2) CO_2, (3) temperature, or (4) DPG.

In addition to pH changes, several other factors are also known to shift the curve. Three of these, all of which shift the curve to the right, are (1) increased carbon dioxide concentration, (2) increased blood temperature, and (3) increased 2,3-diphosphoglycerate, a phosphate compound normally present in the blood but in differing concentrations under different conditions.

A factor that shifts the dissociation curve to the left is the presence of large quantities of *fetal hemoglobin,* a type of hemoglobin present in the fetus before birth and which is different from the normal hemoglobin called *adult hemoglobin.* Shift of the curve to the left when fetal hemoglobin is present is important for oxygen delivery to the fetal tissues under the hypoxic conditions in which the fetus exists. This will be discussed further in Chapter 83.

Importance of the Effect of Carbon Dioxide in Shifting the Hemoglobin Dissociation Curve — the Bohr Effect. Shift of the hemoglobin dissociation curve by changes in the blood CO_2 is important to enhance oxygenation of the blood in the lungs and also to enhance release of oxygen from the blood in the tissues. This is called the *Bohr effect,* and it can be explained as follows: As the blood passes through the lungs, carbon dioxide diffuses from the blood into the alveoli. This reduces the blood P_{CO_2} and also increases the pH; both of these effects shift the hemoglobin dissociation curve to the left and upward. Therefore, the quantity of oxygen that binds with the hemoglobin becomes considerably increased, thus allowing greater oxygen transport to the tissues. Then when the blood reaches the tissue capillaries, exactly the opposite effect occurs. Carbon dioxide entering the blood from the tissues displaces oxygen from the hemoglobin and therefore delivers oxygen to the tissues at a higher P_{O_2} than would otherwise occur. That is, the dissociation curve shifts to the right in the tissues — exactly opposite to the leftward shift in the lungs.

Effect of 2,3-Diphosphoglycerate (DPG). The normal DPG in the blood keeps the hemoglobin dissociation curve shifted slightly to the right all the time. However, in hypoxic conditions that last longer than a few hours, the quantity of DPG in the blood increases considerably, thus shifting the hemoglobin dissociation curve even farther to the right. This causes oxygen to be released to the tissues at as much as 10 mm. Hg higher oxygen pressure than would be the case without this increased DPG. Therefore, it has been taught for the past few years that this might be an important mechanism for adaptation to hypoxia. However, the presence of the excess DPG also makes it difficult for the hemoglobin in the lungs to combine with oxygen in the hypoxic state, thereby often creating as much harm as good. Therefore, it is questionable whether the DPG shift of the dissociation curve is always beneficial in hypoxia.

Shift of the Dissociation Curve During Exercise. In exercise, several factors shift the dissociation curve considerably to the right. The exercising muscles release large quantities of carbon dioxide and this plus acids released by the exercising muscle increases the hydrogen ion concentration in the muscle capillary blood. In addition, the temperature of the muscle often rises as much as 3 to 4 degrees, and, finally, phosphate compounds are also released. All these factors acting together shift the hemoglobin dissociation curve *of the muscle capillary blood* considerably to the right. Therefore, oxygen can sometimes be released to the muscle at P_{O_2}s as great as 40 mm. Hg (the normal resting value) even though as much as 75 to 85 per cent of the oxygen is removed from the hemoglobin. Then, in the lungs, the shift occurs in the opposite direction, thus allowing pickup of extra amounts of oxygen from the alveoli.

TOTAL RATE OF OXYGEN TRANSPORT FROM THE LUNGS TO THE TISSUES

Under resting conditions about 5 ml. of oxygen is transported by each 100 ml. of blood, and the normal cardiac output is approximately 5000 ml. per minute. Therefore, the total quantity of oxygen delivered to the tissues each minute is about 250 ml.

This rate of oxygen transport to the tissues can be increased to about 15 times normal in strenuous exercise and in other instances of excessive need for oxygen (and very rarely to as high as 20 times normal in the best-trained athletes). Oxygen transport can be increased to 3 times normal simply by an increase in utilization coefficient, and it can be increased another five-fold as a result of increased cardiac output, thus accounting for the total 15-fold increase. Therefore, the maximum rate of oxygen transport to the tissues is about 15×250 ml., or 3750 ml., per minute in the normal young adult. Special adaptations in athletic training, especially an increase in maximum cardiac output, can sometimes increase this value to as high as 4.5 to 5 liters per minute.

METABOLIC USE OF OXYGEN BY THE CELLS

Relationship Between Intracellular P_{O_2} and Rate of Oxygen Usage. Only a minute level of oxygen pressure is required in the cells for normal intracellular chemical reactions to take place. The reason for this is that the respiratory enzyme systems of the cell, which will be discussed in Chapter 67, are geared so that when the cellular P_{O_2} is more than 3 to 5 mm. Hg, oxygen availability is no longer a limiting factor in the rates of the chemical reactions. Instead, the main limiting factor then is the *concentration of adenosine diphosphate* (ADP) in the cells, as was explained in Chapter 3. This effect is illustrated in Figure 41–11, which shows the relationship between intracellular P_{O_2} and rate of oxygen usage. Note that whenever the intracellular P_{O_2} is above 3 to 5 mm. Hg the rate of oxygen usage becomes constant for any given concentration of ADP in the cell. On the other hand, when the ADP concentration is altered, the rate of oxygen usage changes in proportion to the change in ADP concentration.

It will be recalled from the discussion in Chapter 3 that when adenosine triphosphate (ATP) is utilized in the cells to provide energy, it is converted

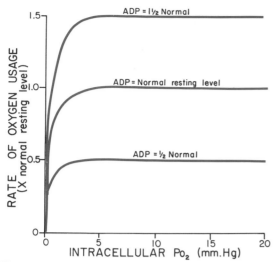

Figure 41–11. Effect of intracellular P_{O_2} on rate of oxygen usage by the cells. Note that increasing the intracellular concentration of adenosine diphosphate (*ADP*) increases the rate of oxygen usage.

into ADP. The increasing concentration of ADP, in turn, increases the metabolic usage of both oxygen and the various nutrients that combine with oxygen to release energy. This energy is used to reform the ATP. Therfore, *under normal operating conditions the rate of oxygen utilization by the cells is controlled by the rate of energy expenditure within the cells — that is, by the rate at which ADP is formed from ATP — and not by the availability of oxygen to the cells.*

Effect of Diffusion Distance from the Capillary to the Cell on Oxygen Utilization. Cells are rarely more than 50 microns away from a capillary, and oxygen normally can diffuse readily enough from the capillary to the cell to supply all the required amounts of oxygen for metabolism. However, occasionally, cells are located farther than this from the capillaries, and the rate of oxygen diffusion to these cells is so low that the intracellular P_{O_2} falls below the critical level of 3 to 5 mm. Hg required to maintain maximum intracellular metabolism. Thus, under these conditions, oxygen utilization by the cells is *diffusion limited* and is no longer determined by the amount of ADP formed in the cells.

Effect of Blood Flow on Metabolic Use of Oxygen. The total amount of oxygen available each minute for use in any given tissue is determined by (1) the quantity of oxygen transported in each 100 ml. of blood and (2) the rate of blood flow. If the rate of blood flow falls to zero, the amount of oxygen available for metabolism obviously also falls to zero. Thus, there are times when the rate of blood flow through a tissue can be so low that the tissue P_{O_2} falls below the critical 3 to 5 mm. Hg required for maximal intracellular metabolism. Under these conditions, the rate of tissue utiliza-

tion of oxygen is *blood flow limited*. However, neither diffusion limited nor blood flow limited usage of oxygen can continue for long, because the cells then often receive less oxygen than is required for continuation of the life of the cells themselves.

TRANSPORT OF OXYGEN IN THE DISSOLVED STATE

At the normal arterial P_{O_2} of 95 mm. Hg, approximately 0.29 ml. of oxygen is dissolved in every 100 ml. of water in the blood. When the P_{O_2} of the blood falls to 40 mm. Hg in the tissue capillaries, 0.12 ml. of oxygen remains dissolved. In other words, 0.17 ml. of oxygen is normally transported in the dissolved state to the tissues by each 100 ml. of blood water. This compares with about 5.0 ml. transported by the hemoglobin. Therefore, the amount of oxygen transported to the tissues in the dissolved state is normally slight, only about 3 per cent of the total as compared with the 97 per cent transported by the hemoglobin. During strenuous exercise, when transport increases another 3-fold, the relative quantity then transported in the dissolved state falls to as little as 1.5 per cent. Yet, if a person breathes oxygen at very high P_{O_2}s, the amount then transported in the dissolved state can become tremendous, so much so that serious excesses of oxygen occur in the tissues and "oxygen poisoning" ensues. This often leads to convulsions and even death, as will be discussed in detail in Chapter 45 in relation to high pressure breathing.

COMBINATION OF HEMOGLOBIN WITH CARBON MONOXIDE

Carbon monoxide combines with hemoglobin at the same point on the hemoglobin molecule as does oxygen. Furthermore, it binds with about 230 times as much tenacity as oxygen, which is illustrated by the carbon monoxide–hemoglobin dissociation curve in Figure 41–12. This curve is almost identical with the oxygen–hemoglobin dissociation curve, except that the pressures of the carbon monoxide shown on the abscissa are

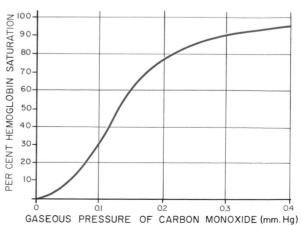

Figure 41–12. The carbon monoxide-hemoglobin dissociation curve.

at a level $1/230$ of those in the oxygen–hemoglobin dissociation curve of Figure 41–8. Therefore, a carbon monoxide pressure of only 0.4 mm. Hg in the alveoli, $1/230$ that of the alveolar oxygen, allows the carbon monoxide to compete equally with the oxygen for combination with the hemoglobin and causes half the hemoglobin in the blood to become bound with carbon monoxide instead of with oxygen. A carbon monoxide pressure of 0.7 mm. Hg (a concentration of about 0.1 per cent) can be lethal.

A patient severely poisoned with carbon monoxide can be advantageously treated by administering pure oxygen, for oxygen at high alveolar pressures displaces carbon monoxide from its combination with hemoglobin far more rapidly than can oxygen at the low pressure of atmospheric oxygen.

The patient can also be benefited by simultaneous administration of a few per cent carbon dioxide because this strongly stimulates the respiratory center, as is discussed in the following chapter. This increases alveolar ventilation and reduces the alveolar carbon monoxide concentration, which allows increased carbon monoxide release from the blood.

With intensive oxygen and carbon dioxide therapy, carbon monoxide can be removed from the blood 10 to 20 times as rapidly as without therapy.

TRANSPORT OF CARBON DIOXIDE IN THE BLOOD

Transport of carbon dioxide by the blood is not nearly so great a problem as transport of oxygen, because even in the most abnormal conditions carbon dioxide can usually be transported in far greater quantities than can oxygen. However, the amount of carbon dioxide in the blood does have much to do with acid-base balance of the body fluids, which was discussed in detail in Chapter 37.

Under normal resting conditions *an average of 4 ml. of carbon dioxide is transported from the tissues to the lungs in each 100 ml. of blood.*

CHEMICAL FORMS IN WHICH CARBON DIOXIDE IS TRANSPORTED

To begin the process of carbon dioxide transport, carbon dioxide diffuses out of the tissue cells in the gaseous form (but not to a significant extent in the bicarbonate form because the cell membrane is far less permeable to bicarbonate than to the dissolved gas). On entering the capillary, the chemical reactions illustrated in Figure 41–13 occur immediately; the quantitative aspects of these can be described as follows:

Transport of Carbon Dioxide in the Dissolved State. A small portion of the carbon dioxide is transported in the dissolved state to the lungs. It will be recalled that the P_{CO_2} of venous blood is 45 mm. Hg and that of arterial blood is 40 mm. Hg. The amount of carbon dioxide dissolved in the

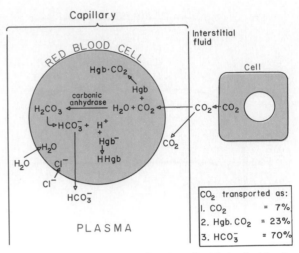

Figure 41–13. Transport of carbon dioxide in the blood.

fluid of the blood at 45 mm. Hg is about 2.7 ml. per 100 ml. (2.7 volumes per cent). The amount dissolved at 40 mm. Hg is about 2.4 ml., or a difference of 0.3 ml. Therefore, only about 0.3 ml. of carbon dioxide is transported in the form of dissolved carbon dioxide by each 100 ml. of blood. This is about 7 per cent of all the carbon dioxide transported.

Transport of Carbon Dioxide in the Form of Bicarbonate Ion. *Reaction of Carbon Dioxide with Water in the Red Blood Cells — Effect of Carbonic Anhydrase.* The dissolved carbon dioxide in the blood reacts with water to form carbonic acid. However, this reaction would occur too slowly to be of importance were it not for the fact that inside the red blood cells the enzyme called *carbonic anhydrase* catalyzes the reaction between carbon dioxide and water, accelerating its rate about 5000-fold. Therefore, instead of requiring many seconds to occur, as is true in the plasma, the reaction occurs so rapidly in the red blood cells that it reaches almost complete equilibrium within a fraction of a second. This allows tremendous amounts of carbon dioxide to react with the red cell water even before the blood leaves the tissue capillaries.

Dissociation of Carbonic Acid into Bicarbonate and Hydrogen Ions. In another small fraction of a second the carbonic acid formed in the red cells dissociates into *hydrogen* and *bicarbonate ions.* Most hydrogen ions then combine with the hemoglobin in the red blood cells because hemoglobin is a powerful acid-base buffer. In turn, many of the bicarbonate ions diffuse into the plasma while chloride ions diffuse into the red cells to take their place. Thus, the chloride content of venous red blood cells is greater than that of arterial cells, a phenomenon called the *chloride shift.*

The reversible combination of carbon dioxide with water in the red blood cells under the influence of carbonic anhydrase accounts for an average of about 70 per cent of all the carbon dioxide transported from the tissues to the lungs. Thus, this means of transporting carbon dioxide is by far the most important of all the methods for transport. Indeed, when a carbonic anhydrase inhibitor (acetazolamide) is administered to an animal to block the action of carbonic anhydrase in the red blood cells, carbon dioxide transport from the tissues becomes very poor — so poor that the tissue P_{CO_2} can as a result rise to as high as 80 mm. Hg instead of the normal 45 mm. Hg.

Transport of Carbon Dioxide in Combination with Hemoglobin and Plasma Proteins — Carbaminohemoglobin. In addition to reacting with water, carbon dioxide also reacts directly with hemoglobin. The combination of carbon dioxide with hemoglobin is a reversible reaction that occurs with a very loose bond. The compound formed by this reaction is known as "carbaminohemoglobin." A small amount of carbon dioxide also reacts in this same way with the plasma proteins, but this is much less significant because the quantity of these proteins in the blood is only one-fourth as great as the quantity of hemoglobin.

The *theoretical* quantity of carbon dioxide that can be carried from the tissues to the lungs in combination with hemoglobin and plasma proteins is approximately 30 per cent of the total quantity transported — that is, about 1.5 ml. of carbon dioxide in each 100 ml. of blood. However, this reaction is much slower than the reaction of carbon dioxide with water inside the red blood cells. Therefore, it is doubtful that this mechanism provides transport of more than 15 to 25 per cent of the total quantity of carbon dioxide.

THE CARBON DIOXIDE DISSOCIATION CURVE

It is apparent that carbon dioxide can exist in blood in many different forms, (1) as free carbon dioxide and (2) in chemical combinations with water, hemoglobin, and plasma protein. The total quantity of carbon dioxide combined with the blood in all these forms depends on the P_{CO_2}. The curve shown in Figure 41–14 depicts this dependence of total blood CO_2 in all its forms on P_{CO_2}; this curve is called the *carbon dioxide dissociation curve*.

Note that the normal blood P_{CO_2} ranges between the limits of 40 mm. Hg in arterial blood and 45 mm. Hg in venous blood, which is a very narrow range. Note also that the normal concentration of carbon dioxide in the blood is about 50 volumes per cent but that only 4 volumes per cent of this is actually exchanged in the process of transporting carbon dioxide from the tissues to the lungs. That is, the concentration rises to about 52 volumes per cent as the blood passes through the

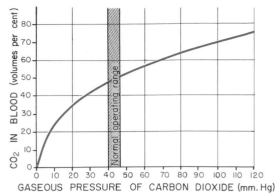

Figure 41–14. The carbon dioxide dissociation curve.

tissues, and falls to about 48 volumes per cent as it passes through the lungs.

EFFECT OF THE OXYGEN-HEMOGLOBIN REACTION ON CARBON DIOXIDE TRANSPORT — THE HALDANE EFFECT

Earlier in the chapter it was pointed out that an increase in carbon dioxide in the blood will cause oxygen to be displaced from the hemoglobin and that this is an important factor in promoting oxygen transport. The reverse is also true: binding of oxygen with hemoglobin tends to displace carbon dioxide from the blood. Indeed, this effect, called the *Haldane effect,* is quantitatively far more important in promoting carbon dioxide transport than is the Bohr effect in promoting oxygen transport.

The Haldane effect results from the simple fact that combination of oxygen with hemoglobin causes the hemoglobin to become a stronger acid. This in turn displaces carbon dioxide from the blood in two ways: (1) The more highly acidic hemoglobin has less tendency to combine with carbon dioxide to form carbaminohemoglobin, thus displacing much of the carbon dioxide that is present in this form in the blood. (2) The increased acidity of the hemoglobin causes a general increase in the acidity of all the fluids both in the red blood cells and in the plasma. The increased hydrogen ions combine with the bicarbonate ions of the blood to form carbonic acid, which then dissociates and releases carbon dioxide from the blood.

Therefore, in the tissue capillaries the Haldane effect causes increased pickup of carbon dioxide because of oxygen removal from the hemoglobin, and in the lungs it causes increased release of carbon dioxide because of oxygen pickup by the hemoglobin.

Figure 41–15 illustrates quantitatively the significance of the Haldane effect on the transport of carbon dioxide from the tissues to the lungs. This figure shows small portions of two separate carbon dioxide dissociation curves, the solid curve when the P_{O_2} is 100 mm. Hg, which is the case in the lungs, and the dashed curve when the P_{O_2} is 40 mm. Hg, which is the case in the tissue capillaries. Point A on the dashed curve shows that the normal P_{CO_2} of 45 mm. Hg in the tissues causes 52 volumes per cent of carbon dioxide to combine with the blood. On entering the lungs the P_{CO_2} falls to 40 mm.

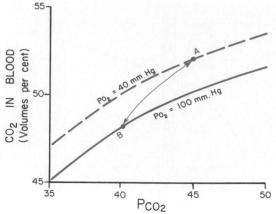

Figure 41–15. Portions of the carbon dioxide dissociation curves when the Po_2 is 100 mm. Hg and 40 mm. Hg, respectively. The arrow represents the Haldane effect on the transport of carbon dioxide, as explained in the text.

Hg while the Po_2 rises to 100 mm. Hg. If the carbon dioxide dissociation curve did not change, the carbon dioxide content of the blood would fall only to 50 volumes per cent, which would be a loss of only 2 volumes per cent of carbon dioxide. However, the increase in Po_2 in the lungs decreases the carbon dioxide dissociation curve from the dashed curve to the solid curve of the figure, so that the carbon dioxide content falls to 48 volumes per cent (point B). This represents an additional 2 volumes per cent loss of carbon dioxide. Thus, the Haldane effect approximately doubles the amount of carbon dioxide released from the blood in the lungs and approximately doubles the pickup of Pco_2 in the tissues.

CHANGE IN BLOOD ACIDITY DURING CARBON DIOXIDE TRANSPORT

The carbonic acid formed when carbon dioxide enters the blood in the tissues decreases the blood pH. Fortunately, though, the reaction of this acid with the buffers of the blood prevents the hydrogen ion concentration from rising greatly. Ordinarily, arterial blood has a pH of approximately 7.40, and, as the blood acquires carbon dioxide in the tissue capillaries, the pH falls to approximately 7.36. In other words, a pH change of 0.04 unit takes place. The reverse occurs when carbon dioxide is released from the blood in the lungs, the pH rising to the arterial value once again. In exercise, in other conditions of high metabolic activity, or when the blood flow through the tissues is sluggish, the decrease in pH in the tissue blood (and in the tissues themselves) can be as much as 0.5, or on occasion even more than this, thus causing tissue acidosis.

THE RESPIRATORY EXCHANGE RATIO

The discerning student will have noted that normal transport of oxygen from the lungs to the tissues by each 100 ml. of blood is about 5 ml., while normal transport of carbon dioxide from the tissues to the lungs is about 4 ml. Thus, under normal resting conditions only about 80 per cent as much carbon dioxide is expired from the lungs as there is oxygen uptake by the lungs. The ratio of carbon dioxide output to oxygen uptake is called the *respiratory exchange ratio* (R). That is,

$$R = \frac{\text{Rate of carbon dioxide output}}{\text{Rate of oxygen uptake}}$$

The value for R changes under different metabolic conditions. When a person is utilizing entirely carbohydrates for body metabolism, R rises to 1.00. On the other hand, when the person is utilizing fats almost entirely for metabolic energy, the level falls to as low as 0.7. The reason for this difference is that when oxygen is metabolized with carbohydrates one molecule of carbon dioxide is formed for each molecule of oxygen consumed, while when oxygen reacts with fats a large share of the oxygen combines with hydrogen atoms to form water instead of carbon dioxide. In other words, the *respiratory quotient of the chemical reactions* in the tissues when fats are metabolized is about 0.70 instead of 1.00 which is the case when carbohydrates are being utilized. The tissue respiratory quotient will be discussed in Chapter 71.

For a person on a normal diet consuming average amounts of carbohydrates, fats, and proteins, the average value for R is considered to be 0.825.

REFERENCES

Adamson, J. W., and Finch, C. A.: Hemoglobin function, oxygen affinity, and erythropoietin. *Annu. Rev. Physiol., 37*:351, 1975.

Bartels, H., and Baumann, R.: Respiratory function of hemoglobin. *Int. Rev. Physiol., 14*:107, 1977.

Bauer, C., *et al.* (eds.): Biophysics and Physiology of Carbon Dioxide. New York, Springer-Verlag, 1980.

Bruley, D. F.: Mathematical considerations for oxygen transport to tissue. *Adv. Exp. Med. Biol., 37*:749, 1973.

Caughey, W. S. (ed.): Oxygen; Biochemical and Clinical Aspects. New York, Academic Press, 1979.

Chance, B.: Regulation of intracellular oxygen. *Proc. Int. Union Physiol. Sci., 6*:13, 1968.

Cherniack, N. S., and Longobardo, G. S.: Oxygen and carbon dioxide gas stores of the body. *Physiol. Rev., 50*:196, 1970.

Crowell, J. W., and Smith, E. E.: Determinants of the optimal hematocrit. *J. Appl. Physiol., 22*:501, 1967.

Forster, R. E.: CO_2: chemical, biochemical, and physiological aspects. *Physiologist, 13*:398, 1970.

Forster, R. E.: Pulmonary ventilation and blood gas exchange. *In* Sodeman, W. A., Jr., and Sodeman, W. A. (eds.): Pathologic Physiology: Mechanisms of Disease, 5th Ed. Philadelphia, W. B. Saunders Company, 1974, p. 371.

Grodins, F. S., and Yamashiro, S. M.: Optimization of the mammalian respiratory gas transport system. *Annu. Rev. Biophys. Bioeng., 4*:115, 1973.

Haldane, J. S., and Priestley, J. G.: Respiration. New Haven, Yale University Press, 1935.

Hayaishi, O.: Molecular Oxygen in Biology. New York, American Elsevier Publishing Co., 1974.

Jöbsis, F. F.: Intracellular metabolism of oxygen. *Am. Rev. Resp. Dis., 110*:58, 1974.

Jones, C. E., *et al.*: Determination of mean tissue oxygen tensions by implanted perforated capsules. *J. Appl. Physiol., 26*:630, 1969.

Jones, N. L.: Blood Gases and Acid-Base Physiology. New York, B. C. Decker, 1980.

Kessler, M., *et al.*: (eds.): Oxygen Supply, Baltimore, University Park Press, 1973.

Kilmartin, J. V., and Rossi-Bernardi, L.: Interaction of hemoglobin with hydrogen ions, carbon dioxide, and organic phosphates. *Physiol. Rev., 53*:836, 1973.

Konigsberg, W.: Protein structure and molecular dysfunction: Hemoglobin. *In* Bondy, P. K., and Rosenberg, L. E. (eds.): Metabolic Control and Disease, 8th Ed. Philadelphia, W. B. Saunders Co., 1980, p. 27.

Lehman, H., and Huntsman, R. G.: Man's Hemoglobins, 2nd Ed. Philadelphia, J. B. Lippincott Co., 1974.

Maren, T. H.: Carbonic anhydrase: Chemistry, physiology, and inhibition. *Physiol. Rev., 47*:595, 1967.

Michel, C. C.: The transport of oxygen and carbon dioxide by the blood. *In* MTP International Review of Science: Physiology. Vol. 2. Baltimore, University Park Press, 1974, p. 67.

Morson, B. C. (ed.): Hypoxia and Ischaemia. London, Journal of Clinical Pathology for the College, 1977.

Oski, F. A., and Gottlieb, A. J.: The interrelationships between red blood cell metabolites, hemoglobin, and the oxygen-equilibrium curve. *Prog. Hematol., 7*:33, 1971.

Paintal, A. S., and Gill-Kumar, P. (eds.): Respiratory Adaptations, Capillary Exchange, and Reflex Mechanisms. Delhi, India, Vallabhbhai Patel Chest Institute, University of Delhi, 1977.

Perutz, M. F.: Hemoglobin structure and respiratory transport. *Sci. Am., 239*(6):92, 1978.

Randall, D. J.: The Evolution of Air Breathing in Vertebrates. New York, Cambridge University Press, 1980.

Robin, E. D., and Simon, L. M.: Oxygen transport and cellular respiration. *In* Frohlich, E. D. (ed.): Pathophysiology, 2nd Ed. Philadelphia, J. B. Lippincott Co., 1976, p. 167.

Root, W. S.: Carbon monoxide. *In* Fenn, W. O., and Rahn, H. (eds.): Handbook of Physiology. Sec. 2, Vol. 2. Baltimore, Williams & Wilkins Co., 1965, p. 1087.

Stainsby, W. N., and Barclay, J. K.: Oxygen uptake by striated muscle. *Muscle Biol., 1*:273, 1972.

Wagner, P. D.: Diffusion and chemical reaction in pulmonary gas exchange. *Physiol. Rev., 57*:257, 1977.

Wagner, P. D.: The oxyhemoglobin dissociation curve and pulmonary gas exchange. *Sem. Hematol., 11*:405, 1974.

West, J. B.: Pulmonary gas exchange. *Int. Rev. Physiol., 14*:83, 1977.

Whalen, W. J.: Intracellular PO_2 in heart and skeletal muscle. *Physiologist, 14*:69, 1971.

Wittenberg, J. B.: Myoglobin-facilitated oxygen diffusion: Role of myoglobin in oxygen entry into muscle. *Physiol. Rev., 50*:559, 1970.

42

Regulation of Respiration

The nervous system adjusts the rate of alveolar ventilation almost exactly to the demands of the body so that the blood oxygen pressure (P_{O_2}) and carbon dioxide pressure (P_{CO_2}) are hardly altered even during strenuous exercise or other types of respiratory stress.

The present chapter describes the operation of this neurogenic system for regulation of respiration.

THE RESPIRATORY CENTER

The so-called "respiratory center" is a widely dispersed group of neurons located bilaterally in the reticular substance of the medulla oblongata and pons, as illustrated in Figure 42–1. It is divided into three major areas: (1) a dorsal medullary group of neurons, which is mainly an *inspiratory area,* (2) a ventral respiratory group of neurons, which is mainly an *expiratory area,* and (3) an area in the pons that helps to control the respiratory rate, called the *pneumotaxic area.* It is the inspiratory area that plays the fundamental role in the control of respiration. Therefore, let us discuss its function first.

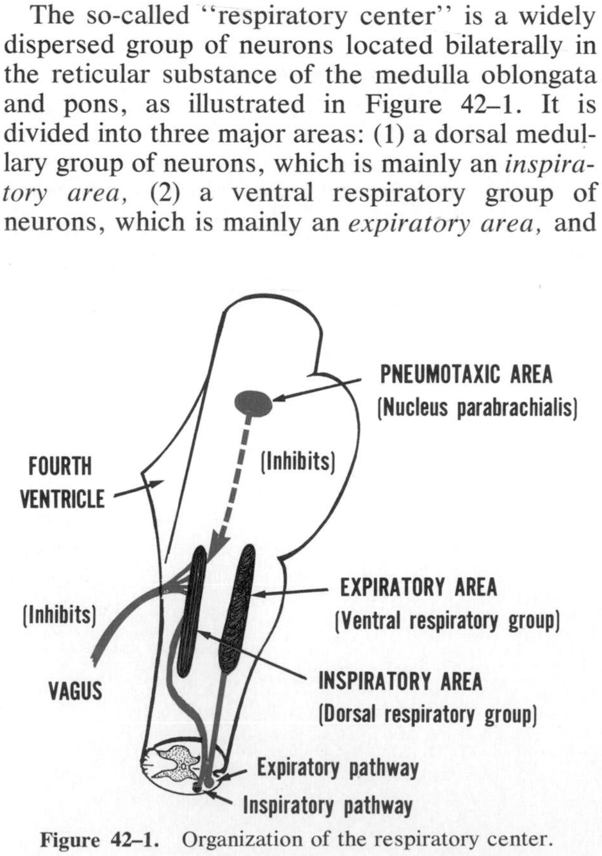

PNEUMOTAXIC AREA
(Nucleus parabrachialis)
(Inhibits)
FOURTH
VENTRICLE
EXPIRATORY AREA
(Ventral respiratory group)
(Inhibits)
VAGUS
INSPIRATORY AREA
(Dorsal respiratory group)
Expiratory pathway
Inspiratory pathway

Figure 42–1. Organization of the respiratory center.

The Inspiratory Area

The inspiratory area lies bilaterally in the dorsal portion of the medulla, extending approximately the entire length of the medulla. Its neurons are very near to and interconnect closely with the tractus solitarius, which is the sensory termination of both the vagal and glossopharyngeal nerves; each of these nerves in turn transmits sensory signals from the peripheral chemoreceptors, in this way helping to control pulmonary ventilation. In addition, the vagi also transmit sensory signals from the lungs that help to control lung inflation and respiratory rate, as we shall discuss in a subsequent section of this chapter.

Rhythmical Oscillation in the Inspiratory Area. The basic rhythm of respiration is generated in the inspiratory area. Even when all incoming nerve fiber connections to this area have been sectioned or blocked, the area still emits repetitive bursts of action potentials that cause rhythmical inspiratory cycles.

During expiration the inspiratory center becomes dormant, but after a few seconds the inspiratory center suddenly and automatically turns on again. This seems to result from an inherent, intrinsic excitability of the inspiratory neurons themselves that causes reverberatory oscillation to occur from neuron to neuron within the dorsal respiratory group; this neuronal activity initiates the inspiratory nervous signals to the diaphragm and other inspiratory muscles. During the next few seconds the intensity of the inspiratory signal increases rapidly, causing the diaphragm and other inspiratory muscles to contract more and more forcefully. In normal respiration this "ramp" increase in inspiratory signal lasts for about two seconds, at the end of which time it suddenly comes to a halt. Then the inspiratory neurons remain dormant again for approximately the next three seconds in normal respiration before the entire cycle repeats itself once more, this repetition continuing on and on throughout the life of the person.

Function of the Pneumotaxic Center to Limit the Duration of Inspiration and to Increase Respiratory Rate. The pneumotaxic center, located in the pons, transmits impulses continuously to the inspiratory area. The primary effect of these is to help turn off the inspiratory signal before the lungs become too full of air. When the pneumotaxic signals are strong, inspiration might last for as short as one-half second. When the pneumotaxic signal is extremely weak the inspiratory ramp signal will continue to ascend the crescendo of strength for perhaps as long as five to seven seconds, thus filling the lungs with a great excess of air.

Thus, the function of the pneumotaxic center is primarily to limit inspiration. However, this has a secondary effect on the rate of breathing because limitation of inspiration shortens the entire period of respiration. Therefore, a new cycle of inspiration begins again at a much earlier time. Thus, a strong pneumotaxic signal can increase the rate of breathing up to 30 to 40 breaths per minute while a weak pneumotaxic signal may reduce the rate of respiration to only a few breaths per minute.

The Apneustic Center in the Lower Pons. In the lower portion of the pons is still another, less well-defined center called the *apneustic center*. This center transmits signals into the inspiratory area in an attempt to *prevent* turn-off of the inspiratory ramp signal. As long as the pneumotaxic center is active, it overrides the apneustic center. However, when the pons is transected at the mid to lower level so that the pneumotaxic center is no longer connected to the inspiratory area, the apneustic center then does indeed either prevent turning off or makes it very difficult for inspiration to turn off. Consequently, the lungs become inflated near to maximum, and only occasional expiratory gasps occur. But this rarely happens in life, even when severe damage to the brain has occurred.

Limitation of Inspiration by Vagal Lung Inflation Signals — The Hering-Breuer Reflex. Located in the walls of the bronchi and bronchioles throughout the lungs are *stretch receptors* that transmit signals through the vagi into the inspiratory center when overstretched. These signals affect inspiration in almost exactly the same way as signals from the pneumotaxic center; that is, they limit the duration of inspiration. Therefore, when the lungs become overly inflated, the stretch receptors activate an appropriate feedback response to limit further inspiration. This is called the *Hering-Breuer reflex*. This reflex also has the same effect

as the pneumotaxic signals in increasing the rate of respiration because of the reduced period of inspiration.

Though in the past it was believed that the Hering-Breuer reflex plays an important role in controlling even normal lung inflation, recent experimental studies have shown that the reflex is not activated until the tidal volume increases to greater than approximately 1.5 liters. Therefore, the reflex appears to be mainly a protective mechanism for preventing excess lung inflation rather than an important ingredient in the normal control of ventilation.

The Expiratory Area

Located bilaterally in the medulla, and also extending the entire length of the medulla, is a ventral respiratory group of neurons which when stimulated excites the expiratory muscles. This expiratory area remains dormant during most normal quiet respiration, because quiet respiration is achieved by contraction only of the inspiratory muscles while expiration results from passive recoil of the elastic structures of the lung and surrounding chest cage. Furthermore, no evidence has been found that the expiratory area participates in the basic rhythmical oscillation that controls respiration.

On the other hand, when the respiratory drive for increased pulmonary ventilation becomes greater than normal, signals then spill over into the expiratory area from the basic oscillating mechanism in the inspiratory area. As a consequence, the expiratory muscles then contribute their powerful contractile forces to the pulmonary ventilatory process. Unfortunately, the neurophysiological basis for this interaction between the inspiratory and expiratory centers is not yet known.

Control of Respiratory Center Activity

It is already clear from the discussions in the preceding few chapters that pulmonary ventilation must change vastly under different conditions, especially when the body's metabolic rate is increased so that oxygen utilization and carbon dioxide formation are increased as much as 20-fold. These vast changes in pulmonary ventilation are achieved in two different ways: (1) by feedback control of respiratory center activity in response to changes in the chemical composition of the blood, especially its concentrations of carbon dioxide, hydrogen ions, and oxygen, and (2) by control from other parts of the nervous system, especially during exercise. The main purpose of the remainder of this chapter is to discuss this control of ventilation in response to the needs of the body.

CHEMICAL CONTROL OF RESPIRATION

The ultimate goal of respiration is to maintain proper concentrations of oxygen, carbon dioxide, and hydrogen ions in the body fluid. It is fortunate, therefore, that respiratory activity is highly responsive to changes in any one of these.

Excess carbon dioxide or hydrogen ions affect respiration mainly by direct excitatory effects on the respiratory center itself, causing greatly increased strength of both the inspiratory and expiratory signals to the respiratory muscles. The resulting increase in ventilation increases the elimination of carbon dioxide from the blood; this also removes hydrogen ions from the blood because of decreased blood carbonic acid.

Oxygen, on the other hand, does not seem to have a significant direct effect on the respiratory center of the brain in controlling respiration. Instead, it probably acts entirely on peripheral chemoreceptors located in the carotid and aortic bodies, and these in turn transmit appropriate neuronal signals to the respiratory center for control of respiration.

Let us discuss first the direct stimulation of the respiratory center itself by carbon dioxide and hydrogen ions.

DIRECT CHEMICAL CONTROL OF RESPIRATORY CENTER ACTIVITY BY CARBON DIOXIDE AND HYDROGEN IONS

The Chemosensitive Area of the Respiratory Center. Thus far, we have discussed mainly three different areas of the respiratory center: the inspiratory area, the expiratory area, and the pneumotaxic area. However, it is believed that none of these are affected directly by changes in blood carbon dioxide concentration or hydrogen ion concentration. Instead, a very sensitive *chemosensitive* area, illustrated in Figure 42–2, is located bilaterally and ventrally in the substance of the medulla, lying only a few microns beneath the surface anterior to the entry of the glossopharyngeal and vagal nerves into the medulla. This area is highly sensitive to changes in either blood CO_2 or hydrogen ion concentration, and it in turn excites the other portions of the respiratory center. It has especially potent effects on increasing the degree of activity of the inspiratory center, increasing both the rate of rise of the inspiratory ramp signal and also the intensity of the signal. This in turn has an automatic secondary effect of increasing the frequency of the respiratory rhythm.

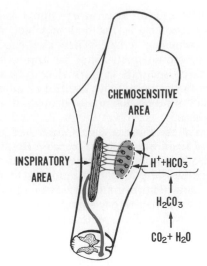

Figure 42–2. Stimulation of the inspiratory area by the *chemosensitive area* located bilaterally in the medulla, lying only a few microns beneath the ventral medullary surface. Note also that hydrogen ions stimulate the chemosensitive area, while mainly carbon dioxide in the fluid gives rise to the hydrogen ions.

Response of the Chemosensitive Neurons to Hydrogen Ions — The Primary Stimulus

The sensor neurons in the chemosensitive area are especially excited by hydrogen ions; in fact, it is believed that hydrogen ions are perhaps the only important direct stimulus for these neurons. Unfortunately, though, hydrogen ions do not easily cross either the blood-brain barrier or the blood–cerebrospinal fluid barrier. For this reason, changes in hydrogen ion concentration in the blood actually have considerably less effect in stimulating the chemosensitive neurons than do changes in carbon dioxide, even though carbon dioxide stimulates these neurons indirectly, as will be explained below.

Effect of Blood Carbon Dioxide on Stimulating the Chemosensitive Area

Though carbon dioxide has very little direct effect on stimulating the neurons in the chemosensitive area, it does have a very potent indirect effect. It does this by reacting with the water of the tissues to form carbonic acid. This in turn dissociates into hydrogen and bicarbonate ions; the hydrogen ions then have a potent direct stimulatory effect. These effects are illustrated in Figure 42–2.

But, why is it that blood CO_2 has a much more potent effect on stimulating the chemosensitive neurons than do blood hydrogen ions? The answer is that hydrogen ions, as noted above, pass through both the blood-brain barrier and the blood–cerebrospinal fluid barrier only very poorly while carbon dioxide passes through both these barriers almost as if they did not exist. Consequently, whenever the blood carbon dioxide concentration increases, so also does the P_{CO_2} increase in both the interstitial fluid of the medulla and also in the cerebrospinal fluid. And, in both of these fluids the carbon dioxide immediately reacts with the water to form hydrogen ions. Thus, paradoxically, more hydrogen ions are released into the respiratory chemosensitive sensory area when the blood carbon dioxide concentration increases than when the blood hydrogen ion concentration increases. For this reason, respiratory center activity is affected considerably more by changes in blood carbon dioxide than by changes in blood hydrogen ions, a fact that we will subsequently discuss quantitatively.

Stimulation of the Chemosensitive Area by Carbon Dioxide in the Cerebrospinal Fluid

On first thought, one would suspect that most stimulation of respiratory center activity would result from changes in carbon dioxide and hydrogen ion concentrations in the interstitial fluid of the respiratory center itself. However, most respiratory physiologists now believe that an increase in carbon dioxide concentration in the cerebrospinal fluid has more effect on stimulating the chemosensitive area than does a change in carbon dioxide in the interstitial fluid. This concept is still controversial but if it is true, the reason is probably a difference in the degree of acid-base buffering of the two fluids. The cerebrospinal fluid has very little protein buffer, while the interstitial fluid is lined on all sides by cells containing high concentrations of acid-base buffers. Therefore, a given change in carbon dioxide concentration in the cerebrospinal fluid will cause far more change in hydrogen ion concentration than in the interstitial fluid. And, since the chemosensitive neurons are located immediately beneath the surface of the medulla, diffusion of hydrogen ions into these neurons from the cerebrospinal fluid seems to be the major factor in the control of respiration.

One of the advantages of this cerebrospinal fluid system for control of respiration is the rapidity with which it can function. The cerebrospinal fluid is in intimate contact with the very rich blood supply of the arachnoid plexus. Therefore, within seconds after the blood P_{CO_2} changes, so also do the P_{CO_2} and hydrogen ion concentration of the cerebrospinal fluid change. On the other hand, a minute or more is required for full change of the P_{CO_2} in the brain interstitial fluid.

Decrease in the Effect of Blood CO_2 on Respira-tory Activity in One to Two Days. The effect of increased blood carbon dioxide on respiration reaches its peak within a minute or so after an increase in blood P_{CO_2}. Thereafter, the effect gradually declines during the next one to two days, decreasing by the end of that time to as little as one-fifth to one-eighth the initial effect. The exact cause of this decreasing stimulation is not known, but it is believed to result from active transport of bicarbonate ions from the blood into the cerebrospinal fluid through the arachnoidal cells lining the cerebrospinal fluid cavity. The bicarbonate ions in turn combine with the excess of hydrogen ions, thus reducing the hydrogen ion concentration and simultaneously reducing the respiratory drive.

Therefore, a change in blood carbon dioxide concentration has a very potent *acute* effect for controlling respiration but only a weak *chronic* effect after a few days' adaptation. On the other hand, the effect of blood hydrogen ion concentration on respiration becomes somewhat more potent with time because of the slow movement of hydrogen ions from the blood into the interstitial and cerebrospinal fluids.

Quantitative Effects of Blood P_{CO_2} and Hydrogen Ion Concentration on Alveolar Ventilation

Figure 42–3 illustrates quantitatively the approximate effects of blood P_{CO_2} and blood pH (which is an inverse measure of hydrogen ion concentration) on alveolar ventilation. Note the marked increase in ventilation caused by *acute* increase in P_{CO_2}. But note also the much smaller effect of increased hydrogen ion concentration (that is, decreased pH).

Finally, note that this *difference* in stimulation of ventilation is especially great in the normal P_{CO_2} and pH ranges: P_{CO_2}s between 40 and 45 mm. Hg and pHs between 7.45 and 7.35. Therefore, from a practical point, changes in blood carbon dioxide play by far the greatest role in the normal minute-by-minute control of pulmonary ventilation.

Value of Carbon Dioxide as a Regulator of Alveolar Ventilation. Since carbon dioxide is one of the end-products of metabolism, its concentration in the body fluids greatly affects the chemical reactions of the cells and also affects the tissue pH. For these reasons, the tissue fluid P_{CO_2} must be regulated exactly. In the preceding chapter, it was pointed out that blood and interstitial fluid P_{CO_2} are determined to a great extent by the rate of alveolar ventilation. Therefore, stimulation of the respiratory center by carbon dioxide provides an important feedback mechanism for regulation of the concentration of carbon dioxide throughout

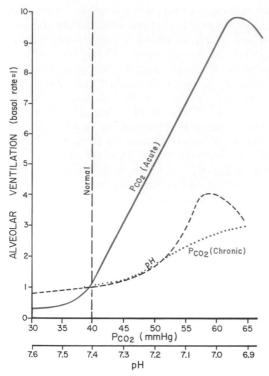

Figure 42–3. Effects of increased arterial P_{CO_2} (both acute and chronic) and decreased arterial pH on the rate of alveolar ventilation.

the body. That is, (1) an increase in P_{CO_2} stimulates the respiratory center; (2) this increases alveolar ventilation and reduces the alveolar carbon dioxide; (3) as a result, the tissue P_{CO_2} returns most of the way back toward normal. In this way, the respiratory center maintains the P_{CO_2} of the tissue fluids at a relatively constant level and, therefore, might well be called a "carbon dioxide pressostat."

THE PERIPHERAL CHEMORECEPTOR SYSTEM FOR CONTROL OF RESPIRATORY ACTIVITY— ROLE OF OXYGEN IN RESPIRATORY CONTROL

Aside from the direct sensitivity of the respiratory center itself to CO_2 and hydrogen ions, special chemical receptors called *chemoreceptors*, located outside the central nervous system, are also responsive to changes in oxygen, carbon dioxide, and hydrogen ion concentrations. These transmit signals to the respiratory center to help regulate respiratory activity. These chemoreceptors are located in the *carotid* and *aortic bodies,* which are illustrated in Figure 42–4 along with

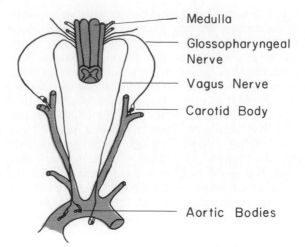

Figure 42–4. Respiratory control by the carotid and aortic bodies.

their afferent nerve connections to the respiratory center. The *carotid bodies* are located bilaterally in the bifurcations of the common carotid arteries, and their afferent nerve fibers pass through Hering's nerves to the glossopharyngeal nerves and thence to the medulla. The *aortic bodies* are located along the arch of the aorta; their afferent nerve fibers pass to the medulla through the vagi. Each of these chemoreceptor bodies receives a special blood supply through a minute artery directly from the adjacent arterial trunk.

Stimulation of the Chemoreceptors by Decreased Arterial Oxygen. Changes in arterial oxygen concentration have *no* direct stimulatory effect on the respiratory center itself, but when the oxygen concentration in the arterial blood falls below normal, the chemoreceptors become strongly stimulated. This effect is illustrated in Figure 42–5, which shows the relationship between *arterial* P_{O_2} and rate of nerve impulse transmission from a carotid body. Note that the impulse rate is particularly sensitive to changes in arterial P_{O_2} in the

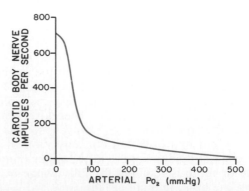

Figure 42–5. Effect of arterial P_{O_2} on impulse rate from the carotid body of a cat. (Curve drawn from data from several sources, but primarily from Von Euler.)

range between 60 and 30 mm. Hg, which is the range in which the arterial hemoglobin saturation with oxygen decreases rapidly.

Effect of Carbon Dioxide and Hydrogen Ion Concentration on Chemoreceptor Activity. An increase in either carbon dioxide concentration or hydrogen ion concentration also excites the chemoreceptors and in this way indirectly increases respiratory activity. However, the direct effects of both of these factors in the respiratory center itself are so much more powerful than their effects mediated through the chemoreceptors that for most practical purposes one can disregard the indirect effects through the chemoreceptors. In the case of oxygen, on the other hand, this is not true, because diminished oxygen in the arterial blood can affect the respiration significantly *only* by acting through the chemoreceptors.

Basic Mechanism of Stimulation of the Chemoreceptors by Oxygen Deficiency. The blood flow through the carotid and aortic bodies is extremely high, the highest that has been found for any tissue in the body. Because of this, the A-V oxygen difference is less than one volume per cent, which means that the venous blood leaving the carotid bodies still has a P_{O_2} nearly equal to that of the arterial blood. It also means that the P_{O_2} of the tissues in the carotid and aortic bodies at all times remains also almost equal to that of the arterial blood. Therefore, it is the *arterial* P_{O_2} that normally determines the degree of stimulation of the chemoreceptors.

Yet, in hypotension, particularly when the mean arterial pressure falls below 80 mm. Hg (and even more so when it falls below 60 mm. Hg), the blood flow through the carotid and aortic bodies then becomes sluggish, and the tissue P_{O_2} falls considerably below the arterial P_{O_2}. At these lower pressures, therefore, the chemoreceptors now do become stimulated in response to the hypotension even when the arterial P_{O_2} is normal. This gives rise to reflexes that enhance respiration and that cause also peripheral vasoconstriction that increases the arterial pressure back toward normal.

The exact means by which low P_{O_2} excites the nerve endings in the carotid and aortic bodies is still unknown. Since both these bodies have two different, highly characteristic glandular-like cells in them, it had been believed that these cells acted as chemoreceptors and that they in turn stimulated the nerve endings. However, recent studies suggest that the nerve endings themselves are directly sensitive to the low P_{O_2}.

Quantitative Effect of Low Blood P_{O_2} on Alveolar Ventilation

Low blood P_{O_2} normally will not increase alveolar ventilation significantly until the alveolar P_{O_2} falls almost to one-half normal. This is illustrated in Figure 42–6. The lowermost curve of this figure shows that changing the alveolar arterial P_{O_2} from slightly more than 100 mm. Hg down to about 60 mm. Hg has an imperceptible effect on ventilation. But, then, as the P_{O_2} falls still further, down to 40 and then to 30 mm. Hg, alveolar ventilation increases 1.5- to 1.7-fold. However, contrast this rather feeble increase in alveolar

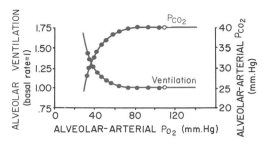

Figure 42–6. Effect of arterial P_{O_2} on alveolar ventilation and on the subsequent decrease in arterial P_{CO_2}. (From Gray: Pulmonary Ventilation and Its Physiological Regulation. Charles C Thomas.)

ventilation to the four-fold increase caused by decreasing blood pH to 7.0 or the ten-fold increase caused by increasing the P_{CO_2} only 50 per cent. Thus, it is clear that the normal effect of changes in blood P_{O_2} on respiratory activity is very slight, especially when compared with the effect of P_{CO_2}.

Cause of the Poor Acute Effect of Low P_{O_2} on Respiration — The "Braking" Effect of the P_{CO_2} and pH Regulatory Mechanisms. The cause of the poor effect of P_{O_2} changes on respiratory control is a "braking" effect caused by *both* the carbon dioxide and the hydrogen ion control mechanisms. This phenomenon can be explained by referring again to Figure 42–6, which shows the slight increase in ventilation as the alveolar P_{O_2} is decreased. The increase in ventilation that does occur blows off carbon dioxide from the blood and therefore decreases the P_{CO_2}, which is also illustrated in the figure; at the same time it also decreases the hydrogen ion concentration. Therefore, two powerful respiratory inhibitory effects are caused by (a) diminished carbon dioxide and (b) diminished hydrogen ions. These two exert an inhibitory "braking" effect that opposes the excitatory effect of the diminished oxygen. As a result, they keep the decreased oxygen from causing a marked increase in ventilation until the P_{O_2} falls to 20 to 40 mm. Hg, a range that is incompatible with life for more than a few minutes. Therefore, the maximum effect of decreased alveolar oxygen on alveolar ventilation, in the range compatible with life, is normally only a 66 per cent increase.

Thus, one can see that for the control of the usual normal respiration the P_{CO_2} and pH feedback control mechanisms are extremely powerful in relation to the P_{O_2} feedback control of respiration. Indeed, under normal conditions the P_{O_2} mechanism is of almost no significance in the control of respiration.

Yet, under some abnormal conditions the P_{CO_2} and hydrogen ion concentrations *increase at the same time that the arterial P_{O_2} decreases.* Under these conditions, all three of the feedback mechanisms support each other, and the P_{O_2} mechanism then exerts its full share of respiratory stimulation, sometimes becoming even more potent as a controller of respiration than are the P_{CO_2} and hydrogen ion mechanisms.

Effect of Diminished Oxygen on Alveolar Ventilation When the Carbon Dioxide and pH Mechanisms Do Not "Brake" the P_{O_2} Effect. When the concentrations of carbon dioxide and hydrogen ions are prevented from changing while decreased blood P_{O_2} increases the ventilation, the effect of diminished P_{O_2} is then 8 to 10 times as great as when the carbon dioxide and hydrogen ions concentrations do change and inhibit the P_{O_2} effect. This is illustrated in Figure 42–7 which compares the effects of acute changes in P_{O_2}, pH, and P_{CO_2} on alveolar ventilation when only one of these factors changes while the other two are kept exactly constant. This figure illustrates that under these conditions changes in P_{O_2} have almost as much effect on alveolar ventilation as changes in the other two factors. The reason that the other two factors dominate normal respiratory control is that the P_{O_2} control mechanism does not exert a significant "braking" effect on the P_{CO_2} and hydrogen ion mechanisms, in contrast to the powerful effect of these other two mechanisms in interfering with the P_{O_2} effect.

Conditions Under Which Diminished Oxygen Does Play a Major Role in the Regulation of Respiration. In pneumonia, emphysema, and other lung ailments in which gases are not readily exchanged between the atmosphere and the pulmonary blood, the oxygen regulatory system *does* then play a major role in the regulation of respiration. Contrary to the normal effect, the increased

ventilation caused by oxygen lack is not followed by reduced arterial P_{CO_2} and hydrogen ion concentration, because the pulmonary disease also diminishes carbon dioxide exchange as well as oxygen exchange. Instead, the CO_2 either remains constant or builds up in the blood, and the hydrogen ion concentration behaves similarly. Therefore, the "braking" effect of these other two control systems on the oxygen lack system is not present. As a result, the oxygen lack system develops its full power and can increase alveolar ventilation as much as five- to seven-fold.

Effects of the Oxygen Lack Mechanism at High Altitudes. When a person first ascends to high altitudes or in any other way is exposed to a rarefied atmosphere, the diminished oxygen in the air stimulates the oxygen lack control system of respiration. The respiration at first increases to a maximum of about two-thirds above normal, which is a comparatively slight increase. Once again, the cause of this slight increase is the tremendous "braking" effect of the carbon dioxide and hydrogen ion control mechanisms on the oxygen lack mechanism.

However, over about a week, the respiratory center gradually becomes "adapted" to the diminished carbon dioxide as explained earlier in the chapter, so that this now depresses the respiratory center very little. Thus, the "braking" effect on the oxygen control is gradually lost, and alveolar ventilation then rises to as high as 5 to 7 times normal. This is part of the acclimatization that occurs as a person slowly ascends a mountain, thus allowing the person to adjust respiration gradually to a level fitted for the higher altitude.

Why Oxygen Regulation of Respiration Is Not Normally Needed. On first thought, it seems strange that oxygen should play so small a role in the normal regulation of respiration, particularly since one of the primary functions of the respiratory center is to provide adequate intake of oxygen. However, oxygen control of respiration is not needed under most normal circumstances for the following reason:

The respiratory system ordinarily maintains an alveolar P_{O_2} actually *higher* than the level needed to saturate almost completely the hemoglobin of the arterial blood. It does not matter whether alveolar ventilation is normal or 10 times normal, the blood will still be essentially fully saturated. Also, alveolar ventilation can decrease to as low as one-half normal, and the blood still remains within 10 per cent of complete saturation. Therefore, one can see that alveolar ventilation can change tremendously without significantly affecting oxygen transport to the tissues.

On the other hand, changes in alveolar ventilation do have a tremendous effect on tissue carbon dioxide concentration, as was explained earlier in

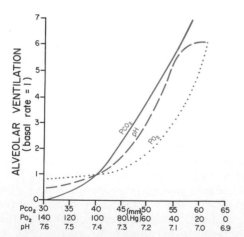

Figure 42–7. Approximate effects on alveolar ventilation of changing the concentrations of carbon dioxide, hydrogen ions, and oxygen in the arterial blood when only one of the humoral factors is changed at a time and the other two are maintained at absolutely normal levels. (Modified from Comroe, et al.: The Lung: Clinical Physiology and Pulmonary Function Tests. Chicago, Year Book Medical Publishers.)

the chapter and illustrated in Figure 42–6. Therefore, it is exceedingly important that carbon dioxide — not oxygen — be the major controller of respiration under normal conditions.

COMPOSITE EFFECTS OF P_{CO_2}, pH, AND P_{O_2} ON RESPIRATORY ACTIVITY

Now that the effects of each of the individual humoral factors on respiratory activity have been discussed, we need to see how all of them function together in controlling respiration. We have already seen that the different factors oppose each other in some instances but at other times support each other. One of the best attempts to show how these factors interact is illustrated in Figure 42–8, which shows two separate "families" of curves relating alveolar P_{CO_2} to pulmonary ventilation. The "family" of curves represented by the solid lines shows the effects of different levels of alveolar P_{CO_2} on alveolar ventilation determined for the average person at different alveolar P_{O_2}s from 40 to 100 mm. Hg and with the pH constant at the normal level of 7.4. The dashed "family" of curves was determined after the same average person had developed metabolic acidosis, with a shift of the blood pH from 7.4 to 7.3. Both these families of curves show that the *slope* of the curve relating P_{CO_2} to alveolar ventilation increases markedly as the P_{O_2} decreases. To express this another way, a decrease in P_{O_2} in the alveoli *multiplies* the effects on alveolar ventilation of *changes* in carbon dioxide concentration.

Figure 42–8 shows also that acidosis shifts the entire family of curves to the left. This means that increased hydrogen ion concentration decreases the carbon dioxide concentration required to stimulate the respiratory center.

The curves in Figure 42–8 are important when one wishes to analyze respiratory control from a quantitative point of view because, in most instances in which respiratory control becomes altered, more than one of the individual humoral factors are changed at the same time.

REGULATION OF RESPIRATION DURING EXERCISE

In strenuous exercise, oxygen utilization and carbon dioxide formation can increase as much as 20-fold (Fig. 42–9). Yet, except in very heavy exercise, alveolar ventilation ordinarily increases almost the same amount so that the blood P_{O_2}, P_{CO_2}, and pH all remain *almost exactly normal.*

In trying to analyze the factors that cause increased ventilation during exercise, one is tempted immediately to ascribe this to the chemical alterations in the body fluids during exercise, including increase of carbon dioxide, increase of hydrogen ions, and decrease of oxygen. However, this is not valid, for measurements of arterial P_{CO_2}, pH, and P_{O_2} show that none of these usually changes significantly and certainly not enough to account for the increase in ventilation. Indeed, even if a very high P_{CO_2} should develop during exercise, this still would be sufficient to account for only two thirds of the increased ventilation of heavy muscular exercise, for, as shown in Figure 42–10, the minute respiratory volume in exercise is about 50 per cent greater than that which can be effected by maximal carbon dioxide stimulation.

Therefore, the question must be asked: What is it during exercise that causes the intense respiration? This question has not been answered, but at least two different effects seem to be predominantly concerned:

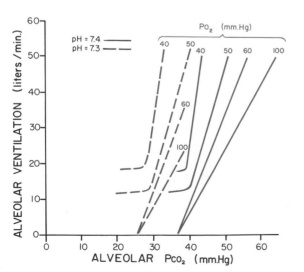

Figure 42–8. A composite diagram showing the interrelated effects of P_{CO_2}, P_{O_2}, and pH on alveolar ventilation. (Drawn from data presented in Cunningham and Lloyd: The Regulation of Human Respiration. F. A. Davis Co.)

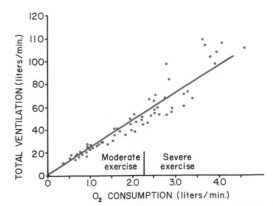

Figure 42–9. Effect of exercise on oxygen consumption and ventilatory rate. (From Gray: Pulmonary Ventilation and Its Physiological Regulation. Charles C Thomas.)

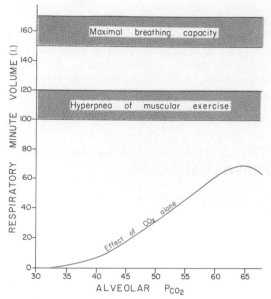

Figure 42–10. Relationship of hyperpnea caused by muscular exercise to that caused by increased alveolar P_{CO_2}. (Modified from Comroe: The Lung: Clinical Physiology and Pulmonary Function Tests. Chicago, Year Book Medical Publishers.)

1. The cerebral cortex, on transmitting impulses to the contracting muscles, is believed to transmit collateral impulses into the reticular substance of the brain stem to excite the respiratory center. This is analogous to the stimulatory effect that causes the arterial pressure to rise during exercise when similar collateral impulses pass to the vasomotor center.

2. During exercise, the body movements, especially of the limbs, are believed to increase pulmonary ventilation by exciting joint proprioceptors that then transmit excitatory impulses to the respiratory center. The reason for believing this is that even passive movements of the limbs often increase pulmonary ventilation several-fold.

It is possible that still other factors are also important in regulating respiration during exercise. For instance, some experiments even suggest that hypoxia developing in the muscles during exercise elicits afferent nerve signals to the respiratory center to excite respiration. However, since the increase in ventilation begins immediately upon the initiation of exercise, most of the increase in respiration probably results from the two neurogenic factors noted above, namely *stimulatory impulses from the higher centers of the brain* and *proprioceptive stimulatory reflexes*.

Interrelationship Between Chemical Factors and Nervous Factors in the Control of Respiration During Exercise. Figure 42–11 illustrates the different factors that operate in the control of respiration during exercise, showing two neurogenic factors: (1) direct stimulation of the respiratory center by the cerebral cortex and (2) indirect stimulation by proprioceptors. It shows also the three humoral factors of carbon dioxide, hydrogen ions, and oxygen.

When a person exercises, usually the nervous factors stimulate the respiratory center almost exactly the proper amount to supply the extra oxygen requirements for the exercise and to blow off the extra carbon dioxide. But, occasionally, the nervous signals are either too strong or too weak in their stimulation of the respiratory center. Then, the chemical factors play a very significant role in bringing about the final adjustment in respiration required to keep the carbon dioxide and hydrogen ion concentrations of the body fluids as nearly normal as possible. This effect is illustrated in Figure 42–12, which shows changes in alveolar ventilation and in arterial P_{CO_2} during a one-minute period of exercise and then for another minute after the exercise is over. Note that the alveolar ventilation increases without an initial increase in arterial P_{CO_2}; this increase is caused by a stimulus that originates in the brain at the same time that the brain excites the muscles. Often, the increase in alveolar ventilation is so great that it actually *decreases* arterial P_{CO_2} below normal, as shown in the figure, even though the exercising muscles are beginning to form large amounts of carbon dioxide. The reason for this is that the ventilation forges ahead of the increase in carbon dioxide formation. Thus, the brain provides an "anticipatory" stimulation of respiration at the

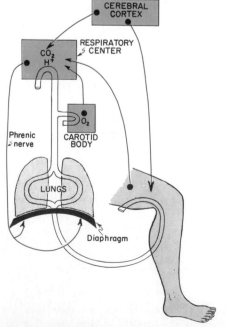

Figure 42–11. The different factors that enter into regulation of respiration during exercise.

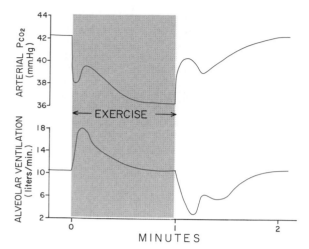

Figure 42–12. Changes in alveolar ventilation and arterial P_{CO_2} during a 1-minute period of exercise and also following termination of the exercise. (Extrapolated to the human being from data in dogs; from Bainton: *J. Appl. Physiol., 33*:778, 1972.)

onset of exercise, causing excessive alveolar ventilation even before this is needed.

Note also in the figure that after a short delay the decrease in arterial P_{CO_2} finally acts on the respiratory center to reduce the alveolar ventilation back toward normal for a fraction of a minute; soon thereafter, the large amount of carbon dioxide formed by the muscles gradually increases the P_{CO_2} up toward the normal level, and the alveolar ventilation increases once again.

The anticipatory signal from the brain thus initiates the increase in alveolar ventilation during exercise; then the chemical feedback factors (mainly arterial P_{CO_2}) make additional adjustments, either upward or downward, to balance the rate of alveolar ventilation with the rate of metabolism in the body. Because of the direct brain stimulus, the final level of P_{CO_2}, hydrogen ion concentration, and P_{O_2} in the arterial blood is almost exactly normal during exercise — sometimes slightly above normal and sometimes slightly below normal.

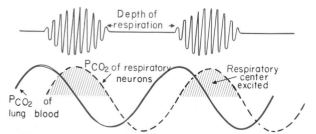

Figure 42–13. Cheyne-Stokes breathing, showing the changing P_{CO_2} in the pulmonary blood (solid line) and the delayed changes in P_{CO_2} of the fluids of the respiratory center (dashed line).

Finally, note in Figure 42–12 that exactly the opposite effects occur when exercise is over. The shutting-off of the brain stimulus returns the alveolar ventilation toward normal instantaneously, but this gets ahead of the decrease in carbon dioxide formation by the muscles so that a sequence of chemical-directed feedback readjustments is necessary before the alveolar ventilation finally becomes stabilized.

Now for a final comment regarding the brain factor for stimulation of respiration during exercise: Many experiments indicate that this brain factor is partly if not entirely a learned response. That is, with repeated periods of exercise of the same degree of strenuousness, the brain seems to become progressively more able to establish the proper amount of brain signal to maintain the chemical factors at their normal levels during exercise. Therefore, there is much reason to believe that some of the higher centers of learning in the brain are important in this brain factor — probably even the cerebral cortex.

OTHER FACTORS THAT AFFECT RESPIRATION

Voluntary Control of Respiration. Thus far, we have discussed the involuntary system for control of respiration. However, we all know that respiration can be controlled voluntarily, and that one can hyperventilate or hypoventilate to such an extent that serious derangements in P_{CO_2}, pH, and P_{O_2} can occur in the blood.

Voluntary control of respiration seems not to be mediated through control of the involuntary respiratory center of the medulla. Instead, the nervous pathway for voluntary control passes directly from the cortex and other higher centers downward through the corticospinal tract to the spinal neurons that drive the respiratory muscles.

Influence of Vasomotor Center Activity on Respiration. The vasomotor center that controls peripheral vasoconstriction and heart activity is closely related to the respiratory center in the medulla. Furthermore, a moderate degree of spillover of nerve signals occurs mutually between the two centers. Therefore, almost any factor that increases the activity of the vasomotor center also has at least a moderate effect on increasing respiration. For instance, a decrease in arterial pressure inhibits the baroreceptors, which increases vasomotor activity and increases pulmonary ventilation as well.

Effect of Body Temperature on Respiration. An increase in body temperature increases the rate of respiration. An *indirect* effect is caused by the action of the increased temperature in increasing cellular metabolism all through the body, which in turn creates a chemical stimulus for increased respiration. Increased body temperature also has a *direct* effect by increasing respiratory center activity. Therefore, in patients with fever, both the rate and depth of ventilation are often greatly increased.

Panting. Many lower animals have very little ability

to lose heat from the surfaces of their bodies for two reasons: First, the surfaces are usually covered with fur, and, second, the skin of most lower animals is not supplied with sweat glands, which prevents evaporative loss of heat. Therefore, a substitute mechanism, the *panting* mechanism, is used by many lower animals as a means for dissipating heat.

The phenomenon of panting is "turned on" by the hypothalamic thermoregulator center that will be discussed in Chapter 72. When the blood becomes overheated, the hypothalamus initiates a series of neurogenic signals to decrease the body temperature. One of these signals is to initiate panting. The actual panting process is then controlled by a *panting center* that is closely related to the pneumotaxic center in the pons.

When an animal pants, it breathes in and out rapidly so that large quantities of new air from the exterior come in contact with the upper portions of its respiratory passages, and this cools both these passages and the blood in the mucosa as a result of water evaporation from the mucosal surfaces. Yet, panting does not increase the alveolar ventilation more than is required for proper control of the blood gases, because each breath is extremely shallow; most of the air that enters the alveoli is dead space air.

Effect of Irritant Receptors in the Airways. The epithelium of the bronchi and bronchioles is supplied with sensory nerve endings that are stimulated by irritants in the respiratory airways. These cause coughing and sneezing, as was discussed in Chapter 39. They possibly also cause bronchial constriction in such diseases as asthma.

Function of Lung "J" Receptors. A few sensory nerve endings occur in the alveolar walls in *juxta*position to the capillaries, from whence comes the name "J" receptors. They are stimulated when irritant chemicals are injected into the pulmonary blood, and they are also believed to be excited when the pulmonary capillaries become engorged with blood or when pulmonary edema occurs in such conditions as congestive heart failure. Though the functional role of the J receptors is not known, their excitation perhaps does give the person a feeling of dyspnea.

ABNORMALITIES OF RESPIRATORY CONTROL

RESPIRATORY CENTER DEPRESSION

Cerebrovascular Disease. Probably the most common cause of long-term respiratory center depression is cerebrovascular disease in older aged patients, especially following vascular occlusions or hemorrhages that damage the respiratory center areas. In such instances, a person may have chronically elevated arterial Pco_2s and depressed Po_2s.

Clinical Measurement of Respiratory Center Depression. In the clinical pulmonary function laboratory, the degree of respiratory center depression is frequently measured by the simple maneuver of having the patient breathe carbon dioxide at successively increasing concentrations. Then, from measurements of alveolar Pco_2 and alveolar ventilation, a *carbon dioxide–alveolar ven-*

tilation stimulation curve, similar to that illustrated in Figure 42–3, is plotted. The degree of respiratory center depression is ascertained by comparing the slope of the carbon dioxide stimulation curve with that of the normal curve, assuming, of course, that all other aspects of pulmonary function are normal. If they are not normal, these also must be factored into the analysis.

Acute Brain Edema. The activity of the respiratory center may be depressed or totally inactivated by acute brain edema resulting from brain concussion. For instance, the head might be struck against some solid object, following which the damaged brain tissues swell, compressing the cerebral arteries against the cranial vault and thus totally or partially blocking the cerebral blood supply. As a result, the neurons of the respiratory center first become inactive and later die. In this manner brain edema may either depress or totally inactivate the respiratory center.

Occasionally, respiratory depression resulting from brain edema can be relieved temporarily by intravenous injection of hypertonic solutions such as highly concentrated glucose solution. These solutions osmotically remove some of the intracellular fluids of the brain, thus relieving intracranial pressure and sometimes reestablishing respiration within a few minutes.

Anesthesia. Perhaps the most prevalent cause of respiratory depression and respiratory arrest is overdosage of anesthetics or narcotics. The best agent to be used for anesthesia is one that depresses the respiratory center the least while depressing the cerebral cortex the most. Ether is among the best of the anesthetics by these criteria, though halothane, cyclopropane, ethylene, nitrous oxide, and a few others have almost the same value. On the other hand, sodium pentobarbital is a poor anesthetic because it depresses the respiratory center considerably more than the above agents. In early days morphine was used as an anesthetic, but this drug is now used only as an adjunct to anesthetics because it greatly depresses the respiratory center.

PERIODIC BREATHING

An abnormality of respiration called periodic breathing occurs in a number of different disease conditions. The person breathes deeply for a short interval of time and then breathes slightly or not at all for an additional interval, the cycle repeating itself over and over again.

The most common type of periodic breathing, *Cheyne-Stokes breathing,* is characterized by slowly waxing and waning respiration, occurring over and over again every 45 seconds to 3 minutes.

Basic Mechanism of Cheyne-Stokes Breathing. Let us assume that the respiration becomes much more rapid and deeper than usual. This causes the Pco_2 in the pulmonary blood to *decrease.* A few seconds later the pulmonary blood reaches the brain, and the decreased Pco_2 inhibits respiration. As a result, the pulmonary blood Pco_2 gradually *increases.* After another few seconds the blood carrying the increased CO_2 arrives at the respiratory center and stimulates respiration again, thus making the person overbreathe once again and initiating a new cycle of depressed respiration; and the cycles thus continue on and on, causing Cheyne-Stokes periodic breathing.

The successive changes in pulmonary and respiratory center P_{CO_2} during Cheyne-Stokes breathing are illustrated in Figure 42–13.

Since the basic feedback mechanism for causing Cheyne-Stokes respiration is present in every person, the real question that must be answered is not why Cheyne-Stokes breathing develops under certain pathological conditions but, instead, why is it not present all the time in everyone? The reason for this is that the feedback is highly "damped," which may be explained as follows:

Damping of the Normal Respiratory Control Mechanism. In order for Cheyne-Stokes oscillation to occur, sufficient time must occur during the hyperpneic phase for the body fluid P_{CO_2} to fall considerably below the mean value. Then this decrease in P_{CO_2} initiates the apneic phase, which also must last long enough for the tissue P_{CO_2} to rise well above the mean value. In the normal person, before significant changes can occur the respiratory center ordinarily readjusts the breathing back toward normal, thereby preventing the extreme excursions of P_{CO_2} in the tissue fluids that are necessary to cause Cheyne-Stokes breathing. Thus, the tremendous storage capacity of the tissues for carbon dioxide functions as an important "damping" factor to prevent Cheyne-Stokes respiration in the normal person.

A number of different conditions can overcome the damping of the feedback mechanism and cause it to oscillate spontaneously. Two of these are: (1) increased delay time in the flow of blood from the lungs to the brain and (2) increased feedback gain of the respiratory center mechanisms for control of respiration.

Cheyne-Stokes Breathing Caused by Delay of Blood Flow From the Lungs to the Brain – Effect of Cardiac Failure. When the heart fails, blood flow from the lungs to the brain is sometimes greatly delayed, partly because the cardiac output decreases but mainly because the left heart sometimes becomes tremendous in size, thus requiring a long period of time for the blood to flow through the left ventricle and thence to the brain. When such a delay occurs, the decrease in blood P_{CO_2} caused by increased pulmonary ventilation is not detected by the brain for many seconds. Therefore, the blood P_{CO_2} continues to fall to very low levels before an appropriate correction is made in the ventilation. When the over-respired blood does reach the brain, the respiratory center then becomes excessively depressed, and this in turn depresses pulmonary ventilation as well. Once again the effect continues long after the blood in the lungs has collected excessive amounts of carbon dioxide.

Thus, the delay in transport of blood from the lungs to the brain is by itself a sufficient cause for the development of Cheyne-Stokes breathing. In patients with chronic heart failure this type of breathing occasionally occurs continuously for months at a time.

Cheyne-Stokes Breathing Caused by Increased Feedback Gain in the Respiratory Center. Occasionally, damage to the brain stem increases the feedback gain of the respiratory center in response to changes in blood P_{CO_2} and blood pH. This does not mean that the sensitivity of these centers to these two chemical factors necessarily increases but instead that a minute concentration *change* in one of the two factors causes a drastic *change*

in ventilation. Indeed, the respiratory center is often severely depressed even though the gain is very high. An example of this would be complete cessation of ventilation when the P_{CO_2} is 30 mm. Hg, but one-half normal ventilation when the P_{CO_2} is 40 mm. Hg. In this case the percentage *change* in ventilation is infinite — from zero to a finite value. Therefore, the change in gain is infinite even though the ventilation is only one-half normal. Consequently, the respiratory control system can oscillate back and forth between apnea and breathing, which is the typical pattern of Cheyne-Stokes breathing. This effect probably explains why many patients with brain damage develop a type of Cheyne-Stokes breathing, and when it does occur in such a patient it often is an ominous sign, foretelling that death is near at hand.

REFERENCES

Bainton, C. R.: Effect of speed versus grade and shivering on ventilation in dogs during active exercise. *J. Appl. Physiol., 33*:778, 1972.

Biscoe, T. J.: Carotid body: Structure and function. *Physiol. Rev., 51*:427, 1971.

Bouverot, P.: Control of breathing in birds compared with mammals. *Physiol. Rev., 58*:604, 1978.

Carroll, D.: Sleep, periodic breathing, and snoring in the aged: Control of ventilation in the aging and diseased respiratory system. *J. Am. Geriatr. Soc., 22*:307, 1974.

Cherniack, N. S., and Longobardo, G. S.: Cheyne-Stokes breathing. An instability in physiologic control. *N. Engl. J. Med., 288*:952, 1973.

Cohen, M. I.: Discharge patterns of brain-stem respiratory neurons in relation to carbon dioxide tension. *J. Neurophysiol., 31*:142, 1968.

Cohen, M. I.: Neurogenesis of respiratory rhythm in the mammal. *Physiol. Rev., 59*:1105, 1979.

Coleridge, H. M., et al.: Thoracic chemoreceptors in the dog: A histological and electrophysiological study of the location, innervation and blood supply of the aortic bodies. *Circ. Res., 26*:235, 1970.

Cunningham, D. J. C.: Integrative aspects of the regulation of breathing: A personal view. *In* MTP International Review of Science: Physiology. Vol. 2. Baltimore, University Park Press, 1974, p. 303.

Cunningham, D. J. C., and Lloyd, B. B. (eds.): The Regulation of Human Respiration. Philadelphia, F. A. Davis Co., 1963.

Fencl, V.: Notes on the centrogenic drive in respiration. *Physiologist, 16*:589, 1973.

Fitzgerald, R., et al. (eds.): The Regulation of Respiration During Sleep and Anesthesia. New York, Plenum Press, 1978.

Guyton, A. C., et al.: Basic oscillating mechanism of Cheyne-Stokes breathing. *Am. J. Physiol., 187*:395, 1956.

Guz, A.: Regulation of respiration in man. *Annu. Rev. Physiol., 37*:303, 1975.

Hechtman, H. B. (ed.): Acute Respiratory Failure: Etiology and Treatment. West Palm Beach, Fla., CRC Press, 1979.

Heinemann, H. O., and Goldring, R. M.: Bicarbonate and the regulation of ventilation. *Am. J. Med., 57*:361, 1974.

Huch, A., et al. (eds.): Continuous Transcutaneous Blood Gas Monitoring. New York, A. R. Liss, 1979.

Karczewski, W. A.: Organization of the brainstem respiratory complex. *In* MTP International Review of Science: Physiology. Vol. 2. Baltimore, University Park Press, 1974. p. 197.

Leusen, I.: Regulation of cerebrospinal fluid composition with reference to breathing. *Physiol. Rev., 52*:1, 1972.

Loeschcke, H. H.: Respiratory chemosensitivity in the medulla oblongata. *Acta Neurobiol. Exp., 33*:97, 1973.

Loeschcke, H. H.: Central nervous chemoreceptors. *In* MTP International Review of Science: Physiology. Vol. 2. Baltimore, University Park Press, 1974, p. 167.

Milhorn, H. T., Jr., and Guyton, A. C.: An analog computer analysis of Cheyne-Stokes breathing, *J. Appl. Physiol., 20*:328, 1965.

Milhorn, H. T., Jr., et al.: A mathematical model of the human respiratory control system. *Biophys. J., 5*:27, 1965.

Mitchell, R. A.: Control of respiration. *In* Frohlich, E. D. (ed.): Pathophysiology, 2nd Ed. Philadelphia, J. B. Lippincott Co., 1976, p. 131.

Mitchell, R. A., and Berger, A. J.: Neural regulation of respiration. *Am. Rev. Resp. Dis., 111*:206, 1975.

Paintal, A. S.: Vagal sensory receptors and their reflex effects. *Physiol. Rev., 53*:159, 1973.

Paintal, A. S., and Gill-Kumar, P. (eds.): Respiratory Adaptations, Capillary Exchange, and Reflex Mechanisms. Delhi, India, Vallabhbhai Patel Chest Institute, University of Delhi, 1977.

Sykes, M. K., *et al.:* Respiratory Failure. Philadelphia, J. B. Lippincott Co., 1974.

Von Euler, C., and Lagercrantz, H. (eds.): Central Nervous Control Mechanisms in Breathing. New York, Pergamon Press, 1980.

Williams, M. H. (ed.): Symposium on Disturbance of Respiratory Control. Philadelphia, W. B. Saunders Co., 1980.

43

Respiratory Insufficiency

The diagnosis and treatment of most respiratory disorders have come to depend highly on an understanding of the basic physiological principles of respiration and gas exchange. Some diseases of respiration result from inadequate ventilation, while others result from abnormalities of diffusion through the pulmonary membrane or of transport from the lungs to the tissues. In each of these instances, the therapy is often entirely different so that it is no longer satisfactory simply to make a diagnosis of "respiratory insufficiency."

Definitions. In describing different respiratory disorders the following descriptive terms are used:

Eupnea means normal breathing; *tachypnea* means rapid breathing; and *bradypnea* means slow breathing.

Hyperpnea means a rate of alveolar ventilation great enough to cause over-respiration. However, this term is commonly used also to indicate simply a very high level of alveolar ventilation without necessarily implying over-respiration. *Hypopnea* is the opposite of hyperpnea, indicating under-respiration.

Anoxia means total lack of oxygen, but this term is more frequently used to mean simply decreased oxygen. A more correct term for decreased oxygen is *hypoxia*.

Anoxemia means lack of oxygen in the blood, but the term is generally used to mean simply reduced oxygen in the blood. A better term for this is *hypoxemia*.

Hypercapnia means excess carbon dioxide in the blood, and *hypocapnia* means depressed carbon dioxide. The term *acapnia* is also frequently used to imply hypocapnia, but, as is true with anoxemia, during life the absolute state of acapnia, with no carbon dioxide at all in the body fluids, never exists.

ADDITIONAL METHODS FOR STUDYING RESPIRATORY ABNORMALITIES

In the previous four chapters we have discussed a number of different methods for studying respiratory abnormalities, including measuring vital capacity, tidal air, functional residual capacity, dead space, physiologic shunt, physiologic dead space, maximum expiratory flow, forced expiratory volume, and respiratory center sensitivity. This array of measurements is only part of the armamentarium of the clinical pulmonary physiologist. Some other interesting tools available will be described.

Use of the Body Plethysmograph to Study Pulmonary

Function. A body plethysmograph, which is illustrated in Figure 43–1, is simply an airtight tank into which a person is placed. It can be used in different ways to measure many different aspects of pulmonary function.

1. The body plethysmograph can be used *as a spirometer*. The person breathes through a tube to the exterior of the tank. The excursions of the chest cause displacement of air volume within the tank, and an appropriate recording device connected to the tank can record this displacement in the form of volume changes, thus giving the usual spirometric recordings.

2. The body plethysmograph can be used to *measure lung volumes*. The person does not breathe to the exterior of the tank but instead breathes the air within the tank while the tank is completely closed, then expires with a forceful expiration against a manometer,

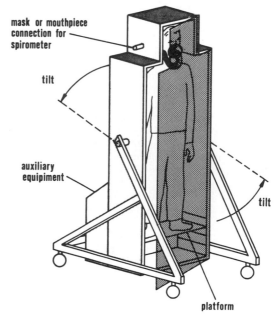

Figure 43–1. A body plethysmograph. (From Ray (ed.): "Instrumentation for Pulmonary Function," in Medical Engineering. © 1974 by Year Book Medical Publishers, Inc., Chicago. Used by permission. Adapted from "Pulmo-Box," Med-Science Electronics, Inc., St. Louis, Missouri.)

and the expiratory pressure is recorded. The increased pressure within the chest decreases the volume of gases in the chest in accordance with Boyle's law: *pressure times volume equals constant,* thus the chest dimensions decrease. Therefore, the person's body requires less volume within the body plethysmograph. An appropriate calibrated pressure recorder connected to the tank records the amount of volume compression of the gases in the lungs. This measure, combined with the recording of pressure change within the chest, can be used to calculate the instantaneous volume of gas in the lung.

3. The body plethysmograph can be used *to determine residual volume* of the lung. The volume of air in the lung is measured by the above method and then as much volume as possible is expired. Subtracting the volume of the expired air from the previously measured lung volume gives the residual volume.

4. The body plethysmograph can be used *to measure airway resistance.* The person is allowed to breathe the air within the tank while volume changes in the tank are recorded. During each expiration the gases in the lung become compressed so that the person's body during this brief period of time actually occupies less space within the chamber. This decreases the tank volume, and the volume decrease is recorded. Knowing the lung volume, as measured above, as well as the *change* in volume during expiration, one can then calculate the pressure of the gas within the lung. This pressure minus the pressure of the gas in the tank gives the pressure difference that forces air through the airways. If the expiratory air flow is measured at the same time, one can then use these two pieces of information to calculate the airway resistance.

5. The body plethysmograph can be used *to measure pulmonary blood flow.* The tank is completely closed from the surrounding atmosphere and the person breathes a gas mixture containing nitrous oxide from a flaccid balloon. As blood flows through the lung the nitrous oxide is absorbed into the pulmonary blood. This reduces the total quantity of gases in the tank (because the balloon itself is inside the tank) and thereby reduces its volume. One can then calculate the rate of pulmonary blood flow from (a) the rate of reduction of the volume in the tank, (b) the concentration of the nitrous oxide in the alveoli, and (c) the solubility coefficient of nitrous oxide in the blood.

Thus, the body plethysmograph is an extremely versatile instrument that has become an important tool for the clinical respiratory physiologist.

STUDY OF BLOOD GASES AND pH

Among the most fundamental of all tests of pulmonary performance are determinations of the blood P_{O_2}, P_{CO_2}, and pH. The discerning physician can usually make reasonable estimates of all of these from careful physical examination of the patient. For instance, a moderate decrease in blood oxygen causes the skin to become bluish because deoxygenated hemoglobin is bluish purple; this condition, called *cyanosis,* is discussed in detail later in the chapter. To estimate approximate concentrations of carbon dioxide and hydrogen ion, the rate and depth of respiration are observed, since these increase greatly as blood carbon dioxide and hydrogen ion concentrations increase.

However, such crude estimates of P_{O_2}, P_{CO_2}, and pH are not sufficient for some purposes and, indeed, can often be very deceiving. Furthermore, in acute situations it is often important to make these measurements extremely rapidly as an aid in determining appropriate therapy for acute respiratory distress or acute abnormalities of acid-base balance. Therefore, several simple and very rapid methods have been developed to make these measurements in only a few minutes, using no more than a few drops of blood. These methods are the following:

Determination of Blood pH. Blood pH is measured using a glass pH electrode of the type used in all chemical laboratories. However, the electrodes used for this purpose are miniaturized so that no more than a drop or so of blood need be used. The voltage generated by the glass electrode is a direct measure of the pH, and this is generally read directly from a voltmeter scale, or it is recorded on a chart.

Determination of Blood P_{CO_2}. A glass electrode pH meter can be used to determine blood P_{CO_2} in the following way: When a weak solution of sodium bicarbonate is exposed to carbon dioxide, the carbon dioxide dissolves in the solution until an equilibrium state is established. In this equilibrium state, the pH of the solution is a function of the carbon dioxide and bicarbonate ion concentrations in accordance with the Henderson-Hasselbalch equation that was discussed in Chapter 37 (that is, $pH = \log \dfrac{HCO_3}{CO_2} + 6.1$). Therefore, with appropriate calibration of the pH meter, the carbon dioxide concentration (or P_{CO_2}) can be read directly from the meter.

When this apparatus is used to measure P_{CO_2} in blood, a miniature glass electrode is coated with a thin solution of sodium bicarbonate, and this is separated from the blood by a very thin plastic membrane that allows carbon dioxide to diffuse from the blood into the solution. Here again, only a drop or so of blood is required.

Determination of Blood P_{O_2}. The concentration of oxygen in a fluid can be measured by a technique called *polarography.* Electrical current is made to flow between two electrodes in the solution. If the voltage of the negative electrode is more than -0.6 volt different from the voltage of the solution, oxygen will deposit on the electrode. Furthermore, the rate of current flow through the electrode will be directly proportional to the concentration of oxygen (and therefore P_{O_2} as well). In practice, a negative platinum electrode with a surface area of about 1 square millimeter is used, and this is separated from the blood by a thin plastic membrane that allows diffusion of oxygen but not diffusion of proteins or other substances that will ''poison'' the electrode. Once again, all parts of the system are miniaturized so that no more than a drop or so of blood is required.

Most often, all three of the measuring devices for pH, P_{CO_2}, and P_{O_2} are built into the same apparatus, and all these measurements can be made within a minute or so using a single, very small sample of blood. Thus, changes in the blood gases and pH can be followed almost moment by moment.

PHYSIOLOGIC TYPES OF RESPIRATORY INSUFFICIENCY

In general, the different types of abnormalities that cause respiratory insufficiency can be divided into three major categories: (1) those that cause inadequate ventilation of the alveoli, (2) those that reduce gaseous diffusion through the respiratory membrane, and (3) those that decrease oxygen transport from the lungs to the tissues.

ABNORMALITIES THAT CAUSE ALVEOLAR HYPOVENTILATION

The different respiratory abnormalities that can cause alveolar hypoventilation mainly have been discussed in the previous few chapters. However, let us summarize briefly some of the clinical causes of alveolar hypoventilation:

Severe hypoventilation frequently results from *paralysis of the respiratory muscles,* resulting from such factors as (1) *bulbar polio,* which often depresses the respiratory center, (2) *cervical transection of the spinal cord,* which prevents transmission of signals from the respiratory center to the respiratory muscles, or (3) *depression of the respiratory center* by anesthetics, drugs, or so forth.

Increased airway resistance is another very common cause of pulmonary hypoventilation; the two classic examples occur in *asthma* and *emphysema.*

Increased tissue resistance also often makes it more difficult to ventilate the lungs. This can be caused by increased viscosity of the lung tissues resulting from such factors as (1) *emphysema,* (2) *pulmonary fibrosis,* (3) *tuberculosis,* (4) *various infections,* or (5) *pulmonary edema.*

Finally, another very common cause of alveolar hypoventilation is *decreased compliance of the lungs and chest wall.* As discussed in Chapter 39, any factor that makes it difficult for the lungs to expand will increase the "work" of ventilation and therefore also often will depress pulmonary ventilation. Some of the diseases of the lungs that do this include (1) *silicosis,* (2) *asbestosis,* (3) *sarcoidosis,* (4) *tuberculosis,* (5) *cancer,* and (6) *pneumonia.* In addition to the lung abnormalities that decrease compliance, restrictive diseases of the chest cage can also greatly decrease the compliance. These include especially *scoliosis* and *kyphosis.*

DISEASES THAT DECREASE LUNG DIFFUSING CAPACITY

Three different types of abnormalities can decrease the diffusing capacity of the lungs. These are: (1) decreased area of the respiratory membrane, (2) increased thickness of the respiratory membrane, called *alveolocapillary block,* and (3) abnormal ventilation-perfusion ratio in some parts of the lungs.

Decreased Area of the Respiratory Membrane. Diseases or abnormalities that decrease the area of the respiratory membrane include *removal of part or all of one lung, tuberculous destruction of the lung, cancerous destruction,* and *emphysema* which causes gradual destruction of alveolar septa.

Also, any acute condition that fills the alveoli with fluid or otherwise prevents air from coming in contact with the alveolar membrane, such as *pneumonia, pulmonary edema,* and *atelectasis,* can temporarily reduce the surface area of the respiratory membrane.

Increased Thickness of the Respiratory Membrane — Alveolocapillary Block. The most common acute cause of increased thickness of the respiratory membrane is *pulmonary edema* resulting from left heart failure or pneumonia. However, *silicosis, tuberculosis,* and *many other fibrotic conditions* can cause progressive deposition of fibrous tissue in the interstitial spaces between the alveolar membrane and the pulmonary capillary membrane, thereby increasing the thickness of the respiratory membrane. This is usually called *alveolocapillary block,* or, occasionally, *interstitial fibrosis.* Since the rate of gaseous diffusion through the respiratory membrane is inversely proportional to the distance that the gas must diffuse, it is readily understood how alveolocapillary block can reduce the diffusing capacity of the lungs.

Abnormal Ventilation-Perfusion Ratio. In Chapter 40 we pointed out that probably the most common cause of decreased lung diffusing capacity is abnormal ventilation-perfusion ratio. That is, in some alveoli there is too little ventilation for the amount of blood flow so that the blood cannot become fully oxygenated. On the other hand, in other alveoli, ventilation is adequate but there is too little blood flow to accept the oxygen. Thus, in either instance, oxygen transfer to the blood becomes greatly compromised. These principles were discussed in detail in Chapter 40. It was pointed out that underventilated alveoli, in which $\dot{V}_A/\dot{Q}$ is less than normal, leads to so-called *physiologic shunt* — that is, blood that is shunted past the lungs without becoming oxygenated. And it was also pointed out that when alveoli are overventilated while blood flow is too little — that is, $\dot{V}_A/\dot{Q}$ is greater than normal — causes *physiologic dead space.* This means that there is ventilation that is not being used by the blood for oxygenation and therefore is *wasted ventilation.* It would be good for the student to return to Chapter 40 and review the principles of abnormal ventilation-perfusion ratio because of the very high prevalence of this abnormality in respiratory disease, especially in the most common of all serious lung diseases, *pulmonary emphysema* caused by smoking.

Diseases that cause abnormal ventilation-perfusion ratios include *thrombosis of a pulmonary artery, excessive airway resistance to some alveoli (emphysema), reduced compliance of one lung without concomitant abnormality of the other lung,* and many other conditions that cause diffuse damage throughout the lungs.

ABNORMALITIES OF OXYGEN TRANSPORT FROM THE LUNGS TO THE TISSUES

Different conditions that reduce oxygen transport from the lungs to the tissues include *anemia,* in which the total amount of hemoglobin available to transport the oxygen is reduced, *carbon monoxide poisoning,* in which a large proportion of the hemoglobin becomes unable to transport oxygen, and *decreased blood flow to the tissues* caused by either low cardiac output or localized tissue ischemia. These have been discussed in Chapters 5, 41, and 28, respectively.

PHYSIOLOGIC PECULIARITIES OF SPECIFIC PULMONARY ABNORMALITIES

CHRONIC EMPHYSEMA

Chronic emphysema is prevalent mainly because of the effects of tobacco smoking. It results from two major pathophysiological changes in the lungs. First, air flow through many of the terminal bronchioles is obstructed. Second, many of the alveolar walls are destroyed.

Many clinicians believe that chronic emphysema begins with chronic infection in the lung that causes *bronchiolitis,* which means inflammation of the small air passages of the lungs. It is likely that this inflammation also involves alveolar septa and destroys many of these or that obstruction to expiration causes excess expiratory pressures in the alveoli and that these pressures rupture the alveolar septa. At any rate, many bronchioles become irreparably obstructed, and the total surface of the respiratory membrane also becomes greatly decreased, as illustrated in Figures 43–2 and 43–3, sometimes to as little as one-tenth to one-quarter normal.

The physiological effects of chronic emphysema are extremely varied, depending on the severity of the disease and on the relative degree of bronchiolar obstruction versus lung parenchymal destruction. However, among the different abnormalities are the following:

First, the bronchiolar obstruction greatly *increases airway resistance* and results in greatly increased work of breathing. It is especially difficult for the person to move air through the bronchioles during expiration because the compressive force on the outside of the lung not only compresses the alveoli but also compresses the bronchioles, which further increases their resistance.

Second, the marked loss of lung parenchyma greatly *decreases the diffusing capacity* of the lung, which reduces the ability of the lungs to oxygenate the blood and to remove carbon dioxide.

Third, the obstructive process is frequently much worse in some parts of the lungs than in other parts so that some portions of the lungs are well ventilated while other portions are poorly ventilated. This often causes an *extremely abnormal ventilation-perfusion ratio,* which also causes *physiologic shunt* resulting in poor aeration of the blood and a greatly expanded *physiologic dead space* resulting in wasted ventilation, both effects occurring in the same lungs.

Fourth, loss of large portions of the lung parenchyma also decreases the number of pulmonary capillaries through which blood can pass. As a result, the pulmonary vascular resistance increases markedly, causing pulmonary hypertension. This in turn overloads the right heart and frequently causes right-heart failure.

Chronic emphysema usually progresses slowly over

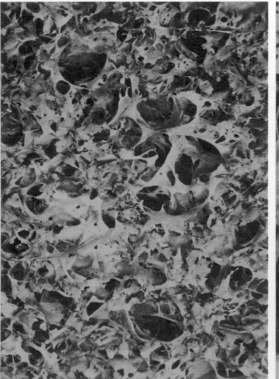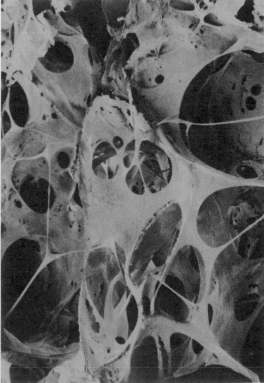

Figure 43–2. Contrast of the emphysematous lung (right) with the normal lung (left), showing extensive alveolar destruction. (Reproduced with permission of Patricia Delaney and the Department of Anatomy, The Medical College of Wisconsin.)

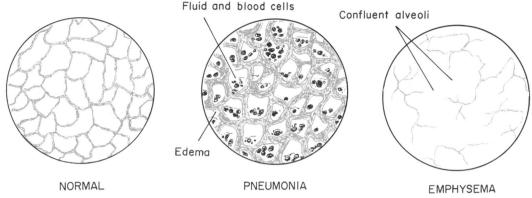

Figure 43–3. Pulmonary changes in pneumonia and emphysema.

many years. The person develops hypoxia and hypercapnia because of hypoventilation of many alveoli and because of loss of lung parenchyma. The net result of all of these effects is severe and prolonged air hunger that can last for years until the hypoxia and hypercapnia cause death — a very high penalty to pay for smoking.

PNEUMONIA

The term pneumonia describes any inflammatory condition of the lung in which the alveoli are usually filled with fluid and blood cells, as shown in Figure 43–3. A common type of pneumonia is *bacterial pneumonia,* caused most frequently by pneumococci. This disease begins with infection in the alveoli; the pulmonary membrane becomes inflamed and highly porous so that fluid and even red and white blood cells pass out of the blood into the alveoli. Thus, the infected alveoli become progressively filled with fluid and cells, and the infection spreads by extension of bacteria from alveolus to alveolus. Eventually, large areas of the lungs, sometimes whole lobes or even a whole lung, become "consolidated," which means that they are filled with fluid and cellular debris.

The pulmonary function of the lungs during pneumonia changes in different stages of the disease. In the early stages, the pneumonia process might well be localized to only one lung, and alveolar ventilation may be reduced even though blood flow through the lung continues normally. This results in two major pulmonary abnormalities: (1) reduction in the total available surface area of the respiratory membrane and (2) decreased ventilation-perfusion ratio. Both these effects cause reduced diffusing capacity, which results in hypoxemia.

Figure 43–4 illustrates the effect of the decreased ventilation-perfusion ratio in pneumonia, showing that the blood passing through the aerated lung becomes 97 per cent saturated while that passing through the un-aerated lung remains only 60 per cent saturated, causing the mean saturation of the aortic blood to be about 78 per cent, which is far below normal. Fortunately, in most stages of pneumonia the blood flow through the diseased areas decreases concurrently with the decrease in ventilation. This gives much less debility than that

resulting when the blood flow through the unventilated lung is normal.

ATELECTASIS

Atelectasis means collapse of the alveoli. It can occur in a localized area of a lung, in an entire lobe, or in an entire lung. Its most common causes are two-fold: (1) obstruction of the airway, or (2) lack of surfactant in the fluids lining the alveoli.

Airway Obstruction. The airway obstruction type of atelectasis usually results from blockage of many small bronchi with mucus or obstruction of a major bronchus by either a large mucous plug or some solid object such as cancer. The air entrapped beyond the block is absorbed within minutes to hours by the blood flowing in the pulmonary capillaries. If the lung tissue is pliable enough, this will lead simply to collapse of the alveoli. However, if the lung tissue cannot collapse, absorption of air from the alveoli creates tremendously negative pressures within the alveoli and pulls fluid out of the pulmonary interstitium into the alveoli, thus causing the alveoli to fill completely with edema fluid. This almost always is the effect that occurs when an entire lung becomes atelectatic, a condition called *massive collapse* of the lung, because the solidity of the chest wall and of the mediastinum allows the lung to decrease only to about one-half normal size rather than collapsing completely. The remainder of the space in the alveoli must then become filled with fluid.

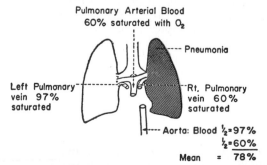

Figure 43–4. Effect of pneumonia on arterial blood oxygen saturation.

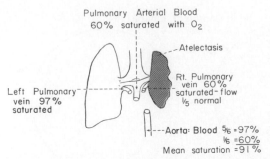

Figure 43–5.   Effect of atelectasis on arterial blood oxygen saturation.

The effects on overall pulmonary function caused by *massive collapse* (atelectasis) of an entire lung are illustrated in Figure 43–5. Collapse of the lung tissue not only occludes the alveoli but also increases the resistance to blood flow through the pulmonary vessels as much as five-fold. This resistance increase occurs partially because of the collapse itself, which compresses and folds the vessels as the volume of the lung decreases. But, in addition, hypoxia in the collapsed alveoli causes additional vasoconstriction, as was explained in Chapter 24.

Because of the vascular constriction, blood flow through the atelectatic lung becomes slight. Most of the blood is routed through the ventilated lung and therefore becomes well aerated. In the situation shown in Figure 43–5, five-sixths of the blood passes through the aerated lung and only one-sixth through the unaerated lung. As a result, the overall ventilation-perfusion ratio is only moderately compromised, so that the aortic blood has only mild oxygen desaturation.

Lack of Surfactant.   In Chapter 39 it was pointed out that the substance surfactant is secreted by the alveolar epithelium into the fluids that line the alveoli. This substance decreases the surface tension in the alveoli 2- to 14-fold. In the normal lung this plays a major role in preventing alveolar collapse. However, in a number of different conditions the quantity of surfactant secreted by the alveoli is greatly decreased. Sometimes this is severe enough to cause atelectasis. For instance, in the condition called *hyaline membrane disease* or *respiratory distress syndrome*, which often occurs in newborn babies, the quantity of surfactant secreted by the alveoli is greatly depressed. This effect causes a serious tendency for the lungs of these babies to collapse or to become filled with fluid as explained above; many of the infants die of suffocation as increasing portions of the lungs become atelectatic.

Atelectasis also frequently results from decreased surfactant secretion following extracorporeal perfusion of the lungs during open-heart surgery. The reason for the reduced surfactant secretion in this instance is not understood.

BRONCHIAL ASTHMA

Bronchial asthma is usually caused by allergic hypersensitivity of the person to foreign substances in the air — especially to plant pollens. The allergic reaction is

believed to occur in the following way: The typically allergic person has a tendency to form an abnormal type of antibody called IgE, and these antibodies cause allergic reactions when they react with their complementary antigens as was explained in Chapter 7. In asthma, the antibodies are believed to become attached to mast cells that lie in the lung interstitium in close association with the bronchioles and small bronchi. Then when the person breathes in a pollen to which he or she is sensitive (that is, to which the person has developed IgE antibodies), the pollen reacts with the mast cell–attached antibodies and causes these cells to release several different substances. Among them are *histamine, slow-reacting substance of anaphylaxis, eosinophilic chemotactic factor,* and *bradykinin.* The combined effects of all these factors are to produce (1) localized edema in the walls of the small bronchioles as well as secretion of thick mucus into the bronchiolar lumens and (2) spasm of the bronchiolar smooth muscle. Obviously, therefore, the airway resistance increases greatly.

As discussed in Chapter 39, the bronchiolar diameter becomes more reduced during expiration than during inspiration in asthma. The reason for this is that the increased intrapulmonary pressure during expiratory effort not only compresses the air in the alveoli but compresses the outsides of the bronchioles as well. Since the bronchioles are already partially occluded, further occlusion resulting from the external pressure creates especially severe obstruction during expiration. Therefore, the asthmatic person usually can inspire quite adequately but has great difficulty expiring. This results in dyspnea, or "air hunger," which is discussed later in the chapter.

The functional residual capacity and the residual volume of the lung become greatly increased during the asthmatic attack because of the difficulty in expiring air from the lungs. Over a long period of time the chest cage becomes permanently enlarged, causing a "barrel chest," and the functional residual capacity and residual volume also become permanently increased.

TUBERCULOSIS

In tuberculosis the tubercle bacilli cause a peculiar tissue reaction in the lungs including, first, invasion of the infected region by macrophages and, second, walling off of the lesion by fibrous tissue to form the so-called "tubercle." This walling-off process helps to limit further transmission of the tubercle bacilli in the lungs and, therefore, is part of the protective process against the infection. However, in approximately 3 per cent of all persons who contract tuberculosis, the walling-off process fails, and tubercle bacilli spread throughout the lungs. Thus, tuberculosis in its late stages causes many areas of fibrosis throughout the lungs, and, secondly, it reduces the total amount of functional lung tissue. These effects cause (1) increased effort on the part of the respiratory muscles causing pulmonary ventilation and therefore *reduced vital capacity and breathing capacity,* (2) *reduced total respiratory membrane surface area* and *increased thickness of the respiratory membrane,* these causing progressively diminished pulmonary diffusing capacity, and (3) *abnor-*

mal ventilation-perfusion ratio in the lungs, further reducing the pulmonary diffusing capacity.

HYPOXIA

Obviously, almost any of the conditions discussed in the past few sections of this chapter can cause serious degrees of cellular hypoxia. In some of these, oxygen therapy is of great value; in others it is of moderate value; in still others it is of almost no value. Therefore, it is important to classify the different types of hypoxia, then we can readily discuss the physiological principles of therapy. The following is a descriptive classification of the different causes of hypoxia:

1. Inadequate oxygenation of the lungs because of extrinsic reasons
   a. Deficiency of oxygen in atmosphere
   b. Hypoventilation (neuromuscular disorders)
2. Pulmonary disease
   a. Hypoventilation due to increased airway resistance or decreased pulmonary compliance
   b. Uneven alveolar ventilation-perfusion ratio (including increased physiologic dead space and physiologic shunt)
   c. Diminished respiratory membrane diffusion
3. Venous-to-arterial shunts
4. Inadequate transport and delivery of oxygen
   a. Anemia, abnormal hemoglobin
   b. General circulatory deficiency
   c. Localized circulatory deficiency (peripheral, cerebral, coronary vessels)
5. Inadequate tissue oxygenation or capability of using oxygen
   a. Tissue edema
   b. Abnormal tissue demand
   c. Poisoning of cellular enzymes

This classification of the different types of hypoxia is mainly self-evident from the discussions earlier in the chapter. Only one of the types of hypoxia in the above classification needs further elaboration; this is the hypoxia caused by inadequate tissue oxygenation or capability of using oxygen.

Inadequate Tissue Oxygenation or Capability of Using Oxygen. The classic cause of inability of the tissues to use oxygen is cyanide poisoning, in which the action of cytochrome oxidase is completely blocked — to such an extent that the tissues simply cannot utilize the oxygen even though plenty is available. This type of hypoxia is frequently also called *histotoxic hypoxia*.

A common type of tissue hypoxia also occurs in tissue edema, which causes increased distances through which the oxygen must diffuse before it can reach the cells. This type of hypoxia can become so severe that the tissues in edematous areas actually die, as is often illustrated by serious ulcers in edematous skin.

Finally, tissues can become hypoxic when the cells themselves demand more oxygen than can be supplied to them by the normal respiratory and oxygen transport systems. For instance, in strenuous exercise, the major limiting factor to the degree of exercise that can be performed is the tissue hypoxia that develops.

Effects of Hypoxia on the Body. Hypoxia, if severe enough, can actually cause death of the cells, but in less severe degrees it results principally in (1) depressed mental activity, sometimes culminating in coma, and (2) reduced work capacity of the muscles. These effects are discussed in the following chapter in relation to high altitude physiology and, therefore, are only mentioned here.

CYANOSIS

The term "cyanosis" means blueness of the skin, and its cause is excessive amounts of deoxygenated hemoglobin in the skin blood vessels, especially in the capillaries. This deoxygenated hemoglobin has an intense dark blue-purple color that is transmitted through the skin.

The presence of cyanosis is one of the most common clinical signs of different degrees of respiratory insufficiency, and for this reason it is important to understand the factors that determine the degree of cyanosis. These are the following:

1. One of the most important factors determining the degree of cyanosis is the *quantity of deoxygenated hemoglobin in the arterial blood*. It is not the percentage deoxygenation of the hemoglobin that causes the bluish hue of the skin, but principally the *concentration of deoxygenated hemoglobin without regard to the concentration of oxygenated hemoglobin*. The reason for this is that the red color of oxygenated blood is weak in comparison with the dark blue color of deoxygenated blood. Therefore, when the two are mixed together, the oxygenated blood has relatively little coloring effect in comparison with that of the deoxygenated blood.

In general, definite cyanosis appears whenever the arterial blood contains more than 5 grams per cent of deoxygenated hemoglobin, and mild cyanosis can frequently be discerned when as little as 3 to 4 grams per cent deoxygenated hemoglobin is present. In polycythemia, cyanosis is very common because of the large amount of hemoglobin in the blood whereas in anemia, cyanosis is rare because it is difficult for there to be enough deoxygenated hemoglobin to produce the blue color — even though anemia is much more likely to cause tissue hypoxia than is polycythemia.

2. Another important factor that affects the degree of cyanosis is the *rate of blood flow through the skin*, for the following reasons: Principally, it is the blood in the capillaries that determines skin color; the blueness of this capillary blood is determined by two factors: (1) the concentration of deoxygenated hemoglobin in the arterial blood entering the capillaries and (2) the amount of deoxygenation that occurs as the blood passes through the capillaries. Ordinarily, the metabolism of the skin is relatively low so that little deoxygenation occurs as the blood passes through the skin capillaries. However, if the blood flow becomes extremely sluggish, even this low metabolism can cause marked desaturation of the blood and therefore can cause cyanosis. This explains the cyanosis that appears in very cold weather, particularly in children who have thin skins.

3. A final factor that affects the blueness of the skin is *skin thickness* because this determines the intensity of the color that is transmitted from the deeper vascular

tissues. For instance, in newborn babies, who have very thin skin, cyanosis occurs readily, particularly in highly vascular portions of the body such as the heels. Also, in adults (as well as babies) the lips and fingernails often appear cyanotic before the remainder of the body shows any blueness.

DYSPNEA

Dyspnea means primarily a desire for air or mental anguish associated with inability to ventilate enough to satisfy the air demand. A common synonym is "air hunger."

At least three different factors often enter into the development of the sensation of dyspnea. These are: (1) abnormality of the respiratory gases in the body fluids, especially hypercapnia (that will be discussed below) and to much less extent hypoxia, (2) the amount of work that must be performed by the respiratory muscles to provide adequate ventilation, and (3) the state of the mind. A person becomes very dyspneic especially from excess buildup of carbon dioxide in body fluids.

At times, however, the levels of both carbon dioxide and oxygen in the body fluids are completely normal, but to attain this normality of the respiratory gases, the person has to breathe forcefully. In these instances the forceful activity of the respiratory muscles gives the person a sensation of air hunger. Indeed, the dyspnea can be so intense, despite normal gas concentrations in the body fluids, that clinicians often especially emphasize this cause of dyspnea.

Finally, the person's respiratory functions may be completely normal, and still dyspnea may be experienced because of an abnormal state of mind. This is called *neurogenic dyspnea* or, sometimes, *emotional dyspnea*. For instance, almost anyone momentarily thinking about the act of breathing may suddenly start taking breaths a little more deeply than ordinarily because of a feeling of mild dyspnea. This feeling is greatly enhanced in persons who have a psychic fear of not being able to receive a sufficient quantity of air. For example, many persons on entering small or crowded rooms immediately experience emotional dyspnea, and patients with "cardiac neurosis" who have heard that dyspnea is associated with heart failure frequently experience severe psychic dyspnea even though the blood gases are completely normal. Neurogenic dyspnea has been known to be so intense that the person over-respires and causes alkalotic tetany.

HYPERCAPNIA

Hypercapnia means excess carbon dioxide in the body fluids.

One might suspect on first thought that any respiratory condition that causes hypoxia would also cause hypercapnia. However, hypercapnia usually occurs in association with hypoxia only when the hypoxia is caused by *hypoventilation* or by *circulatory deficiency*. The reasons for this are the following:

Obviously, hypoxia caused by *too little oxygen in the air,* by *too little hemoglobin,* or by *poisoning of the oxidative enzymes* has to do only with the availability of

oxygen or use of oxygen by the tissues. Therefore, it is readily understandable that hypercapnia is *not* a concomitant of these types of hypoxia.

Also, in hypoxia resulting from poor diffusion through the pulmonary membrane or through the tissues, serious hypercapnia usually does not occur because carbon dioxide diffuses 20 times as rapidly as oxygen. Also, if hypercapnia does begin to occur this immediately stimulates pulmonary ventilation, which corrects the hypercapnia.

However, in hypoxia caused by hypoventilation, carbon dioxide transfer between the alveoli and the atmosphere is affected as much as is oxygen transfer. Therefore, hypercapnia always results along with hypoxia. And in circulatory deficiency, diminished flow of blood decreases the removal of carbon dioxide from the tissues, resulting in tissue hypercapnia. However, the transport capacity of the blood for carbon dioxide is about three times that for oxygen, so that even here the hypercapnia is much less than the hypoxia.

EFFECTS OF HYPERCAPNIA ON THE BODY

When the alveolar P_{CO_2} rises above approximately 60 to 75 mm. Hg, dyspnea usually becomes intolerable, and as the P_{CO_2} rises to 70 to 80 mm. Hg, the person becomes lethargic and sometimes even semicomatose. Total anesthesia and death result when the P_{CO_2} rises to 100 to 150 mm. Hg.

OXYGEN THERAPY IN THE DIFFERENT TYPES OF HYPOXIA

Oxygen can be administered by (1) placing the patient's head in a "tent" that contains air fortified with oxygen, (2) allowing the patient to breathe either pure oxygen or high concentrations of oxygen from a mask, or (3) administering oxygen through an intranasal tube.

Oxygen therapy is of great value in certain types of

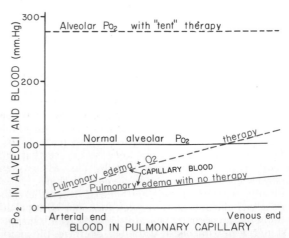

Figure 43–6. Absorption of oxygen into the pulmonary capillary blood in pulmonary edema with and without oxygen therapy.

hypoxia but of almost no value at all in other types. However, recalling the basic physiological principles of the different types of hypoxia, one can readily decide when oxygen therapy is of value and, if so, how valuable. For instance:

In *atmospheric hypoxia,* oxygen therapy can obviously completely correct the depressed oxygen level in the inspired gases and, therefore, provide 100 per cent effective therapy.

In *hypoventilation hypoxia,* a person breathing 100 per cent oxygen can move 5 times as much oxygen into the alveoli with each breath as when breathing normal air. Therefore, here again oxygen therapy can be extremely beneficial, increasing the available oxygen as much as five-fold. (However, this provides no benefit for the hypercapnia also caused by the hypoventilation.)

In *hypoxia caused by impaired diffusion,* essentially the same result occurs as in hypoventilation hypoxia, for oxygen therapy can increase the P_{O_2} in the lungs from a normal value of about 100 mm. Hg to as high as 600 mm. This causes a greatly increased oxygen diffusion gradient between the alveoli and the blood, the gradient rising from a normal value of 60 mm. Hg to as high as 560 mm. Hg, or an increase of more than 800 per cent. This highly beneficial effect of oxygen therapy in diffusion hypoxia is illustrated in Figure 43–6, which shows that the pulmonary blood in a patient with pulmonary edema picks up oxygen up to 8 times as rapidly as it would with no therapy.

In *hypoxia caused by anemia* or *other abnormality of hemoglobin transport,* oxygen therapy is of slight to moderate value because the amount of oxygen transported in the dissolved form in the fluids of the blood can still be increased above normal even though that transported by the hemoglobin is hardly altered. This is illustrated in Figure 43–7, which shows that an increase in alveolar P_{O_2} from a normal value of 100 mm. Hg to 600 mm. Hg increases the total oxygen in the blood (that combined with hemoglobin plus the dissolved oxygen) from 5 volumes per cent to 6.5 volumes per cent. This

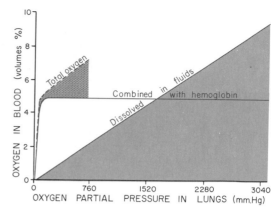

Figure 43–7.   Effects of oxygen therapy in hypohemoglobinemic hypoxia.

represents a 30 per cent increase in the amount of oxygen transported to the tissues, and 30 per cent is often the difference between life and death. However, Figure 43–7 also shows that oxygen therapy in a *hyperbaric pressure chamber* at pressures above atmospheric pressure (760 mm. Hg) can be of even greater value.

In *hypoxia caused by circulatory deficiency,* also called *ischemic hypoxia,* the value of oxygen therapy is usually very slight, because the problem here is sluggish flow of blood and not insufficient oxygen. Figure 43–8 illustrates that normal blood can carry only a small amount of extra oxygen to the tissues (about 10 per cent extra) when the alveolar oxygen concentration is increased to 600 mm. Hg. Yet this 10 per cent difference may save the life of the patient, as, for example, following an acute heart attack which causes the cardiac output to fall very low.

In hypoxia caused by *physiologic shunt,* the value of oxygen therapy is even less than in ischemic hypoxia. Only that blood passing through the ventilated alveoli

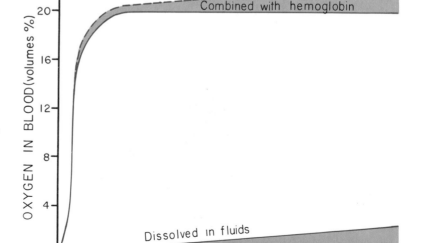

Figure 43–8.   Effects of oxygen therapy in ischemic hypoxia.

can receive additional oxygen. And since the blood passing through these alveoli is already either fully or almost fully saturated, the amount of additional oxygen that can be transported is only the portion that can become dissolved in the fluid of the blood, a very small amount. Yet again, this may be the difference between life and death.

In the different types of *hypoxia caused by inadequate tissue use of oxygen,* there is no abnormality of oxygen pickup by the lungs or of transport to the tissues. Instead, the tissues simply need more oxygen than can reach them or than the enzymes can utilize. Therefore, it is doubtful that oxygen therapy is of any benefit at all.

DANGER OF HYPERCAPNIA DURING OXYGEN THERAPY

Much of the stimulus that helps to maintain ventilation in hypoxia results from hypoxic stimulation of the aortic and carotid chemoreceptors. In the preceding chapter it was noted that in chronic hypoxia, oxygen lack becomes a far more powerful stimulus to respiration than usual, sometimes increasing the ventilation as much as 5- to 7-fold. Therefore, during oxygen therapy, relief of the hypoxia occasionally causes pulmonary ventilation to decrease so low that lethal levels of hypercapnia develop. For this reason, oxygen therapy in hypoxia is sometimes contraindicated, particularly in conditions that otherwise tend to cause hypercapnia, such as uneven ventilation-perfusion ratio or depressed respiratory center activity.

ABSORPTION OF ENTRAPPED AIR

Whenever air becomes entrapped and closed off from the atmosphere anywhere in the body, whether in the alveoli as a result of bronchial obstruction, in the gastrointestinal tract, in one of the nasal or auditory cavities, or simply as air injected beneath the skin, it is usually absorbed in a few hours or days.

Physical Principles of Air Absorption. Figure 43–9 shows the progressive stages of air absorption from a cavity. The cavity to the left is filled with recently injected air which has become humidified almost instantaneously by the fluids surrounding the cavity. Therefore, this cavity contains the normal concentrations of oxygen, nitrogen, and water vapor found in humidified air at sea level. Note that the total pressure of these three gases added together equals 760 mm. Hg. Furthermore, the pressure in the cavity remains almost the same during the entire process of absorption because the atmospheric pressure continually presses against the body, and the body's tissues are flexible enough that the atmospheric pressure is transmitted exactly to most cavities beneath the skin.

At first, the P_{O_2} in the injected air is much greater than the P_{O_2} in the interstitial fluid, so that oxygen begins to be absorbed rapidly. On the other hand, the nitrogen pressure (P_{N_2}) in the cavity is actually less than that in the interstitial fluid because nitrogen, unlike oxygen, is not metabolized by the tissues of the body. As a result, minute amounts of nitrogen actually diffuse into the cavity at first. But after large amounts of oxygen have been absorbed, the P_{O_2} falls to as little as 90 mm. Hg, as shown in the second cavity in Figure 43–9. This absorption of oxygen makes the cavity become smaller, now increasing the P_{N_2} to a value greater than that in the interstitial fluid. Also, carbon dioxide diffuses into the cavity.

The final equilibrium state between the gases in the cavity and in the interstitial fluid is shown in the third cavity; the P_{O_2} has decreased to about 43 mm. Hg, the P_{CO_2} has increased to 40 mm. Hg, and the P_{N_2} has increased to 630 mm. Hg. From then on, both oxygen and nitrogen are absorbed continually because both of these are then at higher pressures in the cavity than in the surrounding tissues. As this reduces the size of the cavity, the carbon dioxide and water pressures increase to slightly higher than the pressures of the surrounding fluids. As a consequence, the carbon dioxide and water vapor are also absorbed. This process continues until all the gases leave the cavity and the cavity collapses totally.

REFERENCES

Avery, M. E., *et al.*: The lung of the newborn infant. *Sci. Am., 228*:74, 1973.

Bass, H.: Assessment of regional pulmonary function with radioactive gases. *Prog. Nucl. Med., 3*:67, 1973.

Bendixen, H., *et al.*:Respiratory Care, 2nd Ed. St. Louis, C. V. Mosby, 1969.

Cohen, A. B., and Gold, W. M.: Defense mechanisms of the lungs. *Annu. Rev. Physiol., 37*:325, 1975.

Comroe, J. H., Jr., *et al.*: The Lung: Clinical Physiology and Pulmonary Function Tests, 2nd Ed. Chicago, Year Book Medical Publishers, 1962.

Crofton, J., and Douglas, A.: Respiratory Diseases. Philadelphia, J. B. Lippincott Co., 1975.

Cumming, G., and Semple, S. J.: Disorders of the Respiratory System. Philadelphia, J. B. Lippincott Co., 1973.

Czerwinski, B. S.: A Manual of Patient Education for Cardiopulmonary Disease. St. Louis, C. V. Mosby, 1980.

Domm, B. M., and Vassallo, C. L.: Pulmonary function testing. *Clin. Anesth., 9*:191, 1973.

Dosman, J. A., and Cotton, D. J. (eds.): Occupational Pulmonary Disease: Focus on Grain Dust and Health. New York, Academic Press, 1979.

Filley, G.: Pulmonary Insufficiency and Respiratory Failure. Philadelphia, Lea & Febiger, 1967.

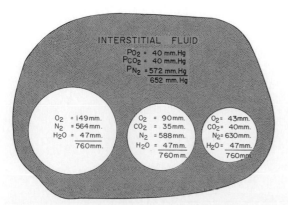

Figure 43–9. Absorption of air from an occluded cavity of the body.

Fisher, A. B., *et al.*: Oxygen toxicity of the lung: Biochemical aspects. *In* Fishman, A. P., and Renkin, E. M. (eds.): Pulmonary Edema. Baltimore, Waverly Press, 1979, p. 207.

Fishman, A. P.: Assessment of Pulmonary Function. New York, McGraw-Hill, 1980.

Fishman, A. P., and Pietra, G. G.: Primary pulmonary hypertension. *Annu. Rev. Med., 31*:421, 1980.

Gelb, A. F., and MacAnally, B. J.: Early detection of obstructive lung disease by analysis of maximal expiratory flow-volume curves. *Chest, 64*:749, 1973.

Gross, N. J.: Bronchial Asthma. Hagerstown, Md., Harper & Row, 1974.

Guenter, C. A., *et al.*: Clinical Aspects of Respiratory Physiology. Philadelphia, J. B. Lippincott Co., 1978.

Guyton, A. C., and Farish, C. A.: A rapidly responding continuous oxygen consumption recorder. *J. Appl. Physiol., 14*:143, 1959.

Hall, W. J., and Douglas, R. G., Jr.: Pulmonary function during and after common respiratory infections. *Annu. Rev. Med., 31*:233, 1980.

Hemingway, A.: Measurement of Airway Resistance with the Body Plethysmograph. Springfield, Ill., Charles C Thomas, 1973.

Hodgkin, J. E. (ed.): Chronic Obstructive Pulmonary Disease: Current Concepts in Diagnosis and Comprehensive Care. Park Ridge, Ill., American College of Chest Physicians, 1979.

Irsigler, G. B., and Severinghaus, J. W.: Clinical problems of ventilatory control. *Annu. Rev. Med., 31*:109, 1980.

Killough, J. H.: Protective mechanisms of the lungs; pulmonary disease; pleural disease. *In* Sodeman, W. A., Jr., and Sodeman, T. M. (eds.): Sodeman's Pathologic Physiology; Mechanisms of Disease, 6th Ed. Philadelphia, W. B. Saunders Co., 1979, p. 451.

Lane, D. J.: Asthma: The Facts. New York, Oxford University Press, 1979.

Macklem, P. T., and Permutt, S. (eds.): The Lung in the Transition Between Health and Disease. New York, Marcel Dekker, 1978.

Middleton, E., Jr.: Autonomic imbalance in asthma with special reference to beta adrenergic blockade. *Adv. Intern. Med., 18*:177, 1972.

Moore, F. D.: Postoperative pulmonary insufficiency: Anoxia, the shunted lung, and mechanical assistance. *Cardiovasc. Clin., 3*:121, 1971.

Moser, K. M. (ed.): Pulmonary Vascular Diseases. New York, Marcel Dekker, 1979.

Nilsson, N. J.: Oximetry. *Physiol. Rev., 40*:1, 1960.

Oxygen Free Radicals and Tissue Damage. Ciba Foundation Symposium. New York, Excerpta Medica, 1979.

Paleček, F.: Control of breathing in diseases of the respiratory system. *Int. Rev. Physiol., 14*:255, 1977.

Petty, T. L. (ed.): Chronic Obstructive Pulmonary Disease. New York, Marcel Dekker, 1978.

Putnam, J. S. (ed.): Advances in Pulmonary Medicine. Philadelphia, W. B. Saunders Co., 1978.

Rochester, D. F., and Enson, Y.: Current concepts in the pathogenesis of the obesity-hypoventilation syndrome. Mechanical and circulatory factors. *Am. J. Med., 57*:402, 1974.

Said, S. I.: Metabolic functions of the lung. *In* Frolich, E. D. (ed.): Pathophysiology, 2nd Ed. Philadelphia, J. B. Lippincott Co., 1976, p. 189.

Schonell, M.: Respiratory Medicine. New York, Churchill Livingstone, 1974.

Secker-Walker, R. H., and Evens, R. G.: The clinical application of computers in ventilation-perfusion studies. *Prog. Nucl. Med., 3*:166, 1973.

Selikoff, I. J., and Lee, D. H. K.: Asbestos and Disease. New York, Academic Press, 1978.

Siegel, B. A., and Potchen, E. J.: Radionuclide studies of pulmonary function. Anatomic and physiologic considerations. *Prog. Nucl. Med., 3*:49, 1973.

Slonim, N. B., *et al.*:Cardiopulmonary Laboratory Basic Methods and Calculations. Springfield, Ill., Charles C Thomas, 1974.

Staub, N. C.: Pulmonary edema. *Physiol. Rev., 54*:678, 1974.

Tisi, G. M.: Pulmonary Physiology in Clinical Medicine. Baltimore, Williams & Wilkins, 1980.

Tysinger, D. S., Jr.: The Clinical Physics and Physiology of Chronic Lung Disease, Inhalation Therapy, and Pulmonary Function Testing. Springfield, Ill., Charles C Thomas, 1973.

Weiss, E. B. (ed.): Status Asthmaticus. Baltimore, University Park Press, 1978.

Wilson, A. F., and McPhillips, J. J.: Pharmacological control of asthma. *Annu. Rev. Pharmacol. Toxicol., 18*:541, 1978.

Wolfe, W. G., and Sabiston, D. C.: Pulmonary Embolism. Philadelphia, W. B. Saunders Co., 1980.

Wright, G. R., and Shepard, R. J.: Physiological effects of carbon monoxide. *In* Robertshaw, D. (ed.): International Review of Physiology: Environmental Physiology III. Vol. 20. Baltimore, University Park Press, 1979, p. 311.

Part VIII

AVIATION, SPACE, AND DEEP SEA DIVING PHYSIOLOGY

44

Aviation, High Altitude, and Space Physiology

As man has ascended to higher and higher altitudes in aviation, in mountain climbing, and in space vehicles, it has become progressively more important to understand the effects of altitude and low gas pressures on the human body. In the early days of aviation only two factors were of concern: (1) the effects of hypoxia on the body and (2) the effects of physical factors of high altitude, such as temperature and ultraviolet radiation. With further development of airplanes, it was soon learned that they could be built to withstand acceleratory forces far greater than the human body can stand. Now, in the space age, all these problems have multiplied to the point where the physical conditions in the spacecraft must be created artificially.

The present chapter deals with all these problems: first, the hypoxia at high altitudes, second, some of the other physical factors affecting the body at high altitudes, and, third, the tremendous acceleratory forces that occur in both aviation and space physiology.

EFFECTS OF LOW OXYGEN PRESSURE ON THE BODY

Barometric Pressures at Different Altitudes. As a prelude to discussing the effects of low oxygen pressure on the body, we must recall that the total pressure of all the gases in the air, the *barometric pressure,* decreases as one rises to progressively higher altitudes. Table 44–1 gives the pressures at different altitudes, showing that at sea level the pressure is 760 mm. Hg, while at 10,000 feet it is only 523 mm. Hg, and at 50,000 feet, 87 mm. Hg. This decrease in barometric pressure is the basic cause of all the hypoxia problems in high altitude physiology, for as the barometric pressure decreases, the oxygen pressure decreases proportionately, remaining at all times slightly less than 21 per cent of the total barometric pressure.

Oxygen Partial Pressures at Different Elevations. Table 44–1 also shows that the partial pressure of oxygen (P_{O_2}) in dry air at sea level is approximately 159 mm. Hg, though this can be decreased as much as 10 mm. when large amounts of water vapor exist in the air. The P_{O_2} at 10,000 feet is approximately 110 mm. Hg; at 20,000 feet, 73 mm. Hg; and at 50,000 feet, only 18 mm. Hg.

ALVEOLAR P_{O_2} AT DIFFERENT ELEVATIONS

Obviously, when the P_{O_2} in the atmosphere decreases at higher elevations, a decrease in alveolar P_{O_2} is also to be expected. At low altitudes the alveolar P_{O_2} does not decrease quite so much as the

TABLE 44–1 EFFECTS ON ALVEOLAR GAS CONCENTRATIONS AND ON ARTERIAL OXYGEN SATURATION OF ACUTE EXPOSURE TO LOW ATMOSPHERIC PRESSURES

Altitude (ft.)	Barometric Pressure (mm. Hg)	P_{O_2} in Air (mm. Hg)	Breathing Air			Breathing Pure Oxygen		
			P_{CO_2} in Alveoli (mm. Hg)	P_{O_2} in Alveoli (mm. Hg)	Arterial Oxygen Saturation (%)	P_{CO_2} in Alveoli (mm. Hg)	P_{O_2} in Alveoli (mm. Hg)	Arterial Oxygen Saturation (%)
0	760	159	40	104	97	40	673	100
10,000	523	110	36	67	90	40	436	100
20,000	349	73	24	40	70	40	262	100
30,000	226	47	24	21	20	40	139	99
40,000	141	29	24	8	5	36	58	87
50,000	87	18	24	1	1	24	16	15

atmospheric P_{O_2} because increased pulmonary ventilation helps to compensate for the diminished atmospheric oxygen. But at higher altitudes the alveolar P_{O_2} decreases even more than atmospheric P_{O_2} for peculiar reasons that are explained as follows:

Effect of Carbon Dioxide and Water Vapor on Alveolar Oxygen. Even at high altitudes carbon dioxide is continually excreted from the pulmonary blood into the alveoli. Also, water vaporizes into the inspired air from the respiratory surfaces. Therefore, these two gases dilute the oxygen and nitrogen already in the alveoli, thus reducing the oxygen concentration.

The presence of carbon dioxide and water vapor in the alveoli becomes exceedingly important at high altitudes because the total barometric pressure falls to low levels while the pressures of carbon dioxide and water vapor do not fall comparably. Water vapor pressure remains 47 mm. Hg as long as the body temperature is normal, regardless of altitude; and during acute exposure to very high altitudes the pressure of carbon dioxide falls from about 40 mm. Hg at sea level to about 24 mm. Hg because of increased respiration.

Now let us see how the pressures of these two gases affect the available space for oxygen. Let us assume that the total barometric pressure falls to 100 mm. Hg; 47 mm. Hg of this must be water vapor, leaving only 53 mm. Hg for all the other gases. Under acute exposure to high altitude, 24 mm. Hg of the 53 mm. Hg must be carbon dioxide, leaving a remaining space of only 29 mm. Hg. If there were no use of oxygen by the body, one-fifth of this 29 mm. Hg would be oxygen, and four-fifths would be nitrogen; or, the P_{O_2} in the alveoli would be 6 mm. Hg. However, by this time the person's tissues would be almost totally anoxic so that even most of this last remaining amount of alveolar oxygen would be absorbed into the blood, leaving not more than 1 mm. Hg oxygen pressure in the alveoli. Therefore, at a barometric pressure of 100 mm. Hg, the person could not possibly survive when breathing air. But the effect is very much different if the person is breathing pure oxygen, as we shall see in the following discussions.

A simple formula for approximating alveolar P_{O_2} is the following:

$$\text{Alveolar } P_{O_2} = \frac{P_B - P_{CO_2} - 47}{5} - P_{O_2} \text{ ABSORPTION}$$

In this formula P_B is barometric pressure, the value 47 is the vapor pressure of water, and P_{O_2} ABSORPTION is the oxygen pressure decrease caused by oxygen uptake into the blood.

Alveolar P_{O_2} at Different Altitudes. Table 44–1 also shows the P_{O_2}s in the alveoli at different altitudes when one is breathing air and when breathing pure oxygen. When breathing air, the alveolar P_{O_2} is 104 mm. Hg at sea level; it falls to approximately 67 mm. Hg at 10,000 feet and to only 1 mm. Hg at 50,000 feet.

Saturation of Hemoglobin with Oxygen at Different Altitudes. Figure 44–1 illustrates arterial oxygen saturation at different altitudes when breathing air and when breathing oxygen, and the actual per cent saturation at each 10,000 foot level is given in Table 44–1. Up to an altitude of approximately 10,000 feet, even

when air is breathed the arterial oxygen saturation remains at least as high as 90 per cent. However, above 10,000 feet the arterial oxygen saturation falls progressively, as illustrated by the left-hand curve of the figure, until it is only 70 per cent at 20,000 feet altitude and still less at higher altitudes.

EFFECT OF BREATHING PURE OXYGEN ON THE ALVEOLAR P_{O_2} AT DIFFERENT ALTITUDES

Referring once again to Table 44–1, note that when a person breathes air at 30,000 feet, the alveolar P_{O_2} is only 21 mm. Hg even though the barometric pressure is 226 mm. Hg. This difference is caused primarily by the fact that a considerable proportion of the alveolar air is nitrogen. But if the person breathes pure oxygen instead of air, most of the space in the alveoli formerly occupied by nitrogen now becomes occupied by oxygen instead. Theoretically, at 30,000 feet the aviator could have an alveolar P_{O_2} of 139 mm. Hg instead of the 21 mm. Hg that one has when breathing air.

The second curve of Figure 44–1 illustrates the arterial oxygen saturation at different altitudes when one is breathing pure oxygen. Note that the saturation remains above 90 per cent until the aviator ascends to approximately 39,000 feet; then it falls rapidly to approximately 50 per cent at about 47,000 feet.

THE "CEILING" WHEN BREATHING AIR AND WHEN BREATHING OXYGEN IN AN UNPRESSURIZED AIRPLANE

Comparing the two arterial oxygen saturation curves in Figure 44–1, one notes that an aviator breathing oxygen can ascend to far higher altitudes than one not breathing oxygen. For instance, the arterial saturation at 47,000 feet when breathing oxygen is about 50 per cent and is equivalent to the arterial oxygen saturation at 23,000 feet when breathing air. And, because a person ordinarily can remain conscious until the arterial oxygen saturation falls to 40 to 50 per cent, the ceiling for an unacclimatized aviator in an unpressurized airplane when breathing air is approximately 23,000 feet and when breathing pure oxygen about 47,000 feet, provided the oxygen-supplying equipment operates perfectly.

EFFECTS OF HYPOXIA

The rate of pulmonary ventilation ordinarily does not increase significantly until one has ascended to about 8000 feet. At this height the arterial oxygen saturation has fallen to approximately 93 per cent, at which level the chemoreceptors respond significantly. Above 8000 feet the chemoreceptor stimulatory mechanism progressively increases the ventilation until one reaches approximately 16,000 to 20,000 feet, at which altitude the ventilation has reached a maximum of approximately 65 per cent above normal if the person is exposed only acutely to the high altitude. (But several days of exposure increases ventilation about 300 per cent.)

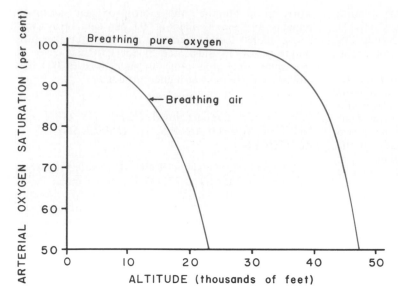

Figure 44–1. Effect of low atmospheric pressure on arterial oxygen saturation when breathing air and when breathing pure oxygen.

Other effects of hypoxia, beginning at an altitude of approximately 12,000 feet, are drowsiness, lassitude, mental fatigue, sometimes headache, occasionally nausea, and sometimes euphoria. Most of these symptoms increase in intensity at still higher altitudes, the headache often becoming especially prominent and the cerebral symptoms sometimes progressing to the stage of twitchings or convulsions and ending, above 23,000 feet in the unacclimatized person, in coma.

One of the most important effects of hypoxia is decreased mental proficiency, which decreases judgment, memory, and the performance of discrete motor movements. Ordinarily these abilities remain absolutely normal up to approximately 9000 feet, and may be completely normal for a short time up to elevations of 15,000 feet. But, if the aviator is exposed to hypoxia for a long time, mental proficiency, as measured by reaction times, handwriting, and psychological tests, may decrease to 80 per cent of normal even at altitudes as low as 11,000 feet. If an aviator stays at 15,000 feet for one hour, mental proficiency ordinarily will have fallen to approximately 50 per cent of normal, and after 18 hours at this level to approximately 20 per cent of normal.

Respiratory Center Depression as the Cause of Death in Hypoxia. When a person develops severe hypoxia — that is, to the stage of coma — the respiratory center itself often becomes depressed within another few minutes because of metabolic deficit in its neuronal cells. This counteracts the stimulatory effect of the chemoreceptor mechanism, and the respiration, instead of increasing further, decreases precipitously until it actually ceases.

Effects of Sudden Exposure to Low Po_2. At times the aviator who has been flying at a very high altitude with special oxygen equipment or in a pressurized cabin suddenly becomes detached from the oxygen supply, or the cabin decompresses. The alveolar Po_2 falls within a few seconds to a low value, but because oxygen is stored in the body fluids (combined with hemoglobin and with various oxygen carriers of the tissues), a short time elapses before the body suffers dire results from oxygen lack.

Figure 44–2 illustrates graphically the time that ordinarily elapses, first, before the aviator shows signs of diminished consciousness, and, second, before actual coma results. Note from this figure that at 38,000 feet diminished consciousness begins in approximately 30 seconds, and coma results in approximately one minute, while at 28,000 feet diminished consciousness begins in approximately one minute, and coma occurs in approximately three minutes.

ACCLIMATIZATION TO LOW Po_2

A person remaining at high altitudes for days, weeks, or years gradually becomes acclimatized to the low Po_2 so that it causes fewer and fewer deleterious effects on the body and also so that it becomes possible for the person to work harder or to ascend to still higher altitudes. The five principal means by which acclimatization comes about are: (1) increased pulmo-

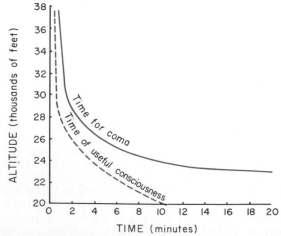

Figure 44–2. Time of exposure to low oxygen concentrations required to cause diminished consciousness or coma. (From Amstrong: Principles and Practice of Aviation Medicine. The Williams & Wilkins Co.)

nary ventilation, (2) increased hemoglobin in the blood, (3) increased diffusing capacity of the lungs, (4) increased vascularity of the tissues, and (5) increased ability of the cells to utilize oxygen despite the low P_{O_2}.

Increased Pulmonary Ventilation. On immediate exposure to low P_{O_2}, the hypoxic stimulation of the chemoreceptors increases alveolar ventilation to a maximum of about 65 per cent. This is an immediate compensation for the high altitude, and it alone allows the person to rise several thousand feet higher than would be possible without the increased ventilation. Then, if the person remains at a very high altitude for several days, the ventilation gradually increases to as much as 3 to 7 times normal. The basic cause of this gradual increase is the following:

The immediate 65 per cent increase in pulmonary ventilation on rising to a high altitude blows off large quantities of carbon dioxide, reducing the P_{CO_2} and increasing the pH of the body fluids. Both of these changes *inhibit* the respiratory center and thereby *oppose the stimulation by the hypoxia.* However, during the ensuing two to five days, this inhibition fades away, allowing the respiratory center now to respond with full force to the chemoreceptor stimuli resulting from hypoxia, and the ventilation increases to about 3 to 7 times normal. The cause of this fading inhibition is unknown, but there is some evidence that it results from reduced CSF bicarbonate ion and perhaps also reduced brain tissue bicarbonate ion. These changes, in turn, decrease the pH in the fluids surrounding the chemosensitive neurons of the respiratory center, thus increasing the activity of the center. (See Chapter 42.)

Rise in Hemoglobin During Acclimatization. It will be recalled from Chapter 5 that hypoxia is the principal stimulus for causing an increase in red blood cell production. Ordinarily, in full acclimatization to low oxygen the hematocrit rises from a normal value of 40 to 45 to an average of 60 to 65, with an average increase in hemoglobin concentration from a normal of 15 gm. per cent to about 22 gm. per cent.

In addition, the blood volume also increases, often by as much as 20 to 30 per cent, resulting in a total increase in circulating hemoglobin of as much as 50 to 90 per cent.

Unfortunately, this increase in hemoglobin and blood volume is a slow one, having almost no effect until after two to three weeks, reaching half development in a month or so, and becoming fully developed only after many months.

Decreased Affinity of the Hemoglobin for Oxygen Under Hypoxic Conditions. Within a few hours after the blood is first exposed to hypoxia at high altitudes, increased quantities of phosphate compounds are formed inside the red blood cells, and some of these combine with the hemoglobin to decrease its affinity for oxygen, that is, to shift the hemoglobin-oxygen saturation curve toward higher P_{O_2}s, as was illustrated in Chapter 41. One of these, 2,3-diphosphoglycerate, commonly called 2,3-DPG, is especially significant. Because of the reduced affinity for oxygen, the hemoglobin delivers the oxygen to the tissue cells at a higher P_{O_2}. At altitudes up to about 15,000 feet this effect increases oxygen delivery as much as 10 to 20 per cent. But at still higher altitudes the decreased

affinity for oxygen also decreases the pickup of oxygen by the hemoglobin in the lungs, and therefore it decreases the overall availability of oxygen to the tissues, which is more harmful than the increased tendency for oxygen release by the hemoglobin at the tissue level is helpful.

Increased Diffusing Capacity During Acclimatization. It will be recalled that the normal diffusing capacity for oxygen through the pulmonary membrane is approximately 21 ml. per mm. Hg pressure gradient per minute, and this diffusing capacity can increase as much as three-fold during exercise. A similar increase in diffusing capacity occurs at high altitude. Part of the increase probably results from greatly increased pulmonary capillary blood volume, which expands the capillaries and increases the surface through which oxygen can diffuse into the blood. Another part results from an increase in lung volume, which expands the surface area of the alveolar membrane. A final part results from an increase in pulmonary arterial pressure; this forces blood into greater numbers of alveolar capillaries than normally — especially in the upper parts of the lungs, which are poorly perfused under usual conditions.

The Circulatory System in Acclimatization — Increased Vascularity. The cardiac output often increases as much as 20 to 30 per cent immediately after a person ascends to a high altitude, but it usually falls back to normal within a few days. In the meantime, though, the blood flow through certain organs, such as the skin and kidneys, decreases, while the flow through the muscles, heart, brain, and other organs that normally require large quantities of oxygen increases. Furthermore, histological studies of animals that have been exposed to low oxygen levels for months or years show *greatly increased vascularity* (increased numbers and sizes of capillaries) of the hypoxic tissues. This helps to explain what happens to the 20 to 30 per cent increase in blood volume, and it means that the blood comes into much closer contact with the tissue cells than normally.

Cellular Acclimatization. In animals native to altitudes of 13,000 to 17,000 feet, mitochondria and certain cellular oxidative enzyme systems are more plentiful than in sea level inhabitants. Therefore, it is presumed that acclimatized human beings as well as these animals can utilize oxygen more effectively than can their sea level counterparts.

NATURAL ACCLIMATIZATION OF NATIVES LIVING AT HIGH ALTITUDES

Many natives in the Andes and in the Himalayas live at altitudes above 13,000 feet — one group in the Peruvian Andes actually lives at an altitude of 17,500 feet and works a mine at an altitude of 19,000 feet. Many of these natives are born at these altitudes and live there all their lives. In all the aspects of acclimatization listed above, the natives are superior to even the best-acclimatized lowlanders, even though the lowlanders might have also lived at high altitudes for ten or more years. This process of acclimatization of the natives begins in infancy. The chest size, especially, is greatly increased while the body size is

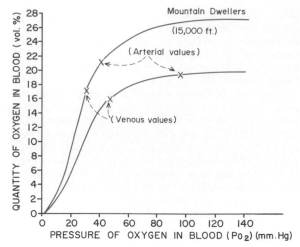

Figure 44–3. Oxygen-dissociation curves for blood of high-altitude and sea-level residents, showing the respective arterial and venous P_{O_2}s and oxygen contents as recorded in their native surroundings. (From "Oxygen-dissociation curves for bloods of high-altitude and sea-level residents." PAHO Scientific Publication No. 140, Life at High Altitudes, 1966.)

somewhat decreased, giving a high ratio of ventilatory capacity to body mass. In addition, their hearts, particularly the right heart which provides a high pulmonary arterial pressure to pump blood through a greatly expanded pulmonary capillary system, are considerably larger than the hearts of lowlanders.

The delivery of oxygen by the blood to the tissues is also highly facilitated in these natives. For instance, Figure 44–3 shows the hemoglobin-oxygen dissociation curves for natives who live at sea level and for their counterparts who live at 15,000 feet. Note that the arterial oxygen P_{O_2} in the natives at higher altitude is only 40 mm. Hg, but because of the greater quantity of hemoglobin the quantity of oxygen in the arterial blood is actually greater than in the blood of the natives at the lower altitude. Note also that the venous P_{O_2} in the high altitude natives is only 10 mm. Hg less than the venous P_{O_2} for the lowlanders, despite the low arterial P_{O_2}, indicating that oxygen transport to the tissues is exceedingly effective in the naturally acclimatized high altitude natives.

WORK CAPACITY AT HIGH ALTITUDES: THE EFFECT OF ACCLIMATIZATION

Aside from the mental depression caused by hypoxia, as discussed earlier, the work capacity of all the muscles is also greatly decreased in hypoxia. This includes not only the skeletal muscles but also the cardiac muscle so that even the maximum level of cardiac output is reduced. In general, the work capacity is reduced in direct proportion to the decrease in maximum rate of oxygen uptake that the body can achieve.

To give an idea of the importance of acclimatization for work capacity, consider this: The work capacities

in per cent of sea level maximum for a normal person at an altitude of 17,000 feet are the following:

	per cent
Unacclimatized	50
Acclimatized for two months	68
Native living at 13,200 feet but working at 17,000 feet	87

Thus, naturally acclimatized natives can achieve a daily work output even at these high altitudes almost equal to that of a normal person at sea level, but even well-acclimatized lowlanders can almost never achieve this result.

To emphasize further the importance of natural acclimatization of natives, the Sherpas of the Himalayas can survive without oxygen for hours at altitudes as high as Mt. Everest — over 29,000 feet!

CHRONIC MOUNTAIN SICKNESS

Occasionally, a person who remains at high altitude too long develops chronic mountain sickness, in which the following effects occur: (1) the red cell mass and hematocrit become exceptionally high, (2) the pulmonary arterial pressure becomes elevated even more than the normal elevation that occurs during acclimatization, (3) the right heart becomes greatly enlarged, (4) the peripheral arterial pressure begins to fall, (5) congestive failure ensues, and (6) death often follows unless the person is removed to a lower altitude. The cause of this sequence of events is almost certainly two-fold: First, the red cell mass becomes so great that the blood viscosity increases several-fold; this now actually *decreases* tissue blood flow to so low a level that oxygen delivery is seriously compromised. Second, the pulmonary arterioles become vasospastic because of the lung hypoxia. This results from the hypoxic vascular constrictor effect that normally operates to divert blood flow from unaerated to aerated alveoli as explained in Chapter 24. But in this instance all the arterioles become constricted, the pulmonary arterial pressure rises excessively, and the right heart fails. Fortunately, most of these persons recover promptly when they are moved to a lower altitude.

EFFECTS OF ACCELERATORY FORCES ON THE BODY IN AVIATION AND SPACE PHYSIOLOGY

Because of rapid changes in velocity and direction of motion in airplanes and space ships, several types of acceleratory forces often affect the body during flight. At the beginning of flight, simple linear acceleration occurs; at the end of flight, deceleration; and, every time the vehicle turns, angular and centrifugal acceleration. In aviation physiology it is usually centrifugal acceleration that demands greatest consideration because the structure of the airplane is capable of withstanding much greater centrifugal acceleration than is the human body.

CENTRIFUGAL ACCELERATORY FORCES

When an airplane makes a turn, the force of centrifugal acceleration is determined by the following relationship:

$$f = \frac{mv^2}{r}$$

in which f is the centrifugal acceleratory force, m is the mass of the object, v is the velocity of travel, and r is the radius of curvature of the turn. From this formula it is obvious that as the velocity increases, the force of centrifugal acceleration increases in proportion to the *square* of the velocity. It is also obvious that the force of acceleration is directly proportional to the sharpness of the turn (the less the radius).

Measurement of Acceleratory Force — "G." When a man is simply sitting in his seat, the force with which he is pressing against the seat results from the pull of gravity, and it is equal to his weight. The intensity of this force is said to be 1 "G" because it is equal to the pull of gravity. If the force with which he presses against the seat becomes 5 times his normal weight during pull-out from a dive, the force acting upon the seat is 5 G.

If the airplane goes through an outside loop so that the man is held down by his seat belt, *negative G* is applied to his body, and, if the force with which he is thrown against his belt is equal to the weight of his body, the negative force is — 1 G.

Effects of Centrifugal Acceleratory Force on the Body. *Effects on the Circulatory System.* The most important effect of centrifugal acceleration is on the circulatory system because blood is mobile and can be translocated by centrifugal forces. Centrifugal forces also tend to displace the tissues, but, because of their more solid structure, they only sag — ordinarily not enough to cause abnormal function.

When the aviator is subjected to *positive G* the blood is centrifuged toward the lower part of the body. Thus, if the centrifugal acceleratory force is 5 G and the person is in an immobilized standing position, the hydrostatic pressure in the veins of the feet is 5 times normal, or approximately 450 mm. Hg; even in the sitting position this pressure is nearly 300 mm. Hg. As the pressure in the vessels of the lower part of the body increases, the vessels passively dilate, and a major proportion of the blood from the upper part of the body is translocated into these lower vessels. Because the heart cannot pump unless blood returns to it, the greater the quantity of blood "pooled" in the lower body the less becomes the cardiac output.

Figure 44–4 illustrates the effect of different degrees of acceleration on systemic arterial pressure in an aviator in the sitting position, showing that when the acceleration rises to 4 G the systemic arterial pressure at the level of the heart falls to approximately 40 mm.

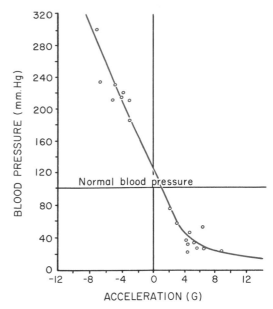

Figure 44–4. Effect of angular acceleratory forces on arterial pressure measured at the heart. (From Armstrong: Principles and Practice of Aviation Medicine. The Williams & Wilkins Co.)

Hg. As a result, positive G diminishes blood flow to the brain. Acceleration greater than 4 to 6 G ordinarily causes "blackout" of vision within a few seconds and unconsciousness shortly thereafter. The time required for these symptoms to appear is shown in Figure 44–5. The upper part of this figure shows that the arterial pressure recovers considerably after about 15 seconds of exposure to 3.3 G of acceleration; this

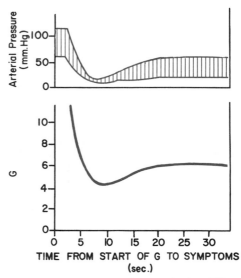

Figure 44–5. *Lower curve:* Time required at different G to cause blackout. *Top curve:* Changes in systolic and diastolic arterial pressures following abrupt and continuing exposure to an angular acceleratory force of 3.3 G. (Modified from Martin and Henry: *J. Aviation Med.*, 22:382, 1951.)

recovery is caused by activation of the baroreceptor reflexes.

Effects on the Vertebrae. Extremely high acceleratory forces for even a fraction of a second can fracture the vertebrae. The degree of positive acceleration that the average person can withstand in the sitting position before vertebral fracture occurs is approximately 20 G. This force also approaches the limit of safety for the structural elements of most airplanes. Therefore, any attempt to utilize such intense positive G as this in maneuvers would be extremely dangerous for reasons other than simply the effects on the body.

Transverse G. The human body can withstand tremendous transverse acceleratory forces (forces applied along the antero-posterior axis of the body — lying down in the airplane, for instance). If the transverse acceleratory forces are applied rather uniformly over large areas of the body, as much as 15 to 25 G can be withstood for many seconds without serious effects other than occasional collapse of a lung, which is not a lethal effect. Therefore, when very large acceleratory forces are to be involved, the aviator or astronaut flies in the semi-reclining or lying position.

Negative G. The effects of negative G on the body are less dramatic acutely but possibly more damaging permanently than the effects of positive G. An aviator can usually go through outside loops up to negative acceleratory forces of −4 to −5 G without harm except for intense momentary hyperemia of the head, though occasionally psychotic disturbances lasting for 15 to 20 minutes occur thereafter as a result of brain edema.

Figure 44–4 shows that the arterial pressure at the level of the heart increases greatly, as would be expected, and the pressure in the head obviously increases far more, to as high as 400 mm. Hg. An interesting but paradoxical effect is often caused by this very high pressure: it elicits such an extreme baroreceptor reflex that severe vagal slowing of the heart occurs, sometimes actually stopping the heart for 5 to 10 seconds.

If the arterial pressure becomes great enough, it can cause some of the small vessels on the surface of the head and in the brain to rupture. However, the vessels inside the cranium show less tendency for rupture than would be expected, for the cerebrospinal fluid is centrifuged toward the head at the same time that blood is centrifuged toward the cranial vessels, and the greatly increased pressure of this fluid acts as a cushioning buffer on the outside of the brain to prevent vascular rupture. Even so, animals exposed to negative acceleratory forces of 20 to 40 G have developed subarachnoid hemorrhages, and undoubtedly such effects could occur in the human being.

Because the eyes are not protected by the cranium, intense hyperemia occurs in them during negative G. As a result, the eyes often become temporarily blinded with "redout."

Protection of the Body Against Centrifugal Acceleratory Forces. Specific procedures and apparatus have been developed to protect aviators against the circulatory collapse that occurs during positive G. First, if the aviator tightens the abdominal muscles to an extreme degree and leans forward to compress the abdomen, some of the pooling of blood in the large vessels

of the abdomen can be prevented, thereby delaying the onset of blackout. Also, special "anti-G" suits have been devised to prevent pooling of blood in the lower abdomen and legs. The simplest of these applies positive pressure to the legs and abdomen by inflating compression bags as the G increases. Theoretically, a pilot submerged in a tank or suit of water could withstand very high degrees of acceleration, both positive and negative, for the pressures developed in the water and pressing on the outside of the body during centrifugal acceleration almost exactly balance the forces acting on the body. Unfortunately, however, the presence of air in the lungs still allows displacement of the heart, the lung tissues, and the diaphragm into seriously abnormal positions despite submersion in water. Therefore, even when this procedure is used, the limits of safety are about 15 to 20 G.

EFFECTS OF LINEAR ACCELERATORY FORCES ON THE BODY

Acceleratory Forces in Space Travel. Unlike an airplane, a spacecraft cannot make rapid turns; therefore, centrifugal acceleration is of little importance except when the spacecraft goes into abnormal gyrations. On the other hand, blast-off acceleration and landing deceleration might be tremendous; both of these are types of linear acceleration.

Figure 44–6 illustrates a typical profile of the acceleration during blast-off in a three-stage spacecraft, showing that the first stage booster causes acceleration as high as 9 G and the second stage booster, as high as 8G. In the standing position the human body could not withstand this much acceleration, but in a semi-reclining position *transverse to the axis of acceleration,* this amount of acceleration can be withstood with ease despite the fact that the acceleratory forces continue for as long as five minutes at a time. Therefore, we see the reason for the reclining seats used by the astronauts.

Problems also occur during deceleration when the spacecraft re-enters the atmosphere. A person travel-

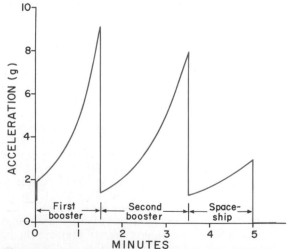

Figure 44–6. Acceleratory forces during the take-off of a spacecraft.

ing at Mach 1 (the speed of sound and of fast airplanes) can be safely decelerated in a distance of approximately 0.12 mile, whereas a person traveling at a speed of Mach 100 (a speed possible in interplanetary space travel) requires a distance of about 10,000 miles for safe deceleration. The principal reason for this difference is that the total amount of energy that must be dispelled during deceleration is proportional to the *square* of the velocity, which alone increases the distance 10,000-fold. But, in addition to this, a human being can withstand far less deceleration if it lasts for a long time than for a short time. Therefore, deceleration must be accomplished much more slowly from the very high velocities than is necessary at the slower velocities.

Deceleratory Forces Associated with Parachute Jumps. When the parachuting aviator leaves the airplane, the velocity of fall is exactly 0 feet per second at first. However, because of the acceleratory force of gravity, within 1 second the velocity of fall is 32 feet per second (if there is no air resistance); in two seconds it is 64 feet per second; and so on. However, as the rate of fall increases, the air resistance tending to slow the fall also increases. Finally, the deceleratory force of the air resistance is equal to the acceleratory force of gravity so that after falling for approximately 12 seconds and a distance of 1400 feet, the person will be falling at a "terminal velocity" of 109 to 119 miles per hour (175 feet per second). However, if the pilot jumps from a very high altitude, where the atmosphere offers little resistance, the terminal velocity is much greater than 175 feet per second, but it slows to 175 feet per second as the body reaches the higher-density atmosphere close to earth.

If the parachutist has already reached the terminal velocity of fall before opening the parachute, an "opening shock load" of approximately 1200 pounds occurs on the parachute shrouds.

The usual-sized parachute slows the fall of the parachutist to approximately one-ninth the terminal velocity. In other words, the speed of landing is approximately 20 feet per second, and the force of impact against the earth is approximately 1/81 the impact force without a parachute. Even so, the force of impact is still great enough to cause considerable damage to the body unless the parachutist is properly trained in landing, because it is not evident as one descends how rapidly the earth is approaching. Actually, the force of impact with the earth is approximately the same as that which would be experienced by jumping from a height of about 7 feet. Unless forewarned, the parachutist will be tricked by the senses into striking the earth with extended legs, and this will result in tremendous deceleratory forces along the skeletal axis of the body, resulting in fracture of the pelvis, of vertebrae, or a leg. Consequently, the trained parachutist strikes the earth with knees bent, but muscles taut, to cushion the shock of landing.

PERCEPTIONS OF EQUILIBRIUM AND TURNING IN BLIND FLYING

The various sensations of equilibrium will be discussed in Chapter 52, but, in essence, whenever the head is not positioned along the vertical axis of forces applied to the body, the otoliths in the utricles of the labyrinths apprise the psyche of a lack of equilibrium. However, the degree of proficiency of equilibrium perceptions has certain limits. For instance, the body must lean forward as much as 5 to 10 degrees before the utricles sense this forward leaning. An aviator not flying blind can perceive forward leaning of the airplane by observing the ground, but when flying blind, it is necessary that appropriate instruments be available to indicate the rate of descent. Also, the utricle may fail to apprise the pilot of ascent up to an error of as much as 24 degrees. Obviously, then, blind flying often results in stall of the airplane unless appropriate instruments are available.

Perhaps the least effective of the perceptive organs in blind flying are those for turning, because the organs that perceive turning — the semicircular canals — are excited only for the first few seconds after a turn begins and only for the first few seconds after the turn ends. Consequently, an aviator might perceive beginning to turn, but once in the turn, all perception of continuing the turn may be lost. Furthermore, if one goes into the turn slowly — at an angular acceleratory rate less than 2 degrees per second^2 — the fact of entering the turn may not even be perceived. Consequently, of especial importance among the instruments for blind flying are those that apprise the aviator of the direction of travel of the airplane and the rate of turn.

PROBLEMS OF TEMPERATURE IN AVIATION AND SPACE PHYSIOLOGY

The cold temperatures of the upper atmosphere involve essentially the same physiological problems as cold temperatures on the surface of the earth; these problems will be discussed in Chapter 72. Therefore, for the present it shall only be pointed out how the temperatures change as one ascends to higher elevations. If the temperature at the earth's surface is approximately 20° C., temperatures at different altitudes are approximately the following:

feet	° C.
0	20
10,000	0
20,000	−22
30,000	−44
40,000	−55

It is obvious from these values that special clothing or special heating apparatus must be designed for flying at high altitudes.

Temperature in Space. Though the temperature of the air falls to about −55° C. several miles above the earth, the temperature rises to very high values in space. In the ionosphere several hundred miles above the earth and in space, the kinetic energy of the few molecules, atoms, and ions that are there is extreme. The reason for this is that any particles that can escape this far away from the gravitational pull of the celestial bodies must have a great velocity. As a result, by the time the spacecraft has reached an altitude above 350 miles the temperature of the surrounding particles is about 3000° C. Yet, strangely enough, this has almost no effect on the temperature of the spacecraft because of the *sparsity* of these particles — that is, they are far too few to impart any significant amount of heat to the spacecraft. Instead, the temperature of the spacecraft is determined by the relative *absorption of radiant energy* from the sun versus the *re-radiation of energy* away from the spacecraft. Different coatings for the spacecraft have different absorptive and radiation characteristics. Therefore, with appropriate coating and orientation of the spacecraft appropriately with respect to the sun, the temperature inside can be made almost any desired value.

RADIATION AT HIGH ALTITUDES AND IN SPACE

Radiation in the Atmosphere. The electromagnetic radiations change considerably between the surface of the earth and at high altitudes because the atmosphere filters some of the radiations from the sun's rays before they reach the earth's surface. Ordinarily most of the ultraviolet light is absorbed before it reaches the earth's surface. Consequently, a person exposed to the sun in the upper atmosphere is many times as likely to develop sunburn as on the earth's surface.

Approximately 18 per cent of the visible light is absorbed by the atmosphere before it reaches the surface of the earth even on a perfectly clear day. Therefore, the brightness of the sun is 1.2 times as great in the upper atmosphere as on earth. This makes the earth less distinct to the aviator than otherwise for two reasons: first, he is looking from an area of greater brightness toward an area of lesser brightness, and, second, light is reflected from the atmosphere into his eyes. This reflected light especially blocks his vision of the horizon.

In aviation there is no significant hazard from gamma and x-rays, but this does become a problem in space physiology.

Radiation Hazards in Space Physiology. Large quantities of cosmic particles are continually bombarding the earth's upper atmosphere, some originating from the sun and some from outer space. The magnetic field of the earth traps many of these cosmic particles in two major belts around the earth called *Van Allen radiation belts.* The inner belt begins at an altitude of about 300 miles and extends to about 3000 miles. The outer belt begins at about 6000 miles and extends to 20,000 miles. However, because of the magnetic field of the earth, these belts occur mainly in the tropical and temperate zones around the earth and not at the two poles.

The types of radiation in the two Van Allen belts are almost entirely high energy electrons and protons. Even with very good shielding, a person traversing these two belts during an interplanetary space trip would be expected to receive as much as 10 roentgens of radiation, which is about one-fortieth the lethal dose; and a person in a spacecraft orbiting the earth within one of these two belts could receive enough radiation in only a few hours to cause death.

Thus, it is important to orbit spacecraft below an altitude of 200 to 300 miles, an altitude at which the radiation hazard is slight. Also, it is possible to minimize the radiation hazard during interplanetary space travel by leaving the earth or returning to earth near one of the poles rather than near the equator.

"ARTIFICIAL CLIMATE" IN THE SEALED SPACECRAFT

Since there is no atmosphere in outer space, an atmosphere and conditions of climate must be provided artificially. The ability of a person to survive in this artificial climate depends entirely on appropriate engineering design.

Most important of all, the oxygen concentration must remain high enough and the carbon dioxide concentration low enough. In some of the space missions a capsule atmosphere containing pure oxygen at about 260 mm. Hg pressure has been used. In others, normal air at 760 mm. Hg pressure has been used. The presence of nitrogen in the mixture greatly diminishes the likelihood of fire and explosion.

For space travel lasting more than several months, it will be impractical to carry along an adequate oxygen supply and enough carbon dioxide absorbent. For this reason, "recycling techniques" have been developed for use over and over again of the same oxygen. These techniques also frequently include re-use of the same food and water. Basically, they involve (1) a method for removing oxygen from carbon dioxide, (2) a method for removing water from the human excreta, and (3) use of the human excreta for resynthesizing or regrowing an adequate food supply.

Large amounts of energy are required for these processes, and the real problem at present is to derive enough energy from the sun's radiation to energize the necessary chemical reactions. Some recycling processes depend on purely physical procedures, such as distillation, electrolysis of water, and capture of the sun's energy by solar batteries, while others depend on biological methods, such as use of algae with its large store of chlorophyll, to generate foodstuffs by photosynthesis. Unfortunately, a completely practical system for recycling is yet to be achieved. The problem is the weight of the equipment that must be carried.

WEIGHTLESSNESS IN SPACE

A person in an orbiting satellite or in any nonpropelled spacecraft experiences weightlessness. That is, the person is not drawn toward the bottom, sides, or top of the spacecraft but simply floats inside its chambers. The cause of this is not failure of gravity to pull on the body, because gravity from any nearby heavenly body is still active. However, the gravity acts on both the spacecraft and the person at the same time, and since there is no resistance to movement in space, both are pulled with exactly the same acceleratory forces and in the same direction. For this reason, the person simply is not attracted toward any wall of the spacecraft.

Weightlessness is mainly an engineering problem — that is, to provide special techniques for eating and drinking (since food and water will not stay on open plates or in glasses), special waste disposal systems, and adequate hand holds or other means for stabilizing the person in the spacecraft for adequately controlling the operation of the ship.

Physiological Problems of Weightlessness. Fortunately, the physiological problems of weightlessness have not proved to be severe. Most of the problems that do occur appear to be related to two effects of the weightlessness: (1) translocation of fluids within the body because of failure of gravity to cause hydrostatic pressures, and (2) diminishment of physical activity because no strength of muscle contraction is required to oppose the force of gravity.

The observed effects of prolonged stay in space are the following: (1) decrease in blood volume, (2) decrease in red cell mass, (3) decreased work capacity, (4) decrease in maximum cardiac output, and (5) loss of calcium from the bones. Essentially these same effects also occur in persons lying in bed for an extended period of time. For this reason an extensive exercise program was carried out during the most recent sojourn by three astronauts in the Space Laboratory, and all of the above effects were greatly reduced. In previous Space Lab expeditions, in which the exercise program had been less vigorous, the astronauts had considerably lower work capacities for the first few days after returning to earth. They also had a tendency to faint when they stood up during the first day or so after return to gravity because of translocation of their diminished blood volume toward their abdomens and legs.

The effects of weightlessness that do occur usually reach their maximum within the first few weeks after the astronaut enters the space environment and fortunately are not progressive thereafter. Therefore, it appears that with an appropriate exercise program the physiological effects of weightlessness will not be a serious problem even during prolonged space voyages.

REFERENCES

Adey, W. R.: The physiology of weightlessness. *Physiologist, 16*:178, 1974.

Andrews, H. L.: Radiation Biophysics, 2nd Ed. Englewood Cliffs, N. J., Prentice-Hall, 1974.

Armstrong, H. G.: Aerospace Medicine. Baltimore, Williams & Wilkins, 1961.

Brown, J. H. U.: Physiology of Man in Space. New York, Academic Press, 1963.

Bullard, R. W.: Physiological problems of space travel. *Annu. Rev. Physiol., 34*:205, 1972.

Burton, R. R. *et al.:* Man at high sustained +Gz acceleration: A review. *Aerospace Med., 45*:1115, 1974.

Cardiovascular Problems Associated with Aviation Safety. Springfield, Va., National Technical Information Service, 1976.

Ciba Foundation Symposia: High Altitude Physiology. New York, Churchill Livingstone, 1971.

Dill, D. B., *et al.:* Hemoconcentration at altitude. *J. Appl. Physiol., 27*:514, 1969.

Frisancho, A. R.: Functional adaptation to high altitude hypoxia. *Science, 187*:313, 1975.

Grover, R. F., *et al.:* High-altitude pulmonary edema. *In* Fishman, A. P., and Renkin, E. M. (eds.): Pulmonary Edema. Baltimore, Waverly Press, 1979, p. 229.

Hempleman, H. V., and Lockwood, A. P. M.: The Physiology of Diving in Man and Other Animals. London, Edward Arnold, 1978.

Hochachka, P. W.: Living Without Oxygen: Closed and Open Systems of Hypoxia Tolerance. Cambridge, Mass., Harvard University Press, 1980.

Hock, R. J.: The physiology of high altitude. *Sci. Am., 222*:52, 1970.

Kellogg, R. H.: Altitude acclimatization, a historical introduction emphasizing the regulation of breathing. *Physiologist, 11*:37, 1968.

Korner, P. I.: Circulatory adaptations in hypoxia. *Physiol. Rev., 39*:687, 1959.

Lahiri, S.: Physiological responses and adaptations to high altitude. *In* MTP International Review of Science: Physiology. Vol. 7. Baltimore, University Park Press, 1974, p. 271.

Lahiri, S.: Physiological responses and adaptations to high altitude. *Int. Rev. Physiol., 15*:217, 1977.

Lee, D. H. K. (ed.): Physiology, Environment, and Man, New York, Academic Press, 1970.

McCally, M.: Hypodynamics and Hypogravics: The Physiology of Inactivity and Weightlessness. New York, Academic Press, 1969.

Pace, N.: Respiration at high altitude. *Fed. Proc., 33*:2126, 1974.

Painter, R. B.: The action of ultraviolet light on mammalian cells. *Photophysiology, 5*:169, 1970.

Reeves, J. T., *et al.:* Physiological effects of high altitude on the pulmonary circulation. *In* Robertshaw, D. (ed.): International Review of Physiology: Environmental Physiology III. Vol. 20. Baltimore, University Park Press, 1979, p. 289.

Sandler, H., and Winter, D. L.: Physiological Responses of Women to Simulated Weightlessness: A Review of the Significant Findings of the First Female Bed-Rest Study. Springfield, Va., National Technical Information Service, 1978.

Sloan, A. W.: Man in Extreme Environments. Springfield, Ill., Charles C Thomas, 1979.

Smith, E. E., and Crowell, J. W.: Influence of hematocrit ratio on survival of unacclimatized dogs at simulated high altitude. *Am. J. Physiol., 205*:1172, 1963.

Smith, E. E., and Crowell, J. W.: Role of the hematocrit in altitude acclimatization. *Aerospace Med., 38*:39, 1966.

Stickney, J. C.: Some problems of homeostasis in high-altitude exposure. *Physiologist, 15*:349, 1972.

45

Physiology of Deep Sea Diving and Other High-Pressure Operations

When a person descends beneath the sea, the pressure around him increases tremendously. To keep his lungs from collapsing, air must be supplied also under high pressure, which exposes the blood in his lungs to extremely high alveolar gas pressures. Beyond certain limits these high pressures can cause tremendous alterations in the physiology of the body, which explains the necessity for the present discussion.

Also exposed to high atmospheric pressures are caisson workers who, in digging tunnels beneath rivers or elsewhere, often must work in a pressurized area to keep the tunnel from caving in. Here again, the same problems of excessively high gas pressures in the alveoli occur.

Before explaining the effects of high alveolar gas pressures on the body, it is necessary to review some physical principles of pressure and volume changes at different depths beneath the sea.

Relationship of Sea Depth to Pressure. A column of fresh water 34 feet high (sea water, 33 feet) exerts the same pressure at its bottom as all the atmosphere above the earth. Therefore, a person 33 feet beneath the ocean surface is exposed to a pressure of 2 atmospheres, 1 atmosphere of pressure caused by the air above the water and the second atmosphere by the weight of the water itself. At 66 feet the pressure is 3 atmospheres, and so forth, in accord with the table in Figure 45–1.

Effect of Depth on the Volume of Gases. Another important effect of depth is the compression of gases to smaller and smaller volumes. Figure 45–1 also illustrates a bell jar at sea level containing 1 liter of air. At 33 feet beneath the sea where the pressure is 2 atmospheres, the volume has been compressed to only one-half liter. At 100 feet, where the pressure is 4 atmospheres, the volume has been compressed to one-fourth liter, and at 8 atmospheres (233 feet) to one-eighth liter. This is an extremely important effect in diving because it can cause the air chambers of the diver's body, including the lungs, to become so small in some instances that serious damage results, as is discussed later in the chapter.

Many times in this chapter it is necessary to refer to *actual volume* versus *sea level volume*. For instance, we

might speak of an actual volume of 1 liter at a depth of 300 feet; this is the same quantity of air as a sea level volume of 10 liters.

EFFECT OF HIGH PARTIAL PRESSURES OF GASES ON THE BODY

The three gases to which a diver breathing air is normally exposed are nitrogen, oxygen, and carbon dioxide. However, helium is often substituted for nitrogen in the diving mixture; therefore, the effects of this gas under high pressure must also be considered.

Nitrogen Narcosis at High Nitrogen Pressures. Approximately four-fifths of the air is nitrogen. At sea level pressure this has no known effect on bodily function, but at high pressures it can cause varying degrees of narcosis. When the diver remains beneath the sea for an hour or more and is breathing compressed air, the depth at which the first symptoms of mild narcosis appear is approximately 120 feet, at which level he begins to exhibit joviality and to lose many of his cares. At 150 to 200 feet, he becomes drowsy. At 200 to 250 feet, his strength wanes considerably, and he often becomes too clumsy to perform the work required of him. Beyond 300 feet (10 atmospheres pressure), the diver usually becomes almost useless as a result of nitrogen narcosis.

Nitrogen narcosis has characteristics very similar to those of alcohol intoxication, and for this reason it has frequently been called "raptures of the depths."

The mechanism of the narcotic effect is believed to be the same as that of essentially all the gas anesthetics. That is, nitrogen dissolves freely in the fats of the body, and it is presumed that it, like most other anesthetic gases, dissolves in the membranes or other lipid structures of the neurons and because of its *physical* effect on altering electrical charge transfer reduces their excitability.

Oxygen Toxicity at High Pressures. *Effect of Extreme-*

Depth (feet)	Atmosphere(s)
Sea level	1
33	2
66	3
100	4
133	5
166	6
200	7
300	10
400	13
500	16

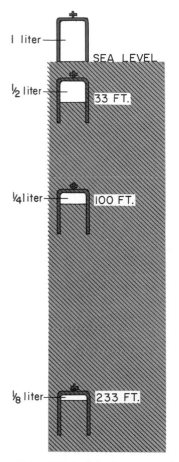

Figure 45–1. Effect of depth on gas volumes.

ly High $P_{O_2}s$ *on Blood Oxygen Transport.* When the P_{O_2} in the blood rises far above 100 mm. Hg, the amount of oxygen dissolved in the water of the blood increases markedly. This is illustrated in Figure 45–2 which depicts the same oxygen hemoglobin dissociation curve as that shown in Chapter 41 except that the alveolar P_{O_2} is now extended to over 3000 mm. Hg. Also depicted is the volume of oxygen dissolved in the fluid of the blood at each P_{O_2} level. Note that in the normal range of alveolar P_{O_2}, almost none of the total oxygen in the blood is accounted for by dissolved oxygen, but as the pressure rises progressively into the thousands of mm. Hg, a large portion of the total oxygen is then dissolved rather than bound with hemoglobin.

Effect of High Alveolar P_{O_2} *on Tissue* P_{O_2}. Let us

assume that the P_{O_2} in the lungs is about 3000 mm. Hg (4 atmospheres pressure). Referring to Figure 45–2, one finds that this represents a total oxygen content in each 100 ml. of blood of about 29 volumes per cent, as illustrated by Point A in the figure. As this blood passes through the tissue capillaries and the tissues utilize their normal amount of oxygen, about 5 ml. from each 100 ml. of blood, the oxygen content on leaving the tissue capillaries is still 24 volumes per cent (Point B in the figure). At this point the P_{O_2} is still about 1200 mm. Hg, which means that oxygen is delivered to the tissues at this extremely high pressure instead of at the normal value of 40 mm. Hg. Thus, once the alveolar P_{O_2} rises above a critical level, the hemoglobin-oxygen buffer mechanism is no longer capable of keeping the tissue P_{O_2} in the normal safe range between 20 and 60 mm. Hg.

Acute Oxygen Poisoning. Because of the extremely high tissue P_{O_2} that occurs when breathing oxygen at a very high alveolar oxygen pressure, one can readily understand that this can be very detrimental to many of the body's tissues. This is especially true of the brain. In fact, exposure to 3 atmospheres pressure of oxygen (P_{O_2} = 2280 mm. Hg) will cause convulsions followed by coma in most persons after about one hour. The convulsions often occur without any warning and, for obvious reasons, are likely to be lethal to divers submerged beneath the sea.

Other symptoms encountered in acute oxygen poisoning include *nausea, muscle twitchings, dizziness, disturbances of vision,* and *irritability.*

Exercise greatly increases a diver's susceptibility to

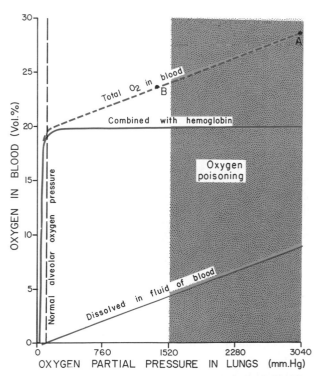

Figure 45–2. Quantity of oxygen dissolved in the fluid of the blood and in combination with hemoglobin at very high $P_{O_2}s$.

oxygen toxicity, causing symptoms to appear much earlier and with far greater severity than in the resting person.

The cause or causes of acute oxygen toxicity are yet unknown, but some of the experimental data and suggested causes are: (1) Following severe oxygen toxicity the concentrations of some of the oxidative enzymes in the tissues are considerably reduced. Therefore, it has been postulated that excess oxygen inactivates some oxidative enzymes and causes toxicity by decreasing the ability of the tissues to form high energy phosphate bonds. (2) Blood flow through the brain decreases 25 to 50 per cent when a person breathes high concentrations of oxygen. This presumably results from the local tissue blood flow control mechanism explained in Chapter 20. That is, the greater the amount of oxygen available in the tissues, the greater is the degree of constriction of the blood vessels, which normally helps to regulate the amount of oxygen delivered to the tissues. Yet, the decrease in blood flow can decrease the availability of other nutrients required by the cerebral tissues or can decrease the removal of cellular waste products such as carbon dioxide and nitrogenous end-products. Therefore, it has been postulated that either lack of certain nutrients or buildup of metabolic end-products can cause the convulsions of oxygen toxicity. (3) Still other experiments have shown that excess oxygen in the tissues can cause the development of large concentrations of oxidizing free radicals that could cause abnormal oxidative destruction of many essential elements of the cells, thereby damaging the cells' metabolic systems.

Chronic Oxygen Poisoning as a Cause of Pulmonary Disability. A person can be exposed to 100 per cent oxygen at normal atmospheric pressure almost indefinitely without developing the *acute* oxygen toxicity described above. However, after 12 hours or so of this exposure, *lung passageway congestion and edema* caused by damage to the linings of the bronchi and alveoli begin to develop. Here again, it appears that these local tissues of the lungs are damaged as a result of oxidative destruction of some of the essential elements of the tissues. The reason this effect occurs in the lungs and not in the other tissues at these lower alveolar Po_2s is that the lungs are directly exposed to the high oxygen pressure, while oxygen is delivered to the other tissues at essentially normal Po_2 because of the hemoglobin-oxygen buffer system described in Chapter 41. When the Po_2 in the air rises above about 1500 mm. Hg, this buffer system fails, which then allows the Po_2 of all tissues to rise and to cause the acute oxygen poisoning described in the previous section.

Carbon Dioxide Toxicity at Great Depths. If the diving gear is properly designed and also functions properly, the diver has no problem from carbon dioxide toxicity, for depth alone does not increase the carbon dioxide partial pressure in the alveoli. This is true because depth does not increase the rate of CO_2 production in the body, and as long as the diver continues to breathe a normal tidal volume, he continues to expire the carbon dioxide as it is formed, maintaining his alveolar carbon dioxide partial pressure at a normal value.

Unfortunately, though, in certain types of diving gear, such as the diving helmet and the different types of rebreathing apparatuses, carbon dioxide can frequently build up in the dead space air of the apparatus and be rebreathed by the diver. Up to a carbon dioxide pressure (Pco_2) of about 80 mm. Hg, 2 times that of normal alveoli, the diver barely tolerates this buildup, his minute respiratory volume increasing up to a maximum of 6- to 10-fold to compensate for the increased carbon dioxide. However, beyond the 80 mm. Hg level the situation becomes intolerable, and eventually the respiratory center begins to be depressed rather than excited; the diver's respiration then begins to fail rather than to compensate. In addition, the diver develops severe respiratory acidosis, and varying degrees of lethargy, narcosis, and finally anesthesia ensue, as was discussed in Chapter 42.

Effects of Helium at High Pressures. In very deep dives, helium is used in place of nitrogen because it has only one-fourth to one-fifth the narcotic effect of nitrogen. Almost no effects of helium are observed until a person descends below about 500 feet. Beyond this level divers begin to experience a so-called "high pressure nervous syndrome" which is characterized by tremor at about 700 feet, drowsiness from 1000 feet onward, and lack of coordination at still deeper levels.

Aside from the much smaller narcotic effect of helium than nitrogen on the nervous system, three other properties also make it desirable in the diving gas mixture under some conditions:

(1) Because of its low atomic weight, its density is slight, which reduces the airway resistance of the diver. (2) Also because of its low atomic weight, helium diffuses through the tissues much more rapidly than nitrogen, which allows more rapid removal of helium than nitrogen from the body fluids under some conditions. (3) Helium is less soluble in the body fluids than nitrogen, which reduces the quantity of bubbles that can form in the tissues when the diver is decompressed after a prolonged dive.

DECOMPRESSION OF THE DIVER AFTER EXPOSURE TO HIGH PRESSURES

When a person breathes air under high pressure for a long time, the amount of nitrogen dissolved in the body fluids becomes great. The reason for this is the following: The blood flowing through the pulmonary capillaries becomes saturated with nitrogen to the same pressure as that in the breathing mixture. Over several hours, enough nitrogen is carried to all the tissues of the body to saturate them also with dissolved nitrogen. And, since nitrogen is not metabolized by the body, it remains dissolved until the nitrogen pressure in the lungs decreases, at which time the nitrogen is then removed by the reverse respiratory process.

Volume of Nitrogen Dissolved in the Body Fluids at Different Depths. At sea level almost 1 liter of nitrogen is dissolved in the entire body. A little less than half of this is dissolved in the water of the body and a little more than half in the fat of the body. This is true despite the fact that fat constitutes only 15 per cent of the normal body, and it is explained by the fact that nitrogen is 5 times as soluble in fat as in water.

After the diver has become totally saturated with nitrogen the *sea level volume of nitrogen* dissolved in the body fluids at the different depths is:

feet	liters
33	2
100	4
200	7
300	10

However, several hours are required for the gas pressures of nitrogen in all the body tissues to come to equilibrium with the gas pressure of nitrogen in the alveoli simply because the blood does not flow rapidly enough and the nitrogen does not diffuse rapidly enough to cause instantaneous equilibrium. The nitrogen dissolved in the water of the body comes to almost complete saturation in about one hour, but the fat, requiring much more nitrogen for saturation and also having a relatively poor blood supply, reaches saturation only after several hours. For this reason, if a person remains at deep levels for only a few minutes not much nitrogen dissolves in his fluids and tissues, while if he remains at a deep level for several hours his fluids and tissues become almost completely saturated with nitrogen.

Decompression Sickness (Synonyms: Compressed Air Sickness, Bends, Caisson Disease, Diver's Paralysis, Dysbarism). If a diver has been beneath the sea long enough so that large amounts of nitrogen have dissolved in his body, and then he suddenly comes back to the surface of the sea, significant quantities of nitrogen bubbles can develop in his body fluids either intracellularly or extracellularly, and these can cause minor or serious damage in almost any area of the body, depending on the amount of bubbles formed; this is decompression sickness.

The principles underlying bubble formation are shown in Figure 45–3. To the left, the diver's tissues have become equilibrated to a very high nitrogen pressure. However, as long as the diver remains deep beneath the sea, the pressure against the outside of his body (5000 mm. Hg) compresses all the body tissues sufficiently to keep the dissolved gases in solution. Then, when the diver suddenly rises to sea level, the pressure on the outside of his body becomes only 1 atmosphere (760 mm. Hg), while the pressure inside the body fluids is the sum of the pressures of water vapor, carbon dioxide, oxygen, and nitrogen, or a total of 4065 mm. Hg, which is far greater than the pressure on the outside of the body. Therefore, the gases can escape from the dissolved state and form actual bubbles inside the tissues. However, the bubbles may not appear for many minutes to hours because the gases can remain dissolved in the "supersaturated" state sometimes for hours before bubbling.

Exercise hastens the formation of bubbles during decompression because of increased agitation of the tissues and fluids. This is an effect analogous to that of shaking an opened bottle of soda pop to release the bubbles.

Symptoms of Decompression Sickness. In persons who have developed decompression sickness, symptoms have occurred with the following frequencies:

	per cent
Local pain in the legs or arms	89
Dizziness	5.3
Paralysis	2.3
Shortness of breath ("the chokes")	1.6
Extreme fatigue and pain	1.3
Collapse with unconsciousness	0.5

From this list it can be seen that the most serious problems are usually related to bubble formation in the nervous system. Bubbles sometimes actually disrupt important pathways in the brain or spinal cord; the bubbles in the peripheral nerves can cause severe pain. Unfortunately, formation of large bubbles in the central nervous system occasionally even leads to permanent paralysis or permanent mental disturbances.

But the nervous system is not the only locus of damage in decompression sickness, for bubbles can also form in the blood and become caught in the capillaries of the lungs; these bubbles block pulmonary blood flow and cause "the chokes," characterized by serious shortness of breath. This is often followed by severe pulmonary edema, which further aggravates the condition and can cause death.

The symptoms of decompression sickness usually appear within a few minutes to an hour after sudden decompression. However, occasional symptoms of decompression sickness develop as long as six or more hours after decompression.

Rate of Nitrogen Elimination from the Body. Decompression Tables. Fortunately, if a diver is brought to the surface slowly, the dissolved nitrogen is eliminated through his lungs rapidly enough to prevent decompression sickness. Approximately two-thirds of the total nitrogen is liberated in one hour and about 90 per cent in six hours. However, some excess nitrogen is still present for many hours, especially in the fatty tissues such as the brain, and the diver is not completely safe for as long as 9 to 12 hours. Therefore, a diver must be decompressed over a period of many hours if he has been deep beneath the sea for a long time. This decom-

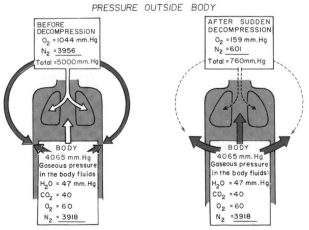

Figure 45–3. Gaseous pressures both inside and outside the body, showing at right the great excess of intrabody pressure that is responsible for bubble formation in the body tissues.

pression is achieved by bringing the diver to the surface slowly, over a period of many minutes to hours.

The rate at which a diver can be brought to the surface depends on, first the *depth* to which he has descended and, second, the *amount of time* he has been there. If he remains at deep levels for only a short time, the body fluids do not become saturated, and, therefore, the decompression time can be reduced accordingly.

Special decompression tables have been prepared by the U.S. Navy that detail procedures for safe decompression. To give the student an idea of the decompression process, a diver who has been breathing air and has been on the bottom for 60 minutes at a depth of 190 feet is decompressed according to the following schedule:

10 minutes at 50 feet depth
17 minutes at 40 feet depth
19 minutes at 30 feet depth
50 minutes at 20 feet depth
84 minutes at 10 feet depth

Thus, for a work period on the bottom of only one hour, the total time for decompression is about three hours.

Decompression in a Tank and Treatment of Decompression Sickness. Another procedure for decompression, used especially in heavily polluted waters and when climatic situations require it, involves bringing the diver to the surface immediately and then placing him in a decompression tank within five minutes after arriving at the surface. Pressure is reapplied, and an appropriate decompression table that prevents bubble formation is used.

A person who begins to develop symptoms of decompression sickness can also be treated by being placed in such a decompression tank for a long time, several times as long as the usual decompression times, and allowing the nitrogen to be released from the body slowly.

Use of Helium-Oxygen Mixtures in Deep Dives. In deep dives helium has advantages over nitrogen, including (1) decreased decompression time, (2) lack of narcotic effect, and (3) decreased airway resistance in the lungs. The decreased decompression time results from two of its properties: (1) Only 40 per cent as much helium dissolves in the body as does nitrogen. (2) Because of its small atomic size it diffuses through the tissues at a velocity about 2.5 times that of nitrogen.

However, helium is not always more advantageous than nitrogen in the breathing mixture. For instance, for short and shallow dives, the rapid rate of diffusion of helium through the tissues allows for more rapid buildup of dissolved helium in the tissue fluids than is true for nitrogen, which has a much slower rate of diffusion. Therefore, the problem of dissolved gas is actually less under these conditions when using nitrogen than when using helium. In general, if the bottom time is one hour, nitrogen is more advantageous at depths less than 100 feet and helium more advantageous at greater depths.

SOME PHYSICAL PROBLEMS OF DIVING

Aside from the effects of high gas pressures on the body, still other physical factors place limitations on diving. For instance:

Volume of Air that Must Be Pumped to the Diver — Relationship to Rate of Carbon Dioxide Elimination. To blow off carbon dioxide from the lungs, the tidal volume of air flowing in and out of the lungs with each breath must remain the same regardless of the depth of the dive. A tidal volume of 0.5 liter at 300 feet depth (10 atmospheres pressure) would be a sea level volume of 5 liters. Therefore, a compressor operating at sea level must pump 5 liters of air to the diver at 300 feet depth for each breath that the diver takes in order to wash the carbon dioxide out of his lungs, and an additional amount must be pumped to wash the carbon dioxide out of the diving gear. Stating this another way, the amount of air that must be pumped to the diver to keep his alveolar carbon dioxide normal is directly proportional to the pressure under which he is operating. At sea level the working diver requires about 1.5 cubic feet of air per minute for adequate carbon dioxide washout from his diving helmet. Therefore, the sea level volumes of air that must be pumped each minute for different depths of operation are the following:

feet	cubic feet
Sea level	1.5
33	3
66	4.5
100	6
200	10.5
300	15

Change in Density of the Air — Effect on Maximum Breathing Capacity. The density of the air increases in proportion to the pressure, which means that the density is 4 times as great at 100 feet depth as at sea level and 7 times as great at 200 feet.

The resistance to air flow through the respiratory passageways increases directly in proportion to the density of the breathing mixture. Therefore, one can readily see that the increased density of the air will increase the work of breathing and, as a corollary, will decrease the maximum breathing capacity (the amount of air that one can breathe each minute). The following table gives the *maximum breathing capacity* in per cent of normal at different depths when breathing air and when breathing a mixture in which helium replaces the nitrogen of the air.

Depth (ft.)	Air (% of normal)	Oxygen-helium (% of normal)
25	75	100+
50	60	100+
100	50	86
200	35	63
400	24	48

Effect of Rapid Descent — "The Squeeze." On rapid descent, the volumes of all gases in the body becomes greatly reduced because of increasing pressure applied to the outside of the body. If additional quantities of air are supplied to the gas cavities during descent — especially to the lungs — no harm is done, but, if the person continues to descend without addition of gas to the cavities, the volume becomes greatly reduced, and serious physical damage results; this is called "the squeeze."

The most damaging effects of the squeeze occur in the lungs, for the smallest volume that the lungs can normally achieve is approximately 1.5 liters. Even if the diver inspires a maximal breath prior to descending, he can go down no further than 100 feet before his chest begins to cave in. Therefore, to prevent lung squeeze, the diver must inspire additional air as he descends.

When air becomes entrapped in the middle ear during descent, the squeeze can cause a ruptured tympanum, and when air is entrapped in one of the nasal sinuses intense pain results. Occasionally, also, when a diver loses air pressure to his helmet his body is literally squeezed upward into the helmet, and, when air volume is lost from the mask of a free diving apparatus, the eyes can actually pop out into the mask and the face can become greatly distorted to fill the mask.

Overexpansion of the Lungs on Rapid Ascent — Air Embolism. Exactly the opposite pulmonary effects occur on rapid ascent if the person fails to expel air from the lungs on the way up. Unfortunately, panic can frequently cause a person to close his glottis spastically, which can result in serious damage to the lungs. When the lungs become expanded to their limit the pressure continues to rise, and above an *excess* alveolar pressure of 80 to 100 mm. Hg, air is forced into the pulmonary capillaries, causing air embolism in the circulation and often resulting in death. Also, increased pressure in the lungs frequently blows out large blebs on the surfaces of the lungs or ruptures the lungs to cause pneumothorax.

In rare instances, in a diver who has been deep below the sea for a long time and in whom large amounts of gas have accumulated in the abdomen, rapid ascent can also cause serious trauma in the gastrointestinal tract.

Rapid ascent can be especially serious when a person is attempting to escape from a submarine and must rise to the surface rapidly without appropriate diving gear. It also occurs frequently when the diver loses control of his diving gear and his suit balloons up so greatly that he "blows up" to the surface.

SCUBA DIVING (SELF-CONTAINED UNDERWATER BREATHING APPARATUS)

In recent years a diving apparatus that does not require connections with the surface has been perfected and is probably best known under the trade name "Aqualung." The two basic types of self-contained underwater breathing apparatuses (SCUBA) from which various modifications have been made are (1) the open circuit demand system and (2) the closed circuit system.

The Open Circuit Demand System. Figure 45–4 illustrates an open circuit demand type of underwater breathing apparatus showing the following components: (1) one or more tanks of compressed air or of some other breathing mixture, (2) a first stage "reducing" valve for reducing the pressure from the tanks to a constant low

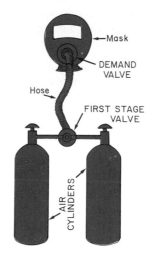

Figure 45–4. A SCUBA apparatus.

pressure level, (3) a combination inhalation "demand" valve and exhalation valve which allows air to be pulled into the lungs with very slight negative pressure and then to be exhausted into the sea, and (4) a mask and tube system with small "dead space."

Basically, the demand system operates as follows: The first stage reducing valve reduces the pressure from the tanks approximately ten-fold, usually to a pressure of about 100 lbs. per square inch. However, the breathing mixture does not flow continually into the mask. Instead, with each inspiration, slight negative pressure in the mask pulls the diaphragm of the demand valve inward, and this automatically releases air from the hose into the mask and lungs. In this way only the amount of air needed for inhalation enters the system. Then, on expiration, the air cannot go back into the tank but instead is expired through the expiration valve.

The most important problem in use of the self-contained underwater breathing apparatus is the time limit that one can remain beneath the surface; only a few minutes are possible at great depths because tremendous airflow from the tanks is required to wash carbon dioxide out of the lungs — the greater the depth, the greater the airflow required, as discussed earlier.

The Closed Circuit System. In the simplest type of closed circuit system a person breathes pure oxygen. This system contains the following elements: (1) a tank of pure oxygen, (2) a rubber bellows into which the diver can breathe back and forth, (3) a valve system for allowing oxygen to flow from the oxygen tank into the bellows as needed to keep it moderately filled, (4) a canister containing soda lime through which the rebreathed air passes to absorb the carbon dioxide, and (5) an appropriate mask system with valves to keep the gaseous mixture flowing through the canister for carbon dioxide removal. Thus, the closed circuit system is similar to a standard anesthetic machine in which oxygen is continually rebreathed — except that no anesthetic is used.

The most important problem in use of this closed circuit system is limitation in depth that a person can remain beneath the sea because of poor oxygen tolerance. A safe working rule is no more than 30 minutes at 30 feet depth.

Several new closed circuit systems are available on an experimental basis. These utilize mixtures of oxygen and helium, and they have separate supply tanks for each gas. As oxygen is utilized from the breathing mixture, an electrical oxygen sensor and control mechanism replenishes the oxygen from the oxygen tank but keeps the oxygen concentration from rising too high; at the same time carbon dioxide is removed by a carbon dioxide absorbent. Since this gaseous mixture can be rebreathed again and again without ever blowing off any of it into the sea, none of the gases are wasted. Therefore, very small tanks of oxygen and helium allow the diver to remain under the sea for many hours.

SPECIAL PHYSIOLOGICAL PROBLEMS IN SUBMARINES

Escape from Submarines. Essentially the same problems as those of deep sea diving are often met in relation to submarines, especially when it is necessary to escape from a submerged submarine. Escape is possible from as deep as 300 feet even without using any special type of apparatus. Proper use of rebreathing devices using helium or hydrogen can theoretically allow escape from as deep as 600 feet or perhaps more.

One of the major problems of escape is prevention of air embolism. As the person ascends, the gases in his lungs expand and sometimes rupture a major pulmonary vessel, allowing the gases to enter into the pulmonary vascular system to cause embolism of the circulation. Therefore, as the person ascends, he must exhale continually.

Expansion and exhalation of gases from the lungs during ascent, even without breathing, is often rapid enough to blow off the accumulating carbon dioxide in the lungs. This keeps the concentration of carbon dioxide from building up in the blood and keeps the person from having the desire to breathe. Therefore, he can hold his breath for an extra long time during ascent.

Health Problems in the Submarine Internal Environment. Except for escape, submarine medicine generally centers around several engineering problems to keep hazards out of the internal environment of the submarine. In atomic submarines there exists the problem of radiation hazards, but, with appropriate shielding, the amount of radiation received by the crew submerged beneath the sea has actually been less than the normal radiation received above the surface of the sea from cosmic rays. Therefore, no essential hazard results from this unless some failure of the apparatus causes unexpected release of radioactive materials.

Second, poisonous gases on occasion escape into the atmosphere of the submarine and must be controlled early. For instance, during several weeks' submergence, cigarette smoking by the crew can liberate sufficient amounts of carbon monoxide, if it is not removed from the air, to cause carbon monoxide poisoning, and on occasion even Freon gas has been found to diffuse through the tubes in refrigeration systems in sufficient quantity to cause toxicity. Finally, it is well known that chlorine and other poisonous gases are released when salt water comes in contact with batteries in the old type of submarines.

A highly publicized factor of submarine medicine has been the possibility of psychological problems caused by prolonged submergence. Fortunately, this has turned out to be more a figment of the public's imagination than truth, for the problems here are the same as those relating (1) to any other confinement or (2) to any other type of danger. Psychological screening has been used to keep such problems almost to zero even in month-long submergence.

REFERENCES

Anderson, H. T.: Physiological adaptations in diving vertebrates. *Physiol. Rev.*, 46:212, 1966

Behnke, A. R., Jr., and Lanphier, E. H.: Underwater physiology. *In* Fenn, W. O., and Rahn, H. (eds.): Handbook of Physiology. Sec. 3, Vol. 2, Baltimore, Williams & Wilkins, 1965, p. 1159.

Bennett, P. B.: The Aetiology of Compressed Air Intoxication and Inert Gas Narcosis. (International Series of Monographs in Pure and Applied Biology, Zoology Div., Vol. 31.) New York, Pergamon Press, 1966.

Bennett, P. B., and Elliott, D. H.: The Physiology and Medicine of Diving and Compressed Air Work, 2nd Ed. Baltimore, Williams & Wilkins, 1975.

Fisher, A. B., *et al.*: Oxygen toxicity of the lung: Biochemical aspects. *In* Fishman, A. P., and Renkin, E. M. (eds.): Pulmonary Edema. Baltimore, Waverly Press, 1979, p. 207.

Gamarra, J. A.: Decompression Sickness. Hagerstown, Md., Harper & Row, 1974.

Gooden, B. A.: Drowning and the diving reflex in man. *Med. J. Aust.*, 2:583, 1972.

Greene, D. G.: Drowning. *In* Fenn, W. O., and Rahn, H. (eds.): Handbook of Physiology. Sec. 3, Vol. 2, Baltimore, Williams & Wilkins, 1965, p. 1195.

Haugaard, N.: Cellular mechanisms of oxygen toxicity. *Physiol. Rev.*, 48:311, 1968.

Hochachka, B. W., and Murphy, B.: Metabolic status during diving and recovery in marine mammals. *In* Robertshaw, D. (ed.): International Review of Physiology: Environmental Physiology III. Vol. 20, Baltimore, University Park Press, 1979, p. 253.

Hochachka, B. W., and Storey, K. B.: Metabolic consequences of diving in animals and man. *Science*, 187:613, 1975.

Lambertsen, C. J.: Proceedings of the Third Symposium on Underwater Physiology. Baltimore, Williams and Wilkins, 1967.

Lanphier, E. H.: Human respiration under increased pressures. *Symp. Soc. Exp. Biol.*, 26:379, 1972.

Miles, S.: Underwater Medicine, 2nd Ed. Philadelphia, J. B. Lippincott Co., 1966.

Miller, J. W.: Vertical Excursions Breathing Air from Nitrogen-Oxygen or Air Saturation Exposures. Washington, D.C., U.S. Government Printing Office, 1976.

Oxygen Free Radicals and Tissue Damage. Ciba Foundation Symposium. New York, Excerpta Medica, 1979.

Shilling, C. W., and Beckett, M. W. (eds.): Underwater Physiology IV. Bethesda, Md., Federation of American Societies for Experimental Biology, 1978.

Sloan, A. W.: Man in Extreme Environments. Springfield, Ill., Charles C Thomas, 1979.

Submarine Medicine Practice, Department of the Navy. Washington, D.C., U.S. Government Printing Office, 1956.

Undersea Medical Society: Glossary of Diving and Hyperbaric Terms. Bethesda, Md., The Society, 1978.

Vail, E. G.: Hyperbaric respiratory mechanics. *Aerospace Med.*, 42:536, 1971.

Wyndam, C. H.: Physiological problems of deep level mining. *Proc. Int. Union Physiol. Sci.*, 6:45, 1968.

Zimmerman, A. M. (ed.): High Pressure Effects on Cellular Processes. New York, Academic Press, 1970.

Part IX

THE NERVOUS SYSTEM

46

Organization of the Nervous System; Basic Functions of Synapses

The nervous system, along with the endocrine system, provides most of the control functions for the body. In general, the nervous system controls the rapid activities of the body, such as muscular contractions, rapidly changing visceral events, and even the rates of secretion of some endocrine glands. The endocrine system, by contrast, regulates principally the metabolic functions of the body.

The nervous system is unique in the vast complexity of the control actions that it can perform. It receives literally thousands of bits of information from the different sensory organs and then integrates all these to determine the response to be made by the body. The purpose of this chapter is to present, first, a general outline of the overall mechanisms by which the nervous system performs such functions. Then we will discuss the function of central nervous system synapses, the basic structures that control the passage of signals into, through, and then out of the nervous system. In succeeding chapters we will analyze in detail the functions of the individual parts of the nervous system. Before beginning this discussion, however, the reader should refer to Chapters 10 and 12, which present, respectively, the principles of membrane potentials and transmission of impulses through neuromuscular junctions.

GENERAL DESIGN OF THE NERVOUS SYSTEM

THE SENSORY DIVISION — SENSORY RECEPTORS

Most activities of the nervous system are initiated by sensory experience emanating from *sensory receptors,* whether these be visual receptors, auditory receptors, tactile receptors on the surface of the body, or other kinds of receptors. This sensory experience can cause an immediate reaction, or its memory can be stored in the brain for minutes, weeks, or years and then can help to determine the bodily reactions at some future date.

Figure 46–1 illustrates a portion of the sensory system, the *somatic* portion that transmits sensory information from the receptors of the entire surface of the body and deep structures. This information enters the nervous system through the spinal nerves and is conducted into (a) the spinal cord at all levels, (b) the reticular substance of the medulla, pons, and mesencephalon, (c) the cerebellum, (d) the thalamus, and (e) the somesthetic areas of the cerebral cortex. But in addition to these "primary sensory" areas, signals are then relayed to essentially all other segments of the nervous system.

THE MOTOR DIVISION — THE EFFECTORS

The most important ultimate role of the nervous system is control of bodily activities. This is achieved by controlling (a) contraction of skeletal muscles throughout the body, (b) contraction of smooth muscle in the internal organs, and (c) secretion by both exocrine and endocrine glands in many parts of the body. These activities are collectively called *motor functions* of the nervous system, and the muscles and glands are called *effectors* because they perform the functions dictated by the nerve signals. That portion of the nervous system directly concerned with transmitting signals to the muscles and glands is called the motor division of the nervous system.

Figure 46–2 illustrates the *motor axis* of the nervous system for controlling skeletal muscle contraction. Operating parallel to this axis is another similar system for control of the smooth

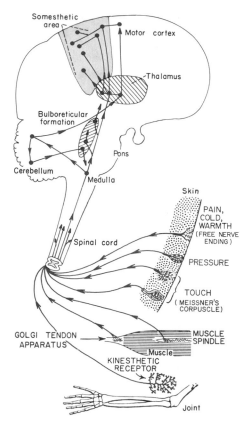

Figure 46–1. The somatic sensory axis of the nervous system.

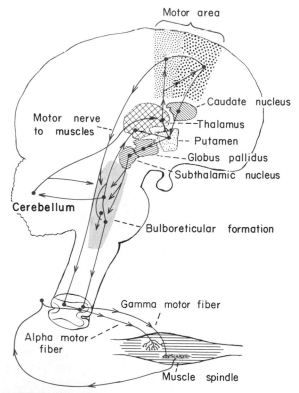

Figure 46–2. The motor axis of the nervous system.

muscles and glands; it is the *autonomic nervous system,* which will be presented in detail in Chapter 57. Note in Figure 46–2 that the skeletal muscles can be controlled from many different levels of the central nervous system, including (a) the spinal cord, (b) the reticular substance of the medulla, pons, and mesencephalon, (c) the basal ganglia, (d) the cerebellum, and (e) the motor cortex. Each of these different areas plays its own specific role in the control of body movements, the lower regions being concerned primarily with automatic, instantaneous responses of the body to sensory stimuli and the higher regions with deliberate movements controlled by the thought processes of the cerebrum.

PROCESSING OF INFORMATION

The nervous system would not be at all effective in controlling bodily functions if each bit of sensory information caused some motor reaction. Therefore, one of the major functions of the nervous system is to process incoming information in such a way that *appropriate* motor responses occur. Indeed, more than 99 per cent of all sensory information is discarded by the brain as irrelevant and unimportant. For instance, one is ordinarily totally unaware of the parts of the body that are in contact with clothing and is also unaware of the seat pressure when sitting. Likewise, attention is drawn only to an occasional object in one's field of vision, and even the perpetual noise of our surroundings is usually relegated to the background.

After the important sensory information has been selected, it is then channeled into proper motor regions of the brain to cause the desired responses. Thus, if a person places a hand on a hot stove, the desired response is to lift the hand, plus other associated responses such as moving the entire body away from the stove and perhaps even shouting with pain. Yet even these responses represent activity by only a small fraction of the total motor system of the body.

Role of Synapses in Processing Information. The synapse is the junction point from one neuron to the next and, therefore, is an advantageous site for control of signal transmission. Later in this chapter we will discuss the details of synaptic function. However, it is important to point out here that the synapses determine the directions that the nervous signals spread in the nervous system. Some synapses transmit signals from one neuron to the next with ease, while others transmit signals only with difficulty. Also, facilitatory and inhibitory signals from other areas in the nervous system can control synaptic activity, sometimes opening the synapses for transmission and other times closing them. In addition, some post-synaptic neurons respond with large numbers of impulses, while others respond with only a few. Thus, the synapses perform a selective action, often blocking the weak signals while allowing the strong signals to pass, often selecting and amplifying certain weak signals, and often channeling the signal in many different directions rather than simply in one direction. The basic principles of this processing of information by the synapses are so important that they are discussed in detail in the latter part of this chapter and in the entire following chapter.

STORAGE OF INFORMATION — MEMORY

Only a small fraction of the important sensory information causes an immediate motor response. Much of the remainder is stored for future control of motor activities and for use in the thinking processes. Most of this storage occurs in the *cerebral cortex,* but not all, for even the basal regions of the brain and perhaps even the spinal cord can store small amounts of information.

The storage of information is the process we call *memory,* and this too is a function of the synapses. That is, each time a particular sensory signal passes through a sequence of synapses, these synapses become more capable of transmitting the same signal the next time, which process is called *facilitation.* After the sensory signal has passed through the synapses a large number of times, the synapses become so facilitated that signals generated within the brain itself can also cause transmission of impulses through the same sequence of synapses even though the sensory input has not been excited. This gives the person a perception of experiencing the original sensation, though in effect it is only a memory of the sensation.

Unfortunately, we do not know the precise mechanism by which facilitation of synapses occurs in the memory process, but what is known about this and other details of the memory process will be discussed in Chapter 55.

Once memories have been stored in the nervous system, they become part of the processing mechanism. The thought processes of the brain compare new sensory experiences with the stored memories; the memories help to select the important new sensory information and to channel this into appropriate storage areas for future use or into motor areas to cause bodily responses.

THE THREE MAJOR LEVELS OF NERVOUS SYSTEM FUNCTION

The human nervous system has inherited specific characteristics from each stage of evolutionary development. From this heritage, there remain three major levels of the nervous system that have special functional significance: (1) the spinal cord level, (2) the lower brain level, and (3) the higher brain or cortical level.

THE SPINAL CORD LEVEL

The spinal cord of the human being still retains many functions of the multisegmental animal. Sensory signals are transmitted through the spinal nerves into each *segment* of the spinal cord, and these signals can cause localized motor responses either in the segment of the body from which the sensory information is received or in adjacent segments. Essentially all the spinal cord motor responses are automatic and occur almost instantaneously in response to the sensory signal. In addition, they occur in specific patterns of response called *reflexes.*

Figure 46–3 illustrates two of the simpler cord reflexes. To the left is the neural control of the *muscle stretch reflex.* If a muscle suddenly becomes stretched, a sensory nerve receptor in the muscle called the *muscle spindle* becomes stimulated and transmits nerve impulses through a sensory nerve fiber into the spinal cord. This fiber synapses directly with a motoneuron in the anterior horn of the cord gray matter, and the motoneuron in turn transmits impulses back to the muscle to cause the muscle, the effector, to contract. The muscle contraction opposes the original muscle stretch. Thus, this reflex acts as a *feedback* mechanism, operating from a receptor to an effector, to prevent sudden change in the length of the muscle. This allows a person to maintain the limbs and other parts of the body in desired positions despite sudden outside forces that tend to move the parts out of position.

To the right in Figure 46–3 is illustrated the neural control of another reflex called the *with-*

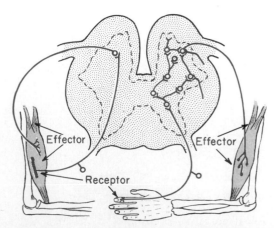

Figure 46–3. *Left:* The simple stretch reflex. *Right:* A withdrawal reflex.

drawal reflex. This is a protective reflex that causes withdrawal of any part of the body from an object that causes pain. For instance, let us assume that the hand is placed on a sharp object. Pain signals are transmitted into the gray matter of the spinal cord, and, after appropriate selection of information by the synapses, signals are channeled to the appropriate motoneurons to cause flexion of the biceps muscle. This obviously lifts the hand away from the sharp object.

We see, then, that the withdrawal reflex is much more complex than the stretch reflex, for it involves many neurons in the gray matter of the cord, and signals are transmitted to many adjacent segments of the cord to cause contraction of the appropriate muscles.

Cord Functions After the Brain is Removed. The many reflexes of the spinal cord will be discussed in Chapter 51; however, the following list of important cord reflex functions that occur even after the brain is removed illustrates the many capabilities of the spinal cord:

1. An animal can under certain conditions be made to stand up. This is caused primarily by reflexes initiated from the pads of the feet. Sensory signals from the pads cause the extensor muscles of the limbs to tighten, which in turn allows the limbs to support the animal's body.

2. An animal with its spinal cord transected, when held in a sling so that its feet hang downward, often begins walking or galloping movements involving one, two, or all its legs. This illustrates that the basic patterns for causing the limb movements of locomotion are present in the spinal cord.

3. A flea crawling on the skin of a spinal animal causes reflex to-and-fro scratching by the paw, and the paw can actually localize the flea on the surface of the body.

4. Cord reflexes exist to cause emptying of the urinary bladder and of the rectum.

5. Segmental temperature reflexes are present throughout the body. Local cooling of the skin causes vasoconstriction, which helps to conserve heat in the body. Conversely, local heating in the skin causes vasodilatation, resulting in loss of heat from the body.

This list of some of the segmental and multisegmental reflexes of the spinal cord demonstrates that many of our day-by-day and moment-by-moment activities are controlled locally by the respective segmental levels of the spinal cord, and the brain plays only a modifying role in these local controls.

THE LOWER BRAIN LEVEL

Many if not most of what we call subconscious activities of the body are controlled in the lower areas of the brain — in the medulla, pons, mesencephalon, hypothalamus, thalamus, cerebellum, and basal ganglia. Subconscious control of arterial blood pressure and respiration is achieved primarily in the reticular substance of the medulla and pons. Control of equilibrium is a combined function of the older portions of the cerebellum and the reticular substance of the medulla, pons, and mesencephalon. The coordinated turning movements of the head, of the entire body, and of the eyes are controlled by specific centers located in the mesencephalon, paleocerebellum, and lower basal ganglia. Feeding reflexes, such as salivation in response to taste of food and licking of the lips, are controlled by areas in the medulla, pons, mesencephalon, amygdala, and hypothalamus. And many emotional patterns, such as anger, excitement, sexual activities, reactions to pain, or reactions of pleasure, can occur in animals without a cerebral cortex.

In short, the subconscious but coordinate functions of the body, as well as many of the life processes themselves — arterial pressure and respiration, for instance — are controlled by the lower regions of the brain, regions that usually, but not always, operate below the conscious level.

THE HIGHER BRAIN OR CORTICAL LEVEL

We have seen from the above discussion that many of the intrinsic life processes of the body are controlled by subcortical regions of the brain or by the spinal cord. What then is the function of the cerebral cortex? The cerebral cortex is primarily a vast information storage area. Approximately three quarters of all the neuronal cell bodies of the entire nervous system are located in the cerebral cortex. It is here that most of the memories of past experiences are stored, and it is here that many of the patterns of motor responses are stored, which information can be called forth at will to control motor functions of the body.

Relation of the Cortex to the Thalamus and Other Lower Centers. The cerebral cortex is actually an outgrowth of the lower regions of the brain, particularly of the thalamus. For each area of the cerebral cortex there is a corresponding and connecting area of the thalamus, and activation of a minute portion of the thalamus activates the corresponding and much larger portion of the cerebral cortex. It is presumed that in this way the thalamus can call forth cortical activities at will. Also, activation of regions in the mesencephalon transmits diffuse signals to the cerebral cortex, partially through the thalamus and partially through other pathways, to activate the entire cortex. This is the process that we call *wakefulness*. On the other hand, when these areas of the mesencephalon become inactive, the thalamic and cortical regions also become inactive, which is the process we call *sleep*.

Function of the Cerebral Cortex in Thought Processes. Some areas of the cerebral cortex are not directly concerned with either sensory or motor functions of the nervous system — for example, the prefrontal lobes and large portions of the temporal and parietal lobes. These areas are set aside for the more abstract processes of thought, but even they also have direct nerve connections with the lower regions of the brain.

Large areas of the cerebral cortex can be destroyed without blocking the subconscious and even many involuntary conscious activities of the body. For instance, destruction of the somesthetic cortex does not destroy one's ability to feel objects touching the skin, but it does destroy the ability to distinguish the shapes of objects, their character, and the precise points on the skin where the objects are touching. Thus, the cortex is not required for perception of sensation, but it does add immeasurably to its depth of meaning. Likewise, destruction of the prefrontal lobe does not destroy one's ability to think, but it does destroy the ability to think in deeply abstract terms. In other words, each time a portion of the cerebral cortex is destroyed, a vast amount of information is lost to the thinking process and some of the mechanisms for processing this information are also lost. Therefore, total loss of the cerebral cortex causes a vegetative type of existence rather than a "living" existence.

Telencephalization. In the process of evolution, the higher regions of the human nervous system have taken over many sensory and motor functions performed by the less developed brain of lower animals. For instance, if the spinal cord of an opossum is cut in the midthoracic region, the opossum can still walk perfectly well on both its forelimbs and hindlimbs, except that the hindlimbs are then unsynchronized with the forelimbs. A similar transection of the spinal cord in the human being causes complete loss of ability to use the legs for locomotion.

This process by which the progressively higher centers of the brain have taken over more and more of the function of the lower centers is called *telencephalization*. Yet, despite telencephalization, some of the very basic functions of the lower centers still remain active or at least partially active in the human being, such as many of the cord reflexes and the stereotyped control systems of the basal regions of the brain that were just discussed.

COMPARISON OF THE NERVOUS SYSTEM WITH AN ELECTRONIC COMPUTER

When electronic computers were first developed in many different laboratories of the world by as many

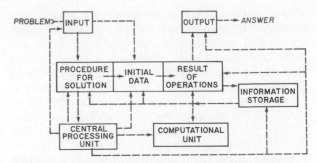

Figure 46–4. Block diagram of a general-purpose electronic computer, showing the basic components and their interrelationships.

different scientists, it soon became apparent that all these machines have many features in common with the nervous system. First, they all have input circuits that are comparable to the sensory portion of the nervous system, and output circuits that are comparable to the motor portion of the nervous system. In the conducting pathway between the inputs and the outputs are the mechanisms for performing the different types of computations.

In simple computers, the output signals are controlled directly by the input signals, operating in a manner similar to that of the simple reflexes of the spinal cord. But, in the more complex computers, the output is determined both by the input signals and by information that has already been stored in memory in the computer, which is analogous to the more complex reflex and processing mechanisms of our higher nervous system. Furthermore, as the computers become even more complex it is necessary to add still another unit, called the *central programming unit*, which determines the sequence of all operations. This unit is analogous to the mechanism in our brain that allows us to direct our attention first to one thought or sensation or motor activity, then to another, and so forth, until complex sequences of thought or action take place.

Figure 46–4 illustrates a simple block diagram of a modern computer. Even a rapid study of this diagram will demonstrate its similarity to the nervous system. Furthermore, general purpose computers designed and built by many different scientific groups have ended up with almost identically the same basic components. The fact that these components are analogous to those of the human nervous system demonstrates that the brain is basically a computer that continuously collects sensory information and uses this along with stored information to compute the daily course of bodily activity.

FUNCTION OF CENTRAL NERVOUS SYSTEM SYNAPSES

Every medical student is aware that information is transmitted in the central nervous system mainly in the form of nerve impulses through a succession of neurons, one after another. However, it is not immediately apparent that each impulse may be (a) blocked in its transmission from one neuron to the next, (b) changed from a single impulse into repetitive impulses, or (c) integrated with impulses from other neurons to cause highly intricate patterns of impulses in successive neurons. All these functions can be classified as *synaptic functions of neurons*.

PHYSIOLOGIC ANATOMY OF THE SYNAPSE

The junction between one neuron and the next is called a *synapse*. Figure 46–5 illustrates a typical *motoneuron* in the anterior horn of the spinal cord. It is comprised of three major parts: the *soma*, which is the main body of the neuron; a single *axon*, which extends from the soma into the peripheral nerve; and the *dendrites*, which are thin projections of the soma that extend up to 1 mm. into the surrounding areas of the cord.

An average of about 6000 small knobs called *synaptic knobs* lie on the surfaces of the dendrites and soma of the motoneuron, approximately 80 to 90 per cent of them on the dendrites. These knobs are the terminal ends of nerve fibrils that originate in many other neurons, and usually not more than a few of the knobs are derived from any single previous neuron. Later it will become evident that many of these synaptic knobs are *excitatory* and

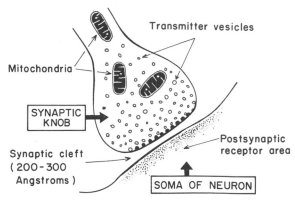

Figure 46–6. Physiologic anatomy of the synapse.

secrete a substance that excites the neuron, while others are *inhibitory* and secrete a substance that inhibits the neuron.

Neurons in other parts of the cord and brain differ markedly from the motoneuron in (1) the size of the cell body, (2) the length, size, and number of dendrites, ranging in length from almost none at all up to as long as one meter (the peripheral sensory nerve fiber), (3) the length and size of the axon, and (4) the number of synaptic knobs, which may range from only a few to more than a hundred thousand. These differences make neurons in different parts of the nervous system react differently to incoming signals and therefore perform different functions, as will be explained in subsequent chapters.

The Synaptic Knobs. Electron microscope studies of the synaptic knobs show that they have varied anatomical forms, but most resemble small round or oval knobs and therefore are frequently called *terminal knobs, boutons, end-feet,* or simply *presynaptic terminals*.

Figure 46–6 illustrates the basic structure of the synaptic knob (presynaptic terminal). It is separated from the neuronal soma by a *synaptic cleft* having a width usually of 200 to 300 Angstroms. The knob has two internal structures important to the excitatory or inhibitory functions of the synapse: the *synaptic vesicles* and the *mitochondria*. The synaptic vesicles contain a *transmitter substance* which, when released into the synaptic cleft, either *excites* or *inhibits* the neurons — excites if the neuronal membrane contains *excitatory receptors,* inhibits if it contains *inhibitory receptors*. The mitochondria provide ATP, which is required to synthesize new transmitter substance. The transmitter must be synthesized extremely rapidly because the amount stored in the vesicles is sufficient to last for only a few seconds to a few minutes of maximum activity.

When an action potential spreads over a presynaptic terminal, the membrane depolarization causes emptying of a small number of vesicles into

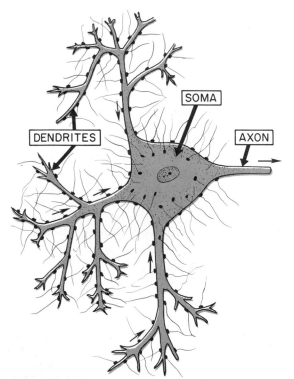

Figure 46–5. A typical motoneuron, showing synaptic knobs on the neuronal soma and dendrites. Note also the single axon.

the cleft; and the released transmitter in turn causes an immediate change in the permeability characteristics of the subsynaptic neuronal membrane, which leads to excitation or inhibition of the neuron, depending on the type of receptor substance.

Mechanism by Which the Synaptic Knob Action Potential Causes Release of Transmitter Vesicles. Unfortunately, we can only guess at the mechanism by which an action potential causes the vesicles to release transmitter substance into the synaptic cleft on reaching the synaptic knob. However, the number of vesicles released with each action potential is greatly reduced (a) when the quantity of calcium ions in the extracellular fluid is diminished, (b) when the quantity of sodium ions in the extracellular fluid is diminished, (c) when the quantity of magnesium ions in the extracellular fluids is increased, or (d) when the membrane of the synaptic knob has already been partially depolarized prior to transmission of the action potential so that the action potential is weaker than usual. On the basis of these characteristics, it has been suggested that the spread of the action potential over the membrane of the knob causes small amounts of calcium ions to leak into the knob. The calcium ions then are believed to cause several submembranal transmitter vesicles to fuse with the membrane and shortly thereafter to rupture, thus allowing spillage of their contents into the synaptic cleft.

At any rate, it is known that each time an action potential travels over the surface of the synaptic knob, a few vesicles of transmitter substance are emptied into the synaptic cleft, the number of vesicles depending upon the factors mentioned above. After emptying its contents, each vesicle then separates from the membrane and is used over and over again for transmitter storage and release.

One transmitter substance that occurs in certain parts of the nervous system is acetylcholine, as is discussed later. It has been calculated that about 3000 molecules of acetylcholine are present in each vesicle, and enough vesicles are present in the synaptic knob on a neuron to transmit a few thousand impulses.

Synthesis of New Transmitter Substance. Fortunately, the synaptic knobs have the capability of continually synthesizing new transmitter substance. Were it not for this ability, synaptic transmission would become completely ineffective within a few minutes. The synthesis usually occurs either partially or totally in the cytoplasm of the synaptic knobs, and then the newly synthesized or partially synthesized transmitter is immediately absorbed into the vesicles and stored or further processed until needed. Thus, each time a vesicle empties its contents into the synaptic cleft,

soon thereafter it becomes filled again with new transmitter.

Acetylcholine is synthesized from acetyl-CoA and choline in the presence of the enzyme *choline acetyltransferase,* an enzyme that is present in abundance in the cytoplasm of the cholinergic type of synaptic knob. When acetylcholine is released from the knob into the synaptic cleft, it is rapidly split again to acetate and choline by the enzyme cholinesterase that is adherent to the outer surface of the knob. Then the choline is actively transported back into the knob to be used once more for synthesis of new acetylcholine. Thus, the vesicles are used again and again. But even so, both the vesicles and the mitochondria that supply the energy for transmitter synthesis eventually disintegrate. Fortunately, new vesicles and mitochondria are continually transported from the cell soma down the axon to the synaptic knob, moving along the axon at a velocity of about 10 cm. per day, thus replenishing the supply in the knobs.

The formation, release, and re-uptake of norepinephrine by synaptic knobs of the sympathetic nervous system will be discussed in detail in Chapter 57 in relation to function of the autonomic nervous system.

Action of the Transmitter Substance on the Postsynaptic Neuron. The membrane of the postsynaptic neuron where a synaptic knob abuts is believed to contain specific receptor molecules that bind the transmitter substance. These receptors are probably proteins that, in response to the binding, change their shapes or activities in such a way that they increase the membrane permeability to most ions when the membrane receptor is excitatory and increase the permeability mainly to chloride ions when the receptor is inhibitory.

The ribosomes and the endoplasmic reticulum are both increased in the area immediately beneath the synapse, as illustrated by the density of the postsynaptic area in Figure 46–6. It is possible that the degree of development of this postsynaptic receptor area increases with the intensity of activity of the synapse; therefore, this is possibly a means by which the synapse can subserve the memory function, as will be discussed in detail in Chapter 55.

CHEMICAL AND PHYSIOLOGICAL NATURES OF THE TRANSMITTER SUBSTANCES

Excitation and Inhibition. Whether a transmitter will cause excitation or inhibition is determined not only by the nature of the transmitter but also by the nature of the receptor in the postsynaptic membrane. To give an example, the same neuron might be excited by a synapse that releases acetylcholine but inhibited by still another synapse that releases glycine. Thus, the neuronal membrane contains an *excitatory receptor* for acetylcholine and an *inhibitory receptor* for glycine. To give another example, norepinephrine released by some synapses in the central nervous system causes inhibition while at other synapses it causes excitation. In the first case, the neuronal membranes contain an inhibitory receptor for norepinephrine while the others contain an excitatory receptor for the same transmitter.

Differences in Function of Different Transmitters. Aside from the fact that transmitter substances sometimes cause excitation and sometimes inhibition, they also have other differences, the most important of which is duration of stimulation. For instance, the excitatory synapses on the anterior motoneurons of the spinal cord develop their excitatory stimulus in 1 to 2 milliseconds, and the excited state then lasts for another 10 to 20 milliseconds. On the other hand, some inhibitory synapses in the brain require as long as 10 to 15 milliseconds to develop full effect, and the effect often lasts for as long as 200 to 300 milliseconds. Still other substances have been isolated from the cerebrospinal fluid which, when infused into the aqueduct of Sylvius, cause brain stem inhibition and make an animal go to sleep for as long as several hours. This suggests that these substances might be synaptic transmitters the effects of which can last for hours.

Another difference among transmitters is that some cause an increase in the rate of firing of neurons while others have no effect at all on the firing rate but do change the neuron's sensitivity to still other transmitter substances. This type of transmitter is called a *modulator*.

One can readily see, therefore, that the brain can call on many different types of transmitters, each having its own special attributes. Obviously, these can be used in many different ways to achieve the varied functions of the nervous system. This is an area of neurophysiological research that is only now beginning to be understood.

Release of Only a Single Type of Transmitter Substance by Each Neuron. A single neuron releases only one type of transmitter, and it releases it at all its nerve terminals. Therefore, release of the different types of transmitters in the central nervous system requires different types of neurons.

In general, there are special neuronal *systems* for secreting the different transmitters. For example, most of the neurons located in the substantia nigra of the brain stem give rise to a system of nerve fibers that projects upward into the basal ganglia and into closely associated areas; all of these fibers release dopamine. The dopamine in

turn has an inhibitory effect on that entire region of the brain. A similar system for release of norepinephrine originates in the reticular substance of the brain stem. And other similar systems exist for other transmitter substances elsewhere in the brain.

Some of the Important Transmitter Substances. There now exists very good evidence for about thirty different nervous system transmitters. Some types of evidence that have been used for discovering the different transmitters include the following: (1) the transmitter substance has been isolated from the neuronal tissue or shown to exist in specific neurons by chemical or immunological techniques; (2) the enzymes required for synthesis of the specific transmitter have been isolated from the nervous tissue or shown to exist specifically in the synaptic knobs; (3) injection of the transmitter substance into the local neuronal areas will cause excitation or inhibition; (4) substances that prevent destruction of the transmitter will potentiate its effect on the local neurons; and (5) specific mechanisms for removing a specific transmitter after it is released have been discovered in the local tissue area. Almost none of the presumed nervous system transmitter substances fulfill all these criteria, but some of the substances that fulfill many of them are the following:

Acetylcholine is secreted by neurons in many areas of the brain. Two places where this has been proved beyond doubt are in the striate region of the basal ganglia and at the endings of the recurrent nerve fibers from the anterior motoneurons where they terminate on the Renshaw cells in the anterior horns of the cord. In addition, histochemical studies suggest that acetylcholine is a widely used transmitter throughout the brain. It probably has an excitatory effect either everywhere or almost everywhere that it is released, even though it is known to have inhibitory effects in some portions of the peripheral parasympathetic nervous system, such as inhibition of the heart by the vagus nerves.

Norepinephrine is secreted by many neurons of which the cell bodies are located in the reticular formation of the brain stem and also in the hypothalamus. These neurons send nerve fibers to widespread areas of the brain. In most instances they probably cause inhibition, though it is probable that they cause excitation in some areas.

Epinephrine is secreted by a fewer number of neurons, but in general they parallel the norepinephrine system.

Dopamine is secreted by neurons that originate in the substantia nigra. The terminations of these neurons are mainly in the striatal region of the basal ganglia. The effect of dopamine is usually inhibition.

Glycine is secreted mainly at synapses in the

spinal cord. It probably always acts as an inhibitory transmitter.

Gamma amino butyric acid (GABA) is secreted by nerve terminals in the spinal cord, the cerebellum, the basal ganglia, and many other areas. It is believed always to cause inhibition.

Glutamic acid is probably secreted by the synaptic knobs in some or many of the sensory pathways. It probably always causes excitation.

Substance P is probably released by pain fiber terminals in the substantia gelatinosa of the spinal cord. In general, it causes excitation.

Enkephalins and endorphins are probably secreted by nerve terminals in the spinal cord, in the brain stem, in the thalamus, and in the hypothalamus. These probably act as excitatory transmitters to excite another system that inhibits the transmission of pain.

Serotonin is secreted by nuclei that originate in the median raphe of the brain stem and project to many brain areas, especially to the dorsal horns of the spinal cord and to the hypothalamus. Serotonin acts as an inhibitor of pain pathways in the cord, and it is also believed to help control the mood of the person, perhaps even to cause sleep.

Despite this long list of transmitter substances, it is only a partial list. Other substances that have been proved or suggested include various peptides, other amino acids, histamine, prostaglandins, cyclic AMP, and many others.

ELECTRICAL EVENTS DURING NEURONAL EXCITATION

The electrical events in neuronal excitation have been studied mainly in the large motoneuron of the anterior horn of the spinal cord. Therefore, the events to be described in the following few sections pertain essentially to these neurons. However, except for some quantitative differences, they apply to most other neurons of the nervous system as well.

The Resting Membrane Potential of the Neuronal Soma. Figure 46–7 illustrates the soma of a motoneuron, showing the resting membrane potential to be about −70 millivolts. This is somewhat less than the −90 millivolts found in large peripheral nerve fibers and in skeletal muscle fibers; the lower voltage is important, however, because it allows both positive and negative control of the degree of excitability of the neuron. That is, decreasing the voltage to a less negative value makes the membrane of the neuron more excitable, while increasing this voltage to a more negative value makes the neuron less excitable. This is the basis of the two modes of function of the neuron — either excitation or inhibition — as we will explain in detail in the following sections.

Concentration Differences of Ions Across the Neuronal Somal Membrane. Figure 46–7 also illustrates the concentration differences across the neuronal somal membrane of the three ions that are most important for neuronal function: sodium ions, potassium ions, and chloride ions.

At the top, the sodium ion concentration is shown to be very great in the extracellular fluid but low inside the neuron. This sodium concentration gradient is caused by a strong sodium pump that continually pumps sodium out of the neuron.

The figure also shows that the potassium ion concentration is large inside the neuronal soma but very low in the extracellular fluid. It illustrates that there is a weak potassium pump that tends to pump potassium to the interior while there is a very high degree of permeability to potassium. The pump is relatively unimportant because potassium ions leak through the neuronal somal pores so readily that this nullifies most of the effectiveness of the pump. Therefore, for most purposes one can consider the membrane to be relatively permeable to potassium and the potassium pump to be of only slight importance.

Figure 46–7 shows the chloride ion to be of high concentration in the extracellular fluid but low concentration inside the neuron. It also shows that the membrane is highly permeable to chloride ions

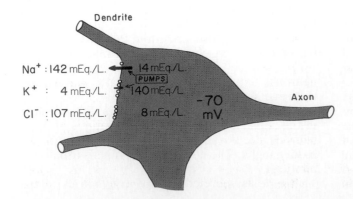

Figure 46–7. Distribution of sodium, potassium, and chloride ions across the neuronal somal membrane; origin of the intrasomal membrane potential.

and that there is no chloride pump. Therefore, the chloride ions become distributed across the membrane passively. The reason for the low concentration of chloride ions inside the neuron is the −70 millivolts in the neuron. That is, this negative voltage repels the negatively charged chloride ions, forcing them outward through the pores until the concentration difference is so great that its tendency to move chloride ions inward exactly balances the tendency of the electrical difference to move them outward. That is, the chloride ions become distributed across the membrane in accordance with the Nernst equation for equilibrium conditions. This equation was discussed in detail in Chapter 10.

Origin of the Resting Membrane Potential of the Neuronal Soma. The basic cause of the −70 millivolt resting membrane potential of the neuronal soma is the sodium-potassium pump. This pump causes the extrusion of more positively charged sodium ions to the exterior than potassium to the interior — 3 sodium ions for each 2 potassium ions. Since there are large numbers of negatively charged ions inside the soma that cannot diffuse through the membrane — protein ions, phosphate ions, and many others — extrusion of the positively charged sodium ions to the exterior leaves all these nondiffusible negative ions unbalanced by positive ions on the inside. Therefore, the interior of the neuron becomes negatively charged as the result of the sodium-potassium pump. This principle was discussed in more detail in Chapter 10 in relation to the resting membrane potential of nerves.

Uniform Distribution of the Potential Inside the Soma. The interior of the neuronal soma contains a very highly conductive electrolytic solution, the intracellular fluid of the neuron. Furthermore, the diameter of the neuronal soma is very large (from 10 to 80 microns in diameter) causing there to be almost no resistance to conduction of electrical current from one part of the somal interior to another part. Therefore, any change in potential in any part of the intrasomal fluid causes an almost exactly equal change in potential at all other points in the soma. This is an important principle because it plays a major role in the summation of signals entering the neuron from multiple sources, as we shall see in subsequent sections of this chapter.

Effect of Synaptic Excitation on the Postsynaptic Membrane — The Excitatory Postsynaptic Potential. Figure 46–8A illustrates the resting neuron with an unexcited synaptic knob resting upon its surface. The resting membrane potential everywhere in the soma is −70 millivolts.

Figure 46–8B illustrates a synaptic knob that has secreted a transmitter into the cleft between the knob and the neuronal somal membrane. This

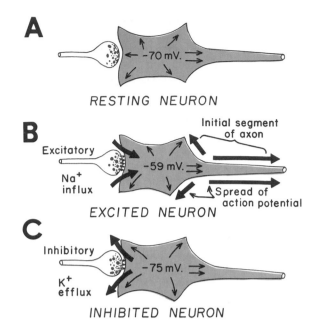

Figure 46–8. Three states of a neuron. A, A resting neuron. B, A neuron in an excited state, with increased intraneuronal potential caused by sodium influx. C, A neuron in an inhibited state, with decreased intraneuronal membrane potential caused by potassium ion efflux.

transmitter acts on a membrane excitatory receptor *to increase the membrane's permeability to all ions.* However, because of the large electrochemical gradient that tends to move sodium inward and because both the potassium and chloride concentration gradients are almost in equilibrium with the electrical potential, this large opening of the membrane pores mainly allows sodium ions to rush to the inside of the membrane.

The rapid influx of the positively charged sodium ions to the interior of the neuron neutralizes part of the negativity of the resting membrane potential. Thus, in Figure 46–8B the resting membrane potential has been increased from −70 millivolts to −59 millivolts. This increase in voltage above the normal resting neuronal potential — that is, to a less negative value — is called the *excitatory postsynaptic potential* (or EPSP) because when this potential rises high enough it will elicit an action potential in the neuron, thus exciting it.

However, we must issue a word of warning at this point. Discharge of a single synaptic knob can never increase the neuronal potential from −70 millivolts up to −59 millivolts. Instead, an increase of this magnitude requires the simultaneous discharge of many knobs — ten or more usually — at the same time or in rapid succession. This occurs by a process called *summation,* which will be discussed in detail in the following sections.

Generation of Action Potentials at the Initial Segment of the Axon — Threshold for Excita-

tion. When the membrane potential inside the neuron rises high enough, there comes a point that this initiates an action potential in the neuron. However, the action potential does not begin on the somal membrane adjacent to the excitatory synapses. Instead, it begins in the initial segment of the axon (the "axon hillock"). This may be explained as follows: Any factor that increases the potential inside the soma at any single point also increases this potential everywhere in the soma at the same time because of the very low resistance inside the soma, as explained earlier. Yet, because of physical differences in the membrane and differences in geometrical arrangement of the membrane in different parts of the neuron, the most excitable part of the neuron by far is the initial segment of the axon — that is, the first 50 to 100 microns of the axon after it leaves the neuronal soma. The excitatory postsynaptic potential that will elicit an action potential at this point on the neuron is approximately 11 millivolts. This is in contrast to approximately 30 millivolts required on the soma itself. Therefore, the new action potential originates in the initial segment of the axon and not on the soma. Once the action potential begins, it travels peripherally along the axon and also travels backward over the soma of the neuron (however, it usually does not travel backward very far into the dendrites).

Thus, in Figure 46–8B, it is shown that under normal conditions the *threshold* for excitation of the neuron is −59 millivolts, which represents an excitatory postsynaptic potential of +11 millivolts — that is, 11 millivolts more positive than the normal resting neuronal potential of −70 millivolts.

ELECTRICAL EVENTS IN NEURONAL INHIBITION

Effect of Inhibitory Synapses on the Postsynaptic Membrane — The Inhibitory Postsynaptic Potential. It was pointed out above that excitatory synapses increase the permeability of the somal membrane to all ions — including sodium, potassium, and chloride. The inhibitory synapses, in contrast, increase the permeability of the postsynaptic membrane only to potassium and chloride ions. And the opened channels are too small to allow the large hydrated sodium ions to pass through. However, potassium efflux does occur, as illustrated in Figure 46–8C. The reason potassium flows outward through the membrane is the following: In the resting state, a weak potassium pump has been pumping a slight excess of potassium ions to the interior of the neuron. Therefore, there are too many potassium ions on the inside to be in an exact equilibrium state as described by the Nernst equation for equilibrium conditions.

Consequently, opening of the pores causes outward diffusion of some of the excess potassium ions, thereby decreasing the positive ions inside the neuron and leaving the nondiffusible negative ions of the neuron (protein ions, phosphate ions, and others). This concentration of negative ions makes the internal potential of the neuron more negative, as illustrated by the −75 millivolt potential inside the neuron in Figure 46–8C. This is called a *hyperpolarized state*. And the 5 millivolt decrease in intraneuronal voltage below the normal resting potential of −70 millivolts caused by the inhibitory transmitter is called the *inhibitory postsynaptic potential*.

Obviously, the increased negativity of the membrane potential (−75 millivolts) makes the neuron less excitable than it is normally. Since the potential must rise to −59 millivolts to excite the neuron, an excitatory postsynaptic potential must now be 16 millivolts instead of the normal 11 millivolts to cause excitation. Thus, in this way the inhibitory transmitter inhibits the neuron.

"Clamping" of the Resting Membrane Potential as a Means to Inhibit Neurons. Sometimes, excitation of the inhibitory synapses causes little or no inhibitory postsynaptic potential, but it still inhibits the neuron. The reason that the potential does not change is that in some neurons the summated concentration differences across the membrane for the potassium and chloride ions are only barely able to create a diffusion potential equal to the normal resting potential. Therefore, when the inhibitory pores open, there is no net flow of ions to cause an inhibitory postsynaptic potential. Yet, both the potassium and the chloride ions do diffuse bidirectionally through the wide-open pores many times as rapidly as normally, and this high flux of these two ions inhibits the neuron in the following way: When excitatory synapses fire and sodium ions flow into the neuron, this raises the intraneuronal voltage far less than usual because any tendency for the membrane potential to change away from the resting potential is immediately opposed by rapid flux of potassium and chloride ions through the inhibitory pores to bring the potential back to the resting value. Therefore, the amount of influx of sodium ions required to cause excitation may be increased to as much as 5 to 20 times normal.

This tendency for the potassium and chloride ions to maintain a membrane potential near the resting value when the inhibitory pores are wide open is called "clamping" of the potential.

To express the phenomenon for clamping more mathematically, one needs to recall the Goldman equation from Chapter 10. This equation indicates that the membrane potential is determined by summation of the tendencies for the different ions to carry electrical charges through the membrane in the two directions. The membrane potential will approach the Nernst equilibrium potential for those ions that permeate the membrane to the greatest extent. When the inhibitory channels are wide open, the chloride and potassium ions permeate the membrane very greatly. Therefore, when the excitatory channels open, it is difficult to raise the neuronal potential up to the threshold value for excitation.

It is also interesting that some recent studies suggest that inhibitory channels are considerably more permeable to chloride ions than to potassium ions. Since chloride permeability inhibits neurons only by the clamping mechanism, this emphasizes the importance of clamping in the inhibitory process.

Presynaptic Inhibition

In addition to the inhibition caused by inhibitory synapses operating at the neuronal membrane, called *postsynaptic inhibition,* another type of inhibition often occurs before the signal reaches the synapse. This type of inhibition, called *presynaptic inhibition,* is believed to occur in the following way:

In presynaptic inhibition, instead of the postsynaptic neuron being inhibited, the presynaptic terminal fibrils and synaptic knobs are inhibited. This inhibition is caused by "presynaptic" synapses that lie on the terminal fibrils and synaptic knobs before the knobs themselves terminate on the following neuron. It is believed that these presynaptic synapses secrete a transmitter substance that in some way not yet understood depresses the voltage of the action potential that occurs at the membrane of the knob and, as has already been pointed out, this greatly decreases the amount of transmitter released by the knob. Therefore, the degree of excitation of the neuron is also greatly suppressed, or inhibited.

The nature of the transmitter substance released at the presynaptic synapses is not known. However, presynaptic inhibition is blocked by *picrotoxin* and also by *bicuculline*. Since these two substances also block the action of the transmitter substance GABA, it has been suggested that the transmitter that causes presynaptic inhibition is GABA. On the other hand, the transmitter substance probably associated with most postsynaptic inhibition in the spinal cord is glycine. The reason for believing this is that strychnine inhibits this type of inhibition, and it also blocks the inhibitory action of glycine, but not of GABA, on neurons.

Presynaptic inhibition is different from postsynaptic inhibition in its time sequence. It requires many milliseconds to develop, but once it does occur, it can last for as long as a half second or perhaps even longer, while postsynaptic inhibition, at least of the anterior motoneurons, lasts for only 10 to 15 milliseconds.

Presynaptic inhibition occurs especially at the more peripheral synapses of the sensory pathways. Shortly after a strong sensory signal enters the sensory pathways, negative feedback automatically causes increasing presynaptic inhibition. This, in turn, reduces the degree of transmission of the sensory signals. Therefore, the greater the intensity of the input signal, the greater also becomes the negative feedback inhibition. In this way presynaptic inhibition acts as a sensitivity control on the sensory input. It also sharpens the boundaries between stimulated and nonstimulated areas of the sensory pathway because it prevents excessive spread of the sensory signals to the unexcited neurons. This process is called "contrast enhancement."

SUMMATION OF POSTSYNAPTIC POTENTIALS

Time Course of Postsynaptic Potentials. When a synapse excites the anterior motoneuron, the neuronal membrane becomes highly permeable for only 1 to 2 milliseconds. During this time sodium ions diffuse rapidly to the interior of the cell to increase the intraneuronal potential, thus creating the *excitatory postsynaptic potential*. This potential then persists for about 15 milliseconds because this is the time required for the sodium ions to be pumped out or for potassium ions to leak out or chloride ions to leak in to re-establish the normal resting membrane potential.

Precisely the same effect occurs for the inhibitory postsynaptic potential. That is, the inhibitory synapse increases the permeability of the membrane to potassium and chloride ions for 1 to 2 milliseconds, and this usually decreases the intraneuronal potential to a more negative value than normal, thereby creating the *inhibitory postsynaptic potential*. This potential also persists for about 15 milliseconds.

However, other types of transmitter substances acting on other neurons can perhaps excite or inhibit for hundreds of milliseconds or even for seconds, minutes, or hours.

Spatial Summation of the Postsynaptic Potentials. It has already been pointed out that excitation of a single synaptic knob on the surface of a neuron will almost never excite the neuron. The reason for this is that sufficient transmitter substance is released by a single knob to cause an excitatory postsynaptic potential usually no more than a millivolt at most, instead of the required 10 or more millivolts that is the usual threshold for excitation. However, during an excitatory state in a neuronal pool of the nervous system, many synaptic knobs are usually stimulated at the same time, and even though these knobs are spread over wide areas of the neuron, their effects can still summate. The reason for this summation is the following: It has already been pointed out that a change in the potential at any single point within the soma will cause the potential to change everywhere in the soma almost exactly equally. Therefore, for each excitatory synapse that discharges simultaneously, the intrasomal potential rises another quantal amount — up to about 1 millivolt. When the excitatory postsynaptic potential becomes great enough, the threshold for firing will be reached, and an action potential will generate at the initial segment of the axon. This effect is illustrated in Figure 46–9, which shows several excitatory postsynaptic potentials. The bottom postsynaptic potential in the figure was caused by stimulation of only four excitatory synapses; then the next higher potential was caused by stimulation of two times as many synapses; finally, a still higher excitatory postsynaptic potential was caused by stimulation of four times as many synapses. This time an action potential was generated at the axon hillock.

This effect of summing simultaneous postsynaptic potentials by excitation of multiple knobs on widely spaced areas of the membrane is called *spatial summation*.

Temporal Summation. Most synaptic knobs can fire repetitively in rapid succession only a few milliseconds apart. Each time a knob fires, the released transmitter substance opens the membrane pores for a millisecond or so. Since the postsynaptic potential lasts for up to 15 milliseconds, a second opening of the same pore can increase the postsynaptic potential to a still greater level so that the more rapid the rate of knob stimulation, the greater the effective postsynaptic potential. Thus, successive postsynaptic potentials of individual synaptic knobs, if they occur rapidly enough, can summate in the same way that postsynaptic potentials can summate

from widely distributed knobs over the surface of the neuron. This summation is called *temporal summation*.

Simultaneous Summation of Inhibitory and Excitatory Postsynaptic Potentials. Obviously, if an inhibitory postsynaptic potential is tending to decrease the membrane potential to a more negative value while an excitatory postsynaptic potential is tending to increase the potential at the same time, these two effects can either completely nullify each other or partially nullify each other. Also, inhibitory "clamping" of the membrane potential can nullify much of an excitatory potential. Thus, if a neuron is currently being excited by an excitatory postsynaptic potential, then an inhibitory signal from another source can easily reduce the postsynaptic potential below the threshold value for excitation, thus turning off the activity of the neuron.

Facilitation of Neurons. Often the summated postsynaptic potential is excitatory in nature but has not risen high enough to reach the threshold for excitation. When this happens the neuron is said to be *facilitated*. That is, its membrane potential is nearer the threshold for firing than normally, but it is not yet to the level of firing. Nevertheless, a signal entering the neuron from some other source can then excite the neuron very easily. Diffuse signals in the nervous system often facilitate large groups of neurons so that they can respond quickly and easily to signals arriving from secondary sources.

SPECIAL FUNCTIONS OF DENDRITES IN EXCITING NEURONS

The Large Spatial Field of Excitation of the Dendrites. The dendrites of the anterior motoneurons extend for 0.5 to 1 millimeter in all directions from the neuronal soma. Therefore, these dendrites can receive signals from a large spatial area around the motoneuron. This provides vast opportunity for summation of signals from many separate presynaptic neurons.

It is also important that between 80 and 90 per cent of all the synaptic knobs terminate on the dendrites of the anterior motoneuron in contrast to only 10 to 20 per cent terminating on the neuronal soma. Therefore, the preponderant share of the excitation of the neuron is provided by signals transmitted over the dendrites.

Failure of Many Dendrites to Transmit Action Potentials. Many dendrites fail to transmit action potentials because their thresholds for excitation are very high. Yet they do transmit *electrotonic current* down the dendrites to the soma. (Transmission of electrotonic current means the direct spread of current by electrical conduction in the

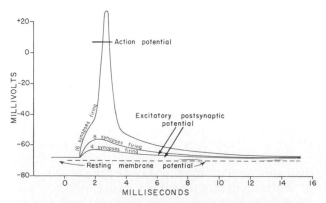

Figure 46–9. Excitatory postsynaptic potentials, showing that simultaneous firing of only a few synapses will not cause sufficient summated potential to elicit an action potential, but that simultaneous firing of many synapses will raise the summated potential to the threshold for excitation and cause a superimposed action potential.

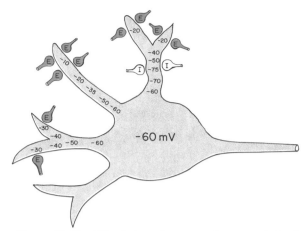

Figure 46-10. Stimulation of a neuron by synaptic knobs located on dendrites, showing, especially, decremental conduction of electrotonic potentials in the two dendrites to the left and inhibition of dendritic excitation in the dendrite that is uppermost.

fluids in the dendrites with no generation of action potentials.) Stimulation of the neuron by this current has special characteristics, as follows:

Decrement of Electrotonic Conduction in the Dendrites — Greater Excitation by Synapses Near the Soma. In Figure 46-10 a number of excitatory and inhibitory synapses are shown stimulating the dendrites of a neuron. On the two dendrites to the left in the figure are shown excitatory effects near the ends of the dendrites; note the high levels of the excitatory postsynaptic potentials at these ends — that is, the less negative potentials. However, a large share of the excitatory postsynaptic potential is lost before it reaches the soma. The reason for this is that the dendrites are long and thin, and their membranes are also very leaky to electrical current. Therefore, before the excitatory potentials can reach the soma, a large share of the potential is lost by leakage through the membrane. This decrease in membrane potential as it spreads along dendrites toward the soma is called *decremental conduction.*

It is also obvious that the nearer the excitatory synapse is to the soma of the neuron, the less will be the decrement of conduction. Therefore, those synapses that lie near the soma have far more excitatory effect than those that lie far away from the soma.

Rapid Re-excitation of the Neuron by the Dendrites After the Neuron Fires. When an action potential is generated in a neuron, this action potential spreads back over the soma but not always over the dendrites. Furthermore, the postsynaptic potentials in the dendrites often remain relatively undisturbed by the action potential. Therefore, just as soon as the action potential is over, the potentials still existing in the dendrites are ready and waiting to re-excite the neuron. Thus, the

dendrites have a "holding capacity" for the excitatory signal from presynaptic sources.

Summation of Excitation and Inhibition in Dendrites. The uppermost dendrite of Figure 46-10 is shown to be stimulated by both excitatory and inhibitory synapses. At the tip of the dendrite is a strong excitatory postsynaptic potential, but much of this is lost by decremental conduction along the dendrite. Then near the soma two inhibitory synapses are stimulating the same dendrite. These inhibitory synapses provide a hyperpolarizing voltage in the dendrite that completely nullifies the excitatory effect and indeed transmits a small amount of inhibition by decremental electrotonic conduction toward the soma. Thus, dendrites can summate excitatory and inhibitory postsynaptic potentials in the same way that the soma can. In those instances when action potentials do develop in dendrites, inhibitory synapses located near the soma can completely block the action potentials and can prevent their ever entering the soma of the neuron. It is interesting that the inhibitory synapses are especially abundant on or near the soma.

RELATION OF STATE OF EXCITATION OF THE NEURON TO THE RATE OF FIRING

The "Central Excitatory State." The "central excitatory state" of a neuron is defined as the degree of excitatory drive to the neuron. If there is a higher degree of excitation than inhibition of the neuron at any given instant, then it is said that there is a *central excitatory state.* On the other hand, if there is more inhibition than excitation, then it is said that there is a *central inhibitory state.*

When the central excitatory state of a neuron rises above the threshold for excitation, then the neuron will fire repetitively as long as the excitatory state remains at this level. However, the rate at which it will fire is determined by how much above threshold it is. To explain this, we must first consider what happens to the neuronal somal potential during and following the action potential.

Changes in Neuronal Somal Potential During and Following the Action Potential. Figure 46-11 illustrates an action potential initiated by an excitatory postsynaptic potential. Following the spike portion of the action potential, there is a very long *positive after-potential* lasting for many milliseconds. During this interval the somal membrane potential falls below the normal resting membrane potential of −70 millivolts. This is probably caused by a very high degree of permeability of the neuronal membrane to potassium ions that persists for many milliseconds after the action potential is over. As a result, potassium ions diffuse from inside the neuron to the exterior carrying positive charges out of the neuron and, therefore, creating an excessively high degree of negativity inside the neuron. This is called *hyperpolarization*.

The importance of this state of hyperpolarization after the spike potential is that the neuron remains in an *inhibited state* during this period of time. Therefore, a far greater central excitatory state is required during this time than normally to cause re-excitation of the neuron.

Relationship of Central Excitatory State to Frequency of Firing. The curve shown at the top of Figure 46–11, labeled "central excitatory state required for re-excitation," depicts the relative level of the central excitatory state required at each instant after an action potential is over to re-excite the neuron. Note that very soon after an action potential is over a very high central excitatory state is required. That is, a very large number of excitatory synapses must be firing simultaneously. Then, after many milliseconds have passed and the state of hyperpolarization of the neuron has begun to disappear, the central excitatory state required becomes greatly reduced.

Therefore, it is immediately evident that when the central excitatory state is high a second action potential will appear very soon after the previous one. Then still a third action potential will appear soon after the second, and this process will continue indefinitely. Thus, at a very high central excitatory state the frequency of firing of the neuron is great.

On the other hand, when the central excitatory state is only barely above threshold, the neuron must recover almost completely from the hyperpolarization, which requires many milliseconds,

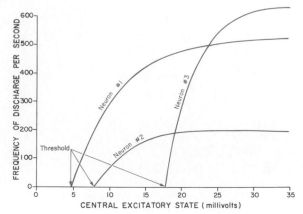

Figure 46–12. Response characteristics of different types of neurons to progressively increasing central excitatory states.

before it will fire again. Therefore, the frequency of neuronal firing is low.

(Note that the increase in the central excitatory state that is required for re-excitation is far greater than the decrease in the neuronal potential during the state of hyperpolarization. The reason for this is that at the same time that the neuron is hyperpolarized its voltage is also "clamped" to a great extent by the high permeability of the membrane to potassium and chloride ions. This clamping mechanism of neuronal inhibition was discussed earlier in the chapter.)

Response Characteristics of Different Neurons to Increasing Levels of Central Excitatory State. Histological study of the nervous system immediately convinces one of the widely varying types of neurons in different parts of the nervous system. And, physiologically, the different types of neurons perform different functions. Therefore, as would be expected, the ability to respond to stimulation by the synapses varies from one type of neuron to another.

Figure 46–12 illustrates theoretical responses of three different types of neurons to varying levels of central excitatory state. Note that neuron number 1 will not discharge at all until the excitatory state rises to 5 millivolts. Then, as the excitatory state rises progressively to 35 millivolts the frequency of discharge rises to slightly over 500 per second.

Neuron number 2 is quite a different type, having a threshold for excitation of 8 millivolts and a maximum rate of discharge of only 190 per second. Finally, neuron number 3 has a high threshold for excitation, about 18 millivolts, but its discharge rate rises rapidly to over 600 per second as the central excitatory state rises only slightly above threshold.

Thus, different neurons respond differently, have different thresholds for excitation, and have

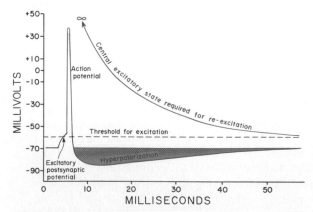

Figure 46–11. A neuronal potential followed by a prolonged period of neuronal hyperpolarization. Also shown is the central excitatory state required for re-excitation of the neuron at given intervals after the action potential is over.

widely differing maximal frequencies of discharge. With a little imagination one can readily understand the importance of having neurons with many different types of response characteristics to perform the widely varying functions of the nervous system.

SOME SPECIAL CHARACTERISTICS OF SYNAPTIC TRANSMISSION

Forward Conduction Through Synapses. From the above discussion, it should be evident by now that impulses are conducted through synapses only from the synaptic knobs to the successive neurons and never in the reverse direction. This is called the *principle of forward conduction*.

Fatigue of Synaptic Transmission. When excitatory synapses are repetitively stimulated at a rapid rate, the number of discharges by the postsynaptic neuron is at first very great, but it becomes progressively less in succeeding milliseconds or seconds. This is called *fatigue* of synaptic transmission.

Fatigue is an exceedingly important characteristic of synaptic function, for when areas of the nervous system become overexcited, fatigue causes them to lose this excess excitability after a while. For example, fatigue is probably the most important means by which the excess excitability of the brain during an epileptic fit is finally subdued so that the fit ceases. Thus, the development of fatigue is a protective mechanism against excess neuronal activity. This will be discussed further in the description of reverberating neuronal circuits in the following chapter.

The mechanism of fatigue is mainly exhaustion of the stores of transmitter substance in the synaptic knobs, particularly since it has been calculated that the excitatory knobs can store enough excitatory transmitter for only 10,000 normal synaptic transmissions, an amount that can be exhausted in only a few seconds to a few minutes. However, part of the fatigue process might also result from progressive inactivation of the membrane receptors.

Post-Tetanic Facilitation. When a rapidly repetitive series of impulses stimulates an excitatory synapse for a period of time and then a rest period is allowed, the postsynaptic neuron will usually be even more responsive to subsequent stimulation than normally. This is called *post-tetanic facilitation*.

Experiments have shown that at least two factors help to cause post-tetanic facilitation. First, for reasons not yet fully understood, a rapidly stimulated synaptic knob develops a state of hyperpolarization (more negative membrane potential than normal), and this causes the action potentials of the synaptic knob to become considerably increased, an effect that also causes increased release of transmitter substance. Second, the transmitter vesicles in the synaptic knob seem to become more mobile, or at least to release their transmitter substances more easily than usual, and this also increases the amount of transmitter substance released. This probably results from enhanced permeability of the synaptic knobs to calcium, the calcium in turn increasing the vesicular release of transmitter. (It is possible that the postsynaptic receptor membrane also becomes more sensitive to the transmitter substance, but this has not yet been proved.) At any rate, post-tetanic facilitation causes the excitatory postsynaptic potential to increase to as much as 2 times normal.

The physiological significance of post-tetanic facilitation is still very doubtful, and it may have no real significance at all. However, since post-tetanic facilitation can last from a few seconds in some neurons to several hours in others, it is immediately apparent that neurons could possibly store information by this mechanism. Therefore post-tetanic facilitation might well be a mechanism of "short-term" memory in the central nervous system. This possibility will be discussed at length in Chapter 55 in relation to the memory function of the cerebral cortex.

Effect of Acidosis and Alkalosis on Synaptic Transmission. The neurons are highly responsive to changes in pH of the surrounding interstitial fluids. *Alkalosis greatly increases neuronal excitability*. For instance, a rise in arterial pH from the normal of 7.4 to about 7.8 often causes cerebral convulsions because of increased excitability of the neurons. This can be demonstrated especially well by having a person who is normally predisposed to epileptic fits overbreathe. The overbreathing elevates the pH of the blood only momentarily, but even this short interval can often precipitate an epileptic convulsive attack.

On the other hand, *acidosis greatly depresses neuronal activity;* a fall in pH from 7.4 to below 7.0 usually causes a comatose state. For instance, in very severe diabetic or uremic acidosis, coma always develops.

Effect of Hypoxia on Synaptic Transmission. Neuronal excitability is also highly dependent on an adequate supply of oxygen. Cessation of oxygen supply for only a few seconds can cause complete inexcitability of the neurons. This is often seen when the cerebral circulation is temporarily interrupted, for within 3 to 5 seconds the person becomes unconscious.

Effect of Drugs on Synaptic Transmission. Many different drugs are known to increase the excitability of neurons, and others are known to decrease the excitability. For instance, caffeine, theophylline, and theobromine, which are found in coffee, tea, and cocoa, respectively, all increase neuronal excitability, presumably by reducing the threshold for excitation of the neurons. However, strychnine, which is one of the best known of all the agents that increase the excitability of neurons, does not reduce the threshold for excitation of the neurons at all but, instead, *inhibits the action of at least some of the inhibitory transmitters* on the neurons, probably especially the inhibitory effect of glycine in the spinal cord. In consequence, the effects of the excitatory transmitters become overwhelming, and the neurons become so excited that they go into rapidly repetitive discharge, resulting in severe convulsions.

Most anesthetics have long been believed to increase membrane threshold for excitation and thereby to decrease neuronal activity throughout the body. This is based principally on the fact that most of the volatile anesthetics are chemically inert compounds but are lipid soluble. Therefore, it has been reasoned that these substances might change the physical characteristics of the neuronal membranes, making them less responsive to excitatory agents.

Synaptic Delay. In transmission of an impulse from a synaptic knob to a postsynaptic neuron, a certain amount of time is consumed in the process of (a) discharge of the transmitter substance by the knob, (b) diffusion of the transmitter to the subsynaptic neuronal membrane, (c) action of the transmitter on the membrane receptor, (d) action of the receptor to increase the membrane permeability, and (e) inward diffusion of sodium to raise the excitatory postsynaptic potential to a high enough value to elicit an action potential. The *minimal* period of time required for all these events to take place, even when large numbers of presynaptic terminals are stimulated simultaneously, is 0.3 to 0.5 millisecond. This is called the *synaptic delay*. It is important for the following reason: Neurophysiologists can measure the *minimal* delay time between an input volley of impulses and an output volley and from this can estimate the number of series neurons in the circuit.

REFERENCES

Adams, D. J., *et al.*: Ionic currents in molluscan soma. *Annu. Rev. Neurosci., 3*:141, 1980.

Baker, P. F., and Reuter, H.: Calcium Movement in Excitable Cells. New York, Pergamon Press, 1975.

Barker, J. L.: Peptides: Roles in neuronal excitability. *Physiol. Rev., 56*:435, 1976.

Bennett, M. V. L.: Electrical transmission: A functional analysis and comparison with chemical transmission. *In* Brookhart, J. M., and Mountcastle, V. B. (eds.): Handbook of Physiology. Sec. 1, Vol. 1. Baltimore, Williams & Wilkins, 1977, p. 357.

Bourne, G. H., *et al.* (eds.): Neuronal Cells and Hormones. New York, Academic Press, 1978.

Ceccarelli, B., and Clementi, F. (eds.): Neurotoxins, Tools in Neurobiology. New York, Raven Press, 1979.

Ceccarelli, B., and Hurlbut, W. P.: Vesicle hypothesis of the release of quanta of acetylcholine. *Physiol. Rev., 60*:396, 1980.

Cobb, W. A., and vanDuijn, H. (eds.): Contemporary Clinical Neurophysiology. New York, Elsevier Scientific Publishing Co., 1978.

Cooper, J. R. *et al.*: The Biochemical Basis of Neuropharmacology. New York, Oxford University Press, 1978.

Cuénod, M., *et al.* (eds.): Development and Chemical Specificity of Neurons. New York, Elsevier Scientific Publishing Co., 1979.

Duggan, A. W.: Pharmacology of mammalian central inhibition: *In* Porter, R. (ed.): International Review of Physiology: Neurophysiology III. Vol. 17. Baltimore, University Park Press, 1978, p. 119.

Eccles, J. C.: My scientific odyssey. *Annu. Rev. Physiol., 39*:1, 1977.

Fuxe, K. (ed.): Peptidergic Neuron. New York, Plenum Press, 1979.

Gotto, A. M., Jr., *et al.* (eds.): Brain Peptides: A New Endocrinology. New York, Elsevier/North-Holland, 1979.

Guroff, G.: Molecular Neurobiology. New York, Marcel Dekker, 1979.

Horn, A. S., *et al.* (eds.): The Neurology of Dopamine. New York, Academic Press, 1979.

Iversen, L. L.: The chemistry of the brain. *Sci. Am. 241*(3):134, 1979.

Johnston, G. A. R.: Neuropharmacology of amino acid inhibitory transmitters. *Annu. Rev. Pharmacol. Toxicol., 18*:269, 1978.

Kelly, R. B., *et al.*: Biochemistry of neurotransmitter release. *Annu. Rev. Neurosci., 2*:399, 1979.

Krogsgaard-Larsen, P., *et al.*: (eds.): GABA-Neuro-Transmitters. New York, Academic Press, 1979.

Kupfermann, I.: Modulatory actions of neurotransmitters. *Annu. Rev. Neurosci., 2*:447, 1979.

Langer, S. Z., *et al.* (eds.): Presynaptic Receptors. New York, Pergamon Press, 1979.

Lynch, G., and Schubert, P.: The use of in vitro brain slices for multidisciplinary studies of synaptic function. *Ann. Rev. Neurosci., 3*:1, 1980.

Maelicke, A., *et al.*: Biochemical aspects of neurotransmitter receptors. *In* Brookhart, J. M., and Mountcastle, V. B. (eds.): Handbook of Physiology. Sec. 1, Vol. 1. Baltimore, Williams & Wilkins, 1977, p. 493.

Mandel, P., and DeFeudis, F. V. (eds.): GABA — Biochemistry and CNS Functions. New York, Plenum Press, 1979.

McLachlan, E. M.: The statistics of transmitter release at chemical synapses. *In* Porter, R. (ed.): International Review of Physiology: Neurophysiology III. Vol. 17. Baltimore, University Park Press, 1978, p. 49.

Moore, R. Y., and Bloom, F. E.: Central catecholamine neuron systems: Anatomy and physiology of the dopamine systems. *Annu. Rev. Neurosci., 1*:129, 1978.

Nicoll, R. A., *et al.*: Substance P as a transmitter candidate. *Ann. Rev. Neurosci., 3*:227, 1980.

Oja, S. S., *et al.*: Amino Acids as Inhibitory Neurotransmitters. New York, Fischer, 1977.

Otsuka, M., and Hall, Z. W. (eds.): Neurobiology of Chemical Transmission. New York, John Wiley & Sons, 1978.

Passonneau, J. V., *et al.*: Cerebral Metabolism and Neural Function. Baltimore, Williams & Wilkins, 1980.

Pepeu, G., *et al.* (eds.): Receptors for Neurotransmitters and Peptide Hormones. New York, Raven Press, 1980.

Redman, S. J.: A quantitative approach to integrative function of dendrites. *Int. Rev. Physiol., 10*:1, 1976.

Rose, F. C. (ed.): Physiological Aspects of Clinical Neurology. Oxford, Blackwell Scientific Publications, 1977.

Schulster, D., and Levitzki, A. (eds.): Cellular Receptors for Hormones and Neurotransmitters. New York, John Wiley & Sons, 1980.

Simon, P. (ed.): Neurotransmitters. New York, Pergamon Press, 1979.

Snyder, S. H., and Bennett, J. P., Jr.: Neurotransmitter receptors in the brain: Biochemical identification. *Annu. Rev. Physiol., 38*:153, 1976.

Spencer, W. A.: The physiology of supraspinal neurons in mammals. *In* Brookhart, J. M., and Mountcastle, V. B. (eds.): Handbook of Physiology. Sec. 1, Vol. 1. Baltimore, Williams & Wilkins, 1977, p. 969.

Stephenson, W. K.: Concepts of Neurophysiology. New York, John Wiley & Sons, 1980.

Stevens, C. F.: The neuron. *Sci. Am. 241*(3):54, 1979.

Tipton, K. F. (ed.): Neurochemistry and Biochemical Pharmacology. Baltimore, University Park Press, 1978.

Vernadakis, A., *et al.* (eds.): Maturation of Neurotransmission: Biochemical Aspects. New York, S. Karger, 1977.

47

Neuronal Mechanisms and Circuits for Processing Information

The primary role of the nervous system is to control a major share of the body's functions. To do this, it gains information from the body's surroundings and from within the body itself and then transmits this information, stores it, changes it, and uses it in any other way that can be advantageous for bodily control. The entire section of this book on the nervous system addresses these many functions of the nervous system.

Throughout the nervous system, certain special neuronal mechanisms and neuronal circuits are used again and again. The purpose of this chapter is to characterize some of these and to preview some of their important uses.

INFORMATION, SIGNALS, AND IMPULSES

The term *information,* as it applies to the nervous system, means a variety of different things, such as knowledge, facts, quantitative values, intensity of pain, intensity of light, temperature, and any other aspect of the body or its immediate surroundings that has meaning. Thus, pain from a pin prick is information, pressure on the bottom of the feet is information, degree of angulation of the joints is information, and a stored memory in the brain is information.

However, information cannot be transmitted in its original form but only in the form of nerve impulses. Thus, a part of the body that is subjected to pain must first convert this information into *nerve impulses;* specific areas of the brain convert abstract thoughts also into nerve impulses that are then transmitted either elsewhere in the brain or into peripheral nerves to motor effectors throughout the body. The retina of the eye converts vision into nerve impulses, the nerve endings of the

joints convert degree of angulation of the joints into nerve impulses, and so forth.

Signals. In the transmission of information, it is frequently not desirable to speak in terms of the individual impulses but instead of the overall pattern of impulses; this pattern is called a *signal.* As an example, when pressure is applied to a large area of skin, impulses are transmitted by large numbers of parallel nerve fibers, and the total pattern of impulses transmitted by all these fibers is called a signal. Thus, we can speak of visual signals, auditory signals, somesthetic sensory signals, motor signals, and so forth.

TRANSMISSION OF SIGNALS IN NERVE TRACTS

SIGNAL STRENGTH

Obviously, one of the characteristics of information that must be conveyed is the quantitative intensity of the information, for instance the intensity of pain. The different gradations of intensity can be transmitted either by utilizing increasing numbers of parallel fibers or by sending more impulses along a single fiber. These two mechanisms are called respectively, spatial summation and temporal summation.

Spatial Summation (Multiple Fiber Summation). Figure 47–1 illustrates the phenomenon of spatial summation, whereby increasing signal strength is transmitted by using progressively greater numbers of fibers. This figure shows a section of skin innervated by a large number of parallel pain nerve fibers. Each of these arborizes into hundreds of minute *free nerve endings* that serve as pain receptors. The entire cluster of fibers from one pain fiber frequently covers an area of skin as large as 5 cm. in diameter, and this area is called the *receptive field*. The number of endings is large in the center of the field but becomes less and less toward the periphery. One can also see from the figure that the arborizing nerve fibrils overlap those from other pain fibers. Therefore, a pin prick of the skin usually stimulates endings from many different pain fibers simultaneously. But, if the pin prick is in the center of the receptive field of a particular pain fiber, the degree of stimulation is far greater than if it is in the periphery of the field. Thus, a mild pin prick might stimulate only the pain fiber receiving the signal from the center of its receptive field, but a strong pin prick would stimulate all the pain fibers whose receptive fields overlap at that point.

Thus, in the lower part of Figure 47–1 is shown three separate views of the cross-section of the nerve bundle leading from the skin area. To the left is shown the effect of a weak stimulus, with only a single nerve fiber in the middle of the bundle stimulated very strongly (represented by the solid fiber) while several adjacent fibers are stimulated weakly (half-solid fibers). The other two views of the nerve cross-section show the effect respectively of a moderate stimulus and a strong stimulus, with progressively more fibers being stimulated. Thus, the signal spreads to more and more fibers. This phenomenon is called *spatial summation,* which means simply that one of the means by which signals of increasing strength are transmitted in the nervous system is by utilizing progressively greater numbers of fibers. The increase in number of fibers as the strength of signal increases is called *recruitment* of the additional fibers.

Temporal Summation. A second means by which signals of increasing strength are transmitted is by increasing the *frequency* of nerve im-

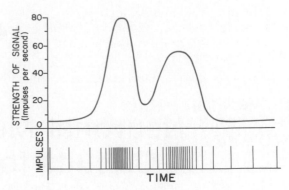

Figure 47–2. Translation of signal strength into a frequency-modulated series of nerve impulses, showing *above* the strength of signal and *below* the separate nerve impulses.

pulses in each fiber, which is called *temporal summation*. Figure 47–2 illustrates this type of summation, showing in the upper part a changing strength of signal and in the lower part the actual impulses transmitted by the nerve fiber.

SPATIAL ORIENTATION OF SIGNALS IN FIBER TRACTS

How does the brain detect the area on the body that is receiving a sensory stimulus, and how does the brain transmit impulses to individual skeletal muscles? It does this by transmitting their signals in spatial patterns through the nerve tracts. All the different nerve tracts, both in the peripheral nerves and in the fiber tracts of the central nervous system, are spatially organized. For instance, in the dorsal columns of the spinal cord the sensory fibers from the feet lie toward the midline, while those fibers entering the dorsal columns at higher levels of the body lie progressively more toward the lateral sides of the dorsal columns. This spatial organization is maintained with precision in this sensory pathway all the way to the somesthetic cortex. Likewise, the fiber tracts within the brain and those extending into motor nerves are spatially oriented in the same way.

As an example, Figure 47–3 illustrates to the left three separate pins stimulating the skin; to the

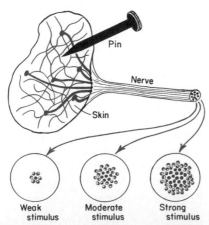

Figure 47–1. Pattern of stimulation of pain fibers in a nerve trunk leading from an area of skin pricked by a pin.

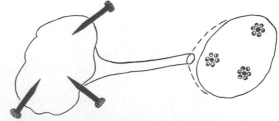

Figure 47–3. Spatial pattern of nerve fiber stimulation in a nerve trunk following stimulation of the skin by three separate but simultaneous pinpricks.

right the spatial orientation of the stimulated fibers in the nerve is shown. Each pin in this example stimulates a single fiber strongly and adjacent fibers less strongly.

TRANSMISSION AND PROCESSING OF SIGNALS IN NEURONAL POOLS

The central nervous system is made up of literally hundreds or even thousands of separate neuronal pools, some of which contain very few neurons, while others hold vast numbers. For instance, the entire cerebral cortex could be considered to be a single large neuronal pool. If all the surface area of the cerebral cortex were flattened out, including the surfaces of the penetrating folds, the total area of this large flat pool would be several square feet. It has many separate fiber tracts coming to it (afferent fibers) and others leaving it (efferent fibers). Furthermore it maintains the same quality of spatial orientation as that found in the nerve bundles, individual points of the cortex connecting with specific points elsewhere in the nervous system or connecting through the peripheral nerves with specific points in the body. However, within this pool of neurons are large numbers of short nerve fibers whereby signals spread horizontally from neuron to neuron within the pool itself.

Other neuronal pools include the different basal ganglia and the specific nuclei in the thalamus, cerebellum, mesencephalon, pons, and medulla. Also, the entire dorsal gray matter of the spinal cord could be considered to be one long pool of neurons, and the entire anterior gray matter another long neuronal pool. Each pool has its own special characteristics of organization which cause it to process signals in its own special way. It is these special characteristics of the different pools that allow the multitude of functions of the nervous system. Yet, despite their differences in function, the pools also have many similarities which are described in the following pages.

RELAYING OF SIGNALS THROUGH NEURONAL POOLS

Organization of Neurons for Relaying Signals. Figure 47–4 is a schematic diagram of several neurons in a neuronal pool, showing "input" fibers to the left and "output" fibers to the right. Each input fiber divides hundreds to thousands of times, providing an average of a thousand or more terminal fibrils that spread over a large area in the pool to synapse with the dendrites or cell bodies of the neurons in the pool. The dendrites of some of

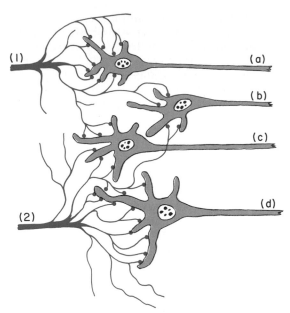

Figure 47–4. Basic organization of a neuronal pool.

the neurons also arborize and spread in the pool. The neuronal area stimulated by each incoming nerve fiber is called its *stimulatory field*. Note that each input fiber arborizes so that large numbers of its synaptic knobs lie on the centermost neurons in its "field," but progressively fewer knobs lie on the neurons farther from the center of the field.

Threshold and Subthreshold Stimuli — Facilitation. In the discussion of synaptic function in the previous chapter, it will be recalled that stimulation of a single excitatory synaptic knob almost never stimulates the postsynaptic neuron. Instead, large numbers of knobs must discharge on the same neuron either simultaneously or in rapid succession to cause excitation. For instance, let us assume that 6 separate knobs must discharge simultaneously or in rapid succession to excite any one of the neurons in Figure 47–4. Note that input fiber 1 contributes a total of 10 synaptic knobs to neuron *a*. Therefore, an incoming impulse in fiber 1 will cause neuron *a* to discharge. This same incoming fiber has two knobs on neuron *b* and three on neuron *c*. Neither of these two neurons will fire.

Incoming nerve fiber 2 has 12 synaptic knobs on neuron *d*, three on neuron *c*, and two on neuron *b*. In this case, an impulse in fiber 2 excites only neuron *d*.

Though fiber 1 fails to stimulate neurons *b* and *c*, discharge of the synaptic knobs changes the membrane potentials of these neurons so that they can be more easily excited by other incoming signals. Thus, a stimulus to a neuron can be either (1) an *excitatory stimulus,* also called a *threshold stimulus* because it is above the threshold required for excitation, or (2) a *subthreshold stimulus*. A subthreshold stimulus fails to excite the neuron but does make the neuron more excitable to impulses from other sources. The neuron that is made more excitable but does not discharge is said to be *facilitated*.

"Convergence" of Subthreshold Stimuli to Cause Excitation. Let us now assume that both fibers 1 and 2 in Figure 47–4 transmit concurrent impulses into the neuronal pool. In this case, neuron *c* is stimulated by six synaptic knobs simultaneously, which is the required number to cause excitation. Thus, subthreshold stimuli can *converge* from several sources and *summate* at a neuron to cause an excitatory stimulus.

Summation of Facilitation. Now, let us consider neuron *b* of Figure 47–4 when both input nerve fibers are stimulated. In this case each fiber excites two synaptic knobs on neuron *b,* but the total of four excited knobs still is not up to the threshold value required to excite the neuron. However, the degree of facilitation does become greatly increased so that now only two additional knobs excited from another source need fire to cause discharge.

The "Field" of Terminals. It must be recognized that Figure 47–4 represents a highly condensed version of the neuronal pool, for each input nerve fiber gives off terminals that spread to as many as 100 or more separate neurons, and the surface of each neuron is covered by many hundred or many thousand synapses. In the central-most portion of this *field* of terminals, almost all the neurons are stimulated by the incoming fiber; while farther toward the periphery of the field, the neurons are facilitated but do not discharge. Figure 47–5 illustrates this effect, showing the field of a single input nerve fiber. The area in the neuronal pool in which all the neurons discharge is called the *discharge* or *excited* or *liminal zone,* and the area to either side in which the neurons are facilitated but do not discharge is called the *facilitated* or *subthreshold* or *subliminal* zone.

Facilitation of the Neuronal Pool by Signals from Accessory Sources. A neuronal pool frequently receives input nerve fibers from many different sources. Thus, in Figure 47–6 a neuron in a given pool receives impulses from a *primary source* and

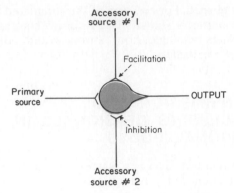

Figure 47–6. Basic neuronal circuit for facilitation or inhibition.

from two *accessory sources*. The nerve fiber from accessory source #1 is an excitatory fiber that stimulates excitatory synapses. If enough synapses from this source are stimulated, the postsynaptic neuron discharges. However, in many parts of the nervous system the accessory sources supply only a few nerve fibers to the pool, not enough usually to cause excitation but yet enough to *facilitate* the neurons. Thus, in Figure 47–6, if a facilitatory signal is entering the pool from accessory source #1, a much weaker than usual signal from the primary source is then able to excite the postsynaptic neuron. In this way the signal from accessory source #1 controls the ease with which signals pass from the primary source to the output, which explains one of the means by which different parts of the nervous system control the degree of activity of other parts.

Inhibition of a Neuronal Pool by Signals from an Accessory Source. It will be recalled from the previous chapter that many synapses in the central nervous system are inhibitory instead of excitatory. Thus, in Figure 47–6, stimulation of the inhibitory fibers from accessory source #2 strongly inhibits the neuronal pool so that a much stronger than usual signal from the primary source is now required to cause even normal output.

Mechanism By Which a Single Input Signal Can Cause Both Excitation and Inhibition — The Inhibitory Circuit. Figure 47–7 illustrates the so-called inhibitory circuit which can change an excitatory signal into an inhibitory signal. In this figure the

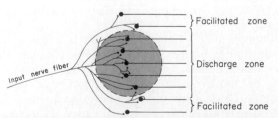

Figure 47–5. "Discharge" and "facilitated" zones of a neuronal pool.

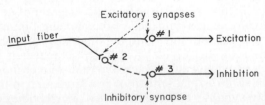

Figure 47–7. Inhibitory circuit. Neuron 2 is an inhibitory neuron.

input fiber divides and provides excitatory synapses at both its endings. This causes excitation of both neurons 1 and 2. However, the synapse between neurons 2 and 3 is inhibitory. Excitation of this neuron therefore inhibits neuron 3.

In short, the usual means to cause inhibition is for a signal *to be transmitted through a neuron that terminates at an inhibitory synapse.*

Convergence. The term *convergence* means control of a single neuron by converging signals from two or more separate input nerve fibers. One type of convergence was illustrated in Figure 47–4, in which two excitatory input nerve fibers from the same source converged upon several separate neurons to stimulate them. This type of convergence from a single source is illustrated again in Figure 47–8A.

However, convergence can also result from input signals (excitatory or inhibitory) from several different sources, which is illustrated in Figure 47–8B. For instance, the interneurons of the spinal cord receive converging signals from (a) peripheral nerve fibers entering the cord, (b) propriospinal fibers passing from one segment of the cord to another, (c) corticospinal fibers from the cerebral cortex, and (d) several other long pathways descending from the brain into the spinal cord. Then the signals from the interneurons converge on the motoneurons to control muscle function.

Such convergence allows summation of information from different sources, and the resulting response is a summated effect of all the different types of information. Obviously, therefore, convergence is one of the important means by which the central nervous system correlates, summates, and sorts different types of information.

Divergence. Divergence means that excitation of a single input nerve fiber stimulates multiple output fibers from the neuronal pool. The two major types of divergence are illustrated in Figure 47–9 and may be described as follows:

An *amplifying* type of divergence often occurs, illustrated in Figure 47–9A. This means simply

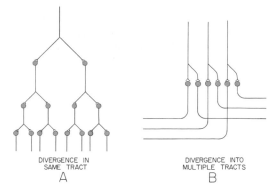

Figure 47–9. "Divergence" in neuronal pathways. A, Divergence within a pathway to cause "amplification" of the signal. B, Divergence into multiple tracts to transmit the signal to separate areas.

that an input signal spreads to an increasing number of neurons as it passes through successive pools of a nervous pathway. This type of divergence is characteristic of the corticospinal pathway in its control of skeletal muscles as follows: Stimulation of a single large pyramidal cell in the motor cortex transmits a single impulse into the spinal cord. Yet, under appropriate conditions of very strong cord facilitation, this impulse can stimulate perhaps several hundred interneurons and anterior motoneurons. Each of the motoneurons then stimulates as many as 100 to 1000 muscle fibers. Thus, there is a total divergence, or amplification, of as much as 10,000-fold.

The second type of divergence, illustrated in Figure 47–9B, is *divergence into multiple tracts*. In this case, the signal is transmitted in two separate directions from the pool, to different parts of the nervous system where it is needed. For instance, information transmitted in the dorsal columns of the spinal cord takes two courses in the lower part of the brain, (1) into the cerebellum and (2) on through the lower regions of the brain to the thalamus and cerebral cortex. Likewise, in the thalamus almost all sensory information is relayed both into deep structures of the thalamus and to discrete regions of the cerebral cortex.

TRANSMISSION OF SPATIAL PATTERNS THROUGH SUCCESSIVE NEURONAL POOLS

Most information is transmitted from one part of the nervous system to another through several successive neuronal pools. For instance, sensory information from the skin passes first through the peripheral nerve fibers, then through second order neurons that originate either in the spinal cord or in the cuneate and gracile nuclei of the medulla, and finally through third order neurons originating in the thalamus to the cerebral cortex. Such a

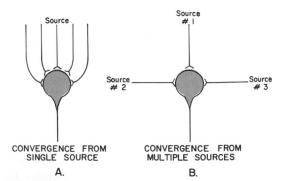

Figure 47–8. "Convergence" of multiple input fibers on a single neuron. A, Input fibers from a single source. B, Input fibers from multiple sources.

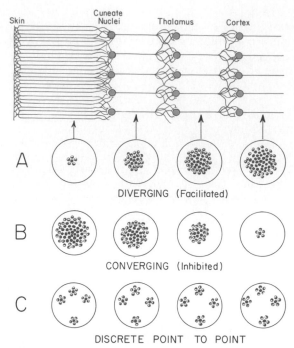

Figure 47–10. Typical organization of a sensory pathway from the skin to the cerebral cortex. *Below:* The patterns of fiber stimulation at different points in the pathway following stimulation by a pinprick when the pathway is (A) facilitated, (B) inhibited, and (C) normally excitable.

pathway is illustrated at the top of Figure 47–10. Note that the sensory nerve endings in the skin overlap each other tremendously; and the terminal fibrils of each nerve fiber, on entering each neuronal pool, spread to many adjacent neurons, innervating perhaps 100 or more separate neurons. On first thought, one would expect signals from the skin to become completely mixed up by this haphazard arrangement of terminal fibrils in each neuronal pool. For statistical reasons, however, this does not occur, which can be explained as follows:

First, if a single point is stimulated in the skin, the nerve fiber with the most nerve endings in that particular spot becomes stimulated to the strongest extent, while the immediately adjacent nerve fibers become stimulated less strongly, and the nerve fibers still farther away become stimulated only weakly. When this signal arrives at the first neuronal pool, the stimulus spreads in many directions in the terminal fibrils of the neuronal pool. Yet the *greatest number* of *excited presynaptic terminals* lies very near the point of entry of the most excited input nerve fiber. Therefore, the neuron closest to this central point is the one that becomes stimulated to the greatest extent. Exactly the same effect occurs in the second neuronal pool in the thalamus and again when the signal reaches the cerebral cortex.

Yet, it is true that a signal passing through *highly facilitated neuronal pools* could diverge so much that the spatial pattern at the terminus of the pathway would be completely obscured. This effect is illustrated in Figure 47–10A, which shows successively expanding spatial patterns of neuron stimulation in such a facilitated, diverging pathway.

However, the degree of facilitation of the different neuronal pools varies from time to time. Under some conditions the degree of facilitation is so low that the pathway becomes converging, as illustrated in Figure 47–10B. In this case, a broad area of the skin is stimulated, but the signal loses part of its fringe stimuli as it passes through each successive pool until the breadth of the stimulus becomes contracted at the opposite end. One can achieve this type of stimulation by pressing ever so lightly with a flat object on the skin. The signal converges to give the person a sensation of almost a point contact.

In Figure 47–10C, four separate points are simultaneously stimulated on the skin, and the degree of excitability in each neuronal pool is exactly that amount required to prevent either divergence or convergence. Therefore, a reasonably true spatial pattern of each of the four points of stimulation is transmitted through the entire pathway.

Centrifugal Control of Neuronal Facilitation in the Sensory Pathways. It is obvious from the above discussion that the degree of facilitation of each neuronal pool must be maintained at exactly the proper level if faithful transmission of the spatial pattern is to occur. Recent discoveries have demonstrated that the degrees of inhibition or facilitation of most — indeed, probably all — neuronal pools in the different sensory pathways are controlled by *centrifugal nerve fibers* that pass from the respective sensory areas of the cortex downward to the separate sensory relay neuronal pools. Thus, these nerve fibers undoubtedly help to control the faithfulness of signal transmission.

Lateral Inhibition to Provide Contrast in the Spatial Pattern. When a single point of the skin or other sensory area is stimulated, not only is a single fiber excited but a number of "fringe" fibers are excited less strongly at the same time, as already explained above. Therefore, the spatial pattern is blurred even before the signal begins to be transmitted through the pathway. However, in many pathways — if not all — such as in the visual pathway and in the somesthetic pathway, lateral *inhibitory circuits* inhibit the fringe neurons and re-establish a truer spatial pattern.

Figure 47–11A illustrates one of the postulated types of lateral inhibitory circuits, showing that the nerve fibers of a pathway give off collateral fibers that excite inhibitory neurons. These inhib-

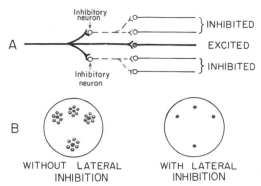

Figure 47–11. A, One type of lateral inhibitory circuit by which an excited fiber of a neuronal pool can cause inhibition of adjacent fibers. B, Increase in contrast of the stimulus pattern caused by the inhibitory circuit.

itory neurons in turn inhibit the less excited fringe neurons in the signal pathway. The effect of this on transmission of the spatial pattern is illustrated in Figure 47–11B, which shows the same point-to-point transmission pattern that was illustrated in Figure 47–10C. The left-hand pattern illustrates four strongly excited fibers (the solid fibers) with penumbras of fringe excitation surrounding each of these. In the illustration to the right, the penumbras have been removed by the lateral inhibitory circuits; obviously, this increases the contrast in the signal and helps in the faithfulness of transmission of the spatial pattern. Unfortunately, we still know too little about the mechanism or mechanisms of this lateral inhibition, but the more we learn the more it becomes clear that this is one of the most useful of all the neuronal circuits of the nervous system to differentiate and analyze sensory signals. We shall discuss this much more fully in the following chapters.

PROLONGATION OF A SIGNAL BY A NEURONAL POOL — "AFTER-DISCHARGE"

Thus far, we have considered signals that are merely relayed through neuronal pools. However, in many instances, a signal entering a pool causes a prolonged output discharge, called *after-discharge,* even after the incoming signal is over and lasting from a few milliseconds to as long as many minutes. The three basic mechanisms by which after-discharge occurs are the following:

Synaptic After-Discharge. When excitatory synapses discharge on the surfaces of dendrites or the soma of a neuron, a postsynaptic potential develops in the neuron and lasts for many milliseconds — in the anterior motoneuron for about 15 milliseconds, though perhaps much longer in other neurons — especially so when some of the long-acting synaptic transmitter substances are in-

volved. As long as this potential lasts it can continue to excite the neuron, causing it to transmit a continuous train of output impulses, as was explained in the previous chapter. Thus, as a result of this synaptic after-discharge mechanism alone, it is possible for a single instantaneous input to cause a sustained signal output (a series of repetitive discharges) lasting as long as 15 or more milliseconds.

The Parallel Circuit Type of After-Discharge. Figure 47–12 illustrates a second type of neuronal circuit that can cause short periods of after-discharge. In this case, the input signal spreads through a series of neurons in the neuronal pool, and from many of these neurons impulses keep converging on an output neuron. It will be recalled that a signal is delayed at each synapse for about 0.5 millisecond, which is called the *synaptic delay.* Therefore, signals that pass through a succession of intermediate neurons reach the output neuron one by one after varying periods of delay. Therefore, the output neuron continues to be stimulated for many milliseconds.

It is doubtful that more than a few dozen successive neurons ordinarily enter into a parallel after-discharge circuit. Therefore, one would suspect that this type of after-discharge circuit could cause after-discharges that last for no more than perhaps 25 to 50 milliseconds. Yet, this circuit does represent a means by which a single input signal, lasting less than 1 millisecond, can be converted into a sustained output signal lasting many milliseconds.

The Reverberatory (Oscillatory) Circuit as a Cause of After-Discharge. Many neurophysiologists believe that one of the most important of all circuits in the entire nervous system is the *reverberatory,* or *oscillatory, circuit.* Such circuits are caused by positive feedback within the neuronal pool. That is, the output of a neuronal circuit feeds back to re-excite the same circuit. Consequently, once stimulated, the circuit discharges repetitively for a long time.

Several different postulated varieties of reverberatory circuits are illustrated in Figure 47–13, the simplest — in Figure 47–13A — involving only a single neuron. In this case, the output

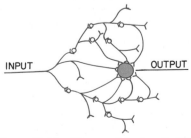

Figure 47–12. The parallel after-discharge circuit.

neuron simply sends a collateral nerve fiber back to its own dendrites or soma to restimulate itself; therefore, once the neuron should discharge, the feedback stimuli could theoretically keep the neuron discharging for a long time thereafter.

Figure 47–13B illustrates a few additional neurons in the feedback circuit, which would give a longer period of time between the initial discharge and the feedback signal. Figure 47–13C illustrates a still more complex system in which both facilitatory and inhibitory fibers impinge on the reverberating pool. A facilitatory signal increases the ease with which reverberation takes place, while an inhibitory signal decreases the ease of reverberation.

Figure 47–13D illustrates that most reverberating pathways are constituted of many parallel fibers, and at each cell station the terminal fibrils diffuse widely. In such a system the total reverberating signal can be either weak or strong, depending on how many parallel nerve fibers are momentarily involved in the reverberation.

Finally, reverberation need not occur only in a single neuronal pool, for it can occur through a circuit involving two or more successive pools in the positive feedback pathway.

Characteristics of After-Discharge from a Rever-

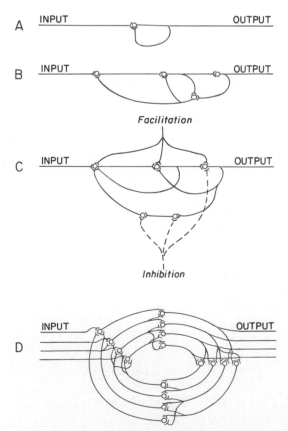

Figure 47–13. Reverberatory circuits of increasing complexity.

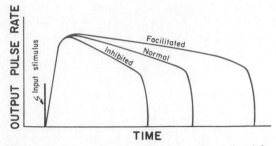

Figure 47–14. Typical pattern of the output signal from a reverberatory circuit following a single input stimulus, showing the effects of facilitation and inhibition.

beratory Circuit. Figure 47–14 illustrates postulated output signals from a reverberatory after-discharge circuit. The input stimulus need last only 1 millisecond or so, and yet the output can last for many milliseconds or even minutes. The figure demonstrates that the intensity of the output signal increases to a reasonably high value early in the reverberation, then decreases to a critical point, and suddenly ceases entirely. Furthermore, the duration of the after-discharge is determined by the degree of inhibition or facilitation of the neuronal pool. In this way, signals from other parts of the brain can control the reaction of the pool to the input stimulus.

Almost these exact patterns of output signals can be recorded from the motor nerves exciting a muscle involved in the flexor reflex (discussed in Chapter 51), which is believed to be caused by a reverberating type of after-discharge following stimulation of pain fibers.

Importance of Synaptic Fatigue in Determining the Duration of Reverberation. It was pointed out in the previous chapter that synapses fatigue if stimulated for prolonged periods of time. Therefore, one of the most important factors that determines the duration of the reverberatory type of after-discharge is probably the rapidity with which the involved synapses fatigue. Rapid fatigue would obviously tend to shorten the period of after-discharge.

Furthermore, the greater the number of neurons in the reverberatory pathway and the greater the number of collateral feedback fibrils, the easier would it be to keep the reverberation going. Therefore, it is to be expected that longer reverberating pathways would in general sustain after-discharges for longer periods of time.

Duration of Reverberation. Even though neurophysiologists are not certain which functions of the nervous system can rightfully be ascribed to reverberatory circuits, some of those that have been postulated have reverberatory durations as short as 10 milliseconds or as long as several minutes, or perhaps even hours. Some examples include: (1) In an animal whose spinal cord is transected in the neck, a sudden painful stimulus to the animal's paw will cause the flexor muscles to contract and remain contracted a fraction of a second to more than a second after the stimulus ends. It is believed that this is caused by an after-discharge of the reverberatory type. (2) During respiration the inspiratory neuronal pool in the medulla becomes excited for about 2 seconds during each respiratory cycle. One theory suggests that this is caused by reverberation within the inspiratory neuronal pool. And (3) a theory of wakefulness is that continual reverberation occurs somewhere within the brain stem to keep a wakefulness area excited during the waking hours. If this be true, then this would represent a 14 to 18 hour period of reverberation.

CONTINUOUS SIGNAL OUTPUT FROM NEURONAL POOLS

Some neuronal pools emit output signals continuously even without excitatory input signals. At least two different mechanisms theoretically can cause this effect: (1) intrinsic neuronal discharge and (2) reverberatory signals.

Continuous Discharge Caused by Intrinsic Neuronal Excitability. Neurons, like other excitable tissues, discharge repetitively if their membrane potentials rise above certain threshold levels. The membrane potentials of many neurons even normally are high enough to cause them to emit impulses continually. This occurs especially in large numbers of the neurons of the cerebellum as well as in most of the interneurons of the spinal cord. The rates at which these cells emit impulses can be increased by facilitatory signals or decreased by inhibitory signals; the latter can sometimes decrease the rate to extinction.

Continuous Signals Emitted from Reverberating Circuits. Obviously, a reverberating circuit that never fatigues to extinction could also be a source of continual impulses. Facilitatory impulses entering the reverberating pool, as illustrated in Figure 47–13C, could increase the output signal, and inhibition could decrease or even extinguish the output signal.

Figure 47–15 illustrates a continual output signal from a pool of neurons, whether it be a pool emitting impulses because of intrinsic neuronal excitability or as a result of reverberation. Note that an excitatory (or facilitatory) input signal greatly increases the output signal, whereas an inhibitory input signal greatly decreases the output. Those students who are familiar with radio transmitters will recognize this to be a *carrier wave* type of information transmission. That is,

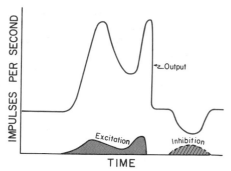

Figure 47–15. Continuous output from either a reverberating circuit or from a pool of intrinsically discharging neurons. This figure also shows the effect of excitatory or inhibitory input signals.

the excitatory and inhibitory control signals are not the *cause* of the output signal, but they do *control* it. Note that this carrier wave system allows decrease in signal intensity as well as increase, whereas, up to this point, the types of information transmission that we have discussed have been only positive information rather than negative information. This type of information transmission is used by the autonomic nervous system to control such functions as vascular tone, gut tone, degree of constriction of the iris, heart rate, and others.

RHYTHMIC SIGNAL OUTPUT

Many neuronal circuits emit rhythmic output signals — for instance, the rhythmic respiratory signal originating in the reticular substance of the medulla and pons. This repetitive rhythmic signal continues throughout life, while other rhythmic signals, such as those that cause scratching movements by the hind leg of a dog or the walking movements in an animal, require input stimuli into the respective circuits to initiate the signals.

Many rhythmic signals are postulated to result from reverberating pathways. One can readily understand that each time a signal passes around a reverberatory loop, collateral impulses could be transmitted into the output pathway. Thus, the rhythmic respiratory signals could result from such a reverberatory mechanism.

Obviously, facilitatory or inhibitory signals can affect rhythmic signal output in the same way that they can affect continual signal outputs. Figure 47–16, for instance, illustrates the rhythmic respiratory signal in the phrenic nerve. However, when the carotid body is stimulated by arterial oxygen deficiency, the frequency and amplitude of the rhythmic signal pattern increase progressively.

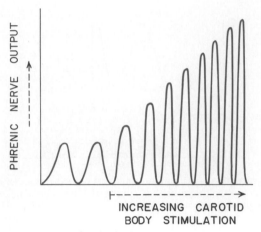

Figure 47–16. The rhythmical output from the respiratory center, showing that progressively increasing stimulation of the carotid body increases both the intensity and frequency of oscillation.

INSTABILITY AND STABILITY OF NEURONAL CIRCUITS

Almost every part of the brain connects either directly or indirectly with every other part, and this creates a serious problem. If the first part excites the second, the second the third, the third the fourth, and so on until finally the signal re-excites the first part, it is clear that an excitatory signal entering any part of the brain would set off a continuous cycle of re-excitation of all parts. If this should occur, the brain would be inundated by a mass of uncontrolled reverberating signals —signals that would be transmitting no information but, nevertheless, would be consuming the circuits of the brain so that none of the informational signals could be transmitted. Such an effect actually occurs in widespread areas of the brain during *epileptic fits*.

How does the central nervous system prevent this from happening all the time? The answer seems to lie in two basic mechanisms that function throughout the central nervous system: (1) inhibitory circuits, and (2) fatigue of synapses.

INHIBITORY CIRCUITS AS A MECHANISM FOR STABILIZING NERVOUS SYSTEM FUNCTION

The phenomenon of lateral inhibition that was discussed earlier in the chapter prevents signals in an informational pathway from spreading diffusely everywhere. In addition, two other types of inhibitory circuits in widespread areas of the brain help to prevent excessive spread of unwanted signals: (a) inhibitory feedback circuits that return from the termini of pathways back to the initial excitatory neurons of the same pathways — these inhib-

it the input neurons when the termini become overly excited, and (b) some neuronal pools that exert gross inhibitory control over widespread areas of the brain — for instance, many of the basal ganglia exert inhibitory influences throughout the motor control system.

SYNAPTIC FATIGUE AS A MEANS FOR STABILIZING THE NERVOUS SYSTEM (DECREMENTAL CONDUCTION)

Synaptic fatigue means simply that the signal becomes progressively weaker the more prolonged the period of excitation. This phenomenon is also frequently called *decremental conduction* because the output signal from a strongly excited pathway *decrements* with time.

Figure 47–17 illustrates typical synaptic fatigue, showing a muscle response during the flexor reflex. That is, stimulation of the pain sensory endings in a limb causes reflex contraction of the flexor muscles of the limb, leading to withdrawal of the limb from the painful stimulus. However, if this stimulus is repeated at close intervals, as was done in Figure 47–17, the response progressively decreases. But how does this fatigue, as illustrated by the decreasing response of the flexor reflex, play a role in stabilizing nervous system function? To answer this, let us assume that a large series of interconnecting neurons in a single pool becomes excessively excitable. As soon as this occurs, positive feedback and reverberation will begin. But so also will fatigue begin, and after a period of time the level of excitability of the entire neuronal pool will automatically return below the threshold level for continued reverberation.

Neuronal Circuits That Exhibit Fatigue; Those That Do Not. Fortunately, some synaptic pathways exhibit very little fatigue, while others exhibit extreme degrees. One type of pathway that exhibits almost no fatigue is the fast-conducting sensory pathways from the peripheral parts of the

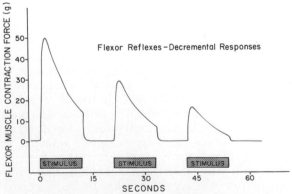

Figure 47–17. Successive flexor reflexes illustrating fatigue of conduction through the reflex pathway.

body to the brain, such as the pathways through the dorsal columns to the thalamus and thence to the somesthetic cortex. This seems to be of purposeful benefit to nervous function because it allows faithful transmission of sensory information into the brain.

On the other hand, the interconnecting links through the short interneurons between the sensory system and the motor system frequently exhibit extreme degrees of fatigue. This was illustrated in Figure 47–17 for the flexor reflex pathway, because it is the interneurons between the sensory input and the motor output of the spinal cord that fatigue. At least in certain parts of the nervous system, this fatigue has been shown to result from exhaustion of transmitter substance in the synaptic knobs.

Recovery From Fatigue. After a major degree of fatigue has occurred in the conduction of a synapse, the conduction recovers if an adequate period of rest between stimuli is allowed. Thus, when a pathway is not used often, it becomes progressively more excitable until full conduction returns.

Automatic Adjustment of Sensitivity of Pathways by the Fatigue Mechanism. It is clear by now that if the excitability of a pathway in the nervous system is too great, then too many impulses will be transmitted and the fatigue mechanism will automatically adjust the sensitivity of the pathway to a lower level. Conversely, if the sensitivity is too low, then too few impulses will be transmitted, and the sensitivity of the system will automatically adjust to a higher level. Therefore, this fatigue mechanism, and its recovery during rest, is probably an extremely important mechanism for maintaining a proper balance among the conductivities of the respective pathways throughout the brain.

REFERENCES

An der Heiden, U.: Analysis of Neural Networks. New York, Springer-Verlag, 1980.

Andreassen, S.: Interval Pattern of Single Motor Units. Aalborg, Denmark, Institute of Electronic Systems, Aalborg University Centre, 1978.

Anderson, H., et al.: Developmental neurobiology of invertebrates. Annu. Rev. Neurosci., 3:97, 1980.

Asanuma, H., and Wilson V. J. (eds.): Integration in the Nervous System. New York, Igaku-Shoin, 1979.

Bennett, M. V. L. (ed.): Synaptic Transmission and Neuronal Interaction. New York, Raven Press, 1975.

Björklund, A., and Stenevi, U.: Regeneration of monoaminergic and cholinergic neurons in the mammalian central nervous system. Physiol. Rev., 59:62, 1979.

Blumenthal, R., et al. (eds.): Dynamic Patterns of Brain Cell Assemblies. Neurosciences Research Program Bulletin. Vol. 12, No. 1. Cambridge, Mass., Massachusetts Institute of Technology, 1974.

Cowan, W. M.: The development of the brain. Sci. Am. 241(3):112, 1979.

Freides, D.: Human information processing and sensory modality: Cross-modal functions, information complexity, memory, and deficit. Psychol. Bull., 81:284, 1974.

Friesen, W. O., and Stent, G. S.: Neural circuits for generating rhythmic movements. Annu. Rev. Biophys. Bioeng., 7:37, 1978.

Grinnell, A. D.: Specificity of neurons and their interconnections. In Brookhart, J. M., and Mountcastle, V. B. (eds.): Handbook of Physiology. Sec. 1, Vol. 1. Baltimore, Williams & Wilkins, 1977, p. 803.

Kandel, E. R.: Small systems of neurons. Sci. Am. 241(3):66, 1979.

Karlin, A., et al. (eds.): Neuronal Information Transfer. New York, Academic Press, 1978.

Kater, S. B., et al.: Identifiable neurons and invertebrate behavior. In MTP International Review of Science: Physiology. Vol. 3. Baltimore, University Park Press, 1974, p. 1.

Kennedy, D., and Davis, W. J.: Organization of invertebrate motor systems. In Brookhart, J. M., and Mountcastle, V. B. (eds.): Handbook of Physiology. Sec. 1, Vol. 1. Baltimore, Williams & Wilkins, 1977, p. 1023.

Levi-Montalcini, R., and Calissano, P.: The nerve-growth factor. Sci. Am. 240(6):68, 1979.

Macagno, E. R., et al.: Three-dimensional computer reconstruction of neurons and neuronal assemblies. Ann. Rev. Biophys. Bioeng. 8:323, 1979.

Martin, A. R.: Synaptic transmission. In MTP International Review of Science: Physiology. Vol. 3. Baltimore, University Park Press, 1974, p. 53.

Nicholls, J. C., and Van Essen, D.: The nervous system of the leech. Sci. Am., 230:38, 1974.

Patterson, P. H., et al.: The chemical differentiation of nerve cells. Sci. Am., 239(1):50, 1978.

Pinsker, H. M., and Willis, W. D., Jr. (eds.): Information Processing in the Nervous System. New York, Raven Press, 1980.

Purves, D., and Lichtman, J. W.: Formation and maintenance of synaptic connections in autonomic ganglia. Physiol. Rev., 58:821, 1978.

Rovainen, C. M.: Neurobiology of lampreys. Physiol. Rev., 59:1007, 1979.

Schmidt, R. F. (ed.): Fundamentals of Neurophysiology. New York, Springer-Verlag, 1978.

Schmidt, R. F. (ed.): Fundamentals of Sensory Physiology. New York, Springer-Verlag, 1978.

Shepherd, B. M.: The Synaptic Organization of the Brain: an Introduction. New York, Oxford University Press, 1974.

Shepherd, G. M.: Microcircuits in the nervous system. Sci. Am., 238(2):92, 1978.

Szentagothai, J., et al. (eds.): Conceptual Models of Neural Organization. Neurosciences Research Program. Cambridge, Mass., Massachusetts Institute of Technology, 1974.

Uttley, A. M.: Information Transmission in the Nervous System. New York, Academic Press, 1979.

Wooldridge, D. E.: Sensory Processing in the Brain: An Exercise in Neuroconnective Modeling. New York, John Wiley & Sons, 1979.

48

Sensory Receptors and Their Basic Mechanisms of Action

Input to the nervous system is provided by the sensory receptors that detect such sensory stimuli as touch, sound, light, pain, cold, warmth, and so forth. The purpose of this chapter is to discuss the basic mechanisms by which these receptors change sensory stimuli into nerve signals and, also, how both the type of sensory stimulus and its strength are detected by the brain.

TYPES OF SENSORY RECEPTORS AND THE SENSORY STIMULI THEY DETECT

The student of medical physiology will have already studied many different anatomical types of nerve endings, and Table 48–1 gives a list and classification of most of the body's sensory receptors. This table shows that there are basically five different types of sensory receptors: (1) *mechanoreceptors,* which detect mechanical deformation of the receptor or of cells adjacent to the receptors; (2) *thermoreceptors,* which detect changes in temperature, some receptors detecting cold and others warmth; (3) *nociceptors,* which detect damage in the tissues, whether it be physical damage or chemical damage; (4) *electromagnetic receptors,* which detect light on the retina of the eye; and (5) *chemoreceptors,* which detect taste in the mouth, smell in the nose, oxygen level in the arterial blood, osmolality of the body fluids, carbon dioxide concentration, and perhaps other factors that make up the chemistry of the body.

This chapter will discuss the function of a few specific types of receptors, primarily the peripheral mechanoreceptors, to illustrate some of the basic principles by which receptors in general operate. Other receptors will be discussed in relation to the sensory systems that they subserve, mainly in the next few chapters. Figure 48–1 illustrates some of the different types of mechanoreceptors found in the skin or in the deep structures of the body, and Table 48–1 gives their respective sensory functions. All these receptors will be discussed in the following chapters in relation to the respective sensory systems. However, the functions of some of these are described here briefly as follows:

Free nerve endings are found in all parts of the body. A large proportion of these detect pain. However, other free nerve endings detect crude touch, pressure, and tickle sensations and possibly warmth and cold.

Several of the more complex receptors listed in Figure 48–1 detect tissue deformation. These include the *Merkel's discs,* the *tactile hairs, pacinian corpuscles, Meissner's corpuscles, Krause's corpuscles,* and *Ruffini's end-organs.* In the skin, these receptors detect the tactile sensations of touch, pressure, and vibration. In the deep tissues, they detect stretch, deep pressure, or any other type of tissue deformation — even the stretch of joint capsules and ligaments to determine the angulation of a joint.

The *Golgi tendon apparatus* detects degree of tension in tendons, and the *muscle spindle* detects relative changes in muscle length. These receptors will be discussed in Chapter 51 in relation to the muscle and tendon reflexes.

DIFFERENTIAL SENSITIVITY OF RECEPTORS

The first question that must be answered is, how do two types of sensory receptors detect different types of sensory stimuli? The answer is by virtue of differential sensitivities. That is, each type of receptor is very highly sensitive to one type of

TABLE 48–1   CLASSIFICATION OF SENSORY RECEPTORS

Mechanoreceptors

Skin tactile sensibilities (epidermis and dermis)
  Free nerve endings
  Expanded tip endings
    Merkel's discs
    Plus several other variants
  Spray endings
    Ruffini's endings
  Encapsulated endings
    Meissner's corpuscles
    Krause's corpuscles
  Hair end-organs
Deep tissue sensibilities
  Free nerve endings
  Expanded tip endings
    Plus a few other variants
  Spray endings
    Ruffini's endings
  Encapsulated endings
    Pacinian corpuscles
    Plus a few other variants
  Muscle endings
    Muscle spindles
    Golgi tendon receptors
Hearing
  Sound receptors of cochlea
Equilibrium
  Vestibular receptors
Arterial pressure
  Baroreceptors of carotid sinuses and aorta

Thermoreceptors

Cold
  Cold receptors
Warmth
  ?

Nociceptors

Pain
  Free nerve endings

Electromagnetic Receptors

Vision
  Rods
  Cones

Chemoreceptors

Taste
  Receptors of taste buds
Smell
  Receptors of olfactory epithelium
Arterial oxygen
  Receptors of aortic and carotid bodies
Osmolality
  Probably neurons of supraoptic nuclei
Blood CO_2
  Receptors in or on surface of medulla and in aortic and
    carotid bodies
Blood glucose, amino acids, fatty acids
  Receptors in hypothalamus

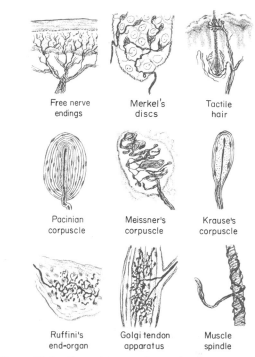

Figure 48–1. Several types of somatic sensory nerve endings.

stimulus for which it is designed and yet is almost nonresponsive to normal intensities of the other types of sensory stimuli. Thus, the rods and cones are highly responsive to light but are almost completely nonresponsive to heat, cold, pressure on the eyeballs, or chemical changes in the blood. The osmoreceptors of the supraoptic nuclei in the hypothalamus detect minute changes in the osmolality of the body fluids but have never been known to respond to sound. Finally, pain receptors in the skin are almost never stimulated by usual touch or pressure stimuli but do become highly active the moment tactile stimuli become severe enough to damage the tissues.

Modality of Sensation — Law of Specific Nerve Energies

Each of the principal types of sensation that we can experience — pain, touch, sight, sound, and so forth — is called a *modality* of sensation. Yet, despite the fact that we experience these different modalities of sensation, nerve fibers transmit only impulses. Therefore, how is it that different nerve fibers transmit different modalities of sensation?

The answer to this is that each nerve tract terminates at a specific point in the central nervous system, and the type of sensation felt when a nerve fiber is stimulated is determined by this specific area in the nervous system to which the fiber leads. For instance, if a pain fiber is stimulated, the person perceives pain regardless of what type of stimulus excites the fiber. This stimulus can be electricity, heat, crushing, or stimulation of the pain nerve ending by damage to the tissue cells. Yet, whatever the means of stimulation, the person still perceives pain. Likewise, if a touch fiber is stimulated by exciting a touch receptor electrically or in any other way, the person perceives touch because touch fibers lead to specific touch areas in the brain. Similarly, fibers from the retina of the eye terminate in the vision areas of the brain, fibers from the ear terminate in the auditory areas of the brain, and temperature fibers terminate in the temperature areas.

This specificity of nerve fibers for transmitting only one modality of sensation is called the *law of specific nerve energies*. It is also called the *labelled line principle*.

TRANSDUCTION OF SENSORY STIMULI INTO NERVE IMPULSES

LOCAL CURRENTS AT NERVE ENDINGS — RECEPTOR POTENTIALS AND GENERATOR POTENTIALS

All sensory receptors have one feature in common. Whatever the type of stimulus that excites the receptor, its immediate effect is to change the potential across the receptor membrane. This change is called a *receptor potential*.

Mechanisms of Receptor Potentials. Different receptors can be excited in several different ways to cause receptor potentials: (1) by mechanical deformation of the receptor, which stretches the membrane and opens pores; (2) by application of a chemical to the membrane, which also opens pores; (3) by change of the temperature of the membrane, which alters the permeability of the membrane; or (4) by the effects of electromagnetic radiation such as light on the receptor, which either directly or indirectly changes the membrane characteristics and allows ions to flow through the membrane pores. It will be recognized that these four different means for exciting receptors correspond in general with the different types of known sensory receptors. In all instances, the basic cause of the change in membrane potential is a change in receptor membrane permeability, which allows ions to diffuse more or less readily through the membrane and thereby change the transmembrane potential.

Excitation of the Terminal Nerve Fiber by the Receptor — the Generator Potential. In many in-

stances, the sensory receptor is the termination of a nerve fiber itself, as is true for all the different types of receptors illustrated in Figure 48–1. However, a few receptors are specialized receptor cells that lie adjacent to the nerve endings. For instance, the hair cells of the cochlea and the taste cells of the taste buds are both specialized epithelial cells that develop receptor potentials, the hair cells when sound enters the cochlea and the taste cells when certain chemical substances are present in the mouth. In each instance, the receptor potential is then transferred to terminal nerve fibrils that synapse at the bases of the receptor cells, and it is this transfer of potential to the terminal nerve fibrils that causes action potentials.

In the case of the nerve ending receptors, it is the receptor potential itself that causes generation of action potentials in the nerve fibers. In the case of the specialized receptor cells, it is the transferred potential from the receptor to the nerve endings that causes the action potentials. In either case, the potential that actually causes generation of action potentials is called a *generator potential*. For the nerve ending receptors, the generator potential is the same as the receptor potential, but for the specialized receptor cells the generator potential is only that portion of the receptor potential that is transferred to the nerve endings.

Relationship of the Generator Potential to Action Potentials. When the generator potential rises above the threshold for eliciting action potentials in the terminal nerve fibril, then action potentials begin to appear. This is illustrated in Figure 48–2. Note also that the more the generator potential rises above the threshold level, the greater becomes the action potential frequency. Thus, the generator potential stimulates the terminal nerve fibril in the same way that the excitatory postsynaptic potential stimulates neurons in the central nervous system neuron.

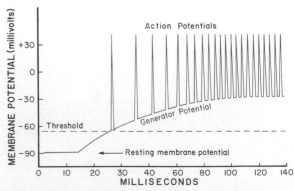

Figure 48–2. Typical relationship between generator potential and action potentials when the generator potential rises above the threshold level.

Illustrative Examples of Receptor Function

The Receptor Potential (Generator Potential) of the Pacinian Corpuscle. The pacinian corpuscle is a very large and easily dissected sensory receptor. For this reason, one can study in detail the mechanism by which tactile stimuli excite it and by which it causes action potentials in the sensory fiber leading from it. Note in Figure 48–1 that the pacinian corpuscle has a central nonmyelinated tip of a nerve fiber extending through its core. Surrounding this fiber are many concentric capsule layers so that compression on the outside of the corpuscle tends to elongate, shorten, indent, or otherwise deform the central core of the fiber, depending on how the compression is applied. The deformation causes a sudden change in membrane potential, as illustrated in Figure 48–3, which is the receptor potential and in this case is also the generator potential that causes generation of action potentials. The receptor potential is believed to result from stretching the nerve fiber membrane, thus increasing its permeability and allowing positively charged sodium ions to leak to the interior of the fiber. This change in local potential causes a local circuit of current flow that spreads along the nerve fiber to its myelinated portion. At the first node of Ranvier, which itself lies inside the capsule of the pacinian corpuscle, the local current flow initiates action potentials in the nerve fiber. That is, the current flow through the node depolarizes it, and this then sets off a typical saltatory transmission of an action potential along the nerve fiber toward the central nervous system, as was explained in Chapter 10.

Relationship Between Stimulus Strength and the Receptor (Generator) Potential. Figure 48–4 illustrates the effect on the amplitude of the receptor (generator) potential caused by progressively stronger mechanical compression applied experimentally to the central core of the pacinian corpuscle. Note that the amplitude increases rapidly at first but then progressively less rapidly at high

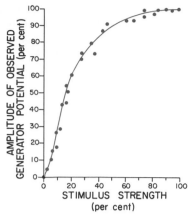

Figure 48–4. Relationship of amplitude of generator potential to strength of a stimulus applied to a pacinian corpuscle. (From Loewenstein: *Ann. N.Y. Acad. Sci., 94*:510, 1961.)

stimulus strengths. The maximum amplitude that can be achieved by receptor potentials is around 100 millivolts. That is, a receptor potential can have almost as high a voltage as an action potential.

Relationship of Amplitude of Receptor Potential to Nerve Impulse Rate. The frequency of action potentials in the nerve fiber (impulse rate) is almost directly proportional to the amplitude of the generator potential. This relationship is illustrated in Figure 48–5, which shows the impulse

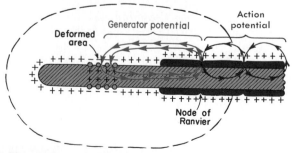

Figure 48–3. Excitation of a sensory nerve fiber by a generator potential produced in a pacinian corpuscle. (Modified from Loewenstein: *Ann. N.Y. Acad. Sci., 94*:510, 1961.)

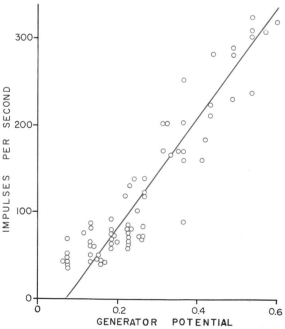

Figure 48–5. Relationship between the generator potential of a muscle spindle and the frequency of sensory impulses transmitted from the spindle. [From Katz: *J. Physiol. (Lond.), 111*:261, 1950.]

rate corresponding to different generator potential amplitudes recorded from a muscle spindle; there is an almost exact proportional relationship. This same relationship between receptor potential and impulse rate is approximately true for most sensory receptors.

ADAPTATION OF RECEPTORS

A special characteristic of all sensory receptors is that they *adapt* either partially or completely to their stimuli after a period of time. That is, when a continuous sensory stimulus is first applied, the receptors respond at a very high impulse rate at first, then progressively less rapidly until finally many of them no longer respond at all.

Figure 48–6 illustrates typical adaptation of certain types of receptors. Note that the pacinian corpuscle adapts extremely rapidly and hair receptors adapt within a second or so, while joint capsule and muscle spindle receptors adapt very slowly.

Furthermore some sensory receptors adapt to a far greater extent than others. For example, the pacinian corpuscles adapt to "extinction" within a few thousandths to a few hundredths of a second, and the hair base receptors adapt to extinction within a second or more. It is probable that all other mechanoreceptors also adapt completely eventually, but some require hours or days to do so for which reason they are frequently called "nonadapting" receptors. The longest measured time for complete adaptation of a mechanoreceptor is about two days for the carotid and aortic baroreceptors.

Some of the other types of receptors, the chemoreceptors and pain receptors for instance, probably never adapt completely. Each of these will be discussed in turn in subsequent chapters.

Mechanisms by Which Receptors Adapt. Adaptation of receptors is an individual property of each type of receptor in much the same way that development of a receptor potential is an individual property. For instance, in the eye, the rods and cones adapt by changing their chemical compositions (which will be discussed in Chapter 59). In the case of the mechanoreceptors, the receptor that has been studied in greatest detail is the pacinian corpuscle. Adaptation occurs in this receptor in two ways. First, the corpuscular structure itself very rapidly adapts to the deformation of the tissue. This can be explained as follows: The pacinian corpuscle is a viscoelastic structure so that when a distorting force is suddenly applied to one side of the corpuscle it is transmitted by the viscous component of the corpuscle directly to the same side of the central core, thus eliciting a receptor potential. However, within a few thousandths to a few hundredths of a second the fluid within the corpuscle redistributes so that the pressure becomes essentially equal all through the corpuscle; this applies an even pressure on all sides of the central core fiber, so that the receptor potential is no longer elicited. Thus, a receptor potential appears at the onset of compression but then disappears within a small fraction of a second.

Then, when the distorting force is removed from the corpuscle, essentially the reverse events occur. The sudden removal of the distortion from one side of the corpuscle allows rapid expansion on that side and a corresponding distortion of the central core occurs once more. Again, within milliseconds, the pressure becomes equalized all through the corpuscle and the stimulus is lost. Thus, the pacinian corpuscle signals the onset of compression and again signals the offset of compression.

The second mechanism of adaptation of the pacinian corpuscle, but a much slower one, results from a process called *accommodation* that occurs in the nerve fiber itself. That is, even if by chance the central core fiber should continue to be excited, as can be achieved after the capsule has been removed and the core is compressed with a stylus, the tip of the nerve fiber itself gradually becomes "accommodated" to the stimulus. This perhaps results from redistribution of ions across the nerve fiber membrane.

Presumably, these same two general mechanisms of adaptation apply to other types of mechanoreceptors. That is, part of the adaptation results from readjustments in the structure of the receptor itself, and part results from accommodation in the terminal nerve fibril.

Function of the Slowly Adapting and Nonadapting Receptors — The "Tonic" Receptors. The slowly adapting receptors (and also the receptors that do not adapt to extinction) continue to trans-

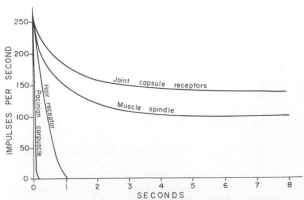

Figure 48–6. Adaptation of different types of receptors, showing rapid adaptation of some receptors and slow adaptation of others.

mit impulses to the brain as long as the stimulus is present (or at least for many minutes or hours). Therefore, they keep the brain constantly apprised of the status of the body and its relation to its surroundings. For instance, impulses from the slowly adapting joint capsule receptors allow the person to "know" at all times the degree of bending of the joints and therefore the positions of the different parts of the body. And impulses from the muscle spindles and Golgi tendon apparatuses allow the central nervous system to know respectively the status of muscle contraction and the load on the muscle tendon at each instant.

Other types of slowly adapting receptors include the receptors of the macula in the vestibular apparatus, the pain receptors, the baroreceptors of the arterial tree, the chemoreceptors of the carotid and aortic bodies, and some of the tactile receptors, such as the Ruffini endings and the Merkel's discs.

Because the slowly adapting receptors can continue to transmit information for many hours, they are also called *tonic* receptors. Many of these slowly adapting receptors would adapt to extinction if the intensity of the stimulus should remain absolutely constant for several hours or days. Fortunately, because of our continually changing bodily state, these receptors can almost never reach a state of complete adaptation.

Function of the Rapidly Adapting Receptors— The Rate Receptors or Movement Receptors or Phasic Receptors. Obviously, receptors that adapt rapidly cannot be used to transmit a continuous signal because these receptors are stimulated only when the stimulus strength changes. Yet they react strongly *while a change is actually taking place.* Furthermore, the number of impulses transmitted is directly related to the *rate at which the change takes place.* Therefore, these receptors are called *rate* receptors, *movement* receptors, or *phasic* receptors. Thus, in the case of the pacinian corpuscle, sudden pressure applied to the skin excites this receptor for a few milliseconds, and then its excitation is over even though the pressure continues. But then it transmits a signal again when the pressure is released. In other words, the pacinian corpuscle is exceedingly important for transmitting information about rapid changes in pressure against the body, but it is useless for transmitting information about constant pressure applied to the body.

Importance of the Rate Receptors–Their Predictive Function. Knowing the rate at which some change in bodily status is taking place, one can predict ahead to the state of the body a few seconds or even a few minutes later. For instance, the receptors of the semicircular canals in the vestibular apparatus of the ear detect the rate at which the head begins to turn when one runs around a curve. Using this information, a person can predict a turn of 10, 30, or some other number of degrees within the next 10 seconds and can adjust the motion of the limbs *ahead of time* to keep from losing balance. Likewise, pacinian corpuscles located in or near the joint capsules help to detect the rates of movement of the different parts of the body. Therefore, when one is running, information from these receptors allows the nervous system to predict ahead of time where the feet will be during any precise fraction of a second, and appropriate motor signals can be transmitted to the muscles of the legs to make any necessary anticipatory corrections in limb position so that the person will not fall. Loss of this predictive function makes it impossible for the person to run.

PSYCHIC INTERPRETATION OF STIMULUS STRENGTH

The ultimate goal of most sensory stimulation is to apprise the psyche of the state of the body and its surroundings. Therefore, it is important that we discuss briefly some of the principles related to the transmission of sensory stimulus strength to the higher levels of the nervous system.

The first question that comes to mind is: How is it possible for the sensory system to transmit sensory experiences of tremendously varying intensities? For instance, the auditory system can detect the weakest possible whisper but can also discern the meanings of an explosive sound only a few feet away, even though the sound intensities of these two experiences can vary almost a trillion-fold; the eyes can see visual images with light intensities that vary as much as a million-fold; or the skin can detect pressure differences of ten thousand- to one hundred thousand-fold.

As a partial explanation of these effects, note in Figure 48–4 the relationship of the generator potential produced by the pacinian corpuscle to the strength of stimulus. At low stimulus strength, very slight changes in stimulus strength increase the potential markedly; while at high levels of stimulus strength, further increases in receptor potential are very slight. Thus, the pacinian corpuscle is capable of accurately measuring extremely minute changes in stimulus strength at low intensity levels and is also capable of detecting much larger changes in stimulus strength at high intensity levels.

The transduction mechanism for detecting sound by the cochlea of the ear illustrates still another method for separating gradations of stimulus intensity. When sound causes vibration at a specific point on the basilar membrane, weak vibration stimulates only those hair cells at the point of maximum vibration. But, as the vibration intensity increases, not only do these hair cells become more intensely stimulated, but still many more hair cells in each direction farther away from the maximum vibratory point also become stimulated. Thus, signals transmitted over progressively increasing numbers of cochlear nerve fibers, as well as increasing intensity of signal strength in each nerve fiber, are two mechanisms by which stimulus strength is transmitted into the central nervous system. These two mechanisms, as well as several others, all of which multiply each other, make it possible for the ear to operate reasonably faithfully at stimulus intensity levels changing almost a trillion-fold.

Importance of the Tremendous Intensity Range of Sensory Reception. Were it not for the tremendous intensity range of sensory reception that we can experience, the various sensory systems would more often than not operate in the wrong range. This is illustrated by the attempts of most persons to adjust the light exposure on a camera without using a light meter. Left to intuitive judgment of light intensity, a person almost always overexposes the film on very bright days and greatly underexposes the film at twilight. Yet, that person's eyes are perfectly capable of discriminating with great detail the surrounding objects in both very bright sunlight and at twilight; the camera cannot do this because of the narrow critical range of light intensity required for proper exposure of film.

JUDGMENT OF STIMULUS STRENGTH

Physiopsychologists have evolved numerous methods for testing one's judgment of sensory stimulus strength, but only rarely do the results from the different methods agree with each other. For instance, one testing method requests selection of a weight that is exactly 100 per cent heavier than another. But the person usually selects a weight that is about 50 per cent heavier instead of 100 per cent heavier. Thus, the weak stimulus is underestimated and the strong stimulus is overestimated. In still another test procedure, a person is given a weight to hold and is then required to select the minimum amount of additional weight that must be added to detect a difference. In this case the person might be holding a 30 gram weight and finds that an additional 1 gram is needed to detect a difference. Then the person holds a 300 gram weight and finds that 10 grams of additional weight are required. In this instance the discriminatory ability is far greater at the low intensity level that at the high intensity level. Thus, the results of these two different types of tests are exactly opposite to each other, which means that the real argument lies in the meaning of the tests themselves. Therefore, at present no real agreement exists as to a proper method for measuring one judgment of stimulus strength. Yet, two principles are widely discussed in the physiopsychology field of sensory interpretation: the *Weber-Fechner principle* and the *power principle*.

The Weber-Fechner Principle — Detection of "Ratio" of Stimulus Strength. In the mid-1800s, Weber first and Fechner later proposed the principle that *gradations of stimulus strength are discriminated approximately in proportion to the logarithm of stimulus strength*. This law is based primarily on one's ability to judge minimal *changes* in stimulus strength. That is, in the second test described in the previous section the person could barely detect a 1 gram increase in weight when holding 30 grams, or a 10 gram increase when holding 300 grams. Thus, the *ratio* of the change in stimulus strength

required for detection of a change remained essentially constant, about 1 to 30, which is what the logarithmic principle means. To express this mathematically,

Interpreted signal strength = log (Stimulus) + Constant

Because the Weber-Fechner principle offers a ready explanation for the tremendous range of stimulus strength that our nervous system can discern, it unfortunately became widely accepted for all types of sensory experience and for all levels of background sensory intensity. More recently it has become evident that this principle applies mainly to higher intensities of visual, auditory, and cutaneous sensory experience and it applies only poorly to most other types of sensory experience.

Yet, the Weber-Fechner principle is still a good one to remember because it emphasizes that the greater the background sensory stimulus, the greater also must be the additional change in stimulus strength in order for the psyche to detect the change.

The Power Law. Another attempt by physiopsychologists to find a good mathematical relationship between actual stimulus strength and interpretation of stimulus strength is the following formula, known as the power law:

Interpreted signal strength = K · (Stimulus − k)^y

In this formula K and k are constants, and y is the power to which the stimulus strength is raised. The exponent y and the constants K and k are different for each type of sensation.

When this power law relationship is plotted on a graph using double logarithmic coordinates, as illustrated in Figure 48-7, a linear curve can be attained between interpreted stimulus strength and actual stimulus strength over a large range for almost any type of sensory perception. However, as illustrated in the fig-

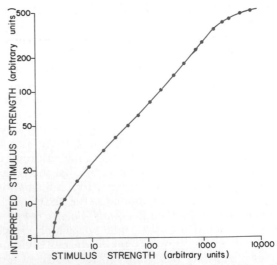

Figure 48-7. Graphical demonstration of the "power law" relationship between actual stimulus strength and strength that the psyche interprets it to be. Note that the power law does not hold at either very weak or very strong stimulus strengths.

ure, even this power law relationship fails to hold satisfactorily at both very low and very high stimulus strengths.

PHYSIOLOGICAL CLASSIFICATION OF NERVE FIBERS

Some sensory signals need to be transmitted to the central nervous system extremely rapidly, or otherwise the information would be useless. An example of this is the sensory signals that apprise the brain of the momentary positions of the limbs at each fraction of a second during running. It is important that this information be transmitted by way of extremely rapidly conducting nerve fibers. At the other extreme, some types of sensory information, such as that depicting prolonged, aching pain, do not need to be transmitted rapidly at all, so that very slowly conducting fibers will suffice. Fortunately, nerve fibers come in all sizes between 0.2 and 20 microns in diameter — the larger the diameter, the greater the conducting velocity. The range of conducting velocities is between 0.5 and 120 meters per second.

Figure 48–8 gives two different classifications of nerve fibers that are in general use. One of these is a general classification that includes both sensory and motor fibers, including the autonomic nerve fibers as well. The other is a classification of sensory nerve fibers that is used primarily by sensory neurophysiologists.

In the general classification, the fibers are divided into types A and C, and the type A fibers are further subdivided into α, β, γ, and δ fibers.

Type A fibers are the typical myelinated fibers of spinal nerves. Type C fibers are the very small, unmyelinated nerve fibers that conduct impulses at low velocities. These constitute more than half the sensory fibers in most peripheral nerves and also all of the post-ganglionic autonomic fibers.

The sizes, velocities of conduction, and functions of the different nerve fiber types are given in the figure. Note that the very large fibers can transmit impulses at velocities as great as 120 meters per second, a distance in one second that is longer than a football field. On the other hand, the smallest fibers transmit impulses as slowly as 0.5 meter per second, requiring about 2 seconds to go from the big toe to the spinal cord.

Over two-thirds of all the nerve fibers in peripheral nerves are type C fibers. Because of their great number, these can transmit tremendous amounts of information from the surface of the body, even though their velocities of transmission are very slow. Utilization of type C fibers for transmitting this great mass of information represents an important economy of space in the nerves, for use of the larger type A fibers for transmitting all information would require peripheral nerves the size of large ropes and a spinal cord almost as large as the body itself.

Alternate Classification Used by Sensory Physiologists. Certain recording techniques have made it possible to separate the type Aα fibers into two subgroups; and, yet, these same recording techniques cannot distin-

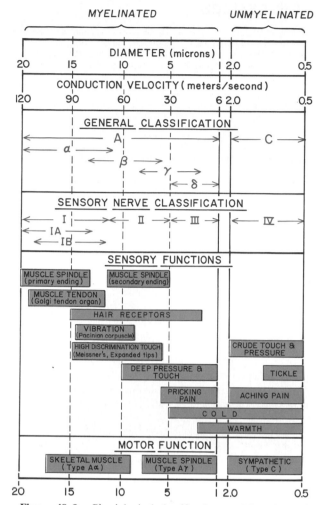

Figure 48–8. Physiological classifications and functions of nerve fibers.

guish easily between Aβ and Aγ fibers. Therefore, the following classification is frequently used by sensory physiologists:

Group Ia. Fibers from the annulospiral endings of muscle spindles. (Average about 17 microns in diameter. These are alpha type A fibers in the general classification.)

Group Ib. Fibers from the Golgi tendon organs. (Average about 16 microns in diameter; these also are alpha type A fibers.)

Group II. Fibers from the discrete cutaneous tactile receptors and also from the flower-spray endings of the muscle spindles. (Average about 8 microns in diameter; these are beta and gamma type A fibers in the other classification.)

Group III. Fibers carrying temperature, crude touch, and pricking pain sensations. (Average about 3 microns in diameter; these are delta type A fibers in the other classification.)

Group IV. Unmyelinated fibers carrying pain, itch, temperature, and crude touch sensations. (0.5 to 2 microns in diameter; called type C fibers in the other classification.)

REFERENCES

Anderson, D. J., *et al.*: Sensory mechanisms in mammalian teeth and their supporting structures. *Physiol. Rev., 50*:171, 1970.

Anstis, S. M., *et al.*: Perception. New York, Springer-Verlag, 1978.

Babel, J., *et al.*: Ultrastructure of the Peripheral Nervous System and Sense Organs. New York, Churchill Livingstone, 1971.

Bate, C. M., *et al.*: Development of Sensory System. New York, Springer-Verlag, 1978.

Bennett, T. L.: The Sensory World: An Introduction to Sensation and Perception. Monterey, Cal., Brooks/Cole Publishing Co., 1978.

Brown, E., and Deffenbacher, K.: Perception and the Senses. New York, Oxford University Press, 1979.

Brown, M.: Touch Will Tell. New York, Franklin Watts, 1979.

Catton, W. T.: Mechanoreceptor function. *Physiol. Rev., 50*:297, 1970.

Coren, S., *et al.*: Sensation and Perception. New York, Academic Press, 1979.

Goldstein, E. B.: Sensation and Perception. Belmont, Cal., Wadsworth Publishing Co., 1980.

Granit, R.: Receptors and Sensory Perception. New Haven, Conn., Yale University Press, 1955.

Gray, J. A. B.: Initiation of impulses at receptors. *In* Magoun, H. W. (ed.): Handbook of Physiology. Sec. 1, Vol. 1. Baltimore, Williams & Wilkins, 1959.

Halata, Z.: The Mechanoreceptors of the Mammalian Skin. New York, Springer-Verlag, 1975.

Loewenstein, W. R.: The generation of electric activity in a nerve ending. *Ann. N.Y. Acad Sci., 81*:367, 1959.

Loewenstein, W. R.: Excitation and inactivation in a receptor membrane. *Ann. N.Y. Acad. Sci., 94*:510, 1961.

Lynn, B.: Somatosensory receptors and their CNS connections. *Annu. Rev. Physiol., 37*:105, 1975.

McCloskey, D. I.: Kinesthetic sensibility. *Physiol. Rev., 58*:763, 1978.

Paintal, A. A. (ed.): Morphology and Mechanisms of Chemoreceptors. Delhi, India, Vallabhbhai Patel Chest Institute, University of Delhi, 1976.

Porter, R. (ed.): Studies in Neurophysiology. New York, Cambridge University Press, 1978.

Schmidt, R. F. (ed.): Fundamentals of Sensory Physiology. New York, Springer-Verlag, 1978.

Somjen, G.: Sensory Coding in the Mammalian Nervous System. New York, Appleton-Century-Crofts, 1972.

Thompson, R. F., and Patterson, M. (eds.): Bioelectric Recording Techniques, Part C. Receptor and Effector Processes. New York, Academic Press, 1974.

Wiederhold, M. L.: Mechanosensory transduction in "sensory" and "motile" cilia. *Annu. Rev. Biophys. Bioeng., 5*:39, 1976.

Wiersma, C. A. G., and Roach, J. L. M.: Principles in the organization of invertebrate sensory systems. *In* Brookhart, J. M., and Mountcastle, V. B. (eds.): *Handbook of Physiology.* Sec. 1, Vol. 1. Baltimore, Williams & Wilkins, 1977, p. 1089.

Zimmermann, M.: Neurophysiology of nociception. *Int. Rev. Physiol., 10*:179, 1976.

49

Somatic Sensations: I. The Mechanoreceptive Sensations

The *somatic senses* are the nervous mechanisms that collect sensory information from the body. These senses are in contradistinction to the *special senses*, which mean specifically sight, hearing, smell, taste, and equilibrium.

CLASSIFICATION OF SOMATIC SENSES

The somatic senses can be classified into three different physiological types: (1) the *mechanoreceptive somatic senses,* stimulated by mechanical displacement of some tissue of the body, (2) the *thermoreceptive senses,* which detect heat and cold, and (3) the *pain sense,* which is activated by any factor that damages the tissues. This chapter deals with the mechanoreceptive somatic senses, and the following chapter discusses the thermoreceptive and pain senses.

The mechanoreceptive senses include *touch, pressure, vibration,* and *tickle* senses (which are frequently called the *tactile senses*) and the *position sense,* which determines the relative positions and rates of movement of the different parts of the body.

Other Classifications of Somatic Sensations. Different types of somatic sensations are also grouped together in special classes that are not necessarily mutually exclusive, as follows:

Exteroreceptive sensations are those from the surface of the body.

Proprioceptive sensations are those having to do with the physical state of the body, including position sensations, tendon and muscle sensations, pressure sensations from the bottom of the feet, and even the sensation of equilibrium, which is generally considered to be a "special" sensation rather than a somatic sensation.

Visceral sensations are those from the viscera of the body; in using this term one usually refers specifically to sensations from the internal organs.

The *deep sensations* are those that come from the deep tissues, such as from the bone, fasciae, and so forth. These include mainly "deep" pressure, pain, and vibration.

DETECTION AND TRANSMISSION OF TACTILE SENSATIONS

Interrelationship Between the Tactile Sensations of Touch, Pressure, and Vibration. Though touch, pressure, and vibration are frequently classified as separate sensations, they are all detected by the same types of receptors. The only differences among these three are (1) touch sensation generally results from stimulation of tactile receptors in the skin or in tissues immediately beneath the skin, (2) pressure sensation generally results from deformation of deeper tissues, and (3) vibration sensation results from rapidly repetitive sensory signals, but some of the same types of receptors as those for touch and pressure are utilized — specifically the very rapidly adapting types of receptors.

The Tactile Receptors. At least six entirely different types of tactile receptors are known, but many more similar to these also exist. Some of these receptors were illustrated in Figure 48–1 of Chapter 48, and their special characteristics are the following:

First, some *free nerve endings,* which are found everywhere in the skin and in many other tissues, can detect touch and pressure. For instance, even light contact with the cornea of the eye, which contains no other type of nerve ending besides free nerve endings, can nevertheless elicit touch and pressure sensations.

Second, a touch receptor of special sensitivity is *Meissner's corpuscle,* an encapsulated nerve ending that excites a large (type Aβ) myelinated sensory nerve fiber. Inside the capsulation are many whorls of terminal nerve filaments. These receptors are present in the nonhairy parts of the skin (called *glabrous skin*) and are particularly abundant in the fingertips, lips, and other areas of the skin where one's ability to discern spatial characteristics of touch sensations is highly developed. They, along with the expanded tip receptors described subsequently, are mainly responsible for the ability to recognize exactly what point of the body is touched and to recognize the texture of objects touched. Meissner's corpuscles adapt in less than a second after they are stimulated, which means that they are particularly sensitive to movement of very light objects over the surface of the skin and also to low frequency vibration.

Third, the fingertips and other areas that contain large numbers of Meissner's corpuscles also contain large numbers of *expanded tip tactile receptors,* one type of which is *Merkel's discs.* The hairy parts of the skin also contain moderate numbers of expanded tip receptors, even though they have almost no Meissner's corpuscles. These receptors differ from Meissner's corpuscles in that they transmit an initially strong but partially adapting signal and then a continuing weaker signal that adapts only slowly. Therefore, they are responsible for giving steady state signals that allow one to determine continuous touch of objects against the skin. Merkel's discs are usually grouped together in a single receptor organ called the *Iggo dome receptor,* which projects upward against the underside of the epithelium of the skin, as illustrated in Figure 49–1. This causes the epithelium at this point to protrude outward, thus creating a dome. Also note that the entire group of

Merkel's discs is innervated by a single large type of myelinated nerve fiber (type Aβ). Each dome receptor has a diameter averaging only 0.2 mm., and these receptors are extremely sensitive. They probably play a very important role in localizing touch sensations to the specific surface areas of the body.

Fourth, slight movement of any hair on the body stimulates the nerve fiber entwining its base. Thus, each hair and its basal nerve fiber, called the *hair end-organ,* are also a type of touch receptor. This receptor adapts readily and, therefore, like Meissner's corpuscles, detects mainly movement of objects on the surface of the body or initial contact with the body.

Fifth, located in the deeper layers of the skin and also in deeper tissues are many *Ruffini's end-organs,* which are multibranched endings, as described and illustrated in the previous chapter. These endings adapt very little and, therefore, are important for signaling continuous states of deformation of the skin and deeper tissues, such as heavy and continuous touch signals and pressure signals. They are also found in joint capsules and signal the degree of joint rotation.

Sixth, many *pacinian corpuscles,* which were discussed in detail in Chapter 48, lie both beneath the skin and also deep in the tissues of the body. These are stimulated only by very rapid movement of the tissues because these receptors adapt in a few thousandths of a second. Therefore, they are particularly important for detecting tissue vibration or other extremely rapid changes in the mechanical state of the tissues.

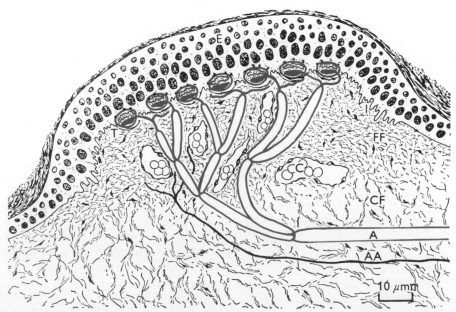

Figure 49–1.   The Iggo dome receptor. Note the multiple numbers of Merkel's discs innervated by a single large myelinated fiber and abutting tightly the undersurface of the epithelium (From Iggo and Muir: *J. Physiol., 200*:763, 1969.)

Transmission of Tactile Sensations in Peripheral Nerve Fibers. Almost all the specialized sensory receptors, such as Meissner's corpuscles, Iggo dome receptors, hair receptors, pacinian corpuscles, and Ruffini's endings, transmit their signals in type Aβ nerve fibers that have transmission velocities of 30 to 70 meters per second. On the other hand, free nerve ending tactile receptors transmit signals mainly via the small type Aδ myelinated fibers that conduct at velocities of 5 to 30 meters per second. Some tactile free nerve endings transmit via type C unmyelinated fibers at velocities from a fraction of a meter up to 2 meters per second; these send signals into the spinal cord and lower brain stem, probably subserving mainly the sensation of tickle. Thus, the more critical types of sensory signals — those that help to determine precise localization on the skin, minute gradations of intensity, or rapid changes in sensory signal intensity — are all transmitted in the rapidly conducting types of sensory nerve fibers. On the other hand, the cruder types of signals, such as crude pressure, poorly localized touch, and especially tickle, are transmitted via much slower nerve fibers, fibers that also require much less space than in the nerves.

DETECTION OF VIBRATION

All the different tactile receptors are involved in detection of vibration, though different receptors detect different frequencies of vibration. Pacinian corpuscles can signal vibrations from 60 to 500 cycles per second, because they respond extremely rapidly to minute and rapid deformations of the tissues, and they also transmit their signals over type Aβ nerve fibers, which can transmit more than 1000 impulses per second.

Low frequency vibrations up to 80 cycles per second, on the other hand, stimulate other tactile receptors — especially Meissner's corpuscles, which are less rapidly adapting than pacinian corpuscles.

THE SUBCONSCIOUS "MUSCLE SENSE"

Highly specialized sensory receptors are found in both skeletal muscles and tendons — *muscle spindles* in muscles and *Golgi tendon organs* in tendons. These receptors transmit their signals into the spinal cord, cerebellum, and cerebral cortex to control reflex contraction of the muscles. But, from what is known at present, these signals operate entirely at a subconscious level or almost so. These specialized receptors and reflexes that they subserve will be considered in detail in relation to spinal cord reflexes and control of muscle contraction in Chapter 51.

However, some psychological effects have pointed to the existence of a conscious "muscle sense." For instance, one is capable of determining how heavy an object is, and this has long been believed to be detected by the specialized muscle spindles or tendon organs; but critical experiments have not proved this. Instead, this type of information is probably signaled in two ways: (1) by usual tactile signals in the deep tissues and (2) by signals from the motor portions of the brain indicating intensity of motor signals to the muscles required to lift an object.

TICKLING AND ITCH

The phenomenon of tickling and itch has often been stated to be caused by very mild stimulation of pain nerve endings, because whenever pain is blocked by local anesthesia of a nerve or by compressing the nerve, the phenomenon of tickling and itch also disappears. However, recent neurophysiological studies have demonstrated the existence of very sensitive, rapidly adapting, free nerve endings that elicit only the tickle and itch sensation. Furthermore, these endings are found almost exclusively in the superficial layers of the skin, which is also the only tissue from which the tickle and itch sensation can usually be elicited. Furthermore, exciting itch receptors in animals initiates scratch reflexes, which contrasts with the effect of exciting pain nerve endings that always causes withdrawal reflexes instead.

Therefore, it seems clear that the itch and tickle sensations are transmitted by very small type C, unmyelinated fibers similar to those that transmit the burning type of pain; these fibers, however, are distinctly separate from the pain fibers. Furthermore, the endings of these fibers are readily excited by light mechanoreceptive stimuli, indicating that they are mechanoreceptors in contradistinction to the probable chemoreceptor nature of pain fibers.

The purpose of the itch sensation is presumably to call attention to mild surface stimuli such as a flea crawling on the skin or a fly about to bite, and the elicited signals then lead to scratching or other maneuvers that rid the host of the irritant.

The relief of itch by the process of scratching occurs only when the irritant is removed or when the scratch is strong enough to elicit pain. The pain signals are believed to suppress the itch signals in the cord by the process of inhibition that will be described in the following chapter.

THE DUAL SYSTEM FOR TRANSMISSION OF MECHANORECEPTIVE SOMATIC SENSORY SIGNALS INTO THE CENTRAL NERVOUS SYSTEM

Either all or almost all sensory information from the somatic segments of the body enters the spinal cord through the posterior roots. Immediately after entering the cord, the nerve fibers separate into two major groups: (1) a *dorsal-lemniscal system* that includes the (a) *dorsal columns* and (b) the *spinocervical* tracts located in the *dorsolateral columns;* and (2) the *anterolateral spinothalamic system* located in the anterior and lateral columns. Closely associated with the dorsal-lemniscal system are the dorsal and ventral spinocerebellar tracts, which operate at a subconscious level and transmit sensory information to the cerebellum. These tracts will be discussed in Chapter 53 along with the discussion of cerebellar function.

Comparison of the Dorsal-Lemniscal System with the Anterolateral Spinothalamic System. The distinguishing difference between the dorsal-lemniscal system and the anterolateral spinothalamic system is that the dorsal-lemniscal system is comprised mainly of large myelinated nerve fibers that transmit signals to the brain at velocities of 30 to 110 meters per second while the anterolateral spinothalamic system is comprised of much smaller myelinated fibers (averaging 4 microns in diameter) that transmit impulses at velocities ranging between 10 and 60 meters per second.

Another difference between the two systems is that the dorsal system has a very high degree of spatial orientation of the nerve fibers with respect to their origin on the surface of the body, while the spinothalamic system has a much smaller degree of spatial orientation, with some fibers seeming to have very little orientation at all.

These differences in the two systems immediately characterize the types of sensory information that can be transmitted by the two systems. First, sensory information that must be transmitted rapidly and with temporal fidelity is transmitted in the dorsal-lemniscal system, while that which does not need to be transmitted rapidly is transmitted mainly in the anterolateral spinothalamic system. Second, those sensations that detect fine gradations of intensity are transmitted in

the dorsal system, while those that lack the fine gradations are transmitted in the spinothalamic system. And, third, sensations that are discretely localized to exact points in the body are transmitted in the dorsal system, while those transmitted in the spinothalamic system can be localized much less exactly. On the other hand, the spinothalamic system has a special capability that the dorsal system does not have: the ability to transmit a broad spectrum of sensory modalities — pain, warmth, cold, and crude tactile sensations; the dorsal system is limited to mechanoreceptive sensations alone. With this differentiation in mind we can now list the types of sensations transmitted in the two systems.

The Dorsal-Lemniscal System

1. Touch sensations requiring a high degree of localization of the stimulus.
2. Touch sensations requiring transmission of fine gradations of intensity.
3. Phasic sensations, such as vibratory sensations.
4. Sensations that signal movement against the skin.
5. Position sensations.
6. Pressure sensations having to do with fine degrees of judgment of pressure intensity.

The Anterolateral Spinothalamic System

1. Pain.
2. Thermal sensations, including both warm and cold sensations.
3. Crude touch and pressure sensations capable of only crude localizing ability on the surface of the body and having little capability for intensity discrimination.
4. Tickle and itch sensations.
5. Sexual sensations.

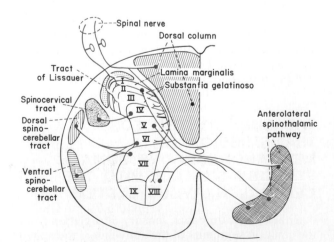

Figure 49–2. Cross-section of the spinal cord showing the anatomical laminae I through IX of the cord gray matter and the ascending sensory tracts in the white columns of the spinal cord.

FUNCTION OF THE SPINAL CORD NEURONS IN TRANSMITTING SENSORY SIGNALS

The gray matter of the spinal cord subserves three major functions. First, it is the locus of the integrating system for the spinal cord reflexes, a function that will be discussed in Chapter 51. Second, it functions as the initial processor of incoming sensory signals and as a relay station to relay most of these signals all the way to the brain; it is with this function that we are concerned in this chapter. Third, it is the area for final processing of motor signals transmitted downward from the brain to the muscles. This will be discussed in Chapter 53.

Relay of Sensory Signals in the Spinal Cord. Upon first entering the spinal cord through the posterior roots, as illustrated in Figure 49–2, most of the large, sensory nerve fibers, mainly type Aβ fibers, turn medially toward the dorsal columns. Then they usually divide into two major branches: one ascends upward through the dorsal columns and the other enters the anterior portion of the dorsal horn of the spinal cord gray matter. Also, a large share of the fibers that ascend in the dorsal columns enters the gray matter a few segments farther up the cord. Approximately 25 per cent of them travel the entire distance to the terminus of the dorsal columns at the dorsal column nuclei in the medulla.

In the dorsal horn gray matter, the terminals of these large fibers synapse with second order neurons of two types: (1) local neurons that play an intricate role in the control of spinal cord reflexes, and (2) relay neurons that give rise to long, ascending fiber tracts that transmit sensory information to the brain. Most of the fibers from the relay neurons ascend in the *spinocervical tract* located in the dorsolateral column, as shown in Figure 49–2.

The smaller fibers entering the spinal cord from the posterior roots, in general, take a more lateral pathway than the larger fibers, and they either enter the gray matter of the dorsal horn immediately or either ascend or descend a few segments and then enter the dorsal horn. Most of these fibers terminate on small neurons in the posterior portions of the dorsal horns, and these often give rise to short fibers that terminate somewhere else in the gray matter before the information is finally relayed to the long ascending pathways. Essentially all the sensory information from the smaller fibers eventually enters the anterolateral spinothalamic tract that crosses to the opposite side of the cord in the anterior commissure and then ascends to the brain in the anterolateral white matter.

Processing of Sensory Signals by Spinal Neurons. Most of the sensory information transmitted through the large sensory fibers is transmitted very faithfully, with either no processing or very little processing in the spinal cord. On the other hand, the signals transmitted through the smaller fibers, especially the pain signals, are often subject to a high degree of processing. This occurs mainly in the posterior portions of the dorsal horns, and it consists principally of (1) control of sensitivity of sensory transmission as a result of inhibitory signals transmitted from the brain to the cord, (2) control of sensitivity as a result of collateral signals from other fibers entering the spinal cord through the posterior roots, and (3) summation of signals from adjacent sensory nerves. This will be discussed in more detail in the following chapter in relation to the processing of pain signals.

The Anatomical Laminae of the Spinal Cord Gray Matter. Figure 49–2 illustrates the spinal cord gray matter divided into nine separate laminae. These divisions are based mainly on the types of neurons found within each separate area of the gray matter. Unfortunately, the exact functions of the neurons in the different areas are not fully known. However, the locus of some of the relay neurons for the ascending sensory tracts have been localized to the following laminae:

Spinothalamic tract	Laminae I, IV, V, and VI
Spinocervical tract	Laminae IV, V, and VI
Dorsal and ventral spinocerebellar tracts	Laminae VI and VII
Spinoreticular tract	Laminae VI, VII, and VIII

TRANSMISSION IN THE DORSAL-LEMNISCAL SYSTEM

ANATOMY OF THE DORSAL-LEMNISCAL SYSTEM

Sensory signals are transmitted to the brain by way of two major pathways in the dorsal-lemniscal system: (1) the *dorsal column pathway,* and (2) the *spinocervical pathway.* Until recently it was believed that all the sensory signals transmitted in this system were conducted only in the dorsal columns. However, complete transection of the dorsal columns is now known to cause only partial loss of sensations normally ascribed to the dorsal-lemniscal system, and it has been discovered that a major share of the remaining signals is transmitted in the spinocervical tracts located in the dorsolateral white columns of the cord, as illustrated in Figure 49–2. The routes that both the dorsal column pathway and the spinocervical pathway take to the brain are illustrated in Figure 49–3.

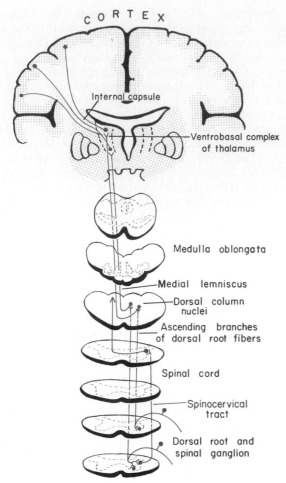

Figure 49–3. The dorsal column and spinocervical pathways for transmitting critical types of tactile signals. (Modified from Ranson and Clark: Anatomy of the Nervous System Philadelphia, W. B. Saunders Company, 1959.)

Anatomy of the Dorsal Column Pathway. Note in Figure 49–3 that the nerve fibers entering the dorsal columns pass up these columns to the medulla, where they synapse in the *dorsal column nuclei* (the *cuneate* and *gracile nuclei*). From here, *second order neurons* decussate immediately to the opposite side and then pass upward to the thalamus through bilateral pathways called the *medial lemnisci*. Each medial lemniscus terminates in a *ventrobasal complex of nuclei* located in the ventral posterolateral nucleus of the *thalamus*. In its pathway through the hindbrain, the medial lemniscus is joined by additional fibers from the *main sensory nucleus of the trigeminal nerve and from the upper portion of its descending nuclei;* these fibers subserve the same sensory functions for the head that the dorsal column fibers subserve for the body.

From the ventrobasal complex, *third order neurons* project, as shown in Figure 49–4, mainly to the *postcentral gyrus* of the *cerebral cortex* called *somatic sensory area I*. But, in addition, neurons also project to closely associated regions of the cortex behind and in front of the postcentral gyrus. Finally, a few fibers project to the lowermost lateral portion of each parietal lobe, an area called *somatic sensory area II*.

Anatomy of the Spinocervical Pathway. The anatomy of the spinocervical pathway is much less well known than that of the dorsal column pathway. However, many of the large sensory fibers that enter the cord in the dorsal roots soon synapse mainly in lamina IV of the cord gray matter but also in adjacent laminae as well, giving rise to second order fibers that enter the *dorsolateral white columns* and ascend in the spinocervical tract to the cervical region of the cord or even to the medulla. At these points these fibers again synapse either in the dorsal horn of the cord or in medullary nuclei either adjacent to or as part of the *dorsal column nuclei*. Then third order neurons decussate to the opposite side and pass along with the second order neurons of the dorsal column pathway upward to the thalamus through the *medial lemnisci*. Thus, the pathway within the brain parallels that of the dorsal column pathway.

Separation of Sensory Modalities Between the Dorsal Column Pathway and the Spinocervical Pathway

The way in which the nervous system distinguishes among the different modalities of sensation is to transmit the signals of different modalities to separate the central nervous system areas. This separation of modality begins in the spinal cord. It has already been noted that only certain mechanoreceptor modalities are transmitted in the dorsal-lemniscal system while all the other somatic modalities are transmitted in the anterolateral spinothalamic system. However, even within the dorsal system there is an additional degree of separation. The dorsal column pathway transmits mainly signals from rapidly adapting sensory receptors. For instance, it is only through the dorsal column pathway that signals are transmitted from the extremely rapidly adapting pacinian corpuscles. Also, most of the signals from the Meissner's corpuscles and from the hair receptors, both

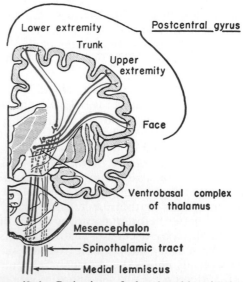

Figure 49–4. Projection of the dorsal-lemniscal system from the thalamus to the somesthetic cortex. (Modified from Brodal: Neurological Anatomy in Relation to Clinical Medicine. New York, Oxford University Press, 1969.)

of which are moderately rapidly adapting receptors, are transmitted through this pathway.

On the other hand, the more slowly adapting signals from the Merkel's discs, from the deep tissue Ruffini end-organs, and from the slowly adapting Ruffini position sense receptors of the joint capsules seem to be transmitted through the spinocervical pathway.

At the level of the cerebral cortex, separation of the mechanoreceptor sensory modalities is also found in somatic sensory area I, as will be discussed later. Therefore, it is presumed that these modalities remain separated throughout their entire pathway from the spinal cord to the cortex.

Spatial Orientation of the Nerve Fibers in the Dorsal-Lemniscal System

One of the distinguishing features of the dorsal-lemniscal system is a distinct spatial orientation of nerve fibers from the individual parts of the body that is maintained throughout. For instance, in the dorsal columns, the fibers from the lower parts of the body lie toward the center, while those that enter the spinal cord at progressively higher segmental levels form successive layers laterally.

The spatial orientation in the spinocervical pathway is less well known. However, stimulation experiments of single fibers within this pathway have shown that the sensory signals diverge very little, indicating a high degree of spatial orientation in this pathway as well.

In the thalamus, the distinct spatial orientation is still maintained, with the tail end of the body represented by the most lateral portions of the ventrobasal complex and the head and face represented in the medial component of the complex. However, because of the crossing of the medial lemnisci in the medulla, the left side of the body is represented in the right side of the thalamus, and the right side of the body is represented in the left side of the thalamus. In a similar manner, the fibers passing to the cerebral cortex also are spatially oriented so that a single part of the cortex receives signals from a discrete area of the body, as is described below.

THE SOMESTHETIC CORTEX

The area of the cerebral cortex to which the sensory signals are projected is called the *somesthetic cortex*. In the human being, this area lies mainly in the anterior portions of the parietal lobes. Two distinct and separate areas are known to receive direct afferent nerve fibers from the relay nuclei of the thalamus; these, called *somatic sensory area I* and *somatic sensory area II,* are illustrated in Figure 49–5. However, somatic sensory area I is so such more important to the sensory functions of the body than is somatic sensory area II that in popular usage, the term

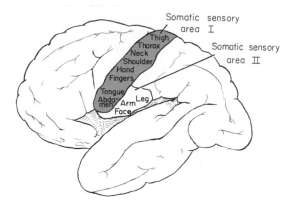

Figure 49–5. The two somesthetic cortical areas, somatic sensory areas I and II.

somesthetic cortex is almost always used to mean area I exclusive of area II. Yet, to keep these two areas separated, we will henceforth refer to them separately as somatic sensory area I and somatic sensory area II.

Projection of the Body in Somatic Sensory Area I. Somatic sensory area I lies in the postcentral gyrus of the human cerebral cortex (in Brodmann areas 3, 1, and 2 immediately behind the central sulcus). A distinct spatial orientation exists in this area for reception of nerve signals from the different areas of the body. Figure 49–6 illustrates a cross-section through the brain at the level of the postcentral gyrus, showing the representations of the different parts of the body in separate regions of somatic sensory area I. Note, however, that

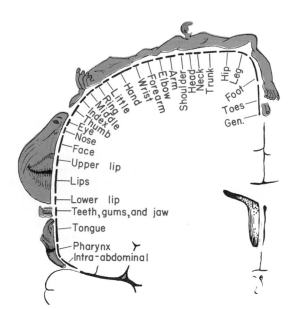

Figure 49–6. Representation of the different areas of the body in the somatic sensory area I of the cortex. (From Penfield and Rasmussen: Cerebral Cortex of Man: A Clinical Study of Localization of Function. New York, Macmillan Company, 1968.)

each side of the cortex receives sensory information exclusively from the opposite side of the body (with the exception of a small amount of sensory information from the same side of the face).

Some areas of the body are represented by large areas in the somatic cortex — the lips by far the greatest of all, followed by the face and thumb — while the entire trunk and lower part of the body are represented by relatively small areas. The sizes of these areas are directly proportional to the number of specialized sensory receptors in each respective peripheral area of the body. For instance, a great number of specialized nerve endings are found in the lips and thumb, while only a few are present in the skin of the trunk.

Note also that the head is represented in the lower or lateral portion of the postcentral gyrus, while the lower part of the body is represented in the medial or upper portion of the postcentral gyrus.

Modality Differentiation in Somatic Sensory Area I. Not only is there spatial projection of the body in somatic sensory area I but also modality separation of different types of mechanoreceptor signals. Tactile signals stimulate the central and anterior portions of the postcentral gyrus, while two other types of signals stimulate respectively the posterior and very anterior portions of somatic sensory area I. The posterior portion receives signals mainly from the joint receptors that provide the brain with information about position of all parts of the body with respect to each other. The very anterior edge of somatic sensory area I, where it dips deeply into the central sulcus and abuts the primary motor cortex anterior to the central sulcus, receives sensory signals mainly from the muscle spindles. This muscle information seemingly operates at a subconscious level and is probably used almost exclusively to help control the muscles from which the signals derive. This separation of modalities is important because it helps us to understand how the brain dissects different types of information from incoming sensory signals.

Somatic Sensory Area II. The second cortical area to which somatic afferent fibers project, somatic sensory area II, is a much smaller area that lies posterior and inferior to the lower end of the postcentral gyrus and on the upper wall of the lateral fissure, as shown in Figure 49–5. The degree of localization of the different parts of the body is very poor in this area compared with somatic sensory area I. The face is represented anteriorly, the arms centrally, and the legs posteriorly.

So little is known about the function of somatic sensory area II that it cannot be discussed intelligently. It is known that signals enter this area from both the dorsal system and the spino-reticulo-thalamic system, and that signals reach this area from both sides of the body. Also, stimulation of somatic sensory area II in some instances causes complex body movements, for

which reason it possibly plays a role in sensory control of motor functions.

Excitation of Vertical Columns of Neurons in the Somesthetic Cortex

The cerebral cortex contains *six* separate layers of neurons, and, as would be expected, the neurons in each layer perform functions different from those in other layers. Also, the neurons are arranged functionally in vertical columns extending all the way through the six layers of the cortex, each column having a diameter of about 0.33 to 1 millimeter and containing about 100,000 neuronal cell bodies. Unfortunately, we still know relatively little about the functions of these columns of cells but we are certain about the following facts:

1. The incoming sensory signal excites mainly neuronal layer IV first; then the signal spreads toward the surface of the cortex and also toward the deeper layers.

2. Layers I and II receive a diffuse, nonspecific input from the reticular activating system that can facilitate the whole brain at once; this system will be described in Chapter 54. This input perhaps controls the overall level of excitability of the cortex.

3. The neurons in layers V and VI send axons to other parts of the nervous system — some to other areas of the cortex, some to deeper structures of the brain, such as the thalamus or brain stem, and some even to the spinal cord.

Similar vertical columns of neurons exist in all other areas of the cortex as well as in the somesthetic cortex. In particular, they have been shown to be very important for function of the visual cortex and the motor cortex.

Each vertical column of neurons seems to be able to decipher a specific quality of information from the sensory signal. For instance, in the visual cortex one specific column will detect a line oriented in a particular direction, while an adjacent column will detect a similar line oriented in a slightly different direction. Presumably, in the somesthetic cortex each column detects separate qualities of signals (angles of orientation of rough spots, lengths of rough spots, perhaps roundness of objects, perhaps sharpness of objects, and so forth) from specific surface areas of the body.

Functions of Somatic Sensory Area I

The functional capabilities of different areas of the somesthetic cortex have been determined by selective excision of the different portions. Widespread excision of somatic sensory area I causes loss of the following types of sensory judgment:

1. The person is unable to localize discretely the different sensations in the different parts of the body. However, he or she can localize these sensations very crudely, such as to a particular hand, which indicates that the thalamus or parts of the cerebral cortex not normally considered to be concerned with somatic sensations can perform some degree of localization.

2. He is unable to judge critical degrees of pressure against his body.

3. He is unable to judge exactly the weights of objects.

4. He is unable to judge shapes or forms of objects. This is called *astereognosis.*

5. He is unable to judge texture of materials, for this type of judgment depends on highly critical sensations caused by movement of the skin over the surface to be judged.

6. He is unable to recognize the relative orientation of the different parts of his body with respect to each other.

Note in the above list that nothing has been said about loss of pain. However, in the absence of somatic sensory area I, the appreciation of pain may be altered either in quality or in intensity. But more important, the pain that does occur is poorly localized, indicating that pain localization probably depends mainly upon simultaneous stimulation of tactile stimuli that use the topographical map of the body in somatic sensory area I to localize the source of the pain.

SOMATIC ASSOCIATION AREAS

Brodmann areas 5 and 7 of the cerebral cortex, which are located in the parietal cortex immediately behind somatic sensory area I and above somatic sensory area II, play important roles in deciphering the sensory information that enters the somatic sensory areas. Therefore, these areas are called the *somatic association areas.*

Electrical stimulation in the somatic association area can occasionally cause a person to experience a complex somatic sensation, sometimes even the "feeling" of an object such as a knife or a ball. Therefore, it seems clear that the somatic association area combines information from multiple points in the somatic sensory area to decipher its meaning. This also fits with the anatomical arrangement of the neuronal tracts that enter the somatic association area, for it receives signals from (a) the primary somatic areas, (b) the ventrobasal complex of the thalamus, and (c) adjacent areas of the thalamus which themselves receive input from the ventrobasal complex.

Effect of Removing the Somatic Association Area — Amorphosynthesis. When the somatic association area is removed, the person especially loses the ability to recognize complex objects and complex forms that are felt. In addition, the person loses most of the sense of form of his or her own body. An especially interesting fact is that loss of the somatic association area on one side of the brain causes the person sometimes to be oblivious of the opposite side of the body — that is, to forget that it is there. Likewise, when feeling objects, the person will tend to feel only one side of the object and to forget that the other side even exists. This complex sensory deficit is called *amorphosynthesis.*

CHARACTERISTICS OF TRANSMISSION IN THE DORSAL-LEMNISCAL SYSTEM

Faithfulness of Transmission. The most important functional characteristic of the dorsal-lemniscal system is its *faithfulness* of transmission. That is, each time a point in the periphery is stimulated, a signal ordinarily is transmitted all the way to the somesthetic cortex. Also, if this peripheral stimulus increases in intensity, the intensity of the signal at the cerebral cortex increases proportionately. And, finally, when a discrete area of the body is stimulated, the signal from this area is transmitted to a discrete area of the cerebral cortex. Thus, the dorsal system is adequately organized for transmission of accurate information from the periphery to the sensorium. Furthermore, the responsiveness of this system can be altered only moderately by stimuli from other areas of the nervous system, and it is not depressed to a significant extent even by moderate degrees of general anesthesia; this is quite different for most portions of the spinothalamic pathway.

Basic Neuronal Circuit and Discharge Pattern in the Dorsal-Lemniscal System. The lower part of Figure 49–7 illustrates the basic organization of the neuronal circuit of the dorsal column pathway, showing that at each synaptic stage divergence occurs. However, the upper part of the figure shows that a single receptor stimulus on the skin does not cause all the cortical neurons with which that receptor connects to discharge at the same rate. Instead, the cortical neurons that discharge to the greatest extent are those in a central part of the cortical "field" for each respective receptor. Thus, a weak stimulus causes only the centralmost neurons to fire. A stronger stimulus causes still more neurons to fire, but those in the center still discharge at a considerably more rapid rate than do those farther away from the center.

Two-Point Discrimination. A method frequently used to test tactile capabilities is to determine a person's so-called "two-point discriminatory ability." In this test, two needles are pressed against the skin, and the subject determines whether two points of stimulus are felt or one point. On the tips

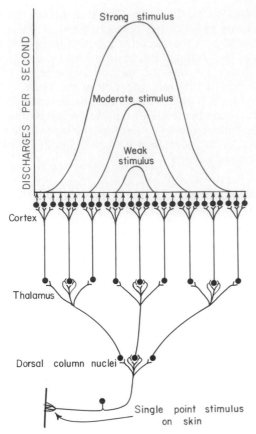

Figure 49–7. Transmission of pinpoint stimulus signal to the cortex.

of the fingers a person can distinguish two separate points even when the needles are as close together as 1 to 2 mm. However, on the person's back, the needles must usually be as far apart as 30 to 70 mm. before one can detect two separate points. The reason for this is that there are many specialized tactile receptors in the tips of the fingers in comparison with a small number in the skin of the back. Referring back to Figure 49–6, we can see also that the portions of the body that have a high degree of two-point discrimination have a correspondingly large cortical representation in somatic sensory area I.

Figure 49–8 illustrates the mechanism by which the dorsal column pathway, and the other sensory pathways as well but to a less critical degree, transmit two-point discriminatory information. This figure shows two adjacent points on the skin that are strongly stimulated, and it also shows the small area of the somesthetic cortex (greatly enlarged) that is excited by signals from the two stimulated points. The solid black curve shows the spatial pattern of cortical excitation when both skin points are stimulated simultaneously. Note that the resultant zone of excitation has two separate peaks. It is these two peaks separated by

a valley that allow the sensory cortex to detect the presence of two stimulatory points rather than a single point. However, the capability of the sensorium to distinguish between two points of stimulation is strongly influenced by another mechanism, the mechanism of lateral inhibition, as explained in the following section.

Increase in Contrast in the Perceived Spatial Pattern Caused by Lateral Inhibition. In Chapter 47 it was pointed out that contrast in sensory patterns is increased by inhibitory signals transmitted laterally in the sensory pathway. This effect was illustrated in Figure 47–11 of that chapter.

In the case of the dorsal column system, an excited sensory receptor in the skin transmits not only excitatory signals to the somesthetic cortex but also inhibitory signals laterally to adjacent fiber pathways. These inhibitory signals help to block lateral spread of the excitatory signal, a process called *lateral inhibition* or *surround inhibition*. As a result, the peak of excitation stands out, and much of the surrounding diffuse stimulation is blocked. This effect is illustrated by the two colored curves in Figure 49–8, showing complete separation of the peaks when the intensity of the surround inhibition is very great. Obviously, this mechanism accentuates the contrast between the areas of peak stimulation and the surrounding areas, thus greatly increasing the contrast or sharpness of the perceived spatial pattern.

Transmission of Rapidly Changing and Repetitive Sensations. The dorsal column system is of particular value for apprising the sensorium of rapidly changing peripheral conditions. This system can "follow" changing stimuli up to at least 400 cycles per seond and can "detect" changes as

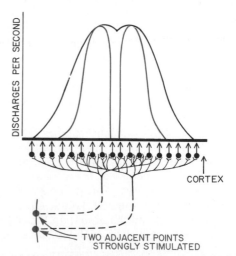

Figure 49–8. Transmission of signals to the cortex from two adjacent pinpoint stimuli. The solid black curve represents the pattern of cortical stimulation without "surround" inhibition, and the two colored curves represent the pattern with "surround" inhibition.

high as 700 cycles. Obviously, very rapid response to a sudden stimulus on the skin allows one to direct attention immediately to any point of contact, which, in turn, allows one to make necessary corrections before damage can be done.

Vibratory Sensation. Because vibratory signals are rapidly repetitive (can be detected as vibration up to 700 cycles per second), they can be transmitted only in the dorsal column pathway and not in the other, more slowly transmitting pathways. For this reason, application of vibration with a tuning fork to different peripheral parts of the body is an important tool used by the neurologist for testing functional integrity of the dorsal columns.

THE POSITION SENSE

The term position sense can be divided into two subtypes: (1) *static position,* which means conscious recognition of the orientation of the different parts of the body with respect to each other, (2) *kinesthesia,* which means conscious recognition of rates of movement of the different parts of the body. The position sensations are transmitted to the sensorium mainly through the dorsal-lemniscal system. However, special features of the position sense require further explanation.

The Position Sense Receptors. Sensory information from many different types of receptors is used to determine both static position and kinesthesia. These include especially the extensive sensory endings in the joint capsules and ligaments but also receptors in the skin and deep tissues near the joints.

Three major types of nerve endings have been described in the joint capsules and ligaments about the joints. (1) By far the most abundant of these are spray-type *Ruffini endings,* one of which was illustrated in Figure 48–1 of Chapter 48. These endings are stimulated strongly when the joint is suddenly moved; they adapt slightly at first but then transmit a steady signal thereafter. (2) A second type of ending resembling the stretch receptors found in muscle tendons (called *Golgi tendon receptors)* is found particularly in the ligaments about the joints. Though far less numerous than the Ruffini endings, they have essentially the same response properties. (3) A few *pacinian corpuscles* are also found in the tissues around the joints. These adapt extremely rapidly and presumably help to detect *rate of rotation* at the joint.

Detection of Static Position by the Joint Receptors. Figure 49–9 illustrates the excitation of seven different nerve fibers leading from separate joint receptors in the capsule of a cat's knee joint. Note that at 180 degrees of joint rotation one of the receptors is stimulated; then at 150 degrees still another is stimulated; at 140 degrees two are stimulated, and so forth. The information from

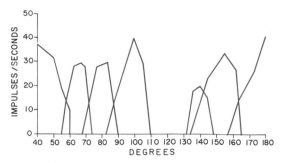

Figure 49–9. Responses of seven different nerve fibers from knee joint receptors in a cat at different degrees of rotation. (Modified from Skoglund: *Acta Physiol. Scand.,* Suppl. 124, *36*:1, 1956.)

these joint receptors continually apprises the central nervous system of the momentary rotation of the joint. That is, the rotation determines *which* receptor is stimulated and how much it is stimulated, and from this the brain knows how far the joint is bent.

Detection of Rate of Movement (Kinesthesia) at the Joint. Because pacinian corpuscles are especially adapted for detecting movement of tissues, it is tempting to suggest that rate of movement at the joints is detected by the pacinian corpuscles. However, the number of pacinian corpuscles in the joint tissues is small, for which reason rate of movement at the joint is probably detected mainly in the following way: The Ruffini and Golgi endings in the joint tissues are stimulated very strongly at first by the joint movement, but within a fraction of a second this strong level of stimulation fades to a lower, steady state rate of firing. Nevertheless, this early overshoot in receptor stimulation is directly proportional to the rate of joint movement and is believed to be the signal used by the brain to discern the rate of movement. However, it is likely that the few pacinian corpuscles also play at least some role in this process.

Transmission of Position Sense Signals in the Dorsal-Lemniscal Pathways. In the past it had been believed that essentially all of the position sense signals were transmitted in the dorsal columns. However, on the basis of studies in animals, there is now much reason to believe that many of these signals are transmitted in the spino-cervical pathway as well. This is probably especially true of the static position signals, that is, those signals that are slowly adapting and that transmit information about the relative positions of the different parts of the body when the person is not moving.

Processing of Position Sense Information in the Dorsal-Lemniscal Pathways. Despite the faithfulness of transmission of signals from the periphery to the sensory cortex in the dorsal-lemniscal system, there is also some processing of sensory

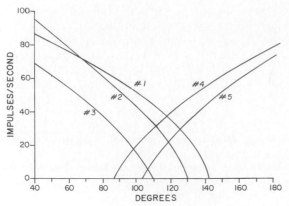

Figure 49–10. Typical responses of five different neurons in the knee joint receptor field of the ventrobasal complex when the knee joint is moved through its range of motion. (The curves were constructed from data in Mountcastle et al.: *J. Neurophysiol.*, 26:807, 1963.)

information at lower synaptic levels before it reaches the cerebral cortex. For instance, the signal pattern from static position receptors changes as it passes progressively up the dorsal column system. Figure 49–9 showed that individual joint receptors are stimulated maximally at specific degrees of rotation of the joint, with the intensity of stimulation decreasing on either side of the maximal point for each receptor. However, the static position signal for joint rotation is quite different at the level of the ventrobasal complex of the thalamus, as can be seen by referring to Figure 49–10. This figure shows that the ventrobasal neurons that respond to the joint rotation signal are of two types: (1) those that are maximally stimulated when the joint is at full rotation and (2) those that are maximally stimulated when the joint is at minimal rotation. In each case, as the degree of rotation changes, the rate of stimulation of the neuron either decreases or increases, depending on the direction in which the joint is being rotated. Furthermore, the intensity of neuronal excitation changes over angles of 40 to 60 degrees of angulation in contrast to 20 to 30 degrees for the individual receptors, as was illustrated in Figure 49–9. Thus, the signals from the individual joint receptors are integrated in the space domain by the thalamic neurons, giving a progressively stronger signal as the joint moves in only one direction rather than giving a peaked signal as occurs in stimulation of individual receptors.

TRANSMISSION IN THE ANTEROLATERAL SPINOTHALAMIC SYSTEM

It was pointed out earlier in the chapter that the anterolateral spinothalamic system transmits sen-

sory signals which do not require highly discrete localization of the signal source and also do not require discrimination of fine gradations of intensity. These include pain, heat, cold, crude tactile, tickle and itch, and sexual sensations. In the following chapter pain and temperature sensations will be discussed, while the present chapter is concerned principally with transmission of the tactile sensations.

ANATOMY OF THE ANTEROLATERAL SPINOTHALAMIC PATHWAY

The anterolateral spinothalamic fibers originate mainly in laminae I, IV, V, and VI in the dorsal horns where the small, peripheral, sensory nerve fibers terminate after entering the cord (see Figure 49–2). Then, as illustrated in Figure 49–11, the fibers immediately cross in the anterior commissure of the cord to the opposite anterolateral white column where they turn upward toward the brain. These fibers ascend rather diffusely throughout the anterolateral columns. However, most anatomists still separate this pathway into a ventral spinothalamic tract and a lateral spinothalamic tract as illustrated in Figure 49–11 even though physiologically it

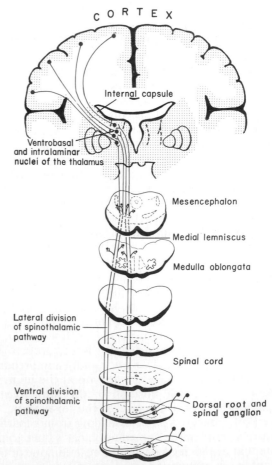

Figure 49–11. The spinothalamic pathways. (Modified from Ranson and Clark: Anatomy of the Nervous System. Philadelphia, W. B. Saunders Company, 1959.)

has been difficult to make this differentiation using electrical recording techniques.

The upper terminus of the anterolateral spinothalamic pathway is mainly two-fold: (1) throughout the *reticular nuclei of the brain stem,* and (2) in two different nuclear complexes of the thalamus, the *ventrobasal complex* and the *intralaminar nuclei*. In general, the tactile signals are transmitted mainly into the ventrobasal complex, and this is probably also true for the temperature signals. On the other hand, only part of the pain signals project to this complex. Instead, most of these enter the reticular nuclei of the brain stem and intralaminar nuclei of the thalamus, as will be discussed in greater detail in the following chapter.

Spinoreticular, Spinotectal, and Paleospinothalamic Pathways. Closely associated with the anterolateral spinothalamic pathway are several other diffuse pathways: (1) the *spinoreticular pathway* that originates in laminae VI, VII, and VIII and terminates in the reticular nuclei of the brain stem; (2) a *spinotectal pathway* that closely parallels the anterolateral spinothalamic pathway but terminates instead in the tectum; and (3) a *paleospinothalamic pathway* that is composed of multiple neurons in series that transmit crude sensory signals from segment to segment of the cord and also upward through the reticular nuclei of the brain stem to the intralaminar nuclei of the thalamus. The functions of all these tracts are not well documented, but they presumably are related to the transmission of crude tactile, thermal, and pain signals and perhaps play either subconscious or mildly conscious roles in controlling the neural functions of the brain stem.

Projection of Spinothalamic Signals from the Thalamus to the Cortex. Spinothalamic signals entering the ventrobasal complex of the thalamus are relayed in association with those from the dorsal-lemniscal system mainly to somatic area I but to a lesser extent to somatic area II. These relayed signals seem to be concerned mainly with tactile sensation, perhaps to a moderate extent with temperature sensation, but probably very little with pain sensation. On the other hand, most of the pain signals seem to be relayed into other areas of the thalamus and into surrounding basal regions of the brain such as the hypothalamus, septal nuclei, and so forth.

Characteristics of Transmission in the Anterolateral Spinothalamic Pathway. In general, the same principles apply to transmission in the anterolateral spinothalamic pathway as in the dorsal-lemniscal system except for the following differences: (a) the velocities of transmission in the spinothalamic pathway are only one-half to one-third those in the dorsal system, ranging between 10 and 60 meters per second; (b) the degree of spatial localization of signals is poor, especially in the pain pathways; (c) the gradations of intensities are also far less acute, most of the sensations being recognized in 10 to 20 gradations of strength rather than as many as 100 gradations for the dorsal system; and (d) the ability to transmit rapidly repetitive sensations is poor.

Thus, it is evident that the anterolateral spinothalamic system is a cruder type of transmission system than the dorsal-lemniscal system. Even so, certain modalities of sensation are transmitted in this system only and not at all in the dorsal-lemniscal system. These are pain, thermal, tickle and itch, and sexual sensations.

SOME SPECIAL ASPECTS OF SENSORY FUNCTION

Function of the Thalamus in Somatic Sensation. When the somesthetic cortex of a human being is destroyed, that person loses most critical tactile sensibilities, but a slight degree of crude tactile sensibility does return. Therefore, it must be assumed that the thalamus has a slight ability to discriminate tactile sensation but functions mainly to relay this type of information to the cortex.

On the other hand, loss of the somesthetic cortex has little effect on one's perception of pain sensation and only a moderate effect on the perception of temperature. Therefore, there is much reason to believe that the thalamus and other associated basal regions of the brain play perhaps the dominant role in discrimination of these sensibilities; it is interesting that these sensibilities appeared very early in the phylogenetic development of animalhood while the critical tactile sensibilities were a late development.

Cortical Control of Sensory Sensitivity. Almost all sensory information that enters the cerebrum, except that from the olfactory system, but including sensory information from the eyes, the ears, the taste receptors, and all the somatic receptors, is relayed through one or another of the thalamic nuclei. Furthermore, the conscious brain is capable of directing its attention to different segments of the sensory system. This function is believed to be mainly achieved through facilitation or inhibition of the cortical receptive areas.

But, in addition, "corticofugal" signals are transmitted from the cortex to the lower relay stations in the sensory pathways to *inhibit* transmission. For instance, corticofugal pathways control the sensitivity of the synapses at all levels of the sensory pathways — in the thalamus, in the brain stem reticular nuclei, in the dorsal column nuclei, and especially in the dorsal horn relay station of the spino-reticulo-thalamic system. Also, similar inhibitory mechanisms are known for the visual, auditory, and olfactory systems, which are discussed in later chapters. Each corticofugal pathway begins in the cortex where the sensory pathway that it controls terminates. Thus, a feedback control loop exists for each sensory pathway.

Obviously, corticofugal control of sensory input could allow the cerebral cortex to alter the threshold for different sensory signals. Also, it might help the brain focus its attention on specific types of information, which is an important and necessary quality of nervous system function.

Automatic Gain Control in the Sensory Pathways, and Enhancement of Contrast. One of the byproducts of corticofugal inhibition is believed to be automatic gain control in many if not most of the sensory pathways.

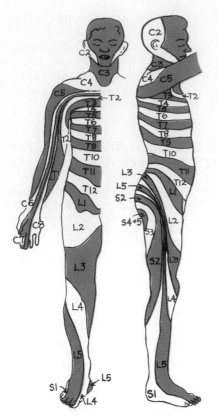

Figure 49–12. The dermatomes. (Modified from Grinker and Sahs: Neurology. Charles C Thomas, 1966.)

That is, when excess sensory signals pass to the brain, recurrent inhibition back to the cord decreases the sensitivity of the sensory pathway. This obviously prevents overloading of the pathway with signals. Also, it helps to preserve contrast in the perception of sensory signals because an excess of signals in the pathway would deluge the cerebral cortex with such a high degree of excitation that it would be impossible to distinguish the highlights from the chaff.

Segmental Fields of Sensation — The Dermatomes. Each spinal nerve innervates a "segmental field" of the skin called a *dermatome*. The different dermatomes are illustrated in Figure 49–12. However, these are shown as if there were distinct borders between the adjacent dermatomes, which is far from true because much overlap exists from segment to segment.

Indeed, because of the great overlap, the posterior roots from an entire segment of the spinal cord can be destroyed without causing significant loss of sensation in the skin.

Figure 49–12 shows that the anal region of the body lies in the dermatome of the most distal cord segment. In the embryo, this is the tail region and is the most distal portion of the body. The legs develop from the lumbar and upper sacral segments rather than from the distal sacral segments, which is evident from the dermatomal map. Obviously, one can use a dermatomal map such as that illustrated in Figure 49–12 to determine the level in the spinal cord at which various cord injuries may have occurred when the peripheral sensations are disturbed.

REFERENCES

Also see bibliographies of Chapters 48 and 50.

Anstis, S. M., *et al.*: Perception. New York, Springer-Verlag, 1978.
Boudreau, J. C., and Tsuchitani, C.: Sensory Neurophysiology: with Special Reference to the Cat. New York, Van Nostrand Reinhold Co., 1973.
Brown, M.: Touch Will Tell. New York, Franklin Watts, 1979.
Coren, S. *et al.*: Sensation and Perception. New York, Academic Press, 1979.
Darian-Smith, I., *et al.*: Posterior parietal cortex: Relations of unit activity to sensorimotor function. *Annu. Rev. Physiol., 41*:141, 1979.
Emmers, R., and Tasker, R. R.: The Human Somesthetic Thalamus. New York, Raven Press, 1975.
Goldstein, E. B.: Sensation and Perception. Belmont, Cal., Wadsworth Publishing Co., 1980.
Gordon, G. (ed.): Active Touch; The Mechanism of Recognition of Objects by Manipulation. New York, Pergamon Press, 1978.
Heath, C. J.: The somatic sensory neurons of pericentral cortex. *In* Porter, R. (ed.): International Review of Physiology: Neurophysiology III. Vol. 17. Baltimore, University Park Press, 1978, p. 193.
Kenshalo, D. R. (ed.): Sensory Function of the Skin of Humans. New York, Plenum Press, 1979.
Lapresle, J., and Haguenau, M.: Anatomico-chemical correlation in focal thalamic lesions. *Z. Neurol., 205*:29, 1973.
Lynn, B.: Somatosensory receptors and their CNS connections. *Annu. Rev. Physiol., 37*:105, 1975.
McCloskey, D. I.: Kinesthetic sensibility. *Physiol. Rev., 58*:763, 1978.
Mountcastle, V. B., et al.: The relation of thalamic cell response to peripheral stimuli varied over an intensive continuum. *J. Neurophysiol., 26*:807, 1963.
Norrsell, U.: Behavioral studies of the somatosensory system. *Physiol. Rev., 60*:327, 1980.
Olton, D. S.: Spatial memory. *Sci. Am., 236*(6):82, 1977.
Perl, E. R., and Boivie, J. G.: Neural substrates of somatic sensations. *In* MTP International Review of Science: Physiology. Vol. 3. Baltimore, University Park Press, 1974, p. 303.
Vallbo, A. B., *et al.*: Somatosensory, proprioceptive, and sympathetic activity in human peripheral nerves. *Physiol. Rev., 59*:919, 1979.
Wall, P. D., and Dubner, R.: Somatosensory pathways. *Annu. Rev. Physiol., 34*:315, 1972.

50

Somatic Sensations: II. Pain,
Visceral Pain, Headache, and
Thermal Sensations

Many, if not most, ailments of the body cause pain. Furthermore, the ability to diagnose different diseases depends to a great extent on a doctor's knowledge of the different qualities of pain, a knowledge of how pain can be referred from one part of the body to another, how pain can spread from the painful site, and, finally, what the different causes of pain are. For these reasons, the present chapter is devoted mainly to pain and to the physiologic basis of some of the associated clinical phenomena.

The Purpose of Pain. Pain is a protective mechanism for the body; it occurs whenever any tissues are being damaged, and it causes the individual to react to remove the pain stimulus. Even such simple activities as sitting for a long time on the ischia can cause tissue destruction because of lack of blood flow to the skin where the skin is compressed by the weight of the body. When the skin becomes painful as a result of the ischemia, the person shifts weight unconsciously. A person who has lost the pain sense, such as after spinal cord injury, fails to feel the pain and therefore fails to shift weight. This eventually results in ulceration at the areas of pressure unless special measures are taken to move the person from time to time.

QUALITIES OF PAIN

Pain has been classified into three different major types: pricking, burning, and aching pain. Other terms used to describe different types of pain include throbbing pain, nauseous pain, cramping pain, sharp pain, electric pain, and others, most of which are well known to almost everyone.

Pricking pain is felt when a needle is stuck into the skin or when the skin is cut with a knife. It is also often felt when a widespread area of the skin is diffusely but strongly irritated.

Burning pain is, as its name implies, the type of pain felt when the skin is burned. It can be excruciating and is the most likely of the pain types to cause suffering.

Aching pain is not felt on the surface of the body, but, instead, is a deep pain with varying degrees of annoyance. Aching pain of low intensity in widespread areas of the body can summate into a very disagreeable sensation.

It is not necessary to describe these different qualities of pain in great detail because they are well known to all persons. The real problem, and one that is only partially solved, is what causes the differences in quality. It is known, however, that pricking pain results from stimulation of delta type A pain fibers, whereas burning and aching pain results from stimulation of the more primitive type C fibers, which will be discussed later in the chapter.

METHODS FOR MEASURING THE PERCEPTION OF PAIN

The intensity of a stimulus necessary to cause pain can be measured in many different ways, but the most used methods have been pricking the skin with a pin at measured pressures, pressing a solid object against a protruding bone with measured force, pinching the skin, or heating the skin with measured amounts of heat. The latter method has proved to be especially accurate from a quantitative point of view.

Figure 50–1 illustrates the basic principles of a heat apparatus used for measuring pain threshold. An intense light is focused by a large condenser lens onto a black spot painted on the forehead of the subject, and the heat intensity delivered by the light is controlled by a rheostat. In determining the subject's threshold for pain, the intensity of the heat is increased in progressive steps,

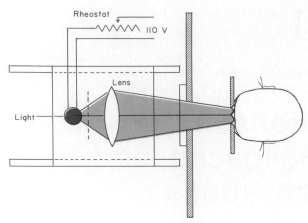

Figure 50–1. Heat apparatus for measuring pain threshold. (From Hardy, Wolff, and Goodell: *J. Clin. Invest*, *19*:649, 1940.)

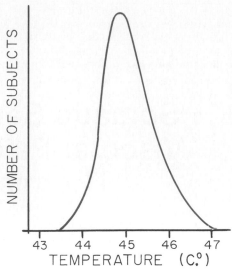

Figure 50–3. Distribution curve obtained from a large number of subjects of the minimal skin temperature that causes pain. (Modified from Hardy: *J. Chronic Dis.*, *4*:22, 1956.)

and the length of time required for the forehead to heat sufficiently to elicit pain is recorded for each heat intensity. These data are then plotted in the form of a "strength-duration curve" to express pain threshold, as follows.

Strength-Duration Curve for Expressing Pain Threshold. Figure 50–2 illustrates a typical strength-duration curve obtained by using the above procedure for measuring pain threshold. This curve is almost identical with the strength-duration curves discussed in Chapter 10 for stimulation of nerve fibers by electrical currents of increasing intensities. Note that a very intense stimulus applied for only a second elicits a sensation of pain while a stimulus of much less intensity may require many seconds. The lowest intensity of stimulus that will excite the sensation of pain when the stimulus is applied for a prolonged period of time is called the *pain threshold.*

The Intensity of Pain — "JNDs." The increase in stimulus intensity that will barely cause a detectable difference in degree of pain is called a *just noticeable difference* (JND). By applying all different stimulus intensities between the level of no pain at all and the most intense pain that a person can distinguish, it has been found that approximately 22 JNDs can be discerned by the average person.

The Pain Threshold. Figure 50–3 shows graphically the skin temperature at which pain is first perceived by

different persons. By far the greatest number perceive pain when the skin temperature reaches almost exactly 45° C., and almost everyone perceives pain before the temperature reaches 47° C. Indeed, measurements in people as widely different as Eskimos, Indians, and whites have shown little differences in their *thresholds for pain*. However, different people do *react* very differently to pain, as is discussed below.

THE PAIN RECEPTORS AND THEIR STIMULATION

Free Nerve Endings as Pain Receptors. The pain receptors in the skin and other tissues are all free nerve endings. They are widespread in the superficial layers of the *skin* and also in certain internal tissues, such as the *periosteum,* the *arterial walls,* the *joint surfaces*, and the *falx* and *tentorium* of the cranial vault. Most of the other deep tissues are not extensively supplied with pain endings but are weakly supplied; nevertheless, any widespread tissue damage can still summate to cause the aching type of pain in these areas.

Types of Stimuli that Excite Pain Receptors — Mechanical, Thermal, and Chemical. Some pain fibers are excited almost entirely by excessive mechanical stress or mechanical damage to the tissues; these are called *mechanosensitive pain receptors*. Others are sensitive to extremes of heat or cold and therefore are called *thermosensitive pain receptors*. And, still others are sensitive to various chemical substances and are called *chemosensitive pain receptors*. Some of the different chemicals that excite the chemosensitive receptors include *bradykinin, serotonin, histamine, po-*

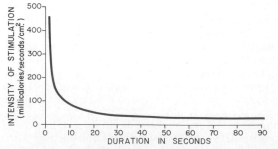

Figure 50–2. Strength-duration curve for depicting pain threshold. (From Hardy: *J. Chronic Dis.* *4*:20, 1956.)

tassium ions, acids, prostaglandins, acetylcholine, and *proteolytic enzymes.*

Though some pain receptors are mainly sensitive to only one of the above types of stimuli, most are sensitive to more than one of the types.

Nonadapting Nature of Pain Receptors. In contrast to most other sensory receptors of the body the pain receptors adapt either not at all or almost not at all. In fact, under some conditions, the threshold for excitation of the pain fibers becomes progressively lower and lower as the pain stimulus continues, thus allowing these receptors to become progressively more activated with time. This increase in sensitivity of the pain receptors is called *hyperalgesia.*

One can readily understand the importance of this failure of pain receptors to adapt, for it allows them to keep the person apprised of a damaging stimulus that causes the pain as long as it persists.

RATE OF TISSUE DAMAGE AS THE CAUSE OF PAIN

The average critical temperature of 45° C. at which a person first begins to perceive pain is also the temperature at which the tissues begin to be damaged by heat; indeed, the tissues are eventually completely destroyed if the temperature remains at this level indefinitely. Therefore, it is immediately apparent that pain resulting from heat is closely correlated with the ability of heat to damage the tissues.

Furthermore, in studying soldiers who had been severely wounded in World War II, it was found that the majority of them felt little or no pain except for a short time after the severe wound had been sustained. This, too, indicates that *pain generally is not felt after damage has been done* but only *while damage is being done.*

Mechanism by Which Tissue Damage Stimulates Pain Receptors

It has already been pointed out that the tissues contain three different types of pain receptors, those that respond to excessive mechanical stress in the tissues, those that respond to extremes of heat, and those that respond to abnormal chemicals in the tissues. Therefore, all the pain receptors can be stimulated by different types of tissue damage.

Special Importance of Chemical Pain Stimuli During Tissue Damage. Injection of certain chemical substances under the skin in human beings can cause the highest degrees of pain. Furthermore, extracts from damaged tissues also cause intense pain when injected beneath the normal skin. Among the substances in such ex-

tracts that are especially painful are *bradykinin, histamine, prostaglandins, acids, excesses of potassium ions, serotonin,* and *proteolytic enzymes.* Obviously, many of these substances could cause direct damage to the pain nerve endings, especially the proteolytic enzymes. But some of the other substances, such as bradykinin and some of the prostaglandins, can cause direct extreme stimulation of pain nerve fibers without necessarily damaging them.

Release of the various substances listed above not only stimulates the chemosensitive pain endings but also greatly decreases the threshold for stimulation of the mechanosensitive and thermosensitive pain receptors as well. A widely known example of this is the extreme pain caused by slight mechanical or heat stimuli following tissue damage by sunburn.

Tissue Ischemia as a Cause of Pain. When blood flow to a tissue is blocked, the tissue becomes very painful within a few minutes. And the greater the rate of metabolism of the tissue, the more rapidly the pain appears. For instance, if a blood pressure cuff is placed around the upper arm and inflated until the arterial blood flow ceases, exercise of the forearm muscles can cause severe muscle pain within 15 to 20 seconds. In the absence of muscle exercise, the pain will not appear for three to four minutes. Cessation of blood flow to the skin, in which the metabolic rate is very low, usually does not cause pain for about 20 to 30 minutes.

The cause of pain in ischemia is yet unknown; however, it is relieved by supplying oxygen to the ischemic tissue. Flow of unoxygenated blood to the tissue will not relieve the pain.

One of the suggested causes of pain in ischemia is accumulation of large amounts of lactic acid in the tissues, formed as a consequence of the anaerobic metabolism (metabolism without oxygen) that occurs during ischemia. However, it is also possible that other chemical agents, such as bradykinin, proteolytic enzymes, and so forth, are formed in the tissues because of cell damage and that these, rather than lactic acid, stimulate the pain nerve endings.

Muscle Spasm as a Cause of Pain. Muscle spasm is also a very common cause of pain, and it is the basis of many clinical pain syndromes. This pain probably results partially from the direct effect of muscle spasm in stimulating mechanosensitive pain receptors. However, it possibly results also from the indirect effect of muscle spasm in causing ischemia and thereby stimulating chemosensitive pain receptors. The muscle spasm not only compresses the blood vessels and diminishes blood flow but also increases the rate of metabolism in the muscle tissue at the same time, thus making the relative ischemia even greater and creating ideal conditions for release of chemical pain-inducing substances.

TRANSMISSION OF PAIN SIGNALS INTO THE CENTRAL NERVOUS SYSTEM

"Fast" Pain Fibers and "Slow" Pain Fibers. Pain signals are transmitted from the periphery to the spinal cord by small delta type A fibers at velocities between 6 and 30 meters per second and also by type C fibers at velocities between 0.5 and 2 meters per second. When the delta type A fibers are blocked without blocking the C fibers by moderate compression of the nerve trunk, the pricking type of pain disappears. On the other hand, when the type C fibers are blocked without blocking the delta fibers by low concentrations of local anesthetic, the burning and aching types of pain disappear.

Because of this double system of pain innervation, a sudden onset of painful stimulus gives a "double" pain sensation: a fast pricking pain followed a second or so later by a slow burning pain. The pricking pain apprises the person very rapidly of a damaging influence and, therefore, plays an important role in making the person react immediately to remove himself from the stimulus. On the other hand, the slow burning sensation tends to become more and more painful over a period of time. It is this sensation that gives one the intolerable suffering of long-continued pain.

Transmission in the Anterolateral Spinothalamic Pathway. Pain fibers enter the cord through the dorsal roots, ascend or descend one to two segments in the *tract of Lissauer,* and then terminate on neurons in the dorsal horns of the cord gray matter, the type Aδ fibers in laminae I and V and the type C fibers in laminae II and III, also called the substantia gelatinosa. Most of the signals then probably pass through one or more additional short-fibered neurons, the last of which gives rise to long fibers that cross immediately to the opposite side of the cord in the *anterior commissure* and pass upward to the brain via the anterolateral spinothalamic pathway as was described in the previous chapter.

We shall see later that the intensity of pain signals can be modified markedly as they pass through the neuronal synapses of the gray matter of the dorsal horns, especially in response to simultaneous signals transmitted by mechanoreceptor sensory nerve fibers and in response to signals entering the dorsal horns from the brain via corticofugal fibers.

As the pain pathways pass into the brain they separate into two pathways: *the pricking pain pathway* and *the burning pain pathway.*

The Pricking Pain Pathway. Figure 50–4 illustrates that the pricking pain pathway terminates in the ventrobasal complex in close association with the areas of termination of the tactile sensation fibers of both the dorsal-lemniscal system and the spinothalamic system. From here signals are transmitted into other areas of the thalamus and to the somatic sensory cortex, mainly to somatic area I. However, the signals to the cortex are probably important mainly for localizing the pain, not for interpreting it.

The Burning Pain Pathway — Stimulation of the Reticular Activating System. Figure 50–4 also shows that the burning and aching pain fibers terminate in the reticular area of the brain stem and in the intralaminar nuclei of the thalamus, which are themselves an upward extension of the reticular formation protruding among the thalamic specific nuclei. Both the reticular area of the brain stem and the intralaminar nuclei are parts of the reticular activating system. In Chapter 54 we will discuss in detail the functions of this system; briefly, it transmits activating signals into essentially all parts of the brain, especially upward through the thalamus to all areas of the cerebral cortex and also laterally into the basal regions of the brain around the thalamus including very importantly the hypothalamus.

Thus, the burning and aching pain fibers, because they do excite the reticular activating system, have a very potent effect for activating essentially the entire nervous system, that is, to arouse one from sleep, to create a state of excitement, to create a sense of urgency, and to promote

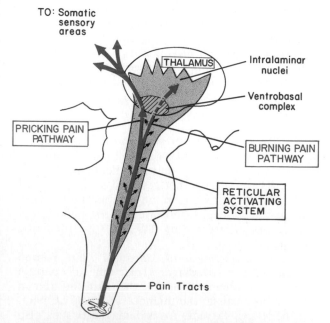

Figure 50–4. Transmission of pain signals into the hindbrain, thalamus, and cortex via the "pricking pain" pathway and the "burning pain" pathway.

defense and aversion reactions designed to rid the person or animal of the painful stimulus.

The signals that are transmitted through the burning pain pathway can be localized only to very gross areas of the body. Therefore, these signals are designed almost entirely for the single purpose of calling one's attention to injurious processes in the body. They create suffering that is sometimes intolerable. Their gradation of intensity is poor; instead, even weak pain signals can summate over a period of time by a process of temporal summation to create an unbearable feeling even though the same pain for short periods of time may be relatively mild.

Function of the Thalamus and Cerebral Cortex in the Appreciation of Pain. Complete removal of the somatic sensory areas of the cerebral cortex does not destroy one's ability to perceive pain. Therefore, it is believed that pain impulses entering only the thalamus and other lower centers cause at least some conscious perception of pain. However, this does not mean that the cerebral cortex has nothing to do with normal pain appreciation; indeed, electrical stimulation of the somesthetic cortical areas causes a person to perceive mild pain in approximately 3 per cent of the stimulations. It is believed that the cortex plays an important role in interpreting the quality of pain even though pain perception might be a function of lower centers.

Localization of Pain in the Body. Most localization of pain probably results from simultaneous stimulation of tactile receptors along with the pain stimulation. However, the pricking type of pain, transmitted through delta type A fibers, can be localized perhaps within 10 to 20 cm. of the stimulated area. On the other hand, the burning and aching types of pain, transmitted through C fibers, are localizable only very grossly, perhaps to a major part of the body such as a limb but certainly not to small areas. This is in keeping with the fact that these fibers terminate extremely diffusely in the hindbrain and thalamus.

Surgical Interruption of Pain Pathways. Often a person has such severe and intractable pain (this often results from rapidly spreading cancer) that it is necessary to relieve the pain. To do this the pain pathway can be destroyed at any one of several different points. If the pain is in the lower part of the body, a *cordotomy* in the upper thoracic region usually relieves the pain at least for a few months. To do this, the spinal cord on the side opposite to the pain is sectioned almost entirely through its anterolateral quadrant, which interrupts the spinothalamic tracts.

Unfortunately, though, the cordotomy is not always successful in relieving the pain for two reasons. First, many of the pain fibers from the upper part of the body do not cross to the opposite side of the spinal cord until they have reached the brain so that the cordotomy does not transsect these fibers. Second, pain frequently returns several months later; indeed, the new pain is often even more objectionable than the original pain. This pain presumably is transmitted in the dorsal half of the cord, but the locus and destination of the fiber pathways are unknown.

An experimental operative procedure to relieve pain is to place lesions in or around the pain-receptive areas of the thalamus to block the pain pathway at this point. It has been claimed that destruction of portions of the intralaminar nuclei in the thalamus can relieve the suffering elicited by pain while still leaving intact one's appreciation of the pricking quality of pain, which remains an important protective mechanism.

THE REACTION TO PAIN AND ITS CONTROL WITHIN THE NERVOUS SYSTEM

It has already been pointed out that the threshold for perceiving pain is determined by the sensitivity of the pain receptors themselves and that this threshold is approximately equal from one person to another. On the other hand, the degree to which each person reacts to pain varies tremendously. Also, the intensity of the pain signals transmitted up the spinal cord to the different pain-receptive areas of the brain can change tremendously under different conditions. This results mainly from activation of a pain inhibiting system, both in the spinal cord and in the brain, that we shall discuss shortly.

Pain causes both reflex motor reactions and psychic reactions. Some of the reflex actions occur in the spinal cord before the pain signals reach the brain, for pain impulses entering the gray matter of the cord can directly initiate "withdrawal reflexes" that remove the body or a portion of the body from the noxious stimulus, as will be discussed in Chapter 51. These primitive spinal cord reflexes, though important in lower animals, are mainly suppressed in the human being by the higher centers of the central nervous system. Yet, in their place, much more complicated and more effective reflexes from the brain are initiated by the pain signals to eliminate the painful stimulus.

The psychic reactions to pain are likely to be far more subtle; they include all the well-known aspects of pain such as anguish, anxiety, crying, depression, nausea, and excess muscular excitability throughout the body. These reactions vary tremendously from one person to another following comparable degrees of pain stimuli.

A PAIN CONTROL ("ANALGESIC") SYSTEM IN THE BRAIN AND SPINAL CORD

Electrical stimulation in several different areas of the brain and spinal cord can greatly reduce or almost block pain signals transmitted in the spinal cord. The most important areas from which this "stimulus-produced analgesia" can be produced include the *periventricular area of the diencephalon* immediately adjacent to the third ventricle, the *periaqueductal gray area* of the brain stem, the *midline raphe nuclei* of the brain stem, and to a lesser extent the *medial forebrain bundle*.

It is believed that this analgesia system operates in the following way: Stimulation either in the periventricular area of the diencephalon or the periaqueductal gray area transmits signals into the midline raphe nuclei. Then, from these nuclei, especially from the *raphe magnus nucleus* in the medulla, fiber tracts pass down the spinal cord to terminate in the dorsal horns, mainly in laminae I, II, and III. It will be recalled that this is the general area of the dorsal horns where the pain sensory fibers from the periphery terminate. It has been shown experimentally that stimulation of the analgesic system will block or suppress transmission of pain impulses through the local neurons of this dorsal horn pain receptive area, most of the block probably occurring in laminae II and III, which is the area called the *substantia gelatinosa*. Stimulation of this analgesic system will suppress pain transmitted by both the type A delta pain fibers and type C fibers as well, but it has almost no effect on the transmission of the other modalities of sensation.

It is very probable that this analgesic system also inhibits brain transmission at other points in the brain pathway, especially in the *reticular nuclei in the brain stem and in the intralaminar nuclei of the thalamus*.

The Brain's Opiate System – the Enkephalins and the Endorphins

A most remarkable research discovery 15 years ago was that injection of morphine in extremely minute quantities into either the periventricular area around the third ventricle of the diencephalon or in the periaqueductal gray area of the brain stem will cause an extreme degree of analgesia. It was immediately assumed that a specific neuronal membrane receptor substance existed in these areas for morphine or morphine-like substances. It was also postulated that substances similar to morphine must exist even normally within a brain analgesic system. This set off a long and intensive search for such morphine-like substances. Finally, only a few years ago two closely related types of compounds with morphine-like actions, called the *enkephalins* and the *endorphins*, were isolated. The enkephalins are found mainly in those areas of the brain associated with pain control, including the periventricular area, the periaqueductal gray, the midline raphe nuclei, the substantia gelatinosa of the dorsal horns in the spinal cord, and the intralaminar nuclei of the thalamus. The endor-

phins have been found in abundance in the hypothalamus and in the pituitary gland.

Therefore, it is now presumed that the enkephalins and the endorphins function as excitatory transmitter substances that activate portions of the brain's analgesic system. Also, infusion of either of these into the cerebrospinal fluid of the third ventricle can lead to analgesia.

Two different types of enkephalins, *met-enkephalin* and *leu-enkephalin*, are found at different points in the analgesic system. Though several different types of endorphins have been isolated, the one that is most potent and presumably most important is *β-endorphin*.

Inhibition of Pain Transmission at the Cord Level by Tactile Signals

Another important landmark in the saga of pain control was the discovery that stimulation of large sensory fibers from the peripheral tactile receptors depresses the transmission of pain signals either from the same area of the body or even from areas sometimes located many segments away. This explains why such simple maneuvers as rubbing the skin near painful areas is often very effective in relieving pain. And it probably also explains why liniments are often useful in the relief of pain. This mechanism and simultaneous psychogenic excitation of the central analgesic system are probably the basis of pain relief by acupuncture.

Treatment of Pain by Electrical Stimulation. Several clinical procedures have been developed recently for suppressing pain by stimulating large sensory nerve fibers. The stimulating electrodes are placed on selected areas of the skin, or on occasion they have been implanted over the spinal cord to stimulate the dorsal sensory columns. And, in a few patients, electrodes have even been placed stereotaxically in the periventricular area of the diencephalon. The patient can then personally control the degree of stimulation. Dramatic relief has been reported in some instances.

The Gating Theory for Pain Control

About a dozen years ago it was proposed that the dorsal horns of the spinal column function as *gates* for controlling entry of pain signals into the pain pathways. And, more specifically, it was proposed that pain signals are constantly balanced against tactile signals, each of these capable of inhibiting the other. It was reasoned that so long as there was excess tactile stimulation, the gates for transmission of pain would be closed, while in the absence of tactile stimulation, pain transmission is unimpeded.

In the intervening years, researchers have now shown that the precise neuronal circuitry proposed for the gating mechanism was mainly incorrect but that slightly different mechanisms for pain sensitivity control do indeed exist. These have been discussed above. Therefore, in a sense, a new concept of pain gating has now been developed, thus supervening the original gating concept but subserving the same function. Yet, it is probably better not to speak in terms of pain gating but instead of pain control.

REFERRED PAIN

Often a person feels pain in a part of his body that is considerably removed from the tissues causing the pain. This pain is called *referred pain*. Usually the pain is initiated in one of the visceral organs and referred to an area of the body surface. Also, pain may originate in a viscus and be referred to another deep area of the body not exactly coincident with the location of the viscus producing the pain. A knowledge of these different types of referred pain is extremely important in diagnosis because many visceral ailments cause no other signs except referred pain.

Mechanism of Referred Pain.   Figure 50–5 illustrates the most likely mechanism by which most pain is referred. In the figure, branches of visceral pain fibers are shown to synapse in the spinal cord with some of the same second order neurons that receive pain fibers from the skin. When the visceral pain fibers are stimulated, pain signals from the viscera are then conducted through at least some of the same neurons that conduct pain signals from the skin, and the person has the feeling that the sensations actually originate in the skin itself.

REFERRED PAIN CAUSED BY REFLEX MUSCULAR SPASM

Some types of referred pain are caused secondarily by reflex muscular spasm. For instance, pain in a ureter can cause reflex spasm of the lumbar muscles. Often the

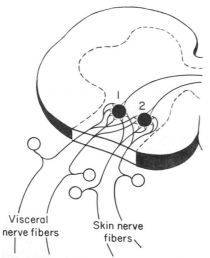

Figure 50–5. Mechanism of referred pain and referred hyperalgesia.

pain from the ureter itself is hardly felt at all, but instead almost all the pain results from spasm of the lumbar muscles.

Many back pains and some types of headache also appear to be caused by muscular spasm, the spasm originating reflexly from much weaker pain signals originating elsewhere in the body.

VISCERAL PAIN

In clinical diagnosis, pain from the different viscera of the abdomen and chest is one of the few criteria that can be used for diagnosing visceral inflammation, disease, and other ailments. In general, the viscera have sensory receptors for no other modalities of sensation besides pain. Also, visceral pain differs from surface pain in several important aspects.

One of the most important differences between surface pain and visceral pain is that highly localized types of damage to the viscera rarely cause severe pain. For instance, a surgeon can cut the gut entirely in two in a patient who is awake without causing significant pain. On the other hand, any stimulus that causes *diffuse stimulation of pain nerve endings* throughout a viscus causes pain that can be extremely severe. For instance, ischemia caused by occluding the blood supply to a large area of gut stimulates many diffuse pain fibers at the same time and can result in extreme pain.

CAUSES OF TRUE VISCERAL PAIN

Any stimulus that excites pain nerve endings in diffuse areas of the viscera causes visceral pain. Such stimuli include ischemia of visceral tissue, chemical damage to the surfaces of the viscera, spasm of the smooth muscle in a hollow viscus, distention of a hollow viscus, or stretching of the ligaments.

Essentially all the true visceral pain originating in the thoracic and abdominal cavities is transmitted through sensory nerve fibers that run in the sympathetic nerves. These fibers are small type C fibers and, therefore, can transmit only burning and aching types of pain. The pathways for transmitting true visceral pain will be discussed in more detail later in this chapter.

Ischemia.   Ischemia causes visceral pain in exactly the same way that it does in other tissues, presumably because of the formation of acidic metabolic end-products or tissue degenerative products, such as bradykinin, proteolytic enzymes, or so forth that stimulate the pain nerve endings.

Chemical Stimuli.   On occasion, damaging substances leak from the gastrointestinal tract into the peritoneal cavity. For instance, proteolytic acidic gastric juice often leaks through a ruptured gastric or duodenal ulcer. This juice causes widespread digestion of the visceral peritoneum, thus stimulating extremely broad areas of pain fibers. The pain is usually extremely severe.

Spasm of a Hollow Viscus. Spasm of the gut, the gallbladder, a bile duct, the ureter, or any other hollow viscus can cause pain in exactly the same way that spasm of skeletal muscle causes pain. This possibly results from mechanical stimulation of the pain endings. Or its cause might be diminished blood flow to the muscle combined with increased metabolic need of the muscle for nutrients. Thus, *relative* ischemia could develop, which causes severe pain.

Often, pain from a spastic viscus occurs in the form of *cramps*, the pain increasing to a high degree of severity and then subsiding, this process continuing rhythmically once every few minutes. The rhythmic cycles result from rhythmic contraction of smooth muscle. For instance, each time a peristaltic wave travels along an overly excitable spastic gut, a cramp occurs. The cramping type of pain frequently occurs in gastroenteritis, constipation, menstruation, parturition, gallbladder disease, or ureteral obstruction.

Overdistension of a Hollow Viscus. Extreme overfilling of a hollow viscus also results in pain, presumably because of overstretch of the tissues themselves. However, overdistension can also collapse the blood vessels that encircle the viscus, or that pass into its wall, thus perhaps promoting ischemic pain.

Insensitive Viscera

A few visceral areas are almost entirely insensitive to pain of any type. These include the parenchyma of the liver and the alveoli of the lungs. Yet the liver *capsule* is extremely sensitive to both direct trauma and stretch, and the *bile ducts* are also sensitive to pain. In the lungs, even though the alveoli are insensitive, the bronchi and the parietal pleura are both very sensitive to pain.

"PARIETAL" PAIN CAUSED BY VISCERAL DAMAGE

In addition to true visceral pain, some pain sensations are also transmitted from the viscera through nerve fibers that innervate the parietal peritoneum, pleura, or pericardium. The parietal surfaces of the visceral cavities are supplied mainly by spinal nerve fibers that penetrate from the surface of the body inward.

Characteristics of Parietal Visceral Pain. When a disease affects a viscus, it often spreads to the parietal wall of the visceral cavity. This wall, like the skin, is supplied with extensive innervation

from the spinal nerves, not from the sympathetic nerves, including the "fast" delta fibers which are different from only "slow" type C fibers in the true visceral pain pathways of the sympathetic nerves. Therefore, pain from the parietal wall of the visceral cavity is frequently very sharp and pricking in quality, though it can also have burning and aching qualities as well if the pain stimulus is diffuse. To emphasize the difference between this pain and true visceral pain, a knife incision through the *parietal* peritoneum is very painful, even though a similar cut through the visceral peritoneum or through a gut is not painful.

LOCALIZATION OF VISCERAL PAIN – THE "VISCERAL" AND THE "PARIETAL" TRANSMISSION PATHWAYS

Pain from the different viscera is frequently difficult to localize for a number of reasons. First, the brain does not know from firsthand experience that the different organs exist, and, therefore, any pain that originates internally can be localized only generally. Second, sensations from the abdomen and thorax are transmitted by two separate pathways to the central nervous system — the *true visceral pathway* and the *parietal pathway*. The true visceral pain is transmitted via sensory fibers of the autonomic nervous system, and the sensations are *referred* to surface areas of the body often far from the painful organ. On the other hand, parietal sensations are conducted *directly* from the parietal peritoneum, pleura, or pericardium, and the sensations are usually *localized directly over the painful area*.

The True Visceral Pathway for Transmission of Pain. Most of the internal organs of the body are supplied by type C pain fibers that pass along the visceral sympathetic nerves into the spinal cord and thence up the anterolateral spinothalamic tract along with the pain fibers from the body's surface. A few visceral pain fibers — those from the distal portion of the colon, from the rectum, and from the bladder — enter the spinal cord through the sacral parasympathetic nerves, and some enter the central nervous system through various cranial nerves. These include fibers in the glossopharyngeal and vagus nerves which transmit pain from the pharynx, trachea, and upper esophagus. Some fibers from the surfaces of the diaphragm as well as from the lower esophagus are also carried in the phrenic nerves.

Localization of Referred Pain Transmitted by the Visceral Pathways. The position in the cord to which visceral afferent fibers pass from each organ depends on the segment of the body from which the organ developed embryologically. For instance, the heart originated in the neck and upper thorax. Consequently, the heart's visceral pain fibers enter the cord all the way from C-3 down to T-5. The stomach had its origin approximately from the seventh to the ninth thoracic segments of the embryo, and consequently the visceral afferents from the stomach enter the spinal cord between these levels. The gallbladder had its origin almost entirely in the ninth thoracic segment, so that the visceral afferents from the gallbladder enter the spinal cord at T-9.

Because the visceral afferent pain fibers are responsible for transmitting referred pain from the viscera, the location of the referred pain on the surface of the body is in the dermatome of the segment from which the visceral organ was originally derived in the embryo. Some of the areas of referred pain on the surface of the body are shown in Figure 50–6.

The Parietal Pathway for Transmission of Abdominal and Thoracic Pain.   A second set of pain fibers penetrates inward from the spinal nerves to innervate the parietal peritoneum, parietal pleura, and parietal pericardium. Also, retroperitoneal visceral organs and perhaps portions of the mesentery are innervated to some extent by parietal pain fibers. The kidney, for instance, is supplied by both visceral and parietal fibers.

Pain from the viscera is frequently localized to two surface areas of the body at the same time because of the dual pathways for transmission of pain. Figure 50–7 illustrates dual transmission of pain from an inflamed appendix. Impulses pass from the appendix through the sympathetic visceral pain fibers into the sympathetic chain and then into the spinal cord at approximately T-10 or T-11; this pain is referred to an area around the umbilicus and is of the aching, cramping type. On the other hand, pain impulses also often originate in

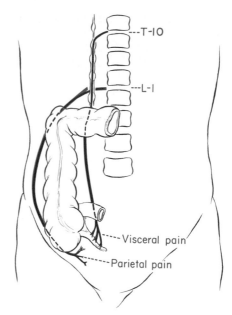

Figure 50–7.   Visceral and parietal transmission of pain from the appendix.

the parietal peritoneum where the inflamed appendix touches the abdominal wall, and these impulses pass through the spinal nerves into the spinal cord at a level of approximately L-1 or L-2. This pain is localized directly over the irritated peritoneum in the right lower quadrant of the abdomen and is of the sharp type.

VISCERAL PAIN FROM VARIOUS ORGANS

Cardiac Pain.   Almost all pain that originates in the heart results from ischemia secondary to coronary sclerosis, which was discussed in Chapter 25. This pain is referred mainly to the base of the neck, over the shoulders, over the pectoral muscles, and down the arms. Most frequently, the referred pain is on the left side rather than on the right — probably because the left side of the heart is much more frequently involved in coronary disease than is the right side — but occasionally referred pain occurs on the right side of the body as well as on the left.

The pain impulses are conducted through sympathetic nerves passing to the middle cervical ganglia, to the stellate ganglia, and to the first four or five thoracic ganglia of the sympathetic chains. Then the impulses pass into the spinal cord through the second, third, fourth and fifth thoracic spinal nerves.

Direct Parietal Pain from the Heart.   When coronary ischemia is extremely severe, such as immediately after a coronary thrombosis, intense cardiac pain frequently occurs directly underneath the sternum simultaneously with pain referred to other areas. This direct pain from underneath the sternum is difficult to explain on the basis of the visceral nerve connections. Therefore, it is highly probable that sensory nerve endings passing from the heart through the pericardial reflections around the great vessels conduct this direct pain.

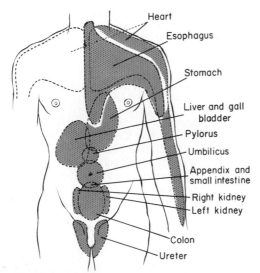

Figure 50–6.   Surface areas of referred pain from different visceral organs.

In addition to pain from the heart, other sensations may accompany coronary thrombosis. One of these is a tight, oppressive sensation beneath the sternum. The exact cause of this is unknown, but a possible cause is reflex spasm of blood vessels, bronchioles, or muscles in the chest region.

Relief of Referred Cardiac Pain by Sympathectomy. To interrupt pain impulses from the heart one can either cut the sympathetic nerves that pass from the heart to the sympathetic chains or, as is usually performed, cut nerve fibers as they pass through the sympathetic chains into the spinal cord. Relief of cardiac pain can frequently be accomplished simply by removing the sympathetic chain from T-2 through T-5 on only the left side, but sometimes it is necessary to remove the fibers on both sides in order to obtain satisfactory results.

Esophageal Pain.   Pain from the esophagus is usually referred to the pharynx, to the lower neck, to the arms, or to midline chest regions beginning at the upper portion of the sternum and ending approximately at the lower level of the heart. Irritation of the gastric end of the esophagus may cause pain directly over the heart, though the pain has nothing whatever to do with the heart. Such pain may be caused by spasm of the cardia, the area where the esophagus empties into the stomach, which causes excessive dilatation of the lower esophagus, or it may result from chemical, bacterial, or other types of inflammatory irritations.

Gastric Pain.   Pain arising in the fundus of the stomach — usually caused by gastritis — is referred to the anterior surface of the chest or upper abdomen, from slightly below the heart to an inch or so below the xyphoid process. This pain is frequently characterized as burning pain; it, or pain from the lower esophagus, causes the condition known as "heartburn."

Most peptic ulcers occur within 1 to 2 inches on either side of the pylorus in the stomach or in the duodenum, and pain from such ulcers is usually referred to a surface point approximately midway between the umbilicus and the xiphoid process. The cause of ulcer pain is almost undoubtedly chemical, because when the acid juices of the stomach are not allowed to reach the pain fibers in the ulcer crater the pain does not exist. This pain is characteristically intensely burning.

Biliary and Gallbladder Pain.   Pain from the bile ducts and gallbladder is localized in the midepigastrium almost coincident with pains caused by peptic ulcers. Also, biliary and gallbladder pain is often burning, like that from ulcers, though cramps often occur too.

Biliary disease, in addition to causing pain on the abdominal surface, frequently refers pain to a small area at the tip of the right scapula. This pain is transmitted through sympathetic afferent fibers that enter the ninth thoracic segment of the spinal cord.

Pancreatic Pain.   Lesions of the pancreas, such as acute or chronic pancreatitis in which the pancreatic enzymes eat away the pancreas and surrounding structures, promote intense pain in areas both anterior to and behind the pancreas. It should be remembered that the pancreas is located beneath the parietal peritoneum and that it receives many parietal sensory fibers from the posterior abdominal wall. Therefore, the pain is usually localized directly behind the pancreas in the back and is severe and burning in character.

Renal Pain.   The kidney, kidney pelvis, and ureters are all retroperitoneal structures and receive most of their pain fibers directly from skeletal nerves. Therefore, pain is usually felt directly behind the ailing structure. However, pain occasionally is referred via visceral afferents to the anterior abdominal wall below and about 2 inches to the side of the umbilicus.

Pain from the bladder is felt directly over the bladder, presumably because the bladder is well innervated by parietal pain fibers. However, pain also is sometimes referred to the groin and testicles because some nerve fibers from the bladder apparently synapse in the cord in association with fibers from the genital areas.

Uterine Pain.   Both parietal and visceral afferent pain may be transmitted from the uterus. The low abdominal cramping pains of dysmenorrhea are mediated through the sympathetic afferents, and an operation to cut the hypogastric nerves between the hypogastric plexus and the uterus will in many instances relieve this pain. On the other hand, lesions of the uterus that spread into the adnexa around the uterus, or lesions of the fallopian tubes and broad ligaments, usually cause pain in the lower back or side. This pain is conducted over parietal nerve fibers and is usually sharper in nature rather than the diffuse cramping pain of true dysmenorrhea.

SOME CLINICAL ABNORMALITIES OF PAIN AND OTHER SENSATIONS

HYPERALGESIA

A pain pathway may become excessively excitable; this gives rise to hyperalgesia, which means hypersensitivity to pain. The basic causes of hyperalgesia are: (1) excessive sensitivity of the pain receptors themselves, which is called *primary hyperalgesia,* or facilitation of sensory transmission, which is called *secondary hyperalgesia.*

An example of primary hyperalgesia is the extreme sensitivity of sunburned skin. Secondary hyperalgesia frequently results from lesions in the spinal cord or in the thalamus. Several of these will be discussed in subsequent sections.

THE THALAMIC SYNDROME

Occasionally the posterolateral branch of the posterior cerebral artery, a small artery supplying the posteroventral portion of the thalamus, becomes blocked by thrombosis so that the nuclei of this area of the thalamus degenerate, while the medial and anterior nuclei of the thalamus remain intact. The patient suffers a series of abnormalities, as follows: First, loss of almost all sensations from the opposite side of the body occurs because of destruction of the relay nuclei. Second, ataxia (inability to control movements precisely) may be evident because of loss of position and kinesthetic signals normally relayed through the thalamus to the cortex. Third, after a few weeks to a few months some sensory perception in the opposite side of the body returns, but strong stimuli are usually necessary to elicit this. When the sensations do occur, they are poorly —if at all— localized, almost always very painful, sometimes lancinating, regardless of the type of stimulus applied to the body. Fourth, the person is likely to perceive many affective sensations of extreme unpleasantness or, rarely, extreme pleasantness; the unpleasant ones are often associated with emotional tirades.

The medial nuclei of the thalamus are not destroyed by thrombosis of the artery. Therefore, it is believed that these nuclei become facilitated and give rise to the enhanced sensitivity to pain transmitted through the reticular system as well as to the affective perceptions.

HERPES ZOSTER

Occasionally a virus that is believed to be identical to the chickenpox virus infects a dorsal root ganglion. This causes severe pain in the dermatomal segment normally subserved by the ganglion, thus eliciting a segmental type of pain that circles halfway around the body. The disease is called herpes zoster, and the pain is commonly called "shingles."

There are two possible causes of the pain of herpes zoster. One is that the disease destroys mainly the large mechanoreceptor afferent sensory fibers. This theoretically could reduce the normal inhibitory effect of these fibers on the pain pathway in the substantia gelatinosa of the dorsal horn. Therefore, pain signals could become exacerbated in the absence of this inhibition.

A second possibility is that the irritated neuronal cell bodies of the root ganglion are stimulated to excessive activity and thereby cause the pain.

TIC DOULOUREUX

Lancinating pains occur in some persons over one side of the face in part of the sensory distribution area of the fifth or ninth nerves; this phenomenon is called tic douloureux (or trigeminal neuralgia or glossopharyngeal neuralgia). The pains feel like sudden electric shocks, and they may appear for only a few seconds at a time or they may be almost continuous. Often, they are set off by exceedingly sensitive "trigger areas" on the surface of the face, in the mouth or in the throat — almost always by a mechanoreceptive stimulus instead of a pain stimulus. For instance, when the patient swallows a bolus of food, as the food touches a tonsil it might set off a severe lancinating pain in the mandibular portion of the fifth nerve.

The pain of tic douloureux can usually be blocked by cutting the peripheral nerve from the hypersensitive area. The sensory portion of the fifth nerve is often sectioned immediately inside the cranium, where the motor and sensory roots of the fifth nerve can be separated so that the motor portions, which are needed for many of the jaw movements, are spared while the sensory elements are destroyed. Obviously, this operation leaves the side of the face anesthetic, which in itself may be annoying. Furthermore, it is sometimes unsuccessful, indicating that the lesion might be more central than the nerves themselves.

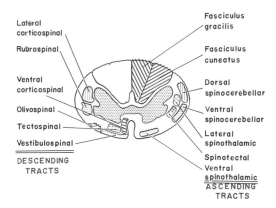

Figure 50–8. Cross-section of the spinal cord, showing principal ascending tracts on the right and principal descending tracts on the left.

THE BROWN-SÉQUARD SYNDROME

Obviously, if the spinal cord is transected entirely, all sensations and motor functions distal to the segments of transection are blocked, but if only one side of the spinal cord is transected, the so-called Brown-Séquard syndrome occurs. The following effects of such a transection occur, and these can be predicted from a knowledge of the cord fiber tracts illustrated in Figure 50–8: All motor functions are blocked on the side of the transection in all segments below the level of the transection; yet, only some of the modalities of sensation are lost on the transected side and others are lost on the opposite side. The sensations of pain, heat, and cold are lost on the opposite side of the body in all dermatomes two to six segments below the level of the transection. The sensations that are transmitted only in the dorsal and dorsolateral columns — kinesthetic and position sensations, vibration sensation, discrete localization, and two-point discrimination — are lost entirely on the side of the transection in all dermatomes below the level of the transection. Touch is impaired on the side of the transection because the principal pathway for transmission of light touch, the dorsal columns, is transected. Yet, "crude touch," which is poorly localized, still persists because of transmission in the opposite ventral spinothalamic tract.

HEADACHE

Headaches are actually referred pain to the surface of the head from the deep structures. Many headaches result from pain stimuli arising inside the cranium, but equally as many probably result from pain arising outside the cranium.

HEADACHE OF INTRACRANIAL ORIGIN

Pain-Sensitive Areas in the Cranial Vault. The brain itself is almost totally insensitive to pain. Even cutting or electrically stimulating the somesthetic centers of the cortex only occasionally causes pain; instead, it causes tactile paresthesias on the area of the body represented by the portion of the somesthetic cortex stimulated. Therefore, it is obvious from the outset that much or most of the pain of headache probably is not caused by damage within the brain itself.

On the other hand, *tugging on the venous sinuses, damaging the tentorium,* or *stretching the dura at the base of the brain* can all cause intense pain that is recognized as headache. Also, almost any type of traumatizing, crushing, or stretching stimulus to the *blood vessels of the dura* can cause headache. A very sensitive structure is the middle meningeal artery, and neurosurgeons are careful to anesthetize this artery specifically when performing brain operations under local anesthesia.

Areas of the Head to Which Intracranial Headache Is Referred. Stimulation of pain receptors in the intracranial vault above the tentorium, including the upper surface of the tentorium itself, initiates impulses in the fifth nerve and, therefore, causes referred headache to the front half of the head in the area supplied by the fifth cranial nerve, as illustrated in Figure 50–9.

On the other hand, pain impulses from beneath the tentorium enter the central nervous system mainly through the second cervical nerve, which also supplies the scalp behind the ear. Therefore, subtentorial pain stimuli cause "occipital headache" referred to the posterior part of the head as shown in Figure 50–9.

Types of Intracranial Headache. *Headache of Meningitis.* One of the most severe headaches of all is that resulting from meningitis, which causes inflammation of all the meninges, including the sensitive areas of the dura and the sensitive areas around the venous sinuses. Such intense damage as this can cause extreme headache pain referred over the entire head.

Headache Resulting from Direct Meningeal Trauma. Following a brain operation the patient ordinarily has intense headache for several days to several weeks. Though part of this headache may result from the brain trauma, experiments indicate that most of it results from meningeal irritation.

Another type of meningeal trauma that almost invariably causes headache is the meningeal irritation resulting from brain tumor. Usually, tumor headache is referred to a localized area of the head, the exact area depending on the portion of the meninges affected by the tumor. Since a tumor above the tentorium is likely to refer its pain to the front half of the head and any tumor below the tentorium to the occipital region of the skull, the general location of an intracranial tumor can often be predicted from the area of the headache.

Headache Caused by Low Cerebrospinal Fluid Pressure. Removing as little as 20 ml. of fluid from the spinal canal, particularly if the person remains in the upright position, often causes intense intracranial headache. Removing this quantity of fluid removes the flotation for the brain that is normally provided by the cerebrospinal fluid. Therefore, the weight of the brain stretches the various dural surfaces and thereby elicits the pain which causes the headache.

Migraine Headache. Migraine headache is a special type of headache that is thought to result from abnormal vascular phenomena, though the exact mechanism is unknown.

Migraine headaches often begin with various prodromal sensations, such as nausea, loss of vision in part of the field of vision, visual aura, or other types of sensory hallucinations. Ordinarily, the prodromal symptoms begin half an hour to an hour prior to the beginning of the headache itself. Therefore, any theory that explains migraine headache must also explain these prodromal symptoms.

One of the theories of the cause of migraine headaches is that prolonged emotion or tension causes reflex vasospasm of some of the arteries of the head, including arteries that supply the brain itself. The vasospasm theoretically produces ischemia of portions of the brain, and this is responsible for the prodromal symptoms. Then, as a result of the intense ischemia, something happens to the vascular wall to allow it to become flaccid and incapable of maintaining vascular tone for 24 to 48 hours. The blood pressure in the vessels causes them to dilate and pulsate intensely, and it is supposedly the excessive stretching of the walls of the arteries — including the extracranial arteries such as the temporal artery — that causes the actual pain of migraine headaches. However, it is possible that diffuse after-effects of ischemia in the brain itself are at least partially responsible for this type of headache.

Alcoholic Headache. As many people have experienced, a headache usually follows an alcoholic binge. It is most likely that alcohol, because it is toxic to tissues, directly irritates the meninges and causes intracranial pain.

Headache Caused by Constipation. Constipation causes headache in many persons. Because it has been shown that constipation headache can occur in persons whose spinal cords have been cut, we know that this headache is not caused by nervous impulses from the colon. Therefore, it possibly results from absorbed toxic products or from changes in the circulatory system. Indeed, constipation sometimes does cause temporary loss of plasma into the wall of the gut, and a resulting poor flow of blood to the head could be the cause of the headache.

EXTRACRANIAL TYPES OF HEADACHE

Headache Resulting from Muscular Spasm. Emotional tension often causes many of the muscles of the

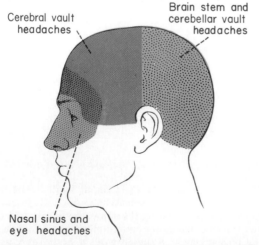

Figure 50–9. Areas of headache resulting from different causes.

Cerebral vault headaches

Brain stem and cerebellar vault headaches

Nasal sinus and eye headaches

head, including especially those muscles attached to the scalp and the neck muscles attached to the occiput, to become spastic, and it is postulated that this is one of the most common causes of headache. The pain of the spastic head muscles supposedly is referred to the overlying areas of the head and gives one the same type of headache as do intracranial lesions.

Headache Caused by Irritation of the Nasal and Accessory Nasal Structures.   The mucous membranes of the nose and also of all the nasal sinuses are sensitive to pain, but not intensely so. Nevertheless, infection or other irritative processes in widespread areas of the nasal structures usually cause headache that is referred behind the eyes or, in the case of frontal sinus infection, to the frontal surfaces of the forehead and scalp, as illustrated in Figure 50–9. Also, pain from the lower sinuses — such as the maxillary sinuses — can be felt in the face.

Headache Caused by Eye Disorders.   Difficulty in focusing one's eyes clearly may cause excessive contraction of the ciliary muscles in an attempt to gain clear vision. Even though these muscles are extremely small, tonic contraction of them can be the cause of retro-orbital headache. Also, excessive attempts to focus the eyes can result in reflex spasm in various facial and extraocular muscles. Tonic spasm in these muscles is also a possible cause of headache.

A second type of headache originating in the eyes occurs when the eyes are exposed to excessive irradiation by ultraviolet light rays. Watching the sun or the arc of an arc-welder for even a few seconds may result in headache that lasts from 24 to 48 hours. The headache sometimes results from "actinic" irritation of the conjunctivae, and the pain is referred to the surface of the head or retro-orbitally. However, focusing intense light from an arc or the sun on the retina can actually burn the retina, and this could result in headache.

THERMAL SENSATIONS

THERMAL RECEPTORS AND THEIR EXCITATION

The human being can perceive different gradations of cold and heat, progressing from *freezing cold* to *cold* to *cool* to *indifferent* to *warm* to *hot* to *burning hot*.

Thermal gradations are discriminated by at least three different types of sensory receptors: the cold receptors, the warmth receptors, and two subtypes of pain receptors, cold-pain receptors and warmth-pain receptors. The two types of pain receptors are stimulated only by extreme degrees of heat or cold and therefore are responsible, along with the cold and warmth receptors, for "freezing cold" and "burning hot" sensations.

The cold and warmth receptors are located immediately under the skin at discrete but separated points, each having a stimulatory diameter of about 1 mm. In most areas of the body there are 3 to 4 times as many cold receptors as warmth receptors, and the number varies from as great as

15 to 25 cold points per square centimeter in the lips, to 3 to 5 cold points per square centimeter in the finger, and to less than 1 cold point per square centimeter in some broad surface areas of the body — and correspondingly fewer numbers of warmth points.

Though it is quite certain on the basis of psychological tests that there are distinctive warmth nerve endings, these have not yet been identified histologically. They are presumed to be free nerve endings because warmth signals are transmitted over type C nerve fibers at transmission velocities of only 0.4 to 2 meters per second.

On the other hand, a definitive cold receptor has been identified. It is a special, small, type A delta myelinated nerve ending that branches a number of times, the tips of which protrude into the bottom surfaces of basal epidermal cells. Signals are transmitted from these receptors via the delta nerve fibers at velocities up to about 20 meters per second. However, cold sensations are also transmitted in type C nerve fibers, which suggests that some free nerve endings also might function as cold receptors.

Stimulation of Thermal Receptors — Sensations of Cold, Cool, Indifferent, Warmth, and Hot.   Figure 50–10 illustrates the effects of different temperatures on the responses of four different types of nerve fibers: (1) a cold-pain fiber, (2) a cold fiber, (3) a warmth fiber, and (4) a heat-pain fiber. Note especially that these fibers respond differently at different levels of temperature. For instance, in the *very* cold region only the cold-pain fibers are stimulated (if the skin becomes even colder so that it nearly freezes or actually does freeze, even these fibers cannot be stimulated). As the temperature rises to 10 to 15° C., pain impulses cease, but the cold receptors begin to be stimulated. Then, above about 30° C. the warmth receptors become stimulated while the cold recep-

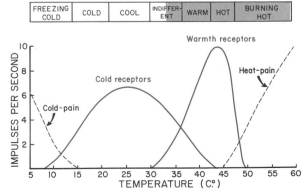

Figure 50–10.   Frequencies of discharge of (1) a cold-pain fiber, (2) a cold fiber, (3) a warmth fiber, and (4) a heat-pain fiber. (The responses of these fibers are drawn from original data collected in separate experiments by Zotterman, Hensel, and Kenshalo.)

tors fade out at about 43° C. Finally, at around 45° C., the heat-pain fibers begin to be stimulated.

One can understand from Figure 50–10, therefore, that a person determines the different gradations of thermal sensations by the relative degrees of stimulation of the different types of endings. For instance, at 20° C. only cold endings are stimulated, whereas at 44° C. only warm endings are stimulated; at 35° C. both cold and warm endings are stimulated, and at 60° C. both cold and pain endings are stimulated. One can understand also from this figure why extreme degrees of cold or heat can be painful and why both these sensations, when intense enough, may give almost exactly the same quality of sensations — that is, freezing cold and burning hot sensations feel almost alike; they are both very painful.

Stimulatory Effects of Rising and Falling Temperature — Adaptation of Thermal Receptors. When a cold receptor is suddenly subjected to an abrupt fall in temperature, it becomes strongly stimulated at first, but this stimulation fades rapidly during the first few seconds and progressively more slowly during the next half hour or more. In other words, the receptor adapts to a very great extent; this is illustrated in Figure 50–11, which shows that the frequency of discharge of a cold receptor rose approximately four-fold when the temperature fell suddenly from 32° to 30° C., but in less than a minute the frequency fell about five sixths of the way back to the original control value. Later, the temperature was suddenly raised from 30° to 32° C. The cold receptor stopped firing entirely for a short time, but after adapting returned to its original control level.

Thus, it is evident that the thermal senses respond markedly to *changes in temperature* in addition to being able to respond to steady states of temperature, as was depicted in Figure 50–10.

This means, therefore, that when the temperature of the skin is actively falling, a person feels much colder than when the temperature remains at the same level. Conversely, if the temperature is actively rising the person feels much warmer than he would at the same temperature if it were constant.

The response to changes in temperature explains the extreme degree of heat that one feels on first entering a tub of hot water and the extreme degree of cold felt on going from a heated room to the out-of-doors on a cold day. The adaptation of the thermal senses explains the ability of the person to become accustomed to the new temperature environment.

Mechanism of Stimulation of the Thermal Receptors. It is believed that the thermal receptors are stimulated by changes in their metabolic rates, these changes resulting from the fact that temperature alters the rates of intracellular chemical reactions about 2 times for each 10° C. change. In other words, thermal detection probably results not from direct physical stimulation but instead from chemical stimulation of the endings as modified by the temperature.

Spatial Summation of Thermal Sensations. Because the number of cold or warmth endings in any one surface area of the body is very slight, it is difficult to judge gradations of temperature when small areas are stimulated. However, when a large area of the body is stimulated all at once, the thermal signals from the entire area summate. Indeed, one reaches the maximum ability to discern minute temperature variations when the entire body is subjected to a temperature change all at once. For instance, rapid changes in temperature as little as 0.01° C. can be detected if this change affects the entire surface of the body simultaneously. On the other hand, temperature changes 100 times this great might not be detected when the skin surface affected is only a square centimeter or so in size.

TRANSMISSION OF THERMAL SIGNALS IN THE NERVOUS SYSTEM

In general, thermal signals are transmitted in almost exactly the same pathways as pain signals. On entering the spinal cord, the signals travel for a few segments upward or downward in the *tract of Lissauer* and then terminate mainly in laminae I, II, and III of the dorsal horns. After a small amount of processing by one or more cord neurons, the signals enter long, ascending thermal fibers that cross to the opposite anterolateral spinothalamic tract and terminate in (a) the reticular areas of the brain stem, (b) the ventrobasal complex of the thalamus, and perhaps (c) in the intralaminar nuclei of the thalamus along with pain

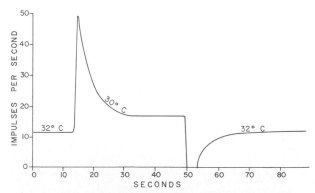

Figure 50–11. Response of a nerve fiber from a cold receptor following, first, instantaneous change in skin temperature from 32° to 30°C. and, second, instantaneous change back to 32°C. Note the adaptation of the receptor and also the higher steady state level of discharge at 30° than at 32°.

signals. A few thermal signals are also relayed to the somesthetic cortex from the ventrobasal complex. Occasionally, a neuron in somatic sensory area I has been found by microelectrode studies to be directly responsive to either cold or warm stimuli in specific areas of the skin. Furthermore, it is known that removal of the postcentral gyrus in the human being reduces the ability to distinguish different gradations of temperature.

REFERENCES

Also see bibliographies following Chapters 47, 48, and 49.

Beaumont, A., and Hughes, J.: Biology of opioid peptides. *Annu. Rev. Pharmacol. Toxicol., 19*:245, 1979.

Bonica, J. J. (ed.): International Symposium on Pain. New York, Raven Press, 1974.

Bonica, J. J., *et al.*: Recent Advances in Pain. Springfield, Ill., Charles C Thomas, 1974.

Bonica. J. J., *et al.* (eds.): Proceedings of the Second World Congress on Pain. New York, Raven Press, 1979.

Brainard, J. B.: Control of Migraine. New York, W. W. Norton & Co., 1979.

Bresler, D. E., and Trubo, R.: Freedom from Pain. New York, Simon and Schuster, 1979.

Casey, K. L.: Pain: A current view of neural mechanisms. *Am. Sci., 61*:194, 1973.

Coren, S., *et al.*: Sensation and Perception. New York, Academic Press, 1979.

Currie, D. J.: Abdominal Pain. Washington, D.C., Hemisphere Publishing Corporation, 1979.

Dalessio, D. J.: Wolff's Headache and Other Head Pain, 3rd Ed. New York, Oxford University Press, 1972.

Fairley, P.: The Conquest of Pain. New York, Charles Scribner's Sons, 1979.

Fields, H. L., and Bashbaum, A. I.: Brainstem control of spinal pain-transmission neurons. *Annu. Rev. Physiol., 40*:217, 1978.

Goldstein, E. B.: Sensation and Perception. Belmont, Cal., Wadsworth Publishing Co., 1980.

Greep, J. M., *et al.* (eds): Pain in Shoulder and Arm. Hingham, Mass., Kluwer Boston, 1979.

Guyton, A. C., and Reeder, R. C.: Pain and contracture in poliomyelitis. *Arch. Neurol. Psychiatr., 63*:954, 1950.

Hardy, J. D.: The nature of pain. *J. Chronic Dis., 4*:22, 1956.

Hardy, J. D., *et al.*: Pain Sensations and Reactions. Baltimore, Williams & Wilkins, 1952.

Herz, A. (ed.): Developments in Opiate Research. New York, Marcel Dekker, 1978.

Hewer, C. L., and Atkinson, R. S. (eds.): Recent Advances in Anaesthesia and Analgesia. Boston, Little, Brown, 1978.

Jacob, J. (ed.): Receptors, New York, Pergamon Press, 1979.

Kenshalo, D. R. (ed.): Sensory Function of the Skin of Humans. New York, Plenum Press, 1979.

Leroy, P. L., *et al.* (eds.): Current Concepts in the Management of Chronic Pain. Miami, Symposia Specialists, 1977.

Lipton, S. (ed.): Persistent Pain: Modern Methods of Treatment. New York, Grune & Stratton, 1977.

Livingston, W. K.: Pain Mechanisms: A Physiologic Interpretation of Causalgia and Its Related States. Wilmington, Del., International Academic Publishers, 1979.

Newman, P. O.: Visceral Afferent Functions of the Nervous System. Monograph of the Physiological Society, No. 25. Baltimore, Williams & Wilkins, 1974.

Perl, E. R., and Boivie, J. G.: Neural substrates of somatic sensation. *In* MTP International Review of Science: Physiology. Vol. 3. Baltimore, University Park Press, 1974, p. 303.

Raskin, N. H., and Appenzeller, O.: Headache. Philadelphia, W. B. Saunders Co., 1980.

Ryan, R. E., and Ryan, R. E., Jr. (eds.): Headache and Head Pain: Diagnosis and Treatment. St. Louis, C. V. Mosby, 1978.

Seltzer, S.: Pain in Dentistry: Diagnosis and Management. Philadelphia, J. B. Lippincott Co., 1978.

Silen, W. (rev.): Cope's Early Diagnosis of the Acute Abdomen. New York, Oxford University Press, 1979.

Simon, E. J., and Hiller, J. M.: The opiate receptors. *Annu. Rev. Pharmacol. Toxicol., 18*:371, 1978.

Snyder, S. H.: Opiate receptors and internal opiates. Sci. Am. *236*(3): 44, 1977.

Snyder, S. H., and Childres, S. R.: Opiate receptors and opioid peptides. *Annu. Rev. Neurosci., 2*:35, 1979.

Swerdlow, M.: Relief of Intractable Pain. New York, American Elsevier Publishing Co., 1974.

Wilson, M. E.: The neurological mechanisms of pain. A review. *Anaesthesia, 29*:407, 1974.

Zimmerman, M.: Neurophysiology of nociception. *Int. Rev. Physiol., 10*:179, 1976.

51

Motor Functions of the Spinal Cord and the Cord Reflexes

In the discussion of the nervous system thus far, we have considered principally the input of sensory information. In the following chapters we will discuss the origin and output of motor signals, the signals that cause muscle contraction and other motor effects throughout the body. Sensory information is integrated at all levels of the nervous system and causes appropriate motor responses, beginning in the spinal cord with relatively simple reflexes, extending into the brain stem with still more complicated responses, and, finally, extending to the cerebrum where the most complicated responses are controlled. The present chapter discusses the control of motor function especially in response to spinal cord reflexes.

Experimental Preparations for Studying Cord Reflexes — The Spinal Animal and the Decerebrate Animal. Normally the functions of the spinal cord are strongly controlled by signals from the brain. Therefore, to study the isolated cord reflexes it is necessary to separate the cord from the higher centers. This is usually done in one of two different types of preparations: (1) the *spinal animal* or (2) the *decerebrate animal.*

The spinal animal is prepared by cutting the spinal cord at any level above the region in which the cord reflexes are to be studied. For a variable period of time, depending upon the phylogenetic level of the animal, as will be discussed later in the chapter, the cord reflexes are deeply depressed immediately after removal of the signals from the brain. However, over a period of minutes, hours, or days in animals and over a period of weeks in the human being, the cord reflexes become progressively more active and can then be studied independently of control by the upper levels of the nervous system. Often the cord reflexes become even more excitable under these conditions than normally.

In the decerebrate preparation, the brain stem is usually transected between the superior and inferior colliculi. Transecting the brain stem at this level removes the voluntary control centers of the forebrain, and it also removes inhibitory influences from the basal ganglia that normally suppress the activities of the lower brain stem and spinal cord; this will be discussed in detail in the following chapter. Removal of this inhibi-

tion causes an immediate increase in the activity of many of the cord reflexes, especially the extensor reflexes that help the animal hold itself up against gravity. This preparation has an advantage over the spinal animal in that many of the cord reflexes are exacerbated even in the acute preparation and, therefore, can be studied immediately, though, of course, lower brain stem reflexes may still interfere with the cord functions.

ORGANIZATION OF THE SPINAL CORD FOR MOTOR FUNCTIONS

The cord gray matter is the integrative area for the cord reflexes and other motor functions. Figure 51–1 shows the typical organization of the cord gray matter in a single cord segment. Sensory signals enter the cord through the sensory roots. After entering the cord, every sensory signal travels to two separate destinations. First, either in the same segment of the cord or in nearby segments, the sensory nerve or its collaterals terminate in the gray matter of the cord and elicit local segmental

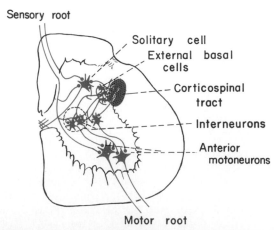

Figure 51–1. Connections of the sensory fibers and corticospinal fibers with the interneurons and anterior motoneurons of the spinal cord.

responses — local excitatory effects, facilitatory effects, reflexes, and so forth. Second, the signals travel to higher levels of the nervous system — to higher levels in the cord itself, to the brain stem, or even to the cerebral cortex. It is these sensory signals that cause the conscious sensory experiences described in the past few chapters.

Each segment of the spinal cord has several million neurons in its gray matter. Aside from the sensory relay neurons already discussed, these neurons are divided into two separate types, the *anterior motoneurons* and the *interneurons* (also called *internuncial cells*).

The Anterior Motoneurons. Located in each segment of the anterior horns of the cord gray matter are several thousand neurons that are 50 to 100 per cent larger than most of the others and are called *anterior motoneurons*. These give rise to the nerve fibers that leave the cord via the anterior roots and innervate the skeletal muscle fibers. The fibers are of two types, the *alpha motoneurons* and the *gamma motoneurons*.

The Alpha Motoneurons. The alpha motoneurons give rise to large, type A alpha (Aα) nerve fibers ranging from 9 to 20 microns in diameter and passing through the spinal nerves to innervate the skeletal muscle fibers. Stimulation of a single nerve fiber excites from 3 to 2000 skeletal muscle fibers which are collectively called the *motor unit*. Transmission of nerve impulses into skeletal muscles and their stimulation of the muscle fibers were discussed in Chapters 10 through 12.

The Gamma Motoneurons. In addition to the alpha motoneurons that excite contraction of the skeletal muscle fibers, about one-half as many much smaller motoneurons, located along with the alpha motoneurons also in the anterior horns, transmit impulses through type A gamma (Aα) fibers, averaging 5 microns in diameter, to special skeletal muscle fibers called *intrafusal fibers*. These are part of the *muscle spindle,* which is discussed at length later in the chapter.

The Interneurons. The interneurons are present in all areas of the cord gray matter — in the dorsal horns, in the anterior horns, and in the intermediate areas between these two. These cells are numerous — approximately 30 times as numerous as the anterior motoneurons. They are small and highly excitable, often exhibiting spontaneous activity and capable of firing as rapidly as 1500 times per second. They have many interconnections one with the other, and many of them directly innervate the anterior motoneurons as illustrated in Figure 51–1. The interconnections among the interneurons and anterior motoneurons are responsible for many of the integrative functions of the spinal cord that are discussed in the remainder of this chapter.

Essentially all the different types of neuronal circuits described in Chapter 47 are found in the interneuron pool of cells of the spinal cord, including the *diverging, converging,* and *repetitive-discharge* circuits. In the following sections of this chapter we will see the many applications of these different circuits to the performance of specific reflex acts by the spinal cord.

Only a few incoming sensory signals to the spinal cord or signals from the brain terminate directly on the anterior motoneurons. Instead, most of them are transmitted first through interneurons where they are appropriately processed before stimulating the anterior motoneurons. Thus, in Figure 51–1, it is shown that the corticospinal tract terminates almost entirely on interneurons, and it is through these that the cerebrum transmits most of its signals for control of muscular function.

The Renshaw Cell Inhibitory System. Located also in the ventral horns of the spinal cord in close association with the motoneurons are a large number of small interneurons called *Renshaw cells*. Almost immediately after the motor axon leaves the motoneuron, collateral branches from the axon pass to the adjacent Renshaw cells. These in turn are inhibitory cells that transmit inhibitory signals to the nearby motoneurons. Thus, stimulation of each motoneuron tends to inhibit the surrounding motoneurons, an effect called recurrent inhibition. This effect is probably important for two major reasons:

1. It shows that the motor system utilizes the principle of lateral inhibition to focus, or sharpen, its signals — that is, to allow unabated transmission of the primary signal while suppressing the tendency for signals to spread to adjacent neurons.

2. The peripheral axon of the anterior motoneuron is known to secrete acetylcholine at its nerve endings. Therefore, the ability of collateral nerve endings from the same axon to stimulate the Renshaw cells proves that acetylcholine is one of the central nervous system excitatory transmitters, acting in the same way that it acts as an excitatory transmitter at the neuromuscular junction.

Sensory Input to the Motoneurons and Interneurons

Most of the sensory fibrils entering each segment of the spinal cord terminate on interneurons, but a very small number of large sensory fibers from the muscle spindles terminate directly on the anterior motoneurons. Thus, there are two pathways in the spinal cord that cord reflexes can take: either directly to the anterior motoneuron itself, utilizing a *monosynaptic pathway;* or through one or more interneurons first, before passing to the anterior motoneuron, thus utilizing a *polysynaptic pathway.* The monosynaptic pathway provides an extremely rapid reflex feedback system and is the basis of the very important muscle *stretch reflex* that will be discussed later in the chapter. All other cord reflexes utilize polysynaptic pathways, pathways that can modify the signals tremendously and that can cause complex reflex patterns. For instance, the very important protective reflex called the *withdrawal reflex* (flexor reflex) utilizes this type of pathway.

Multisegmental Connections in the Spinal Cord — The Propriospinal Fibers. More than half of all the nerve fibers ascending and descending in the spinal cord are *propriospinal fibers*. These are fibers that run from one segment of the cord to another. In addition, the terminal fibrils of sensory fibers as they enter the cord branch both up and down the spinal cord, some of the branches transmitting signals only a segment or two in each direction, while others transmit signals many segments. These ascending and descending fibers of the cord provide pathways for the multisegmental reflexes that will be described later in this chapter, including many reflexes that coordinate movements in both the forelimbs and hindlimbs simultaneously.

ROLE OF THE MUSCLE SPINDLE IN MOTOR CONTROL

Muscles and tendons have an abundance of two special types of receptors: (1) *muscle spindles* that detect (a) change in length of muscle fibers and (b) rate of this change in length, and (2) *Golgi tendon organs* that detect the tension applied to the muscle tendon during muscle contraction or muscle stretch.

The signals from these two receptors operate entirely at a subconscious level, causing no sensory perception at all. But they do transmit tremendous amounts of information into the spinal cord, cerebellum, and even the cerebral cortex, thereby helping all these portions of the nervous system to perform their functions for controlling muscle contraction.

RECEPTOR FUNCTION OF THE MUSCLE SPINDLE

Structure and Innervation of the Muscle Spindle. The physiologic organization of the muscle spindle is illustrated in Figure 51–2. Each spindle is built around three to ten small *intrafusal muscle fibers* that are pointed at their ends and are attached to the sheaths of the surrounding *extrafusal* skeletal muscle fibers. The intrafusal fiber is a very small skeletal muscle fiber. However, the central region of each of these fibers has either no or few actin and myosin filaments. Therefore, this central portion does not contract when the ends do. The end portions are excited by the small *gamma efferent* motor nerve fibers described earlier.

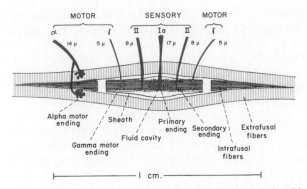

Figure 51–2. The muscle spindle, showing its relationship to the large extrafusal skeletal muscle fibers. Note also both the motor and the sensory innervation of the muscle spindle.

Excitation of the Spindle Receptors

The receptor portion of the muscle spindle is located midway between its two ends where the intrafusal muscle fibers have no contractile elements. As illustrated in Figure 51–2, sensory fibers originate in this area and are stimulated by stretch of this mid-portion of the spindle. One can readily see that there are two different ways in which the muscle spindle receptor can be excited:

1. Stretching the whole muscle in which the muscle spindle is located will obviously stretch the mid-portion of the spindle and therefore excite the receptor.

2. Even if the length of the entire muscle does not change, contraction of the end-portions of the intrafusal fibers will also stretch the mid-portions and therefore excite the receptor.

Two types of sensory endings are found in the receptor area of the muscle spindle. These are:

The Primary Ending. In the very center of the receptor area a large, type Ia fiber encircles the central portions of the intrafusal fibers, forming the so-called *primary ending,* also called the *annulospiral ending.* This nerve fiber averages 17 microns in diameter and transmits sensory signals to the spinal cord at a velocity of about 100 meters per second, as rapidly as any type of sensory nerve fiber in the entire body.

The Secondary Endings. Usually two smaller sensory nerve fibers, type II fibers with an average diameter of 8 microns, innervate the receptor regions on either side of the primary ending, as illustrated in Figure 51–2. These sensory endings are called *secondary endings,* or sometimes they are called *flower spray endings* because in some preparations they look like flower sprays even though they mainly encircle the intrafusal fibers in the same way that the type Ia fiber does.

Division of the Intrafusal Fibers into Nuclear Bag and Nuclear Chain Fibers — Dynamic and Static

Responses of the Muscle Spindle. There are also two different types of intrafusal fibers: (1) *nuclear bag fibers* (one to three), in which a large number of nuclei are congregated into a bag in the central portion of the receptor area, and (2) *nuclear chain fibers* (three to seven), which are about half as large in diameter and half as long as the nuclear bag fibers and have nuclei spread in a chain throughout the receptor area. The primary ending innervates both the nuclear bag intrafusal fibers *and* the nuclear chain fibers. On the other hand, the secondary endings are located almost entirely on the nuclear chain fibers. These relationships are illustrated in Figure 51–3.

Static Response of Both the Primary and the Secondary Receptors. When the receptor portion of the muscle spindle is stretched slowly, the number of impulses transmitted from both the primary and the secondary endings increases almost directly in proportion to the degree of stretch, and the endings continue to transmit these impulses for many minutes. This effect is called the *static response* of the spindle receptors, meaning simply that the receptors continue to transmit their signals for a prolonged period of time.

Dynamic Response of the Primary Endings. The primary ending also exhibits a very strong *dynamic response,* which means that it responds extremely actively to a rapid *rate of change* in length. When the length of a spindle receptor increases only a fraction of a micron, if this increase occurs in a fraction of a second, the primary receptor transmits tremendous numbers of excess impulses into the Ia fiber, but only *while the length is actually increasing*. As soon as the length has stopped increasing, the rate of impulse discharge returns almost back to its original level, except for the small static response that is still present in the signal.

Conversely, when the spindle receptor shortens, this change momentarily decreases the impulse output from the primary ending; as soon as the receptor area has reached its new shortened length, the impulses reappear in the Ia fiber within a fraction of a second. Thus, the primary ending sends extremely strong signals to the central nervous system to apprise it of any change in length of the spindle receptor area.

Since both the primary and the secondary endings innervate the nuclear chain intrafusal fibers, it is assumed that it is these nuclear chain fibers that are responsible for the static response of both the primary and the secondary endings. On the other hand, only the primary endings innervate the nuclear bag fibers. Therefore, it is assumed that the nuclear bag fibers are responsible for the powerful dynamic response.

Control of the Static and Dynamic Responses by the Gamma Efferent Nerves. Some physiologists believe that the gamma efferent nerves to the muscle spindle can be divided into two different types: gamma-dynamic (gamma-d) and gamma-static (gamma-s). The first of these excites mainly the nuclear bag intrafusal fibers and the second excites mainly the nuclear chain intrafusal fibers. When the gamma-d fibers excite the nuclear bag fibers, the dynamic response of the muscle spindle becomes tremendously enhanced, whereas the static response remains very weak or even none at all. On the other hand, stimulation of the gamma-s fibers, which excite mainly the nuclear chain fibers, supposedly enhances the static response while having little influence on the dynamic response. We shall see in subsequent paragraphs that these two different types of responses of the muscle spindle are exceedingly important in different types of muscle control.

Continuous Discharge of the Muscle Spindles Under Normal Conditions. Normally, particularly when there is a slight amount of gamma efferent excitation, the muscle spindles emit sensory nerve impulses all of the time. Stretching the muscle spindles increases the rate of firing, whereas shortening the spindle decreases this rate of firing. Thus, the degree of excitation of the spindles can be either increased or decreased.

THE STRETCH REFLEX (ALSO CALLED MUSCLE SPINDLE REFLEX OR MYOTATIC REFLEX)

From the above description of all the intricacies of the muscle spindle, one can readily see that the spindle is a very complex organ. Its function is manifested in the form of the muscle *stretch reflex* — that is, whenever a muscle is stretched, excitation of the spindles causes reflex contraction of the muscle. And, because of the static and the dynamic types of muscle spindle response, the stretch reflex also consists of a *static reflex* and a *dynamic reflex.*

Neuronal Circuitry of the Stretch Reflex. Figure 51–4 illustrates the basic circuit of the muscle

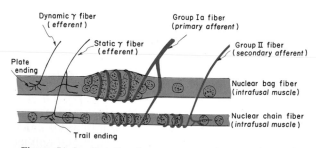

Figure 51–3. Details of nerve connections to the nuclear bag and nuclear chain muscle spindle fibers. (Modified from Stein: *Physiol. Rev.,* 54:225, 1974, and Boyd: *Philos. Trans. R. Soc. Lond.* [*Biol. Sci.*], 245:81, 1962.)

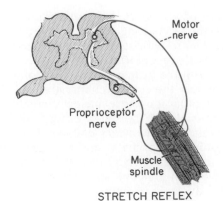

STRETCH REFLEX

Figure 51–4.   Neuronal circuit of the stretch reflex.

spindle reflex, showing a type Ia nerve fiber originating in a muscle spindle and entering the dorsal root of the spinal cord. It then synapses directly with an anterior motoneuron whose motor nerve fiber transmits an appropriate reflex signal, with almost negligible delay, back to the same muscle containing the muscle spindle.

In addition to the monosynaptic pathway of the Ia muscle spindle nerve fiber synapsing with the anterior motoneuron, collaterals from this fiber, as well as the type II fibers from the secondary spindle endings, terminate on interneurons in the gray matter of the cord, and these in turn transmit more delayed signals to the anterior motoneuron.

The Dynamic Stretch Reflex.   The dynamic stretch reflex is caused by the potent dynamic signal transmitted via the primary endings of the muscle spindles. That is, when a muscle is suddenly stretched, a strong signal is transmitted to the spinal cord, and this causes an instantaneous, very strong reflex contraction of the same muscle from which the signal originated. Thus, the reflex functions to oppose sudden changes in the length of the muscle, because the muscle contraction opposes the stretch.

The Static Stretch Reflex.   Though the dynamic stretch reflex is over within a fraction of a second after the muscle has been stretched to its new length, a much weaker static stretch reflex continues for a prolonged period of time thereafter. This reflex is elicited by the continuous static receptor signals transmitted by both the primary and secondary endings. The importance of the static stretch reflex is that it continues to cause muscle contraction as long as the muscle is maintained at an excessive length (for as long as several hours, but not for days). The muscle contraction in turn opposes the force that is causing the excess length.

The Negative Stretch Reflex.   When a muscle is suddenly shortened, exactly opposite effects occur. If the muscle is already taut, any sudden release of the load on the muscle that allows it to shorten will elicit both dynamic and static reflex muscle inhibition rather than reflex excitation. Thus, *this negative stretch reflex* opposes the shortening of the muscle in the same way that the positive stretch reflex opposes lengthening of the muscle. Therefore, one can begin to see that the stretch reflex tends to maintain the status quo for the length of a muscle.

Feedback Control of Muscle Length — The Load Reflex

Let us assume that a person's biceps is contracted so that the forearm is horizontal to the earth. Then assume that a five-pound weight is put in the hand. The hand will immediately drop. However, the amount that the hand will drop is determined to a great extent by the degree of activity of the muscle spindle reflex. If the gamma-s fibers to the muscle spindles are strongly stimulated so that the static reflex is very active, even slight lengthening of the biceps, and therefore also of the muscle spindles in the biceps, will cause a strong feedback contraction of the extrafusal skeletal muscle fibers of the biceps. This contraction in turn will limit the degree of fall of the hand, thus automatically maintaining the forearm in a nearly horizontal position. This response is called a *load reflex*.

Effect of Gamma Efferent Stimulation on Increasing the Effectiveness of the Load Reflex.   If the gamma efferent nerves to the static portion of the muscle spindle (the gamma-s fibers) are not stimulated at all, then the spindle sensitivities are greatly depressed, and the effectiveness of the load reflex is almost zero. On the other hand, if these gamma efferents are strongly stimulated so that the static muscle spindle reflex is an extremely potent one, then one would expect the length of the muscle to remain almost exactly constant regardless of the change in load. In many of our muscle activities a particular part of our body often is fixed in a given position, and any attempt to move that part of the body from the position is met by instantaneous reflex resistance. This probably results to a great extent from the load reflex. Furthermore, the fact that we can change the sensitivity of this load reflex by changing the intensity of gamma-s stimulation allows us to make a particular part of the body be either flail or tightly locked in place. To give an example, when a person is threading a needle he tends to fix both of his hands into precise positions, and any extra load applied to either of the two hands will hardly change its position.

The Damping Function of the Dynamic and Static Stretch Reflexes

Another extremely important function of the stretch reflex — indeed, probably just as important as, if not more important than, the load reflex — is the ability of the stretch reflex to prevent some types of oscillation and jerkiness of the body movements. This is a damping, or smoothing, function. An example is the following:

Use of the Damping Mechanism in Smoothing Muscle Contraction. Occasionally, signals from other parts of the nervous system are transmitted to a muscle in a very unsmooth form, first increasing in intensity for a few milliseconds, then decreasing in intensity, then changing to another intensity level, and so forth. When the muscle spindle apparatus is not functioning satisfactorily, the degree of muscle contraction changes jerkily during the course of such a signal. This effect is illustrated in Figure 51–5, which shows an experiment in which a sensory nerve signal entering one side of the cord is transmitted to a motor nerve on the other side of the cord to excite a muscle. In curve A the muscle spindle reflex of the excited muscle is intact. Note that the contraction is relatively smooth even though the sensory nerve is excited at a very slow frequency of 8 per second. Curve B, on the other hand, is the same experiment in an animal whose muscle spindle sensory nerves from the excited muscle had been sectioned three months earlier. Note the very unsmooth muscle contraction. Thus, curve A illustrates very graphically the ability of the damping mechanism of the muscle spindle to smooth muscle contractions even though the input signals to the muscle motor system may themselves be very jerky. This effect can also be called a *signal averaging* function of the muscle spindle reflex.

Function of the Gamma Efferent System in Controlling the Degree of Damping. In the same way that the gamma efferent system can play a potent role in determining the effectiveness of the load reflex, so also can the gamma efferent system determine the degree of damping. For instance, there are times when a person wishes his limbs to move extremely rapidly in response to rapidly changing input signals. Under such conditions, one would wish less damping. On the other hand, there are other times when it is very important that the muscle contractions be very smooth. Under these conditions one would like a high degree of damping. Therefore, even though we are not completely sure how the nervous system can make such changes as this, psychophysiological tests do tell us that our motor system does have such capability to increase or decrease the degree of damping in any given muscle response.

ROLE OF THE MUSCLE SPINDLE IN VOLUNTARY MOTOR ACTIVITY

To emphasize the importance of the gamma efferent system, one needs to recognize that 31 per cent of all the motor nerve fibers to the muscle are gamma efferent fibers rather than large, type A alpha motor fibers. Whenever signals are transmitted from the motor cortex or from any other area of the brain to the alpha motoneurons, almost invariably the gamma motoneurons are stimulated simultaneously, an effect called *co-activation* of the alpha and gamma motor neurons. This causes both the extrafusal and the intrafusal muscle fibers to contract at the same time.

The purpose of contracting the muscle spindle fibers at the same time that the large skeletal muscle fibers contract is probably two-fold: First, it keeps the muscle spindle from opposing the muscle contraction. Second, it also maintans proper damping and proper load responsiveness of the muscle spindle regardless of change in muscle length. For instance, if the muscle spindle should not contract and relax along with the large muscle fibers, the receptor portion of the spindle would sometimes be flail and at other times be overstretched, in neither instance operating under optimal conditions for spindle function.

Possible "Servo" Control of Muscle Contraction by Exciting the Gamma Motor Neurons. Several physiologists have suggested that the muscle spindle reflex can also operate as a servo controller of muscle contraction. A "servo" control system is one in which the major function is not activated directly but instead is activated indirectly in the following way: Suppose that a gamma efferent nerve fiber to a muscle spindle were stimulated.

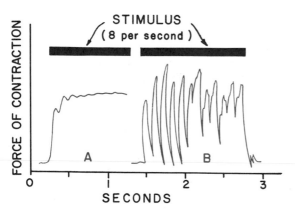

Figure 51–5. Muscle contraction caused by a central nervous system signal under two different conditions: (A) in a normal muscle, and (B) in a muscle whose muscle spindles had been denervated by section of the posterior roots of the cord 82 days previously. Note the smoothing effect of the muscle spindles in Part A.

This would contract the intrafusal fibers and excite the muscle spindle receptors. The signals from these receptors would then pass to the cord and thence back to the large skeletal muscle fibers to cause them to contract also. Thus, an initial contraction of the spindle intrafusal fibers would cause a similar contraction of the surrounding large skeletal muscle fibers.

If such a servo system as this should exist, signals transmitted from the brain downward through the gamma efferent system could cause secondary contractions of the muscles in accordance with the dictates of the muscle spindles. Unfortunately, most of the data indicate that a mechanism such as this probably does not occur to a significant extent. For a system of this type to function properly, it would be necessary to have a static response of the muscle spindles that is very potent and also unchanging. Unfortunately, the static response of the muscle spindles is usually very weak — that is, it has a low gain. Also, it decays very slowly over a period of many minutes to an hour so that changes in length of a muscle spindle could cause only temporary changes in length of the surrounding large skeletal muscle fibers. Therefore, despite the attractiveness of this possible type of muscle control, it is doubtful that it does actually occur in normal function.

Probable Servo-Assist Function of the Muscle Spindle Reflex. It is much more likely that the muscle spindle reflex acts as a servo-assist device than as a pure servo controller. But, first, let us explain what is meant by a "servo-assist mechanism."

When both the alpha and gamma motoneurons are stimulated simultaneously, if the intra- and extrafusal fibers contract equal amounts, the degree of stimulation of the muscle spindles will not change at all — neither increase nor decrease. However, in case the extrafusal muscle fibers contract less than the intrafusal fibers (as might occur when the muscle is contracting against a great load), this mismatch will stretch the receptor portions of the spindles and, therefore, elicit a stretch reflex that will further excite the extrafusal fibers. Thus, failure of the extrafusal fibers to contract to the same degree as the intrafusal fibers contract causes an accessory neuronal signal that increases the degree of stimulation of the extrafusal fibers. This is exactly the same mechanism as that employed by power steering in an automobile. That is, the steering wheel directly turns the front wheels and will do this even when the power steering is not effective. However, if the wheels fail to follow even the slightest extra force applied to the steering wheel, a servo-assist device is activated that applies additional power to turn the wheels.

The servo-assist type of motor function has several important advantages, as follows:

1. It allows the brain to cause a muscle contraction against a load without the brain having to expend much extra nervous energy — instead, the spindle reflex provides most of the nervous energy.

2. It makes the muscle contract almost the desired length even when the load is increased or decreased between successive contractions. In other words, it makes the length of contraction less load-insensitive.

3. It compensates for fatigue or other abnormalities of the muscle itself because any failure of the muscle to provide the proper contraction elicits an additional

muscle spindle reflex stimulus to make the contraction occur.

But, unfortunately, we still do not know how important this probable function of the muscle spindle reflex actually is.

Brain Areas for Control of the Gamma Efferent System

The gamma efferent system is excited primarily by the *bulboreticular facilitatory* region of the brain stem, and secondarily by impulses transmitted into this area from (a) the *cerebellum,* (b) the *basal ganglia,* and even (c) the *cerebral cortex.* Unfortunately, little is known about the precise mechanisms of control of the gamma efferent system. However, since the bulboreticular facilitatory area is particularly concerned with postural contractions, emphasis is given to the possible or probable important role of the gamma efferent mechanism in controlling muscle contraction for positioning the different parts of the body and for damping the movements of the different parts.

In addition to transmitting signals into the spinal cord, the muscle spindle also transmits signals up the cord to the cerebellum and thence into the bulboreticular areas and also as far as the cerebral cortex, as will be discussed in greater detail in Chapter 53. However, both the signals that operate in the spinal cord and those that pass to the brain are entirely subconscious, so that the conscious portion of the brain is never apprised of the immediate changes in length of the muscles. On the other hand, as has already been pointed out in Chapter 49, signals from the position receptors constantly apprise the conscious brain of the positions of the different parts of the body, even though the muscle receptors do not.

CLINICAL APPLICATIONS OF THE STRETCH REFLEX

The Knee Jerk and Other Muscle Jerks. Clinically, a method used to determine the functional integrity of the stretch reflexes is to elicit the knee jerk and other muscle jerks. The knee jerk can be elicited by simply striking the patellar tendon with a reflex hammer; this stretches the quadriceps muscle and initiates a *dynamic stretch reflex* to cause the lower leg to jerk forward. The upper part of Figure 51–6 illustrates a myogram from the quadriceps muscle recorded during a knee jerk.

Similar reflexes can be obtained from almost any muscle of the body either by striking the tendon of the muscle or by striking the belly of the muscle itself. In other words, sudden stretch of muscle spindles is all that is required to elicit a stretch reflex.

The muscle jerks are used by neurologists to assess the degree of facilitation of spinal cord centers. When large numbers of facilitatory impulses are being transmitted from the upper regions of the central nervous system into the cord, the muscle jerks are greatly

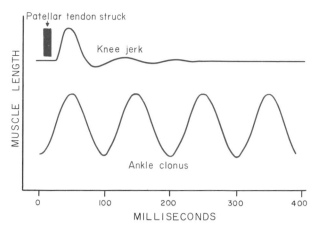

Figure 51–6. Myograms recorded from the quadriceps muscle during elicitation of the knee jerk and from the gastrocnemius muscle during ankle clonus.

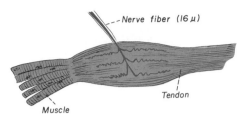

Figure 51–7. Golgi tendon organ.

exacerbated. On the other hand, if the facilitatory impulses are depressed or abrogated, the muscle jerks are considerably weakened or completely absent. These reflexes are used most frequently to determine the presence or absence of muscle spasticity following lesions in the motor areas of the brain. Ordinarily, large lesions in the contralateral motor areas of the cerebral cortex, especially those caused by strokes or brain tumors, cause greatly exacerbated muscle jerks.

Clonus. Under appropriate conditions, the muscle jerks can oscillate, a phenomenon called clonus (see lower myogram, Fig. 51–6). Oscillation can be explained particularly well in relation to ankle clonus, as follows:

If a man standing on his tiptoes suddenly drops his body downward to stretch one of his gastrocnemius muscles, impulses are transmitted from the muscle spindles into the spinal cord. These reflexly excite the stretched muscle, which lifts the body back up again. After a fraction of a second, the reflex contraction of the muscle dies out and the body falls again, thus stretching the spindles a second time. Again a dynamic stretch reflex lifts the body, but this too dies out after a fraction of a second, and the body falls once more to elicit still a new cycle. In this way, the stretch reflex of the gastrocnemius muscle continues to oscillate, often for long periods of time; this is clonus.

Clonus ordinarily occurs only if the stretch reflex is highly sensitized by facilitatory impulses from the brain. For instance, in the decerebrate animal, in which the stretch reflexes are highly facilitated, clonus develops readily. Therefore, to determine the degree of facilitation of the spinal cord, neurologists test patients for clonus by suddenly stretching a muscle and keeping a steady stretching force applied to the muscle. If clonus occurs, the degree of facilitation is certain to be very high.

THE TENDON REFLEX

The Golgi Tendon Organ and Its Excitation. Golgi tendon organs, illustrated in Figure

51–7, lie within muscle tendons immediately beyond their attachments to the muscle fibers. An average of 10 to 15 muscle fibers are usually connected in series with each Golgi tendon organ, and the organ is stimulated by the tension produced by this small bundle of muscle fibers. Thus, the major difference between the function of the Golgi tendon organ and the muscle spindle is that the spindle detects relative muscle length, and the tendon organ detects muscle *tension.*

The tendon organ, like the primary receptor of the muscle spindle, has both a *dynamic response* and a *static response,* responding very intensely when the muscle tension suddenly increases (the dynamic response), but within a small fraction of a second it settles down to a lower level of steady state firing that is almost directly proportional to the muscle tension (the static response). Thus, the Golgi tendon organs provide the nervous system with instantaneous information of the degree of tension on each small segment of each muscle.

Transmission of Impulses from the Tendon Organ into the Central Nervous System. Signals from the tendon organ are transmitted through large, rapidly conducting type Ib nerve fibers, fibers averaging 16 microns in diameter, only slightly smaller than those from the primary receptors of the muscle spindle. These fibers, like those from the primary receptors, transmit signals both into local areas of the cord and through the spinocerebellar tracts into the cerebellum and through still other tracts to the cerebral cortex. The local signal is believed to excite a single inhibitory interneuron that in turn inhibits the anterior alpha motoneuron. Thus, this local circuit directly inhibits the individual muscle without affecting adjacent muscles. The signals to the brain will be discussed in Chapter 53.

Inhibitory Nature of the Tendon Reflex. When the Goli tendon organs of a muscle are stimulated by increased muscle tension, signals are transmitted into the spinal cord to cause reflex effects in the respective muscle. However, this reflex is entirely inhibitory, the exact opposite to the muscle spindle reflex. Thus, this reflex provides a negative feedback mechanism that prevents the development of too much tension on the muscle.

When tension on the muscle and, therefore, on the tendon becomes extreme, the inhibitory effect from the tendon organ can be so great that it causes sudden relaxation of the entire muscle. This effect is called the *lengthening reaction;* it is probably a protective mechanism to prevent tearing of the muscle or avulsion of the tendon from its attachments to the bone. We know, for instance, that direct electrical stimulation of muscles in the laboratory can frequently cause such destructive effects.

However, possibly as important as this protective reaction is function of the tendon reflex as a part of the overall servo control of muscle contraction in the following manner:

The Tendon Reflex as a Possible Servo Control Mechanism for Muscle Tension.   In the same way that the strech reflex possibly operates as a feedback mechanism to control muscle length, the tendon reflex theoretically can operate as a servo feedback mechanism to control muscle tension. That is, if the tension on the muscle becomes too great, inhibition from the tendon organ decreases this tension back to a lower value. On the other hand, if the tension becomes too little, impulses from the tendon organ cease; and the resulting loss of inhibition allows the alpha motoneurons to become active again, thus increasing muscle tension toward a higher level.

Very little is known at present about the function of or control of this tension servo feedback mechanism, but it is postulated to operate in the following basic manner: Signals from the brain are presumably transmitted to the cord centers to set the gain of the tendon feedback system. This can be done by changing the degree of facilitation of the neurons in the feedback loop. If the gain is high, then this system will be extremely sensitive to signals coming from the tendon organs; on the other hand, lack of excitatory signals from the brain could make the system very insensitive to the signals from the tendon organ. In this way, control signals from higher nervous centers could automatically set the level of tension at which the muscle would be maintained.

Yet, a word of caution: Despite this speculation, the measured feedback gains of the tendon-muscle feedback system have been far too little to prove that it can be very useful as a tension servo controller.

Value of a Servo Mechanism for Control of Tension.   An obvious value of a mechanism for setting the degree of muscle tension would be to allow the different muscles to apply a desired amount of force irrespective of how far the muscles contract. An example of this is paddling a boat, in which a person sets the amount of force that he pulls backward on the paddle and maintains that degree of force throughout the entire movement. Were it not for some type of tension feedback mechanism, the same amount of force would not be maintained throughout the stroke because uncontrolled muscles change their force of contraction as their lengths change. One can imagine hundreds of different patterns of muscle function that might require maintenance of constant tension rather than maintenance of constant lengths of the muscles.

THE FLEXOR REFLEX (THE WITHDRAWAL REFLEXES)

In the spinal or decerebrate animal, almost any type of cutaneous sensory stimulus on a limb is likely to cause the flexor muscles of the limb to contract strongly, thereby withdrawing the limb from the stimulus. This is called the flexor reflex.

In its classic form the flexor reflex is elicited most frequently by stimulation of pain endings, such as by pinprick, heat, or some other painful stimulus, for which reason it is also frequently called a *nociceptive reflex,* or simply *pain reflex.* However, even stimulation of the touch receptors can also occasionally elicit a weaker and less prolonged flexor reflex.

If some part of the body besides one of the limbs is painfully stimulated, this part, in a similar manner, will be withdrawn from the stimulus, but the reflex may not be confined entirely to flexor muscles even though it is basically the same type of reflex. Therefore, the many patterns of reflexes of this type in the different areas of the body are called the *withdrawal reflexes.*

Neuronal Mechanism of the Flexor Reflex.   The left-hand portion of Figure 51–8 illustrates the neuronal pathways for the flexor reflex. In this instance, a painful stimulus is applied to the hand; as a result, the flexor muscles of the upper arm become reflexly excited, thus withdrawing the hand from the painful stimulus.

The pathways for eliciting the flexor reflex do not pass directly to the anterior motoneurons but, instead, pass first into the interneuron pool of neurons and then to the motoneurons. The shortest possible circuit is a three- or four-neuron arc; however, most of the signals of the reflex traverse many more neurons than this and involve the following basic types of circuits: (1) diverging circuits to spread the reflex to the necessary

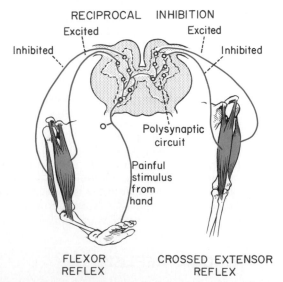

Figure 51–8.   The flexor reflex, the crossed extensor reflex, and reciprocal inhibition.

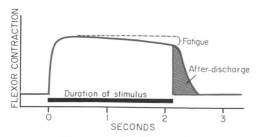

Figure 51–9. Myogram of a flexor reflex, showing rapid onset of the reflex, an interval of fatigue, and finally after-discharge after the stimulus is over.

muscles for withdrawal, (2) circuits to inhibit the antagonist muscles, called *reciprocal inhibition circuits,* and (3) circuits to cause a prolonged repetitive after-discharge even after the stimulus is over.

Figure 51–9 illustrates a typical myogram from a flexor muscle during a flexor reflex. Within a few milliseconds after a pain nerve is stimulated, the flexor response appears. Then, in the next few seconds the reflex begins to *fatigue,* which is characteristic of essentially all of the more complex integrative reflexes of the spinal cord. Then, soon after the stimulus is over, the contraction of the muscle begins to return toward the base line, but, because of *after-discharge,* will not return all the way for many milliseconds. The duration of the after-discharge depends on the intensity of the sensory stimulus that had elicited the reflex; a weak stimulus causes almost no after-discharge in contrast to an after-discharge lasting for several seconds following a very strong stimulus. Furthermore, a flexor reflex initiated by nonpainful stimuli and transmitted through the large sensory fibers causes essentially no after-discharge, whereas nociceptive impulses transmitted through the small type A fibers and type C fibers cause prolonged after-discharge.

The after-discharge that occurs in the flexor reflex almost certainly results from all three types of repetitive-discharge circuits that were discussed in Chapter 47. Electrophysiological studies indicate that the immediate after-discharge, lasting for about 6 to 8 milliseconds, results from the interneuron repetitive firing mechanism and from the parallel type of circuit, with impulses being transmitted from one interneuron to another to another and all these in turn transmitting their signals successively to the anterior motoneurons. However, the prolonged after-discharge that occurs following strong pain stimuli presumably involves reverberating circuits in the interneurons, these transmitting impulses to the anterior motoneurons sometimes for several seconds after the incoming sensory signal is completely over.

Thus, the flexor reflex is appropriately organized to withdraw a pained or otherwise irritated part of the body away from the stimulus. Furthermore, because of the after-discharge the reflex can hold the irritated part away from the stimulus for as long as 1 to 3 seconds even after the irritation is over. During this time, other reflexes and actions of the central nervous system can move the entire body away from the painful stimulus.

The Pattern of Withdrawal. The pattern of withdrawal that results when the flexor reflex (or the many other types of withdrawal reflexes) is elicited depends on the sensory nerve that is stimulated. Thus, a painful stimulus on the inside of the arm not only elicits a flexor reflex in the arm but also contracts the abductor muscles to pull the arm outward. In other words, the integrative centers of the cord cause those muscles to contract that can most effectively remove the pained part of the body from the object that causes pain. This same principle applies for any part of the body but especially to the limbs, because they have highly developed flexor reflexes.

THE CROSSED EXTENSOR REFLEX

Approximately 0.2 to 0.5 second after a stimulus elicits a flexor reflex in one limb, the opposite limb begins to extend. This is called the *crossed extensor reflex.* Extension of the opposite limb obviously can push the entire body away from the object causing the painful stimulus.

Neuronal Mechanism of the Crossed Extensor Reflex. The right-hand portion of Figure 51–8 illustrates the neuronal circuit responsible for the crossed extensor reflex, showing that signals from the sensory nerves cross to the opposite side of the cord to cause exactly opposite reactions to those that cause the flexor reflex. Because the crossed extensor reflex usually does not begin until 200 to 500 milliseconds following the initial pain stimulus, it is certain that many internuncial neurons are in the circuit between the incoming sensory neuron and the motoneurons of the opposite side of the cord responsible for the crossed extension. Furthermore, after the painful stimulus is removed, the crossed extensor reflex continues for an even longer period of after-discharge than that for the flexor reflex. Therefore, again, it is presumed that this prolonged after-discharge results from reverberatory circuits among the internuncial cells.

Figure 51–10 illustrates a myogram recorded from a muscle involved in a crossed extensor reflex. This shows the relatively long latency before the reflex begins and also the long after-discharge following the end of the stimulus. The prolonged after-discharge obviously would be of

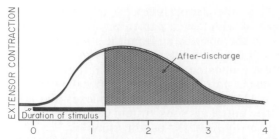

Figure 51–10. Myogram of a crossed extensor reflex, showing slow onset but prolonged after-discharge.

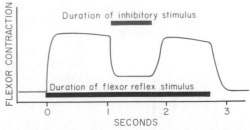

Figure 51–11. Myogram of a flexor reflex, illustrating reciprocal inhibition caused by a stronger flexor reflex in the opposite limb.

benefit in holding the body away from a painful object until other neurogenic reactions could cause the body to move away.

RECIPROCAL INHIBITION AND RECIPROCAL INNERVATION

In the foregoing paragraphs we have pointed out several times that excitation of one group of muscles is often associated with inhibition of another group. For instance, when a stretch reflex excites one muscle, it simultaneously inhibits the antagonist muscles. This is the phenomenon of *reciprocal inhibition*, and the neuronal mechanism that causes this reciprocal relationship is called *reciprocal innervation*. Likewise, reciprocal relationships exist between the two sides of the cord, as exemplified by the flexor and extensor reflexes described above.

Figure 51–11 illustrates a typical example of reciprocal inhibition. In this instance, a moderate but prolonged flexor reflex is elicited from one limb of the body, and while this reflex is still being elicited a still stronger flexor reflex is elicited in the opposite limb, causing reciprocal inhibition of the first limb. Then removal of the strong flexor reflex allows the original reflex to reassume its previous intensity.

FATIGUE OF REFLEXES; REBOUND

Figure 51–9 illustrated that the flexor reflex begins to *fatigue* within a few seconds after its initiation. This is a common effect in most of the cord reflexes as well as many other reflexes of the central nervous system, and it presumably results from progressive fatigue of synaptic transmission in the reflex circuits, a subject that was discussed more fully in Chapter 46.

Another effect closely allied to fatigue is *rebound*. This means that, immediately after a reflex is over, a second reflex of the same type is much more difficult to elicit for a given time thereafter.

However, because of reciprocal innervation, reflexes of the antagonist muscles become even more easily elicited during that same period of time. For instance, if a flexor reflex occurs in a left limb, a second flexor reflex is more difficult to establish for a few seconds thereafter, but a crossed extensor reflex in this same limb will be greatly exacerbated. Rebound is probably one of the important mechanisms by which the rhythmic to-and-fro movements required in locomotion are effected, which are described in more detail later in the chapter.

THE REFLEXES OF POSTURE AND LOCOMOTION

THE POSTURAL AND LOCOMOTIVE REFLEXES OF THE CORD

The Positive Supportive Reaction. Pressure on the footpad of a decerebrate animal causes the limb to extend against the pressure being applied to the foot. Indeed, this reflex is so strong that an animal whose spinal cord has been transected for several months can often be placed on its feet, and the reflex will stiffen the limbs sufficiently to support the weight of the body — the animal will stand in a rigid position. This reflex is called the *positive supportive reaction*.

The positive supportive reaction involves a complex circuit in the interneurons similar to those responsible for the flexor and the crossed-extensor reflexes. Furthermore, the locus of the pressure on the pad of the foot determines the position to which the limb will extend; pressure on one side causes extension in that direction, an effect called the *magnet reaction*. This obviously helps to keep an animal from falling to one side.

The Cord "Righting" Reflexes. When a spinal cat or even a well-recovered young spinal dog is laid on its side, it will make incoordinate movements that indicate that it is trying to raise itself to the standing position. This is called a *cord righting reflex,* and it illustrates that relatively complicated reflexes associated with posture are at least partially integrated in the spinal cord. Indeed, a puppy with a well-healed transected thoracic cord caudal to the cord level for the forelegs can completely right itself from the lying position and can even walk on its hindlimbs. And, in the case of the opossum with a similar transection of the thoracic cord, the walking movements of the hindlimbs are hardly different from those in the normal opossum — except that the hindlimb movements are not synchronized with those of the forelimbs as is normally the case.

Rhythmic Stepping Movements of a Single Limb. Rhythmic stepping movements are frequently observed in the limbs of spinal animals. Indeed, even when the lumbar portion of the spinal cord is separated from the remainder of the cord and a longitudinal section is made down the center of the cord to block neuronal connections between the two limbs, each hind limb can still perform stepping functions. Forward flexion of the limb is followed a second or so later by backward extension. Then flexion occurs again, and the cycle is repeated over and over.

This oscillation back and forth between the flexor and extensor muscles occurs even after the sensory nerves have been cut, and it seems to result mainly from reciprocal inhibition and rebound. That is, the forward flexion of the limb causes reciprocal inhibition of the cord center controlling the extensor muscles, but shortly thereafter the flexion begins to die out; as it does so, *rebound* inhibition of the flexors and reciprocal *rebound* excitation of the extensors cause the leg to move downward and backward. After extension has continued for a time, it, too, begins to die and is followed by reciprocal rebound excitation of the flexor muscles. And the oscillating cycle continues again and again.

The sensory signals from the footpads and from the position sensors around the joints play a strong role in controlling foot pressure and rate of stepping when the foot is allowed to walk along a surface.

Reciprocal Stepping of Opposite Limbs. If the lumbar spinal cord is not sectioned down its center as noted above, every time stepping occurs in the forward direction in one limb, the opposite limb ordinarily steps backward. This effect results from reciprocal innervation between the two limbs.

Diagonal Stepping of All Four Limbs — The "Mark Time" Reflex. If a well-healed spinal animal is held up from the table and its legs are allowed to fall downward as illustrated in Figure 51–12, the stretch on the limbs occasionally elicits stepping reflexes that involve all four limbs. In general, stepping occurs diagonally between the fore- and hindlimbs. That is, the right hindlimb and the left forelimb move backward together while the right forelimb and left hindlimb move forward. This diagonal response is another manifestation of reciprocal innervation, this time oc-

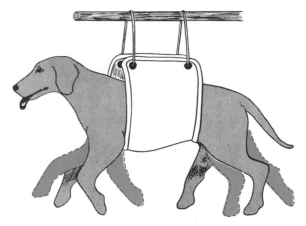

Walking movement

Figure 51–12. Diagonal stepping movements exhibited by a spinal animal.

curring the entire distance up and down the cord between the fore- and hindlimbs. Such a walking pattern is often called a *mark time reflex.*

The Galloping Reflex. Another type of reflex that occasionally develops in the spinal animal is the galloping reflex, in which both forelimbs move backward in unison while both hindlimbs move forward. If stretch or pressure stimuli are applied almost exactly equally to opposite limbs at the same time, a galloping reflex will likely result, whereas unequal stimulation of one side versus the other elicits the diagonal walking reflex. This is in keeping with the normal patterns of walking and of galloping, for, in walking, only one limb at a time is stimulated, and this would predispose to continued walking. Conversely, when the animal strikes the ground during galloping, the limbs on both sides are stimulated approximately equally; this obviously would predispose to further galloping and, therefore, would continue this pattern of motion in contradistinction to the walking pattern.

THE SCRATCH REFLEX

An especially important cord reflex in lower animals is the scratch reflex, which is initiated by the *itch and tickle sensation.* It actually involves two different functions: (1) a *position sense* that allows the paw to find the exact point of irritation on the surface of the body and (2) a *to-and-fro scratching movement.*

Obviously, the scratch reflex, like the stepping movements of locomotion, involves reciprocal innervation circuits that cause oscillation. One of the important discoveries in relation to the to-and-fro movement of the scratch reflex (and also of the stepping movements as described above) is that it can still occur even when all the sensory roots from the oscillating limb are sectioned. In other words, feedback from the limb itself is not necessary to maintain the neuronal oscillation, which means that the oscillation can occur intrinsically as a result of oscillating circuits within the spinal interneurons themselves.

The *position sense* of the scratch reflex is also a highly developed function, for even though a flea might be crawling as far forward as the shoulder of a spinal animal, the hind paw can often find its position. Furthermore, this can be accomplished even though 19 different muscles in the limb must be contracted simultaneously in a precise pattern to bring the paw to the position of the crawling flea. To make the reflex even more complicated, when the flea crosses the midline, the paw stops scratching, but the opposite paw begins the to-and-fro motion and eventually finds the flea.

THE SPINAL CORD REFLEXES THAT CAUSE MUSCLE SPASM

In human beings, local muscle spasm is often observed. The mechanism of this has not been elucidated to complete satisfaction even in experimental animals, but it is known that pain stimuli can cause reflex spasm of local muscles, which presumably is the cause of much if not most of the muscle spasm observed at localized regions of the human body.

Muscle Spasm Resulting from a Broken Bone. One type of clinically important spasm occurs in muscles surrounding a broken bone. This seems to result from the pain impulses initiated from the broken edges of the bone, which cause the muscles surrounding the area to contract powerfully and tonically. Relief of the pain by injection of a local anesthetic relieves the spasm; a general anesthetic also relieves the spasm. One of these procedures is often necessary before the spasm can be overcome sufficiently for the two ends of the bone to be set back into appropriate positions.

Abdominal Muscle Spasm in Peritonitis. Another type of local spasm caused by a cord reflex is the abdominal spasm resulting from irritation of the parietal peritoneum by peritonitis. Here, again, relief of the pain caused by the peritonitis allows the spastic muscle to relax. Almost the same type of spasm often occurs during surgical operations; pain impulses from the parietal peritoneum cause the abdominal muscles to contract extensively and sometimes actually to extrude the intestines through the surgical wound. For this reason deep surgical anesthesia is usually required for intra-abdominal operations.

Muscle Cramps. Still another type of local spasm is the typical muscle cramp. Electromyographic studies indicate that the cause of at least some muscle cramps is the following:

Any local irritating factor or metabolic abnormality of a muscle — such as severe cold, lack of blood flow to the muscle, or overexercise of the muscle — can elicit pain or other types of sensory impulses that are transmitted from the muscle to the spinal cord, thus causing reflex muscle contraction. The contraction in turn stimulates the same sensory receptors still more, which causes the spinal cord to increase the intensity of contraction still further. Thus, a positive feedback mechanism occurs so that a small amount of initial irritation causes more and more contraction until a full-blown muscle cramp ensues. Reciprocal inhibition

of the muscle can sometimes relieve the cramp. That is, if a person purposefully contracts the muscle on the opposite side of the joint from the cramped muscle while at the same time using another hand or leg to prevent movement of the joint, the reciprocal inhibition that occurs in the cramped muscle can at times relieve the cramp immediately.

THE AUTONOMIC REFLEXES IN THE SPINAL CORD

Many different types of segmental autonomic reflexes occur in the spinal cord, most of which are discussed in other chapters. Briefly, these include: (1) changes in vascular tone, resulting from heat and cold (Chap. 72), (2) sweating, which results from localized heat at the surface of the body (Chap. 72), (3) intestino-intestinal reflexes that control some motor functions of the gut (Chap. 63), (4) peritoneointestinal reflexes that inhibit gastric motility in response to peritoneal irritation (Chap. 66), and (5) evacuation reflexes for emptying the bladder (Chap. 38) and the colon (Chap. 63). In addition, all the segmental reflexes can at times be elicited simultaneously in the form of the so-called mass reflex as follows:

The Mass Reflex. In a spinal animal or human being, the spinal cord sometimes suddenly becomes excessively active, causing massive discharge of large portions of the cord. The usual stimulus that causes this is a strong nociceptive stimulus to the skin or excessive filling of a viscus, such as overdistention of the bladder or of the gut. Regardless of the type of stimulus, the resulting reflex, called the mass reflex, involves large portions or even all of the cord, and its pattern of reaction is the same. The effects are: (1) a major portion of the body goes into strong flexor spasm, (2) the colon and bladder are likely to evacuate, (3) the arterial pressure often rises to maximal values — sometimes to a mean pressure well over 200 mm. Hg, and (4) large areas of the body break out into profuse sweating. The mass reflex might be likened to the epileptic attacks that involve the central nervous system in which large portions of the brain become massively activated.

The precise neuronal mechanism of the mass reflex is unknown. However, since it lasts for minutes, it presumably results from activation of great masses of reverberating circuits that excite large areas of the cord at once.

SPINAL CORD TRANSECTION AND SPINAL SHOCK

When the spinal cord is suddenly transected, essentially all cord functions, including the cord reflexes, immediately become depressed to the point of oblivion, a reaction called *spinal shock*. The reason for this is that normal activity of the cord neurons depends to a great extent on continual tonic discharges from higher centers, particularly discharges transmitted through the reticulospinal tracts and corticospinal tracts.

After a few hours to a few days or weeks of spinal shock, the spinal neurons gradually regain their excitability. This seems to be a natural characteristic of neurons everywhere in the nervous system — that is, after they lose their source of facilitatory impulses, they increase their own natural degree of excitability to make up for the loss. In most nonprimates, the excitability of the cord centers returns essentially to normal within a few hours to a day or so, but in human beings the return is often delayed for several weeks and occasionally is never complete; or, on the other hand, recovery is sometimes excessive, with resultant hyperexcitability of all or most cord functions.

Some of the spinal functions specifically affected during or following spinal shock are: (1) The arterial blood pressure falls immediately — sometimes to as low as 40 mm. Hg — thus illustrating that sympathetic activity becomes blocked almost to extinction. However, the pressure ordinarily returns to normal within a few days. (2) All skeletal muscle reflexes integrated in the spinal cord are completely blocked during the initial stages of shock. In lower animals, a few hours to a few days are required for these reflexes to return to normal, and in human beings two weeks to several months are usually required. Sometimes, both in animals and people, some reflexes eventually become hyperexcitable, particularly if a few facilitatory pathways remain intact between the brain and the cord while the remainder of the spinal cord is transected. The first reflexes to return are the stretch reflex, followed in order by the progressively more complex reflexes, the flexor reflexes, the postural antigravity reflexes, and remnants of stepping reflexes. (3) The sacral reflexes for control of bladder and colon evacuation are completely suppresed in human beings for the first few weeks following cord transection, but they eventually return. These effects are discussed in Chapters 38 and 66.

REFERENCES

Burke, R. E., and Rudonmin, P.: Spinal neurons and synapses. *In* Brookhart, J. M., and Mountcastle, V. B. (eds.): Handbook of Physiology. Sec. 1, Vol. 1. Baltimore, Williams & Wilkins, 1977, p. 877.

Creed, R. S. *et al.*: Reflex Activity of the Spinal Cord. New York, Oxford University Press, 1932.

Crowley, W. J.: Neural control of skeletal muscle. *In* Frohlich, E. D. (ed.): Pathophysiology, 2nd Ed. Philadelphia, J. B. Lippincott Co., 1976, p. 735.

Daube, J. R., *et al.*: Medical Neurosciences: An Approach to Anatomy, Pathology, and Physiology by Systems and Levels. Boston, Little, Brown, 1978.

Desmedt, J. E. (ed.): Physiological Tremor, Pathological Tremors and Clonus. New York, S. Karger, 1978.

Easton, T. A.: On the normal use of reflexes. *Am. Sci., 60*:591, 1972.

Gallistel, C. R.: The Organization of Action: A New Synthesis. New York, Halsted Press, 1979.

Granit, R.: Muscular tone, *J. Sport Med., 2*:46, 1962.

Granit, R., and Pompeiano, O. (eds.): Reflex Control of Posture and Movement. New York, Elsevier Scientific Publishing Co., 1979.

Granit, R., *et al.*: First supraspinal control of mammalian muscle spindles: extra- and intrafusal co-activation. *J. Physiol. (Lond.), 147*:385, 1959.

Houk, J. C.: Regulation of stiffness in skeletomotor reflexes. *Annu. Rev. Physiol., 41*:99, 1979.

Hughes, J. T.: Pathology of the Spinal Cord. Philadelphia, W. B. Saunders Co., 1978.

Hunt, C. C., and Perl, E. R.: Spinal reflex mechanisms concerned with skeletal muscle. *Physiol. Rev., 40*:538, 1960.

Kostyuk, P. G., and Vasilenko, D.: Spinal interneurons. *Annu. Rev. Physiol., 41*:115, 1979.

Matthews, P. B. C.: Mammalian Muscle Receptors and Their Central Actions. Baltimore, Williams & Wilkins, 1972.

Merton, P. A.: How we control the contraction of our muscles. *Sci. Am., 226*:30, 1972.

Orlovsky, G. N., and Shik, M. L.: Control of locomotion: A neurophysiological analysis of the cat locomotor system. *Int. Rev. Physiol., 10*:281, 1976.

Pearson, K.: The control of walking. *Sci. Am. 235*(6):72, 1976.

Peterson, B. W.: Reticulospinal projections to spinal motor nuclei. *Annu. Rev. Physiol., 41*:127, 1979.

Porter, R.: The neurophysiology of movement performance. *In* MTP International Review of Science: Physiology. Vol. 3. Baltimore, University Park Press, 1974, p. 151.

Purves, D.: Long-term regulation in the vertebrate peripheral nervous system. *Int. Rev. Physiol., 10*:125, 1976.

Roaf, R., and Hodkinson, L. J.: The Paralysed Patient. Philadelphia, J. B. Lippincott Co., 1977.

Rogers, M. A.: Paraplegia. Boston, Faber and Faber, 1978.

Sherrington, C. S.: The Integrative Action of the Nervous System. New Haven, Conn., Yale University Press, 1911.

Shik, M. L., and Orlovsky, G. N.: Neurophysiology of Locomotor Automatism. *Physiol. Rev., 56*:465, 1976.

Stein, P. S. G.: Motor systems with specific reference to the control of locomotion. *Annu. Rev. Neurosci., 1*:61, 1978.

Stein, R. B.: Peripheral control of movement. *Physiol. Rev., 54*:215, 1974.

Trieschmann, R. B.: Spinal Cord Injuries. New York, Pergamon Press, 1979.

52

Motor Functions of the Brain Stem and Basal Ganglia — Reticular Formation, Vestibular Apparatus, Equilibrium, and Brain Stem Reflexes

The brain stem is a complex extension of the spinal cord. Collected in it are numerous neuronal circuits to control respiration, cardiovascular function, gastrointestinal function, eye movement, equilibrium, support of the body against gravity, and many special stereotyped movements of the body. Some of these functions — such as control of respiration and cardiovascular functions — are described in special sections of this text. The present chapter deals primarily with the control of whole body movement and equilibrium.

THE RETICULAR FORMATION, AND SUPPORT OF THE BODY AGAINST GRAVITY

Throughout the entire extent of the brain stem — in the medulla, pons, mesencephalon, and even in portions of the diencephalon — are areas of diffuse neurons collectively known as the *reticular formation*. Figure 52–1 illustrates the extent of the reticular formation, showing it to begin at the upper end of the spinal cord and to extend upward through the medulla, pons, and mesencephalon into (a) the central portions of the thalamus, (b) the hypothalamus, and (c) other areas adjacent to the thalamus. The lower end of the reticular formation is continuous with the interneurons of the spinal cord, and, indeed, the reticular formation of the brain stem functions in a

manner quite analogous to many of the functions of the interneurons in the cord gray matter.

Interspersed throughout the reticular formation are both motor and sensory neurons; these vary in size from very small to very large. The small

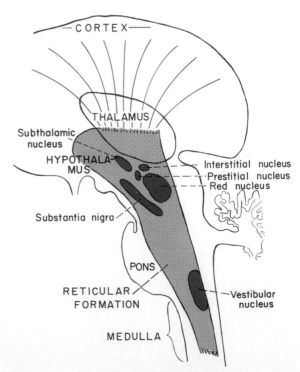

Figure 52–1. The reticular formation and associated nuclei.

640

neurons, which constitute the greater number, have short axons that make multiple connections within the reticular formation itself. The large neurons are mainly motor in function, and their axons often bifurcate almost immediately, with one division extending downward to the spinal cord and the other extending upward to the thalamus or other basal regions of the diencephalon or cerebrum.

The sensory input to the reticular formation is from multiple sources, including: (1) the spinoreticular tracts and collaterals from the spinothalamic tracts, (2) the vestibular nuclei, (3) the cerebellum, (4) the basal ganglia, (5) the cerebral cortex, especially the motor regions, and (6) the hypothalamus and other nearby associated areas.

Though most of the neurons in the reticular formation are evenly dispersed, some of them are collected into *specific nuclei*, some of which are labeled in Figure 52–1. In general, these specific nuclei are not considered to be part of the reticular formation per se even though they do operate in association with it. In most instances they are the loci of "preprogrammed" control of stereotyped movements. As an example, the vestibular nuclei provide preprogrammed attitudinal contractions of the muscles for maintenance of equilibrium, as will be discussed later in the chapter.

EXCITATORY FUNCTION OF THE RETICULAR FORMATION

The reticular formation and the vestibular nuclei, which function in very close association with each other, are both intrinsically excitable, but this excitability is usually held in check by inhibitory signals that flow into this area mainly from the basal ganglia. Also, some areas of the cerebral cortex send inhibitory signals to the reticular formation mainly by way of the basal ganglia. As will be noted later in the chapter when we discuss the decerebrate animal, destruction of the higher portions of the nervous system, the basal ganglia and the cortex, removes this inhibition and allows the reticular formation and vestibular nuclei to become tonically active; this causes rigidity of the antigravity skeletal muscles throughout the body.

On the other hand, when the brain stem is sectioned at a level slightly below the vestibular nuclei, loss of the motor signals normally transmitted from the reticular formation and vestibular nuclei to the spinal cord causes almost all the musculature to become totally flaccid.

Reciprocal Excitation or Inhibition of Antagonist Muscles by the Reticular Formation. Though very little is known about the function of specific areas in the reticular formation, in general stimulation near the midline of the reticular formation causes the flexor muscles on the same side of the body to contract and the extensors to relax. Stimulation in the lateral portions of the reticular formation tends to cause opposite effects, excitation of the extensor muscles and inhibition of the flexors. And at the same time that contraction occurs on one side of the body the same muscle on the opposite side tends to relax while its antagonist contracts. Thus, the phenomenon of reciprocal inhibition is strongly expressed in the reticular formation, as it is in the spinal cord, both for reciprocal control of antagonist pairs of muscles as well as for control of muscles between the two sides of the body.

SUPPORT OF THE BODY AGAINST GRAVITY

When a person or an animal stands, the vestibular nuclei and the closely related nuclei in the reticular formation transmit continuous impulses into the spinal cord and thence to the extensor muscles to stiffen the limbs. This allows the limbs to support the body against gravity. These impulses are transmitted mainly by way of the vestibulospinal and reticulospinal tracts.

However, the degree of activity in the individual extensor muscles is determined by the equilibrium mechanisms. Thus, if an animal begins to fall to one side, the extensor muscles on that side stiffen while those on the opposite side relax. And analogous effects occur when it tends to fall forward or backward.

In essence, then, the vestibular nuclei and the reticular formation provide the nervous energy to support the body against gravity. But other factors, particularly the vestibular apparatuses, control the relative degree of extensor contraction in the different parts of the body, which provides the function of equilibrium.

VESTIBULAR SENSATIONS AND THE MAINTENANCE OF EQUILIBRIUM

THE VESTIBULAR APPARATUS

The vestibular apparatus is the sensory organ that detects sensations concerned with equilibrium. It is composed of a *bony labyrinth* containing the *membranous labyrinth* which is the functional part of the apparatus. The top of Figure 52–2 illustrates the membranous labyrinth; it is composed mainly of the *cochlear duct*, the three *semicircular canals*, and the two large chambers known as the *utricle* and the *saccule*. The cochlear duct is the major sensory area for hearing and has nothing to do with equilibrium. However, the

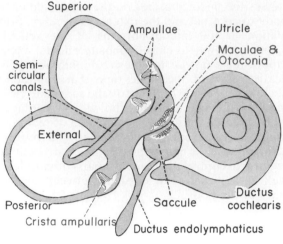

MEMBRANOUS LABYRINTH

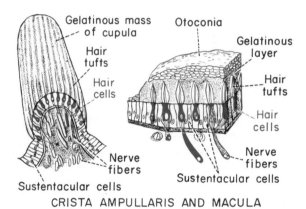

CRISTA AMPULLARIS AND MACULA

Figure 52–2. The membranous labyrinth, and organization of the crista ampullaris and the macula. (From Goss: Gray's Anatomy of the Human Body. Lea & Febiger; modified from Kolmer by Buchanan: Functional Neuroanatomy. Lea & Febiger.)

utricle, the saccule, and the semicircular canals are all integral parts of the equilibrium mechanism.

The Utricle and the Saccule. Located on the wall of both the utricle and the saccule is a small area slightly over 2 mm. in diameter called a macula. Each of these maculae is a sensory area for detecting the orientation of the head with respect to the direction of gravitational pull or of other acceleratory forces, as will be explained in subsequent sections of this chapter. Each macula is covered by a gelatinous layer in which many small calcium carbonate crystals called otoconia are imbedded. Also, in the macula are thousands of hair cells, which project cilia up into the gelatinous layer. The bases and sides of the hair cells synapse with sensory axons of the vestibular nerve.

Even under resting conditions, most of the nerve fibers leading from the hair cells transmit a continuous series of nerve impulses. Bending the cilia of a hair cell to one side causes the impulse traffic in its nerve fibers to increase markedly; bending the cilia to the opposite side decreases the impulse traffic, often turning it off completely. Therefore, as the orientation of the head in space changes and the weight of the otoconia (whose specific gravity is about 3 times that of the surrounding tissues) bends the cilia, appropriate signals are transmitted to the brain to control equilibrium.

In each macula the different hair cells are oriented in different directions so that some of them are stimulated when the head bends forward, some when it bends backward, others when it bends to one side, and so forth. Therefore, a different pattern of excitation occurs in the macula for each position of the head; it is this "pattern" that apprises the brain of the head's orientation.

The Semicircular Canals. The three semicircular canals in each vestibular apparatus, known respectively as the superior, posterior, and external (or horizontal) semicircular canals, are arranged at right angles to each other so that they represent all three planes in space. When the head is bent forward approximately 30 degrees, the two external semicircular canals are located approximately horizontal with respect to the surface of the earth. The superior canals are then located in vertical planes that project forward and 45 degrees outward, and the posterior canals are also then in vertical planes but project backward and 45 degrees outward. Thus, the superior canal on each side of the head is in a plane parallel to that of the posterior canal on the opposite side of the head, whereas the two external canals on the two sides are located in approximately the same plane.

Each semicircular canal has an enlargement at one of its ends called the ampulla, and the canals are filled with fluid called endolymph. Flow of this fluid in the canals excites a sensory organ in the ampulla. As illustrated in Figure 52–2, located in each ampulla is a small crest called a crista ampullaris, and on top of the crista is a gelatinous mass similar to that in the utricle and known as the cupula. Into the cupula are projected hairs from hair cells located along the ampullary crest, and these hair cells in turn are connected to sensory nerve fibers that pass into the vestibular nerve. Bending the cupula to one side, caused by flow of fluid in the canal, stimulates the hair cells, while bending in the opposite direction inhibits them. Thus, appropriate signals are sent through the vestibular nerve to apprise the central nervous system of fluid movement in the respective canal.

Directional Sensitivity of the Hair Cells — The Kinocilium. Each hair cell, whether in a macula or a cupula, has a large number of very small cilia plus one very large cilium called the kinocilium.

This kinocilium is located to one side of the hair cell, always on the same side of the cell with respect to its orientation on the ampullary crest. This is the cause of the directional sensitivity of the hair cells: namely, stimulation when the cilia are bent toward the kinocilium side and inhibition when bent in the opposite direction.

Neuronal Connections of the Vestibular Apparatus with the Central Nervous System. Figure 52–3 illustrates the central connections of the vestibular nerve. Most of the vestibular nerve fibers end in the vestibular nuclei, which are located approximately at the junction of the medulla and the pons, but some fibers pass without synapsing to the fastigial nuclei, uvula, and flocculonodular lobes of the cerebellum. The fibers that end in the vestibular nuclei synapse with second order neurons that also send fibers into these areas of the cerebellum as well as to the cortex of other portions of the cerebellum, into the vestibulospinal tract, into the medial longitudinal fasciculus, and to other areas of the brain stem, particularly the reticular formation.

Note especially the very close association between the vestibular apparatus, the vestibular nuclei, and the cerebellum. The primary pathway for the reflexes of equilibrium begins in the vestibular nerves and passes next to both the vestibular nuclei and the cerebellum. Then, after much two-way traffic of impulses between these two, signals are sent into the reticular nuclei of the brain stem as well as down the spinal cord via vestibulospinal and reticulospinal tracts. In turn, the signals to the cord control of the interplay between facilitation and inhibition of the extensor muscles, thus automatically controlling equilibrium.

The *flocculonodular lobes* of the cerebellum seem to be especially concerned with equilibrium functions of the semicircular canals because destruction of these lobes gives almost exactly the same clinical symptoms as destruction of the semicircular canals themselves. That is, severe injury to either of these structures causes loss of equilibrium during *rapid changes in direction of motion* but does not seriously disturb equilibrium under static conditions, as will be discussed in subsequent sections. It is also believed that the uvula of the cerebellum plays an equally important role in static equilibrium.

Signals transmitted upward in the brain stem from both the vestibular nuclei and the cerebellum via the *medial longitudinal fasciculus* cause corrective movements of the eyes every time the head rotates so that the eyes can remain fixed on a specific visual object. Signals also pass upward (either through this same tract or through reticular tracts) to the cerebral cortex, probably terminating in a primary cortical center for equilibrium located in the parietal lobe deep in the Sylvian fissure, on the opposite side of the fissure wall from the auditory area of the superior temporal gyrus. These signals apprise the psyche of the equilibrium status of the body.

FUNCTION OF THE UTRICLE AND THE SACCULE IN THE MAINTENANCE OF STATIC EQUILIBRIUM

It is especially important that the different hair cells are oriented in all different directions in the maculae of the utricles and saccules so that at different positions of the head, different hair cells become stimulated. The "patterns" of stimulation of the different hair cells apprise the nervous system of the position of the head with respect to the pull of gravity. In turn, the vestibular, cerebellar, and reticular motor systems reflexly excite the appropriate muscles to maintain proper equilibrium.

The maculae in the utricle and saccule function extremely effectively for maintaining equilibrium especially when the head is in the near-vertical position. That is, a person can determine as little as a half-degree of mal-equilibrium when the head leans from the precise upright position. On the other hand, as the head is leaned further and further from the upright, the determination of head orientation by the vestibular sense becomes poorer and poorer. Obviously, extreme sensitivity in the upright position is of major importance for maintenance of precise vertical static equilibrium, which is the most essential function of the vestibular apparatus.

Detection of Linear Acceleration by the Maculae. When the body is suddenly thrust forward — that is, when the body accelerates — the otoconia, which have greater inertia than the surrounding fluids, fall backward on the hair cell cilia, and information of mal-equilibrium is sent into the nervous centers, causing the individual to feel as if he were falling backward. This automatically causes him to lean his body forward until the anterior shift of the otoconia caused by leaning exactly equals the tendency for the otoconia to fall backward. At this point, the nervous system detects a state of proper equilibrium and therefore leans the body no farther forward. As long as the degree of linear acceleration remains constant and

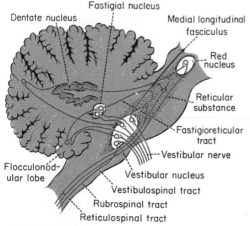

Figure 52–3. Connections of vestibular nerves in the central nervous system.

the body is maintained in this forward leaning position, the person falls neither forward nor backward. Thus, the maculae operate to maintain equilibrium during linear acceleration in exactly the same manner as they operate in static equilibrium.

The otoconia *do not* operate for the detection of linear *velocity*. When a runner first begins to run, he must lean far forward to keep from falling over backward because of *acceleration,* but once he has achieved running speed, he would not have to lean forward at all if he were running in a vacuum. When running in air he leans forward to maintain equilibrium only because of the air resistance against his body, and in this instance it is not the otoconia that make him lean but the pressure of the air acting on pressure end-organs in the skin, which initiate the appropriate equilibrium adjustments to prevent falling.

THE SEMICIRCULAR CANALS AND THEIR DETECTION OF ANGULAR ACCELERATION AND ANGULAR VELOCITY

When the head suddenly *begins* to rotate in any direction, the endolymph in the membranous semicircular canals, because of its inertia, tends to remain stationary while the semicircular canals themselves turn. This causes relative fluid flow in the canals in a direction opposite to the rotation of the head.

Figure 52–4 illustrates an *ampulla* of one of the semicircular canals, showing the *cupula* and its embedded hairs bending in the direction of fluid movement. And Figure 52–5 illustrates the discharge signal from a single hair cell in the crista ampullaris when an animal is rotated for 40 sec-

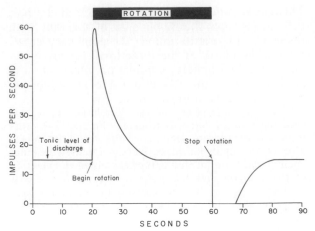

Figure 52–5. Response of a hair cell when a semicircular canal is stimulated first by rotation and then by stopping rotation.

onds, showing that (1) even when the cupula is in its resting position the hair cell emits a tonic discharge of approximately 15 impulses per second; (2) when the animal is rotated, the hairs bend to one side and the rate of discharge increases greatly; and (3) with continued rotation, the excess discharge of the hair cell gradually subsides back to the resting level in about 20 seconds.

The reason for this adaptation of the receptor is that within a second or more of rotation, friction in the semicircular canal causes the endolymph to rotate as rapidly as the semicircular canal itself; then in an additional 15 to 20 seconds the cupula slowly returns to its resting position in the middle of the ampulla because of its own elastic recoil.

When the rotation suddenly stops, exactly the opposite effects take place: the endolymph continues to rotate while the semicircular canal stops. This time the cupula is bent in the opposite direction, causing the hair cell to stop discharging entirely. After another few seconds, the endolymph stops moving, and the cupula returns gradually to its resting position in about 20 seconds, thus allowing the discharge of the hair cell to return to its normal tonic level as shown to the right in Figure 52–5.

Thus, the semicircular canal transmits a positive signal when the head *begins* to rotate and a negative signal when it *stops* rotating. Furthermore, at least some hair cells will always respond to rotation in any plane — horizontal, sagittal, or coronal — for fluid movement always occurs in at least one semicircular canal.

Detection of Angular Acceleration and Angular Velocity. Angular velocity means the rate of rotation; that is, it is the number of revolutions about an axis in a given period of time. Angular acceleration means the rate at which the angular velocity is changing.

The semicircular canals always give a person a feeling of increasing rotation when he begins to rotate or a

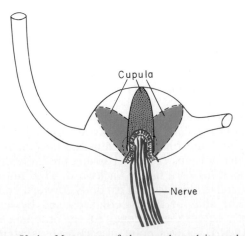

Figure 52–4. Movement of the cupula and its embedded hairs during rotation first in one direction and then in the opposite direction. (Often, the top of the cupula is attached so that it merely bulges in one direction or the other.)

feeling of decreasing rotation when he stops rotating — that is, they give a signal for angular acceleration. Therefore, it is commonly stated that the semicircular canals are primarily a detector of angular acceleration. However, a person also continues to feel that he is rotating for 5 to 15 seconds after he has begun to rotate. That is, he feels angular velocity during this period of time. Therefore, the semicircular canals actually detect a combination of angular acceleration and angular velocity, although the angular velocity effect may last only for 5 to 15 seconds.

Rate of Angular Acceleration Required to Stimulate the Semicircular Canals. The angular acceleration required to stimulate the semicircular canals in the human being averages about 1 degree per second per second. In other words, when one begins to rotate, the velocity of rotation must be as much as 1 degree per second by the end of the first second, 2 degrees per second by the end of the second second, 3 degrees per second by the end of the third second, and so forth, in order for the person barely to detect that the rate of rotation is increasing.

"Predictive" Function of the Semicircular Canals in the Maintenance of Equilibrium. Since the semicircular canals do not detect that the body is off balance in the forward direction, in the side direction, or in the backward direction, one might at first ask: What is the function of the semicircular canals in the maintenance of equilibrium? All they detect is that the person's head is rotating or beginning to rotate or stopping rotation in one direction or another. Therefore, the function of the semicircular canals is not likely to be to maintain static equilibrium or to maintain equilibrium during linear acceleration or when the person is exposed to steady centrifugal forces. Yet loss of function of the semicircular canals causes a person to have very poor equilibrium when attempting to perform *rapid* and *intricate* body movements.

We can explain the function of the semicircular canals best by the following illustration. If a person is running forward rapidly, and then suddenly begins to turn to one side, he falls off balance a second or so later unless appropriate corrections are made *ahead of time*. But, unfortunately, the utricle cannot detect that he is off balance until *after* this has occurred. On the other hand, the semicircular canals will have already detected that the person is turning, and this information can easily apprise the central nervous system of the fact that the person *will* fall off balance within the next second or so unless some correction is made. In other words, the semicircular canal mechanism *predicts ahead of time* that mal-equilibrium is going to occur even before it does occur and thereby causes the equilibrium centers to make appropriate preventive adjustments. In this way, the person need not fall off balance before he begins to correct the situation.

Removal of the flocculonodular lobes of the cerebellum prevents normal function of the semicircular canals but does not prevent normal func-

tion of the macular receptors. It is especially interesting in this connection that the cerebellum serves as a "predictive" organ for most of the other rapid movements of the body as well as those having to do with equilibrium. These other functions of the cerebellum are discussed in the following chapter.

VESTIBULAR POSTURAL REFLEXES

Sudden changes in the orientation of an animal in space elicit reflexes that help to maintain equilibrium and posture. For instance, if an animal is suddenly pushed to the right, even before it can fall more than a few degrees its right legs extend instantaneously. In other words, this mechanism *anticipates* that the animal will be off balance in a few seconds and makes appropriate adjustments to prevent this.

Another type of vestibular postural reflex occurs when the animal suddenly falls forward. When this occurs, the forepaws extend forward, the extensor muscles tighten, and the muscles in the back of the neck stiffen to prevent the animal's head striking the ground. This reflex is probably also of importance in locomotion, for, in the case of the galloping horse, the downward thrust of the head can automatically provide reflex thrust of the forelimbs to move the animal forward for the next gallop.

VESTIBULAR MECHANISM FOR STABILIZING THE EYES AND FOR NYSTAGMUS

When a person changes his direction of movement rapidly, or even leans his head sideways, forward, or backward, it would be impossible for him to maintain a stable image on the retinae of his eyes unless he had some automatic control mechanism to stabilize the direction of gaze of the eyes. In addition, the eyes would be of little use in detecting an image unless they remained "fixed" on each object long enough to gain a clear image. Fortunately, each time the head is suddenly angulated, signals from the semicircular canals cause the eyes to angulate in an equal and opposite direction to the angulation of the head. This results from reflexes transmitted from the canals through the *vestibular nuclei*, the *cerebellum*, and the *medial longitudinal fasciculus* to the *ocular nuclei*.

A special example of this stabilization of the eyes occurs when a person begins to rotate. At first the eyes remain glued to the object. But after the head has rotated far to one side, they jump suddenly in the direction of rotation of the head and "fix" on a new object; then they deviate slowly backward again as the rotation proceeds. This sudden jumping motion forward and then slow backward motion is called *nystagmus*. The jumping motion is called the *fast component* of the nystagmus, and the slow movement, the *slow component*.

Nystagmus always occurs automatically when the semicircular canals are stimulated. For instance, if a person's head begins to rotate to the right, backward movement of fluid in the left horizontal canal and forward movement in the right horizontal canal cause the eyes to move slowly to the left; thus, the slow component of nystagmus is initiated by the vestibular apparatuses. But, when the eyes have moved as far to the left as they reasonably can, centers located in the brain stem in close approximation to the nuclei of the abducens nerves cause the eyes to jump suddenly to the right; then the vestibular apparatuses take over once more to move the eyes again slowly to the left.

CLINICAL TESTS FOR INTEGRITY OF VESTIBULAR FUNCTION

Balancing Test. One of the simplest clinical tests for integrity of the equilibrium mechanism is to have the individual stand perfectly still with eyes closed. If there is no longer a functioning static equilibrium system of the utricles, the person will waver to one side or the other and possibly even fall. However, as noted below, some of the proprioceptive mechanisms of equilibrium are occasionally sufficiently well developed to maintain balance even with the eyes closed.

Barany Test. A second test that is frequently performed determines the integrity of the semicircular canals. In this instance the individual is placed in a "Barany chair" and rotated rapidly while placing the head respectively in various planes — first forward, then angulated to one side or the other. By such positioning, each pair of semicircular canals is successively placed in the horizontal plane of rotation. When the rotation of the chair is stopped suddenly, the endolymph, because of its momentum, continues to rotate in the pair of semicircular canals that has been placed in the horizontal plane, causing the cupula to bend in the direction of rotation. As a result, nystagmus occurs, with the slow component in the direction of rotation and the fast component in the opposite direction. Also, as long as the nystagmus lasts (about 15 to 20 seconds) the individual has the sensation that he is rotating in the direction *opposite* to that in which he was actually rotated in the chair. Obviously, this test checks the semicircular canals on both sides of the head at the same time.

Ice Water Test. A clinical test of one vestibular apparatus separately from the other depends on placing ice water in one ear. The external semicircular canal lies adjacent to the ear, and cooling the ear can transfer a sufficient amount of heat from this canal to cool the endolymph. This increases the density of the endolymph, thereby causing it to sink downward and resulting in slight movement of fluid around the semicircular canal. This stimulates the canal, giving the individual a sensation of rotating and also initiating nystagmus. From these two findings one can determine whether the respective semicircular canals are functioning properly. When the semicircular canals are normal, the utricles and saccules are usually normal also, for disease usually destroys the function of all of these components at the same time.

OTHER FACTORS CONCERNED WITH EQUILIBRIUM

The Neck Proprioceptors. The vestibular apparatus detects the orientation and movements *only of the head*.

Therefore, it is essential that the nervous centers also receive appropriate information depicting the orientation of the head with respect to the body as well as the orientation of the different parts of the body with respect to each other. This information is transmitted from the proprioceptors of the neck and body directly into the vestibular and reticular nuclei of the brain stem and also indirectly by way of the cerebellum.

By far the most important proprioceptive information needed for the maintenance of equilibrium is that derived from the *joint receptors of the neck,* for this apprises the nervous system of the orientation of the head with respect to the body. When the head is bent in one direction or the other by bending the neck, impulses from the neck proprioceptors keep the vestibular apparatuses from giving the person a sense of malequilibrium. They do this by transmitting signals that exactly oppose the signals transmitted from the vestibular apparatuses. However, *when the entire body* is changed to a new position with respect to gravity, the impulses from the vestibular apparatuses *are not opposed* by the neck proprioceptors; therefore, the person in this instance does perceive a change in equilibrium status.

The Neck Reflexes. In an animal *whose vestibular apparatuses have been destroyed,* bending the neck causes immediate muscular reflexes called *neck reflexes* occurring especially in the forelimbs. For instance, bending the head forward causes both forelimbs to relax. However, when the vestibular apparatuses are *intact,* this effect does *not* occur because the vestibular reflexes function almost exactly oppositely to the neck reflexes. Thus, if the head is flexed downward, the vestibular reflex tends to extend the forelimbs, while the neck reflexes tend to relax them. Since the equilibrium of the entire body and not of the head alone must be maintained, it is easy to understand that the vestibular and neck reflexes must function oppositely. Otherwise, each time the neck should bend, the animal would immediately fall off balance.

Proprioceptive and Exteroceptive Information from Other Parts of the Body. Proprioceptive information from other parts of the body besides the neck is also necessary for maintenance of equilibrium because appropriate equilibrium adjustments must be made whenever the body is angulated in the chest or abdomen region or elsewhere. Presumably, all this information is algebraically added in the cerebellum, vestibular nuclei, and reticular nuclei of the brain stem, thus causing appropriate adjustments in the postural muscles.

Also important in the maintenance of equilibrium are several types of exteroceptive sensations. For instance, pressure sensations from the footpads can tell one (a) whether his weight is distributed equally between his two feet and (b) whether his weight is more forward or backward on his feet.

Another instance in which exteroceptive information is necessary for maintenance of equilibrium occurs when a person is running. The air pressure against the front of the body signals that a force is opposing the body in a direction different from that caused by gravitational pull; as a result, the person leans forward to oppose this.

Importance of Visual Information in the Maintenance of Equilibrium. After complete destruction of the vestibular apparatuses, and even after loss of most proprioceptive information from the body, a person can still use the visual mechanisms effectively for maintaining equilibrium. Even slight linear or angular movement of the body instaneously shifts the visual images on the retina, and this information is relayed to the equilibrium centers. Many persons with complete destruction of the vestibular apparatus have almost normal equilibrium as long as their eyes are open and as long as they perform all motions slowly. But, when moving rapidly or when the eyes are closed, equilibrium is immediately lost.

Conscious Perception of Equilibrium. A cortical center for conscious perception of the state of equilibrium is believed to lie in the lowest part of the parietal cortex deep in the sylvian fissue, on the medial wall of this fissure opposite to the primary cortical area for hearing. The sensations from the vestibular apparatuses, from the neck proprioceptors, and from most of the other proprioceptors are probably first integrated in the equilibrium centers of the brain stem before being transmitted to the cerebral cortex. Various pathological processes in the vestibular apparatuses or in the vestibular neuronal circuits often affect the equilibrium sensations. Thus, loss of one or both flocculonodular lobes or one or both fastigial nuclei of the cerebellum gives the person a sensation of constant mal-equilibrium for the first few weeks or months after the loss. This probably results because impulses from these cerebellar areas are normally integrated into the subconscious sense of equilibrium by the vestibular and reticular nuclei even before equilibrium information is sent to the cerebral cortex.

FUNCTIONS OF THE RETICULAR FORMATION AND SPECIFIC BRAIN STEM NUCLEI IN CONTROLLING SUBCONSCIOUS, STEREOTYPED MOVEMENTS

Rarely, a child called an *anencephalic monster* is born without brain structures above the mesencephalic region, and some of these children have been kept alive for many months. They are able to perform essentially all the functions of feeding, such as suckling, extrusion of unpleasant food from the mouth, and moving the hands to the mouth to suck the fingers. In addition, they can yawn and stretch. They can cry and follow objects with the eyes and by movements of the head. Also, placing pressure on the upper anterior parts of their legs will cause them to pull to the sitting position.

Therefore, it is obvious that many of the stereotyped motor functions of the human being are integrated in the brain stem. Unfortunately, the loci of most of the different motor control systems have not been found except for the following:

Stereotyped Body Movements. Most movements of the trunk and head can be classified into several simple movements, such as forward flexion, extension, rotation, and turning movements of the entire body. These types of movements are controlled by special nuclei located mainly in the mesencephalic and lower dien-

cephalic region. For instance, *rotational movements* of the head and eyes are controlled by the *interstitial nucleus,* which is illustrated in Figure 52–1. This nucleus lies in close approximation to the *medial longitudinal fasciculus,* through which it transmits a major portion of its control impulses. The *raising movements* of the head and body are controlled by the *prestitial nucleus,* which is located approximately at the juncture of the diencephalon and mesencephalon. On the other hand, the *flexing movements* of the head and body are controlled by the *nucleus precommissuralis* located at the level of the posterior commissure. Finally, the *turning movements* of the entire body, which are much more complicated, involve both the pontile and mesencephalic reticular formation. However, for full expression of the turning movements, the caudate nucleus and the cingulate gyrus of the cerebral cortex are also required. The turning movements can cause an animal or person to continue circling around and around in one direction or the other.

Function of the Red Nucleus. The red nucleus, illustrated in Figure 52–6, is composed of two distinct parts, the *magnocellular portion* and a *small cellular portion*.

The magnocellular nuclei receive impulses both from the *basal ganglia* and from the *prestitial nucleus* and *nucleus commissuralis*. In turn they transmit signals by way of the *rubrospinal tract* into the spinal cord and by way of collaterals from the rubrospinal tract into the *reticular nuclei* of the brain stem. Stimulation of this portion of the red nucleus causes the head and upper trunk to extend backward. Thus, this portion of the red nucleus enters into a particular type of gross body movement concerned with forward and backward deviation of the body axis.

The small cells of the red nucleus are excited principally by impulses from the *cerebellum* through the *cerebellorubral tracts*. In turn, they send impulses mainly into the reticular formation of the brain stem. These tracts are part of the overall cerebellar mechanism for control of motor function, which will be discussed in the following chapter.

Function of the Subthalamic Areas — Forward Progression. Much less is known about function of higher brain stem centers in posture and locomotion than of the lower centers, principally because of the complexity of the neuronal connections. However, it is known that stimulation of centers in or around the subthalamic nuclei can cause rhythmic limb motions, including forward walking reflexes. This does not mean that the individual muscles of walking are controlled from this region but simply that excitation of this region sends "command" signals to activate the cord centers where *pre-programmed* neuronal circuits then cause the actual walking movements.

A cat with its brain transected beneath the thalamus but above the subthalamus can walk in an almost completely normal fashion — so much so that the observer often cannot tell the difference. However, when the animal comes to an obstruction it simply butts its head against the obstruction and tries to keep on walking. Thus, it lacks *purposefulness of locomotion*.

The function of the subthalamic region in walking is frequently described as that of controlling *forward progression*. In the previous chapter we have already noted that a decerebrate animal can stand perfectly well, and also it can make attempts to right itself from the lying position. Unfortunately, though, the decerebrate animal with the brain stem sectioned below the subthalamus cannot force itself to move forward in a normal walking pattern. This seems to be the function of areas either in or somewhere close to the subthalamic nuclei.

MOTOR FUNCTIONS OF THE BASAL GANGLIA

Physiologic Anatomy of the Basal Ganglia. The anatomy of the basal ganglia is so complex and so poorly known in its details that it would be pointless to attempt a complete description at this time. However, Figure 52–6 illustrates the principal structures of the basal ganglia and their multitude of neural connections with other parts of the nervous system. Physiologically, the basal ganglia are considered to be the *caudate nucleus, putamen, globus pallidus, substantia nigra,* and *subthalamus.* However, major portions of the *thalamus, reticular formation,* and *red nuclei* operate in close association with these and are therefore part of the basal ganglial system for motor control.

Some important features of the different pathways illustrated in Figure 52–6 are the following:

1. Numerous nerve pathways pass from the motor portion of the cerebral cortex, particularly from the *motor association areas* in front of the primary motor cortex (also called the premotor areas), to the caudate nucleus and putamen, which together are called the *striate body.* In turn, the caudate nucleus and putamen send numerous fibers to the globus pallidus, substantia nigra, and subthalamus. Then these nuclei send fibers to the ventrolateral and ventroanterior nuclei of the thalamus, and still another set of fibers from these nuclei pass

back to the motor areas of the cerebral cortex. Thus, circular pathways are established from the motor cortical regions to the basal ganglia, the thalamus, and back to the motor regions from which the pathways begin. These circuits obviously could operate as a feedback system of the servo control type, as will be discussed later.

2. Still another circular pathway exists among the basal ganglia themselves, in which signals are transmitted downward from the caudate nucleus and putamen through the globus pallidus to the substantia nigra; then signals return upward from the substantia nigra back to the putamen and caudate nucleus. A special feature of this circuit is that both its downward and upward limbs are probably inhibitory, the downward pathway secreting the transmitter substance *GABA* and the upward pathway from the substantia nigra to the putamen and caudate nucleus secreting *dopamine.* As we shall see, these two pathways play a major role in controlling the intrinsic activity level in the basal ganglia themselves.

3. In the following chapter it will be noted that signals also pass from the motor regions of the cerebral cortex to the pons and then into the cerebellum. In turn, signals from the cerebellum are transmitted back to the motor cortex by way of the ventrolateral nucleus of the thalamus, a nucleus through which many of the basal ganglial signals returning to the cortex also are transmitted. This circuit could allow integration between the basal ganglial feedback signals and the feedback signals from the cerebellum.

4. The basal ganglia have numerous short neuronal connections among themselves. Also, the lower basal ganglia, such as the globus pallidus, substantia nigra, and subthalamus send signals into the lower brain stem, projecting especially onto the reticular nuclei, the red nucleus, and the inferior olive. It is presumably through these pathways that many of the so-called extrapyramidal signals for motor control are transmitted.

5. Large numbers of nerve fibers pass directly from the motor cortex to the reticular nuclei and other nuclei of the brain stem. These bypass the basal ganglia but converge on the same brain stem nuclei that are innervated by the basal ganglia.

FUNCTIONS OF THE DIFFERENT BASAL GANGLIA

Before attempting to discuss the functions of the basal ganglia in human beings, we should speak briefly of the better known functions of these ganglia in lower animals. In birds, for instance, the cerebral cortex is poorly developed while the basal ganglia are highly developed. These ganglia perform essentially all the motor functions, even controlling the voluntary movements in much the same manner that the motor cortex of the human being controls voluntary movements. Furthermore, in the cat, and to a lesser extent in the dog, decortication removes only the discrete types of motor functions and does not interfere with the animal's ability to walk, eat, fight, develop rage, have periodic sleep and wakefulness, and even participate naturally in sexual activities. However, if a major portion of the basal ganglia is

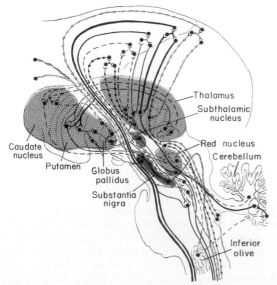

Figure 52–6. Pathways through the basal ganglia and related structures of the brain stem, thalamus, and cerebral cortex. (From Jung and Hassler: Handbook of Physiology, Sec. I, Vol. II. Baltimore, Williams & Wilkins Company, 1960.)

destroyed, only gross stereotyped movements remain, which were discussed above in relation to the mesencephalic animal.

Finally, in the human being, cortical lesions in very young individuals destroy the discrete movements of the body, particularly of the hands and distal portions of the lower limbs, but do not destroy the person's ability to walk crudely, to control equilibrium, or to perform many other subconscious types of movements. However, simultaneous destruction of a major portion of the caudate nuclei almost totally paralyzes the opposite side of the body except for a few stereotyped reflexes integrated in the cord or brain stem.

With this brief background of the overall function of the basal ganglia, we can attempt to dissect the functions of the individual portions of the basal ganglia system, realizing that the system actually operates, along with the motor cortex and cerebellum, as a total unit and that individual functions cannot be ascribed completely to the different parts of the basal ganglia.

Inhibition of Motor Tone by the Basal Ganglia. Though it is wrong to ascribe a single function to all the basal ganglia, nevertheless, one of the general effects of diffuse basal ganglia excitation is to inhibit muscle tone throughout the body. This effect results from inhibitory signals transmitted from the basal ganglia to both the motor cortex and the lower brain stem. Therefore, whenever widespread destruction of the basal ganglia occurs, this causes muscle rigidity throughout the body. For instance, when the brain stem is transected at the mesencephalic level, which removes the inhibitory effects of the basal ganglia, the phenomenon of decerebrate rigidity occurs.

Yet, despite this general inhibitory effect of the basal ganglia, stimulation of specific areas within the basal ganglia can at times elicit positive muscle contractions and at times even complex patterns of movements.

Function of the Caudate Nucleus and Putamen — The Striate Body. The caudate nucleus and putamen, because of their gross appearance on sections of the brain, are together called the *striate body*. They seem to function together to initiate and regulate gross intentional movements of the body. To perform this function they transmit inpulses through two different pathways: (1) into the *globus pallidus* and *substantia nigra*, thence by way of the *thalamus* to the *cerebral cortex*, and finally downward into the spinal cord through the *corticospinal* and *extracorticospinal pathways;* (2) downward through the *globus pallidus* and the *substantia nigra* by way of short axons into the *reticular formation* and finally into the spinal cord mainly through the *reticulospinal tracts*.

In summary, the striate body helps to control gross intentional movements that we normally perform unconsciously. However, this control also involves the motor cortex, with which the striate body is very closely connected.

Function of the Globus Pallidus. It has been suggested that the principal function of the globus pallidus is to provide background muscle tone for intended movements, whether these be initiated by impulses from the cerebral cortex or from the striate body. That is, a person wishing to perform an exact function with one hand, first positions the body, next the legs and arms, and also tenses the muscles of the upper arm. These associated tonic contractions are supposedly initiated by circuits that strongly involve the globus pallidus. Destruction of the globus pallidus removes these attitudinal movements that are necessary to position the hand and, therefore, makes it difficult or impossible for the distal portions of the limbs to perform their more discrete activities.

The globus pallidus is believed to function through two pathways: first, through feedback circuits to the thalamus, to the cerebral cortex, and thence by way of corticospinal and extracorticospinal tracts to the spinal cord; and, second, by way of short axons to the reticular formation of the brain stem and thence mainly by way of the reticulospinal tracts into the spinal cord.

Electrical stimulation of the globus pallidus while an animal is performing a gross body movement will often stop the movement in a static position, the animal holding that position for many seconds while the stimulation continues. This fits with the concept that the globus pallidus is involved in some type of servo feedback motor control system that is capable of locking the different parts of the body into specific positions. Obviously, such a circuit could be extremely important in providing the background body movements and upper limb movements when a person performs delicate tasks with the hands.

CLINICAL SYNDROMES RESULTING FROM DAMAGE TO THE BASAL GANGLIA

Much of what we know about the function of the basal ganglia comes from study of patients with basal ganglia lesions whose brains have undergone pathologic studies after death. Among the different clinical syndromes are:

Chorea. Chorea is a disease in which random uncontrolled movement patterns occur continuously. Normal progression of movements cannot occur; instead, the person may perform one pattern of movement for a few seconds and then suddenly begin another pattern of movement; then still another pattern begins after a few seconds. Because of this peculiar progression of movements, one type of chorea is frequently called *St. Vitus' dance.*

Pathologically, chorea results from *diffuse and widespread damage in the caudate and putamen nuclei.* Especially, the neurons that secrete the inhibitory transmitter GABA are greatly reduced in number or activity. Therefore, the normal inhibitory signals transmitted from the caudate nucleus and putamen into the lower basal ganglia — especially into the globus pallidus and substantia nigra — can no longer hold these lower areas in check. As a result, recurrent signals from these lower areas, probably mainly from the globus pallidus to the thalamus and thence to the motor cortex, elicit trains of successive choreiform movements.

Athetosis. In this disease, slow, writhing movements of a hand, the neck, the face, the tongue, or some other part of the body occur continually. The movements are likely to be wormlike, first with overextension of the hands and fingers, then flexion, then rotary twisting to the side — all these continuing in a slow, rhythmic, repetitive writhing pattern. The contracting muscles exhibit a high degree of spasm, and the movements are enhanced by emotions or by excessive signals from the sensory organs. Furthermore, voluntary movements in the affected area are greatly impaired or sometimes even impossible.

The damage in athetosis is usually found in the *outer portion of the globus pallidus* or in this area and the *striate body*. Athetosis is usually attributed to the interruption of feedback circuits among the basal ganglia, thalamus, and cerebral cortex. The normal feedback circuits presumably allow a constant and rapid interplay between antagonistic muscle groups so that finely controlled static or progressive movements can take place. However, if the feedback circuits are blocked, it is supposed that the detouring impulses may take devious routes through the basal ganglia, thalamus, and motor cortex, causing a succession of abnormal movements.

Hemiballismus. Hemiballismus is an uncontrollable succession of violent movements of large areas of the body. These may occur once every few seconds or sometimes only once in many minutes. For instance, an entire leg might suddenly jerk uncontrollably to full flexion, or the entire trunk might go through an extreme, sudden torsion movement, or an arm might be pulled upward suddenly with great force. Hemiballismus of the legs or trunk causes the person to fall to the ground if walking, and even in bed, the person tosses violently when affected by these powerful and strong intermittent movements. Furthermore, attempts to perform voluntary movements frequently invoke ballistic movements in place of the normal movements.

Hemiballismus of one side of the body results from a *large lesion in the opposite subthalamus*. The smooth, progressive or rhythmic movements of the limbs or other parts of the body normally integrated in this area can no longer occur, but excitatory impulses attempting to evoke such movements elicit instead uncontrollable ballistic movements.

Parkinson's Disease. Parkinson's disease, which is also known as *paralysis agitans,* results almost invariably from *widespread destruction of the substantia nigra,* but often associated with lesions of the *globus pallidus* and other related areas. It is characterized by (1) *rigidity* of the musculature either in widespread areas of the body or in isolated areas, (2) *tremor at rest* of the involved areas in most but not all instances, and (3) a serious *inability to initiate movement,* called *akinesia*.

The cause or causes of these abnormal motor activi-

ties are almost entirely unknown. However, they are almost certainly related to *loss of dopamine secretion* in the caudate nucleus and putamen by the nerve endings of the *nigrostriatal tract*. Destruction of the substantia nigra causes this tract to degenerate and the dopamine normally secreted in the caudate nucleus and putamen no longer to be present. But still present are large numbers of neurons that secrete acetylcholine. These are believed to transmit excitatory signals throughout the basal ganglia. It is also believed that the dopamine from the nigrostriatal pathway normally acts to inhibit these acetylcholine-producing neurons or in some other way to counter their activity. But, in the absence of dopamine secretion, the acetylcholine pathways become overly active, which presumably is the basis for the motor symptoms in Parkinson's disease.

The rigidity in this disease is somewhat different from that which occurs in decerebrate rigidity, for decerebrate rigidity results mainly from hyperactivity of the muscle spindle system, and even slight movement of a muscle is met with rather extreme reflex resistance resulting from feedback through the stretch reflex mechanism. Parkinsonian rigidity, on the other hand, is of a "plastic" type. That is, sudden movement is not met by intense resistance from the stretch reflexes as in the decerebrate type of rigidity, but instead both protagonist and antagonist muscles remain tightly contracted throughout the movement. Therefore, it is believed that the rigidity of Parkinson's disease results to a great extent from excess impulses transmitted in the corticospinal system, thus activating the alpha motor fibers to the muscles, in addition to probable excess activation of the gamma efferent system as well.

Tremor usually, but not always, occurs in Parkinson's disease. Its frequency is normally four to six cycles per second. When the parkinsonian patient performs voluntary movements, the tremor becomes temporarily or partially blocked, presumably because other motor control signals —perhaps from the cerebral cortex and cerebellum —override the abnormal basal ganglial signals.

The mechanism of the tremor in Parkinson's disease is not known. However, a type of treatment that frequently relieves the tremor is surgical destruction of the *ventrolateral nucleus of the thalamus* which is one of the feedback pathways from the basal ganglia to the motor cortex. Therefore, it is presumed that loss of the inhibitory influence of dopamine in the basal ganglia leads to enhanced activity of the cortical-basal ganglial-thalamic-cortical feedback circuit, leading to an oscillation that produces the muscle tremor.

Though the muscle rigidity and the tremor are both distressing to the parkinsonian patient, even more serious is the *akinesia* that occurs in the final stages of the disease. To perform even the simplest of movement, the person must exert the highest degree of concentration, and the mental effort, even mental anguish, that is necessary to make the movement "go" is often almost beyond the patient's willpower. Then, when the movement does occur, it is stiff and often staccato in character instead of occurring with smooth progression. For instance, when a patient begins to perform a discrete voluntary movement with his hands, the automatic "associated" adjustments of the trunk of his body and the upper arm segments do not occur. Instead, he must voluntarily adjust these segments before he can use his hands. Furthermore, a tremendous amount of nervous effort must be made by the voluntary motor control system to overcome "motor stiffness" of his musculature. Thus, the person with Parkinson's disease has a masklike face, showing almost no automatic emotional facial expressions; he is usually bent forward because of his muscle rigidity; and all his movements of necessity are highly deliberate rather than the many casual subconscious movements that are normally a part of our everyday life.

The cause of the akinesia in Parkinson's disease is not known, and again we must resort to theory. It is presumed that *loss of dopamine secretion in the caudate nucleus and putamen by the nigrostriatal fibers* allows excessive activity of the acetylcholine-producing neurons. But normal operation of the basal ganglia requires a balance between both excitatory and inhibitory activities, and loss of this balance, in effect, leads to a functionless basal ganglia system. We have already pointed out that the basal ganglia are responsible for many of the subconscious stereotyped movements of the body, and also responsible even for the background movements of the trunk, legs, neck, and upper arms that are required preliminary to performing the more discrete movements of the hands. If the subconscious and the background movements cannot occur, then other neural mechanisms must be substituted, especially those of the motor cortex and cerebellum. Unfortunately, though, these cannot replace the movements normally controlled by the basal ganglia and certainly cannot function at a subconscious level.

Treatment with L-Dopa and Anticholinergic Drugs. Administration of L-dopa to patients with Parkinson's disease ameliorates most of the symptoms, especially the rigidity and the akinesia, in about two-thirds of the patients. The reason for this seems to be the following: The dopamine secreted in the caudate nucleus and putamen by the nigrostriated fibers is a derivative of L-dopa. When the substantia nigra is destroyed and the person develops Parkinson's disease, the administered L-dopa is believed to substitute for the dopamine no longer secreted by the destroyed neurons. This causes more or less normal inhibition of the basal ganglia and relieves much or most of the akinesia and rigidity.

The same basal ganglia that are inhibited by dopamine are excited by acetylcholine-secreting neurons, as was discussed earlier. Therefore, as would be expected, administration of anticholinergic drugs such as scopolamine can also decrease the level of activity in the basal ganglia and therefore benefit the parkinsonian patient.

Coagulation of the Ventrolateral Nucleus of the Thalamus for Treatment of Parkinson's Disease. In recent years neurosurgeons have treated Parkinson's disease patients, with varying success, by destroying portions of the basal ganglia or of the thalamus. The most prevalent treatment has been widespread destruction of the ventrolateral nucleus of the thalamus, usually by electrocoagulation. Most fiber pathways from the basal ganglia and cerebellum to the cerebral cortex pass through this nucleus so that its destruction blocks many or most of the feedback functions of the basal ganglia and cerebellum. It is presumed that blockage of some of these feedbacks limits the functions of at least certain basal ganglia and thereby removes the factors that cause especially the tremor of Parkinson's disease.

REFERENCES

Baldessarini, R. J., and Tarsey, D.: Dopamine and the pathophysiology of dyskinesias induced by antipsychotic drugs. *Annu. Rev. Neurosci.,* 3:23. 1980.

Bizzi, E.: The coordination of eye-head movements. *Sci. Am.,* 231(4):100, 1974.

Bobath, B., and Bobath, K.: Motor Development in the Different Types of Cerebral Palsy. London, W. Heinemann Medical Books, 1975.

Brodal, A. (ed.): Basic Aspects of Central Vestibular Mechanisms. New York, American Elsevier Publishing Co., 1971.

Chase, T. N., *et al.* (eds.): Huntington's Disease. New York, Raven Press, 1979.

Crowley, W. J.: Neural control of skeletal muscle. *In* Frohlich, E. D. (ed.): Pathophysiology, 2nd Ed. Philadelphia, J. B. Lippincott Co., 1976, p. 735.

Denny-Brown, D.: The Basal Ganglia: Their Relation to Disorders of Movement. New York, Oxford University Press, 1962.

Divac, I., and Öberg, R. G. E. (eds.): The Neostriatum. New York, Pergamon Press, 1979.

Duvoisin, R. C.: Parkinson's Disease. New York, Raven Press, 1978.

Elder, H. Y., and Trueman, E. R. (eds.): Aspects of Animal Movement. New York, Cambridge University Press, 1980.

Evarts, E. V.: Brain mechanisms of movement. *Sci. Am.* 241(3):164, 1979.

Fuxe, K., and Calne, D. B. (eds.): Dopaminergic Ergot Derivatives and Motor Function. New York, Pergamon Press, 1979.

Goldberg, J. M., and Fernandez, C.: Vestibular mechanisms. *Annu. Rev. Physiol.,* 37:129, 1975.

Granit, R., and Pompeiano, O. (eds.): Reflex Control of Posture and Movement. New York, Elsevier Scientific Publishing Co., 1979.

Hobson, J. A., and Brazier, M. A. B. (eds.): The Reticular Formation Revisited: Specifying Function for A Nonspecific System. New York, Raven Press, 1980.

Hood, J. D. (ed.): Vestibular Mechanisms in Health and Disease. New York, Academic Press, 1978.

Horn, A. S., *et al.* (eds.): The Neurology of Dopamine. New York, Academic Press, 1979.

Nauta, W. J. H., and Freitag, M.: The organization of the brain. *Sci. Am.,* 241(3).88, 1979.

Oosterveld, W. J. (ed.): Audio-Vestibular System and Facial Nerve. New York, S. Karger, 1977.

Orlovsky, G. N., and Shik, M. L.: Control of locomotion: A neurophysiological analysis of the cat locomotor system. *Int. Rev. Physiol.,* 10:281, 1976.

Pearson, K.: The control of walking. *Sci. Am.,* 235(6):72, 1976.

Peterson, B. W.: Reticulospinal projections of spinal motor nuclei. *Annu. Rev. Physiol.,* 41:127, 1979.

Precht, W.: Vestibular system. *In* MTP International Review of Science: Physiology. Vol. 3. Baltimore, University Park Press, 1974, p. 81.

Precht, W.: Vestibular mechanisms. *Annu. Rev. Neurosci.,* 2:265, 1979.

Riklan, M.: L-Dopa and Parkinsonism. Springfield, Ill., Charles C Thomas, 1973.

Sarno, J. E., and Sarno, M. T.: Stroke; The Condition and the Patient. New York, McGraw-Hill, 1979.

Sherrington, C. S.: Decerebrate ridigity and reflex coordination of movements. *J. Physiol. (Lond.),* 22:319, 1898.

Shik, M. L., and Orlovsky, G. N.: Neurophysiology of locomotor automatism. *Physiol. Rev.,* 56:465, 1976.

Stein, P. S. G.: Motor systems with specific reference to the control of locomotion. *Annu. Rev. Neurosci.* 1:61, 1978.

Talbott, R. E., and Humphrey, D. R. (eds.): Posture and Movement. New York, Raven Press, 1979.

Valentinuzzi, M.: The Organs of Equilibrium and Orientation as a Control System. New York, Harwood Academic Publishers, 1980.

Wilson, V., and Jones. G. M.: Mammalian Vestibular Physiology. New York, Plenum Press, 1979.

Wilson, V. J., and Peterson, B. W.: Peripheral and central substrates of vestibulospinal reflexes. *Physiol. Rev.,* 58:80, 1978

Young, L. R.: Cross coupling between effects of linear and angular acceleration on vestibular nystagmus. *Bibl. Ophthalmol.,* 82:116, 1972.

53

Cortical and Cerebellar Control of Motor Functions

In preceding chapters we have been concerned with many of the subconscious motor activities integrated in the spinal cord and brain stem, especially those responsible for locomotion. In the present chapter we will discuss the control of motor function by the cerebral cortex and cerebellum as well as their relationship to the basal ganglia and the other lower centers. Much of this control is "voluntary" in contradistinction to the subconscious control effected by the lower centers. Yet we will also see that at least some motor functions of the cerebral cortex and cerebellum are not entirely "voluntary."

PHYSIOLOGIC ANATOMY OF THE MOTOR AREAS OF THE CORTEX AND THEIR PATHWAYS TO THE CORD

Figure 53–1 illustrates a broad area of the cerebral cortex called the *sensorimotor cortex* that is concerned either with sensation from the somatic areas of the body or with control of body movement. We have discussed in previous chapters the somesthetic cortex, which constitutes the posterior portion of the sensorimotor cortex. This area plays a major role in controlling the functions of the motor cortex, as we shall see in more detail later in the chapter.

The *motor cortex* lies anterior to the central sulcus. Its posterior portion, as shown in the figure, is characterized by large pyramid-shaped cells. This area is usually referred to simply as area IV which is the number of the area in the Broadmann classification of the cortical areas. It is frequently also called the *primary motor cortex* or the *pyramidal area*. The anterior portion of the motor cortex is usually referred to as area VI because it is mainly composed of area VI in Broadmann's classification. It is also frequently referred to as either the *motor association area* or the *premotor area*.

The Pyramidal Area and the Pyramidal Tract (Corti- cospinal Tract). The area designated in Figure 53–1 by the darker shading contains in each hemisphere of the cortex approximately 34,000 *giant Betz cells,* or *giant pyramidal cells,* for which reason it is called the "pyramidal area." This area also causes motor movements with the least amount of electrical excitation, which explains its other name, the "primary motor cortex."

One of the major pathways by which motor signals are transmitted from all the motor areas of the cortex to the anterior motoneurons of the spinal cord is through the *pyramidal tract,* or *corticospinal tract,* which is illustrated in Figure 53–2. This tract originates in all the shaded areas in Figure 53–1, including both the motor and the somesthetic areas, about 80 per cent from the motor areas and about 20 per cent from the somesthetic regions posterior to the central sulcus. The function of the fibers from the somesthetic cortex is probably not motor, but instead to cause feedback control of sensory input to the nervous system, as was discussed in Chapter 49.

The most impressive fibers in the pyramidal tract are the large myelinated fibers that originate in the giant Betz cells of the pyramidal area. These account for approximately 34,000 large fibers (mean diameter of about 16 microns) in the pyramidal tract from *each* side of the cortex. However, since each corticospinal tract contains more than a million fibers, only 3 per cent of the total number of corticospinal fibers are of this large type. The other 97 per cent are mainly fibers smaller than 4 microns in diameter.

The pyramidal tract passes downward through the *brain stem;* then it decussates mainly to the opposite side to form the *pyramids of the medulla.* By far the majority of the pyramidal fibers then descend in the *lateral corticospinal tracts* of the cord and terminate principally on interneurons at the bases of the dorsal horns of the cord gray matter. A few fibers, however, do not cross to the opposite side but pass ipsilaterally down the cord in the *ventral corticospinal tracts* and then mainly decussate to the opposite side farther down the cord.

Collaterals from the Pyramidal Tract in the Brain. Even before the pyramidal tract leaves the brain, many collaterals are given off as follows:

1. The axons from the giant Betz cells send short

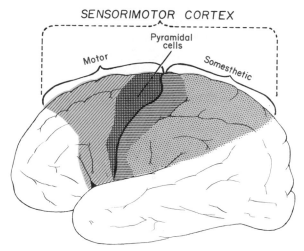

Figure 53–1. Relationship of the motor cortex to the somesthetic cortex.

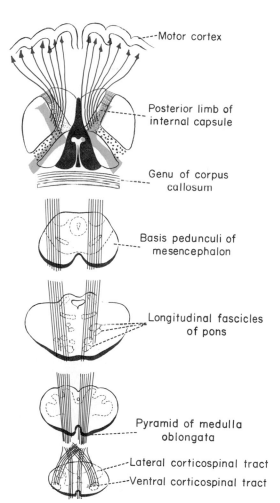

Figure 53–2. The pyramidal tract. (From Ranson and Clark: Anatomy of the Nervous System. Philadelphia, W. B. Saunders Company, 1959.)

collaterals back to the cortex itself. It is believed that these collaterals mainly inhibit adjacent regions of the cortex when the Betz cells discharge, thereby "sharpening" the boundaries of the excitatory signal.

2. A large body of collateral fibers pass into the *striate body,* and *putamen.* From here additional pathways extend through several neurons down the brain stem and then into the spinal cord through the *extrapyramidal tracts,* which are discussed later in the chapter.

3. A moderate number of collaterals pass from the pyramidal tract into the *red nuclei.* From these, additional pathways pass down the cord through the *rubrospinal tract.*

4. A moderate number of collaterals also deviate from the pyramidal tract into the *reticular substance* of the brain stem; from here signals go to the cord via *reticulospinal tracts* and others go to the cerebellum via *reticulocerebellar tracts.*

5. A tremendous number of collaterals synapse in the pontile nuclei, which give rise to the *pontocerebellar fibers.* Thus, whenever signals are transmitted from the motor cortex through the pyramidal tract, simultaneous signals are transmitted into the cerebellar hemispheres.

6. Many collaterals also terminate in the *inferior olivary nuclei,* and from here secondary *olivocerebellar fibers* transmit signals to most areas of the cerebellum.

Thus, the basal ganglia, the brain stem, and the cerebellum all receive strong signals from the pyramidal tract every time a signal is also transmitted down the spinal cord to cause a motor activity.

Importance of the Area Pyramidalis. The giant pyramidal cells of the *area pyramidalis* are very excitable. A single weak electrical shock applied to a focal point in this area is sometimes strong enough to cause a definite motor movement somewhere in the body — for instance, a flick of a finger, deviation of the lip, or some other discrete movement; and a short train of stimuli almost always elicits a movement in some discrete part of the body. In almost no other region of the cortex can one be certain of achieving a motor response with such weak stimuli.

The Extrapyramidal Tracts. The extrapyramidal tracts are, collectively, all the tracts besides the pyramidal tract itself that transmit motor signals from the cortex to the spinal cord. In the above discussion of the collaterals from the pyramidal tract, we have noted some of these extrapyramidal pathways that pass through the striate nuclei, red nuclei, and reticular nuclei of the brain stem. In addition to these collateral fibers from the pyramidal tract, still other fibers course directly from the cortex to these intermediate nuclei. For instance, large numbers of neurons project directly from area VI of the motor cortex into the *caudate nucleus* and then through the *putamen, globus pallidus, subthalamic nucleus, red nucleus, substantia nigra,* and *reticular nuclei of the brain stem* before passing into the spinal cord. The multiplicity of connections within these intermediate nuclei of the basal ganglia and reticular substance was presented in the preceding chapter.

The final pathways for transmission of extrapyramidal signals into the cord are the *reticulospinal tracts,* which lie in both the ventral and lateral columns of the cord, and the *rubrospinal, tectospinal,* and *vestibulospinal tracts.*

THE PRIMARY MOTOR CORTEX OF THE HUMAN BEING

Figure 53–3 illustrates a topographical map of motor and sensory areas in the monkey cortex, showing the left side of the cerebral cortex and, above, a folded-out view of the medial side of the hemisphere as seen from the longitudinal fissure. In the posterior part of this map one sees the two somatic sensory areas, Sm I and Sm II, which were discussed in Chapter 49. Anterior to the central sulcus are illustrated two motor areas. The darkly shaded area, labeled Ms I, is the *primary motor cortex,* while the lightly shaded motor area located on the medial aspect of the frontal lobe, labeled Ms II, is called the *supplemental area,* the function of which is doubtful as we shall discuss later. Note especially in the primary motor cortex the topographical areas from which muscles in different parts of the body can be activated. Note also the very high degree of representation of the hands, feet, and face — especially the digits of the hands and feet. In the human being, the areas representing the hand and also the mouth and tongue are especially enlarged, as we shall see in subsequent figures. Figure 53–4 gives an approximate map of the human brain, showing the points in the primary motor cortex that cause muscle contractions in different parts of the body when

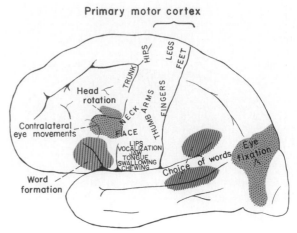

Figure 53–4. Representation of the different muscles of the body in the motor cortex and location of other cortical areas responsible for certain types of motor movements.

electrically stimulated. These have been determined by electrical stimulation of the human brain in patients having brain operations under local anesthesia. Indeed, such stimulation is commonly used by the neurosurgeon to determine the location of the motor cortex, thus permitting avoidance of this area, if possible, during the operation.

Note in the figure that stimulation of the most lateral portions of the motor cortex causes muscular contractions related to swallowing, chewing, and facial movements, while stimulation of the midline portion of the motor cortex where it bends over into the longitudinal fissure causes contraction of the legs, feet, or toes. The spatial organization is similar to that of the somatic sensory cortex I, which was shown in Figure 49–5 of Chapter 49.

Afferent Fibers That Excite the Primary Motor Cortex. The primary motor cortex is stimulated to action by signals from many different sources, including:

1. Subcortical fibers from adjacent regions of the cortex, especially from the somatic sensory areas and from the frontal areas — also subcortical fibers from the visual and auditory cortices.

2. Subcortical fibers that pass through the corpus callosum from the opposite cerebral hemisphere. These fibers connect corresponding areas of the motor cortices in the two sides of the brain.

3. Somatic sensory fibers derived directly from the ventrobasal complex of the thalamus. These transmit mainly cutaneous tactile signals and joint and muscle signals.

4. Tracts from the ventrolateral and ventroanterior nuclei of the thalamus, which in turn receive tracts from cerebellum and the basal ganglia. These tracts provide signals that are necessary for coordination between the functions of the motor cortex, the basal ganglia, and the cerebellum.

5. Fibers from the nonspecific nuclei of the thalamus. These fibers control the general level of excitabili-

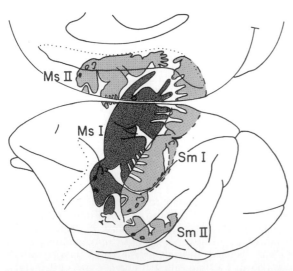

Figure 53–3. The different motor areas of the monkey cortex, illustrating especially the primary motor area (Ms I) located in the precentral gyrus. (The upper portion of the figure illustrates the medial surface of the cerebral hemisphere.) (From Woolsey, *in* Harlow and Woolsey (eds.): Biological and Biochemical Bases of Behavior. University of Wisconsin Press, 1958.)

ty of the motor cortex in the same manner that they also control the general level of excitability of most other regions of the cerebral cortex.

Vertical Columnar Arrangement of the Cells in the Motor Cortex. In Chapters 49 and 60 it is pointed out that the cells in the somatic sensory cortex and the visual cortex — and perhaps in all other parts of the brain as well — are organized in vertical columns of cells. In a like manner, the cells of the motor cortex are also organized in vertical columns — columns 0.5 to 1 mm. in diameter and having perhaps 10,000 to 50,000 neurons in each column.

The columns of cells, in turn, are themselves arranged in six distinct layers, as is the arrangement throughout the cerebral cortex. The functions of these layers from the surface inward are:

Layers I through IV are mainly sensory; that is, they receive afferent fibers from other sources. The fibers from the nonspecific nuclei of the thalamus (from the "reticular activating system" that will be discussed in the following chapter) terminate in layers I and II. These fibers provide general enhancement of the degree of excitability of the cortex when the reticular activating system is stimulated.

Layer IV is the main terminus and layer III the secondary terminus of the specific sensory signals from the thalamus, the signals from the sensory areas of the cortex, the signals from the ventrolateral and ventroanterior nuclei of the thalamus that transmit signals from the cerebellum and the basal ganglia, and the corpus callosal signals from the opposite cerebral hemisphere.

Layers V and VI are the origins of the efferent fibers or "output fibers" from the motor cortex. The large Betz cells that give rise to the long pyramidal tract fibers lie in layer V. Efferent fibers also pass from both layers V and VI to other areas of the brain.

Function of the Columns of Cells. Each column of cells seems to perform a specific motor function, such as stimulating a particular muscle or perhaps stimulating several synergistic muscles. The cells in each column operate both as an integrative and an amplifying system. The integrative function of the column is its ability to combine signals from many different sources and to determine the appropriate output response. Specific cells within a column seem to be responsive to different types of input signals. For instance, some cells respond to somatic sensory signals related to joint movement, others to tactile stimuli, and still others to feedback signals from the cerebellum and basal ganglia. Also, the general level of activation of the motor cortex is provided by signals from the nonspecific nuclei of the thalamus.

The amplifying function of the column of cells allows weak input signals to cause stimulation of large numbers of output neurons. Direct electrical stimulation of a signal output neuron in the motor cortex by means of a micropipette electrode will almost never excite a muscle contraction. Usually as many as 100 pyramidal cells need to be excited simultaneously. Therefore, divergence of the signal within the column provides this needed motive power.

Degree of Representation of Different Muscle Groups in the Primary Motor Cortex. The different muscle groups of the body are not repre-

sented equally in the motor cortex. In general, the degree of representation is proportional to the discreteness of movement required of the respective part of the body. Thus, the thumb and fingers have large representations, as is true also of the lips, tongue, and vocal cords. The relative degrees of representation of the different parts of the body are illustrated in Figure 53–5, a figure constructed by Penfield and Rasmussen on the basis of stimulatory charts made of the human motor cortex during hundreds of brain operations.

When barely threshold stimuli are used, only small segments of the peripheral musculature ordinarily contract at a time. In the "finger" and "thumb" regions, which have tremendous representation in the cerebral cortex, threshold stimuli can sometimes cause single muscles or, at times, even single fasciculi of muscles to contract, thus illustrating that a high degree of control is exercised by this portion of the motor cortex over discrete muscular movement.

On the other hand, threshold stimuli in the trunk region of the body might cause as many as 30 to 50 small, closely allied trunk muscles to contract simultaneously, thus illustrating that the motor cortex does not control discrete trunk muscles but instead controls *groups* of muscles. Similarly, threshold stimuli in the lip, tongue, and vocal cord regions of the motor cortex cause contraction of minute muscular areas, whereas stimulation in the leg region ordinarily excites several synergistic muscles at a time, causing some gross movement of the leg. Also, it should be recognized that the

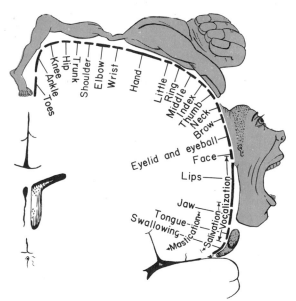

Figure 53–5. Degree of representation of the different muscles of the body in the motor cortex. (From Penfield and Rasmussen: The Cerebral Cortex of Man: A Clinical Study of Localization of Function, New York, Macmillan Company, 1968.)

finger muscles have far more representation in the motor cortex than do the muscles of the upper arm.

COMPLEX MOVEMENTS ELICITED BY STIMULATING THE CORTEX ANTERIOR TO THE MOTOR CORTEX — THE CONCEPT OF A MOTOR ASSOCIATION AREA

Electrical stimulation of the cerebral cortex for distances 1 to 3 centimeters in front of the primary motor cortex will often elicit complex contractions of groups of muscles. Occasionally, vocalization occurs, or rhythmic movements such as alternate thrusting of a leg forward and backward, coordinate moving of the eyes, chewing, swallowing, or contortions of parts of the body into different postural positions.

Some neurophysiologists have called this area the *motor association area* or *premotor cortex* and have ascribed special capabilities to it to control coordinated movements involving many muscles simultaneously. In fact, it is peculiarly organized to perform such a function for the following reasons: (1) It has long subcortical neuronal connections with the sensory association areas of the parietal lobe; (2) It has direct subcortical connections with the primary motor cortex; (3) It connects with areas in the thalamus contiguous with the thalamic areas that connect with the primary motor cortex; and (4) most important of all, it has abundant direct connections with the basal ganglia.

Still another reason for the belief that there is a motor association area (a "premotor cortex") is that damage to this area causes loss of certain coordinate skills, as follows:

Broca's Area and Speech. Note in Figure 53–4 that immediately anterior to the primary motor cortex and immediately above the sylvian fissure is an area labeled "word formation." This region is called *Broca's area*. Damage to it does not prevent a person from vocalizing, but it does make it impossible for the person to speak whole words other than simple utterances such as "no" or "yes." A closely associated cortical area also causes appropriate respiratory function so that respiratory activation of vocal cords can occur simultaneously with the movements of the mouth and tongue during speech. Thus, the activities that are related to Broca's area are highly complex.

The Voluntary Eye Movement Field. Immediately above Broca's area is a locus for controlling eye movements. Damage to this area prevents a person from voluntarily moving the eyes toward different objects. Instead, the eyes tend to lock on specific objects, an effect controlled by signals from the occipital region, as explained in Chapter 60. This frontal area also controls eyelid movements such as blinking.

Head Rotation Area. Still slightly higher in the motor association area, electrical stimulation will elicit head rotation. This area is closely associated with the eye movement field and is presumably related to directing the head toward different objects.

Area for Hand Skills. In the frontal area immediately anterior to the primary motor cortex for the hands and fingers is a region neurosurgeons have called an area for hand skills. That is, when tumors or other lesions cause destruction in this area, the hand movements become incoordinate and nonpurposeful, a condition called *motor apraxia*.

Summary of the Motor Association Area Concept. It is clear that at least some areas anterior to the primary motor cortex can cause complex coordinate movements, such as for speech, for eye movements, for head movements, and perhaps even for hand skills. However, it should be remembered that all these areas are closely connected with corresponding areas in the primary motor cortex, the thalamus, and the basal ganglia. Therefore, the complex coordinate movements almost certainly result from a cooperative effort of all these structures.

EFFECTS OF LESIONS IN THE PRIMARY MOTOR AND MOTOR ASSOCIATION CORTEX

The motor cortex is frequently damaged — especially by the common abnormality called a "stroke," which is caused by loss of blood supply to the cortex. Also, experiments have been performed in animals to remove selectively different parts of the motor cortex.

Ablation of the Area Pyramidalis. Removal of a very small portion of the area pyramidalis (the primary motor cortex — area IV) in a monkey causes paralysis of the represented muscles. If the sublying caudate nucleus and the adjacent motor association area (area VI) are not damaged, gross postural and limb "fixation" movements can still be performed, but the animal loses voluntary control of discrete movements of the distal segments of the limbs — of the hands and fingers, for instance. This does not mean that the muscles themselves cannot contract, but that the animal's ability to control the fine movements is gone.

From these results one can conclude that the area pyramidalis is concerned mainly with voluntary initiation of finely controlled movements. On the other hand, area VI and the deeper motor areas, particularly the basal ganglia, seem to be responsible mainly for the involuntary and postural body movements.

Muscle Spasticity Caused by Ablation of Large Areas of the Motor Cortex — Extrapyramidal Lesions as the Basis of Spasticity. It should be recalled that the motor cortex gives rise to tracts that descend to the spinal cord through both the pyramidal tract and the extrapyramidal tract. These two tracts have opposing effects on the tone of the body muscles. The pyramidal tract causes continuous facilitation and, therefore, a tendency to increase muscle tone throughout the body. On the other hand, the extrapyramidal system transmits inhibitory signals through the basal ganglia and the reticular formation of the brain stem, with resultant inhibition of muscle action. When the motor cortex is destroyed, the balance between these two opposing effects may be altered. If the lesion is located discretely in the area pyramidalis where the large Betz cells lie, there may be a slight degree of muscle flaccidity, but most often very little change in muscle tone results because both pyramidal and extrapyramidal elements are affected about equally. On the other hand, the usual lesion is very large and involves large portions of area VI (the motor association area) and the sensory cortex as well, both anterior and posterior to the area pyramidalis, and both of these regions normally transmit inhibitory signals through the extrapyramidal tracts. Therefore, loss of extrapyramidal inhibition is the dominant feature, thus leading to muscle spasm.

If the lesion involves the basal ganglia as well as the motor cortex itself, the spasm is even more intense because the basal ganglia normally provide very strong inhibition of the brain stem motor centers, and loss of this inhibition further exacerbates the extrapyramidal excitation of the muscles. In patients with strokes, the lesion almost invariably affects both the motor cortex itself and the sublying basal ganglia, so that very intense spasm normally occurs in the muscles of the opposite side of the body.

The muscle spasm that results from damage to the extrapyramidal system is different from the spasticity that occurs in decerebrate rigidity. In decerebrate rigidity the spasm is of the antigravity muscles and is called *extensor spasm.* On the other hand, the spasm resulting from lesions of the extrapyramidal system usually involves the flexors as well, causing intense stiffening of the limbs and other parts of the body.

The Babinski Sign. Destruction of the foot region of the area pyramidalis or transection of the foot portion of the pyramidal tract causes a peculiar response of the foot called the *Babinski sign.* This response is demonstrated when a firm tactile stimulus is applied to the sole of the foot: The great toe extends upward and the other toes fan outward. This is in contradistinction to the normal effect in which all the toes bend downward. Also, the Babinski sign does not occur when the damage is only in the extrapyramidal system. Therefore, the sign is often used clinically to detect damage specifically in the pyramidal portion of the motor control system.

The cause of the Babinski sign is believed to be the following: The pyramidal tract is a major controller of muscle activity for performance of voluntary, purposeful activity. On the other hand, the extrapyramidal system, a much older motor control system, is concerned to a great extent with protection. Therefore, when only the extrapyramidal system is functional, stimuli to the bottom of the feet cause a typical withdrawal protective type of reflex which is expressed ty the upturned great toe and fanning of the other toes. But, when the pyramidal system is also fully functional, it suppresses the protective reflex and instead excites the postural reflexes, including the normal effect of causing downward bending of the toes in response to sensory stimuli from the bottom of the feet.

THE SUPPLEMENTAL MOTOR AREA

Referring again to Figure 53–3, one sees a small area located on the medial side of the frontal lobe, called the *supplemental area.* This area requires considerably stronger electrical stimuli to cause motor activity than does the primary motor area. Also, the movements involve coordinate contraction of many muscles in contradistinction to much more discrete movements elicited from the primary area. Normal function of this area is not known, but strong stimulation is likely to elicit such effects as movement of the head and eyes, vocalization, or yawning. Also, movements are sometimes bilateral rather than unilateral.

STIMULATION OF THE SPINAL MOTONEURONS BY MOTOR SIGNALS FROM THE BRAIN

Figure 53–6 shows several different motor tracts entering a segment of the spinal cord from the brain. The corticospinal tract (pyramidal tract) terminates mainly on small interneurons in the base of the dorsal horns, although as many as 8 per cent may also terminate directly on the anterior motoneurons themselves. From the primary interneurons, most of the motor signals are transmitted through still other interneurons before finally exciting the anterior motoneurons.

Figure 53–6 shows several other descending tracts from the brain, collectively called the *extrapyramidal tracts,* which also carry signals to the anterior motoneurons: (1) the *rubrospinal tract,* (2) the *reticulospinal tracts,* (3) the *tectospinal*

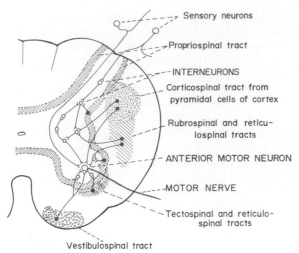

Figure 53–6. Convergence of all the different motor pathways on the anterior motoneurons.

tract, and (4) the *vestibulospinal tract.* In addition, sensory signals arriving through the *dorsal sensory roots,* as well as signals transmitted from segment to segment of the spinal cord via the *propriospinal tracts,* also stimulate the anterior motoneurons. In general, the extrapyramidal tracts terminate farther anteriorly in the gray matter of the spinal cord than does the corticospinal tract; a larger share of these fibers terminate directly on the anterior motoneurons. Also, some terminate on *inhibitory neurons,* which in turn inhibit the anterior motoneurons, an effect opposite to the excitatory effect of the corticospinal tract.

The corticospinal tract seems to cause very specific muscle contractions; neurophysiological recordings have demonstrated that the intensity of muscle contraction — everything else being equal — is directly proportional to the nervous energy transmitted through the corticospinal tract. On the other hand, some of the extrapyramidal tracts provide less specific muscle contractions. Instead, they provide such effects as general facilitation, general inhibition, or gross postural signals, all of which provide the background against which the corticospinal system operates. Finally, the sensory and propriospinal signals also add their input to the total melee of spinal integration.

Patterns of Movement Elicited by Spinal Cord Centers. From Chapter 51, recall that the spinal cord can provide specific reflex patterns of movement in response to sensory nerve stimulation. Many of these patterns are also important when the anterior motoneurons are excited by signals from the brain. For instance, the stretch reflex is functional at all times, helping to damp the motor movements initiated from the brain and probably providing at least part of the motive power required to cause the muscle contractions employing the servo-assist mechanism that was described in Chapter 51.

Also, when a brain signal excites an agonist muscle, it is not necessary to transmit an inverse signal to the antagonist at the same time; this transmission will be achieved by the reciprocal innervation circuit that is always present in the cord for coordinating the functions of antagonistic pairs of muscles.

Finally, parts of the other reflex mechanisms such as withdrawal, stepping and walking, scratching, postural mechanisms, and so forth, can be activated by signals from the brain. Thus, very simple signals from the brain can lead, at least theoretically, to many of our normal motor activities, particularly for such functions as walking and the attainment of different postural attitudes of the body.

Inhibition of Sensory Input Signals by Corticospinal Signals. At times it is important to suppress sensory input from the periphery to prevent its interference with motor activity. Experimental studies have shown that the same corticospinal signals that activate the muscles or muscle groups at the same time inhibit the local cord sensory input to the same muscles. This results from specific inhibitory circuits located in the cord itself at the base of the dorsal horns, through which most of the sensory information feeds. For instance, if pain signals are arriving in the cord from the bottom of the feet and are attempting to cause withdrawal of the feet from a source of pain, nevertheless the cortical signals can still suppress these and allow continued use of the feet for running or other purposes.

THE CEREBELLUM AND ITS MOTOR FUNCTIONS

The cerebellum has long been called a *silent area* of the brain principally because electrical excitation of this structure does not cause any sensation and rarely any motor movement. However, as we shall see, removal of the cerebellum does cause the motor movements to become highly abnormal. The cerebellum is especially vital to the control of very rapid muscular activities such as running, typing, playing the piano, and even talking. Loss of this area of the brain can cause almost total incoordination of these activities even though its loss causes paralysis of no muscles.

But how is it that the cerebellum can be so important when it has no direct control over muscle contraction? The answer to this is that it *monitors and makes corrective adjustments in the motor activities elicited by other parts of the brain*. It receives continuously updated information from the peripheral parts of the body to determine the instantaneous status of each part of the body — its position, its rate of movement, forces acting on it, and so forth. And it is believed that the cerebellum *compares* the actual instantaneous status of each part of the body as depicted by the sensory information with the status that is intended by the motor system. If the two do not compare favorably, then appropriate corrective signals are transmitted instantaneously back into the motor system to increase or decrease the levels of activation of the specific muscles.

Since the cerebellum must make major motor corrections extremely rapidly *during the course of motor movements*, a very extensive and rapidly acting cerebellar input system is required both from the peripheral parts of the body and from the cerebral motor areas. Also, an extensive output system feeding equally as rapidly into the motor system is necessary to provide the necessary corrections of the motor signals.

THE INPUT SYSTEM TO THE CEREBELLUM

For purposes of discussion, the cerebellum is generally divided into three lobes, as illustrated in Figure 53–7: (1) the *flocculonodular lobe;* (2) the *anterior lobe;* and (3) the *posterior lobe.*

The posterior lobe is greatly enlarged in primates, especially so in the human being, forming bilateral protrusions called the *cerebellar hemispheres* (which are illustrated in Figures 53–9 and 53–10). These hemispheres are also known as the *neocerebellum* (new cerebellum) because they represent a phylogenetically new portion of the cerebellum.

The oldest part of the cerebellum is the flocculo-

nodular lobe, which developed in association with the equilibrium apparatus and vestibular nuclei. In the human being, this part of the cerebellum still functions almost entirely in relation to equilibrium, as was described in the previous chapter.

Another part of the cerebellum that is also very old is the entire midline area, 2 to 3 centimeters wide, in both the anterior and posterior lobes, called the *vermis.* In this area most of the nerve signals from the somatic areas of the body terminate. Therefore, this midline structure (the vermis) plays an important role in the integration of subconscious postural mechanisms.

On the other hand, the cerebellar hemispheres are the terminus of most of the signals arriving from the higher levels of the brain, especially from the motor areas of the cerebral cortex.

The Afferent Pathways. The basic afferent pathways to the cerebellum are also illustrated in Figure 53–7. An extensive and important afferent pathway is the *corticocerebellar pathway,* which originates mainly in the *motor cortex* (but to a lesser extent in the sensory cortex as well) and then passes by way of the *pontile nuclei* and *pontocerebellar tracts* to the contralateral cortex of the cerebellum. In addition, important afferent tracts originate in the brain stem; they include: (a) an extensive *olivocerebellar* tract, which passes from the *inferior olive* to all parts of the cerebellum; this tract is excited by fibers from the *motor cortex,* the *basal ganglia,* widespread areas of the *reticular formation,* and the *spinal cord;* (b) *vestibulocerebellar fibers,* some of which originate in the vestibular apparatus itself and others from the vestibular nuclei; most of these terminate in the *flocculonodular lobe* and *fastigial nucleus* of the cerebellum; and (c) *reticulocerebellar fibers,* which originate in different portions of the reticular formation and terminate mainly in the midline structures (the vermis).

The cerebellum also receives important sensory signals directly from the peripheral parts of the body, which reach the cerebellum by way of the *ventral* and *dorsal spinocerebellar* tracts (which pass ipsilaterally up to the cerebellum, as illustrated in Figure 53–8. The signals transmitted in these tracts originate in the muscle spindles, the Golgi tendon organs, the large tactile receptors of the skin, and the joint receptors, and they apprise the cerebellum of the momentary status of muscle contraction, degree of tension on the muscle tendons, positions and rates of movement of the parts of the body, and forces acting on the surfaces of the body. All this information keeps the cerebellum constantly apprised of the instantaneous physical status of the body.

The spinocerebellar pathways can transmit impulses at velocities over 100 meters per second, which is the most rapid conduction of any pathway in the entire central nervous system. This extremely rapid conduction is also important for the instantaneous apprisal of the cerebellum of changes that take place in the status of the body.

In addition to the signals in the spinocerebellar tracts, other signals are transmitted through the dorsal and dorsolateral columns to the medulla and then relayed from there to the cerebellum. Likewise, signals are transmitted through the *spinoreticular pathway* to the reticular substance of the brain stem and through the

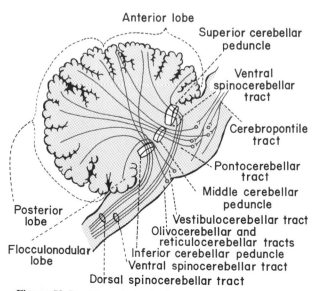

Figure 53–7. The principal afferent tracts to the cerebellum.

Anterior lobe
Superior cerebellar peduncle
Ventral spinocerebellar tract
Cerebropontile tract
Pontocerebellar tract
Middle cerebellar peduncle
Vestibulocerebellar tract
Olivocerebellar and reticulocerebellar tracts
Inferior cerebellar peduncle
Ventral spinocerebellar tract
Dorsal spinocerebellar tract
Posterior lobe
Flocculonodular lobe

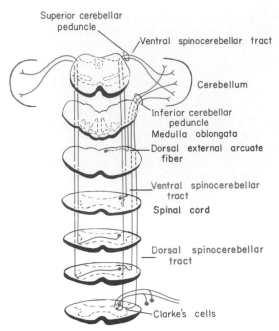

Figure 53–8. The spinocerebellar tracts.

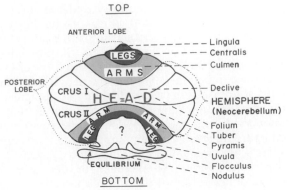

Figure 53–10. Localization of somatic sensory signals in the cerebellum of the human being.

spino-olivary pathway to the inferior olivary nucleus and then relayed to the cerebellum. Thus, the cerebellum collects continual information about all parts of the body even though it is operating at a subconscious level.

Topographical Localization of Sensory Input to the Cerebellar Cortex. Because the cerebellum operates in close association both with the motor cortex and with the muscles that are being activated, one would expect separate parts of the cerebellum to be concerned specifically with separate parts of the body. However, it has not been possible to localize the different parts of the body in most areas of the cerebellum. Instead, only a gross representation has been found in the midline

structures. Figure 53–9 illustrates the loci of two different "homunculi" that are believed to have been found in the monkey cerebellar cortex, representing the terminations of somatic sensory fibers from the different parts of the body. Figure 53–10 illustrates the human cerebellum, showing approximate sensory terminations based on anatomical data and extrapolation from animals. However, it should be recognized that these are only gross representations and are based on uncertain data. It should also be noted that no specific somatic representations have been found in the very large cerebellar hemispheres. Even so, many neurophysiologists still believe that point-to-point spatial relationships between these hemispheres and specific stimulatory areas of the cerebral cortex are likely because of the well-known fact that whenever a motor signal is transmitted to the periphery, a signal is transmitted into the cerebral hemispheres at the same time.

OUTPUT SIGNALS FROM THE CEREBELLUM

The Deep Cerebellar Nuclei and the Efferent Pathways. Located deep in the cerebellar mass are three *deep cerebellar nuclei* — the *dentate, interpositus,* and *fastigial nuclei.* The vestibular nuclei in the medulla also function in some respects as if they were deep cerebellar nuclei because of their direct connections with the cortex of the flocculonodular lobe. These nuclei receive signals from two different sources: (1) the cerebellar cortex, and (2) the sensory afferent tracts to the cerebellum. Each time an input signal arrives in the cerebellum, it goes both to the cerebellar cortex and also directly to the deep nuclei (or the vestibular nuclei) through collateral fibers; then, a short time later, the cerebellar cortex relays its output signals also to the deep nuclei. Thus, all the input signals that enter the cerebellum eventually end in the deep nuclei. We shall discuss this circuit in greater detail below.

Three major efferent pathways lead out of the cerebellum, as illustrated in Figure 53–11:

(1) A pathway that begins in the *cortex of the two cerebellar hemispheres,* then passes to the *dentate nucleus,* next to the *ventrolateral nucleus of the thalamus,* and finally to the *motor cortex.* This pathway plays an important role in helping to coordinate "voluntary"

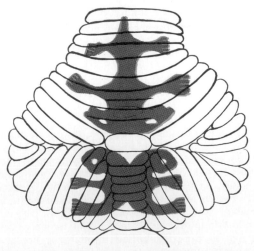

Figure 53–9. The sensory projection areas, called "homunculi," on the cortex of the cerebellum. (From Snider: *Sci. Amer., 199:*4, 1958. © 1958 by Scientific American, Inc. All rights reserved.)

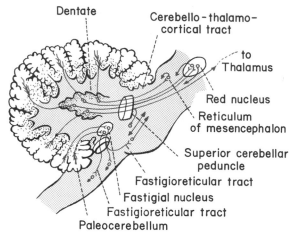

Figure 53–11. Principal efferent tracts from the cerebellum.

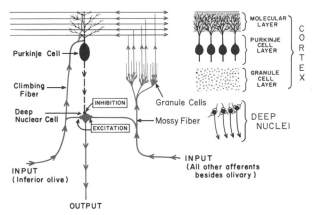

Figure 53–12. Basic neuronal circuit of the cerebellum, showing excitatory pathways in color. At right are the three major layers of the cerebellar cortex and also the deep nuclei.

motor activities initiated by the motor cortex and its associated structures.

(2) A pathway that originates in the *midline structures of the cerebellum* (the *vermis*) and then passes through the *fastigial nuclei* into the *medullary* and *pontile regions of the brain stem.* This circuit functions in close association with the equilibrium apparatus to help control equilibrium and also to control the postural attitudes of the body. It was discussed in detail in the previous chapter in relation to equilibrium.

(3) A pathway that originates in the *intermediate areas* on each side of the cerebellum, between the vermis and the cerebellar hemispheres, then passes (a) through the *nucleus interpositus* to the *ventrolateral nucleus of the thalamus,* and thence to the *motor cortex,* (b) to several *midline structures* of the *thalamus* and thence to the *basal ganglia,* and (c) to the *red nucleus* and *reticular formation* of the upper portion of the brain stem. This circuit functions to coordinate activities between the first two cerebellar output pathways noted above — that is, to help coordinate the interrelationships between subconscious body postural control and conscious command control from the motor cortex.

THE NEURONAL CIRCUIT OF THE CEREBELLUM

The structure of the cerebellar cortex is entirely different from that of the cerebral cortex. Furthermore, each part of the cerebellar cortex has a neuronal organization almost precisely the same as that in all other parts.

The human cerebellum is actually a large folded sheet, approximately 17 cm. wide by 120 cm. long, the folds lying crosswise, as illustrated in Figures 53–9 and 53–10. Each fold is called a *folium.*

The Functional Unit of the Cerebellar Cortex — the Purkinje Cell. The cerebellum has approximately 30 million nearly identical functional units, one of which is illustrated in Figure 53–12. This functional unit centers on the Purkinje cell, of

which there are also 30 million in the cerebellar cortex.

Note to the far right in Figure 53–12 the three major layers of the cerebellar cortex: the *molecular layer,* the *Purkinje cell layer,* and the *granular cell layer.* Then, beneath these layers, the *deep nuclei* are located far within the center of the cerebellar mass.

The Neuronal Circuit of the Functional Unit. As illustrated in Figure 53–12, the output from the functional unit is from a deep nuclear cell. However, this cell is continually under the influence of both excitatory and inhibitory influences. The excitatory influences arise from direct connections with the afferent fibers that enter the cerebellum. The inhibitory influences arise entirely from the Purkinje cells in the cortex of the cerebellum.

The afferent inputs to the cerebellum are of two types, one called the *climbing fiber type* and the other called the *mossy fiber type.* There is one climbing fiber for about 10 Purkinje cells. After sending collaterals to several deep nuclear cells, the climbing fiber projects all the way to the molecular layer of the cerebellar cortex where it makes about 300 synapses with the dendrites of each Purkinje cell. This climbing fiber is distinguished by the fact that a single impulse in it will always cause a single, very prolonged and peculiar oscillatory type of action potential in each Purkinje cell with which it connects. Another distinguishing feature of the climbing fibers is that they all originate in the inferior olive of the medulla.

The mossy fibers also send collaterals to excite the deep nuclear cells. Then these fibers proceed to the granular layer of the cortex where they synapse with hundreds of *granule cells.* These in turn send their very small axons, less than 1 micron in diameter, up to the outer surface of the cerebellar cortex to enter the molecular layers.

Here the axons divide into two branches that extend 1 to 2 millimeters in each direction parallel to the folia. There are literally millions of these *parallel nerve fibers* in each small segment of the cerebellar cortex (there are about 1000 granule cells for every Purkinje cell). It is into this molecular layer that the dendrites of the Purkinje cells project, and 80,000 to 200,000 of these parallel fibers synapse with each Purkinje cell. Yet, the mossy fiber input to the Purkinje cell is quite different from the climbing fiber input because stimulation of a single mossy fiber will never elicit an action potential in the Purkinje cell; instead, large numbers of mossy fibers must be stimulated simultaneously to activate the Purkinje cell. Furthermore, this activation usually takes the form of prolonged facilitation or excitation that, when it reaches threshold for stimulation, causes repetitive Purkinje cell firing of normal, short-duration action potentials rather than the single prolonged action potential occurring in response to the climbing fiber input.

Thus, the Purkinje cells are stimulated by two types of input circuits — one that causes a highly specific output in response to the incoming signal and the other that causes a less specific but tonic type of response. It should be noted that the greater proportion of the afferent input to the cerebellum is of the mossy fiber type because this represents the afferent input from all the cerebellar afferent tracts besides those from the inferior olive.

Balance Between Excitation and Inhibition in the Deep Cerebellar Nuclei. The output signals from the Purkinje cells to the deep nuclei are entirely inhibitory. Therefore, referring again to the circuit of Figure 53–12, *one should note that direct stimulation of the deep nuclear cells by both the climbing and the mossy fibers excites them, whereas the signals arriving from the Purkinje cells inhibit them.* Normally, there is a continual balance between these two effects so that the degree of output from the deep nuclear cell remains relatively constant. On the other hand, in the execution of rapid motor movements, the *timing* of the two effects on the deep nuclei is such that the excitation appears before the inhibition. Then a few milliseconds later inhibition occurs. In this way, very rapid excitatory and inhibitory transient signals can be fed back into the motor pathways to modify motor movements. The inhibitory portions of these signals resemble "delay-line" negative feedback signals of the type that are very effective in providing *damping*. That is, when the motor system is excited, a negative feedback signal occurs after a short delay to stop the muscle movement from overshooting its mark, which is the usual cause of oscillation.

Other Inhibitory Cells in the Cerebellar Cortex. In addition to the granule cells and Purkinje cells, three other types of neurons are also located in the cerebellar cortex: *basket cells, stellate cells,* and *Golgi cells.* All of these are inhibitory cells with very short axons. Both the basket cells and the stellate cells are excited by the parallel fibers in the molecular layer of the cerebellar cortex. These cells in turn cause lateral inhibition of the adjacent Purkinje cells, thus sharpening the signal in the same manner that lateral inhibition sharpens the contrast of signals in many other areas of the nervous system. The Golgi cells are also stimulated by the parallel fibers of the molecular layer, but they inhibit the granule cells instead of the Purkinje cells, possibly functioning as an automatic feedback gain control circuit to prevent overamplification of the input signals.

Special Features of the Cerebellar Neuronal Circuit. A special feature of the cerebellum is that there are no reverberatory pathways in the cerebellar neuronal circuits, so that the input-output signals of the cerebellum are very rapid transients that never persist for long periods of time.

Another special feature is that many of the cells of the cerebellum are constantly active. This is especially true of the deep nuclear cells; they continually send output signals to the other areas of the motor system. The importance of this is that decrease of the nuclear cell firing rate can provide an inhibitory output signal from the cerebellum, while an increase in firing rate can provide an excitatory output signal.

FUNCTION OF THE CEREBELLUM IN CONTROLLING MOVEMENTS

The cerebellum functions in motor control only in association with motor activities initiated elsewhere in the nervous system. These activities may originate in the spinal cord, in the reticular formation, in the basal ganglia, or in motor areas of the cerebral cortex. We will discuss, first, the operation of the cerebellum in association with the spinal cord and lower brain stem for control of postural movements and equilibrium and then discuss its function in association with the motor cortex for control of voluntary movements.

FUNCTION OF THE CEREBELLUM WITH THE SPINAL CORD AND LOWER BRAIN STEM TO CONTROL POSTURAL AND EQUILIBRIUM MOVEMENTS

The cerebellum originated phylogenetically at about the same time that the vestibular apparatus developed. Furthermore, as was discussed in the previous chapter, loss of the flocculonodular lobes of the cerebellum causes extreme disturbance of equilibrium. Yet, we still must ask the question, what role does the cerebellum play in equilibrium that cannot be provided by the other neuronal machinery of the brain stem? A clue is the fact that equilibrium is far more disturbed during performance of rapid motions than during stasis. This suggests that the cerebellum is especially important in controlling the balance between agonist and antagonist muscle contractions during rapid changes in body positions as dictated by the vestibular apparatuses. One of the major problems in controlling this balance is the time required to transmit position signals and kinesthetic signals from the different parts of the body to the brain. Even when utilizing the most rapidly conducting sensory pathways at 120 meters per second, which are indeed utilized by the spinocerebellar system, the delay for transmission from the feet to the brain is still 15 to 20 milliseconds. The feet of a person running rapidly can move as much as 10 inches during this time. Therefore, it is impossible for the brain to know at any given instant during rapid motion the exact position of the different parts of the body.

On the other hand, with appropriate neuronal circuitry, it would be possible for the cerebellum or some other portion of the brain to know how rapidly and in what direction a part of the body was moving 15 to 20 milliseconds previously and then to predict from this information where the parts of the body should be at the present time. And this seems to be one of the major functions of the cerebellum.

As we have already discussed in relation to the neuronal circuitry of the cerebellum, there are abundant sensory pathways from the somatic areas of the body, especially from the muscles, joints, and skin surface, that feed both into the brain stem and into the older areas of the cerebellum — into the flocculonodular lobes through the vestibular nuclei and into the vermis and intermediate areas of the cerebellum through the dorsal and ventral spinocerebellar tracts and reticulocerebellar tracts. Also, the vestibular apparatus is located within a few centimeters of the flocculonodular lobes, allowing no more than a millisecond or so of delay in transmission of the vestibular information.

Therefore, during the control of equilibrium, it is presumed that the extremely rapidly conducted vestibular apparatus information is used in a typical feedback control circuit to provide almost instantaneous correction of postural motor signals as necessary for maintaining equilibrium even during extremely rapid motion, including rapidly changing directions of motion. The feedback signals from the peripheral areas of the body help in this process, but their help is presumably contingent upon some function of the cerebellum to compute positions of the respective parts of the body at any given time, despite the long delay time from the periphery to the cerebellum.

Relationship of Cerebellar Function to the Spinal Cord Stretch Reflex

One major component of cerebellar control of posture and equilibrium is an extreme amount of information transmitted from the muscle spindles to the cerebellum through the dorsal spinocerebellar tracts. In turn, signals are transmitted into the brain stem through the cerebellar fastigial nuclei to stimulate the gamma efferent fibers that innervate the muscle spindles themselves. Therefore, a cerebellar stretch reflex occurs that is similar to but more complex than the spinal cord stretch reflex. It utilizes signals that pass all the way to the cerebellum and back again to the muscles. In general, this reflex adds additional support to the cord stretch reflex, but its feedback time is considerably longer, thus prolonging the effect. It is through this feedback pathway that many of the postural adjustments of the body are believed to occur.

Function of the Cerebellum in Voluntary Muscle Control

In addition to the feedback circuitry between the body periphery and the cerebellum, an almost entirely independent feedback circuitry exists between the motor cortex and the cerebellum. This is illustrated in its simplest form in Figure 53–13 and in a much more complex form, involving the basal ganglia also in the control circuit, in Figure 53–14. Most of the signals of this circuit pass from the motor cortex to the cerebellar hemispheres and then back to the motor cortex again, successively, through the dentate nuclei and ventrolateral nuclei of the thalamus. These circuits are not involved in the control of posture. However, a small component of the feedback to the cortex involves the intermediate areas of the cerebellum lying between the cerebellar hemispheres and the

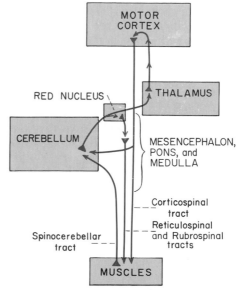

Figure 53–13. Pathways for cerebellar control of voluntary movements.

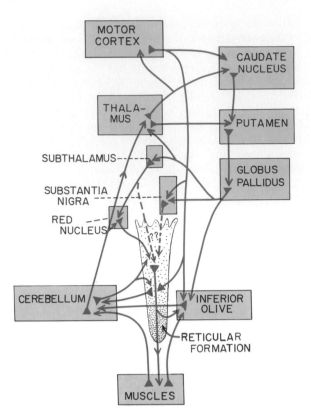

Figure 53–14. Pathways for cerebellar "error" control of involuntary movements.

vermis. This cerebellar area does receive feedback signals from the periphery. Therefore, there is reason to believe that the cerebellum functions in relation to cortical motor control in two ways: (1) by direct feedback circuitry from the motor cortex to the cerebellum, then back to the motor cortex without involving peripheral feedback, and (2) by similar feedback but with the return signals from the cerebellum being modified by conditioning information received from the peripheral parts of the body. Let us discuss the second of these first.

Cerebellar Feedback Control of Cortical Motor Function

There is much reason to believe that the cerebellum operates to provide so-called "error control" in much the same manner as servomechanisms used in (a) various industrial control systems, (b) the control system of the antiaircraft gun, or (c) the control system of the automatic pilot. That is, the motor cortex transmits signals to the periphery to cause a motor function, but at the same time it transmits the same information into the cerebellum. Then the cerebellum seems to compare the "intentions" of the cortex with the "performance" by the parts of the body, and, if the parts

are not moving according to the intentions of the cortex, the "error" between these two is calculated by the cerebellum so that appropriate and immediate corrections can be made. Thus, if the cortex has transmitted a signal intending the limb to move to a particular point, but the limb begins to move too fast and obviously is going to overshoot the point of intention, the cerebellum can initiate "braking impulses" to slow down the movement of the limb and stop it at the point of intention.

Ordinarily, during rapid movements, the motor cortex transmits far more impulses than are needed to perform each intended movement, and the cerebellum therefore must act to inhibit the motor cortex at the appropriate time after the muscle has begun to move. The cerebellum is believed to assess the rate of movement and calculate the length of time that will be required to reach the point of intention. Then appropriate inhibitory impulses are transmitted to the motor cortex to inhibit the agonist muscle and to excite the antagonist muscle. In this way, appropriate "brakes" are applied to stop the movement at the precise point of intention.

Experiments have demonstrated two important characteristics of the cerebellar feedback system:

1. When a person performs a particular act, such as moving a heavy weight rapidly to a new position, at first he is unable to judge the degree of inertia that will be involved in the movement. Therefore, he almost always overshoots the point of intention on his first trial. However, after several trials, he can stop the movement at the precise point. This illustrates that learned knowledge of the inertia of the system is an important feature of the cerebellar feedback mechanism, though it is possible that this learning occurs in the cerebral cortex rather than in the cerebellum.

2. When a rapid movement is made toward a point of intention, the agonist muscle contracts strongly throughout the early course of movement. Then, suddenly, shortly before the point of intention is reached, the agonist muscle becomes completely inhibited while the antagonist muscle becomes strongly excited. Furthermore, the point at which this reversal of excitation occurs depends on the rate of movement and on the previously learned knowledge of the inertia of the system. The faster the movement and the greater the inertia, the earlier the reversal point appears in the course of movement.

Since all these events transpire much too rapidly for the motor cortex to reverse the excitation "voluntarily," it is evident that the excitation of the antagonist muscle toward the end of a movement is an entirely automatic and subconscious function and is not a "willed" contraction of the same nature as the original contraction of the agonist muscle. We shall see below that in patients with serious cerebellar damage, excitation of the antagonist muscles does not occur at the appropriate time but instead always too late. Therefore, it is almost certain that one of the major functions of the cerebellum is automatic excitation of antagonist muscles at the end of a movement while at the same time inhibiting agonist muscles that have started the movement.

The "Damping" Function of the Cerebellum. One of the by-products of the cerebellar feedback mechanism is its ability to "damp" muscular movements. To explain the meaning of "damping" we must first point out that essentially all movements of the body are "pendular." For instance, when an arm is moved, momentum develops, and the momentum must be overcome before the movement can be stopped. And, because of the momentum, all pendular movements have a tendency to overshoot. If overshooting does occur in a person whose cerebellum has been destroyed, the conscious centers of the cerebrum eventually recognize this and initiate a movement in the opposite direction to bring the arm to its intended position. But again the arm, by virtue of its momentum, overshoots, and appropriate corrective signals must again be instituted. Thus, the arm oscillates back and forth past its intended point for several cycles before it finally fixes on its mark. This effect is called an *action tremor,* or *intention tremor.*

However, if the cerebellum is intact, appropriate subconscious signals stop the movement precisely at the intended point, thereby preventing the overshoot and also the tremor. This is the basic characteristic of a damping system. All servocontrol systems regulating pendular elements that have inertia must have damping circuits built into the servomechanisms. In the motor control system of our central nervous system, the cerebellum seems to provide much of this damping function.

Function of the Cerebellum in Prediction. Another important by-product of the cerebellar feedback mechanism is its ability to help the central nervous system predict the future positions of moving parts of the body. Without the cerebellum a person "loses" his or her limbs *when they move rapidly,* indicating that feedback information from the periphery probably must be analyzed by the cerebellum if the brain is to keep up with the motor movements. Thus, the cerebellum detects from the incoming proprioceptive signals the rapidity with which the limb is moving and then seems to predict from this the projected time course of movement. This allows the cerebellum, operating through the cerebellar output circuits to inhibit the agonist muscles and to excite the an-

tagonist muscles when the movement approaches the point of intention.

Without the cerebellum this predictive function is so deficient that *rapidly* moving parts of the body move much farther than the point of intention. This failure to control the distance that the parts of the body move is called *dysmetria,* which simply means poor control of the distance of movement. As would be expected, dysmetria is much more pronounced in rapid movements than in slow movements.

Failure of Smooth "Progression of Movements." One of the most important features of normal motor function is one's ability to progress from one movement to the next in an orderly succession. When the cerebellum becomes dysfunctional and the subconscious ability to predict ahead of time how far the different parts of the body will move in a given time is lost, the person becomes unable also to control the beginning of the next movement. As a result, the succeeding movement may begin much too early or much too late. Therefore, movements such as those required for writing, running, or even talking all become completely incoordinate, lacking completely in the ability to progress in an orderly sequence from one movement to the next.

Extramotor Predictive Functions of the Cerebellum. The cerebellum also plays a role in predicting other events besides movements of the body. For instance, the rates of progression of both auditory and visual phenomena can be predicted. As an example, a person can predict from the changing visual scene how rapidly he is approaching an object. A striking experiment that demonstrates the importance of the cerebellum in this ability is the effect of removing the "head" portion of the cerebellar vermis in monkeys. Such a monkey occasionally charges the wall of a corridor and literally bashes its brains out because it is unable to predict when it will reach the wall.

Unfortunately, we are only now beginning to learn about these extramotor predictive functions of the cerebellum. It is quite possible that the cerebellum provides a "time base", perhaps utilizing time-delay circuits, against which signals from other parts of the central nervous system can be compared. It is often stated that the cerebellum is especially important in interpreting *spatiotemporal relationships* in sensory information.

Cerebellar Control of Ballistic Movements

We must emphasize again that *feedback* control of motor movements in response to signals from the periphery involves a very long delay time, as much as 15 to 25 milliseconds from the periphery to the motor cortex and at least that much longer back to the periphery, giving a total time interval of 1/30 to 1/20 of a second — far too long for any effective control of extremely rapid movements. The cerebellum does utilize its predictive capability to help overcome this problem, but even so some motor movements are entirely over before any feedback whatsoever can return to the motor cortex. These are called *ballistic* movements. An important example is the saccadic movements of the eyes, in which the eyes jump from one position to the next when reading or when looking at the side of the road while moving in a car, or so forth. The movements are entirely over before any feedback can reach the cerebellum, much less reach the motor cortex. And the same is true of many of the limb movements as well.

Yet, for proper performance of these ballistic movements, the cerebellum is essential. In fact, they are usually disturbed even more than the slower movements when the cerebellum is impaired.

It is believed that mainly the hemispheres of the cerebellum are concerned with these ballistic movements, because the two-way circuits between the motor cortex and these hemispheres provide the most direct and most rapidly functioning pathway between the cortex and cerebellum. Furthermore, this pathway begins mainly in area VI of the motor cortex and returns to area IV. Therefore, there is very little, if any, feed*back* to the point of origin of the signal; instead, this is a feed*forward* circuit.

When the cerebellum is removed, three major changes occur in the ballistic movements: (1) they are slow to begin, (2) the force development is weak, and (3) the movements are slow to turn off. Therefore, it becomes very difficult to perform the very rapid ballistic movements. Furthermore, it is almost impossible to control how far the movement will go because of the difficulty of turning the movement off once it is begun. Thus, in the absence of the cerebellar circuit, the motor cortex has to think very hard to turn ballistic movements on and again has to think hard and take extra time to turn the movement off. Thus, the automatism of ballistic movements is lost.

But, how does the cerebellum function in the control of ballistic movements? We do not know the answer to this. The supposition is: When the motor cortex first initiates the movement, it immediately sends signals to the cerebellum at the same time. The first effect of the signals is to excite the deep cerebellar dentate nuclei, and these immediately send an excitatory signal back to the motor cortex or other motor nuclei to reinforce strongly the onset of the ballistic movement. A few milliseconds later, the signal entering the cerebellum will have had time to go through the delay circuits of the cerebellar cortex and to return by way of the Purkinje cells back to the dentate nuclei, but this

time inhibiting these rather than exciting them. Therefore, after this given delay time, this automatic delayed inhibitory signal presumably stops the ballistic movement by turning off the agonist muscle and, because of reciprocal innervation, turning on the antagonist at the same time.

If you will reconsider once again the circuitry of the cerebellum as described earlier in the chapter, you will see that it is beautifully organized to perform this biphasic, first excitatory and then delayed, inhibitory function that is required for ballistic movements. You will also see that the time delay circuits of the cerebellar cortex almost undoubtedly are fundamental to this particular ability of the cerebellum.

Function of the Cerebellum in Relation to the Basal Ganglia

It is already clear that proper function of the motor control system requires not only active participation of the motor cortex but also of the cerebellum and basal ganglia. Furthermore, both of these can provide either feed*back* or feed*forward* signals. Yet, their functions are entirely different. Some of the differences are the following:

First, the cerebellum can provide feedback or feedforward signals that last no longer than a few milliseconds at a time. Furthermore, the cerebellum cannot originate motor signals. On the other hand, the basal ganglia can originate signals; they can provide long-term continuous signals to the motor control pathway; they can provide rhythmical patterns of activities; and they can transmit motor control signals into the brain stem and to the muscles sometimes without involving the motor cortex at all.

Second, the basal ganglia are concerned especially with the balance between agonist and antagonist muscle contractions during either static contractions or during smooth progressive contractions. In contrast, the cerebellum is concerned principally with the onset and offset of rapid movements.

Third, the basal ganglia seem to be harbingers of many subconscious stereotyped or even learned programs of movement, either functioning alone or functioning in close association with the cerebral cortex. The cerebellum does not provide such programs of movement. However, whenever the basal ganglial programs involve rapid movements, the cerebellum plays the same role in relation to these movements as it does to the more direct signals from the motor cortex.

CLINICAL ABNORMALITIES OF THE CEREBELLUM

Destruction of small portions of the cerebellar cortex causes no detectable abnormality in motor function. In fact, several months after as much as half the cerebellar cortex has been removed, the motor functions appear to be almost entirely normal — but only as long as the person performs all movements slowly. Thus, the remaining areas of the cerebellum compensate tremendously for loss of part of the cerebellum.

To cause serious and continuous dysfunction of the cerebellum, the cerebellar lesion must usually involve the deep cerebellar nuclei — the *dentate, interpositus,* and *fastigial nuclei* — as well as the cerebellar cortex.

Dysmetria and Ataxia. Two of the most important symptoms of cerebellar disease are dysmetria and ataxia. It was pointed out above that in the absence of the cerebellum a person cannot predict ahead of time how far movements will go. Therefore, the movements ordinarily overshoot their intended mark. This effect is called *dysmetria,* and it results in incoordinate movements which are called *ataxia.*

Dysmetria and ataxia can also result from lesions in the spinocerebellar tracts, for the feedback information from the moving parts of the body is essential for accurate control of the muscular movements.

Past Pointing. Past pointing means that in the absence of the cerebellum a person ordinarily moves the hand or some other moving part of the body considerably beyond the point of intention. This probably results from the following effect: The motor cortex normally transmits more impulses to the muscles to perform a given motor function than are actually needed. The cerebellum automatically corrects this by inhibiting the movement after it has begun. However, if the cerebellum is not available to cause this inhibition, the movement ordinarily goes beyond the intended point. Therefore, past pointing is actually a manifestation of dysmetria.

Failure of Progression. *Dysdiadochokinesia.* When the motor control system fails to predict ahead of time where the different parts of the body will be at a given time, it temporarily ''loses'' the parts during rapid motor movements. As a result, the succeeding movement may begin much too early or much too late so that no orderly ''progression of movement'' can occur. One can demonstrate this readily by having a patient with cerebellar damage turn one hand upward and downward at a rapid rate. The patient rapidly ''loses'' the hand and does not know its position during any portion of the movement. As a result, a series of jumbled movements occurs instead of the normal coordinate upward and downward motions. This is called dysdiadochokinesia.

Dysarthria. Another instance in which failure of progression occurs is in talking, for the formation of words depends on rapid and orderly succession of individual muscular movements in the larynx, mouth, and respiratory system. Lack of coordination between these and inability to predict either the intensity of the sound or the duration of each successive sound cause jumbled vocalization, with some syllables loud, some weak, some held long, some held for a short interval, and resultant speech that is almost completely unintelligible. This is called dysarthria.

Intention Tremor. When a person who has lost the cerebellum performs a voluntary act, the muscular movements are jerky; this reaction is called an *intention tremor* or an *action tremor,* and it results from failure of the cerebellar system to damp the motor movements. Tremor is particularly evident when the dentate nuclei or the superior cerebellar peduncle is destroyed, but it is not present when the spinocerebellar tracts from the periphery to the cerebellum are destroyed. This indicates that the feedback pathway from the cerebellum to the motor cortex is a principal pathway for damping of muscular movements.

Cerebellar Nystagmus. Cerebellar nystagmus is a tremor of the eyeballs that occurs usually when one attempts to fixate the eyes on a scene to the side of the head. This off-center type of fixation results in rapid, tremulous movements of the eyes rather than a steady fixation, and it is probably another manifestation of the failure of damping by the cerebellum. It occurs especially when the flocculonodular lobes are damaged; in this instance it is associated with loss of equilibrium, presumably because of dysfunction of the pathways through the cerebellum from the semicircular canals. Nystagmus resulting from damage to the semicircular canals was discussed in Chapter 52.

Rebound. If a person with cerebellar disease is asked to contract an arm tightly while the physician holds it back at first and then lets go, the arm will fly back until it strikes the face instead of being automatically stopped. This is called rebound, and it results from loss of the cerebellar component of the stretch reflex. That is, the normal cerebellum ordinarily instantaneously and powerfully sensitizes the spinal cord reflex mechanism whenever a portion of the body begins to move unexpectedly in an unwilled direction. But, without the cerebellum, activation of the antagonist muscles fails to occur, thus allowing overmovement of the limb.

Hypotonia. Loss of the deep cerebellar nuclei, particularly the dentate nuclei, causes moderate decrease in tone of the peripheral musculature on the side of the lesion, though after several months the motor cortex usually compensates for this by an increase in its intrinsic activity. The hypotonia results from loss of facilitation of the motor cortex and brain stem nuclei by the tonic discharge of the deep nuclei.

SENSORY FEEDBACK CONTROL OF MOTOR FUNCTIONS

Everyone who has ever studied the relationship of the somatic sensory areas to the motor areas of the cortex has been impressed by the close functional interdependence of the two areas. Anatomically, the characteristic pyramidal cells of the motor cortex extend backward into the anterior lip of the postcentral gyrus where they intermingle with large numbers of granule cells of the somatic sensory cortex. Likewise, the granule cells of the sensory cortex extend anteriorly into the precentral gyrus, the *area pyramidalis.* Thus the two areas fade into each other with many somatic sensory fibers actually terminating directly in the motor cortex and some motor signals originating in the sensory cortex.

Furthermore, when a portion of the somatic sensory cortex in the postcentral gyrus is removed, the muscles controlled by the motor cortex immediately anterior to the removed area often lose much of their coordination. This observation illustrates that the somatic sensory area plays a major role in the control of motor functions.

THE SENSORY ENGRAM FOR MOTOR CONTROL

A person performs a motor movement mainly to achieve a purpose. Primarily in the sensory and sensory association areas the person experiences effects of motor movements and records "memories" of the different patterns of motor movements. These are called sensory engrams of the motor movements. Wishing to achieve some purposeful act, one presumably calls forth one of these engrams and then sets the motor system of the brain into action to reproduce the sensory pattern that is laid down in the engram.

The Proprioceptor Feedback Servomechanism for Reproducing the Sensory Engram. In addition to the feedback pathways through the cerebellum, feedback pathways also pass from proprioceptors to the sensory areas of the cerebral cortex and thence back to the motor cortex. And these feedback pathways are capable of modifying the motor response. For instance, as a person learns to cut out a paper doll with scissors, the movements involved in this process cause a particular sequential pattern of proprioceptive impulses to pass to the somatic sensory area. Once this pattern has been "learned" by the sensory cortex, the memory engram of the pattern can be used to activate the motor system to perform the same sequential pattern whenever it is required — even when the person is blindfolded.

To do this, the proprioceptor signals from the fingers, hands, and arms are compared with the engram, and if the two do not match each other, the difference, called the "error," supposedly initiates additonal motor signals that automatically activate appropriate muscles to bring the fingers, hands, and arms into the necessary sequential attitudes for performance of the task. Each successive portion of the engram presumably is projected according to a time sequence, and the motor control system automatically follows from one point to the next so that the fingers go through the precise motions necessary to duplicate the sensory engram of the motor activity.

Thus, one can see that the motor system in this case acts as a *servomechanism,* for it is not the motor cortex itself that controls the pattern of activity to be accomplished. Instead, the pattern is located in the sensory part of the brain, and the motor system merely "follows" the pattern, which is the definition of a servomechanism. If ever the motor system fails to follow the pattern, sensory signals are fed back to the cerebral cortex to apprise the sensorium of this failure, and appropriate corrective signals are transmitted to the muscles.

Other sensory signals besides somesthetic signals are also involved in motor control, particularly visual signals. However, these other sensory systems are often slower to recognize error than is the somatic proprioceptor system. Therefore, when the sensory engram depends on visual feedback for control purposes, the motor movements are usually considerably slowed in comparison with those that depend on somatic feedback.

An extremely interesting experiment that demonstrates the importance of the sensory engram for control of motor movements is one in which a monkey has been trained to perform some complex but slow-moving activity, and then various portions of its cortex are removed. Removal of small portions of the motor cortex that control the muscles normally used for the activity does not prevent the monkey from performing the activity. Instead it automatically uses other muscles in place of the paralyzed ones to perform the same activity. On the other hand, if the corresponding somatic sensory cortex is removed while the motor cortex is left intact, the monkey loses all ability to perform the activity. Thus, this experiment demonstrates that the motor system acts automatically as a servomechanism to use whatever muscles are available to follow the pattern of the sensory engram, and if some muscles are missing, other muscles are substituted automatically. The experiment also demonstrates forcefully that the somatic sensory cortex is essential for performance of some types of "learned" motor activities.

ESTABLISHMENT OF RAPID MOTOR PATTERNS

Many motor activities are performed so rapidly that there is insufficient time for sensory feedback signals to control these activities. For instance, the movements of the fingers during typing occur much too rapidly for somatic sensory signals to be transmitted either to the somatic sensory cortex or even directly to the motor cortex and for these then to control each discrete movement. It is believed that the patterns for control of these rapid coordinate muscular movements are established in the motor system itself, probably involving complex circuitry in the primary motor cortex, in the motor association cortex, the basal ganglia, and even the cerebellum. Indeed, lesions in any of these areas can destroy one's ability to perform rapid coordinated muscular contractions, such as those required during the act of typing, talking, or writing by hand.

Role of Sensory Feedback During Establishment of the Rapid Motor Patterns. Even a highly skilled motor activity can be performed the very first time if it is performed extremely slowly — slowly enough for sensory feedback to guide the movements through each step. However, to be really useful, many skilled motor activities must be performed rapidly. This probably is achieved by successive performance of the same skilled activity until finally an engram for the skilled activity is laid down in the motor control areas of the cortex as well as in the sensory system. This motor engram causes a precise set of muscles to go through a specific sequence of movements required to perform the skilled activity. Therefore, such an engram is called a *pattern of skilled motor function,* and the motor areas are primarily concerned with this.

After a person has performed a skilled activity many times, the motor pattern of this activity can thereafter cause the hand or arm or other part of the body to go through the same pattern of activity again and again, now entirely *without* sensory feedback control. However, even though sensory feedback control is no longer present, the sensory system still determines whether or not the act has been performed correctly. This determination is made in retrospect rather than while the act is being performed. If the pattern has not been performed correctly, information from the sensory system can help to correct the pattern the next time it is performed.

Thus, eventually, hundreds of patterns of different coordinate movements are believed to be laid down in the motor system, and these can be called upon one at a time in different sequential orders to perform literally thousands of complex motor activities.

An interesting experiment that demonstrates the applicability of these theoretical methods of muscular control is one in which the eyes are made to "follow" an object that moves around and around in a circle. At first, the eyes can follow the object only when it moves around the circle slowly, and even then the movements of the eyes are extremely jerky. Thus, sensory feedback is being utilized to control the eye movements for following the object. However, after a few seconds, the eyes begin to follow the moving object rather faithfully, and the rapidity of movement around the circle can be increased to many times per second, and still, the eyes continue to follow the object. Sensory feedback control of each stage of the eye movements at these rapid rates would be completely impossible. Therefore, by this time, the eyes have developed a pattern of movement that is not dependent upon step-by-step sensory feedback. Nevertheless, if the eyes should fail to follow the object around the circle, the sensory system would immediately become aware of this and presumably could make corrections in the pattern of movement.

INITIATION OF VOLUNTARY MOTOR ACTIVITY

Because of the spectacular properties of the primary motor cortex (the area pyramidalis) and of the instantaneous muscle contractions that can be achieved by stimulating this area, it has become customary to think that the initial brain signals that elicit voluntary muscle contractions begin in the primary motor cortex. However, this almost certainly is far from the truth. Indeed, experiments have shown that the cerebellum and the basal ganglia are activated at almost exactly the same time that the motor cortex is activated, and all of these are activated even *before* the voluntary movement occurs. Furthermore, there is no known mechanism by which the motor cortex can conceive the entire sequential pattern that is to be achieved by the motor movements.

Therefore, we are left with an unanswered question: What is the locus of the initiation of voluntary motor activity? In the following chapter we will learn that the reticular formation of the brain stem and much of the thalamus play essential roles in activating all other parts of the brain. Therefore, it is very likely that these areas provide at least part of the initial signals that lead to subsequent activity in the motor cortex, the basal ganglia, and the cerebellum at the onset of voluntary movement. Here again, we come to a circular question: What is it that initiates the activity in the brain stem and the thalamus? We do know part of the answer to this: The activity is initiated by continual sensory input into these areas, including sensory input from the peripheral receptors of the body and from the cortical sensory storage, or memory, areas. Much of motor activity is almost reflex in nature, occurring instantly after an incoming sensory signal from the periphery. But so-called voluntary activity occurs minutes, hours, or even days after the initiating sensory input — after analysis, storage of memories, recall of the memories, and finally initiation of a motor response. We shall see in Chapter 55 that the control of motor activity is strongly influenced by these prolonged procedures of cerebration. Furthermore, damage to the essential areas of the cerebral cortex for analysis of sensory information leads to serious deficits and abnormalities of voluntary muscle control.

Therefore, for the present, let us conclude that the immediate energy for eliciting voluntary motor activity probably comes from the middle regions of the brain. These, in turn, are under the control of the different sensory inputs, the memory storage areas, and the associated areas of the brain that are devoted to the processes of analysis, a mechanism frequently called *cerebration*.

REFERENCES

Allen, G. I., and Tsukahara, N.: Cerebrocerebellar communication systems. *Physiol. Rev., 54*:957, 1974.

Armstrong, D. M.: The mammalian cerebellum and its contribution to movement control. *In* Porter, R. (ed.): International Review of Physiology: Neurophysiology III. Vol. 17. Baltimore, University Park Press, 1978, p. 239.

Asanuma, H.: Cerebral cortical control of movement. *Physiologist, 16*:143, 1973.

Asanuma, H.: Recent developments in the study of the columnar arrangement of neurons within the motor cortex. *Physiol. Rev., 55*:143, 1975.

Brooks, V. B.,, and Stoney, S. D., Jr.: Motor mechanisms: The role of the pyramidal system in motor control. *Annu. Rev. Physiol., 33*:337, 1971.

Cooper, I. S., *et al.* (eds.): The Cerebellum, Epilepsy, and Behavior. New York, Plenum Press, 1974.

Crowley, W. J.: Neural control of skeletal muscle. *In* Frohlich, E. D. (ed.): Pathophysiology, 2nd Ed. Philadelphia, J. B. Lippincott Co., 1976, p. 735.

Desmedt, J. E. (ed.): Cerebral Motor Control in Man: Long Loop Mechanisms. New York, S. Karger, 1978.

Eccles, J. C.: The Understanding of the Brain. New York, McGraw-Hill, 1973.

Evarts, E. V.: Brain mechanisms in movement. *Sci. Am., 229*:96, 1973.

Evarts, E. V.: Brain mechanisms of movement. *Sci. Am. 241*(3):164, 1979.

Evarts, E. V., and Thach, W. T.: Motor mechanisms of the CNS: Cerebrocerebellar interrelations. *Annu. Rev. Physiol., 31*:451, 1969.

Gallistel, C. R.: The Organization of Action: A New Synthesis. New York, Halsted Press, 1979.

Granit, R.: The Basis of Motor Control. New York, Academic Press, 1970.

Grillner, S.: Locomotion in vertebrates: Central mechanisms and reflex interaction. *Physiol. Rev., 55*:247, 1975.

Llinas, R.: Eighteenth Bowditch lecture. Motor aspects of cerebellar control. *Physiologist, 17*:19, 1974.

Massion, J., and Sasaki, K. (eds.): Cerebro-Cerebellar Interactions. New York, Elsevier/North-Holland, 1979.

McHenry, L. C., Jr.: Cerebral Circulation and Stroke. St. Louis, W. H. Green, 1978.

O'Connell, A. L., and Gardner, E. B.: Understanding the Scientific Bases of Human Movement. Baltimore, Williams & Wilkins. 1972.

Orlovsky, G. N., and Shik, M. L.: Control of locomotion: A neurophysiological analysis of the cat locomotor system. *10*:281, 1976.

Paillard, J.: The patterning of skilled movements. *In* Magoun, H. W. (ed.): Handbook of Physiology. Sec. 1, Vol. 3. Baltimore. Williams & Wilkins, 1960, p. 1679.

Pearson, K.: The control of walking. *Sci. Am., 235*(6):72, 1976.

Penfield, W., and Rasmussen, T.: The Cerebral Cortex of Man. New York, The Macmillan Co., 1950.

Pepper, R. L., and Herman, L. M.: Decay and Interference Effects in the Short-Term Retention of a Discrete Motor Act. Washington, D.C., American Physiological Association, 1970.

Porter, R.: Functions of the mammalian cerebral cortex in movement. *Prog. Neurobiol., 1*:3, 1973.

Porter, R.: The neurophysiology of movement performance. *In* MTP International Review of Science: Physiology. Vol. 3. Baltimore, University Park Press, 1974, p. 151.

Porter, R.: Influences of movement detectors on pyramidal tract neurons in primates. *Annu. Rev. Physiol., 38*:121, 1976.

Shik, M. L., and Orlovsky, G. N.: Neurophysiology of Locomotor Automatism. *Physiol. Rev., 56*:465, 1976.

Stein, P. S. G.: Motor systems with specific reference to the control of locomotion. *Annu. Rev. Neurosci., 1*:61, 1978.

Wilson, D. M.: The flight-control system of the locust. *Sci. Am., 218*:(5):83, 1968.

54

Activation of the Brain – The Reticular Activating System; The Generalized Thalamocortical System; Brain Waves; Epilepsy; Wakefulness and Sleep

We will now consider the many brain functions related to the *reticular activating system,* a system that controls the overall degree of central nervous system activity, including control of wakefulness and sleep, and control of at least part of our ability to direct attention toward specific areas of our conscious minds.

Figure 54–1A illustrates the extent of this system, showing that it begins in the lower brain stem and extends upward through the mesencephalon and thalamus to be distributed throughout the cerebral cortex. Impulses are transmitted from the ascending reticular activating system to the cortex by two different pathways. One pathway passes upward from the brain stem portion of the reticular formation — especially from the mesencephalic region — to the intralaminar, midline, and reticular nuclei of the thalamus and thence through diverse pathways to essentially all parts of the cerebral cortex as well as the basal ganglia. A second and probably much less important pathway is through the subthalamic, hypothalamic, and adjacent areas.

FUNCTION OF THE RETICULAR ACTIVATING SYSTEM IN WAKEFULNESS

Diffuse electrical stimulation in the *mesencephalic, pontile, and upper medullary portions of the reticular formation* — an area discussed in Chapter 52 in relation to the motor functions of the nervous system — causes immediate and marked activation of the cerebral cortex and even causes a sleeping animal to awaken instantaneously. Furthermore, when this mesencephalic portion of the reticular formation is damaged severely, as occurs (a) when a *brain tumor* develops in this region, (b) when serious *hemorrhage* occurs, or (c) in diseases such as *encephalitis lethargica* (sleeping sickness), the person passes into coma and is completely nonsusceptible to normal awakening stimuli.

Function of the Mesencephalic Portion of the Reticular Activating System. Electrical stimuli applied to different portions of the reticular activating system have shown that the mesencephalic portion functions quite differently from the thalamic portion. Electrical stimulation of the mesencephalic portion causes generalized activation of the entire brain, including activation of the cerebral cortex, thalamic nuclei, basal ganglia, hypothalamus, other portions of the brain stem, and even the spinal cord. Furthermore, once the mesencephalic portion is stimulated, the degree of activation throughout the nervous system remains high for as long as a half minute or more after the stimulation is over. Therefore, *it is believed that the mesencephalic portion of the reticular activating system is basically responsible for normal wakefulness of the brain.*

Function of the Thalamic Portion of the Activating System. Electrical stimulation in different areas of the thalamic portion of the activating system (if the stimulation is not too strong) activates specific regions of the cerebral cortex more than others. This is distinctly different from stimu-

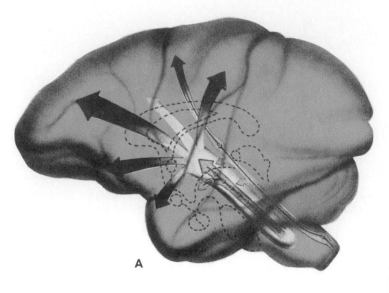

A

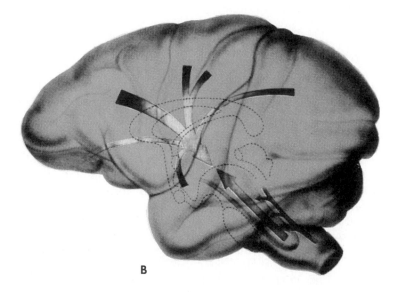

B

Figure 54–1. (A) The ascending reticular activating system schematically projected on a monkey
brain. (From Lindsley: Reticular Formation of the
Brain. Little, Brown, and Co.) (B) Convergence of
pathways from the cerebral cortex and from the
spinal afferent systems on the reticular activating
system. (From French, Hernandez-Peon, and
Livingston: *J. Neurophysiol., 18*:74, 1955.)

lation in the mesencephalic portion, which activates all the brain at the same time. Therefore, it is
believed that the thalamic portion of the activating
system has two specific functions: first, it relays
most of the diffuse facilitatory signals from the
mesencephalon to all parts of the cerebral cortex
to cause generalized activation of the cerebrum;
second, stimulation of selected points in the thalamic activating system causes specific activation of
certain areas of the cerebral cortex in distinction
to the other areas. This selective activation of
specific cortical areas possibly or probably plays
an important role in our ability to direct our
attention to certain parts of our mental activity,
which is discussed later in the chapter.

THE AROUSAL REACTION — ACTIVATION OF THE RETICULAR ACTIVATING SYSTEM BY SENSORY SIGNALS

When an animal is asleep, the level of activity of
the reticular activating system is greatly decreased; yet almost any type of sensory signal can
immediately activate the system. For instance,
proprioceptive signals from the joints and muscles, pain impulses from the skin, visual signals
from the eyes, auditory signals from the ears, or
even visceral sensations from the gut can all cause
sudden activation of the reticular activating system and therefore arouse the animal. This is called
the *arousal reaction*.

Some types of sensory stimuli are more potent than others in eliciting the arousal reaction; the most potent are pain and proprioceptive somatic impulses.

Anatomically, the reticular formation of the brain stem is admirably constructed to perform the arousal functions. It receives tremendous numbers of signals either directly or indirectly from the *spinoreticular tracts,* the *spinothalamic tracts,* the *spinotectal tracts,* the *auditory tracts,* the *visual tracts,* and others, so that almost any sensory stimulus in the body can activate it. The reticular formation in turn can transmit signals both upward into the brain and downward into the spinal cord. Indeed, many of the fibers originating from cells in the reticular formation divide, with one branch of the fiber passing upward and the other branch passing downward, as was explained in Chapter 52.

STIMULATION OF THE RETICULAR ACTIVATING SYSTEM BY THE CEREBRAL CORTEX

In addition to activation of the reticular activating system by sensory impulses, the cerebral cortex can also stimulate this system. Direct fiber pathways pass into the reticular formation, as shown in Figure 54–1B, from almost all parts of the cerebrum but particularly from (1) the *sensorimotor cortex* of the pre- and postcentral gyri, (2) the *frontal cortex,* (3) the *cingulate gyrus,* (4) the *hippocampus* and *other limbic structures,* (5) the *hypothalamus,* and (6) the *basal ganglia.* Because of an exceedingly large number of nerve fibers that pass from the motor regions of the cerebral cortex to the reticular formation, motor activity in particular is associated with a high degree of wakefulness, which partially explains the importance of movement to keep a person awake. However, intense activity of any other part of the cerebrum can also activate the reticular activating system and consequently can cause a high degree of wakefulness.

THE GENERALIZED THALAMOCORTICAL SYSTEM

At this point it is important that we explain some of the interrelationships between the thalamus and the cerebral cortex. The thalamus is the entryway for essentially all sensory nervous signals to the cortex, with the single exception of signals from the olfactory system. In Chapters 49 and 50 we have already discussed the relay of the somatic signals through the *ventrobasal nuclei* of the thalamus to the cerebral cortex, and in Chapters 52 and 53 we traced signals from the cerebellum and basal ganglia through the *ventrolateral* and *ventroanterior thalamic nuclei.* Also, we shall see in Chapters 60, 61, and 62 that all the signals from the visual, auditory, and taste systems are also relayed in the thalamus on their way to the cortex. All these thalamic relay nuclei are called the *specific nuclei of the thalamus;* the combination of these nuclei and their connecting areas in the cortex is called the *specific thalamocortical system.* It is this system of which we usually speak when we discuss the transmission of sensory information to the cerebral cortex.

In addition to the specific thalamocortical system, there is a separate system, partially separated from the specific system, called the *generalized thalamocortical system,* or the *diffuse thalamocortical system,* or the *nonspecific thalamocortical system.* This system is subserved by *diffuse thalamic nuclei* and also by *diffuse fiber projections* from the thalamus to the cortex. The thalamic neurons of this system are, in reality, the uppermost end of the reticular formation, and they are not collected into discrete nuclei as are the neurons of the specific system. Instead, they lie mainly between the specific nuclei or on the outer surface of the thalamus; they are divided into three separate groups: (1) the *intralaminar nuclei,* which are diffuse collections of neurons that lie within the thalamic mass among the specific nuclei; (2) the *periventricular nuclei,* which lie adjacent to the third ventricle and have little organization; and (3) the *reticular nuclei* of the thalamus, which form a thin shell of diffuse neurons that lies over the entire dorsal, anterior, and lateral surfaces of the thalamus and through which all the pathways from the thalamus to the cerebral cortex must pass. All these nuclei make multiple connections with the specific thalamic nuclei. They also project very small fibers to all parts of the cerebral cortex, with the exception of small portions of the temporal lobes that are associated with the limbic system, a very old part of the brain that plays major roles in behavior, emotions, and so forth as will be discussed in Chapter 56.

The generalized system of the thalamus is continuous with the upper end of the reticular formation in the brain stem and receives much of its input from this source. It is also an important terminus of one of the very important somatic sensory pathways, the *paleospinothalamic pathway for pain* — the pathway that transmits the intolerable type of burning, aching pain. This is the same pathway that transmits the major sensory signals into much of the reticular formation of the lower brain stem as well.

Control Functions of the Generalized Thalamocortical System. One can well understand from the above description of the diffuse thalamic nu-

clei that they lie in propitious locations for affecting the levels of activity of all of the specific nuclei
of the thalamus. It is also clear that they are
properly located to relay control functions from
the mesencephalic reticular formation to the cerebrum, and that they in turn can receive much
helpful input information from the specific nuclei
of the thalamus. Perhaps most important of all,
however, is the fact that these nuclei have a
specific capability to control the level of activity in
the entire cerebral cortex; this is mediated by
signals that play on the cerebral cortex through the
diffuse thalamocortical projections.

**Effect of the Generalized Thalamic System on
Cortical Activity.**   Activation of the cortex by the
generalized thalamic system is entirely different
from activation by the specific system. Some of
the differences are the following:

1. Stimulation of a specific thalamic nucleus — such as the ventrobasal complex that
transmits somatic signals to the somesthetic cortex — activates the cortex within 1 to 2 milliseconds, whereas stimulation of the generalized system causes no activation for approximately the
first 25 milliseconds. The activation level builds up
over a period of many milliseconds. These differences are illustrated in Figure 54–2.

2. At the end of stimulation, the activation of
the cortex by the specific nuclei dies away within
another few milliseconds, whereas the activation
by the generalized system sometimes continues as
an ''after-discharge'' for as long as 30 seconds.

3. Signals from the specific nuclei to the cortex
activate mainly layer IV of the cortex, as was
explained in Chapter 49, whereas activation of the
generalized thalamic system activates mainly
layers I and II of the cortex. Since this latter
activation is prolonged and because layers I and II
are the loci of many of the dendrites of the deeper

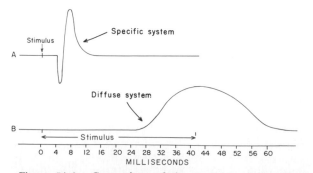

Figure 54–2. Comparison of the response of the visual
cortex following stimulation of the lateral geniculate
body (record A), which is part of the specific thalamocortical system, with the response of the visual cortex following stimulation of the visual portion of the generalized
thalamocortical system (record B). Note in record B the
long latent period before appearance of the diffuse response and also the prolonged duration of the response.

cortical neurons, it is supposed that the stimulation by the generalized thalamic system mainly
causes partial depolarization of large numbers of
dendrites near the surface of the cortex; this in
turn causes a generalized increase in the degree of
facilitation of the cortex. When the cortex is thus
facilitated, specific signals that enter the cortex
from other sources are exuberantly received.

4. Stimulation in the generalized thalamic system *facilitates* an *area* of several square centimeters in the cortex, while stimulation at a point in a
specific thalamic nucleus *excites* a specific point in
the cortex.

In summary, the generalized thalamocortical
system controls the overall degree of activity of
the cortex. It can at times facilitate activity in
regional areas of the cortex distinct from the
remainder of the cortex. Collateral signals from
this system also control the level of activity in the
specific nuclei of the thalamus, the basal ganglia,
the hypothalamus, and other structures of the
cerebrum as well.

ATTENTION

So long as a person is awake, he has the ability
to direct his attention to specific aspects of his
mental environment. Furthermore, his degree of
attention can change remarkably from (a) almost
no attention at all to (b) broad attention to almost
everything that is going on, or to (c) intense
attention to a minute facet of his momentary
mental experience.

Unfortunately, the basic mechanisms by which
the brain accomplishes its diverse acts of attention
are not known. However, a few clues are beginning to fall into place, as follows:

**Reticular Activating System Control of Overall
Attentiveness.**   In exactly the same way that a
person can change from a state of sleep to a state
of wakefulness, there can be all degrees of wakefulness, from wakefulness in which a person is
nonattentive to almost all the surroundings to an
extremely high degree of wakefulness in which the
person reacts instantaneously to almost any sensory experience. These changes in degree of *overall attentiveness* seem to be caused primarily by
changes in activity of the mesencephalic portion of
the reticular activating system. Thus, control of
the general level of attentiveness is probably exerted by the same mechanism that controls wakefulness, the control center for which is located in
the mesencephalon and upper pons.

Function of the Thalamus in Attention.   Earlier
in the chapter it was pointed out that stimulation
of a single specific area in the thalamic portion of
the reticular activating system, when the stimulus
intensity is not too strong, activates only a small

area of the cerebral cortex. Since the cerebral cortex is one of the most important areas of the brain for conscious awareness of our surroundings, one can surmise that the ability of specific thalamic areas to excite specific cortical regions might be one of the mechanisms by which a person can direct attention to specific aspects of the mental environment, whether these be immediate sensory experiences or stored memories.

Relation of Centrifugal Control of Sensory Information to Attention. Nervous pathways extend centrifugally from all sensory areas of the brain toward the lower centers to control the intensity of sensory input to the brain. For instance, the auditory cortex can either inhibit or facilitate signals from the cochlea, the visual cortex can control the signals from the retina in the same way, and the somesthetic cortex can control the intensities of signals from the somatic areas of the body.

Thus, it is likely that activated regions of the cortex control their own sensory input. This is another means by which the brain might direct its attention to specific phases of its mental activity.

POSSIBLE "SEARCHING" AND "PROGRAMMING" FUNCTIONS OF THE BRAIN'S ACTIVATING SYSTEM

At this point it is important to refer back to the analogy drawn in Chapter 46 between the functions of the central nervous system and of a general purpose computer. Almost all complex computers, particularly those that are capable of storing information and then recalling this information at later times, have a control center called a "programming unit" that directs the attention of different parts of the computer to specific information stored in other parts of the computer. It is quite likely that the activating system of the brain operates in much the same way as the programming unit of a computer for the following reasons:

1. The probable ability of the activating system to direct one's attention to specific mental activities is analogous to the ability of the programming unit to call forth information that has been stored in the memory of the computer or to call forth information that is given to the computer at its input, information that is comparable to sensory information in the human being.

2. The programming unit can "find information" when its exact locus is not known. In the computer, this is achieved by dictating certain qualities required of the information and then searching through hundreds or thousands of stored bits of information until the appropriate information is found, a function called *searching*. We know from psychological tests that our own brain can perform this same function, though we do not know how it does so. Yet, since the thalamus perhaps plays a role in directing our attention to stored information in specific areas of the cortex, it is reasonable to postulate that the thalamus, probably operating in conjunction with other basal areas of the brain, plays a major role in this searching operation.

3. The programming unit of a computer determines the sequence of processing of information. Our brain can also control its thoughts in orderly sequence. Here again, mainly on the basis of anatomical considerations and on the basis of the fact that the reticular activating system is the primary controller of cerebral activity, it can be surmised, until we know more about the subject, that the reticular activating system is at least one of the controllers of our sequence of thoughts.

EFFECT OF BARBITURATE ANESTHESIA ON THE RETICULAR ACTIVATING SYSTEM

The barbiturates have a specific depressant effect on the brain stem portion of the reticular activating system. Therefore, barbiturates obviously can either depress brain activity or even cause sleep. Yet it is especially interesting that barbiturate anesthesia does not block transmission in most of the specific sensory systems and also does not entirely block function of the thalamic portion of the reticular activating system. It is probable that many other clinically used anesthetics also have specific depressant effects on the midbrain portion of the reticular activating system and in this way cause general anesthesia.

BRAIN WAVES

Electrical recordings from the surface of the brain or from the outer surface of the head demonstrate continuous electrical activity in the brain. Both the intensity and patterns of this electrical activity are determined to a great extent by the overall level of excitation of the brain resulting from functions in the reticular activating system. The undulations in the recorded electrical potentials, shown in Figure 54–3, are called *brain waves,* and the entire record is called an *electroencephalogram* (EEG).

The intensities of the brain waves on the surface of the scalp range from 0 to 300 microvolts, and their frequencies range from once every few seconds to 50 or more per second. The character of the waves is highly dependent on the degree of activity of the cerebral cortex, and the waves change markedly between the states of wakefulness and sleep.

Much of the time, the brain waves are irregular, and no general pattern can be discerned in the EEG. However, at other times, distinct patterns do appear. Some of these are characteristic of specific abnormalities of the brain, such as epilepsy, which is discussed later. Others occur even in normal persons and can be classified into *alpha, beta, theta,* and *delta waves,* which are all illustrated in Figure 54–3.

Alpha waves are rhythmic waves occurring at a frequency between 8 and 13 per second and are found in the EEG's of almost all normal adult persons when they are awake in a quiet, resting state of cerebration. These waves occur most intensely in the occipital region but can also be recorded at times from the parietal and frontal regions of the scalp. Their voltage usually is about 50 microvolts. During sleep the alpha waves disappear entirely, and when the awake person's atten-

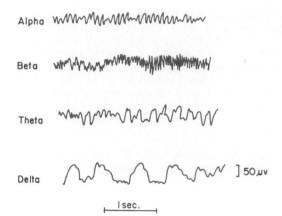

Figure 54–3. Different types of normal electroencephalographic waves.

tion is directed to some specific type of mental activity, the alpha waves are replaced by asynchronous, higher frequency but lower voltage beta waves. Figure 54–4 illustrates the effect on the alpha waves of simply opening the eyes in bright light and then closing the eyes again. Note that the visual sensations cause immediate cessation of the alpha waves and that these are replaced by low voltage, asynchronous beta waves.

Beta waves occur at frequencies of more than 14 cycles per second and as high as 25 and rarely 50 cycles per second. These are most frequently recorded from the parietal and frontal regions of the scalp, and they can be divided into two major types, *beta I* and *beta II*. The beta I waves, which are illustrated in Figure 54–3, have a frequency about twice that of the alpha waves, and these are affected by mental activity in very much the same way as the alpha waves — that is, they disappear and in their place appears an asynchronous but low voltage recording. The beta II waves, on the contrary, appear during activation of the central nervous system or during tension. Thus, one type of beta wave is inhibited by cerebral activity while the other is elicited.

Theta waves have frequencies between 4 and 7 cycles per second. These occur mainly in the parietal and temporal regions in children, but they also occur during emotional stress in some adults, particularly during disappointment and frustration. They can often be brought out in the EEG of a frustrated person by allowing enjoyment of some pleasant experience and then suddenly removing this element of pleasure; this causes approximately 20 seconds of theta waves. These same waves also occur in many brain disorders.

Delta waves include all the waves of the EEG below 3.5 cycles per second and sometimes as low as 1 cycle every 2 to 3 seconds. These occur in deep sleep, in infancy, and in very serious organic brain disease. And

they occur in the cortex of animals that have had subcortical transections separating the cerebral cortex from the thalamus. Therefore, delta waves can occur strictly in the cortex independently of activities in lower regions of the brain.

ORIGIN OF THE DIFFERENT TYPES OF BRAIN WAVES

The discharge of a single neuron or single nerve fiber in the brain cannot be recorded from the surface of the head. Instead, for an electrical potential to be recorded all the way through the skull, large portions of nervous tissue must emit electrical current simultaneously. There are two ways in which this can occur. First, tremendous numbers of nerve fibers can discharge in synchrony with each other, thereby generating very strong electrical currents. Second, large numbers of neurons can partially discharge, though not emit action potentials; furthermore, these partially discharged neurons can give periods of current flow that can undulate with changing degrees of excitability of the neurons. Simultaneous electrical measurements within the brain while recording brain waves from the scalp indicate that it is the second of these that causes the usual brain waves.

To be more specific, the surface of the cerebral cortex is composed almost entirely of a mat of dendrites extending to the surface from neuronal cells in the lower layers of the cortex. When signals impinge on these dendrites, the dendrites become partially discharged. This partially discharged state makes the neurons of the cortex highly excitable — that is, facilitates them, and the negative potential is simultaneously recorded from the surface of the scalp, indicating this high degree of excitability.

One of the important sources of signals to excite the outer dendritic layer of the cerebral cortex is the ascending reticular activating system. Therefore, brain wave intensity is closely related to the degree of activity in either the brain stem or the thalamic portions of the reticular activating system.

Origin of Delta Waves. Transection of the fiber tracts from the thalamus to the cortex, which blocks the reticular activating system fibers, causes delta waves in the cortex. This indicates that some synchronizing mechanism can occur in the cortical neurons themselves — entirely independently of lower structures in the brain — to cause the delta waves.

Delta waves also occur in very deep "slow wave" sleep; and this suggests that the cortex is then released from the activating influences of the reticular activating system, as was explained earlier in the chapter.

Origin of the Alpha Waves. Alpha waves will *not* occur in the cortex without connections with the thalamus. Also, stimulation in the generalized thalamic nuclei often sets up waves in the generalized thalamocortical system at a frequency of between 8 and 13 per second, the natural frequency of the alpha waves. Therefore, it is assumed that the alpha waves result from spontaneous activity in the generalized thalamocortical system, which causes both the periodicity of the alpha waves and the synchronous activation of literally millions of cortical neurons during each wave.

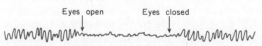

Figure 54–4. Replacement of the alpha rhythm by an asynchronous discharge on opening the eyes.

EFFECT OF VARYING DEGREES OF CEREBRAL ACTIVITY ON THE BASIC RHYTHM OF THE ELECTROENCEPHALOGRAM

There is a general relationship between the degree of cerebral activity and the average frequency of the electroencephalographic rhythm, the average frequency increasing progressively with higher and higher degrees of activity. This is illustrated in Figure 54–5, which shows the existence of delta waves in stupor, surgical anesthesia, and sleep; theta waves in psychomotor states and in infants; alpha waves during relaxed states; and beta waves during periods of intense mental activity. However, during periods of mental activity the waves usually become asynchronous rather than synchronous so that the voltage falls considerably, despite increased cortical activity, as illustrated in Figure 54–4.

CLINICAL USE OF THE ELECTROENCEPHALOGRAM

One of the most important uses of the EEG is to diagnose different types of epilepsy and to find the focus in the brain causing the epilepsy. This is discussed further below. But, in addition, the EEG can be used to localize brain tumors or other space-occupying lesions of the brain and to diagnose certain types of psychopathic disturbances.

Localization of Brain Tumors.   There are two means by which brain tumors can be localized. Some brain tumors are so large that they block electrical activity from a given portion of the cerebral cortex, and when this occurs the voltage of the brain waves is considerably reduced in the region of the tumor. However, more frequently a brain tumor compresses the surrounding neuronal tissue and thereby causes abnormal electrical excitation of these surrounding areas; this in turn leads to synchronous discharges of very high voltage waves in the EEG, as shown in the middle two records of Figure 54–6. Localization of the origin of these spikes on the surface of the scalp is a valuable means for locating the brain tumor.

The upper part of Figure 54–6 shows the placement of 16 different electrodes on the scalp, and the lower part of the figure shows the brain waves from four of these electrodes marked in the figure by Xs. Note that in two of these, intense brain waves are recorded and, furthermore, that the two waves are essentially of reverse polarity to each other. This reverse polarity means that the origin of the spikes is somewhere in the area *between* the two respective electrodes. Thus, the excessively excitable area of the brain has been located, and this is a lead to the location of the brain tumor.

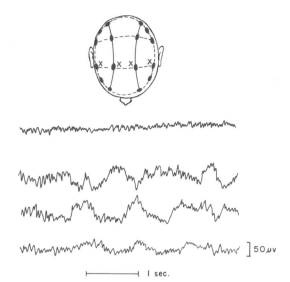

Figure 54–6. Localization of a brain tumor by means of the EEG, illustrating (above) the placement of electrodes and (below) the records from the four electrodes designated by Xs.

Diagnosing Psychopathic Disturbances.   Use of brain waves in diagnosing psychopathic abnormalities is generally not very satisfactory because only a few of these cause distinct brain wave patterns. Yet by observing combinations of different types of basic rhythms, reactions of the rhythms to attention, changes of the rhythms during alkalosis caused by forced breathing, the appearance of particular characteristics in the brain waves (such as "spindles" of alpha waves), and so forth, an experienced electroencephalographer can detect at least certain types of psychopathic disturbances. Also, theta waves are frequently found in persons with brain abnormalities.

EPILEPSY

Epilepsy is characterized by uncontrolled excessive activity of either a part or all of the central nervous system. A person who is predisposed to epilepsy has attacks when the basal level of excitability of the nervous system (or of the part that is susceptible to the epileptic state) rises above a certain critical threshold. But, as long as the degree of excitability is held below this threshold, no attack occurs.

Basically, epilepsy can be classified into three major types: *grand mal epilepsy, petit mal epilepsy,* and *focal epilepsy.*

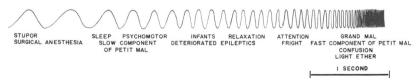

| STUPOR SURGICAL ANESTHESIA | SLEEP PSYCHOMOTOR SLOW COMPONENT OF PETIT MAL | INFANTS DETERIORATED EPILEPTICS | RELAXATION | ATTENTION FRIGHT | GRAND MAL FAST COMPONENT OF PETIT MAL CONFUSION LIGHT ETHER |

I SECOND

Figure 54–5. Effect of varying degrees of cerebral activity on the basic rhythm of the EEG. (From Gibbs and Gibbs: Atlas of Electroencephalography. Addison-Wesley, 1974.)

GRAND MAL EPILEPSY

Grand mal epilepsy is characterized by extreme neuronal discharges in all areas of the brain — in the cortex, in the deeper parts of the cerebrum, and even in all areas of the reticular activating system. Also, discharges are transmitted from the reticular formation into the spinal cord to cause generalized *tonic convulsions* of the entire body, followed toward the end of the attack by alternating muscular contractions called *tonic-clonic convulsions*. Often the person bites or "swallows" the tongue and usually has difficulty in breathing, sometimes to the extent of developing cyanosis. Also, signals to the viscera frequently cause urination and defecation.

The grand mal seizure lasts from a few seconds to as long as three to four minutes and is characterized by post-seizure depression of the entire nervous system; the person remains in stupor for one to many minutes after the attack is over and then often remains severely fatigued for many hours thereafter.

The middle recording of Figure 54–7 illustrates a typical electroencephalogram from almost any region of the cortex during the tonic phase of a grand mal attack. This illustrates that high voltage, synchronous discharges occur over the entire cortex, having almost the same periodicity as the normal alpha waves. Furthermore, the same type of discharge occurs on both sides of the brain at the same time, illustrating that the abnormal neuronal circuitry responsible for the attack strongly involves the reticular activating system itself.

In experimental animals or even in human beings, grand mal attacks can be initiated by administering neuronal stimulants, such as the well-known drug Metrazol, or they can be caused by insulin hypoglycemia or by the passage of alternating electrical current directly through the brain. Electrical recordings from the thalamus and also from the reticular formation of the brain stem during the grand mal attack show typical high voltage activity in both of these areas similar to that recorded from the cerebral cortex. Furthermore, in an experimental animal, even after transecting the brain stem immediately above the mesencephalon, a typical grand mal seizure can still be induced in the portion of the brain stem beneath the transection.

Presumably, therefore, a grand mal attack is caused by abnormal activation of the reticular activating system itself. The synchronous discharges from this region could result from local reverberating circuits or from reverberation back and forth between the reticular activating system and the cortex.

What Initiates a Grand Mal Attack? Most persons who have grand mal attacks have a hereditary predisposition to epilepsy, a predisposition that occurs in about 1 of every 50 people. In such persons, some of the factors that can increase the excitability of the abnormal "epileptogenic" circuitry enough to precipitate attacks are (1) strong emotional stimuli, (2) alkalosis caused by overbreathing, (3) drugs, (4) fever, or (5) loud noises or flashing lights. Also, even in persons not genetically predisposed, traumatic lesions in almost any part of the brain can cause excess excitability of local brain areas as we shall discuss shortly, and these too can transmit signals into the reticular activating system to elicit grand mal seizures.

What Stops the Grand Mal Attack? The cause of the extreme neuronal overactivity during a grand mal attack is presumed to be massive activation of many reverberating pathways throughout the brain. It is presumed that the major factor, or at least one of the major factors, that stops the attack after a few minutes is the phenomenon of neuronal *fatigue*. However, a second factor is probably *active inhibition* by certain structures of the brain. The change from the tonic type of convulsion to the clonic type during the latter part of the grand mal attack has been suggested to result from partial fatigue of the neuronal system so that some of the excited neurons fade out for a moment, then return to activity after a brief rest, only to fatigue a second time and then a third time, and so forth, until the entire seizure is over. The stupor and fatigue that occur after a grand mal seizure is over are believed to result from the intense fatigue of the neurons following their intensive activity during the grand mal attack.

The *active inhibition* that helps to stop the grand mal attack is believed to result from feedback circuits through inhibitory areas of the brain. The grand mal attack undoubtedly excites such areas as the basal ganglia, which in turn emit many inhibitory impulses into the reticular formation of the brain stem. But the nature of such active inhibition is very much a matter of speculation.

PETIT MAL EPILEPSY

Petit mal epilepsy is closely allied to grand mal epilepsy in that it too almost certainly involves the reticular activating system, but probably the thalamocortical portion rather than the mesencephalic portion. It is characterized by 3 to 30 seconds of unconsciousness during which the person has several twitchlike contractions of the muscles, usually in the head region — especially blinking of the eyes; this is followed by return of consciousness and resumption of previous activities. The patient may have one such attack in many months or in rare instances may have a rapid series of attacks, one following the other. However, the usual course is for the petit mal attacks to appear in late childhood and then to disappear entirely by the age of 30. On occasion, a petit mal epileptic attack will initiate a grand mal attack.

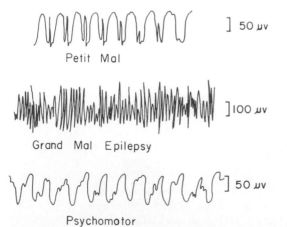

Figure 54–7. Electroencephalograms in different types of epilepsy.

The brain wave pattern in petit mal epilepsy is illustrated by the first record of Figure 54–7, which is typified by a *spike and dome pattern*. The spike portion of this recording is almost identical to the spikes that occur in grand mal epilepsy, but the dome portion is distinctly different. The spike and dome can be recorded over most or all of the cerebral cortex, illustrating that the seizure involves the reticular activating system of the brain.

However, the petit mal attack can sometimes also cause motor twitches or at other times short intervals of psychomotor disturbances. These specific effects suggest that at least some petit mal attacks originate in focal regions of the diffuse thalamic nuclei, possibly in the thalamic intralaminar nuclei.

FOCAL EPILEPSY

Focal epilepsy can involve almost any part of the brain, either localized regions of the cerebral cortex or deeper structures of both the cerebrum and brain stem. And almost always, focal epilepsy results from some localized organic lesion or functional abnormality, such as a scar that pulls on the neuronal tissue, a tumor that compresses an area of the brain, a destroyed area of brain tissue, or congenitally deranged local circuitry. Lesions such as these can promote extremely rapid discharges in the local neurons, and when the discharge rate rises above approximately 1000 per second, synchronous waves begin to spread over the adjacent cortical regions. These waves presumably result from *localized reverberating circuits* that gradually recruit adjacent areas of the cortex into the discharge zone. The process spreads to adjacent areas at a rate as slow as a few millimeters a minute to as fast as several centimeters per second. When such a wave of excitation spreads over the motor cortex, it causes a progressive ''march'' of muscular contractions throughout the opposite side of the body, beginning most characteristically in the mouth region and marching progressively downward to the legs, but at other times marching in the opposite direction. This is called *jacksonian epilepsy*.

A focal epileptic attack may remain confined to a single area of the brain, but in many instances the strong signals from the convulsing cortex or other part of the brain excite the mesencephalic portion of the reticular activating system so greatly that a grand mal epileptic attack ensues as well.

One type of focal epilepsy is the so-called *psychomotor seizure,* which may cause (1) a short period of amnesia, (2) an attack of abnormal rage, (3) sudden anxiety, discomfort, or fear, (4) a moment of incoherent speech or mumbling of some trite phrase, or (5) a motor act to attack someone, to rub the face with the hand, or so forth. Sometimes the person cannot remember his activities during the attack, but at other times he will have been conscious of everything that he had been doing but unable to control it. Attacks of this type characteristically involve part of the limbic portion of the brain, such as the hippocampus, the amygdala, the septum, or the temporal cortex.

The lower tracing of Figure 54–7 illustrates a typical electroencephalogram during a psychomotor attack, showing a low frequency rectangular wave with a frequency between 2 and 4 per second and with superimposed 14 per second waves.

The electroencephalogram can often be used to localize abnormal spiking waves originating in areas of organic brain disease that might predispose to focal epileptic attacks. Once such a focal point is found, surgical excision of the focus frequently prevents future epileptic attacks.

SLEEP AND WAKEFULNESS

Sleep is defined as a state of unconsciousness from which a person can be aroused by appropriate sensory or other stimuli. Therefore, the unconsciousness caused by deep anesthesia, by total inactivity of the reticular activating system in diseased states (coma), and by excessive activity of the reticular activating system as occurs in grand mal epilepsy would not be considered to be sleep. However, coma and anesthesia do have many characteristics similar to those of deep sleep.

Two Different Types of Sleep. There are two different ways in which sleep can occur. First, it can result from decreased activity in the reticular activating system; this is called *slow wave sleep* because the brain waves are very slow. Second, sleep can result from abnormal channeling of signals in the brain even though brain activity may not be significantly depressed; this is called *paradoxical sleep* or *desynchronized sleep*.

Most of the sleep during each night is of the slow wave variety; this is the deep, restful type of sleep that the person experiences after having been kept awake for the previous 24 to 48 hours. On the other hand, short episodes of paradoxical sleep usually occur at intervals during each night, and this type of sleep may have purposeful functions that will be discussed later.

SLOW WAVE SLEEP

Slow wave sleep is frequently called by different names, such as *deep restful sleep, dreamless sleep, delta wave sleep,* or *normal sleep*. However, we shall see later that paradoxical sleep is also normal and that it has some characteristics of deep sleep.

Electroencephalographic Changes as a Person Falls Asleep. Beginning with wakefulness and proceeding to deep slow wave sleep, the electroencephalogram changes as follows:

1. Alert wakefulness — low voltage, high frequency beta waves showing desynchrony, as illustrated by the second record in Figure 54–3;

2. Quiet restfulness — predominance of alpha waves; a type of ''synchronized'' brain waves;

3. Light sleep — slowing of the brain waves mainly to theta or low-voltage variety, but interspersed with spindles of alpha waves called *sleep spindles* that last for a few seconds at a time;

4. Deep slow wave sleep — high voltage delta waves occurring at a rate of 1 to 2 per second, as illustrated by the fourth record of Figure 54–3.

In paradoxical sleep the brain waves change to still a different pattern approaching that of normal wakefulness, as will be discussed below.

Origin of the Delta Waves in Sleep. When the fiber tracts between the thalamus and the cortex are transected, delta waves are generated in the isolated cortex, indicating that this type of wave probably occurs intrinsically in the cortex when it is not being driven from below. Therefore, it is assumed that the degree of activity in the reticular activating system has fallen to a level too low to maintain normal excitability of the cortex so that the cortex then becomes its own pacemaker.

Characteristics of Deep Slow Wave Sleep. Most of us can understand the characteristics of deep slow wave sleep by remembering the last time that we were kept awake for more than 24 hours and then the deep sleep that occurred within 30 minutes to an hour after going to sleep. This sleep is dreamless, exceedingly restful, and is associated with a decrease in both peripheral vascular tone and also most of the other vegetative functions of the body as well. There is also a 10 to 30 per cent decrease in blood pressure, respiratory rate, and basal metabolic rate.

PARADOXICAL SLEEP (REM SLEEP)

In a normal night of sleep, bouts of paradoxical sleep lasting 5 to 20 minutes usually appear on the average every 90 minutes, the first such period occurring 80 to 100 minutes after the person falls asleep. When the person is extremely tired, the duration of each bout of paradoxical sleep is very short, and it may even be absent. On the other hand, as the person becomes more rested through the night, the duration of the paradoxical bouts greatly increases.

There are several very important characteristics of paradoxical sleep:

1. It is usually associated with active dreaming.

2. The person is even more difficult to arouse than during deep slow wave sleep.

3. The muscle tone throughout the body is exceedingly depressed, indicating strong inhibition of the spinal projections from the reticular activating system.

4. The heart rate and respiration usually become irregular, which is characteristic of the dream state.

5. Despite the extreme inhibition of the peripheral muscles, a few irregular muscle movements occur. These include, in particular, rapid movements of the eyes; consequently, paradoxi-cal sleep has often been called *REM sleep,* for "rapid eye movements."

6. The electroencephalogram shows a desynchronized pattern of low voltage beta waves similar to those that occur during wakefulness. Therefore, this type of sleep is also frequently called *desynchronized sleep,* meaning desynchronized brain waves.

In summary, paradoxical sleep is a type of sleep in which the brain is quite active. However, the brain activity is not channeled in the proper direction for the persons to be aware of their surroundings and therefore to be awake.

BASIC THEORIES OF SLEEP AND WAKEFULNESS

Role of the Reticular Activating System in Sleep and Wakefulness. From the discussion of the reticular activating system earlier in the chapter, it should already be clear that stimulation of this system produces the state of wakefulness. But, what causes sleep? In deep, slow wave sleep, transmission of signals from the reticular activating system to the cortex is greatly diminished, indeed almost absent. Therefore, slow wave sleep presumably results from decreased activity of the reticular activating system.

On the other hand, paradoxical sleep is, as its name implies, paradoxical because some areas of the cerebrum are quite active despite the state of sleep. Therefore, it is assumed that this type of sleep results from a curious mixture of activation of some brain regions while other regions are still suppressed.

Yet, the questions that must be answered before we will understand wakefulness and sleep are: (1) What are the mechanisms that activate the reticular activating system during wakefulness? And (2) what suppresses this system during sleep?

Neuronal Centers, Transmitters, and Mechanisms That Cause Wakefulness

Most of the factors that can cause wakefulness have already been discussed in relation to the reticular activating system. However, let us quickly review these:

1. Stimulation of the medial portion of the reticular formation, especially in the mesencephalon and upper pons, will cause intense wakefulness.

2. Widespread stimulation of sensory nerves throughout the body will also cause wakefulness. These nerves transmit strong signals into the mesencephalic portion of the reticular activating system.

3. Stimulation of most areas of the cerebral

cortex will also cause a high level of wakefulness. These areas also transmit strong signals into both the mesencephalic and thalamic portions of the reticular activating system.

4.   Stimulation in certain regions of the hypothalamus, especially in the lateral regions, can also cause extreme degrees of wakefulness. Here again, strong signals are known to be transmitted into the reticular activating system.

5.   It is believed that excitation of the area in the pons called the *locus ceruleus* is especially important in maintaining activity in the reticular activating system. The locus ceruleus lies bilaterally immediately beneath the floor of the fourth ventricle. It is a collection of neurons that is considered to be part of the reticular formation though these neurons perform specific functions as follows: The nerve fibers from this area are distributed widely throughout other portions of the reticular formation and throughout almost all areas of the diencephalon and cerebrum as well. They all secrete *norepinephrine* at their endings. It is believed that this norepinephrine in some way plays a role in the wakefulness process. It has been suggested that dopamine and epinephrine, both of which are very similar to norepinephrine, might also contribute to wakefulness because neurons in closely allied regions of the brain stem secrete these transmitter substances and seem to be activated, in many instances, along with the norepinephrine system.

Effect of Lesions in the Wakefulness Areas.   Lesions in the mesencephalic portion of the reticular activating system or in fiber pathways leading upward from this area through the diencephalon, if large enough, will invariably lead to coma from which the person cannot be aroused with any type of stimuli.

Also, very discrete lesions located bilaterally in the locus ceruleus will cause a type of sleep that closely resembles natural sleep.

Neuronal Centers, Transmitters, and Mechanisms That Can Cause Sleep

Stimulation of several specific areas of the brain can produce sleep with characteristics very near to those of natural sleep. Some of these are the following:

1.   The most conspicuous stimulation area for causing almost natural sleep is the raphe nuclei in the pons and medulla. These are a thin sheet of nuclei located in the midline. Nerve fibers from these nuclei spread widely in the reticular formation and also upward into the thalamus, hypothalamus, and most areas of the limbic cortex. In addition, they extend downward into the spinal cord, terminating in the posterior horns where they can inhibit incoming pain signals, as was discussed in Chapter 50. It is also known that the endings of fibers from these raphe neurons secrete *serotonin*. Therefore, it is assumed that serotonin is the major transmitter substance associated with production of sleep.

2.   Stimulation of several other regions in the lower brain stem and diencephalon can also lead to sleep, including (a) the rostral part of the hypothalamus, mainly in the suprachiasmal area, and (b) an occasional area in the diffuse nuclei of the thalamus.

Effect of Lesions in the Sleep-Promoting Centers.   Discrete lesions in the raphe nuclei lead to a high state of wakefulness. This is also true of bilateral lesions in the mediorostral suprachiasmal portion of the anterior hypothalamus. In both instances, the reticular activating system seems to become released from inhibition. Indeed, the lesions of the anterior hypothalamus can sometimes cause such intense wakefulness that the animal actually dies of exhaustion.

Other Possible Transmitter Substances Related to Sleep.   Experiments have shown that the cerebrospinal fluid and also the blood of animals that have been kept awake for several days contain a substance or substances that cause sleep when injected into the ventricular system of an animal. One of these substances is a small polypeptide with a molecular weight of less than 500. When cerebrospinal fluid containing this sleep-producing substance or substances is injected into the third ventricle, almost natural sleep occurs within a few minutes, and the animal may then stay asleep for several hours. Therefore, it is possible that prolonged wakefulness causes progressive accumulation of a sleep factor in the brain stem or in the cerebrospinal fluid that leads to sleep.

The Cycle Between Sleep and Wakefulness

The above discussions have merely identified neuronal areas, transmitters, and mechanisms that are related either to wakefulness or to sleep. However, they have not explained the cyclic, reciprocal operation of the sleep-wakefulness cycle. It is quite possible that this is caused by a free running intrinsic oscillator within the brain stem that cycles back and forth between the sleep and wakefulness centers, the wakefulness centers presumably activating the reticular activating system, and the sleep centers inhibiting this system.

However, there is much reason to believe that feedback signals from the cerebral cortex and also from the peripheral nerve receptors might also play a very important role in causing the sleep-wakefulness rhythm. One reason for believing this is that interruption of all the sensory nerve tracts leading from the periphery to the reticular activating system will cause the animal to go to sleep permanently. Also, as explained above, stimulation of the cerebral cortex will cause powerful activation of the reticular activating system.

Therefore, a very likely mechanism for causing the rhythmicity of the sleep-wakefulness cycle is the following:

When the reticular activating system is completely rested and the sleep centers are not activated, the wakefulness centers then presumably begin spontaneous activity. This in turn excites both the cerebral cortex and the peripheral nervous system. Next, positive feedback signals come from both of these areas back to the reticular activating system to activate it still further. Thus, once the wakefulness state begins, it has a natural tendency to sustain itself.

However, after the brain remains activated for many hours, even the neurons within the activating system presumably will fatigue, or other factors might activate the sleep centers. Consequently, the positive feedback cycle between the reticular activating system and the cortex, and also that between the reticular activating system and the periphery, will begin to fade. As soon as a few of the neurons in the reticular activating system become inactive, this also eliminates part of the feedback stimulus to the other neurons as well. Therefore, these too become inactive, the process spreading rapidly through the neurons and leading to rapid transition from the wakefulness state to the sleep state.

Then, one could postulate that during sleep the excitatory neurons of the reticular activating system gradually become more and more excitable because of the prolonged rest, while the inhibitory neurons of the sleep centers become less excitable, thus leading to a new cycle of wakefulness.

This theory obviously can explain the rapid transitions from sleep to wakefulness and from wakefulness to sleep. It can also explain arousal, the insomnia that occurs when a person's mind becomes preoccupied with a thought, the wakefulness that is produced by bodily activity, and many other conditions that affect the person's state of sleep or wakefulness.

PHYSIOLOGICAL EFFECTS OF SLEEP

Sleep causes two major types of physiological effects: first, effects on the nervous system itself and, second, effects on other structures of the body. The first of these seems to be by far the more important, for any person who has a transected spinal cord in the neck shows no physiological effects in the body beneath the level of transection that can be attributed to a sleep and wakefulness cycle; that is, lack of sleep and wakefulness causes neither significant harm to the

bodily organs nor even any deranged function. On the other hand, lack of sleep certainly does affect the functions of the central nervous system.

Prolonged wakefulness is often associated with progressive malfunction of the mind and behavioral activities of the nervous system. We are all familiar with the increased sluggishness of thought that occurs toward the end of a prolonged wakeful period, but in addition, a person can become irritable or even psychotic following forced wakefulness for prolonged periods of time. Therefore, one can assume that sleep in some way not presently understood restores both normal sensitivities of and normal "balance" among the different parts of the central nervous system. This might be likened to the "rezeroing" of electronic analog computers after prolonged use, for all computers of this type gradually lose their "base line" of operation; it is reasonable to assume that the same effect occurs in the central nervous system, because overuse of some neurons during wakefulness could easily throw all these out of balance with the remainder of the nervous system. Therefore, in the absence of any definitely demonstrated functional value of sleep, we might postulate, on the basis of known psychological changes that occur with prolonged wakefulness, that sleep performs this rezeroing function for the nervous system to re-establish the natural balance among the neuronal centers.

Even though, as pointed out above, wakefulness and sleep have not been shown to be necessary for somatic functions of the body, the cycle of enhanced and depressed nervous excitability that follows the cycle of wakefulness and sleep does have moderate effects on the peripheral body. For instance, there is enhanced sympathetic activity during wakefulness and also enhanced numbers of impulses to the skeletal musculature to increase muscle tone. Conversely, during sleep, sympathetic activity decreases while parasympathetic activity occasionally increases, and the muscle tone becomes almost nil. Therefore, arterial blood pressure falls, pulse rate decreases, skin vessels dilate, activity of the gastrointestinal tract sometimes increases, muscles fall into a completely relaxed state, and overall basal metabolic rate of the body falls by 10 to 30 per cent.

REFERENCES

Arkin, J. S., et al. (eds.): The Mind in Sleep. New York, Halsted Press, 1978.

Begleiter, H. (ed.): Evoked Brain Potentials and Behavior. New York, Plenum Press, 1979.

Block, G. D., and Page, T. L.: Circadian pacemakers in the nervous system. Annu. Rev. Neurosci., 1:19, 1978.

Buser, P.: Higher functions of the nervous system. Annu. Rev. Physiol., 38:217, 1978.

Buser, P. A., and Rougeul-Buser, A. (eds.): Cerebral Correlates of Conscious Experience. New York, Elsevier/North-Holland, 1978.

Cartwright, R. D.: A Primer on Sleep and Dreaming. Reading, Mass., Addison-Wesley Publishing Co., 1978.

Diehl, L. W.: Treatment of Complicated Epilepsies in Adults: A Clinical-Statistical Study. New York, S. Karger, 1978.

Drucker-Colín, R., et al. (eds.): The Functions of Sleep. New York, Academic Press, 1979.

Edelman, G. M., and Mountcastle, V. B.: The Mindful Brain. Cambridge, Mass., MIT Press, 1978.

Enright, J. T.: The Timing of Sleep and Wakefulness: On the Substructure and Dynamics of the Circadian Pacemakers Underlying the Wake-Sleep Cycle. New York, Springer-Verlag, 1979.

Fröscher, W.: Treatment of Status Epilepticus. Baltimore, University Park Press, 1978.

Gillin, J. C., et al.: The neuropharmacology of sleep and wakefulness. Annu. Rev. Pharmacol. Toxicol., 18:563, 1978.

Glaser, G. H., et al. (eds.): Antiepileptic Drugs. New York, Raven Press, 1980.

Goldberg, P., and Kaufman, D.: Natural Sleep. Emmaus, Pa., Rodale Press, 1978.

Guilleminault, C., and Dement, W. C. (eds.): Sleep Apnea Syndromes. New York, A. R. Liss, 1978.

Hector, M. L.: EEG Recording. Boston, Butterworths, 1979.

Hobson, J. A., and Brazier, M. A. B. (eds.): The Reticular Formation Revisited: Specifying Function for a Nonspecific System. New York, Raven Press, 1980.

Ito, M., et al. (eds.): Integrative Control Functions of the Brain. New York, Elsevier/North-Holland, 1978.

Klass, D. W., and Daly, D. C. (eds.): Current Practice of Clinical Electroencephalography. New York, Raven Press, 1979.

Kool, K. A., et al.: Fundamentals of Electroencephalography. Hagerstown, Md., Harper & Row, 1978.

Livingston, R. B.: Neural integration. In Frohlich, E. D. (ed.): Pathophysiology, 2nd Ed. Philadelphia, J. B. Lippincott Co., 1976, p. 681.

Livingston, R. B.: Sensory Processing, Perception, and Behavior. New York, Raven Press, 1978.

Moore, R. Y., and Bloom, F. E.: Central catecholamine neuron systems: Anatomy and physiology of the norepinephrine and epinephrine systems. Annu. Rev. Neurosci., 2:113, 1979.

Newmark, M. E., and Penry, J. K.: Photosensitivity and Epilepsy: A Review. New York, Raven Press, 1979.

Newmark, M. E., and Penry, J. K.: Genetics of Epilepsy: A Review. New York, Churchill Livingstone, 1980.

O'Keefe, J., and Nadel, L.: The Hippocampus as a Cognitive Map. New York, Oxford University Press, 1978.

Pappenheimer, J. R.: The sleep factor. Sci. Am., 235(2):24, 1976.

Passouant, P., and Oswald, I. (eds.): Pharmacology of the States of Alertness. New York, Pergamon Press, 1979.

Plum, F., and Posner, J. B.: The Diagnosis of Stupor and Coma, 3rd Ed. Philadelphia, F. A. Davis, 1980.

Prince, D. A.: Neurophysiology of epilepsy. Annu. Rev. Neurosci., 1:395, 1978.

Regan, D.: Electrical responses evoked from the human brain. Sci. Am., 241(6):134, 1979.

Roth, S. H.: Physical mechanisms of anesthesia. Annu. Rev. Pharmacol. Toxicol., 19:159, 1979.

Schneider, A. M., and Tarshis, B.: An Introduction to Physiological Psychology. New York, Random House, 1979.

Singer, W.: Control of thalamic transmission by corticofugal and ascending reticular pathways in the visual system. Physiol. Rev., 57:386, 1977.

Spiegel, R.: Sleep and Sleeplessness in Advanced Age. Jamaica, N.Y., Spectrum Publications, 1979.

Stern, R. M., et al.: Psychophysiological Recording. New York, Oxford University Press, 1980.

Wada, J. A. (ed.): Modern Perspectives in Epilepsy: Proceedings of the Inaugural Symposium of the Canadian League Against Epilepsy. St. Albans, Vt., Eden Press, 1978.

55

The Cerebral Cortex and Intellectual Functions of the Brain

It is ironic that we know least about the mechanisms of the cerebral cortex of almost all parts of the brain, even though it is by far the largest portion of the nervous system. Yet, we do know the effects of destruction or of specific stimulation of various portions of the cortex, and still more is known from electrical recordings from the cortex or from the surface of the scalp. In the early part of the present chapter the facts known about cortical functions are discussed and then some basic theories of the neuronal mechanisms involved in thought processes, memory, analysis of sensory information, and so forth are presented briefly.

PHYSIOLOGIC ANATOMY OF THE CEREBRAL CORTEX

The functional part of the cerebral cortex is comprised mainly of a thin layer of neurons 2 to 5 mm. in thickness, covering the surface of all the convolutions of the cerebrum and having a total area of about one-quarter square meter. The total cerebral cortex probably contains about 100 billion neurons.

Figure 55–1 illustrates the typical structure of the cerebral cortex, showing successive layers of different types of cells. Most of the cells are of three types: *granular, fusiform,* and *pyramidal,* the latter named for their characteristic pyramidal shape. To the right in Figure 55–1 is illustrated the typical organization of nerve fibers within the different layers of the cortex. Note particularly the large number of horizontal fibers extending between adjacent areas of the cortex, but note also the vertical fibers that extend to and from the cortex to lower areas of the brain stem or to distant regions of the cerebral cortex through long association bundles of fibers.

Neurohistologists have divided the cerebral cortex into almost 100 different areas which have slightly different architectural characteristics. Yet in all these different areas except the hippocampal region there still persist representations of all the six major layers of the cortex. To the untrained histologist, only five major architectural types of cortex can be distinguished as follows: Type 1 contains large numbers of pyramidal cells with few granular cells and, therefore, is often called the *agranular cortex.* At the other extreme, type 5 contains almost no pyramidal cells but is filled with closely packed granule cells and is called the *granular cortex.* Types 2, 3, and 4 have intermediate characteris-

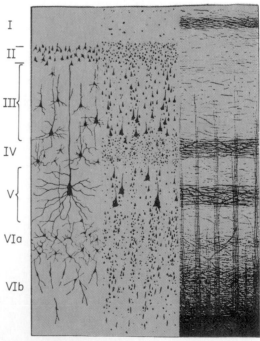

Figure 55–1. Structure of the cerebral cortex, illustrating: *I,* molecular layer; *II,* external granular layer; *III,* layer of pyramidal cells; *IV,* internal granular layer; *V,* large pyramidal cell layer; *VI,* layer of fusiform or polymorphic cells. (From Ranson and Clark (after Brodmann): Anatomy of the Nervous System. Philadelphia, W. B. Saunders Company, 1959.)

684

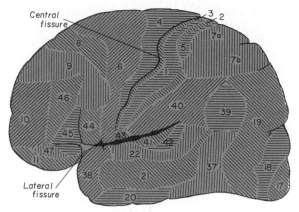

Figure 55–2. Structurally distinct areas of the human cerebral cortex. (From Brodmann, modified by Buchanan: Functional Neuroanatomy. Lea & Febiger, 1966.)

tics containing gradated proportions of pyramidal and agranular cells.

The agranular cortex, which contains large numbers of pyramidal cells, is characteristic of the motor areas of the cerebral cortex, while the granular cortex, containing almost no pyramidal cells, is characteristic of the primary sensory areas. The other three intermediate types of cortex are characteristic of the association areas between the primary sensory and motor regions.

Figure 55–2 shows a map of cortical areas classified on the basis of histological studies by Brodmann. This classification provides a basis for discussion of functional areas of the brain, even though different architectural types of cortex often have similar functions.

Anatomical Relationship of the Cerebral Cortex to the Thalamus and Other Lower Centers. All areas of the cerebral cortex have direct afferent and efferent connections with the thalamus. Figure 55–3 shows the areas of the cerebral cortex connected with specific parts of the thalamus. These connections are in *two* directions, both from the thalamus to the cortex and then from the cortex back to essentially the same area of the thalamus. Furthermore, when the thalamic connections are cut, the functions of the corresponding cortical area become

entirely or almost entirely abrogated. Therefore, the cortex operates in close association with the thalamus and can almost be considered both anatomically and functionally to be a large outgrowth of the thalamus; for this reason the thalamus and the cortex together are called the *thalamocortical system,* as was explained in the previous chapter. Also, all pathways from the sensory organs to the cortex pass through the thalamus, with the single exception of the sensory pathways of the olfactory tract.

FUNCTIONS OF CERTAIN SPECIFIC CORTICAL AREAS

Studies in human beings by neurosurgeons have shown that some specific functions are localized to certain general areas of the cerebral cortex. Figure 55–4 gives a map of some of these areas as determined by Penfield and Rasmussen from electrical stimulation of the cortex or by neurological examination of patients after portions of the cortex had been removed. The lightly shaded areas are *primary sensory areas,* while the darkly shaded area is the *voluntary motor area* (also called *primary motor area*) from which muscular movements can be elicited with relatively weak electrical stimuli. These primary sensory and motor areas have highly specific functions, while other areas of the cortex perform more general functions that we call association or cerebration.

SPECIFIC FUNCTIONS OF THE PRIMARY SENSORY AREAS

The primary sensory areas all have certain functions in common. For instance, somatic sensory areas, visual sensory areas, and auditory sensory areas all have spatial localizations of

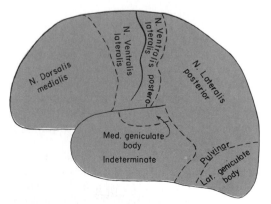

Figure 55–3. Areas of the cerebral cortex that connect with specific portions of the thalamus. (Modified from Elliott: Textbook of the Nervous System. J. B. Lippincott Co.)

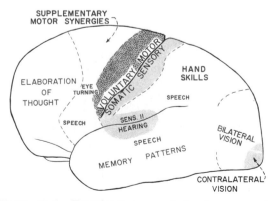

Figure 55–4. Functional areas of the human cerebral cortex as determined by electrical stimulation of the cortex during neurosurgical operations and by neurological examinations of patients with destroyed cortical regions. (From Penfield and Rasmussen: The Cerebral Cortex of Man: A Clinical Study of Localization of Function. New York, Macmillan Company, 1968.)

signals from the peripheral receptors (which are discussed in detail in Chapters 49, 60, and 61).

Electrical stimulation of the primary sensory areas in the parietal lobes in awake patients elicits relatively uncomplicated sensations. For instance, in the somatic sensory area the patient expresses feelings of tingling in the skin, numbness, mild "electric" feeling, or, rarely, mild degrees of temperature sensations. And these sensations are localized to discrete areas of the body in accord with the spatial representation in the somatic sensory cortex, as described in Chapter 49. Therefore, it is believed that the primary somatic sensory cortex analyzes only the simple aspects of sensations and that analysis of intricate patterns of sensory experience requires also adjacent parts of the parietal lobes called the *sensory association areas*.

Electrical stimulation of the primary visual cortex in the occipital lobes causes the person to see flashes of light, bright lines, colors, or other simple visions. Here again, the visual images are localized to specific regions of the visual fields in accord with the portion of the primary visual cortex stimulated, as described in Chapter 60. But the visual cortex alone is not capable of complete analysis of complicated visual patterns; for this, the visual cortex must operate in association with adjacent regions of the occipital cortex, the *visual association areas*.

Electrical stimulation of the auditory cortex in the temporal lobes causes a person to hear a simple sound which may be weak or loud, have low or high frequency, or have other uncomplicated characteristics, such as a squeak or even an undulation. But never are words or any other fully intelligible sound heard. Thus, the primary auditory cortex, like the other primary sensory areas, can detect the individual elements of auditory experience but cannot analyze complicated sounds. Therefore, the primary auditory cortex alone is not sufficient to give one even the usual auditory experiences; these can be achieved, however, when the primary area operates together with the *auditory association areas* in adjacent regions of the temporal lobes.

Despite the inability of the primary sensory areas to analyze the incoming sensations fully, when these areas are destroyed, the ability of the person to utilize the respective sensations usually suffers drastically. For instance, loss of the primary visual cortex in one occipital lobe causes a person to become blind in the ipsilateral halves of both retinae, and loss of the primary visual cortices in both hemispheres causes total blindness. Likewise, loss of both primary auditory cortices causes almost total deafness. Loss of the postcentral gyri causes *depression* of somatic sensory sensations — though not total loss — presumably

because of additional cortical representation of these sensations in other cortical areas: somatic sensory area II and the motor cortex, for example. (In animals far down the phylogenetic scale, loss of the visual and auditory cortices may have little effect on vision and hearing, and even an anencephalic human infant having no brain above the mesencephalon detects some visual scenes in the absence of the visual cortex, for such a child can observe a moving object and even follow it by movement of the eyes and head.)

Therefore, we can summarize the functions of the primary sensory areas of the human cerebral cortex in the following way: The lower centers of the brain relay a large part of the sensory signals to the cerebral cortex for analysis. In turn, the primary sensory areas transmit the results of their analyses back to the lower centers and to other regions of the cerebral cortex, as is discussed later in the chapter.

THE SENSORY ASSOCIATION AREAS

Around the borders of the primary sensory areas are regions called *sensory association areas* or *secondary sensory areas*. In general, these areas extend 1 to 5 centimeters in one or more directions from the primary sensory areas; each time a primary area receives a sensory signal, secondary signals spread, after a delay of a few milliseconds, into the respective association area as well. Part of this spread occurs directly from the primary area through subcortical fiber tracts, but a major part also occurs in the thalamus, beginning in the sensory relay nuclei, passing next to corresponding *thalamic association areas,* and then traveling to the association cortex.

The general function of the sensory association areas is to provide a higher level of interpretation of the sensory experiences. The general areas for the interpretative functions for somatic, visual, and auditory experiences are illustrated in Figure 55–5.

Destruction of the sensory association areas greatly reduces the capability of the brain to analyze different characteristics of sensory experiences. For instance, damage in the temporal lobe below and behind the primary auditory area in the "dominant hemisphere" of the brain often causes a person to lose his ability to understand words or other auditory experiences even though he hears them.

Likewise, destruction of the visual association area in Brodmann's areas 18 and 19 of the occipital lobe in the dominant hemisphere (see Figure 55–2), or the presence of a brain tumor or other lesion in these areas, does not cause blindness or prevent normal activation of the primary visual cortex but does greatly reduce the person's ability to inter-

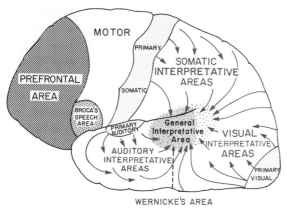

Figure 55–5. Organization of the somatic, auditory, and visual association areas into a general mechanism for interpretation of sensory experience. All of these feed into the *general interpretative area* located in the postero-superior portion of the temporal lobe and the angular gyrus. Note also the prefrontal area, and Broca's speech area.

pret what is seen. Such a person often loses the ability to recognize the meanings of words, a condition that is called *word blindness* or *dyslexia*.

Finally, destruction of the somatic sensory association area in the parietal cortex posterior to primary somatic area I causes the person to lose spatial perception for location of the different parts of the body. In the case of the hand that has been "lost," the skills of the hand are greatly reduced. Thus, this area of the cortex seems to be necessary for interpretation of somatic sensory experiences.

The functions of the association areas are described in more detail in Chapter 49 for somatic, in Chapter 60 for visual, and in Chapter 61 for auditory experiences.

INTERPRETATIVE FUNCTION OF THE POSTERIOR SUPERIOR TEMPORAL LOBE — THE GENERAL INTERPRETATIVE AREA (OR WERNICKE'S AREA)

The somatic, visual, and auditory association areas, which can actually be called "interpretative areas," all meet one another in the posterior part of the superior temporal lobe and in the anterior part of the angular gyrus where the temporal, parietal, and occipital lobes all come together. This area of confluence of the different sensory interpretative areas is especially highly developed in the dominant side of the brain — the *left side* in right-handed persons — and it plays the greatest single role of any part of the cerebral cortex in the higher levels of brain function that we call *cerebration*. Therefore, this region has frequently been

called by different names suggestive of the area having almost global importance: the *general interpretative area*, the *gnostic area*, the *knowing area*, the *tertiary association area*, and so forth. The temporal portion of the general interpretative area is also called *Wernicke's area* in honor of the neurologist who first described its special significance in intellectual processes.

Following severe damage in the general interpretative area, a person might hear perfectly well and even recognize different words but still might be unable to arrange these words into a coherent thought. Likewise, the person may be able to read words from the printed page but be unable to recognize the thought that is conveyed. In addition, the person has similar difficulties in understanding the higher levels of meaning of somatic sensory experiences, even though there is no loss of sensation itself.

Electrical stimulation in the posterior superior temporal lobe of the conscious patient occasionally causes a highly complex thought. This is particularly true when the stimulatory electrode is passed deep enough into the brain to approach the corresponding connecting areas of the thalamus. The types of thoughts that might be experienced include memories of complicated visual scenes that one might remember from childhood, auditory hallucinations such as a specific musical piece, or even a discourse by a specific person. For this reason it is believed that complicated memory patterns involving more than one sensory modality are stored at least partially in the temporal lobe. This belief is in accord with the importance of the general interpretative area in interpretation of the complicated meanings of different sensory experiences.

The angular gyrus portion of the general interpretative area is especially important for interpretation of visual information. If this region is destroyed while the temporal lobe portion of this interpretative area (Wernicke's area) is still intact, the person can still interpret both auditory and somatic experiences as usual, but the stream of visual experiences passing into the general interpretative area from the visual cortex is mainly blocked. Therefore, the person may be able to see words and even know they are words but, nevertheless, not be able to interpret their meanings. This is the condition called *dyslexia*.

Let us again emphasize the global importance of the general interpretative area for most intellectual functions of the brain. Loss of this area in an adult usually leads thereafter to a lifetime of almost demented existence.

The Dominant Hemisphere. The general interpretative functions of Wernicke's area and of the angular gyrus, and also the functions of the speech and motor control areas, are usually much more highly developed in one cerebral hemisphere than in the other. This is called the *dominant hemisphere*. In at least 9 of 10 persons the left hemisphere is the dominant one. At birth, Wernicke's area of the brain is often as much as 50 per cent larger in the left hemisphere than in the right. Therefore, it is easy to understand why the left side of the brain might become dominant over the right side. However, if for some reason the dominant Wernicke's area is removed in early childhood, the opposite side of the brain can develop full dominant characteristics.

A theory that can explain the capability of one hemisphere to dominate the other hemisphere is the following:

The attention of the "mind" seems to be directed to one portion of the brain at a time, as was explained in the preceding chapter. Presumably, because its size is usually larger at birth, the left temporal lobe normally begins to be used to a greater extent than the right, and, thenceforth, because of the tendency to direct one's attention to the better developed region, the rate of learning in the cerebral hemisphere that gains the first start increases rapidly while that in the opposite side remains slight. Therefore, in the normal human being, one side becomes dominant over the other.

In more than 9 of 10 persons the left temporal lobe and angular gyrus become dominant, and in the remaining one tenth of the population either both sides develop simultaneously to have dual dominance, or, more rarely, the right side alone becomes highly developed.

Usually associated with the dominant temporal lobe and angular gyrus is dominance of certain portions of the somesthetic cortex and motor cortex for control of voluntary motor functions. For instance, as is discussed later in the chapter, the premotor speech area (Broca's area), located far laterally in the intermediate frontal area, is almost always dominant also on the left side of the brain. This speech area causes the formation of words by exciting simultaneously the laryngeal muscles, the respiratory muscles, and the muscles of the mouth.

Though the interpretative areas of the temporal lobe and angular gyrus, as well as many of the motor areas, are highly developed in only a single hemisphere, they are capable of receiving sensory information from both hemispheres and are also capable of controlling motor activities in both hemispheres, utilizing mainly fiber pathways in the *corpus callosum* for communication between the two hemispheres. This unitary, cross-feeding organization prevents interference between the two sides of the brain; such interference, obviously, could create havoc with both thoughts and motor responses.

Role of Language in Function of the General Interpretative Area and in Intellectual Functions. A major share of our sensory experience is converted into their language equivalent before being stored in the memory areas of the brain and before being processed for other intellectual purposes. For instance, when we read a book, we do not store the visual images of the words but, instead, store the words themselves in language form. Also, the information is usually converted to language form before its meaning is discerned.

The sensory area of the dominant hemisphere for interpretation of language is closely associated with the primary hearing area located in the auditory association areas of the temporal lobe. This locus probably results from the fact that the first introduction to language is by way of hearing. Later in life, when visual perception of language through the medium of reading develops, the visual information is then presumably channeled into the already developed language regions of the dominant temporal lobe. This probably also explains why the general interpretative area of the brain is more closely allied with the auditory association areas than with other sensory areas of the cortex.

Wernicke's Area in the Nondominant Hemisphere. When Wernicke's area in the dominant hemisphere is destroyed, the person normally loses almost all intellectual functions associated with language or symbolism, such as ability to read, ability to perform mathematical operations, and even the ability to think through logical problems. However, other types of interpretative capabilities, some of which undoubtedly utilize the temporal lobe and angular gyrus regions of the opposite hemisphere, are retained. Psychological studies in patients with damage to their nondominant hemispheres have suggested that this hemisphere may be especially important for understanding and interpreting music, nonverbal visual experiences, spatial relationships between the person and the surroundings, and probably also interpreting many somatic experiences related to use of the limbs and hands. Thus, even though we speak of the "dominant" hemisphere, this dominance is primarily for language- or symbolism-related intellectual functions; the opposite hemisphere can actually be dominant for some other types of intelligence.

An Area for Recognition of Faces

An interesting type of brain abnormality called *prosophenosia* is the inability to recognize faces. This occurs in persons who have extensive damage on the medial undersides of both occipital lobes and along the medioventral surfaces of the temporal lobes as illustrated in Figure 55–6. Loss of these face recognition areas, strangely enough, results in very little other abnormality of brain function.

One wonders why so much of the cerebral cortex should be reserved for the simple task of face recognition. However, when it is remembered that most of our daily tasks involve associations with other people, one can see the importance of this intellectual function.

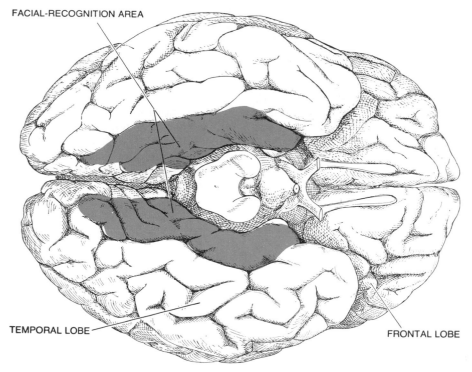

FACIAL-RECOGNITION AREA

TEMPORAL LOBE

FRONTAL LOBE

Figure 55–6. Facial-recognition areas located on the underside of the brain in the medial occipital and temporal lobes. (From Geschwind: *Sci. Amer. 241*:180, 1979. © 1970 by Scientific American, Inc. All rights reserved.)

The occipital portion of this area is contiguous with the primary visual cortex, and the temporal portion is closely associated with the limbic system that has to do with emotions, brain activation, and control of one's behavioral response to the environment.

THE PREFRONTAL AREAS

The prefrontal areas are those portions of the frontal lobes that lie anterior to the motor regions, as shown in Figure 55–5. For years, this part of the cortex has been considered to be the locus of the higher intellect of the human being, principally because the main difference between the brain of monkeys and that of human beings is the great prominence of the human prefrontal areas. Yet efforts to show that the prefrontal cortex is more important in higher intellectual functions than other portions of the cortex have not been successful. Indeed, destruction of the posterior temporal lobe and angular gyrus region in the dominant hemisphere causes infinitely more harm to the intellect than does destruction of both prefrontal areas.

Psychological studies in lower animals have shown that all portions of the cortex not immediately associated with either sensory or motor functions are important in the ability of an animal to learn complicated information — such as a rat's learning to run through a maze — and that all the different areas are approximately equipotent in this regard. Therefore, the importance of the prefrontal areas to the human being is perhaps mainly that they supply much additional cortical area in which cerebration can take place.

Yet, the prefrontal areas do seem to have some specific functions that are all their own. One of these has to do with control of some types of behavior, especially choice of behavioral options for each social or physical situation. For this purpose, the prefrontal areas transmit signals into the limbic areas of the brain, as will be described in the following two chapters. Some of the other possible functions of the prefrontal lobes include:

Prevention of Distractibility by the Prefrontal Areas — Importance for Sequencing Thoughts. One of the outstanding characteristics of a person who has lost the prefrontal areas is the ease with which he can be *distracted* from a sequence of thoughts. Likewise, in lower animals whose prefrontal areas have been removed, the ability to concentrate on psychological tests is almost completely lost. The human being without prefrontal areas is still capable of performing many intellectual tasks, such as answering short questions and performing simple arithmetic computations (such as $9 \times 6 = 54$), thus illustrating that the basic intellectual activities of the cerebral cortex are still intact without the prefrontal areas.

Yet if concerted *sequences* of cerebral functions are required of the person, he becomes completely disorganized. Therefore, the prefrontal areas seem to be important in keeping the mental functions directed toward goals.

Elaboration of Thought, Prognostication, and Performance of Higher Intellectual Functions by the Prefrontal Areas. Another function that has been ascribed to the prefrontal areas by psychologists and neurologists is *elaboration of thought*. This means simply an increase in depth and abstractness of the different thoughts. Psychological tests have shown that prefrontal lobectomized lower animals presented with successive bits of sensory information fail to store these bits in memory — probably because they are distracted so easily that they cannot hold thoughts long enough for storage to take place. If the prefrontal areas are intact, many such successive bits of information can be stored in all areas of the brain and can be called forth again and again during the subsequent periods of cerebration. This postulated ability of the prefrontal areas to cause storage — even though it be temporary — of many types of information simultaneously, and then perhaps also to cause recall of this information, could well explain the many functions of the brain that we associate with higher intelligence, such as the abilities to (1) prognosticate, (2) plan for the future, (3) delay action in response to incoming sensory signals so that the sensory information can be weighed until the best course of response is decided, (4) consider the consequences of motor actions even before these are performed, (5) solve complicated mathematical, legal, or philosophical problems, (6) correlate all avenues of information in diagnosing rare diseases, and (7) control one's activities in accord with moral laws.

Effects of Destruction of the Prefrontal Areas

The person without prefrontal areas ordinarily acts precipitously in response to incoming sensory signals, such as reacting angrily to slight provocations. Also, he is likely to lose many or most of his morals; he has little embarrassment in relation to his excretory, sexual, and social activities; and he is prone to quickly changing moods of sweetness, hate, joy, sadness, exhilaration, and rage. In short, he is a highly *distractible* person with lack of ability to pursue long and complicated thoughts.

THOUGHTS, CONSCIOUSNESS, AND MEMORY

Our most difficult problem in discussing consciousness, thoughts, memory, and learning is that we do not know the neural mechanism of a thought. We know that destruction of large portions of the cerebral cortex does not prevent a person from having thoughts, but it does reduce the *degree* of awareness of the surroundings. On the other hand, destruction of far smaller portions of the thalamus and especially of the mesencephalic portion of the reticular activating system can cause tremendously decreased awareness or even complete unconsciousness.

Each thought almost certainly involves simultaneous signals in portions of the cerebral cortex, thalamus, limbic system, and reticular formation of the brain stem. Some crude thoughts probably depend almost entirely on lower centers; the thought of pain is probably a good example, for electrical stimulation of the human cortex rarely elicits anything more than the mildest degrees of pain, while stimulation of certain areas of the hypothalamus and mesencephalon in animals apparently causes excruciating pain. On the other hand, a type of thought pattern that requires mainly the cerebral cortex is that involving vision, because loss of the visual cortex causes complete inability to perceive visual form or color.

Therefore, we might formulate a definition of a thought in terms of neural activity as follows: A thought probably results from the momentary "pattern" of stimulation of many different parts of the nervous system at the same time, probably involving most importantly the cerebral cortex, the thalamus, the limbic system, and the upper reticular formation of the brain stem. This is called the *holistic theory* of thoughts. The stimulated areas of the limbic system, thalamus, and reticular formation perhaps determine the general nature of the thought, giving it such qualities as pleasure, displeasure, pain, comfort, crude modalities of sensation, localization to gross areas of the body, and other general characteristics. On the other hand, the stimulated areas of the cortex probably determine the discrete characteristics of the thought (such as specific localization of sensations on the body and of objects in the fields of vision), discrete patterns of sensation (such as the rectangular pattern of a concrete block wall or the texture of a rug), and other individual characteristics that enter into the overall awareness of a particular instant.

MEMORY AND TYPES OF MEMORY

If we accept the above approximation of what constitutes a thought, we can see immediately that the mechanism of memory must be equally as complex as the mechanism of a thought, for, to provide memory, the nervous system must recreate the same spatial and temporal pattern (the "holistic" pattern) of stimulation in the central nervous system at some future date. Though we cannot explain in detail what a memory is, we do know some of the basic psychological and neuronal processes that probably lead to the process of memory.

All of us know that all degrees of memory occur, some memories lasting a few seconds and others lasting hours, days, months, or years. Possibly all these types of memory are caused by the same mechanism operating to different degrees of fulfillment. Yet, it is also possible that different mechanisms of memory do exist. Indeed, most physiologists classify memory into from two to four different types. For the purpose of the present discussion, we will use the following classification:

1. *Sensory memory;*
2. *Short-term memory* or *primary memory;*
3. *Long-term memory,* which itself can be divided into *secondary memory* and *tertiary memory.*

The basic characteristics of these types of memory are the following:

Sensory Memory. Sensory memory means the ability to retain sensory signals in the sensory areas of the brain for a very short interval of time following the actual sensory experience. Usually these signals remain available for analysis for several hundred milliseconds but are replaced by new sensory signals in less than one second. Nevertheless, during the short interval of time that the instantaneous sensory information remains in the brain it can continue to be used for further processing; most important, it can be "scanned" to pick out the important points. Thus, this is the initial stage of the memory process.

Short-Term Memory (Primary Memory). Short-term memory (or primary memory) is the memory of a few facts, words, numbers, letters, or other bits of information for a few seconds to a minute or more at a time. This is typified by a person's memory of the digits in a telephone number for a short period of time after he has looked up the number in the telephone directory. This type of memory is usually limited to about seven bits of information, and when new bits of information are put into this *short-term store,* older information is displaced. Thus, if a person looks up a second telephone number, the first is usually lost. One of the most important features of short-term memory is that the information in this memory store is instantaneously available so that the person does not have to search through his mind for it as he does for information that has been put away in the long-term memory stores.

Long-Term Memory. Long-term memory is the storage in the brain of information that can be recalled at some later time — many minutes, hours, days, months, or years later. This type of memory has been called *fixed memory, permanent memory,* and several other names. Long-term memory is also usually divided into two different types, *secondary memory* and *tertiary memory,* the characteristics of which are the following:

A *secondary memory* is a long-term memory that is stored with either a weak or only a moderately strong memory trace. For this reason it is easy to forget and it is sometimes difficult to recall. Furthermore, the time required to search for the information is relatively long. This type of memory can last from several minutes to several years. When the memories are so weak that they will last for only a few minutes to a few days, they are also frequently called *recent memory.*

A *tertiary memory* is a memory that has become so well ingrained in the mind that the memory can usually last the lifetime of the person. Furthermore, the very strong memory traces of this type of memory make the stored information available within a split second. This type of memory is typified by one's knowledge of his own name, by his ability to recall immediately the numbers from 1 to 10, the letters of the alphabet, and the words that he uses in speech, and also by the memory of his own precise physical structure and of his very familiar immediate surroundings.

PHYSIOLOGICAL BASIS OF MEMORY

Despite the many advances in neurophysiology during the past half century, we still cannot explain what is perhaps the most important function of the brain: its capability for memory. Yet, physiological experiments are beginning to generate conceptual theories of the means by which memory could occur. Some of these are discussed in the following few sections.

Possible Mechanisms for Short-Term Memory. Short-term memory requires a neuronal mechanism that can hold specific information signals for a few seconds to at most a minute or more. Several such mechanisms are the following:

Reverberating Circuit Theory of Short-Term Memory. When a tetanizing electrical stimulus is applied directly to the surface of the cerebral cortex and then is removed after a second or more, the local area excited by this stimulus continues to emit rhythmic action potentials for short periods of time. This effect is believed to result from local reverberating circuits, the signals passing through a multistage circuit of neurons in the local area of the cortex itself or perhaps back and forth between the cortex and the thalamus.

It is postulated that sensory signals reaching the cerebral cortex can set up similar reverberating oscillations and that these could be the basis for short-term memory. Then, as the reverberating circuit fatigues, or as new signals interfere with the reverberations, the short-term memory fades away.

One of the principal observations in support of this theory is that any factor that causes a general disturbance of brain function, such as sudden fright, a very loud noise, or any other sensory experience that attracts the person's undivided attention immediately erases the short-term memory. The memory cannot be recalled when the disturbance is over unless a portion of this memory had already been placed into the long-term memory store, as will be discussed in subsequent sections.

Post-Tetanic Potentiation Theory of Short-Term Memory. In most parts of the nervous system, including even the anterior motoneurons of the spinal cord, tetanic stimulation of a synapse for a few seconds causes increased excitability of the synapse for a few seconds to a few hours.

If during this time the synapse is stimulated again, the stimulated neuron responds much more vigorously than normally, a phenomenon called *post-tetanic potentiation*. This is obviously a type of memory that depends on change in the excitability of the synapse, and it could be the basis for short-term memory. Recent experiments suggest that this phenomenon might result from increased permeability of the synaptic knob membranes to calcium ions, which will be discussed in a later section.

DC Potential Theory of Short-Term Memory. Another change that often occurs in neurons following a period of excitation is a prolonged decrease in the membrane potential of the neuron lasting for from seconds to minutes. Because this changes the excitability of the neuron, it could be the basis for short-term memory. Such changes in neuronal potentials are called *DC potentials* or sometimes *electrotonic potentials*. Measurements in the cerebral cortex show that such potentials occur especially in the superficial dendritic layers of the cortex, indicating that the process of short-term memory could result from changes in dendritic membrane potentials.

Mechanism of Long-Term Memory, Enhancement of Synaptic Transmission Facility

Long-term memory means the ability of the nervous system to recall thoughts long after initial elicitation of the thoughts is over. We know that long-term memory does not depend on continued activity of the nervous system, because the brain can be totally inactivated by cooling, by general anesthesia, by hypoxia, by ischemia, or by any other method and yet memories that have been previously stored are still retained when the brain becomes active once again. Therefore, it is assumed that long-term memory must result from

some actual alterations of the synapses, either physical or chemical.

Many different theories have been offered to explain the synaptic changes that cause long-term memory. Among the most important of these are:

1. Anatomical Changes in the Synapses. Cajal, almost a century ago, discovered that the number of terminal fibrils ending on neuronal cells and dendrites in the cerebral cortex increases with age. Conversely, physiologists have shown that inactivity of regions of the cortex causes thinning of the cortex: for instance, thinning of the primary visual cortex in animals that have lost their eyesight. Also, intense activity of a particular part of the cortex can cause excessive thickening of the cortical shell in that area alone. This has been demonstrated especially in the visual cortex of animals subjected to repeated visual experiences. Finally, some neuroanatomists have observed electron micrographic changes in presynaptic terminals that have been subjected to intense and prolonged activity.

All these observations have led to a widely held belief that fixation of memories in the brain results from anatomical changes in the synapses themselves: perhaps changes in numbers of presynaptic terminals, perhaps in sizes of the terminals, or perhaps in the sizes and conductivities of the dendrites. Such anatomical changes could cause permanent or semipermanent increase in the degree of facilitation of specific neuronal circuits, thus allowing signals to pass through the circuits with progressive ease the more often the memory trace is used. This obviously would explain the tendency for memories to become more and more deeply fixed in the nervous system the more often they are recalled or the more often the person repeats the sensory experience that leads to the memory trace.

Another variant of the anatomical theory is that in early life an excess of synaptic connections is made, and those synapses that are active continue to remain functional, while the others eventually degenerate. Thus, the pattern of synaptic transmission could be established by destroying unused synapses rather than by enhancing the used ones. This theory fits the finding that unused synapses and even unused neurons often disappear completely.

2. Physical or Chemical Changes in the Presynaptic Terminal or the Postsynaptic Membrane. Recent studies in the large snail *Aplysia* have uncovered a mechanism of memory that results from either a physical or chemical change of the presynaptic terminal. This mechanism, illustrated in Figure 55–7, works in the following way: There are two separate presynaptic terminals, one of which ends on the subsequent neuron

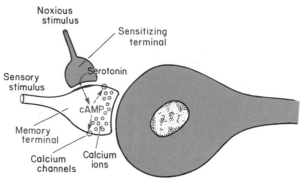

Figure 55–7. A memory system that has been discovered in the snail *Aplysia*.

and the other of which ends on the first presynaptic terminal. In the figure these are called the *memory terminal* and the *sensitizing terminal*. When the memory terminal is stimulated repeatedly but without stimulating the sensitizing terminal, signal transmission at first is very great, but this becomes less and less intense with repeated stimulation until transmission almost ceases. This phenomenon is called *habituation*. It is a type of memory that causes the neuronal circuit to lose its response to repeated events that are insignificant. On the other hand, if a noxious stimulus excites the sensitizing terminal at the same time that the memory terminal is stimulated, then, instead of the transmitted signal becoming progressively weaker, the ease of transmission becomes much stronger and will remain strong for hours, days, or even weeks, even without further stimulation of the sensitizing terminal. Thus, the noxious stimulus causes the memory pathway to become facilitated for weeks thereafter. It is especially interesting that only a few repeated action potentials are required to cause either habituation or sensitization. However, once habituation has occurred, the synapses can become sensitized very rapidly with only a few sensitizing signals.

At the molecular level, the habituation effect in the memory terminal results from progressive closure of calcium channels of the terminal membrane. As a result, much smaller than normal amounts of calcium diffuse into this terminal when action potentials occur, and much less transmitter is therefore released because calcium entry is the stimulus for transmitter release (as was discussed in Chapter 46).

In the case of sensitization, excess amounts of cyclic AMP are believed to develop inside the memory terminal and this opens increased numbers of calcium channels, which correspondingly enhances signal transmission.

Other variants of sensitized synaptic transmission have also been suggested, such as increased amounts of receptor protein in the postsynaptic membrane, decreased rate of destruction of transmitter substance, and so on.

3. Theoretical Function of RNA in Memory. The discovery that DNA and RNA can act as codes to control reproduction, which in itself is a type of memory from one generation to another, plus the fact that these substances, once formed in a cell, tend to persist for the lifetime of the cell, has led to the theory that nucleic acids might be involved in memory changes of the neurons, changes that could last for the lifetime of the person. In addition, biochemical studies have shown an increase in RNA in some active neurons. Yet, a mechanism by which RNA could cause facilitation of synaptic transmission has not been found. Therefore, this theory seems to be based mainly on analogy rather than on factual evidence. The only possible supporting evidence is that several research workers have claimed that long-term memory will not occur to a significant extent when the chemical processes for formation of RNA are blocked. Since RNA is required to promote most anatomical or physical changes in the synapses, however, these observations could support almost any type of theory for long-term memory.

Summary. The theory that seems most likely at present for explaining long-term memory is that some actual anatomical, physical, or chemical change occurs in the synaptic knobs themselves or in the postsynaptic neurons, these changes permanently facilitating the transmission of impulses at the synapses. If all the synapses are thus facilitated in a thought circuit, this circuit can be re-excited by any one of many diverse incoming signals at later dates, thereby causing memory. The overall facilitated circuit is called a *memory engram* or a *memory trace*.

Consolidation of Long-Term Memory

If a memory is to last in the brain so that it can be recalled days later, it must become "consolidated" — that is, the synapses must become permanently facilitated. This process requires 5 to 10 minutes for minimal consolidation and an hour or more for maximal consolidation. For instance, if a strong sensory impression is made on the brain but is then followed within a minute or so by an electrically induced brain convulsion, the sensory experience will not be remembered at all. Likewise, brain concussion, sudden application of deep general anesthesia, and other effects that temporarily block the dynamic function of the brain can prevent consolidation.

However, if the same sensory stimulus is impressed on the brain and the strong electrical shock is delayed for more than 5 to 10 minutes, at least part of the memory trace will have become established. If the shock is delayed for an hour, the memory will have become fully consolidated.

The process of consolidation and the time required for consolidation can probably be explained by the phenomenon of *rehearsal* of the short-term memory as follows:

Role of Rehearsal in Transference of Short-Term Memory into Long-Term Memory. Psychological studies have shown that rehearsal of the same information again and again accelerates and potentiates the degree of transfer of short-term memory into long-term memory, and therefore also accelerates and potentiates the process of consolidation. The brain has a natural tendency to rehearse newfound information, and especially to rehearse newfound information that catches the mind's attention. Therefore, over a period of time the important features of sensory experiences become progressively more and more fixed in the long-term memory stores. This explains why a person can remember small amounts of information studied in depth far better than large amounts of information studied only superficially. And it also explains why a person who is wide-awake will consolidate memories far better than will a person who is in a state of mental fatigue.

Codifying of Memories During the Process of Consolidation. One of the most important features of the process of consolidation is that memories to be placed permanently into the long-term memory storehouse are first codified into different classes of information. During this process similar information is recalled from the long-term storage bins and is used to help process the new information. The new and old are compared for similarities and for differences, and part of the storage process is to store the information about these similarities and differences rather than simply to store the information unprocessed. Thus, during the process of consolidation, the new memories are not stored randomly in the brain, but instead are stored in direct association with other memories of the same type. This is obviously necessary if one is to be able to scan the memory store at a later date to find the required information.

Change of Long-Term Secondary Memory into Long-Term Tertiary Memory — Role of Rehearsal. Rehearsal also plays an extremely important role in changing the weak trace type of long-term memory, called secondary memory, into the strong trace type, called tertiary memory. That is, each time a memory is recalled or each time the same sensory experience is repeated, a more and more indelible memory trace develops in the brain. The memory finally becomes so deeply fixed in the brain that it can be recalled within a fraction of a second and it will also last for a lifetime, both of which are the characteristics of long-term tertiary memory.

Role of Specific Parts of the Brain in the Memory Process

Role of the Hippocampus for Rehearsal, Codification, and Consolidation of Memories — Anterograde Amnesia. Persons who have had both hippocampi (or other associated limbic structures) removed have essentially normal memory for information stored in the brain prior to removal of the hippocampi. However, after loss of these structures, these persons have very little capability for transferring short-term memory into long-term memory. That is, they do not have the ability to separate out the important information, to codify it, to rehearse it, and to consolidate it in the long-term memory store. Therefore, these persons develop serious *anterograde amnesia,* meaning simply inability to establish new memories.

Other types of lesions that frequently give anterograde amnesia include lesions of (1) the mammillary bodies, (2) the anterior nuclei of the thalamus, (3) the anterior columns of the fornix, or (4) the dorsal medial nuclei of the thalamus.

Retrograde Amnesia. Retrograde amnesia means inability to recall memories from the long-term memory storage bins even though the memories are known to be still there. Usually when a person develops retrograde amnesia, the degree of amnesia for recent events is likely to be much greater than for events of the distant past, though even the distant memories are usually affected at least to some extent.

In most persons who have hippocampal lesions and resultant anterograde amnesia, there is also some degree of retrograde amnesia, which suggests that these two types of amnesia are at least partially related and that hippocampal lesions can cause both. However, it has also been claimed that damage in some thalamic areas can lead specifically to retrograde amnesia without significant anterograde amnesia. This perhaps can be explained by the fact that the thalamus probably plays a major role in directing the person's attention to information in different parts of the memory storehouse, as was discussed in the previous chapter. Thus, the thalamus could easily play important roles in codifying, storing, and recalling memories.

ANALYTICAL OPERATIONS OF THE BRAIN

Thus far, we have considered the approximate nature of thoughts and possible mechanisms by which memory and learning can occur. Now we need to consider the mechanisms by which the brain performs complex intellectual operations, such as the analysis of sensory information and the establishment of abstract thoughts. About these mechanisms we know almost nothing, but experiments along these lines have established the following important facts: First, the brain can focus its attention on specific types of information (which was discussed in the preceding chapter). Second, the different qualities of each set of information signals are split away from the central signal and are transmitted to multiple areas of the brain. Third, the brain compares new information with old information in its memory loci. And, fourth, the brain determines patterns of stimulation.

Analysis of Information by Splitting its Qualities. When sensory information enters the nervous system, one of the first steps in its analysis is to transmit the signal to separate parts of the brain that are selectively adapted for detecting specific qualities. For instance, if the hand is placed on a hot stove, (a) pain information is sent through the spinoreticular and spinothalamic tracts into the reticular formation of the brain stem and into certain thalamic and hypothalamic nuclei, (b) tactile information is sent through the dorsal and dorsolateral column systems to the somesthetic cortex, giving the cortex a detailed description of the part of the hand that is touching the hot stove, and (c) position sense information from the muscles and joints is sent to the cerebellum, the reticular formation, and to higher centers of the brain to give momentary information of the position of the hand. Thus, the different qualities of the overall information are dissected and transmitted to different parts of the brain. On the basis of analyses in all these parts of the brain, motor responses to the incoming sensory information are formulated.

One can see that different centers of the brain react to specific qualities of information — the reticular formation and certain regions of the thalamus and hypothalamus to pain, regions of the mesencephalon and hypothalamus to the affective nature of the sensation (that is, whether it is pleasant or unpleasant), the somatic cortex to the localization of sensation, other areas to kinesthetic activities, and others to visual, auditory, gustatory, olfactory, and vestibular information.

Analysis of New Information by Comparison with Memories. We all know that new sensory experiences are immediately compared with previous experiences of the same or similar types. For instance, this is the way we recognize a person whom we know. Yet, it still remains a mystery how we can make these comparisons. One theory suggests that if the new pattern of stimulation matches the memory trace, some interaction between the two gives the person a sense of recognition.

Analysis of Patterns. In analyzing information, the brain depends to a great extent on "patterns" of stimulation. For instance, a square is detected

as a square regardless of its position or angle of rotation in the visual field. Likewise, a series of parallel bars is detected as parallel bars regardless of their orientation, or a fly is detected as a fly whether it is seen in the peripheral or central field of vision. We can extend the same logic to the somatic sensory areas, for a person can detect a cube simply by feeling whether it be in an upright, horizontal, or angulated position. Also, it can be detected even by one's feet even though one's feet may have never felt a cube before.

Processing of Information so That Patterns Can Be Determined. Beautiful examples of the ability of the brain to process information for determination of patterns have been discovered in relation to the visual system. In Chapter 60 we shall point out that the retina itself processes visual information to a great extent even before it is transmitted into the brain. For instance, the neuronal circuits in the retina are organized to detect lines or boundaries between light and dark areas. That is, the visual scene is "differentiated" by this mechanism, bringing out strongly the contrasts of the scene but de-emphasizing the flat areas. This explains why a few lines drawn on a paper can give one the impression of a person's appearance. Certainly the lines do not represent the actual picture of a person, but they do give the visual cortex the same pattern of contrasts that one's own visual system would give.

One can see, therefore, that the visual cortex utilizes transformed types of information to point up the most striking characteristics of the visual scene while neglecting unimportant information. Then, the patterns of the visual scene can be extracted from this "preselected" information.

FUNCTION OF THE BRAIN IN COMMUNICATION

One of the most important differences between the human being and lower animals is the facility with which human beings can communicate with one another. Furthermore, because neurological tests can easily assess the ability of a person to communicate with others, we know perhaps more about the sensory and motor systems related to communication than about any other segment of cortical function. Therefore, we will review rapidly the function of the cortex in communication, and from this one can see immediately how the principles of sensory analysis and motor control apply to this art.

There are two aspects to communication: first, the *sensory aspect,* involving the ears and eyes, and, second, the *motor aspect,* involving vocalization and its control.

Sensory Aspects of Communication. We noted earlier in the chapter that destruction of portions

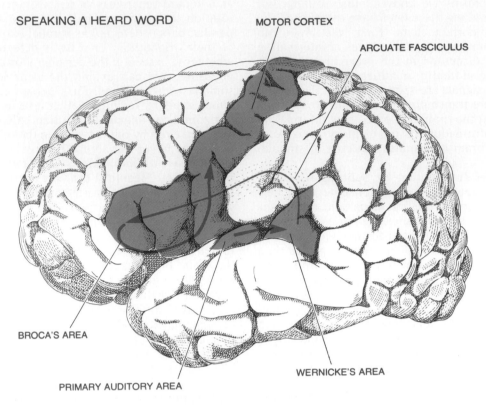

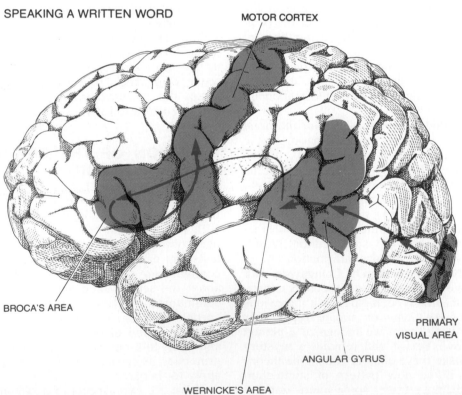

Figure 55–8. Brain pathways for (a) perception of the heard word and then speaking the same word, and (b) perception of the written word and then speaking the same word. (From Geschwind: *Sci. Amer. 241*:180, 1979. © 1979 by Scientific American, Inc. All rights reserved.)

of the *auditory* and *visual association areas* of the cortex can result in inability to understand the spoken word or the written word. These effects are called, respectively, *auditory receptive aphasia* and *visual receptive aphasia* or, more commonly, *word deafness* and *word blindness* (also called *dyslexia*). On the other hand, some persons are perfectly capable of understanding either the spoken word or the written word but are unable to interpret the thought that each expresses. This results most frequently when *Wernicke's area* in the *posterior portion of the superior temporal gyrus* in the dominant hemisphere is damaged or destroyed. This is considered to be *general sensory aphasia* or *general agnosia*.

Motor Aspects of Communication. *Sensory Aphasia.* The process of speech involves (a) formation in the mind of thoughts to be expressed and choice of the words to be used, then (b) the actual act of vocalization. The formation of thoughts and choice of words is principally the function of the sensory areas of the brain, for a person who has a destructive lesion in the same *Wernicke's area* in the *posterior superior temporal lobe* that is involved in general sensory aphasia also has inability to formulate intelligible thoughts to be communicated. And at other times the thoughts can be formulated, but the person is unable to put together the appropriate words to express the thought. These inabilities to formulate thoughts and word sequences are called *sensory aphasias* or, in honor of the neurologist who first delimited the brain area responsible, *Wernicke's aphasia*.

Motor Aphasia. Often a person is perfectly capable of deciding what he wishes to say, and he is capable of vocalizing, but he simply cannot make his vocal system emit words instead of noises. This effect, called *motor aphasia,* almost always results from damage to *Broca's speech area,* which lies in the *premotor* facial region of the cortex — about 95 per cent of the time in the left hemisphere, as illustrated in Figure 55–5. Therefore, we assume that the *skilled motor patterns* for control of the larynx, lips, mouth, respiratory system, and other accessory muscles of articulation are all controlled in this area.

Articulation. Finally, we have the act of articulation itself, which means the muscular movements of the mouth, tongue, larynx, and so forth, that are responsible for the actual emission of sound. The *facial and laryngeal regions of the motor cortex* activate these muscles, and the *cerebellum, basal ganglia,* and *sensory cortex* all help to control the muscle contractions by feedback mechanisms described in Chapter 53. Destruction of these regions can cause either total or partial inability to speak distinctly.

Summary. Figure 55–8 illustrates two principal pathways for communication. The upper half of the figure shows the pathway involved in speaking a heard word. This sequence is the following: (1) reception in the primary auditory area of the sound signals that encode the word; (2) interpretation of the word in Wernicke's area; (3) determination also in Wernicke's area that the word is to be spoken; (4) transmission of signals from Wernicke's area to Broca's area via the *arcuate fasciculus;* (5) activation of the skilled motor programs in Broca's area for control of word formation; and (6) transmission of appropriate signals into the motor cortex to control the speech muscles.

The lower figure illustrates the comparable steps in speaking a written word. The initial receptive area for the word is in the primary visual area rather than in the primary auditory area. Then the word information passes through early stages of interpretation in the *angular gyrus region* and finally reaches its full level of recognition in Wernicke's area. From here, the sequence is the same as for speaking a heard word.

FUNCTION OF THE CORPUS CALLOSUM AND ANTERIOR COMMISSURE TO TRANSFER THOUGHTS, MEMORIES, AND OTHER INFORMATION TO THE OPPOSITE HEMISPHERE

Fibers in the *corpus callosum* connect the respective cortical areas of the two hemispheres with each other except for the anterior portions of the temporal lobes; these temporal areas, including especially the *amygdala,* are interconnected by fibers that pass through the *anterior commissure.* Because of the tremendous number of fibers in the corpus callosum, it obviously was assumed that this massive structure must have some important function to correlate activities of the two cerebral hemispheres. However, after cutting the corpus callosum in experimental animals, it was difficult to discern changes in brain function. Therefore, for a long time the function of the corpus callosum was a mystery.

Yet, properly designed psychological experiments have demonstrated the extremely important functions of the corpus callosum and anterior commissure. These can be explained best by recounting one of the experiments. A monkey is first prepared by cutting the corpus callosum and splitting the optic chiasm longitudinally. Then it is taught to recognize different types of objects with its right eye while its left eye is covered. Next, the right eye is covered and the monkey is tested to determine whether or not its left eye can recognize the same object. The answer to this is that the left eye *cannot* recognize the object. Yet, on repeating the same experiment in another monkey with the optic chiasm split but the corpus callosum intact, it is found invariably that recognition in one hemisphere of the brain creates recognition also in the opposite hemisphere.

Thus, one of the functions of the corpus callosum and the anterior commissure is to make information stored in the cortex of one hemisphere available to the corresponding cortical area of the opposite hemisphere. Two important examples of such cooperation between the two hemispheres are:

1. Cutting of the corpus callosum blocks transfer of information from the general interpretative area of the dominant hemisphere to the motor cortex on the opposite side of the brain. Therefore, the intellectual functions of the brain, located primarily in the dominant hemisphere, lose their control over the right motor cortex and therefore also of the voluntary motor functions of the left hand and arm even though the usual subconscious movements of the left hand and arm are completely normal.

2. Cutting of the corpus callosum prevents transfer of somatic and visual information from the right hemisphere into the general interpretative area of the dominant hemisphere. Therefore, somatic and visual information from the left side of the body frequently fails to reach the general interpretative area of the brain and, therefore, cannot be used for decision making.

3. Finally, a teen-aged boy whose corpus callosum had been completely sectioned, but whose anterior commissure was not sectioned, recently was found to have two entirely separate conscious portions to his brain. Only the left half of his brain could understand the spoken word, because it was the dominant hemisphere. On the other hand, the right side of the brain could understand the written word and could elicit a motor response to it without the left side of the brain ever knowing why the response was performed. Yet, the effect was somewhat different when an emotional response was evoked in the right side of the brain, because in this case a subconscious emotional response was felt in the left side of the brain as well. This undoubtedly occurred because the limbic areas in the two sides of the brain were still communicating with each other through anterior commissural fibers interconnecting the amygdalas. For instance, when the command ''kiss'' was written for the right half of his brain to see, the boy immediately and with full emotion, said ''No way!'' But, when questioned why he said this, he could not explain. Thus, the two halves of the brain have independent capabilities for consciousness, memory storage, communication, and control of motor activities. The corpus callosum is required for the two sides to operate cooperatively, and the anterior commissure plays an important role in unifying the emotional responses of the two brain sides.

REFERENCES

Bannister, R. (rev.): Brain's Clinical Neurology. New York, Oxford University Press, 1978.
Bekhtereva, N. P.: The Neurophysiological Aspects of Human Mental Activity. New York, Oxford University Press, 1978.
Benson, D. F.: Aphasia, Alexia, and Agraphia. New York, Churchill Livingstone, 1979.
Benton, A. L., and Pearl, D.: Dyslexia: An Appraisal of Current Knowledge. New York, Oxford University Press, 1978.
Bloodstein, O.: Speech Pathology: An Introduction. Boston, Houghton Mifflin, 1979.

Buser, P.: Higher functions of the nervous system. Annu. Rev. Physiol., 38:217, 1978.
Corning, W. C., et al. (eds.): Invertebrate Learning. Vol. 3. New York, Plenum Press, 1974.
Daniloff, R., et al.: The Physiological Bases of Verbal Communication. Englewood Cliffs, N.J., Prentice-Hall, 1980.
Darley, F. L., et al. (eds.): Evaluation of Appraisal Techniques in Speech and Language Pathology. Reading, Mass., Addison-Wesley Publishing Co., 1979.
De Silva, F. H. L., and Arnolds, D.E.A.T.: Physiology of the hippocampus and related structures. Annu. Rev. Physiol., 40:185, 1978.
Desmedt, J. E. (ed.): Language and Hemispheric Specialization in Man; Cerebral Event-Related Potentials. New York, S. Karger, 1977.
Deutsch, J. A. (ed.): The Physiological Basis of Memory. New York, Academic Press, 1972.
DiCara, L. V.: Learning mechanisms. In Frohlich, E. D. (ed.): Pathophysiology, 2nd Ed. Philadelphia, J. B. Lippincott Co., 1976, p. 757.
Dimond, S. J.: Neuropsychology: A Textbook of Systems and Psychological Functions of the Human Brain. Boston, Butterworths, 1979.
Edelman, G. M., and Mountcastle, V. B.: The Mindful Brain: Cortical Organization and the Group-Selective Theory of Higher Brain Function. Cambridge, Mass., MIT Press, 1978.
Eggert, G. H.: Wernicke's Works on Aphasia: A Sourcebook and Review. The Hague, Mouton, 1977.
Gazzaniga, M. (ed.): Neuropsychology. New York, Plenum Press, 1978.
Geschwind, N.: The apraxias: Neural mechanisms of disorders of learned movement. Am. Sci., 63:188, 1975.
Geschwind, N.: Specializations of the human brain. Sci. Am., 241(3):180, 1979.
Herron, J. (ed.): Neuropsychology of Left-Handedness. New York, Academic Press, 1979.
Hixon, T. J., et al. (eds.): Introduction to Communicative Disorders. Englewood Cliffs, N. J., Prentice-Hall, 1980.
Hubel, D. H.: The brain. Sci. Am., 241(3):44, 1979.
Johns, D. F. (ed.): Clinical Management of Neurogenic Communicative Disorders. Boston, Little, Brown, 1978.
Kandel, E. R.: Neuronal plasticity and the modification of behavior. In Brookhart, J. M., and Mountcastle, V. B. (eds.): Handbook of Physiology. Sec. 1, Vol. 1. Baltimore, Williams & Wilkins, 1977, p. 1137.
Kandel, E. R.: A Cell-Biological Approach to Learning. Bethesda, Md., Society for Neuroscience, 1978.
Lehmann, D., and Callaway, E. (eds.): Event-Related Potentials in Man: Applications and Probes. New York, Plenum Press, 1979.
Levinthal, C. F.: The Physiological Approach in Psychology. Englewood Cliffs, N.J., Prentice-Hall, 1979.
Moskowitz, B. A.: The acquisition of language. Sci. Am., 239(5):92, 1978.
Nebes, R. D.: Hemispheric specialization in commissurotomized man. Psychol. Bull., 81:1, 1974.
Olton, D. S.: Spatial memory. Sci. Am., 236(6):82, 1977.
Ordy, J. M., and Brizzee, K. R. (eds.): Sensory Systems and Communication in the Elderly. New York, Raven Press, 1979.
Russell, I. S., and vanHof, M. W. (eds.): Structure and Function of Cerebral Commissures. Baltimore, University Park Press, 1978.
Schiefelbusch, R. L., et al. (eds.): Language Intervention Strategies. Baltimore, University Park Press, 1978.
Siegel, R. K.: Hallucinations. Sci. Am., 237(4):132, 1977.
Smith, J. H. (ed.): Psychoanalysis and Language. New Haven, Conn., Yale University Press, 1978.
Sperry, R. W.: Changing concepts of consciousness and free will. Perspect. Biol. Med., 20:9, 1976.
Stein, D. G., and Rosen, J. J.: Learning and Memory. New York, The Macmillan Co., 1974.
Steklis, H. D., and Raleigh, M. J. (eds.): Neurobiology of Social Communication in Primates: An Evolutionary Perspective. New York, Academic Press, 1979.
Terry, R. D., and Davies, P.: Dementia of the Alzheimer type. Annu. Rev. Neurosci., 3:77, 1980.
Uttal, W. R.: The Psychobiology of Mind. New York, Halsted Press, 1978.
Vellutino, F. R.: Dyslexia: Theory and Research. Cambridge, Mass., MIT Press, 1979.

56

Behavioral Functions of the Brain: The Limbic System, Role of the Hypothalamus, and Control of Vegetative Functions of the Body

Behavior is a function of the entire nervous system, not of any particular portion. Even the discrete cord reflexes are an element of behavior, and the wakefulness and sleep cycle discussed in Chapter 54 is certainly one of the most important of our behavioral patterns. However, in this chapter we will deal with those special types of behavior associated with emotions, subconscious motor and sensory drives, and the intrinsic feelings of punishment and pleasure. These functions of the nervous system are performed mainly by subcortical structures located in the basal regions of the brain.

The word "limbic" means "border," and the original usage of the term *limbic system* was to describe the brain structures that lie in the border region between the hypothalamus and its related structures, on the one hand, and the cerebral cortex. However, as we have learned more about the functions of the hypothalamus and the limbic system, it has become clear that they function together as a total system. Therefore, the term limbic system has now been expanded in common usage to mean this entire basal system of the brain that mainly controls the person's emotional behavior and drive.

The hypothalamus and its related structures also control many internal conditions of the body as well as aspects of behavior — such conditions as body temperature, osmolality of the body fluids, the drives to eat and drink, body weight, and so forth. These internal functions are collectively called *vegetative functions* of the body, and their control is obviously closely related to behavior.

FUNCTIONAL ANATOMY OF THE LIMBIC SYSTEM; ITS RELATION TO THE HYPOTHALAMUS

Figure 56–1 illustrates the anatomical structures of the limbic system and its relationship to the hypothalamus, showing these to be an interconnected complex of basal brain elements. Located in the midst of all these is the hypothalamus, which is considered by many anatomists to be a separate structure from the remainder of the limbic system but which, from a physiological point of view, is one of the central elements of the system. Therefore, Figure 56–2 illustrates schematically this key position of the hypothalamus in the limbic system and shows that surrounding it are the other subcortical structures of the limbic system, including the *preoptic area,* the *septum,* the *paraolfactory area,* the *epithalamus,* the *anterior nuclei of the thalamus, portions of the basal ganglia,* the *hippocampus,* and the *amygdala.*

Surrounding the subcortical limbic areas is the *limbic cortex* composed of a ring of cerebral cortex (a) beginning in the *orbitofrontal area* on the ventral surface of the frontal lobes, (b) extending upward in front of and over the corpus callosum onto the medial aspect of the cerebral hemisphere to the *cingulate gyrus,* and finally (c) passing behind the corpus callosum and downward onto the ventromedial surface of the temporal lobe to the *hippocampal gyrus, pyriform area,* and *uncus.* Thus, on the medial and ventral surfaces of each cerebral hemisphere is a ring of *paleocortex* that surrounds a group of deep structures intimately associated with overall behavior and with emotions. In turn, this ring of paleocortex functions as a two-way communication and association linkage between the *neocortex* and the lower limbic structures.

It is also important to recognize that many of the

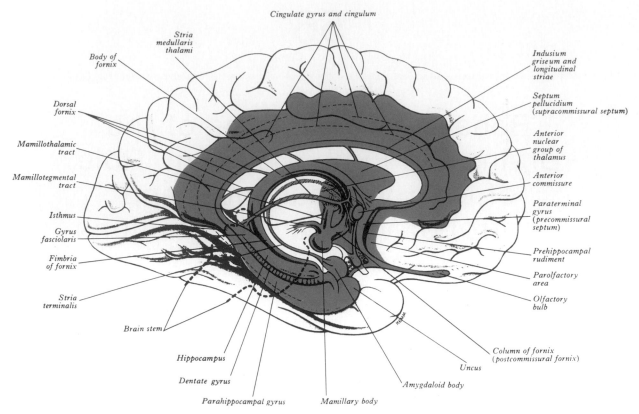

Figure 56–1. Anatomy of the limbic system illustrated by the shaded areas of the figure. (From Warwick and Williams: Gray's Anatomy, 35th Brit. Ed. London, Longman Group Ltd., 1973.)

behavioral functions elicited from the hypothalamus and from other limbic structures are mediated through the reticular formation of the brain stem. It was pointed out in Chapters 52 and 54 that stimulation of the excitatory portion of the reticular formation can cause high degrees of somatic excitability; in Chapter 57 we will see that most of the signals for control of the autonomic nervous system either originate in or are transmitted from higher centers through nuclei located in the brain stem reticular formation.

Therefore, from a physiological point of view the reticular formation is a very important part of the limbic system even though anatomically it is considered to be a separate entity. A very important route of communication between the limbic system and the reticular formation is the *medial forebrain bundle* that extends from the septal and orbitofrontal cortical regions downward through the hypothalamus to the reticular formation. This bundle carries fibers in both directions, forming a trunk-line communication system. A second route of communication is through short pathways among the reticular formation, the thalamus, the hypothalamus, and most of the other contiguous areas of the basal brain.

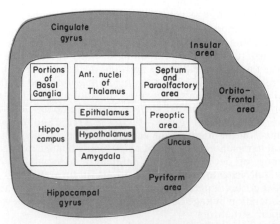

Figure 56–2. The limbic system.

THE HYPOTHALAMUS, THE MAJOR OUTPUT PATHWAY OF THE LIMBIC SYSTEM

Note in Figure 56–2 that the hypothalamus lies in the very middle of the limbic system. It also has communicating pathways with all levels of this system. In turn, it and its closely allied structures, the *septum* and the *mammillary bodies,* send output signals in two directions, (1) downward through the brain stem mainly into the reticular formation of the mesencephalon, pons, and medulla, and (2) upward toward many areas of the cerebrum, especially the anterior thalamus and the limbic cortex. In addition, the hypothalamus indirectly affects cerebral cortical function very dramatically through activation or inhibition of the reticular activating system that originates in the brain stem.

Thus, the hypothalamus (along with its closely related structures) is the principal motor output pathway of the limbic system. It in turn controls most of the vegetative functions of the body as well as many aspects of emotional behavior. Let us discuss first the vegetative control functions and then return to the behavioral control functions to see how these operate together.

VEGETATIVE CONTROL FUNCTIONS OF THE HYPOTHALAMUS

The different hypothalamic mechanisms for controlling the vegetative functions of the body are so important that they, along with the functions that they control, are discussed in many different chapters throughout this text. For instance, the role of the hypothalamus in arterial pressure regulation is discussed in Chapter 21, thirst and water conservation in Chapter 36, and temperature regulation in Chapter 72. However, to illustrate the organization of the hypothalamus as a functional unit, let us summarize the more important of its vegetative functions here as well.

Figure 56–3 shows an enlargement of the hypothalamus, which represented only a small area in Figure 56–1. Please take a few minutes to study this diagram, especially to read the multiple vegetative activities that are stimulated or inhibited when respective centers are stimulated. In addi-

tion to those centers that are illustrated, a large *lateral hypothalamic* area overlies the illustrated areas on each side of the hypothalamus. The lateral areas are especially important in controlling thirst, hunger, and many of the emotional drives.

A word of caution must be issued for studying this diagram, however, for the areas that cause specific activities are not nearly so discrete as shown in the figure. The illustrated areas are merely those from which the functions are likely to be elicited. Also, it is not known whether the effects noted in the figure result from activation of specific control nuclei or whether they result merely from activation of fiber tracts leading from control nuclei located elsewhere. For instance, stimulation of the posterior hypothalamus, through which many of the fiber pathways from other portions of the hypothalamus pass, can at times elicit many functions believed to be controlled primarily by other hypothalamic nuclei. With this caution in mind, we can give the following general description of the vegetative control functions of the hypothalamus.

Cardiovascular Regulation. Stimulation of different areas throughout the hypothalamus can cause every known type of neurogenic effect in the cardiovascular system, including increased arterial pressure, decreased arterial pressure, increased heart rate, and decreased heart rate. In general, stimulation in the *posterior* and *lateral hypothalamus* increases the arterial pressure and heart rate, while stimulation in the *preoptic area* often has opposite effects, causing a decrease in both heart rate and arterial pressure. These effects are transmitted mainly through the cardiovascular control centers in the reticular substance of the medulla and pons.

Regulation of Body Temperature. Large areas in the anterior portion of the hypothalamus, especially in the *preoptic area*, are concerned with regulation of body

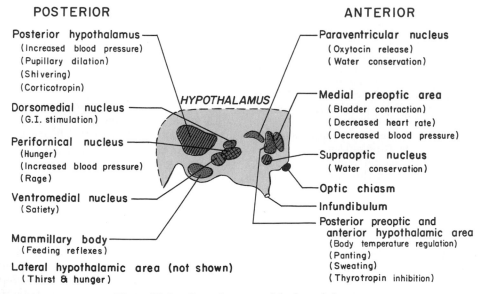

Figure 56–3. Control centers of the hypothalamus.

temperature. An increase in the temperature of the blood flowing through these areas increases their activity, while a decrease in temperature decreases their activity. In turn, these areas control the mechanisms for increasing or decreasing body temperature, as discussed in Chapter 72.

Regulation of Body Water. The hypothalamus regulates body water in two separate ways, by (1) creating the sensation of thirst, which makes an animal drink water, and (2) controlling the excretion of water into the urine. An area called the *thirst center* is located in the lateral hypothalamus. When the electrolytes inside the neurons of this center become too concentrated, the animal develops an intense desire to drink water; it will search out the nearest source of water and drink enough to return the electrolyte concentration of the thirst center to normal.

Control of renal excretion of water is vested mainly in the *supraoptic nuclei*. When the body fluids become too concentrated, the neurons of this area become stimulated. The nerve fibers from these neurons project into the posterior pituitary gland where they secrete a hormone called *antidiuretic hormone*. This hormone is then absorbed into the blood and acts on the collecting ducts of the kidneys to cause massive reabsorption of water, thereby decreasing the loss of water into the urine. These functions were presented in detail in Chapter 36.

Regulation of Uterine Contractility and of Milk Ejection by the Breasts. Stimulation of the *paraventricular nuclei* causes its neuronal cells to secrete the hormone *oxytocin*. This in turn causes increased contractility of the uterus and also contraction of the myoepithelial cells that surround the alveoli of the breasts, which then causes the alveoli to empty the milk through the nipples. At the end of pregnancy, especially large quantities of oxytocin are secreted, and this secretion helps to promote labor contractions that expel the baby. Also, when a baby suckles the mother's breast, a reflex signal from the nipple to the hypothalamus causes oxytocin release, and the oxytocin then performs the necessary function of expelling the milk through the nipples so that the baby can nourish itself. These functions are discussed in Chapter 82.

Gastrointestinal and Feeding Regulation. Stimulation of several areas of the hypothalamus causes an animal to experience extreme hunger, a voracious appetite, and an intense desire to search for food. The area most associated with hunger is the *lateral hypothalamic area.* On the other hand, damage to this causes the animal to lose desire for food, sometimes causing lethal starvation.

A center that opposes the desire for food, called the *satiety center,* is located in the *ventromedial nucleus.* When this center is stimulated, an animal that is eating food suddenly stops eating and shows complete indifference to the food. On the other hand, if this area is destroyed, the animal cannot be satiated, but instead, its hypothalamic hunger centers become overactive so that it has a voracious appetite, resulting in tremendous obesity.

Another area of the hypothalamus that enters into the overall control of gastrointestinal activity is the *mammillary bodies* that activate many feeding reflexes, such as licking the lips and swallowing.

Hypothalamic Control of the Anterior Pituitary Gland. Stimulation of certain areas of the hypothalamus also causes the anterior pituitary gland to secrete its hormones. This subject will be discussed in detail in Chapter 75 in relation to the overall control of the endocrine glands, but, briefly, the basic mechanism of the control of the anterior pituitary is the following:

The adenohypophysis receives its blood supply mainly from veins that flow into the anterior pituitary sinuses from the lower part of the hypothalamus. As the blood courses through the hypothalamus before reaching the anterior pituitary, *neurosecretory substances* are secreted into the blood by various hypothalamic nuclei. They are then transported in the blood to the anterior pituitary where they act on the glandular cells to cause release of the anterior pituitary hormones.

Stimulation of the following areas of the hypothalamus has been said to cause secretion of specific anterior pituitary hormones: (1) preoptic and anterior hypothalamic nuclei, thyrotropin that controls secretion of the thyroid hormones; (2) posterior hypothalamus and median eminence of the infundibulum, corticotropin that causes secretion of adrenocortical hormones; (3) median eminence, follicle-stimulating hormone, and anterior hypothalamus, luteinizing hormone, the two of which cause secretion of the gonadotropic hormones. Other nuclei of the hypothalamus control the secretion of growth hormone and prolactin by the anterior pituitary gland, the first of these accelerating body growth and the second causing lactation.

Summary. A number of discrete areas of the hypothalamus have now been found that control specific vegetative functions. However, these areas are still poorly delimited, so much so that the above separation of different areas for different hypothalamic functions is partially artificial.

BEHAVIORAL FUNCTIONS OF THE HYPOTHALAMUS AND ASSOCIATED LIMBIC STRUCTURES

Aside from the vegetative functions of the hypothalamus, experiments have shown that stimulation of or lesions in the hypothalamus often have profound effects on emotional behavior of animals or human beings.

In animals, the effects of stimulation are the following:

1. Stimulation in the *lateral hypothalamus* not only causes thirst and eating as discussed above in relation to the vegetative functions but also increases the general level of activity of the animal, sometimes leading to overt rage and fighting as will be discussed subsequently.

2. Stimulation of the *most medial portion of the medial hypothalamus* immediately adjacent to the third ventricle (or also stimulation of the central gray area of the mesencephalon that is continuous with this portion of the hypothalamus) usually leads to fear and punishment reactions.

3. Stimulation of the *lateral portion of the medial hypothalamus,* the area lying between the two

areas discussed above, can elicit a number of different hypothalamic drives. In the most anterior portion, stimulation is most likely to increase sexual drive. Immediately behind this area, in the *preoptic hypothalamus,* is the area for control of body temperature. Next, in the anterior hypothalamus is an area that increases the drive mainly for drinking, but to some extent for eating. A little further posteriorly, the drive converts mainly to excessive eating, and, finally, at the very posterior of this mid-zone of the hypothalamus, the sex drive again becomes most prominent.

Lesions in the hypothalamus, in general, cause the opposite effects. For instance:

1. Bilateral lesions in the lateral hypothalamus will decrease drinking and eating almost to zero, often leading to lethal starvation, as was discussed above in relation to the vegetative functions of the lateral hypothalamus. But these lesions cause extreme *passivity* of the animal as well, with loss of most of its overt activities.

2. Bilateral lesions of the ventromedial areas of the hypothalamus cause exactly opposite effects: excessive drinking and eating, as well as hyperactivity and often contiguous savagery along with frequent bouts of extreme rage on the slightest provocation.

Lesions or stimulation in other regions of the limbic system, especially the amygdala, the septal area, and areas in the mesencephalon, often also cause effects similar to those elicited directly from the hypothalamus. We will discuss some of these in more detail later.

The Reward and Punishment (or Pleasure and Pain) Function of the Limbic System

From the above discussion it is already clear that the hypothalamus and other limbic structures are particularly concerned with the affective nature of sensory sensations — that is, whether the sensations are *pleasant* or *painful.* These affective qualities are also called *reward* and *punishment* or *satisfaction* and *aversion.* Electrical stimulation of certain regions pleases or satisfies the animal, whereas electrical stimulation of other regions causes terror, pain, fear, defense, escape reactions, and all the other elements of punishment. Obviously, these two oppositely responding systems greatly affect the behavior of the animal.

Reward Centers. Figure 56–4 illustrates a technique that has been used for localizing the specific reward and punishment areas of the brain. In this figure a lever is placed at the side of the cage and is arranged so that depressing the lever makes electrical contact with a stimulator. Electrodes are placed successively at different areas in the brain so that the animal can stimulate the area by pressing the lever. If stimulating the particular

Figure 56–4. Technique for localizing reward and punishment centers in the brain of a monkey.

area gives the animal a sense of reward, then it will press the lever again and again, sometimes as much as 10,000 to 15,000 times per hour. Furthermore, when offered the choice of eating some delectable food as opposed to the opportunity to stimulate the reward center, it often chooses the electrical stimulation.

By using this procedure, the major reward centers have been found to be located in the septum and hypothalamus, primarily *along the course of the medial forebrain bundle* and in both the *lateral and the ventromedial nuclei of the hypothalamus.* It is strange that the lateral nuclei should be included among the reward areas — indeed it is one of the most potent reward areas of all — because even stronger stimuli in this area can cause rage. But this is true in many areas, with weaker stimuli giving a sense of reward and stronger ones, a sense of punishment.

Less potent reward centers, which are probably secondary to the major ones in the septum and hypothalamus, are found in the amygdala, certain areas of the thalamus and basal ganglia, and finally extending downward into the basal tegmentum of the mesencephalon.

Punishment Centers. The apparatus illustrated in Figure 56–4 can also be connected so that pressing the lever turns off rather than turns on an electrical stimulus. In this case, the animal will not turn the stimulus off when the electrode is in one of the reward areas, but when it is in certain other areas it immediately learns to turn it off. Stimulation in these areas causes the animal to show all the signs of displeasure, fear, terror, and punishment. Furthermore, prolonged stimulation for 24 hours or more causes the animal to become severely sick and actually leads to death.

By means of this technique, the principal centers for pain, punishment, and escape tendencies have been found in the *central gray area surrounding the aqueduct of Sylvius in the mesencephalon* and extending upward into the *periventricular structures of the hypothalamus and thalamus.*

The extent of the punishment areas in the limbic system is only one-seventh as great as the reward areas. Even so, it is particularly interesting that stimulation in the punishment centers can frequently inhibit the reward and pleasure centers completely, illustrating that punishment and fear can take precedence over pleasure and reward.

Importance of Reward and Punishment in Behavior. Almost everything that we do depends on reward and punishment. If we are doing something that is rewarding, we continue to do it; if it is punishing, we cease to do it. Therefore, the reward and punishment centers undoubtedly constitute one of the most important of all the controllers of our bodily activities, our motivations, and so forth.

Importance of Reward and Punishment in Learning and Memory — Habituation or Reinforcement. Animal experiments have shown that a sensory experience causing neither reward nor punishment is remembered hardly at all. Electrical recordings have shown that new and novel sensory stimuli always excite the cerebral cortex. But repetition of the stimulus over and over leads to almost complete extinction of the cortical response if the sensory experience does not elicit either a sense of reward or punishment. Thus, the animal becomes *habituated* to the sensory stimulus, and thereafter ignores this stimulus.

If the stimulus causes either reward or punishment rather than indifference, however, the cortical response becomes progressively more and more intense with repetitive stimulation instead of fading away, and the response is said to be *reinforced.* Thus, an animal builds up strong memory traces for sensations that are either rewarding or punishing but, on the other hand, develops complete habituation to indifferent sensory stimuli. Therefore, it is evident that the reward and punishment centers of the midbrain have much to do with selecting the information that we learn.

Effect of Tranquilizers on the Reward and Punishment Centers. Administration of a tranquilizer, such as chlorpromazine, inhibits both the reward and punishment centers, thereby greatly decreasing the affective reactivity of the animal. Therefore, it is presumed that tranquilizers function in psychotic states by suppressing many of the important behavioral areas of the hypothalamus and its associated regions of the brain.

An emotional pattern that involves the hypothalamus and has been well characterized is the *rage pattern.* This can be described as follows:

Strong stimulation of the punishment centers of the brain, especially the *periventricular areas of the hypothalamus,* causes the animal to (1) develop a defense posture, (2) extend its claws, (3) lift its tail, (4) hiss, (5) spit, (6) growl, and (7) develop pilo-erection, wide-open eyes, and dilated pupils. Furthermore, even the slightest provocation causes an immediate savage attack. This is approximately the behavior that one would expect from an animal being severely punished, and it is a pattern of behavior that is called *rage.*

Stimulation of the more rostral areas of the punishment areas — in the midline preoptic areas — causes mainly fear and anxiety, associated with a tendency for the animal to run away.

In the normal animal the rage phenomenon is held in check mainly by counterbalancing activity of the lateral portions of the ventromedial nuclei of the hypothalamus. This is the area of the hypothalamus that lies between the punishment centers adjacent to the third ventricle, on the one hand, and, on the other hand, the lateral nuclei of the hypothalamus that provides some of the very strong behavioral drives.

In addition, the hippocampus, the amygdala, and the anterior portions of the limbic cortex, especially the anterior limbic cortex of the subcallosal area and of the anterior temporal lobe, transmit signals that help to suppress the rage phenomenon. Conversely, if these portions of the limbic system are damaged or destroyed, the animal (also the human being) becomes far more susceptible to bouts of rage.

Placidity and Tameness. Exactly the opposite emotional behavioral patterns occur when the reward centers are stimulated: placidity and tameness.

Function of the Hypothalamus and Other Limbic Areas in Sleep, Wakefulness, Alertness, and Excitement

Stimulation of regions of the hypothalamus dorsal to the mammillary bodies greatly excites the reticular activating system and therefore causes wakefulness, alertness, and excitement. In addition, the sympathetic nervous system becomes excited in general, increasing the arterial pressure, causing pupillary dilatation, and enhancing other activities associated with sympathetic activity.

On the other hand, stimulation in the septum, in the anterior hypothalamus, or in isolated points of the thalamic portions of the reticular activating system often inhibits the mesencephalic portion of the reticular activating system, causing somnolence and sometimes actual sleep. Thus, the hypothalamus and other limbic structures indirectly contribute much to the control of the degree of excitement and alertness.

SPECIFIC FUNCTIONS OF OTHER PARTS OF THE LIMBIC SYSTEM

FUNCTIONS OF THE AMYGDALA

The amygdala is a complex of nuclei located immediately beneath the medial surface of the cerebral cortex in the pole of each temporal lobe, and it has abundant direct connections with the hypothalamus. In lower animals, this complex is concerned primarily with association of olfactory stimuli with stimuli from other parts of the brain. Indeed, it is pointed out in Chapter 62 that one of the major divisions of the olfactory tract leads directly to a portion of the amygdala called the *corticomedial nuclei* that lie immediately beneath the cortex in the pyriform area of the temporal lobe. However, in the human being, another portion of the amygdala, the *basolateral nuclei,* has become much more highly developed than the olfactory portion and plays very important roles in many behavioral activities not generally associated with olfactory stimuli.

The amygdala receives impulses from all portions of the limbic cortex, from the orbital surfaces of the frontal lobes, from the cingulate gyrus, from the hippocampal gyrus, and from the neocortex of the temporal, parietal, and occipital lobes, especially from the auditory and visual association areas. Because of these multiple connections, the amygdala has been called the "window" through which the limbic system sees the place of the person in the world. In turn, the amygdala transmits signals (a) back into these same cortical areas, (b) into the hippocampus, (c) into the septum, (d) into the thalamus, and (e) especially into the hypothalamus.

Effects of Stimulating the Amygdala. In general, stimulation in the amygdala can cause almost all the same effects as those elicited by stimulation of the hypothalamus, plus still other effects. The effects that are mediated through the hypothalamus include (1) increases or decreases in arterial pressure, (2) increases or decreases in heart rate, (3) increases or decreases in gastrointestinal motility and secretion, (4) defecation and micturition, (5) pupillary dilatation or, rarely, constriction, (6) pilo-erection, (7) secretion of the various anterior pituitary hormones, including especially the gonadotropins and corticotropin.

Aside from these effects mediated through the hypothalamus, amygdala stimulation can also cause different types of involuntary movement. These include (1) tonic movements, such as raising the head or bending the body, (2) circling movements, (3) occasionally clonic, rhythmic movements, and (4) different types of movements associated with olfaction and eating, such as licking, chewing, and swallowing.

In addition, stimulation of certain amygdaloid nuclei can, rarely, cause a pattern of rage, escape, punishment, and fear similar to the rage pattern elicited from the hypothalamus as described above. And stimulation of other nuclei can give reactions of reward and pleasure.

Finally, excitation of still other portions of the amygdala can cause sexual activities that include erection, copulatory movements, ejaculation, ovulation, uterine activity, and premature labor.

Effects of Bilateral Ablation of the Amygdala — the Klüver-Bucy Syndrome. When the anterior portions of both temporal lobes are destroyed in a monkey, this not only removes the temporal cortex but also the amygdalas that lie deep in these parts of the temporal lobes. This causes a combination of changes in behavior called the Klüver-Bucy syndrome, which includes (1) excessive tendency to examine objects orally, (2) loss of fear, (3) decreased aggressiveness, (4) tameness, (5) changes in dietary habits, even to the extent that a herbivorous animal frequently becomes carnivorous, (6) sometimes psychic blindness, and (7) often excessive sex drive. The characteristic picture is of an animal that is not afraid of anything, has extreme curiosity about everything, forgets very rapidly, has a tendency to place everything in its mouth and sometimes even tries to eat solid objects, and, finally, often has a sex drive so strong that it attempts to copulate with immature animals, animals of the wrong sex, or animals of a different species.

Though similar lesions in human beings are rare, afflicted persons respond in a mannner not too different from that of the monkey.

Overall Function of the Amygdala. The amygdala seems to be a behavioral awareness area that operates at a semiconscious level. It also seems to project into the limbic system one's present status in relation both to surroundings and thoughts. Based on this information, the amygdala is believed to help pattern the person's behavioral response so that it is appropriate for each occasion.

FUNCTIONS OF THE HIPPOCAMPUS

The hippocampus is an elongated structure composed of a modified type of cerebral cortex. In fact it is that portion of the temporal lobe cortex that folds inward to form the ventral surface of the inferior horn of the lateral ventricle. One end of the hippocampus abuts the amygdaloid nuclei, and it also fuses along one of its borders with the hippocampal gyrus, which is the cortex of the ventromedial surface of the temporal lobe.

The hippocampus has numerous connections with most portions of the sensory cortex as well as with the basic structures of the limbic system — the amygdala, the hypothalamus, the septum, and the mammillary bodies. Almost any type of sensory experience causes instantaneous activation of different parts of the hippocampus, and the hippocampus in turn distributes many outgoing signals to the hypothalamus and other parts of the limbic system, especially through the *fornix.* Thus, the hippocampus, like the amygdala, is an additional channel through which incoming sensory signals can lead to appropriate limbic reactions, but perhaps for different purposes as we shall see later.

Like other limbic structures, stimulation of different

areas in the hippocampus can cause almost any one of different behavioral patterns, such as rage, passivity, excess sex drive, or so forth.

Another feature of the hippocampus is that very weak electrical stimuli can cause local epileptic seizures that persist for many seconds after the stimulation is over, suggesting that the hippocampus can perhaps give off prolonged output signals even under normal functioning conditions. During the hippocampal seizures, the person experiences various psychomotor effects, including olfactory, visual, auditory, tactile, and other types of hallucinations that cannot be suppressed even though the person has not lost consciousness and knows these hallucinations to be unreal. Probably one of the reasons for this hyperexcitability of the hippocampus is that it is composed of a different type of cortex from that elsewhere in the cerebrum, having only three neuronal layers instead of the six layers found elsewhere.

Effect of Bilateral Removal of the Hippocampi — Inability to Learn. The hippocampi have been removed or destroyed bilaterally in a few human beings. These persons can perform most previously learned activities satisfactorily. However, they can learn essentially nothing new. In fact, they cannot even learn the names or the faces of persons with whom they come in contact every day. Yet, they can remember for a moment or so what transpires during the course of their activities. Thus, they do have a type of short-term memory even though their ability to establish new long-term memories is either completely or almost completely abolished, which is the phenomenon called *anterograde amnesia* that was discussed in the previous chapter.

Destruction of the hippocampi also causes some deficit in previously learned memories (retrograde amnesia), a little more so for recent memories than for memories of the distant past.

Theoretical Function of the Hippocampus in Learning. The hippocampus originated as part of the olfactory cortex. In the very lowest animals it plays essential roles in determining whether the animal will eat a particular food, whether the smell of a particular object suggests danger, whether the odor is sexually inviting, and in making other decisions that are of life and death importance. Thus, very early in the development of the brain, the hippocampus presumably became the critical decision-making neuronal mechanism, determining the importance and type of importance of the incoming sensory signals. Presumably, as the remainder of the brain developed, the connections from the other sensory areas into the hypothalamus have continued to utilize this decision-making capability.

Earlier in this chapter (and also in the previous chapter), it was pointed out that reward and punishment play a major role in determining the importance of information and especially whether or not the information will be stored in memory. A person rapidly becomes habituated to indifferent stimuli but learns assiduously any sensory experience that causes either pleasure or punishment. Yet, what is the mechanism by which this occurs? It has been suggested that the hippocampus acts as the encoding mechanism for translating short-term memory into long-term memory — that is, it translates the short-term memory signals into an appropriate form so that they can be stored in long-term memory and also transmits an additional signal to the long-term memory storage area directing that storage shall take place.

Whatever the mechanism, without the hippocampi *consolidation* of long-term memories does not take place. This is especially true for verbal information, perhaps because the temporal lobes, in which the hippocampi are located, are particularly concerned with verbal information.

FUNCTION OF THE LIMBIC CORTEX

Probably the most poorly understood portion of the entire limbic system is the ring of cerebral cortex called the *limbic cortex* that surrounds the subcortical limbic structures. This cortex functions as a transitional zone through which signals are transmitted from the remainder of the cortex into the limbic system. Therefore, it is presumed that the limbic cortex functions as a cerebral *association area for control of behavior*.

Stimulation of the different regions of the limbic cortex has failed to give any real idea of their functions. However, as is true of so many other portions of the limbic system, essentially all the behavioral patterns that have already been described can also be elicited by stimulation in different portions of the limbic cortex. Likewise, ablation of a few limbic cortical areas can cause persistent changes in an animal's behavior, though ablation in most areas has little effect. Some of the areas that have been noted to cause specific changes include:

Ablation of the Temporal Cortex. When the temporal cortex is ablated bilaterally, the amygdala is almost invariably damaged as well. This was discussed earlier, and it was pointed out that the Klüver-Bucy syndrome occurs. The animal develops consummatory behavior especially, investigates any and all objects, has intense sex drives toward inappropriate animals or even inanimate objects, and loses all fear — thus develops tameness as well.

Ablation of the Posterior Orbital Frontal Cortex. Bilateral removal of the posterior portion of the orbital frontal cortex often causes an animal to develop insomnia and an intense degree of motor restlessness, becoming unable to sit still but moving about continually.

Ablation of the Subcallosal Gyri. The subcallosal gyri are the portions of the limbic cortex that turn downward and under the corpus callosum on the medial portions of the frontal lobes. Destruction of these gyri bilaterally releases the rage centers of the septum and hypothalamus from an inhibitory influence. Therefore, the animal can become vicious and much more subject to fits of rage than normally.

Summary. Until further information is available, it is perhaps best to state that the cortical regions of the limbic system occupy intermediate associative positions between the functions of the remainder of the cerebral cortex and the functions of the subcortical limbic structures for control of behavioral patterns. Thus, in the insular and anterior temporal cortex one especially finds gustatory and olfactory associations. In the hippocampal gyri there is a tendency for complex auditory associations and also complex thought associations derived from the general interpretative areas of the posterior temporal lobes. In the cingulate cortex, there is reason to believe that sensorimotor associations occur. And, finally, the orbitofrontal cortex presumably aids in the analytical functions of the prefrontal lobes.

FUNCTION OF SPECIFIC CHEMICAL TRANSMITTER SYSTEMS FOR BEHAVIOR CONTROL

Recently it has become apparent that some of the synaptic chemical transmitter substances of the brain play especially important roles in behavior. Though this field is only beginning to be explored, the chemical transmitter systems that seem to be especially important are the following:

The Norepinephrine System. In Chapter 54 it was pointed out that large numbers of norepinephrine-secreting neurons are located in the reticular formation, especially in the locus ceruleus, and that they send fibers upward through the reticular activating system to essentially all parts of the diencephalon and cerebrum. One area of the hypothalamus that is especially activated by the norepinephrine system is the ventromedial nuclei that can block many of the drives such as eating, drinking, sex, and so forth when stimulated. Therefore, at the same time that the norepinephrine system activates the cerebral cortex, it also seems to block these other drives that might compete with the higher levels of cerebration.

It is believed that overstimulation of the norepinephrine system is the cause of the manic phase of the manic-depressive psychosis. And, conversely, there is much reason to believe that depressed activity of this system is the basis of the depression stage of the manic-depressive syndrome. Drugs that decrease norepinephrine removal from the synapses are frequently very effective both in increasing the activity of the norepinephrine system and also in the treatment of depression psychoses.

The Dopamine System. In Chapter 52 the dopaminergic neurons located in the substantia nigra and projecting to the striate portion of the basal ganglia were discussed. This system exerts an important continuous restraint on basal ganglia activity.

In addition to projecting to the basal ganglia, these dopaminergic fibers also project to other basal regions of the brain, especially into two areas of the hypothalamus — the area immediately adjacent to the third ventricle and also the lateral hypothalamus. Stimulation of this system especially increases the activity of the lateral hypothalamus, producing enhanced drives for eating and some other types of activity such as fighting. Conversely, destruction of the dopamine system has almost exactly the same effects as destroying the two lateral hypothalami, causing the animal to lose its desire to eat and also decreasing its aversive drives.

It has already been pointed out in Chapter 52 that destruction of the substantia nigral dopamine system causes Parkinson's disease. However, during treatment of Parkinson's disease with L-dopa, which in turn releases dopamine, patients sometimes develop schizophrenic symptoms, indicating that excess dopaminergic activity can cause dissociation of a person's drives and thought patterns. The phenothiazine tranquilizers depress the dopamine system and also are effective in the treatment of many patients with schizophrenia.

The Serotonin System. Located in the medial raphe nuclei of the medulla is a system of serotonin-secreting neurons. Many of the nerve fibers of this system spread downward into the cord where they reduce the input level of pain signals, as was explained in Chapter 50. However, the fibers also spread upward into the reticular formation and into other basal areas of the brain to suppress the activity of the reticular activating system, as well as other brain activity. Thus, this system can promote sleep, as was discussed in Chapter 54. The drug p-chlorophenylalanine inhibits the formation of serotonin, and it can cause prolonged wakefulness in animals. The drug LSD (lysergic acid diethylamide) is also known to act as an antagonist against some functions of serotonin. Its propensity to cause dissociation of both behavioral and mental activities is well known.

The Enkephalin-Endorphin System. Very little is yet known about the behavioral functions of the enkephalin-endorphin system. However, the probable role of this system for suppression of pain was discussed in Chapter 50. It presumably has many other activities as well, because it is secreted in many areas of the brain stem and thalamus.

Effects of the General Hormones on Behavior. Several of the general hormones have specific effects on behavior. Some of these are:

Thyroid hormone increases the overall *metabolism of the body* and increases overall activity of the nervous system as well.

Both *testosterone* in the male and estrogen in the female increase *libido*. Also, injection of a minute quantity of estrogen into the anterior hypothalamus of some lower animals will cause estrous behavior (heat). The sex drive of the human female seems to be increased near the time of ovulation when estrogen secretion is high.

A small, 6 or 7 amino acid *fragment of ACTH,* when injected into the brain of an animal, can cause *intense fear.*

Injection of *norephinephrine,* or excess secretion of norepinephrine or epinephrine by the adrenal medullae, increases the overall activity of the brain, which is essentially the same effect as stimulating the reticular activating system.

Thus, it is becoming clear that various chemical and transmitter systems play important roles in determining an animal's behavior. Furthermore, abnormalities of some of the brain's chemical transmitter systems seem to be the bases of some of the common psychoses.

PSYCHOSOMATIC EFFECTS OF THE BEHAVIORAL SYSTEM

We are all familiar with the fact that abnormal function in the central nervous system can frequently lead to serious dysfunction of the different somatic organs of the body. This is also true in experimental animals. Indeed, prolonged electrical stimulation in the punishment regions of the brain can actually lead to severe sickness of the animal, culminating in death within 24 to 48 hours. We need, therefore, to discuss briefly the mechanisms by which stimulatory effects in the brain can affect the peripheral organs. Ordinarily, this occurs through three routes: (1) through the motor nerves to the skeletal muscles throughout the body, (2) through the autonomic nerves to the different internal organs of the body, and (3) through the hormones secreted by the pituitary gland in response to nervous activity in the hypothalamus, as will be explained in Chapter 75.

Psychosomatic Disorders Transmitted Through the Skeletal Nervous System. Abnormal psychic states can greatly alter the degree of nervous stimulation to the skeletal musculature throughout the body and thereby increase or decrease the skeletal muscular tone. During states of excitement the general skeletal muscular tone as well as sympathetic tone normally increases, whereas during somnolent states skeletal muscular and sympathetic activity both decrease greatly. In neurotic and psychotic states, such as anxiety, tension, and mania, generalized overactivity of both the muscles and sympathetic system often occurs throughout the body. This in turn results in intense feedback from the muscle proprioceptors to the reticular activating system, and the norepinephrine and epinephrine circulating in the blood as a result of the sympathetic activity also excite the reticular activating system, all of which undoubtedly help to maintain an extreme degree of wakefulness and alertness that characterizes these emotional states. Unfortunately, though, the wakefulness prevents adequate sleep and also leads to progressive bodily fatigue and mental dissociations.

Transmission of Psychosomatic Effects Through the Autonomic Nervous System. Many psychosomatic abnormalities result from hyperactivity of either the sympathetic or parasympathetic system. In general, hyperactivity of the sympathetic system occurs in many areas of the body at the same time rather than in focal areas, and the usual effects are (1) increased heart rate — sometimes with palpitation of the heart, (2) increased arterial pressure, (3) constipation, and (4) increased metabolic rate. On the other hand, parasympathetic signals are likely to be much more focal. For instance, stimulation of specific areas in the dorsal motor nuclei of the vagus nerves can cause more or less specifically (1) increased or decreased heart rate or palpitation of the heart, (2) esophageal spasm, (3) increased peristalsis in the upper gastrointestinal tract, or (4) increased hyperacidity of the stomach with resultant development of peptic ulcer. Stimulation of the sacral region of the parasympathetic system, on the other hand, is likely to cause extreme colonic glandular secretion and peristalsis with resulting diarrhea. One can readily see, then, that emotional patterns controlling the sympathetic and parasympathetic centers of the hypothalamus and lower brain stem can cause wide varieties of peripheral psychosomatic effects.

Psychosomatic Effects Transmitted Through the Anterior Pituitary Gland. Electrical stimulation of the posterior hypothalamus increases the secretion of corticotropin by the anterior pituitary gland (as explained earlier in the chapter) and therefore indirectly increases the output of adrenocortical hormones. One of the effects of this is a gradual increase in stomach hyperacidity because of the effect of glucocorticoids on stomach secretion. Over a prolonged period of time this obviously could lead to peptic ulcer, which is a well-

known effect of hypersecretion by the adrenal cortex. Likewise, activity in the anterior hypothalamus increases the pituitary secretion of thyrotropin, which in turn increases the output of thyroxine and leads to an elevated basal metabolic rate. It is well known that different types of emotional disturbances can lead to thyrotoxicosis (as will be explained in Chapter 76), this presumably resulting from overactivity in the anterior hypothalamus.

From these examples, therefore, it is evident that many types of psychosomatic diseases of the body can be caused by abnormal control of anterior pituitary secretion. These will be discussed more extensively in the section on endocrinology later in this text.

REFERENCES

Akiskal, H. (ed.): Affective Disorders: Special Clinical Forms. Philadelphia, W. B. Saunders Co., 1979

Bargmann, W., *et al.* (eds.): Neurosecretion and Neuroendocrine Activity: Evolution, Structure and Function. New York, Springer-Verlag, 1978.

Belmaker, R. H., and vanPragg, H. (eds.): Mania, An Evolving Concept. Jamaica, N. Y., Spectrum Publications, 1980.

Bentley, D., and Konishi, M.: Neural control of behavior. *Annu. Rev. Neurosci., 1*:35, 1978.

Bourne, G. H., *et al.* (eds.): Neuronal Cells and Hormones. New York, Academic Press, 1978.

Bowsher, D.: Mechanisms of Nervous Disorder: An Introduction. Philadelphia, J. B. Lippincott Co., 1978.

CIBA Foundation Symposium: Functions of the Septo-Hippocampal System. New York, Elsevier/Excerpta Medica/North-Holland, 1978.

Cooper, D. G.: The Language of Madness. London, Allen Lane, 1978.

Cotman, C. W., and McGaugh, J. L.: Behavioral Neuroscience. New York, Academic Press, 1979.

DeFeudis, F. V., and DeFeudis, P. A. F.: Elements of the Behavioral Code. New York, Academic Press, 1977.

Depue, R. A. (ed.): The Psychobiology of the Depressive Disorders: Implications For the Effects of Stress. New York, Academic Press, 1979.

De Silva, F. H. L., and Arnolds, D. E. A. T.: Physiology of the hippocampus and related structures. *Annu. Rev. Physiol., 40*:185, 1978.

DiCara, L. V.: Learning mechanisms. *In* Frohlich, E. D. (eds.): Pathophysiology, 2nd Ed. Philadelphia, J. B. Lippincott Co., 1976, p. 757.

Ehrlich, Y. H. (ed.): Modulators, Medicators, and Specifiers in Brain Function. New York, Plenum Press, 1979.

Elde, R., and Hökfelt, T.: Localization of hypophysiotropic peptides and other biologically active peptides within the brain. *Annu. Rev. Physiol., 41*:587, 1979.

Ferrendelli, J. A., and Gurvitch, G. (eds.): Aspects of Behavioral Neurobiology. Bethesda, Md., Society of Neuroscience, 1977.

Fink, G., and Geffen, L. B.: The hypothalamo-hypophysial system: Model for central peptidergic and monoaminergic transmission. *In* Porter, R. (ed.): International Review of Physiology: Neurophysiology III. Vol. 17. Baltimore, University Park Press, 1978, p. 1.

Fink, M.: Convulsive Therapy: Theory and Practice. New York, Raven Press, 1979.

Fuxe, K. (ed.): Peptidergic Neuron. New York, Plenum Press, 1979.

Gershon, M. D.: Biochemistry and physiology of serotonergic transmission. *In* Brookhart, J. M., and Mountcastle, V. B. (eds.): Handbook of Physiology. Sec. 1, Vol. 1. Baltimore, Williams & Wilkins, 1977, p. 573.

Greist, J. H., and Greist, T. H.: Antidepressant Treatment in Primary Care. Baltimore, Williams & Wilkins, 1979.

Guillemin, R.: Neuroendocrine interrelations. *In* Bondy, P. K., and Rosenberg, L. E. (eds.): Metabolic Control and Disease, 8th Ed. Philadelphia, W. B. Saunders Co., 1980, p. 1155.

Guroff, G.: Molecular Neurobiology. New York, Marcel Dekker, 1979.

Jeffcoate, S. L., and Hutchinson, J. S. M. (eds.): The Endocrine Hypothalamus. New York, Academic Press, 1978.

Jones, N. B., and Reynolds, V.: Human Behaviour and Adaptation. New York, Halsted Press, 1978.

Kelly, D.: Anxiety and Emotions: Physiological Basis and Treatment. Springfield, Ill., Charles C Thomas, 1979.

Lederis, K., and Veale, W. L. (eds.): Current Studies of Hypothalamic Function, 1978. New York, S. Karger, 1978.

Legg, N. J. (ed.): Neurotransmitter Systems and Their Clinical Disorders. New York, Academic Press, 1978.

Livingston, K. E., and Hornykiewicz, O. (eds.): Limbic Mechanism. New York, Plenum Press, 1978.

Livingston, R. B.: Neural integration. *In* Frohlich, E. D. (ed.): Pathophysiology, 2nd Ed. Philadelphia, J. B. Lippincott Co., 1976, p. 681.

Mandel, P., and DeFeudis, F. V. (eds.): GABA — Biochemistry and CNS Functions. New York, Plenum Press, 1979.

McFadden, D. (ed.): Neural Mechanisms in Behavior. New York, Springer-Verlag, 1980.

McGuigan, F. J.: Psychophysiological Measurement of Covert Behavior. New York, Halsted Press, 1979.

Morgane, P. J., and Panksepp, J. (eds.): Anatomy of the Hypothalamus. New York, Marcel Dekker, 1979.

Morgane, P. J., and Panksepp, J. (eds.): Handbook of the Hypothalamus. New York, Marcel Dekker, 1979.

Norrsel, U.: Behavioral studies of the somatosensory system. *Physiol. Rev., 60*:327, 1980.

Obata, K.: Biochemistry and physiology of amino acid transmitters. *In* Brookhart, J. M., and Mountcastle, V. B. (eds.): Handbook of Physiology. Sec. 1, Vol. 1. Baltimore, Williams & Wilkins, 1977, p. 625.

Pepeu, G., *et al.* (eds.): Receptors for Neurotransmitters and Peptide Hormones, New York, Raven Press, 1980.

Plotnik, R., and Mollenauer, S.: Brain and Behavior: An Introduction to Physiological Psychology. San Francisco, Canfield Press, 1978.

Rimm, D. C., and Masters, J. C.: Behavior Therapy: Techniques and Empirical Findings, 2nd Ed. New York, Academic Press, 1979.

Routtenberg, A.: The reward system of the brain. *Sci. Am., 239* (5):154, 1978.

Sachar, E. J., and Baron, M.: The biology of affective disorders. *Annu. Rev. Neurosci., 2*:505, 1979.

Sadoul, P. (ed.): Muscular Exercise in Chronic Lung Disease. New York, Pergamon Press, 1979.

Schou, M., and Strömgren, E. (eds.): Origin, Prevention, and Treatment of Affective Disorders. New York, Academic Press, 1979.

Shopsin, B. (ed.): Manic Illness. New York, Raven Press, 1979.

Snyder, S.: Biological Aspects of Mental Disorder. New York, Oxford University Press, 1980.

Sowers, J. R. (ed.): Hypothalamic Hormones. Stroudsburg, Pa., Dowden, Hutchinson & Ross, 1980.

Tipton, K. F. (ed.): Neurochemistry and Biochemical Pharmacology. Baltimore, University Park Press, 1978.

Tolis, G., *et al.* (eds.): Clinical Neuroendocrinology. New York, Raven Press, 1979.

Vale, W., Rivier, C., and Brown, M.: Regulatory peptides of the hypothalamus. *Annu. Rev. Physiol., 39*:473, 1977.

Van Ree, J. M., and Terenius, L. (eds.): Characteristics and Function of Opioids. New York, Elsevier/North Holland, 1978.

Wassermann, G. D.: Neurobiological Theory of Psychological Phenomena. Baltimore, University Park Press, 1978.

Way, E. L. (ed.): Endogenous and Exogenous Opiate Agonists and Antagonists. New York, Pergamon Press, 1979.

Weiner, R. I., and Ganong, W. F.: Role of brain monoamines and histamine in regulation of anterior pituitary secretion. *Physiol. Rev., 58*:905, 1978.

The Autonomic Nervous System;
The Adrenal Medulla

The portion of the nervous system that controls the visceral functions of the body is called the *autonomic nervous system.* This system helps to control arterial pressure, gastrointestinal motility and secretion, urinary bladder emptying, sweating, body temperature, and many other activities, some of which are controlled almost entirely and some only partially by the autonomic nervous system.

GENERAL ORGANIZATION OF THE AUTONOMIC NERVOUS SYSTEM

The autonomic nervous system is activated mainly by centers located in the *spinal cord, brain stem,* and *hypothalamus.* Also, portions of the cerebral cortex and especially of the limbic system can transmit impulses to the lower centers and in this way influence autonomic control. Often the autonomic nervous system operates by means of *visceral reflexes.* That is, sensory signals enter the centers of the cord, brain stem, or hypothalamus, and these in turn transmit appropriate reflex responses back to the visceral organs to control their activities.

The autonomic impulses are transmitted to the body through two major subdivisions called the *sympathetic* and *parasympathetic systems,* the characteristics and functions of which follow.

PHYSIOLOGIC ANATOMY OF THE SYMPATHETIC NERVOUS SYSTEM

Figure 57–1 illustrates the general organization of the sympathetic nervous system, showing one of the two *paravertebral sympathetic chains* to the side of the spinal column and nerves extending to the different internal organs. The sympathetic nerves originate in the spinal cord between the segments T-1 and L-2 and pass from here first into the sympathetic chain, thence to the

tissues and organs that are stimulated by the sympathetic nerves.

Preganglionic and Postganglionic Sympathetic Neurons. The sympathetic nerves are different from skeletal motor nerves in the following way: Each motor pathway from the cord to a skeletal muscle is comprised of a single fiber. Each sympathetic pathway, on the other hand, is comprised of two fibers, a *preganglionic*

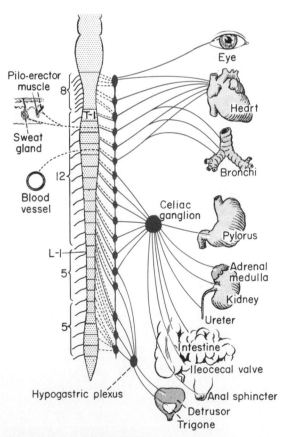

Figure 57–1. The sympathetic nervous system. Dashed lines represent postganglionic fibers in the gray rami leading into the spinal nerves for distribution to blood vessels, sweat glands, and pilo-erector muscles.

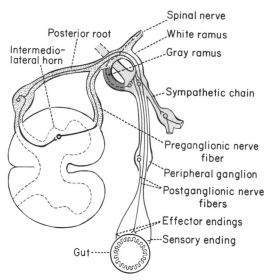

Figure 57–2. Nerve connections between the spinal cord, sympathetic chain, spinal nerves, and peripheral sympathetic nerves.

neuron and a *postganglionic neuron.* The cell body of a preganglionic neuron lies in the *intermediolateral horn* of the spinal cord, and its fiber passes, as illustrated in Figure 57–2, through an *anterior root* of the cord into a *spinal nerve.*

Immediately after the spinal nerve leaves the spinal column, the preganglionic sympathetic fibers leave the nerve and pass through the *white ramus* into one of the *ganglia* of the *sympathetic chain.* Then the course of the fibers can be one of the following three: (1) It can synapse with postganglionic neurons in the ganglion that it enters. (2) It can pass upward or downward in the chain and synapse in one of the other ganglia of the chain. Or (3) it can pass for variable distances through the chain and then through one of the nerves radiating outward from the chain, finally terminating in an *outlying sympathetic ganglion.*

The postganglionic neuron, therefore, can originate either in one of the sympathetic chain ganglia or in one of the outlying ganglia. From either of these two sources, the postganglionic fibers then travel to their destinations in the various organs.

Sympathetic Nerve Fibers in the Skeletal Nerves. Many of the postganglionic fibers pass back from the sympathetic chain into the spinal nerves through *gray rami* (see Fig. 57–2) at all levels of the cord. These pathways are made up of type C fibers that extend to all parts of the body in the skeletal nerves. They control the blood vessels, sweat glands, and piloerector muscles of the hairs. Approximately 8 per cent of the fibers in the average skeletal nerve are sympathetic fibers, a fact that indicates their importance.

Segmental Distribution of Sympathetic Nerves. The sympathetic pathways originating in the different segments of the spinal cord are not necessarily distributed to the same part of the body as the spinal nerve fibers from the same segments. Instead, the *sympathetic fibers from T-1 generally pass up the sympathetic chain into the head; from T-2 into the neck; T-3, T-4, T-5, and T-6*

into the thorax; T-7, T-8, T-9, T-10, and T-11 into the abdomen; T-12, L-1, and L-2 into the legs. This distribution is only approximate and overlaps greatly.

The distribution of sympathetic nerves to each organ is determined partly by the position in the embryo at which the organ originates. For instance, the heart receives many sympathetic nerves from the neck portion of the sympathetic chain because the heart originates in the neck of the embryo. Likewise, the abdominal organs receive their sympathetic innervation from the lower thoracic segments because the primitive gut originates in this area.

Special Nature of the Sympathetic Nerve Endings in the Adrenal Medullae. Preganglionic sympathetic nerve fibers pass, without synapsing, all the way from the intermediolateral horn cells of the spinal cord, through the sympathetic chains, through the splanchnic nerves, and finally into the adrenal medullae. There they end directly on special cells that secrete epinephrine and norepinephrine directly into the circulatory blood. These secretory cells embryologically are derived from nervous tissue and are analogous to postganglionic neurons; indeed, they even have rudimentary nerve fibers, and it is these fibers that secrete the hormones.

PHYSIOLOGIC ANATOMY OF THE PARASYMPATHETIC NERVOUS SYSTEM

The parasympathetic nervous system is illustrated in Figure 57–3, showing that parasympathetic fibers leave the central nervous system through several of the crani-

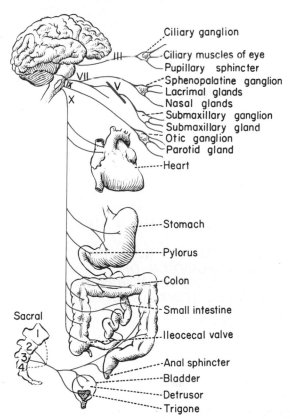

Figure 57–3. The parasympathic nervous system.

al nerves, the second and third sacral spinal nerves, and occasionally the first and fourth sacral nerves. About 75 per cent of all parasympathetic nerve fibers are in the vagus nerves, passing to the entire thoracic and abdominal regions of the body. Therefore, a physiologist speaking of the parasympathetic nervous system is thinking mainly of the two vagus nerves. The vagus nerves supply parasympathetic nerves to the heart, the lungs, the esophagus, the stomach, the small intestine, the proximal half of the colon, the liver, the gallbladder, the pancreas, and the upper portions of the ureters.

Parasympathetic fibers in the *third nerve* flow to the pupillary sphincters and ciliary muscles of the eye. Fibers from the *seventh nerve* pass to the lacrimal, nasal, and submaxillary glands, and fibers from the *ninth nerve* pass to the parotid gland.

The sacral parasympathetic fibers congregate in the form of the *nervi erigentes,* also called the *pelvic nerves,* which leave the sacral plexus on each side of the cord and distribute their peripheral fibers to the descending colon, rectum, bladder, and lower portions of the ureters. Also, this sacral group of parasympathetics supplies fibers to the external genitalia to cause various sexual reactions.

Preganglionic and Postganglionic Parasympathetic Neurons. The parasympathetic system, like the sympathetic, has both preganglionic and postganglionic neurons, but, except in the case of a few cranial parasympathetic nerves, the *preganglionic fibers* pass uninterrupted to the organ that is to be excited by parasympathetic impulses. In the wall of the organ are located the *postganglionic neurons* of the parasympathetic system. The preganglionic fibers synapse with these; then short postganglionic fibers, 1 millimeter to several centimeters in length, leave the neurons to spread in the substance of the organ. This location of the parasympathetic postganglionic neurons in the visceral organ itself is quite different from the arrangement of the sympathetic ganglia, for the cell bodies of the sympathetic postganglionic neurons are almost always located in ganglia of the sympathetic chain or in various other discrete ganglia in the abdomen rather than in the excited organ itself.

BASIC CHARACTERISTICS OF SYMPATHETIC AND PARASYMPATHETIC FUNCTION

CHOLINERGIC AND ADRENERGIC FIBERS — SECRETION OF ACETYLCHOLINE OR NOREPINEPHRINE BY THE POSTGANGLIONIC NEURONS

The sympathetic and parasympathetic nerve endings secrete one of the two synaptic transmitter substances, *acetylcholine* or *norepinephrine.* Those fibers that secrete acetylcholine are said to be *cholinergic.* Those that secrete norepinephrine are said to be *adrenergic,* a term derived from *adrenalin,* which is the British name for epinephrine.

All preganglionic neurons are cholinergic in both the sympathetic and parasympathetic nervous systems. Therefore, acetylcholine or acetylcholine-like substances, when applied to the ganglia, will excite both sympathetic and parasympathetic postganglionic neurons.

The postganglionic neurons of the parasympathetic system are also all cholinergic. On the other hand, most of the postganglionic sympathetic neurons are adrenergic, though this is not entirely so because the sympathetic nerve fibers to the sweat glands and to a few blood vessels are cholinergic.

Thus, in general, the terminal nerve endings of the parasympathetic system secrete acetylcholine, and most of the sympathetic nerve endings secrete norepinephrine. These hormones, in turn, act on the different organs to cause the respective parasympathetic and sympathetic effects. Therefore, these substances are often called respectively *parasympathetic* and *sympathetic mediators.* These are molecular structures of acetylcholine and norepinephrine:

$$CH_3-\underset{\underset{O}{\|}}{C}-O-CH_2-CH_2-\overset{+}{N}\underset{CH_3}{\overset{CH_3}{\diagup}}CH_3$$

Acetylcholine

Norepinephrine

Mechanism of Secretion of Acetylcholine and Norepinephrine by Autonomic Nerve Endings. Some of the autonomic nerve endings, especially those of the parasympathetic nerves, are similar to but much smaller in size than those of the skeletal neuromuscular junction. However, most of the sympathetic nerve fibers merely touch the effector cells of the organs that they innervate as they pass by; in some instances they terminate in connective tissue located adjacent to the cells that are to be stimulated. Where these filaments pass over or near the effector cells, they usually have bulbous enlargements called *varicosities,* and it is in these varicosities that the transmitter vesicles of acetylcholine and norepinephrine are found. Also in the varicosities are large numbers of mitochondria to supply the ATP required to energize acetylcholine or norepinephrine synthesis.

When an action potential spreads over the terminal fibers, the depolarization process increases the permeability of the fiber membrane to calcium ions, thus allowing these to diffuse in moderate numbers into the nerve terminals. There they interact with the vesicles adjacent to the membrane, causing them to fuse with the nerve membrane and to empty their contents to the exterior. Thus, the transmitter substance is secreted.

Synthesis of Acetylcholine, Its Destruction After Secretion, and Duration of Action. Acetylcholine is synthesized in the terminal endings of cholinergic nerve fibers. Most of this synthesis probably occurs in the axoplasm, and the acetylcholine is then transported to the interior of the vesicles. The basic chemical reaction of this synthesis is the following:

$$\text{Acetyl-CoA} + \text{Choline} \xrightarrow{\text{choline acetyl transferase}} \text{Acetylcholine}$$

Once the acetylcholine has been secreted by the cholinergic nerve ending, most of it is split into acetate ion and choline by the enzyme *acetylcholinesterase* that is present both in the terminal nerve ending itself and on the surface of the receptor organ. Thus, this is the same mechanism of acetylcholine destruction that occurs at the neuromuscular junctions of skeletal nerve fibers. The choline that is formed is in turn transported back into the terminal nerve ending where it is used again for synthesis of new acetylcholine. Though most of the acetylcholine is usually destroyed within a fraction of a second after its secretion, it sometimes persists for as long as several seconds, and a small amount also diffuses into the surrounding fluids. These fluids contain a different type of cholinesterase called *serum cholinesterase* that destroys the remaining acetylcholine within another few seconds. Therefore, the action of acetylcholine released by cholinergic nerve fibers usually lasts for a few seconds at most.

Synthesis of Norepinephrine, Its Removal, and Duration of Action. Synthesis of norepinephrine begins in the axoplasm of the terminal nerve endings of adrenergic nerve fibers but is completed inside the vesicles. The basic steps are the following:

1. Tyrosine $\xrightarrow{\text{hydroxylation}}$ DOPA

2. DOPA $\xrightarrow{\text{decarboxylation}}$ Dopamine

3. Transport of dopamine into the vesicles

4. Dopamine $\xrightarrow{\text{hydroxylation}}$ Norepinephrine

In the adrenal medulla this reaction goes still one step further to form epinephrine, as follows:

5. Norepinephrine $\xrightarrow{\text{methylation}}$ Epinephrine

Following secretion of norepinephrine by the terminal nerve endings, it is removed from the secretory site in three different ways: (1) re-uptake into the adrenergic nerve endings themselves by an active transport process — accounting for removal of 50 to 80 per cent of the secreted norepinephrine; (2) diffusion away from the nerve endings into the surrounding body fluids and thence into the blood — accounting for removal of most of the remainder of the norepinephrine; and (3) destruction by enzymes to a slight extent (one of these enzymes is *monoamine oxidase,* which is found in the nerve endings themselves, and another is *catechol-O-methyl transferase,* which is present diffusely in all tissues).

Ordinarily, the norepinephrine secreted directly in a tissue by adrenergic nerve endings remains active for only a few seconds, illustrating that its re-uptake and diffusion away from the tissue are rapid. However, the norepinephrine and epinephrine secreted into the body by the adrenal medullae remain active until they diffuse into some tissue where they are destroyed by catechol-O-methyl transferase; this occurs mainly in the liver. Therefore, when secreted into the blood, both norepinephrine and epinephrine remain very active for 10 to 30 seconds, followed by decreasing activity thereafter for one to several minutes.

RECEPTOR SUBSTANCES OF THE EFFECTOR ORGANS

The acetylcholine, norepinephrine, and epinephrine secreted by the autonomic nervous system all stimulate the effector organs by first reacting with receptor substances in the effector cells. The receptor in most instances is in the cell membrane and is a protein molecule. The most likely mechanism for function of the receptor is that the transmitter substance first binds with the receptor and this causes a basic change in the molecular structure of the receptor compound. Because the receptor is in integral part of the cell membrane, this structural change often alters the permeability of the cell membrane to various ions — for instance, to allow rapid influx of sodium, chloride, or calcium ions into the cell or to allow rapid efflux of potassium ions out of the cell. These ionic changes then usually alter the membrane potential, sometimes eliciting action potentials (as occurs in smooth muscle cells) and at other times causing electrotonic effects on the cells (as occurs on glandular cells) to produce the responses. The ions themselves have direct effects within the receptor cells, such as the effect of calcium ions to promote smooth muscle contraction.

Another way that the receptor can function, besides changing the membrane permeability, is to activate an enzyme in the cell membrane and this enzyme in turn promotes chemical reactions within the cell. For instance, epinephrine increases the activity of *adenyl cyclase* in some cell membranes, and this then causes the formation of cyclic AMP at the inner surfaces of the membranes; the cyclic AMP then initiates many intracellular activities.

The Acetylcholine Receptors — "Muscarinic" and "Nicotinic" Receptors. Acetylcholine activates two different types of receptors. These are called *muscarinic* and *nicotinic* receptors. The reason for these names is that muscarine, a poison from toadstools, also activates the muscarinic receptors but will not activate the nicotinic receptors, whereas nicotine will activate only the other receptors; acetylcholine activates both of them.

The muscarinic receptors are found in all the effector cells stimulated by the postganglionic neurons of the parasympathetic nervous system as well as those stimulated by the postganglionic cholinergic neurons of the sympathetic system.

The nicotinic receptors are found in the synapses between the pre- and postganglionic neurons of both the sympathetic and parasympathetic systems and also in the membranes of skeletal muscle fibers at the neuromuscular junction (discussed in Chapter 12).

An understanding of the two different types of receptors is especially important because specific drugs are frequently used in the practice of medicine to stimulate or to block one or the other of the two types of receptors.

The Adrenergic Receptors — "Alpha" and "Beta" Receptors. Research experiments using different drugs (called *sympathomimetic drugs)* to mimic the action of norepinephrine on sympathetic effector organs have shown that there are two major types of adrenergic receptors, called *alpha receptors* and *beta receptors*. (The beta receptors in turn are divided into *beta_1* and *beta_2* receptors because certain drugs affect some beta receptors but not all.)

Norepinephrine and epinephrine, both of which are secreted by the adrenal medulla, have somewhat different effects in exciting the alpha and beta receptors. Norepinephrine excites mainly alpha receptors but excites the beta receptors to a very slight extent as well. On the other hand, epinephrine excites both types of receptors approximately equally. Therefore, the relative effects of norepinephrine and epinephrine on different effector organs is determined by the types of receptors in the organs. Obviously, if they are all beta receptors, epinephrine will be the more effective excitant.

Table 57–1 gives the distribution of alpha and beta receptors in some of the organs and systems controlled by the sympathetics. Note that certain alpha functions are excitatory while others are inhibitory. Likewise, certain beta functions are excitatory and others are inhibitory. Therefore, alpha and beta receptors are not necessarily associated with excitation or inhibition but simply with the affinity of the hormone for the receptors in a given effector organ.

A synthetic hormone chemically similar to epinephrine and norepinephrine, *isopropyl norepinephrine,* has an extremely strong action on beta receptors but essentially no action on alpha receptors. Later in the chapter we will discuss various drugs that can mimic the adrenergic actions of epinephrine and norepinephrine or that will block specifically the alpha or beta receptors.

EXCITATORY AND INHIBITORY ACTIONS OF SYMPATHETIC AND PARASYMPATHETIC STIMULATION

Table 57–2 lists the effects on different visceral functions of the body caused by stimulating the parasympathetic and sympathetic nerves. From this table it can be seen that *sympathetic stimulation causes excitatory effects in some organs but inhibitory effects in others. Likewise, parasympathetic stimulation causes excitation in some organs but inhibition in others.* Also, when sympathetic stimulation excites a particular organ, parasympathetic stimulation often inhibits it, illustrating that the two systems occasionally act reciprocally to each other. However, most organs are dominantly controlled by one or the other of the two systems.

There is no generalization one can use to explain whether sympathetic or parasympathetic stimulation will cause excitation or inhibition of a particular organ. Therefore, to understand sympathetic and parasympathetic function, one must learn the functions of these two nervous systems as listed in Table 57–2. Some of these functions need to be clarified in still greater detail as follows:

EFFECTS OF SYMPATHETIC AND PARASYMPATHETIC STIMULATION ON SPECIFIC ORGANS

The Eye. Two functions of the eye are controlled by the autonomic nervous system: These are the pupillary opening and the focus of the lens. Sympathetic stimulation contracts the meridional *fibers of the iris* and, therefore, dilates the pupil, while parasympathetic stimulation contracts the *circular muscle of the iris* to constrict the pupil. The parasympathetics that control the pupil are reflexly stimulated when excess light enters the eyes; this reflex reduces the pupillary opening and decreases the amount of light that strikes the retina. On the other hand, the sympathetics become stimulated during periods of excitement and, therefore, increase the pupillary opening at these times.

Focusing of the lens is controlled almost entirely by the parasympathetic nervous system. The lens is normally held in a flattened state by tension of its radial

TABLE 57–1. ADRENERGIC RECEPTORS AND FUNCTION

Alpha Receptor	Beta Receptor
Vasoconstriction	Vasodilatation (β_2)
Iris dilatation	Cardioacceleration (β_1)
Intestinal relaxation	Increased myocardial strength (β_1)
Intestinal sphincter contraction	Intestinal relaxation (β_2)
Pilomotor contraction	Uterus relaxation (β_2)
Bladder sphincter contraction	Bronchodilatation (β_2)
	Calorigenesis (β_2)
	Glycogenolysis (β_2)
	Lipolysis (β_1)
	Bladder relaxation (β_2)

TABLE 57–2.   AUTONOMIC EFFECTS ON VARIOUS ORGANS OF THE BODY

Organ	Effect of Sympathetic Stimulation	Effect of Parasympathetic Stimulation
Eye: Pupil	Dilated	Constricted
Ciliary muscle	Slight relaxation	Contracted
Glands: Nasal	Vasoconstriction and slight	Stimulation of thin, copious
Lacrimal	secretion	secretion (containing many enzymes
Parotid		for enzyme-secreting glands)
Submaxillary		
Gastric		
Pancreatic		
Sweat glands	Copious sweating (cholinergic)	None
Apocrine glands	Thick, odoriferous secretion	None
Heart: Muscle	Increased rate	Slowed rate
	Increased force of contraction	Decreased force of atrial contraction
Coronaries	Dilated (β_2); constricted (α)	Dilated
Lungs: Bronchi	Dilated	Constricted
Blood vessels	Mildly constricted	? Dilated
Gut: Lumen	Decreased peristalsis and tone	Increased peristalsis and tone
Sphincter	Increased tone	Relaxed
Liver	Glucose released	Slight glycogen synthesis
Gallbladder and bile ducts	Relaxed	Contracted
Kidney	Decreased output	None
Bladder: Detrusor	Relaxed	Excited
Trigone	Excited	Relaxed
Penis	Ejaculation	Erection
Systemic blood vessels:		
Abdominal	Constricted	None
Muscle	Constricted (adrenergic α)	None
	Dilated (adrenergic β)	
	Dilated (cholinergic)	
Skin	Constricted	None
Blood: Coagulation	Increased	None
Glucose	Increased	None
Basal metabolism	Increased up to 100%	None
Adrenal cortical secretion	Increased	None
Mental activity	Increased	None
Piloerector muscles	Excited	None
Skeletal muscle	Increased glycogenolysis	None
	Increased strength	

ligaments. Parasympathetic excitation contracts the *ciliary muscle,* which releases this tension and allows the lens to become more convex. This causes the eye to focus on objects near at hand. The focusing mechanism is discussed in Chapters 58 and 60 in relation to function of the eyes.

The Glands of the Body.   The *nasal, lacrimal, salivary,* and many *gastrointestinal glands* are all strongly stimulated by the parasympthetic nervous system, resulting in copious quantities of secretion. The glands of the alimentary tract most strongly stimulated by the parasympathetics are those of the upper tract, especially those of the mouth and stomach. The glands of the small and large intestines are controlled principally by local factors in the intestinal tract itself and not by the autonomic nerves.

Sympathetic stimulation has a slight direct effect on glandular cells in causing formation of a concentrated secretion. However, it also causes vasoconstriction of the blood vessels supplying the glands and in this way often reduces their rates of secretion.

The *sweat glands* secrete large quantities of sweat when the sympathetic nerves are stimulated, but no effect is caused by stimulating the parasympathetic nerves. However, the sympathetic fibers to most sweat glands are *cholinergic* (except for a few adrenergic fibers to palms of the hand and the soles of the feet), in contrast to most other sympathetic fibers, which are adrenergic. Furthermore, the sweat glands are stimulated primarily by centers in the hypothalamus that are usually considered to be parasympathetic centers. Therefore, sweating could be called a parasympathetic function.

The *apocrine glands* secrete a thick, odoriferous secretion as a result of sympathetic stimulation, but they do not react to parasympathetic stimulation. Furthermore, the apocrine glands, despite their close embryological relationship to sweat glands, are controlled by adrenergic fibers rather than by cholinergic fibers and are controlled by the sympathetic centers of the central nervous system rather than by the parasympathetic centers.

The Gastrointestinal System.   The gastrointestinal system has its own intrinsic set of nerves known as the *intramural plexus.* However, both parasympathetic and sympathetic stimulation can affect gastrointestinal activity — parasympathetic especially. Parasympathetic stimulation, in general, increases the overall degree of activity of the gastrointestinal tract by promoting peristalsis and relaxing the sphincters, thus allowing rapid propulsion of contents along the tract. This propulsive effect is associated with simultaneous increases in rates of secretion by many of the gastrointestinal glands, which was described above.

Normal function of the gastrointestinal tract is not very dependent on sympathetic stimulation. However, in some diseases, strong sympathetic stimulation inhibits peristalsis and increases the tone of the sphincters. The net result is greatly slowed propulsion of food through the tract.

The Heart. In general, sympathetic stimulation increases the overall activity of the heart. This is accomplished by increasing both the rate and force of the heartbeat. Parasympathetic stimulation causes mainly the opposite effects, decreasing the overall activity of the heart. To express these effects in another way, sympathetic stimulation increases the effectiveness of the heart as a pump, while parasympathetic stimulation decreases its pumping capability. However, sympathetic stimulation unfortunately also greatly increases the metabolism of the heart while parasympathetic stimulation decreases its metabolism and allows the heart a degree of rest.

Systemic Blood Vessels. Most blood vessels, especially those of the abdominal viscera and the skin of the limbs, are constricted by sympathetic stimulation. Parasympathetic stimulation generally has almost no effects on blood vessels but does dilate vessels in certain restricted areas such as in the blush area of the face. Under some conditions, the beta stimulatory function of the sympathetics causes vascular dilatation, especially when drugs have paralyzed the sympathetic alpha effects.

Effect of Sympathetic and Parasympathetic Stimulation on Arterial Pressure. The arterial pressure is caused by propulsion of blood by the heart and by resistance to flow of this blood through the blood vessels. In general, sympathetic stimulation increases both propulsion by the heart and resistance to flow, which can cause the pressure to increase greatly.

On the other hand, parasympathetic stimulation decreases the pumping by the heart, which lowers the pressure a moderate amount, though not nearly so much as the sympathetics can increase the pressure.

The Lungs. In general, the structures in the lungs do not have extensive sympathetic or parasympathetic innervation. Therefore, most effects of stimulation are mild. Sympathetic stimulation can dilate the bronchi and mildly constrict the blood vessels. On the contrary, parasympathetic stimulation can cause mild constriction of the bronchi and can perhaps mildly dilate the vessels.

Effects of Sympathetic and Parasympathetic Stimulation on Other Functions of the Body. Because of the great importance of the sympathetic and parasympathetic control systems, these are discussed many times in this text in relation to a myriad of body functions that are not considered in detail here. In general, most of the entodermal structures, such as the ducts of the liver, the gallbladder, the ureter, and the bladder, are inhibited by sympathetic stimulation but excited by parasympathetic stimulation. Sympathetic stimulation also has metabolic effects, causing release of glucose from the liver, increase in blood glucose concentration, increase in glycogenolysis in muscle, increase in muscle strength, increase in basal metabolic rate, and increase in mental activity. Finally, the sympathetics and parasympathetics are involved in the execution of the male and female sexual acts, as will be explained in Chapters 80 and 81.

FUNCTION OF THE ADRENAL MEDULLAE

Stimulation of the sympathetic nerves to the adrenal medullae causes large quantities of epinephrine and norepinephrine to be released into the circulating blood, and these two hormones in turn are carried in the blood to all tissues of the body. On the average, approximately 80 per cent of the secretion is epinephrine and 20 per cent is norepinephrine, though the relative proportions of these change considerably under different physiological conditions.

The circulating epinephrine and norepinephrine have almost the same effects on the different organs as those caused by direct sympathetic stimulation, except that *the effects last about 10 times as long* because these hormones are removed from the blood slowly.

The circulating norepinephrine causes constriction of essentially all the blood vessels of the body; it causes increased activity of the heart, inhibition of the gastrointestinal tract, dilation of the pupils of the eyes, and so forth.

Epinephrine causes almost the same effects as those caused by norepinephrine, but the effects differ in the following respects: First, epinephrine has a greater effect on cardiac activity than norepinephrine. Second, epinephrine causes only weak constriction of the blood vessels of the muscles in comparison with a much stronger constriction that results from norepinephrine. Since the muscle vessels represent a major segment of all vessels of the body, this difference is of special importance because norepinephrine greatly increases the total peripheral resistance and thereby greatly elevates arterial pressure, whereas epinephrine raises the arterial pressure to a lesser extent but increases the cardiac output considerably more because of its effect on the heart and veins.

A third difference between the action of epinephrine and norepinephrine relates to their effects on tissue metabolism. Epinephrine probably has several times as great a metabolic effect as norepinephrine. Indeed, the epinephrine secreted by the adrenal medullae increases the metabolic rate of the body often to as much as 100 per cent above normal, in this way increasing the activity and excitability of the whole body. It also increases the rate of other metabolic activities, such as glycogenolysis in the liver and muscle and glucose release into the blood.

In summary, stimulation of the adrenal medullae causes the release of hormones that have almost the same effects throughout the body as direct sympathetic stimulation, except that the effects are greatly prolonged, up to a minute or two after the stimulation is over. The only significant differences are caused by the epinephrine in the secretion, which increases the rate of metabolism and cardiac output to a greater extent than is caused by direct sympathetic stimulation.

Value of the Adrenal Medullae to the Function of the Sympathetic Nervous System. Usually, when any part of the sympathetic nervous system is stimulated, the entire system, or at least major portions of it, is stimulated at the same time. Therefore, norepinephrine and epinephrine are almost always released by the adrenal medullae at the same time that the different organs are being stimulated directly by the sympathetic nerves. Therefore, the organs are actually stimulated in two different ways simultaneously, directly by the sympathetic nerves and indirectly by the medullary hormones. The two means of stimulation support each other, and either can usually substitute for the other. For instance, destruction of the direct sympathetic pathways to the organs does not abrogate excitation of the organs because norepinephrine and epinephrine are still released into the circulating fluids and indirectly cause stimulation. Likewise, total loss of the two adrenal medullae usually has little significant effect on the operation of the sympathetic nervous system because the direct pathways can still perform almost all the necessary duties. Thus, the dual mechanism of sympathetic stimulation provides a safety factor, one mechanism substituting for the other when the second is missing.

Another important value of the adrenal medullae is the capability of epinephrine and norepinephrine to stimulate structures of the body that are not innervated by direct sympathetic fibers. For instance, the metabolic rate of every cell of the body is increased by these hormones, especially by epinephrine, even though only a small proportion of all the cells in the body are innervated directly by sympathetic fibers.

RELATIONSHIP OF STIMULUS RATE TO DEGREE OF SYMPATHETIC AND PARASYMPATHETIC EFFECT

A special difference between the autonomic nervous system and the skeletal nervous system is the low frequency of stimulation required for full activation of autonomic effectors. In general, only one impulse every second or so suffices to maintain normal sympathetic or parasympathetic effect, and full activation occurs when the nerve fibers discharge 10 to 20 times per second. This compares with full activation in the skeletal nervous system at about 75 to 200 impulses per second.

SYMPATHETIC AND PARASYMPATHETIC "TONE"

The sympathetic and parasympathetic systems are continually active, and the basal rates of activity are known, respectively, as *sympathetic tone* or *parasympathetic tone*.

The value of tone is that it allows a single nervous system to increase or to decrease the activity of a stimulated organ. For instance, sympathetic tone normally keeps almost all the blood vessels of the body constricted to approximately half their maximum diameter. By increasing the degree of sympathetic stimulation, the vessels can be constricted even more; but, on the other hand, by inhibiting the normal tone, the vessels can be dilated. If it were not for the continual sympathetic tone, the sympathetic system could cause only vasoconstriction, never vasodilatation.

Another interesting example of tone is that of the parasympathetics in the gastrointestinal tract. Surgical removal of the parasympathetic supply to the gut by cutting the vagi can cause serious and prolonged gastric and intestinal "atony," thus illustrating that in normal function the parasympathetic tone to the gut is strong. This tone can be decreased by the brain, thereby inhibiting gastrointestinal motility, or it can be increased, thereby promoting increased gastrointestinal activity.

Tone Caused by Basal Secretion of Norepinephrine and Epinephrine by the Adrenal Medullae. The normal resting rate of secretion by the adrenal medullae is about 0.2 μgm./kg./min. of epinephrine and about 0.05 μgm./kg./min. of norepinephrine. These quantities are considerable — indeed, enough to maintain the blood pressure almost up to the normal value even if all direct sympathetic pathways to the cardiovascular system are removed. Therefore, it is obvious that much of the overall tone of the sympathetic nervous system results from basal secretion of epinephrine and norepinephrine in addition to the tone that results from direct sympathetic stimulation.

Effect of Loss of Sympathetic or Parasympathetic Tone Following Denervation. Immediately after a sympathetic or parasympathetic nerve is cut, the innervated organ loses its sympathetic or parasympathetic tone. In the case of the blood vessels, for instance, cutting the sympathetic nerves results immediately in almost maximal vasodilatation. However, over several days or weeks, the *intrinsic tone* in the smooth muscle of the vessels increases, usually restoring almost normal vasoconstriction.

Essentially the same events occur in most effector organs whenever sympathetic or parasympathetic tone is lost. That is, intrinsic compensation soon develops to return the function of the organ almost to its normal basal level. However, in the parasympathetic system, the compensation sometimes requires many months. For instance, loss of parasympathetic tone to the heart increases the heart rate from 90 to 160 beats per minute in a dog, and this will still be about 120 beats six months later.

DENERVATION SUPERSENSITIVITY OF SYMPATHETIC AND PARASYMPATHETIC ORGANS FOLLOWING DENERVATION

During the first week or so after a sympathetic or parasympathetic nerve is destroyed, the innervated organ becomes more and more sensitive to injected norepinephrine or acetylcholine, respectively. This effect is illustrated in Figure 57–4; the blood flow in the forearm before removal of the sympathetics was 200 ml. per minute, and a test dose of norepinephrine caused only a slight depression in flow. Then the stellate ganglion was removed, and normal sympathetic tone was lost. At first, the blood flow rose markedly because of the lost vascular tone, but over a period of days to weeks the blood flow returned almost to normal because of progressive increase in intrinsic tone of the vascular musculature itself, thus compensating for the loss of sympathetic tone. Another test dose of norepinephrine was then administered and the blood flow decreased much more than before, illustrating that the blood vessels had become about two to four times as responsive to norepinephrine as previously. This phenomenon is called *denervation supersensitivity*. It occurs in both sympathetic and parasympathetic organs and to a far greater extent in some organs than in others, often increasing the response as much as 10- to 50-fold.

Mechanism of Denervation Supersensitivity. The cause of denervation supersensitivity is only partially known. Part of the answer is probably that the number of receptors in the postsynaptic membranes of the effector cells increases — sometimes many-fold — when norepinephrine or acetylcholine are no longer released at the synapses. Therefore, when these hor-

mones appear in the circulating blood, the effector reaction is vastly enhanced.

THE AUTONOMIC REFLEXES

It is mainly by means of autonomic reflexes that the autonomic nervous system regulates visceral functions. Throughout this text the functions of these reflexes are discussed in relation to individual organ systems, but, to illustrate their importance, a few are presented here briefly.

Cardiovascular Autonomic Reflexes. Several reflexes in the cardiovascular system help to control the arterial blood pressure, cardiac output, and heart rate. One of these is the *baroreceptor reflex,* which was described in Chapter 22 along with other cardiovascular reflexes. Briefly, stretch receptors called *baroreceptors* are located in the walls of the major arteries, including the carotid arteries and the aorta. When these become stretched by high pressure, signals are transmitted to the brain stem, where they inhibit the sympathetic impulses to the heart and blood vessels, which allows the arterial pressure to fall back toward normal.

The Gastrointestinal Autonomic Reflexes. The uppermost part of the gastrointestinal tract and also the rectum are controlled principally by autonomic reflexes. For instance, the smell of appetizing food initiates signals from the nose to the vagal, glossopharyngeal, and salivary nuclei of the brain stem. These in turn transmit signals through the parasympathetic nerves to the secretory glands of the mouth and stomach, causing secretion of digestive juices even before food enters the mouth. And when fecal matter fills the rectum at the other end of the alimentary canal, sensory impulses inititated by stretching the rectum are sent to the sacral portion of the spinal cord, and a reflex signal is retransmitted through the parasympathetics to the distal parts of the colon; these result in strong peristaltic contractions that empty the bowel.

Other Autonomic Reflexes. Emptying of the bladder is controlled in the same way as emptying the rectum; stretching of the bladder sends impulses to the sacral cord, and this in turn causes contraction of the bladder as well as relaxation of the urinary sphincters, thereby promoting micturition.

Also important are the sexual reflexes which are initiated both by psychic stimuli from the brain and stimuli from the sexual organs. Impulses from these sources converge on the sacral cord and, in the male, result, first, in erection, mainly a parasympathetic function, and then in ejaculation, a sympathetic function.

Other autonomic reflexes include reflex contributions to the regulation of pancreatic secre-

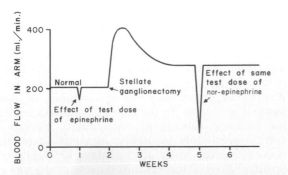

Figure 57–4. Effect of sympathectomy on blood flow in the arm, and the effect of a test dose of norepinephrine before and after sympathectomy, showing sensitization of the vasculature to norepinephrine.

tion, gallbladder emptying, urinary excretion, sweating, blood glucose concentration, and many other visceral functions, all of which are discussed in detail at other points in this text.

MASS DISCHARGE OF THE SYMPATHETIC SYSTEM VERSUS DISCRETE CHARACTERISTICS OF PARASYMPATHETIC REFLEXES

Large portions of the sympathetic nervous system often become stimulated simultaneously, a phenomenon called *mass discharge*. This characteristic of sympathetic action is in keeping with the usually diffuse nature of sympathetic function, such as overall regulation of arterial pressure or of metabolic rate.

However, in a few instances sympathetic activity does occur in isolated portions of the system. The most important of these are: (1) In the process of heat regulation, the sympathetics control sweating and blood flow in the skin without affecting other organs innervated by the sympathetics. (2) During muscular activity in some animals, cholinergic vasodilator fibers of the skeletal muscles are stimulated independently of all the remainder of the sympathetic system. (3) Many "local reflexes" involving the spinal cord but usually not the higher nervous centers affect local areas. For instance, heating a local skin area causes local vasodilatation and enhanced local sweating, while cooling causes the opposite effects.

In contrast to the sympathetic system, most reflexes of the parasympathetic system are very specific. For instance, parasympathetic cardiovascular reflexes usually act only on the heart to increase or decrease its rate of beating. Likewise, parasympathetic reflexes frequently cause secretion mainly in the mouth or, in other instances, secretion mainly in the stomach glands. Finally, the rectal emptying reflex does not affect other parts of the bowel to a major extent.

Yet there is often association between closely allied parasympathetic functions. For instance, though salivary secretion can occur independently of gastric secretion, these two often also occur together, and pancreatic secretion frequently occurs at the same time. Also, the rectal emptying reflex often initiates a bladder emptying reflex, resulting in simultaneous emptying of both the bladder and rectum. Conversely, the bladder emptying reflex can help to initiate rectal emptying.

"ALARM" OR "STRESS" FUNCTION OF THE SYMPATHETIC NERVOUS SYSTEM

From the above discussions of the sympathetic nervous system, one can already see that mass sympathetic discharge in many different ways increases the capability of the body to perform vigorous muscle activity. Let us quickly summarize these ways:

1. increased arterial pressure
2. increased blood flow to active muscles concurrent with decreased blood flow to organs that are not needed for rapid activity
3. increased rates of cellular metabolism throughout the body
4. increased blood glucose concentration
5. increased glycolysis in muscle
6. increased muscle strength
7. increased mental activity
8. increased rate of blood coagulation

The sum of these effects permits the person to perform far more strenuous physical activity than would otherwise be possible. Since it is physical *stress* that usually excites the sympathetic system, it is frequently said that the purpose of the sympathetic system is to provide extra activation of the body in states of stress: this is often called the sympathetic stress *reaction*.

The sympathetic system is also strongly activated in many emotional states. For instance, in the state of *rage*, which is elicited mainly by stimulating the hypothalamus, signals are transmitted downward through the reticular formation and spinal cord to cause massive sympathetic discharge, and all of the sympathetic events listed above ensue immediately. This is called the sympathetic *alarm reaction*. It is also frequently called the *fight or flight reaction* because an animal in this state decides almost instantly whether to stand and fight or to run. In either event, the sympathetic alarm reaction makes the animal's subsequent activities extremely vigorous.

MEDULLARY, PONTINE, AND MESENCEPHALIC CONTROL OF THE AUTONOMIC NERVOUS SYSTEM

Many areas in the reticular substance of the medulla, pons, and mesencephalon, as well as many special nuclei (see Fig. 57–5), control different autonomic functions such as arterial pressure, heart rate, glandular secretion in the upper part of the gastrointestinal tract, gastrointestinal peristalsis, the degree of contraction of the urinary bladder, and many others. The control of each of these is discussed in detail at appropriate points in this text. Suffice it to point out here that the most important factors controlled in the lower brain stem are arterial pressure, heart rate, and respiration. Indeed, transection of the brain stem at the midpontine level allows normal basal control of arterial pressure to continue as before but prevents its modulation by higher nervous centers, particularly the hypothalamus. On the other hand, transection immediately below the medulla causes

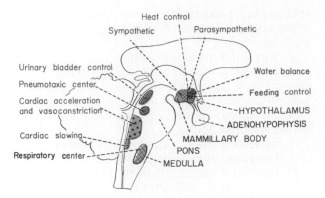

Figure 57–5. Autonomic control centers of the brain stem.

the arterial pressure to fall to about one-half normal for several hours or several days after the transection.

Closely associated with the cardiovascular regulatory centers in the medulla is the medullary center for regulation of respiration discussed in detail in Chapter 42. Though this is not considered to be an autonomic function, it is one of the *involuntary* functions of the body.

Control of Lower Brain Stem Autonomic Centers by Higher Areas. Signals from the hypothalamus and even from the cerebrum can affect the activities of almost all the lower brain stem autonomic control centers. For instance, stimulation in appropriate areas of the hypothalamus can activate the medullary cardiovascular control centers strongly enough to increase the arterial pressure to more than double normal. Likewise, other hypothalamic centers can control body temperature, increase or decrease salivation and gastrointestinal activity, or cause bladder emptying. To a great extent the autonomic centers in the lower brain stem are relay stations for control activities initiated at higher levels of the brain.

In the previous chapter it was also pointed out that many of the behavioral responses of an animal are mediated through the hypothalamus, reticular formation, and the autonomic nervous system. Indeed, the higher areas of the brain can alter the function of the whole autonomic nervous system or of portions of it strongly enough to cause severe autonomic-induced disease, such as peptic ulcer, constipation, heart palpitation, and even heart attacks.

PHARMACOLOGY OF THE AUTONOMIC NERVOUS SYSTEM

DRUGS THAT ACT ON ADRENERGIC EFFECTOR ORGANS — THE SYMPATHOMIMETIC DRUGS

From the foregoing discussion, it is obvious that intravenous injection of norepinephrine causes essentially the same effects throughout the body as sympathetic stimulation. Therefore, norepinephrine is called a *sympathomimetic*, or *adrenergic*, *drug*. Other sympathomimetic drugs are *epinephrine, methoxamine,* and many others. These differ from each other in the degree to which they stimulate different sympathetic effector organs and in their duration of action. Norepinephrine and epinephrine have actions as short as 1 to 2 minutes, while the actions of most other commonly used sympathomimetic drugs last 30 minutes to 2 hours.

Important drugs that stimulate specific adrenergic receptors but not the others are:

phenylephrine — α receptors
isoproterenol — β receptors
salbutamol — only β_2 receptors

Drugs That Cause Release of Norepinephrine from Nerve Endings. Certain drugs have a sympathomimetic action in an indirect manner rather than by directly exciting adrenergic effector organs. These drugs include ephedrine, tyramine, and amphetamine. Their effect is to cause release of norepinephrine from its storage vesicles in the sympathetic nerve endings. The norepinephrine in turn causes the sympathetic effects.

Drugs That Block Adrenergic Activity. Adrenergic activity can be blocked at several different points in the stimulatory process as follows:

1. The synthesis and storage of norepinephrine in the sympathetic nerve endings can be prevented. The best known drug that causes this effect is *reserpine*.

2. Release of norepinephrine from the sympathetic endings can be blocked. This is caused by *guanethidine*.

3. The *alpha* receptors can be blocked. Two drugs that cause this effect are *phenoxybenzamine* and *phentolamine*.

4. The beta receptors can be blocked. A drug that blocks all beta receptors is *propranolol*. One that blocks only beta$_1$ receptors is *practolol*.

5. Sympathetic activity can be blocked by drugs that block transmission of nerve impulses through the autonomic ganglia. These are discussed in the following section, but the most important drug for blockade of both sympathetic and parasympathetic transmission through the ganglia is *hexamethonium*.

DRUGS THAT ACT ON CHOLINERGIC EFFECTOR ORGANS

Parasympathomimetic Drugs (Muscarinic Drugs). Acetylcholine injected intravenously usually does not cause exactly the same effects throughout the body as parasympathetic stimulation because the acetylcholine is destroyed by cholinesterase in the blood and body fluids before it can reach all the effector organs. Yet a number of other drugs that are not so rapidly destroyed can produce typical parasympathetic effects, and these are called parasympathomimetic drugs.

Two commonly used parasympathomimetic drugs are *pilocarpine* and *methacholine*. These act directly on the muscarinic type of cholinergic receptors.

Parasympathomimetic drugs act on the effector organs of cholinergic *sympathetic* fibers also. For instance, these drugs cause profuse sweating. Also, they cause vascular dilatation in some organs, this effect occurring even in vessels not innervated by cholinergic fibers.

Drugs That Have a Parasympathetic Potentiating Effect. Some drugs do not have a direct effect on parasympathetic effector organs but do potentiate the effects of the naturally secreted acetylcholine at the parasympathetic endings. These are the same drugs as those listed in Chapter 12 that potentiate the effect of acetylcholine at the neuromuscular junction — that is, *neostigmine, physostigmine,* and *diisopropyl fluorophosphate.* These inhibit acetylcholinesterase, thus preventing rapid destruction of the acetylcholine liberated by the parasympathetic nerve endings. As a consequence, the quantity of acetylcholine acting on the effector organs progressively increases with successive stimuli, and the degree of action also increases.

Drugs That Block Cholinergic Activity at Effector Organs. *Atropine* and similar drugs, such as *homatropine* and *scopolamine,* block the action of acetylcholine on the muscarinic type of cholinergic effector organs. However, these drugs do not affect the nicotinic action of acetylcholine on the postganglionic neurons or on skeletal muscle.

DRUGS THAT STIMULATE THE POSTGANGLIONIC NEURONS – "NICOTINIC DRUGS"

The preganglionic neurons of both the parasympathetic and sympathetic nervous systems secrete acetylcholine at their endings, and the acetylcholine in turn stimulates the postganglionic neurons. Therefore, injected acetylcholine can also stimulate the postganglionic neurons of both systems, thereby causing both sympathetic and parasympathetic effects in the body. *Nicotine* is a drug that can also stimulate postganglionic neurons in the same manner as acetylcholine because the neuronal membranes contain *nicotinic receptors,* but it cannot directly stimulate the autonomic effector organs, which have muscarinic receptors as explained earlier in the chapter. Therefore, drugs that cause autonomic effects by stimulating the postganglionic neurons are frequently called nicotinic drugs. Some drugs, such as *acetylcholine* itself and *methacholine,* have both nicotinic and muscarinic actions, whereas pilocarpine has only muscarinic actions.

Nicotine excites both the sympathetic and parasympathetic systems at the same time, resulting in strong sympathetic vasoconstriction in the abdominal organs and limbs, but at the same time resulting in parasympathetic effects, such as increased gastrointestinal activity and, sometimes, slowing of the heart.

Ganglionic Blocking Drugs. Many important drugs block impulse transmission from the preganglionic neurons to the postganglionic neurons, including *tetraethyl ammonium ion, hexamethonium ion,* and *pentolinium.* These inhibit impulse transmission in both the sympathetic and parasympathetic systems simultaneously. They are often used for blocking sympathetic activity but rarely for blocking parasympathetic activity, because the sympathetic blockade usually far overshadows the effects of parasympathetic blockade. The ganglionic blocking drugs have been especially important in reducing arterial pressure in patients with hypertension.

REFERENCES

Aviado, D. M., et al. (eds.): Pharmacology of Ganglionic Transmission. New York, Springer-Verlag, 1979.

Axelrod, J.: Neurotransmitters. *Sci. Am., 230(6)*:58, 1974.

Bhagat, B. D.: Mode of Action of Autonomic Drugs. Flushing, N.Y., Graceway Publishing Company, 1979.

Black, I. B.: Regulation of autonomic development. *Annu. Rev. Neurosci., 1*:183, 1978.

Brooks, C. M., et al. (eds.): Integrative Functions of the Autonomic Nervous System. Tokyo, University of Tokyo Press, 1979.

Carrier, O., Jr.: Pharmacology of the Peripheral Autonomic Nervous System. Chicago, Year Book Medical Publishers, 1972.

Collier, B.: Biochemistry and physiology of cholinergic tranmission. *In* Brookhart, J. M., and Mountcastle, V. B. (eds.): Handbook of Physiology. Sec. 1, Vol. I. Baltimore, Williams & Wilkins, 1977, p. 463.

DeQuattro, V., et al.: Anatomy and biochemistry of the sympathetic nervous system. *In* DeGroot, L. J., et al. (eds.): *Endocrinology.* Vol. 2. New York, Grune & Stratton, 1979, p. 1241.

Eränkö, O.: Small intensely fluorescent (SIF) cells and nervous transmission in sympathetic ganglia. *Annu. Rev. Pharmacol. Toxicol., 18*:417, 1978.

Geffen, L. B., and Jarrott, B.: Cellular aspects of catecholaminergic neurons. *In* Brookhart, J. M., and Mountcastle, V. B. (eds.): Handbook of Physiology. Sec. 1, Vol. 1. Baltimore, Williams & Wilkins, 1977, p. 521.

Guyton, A. C., and Gillespie, W. M., Jr.: Constant infusion of epinephrine: Rate of epinephrine secretion and destruction in the body. *Am. J. Physiol., 165*:319, 1951.

Guyton, A. C., and Reeder, R. C.: Quantitative studies on the autonomic actions of curare. *J. Pharmacol. Exp. Ther., 98*:188, 1950.

Haber, E., and Wrenn, S.: Problems in identification of the beta-adrenergic receptor. *Physiol. Rev., 56*:317, 1976.

Hayward, J. N.: Functional and morphological aspects of hypothalamic neurons. *Physiol. Rev., 57*:574, 1977.

Kalsner, S. (ed.): Trends in Autonomic Pharmacology. Baltimore, Urban & Schwarzenberg, 1979.

Kunos, G.: Adrenoceptors. *Annu. Rev. Pharmacol. Toxicol., 18*:291, 1978.

Landsberg, L., and Young, J. B.: Catecholamines and the adrenal medulla. *In* Bondy, P. K., and Rosenberg, L. E. (eds.): Metabolic Control and Disease, 8th Ed. Philadelphia, W. B. Suunders Co., 1980, p. 1621.

Levitzki, A.: Catecholamine receptors. *In* Adrian, R. H., et al. (eds.): Reviews of Physiology, Biochemistry, and Pharmacology. New York, Springer-Verlag, 1978, p. 1.

Mason, C. A., and Bern, H. A.: Cellular biology of the neurosecretory neuron. *In* Brookhart, J. M., and Mountcastle, V. B. (eds.): Handbook of Physiology. Sec. 1, Vol. 1. Baltimore, Williams & Wilkins, 1977, p. 651.

Miyahara, M., and Simpson, F. O.: Adrenoreceptor blockers in cardiovascular therapy. *In* Hayase, S., and Murao, S. (eds.): Cardiology: Proceedings of the VIII World Congress of Cardiology, Tokyo, 1978. New York, Elsevier/North-Holland, 1979, p. 460.

Moore, R. Y., and Bloom, F. E.: Central catecholamine neuron systems: Anatomy and physiology of the dopamine system. *Annu. Rev. Neurosci., 1*:129, 1978.

Morgane, P. J., and Panksepp, J. (eds.): Handbook of the hypothalamus. New York, Marcel Dekker, 1979.

Paton, D. M. (ed.): The Release of Catecholamines from Adrenergic Neurons. New York, Pergamon Press, 1979.

Patterson, P. H.: Environmental determination of autonomic neurotransmitter functions. *Annu. Rev. Neurosci., 1*:1, 1978.

Rémond, A., and Izard, C. (eds.): Electrophysiological Effects of Nicotine. New York, Elsevier/North Holland, 1979.

Robinson, R.: Tumours That Secrete Catecholamines: A Study of Their Natural History and Their Diagnosis. New York, John Wiley & Sons, 1980.

Simon, P. (ed.): Neurotransmitters. New York, Pergamon Press, 1979.

Szabadi, E., *et al.*: Recent Advances in the Pharmacology of Adrenoceptors. New York, Elsevier/North-Holland, 1978.

Tŭcek, S. (eds.): The Cholinergic Synapse. New York, Elsevier/North-Holland, 1979.

Usdin, E., *et al.* (eds.): Catecholamines: Basic and Clinical Frontiers. New York, Pergamon Press, 1979.

von Euler, U. S.: Noradrenaline. Springfield, Ill., Charles C Thomas, 1956.

Westfall, T. C.: Local regulation of adrenergic neurotransmission. *Physiol. Rev., 57*:659, 1977.

Wolf, S. G.: Neural control of visceral function. *In* Frohlich, E. D. (ed.): Pathophysiology, 2nd Ed. Philadelphia, J. B. Lippincott Co., 1976, p. 711.

Part X
THE SPECIAL SENSES

58

The Eye:
I. Optics of Vision

PHYSICAL PRINCIPLES OF OPTICS

Before it is possible to understand the optical system of the eye, the student must be thoroughly familiar with the basic physical principles of optics, including the physics of refraction, a knowledge of focusing, depth of focus, and so forth. Therefore, in the present study of the optics of the eye, a brief review of these physical principles is first presented, and then the optics of the eye is discussed.

REFRACTION OF LIGHT

The Refractive Index of a Transparent Substance. Light rays travel through air at a velocity of approximately 300,000 kilometers per second but much slower through transparent solids and liquids. The refractive index of a transparent substance is the *ratio* of the velocity of light in air to that in the substance. Obviously, the refractive index of air itself is 1.00.

If light travels through a particular type of glass at a velocity of 200,000 kilometers per second, the refractive index of this glass is 300,000 divided by 200,000, or 1.50.

Refraction of Light Rays at an Interface Between Two Media with Different Refractive Indices. When light waves traveling forward in a beam, as shown in the upper part of Figure 58–1, strike an interface that is perpendicular to the beam, the waves enter the second refractive medium without deviating in their course. The only effect that occurs is decreased velocity of transmission. On the other hand, as illustrated in the lower part of Figure 58–1, if the light waves strike an angulated interface, the light waves bend if the refractive indices of the two media are different from each other. In this particular figure the light waves are leaving air, which has a refractive index of 1.00, and are entering a block of glass having a refractive index of 1.50. When the beam first strikes the angulated interface, the lower edge of the beam enters the glass ahead of the upper edge. The wave front in the upper portion

of the beam continues to travel at a velocity of 300,000 kilometers per second while that which has entered the glass travels at a velocity of 200,000 kilometers per second. This causes the upper portion of the wave front to move ahead of the lower portion so that it is no longer vertical but is angulated to the right. Because *the direction in which light travels is always perpendicular to the wave front,* the direction of travel of the light beam now bends downward.

The bending of light rays at an angulated interface is known as *refraction*. Note particularly that the degree of refraction increases as a function of (1) the ratio of the two refractive indices of the two transparent media and (2) the degree of angulation between the interface and the entering wave front.

APPLICATION OF REFRACTIVE PRINCIPLES TO LENSES

The Convex Lens. Figure 58–2 shows parallel light rays entering a convex lens. The light rays passing

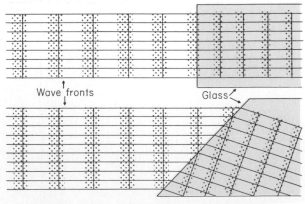

Figure 58–1. Wave fronts entering (*top*) a glass surface perpendicular to the light rays and (*bottom*) a glass surface angulated to the light rays. This figure illustrates that the distance between waves after they enter the glass is shortened to approximately two-thirds that in air. It also illustrates that light rays striking an angulated glass surface are refracted.

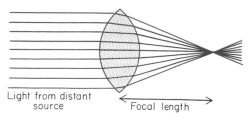

Figure 58-2. Bending of light rays at each surface of a convex spherical lens, showing that parallel light rays are focused to a point focus.

through the center of the lens strike the lens exactly perpendicular to the lens surfaces and therefore pass through the lens without being refracted at all. Toward either edge of the lens, however, the light rays strike a progressively more angulated interface. Therefore, the outer rays bend more and more toward the center. Half of the bending occurs when the rays enter the lens and half as they exit from the opposite side.

Thus, parallel light rays entering an appropriately formed convex lens come to a single point focus at some distance beyond the lens.

The Concave Lens. Figure 58-3 shows the effect of a concave lens on parallel light rays. The rays that enter the very center of the lens strike an interface that is absolutely perpendicular to the beam and, therefore, do not refract at all. The rays at the edge of the lens enter the lens ahead of the rays toward the center. This is opposite to the effect in the convex lens, and it causes the peripheral light rays to *diverge* away from the light rays that pass through the center of the lens.

Thus, the concave lens *diverges* light rays, while the convex lens *converges* light rays.

Spherical Versus Cylindrical Lenses. Figure 58-4 illustrates both a convex *spherical* lens and a convex *cylindrical* lens. Note that the convex cylindrical lens bends light rays from the two sides of the lens but not from either the top or the bottom. Therefore, parallel light rays are bent to a focal *line*. On the other hand, light rays that pass through the spherical lens are refracted at all edges of the lens toward the central ray, and all the rays come to a focal *point*.

The cylindrical lens is well illustrated by a test tube full of water. If the test tube is placed in a beam of sunlight and a piece of paper is brought progressively

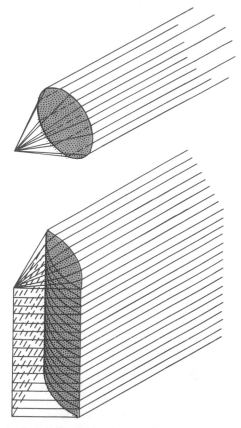

Figure 58-4. *Top*: Point focus of parallel light rays by a spherical convex lens. *Bottom*: Line focus of parallel light rays by a cylindrical convex lens.

closer to the tube, a certain distance will be found at which the light rays come to a focal line. On the other hand, the spherical lens is illustrated by an ordinary magnifying glass. If such a lens is placed in a beam of sunlight and a piece of paper is brought progressively closer to the lens, the light rays will impinge on a common focal point at an appropriate distance.

Concave cylindrical lenses *diverge* light rays in only one plane in the same manner that *convex* cylindrical lenses *converge* light rays in one plane.

Figure 58-5 shows two convex cylindrical lenses at right angles to each other. The vertical cylindrical lens causes convergence of the light rays that pass through the two sides of the lens. The horizontal lens converges the top and bottom rays. Thus, all the light rays come to a single point focus. In other words, *two cylindrical lenses crossed at right angles to each other perform the same function as one spherical lens of the same refractive power.*

FOCAL LENGTH OF THE CONVEX LENS

The distance from a convex lens at which parallel rays converge to a common focal point is the *focal length* of the lens. The diagram at the top of Figure 58-6 illustrates this focusing of parallel light rays. In the middle diagram of Figure 58-6, the light rays that enter the convex lens are not parallel but are diverging because

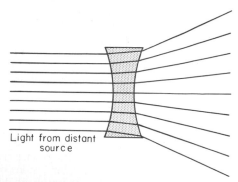

Figure 58-3. Bending of light rays at each surface of a concave spherical lens, illustrating that parallel light rays are diverged by a concave lens.

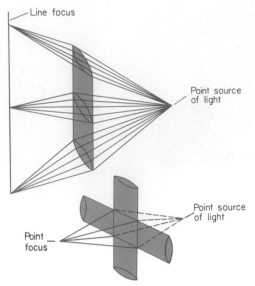

Figure 58–5. Two cylindrical convex lenses at right angles to each other, illustrating that one lens converges light rays in one plane and the other lens converges light rays in the plane at right angles. The two lenses combined give the same point focus as that obtained with a spherical convex lens.

the origin of the light is a point source not far away from the lens itself. The rays striking the center of the lens pass through the lens without any refraction as pointed out above. And the rays striking the edges are refracted toward the center. However, because these rays are diverging outward from the point source, it can be seen from the diagram that they do not come to a point focus at the same distance away from the lens as do parallel

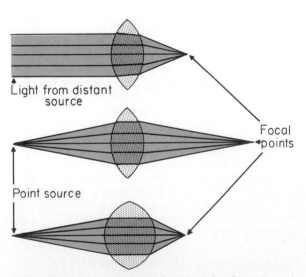

Figure 58–6. The upper two lenses of this figure have the same strength, but the light rays entering the top lens are parallel, while those entering the second lens are diverging; the effect of parallel versus diverging rays on the focal distance is illustrated. The bottom lens has far more refractive power than either of the other two lenses, illustrating that the stronger the lens the nearer to the lens is the point focus.

rays. In other words, when rays of light that are already diverging enter a convex lens, the distance of focus on the other side of the lens is farther from the lens than is the case when the entering rays are parallel to each other.

In the lower diagram of Figure 58–6 are shown light rays that also are diverging toward a convex lens that has far greater curvature than that of the upper two lenses of the figure. In this diagram the distance from the lens at which the light rays come to a focus is exactly the same as that from the lens in the first diagram, in which the lens was less convex but the rays entering it were parallel. This illustrates that both parallel rays and diverging rays can be focused at the same distance behind a lens provided that the lens changes its convexity. It is obvious that the nearer the point source of light is to the lens, the greater is the divergence of the light rays entering the lens and the greater must be the curvature of the lens to cause focusing at the same distance.

The relationship of focal length of the lens, distance of the point source of light, and distance of focus is expressed by the following formula:

$$\frac{1}{f} = \frac{1}{a} + \frac{1}{b}$$

in which f is the focal length of the lens, a the distance of the point source of light from the lens, and b the distance of focus from the lens.

FORMATION OF AN IMAGE BY A CONVEX LENS

The upper drawing of Figure 58–7 illustrates a convex lens with two point sources of light to the left. Because light rays from any point source pass through the center of a convex lens without being refracted in either direction, the light rays from each point source of light are shown to come to a point focus behind the lens *directly in line with the point source and the center of the lens.*

Any object in front of the lens is in reality a mosaic of point sources of light. Some of these points are very bright, some are very weak, and they vary in color. The light rays from each point source of light that enter the very center of the convex lens pass directly through this lens without any of the rays bending. Furthermore, the light rays that enter the edges of the lens come to focal points behind the lens in line with the rays that pass through the center. Therefore, every point source of light on the object comes to a separate point focus on the opposite side of the lens. If all portions of the object are the same distance in front of the lens, all the focal points behind the lens will fall in a common plane a certain distance behind the lens. If a white piece of paper is placed at this distance, one can see an image of the object, as is illustrated in the lower portion of Figure 58–7. However, this image is upside down with respect to the original object, and the two lateral sides of the image are reversed with respect to the original. This is the method by which the lens of a camera focuses light rays on the camera film.

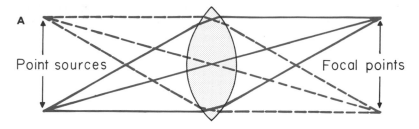

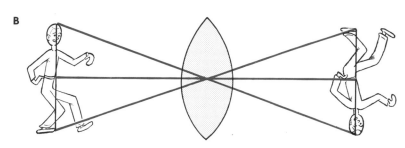

Figure 58–7. A, Two point sources of light focused at two separate points on the opposite side of the lens. B, Formation of an image by a convex spherical lens.

MEASUREMENT OF THE REFRACTIVE POWER OF A LENS – THE DIOPTER

The more a lens bends light rays, the greater is its "refractive power." This refractive power is measured in terms of *diopters*. The refractive power of a convex lens is equal to 1 meter divided by its focal length. Thus a spherical lens has a refractive power of +1 diopter if it converges parallel light rays to a focal point 1 meter beyond the lens, as illustrated in Figure 58–8. If the lens is capable of bending parallel light rays twice as much as a lens with a power of +1 diopter, it is said to have a strength of +2 diopters, and, obviously, the light rays come to a focal point 0.5 meter beyond the lens. A lens capable of converging parallel light rays to a focal point only 10 cm. (¹⁄₁₀ meter) beyond the lens has a refractive power of +10 diopters.

The refractive power of concave lenses cannot be stated in terms of the focal distance beyond the lens because the light rays diverge rather than focusing to a point. Therefore, the power of a concave lens is stated in terms of its ability to diverge light rays in comparison with the ability of convex lenses to converge light rays. That is, if a concave lens diverges light rays the same amount that a 1 diopter convex lens converges them, the concave lens is said to have a dioptric strength of −1. Likewise, if the concave lens diverges the light rays as much as a +10 diopter lens converges them, it is said to have a strength of −10 diopters.

Note particularly that concave lenses can "neutralize" the refractive power of convex lenses. Thus, placing a 1 diopter concave lens immediately in front of a 1 diopter convex lens results in a lens system with zero refractive power.

The strengths of cylindrical lenses are computed in the same manner as the strengths of spherical lenses. If a cylindrical lens focuses parallel light rays to a line focus 1 meter beyond the lens, it has a strength of +1 diopter. On the other hand, if a cylindrical lens of a concave type *diverges* light rays as much as a +1 diopter cylindrical lens *converges* them, it has a strength of −1 diopter.

THE OPTICS OF THE EYE

THE EYE AS A CAMERA

The eye, as illustrated in Figure 58–9, is optically equivalent to the usual photographic camera, for it has a lens system, a variable aperture system (the

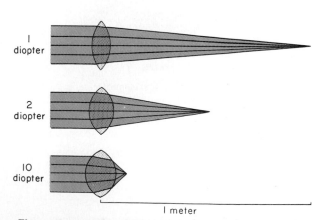

Figure 58–8. Effect of lens strength on the focal distance.

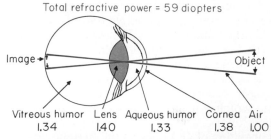

Figure 58–9. The eye as a camera. The numbers are the refractive indices.

pupil), and a retina that corresponds to the film. The lens system of the eye is composed of (1) the interface between air and the anterior surface of the cornea, (2) the interface between the posterior surface of the cornea and the aqueous humor, (3) the interface between the aqueous humor and the anterior surface of the lens, and (4) the interface between the posterior surface of the lens and the vitreous humor. The refractive index of air is 1; the cornea, 1.38; the aqueous humor, 1.33; the lens (on the average), 1.40; and the vitreous humor, 1.34.

The Reduced Eye. If all the refractive surfaces of the eye are algebraically added together and then considered to be one single lens, the optics of the normal eye may be simplified and represented schematically as a "reduced eye." This is useful in simple calculations. In the reduced eye, a single lens is considered to exist with its central point 17 mm. in front of the retina and to have a total refractive power of approximately 59 diopters when the lens is accommodated for distant vision.

The anterior surface of the cornea provides about 48 diopters of the eye's total dioptric strength for three reasons: (1) the refractive index of the cornea is markedly different from that of air; (2) the surface of the cornea is farther away from the retina than are the surfaces of the eye lens; and (3) the curvature of the cornea is reasonably great.

The posterior surface of the cornea is concave and actually acts as a concave lens, but, because the difference in refractive index of the cornea and the aqueous humor is slight, this posterior surface of the cornea has a refractive power of only about −4 diopters, which neutralizes only a small part of the refractive power of the other refractive surfaces of the eye.

The total refractive power of the crystalline lens of the eye when it is surrounded by fluid on each side is only 15 diopters of the total refractive power of the eye's lens system. If this lens were removed from the eye and then surrounded by air, its refractive power would be about 150 diopters. Thus, it can be seen that the lens inside the eye is not nearly so powerful as it would be outside the eye. The reason for this is that the fluids surrounding the lens have refractive indices not greatly different from the refractive index of the lens itself, the smallness of the differences greatly decreasing the amount of light refraction at the lens interfaces. But the importance of the crystalline lens is that its curvature, and therefore also its strength, can be changed to provide accommodation, which will be discussed later in the chapter.

Formation of an Image on the Retina. In exactly the same manner that a glass lens can focus an image on a sheet of paper, the lens system of the eye can also focus an image on the retina. The image is inverted and reversed with respect to the object. However, the mind perceives objects in the upright position despite the upside-down orientation on the retina because the brain is trained to consider an inverted image as the normal.

THE MECHANISM OF ACCOMMODATION

The refractive power of the crystalline lens of the eye can be voluntarily increased from 15 diopters to approximately 29 diopters in young children; this is a total "accommodation" of 14 diopters. To do this, the shape of the lens is changed from that of a moderately convex lens to that of a very convex lens. The mechanism of this is the following:

Normally, the lens is composed of a strong elastic capsule filled with viscous, proteinaceous, but transparent fibers. When the lens is in a relaxed state, with no tension on its capsule, it assumes a spherical shape, owing entirely to the elasticity of the lens capsule. However, as illustrated in Figure 58–10, approximately 70 ligaments attach radially around the lens, pulling the lens edges toward the edge of the choroid. These ligaments are constantly tensed by the elastic pull of their attachments to the choroid, and the tension on the ligaments causes the lens to remain relatively flat under normal resting conditions of the eye. At the insertions of the ligaments in the choroid is the ciliary muscle, which has two sets of smooth muscle fibers, the *meridional fibers* and the *circular fibers*. The meridional fibers extend from the corneoscleral junction to the insertions of the ligaments in the choroid approximately 2 to 3 mm. behind the corneoscleral junction. When

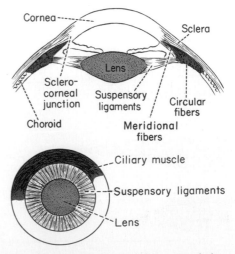

Figure 58–10. Mechanism of accommodation.

these muscle fibers contract, the insertions of the ligaments are pulled forward, thereby releasing a certain amount of tension on the crystalline lens. The circular fibers are arranged circularly all the way around the eye so that when they contract a sphincter-like action occurs, decreasing the diameter of the circle of ligament attachments and allowing the ligaments to pull less on the lens capsule.

Thus, contraction of both sets of smooth muscle fibers in the ciliary muscle relaxes the ligaments to the lens capsule, and the lens assumes a more spherical shape, like that of a balloon, because of elasticity of its capsule. When the ciliary muscle is completely relaxed, the dioptric strength of the lens is as weak as it can become. On the other hand, when the ciliary muscle contracts as strongly as possible, the dioptric strength of the lens becomes maximal.

Autonomic Control of Accommodation. The ciliary muscle is controlled mainly by the parasympathetic nervous system but also to a slight extent by the sympathetic system, as will be discussed in Chapter 60. Stimulation of the parasympathetic fibers to the eye contracts the ciliary muscle, which in turn relaxes the ligaments of the lens and increases its refractive power. With an increased refractive power, the eye is capable of focusing on objects that are nearer to it than is an eye with less refractive power. Consequently, as a distant object moves toward the eye, the number of parasympathetic impulses impinging on the ciliary muscle must be progressively increased for the eye to keep the object constantly in focus. (Sympathetic stimulation has a weak effect in relaxing the ciliary muscle but this plays almost no role in the normal accommodation mechanism, the neurology of which will be discussed in Chapter 60.)

Presbyopia. As a person grows older, the lens loses its elastic nature and becomes a relatively solid mass, probably because of progressive denaturation of the lens proteins. Therefore, the ability of the lens to assume a spherical shape progressively decreases, and the power of accommodation decreases from approximately 14 diopters shortly after birth to approximately 2 diopters at the age of 45 to 50. Thereafter, the lens may be considered to be almost totally nonaccommodating, a condition known as "presbyopia."

Once a person has reached the state of presbyopia, each eye remains focused permanently at an almost constant distance; this distance depends on the physical characteristics of each individual's eyes. Obviously, the eyes can no longer accommodate for both near and far vision. Therefore, to see clearly both in the distance and nearby, an older person must wear bifocal glasses with the upper segment normally focused for far-seeing and the lower segment focused for near-seeing.

THE PUPILLARY APERTURE

A major function of the iris is to increase the amount of light that enters the eye during darkness and to decrease the light in bright light. The reflexes for controlling this mechanism will be considered in the discussion of the neurology of the eye in Chapter 60. The amount of light that enters the eye through the pupil is proportional to the area of the pupil or to the *square of the diameter* of the pupil. The pupil of the human eye can become as small as approximately 1.5 mm. and as large as 8 mm. in diameter. Therefore, the quantity of light entering the eye may vary approximately 30 times as a result of changes in pupillary aperture.

Depth of Focus of the Lens System of the Eye. Figure 58–11 illustrates two separate eyes that are exactly alike except for the diameters of the pupillary apertures. In the upper eye the pupillary aperture is small, and in the lower eye the aperture is large. In front of each of these two eyes are two small point sources of light, and light from each passes through the pupillary aperture and focuses on the retina. Consequently, in both eyes the retina sees two spots of light in perfect focus. It is evident from the diagrams, however, that if the retina is moved forward or backward to an out-of-focus position, the size of each spot will not change much in the upper eye, but in the lower eye the size of each spot will increase greatly, becoming a "blur circle." In other words, the upper lens system has far greater *depth of focus* than the bottom lens system. When a lens system has great depth of focus, the retina can be considerably displaced from the focal plane and still the image remains in sharp focus; whereas, when a lens system has a shallow depth of focus, moving the retina only slightly away from the focal plane causes extreme blurring.

The greatest possible depth of focus occurs when the pupil is extremely small. The reason for

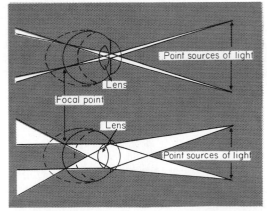

Figure 58–11. Effect of small and large pupillary apertures on the depth of focus.

this is that with a very small aperture the eye acts more or less as a pinhole camera. It will be recalled from the study of physics that light rays coming from an object and passing through a pinhole are in focus on any surface at all distances beyond the pinhole. In other words, the depth of focus of a pinhole camera is infinite.

"NORMAL" ABERRATIONS OF VISION

Spherical Aberration.   The crystalline lens of the eye is not nearly so regularly formed as lenses made by good opticians. Indeed, the light rays passing through the peripheral edges of the eye lens are not brought to really sharp focus with the other light rays, as illustrated in Figure 58–12. This effect is known as "spherical aberration," and the lens system of the human eye is quite subject to such an error. Therefore, increasing the aperture of the pupil progressively decreases the sharpness of focus. This partially explains why visual acuity is decreased at low illumination levels, which cause the pupil to dilate.

Chromatic Aberration.   The lens of the eye is also subject to "chromatic aberration," which is illustrated in Figure 58–13. This means that the refractive power of the lens is different for the different colors, so that different colors focus at different distances behind the lens. Furthermore, the greater the aperture of the lens, the greater are the errors of chromatic aberration, for those light rays passing near the center of the lens are proportionately less affected than are the rays passing through the periphery of the lens.

Diffractive Errors of the Eye.   Still another error in the optical system of all eyes is "diffraction" of light rays. Diffraction means bending of light rays as they pass over sharp edges; this obviously occurs as the rays pass over the edges of the pupil. Diffractive errors become especially important when the pupil becomes very small, because "interference" patterns then appear on the retina. A thorough consideration of diffraction at this point is impossible, and this phenomenon is noted simply to explain that as the pupil becomes 1.5 mm. in size the sharpness of vision becomes less than when the pupil is approximately 2.5 mm. in size, despite the fact that the depth of focus is better the smaller the pupillary diameter.

Cataracts.   Cataracts are an especially common eye abnormality that occurs in older people. A cataract is a cloudy or opaque area in the lens. In the early stage of cataract formation the proteins in the lens fibers immediately beneath the capsule become denatured. Later, these same proteins coagulate to form opaque areas in place of the normal transparent protein fibers of the

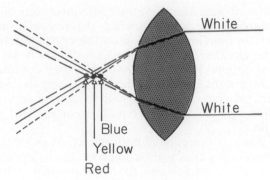
Figure 58–13.   Chromatic aberration.

lens. Finally, in still later stages, calcium is often deposited in the coagulated proteins, thus further increasing the opacity.

When a cataract has obscured light transmission so greatly that it seriously impairs vision, the condition can be corrected by surgical removal of the entire lens. When this is done, however, the eye loses a large portion of its refractive power, which must be replaced by a powerful convex lens (about +15 diopters) in front of the eye, as will be explained in the following sections.

ERRORS OF REFRACTION

Emmetropia.   As shown in Figure 58–14, the eye is considered to be normal or "emmetropic" if, when the ciliary muscle is completely relaxed, parallel light rays from distant objects are in sharp focus on the retina. This means that the emmetropic eye can, with its ciliary muscle completely relaxed, see all distant objects clearly, but to focus objects at close range it must contract its

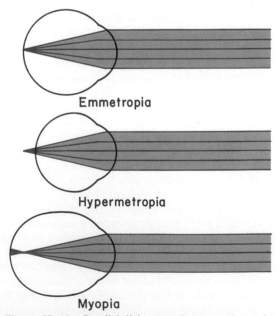
Figure 58–14.   Parallel light rays focus on the retina in emmetropia, behind the retina in hypermetropia, and in front of the retina in myopia.

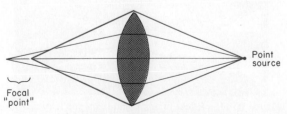
Figure 58–12.   Spherical aberration.

ciliary muscle and thereby provide various degrees of accommodation.

Hypermetropia (Hyperopia). Hypermetropia, which is also known as "far-sightedness," is usually due either to an eyeball that is too short or occasionally to a lens system that is too weak when the ciliary muscle is completely relaxed. In this condition, parallel light rays are not bent sufficiently by the lens system to come to a focus by the time they reach the retina. In order to overcome this abnormality, the ciliary muscle may contract to increase the strength of the lens. Therefore, the far-sighted person is capable, by using the mechanism of accommodation, of focusing distant objects on the retina. If he has used only a small amount of strength in his ciliary muscle to accommodate for the distant objects, then he still has much accommodative power left, and objects closer and closer to the eye can also be focused sharply until the ciliary muscle has contracted to its limit. The distance of the object away from the eye at this point is known as the "near point" of vision.

In old age, when the lens becomes presbyopic, the far-sighted person often is not able to accommodate his lens sufficiently to focus even distant objects, much less to focus near objects.

Myopia. In myopia, or "near-sightedness," when the ciliary muscle is completely relaxed, the light rays coming from distant objects are focused in front of the retina. This is usually due to too long an eyeball but it can occasionally result from too much power of the lens system of the eye.

No mechanism exists by which the eye can decrease the strength of its lens to less than that which exists when the ciliary muscle is completely relaxed. Therefore, the myopic person has no mechanism by which he can ever focus distant objects sharply on his retina. However, as an object comes nearer to his eye it finally comes near enough that its image will focus on the retina. Then, when the object comes still closer to the eye, the person can use his mechanism of accommodation to keep the image focused clearly. Therefore, a myopic person has a definite limiting "far point" for clear vision as well as a "near point"; when an object comes inside the "far point," he can use his mechanism of accommodation to keep the object in focus until the object reaches the "near point."

Correction of Myopia and Hypermetropia by Use of Lenses. It will be recalled that light rays passing through a concave lens diverge. Therefore, if the refractive surfaces of the eye have too much refractive power, as in myopia, some of this excessive refractive power can be neutralized by placing in front of the eye a concave spherical lens, which will diverge rays. On the other hand, in a person who has hypermetropia — that is, someone who has too weak a lens for the distance of the retina from the lens — the abnormal vision can be corrected by adding refractive power with a convex lens in front of the eye. These corrections are illustrated in Figure 58–15. One usually determines the strength of the concave or convex lens needed for clear vision by "trial and error" — that is, by trying first a strong lens and then a stronger or weaker lens until the one that gives the best visual acuity is found.

Astigmatism. Astigmatism is a refractive error of the lens system of the eye caused usually by an oblong shape of the cornea or, rarely, by an oblong shape of the

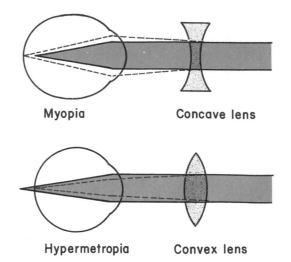

Figure 58–15. Correction of myopia with a concave lens, and correction of hypermetropia with a convex lens.

lens. A lens surface like the side of an egg lying edgewise to the incoming light would be an example of an astigmatic lens. The degree of curvature in a plane through the long axis of the egg is not nearly so great as the degree of curvature in a plane through the short axis. The same is true of an astigmatic lens. Because the curvature of the astigmatic lens along one plane is less than the curvature along the other plane, light rays striking the peripheral portions of the lens in one plane are not bent nearly so much as are rays striking the peripheral portions of the other plane.

This is illustrated in Figure 58–16, which shows what happens to rays of light emanating from a point source and passing through an oblong astigmatic lens. The light rays in the vertical plane, which is indicated by plane BD, are refracted greatly by the astigmatic lens because of the greater curvature in the vertical direction than in the horizontal direction. However, the light rays in the horizontal plane, indicated by plane AC, are bent not nearly so much as the light rays in the vertical plane. It is obvious, therefore, that the light rays passing through an astigmatic lens do not all come to a common focal

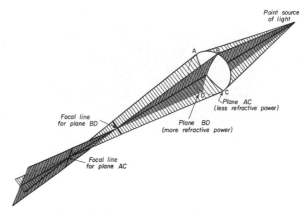

Figure 58–16. Astigmatism, illustrating that light rays focus at one focal distance in one focal plane and at another focal distance in the plane at right angles.

point because the light rays passing through one plane of the lens focus far in front of those passing through the other plane.

Placing an appropriate *spherical* lens in front of an astigmatic eye can bring the light rays that pass through *one plane* of the lens into focus on the retina, but spherical lenses can never bring *all* the light rays into complete focus at the same time. This is the reason why astigmatism is a very undesirable refractive error of the eyes. Furthermore, the accommodative powers of the eyes cannot compensate for astigmatism for the same reasons that spherical lenses placed in front of the eyes cannot correct the condition.

Correction of Astigmatism with a Cylindrical Lens. In correcting astigmatism with lenses it must always be remembered that two cylindrical lenses of equal strength may be crossed at right angles to give the same refractive effects as those of a spherical lens of the same strength. However, if one of the crossed cylindrical lenses has a different strength from that of the second lens, the light rays in one plane may not be brought to a common focal point with the light rays in the opposite plane. This is the situation that one usually finds in the astigmatic eye. In other words, one may consider an astigmatic eye as having a lens system made up of two cylindrical lenses of slightly different strengths. Another way of looking at the astigmatic lens system of the eye is that this system is a spherical lens with a superimposed cylindrical lens.

To correct the focusing of an astigmatic lens system, it is necessary to determine both the *strength* of the cylindrical lens needed to neutralize the excess cylindrical power of the eye lens and the *axis* of this abnormal cylindrical lens.

There are several methods for determining the axis of the abnormal cylindrical component of the lens system of an eye. One of these methods is based on the use of parallel black bars as shown in Figure 58–17. Some of these parallel bars are vertical, some horizontal, and some at various angles to the vertical and horizontal axes. After placing, by trial and error, various spherical lenses in front of the astigmatic eye, a strength of lens will usually be found that will cause sharp focus on one set of these parallel bars on the retina of the astigmatic eye. If there is no accompanying myopia or hypermetropia, the spherical lens may not be necessary.

It can be shown from the physical principles of optics discussed earlier in this chapter that the axis of the *out-of-focus* cylindrical component of the optical system is parallel to the black bars that are fuzzy in appearance. Once this axis is found, the examiner tries progressively stronger and weaker positive or negative cylindrical lenses, the axes of which are placed parallel to the out-of-focus bars, until the patient sees all the crossed bars with equal clarity. When this has been accomplished, the examiner directs the optician to grind a special lens having both the spherical correction plus the cylindrical correction at the appropriate axis.

Correction of Optical Abnormalities by Use of Contact Lenses. In recent years, either glass or plastic contact lenses have been fitted snugly against the anterior surface of the cornea. These lenses are held in place by a thin layer of tears that fill the space between the contact lens and the anterior eye surface.

A special feature of the contact lens is that it nullifies almost entirely the refraction that normally occurs at the anterior surface of the cornea. The reason for this is that the tears between the contact lens and the cornea have a refractive index almost equal to that of the cornea so that no longer does the anterior surface of the cornea play a significant role in the eye's optical system. Instead, the anterior surface of the contact lens now plays the major role and its posterior surface a minor role. Thus, the refraction of this lens now substitutes for the cornea's usual refraction. This is especially important in persons whose eye refractive errors are caused by an abnormally shaped cornea, such as persons who have an odd-shaped, bulging cornea — a condition called *keratoconus*. Without the contact lens the bulging cornea causes such severe abnormality of vision that almost no glasses can correct the vision satisfactorily; when a contact lens is used, however, the corneal refraction is neutralized, and normal refraction by the contact lens is substituted in its place.

The contact lens has several other advantages as well, including (1) the lens turns with the eye and gives a broader field of clear vision than do usual glasses, and (2) the contact lens has little effect on the size of the object that the person sees through the lens; on the other hand, lenses placed several centimeters in front of the eye do affect the size of the image even though they correct the focus.

SIZE OF THE IMAGE ON THE RETINA AND VISUAL ACUITY

If the distance from an object to the eye lens is 17 meters and the distance from the center of the lens to the image is 17 millimeters, the ratio of the object size to image size is 1000 to 1. Therefore, an object 17 meters in front of the eye and 1 meter in size produces an image on the retina 1 millimeter in size.

Theoretically, a point of light from a distant

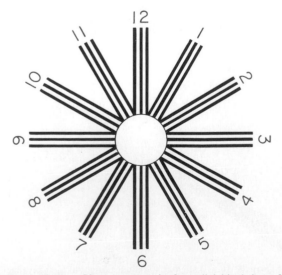

Figure 58–17. Chart composed of parallel black bars for determining the axis (meridian) of astigmatism.

point source, when focused on the retina, should be infinitely small. However, since the lens system of the eye is not perfect, such a retinal spot ordinarily has a total diameter of about 11 microns even with maximal resolution of the optical system. However, it is brightest in its very center and shades off gradually toward the edges, as illustrated by the two point images in Figure 58–18.

The average diameter of cones *in the fovea* of the retina, the central part of the retina where vision is most highly developed, is approximately 1.5 microns, which is one-seventh the diameter of the spot of light. Nevertheless, since the spot of light has a bright center point and shaded edges, a person can distinguish two separate points if their centers lie approximately 2 microns apart on the retina, which is slightly greater than the width of a foveal cone. This discrimination between points is illustrated in Figure 58–18.

The maximum visual acuity of the human eye for discriminating between point sources of light is 26 seconds. That is, when light rays from two separate points strike the eye with an angle of at least 26 seconds between them, they can usually be recognized as two points instead of one. This means that a person with maximal acuity looking at two bright pinpoint spots of light 10 meters away can barely distinguish the spots as separate entities when they are 1 millimeter apart.

The fovea is less than a millimeter in diameter, which means that maximum visual acuity occurs in only 3 degrees of the visual field. Outside this foveal area the visual acuity is reduced 5- to 10-fold, and it becomes progressively poorer as the periphery is approached. This is caused by the connection of many rods and cones to the same nerve fiber, as discussed in Chapter 60.

Clinical Method for Stating Visual Acuity. Usually the test chart for testing eyes is placed 20 feet away from the tested person, and if the person can see the letters of the size that he should be able to see at 20 feet, he is said to have 20/20 vision: that is, normal vision. If he can see

only letters that he should be able to see at 200 feet, he is said to have 20/200 vision. On the other hand, if he can see at 20 feet letters that he should be able to see only at 15 feet, he is said to have 20/15 vision. In other words, the clinical method for expressing visual acuity is to use a mathematical fraction that expresses the ratio of two distances, which is also the ratio of one's visual acuity to that of the normal person.

DETERMINATION OF DISTANCE OF AN OBJECT FROM THE EYE – DEPTH PERCEPTION

There are three major means by which the visual apparatus normally perceives distance, a phenomenon that is known as depth perception. These are (1) relative sizes of objects, (2) moving parallax, and (3) stereopsis.

Determination of Distance by Relative Sizes. If a person knows that a man is six feet tall and then he sees this man even with only one eye, he can determine how far away the man is simply by the size of the man's image on his retina. He does not consciously think about the size of this image, but his brain has learned to determine automatically from image sizes the distances of objects from the eye when the dimensions of these objects are already known.

Determination of Distance by Moving Parallax. Another important means by which the eyes determine distance is that of moving parallax. If a person looks off into the distance with his eyes completely still, he perceives no moving parallax, but, when he moves his head to one side or the other, the images of objects close to him move rapidly across his retinae while the images of distant objects remain rather stationary. For instance, if he moves his head 1 inch and an object is only 1 inch in front of his eye, the image moves almost all the way across his retinae, whereas the image of an object 200 feet away from his eyes does not move perceptibly. Thus, by this mechanism of moving parallax, one can tell the *relative distances* of different objects even though only one eye is used.

Determination of Distance by Stereopsis. Another method by which one perceives parallax is that of binocular vision. Because one eye is a little more than 2 inches to one side of the other eye, the images on the two retinae are different one from the other — that is, an object that is 1 inch in front of the bridge of the nose forms an image on the temporal portion of the retina of each eye, while a small object 20 feet in front of the nose has its image at closely corresponding points in the middle of each retina. This type of parallax is illustrated in Figure 58–19, which shows the images of a black spot and a square actually reversed on the

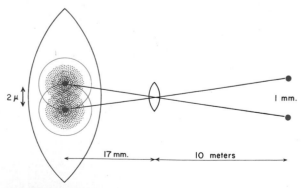

Figure 58–18. Maximum visual acuity for two point sources of light.

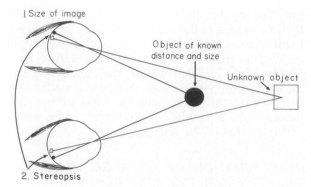

Figure 58–19.   Perception of distance (1) by the size of the image on the retina, and (2) as a result of stereopsis.

retinae because they are at different distances in front of the eyes. This gives a type of parallax that is present all the time when both eyes are being used. It is almost entirely this binocular parallax (or stereopsis) that gives a person with two eyes far greater ability to judge relative distances *when objects are nearby* than a person who has only one eye. However, stereopsis is virtually useless for depth perception at distances beyond 200 feet.

OPTICAL INSTRUMENTS

THE OPHTHALMOSCOPE

The ophthalmoscope is an instrument designed so that an observer can look through it into another person's eye and see the retina with clarity. Though the ophthalmoscope appears to be a relatively complicated instrument, its principles are simple. The basic components are illustrated in Figure 58–20 and may be explained as follows.

If a bright spot of light is on the retina of an emmetropic eye, light rays from the spot diverge toward the lens system of the eye, and, after they pass through the lens system, they are parallel with each other because the retina is located in the focal plane of the lens. When these parallel rays pass into an emmetropic eye of another person, they focus to a point on the retina of the second person because his retina is in the focal plane of his lens. Therefore, any spot of light on the

retina of the observed eye comes to a focal spot on the retina of the observing eye. Likewise, when the bright spot of light is moved to different points on the observed retina, the focal spot on the retina of the observer also moves an equal amount. Thus, if the retina of one person is made to emit light, the image of his retina will be focused on the retina of the observer provided the two eyes are simply looking into each other. These principles, of course, apply only to completely emmetropic eyes.

To make an ophthalmoscope, one need only devise a means for illuminating the retina to be examined. Then, the reflected light from that retina can be seen by the observer simply by putting the two eyes close to each other. To illuminate the retina of the observed eye, an angulated mirror or a segment of a prism is placed in front of the observed eye in such a manner that light from a bulb is reflected into the observed eye. Thus, the retina is illuminated through the pupil, and the observer sees into the subject's pupil by looking over the edge of the mirror or prism, or preferably *through* an appropriately designed prism so that the light will not have to enter the pupil at an angle.

It was noted above that these principles apply only to persons with completely emmetropic eyes. If the refractive power of either eye is abnormal, it is necessary to correct this refractive power in order for the observer to see a sharp image of the observed retina. Therefore, the usual ophthalmoscope has a series of about 20 lenses mounted on a turret so that the turret can be rotated from one lens to another, and the correction for abnormal refractive power of either or both eyes can be made at the same time by selecting a single lens of appropriate strength. In normal young adults, when the two eyes come close together, a natural accommodative reflex occurs that causes approximately +2 diopters increase in the strength of the lens of each eye. To correct for this, it is necessary that the lens turret be rotated to approximately −4 diopters correction.

THE RETINOSCOPE

The retinoscope, illustrated in Figure 58–21, is an instrument that can be used to determine the refractive power of an eye even though the subject cannot converse with the observer. Such a procedure is valuable for fitting glasses to an infant.

To use the retinoscope, one places a bright spot of light behind and to one side of the observed eye, and stands 1 meter away, looking through a hole in the

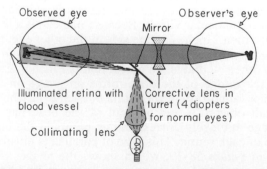

Figure 58–20.   The optical system of the ophthalmoscope.

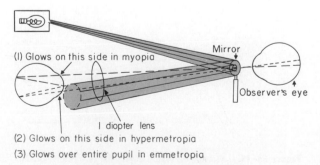

Figure 58–21.   The retinoscope.

middle of a mirror. The observer then rotates this mirror from side to side, making a reflected beam of light travel across the pupil of the observed eye while the subject keeps his gaze intently on the observer's eye. If the observed eye is normal, when the edge of this beam of light first enters the pupil, the entire pupil suddenly glows red. If the eye has abnormal refractive power, the red glow appears either on the side of the pupil into which the light edge first shines or on the opposite side of the pupil — one or the other. *In hypermetropia the first glow appears on the side of the pupil from which the light beam is being moved. In myopia the first glow appears on the opposite side of the pupil.* The cause of this difference is that the lens of the hyperopic eye is too weak and that of the myopic eye too strong to focus the light from the edge of the pupil on the retina exactly in line with the observer's eye. For a fuller understanding of this effect, however, the student is referred to texts on physiologic optics.

One can fit glasses to a patient by placing selected lenses in front of the observed eye one at a time until the glow suddenly covers the pupil over its entire extent rather than spreading from one side of the pupil to the other. However, it should be noted that in retinoscopy one tests an eye that is focused on the observer's eye at 1 meter's distance. This must be taken into consideration in prescribing glasses; 1 diopter strength must be subtracted for far vision.

REFERENCES

Allen, E. W.: Essentials of Ophthalmic Optics. New York, Oxford University Press, 1979.

Campbell, C. J., *et al.*: Physiological Optics. Hagerstown, Md., Harper & Row, 1974.

Collins, R., and Van der Werff, T. J.: Mathematical Models of the Dynamics of the Human Eye. New York, Springer-Verlag, 1980.

Davson, H.: The Physiology of the Eye, 3rd Ed. New York. Academic Press, 1972.

Davson, H., and Graham, L. T., Jr.: The Eye. Vols. 1–6. New York, Academic Press, 1969–1974.

Dick, G. L.: Studies in Ocular Anatomy and Physiology. Kensington, N. S. W., New South Wales University Press, 1976.

Hartstein, J.: Basics of Contact Lenses, 3rd Ed. San Francisco, American Academy of Ophthalmology, 1979.

Kavner, R. S., and Dusky, L.: Total Vision. New York, A & W Publishers, 1980.

Kolder, H. E. J. W. (ed.): Cataracts. Boston, Little, Brown, 1978.

Lerman, S.: Radiant Energy and the Eye. New York, The Macmillan Co., 1979.

Miller, D.: Ophthalmology; The Essentials. Boston, Houghton Mifflin, 1979.

Morgan, M. W.: The Optics of Ophthalmic Lenses. Chicago, Professional Press, 1978.

Polyak, S.: The Vertebrate Visual System. Chicago, University of Chicago Press, 1957.

Records, R. E.: Physiology of the Human Eye and Visual System. Hagerstown, Md., Harper & Row, 1979.

Regan, D., *et al.*: The visual perception of motion in depth. *Sci. Am., 241*(1):136, 1979.

Roth, H. W., and Roth-Wittig, M.: Contact Lenses. Hagerstown, Md., Harper & Row, 1980.

Ruben, M.: Understanding Contact Lenses (and the Correction of the Abnormal Eye). London, Heinemann Health Books, 1975.

Safir, A. (ed.): Refraction and Clinical Optics. Hagerstown, Md., Harper & Row, 1980.

Sloane, A. E. (ed.): Manual of Refraction. Boston, Little, Brown, 1979.

Soft Contact Lenses. New York, Elsevier/North-Holland, 1977.

Toates, F. M.: Accommodation function of the human eye. *Physiol. Rev., 52*:828, 1972.

Van Heyningen, R.: What happens to the human lens in cataract. *Sci. Am., 233*(6):70, 1975.

Whitnall, S. E.: The Anatomy of the Human Orbit and Accessory Organs of Vision. Huntington, N.Y., R. E. Krieger Publishing Co., 1979.

Whitteridge, D.: Binocular vision and cortical function. *Proc. R. Soc. Med., 65*:947, 1972.

Witkovsky, P.: Peripheral mechanisms of vision. *Annu. Rev. Physiol., 33*:257, 1971.

The Eye: II. Receptor Functions of the Retina

The retina is the light-sensitive portion of the eye, containing the cones which are responsible for color vision, and the rods, which are mainly responsible for vision in the dark. When the rods and cones are excited, signals are transmitted through successive neurons in the retina itself and finally into the optic nerve fibers and cerebral cortex. The purpose of the present chapter is to explain specifically the mechanisms by which the rods and cones detect both white and colored light.

ANATOMY AND FUNCTION OF THE STRUCTURAL ELEMENTS OF THE RETINA

The Layers of the Retina. Figure 59–1 shows the functional components of the retina arranged in layers from the outside to the inside as follows: (1) pigment layer, (2) layer of rods and cones projecting into the pigment, (3) outer lining membrane, (4) outer nuclear layer, (5) outer plexiform layer, (6) inner nuclear layer, (7) inner plexiform layer, (8) ganglionic layer, (9) layer of optic nerve fibers, and (10) inner limiting membrane.

After light passes through the lens system of the eye and then through the vitreous humor, it enters the retina from the bottom (see Figure 59–1); that is, it passes through the ganglion cells, the plexiform layer, the nuclear layer, and the limiting membranes before it finally reaches the layer of rods and cones located all the way on the opposite side of the retina. This distance is a thickness of several hundred microns; visual acuity is obviously decreased by this passage through such non-homogeneous tissue. However, in the central region of the retina, as will be discussed below, the initial layers are pulled aside to prevent this loss of acuity.

The Foveal Region of the Retina and Its Importance in Acute Vision. A minute area in the center of the retina, illustrated in Figure 59–2, called the *macula* and occupying a total area of less than 1 square millimeter, is especially capable of acute and detailed vision. This area is composed entirely of cones, but the cones are very much elongated and have a diameter of only 1.5 microns in contradistinction to the very large cones located farther peripherally in the retina. The central portion of the macula, only 0.4 mm. in diameter, is called the *fovea;* in this region the blood vessels, the ganglion cells, the inner nuclear layer of cells, and the plexiform layers are all displaced to one side rather than resting directly on top of the cones. This allows light to pass unimpeded to the cones rather than through several layers of retina, which aids immensely in the acuity of visual perception by this region of the retina.

The Rods and Cones. Figure 59–3 is a diagrammatic representation of a photoreceptor (either a rod or a cone) though the cones are distinguished by having a conical upper end (the outer segment), as shown in Figure 59–4. In general, the rods are narrower and longer than the cones, but this is not always the case. In the peripheral portions of the retina the rods are 2 to 5

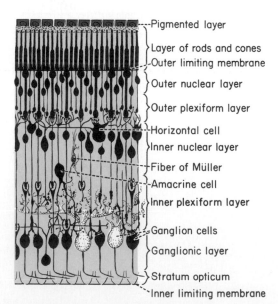

Figure 59–1. Plan of the retinal neurons. (From Polyak: The Retina. University of Chicago Press.)

The following labels appear in Figure 59–1:
- Pigmented layer
- Layer of rods and cones
- Outer limiting membrane
- Outer nuclear layer
- Outer plexiform layer
- Horizontal cell
- Inner nuclear layer
- Fiber of Müller
- Amacrine cell
- Inner plexiform layer
- Ganglion cells
- Ganglionic layer
- Stratum opticum
- Inner limiting membrane

Figure 59–2. Photomicrograph of the macula and of the fovea in its center. Note that the inner layers of the retina are pulled to the side to decrease the interference with light transmission. (From Bloom and Fawcett: A Textbook of Histology, 10th Ed. Philadelphia, W. B. Saunders Company, 1975; courtesy of H. Mizoguchi.)

microns in diameter while the cones are 5 to 8 microns in diameter; in the central part of the retina, in the fovea, the cones have a diameter of only 1.5 microns.

To the right in Figure 59–3 are labeled the four major functional segments of either a rod or a cone: (1) the outer segment, (2) the inner segment, (3) the nucleus, and (4) the synaptic body. In the outer segment the light-sensitive photochemical is found. In the case of the rods, this is *rhodopsin,* and in the cones it is one of several photochemicals almost exactly the same as rhodopsin except for a difference in spectral sensitivity. The cross-marks in the outer segment represent discs,

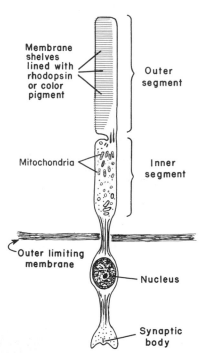

Figure 59–3. Schematic drawing of the functional parts of the rods and cones.

or "shelves," formed by infoldings of the cell membrane. The photosensitive pigments are incorporated into these membranes as part of their structures. They are concentrated at each surface of the membrane — both the inside and the outside surfaces. In this outer segment the concentration of the photosensitive pigment is approximately 40 per cent.

The inner segment contains the usual cytoplasm of the cell with the usual cytoplasmic organelles. Particularly important are the mitochondria, for we shall see later that the mitochondria in this segment play an important role in providing most of the energy for function of the photoreceptors.

The synaptic body is the portion of the rod and cone that connects with the subsequent neuronal cells, the horizontal and bipolar cells, that represent the next stages in the vision chain.

Figure 59–4 shows in more detail the photoreceptor discs in the outer segments of the rods and cones. Note especially their relationships to the cell membrane from which they originated.

Regeneration of Photoreceptors. The inner segments of the rods continually synthesize new rhodopsin. This in turn migrates to the base of the outer segment, where the cell membrane is continually forming new folds that become the discs of the rod and to which the rhodopsin attaches. Several new discs are formed each hour, and these migrate outward along the rod toward the tip, carrying the new rhodopsin with them. When the discs reach the top they degenerate and are dissoluted by the pigment epithelium. Thus, the light receptor portion of the rod is continually being replaced.

On the other hand, new disc formation does not seem to occur in the cones. However, new visual pigment does migrate throughout the outer segment, and local repair of the disc system does seem to occur.

These processes of disc generation and repair are probably necessary to provide continually viable photoreceptor chemical mechanisms in the photoreceptors.

The Pigment Layer of the Retina. The black pigment

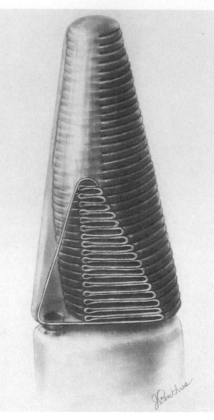

Figure 59–4. Membranous structures of the outer segments of a rod (left) and a cone (right). (Courtesy of Dr. Richard Young.)

melanin in the pigment layer, and still more melanin in the choroid, prevent light reflection throughout the globe of the eyeball, and this is extremely important for clear vision. This pigment performs the same function in the eye as does the black paint inside the bellows of a camera. Without it, light rays would be reflected in all directions within the eyeball and would cause diffuse lighting of the retina rather than the contrast between dark and light spots required for formation of precise images.

The importance of melanin in the pigment layer and choroid is well illustrated by its absence in *albinos,* persons hereditarily lacking in melanin pigment in all parts of their bodies. When an albino enters a bright area, light that impinges on the retina is reflected in all directions by the white surface of the unpigmented choroid so that a single discrete spot of light that would normally excite only a few rods or cones is reflected everywhere and excites many of the receptors. Also, a large amount of diffuse light enters the eye through the unpigmented iris. Thus, the lack of melanin causes the visual acuity of albinos, even with the best of optical correction, to be rarely better than 20/100 to 20/200. The pigment layer also stores large quantities of *vitamin A.* This vitamin A is exchanged back and forth through the membranes of the outer segments of the rods and cones, which themselves are embedded in the pigment layer. We shall see later that vitamin A is an important precursor of the photosensitive pigments and that this interchange of vitamin A is very important for adjustment of the light sensitivity of the receptors.

The Blood Supply of the Retina — the Retinal Arterial System and the Choroid.   The nutrient blood supply for the inner layers of the retina is derived from the central retinal artery, which enters the inside of the eye along with the optic nerve and then divides to supply the entire inner retinal surface. Thus, to a great extent, the retina has its own blood supply independent of the other structures of the eye.

However, the outer layer of the retina is adherent to the *choroid,* which is a highly vascular tissue between the retina and the sclera. The outer layers of the retina, including the outer segments of the rods and cones, depend mainly on diffusion from the choroid vessels for their nutrition, especially for their oxygen.

Retinal Detachment.   The neural retina occasionally detaches from the pigment epithelium. In a few instances the cause of such detachment is injury to the eyeball that allows fluid or blood to collect between the retina and the pigment epithelium, but more often it is caused by contracture of fine collagenous fibrils in the vitreous humor, which pull the retina unevenly toward the interior of the globe.

Fortunately, partly because of diffusion across the detachment gap and partly because of the independent blood supply to the retina through the retinal artery, the retina can resist degeneration for a number of days and can become functional once again if surgically replaced in its normal relationship with the pigment epithelium. But, if not replaced soon, the retina finally degenerates and then is unable to function even after surgical repair.

PHOTOCHEMISTRY OF VISION

Both the rods and cones contain chemicals that decompose on exposure to light and, in the process, excite the nerve fibers leading from the eye. The chemical in the *rods* is called *rhodopsin,* and the light-sensitive chemicals in the *cones* have compositions only slightly different from that of rhodopsin.

In the present section we will discuss principally the photochemistry of rhodopsin, but we can apply almost exactly the same principles to the photochemistry of the color-sensitive substances of the cones.

THE RHODOPSIN-RETINAL VISUAL CYCLE, AND EXCITATION OF THE RODS

Rhodopsin and Its Decomposition by Light Energy. The outer segment of the rod that projects into the pigment layer of the retina has a concentration of about 40 per cent of the light-sensitive pigment called *rhodopsin* or *visual purple.* This substance is a combination of the protein *scotopsin* and the carotenoid pigment *retinal* (also called "retinene"). Furthermore, the retinal is a particular type called 11-*cis* retinal. This *cis* form of the retinal is important because only this form can combine with scotopsin to synthesize rhodopsin.

When light energy is absorbed by rhodopsin, the rhodopsin immediately begins to decompose, as shown in Figure 59–5. The cause of this is photoactivation of electrons in the retinal portion of

the rhodopsin, which leads to an instantaneous change of the *cis* form of retinal into an all-*trans* form, which still has the same chemical structure as the *cis* form but has a different physical structure — a straight molecule rather than a curved molecule. Because the three-dimensional orientation of the reactive sites of the all-*trans* retinal no longer fits with that of the reactive sites on the protein scotopsin, it begins to pull away from the scotopsin. The immediate product is *bathorhodopsin* (also called "prelumirhodopsin"), which is a partially split combination of the all-*trans* retinal and scotopsin. However, bathorhodopsin is an extremely unstable compound and decays in nanoseconds to *lumirhodopsin.* This decays in microseconds to *metarhodopsin I,* then in about a millisecond to *metarhodopsin II,* and finally, much more slowly (in about a minute), to *pararhodopsin.* All of these are loose combinations of the all-*trans* retinal and scotopsin, but the pararhodopsin, too, is unstable, and it decomposes during the next few minutes into completely split products — scotopsin and all-*trans* retinal. During the first stages of splitting, the rods are excited, and signals are transmitted into the central nervous system.

Reformation of Rhodopsin. The first stage in reformation of rhodopsin, as shown in Figure 59–5, is to reconvert the all-*trans* retinal into 11-*cis* retinal. This process is catalyzed by the enzyme *retinal isomerase.* However, this process also requires metabolic energy from the rods and cones. Once the 11-*cis* retinal is formed, it automatically recombines with the scotopsin to reform rhodopsin, an exergonic process (which means that it gives off energy). The product, rhodopsin, is a stable compound until its decomposition is again triggered by absorption of light energy.

Excitation of the Rods When Rhodopsin Decomposes. Although the exact way in which rhodopsin decomposition excites the rod is still speculative, it is believed to occur in the following general way: The *retinal* portion of rhodopsin is a photosensitive compound that becomes activated in the presence of light. The initial effect of the light is to cause photoexcitation of electrons in the retinal molecule, raising these electrons to higher energy states. During this excited state, the internal physical structure (but not yet its chemical structure) begins to change, and the ultimate effect is conversion of the 11-*cis*-retinal into all-*trans*-retinal. All this occurs within nanoseconds or microseconds, and at the same time the retinal portion of the rhodopsin molecule undergoes a conformational change, from an angulated molecular structure of the *cis* form to a straight molecular structure of the *trans* form. Then, in rapid succession, the rhodopsin decomposes chemically through the several

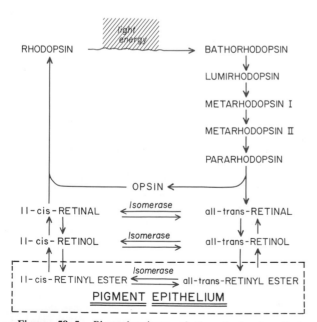

Figure 59–5. Photochemistry of the rhodopsin-retinal-vitamin A visual cycle.

stages of decomposition products as described above.

But, how does this photoexcitation and decomposition of the rhodopsin molecule excite the rod? Rod stimulation is believed to occur at the very instant that the rhodopsin molecule becomes excited by the light. The instantaneous effect on the rod is a sudden *decrease* in electrical conductance of the rod membrane, which leads to a change in the membrane potential, as we shall discuss subsequently.

Because stimulation of the rod takes place at the very outset of rhodopsin decomposition, it is believed that the initial photoexcitation of the retinal portion of the rhodopsin molecule is the prime cause of the stimulation. This could occur in one of several ways: One of the suggestions has been that the excited electrons themselves in some way lead directly to a change in the membrane conductance. A second suggestion has been that the excited electrons cause a change in a second chemical substance that in turn affects membrane conductance. Still a third suggestion has been that the rhodopsin molecule itself undergoes an instantaneous conformational change that alters the actual physical structure of the membrane, in this way changing its conductance at the same time; this theory is based principally on the fact that the rhodopsin molecules are actually a physical part of the membrane structure itself.

The physical construction of the outer segment of the rod is compatible with all these concepts because the cell membrane folds inward to form tremendous numbers of shelf-like discs lying one on top of the other, each one lined on both the inside and outside of the membrane with aggregates of rhodopsin. This extensive relationship between rhodopsin and the cell membrane could explain why the rod is so exquisitely sensitive to light, being capable of detectable excitation following absorption of only one quantum of light energy.

Generation of the Receptor Potential. Regardless of the precise mechanism by which the membrane of the outer segment becomes altered in the presence of decomposing rhodopsin, it is known that this excitation process *decreases* the conductance of this membrane for sodium ions. Based on this fact, the following theory for development of the receptor potential has been proposed:

Figure 59–6 illustrates movement of sodium ions in a complete electrical circuit through the inner and outer segments of the rod. The inner segment continually pumps sodium from inside the rod to the outside, thereby creating a negative potential on the inside of the entire cell. However, the membrane of the outer segment of the rod, in the unexcited condition, is very leaky to sodium. Therefore, sodium continually leaks back to the

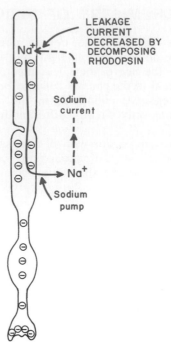

Figure 59–6. Theoretical basis for the generation of a hyperpolarization receptor potential caused by rhodopsin decomposition.

inside of the rod and thereby neutralizes much of the negativity on the inside of the entire cell. Thus, under normal conditions, when the rod is not excited, there is a reduced amount of electronegativity inside the membrane of the rod.

When the rhodopsin in the outer segment of the rod is exposed to light and begins to decompose, however, this *decreases* the leakage of sodium to the interior of the rod even though sodium continues to be pumped out. Thus, a net loss of sodium from the rod creates increased negativity inside the membrane, and the greater the amount of light energy striking the rod, the greater the electronegativity. This process is called *hyperpolarization*. Furthermore, it is exactly opposite to the effect that occurs in almost all other sensory receptors in which the degree of negativity is generally reduced during stimulation rather than increased, producing a state of depolarization rather than hyperpolarization.

Characteristics of the Receptor Potential. The receptor potential has two phases: an instantaneous, very low voltage phase, the electrical potential of which is directly proportional to the light energy; and a secondary phase of much greater voltage, the potential of which is proportional to the logarithm of the light energy. The first potential is believed to result from the immediate interaction of light with the rhodopsin. The second is believed to result from the sodium current mechanism described above.

This transduction of light energy into a logarithmic type of receptor potential seems to be an exceedingly important one for function of the eye, because it allows the eye to discriminate light intensities through a range many thousand times as great as would otherwise be possible.

Relationship Between Retinal and Vitamin A. The lower part of the scheme in Figure 59–5 illustrates that each of the two types of retinal can be converted into corresponding types of *retinol* and *retinyl ester,* both of which are forms of vitamin A. In turn, both of these can be reconverted into the two types of retinal. Thus, the two retinals are in dynamic equilibrium with vitamin A. Most of the vitamin A of the retina is stored in the pigment layer of the retina rather than in the rods themselves, but this vitamin A is readily available to the rods.

An important feature of the conversion of retinal into vitamin A, or the converse conversion of vitamin A into retinal, is that these processes require a longer time to approach equilibrium than it takes for conversion of retinal and scotopsin into rhodopsin, or for conversion (under the influence of strong light energy) of rhodopsin into retinal and scotopsin. Therefore, all the reactions of the upper part of Figure 59–5 can take place relatively rapidly in comparison with the slower interconversions between retinal and vitamin A.

Yet if the retina remains exposed to strong light for a long time, most of the stored rhodopsin will be converted eventually into vitamin A, thereby decreasing the concentration of all the photochemicals in the rods much more than would be true were it not for this subsequent conversion.

Conversely, during total darkness, essentially all the retinal already in the rods becomes converted into rhodopsin within a few minutes. This then allows much of the vitamin A to be converted into still additional retinal, which also becomes rhodopsin within another few minutes. Thus, when a person remains in complete darkness for a prolonged period of time, not only is almost all the retinal of his rods converted into rhodopsin but also much of the vitamin A stored in the pigment layer of the retina is absorbed by the rods and also converted into rhodopsin.

Night Blindness. Night blindness occurs in severe vitamin A deficiency. When the total quantity of vitamin A in the blood becomes greatly reduced, the quantities of vitamin A, retinal, and rhodopsin in the rods, as well as the color photosensitive chemicals in the cones, are all depressed, thus decreasing the sensitivities of the rods and cones. This condition is called night blindness because at night the amount of available light is far too little to permit adequate vision, though in daylight, sufficient light is available to excite the rods and cones despite their reduction in photochemical substances.

For night blindness to occur, a person often must remain on a vitamin A–deficient diet for months, because large quantities of vitamin A are normally stored in the liver and are made available to the rest of the body in times of need. However, once night blindness does develop it can sometimes be completely cured in a half hour or more by intravenous injection of vitamin A. This results from the ready conversion of vitamin A into retinal and thence into rhodopsin.

PHOTOCHEMISTRY OF COLOR VISION BY THE CONES

It was pointed out at the outset of this discussion that the photochemicals in the cones have almost exactly the same chemical composition as that of rhodopsin in the rods. The only difference is that the protein portions, the opsins, called *photopsins* in the cones, are different from the scotopsin of the rods. The retinal portions are exactly the same in the cones as in the rods. The color-sensitive pigments of the cones, therefore, are combinations of retinal and photopsins.

In the discussion of color vision later in the chapter, it will become evident that three different types of photochemicals are present in different cones, thus making these cones selectively sensitive to the different colors of blue, green, and red. These photochemicals are called respectively *blue-sensitive pigment, green-sensitive pigment,* and *red-sensitive pigment.* The absorption characteristics of the pigments in the three types of cones show peak absorbancies at light wavelengths, respectively, of 430, 535, and 575 millimicrons. These are also the wavelengths for peak light sensitivity for each type of cone, which begins to explain how the retina differentiates the colors. The approximate absorption curves for these three pigments are shown in Figure 59–7. The peak absorption for the rhodopsin of the rods, on the other hand, occurs at 505 millimicrons.

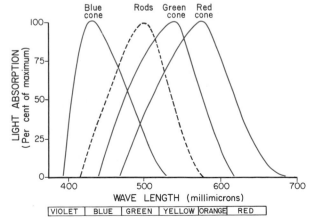

Figure 59–7. Light absorption by the respective pigments of the three color-receptive cones of the human retina. (Drawn from curves recorded by Marks, Dobelle, and MacNichol, Jr.: *Science, 143*:1181, 1964, and by Brown and Wald: *Science, 144*:45, 1964.)

Note that the peak absorption of the so-called "red" cone is actually in the orange color band. It is called the "red" cone because it responds to red more intensely than do any of the other cones.

AUTOMATIC REGULATION OF RETINAL SENSITIVITY — DARK AND LIGHT ADAPTATION

Relationship of Sensitivity to Pigment Concentration. The sensitivity of rods is approximately proportional to the antilogarithm of the rhodopsin concentration, and it is assumed that this relationship also holds true in the cones. Therefore, the sensitivity of the rods and cones can be altered up or down tremendously by only slight changes in concentrations of the photosensitive chemicals.

Light and Dark Adaptation. If a person has been in bright light for a long time, large proportions of the photochemicals in both the rods and cones have been reduced to retinal and opsins. Furthermore, most of the retinal of both the rods and cones has been converted into vitamin A. Because of these two effects, the concentrations of the photosensitive chemicals are considerably reduced, and the sensitivity of the eye to light is even more reduced. This is called *light adaptation*.

On the other hand, if the person remains in darkness for a long time, essentially all the retinal and opsins in the rods and cones become converted into light-sensitive pigments. Furthermore, large amounts of vitamin A are converted into retinal, which is then changed into additional light-sensitive pigments, the final limit being determined by the amount of opsins in the rods and cones. Because of these two effects, the visual receptors gradually become so sensitive that even the minutest amount of light causes excitation. This is called *dark adaptation*.

Figure 59–8 illustrates the course of dark adaptation when a person is exposed to total darkness after having been exposed to bright light for several hours. Note that sensitivity of the retina is very low when one first enters the darkness, but within 1 minute the sensitivity has increased 10-fold — that is, the retina can respond to light of one-tenth intensity. At the end of 20 minutes the sensitivity has increased about 6000-fold, and at the end of 40 minutes it has increased about 25,000-fold.

The resulting curve of Figure 59–8 is called the *dark adaptation curve*. Note, however, the inflection in the curve. The early portion of the curve is caused by adaptation of the cones, for these adapt much more rapidly than the rods because of a basic difference in the rate at which they resynthesize their photosensitive pigments. On the other hand, the cones do not achieve anywhere near the same degree of sensitivity as the rods. Therefore,

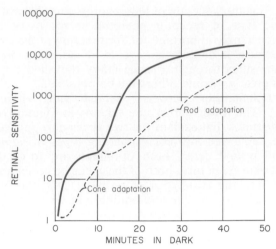

Figure 59–8. Dark adaptation, illustrating the relationship of cone adaptation to rod adaptation.

despite rapid adaptation by the cones, they cease adapting after only a few minutes, while the slowly adapting rods continue to adapt for many minutes or even hours, their sensitivity increasing tremendously. However, a large share of the greater sensitivity of the rods is also caused by convergence of as many as 100 rods onto a single ganglion cell in the retina; these rods summate to increase their sensitivity, as will be discussed in the following chapter.

Figure 59–9 illustrates several additional dark adaptation curves, but in this figure it is shown that the rate of dark adaptation differs depending on the previous degree of exposure of the retina to light. Curve number 1 shows the dark adaptation after a person had been in bright light for approximately 20 minutes, curve number 2 shows the rate after he had been in bright light for only a minute or so, and curve number 3 after he had been in bright light for many hours. Note that the eye exposed to bright light for a long time adapts slowly in comparison with the eye that has been in bright light for only a few minutes. The cause of this difference is probably the fact that long exposure to light converts most of the retinal into vitamin A that is then stored in the pigment

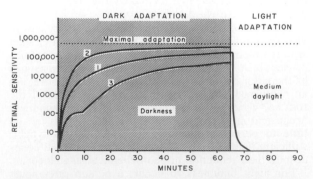

Figure 59–9. Dark and light adaptation. (The retina is considered to have a sensitivity of 1 when maximally light-adapted.)

epithelium, and this is slow to return to the rods and cones.

To the right in Figure 59–9 is shown a *light adaptation curve,* illustrating that light adaptation occurs much more rapidly than dark adaptation. Also, the slow component caused by retinal–vitamin A interconversion does not play as prominent a role in light adaptation as in dark adaptation. Therefore, a person exposed to extremely bright light after having been in prolonged darkness becomes adjusted to the new light conditions in only a few minutes. This is in contrast to dark adaptation, which requires 10 to 20 minutes for moderate adaptation and 10 to 18 hours for maximal adaptation.

Other Mechanisms of Light and Dark Adaptation. In addition to adaptation caused by changes in concentrations of rhodopsin or color photochemicals, the eye has two other mechanisms for light and dark adaptation. The first of these is the *change in pupillary size,* which was discussed in the previous chapter. This can cause a degree of adaptation of approximately 30-fold because of changes in the amount of light allowed through the pupillary opening.

The other mechanism is *neural adaptation,* involving the neurons in the successive stages of the visual chain in the retina itself. That is, when the light intensity first increases, the intensities of the signals transmitted by the bipolar cells, the horizontal cells, the amacrine cells, and the ganglion cells are all very intense. However, the intensities of these signals all decrease rapidly. Although the degree of this adaptation is only a few-fold rather than the many thousand-fold that occurs during adaptation of the photochemical system, this neural adaptation occurs in a fraction of a second, in contrast to the many minutes required for full adaptation by the photochemicals.

Value of Light and Dark Adaptation in Vision. Between the limits of maximal dark adaptation and maximal light adaptation, the eye can change its sensitivity to light by as much as 500,000 to 1,000,000 times, the sensitivity automatically adjusting to changes in illumination.

Since the registration of images by the retina requires detection of both dark and light spots in the image, it is essential that the sensitivity of the retina always be adjusted so that the receptors respond to the lighter areas and not to the darker areas. An example of maladjustment of the retina occurs when a person leaves a movie theater and enters the bright sunlight, for even the dark spots in the images then seem exceedingly bright, and, as a consequence, the entire visual image is bleached, having little contrast between its different parts. Obviously, this is poor vision, and it remains poor until the retina had adapted sufficiently that the dark spots of the image no longer stimulate the receptors excessively.

Conversely, when a person enters darkness, the sensitivity of the retina is usually so slight that even the light spots in the image cannot excite the retina. But, after dark adaptation, the light spots begin to register. As an example of the extremes of

light and dark adaptation, the light intensity of the sun is approximately 30,000 times that of the moon; yet the eye can function well both in bright sunlight and in bright moonlight.

Negative After-Images. If one looks steadily at a scene for a while, the bright portions of the image cause light adaptation of the retina while the dark portions of the image cause dark adaptation. In other words, areas of the retina that are stimulated by light become less sensitive while areas that are exposed only to darkness gain in sensitivity. If the person then moves his eyes away from the scene and looks at a bright white surface, he sees exactly the same scene that he had been viewing, but the light areas of the scene now appear dark, and the dark areas appear light. This is known as the negative after-image, and it is a natural consequence of light and dark adaptation.

The negative after-image persists as long as any degree of light and dark adaptation remains in the respective portions of the retina. Referring back to Figure 59–9, one can see from the dark adaptation curves that a negative after-image could possibly persist as long as an hour under favorable conditions.

Photopic versus Scotopic Vision. *Photopic vision* means vision capable of discriminating color, while *scotopic vision* means vision capable of discriminating only between shades of black and white. In bright light one's vision is photopic, while below a critical light intensity, vision is scotopic. The reason for this difference is the following: In very dim light, only rods are capable of becoming dark-adapted to a sensitivity level required for light detection. Therefore, in dim light the retina is capable only of scotopic vision. On the other hand, in very bright light, the rods become light-adapted to the point that they either become inoperative or overshadowed by the signals from the cones; in contrast, the cones find bright light especially suitable for optimal function. Some physiologists believe that the cones in bright light inhibit rod function by transmitting inhibiting signals through the horizontal cells to the synaptic bodies of the rods. At any rate, in bright light, function of the retina appears to be based almost entirely on cone detection of the light signals.

FUSION OF FLICKERING LIGHTS BY THE RETINA

In a flickering light, the intensity alternately increases and decreases rapidly. An instantaneous flash of light excites the visual receptors for as long as $\frac{1}{10}$ to $\frac{1}{5}$ second, and because of the *persistence* of excitation, rapidly successive flashes of light become *fused* together to give the appearance of being continuous. This effect is well known when one observes motion pictures or television. The images on the motion picture screen are flashed at a rate of 24 frames per second, while those of the television screen are flashed at a rate of 60 frames per second. As a result, the images fuse together, and continuous motion is observed.

The frequency at which flicker fusion occurs,

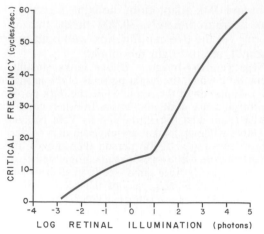

Figure 59-10. Relationship of intensity of illumination to the critical frequency for fusion.

called the *critical frequency for fusion,* varies with the light intensity. Figure 59–10 illustrates the effect of intensity of illumination on the critical fusion frequency. At a low intensity, fusion results even when the rate of flicker is as low as 2 to 6 per second. However, in bright illumination, the critical frequency for fusion rises to as great as 60 flashes per second. This difference results at least partly from the fact that the cones, which operate mainly at high levels of illumination, can detect much more rapid alterations in illumination than can the rods, which are the important receptors in dim light.

COLOR VISION

From the preceding sections, we know that different cones are sensitive to different colors of light. The present section is a discussion of the mechanisms by which the retina detects the different gradations of color in the visual spectrum.

THE TRI-COLOR THEORY OF COLOR PERCEPTION

Many different theories have been proposed to explain the phenomenon of color vision, but they are all based on the well-known observation that the human eye can detect almost all gradations of colors when red, green, and blue monochromatic lights are appropriately mixed in different combinations.

The first important theory of color vision was that of Young, which was later expanded and given a more experimental basis by Helmholtz. Therefore, the theory is known as the *Young-Helmholtz theory*. According to this theory, there are three different types of cones, each of which responds maximally to a different color.

As time has gone by, the Young-Helmholtz theory has been expanded, and more details have been worked out. It now is generally accepted as the mechanism of color vision.

Spectral Sensitivities of the Three Types of Cones. On the basis of color vision tests, the spectral sensitivities of the three different types of cones in human beings are essentially the same as the light absorption curves for the three types of pigment found in the respective cones. These were illustrated in Figure 59–7 and are also shown in Figure 59–11. These curves can readily explain almost all the phenomena of color vision.

Interpretation of Color in the Nervous System. Referring to Figure 59–11, one can see that an orange monochromatic light with a wavelength of 580 millimicrons stimulates the red cones to a stimulus value of approximately 99 (99 per cent of the peak stimulation at optimum wavelength), while it stimulates the green cones to a stimulus value of approximately 42 and the blue cones not at all. Thus, the ratios of stimulation of the three different types of cones in this instance are 99:42:0. The nervous system interprets this set of ratios as the sensation of orange. On the other hand, a monochromatic blue light with a wavelength of 450 millimicrons stimulates the red cones to a stimulus value of 0, the green cones to a value of 0 and the blue cones to a value of 97. This set of ratios — 0:0:97 — is interpreted by the nervous system as blue. Likewise, ratios of 83:83:0 are interpreted as yellow and 31:67:36, as green.

This scheme also shows how it is possible for a person to perceive a sensation of yellow when a red light and a green light are shone into the eye at the same time, for this stimulates the red and green cones approximately equally, which gives a sensation of yellow even though no wavelength of light corresponding to yellow is present.

Perception of White Light. Approximately

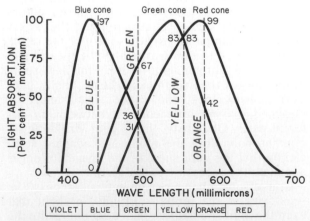

Figure 59-11. Demonstration of the degree of stimulation of the different color-sensitive cones by monochromatic lights of four separate colors: blue, green, yellow, and orange.

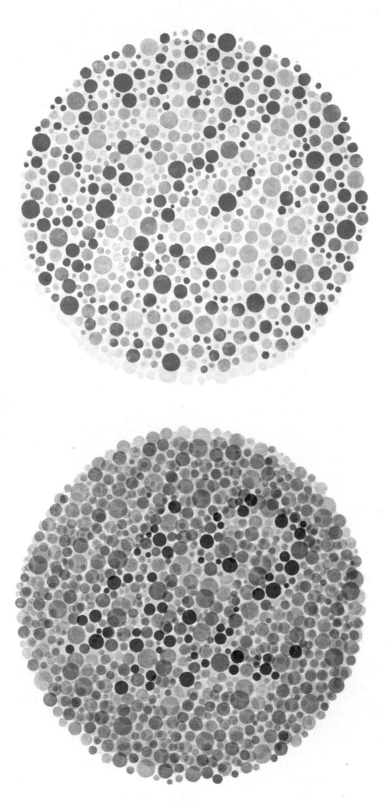

Figure 59–12.   Two Ishihara charts. *Upper:* In this chart, the normal person reads "74," whereas the red-green color blind person reads "21." *Lower:* In this chart, the red-blind person (protanope) reads "2," while the green-blind person (deuteranope) reads "4." The normal person reads "42." (From Ishihara: Tests for Colour-Blindness. Tokyo, Kanehara and Co.)

equal stimulation of all the red, green, and blue cones gives one the sensation of seeing white. Yet there is no wavelength of light corresponding to white; instead, white is a combination of all the wavelengths of the spectrum. Furthermore, the sensation of white can be achieved by stimulating the retina with a proper combination of only three chosen colors that stimulate the respective types of cones approximately equally.

Integration of Color Sensations by the Retina and the Brain. From psychological studies, we know that the interpretation of color is performed partly by the retina and partly by the brain. If a monochromatic green filter is placed in front of the left eye and a monochromatic red filter in front of the right eye, the visual object appears mainly yellow. This integration of color sensations obviously could not be occurring in the retina because one retina is allowed to respond only to green light and the other only to red light. However, sensations perceived in this way are not as clearly mixed as those perceived when the two monochromatic lights are mixed in the same retina. Therefore, at least some degree of the interpretation of color occurs in the retina itself even before the light information is transmitted into the brain. Some of the known features of color signal processing in the neuronal cells of the retina will be presented in the following chapter.

COLOR BLINDNESS

Red-Green Color Blindness. When a single group of color receptive cones is missing from the eye, the person is unable to distinguish some colors from others. As can be observed by studying Figure 59-9 or 59-11, if the red cones are missing, light of 525 to 625 millimicrons' wavelength can stimulate only the green-sensitive cones, so that the *ratio* of stimulation of the different cones does not change as the color changes from green all the way through the red spectrum. Therefore, within this wavelength range, all colors appear to be the same to this "color blind" person.

On the other hand, if the green-sensitive cones are missing, the colors in the range from green to red can stimulate only the red-sensitive cones, and the person also perceives only one color within these limits. Therefore, when either the red or green types of cones are lacking, the person is said to be "red-green" color blind. However, when one or more types of cones are abnormal but still function partially, a person has "color weakness," instead of color blindness.

The person with loss of red cones is called a *protanope;* his overall visual spectrum is noticeably shortened at the long wavelength end because of lack of the red cones. The color blind person who lacks green cones is called a *deuteranope;* this person has a perfectly normal visual spectral width because the absent green cones operate in the middle of the spectrum where red or blue cones also operate.

Blue Weakness. Occasionally, a person has "blue weakness," which results from diminished or absent blue receptors. If we observe Figure 59-11 once again, we can see that the blue cones are sensitive to a spectral range almost entirely different from that of both the red and green cones. Therefore, if the blue receptors are completely absent, the person has a greater preponderance of green, yellow, orange, and red in the visual spectrum than of blue, thus producing this rarely observed type of color weakness or blindness.

Tests for Color Blindness. Tests for color blindness depend on the person's ability to distinguish various colors from each other and also on his ability to judge correctly the degree of contrast between colors. For instance, to determine whether or not a person is red-green color blind, he may be given many small tufts of wool whose colors encompass the entire visual spectrum. He is then asked to place those tufts that have the same colors in the same piles. If he is not color blind, he recognizes immediately that all the tufts have slightly different colors; however, if he is red-green color blind, he places the red, orange, yellow, and yellow-green colors all together as having essentially the same color.

Stilling and Ishihara Test Charts. A rapid method for determining color blindness is based on the use of spot-charts such as those illustrated in Figure 59-12. These charts are arranged with a confusion of spots of several different colors. In the top chart, the normal person reads "74," while the red-green color blind person reads "21." In the bottom chart, the normal person reads "42," while the red blind protanope reads "2," and the green blind deuteranope reads "4."

If one will study these charts while at the same time observing the spectral sensitivity curves of the different cones in Figure 59-11, it can be readily understood how excessive emphasis can be placed on spots of certain colors by color blind persons in comparison with normal persons.

Genetics of Color Blindness. Color blindness is sex-linked and results from absence of appropriate color genes in the X chromosomes. This lack of color genes is a recessive trait; therefore, color blindness will not appear as long as another X chromosome carries the genes necessary for development of the respective color-receptive cones.

Because the male human being has only one X chromosome, all three color genes must be present in this single chromosome if he is not to be color blind. In approximately 1 of every 50 times, the X chromosome lacks the red gene; in approximately 1 of every 16, it lacks the green gene; and, rarely, it lacks the blue gene. This means, therefore, that 2 per cent of all men are red color blind (protanopes) and 6 per cent are green color blind (deuteranopes), making a total of approximately 8 per cent who are red-green color blind. Because a female has two X chromosomes, red-green color blindness is a rare abnormality in the female.

REFERENCES

Benson, W. E.: Retinal Detachment: Diagnosis and Management. Philadelphia, Harper & Row, 1980.
Brindley, G. S.: Physiology of the Retina and Visual Pathway, 2nd Ed. Baltimore, Williams & Wilkins, 1970.

Callender, R., and Honig, B.: Resonance Raman studies of visual pigments. *Annu. Rev. Biophys. Bioeng., 6*:33, 1977.

Cervetto, L., and Fuortes, M. G. F.: Excitation and interactions in the retina. *Annu. Rev. Biophys. Bioeng., 7*:229, 1978.

Chignell, A. H.: Retinal Detachment Surgery. New York, Springer-Verlag, 1979.

Cunha-Vaz, J. G. (ed.): The Blood-Retinal Barriers. New York, Plenum Press, 1980.

Davson, H.: The Physiology of the Eye, 3rd Ed. New York, Academic Press, 1972.

Daw, N. W.: Neurophysiology of color vision. *Physiol. Rev., 53*:571, 1973.

Dick, G. L.: Studies in Ocular Anatomy and Physiology. Kensington, N. S. W., New South Wales University Press, 1976.

Fatt, I.: Physiology of the Eye: An Introduction to the Vegetative Function. Boston, Butterworths, 1978.

Favreau, O. E., and Corballis, M. C.: Negative aftereffects in visual perception. *Sci. Am., 235*(6):42, 1976.

Fine, B. S., and Yanoff, M.: Ocular Histology: A Text and Atlas. Hagerstown, Md., Harper & Row, 1979.

Freeman, R. D. (ed.): Developmental Neurobiology of Vision. New York, Plenum Press, 1979.

Friedlaender, M. H.: Allergy and Immunology of the Eye. Hagerstown, Md., Harper & Row, 1979.

Kaneko, A.: Physiology of the retina. *Annu. Rev. Neurosci., 2*:169, 1979.

Klein, R. M., and Katzin, H. M.: Cellular and Biochemical Aspects in Diabetic Retinopathy. Baltimore, Williams & Wilkins, 1978.

Kohner, E. M. (ed.): Diabetic Retinopathy. Boston, Little, Brown, 1978.

Land, E. H.: The retinex theory of color vision. *Sci. Am., 237*(6):108, 1977.

MacNichol, E. F., Jr.: Three-pigment color vision. *Sci. Am., 211*:48, 1964.

Marks, W. B., *et al.*: Visual pigments of single primate cones. *Science, 143*:1181, 1964.

Michael, C. R.: Color vision. *N. Engl. J. Med., 288*:724, 1973.

Michaelson, I. C.: Textbook of the Fundus of The Eye. New York, Churchill Livingstone, 1980.

Ming, A. L. S., and Constable, I. J.: Colour Atlas of Ophthalmology. Boston, Houghton Mifflin, 1979.

Padgham, C. A., and Saunders, J. E.: The Perception of Light and Color. New York, Academic Press, 1975.

Retina and Vitreous. Rochester, Minn., American Academy of Ophthalmology, 1978.

Rushton, W. A. H.: Visual pigments and color blindness. *Sci. Am., 232*(3):64, 1975.

Schepens, C. L.: Retinal Detachment and Allied Diseases. Philadelphia, W. B. Saunders Co., 1981.

Yannuzzi, L. A., *et al.* (eds.): The Macula: A Comprehensive Text and Atlas. Baltimore, Williams & Wilkins, 1978.

Young, R. W.: Proceedings: Biogenesis and renewal of visual cell outer segment membranes. *Exp. Eye Res., 18*:215, 1974.

Zinn, K. M., and Marmor, M. F. (eds.): The Retinal Pigment Epithelium. Cambridge, Mass., Harvard University Press, 1979.

60

The Eye:
III. Neurophysiology
of Vision

THE VISUAL PATHWAY

Figure 60–1 illustrates the visual pathway from the two retinae back to the *visual cortex.* After impulses leave the retinae they pass backward through the *optic nerves.* At the *optic chiasm* all the fibers from the nasal halves of the retinae cross to the opposite side where they join the fibers from the opposite temporal retinae to form the *optic tracts.* The fibers of each optic tract synapse in the *lateral geniculate body* and from here, the *geniculocalcarine fibers* pass through the *optic radiation,* or *geniculocalcarine tract,* to the *primary visual cortex* in the calcarine area of the occipital lobe.

In addition, visual fibers also pass to other areas of the brain: (1) from the optic tracts to the *suprachiasmatic nucleus of the hypothalamus,* presumably for controlling circadian rhythms; (2) from the optic tracts to the *brain stem motor nuclei,* for control of eye movements and head turning; (3) into the *pretectal nuclei,* for control of fixation of the eyes on objects of importance and also for activating the pupillary light reflex; (4) into the *superior colliculus,* for control of bilateral simultaneous movement of the two eyes; (5) into the *pulvinar* as a secondary visual pathway either directly from the optic tracts or indirectly from the superior colliculus — then from the pulvinar to the secondary visual areas, areas 18 and 19, of the occipital lobes; and (6) into other areas of the thalamus and brain stem, possibly partly for perception of light intensity.

NEURAL FUNCTION OF THE RETINA

NEURAL ORGANIZATION OF THE RETINA

The detailed anatomy of the retina was illustrated in Figure 59–1 of the preceding chapter, and Figure 60–2 illustrates the basic essentials of the retina's neural connections; to the left is the general organization of the neural elements in a peripheral retinal area and to the right the organization of the foveal area. Note that in the peripheral region both rods and cones converge on *bipolar cells* which in turn converge on *ganglion cells.* In the fovea, where only cones exist, there is little convergence; instead, the cones are represented by approximately equal numbers of bipolar and ganglion cells.

Not emphasized in the figure is the fact that in

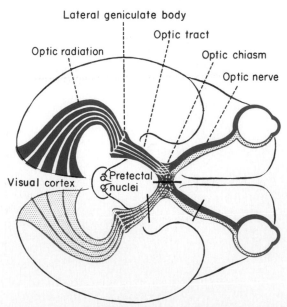

Figure 60–1. The visual pathways from the eyes to the visual cortex. (Modified from Polyak: The Retina. University of Chicago Press.)

748

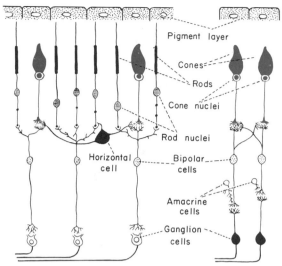

Figure 60–2. Neural organization of the retina: peripheral area to the left, foveal area to the right.

both the outer and inner plexiform layers (where the neurons synapse with each other) are many lateral connections between the different neural elements. Many of these lateral connections are collateral branches of the dendrites and axons of the bipolar and ganglion cells. But, in addition, two special types of cells are present in the inner nuclear layer that transmits signals laterally: (1) the *horizontal cells*, and (2) the *amacrine cells*.

Each retina contains about 125 million rods and 5.5 million cones; yet, as counted with the light microscope, only 900,000 optic nerve fibers lead from the retina to the brain. Thus, an average of 140 rods plus 6 cones converge on each optic nerve fiber. However, there are major differences between the peripheral retina and the central retina, for nearer the fovea fewer and fewer rods and cones converge on each optic fiber, and the rods and cones both become slenderer. These two effects progressively increase the acuity of vision toward the central retina. In the very central portion, in the *fovea*, there are no rods at all. Also, the number of optic nerve fibers leading from this part of the retina is almost equal to the number of cones, as shown to the right in Figure 60–2. This mainly explains the high degree of visual acuity in the central portion of the retina in comparison with the very poor acuity in the peripheral portions.

Another difference between the peripheral and central portions of the retina is the considerably greater sensitivity of the peripheral retina to weak light. This results partly from the fact that as many as 300 rods converge on the same optic nerve fiber in the most peripheral portions of the retina; the signals from the rods summate to give intense stimulation of the peripheral ganglion cells. But, in addition, rods are more sensitive to weak light than are the cones because of intrinsic differences in the receptors themselves.

STIMULATION OF THE RODS AND CONES — THE RECEPTOR POTENTIAL

Stimulation of the rods and cones was discussed at length in the preceding chapter. Briefly, when light strikes either of these receptors, it decomposes a photochemical that in turn acts on the membrane of the *outer segment* of the receptor (the part that protrudes into the pigment layer) to cause a sustained *receptor potential* that lasts as long as the light continues. However, as pointed out in the previous chapter, this receptor potential is different from all other receptor potentials that have been recorded, for it is a *hyperpolarization* signal rather than a depolarization signal. Also, its intensity is proportional to the logarithm of light energy, in contrast to the more linear response of many other receptors. This logarithmic response is very important to vision, because it allows the eyes to detect contrasts in the image even when light intensities vary many thousand-fold, an effect that would not be possible if there were linear transduction of the signal.

Receptor potentials are transmitted unchanged through the bodies of the rods and cones. Neither the rods nor the cones generate action potentials. Instead, the receptor potentials themselves, acting at the synaptic bodies, induce signals in the successive neurons, the bipolar and horizontal cells.

STIMULATION OF THE BIPOLAR AND HORIZONTAL CELLS

The synaptic bodies of the rods and cones make intimate contact with the dendrites of both the bipolar and horizontal cells. For this reason, it has been believed that signals were transmitted from the rods and cones to the bipolar and horizontal cells by electrotonic conduction, that is, by flow of electrical current directly from the rods and cones into the bipolar and horizontal cells. However, it is now almost certain that the rods and cones secrete a transmitter substance that then induces the potential changes in the bipolar and horizontal cells. Unfortunately, the exact nature of the transmitter is yet unknown.

The transmitter from the rods and cones causes *depolarization of the horizontal cells* and *hyperpolarization of the bipolar cells*. Therefore, when the rods and cones are excited, the bipolar cells become excited, while the horizontal cells become inhibited.

Neither the bipolar cells nor the horizontal cells transmit action potentials. Instead, their signals, like those of the rods and cones, result from electrotonic depolarization or hyperpolarization of the entire neuronal membrane all at once.

Function of the Bipolar Cells. The bipolar cells are the main transmitting link for the visual signal from the rods and cones to the ganglion cells. They are entirely excitatory cells; that is, they can only excite the ganglion cells.

Function of the Horizontal Cells. The horizontal cells lie in the inner nuclear layer and spread widely in the retina, transmitting signals laterally as far as several hundred microns. They are inhibited by the synaptic bodies of the rods and cones and in turn transmit their inhibitory signals mainly to bipolar cells located in areas lateral to the excited rods and cones.

The horizontal cells, therefore, function as inhibitory pathways, in contrast to the bipolar cells, which are excitatory. Therefore, the signals transmitted by the horizontal cells from the excited rods and cones cause inhibition of the bipolar cells lateral to the excited point in the retina. That is, excited rods and cones transmit excitatory signals in a direct line through the bipolar cells in the area of excitation but transmit inhibitory signals through the surrounding bipolar cells.

The horizontal cells are of major importance for enhancing and helping to detect contrasts in the visual scene. They are also of importance in helping to differentiate colors. Both of these functions will be discussed in subsequent sections of this chapter.

Stimulation and Function of the Amacrine Cells. The amacrine cells, also located in the inner nuclear layer, are excited mainly by the bipolar cells but possibly also occasionally directly by the synaptic bodies of the rods and cones. These cells in turn synapse with the ganglion cells. However, their response is a *transient* one rather than the steady, continuous response of the bipolar and horizontal cells. That is, when the photoreceptors are first stimulated, the signal transmitted by the amacrine cells is very intense; this signal dies away to almost nothing in a fraction of a second.

Another difference between the amacrine cells and the bipolar and horizontal cells is that the amacrine cells often transmit action potentials (and can transmit signals by electrotonic conduction as well), but the significance of this difference is not known.

Bipolar cells, horizontal cells, and amacrine cells are all believed to secrete transmitter substances at their nerve terminations, but the types of transmitter substances are yet unknown. Some of the different transmitter substances that have been found in the retina include acetylcholine, dopamine, GABA, glutamate ions, and aspartate ions.

EXCITATION OF THE GANGLION CELLS

Spontaneous, Continuous Discharge of the Ganglion Cells. The ganglion cells transmit their signals through the optic nerve fibers to the brain in the form of action potentials. These cells, even when unstimulated, transmit continuous nerve impulses at an average rate of about 5 per second. The visual signal is superimposed onto this basic level of ganglion cell stimulation. It can be either an excitatory signal, with the number of impulses increasing to greater than 5 per second, or an inhibitory signal, with the number of nerve impulses decreasing to below 5 per second — often all the way to zero.

Summation at the Ganglion Cells of Signals from the Bipolar Cells, the Horizontal Cells, and the Amacrine Cells. The bipolar cells transmit the main direct *excitatory* information from the rods and the cones to the ganglion cells: the horizontal cells transmit *inhibitory* information from laterally displaced rods and cones to surrounding bipolar cells and then to the ganglion cells; the amacrine cells seem to transmit direct but short-lived transient signals that signal a *change* in the level of illumination of the retina. Thus, each of these three types of cells performs a separate function in stimulating the ganglion cells.

DIFFERENT TYPES OF SIGNALS TRANSMITTED BY THE GANGLION CELLS THROUGH THE OPTIC NERVE

Transmission of Luminosity Signals. The ganglion cells are of several different types, and they also have different patterns of stimulation by the bipolar, horizontal, and amacrine cells. A very few of the ganglion cells respond to the intensity *(luminosity)* of the light falling on the photoreceptors. The rate of impulses from these cells remains at a level greater than the natural rate of firing as long as the luminosity is high. It is the signals from these cells that are believed to apprise the brain of the overall level of light intensity of the observed scene. However, by far the greater number of ganglion cells are concerned with transmitting information concerning the form of the scene in terms of contrast borders, as will be discussed below.

Transmission of Signals Depicting Contrasts in the Visual Scene — The Process of Lateral Inhibition. Most of the ganglion cells do not respond to the actual level of illumination of the scene; instead they respond only to contrast borders in the scene. Since it seems that this is the major means by which the form of the scene is transmitted to the brain, let us explain how this process occurs.

When flat light is applied to the entire retina —

that is, when all the photoreceptors are stimulated equally by the incident light — the contrast type of ganglion cell is neither stimulated nor inhibited. The reason for this is that the signals transmitted *directly* from the photoreceptors through the bipolar cells are excitatory, whereas the signals transmitted *laterally* through the horizontal cells are inhibitory. These two effects neutralize each other. Now, let us examine what happens when a contrast border occurs in the visual scene. Figure 60-3 shows a spot of light shining on the retina and exciting the photoreceptors. The signal is transmitted directly through the bipolar cells and thence to the ganglion cells in the same area to cause excitation. At the same time these same excited photoreceptors transmit *lateral inhibitory* signals through the horizontal cells to the surrounding bipolar and ganglion cells. Thus, when there is a contrast border, with stimulated photoreceptors on one side of the border and unstimulated photoreceptors on the opposite side of the border, the mutual cancellation of the excitatory signals through the bipolar cells and the inhibitory signals through the horizontal cells no longer occurs. Consequently, the ganglion cells on either side of this border become either excited or inhibited — excited on the light side of the border and inhibited on the dark side.

A linear contrast border acts almost identically as does a spot of light; the ganglion cells on the light side of the border are stimulated and on the dark side are inhibited. On the other hand, all of the contrast type of ganglion cells that lie in the flatly illuminated portion of the scene still remain unstimulated.

This process of contrasting an excitatory signal with an inhibitory signal is essentially the same as the process of *lateral inhibition* that occurs in

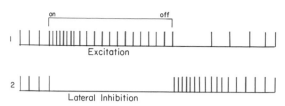

Figure 60-4.   Responses of ganglion cells to light in (1) an area excited by a spot of light and (2) an area immediately adjacent to the excited spot; the ganglion cells in this area are inhibited by the mechanism of lateral inhibition. (Modified from Granit: Receptors and Sensory Perception: A Discussion of Aims, Means, and Results of Electrophysiological Research into the Process of Reception. Yale University Press, 1955.)

most other types of sensory signal transmission. The process is a mechanism used by the nervous system for contrast detection and enhancement.

When impulses are recorded from ganglion cells, it is possible to observe the effects of contrast borders on the signals transmitted in optic nerve fibers. Figure 60-4 illustrates this effect. The upper recording is from a ganglion cell located in the area excited by a spot of light, whereas the lower recording is from a ganglion cell located a few hundred microns to the side of the spot. Note that when the spot of light is turned on, the ganglion cell in the center of the light becomes excited while the ganglion cell located laterally becomes inhibited.

Detection of Instantaneous Changes in Light Intensity — The On-Off Response.   Many of the ganglion cells are especially excited by *change* in light intensity; this effect most often occurs in the same ganglion cells that transmit contrast border signals. For instance, the upper tracing of Figure 60-4 shows that when the light was first turned on the ganglion cell became strongly excited for a fraction of a second, and then the level of excitation diminished. The bottom tracing shows that a second ganglion cell located in the dark area lateral to the spot of light was markedly inhibited at the same time. Then when the light was turned off, exactly the opposite effects occurred. Thus, the responses of these two types of cells are called the "on-off" and the "off-on" responses.

This ability of the retina to detect and transmit signals related to *change* in light intensity is caused by a rapid phase of "adaptation" of some of the neurons in the visual chain. Since this effect is known to be extremely marked in the amacrine cells, it has been suggested that the amacrine cells are peculiarly adapted to detecting light intensity change.

This capability to detect change in light intensity is especially well developed in the peripheral retina. For instance, a minute gnat flying across the peripheral field of vision is instantaneously

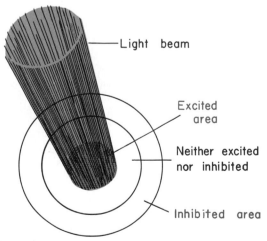

Figure 60-3.   Excitation and inhibition of a retinal area caused by a small beam of light.

detected. On the other hand, the same gnat sitting quietly in the peripheral field of vision remains entirely below the threshold of visual detection.

Transmission of Color Signals by the Ganglion Cells. A single ganglion cell may be stimulated by a number of cones or by only a very few. When all three types of cones — the red, blue, and green types — all stimulate the same ganglion cell, the signal transmitted through the ganglion cell is the same for any color of the spectrum. Therefore, this signal plays no role in the detection of the different colors. Instead, it is a "luminosity" signal.

On the other hand, many of the ganglion cells are excited by only one color type of cone but inhibited by a second type. For instance, this frequently occurs for the red and green cones, red causing excitation and green causing inhibition — or, vice versa, that is, green causing excitation and red, inhibition. The same type of reciprocal effect also occurs between blue cones and red or green cones.

The mechanism of this opposing effect of colors is the following: One color type cone excites the ganglion cell by the direct excitatory route through a bipolar cell, while the other color type inhibits the ganglion cell by the indirect inhibitory route through a horizontal cell.

The importance of these color-contrast mechanisms is that they represent a mechanism by which the retina itself differentiates colors. Thus each color-contrast type of ganglion cell is excited by one color but inhibited by the "opponent." Therefore, the process of color analysis begins in the retina and is not entirely a function of the brain.

FUNCTION OF THE LATERAL GENICULATE BODY

ANATOMICAL ORGANIZATION OF THE LATERAL GENICULATE NUCLEI

Each lateral geniculate body is composed of six nuclear layers. Layers 2, 3, and 5 (from the inside outward) receive signals from the temporal portion of the ipsilateral retina, while layers 1, 4, and 6 receive signals from the nasal retina of the opposite eye.

All layers of the lateral geniculate body relay visual information to the *visual cortex* through the *geniculocalcarine tract.*

Though it has not been proved, the pairing of layers from the two eyes probably plays a role in *fusion of vision* from the two eyes, because corresponding retinal fields in the two eyes connect with respective neurons that are approximately superimposed over each other in the successive layers. Also, with a little imagination, one can postulate that interaction between the successive layers could be part of the mechanism by which *stereoscopic visual depth perception* occurs, because this depends on comparing the visual images of the two eyes and determining their slight differences, as was discussed in Chapter 58. However, both fusion of vision and stereopsis also require the visual cortex, as we shall discuss in more detail later in the chapter.

Characteristics of the Visual Signals in the Lateral Geniculate Body

The signals recorded in the relay neurons of the lateral geniculate body are similar to those recorded in the ganglion cells of the retina. A few of the neurons transmit luminosity signals, while the majority transmit signals depicting only contrast borders or opponent colors in the visual image; also, many of the neurons are particularly responsive to movement of objects across the visual scene. However, the signals of the geniculate neurons are different from those in the ganglion cells in that a greater number of the complex interactions are found. That is, a much higher percentage of the neurons respond to contrast in the visual scene or to movement. These more complex reactions presumably result from convergence of excitatory and inhibitory signals from two or more ganglion cells on the relay neurons of the lateral geniculate body.

Color Signals in the Lateral Geniculate Body. Signals related to black and white vision are mainly found in layers 1 and 2 of the lateral geniculate body, while color signals occur mainly in layers 3 through 6. A few of the neurons in these last four layers respond to all colors and therefore transmit "white light" information. However, about three-quarters of the neurons respond to "opponent colors;" that is, a given cell may be excited by red but be inhibited by green. Another cell will operate exactly oppositely. These are called the "red-green" cells. Still other lateral geniculate body cells respond to the opponent colors blue and yellow, with one of these exciting the cell and the other inhibiting it. The mechanism of this effect is the same as in the retina; that is, it is caused by convergence of signals from red and green cones onto the same neuron, one of these signals causing inhibition while the other causes excitation.

By various interactions of this type, the color information in the visual scene is gradually dissected, and the information is then transmitted to other areas of the brain as signals that express the *ratios* of different colors in the different areas of the visual scene.

FUNCTION OF THE PRIMARY VISUAL CORTEX

The ability of the visual system to detect spatial organization of the visual scene — that is, to detect the forms of objects, brightness of the individual parts of the objects, shading, and so forth — depends on function of the *primary visual cortex,* the anatomy of which is illustrated in Figure 60–5. This area lies mainly in the calcarine fissure located bilaterally on the medial aspect of each occipital cortex. Specific points of the retina connect with specific points of the visual cortex, the right halves of the two respective retinae connecting with the right visual cortex and the left halves connecting with the left visual cortex. The macula is represented at the occipital pole of the visual cortex and the peripheral regions of the retina are represented in concentric circles farther and farther forward from the occipital pole. The upper portion of the retina is represented superiorly in the visual cortex, and the lower portion, inferiorly. Note the large area of cortex receiving signals from the macular region of the retina. It is in this region that the fovea is represented, which gives the highest degree of visual acuity. Based on area, the fovea has 35 times as much representation in the primary visual cortex as do the peripheral portions of the retina.

DETECTION OF LINES AND BORDERS BY THE PRIMARY VISUAL CORTEX

If a person looks at a blank wall, only a few neurons of the primary visual cortex will be stimulated, whether the illumination of the wall is bright or weak. Therefore, the question must be asked: What does the visual cortex do? To answer this question, let us now place on the wall a large black cross such as that illustrated to the left in Figure 60–6. To the right is illustrated the spatial pattern of the greater majority of the excited neurons that one finds in the visual cortex. *Note that the areas*

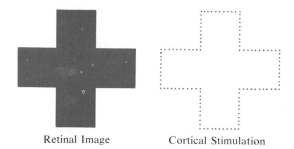

Retinal Image   Cortical Stimulation

Figure 60–6. Pattern of excitation occurring in the visual cortex in response to a retinal image of a dark cross.

of excitation occur along the sharp borders of the visual pattern. Thus, by the time the visual signal is recorded in the primary visual cortex, it is concerned mainly with the *contrasts* in the visual scene rather than with the flat areas. At each point in the visual scene where there is a change from dark to light or light to dark, the corresponding area of the primary visual cortex becomes stimulated. The intensity of stimulation is determined by the *gradient of contrast.* That is, the greater the sharpness in the contrast border and the greater the difference in intensities between the light and dark areas, the greater is the degree of stimulation.

Thus, the *pattern of contrasts* in the visual scene is impressed upon the neurons of the visual cortex, and this pattern has a spatial orientation roughly the same as that of the retinal image.

Detection of Orientation of Lines and Borders — Neuronal Columns. Not only does the visual cortex detect the *existence* of lines and borders in the different areas of the retinal image, but it also detects the *orientation* of each line or border — that is, whether it is a vertical or horizontal line or border, or lies at some degree of inclination. The mechanism of this effect is the following:

In each small area of the visual cortex the neurons are arranged in columns, each column containing about 100,000 neurons and having a diameter of about 1 mm., extending downward from the surface of the cortex through its six layers. The signals arriving in the cortex from the lateral geniculate body terminate in layer 4, and from here secondary signals spread upward or downward in the column. The neurons in the deeper portions of layer 4 respond in almost exactly the same way as the ganglion cells of the retina and the cells of the lateral geniculate body — that is, they respond to spots of light, borders, or lines. However, as the signal travels outward toward the surface of the cortex (into the outer portion of layer 4 and then into layers 3 and 2) or inward (into layers 5 and 6), the neurons

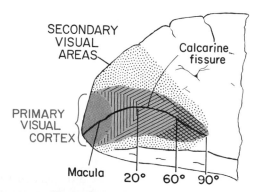

SECONDARY VISUAL AREAS

Calcarine fissure

PRIMARY VISUAL CORTEX

Macula   20°   60°   90°

Figure 60–5. The visual cortex.

respond no longer to just any spot, line, or border but instead only to lines or borders oriented in certain directions. Some respond to vertical lines or borders, others to lines or borders slightly angulated, still others to lines or borders more angulated, and so forth. This process of selective stimulation is presumably caused by convergence of signals from a number of layer 4 neurons. The effect is similar to that which occurs in the retina, some of the signals causing inhibition and others, excitation.

If the orientation of the border between the dark area and the light area in the visual scene corresponds with the orientation of the neural connections of inhibitory and excitatory input fibers to a given neuron of the visual cortex, the neuron would be excited. By having many different cortical neurons with slightly different synaptic connections, such interactions could lead to selective stimulation of one neuron for one orientation of a border, another neuron for another orientation, and so forth. This represents another stage in the mechanism for deciphering information in the visual scene.

Detection of Line Orientation Even When the Line Is Displaced Laterally. Still other neurons in the columns are capable of detecting lines oriented in one direction even when the lines are moved to slightly different areas on the retina. These neurons presumably are excited by signals from many separate line-oriented cells in adjacent columns. The cells that detect line orientation when the line is not displaced are called *simple cells*, while those neurons that detect line orientation when the line is displaced are called *complex cells*.

Detection of Length of Lines. Still a higher order for deciphering information from the visual scene is the ability of yet other neuronal cells to be stimulated by lines or borders of specific lengths. This type of detection presumably results from convergence of signals within the neuronal columns and between adjacent columns. The neurons that detect lines or borders of specific length are called *hypercomplex cells*.

Analysis of Color by the Visual Cortex. One finds in the primary visual cortex specific cells that are stimulated by color intensity or by contrasts of the opponent colors, the red-green opponent colors and the blue-yellow opponent colors. These effects are almost identical to those found in the lateral geniculate body. However, the proportion of cells excited by the opponent color contrasts is vastly reduced from the proportion found in the lateral geniculate body. Since neuronal excitation by color contrasts is a means of deciphering color, it is believed that the primary visual cortex is concerned with an even higher order of detection of color than simply the deciphering of color itself.

Full appreciation of color, however, does not seem to be achieved in the primary visual cortex, for it has recently been found in a number of human beings that lesions of the ventral surface of the occipital lobe can cause loss of color perception. It is suggested, therefore, that this area might be the final region for color signal processing.

Interaction of Visual Signals from the Two Different Eyes in the Primary Visual Cortex. From our earlier description of the lateral geniculate body, it will be recalled that the visual signals from the two separate eyes are relayed through separate lateral geniculate body neuronal layers. These signals from the two eyes are still separated from each other when they arrive in the primary visual cortex. The cortex is interlaced with zebra-like stripes of neurons, each stripe about 0.5 mm. wide; the signals from one eye enter every other stripe, and the signals from the other eye the opposite set of stripes. Each neuronal column of cells lies approximately one half in one of the stripes and the other half in the next stripe from the opposite eye. The signals also remain separated from each other when they excite the layer 4 neurons; this is also true for excitation of almost all the simple cells that detect simple line orientation. However, by the time the complex, and especially the hypercomplex, cells are excited, many of the signals from the two separate eyes have by then begun to converge. As we shall see later in the chapter, when these signals are not "in register" — that is, not both excited or inhibited together — *interference patterns* develop and excite still other cells. It is presumed that these patterns are the basis for controlling fusion of the visual images from the two eyes, and they are likely also the basis of stereopsis.

PERCEPTION OF LUMINOSITY

Although most of the neurons in the visual cortex are mainly responsive to contrasts caused by lines, borders, moving objects, or opponent colors in the visual scene, a few are directly responsive to the levels of luminosity in the different areas of the visual scene. Presumably, it is these cells that detect flat areas in the scene and also the overall level of luminosity. It is likely, too, that some of the visual signals entering the brain stem and thalamus help to determine luminosity.

EFFECT OF REMOVING THE PRIMARY VISUAL CORTEX

Removal of the primary visual cortex in the human being always causes loss of conscious vision. However, psychological studies have demonstrated that such persons can still react subconsciously to many aspects of vision, especially to changes in light intensity, and to crude aspects of form as well. The reactions include turning the eyes, turning the head, avoidance, and so on. In some low-level mammals a large share of the vision is still preserved even when the primary visual cortex is completely removed. This vision is believed to be subserved by neuronal pathways from the optic tracts to the superior colliculi and thence to the pulvinar and secondary visual cortical areas, areas 18 and 19 of the occipital lobe.

TRANSMISSION OF VISUAL INFORMATION INTO OTHER REGIONS OF THE CEREBRAL CORTEX

The primary visual cortex is located in the *calcarine fissure area*, mainly on the medial aspect of each hemisphere but also spreading slightly over the occipital pole. It is known as *Brodmann area 17* of the cortex, and it is also called the *striate area* because of its striated appearance to the naked eye.

Signals from the striate area project laterally in the occipital cortex into area 18 and then into area 19, as illustrated in Figure 60–7. These areas are called the *secondary visual cortex*, and they are also known as *visual association areas*. In reality, they are simply the loci for additional processing of visual information.

The visual projection images in areas 18 and 19 are organized into columns of cells in the same manner as described earlier for the primary visual cortex. There are also topographic representations of the visual field as found in the primary visual cortex. However, the neuronal cells respond to more complex patterns than do those in the primary visual cortex. For instance, some cells are stimulated by simple geometric patterns, such as curving borders, angles, and so forth. Presumably, these progressively more complex interpretations eventually decode the visual information, giving the overall impression of the visual scene that the person is observing.

Projection of Visual Information into the Temporal Cortex. From areas 18 and 19, visual signals next proceed to the posterior portion of the temporal lobes, Brodmann areas 20 and 21. In these areas simple visual patterns such as lines, borders, angles, and so forth fail to cause excitation of specific neurons. The degree of integration of visual information at this level, such as interpretation of letters or words, is presumably much higher.

Effects of Destruction of the Secondary Visual Cortex. Human beings who have destructive lesions of any of the visual association areas, areas 18 and 19 in the occipital cortex or areas 20 and 21 in the temporal cortex, have difficulty with certain types of visual perception and visual learning. For instance, destruction in areas 18 and 19 generally makes it difficult to perceive form, such as shapes of objects, their sizes, and their meanings. It is believed that this type of lesion in the dominant hemisphere can cause the abnormality known as *dyslexia* or *word blindness,* which means that the person has difficulty understanding the meanings of words that are seen.

Destruction of the temporal projections of the visual system makes it especially difficult for animals, and presumably for human beings as well, to learn tasks that are based upon visual perceptions. For instance, a person might be able to see a plate of food perfectly well but be unable to utilize this visual information to direct a fork toward the food. Yet, when the person feels the plate with the other hand, the fork can be directed accurately by stereotactic information from the somesthetic cortex.

Effects of Electrical Stimulation of the Visual Areas. Electrical stimulation in the primary visual cortex or in the parietal association areas, areas 18 and 19, causes a person to have simple optic auras — that is, flashes of light, simple colors, or simple forms such as lines, stars, or so forth — but complicated forms are not seen.

Stimulation of the visual association areas of the temporal cortex, areas 20 and 21, occasionally elicits complicated visual perceptions, sometimes causing the person to see a scene that had been known many years before. This is in keeping with the interpretative function of this area for visual function.

THE FIELDS OF VISION; PERIMETRY

The *field of vision* is the area seen by an eye at a given instant. The area seen to the nasal side is called the *nasal field of vision*, and the area seen to the lateral side is called the *temporal field of vision.*

To diagnose blindness in specific portions of the retinae, one charts the field of vision for each eye by a process known as *perimetry*. This is done by having the subject look with one eye toward a central spot directly in front of the eye. Then a small dot of light or a small object is moved back and forth in all areas of the field of vision, both laterally and nasally and upward and downward, and the person indicates when the spot of light or object can be seen and when it cannot. At the same time, a chart (Fig. 60–8) is made for the eye, showing the areas in which the subject can and cannot see the spot. Thus, the field of vision is plotted.

In all perimetry charts, a blind spot caused by lack of rods and cones in the retina over the optic disc is found approximately 15 degrees lateral to the central point of vision, as illustrated in the figure.

Abnormalities in the Fields of Vision. Occasionally

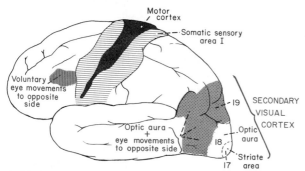

Figure 60–7. The visual association fields and the cortical areas for control of eye movements.

Right 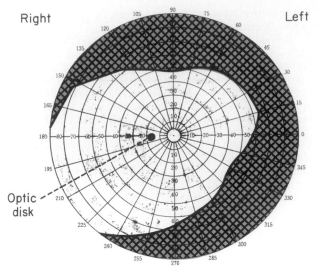 Left

Figure 60–8.   A perimetry chart, showing the field of vision for the left eye.

blind spots are found in other portions of the field of vision besides the optic disc area. Such blind spots are called *scotomata;* they frequently result from allergic reactions in the retina or from toxic conditions, such as lead poisoning or even excessive use of tobacco.

Still another condition that can be diagnosed by perimetry is *retinitis pigmentosa*. In this disease, portions of the retina degenerate and excessive melanin pigment deposits in the degenerated areas. Retinitis pigmentosa generally causes blindness in the peripheral field of vision first and then gradually encroaches on the central areas.

Effect of Lesions in the Optic Pathway on the Fields of Vision.   One of the most important uses of perimetry is for localization of lesions in the visual nervous pathway. Lesions in the optic nerve, in the optic chiasm, in the optic tract, and in the geniculocalcarine tract all cause blind areas in the visual fields, and the ''patterns'' of these blind areas indicate the location of the lesion.

Destruction of an entire *optic nerve* obviously causes blindness of the respective eye. Destruction of the *optic chiasm,* as shown by the longitudinal line across the chiasm in Figure 60–1, prevents the passage of impulses from the nasal halves of the two retinae to the opposite optic tracts. Therefore, the nasal halves are both blinded, which means that the person is blind in both temporal fields of vision; this condition is called *bitemporal hemianopsia.* Such lesions frequently result from tumors of the adenohypophysis pressing upward on the optic chiasm.

Interruption of an *optic tract,* which is also shown by a line in Figure 60–1, denervates the corresponding half of each retina on the same side as the lesion, and, as a result, neither eye can see objects to the opposite side. This condition is known as *homonymous hemianopsia.* Destruction of the *optic radiation* or the *visual cortex* of one side also causes homonymous hemianopsia. A common condition that destroys the visual cortex is thrombosis of the posterior cerebral artery, which infarcts the occipital cortex except for the foveal area, thus often sparing central vision.

One can differentiate a lesion in the optic tract from a lesion in the geniculocalcarine tract or visual cortex by determining whether impulses can still be transmitted into the pretectal nuclei to initiate a pupillary light reflex. To do this, light is shown onto the blinded half of one of the retinae, and, if a pupillary light reflex can still occur, it is known that the lesion is at or beyond the lateral geniculate body, that is, in the geniculocalcarine tract or visual cortex. However, if the light reflex is also lost, then the lesion is in the optic tract itself.

EYE MOVEMENTS AND THEIR CONTROL

To make use of the abilities of the eye, almost equally as important as the system for interpretation of the visual signals from the eyes is the cerebral control system for directing the eyes toward the object to be viewed.

Muscular Control of Eye Movements.   The eye movements are controlled by three separate pairs of muscles, shown in Figure 60–9: (1) the *medial* and *lateral recti,* (2) the *superior* and *inferior recti,* and (3) the *superior* and *inferior obliques.* The medial and lateral recti contract reciprocally to move the eyes from side to side. The superior and inferior recti contract reciprocally to move the eyes upward or downward. And the oblique muscles function mainly to rotate the eyeballs to keep the visual fields in the upright position.

Neural Pathways for Control of Eye Movements.   Figure 60–9 also illustrates the nuclei of the third, fourth, and sixth cranial nerves and their innervation of the ocular muscles. Shown, too, are the interconnections among these three nuclei through the *medial longitudinal fasciculus.* Either by way of this fasciculus or by way of other closely associated pathways, each of the three sets of muscles to each eye is *reciprocally* innervated so that one muscle of the pair relaxes while the other contracts.

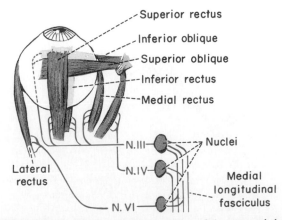

Figure 60–9.   The extraocular muscles of the eye and their innervation.

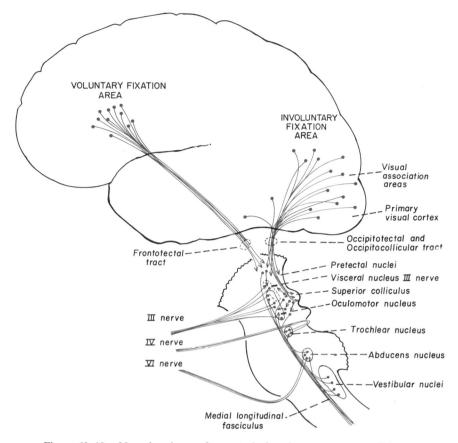

Figure 60–10.   Neural pathways for control of conjugate movement of the eyes.

Figure 60–10 illustrates cortical control of the oculomotor apparatus, showing spread of signals from the occipital visual areas through occipito-tectal and occipitocollicular tracts into the pretectal and superior colliculus areas of the brain stem. In addition, a frontotectal tract passes from the frontal cortex into the pretectal area. From both the pretectal and the superior colliculus areas, the oculomotor control signals then pass to the nuclei of the oculomotor nerves. Strong signals are also transmitted into the oculomotor system from the vestibular nuclei by way of the medial longitudinal fasciculus.

CONJUGATE MOVEMENT OF THE EYES

Simultaneous movement of both eyes in the same direction is called *conjugate movement* of the eyes. The signals to cause these movements probably originate in the reticular nuclei of the mesencephalon and pons in close association with the oculomotor nuclei. However, the movements are under strong control of the cerebral cortex, which transmits signals into the reticular nuclei by way of the pretectal nuclei and the superior colliculi. Signals from the vestibular nuclei also elicit conjugate movement of the eyes, mainly by sending signals through the medial longitudinal fasciculi.

FIXATION MOVEMENTS OF THE EYES

Perhaps the most important movements of the eyes are those that cause the eyes to "fix" on a discrete portion of the field of vision.

Fixation movements are controlled by two different neuronal mechanisms. The first of these allows the person to move his eyes voluntarily to find the object upon which he wishes to fix his vision; this is called the *voluntary fixation mechanism*. The second is an in-voluntary mechanism that holds the eyes firmly on the object once it has been found; this is called the *involun-tary fixation mechanism*.

The voluntary fixation movements are controlled by a small cortical field located bilaterally in the premotor cortical regions of the frontal lobes, as illustrated in Figure 60–10. Bilateral dysfunction or destruction of these areas makes it difficult or almost impossible for the person to "unlock" his eyes from one point of fixation and then move them to another point. It is usually necessary for him to blink his eyes or put his hand over his eyes for a short time, which then allows him to move the eyes.

On the other hand, the fixation mechanism that causes the eyes to "lock" on the object of attention once it is found is controlled by the *secondary visual area of the occipital cortex* — mainly area 19 — which is also illustrated in Figure 60–10. When this area is destroyed bilaterally, the person has difficulty or becomes completely unable to keep his eyes directed toward a given fixation point.

To summarize, the posterior eye fields automatically "lock" the eyes on a given spot of the visual field and thereby prevent movement of the image across the retina. To unlock this visual fixation, voluntary impulses must be transmitted from the "voluntary" eye fields located in the frontal areas.

Mechanism of Fixation. Visual fixation results from a negative feedback mechanism that prevents the object of attention from leaving the foveal portion of the retina. The eyes even normally have three types of continuous but almost imperceptible movements: (1) a *continuous tremor* at a rate of 30 to 80 cycles per second caused by successive contractions of the motor units in the ocular muscles, (2) a *slow drift* of the eyeballs in one direction or another, and (3) sudden *flicking movements* which are controlled by the involuntary fixation mechanism. When a spot of light has become fixed on the foveal region of the retina, the tremorous movements cause the spot to move back and forth at a rapid rate across the cones, and the drifting movements cause it to drift slowly across the cones. However, each time the spot of light drifts as far as the edge of the fovea, a sudden reflex reaction occurs, producing a flicking movement that moves the spot away from this edge back toward the center. Thus, whenever the image drifts away from the point of fixation, an automatic response moves the image back toward the central portion of the fovea. These drifting and flicking motions are illustrated in Figure 60–11, which shows by the dashed lines the slow drifting across the retina and by the solid lines the flicks that keep the image from leaving the foveal region.

Fixation of the Eyes on Important Visual Highlights — Role of the Superior Colliculi. The eyes have an automatic capability for instantaneously *fixing* on an important highlight in the visual field. However, this capability is mostly lost when the superior colliculi are destroyed even though most other conjugate movements of the eyes occur normally in the absence of the

superior colliculi. The signals for fixation originate in the visual areas of the occipital cortex, primarily in the secondary visual areas, then pass to the superior colliculi, from there to reticular areas around the oculomotor nuclei, and thence into the motor nuclei themselves.

The superior colliculi are also important in causing sudden turning of the eyes to the side when a flash of light or some other sudden visual disturbance occurs on that side. In addition, signals are transmitted through the superior colliculi into the medial longitudinal fasciculus and other areas of the brain stem to cause turning of the whole head and perhaps even of the whole body toward the direction of the light. Other types of disturbances besides visual disturbances, such as strong sounds or even stroking the side of the body, will cause similar turning of the eyes, head, and body if the superior colliculi are intact. This effect, however, is absent or severely disturbed when the superior colliculi are destroyed. Therefore, it is frequently said that the superior colliculi play an important role in orienting the eyes, the head, and the body with respect to external signals — visual signals, auditory signals, somatic signals, and perhaps even others.

Saccadic Movement of the Eyes. When the visual scene is moving continually before the eyes, such as when a person is riding in a car or turning around, the eyes fix on one highlight after another in the visual field, jumping from one to the next at a rate of two to three jumps per second. The jumps are called *saccades*, and the movements are called *opticokinetic movements*. The saccades occur so rapidly that not more than 10 per cent of the total time is spent in moving the eyes, 90 per cent of the time being allocated to the fixation sites. Also, the brain suppresses the visual image during the saccades so that one is completely unconscious of the movements from point to point.

Saccadic Movements During Reading. During the process of reading, a person usually makes several saccadic movements of the eyes for each line. In this case the visual scene is not moving past the eyes, but the eyes are trained to scan across the visual scene to extract the important information. Similar saccades occur when a person observes a painting, except that the saccades occur in one direction after another from one highlight of the painting to another, then another, and so forth.

Fixation on Moving Objects — "Pursuit Movements." The eyes can also remain fixed on a moving object, which is called *pursuit movement*. A highly developed cortical mechanism automatically detects the course of movement of an object and then gradually develops a similar course of movement of the eyes. For instance, if an object is moving up and down in a wavelike form at a rate of several times per second, the eyes at first may be completely unable to fixate on it. However, after a second or so the eyes begin to jump coarsely in approximately the same pattern of movement as that of the object. Then after a few more seconds, the eyes develop progressively smoother and smoother movements and finally follow the course of movement almost exactly. This represents a high degree of automatic, subconscious computational ability by the cerebral cortex.

Vestibular Control of Eye Movements. Another type of eye movement is elicited by stimulation of the

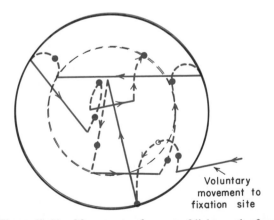

Voluntary movement to fixation site

Figure 60–11. Movements of a spot of light on the fovea, showing sudden "flicking" movements to move the spot back toward the center of the fovea whenever it drifts to the foveal edge. (The dashed lines represent slow drifting movements, and the solid lines represent sudden flicking movements.) (Modified from Whitteridge: Handbook of Physiology, Vol. 2, Sec. 1. Baltimore, The Williams & Wilkins Company, 1960, p. 1089.)

vestibular apparatus. The vestibular nuclei are connected directly with the brainstem nuclei that control ocular movements, and, whenever the head is accelerated in a vertical, longitudinal, lateral or angular direction, an immediate compensatory motion of the eyes occurs in the opposite direction. This allows the eyes to remain fixed on an object of attention despite rapid movements of the body or head.

Vestibular control of the eyes is especially valuable when a person is subjected to jerky motions of the body. For instance, when a person with bilateral destruction of the vestibular apparatuses rides over rough roads, fixing the eyes on the road or on any horizontal scene is extremely difficult. The opticokinetic type of movement is not capable of keeping the eyes fixed under such conditions because it has a latent period of about one-fifth second before the direction of movement of the visual scene can be detected and followed by the eyes. On the other hand, the vestibular type of eye movement has a short latent period, measured in thousandths of a second rather than one-fifth second.

Also when a person begins to rotate, the vestibular apparatuses cause a vestibular type of saccadic movements called *vestibular nystagmus*. That is, the eyes lag behind the rotating head, then jump forward, then lag behind again, jump again, and so forth. This type of nystagmus is discussed in Chapter 52 in relation to vestibular function.

Pathologic Types of Nystagmus. Occasionally, abnormalities occur in the control system for eye movements that cause continuous nystagmus (back and forth movements) despite the fact that neither the visual scene nor the body is moving. This is likely to occur when one of the vestibular apparatuses is damaged or when severe damage is sustained in the deep nuclei of the cerebellum. This is discussed further in Chapter 53.

Another pathologic type of eye movement that is sometimes called nystagmus occurs when the foveal regions of the two eyes have been destroyed or when the vision in these areas is greatly weakened. In such a condition the eyes attempt to fix the object of attention on the foveae but always overshoot the mark because of foveal insensitivity. Therefore, they oscillate back and forth but never achieve foveal fixation. Even though this condition is known clinically as a type of nystagmus, physiologically it is completely different from the nystagmus that keeps the eyes fixed on a moving scene.

FUSION OF THE VISUAL IMAGES

To make the visual perceptions more meaningful and also to aid in depth perception by the mechanism of stereopsis, which was discussed in Chapter 58, the visual images in the two eyes normally *fuse* with each other on "corresponding points" of the two retinae.

The visual cortex especially, and perhaps the lateral geniculate body as well, plays a very important role in fusion. It was pointed out earlier in the chapter that corresponding points of the two retinae transmit visual signals to different neuronal layers of the lateral geniculate body, and these signals in turn are relayed to parallel stripes of neurons in the visual cortex. Interactions occur between the stripes of cortical neurons; these cause *interference patterns of excitation* in still

other neuronal cells when the two visual images are not precisely "in register" — that is, not precisely fused. This excitation presumably provides the signal that is transmitted to the oculomotor apparatus to cause convergence or divergence or rotation of the eyes so that fusion can be re-established. Once the corresponding points of the retinae are precisely in register with each other, the excitation of the specific cells in the visual cortex disappears.

The Neural Mechanism for Stereopsis. The visual images that appear on the retina during the process of stereopsis were discussed in Chapter 58. It was pointed out that because the two eyes are a little more than 2 inches apart the images on the two retinae are not exactly the same. The closer the object is to the eye the greater is the disparity between the two images. Consequently, it is impossible for all corresponding points in the visual image to be in complete register at the same time. Furthermore, the nearer the object is to the eye the less is the degree of register. Here again, specific cells in the primary visual cortex become excited in the areas of the visual field where highlights are out of register. Presumably, this excitation is the source of the signals for detection of the distance of the object in front of the eyes; this mechanism is called *stereopsis*.

Strabismus. Strabismus, which is also called *squint* or *cross-eyedness*, means lack of fusion of the eyes in one or more of the coordinates described above. Three basic types of strabismus are illustrated in Figure 60–12: *horizontal strabismus, vertical strabismus,* and *torsional strabismus.* However, combinations of two or even of all three of the different types of strabismus often occur.

Strabismus is caused by an abnormal "set" of the fusion mechanism of the visual system. That is, in the early efforts of the child to fixate the two eyes on the same object, one of the eyes fixates satisfactorily while the other fails to fixate, or they both fixate satisfactorily but never simultaneously. Soon, the patterns of conjugate movements of the eyes become abnormally "set" so that the eyes never fuse.

Frequently, some abnormality of the eyes contributes to the failure of the two eyes to fixate on the same point. For instance, if at birth one eye has poor vision in comparison with the other, the good eye tends to fixate on the object of attention while the poor eye might never do so. Also, in hypermetropic infants, intense impulses must be transmitted to the ciliary muscles to focus the eyes, and some of these impulses overflow into the oculomotor nuclei to cause simultaneous convergence of the eyes, as will be discussed below. As a result, the child's fusion mechanism becomes "set" for continual inward deviation of the eyes.

Suppression of Visual Image from a Repressed Eye. In

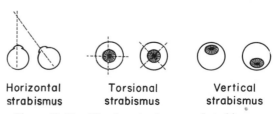

Horizontal strabismus   Torsional strabismus   Vertical strabismus

Figure 60–12. The three basic types of strabismus.

many patients with strabismus the eyes alternate in fixing on the object of attention. However, in other patients, one eye alone is used all the time while the other eye becomes repressed and is never used for vision. The vision in the repressed eye develops only slightly, usually remaining 20/400 or less. If the dominant eye then becomes blinded, vision in the repressed eye can develop only to a slight extent in the adult but far more in young children. This illustrates that visual acuity is highly dependent on proper development of the central synaptic connections from the eyes.

AUTONOMIC CONTROL OF ACCOMMODATION AND PUPILLARY APERTURE

The Autonomic Nerves to the Eyes. The eye is innervated by both parasympathetic and sympathetic fibers, as illustrated in Figure 60–13. The parasympathetic preganglionic fibers arise in the *Edinger-Westphal nucleus* (the visceral nucleus of the third nerve) and then pass in the *third nerve* to the *ciliary ganglion,* which lies about 1 cm. behind the eye. Here the preganglionic fibers synapse with postganglionic parasympathetic neurons that, in turn, send fibers through the *ciliary nerves* into the eyeball. These nerves excite the ciliary muscle and the sphincter of the iris.

The sympathetic innervation of the eye originates in the *intermediolateral horn cells* of the first thoracic segment of the spinal cord. From here, sympathetic fibers enter the sympathetic chain and pass upward to the *superior cervical ganglion* where they synapse with postganglionic neurons. Fibers from these spread along the carotid artery and successively smaller arteries until they reach the eye. There the sympathetic fibers innervate the radial fibers of the iris as well as several extraocular structures around the eye, which are discussed shortly in relation to Horner's syndrome. Also, they supply very weak innervation to the ciliary muscle.

CONTROL OF ACCOMMODATION

The accommodation mechanism — that is, the mechanism which focuses the lens system of the eye — is essential for a high degree of visual acuity. Accommodation results from contraction or relaxation of the ciliary muscle, contraction causing increased strength of the lens system, as explained in Chapter 58, and relaxation causing decreased strength. The question that must be answered now is: How does a person adjust accommodation to keep the eyes in focus all the time?

Accommodation of the lens is regulated by a negative feedback mechanism that automatically adjusts the focal power of the lens for the highest degree of visual acuity. When the eyes have been fixed on some far object and then suddenly fix on a near object, the lens accommodates for maximum acuity of vision usually within one second.

Though the precise control mechanism that causes this rapid and accurate focusing of the eye is still unclear, some of the known features of the mechanism are the following:

First, when the eyes suddenly change the distance of the fixation point, the lens always changes its strength in the proper direction to achieve a new state of focus. In other words, the lens *does not hunt* back and forth on the two sides of focus in an attempt to find the focus.

Second, different types of clues that can help the lens change its strength in the proper direction include the following: (1) *Chromatic aberration* appears to be important. That is, the red light rays focus slightly posteriorly to the blue light rays. The eyes appear to be able to detect which of these two types of rays is in better focus, and this clue relays information to the accommodation mechanism whether to make the lens stronger or weaker. (2) When the eyes fixate on a near object they also converge toward each other. The neural mechanisms for *convergence cause a simultaneous signal to strengthen the lens of the eye.* (3) *Since the fovea is a depressed area, the clarity of focus in the depth of the fovea versus the clarity of focus on the edges will be different.* It has been suggested that this also gives clues as to which way the strength of the lens needs to be changed. (4) It has been found that *the degree of accommodation of the lens oscillates slightly* all of the time, at a frequency up to 2 times per second. It has been suggested that the visual image becomes clearer when the oscillation of the lens strength is in the appropriate direction and poorer when the lens strength is in the wrong direction. This could give a rapid cue as to which way the strength of the lens needs to change to provide appropriate focus.

It is presumed that the cortical areas that control accommodation closely parallel those that control fixation movements of the eyes, with final integration of the visual signals in areas 18 and 19 and transmission of motor signals to the ciliary muscle through the pretectal area and Edinger-Westphal nucleus.

CONTROL OF THE PUPILLARY APERTURE

Stimulation of the parasympathetic nerves excites the pupillary sphincter, thereby decreasing the pupillary aperture; this is called *miosis.* On the other hand, stimulation of the sympathetic nerves excites the radial fibers of the iris and causes pupillary dilatation, which is called *mydriasis.*

The Pupillary Light Reflex. When light is shone into the eyes the pupils constrict, a reaction that is called the pupillary light reflex. The neuronal pathway for this reflex is illustrated in Figure

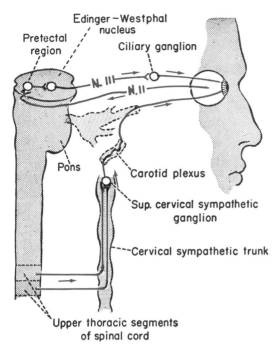

Figure 60–13. Autonomic innervation of the eye, showing also the reflex arc of the light reflex. (Modified from Ranson and Clark: Anatomy of the Nervous System. Philadelphia, W. B. Saunders Company, 1959.)

60–13. When light impinges on the retina, the resulting impulses pass through the optic nerves and optic tracts to the pretectal nuclei. From here, impulses pass to the *Edinger-Westphal nucleus* and finally back through the *parasympathetic nerves* to constrict the sphincter of the iris. In darkness, the reflex becomes inhibited, which results in dilatation of the pupil.

The function of the light reflex is to help the eye adapt extremely rapidly to changing light conditions, the importance of which was explained in relation to retinal adaptation in the previous chapter. The limits of pupillary diameter are about 1.5 mm. on the small side and 8 mm. on the large side. Therefore, the range of light adaptation that can be effected by the pupillary reflex is about 30 to 1.

Pupillary Reflexes in Central Nervous System Disease. Certain central nervous system diseases block the transmission of visual signals from the retinae to the Edinger-Westphal nucleus. Such blocks frequently occur as a result of *central nervous system syphilis, alcoholism, encephalitis,* or so forth. The block usually occurs in the pretectal region of the brain stem, though it can also result from destruction of the small afferent fibers in the optic nerves.

The light signals to the Edinger-Westphal nucleus are of the inhibitory type. Therefore, when they are blocked, the nucleus becomes chronically active, causing the pupils thereafter to remain partially constricted in addition to their failure to respond to light.

When the light reflex is lost, the pupils can still constrict an additional amount if the Edinger-Westphal nucleus is stimulated through some other pathways. For instance, when the eyes fixate on a near object, the signals that cause accommodation of the lens (and also those that cause convergence of the two eyes) cause a mild degree of pupillary constriction at the same time. This is called the *accommodation reflex.* Such a pupil that fails to respond to light but does respond to accommodation and also is very small (an *Argyll Robertson pupil*) is an important diagnostic sign of central nervous system disease — very often syphilis.

Horner's Syndrome. The sympathetic nerves to the eye are occasionally interrupted, and this interruption frequently occurs in the cervical sympathetic chain. This results in *Horner's syndrome*, which consists of the following effects: First, because of interruption of fibers to the pupillary dilator muscle, the pupil remains persistently constricted to a smaller diameter than that of the pupil of the opposite eye. Second, the superior eyelid droops because this eyelid is normally maintained in an open position during the waking hours partly by contraction of a smooth muscle, the *superior palpebral muscle,* which is innervated by the sympathetics. Therefore, destruction of the sympathetics makes it impossible to open the superior eyelid nearly as widely as normally. Third, the blood vessels on the corresponding side of the face and head become persistently dilated. And, fourth, sweating cannot occur on the side of the face and head affected by Horner's syndrome.

REFERENCES

Armington, J. C.: The Electroretinogram. New York, Academic Press, 1974.

Bizzi, E.: The coordination of eye-head movements. *Sci. Am., 231*(4):100, 1974.

Critchley, M., and Critchley, E. A.: Dysplexia Defined. Springfield, Ill., Charles C Thomas, 1978.

Dunn-Rankin, P.: The visual characteristics of words. *Sci. Am., 238*(1):122, 1978.

Ewert, J. P.: The neural basis of visually guided behavior. *Sci. Am., 230*(3):34, 1974.

Fatt, I.: Physiology of the Eye: An Introduction to the Vegetative Function. Boston, Butterworths, 1978.

Fraser, S. E., and Hunt, R. K.: Retinotectal specificity: Models and experiments in search of a mapping function. *Annu. Rev. Neurosci., 3*:319, 1980.

Goldchrist, A. L.: The perception of surface blacks and whites. *Sci. Am., 240*(3):112, 1979.

Glickstein, M., and Gibson, A. R.: Visual cells in the pons of the brain. *Sci. Am., 235*(5):90, 1976.

Gogel, W. C.: The adjacency principle in visual perception. *Sci. Am., 238*(5):126, 1978.

Guyton, J. S., and Kirkman, N.: Ocular movement: I. Mechanics, pathogenesis, and surgical treatment of alternating hypertropia (dissociated vertical divergence, double hypertropia) and some related phenomena. *Am. J. Ophthalmol., 41*:438, 1956.

Hess, E. H.: The role of pupil size in communication. *Sci. Am., 235*(5):110, 1975.

Hubbell, W. L., and Bownds, M. D.: Visual transduction in vertebrate photoreceptors. *Annu. Rev. Neurosci., 2*:17, 1979.

Hubel, D. H.: Eleventh Bowditch Lecture. Effects of distortion of sensory input on the visual system of kittens. *Physiologist, 10*:17, 1967.

Hubel, D. H., and Wiesel, T. N.: Receptive fields of cells in striate cortex of very young, visually inexperienced kittens. *J. Neurophysiol., 26*:994, 1963.

Hubel, D. H., and Wiesel, T. N.: Cortical and callosal connections concerned with vertical meridian of visual fields in the cat. *J. Neurophysiol., 30*:1561, 1967.

Hubel, D. H., and Wiesel, T. N.: Brain mechanisms of vision. *Sci. Am., 241*(3):150, 1979.

Kuffler, S. W.: The single-cell approach in the visual system and the study of receptive fields. *Invest. Ophthalmol., 12*:794, 1973.

Johansson, G.: Visual motion perception. *Sci. Am., 232*(6):1975.

Julesz, B.: Experiments in the visual perception of texture. *Sci. Am., 232*(4):34, 1975.

Kool, K. A., and Marshall, R. E.: Visual Evoked Potentials in Central Disorders of the Visual System. New York, Harper & Row, 1979.

McIlwain, J. T.: Large receptive fields and spatial transformations in the visual system. *Int. Rev. Physiol., 10*:223, 1976.

Neuroanatomy and Neuro-Ophthalmology, May–June 1978. Rochester, Minn., American Academy of Ophthalmology, 1978.

Ordy, J. M., and Brizzee, K. R. (eds.): Sensory Systems and Communication in the Elderly. New York, Raven Press, 1979.

Polyak, S.: The Retina. Chicago, University of Chicago Press, 1941.

Raphan, T., and Cohen, B.: Brainstem mechanisms for rapid and slow eye movements. *Annu. Rev. Physiol., 40*:527, 1978.

Regan, D., *et al.*: The visual perception of motion in depth. *Sci. Am. 241*(1):136, 1979.

Reinecke, R. D. (ed.): Strabismus. New York, Grune & Stratton, 1978.

Rodieck, R. W.: Visual pathways. *Annu. Rev. Neurosci., 2*:193, 1979.

Sekuler, R., and Levinson, E.: The perception of moving targets. *Sci. Am., 236*(1):60, 1977.

Singer, W.: Control of thalamic transmission by corticofugal and ascending reticular pathways in the visual system. *Physiol. Rev., 57*:386, 1977.

Smith, J. L.: Neuro-Ophthalmology, Focus 1980. New York, Masson Publishing USA, 1979.

Toates, F. M.: Accommodation function in the human eye. *Physiol. Rev., 52*:828, 1972.

Van Essen, D. C.: Visual areas of the mammalian cerebral cortex. *Annu. Rev. Neurosci., 2*:227, 1979.

Walsh, T. J.: Neuro-Ophthalmology: Clinical Signs and Symptoms. Philadelphia, Lea & Febiger, 1978.

Wurtz, R. H., and Albano, J. E.: Visual-motor function of the primate superior colliculus. *Annu. Rev. Neurosci., 3*:189, 1980.

Yannuzzi, L. A., *et al.* (eds.): The Macula: A Comprehensive Text and Atlas. Baltimore, Williams & Wilkins, 1978.

61

The Sense of Hearing

Hearing, like many somatic senses, is a mechanoreceptive sense, for the ear responds to mechanical vibration of the sound waves in the air. The purpose of the present chapter is to describe and explain the mechanism by which the ear receives sound waves, discriminates their frequencies, and finally transmits auditory information into the central nervous system.

THE TYMPANIC MEMBRANE AND THE OSSICULAR SYSTEM

TRANSMISSION OF SOUND FROM THE TYMPANIC MEMBRANE TO THE COCHLEA

Figure 61–1 illustrates the *tympanic membrane* (commonly called the *eardrum)* and the *ossicular system,* which transmits sound through the middle ear. The tympanic membrane is cone-shaped, with its concavity facing downward and outward toward the auditory canal. Attached to the very center of the tympanic membrane is the *handle* of the *malleus.* At its other end the malleus is tightly bound to the *incus* by ligaments so that whenever the malleus moves the incus moves with it. The opposite end of the incus in turn articulates with the stem of the *stapes,* and the *faceplate* of the stapes lies against the membranous labyrinth in the opening of the oval window where sound waves are transmitted into the inner ear, the *cochlea.*

The ossicles of the middle ear are suspended by ligaments in such a way that the combined malleus and incus act as a single lever having its fulcrum approximately at the border of the tympanic membrane. The large *head* of the malleus, which is on the opposite side of the fulcrum from the handle, almost exactly balances the other end of the lever so that changes in position of the body will not increase or decrease the tension on the tympanic membrane.

The articulation of the incus with the stapes causes the stapes to push forward on the cochlear fluid every time the handle of the malleus moves inward and to pull backward on the fluid every time the malleus moves outward, which promotes inward and outward motion of the faceplate at the oval window.

The handle of the malleus is constantly pulled inward by ligaments and by the tensor tympani muscle, which keeps the tympanic membrane tensed. This allows sound vibrations on *any* portion of the tympanic membrane to be transmitted to the malleus, which would not be true if the membrane were lax.

Impedance Matching by the Ossicular System. The amplitude of movement of the stapes faceplate with each sound vibration is only three-fourths as much as the amplitude of the handle of the malleus. Therefore, the ossicular lever system does not amplify the movement distance of the stapes, as is commonly believed. Instead, the system actually reduces the amplitude but increases the *force* of movement about 1.3 times. However, the surface area of the tympanic membrane is approximately 55 sq. mm., while the surface area of the stapes averages 3.2 sq. mm. This 17-fold difference times the 1.3-fold ratio of the lever system allows all the energy of a sound wave impinging on the tympanic membrane to be

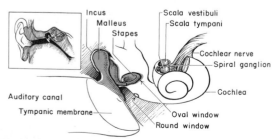

Figure 61–1. The tympanic membrane, the ossicular system of the middle ear, and the inner ear.

applied to the small faceplate of the stapes, causing approximately 22 times as much *pressure* on the fluid of the cochlea as is exerted by the sound wave against the tympanic membrane. Since fluid has far greater inertia than air, it is easily understood that increased amounts of pressure are needed to cause vibration in the fluid. Therefore, the tympanic membrane and ossicular system provide *impedance matching* between the sound waves in air and the sound vibrations in the fluid of the cochlea. Indeed, the impedance matching is about 50 to 75 per cent of perfect for sound frequencies between 300 to 3000 cycles per second, which allows almost full utilization of the energy in the incoming sound waves.

In the absence of the ossicular system and tympanum, sound waves can travel directly through the air of the middle ear and can enter the cochlea at the oval window. However, the sensitivity for hearing is then 30 decibels less than for ossicular transmission — equivalent to a decrease from a very loud voice to a barely audible voice level.

Attenuation of Sound by Contraction of the Stapedius and Tensor Tympani Muscles. When loud sounds are transmitted through the ossicular system into the central nervous system, a reflex occurs after a latent period of 40 to 80 milliseconds to cause contraction of both the stapedius and tensor tympani muscles. The tensor tympani muscle pulls the handle of the malleus inward while the stapedius muscle pulls the stapes outward. These two forces oppose each other and thereby cause the entire ossicular system to develop a high degree of rigidity, thus greatly reducing the ossicular transmission of low frequency sound, mainly frequencies below 1000 cycles per second.

This *attenuation reflex* can reduce the intensity of sound transmission by as much as 30 to 40 decibels, which is about the same difference as that between a whisper and the sound emitted by a loud voice. The function of this mechanism is probably two-fold:

1. To *protect* the cochlea from damaging vibrations caused by excessively loud sound. It is mainly low frequency sounds (the ones that are attenuated) that are frequently loud enough to damage the basilar membrane of the cochlea. Unfortunately, because of the 40 or more millisecond latency for reaction of the reflex, the sudden, loud, thunderous sounds that result from explosions can still cause extensive cochlear damage.

2. To *mask* low frequency sounds in loud environments. This usually removes a major share of the background noise and allows a person to concentrate on sounds above 1000 cycles per second frequency. It is in this upper frequency range that most of the pertinent information in voice communication is transmitted.

Another function of the tensor tympani and stapedius muscles is to decrease a person's hearing sensitivity to his or her own speech. This effect is activated by collateral signals transmitted to these muscles at the same time that the brain activates the voice mechanism.

TRANSMISSION OF SOUND THROUGH BONE

Because the inner ear, the *cochlea,* is embedded in a bony cavity in the temporal bone called the bony labyrinth, vibrations of the entire skull can cause fluid vibrations in the cochlea itself. Therefore, under appropriate conditions, a tuning fork or an electronic vibrator placed on any bony protuberance of the skull causes the person to hear the sound if it is intense enough. Unfortunately, the energy available even in very loud sound in the air is not sufficient to cause hearing through the bone except when a special electromechanical sound-transmitting device is applied directly to the bone, usually to the mastoid process.

THE COCHLEA

FUNCTIONAL ANATOMY OF THE COCHLEA

The cochlea is a system of coiled tubes, shown in Figure 61–1 and in cross-section in Figures 61–2 and 61–3, with three different tubes coiled side by side: the *scala vestibuli,* the *scala media,* and the *scala tympani.* The scala vestibuli and scala media are separated from each other by *Reissner's membrane* (also called the *vestibular membrane*), and

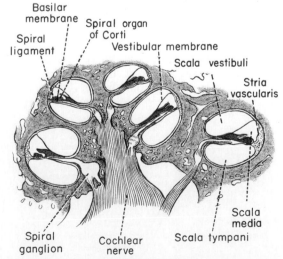

Figure 61–2. The cochlea. (From Goss, C. M. (ed.): Gray's Anatomy of the Human Body. Lea & Febiger.)

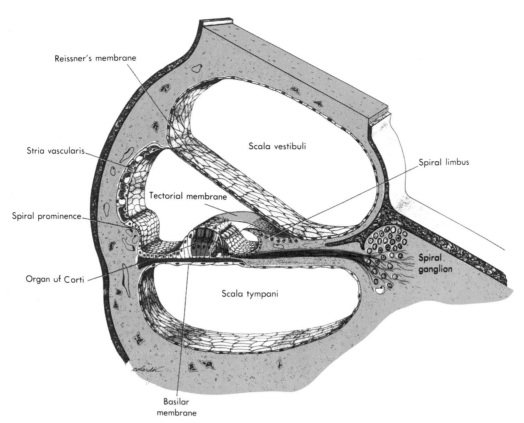

Figure 61–3. A section through one of the turns of the cochlea. (Drawn by Sylvia Colard Keene. From Bloom and Fawcett: A Textbook of Histology. Philadelphia, W. B. Saunders Company, 1975.)

the scala tympani and scala media are separated from each other by the *basilar membrane*. On the surface of the basilar membrane lies a structure, the *organ of Corti,* which contains a series of mechanically sensitive cells, the *hair cells.* These are the receptive end-organs that generate nerve impulses in response to sound vibrations.

Figure 61–4 diagrams the functional parts of the uncoiled cochlea for transmission of sound vibrations. First, note that Reissner's membrane is missing from this figure. This membrane is so thin and so easily moved that it does not obstruct the passage of sound vibrations from the scala vestibuli into the scala media at all. Therefore, so far

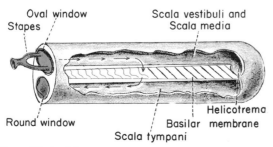

Figure 61–4. Movement of fluid in the cochlea following forward thrust of the stapes.

as the transmission of sound is concerned, the scala vestibuli and scala media are considered to be a single chamber. The importance of Reissner's membrane is to maintain a special fluid in the scala media that is required for normal function of the sound receptive hair cells, as discussed later in the chapter.

Sound vibrations enter the scala vestibuli from the faceplate of the stapes at the oval window. The faceplate covers this window and is connected with the window's edges by a relatively loose annular ligament so that it can move inward and outward with the sound vibrations. Inward movement causes the fluid to move into the scala vestibuli and scala media, which immediately increases the pressure in the entire cochlea and causes the round window to bulge outward.

Note in Figure 61–4 that the distal ends of the scala vestibuli and scala tympani are continuous with each other through an opening at the end, the *helicotrema.* If the stapes moves inward *very slowly,* fluid from the scala vestibuli is pushed through the helicotrema into the scala tympani, and this causes the round window to bulge outward. However, if the stapes vibrates inward and outward rapidly, the fluid simply does not have time to pass all the way to the helicotrema, then to the round window, and back again to the oval window between each two successive vibrations. Instead, the fluid wave takes a shortcut through the basilar membrane, causing it to bulge back and forth with each sound vibration. We shall see later that each frequency of sound causes a different "pattern" of vibration in the basilar membrane and that this is one of the important means by which the sound frequencies are discriminated from each other.

The Basilar Membrane and Resonance in the Cochlea. The basilar membrane contains 20,000 to 30,000 *basilar fibers* that project from the bony center of the cochlea, the *modiolus,* toward the outer wall. These fibers are stiff, elastic, reedlike structures that are fixed at their basal ends but not at their distal ends except that the distal ends are embedded in the loose basilar membrane. Because the fibers are stiff and also free at one end, they can vibrate like reeds of a harmonica.

The lengths of the basilar fibers increase progressively as one goes from the oval and round windows toward the helicotrema, from a length of approximately 0.04 mm. near the windows to 0.5 mm. at the helicotrema, a 12-fold increase in length.

The diameters of the fibers, on the other hand, decrease from the base to the helicotrema, so that their overall stiffness decreases more than 100-fold. As a result, the stiff, short fibers near the base of the cochlea have a tendency to vibrate at a high frequency, while the long, limber fibers near the helicotrema have a tendency to vibrate at a low frequency.

In addition to the differences in stiffness of the basilar fibers, they are also differently "loaded" by the fluid mass of the cochlea. That is, when a fiber vibrates back and forth, all the fluid between the vibrating fiber and the oval and round windows must also move back and forth at the same time. For a fiber vibrating near the base of the cochlea (near the two windows), the total mass of moving fluid is slight in comparison with that for a fiber vibrating near the helicotrema. This difference, too, favors high frequency vibration near the windows and low frequency vibration near the tip of the cochlea.

Thus, high frequency resonance of the basilar membrane occurs near the base, and low frequency resonance occurs near the apex because of (1) difference in stiffness of the fibers and (2) difference in "loading."

TRANSMISSION OF SOUND WAVES IN THE COCHLEA— THE "TRAVELING WAVE"

If the foot of the stapes moves inward instantaneously, the round window must also bulge outward instantaneously because the cochlea is bounded on all sides by bony walls. Since the fluid wave will not have time to move all the way from the oval window to the helicotrema and back to the round window, the initial effect is to cause the basilar membrane at the very base of the cochlea to bulge in the direction of the round window. However, the elastic tension that is built up in the basilar fibers as they bend toward the round window initiates a wave that "travels" along the basilar membrane toward the helicotrema, as illustrated in Figure 61–5. Figure 61–5A shows movement of a high frequency wave down the basilar membrane; Figure 61–5B, a medium frequency wave; and Figure 61–5C, a very low frequency wave. Movement of the wave along the basilar membrane is comparable to the movement of a pressure wave along the arterial walls, which was discussed in Chapter 19, or it is also comparable to the wave that travels along the surface of a pond.

Pattern of Vibration of the Basilar Membrane for Different Sound Frequencies. Note in Figure 61–5 the different patterns of transmission for sound waves of different frequencies. Each wave is relatively weak at the outset but becomes strong when it reaches that portion of the basilar membrane that has a natural resonant frequency equal to the respective sound frequency. At this point the basilar membrane can vibrate back and forth with such great ease that the energy in the wave is completely dissipated. Consequently, the wave ceases at this point and fails to travel the remaining distance along the basilar membrane. Thus, a high frequency sound wave travels only a short

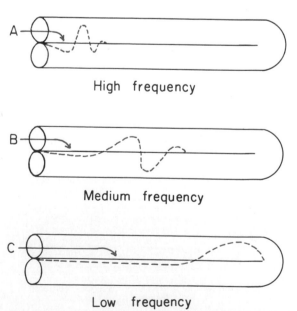

Figure 61–5. "Traveling waves" along the basilar membrane for high, medium, and low frequency sounds.

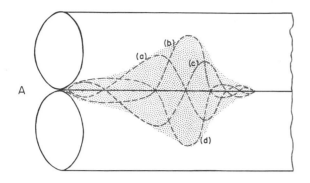

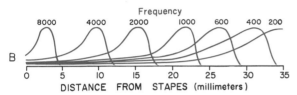

Figure 61–6. (A) Amplitude pattern of vibration of the basilar membrane for a medium frequency sound. (B) Amplitude patterns for sounds of all frequencies between 50 and 8000 per second, showing the points of maximum amplitude (the resonance points) on the basilar membrane for the different frequencies.

distance along the basilar membrane before it reaches its resonant point and dies out; a medium frequency sound wave travels about halfway and then dies out; and, finally, a very low frequency sound wave travels the entire distance along the membrane.

Another feature of the traveling wave is that it travels fast along the initial portion of the basilar membrane but progressively more slowly as it goes farther and farther into the cochlea. The cause of this is the high coefficient of elasticity of the basilar fibers near the stapes but a progressively decreasing coefficient farther along the membrane. This rapid initial transmission of the wave allows the high frequency sounds to travel far enough into the cochlea to spread out and separate from each other on the basilar membrane. Without this spread, all the high frequency waves would be bunched together within the first millimeter or so of the basilar membrane, and their frequencies could not be discriminated one from the other.

Amplitude Pattern of Vibration of the Basilar Membrane. The dashed curves of Figure 61–6A show the position of a sound wave on the basilar membrane when the stapes (*a*) is all the way inward, (*b*) has moved back to the neutral point, (*c*) is all the way outward, and (*d*) has moved back again to the neutral point but is moving inward. The shaded area around these different waves shows the maximum extent of vibration of the basilar membrane during a complete vibratory cycle. This is the amplitude pattern of vibration of

the basilar membrane for this particular sound frequency.

Figure 61–6B shows the amplitude patterns of vibration for different frequencies, showing that the maximum amplitude for 8000 cycles occurs near the base of the cochlea, while that for frequencies less than 400 to 500 cycles per second is all the way at the tip of the basilar membrane near the helicotrema.

Note in Figure 61–6B that the basal end of the basilar membrane vibrates at least weakly for all frequencies. However, beyond the resonant area for each given frequency, the vibration of the basilar membrane cuts off sharply. The principal method by which sound frequencies, especially those above 400 to 500 cycles per second, are discriminated from each other is based on the "place" of maximum stimulation of the nerve fibers from the organ of Corti lying on the basilar membrane, as will be explained in the following section.

FUNCTION OF THE ORGAN OF CORTI

The organ of Corti, illustrated in Figures 61–2, 61–3, and 61–7, is the receptor organ that generates nerve impulses in response to vibration of the basilar membrane. Note that the organ of Corti lies on the surface of the basilar fibers and basilar membrane. The actual sensory receptors in the organ of Corti are two types of *hair cells,* a single row of *internal hair cells,* numbering about 3500 and measuring about 12 microns in diameter, and three to four rows of *external hair cells,* numbering about 20,000 and having diameters of only about 8 microns. The bases and sides of the hair cells synapse with a network of cochlear nerve endings. These lead to the *spiral ganglion of Corti,* which lies in the modiolus (the center) of the cochlea. The spiral ganglion in turn sends axons into the *cochlear nerve* and thence into the central nervous system at the level of the upper medulla. The relationship of the organ of Corti to the spiral ganglion and to the cochlear nerve is illustrated in Figure 61–2.

Excitation of the Hair Cells. Note in Figure 61–7 that minute hairs, or cilia, project upward from the hair cells and either touch or are embedded in the surface gel coating of the *tectorial membrane* which lies above the cilia in the scala media. These hair cells are similar to the hair cells found in the macula and cristae ampullaris of the vestibular apparatus, which were discussed in Chapter 52. Bending of the hairs in one direction depolarizes the hair cells, and this in turn excites the nerve fibers synapsing with their bases.

Figure 61–8 illustrates the mechanism by which vibration of the basilar membrane excites the hair endings. This shows that the upper ends of the hair

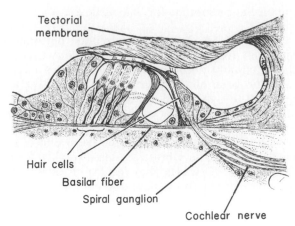

Figure 61-7.   The organ of Corti, showing especially the hair cells and the tectorial membrane against the projecting hairs.

cells are fixed tightly in a structure called the *reticular lamina*. Furthermore, the reticular lamina is very rigid and is continuous with a rigid triangular structure called the *rods of Corti* that rests on the basilar fibers. Therefore, the basilar fiber, the rods of Corti, and the reticular lamina all move as a unit.

Upward movement of the basilar fiber rocks the reticular lamina upward and *inward*. Then, when the basilar membrane moves downward, the reticular lamina rocks downward and *outward*. The inward and outward motion causes the hairs to shear back and forth against the tectorial membrane, thus exciting the cochlear nerve fibers whenever the basilar membrane vibrates.

Mechanism by Which the Hair Cells Excite the Nerve Fibers — Receptor Potentials.   Back-and-forth bending of the hairs causes alternate changes in the electrical potential across the hair cell membrane. This alternating potential is the *receptor potential* of the hair cell, and it in turn stimulates the cochlear nerve endings that terminate on the hair cells. Most physiologists believe that the receptor potential stimulates the endings by direct electrical excitation.

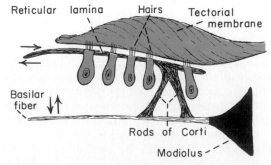

Figure 61-8.   Stimulation of the hair cells by the to-and-fro movement of the hairs in the tectorial membrane.

When the basilar fiber bends toward the scala vestibuli (in the upward direction in Figure 61-8), the hair cell becomes depolarized, and it is this depolarization that excites an increased number of action potentials in the nerve fiber. When the basilar fiber moves in the opposite direction, the hair cell becomes hyperpolarized, and the number of action potentials decreases.

The Endocochlear Potential.   To explain even more fully the electrical potentials generated by the hair cells, we need to explain still another electrical phenomenon called the endocochlear potential: The scala media is filled with a fluid called *endolymph* in contradistinction to the *perilymph* present in the scala vestibuli and scala tympani. The scala vestibuli and scala tympani in most young children and in some adults communicate directly with the subarachnoid space around the brain, so that the perilymph is almost identical with cerebrospinal fluid. On the other hand, the endolymph that fills the scala media is an entirely different fluid probably secreted by the *stria vascularis,* a highly vascular area on the outer wall of the scala media. Endolymph contains a very high concentration of potassium and a very low concentration of sodium, which is exactly opposite to the perilymph.

An electrical potential of approximately 80 mv. exists all the time between the endolymph and the perilymph, with positivity inside the scala media and negativity outside. This is called the *endocochlear potential,* and it is believed to be generated by continual secretion of positive potassium ions into the scala media by the stria vascularis.

The importance of the endocochlear potential is that the tops of the hair cells project through the reticular lamina into the endolymph of the scala media while perilymph bathes the lower bodies of the hair cells. Furthermore, the hair cells have a negative intracellular potential of -70 millivolts with respect to the perilymph, but -150 millivolts with respect to the endolymph at their upper surfaces where the hairs project into the endolymph. It is believed that this high electrical potential at the hair border of the cell greatly sensitizes the cell, thereby increasing its ability to respond to slight movement of the hairs.

DETERMINATION OF PITCH – THE "PLACE" PRINCIPLE

In the minds of most persons, sound pitch and sound frequency are the same thing. However, the two are slightly different in the following way: Pitch is the conscious perception of the sound frequency and may not be the same as the true sound frequency. The pitch is especially likely to deviate from the true sound frequency at both very high and very low frequency; also, the pitch usually changes slightly as the intensity of the sound changes even though the frequency remains constant. Yet, from the point of view of the following discussion, we can still consider the two to be essentially the same.

From earlier discussions in this chapter it is already apparent that low pitch (or low frequency) sounds cause maximal activation of the basilar membrane near the apex of the cochlea, sounds of high pitch (or high frequency) activate the basilar membrane near the base of the cochlea, and intermediate frequencies activate the membrane at intermediate distances between these two extremes. Furthermore, there is spatial organization of the cochlear nerve fibers from the cochlea to the cochlear nuclei in the brain stem, the fibers from each respective area of the basilar membrane terminating in a corresponding area in the cochlear nuclei. We shall see later that this spatial organization continues all the way up the brain stem to the cerebral cortex. The recording of signals from the auditory tracts in the brain stem and from the auditory receptive fields in the cerebral cortex shows that specific neurons are activated by specific pitches. Therefore, the major method used by the nervous system to detect different pitches is to determine the position along the basilar membrane that is most stimulated. This is called the *place principle* for determination of pitch.

Yet, referring again to Figure 61–6, one can see that the distal end of the basilar membrane at the helicotrema is stimulated by all sound frequencies below 400 to 500 cycles per second. Therefore, it has been difficult to understand from the place principle how one can differentiate between very low sound frequencies. It is postulated that these low frequencies are discriminated mainly by the so-called *frequency principle*. That is, the low frequency sounds cause volleys of impulses at the same low frequencies to be transmitted by the cochlear nerve into the cochlear nuclei. And it is believed that the cochlear nuclei then distinguish the different frequencies. In fact, destruction of the entire apical half of the cochlea, which destroys the basilar membrane where all of the lower frequency sounds are normally detected, still does not completely eliminate the discrimination of low frequency sounds.

DETERMINATION OF LOUDNESS

Loudness is determined by the auditory system in at least three different ways: First, as the sound becomes louder, the amplitude of vibration of the basilar membrane and hair cells also increases so that the hair cells excite the nerve endings at more rapid rates. Second, as the amplitude of vibration increases, it causes more and more of the hair cells on the fringes of the vibrating portion of the basilar membrane to become stimulated, thus causing *spatial summation* of impulses — that is, transmission through many nerve fibers, rather than through a few. Third, certain hair cells do not become stimulated until the vibration of the basilar membrane reaches a relatively high intensity, and it is believed that stimulation of these cells in some way apprises the nervous system that the sound is then very loud.

Detection of Changes in Loudness — The Power Law. It was pointed out in Chapter 48 that a person interprets changes in intensity of sensory stimuli approximately in proportion to a power function of the actual intensity. In the case of sound, the interpreted sensation changes approximately in proportion to the cube root of the actual sound intensity. To express this another way, the ear can discriminate differences in sound intensity from the softest whisper to the loudest possible noise, representing an *approximate one trillion times* increase in sound energy. Yet the ear interprets this much difference in sound level as approximately a 10,000-fold change. Thus, the scale of intensity is greatly "compressed" by the sound perception mechanisms of the auditory system. This obviously allows a person to interpret differences in sound intensities over an extremely wide range, a far broader range than would be possible were it not for compression of the scale.

The Decibel Unit. Because of the extreme changes in sound intensities that the ear can detect and discriminate, sound intensities are usually expressed in terms of the logarithm of their actual intensities. A 10-fold increase in sound energy (or a $\sqrt{10}$-fold increase in sound pressure, because energy is proportional to the square of pressure) is called 1 *bel*, and one-tenth bel is called 1 *decibel*. One decibel represents an actual increase in intensity of 1.26 times.

Another reason for using the decibel system in expressing changes in loudness is that, in the usual sound intensity range for communication, the ears can detect approximately a 1 decibel change in sound intensity.

The "Zero" Decibel Reference Level. The usual method for expressing the intensity of sound is to state the pressure difference between the peak of the sound compression wave and the trough of the wave. A pressure of *0.0002 dyne per square centimeter* is considered by audiologists to be unit intensity, and this is expressed as *zero decibels* when converted to the decibel scale because the logarithm of unity is zero. This sound level is also approximately the minimum that can be detected by the normal ear at the optimal frequency of 2000 cycles per second. (On the other hand, sound engineers frequently consider a sound pressure of 1 dyne per square centimeter to be zero decibels.)

Threshold for Hearing Sound at Different Frequencies. Figure 61–9 shows the energy threshold at which sounds of different frequencies can barely be heard by the ear. This figure illustrates that a 2000 cycle per second sound can be heard even when its intensity is as low as 70 decibels below 1 dyne/cm.^{2} sound pressure level, which is one ten-millionth microwatt/cm.^{2}. On the other hand, a 100 cycle per second sound can be detected only if its intensity is 10,000 times as great as this.

Frequency Range of Hearing. The frequencies of sound that a young person can hear, before aging has occurred in the ears, is generally stated to be between 30 and 20,000 cycles per second. However, referring again to Figure 61–9, we see that the sound range depends to a great extent on intensity. If the intensity is 60 decibels

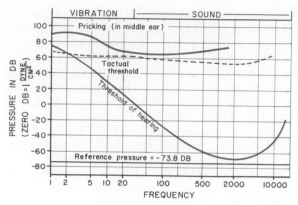

Figure 61–9. Relationship of the threshold of hearing and the threshold of somesthetic perception to the sound energy level at each sound frequency. (Modified from Stevens and Davis: Hearing. John Wiley & Sons.)

below the 1 dyne/cm.² sound pressure level, the sound range is 500 to 5000 cycles per second, and only with intense sounds can the complete range of 30 to 20,000 cycles be achieved. In old age, the frequency range falls to 50 to 8,000 cycles per second or less, as is discussed later in the chapter.

CENTRAL AUDITORY MECHANISMS

THE AUDITORY PATHWAY

Figure 61–10 illustrates the major auditory pathways. It shows that nerve fibers from the *spiral ganglion of the organ of Corti* enter the *cochlear nuclei* located in the

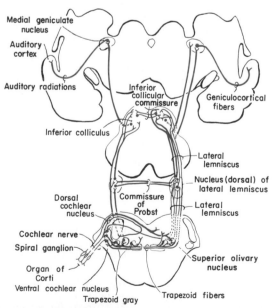

Figure 61–10. The auditory pathway. (Modified from Crosby, Humphrey, and Lauer: Correlative Anatomy of the Nervous System. The Macmillan Co.)

upper part of the medulla. At this point, all the fibers synapse, and second order neurons pass mainly to the opposite side of the brain stem through the *trapezoid body* to the *superior olivary nucleus*. However, some of the second order fibers pass ipsilaterally to the superior olivary nucleus on the same side. Most of the fibers entering the superior olivary nucleus on either side terminate here, but some pass on through this nucleus. From the superior olivary nucleus the auditory pathway then passes upward through the *lateral lemniscus,* and many of the fibers terminate in the *nucleus of the lateral lemniscus,* but many also bypass this nucleus and pass on to the inferior colliculus where most terminate; a few pass on without terminating to higher levels. A few fibers cross from the nucleus of the lateral lemniscus through the *commissure of Probst* to the contralateral nucleus, and still other fibers cross through the *inferior collicular commissure* from one inferior colliculus to the other. From the inferior colliculus, the pathway passes through the *peduncle of the inferior colliculus* to the *medial geniculate nucleus,* where all the fibers synapse. From here, the auditory tract spreads by way of the *auditory radiation* to the *auditory cortex* located mainly in the superior temporal gyrus.

Several points of importance in relation to the auditory pathway should be noted. First, impulses from either ear are transmitted through the auditory pathways of both sides of the brain stem with only slight preponderance of transmission in the contralateral pathway. In at least three different places in the brain stem crossing-over occurs between the two pathways: (a) in the trapezoid body, (b) in the commissure of Probst, and (c) in the commissure connecting the two inferior colliculi.

Second, many collateral fibers from the auditory tracts pass directly into the reticular activating system of the brain stem. This system projects diffusely upward into the cerebral cortex and downward into the spinal cord.

Third, the pathway for transmission of sound impulses from the cochlea to the cortex consists of at least four neurons and sometimes as many as six. Neurons *may* or *may not* synapse in the superior olivary nuclei, in the nuclei of the lateral lemniscus, and in the inferior colliculi. Therefore, some of the tracts are more direct than others, which means that some impulses arrive at the cortex well ahead of others even though they might have originated at exactly the same time.

Fourth, several important pathways also exist from the auditory system into the cerebellum: (a) directly from the cochlear nuclei, (b) from the inferior colliculi, (c) from the reticular substance of the brain stem, and (d) from the cerebral auditory areas. These activate the *cerebellar vermis* instantaneously in the event of a sudden noise.

Fifth, a high degree of spatial orientation is maintained in the fiber tracts from the cochlea all the way to the cortex. In fact, there are three different spatial representations of sound frequencies in the cochlear nuclei, two representations in the inferior colliculi, one precise representation for discrete sound frequencies in the auditory cortex, and several less precise representations in the auditory association areas.

Firing Rates at Different Levels of the Auditory Tract. Single nerve fibers entering the cochlear nuclei from the eighth nerve can fire at rates up to 1000 per second, the rate being determined mainly by the loudness of the sound. At low sound frequencies, the nerve impulses are usually synchronized with the sound waves but they do not necessarily occur with every wave.

In the auditory tracts of the brain stem, the firing is usually no longer synchronized with the sound frequency except at sound frequencies below 200 cycles per second. And above the level of the inferior colliculi, even this synchronization is mainly lost. These findings demonstrate that the sound signals are not transmitted unchanged directly from the ear to the higher levels of the brain; instead, information from the sound signals begins to be dissected from the impulse traffic at levels as low as the cochlear nuclei. We will have more to say about this later, especially in relation to perception of direction from which sound comes.

Another significant feature of the auditory pathways is that low rates of impulse firing continue even in the absence of sound all the way from the cochlear nerve fibers to the auditory cortex. When the basilar membrane moves toward the scala vestibuli, the impulse traffic increases; and when the basilar membrane moves toward the scala tympani, the impulse traffic decreases. Thus, the presence of this background signal allows information to be transmitted from the basilar membrane when the membrane moves in either direction: positive information in one direction and negative information in the opposite direction. Were it not for the background signal, only the positive half of the information could be transmitted. This type of so-called "carrier wave" method for transmitting information is utilized in many parts of the brain, as has been discussed in several of the preceding chapters.

Function of the Auditory Relay Nuclei. Very little is known about the function of the different nuclei in the auditory pathway. However, cats and even monkeys can still detect very low intensity sound when the cerebral cortex is removed bilaterally, which indicates that the nuclei in the brain stem and thalamus can perform many auditory functions even without the cerebral cortex. However, discrimination of tonal patterns and sound sequences is considerably impaired. In human beings, *bilateral* destruction of the cortical auditory centers may give a different picture: it is said to cause severe hearing loss; but, unfortunately, this has not been studied adequately.

FUNCTION OF THE CEREBRAL CORTEX IN HEARING

The projection of the auditory pathway to the cerebral cortex is illustrated in Figure 61–11, which shows that the auditory cortex lies princi-

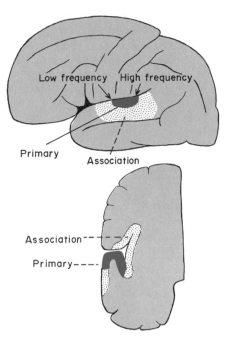

Figure 61–11. The auditory cortex.

pally on the *supratemporal plane of the superior temporal gyrus* but also extends over the *lateral border of the temporal lobe,* over much of the *insular cortex,* and even into the most lateral portion of the *parietal operculum.*

Two separate areas are shown in Figure 61–11: the *primary auditory cortex* and the *auditory association cortex* (also called the *secondary auditory cortex*). The primary auditory cortex is directly excited by projections from the medial geniculate body, while the auditory association areas are usually excited secondarily by impulses from the primary auditory cortex and by projections from thalamic association areas adjacent to the medial geniculate body.

Locus of Sound Frequency Perception in the Primary Auditory Cortex. Certain parts of the primary auditory cortex are known to respond to high frequencies and other parts to low frequencies. In monkeys, the posteromedial part of the supratemporal plane responds to high frequencies, while the anterolateral part responds to low frequencies. Presumably, the same frequency localization occurs in the human cortex, but this is yet unproven.

The frequency range to which each individual neuron in the auditory cortex responds is much narrower than that in the cochlear and brain stem relay nuclei. Referring back to Figure 61–6B, we note that the basilar membrane near the base of the cochlea is stimulated by all frequency sounds, and in the cochlear nuclei this same breadth of sound representation is found. Yet by the time the excitation has reached the cerebral cortex, each sound-responsive neuron responds to only a narrow range of frequencies rather than to a broad range. Therefore, somewhere along the pathway, processing mechanisms in some way "sharpen" the frequency response. It is believed that this sharpening effect is caused mainly by the phenomenon of lateral inhibition, which was discussed in Chapter 47 in relation to mechanisms for transmitting information in nerves. That is, stimulation of the cochlea at one frequency causes inhibition of signals caused by sound frequencies on either side of the stimulated frequency, this effect resulting from collateral fibers angling off the primary signal pathway and exerting inhibitory influences on adjacent pathways. The same effect has also been demonstrated to be important in sharpening patterns of somesthetic images, visual images, and other types of sensations.

A large share of the neurons in the auditory cortex, especially in the auditory association cortex, do not respond to specific sound frequencies in the ear. It is believed that these neurons "associate" different sound frequencies with each other or associate sound information with information from other sensory areas of the cortex. Indeed, the parietal portion of the auditory association cortex partly overlaps somatic sensory area II, which could provide easy opportunity for association of auditory information with somatic sensory information.

Discrimination of Sound "Patterns" by the Auditory Cortex. Complete bilateral removal of the auditory cortex does not prevent an animal from detecting sounds or reacting in a crude manner to the sounds. However, it does greatly reduce or sometimes even abolish its ability to discriminate different sound pitches and especially *patterns of sound*. For instance, an animal that has been trained to recognize a combination or sequence of tones, one following the other in a particular pattern, loses this ability when the auditory cortex is destroyed, and, furthermore, it cannot relearn this type of response. Therefore, the auditory cortex is important in the discrimination of *tonal* and *sequential sound patterns*.

In the human being, lesions affecting the auditory association areas but not affecting the primary auditory cortex will allow the person full capability to hear and differentiate sound tones as well as to interpret at least a few simple patterns of sound. However, he will often be unable to interpret the *meaning* of the sound that he hears. For instance, lesions in the posterior portion of the superior temporal gyrus often make it impossible for the person to interpret the meanings of words even though he hears them perfectly well and can even repeat them. These functions of the auditory association areas and their relationship to the overall intellectual functions of the brain were discussed in detail in Chapter 55.

DISCRIMINATION OF DIRECTION FROM WHICH SOUND EMANATES

A person determines the direction from which sound emanates by two principal mechanisms: (1) by the time lag between the entry of sound into one ear and into the opposite ear and (2) by the difference between the intensities of the sounds in the two ears. The first mechanism functions best for frequencies below 3000 cycles per second, and the intensity mechanism operates best at higher frequencies because the head acts as a sound barrier at these frequencies. The time lag mechanism discriminates direction much more exactly than the intensity mechanism, for the time lag mechanism does not depend on extraneous factors but only on an exact interval of time between two acoustical signals. If a person is looking straight toward the sound, the sound reaches both ears at exactly the same instant, while, if the right ear is closer to the sound than the left ear, the sound signals from the right ear are perceived ahead of those from the left ear.

If a sound emanates from the right side 45 degrees from the frontal direction and if still a second sound emanates also from the right but 45 degrees from behind, the difference in time of arrival of the two sounds at the two ears will be exactly the same. For this reason, it is difficult to distinguish whether the sound is originating from the frontal or the posterior quadrant. The principal method by which a person determines this is to rotate his head quickly. If he turns his head toward the right when the sound is coming from the frontal quadrant, the time lag between the two ears becomes smaller. If the sound is originating from the posterior quadrant, the time lag becomes greater. A sudden sound that occurs so rapidly that the person does not have time to move his head often cannot be properly localized to the frontal or posterior quadrant.

Neural Mechanisms for Detecting Sound Direction. Destruction of the auditory cortex on both sides of the brain, in either human beings or lower mammals, causes loss of almost all ability to detect the direction from which sound comes. Yet, the mechanism for this detection process begins in the superior olivary nuclei, even though it requires the neural pathways all the way from these nuclei to the cortex for interpretation of the signals. The mechanism is believed to be the following:

Some neurons in the superior olivary nucleus are inhibited by signals from one ear but excited by signals from the other ear. Therefore, if the signals from both ears reach these cells at the same time, they cancel each other and no subsequent signal is transmitted to the auditory cortex. However, if the sound reaches one ear before the other, then the two signals do not cancel each other, and the time interval between the sounds in the two ears determines the degree of cancellation, thus providing the clue required to discriminate the direction of the sound.

Let us add still another feature to the mechanism: Because of multiple neuronal relays between the cochlear nuclei and the superior olivary nuclei, the signals reaching some of the olivary neurons arrive later than others. Therefore, the signals for some of these neurons are cancelled when the time interval between the sounds in the two ears is very short while for other neurons the signals are cancelled when the time interval is very long. Thus, a spatial pattern of neuronal stimulation develops, with sound from directly in front of the head stimulating one set of olivary neurons maximally and sounds from different side angles stimulating other sets of neurons maximally. This spatial orientation of signals is then transmitted all the way to the auditory cortex where sound direction is determined by the locus in the cortex that is stimulated maximally. It is believed that the signals for determining sound direction are transmitted through a slightly different pathway and that this pathway terminates in the cerebral cortex in a slightly different locus from the transmission pathway and the termination locus of the tonal patterns of sound.

This mechanism for detection of sound direction indicates again how information in sensory signals is dissected out as the signals pass through different levels of neuronal activity. In this case, the "quality" of sound direction is separated from the "quality" of sound tones at the level of the superior olivary nuclei.

CENTRIFUGAL CONDUCTION OF IMPULSES FROM THE CENTRAL NERVOUS SYSTEM

Retrograde pathways have been demonstrated at each level of the central nervous system all the way from the auditory cortex to the cochlea. The final pathway is mainly from the superior olivary nucleus to the organ of Corti.

These retrograde fibers are inhibitory. Indeed, direct stimulation of discrete points in the olivary nucleus have been shown to inhibit specific areas of the organ of Corti, reducing their sound sensitivities as much as 15 to 20 decibels. One can readily understand how this could allow a person to direct attention to sounds of particular qualities while rejecting sounds of other qualities. This is readily demonstrated when one listens to a single instrument in a symphony orchestra.

HEARING ABNORMALITIES

TYPES OF DEAFNESS

Deafness is usually divided into two types; first, that caused by impairment of the cochlea or auditory nerve, which is usually classed under the heading "nerve deafness," and, second, that caused by impairment of the middle ear mechanisms for transmitting sound into the cochlea, which is usually called "conduction deafness." Obviously, if either the cochlea or the auditory nerve is completely destroyed the person is permanently deaf. However, if the cochlea and nerve are still intact but the ossicular system has been destroyed or ankylosed ("frozen" in place by fibrosis or calcification), sound waves can still be conducted into the cochlea by means of bone conduction.

Tuning Fork Test for Differentiation Between Nerve Deafness and Conduction Deafness. To test an ear for bone conduction with a tuning fork, one places a vibrating fork in front of the ear, and the subject listens to the tone of the fork until he can no longer hear it. Then the butt of the still weakly vibrating fork is immediately placed against the mastoid process. If his bone conduction is better than his air conduction, he will again hear the sound of the tuning fork. If this occurs, his deafness may be considered to be conduction deafness. However, if on placing the butt of the fork against the mastoid process he cannot hear the sound of the tuning fork, his bone conduction is probably decreased as much as his air conduction, and the deafness is presumably due to damage in the cochlea or in the nervous system rather than in the ossicular system — that is, it is nerve deafness.

The Audiometer. To determine the nature of hearing disabilities more exactly than can be accomplished by the above method, the audiometer is used. This is simply an earphone connected to an electronic oscillator capable of emitting pure tones ranging from low frequencies to high frequencies. Based on previous studies of normal persons, the instrument is calibrated so that the zero intensity level of sound at each frequency is the loudness that can barely be heard by the normal person. However, a calibrated volume control can increase or decrease the loudness of each tone above or below the zero level. If the loudness of a tone must be increased to 30 decibels above normal before it can be heard, the subject is said to have a *hearing loss* of 30 decibels for that particular tone.

In performing a hearing test using an audiometer, one tests approximately 8 to 10 frequencies covering the auditory spectrum, one at a time, and the hearing loss is determined for each of these frequencies. Then the

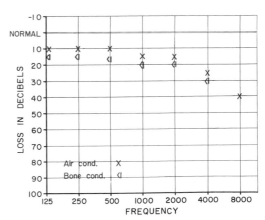

Figure 61–12. Audiogram of the old-age type of nerve deafness.

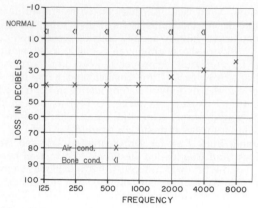

Figure 61–13. Audiogram of deafness resulting from middle ear sclerosis.

so-called "audiogram" is plotted as shown in Figures 61–12 and 61–13, depicting the hearing loss for each of the frequencies in the auditory spectrum.

The audiometer, in addition to being equipped with an earphone for testing air conduction by the ear, is also equipped with an electronic vibrator for testing bone conduction from the mastoid process into the cochlea.

The Audiogram in Nerve Deafness. In nerve deafness — this term including damage to the cochlea, to the auditory nerve, or to the central nervous system circuits from the ear — the person has lost the ability to hear sound as tested by both the air conduction apparatus and the bone conduction apparatus. An audiogram depicting nerve deafness is illustrated in Figure 61–12. In this figure the deafness is mainly for high frequency sound. Such deafness could be caused by damage to the base of the cochlea. This type of deafness occurs to some extent in almost all older persons.

Other patterns of nerve deafness frequently occur as follows: (1) deafness usually for low frequency sounds caused by excessive and prolonged exposure to very loud sounds (the rock band or the jet airplane engine) because low frequency sounds are usually louder and more damaging to the organ of Corti, and (2) deafness for all frequencies caused by drug sensitivity of the organ of Corti, especially sensitivity to some antibiotics such as streptomycin, kanamycin, and chloramphenicol.

The Audiogram in Conduction Deafness. A second and frequent type of deafness is that caused by fibrosis of the middle ear following repeated infection in the middle ear or in the hereditary disease called *otosclerosis.* In this instance the sound waves cannot be transmitted easily to the oval window. Figure 61–13 illustrates an audiogram from a person with "middle ear deafness" of this type. In this case the bone conduction is essentially normal, but air conduction is greatly depressed at all frequencies, more so at the low frequencies. In this

type of deafness, the faceplate of the stapes frequently becomes "ankylosed" by bony overgrowth to the edges of the oval window. In this case, the person becomes totally deaf for air conduction, but can be made to hear again almost normally by removing the stapes and replacing it with a minute Teflon or metal prosthesis that transmits the sound from the incus to the oval window.

REFERENCES

Aitkin, L. M.: Tonotopic organization at higher levels of the auditory pathway. *Int. Rev. Physiol., 10*:249, 1976.

Beagley, H. A. (ed.): Auditory Investigation: The Scientific and Technological Basis. New York, Oxford University Press, 1979.

Bench, J., and Bamford, J. (eds.): Speech-Hearing Tests and the Spoken Language of Hearing-Impaired Children. New York, Academic Press, 1979.

Brugge, J. F., and Geisler, C. D.: Auditory mechanisms of the lower brainstem. *Annu. Rev. Neurosci., 1*:363, 1978.

Busnel, R., and Fish, J. F. (eds.): Animal Sonar Systems. New York, Plenum Press, 1979.

Carterette, E. C., and Friedman, M. P.: Hearing. New York, Academic Press, 1978.

Dallos, P.: The Auditory Periphery. New York, Academic Press, 1973.

Davis, H.: Biophysics and physiology of the inner ear. *Physiol. Rev., 37*:1, 1957.

Egerton, B. J., and Danhauer, J. L.: Clinical Implications of Speech Discrimination Testing Using Nonsense Stimuli. Baltimore, University Park Press, 1979.

Evans, E. F., and Wilson, J. P. (eds.): Psychophysics and Physiology of Hearing. New York, Academic Press, 1977.

Gerber, S. E.: Introductory Hearing Science, Philadelphia, W. B. Saunders Co., 1974.

Harris, J. D.: The Electrophysiology and Layout of the Auditory Nervous System. Indianapolis, The Bobbs-Merrill Co., 1974.

Heasley, B. E.: Auditory Processing Disorders and Remediation, 2nd Ed. Springfield, Ill., Charles C Thomas, 1980.

Jarvis, J. F.: An Introduction to the Anatomy and Physiology of Speech and Hearing. Cape Town, South Africa, Juta & Co., 1978.

Lipscomb, D. M. (ed.): Noise and Audiology. Baltimore, University Park Press, 1978.

Naunton, R. F., and Fernandez, C. (eds.): Evoked Electrical Activity in the Auditory Nervous System. New York, Academic Press, 1978.

Rintelmann, W. F. (ed.): Hearing Assessment. Baltimore, University Park Press, 1979.

Rosenblum, E. H.: Fundamentals of Hearing for Health Professionals. Boston, Little, Brown, 1979.

Sataloff, J., *et al.*: Hearing Loss, 2nd Ed. Philadelphia, J. B. Lippincott Co., 1980.

Scheich, O. C. H., and Schreiner, C. (eds.): Hearing Mechanisms and Speech. New York, Springer-Verlag, 1979.

Singh, R. P.: Anatomy of Hearing and Speech. New York, Oxford University Press, 1980.

Skinner, P. H., and Shelton, R. L.: Speech, Language, and Hearing: Normal Processes and Disorders. Reading, Mass., Addison-Wesley Publishing Co., 1978.

Starr, A.: Neurophysiological mechanisms of sound localization. *Fed. Proc., 33*:1911, 1974.

Tobias, J. V. (ed.): Foundations of Modern Auditory Theory. Vols. 1 and 2. New York, Academic Press, 1970.

Van Hattum, R. J.: Communication Disorders. New York, The Macmillan Co., 1980.

Wever, E. G., and Lawrence, M.: Physiological Acoustics. Princeton, Princeton University Press, 1954.

Yanick, P., Jr., and Freifeld, S. (eds.): The Application of Signal Processing Concepts of Hearing Aids. New York, Grune & Stratton, 1978.

62

The Chemical Senses— Taste and Smell

THE SENSE OF TASTE

Taste is mainly a function of the *taste buds* in the mouth, but it is also common experience that one's sense of smell also contributes strongly to taste perception. In addition, the texture of food, as detected by tactual senses of the mouth, and the presence of such elements in the food as pepper, which stimulate pain endings, greatly condition the taste experience. The importance of taste lies in the fact it allows a person to select food in accord with desires and perhaps also in accord with the needs of the tissues for specific nutritive substances.

On the basis of psychological studies, there are generally believed to be four *primary* sensations of taste: *sour, salty, sweet, and bitter.* Yet we know that a person can perceive literally hundreds of different tastes. These are all supposed to be combinations of the four primary sensations in the same manner that all the colors of the spectrum are combinations of three primary color sensations, as described in Chapter 59. There might be other, less conspicuous classes or subclasses of primary sensations; nevertheless, the following discussion is based on the usual classification of only four primary tastes.

THE PRIMARY SENSATIONS OF TASTE

The Sour Taste. The sour taste is caused by acids, and the intensity of the taste sensation is approximately proportional to the logarithm of the *hydrogen ion concentration.* That is, the more acidic the acid, the stronger becomes the sensation.

The Salty Taste. The salty taste is elicited by ionized salts. The quality of the taste varies somewhat from one salt to another because the salts also elicit other taste sensations besides saltiness. The cations of the salts are mainly responsible for the salty taste, but the anions also contribute at least to some extent.

The Sweet Taste. The sweet taste is not caused by any single class of chemicals. A list of some of the types of chemicals that cause this taste includes: sugars, glycols, alcohols, aldehydes, ketones, amides, esters, amino acids, sulfonic acids, halogenated acids, and inorganic salts of lead and beryllium. Note specifically that most of the substances that cause a sweet taste are organic chemicals. It is especially interesting that very slight changes in the chemical structure, such as addition of a simple radical, can often change the substance from sweet to bitter.

The third column of Table 62–1 shows the relative intensities of taste of certain substances that cause the sweet taste. *Sucrose,* which is common table sugar, is considered to have an index of 1. Note that one of the substances has a sweet index 5000 times as great as that of sucrose. However, this extremely sweet substance, known as *P-4000,* is unfortunately extremely toxic and therefore cannot be used as a sweetening agent. *Saccharin,* on the other hand, is also more than 600 times as sweet as common table sugar, and since it is not toxic (except that it might be mildly carcinogenic), it can be used with impunity as a sweetening agent.

The Bitter Taste. The bitter taste, like the sweet taste, is not caused by any single type of chemical agent, but, here again, the substances that give the bitter taste are almost entirely organic substances. Two particular classes of substances are especially likely to cause bitter taste sensations: (1) long chain organic substances and (2) alkaloids. The alkaloids include many of the drugs used in medicines such as quinine, caffeine, strychnine, and nicotine.

Some substances that at first taste sweet have a bitter after-taste. This is true of saccharin, which makes this substance objectionable to some people. Some substances have a sweet taste on the

TABLE 62–1 RELATIVE TASTE INDICES OF DIFFERENT SUBSTANCES

Sour Substances	Index	Bitter Substances	Index	Sweet Substances	Index	Salty Substances	Index
Hydrochloric acid	1	Quinine	1	Sucrose	1	NaCl	1
Formic acid	1.1	Brucine	11	1-propoxy-2-amino-		NaF	2
Chloracetic acid	0.9	Strychnine	3.1	4-nitrobenzene	5000	CaCl$_2$	1
Acetyllactic acid	0.85	Nicotine	1.3	Saccharin	675	NaBr	0.4
Lactic acid	0.85	Phenylthiourea	0.9	Chloroform	40	NaI	0.35
Tartaric acid	0.7	Caffeine	0.4	Fructose	1.7	LiCl	0.4
Malic acid	0.6	Veratrine	0.2	Alanine	1.3	NH$_4$Cl	2.5
Potassium H tartrate	0.58	Pilocarpine	0.16	Glucose	0.8	KCl	0.6
Acetic acid	0.55	Atropine	0.13	Maltose	0.45		
Citric acid	0.46	Cocaine	0.02	Galactose	0.32		
Carbonic acid	0.06	Morphine	0.02	Lactose	0.3		

From Derma: *Proc. Oklahoma Acad. Sc.*, 27:9, 1947; and Pfaffman: Handbook of Physiology, Sec. I, Vol. I, p. 507, 1959. Baltimore, The Williams & Wilkins Co.

front of the tongue, where taste buds with special sensitivity to the sweet taste are principally located, and a bitter taste on the back of the tongue, where taste buds more sensitive to the bitter taste are located.

The bitter taste, when it occurs in high intensity, usually causes the person or animal to reject the food. This is undoubtedly an important purposive function of the bitter taste sensation because many of the deadly toxins found in poisonous plants are alkaloids, and these all cause intensely bitter taste.

Threshold for Taste

The threshold for stimulation of the sour taste by hydrochloric acid averages 0.0009 N; for stimulation of the salty taste by sodium chloride, 0.01 M; for the sweet taste by sucrose, 0.01 M; and for the bitter taste by quinine, 0.000008 M. Note especially how much more sensitive is the bitter taste sense to stimuli than all the others, which would be expected since this sensation provides an important protective function.

Table 62–1 gives the relative taste indices (the reciprocals of the taste thresholds) of different substances. In this table, the intensities of the four different primary sensations of taste are referred, respectively, to the intensities of taste of hydrochloric acid, quinine, sucrose, and sodium chloride, each of which is considered to have a taste index of 1.

Taste Blindness. Many persons are taste blind for certain substances, especially for different types of thiourea compounds. A substance used frequently by psychologists for demonstrating taste blindness is *phenylthiocarbamide,* for which approximately 15 to 30 per cent of all people exhibit taste blindness, the exact percentage depending on the method of testing and the concentration of the substance.

THE TASTE BUD AND ITS FUNCTION

Figure 62–1 illustrates a taste bud, which has a diameter of about $^1/_{30}$ millimeter and a length of about $^1/_{16}$ millimeter. The taste bud is composed of about 40 modified epithelial cells, some of which

are supporting cells and others are *taste cells.* The taste cells are continually being replaced by mitotic division from the surrounding epithelial cells so that some are young cells and others are mature cells that lie toward the center of the bud and soon dissolute. The life span of each taste cell is about ten days.

The outer tips of the taste cells are arranged around a minute *taste pore,* shown in Figure 62–1. From the tip of each cell, several *microvilli,* or *taste hairs,* about 2 to 3 microns in length and 0.1 to 0.2 micron in width, protrude outward through the taste pore to approach the cavity of the mouth. These microvilli are believed to provide the receptor surface for taste.

Interwoven among the taste cells is a branching terminal network of several *taste nerve fibers* that are stimulated by the taste cells. These fibers invaginate into folds of the taste cell membranes, so that there is extremely intimate contact between the taste cells and the nerves. Several taste buds can be innervated by the same taste fiber.

An interesting feature of the taste buds is that they completely degenerate when the taste nerve fibers are destroyed. Then, if the taste fibers regrow to the epithelial surface of the mouth, the local epithelial cells regroup themselves to form

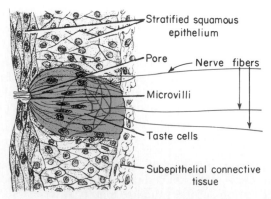

Figure 62–1. The taste bud.

new taste buds. This illustrates the important principle of "trophic" function of nerve fibers in certain parts of the body. The cause of the tropism is unknown, but it has been postulated to be a protein trophic factor secreted by the nerve endings.

Location of the Taste Buds. The taste buds are found on three of four different types of papillae of the tongue, as follows: (1) A large number of taste buds are on the walls of the troughs that surround the circumvallate papillae, which form a V line toward the posterior of the tongue. (2) Moderate numbers of taste buds are on the fungiform papillae over the front surface of the tongue. (3) Moderate numbers are on the foliate papillae located in the folds along the lateral surfaces of the tongue. Additional taste buds are located on the palate and a few on the tonsillar pillars and at other points around the nasopharynx. Adults have approximately 10,000 taste buds, and children a few more. Beyond the age of 45 many taste buds rapidly degenerate, causing the taste sensation to become progressively less critical.

Especially important in relation to taste is the tendency for taste buds subserving particular primary sensations of taste to be localized in special areas. The sweet taste is localized *principally* on the anterior surface and tip of the tongue, the salty and sour tastes on the two lateral sides of the tongue, and the bitter taste on the circumvallate papillae on the posterior of the tongue.

Specificity of Taste Buds for the Primary Taste Stimuli. Psychological tests using different types of taste stimuli carefully applied to individual taste buds, one at a time, have suggested that we have four distinctly different varieties of taste buds, each sensitive for only one type of taste. Yet, microelectrode studies from single taste buds while they are stimulated successively by the four different primary taste stimuli have shown that most of them can be excited by two, three, or even four of the primary taste stimuli, though usually with one or two of these predominating. Thus, at present, there are conflicting beliefs about the degree of specificity of taste buds, some physiologists believing them always to be highly specific for only one primary taste stimulus and other physiologists believing this to be true only in a statistical sense because of dominance of one taste perception over the others.

Regardless of which of the two theories is correct, one can well understand that the hundreds of different types of tastes that we experience result from different quantitative degrees of stimulation of the four primary sensations of taste, as well as simultaneous stimulation of smell in the nose and tactile and pain nerve endings in the mouth.

Mechanism of Stimulation of Taste Buds. *The Receptor Potential.* The membrane of the taste cell, like that of other sensory receptor cells, normally is negatively charged on the inside with respect to the outside. Application of a taste substance to the taste hairs causes partial loss of this negative potential. The decrease in potential, within a wide range, is approximately proportional to the logarithm of concentration of the stimulating substance. This change in potential in the taste cell is the *receptor potential* for taste.

The mechanism by which the stimulating substance reacts with the taste hairs to initiate the receptor potential is unknown. It is believed by some physiologists that the substance is simply adsorbed to receptors on the membrane surface of the taste hair and that this adsorption changes the physical characteristics of the hair membrane. This in turn makes the taste cell more permeable to sodium ions and thus depolarizes the cell. The substance is gradually washed away from the taste hair by the saliva, thus removing the taste stimulus. Supposedly, the type of receptor substance or substances in each taste hair determines the types of taste substances that will elicit responses.

Generation of Nerve Impulses by the Taste Bud. The taste nerve fiber endings are encased by folds of the taste cell membranes. In some way not understood the receptor potentials of the taste cells generate impulses in the taste fibers. On first application of the taste stimulus, the rate of discharge of the nerve fibers rises to a peak in a small fraction of a second, but then it adapts within the next 2 seconds back to a much lower steady level. Thus, a strong immediate signal is transmitted by the taste nerve, and a much weaker continuous signal is transmitted as long as the taste bud is exposed to the taste stimulus.

TRANSMISSION OF TASTE SIGNALS INTO THE CENTRAL NERVOUS SYSTEM

Figure 62–2 illustrates the neuronal pathways for transmission of taste signals from the tongue and pharyngeal region into the central nervous system. Taste impulses from the anterior two thirds of the tongue pass first into the *fifth nerve*

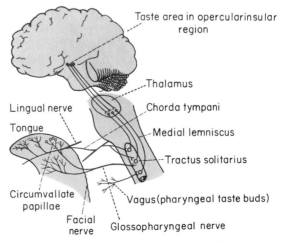

Figure 62–2. Transmission of taste impulses into the central nervous system.

and then through the *chorda tympani* into the *seventh nerve,* thence into the *tractus solitarius* in the brain stem. Taste sensations from the circumvallate papillae on the back of the tongue and from other posterior regions of the mouth are transmitted through the *ninth nerve* also into the *tractus solitarius* but at a slightly lower level. Finally, a few taste impulses are transmitted into the *tractus solitarius* from the base of the tongue and other parts of the pharyngeal region by way of the *vagus nerve.*

All taste fibers synapse in the *nuclei of the tractus solitarius* and send second order neurons to a small area of the *thalamus* located slightly medial to the thalamic terminations of the facial regions of the dorsal column–medial lemniscal system. From the thalamus, third order neurons are transmitted to the *lower tip of the postcentral gyrus in the parietal cortex* where it curls deep into the sylvian fissure. This lies in close association with, or even superimposed on, the tongue area of somatic area I. Third order neurons also project to the nearby *opercular-insular area,* located deep in the sylvian fissure as well.

From this description of the taste pathways, it immediately becomes evident that they parallel closely the somatic pathways from the tongue.

Taste Reflexes.   From the tractus solitarius a large number of impulses are transmitted directly into the *superior* and *inferior salivatory nuclei,* and these in turn transmit impulses to the submaxillary and parotid glands to help control the secretion of saliva during the ingestion of food.

Adaptation of Taste.   Everyone is familiar with the fact that taste sensations adapt rapidly, often with almost complete adaptation within 1 to 5 minutes of continuous stimulation. Yet, from electrophysiological studies of taste nerve fibers, it seems that the taste buds themselves do not adapt enough to account for all or even most of the taste adaptation. They have a rapid period of adaptation during the first 2 to 3 seconds after contact with the taste stimulus. The first burst of impulses from the taste bud allows one to detect extremely minute concentrations of taste substances, but normal concentrations of taste substances cause prolonged discharge of the taste fibers. Therefore, the progressive adaptation that occurs in the sensation of taste has been postulated to occur in the central nervous system itself, though the mechanism and site of this is not known. If this is true, it is a mechanism that is different from that of other sensory systems, which adapt almost entirely in the receptors.

SPECIAL ATTRIBUTES OF THE TASTE SENSE

Affective Nature of Taste.   *Pleasantness* and *unpleasantness* are called the "affective" attributes of a sensa-

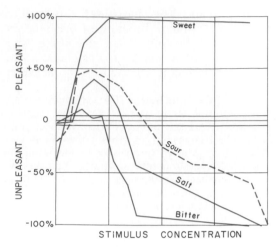

Figure 62–3.   The affective nature of the different primary sensations of taste at progressively increasing degrees of taste stimulus. (From Engel in Woodworth and Schlosberg: Experimental Psychology. Holt, Rinehart & Winston, Inc.)

tion. Figure 62–3 illustrates the affective effects of different types of taste at different concentrations of the stimulating substances, showing, strangely enough, that the sweet taste is likely to be unpleasant at a very low concentration but very pleasant at high concentrations. The other types of taste, on the other hand, are likely to be pleasant at low concentrations but exceedingly unpleasant at high concentrations. This is particularly true of the bitter taste.

Importance of the Sense of Smell in Taste.   Persons with severe colds frequently state that they have lost their sense of taste. However, on testing, the taste sensations are found to be completely normal. This illustrates that much of what we call taste is actually smell. Odors from the food can pass upward into the nasopharynx, often stimulating the olfactory system thousands of times as strongly as the taste system. For instance, if the olfactory system is intact, alcohol can be "tasted" in 1/25,000 the concentration required when the olfactory system is not intact.

Taste Preference and Control of the Diet.   Taste preferences mean simply that an animal will choose certain types of food in preference to others, and it automatically uses this to help control the type of diet it eats. Furthermore, its taste preferences often change in accord with the needs of the body for certain specific substances. The following experimental studies illustrate this ability of an animal to choose food in accord with the need of its body: First, adrenalectomized animals automatically select drinking water with a high concentration of sodium chloride in preference to pure water, and this in many instances is sufficient to supply the needs of the body and prevent death as a result of salt depletion. Second, an animal injected with excessive amounts of insulin develops a depleted blood sugar, and it automatically chooses the sweetest food from among many samples. Third, parathyroidectomized animals automatically choose drinking water with a high concentration of calcium chloride.

These same phenomena are also observed in many instances of everyday life. For instance, the salt licks of the desert region are known to attract animals from far and wide, and even the human being rejects any food that has an unpleasant affective sensation, which certainly in many instances protects our bodies from undesirable substances.

The phenomenon of taste preference almost certainly results from some mechanism located in the central nervous system and not from a mechanism in the taste buds themselves, because many experiments have demonstrated that taste preference can occur in animals even in the absence of changes in stimulus threshold of the taste buds for the substances involved. Another reason for believing this to be a central phenomenon is that previous experience with unpleasant or pleasant tastes plays a major role in determining one's different taste preferences. For instance, if a person becomes sick immediately after eating a particular type of food, the person generally develops a negative taste preference, or taste aversion, for that particular food thereafter; the same effect can be demonstrated in animals.

THE SENSE OF SMELL

Smell is the least understood sense. This results partly from the location of the olfactory membrane high in the nose where it is difficult to study and partly from the fact that the sense of smell is a subjective phenomenon that cannot be studied with ease in lower animals. Still another complicating problem is the fact that the sense of smell is almost rudimentary in the human being in comparison with that of some lower animals.

THE OLFACTORY MEMBRANE

The olfactory membrane lies in the superior part of each nostril, as illustrated in Figure 62–5. Medially it folds downward over the surface of the septum, and laterally it folds over the superior turbinate and even over a small portion of the upper surface of the middle turbinate. In each nostril the olfactory membrane has a surface area of approximately 2.4 square centimeters.

The Olfactory Cells. The receptor cells for the smell sensation are the *olfactory cells,* which are actually bipolar nerve cells derived originally from the central nervous system itself. There are about 100 million of these cells in the olfactory epithelium interspersed among *sustentacular cells,* as shown in Figure 62–4. The mucosal end of the olfactory cell forms a knob called the *olfactory rod* from which 6 to 12 *olfactory hairs,* or *cilia,* 0.3 micron in diameter and several microns in length, project into the mucus that coats the inner surface of the nasal cavity. These projecting olfactory hairs are believed to react to odors in the air and then to stimulate the olfactory cells, as discussed below. Spaced among the olfactory cells in the olfactory membrane are many small *glands of Bowman* that secrete mucus onto the surface of the olfactory membrane.

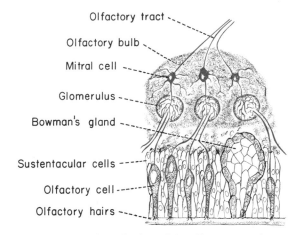

Figure 62–4. Organization of the olfactory membrane.

STIMULATION OF THE OLFACTORY CELLS

The Necessary Stimulus for Smell. We do not know what it takes chemically to stimulate the olfactory cells. Yet we do know the physical characteristics of the substances that cause olfactory stimulation: First, the substance must be volatile so that it can be sniffed into the nostrils. Second, it must be at least slightly water-soluble so that it can pass through the mucus to the olfactory cells. And, third, it must also be lipid-soluble, presumably because the olfactory cilia and outer tips of the olfactory cells are composed principally of lipid materials.

Regardless of the basic mechanism by which the olfactory cells are stimulated, it is known that they become stimulated only when air blasts upward into the superior region of the nose. Therefore, smell occurs in cycles along with the inspirations, which indicates that the olfactory receptors respond in milliseconds to the volatile agents. Because smell intensity is exacerbated by blasting air through the upper reaches of the nose, the sensitivity of smell can be greatly increased by the well-known sniffing technique.

Receptor Potentials in Olfactory Cells. The olfactory cells are believed to react to olfactory stimuli in the same manner that most other sensory receptors react to their specific stimuli; that is, by generating a receptor potential, which in turn initiates nerve impulses in the olfactory nerve fibers. An experiment that probably demonstrates this property of the olfactory receptors is the following: An electrode is placed on the surface of the olfactory membrane, and its electrical potential with respect to the remainder of the body is recorded. When an odorous substance is blown into the nostril, the potential becomes negative and remains negative as long as the odorous air continues to pass through the nostril. This electrical recording is called the *electro-olfactogram,* and it is believed to result from summation of receptor potentials developed in the receptor olfactory cells and perhaps also in the sustentacular cells that seem to be electrically connected to the olfactory cells.

Over a wide range, both the amplitude of the electro-olfactogram and the rate of olfactory nerve impulses are approximately proportional to the logarithm of the stimulus strength, which illustrates that the olfactory receptors tend to obey principles of transduction similar to those of other sensory receptors.

Adaptation. The olfactory receptors adapt approximately 50 per cent in the first second or so after stimulation. Thereafter, they adapt further very slowly. Yet we all know from our own experience that smell sensations adapt almost to extinction within a minute or so after one enters a strongly odorous atmosphere. Since the psychological adaptation is far more rapid than the adaptation of the receptors, it has been suggested that most of this adaptation occurs in the central nervous system, as has also been postulated for adaptation of taste sensations.

Search for the Primary Sensations of Smell. Most physiologists are convinced that the many smell sensations are subserved by a few rather discrete primary sensations, in the same way that taste is subserved by sour, sweet, bitter, and salty sensations. But, thus far, only minor success has been achieved in classifying the primary sensations of smell. Yet, on the basis of psychological tests and action potential studies from various points in the olfactory nerve pathways, it has been postulated that about seven different primary classes of olfactory stimulants preferentially excite separate olfactory cells. These classes of olfactory stimulants may be characterized as follows.

1. Camphoraceous
2. Musky
3. Floral
4. Pepperminty
5. Ethereal
6. Pungent
7. Putrid

However, it is unlikely that this list actually represents the true primary sensations of smell, even though it does illustrate the results of one of the many attempts to classify them. Indeed, several clues in recent years have indicated that there may be as many as *50* or more primary sensations of smell — a marked contrast to only *three* primary sensations of color detected by the eyes and only *four* primary sensations of taste detected by the tongue. For instance, persons have been found who have *odor blindness* for single substances; and such discrete odor blindness has been identified for more than 50 different substances. Since it is presumed that odor blindness for each substance represents a lack of the appropriate receptor cell for that substance, it is postulated that the sense of smell might be subserved by 50 or more primary smell sensations.

Two basic theories have been postulated to explain the abilities of different receptors to respond selectively to different types of olfactory stimulants: the *chemical theory* and the *physical theory*. The chemical theory assumes that *receptor chemicals* in the membranes of the olfactory cilia react specifically with the different types of olfactory stimulants. The type of receptor chemical determines the type of stimulant that will elicit a response in the olfactory cell. The reaction between the stimulant and the receptor substance supposedly increases the permeability of the olfactory ciliary membrane, and this in turn creates the receptor potential in the olfactory cell that generates impulses in the olfactory nerve fibers.

The physical theory assumes that differences in *physical receptor sites* on the olfactory ciliary membranes of separate olfactory cells allow specific olfactory stimulants to adsorb to the membranes of different olfactory cells. A fact supporting this theory is that some substances with very different chemical properties, but with almost identical molecular shapes, have the same odor. Yet, this principle does not hold for most odoriferous substances so that the physical theory is probably of historical rather than factual importance.

Affective Nature of Smell. Smell, equally as much as taste, has the affective qualities of either pleasantness or unpleasantness. Because of this, smell is as important as, if not more important than, taste in the selection of food. Indeed, a person who has previously eaten food that has disagreed with him is often nauseated by the smell of that same type of food on a second occasion. Other types of odors that have proved to be unpleasant in the past may also provoke a disagreeable feeling; on the other hand, perfume of the right quality can wreak havoc with masculine emotions. In addition, in some lower animals odors are the primary excitant of sexual drive.

Threshold for Smell. One of the principal characteristics of smell is the minute quantity of the stimulating agent in the air often required to effect a smell sensation. For instance, the substance *methyl mercaptan* can be smelled when only 1/25,000,000,000 milligram is present in each milliliter of air. Because of this low threshold, this substance is mixed with natural gas to give it an odor that can be detected when it leaks from a gas pipe.

Measurement of Smell Threshold. One of the problems in studying smell has been difficulty in obtaining accurate measurements of the threshold stimulus required to induce smell. The simplest technique is simply to allow a person to sniff different substances in the usual manner of smelling. Indeed, some investigators feel that this is as satisfactory as almost any other procedure. However, to eliminate variations from person to person, more objective methods have been developed: One of these has been to place a box containing the volatilized agent over the subject's head. Appropriate precautions are taken to exclude odors from the person's own body. The person is allowed to breathe naturally, but the volatilized agent is distributed evenly in the air that is breathed.

Gradations of Smell Intensities. Though the threshold concentrations of substances that evoke smell are extremely slight, concentrations only 10 to 50 times above the threshold values evoke maximum intensity of smell. This is in contrast to most other sensory systems of the body, in which the ranges of detection are tremendous — for instance, 500,000 to 1 in the case of the eyes and 1,000,000,000,000 to 1 in the case of the ears. This perhaps can be explained by the fact that smell is concerned more with detecting the presence or absence of odors than with quantitative detection of their intensities.

TRANSMISSION OF SMELL SIGNALS INTO THE CENTRAL NERVOUS SYSTEM

The function of the central nervous system in olfaction is almost as vague as the function of the peripheral receptors. However, Figures 62–4 and 62–5 illustrate the general plan for transmission of olfactory signals into the central nervous system. Figure 62–4 shows a number of separate *olfactory cells* sending axons into the *olfactory bulb* to end on *dendrites from mitral cells* in a structure called the *glomerulus.* Approximately 25,000 axons enter each glomerulus and synapse with about 25 mitral cells that in turn send signals into the brain. Also in the glomerulus are smaller cells, the *tufted cells,* that send signals to the brain as well — perhaps to the older portions of the olfactory nervous system.

Figure 62–5 shows the major pathways for transmission of olfactory signals from the mitral and tufted cells into the brain. The fibers from the cells travel through the olfactory tract and terminate either primarily or after relay neurons in two principal areas of the brain called the *medial*

olfactory area and the *lateral olfactory area,* respectively. The medial olfactory area is composed of a large group of nuclei located in the midportion of the brain superiorly and anteriorly to the hypothalamus. This group includes the *olfactory trigone,* the *medial part of the anterior perforated substance,* the *paraolfactory area,* and the *nucleus of the stria terminalis.*

The lateral olfactory area is composed of the *prepyriform area,* the *lateral part of the anterior perforated substance,* and part of the *amygdaloid nuclei.*

Secondary olfactory tracts pass from the nuclei of both the medial olfactory area and the lateral olfactory area into the *hypothalamus, septum, thalamus, hippocampus, gyrus subcallosus,* and *brain stem nuclei.* These secondary areas control the automatic responses of the body to olfactory stimuli, including automatic feeding activities and also emotional responses, such as fear, excitement, and pleasure and sexual drives.

Secondary olfactory tracts also spread from the lateral olfactory area into the *uncus,* the lateroposterior orbitofrontal cortex, and the prefrontal cortex. It is probably in this lateral olfactory area, especially in the amygdala, its associated cortical regions in the temporal cortex, and the lateroposterior orbitofrontal cortex, that the more complex aspects of olfaction are integrated, such as association of olfactory sensations with somatic, visual, tactile, and other types of sensation. However, complete removal of the lateral olfactory area hardly affects the primitive responses to olfaction, such as licking the lips, salivation, and other feeding responses caused by the smell of food or such as the various emotions associated with smell. On the other hand, its removal does abolish the more complicated conditioned reflexes depending on olfactory stimuli. Therefore, cortical portions of this region are often considered to be the *primary olfactory cortex* for smell. In human beings, tumors in the region of the uncus, amygdala, or lateroposterior orbitofrontal cortex frequently cause the person to perceive abnormal smells.

Centrifugal Control of the Olfactory Bulb by the Central Nervous System. The central nervous system also transmits impulses in a backward direction to the olfactory bulb. These originate mainly from undetermined areas in the brain stem and olfactory portion of the cerebrum and terminate on small *granule* cells in the center of the bulb, which in turn send axons to the mitral and tufted cells to inhibit these.

Mechanism of Function of the Olfactory Tracts. Electrophysiological studies of the olfactory system show that the mitral and tufted cells are continually active, and superimposed on this background are increases or decreases in impulse traffic caused by different odors. Thus, the olfactory stimuli presumably *modulate* the frequency of impulses in the olfactory system and in this way transmit the olfactory information.

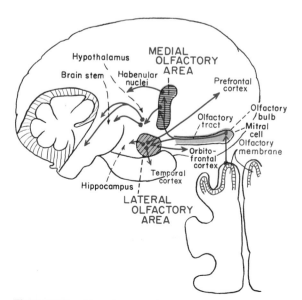

Figure 62–5. Neural connections of the olfactory system.

REFERENCES

Alberts, J. R.: Producing and interpreting experimental olfactory deficits. *Physiol. Behav., 12*:657, 1974.

Bradley, R. M., and Mistretta, C. M.: Investigations of taste function and swallowing in fetal sheep. *Symp. Oral Sens. Percept., 4*:185, 1973.

Dastoli, F. R.: Taste receptor proteins. *Life Sci., 14*:1417, 1974.

Denton, D. A.: Salt appetite. *In* Code, C. F., and Heidel, W. (eds.): Handbook of Physiology. Sec. 6, Vol. 1. Baltimore, Williams & Wilkins, 1967, p. 433.

Douek, E.: The Sense of Smell and Its Abnormalities. New York, Churchill Livingstone, 1974.

Forss, D. A.: Odor and flavor compounds from lipids. *Prog. Chem. Fats Other Lipids., 13*:177, 1972.

Hodgson, E. S.: Taste receptors. *Sci. Am., 204*:135, 1961

Kare, M. R., and Maller, O.: The Chemical Sense and Nutrition. Baltimore, The Johns Hopkins Press, 1967.

Lat, J.: Self-selection of dietary components. *In* Code, C. F., and Heidel, W. (eds.): Handbook of Physiology. Sec. 6, Vol. 1. Baltimore, Williams & Wilkins, 1967, p. 367.

Moulton, D. G.: Spatial patterning of response to odors in the peripheral olfactory system. *Physiol. Rev., 56*:578, 1976.

Moulton, D. G., and Beidler, L. M.: Structure and function in the peripheral olfactory system. *Physiol. Rev., 47*:1, 1967.

Norsiek, F. W.: The sweet tooth. *Am. Sci., 60*:41, 1972.

Oakley, B., and Benjamin, R. M.: Neural mechanisms of taste. *Physiol. Rev., 46*:173, 1966.

Ohloff, G., and Thomas, A. F. (eds.): Gustation and Olfaction. New York, Academic Press, 1971.

Schultz, E. F., and Tapp, J. T.: Olfactory control of behavior in rodents. *Psychol. Bull., 79*:21, 1973.

Shepherd, G. M.: Synaptic organization of the mammalian olfactory bulb. *Physiol. Rev., 52*:864, 1972.

Shepherd, G. M.: The olfactory bulb: A simple system in the mammalian brain. *In* Brookhart, J. M., and Mountcastle, V. B. (eds.): Handbook of Physiology. Sec. 1, Vol. 1. Baltimore, Williams & Wilkins, 1977, p. 945.

Todd, J. H.: The chemical languages of fishes. *Sci. Am., 224*:98, 1971.

Weiffenbach, J. M. (ed.): Taste and Development: The Genesis of Sweet Preference. Washington, D.C., U.S. Government Printing Office, 1977.

Wenzel, B. M., and Sieck, M. H.: Olfaction. *Annu. Rev. Physiol., 28*:381, 1966.

Zotterman, Y.: Olfaction and Taste. New York, The Macmillan Co., 1963.

Part XI

THE GASTROINTESTINAL TRACT

63

Movement of Food Through
the Alimentary Tract

The primary function of the alimentary tract is to provide the body with a continual supply of water, electrolytes, and nutrients, but before this can be achieved food must be moved along the alimentary tract at an appropriate rate for the digestive and absorptive functions to take place. Therefore, discussion of the alimentary system is presented in three different phases in the next three chapters: (1) movement of food through the alimentary tract, (2) secretion of the digestive juices, and (3) absorption of the digested foods, water, and the various electrolytes.

Figure 63–1 illustrates the entire alimentary tract, showing major anatomical differences between its parts. Each part is adapted for specific functions, such as: (1) simple passage of food from one point to another, as in the esophagus, (2) storage of food in the body of the stomach or fecal matter in the descending colon, (3) digestion of food in the stomach, duodenum, jejunum, and ileum, and (4) absorption of the digestive end-products and fluids in the entire small intestine and proximal half of the colon. One of the most important features of the gastrointestinal tract that is discussed in the present chapter is the myriad of autoregulatory processes in the gut that keeps the food moving at an appropriate pace — slow enough for digestion and absorption to take place but fast enough to provide the nutrients needed by the body.

GENERAL PRINCIPLES OF INTESTINAL MOTILITY

CHARACTERISTICS OF THE INTESTINAL WALL

Figure 63–2 illustrates a typical section of the intestinal wall, showing the following layers from the outer surface inward: (1) the *serosa,* (2) a *longitudinal muscle layer,* (3) a *circular muscle layer,* (4) the *submucosa,* and (5) the *mucosa.* In addition, a sparse layer of smooth muscle fibers, the *muscularis mucosae,* lies in the deeper layers of the mucosa. The motor functions of the gut are performed by the different layers of smooth muscle.

Characteristics of Intestinal Smooth Muscle — Electrical Activity and Contraction. The general characteristics of smooth muscle and its function were discussed in Chapter 12. However, the specific characteristics of smooth muscle in the gut are the following:

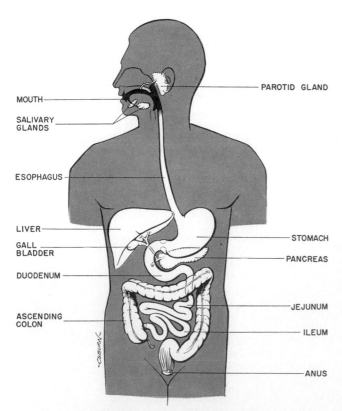

MOUTH

SALIVARY GLANDS

ESOPHAGUS

LIVER

GALL BLADDER

DUODENUM

ASCENDING COLON

PAROTID GLAND

STOMACH

PANCREAS

JEJUNUM

ILEUM

ANUS

Figure 63–1.   The alimentary tract.

784

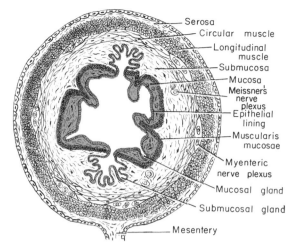

Figure 63–2. Typical cross-section of the gut.

The Functional Syncytium. The individual smooth muscle fibers in most areas of the gastrointestinal tract abut each other extremely closely. About 12 per cent of their membrane surfaces are actually fused with the membranes of adjacent muscle fibers in the form of *nexus,* and most of the remainder of the cell membranes of adjacent fibers lie in extremely close apposition. Measurements of ionic transport through these areas of close contact demonstrate extremely low electrical resistance — so little that intracellular electrical current can travel very easily from one smooth muscle fiber to another. Therefore, the smooth muscle of the gastrointestinal tract performs as a *functional syncytium,* which means that electrical signals originating in one smooth muscle fiber are generally propagated from fiber to fiber.

Electrical Activity of Gastrointestinal Smooth Muscle. The smooth muscle of the gastrointestinal tract undergoes electrical activity almost continuously. This activity is sometimes very bizarre, but it tends to have two basic types of waves, *slow waves* and *spikes,* both of which are illustrated in Figure 63–3. The slow waves occur at frequencies between 3 and 12 per minute, and they represent a basic continuous oscillation that occurs at the membranes of some smooth muscle in the gastrointestinal tract, especially the muscle in the longitudinal layer. The slow waves are not all-or-nothing action potentials but, instead, can be of any graded degree of intensity.

When the muscle is stimulated by stretch, by acetylcholine, or by parasympathetic excitation, the intracellular resting membrane potential of the muscle fibers becomes more positive, which raises the entire potential level of the slow waves from their normal mean voltage of −50 to −40 millivolts to some less negative value. This effect is called *depolarization.* As the depolarization rises above −40 millivolts, spike potentials begin to occur on the peaks of the slow waves; the frequency of spikes increases progressively as the resting potential rises still further. However, with very strong stimulation, when the resting membrane potential rises to a value of −20 to −15 millivolts, the spikes are likely to disappear because the membrane now remains totally depolarized all the time.

Figure 63–3 also illustrates that stimulation of the smooth muscle by either epinephrine or sympathetic excitation will *decrease* the resting membrane potential to a "hyperpolarized" value approaching −70 millivolts, and the electrical activity as well as the mechanical activity of the smooth muscle approaches zero.

Excitation of Muscle Contraction. Most gut contraction occurs in response to the spike potentials. Indeed, there is ordinarily no contraction at all in response to slow waves when these have no superimposed spikes. Thus, the spikes are comparable to action potentials in skeletal muscle and are responsible for the membrane changes that excite contraction. The contraction results mainly from entry of calcium through the cell membrane to the interior of the smooth muscle where it initiates a reaction between the myosin and actin of the smooth muscle, as was discussed in Chapter 12.

The smooth muscle of the gastrointestinal tract exhibits both *tonic contraction* and *rhythmic contraction,* both of which are characteristic of most types of smooth muscle, as discussed in Chapter 12.

Tonic contraction is continuous, lasting minute after minute or even hour after hour, sometimes increasing or decreasing in intensity but, nevertheless, continuing. It is believed to be caused by a series of spike potentials, the frequency of these determining the degree of tonic contraction. The intensity of tonic contraction in each segment of

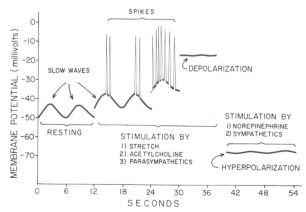

Figure 63–3. Membrane potentials in intestinal smooth muscle. Note the slow waves, the spike potentials, total depolarization, and hyperpolarization, all of which occur under different physiological conditions of the intestine.

the gut determines the amount of steady pressure in the segment, and tonic contraction of the sphincters determines the amount of resistance offered at the sphincters to the movement of intestinal contents. In this way the *pyloric,* the *ileocecal,* and the *anal sphincters* all help to regulate food movement in the gut.

In different parts of the gut the rhythmic contractions of the gastrointestinal smooth muscle occur at rates as rapid as 12 times per minute or as slow as 3 times per minute. These are also the frequencies of the respective slow waves in these segments; it is these slow waves that set the frequency. These contractions are responsible for the phasic functions of the gastrointestinal tract, such as mixing of the food and peristaltic propulsion of food as discussed below.

INNERVATION OF THE GUT — THE INTRINSIC NERVOUS SYSTEM

The gastrointestinal tract has an *intrinsic nervous system* of its own that begins in the esophagus and extends all the way to the anus. This system controls most gastrointestinal functions, especially gastrointestinal movements and secretion. On the other hand, both parasympathetic and sympathetic nervous signals to the gastrointestinal tract from the brain can strongly alter the degree of activity of this intrinsic nervous system, parasympathetic activity in general increasing its activity and sympathetic stimulation decreasing the activity.

The intrinsic nervous system is composed principally of two layers of neurons and appropriate connecting fibers: the outer layer, called the *myenteric plexus* or *Auerbach's plexus,* lies between the longitudinal and circular muscular layers; the inner layer, called the *submucosal plexus* or *Meissner's plexus,* lies in the submucosa. The myenteric plexus controls mainly the *gastrointestinal movements,* while the submucosal plexus is important in controlling *secretion* and also subserves many *sensory functions,* receiving signals principally from the gut epithelium and from stretch receptors in the gut wall.

In general, stimulation of the myenteric plexus increases the activity of the gut, causing four principal effects: (1) increased tonic contraction, or "tone," of the gut wall, (2) increased intensity of the rhythmic contractions, (3) increased rate of rhythmic contraction, and (4) increased velocity of conduction of excitatory waves along the gut wall. On the other hand, some myenteric plexus fibers are inhibitory rather than excitatory; these fibers are believed to be *purinergic* — that is, they secrete ATP or some similar purine-based transmitter substance. The excitatory fibers are mainly *cholinergic* — that is, they secrete acetylcholine,

though some probably secrete one or more other excitatory transmitters as well.

The intrinsic nervous system, including both the submucosal sensory plexus and the myenteric motor plexus, is especially responsible for many neurogenic reflexes that occur locally in the gut, such as reflexes from the mucosal epithelium to increase the activity of the gut muscle or to cause localized secretion of digestive juices by the submucosal glands.

Autonomic Control of the Gastrointestinal Tract.   The gastrointestinal tract receives extensive parasympathetic and sympathetic innervation that is capable of altering the overall activity of the entire gut or of specific parts of it, particularly of its upper end down to the stomach and its distal end from the mid-colon region to the anus.

Parasympathetic Innervation. The parasympathetic supply to the gut is divided into *cranial* and *sacral divisions,* which were discussed in Chapter 57. Except for a few parasympathetic fibers to the mouth and pharyngeal regions of the alimentary tract, the cranial parasympathetics are transmitted almost entirely in the *vagus nerves.* These fibers provide extensive innervation to the esophagus, stomach, pancreas, and first half of the large intestine (but rather little innervation to the small intestine). The sacral parasympathetics originate in the second, third, and fourth sacral segments of the spinal cord and pass through the *nervi erigentes* to the distal half of the large intestine. The sigmoidal, rectal, and anal regions of the large intestine are considerably better supplied with parasympathetic fibers than are the other portions. These fibers function especially in the defecation reflexes, which are discussed later in the chapter.

The postganglionic neurons of the parasympathetic system are probably part of the myenteric plexus, so that stimulation of the parasympathetic nerves causes a general increase in activity of this plexus. This in turn excites the gut wall and facilitates most of the intrinsic excitatory nervous reflexes of the gastrointestinal tract.

Sympathetic Innervation. The sympathetic fibers to the gastrointestinal tract originate in the spinal cord between the segments T-8 and L-3. The preganglionic fibers, after leaving the cord, enter the sympathetic chains and pass through the chains to outlying ganglia, such as the *celiac ganglion* and various *mesenteric ganglia.* Here, the postganglionic neuron bodies are located, and postganglionic fibers spread from them along with the blood vessels to all parts of the gut. The sympathetics innervate essentially all portions of the gastrointestinal tract rather than being more extensively supplied to the most orad and most analward portions as is true of the parasympathetics. The sympathetic nerve endings secrete *norepinephrine.*

In general, stimulation of the sympathetic nervous system inhibits activity in the gastrointestinal tract, causing effects essentially opposite to those of the parasympathetic system. It exerts its effects in two different ways: (1) by direct effect of the norepinephrine on the smooth muscle to inhibit this, and (2) by an inhibitory effect of the norepinephrine on the neurons of the intrinsic nervous system. Thus, strong stimulation of the sympathetic system can totally block movement of food through the gastrointestinal tract.

Afferent Nerve Fibers from the Gut. Many afferent nerve fibers arise in the gut. Some of these have their cell bodies in the submucosal plexus and terminate mainly in the myenteric plexus. These nerves can be stimulated by (a) irritation of the gut mucosa, (b) excessive distension of the gut, or (c) the presence of specific chemical substances in the gut. Signals transmitted through these fibers can cause excitation or, under some conditions, inhibition of intestinal movements or intestinal secretion.

In addition to the afferent fibers that terminate in the intramural plexus, other afferent fibers whose cell bodies lie in the dorsal root ganglia of the spinal cord or in the cranial nerve ganglia transmit signals all the way to the central nervous system, traveling along the sympathetic or parasympathetic nerves. For example, pain nerve fibers, except those from the esophagus, course along the sympathetic nerves to the spinal cord. Also, about 80 per cent of the nerve fibers in the vagus nerves are afferent rather than efferent. These fibers transmit signals into the medulla to help initiate vagal signals that in turn control several important functions of the gastrointestinal tract, as described in this and the following chapter.

HORMONAL CONTROL OF GASTROINTESTINAL MOTILITY

In the following chapter we shall discuss in detail the extreme importance of several hormones for controlling gastrointestinal secretion. Most of these same hormones also affect the motility of some parts of the gastrointestinal tract. Though these effects are much less important than the secretory effects of the hormones, some of the more important of them are the following:

Gastrin, which is secreted by the mucosa of the stomach antrum when food enters the stomach, increases stomach motility. Also, it increases the degree of constriction of the lower esophageal sphincter, thus helping to prevent reflux of the stomach contents into the esophagus. Gastrin also has a mild effect in increasing the motility of the small intestine and also of the gallbladder.

Cholecystokinin, which is secreted mainly by the mucosa of the jejunum in response to the presence of fatty substances in the intestinal contents, has an extremely potent effect in increasing contractility of the gallbladder, thus expelling bile into the small intestine where the bile then plays important roles in emulsifying fatty substances so that they can be digested and in causing fat absorption.

Secretin, which is secreted by the mucosa of the duodenum in response to acidic gastric juice emptied from the stomach through the pylorus, has a mild inhibitory effect on the motility of most of the gastrointestinal tract.

Gastric inhibitory peptide, which is secreted by the mucosa of the upper small intestine, mainly in response to fat but to a lesser extent in response to carbohydrate, has a moderate effect in decreasing motor activity of the stomach and therefore slowing the emptying of gastric contents into the duodenum when the upper small intestine is already oversupplied with food products.

FUNCTIONAL TYPES OF MOVEMENTS IN THE GASTROINTESTINAL TRACT

Two basic types of movements occur in the gastrointestinal tract: (1) *mixing movements*, which keep the intestinal contents thoroughly mixed at all times, and (2) *propulsive movements*, which cause food to move forward along the tract at an appropriate rate for digestion and absorption.

THE MIXING MOVEMENTS

In most parts of the alimentary tract, the mixing movements are caused by either *peristaltic contractions* or *local constrictive contractions of small segments of the gut wall*. These movements are modified in different parts of the gastrointestinal tract for proper performance of the respective activities of each part, as discussed separately later in the chapter.

THE PROPULSIVE MOVEMENTS — PERISTALSIS

The basic propulsive movement of the gastrointestinal tract is peristalsis, which is illustrated in Figure 63–4. A contractile ring appears around the gut and then moves forward; this is analogous to putting one's fingers around a thin distended tube,

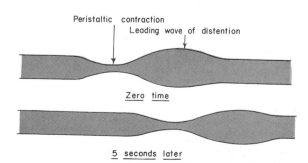

Figure 63–4. Peristalsis.

then constricting the fingers and moving forward along the tube. Obviously, any material in front of the contractile ring is moved forward.

Peristalsis is an inherent property of any syncytial smooth muscle tube, and stimulation at any point causes a contractile ring to spread in both directions. Thus, peristalsis occurs in (a) the gastrointestinal tract, (b) the bile ducts, (c) other glandular ducts throughout the body, (d) the ureters, and (e) most other smooth muscle tubes of the body.

The usual stimulus for peristalsis is *distension*. That is, if a large amount of food collects at any point in the gut, the distension stimulates the gut wall 2 to 3 cm. above this point, and a contractile ring appears that initiates a peristaltic movement.

Function of the Myenteric Plexus in Peristalsis. Even though peristalsis is a basic characteristic of all tubular smooth muscle structures, it occurs only weakly in portions of the gastrointestinal tract that have congenital absence of the myenteric plexus. Also, it is greatly depressed or completely blocked in the entire gut when the person is treated with atropine to paralyze the cholinergic nerve endings of the myenteric plexus. Furthermore, since the myenteric plexus is strongly affected in most parts of the gastrointestinal tract, except in the small intestine, by the parasympathetic nerves, the intensity of peristalsis in these areas and its velocity of conduction can be altered by parasympathetic stimulation.

Therefore, *even though the basic phenomenon of peristalsis is not dependent on the myenteric nerve plexus, effectual* peristalsis does require an active myenteric plexus.

Analward Peristaltic Movements. Peristalsis, theoretically, can occur in either direction from a stimulated point, but it normally dies out rapidly in the orad direction while continuing for a considerable distance analward. The cause of this directional transmission of peristalsis has never been ascertained, though several suggestions have been offered, as follows:

Receptive Relaxation and the "Law of the Gut." Some physiologists believe the directional movement of peristalsis to be caused by a special organization of the myenteric plexus, which allows preferential transmission of analward signals. One reason for believing this is the effect of electrical stimulation on the gut: An electrical stimulus causes a contractile ring to appear near the point of the stimulus but at the same time sometimes causes relaxation, called "receptive relaxation," several centimeters down the gut toward the anus. It is believed that this relaxation could occur only as a result of conduction in the myenteric plexus. Obviously, a leading wave of receptive relaxation could allow food to be pro-

pelled more easily analward than in the orad direction.

This response to electrical stimulation is also called the "law of the gut" or sometimes simply the "myenteric reflex." It can be particularly well demonstrated in the esophagus and in the pyloric region of the stomach.

The Gradient Theory for Forward Propulsion. Another theory for forward propulsion of intestinal contents is the gradient theory. This is based on the fact that the upper part of the gastrointestinal tract usually displays a higher degree of activity than the lower part. For instance, the basic contractile rhythm in the duodenum is approximately 12 contractions per minute, while that in the ileum is 7 to 9 contractions per minute. Also, in the upper part of the gastrointestinal tract far greater quantities of secretions are formed, and these, by distending the gut, initiate far more peristaltic waves than are initiated in the lower gut. Therefore, it is postulated that a greater number of peristaltic impulses originate in the orad portions of the gut than in the more distal regions, thereby causing peristaltic waves generally to travel down rather than up the gut. This is analogous to the *pacemaker* function of the S-A node in the heart.

INGESTION OF FOOD

The amount of food that a person ingests is determined principally by the intrinsic desire for food called *hunger*. The type of food that a person preferentially seeks is determined by *appetite*. These mechanisms in themselves are extremely important automatic regulatory systems for maintaining an adequate nutritional supply for the body, and they will be discussed in detail in Chapter 73 in relation to nutrition of the body. The present discussion is confined to the actual mechanical aspects of food ingestion, including especially *mastication* and *swallowing*.

MASTICATION (CHEWING)

The teeth are admirably designed for chewing, the anterior teeth (incisors) providing a strong cutting action and the posterior teeth (molars) a grinding action. All the jaw muscles working together can close the teeth with a force as great as 55 pounds on the incisors and 200 pounds on the molars. When this is applied to a small object, such as a small seed between the molars, the actual force *per square inch* may be several thousand pounds.

Most of the muscles of chewing are innervated by the motor branch of the 5th cranial nerve, and the chewing process is controlled by nuclei in the hindbrain. Stimulation of the reticular formation near the hindbrain centers for taste can cause continual rhythmic chewing movements. Also, stimulation of areas in the hypothalamus, amygdaloid nuclei, and even in the cerebral cortex near the sensory areas for taste and smell can cause chewing.

Much of the chewing process is caused by the *chewing reflex,* which may be explained as follows: The presence of a bolus of food in the mouth causes reflex inhibition of the muscles of mastication, which allows the lower jaw to drop. The drop in turn initiates a stretch reflex of the jaw muscles that leads to *rebound* contraction. This automatically raises the jaw to cause closure of the teeth, but it also compresses the bolus again against the linings of the mouth, which inhibits the jaw muscles once again, allowing the jaw to drop and rebound another time, and this is repeated again and again.

Chewing of the food is important for digestion of all foods, but it is especially important for most fruits and raw vegetables, because these have undigestible cellulose membranes around their nutrient portions which must be broken before the food can be utilized. Chewing aids in the digestion of food for the following simple reason: Since the *digestive enzymes act only on the surfaces of food particles,* the rate of digestion is highly dependent on the total surface area exposed to the intestinal secretions. Also, grinding the food to a very fine particulate consistency prevents excoriation of the gastrointestinal tract and increases the ease with which food is emptied from the stomach into the small intestine and thence into all succeeding segments of the gut.

SWALLOWING (DEGLUTITION)

Swallowing is a complicated mechanism, principally because the pharynx most of the time subserves several other functions besides swallowing and is converted for only a few seconds at a time into a tract for propulsion of food. Especially is it important that respiration not be seriously compromised during swallowing.

In general, swallowing can be divided into: (1) the *voluntary stage,* which initiates the swallowing process, (2) the *pharyngeal stage,* which is involuntary and constitutes the passage of food through the pharynx into the esophagus, and (3) the *esophageal stage,* another involuntary phase which promotes passage of food from the pharynx to the stomach.

Voluntary Stage of Swallowing. When the food is ready for swallowing, it is "voluntarily" squeezed or rolled posteriorly in the mouth by pressure of the tongue upward and backward against the palate, as shown in Figure 63–5. Thus, the tongue forces the bolus of food into the pharynx. From here on, the process of swallowing becomes entirely, or almost entirely, automatic and ordinarily cannot be stopped.

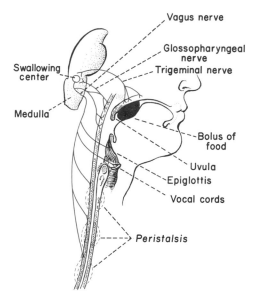

Figure 63–5. The swallowing mechanism.

Pharyngeal Stage of Swallowing. When the bolus of food is pushed backward in the mouth, it stimulates *swallowing receptor areas* all around the opening of the pharynx, especially on the tonsillar pillars, and impulses from these pass to the brain stem to initiate a series of automatic pharyngeal muscular contractions as follows:

1. The soft palate is pulled upward to close the posterior nares, in this way preventing reflux of food into the nasal cavities.

2. The palatopharyngeal folds on either side of the pharynx are pulled medialward to approximate each other. In this way these folds form a sagittal slit through which the food must pass into the posterior pharynx. This slit performs a selective action, allowing food that has been masticated properly to pass with ease while impeding the passage of large objects. Since this stage of swallowing lasts only 1 second, any large object is usually impeded too much to pass through the pharynx into the esophagus.

3. The vocal cords of the larynx are strongly approximated, and the epiglottis swings backward over the superior opening of the larynx. Both of these effects prevent passage of food into the trachea. Especially important is the approximation of the vocal cords, but the epiglottis helps to prevent food from ever getting as far as the vocal cords. Destruction of the vocal cords or of the muscles that approximate them can cause strangulation. On the other hand, removal of the epiglottis usually does not cause serious debility in swallowing.

4. The entire larynx is pulled upward and forward by muscles attached to the hyoid bone; this movement of the larynx stretches the opening of the esophagus. At the same time, the upper 3 to

4 centimeters of the esophagus, an area called the *upper esophageal sphincter* or the *pharyngoesophageal sphincter*, relaxes, thus allowing food to move easily and freely from the posterior pharynx into the upper esophagus. This sphincter, between swallows, remains tonically and strongly contracted, thereby preventing air from going into the esophagus during respiration. The upward movement of the larynx also lifts the glottis out of the main stream of food flow so that the food usually passes on either side of the epiglottis rather than over its surface; this adds still another protection against passage of food into the trachea.

5. At the same time that the larynx is raised and the pharyngoesophageal sphincter is relaxed, the superior constrictor muscle of the pharynx contracts, giving rise to a rapid peristaltic wave passing downward over the pharyngeal muscles and into the esophagus, which also propels the food into the esophagus.

To summarize the mechanics of the pharyngeal stage of swallowing — the trachea is closed, the esophagus is opened, and a fast peristaltic wave originating in the pharynx then forces the bolus of food into the upper esophagus, the entire process occurring in 1 to 2 seconds.

Nervous Control of the Pharyngeal Stage of Swallowing. The most sensitive tactile areas of the pharynx for initiation of the pharyngeal stage of swallowing lie in a ring around the pharyngeal opening, with greatest sensitivity in the tonsillar pillars. Impulses are transmitted from these areas through the sensory portions of the trigeminal and glossopharyngeal nerves into a region of the medulla oblongata closely associated with the *tractus solitarius* which receives essentially all sensory impulses from the mouth.

The successive stages of the swallowing process are then automatically controlled in orderly sequence by neuronal areas distributed throughout the reticular substance of the medulla and lower portion of the pons. The sequence of the swallowing reflex remains the same from one swallow to the next, and the timing of the entire cycle also remains constant from one swallow to the next. The areas in the medulla and lower pons that control swallowing are collectively called the *deglutition* or *swallowing center*.

The motor impulses from the swallowing center to the pharynx and upper esophagus that cause swallowing are transmitted by the 5th, 9th, 10th, and 12th cranial nerves and even a few of the superior cervical nerves.

In summary, the pharyngeal stage of swallowing is principally a reflex act. It is almost never initiated by direct stimuli to the swallowing center from higher regions of the central nervous system. Instead, it is almost always initiated by voluntary movement of food into the back of the mouth, which, in turn, elicits the swallowing reflex.

Effect of the Pharyngeal Stage of Swallowing on Respiration. The entire pharyngeal stage of swallowing occurs in less than 1 to 2 seconds, thereby interrupting respiration for only a fraction of a usual respiratory cycle. The swallowing center specifically inhibits the respiratory center of the medulla during this time, halting respiration at any point in its cycle to allow swallowing to proceed. Yet, even while a person is talking, swallowing interrupts respiration for such a short time that it is hardly noticeable.

Esophageal Stage of Swallowing. The esophagus functions primarily to conduct food from the pharynx to the stomach, and its movements are organized specifically for this function.

Normally the esophagus exhibits two types of peristaltic movements — *primary peristalsis* and *secondary peristalsis*. Primary peristalsis is simply a continuation of the peristaltic wave that begins in the pharynx and spreads into the esophagus during the pharyngeal stage of swallowing. This wave passes all the way from the pharynx to the stomach in approximately 5 to 10 seconds. However, food swallowed by a person who is in the upright position is usually transmitted to the lower end of the esophagus even more rapidly than the peristaltic wave itself, in about 4 to 8 seconds, because of the additional effect of gravity pulling the food downward. If the primary peristaltic wave fails to move all the food that has entered the esophagus into the stomach, secondary peristaltic waves result from distension of the esophagus by the retained food. These waves are essentially the same as the primary peristaltic waves, except that they originate in the esophagus itself rather than in the pharynx. Secondary peristaltic waves continue to be initiated until all the food has emptied into the stomach.

The peristaltic waves of the esophagus are controlled almost entirely by vagal reflexes that are part of the overall swallowing mechanism. These reflexes are transmitted through *vagal afferent fibers* from the esophagus to the medulla and then back again to the esophagus through *vagal efferent fibers*.

The musculature of the pharynx and the upper quarter of the esophagus is skeletal muscle, and, therefore, the peristaltic waves in these regions are controlled only by skeletal nerve impulses. In the lower two thirds of the esophagus, the musculature is smooth, but even this portion of the esophagus is normally under the control of the vagus nerve. However, when the vagus nerves to the esophagus are sectioned, the myenteric nerve plexus of the esophagus becomes excitable enough after several days to cause secondary peristaltic waves even without support from the vagal reflexes. Therefore, following paralysis of the swallowing reflex, food forced into the upper esophagus and then pulled by gravity to the lower esophagus still passes readily into the stomach.

Receptive Relaxation of the Stomach. As the esophageal peristaltic wave passes toward the stomach, a wave of relaxation precedes the constriction. Furthermore, the entire stomach and, to a lesser extent, even the duodenum become relaxed as this wave reaches the lower end of the esophagus. Especially important, also, is relaxation of the gastroesophageal sphincter at the juncture between the esophagus and the stomach. In other words, the constrictor and the stomach are prepared well ahead of time to receive food being propelled down the esophagus during the swallowing act.

FUNCTION OF THE LOWER ESOPHAGEAL SPHINCTER (GASTROESOPHAGEAL SPHINCTER)

At the lower end of the esophagus, about 2 to 5 cm. above its juncture with the stomach, the circular muscle functions as a so-called *lower esophageal sphincter* or *gastroesophageal sphincter*. Anatomically this sphincter is no different from the remainder of the esophagus. However, physiologically, it remains tonically constricted, in contrast to the mid- and upper portions of the esophagus which normally remain completely relaxed. However, when a peristaltic wave of swallowing passes down the esophagus, "receptive relaxation" relaxes the lower esophageal sphincter ahead of the peristaltic wave, and allows easy propulsion of the swallowed food into the stomach. Rarely, the sphincter does not relax satisfactorily, resulting in a condition called *achalasia,* which will be discussed in detail in Chapter 66.

A principal function of the lower esophageal sphincter is to prevent reflux of stomach contents into the upper esophagus. The stomach contents are highly acidic and contain many proteolytic enzymes. The esophageal mucosa, except in the lower eighth of the esophagus, is not capable of resisting for long the digestive action of gastric secretions. Fortunately, the tonic constriction of the lower esophageal sphincter prevents significant reflux of stomach contents into the esophagus except under abnormal conditions. Indeed, increased intragastric pressure, except during vomiting, causes a vagal reflex that further constricts the sphincter to add extra insurance against reflux.

Prevention of Reflux by Flutter-Valve Closure of the Distal End of the Esophagus. Another factor that prevents reflux is a valvelike mechanism of that portion of the esophagus that lies immediately beneath the diaphragm. Greatly increased intra-abdominal pressure caves the esophagus inward at this point at the same time that the abdominal pressure also increases the intragastric pressure. This flutter-valve closure of the lower esophagus therefore prevents the high pressure in the stomach from forcing stomach contents into the esophagus. Otherwise, every time we should walk, cough, or breathe hard, we might expel acid into the esophagus.

MOTOR FUNCTIONS OF THE STOMACH

The motor functions of the stomach are threefold: (1) storage of large quantities of food until it can be accommodated in the lower portion of the gastrointestinal tract, (2) mixing of this food with gastric secretions until it forms a semifluid mixture called *chyme,* and (3) slow emptying of the food from the stomach into the small intestine at a rate suitable for proper digestion and absorption by the small intestine.

Figure 63–6 illustrates the basic anatomy of the stomach. Physiologically, the stomach can be divided into two major parts: (1) the *corpus,* or *body,* and (2) the *antrum.* The *fundus,* located at the upper end of the body of the stomach, is considered by anatomists to be a separate entity from the body, but physiologically the fundus functions mainly as part of the body.

STORAGE FUNCTION OF THE STOMACH

As food enters the stomach, it forms concentric circles in the body and fundus of the stomach, the

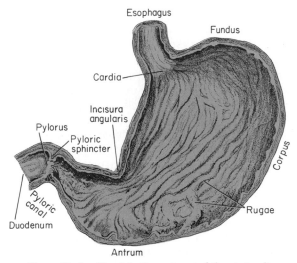

Figure 63–6. Physiologic anatomy of the stomach.

newest food lying closest to the esophageal opening and the oldest food lying nearest the wall of the stomach. Normally, the body of the stomach has relatively little tone in its muscular wall so that it can bulge progressively outward, thereby accommodating greater and greater quantities of food up to a limit of almost 1 liter. The pressure in the stomach remains low until this limit is approached for three reasons: First, the smooth muscle in the wall of the stomach exhibits a quality of *plasticity,* which means that it can increase its length greatly without significantly changing its tone. Second, the greater the diameter of the stomach, the greater also becomes the radius of curvature of the walls. The stretching force on the walls increases in direct proportion to this radius of curvature (an effect called the *law of Laplace*); therefore, the pressure inside the stomach increases only slightly despite marked distension. Third, stretch of the stomach also causes a *vagal reflex* that inhibits muscle activity in the body of the stomach.

MIXING IN THE STOMACH— THE BASIC ELECTRICAL RHYTHM OF THE STOMACH

The digestive juices of the stomach are secreted by the *gastric glands,* which cover almost the entire outer wall of the body of the stomach. These secretions come immediately into contact with that portion of the stored food lying against the mucosal surface of the stomach; when the stomach is filled, weak *constrictor waves,* also called *mixing waves,* move toward the antrum along the stomach wall approximately once every 20 seconds. These waves are caused by a *basic electrical rhythm* (called *BER*) consisting of electrical "slow waves" that occur spontaneously in the longitudinal muscle of the stomach wall and then spread by conduction to the circular muscle. As the waves move along the stomach wall, they not only cause the secretions to mix with the stored food, but they also provide weak propulsion to move these mixed contents into the antrum. When the stomach is full, these mixing waves usually begin near the midpoint of the stomach, but, as the stomach empties, the waves originate farther back up the stomach wall, thus propelling the last vestiges of stored food into the stomach antrum.

As the constrictor waves progress from the body of the stomach into the antrum, they usually become more intense, some becoming extremely intense and providing powerful *peristaltic constrictor rings* that force the antral contents under high pressure toward the pylorus. These constrictor rings also play an exceedingly important role in mixing the stomach contents in the following way:

Each time a peristaltic wave passes over the antrum toward the pylorus, it digs deeply into the contents of the antrum. Yet the opening of the pylorus is small enough that only a few milliliters of antral contents are expelled into the duodenum with each peristaltic wave. Instead, most of the antral contents are squirted backward through the peristaltic ring toward the body of the stomach. Thus, the moving peristaltic constrictive ring, combined with the squirting action, is an exceedingly important mixing mechanism of the stomach.

Chyme. After the food has become mixed with the stomach secretions, the resulting mixture that passes on down the gut is called *chyme.* The degree of fluidity of chyme depends on the relative amounts of food and stomach secretions and on the degree of digestion that has occurred. The appearance of chyme is that of a murky, milky semifluid or paste.

Propulsion of Food Through the Stomach. Strong peristaltic waves occur about 20 per cent of the time in the antrum of the stomach. These waves, like the mixing waves, occur about once every 20 seconds and, in fact, are extensions of mixing waves that have become especially potentiated as they spread from the body of the stomach into the antrum. They become intense approximately at the incisura angularis, from which they spread, no longer as weak mixing waves but instead as strong, peristaltic, ringlike constrictions, through the antrum. As the stomach becomes progressively more and more empty, these intense waves begin farther and farther up the body of the stomach, gradually pinching off the lowermost portions of stored food, adding this food to the chyme in the antrum.

The peristaltic waves often exert as much as 50 to 70 cm. of water pressure, which is about 6 times as powerful as the usual mixing waves.

Hunger Contractions. In addition to the mixing and peristaltic contractions of the stomach, a third type of intense contractions, called hunger contractions, often occurs when the stomach has been empty for a long time. These are usually rhythmic peristaltic contractions probably representing exacerbated mixing waves in the *body* of the stomach. However, when they become extremely strong, they often fuse together to cause a continuing tetanic contraction lasting for as long as two to three minutes.

Hunger contractions are usually most intense in young healthy persons with high degrees of gastrointestinal tonus, and they are also greatly increased by a low level of blood sugar.

When hunger contractions occur in the stomach, the person sometimes experiences a sensation of pain in the pit of the stomach, called *hunger pangs.* Hunger pangs usually do not begin until 12 to 24 hours after the last ingestion of food; in starvation they reach their greatest intensity in three to four days, and then gradually weaken in succeeding days.

Hunger contractions are often associated with a feeling of hunger and therefore are perhaps an important means by which the alimentary tract intensifies the desire for food when a person is in a state of incipient starvation.

Reflex Regulation of Stomach Contractions. Distension of the stomach by food initiates vagal afferent signals that pass to the medulla oblongata and reflexly inhibit the tone in the storage area of the stomach. At the same time these signals increase the rate of stomach secretion and the intensity of both the mixing and peristaltic waves. Thus, the rate of digestion and removal of the stored food is accelerated.

EMPTYING OF THE STOMACH

Basically, stomach emptying is opposed by resistance of the pylorus to the passage of food, and it is promoted by peristaltic waves in the antrum of the stomach. Usually, these two are reciprocally related to each other, with those factors that increase antral peristalsis usually decreasing the tone of the pyloric muscle.

Role of the Pylorus in Stomach Emptying. The pylorus normally remains almost, but not completely, closed because of tonic contraction of the pyloric muscle. The closing force is weak enough that water and other fluids empty from the stomach with ease. On the other hand, it is great enough to prevent movement of semi-solid chyme into the duodenum except when a strong antral peristaltic wave forces the chyme through. On the other hand, the degree of constriction of the pyloric sphincter can increase or decrease under the influence of signals both from the stomach and from the duodenum, as we shall discuss subsequently. In this way, the pylorus plays an important role in the control of stomach emptying.

Role of Antral Peristalsis in Stomach Emptying — The Pyloric Pump. The intensity of antral peristalsis changes markedly under different conditions, especially in response to signals both from the stomach and from the duodenum. Therefore, the intensity of antral peristalsis is the other principal factor determining the rate of stomach emptying.

When pyloric tone is normal, each strong antral peristaltic wave forces several milliliters of chyme into the duodenum. Thus, the peristaltic waves provide a pumping action that is frequently called the "pyloric pump."

Regulation of Stomach Emptying

The rate at which the stomach empties is regulated by signals both from the stomach and from the duodenum. The stomach signals are mainly two-fold: (1) nervous signals caused by distension of the stomach by food, and (2) the hormone *gastrin* released from the antral mucosa in response to the presence of certain types of food in the stomach. Both these signals increase pyloric pumping force and at the same time inhibit the pylorus, thus promoting stomach emptying.

On the other hand, signals from the duodenum depress the pyloric pump and usually increase pyloric tone at the same time. In general, when an excess volume of chyme or excesses of certain types of chyme enter the duodenum, strong *negative* feedback signals, both nervous and hormonal, depress the pyloric pump and enhance pyloric sphincter tone. Obviously, these feedback signals allow chyme to enter the duodenum only as rapidly as it can be processed by the small intestine.

Effect of Gastric Food Volume on Rate of Emptying. It is very easy to see how increased food volume in the stomach could promote increased emptying from the stomach. However, this increased emptying does not occur for the reasons that one would expect. It is not increased pressure in the stomach that causes the increased emptying because, in the usual normal range of volume, the increase in volume does not increase the pressure significantly, as was discussed in an earlier section. On the other hand, stretch of the stomach wall does elicit vagal and local myenteric reflexes in the wall that increase the activity of the pyloric pump and at the same time inhibit the pylorus. In general, the rate of food emptying from the stomach is approximately proportional to the *square root* of the volume of food remaining in the stomach at any given time.

Effect of the Hormone Gastrin on Stomach Emptying. In the following chapter we shall see that stretch, as well as the presence of certain types of foods in the stomach — particularly meat — elicits release of a hormone called *gastrin* from the antral mucosa, and this has potent effects on causing secretion of highly acidic gastric juice by the stomach fundic glands. However, gastrin also has potent stimulatory effects on motor functions of the stomach. Most important, it enhances the activity of the pyloric pump while at the same time relaxing the pylorus. Thus, it is a strong influence for promoting stomach emptying. It also has a constrictor effect on the gastroesophageal sphincter at the lower end of the esophagus for preventing reflux of gastric contents into the esophagus during the enhanced gastric activity.

The Inhibitory Effect of the Enterogastric Reflex from the Duodenum on Pyloric Activity. Reflex nervous signals are transmitted from the duodenum back to the stomach most of the time when the stomach is emptying food into the duodenum. These signals probably play an especially important role in controlling both the pyloric pump and the pylorus and, therefore, also in determining the rate of emptying of the stomach. The nervous reflexes probably are mediated mainly by way of afferent nerve fibers in the vagus nerve to the brain stem and then back through efferent nerve fibers to the stomach, also by way of the vagi. However, some of the signals are probably transmitted directly by way of the myenteric plexus as well.

The types of factors that are continually monitored in the duodenum and that can elicit the enterogastric reflex include:

1.   The degree of distension of the duodenum.
2.   The presence of any degree of irritation of the duodenal mucosa.
3.   The degree of acidity of the duodenal chyme.
4.   The degree of osmolality of the chyme.
5.   The presence of certain breakdown products in the chyme, especially breakdown products of proteins and perhaps to a lesser extent of fats.

The enterogastric reflex is especially sensitive to the presence of irritants and acids in the duodenal chyme. For instance, whenever the pH of the chyme in the duodenum falls below approximately 3.5 to 4, this reflex is immediately elicited, which inhibits the pyloric pump and increases pyloric constriction, thus reducing or even blocking further release of acidic stomach contents into the duodenum until the duodenal chyme can be neutralized by pancreatic and other secretions.

Breakdown products of protein digestion will also elicit this reflex; by slowing the rate of stomach emptying, sufficient time is insured for adequate protein digestion in the upper portion of the small intestine.

Finally, either hypo- or hypertonic fluids (especially hypertonic) will elicit the enterogastric reflex. This effect prevents too rapid flow of nonisotonic fluids into the small intestine, thereby preventing rapid changes in electrolyte balance of the body fluids during absorption of the intestinal contents.

Hormonal Feedback from the Duodenum in Inhibiting Gastric Emptying — Role of Fats. When excess chyme empties into the duodenum from the stomach, especially chyme containing fats, the activity of the pyloric pump is depressed while the pyloric sphincter is slightly constricted, and stomach emptying is correspondingly slowed. This plays an important role in allowing slow digestion of the fats before they proceed into the deeper recesses of the intestine.

Unfortunately, the precise mechanism by which fats cause this effect of slowing the emptying of the stomach is not completely known. Most of the effect still occurs even after the enterogastric reflex has been blocked. Therefore, it is presumed that the effect results from some hormonal feedback mechanism elicited by the presence of fats in the duodenum. In the past, this hormone has been called *enterogastrone,* but such a hormone has never yet been identified as a specific entity. On the other hand, several different hormones released by the mucosa of the upper small intestine are known to inhibit stomach emptying. One of these is *cholecystokinin,* which is released from the mucosa of the jejunum in response to fatty substances in the chyme. This hormone acts as a competitive inhibitor to block the increased stomach motility caused by gastrin. Another is the hormone *secretin,* which is released mainly from the duodenal mucosa in response to gastric acid released from the stomach through the pylorus. This hormone has the general effect of decreasing gastrointestinal motility. Finally, a hormone called *gastric inhibitory peptide,* which is released from the upper small intestine in response mainly to fat in the chyme but to carbohydrates as well, is known also to inhibit gastric motility under some conditions. (However, its effect at physiological concentrations is probably mainly to stimulate the secretion of insulin by the pancreas.) All these hormones will be discussed at greater length elsewhere in this text, especially in the following chapter where both cholecystokinin and secretin will be discussed in detail.

In summary, several different hormones are known that could serve as hormonal mechanisms for inhibiting gastric emptying when excess quantities of chyme, especially acidic or fatty chyme, enter the duodenum from the stomach.

Summary.   Emptying of the stomach is controlled to a moderate degree by stomach factors, such as the degree of filling in the stomach and the excitatory effect of gastrin on antral peristalsis. However, probably the more important control of stomach emptying resides in feedback signals from the duodenum, including both the enterogastric reflex and hormonal feedback. These two feedback signals work together to slow the rate of emptying when (a) too much chyme is already in the small intestine or (b) the chyme is excessively acid, contains too much protein or fat, is hypotonic or hypertonic, or is irritating. In this way the rate of stomach emptying is limited to that amount of chyme that the small intestine can process.

MOVEMENTS OF THE SMALL INTESTINE

The movements of the small intestine, as elsewhere in the gastrointestinal tract, can be divided into the *mixing contractions* and the *propulsive contractions.* However, to a great extent this separation is artificial because essentially all movements of the small intestine cause at least some degree of both mixing and propulsion. Yet, the usual classification of these processes is the following:

MIXING CONTRACTIONS (SEGMENTATION CONTRACTIONS)

When a portion of the small intestine becomes distended with chyme, the stretch of the intestinal wall elicits localized concentric contractions spaced at intervals along the intestine. The longitudinal length of each one of the contractions is only about 1 cm. so that each set of contractions causes "segmentation" of the small intestine, as illustrated in Figure 63–7, dividing the intestine at times into regularly spaced segments that have the appearance of a chain of sausages. As one set of segmentation contractions relaxes a new set begins, but the contractions this time occur at new points between the previous contractions. These segmentation contractions "chop" the chyme as often as 7 to 12 times a minute, in this way promoting progressive mixing of the solid food particles with the secretions of the small intestine.

The mixing movements depend mainly on the reflex signals generated in the myenteric plexus of the gut in response to the intestinal distension, though very weak concentric contractions can still occur even when the myenteric plexus is blocked by atropine.

PROPULSIVE MOVEMENTS

Chyme is propelled through the small intestine by *peristaltic waves*. These can occur in any part of the small intestine, and they move analward at a velocity of 0.5 to 2 cm. per second, much faster in the proximal intestine and much slower in the terminal intestine. However, they are normally very weak and usually die out after traveling only a few centimeters, so that movement of the chyme is much slower. As a result, the *net* movement of the chyme along the small intestine averages only 1 cm. per minute. This means that 3 to 5 hours are normally required for passage of chyme from the pylorus to the ileocecal valve.

Peristaltic activity of the small intestine is greatly increased after a meal. This is caused partly by the beginning entry of chyme into the duodenum but also by the so-called *gastroenteric reflex* that is initiated by distension of the stomach and conducted principally through the myenteric plexus from the stomach down along the wall of the small intestine. This reflex increases the overall degree of excitability of the small intestine, including both increased motility and secretion.

The Peristaltic Reflex. The usual cause of peristalsis in the small intestine is distension. Circumferential stretch of the intestine excites receptors in the gut wall, and these elicit a local myenteric reflex that begins with contraction of the *longitudinal muscle* over a distance of several centimeters, followed by contraction of the circular muscle. Simultaneously, the contractile process spreads in an analward direction by the process of peristalsis. Movement of the peristaltic contraction down the gut is probably a function mainly of the myenteric plexus because it occurs only weakly and more slowly when this plexus has been blocked by drugs or when the plexus has degenerated.

Very intense irritation of the intestinal mucosa, such as occurs in some infectious processes, can elicit a so-called *peristaltic rush*, which is a powerful peristaltic wave that travels long distances in the small intestine in a few minutes. These waves can sweep the contents of the intestine into the colon and thereby relieve the small intestine of either irritative chyme or excessive distension.

The function of the peristaltic waves in the small intestine is not only to cause progression of the chyme toward the ileocecal valve but also to spread out the chyme along the intestinal mucosa. As the chyme enters the intestine from the stomach and causes initial distension of the proximal intestine, the elicited peristaltic waves begin immediately to spread the chyme along the intestine, and this process intensifies as additional chyme enters the intestine. On reaching the ileocecal valve the chyme is sometimes blocked for several hours until the person eats another meal, when a new *gastroenteric* (also called *gastroileal*) reflex intensifies the peristalsis in the ileum and forces the remaining chyme through the ileocecal valve into the cecum.

The Propulsive Effect of the Segmentation Movements. The segmentation movements, though they last for only a few seconds, also travel in the analward direction and help to propel the food down the intestine. Therefore, the difference between the segmentation and the peristaltic movements is not as great as might be implied by their separation into these two classifications.

The Basic Electrical Rhythm of the Small Intestine and Its Control of Intestinal Contractions. The basic electrical rhythm (BER) of the small intestine is generated in the longitudinal

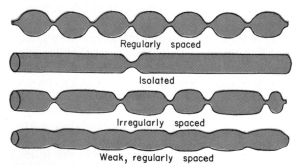

Regularly spaced

Isolated

Irregularly spaced

Weak, regularly spaced

Figure 63–7. Segmentation movements of the small intestine.

muscle. Slow oscillatory electrical waves, as described earlier in this chapter, occur at the membranes of the longitudinal smooth muscle, at rhythmical frequencies of 11 to 12 per minute in the duodenum, decreasing to 7 to 9 per minute in the terminal ileum. However, these electrical waves do not in themselves produce intestinal contractions. They merely set the background conditions for the contractions.

When the intestinal tract becomes overly distended or when the mucosa becomes irritated, this superimposes an additional stimulus, leading to myenteric reflexes that enhance the electrical activity of the gut. Now, during the positive phases of the slow waves, spike potentials occur. These spike potentials in turn spread through both the longitudinal muscle and then the circular muscle to elicit the contractions. However, it is the basic rhythm of the slow waves that still determines the frequency of the contractions.

Furthermore, the slow waves travel in an analward direction. This property of the slow waves is presumably the reason, or at least part of the reason, for the analward progression of both the segmentation and the peristaltic contractions of the intestines.

The most rapid rate of rhythm of the slow waves occurs in the intestinal muscle surrounding the point of entry of the bile duct into the duodenum. Since the slow wave travels downward along the intestine, the rate of rhythm of this point in the intestine tends also to be transmitted along the intestinal wall, thus acting as a "pacemaker" for the intestine. This is analogous to the pacemaker action of the S-A node of the heart, which controls the rate of rhythm of the entire heart. However, when this pacemaker slow wave fails to pass the entire distance down the small intestine, naturally occurring slow waves at slower frequencies in the more distal portions of the intestine then become local pacemakers.

Movements Caused by the Muscularis Mucosae and Muscle Fibers of the Villi. The muscularis mucosae, which is stimulated by local nervous reflexes in the submucosal plexus, can cause short or long folds to appear in the intestinal mucosa. Also, individual fibers from this muscle extend into the intestinal villi and cause them to contract intermittently. The mucosal folds increase the surface area exposed to the chyme, thereby increasing the rate of absorption. The contractions of the villi — shortening, elongating, and shortening again — "milk" the villi so that lymph flows freely from the central lacteals into the lymphatic system. Both these types of contraction also agitate the fluids surrounding the villi so that progressively new areas of fluid become exposed to absorption.

These mucosal and villous contractions are initiated by local nervous reflexes that occur in response to chyme in the small intestine. It is also believed that the villi are stimulated by a hormone called *villikinin*. The chyme in the intestine extracts villikinin from the mucosa, and this in turn is absorbed into the blood to excite the villi.

CONTRACTION AND EMPTYING OF THE GALLBLADDER — CHOLECYSTOKININ

The functions of the gallbladder will be discussed specifically in relation to overall function of the liver in Chapter 70. However, its activity is closely associated with that of the small intestine.

The liver continually secretes bile, which is then stored and concentrated in the gallbladder. The bile does not enter the small intestine until a specific stimulus causes the gallbladder to contract; this stimulus in general is initiated by the presence of fat in the small intestine as follows:

When fat and protein digestates enter the small intestine, within a few minutes these extract the hormone *cholecystokinin* from the mucosa. This in turn passes by way of the blood to the gallbladder and causes it to contract rhythmically. However, these contractions alone are not sufficient to cause emptying of the gallbladder, because the sphincter of Oddi, which constricts the common bile duct where it empties into the small intestine, normally remains tonically contracted despite gallbladder contraction. However, cholecystokinin also relaxes the sphincter of Oddi. In addition, every time a peristaltic wave passes over the duodenum, a wave of relaxation immediately ahead of the peristaltic contraction helps to relax the sphincter of Oddi, thus allowing small squirts of bile to flow into the duodenum. Therefore, the emptying of the gallbladder results from a combined action of cholecystokinin and intestinal peristalsis.

Regulation of gallbladder emptying by fats is a purposive effect, for the bile salts are stored in the gallbladder during the interdigestive period and are emptied only when needed to emulsify the fats in the gut so that they can be digested by the intestinal lipase. Bile salts also help to transport the digested fats to the intestinal epithelium.

FUNCTION OF THE ILEOCECAL VALVE

A principal function of the ileocecal valve is to prevent backflow of fecal contents from the colon into the small intestine. As illustrated in Figure 63–8, the lips of the ileocecal valve protrude into the lumen of the cecum and therefore are forcefully closed when the cecum fills. Usually the valve can resist reverse pressure of as much as 50 to 60 cm. water.

The wall of the ileum for several centimeters immediately preceding the ileocecal valve has a thickened muscular coat called the *ileocecal sphincter*. This normally remains mildly constricted and slows the emptying of ileal contents into the cecum except immediately following a meal when a gastroileal reflex (described above) intensifies the peristalsis in the ileum. Also, the hormone *gastrin*, which is liberated from the stomach mucosa in response to food in the stomach, has a direct relaxant effect on the ileocecal sphincter,

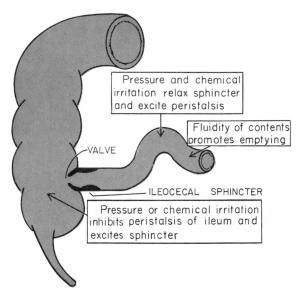

Figure 63–8. Emptying at the ileocecal valve.

thus allowing increased emptying. Even so, only about 750 ml. of chyme empty into the cecum each day. The resistance to emptying at the ileocecal valve prolongs the stay of chyme in the ileum and, therefore, facilitates absorption.

Feedback Control of the Ileocecal Sphincter. The degree of contraction of the ileocecal sphincter is also controlled strongly by reflexes from the cecum. Whenever the cecum is distended, the degree of contraction of the ileocecal sphincter is intensified, which greatly delays emptying of additional chyme from the ileum. Also, any irritant in the cecum causes constriction of the ileocecal sphincter. For instance, when a person has an inflamed appendix, the irritation of this vestigial remnant of the cecum can cause such intense spasm of the ileocecal sphincter that it completely blocks emptying of the ileum. These reflexes from the cecum to the ileocecal sphincter are mediated by way of the myenteric plexus.

MOVEMENTS OF THE COLON

The functions of the colon are (1) absorption of water and electrolytes from the chyme and (2) storage of fecal matter until it can be expelled. The proximal half of the colon, illustrated in Figure 63–9, is concerned principally with absorption, and the distal half with storage. Since intense movements are not required for these functions, the movements of the colon are normally sluggish. Yet in a sluggish manner, the movements still have characteristics similar to those of the small intestine and can be divided once again into mixing movements and propulsive movements.

Mixing Movements — Haustrations. In the

same manner that segmentation movements occur in the small intestine, large circular constrictions also occur in the large intestine. At each of these constriction points, about 2.5 cm. of the circular muscle contracts, sometimes constricting the lumen of the colon to almost complete occlusion. At the same time, the longitudinal muscle of the colon, which is aggregated into three longitudinal strips called the *tineae coli,* contract. These combined contractions of the circular and longitudinal smooth muscle cause the unstimulated portion of the large intestine to bulge outward into baglike sacs called *haustrations.* The haustral contractions, once initiated, usually reach peak intensity in about 30 seconds and then disappear during the next 60 seconds. They also at times move slowly analward during their period of contraction, especially in the cecum and ascending colon. After another few minutes, new haustral contractions occur in other areas nearby. Therefore, the fecal material in the large intestine is slowly *dug into and rolled over* in much the same manner that one spades the earth. In this way, all the fecal material is gradually exposed to the surface of the large intestine, and fluid is progressively absorbed until only 80 to 150 ml. of the 750 ml. daily load of chyme is lost in the feces.

Propulsive Movements — "Mass Movements." Peristaltic waves of the type seen in the small intestine only rarely occur in most parts of the colon. Instead, most propulsion occurs by (1) the *haustral contractions* discussed above and (2) *mass movements.*

Most of the propulsion in the cecum and ascending colon results from the slow but persistent haustral contractions, requiring as many as 8 to 15 hours to move the chyme only from the ileocecal valve to the transverse colon, while the chyme itself becomes fecal in quality and also becomes a semi-solid slush instead of a semi-fluid.

From the transverse colon to the sigmoid, mass

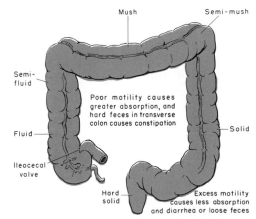

Figure 63–9. Absorptive and storage functions of the large intestine.

movements take over the propulsive role. These movements usually occur only a few times each day, most abundantly for about 15 minutes during the first hour after eating breakfast.

A mass movement is characterized by the following sequence of events: First, a constrictive ring occurs at a distended or irritated point in the colon, usually in the transverse colon, and then rapidly thereafter the 20 or more centimeters of colon *distal* to the constriction contract almost as a unit, forcing the fecal material in this segment *en masse* down the colon. During this process, the haustrations disappear completely. The initiation of contraction is complete in about 30 seconds, and relaxation then occurs during the next 2 to 3 minutes before another mass movement occurs. But the whole series of mass movements will usually persist for only 10 minutes to half an hour, and they will then return perhaps a half day or even a day later.

Mass movements can occur in any part of the colon, though most often they occur in the transverse or descending colon. When they have forced a mass of feces into the rectum, the desire for defecation is felt.

Initiation of Mass Movements by the Gastrocolic and Duodenocolic Reflexes. The appearance of mass movements after meals is caused at least partially by *gastrocolic* and *duodenocolic reflexes*. These reflexes result from distension of the stomach and duodenum. They can take place, though with decreased intensity, when the autonomic nerves are removed; therefore, it is probable that the reflexes are basically transmitted through the myenteric plexus, though reflexes conducted through the autonomic nervous system probably reinforce this direct route of transmission.

It is likely that the hormone gastrin, which is secreted by the stomach antral mucosa in response to distension, also plays some role in this effect, because gastrin has an excitatory effect on the colon and an inhibitory effect on the ileocecal valve, thus allowing rapid emptying of ileal contents into the cecum. This, in turn, elicits increased colonic activity.

Irritation in the colon can also initiate intense mass movements. For instance, a person who has an ulcerated condition of the colon (*ulcerative colitis*) frequently has mass movements that persist almost all of the time.

Mass movements are also initiated by intense stimulation of the parasympathetic nervous system or simply by overdistension of a segment of the colon.

DEFECATION

Most of the time, the rectum is empty of feces. This results partly from the fact that a weak functional sphincter exists approximately 20 cm. from the anus at the juncture between the sigmoid and the rectum. However, when a mass movement forces feces into the rectum the desire for defecation is normally initiated, including reflex contraction of the rectum and relaxation of the anal sphincters.

Continual dribble of fecal matter through the anus is prevented by tonic constriction of (1) the *internal anal sphincter,* a circular mass of smooth muscle that lies immediately inside the anus, and (2) the *external anal sphincter,* composed of striated voluntary muscle that both surrounds the internal sphincter and also extends distal to it; the external sphincter is controlled by the somatic nervous system and therefore is under voluntary control.

The Defecation Reflexes. Ordinarily, defecation results from the *defecation reflexes* which can be described as follows: When the feces enter the rectum, distension of the rectal wall initiates afferent signals that spread through the *myenteric plexus* to initiate peristaltic waves in the descending colon, sigmoid, and rectum, forcing feces toward the anus. As the peristaltic wave approaches the anus, the internal anal sphincter is inhibited by the usual phenomenon of *receptive relaxation,* and if the external anal sphincter is relaxed, defecation will occur. This overall effect is the *intrinsic defecation reflex* of the colon itself.

However, the intrinsic defecation reflex itself is usually weak, and to be effective in causing defecation it must be fortified by another type of defecation reflex, a *parasympathetic defecation reflex* that involves the sacral segments of the spinal cord, as illustrated in Figure 63–10. When the afferent fibers in the rectum are stimulated, signals are transmitted into the spinal cord and thence, reflexly, back to the descending colon,

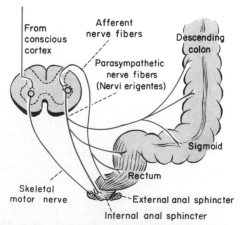

Figure 63–10. The afferent and efferent pathways of the parasympathetic mechanism for enhancing the defecation reflex.

sigmoid, rectum, and anus by way of parasympathetic nerve fibers in the *nervi erigentes*. These parasympathetic signals greatly intensify the peristaltic waves and convert the intrinsic defecation reflex from an ineffectual weak movement into a powerful process of defecation that is sometimes effective in emptying the large bowel in one movement all the way from the splenic flexure to the anus. Also, the afferent signals entering the spinal cord initiate other effects, such as taking a deep breath, closure of the glottis, and contraction of the abdominal muscles to force downward on the fecal contents of the colon while at the same time causing the pelvic floor to pull outward and upward on the anus to evaginate the feces downward.

However, despite the defecation reflexes other effects are also necessary before actual defecation occurs because relaxation of the internal sphincter and forward movement of feces toward the anus normally initiates an instantaneous contraction of the external sphincter, which still temporarily prevents defecation. Except in babies and mentally inept persons, the conscious mind then takes over voluntary control of the external sphincter and either inhibits it to allow defecation to occur or further contracts it if the moment is not socially acceptable for defecation. When the contraction is maintained, the defecation reflexes die out after a few minutes and usually will not return until an additional amount of feces enters the rectum, which may not occur until several hours thereafter.

When it becomes convenient for the person to defecate, the defecation reflexes can sometimes be excited by taking a deep breath to move the diaphragm downward and then contracting the abdominal muscles to increase the pressure in the abdomen, thus forcing fecal contents into the rectum to elicit new reflexes. Unfortunately, reflexes initiated in this way are never as effective as those that arise naturally, for which reason people who too often inhibit their natural reflexes become severely constipated.

In the newborn baby and in some persons with transected spinal cords, the defecation reflexes cause automatic emptying of the lower bowel without the normal control exercised through contraction of the external anal sphincter.

OTHER AUTONOMIC REFLEXES AFFECTING BOWEL ACTIVITY

Aside from the duodenocolic, gastrocolic, gastroileal, enterogastric, and defecation reflexes which have been discussed in this chapter, several other important nervous reflexes can affect the overall degree of bowel activity. These are the intestino-intestinal reflex, peritoneo-intestinal reflex, reno-intestinal reflex, vesico-intestinal reflex, and somato-intestinal reflex. All these reflexes are initiated by sensory signals that pass to the spinal cord and then are transmitted through the sympathetic nervous system back to the gut. And they all *inhibit* gastrointestinal activity. Thus, the *intestino-intestinal reflex* occurs when one part of the intestine becomes overdistended or its mucosa becomes excessively irritated; this blocks activity in other parts of the intestine while the local distension or irritability increases activity in the localized region and moves the intestinal contents away from the distended or irritated area.

The *peritoneo-intestinal reflex* is very much like the intestino-intestinal reflex, except that it results from irritation of the peritoneum; it causes intestinal paralysis. The *reno-intestinal* and *vesico-intestinal reflexes* inhibit intestinal activity as a result of kidney or bladder irritation. Finally, the *somato-intestinal reflex* causes intestinal inhibition when the skin over the abdomen is irritatingly stimulated. Some of these reflexes will be discussed further in Chapter 66 in relation to abnormalities of gastrointestinal function.

REFERENCES

Atanassova, E., and Papasova, M.: Gastrointestinal motility. *Int. Rev. Physiol., 12*:35, 1977.

Barrett, R. H., and Hanson, M. L.: Oral Myofunctional Disorders. St. Louis, C. V. Mosby, 1978.

Barry, R. J., and Eggenton, J.: Electrical activity of the intestine. *Biomembranes, 4B*:917, 1974.

Bofak-Gyovai, L. A., and Manzione, J. V.: Oral Medicine: Patient Evaluation and Management. Baltimore, Williams & Wilkins, 1980.

Bortoff, A.: Myogenic control of intestinal motility. *Physiol. Rev., 56*:418, 1976.

Brooks, F. P. (ed.): Gastrointestinal Pathophysiology. New York, Oxford University Press, 1978.

Brooks, F. P., and Evers, P. W. (eds.): Nerves and the gut. Thorofare, N.J., C. B. Slack, 1977.

Cohen, S., *et al.*: Gastrointestinal motility. In Crane, R. K. (ed.): *International Review of Physiology: Gastrointestinal Physiology III.* Vol. 19. Baltimore, University Park Press, 1979, p. 107.

Davenport, H. W.: A Digest of Digestion, 2nd Ed. Chicago, Year Book Medical Publishers, 1978.

Dickray, G.: Comparative biochemistry and physiology of gut hormones. *Annu. Rev. Physiol., 41*:83, 1979.

Dubner, A., *et al.*: The Neural Basis of Oral and Facial Function. New York, Plenum Press, 1978.

Duthie, H. S. (ed.): Gastrointestinal Motility in Health and Disease. Baltimore, University Park Press, 1978.

Furness, J. B., and Costa, M.: The adrenergic innervation of the gastrointestinal tract. *Ergeb. Physiol., 69*:2, 1974.

Grossman, M. I.: Neural and hormonal regulation of gastrointestinal function: An overview. *Annu. Rev. Physiol., 41*:27, 1979.

Hunt, J. N., and Knox, M. T.: Regulation of gastric emptying. In Code, C. F., and Heidel, W. (eds.): Handbook of Physiology. Sec. 6, Vol. 4. Baltimore, Williams & Wilkins, 1968, p. 1917.

Hurwitz, A. L., *et al.*: Disorders of Esophageal Motility. Philadelphia, W. B. Saunders Co., 1979.

Jenkins, G. N.: The Physiology and Biochemistry of the Mouth. Philadelphia, J. B. Lippincott Co., 1978.

Johnson, L. R.: Gastrointestinal hormones and their functions. *Annu. Rev. Physiol., 39*:135, 1977.

Kirsner, J. B., and Winans, C. S.: The stomach. In Sodeman, W. A., Jr., and Sodeman, T. M. (eds.): Sodeman's Pathologic Physiology; Mechanisms of Disease, 6th Ed. Philadelphia, W. B. Saunders Co., 1979, p. 798.

Levitt, M. D., and Bond, J. H.: Flatulence. *Annu. Rev. Med., 31*:127, 1980.

Makhlouf, G. M.: The neuroendocrine design of the gut. The play of chemicals in a chemical playground. *Gastroenterology, 67*:159, 1974.

Myren, J., and Vatn, M. H. (eds.): Gastric Inhibition. Oslo, Universitetsforl, 1976.

Phillips, S. F., and Devroede, G. J.: Functions of the large intestine. *In* Crane, R. K. (ed.): International Review of Physiology: Gastrointestinal Physiology III. Vol. 19. Baltimore, University Park Press, 1979, p. 263.

Rehfeld, J. F.: Gastrointestinal hormones. *In* Crane, R. K. (ed.): International Review of Physiology: Gastrointestinal Physiology III. Vol. 19. Baltimore, University Park Press, 1979, p. 291.

Sernka, T. J., and Jacobson, E. D.: Gastrointestinal Physiology; The Essentials. Baltimore, Williams & Wilkins, 1979.

Sodeman, W. A., Jr., and Watson, D. W.: The large intestine. *In*

Sodeman, W. A., Jr., and Sodeman, T. M. (eds.): Sodeman's Pathologic Physiology; Mechanisms of Disease, 6th Ed. Philadelphia, W. B. Saunders Co., 1979, p. 860.

Van Der Reis, L. (ed.): The Esophagus. New York, S. Karger, 1978.

Weisbrodt, N. W.: Gastrointestinal motility. *In* MTP International Review of Science: Physiology. Vol. 4. Baltimore, University Park Press, 1974, p. 105.

Wood, J. D.: Neurophysiology of Auerbach's plexus and control of intestinal motility. *Physiol. Rev., 55*:307, 1975.

Young, J. A., and vanLennep, E. W.: The Morphology of Salivary Glands. New York, Academic Press, 1978.

64

Secretory Functions of the Alimentary Tract

Throughout the gastrointestinal tract secretory glands subserve two primary functions: First, digestive enzymes are secreted in most areas from the mouth to the distal end of the ileum. Second, mucous glands, present from the mouth to the anus, provide mucus for lubrication and protection of all parts of the alimentary tract.

Most digestive secretions are formed only in response to the presence of food in the alimentary tract, and the quantity secreted in each segment of the tract is almost exactly the amount needed for proper digestion. Furthermore, in some portions of the gastrointestinal tract even the types of enzymes and other constituents of the secretions are varied in accordance with the types of food present. The purpose of the present chapter, therefore, is to describe the different alimentary secretions, their functions, and regulation of their production.

GENERAL PRINCIPLES OF GASTROINTESTINAL SECRETION

ANATOMICAL TYPES OF GLANDS

Several types of glands provide the different types of secretions in the gastrointestinal tract. First, on the surface of the epithelium in most parts of the gastrointestinal tract are literally billions of *single cell mucous glands* called *goblet cells*. These function entirely by themselves without the necessity of coordination with other goblet cells, and they simply extrude their mucus directly into the lumen of the gastrointestinal tract.

Second, most surface areas of the gastrointestinal tract are lined by pits which represent invaginations of the epithelium into the submucosa. In the small intestine these pits, called *crypts of Lieberkühn,* are deep and contain specialized secretory cells. One of these is illustrated in Figure 64–11. They are lined with goblet cells that produce mucus and with other epithelial cells that produce mainly serous fluids.

Third, in the stomach and upper duodenum are found large numbers of deep *tubular glands*. A typical tubular gland is illustrated in Figure 64–4, which shows an acid- and peptic-secreting gland of the stomach.

Fourth, also associated with the gastrointestinal tract are several complex glands — the *salivary glands,* the *pancreas,* and the *liver* — which provide secretions for digestion or emulsification of food. The liver has a highly specialized structure that will be discussed in Chapter 70. The salivary glands and the pancreas are compound acinous glands of the type illustrated in Figure 64–2. These glands lie completely outside the walls of the gastrointestinal tract and, in this, differ from all other gastrointestinal glands. The major secreting structures, called *acini,* are lined with secreting glandular cells; these feed into a system of ducts that finally empty through one or more portals into the intestinal tract itself.

BASIC MECHANISMS OF STIMULATION OF THE GASTROINTESTINAL GLANDS

Effect of Local Stimuli. The mechanical presence of food in a particular segment of the gastrointestinal tract usually causes the glands of that region, and often of adjacent regions, to secrete moderate to large quantities of digestive juices. Part of this local effect results from direct stimulation of the surface glandular cells by contact with the food. For instance, the goblet cells on the surface of the epithelium are stimulated mainly in this way. However, most local stimulation of the intestinal glands results from one of the following three methods of stimulation:

(1) Tactile stimulation or chemical irritation of the mucosa can elicit reflexes that pass through the intrinsic nervous plexus of the intestinal wall to stimulate either the mucous cells on the surface or the deeper glands of the mucosa. (2) Distension of the gut can also elicit nervous reflexes that

stimulate secretion. (3) Tactile or chemical stimuli or distension can result in increased motility of the gut, as described in the preceding chapter, and the motility in turn can then increase the rate of secretion.

Autonomic Stimulation of Secretion. *Parasympathetic Stimulation.* Stimulation of the parasympathetic nerves to the alimentary tract almost invariably increases the rates of glandular secretion. This is especially true of the glands in the upper portion of the tract, innervated by the vagus and other cranial parasympathetic nerves — including the salivary, esophageal, and gastric glands, the pancreas, and Brunner's glands in the duodenum — and also of the distal portion of the large intestine, innervated by the pelvic parasympathetic nerves. Secretion in the remainder of the small intestine and in the first two-thirds of the large intestine occurs almost entirely in response to local stimuli.

Sympathetic Stimulation. Stimulation of sympathetic nerves in some parts of the gastrointestinal tract causes a slight increase in secretion by the respective glands. On the other hand, sympathetic stimulation also results in constriction of the blood vessels supplying the glands. Therefore, sympathetic stimulation can have a dual effect: First, sympathetic stimulation alone can slightly increase secretion. But, second, if parasympathetic or hormonal stimulation is causing copious secretion by the glands, superimposed sympathetic stimulation usually reduces the secretion mainly because of reduced blood supply.

Regulation of Glandular Secretion by Hormones. In the stomach and intestine several different *gastrointestinal hormones* help to regulate the volume and character of the secretions. These hormones are liberated from the gastrointestinal mucosa in response to the presence of foods in the lumen of the gut. They then are absorbed into the blood and are carried to glands where they stimulate secretion. This type of stimulation is particularly valuable in increasing the output of gastric juice and pancreatic juice when food enters the stomach or duodenum. Also, hormonal stimulation of the gallbladder wall causes it to empty its stored bile into the duodenum, as discussed in the preceding chapter. Other hormones that are still of doubtful value have also been postulated to stimulate secretion by the glands of the small intestine.

Chemically, the gastrointestinal hormones are polypeptides or polypeptide derivatives.

BASIC MECHANISM OF SECRETION BY GLANDULAR CELLS

Secretion of Organic Substances. Though all the basic mechanisms by which glandular cells form different secretions and then extrude these to the exterior are not known, experimental evidence

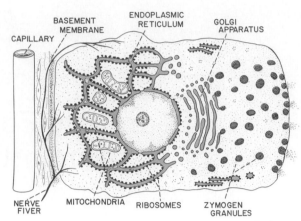

Figure 64–1. Typical function of a glandular cell in formation and secretion of enzymes or other secretory substances.

points to the following basic principles of secretion by glandular cells, as illustrated in Figure 64–1.

(1) The nutrient material needed for formation of the secretion must diffuse or be actively transported from the capillary into the base of the glandular cell.

(2) Many *mitochondria* located inside the cell near its base provide oxidative energy for formation of adenosine triphosphate.

(3) Energy from the adenosine triphosphate (see Chapter 67), along with appropriate nutrients, is then used for synthesis of the organic substances; this synthesis occurs almost entirely on the *endoplasmic reticulum*. The *ribosomes* adherent to this reticulum are specifically responsible for formation of proteins that are to be secreted.

(4) The secretory materials are transported into and through the tubules of the endoplasmic reticulum, passing in about 20 minutes all the way to the vesicles of the Golgi complex, which lies near the secretory ends of the cells.

(5) The materials then are modified, added to, concentrated, and discharged into the cytoplasm in the form of *secretory vesicles* that are stored in the apical ends of the secretory cells.

(6) These vesicles remain stored until nervous or hormonal control signals cause them to extrude their contents through the cell's surface. This probably occurs in the following way: The control signal first increases the cell membrane permeability to calcium, and calcium enters the cell. The calcium in turn causes many of the vesicles to fuse with the cell membrane and then to break open on their outer surfaces, thus emptying their contents to the exterior; this process is called *exocytosis*.

Water and Electrolyte Secretion. A second necessity for glandular secretion is sufficient water and electrolytes to be secreted along with the organic substances. The following is a postulated method by which nervous stimulation causes water and salts to pass through the glandular cells in great profusion, which washes the organic substances through the secretory border of the cells at the same time:

(1) Nerve stimulation has a specific effect on the *basal* portion of the cell membrane to cause active transport of chloride ions to the interior. (2) The resulting increase in electronegativity inside the cell then causes positive ions also to move to the interior of the cell. (3) The excess of both of these ions inside the cell creates an osmotic force which pulls water to the interior, thereby increasing the hydrostatic pressure inside the cell and causing the cell itself to swell. (4) The pressure in the cell then results in minute ruptures of the secretory border of the cell and causes flushing of water, electrolytes, and organic materials out of the secretory end of the glandular cell and into the lumen of the gland.

In support of this theory have been the following findings: First, the nerve endings on glandular cells are principally on the bases of the cells. Second, microelectrode studies show that the electrical potential across the membrane at the base of the cell is between 30 and 40 mv., with negativity on the interior and positivity on the exterior. Parasympathetic stimulation increases this polarization voltage to values some 10 and 20 millivolts more negative than normal. This increase in polarization occurs a second or more after the nerve signal has arrived, indicating that it is caused by movement of negative ions through the membrane to the interior of the cell.

Though this mechanism for secretion is still partly theoretical, it does explain how it would be possible for nerve impulses to regulate secretion. Obviously, hormonal effects on the cell membrane could cause similar results.

LUBRICATING AND PROTECTIVE PROPERTIES OF MUCUS AND ITS IMPORTANCE IN THE GASTROINTESTINAL TRACT

Mucus is a thick secretion composed mainly of water, electrolytes, and a mixture of several glycoproteins, which themselves are composed of large polysaccharides bound with much smaller quantities of protein. Mucus is slightly different in different parts of the gastrointestinal tract, but everywhere it has several important characteristics that make it both an excellent lubricant and a protectant for the wall of the gut. *First,* mucus has adherent qualities that make it adhere tightly to the food or other particles and also to spread as a thin film over the surfaces. *Second,* it has sufficient *body* that it coats the wall of the gut and prevents actual contact of food particles with the

mucosa. *Third,* mucus has a low resistance to slippage so that the particles can slide along the epithelium with great ease. *Fourth*, mucus causes fecal particles to adhere to each other to form the fecal masses that are expelled during a bowel movement. *Fifth,* mucus is strongly resistant to digestion by the gastrointestinal enzymes. And, *sixth,* the glycoproteins of mucus have amphoteric properties and are therefore capable of buffering small amounts of either acids or alkalies; also, mucus often contains moderate quantities of bicarbonate ions, which specifically neutralize acids.

In summary, mucus has the ability to allow easy slippage of food along the gastrointestinal tract and also to prevent excoriative or chemical damage to the epithelium. A person becomes acutely aware of the lubricating qualities of mucus when the salivary glands fail to secrete saliva, for under these circumstances it is extremely difficult to swallow solid food even when it is taken with large amounts of water.

SECRETION OF SALIVA

The Salivary Glands; Characteristics of Saliva. The principal glands of salivation are the *parotid, submaxillary* (also called *submandibular*), and *sublingual* glands; in addition, there are many small *buccal* glands. The daily secretion of saliva normally ranges between 1000 and 1500 milliliters, as shown in Table 64–1.

Saliva contains two major types of protein secretion: (1) a *serous secretion* containing *ptyalin* (an α-amylase), which is an enzyme for digesting starches, and (2) *mucous secretion* containing *mucin* for lubricating purposes. The parotid glands secrete entirely the serous type, and the submaxillary glands secrete both the serous type and mucus. The sublingual and buccal glands secrete only mucus. Saliva has a pH between 6.0 and 7.4, a favorable range for the digestive action of ptyalin.

TABLE 64–1 DAILY SECRETION OF INTESTINAL JUICES

	Daily Volume (ml.)	*pH*
Saliva	1200	6.0–7.0
Gastric secretion	2000	1.0–3.5
Pancreatic secretion	1200	8.0–8.3
Bile	700	7.8
Succus entericus	2000	7.8–8.0
Brunner's gland secretion	50(?)	8.0–8.9
Large intestinal secretion	60	7.5–8.0
Total	7210	

Secretion of Ions in the Saliva. Saliva contains especially large quantities of potassium and bicarbonate ions. On the other hand, the concentrations of both sodium and chloride ions are considerably less in saliva than in plasma. One can understand these special concentrations of ions in the saliva from the following description of the mechanism for secretion of saliva.

Figure 64–2 illustrates secretion by the submaxillary gland, a typical compound gland containing both *acini* and *salivary ducts*. Salivary secretion is a two-stage operation; the first stage involves the acini and the second, the salivary ducts. The acini secrete a *primary secretion* that contains ptyalin and/or mucin in a solution of ions in concentrations not greatly different from those of typical extracellular fluid. However, as the primary secretion flows through the ducts, two major active transport processes take place that markedly modify the ionic composition of the saliva. First, in the initial portion of the salivary ducts, near their junctions with the acini, *bicarbonate ion* is actively secreted. This process is catalyzed by *carbonic anhydrase* in the epithelial cells, which convert carbon dioxide and water to carbonic acid, then secrete the bicarbonate ion while reabsorbing *chloride ions*. Second, *sodium ions* are actively reabsorbed from all the salivary ducts, and *potassium ions* are actively secreted, but at a slower rate, in exchange for the sodium. Therefore, the sodium concentration of the saliva becomes greatly reduced while the potassium ion concentration becomes increased. The great excess of sodium reabsorption over potassium secretion creates negativity of about -70 mV. in the salivary ducts, and this causes still far more chloride ions to be reabsorbed passively.

The net result of these active transport processes is that, under resting conditions, the concentrations of sodium and chloride ions in the saliva are only about 15 mEq./liter each, approximately 1/7 to 1/10 their concentrations in plasma. On the other hand, the concentration of potassium ions is about 30 mEq./liter, and 7 times as great as its concentration in plasma, and the concentration of bicarbonate ions is 50 to 90 mEq./liter, about 2 to 4 times that of plasma.

During maximal salivation, the salivary ionic concentrations change considerably because the rate of formation of primary secretion by the acini can increase as much as 20-fold. As a result this secretion then flows through the ducts so rapidly that the ductal reconditioning of the secretion is considerably reduced. Therefore, when copious quantities of saliva are secreted, the sodium chloride concentration rises to about 1/2 to 2/3 that of plasma, while the potassium concentration falls to only 4 times that of plasma.

In the presence of excess aldosterone secretion, the sodium and chloride reabsorption and the potassium secretion become greatly increased so that the sodium chloride concentration in the saliva is sometimes reduced almost to zero, while the potassium concentration increases still more.

Because of the high potassium ion concentration of saliva, in any abnormal state in which the saliva is lost to the exterior of the body for long periods of time a person can develop serious depletion of potassium ions in the body, leading eventually to serious hypokalemia and paralysis, as was discussed in Chapter 10.

Function of Saliva for Oral Hygiene. Under basal conditions, between 0.5 and 1 ml./min. of saliva, almost entirely of the mucous type, is secreted all the time. This secretion plays an exceedingly important role in maintaining healthy oral tissues. The mouth is loaded with pathogenic bacteria that can easily destroy tissues and can also cause dental caries. However, saliva helps to prevent the deteriorative processes in several ways: First, the flow of saliva itself helps to wash away the pathogenic bacteria, as well as the food particles that provide their metabolic support. Second, the saliva also contains several factors that actually destroy bacteria. One of these is *thiocyanate ions* and another is several *proteolytic enzymes* that (a) attack the bacteria, (b) aid the thiocyanate ions in entering the bacteria where they in turn become bactericidal, and (c) digest food particles, thus helping further to remove the bacterial metabolic support. And, third, saliva often contains significant amounts of protein antibodies that can destroy the oral bacteria, including those that cause dental caries.

Therefore, in the absence of salivation, the oral tissues become ulcerated and otherwise infected, and caries of the teeth becomes rampant.

Nervous Regulation of Salivary Secretion. Figure 64–3 illustrates the nervous pathways for regulation of salivation, showing that the salivary glands are controlled by *parasympathetic nervous signals* from the *salivatory nuclei*. The salivatory nuclei are located approximately at the juncture of the medulla and pons and are excited by both taste

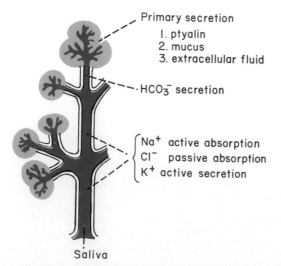

Figure 64–2. Formation and secretion of saliva by a salivary gland.

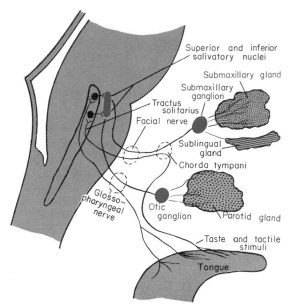

Figure 64–3. Nervous regulation of salivary secretion.

and tactile stimuli from the tongue and other areas of the mouth. Many taste stimuli, especially the sour taste, elicit copious secretion of saliva — often as much as 5 to 10 ml. per minute or 8 to 20 times the basal rate of secretion. Also, certain tactile stimuli, such as the presence of smooth objects in the mouth (a pebble, for instance), cause marked salivation, while rough objects cause less salivation and occasionally even inhibit salivation.

Salivation can also be stimulated or inhibited by impulses arriving in the salivatory nuclei from higher centers of the central nervous system. For instance, when a person smells or eats favorite foods, salivation is far greater than when disliked food is smelled or eaten. The *appetite area* of the brain that partially regulates these effects is located in close proximity to the parasympathetic centers of the anterior hypothalamus, and it functions to a great extent in response to signals from the taste and smell areas of the cerebral cortex or amygdala.

Finally, salivation also occurs in response to reflexes originating in the stomach and upper intestines — particularly when very irritating foods are swallowed or when a person is nauseated because of some gastrointestinal abnormality. The swallowed saliva presumably helps to remove the irritating factor in the gastrointestinal tract by diluting or neutralizing the irritant substances.

ESOPHAGEAL SECRETION

The esophageal secretions are entirely mucoid in character and principally provide lubrication for swallowing. The main body of the esophagus is lined with many simple mucous glands, but at the gastric end, and to a lesser extent in the initial portion of the esophagus, there are many compound mucous glands. The mucus secreted by the compound glands in the upper esophagus prevents mucosal excoriation by the newly entering food, while the compound glands near the esophagogastric juncture protect the esophageal wall from digestion by gastric juices that reflux into the lower esophagus. Despite this protection, a peptic ulcer at times may occur at the gastric end of the esophagus.

GASTRIC SECRETION

CHARACTERISTICS OF THE GASTRIC SECRETIONS

In addition to the mucus-secreting cells that line the surface of the stomach, the stomach mucosa has two different types of tubular glands: the *oxyntic* or *gastric glands* and the *pyloric glands.* The oxyntic glands secrete *hydrochloric acid, pepsinogen,* and *mucus,* and the pyloric glands secrete mainly *mucus* for protection of the pyloric mucosa but also a different type of *pepsinogen* as well as the hormone *gastrin.* The oxyntic glands are located everywhere in the mucosa of the body and fundus of the stomach except along the lesser curvature, and the pyloric glands are located in the antral portion of the stomach. In addition, a few mucus-secreting *cardiac glands,* which are almost identical to the pyloric glands, are located about 1 cm. immediately surrounding the entry point of the esophagus.

The Secretions from the Gastric Glands. A typical oxyntic gland is shown in Figure 64–4. It is composed of three different types of cells: the *mucous neck cells,* which secrete mainly mucus

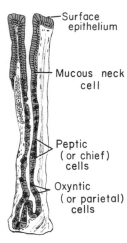

Figure 64–4. An oxyntic gland from the body or fundus of the stomach.

but also some pepsinogen; the *peptic* (or *chief*) *cells,* which secrete large quantities of pepsinogen; and the *oxyntic* (or *parietal*) *cells*, which secrete hydrochloric acid and which lie mainly behind the mucous neck cells or, less often, behind the chief cells. A postulated mechanism for secretion of mucus and pepsinogen was given earlier in the chapter and was depicted in Figure 64–1. However, secretion of hydrochloric acid by the parietal cells involves particular problems which require further consideration as follows:

Basic Mechanism of Hydrochloric Acid Secretion. The parietal cells secrete an electrolytic solution containing 160 millimols of hydrochloric acid per liter, which is almost exactly isotonic with the body fluids. The pH of this acid solution is approximately 0.8, thus illustrating its extreme acidity. At this pH the hydrogen ion concentration is about 3 million times that of the arterial blood. And to concentrate the hydrogen ions this tremendous amount requires over 1500 calories of energy per liter of gastric juice, as discussed in Chapter 4 in relation to membrane transport mechanisms.

Figure 64–5 illustrates the basic structure of an oxyntic cell, showing that it contains a system of *intracellular canaliculi.* The hydrochloric acid is formed at the membranes of these canaliculi and then conducted through openings to the exterior.

Different suggestions for the precise mechanism of hydrochloric acid formation have been offered. One of these is illustrated in Figure 64–6 and consists of the following steps:

1. Chloride ion is actively transported from the cytoplasm of the oxyntic cell into the lumen of the canaliculus. This creates a negative potential of -40 to -70 millivolts in the canaliculus, which in turn causes passive diffusion of positively charged potassium ions from the cell cytoplasm also into the canaliculus. Thus, in effect, potassium chloride enters the canaliculus.

2. Water is dissociated into hydrogen ions and

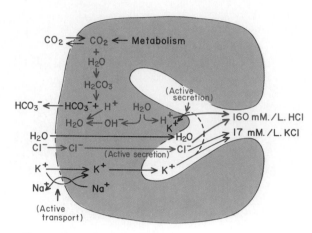

Figure 64–6. Postulated mechanism for the secretion of hydrochloric acid.

hydroxyl ions in the cell cytoplasm. The hydroxyl ion is then actively secreted into the canaliculus in exchange for potassium ions, this active exchange process being catalyzed by H^+-K^+ ATPase. Thus, most of the potassium ions that had been secreted along with the chloride ions are reabsorbed, and hydrogen ions take their place in the canaliculus.

3. Water passes through the cell and into the canaliculus by osmosis. Thus the final secretion from the canaliculus is a solution containing hydrochloric acid in a concentration of 160 millimoles per liter and potassium chloride in a concentration of 17 millimoles per liter.

4. Finally, carbon dioxide, either formed during metabolism in the cell or entering the cell from the blood, combines with water under the influence of carbonic anhydrase to form carbonic acid. This, in turn, dissociates into bicarbonate ion and hydrogen ion. The hydrogen ion combines with the hydroxyl ion released in Step one to form water. The bicarbonate ion, in turn, diffuses out of the cell into the blood. The importance of carbon dioxide in the chemical reactions for formation of hydrochloric acid is illustrated by the fact that carbonic anhydrase inhibition by the drug *acetazolamide* almost completely blocks the formation of hydrochloric acid.

Activation of Pepsinogen. Several different types of pepsinogen are secreted by the peptic and mucous cells of the gastric glands. Even so, all the pepsinogens perform essentially the same functions. When the pepsinogens are first secreted, they have no digestive activity. However, as soon as they come in contact with hydrochloric acid, and especially when they come in contact with previously formed pepsin as well as hydrochloric acid, they are immediately activated to form active *pepsin.* In this process, the pepsinogen molecule, having a molecular weight of about 42,500, is split to the pepsin molecule, having a molecular weight of about 35,000.

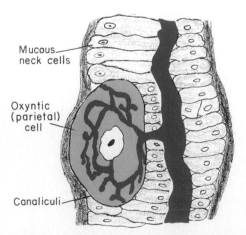

Figure 64–5. Anatomy of the canaliculi in an oxyntic (parietal) cell.

Pepsin is an active proteolytic enzyme in a highly acid medium (optimum pH = 2.0), but above a pH of about 5 it has little proteolytic activity and soon becomes completely inactivated. Therefore, hydrochloric acid secretion is equally as necessary as pepsin secretion for protein digestion in the stomach; this will be discussed in the following chapter.

Secretion of Other Enzymes. Small quantities of other enzymes are also secreted in the stomach juices, including *gastric lipase* and *gastric amylase*. Gastric lipase is of little quantitative importance and is actually a *tributyrase,* for its principal activity is on tributyrin, which is butterfat; it has almost no lipolytic activity on the other fats. Gastric amylase plays a very minor role in digestion of starches.

Secretion of Intrinsic Factor. A substance called *intrinsic factor,* that is essential for absorption of vitamin B_{12} in the ileum is secreted by the *oxyntic cells* along with the secretion of hydrochloric acid. Therefore, when the acid-producing cells of the stomach are destroyed, which frequently occurs in chronic gastritis, the person not only develops *achlorhydria* but also develops *pernicious anemia* because of failure of maturation of the red blood cells in the absence of vitamin B_{12} stimulation of the bone marrow. This was discussed in detail in Chapter 5.

Secretion of Mucus in the Stomach. The pyloric glands are structurally similar to the oxyntic glands, but contain few peptic and oxyntic cells. Instead, they contain almost entirely mucous cells that are identical with the mucous neck cells of the gastric glands. These cells secrete a small amount of pepsinogen, as discussed above, and an especially large amount of thin mucus which protects the stomach wall from digestion by the gastric enzymes.

In addition, the surface of the stomach mucosa between glands has a continuous layer of mucous cells that secrete large quantities of a far more *viscid and alkaline mucus* that coats the mucosa with a mucous gel layer often more than 1 mm. thick, thus providing a major shell of protection for the stomach wall as well as contributing to lubrication of food transport. Even the slightest irritation of the mucosa directly stimulates the mucous cells to secrete copious quantities of this thick, viscid mucus.

REGULATION OF GASTRIC SECRETION BY NERVOUS AND HORMONAL MECHANISMS

Gastric secretion is regulated by both nervous and hormonal mechanisms, nervous regulation being effected through the parasympathetic fibers of the vagus nerves as well as through local intrinsic nerve plexus reflexes, and hormonal regulation taking place by means of the hormone *gastrin.* Thus, regulation of gastric secretion is different from the regulation of salivary secretion, which is effected entirely by nervous mechanisms.

Vagal Stimulation of Gastric Secretion

Nervous signals to cause gastric secretion originate in the dorsal motor nuclei of the vagi and pass via the vagus nerves to the intrinsic nerve plexus of the stomach and thence to the oxyntic glands. In response, these glands secrete vast quantities of both pepsin and acid, but with a higher proportion of pepsin than in gastric juice elicited in other ways. Also, vagal signals to all the mucus-secreting glands and epithelial lining cells cause increased mucus secretion as well.

Still another effect of vagal stimulation is to cause the antral part of the stomach mucosa to secrete the hormone *gastrin.* As will be explained in the following paragraphs, this hormone then acts on the gastric glands to cause additional flow of highly acidic gastric juice. Thus, vagal stimulation excites stomach secretion both directly by stimulation of the gastric glands and indirectly through the gastrin mechanism.

Stimulation of Gastric Secretion by Gastrin

When food enters the stomach, it causes the antral portion of the stomach mucosa to secrete the hormone gastrin. This hormone is secreted by *gastrin cells*, also called *G cells,* in the pyloric glands and to a lesser extent in the proximal glands of the duodenum. Gastrin is a large peptide secreted in two forms, a large form called *G-34*, containing 34 amino acids, and a smaller form, *G-17*, containing 17 amino acids. Though both of these are important, the smaller form is more abundant.

The food causes release of this hormone in two ways: (1) The actual bulk of the food distends the stomach, and this causes the hormone gastrin to be released from the antral mucosa. (2) Certain substances called secretagogues — such as food extractives, partially digested proteins, alcohol (in low concentration), caffeine, and so forth — also cause gastrin to be liberated from the antral mucosa.

Both of these stimuli — the distension and the chemical action of the secretagogues — elicit gastrin release by means of a local nerve reflex. That is, they stimulate sensory nerve fibers in the stomach epithelium which in turn synapse with the intrinsic nerve plexus. This then transmits efferent signals to the gastrin cells, causing them to secrete the gastrin. Therefore, any factor that blocks this reflex will also block the formation of gastrin. For instance, anesthetization of the gastric mucosa to block the sensory stimuli will prevent gastrin release; administration of atropine, which blocks the action on the gastrin cells of the acetylcholine released by the intrinsic nerves and also by the vagus nerves, will also prevent gastrin release.

Gastrin is absorbed into the blood and carried to the gastric glands where it stimulates mainly the oxyntic cells and to much less extent the peptic cells, also. The oxyntic cells increase their rate of hydrochloric acid secretion as much as eight-fold, and the chief cells increase their rate of enzyme secretion two- to four-fold.

The rate of secretion in response to gastrin is somewhat less than to vagal stimulation, 200 ml. per hour in contrast to about 500 ml. per hour, indicating that the gastrin mechanism is a less potent mechanism for stimulation of stomach secretion than is vagal stimulation. However, the gastrin mechanism usually continues for several hours in contrast to a much shorter period of time for vagal stimulation. Therefore, as a whole, it is likely that the gastrin mechanism is equally as important as, if not more important than, the vagal mechanism for control of gastric secretion. Yet, when both of these work together, the total secretion is much greater than the sum of the individual secretions caused by each of the two mechanisms. In other words, *the two mechanisms multiply each other rather than simply add to each other*.

Histamine, an amino acid derivative, also stimulates gastric secretion. When the H-2 type of histamine receptors (but not the H-1 receptors) of the gastric glands is blocked with the histamine antagonist drug *cimetidine*, not only does this block the stimulatory effect of histamine on gastric secretion but the effect of gastrin as well. Therefore, it is believed that local gastric mucosal histamine is a prerequisite for the stimulatory function of gastrin.

Below are illustrated the amino. acid compositions of *gastrin-17* and also of *cholecystokinin* and *secretin*, which will be discussed later in the chapter. Note that all are polypeptides and that the last five amino acids in the gastrin and cholecystokinin molecular chains are exactly the same. The activity of gastrin resides in the terminal four amino acids and in the terminal eight amino acids for cholecystokinin; all the amino acids are essential in secretin.

Feedback Inhibition of Gastric Acid Secretion. When the acidity of the gastric juices increases to a pH of 2.0, the gastrin mechanism for stimulating gastric secretion becomes totally blocked. This effect probably results from two different factors. First, greatly enhanced acidity depresses or blocks the extraction of gastrin itself from the antral mucosa. Second, the acid seems to extract an inhibitory hormone from the gastric mucosa or to cause an inhibitory reflex that inhibits gastric acid secretion.

Obviously, this feedback inhibition of the gastric glands plays an important role in protecting the stomach against excessively acid secretions, which would readily cause peptic ulceration. However, in addition to this protective effect, the feedback mechanism is also important in maintaining optimal pH for function of the peptic enzymes in the digestive process because whenever the pH rises gastrin begins to be secreted again.

The Three Phases of Gastric Secretion

Gastric secretion is said to occur in three separate phases (as illustrated in Fig. 64–7): a *cephalic phase*, a *gastric phase*, and an *intestinal phase*. However, as will be apparent in the following discussion, these three phases in reality fuse together.

The Cephalic Phase. The cephalic phase of gastric secretion occurs even before food enters the stomach. It results from the sight, smell, thought, or taste of food;

Glu- Gly- Pro- Trp- Leu- Glu- Glu- Glu- Glu- Glu- Ala- Tyr- Gly- Trp- Met- Asp- Phe- NH$_2$

$\qquad\qquad\qquad\qquad\qquad\qquad\qquad\qquad\qquad$ HSO$_3$

GASTRIN

Lys- (Ala, Gly, Pro, Ser)- Arg- Val- (Ile, Met, Ser)- Lys- Asn- (Asn, Gln, His, Leu$_2$,

Pro, Ser$_2$)- Arg- Ile- (Asp, Ser)- Arg- Asp- Tyr- Met- Gly- Trp- Met- Asp- Phe- NH$_2$

$\qquad\qquad\qquad\qquad\qquad$ HSO$_3$

CHOLECYSTOKININ

His- Ser- Asp- Gly- Thr- Phe- Thr- Ser- Glu- Leu- Ser- Arg- Leu- Arg- Asp- Ser-

$\qquad\qquad\qquad$ Ala- Arg- Leu- Gln- Arg- Leu- Leu- Gln- Gly- Leu- Val- NH$_2$

SECRETIN

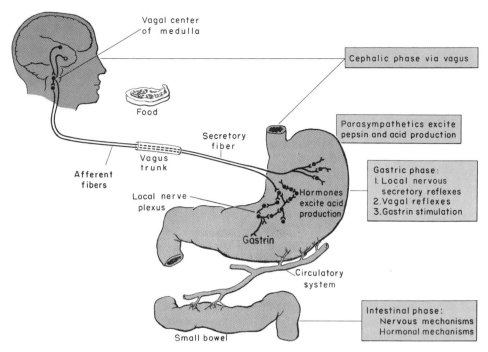

Figure 64–7. The phases of gastric secretion and their regulation.

and the greater the appetite, the more intense is the stimulation. Neurogenic signals causing the cephalic phase of secretion can originate in the cerebral cortex or in the appetite centers of the amygdala or hypothalamus. They are transmitted through the dorsal motor nuclei of the vagi to the stomach. This phase of secretion probably accounts for less than one-tenth of the gastric secretion normally associated with eating a meal.

The Gastric Phase.   Once the food enters the stomach, it excites the gastrin mechanism, which in turn causes secretion of gastric juice that continues throughout the several hours that the food remains in the stomach.

In addition, the presence of food in the stomach also causes (a) local reflexes in the intrinsic nerve plexus of the stomach and (b) vagovagal reflexes that pass all the way to the brain stem and back to the stomach; both these reflexes cause parasympathetic stimulation of the gastric glands and add to the secretion caused by the gastrin mechanism.

The gastric phase of secretion accounts for more than two-thirds of the total gastric secretion associated with eating a meal and, therefore, accounts for most of the total daily gastric secretion of about 2000 ml.

The Intestinal Phase.   Presence of food in the upper portion of the small intestine, particularly in the duodenum, also causes the stomach to secrete small amounts of gastric juice. This probably results partly from the fact that small amounts of gastrin, called *enteric gastrin*, are also released by the duodenal mucosa in response to distension or chemical stimuli of the same type as those that stimulate the stomach gastrin mechanism. However, it is probable that several other hormones or reflexes also play minor roles in causing secretion of gastric juice.

Inhibition of Gastric Secretion by Intestinal Factors

Though chyme stimulates gastric secretion during the intestinal phase of secretion, it paradoxically often partially inhibits secretion during the gastric phase. This results from at least two different influences:

1. The presence of food in the small intestine initiates an *enterogastric reflex*, transmitted through the intrinsic nerve plexus, the sympathetic nerves, and the vagus nerves, that inhibits stomach secretion. This reflex can be initiated by distension of the small bowel, the presence of acid in the upper intestine, the presence of protein breakdown products, or irritation of the mucosa. This is part of the complex mechanism discussed in the preceding chapter for slowing down stomach emptying when the intestines are already filled.

2. The presence of acid, fat, protein breakdown products, hyper- or hypo-osmotic fluids, or any irritating factor in the upper small intestine causes the release of several intestinal hormones. Two of these are *secretin* and *cholecystokinin*. Both of these are especially important for control of pancreatic secretion, and cholecystokinin is also important in promoting emptying of the gallbladder. However, in addition to these effects, they both oppose the stimulatory effects of gastrin. Also, another hormone, *gastric inhibitory peptide,* has a moderate effect on inhibiting gastric secretion.

The functional purpose of the inhibition of gastric secretion by intestinal factors is probably to slow the release of chyme from the stomach when the small intestine is already filled. In fact, the enterogastric reflex and also these inhibitory hormones reduce stomach motility at the same time that they reduce gastric secretion, as was discussed in the previous chapter.

Secretion During the Interdigestive Period.   The

stomach secretes only a few milliliters of gastric juice per hour during the "interdigestive period" when little or no digestion is occurring anywhere in the gut. Furthermore, the secretion that does occur is almost entirely of the so-called nonoxyntic type, meaning that it is composed mainly of mucus containing very little pepsin and either no or very little acid. In fact, on occasion, the fluid is actually slightly alkaline, containing moderate quantities of sodium bicarbonate. However, strong emotional stimuli frequently increase the interdigestive secretion to 50 ml. or more of highly peptic and highly acidic gastric juice per hour, in very much the same manner that the cephalic phase of gastric secretion excites secretion at the onset of a meal. This increase of secretion during the presence of emotional stimuli is believed to be one of the factors in the development of peptic ulcers, as will be discussed in Chapter 66.

PANCREATIC SECRETION

The pancreas, which lies parallel to and beneath the stomach, is a very large compound gland with a structure almost identical to that of the salivary glands. Digestive enzymes are secreted by the acini, and large volumes of sodium bicarbonate solution are secreted by the small ductules leading from the acini. The combined product then flows through a long pancreatic duct that usually joins the hepatic duct immediately before it empties into the duodenum through the sphincter of Oddi. However, on occasion this duct empties separately into the duodenum. Pancreatic juice is secreted mainly in response to the presence of chyme in the upper portions of the small intestine, and the characteristics of the pancreatic juice are determined to a great extent by the types of food in the chyme.

Characteristics of Pancreatic Juice. Pancreatic juice contains enzymes for digesting all three major types of food: proteins, carbohydrates, and fats. It also contains large quantities of bicarbonate ions which play an important role in neutralizing the acid chyme emptied by the stomach into the duodenum.

The proteolytic enzymes are *trypsin, chymotrypsin, carboxypolypeptidase, ribonuclease,* and *deoxyribonuclease.* By far the most abundant of these is trypsin. The first three split whole and partially digested proteins into small peptides or amino acids, while the nucleases split the two types of nucleic acids: ribonucleic and deoxyribonucleic acids.

The digestive enzyme for carbohydrates is *pancreatic amylase*, which hydrolyzes starches, glycogen, and most other carbohydrates except cellulose to form disaccharides.

The main enzymes for fat digestion are *pancreatic lipase,* which is capable of hydrolyzing neutral fat into glycerol and fatty acids, and *cholester-*

ol esterase, which causes hydrolysis of cholesterol esters.

The proteolytic enzymes when synthesized in the pancreatic cells are in the inactive forms *trypsinogen, chymotrypsinogen,* and *procarboxypolypeptidase,* which are all enzymatically inactive. These become activated only after they are secreted into the intestinal tract. Trypsinogen is activated by an enzyme called *enterokinase,* which is secreted by the intestinal mucosa when chyme comes in contact with the mucosa. Also, trypsinogen can be activated by trypsin that has already been formed. Chymotrypsinogen is activated by trypsin to form chymotrypsin, and procarboxypolypeptidase is activated in a similar manner.

Secretion of Trypsin Inhibitor. It is important that the proteolytic enzymes of the pancreatic juice not become activated until they have been secreted into the intestine, for the trypsin and the other enzymes would digest the pancreas itself. Fortunately, the same cells that secrete the proteolytic enzymes into the acini of the pancreas secrete simultaneously another substance called *trypsin inhibitor*. This substance is stored in the cytoplasm of the glandular cells surrounding the enzyme granules, and it prevents activation of trypsin both inside the secretory cells and in the acini and ducts of the pancreas. Since it is trypsin that activates the other pancreatic proteolytic enzymes, trypsin inhibitor also prevents the subsequent activation of all these.

However, when the pancreas becomes severely damaged or when a duct becomes blocked, large quantities of pancreatic secretion become pooled in the damaged areas of the pancreas. Under these conditions, the effect of trypsin inhibitor is sometimes overwhelmed, in which case the pancreatic secretions rapidly become activated and literally digest the entire pancreas within a few hours, giving rise to the condition called *acute pancreatitis.* This often is lethal because of accompanying shock, and even if not lethal it leads to a lifetime of pancreatic insufficiency.

Secretion of Bicarbonate Ions. The enzymes of the pancreatic juice are secreted entirely by the acini of the pancreatic glands. On the other hand, two other important components of pancreatic juice, water and bicarbonate ion, are secreted mainly by the epithelial cells of the small ductules leading from the acini. We shall see below that the stimulatory mechanisms for (a) enzyme production and (b) production of water and bicarbonate ions are also quite different. When the pancreas is stimulated to secrete copious quantities of pancreatic juice — that is, copious quantities of the water and bicarbonate ions — the bicarbonate ion concentration can rise to as high as 145 mEq./liter, a value approximately 5 times that of bicarbonate

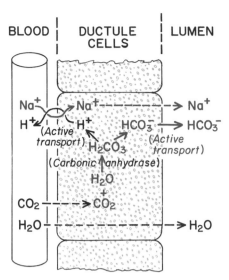

Figure 64-8. Secretion of iso-osmotic sodium bicarbonate solution by the pancreatic ductules.

ions in the plasma. Obviously, this provides a large quantity of alkaline ion in the pancreatic juice that serves to neutralize acid emptied into the duodenum from the stomach.

The cellular mechanism for secreting sodium bicarbonate solution into the pancreatic ductules is illustrated in Figure 64-8. The major steps in this process are:

1. Carbon dioxide diffuses to the interior of the cell from the blood and combines with water under the influence of carbonic anhydrase to form carbonic acid. The carbonic acid in turn dissociates into bicarbonate ion and hydrogen ion. Then the bicarbonate ion is actively transported through the *luminal* border of the cell into the lumen of the duct.

2. The hydrogen ion formed by dissociation of carbonic acid inside the cell is exchanged for sodium ion through the *blood* border of the cell, also by an active transport process. The sodium ion in turn diffuses through the *luminal* border into the pancreatic duct to provide electrical neutrality for the bicarbonate ion.

3. The movement of sodium and bicarbonate ions from the blood to the lumen creates an osmotic gradient that causes osmosis of water also into the pancreatic duct, thus forming the bicarbonate solution.

REGULATION OF PANCREATIC SECRETION

Pancreatic secretion, like gastric secretion, is regulated by both nervous and hormonal mechanisms. However, in this case, hormonal regulation is by far the more important.

Nervous Regulation. When the cephalic and gastric phases of stomach secretion occur, parasympathetic impulses are simultaneously transmitted along the vagus nerves to the pancreas, resulting in *acetylcholine* release followed by secretion of moderate amounts of enzymes into the pancreatic acini. However, little secretion flows through the pancreatic ducts to the intestine because only small amounts of water and electrolytes are secreted along with the enzymes. Therefore, most of the enzymes are temporarily stored in the acini.

Hormonal Regulation. After chyme enters the small intestine, pancreatic secretion becomes copious, mainly in response to the hormone *secretin*. In addition, a second hormone, *cholecystokinin*, causes greatly increased secretion of enzymes.

Stimulation of Secretion of Copious Quantities of Bicarbonate by Secretin – Neutralization of the Acidic Chyme. Secretin is a polypeptide containing 27 amino acids (molecular weight of about 3400) that is present in the mucosa of the upper small intestine in an inactive form *prosecretin*. When chyme enters the intestine, it causes the release and activation of secretin, which is subsequently absorbed into the blood. The one constituent of chyme that causes greatest secretin release is hydrochloric acid, though almost any type of food will cause at least some release.

Secretin causes the pancreas to secrete large quantities of fluid containing a high concentration of bicarbonate ion (up to 145 mEq./liter) but a low concentration of chloride ion. However, this fluid contains almost no enzymes when the pancreas is stimulated only by secretin because secretin does not stimulate the acinar cells.

The secretin mechanism is especially important for two reasons: *First*, secretin is released in especially large quantities from the mucosa of the small intestine whenever the pH of the duodenal contents falls below 4.0 to 5.0. This immediately causes large quantities of pancreatic juice containing abundant amounts of sodium bicarbonate to be secreted, which results in the following reaction in the duodenum:

$$HCl + NaHCO_3 \rightarrow NaCl + H_2CO_3$$

The carbonic acid immediately dissociates into carbon dioxide and water, and the carbon dioxide is absorbed into the blood and expired through the lungs, thus leaving a neutral solution of sodium chloride in the duodenum. In this way, the acid contents emptied into the duodenum from the stomach become neutralized, and the peptic activity of the gastric juices is immediately blocked. Since the mucosa of the small intestine cannot withstand the intense digestive action of gastric juice, this is a highly important and even essential protective mechanism against the development of

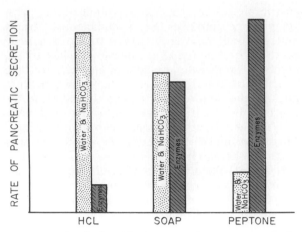

Figure 64–9. Sodium bicarbonate and enzyme secretion by the pancreas caused respectively by the presence of acid, fat (soap), or peptone solutions in the duodenum.

duodenal ulcers, which will be discussed in further detail in Chapter 66.

A *second* importance of bicarbonate secretion by the pancreas is to provide an appropriate pH for action of the pancreatic enzymes. All these function optimally in a slightly alkaline or neutral medium. The pH of the sodium bicarbonate secretion averages 8.0.

Cholecystokinin – Control of Enzyme Secretion by the Pancreas. The presence of food in the upper small intestine also causes a second hormone, cholecystokinin, a polypeptide containing 33 amino acids, to be released from the mucosa. This results especially from the presence of proteoses and peptones, which are products of partial protein digestion, and of fats; however, acids will also cause its release in smaller quantities. Cholecystokinin, like secretin, passes by way of the blood to the pancreas but, instead of causing sodium bicarbonate secretion, causes secretion of large quantities of digestive enzymes, which is similar to the

effect of vagal stimulation but even more pronounced.

The differences between the stimulatory effects of secretin and cholecystokinin are shown in Figure 64–9, which illustrates (a) intense sodium bicarbonate secretion in response to acid in the duodenum, (b) a dual effect in response to soap (a fat), and (c) intense enzyme secretion in response to peptones.

Figure 64–10 summarizes the overall regulation of pancreatic secretion. The total amount secreted each day is about 1200 ml.

SECRETION OF BILE BY THE LIVER

The secretion of bile by the liver will be discussed in more detail in relation to the overall function of the liver in Chapter 70. Basically, bile contains no digestive enzyme and is important for digestion only because of the presence of bile salts which (1) help to emulsify fat globules so that they can be digested by the intestinal lipases and (2) transport the end-products of fat digestion to the intestinal villi so that they can be absorbed into the lymphatics.

Bile is secreted continually by the liver rather than intermittently as in the case of most other gastrointestinal secretions, but the bile is stored in the gallbladder until it is needed in the gut. The gallbladder then empties the bile into the intestine in response to *cholecystokinin,* the same hormone that causes enzyme secretion by the pancreas. Therefore, bile becomes available to aid in the processes of fat digestion. This mechanism of gallbladder emptying was discussed in the preceding chapter.

The rate of bile secretion can be altered in response to four different effects: (1) Vagal stimulation can sometimes more than double the secretion. (2) Secretin can increase bile secretion as much as 80 per cent, mainly by stimulating the small bile ducts to secrete sodium bicarbonate solution — the same effect that secretin has in the pancreas. The sodium bicarbonate, in turn, helps to keep the bile salts in solution in the bile. (3) The greater the liver blood flow (up to a point), the greater the

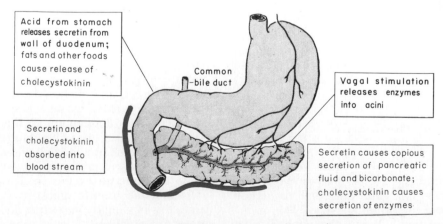

Figure 64–10.   Regulation of pancreatic secretion.

secretion. And (4) the presence of large amounts of bile salts in the blood increases the rate of liver secretion proportionately.

Most of the bile salts secreted in the bile are reabsorbed by the ileum and then resecreted by the liver over and over again, performing their actions in relation to fat digestion and absorption about 20 times before finally being lost in the feces. This continual recycling of bile salts is important for maintaining even the normal daily flow of bile because the liver otherwise would not be able to synthesize nearly enough of the salts to perform their required functions. When the salts are lost to the exterior through a bile fistula rather than being reabsorbed, the rate of bile secretion becomes reduced to as little as one-third normal. And, in the absence of bile in the intestines, as much as one-half of the fat in the food fails to be digested and absorbed, thus causing *steatorrhea* (fatty stools).

The daily volume of bile production averages 600 to 1000 ml.

SECRETIONS OF THE SMALL INTESTINE

SECRETION OF MUCUS BY BRUNNER'S GLANDS AND BY MUCOUS CELLS OF THE INTESTINAL SURFACE

An extensive array of compound mucous glands, called *Brunner's glands*, is located in the first few centimeters of the duodenum, mainly between the pylorus and the papilla of Vater where the pancreatic juices and bile empty into the duodenum. These glands secrete mucus in response to: (a) direct tactile stimuli or irritating stimuli of the overlying mucosa, (b) vagal stimulation, which causes secretion concurrently with increase in stomach secretion, and (c) intestinal hormones, especially secretin. The function of the mucus secreted by Brunner's glands is to protect the duodenal wall from digestion by the gastric juice, and their rapid and intense response to irritating stimuli is especially geared to this purpose.

Brunner's glands are inhibited by sympathetic stimulation; therefore, such stimulation is likely to leave the duodenal bulb unprotected and is perhaps one of the factors that cause this area of the gastrointestinal tract to be the site of peptic ulcers in about 50 per cent of the cases.

Mucus is also secreted in large quantities by goblet cells located extensively over the surface of the intestinal mucosa. This secretion results principally from direct tactile or chemical stimulation of the mucosa by the chyme. Additional mucus is also secreted by the goblet cells in the intestinal pits, the crypts of Lieberkühn. This secretion is controlled mainly by local nervous reflexes.

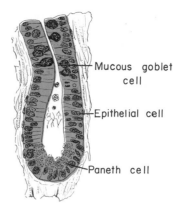

Figure 64–11. A crypt of Lieberkühn, found in all parts of the small intestine between the villi, which secretes almost pure extracellular fluid.

SECRETION OF THE INTESTINAL DIGESTIVE JUICES – THE CRYPTS OF LIEBERKÜHN

Located on the entire surface of the small intestine, with the exception of the Brunner's gland area of the duodenum, are small pits called *crypts of Lieberkühn,* one of which is illustrated in Figure 64–11. The intestinal secretions are formed by the epithelial cells in these crypts at a rate of about 2000 ml. per day. The secretions are almost pure extracellular fluid, and they have a neutral pH in the range of 6.5 to 7.5. They are rapidly reabsorbed by the villi. This circulation of fluid from the crypts to the villi obviously supplies a watery vehicle for absorption of substances from the chyme as it comes in contact with the villi, which is one of the primary functions of the small intestine.

Enzymes in the Small Intestinal Secretion. When secretions of the small intestine are collected without cellular debris, they have almost no enzymes. However, the epithelial cells of the mucosa do contain digestive enzymes that are believed to digest food substances *while* they are being absorbed through the epithelium. These enzymes are the following: (1) several different *peptidases* for splitting small peptides into amino acids, (2) four enzymes for splitting disaccharides into monosaccharides — *sucrase, maltase, isomaltase,* and *lactase,* and (3) small amounts of *intestinal lipase* for splitting neutral fats into glycerol and fatty acids. Most, if not all of these enzymes, are mainly in the brush border of the epithelial cells. Therefore, they presumably catalyze hydrolysis of the foods on the outside surfaces of the microvilli prior to absorption of the end-products of digestion.

The epithelial cells deep in the crypts of Lieberkühn continually undergo mitosis, and the new cells gradually migrate along the basement membrane upward out of the crypts toward the tips of the villi where they are finally shed into the intestinal secretions. The life cycle of an intestinal epithelial cell is approximately 5 days. This rapid growth of new cells allows rapid repair of any excoriation that occurs in the mucosa.

REGULATION OF SMALL INTESTINAL SECRETION

Local Stimuli. By far the most important means for regulating small intestinal secretion are various local nervous reflexes, especially reflexes initiated by tactile or irritative stimuli. Therefore, for the most part, secretion in the small intestine occurs simply in response to the presence of chyme in the intestine — the greater the amount of chyme, the greater the secretion.

Hormonal Regulation. Some of the same hormones that promote secretion elsewhere in the gastrointestinal tract also increase small intestinal secretion, especially secretin and cholecystokinin. Also, some experiments suggest that other hormonal substances extracted from the small intestinal mucosa by the chyme might help to control secretion. However, in general, the reflex mechanisms probably play the dominant role.

SECRETIONS OF THE LARGE INTESTINE

Mucus Secretion. The mucosa of the large intestine, like that of the small intestine, is lined with crypts of Lieberkühn, but the epithelial cells contain almost no enzymes. Instead, they are lined almost entirely by goblet cells that secrete mucus. Also, on the surface epithelium of the large intestine are large numbers of goblet cells dispersed among the other epithelial cells.

Therefore, the great preponderance of secretion in the large intestine is mucus. Its rate of secretion is regulated principally by direct, tactile stimulation of the goblet cells on the surface of the mucosa and by local myenteric reflexes to the goblet cells in the crypts of Lieberkühn. However, stimulation of the nervi erigentes, which carry the parasympathetic innervation to the distal one-half to two-thirds of the large intestine, also causes marked increase in the secretion of mucus. This occurs along with an increase in motility, which was discussed in the preceding chapter. Therefore, during extreme parasympathetic stimulation, often caused by emotional disturbances, so much mucus may be secreted into the large intestine that the person has a bowel movement of ropy mucus as often as every 30 minutes; the mucus contains little or no fecal material.

Mucus in the large intestine obviously protects the wall against excoriation, but, in addition, it provides the adherent medium for holding fecal matter together. Furthermore, it protects the intestinal wall from the great amount of bacterial activity that takes place inside the feces, and it, plus the alkalinity of the secretion (pH of 8.0 caused by large amounts of sodium bicarbonate), also provides a barrier to keep acids formed deep in the feces from attacking the intestinal wall.

Secretion of Water and Electrolytes in Response to Irritation. Whenever a segment of the large intestine becomes intensely irritated, as occurs when bacterial infection becomes rampant during *enteritis*, the mucosa then secretes large quantities of water and electrolytes in addition to the normal viscid solution of mucus. This acts to dilute the irritating factors and to cause rapid movement of the feces toward the anus. The usual result is *diarrhea* with loss of large quantities of water and electrolytes but also earlier recovery from the disease than would otherwise occur.

REFERENCES

Banks, P. A.: Pancreatitis. New York, Plenum Press, 1978
Baron, J. H.: Clinical tests of Gastric Secretion: History, Methodology, and Interpretation. New York, Oxford University Press, 1978.
Bennett, A.: Prostaglandins and the Gut, 1976. Montreal, Eden Press, 1976.
Binder, H. J. (ed.): Mechanisms of Intestinal Secretion. New York, A. R. Liss, 1979.
Bloom, S. R.: Gastrointestinal hormones. *Int. Rev. Physiol., 12*:71, 1977.
Bloom, S. R. (ed.): Gut Hormones. New York, Longman, Inc., 1978.
Bonfils, S., *et al.*: Vagal control of gastric secretion. *In* Crane, R. K. (ed.): International Review of Physiology: Gastrointestinal Physiology III. Vol. 19. Baltimore, University Park Press, 1979, p. 59.
Brooks, F.: Diseases of the Exocrine Pancreas. Philadelphia, W. B. Saunders Co., 1980.
Burland, W. L., and Simkins, M. A. (eds.): Cimetidine. New York, Elsevier/North-Holland, 1977.
Davenport, H. W.: Mechanisms of gastric and pancreatic secretion. *In* Frohlich, E. D., (ed.): Pathophysiology, 2nd Ed. Philadelphia, J. B. Lippincott Co., 1976, p. 481.
Forker, E. L.: Mechanisms of hepatic bile formation. *Annu. Rev. Physiol., 39*:323, 1977.
Gardner, J.: Regulation of pancreatic exocrine function in vitro: Initial steps in the actions of secretagogues. *Annu. Rev. Physiol., 41*:55, 1979.
Gerolami, A., and Sarles, J. C.: Biliary secretion and motility. *Int. Rev. Physiol., 12*:223, 1977.
Grayson, J.: The gastrointestinal circulation. *In* MTP International Review of Science: Physiology. Vol. 4. Baltimore, University Park Press, 1974, p. 105.
Grossman, M. I.: Neural and hormonal regulation of gastrointestinal function: An overview. *Annu. Rev. Physiol., 41*:27, 1979.
Hendrix, T. R., and Paulk, H. T.: Intestinal secretion. *Int. Rev. Physiol., 12*:257, 1977.
Hirschowitz, B. I.: H–2 histamine receptors. *Annu. Rev. Pharmacol. Toxicol., 19*:203, 1979.
Jaffe, B. M.: Hormones of the gastrointestinal tract. *In* DeGroot, L. J., *et al.* (eds.): Endocrinology. Vol. 3. New York, Grune & Stratton, 1979, p. 1669.
Jerzy Glass, G. B.: Gastrointestinal Hormones. New York, Raven Press, 1980.
Johnson, L. R.: Gastrointestinal hormones and their functions. *Annu. Rev. Physiol., 39*:135, 1977.
Jones, R. S., and Myers, W. C.: Regulation of hepatic biliary secretion. *Annu. Rev. Physiol., 41*:67, 1979.
Konturek, S. J.: Gastric secretion. *In* MTP International Review of Science: Physiology. Vol. 4. Baltimore, University Park Press, 1974, p. 227.

Mason, D. K.: Salivary Glands in Health and Disease. Philadelphia, W. B. Saunders Company, 1975.

Miyoshi, A. (ed.): Gut Peptides: Secretion, Function, and Clinical Aspects. New York, Elsevier/North-Holland, 1979.

Myren, J., and Vatn, M. H. (eds.): Gastric Inhibition. Oslo, Universitetsforl, 1976.

Petersen, O. H.: Electrophysiology of mammalian gland cells. *Physiol. Rev., 56*:535, 1976.

Picazo, J. (ed.): Glucagon in Gastroenterology. Baltimore, University Park Press, 1978.

Preshaw, R. M.: Pancreatic exocrine secretion. *In* MTP International Review of Science: Physiology. Vol. 4, Baltimore, University Park Press, 1974, p. 265.

Rehfeld, J. F.: Gastrointestinal hormones. *In* Crane, R. K. (ed.): International Review of Physiology: Gastrointestinal Physiology III. Vol. 19. Baltimore, University Park Press, 1979, p. 291.

Rehfeld, J. F., and Amdrup, E. (eds.): Gastrins and the Vagus. New York, Academic Press, 1979.

Reichen, J., and Paumgartner, G.: Excretory function of the liver. *In* Javitt, N. B. (ed.): International Review of Physiology: Liver and Biliary Tract Physiology I. Vol. 21. Baltimore, University Park Press, 1980, p. 103.

Rosselin, G., et al. (eds.): Hormone Receptors in Digestion and Nutrition. New York, Elsevier/North-Holland, 1979.

Rothman, S. S.: The digestive enzymes of the pancreas: A mixture of inconstant proportions. *Annu. Rev. Physiol., 39*:373, 1977.

Sachs, G., et al.: H^+ transport: Regulation and mechanism in gastric mucosa and membrane vesicles. *Physiol. Rev., 58*:106, 1978.

Sachs, G., et al.: Gastric secretion. *Int. Rev. Physiol., 12*:127, 1977.

Sarles, H.: The exocrine pancreas. *Int. Rev. Physiol., 12*:173, 1977.

Satir, B.: The final steps in secretion. *Sci. Am. 233*(4): 28, 1975.

Schneyer, L. H.: Salivary secretion. *In* MTP International Review of Science: Physiology. Vol. 4. Baltimore, University Park Press, 1974, p. 183.

Soll, A., and Walsh, J. H.: Regulation of gastric acid secretion. *Annu. Rev. Physiol., 41*:35, 1979.

Stroud, R. M., et al.: Mechanisms of zymogen activation. *Annu. Rev. Biophys. Bioeng., 6*:177, 1977.

Walsh, J. H.: Circulating gastrin. *Ann. Rev. Physiol., 37*:81, 1975.

Young, J. A.: Salivary secretion of inorganic electrolytes. *In* Crane, R. K. (ed.): International Review of Physiology: Gastrointestinal Physiology III. Vol. 19. Baltimore, University Park Press, 1979, p. 1.

Young, J. A., and van Lennep, E. W.: Morphology and physiology of salivary myoepithelial cells. *Int. Rev. Physiol., 12*:105, 1977.

65

Digestion and Absorption in the Gastrointestinal Tract

The foods on which the body lives, with the exception of small quantities of substances such as vitamins and minerals, can be classified as carbohydrates, fats, and proteins. However, these generally cannot be absorbed in their natural forms through the gastrointestinal mucosa and, for this reason, are useless as nutrients without the preliminary process of digestion. Therefore, the present chapter discusses, first, the processes by which carbohydrates, fats, and proteins are digested into small enough compounds for absorption and, second, the mechanisms by which the digestive end-products, as well as water, electrolytes, and other substances, are absorbed.

DIGESTION OF THE VARIOUS FOODS

Hydrolysis as the Basic Process of Digestion. Almost all the carbohydrates of the diet are large *polysaccharides* or *disaccharides,* which are combinations of *monosaccharides* bound to each other by the process of *condensation*. This means that a hydrogen ion has been removed from one of the monosaccharides, while a hydroxyl ion has been removed from the next one; the two monosaccharides then are combined with each other at these sites of removal, and the hydrogen and hydroxyl ions combine to form water. When the carbohydrates are digested back into monosaccharides, specific enzymes return the hydrogen and hydroxyl ions to the polysaccharides and thereby separate the monosaccharides from each other. This process, called *hydrolysis,* is the following:

$$R'' - R' + H_2O \xrightarrow[enzyme]{digestive} R''OH + R'H$$

Almost the entire fat portion of the diet consists of triglycerides (neutral fats), which are combinations of three *fatty acid* molecules condensed with a single *glycerol* molecule. In the process of condensation, three molecules of water had been removed. Digestion of the triglycerides consists of the reverse process, the fat-digesting enzymes returning the three molecules of water to each molecule of neutral fat and thereby splitting the fatty acid molecules away from the glycerol. Here again, the process is one of hydrolysis.

Finally, proteins are formed from *amino acids* that are bound together by means of *peptide linkages.* In this linkage a hydroxyl ion is removed from one amino acid, while a hydrogen ion is removed from the succeeding one; thus, the amino acids also combine together by a process of condensation while losing a molecule of water. Digestion of proteins, therefore, also involves a process of hydrolysis, the proteolytic enzymes returning the water to the protein molecules to split them into their constituent amino acids.

Therefore, the chemistry of digestion is really simple, for in the case of all three major types of food, the same basic process of *hydrolysis* is involved. The only difference lies in the enzymes required to promote the reactions for each type of food.

All the digestive enzymes are proteins. Their secretion by the different gastrointestinal glands is discussed in the preceding chapter.

DIGESTION OF CARBOHYDRATES

The Carbohydrate Foods of the Diet. Only three major sources of carbohydrates exist in the normal human diet. These are sucrose, which is the disaccharide known popularly as cane sugar; lactose, which is a disaccharide in milk; and starches, which are large polysaccharides present in almost all foods and particularly in the grains. Other types of carbohydrates ingested to a slight extent are glycogen, alcohol, lactic acid, pyruvic acid, pectins, dextrins, and minor quantities of other carbohydrate derivatives in meats. The diet also contains a large amount of cellulose, which is a carbohydrate. However, no enzymes capable of hydrolyzing cellulose are secreted by the human digestive tract. Consequently, cellulose cannot be considered to be a food for the human being, though it can be utilized by some lower animals.

Digestion of Carbohydrates in the Mouth. When food is chewed, it is mixed with the saliva, which contains the enzyme *ptyalin* (α-amylase) secreted mainly by the

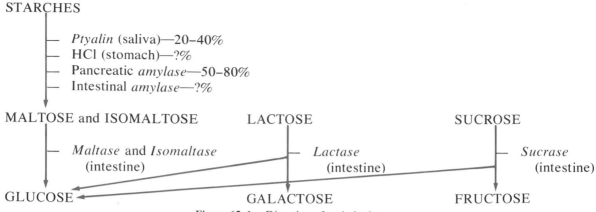

Figure 65–1. Digestion of carbohydrates.

parotid glands. This enzyme hydrolyzes starch into the disaccharides *maltose* and *isomaltose,* as shown in Figure 65–1; but the food remains in the mouth only a short time, and probably not more than 3 to 5 per cent of all the starches that are eaten will have become hydrolyzed into maltose and isomaltose by the time the food is swallowed. One can demonstrate the digestive action of ptyalin in the mouth by chewing a piece of bread for several minutes; after this time, the bread tastes sweet because of the maltose and isomaltose that has been liberated from the starches of the bread.

Most starches in their natural state, unfortunately, are present in the food in small globules, each of which has a thin protective cellulose covering. Therefore, most naturally occurring starches are digested only poorly by ptyalin unless the food is cooked to destroy the protective membrane.

Digestion of Carbohydrates in the Stomach. Even though food does not remain in the mouth long enough for ptyalin to complete the breakdown of starches into maltose, the action of ptyalin continues for as long as several hours after the food has entered the stomach; that is, until the contents of the fundus are mixed with the stomach secretions. Then the activity of the salivary amylase is blocked by the acid of the gastric secretions, for it is essentially nonactive as an enzyme once the pH of the medium falls below approximately 4.0. Nevertheless, on the average, before the food becomes completely mixed with the gastric secretions, as much as 30 to 40 per cent of the starches will have been changed into maltose and isomaltose.

Digestion of Carbohydrates in the Small Intestine. *Digestion by Pancreatic Amylase.* Pancreatic secretion, like saliva, contains a large quantity of α-amylase which is almost identical in its function with the α-amylase of saliva and is capable of splitting starches into *maltose* and *isomaltose.* Therefore, immediately after the chyme empties from the stomach into the duodenum and mixes with pancreatic juice, the starches that have not already been split are digested by amylase. In general, the starches are almost totally converted into maltose and isomaltose before they have passed beyond the jejunum.

Hydrolysis of Disaccharides into Monosaccharides by the Intestinal Epithelial Enzymes. The epithelial cells lining the small intestine contain the four enzymes *lactase, sucrase, maltase,* and *isomaltase,* which are capable of splitting the disaccharides lactose, sucrose, maltose, and isomaltose, respectively, into their constituent monosaccharides. These enzymes are located in the brush border of the cells lining the lumen of the intestine, and the disaccharides are digested as they come in contact with this border. The digested products, the monosaccharides, are then immediately absorbed into the portal blood. Lactose splits into a molecule of *galactose* and a molecule of *glucose.* Sucrose splits into a molecule of *fructose* and a molecule of *glucose.* Maltose and isomaltose each split into *two molecules of glucose.* Thus, the final products of carbohydrate digestion that are absorbed into the blood are all monosaccharides.

In the ordinary diet, which contains far more starches than either sucrose or lactose, glucose represents about 80 per cent of the final products of carbohydrate digestion, and galactose and fructose each represent, on the average, about 10 per cent of the products of carbohydrate digestion.

DIGESTION OF FATS

The Fats of the Diet. By far the most common fats of the diet are the neutral fats, also known as *triglycerides,* each molecule of which is composed of a glycerol nucleus and three fatty acids, as illustrated in Figure 65–2. Neutral fat is found in food of both animal origin and plant origin.

In the usual diet are also small quantities of phospholipids, cholesterol, and cholesterol esters. The phospholipids and cholesterol esters contain fatty acid, and, therefore, can be considered to be fats themselves. Cholesterol, on the other hand, is a sterol compound containing no fatty acid, but it does exhibit some of the physical and chemical characteristics of fats; it is derived from fats, and it is metabolized similarly to fats. Therefore, cholesterol is considered from a dietary point of view to be a fat.

Digestion of Fats in the Intestine. A small amount of short chain triglycerides of butter fat origin is digested in the stomach by gastric lipase *(tributyrase).* However, the amount of digestion is so slight that it is unimportant. Instead, essentially all fat digestion occurs in the small intestine as follows:

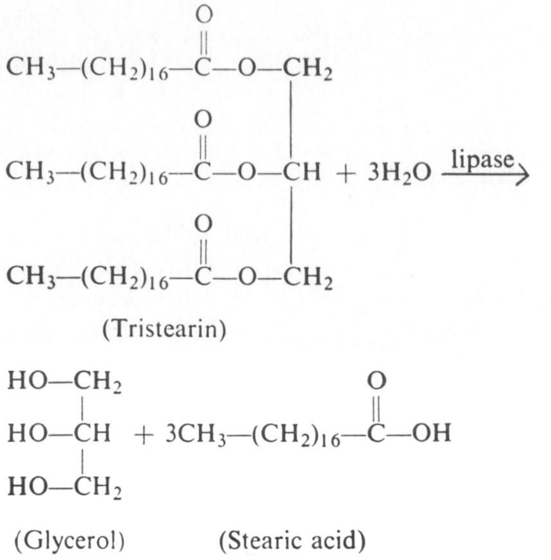

$$CH_3-(CH_2)_{16}-\overset{\overset{O}{\parallel}}{C}-O-CH_2$$

$$CH_3-(CH_2)_{16}-\overset{\overset{O}{\parallel}}{C}-O-CH + 3H_2O \xrightarrow{\text{lipase}}$$

$$CH_3-(CH_2)_{16}-\overset{\overset{O}{\parallel}}{C}-O-CH_2$$

(Tristearin)

$$\begin{matrix} HO-CH_2 \\ | \\ HO-CH \\ | \\ HO-CH_2 \end{matrix} + 3CH_3-(CH_2)_{16}-\overset{\overset{O}{\parallel}}{C}-OH$$

(Glycerol)   (Stearic acid)

Figure 65–2. Hydrolysis of neutral fat catalyzed by lipase.

Emulsification of Fat by Bile Acids. The first step in fat digestion is to break the fat globules into small sizes so that the water-soluble digestive enzymes can act on the globule surfaces. This process is called emulsification of the fat, and it is achieved under the influence of *bile*, the secretion of the liver that does not contain any digestive enzymes. However, bile does contain a large quantity of *bile salts*, mainly in the form of ionized sodium salts, which are extremely important for the emulsification of fat. The carboxyl (or polar) part of the bile salt is highly soluble in water, whereas the sterol portion of the bile salt is highly soluble in fat. Therefore, the fat-soluble portion of the bile salt dissolves in the surface layer of the fat globule but with the carboxyl portion of the salt projecting outward and soluble in the surrounding fluids; this effect greatly decreases the interfacial tension of the fat.

When the interfacial tension of a globule of nonmiscible fluid is low, this nonmiscible fluid, on agitation, can be broken up into many minute particles far more easily than it can when the interfacial tension is great. Consequently, a major function of the bile salts is to make the fat globules readily fragmentable by agitation in the small bowel. This action is the same as that of many detergents that are used widely in most household cleansers for removing grease.

Each time the diameters of the fat globules are decreased by a factor of 2 as a result of agitation in the small intestine, the total surface area of the fat increases 2 times. In other words, the total surface area of the fat particles in the intestinal contents is inversely proportional to the diameters of the particles.

The lipases are water-soluble compounds and can attack the fat globules only on their surfaces. Consequently, it can be readily understood how important this detergent function of bile salts is for the digestion of fats.

Digestion of Fats by Pancreatic Lipase. By far the most important enzyme for the digestion of fats is *pancreatic lipase* in the pancreatic juice. However, the epithelial cells of the small intestine also contain a small quantity of lipase known as *enteric lipase*. Both of these act alike to cause hydrolysis of fat.

End-Products of Fat Digestion. Most of the triglycerides of the diet are finally split into *monoglycerides, free fatty acids,* and *glycerol,* as illustrated in Figure 65–3. However, small portions are not digested at all or remain in a diglyceride state.

Role of Bile Salts in Accelerating Fat Digestion — Formation of Micelles. The hydrolysis of triglycerides is a highly reversible process; therefore, accumulation of monoglycerides and free fatty acids in the vicinity of digesting fats very quickly blocks further digestion. Fortunately, the bile salts play an important role in removing the monoglycerides and free fatty acids from the vicinity of the digesting fat globules almost as rapidly as these end-products of digestion are formed. This occurs in the following way:

Bile salts have the propensity to form *micelles,* which are small spherical globules about 25 Angstroms in diameter and composed of 20 to 50 molecules of bile salt. These develop because each bile salt molecule is composed of a sterol nucleus that is highly fat-soluble and a polar group that is highly water-soluble. The sterol nuclei of the 20 to 50 bile salt molecules of the micelle aggregate together to form a small fat globule in the middle of the micelle. This aggregation causes the polar groups to project outward to cover the surface of the micelle. Since these polar groups are negatively charged, they allow the entire micelle globule to become dissolved in the water of the digestive fluids and to remain in stable solution despite the very large size of the micelle.

During triglyceride digestion, as rapidly as the monoglycerides and free fatty acids are formed they become dissolved in the fatty portion of the micelles, which immediately removes these end-products of digestion from the vicinity of the digesting fat globules. Consequently, the digestive process can proceed unabated.

The bile salt micelles also act as a transport medium to carry the monoglycerides and the free fatty acids, both of which would otherwise be almost completely insoluble, to the brush borders of the epithelial cells. There the monoglycerides and free fatty acids are absorbed, as will be discussed later. On delivery of these substances to the brush border, the bile salts are again released back into the chyme to be used again and again for this "ferrying" process.

$$\text{Fat} \xrightarrow{\text{(Bile + Agitation)}} \text{Emulsified fat}$$

$$\text{Emulsified fat} \xrightarrow{\textit{Pancreatic lipase}} \begin{cases} \text{Fatty acids} \\ \text{Glycerol} \end{cases} 40\% \ (?) \\ \text{Glycerides} \ 60\% \ (?)$$

Figure 65–3. Digestion of fats.

Digestion of Cholesterol Esters. Most of the cholesterol in the diet is in the form of cholesterol esters, which cannot be absorbed in this form, though free cholesterol is readily absorbed. A *cholesterol esterase* in the pancreatic juice hydrolyzes the esters and thus frees the cholesterol. The bile salt micelles play identically the same role in "ferrying" cholesterol as they play in "ferrying" monoglycerides and free fatty acids. Indeed, this role of the bile salt micelles is absolutely essential to the absorption of cholesterol because essentially no cholesterol is absorbed without the presence of bile salts. On the other hand, as much as 60 per cent of the triglycerides can be digested and absorbed even in the absence of bile salts.

DIGESTION OF PROTEINS

The Proteins of the Diet. The dietary proteins are derived almost entirely from meats and vegetables. These proteins in turn are formed of long chains of amino acids bound together by *peptide linkages*. A typical linkage is the following:

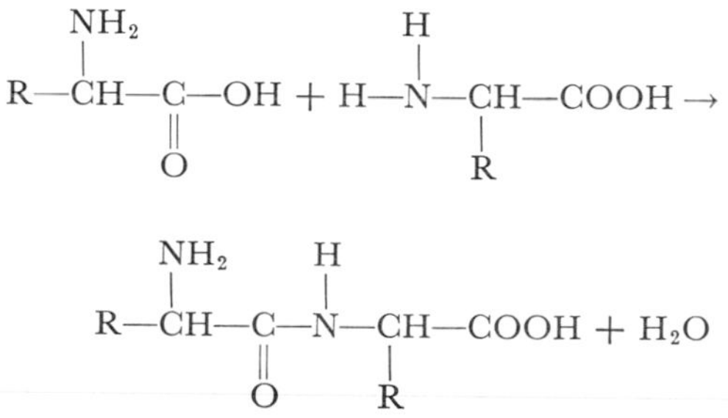

$$R-CH-C-OH + H-N-CH-COOH \rightarrow$$

with NH_2, O on the left group and H, R on the right group

$$R-CH-C-N-CH-COOH + H_2O$$

with NH_2, O, H, R

The characteristics of each type of protein are determined by the types of amino acids in the protein molecule and by the arrangement of these amino acids. The physical and chemical characteristics of the different proteins will be discussed in Chapter 69.

Digestion of Proteins in the Stomach. *Pepsin,* the important peptic enzyme of the stomach, is most active at a pH of about 2 and is completely inactive at a pH above approximately 5. Consequently, for this enzyme to cause any digestive action on protein, the stomach juices must be acidic. It will be recalled from Chapter 64 that the gastric glands secrete a large quantity of hydrochloric acid. This hydrochloric acid is secreted by the oxyntic (parietal) cells at a pH of about 0.8, but, by the time it is mixed with the stomach contents and with the secretions from the nonoxyntic glandular cells of the stomach, the pH ranges around 2 to 3, a highly favorable range of acidity for pepsin activity.

Pepsin is capable of digesting essentially all the different types of proteins in the diet. One of the important features of pepsin digestion is its ability to digest collagen, an albuminoid that is affected little by other digestive enzymes. Collagen is a major constituent of the intercellular connective tissue of meats, and for the digestive enzymes of the digestive tract to penetrate meats and digest the cellular proteins it is first necessary that the collagen fibers be digested. Consequently, in persons lacking peptic activity in the stomach, the ingested meats are poorly penetrated by the digestive enzymes and, therefore, are poorly digested.

As illustrated in Figure 65–4, pepsin usually only

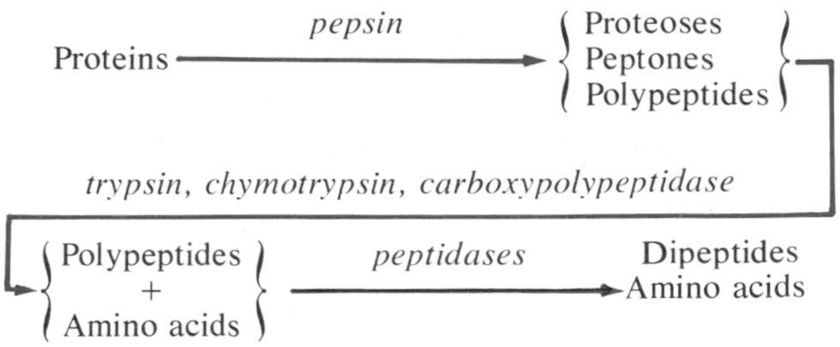

Figure 65–4. Digestion of proteins.

begins the process of protein digestion, simply splitting the proteins into proteoses, peptones, and large polypeptides. This splitting of proteins is a process of hydrolysis occurring at the peptide linkages between the amino acids.

Digestion of Proteins by Pancreatic Secretions. When the proteins leave the stomach, they ordinarily are mainly in the form of proteoses, peptones, and large polypeptides. Immediately upon entering the small intestine, the partial breakdown products are attacked by the pancreatic enzymes *trypsin, chymotrypsin,* and *carboxypolypeptidase*. As illustrated in Figure 65–4, these enzymes are capable of hydrolyzing all the partial breakdown products of protein to peptides and many also to the final stage of amino acids.

Digestion of Peptides by the Epithelial Peptidases of the Small Intestine. The epithelial cells of the small intestine contain several different enzymes for hydrolyzing the final peptide linkages of the remaining dipeptides and other small polypeptides as they come in contact with the epithelium of the villi. The enzymes responsible for final hydrolysis of the peptides into amino acids are *amino-polypeptidase* and the *dipeptidases*.

All the proteolytic enzymes — including those of the gastric juice, the pancreatic juice, and the brush border of the intestinal epithelial cells — are very specific for hydrolyzing individual types of peptide linkages. The linkages between certain pairs of amino acids differ in their bond energy and other physical characteristics from the linkages between other pairs. Therefore, a specific enzyme is required for each specific type of linkage. This accounts for the multiplicity of proteolytic enzymes as well as for the fact that no one single enzyme can usually digest protein all the way to all its constituent amino acids.

When food has been properly masticated and is not eaten in too large a quantity at any one time, about 98 per cent of all the proteins finally become either amino acids or very small peptides, mainly dipeptides. A few molecules of protein are never digested at all, and some remain in the stages of proteoses, peptones, and varying sizes of polypeptides.

BASIC PRINCIPLES OF GASTROINTESTINAL ABSORPTION

ANATOMICAL BASIS OF ABSORPTION

The total quantity of fluid that must be absorbed each day is equal to the ingested fluid (about 1.5

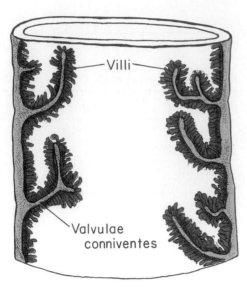

Figure 65–5. A longitudinal section of the small intestine, showing the valvulae conniventes covered by villi.

sorptive surface of the intestinal mucosa, showing many folds called *valvulae conniventes* (or *folds of Kerckring*), which increase the surface area of the absorptive mucosa about three-fold. These folds extend circularly most of the way around the intestine and are especially well developed in the duodenum and jejunum, where they often protrude as much as 8 mm. into the lumen.

Located over the entire surface of the small intestine, from approximately the point at which the common bile duct empties into the duodenum down to the ileocecal valve, are literally millions of small *villi*, which project about 1 mm. from the surface of the mucosa, as shown on the surfaces of the valvulae conniventes in Figure 65–5 and in detail in Figure 65–6. These villi lie so close to each other in the upper small intestine that they actually touch in most areas, but their distribution is less profuse in the distal small intestine. The presence of villi on the mucosal surface enhances the absorptive area another 10-fold.

The intestinal epithelial cells are characterized by a brush border, consisting of about 600 *microvilli* 1 μm in length and 0.1 μm in diameter protruding from each cell; these are illustrated in the electron micrograph in Figure 65–7. This increases the surface area exposed to the intestinal materials another 20-fold. Thus, the combination of the folds of Kerckring, the villi, and the microvilli increases the absorptive area of the mucosa about 600-fold, making a tremendous total area of about 250 square meters for the entire small intestine — about the surface area of a tennis court.

Figure 65–6A illustrates the general organization of a villus, emphasizing especially the advantageous arrangement of the vascular system for

liters) plus that secreted in the various gastrointestinal secretions (about 7.5 liters). This comes to a total of approximately 9 liters. About 8 to 8.5 liters of this is absorbed in the small intestine, leaving only 0.5 to 1 liter to pass through the ileocecal valve into the colon each day.

The stomach is a poor absorptive area of the gastrointestinal tract because it lacks the typical villus type of absorptive membrane and also because the junctions between the epithelial cells are tight junctions. Only a few highly lipid-soluble substances, such as alcohol and some drugs such as aspirin, can be absorbed in small quantities.

The Absorptive Surface of the Intestinal Mucosa — The Villi. Figure 65–5 illustrates the ab-

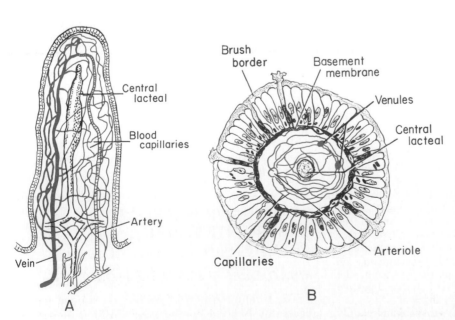

Figure 65–6. Functional organization of the villus: (A) Longitudinal section. (B) Cross-section showing the epithelial cells and basement membrane.

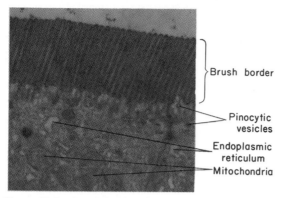

Figure 65–7. Brush border of the gastrointestinal epithelial cell, showing, also, pinocytic vesicles, mitochondria, and endoplasmic reticulum lying immediately beneath the brush border. (Courtesy of Dr. William Lockwood.)

absorption of fluid and dissolved material into the portal blood, and the arrangement of the *central lacteal* for absorption into the lymph. Figure 65–6B shows the cross-section of a villus, and Figure 65–7 shows many small *pinocytic* vesicles, which are pinched-off portions of infolded epithelium surrounding extracellular materials that have been entrapped inside the cells. Small amounts of substances are absorbed by this physical process of *pinocytosis,* though, as noted later in the chapter, most absorption occurs by means of single molecular transfer. Located near the brush border of the epithelial cell are many *mitochondria,* which supply the cell with oxidative energy needed for *active transport* of materials through the intestinal epithelium; this also is discussed later in the chapter.

BASIC MECHANISMS OF ABSORPTION

Absorption through the gastrointestinal mucosa occurs by *active transport* and by *diffusion,* as is also true for other membranes. The physical principles of these processes were explained in Chapter 4.

Briefly, active transport imparts energy to the substance as it is being transported for the purpose of concentrating it on the other side of the membrane or for moving it against an electrical potential. On the other hand, the term diffusion means simply transport of substances through the membrane as a result of molecular movement *along,* rather than against, an electrochemical gradient.

ABSORPTION IN THE SMALL INTESTINE

Normally, absorption from the small intestine each day consists of several hundred grams of

carbohydrates, 100 or more grams of fat, 50 to 100 grams of amino acids, 50 to 100 grams of ions, and 8 or 9 liters of water. However, the absorptive *capacity* of the small intestine is far greater than this: as much as several kilograms of carbohydrates per day, 500 to 1000 grams of fat per day, 500 to 700 grams of amino acids per day, and 20 or more liters of water per day. In addition, the large intestine can absorb still more water and ions, though almost no nutrients.

ABSORPTION OF WATER

Isosmotic Absorption. Water is transported through the intestinal membrane entirely by the process of *diffusion*. Furthermore, this diffusion obeys the usual laws of osmosis. Therefore, when the chyme is dilute, water is absorbed through the intestinal mucosa into the blood of the villi by osmosis.

Water can also be transported by osmosis from the plasma into the chyme. This occurs whenever hyperosmotic solutions are discharged from the stomach into the duodenum. Usually within minutes, sufficient water is transferred by osmosis to make the chyme isosmotic with the plasma. On the other hand, if there is excess water in the chyme, the osmosis into the plasma also causes an isosmotic state within a few minutes. Thereafter, the chyme remains almost exactly isosmotic throughout its total passage through the small and large intestine.

As dissolved substances are absorbed from the lumen of the gut into the blood the absorption tends to decrease the osmotic pressure of the chyme, but water diffuses so readily through the intestinal membrane (because of large 7 to 15 A intercellular pores) that it almost instantaneously "follows" the absorbed substances into the blood. Therefore, as ions and nutrients are absorbed, so also is an isosmotic equivalent of water absorbed. In this way not only are the ions and nutrients almost entirely absorbed before the chyme passes through the small intestine, but so also is almost 95 per cent of the water absorbed.

ABSORPTION OF IONS

Active Transport of Sodium. Twenty to 30 grams of sodium are secreted into the intestinal secretions each day. In addition, the normal person eats 5 to 8 grams of sodium each day. Combining these two, the small intestine absorbs 25 to 35 grams of sodium each day, which amounts to about one-seventh of all the sodium that is present in the body. One can well understand that whenever the intestinal secretions are lost to the exterior, as in extreme diarrhea, the sodium reserves of the body can be depleted to a lethal level within hours. Normally, this sodium is secreted and reabsorbed continually with only about 1 milliequivalent lost in the feces each day. The sodium plays an important role in the absorption of sugars and amino acids, as we shall see in subsequent discussions.

The basic mechanism of sodium absorption from the intestine is illustrated in Figure 65–8. The principles of this mechanism, which were discussed in Chapter 4, are also essentially the same as those for absorption of sodium from the renal tubules, as discussed in Chapter 34. The motive power for the sodium absorption is provided by active transport of sodium from inside the epithelial cells through the side walls of these cells into the intercellular spaces. This is illustrated by the heavy black arrows in Figure 65–8. This active transport obeys the usual laws of active transport: it requires a carrier, it requires energy, and it is catalyzed by appropriate ATPase carrier enzymes in the cell membrane. Part of the sodium is transported along with chloride ions, part is transported in exchange for potassium ions, and part is transported without either of these.

The active transport of sodium reduces its concentration in the cell to a low value (about 50 mEq./liter), as also illustrated in Figure 65–8. Since the sodium concentration in the chyme is normally about 142 mEq./liter (that is, approximately equal to that in the plasma), sodium diffuses from the chyme through the brush border of the epithelial cell into the epithelial cell cytoplasm. This replaces the sodium that is actively transported out of the epithelial cells into the intercellular spaces.

The next step in the transport process is osmosis of water out of the epithelial cell into the intercellular spaces. This movement is caused by the osmotic gradient created by the reduced concentration of sodium inside the cell and the elevated concentration in the intercellular space. The osmotic movement of water creates a flow of fluid

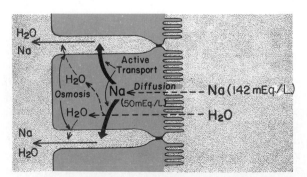

Figure 65–8. Absorption of sodium through the intestinal epithelium. Note also the osmotic absorption of water—that is, the water "follows" the sodium through the epithelial membrane.

into the intercellular space, then through the basement membrane of the epithelium, and finally into the circulating blood of the villi. New water diffuses along with sodium through the brush border of the epithelial cell to replenish the water that osmoses into the intercellular spaces.

Transport of Chloride. In the upper part of the small intestine chloride transport is mainly by passive diffusion. The transport of sodium ions through the epithelium creates electronegativity in the chyme and electropositivity on the basal side of the epithelial cells. Then chloride ions move along this electrical gradient to "follow" the sodium ions.

Active Absorption of Chloride Ions and Active Secretion of Bicarbonate Ions in the Lower Ileum and in the Large Intestine. The epithelial cells of the distal ileum and of the large intestine have the special capability of actively absorbing chloride ions by means of a tightly coupled active transport mechanism in which an equivalent number of bicarbonate ions are secreted. The functional role of this mechanism is to provide bicarbonate ions for neutralization of acidic products formed by bacteria — especially in the large intestine.

Various bacterial toxins, particularly those of cholera, colon bacilli, and staphylococci, can strongly stimulate this chloride-bicarbonate exchange mechanism. The secreted bicarbonate ion carries with it sodium ions, and the two of these together carry an isosmotic equivalent of water as well. This results in rapid flow of fluid from the distal part of the gut, thus causing diarrhea. In cholera, especially, the diarrhea can be so severe that it can cause death within 24 hours.

Absorption of Other Ions. Calcium ions are actively absorbed, especially from the duodenum, and calcium ion absorption is exactly controlled in relation to the need of the body for calcium. One important factor controlling calcium absorption is parathyroid hormone secreted by the parathyroid glands, and another is vitamin D. The parathyroid hormone activates vitamin D in the kidneys, and the activated vitamin D in turn greatly enhances calcium absorption. These effects are discussed in Chapter 79.

Iron ions are also actively absorbed from the small intestine. The principles of iron absorption and the regulation of its absorption in proportion to the body's need for iron were discussed in Chapter 5.

Potassium, magnesium, phosphate, and probably still other ions can also be actively absorbed through the mucosa. In general, the monovalent ions are absorbed with ease and in great quantities. On the other hand, the bivalent ions are normally absorbed in only small amounts; fortunately, only small quantities of these are normally needed by the body.

ABSORPTION OF NUTRIENTS

Absorption of Carbohydrates

Essentially all the carbohydrates are absorbed in the form of monosaccharides, only a small fraction of a per cent being absorbed as disaccharides and almost none as larger carbohydrate compounds. Furthermore, little carbohydrate absorption results from simple diffusion, for the pores of the mucosa through which diffusion occurs are essentially impermeable to water-soluble solutes with molecular weights greater than 100.

That the transport of most monosaccharides through the intestinal membrane is an active process is demonstrated by several important experimental observations:

1. Transport of most of them, especially glucose and galactose, can be blocked by metabolic inhibitors, such as iodoacetic acid, cyanides, and phlorhizin.

2. The transport is selective, specifically transporting certain monosaccharides without transporting others. The order of preference for transporting different monosaccharides and their relative rates of transport in comparison with glucose are:

Galactose	1.1
Glucose	1.0
Fructose	0.4
Mannose	0.2
Xylose	0.15
Arabinose	0.1

3. There is a maximum rate of transport for each type of monosaccharide. The most rapidly transported monosaccharide is galactose, with glucose running a close second. Fructose, which is also one of the three important monosaccharides for nutrition, is absorbed less than half as rapidly as either galactose or glucose; also, its mechanism of absorption is different, as will be explained below.

4. There is competition between certain sugars for the respective carrier system. For instance, if large amounts of galactose are being transported, the amount of glucose that can be transported simultaneously is considerably reduced.

Mechanism of Glucose and Galactose Absorption. Glucose and galactose transport ceases whenever active sodium transport is blocked. Therefore, it is assumed that the energy required for transport of these two monosaccharides is actually provided by the sodium transport system. A theory that attempts to explain this is the following: It is known that the carrier for transport of glucose (which is the carrier for galactose as well) is present in the brush border of the epithelial cell. However, this carrier will not transport the glucose in the absence of sodium transport. Therefore, it is believed that the carrier has receptor sites for both a glucose molecule and a sodium ion, and that it will not transport either of these to the interior of the epithelial cell until both receptor sites are simultaneously filled. The energy to cause movement of the carrier from the exterior of the membrane to the interior is derived from the difference in sodium concentration between the outside and inside. That is, as sodium diffuses to the inside of the cell it "drags" the carrier, and therefore the glucose as well, along with it, thus providing the energy for transport of the glucose. For obvious reasons, this explanation is called the *sodium co-transport theory* for glucose transport.

Subsequently, we will see that sodium transport is also required for transport of amino acids, suggesting a similar "carrier-drag" mechanism for amino acid transport.

Absorption of Fructose. Transport of fructose is slightly different from that of most other monosaccharides. It is not blocked by some of the same metabolic poisons — specifically, phlorhizin — and it does not require metabolic energy for transport, even though it does require a specific carrier. Therefore, it is transported by *facilitated diffusion* rather than active transport. Also, it is mainly converted into glucose inside the epithelial cell before entering the portal blood, the fructose first becoming phosphorylated, then converted to glucose, and finally released from the epithelial cell into the blood.

Absorption of Proteins

Most proteins are absorbed in the form of amino acids. However, small quantities of dipeptides and even tripeptides are also absorbed, and extremely minute quantities of whole proteins can at times be absorbed by the process of pinocytosis, though not by the usual absorptive mechanisms.

The absorption of amino acids also obeys the principles listed above for active absorption of glucose; that is, the different types of amino acids are absorbed selectively and certain ones interfere with the absorption of others, illustrating that common carrier systems exist. Finally, metabolic poisons block the absorption of amino acids in the same way that they block the absorption of glucose.

Absorption of amino acids through the intestinal mucosa can occur far more rapidly than can protein digestion in the lumen of the intestine. As a result, the normal rate of absorption is determined not by the rate at which they can be absorbed but by the rate at which they can be released from the proteins during digestion. For these reasons, essentially no free amino acids can be found in the intestine during digestion — that is, they are absorbed as rapidly as they are formed. Since most protein digestion occurs in the upper small intestine, most protein absorption occurs in the duodenum and jejunum.

Basic Mechanisms of Amino Acid Transport. As is true for monosaccharide absorption, very little is known about the basic mechanisms of amino acid transport. However, at least four different carrier systems transport different amino acids — one transports *neutral amino acids,* a second transports *basic amino acids,* a third transports *acidic amino acids,* and a fourth has specificity for the two imino acids *proline* and *hydroxyproline.* Also, the transport mechanisms have far greater affinity for transporting L-stereoisomers of amino acids than D-stereoisomers. And experiments have demonstrated that *pyridoxal phosphate,* a derivative of the vitamin pyridoxine, is required for transport of many amino acids.

Amino acid transport, like glucose transport, occurs only in the presence of simultaneous sodium transport. Furthermore, the carrier systems for amino acid transport, like those for glucose transport, are in the brush border of the epithelial cell. It is believed that amino acids are transported by the same *sodium co-transport* mechanism as that explained above for glucose transport. That is, the theory postulates that the carrier has receptor sites for both an amino acid molecule and a sodium ion. Only when both of the sites are filled will the carrier move to the interior of the cell. Because of the sodium gradient across the brush border, the sodium diffusion to the cell interior pulls the carrier and its attached amino acid to the interior where the amino acid becomes trapped. Therefore, amino acid concentrations increase within the cell, and they then diffuse through the sides or base of the cell into the portal blood.

Absorption of Fats

Earlier in this chapter it was pointed out that as fats are digested to form monoglycerides and free fatty acids, both of these digestive end-products become dissolved in the lipid portion of the bile acid micelles. Because of the molecular dimensions of these micelles and also because of their highly charged exterior, they are soluble in the chyme. In this form the monoglycerides and the fatty acids are transported to the surfaces of the epithelial cells. On coming in contact with these surfaces, both the monoglycerides and the fatty acids immediately diffuse through the epithelial membrane, leaving the bile acid micelles still in the chyme. The micelles then diffuse back into the chyme and absorb still more monoglycerides and

fatty acids, and similarly transport these also to the epithelial cells. Thus, the bile acids perform a "ferrying" function, which is highly important for fat absorption. In the presence of an abundance of bile acids, approximately 97 per cent of the fat is absorbed; in the absence of bile acids, only 50 to 60 per cent is normally absorbed.

The mechanism for absorption of the monoglycerides and fatty acids through the brush border is based on the fact that both of these substances are highly lipid-soluble. Therefore, they become dissolved in the membrane and simply diffuse to the interior of the cell.

The undigested triglycerides and the diglycerides are both also highly soluble in the lipid membrane of the epithelial cell. However, only small quantities of these are normally absorbed because the bile acid micelles will not dissolve either triglycerides or diglycerides and therefore will not ferry them to the epithelial membrane.

During entry into the epithelial cell, many of the monoglycerides are further digested into glycerol and fatty acids by an epithelial cell lipase. Then, the free fatty acids are reconstituted by the smooth endoplasmic reticulum into triglycerides. Almost all of the glycerol that is utilized for this purpose is synthesized *de novo* from alpha-glycerophosphate, this synthesis requiring both energy from ATP and a complex of enzymes to catalyze the reactions.

Once formed, the triglycerides aggregate within the endoplasmic reticulum into globules along with absorbed cholesterol, absorbed phospholipids, and newly synthesized cholesterol and phospholipids. Each of these is then encased in a protein coat, utilizing protein also synthesized by the endoplasmic reticulum. This globular mass, along with its protein coat, is extruded from the sides of the epithelial cells into the intercellular spaces, and from here it passes into the central lacteal of the villi. Such globules are called *chylomicrons.*

The protein coat of the chylomicrons makes them hydrophilic, allowing a reasonable degree of suspension stability in the extracellular fluids. Poisons or genetic disorders that prevent formation of the protein for coating the chylomicrons cause the fat to accumulate in the epithelial cell and not to be extruded into the extracellular fluid.

Transport of the Chylomicrons in the Lymph. From the sides of the epithelial cells the chylomicrons wend their way through the basement membrane and into the central lacteal of the villi and from here are propelled, along with the lymph, by the lymphatic pump upward through the thoracic duct to be emptied into the great veins of the neck. Between 80 and 90 per cent of all fat absorbed from the gut is absorbed in this manner and is transported to the blood by way of the thoracic lymph in the form of chylomicrons.

Direct Absorption of Fatty Acids into the Portal Blood. Small quantities of short chain fatty acids, such as those from butterfat, are absorbed directly into the portal blood rather than being converted into triglycerides and absorbed into the lymphatics. The cause of this difference between short and long chain fatty acid absorption is presumably that the shorter chain fatty acids are more water-soluble, which allows direct diffusion of fatty acids from the epithelial cells into the capillary blood of the villus.

Absorption of Bile Salts. In the upper portion of the small intestine the bile salts are not absorbed; this failure to be absorbed requires them to remain in the chyme and to continue their function of "ferrying" free fatty acids and monoglycerides to the intestinal mucosa through the entire extent of the small intestine. Once the processes of fat digestion and fat absorption have been accomplished in the upper and mid-intestinal levels, however, the bile salts themselves are then absorbed from the distal ileum before the chyme empties into the large intestine. This absorption is an active process and is carrier-mediated.

After being absorbed from the distal ileum, the bile salts are again secreted in the bile by the liver and returned once more to the upper intestine. Thus, the same bile salts are re-secreted several times each day and are used again and again in the process of fat absorption. Only a small portion of the bile salts (approximately 5 per cent) is lost during each cycle of this "bile salt circulation."

On occasion, the bile salts fail to be absorbed in the ileum and instead empty with the chyme into the large intestine; this occurs especially in patients whose distal ileum has been removed because of ileitis. The presence of bile salts in the large intestine frequently causes severe diarrhea, presumably because of the detergent effect of these salts acting on and irritating the large intestinal mucosa.

ABSORPTION IN THE LARGE INTESTINE: FORMATION OF THE FECES

Approximately 500 to 1000 ml. of chyme passes through the ileocecal valve into the large intestine each day. Most of the water and electrolytes in this are absorbed in the colon, usually leaving less than 100 ml. of fluid to be excreted in the feces. Also, essentially all the ions are also absorbed, leaving less than 1 milliequivalent each of sodium and chloride ions to be lost in the feces.

Most of the absorption in the large intestine occurs in the proximal half of the colon, giving this portion the name *absorbing colon,* while the distal colon functions principally for storage and is therefore called the *storage colon.*

Absorption and Secretion of Electrolytes and Water. The mucosa of the large intestine, like that of the small intestine, has a very high capacity for active absorption of sodium, and the electrical potential created by the absorption of the sodium causes chloride absorption as well. In addition, as in the distal portion of the small intestine, the mucosa of the large intestine actively secretes bicarbonate ions while it simultaneously actively absorbs an equal amount of chloride ions in an exchange transport process. The bicarbonate helps to neutralize the acidic end-products of bacterial action in the colon.

The absorption of sodium and chloride ions creates an osmotic gradient across the large intestinal mucosa, which in turn causes absorption of water.

Bacterial Action in the Colon. Numerous bacteria, especially colon bacilli, are present in the absorbing colon. These are capable of digesting small amounts of cellulose, in this way providing a few calories of nutrition to the body each day. In herbivorous animals this source of energy is very significant, though it is of negligible importance in the human being. Other substances formed as a result of bacterial activity are vitamin K, vitamin B_{12}, thiamin, riboflavin, and various gases that contribute to *flatus* in the colon. Vitamin K is especially important, for the amount of this vitamin in the ingested foods is normally insufficient to maintain adequate blood coagulation.

Composition of the Feces. The feces normally are about three-fourths water and one-fourth solid matter composed of about 30 per cent dead bacteria, 10 to 20 per cent fat, 10 to 20 per cent inorganic matter, 2 to 3 per cent protein, and 30 per cent undigested roughage of the food and dried constituents of digestive juices, such as bile pigment and sloughed epithelial cells. The large amount of fat derives from unabsorbed fatty acids from the diet, fat formed by bacteria, and fat in the sloughed epithelial cells.

The brown color o. feces is caused by *stercobilin* and *urobilin,* which are derivatives of bilirubin. The odor is caused principally by the products of bacterial action; these vary from one person to another, depending on each person's colonic bacterial flora and on the type of food eaten. The actual odoriferous products include *indole, skatole, mercaptans,* and *hydrogen sulfide.*

REFERENCES

Beck, I. T.: The role of pancreatic enzymes in digestion. *Am. J. Clin. Nutr.,* 26:311, 1973.

Borgström, B.: Fat digestion and absorption. *Biomembranes,* 4B:555, 1974.

Brindley, D. N.: The intracellular phase of fat absorption. *Biomembranes,* 4B:621, 1974.

Cluysenaer, O. J. J., and van Tongeren, J. H. M.: Malabsorption in Coeliac Sprue. The Hague, M. Nijhoff Medical Division, 1977.

Crane, R. K.: Intestinal absorption of glucose. *Biomembranes,* 4A:541, 1974.

Creamer, B.: Intestinal structure in relation to absorption. *Biomembranes, 4A*:1, 1974.

Davenport, H. W.: A Digest of Digestion, 2nd Ed. Chicago, Year Book Medical Publishers, 1978.

Forth, W., and Rummel, W.: Iron absorption. *Physiol. Rev., 53*:724, 1973.

Frizzell, R. A., and Schultz, S. G.: Models of electrolyte absorption and secretion by gastrointestinal epithelia. *In* Crane, R. K. (ed.): International Review of Physiology: Gastrointestinal Physiology III. Vol. 19. Baltimore, University Park Press, 1979, p. 205.

Holzer, H., and Tschesche, H. (eds.): Biological Functions of Proteinases. New York, Springer-Verlag, 1979.

Jackson, M. J.: Treatment of short chain fatty acids. *Biomembranes, 4B*:673, 1974.

Kim, Y. S. *et al.*: Intestinal peptide hydrolases: Peptide and amino acid absorption. *Med. Clin. North Am., 58*:1397, 1974.

Kotyk, A.: Mechanisms of nonelectrolyte transport. *Biochim. Biophys. Acta, 300*:183, 1973.

Levitan, R., and Wilson, D. E.: Absorption of water soluble substances. *In* MPT International Review of Physiology. Vol. 4. Baltimore, University Park Press, 1974, p. 293.

Matthews, D. M.: Intestinal absorption of amino acids and peptides. *Proc. Nutr. Soc., 31*:171, 1972.

Matthews, D. M.: Absorption of amino acids and peptides from the intestine. *Clin. Endocrinol. Metabol., 3*:3, 1974.

Matthews, D. M.: Absorption of water-soluble vitamins. *Biomembranes, 4B*:847, 1974.

Matthews, D. M.: Intestinal absorption of peptides. *Physiol. Rev., 55*:537, 1975.

Ockner, R. K., and Isselbacher, K. J.: Recent concepts of intestinal fat absorption. *Rev. Physiol. Biochem. Pharmacol., 71*:107, 1974.

Okuda, K.: Intestinal mucosa and vitamin B_{12} absorption. *Digestion, 6*:173, 1972.

Olsen, W. A.: Carbohydrate absorption. *Med. Clin. North Am. 58*:1387, 1974.

Schultz, S. G.: Principles of electrophysiology and their application to epithelial tissues. *In* MPT International Review of Science: Physiology. Vol. 4, Baltimore, University Park Press, 1974, p. 69.

Schultz, S. G., *et al.*: Ion transport by mammalian small intestine. *Annu. Rev. Physiol., 36*:51, 1974.

Silk, D. B. A., and Dawson, A. M.: Intestinal absorption of carbohydrate and protein in man. *In* Crane, R. K. (ed.): International Review of Physiology: Gastrointestinal Physiology III. Vol. 19. Baltimore, University Park Press, 1979, p. 151.

Simmonds, W. J.: Absorption of lipids. *In* MPT International Review of Science: Physiology. Vol. 4. Baltimore, University Park Press, 1974, p. 343.

Smyth, D. H. (ed.): Intestinal Absorption. Vols. 4A and 4B. New York, Plenum Press, 1974.

Soergel, K. H., and Hofmann, A. F.: Absorption. *In* Frohlich, E. D. (ed.): Pathophysiology, 2nd Ed. Philadelphia, J. B. Lippincott Co., 1976, p. 499.

Turnberg, L. A.: Absorption and secretion of salt and water by the small intestine. *Digestion, 9*:357, 1973.

Ugolev, A. M.: Membrane (contact) digestion. *Biomembranes, 4A*:285, 1974.

Van Campen, D.: Regulation of iron absorption. *Fed. Proc. 33*:100, 1974.

Watson, D. W., and Sodeman, W. A., Jr.: The small intestine. *In* Sodeman, W. A., Jr. and Sodeman, T. M. (eds.): Pathologic Physiology: Mechanisms of Disease, 6th Ed. Philadelphia, W. B. Saunders Company, 1979, p. 824.

Wiseman, G.: Absorption of protein digestion products. *Biomembranes, 4A*:363, 1974

66

Physiology of Gastrointestinal Disorders

The logical treatment of most gastrointestinal disorders depends on a basic knowledge of gastrointestinal physiology. The purpose of this chapter, therefore, is to discuss a few representative types of malfunction that have special physiological bases or consequences.

DISORDERS OF SWALLOWING AND OF THE ESOPHAGUS

Paralysis of the Swallowing Mechanism. Damage to the 5th, 9th, or 10th nerve can cause paralysis of significant portions of the swallowing mechanism. Also, a few diseases, such as poliomyelitis or encephalitis, can prevent normal swallowing by damaging the swallowing center in the brain stem. Finally, malfunction of the swallowing muscles, as occurs in *muscle dystrophy* or in failure of neuromuscular transmission in *myasthenia gravis* or *botulism*, can also prevent normal swallowing.

When the swallowing mechanism is partially or totally paralyzed, the abnormalities that can occur include: (1) complete abrogation of the swallowing act so that swallowing cannot occur at all, (2) failure of the glottis to close so that food passes into the lungs instead of the esophagus, (3) failure of the soft palate and uvula to close the posterior nares so that food refluxes into the nose during swallowing, or (4) failure of the cricopharyngeal sphincter to remain closed during normal breathing, thus allowing large quantities of air to be sucked into the esophagus and thence swallowed.

One of the most serious instances of paralysis of the swallowing mechanism occurs when patients are under deep anesthesia. Often they vomit large quantities of materials from the stomach into the pharynx; then, instead of swallowing the materials again, they simply suck them into the trachea because the anesthetic has blocked the reflex mechanism of swallowing. As a result, such patients often choke to death on their own vomitus.

Achalasia and Megaesophagus. Achalasia is a condition in which the lower few centimeters of the esophagus fail to relax during the swallowing mechanism. Also, the lower two-thirds of the esophagus tends to contract in unison rather than peristaltically. As a result, food transmission from the esophagus into the stomach is impeded or prevented. Pathological studies have shown the physiological basis of this condition to be either damage to or absence of the myenteric plexus in the lower two-thirds of the esophagus. The musculature of the lower esophagus instead remains incoordinately contracted, and the myenteric plexus has lost the ability to transmit a signal to cause "receptive relaxation" of the gastroesophageal sphincter as food approaches this area during the swallowing process.

When achalasia becomes severe, the esophagus may not empty the swallowed food into the stomach for many hours, instead of within a few seconds which is the normal time. Over months and years, the esophagus becomes tremendously enlarged until it often can hold as much as one liter of food, which becomes putridly infected during the long periods of esophageal stasis. The infection may also cause ulceration of the esophageal mucosa, sometimes leading to severe substernal pain or even rupture and death. The food often refluxes into the pharynx and sometimes is then aspirated into the lungs, causing aspiration pneumonia. Fortunately, in most patients, considerable benefit or even cure can be achieved by stretching the lower end of the esophagus by means of a balloon inflated on the end of a swallowed esophageal tube.

DISORDERS OF THE STOMACH

Gastritis. Gastritis means inflammation of the gastric mucosa. This can result from (a) action of irritant foods on the gastric mucosa, (b) excessive excoriation of the stomach mucosa by the stomach's own peptic secretions, or (c) occasionally, bacterial inflammation. One of the most frequent causes of gastritis is irritation of the mucosa by alcohol or drugs such as aspirin.

The inflamed mucosa in gastritis is often painful, causing a diffuse burning pain referred to the high epigastrium. Reflexes initiated in the stomach mucosa cause the salivary glands to salivate intensely, and the frequent swallowing of foamy saliva makes air accumulate in the stomach. As a result, the person usually belches profusely, a burning sensation often occurring in the throat with each belch.

The Gastric Barrier and Its Penetration in Gastritis. Absorption from the stomach is normally very slight. This low level of absorption is probably caused by two specific features of the gastric mucosa: (1) it is lined with highly resistant mucus cells, and (2) it has very tight junctions between the adjacent epithelial cells. Normally this barrier is so resistant to diffusion that even the highly concentrated hydrogen ions of the gastric juice, averaging about 100,000 times the concentration of the hydrogen ions in the plasma, barely diffuse through the epithelial membrane. However, in gastritis, this barrier becomes inflamed and its permeability is greatly increased. The hydrogen ions do then diffuse into the stomach epithelium, creating additional havoc and leading to progressive stomach mucosal damage and atrophy. It also makes the mucosa susceptible to peptic digestion, thus frequently resulting in gastric ulcer.

Gastric Atrophy. In many persons who have chronic gastritis, the mucosa gradually becomes atrophic until little or no gastric gland activity remains. It is also believed that some persons develop autoimmunity against the gastric mucosa, this leading eventually to gastric atrophy. Loss of the stomach secretions in gastric atrophy leads to achlorhydria and, occasionally, to *pernicious anemia*.

Achlorhydria (and Hypochlorhydria). Achlorhydria means simply that the stomach fails to secrete hydrochloric acid, and hypochlorhydria means diminished acid secretion. Usually, when acid is not secreted, pepsin also is not secreted, and, even if it is, the lack of acid prevents it from functioning because pepsin requires an acid medium for activity. Obviously, then, essentially all digestive function in the stomach is lost when achlorhydria is present.

The method usually used to determine the degree of hypochlorhydria is to inject 0.5 milligram of histamine and then to aspirate the stomach secretions through a tube for the following hour. Each 10-minute sample of stomach secretion is titrated against sodium hydroxide to a pH of 8.5, which gives a measure of *total acid* secreted. The maximum rate of acid secretion during any 10-minute interval in the normal person rises to almost exactly 1 milliequivalent of hydrochloric acid per minute, and the total hydrochloric acid secreted during the entire hour after the injection averages about 18 milliequivalents. All degrees of hypochlorhydria — down to no acid whatsoever — occur.

The acid secretion is frequently divided into *free acid* and *combined acid*. The amount of free acid is determined by titrating the gastric secretions to a pH of 3.5, using *dimethylaminoazobenzene* as an indicator. After this titration has been performed, the same secretions are titrated to a pH of 8.5, using *phenolphthalein* as an indicator; this measures the combined acid. Gastric secretions mixed with food in the stomach usually show little or no free acid but a large amount of combined acid. On the other hand, when the stomach secretes large quantities of gastric juice while it is almost empty of food, the larger portion of the acid may then be free acid and only a small amount combined.

Though achlorhydria is associated with depressed or even no digestive capability by the stomach, the overall digestion of food in the entire gastrointestinal tract is still almost normal. This is because trypsin and other enzymes secreted by the pancreas are capable of digesting most of the protein in the diet — particularly if the food is well chewed so that no portion of the protein is protected by collagen fibers, which need peptic activity for most effective digestion.

Pernicious Anemia in Gastric Atrophy. Pernicious anemia is a common accompaniment of gastric atrophy. The normal gastric secretions contain a glycoprotein called *intrinsic factor*, which is secreted by the oxyntic (parietal) cells (the HCl-producing cells) and which must be present for adequate absorption of vitamin B_{12} from the ileum. The instrinsic factor combines with vitamin B_{12}, and the complex then binds with receptors on the ileal epithelial surface. In some yet unknown way this makes it possible for the vitamin B_{12} to be absorbed. In the absence of intrinsic factor, an adequate amount of vitamin B_{12} is not made available from the foods. As a result, maturation failure occurs in the bone marrow, resulting in pernicious anemia. This subject was discussed in more detail in Chapter 5.

Pernicious anemia also occurs frequently when most of the stomach has been removed for treatment of either stomach ulcer or gastric cancer or when the ileum, where vitamin B_{12} is almost entirely absorbed, is removed.

PEPTIC ULCER

A peptic ulcer is an excoriated area of the mucosa caused by the digestive action of gastric juice. Figure 66–1 illustrates the points in the gastrointestinal tract at which peptic ulcers frequently occur, showing that by far the most frequent site is in the first few centimeters of the duodenum. In addition, peptic ulcers frequently occur along the lesser curvature of the antral end of the stomach or, more rarely, in the lower end of the esophagus where stomach juices frequently reflux. A peptic ulcer called a *marginal ulcer* also frequently occurs wherever an abnormal opening, such as a gastrojejunostomy, is made between the stomach and some portion of the small intestine.

Basic Cause of Peptic Ulceration. The usual cause of peptic ulceration is too much secretion of gastric juice in relation to the degree of protection afforded by the gastroduodenal mucosal barrier and the neutralization of the gastric acid by duodenal juices. It will be recalled that all areas normally exposed to gastric juice are well supplied with mucous glands, beginning with the compound mucous glands of the lower esophagus, then

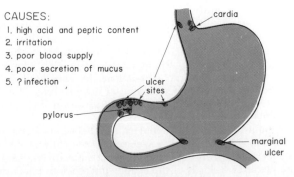

CAUSES:
1. high acid and peptic content
2. irritation
3. poor blood supply
4. poor secretion of mucus
5. ? infection

Figure 66–1. Peptic ulcer.

including the mucous cell coating of the stomach mucosa, the mucous neck cells of the gastric glands, the deep pyloric glands that secrete mainly mucus, and, finally, the glands of Brunner of the upper duodenum, which secrete a highly alkaline mucus.

In addition to the mucus protection of the mucosa, the duodenum is also protected by the alkalinity of the small intestinal secretions. Especially important is pancreatic secretion, which contains large quantities of sodium bicarbonate that neutralize the hydrochloric acid of the gastric juice, thus inactivating the pepsin to prevent digestion of the mucosa. Two additional mechanisms insure that this neutralization of gastric juices is complete:

1. When excess acid enters the duodenum, it reflexly inhibits gastric secretion and peristalsis in the stomach, both nervously and hormonally, thereby decreasing the rate of gastric emptying. This allows increased time for pancreatic secretion to enter the duodenum to neutralize the acid already present. After neutralization has taken place, the reflex subsides and more stomach contents are emptied.

2. The presence of acid in the small intestine liberates secretin from the intestinal mucosa, which then passes by way of the blood to the pancreas to promote rapid secretion of pancreatic juice containing a high concentration of sodium bicarbonate, thus making more sodium bicarbonate available for neutralization of the acid. These mechanisms were discussed in detail in Chapters 63 and 64 in relation to gastrointestinal motility and secretion.

Experimental Peptic Ulcer. Experimental peptic ulcers have been created in dogs and other animals in the following ways: (a) *Feeding ground glass* to an animal causes excoriation of the pyloric wall and allows the peptic juices to begin digesting the deeper layers of the mucosa. (b) *Transplantation of the pancreatic duct to the ileum* removes the normal neutralizing effect of pancreatic secretion in the duodenum and, therefore, allows the gastric juice to attack the mucosa of the upper duodenum. (c) *Repeated injection of histamine* causes excessive secretion of gastric juice. (d) *Continual infusion of hydrochloric acid* through a tube into the stomach causes direct irritation of the stomach mucosa and also prevents full neutralization of the gastric juices by the pancreatic and other secretions of the small intestine. Therefore, peptic activity is enhanced, and ulcers appear in either the stomach or duodenum. (e) A portion of the *stomach is anastomosed directly to the small intestine* so that gastric juice can pass directly and rapidly into the small intestine. The mucosa of the small intestine, except at the uppermost part of the duodenum, is not sufficiently resistant to gastric juice to prevent peptic digestion. (f) *Obstruction of the blood flow* or even reduction of the blood flow to an area of the stomach or upper duodenum will cause ulcers to develop because the local epithelium degenerates or cannot produce appropriate protective secretions.

In summary, any factor that (1) *increases the rate of production of gastric juice* or (2) *blocks the normal protective mechanisms* against this juice can produce peptic ulcers. The same general principles apply to the development of peptic ulcers in the human being.

Causes of Peptic Ulcer in the Human Being. Peptic ulcer occurs much more frequently in the white collar worker than in the laborer, and persons subjected to extreme anxiety for a long time seem particularly prone to peptic ulcer. For instance, the number of persons who developed peptic ulcer increased greatly during the air raids of London in World War II. Therefore, it is believed that many if not most instances of duodenal peptic ulcer in the human being result from excessive stimulation of the dorsal motor nucleus of the vagus by impulses originating in the cerebrum. The duodenal ulcers develop in the area of the duodenum where the gastric juices still have not been fully neutralized, immediately beyond the pylorus, as illustrated in Figure 66–1. Supporting this theory is the fact that duodenal ulcer patients have a high rate of gastric secretion during the interdigestive period between meals when the stomach is empty. The normal stomach secretes a total of approximately 18 milliequivalents of hydrochloric acid during the 12-hour interdigestive period through the night, while duodenal ulcer patients occasionally secrete as much as 300 milliequivalents of hydrochloric acid during this same time.

Paradoxically, gastric ulcers, in contradistinction to duodenal ulcers, often occur in patients who have normal or low secretion of hydrochloric acid. However, these patients almost invariably have an associated, often atrophic, gastritis, indicating that ulceration in the stomach almost certainly results from reduced resistance of the stomach mucosa to digestion rather than from excess secretion of gastric juice. Stomach ulceration frequently occurs in patients who have ingested substances such as aspirin or alcohol that reduce the mucosal resistance. Also, reflux of duodenal contents into the stomach often leads to gastric ulceration; this results mainly from the bile acids in the refluxed chyme because these acids have a detergent effect that reduces the mucosal resistance.

Physiology of Treatment. The usual medical treatment for peptic ulcer is a combination of (1) reduction of stressful situations that might lead to excessive acid secretion, (2) administration of antacid drugs to neutralize much of the acid in the stomach secretions, (3) administration of the drug *cimetidine* which blocks the effect of histamine on H_2-receptors and thereby blocks the action of gastrin in stimulating gastric juice secretion, (4) prescription of a bland diet and sometimes small meals many times a day rather than three meals a day (though there is not yet proof that spreading meals is truly effective), (5) interdiction of smoking because statistical studies have shown that smokers are several times as prone to have peptic ulcers as are nonsmokers, and (6) removal of such ulcer-causing factors as alcohol, aspirin, or other substances that might irritate the gastroduodenal mucosa.

Surgical treatment of peptic ulcer is effected by one or both of two procedures: (1) removal of a large portion of the stomach or (2) vagotomy. When the stomach is surgically removed, usually at least the lower three-fourths to four-fifths of the stomach is removed, and the upper stump of the stomach is then anastomosed to the jejunum. If less of the stomach than this is removed, far too much gastric juice continues to be secreted, and a marginal ulcer soon develops where the stomach is anastomosed to the intestine. Since a marginal ulcer is frequently equally as debilitating as, or sometimes even more debilitating than, the original ulcer, nothing will have been accomplished. However, if only one-fifth of the stomach remains, the rate of gastric secretion is usually low enough that a marginal ulcer does not develop, and yet the stomach is still large enough to hold small meals. Furthermore, the presence of at least some functional stomach mucosa usually prevents the development of pernicious anemia.

Vagotomy temporarily blocks almost all secretion by the stomach and cures a peptic ulcer within less than a week after the operation is performed. The vagus nerves may be cut in the chest as they pass downward over the surface of the esophagus, or they may be cut as they pass through the diaphragm to spread over the surface of the stomach. Unfortunately, though, a large amount of basal stomach secretion returns a few months after the vagotomy, and in many patients the ulcer itself also returns. Even more distressing is the *gastric atony* that usually follows vagotomy, for the motility of the stomach is often reduced to such a low level that almost no gastric emptying occurs after vagotomy. For this reason, vagotomy alone is rarely performed in the treatment of peptic ulcer. However, it is performed frequently in association with simultaneous removal of the stomach antrum or a plastic procedure to increase the size of the opening of the pylorus from the stomach into the small intestine.

DISORDERS OF THE SMALL INTESTINE

Abnormal Digestion of Food in the Small Intestine; Pancreatic Failure. By far the greater portion of all digestion in the alimentary tract occurs in the upper third of the small intestine. Only rarely is digestion impaired enough to cause malnutrition. Indeed, as much as three-fourths of the small intestine has been removed without patients suffering serious malnutrition.

Perhaps the commonest cause of abnormal digestion is failure of the pancreas to secrete its juice into the small intestine. Lack of pancreatic secretion frequently occurs (a) in *pancreatitis,* which is discussed below, (b) when the *pancreatic duct is blocked* by a gallstone at the papilla of Vater, or (c) after the *head of the pancreas has been removed* because of malignancy. Loss of pancreatic juice means loss of trypsin, chymotrypsin, carboxypolypeptidase, pancreatic amylase, pancreatic lipase, and still a few other digestive enzymes. Without these enzymes, as much as one-half of the fat entering the small intestine may go unabsorbed and as much as one-third of the proteins and starches. As a result, large portions of the ingested food are not utilized for nutrition; and copious, fatty feces are excreted.

Pancreatitis. Pancreatitis means inflammation of the pancreas, and this can occur in the form of either *acute pancreatitis* or *chronic pancreatitis*. The commonest cause of acute pancreatitis is blockage of the papilla of Vater by a gallstone; this blocks the main secretory duct from the pancreas as well as the common bile duct. The pancreatic enzymes are then dammed up in the ducts and acini of the pancreas. Eventually, so much trypsinogen accumulates that it *overcomes the trypsin inhibitor* in the secretions, and a small quantity of trypsinogen becomes actived to form trypsin. Once this happens the trypsin activates still more trypsinogen as well as chymotrypsinogen and carboxypolypeptidase, resulting in a vicious cycle until most of the proteolytic enzymes in the pancreatic ducts and acini become activated. These rapidly digest large portions of the pancreas itself, sometimes completely and permanently destroying the ability of the pancreas to secrete digestive enzymes.

Often in acute pancreatitis the proteolytic enzymes eat their way all the way to the surface of the pancreas, and pancreatic juice then empties into the peritoneal cavity where additional proteolytic activity occurs, causing *chemical peritonitis*. However, as soon as large portions of the pancreas have been destroyed, the acute stage of pancreatitis subsides, leaving the person with diminished or sometimes totally absent pancreatic secretion into the gut.

Chronic pancreatitis most often results from chronic alcoholism, beginning with acute episodes of pancreatic inflammation usually with severe pain, and progressing to pancreatic fibrosis until pancreatic function becomes greatly diminished.

Ordinarily, the islets of Langerhans are not seriously affected by pancreatitis, so that insulin still continues to be secreted by the pancreas even though secretion of pancreatic juice into the intestine is markedly reduced.

Malabsorption by the Small Intestinal Mucosa — "Sprue." Occasionally, nutrients are not adequately absorbed from the small intestine even though the food is well digested. Several different diseases can cause decreased absorbability by the mucosa; these are often classified together under the general heading of *sprue*. Obviously, also, malabsorption can occur when large portions of the small intestine have been removed.

Nontropical Sprue. One type of sprue, called variously by the names *idiopathic sprue, celiac disease* (in children), or *gluten enteropathy*, results from the toxic effects of *gluten* present in certain types of grains, especially wheat and rye. It is the gliadin fraction that is responsible for this toxic effect, perhaps because of a direct destructive effect on the intestinal epithelial cells, or perhaps as the result of an immunological or allergic reaction to the gliadin. In milder forms of the disease, the microvilli of the absorbing cells on the villi are destroyed, thus decreasing the absorptive surface area as much as 2-fold. In the more severe forms, the gliadin causes early and total destruction and dissolution of newly forming epithelial cells in the crypts of Lieberkühn. Therefore, these cells fail to migrate upward onto the villi. As a result, the villi become blunted or disappear altogether, thus still further reducing the absorptive area of the gut. Removal of wheat and rye flour from the diet frequently results in a miraculous cure within weeks, especially in children with this disease.

Tropical Sprue. A different type of sprue called *tropical sprue* frequently occurs in the tropics and can often be treated with antibacterial agents. Even though no specific bacterium has ever been implicated as the cause, it is believed that this variety is often caused by inflammation of the intestinal mucosa resulting from yet unidentified infectious agents.

Malabsorption in Sprue. In the early stages of sprue, the absorption of fats is more impaired than the absorption of other digestive products. The fat appearing in the stools is almost entirely in the form of soaps rather than undigested neutral fat, illustrating that the problem is one of absorption and not of digestion. In this stage of sprue, the conditon is frequently called *idiopathic steatorrhea*, which means simply excess fats in the stools as a result of unknown causes.

In more severe cases of sprue the absorption of proteins, carbohydrates, calcium, vitamin K, folic acid, and vitamin B_{12}, as well as many other important substances, becomes greatly impaired. As a result, the person suffers (1) severe nutritional deficiency, often developing severe wasting of the tissues, (2) osteomalacia (demineralization of the bones because of calcium lack), (3) inadequate blood coagulation due to lack of vitamin K, and (4) macrocytic anemia of the pernicious anemia type, owing to diminished vitamin B_{12} and folic acid absorption.

Regional Enteritis and Appendicitis. Regional enteritis means an inflammatory condition of the terminal portion of the ileum, and appendicitis means inflammation of the appendix. Appendicitis is usually caused by an actual infection in the appendix, while the cause of regional enteritis is not known but in some instances might also result from a low grade infectious process, or perhaps an allergic or autoimmune reaction. Either appendicitis or ileitis cause crampy pain that is referred to the mid-abdomen. Simultaneously, each also elicits intense enterointestinal reflexes resulting in severe inhibition of gastrointestinal mobility. As a result, functional obstruction often occurs in the small bowel, causing symptoms of acute intestinal obstruction, mainly distension of the upper small intestine associated with protracted cramping pains and vomiting. These will be discussed later in the chapter.

DISORDERS OF THE LARGE INTESTINE

CONSTIPATION

Constipation means slow movement of feces through the large intestine, and it is often associated with large quantities of dry, hard feces in the descending colon which accumulate because of the long time available for absorption of fluid.

A frequent cause of constipation is irregular bowel habits that have developed through a lifetime of inhibition of the normal defecation reflexes. Newborn children are rarely constipated, but part of their training in the early years of life requires that they learn to control defecation, and this control is effected by inhibiting the natural defecation reflexes. Clinical experience shows that if one fails to allow defecation to occur when the defecation reflexes are excited or if one overuses laxatives to take the place of natural bowel function, the reflexes themselves becomes progressively less strong over a period of time and the colon becomes *atonic*. For this reason, if a person establishes regular bowel habits early in life, usually defecating in the morning after breakfast when the gastrocolic and duodenocolic reflexes cause mass movements in the large intestine, the development of constipation in later life can generally be prevented.

Constipation can also result from spasm of a small segment of the sigmoid. It should be recalled that motility, even normally, is weak in the large intestine, so that even a slight degree of spasm is often capable of causing serious constipation. This effect frequently occurs in the so-called *irritable colon syndrome* in which bowel spasm causes constipation and consequent development of small and hard feces, a condition also often associated with crampy abdominal pains. After several days, the constipation is relieved by a bowel movement that often contains a moderate to large amount of mucus. The period of constipation may then be followed by a day or so of diarrhea. Following this, the cycle begins again, with repeated bouts of alternating constipation and diarrhea.

Megacolon. Occasionally, a person develops constipation which is so severe that bowel movements occur only once every week or so. Obviously, this allows tremendous quantities of fecal matter to accumulate in the colon, causing the colon sometimes to distend to a diameter as great as 3 to 4 inches. The condition is called megacolon, or *Hirschsprung's disease*.

The most frequent cause of megacolon is lack of ganglion cells of the myenteric plexus in a segment of the sigmoid. As a consequence, neither defecation reflexes nor peristaltic motility can occur through this area of the large intestine. And the sigmoid itself becomes small and almost spastic while feces accumulate proximal to this area, causing megacolon. Thus, the diseased portion of the large intestine appears normal by x-ray, while the originally normal portion of the large intestine appears greatly enlarged.

DIARRHEA

Diarrhea, which is the opposite of constipation, results from rapid movement of fecal matter through the large intestine. Several causes of diarrhea with important physiological overtones are:

Enteritis. Enteritis means infection caused either by a virus or by bacteria in the intestinal tract. In usual infectious diarrhea, the infection is most extensive in the large intestine and the distal end of the ileum. Everywhere that the infection is present, the mucosa becomes extensively irritated, and its rate of secretion becomes greatly enhanced. In addition, the motility of the intestinal wall usually increases many fold. As a result, large quantities of fluid are made available for washing the infectious agent toward the anus, and at the same time strong propulsive movements propel this fluid forward. Obviously, this is an important mechanism for ridding the intestinal tract of the debilitating infection.

Of special interest is the diarrhea caused by *cholera* (and sometimes by other bacteria such as pathogenic colon bacilli). The cholera toxin directly stimulates excessive secretion of electrolytes and fluid from the crypts of Lieberkühn in the distal ileum and colon, and it specifically enhances the bicarbonate-chloride exchange mechanism, causing extreme quantities of bicarbonate ions to be secreted into the intestinal tract. The loss of fluid and electrolytes can be so debilitating within a day or so that death ensues. Therefore, the most important physiological basis of therapy is simply to replace the fluid and electrolytes as rapidly as they are lost. With proper therapy of this type, along with the use of antibiotics, almost no cholera patients die, but without therapy 50 per cent or more do.

Psychogenic Diarrhea. Everyone is familiar with the diarrhea that accompanies periods of nervous tension, such as during examination time or when a soldier is about to go into battle. This type of diarrhea, called psychogenic or emotional diarrhea, is caused by excessive stimulation of the parasympathetic nervous system, which greatly excites both motility and secretion of mucus in the distal colon. These two effects added together can cause marked diarrhea.

The *irritable colon syndrome*, which was discussed above in relation to constipation and which often causes alternating constipation and diarrhea, is also generally believed to be at least partially a psychogenic (or emotional) disorder; it usually occurs in anxiety states.

Ulcerative Colitis. Ulcerative colitis is a disease in which extensive areas of the walls of the large intestine become ulcerated. This, like psychogenic diarrhea, is frequently associated with different states of nervous tension. The motility of the ulcerated colon is often so great that mass movements occur most of the time, rather than the usual 10 to 20 minutes per day. Also, the colon's secretions are greatly enhanced. As a result, the patient has repeated diarrheal bowel movements.

Though ulcerative colitis is associated with psychogenic disorders, the precise cause of the actual ulcers is unknown. Some clinicians believe that these are caused by specific infectious bacteria invading the mucosa; others believe that they result from an allergic or immune destructive effect. Regardless of the cause, if the condition has not progressed too far, removal of factors that have been causing nervous tension frequently leads to a cure. On the other hand, if the condition has progressed extremely far, the ulcers usually will not heal until an ileostomy is performed to allow the intestinal contents to drain to the exterior rather than to flow through the colon. Even then the ulcers sometimes fail to heal, and the only solution is removal of the colon itself.

PARALYSIS OF DEFECATION IN SPINAL CORD INJURIES

From Chapter 63 it will be recalled that defecation is normally initiated by the movement of feces into the rectum, which causes a cord-mediated defecation reflex passing from the rectum to the spinal cord and then back to the descending colon, sigmoid, rectum, and anus. This reflex adds greatly to the activity of the intrinsic defecation reflex mediated through the myenteric plexus of the sigmoid and colon wall itself. Frequently, this *cord* defecation reflex is blocked or altered by spinal cord injuries. For instance, destruction of the *conus medullaris* of the spinal cord destroys the sacral centers in which the cord reflex is integrated and therefore almost completely paralyzes defecation. In such instances defecation requires extensive supportive measures, such as cathartics and large enemas.

But, more frequently, the spinal cord is injured somewhere between the conus medullaris and the brain, in which case the voluntary portion of the defecation act is blocked while the basic cord reflexes for defecation are still intact. Nevertheless, loss of the voluntary aid to defecation — that is, loss of the increased abdominal pressure, the lifting of the pelvic floor, and the stretching of the anal ring by the pelvic muscles — often makes defecation a difficult process in the person with this type of cord injury. Yet, since cord defecation reflexes can still occur, a small enema to potentiate the action of these reflexes, usually given in the morning shortly after a meal, can often cause adequate defecation. In this way persons with spinal cord injuries can usually control their bowel movements each day.

GENERAL DISORDERS OF THE GASTROINTESTINAL TRACT

VOMITING

Vomiting is the means by which the upper gastrointestinal tract rids itself of its contents when almost any part of the gastrointestinal tract becomes excessively irritated, overdistended, or even overexcitable. Distention or irritation of the stomach or duodenum provides the strongest stimulus for vomiting. Impulses are transmitted, as illustrated in Figure 66–2, by both vagal and sympathetic afferents to the bilateral *vomiting center* of the medulla, which lies near the tractus solitarius at approximately the level of the dorsal motor nucleus of

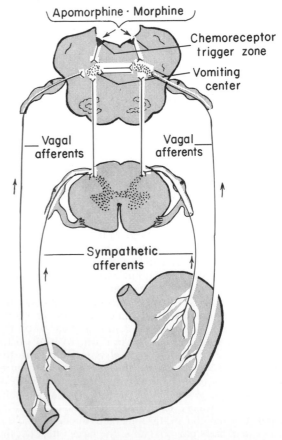

Figure 66–2. The afferent connections of the vomiting center.

the vagus. Appropriate motor reactions are then instituted to cause the vomiting act, and the motor impulses that cause the actual vomiting are transmitted from the vomiting center through the fifth, seventh, ninth, tenth, and twelfth cranial nerves to the upper gastrointestinal tract and through the spinal nerves to the diaphragm and abdominal muscles.

The Vomiting Act. Once the vomiting center has been sufficiently stimulated and the vomiting act instituted, the first effects are (1) a deep breath, (2) raising of the hyoid bone and the larynx to pull the crico-esophageal sphincter open, (3) closing of the glottis, and (4) lifting of the soft palate to close the posterior nares. Next comes a strong downward contraction of the diaphragm along with simultaneous contraction of all the abdominal muscles. This obviously squeezes the stomach between the two sets of muscles, building the intragastric pressure to a high level. Finally, the gastro-esophageal sphincter relaxes, allowing expulsion of the gastric contents upward through the esophagus.

Thus, the vomiting act results from a squeezing action of the muscles of the abdomen associated with opening of the esophageal sphincters so that the gastric contents can be expelled.

The Chemoreceptor Trigger Zone of the Medulla for Initiation of Vomiting by Drugs or by Motion Sickness. Aside from the vomiting initiated by irritative stimuli in the gastrointestinal tract itself, vomiting can also be caused by impulses arising in areas of the brain outside the vomiting center. This is particularly true of a small area located bilaterally on the floor of the fourth ventricle in or above the area postrema and called the *chemoreceptor trigger zone*. Electrical stimulation of this area will initiate vomiting, but, more importantly, administration of certain drugs, including apomorphine, morphine, and some of the digitalis derivatives, can directly stimulate the chemoreceptor trigger zone and initiate vomiting. Destruction of this area blocks this type of vomiting but does not block vomiting resulting from irritative stimuli in the gastrointestinal tract itself.

Also, it is well known that rapidly changing motions of the body cause certain people to vomit. The mechanism for this is the following: The motion stimulates the receptors of the labyrinth, and impulses are transmitted mainly by way of the vestibular nuclei into the cerebellum. After passing through the uvula and nodule of the cerebellum, the signals are believed to be transmitted to the chemoreceptor trigger zone and thence to the vomiting center to cause vomiting.

Cortical Excitation of Vomiting. Various psychic stimuli, including disquieting scenes, noisome odors, and other similar psychological factors, can also cause vomiting. Stimulation of certain areas of the hypothalamus also causes vomiting. The precise neuronal connections for these effects are not known, though it is probable that the impulses pass directly to the vomiting center and do not involve the chemoreceptor trigger zone.

NAUSEA

Everyone has experienced the sensation of nausea and knows that it is often a prodrome of vomiting.

Nausea is the conscious recognition of subconscious excitation in an area of the medulla closely associated with or part of the vomiting center, and it can be caused by irritative impulses coming from the gastrointestinal tract, impulses originating in the lower brain associated with motion sickness, or impulses from the cerebral cortex to initiate vomiting. However, vomiting occasionally occurs without the prodromal sensation of nausea, indicating that only certain portions of the vomiting centers are associated with the sensation of nausea.

A common cause of nausea is distension or irritation of the duodenum or lower small intestine. When this occurs, the intestine contracts forcefully while the stomach relaxes, often allowing the intestinal contents to reflux into the stomach. This is preliminary to the vomiting that often follows.

GASTROINTESTINAL OBSTRUCTION

The gastrointestinal tract can become obstructed at almost any point along its course, as illustrated in Figure 66–3; some common causes of obstruction are: *cancer, fibrotic constriction resulting from ulceration or from peritoneal adhesions, spasm of a segment of the gut,* or *paralysis of a segment of the gut.*

The abnormal consequences of obstruction depend on the point in the gastrointestinal tract that becomes obstructed. If the obstruction occurs at the pylorus, which results often from fibrotic constriction following peptic ulceration, persistent vomiting of stomach contents occurs. This obviously depresses bodily nutrition, and it also causes excessive loss of hydrogen ions from the body and can result in various degrees of alkalosis.

If the obstruction is beyond the stomach, reflux from the small intestine causes the intestinal juices to flow backward into the stomach, and these are vomited along with the stomach secretions. In this instance the person loses large amounts of water and electrolytes so that he becomes severely dehydrated, but the loss of acids and bases may be approximately equal so that little change in acid-base balance occurs. If the obstruction is near the lower end of the small intestine, then it is actually possible to vomit more basic than acidic substances; in this case acidosis may result. Also, the vomitus, after a

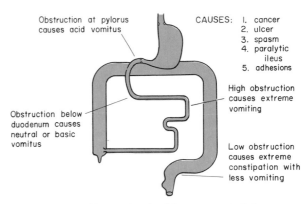

Figure 66–3. Obstruction in different parts of the gastrointestinal tract.

few days of obstruction, becomes fecal in character. Some physiologists believe that this reflux of intestinal contents all the way to the stomach is caused by reverse peristalsis, an effect that is not normal to the intestinal tract.

Also important in obstruction of the small intestine is marked distension of the intestine proximal to the obstructed point. Large quantities of fluid and electrolytes continue to be secreted into the lumen of the small intestine and even large amounts of proteins are lost from the blood stream, partly into the intestinal lumen and partly into the gut wall, which becomes edematous as a result of the excessive distention. The plasma volume diminishes because of the protein loss, and severe circulatory shock often ensues. One might immediately ask: Why does the small intestine not absorb these fluids and electrolytes? The answer is that distension of the gut usually stimulates the secretory activity of the gut but does not correspondingly increase the rate of absorption. Normally this would act to flush the chyme farther down the small intestine and therefore relieve the distension. But, if obstruction is present, obviously this normal mechanism backfires and simply causes a vicious cycle of more and more distension.

If the obstruction is near the distal end of the large intestine, feces can accumulate in the colon for several weeks. The patient develops an intense feeling of constipation, but in the first stages of the obstruction vomiting is not severe. After the large intestine has become completely filled and it finally becomes impossible for additional chyme to move from the small intestine into the large intestine, vomiting does then begin severely. Prolonged obstruction of the large intestine will finally cause rupture of the intestine itself or dehydration and circulatory shock resulting from the severe vomiting.

GASES IN THE GASTROINTESTINAL TRACT AND FLATUS

Gases can enter the gastrointestinal tract from three different sources: (1) swallowed air, (2) gases formed as a result of bacterial action, and (3) gases that diffuse from the blood into the gastrointestinal tract.

Most gases in the stomach are nitrogen and oxygen derived from swallowed air, and in the normal person most of these are expelled by belching.

Only small amounts of gas are usually present in the small intestine, and much of these are air that passes from the stomach into the intestinal tract. In addition, considerable carbon dioxide also often occurs because the reaction between acidic gastric juice and bicarbonate in pancreatic juice sometimes is too rapid for the liberated CO_2 to be absorbed.

In the large intestine, the greater proportion of the gases is derived from bacterial action; these gases include especially *carbon dioxide, methane*, and *hydrogen*. When the methane and hydrogen become suitably mixed with oxygen from swallowed air, an actual explosive mixture is occasionally formed; and use of the electric cautery during sigmoidoscopy has rarely caused colon explosions.

Essentially all the gases in the large intestine are highly diffusible through the intestinal mucosa. There-fore, if the gases remain in the large intestine for many hours the final mixture contains approximately 75 per cent or more of nitrogen and little of the other gases. The reason for this is that nitrogen in the gut cannot easily be absorbed into the blood because of the high P_{N_2} already in the blood, as explained in Chapter 43, even though essentially all other gases are easily absorbed. Indeed, the high blood P_{N_2} can actually cause nitrogen to diffuse *into* the colon. However, if the gases are passed on through the colon rapidly, the composition of the expelled *flatus* may be as little as 20 per cent nitrogen, with the remaining 80 per cent composed mainly of carbon dioxide, methane, and hydrogen, all of which are formed locally by bacterial action.

Certain foods are known to cause greater expulsion of flatus from the large intestine than others — beans, cabbage, onion, cauliflower, corn, and certain highly irritant foods such as vinegar. Some of these foods serve as a suitable medium for gas-forming bacteria, especially because of unabsorbed fermentable types of carbohydrates (for instance, beans contain an undigestible sugar that passes on into the colon and becomes a superior food for the colonic bacteria), but in other instances excess gas results from irritation of the large intestine, which promotes rapid expulsion of the gases before they can be absorbed.

The amount of gases entering or forming in the large intestine each day averages 7 to 10 liters, whereas the average amount expelled is usually only about 0.6 liter. The remainder is absorbed through the intestinal mucosa. Most often, a person expels large quantities of gases not because of excessive bacterial activity, but because of excessive motility of the large intestine, the gases being moved on through the large intestine before they can be absorbed.

REFERENCES

Blair, E. (ed.): Markowitz's Experimental Surgery, 6th Ed. Baltimore, Williams & Wilkins, 1975.

Bockus, H. L., et al. (eds.): Gastroenterology, 3rd Ed. Philadelphia, W. B. Saunders Co., 1975.

Boley, S. J., et al.: Ischemic Disorders of the Intestines. Chicago, Year Book Medical Publishers, 1978.

Brooke, B. N.: Crohn's Disease: Aetiology, Clinical Manifestations and Management. New York, Oxford University Press, 1977.

Brooks, F. P. (ed.): Gastrointestinal Pathophysiology. New York, Oxford University Press, 1974.

Cook, G. C.: Tropical Gastroenterology. New York, Oxford University Press, 1980.

Csaky, T. Z. (ed.): Intestinal Absorption and Malabsorption. New York, Raven Press, 1975.

Duthie, H. L. (ed.): Gastrointestinal Motility in Health and Disease. Baltimore, University Park Press, 1978.

Dworken, H. J.: The Alimentary Tract. Philadelphia, W. B. Saunders Co., 1974.

Eisenberg, M. M.: Physiologic Approach to the Surgical Management of Duodenal Ulcer. Chicago, Year Book Medical Publishers, 1977.

Eisenberg, M. M.: Ulcers. Boston, G. K. Hall, 1979.

Goodman, M. J., and Sparberg, M.: Ulcerative Colitis. New York, John Wiley & Sons, 1978.

Halsted, C. H.: Intestinal absorption and malabsorption of folates. Annu. Rev. Med., 31:79, 1980.

Jones, F. A., et al. (eds.): Peptic Ulcer Healing: Recent Studies on Carbenoxolone. Baltimore, University Park Press, 1978.

Kirsner, J. B., and Shorter, R. G. (eds.): Inflammatory Bowel Disease, 2nd Ed. Philadelphia, Lea & Febiger, 1980.

Kirsner, J. B., and Winans, C. S.: The stomach. *In* Sodeman, W. A., Jr., and Sodeman, T. M. (eds.): Sodeman's Pathologic Physiology; Mechanisms of Disease, 6th Ed. Philadelphia, W. B. Saunders Co., 1979, p. 798.

Lebenthal, E. (ed.): Digestive Disease in Children. New York, Grune & Stratton, 1978.

Maingot, R.: Abdominal Operations. New York, Appleton-Century-Crofts, 1979.

McNicholl, B., *et al.* (eds.): Perspectives in Coeliac Disease. Baltimore, University Park Press, 1978.

Money, K. E.: Motion sickness. *Physiol. Rev., 50*:1, 1970.

Najarian, J. S., and Delaney, J. P. (eds.): Gastrointestinal Surgery. Chicago, Year Book Medical Publishers, 1979.

Payne, W. S., and Olsen, A. M.: The Esophagus. Philadelphia, Lea & Febiger, 1974.

Powell, L. W., and Piper, D. W. (eds.): Fundamentals of Gastroenterology, 3rd Ed. Baltimore, University Park Press, 1979.

Rankow, R. M., and Polayes, I. M.: Diseases of the Salivary Glands. Philadelphia, W. B. Saunders Co., 1976.

Silen, W. (rev.): Cope's Early Diagnosis of the Acute Abdomen. New York, Oxford University Press, 1979.

Sleisenger, M. H., and Fordtran, J. S.: Gastrointestinal Disease: Pathophysiology, Diagnosis, Management. Philadelphia, W. B. Saunders Co., 1977.

Welch, C. E., *et al.*: Manual of Lower Gastrointestinal Surgery. New York, Springer-Verlag, 1979.

Westergaard, H., and Dietschy, J. M.: Normal mechanisms of fat absorption and derangements induced by various gastrointestinal diseases. *Med. Clin. North Am., 58*:1413, 1974.

Part XII

METABOLISM AND TEMPERATURE REGULATION

67

Metabolism of Carbohydrates and Formation of Adenosine Triphosphate

The next few chapters deal with metabolism in the body, which means the chemical processes that make it possible for the cells to continue living. It is not the purpose of this textbook, however, to present the chemical details of all the various cellular reactions, for this lies in the discipline of biochemistry. Instead, these chapters are devoted to (1) a review of the principal chemical processes of the cell, and (2) an analysis of their physiological implications, especially in relation to the manner in which they fit into the overall concept of homeostasis.

RELEASE OF ENERGY FROM FOODS AND THE CONCEPT OF "FREE ENERGY"

A great proportion of the chemical reactions in the cells is concerned with making the energy in foods available to the various physiological systems of the cell. For instance, energy is required for (a) muscular activity, (b) secretion by the glands, (c) maintenance of membrane potentials by the nerve and muscle fibers, (d) synthesis of substances in the cells, and (e) absorption of foods from the gastrointestinal tract.

Coupled Reactions. All the energy foods — carbohydrates, fats and proteins — can be oxidized with oxygen in the cells, and in this process large amounts of energy are released. These same foods can also be burned with pure oxygen outside the body in an actual fire, again releasing large amounts of energy. However, this time the energy is released suddenly, all in the form of heat. The energy needed by the physiological processes of the cells is not heat but instead energy (a) to cause mechanical movement in the case of muscle function, (b) to concentrate solutes in the case of glandular secretion, and (c) to effect other functions. To provide this energy, the chemical reactions must be "coupled" with the systems responsible for these physiological functions. This coupling is accomplished by special cellular enzyme and energy transfer systems, some of which will be explained in this and subsequent chapters.

"Free Energy." The amount of energy liberated by complete oxidation of a food is called *the free energy of the food,* and this is generally represented by the symbol ΔF. Free energy is usually expressed in terms of calories per mole of food substance. For instance, the amount of free energy liberated by oxidation of 1 mole of glucose (180 grams of glucose) is 686,000 calories.

ROLE OF ADENOSINE TRIPHOSPHATE (ATP) IN METABOLISM

ATP is a labile chemical compound that is present in all cells and has the chemical structure shown in the formula on the opposite page.

From this formula it can be seen that ATP is a combination of adenine, ribose, and three phosphate radicals. The last two phosphate radicals are connected with the remainder of the molecule by so-called *high energy bonds,* which are indicated by the symbol $\sim$. The amount of free energy in each of these high energy bonds per mole of ATP is approximately 7300 calories under standard conditions and 8000 calories under the conditions of temperature and concentrations of the reactants in the body. Therefore, removal of each phosphate radical liberates 8000 calories of energy. After loss of one phosphate radical from ATP, the compound becomes *adenosine diphosphate* (ADP), and after loss of the second phosphate radical the compound becomes *adenosine monophosphate* (AMP). The interconversions between ATP, ADP, and AMP are the following:

$$\text{ATP} \underset{\text{+8000 cal.}}{\overset{-8000 \text{ cal.}}{\rightleftarrows}} \begin{Bmatrix} \text{ADP} \\ + \\ \text{PO}_4 \end{Bmatrix} \underset{\text{+8000 cal.}}{\overset{-8000 \text{ cal.}}{\rightleftarrows}} \begin{Bmatrix} \text{AMP} \\ + \\ 2\text{PO}_4 \end{Bmatrix}$$

ATP is present everywhere in the cytoplasm and nucleoplasm of all cells, and essentially all the physiological mechanisms that require energy for operation obtain this directly from the ATP (or some other similar high energy compound – guanosine triphos-

$$\left.\begin{array}{l} \text{ADENINE} \end{array}\right\{ \quad \text{(adenine ring structure with } NH_2)$$

ADENINE $\{$ structure

RIBOSE $\{$ structure

TRIPHOSPHATE

$$CH_2-O-\overset{\overset{\displaystyle O}{\|}}{\underset{\underset{\displaystyle O^-}{}}{P}}-O\sim\overset{\overset{\displaystyle O}{\|}}{\underset{\underset{\displaystyle O^-}{}}{P}}-O\sim\overset{\overset{\displaystyle O}{\|}}{\underset{\underset{\displaystyle O^-}{}}{P}}-O^-$$

phate, GTP, for example). In turn, the food in the cells is gradually oxidized, and the released energy is used to re-form the ATP, thus always maintaining a supply of this substance; all of these energy transfers take place by means of coupled reactions.

In summary, ATP is an intermediary compound that has the peculiar ability of entering into many coupled reactions — reactions with the food to extract energy, and reactions in relation to many physiological mechanisms to provide energy for their operation. For this reason, ATP has frequently been called the energy *currency* of the body that can be gained and spent again and again.

The principal purpose of the present chapter is to explain how the energy from carbohydrates can be used to form ATP in the cells. At least 99 per cent of all the carbohydrates utilized by the body is used for this purpose.

CENTRAL ROLE OF GLUCOSE IN CARBOHYDRATE METABOLISM

From Chapter 65 it will be recalled that the final products of carbohydrate digestion in the alimentary tract are almost entirely glucose, fructose, and galactose — with glucose representing, on the average, about 80 per cent of these. After absorption from the intestinal tract, most of the fructose and galactose are also almost immediately converted into glucose. Therefore, very little fructose and galactose are present in the circulating blood. Glucose thus becomes the final common pathway for transport of almost all carbohydrates to the tissue cells.

In the case of fructose, much of this is converted into glucose as it is absorbed through the intestinal epithelial cells into the portal blood. This conversion was discussed in Chapter 65. Most of the remaining fructose, along with galactose, is then converted into glucose by the liver as explained below.

In liver cells, appropriate enzymes are available to promote interconversions among the monosaccharides, as shown in Figure 67–1. Furthermore, the dynamics of the reactions are such that when the liver

releases the monosaccharides back into the blood the final product is almost entirely glucose. Note that all three of the monosaccharides, galactose, glucose, and fructose, on entering the liver cells bind with phosphate, and reversible reactions allow interconversions among these three phosphorylated monosaccharides. Then, because the liver cells contain large amounts of *glucose phosphatase*, the glucose 6-phosphate can be degraded back to glucose and phosphate, and the glucose transported back into the blood.

Therefore, once again it should be emphasized that essentially all the monosaccharides that circulate in the blood are the final conversion product, glucose.

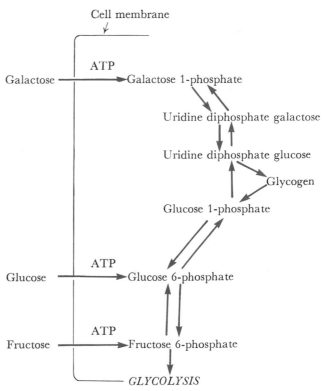

Figure 67–1. Interconversions of the three major monosaccharides—glucose, fructose, and galactose—in liver cells.

TRANSPORT OF GLUCOSE THROUGH THE CELL MEMBRANE

Before glucose can be used by the cells, it must be transported through the cell membrane into the cellular cytoplasm. However, glucose cannot diffuse through the pores of the cell membrane because the maximum molecular weight of particles that can do this is about 100, while glucose has a molecular weight of 180. Yet, glucose does pass to the interior of the cells with a reasonable degree of freedom but is transported through the membrane by the mechanism of *facilitated diffusion*. The principles of this type of transport were discussed in Chapter 4. Basically, they are the following: In the lipid matrix of the cell membrane are large numbers of protein *carrier* molecules that can bind with glucose. In this bound form the glucose can be transported by the carrier from one side of the membrane to the other side and then released. Therefore, if the concentration of glucose is greater on one side of the membrane than on the other side, more glucose will be transported from the high concentration area than in the opposite direction.

It should be noted that this transport of glucose through the membranes of most tissue cells is quite different from that which occurs through the gastrointestinal membrane or through the epithelium of the renal tubules. In both these cases, the glucose is transported by the mechanism of active sodium co-transport, in which active transport of sodium provides energy for absorbing glucose against a concentration difference. This sodium co-transport mechanism functions only in certain special epithelial cells that are specifically adapted for active absorption of glucose. Instead, at other cell membranes, glucose is transported only from higher concentration toward lower concentration by the process of facilitated diffusion, made possible by the special binding properties of the glucose carrier.

FACILITATION OF GLUCOSE TRANSPORT BY INSULIN

The rate of glucose transport, and also transport of some other monosaccharides, is greatly increased by insulin. When large amounts of insulin are secreted by the pancreas, the rate of glucose transport into some cells increases to as much as 10 times the rate of transport when no insulin at all is secreted. And, for practical considerations, the amounts of glucose that can diffuse to the insides of most cells of the body in the absence of insulin, with the exceptions of the liver and the brain, are far too little to supply anywhere near the amount of glucose normally required for energy metabolism. Therefore, in effect, the rate of carbohydrate utilization by the cells is controlled by the rate of insulin secretion in the pancreas. The functions of insulin and its control of carbohydrate metabolism will be discussed in detail in Chapter 78.

Failure of Disaccharides to Be Transported. Very minute amounts of disaccharides are absorbed into the blood from the gastrointestinal tract, but essentially none of these can be transported into the cells. There-

fore, no disaccharides or larger polysaccharides are utilized for cellular metabolism. Instead, they are excreted completely in the urine.

PHOSPHORYLATION OF GLUCOSE

Immediately upon entry into the cells, glucose combines with a phosphate radical in accordance with the following reaction:

$$\text{Glucose} \xrightarrow[\text{+ATP}]{\text{glucokinase}} \text{Glucose 6-phosphate}$$

This phosphorylation is promoted by the enzyme *glucokinase*. Similarly, *fructokinase* promotes fructose phosphorylation and *galactokinase* promotes galactose phosphorylation in those cells into which they are transported, principally the liver cells.

The phosphorylation of glucose is almost completely irreversible except in the liver cells, the renal tubular epithelium, and the intestinal epithelial cells in which glucose phosphatase is available for reversing the reaction. Therefore, in most tissues of the body phosphorylation serves to *capture* the glucose in the cell — once *in* the cell the glucose will not diffuse back out except from those special cells listed above that have the necessary phosphatase.

STORAGE OF GLYCOGEN IN LIVER AND MUSCLE

After absorption into the cells, glucose can be used immediately for release of energy to the cells or it can be stored in the form of *glycogen,* which is a large polymer of glucose.

All cells of the body are capable of storing at least some glycogen, but certain cells can store large amounts, especially the liver cells, which can store up to 5 to 8 per cent of their weight as glycogen, and muscle cells, which can store up to 1 per cent glycogen. The glycogen molecules can be polymerized to almost any molecular weight, the average molecular weight being 5,000,000 or greater; most of the glycogen precipitates in the form of solid granules. This conversion of the monosaccharides into a high molecular weight, precipitated compound makes it possible to store large quantities of carbohydrates without significantly altering the osmotic pressure of the intracellular fluids. Obviously, high concentrations of low molecular weight, soluble monosaccharides would play havoc with the osmotic relationships between intracellular and extracellular fluids.

GLYCOGENESIS

Glycogenesis is the process of glycogen formation, the chemical reactions for which are illustrated in Figure 67–2. From this figure it can be seen that *glucose 6-phosphate* first becomes *glucose 1-phosphate;* then this is converted to *uridine diphosphate glucose,* which is then converted into glycogen. Several specific enzymes are required to cause these conversions,

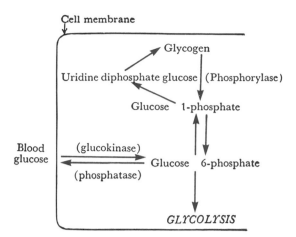

Figure 67–2. The chemical reactions of glycogenesis and glycogenolysis, showing also the interconversions between blood glucose and liver glycogen. (The phosphatase required for release of glucose from the cell is absent in muscle cells.)

and any monosaccharide that can be converted into glucose obviously can enter into the reactions. Certain smaller compounds, including *lactic acid, glycerol, pyruvic acid,* and *some deaminated amino acids,* can also be converted into glucose or closely allied compounds and thence into glycogen.

GLYCOGENOLYSIS

Glycogenolysis means the breakdown of glycogen to re-form glucose in the cells. Glycogenolysis does not occur by reversal of the same chemical reactions that serve to form glycogen; instead, each succeeding glucose molecule on each branch of the glycogen polymer is split away by a process of *phosphorylation,* catalyzed by the enzyme *phosphorylase* (several other enzymes split the glycogen molecule at the branching points).

Under resting conditions, the phosphorylase is in an inactive form so that glycogen can be stored but not reconverted into glucose. When it is necessary to re-form glucose from glycogen, therefore, the phosphorylase must first be activated. This is accomplished in the following two ways:

Activation of Phosphorylase by Epinephrine and Glucagon. Two hormones, epinephrine and glucagon, can specifically activate phosphorylase and thereby cause rapid glycogenolysis. The initial effect of each of these hormones is to increase the formation of *cyclic adenosine monophosphate (cAMP)* in the cells. This substance then initiates a cascade of chemical reactions that activates the phosphorylase. This will be discussed in more detail in Chapter 78.

Epinephrine is released by the adrenal medullae when the sympathetic nervous system is stimulated. Therefore, one of the functions of the sympathetic nervous system is to increase the availability of glucose for rapid metabolism. This function of epinephrine occurs markedly both in liver cells and in muscle, thereby contributing, along with other effects of sympathetic stimulation, to preparing the body for action, as was discussed in Chapter 57.

Glucagon is a hormone secreted by the *alpha cells* of the pancreas when the blood glucose concentration falls too low. It stimulates the formation of cAMP mainly in the liver. Therefore, its effect is primarily to convert liver glycogen into glucose and to release this into the blood, thereby elevating blood glucose concentration. The function of glucagon in blood glucose regulation is discussed in Chapter 78.

RELEASE OF ENERGY FROM THE GLUCOSE MOLECULE BY THE GLYCOLYTIC PATHWAY

Since complete oxidation of 1 gram-mole of glucose releases 686,000 calories of energy, and only 8000 calories of energy are required to form 1 gram-mole of adenosine triphosphate (ATP), it would be extremely wasteful of energy if glucose should be decomposed all the way into water and carbon dioxide at once while forming only a single ATP molecule. Fortunately, cells contain an extensive series of different protein enzymes that cause the glucose molecule to split a little at a time in many successive steps, with its energy released in small packets to form 1 molecule of ATP at a time, forming a total of 38 moles of ATP for each mole of glucose utilized by the cells.

The purpose of the present section is to describe the basic principles by which the glucose molecule is progressively dissected and its energy released to form ATP.

GLYCOLYSIS AND THE FORMATION OF PYRUVIC ACID

By far the most important means by which energy is released from the glucose molecule is by the process of *glycolysis* and then *oxidation of the end-products of glycolysis.* Glycolysis means splitting of the glucose molecule to form 2 molecules of pyruvic acid. This occurs by 10 successive steps of chemical reactions, as illustrated in Figure 67–3. Each step is catalyzed by at least 1 specific protein enzyme. Note that glucose is first converted into fructose 1,6-phosphate and then split into 2 three-carbon atom molecules, each of which is then converted through 5 successive steps into pyruvic acid.

Formation of Adenosine Triphosphate (ATP) During Glycolysis. Despite the many chemical reactions in the glycolytic series, little energy is released, and much of the energy that is released simply becomes heat and is lost from the metabolic systems of the cells. However, between the 1,3-diphosphoglyceric acid and the 3-phosphoglyceric acid stages, and again between the phosphopyruvic acid and the pyruvic acid stages, the packets of energy released are greater than 8000 calories per mole, the amount required to form ATP, and the reactions are coupled in such a way that ATP is formed. Thus, a total of 4 moles of ATP is formed for each mole of fructose 1,6-phosphate that is split into pyruvic acid.

Yet 2 moles of ATP had been required to phosphorylate the original glucose to form fructose 1,6-

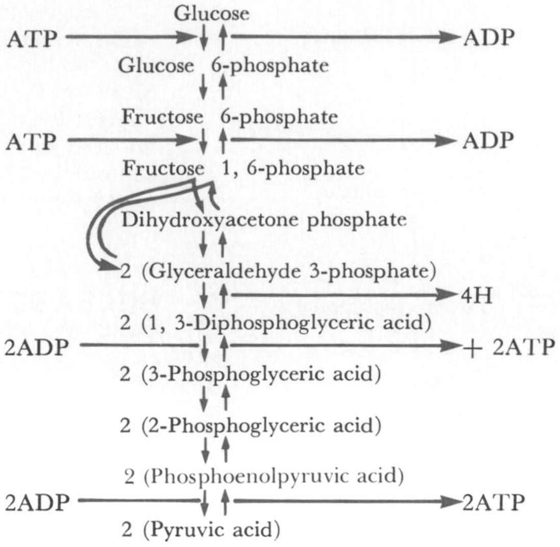

Glucose

$ATP \longrightarrow \downarrow \uparrow \longrightarrow ADP$

Glucose 6-phosphate

Fructose 6-phosphate

$ATP \longrightarrow \downarrow \uparrow \longrightarrow ADP$

Fructose 1, 6-phosphate

Dihydroxyacetone phosphate

2 (Glyceraldehyde 3-phosphate)

$\longrightarrow 4H$

2 (1, 3-Diphosphoglyceric acid)

$2ADP \longrightarrow \downarrow \uparrow \longrightarrow + 2ATP$

2 (3-Phosphoglyceric acid)

2 (2-Phosphoglyceric acid)

2 (Phosphoenolpyruvic acid)

$2ADP \longrightarrow \downarrow \uparrow \longrightarrow 2ATP$

2 (Pyruvic acid)

Net reaction:

$$Glucose + 2ADP + 2PO_4^{---} \longrightarrow 2 \text{ Pyruvic acid} + 2ATP + 4H$$

Figure 67–3.   The sequence of chemical reactions responsible for glycolysis.

phosophate before glycolysis could begin. Therefore, the net gain in ATP molecules by the entire glycolytic process is only 2 moles for each mole of glucose utilized. This amounts to 16,000 calories of energy stored in the form of ATP, but during glycolysis a total of 56,000 calories of energy is lost from the original glucose, giving an overall *efficiency* for ATP formation of 29 per cent. The remaining 71 per cent of the energy is lost in the form of heat.

Release of Hydrogen Atoms During Glycolysis.   Note in Figure 67–3 that 2 hydrogen atoms are released from each molecule of glyceraldehyde during its conversion to 1,3-diphosphoglyceric acid. And, since 2 molecules of glyceraldehyde are formed from each glucose molecule, a total of 4 hydrogen atoms is released. Later in the chapter we will see that still many more hydrogen atoms are released when pyruvic acid is split into its component parts and that these hydrogen atoms plus the 4 released during glycolysis are oxidized to provide most of the energy used in the synthesis of ATP.

CONVERSION OF PYRUVIC ACID TO ACETYL COENZYME A

The next stage in the degradation of glucose is (1) facilitated transport of the 2 derivative pyruvic acid molecules into the matrix of the mitochondria, and then (2) conversion of these into 2 molecules of *acetyl coenzyme A* (acetyl Co-A) in accordance with the following reaction:

$$2 \text{ CH}_3\text{—}\overset{\overset{\displaystyle O}{\|}}{C}\text{—COOH} + 2 \text{ Co-A—SH} \longrightarrow$$
(Pyruvic acid)   (Coenzyme A)

$$2 \text{ CH}_3\text{—}\overset{\overset{\displaystyle O}{\|}}{C}\text{—S—Co-A} + 2\text{CO}_2 + 4H$$
(Acetyl Co-A)

From this reaction it can be seen that 2 carbon dioxide molecules and 4 hydrogen atoms are released, while the remainders of the 2 pyruvic acid molecules combine with coenzyme A, a derivative of the vitamin pantothenic acid, to form 2 molecules of acetyl Co-A. In this conversion, no ATP is formed, but up to 6 molecules of ATP are formed when the 4 hydrogen atoms are later oxidized, as will be discussed in a later section.

THE CITRIC ACID CYCLE

The next stage in the degradation of the glucose molecule is called the *citric acid cycle* (also called the *tricarboxylic acid cycle* or *Krebs cycle*). This is a sequence of chemical reactions in which the acetyl portion of acetyl Co-A is degraded to carbon dioxide and hydrogen atoms. These reactions all occur in the matrix of the mitochondrion. The released hydrogen atoms are subsequently oxidized, as will be discussed later, releasing tremendous amounts of energy to form ATP.

Figure 67–4 shows the different stages of the chemical reactions in the citric acid cycle. The substances to the left are added during the chemical reactions, and the products of the chemical reactions are shown to the right. Note at the top of the column that the cycle begins with *oxaloacetic acid,* and then at the bottom of the chain of reactions, *oxaloacetic acid* is formed once again. Thus, the cycle can continue over and over.

In the initial stage of the citric acid cycle, *acetyl Co-A* combines with *oxaloacetic acid* to form *citric acid.* The coenzyme A portion of the acetyl Co-A is released and can be used again and again for the formation of still more quantities of acetyl Co-A from pyruvic acid. The acetyl portion, however, becomes an integral part of the citric acid molecule. During the successive stages of the citric acid cycle, several mol-

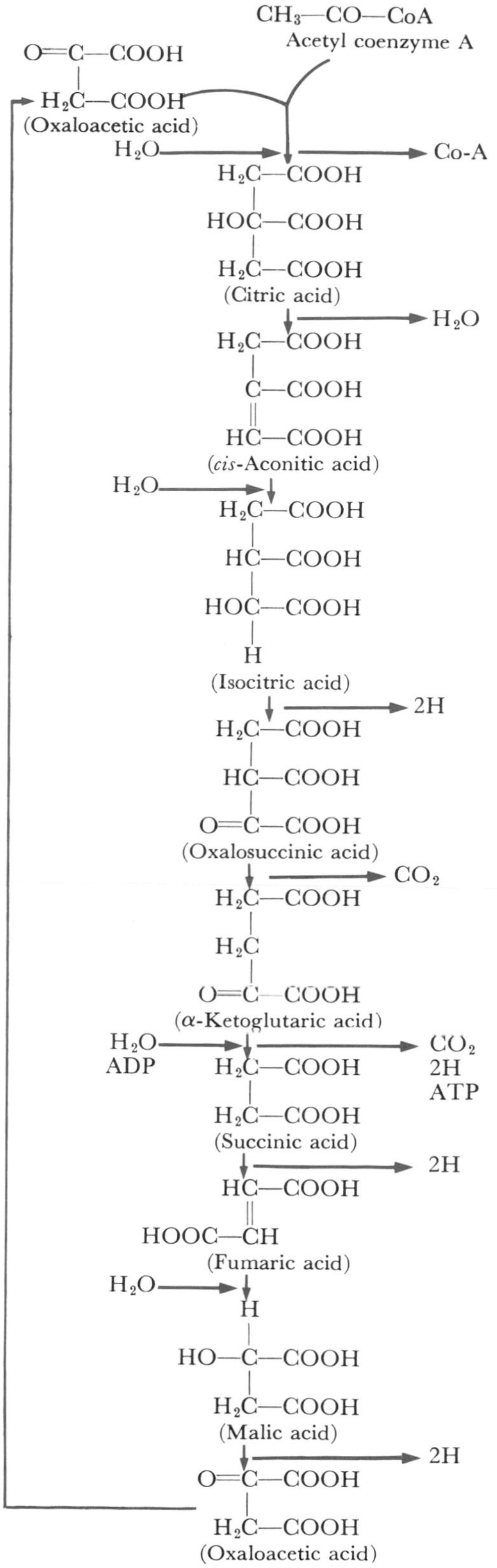

CH₃—CO—CoA
Acetyl coenzyme A

O=C—COOH
|
H₂C—COOH
(Oxaloacetic acid)
H₂O

H₂C—COOH → Co-A
|
HOC—COOH
|
H₂C—COOH
(Citric acid)

→ H₂O

H₂C—COOH
|
C—COOH
‖
HC—COOH
(cis-Aconitic acid)
H₂O

H₂C—COOH
|
HC—COOH
|
HOC—COOH
|
H
(Isocitric acid)

→ 2H

H₂C—COOH
|
HC—COOH
|
O=C—COOH
(Oxalosuccinic acid)

→ CO₂

H₂C—COOH
|
H₂C
|
O=C—COOH
(α-Ketoglutaric acid)

H₂O
ADP

→ CO₂
2H
ATP

H₂C—COOH
|
H₂C—COOH
(Succinic acid)

→ 2H

HC—COOH
‖
HOOC—CH
(Fumaric acid)
H₂O

→ H
|
HO—C—COOH
|
H₂C—COOH
(Malic acid)

→ 2H

O=C—COOH
|
H₂C—COOH
(Oxaloacetic acid)

Figure 67–4. The chemical reactions of the citric acid cycle, showing the release of carbon dioxide and an especially large number of hydrogen atoms during the cycle.

Net reaction per molecule of glucose:

2 Acetyl Co-A + 6H₂O + 2ADP ⟶
4CO₂ + 16H + 2Co-A + 2ATP

ecules of water are added; and *carbon dioxide* and *hydrogen atoms* are released at various stages in the cycle, as shown on the right in the figure.

The net results of the entire citric acid cycle are given in the legend of Figure 67–4, illustrating that for each molecule of glucose originally metabolized, 2 acetyl Co-A molecules enter into the citric acid cycle along with 6 molecules of water. These then are degraded into 4 carbon dioxide molecules, 16 hydrogen atoms, and 2 molecules of coenzyme A.

Formation of ATP in the Citric Acid Cycle. Not a great amount of energy is released during the citric acid cycle itself; in only one of the chemical reactions — during the change from α-ketoglutaric acid to succinic acid — is a molecule of ATP formed. Thus, for each molecule of glucose metabolized, 2 acetyl Co-A molecules pass through the citric acid cycle, each forming a molecule of ATP; or, a total of 2 molecules of ATP is formed.

Function of the Dehydrogenases and Nicotinamide Adenine Dinucleotide (NAD⁺) for Causing Release of Hydrogen Atoms. As noted at several points in this discussion, hydrogen atoms are released during different chemical reactions — 4 hydrogen atoms during glycolysis, 4 during the formation of acetyl Co-A from pyruvic acid, and 16 in the citric acid cycle; this makes a total of 24 hydrogen atoms. However, the hydrogen atoms are not simply turned loose in the intracellular fluid. Instead, they are released in packets of two, and in each instance the release is catalyzed by a specific protein enzyme called a *dehydrogenase*. Twenty of the 24 hydrogen atoms immediately combine with nicotinamide adenine dinucleotide (NAD⁺), a derivative of the vitamin niacin, in accordance with the following reaction:

$$\text{Substrate} \begin{array}{c} \text{H} \\ \diagup \\ \diagdown \\ \text{H} \end{array} + \text{NAD}^+ \xrightarrow{\textit{dehydrogenase}}$$

$$\text{NADH} + \text{H}^+ + \text{Substrate}$$

This reaction will not occur without the initial intermediation of the dehydrogenase nor without the availability of NAD⁺ to act as a hydrogen carrier. Both the free hydrogen ion and the hydrogen bound with NAD⁺ subsequently enter into the oxidative chemical reactions that form tremendous quantities of ATP, as will be discussed later.

The remaining 4 hydrogen atoms released during the breakdown of glucose — the four released during the citric acid cycle between the succinic and fumaric acid stages — combine with a specific dehydrogenase but are not then subsequently released to NAD⁺. Instead, they pass directly from the dehydrogenase into the oxidative processes.

Function of Decarboxylases for Causing Release of Carbon Dioxide. Referring again to the chemical reactions of the citric acid cycle and also to those for formation of acetyl Co-A from pyruvic acid, we find that there are three stages in which carbon dioxide is released. To cause the carbon dioxide release, other specific protein enzymes, called *decarboxylases*, split the carbon dioxide away from the substrate. The carbon dioxide in turn becomes dissolved in the fluids of the cells and is transported to the lungs where it is expired from the body (see Chapter 41).

FORMATION OF ATP BY OXIDATIVE PHOSPHORYLATION

Despite all the complexities of (a) glycolysis, (b) the citric acid cycle, (c) dehydrogenation, and (d) decarboxylation, pitifully small amounts of ATP are formed during all these processes, only 2 ATP molecules in the glycolysis scheme and another two in the citric acid cycle. Instead, about 90 per cent of the final ATP is formed during subsequent oxidation of the hydrogen atoms that are released during these earlier stages of glucose degradation. Indeed, the principal function of all these earlier stages is to make the hydrogen of the glucose molecule available in a form that can be utilized for oxidation.

Oxidation of hydrogen is accomplished by a series of enzymatically catalyzed reactions that (a) change the hydrogen atoms into hydrogen ions and electrons, and (b) use the electrons eventually to change the dissolved oxygen of the fluids into hydroxyl ions. Then the hydrogen and hydroxyl ions combine with each other to form water. During the sequence of oxidative reactions, tremendous quantities of energy are released to form ATP. Formation of ATP in this manner is called *oxidative phosphorylation.*

Unfortunately, even though oxidative phosphorylation is the basis of about 95 per cent of all energy utilization in the body, we still do not know the precise means by which the energy released from oxidation of hydrogen atoms is utilized to form ATP. However, a theory for which there is much support, called the *chemiosmotic theory,* is illustrated in Figure 67–5, and its essentials are described in the following sections.

Ionization of Hydrogen, the Electron Transport Chain, and Formation of Water. The first step in oxidative phosphorylation is to ionize the hydrogen atoms that are removed from the food substrates. As described above, these hydrogen atoms are removed in pairs: one immediately becomes a hydrogen ion, H^+ and the other combines with NAD^+ to form NADH. The upper portion of Figure 67–5 shows in color the subsequent fate of the NADH and H^+. The initial effect is to release the other hydrogen atom bound with NAD to form another hydrogen ion, H^+; this process also reconstitutes NAD^+ that will be reused again and again.

During these changes, the electrons that are removed from the hydrogen atoms to cause the ionization immediately enter an *electron transport chain* that is an integral part of the inner membrane (the shelf membrane) of the mitochondrion. This transport chain consists of a series of electron acceptors that can be reversibly reduced or oxidized by accepting or giving up electrons. The important members of this electron transport chain include *flavoprotein, several iron sulfide proteins, ubiquinone,* and *cytochromes B, C_1, C, A, and A_3.* Each electron is shuttled from one of these acceptors to the next until it finally reaches cytochrome A_3, which is called *cytochrome oxidase* because it is capable, by giving up two electrons, of

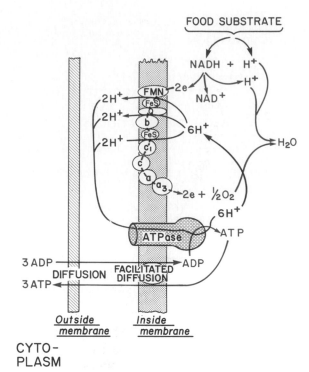

Figure 67–5. The chemiosmotic theory of oxidative phosphorylation for forming great quantities of ATP.

causing elemental oxygen to combine with hydrogen ions to form water.

Thus, Figure 67–5 illustrates transport of electrons through the electron chain and then ultimate use of these by cytochrome oxidase to cause the formation of water molecules. During the transport of these electrons through the electron transport chain energy is released that is later used to cause synthesis of ATP.

Hydrogen Pumping by the Electron Transport Chain. Exactly what happens to the energy released as the electrons pass through the electron transport chain is not known. However, experiments have suggested that this energy is used to pump hydrogen ions from the inner matrix of the mitochondrion into the space between the inner and outer mitochondrial membranes. This creates a high concentration of hydrogen ions in this space, and it also creates a strong negative electrical potential in the inner matrix.

Formation of ATP. The next step in oxidative phosphorylation is to convert ADP into ATP. This occurs in conjunction with a large protein molecule that protrudes all the way through the inner mitochondrial membrane and projects with a knoblike head into the inner matrix. This molecule is an ATPase, the physical nature of which is illustrated in Figure 67–5. It is called *ATP synthetase.* It is postulated that the high concentration of hydrogen ions in the space between the two mitochondrial membranes and the large electrical potential difference across the inner membrane cause the hydrogen ions to flow into the mitochondrial matrix *through the substance of the ATPase molecule.* In doing so, energy derived from this hydrogen ion flow is utilized by the ATPase to convert ADP into ATP by combining an ADP with a phosphate radical, at the same time forming an additional high energy phosphate bond.

The final step in the process is transfer of the ATP from the inside of the mitochondrion back to the cytoplasm. This occurs by facilitated diffusion outward through the inner membrane and then by simple diffusion through the very permeable outer mitochondrial membrane. In turn, ADP is continually transferred in a like manner back into the mitochondrial matrix for continual conversion into ATP.

For each 2 electrons that pass through the entire electron transport chain (representing the ionization of 2 hydrogen atoms), up to 3 ATP molecules are synthesized.

SUMMARY OF ATP FORMATION DURING THE BREAKDOWN OF GLUCOSE

We can now determine the total number of ATP molecules formed by the energy from 1 molecule of glucose.

First, during glycolysis, 4 molecules of ATP are formed while two are expended to cause the initial phosphorylation of glucose to start the process going. This gives a net gain of *2 molecules of ATP.*

Second, during each revolution of the citric acid cycle, 1 molecule of ATP is formed. However, because each glucose molecule splits into 2 pyruvic acid molecules, there are two revolutions of the cycle for each molecule of glucose metabolized, giving a net production of *2 molecules of ATP.*

Third, during the entire schema of glucose breakdown, a total of 24 hydrogen atoms is released during glycolysis and during the citric acid cycle. Twenty of these atoms are oxidized by the schema of oxidative reactions shown in Figure 67–5, with the release of up to 3 ATP molecules per 2 atoms of hydrogen metabolized. This gives a production of *30 ATP molecules.*

Fourth, the remaining 4 hydrogen atoms are released by their dehydrogenase into the oxidative schema beyond the first stage of Figure 67–5, so that for these 4 hydrogen atoms only 2 ATP molecules are usually released for each 2 hydrogen atoms oxidized, giving a total of *4 ATP molecules.*

Now, adding all the ATP molecules formed, we find a maximum of *38 ATP molecules* formed for each molecule of glucose degraded to carbon dioxide and water. Thus, 304,000 calories of energy can be stored in the form of ATP, while 686,000 calories are released during the complete oxidation of each grammole of glucose. This represents an overall maximum *efficiency* of energy transfer of 44 per cent. The remaining 56 per cent of the energy becomes heat and therefore cannot be used by the cells to perform specific functions.

CONTROL OF GLYCOLYSIS AND OXIDATION BY ADENOSINE DIPHOSPHATE (ADP)

Continual release of energy from glucose when the energy is not needed by the cells would be an extremely wasteful process. Fortunately, glycolysis and the subsequent oxidation of hydrogen atoms are continually controlled in accordance with the needs of the cells for ATP. This control is accomplished by a multitude of feedback control mechansims within the chemical schemas. However, probably the most important of these is the following:

Referring back to the various chemical reactions of the glycolytic series, the citric acid cycle, and the oxidation of hydrogen, we see that at different stages *ADP is converted into ATP. If ADP is not available at each of these stages, the reactions cannot occur, thus stopping the degradation of the glucose molecule.* Therefore, once all the ADP in the cells has been converted to ATP, the entire glycolytic and oxidative process stops. Then, when more ATP is used to perform different physiological functions in the cell, new ADP is formed, which automatically turns on glycolysis and oxidation once more. In this way, essentially a full store of ATP is automatically maintained all the time, except when ATP is used up more rapidly than it can be formed.

ANAEROBIC RELEASE OF ENERGY – "ANAEROBIC GLYCOLYSIS"

Occasionally, oxygen becomes either unavailable or insufficient so that oxidative phosphorylation cannot take place. Yet, even under these conditions, a small amount of energy can still be released to the cells by glycolysis, for the chemical reactions in the glycolytic breakdown of glucose to pyruvic acid do not require oxygen. Unfortunately, this process is extremely wasteful of glucose because only 16,000 calories of energy are used to form ATP for each molecule of glucose utilized, which represents only a little over 2 per cent of the total energy in the glucose molecule. Nevertheless, this release of glycolytic energy to the cells can be a lifesaving measure for a few minutes when oxygen becomes unavailable.

Formation of Lactic Acid During Anaerobic Glycolysis. The *law of mass action* states that as the endproducts of a chemical reaction build up in the reacting medium, the rate of the reaction approaches zero. The two end-products of the glycolytic reactions (see Fig. 67–3) are (1) pyruvic acid and (2) hydrogen atoms, which are combined with NAD^+ to form NADH and H^+. The build-up of either or both of these would stop the glycolytic process and prevent further formation of ATP. Fortunately, when their quantities begin to be excessive these two endproducts react with each other to form lactic acid in accordance with the following equation:

$$
\underset{\text{(Pyruvic acid)}}{CH_3-\overset{\overset{\textstyle O}{\|}}{C}-COOH} + NADH + H^+ \xrightarrow[\longleftarrow]{\text{lactic dehydrogenase}}
$$

$$
\underset{\text{(Lactic acid)}}{CH_3-\overset{\overset{\textstyle OH}{|}}{\underset{\underset{\textstyle H}{|}}{C}}-COOH} + NAD^+
$$

Thus, under anaerobic conditions, by far the major proportion of the pyruvic acid is converted into lactic acid, which diffuses readily out of the cells into the extracellular fluids and even into the intracellular fluids of other less active cells. Therefore, lactic acid represents a type of "sinkhole" into which the glycolytic end-products can disappear, thus allowing glycolysis to proceed far longer than would be possible if the pyruvic acid and hydrogen were not removed from the reacting medium. Indeed, glycolysis could proceed for only a few seconds without this conversion. Instead, it can proceed for several minutes, supplying the body with considerable quantities of ATP even in the absence of respiratory oxygen.

Reconversion of Lactic Acid and Pyruvic Acid to Glucose in the Presence of Oxygen. The entire schema of glycolysis, all the way from glucose to lactic acid, is theoretically reversible, but in practice the reverse change of pyruvic acid into phosphoenolpyruvic acid is extremely slow. However, other enzymes cause the pyruvic acid to be converted first into oxaloacetic acid and then into phosphoenolpyruvic acid, after which the scheme of reconversion can proceed. Unfortunately, however, this reverse process consumes 6 ATP molecules in comparison with the 2 ATP molecules formed during glycolysis.

When a person begins to breathe oxygen again after a period of anaerobic metabolism, the extra NADH and H+ as well as the extra pyruvic acid that have built up in the body fluids are rapidly oxidized, thereby greatly reducing their concentrations. As a result, the chemical reaction for formation of lactic acid immediately reverses itself, the lactic acid once again becoming pyruvic acid. Large portions of this are immediately utilized by the citric acid cycle to provide additional oxidative energy, and large quantities of ATP are formed. This excess ATP then causes as much as three-fourths of the remaining excess pyruvic acid to be converted back into glucose.

Thus, the great amount of lactic acid that forms during anaerobic glycolysis does not become lost from the body, for when oxygen is again available, the lactic acid either can be reconverted to glucose or can be used directly for energy. By far the greater proportion of this reconversion occurs in the liver, but a small amount can also occur in other tissues.

Use of Lactic Acid by the Heart for Energy. Heart muscle is especially capable of converting lactic acid to pyruvic acid and then utilizing this for energy. This occurs especially in very heavy exercise, during which large amounts of lactic acid are released into the blood from the skeletal muscles.

RELEASE OF ENERGY FROM GLUCOSE BY THE PHOSPHOGLUCONATE PATHWAY

Though essentially all the carbohydrates utilized by the muscles are degraded to pyruvic acid by glycolysis and then oxidized, the glycolytic schema is not the only means by which glucose can be degraded and then oxidized to provide energy. A second important schema for breakdown and oxidation of glucose is called the *phosphogluconate pathway,* which is responsible for as much as 30 per cent of the glucose breakdown *in the liver and even more than this in fat cells.* It is especially important because it can provide energy independently of all the enzymes of the tricarboxylic acid cycle and therefore is an alternate pathway for energy metabolism in case of some enzymatic abnormality of the cells.

Release of Carbon Dioxide and Hydrogen by Means of the Phosphogluconate Pathway. Figure 67–6 illustrates most of the basic chemical reactions in the phosphogluconate pathway. This shows that glucose, after several stages of conversion, releases 1 molecule of carbon dioxide and 4 atoms of hydrogen, with resultant formation of a five-carbon sugar, d-ribulose. This substance in turn can change progressively into several other five-, four-, seven-, and three-carbon sugars. Finally, various combinations of these sugars can resynthesize glucose. However, *only 5 molecules of glucose are resynthesized for every 6 molecules of glucose that initially enter into the reactions.* That is, the phosphogluconate pathway is a cyclic process in which 1 molecule of glucose is metabolized for each revolution of the cycle. As illustrated at the bottom of Figure 67–6, the net reaction is conversion of the single molecule of glucose plus 6 molecules of water into 6 carbon dioxide molecules and 24 hydrogen atoms. Thus, by revolution of the cycle again and again, all the glucose can eventually be converted into carbon dioxide and hydrogen, and the hydrogen in turn can enter the oxidative phosphorylation pathway to form ATP, or more often it is used for synthesis of fat as follows:

Use of Hydrogen to Synthesize Fat; The Function of Nicotinamide Adenine Dinucleotide Phosphate

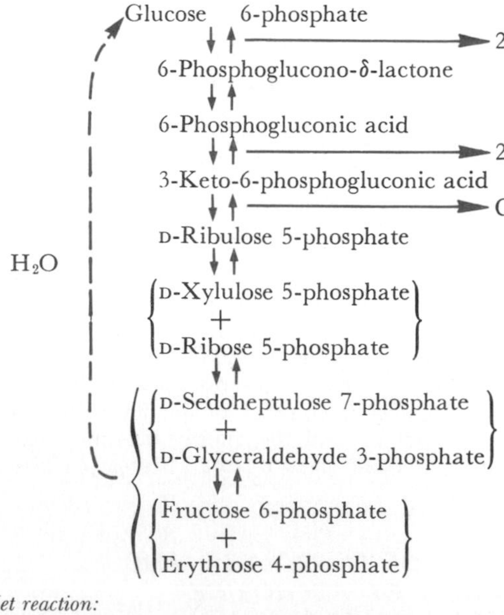

Net reaction:

Glucose + 12NADP+ + 6H₂O ⟶ 6CO₂ + 12H + 12NADPH

Figure 67–6. The phosphogluconate pathway for glucose metabolism.

(NADP$^+$). The hydrogen released during the phosphogluconate cycle does not combine with NAD$^+$ as in the glycolytic pathway, but combines with NADP$^+$, which is almost identical with NAD$^+$ except for an extra phosphate radical. Yet, this difference is extremely significant because only hydrogen bound with NADP$^+$ in the form of NADPH can be used for synthesis of fats from carbohydrates, which is discussed in the following chapter. When glycolysis becomes slowed because of cellular inactivity, the phosphogluconate pathway still remains operative (mainly in the liver) to break down any excess glucose that continues to be transported into the cells, and NADPH becomes abundant to help convert acetyl radicals into long fatty acid chains.

GLUCOSE CONVERSION TO GLYCOGEN OR FAT

When glucose is not immediately required for energy, the extra glucose that continually enters the cells either is stored as glycogen or is converted into fat. Glucose is preferentially stored as glycogen until the cells have stored as much glycogen as they can — an amount sufficient to supply the energy needs of the body for only a few hours. When the cells (primarily liver and muscle cells) approach saturation with glycogen, the additional glucose is then converted into fat in the liver and in the fat cells and then is stored in the fat cells. Other steps in the chemistry of this conversion are discussed in the following chapter.

FORMATION OF CARBOHYDRATES FROM PROTEINS AND FATS— "GLUCONEOGENESIS"

When the body's stores of carbohydrates decrease below normal, moderate quantities of glucose can be formed from *amino acids* and from the *glycerol* portion of fat. This process is called *gluconeogenesis.* Approximately 60 per cent of the amino acids in the body proteins can be converted easily into carbohydrates, while the remaining 40 per cent have chemical configurations that make this difficult. Each amino acid is converted into glucose by a slightly different chemical process. For instance, alanine can be converted directly into pyruvic acid simply by deamination; the pyruvic acid then is converted into glucose, as explained previously. Several of the more complicated amino acids can be converted into different sugars containing three-, four-, five-, or seven-carbon atoms; these can then enter the phosphogluconate pathway and eventually form glucose. Thus, by means of deamination plus several simple interconversions, many of the amino acids become glucose. Similar interconversions can change glycerol into carbohydrates.

Regulation of Gluconeogenesis. Diminished carbohydrates in the cells and decreased blood sugar are the basic stimuli that set off an increase in the rate of gluconeogenesis. The diminished carbohydrates can directly cause reversal of many of the glycolytic and phosphogluconate reactions, thus allowing conversion of deaminated amino acids and glycerol into carbohydrates. However, in addition, several of the hormones secreted by the endocrine glands are especially important in this regulation.

Effect of Corticotropin and Glucocorticoids on Gluconeogenesis. When normal quantities of carbohydrates are not available to the cells, the adenohypophysis, for reasons not completely understood at present, begins to secrete increased quantities of *corticotropin,* which stimulate the adrenal cortex to produce large quantities of *glucocorticoid hormones,* especially *cortisol.* In turn, cortisol mobilizes proteins from essentially all cells of the body, making these available in the form of amino acids in the body fluids. A high proportion of these immediately become deaminated in the liver and therefore provide ideal substrates for conversion into glucose. Thus, one of the most important means by which gluconeogenesis is promoted is through the release of glucocorticoids from the adrenal cortex.

Effect of Thyroxine on Gluconeogenesis. Thyroxine, secreted by the thyroid gland, also increases the rate of gluconeogenesis. This, too, is believed to result principally from mobilization of proteins from the cells. However, it might result to some extent also from the mobilization of fats from the fat depots, the glycerol portion of the fats being converted into glucose.

BLOOD GLUCOSE

The normal blood glucose concentration in a person who has not eaten a meal within the past three to four hours is approximately 90 mg./dl. and, even after a meal containing large amounts of carbohydrates, this rarely rises above 140 mg./dl. unless the person has diabetes mellitus, which will be discussed in Chapter 78.

The regulation of blood glucose concentration is so intimately related to insulin and glucagon that this subject will be discussed in detail in Chapter 78 in relation to the functions of insulin and glucagon.

REFERENCES

Albrink, M. J.: Overnutrition and the fat cell. *In* Bondy, P. K., and Rosenberg, L. E. (eds.): Duncan's Diseases of Metabolism, 7th Ed. Philadelphia, W. B. Saunders Co., 1974, p. 417.

Aoki, T. R., and Cahill, G. F., Jr.: Metabolic effects of insulin, glucagon, and glucose in man: Clinical applications. *In* DeGroot, L. J., *et al.* (eds.): Endocrinology. Vol. 3. New York, Grune & Stratton, 1979, p. 1843.

Baer, H. P., and Drummond, G. I. (eds.): Physiological and Regulatory Functions of Adenosine and Adenine Nucleotides. New York, Raven Press, 1979.

Baltcheffsky, H., and Baltscheffsky, M.: Electron transport phosphorylation. *Annu. Rev. Biochem., 43*:871, 1974.

Butler, T. M., and Davies, R. E.: High-energy phosphates in smooth muscle. *In* Bohr, D. F., *et al.* (eds.): Handbook of Physiology. Sec. 2, Vol. II. Baltimore, Williams & Wilkins, 1980, p. 237.

Careri, G., *et al.*: Enzyme dynamics: The statistical physics approach. *Annu. Rev. Biophys. Bioeng. 8*:69, 1979.

Cornish-Bowden, A.: Fundamentals of Enzyme Kinetics. Boston, Butterworths, 1979.

Desnick, R. J., *et al.*: Toward enzyme therapy for lysosomal storage diseases. *Physiol. Rev., 56*:57, 1978.

Dickerson, R. E.: Cytochrome C and the evolution of energy metabolism. *Sci. Am., 242*(3):136, 1980.

Esmann, V. (ed.): Regulatory Mechanisms of Carbohydrate Metabolism. New York, Pergamon Press, 1978.

Felig, P.: Disorders of carbohydrate metabolism. *In* Bondy, P. K., and Rosenberg, L. E. (eds.): Metabolic Control and Disease, 8th Ed. Philadelphia, W. B. Saunders Co., 1980, p. 276.

Felig, P., and Wahren, J.: Protein turnover and amino acid metabolism in the regulation of gluconeogenesis. *Fed. Proc., 33*:1092, 1974.

Foster, R. L.: The Nature of Enzymology. New York, John Wiley & Sons, 1979.

Friedmann, H. C. (ed.): Enzymes. Stroudsburg, Pa., Dowden, Hutchinson & Ross, 1980.

Hasselberger, F. X.: Uses of Enzymes and Immobilized Enzymes. Chicago, Nelson-Hall, 1978.

Hems, D. A., and Whitton, P. D.: Control of hepatic glycogenolysis. *Physiol. Rev., 60*:1, 1980.

Homsher, E., and Kean, C. J.: Skeletal muscle energetics and metabolism. *Annu. Rev. Physiol., 40*:93, 1978.

Hoolaway, M. R.: The Mechanism of Enzyme Action. London, Oxford University Press, 1976.

Huijing, F.: Glycogen metabolism and glycogen-storage diseases. *Physiol. Rev., 55*:609, 1975.

Jeffery, J. (ed.): Dehydrogenases Requiring Nicotinamide Coenzymes. Boston, Birkhauser, 1979.

Kappas, A., and Alvares, A. P.: How the liver metabolizes foreign substances. *Sci. Am., 232*(6): 22, 1975.

Klachko, D. M., *et al.* (eds.): Hormones and Energy Metabolism. New York, Plenum Press, 1978.

Krebs, H. A.: The tricarboxylic acid cycle. *Harvey Lectures, 44*:165, 1948–1949.

Lieber, C. S.: The metabolism of alcohol. *Sci. Am., 234*(3):25, 1976.

Lund-Andersen, H.: Transport of glucose from blood to brain. *Physiol. Rev., 59*:305, 1979.

Martin, D. B.: Metabolism and energy mechanisms. *In* Frohlich, E. D. (ed.): Pathophysiology, 2nd Ed. Philadelphia, J. B. Lippincott Co., 1976, p. 365.

McCarty, R. E.: How cells make ATP. *Sci. Am., 238*(3):104, 1978.

McGilvery, R. W.: Biochemistry: A Functional Approach, 2nd Ed. Philadelphia, W. B. Saunders Co., 1979.

Mommaerts, W. F.: Energetics of muscular contraction. *Physiol. Rev., 49*:427, 1969.

Purich, D. L. (ed.): Enzyme Kinetics and Mechanism. New York, Academic Press, 1979.

Racker, E. (ed.): Energy Transducing Mechanisms. Baltimore, University Park Press, 1975.

Singer, T. P., and Ondarza, P. N. (eds.): Mechanisms of Oxidizing Enzymes. New York, Elsevier/North-Holland, 1978.

Wynn, C. H.: The Structure and Function of Enzymes, 2nd Ed. Baltimore, University Park Press, 1979.

68

Lipid Metabolism

A number of different chemical compounds in the food and in the body are classified as *lipids*. These include (1) *neutral fat*, known also as *triglycerides*, (2) the *phospholipids*, (3) *cholesterol*, and (4) a few others of less importance. These substances have certain similar physical and chemical properties, especially the fact that they are miscible with each other. Chemically, the basic lipid moiety of both the triglycerides and the phospholipids is *fatty acids*, which are simply long chain hydrocarbon organic acids. A typical fatty acid, palmitic acid, is the following:

$$CH_3(CH_2)_{14}COOH$$

Though cholesterol does not contain fatty acid, its sterol nucleus, as pointed out later in the chapter, is synthesized from degradation products of fatty acid molecules, thus giving it many of the physical and chemical properties of other lipid substances.

The triglycerides are used in the body mainly to provide energy for the different metabolic processes; this function they share almost equally with the carbohydrates. However, some lipids, especially cholesterol, the phospholipids, and derivatives of these, are used throughout the body to provide other intracellular functions.

Basic Chemical Structure of Triglycerides (Neutral Fat). Since most of this chapter deals with utilization of triglycerides for energy, the following basic structure of the triglyceride molecule must be understood:

$$CH_3-(CH_2)_{16}-COO-CH_2$$
$$CH_3-(CH_2)_{16}-COO-CH$$
$$CH_3-(CH_2)_{16}-COO-CH_2$$
Tri-stearin

Note that three long chain fatty acid molecules are bound with one molecule of glycerol. In the human body, the three fatty acids most commonly present in neutral fat are (1) *stearic acid*, which has an 18-carbon chain and is fully saturated with hydrogen atoms, (2) *oleic acid*, which also has an 18-carbon chain but has one double bond in the middle of the chain, and (3) *palmitic acid*, which has 16 carbon atoms and is fully saturated. These or closely similar fatty acids are also the major constituents of the fats in food except for shorter chain fatty acids in milk products.

TRANSPORT OF LIPIDS IN THE BLOOD

TRANSPORT FROM THE GASTROINTESTINAL TRACT — THE CHYLOMICRONS

It will be recalled from Chapter 65 that essentially all the fats of the diet are absorbed into the lymph (with the exception of the short chain fatty acids that can be absorbed directly into the portal blood but that normally represent only a small fraction of the fat in the diet). In the digestive tract, most triglycerides are split into glycerol, monoglycerides, and fatty acids. Then, on passing through the intestinal epithelial cells, these are resynthesized into new molecules of triglycerides which aggregate and enter the lymph as minute, dispersed droplets called *chylomicrons*, having sizes between 0.03 and 0.5 micron. A small amount of the protein apoprotein B adsorbs to the outer surfaces of the chylomicrons; this increases their suspension stability in the fluid of the lymph and prevents their adherence to the lymphatic vessel walls.

Most of the cholesterol and phospholipids absorbed from the gastrointestinal tract, as well as small amounts of phospholipids that are continually synthesized by the intestinal mucosa, also enter the chylomicrons. Thus, chylomicrons are composed principally of triglycerides, but they also contain approximately 9 per cent phospholipids, 3 per cent cholesterol, and 1 per cent apoprotein B as well. The chylomicrons are then transported up the thoracic duct and emptied into the venous blood at the juncture of the jugular and subclavian veins.

Removal of the Chylomicrons from the Blood. Immediately after a meal that contains large quantities of fat, the chylomicron concentration in the plasma may rise to as high as 1 to 2 per cent, and, because of the large sizes of the chylomicrons, the plasma appears turbid and sometimes yellow. However, the chylomicrons (with a half-life of less than an hour) are removed within a few hours and the plasma becomes clear once again. The fat of the chylomicrons is removed mainly in the following way:

Hydrolysis of the Chylomicron Triglycerides by Lipoprotein Lipase; Fat Storage in the Fat and Liver Cells. Most of the chylomicrons are removed from the circulating blood as they pass through the capillaries of adipose tissue and the liver, but to a lesser extent other tissues as well. Both the adipose tissue and the liver contain large quantities of an enzyme called *lipoprotein lipase*. This enzyme migrates to the capillary endothelium where it hydrolyzes the triglycerides of chylomicrons that stick to the endothelial wall, releasing fatty acids and glycerol. The fatty acids, being highly miscible with the membranes of the cells, immediately diffuse into the fat and liver cells. Once within these cells, the fatty acids are resynthesized into triglycerides, new glycerol being supplied by the metabolic processes of the cells, as will be discussed later in the chapter. The lipase also causes hydrolysis of phospholipids, this too releasing fatty acids to be stored in the cells in the same way. Thus, essentially all of the mass of the chylomicrons is removed from the circulating blood, and the fat absorbed from the gastrointestinal tract only a few minutes previously becomes rapidly stored where it waits to be used for other purposes at a later time.

TRANSPORT OF FATTY ACIDS IN COMBINATION WITH ALBUMIN — FREE FATTY ACID

When the fat that has been stored in the fat cells is to be used elsewhere in the body, usually to provide energy, it must first be transported to the other tissues. It is transported almost entirely in the form of *free fatty acid*. This is achieved by hydrolysis of the triglycerides once again into fatty acids and glycerol. Although all the factors that initiate this hydrolysis are not completely understood, it is known that at least two classes of stimuli play important roles. First, decreased availability of glucose to the cell, as occurs in the period between meals, *decreases the quantity of α-glycerophosphate,* and this removes one of the most important stimuli for new synthesis of triglycerides, as will be discussed later in the chapter, thus allowing the equilibrium to shift in favor of hydrolysis. Second, a *hormone-sensitive cellular lipase* that can be activated by a multitude of different hormones is activated at the same time, and this promotes rapid hydrolysis of the triglyceride. This also will be discussed later in the chapter.

On leaving the fat cells, the fatty acids ionize strongly in the plasma and immediately combine with albumin of the plasma proteins. The fatty acid bound with proteins in this manner is called *free fatty acid* or *nonesterified fatty acid* (or simply *FFA* or *NEFA*) to distinguish it from other fatty acids in the plasma that exist in the form of esters of glycerol, cholesterol, or other substances.

The concentration of free fatty acid in the plasma under resting conditions is about 15 mg. per 100 ml. of blood, which is a total of only 0.75 gram of fatty acids in the entire circulatory system. Yet, strangely enough, even this small amount accounts for almost all the transport of fatty acids from one part of the body to another for the following reasons:

(1) Despite the minute amount of free fatty acid in the blood, its rate of "turnover" is extremely rapid; *half the plasma fatty acid is replaced by new fatty acid every two to three minutes*. One can calculate that at this rate over half of all the energy required by the body can be provided by oxidation of the transported free fatty acid

even without increasing the free fatty acid concentration. (2) All conditions that increase the rate of utilization of fat for cellular energy also increase the free fatty acid concentration in the blood; this sometimes increases as much as 5- to 8-fold. Especially does this occur in starvation and in diabetes, when a person is not or cannot be using carbohydrates for energy.

Under normal conditions, about three molecules of fatty acid combine with each molecule of albumin, but as many as 30 fatty acid molecules can combine with a single molecule of albumin when the need for fatty acid transport is extreme. This shows how variable the rate of lipid transport can be under different physiological needs.

THE LIPOPROTEINS

In the postabsorptive state — that is, when no chylomicrons are in the blood — over 95 per cent of all the lipids in the plasma (in terms of mass, but *not* in terms of rate of transport) are in the form of lipoproteins, which are small particles much smaller than chylomicrons but qualitatively similar in composition, containing mixtures of *triglycerides, phospholipids, cholesterol,* and *protein*. The protein in the mixture averages about one fourth to one third of the total constituents, and the remainder is lipids. The total concentration of lipoproteins in the plasma averages about 700 mg. per 100 ml., and this can be broken down into the following average concentrations of the individual constituents:

	mg./100 ml. of plasma
Cholesterol	180
Phospholipids	160
Triglycerides	160
Lipoprotein protein	200

Types of Lipoproteins. Chylomicrons are sometimes classified as lipoproteins because they contain both lipids and protein. In addition to the chylomicrons, however, there are three other major classes of lipoprotein based on their densities, as measured in the ultracentrifuge: (1) *very low density lipoproteins,* which contain high concentrations of triglycerides and moderate concentrations of both phospholipids and cholesterol; (2) *low density lipoproteins,* which contain relatively little triglyceride but a very high percentage of cholesterol; and (3) *high density lipoproteins,* which contain about 50 per cent protein with smaller concentrations of the lipids.

Formation of the Lipoproteins. The lipoproteins are formed almost entirely in the liver, which is in keeping with the fact that most plasma phospholipids, cholesterol, and triglycerides (except those in the chylomicrons) are synthesized in the liver. However, small quantities of high density lipoproteins are synthesized in the intestinal epithelium during the absorption of fatty acids from the intestines.

Function of the Lipoproteins. The functions of the lipoproteins in the plasma are poorly known, though they are known to be a means by which lipid substances can be transported throughout the body. For instance, the turnover of triglycerides in the lipoproteins is as much as 1.5 grams per hour, which could account for as much as 10 to 20 per cent of the total lipids utilized by the body under resting conditions.

Triglycerides, and to a lesser extent other lipids, that are synthesized mainly from carbohydrates in the liver are transported to the adipose tissue and other peripheral tissues in *very low density lipoproteins*. Also important is the transport of cholesterol and phospholipids by the lipoproteins, because these substances are not known to be transported to any significant extent in any other form. The *high density lipoproteins*, in particular, are known to transport cholesterol away from the peripheral tissues and to the liver. Therefore, this type of lipoprotein probably plays a very significant role in preventing the development of atherosclerosis, as we shall discuss later in the chapter.

THE FAT DEPOSITS

ADIPOSE TISSUE

Large quantities of fat are frequently stored in two major tissues of the body, the adipose tissue and the liver. The adipose tissue is usually called the *fat deposits,* or simply the *fat depots*.

The major function of adipose tissue is storage of triglycerides until these are needed to provide energy elsewhere in the body. However, a subsidiary function is to provide heat insulation for the body, as will be discussed in Chapter 72.

The Fat Cells. The fat cells of adipose tissue are modified fibroblasts that are capable of storing almost pure triglycerides in quantities equal to 80 to 95 per cent of their volume. The triglycerides are generally in a liquid form, and when the tissues of the skin are exposed to prolonged cold, the fatty acid chains of the triglycerides, over a period of weeks, become either shorter or more unsaturated to decrease their melting point, thereby always allowing the fat in the fat cells to remain in a liquid state. This is particularly important because only liquid fat can be hydrolyzed and then transported from the cells.

Fat cells can also synthesize fatty acids and triglycerides from carbohydrates, this function supplementing the synthesis of fat in the liver (though only to a small extent in people), as discussed later in the chapter.

Exchange of Fat Between the Adipose Tissue and the Blood — Tissue Lipases. As discussed above, large quantities of lipases are present in adipose tissue. Some of these enzymes catalyze the deposition of triglycerides from the chylomicrons and other lipoproteins. Others, when activated by hormones, cause splitting of the triglycerides of the fat cells to release free fatty acids. Because of rapid exchanges of the fatty acids, the triglycerides in the fat cells are renewed approximately once every two to three weeks, which means that the fat stored in the tissues today is not the same fat that was stored last month, thus emphasizing the dynamic state of the storage fat.

THE LIVER LIPIDS

The principal functions of the liver in lipid metabolism are: (1) to degrade fatty acids into small compounds that can be used for energy, (2) to synthesize triglycerides mainly from carbohydrates and, to a lesser extent, from proteins, and (3) to synthesize other lipids from fatty acids, especially cholesterol and phospholipids.

Large quantities of triglycerides appear in the liver (a) during starvation, (b) in diabetes mellitus, or (c) in any other condition in which fat is being utilized rapidly for energy. In these conditions, the triglycerides are mobilized from the adipose tissue, transported as free fatty acids in the blood, and then redeposited as triglycerides in the liver, where the initial stages of much of the fat degradation begin. Thus, under normal physiological conditions the total amount of triglycerides in the liver is controlled to a great extent by the overall rate at which lipids are being utilized for energy.

Excess triglycerides can accumulate in the liver also when a deficiency of lipoproteins prevents transport of newly synthesized triglycerides to the peripheral tissues.

The liver cells, in addition to containing triglycerides, contain large quantities of phospholipids and cholesterol, which are continually synthesized by the liver. Also, the liver cells are much more capable than other tissues of desaturating fatty acids so that the liver triglycerides normally are much more unsaturated than the triglycerides of the adipose tissue. This capability of the liver to desaturate fatty acids seems to be functionally important to all the tissues of the body, because many of the structural members of all cells contain reasonable quantities of unsaturated fats, and their principal source seems to be the liver. This desaturation is accomplished by a dehydrogenase in the liver cells.

USE OF TRIGLYCERIDES FOR ENERGY, AND FORMATION OF ADENOSINE TRIPHOSPHATE (ATP)

Approximately 40 to 45 per cent of the calories in the normal American diet are derived from fats, which is about equal to the calories derived from carbohydrates. Therefore, the use of fats by the body for energy is equally as important as the use of carbohydrates. In addition, much of the carbohydrates ingested with each meal is converted into triglycerides, then stored, and later utilized as triglycerides for energy. For these reasons it is equally as important to understand the principles of triglyceride oxidation as to understand carbohydrate oxidation.

Hydrolysis of the Triglycerides. The first stage in the utilization of triglycerides for energy is hydrolysis of these compounds into fatty acids and glycerol and subsequent transport of both products of hydrolysis to the active tissues where they are oxidized to give energy. Almost all cells, with the notable exception of brain tissue, can use fatty acids almost interchangeably with glucose for energy.

The glycerol, on entering the active tissue, is immediately changed by intracellular enzymes into glycerol 3-phosphate, which enters the glycolytic pathway for glucose breakdown, and in this way is used for energy. However, before the fatty acids can be used for energy, they must be processed further in the following way:

Entry of Fatty Acids into the Mitochondria. The degradation and oxidation of fatty acids occur only in the mitochondria. Therefore, the first step for utilization of the fatty acids is their transport into the mitochondria. This is an enzyme-catalyzed process that employs *carnitine* as a carrier substance. Once inside the mitochondria, the fatty acid splits away from the carnitine and is then oxidized.

Degradation of Fatty Acid to Acetyl Coenzyme-A by Beta Oxidation. The fatty acid molecule is degraded in the mitochondria by progressive release of 2-carbon segments in the form of acetyl coenzyme-A (acetyl Co-A). This process, which is illustrated in Figure 68–1, is called the *beta oxidation* process for degradation of fatty acids. The successive stages are the following:

1. The fatty acid molecule first combines with coenzyme A to form a *fatty acyl Co-A* molecule. This step is energized by the breakdown of an ATP molecule to AMP, a loss of two high energy phosphate bonds.

2. Fatty acyl Co-A loses two hydrogen atoms from the alpha and beta carbons, leaving a double bond at this point. The hydrogen atoms that are removed become attached to a flavoprotein (FAD) and later are oxidized, as discussed below.

3. A water molecule reacts at the site of the double bond so that a hydrogen atom from the water attaches to the alpha carbon and the remaining hydroxyl radical attaches to the beta carbon.

4. Two additional hydrogen atoms are removed, one from the beta carbon and one from the hydroxyl radical. The hydrogen atoms removed in this process combine with NAD$^+$ and also are later oxidized, as discussed below.

5. The compound splits between the alpha and beta carbons, the long portion of the chain combining with a new molecule of coenzyme A, while the shorter acetyl portion remains combined with the original coenzyme A in the form of *acetyl Co-A*.

The new fatty acyl Co-A, which now has two carbon atoms fewer than the original fatty acyl Co-A, re-enters reaction number 2 in Figure 68–1 and proceeds through the four stages of chemical reactions until another acetyl Co-A molecule is released. This process is repeated again and again until the entire fatty acid molecule is split into acetyl Co-A. For instance, from each molecule of stearic acid, nine molecules of acetyl Co-A are formed.

Oxidation of Acetyl Co-A. The acetyl Co-A molecules formed by beta oxidation of fatty acids enter into the citric acid cycle as explained in the preceding chapter, combining first with oxaloacetic acid to form citric acid, which then is degraded into carbon dioxide and hydrogen atoms. The hydrogen is subsequently oxidized by the oxidative enzyme system of the cells, which was also explained in the preceding chapter. The net reaction for each molecule of acetyl Co-A is the following:

$$CH_3COCo\text{-}A + \text{Oxaloacetic acid} + 3H_2O + ADP$$

$$\xrightarrow{\textit{Citric acid cycle}}$$

$$2CO_2 + 8H + HCo\text{-}A + ATP + \text{Oxaloacetic acid}$$

Thus, after the initial degradation of fatty acids to acetyl Co-A, their final breakdown is precisely the same as that of the acetyl Co-A formed from pyruvic acid during the metabolism of glucose.

ATP Formed by Oxidation of Fatty Acid. In Figure 68–1 note also that 4 hydrogen atoms are released each time a molecule of acetyl Co-A is formed from the fatty acid chain. Therefore, for every stearic acid molecule that is split, a total of 32 hydrogen atoms are removed. In addition, for each acetyl Co-A degraded by the citric acid cycle, 8 hydrogen atoms are removed, making an additional 72 hydrogens for each molecule of stearic acid metabolized. This added to the above 32 hydrogen atoms makes a total of 104 hydrogen atoms. Of this group 34 are removed from the degrading fatty acid by flavoproteins and 70 are removed by NAD$^+$ as NADH and H$^+$. These two groups of hydrogen atoms are oxidized by the cells, as discussed in the preceding chapter, but they enter the oxidative system at different points, so that up to 1 molecule of ATP is synthesized for each of the 34 flavoprotein hydrogens and up to 1.5 molecules of ATP are synthesized for each of the 70 NADH and H$^+$ hydrogens. This makes 34 plus 105, or a total of 139 molecules of ATP formed by the oxidation of hydrogen derived from each molecule of stearic acid. And another 9 molecules of ATP are formed in the citric acid cycle, one for each of the 9 acetyl Co-A molecules metabolized. Thus, a total of 148 molecules of ATP is formed during the complete oxidation of 1 molecule of stearic acid. However, 2 high energy bonds are con-

(1) $RCH_2CH_2CH_2COOH + Co\text{-}A + ATP \underset{}{\overset{\text{thiokinase}}{\rightleftharpoons}} RCH_2CH_2CH_2COCo\text{-}A + AMP + \text{Pyrophosphate}$
  (Fatty acid)   (Fatty acyl Co-A)

(2) $RCH_2CH_2CH_2COCo\text{-}A + FAD \xrightarrow{\text{acyl dehydrogenase}} RCH_2CH=CHCOCo\text{-}A + FADH_2$
  (Fatty acyl Co-A)

(3) $RCH_2CH=CHCOCo\text{-}A + H_2O \underset{}{\overset{\text{enoyl hydrase}}{\rightleftharpoons}} RCH_2CHOHCH_2COCo\text{-}A$

(4) $RCH_2CHOHCH_2COCo\text{-}A + NAD^+ \underset{\text{dehydrogenase}}{\overset{\beta\text{-hydroxyacyl}}{\rightleftharpoons}} RCH_2COCH_2COCo\text{-}A + NADH + H^+$

(5) $RCH_2COCH_2COCo\text{-}A + Co\text{-}A \underset{}{\overset{\text{thiolase}}{\rightleftharpoons}} RCH_2COCo\text{-}A + CH_3COCo\text{-}A$
                                                          (Fatty acyl Co-A)(Acetyl Co-A)

Figure 68–1. Beta oxidation of fatty acids to yield acetyl coenzyme A.

sumed in the initial combination of coenzyme A with the fatty acid molecule, making a net gain of 146 molecules of ATP.

FORMATION OF ACETOACETIC ACID IN THE LIVER AND ITS TRANSPORT IN THE BLOOD

A large share of the initial degradation of fatty acids occurs in the liver, especially when excessive amounts of lipids are mobilized to be used for energy. However, the liver uses only a small proportion of the fatty acids for its own intrinsic metabolic processes. Instead, when the fatty acid chains have been split into acetyl Co-A, two molecules of acetyl Co-A condense to form one molecule of acetoacetic acid, as follows:

$$2CH_3COCo\text{-}A + H_2O \underset{\text{other cells}}{\overset{\text{liver cells}}{\rightleftarrows}}$$

$$\underset{Acetyl\ Co\text{-}A}{}$$

$$CH_3COCH_2COOH + 2HCo\text{-}A$$
$$\underset{Acetoacetic\ acid}{}$$

Then, a large part of the acetoacetic acid is converted into *β-hydroxybutyric acid* and minute quantities to *acetone* in accord with the following reactions:

The acetoacetic acid and β-hydroxybutyric acid then freely diffuse through the liver cell membranes and are transported by the blood to the peripheral tissues. Here they again diffuse into the cells where reverse reactions occur and acetyl Co-A molecules are formed. These in turn enter the citric acid cycle and are oxidized for energy, as explained above.

Normally, the acetoacetic acid and β-hydroxybutyric acid that enter the blood are transported so rapidly to the tissues that their combined concentration in the plasma rarely rises above 3 mg. per cent. Yet despite the small quantities in the blood, large amounts are actually transported; this is analogous to the high rate of free fatty acid transport. The rapid transport of both these substances depends on their high degree of lipid solubility, which allows rapid diffusion through the cell membranes.

Ketosis and Its Occurrence in Starvation, Diabetes, and Other Diseases. Large quantities of acetoacetic acid, β-hydroxybutyric acid, and acetone occasionally accumulate in the blood and interstitial fluids; this condition is called *ketosis* because acetoacetic acid is a keto acid, and the three compounds are called *ketone bodies*. Ketosis occurs especially in starvation, in diabetes

mellitus, or sometimes even when a person's diet is composed almost entirely of fat. In all these states, essentially no carbohydrates are metabolized — in starvation and following a high fat diet because carbohydrates are not available and in diabetes because insulin is not available to cause glucose transport into the cells.

When carbohydrates are not utilized for energy, almost all the energy of the body must come from metabolism of fats. We shall see later in the chapter that lack of availability of carbohydrates automatically increases the rate of removal of fatty acids from adipose tissues, and in addition, several hormonal responses — such as increased secretion of corticotropin by the adenohypophysis, increased secretion of glucocorticoids by the adrenal cortex, and decreased secretion of insulin by the pancreas — all further enhance the removal of fatty acids from the fat tissues. As a result, tremendous quantities of fatty acids become available to the liver for degradation. The ketone bodies in turn pour out of the liver to be carried to the cells. Yet the cells are limited in the amount of ketone bodies that can be oxidized for two reasons: First, they can be oxidized only as rapidly as adenosine diphosphate is formed in the tissues to initiate the oxidative reactions. Second, lack of carbohydrate intermediates of the glycolytic series depresses the activity of citrate lyase, one of the important rate-limiting enzymes controlling the citric acid cycle, which is mainly responsible for degradation of acetoacetic acid, as explained previously. Thus, because of this limit and because of the simultaneous outpouring of tremendous quantities of acetoacetic acid and the other ketone bodies from the liver, the blood concentration of acetoacetic acid and β-hydroxybutyric acid sometimes rises to as high as 30 or more times normal, thus leading to extreme acidosis, as explained in Chapter 37.

The acetone that is formed during ketosis is a volatile substance that is blown off in small quantities in the expired air of the lungs, often giving the breath an acetone smell. This smell is frequently used as a diagnostic criterion of ketosis.

Adaptation to a High Fat Diet. Upon changing slowly from a carbohydrate diet to an almost completely fat diet, a person's body adapts to the utilization of far more acetoacetic acid than usual, and, in this instance, ketosis does not occur. For instance, the Eskimos, who sometimes live almost entirely on a fat diet, do not develop ketosis. Presumably some factor in arctic inhabitants enhances the rate of acetoacetic acid metabolism by the cells.

SYNTHESIS OF TRIGLYCERIDES FROM CARBOHYDRATES

Whenever a greater quantity of carbohydrates enters the body than can be used immediately for energy or stored in the form of glycogen, the excess is rapidly converted into triglycerides and then stored in this form in the adipose tissue. In human beings, most triglyceride synthesis occurs in the liver, but small quantities are also synthesized in the adipose tissue. The triglycerides that are formed in the liver are then transported by the

Step 1:

$$CH_3COCo\text{-}A + CO_2 + ATP$$

(*Acetyl Co-A carboxylase*)

$$\begin{array}{l} COOH \\ \| \\ CH_2 \qquad + ADP + PO_4^{---} \\ | \\ O{=}C{-}CO\text{-}A \end{array}$$

Malonyl Co-A

Step 2:

$$1 \text{ Acetyl Co-A} + 8 \text{ Malonyl Co-A} + 16NADPH + 16H^+ \longrightarrow$$

$$1 \text{ Stearic Acid} + 8CO_2 + 9Co\text{-}A + 16NADP^+ + 7H_2O$$

Figure 68–2.   Synthesis of fatty acids.

very low density lipoproteins to the adipose tissue where they too are stored until needed for energy.

Conversion of Acetyl Co-A into Fatty Acids.   The first step in the synthesis of triglycerides is conversion of carbohydrates into acetyl Co-A. It will be recalled from the preceding chapter that this occurs during the normal degradation of glucose by the glycolytic system. It will also be remembered from earlier in this chapter that fatty acids are actually large polymers of acetic acid. Therefore, it is easy to understand how acetyl Co-A could be converted into fatty acids.

However, synthesis of by far the largest proportion of the triglycerides from acetyl Co-A is not achieved by simply reversing the oxidative degradation that was described above. Instead, its first step is conversion of acetyl Co-A into *malonyl Co-A* in accordance with step 1 of Figure 68–2. A large amount of energy is transferred from ATP to malonyl Co-A, and it is this energy that is utilized to cause the subsequent reactions required in the formation of the fatty acid molecule. Step 2 in Figure 68–2 gives the net reaction in the formation of a stearic acid molecule, showing that one acetyl Co-A molecule and eight malonyl Co-A molecules combine with NADPH and hydrogen ions to form the fatty acid molecule. Carbon dioxide and Co-A are both liberated, and these are utilized again and again in the formation of malonyl Co-A.

The acetyl Co-A that is converted into fatty acid molecules is derived mainly from the *glycolytic* breakdown of glucose, and the NADPH required for fatty acid synthesis is a by-product of the *phosphogluconate pathway* of glucose degradation, which emphasizes the importance of both these pathways in fat synthesis.

These two mechanisms of glucose degradation occur side-by-side in the liver cells and to a much smaller extent in the fat cells, contributing the appropriate proportions of acetyl Co-A and NADPH required for fatty acid synthesis.

Combination of Fatty Acids with α-Glycerophosphate to Form Triglycerides.   Once the synthesized fatty acid chains have grown to contain 14 to 18 carbon atoms, they are then bound to glycerol to form triglycerides. The enzymes that cause this conversion are highly specific for fatty acids with chain lengths of 14 carbon atoms or greater, a factor that actually controls the physical quality of the triglycerides stored in the body.

As illustrated in Figure 68–3, the glycerol portion of the triglyceride is furnished by α-glycerophosphate, which is a product derived from the glycolytic schema of glucose degradation. The mechanism of this was discussed in Chapter 67.

The real importance of this mechanism for formation of triglycerides is that the whole process is controlled to a great extent by the concentration of α-glycerophosphate. When carbohydrates are available to form large quantities of α-glycerophosphate, the equilibrium shifts to promote formation and storage of triglycerides.

Efficiency of Carbohydrate Conversion into Fat.   During triglyceride synthesis, only about 15 per cent of the original energy in the glucose is lost in the form of heat, while the remaining 85 per cent is transferred to the stored triglycerides.

Importance of Fat Synthesis and Storage.   Fat synthesis from carbohydrates is especially important for two

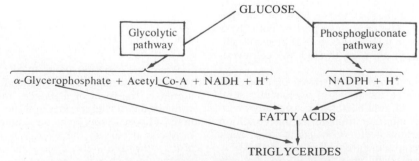

Figure 68–3.   An overall schema for synthesis of triglycerides from glucose.

reasons: (1) The ability of the different cells of the body to store carbohydrates in the form of glycogen is generally slight; a maximum of only a few hundred grams of glycogen can be stored in the liver, the skeletal muscles, and all other tissues of the body put together. In contrast, many kilograms of fat can be stored. Therefore, fat synthesis provides a means by which the energy of excess ingested carbohydrates (and proteins, too) can be stored for later use. Indeed, the average person has almost 150 times as much energy stored in the form of fat as stored in the form of carbohydrate. (2) Each gram of fat contains approximately 2.25 times as many calories of energy as each gram of glycogen. Therefore, for a given weight gain, a person can store far more energy in the form of fat than in the form of carbohydrate, which would be important when an animal must be highly motile to survive.

Failure to Synthesize Fats from Carbohydrates in the Absence of Insulin. When insulin is not available, as in diabetes mellitus, fats are poorly if at all synthesized. This results from the following effects: First, when insulin is not available, glucose does not enter the fat and liver cells satisfactorily, so that little of the acetyl Co-A and NADPH needed for fat synthesis can be derived from glucose. Second, lack of glucose in the fat cells greatly reduces the availability of α-glycerophosphate, which also makes it difficult for the tissues to form triglycerides.

SYNTHESIS OF TRIGLYCERIDES FROM PROTEINS

Many amino acids can be converted into acetyl Co-A, as will be discussed in the following chapter. Obviously, this can be synthesized into triglycerides. Therefore, when people have more proteins in their diets than their tissues can use as proteins, a large share of the excess is stored as fat.

REGULATION OF ENERGY RELEASE FROM TRIGLYCERIDES

Regulation of Energy Release by Formation of Adenosine Diphosphate (ADP) in the Tissues. One of the primary factors that causes energy release from all foodstuffs is the concentration of ADP in the tissues. As explained at many points in this text, essentially all functions of the body are energized by the high energy phosphate bonds of ATP. In the process of liberating this energy the ATP becomes ADP, which in turn is a necessary substrate for the reactions responsible for energy release from essentially all foods. Therefore, when the activity of the tissues accelerates so that increased quantities of ADP are formed, all the oxidative and other energy-releasing processes accelerate, and the degradation and utilization of all foodstuffs, including fats, for energy proceeds apace.

Fat-Sparing Effect of Carbohydrates; Control of Fat Synthesis from Carbohydrates. When adequate quantities of carbohydrates are available in the body, the utilization of triglycerides for energy is greatly depressed even though large quantities of ADP are formed. In place of fat utilization, the carbohydrates are utilized preferentially. There are several different reasons for this "fat-sparing" effect of carbohydrates. One of the most important is the following: The fats in adipose tissue cells are present in two different forms, triglycerides and small quantities of free fatty acids. These are in constant equilibrium with each other. When excess quantities of α-glycerophosphate and NADPH are present, the equilibrium between free fatty acids and triglycerides shifts toward the triglycerides, as explained earlier in the chapter; as a result, only minute quantities of fatty acids are then available to be utilized for energy. Since α-glycerophosphate and NADPH are both important products of glucose metabolism, the availability of large amounts of glucose automatically inhibits the use of fatty acids for energy.

Second, when carbohydrates are available in excess, fats are synthesized more rapidly than they are degraded. This effect is caused partially by the large quantities of acetyl Co-A formed from the carbohydrates and by the low concentration of free fatty acids in the adipose tissue, thus creating conditions appropriate for conversion of acetyl Co-A into fatty acids. However, an even more important effect that promotes conversion of carbohydrates to fats is the following: The first step, and the rate-limiting step, in the synthesis of fatty acids is carboxylation of acetyl Co-A to form malonyl Co-A, which was discussed above. The rate of this reaction is controlled primarily by the enzyme *acetyl Co-A carboxylase,* the activity of which is accelerated in the presence of the intermediates of the citric acid cycle. When excess carbohydrates are being utilized, these intermediates increase, thus automatically causing increased synthesis of fatty acids. Thus, an excess of carbohydrates in the diet not only spares but also increases the fat already in the fat stores.

Acceleration of Fat Utilization for Energy in the Absence of Carbohydrates. All the fat-sparing effects of carbohydrates are lost and are actually reversed when carbohydrates are not available or are available in short supply.

First, pyruvate formed by glycolytic breakdown of glucose is required for formation of oxaloacetic acid, the starting point of the citric acid cycle. Therefore, when diminished amounts of carbohydrates are available, all the citric acid cycle intermediates become depressed in quantity. Some of these are required for activation of the rate-limiting enzyme for fat synthesis, *acetyl Co-A carboxylase.* As a consequence, the acetyl Co-A that is available to the metabolic pool is now shifted away from fatty acid synthesis.

Second, lack of glucose availability to the fat cells makes it impossible for these to store triglycerides, as was explained above, because of the unavailability of α-glycerophosphate. Yet, some of the triglycerides of the fat cells are continually being hydrolyzed all the time and releasing fatty acids to the metabolic pool.

Third, in the absence of carbohydrates, several hormonal changes take place that promote rapid fatty acid mobilization from adipose tissue. Among the most important of these hormonal effects is the marked decrease in insulin secretion caused by absence of carbohydrates. This not only reduces the rate of glucose utilization by the tissues but also decreases fat synthesis and indirectly mobilizes large amounts of fatty acids.

Hormonal Regulation of Fat Utilization. At least seven of the hormones secreted by the endocrine glands have marked effects on fat utilization. In addition to the important effect of *insulin lack* to mobilize fatty acids, as discussed above, some other important hormonal effects on fat metabolism are listed here.

Probably the most dramatic increase that occurs in fat utilization is that observed during heavy exercise. This results almost entirely from rapid release of *epinephrine* and *norepinephrine* by the adrenal medullae during exercise, as a result of sympathetic stimulation. These two hormones directly activate *hormone-sensitive triglyceride lipase* that is present in abundance in the fat cells, and this causes very rapid breakdown of triglycerides and mobilization of fatty acids. Sometimes the free fatty acid concentration in the blood rises as much as 10- to 15-fold. Other types of stress that activate the sympathetic nervous system will increase fatty acid mobilization and utilization in a similar manner.

Stress also causes large quantities of *corticotropin* to be released by the anterior pituitary gland, and this, in turn, causes the adrenal cortex to secrete excessive quantities of *glucocorticoids*. Both the corticotropin and glucocorticoids activate either the same hormone-sensitive triglyceride lipase as that activated by epinephrine and norepinephrine or a similar lipase. Therefore, this is still another mechanism for increasing the release of fatty acids from fat tissue. When corticotropin and glucocorticoids are secreted in excessive amounts for long periods of time, as occurs in endocrine diseases such as Cushing's disease, fats are frequently mobilized to such a great extent that ketosis results. Corticotropin and glucocorticoids are said then to have a *ketogenic effect.*

Growth hormone has an effect similar to but less than that of corticotropin and glucocorticoids in activating the hormone-sensitive lipase. Therefore, growth hormone can also have a mild ketogenic effect.

Finally, *thyroid hormone* causes rapid mobilization of fat, which is believed to result indirectly from an increased rate of energy metabolism in all cells of the body under the influence of this hormone. The resulting reduction in acetyl Co-A and other intermediates of fat metabolism in the cells would then be a stimulus to cause fat mobilization.

The effects of the different hormones on metabolism are discussed further in the chapters dealing with each of them.

OBESITY

Obesity means deposition of excess fat in the body. This subject will be discussed in detail in relation to dietary balances in Chapter 73, but briefly it is caused by ingestion of greater amounts of food than can be utilized by the body for energy. The excess food, whether fats, carbohydrates, or proteins, is then stored as fat in the adipose tissue to be used later for energy. Strains of rats have been found in which *hereditary obesity* occurs. In at least one of these strains, the obesity is caused by ineffective mobilization of fat from the adipose tissue while synthesis and storage of fat continue normally. Obviously, such a one-way process causes progressive enhancement of the fat stores, resulting in severe obesity. The problem of obesity will be discussed further in Chapter 73.

PHOSPHOLIPIDS AND CHOLESTEROL

PHOSPHOLIPIDS

The three major types of body phospholipids are the *lecithins,* the *cephalins,* and the *sphingomyelins,* typical examples of which are shown in Figure 68–4.

Phospholipids always contain one or more fatty acid molecules and one phosphoric acid radical, and they usually contain a nitrogenous base. Though the chemical structures of phospholipids are somewhat variant, their physical properties are similar, for they are all lipid soluble, are transported together in lipoproteins in the blood, and are utilized in similar ways throughout the body for various structural purposes.

Formation of Phospholipids. Phospholipids are formed in essentially all cells of the body, though certain cells have a special ability to form them. Probably 90 per

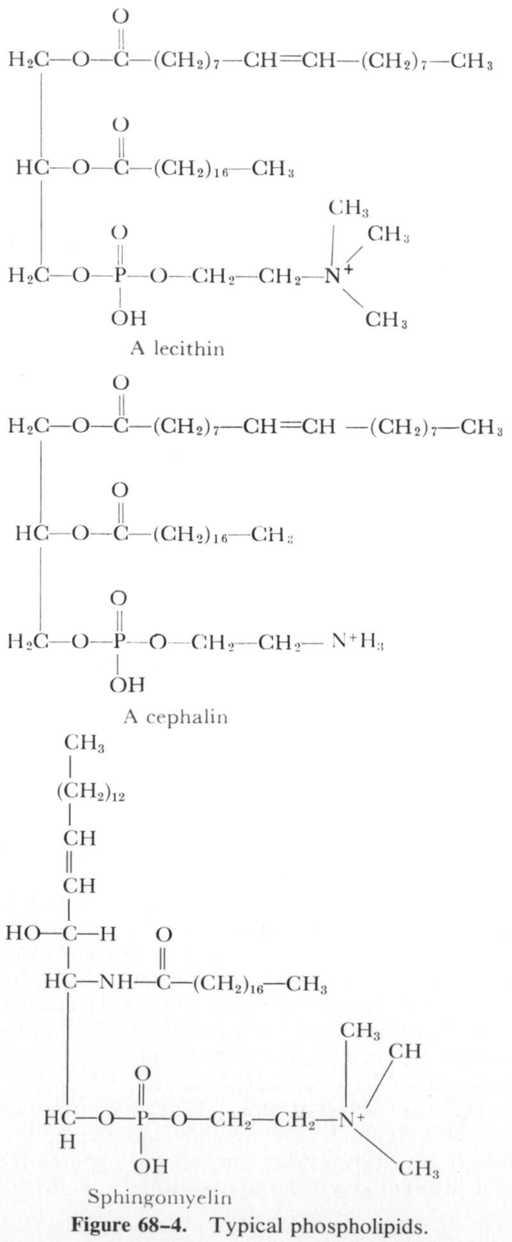

Figure 68–4. Typical phospholipids.

are formed in the liver cells, though reasonable quantities can also be formed by the intestinal epithelial cells during lipid absorption from the gut.

The rate of phospholipid formation is governed to some extent by the usual factors that control the rate of fat metabolism, for when triglycerides are deposited in the liver, the rate of phospholipid formation increases. Also, certain specific chemical substances are needed for formation of some phospholipids. For instance, *choline*, either in the diet or synthesized in the body, is needed for the formation of lecithin because choline is the nitrogenous base of the lecithin molecule. Also, *inositol* is needed for the formation of some cephalins.

Specific Uses of Phospholipids.   Several isolated functions of the phospholipids are the following: (1) Phospholipids are an important constituent of lipoproteins and are essential for the formation of some of these; therefore, in their absence serious abnormalities of fat transport can occur. (2) Thromboplastin, which is necessary to initiate the clotting process, is composed mainly of one of the cephalins. (3) Large quantities of sphingomyelin are present in the nervous system; this substance acts as an insulator in the myelin sheath around nerve fibers. (4) Phospholipids are donors of phosphate radicals when these are needed for different chemical reactions in the tissues. (5) Perhaps the most important of all the functions of phospholipids is participation in the formation of structural elements — mainly membranes — within the cells throughout the body, as is discussed below in connection with cholesterol.

CHOLESTEROL

Cholesterol, the formula of which is illustrated, is present in the diet of all persons, and it can be absorbed slowly from the gastrointestinal tract into the intestinal lymph. It is highly fat soluble, but only slightly soluble in water, and it is capable of forming esters with fatty acids. Indeed, approximately 70 per cent of the cholesterol in the lipoproteins of the plasma is in the form of cholesterol esters.

Formation of Cholesterol.   Besides the cholesterol absorbed each day from the gastrointestinal tract, which is called *exogenous cholesterol*, a large quantity is formed in the cells of the body, called *endogenous cholesterol*. Essentially all the endogenous cholesterol that circulates in the lipoproteins of the plasma is formed by the liver, but all the other cells of the body form at least some cholesterol, which is consistent with the fact that many of the membranous structures of all cells are partially composed of this substance.

As illustrated by the formula of cholesterol, its basic structure is a sterol nucleus. This is synthesized entirely from acetyl Co-A. In turn, the sterol nucleus can be modified by means of various side chains to form (a) cholesterol, (b) cholic acid, which is the basis of the bile acids formed in the liver, and (c) many important steroid hormones secreted by the adrenal cortex, the ovaries, and the testes (these are discussed in later chapters).

Factors That Affect the Plasma Cholesterol Concentration — Feedback Control of Body Cholesterol.   Among the important factors that affect plasma cholesterol concentration are the following:

1. An increase in the amount of cholesterol ingested each day increases the plasma concentration slightly. However, when cholesterol is ingested, the rising concentration of cholesterol inhibits one of the essential enzymes for endogenous synthesis of cholesterol, thus providing an intrinsic feedback control system to regulate plasma cholesterol concentration. As a result, plasma cholesterol concentration *usually* is not changed upward or downward more than ± 15 per cent by altering the amount of cholesterol in the diet, though extremes of cholesterol in the diet can probably alter the level by as much as ± 30 per cent.

2. A *saturated* fat diet increases blood cholesterol concentration as much as 15 to 25 per cent. This results from increased fat deposition in the liver, which then provides increased quantities of acetyl Co-A in the liver cells for production of cholesterol. Therefore, to decrease the blood cholesterol concentration, it is equally as important to maintain a diet low in saturated fat as to maintain a diet low in cholesterol concentration.

3. Ingestion of fat containing highly unsaturated fatty acids usually depresses the blood cholesterol concentration a slight to moderate amount. Unfortunately, the mechanism of this effect is unknown despite the fact that this observation is the basis of much of present-day dietary strategy.

4. Lack of thyroid hormone increases the blood cholesterol concentration, whereas excess thyroid hormone decreases the concentration. This effect is believed to be related to the increased metabolism of all lipid substances under the influence of thyroxine.

5. The blood cholesterol also rises greatly in diabetes mellitus. This is believed to result from the general increase in lipid mobilization in this condition.

6. The female sex hormones, the *estrogens,* decrease blood cholesterol, whereas the male sex hormones, the *androgens,* increase blood cholesterol. Unfortunately, the mechanisms of these effects are unknown, but the sex effects are very important because the higher cholesterol in the male is associated with a higher incidence of heart attacks.

Specific Uses of Cholesterol.   By far the most abundant use of cholesterol in the body is to form cholic acid in the liver. As much as 80 per cent of the cholesterol is converted into cholic acid. As explained in Chapter 70, this is conjugated with other substances to form bile salts, which promote digestion and absorption of fats as has already been discussed in connection with fat digestion.

A small quantity of cholesterol is used (a) by the adrenal glands to form adrenocortical hormones, (b) by the ovaries to form progesterone and estrogen, and (c) by the testes to form testosterone. However, these glands can also synthesize their own sterols and then

Cholesterol

form their hormones from these, as is discussed in the chapters on endocrinology later in the text.

A large amount of cholesterol is precipitated in the corneum of the skin. This, along with other lipids, makes the skin highly resistant to the absorption of water-soluble substances and also to the action of many chemical agents, for cholesterol and the other lipids are highly inert to such substances as acids and different solvents that might otherwise easily penetrate the body. Also, these lipid substances help to prevent water evaporation from the skin; without this protection the amount of evaporation (as occurs in burn patients who have lost their skin) is as much as 5 to 10 liters per day instead of the usual 300 to 400 ml.

STRUCTURAL FUNCTIONS — ESPECIALLY FOR MEMBRANES — OF PHOSPHOLIPIDS AND CHOLESTEROL

The specific uses of phospholipids and cholesterol are probably of only minor importance in comparison with their importance for general structural purposes throughout the cells of the body, mainly for formation of membranes.

In Chapter 2 it was pointed out that large quantities of phospholipids and cholesterol are present in the cell membrane and in the membranes of the internal organelles of all cells. It is also known that both cholesterol and the phospholipids have controlling effects on the permeability of cell membranes.

For membranes to be formed, substances that are not soluble in water must be available, and, in general, the only substances in the body that are not soluble in water (besides the inorganic substances of bone) are the lipids and some proteins. Thus, the physical integrity of cells throughout the body is based mainly on phospholipids, cholesterol, and to a lesser extent triglycerides and certain insoluble proteins. The polar charges on the phospholipids give them the important property of helping to decrease the interfacial tension between the membranes and the surrounding fluids.

Another fact that indicates phospholipids and cholesterol to be mainly concerned with the formation of structural elements of the cells is the slow turnover rate of these substances in most nonhepatic tissues. For instance, radioactive phospholipids formed in the brain of mice remain in the brain several months after they are formed. Thus, these phospholipids are only slowly metabolized, and the fatty acid is not split away from them to any major extent. Consequently, the purpose of their being in the cells of the brain is presumably related to their indestructible physical properties rather than to their chemical properties — in other words, for the formation of actual physical structures within the cells of the brain.

ATHEROSCLEROSIS

Atherosclerosis is a disease of the intima of the arteries, especially of the large arteries, that leads to fatty lesions called *atheromatous plaques* on the inner surfaces of the arteries. The earliest stage in the devel-opment of these lesions is believed to be damage to the endothelial cells and sublying intima. The damage can be caused by physical abrasion of the endothelium, by abnormal substances in the blood, or even by the effect of the pulsating arterial pressure on the vessel wall. Once the damage has occurred, smooth muscle cells proliferate and migrate from the media of the arteries into the lesions. Soon thereafter lipid substances, especially cholesterol, begin to deposit from the blood in the proliferating muscle cells, forming the atheromatous plaques. Because these plaques contain so much cholesterol, they are frequently called simply *cholesterol deposits*. In the later stages of the lesions, fibroblasts infiltrate the degenerative areas and cause progressive sclerosis (fibrosis) of the arteries. In addition, calcium often precipitates with the lipids to develop *calcified plaques*. When these two processes occur, the arteries become extremely hard, and the disease is then called *arteriosclerosis*, or simply "hardening of the arteries."

Obviously, arteriosclerotic arteries lose most of their distensibility, and because of the degenerative areas, they are easily ruptured. Also, the atheromatous plaques often protrude through the intima into the flowing blood, and the roughness of their surfaces causes blood clots to develop, with resultant thrombus or embolus formation (see Chapter 9). Almost half of all human beings die of arteriosclerosis. Approximately two thirds of these deaths are caused by thrombosis of one or more coronary arteries, and the remaining one third by thrombosis or hemorrhage of vessels in other organs of the body — especially the brain, to cause strokes, but also in the kidneys, liver, gastrointestinal tract, limbs, and so forth.

Despite the extreme prevalence of atherosclerosis, little is known about its cause. Therefore, it is necessary to outline the general trends of the experimental studies rather than to present a definitive description of the mechanisms that cause atherosclerosis.

EXPERIMENTAL PRODUCTION OF ATHEROSCLEROSIS IN ANIMALS

In the past, it had been believed that all that was necessary to cause atherosclerosis was to increase the amount of cholesterol circulating in the blood. Indeed, very early it was demonstrated that feeding large amounts of cholesterol to rabbits will cause large numbers of cholesterol deposits in the intimal layer of the arteries. This experiment has often been cited as proof of the hypercholesterolemic theory of atherogenesis. However, these deposits do not lead to subsequent arterial wall fibrosis and death of the animal. Furthermore, it has been difficult to achieve such deposits in carnivorous animals such as the dog, except by feeding extreme amounts of cholesterol and also removing the thyroid gland to prevent normal utilization of cholesterol.

Therefore, in recent years much more emphasis has been placed on the initial endothelial and intimal lesions as the primary cause of the ultimate atherosclerotic plaques. Almost any factor that can cause damage to the endothelial cells will lead to the following sequence of events: First, platelets adhere to the endothelium. Second, the platelets dissolute, and some factor from the platelets causes proliferation of the sublying smooth muscle cells; these then infiltrate the damaged region. Subsequently, cholesterol (and to much lesser extent other lipids) infiltrates the lesion until eventually lesions typical of human atherosclerotic plaques develop. The severity of these lesions is enhanced in the presence of hypercholesterolemia.

ATHEROSCLEROSIS IN THE HUMAN BEING

Effect of Age, Sex, and Heredity on Atherosclerosis. Atherosclerosis is mainly a disease of old age, but small atheromatous plaques can almost always be found in the arteries of young adults. Therefore, the full-blown disease is a culmination of a lifetime of vascular damage and lipid deposition rather than deposition over a few years.

Far more men than woman die of atherosclerotic heart disease. This is especially true of men younger than 50 years of age. For this reason, it is possible that the male sex hormone accelerates the development of atherosclerosis, or that the female sex hormone *protects* a person from atherosclerosis. Indeed, administration of estrogens to men who have already had coronary thromboses has decreased the number of secondary coronary attacks in some clinical trials. Furthermore, administration of estrogens to chickens with atheromatous plaques in their coronaries has in some instances actually caused the disease to regress.

Atherosclerosis and atherosclerotic heart disease are highly hereditary in some families. In some instances, this is related to an inherited *familial hypercholesterolemia*, the excess cholesterol occurring almost entirely in the *low density lipoproteins*. This probably results from lack of a *lipoprotein receptor substance* on the liver cell membranes that recognizes the low density lipoproteins and causes them to adhere to the cells. Normally, this adherence is required before the lipoproteins can deliver their load of cholesterol to the liver cells. In the absence of this, cholesterol can only leave the liver cells, and the internal feedback mechanism of the liver cells causes prolific production of cholesterol, adding to that already in the low density lipoproteins. Homozygous persons with this disease rarely live beyond the age of 20.

In other persons with hereditary atherosclerosis, the blood cholesterol level is completely normal. Inheritance of the tendency to atherosclerosis is sometimes caused by dominant genes, which means that once this dominant trait enters a family a high incidence of the disease occurs among the offspring.

Other Diseases That Predispose to Atherosclerosis. Human beings with severe *diabetes* or severe *hypothyroidism* frequently develop premature and severe atherosclerosis. In both these conditions the blood cholesterol is greatly elevated, which is at least part of the cause of the atherosclerosis.

Another disease associated with atherosclerosis, in human beings as well as in experimental animals, is *hypertension;* the incidence of atherosclerotic coronary heart disease is about twice as great in hypertensive people as in normal persons. Though the cause is not known, it possibly results from pressure damage to the arterial walls, with subsequent deposition of cholesterol plaques.

Relationship of Dietary Fat to Atherosclerosis in the Human Being. A high fat diet, especially one containing cholesterol and saturated fats, increases one's chances of developing atherosclerosis. Therefore, decreasing the fat can help greatly in protecting against atherosclerosis, and some experiments indicate that this can benefit even patients who have already had coronary heart attacks. Also, life insurance statistics show that the rate of mortality — mainly from coronary disease — of normal weight middle and older aged persons is about half the mortality rate of overweight subjects of the same age.

As pointed out above, eating unsaturated fats instead of saturated fats, for reasons not known, decreases the blood cholesterol concentration, and this seems to be especially beneficial in preventing atherosclerosis.

SUMMARY OF FACTORS CAUSING ATHEROSCLEROSIS

Atherosclerosis is highly associated with abnormalities of lipid metabolism, but it is also exacerbated by almost any factor that injures the arterial wall. In particular, elevated blood cholesterol is often related to atherosclerosis. But, perhaps equally as important might be some undiscovered third factor that is inherited from generation to generation which causes increased arterial intimal degeneration or increased rate of cholesterol deposition in the arterial walls irrespective of the blood cholesterol concentration.

REFERENCES

Benditt, E. P.: The origin of atherosclerosis. *Sci. Am., 236*(2):74, 1977.

Brady, R. O.: The sphingolipidoses. *In* Bondy, P. K., and Rosenberg, L. E. (eds.): Metabolic Control and Disease, 8th Ed. Philadelphia, W. B. Saunders Co., 1980, p. 523.

Brady, G. A., and York, D. A.: Hypothalamic and genetic obesity in experimental animals: An autonomic and endocrine hypothesis. *Physiol. Rev., 59*:719, 1979.

Chandler, A. B., *et al.* (eds.): The Thrombotic Process in Atherogenesis. New York, Plenum Press, 1978.

Coleman, J. E.: Metabolic interrelationships between carbohydrates, lipids and proteins. *In* Bondy, P. K., and Rosenberg, L. E. (eds.): Metabolic Control and Disease, 8th Ed. Philadelphia, W. B. Saunders Co., 1980, p. 161.

Crepaldi, G., *et al.* (eds.): Diabetes, Obesity, and Hyperlipidemias. New York, Academic Press, 1978.

Dils, R., and Knudsen, J. (eds.): Regulation of Fatty Acid and Glycerolipid Metabolism. New York, Pergamon Press, 1978.

Gennis, R. B., and Jonas, A.: Protein-Lipid Interactions. *Annu. Rev. Biophys. Bioeng., 6*:195, 1977.

Goldfarb, S.: Regulation of hepatic cholesterogenesis. *In* Javitt, N. B. (ed.): International Review of Physiology: Liver and Biliary Tract Physiology I. Vol. 21. Baltimore, University Park Press, 1980, p. 317.

Goto, Y.: New aspects of lipid metabolism and arteriosclerosis. *In* Hayase, S., and Murao, S. (eds.): Cardiology: Proceedings of the VIII World Congress of Cardiology, Tokyo, 1978. New York, Elsevier/North-Holland, 1979, p. 197.

Gotto, A. M., Jr., *et al.* (eds.): High Density Lipoproteins and Atherosclerosis. New York, Elsevier/North-Holland, 1978.

Gresham, G. A.: Reversing Atherosclerosis. Springfield, Ill., Charles C Thomas, 1979.

Hashimoto, S., and Dayton, S.: Lipid metabolism of arterial smooth muscle. *In* Bohr, D. F., *et al.* (eds.): Handbook of Physiology. Sec. 2, Vol. II. Baltimore, Williams & Wilkins, 1980, p. 161.

Havel, R. J., *et al.:* Lipoproteins and lipid transport. *In* Bondy, P. K., and Rosenberg, L. E. (eds.): Metabolic Control and Disease, 8th Ed. Philadelphia, W. B. Saunders Co., 1980, p. 393.

Hessel, L. W., and Krans, H. M. J. (eds.): Lipoprotein Metabolism and Endocrine Regulation. New York, Elsevier/Noth-Holland, 1979.

Jackson, R. L., *et al.:* Lipoprotein structure and metabolism. *Physiol. Rev., 56*:259, 1976.

Jellinek, H. (ed.): Arterial Lesions and Arteriosclerosis. New York, Plenum Press, 1974.

Levy, R. I. (eds.): Nutrition, Lipids, and Coronary Heart Disease. New York, Raven Press, 1979.

Masoro, E. J.: Lipids and lipid metabolism. *Annu. Rev. Physiol., 39*:301, 1977.

McGill, H. C., Jr.: Morphologic development of the atherosclerotic plaque. *In* Lauer, R. M., and Shekelle, R. B. (eds.): Childhood Prevention of Atherosclerosis and Hypertension. New York, Raven Press, 1980.

Miller, G. J.: High density lipoproteins and atherosclerosis. *Annu. Rev. Med., 31*:97, 1980.

Nelson, G. J. (ed.): Blood Lipids and Lipoproteins: Quantitation, Composition and Metabolism. Huntington, N.Y., R. E. Krieger Publishing Co., 1979.

Robertson, A. L., Jr., and Rosen, L. A.: The endothelial lining of arteries in disease. *In* Kaley, G., and Altura, B. M. (eds.): Microcirculation. Vol. III. Baltimore, University Park Press, 1977.

Robinson, A. M., and Williamson, D. H.: Physiological roles of ketone bodies as substrates and signals in mammalian tissues. *Physiol. Rev., 60*:143, 1980.

Rosell, S., and Belfrage, E.: Blood circulation in adipose tissue. *Physiol. Rev., 59*:1078, 1979.

Ross, R.: Endothelial integrity, cell proliferation, and atherosclerosis. *In* Lauer, R. M., and Shekelle, R. B. (eds.): Childhood Prevention of Atherosclerosis and Hypertension. New York, Raven Press, 1980.

Ross, R., and Kariya, B.: Morphogenesis of vascular smooth muscle in atherosclerosis and cell structure. *In* Bohr, D. R. *et al.* (eds.): Handbook of Physiology. Sec. 2, Vol. II. Baltimore, Williams & Wilkins, 1980, p. 69.

Salans, L. B.: Obesity and the adipose cell. *In* Bondy, P. K., and Rosenberg, L. E. (eds.): Metabolic Control and Disease, 8th Ed. Philadelphia, W. B., Saunders Co., 1980, p. 495.

Scanu, A. M., *et al.* (eds.): The Biochemistry of Atherosclerosis. New York, Marcel Dekker, 1978.

Schonfeld, G.: Hormonal control of lipoprotein metabolism. *In* De-Groot, L. J. *et al.* (eds.): Endocrinology. Vol. 3. New York, Grune & Stratton, 1979, p. 1855.

Texon, M.: The Hemodynamic Basis of Atherosclerosis. Washington, D. C., Hemisphere Publishing Corp., 1979.

Verger, R., and Haas, G. H.: Interfacial enzyme kinetics of lipolysis. *Annu. Rev. Biophys. Bioeng., 5*:77, 1976.

Volpe, J. J., and Vagelos, P. R.: Mechanisms and regulation of biosynthesis of saturated fatty acids. *Physiol. Rev., 56*:339, 1976.

Whayne, T. F., Jr.: Atherogenic mechanisms. *In* Frohlich, E. D. (ed.): Pathophysiology, 2nd Ed. Philadelphia, J. B. Lippincott Co., 1976, p. 103.

Wissler, R. W.: Nutrition, plasma lipids, and atherosclerosis. *In* Lauer, R. M., and Shekelle, R. B. (eds.): Childhood Prevention of Atherosclerosis and Hypertension. New York, Raven Press, 1980.

69

Protein Metabolism

About three quarters of the body solids are proteins. These include *structural proteins, enzymes, genes, proteins that transport oxygen, proteins of the muscle that cause contraction,* and many other types that perform specific functions both intracellularly and extracellularly throughout the body.

The basic chemical properties of proteins that explain their diverse functions are so extensive that they are a major portion of the entire discipline of biochemistry. For this reason, the present discussion is confined to a few specific aspects of protein metabolism that are important as background for other discussions in the text.

BASIC PROPERTIES OF PROTEINS

THE AMINO ACIDS

The principal constituents of proteins are amino acids, 20 of which are present in the body proteins in significant quantities. Figure 69-1 illustrates the chemical formulas of these 20 amino acids, showing that they all have two features in common: Each amino acid has an acidic group (—COOH) and a nitrogen radical attached to the molecule near the acidic radical, usually represented by the amino group (—NH$_2$).

Peptide Linkages and Peptide Chains. In proteins, the amino acids are aggregated into long chains by means of *peptide linkages,* one of which is illustrated by the following reaction:

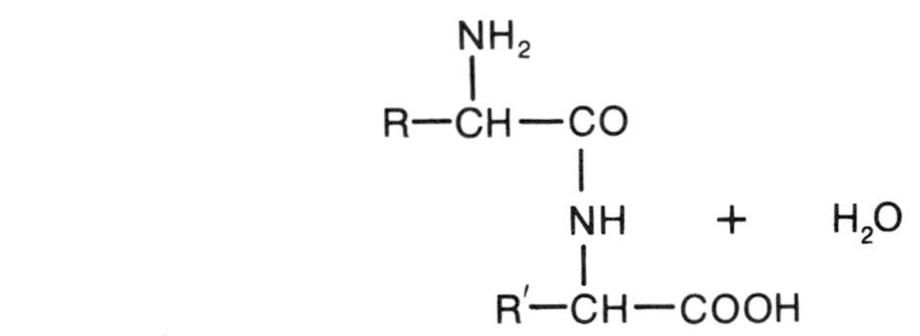

Note in this reaction that the amino radical of one amino acid combines with the carboxyl radical of the other amino acid. A hydrogen atom is released from the amino radical, and a hydroxyl radical is released from the carboxyl radical; these two combine to form a molecule of water. Note that after the peptide linkage has been formed, an amino radical and a carboxyl radical are still in the new molecule, both of which are capable of combining with additional amino acids to form a *peptide chain.* Some complicated protein molecules have as many as a hundred thousand amino acids combined together principally by peptide linkages, and even the smallest protein usually has more than 20 amino acids combined together by peptide linkages.

Other Linkages in Protein Molecules. Some protein molecules are composed of several peptide chains rather than a single chain, and these in turn are bound with each other by other linkages, often by *hydrogen bonds* between the CO and NH radicals of the peptides as follows:

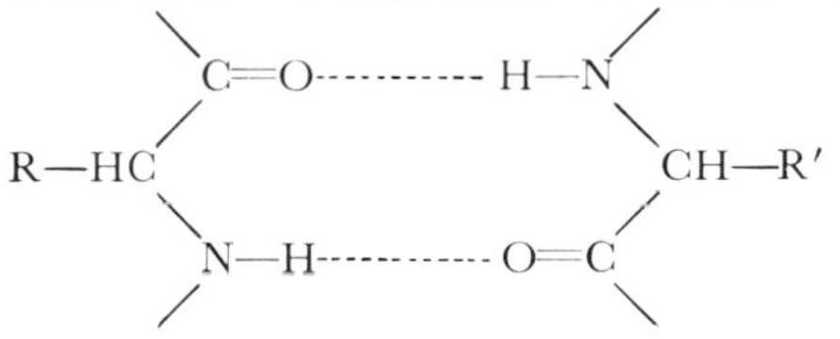

Also, many peptide chains are coiled or folded, and the successive coils or folds are held in a tight spiral or in other shapes by similar hydrogen bonding. But, in addition to hydrogen bonds, separate peptide chains can be held together by hydrophobic bonds, electrostatic forces, and sulfhydryl, phenolic, and salt linkages, as well as by others.

PHYSICAL CHARACTERISTICS OF PROTEINS

GLOBULAR PROTEINS

With the exception of the fibrous proteins, which are discussed subsequently, most proteins of the body assume either a globular or an elliptical shape and are called *globular proteins.* These, in general, are soluble in water or salt solutions, and they are held in a globular shape by coiling and folding of the peptide chains.

Important Types of Globular Proteins in the Body. There is no simple functional classification of the globular proteins for two reasons: First, these

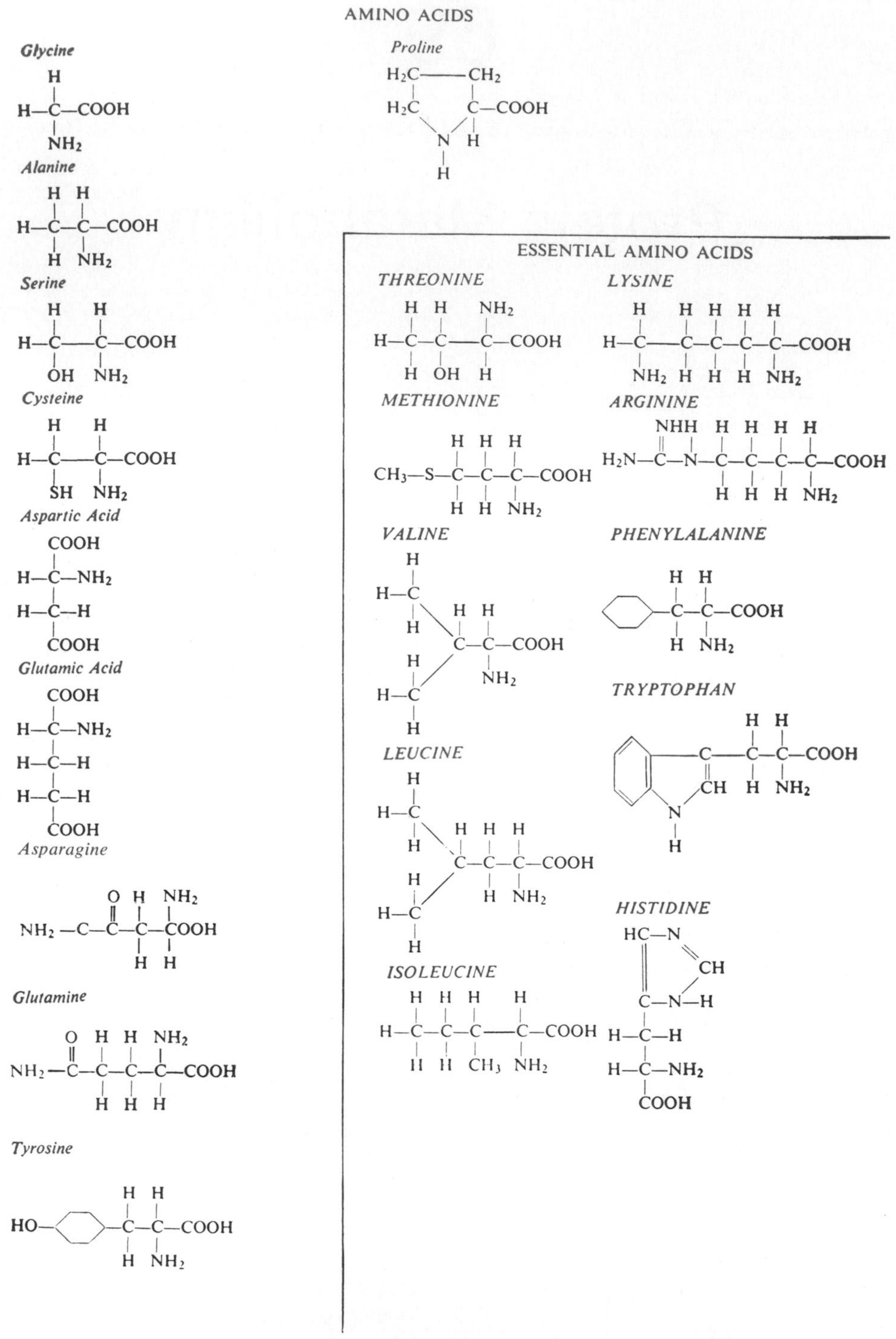

AMINO ACIDS

Glycine · Alanine · Serine · Cysteine · Aspartic Acid · Glutamic Acid · Asparagine · Glutamine · Tyrosine · Proline

ESSENTIAL AMINO ACIDS

THREONINE · LYSINE · METHIONINE · ARGININE · VALINE · PHENYLALANINE · LEUCINE · TRYPTOPHAN · ISOLEUCINE · HISTIDINE

Figure 69–1. The amino acids, showing the 10 essential amino acids, which cannot be synthesized at all or in sufficient quantity in the body.

proteins perform literally thousands of different functions in the body. Second, proteins of widely varying chemical and physical characteristics many times perform very much the same function. Some important examples of globular proteins are the *albumin, globulins,* and *fibrinogen* that constitute the plasma proteins. Also, *hemoglobin, the cytochromes,* and most of the *cellular enzymes* are globular proteins.

FIBROUS PROTEINS

Many of the highly complex proteins are fibrillar and are called fibrous proteins. In these the peptide chains are elongated, and many separate chains are held together in parallel bundles by cross-linkages. Major types of fibrous proteins are (1) *collagens,* which are the basic structural proteins of connective tissue, tendons, cartilage, and bone; (2) *elastins,* which are the elastic fibers of tendons, arteries, and connective tissue; (3) *keratins,* which are the structural proteins of hair and nails; and (4) *actin* and *myosin,* the contractile proteins of muscle.

Since fibrillar proteins are the principal structural proteins of the body, it is important to know about their physical characteristics. In general, the fibrils are extremely strong and are capable of being stretched, and then they recoil to their natural length; these are properties of typical *elastomers.* Another elastomeric characteristic of the fibrillar proteins is their tendency to *creep;* that is, if stretched for a long time, their basic length gradually becomes the stretched length, but, on the other hand, if the tension on the two ends of the fibrils is relaxed, the fibrils creep to a shorter and shorter length. For instance, a large scar that forms shortly after a severe wound gradually creeps to a smaller and smaller size if tension is not placed on the scar. On the other hand the scar can creep to a larger size if it is in an area where tension is high, as occurs frequently in the skin of a person who becomes progressively more obese.

CONJUGATED PROTEINS

In addition to the simple globular and fibrous proteins, many proteins are combined as conjugated proteins with nonprotein substances. These include the following:

Nucleoproteins are combinations of simple proteins, most often the highly basic proteins called histones and protamines, and nucleic acid. The deoxyribose nucleoproteins are the principal constituents of the chromosomes, and the DNA portions of these constitute the genes.

Proteoglycans, also called *mucoproteins,* are composed of protein and extremely large amounts of *glycosaminoglycans,* which are large, negatively charged polysaccharides. They are major components of all tissues, composing as much as 30 per cent of the dry weight of connective tissue. They also serve as a lubricant in joints, and they are responsible for the gel-like consistency of the vitreous humor of the eyes.

In addition to these important types of proteins are the *lipoproteins,* which contain lipid materials; *chromoproteins* (such as hemoglobin and cytochromes), which contain coloring agents; *phosphoproteins,* which contain phosphorus; and *metalloproteins,* which contain magnesium, copper, iron, zinc, or other metallic ions and which constitute many of the enzymes.

TRANSPORT AND STORAGE OF AMINO ACIDS

THE BLOOD AMINO ACIDS

The normal concentration of amino acids in the blood is between 35 and 65 mg. per cent. This is an average of about 2 mg. per cent for each of the 20 amino acids, though some are present in far greater concentrations than others. Since the amino acids are relatively strong acids, they exist in the blood principally in the ionized state and account for 2 to 3 milliequivalents of the negative ions in the blood. The precise distribution of the different amino acids in the blood depends to some extent on the types of proteins ingested, but the concentrations of at least some individual amino acids are regulated by (a) selective synthesis in the different cells and (b) selective excretion by the kidneys.

Fate of Amino Acids Absorbed from the Gastrointestinal Tract.   It will be recalled from Chapter 65 that the end-products of protein digestion in the gastrointestinal tract are almost entirely amino acids and that only rare polypeptide or protein molecules are absorbed. Immediately after a meal, the amino acid concentration in the blood rises, but the rise is usually only a few milligrams per cent for two reasons: First, protein digestion and absorption is usually extended over two to three hours, which allows only small quantities of amino acids to be absorbed at a time. Second, after entering the blood, the excess amino acids are absorbed within 5 to 10 minutes by cells throughout the entire body, especially by the liver. Therefore, almost never do large concentrations of amino acids accumulate in the blood. Nevertheless, the turnover rate of the amino acids is so rapid that many grams of proteins can be carried from one part of the body to another in the form of amino acids each hour.

Active Transport of Amino Acids into the Cells.   The molecules of essentially all the amino acids are much too large to diffuse through the pores of the cell membranes. Therefore, significant quantities of amino acids can be transported through the membrane only by facilitated or active transport utilizing carrier mechanisms. The nature of the carrier mechanisms is still poorly understood, but some of the theories are discussed in Chapter 4.

Renal Threshold for Amino Acids.   One of the special functions of carrier transport of amino acids is to prevent loss of these in the urine. All the different amino acids can be *actively transported* through the proximal tubular epithelium, thus removing them from the glomerular filtrate and returning them to the blood. However, as is true of other active transport mechanisms in the renal tubules, there is an upper limit to the rate at which each type of amino acid can be transported. For this reason, when a particular type of amino acid rises to too high a concentration in the plasma and glomerular filtrate, the excess above that which can be actively reabsorbed is lost into the urine.

In Chapter 38 it was pointed out that appropriate carrier systems for active reabsorption of certain amino acids are often deficient or lacking from the renal tubular epithelium. In these conditions the plasma threshold for the respective amino acids is greatly reduced. However, in the normal person, the loss of amino acids in the urine each day is insignificant.

STORAGE OF AMINO ACIDS AS PROTEINS IN THE CELLS

Almost immediately after entry into the cells, amino acids are conjugated under the influence of intracellular enzymes into cellular proteins so that the concentrations of the amino acids inside the cells probably always remains low. Thus, so far as is known, storage of large quantities of amino acids as such probably does not occur in the cells; instead, they are mainly stored in the form of actual proteins. Yet many intracellular proteins can be rapidly decomposed again into amino acids under the influence of intracellular lysosomal digestive enzymes, and these amino acids in turn can be transported back out of the cell into the blood. The proteins that can be thus decomposed include many cellular enzymes as well as some other functioning proteins. However, the genes of the nucleus and the structural proteins such as collagen and muscle contractile proteins do not participate significantly in this reversible storage of amino acids.

Some tissues of the body participate in the storage of amino acids to a greater extent than others. For instance, the liver, which is a large organ and also has special systems for processing amino acids, stores large quantities of proteins; this is also true of the kidney and the intestinal mucosa.

Release of Amino Acids from the Cells and Regulation of Plasma Amino Acid Concentration. Whenever the plasma amino acid concentrations fall below their normal levels, amino acids are transported out of the cells to replenish the supply in the plasma. Simultaneously, intracellular proteins are degraded back into amino acids.

The plasma concentration of each type of amino acid is maintained at a reasonably constant value. Later it will be noted that the various hormones secreted by the endocrine glands are able to alter the balance between tissue proteins and circulating amino acids; growth hormone and insulin increase the formation of tissue proteins, while the adrenocortical glucocorticoid hormones increase the concentration of circulating amino acids.

Reversible Equilibrium Between the Proteins of Different Parts of the Body. Since cellular proteins can be synthesized rapidly from plasma amino acids and many of these in turn can be degraded and returned to the plasma almost equally as rapidly, there is constant equilibrium between the plasma amino acids and most of the proteins in the cells of the body. Therefore, it follows that there is also equilibrium between the proteins from one type of cell to the next. For instance, if any particular tissue loses proteins, it can synthesize new proteins from the amino acids of the blood; in turn, these are replenished by degradation of proteins from other cells of the body. These effects are particularly noticeable in relation to protein synthesis in cancer cells. Cancer cells are prolific users of amino acids, and, simultaneously, the proteins of the other tissues become markedly depleted.

Upper Limit to the Storage of Proteins. Each particular type of cell has an upper limit to the amount of proteins that it can store. After all the cells have reached their limits, the excess amino acids in the circulation are then degraded into other products and used for energy, as is discussed subsequently, or they are converted to fat or glycogen and stored in these forms.

THE PLASMA PROTEINS

The three major types of protein present in the plasma are *albumin, globulin,* and *fibrinogen.* The principal function of albumin is to provide *colloid osmotic pressure,* which in turn prevents plasma loss from the capillaries, as discussed in Chapter 30. The globulins perform a number of enzymatic functions in the plasma itself, but, more important than this, they are principally responsible for both the natural and acquired immunity that a person has against invading organisms, which was discussed in Chapter 7. The fibrinogen polymerizes into long fibrin threads during blood coagulation, thereby forming blood clots that help to repair leaks in the circulatory system, which was discussed in Chapter 9.

Formation of the Plasma Proteins. Essentially all the albumin and fibrinogen of the plasma proteins, as well as 60 to 80 per cent of the globulins, are formed in the liver. The remainder of the globulins are formed in the lymphoid tissues and other cells of the reticuloendothelial system. These are mainly the gamma globulins that constitute the antibodies.

The rate of plasma protein formation by the liver can be extremely high, as great as 2 grams per hour or as much as 50 grams per day. Certain disease conditions often cause rapid loss of plasma proteins; severe burns that denude large surface areas cause loss of many liters of plasma through the denuded areas each day. The rapid production of plasma proteins by the liver is obviously valuable in preventing death in such states. Furthermore, occasionally, a person with severe renal disease loses as much as 20 grams of plasma protein in the urine each day for months. In some of these patients the plasma protein concentration may remain almost normal throughout the entire illness.

Use of Plasma Proteins by the Tissues. When the tissues become depleted of proteins, the plasma proteins can act as a source for rapid replacement of the tissue proteins. Indeed whole plasma proteins can be imbibed *in toto* by the reticuloendothelial cells by the process of pinocytosis; then, once in the cells, these are split into amino acids that are transported back into the blood and utilized throughout the body to build cellular proteins. In this way, then, the plasma proteins function as a labile protein storage medium and represent a rapidly available source of amino acids whenever a particular tissue requires these.

Reversible Equilibrium Between the Plasma Proteins and the Tissue Proteins. The rate of synthesis of plasma proteins by the liver depends on the concentration of amino acids in the blood, which means that the concentration of plasma proteins becomes reduced whenever an appropriate supply of amino acids is not available. On the other hand, whenever excess proteins are available in the plasma but insufficient proteins are present in the cells, the plasma proteins are used to form tissue proteins. Thus, there is a constant state of equilibrium, as illustrated in Figure 69–2, between the plasma proteins, the amino acids of the blood, and the tissue proteins. It has been estimated from radioactive tracer studies that about 400 grams of body protein are synthesized and degraded each day as part of the continual state of flux of amino acids. This illustrates once again the general principle of reversible exchange of amino acids among the different proteins of the body. Even during starvation or during severe debilitating diseases, the ratio of total tissue proteins to total plasma proteins in the body remains relatively constant at about 33 to 1.

Because of this reversible equilibrium between plasma proteins and the other proteins of the body, one of the most effective of all therapies for severe acute

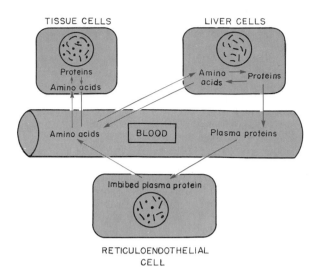

Figure 69-2. Reversible equilibrium between the tissue proteins, plasma proteins, and plasma amino acids.

protein deficiency is intravenous administration of plasma protein. Within hours, the amino acids of the administered protein become distributed throughout the cells of the body to form proteins where they are needed.

CHEMISTRY OF PROTEIN SYNTHESIS

Proteins are synthesized in all cells of the body, and the functional characteristics of each cell depend upon the types of protein that it can form. Basically, the genes of the cells control the protein types and thereby control the functions of the cell. This regulation of cellular function by the genes was discussed in detail in Chapter 3. Chemically, two basic processes must be accomplished for the synthesis of proteins; these are: (1) synthesis of the amino acids and (2) appropriate conjugation of the amino acids to form the respective types of whole proteins in each individual cell.

Essential and Nonessential Amino Acids. Ten of the amino acids normally present in animal proteins can be synthesized in the cells, while the other 10 either cannot be synthesized at all or are synthesized in quantities too small to supply the body's needs. The first group of amino acids is called *nonessential*, while the second group is called *essential amino acids*. The essential amino acids obviously must be present in the diet if protein formation is to take place in the body. Use of the word "essential" does not mean that the other 10 amino

acids are not equally as essential for formation of the proteins, but only that these others are not essential in the diet.

Synthesis of the nonessential amino acids depends on the formation first of appropriate α-keto acids, which are the precursors of the respective amino acids. For instance, *pyruvic acid*, which is formed in large quantities during the glycolytic breakdown of glucose, is the keto acid precursor of the amino acid *alanine*. Then, by the simple process of *transamination*, an amino radical is transferred to the α-keto acid while the keto oxygen is transferred to the donor of the amino radical. This reaction is illustrated in Figure 69-3. Note in this reaction that the amino radical is transferred to the pyruvic acid from another amino acid, *glutamine*. Glutamine is present in the tissues in large quantities, and it functions specifically as an amino radical storehouse. In addition, amino radicals can be transferred from *asparagine, glutamic acid,* and *aspartic acid.*

Transamination is promoted by enzymes called *transaminases,* all of which are derivatives of pyridoxine, one of the B vitamins. Without this vitamin, the nonessential amino acids cannot be synthesized, and, therefore, protein formation cannot proceed normally.

Formation of Proteins from Amino Acids. Once the appropriate amino acids are present in a cell, whole proteins are synthesized rapidly. However, each peptide linkage requires from 500 to 4000 calories of energy, and this must be supplied from ATP and GTP (guanosine triphosphate) in the cell. Protein formation proceeds through two steps: (1) "activation" of each amino acid, during which the amino acid is "energized" by energy derived from ATP and GTP, and (2) alignment of the amino acids into the peptide chains, a function that is under control of the genetic system of each individual cell. Both these processes were discussed in detail in Chapter 3. Indeed, the formation of cellular proteins is the basis of life itself and is so important that the reader would find it valuable to review Chapter 3.

USE OF PROTEINS FOR ENERGY

It was pointed out earlier that there is an upper limit to the amount of protein that can accumulate in each particular type of cell. Once the cells are filled to their limits, any additional amino acids in the body fluids are degraded and used for energy or stored as fat. This degradation occurs almost entirely in the liver, and it begins with the process known as deamination. To help in this process, the excess amino acids in the cells, especially in the liver, induce the production of large

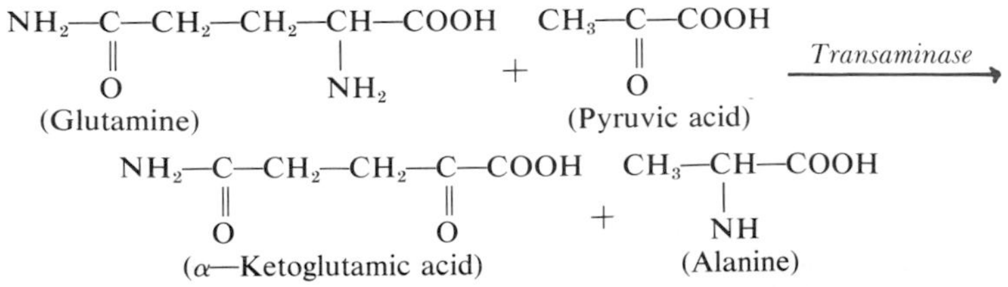

$$NH_2-\underset{\underset{O}{\|}}{C}-CH_2-CH_2-\underset{\underset{NH_2}{|}}{CH}-COOH \quad + \quad CH_3-\underset{\underset{O}{\|}}{C}-COOH \quad \xrightarrow{\textit{Transaminase}}$$

(Glutamine)   (Pyruvic acid)

$$NH_2-\underset{\underset{O}{\|}}{C}-CH_2-CH_2-\underset{\underset{O}{\|}}{C}-COOH \quad + \quad CH_3-\underset{\underset{NH}{|}}{CH}-COOH$$

(α—Ketoglutamic acid)   (Alanine)

Figure 69-3. Synthesis of alanine from pyruvic acid by transamination.

quantities of *aminotransferases,* the enzymes responsible for initiating most deamination.

Deamination. Deamination means removal of the amino groups from the amino acids. This can occur by several different means, two of which are especially important: (1) transamination, which means transfer of the amino group to some acceptor substance as explained above in relation to the synthesis of amino acids, and (2) oxidative deamination.

The greatest amount of deamination occurs by the following transamination schema:

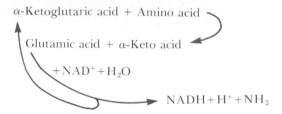

Note from this schema that the amino group from the amino acid is transferred to α-ketoglutaric acid, which then becomes glutamic acid. The glutamic acid can then transfer the amino group to still other substances or can release it in the form of ammonia. In the process of losing the amino group, the glutamic acid once again becomes α-ketoglutaric acid, so that the cycle can be repeated again and again.

Oxidative deamination occurs to much less extent and is catalyzed by amino acid oxidases. In this process the amino acid is oxidized at the point where the amino radical attaches, which causes the amino radical to be released.

Urea Formation by the Liver. The ammonia released during deamination is removed from the blood almost entirely by conversion into urea, two molecules of ammonia and one molecule of carbon dioxide combining in accordance with the following net reaction:

$$2NH_3 + CO_2 \rightarrow H_2N-\underset{\underset{O}{\|}}{C}-NH_2 + H_2O$$

Essentially all urea formed in the human body is synthesized in the liver. In the absence of the liver or in serious liver disease, ammonia accumulates in the blood. This in turn is extremely toxic, especially to the brain, often leading to a state called *hepatic coma.*

The stages in the formation of urea are essentially the following:

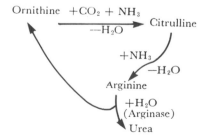

The reaction begins with the amino acid derivative *ornithine,* which combines (by means of a series of

reactions) with one molecule of carbon dioxide and one molecule of ammonia to form a second substance, *citrulline.* This in turn combines with still another molecule of ammonia to form *arginine,* which then splits into *ornithine* and *urea.* The urea diffuses from the liver cells into the body fluids and is excreted by the kidneys, while the ornithine is reused in the cycle again and again.

Oxidation of Deaminated Amino Acids. Once amino acids have been deaminated, the resulting keto acid products can in most instances be oxidized to release energy for metabolic purposes. This usually involves two successive processes: (1) the keto acid is changed into an appropriate chemical substance that can enter the citric acid cycle, and (2) this substance is then degraded by this cycle and used for energy in the same manner that acetyl Co-A derived from carbohydrate and lipid metabolism is used.

In general, the amount of adenosine triphosphate formed for each gram of protein that is oxidized is slightly less than that formed for each gram of glucose oxidized.

Gluconeogenesis and Ketogenesis. Certain deaminated amino acids are similar to the breakdown products that result from glucose and fatty acid metabolism. For instance, deaminated alanine is pyruvic acid. Obviously, this can be converted into glucose or glycogen. Or it can be converted into acetyl Co-A, which can then be polymerized into fatty acids. Also, two molecules of acetyl Co-A can condense to form acetoacetic acid, which is one of the ketone bodies, as explained in the preceding chapter.

The conversion of amino acids into glucose or glycogen is called *gluconeogenesis,* and the conversion of amino acids into keto acids or fatty acids is called *ketogenesis.* Eighteen of 20 of the deaminated amino acids have chemical structures that allow them to be converted into glucose, and 19 can be converted into fats — 5 directly and the other 14 by becoming carbohydrate first and then becoming fat.

OBLIGATORY DEGRADATION OF PROTEINS

When a person eats no proteins, a certain proportion of that person's own body proteins continues to be degraded into amino acids, then deaminated and oxidized. This involves 20 to 30 grams of protein each day, which is called the *obligatory loss* of proteins. Therefore, to prevent a net loss of protein from the body, one must ingest at least 20 to 30 grams of protein each day, and to be on the safe side as much as 75 grams is usually recommended.

The ratios of the different amino acids in the dietary protein must be about the same as the ratios in the body tissues if the entire protein is to be usable. If one particular type of essential amino acid is low in concentration, the others become unusable as well because cells form either whole proteins or none at all, as explained in Chapter 3 in relation to protein synthesis. The unusable amino acids are then deaminated and oxidized. A protein that has a ratio of amino acids different from that of the average body protein is called a *partial protein* or *incomplete protein,* and such a protein is obviously less valuable for nutrition than is the *complete protein*.

Effect of Starvation on Protein Degradation. Except for the excess protein in the diet or the 20 to 30 grams of obligatory protein degradation each day, the body uses almost entirely carbohydrates or fats for energy as long as these are available. However, after several weeks of starvation, when the quantity of stored fats begins to run out, the amino acids of the blood begin to be rapidly deaminated and oxidized for energy. From this point on, the proteins of the tissues degrade rapidly — as much as 125 grams daily — and the cellular functions deteriorate precipitously.

Because carbohydrate and fat utilization for energy occurs in preference to protein utilization, carbohydrates and fats are called *protein sparers*.

HORMONAL REGULATION OF PROTEIN METABOLISM

Growth Hormone. Growth hormone increases the rate of synthesis of cellular proteins, causing the tissue proteins to increase. The precise mechanism by which growth hormone increases the rate of protein synthesis is not known, but growth hormone is believed to act in some direct manner to enhance the transport of amino acids through the cell membranes and/or to accelerate the DNA and RNA processes of protein synthesis. Part of the action might also result from the effect of growth hormone on fat metabolism, for this hormone causes increased rate of fat liberation from the fat depots, making this available for energy. This in turn reduces the rate of oxidation of amino acids and consequently makes increased quantities of amino acids available to the tissues to be synthesized into proteins.

Insulin. Lack of insulin reduces protein synthesis almost to zero. The mechanism by which this hormone affects protein metabolism is also unknown, but insulin does accelerate amino acid transport into the cells, which could be a stimulus to protein synthesis. Also, insulin increases the availability of glucose to the cells so that the use of amino acids for energy becomes correspondingly reduced. This, too, undoubtedly makes far larger quantities of amino acids available to the tissues for protein synthesis.

Glucocorticoids. The glucocorticoids secreted by the adrenal cortex *decrease* the quantity of protein in most tissues while increasing the amino acid concentration in the plasma. However, contrary to elsewhere in the body, these hormones *increase* both the liver proteins and the plasma proteins. It is believed that the glucocorticoids act by increasing the rate of breakdown of extrahepatic proteins, thereby making increased quantities of amino acids available in the body fluids. This in turn supposedly allows the liver to synthesize increased quantities of hepatic cellular proteins and plasma proteins.

The effects of glucocorticoids on protein metabolism are especially important in ketogenesis and gluconeogenesis, for in the absence of these hormones insufficient quantities of amino acids are usually available in the plasma to allow either significant gluconeogenesis or ketogenesis from proteins.

Testosterone. Testosterone, the male sex hormone, causes increased deposition of protein in the tissues throughout the body, including especially increase in the contractile proteins of the muscles. The mechanism by which this effect comes about is unknown, but it is definitely different from the effect of growth hormone in the following way: Growth hormone causes tissues to continue growing almost indefinitely, while testosterone causes the muscles and other protein tissues to enlarge only for several months; beyond that time, despite continued administration of testosterone, further protein deposition ceases.

Estrogen, the principal female sex hormone, also causes slight deposition of protein, but its effect is relatively insignificant in comparison with that of testosterone.

Thyroxine. Thyroxine increases the rate of metabolism of all cells, and, as a result, indirectly affects protein metabolism. If insufficient carbohydrates and fats are available for energy, thyroxine causes rapid degradation of proteins to be used for energy. On the other hand, if adequate quantities of carbohydrates and fats are available and excesses of amino acids are also available in the extracellular fluid, thyroxine can actually increase the rate of protein synthesis. Conversely, in growing animals deficiency of thyroxine causes growth to be greatly inhibited because of lack of protein synthesis. In essence, it is believed that thyroxine has little specific direct effect on protein metabolism but does have an important general effect in increasing the rates of both normal anabolic and normal catabolic protein reactions.

REFERENCES

Adler-Nissen, J., *et al.* (eds.): Biochemical Aspects of New Protein Food. New York, Pergamon Press, 1978.

Arnstein, H. R. V. (ed.): Amino Acid and Protein Biosynthesis II. Baltimore, University Park Press, 1978.

Atkinson, D. E., and Fox, C. F. (eds.): Modulation of Protein Function. New York, Academic Press, 1979.

Banga, I.: Structure and Function of Elastin and Collagen. Budapest, Akadémiai Kiadó, 1966.

Bender, D. A.: Amino Acid Metabolism. New York, John Wiley & Sons, 1978.

Borasky, R.: Ultrastructure of Protein Fibers. New York, Academic Press, 1963.

Bremer, H. J., *et al.*: Clinical Chemistry and Diagnosis of Amino Acid Disturbances. Baltimore, Urban & Schwarzenberg, 1979.

Coleman, J. E.: Metabolic interrelationships between carbohydrates, lipids and proteins. *In* Bondy, P. K., and Rosenberg, L. E. (eds.): Metabolic Control and Disease, 8th Ed. Philadelphia, W. B. Saunders Co., 1980, p. 161.

Crim, M. C., and Munro, H. N.: Protein-energy malnutrition and endocrine function. *In* DeGroot, L. J., *et al.* (eds.): Endocrinology. Vol. 3. New York, Grune & Stratton, 1979, p. 1987.

Croft, L. R.: Handbook of Protein Sequence Analysis. New York, John Wiley & Sons, 1980.

Devenyi, T., and Gergely, J.: Amino Acids, Peptides, and Proteins. New York, American Elsevier Publishing Co., 1974.

Foster, R. L.: The Nature of Enzymology. New York, John Wiley & Sons, 1979.

Franks, F. (ed.): Characterisation of Protein Conformation and Function. London, Symposium Press, 1978.

Friedmann, H. C. (ed.): Enzymes. Stroudsburg, Pa., Dowden, Hutchinson & Ross, 1980.

Galjaard, H.: Genetic Metabolic Diseases: Early Diagnosis and Prenatal Analysis. New York, Elsevier/North Holland, 1980.

Gross, E., and Meienhofer, J. (eds.): The Peptides. New York, Academic Press, 1979.

Gross, E., and Meienhofer, J. (eds.): Peptides; Structure and Biological Function. Rockford, Ill., Pierce Chemical Co., 1979.

Harper, A. E., *et al.*: Effects of ingestion of disproportionate amounts of amino acids. *Physiol. Rev., 50*:428, 1970.

Jackson, R. J.: Protein Biosynthesis. Burlington, N.C., Carolina Biological Supply Co., 1978.

Kostyo, J. L., and Nutting, D. F.: Growth hormone and protein metabolism. *In* Greep, R. O., and Astwood, E. B. (eds.): Handbook of Physiology. Sec. 7, Vol. 4. Baltimore, Williams & Wilkins, 1974, p. 187.

Morgan, H. E., *et al.*: Protein metabolism of the heart. *In* Berne, R. M., *et al.* (eds.): Handbook of Physiology. Sec. 2, Vol. 1. Baltimore, Williams & Wilkins, 1979, p. 845.

Morris, D. R., and Fillingame, R. H.: Regulation of amino acid decarboxylation. *Annu. Rev. Biochem., 43*:303, 1974.

Richards, F. M.: Areas, volumes, packing, and protein structure. *Annu. Rev. Biophys. Bioeng., 6*:151, 1977.

Rosenberg, L. E., and Scriver, C. R.: Disorders of amino acid metabolism. *In* Bondy, P. K., and Rosenberg, L. E. (eds.): Metabolic Control and Disease, 8th Ed. Philadelphia, W. B. Saunders Co., 1980, p. 583.

Ross, R., and Bornstein, P.: Elastic fibers in the body. *Sci. Am., 224*:44, 1971.

Rothschild, M. A.: Albumin synthesis. *In* Javitt, N. B. (ed.): International Review of Physiology: Liver and Biliary Tract Physiology I. Vol. 21. Baltimore, University Park Press, 1980, p. 249.

Rothschild, M. A., *et al.*: Effect of albumin concentration on albumin synthesis in the perfused liver. *Am. J. Physiol., 216*:1127, 1969.

Seegmiller, J. E.: Diseases of purine and pyrimidine metabolism. *In* Bondy, P. K., and Rosenberg, L. E. (eds.): Metabolic Control and Disease, 8th Ed. Philadelphia, W. B. Saunders Co., 1980, p. 777.

Truffa-Bachi, P., and Cohen, G. N.: Amino acid metabolism. *Annu. Rev. Biochem., 42*:113, 1973.

Waterlow, J. C., *et al.*: Protein Turnover in Mammalian Tissues and in the Whole Body. New York, Elsevier/North-Holland, 1978.

Wiessbach, H., and Brot, N.: The role of protein factors in the biosynthesis of proteins. *Cell, 2*:137, 1974.

Williams-Ashman, H. G.: Metabolic effects of testicular androgens. *In* Greep, R. O., and Astwood, E. B. (eds.): Handbook of Physiology. Sec. 7, Vol. 5. Baltimore, Williams & Wilkins, 1975, p. 473.

Wynn, C. H.: The Structure and Function of Enzymes, 2nd Ed. Baltimore, University Park Press, 1979.

Zak, R., *et al.*: Assessment of protein turnover by use of radioisotopic tracers. *Physiol. Rev., 59*:407, 1979.

70

The Liver and Biliary System

Different functions of the liver have been presented at many points in this text because the liver has so many and such varied functions that it is impossible to separate its actions from those of other organ systems. Therefore, the purpose of the present chapter is to summarize the different functions of the liver and to show how the liver operates as an individual organ.

The basic functions of the liver can be divided into: (1) its vascular functions for storage and filtration of blood, (2) its secretory function for secreting bile into the gastrointestinal tract, and (3) its metabolic functions concerned with the majority of the metabolic systems of the body.

PHYSIOLOGIC ANATOMY OF THE LIVER

The basic functional unit of the liver is the liver lobule, which is a cylindrical structure several millimeters in length and 0.8 to 2 mm. in diameter. The human liver contains 50,000 to 100,000 individual lobules.

The liver lobule is constructed around a *central vein* that empties into the hepatic veins and thence into the vena cava. The lobule itself is composed principally of many *hepatic cellular plates* (two of which are illustrated in Figure 70–1) that radiate centrifugally from the central vein like spokes in a wheel. Each hepatic plate is usually two cells thick, and between the adjacent cells lie small *bile canaliculi* that empty into *terminal bile*

ducts that originate in the septa between the adjacent liver lobules.

Also in the septa are small *portal venules* that receive their blood from the portal veins. From these venules blood flows into flat, branching *hepatic sinusoids* that lie between the hepatic plates, and thence into the central vein. Thus, the hepatic cells are exposed continuously to portal venous blood.

In addition to the portal venules, *hepatic arterioles* are also present in the interlobular septa. These arterioles supply arterial blood to the septal tissues, and many of the small arterioles also empty directly into the hepatic sinusoids, most frequently emptying into these about one third of the distance away from the interlobular septa, as shown in Figure 70–1.

The venous sinusoids are lined by two types of cells: (1) typical *endothelial cells* and (2) large *Kupffer cells*, which are reticuloendothelial cells capable of phagocytizing bacteria and other foreign matter in the blood. The endothelial lining of the venous sinusoids has extremely large pores, some of which are almost 1 micron in diameter. Beneath this lining, between the endothelial cells and the hepatic cells, is a very narrow space called the *space of Disse*. Because of the large pores in the endothelium, substances in the plasma move freely into the space of Disse. Even large portions of the plasma proteins diffuse freely into this space.

In the interlobular septa are also vast numbers of *terminal lymphatics*. The spaces of Disse connect directly with the lymphatics so that excess fluid in the spaces is removed through the lymphatics.

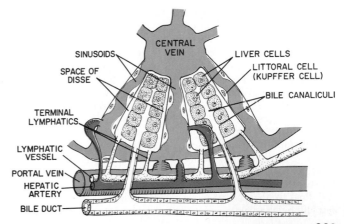

Figure 70–1. Basic structure of a liver lobule showing the hepatic cellular plates, the blood vessels, the bile-collecting system, and the lymph flow system comprised of the spaces of Disse and the interlobular lymphatics. (Reprinted from Guyton, Taylor, and Granger, as modified from Elias: Dynamics of the Body Fluids, 1975.)

FUNCTION OF THE HEPATIC VASCULAR SYSTEM

The function of the hepatic vascular system was discussed in Chapter 29 in connection with the portal veins. Briefly, this can be summarized as follows:

Blood Flow Through the Liver. About 1000 ml. of blood flows from the portal vein into the liver sinusoids each minute, and approximately an additional 400 ml. flows into the sinusoids from the hepatic artery, the total averaging about 1400 ml. per minute.

Total hepatic blood flow per minute can be measured by a modified Fick procedure in which the dye *indocyanine green,* a dye that is removed from the blood only by the liver, is injected into the circulatory system, and its concentration is measured in the arterial blood and also in venous blood collected from the hepatic vein by means of a catheter. From these measurements, one can calculate the arteriovenous difference of the dye. In addition, the rate at which the dye is removed by the liver is determined by continuously infusing the dye until its concentration in the blood remains constant, signifying equal rate of dye infusion and dye removal by the liver. One can then calculate liver blood flow by the usual Fick formula, as follows:

$$\text{Hepatic blood flow} = \frac{\text{Rate of blood clearance of dye}}{\text{A-V difference in dye}}$$

Pressures and Resistance in the Hepatic Vessels. The pressure in the hepatic vein leading from the liver into the vena cava averages almost exactly 0 mm. Hg, while the pressure in the portal vein leading into the liver averages 9 mm. Hg. This shows that the resistance to blood flow through the liver sinusoids is normally low, especially when one considers that about 1.4 liters of blood flow by this route each minute. However, various pathological conditions can cause the resistance to rise markedly, sometimes increasing the portal venous pressure to as high as 20 to 30 mm. Hg. The most common cause of increased hepatic vascular resistance is the disease *liver cirrhosis,* in which many of the vascular channels through the liver are severely restricted by fibrotic constriction of the sinusoids or even complete blockage or destruction.

Storage of Blood in the Liver; Hepatic Congestion. An increase in pressure in the veins draining the liver dams blood in the liver sinusoids and thereby causes the entire liver to swell markedly. The liver can store 200 to 400 ml. of blood in this way as the result of only a 4 to 8 mm. Hg rise in hepatic venous pressure. For this reason, the liver is one of the major *blood reservoirs.* Conversely, if a person hemorrhages so that large amounts of blood are lost from the circulatory system, much of the normal blood in the liver sinusoids drains into the remainder of the circulation to help replace the lost blood.

The most common cause of hepatic congestion is cardiac failure, which often increases the central venous pressure to as high as 10 to 15 mm. Hg. The continual stretching of the liver sinusoids that results and the stasis of blood caused by the hepatic congestion gradually lead to necrosis of many of the hepatic cells in the hepatic cellular plates.

Lymph Flow from the Liver

Because the pores in the hepatic sinusoids allow ready passage of proteins, the lymph draining from the liver usually has a protein concentration of about 6 grams per 100 milliliters, which is only slightly less than the protein concentration of plasma. Also, the extreme permeability of the liver sinusoids allows large quantities of lymph to form. Indeed, between one third and one half of all the lymph formed in the body under resting conditions arises in the liver.

Effects of High Hepatic Vascular Pressures on Fluid Transudation from the Liver Sinusoids and Portal Capillaries. When the hepatic venous pressure rises only 3 to 7 mm. Hg above normal, excessive amounts of fluid begin to transude into the lymph and also to leak through the outer surface of the liver capsule directly into the abdominal cavity. This fluid is almost pure plasma, containing 80 to 90 percent as much protein as normal plasma. At still higher hepatic venous pressures, 10 to 15 mm. Hg, liver lymph flow increases to as much as 20 times normal, and the "sweating" from the surface of the liver can be so great that it causes severe *ascites* (free fluid in the abdominal cavity).

Blockage of portal flow into or through the liver causes very high capillary pressures in the entire gastrointestinal tract, resulting in edema of the gut wall and transudation of fluid through the serosa of the gut into the abdominal cavity. This, too, can cause ascites but is less likely to do so than is sweating from the liver surface, because collateral vascular channels develop rapidly from the portal veins to the systemic veins, decreasing the intestinal capillary pressure back to a safe value.

The Hepatic Reticuloendothelial System

The inner surfaces of all the liver sinusoids are loaded with many *Kupffer cells,* which protrude into the flowing blood as illustrated in Figure 70–1 and as shown in detail in Figure 6–4 of Chapter 6. These cells are highly phagocytic, so much so that they can remove 99 per cent (or more) of bacteria in the portal venous blood before they can pass all the way through the liver sinusoids. Since the portal blood drains from the intestines, it almost always contains a reasonable number of colon bacilli. Therefore, the importance of the Kupffer cell filtration system is readily apparent. The number of Kupffer cells in the sinusoids increases markedly when increased quantities of particulate matter or other debris are present in the blood.

SECRETION OF BILE AND FUNCTIONS OF THE BILIARY TREE

PHYSIOLOGIC ANATOMY OF BILIARY SECRETION

All the hepatic cells continually form a small amount of secretion called *bile.* This is secreted into the minute *bile canaliculi* that lie between the hepatic cells in the hepatic plates, and the bile then flows peripherally toward the interlobular septa where the canaliculi empty into *terminal bile ducts,* then into progressively larger ducts, finally reaching the *hepatic duct* and *common bile duct* from which the bile either empties directly into the duodenum or is diverted into the gallbladder.

Storage of Bile in the Gallbladder. In the discussion of bile and its function in relation to gastrointestinal secretions in Chapter 64, it was pointed out that bile secreted continually by the liver cells is normally stored in the gallbladder until needed in the duodenum. The total secretion of bile each day is some 600 to 1000 ml., and the maximum volume of the gallbladder is only 40 to 70 ml. Nevertheless, as much as 12 hours' bile secretion can be stored in the gallbladder because water, sodium, chloride, and most other small electrolytes are continually absorbed by the gallbladder mucosa, concentrating the other bile constituents, including the bile salts, cholesterol, and bilirubin. Most of this absorption is caused by active transport of sodium through the gallbladder epithelium. Bile is normally concentrated about five-fold, but it can be concentrated up to a maximum of 12- to 18-fold.

Emptying of the Gallbladder. Two basic conditions are necessary for the gallbladder to empty: (1) The sphincter of Oddi must relax to allow bile to flow from the common bile duct into the duodenum and (2) the gallbladder itself must contract to provide the force required to move the bile along the common duct. After a meal, particularly one that contains a high concentration of fat, both these effects take place in the following manner:

First, the fat (also partially digested protein) in the food entering the small intestine causes release of a hormone called *cholecystokinin* from the intestinal mucosa, especially from the upper regions of the small intestine. The cholecystokinin in turn is absorbed into the blood and, on passing to the gallbladder, causes specific contraction of the gallbladder muscle. This provides the pressure that forces bile toward the duodenum.

Second, vagal stimulation associated with the cephalic phase of gastric secretion or with various intestino-intestinal reflexes causes an additional weak contraction of the gallbladder.

Third, when the gallbladder contracts, the sphincter of Oddi becomes inhibited, this effect resulting from either a neurogenic or a myogenic reflex from the gallbladder to the sphincter of Oddi. This inhibition may also, to some extent, be a direct effect of cholecystokinin on the sphincter, causing relaxation.

Fourth, the presence of food in the duodenum causes the degree of peristalsis in the duodenal wall to increase. Each time a peristaltic wave travels toward the sphincter of Oddi, this sphincter, along with the adjacent intestinal wall, momentarily relaxes because of the phenomenon of "receptive relaxation" that travels ahead of the peristaltic contraction wave, and, if the bile in the common bile duct is under sufficient pressure, a small quantity of the bile squirts into the duodenum.

In summary, the gallbladder empties its store of concentrated bile into the duodenum mainly in response to the cholecystokinin stimulus. When fat is not in the meal, the gallbladder empties poorly, but, when adequate quantities of fat are present, the gallbladder empties completely in about one hour.

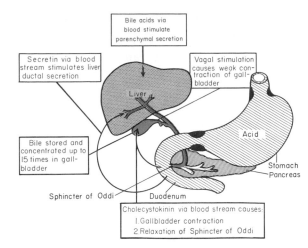

Figure 70–2. Mechanisms of liver secretion and gallbladder emptying.

Figure 70–2 summarizes the secretion of bile, its storage in the gallbladder, and its release from the bladder to the gut.

Composition of Bile. Table 70–1 gives the composition of bile when it is first secreted by the liver and then after it has been concentrated in the gallbladder. This table shows that the most abundant substance secreted in the bile is the *bile salts,* but also secreted or excreted in large concentrations are *bilirubin, cholesterol, lecithin,* and the usual *electrolytes* of plasma. In the concentrating process in the gallbladder, water and large portions of the electrolytes are reabsorbed by the gallbladder mucosa, but essentially all the other constituents, including especially the bile salts and the lipid substances cholesterol and lecithin, are not reabsorbed and therefore become highly concentrated in the gallbladder bile.

THE BILE SALTS AND THEIR FUNCTION

The liver cells form about 0.5 gram of *bile salts* daily. The precursor of the bile salts is *cholesterol,* which is either supplied in the diet or synthesized in the liver cells during the course of fat metabolism and then converted to *cholic acid* or *chenodeoxycholic acid* in about equal quantities. These acids then combine princi-

TABLE 70–1 COMPOSITION OF BILE

	Liver Bile		*Gallbladder Bile*	
Water	97.5	gm. %	92	gm. %
Bile salts	1.1	gm. %	6	gm. %
Bilirubin	0.04	gm. %	0.3	gm. %
Cholesterol	0.1	gm. %	0.3 to 0.9	gm. %
Fatty acids	0.12	gm. %	0.3 to 1.2	gm. %
Lecithin	0.04	gm. %	0.3	gm. %
Na$^+$	145	mEq./l.	130	mEq./l.
K$^+$	5	mEq./l.	12	mEq./l.
Ca$^+$	5	mEq./l.	23	mEq./l.
Cl$^-$	100	mEq./l.	25	mEq./l.
HCO$_3^-$	28	mEq./l.	10	mEq./l.

pally with glycine and to a lesser extent with taurine to form *glyco-* and *tauro-conjugated acids*. The salts of these acids are secreted in the bile.

The bile salts have two important actions in the intestinal tract. First, they have a detergent action on the fat particles in the food, which decreases the surface tension of the particles and allows the agitation in the intestinal tract to break the fat globules into minute sizes. This is called the *emulsifying* or *detergent function* of bile salts. Second, and even more important than the emulsifying function, bile salts help in the absorption of fatty acids, monoglycerides, cholesterol, and other lipids from the intestinal tract. They do this by forming minute complexes with the fatty acids and monoglycerides; the complexes are called *micelles,* and they are highly soluble because of the electrical charges of the bile salts. The lipids are "ferried" in this form to the mucosa where they are then absorbed; this mechanism was described in detail in Chapter 65. Without the presence of bile salts in the intestinal tract, up to 40 per cent of the lipids are lost into the stools, and the person often develops a metabolic deficit due to this nutrient loss.

Also, when fats are not absorbed adequately, the fat-soluble vitamins A, D, E, and K are not absorbed satisfactorily. Though large quantities of the first three of these vitamins are usually stored in the body, this is not true of vitamin K. Within only a few days after bile secretion ceases, the person usually develops a deficiency of vitamin K. This in turn results in deficient formation by the liver of several blood coagulation factors — prothrombin, and factors VII, IX, and X — thus resulting in serious impairment of blood coagulation.

Enterohepatic Circulation of Bile Salts. Approximately 94 per cent of the bile salts are reabsorbed by an active transport process through the intestinal mucosa in the distal ileum. They enter the portal blood and pass to the liver. On reaching the liver the bile salts are absorbed from the venous sinusoids into the hepatic cells and then resecreted into the bile. In this way about 94 per cent of all the bile salts are recirculated into the bile, so that on the average these salts make the entire circuit some 18 times before being carried out in the feces. The small quantities of bile salts lost into the feces are replaced by new amounts formed continually by the liver cells. This recirculation of the bile salts is called the *enterohepatic circulation*.

The quantity of bile secreted by the liver each day is highly dependent on the availability of bile salts — the greater the quantity of bile salts in the enterohepatic circulation (usually a total of about 4 gm.), the greater is the rate of bile secretion. Indeed, ingestion of an excess of bile salts can increase bile secretion by several hundred milliliters per day. When a bile fistula forms so that bile is lost directly from the common bile duct to the exterior, the bile salts cannot be reabsorbed. Therefore, the total quantity of bile salts in the enterohepatic circulation becomes greatly depressed, and concurrently the volume of liver secretion is also depressed.

However, if a bile fistula continues to empty the bile salts to the exterior for several days to several weeks, the liver increases its production of bile salts as much as 10-fold, which increases the rate of bile secretion approximately back to normal. This also demonstrates that the daily rate of bile salt secretion is actively controlled, though the mechanism of this control is unknown.

EXCRETION OF BILIRUBIN IN THE BILE

In addition to secreting substances synthesized by the liver itself, the liver cells also *excrete* a number of substances formed elsewhere in the body. Among the most important of these is *bilirubin,* which is one of the major end-products of hemoglobin decomposition, as was pointed out in Chapter 5.

Briefly, when the red blood cells have lived out their life span, averaging 120 days, and have become too fragile to exist longer in the circulatory system, their cell membranes rupture, and the released hemoglobin is phagocytized by reticuloendothelial cells throughout the body. Here, the hemoglobin is first split into *globin* and *heme,* and the heme ring is opened to give a straight chain of four pyrrole nuclei that is the substrate from which the bile pigments are formed. The first pigment formed is *biliverdin,* but this is rapidly reduced to *free bilirubin,* which is gradually released into the plasma. However, the free bilirubin immediately combines very strongly with the plasma albumin and is transported in this combination throughout the blood and interstitial fluids. Even when bound with the plasma protein, this bilirubin is still called "free bilirubin" to distinguish it from "conjugated bilirubin," which will be discussed below. Within hours, the free bilirubin is absorbed through the hepatic cell membrane, in this process being released from the plasma albumin but almost instantly being combined with another protein (called "Y" protein) inside the hepatic cells that traps the bilirubin inside the cells. Soon thereafter, however, the bilirubin is also removed from this protein and conjugated with other substances. About 80 per cent of it conjugates with glucuronic acid to form *bilirubin glucuronide;* an additional 10 per cent conjugates with sulfate to form *bilirubin sulfate,* and the final 10 per cent conjugates with a multitude of other substances. It is in these forms that the bilirubin is excreted by an active transport process into the bile canaliculi.

A small portion of the conjugated bilirubin formed by the hepatic cells returns to the plasma, either directly into the liver sinusoids or indirectly by absorption into the blood from the bile ducts or lymphatics. Regardless of the exact mechanism by which it re-enters the blood, this causes a small portion of the bilirubin in the extracellular fluids always to be of the conjugated type rather than of the free type.

Formation and Fate of Urobilinogen. Once in the intestine, bilirubin is converted by bacterial action mainly into the substance *urobilinogen,* which is highly soluble. Some of the urobilinogen is reabsorbed through the intestinal mucosa into the blood. Most of this is re-excreted by the liver back into the gut, but about 5 per cent of it is excreted by the kidneys into the urine. After exposure to air in the urine, the urobilinogen becomes oxidized to *urobilin,* or in the feces it becomes altered and oxidized to form *stercobilin*. These interrelationships of bilirubin and the other bile pigments are illustrated in Figure 70–3.

Jaundice. The word "jaundice" means a yellowish

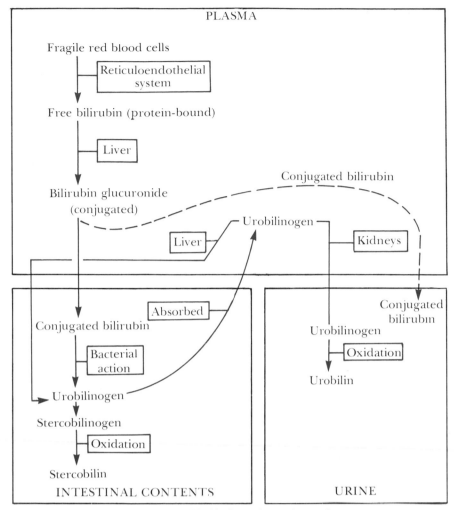

Figure 70–3.   Bilirubin formation and excretion.

tint to the body tissues, including yellowness of the skin and also of the deep tissues. The usual cause of jaundice is large quantities of bilirubin in the extracellular fluids, either free bilirubin or conjugated bilirubin. The normal plasma concentration of bilirubin, including both the free and the conjugated forms, averages 0.5 mg. per 100 ml. of plasma. However, in certain abnormal conditions this can rise to as high as 40 mg. per 100 ml. The skin usually begins to appear jaundiced when the concentration rises to about three times normal — that is, above 1.5 mg. per 100 ml.

The common causes of jaundice are: (1) increased destruction of red blood cells with rapid release of bilirubin into the blood or (2) obstruction of the bile ducts or damage to the liver cells so that even the usual amounts of bilirubin cannot be excreted into the gastrointestinal tract. These two types of jaundice are called, respectively, *hemolytic jaundice* and *obstructive jaundice*. They differ from each other in the following ways:

Hemolytic Jaundice.   In hemolytic jaundice, the excretory function of the liver is not impaired in the least, but red blood cells are hemolyzed rapidly and the hepatic cells simply cannot excrete the bilirubin as

rapidly as it is formed. Therefore, the plasma concentration of *free bilirubin* rises especially high. Likewise, the rate of formation of *urobilinogen* in the intestine is greatly increased, and much of this is absorbed into the blood and later excreted in the urine.

Obstructive   Jaundice.   In   obstructive   jaundice, caused either by obstruction of the bile ducts or by damage to the liver cells, the rate of bilirubin formation is normal, but the bilirubin formed simply cannot pass from the blood into the intestines. However, the free bilirubin usually does still enter the liver cells and becomes conjugated in the usual way. This conjugated bilirubin is then returned to the blood probably by rupture of the congested bile canaliculi and direct emptying of the bile into the lymph leaving the liver. Thus, *most of the bilirubin in the plasma becomes the conjugated type* rather than the free type.

Diagnostic Differences Between Hemolytic and Obstructive Jaundice.   A simple test called the *van den Bergh test* can be used to differentiate between free and conjugated bilirubin in the plasma. If an immediate reaction occurs with the van den Bergh reagent, the bilirubin is of the conjugated type, and the reaction is called a *direct* van den Bergh reaction. However, to demonstrate the presence of free bilirubin, one must first add alcohol to the plasma. This precipitates the protein and "frees" the free bilirubin from its protein complex so that it can then combine with the van den Bergh reagent. This result is called the *indirect* van den Bergh reaction. Thus, *in hemolytic jaundice an indirect van den Bergh reaction occurs (increased free bilirubin) and in obstructive jaundice a direct van den Bergh reaction occurs (increased conjugated bilirubin).*

When there is total obstruction of bile flow, no bilirubin at all can reach the intestines to be converted into urobilinogen by bacteria. Therefore, urobilinogen is not reabsorbed into the blood and is not excreted by the kidneys into the urine. Consequently, in *total* obstructive jaundice, tests for urobilinogen in the urine are completely negative. Also, the stools become clay colored for lack of stercobilin and other bile pigments.

Another major difference between free and conjugated bilirubin is that kidneys can excrete conjugated bilirubin but not free bilirubin. Therefore, in severe obstructive jaundice, large quantities of conjugated bilirubin appear in the urine. This can be demonstrated simply by shaking the urine and observing the foam, which becomes colored an intense yellow.

Thus, by understanding the physiology of bilirubin excretion by the liver and by use of a few simple tests, it is possible to differentiate between obstructive and hemolytic jaundice and, often, also to determine the severity of the disease.

SECRETION OF CHOLESTEROL; GALLSTONE FORMATION

Bile salts are formed in the hepatic cells from cholesterol, and in the process of secreting the bile salts about one-tenth as much cholesterol is also secreted into the bile. No specific function is known for the cholesterol in the bile, and it is presumed that it is simply a by-product of bile salt formation and secretion.

Cholesterol is almost insoluble in pure water, but the bile salts and lecithin in bile combine physically with the cholesterol to form ultramicroscopic *micelles* that are soluble. When the bile becomes concentrated in the gallbladder, the bile salts and lecithin become concentrated along with the cholesterol, which keeps the cholesterol in solution. Also, the hormone *secretin*, which is released into the blood from the duodenal mucosa in response to acid in the chyme (see Chapter 64), causes the bile ducts to secrete about 200 milliliters of highly alkaline fluid each day. This alkalinity is essential for formation of the micelles, for they will not develop in an acid medium.

Nevertheless, under abnormal conditions the cholesterol may precipitate, resulting in the formation of *gallstones,* as shown in Figure 70–4. The different conditions that can cause cholesterol precipitation are (1) too much absorption of water from the bile, (2) too much absorption of bile salts and lecithin from the bile, (3) too much secretion of cholesterol in the bile, and (4) inflammation of the epithelium of the gallbladder. The latter two of these require the following special explanation.

The amount of cholesterol in the bile is determined partly by the quantity of fat that the person eats, for the

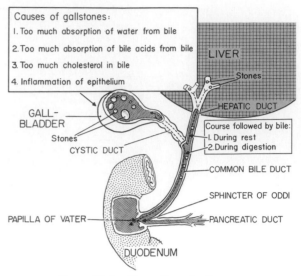

Figure 70–4.　Formation of gallstones.

hepatic cells synthesize cholesterol as one of the products of fat metabolism in the body. For this reason, persons on a high fat diet over a period of many years are prone to the development of gallstones.

Inflammation of the gallbladder epithelium often results from low grade chronic infection; this changes the absorptive characteristics of the gallbladder mucosa, sometimes allowing excessive absorption of water, bile salts, or other substances that are necessary to keep the cholesterol in solution. As a result, cholesterol begins to precipitate, usually forming many small crystals of cholesterol on the surface of the inflamed mucosa. These, in turn, act as nidi for further precipitation of cholesterol, and the crystals grow larger and larger. Occasionally tremendous numbers of sandlike stones develop, but much more frequently they coalesce to form a few large gallstones, or even a single stone that fills the entire gallbladder.

Medical Therapy for Dissolving Gallstones.　Simple cholesterol gallstones in most patients can be dissolved over a period of one to two years by feeding the patient 1 to 1.5 grams of *chenodeoxycholic acid* daily. This is one of the naturally secreted bile acids, and its exogenous administration adds greatly to the enterohepatic pool of bile acids. This causes dissolution and reabsorption of the gallstones in the following ways: (1) The increased quantity of bile acids increases the volume of bile formed and therefore decreases the concentration of cholesterol in the bile. (2) The increase of bile acids in the bile makes the cholesterol that is present more soluble. (3) The exogenous administration of bile acids decreases the formation of bile acids by the liver, which at the same time reduces the formation of cholesterol.

METABOLIC FUNCTIONS OF THE LIVER

The metabolic functions of the liver are so numerous and intricate that they could not possibly be presented completely in this chapter. Therefore, for details, refer to the preceding chapters on the metabolism of carbohydrates, fats, and proteins. Briefly, the specific roles of the liver in the different metabolic processes are described as follows:

CARBOHYDRATE METABOLISM

In carbohydrate metabolism the liver performs the following specific functions: (1) storage of glycogen, (2) conversion of galactose and fructose to glucose, (3) gluconeogenesis, and (4) formation of many important chemical compounds from the intermediate products of carbohydrate metabolism.

The liver is especially important for maintaining a normal blood glucose concentration. For instance, storage of glycogen allows the liver to remove excess glucose from the blood, store it, and then return it to the blood when the blood glucose concentration begins to fall too low. This is called the *glucose buffer function* of the liver. As an example, immediately after a meal containing large amounts of carbohydrates, the blood glucose concentration rises about three times as much in a person with a nonfunctional liver as in a person with a normal liver.

Gluconeogenesis in the liver is also concerned with maintaining a normal blood glucose concentration, for gluconeogenesis occurs to a significant extent only when the glucose concentration begins to fall below normal. In such a case, large amounts of amino acids are converted into glucose, thereby helping to maintain a relatively normal blood glucose concentration.

FAT METABOLISM

Though fat metabolism can take place in almost all cells of the body, certain aspects of fat metabolism occur much more rapidly in the liver than in the other cells. Some specific functions of the liver in fat metabolism are (1) very high rate of beta oxidation of fatty acids and formation of acetoacetic acid, (2) formation of the lipoproteins, (3) formation of large quantities of cholesterol and phospholipids, and (4) conversion of large quantities of carbohydrates and proteins to fat.

To derive energy from neutral fats, the fat is first split into glycerol and fatty acids; then the fatty acids are split by *beta oxidation* into two-carbon acetyl radicals which form *acetyl coenzyme A* (acetyl Co-A). This in turn can then enter the citric acid cycle and be oxidized to liberate tremendous amounts of energy. Beta oxidation can take place in all cells of the body, but it occurs especially rapidly in the hepatic cells. Yet the liver itself cannot utilize all the acetyl Co-A that is formed; instead, this is converted by condensation of two molecules of acetyl Co-A into *acetoacetic acid,* which is a highly soluble acid that passes from the liver cells into the extracellular fluids and then is transported throughout the body to be absorbed by the other tissues. These tissues in turn reconvert the acetoacetic acid into acetyl Co-A and then oxidize it in the usual manner. In this way, therefore, the liver is responsible for a major part of the metabolism of fats.

About 80 per cent of the cholesterol synthesized is converted into bile salts; the remainder enters the blood, transported in the lipoproteins. The phospholipid lecithin likewise is synthesized in the liver and is transported principally in the lipoproteins. It is possible that both these substances are absorbed by cells everywhere in the body to help form the cell membranes and intracellular structures, for it is well known that most membranous structures throughout the body contain major amounts of cholesterol and phospholipids.

Most of the fat synthesis in the body from carbohydrates and proteins also occurs in the liver. After fat is synthesized in the liver it is transported in the lipoproteins to the adipose tissue to be stored.

PROTEIN METABOLISM

Even though a large proportion of the metabolic processes for carbohydrates and fat metabolism occurs in the liver, the body could probably dispense with these functions of the liver and still survive. On the other hand, the body cannot dispense with the services of the liver in protein metabolism for more than a few days without death ensuing. The most important functions of the liver in protein metabolism are (1) deamination of amino acids, (2) formation of urea for removal of ammonia from the body fluids, (3) formation of plasma proteins, and (4) interconversions among the different amino acids and other compounds important to the metabolic processes of the body.

Deamination of the amino acids is required before these can be used for energy or before they can be converted into carbohydrates or fats. A small amount of deamination can occur in the other tissues of the body, especially in the kidney, but the percentage of deamination occurring extrahepatically is so small that it is almost completely unimportant.

Formation of urea by the liver removes ammonia from the body fluids. Moderate amounts of ammonia are continually formed in the gut by bacteria and are then absorbed into the blood; therefore, without this function of the liver the plasma ammonia concentration rises rapidly and results in *hepatic coma* and death. Indeed, any failure of portal blood to flow through the liver — as occurs occasionally when a shunt develops between the portal vein and the vena cava — can also cause excessive ammonia in the blood, an exceedingly toxic condition.

Essentially all the plasma proteins, with the exception of part of the gamma globulins, are formed by the hepatic cells. This accounts for about 90 to 95 per cent of all the plasma proteins. The remaining gamma globulins are the antibodies formed mainly by plasma cells in the lymph tissue of the body. The liver can probably form plasma proteins at a maximum rate of 30 to 60 grams per day. Therefore, after loss of as much as half the plasma proteins from the body, these can be replenished in approximately four to seven days. It is particularly interesting that plasma protein depletion causes rapid mitosis of the hepatic cells and actual growth of the liver to a larger size; these effects are coupled with rapid output of plasma proteins until the plasma concentration returns to normal.

Among the most important functions of the liver is its ability to synthesize certain amino acids and also to synthesize other important chemical compounds from amino acids. For instance, the so-called nonessential amino acids can all be synthesized in the liver. To do this, a keto acid having the same chemical composition (except at the keto oxygen) as that of the amino acid to be formed is first synthesized. Then an amino radical is transferred through several stages of *transamination* from an available amino acid to the keto acid to take the place of the keto oxygen.

MISCELLANEOUS METABOLIC FUNCTIONS OF THE LIVER

Storage of Vitamins. The liver has a particular propensity for storing vitamins and has long been known as an excellent source of certain vitamins in treating patients. The single vitamin stored to the greatest extent in the liver is vitamin A, but large quantities of vitamin D and vitamin B_{12} are normally stored as well. Sufficient quantities of vitamin A can be stored to prevent vitamin A deficiency for as long as one to two years, and sufficient vitamin D and vitamin B_{12} can be stored to prevent deficiency for as long as one to four months.

Relationship of the Liver to Blood Coagulation. The liver forms a large proportion of the blood substances utilized in the coagulation process. These are fibrinogen, prothrombin, accelerator globulin, factor VII, and several other less important coagulation factors. Vitamin K is required by the metabolic processes of the liver for the formation of prothrombin and factors VII, IX, and X. In the absence of vitamin K the concentrations of these substances fall very low, and this almost prevents blood coagulation.

Storage of Iron. Except for the iron in the hemoglobin of the blood, by far the greater proportion of the iron in the body is usually stored in the liver in the form of *ferritin*. The hepatic cells contain large amounts of a protein called *apoferritin,* which is capable of combining with either small or large quantities of iron. Therefore, when iron is available in the body fluids in extra quantities, it combines with the apoferritin to form ferritin and is stored in this form until needed by the body. When the iron in the circulating body fluids reaches a low level, the ferritin releases the iron. Thus, the apoferritin-ferritin system of the liver acts as a *blood iron buffer* and also as an iron storage medium. Other functions of the liver in relation to iron metabolism are considered in Chapter 5.

Removal or Excretion of Drugs, Hormones, and Other Substances by the Liver. The very active chemical medium of the liver is well known for its capability of detoxifying or excreting into the bile many different drugs, including sulfonamides, penicillin, ampicillin, and erythromycin. In a similar manner, several of the different hormones are either chemically altered or excreted, including thyroxine and essentially all of the steroid hormones such as estrogen, cortisol, aldos-

terone, and so forth. Finally, one of the major routes for excreting calcium from the blood is into the bile and then into the feces.

REFERENCES

Alagille, D., and Odievre, M.: Liver and Bilary Tract Disease in Children. New York, John Wiley & Sons, 1979.

Berk, P. D., *et al.*: Disorders of bilirubin metabolism. *In* Bondy, P. K., and Rosenberg, L. E. (eds.): Metabolic Control and Disease, 8th Ed. Philadelphia, W. B. Saunders Co., 1980, p. 1009.

Boucheir, I. A. D.: The medical treatment of gallstones. *Annu. Rev. Med., 31*:59, 1980.

Boyer, J. L.: New concepts of mechanisms of hepatocyte bile formation. *Physiol. Rev., 60*:303, 1980.

Brown, H., and Hardwick, D. F. (eds.): Intermediary Metabolism of the Liver. Springfield, Ill., Charles C Thomas, 1973.

Davidson, C. S. (ed.): Problems in Liver Diseases. New York, Stratton Intercontinental Medical Book Corp., 1979.

Erlinger, S., and Dhumeaux, D.: Mechanisms and control of secretion of bile water and electrolytes. *Gastroenterology, 66*:281, 1974.

Faloon, W. W.: Hepatic mechanisms. *In* Frohlich, E. D. (ed.): Pathophysiology, 2nd Ed. Philadelphia, J. B. Lippincott Co., 1976, p. 531.

Farber, E., and Fisher, M. M. (eds.): Toxic Injury of the Liver. New York, Marcel Dekker, 1979.

Fevery, J., and Heirwegh, K. P. M.: Bilirubin metabolism. *In* Javitt, N. B. (ed.): International Review of Physiology: Liver and Biliary Tract Physiology I. Vol. 21. Baltimore, University Park Press, 1980, p. 171.

Fisher, M. M., *et al.* (ed.): Gall Stones. New York, Plenum Press, 1979.

Forker, E. L.: Mechanisms of hepatic bile formation. *Annu. Rev. Physiol., 39*:323, 1977.

Frizzell, R. A., and Heintze, K.: Transport functions of the gallbladder. *In* Javitt, N. B. (ed.): International Review of Physiology: Liver and Biliary Tract Physiology I. Vol. 21. Baltimore, University Park Press, 1980, p. 221.

Galambos, J. T.: Cirrhosis. Philadelphia, W. B. Saunders Co., 1979.

Gall, E. A., and Mostofi, F. K. (eds.): The Liver. Huntington, N.Y., R. E. Krieger Publishing Co., 1980.

Gerolami, A., and Sarles, J. C.: Biliary secretion and motility. *Int. Rev. Physiol., 12*:223, 1977.

Gitnick, G. L. (ed.): Current Gastroenterology and Hepatology. Boston, Houghton Mifflin, 1979.

Jones, R. S., and Myers, W. C.: Regulation of hepatic biliary secretion. *Annu. Rev. Physiol., 41*:67, 1979.

Kappas, A., and Alvares, A. P.: How the liver metabolizes foreign substances. *Sci. Am., 232*(6): 22, 1975.

MacSween, R. N. M., *et al.* (eds.): Pathology of the Liver. New York, Longman, Inc., 1979.

McGarry, J. D., and Foster, D. W.: Regulation of hepatic ketogenesis. *In* DeGroot, L. J., *et al.* (eds.): Endocrinology. Vol. 2. New York, Grune & Stratton, 1979, p. 997.

Paumgartner, G., *et al.* (eds.): Biological Effects of Bile Acids. Baltimore, University Park Press, 1979.

Presig, R., *et al.*: Physiologic and pathophysiologic aspects of the hepatic hemodynamics. *Prog. Liver Dis., 4*:201, 1972.

Rappaport, A. M.: Hepatic blood flow: Morphologic aspects and physiologic regulation. *In* Javitt, N. B. (ed.): International Review of Physiology: Liver and Biliary Tract Physiology I. Vol. 21. Baltimore, University Park Press, 1980, p. 1.

Reichen, J., and Paumgartner, G.: Excretory function of the liver. *In* Javitt, N. B. (ed.): International Review of Physiology: Liver and Biliary Tract Physiology I. Vol. 21. Baltimore, University Park Press, 1980, p. 103.

Sherlock, S.: The Human Liver. Burlington, N. C., Carolina Biological Supply Co., 1978.

Tanikawa, K.: Ultrastructural Aspects of the Liver and Its Disorders. New York, Igaku-Shoin, 1979.

71

Energetics and Metabolic Rate

IMPORTANCE OF ADENOSINE TRIPHOSPHATE (ATP) IN METABOLISM

In the last few chapters it has been pointed out that carbohydrates, fats and proteins can all be used by cells to synthesize large quantities of ATP, and that in turn the ATP can be used as an energy source for many other cellular functions. For these reasons, ATP has been called an energy "currency" that can be created and expended. Indeed, the cells can transfer energy from the different foodstuffs to most functional systems of the cells only through this medium of ATP [or the very similar nucleotide guanosine triphosphate (GTP)]. Many of the attributes of ATP were presented in Chapter 2, but others require discussion at this point.

An attribute of ATP that makes it highly valuable as a means of energy currency is the large quantity of free energy (about 8000 calories per mole under physiological conditions) vested in each of its two high energy phosphate bonds. The amount of energy in each bond, when liberated by decomposition of one molecule of ATP, is enough to cause almost any step of any chemical reaction in the body to take place if appropriate transfer of the energy is achieved. Some chemical reactions that require ATP energy use only a few hundred of the available 8000 calories, and the remainder of this energy is then lost in the form of heat. Yet even this inefficiency in the utilization of energy is better than not being able to energize the necessary chemical reactions at all.

The precise methods by which ATP is used to cause the many physical, chemical, and other types of functions in the cells lie principally in the province of biochemistry. Therefore, only the more important functions of ATP are listed here.

Use of ATP for Synthesis of Important Cellular Components. Probably by far the most important intracellular process that requires ATP is formation of peptide linkages between amino acids during the synthesis of proteins. The energy from two high energy bonds of ATP and one from GTP is used at different steps in the formation of each peptide linkage. Since an occasional protein of the body has as many as 100,000 amino acids linked together in this manner, it can be understood readily how much ATP (or its equivalent of GTP) is required for this cellular function. The different peptide linkages, depending on which types of amino acids are linked together, require from 500 to 4000 calories of energy per mole. In each instance, the amount of energy available from the three high energy bonds, 24,000 calories, is always more than sufficient.

Also, it will be recalled from the preceding chapters that ATP is utilized in the synthesis of glucose from lactic acid and in the synthesis of fatty acids from acetyl Co-A. In addition, ATP is utilized for synthesis of cholesterol, phospholipids, the hormones, and almost all other substances of the body. Even the urea excreted by the kidneys requires ATP to cause its formation. One might wonder at the advisability of expending energy to form urea which then is simply thrown away from the body. However, remembering the extreme toxicity of ammonia in the body fluids, one can see the value of this reaction, which keeps the ammonia concentration of the body fluids always at a low level.

Use of ATP for Muscular Contraction. Muscular contraction will not occur without energy from ATP. Myosin, one of the important contractile proteins of the muscle fiber, acts as an enzyme to cause breakdown of ATP into adenosine diphosphate (ADP), thus causing the release of energy. However, the means by which this energy is coupled to the contractile process of the muscle fiber is still conjectural. Only a small amount of ATP is normally degraded in muscles when muscular contraction is not occurring, but this rate of ATP usage can rise to more than 100 times the resting level during maximal contraction. The postulated mechanism by which ATP is utilized to cause muscle contraction was discussed in Chapter 11.

Use of ATP for Active Transport Across Membranes. In Chapters 4, 34, and 65, active transport of electrolytes and various nutrients across cell membranes and from the renal tubules and gastrointestinal tract was discussed. In each instance, it was noted that active transport of most electrolytes and other substances such as glucose, amino acids, and acetoacetate can occur against an electrochemical gradient, even though the natural diffusion of the substances would be in the opposite direction. Obviously, to oppose the electrochemical gradient requires energy, as was discussed in Chapter 4. This energy is provided by ATP.

Energy for Glandular Secretion. The same principles apply to glandular secretion as to the absorption of substances against concentration gradients, for energy is required to concentrate substances as they are secreted by the glandular cell. In addition, energy is also required to synthesize the organic compounds to be secreted.

Energy for Nerve Conduction. The energy utilized during propagation of a nerve impulse is derived from the potential energy stored in the form of concentration differences of ions across the membranes. That is, a high concentration of potassium inside the fiber and a low concentration outside the fiber constitutes a type of energy storage. Likewise, a high concentration of sodium on the outside of the membrane and a low concentration on the inside represents another store of energy. The energy needed to pass each impulse along the fiber membrane is derived from this energy storage, with small amounts of potassium transferring out of the cell and sodium into the cell. However, active transport systems then retransport the ions back through the membrane to their former positions. Here, ATP is utilized in abundance to retransfer the sodium and potassium ions after nerve impulses have been conducted, and the rate of ATP usage increases in proportion to the number of nerve impulses transmitted.

CREATINE PHOSPHATE AS A STORAGE DEPOT FOR ENERGY

Despite the paramount importance of ATP as a coupling agent for energy transfer, this substance is not the most abundant store of high energy phosphate bonds in the cells. On the contrary, *creatine phosphate*, which also contains high energy phosphate bonds, is several times as abundant, at least in muscle. The high energy bond of creatine phosphate contains about 8500 calories per mole under standard conditions, or 9500 calories per mole under conditions in the body (38° C. and low concentrations of the reactants). This is not greatly different from the 8000 calories per mole in each of the two high energy phosphate bonds of ATP. The formula for creatine phosphate is the following:

$$\text{HOOC}-\text{CH}_2-\overset{\overset{\text{CH}_3}{|}}{\text{N}}-\overset{\overset{\text{NH}}{\|}}{\text{C}}-\overset{\overset{\text{H}}{|}}{\text{N}}\sim\overset{\overset{\text{O}}{\|}}{\underset{\underset{\text{H}}{|}}{\underset{\text{O}}{|}}{\text{P}}}-\text{OH}$$

Creatine phosphate cannot act in the same manner as ATP as a coupling agent for transfer of energy between the foods and the functional cellular systems. But it can transfer energy interchangeably with ATP. When extra amounts of ATP are available in the cell, much of its energy is utilized to synthesize creatine phosphate, thus building up this storehouse of energy. Then when the ATP begins to be used up, the energy in the creatine phosphate is transferred rapidly back to ATP and from this to the functional systems of the cells. This reversible interrelationship between ATP and creatine phosphate is illustrated by the following equation:

$$\text{Creatine phosphate} + \text{ADP}$$
$$\Updownarrow$$
$$\text{ATP} + \text{Creatine}$$

Note particularly that the higher energy level of the high energy phosphate bond in creatine phosphate, 9500 in comparison with 8000 calories per mole, causes the reaction between creatine phosphate and ATP to proceed to an equilibrium state very much in favor of ATP. Therefore, the slightest utilization of ATP by the cells calls forth the energy from the creatine phosphate to synthesize new ATP. This effect keeps the concentration of ATP at an almost constant level as long as any creatine phosphate remains. For this reason we can actually call the ATP–creatine phosphate system an ATP "buffer" system. Indeed, one can readily understand the importance of keeping the concentration of ATP very nearly constant, because the rates of almost all the reactions in the body depend on this constancy.

ANAEROBIC VERSUS AEROBIC ENERGY

Anaerobic energy means energy that can be derived from foods without the simultaneous utilization of oxygen; *aerobic energy* means energy that can be derived from foods only by oxidative metabolism. In the discussions in the preceding three chapters it was noted that carbohydrates, fats, and proteins can all be oxidized to cause synthesis of ATP. However, carbohydrates are the only significant foods that can be utilized to provide energy without utilization of oxygen; this energy release occurs during glycolytic breakdown of glucose or glycogen to pyruvic acid. For each mole of glucose that is split into pyruvic acid, 2 moles of ATP are formed. However, when glycogen is split to pyruvic acid, each mole of glucose in the glycogen gives rise to 3 moles of ATP. The reason for this difference is that free glucose entering the cell must be phosphorylated by 1 mole of ATP before it can begin to be split, whereas this is not true of glucose derived from glycogen because it comes from the glycogen in the phosphorylated state without the expenditure of ATP. Thus, the best source of energy under anaerobic conditions is the stored glycogen of the cells.

Anaerobic Energy During Hypoxia. One of the prime examples of anaerobic energy utilization occurs in acute hypoxia. When a person stops breathing, there is already a small amount of oxygen stored in the lungs and an additional amount stored in the hemoglobin of the blood. However, these are sufficient to keep the metabolic processes functioning only for about two minutes. Continued life beyond this time requires an additional source of energy. This can be derived for another minute or so from glycolysis, the glycogen of the cells splitting into pyruvic acid and the pyruvic acid in turn becoming lactic acid, which diffuses out of the cells as described in Chapter 67.

Anaerobic Energy Usage in Strenuous Bursts of Activity. It is common knowledge that muscles can perform extreme feats of strength for a few seconds but are much less capable during prolonged activity. The energy used during strenuous activity is derived from (1) ATP already present in the muscle cells, (2) stored creatine phosphate in the cells, (3) anaerobic energy released by glycolytic breakdown of glycogen to lactic acid, and (4) oxidative energy released continuously by oxidative processes in the cells. The speed of the oxidative processes cannot approach that which is required to supply all the energy demands during strenuous bursts of activity. Therefore, the other three sources of energy are called upon to their maximum extent.

The maximum amount of ATP in muscle is only about 5 millimoles per liter of intracellular fluid, and this amount can maintain maximum muscle contraction for not more than a few seconds. The amount of creatine phosphate in the cells may be several times this amount, but even by utilization of all the creatine phosphate, the amount of time that maximum contraction can be maintained is still only a few more seconds. Release of energy by glycolysis can occur much more rapidly than can oxidative release of energy. Consequently, most of the extra energy required during strenuous activity that lasts for more than a few seconds but less than one to two minutes is derived from anaerobic glycolysis. As a result, the glycogen content of muscles after strenuous bouts of exercise becomes greatly reduced, while the lactic acid concentration of the blood rises. Then, immediately after the exercise is over, oxidative metabolism is used to reconvert about four fifths of the lactic acid into glucose, while the remainder becomes pyruvic acid and is degraded and oxidized in the citric acid cycle. The reconversion to glucose occurs principally in the liver cells, and the glucose is then transported in the blood back

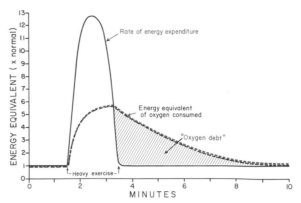

Figure 71–1. Oxygen debt incurred by a bout of strenuous exercise.

to the muscles where it is stored once more in the form of glycogen.

Oxygen Debt. After a period of strenuous exercise, a person continues to breathe hard and to consume excessive amounts of oxygen for many minutes thereafter. This excess oxygen is used (1) to reconvert the lactic acid that has accumulated during exercise back into glucose, (2) to reconvert the decomposed ATP and creatine phosphate to their original states, (3) to re-establish normal concentrations of oxygen bound with hemoglobin and myoglobin, and (4) to raise the concentration of oxygen in the lungs up to its normal level. This excess consumption of oxygen after the exercise is over is called the *oxygen debt.*

The oxygen debt is illustrated by the shaded area in Figure 71–1. This figure shows, first, a period of excess energy expenditure during a bout of heavy exercise, and it also shows the rate of oxygen consumption. Though part of the energy expended during the exercise was provided by oxygen consumption at the time of the exercise, it is clear that a considerable oxygen debt developed and that this was repaid by several minutes of excess oxygen consumption after the exercise ended.

SUMMARY OF ENERGY UTILIZATION BY THE CELLS

With the background of the past few chapters and of the preceding discussion, we can now synthesize a composite picture of overall energy utilization by the cells as illustrated in Figure 71–2. This figure shows the anaerobic utilization of glycogen and glucose to form ATP and also the aerobic utilization of compounds derived from carbohydrates, fats, proteins, and other substances for the formation of still additional ATP. In turn, ATP is in reversible equilibrium with creatine phosphate in the cells, and, since large quantities of creatine phosphate are present in the

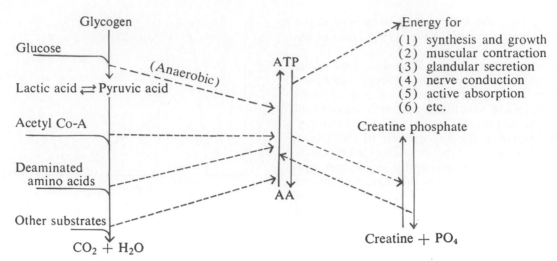

Figure 71-2. Overall schema of energy transfer from foods to the adenylic acid system and then to the functional elements of the cells. (Modified from Soskin and Levine: Carbohydrate Metabolism. University of Chicago Press.)

cell, much of the stored energy of the cell is in this energy storehouse.

Energy from ATP can be utilized by the different functioning systems of the cells to provide for synthesis and growth, muscular contraction, glandular secretion, impulse conduction, active absorption, and other cellular activities. If greater amounts of energy are called forth for cellular activities than can be provided by oxidative metabolism, the creatine phosphate storehouse is first utilized, and this is followed rapidly by anaerobic breakdown of glycogen. Thus, oxidative metabolism cannot deliver energy to the cells nearly so rapidly as can the anaerobic processes, but in contrast it is quantitatively almost inexhaustible.

CONTROL OF ENERGY RELEASE IN THE CELL

Rate Control of Enzyme-Catalyzed Reactions. Before it is possible to discuss the control of energy release in the cell, it is necessary to consider the basic principles of *rate control* of enzymatically catalyzed chemical reactions, which are the type of reactions that occur almost universally throughout the body.

The mechanism by which an enzyme catalyzes a chemical reaction is for the enzyme first to combine loosely with one of the substrates of the reaction. This alters the bonding forces on the substrate sufficiently that it can then react with other substances. Therefore, the rate of the overall chemical reaction is determined by both the concentration of the enzyme and the concentration of the substrate that binds with the enzyme. The basic equation expressing this concept is:

$$\text{Rate of reaction} = \frac{K_1 \cdot \text{Enzyme} \cdot \text{Substrate}}{K_2 + \text{Substrate}}$$

This is called the *Michaelis-Menten equation.* Figure 71-3 illustrates the application of this equation.

Role of Enzyme Concentration in the Regulation of Metabolic Reactions. Figure 71-3 illustrates that *when the substrate is present in excess,* as shown by the curves in the right half of the figure, the rate of a chemical reaction is determined almost entirely by the concentration of the enzyme. Thus, as the enzyme concentration increases from an arbitrary value of 1 up to 2, 4, or 8, the rate of the reaction increases proportionately. As an example, when large quantities of glucose enter the renal tubules in diabetes mellitus, the rate of reabsorption of the glucose is determined almost entirely by the concentration of the glucose transport enzymes in the proximal tubular cells because the substrate is then present in excess.

Role of Substrate Concentration in Regulation of Metabolic Reactions. Note also in Figure 71-3

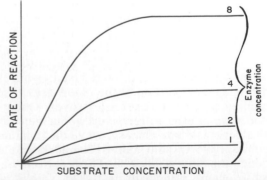

Figure 71-3. Effect of substrate and enzyme concentrations on the rate of enzyme-catalyzed reaction.

that when the substrate concentration becomes low enough that only a small portion of the enzyme is required in the reaction, the rate of the reaction is directly proportional to the substrate concentration as well as enzyme concentration. This is the effect seen in the absorption of substances from the intestinal tract and renal tubules when their concentrations are very low.

Rate Limitation in a Series of Reactions. It has become evident from the discussions in the preceding chapters that almost all chemical reactions of the body occur in series, the product of one reaction acting as a substrate of the next reaction, and so on. It is immediately obvious also that the overall rate of a complex series of chemical reactions is determined by the rate of reaction of the slowest step in the series. This is called the *rate-limiting reaction* for the entire series.

Adenosine Diphosphate (ADP) Concentration as a Rate-Controlling Factor in Energy Release. Under *resting* conditions, the concentration of ADP in the cells is extremely slight so that the chemical reactions that depend on ADP as one of the substrates likewise are very slow. These include all the oxidative metabolic pathways as well as essentially all other pathways for release of energy in the body. Thus, *ADP is a major rate-limiting factor* for almost all energy metabolism of the body.

When the cells become active, regardless of the type of activity, ATP is converted into ADP, increasing the concentration of ADP in direct proportion to the degree of activity of the cell. This automatically increases the rates of reactions of the ADP–rate-limiting steps in the metabolic release of energy. Thus, by this simple process, the amount of energy released in the cell is controlled by the degree of activity of the cell. In the absence of cellular activity, the release of energy stops.

THE METABOLIC RATE

The *metabolism* of the body means simply all the chemical reactions in all the cells of the body, and the *metabolic rate* is normally expressed in terms of the rate of heat liberation during the chemical reactions.

Heat as the Common Denominator of All the Energy Released in the Body. In discussing many of the metabolic reactions of the preceding chapters, we have noted that not all the energy in the foods is transferred to ATP; instead, a large portion of this energy becomes heat. On the average, about 55 per cent of the energy in the foods becomes heat during ATP formation. Then still more energy becomes heat as it is transferred from ATP to the functional systems of the cells, so that even under the best of conditions not more than about 25 per cent of all the energy from the food is finally utilized by the functional systems.

Even though 25 per cent of the energy should reach the functional systems of the cells, almost all of this also becomes heat for the following reasons: We might first consider the synthesis of protein and other growing elements of the body. When proteins are synthesized, large portions of ATP are used to form the peptide linkages, and this stores energy in these linkages. But we also noted in our discussions of proteins in Chapter 69 that there is continuous turnover of proteins, some being degraded while others are being formed. When the proteins are degraded, the energy stored in the peptide linkages is released in the form of heat into the body.

Now let us consider the energy used for muscle activity. Much of this energy simply overcomes the viscosity of the muscles themselves or of the tissues so that the limbs can move. The viscous movement in turn causes friction within the tissues, which generates heat.

We might also consider the energy expended by the heart in pumping blood. The blood distends the arterial system, the distention in itself representing a reservoir of potential energy. However, as the blood flows through the peripheral vessels, the friction of the different layers of blood flowing over each other and the friction of the blood against the walls of the vessels turns this energy into heat.

Therefore, essentially all the energy expended by the body is eventually converted into heat. The only significant exception to this occurs when the muscles are used to perform some form of work outside the body. For instance, when the muscles elevate an object to a height or carry the person's body up steps, a type of potential energy is thus created by raising a mass against gravity. Also, if muscular energy is used to turn a flywheel so that kinetic energy is developed in the flywheel, this, too, would be external expenditure of energy. But, when external expenditure of energy is not taking place, it is safe to consider that all the energy released by the metabolic processes eventually becomes body heat.

The Calorie. To discuss the metabolic rate and related subjects intelligently, it is necessary to use some unit for expressing the quantity of energy released from the different foods or expended by the different functional processes of the body. Most often, the *Calorie* is the unit used for this purpose. It will be recalled that 1 *calorie*, spelled with a small "c" and often called the *gram-calorie*, is the quantity of heat required to raise the temperature of 1 gram of water 1° C. The calorie is much too small a unit for ease of expression in speaking of energy in the body. Consequently the large Calorie, spelled with a capital "C" and often called the *kilogram-calorie*, which is equivalent to 1000 calories, is the unit ordinarily used in discussing energy metabolism.

MEASUREMENT OF THE METABOLIC RATE

Direct Calorimetry. As pointed out above, it is only when the body performs some external work that energy expended within the body does not become heat. Since a person is ordinarily not performing any external work, the metabolic rate can be determined by simply measuring the total quantity of heat liberated from the body in a given time. This method is called *direct calorimetry*.

In determining the metabolic rate by direct calorimetry, one measures the quantity of heat liberated from the body in a large, specially- constructed *calorimeter*. The subject is placed in an air chamber that is so well insulated that no heat can leak through the walls of the chamber. Heat formed by the subject's body warms the air of the chamber. However, the air temperature within the chamber is maintained at a constant level by forcing the air through pipes in a cool water bath. The rate of heat gain by the water bath, a factor that can be measured with an accurate thermometer, is equal to the rate at which heat is liberated by the subject's body.

Obviously, direct calorimetry is physically difficult to perform and, therefore, is used only for research purposes.

Indirect Calorimetry.   Since more than 95 per cent of the energy expended in the body is derived from reaction of oxygen with the different foods, the metabolic rate can also be calculated with a high degree of accuracy from the rate of oxygen utilization. When 1 liter of oxygen is metabolized with glucose, 5.01 Calories of energy are released; when metabolized with starches, 5.06 Calories are released; with fat, 4.70 Calories; and with protein, 4.60 Calories.

From these figures it is striking how nearly equivalent are the quantities of energy liberated per liter of oxygen regardless of the type of food that is being burned. For the average diet, the *quantity of energy liberated per liter of oxygen utilized in the body averages approximately 4.825 Calories*. Using this *energy equivalent* of oxygen, one can calculate approximately the rate of heat liberation in the body from the quantity of oxygen utilized in a given period of time.

If a person should metabolize only carbohydrates during the period of the metabolic rate determination, the calculated quantity of energy liberated based on the value for the average energy equivalent of oxygen (4.825 Calories per liter), would be approximately 4 per cent too little. On the other hand, if the person were obtaining most of his energy from fat, the calculated value would be approximately 4 per cent too great, and, if he were burning almost entirely protein during the test, the error would be insignificant.

The Metabolator.   Figure 71–4 illustrates the metabolator usually used for indirect calorimetry. This apparatus contains a floating drum, under which is an oxygen chamber connected to a mouthpiece through two rubber tubes. A valve in one of these rubber tubes allows air to pass from the oxygen chamber into the mouth, while air passing from the mouth back to the chamber is directed by means of another valve through the second tube. Before the expired air from the mouth enters the upper portion of the oxygen chamber, it flows through a lower chamber containing pellets of soda lime, which combine chemically with the carbon dioxide in the expired air. Therefore, as oxygen is used by the person's body and the carbon dioxide is absorbed by the soda lime, the floating oxygen chamber, which is precisely balanced by a weight, gradually sinks in the water, owing to the oxygen loss. This chamber is coupled to a pen that records on a moving drum the rate at which the chamber sinks in the water and thereby records the rate at which the body utilizes oxygen.

FACTORS THAT AFFECT THE METABOLIC RATE

Factors that increase the chemical activity in the cells also increase the metabolic rate. Some of these are the following:

Exercise.   The factor that causes by far the most dramatic effect on metabolic rate is strenuous exercise. Short bursts of maximal muscle contraction in any single muscle can liberate as much as a hundred times its normal resting amount of heat for a few seconds at a time. In considering the entire body, however, maximal muscle exercise can increase the overall heat production of the body for a few seconds to about 50 times normal or sustained for several minutes to about 20 times normal in the well-trained athlete, which is an increase in metabolic rate to 2000 per cent of normal.

Energy Requirements for Daily Activities.   When an average man of 70 kilograms lies in bed all day, he utilizes approximately 1650 Calories of energy. The process of eating increases the amount of energy utilized each day by an additional 200 or more Calories so that the same man lying in bed and also eating a reasonable diet requires a dietary intake of approximately 1850 Calories per day. If he sits in a chair all day, his total energy requirement reaches 2000 to 2250 Calories. Therefore, in round figures, it can be assumed that the daily energy requirements simply for existing (that is, performing essential functions only) is about 2000 Calories.

Effects of Different Types of Work on Daily Energy Requirements.   Table 71–1 illustrates the rates of energy utilization while one performs different types of activities. Note that walking up stairs requires approximately 17 times as much energy as lying in bed asleep. In general, over a 24-hour period a laborer can achieve a maximum rate of energy utilization as great as 6000 to 7000 Calories — in other words, as much as 3.5 times the basal rate of metabolism.

Specific Dynamic Action of Protein.   After a meal is ingested, the metabolic rate increases. This is believed to result to a very slight extent from the different chemical reactions associated with digestion, absorption, and storage of food in the body. However, it mainly results from the action of certain of the amino acids derived from the proteins of the ingested food to stimulate directly the cellular chemical processes.

After a meal containing a large quantity of carbohydrates or fats, the metabolic rate usually increases only

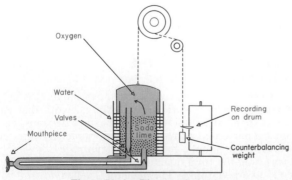

Figure 71–4.   The metabolator.

TABLE 71–1 ENERGY EXPENDITURE PER HOUR DURING DIFFERENT TYPES OF ACTIVITY FOR A 70 KILOGRAM MAN

Form of Activity	Calories per Hour
Sleeping	65
Awake lying still	77
Sitting at rest	100
Standing relaxed	105
Dressing and undressing	118
Tailoring	135
Typewriting rapidly	140
"Light" exercise	170
Walking slowly (2.6 miles per hour)	200
Carpentry, metal working, industrial painting	240
"Active" exercise	290
"Severe" exercise	450
Sawing wood	480
Swimming	500
Running (5.3 miles per hour)	570
"Very severe" exercise	600
Walking very fast (5.3 miles per hour)	650
Walking up stairs	1100

Extracted from data compiled by Professor M. S. Rose.

about 4 per cent. However, after a meal containing large quantities of protein, the metabolic rate usually begins rising within one hour, reaches a maximum about 30 per cent above normal, and lasts for as long as 3 to 12 hours. This effect of protein on the metabolic rate is called the *specific dynamic action* of protein.

Age. The metabolic rate of a young child in relation to its size is almost 2 times that of an old person. This is illustrated in Figure 71–5, which shows the declining metabolic rates of both males and females from birth until very old age. The high metabolic rate of young children results from high rates of cellular reactions, including partly the rapid synthesis of cellular materials and growth of the body, which require moderate quantities of energy.

Thyroid Hormone. When the thyroid gland secretes maximal quantities of thyroxine, the metabolic rate sometimes rises to as high as 100 per cent above normal. On the other hand, total loss of thyroid secretion decreases the metabolic rate to as low as 50 to 60 per cent of normal. These effects can readily be explained by the basic function of thyroxine to increase the rates of activity of almost all the chemical reactions in all cells of the body. This relationship between thyroxine and metabolic rate will be discussed in much greater detail in Chapter 76 in relation to thyroid function, because one of the useful methods for diagnosing abnormal rates of thyroid secretion is to determine the basal metabolic rate of the patient. A normal person usually has a basal metabolic rate within 10 to 15 per cent of normal, while the hyperthyroid person often has a basal metabolic rate as high as 40 to 80 per cent above normal, and a hypothyroid person can have a basal metabolic rate as low as 40 to 50 per cent below normal.

Sympathetic Stimulation. Stimulation of the sympathetic nervous system, with liberation of norepinephrine and epinephrine, increases the metabolic rate of many tissues of the body. These hormones have a direct effect on muscle and liver cells in causing glycogenolysis, and this, along with other intracellular effects, increases cellular activity. However, even more important is the effect of sympathetic stimulation on a certain type of fat tissue called *brown fat* to cause marked liberation of heat. This type of fat contains large numbers of mitochondria and many small globules of fat instead of one large fat globule. In these cells, the process of oxidative phosphorylation is mainly "uncoupled," so that when they are stimulated by the sympathetic nerves their mitochondria produce a large amount of heat but very little ATP. The newborn child has a considerable number of such fat cells, and maximal sympathetic stimulation can increase the child's metabolism more than 100 per cent. This is called *nonshivering thermogenesis*. The magnitude of this effect in the adult human being is in question — probably less than 10 to 15 per cent — though this might be increased following cold adaptation.

Male Sex Hormone. The male sex hormone can increase the basal metabolic rate about 10 to 15 per cent, and the female sex hormone perhaps a few per cent but usually not enough to be of great significance. The difference in metabolic rates of males and females is illustrated in Figure 71–5.

Growth Hormone. Growth hormone can increase the basal metabolic rate as much as 15 to 20 per cent as a result of direct stimulation of cellular metabolism.

Fever. Fever, regardless of its cause, increases the metabolic rate. This is because all chemical reactions, either in the body or in the test tube, increase their rates of reaction an average of about 100 per cent for every 10° C. rise in temperature. However, the body's temperature control system diminishes this effect somewhat.

Climate. Studies of metabolic rates of persons living in the different geographic zones have shown metabolic rates as much as 10 to 20 per cent lower in tropical regions than in arctic regions. This difference is caused to some extent by adaptation of the thyroid gland, with increased secretion in cold climates and decreased secretion in hot climates. Indeed, far more persons develop hyperthyroidism in cold regions of the earth than in tropical regions.

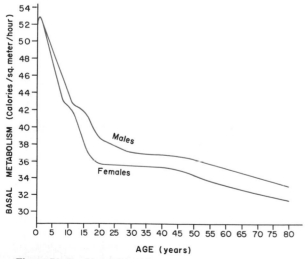

Figure 71–5. Normal basal metabolic rates at different ages for each sex.

Sleep. The metabolic rate falls approximately 10 to 15 per cent below normal during sleep. This fall is presumably due to two principal factors: (1) decreased tone of the skeletal musculature during sleep and (2) decreased activity of the sympathetic nervous system.

Malnutrition. Prolonged malnutrition can decrease the metabolic rate as much as 20 to 30 per cent; this decrease is presumably caused by the paucity of necessary food substances in the cells.

In the final stages of many disease conditions, the inanition that accompanies the disease frequently causes marked premortem decrease in metabolic rate, even to the extent that the body temperature may fall a number of degrees shortly before death.

THE BASAL METABOLIC RATE

The Basal Metabolic Rate as a Method for Comparing Metabolic Rates Between Individuals. It has been extremely important to establish a procedure that will measure the inherent activity of the tissues independently of exercise and other extraneous factors that would make it impossible to compare one person's metabolic rate with that of another person. To do this, the metabolic rate is usually measured under so-called *basal conditions,* and the metabolic rate then measured is called the *basal metabolic rate* (BMR).

Basal Conditions. The basal metabolic rate means the rate of energy utilization in the body during absolute rest but while the person is awake. The following basal conditions are necessary for measuring the basal metabolic rate:

1. The person must not have eaten any food for at least 12 hours because of the specific dynamic action of foods.

2. The basal metabolic rate is determined after a night of restful sleep, for rest reduces the activities of the sympathetic nervous system and other metabolic excitants to their minimal level.

3. No strenuous exercise is performed after the night of restful sleep, and the person must remain at complete rest in a reclining position for at least 30 minutes prior to actual determination of the metabolic rate. This is perhaps the most important of all the conditions for attaining the basal state because of the extreme effect of exercise on metabolism.

4. All psychic and physical factors that cause excitement must be eliminated, and the subject must be made as comfortable as possible. These conditions, obviously, help to reduce the degree of sympathetic activity to as little as possible.

5. The temperature of the air must be comfortable and be somewhere between the limits of 68° and 80° F. Below 68° F., the sympathetic nervous system becomes progressively more activated to help maintain body heat, and above 80° F., discomfort, sweating, and other factors increase the metabolic rate.

Usual Technique for Determining the Basal Metabolic Rate. The usual method for determining basal metabolic rate is first to establish the subject under basal conditions and then to measure the rate of oxygen utilization using a metabolator of the type illustrated in Figure 71–4. Then the basal metabolic rate is calculated as shown in Figure 71–6.

15 liters   —   O$_2$ at standard conditions consumed in 1 hr.

×4.825   —   Calories liberated per liter of O$_2$ burned

72.4   —   Calories liberated per hour

÷1.5   —   Body surface area in square meters

48.3   —   Calories per square meter per hour

−38.5   —   Normal value for 20-year-old man

9.8   —   Excess Calories above normal

$$\frac{9.8 \times 100}{38.5} = 25.5 \text{ per cent above normal}$$

BMR = +25.5

Figure 71–6. Calculation of basal metabolic rate from the rate of oxygen consumption.

In the first few lines of Figure 71–6, the quantity of heat liberated in the body of the person is calculated from the quantity of oxygen utilized. In this figure, note that 15 liters of oxygen (corrected to standard conditions) are consumed in one hour. Multiplying this by the energy equivalent for 1 liter of oxygen (4.825 Calories), the total quantity of energy liberated in the body during the hour is 72.4 Calories.

Expressing the Basal Metabolic Rate in Terms of Surface Area. Obviously, if one subject is much larger than another, the total amount of energy utilized by the two subjects will be considerably different simply because of differences in body size. Experimentally, among normal persons, the average basal metabolic rate varies approximately *in proportion to the body surface area.*

Note from Figure 71–6 that the total number of Calories liberated by the person per hour is divded by the total body surface area, 1.5 square meters. This means that the basal metabolic rate is 48.3 Calories per square meter per hour.

Method for Calculating the Total Surface Area. The surface area of the body varies approximately in propor-

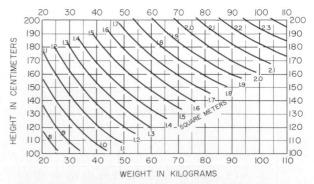

Figure 71–7. Relationship of height and weight to body surface area. (From DuBois: Metabolism in Health and Disease. Lea & Febiger.)

tion to *weight*[0.67]. However, measurements of the body surface area have shown that it can be calculated more accurately by a complicated formula based on the weight and height of the subject as follows:

Body surface area =

$$\text{Weight}^{0.425} \times \text{Height}^{0.725} \times 0.007184$$

Figure 71–7 presents a graph based on this formula. In the formula and in the figure, body surface area is expressed in *square meters,* weight in *kilograms,* and height in *centimeters*. The surface area of the average 70-kilogram adult is 1.73 square meters.

Expressing the Basal Metabolic Rate in Percentage Above or Below Normal.   To compare the basal metabolic rate of any one subject with the normal basal metabolic rate, it is necessary to refer to a chart such as that in Figure 71–5, which gives the normal basal metabolic rate per square meter at each age for each sex. For example, if the person represented in the calculation of Figure 71–6 is a 20-year-old male, then we find from Figure 71–5 that his normal metabolic rate is 38.5 Calories per square meter per hour. But his actual metabolic rate, 48.3 Calories, is 9.8 Calories per square meter per hour *above* the normal mean value. It is then determined that this is 25.5 per cent above normal. Therefore, the basal metabolic rate is expressed as *plus* 25.5. Similarly, basal metabolic rates below normal are expressed as minus values.

Constancy of the Metabolic Rate in the Same Person.   Basal metabolic rates have been measured in many subjects at repeated intervals for as long as 20 or more years. As long as a subject remains healthy, almost invariably the basal metabolic rate when expressed as a percentage of normal does not vary more than 5 to 10 per cent.

Constancy of the Basal Metabolic Rate from Person to Person.   When the basal metabolic rate is measured in a wide variety of different persons and comparisons are made within single age, weight, and sex groups, 85 per cent of normal persons have been found to have basal metabolic rates within 10 per cent of the mean. Thus, it is obvious that measurements of metabolic rates performed under basal conditions offer an excellent means for comparing the rates of metabolism from one person to another.

REFERENCES

Alexander, G.: Cold thermogenesis. *In* Robertshaw, D. (ed.): International Review of Physiology: Environmental Physiology III. Vol. 20. Baltimore, University Park Press, 1979, p. 43.

Ashwell, M. (ed.): Clinical and Scientific Aspects of the Regulation of Metabolism. Boca Raton, Fla., CRC Press, 1979.

Bennett, A. F.: Activity metabolism of the lower vertebrates. *Annu. Rev. Physiol, 40*:447, 1978.

Chance, B.: Enzymes in action in living cells: The steady state of reduced pyridine nucleotides. *Harvey Lectures, 49*:145, 1953–1954.

Christensen, H. N., and Palmer, G. A.: Enzyme Kinetics. Philadelphia, W. B. Saunders Co., 1974.

Consolazio, C. F., *et al.*: Energy requirement and metabolism during exposure to extreme environments. *World Rev. Nutr. Diet, 18*:177, 1973.

Energy production during exercise. *Nutr. Rev., 31*:11, 1973.

Garfinkel, D.: Computer simulation of biologically realistic metabolic networks. *Acta Biol. Med. Ger., 31*:339, 1973.

Guyton, A. C., and Farrish, C. A.: A rapidly responding continuous oxygen consumption recorder. *J. Appl. Physiol., 14*:143, 1959.

Havel, R. J.: Caloric homeostasis and disorders of fuel transport. *N. Engl. J. Med., 287*:1186, 1972.

Hegsted, D. M.: Energy needs and energy utilization. *Nutr. Rev., 32*:33, 1974.

Herman, R. H., *et al.* (eds.): Metabolic Control in Mammals. New York, Plenum Press, 1979.

Hoch, F. L.: Metabolic effects of thyroid hormones. *In* Greep, R. O., and Astwood, E. B. (eds.): Handbook of Physiology. Sec. 7, Vol. 3. Baltimore, Williams & Wilkins, 1974, p. 391.

Homsher, E., and Kean, C. J.: Skeletal muscle energetics and metabolism. *Annu. Rev. Physiol., 40*:93, 1978.

Kagawa, Y.: Reconstitution of oxidative phosphorylation. *Biochim. Biophys. Acta, 265*:297, 1972.

Klachko, D. M., *et al.* (eds.): Hormones and Energy Metabolism. New York, Plenum Press, 1978.

Kleiber, M.: Respiratory exchange and metabolic rate. *In* Fenn, W. O., and Rahn, H. (eds.): Handbook of Physiology. Sec. 3, Vol. 2. Baltimore, Williams & Wilkins Co., 1965, p. 927.

Martin, D. B.: Metabolism and energy mechanisms. *In* Frohlich, E. D. (ed.): Pathophysiology, 2nd Ed. Philadelphia, J. B. Lippincott Co., 1976, p. 365.

Mildvan, A. S.: Mechanism of enzyme action. *Annu. Rev. Biochem., 43*:357, 1974.

Oster, G. F., *et al.*: Network thermodynamics: dynamic modelling of biophysical systems. *Q. Rev. Biophys., 6*:1, 1973.

Roe, C. F.: Temperature regulation and energy metabolism in surgical patients. *Prog. Surg., 12*:96, 1973.

Sinclair, J. C.: Metabolic rate and body size of the newborn. *Clin. Obstet. Gynecol., 14*:840, 1971.

Wyndham, C. H., and Loots, H.: Responses to cold during a year in Antarctica. *J. Appl. Physiol., 27*:696, 1969.

72

Body Temperature, Temperature Regulation, and Fever

"Core" Temperature and Surface Temperature. The temperature of the inside of the body — the "core" — remains almost exactly constant, within ±1° F., day in and day out except when a person develops a febrile illness. Indeed, the nude person can be exposed to temperatures as low as 55° F. or as high as 140° F. in dry air and still maintain an almost constant internal body temperature. Obviously, the mechanisms for control of body temperature represent a beautifully designed control system. It is the purpose of this chapter to discuss this system as it operates in health and in disease.

When speaking of the body temperature regulation, one usually means the temperature in the interior, called the *core temperature,* and not the temperature of the skin or tissues immediately underlying the skin. The *surface temperature,* in contrast to the core temperature, rises and falls with the temperature of the surroundings. This is the temperature that is important when we refer to the ability of the skin to lose heat to the surroundings.

The Normal Body Temperature. No single temperature level can be considered to be normal, for measurements on many normal persons have shown a *range* of normal temperatures, as illustrated in Figure 72–1, from approximately 97° F. to over 99° F. When measured by rectum, the values are approximately 1° F. greater than the oral temperatures. The average normal temperature is generally considered to be 98.6° F. (37° C.) when measured orally and approximately 1° F. or 0.6° C. higher when measured rectally.

The body temperature varies somewhat with exercise and with extremes of temperature of the surroundings, because the temperature regulatory mechanisms are not 100 per cent effective. When excessive heat is produced in the body by strenuous exercise, the rectal temperature can rise to as high as 101° to 104° F. On the other hand, when the body is exposed to extreme cold, the rectal temperature can often fall to values considerably below 98° F.

THE INSULATOR SYSTEM OF THE BODY

The skin, the subcutaneous tissues, and the fat of the subcutaneous tissues are a heat insulator for the body. The fat is especially important because it conducts heat only *one-third* as readily as other tissues. When no blood is flowing from the heated internal organs to the skin, the insulating properties of the normal male body are approximately equal to three-quarters the insulating properties of a usual suit of clothes. In women this insulation is still better. Obviously, the degree of insulation varies from one person to another, depending on the quantity of adipose tissue.

Because most body heat is produced in the deeper portions of the body, the insulation beneath the skin is an effective means for maintaining normal internal temperatures, even though it allows the temperature of the skin to approach the temperature of the surroundings.

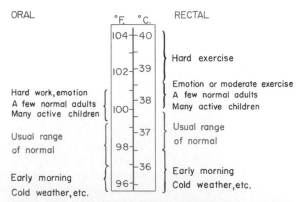

Figure 72–1. Estimated range of body temperature in normal persons. (From DuBois: Fever. Charles C Thomas.)

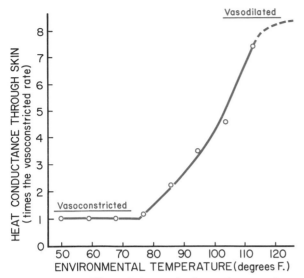

Figure 72–2. Effect of changes in the environmental temperature on the heat conductance from the body core to the skin surface. (Modified from Benzinger: Heat and Temperature; Fundamentals of Medical Physiology. Dowden, Hutchinson and Ross, 1980.)

FLOW OF BLOOD TO THE SKIN AND HEAT TRANSFER FROM THE BODY CORE

Blood vessels penetrate the subcutaneous insulator tissues and are distributed profusely in the subpapillary portions of the skin. Indeed, immediately beneath the skin is a continuous venous plexus that is supplied by inflow of blood. In the most exposed areas of the body — the hands, feet, and ears — blood is supplied through direct *arteriovenous anastomoses* from the arterioles to the veins. The rate of blood flow into this venous plexus can vary tremendously — from barely above zero to as great as 30 per cent of the total cardiac output. A high rate of blood flow causes heat to be conducted from the internal portions of the body to the skin with great efficiency, whereas reduction in the rate of blood flow decreases the efficiency of heat conduction from the internal portions of the body. Figure 72–2 shows quantitatively the effect of skin blood flow on conductance of heat from the body core to the skin surface, illustrating an approximate 8-fold increase in heat conductance between the fully vasoconstricted state and the fully vasodilated state.

Obviously, therefore, the skin is an effective "radiator" system, and the flow of blood to the skin is the most effective mechanism of heat transfer from the body core to the skin. If blood flow from the internal structures to the skin is depressed, the only means by which heat produced internally can be lost to the exterior is by heat diffusion through the insulator tissues of the skin and subcutaneous areas. This form of heat transfer is inadequate to provide the needed heat dissipation in warm and hot weather and also when the body produces excess heat, as occurs during exercise.

Control of Heat Conduction to the Skin. Heat conduction to the skin by the blood is controlled by the degree of vasoconstriction of the arterioles and arteriovenous anastomoses that supply blood to the venous plexus of the skin. This vasoconstriction in turn is controlled almost entirely by the sympathetic nervous system in response to changes in internal body temperature or environmental temperature. This will be discussed later in the chapter in connection with the control of body temperature by the hypothalamus.

BALANCE BETWEEN HEAT PRODUCTION AND HEAT LOSS

Heat is continually being produced in the body as a by-product of metabolism, and body heat is also continually being lost to the surroundings. When the rate of heat production is exactly equal to the rate of loss, the person is said to be in heat balance. But when the two are out of equilibrium, the body heat, and the body temperature as well, will obviously be either increasing or decreasing.

The important factors that play major roles in determining the rate of *heat production* were discussed in detail in Chapter 71; they may be listed as follows: (1) basal rate of metabolism of all the cells of the body; (2) increase in rate of metabolism caused by muscle activity, including that caused by shivering; (3) increase in metabolism caused by the effect of thyroxine on cells; (4) increase in metabolism caused by the effect of norepinephrine and sympathetic stimulation on cells; and (5) increase in metabolism caused by increased temperature of the body cells.

HEAT LOSS

The various methods by which heat is lost from the body are pictured in Figure 72–3. These in-

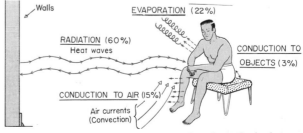

Figure 72–3. Mechanisms of heat loss from the body.

clude *radiation, conduction,* and *evaporation* and may be explained as follows:

Radiation. As illustrated in Figure 72–3, a nude person in a room at normal room temperature loses about 60 per cent of the total heat loss by radiation.

Loss of heat by radiation means loss in the form of infrared heat rays, a type of electromagnetic waves. Most infrared heat rays radiating from the body have wavelengths of 5 to 20 microns, 10 to 30 times the wavelengths of light rays. All objects that are not at absolute zero temperature radiate such rays. Therefore, the human body radiates heat rays in all directions. However, heat rays are also being radiated from the walls and other objects toward the body. If the temperature of the body is greater than the temperature of the surroundings, a greater quantity of heat is radiated from the body than is radiated to the body.

Conduction. Usually, as illustrated in Figure 72–2, only minute quantities of heat are lost from the body by direct conduction from the surface of the body *to other objects,* such as a chair or a bed. When one first sits on a chair while nude, heat is conducted from the body to the chair rapidly, but within a few minutes the temperature of the chair rises almost to equal the temperature of the body, and thereafter the chair actually becomes an insulator to prevent much additional loss of heat.

On the other hand, loss of heat by *conduction to air* does represent a sizeable proportion of the body's heat loss even under normal conditions. It will be recalled that heat is actually the kinetic energy of molecular motion, and the molecules that comprise the skin of the body are continually undergoing vibratory motion. The energy of this motion can be transferred to the air if the air is colder than the skin, thus increasing the velocity of motion of the air molecules. However, once the temperature of the air immediately adjacent to the skin equals the temperature of the skin, no further loss of heat from the body to the air can occur. Therefore, conduction of heat from the body to the air is self-limited unless the heated air moves away from the skin so that new, unheated air is continually brought in contact with the skin, a phenomenon called convection.

Convection. The removal of heat from the body by convection air currents is commonly called "heat loss by convection." Actually, the heat must first be *conducted* to the air and then carried away by the convection currents.

A small amount of convection almost always occurs around the body because of the tendency for the air adjacent to the skin to rise as it becomes heated. Therefore, a nude person seated in a comfortable room without gross air movement still loses about 12 per cent of his heat by conduction to the air and then by convection away from the body.

Cooling Effect of Wind. When the body is exposed to wind, the layer of air immediately adjacent to the skin is replaced by new air much more rapidly than normally, and heat loss by convection increases accordingly. The cooling effect of wind at low velocities is approximately proportional to the square root of the wind velocity. For instance, a wind of 4 miles per hour is about 2 times as effective for cooling as a wind of 1 mile per hour. However, when the wind velocity rises beyond a few miles per hour, additional cooling does not occur to a great extent. This is because once the wind has cooled the skin to the temperature of the air itself, the rate of heat loss cannot increase further, regardless of the wind velocity. Instead, the rate at which heat can flow from the core of the body to the skin is then the factor that determines the rapidity with which heat can be lost.

Conduction and Convection of Heat from a Body Exposed to Water. Water has a specific heat several thousand times as great as that of air so that each unit portion of water adjacent to the skin can absorb far greater quantities of heat than can air. Also, the conductivity of heat through water is marked in comparison with that in air. Consequently, it is impossible to heat a thin layer of water next to the body to form an "insulator zone" as occurs in air. Therefore, the rate of heat loss to water at moderate temperatures is many times as great as the rate of heat loss to air of the same temperature. However, when both the water and air are extremely cold, the rate of heat loss to air becomes almost as great as to water, for both water and air are then capable of carrying away essentially all the heat that can reach the skin from the body core.

Evaporation. When water evaporates from the body surface, 0.58 Calorie of heat is lost for each gram of water that evaporates. Water evaporates *insensibly* from the skin and lungs at a rate of about 600 ml. per day. This causes continual heat loss at a rate of 12 to 16 Calories per hour. Unfortunately, this insensible evaporation of water directly through the skin and lungs cannot be controlled for purposes of temperature regulation because it results from continual diffusion of water molecules through the skin and respiratory surfaces regardless of body temperature. However, loss of heat by evaporation of sweat can be controlled by regulating the rate of sweating, which is discussed later in the chapter.

Evaporation as a Necessary Refrigeration Mechanism at Very High Air Temperatures. In the preceding discussions of radiation and conduction it was noted that as long as skin temperature is greater than that of the surroundings, heat is lost by radiation and conduction, but when the temperature of the surroundings is greater than that of the skin, instead of losing heat the body gains heat by radiation and conduction. Under these conditions, *the only means by which the body can rid itself of heat is by evaporation.* Therefore, any factor that prevents adequate evaporation when the surrounding temperatures are higher than skin temperature permits the body temperature to rise. This occurs occasionally in human beings who are born with congenital absence of sweat glands.

These persons can withstand cold temperatures as well as can normal persons, but they are likely to die of heat stroke in tropical zones, for without the evaporative refrigeration system their skin temperatures often remain at values greater than those of the surroundings.

Effect of Convection Air Currents on Evaporation. Lack of air movement reduces effective evaporation in the same manner that it reduces effective cooling by conduction of heat to the air — that is, the local air becomes saturated with water vapor, and further evaporation cannot occur. Convection currents cause air that has become saturated with moisture to move away from the skin while unsaturated air replaces it. Indeed, convection is even more important for heat loss from the body by evaporation than by conduction, for the instances in which one especially needs to lose heat from the body, such as on hot days, are the same times when evaporative loss of heat from the body is far greater than conductive loss.

Effect of Clothing on Heat Loss. *Effect on Conductive Heat Loss.* Clothing entraps air next to the skin and in the weave of the cloth, thereby increasing the thickness of the so-called "private zone" of air adjacent to the skin and decreasing the flow of convection air currents. Consequently, the rate of heat loss from the body by conduction is greatly depressed. A usual suit of clothes decreases the rate of heat loss to about half that from a nude body, while arctic-type clothing can decrease this heat loss to as little as one sixth that of the nude state.

About half the heat transmitted from the skin to the clothing is radiated to the clothing instead of being conducted across the small intervening space. Therefore, coating the inside of clothing with a thin layer of gold which reflects radiant heat back to the body, makes the insulating properties of clothing far more effective than otherwise. Using this technique, clothing for use in the arctic can be decreased in weight by about half.

The effectiveness of clothing in preventing heat loss is almost completely lost when it becomes wet, because the high conductivity of water increases the rate of heat transmission as much as 20-fold or more. One of the most important factors for protecting the body against cold in arctic regions is extreme caution against wet clothing. Indeed, one must be careful not to overheat oneself even temporarily, for sweating in one's clothes makes them much less effective thereafter as an insulator.

SWEATING AND ITS REGULATION BY THE AUTONOMIC NERVOUS SYSTEM

Stimulation of the preoptic area in the anterior part of the hypothalamus either electrically or by excess heat causes sweating. The impulses from this area that cause sweating are transmitted in the autonomic pathways to the cord and thence through the sympathetic outflow to the skin everywhere in the body.

It should be recalled from the discussion of the autonomic nervous system in Chapter 57 that the sweat glands are innervated by sympathetic *cholinergic* nerve fibers. However, these glands can also be stimulated by epinephrine or norepinephrine circulating in the blood even though the glands themselves, in most parts of the body, do not have adrenergic innervation. It is possible that the sweat glands of the hands and feet do have some adrenergic innervation as well as cholinergic innervation, for many emotional states that excite the adrenergic portions of the sympathetic nervous system are known also to cause local sweating in these areas. Also, during exercise, which normally excites adrenergic activity, localized sweating of the hands and feet occurs. In this instance, the moisture from the sweat helps the surfaces of the hands and feet to gain traction against smooth surfaces and also prevents drying of the thick cornified layers of skin.

Mechanism of Sweat Secretion. The sweat glands are tubular structures consisting of two parts: (1) a deep subdermal *coiled portion* that secretes the sweat, and (2) a *duct portion* passing outward through the dermis and epidermis of the skin. As is true of so many other glands, the secretory portion of the sweat gland secretes a fluid called the *precursor secretion;* then certain constituents of the fluid are reabsorbed as it flows through the duct.

The precursor secretion is an active secretory product of the epithelial cells lining the coiled portion of the sweat gland. Cholinergic sympathetic nerve fibers ending on or near the glandular cells elicit the secretion.

Since large amounts of sodium chloride are lost in the sweat, it is important especially to know how the sweat glands handle sodium and chloride during the secretory process. When the rate of sweat secretion is very low, the sodium and chloride concentrations of the sweat are also very low, because most of these ions are reabsorbed from the precursor secretion before it reaches the surface of the body; their concentrations are sometimes as low as 5 mEq./liter each. On the other hand, when the rate of secretion becomes progressively greater, the rate of sodium chloride reabsorption does not increase commensurately, so that then their concentrations in the sweat usually rise in the unacclimatized person to maximum levels of about 60 mEq/liter, or slightly less than half those in plasma.

Other substances lost in reasonable quantities in the sweat include urea, lactic acid, and potassium ions. At low rates of sweat secretion, the concentrations of all these can be extremely high, but at high rates of secretion the concentration of urea is only about 2 times that in the plasma; lactic acid, about 4 times; and potassium, about 1.2 times.

Acclimatization of the Sweating Mechanism. Though a normal, unacclimatized person can rarely produce more than about 700 ml. of sweat per hour, when exposed to hot weather for one to six weeks, the person sweats progressively more profusely, often increasing the maximum sweat production to as much as 1.5 liters per hour. Evaporation of this much sweat can remove heat from the body at a rate *more than 10 times* the normal basal rate of heat production. This increased effectiveness of the sweating mechanism is caused by a direct increase in sweating capability of the sweat glands themselves.

Associated with the acclimatization is also decreased concentration of sodium chloride in the sweat, which allows progressively better conservation of salt. Most of this effect is caused by *increased secretion of aldosterone,* which, in turn, results from decreasing sodium chloride in the body. An unacclimatized person who sweats profusely often loses as much as 15 to 30 grams of salt each day for the first few days. But, after four to six weeks of acclimatization, the loss may be as little as 3 to 5 grams per day.

Panting as a Means of Evaporative Heat Loss. Many lower animals do not have sweat glands, but to offset this they lose large amounts of heat by the panting mechanism. During panting, only small volumes of air pass in and out of the lungs with each breath so that mainly dead space air enters the alveoli. Because of this and because of the very rapid breathing rate, tremendous amounts of air are moved over the surfaces of the tongue, mouth, and trachea but without causing hypocapnia because of excess CO_2 loss from the alveoli. Evaporation from these respiratory surfaces, especially of saliva on the tongue, is an important mechanism for body heat control.

A special nervous center in the pons controls panting. This center modifies the normal respiratory pattern to provide the rapid and shallow breathing required for the panting mechanism.

REGULATION OF BODY TEMPERATURE—THE "HYPOTHALAMIC THERMOSTAT"

Figure 72–4 illustrates approximately what happens to the temperature of the nude body after a few hours' exposure to dry air ranging from 30° to 170° F. Obviously, the precise dimensions of this curve vary, depending on the movement of air, the amount of moisture in the air, and even the nature of the surroundings. However, in general, between approximately 60° and 130° F. in dry air, the nude body is capable of maintaining indefinitely a normal body core temperature somewhere between 98° and 100° F.

The temperature of the body is regulated almost entirely by nervous feedback mechanisms, and almost all of these operate through *temperature regulating centers* located in the *hypothalamus.* However, for these feedback mechanisms to operate, there must also exist temperature detectors, discussed next, to determine when the body temperature becomes either too hot or too cold.

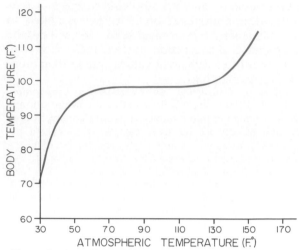

Figure 72–4. Effect of high and low atmospheric temperatures for several hours' duration on the internal body temperature, showing that the internal body temperature remains stable despite wide changes in atmospheric temperature.

Thermostatic Detection of Excess Body Temperature — Role of the Preoptic Area of the Hypothalamus

In recent years, experiments have been performed in which minute areas in the brain have been either heated or cooled by use of a so-called *thermode.* This small, needle-like device is heated by electrical means or by passing hot water through it, or it is cooled by cold water. The principal area in the brain in which heat from a thermode affects body temperature control is the preoptic area of the hypothalamus, and to a lesser extent the adjacent regions of the anterior hypothalamus.

Using the thermode, the preoptic area of the hypothalamus has been found to contain large numbers of heat-sensitive neurons that seem to function as temperature sensors for controlling body temperature. These neurons increase their firing rate as the temperature rises, the rate sometimes increasing as much as 10-fold with an increase in body temperature of 10° C.

In addition to the heat-sensitive neurons, a few cold-sensitive neurons have also been found in other parts of the hypothalamus, in the septum, and in the reticular substance of the mid-brain, all of which increase their rate of firing when exposed to cold. However, the number of these are few, and it is likely that they play a much smaller, if any, role in regulating body temperature.

When the preoptic area is heated, the body immediately breaks out into a profuse sweat while at the same time the skin blood vessels over the entire body become greatly vasodilated. Thus, there is the immediate reaction to cause the body to lose heat, thereby helping to return the body temperature toward the normal level. In addition, excess body heat production is inhibited. Therefore, it is clear that the preoptic area of the hypothalamus has the capability of serving as a thermostatic body temperature control center.

Detection of Cold — Role of Skin and Spinal Cord Receptors

One of the ways in which the body detects cold is by reduced rates of discharge of the heat-sensitive neurons in the preoptic area. However, by the time the internal body temperature has fallen a few tenths of a degree below normal, these neurons have generally become completely inactive so that their signal level cannot be decreased any further. Therefore, detection of cold temperature in the body relies on other temperature receptors; mainly the cold receptors of the skin but to some extent cold receptors also in the spinal cord, abdomen, and possibly other internal structures of the body.

It will be recalled from the discussion of the temperature receptors in Chapter 49 that the skin is endowed with both *cold* and *warmth* receptors. However, there are far more cold receptors than warmth receptors; in fact, 10 times as many in some parts of the body. Therefore, peripheral detection of temperature mainly concerns detecting cool and cold instead of warm temperatures.

When the skin is chilled over the entire body, immediate reflex effects are invoked to increase the temperature of the body in several ways: (1) by providing a strong stimulus to cause shivering, with resultant increase in the rate of body heat production, (2) by inhibiting the process of sweating if this should be occurring, and (3) by promoting skin vasoconstriction to diminish the transfer of body heat to the skin.

Thus, it is clear that the heat center in the preoptic area of the hypothalamus plays a major role in preventing overheating of the body, while the cold receptors in many areas of the body — especially the skin — play very important roles in preventing low body temperatures.

Integration of the Heat and Cold Temperature Signals in the Posterior Hypothalamus

Even though a large share of the signals for cold detection arise in peripheral receptors, these signals help to control body temperature mainly through the hypothalamus. However, the area of the hypothalamus that they stimulate is not the preoptic area but instead an area located bilaterally in the posterior hypothalamus approximately at the level of the mammary bodies. The thermostatic signals from the preoptic area are also transmitted into this posterior hypothalamus area. It is here that the signals from the preoptic area and the signals from the body periphery are combined to provide the heat-producing or heat-losing reactions of the body.

The overall heat-controlling mechanism of the hypothalamus is called the *hypothalamic thermostat*.

NEURONAL EFFECTOR MECHANISMS TO INCREASE OR DECREASE BODY TEMPERATURE

When the hypothalamic thermostat detects that the body temperature is either too hot or too cold, it institutes appropriate heat-decreasing or heat-increasing procedures. The student is familiar with most of these from personal experience, but special features are the following:

Mechanisms for Reducing Body Temperature

The thermostatic system employs three important mechanisms to reduce body heat when the temperature becomes too great:

1. In almost all areas of the body *the skin blood vessels are intensely dilated.* This is caused by *inhibition of the sympathetic centers in the posterior hypothalamus that cause vasoconstriction.* Full vasodilatation can increase the rate of heat transfer to the skin as much as 8-fold.

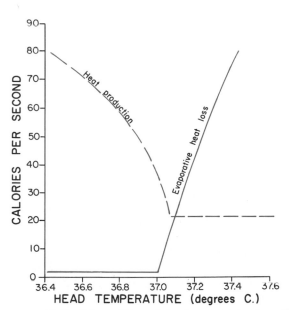

Figure 72-5. Effect of hypothalamic temperature on: (1) evaporative heat loss from the body and (2) heat production caused primarily by muscular activity and shivering. This figure demonstrates the extremely critical temperature level at which increased heat loss begins and increased heat production stops. (Drawn from data in Benzinger, Kitzinger, and Pratt, in Hardy (ed.): Temperature, Part 3, p. 637. Reinhold Publishing Corp.)

2. *Sweating is strongly stimulated.* This effect is illustrated by the solid curve in Figure 72–5, which shows a sharp increase in the rate of evaporative heat loss resulting from sweating when the body core temperature rises above the critical temperature level of 37° C. (98.6° F.) An additional 1° C. increase in body temperature causes enough sweating to remove 10 times the basal rate of body heat production.

3. Heat production by such mechanisms as shivering and chemical thermogenesis is strongly inhibited.

Mechanisms for Increasing Body Heat When the Temperature Is Too Low

In the cold state, the thermostatic mechanism institutes exactly opposite procedures. These are:

1. *Skin vasoconstriction throughout the body* caused by stimulation of the posterior hypothalamic sympathetic centers.

2. *Pilo-erection* — that is, hairs "standing on end." Obviously, this effect is not important in the human being, but in lower animals the upright projection of the hairs entraps a thick layer of insulator air next to the skin so that the transfer of heat to the surroundings is greatly depressed.

3. *Increase in heat production* by causing (a) shivering, (b) sympathetic excitation of heat production, and (c) thyroxine secretion. These require additional explanation, as follows:

Hypothalamic Stimulation of Shivering. Located in the dorsomedial portion of the posterior hypothalamus near the wall of the third ventricle is an area called the *primary motor center for shivering.* This area is normally inhibited by heat signals from the preoptic heat thermostatic area but is driven by cold signals from the skin and spinal cord. Therefore, as illustrated by the dashed curve in Figure 72–5, in response to cold, this center becomes activated and transmits impulses through bilateral tracts down the brain stem, into the lateral columns of the spinal cord, and, finally, to the anterior motoneurons. These impulses are nonrhythmic and do not cause the actual muscle shaking. Instead, they increase the tone of the skeletal muscles throughout the body. When the tone rises above a certain critical level, shivering begins. This probably results from feedback oscillation of the muscle spindle stretch reflex mechanism. During maximum shivering, body heat production can rise to as high as 4 to 5 times normal.

Sympathetic "Chemical" Excitation of Heat Production. It was pointed out in Chapter 71 that either sympathetic stimulation or circulating norepinephrine and epinephrine in the blood can cause an immediate increase in the rate of cellular metabolism; this effect is called *chemical thermogenesis,* and it is believed to result at least partially from the ability of norepinephrine and epinephrine to uncouple oxidative phosphorylation, as a result of which more oxidation of foodstuffs must occur to produce the energy required for normal function of the body. Therefore, the rate of cellular metabolism increases.

The degree of chemical thermogenesis that occurs in an animal is almost directly proportional to the amount of *brown* fat that exists in the animal's tissues. This is a type of fat that contains large numbers of mitochondria in its cells, and these cells are supplied by a strong sympathetic innervation. Upon sympathetic stimulation, the oxidative metabolism of the mitochondria is greatly stimulated, but this probably occurs in an uncoupled manner so that only small amounts of ATP are formed. Regardless of the chemical mechanism, the energy release caused by the oxidative metabolism provides a very important source of heat to the body.

The process of acclimatization greatly affects the intensity of chemical thermogenesis; some animals that have been exposed for several weeks to a very cold environment exhibit as much as a 100 to 500 per cent increase in metabolism when acutely exposed to cold, in contrast to the unacclimatized animal, which responds with an increase in metabolism of perhaps one-third as much.

In adult human beings, who have almost no brown fat, it is rare that chemical thermogenesis increases the rate of heat production more than 10 to 15 per cent. However, in infants, who *do* have a small amount of brown fat in the interscapular space, chemical thermogenesis can increase the rate of heat production as much as 100 per cent, which is probably a very important factor in maintaining normal body temperature in the neonate.

Increased Thyroxine Output as a Cause of Increased Heat Production. Cooling the preoptic area of the hypothalamus also increases the production of the neurosecretory hormone *thyrotropin-releasing hormone* by the hypothalamus. This hormone is carried by way of the hypothalamic portal veins to the adenohypophysis where it stimulates the secretion of thyroid-stimulating hormone. Thyroid-stimulating hormone, in turn, stimulates increased output of thyroxine by the thyroid gland, as will be explained in Chapter 76. The increased thyroxine increases the rate of cellular metabolism throughout the body. However, this increase in metabolism does not occur immediately but requires several weeks for the thyroid gland to hypertrophy before it reaches its new level of thyroxine secretion.

Exposure of animals to extreme cold for several weeks can cause their thyroid gland to increase in size as much as 20 to 40 per cent. However, human beings rarely allow themselves to be exposed to the same degree of cold as that to which animals have been subjected. Therefore, we still do not know, quantitatively, how important the thyroid method of adaptation to cold is in the human being. Yet, isolated measurements have shown that military personnel residing for several months in the Arctic develop increased metabolic rates; Eskimos also have abnormally high basal metabolic rates. Also, the continuous stimulatory effect of cold on the thyroid gland can probably explain the much higher incidence of toxic thyroid goiters in persons living in colder climates than in those living in warmer climates.

THE CONCEPT OF A "SET-POINT" FOR TEMPERATURE CONTROL

In the example of Figure 72–5, it is clear that at a very critical body core temperature, between the level of 37.0° and 37.1° C., drastic changes occur in both the rate of heat loss and heat production. This critical temperature level is called the "set-point" of the temperature control mechanism. That is, all the temperature control mechanisms continually attempt to bring the body temperature back to this "set-point" level whenever it deviates in either direction.

The Feedback Gain for Body Temperature Control. Let us recall for a moment the discussion of feedback gain of control systems that was presented in Chapter 1. Feedback gain is a measure of the effectiveness of a control system. In the case of body temperature control, it is important for the internal body temperature to change as little as possible despite marked changes in the environmental temperature, and the gain is approximately equal to the ratio of change in environmental temperature to change in body temperature. Experiments have shown that the body temperature changes about 1° C. for each 25° to 30° C. change in environmental temperature. Therefore, the feedback gain of the hypothalamic thermostat mechanism for control of body temperature averages about 27, which is an extremely high gain for a biological control system (the baroreceptor arterial pressure control system, for instance, has a gain less than 2).

Effect of Skin Temperature Changes in Altering the Set-Point

The critical temperature set-point in the hypothalamus above which sweating begins and below which shivering begins is determined mainly by the degree of activity of the heat temperature

receptors in the preoptic area of the hypothalamus. However, we have already pointed out that cold temperature signals from the peripheral areas of the body, especially from the skin, also contribute very significantly to body temperature regulation. But, how do they contribute? The answer to this is that they alter the set-point of the hypothalamic thermostat. This effect is illustrated in Figures 72–6 and 72–7.

Figure 72–6 illustrates the effect of different skin temperatures on the set-point for sweating. In the person represented in this figure, the hypothalamic set-point changed from 36.7° C. when the skin temperature was higher than 33° C. to a set-point of 37.4° C. when the skin temperature had fallen to 29° C. Thus, when the skin temperature was very high, sweating began to occur at a much lower hypothalamic temperature than when the skin temperature was low. One can readily understand the logic of a system such as this, for it is important that sweating be inhibited when the skin temperature is low; otherwise, the combined effect of a low skin temperature and sweating at the same time could cause far too much loss of body heat.

A similar but opposite effect occurs to shift the set-point for shivering. Note in Figure 72–7 that shivering begins at a hypothalamic temperature of approximately 37.1° C. when the skin temperature is a cold 20° C.; but when the skin is at a warm

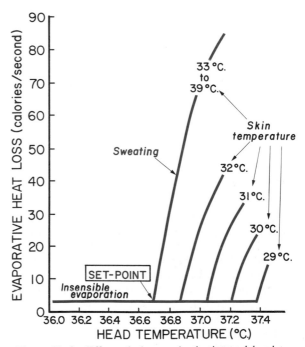

Figure 72–6. Effect of changes in the internal head temperature on the rate of evaporative heat loss from the body. Note also that the skin temperature determines the exact "set-point" level at which sweating begins. (Courtesy of Dr. T. H. Benzinger.)

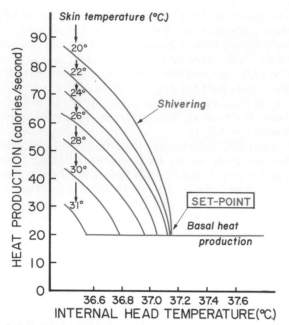

Figure 72–7. Effect of changes in the internal head temperature on the rate of heat production by the body. Note also that the skin temperature determines the exact "set-point" level at which shivering begins. (Courtesy of Dr. T. H. Benzinger.)

level of 31° C., the hypothalamic temperature must fall almost to 36.5° C. before shivering begins. That is, when the skin becomes cold it drives the hypothalamic thermostat to the shivering threshold even when the hypothalamic temperature itself is still quite hot. Here again, one can well understand the logic of this mechanism, because a cold skin temperature will soon lead to deeply depressed body temperature unless heat production is increased. Thus, this mechanism of cold skin temperature increasing heat production actually "anticipates" a possible fall in internal body temperature and prevents its occurrence.

BEHAVIORAL CONTROL OF BODY TEMPERATURE

Aside from the thermostatic mechanism for body temperature control, the body has still another temperature-controlling mechanism that is even more potent than all the above mechanisms together. This is behavioral control of temperature, which can be explained as follows: Whenever the internal body temperature becomes too high, signals from the thermostatic brain areas give the person a psychic sensation of being overheated. Conversely, whenever the body becomes too cold, signals from the skin and probably also from other receptors elicit the feeling of cold discomfort. Therefore, the person makes appropriate environmental adjustments to re-establish comfort. This is a much more powerful system of

body temperature control than most physiologists have recognized in the past. Indeed, for people, this is the only really effective mechanism for body heat control in severely cold environs.

The obvious types of behavioral adjustments include selecting appropriate clothing, moving the body to a different environmental setting, increasing the delivery of heat or cold from appropriate heaters or air conditioners, and so forth.

It is important to note that many other of our body's control systems utilize similar behavior mechanisms to achieve highly refined degrees of control. For instance, even respiration is probably controlled to a great extent in this way — that is, when a person perceives that he is being subjected to air hunger, he consciously breathes more to make up the deficit. Therefore, it is not valid to think of the body's homeostatic control systems as operating only in the subconscious portions of the brain.

LOCAL SKIN REFLEXES

When a person places a foot under a hot lamp and leaves it there for a short time, *local vasodilatation* and mild *local sweating* occur. Conversely, placing the foot in cold water causes vasoconstriction and cessation of sweating. These reactions are caused both by local effects of temperature changes directly on the blood vessels and sweat glands and by local cord reflexes conducted from the skin receptors to the spinal cord and back to the same skin area. However, their *intensity* is controlled by the hypothalamic thermostat, so that the overall effect is approximately proportional to the hypothalamic heat control signal *times* the local signal. Such reflexes can help to prevent excessive heat exchange from locally cooled or heated portions of the body.

Regulation of Internal Body Temperature After Cutting the Spinal Cord. After cutting the spinal cord in the neck above the sympathetic outflow from the cord, regulation of body temperature becomes extremely poor, for the hypothalamus can then no longer control either skin blood flow or the degree of sweating anywhere in the body. On the other hand, the local temperature reflexes originating in the skin, spinal cord, and intra-abdominal receptors still exist. Unfortunately, these reflexes are not powerful. In persons with this condition, body temperature must be regulated principally by the patient's psychic response to cold and hot sensations in the head region — that is, by behavioral control. If he feels himself becoming too hot or if he develops a headache from the heat, he knows to select cooler surroundings, and, conversely, if he has cold sensations, he selects warmer surroundings.

ABNORMALITIES OF BODY TEMPERATURE REGULATION

FEVER

Fever, which means a body temperature above the usual range of normal, may be caused by

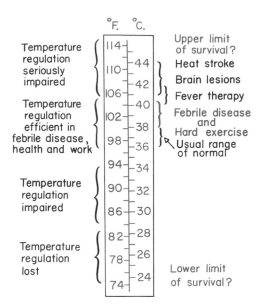

Figure 72–8. Body temperatures under different conditions. (From DuBois: Fever. Charles C Thomas.)

abnormalities in the brain itself or by toxic substances that affect the temperature regulating centers. Some causes of fever are presented in Figure 72–8. These include bacterial diseases, brain tumors, and environmental conditions that may terminate in heat stroke.

Resetting the Hypothalamic Thermostat in Febrile Diseases — Effect of Pyrogens

Many proteins, breakdown products of proteins, and certain other substances, such as lipopolysaccharide toxins secreted by bacteria, can cause the set-point of the hypothalamic thermostat to rise. Substances that cause this effect are called *pyrogens*. It is pyrogens secreted by toxic bacteria or pyrogens released from degenerating tissues of the body that cause fever during disease conditions. When the set-point of the hypothalamic thermostat becomes increased to a higher level than normal, all the mechanisms for raising the body temperature are brought into play, including heat conservation and increased heat production. Within a few hours after the thermostat has been set to a higher level, the body temperature also approaches this level.

Mechanism of Action of Pyrogens in Causing Fever — Role of "Endogeneous Pyrogen." Some physiologists believe that exogenous pyrogenic substances such as bacterial toxins act directly on the hypothalamic thermostat to increase its setting. However, there is reason to believe that most pyrogens affect the hypothalamic thermostat indirectly in the following manner:

When bacterial or other pyrogens are injected into a person they usually do not affect the hypo-

thalamic thermostat immediately; instead, the effect is delayed for many minutes to several hours. This delay is believed to result from the fact that the pyrogens must first react with the polymorphonuclear leukocytes, monocytes, and certain of the reticuloendothelial cells to form still another substance called *endogenous pyrogen*. The endogenous pyrogen is then transmitted in the blood to the hypothalamus and has an extremely powerful direct effect on the preoptic heat-sensitive neurons in increasing the set-point of the hypothalamic thermostat to febrile levels.

To give one an idea of the extremely powerful effect of pyrogens in resetting the hypothalamic thermostat, only a few *nanograms* of purified endogenous pyrogen injected into an animal can cause severe fever.

Effect of Dehydration on the Hypothalamic Thermostat. Dehydration is another factor that can cause the body temperature to rise considerably. Part of this elevation of temperature probably results from lack of available fluid for sweating, but dehydration can also cause temperature elevation even in a cold environment. Consequently, dehydration almost certainly affects the hypothalamic centers directly to set the hypothalamic thermostat to febrile levels.

Characteristics of Febrile Conditions

Chills. When the setting of the thermostat is suddenly changed from the normal level to a higher-than-normal value as a result of tissue destruction, pyrogenic substances, or dehydration, the body temperature usually takes several hours to reach the new temperature setting. For instance, the temperature setting of the hypothalamic thermostat, as illustrated in Figure 72–9, might suddenly rise to 103° F. Because the blood temperature is less than the temperature setting of the hypothalamic thermostat, the usual responses to cause elevation of body temperature occur. During this period the person experiences chills and feels extremely cold, even though body temperature may already be

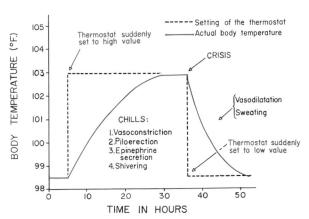

Figure 72–9. Effects of changing the setting of the "hypothalamic thermostat."

above normal. Also, the skin is cold because of vaso-constriction, and the person shakes all over because of shivering. Chills continue until the body temperature rises to the hypothalamic setting of 103° F. Then, when the temperature of the body reaches this value, the person no longer experiences chills but instead feels neither cold nor hot. As long as the factor that is causing the hypothalamic thermostat to be set at this high value continues its effect, the body temperature is regulated more or less in the normal manner but at the high temperature level.

The Crisis of "Flush." If the factor that is causing the high temperature is suddenly removed, the hypotha-lamic thermostat is suddenly set at a lower value — perhaps even back to the normal level, as illustrated in Figure 72–9. In this instance, the body temperature is still 103° F., but the hypothalamus is attempting to regulate the temperature to 98.6° F. This situation is analogous to excessive heating of the preoptic area, which causes intense sweating and sudden development of a hot skin because of vasodilatation everywhere. This sudden change of events in a febrile disease is known as the "crisis" or, more appropriately, the "flush." In olden days, before the advent of antibiotics, the crisis was always awaited, for once this occurred the doctor knew immediately that the patient's temperature would soon be falling.

Heat Stroke

The limits of extreme heat that one can stand depend almost entirely on whether the heat is dry or wet. If the air is completely dry and sufficient convection air cur-rents are flowing to promote rapid evaporation from the body, a person can withstand several hours of air temperature at 150° F. with no apparent ill effects. On the other hand, if the air is 100 per cent humidified or if the body is in water, the body temperature begins to rise whenever the environmental temperature rises above approximately 94° F. If the person is performing very heavy work, this critical temperature level may fall to 85° to 90° F.

Unfortunately, there is a limit to the rate at which the body can lose heat even with maximal sweating. Fur-thermore, when the hypothalamus becomes excessively heated, its heat-regulating ability becomes greatly de-pressed and sweating diminishes. As a result, a high body temperature tends to perpetuate itself unless measures are taken specifically to decrease body heat.

When the body temperature rises beyond a critical temperature, into the range of 106° to 108° F., the person is likely to develop *heat stroke*. The symptoms include dizziness, abdominal distress, sometimes de-lirium, and eventually loss of consciousness if the body temperature is not soon decreased. Many of these symptoms probably result from a mild degree of *circu-latory shock* brought on by excessive loss of fluid and electrolytes in the sweat before the onset of symptoms. However, the hyperpyrexia itself is also exceedingly damaging to the body tissues, especially to the brain, and therefore is undoubtedly responsible for many of the effects. In fact, even a few minutes of very high body temperature can sometimes be fatal. For this reason, many authorities recommend immediate treat-ment of heat stroke by placing the person in an ice water bath. However, because this often induces uncontroll-able shivering with considerable increase in rate of heat production, others have suggested that sponge cooling of the skin is likely to be more effective for rapidly decreasing the body core temperature.

Harmful Effects of the High Temperature. When the body temperature rises above 106° to 108° F., the paren-chyma of many cells begins to be damaged. The patho-logical findings in a person who dies of hyperpyrexia are local hemorrhages and parenchymatous degeneration of cells throughout the entire body, but especially in the brain. Unfortunately, once neuronal cells are destroyed, they can never be replaced. Damage to the liver, kidneys, and other body organs can often be great enough that failure of one or more of these eventually causes death, sometimes not till several days after the heat stroke.

Acclimatization to Heat. It is often extremely impor-tant to acclimatize persons to extreme heat; some examples are acclimatization of soldiers for tropical duty and acclimatization of miners for work in the 2-mile-deep gold mines of South Africa where the temperature approaches body temperature and the hu-midity approaches 100 per cent. Exposure of a person to heat for several hours each day while working a reason-ably heavy work load will develop increased tolerance to hot and humid conditions in about one week; the tolerance will then increase, to a lesser extent, for several additional weeks. Probably the most important physiological changes that occur during this acclima-tization process are an increase in the maximum rate of sweating, increase in the plasma volume, and dimin-ished loss of salt in the sweat and urine; the latter two of these effects result from increased secretion of aldos-terone.

Effects of Chemicals and Drugs on Body Temperature

An extremely large number of foreign substances when injected into the body fluids can cause the body temperature to rise — that is, they are pyrogenic. Bac-teria, pollens, dust, and vaccines are almost all pyrogen-ic. Ordinarily, human plasma is not pyrogenic, but it is possible for the plasma proteins and other proteins that normally do not cause pyrogenic reactions to be changed chemically by degradation or denaturization and thereafter to cause pyrogenic reactions. Some non-protein chemicals are also known to elevate the body temperature; these include especially certain polysac-charides and some nitrated phenol compounds.

Antipyretics. Aspirin, antipyrine, aminopyrine, and a number of other substances known as *antipyretics* have an effect on the hypothalamic thermostat opposite to that of the pyrogens. In other words, they cause the setting of the thermostat to be lowered so that the body temperature falls, though usually not more than a degree or so. Aspirin is especially effective in lowering the hypothalamic setting when pyrogens have raised the setting, but aspirin will not lower the normal tempera-ture. On the other hand, aminopyrine will decrease even the normal body temperature. Obviously, these drugs can be used to prevent damage to the body from excessively high body temperature.

EXPOSURE OF THE BODY TO EXTREME COLD

Unless treated immediately, a person exposed to ice water for approximately 20 to 30 minutes ordinarily dies because of heart standstill or heart fibrillation. By that time, the internal body temperature will have fallen to about 77° F. Yet, if warmed rapidly by application of external heat, the person's life can often be saved.

Treatment of a person whose body temperature has fallen into the 70s usually consists of application of wet heat either in the form of tub treatment or hot packs, with the water at approximately 110° F. If the temperature of the water is less than this, the rate at which heat is returned to the body is too slow for maximal benefit, and, if it is greater than this, the skin might be severely damaged because it becomes overheated while not receiving a satisfactory blood supply.

Loss of Temperature Regulation at Low Temperatures. As noted in Figure 72–8, once the body temperature has fallen below 85° F., the ability of the hypothalamus to regulate temperature is completely lost, and it is greatly impaired even when the body temperature falls below approximately 94° F. Part of the reason for this loss of temperature regulation is that the rate of heat production in each cell is greatly depressed by the low temperature. Also, sleepiness and even coma are likely to develop, which depress the activity of the central nervous system heat-control mechanisms and prevent shivering.

Frostbite. When the body is exposed to extremely low temperatures, surface areas can actually freeze; the freezing is called *frostbite*. This occurs especially in the lobes of the ears and in the digits of the hands and feet. If the parts are thawed immediately, especially with water that is not above approximately 110° F., no permanent damage may result. On the other hand, prolonged freezing causes permanent circulatory impairment as well as local tissue damage. Often gangrene follows thawing, and the frostbitten areas are lost.

Cold-Induced Vasodilation as a Protection Against Frostbite. When the temperature of tissues falls almost to freezing, the smooth muscle in the vascular wall becomes paralyzed because of the cold itself, and sudden vasodilatation occurs, often manifested by a flush of the skin. This mechanism, fortunately, helps to prevent frostbite by delivering warm blood to the skin. Unfortunately, it is a far less well-developed mechanism in humans than in most lower animals that live in the cold all the time.

Artificial Hypothermia. It is very easy to decrease the temperature of a person by giving him a sedative to depress the hypothalamic thermostat and then cooling him with ice, cooling blankets, or otherwise until the temperature falls. The temperature can then be maintained below 90° F. for several days to a week or more by continual sprinkling of cool water or alcohol on the body. Such artificial cooling is often used during heart surgery so that the heart can be stopped artificially for many minutes at a time. Cooling to this extent does not cause severe physiological results. It does slow the heart and greatly depresses body metabolism.

REFERENCES

Benzinger, T. H.: Heat regulation: Homeostasis of central temperature in man. *Physiol. Rev., 49*:671, 1969.

Bligh, J.: Temperature Regulation in Mammals and Other Vertebrates. New York, American Elsevier Publshing Co., 1973.

Blix, A. S., and Steen, J. B.: Temperature regulation in newborn polar homeotherms. *Physiol. Rev., 59*:285, 1979.

Cabanac, M.: Thermoregulatory behavior. *In* MPT International Review of Science: Physiology. Vol. 7. Baltimore, University Park Press, 1974, p. 231.

Cabanac, M.: Temperature regulation. *Annu. Rev. Physiol., 37*:415, 1975.

Cena, K., and Clark, J. A.: Transfer of heat through animal coats and clothing. *In* Robertshaw, D. (ed.): Internal Review of Physiology: Environmental Physiology III. Vol. 20. Baltimore, University Park Press, 1979, p. 1.

Elizondo, R.: Temperature regulation in primates. *Int. Rev. Physiol., 15*:71, 1977.

Hales, J. R. S.: Physiological responses to heat. *In* MTP International Review of Science: Physiology. Vol. 7. Baltimore, University Park Press, 1974, p. 107.

Hardy, J. D.: Physiology of temperature regulation. *Physiol. Rev., 41*:521, 1961.

Hardy, R. N.: Temperature and Animal Life. Baltimore, University Park Press, 1979.

Hazel, J. R., and Prosser, C. L.: Molecular mechanisms of temperature compensation in poikilotherms. *Physiol. Rev., 54*:620, 1974.

Heller, H. C., *et al.*: The thermostat of vertebrate animals. *Sci. Am., 239*(2):102, 1978.

Hensel, H.: Neural processes in thermoregulation. *Physiol. Rev., 53*:948, 1973.

Hensel, H.: Thermoreceptors. *Annu. Rev. Physiol., 36*:233, 1974.

Himms-Hagen, J.: Cellular thermogenesis. *Annu. Rev. Physiol., 38*:315, 1976.

Jansky, L.: Non-shivering thermogenesis and its thermoregulatory significance. *Biol. Rev., 48*:85, 1973.

Kerslake, D. M.: The Stress of Hot Environments. (Physiological Society Monograph, No. 29.) Cambridge, England, Cambridge University Press, 1972.

Kluger, M. J.: Temperature regulation, fever, and disease. *In* Robertshaw, D. (ed.): International Review of Physiology: Environmental Physiology III. Vol. 20. Baltimore, University Park Press, 1979, p. 209.

Maclean, D., and Emslie-Smith, D.: Accidental Hypothermia. Oxford, Blackwell Scientific Publications, 1977.

Mitchell, D.: Physical basis of thermoregulation. *In* MTP International Review of Science: Physiology. Vol. 7. Baltimore, University Park Press, 1974, p. 1.

Mitchell, D.: Physical basis of thermoregulation. *Int. Rev. Physiol., 15*:1, 1977.

Precht, H., *et al.*: Temperature and Life. New York, Springer-Verlag, 1973.

Reeves, R. B.: The interaction of body temperature and acid-base balance in ectothermic vertebrates. *Annu. Rev. Physiol., 39*:559, 1977.

Robertshaw, D.: Role of the adrenal medulla in thermoregulation. *Int. Rev. Physiol:, 15*:189, 1977.

Satinoff, E. (ed.): Thermoregulation. Stroudsburg, Pa., Dowden, Hutchinson & Ross; New York, Academic Press, 1979.

Schmidt-Nielsen, K.: Animal Physiology: Adaptation and Environment. London, Cambridge University Press, 1975.

Sloan, A. W.: Man in Extreme Environments. Springfield, Ill., Charles C Thomas, 1979.

Smith, R. E., and Horwitz, B. A.: Brown fat and thermogenesis. *Physiol. Rev., 49*:330, 1969.

Svanes, K.: Effects of temperature on blood flow. *In* Kaley, G., and Altura, B. M. (eds.): Microcirculation. Vol. III. Baltimore, University Park Press, 1977.

Swan, H.: Thermoregulation and Bioenergetics: Patterns for Vertebrate Survival. New York, American Elsevier Publishing Co., 1974.

Thompson, G. E.: Physiological effects of cold exposure. *Int. Rev. Physiol., 15*:29, 1977.

Underwood, L. S., and Tieszen, L. L. (eds.): Comparative Mechanisms of Cold Adaptation. New York, Academic Press, 1979.

Wang, L. C. H., and Hudson, J. W. (eds.): Strategies in Cold: Natural Torpidity and Thermogenesis. New York, Academic Press, 1978.

Webster, A. J. F.: Physiological effects of cold exposure. *In* MTP International Review of Science: Physiology. Vol. 7. Baltimore, University Park Press, 1974, p. 33.

Wyndham, C. H.: The physiology of exercise under heat stress. *Annu. Rev. Physiol., 35*:193, 1973.

73

Dietary Balances, Regulation of Feeding; Obesity and Starvation

The intake of food must always be sufficient to supply the metabolic needs of the body and yet not enough to cause obesity. Also, since different foods contain different proportions of proteins, carbohydrates, and fats, appropriate balance must be maintained among these different types of food so that all segments of the body's metabolic systems can be supplied with the requisite materials. This chapter, therefore, discusses the problems of balance among the three major types of food and also the mechanisms by which the intake of food is regulated in accordance with the metabolic needs of the body.

DIETARY BALANCES

ENERGY AVAILABLE IN FOODS

The energy liberated from each gram of carbohydrate as it is oxidized to carbon dioxide and water is 4.1 Calories, and that liberated from fat is 9.3 Calories. The energy liberated from metabolism of the average protein of the diet as each gram is oxidized to carbon dioxide, water, and urea is 4.35 Calories. Also, these different substances vary in the average percentages that are absorbed from the gastrointestinal tract: approximately 98 per cent of the carbohydrate, 95 per cent of the fat, and 92 per cent of the protein. Therefore, in round figures the average *physiologically available energy* in each gram of the three different foodstuffs in the diet is:

	Calories
Carbohydrates	4.0
Fat	9.0
Protein	4.0

AVERAGE COMPOSITION OF THE DIET

Average Americans receive approximately 15 per cent of their energy from protein, about 40 per cent from fat, and 45 per cent from carbohydrates. In most other parts of the world the quantity of energy derived from carbohydrates far exceeds that derived from both proteins and fats. Indeed, in Mongolia the energy received from fats and proteins combined is said to be no greater than 15 to 20 per cent.

Daily Requirement for Protein. Twenty to 30 grams of the body proteins are degraded and used for producing other body chemicals daily. Therefore, all cells must continue to form new proteins to take the place of those that are being destroyed, and a supply of protein is needed in the diet for this purpose. An average person can maintain normal stores of protein provided the *daily intake is above 30 to 55 grams.*

Partial Proteins. Another factor that must be considered in analyzing the proteins of the diet is whether the dietary proteins are *complete* proteins or *partial* proteins. Complete proteins have compositions of amino acids in appropriate proportion to each other so that all the amino acids can be properly used by the human body. In general, proteins derived from animal foodstuffs are more nearly complete than are proteins derived from vegetable and grain sources. This subject is more completely discussed in Chapter 69; therefore, for the present, let us note that when partial proteins are in the diet an increased minimal quantity of protein is necessary in the daily rations to maintain protein balance. An example is the diet of many African natives, who subsist primarily on corn meal. The protein of corn is almost totally lacking in tryptophan, and this means that this diet, in effect, is almost completely protein-deficient. As a result, the Africans, especially the children, develop the protein deficiency syndrome called *kwashiorkor,* which consists of failure to grow, lethargy, depressed mentality, and hypoprotein edema.

Necessity for Fat in the Diet. The human body is capable of synthesizing some unsaturated fatty acids, especially in the liver and to a lesser extent in fat cells. However, the body cannot form others such as *linoleic, linolenic,* and *arachidonic acids,* which are essential constituents of the diet for normal operation of the body. In lower animals, scaling and exudative skin lesions appear when these unsaturated fatty acids are totally absent from the diet, and the animals fail to grow. Also, arachidonic acid (or linoleic acid from which arachidonic acid can be synthesized) must be present to form the prostaglandins which are ubiquitous cellular hormones with many local functions.

How much of the essential unsaturated fat is needed in the diet has not been determined; but because Mongolians exist on diets with fat contents as low as 10 per cent, it is presumed that only small quantities of unsaturated fats are needed. However, it was pointed out in Chapter 68 that a high *ratio* of saturated to unsaturated fats in the diet predisposes to atherogenesis, which often leads to heart attacks. Therefore, most nutritionists recommend that the fat of the diet contain a high proportion of unsaturated fatty acids.

Necessity for Carbohydrates in the Diet. In the discussion of carbohydrate metabolism in Chapter 67, it was pointed out that failure to metabolize carbohydrates in sufficient quantity causes rapid metabolism of fats and that ketosis is likely to develop. Also, fats and proteins alone usually cannot supply sufficient energy for all operations of the human body. Consequently, carbohydrates appear to be essential to prevent weakness. This is especially true when the body undertakes a considerable work load. Finally, as already discussed, carbohydrate is a *protein sparer;* that is, carbohydrate is burned in preference to the burning of protein, which is especially important for preserving the functional proteins in the cells.

Composition of Different Foods. Table 73–1 presents the compositions of selected foods, illustrating especially the high proportions of fats and proteins in meat products and the high proportions of carbohydrates in most vegetable and grain products.

Fat is deceptive in the diet, for it often exists as 100 per cent fat undiluted by any other substances, whereas essentially all the proteins and carbohydrates of foods are mixed in watery media and often represent less than 25 per cent of the weight of the food. Thus, the fat of one pat of butter mixed with an entire helping of potato may contain as much energy as all the potato itself. This is one of the reasons why fats are assiduously avoided in the prescription of diets for weight reduction.

STUDY OF ENERGY BALANCES

It is important in many physiological studies to compute the intake of food and to compute at the same time the utilization of the different types of foods by the body. Special procedures are available for determining *utilization* of the different types of foods for energy, as follows:

Determination of the Rate of Protein Metabolism in the Body. The average protein of the diet contains approximately 16 per cent nitrogen, and the remaining 84 per cent is composed of carbon, hydrogen, oxygen, and sulfur. Numerous experiments have shown that when protein is metabolized in the body, the average person excretes about 90 per cent of the protein nitrogen in the urine in the form of urea, uric acid, creatinine, and other less important nitrogen products. The remaining 10 per

TABLE 73–1 PROTEIN, FAT, AND CARBOHYDRATE CONTENT OF DIFFERENT FOODS

Food	Protein %	Fat %	Carbohydrate %	Fuel Value per 100 Grams, Calories
Apples	0.3	0.4	14.9	64
Asparagus	2.2	0.2	3.9	26
Bacon, fat	6.2	76.0	0.7	712
broiled	25.0	55.0	1.0	599
Beef, medium	17.5	22.0	1.0	268
Beets, fresh	1.6	0.1	9.6	46
Bread, white	9.0	3.6	49.8	268
Butter	0.6	81.0	0.4	733
Cabbage	1.4	0.2	5.3	29
Carrots	1.2	0.3	9.3	45
Cashew nuts	19.6	47.2	26.4	609
Cheese, Cheddar, American	23.9	32.3	1.7	393
Chicken, total edible	21.6	2.7	1.0	111
Chocolate	(5.5)	52.9	(18.)	570
Corn (maize), entire	10.0	4.3	73.4	372
Haddock	17.2	0.3	0.5	72
Lamb, leg, intermediate	18.0	17.5	1.0	230
Milk, fresh whole	3.5	3.9	4.9	69
Molasses, medium	0.0	0.0	(60.)	240
Oatmeal, dry, uncooked	14.2	7.4	68.2	396
Oranges	0.9	0.2	11.2	50
Peanuts	26.9	44.2	23.6	600
Peas, fresh	6.7	0.4	17.7	101
Pork, ham, medium	15.2	31.0	1.0	340
Potatoes	2.0	0.1	19.1	85
Spinach	2.3	0.3	3.2	25
Strawberries	0.8	0.6	8.1	41
Tomatoes	1.0	0.3	4.0	23
Tuna, canned	24.2	10.8	0.5	194
Walnuts, English	15.0	64.4	15.6	702

cent is ordinarily excreted in the feces. It is possible, therefore, to estimate relatively accurately the total quantity of protein metabolized by the body in a given time by analyzing the amount of nitrogen excreted in the urine. For instance, if 8 grams of nitrogen are excreted into the urine each day, then the total excretion, including that in the feces (an additional 10 per cent), will be about 8.8 grams. This value multiplied by 100/16 gives a calculated total protein metabolism of 55 grams per day.

"Nitrogen Balance" in the Body. Nitrogen balance experiments are frequently performed to determine the rate of protein increase or decrease in the body. These balance studies are performed by measuring for a week or more (a) the rate of protein intake and (b) the rate of protein utilization, the total protein utilization being estimated as was just described.

A *negative nitrogen balance* implies greater protein utilization than protein intake, causing loss of protein from the body, and a *positive nitrogen* balance implies net gain of protein in the body. Factors that cause negative balance include malnutrition, debilitating diseases, and glucocorticoid hormones. Factors that cause positive nitrogen balance include exercise, growth hormone, insulin, and testosterone.

Relative Utilization of Fat and Carbohydrates – The Respiratory Quotient. When one molecule of glucose is oxidized, the number of molecules of carbon dioxide liberated is exactly equal to the number of oxygen molecules used in the oxidative process. This fact is illustrated in Figure 73–1. Therefore the *respiratory quotient,* which is defined as the *ratio of carbon dioxide output to oxygen usage* is 1.00. On the other hand, oxidation of triolein (the most abundant fat in the body) liberates 57 carbon dioxide molecules while 80 oxygen molecules are being utilized. Consequently, the respiratory quotient in this instance is 0.71. Finally, oxidation of alanine liberates five carbon dioxide molecules for every six oxygen molecules entering into the reaction, giving a respiratory quotient of 0.83.

Thus, the respiratory quotient for utilization of carbohydrates is 1.00 because oxidation of each carbon atom in the molecule requires one molecule of respiratory oxygen. On the other hand, the respiratory quotient of dietary fat has been found to average 0.707, while the respiratory quotient of the protein of the diet averages 0.801.

Use of the Respiratory Quotient for Estimating Relative Rates of Carbohydrate and Fat Metabolism. As pointed out, the average person receives only 15 per cent of his total energy from protein metabolism. Furthermore, the respiratory quotient of protein is approximately midway between the respiratory quotients of fat and carbohydrate (see preceding paragraph). Consequently, when the overall respiratory quotient of a person (which, in studies lasting an hour or more, is equal to the respiratory exchange ratio that was discussed in Chapter 40) is measured by determining the total respiratory intake of oxygen and the total output of carbon dioxide from the lungs, one has a reasonable measure of the relative quantities of fat and carbohydrate being metabolized by the body during that period of time. For instance, if the respiratory quotient is approximately 0.71, the body is burning almost entirely fat to the exclusion of carbohydrates and proteins. If, on the other hand, the respiratory quotient is 1.00, the body is probably metabolizing almost entirely carbohydrate to the exclusion of fat and protein. Finally, a respiratory quotient of 0.85 indicates approximately equal utilization of carbohydrate and fat.

To be still more accurate in the calculation, one must consider the portion of the respiratory quotient that represents protein metabolism. To do this the rate of protein utilization in the body is first determined as discussed above; then the quantities of oxygen utilized and carbon dioxide released as a result of the protein metabolism are calculated and subtracted from the intake of oxygen and output of carbon dioxide, respectively, in order to establish the net respiratory quotient as it applies to fat and carbohydrate.

Variations in the Respiratory Quotient. Shortly after a meal, almost all the food metabolized is carbohydrates. Consequently, the respiratory quotient approaches 1.00 at this time. Approximately 8 to 10 hours following a meal, the quantity of carbohydrate being metabolized is relatively slight, and the respiratory quotient approaches that for fat metabolism; that is, approximately 0.71.

In diabetes mellitus, little carbohydrate is utilized by the body, and consequently, most of the energy is derived from fat. Therefore, most persons with severe diabetes have respiratory quotients approaching the value for fat metabolism, 0.71.

In the human being, the respiratory quotient rarely becomes greater than 1.00, but such values have been recorded in animals in which carbohydrate is being converted rapidly into fat. When carbohydrate is being converted into fat, considerable oxygen is liberated from the carbohydrate molecule, and this oxygen becomes available for oxidation of other foodstuffs; this lessens the quantity of oxygen that must be brought into the body through the lungs. Consequently, if the animal

Respiratory Quotient:

$$C_6H_{12}O_6 + 6\ O_2 \rightarrow 6\ CO_2 + 6\ H_2O \qquad \frac{6}{6} = 1.00$$
Glucose

$$C_{57}H_{104}O_6 + 80\ O_2 \rightarrow 57\ CO_2 + 52\ H_2O \qquad \frac{57}{80} = 0.71$$
Triolein

$$2\ C_3H_7O_2N + 6\ O_2 \rightarrow (NH_2)_2CO + 5\ CO_2 + 5\ H_2O \qquad \frac{5}{6} = 0.83$$
Alanine

Figure 73–1. Utilization of oxygen and release of carbon dioxide during the oxidation of carbohydrate, fat, and protein. The respiratory quotient for each of these reactions is calculated.

is at the same time metabolizing only carbohydrates, the amount of carbon dioxide excreted through the lungs is greater than the intake of oxygen so that the respiratory quotient can rise to values as high as perhaps 1.10.

REGULATION OF FOOD INTAKE

Hunger. The term hunger means a craving for food, and it is associated with a number of objective sensations. For instance, as pointed out in Chapter 63, in a person who has not had food for many hours, the stomach undergoes intense rhythmic contractions called *hunger contractions.* These cause a tight or gnawing feeling in the pit of the stomach and sometimes actually cause pain called *hunger pangs.* In addition to the hunger pangs, the hungry person also becomes more tense and restless than usual, and often has a strange feeling throughout the entire body that might be described by the nonphysiological term "twitterness."

Some physiologists actually define hunger as the tonic contractions of the stomach. However, even after the stomach is completely removed, the psychic sensations of hunger still occur, and craving for food still makes the person search for an adequate food supply.

Appetite. The term appetite is often used in the same sense as hunger except that it usually implies desire for specific types of food instead of food in general. Therefore, appetite helps a person choose the quality of food to eat.

Satiety. Satiety is the opposite of hunger. It means a feeling of fulfillment in the quest for food. Satiety usually results from a filling meal, particularly when the person's nutritional storage depots, the adipose tissue and the glycogen stores, are already filled.

NEURAL CENTERS FOR REGULATION OF FOOD INTAKE

Hunger and Satiety Centers. Stimulation of the *lateral hypothalamus* causes an animal to eat voraciously, while stimulation of the *ventromedial nuclei of the hypothalamus* causes complete satiety, and, even in the presence of highly appetizing food, the animal will still refuse to eat. Conversely, destructive lesions of the two respective areas cause results exactly opposite to those caused by stimulation. That is, ventromedial lesions cause voracious and continued eating until the animal becomes extremely obese, sometimes as large as 4 times normal in size. And lesions of the lateral hypothalamic nuclei cause complete lack of desire for food and progressive inanition of the animal. Therefore, we can label the lateral nuclei of the hypothalamus as the *hunger center* or

feeding center, while we can label the ventromedial nuclei of the hypothalamus as a *satiety center.*

The feeding center operates by directly exciting the emotional drive to search for food (while also stimulating other emotional drives as well — see Chapters 56 and 57). On the other hand, it is believed that the satiety center operates primarily by inhibiting the feeding center.

Other Neural Centers That Enter into Feeding. If the brain is sectioned between the hypothalamus and the mesencephalon, the animal can still perform the basic mechanical features of the feeding process. It can salivate, lick its lips, chew food, and swallow. Therefore, the actual mechanics of feeding are all controlled by centers in the brain stem. The function of the hypothalamus in feeding, then, is to control the quantity of food intake and to excite the lower centers to activity.

Higher centers than the hypothalamus also play important roles in the control of feeding, particularly in the control of appetite. These centers include especially the amygdala and the cortical areas of the limbic system, all of which are closely coupled with the hypothalamus. It will be recalled from the discussion of the sense of smell that the amygdala is one of the major parts of the olfactory nervous system. Destructive lesions in the amygdala have demonstrated that some of its areas greatly increase feeding, while others inhibit feeding. In addition, stimulation of some areas of the amygdala elicits the mechanical act of feeding. However, the most important effect of destruction of the amygdala on both sides of the brain is a "psychic blindness" in the choice of foods. In other words, the animal (and presumably the human being as well) loses or at least partially loses the mechanism of appetite control of type and quality of food that it eats.

The cortical regions of the limbic system, including the infraorbital regions, the hippocampal gyrus, and the cingulate gyrus, all have areas that when stimulated can either increase or decrease feeding activities. These areas seem especially to play a role in the animal's drive to search for food when it is hungry. It is presumed that these centers are also responsible, probably operating in association with the amygdala and hypothalamus, for determining the quality of food that is eaten. For instance, a previous unpleasant experience with almost any type of food often destroys a person's appetite for that food thenceforth.

FACTORS THAT REGULATE FOOD INTAKE

We can divide the regulation of food into (1) *nutritional regulation,* which is concerned primarily with maintenance of normal quantities of nutrient stores in the body, and (2) *alimentary regulation,* which is concerned primarily with the immediate effects of feeding on the alimentary tract and is sometimes called *peripheral regulation* or *short-term regulation.*

Nutritional Regulation. An animal that has been starved for a long time and is then presented with unlimited food eats a far greater quantity than does an animal that has been on a regular diet. Conversely, an animal that has been force-fed for several weeks eats little when allowed to eat according to its own desires. Thus, the feeding center in the hypothalamus is geared to the nutritional status of the body. Some of the nutritional factors that control the degree of activity of the feeding center are the following:

Availability of Glucose to the Body Cells – The Glucostatic Theory of Hunger and of Feeding Regulation. It has long been known that a decrease in blood glucose concentration is associated with development of hunger, which has led to the so-called glucostatic theory of hunger and of feeding regulation: When the blood glucose level falls too low, this automatically causes the animal to increase his feeding, which eventually returns the glucose concentration back toward normal. Two other observations also support the glucostatic theory: (1) An increase in blood glucose level increases the measured electrical activity in the satiety center in the ventromedial nuclei of the hypothalamus and simultaneously decreases the electrical activity in the feeding center of the lateral nuclei. (2) Chemical studies show that the ventromedial nuclei (the satiety center) concentrate glucose while other areas of the hypothalamus fail to concentrate glucose; therefore, it is assumed that glucose acts by increasing the degree of satiety.

Effect of Blood Amino Acid Concentration on Feeding. An increase in amino acid concentration in the blood also reduces feeding, and a decrease enhances feeding. In general, though, this effect is not as powerful as the glucostatic mechanism.

Effect of Fat Metabolites on Feeding — Long-Term Regulation. The overall degree of feeding varies almost inversely with the amount of adipose tissue in the body. That is, as the quantity of adipose tissue increases, the rate of feeding decreases. Therefore, many physiologists believe that *long-term regulation* of feeding is controlled mainly by fat metabolites of undiscovered nature. This is called the "lipostatic" theory of feeding regulation. In support of this is the fact that the long-term average concentration of free fatty acids in the blood is directly proportional to the quantity of adipose tissue in the body. Therefore, it is likely that the free fatty acids or some other similar fat metabolites act in the same manner as glucose and amino acids to cause a negative feedback regulatory effect on feeding. It is also possible, if not probable, that this is the most important long-term regulator of feeding.

Interrelationshp Between Body Temperature and Food Intake. When an animal is exposed to cold, it tends to overeat; when exposed to heat, it tends to undereat. This is caused by interaction within the hypothalamus between the temperature-regulating system (see Chap. 72) and the food intake–regulating system. It is important because increased food intake in the cold animal (a) increases its metabolic rate and (b) provides increased fat for insulation, both of which tend to correct the cold state.

Summary of Long-Term Regulation. Even though our information on the different feedback factors in long-term feeding regulation is imprecise, we can make the following general statement: When the nutrient stores of the body fall below normal, the feeding center of the hypothalamus becomes highly active, and the person exhibits increased hunger; on the other hand, when the nutrient stores are abundant, the person loses the hunger and develops a state of satiety.

Alimentary Regulation (Short-Term, Nonmetabolic Regulation). The degree of hunger or satiety at different times of the day depends to a great extent on habit. For instance, normally people have the habit of eating three meals a day, and, if they miss one, they are likely to develop a state of hunger at mealtime despite completely adequate nutritional stores in their tissues. But, in addition to habit, several other short-term physiological stimuli — mainly related to the alimentary tract — can alter one's desire for food for several hours at a time, as follows:

Gastrointestinal Filling. When the gastrointestinal tract becomes distended, especially the stomach or the duodenum, inhibitory signals temporarily suppress the feeding center, thereby reducing the desire for food. This effect probably depends mainly on sensory signals transmitted through the vagi, but part of the effect still persists after the vagi and the sympathetic nerves from the upper gastrointestinal tract have been severed. Therefore, somatic sensory signals from the stretched abdomen might also play a role. And, recently it has been found that hormonal feedback also suppresses feeding, for *cholecystokinin* that is released in response mainly to fat entering the duodenum has a strong effect on inhibition of further eating.

Obviously, these mechanisms are of particular importance in bringing one's feeding to a halt during a heavy meal.

Metering of Food by Head Receptors. When a person with an esophageal fistula is fed large quantities of food, even though this food is immediately lost again to the exterior the degree of hunger is decreased after a reasonable quantity of food has passed through the mouth. This effect occurs despite the fact that the gastrointestinal tract does not become the least bit filled. Therefore, it is postulated that various "head factors" relating to feeding, such as chewing, salivation, swallowing, and tasting, "meter" the food as it passes through the mouth, and after a certain amount has passed through, the hypothalamic feeding center becomes inhibited. However, the inhibition caused by this metering mechanism is considerably less intense and less lasting, usually lasting only 20 to 40 minutes, than is the inhibition caused by gastrointestinal filling.

Importance of Having Both Long- and Short-term Regulatory Systems for Feeding. The long-term regulatory system, especially the lipostatic feedback mechanism, obviously helps an animal to maintain constant stores of nutrients in its tissues, preventing these from becoming too low or too high. On the other hand, the short-term regulatory stimuli make the animal feed only when the gastrointestinal tract is receptive to food. Thus, food passes through its gastrointestinal tract fairly continuously so that its digestive, absorptive, and storage mechanisms can all work at a steady pace rather than only when the animal needs food for energy. Indeed, the digestive, absorptive, and storage mechanisms can increase their rates of activity above normal only four- to five-fold, while the rate of usage of stored nutrients for energy sometimes increases to 20 times normal.

It is important, then, that feeding occur rather continuously (but at a rate that the gastrointestinal tract can accommodate), regulated principally by the short-term regulatory mechanisms. However, it is also important that the intensity of the daily rhythmic feeding habits be modulated up or down by the long-term regulatory system, based principally on the level of nutrient stores in the body.

OBESITY

Energy Input Versus Energy Output. When greater quantities of energy (in the form of food) enter the body than are expended, the body weight increases. Therefore, obesity is obviously caused by excess energy input over energy output. For each 9.3 Calories of excess energy entering the body, 1 gram of fat is stored.

Excess energy input occurs *only during the developing phase of obesity,* and once a person has become obese, all that is required to remain obese is that the energy input equal the energy output. For the person to reduce, the input must be *less* than the output. Indeed, studies of obese persons have shown that their intake of food in the static stage of obesity (after the obesity has already been attained) is approximately the same as that for normal persons.

Effect of Muscular Activity on Energy Output. About one-third the energy used each day by the normal person goes into muscular activity, and in the laborer as much as two-thirds or occasionally three-fourths is used in this way. Since muscular activity is by far the most important means by which energy is expended in the body, it is frequently said that obesity results from *too high a ratio of food intake to daily exercise.*

Abnormal Feeding Regulation as a Pathological Cause of Obesity

The preceding discussion of the mechanisms that regulate feeding emphasized that the rate of feeding is normally regulated in proportion to the nutrient stores in the body. When these stores begin to approach an optimal level in a normal person, feeding is automatically reduced to prevent overstorage. However, in many obese persons this is not true, for feeding does not slacken until body weight is far above normal. Therefore, in effect, obesity is often caused by an abnormality of the feeding regulatory mechanism. This can result from either psychogenic factors that affect the regulation or actual abnormalities of the hypothalamus itself.

Psychogenic Obesity. Studies of obese patients show that a large proportion of obesity results from psychogenic factors. Perhaps the most common psychogenic factor contributing to obesity is the prevalent idea that healthy eating habits require three meals a day and that each meal must be filling. Many children are forced into this habit by overly solicitous parents, and the children continue to practice it throughout life. In addition, persons are known often to gain large amounts of weight during or following stressful situations, such as the death of a parent, a severe illness, or even mental depression. It seems that eating is often a means of release from tension.

Hypothalamic Abnormalities as a Cause of Obesity. In the preceding discussion of feeding regulation, it was pointed out that lesions in the ventromedial nuclei of the hypothalamus cause an animal to eat excessively and become obese. It has also been discovered that these lesions are associated with excess insulin production, which in turn increases fat depositon. Also, many persons with hypophysial tumors that encroach on the hypothalamus develop progressive obesity, illustrating that obesity in the human being, too, can definitely result from damage to the hypothalamus. Though in the normal obese person hypothalamic damage is almost never found, it is possible that the functional organization of the feeding center is different in the obese person from that of the nonobese person. For instance, an obese person who has reduced to normal weight usually develops hunger that is demonstrably far greater than that of the normal person. This indicates that the "setting" of the obese person's feeding center is at a much higher level of nutrient storage than is that of the normal person.

Genetic Factors in Obesity. Obesity definitely runs in families. Furthermore, identical twins usually maintain weight levels within 2 pounds of each other throughout life if they live under similar conditions, or within 5 pounds of each other if their conditions of life differ markedly. This might result partly from eating habits engendered during childhood, but it is generally believed that this close similarity between twins is genetically controlled.

The genes can direct the degree of feeding in several different ways, including (1) a genetic abnormality of the feeding center to set the level of nutrient storage high or low and (2) abnormal hereditary psychic factors that either whet the appetite or cause the person to eat as a "release" mechanism.

Genetic abnormalities in the *chemistry of fat storage* are also known to cause obesity in certain strains of rats and mice. In one strain of rats, fat is easily stored in the adipose tissue, but the quantity of hormone-sensitive lipase in the adipose tissue is greatly reduced, so that little of the fat can be removed. In addition, the rats develop hyperinsulinism which promotes fat storage. This combination obviously results in a one-way path, the fat continually being deposited but never released. In a strain of mice there is excess of fatty acid synthetase, which causes excess synthesis of fatty acids. These, too, are other possible mechanisms of obesity in some human beings.

Childhood Overnutrition as a Possible Cause of Obesity. The number of fat cells in the adult body is determined almost entirely by the amount of fat stored in the body during early life. The rate of formation of new fat cells is especially rapid in obese infants before they are weaned, and this continues at a lesser rate in obese children until adolescence, occurring most markedly about the time of puberty. Thereafter, the number of fat cells remains almost identically the same throughout the remainder of life. Thus, mainly on the basis of experiments in lower animals, it is believed that overfeeding children, especially in infancy and to a lesser extent during the older years of childhood, can lead to a lifetime of obesity. The person who has excess fat cells is thought to have a higher setting of the hypothalamic feedback autoregulatory mechanism for control of adipose tissues. In support of this belief is the fact that most extremely obese people have far more fat cells than normal people — often as much as 3 or more times as many.

In less obese persons, especially those who become obese in middle or old age, most of the obesity results from hypertrophy of already existing fat cells. This type of obesity is far more susceptible to treatment than is the life-long type.

TREATMENT OF OBESITY

Treatment of obesity depends simply on decreasing energy input below energy expenditure. In other words, this means partial starvation. For this purpose, most diets are designed to contain large quantities of "bulk" which, in general, are made up of non-nutritive cellulose substances. This bulk distends the stomach and thereby partially appeases the hunger. In most lower animals such a procedure simply makes the animal increase its food intake still further, but human beings can often fool themselves because their food intake is sometimes controlled as much by habit as by hunger. As pointed out below in connection with starvation, it is important to prevent vitamin deficiencies during the dieting period.

Various *drugs for decreasing the degree of hunger* have been used in the treatment of obesity. The most important of these is *amphetamine* (or amphetamine derivatives), which directly inhibits the feeding center in the lateral nuclei of the hypothalamus. However, there is danger in using this drug because it simultaneously overexcites the central nervous system, making the person nervous and elevating the blood pressure. Also, a person soon adapts to the drug so that weight reduction is usually no greater than 10 per cent.

Finally, the more exercise one takes, the greater is the daily energy expenditure and the more rapidly the obesity disappears. Therefore, forced exercise is often an essential part of the treatment for obesity.

INANITION

Inanition is the exact opposite of obesity. In addition to inanition caused by inadequate availability of food, both psychogenic and hypothalamic abnormalities can on occasion cause greatly decreased feeding. One such condition, *anorexia nervosa,* is an abnormal psychic state in which a person loses all desire for food and even becomes nauseated by food; as a result, severe inanition occurs. Also, destructive lesions of the hypothalamus, particularly caused by vascular thrombosis, frequently cause a condition called *cachexia;* the term simply means severe inanition.

STARVATION

Depletion of Food Stores in the Body Tissues During Starvation. Even though the tissues preferentially use carbohydrate for energy over both fat and protein, the quantity of carbohydrate stores of the body is only a few hundred grams (mainly glycogen in the liver and muscles), and it can supply the energy required for body function for perhaps half a day. Therefore, except for the first few hours of starvation, the major effects are progressive depletion of tissue fat and protein. Since fat is the prime source of energy, its rate of depletion continues unabated, as illustrated in Figure 73–2, until most of the fat stores in the body are gone.

Protein undergoes three different phases of depletion:

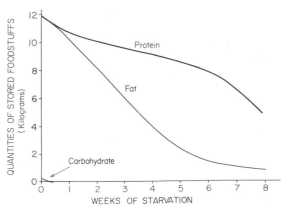

Figure 73–2. Effect of starvation on the food stores of the body.

rapid depletion at first, then greatly slowed depletion, and, finally rapid depletion again shortly before death. The initial rapid depletion is caused by conversion of protein to glucose in the liver by the process of gluconeogenesis. The glucose thus formed (about two-thirds of it, that is) is used mainly to supply energy to the brain which, under normal circumstances, utilizes almost no other metabolic substrate for energy besides glucose. However, after the readily mobilizable protein stores have been depleted during the early phase of starvation, the remaining protein is not so easily removed from the tissues. At this time, the rate of gluconeogenesis decreases to one-third to one-fifth its previous rate, and the rate of depletion of protein becomes greatly decreased. The lessened availability of glucose then initiates a series of events that often lead to *ketosis,* which was discussed in Chapter 68. Fortunately, the ketone bodies, like glucose, can cross the blood-brain barrier and can be utilized by the brain cells for energy. Therefore, approximately two-thirds of the brain's energy now is derived from these ketone bodies, principally from beta-hydroxybutyrate. This sequence of events thus leads to at least partial preservation of the protein stores of the body.

However, there finally comes a time when the fat stores also are almost totally depleted, and the only remaining source of energy is proteins. At that time, the protein stores once again enter a stage of rapid depletion. Since the proteins are essential for maintenance of cellular function, death ordinarily ensues when the proteins of the body have been depleted to approximately one-half their normal level.

Vitamin Deficiencies in Starvation. The stores of some of the vitamins, especially the water-soluble vitamins — the vitamin B group and vitamin C — do not last long during starvation. Consequently, after a week or more of starvation mild vitamin deficiencies usually begin to appear, and after several weeks severe vitamin deficiencies can occur. Obviously, these can add to the debility that leads to death.

REFERENCES

Albrink, M. J.: Overnutrition and the fat cell. *In* Bondy, P. K., and Rosenberg, L. E. (eds.): Duncan's Diseases of Metabolism, 7th Ed. Philadelphia, W. B. Saunders Co., 1974, p. 417.

Alfin-Slater, R. B., and Kritchevsky, D. (eds.): Nutrition and the Adult: Macronutrients. New York, Plenum Press, 1979.

Baile, C. A., and Forbes, J. M.: Control of feed intake and regulation of energy balance in ruminants. *Physiol. Rev., 54*:160, 1974.

Booth, D. A. (ed.): Hunger Models: Computable Theory of Feeding Control. New York, Academic Press, 1978.

Bray, G. A.: Endocrine factors in the control of food intake. *Fed. Proc., 33*:1140, 1974.

Bray, G. A., and Campfield, L. A.: Metabolic factors in the control of energy stores. *Metabolism, 24*:99, 1975.

Bray, G. A., and York, D. A.: Hypothalamic and genetic obesity in experimental animals: An autonomic and endocrine hypothesis. *Physiol. Rev., 59*:719, 1979.

Chaney, M. S., et al.: Nutrition. Boston, Houghton Mifflin, 1979.

Collipp, P. J. (ed.): Childhood Obesity. Littleton, Mass., PSG Publishing Co., 1979.

Crisp, A. H.: Anorexia Nervosa. New York, Academic Press, 1979.

Felig, P.: Starvation. *In* DeGroot, L. J., et al. (eds.): Endocrinology. Vol. 3. New York, Grune & Stratton, 1979, p. 1927.

Festing, M. F. W. (ed.): Animal Models of Obesity. New York, Oxford University Press, 1979.

Garrow, J. S.: Energy Balance and Obesity in Man. New York, Elsevier/North-Holland, 1978.

Havel, R. J.: Caloric homeostasis and disorders of fuel transport. *N. Engl. J. Med., 287*:1186, 1972.

Heird, W. C., and Winters, R. W.: Total parenteral nutrition. The state of the art. *J. Pediatr., 86*:2, 1975.

Hodges, R. E.: Nutrition in Medical Practice. Philadelphia, W. B. Saunders Co., 1979.

Hoebel, B. G.: Feeding: Neural control of intake. *Annu. Rev. Physiol., 33*:533, 1971.

Hunt, S. M., et al.: Nutrition: Principles and Clinical Practice. New York, John Wiley & Sons, 1980.

Jarrett, R. J. (ed.): Nutrition and Disease. Baltimore, University Park Press, 1978.

Keesey, R. E.: Neurophysiologic control of body fatness. *In* Lauer, R. M., and Shekelle, R. B. (eds.): Childhood Prevention of Atherosclerosis and Hypertension. New York, Raven Press, 1980.

LeBow, M. D.: Weight Control: The Behavioural Strategies. New York, John Wiley & Sons, 1980.

Lepkovsky, S.: Newer concepts in the regulation of food intake. *Am. J. Clin. Nutr., 26*:271, 1973.

Mancini, M., and London, B. L. (eds.): Medical Complications of Obesity. New York, Academic Press, 1979.

Olson, R. E. (ed.): Protein-Calorie Malnutrition. New York, Academic Press, 1975.

Oscai, L. B.: Recent progress in the possible prevention of obesity. *In* Lauer, R. M., and Shekelle, R. B. (eds.): Childhood Prevention of Atherosclerosis and Hypertension. New York, Raven Press, 1980.

Panksepp, J.: Hypothalamic regulation of energy balance and feeding behavior. *Fed. Proc., 33*:1150, 1974.

Salans, L. B.: Obesity and the adipose cell. *In* Bondy, P. K., and Rosenberg, L. E. (eds.): Metabolic Control and Disease, 8th Ed. Philadelphia, W. B. Saunders Co., 1980, p. 495.

Shils, M. E., and Goodhart, R. S. (eds.): Modern Nutrition in Health and Disease. Philadelphia, Lea & Febiger, 1979.

Shoden, R. J., and Griffin, W. S.: Fundamentals of Clinical Nutrition. New York, McGraw-Hill, 1980.

Sims, E. A. H.: The syndromes of obesity. *In* DeGroot, L. J., et al. (eds.): Endocrinology. Vol. 3. New York, Grune & Stratton, 1979, p. 1941.

Stamler, J.: The fat-modified diet: Its nature, effectiveness, and safety. *In* Lauer, R. M., and Shekelle, R. B. (eds.): Childhood Prevention of Atherosclerosis and Hypertension. New York, Raven Press, 1980.

Stern, J. S., and Greenwood, M. R. C.: A review of development of adipose cellularity in man and animals. *Fed. Proc., 33*:1952, 1974.

Suitor, C. W., and Hunter, M. F.: Nutrition; Principles and Application in Health Promotion. Philadelphia, J. B. Lippincott Co., 1980.

Thompson, C. I.: Controls of Eating. New York, Spectrum Publications, 1979.

Wurtman, R. J., and Wurtman, J. J. (eds.): Disorders of Eating. New York, Raven Press, 1979.

Young, V. R., and Scrimshaw, N. S.: The physiology of starvation. *Sci. Am. 225*:14, 1971.

74

Vitamin and Mineral Metabolism

The study of vitamin and mineral metabolism rightfully falls in the province of biochemistry. Therefore, the present discussion of this subject is greatly abbreviated. It is the purpose of this chapter only to present the major aspects of vitamin and mineral metabolism as they relate to the overall physiology of the body.

VITAMINS

A vitamin is an organic compound needed in small quantities for operation of normal bodily metabolism and that cannot be manufactured in the cells of the body. Probably hundreds of such substances exist, most of which have not been discovered. However, a few have been studied extensively because they are present in the foods in relatively small quantities, and, as a result, dietary deficiency of one or more of them often occurs. Therefore, from a clinical point of view the agents that are generally considered to be vitamins are those organic compounds that occur in the diet in small quantities and, when lacking, can cause specific metabolic deficits.

Daily Requirements of Vitamins. Table 74–1 lists the amounts of important vitamins required daily by the average adult male. Smaller persons require proportionately less quantities. These requirements vary considerably. For instance, the greater the person's size, the greater is the vitamin requirement. Second, growing persons usually require larger quantities of vitamins than do others. Third, when the person performs exercise, the vitamin requirements are increased. Fourth, during disease and fevers, the vitamin requirements are ordinarily

increased. Fifth, when larger than normal quantities of carbohydrates are metabolized, the requirements for thiamine and perhaps some of the other vitamins of the B complex are increased. Sixth, during pregnancy and lactation the requirement for vitamin D by the mother is greatly increased, and the requirement for vitamin D is considerable during the period of growth in children. Finally, a number of metabolic deficits occur pathologically in which the vitamins themselves cannot be utilized properly in the body; in such conditions the requirement for one or more specific vitamins may be extreme.

Storage of Vitamins in the Body. Vitamins are stored to a slight extent in all the cells. However, some vitamins are stored to a major extent in the liver. For instance, the quantity of vitamin A stored in the liver may be sufficient to maintain a person without any intake of vitamin A for up to ten months, and ordinarily the quantity of vitamin D stored in the liver is sufficient to maintain a person for two to four months without any additional intake of vitamin D.

The storage of vitamin K as well as of most water-soluble vitamins is relatively slight; this applies especially to the vitamin B compounds, for when a person's diet is deficient in vitamin B compounds, clinical symptoms of the deficiency can sometimes be recognized within a few days (except for vitamin B_{12} which can last in the liver for a year or longer). Absence of vitamin C, another water-soluble vitamin, can cause symptoms within a few weeks and can cause death from scurvy in 20 to 30 weeks.

VITAMIN A

Vitamin A occurs in animal tissues as *retinol*, the formula of which is illustrated. This vitamin does not occur in foods of vegetable origin, but *provitamins* for the formation of vitamin A do occur in abundance in many different vegetable foods. These are the yellow and red *carotenoid pigments* which, since they have chemical structures similar to that of vitamin A, can be changed into vitamin A in the liver.

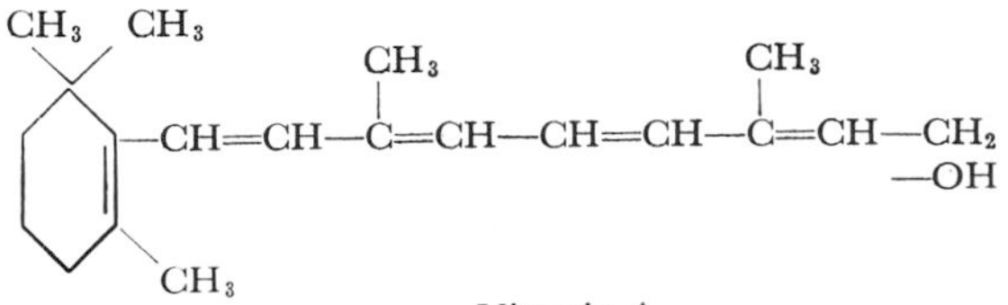

TABLE 74–1 REQUIRED DAILY AMOUNTS OF THE VITAMINS

A	5000 IU
Thiamine	1.5 mg.
Riboflavin	1.8 mg.
Niacin	20 mg.
Ascorbic acid	45 mg.
D	400 IU
E	15 IU
K	none
Folic acid	0.4 mg.
B_{12}	3 μg.
Pyridoxine	2 mg.
Pantothenic acid	unknown

IU, international units.

Vitamin A_1

907

The basic function of vitamin A in the metabolism of the body is not known except in relation to its use in the formation of retinal pigments, which was dicussed in Chapter 59. Nevertheless, some of the other physiological results of vitamin A lack have been well documented. In addition to the need for vitamin A to form the visual pigments and therefore to prevent night blindness, it is also necessary for normal growth of most cells of the body and especially for normal growth and proliferation of the different types of epithelial cells. When vitamin A is lacking, the epithelial structures of the body tend to become stratified and keratinized. Vitamin A deficiency manifests itself by (1) scaliness of the skin and sometimes acne, (2) failure of growth of young animals, including cessation of skeletal growth, (3) failure of reproduction, associated especially with atrophy of the germinal epithelium of the testes and sometimes with interruption of the female sexual cycle, and (4) keratinization of the cornea with resultant corneal opacity and blindness.

Also, the damaged epithelial structures often become infected, for example, in the eyes, the kidneys, or the respiratory passages. Therefore, vitamin A has been called an "anti-infection" vitamin. Vitamin A deficiency also frequently causes kidney stones, probably owing to infection in the renal pelvis.

THIAMINE (VITAMIN B₁)

Thiamine operates in the metabolic systems of the body principally as *thiamine pyrophosphate*; this compound functions as a *cocarboxylase,* operating mainly in conjunction with a protein decarboxylase for decarboxylation of pyruvic acid and other α-keto acids, as discussed in Chapter 67.

Thiamine chloride

Thiamine deficiency causes decreased utilization of pyruvic acid and some amino acids by the tissues but increased utilization of fats. Thus, thiamine is specifically needed for final metabolism of carbohydrates and many amino acids. Probably the decreased utilization of these nutrients is the responsible factor for the debilities associated with thiamine deficiency.

Thiamine Deficiency and the Nervous System. The central nervous system depends almost entirely on the metabolism of carbohydrates for its energy. In thiamine deficiency the utilization of glucose by nervous tissue may be decreased as much as 50 to 60 per cent. Therefore, it is readily understandable how thiamine deficiency could greatly impair function of the central nervous system. The neuronal cells of the central nervous system frequently show chromatolysis and swelling during thiamine deficiency, changes that are characteristic of neuronal cells with poor nutrition. Obviously, such changes as these can disrupt communication in many different portions of the central nervous system.

Also, thiamine deficiency can cause *degeneration of myelin sheaths* of nerve fibers both in the peripheral nerves and in the central nervous system. The lesions in the peripheral nerves frequently cause these nerves to become extremely irritable, resulting in "polyneuritis" characterized by pain radiating along the course of one or more peripheral nerves. Also, in severe thiamine deficiency, the peripheral nerve fibers and fiber tracts in the cord can degenerate to such an extent that *paralysis* occasionally results; and, even in the absence of paralysis, the muscles atrophy, with resultant severe weakness.

Thiamine Deficiency and the Cardiovascular System. Thiamine deficiency also weakens the heart muscle, so that a person with severe thiamine deficiency sometimes develops *cardiac failure*. In general, the right side of the heart becomes greatly enlarged in thiamine deficiency. Furthermore, the return of blood to the heart may be increased to as much as 3 times normal. This indicates that thiamine deficiency causes *peripheral vasodilation* throughout the circulatory system, possibly as a result of metabolic deficiency in the smooth muscle of the vascular system itself. Therefore, the cardiac effects of thiamine deficiency are due partly to excessive return of blood to the heart and partly to primary weakness of the cardiac muscle. *Peripheral edema* and *ascites* also occur to a major extent in some persons with thiamine deficiency because of the cardiac failure.

Thiamine Deficiency and the Gastrointestinal Tract. Among the gastrointestinal symptoms of thiamine deficiency are indigestion, severe constipation, anorexia, gastric atony, and hypochlorhydria. All these effects possibly result from failure of the smooth muscle and glands of the gastrointestinal tract to derive sufficient energy from carbohydrate metabolism.

The overall picture of thiamine deficiency, including polyneuritis, cardiovascular symptoms, and gastrointestinal disorders, is frequently referred to as "beriberi" — especially when the cardiovascular symptoms predominate.

NIACIN

Niacin, also called *nicotinic acid*, functions in the body as coenzymes in the forms of nicotinamide adenine dinucleotide (NAD) and nicotinamide adenine dinucleotide phosphate (NADP), which are also known as DPN and TPN. These coenzymes are hydrogen acceptors; they combine with hydrogen atoms as they are removed from food substrates by many different types of dehydrogenases. Typical operation of both of them is presented in Chapter 67. When a deficiency of niacin exists, the normal rate of dehydrogenation presumably cannot be maintained, and, therefore, oxidative delivery of energy from the foodstuffs to the functioning elements of the cells likewise cannot occur at normal rates.

Niacin

Because NAD and NADP operate in all cells of the body, it is readily understood how lack of niacin could cause multiple symptoms, even though specific metabolic reaction deficiencies have not yet been pinpointed. Clinically, niacin deficiency causes mainly gastrointestinal symptoms, neurologic symptoms, and a characteristic dermatitis. However, it is probably much more proper to say that essentially all functions of the body are depressed or altered.

In the early stages of niacin deficiency simple physiological changes, such as muscular weakness of all the different types of muscles and poor glandular secretion, may occur, but in severe niacin deficiency actual death of tissues ensues. Pathological lesions appear in many parts of the central nervous system, and permanent dementia or any of many different types of psychoses may result. Also, the skin develops a cracked, pigmented scaliness in areas that are exposed to mechanical irritation or sun irradiation; thus, it seems as though the skin were unable to repair the different types of irritative damage.

Niacin deficiency causes intense irritation and inflammation of the mucous membranes of the mouth and other portions of the gastrointestinal tract, thus instituting many digestive abnormalities, leading in severe cases to widespread gastrointestinal hemorrhage. It is possible that this results from generalized depression of metabolism in the gastrointestinal epithelium and failure of appropriate epithelial repair.

The clinical entity called *pellagra* and the canine disease called black tongue are caused mainly by niacin deficiency. Pellagra is greatly exacerbated in persons on a corn diet (such as many of the natives of Africa) because corn is very deficient in the amino acid tryptophan, which can be converted in limited quantities to niacin in the body.

RIBOFLAVIN (VITAMIN B₂)

Riboflavin normally combines in the tissues with phosphoric acid to form two coenzymes, *flavin mononucleotide (FMN)*, and *flavin adenine dinucleotide (FAD)*. These in turn operate as hydrogen carriers in several of the important oxidative systems of the body. Usually, NAD, operating in association with specific dehydrogenases, accepts hydrogen removed from various food substrates and then passes the hydrogen to FMN or FAD; finally, the hydrogen is released as an ion into the surrounding fluids to become oxidized by nascent oxygen, the system for which is described in Chapter 67.

Riboflavin

Deficiency of riboflavin in lower animals causes severe *dermatitis, vomiting, diarrhea, muscular spasticity* which finally becomes muscular weakness, and then *death* preceded by coma and decline in body temperature. Thus, severe riboflavin deficiency can cause many of the same effects as lack of niacin in the diet; presumably the debilities that result in each instance are due to generally depressed oxidative processes within the cells.

In the human being riboflavin deficiency has never been known to be severe enough to cause the marked debilities noted in animal experiments, but mild riboflavin deficiency is probably common. Such deficiency causes digestive disturbances, burning sensations of the skin and eyes, cracking at the corners of the mouth, headaches, mental depression, forgetfulness, and so on. Perhaps the most common characteristic lesion of riboflavin deficiency is *cheilosis*, which is inflammation and cracking at the angles of the mouth. In addition, a fine, scaly dermatitis often occurs at the angles of the nares, and keratitis of the cornea may occur with invasion of the cornea by small blood vessels.

Though the manifestations of riboflavin deficiency are usually relatively mild, this deficiency frequently occurs in association with deficiency of thiamine or niacin. Therefore, many deficiency syndromes, including pellagra, beriberi, sprue, and kwashiorkor, are probably due to a combined deficiency of a number of the vitamins, as well as to other aspects of malnutrition.

VITAMIN B₁₂

Several different *cobalamin* compounds which possess the common prosthetic group illustrated below exhibit so-called "vitamin B₁₂" activity.

Note that this prosthetic group contains cobalt that has coordination bonds similar to those found in relation to the iron of the hemoglobin molecule. It is likely that the cobalt atom functions in much the same way that the iron atom functions.

Vitamin B₁₂ performs several metabolic functions, acting as a hydrogen acceptor coenzyme. Its most important function is probably to act as a coenzyme for reducing ribonucleotides to deoxyribonucleotides, a step that is important in the formation of genes. This could explain the two major functions of vitamin B₁₂, (1) promotion of growth and (2) red blood cell maturation. This latter function was described in detail in Chapter 5 in relation to pernicious anemia, a type of anemia caused by failure of red blood cell maturation when vitamin B₁₂ is deficient.

A special effect of vitamin B₁₂ deficiency is often demyelination of the large nerve fibers of the spinal cord, especially of the posterior columns and occasionally of

Folic acid (pteroylglutamic acid)

the lateral columns. As a result, persons with pernicious anemia frequently have much loss of peripheral sensation and, in severe cases, even become paralyzed.

The usual cause of vitamin B_{12} deficiency is not lack of this vitamin in the food but instead deficiency of formation of *intrinsic factor*, which is normally secreted by the parietal cells of the gastric glands and is essential for absorption of vitamin B_{12} by the ileal mucosa. This was discussed in Chapter 5 and in Chapter 66.

FOLIC ACID (PTEROYLGLUTAMIC ACID)

Several different pteroylglutamic acids, one of which is illustrated above, exhibit the "folic acid effect." Folic acid functions as a carrier of hydroxymethyl and formyl groups. Perhaps its most important use in the body is in the synthesis of purines and thymine, which are required for formation of deoxyribonucleic acid. Therefore, folic acid is required for reproduction of the cellular genes. This perhaps explains one of the most important functions of the folic acid — that is, to promote growth. Indeed, when it is absent from the diet an animal will in fact grow very little.

Folic acid is an even more potent growth promoter than vitamin B_{12}, and, like vitamin B_{12}, is also important for the maturation of red blood cells, as discussed in Chapter 5. However, vitamin B_{12} and folic acid each perform specific and different functions in promoting growth and maturation of red blood cells. One of the very significant effects of folic acid deficiency is development of macrocytic anemia. This often occurs in sprue, which can often be treated effectively with folic acid alone.

PYRIDOXINE (VITAMIN B_6)

Pyridoxine exists in the form of *pyridoxal phosphate* in the cells and functions as a coenzyme for many different chemical reactions relating to amino acid and protein metabolism. Its most important role is that of coenzyme in transamination for the synthesis of amino acids. As a result, pyridoxine plays many key roles in metabolism — especially in protein metabolism. Also, it is believed to act in the transport of some amino acids across cell membranes.

Pyridoxine

Dietary lack of pyridoxine in lower animals can cause dermatitis, decreased rate of growth, development of fatty liver, anemia, and evidence of mental deterioration. In the human being, pyridoxine deficiency has been known to cause convulsions, dermatitis, and gastrointestinal disturbances such as nausea and vomiting in children. However, this deficiency is rare.

PANTOTHENIC ACID

Pantothenic acid mainly is incorporated in the body into *coenzyme A,* which has many metabolic roles in the cells. Two of these discussed at length in Chapters 67 and 68 are (1) conversion of decarboxylated pyruvic acid into acetyl Co-A prior to its entry into the tricarboxylic acid cycle, and (2) degradation of fatty acid molecules into multiple molecules of acetyl Co-A. Thus, lack of pantothenic acid can lead to depressed metabolism of both carbohydrates and fats.

Pantothenic acid

Deficiency of pantothenic acid in lower animals can cause retarded growth, failure of reproduction, graying of the hair, dermatitis, fatty liver, and hemorrhagic adrenal cortical necrosis. In the human being no definite deficiency syndrome has been proved, presumably because of the wide occurrence of this vitamin in almost all foods and because small amounts of the vitamin can probably be synthesized in the body. Nevertheless, this does not mean that pantothenic acid is not of value in the metabolic systems of the body; indeed, it is perhaps as necessary as any other vitamin.

ASCORBIC ACID (VITAMIN C)

Ascorbic acid is essential for many oxidation reactions in the body. For instance, oxidation of tyrosine and phenylalanine requires an adequate supply of ascorbic acid, and ascorbic acid plays a role in formation of hydroxyproline, an integral constituent of collagen, which in turn is essential for growth of subcutaneous tissue, cartilage, and bone.

Though the mechanisms are not clear, ascorbic acid enhances the removal of iron from cellular ferritin, thereby increasing the concentration of iron in the body fluids. Also, in some way not clarified, ascorbic acid potentiates the effects of folic acid in at least some metabolic processes.

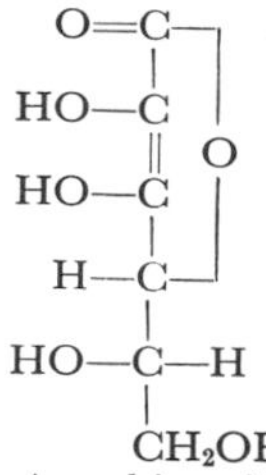

O=C ⌐
HO—C |
        ‖ O
HO—C |
H—C |
HO—C—H
CH₂OH
Ascorbic acid
(vitamin C)

Physiologically, the major function of ascorbic acid appears to be maintenance of normal intercellular substances throughout the body. This includes the formation of collagen, probably because of the action of ascorbic acid in synthesis of hydroxyproline. It also enhances the intercellular cement substance between the cells, the formation of bone matrix, and the formation of tooth dentin.

Deficiency of ascorbic acid for 20 to 30 weeks, as occurred frequently during long sailing voyages in olden days, causes *scurvy*, some effects of which are the following:

One of the most important effects of scurvy is *failure of wounds to heal*. This is caused by failure of the cells to deposit collagen fibrils and intercellular cement substances. As a result, healing of a wound may require several months instead of the several days ordinarily necessary.

Lack of ascorbic acid causes *cessation of bone growth*. The cells of the growing epiphyses continue to proliferate, but no new matrix is laid down between the cells, and the bones fracture easily at the point of growth because of failure to ossify. Also, when an already ossified bone fractures in a person with ascorbic acid deficiency, the osteoblasts cannot secrete a new matrix for the deposition of new bone. Consequently, the fractured bone does not heal.

The *blood vessel walls become extremely fragile* in scurvy, presumably because of failure of the endothelial cells to be cemented together properly and failure to form the collagen fibrils normally present in vessel walls. The capillaries especially are likely to rupture, and as a result many small petechial hemorrhages occur throughout the body. The hemorrhages beneath the skin cause purpuric blotches, sometimes over the entire body. To test for ascorbic acid deficiency, one can produce such petechial hemorrhages by inflating a blood pressure cuff over the upper arm; this occludes the venous return of blood, the capillary pressure rises, and red blotches occur in the skin immediately if there is a sufficiently severe ascorbic acid deficiency.

In extreme scurvy the muscle cells sometimes fragment; lesions of the gums with loosening of the teeth occur; infections of the mouth develop; vomiting of blood, bloody stools, and cerebral hemorrhage can all occur; and, finally, high fever often develops before death.

VITAMIN D

Vitamin D increases calcium absorption from the gastrointestinal tract and also helps to control calcium deposition in the bone. The mechanism by which vitamin D increases calcium absorption is to promote active transport of calcium through the epithelium of the ileum. It especially increases the formation of a calcium-binding protein in the epithepial cells that aids in calcium absorption. The specific functions of vitamin D in relation to overall body calcium metabolism and to bone formation are presented in Chapter 79.

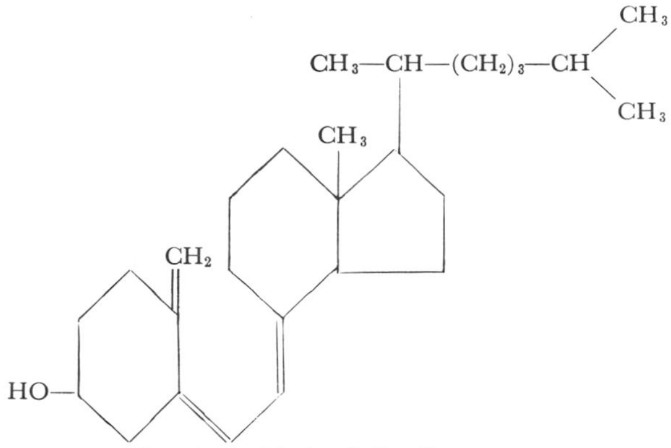

CH₃—CH—(CH₂)₃—CH⟨ CH₃ / CH₃
CH₃
CH₂
HO—
Vitamin D₃ (cholecalciferol)

Several vitamin D compounds exist, one of which is illustrated. This is the natural vitamin D, *cholecalciferol*, which results from ultraviolet irradiation of 7-dehydrocholesterol in the skin. A synthetic vitamin D compound, *ergocalciferol*, is formed by irradiation of ergosterol and is used to a major extent in vitamin D therapy. Both of these vitamin Ds are further altered, first in the liver and then in the kidneys, finally forming *1,25-dihydroxycholecalciferol*, which is the active form of the vitamin. This will be discussed in more detail in Chapter 79.

VITAMIN E

Several related compounds, one of which is illustrated below, exhibit so-called "vitamin E activity." Only rare instances of vitamin E deficiency have occurred in human beings. In lower animals, lack of vitamin E can cause degeneration of the germinal epithelium in the testis and therefore can cause male sterility. Lack of vitamin E can also cause resorption of a fetus after conception in the female. Because of these effects of vitamin E deficiency, vitamin E is sometimes called the "anti-sterility vitamin."

Vitamin E deficiency in animals can also cause paralysis of the hindquarters, and pathological changes occur in the muscles similar to those found in the disease entity muscular dystrophy of the human being. However, administration of vitamin E to patients with muscular dystrophy has not proved to be of any benefit.

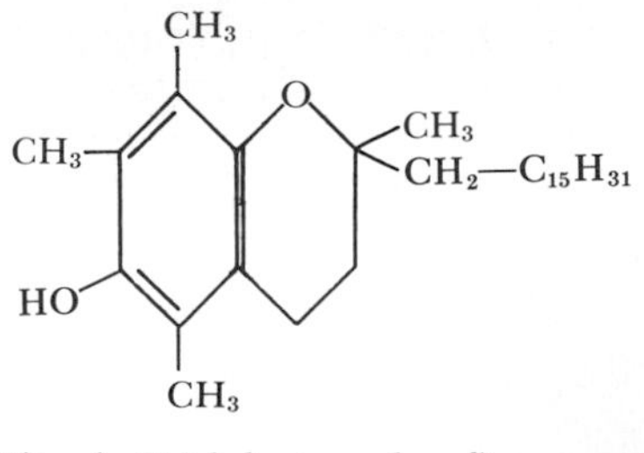

CH₃
CH₃—   O
             CH₃
             CH₂—C₁₅H₃₁
HO—
CH₃

Vitamin E (alpha-tocopherol)

Vitamin K₁ (2-methyl-3-phytyl-1,4-naphthoquinone)

Finally, as is true of almost all the vitamins, deficiency of vitamin E prevents normal growth, and it sometimes causes degeneration of the renal tubular cells.

Vitamin E is believed to function mainly in relation to unsaturated fatty acids, providing a protective role to prevent oxidation of the unsaturated fats. In the absence of vitamin E, the quantity of unsaturated fats in the cells becomes diminished, causing abnormal structure and function of such cellular organelles as the mitochondria, the lysosomes, and even the cell membrane. Indeed, the muscular dystrophy–like syndrome that occurs in the vitamin E deficiency probably results from continual rupture of lysosomes and subsequent autodigestion of the muscle.

VITAMIN K

Vitamin K is necessary for the formation by the liver of prothrombin, factor VII (proconvertin), factor IX, and factor X, all of which are important in blood coagulation. Therefore, when vitamin K deficiency occurs, blood clotting is retarded. The function of this vitamin and its relationships with some of the anticoagulants, such as Dicumarol, have been presented in greater detail in Chapter 9.

Several different compounds, both natural and synthetic, exhibit vitamin K activity. The chemical formula for one of the natural vitamin K compounds is illustrated above. Because vitamin K is synthesized by bacteria in the colon, a dietary source for this vitamin is not usually necessary, but, when the bacteria of the colon are destroyed by administration of large quantities of antibiotic drugs, vitamin K deficiency occurs rapidly because of the paucity of this compound in the normal diet. (See below.)

MINERAL METABOLISM

The functions of many of the minerals, such as sodium, potassium, chloride, and so forth, have been presented at appropriate points in the text. Therefore, only specific functions of minerals not covered elsewhere are mentioned here.

The body content of the most important minerals is listed in Table 74–2, and the daily requirements of these are given in Table 74–3.

Magnesium. Magnesium is approximately one-sixth as plentiful in cells as potassium, and it undoubtedly performs at least some of the same intracellular functions. But magnesium is also required as a catalyst for many intracellular enzymatic reactions, particularly those relating to carbohydrate metabolism.

The extracellular magnesium concentration is slight, only 1.8 to 2.5 mEq./liter. Increased extracellular concentration of magnesium depresses activity in the nervous system and also depresses skeletal muscle contraction. This latter effect can be blocked by administration of calcium. Low magnesium concentration causes increased irritability of the nervous system, peripheral vasodilatation, and cardiac arrhythmias.

Calcium. Calcium is present in the body mainly in the form of calcium phosphate in the bone. This subject will be discussed in detail in Chapter 79, as is also the calcium content of the extracellular fluid.

Excess quantities of calcium ion in the extracellular fluids can cause the heart to stop in systole and can act as a mental depressant. On the other hand, low levels of calcium can cause spontaneous discharge of nerve fibers, resulting in tetany. This, too, will be discussed in Chapter 79.

Phosphorus. Phosphate is the major anion of intracellular fluids. Phosphates have the ability to combine reversibly with a multitude of coenzyme systems and also with a multitude of other compounds that are necessary for operation of the metabolic processes. Many important reactions of phosphates have been catalogued at other points in this text, especially in relation to the functions of ATP, ADP, creatine phosphate, and so forth. Phosphates are perhaps the single most important mineral constituent required for cellular activity. Also, bone contains a tremendous amount of calcium phosphate, which will be discussed in Chapter 79.

TABLE 74–2　CONTENT IN GRAMS OF A 70 KILOGRAM ADULT MAN

Water	41,400	Mg	21
Fat	12,600	Cl	85
Protein	12,600	P	670
Carbohydrate	300	S	112
Na	63	Fe	3
K	150	I	0.014
Ca	1,160		

TABLE 74–3　REQUIRED DAILY AMOUNTS OF MINERALS

Na	3.0 grams	I	150.0 μg.
K	1.0 gram	Mg	0.4 gram
Cl	3.5 grams	Co	unknown
Ca	1.2 grams	Cu	unknown
P	1.2 grams	Mn	unknown
Fe	18.0 mg.	Zn	15 mg.

Iron. The function of iron in the body, especially in relation to the formation of hemoglobin, was discussed in Chapter 5. The major proportion of iron in the body is in the form of hemoglobin, though smaller quantities are present in other forms, especially in the liver and in the bone marrow. Electron carriers containing iron (especially the cytochromes) are present in all the cells of the body and are essential for most of the oxidation that occurs in the cells. Therefore, iron is absolutely essential both for transport of oxygen to the tissues and for maintenance of oxidative systems within the tissue cells, without which life would cease within a few seconds.

Important Trace Elements in the Body. A few elements are present in the body in such small quantities that they are called *trace elements*. Usually, the amounts of these in the foods are also minute. Yet without any one of them a specific deficiency syndrome is likely to develop.

Iodine. The best known of the trace elements is iodine. This element is discussed in Chapter 76 in connection with thyroid hormone; as illustrated in Table 74–2, the entire body contains an average of only 14 milligrams. Iodine is essential for the formation of *thyroxine*, which in turn is essential for maintenance of normal metabolic rates in all the cells.

Copper. Copper deficiency has been observed on rare occasions in infants receiving only a milk diet. The effect of the deficiency is a microcytic, normochromic anemia caused by failure of the intestines to absorb adequate quantities of iron. Lack of copper also decreases the quantities of certain copper-containing enzymes in the tissue cells; especially important are cytochrome oxidase and cytochrome c.

Zinc. Zinc is an integral part of many enzymes, one of the most important of which is *carbonic anhydrase*, present in especially high concentration in the red blood cells. This enzyme is responsible for rapid combination of carbon dioxide with water in the red blood cells of the peripheral capillary blood and for rapid release of carbon dioxide from the pulmonary capillary blood into the alveoli. Carbonic anhydrase is also present to a major extent in the gastrointestinal mucosa, in the tubules of the kidney, and in the epithelial cells of many glands of the body. Consequently, zinc in small quantities is essential for the performance of many reactions relating to carbon dioxide metabolism.

Zinc is also a component of *lactic dehydrogenase* and, therefore, is important for the interconversions between pyruvic acid and lactic acid. Finally, zinc is a component part of some *peptidases* and therefore is important for digestion of proteins in the gastrointestinal tract.

Cobalt. Cobalt is an essential part of vitamin B_{12}, and vitamin B_{12} is essential for maturation of red blood cells, as discussed earlier in this chapter.

Excess cobalt in the diet causes the opposite of anemia; i.e., polycythemia. However, the polycythemic cells formed contain relatively small concentrations of hemoglobin, so that the total quantity of hemoglobin in the circulating blood remains about normal. Thus, it appears that cobalt is concerned principally with formation of the red blood cell structure and not with formation of hemoglobin.

Manganese. Lack of manganese in the diet of animals causes testicular atrophy, though the cause of this is not known. Also, manganese is necessary in the body to activate arginase, which is the principal enzyme necessary for the formation of urea. Consequently, lack of manganese in the diet might prevent the conversion of ammonium ions into urea, and excessive quantites of ammonium compounds developing in the body fluids could cause toxicity. Finally, manganese activates many of the other metabolic enzymes, including cholinesterase, muscle ATPase, and enzymes required for glycolysis.

Fluorine. Fluorine does not seem to be a necessary element for metabolism, but the presence of a small quantity of fluorine in the body during the period of life when the teeth are being formed subsequently protects against carious teeth. Fluorine does not make the teeth themselves stronger but, instead, has some yet unknown effect in suppressing the cariogenic process. It has been suggested that fluorine in the teeth combines with various trace metals that are necessary for activation of the bacterial enzymes, and, because the enzymes are deprived of these trace metals, they remain inactive and cause no caries.

Excessive intake of fluorine causes *fluorosis*, which is manifest in its mild state by mottled teeth and in a more severe state by enlarged bones. It has been postulated that in this condition fluorine combines with trace metals in some metabolic enzymes, including the phosphatases, so that various metabolic systems become partially inactivated. According to this theory, the mottled teeth and enlarged bones are due to abnormal enzyme systems in the odontoblasts and osteoblasts. Even though mottled teeth are highly resistant to the development of caries, the structural strength of these teeth is considerably lessened by the mottling process.

REFERENCES

Alfin-Slater, R. B., and Kritchevsky, D. (eds.): Human Nutrition: A Comprehensive Treatise. New York, Plenum Press, 1979.

Alfin-Slater, R. B., and Kritchevsky, D. (eds.): Nutrition and the Adult: Micronutrients. New York, Plenum Press, 1979.

Benowicz, R. J.: Vitamins and You. New York, Grosset & Dunlap, 1979.

Bourne, G. H. (ed.): Some Aspects of Human and Veterinary Nutrition. New York, S. Karger, 1978.

Christensen, S.: The biological fate of riboflavin in mammals. A survey of literature and own investigations. *Acta Pharmacol. Toxicol. (Kbh) (Suppl.)*, 32:3, 1973.

Coughlan, M. P. (ed.): Molybdenum and Molybdenum-Containing Enzymes. New York, Pergamon Press, 1980.

Cudlipp, E.: Vitamins. New York, Grosset & Dunlap, 1978.

Davies, I. J. T.: The Clinical Significance of the Essential Biological Metals. Springfield, Ill., Charles C Thomas, 1972.

Deluca, H. F.: Metabolism and function of vitamin D. *In* Greep, R. O., and Astwood, G. D. (eds.): Handbook of Physiology. Sec. 7, Vol. 7. Baltimore, Willams & Wilkins, 1976, p. 265.

DeLuca, H. F. (ed.): The Fat-Soluble Vitamins. New York, Plenum Press, 1978.

DeLuca, H. F., and Holick, M. F.: Vitamin D: Biosynthesis, metabolism, and mode of action. *In* DeGroot, L. J., *et al.* (eds.): Endocrinology. Vol. 2. New York, Grune & Stratton, 1979, p. 653.

Fernstrom, J. D., and Wurtman, R. J.: Nutrition and the brain. *Sci. Am.*, 230:84, 1974.

Fraser, D., and Scriver, C. R.: Disorders associated with hereditary or acquired abnormalities in vitamin D function: Hereditary disorders associated with vitamin D resistance or defective phosphate metabolism. *In* DeGroot, L. J., *et al.* (eds.): Endocrinology. Vol. 2. New York, Grune & Stratton, 1979, p. 797.

Frieden, E.: The biochemistry of copper. *Sci. Am.*, 218(5):102, 1968.

Frieden, E.: The chemical elements of life. *Sci. Am.*, 227:52, 1972.

Gardner, L. I., and Amacher, P. (eds.): Endocrine Aspects of Malnutrition: Marasmus, Kwashiorkor, and Psychosocial Deprivation. New York, Raven Press, 1973.

Hamilton, E. I.: The Chemical Elements and Man: Measurements — Perspectives — Applications. Springfield, Ill., Charles C Thomas, 1978.

Hanck, A., and Ritzel, G. (eds.): Re-Evaluation of Vitamin C. Bern, H. Huber, 1977.

Hodges, R. E.: Nutrition in Medical Practice. Philadelphia, W. B. Saunders Co., 1979.

Holwerda, R. A., et al.: Electron transfer reactions of copper proteins. Annu. Rev. Biophys. Bioeng., 5:363, 1976.

Hunt, S. M., et al.: Nutrition: Principles and Clinical Practice. New York, John Wiley & Sons, 1980.

Karcioglu, Z. A., and Sarper, R. M. (eds.): Zinc and Copper in Medicine. Springfield, Ill., Charles C Thomas, 1980.

Kharasch, N. (ed.): Trace Metals in Health and Disease. New York, Raven Press, 1979.

Lawson, D. E. M.: Vitamin D. New York, Academic Press, 1978.

Pories, W. J. et al. (eds.): Clinical Applications of Zinc Metabolism. Springfield, Ill., Charles C Thomas, 1974.

Prasad, A. S.: Zinc in Human Nutrition. Boca Raton, Fla., CRC Press, 1979.

Shoden, R. J., and Griffin, W. S.: Fundamentals of Clinical Nutrition. New York, McGraw-Hill, 1980.

Suitor, C. W., and Hunter, M. F.: Nutrition; Principles and Application in Health Promotion. Philadelphia, J. B. Lippincott Co., 1980.

Suttie, J. W. (ed.): Vitamin K Metabolism and Vitamin K–Dependent Proteins. Baltimore, University Park Press, 1979.

Vitamin Compendium: The Properties of the Vitamins and Their Importance in Human and Animal Nutrition. Basle, F. Hoffmann–La Roche, Vitamins and Chemicals Department, 1976.

Williams, R. J.: Physician's Handbook of Nutrition. Springfield, Ill., Charles C Thomas, 1974.

Zagalak, B., and Friedrich, W. (eds.): Vitamin B12. New York, Walter de Gruyter, 1979.

Part XIII
ENDOCRINOLOGY AND REPRODUCTION

75

Introduction to Endocrinology; and the Pituitary Hormones

The functions of the body are regulated by two major control systems: (1) the nervous system, which has been discussed, and (2) the hormonal, or endocrine, system. In general, the hormonal system is concerned principally with control of the different metabolic functions of the body such as controlling the rates of chemical reactions in the cells or the transport of substances through cell membranes or other aspects of cellular metabolism like growth and secretion. Some hormonal effects occur in seconds, while others require several days simply to start and then continue for weeks, months, or even years.

Many interrelationships exist between the hormonal and nervous systems. For instance, at least two glands secrete their hormones only in response to appropriate nerve stimuli, the *adrenal medullae* and the *posterior pituitary gland,* and few of the adenohypophyseal hormones are secreted to a significant extent except in response to nervous activity in the hypothalamus, as is detailed later in this chapter.

NATURE OF A HORMONE

A hormone is a chemical substance that is secreted into the body fluids by one cell or a group of cells and that exerts a physiological *control* effect on other cells of the body.

At many points in this text we have already discussed different hormones. Some are *local hormones* and others are *general hormones.* Examples of local hormones are *acetylcholine* released at the parasympathetic and skeletal nerve endings, *secretin* released by the duodenal wall and transported in the blood to the pancreas to cause a watery pancreatic secretion, *cholecystokinin* released in the small intestine and transported to the gallbladder to cause contraction and to the pancreas to cause enzyme secretion, and many others. These hormones obviously have specific local effects, whence comes the name local hormones.

The general hormones are secreted by specific *endocrine glands* and are transported in the blood to cause physiologic actions at distant points in the body. A few of the general hormones affect all, or almost all, cells of the body; examples are *growth hormone* from the adenohypophysis and *thyroid hormone* from the thyroid gland. Other general hormones, however, affect only specific tissues. For instance, *adrenocorticotropin* from the anterior pituitary gland specifically stimulates the adrenal cortex, and the *ovarian hormones* have specific effects on the sex organs. The tissues affected specifically in this way are called *target tissues.* Many examples of target tissues will become apparent in the following chapters on endocrinology.

The following general hormones have proved to be of major significance and are discussed in detail in this and the following chapters:

Anterior pituitary hormones: *growth hormone, adrenocorticotropin, thyroid-stimulating hormone, follicle-stimulating hormone, luteinizing hormone, prolactin,* and *melanocyte-stimulating hormone.*
Posterior pituitary hormones: *antidiuretic hormone (vasopressin)* and *oxytocin.*
Adrenocortical hormones: especially *cortisol* and *aldosterone.*
Thyroid hormones: *thyroxine, triiodothyronine,* and *calcitonin.*
Pancreatic hormones: *insulin* and *glucagon.*
Ovarian hormones: *estrogens* and *progesterone.*
Testicular hormone: *testosterone.*
Parathyroid hormone: *parathormone.*
Placental hormones: *human chorionic gonadotropin, estrogens, progesterone,* and *human somatomammotropin.*

Negative Feedback in the Control of Hormonal Secretion. At many points both in this chapter and in the succeeding few chapters we will see that as a hormone accomplishes its physiological function, its rate of secretion is prevented from increasing further and at times is even decreased. This is caused by negative feedback, a phenomenon we have seen to be important in many nervous control systems as well. In general, each gland has a basic tendency to *oversecrete* its particular hormone, but, once the normal physiological effect of the hormone has been achieved, information is transferred either directly or indirectly back to the producing gland to inhibit further secretion. On the other hand, if the gland undersecretes, the physiological effects of the hormone diminish, and the feedback decreases, thus allowing the gland to begin secreting adequate quantities of the hormone once again. In this way, the rate of secretion of each hormone is controlled in accord with the need for the hormone. The specific negative feedback mechanisms are discussed in relation to the different individual hormones.

Chemistry of the Hormones. Chemically, the basic types of hormones are (1) proteins or derivatives of proteins or amino acids, and (2) steroid hormones. For example, the hormones of the pancreas and anterior pituitary are proteins, the hormones of the posterior pituitary are peptides, and those of the thyroid and the adrenal medulla are derivatives of amino acids. The steroids are secreted by the glands derived from the mesenchymal zone of the embryo, including the adrenal cortex, the ovary, and the testis.

Measurement of Hormone Concentrations

Most hormones are present in the circulating body fluids and tissues in extremely minute quantities, some in concentrations as low as one-millionth of a milligram (one picogram) per milliliter. Therefore, except in a few instances, it has been almost impossible to measure these concentrations by usual chemical means. Two important methods have been employed for this purpose: (1) bioassay and (2) radioactive competitive binding methods.

Bioassay. Bioassay means use of an appropriate animal preparation in which one can test the action of the hormone. For instance, an appropriate bioassay for antidiuretic hormone is to measure the degree of water conservation caused by injecting plasma or a concentrated extract of the plasma from an experimental animal or human being into a test animal that does not itself secrete antidiuretic hormone and to compare the animal's response with the response to a known quantity of pure antidiuretic hormone. In a similar manner, bioassay for growth hormone is based on stimulation of growth, usually of rats. Bioassays for gonadotropic hormones are based on their effects on the ovaries or other gonadotropic target tissues. By appropriate titration, a reasonable degree of accuracy can be achieved for most hormones but not for all, because appropriate animal models have not been achieved for all hormones.

Competitive Binding Assays — Radioimmunoassay. For measuring extremely low concentrations of hormones, a substance that specifically binds with the hormone is first found. For instance, antibodies can usually be developed that will bind specifically with a given hormone. Then a mixture is made of three different elements: (1) a fluid from the animal to be assayed, (2) the antibody, and (3) an approximate equivalent amount of purified hormone of the type to be measured but that has been tagged with a radioactive isotope. However, one specific condition must be met: There must be too little antibody for both of the hormones from the two separate sources to combine completely. Therefore, the natural hormone and the radioactive hormone *compete for the binding sites* on the antibody; the quantity of each of the two hormones, the natural and the radioactive, that will bind is proportional to its concentration. After binding is complete, the antibody-hormone complex is separated from the remainder of the solution, and the quantity of radioactive hormone that has bound with the antibody is measured by means of radioactive counting techniques. If a large amount of radioactive hormone has bound, then it is clear that there was only a small amount of natural hormone to compete. Conversely, if only a small amount of radioactive hormone has bound, it is clear that there was a very large amount of natural hormone to compete for the binding sites. Thus, by the use of an appropriate calibration curve, a very precise measurement of the quantity of the natural hormone in the test fluid can be achieved. As little as a fraction of a picogram (one-trillionth of a gram) of vasopressin per milliliter of assay fluid has been measured in this way.

Several other competitive binding techniques for assay of minute quantities of hormones have also been employed. One of these is to use — in place of the antibody — the specific carrier globulins of plasma that are natural binding agents for some hormones. The carrier globulin is substituted for the antibody in the assay process, and then the assay is carried out in exactly the same way as the radioimmunoassay procedure. This technique is used mainly for assay of cortisol and thyroxine.

MECHANISMS OF HORMONAL ACTION

The function of the different hormones is to *control* the activity levels of target tissues. To provide this control function they may alter the chemical reactions within the cells, alter the permeability of the cell membrane to specific substances, or activate some other specific cellular mechanism. The different hormones achieve these effects in many different ways. However, two important general mechanisms by which many of the hormones function are: (1) activation of the cyclic AMP system of cells, which in turn elicits the specific cellular functions, or (2) activation of the genes of the cells which causes the formation of intracellular proteins that initiate specific cellular functions. These mechanisms are described as follows:

INTRACELLULAR HORMONAL MEDIATORS – CYCLIC AMP

Many hormones exert their effects on cells by first causing the substance *cyclic 3′,5′-adenosine monophosphate* (cyclic AMP) to be formed in the cell. Once formed, the cyclic AMP causes the hormonal effects inside the cell. Thus, *cyclic AMP is an intracellular hormonal mediator*. It is also frequently called the *second messenger* for hormone mediation — the "first messenger" being the original stimulating hormone.

The cyclic AMP mechanism has been shown to be the way in which all of the following hormones (and many more) stimulate their target tissues:

1. Adrenocorticotropin
2. Thyroid-stimulating hormone
3. Luteinizing hormone
4. Follicle-stimulating hormone
5. Vasopressin
6. Parathyroid hormone
7. Glucagon
8. Catecholamines
9. Secretin
10. The hypothalamic releasing factors

Figure 75–1 illustrates the function of the cyclic AMP mechanism in more detail. It is believed that the stimulating hormone combines with a specific "receptor" for that hormone on the surface of the target cell. The specificity of the receptor determines which hormone will affect the target cell.

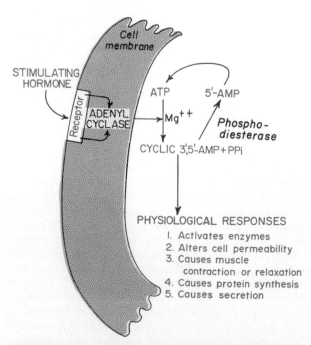

Figure 75–1. The cyclic AMP mechanism by which many hormones exert their control of cell function.

After binding with the receptor, the combination of hormone and receptor activates the enzyme *adenyl cyclase* in the membrane, and the portion of the adenyl cyclase that is exposed to the cytoplasm causes immediate conversion of cytoplasmic ATP into cyclic AMP. The cyclic AMP then initiates any number of cellular functions before it itself is destroyed — functions such as activating enzymes in the cell, altering the cell permeability, initiating synthesis of specific intracellular proteins, causing muscle contraction or relaxation, initiating secretion, and many other possible effects. The types of effects that will occur inside the cell are determined by the character of the cell itself. Thus, a thyroid cell stimulated by cyclic AMP forms thyroid hormones, whereas an adrenocortical cell forms adrenocortical hormones. On the other hand, cyclic AMP affects epithelial cells of the renal tubules by increasing their permeability to water.

Other Intracellular Hormonal Mediators. Another intracellular mediator (or "second messenger") is *cyclic guanosine monophosphate,* which is a nucleotide almost exactly like cyclic AMP except that it contains the base guanine rather than adenine. It and cyclic AMP are activated similarly. However, it usually promotes intracellular reactions that are different from those promoted by cyclic AMP. This allows two separate second messenger control mechanisms to operate simultaneously in the same cell but functioning separately from each other.

Still other types of intracellular hormonal mediators probably also exist. One of these likely is the *prostaglandins,* which are a series of lipid compounds closely related to each other and widely present in cells throughout the body. Hundreds of different cellular control functions have been postulated for the different prostaglandins. However, most of these are local rather than subserving a second messenger function stimulated by one of the general hormones.

ACTION OF STEROID HORMONES ON THE GENES TO CAUSE PROTEIN SYNTHESIS

A second major means by which hormones — specifically the steroid hormones secreted by the adrenal cortex, the ovaries, and the testes — act is to cause synthesis of proteins in the target cells; these proteins then function as enzymes or carrier proteins that in turn activate other functions of the cells.

The sequence of events in steroid function is the following:

1. The steroid hormone enters the cytoplasm of the cell, where it binds with a specific *receptor protein*.

2. The combined receptor protein/hormone

then diffuses into or is transported into the nucleus.

3. Somewhere along this route the receptor protein is altered to form a smaller molecular weight protein, or the steroid hormone is transferred to a second smaller protein.

4. The combination of the small protein and hormone is now the active factor that activates specific genes to form messenger RNA.

5. The messenger RNA diffuses into the cytoplasm where it promotes the translation process at the ribosomes to form new proteins.

To give an example, aldosterone, one of the hormones secreted by the adrenal cortex, enters the cytoplasm of renal tubular cells, which contain its specific receptor protein. Therefore, in these cells the above sequence of events ensues. After about 45 minutes, proteins begin to appear in the renal tubular cells that promote sodium reabsorption from the tubules and potassium secretion into the tubules. Thus, there is a characteristic delay in the beginning action of the steroid hormone of 45 minutes and up to several hours or even days for full action, which is in marked contrast to the almost instantaneous action of some of the peptide and peptide-derived hormones.

OTHER MECHANISMS OF HORMONE FUNCTION

Hormones can have other direct effects on cells, though in most instances the precise mechanisms of these effects are not known. For instance, insulin increases the permeability of the cells to glucose, and growth hormone increases the transport of amino acids into cells. Also, growth hormone stimulates synthesis of proteins in several ways in addition to its effect of increasing intracellular amino acids. In addition, several hormones, such as the catecholamines and acetylcholine, directly affect cell membranes by changing their permeabilities to ions and thereby exciting muscular contraction or causing other effects.

THE PITUITARY GLAND AND ITS RELATIONSHIP TO THE HYPOTHALAMUS

The *pituitary gland* (Fig. 75–2), also called the *hypophysis*, is a small gland — about 1 cm. in diameter and 0.5 to 1 gram in weight — that lies in the *sella turcica* at the base of the brain and is connected with the hypothalamus by the *pituitary* (or *hypophysial*) *stalk*. Physiologically, the pituitary gland is divisible into two distinct portions: the *anterior pituitary*, also known as the *adenohypophysis*, and the *posterior pituitary*, also known as the *neurohypophysis*. Between these is a

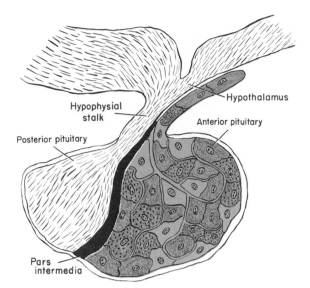

Figure 75–2. The pituitary gland.

small, relatively avascular zone called the *pars intermedia*, which is almost absent in the human being while much larger and much more functional in some lower animals.

Embryologically, the two portions of the pituitary originate from different sources, the anterior pituitary from *Rathke's pouch*, which is an embryonic invagination of the pharyngeal epithelium, and the posterior pituitary from an outgrowth of the hypothalamus. The origin of the anterior pituitary from the pharyngeal epithelium explains the epithelioid nature of its cells, while the origin of the posterior pituitary from neural tissue explains the presence of large numbers of glial-type cells in this gland.

Six very important hormones plus several less important ones are secreted by the *anterior* pituitary, and two important hormones are secreted by the *posterior* pituitary. The hormones of the anterior pituitary play major roles in the control of metabolic functions throughout the body, as shown in Figure 75–3. (1) *Growth hormone* promotes growth of the animal by affecting many metabolic functions throughout the body, especially protein formation. (2) *Adrenocorticotropin* controls the secretion of some of the adrenocortical hormones, which in turn affect the metabolism of glucose, proteins, and fats. (3) *Thyroid-stimulating hormone* controls the rate of secretion of thyroxine by the thyroid gland, and thyroxine in turn controls the rates of most chemical reactions of the entire body. (4) *Prolactin* promotes mammary gland development and milk production. And two separate gonadotropic hormones, (5) *follicle-stimulating hormone* and (6) *luteinizing hormone,* control growth of the gonads as well as their reproductive activities.

The two hormones secreted by the posterior pituitary play other roles. (1) *Antidiuretic hor-*

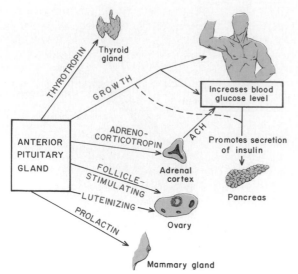

Figure 75–3. Metabolic functions of the anterior pituitary hormones.

mone (also called *vasopressin*) controls the rate of water excretion into the urine and in this way helps to control the concentration of water in the body fluids. (2) *Oxytocin* (a) helps to deliver milk from the glands of the breast to the nipples during suckling, and (b) probably helps in the delivery of the baby at the end of gestation.

CONTROL OF PITUITARY SECRETION BY THE HYPOTHALAMUS

Almost all secretion by the pituitary is controlled by either hormonal or nervous signals from the hypothalamus. Indeed, when the pituitary gland is removed from its normal position beneath the hypothalamus and transplanted to some other part of the body, its rates of secretion of the different hormones (except for prolactin) fall to low levels — in the case of some of the hormones, almost to zero.

Secretion from the posterior pituitary is controlled by nerve fibers originating in the hypothalamus and terminating in the posterior pituitary. In contrast, secretion by the anterior pituitary is controlled by hormones called *hypothalamic releasing* and *inhibitory hormones* (or *factors*) secreted within the hypothalamus itself and then conducted to the anterior pituitary through minute blood vessels called *hypothalamic-hypophysial portal vessels*. In the anterior pituitary these releasing and inhibitory hormones act on the glandular cells to control their secretion. This system of control will be discussed in detail later in the chapter.

The hypothalamus receives signals from almost all possible sources in the nervous system. Thus, when a person is exposed to pain, a portion of the pain signal is transmitted into the hypothalamus. Likewise, when a person experiences some powerful depressing or exciting thought, a portion of the signal is transmitted into the hypothalamus. Olfactory stimuli denoting pleasant or unpleasant smells transmit strong signal components directly and through the amygdaloid nuclei into the hypothalamus. *Even the concentrations of nutrients, electrolytes, water, and various hormones* in the blood excite or inhibit various portions of the hypothalamus. Thus, the hypothalamus is a collecting center for information concerned with the well-being of the body, and in turn much of this information is used to control secretions of the many globally important pituitary hormones.

THE ANTERIOR PITUITARY GLAND AND ITS REGULATION BY HYPOTHALAMIC RELEASING FACTORS

CELL TYPES OF THE ANTERIOR PITUITARY

The anterior pituitary gland is composed of several different types of cells. In general, there is one type of cell for each type of hormone that is formed in this gland; with special staining techniques these various cell types can be differentiated from one another. The only major exception to this is that the same cell type seems to secrete both luteinizing hormone and follicle-stimulating hormone.

However, through the use of usual acid-base histological stains, the cell types can be separated into only three types, commonly known as (1) *acidophils*, which stain strongly with acidic dyes; (2) *basophils*, which stain strongly with basic dyes; and (3) *chromophobes*, which do not stain with either. The *acidophils* produce *growth hormone* and *prolactin;* the *basophils* produce *luteinizing hormone, follicle-stimulating hormone, adrenocorticotropin,* and *thyroid-stimulating hormone;* and the *chromophobes* are believed to be cells mainly in nonsecreting stages of development.

THE HYPOTHALAMIC-HYPOPHYSIAL PORTAL SYSTEM

The anterior pituitary is a highly vascular gland with extensive capillary sinuses among the glandular cells. Almost all the blood that enters these sinuses passes first through a capillary bed in the tissue of the lower tip of the hypothalamus and then through small *hypothalamic-hypophysial portal vessels* into the anterior pituitary sinuses. Thus, Figure 75–4 illustrates a small artery supplying the lowermost portion of the hypothalamus called the *median eminence*. Small vascular tufts project into the substance of the median eminence and then return to its surface, coalescing to form the hypothalamic-hypophysial portal vessels.

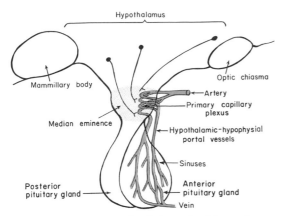

Figure 75-4. The hypothalamic-hypophysial portal system.

These in turn pass downward along the pituitary stalk to supply the anterior pituitary sinuses.

Secretion of Hypothalamic Releasing and Inhibitory Hormones into the Median Eminence. Special neurons in the hypothalamus synthesize and secrete hormones called *hypothalamic releasing* and *inhibitory hormones (or releasing* and *inhibitory factors)* that control the secretion of the anterior pituitary hormones. These neurons originate in various parts of the hypothalamus and send their nerve fibers into the median eminence and the tuber cinereum, the hypothalamic tissue that extends into the pituitary stalk. The endings of these fibers are different from most endings in the central nervous system in that their function is not to transmit signals from one neuron to another but merely to secrete the hypothalamic releasing and inhibitory hormones (factors) into the tissue fluids. These hormones are immediately absorbed into the capillaries of the hypothalamic-hypophysial portal system and carried directly to the sinuses of the anterior pituitary gland.

Function of the Releasing and Inhibitory Hormones. The function of the releasing and inhibitory hormones is to control the secretion of the anterior pituitary hormones. For each type of anterior pituitary hormone there is a corresponding hypothalamic releasing hormone; for some of the anterior pituitary hormones there is also a corresponding hypothalamic inhibitory factor. For most of the anterior pituitary hormones it is the releasing hormone that is important; but, for prolactin, an inhibitory hormone probably exerts most control. The hypothalamic releasing and inhibitory hormones that are of major importance are:

1. *Thyroid-stimulating hormone releasing hormone* (TRH), which causes release of thyroid-stimulating hormone

2. *Corticotropin releasing hormone* (CRH), which causes release of adrenocorticotropin

3. *Growth hormone releasing hormone* (GHRH), which causes release of growth hormone, and *growth hormone inhibitory hormone* (GHIH), which is the same as the hormone *somatostatin* and which inhibits the release of growth hormone

4. *Luteinizing hormone releasing hormone* (LRH), which causes release of both luteinizing hormone and follicle-stimulating hormone — this hormone is also called *gonadotropin releasing hormone* (GHRH)

5. *Prolactin inhibitory hormone* (PIH), which causes inhibition of prolactin secretion

In addition to these more important hypothalamic hormones, still another excites the secretion of prolactin and several hypothalamic inhibitory hormones inhibit some of the other anterior pituitary hormones. Each of the more important hypothalamic hormones will be discussed in detail at the time that the specific hormonal system controlled by them is presented in this and subsequent chapters.

Specific Areas in the Hypothalamus That Control Secretion of Specific Hypothalamic Releasing and Inhibitory Factors. It is believed that all or most of the hypothalamic hormones are secreted at nerve endings in the median eminence before being transported to the anterior pituitary gland. Electrical stimulation of this region excites these nerve endings and therefore causes release of essentially all the hypothalamic factors. However, the neuronal cell bodies that give rise to these median eminence nerve endings are located in other discrete areas of the hypothalamus or in closely related areas of the basal brain. Unfortunately, the specific loci of the neurons that secrete the different hypothalamic hormones are so incompletely known that it would be misleading to attempt a delineation here.

PHYSIOLOGICAL FUNCTIONS OF THE ANTERIOR PITUITARY HORMONES

All the major anterior pituitary hormones besides growth hormone exert their effects by stimulating target glands — the thyroid gland, the adrenal cortex, the ovaries, the testicles, and the mammary glands. The functions of each of these pituitary hormones are so intimately concerned with the functions of the respective target glands that except for growth hormone, their functions will be discussed in subsequent chapters along with the functions of the target glands. Growth hormone, in contrast to other hormones, does not function through a target gland but instead exerts an effect on all or almost all tissues of the body.

GROWTH HORMONE

Growth hormone (GH), also called *somatotrophic hormone* (SH) or *somatotropin*, is a small protein molecule containing 191 amino acids in a

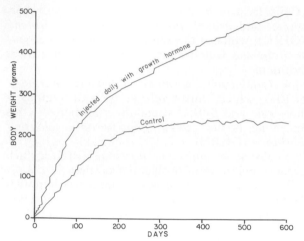

Figure 75–5. Comparison of weight gain of a rat injected daily with growth hormone with that of a normal rat of the same litter.

single chain and having a molecular weight of 22,005. It causes growth of all tissues of the body that are capable of growing. It promotes both increased sizes of the cells and increased mitosis with development of increased numbers of cells.

Figure 75–5 illustrates weight charts of two growing rats, one of which received daily injections of growth hormone, compared with a littermate that did not receive growth hormone. This figure shows marked exacerbation of growth by growth hormone — both in the early days of life and even after the two rats had reached adulthood. In the early stages of development, all organs of the treated rat increased proportionately in size, but, after adulthood was reached, most of the bones ceased growing while the soft tissues continued to grow. This results from the fact that, once the epiphyses of the long bones have united with the shafts, further growth of the bones cannot occur even though most other tissues of the body can continue to grow throughout life.

Stimulation of Growth of Cartilage and Bone – Role of Somatomedin

Growth hormone does not have a *direct* effect on the growth of cartilage and bone, both of which must grow if the overall structure of the animal is to increase. However, growth hormone does indirectly stimulate their growth by causing several small proteins, collectively called *somatomedin,* to be formed in the liver and perhaps the muscles and kidneys as well; somatomedin in turn acts directly on the cartilage and bone to promote their growth. Somatomedin is required for deposition of chondroitin sulfate and collagen, both of which are necessary for growth of the cartilage and bone.

Once the epiphyses of the long bones have

united with the shafts, the bones can no longer increase in length, but they can continue to increase in thickness. Therefore, excess growth hormone after adolescence cannot cause further increase in height of a person but can cause disproportionate growth of the membranous bones and excessive thickening of all bones.

Other Growth Functions of Somatomedin. At least three different somatomedins, and probably as many as five, have been isolated. These have molecular weights ranging between 5,000 and 10,000, and the different somatomedins have somewhat different metabolic stimulating functions. Some physiologists believe that most of the important functions of growth hormone are achieved by the mediation of one of the somatomedins. The reason for believing this is that many of the known actions of growth hormone will not occur when growth hormone, in the absence of somatomedin, is added directly to tissues removed from the body. And when growth hormone does cause direct effects in isolated tissues, this often requires growth hormone concentrations as much as 1000 times as great as those known to exist in the body. And, finally, preliminary studies show that preparations of somatomedin can cause growth of muscle and other tissues similar to its growth-promoting effect on bone.

Metabolic Effects of Growth Hormone

Aside from its specific effect in causing growth, growth hormone has many generalized metabolic effects as well, including especially:

1. Increased rate of protein synthesis in all cells of the body;

2. Increased mobilization of fatty acids from adipose tissue, and increased use of the fatty acids for energy; and

3. Decreased rate of glucose utilization throughout the body.

Thus, in effect, growth hormone enhances the body protein, uses up the fat stores, and conserves carbohydrate. It is probable that the increased rate of growth results mainly from the increased rate of protein synthesis.

Role of Growth Hormone in Promoting Protein Deposition

Although the most important cause of the increased protein deposition caused by growth hormone is not known, a series of different effects are known, all of which can lead to enhanced protein. These effects are:

1. Enhancement of Amino Acid Transport Through the Cell Membranes. Growth hormone directly enhances transport of at least some and perhaps most amino acids through the cell membranes to the interior of the cells. This increases the concentrations of the amino acids in the cells and is presumed to be at least partly responsible for the increased protein synthesis. This control of amino acid transport is similar to the effect of insulin in controlling glucose transport through the membrane, as discussed in Chapters 67 and 78.

2. Enhancement of Protein Synthesis by the Ribosomes. Even when the amino acids are not increased in the cells, growth hormone still causes protein to be synthesized in increased amounts in the cells. This is believed to be caused partly by a direct effect on the ribosomes, making them produce greater numbers of protein molecules; the mechanism by which this effect occurs is as yet unknown.

3. Increased Formation of RNA. Over more prolonged periods of time, growth hormone also stimulates the transcription process in the nucleus, causing formation of increased quantities of RNA. This in turn promotes protein synthesis and also promotes growth if sufficient energy, amino acids, vitamins, and other necessities for growth are available.

4. Decreased Catabolism of Protein and Amino Acid. In addition to the increase in protein synthesis, there is a decrease in the breakdown of cell protein. A possible if not probable reason for this effect is that growth hormone also mobilizes large quantities of free fatty acids from the adipose tissue, and these in turn are used to supply most of the energy for the body cells, thus acting as a potent "protein sparer." And this easy availability of fats for energy also acts as a "carbohydrate sparer," thereby decreasing the necessity to use proteins for gluconeogenesis — another factor that diminishes protein catabolism.

Summary. Growth hormone enhances almost all facets of amino acid uptake and protein synthesis by cells, while at the same time reducing the breakdown of proteins.

Effect of Growth Hormone in Enhancing Fat Utilization for Energy

Growth hormone has a specific effect, but one that takes several hours to develop, in causing release of fatty acids from adipose tissue and, therefore, increasing the fatty acid concentration in the body fluids. In addition, in the tissues it enhances the conversion of fatty acids to acetyl-CoA with subsequent utilization of this for energy. Therefore, under the influence of growth hormone, fat is utilized for energy in preference to both carbohydrates and proteins.

Some research workers have considered this fat mobilization effect of growth hormone to be its most important function and have also considered the protein-sparing effect to be the major factor that promotes protein deposition and growth. However, growth hormone mobilization of fat requires hours to occur, whereas enhancement of cellular protein synthesis can begin in less than a minute under the influence of growth hormone.

Ketogenic Effect of Growth Hormone. Occasionally, fat mobilization under the influence of excessive amounts of growth hormone is so great that excessive quantities of acetoacetic acid are formed by the liver and are released into the body fluids, thus causing ketosis. This excessive mobilization of fat from the adipose tissue also frequently causes a fatty liver.

Effect of Growth Hormone on Carbohydrate Metabolism

Growth hormone has three major effects on cellular metabolism of glucose. These effects are (1) decreased utilization of glucose for energy, (2) enhancement of glycogen deposition in the cells, and (3) diminished uptake of glucose by the cells.

Decreased Glucose Utilization for Energy. Unfortunately, we do not know the precise mechanism by which growth hormone decreases glucose utilization by the cells. However, the decrease probably results partially from the increased mobilization and utilization of fatty acids for energy caused by growth hormone. That is, the fatty acids form large quantities of acetyl-CoA that in turn initiate feedback effects to block the glycolytic breakdown of glucose and glycogen.

Enhancement of Glycogen Deposition. Since glucose and glycogen cannot be utilized for energy, the glucose that does enter the cells is rapidly polymerized into glycogen and deposited. Therefore, the cells rapidly become saturated with glycogen and can store no more.

Diminished Uptake of Glucose by the Cells and Increased Blood Glucose Concentration. When growth hormone is first administered to an animal, the cellular uptake of glucose is enhanced, and the blood glucose concentration falls slightly. However, this effect lasts for only 30 minutes to an hour or so and is then superseded by exactly the opposite effect, namely decreased transport of glucose through the cell membrane — and, indeed, also decreased phosphorylation of the glucose that does cross the membrane. Without normal cellular uptake, the blood concentration of glucose increases, sometimes to as high as 50 to 100 per cent above normal.

Thus, growth hormone seems actually to enhance membrane transport of glucose, though failure of glucose utilization eventually leads to greatly diminished uptake. However, some physiologists believe that there is also a direct effect of growth hormone in reducing glucose uptake — an effect as yet unexplained.

Necessity of Insulin and Carbohydrate for the Growth-Promoting Action of Growth Hormone. Growth hormone fails to cause growth in an animal lacking a pancreas, and it also fails to cause growth if carbohydrates are excluded from the diet. This shows that adequate insulin activity as well as adequate availability of carbohydrates is necessary for growth hormone to be effective. Part of this requirement for carbohydrates and insulin is to provide the energy needed for the metabolism of growth. But there seem to be other effects as well. Especially important is one effect of insulin on enhancing amino acid transport into cells in the same way that it enhances glucose transport.

Diabetogenic Effect of Growth Hormone. We have already pointed out that growth hormone leads to moderately increased blood glucose concentration. This in turn stimulates the beta cells of the islets of Langerhans to secrete extra insulin. In addition to this effect, growth hormone has a moderate, direct, stimulatory effect on the beta cells as well. The combination of these two effects sometimes so greatly overstimulates insulin secretion by the beta cells that they literally "burn out." When this occurs the person develops diabetes mellitus, a disease that will be discussed in detail in Chapter 78. Therefore, growth hormone is said to have a *diabetogenic effect.*

Diabetogenic Effects of Other Anterior Pituitary Hormones. Growth hormone is not the only anterior pituitary hormone that increases the blood glucose concentration. At least three others can do the same: adrenocorticotropin, thyroid-stimulating hormone, and prolactin. Especially important is adrenocorticotropin, which increases the rate of cortisol secretion by the adrenal cortex. Cortisol then increases the blood glucose concentration by increasing the rate of gluconeogenesis. This effect, quantitatively, is probably equally as diabetogenic as the effect of growth hormone.

Pituitary Diabetes. From the above discussion it is readily apparent that specific increase in secretion of growth hormone or generalized increase in secretion of all the anterior pituitary hormones causes elevated blood glucose concentration; this condition is called *pituitary diabetes,* and it differs from *diabetes mellitus,* which results from insulin lack, in the following ways: First, in pituitary diabetes the rate of glucose utilization by the cells is only moderately depressed, in comparison with almost no utilization in diabetes mellitus. Second, the blood glucose concentration is relatively *refractory to insulin* — that is, decreases very little — because adequate insulin is already available in the body; the problem instead is the anti-insulin effect of growth hormone and other pituitary hormones that blocks insulin stimulation of glucose transport into the cells. Third, many of the side effects that result from reduced carbohydrate metabolism in diabetes mellitus are absent in pituitary diabetes.

Regulation of Growth Hormone Secretion

For many years it was believed that growth hormone was secreted primarily during the period of growth but then disappeared from the blood at adolescence. However, this has proved to be very far from the truth, because after adolescence, secretion continues at a rate almost as great as that in childhood. Furthermore, the rate of growth hormone secretion increases and decreases within minutes, sometimes for reasons that are not at all understood, but at other times definitely in relation to the person's state of nutrition or stress, such as during starvation, hypoglycemia, exercise, excitement, and trauma. And it characteristically increases during the first two hours of sleep.

The normal concentration of growth hormone in the plasma of an adult is about 3 millimicrograms per milliliter and in the child about 5 millimicrograms per milliliter. However, these values often increase to as high as 50 millimicrograms per milliliter after depletion of the body stores of proteins or carbohydrates. Under acute conditions, hypoglycemia is a far more potent stimulator of growth hormone secretion than is a decrease in the amino acid concentration in the blood. In fact, acute administration of some amino acids, especially arginine, can cause acute increase in growth hormone secretion; this hormone then increases the rate of protein synthesis.

On the other hand, in chronic conditions the degree of cellular protein depletion seems to be more correlated with the level of growth hormone secretion than is the availability of glucose. For instance, the extremely high levels of growth hormone that occur during starvation are very closely related to the amount of protein depletion. Figure 75–6 illustrates this relationship: the first

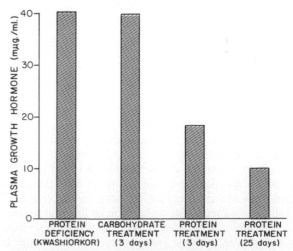

Figure 75–6. Effect of extreme protein deficiency on the concentration of growth hormone in the plasma in the disease kwashiorkor. The figure also shows the failure of carbohydrate treatment but the effectiveness of protein treatment in lowering growth hormone concentration. (Drawn from data in Pimstone: *Am. J. Clin. Nutr., 21*:482, 1968.)

column shows growth hormone levels in children with extreme protein deficiency during the malnutrition disease called *kwashiorkor*; the second column shows the levels in the same children after 3 days of treatment with more than adequate quantities of carbohydrates in their diets, illustrating that the carbohydrates did not lower the plasma growth hormone concentration; the third and fourth columns show the levels after treatment with protein supplement in their diet for 3 and 25 days, with concomitant decrease in the hormone. These results demonstrate that under very severe conditions of protein malnutrition, adequate calories alone are not sufficient to correct the excess production of growth hormone. Instead, the protein deficiency must also be corrected before the growth hormone concentration will return to normal.

Role of the Hypothalamus, Growth Hormone Releasing Hormone, and Somatostatin in the Control of Growth Hormone Secretion. From the above description of the many different factors that can affect growth hormone secretion, one can readily understand the perplexity that has faced physiologists in their attempt to unravel the mysteries of the regulation of growth hormone secretion. However, it is known that growth hormone secretion is controlled almost entirely in response to two factors secreted in the hypothalamus and then transported to the anterior pituitary gland through the hypothalamic-hypophysial portal vessels. These are *growth hormone releasing hormone* (GHRH) and *growth hormone inhibitory hormone* (GHIH), also called *somatostatin.*

The hypothalamic nucleus that causes secretion of growth hormone releasing hormone is the ventromedial nucleus, the same area of the hypothalamus that is known to be sensitive to hypoglycemia and to cause hunger in hypoglycemic states. The secretion of somatostatin is controlled by other nearby areas of the hypothalamus. Therefore, it is reasonable to believe that some of the same signals that modify a person's behavioral feeding instincts also alter the rate of growth hormone secretion. In a similar manner, hypothalamic signals depicting emotions, stress, and trauma can all affect hypothalamic control of growth hormone secretion. In fact, definitive experiments have shown that catecholamines, dopamine, and serotonin, each of which is released by a different neuronal system in the hypothalamus, will increase the rate of growth hormone secretion.

Most of the control of growth hormone secretion is probably mediated through growth hormone releasing hormone rather than through the inhibitory hormone somatostatin. However, it is important to note that somatostatin is also secreted by the delta cells of the islets of Langerhans in the pancreas, and that it can inhibit the secretion of insulin and glucagon by the beta and alpha cells in the islets of Langerhans in the same way that it inhibits the secretion of growth hormone. For this reason, somatostatin might have a widespread role in modulating the functions of multiple hormonal systems.

When growth hormone is administered to an animal over a period of hours, the rate of endogenous growth hormone secretion becomes decreased. This illustrates that growth hormone secretion, as is true of essentially all other hormones, is subject to typical negative feedback control. However, the nature of this feedback mechanism and whether it is mediated through inhibition of growth hormone releasing hormone or enhancement of somatostatin are still unknown.

In summary, our present knowledge of the regulation of growth hormone secretion is not sufficient to describe a composite picture. Yet, because of its undoubted long-term effect on promoting protein synthesis and growth of tissues, one would like to propose that the major long-term controller of growth hormone secretion is the state of nutrition of the tissues themselves, especially their level of protein nutrition. That is, nutritional deficiency would in some way increase the rate of growth hormone secretion. The growth hormone in turn would promote synthesis of new proteins while at the same time conserving the protein already present in the cells.

ABNORMALITIES OF GROWTH HORMONE SECRETION

Panhypopituitarism. This term means decreased secretion of all the anterior pituitary hormones. The decrease in secretion may be congenital (present from birth), or it may occur suddenly or slowly at any time during the life of the individual.

Dwarfism. Some instances of dwarfism result from deficiency of anterior pituitary secretion during childhood. In general, the features of the body develop in appropriate proportion to each other, but the rate of development is greatly decreased. A child who has reached the age of 10 may have the bodily development of a child of 4 to 5, whereas the same person on reaching the age of 20 may have the bodily development of a child of 7 to 10.

The dwarf usually does not exhibit specific thyroid deficiency or adrenocortical deficiency, for the entire body remains so small that only small quantities of thyroid-stimulating and adrenocorticotropic hormones are needed. Also, there is no mental retardation. On the other hand, the panhypopituitary dwarf does not pass through puberty and never secretes a sufficient quantity of gonadotropic hormones to develop adult sexual functions. In one-third of the dwarfs, however, the deficiency is of growth hormone alone; these individuals do mature sexually and occasionally do reproduce. In a very rare condition (the Laron dwarf), the rate of growth hormone secretion is actually high but there is a hereditary inability to form somatomedin in response to the growth hormone.

Panhypopituitarism in the Adult. Panhypopituitarism occurring in adulthood frequently results from one of three common abnormalities: Two tumorous conditions, craniopharyngiomas and chromophobe tumors, may compress the pituitary gland until the functioning anterior pituitary cells are totally or almost totally destroyed. The third cause is thrombosis of the pituitary blood vessels. This occurs occasionally when a mother develops circulatory shock following birth of a baby.

The effects of panhypopituitarism, in general, are (1) hypothyroidsim, (2) depressed production of glucocorticoids by the adrenal glands, and (3) suppressed secretion of the gonadotropic hormones to the point that sexual functions are lost. Thus, the picture is that of a lethargic person (from lack of thyroxine) who is gaining weight because of lack of fat mobilization by growth, adrenocorticotropic, adrenocortical, and thyroid hormones and who has lost all sexual functions. Except for the abnormal sexual functions, the patient can usually be treated satisfactorily by administration of adrenocortical and thyroid hormones.

Giantism. Occasionally, the acidophilic, growth-hormone producing cells of the anterior pituitary become excessively active, and sometimes even acidophilic tumors occur in the gland. As a result, large quantities of growth hormone are produced. All body tissues grow rapidly, including the bones, and if the epiphyses of the long bones have not become fused with the shafts before the development of the anterior pituitary acidophilia, height increases so that the person becomes a giant with heights as great as 8 to 9 feet. Thus, for gigantism to occur, the acidophilia must occur prior to adolescence.

The giant ordinarily has hyperglycemia, and the beta cells of the islets of Langerhans in the pancreas are prone to degenerate, partially because they become overactive owing to the hyperglycemia and partially because of a direct overstimulating effect of growth hormone on the islet cells. Consequently, about 10 per cent of the giants finally develop full-blown diabetes mellitus.

Most giants, unfortunately, eventually develop panhypopituitarism if they remain untreated, because the gigantism is usually caused by a tumor of the pituitary gland that grows until the gland itself is destroyed. This general deficiency of pituitary hormones usually causes death in early adulthood. However, once giantism is diagnosed, further development can often be blocked by microsurgical removal of the tumor from the pituitary gland or, if not in this way, by irradiation of the gland.

Acromegaly. If an acidophilic tumor occurs after adolescence — that is, after the epiphyses of the long bones have fused with the shafts — the person cannot grow taller; but the soft tissues can continue to grow, and the bones can grow in thickness. This condition, which is illustrated in Figure 75–7, is known as *acromegaly.* Enlargement is especially marked in the small bones of the hands and feet and in the *membranous bones,* including the cranium, the nose, the bosses on the forehead, the supraorbital ridges, the lower jawbone, and portions of the vertebrae, for their growth does not cease at adolescence. Consequently, the jaw protrudes forward, sometimes as much as a half inch, the forehead slants forward because of excess development of the supra-orbital ridges, the nose increases to as much as twice normal size, the foot requires a size 14 or larger shoe, and the fingers become extremely thickened so that the hand develops a size almost twice normal. In addition to these effects, changes in the vertebrae ordinarily cause a hunched back, which is known clinically as kyphosis.

 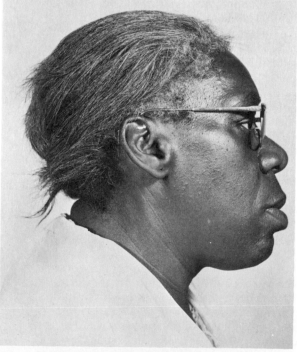

Figure 75–7. An acromegalic patient. (Courtesy of Dr. Herbert Langford.)

Finally, many soft tissue organs, such as the tongue, liver, and especially the kidneys, become greatly enlarged.

THE POSTERIOR PITUITARY GLAND AND ITS RELATION TO THE HYPOTHALAMUS

The *posterior pituitary gland,* also called the *neurohypophysis,* is composed mainly of glial-like cells called *pituicytes.* However, the pituicytes do not secrete hormones; they act simply as a supporting structure for large numbers of *terminal nerve fibers* and *terminal nerve endings* from nerve tracts that originate mainly in the *supraoptic* and *paraventricular nuclei* of the hypothalamus, as shown in Figure 75–8. These tracts pass to the neurohypophysis through the *pituitary stalk* (hypophysial stalk). The nerve endings are bulbous knobs that lie on the surfaces of capillaries onto which they secrete the two posterior pituitary hormones: (1) *antidiuretic hormone* (ADH), also called *vasopressin,* and (2) *oxytocin.*

If the pituitary stalk is cut near the pituitary gland, leaving the entire hypothalamus intact, the posterior pituitary hormones continue, after a transient decrease for a few days, to be secreted almost normally, but they are then secreted by the cut ends of the fibers within the hypothalamus and not by the nerve endings in the posterior pituitary. The reason for this is that the hormones are initially synthesized in the cell bodies of the supraoptic and paraventricular nuclei and are then transported in combination with ''carrier'' proteins called *neurophysins* down to the nerve endings in the posterior pituitary gland, requiring several days to reach the gland.

ADH is formed primarily in the supraoptic nuclei, while *oxytocin is formed primarily in the paraventricular nuclei.* However, each of these two nuclei can synthesize approximately one-sixth as much of the second hormone as of its primary hormone.

Under resting conditions, large quantities of both ADH and oxytocin accumulate in large secretory granules in the nerve endings of the posterior pituitary gland, still loosely bound with their respective neurophysins. Then, when nerve impulses are transmitted downward along the fibers from the supraoptic and paraventricular nuclei, the hormones are immediately released from the nerve endings by the usual secretory mechanism of *exocytosis* and are absorbed into adjacent capillaries. Both the neurophysin and the hormone are secreted together, but since they are only loosely bound to each other, it is believed that the hormone separates almost immediately.

PHYSIOLOGICAL FUNCTIONS OF ANTIDIURETIC HORMONE (VASOPRESSIN)

Extremely minute quantities of antidiuretic hormone (ADH) — as small as 2 millimicrograms — when injected into a person can cause antidiuresis, that is, decreased excretion of water by the kidneys. This antidiuretic effect was discussed in detail in Chapter 36. Briefly, in the absence of ADH, the collecting ducts and distal tubules are almost totally impermeable to water, which prevents significant reabsorption of water and therefore allows extreme loss of water into the urine. On the other hand, in the presence of ADH the permeability of the collecting ducts to water increases greatly and allows most of the water to be reabsorbed as the tubular fluid passes through the collecting ducts, thereby conserving water in the body.

The precise mechanism by which ADH acts on the ducts to increase their permeability is unknown. However, the hormone first becomes fixed to specific receptors on the basal side of the collecting duct epithelial cells, causing them to form large quantities of cyclic AMP in their cytoplasm. This in turn opens many pores in the cell membranes and allows free diffusion of water between the tubular and peritubular fluids, though the manner in which cyclic AMP acid causes its effect on the pores is mainly unknown. The water is then absorbed by osmosis, as explained in relation to the concentrating mechanism of the kidney in Chapter 36.

Regulation of ADH Production

Osmotic Regulation. When a concentrated electrolyte solution is injected into the artery supplying the hypothalamus, the ADH neurons in the supraoptic and paraventricular nuclei immediately transmit impulses into the posterior pituitary

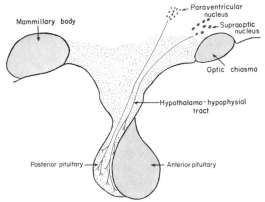

Figure 75–8. Hypothalamic control of the posterior pituitary.

to release large quantities of ADH into the circulating blood. Conversely, injection of pure water into this artery causes complete cessation of the impulses and essentially complete cessation of ADH secretion. The ADH that is already in the blood is destroyed by the tissues at a rate of approximately one-half every 15 minutes. Thus, the concentration of ADH in the body fluids can change from small amounts to large amounts, or vice versa, in only a few minutes.

It has been postulated that neurons that are separate from the ADH neurons but are located near them function as *osmoreceptors*. These presumably could increase and decrease in size in relation to the degree of concentration of the extracellular fluids. When the extracellular fluids are highly dilute, osmosis of water into the cell supposedly increases their volume, while concentrated extracellular fluids supposedly reduce the volume. Regardless of the mechanism, concentrated body fluids do stimulate the supraoptic nuclei, while dilute body fluids inhibit them. Therefore, a feedback control system is available to control the total osmotic pressure of the body fluids, operating as follows:

When the body fluids become highly concentrated, the supraoptic nuclei become excited, impulses are transmitted to the posterior pituitary, and ADH is secreted. This passes by way of the blood to the kidneys, where it increases the permeability of the collecting ducts to water. As a result, most of the water is reabsorbed from the tubular fluid, while electrolytes continue to be lost into the urine. This dilutes the extracellular fluids, returning them to a reasonably normal osmotic composition.

Role of ADH in Controlling Extracellular Fluid Sodium Ion Concentration. Under normal conditions, about 95 per cent of the total osmotic pressure of the extracellular fluids is determined by the sodium ion concentration of these fluids (because of the high concentration of sodium and because the anion concentration "follows" the sodium concentration). Therefore, in effect, when one says that ADH controls the osmolality of the extracellular fluids, this also means that ADH controls the sodium ion concentration of these fluids at the same time. Furthermore, recent experiments have shown that the major change that occurs in the composition of the extracellular fluid following either increased or decreased ADH secretion is a change in sodium ion concentration, with much less change in volume of water in the extracellular fluids. Therefore, it is becoming clear that ADH is a very potent controller of sodium ion concentration, a fact that was discussed in detail in Chapter 36.

Stimulation of ADH Secretion by Low Blood Volume — Pressor Effect of ADH. ADH in moderate to high concentrations has a very potent effect of constricting the arterioles and therefore of increasing the arterial pressure. Also, one of the most powerful stimuli of all for increasing the secretion of ADH is severe loss of blood volume. As little as 10 per cent loss of blood will promote a moderate increase in ADH secretion, and 25 per cent or more blood loss can cause as much as 50 to 100 times normal rates of secretion.

The increased secretion is believed to result mainly from the low pressure caused in the atria of the heart by the low blood volume. The relaxation of the atrial stretch receptors supposedly elicits the increase in ADH secretion. However, the baroreceptors of the carotid, aortic, and pulmonary regions also participate in the control of ADH secretion.

It is generally stated that the normal concentration of ADH in the body fluids, between 1 and 2 picograms per milliliter (trillionths of a gram per milliliter), is too small to cause any significant pressor effect. However, recent studies by Cowley in which the compensatory role of the nervous reflexes was eliminated showed that between 5 and 10 mm. Hg of the normal arterial pressure is maintained by this normal ADH. Also, he showed that, when the arterial pressure falls, the amount of ADH secreted is enough to bring the pressure three-fourths of the way back to normal. This shows that ADH plays a much greater role in arterial pressure homeostasis than has previously been believed.

Because ADH has this potent pressor effect, it is also called *vasopressin*.

Other Factors That Affect ADH Production. Other factors that frequently increase the output of ADH include *trauma* to the body, *pain, anxiety,* and drugs such as *morphine, nicotine, tranquilizers,* and some *anesthetics*. Each of these factors can cause retention of water in the body. This explains the frequent accumulation of water in many emotional states, and it also explains the diuresis that occurs when the state is over.

A substance that inhibits ADH secretion is *alcohol*. Therefore, during an alcoholic bout, lack of ADH allows marked diuresis. Alcohol probably also dilates the afferent arterioles of the nephrons, which adds to the diuretic effect.

Diabetes Insipidus. Diabetes insipidus is the disease that occurs when the supraoptico-hypophysial system secreting ADH fails. This will occur only when the ADH neurons in the supraoptic and paraventricular nuclei or their nerve fibers near these nuclei are destroyed. It will not occur when the posterior pituitary gland alone is destroyed or when the pituitary stalk is cut below the median eminence, because the cut nerve fibers can continue to secrete ADH. In a person with full-blown diabetes insipidus, lack of ADH makes the urine almost always very dilute except in very severe dehydration states. The urine specific gravity remains almost constantly between 1.002 and 1.006; the urine output is usually 4 to 6 liters per day but can be as great as 12 to 15 liters per day, depending principally on how much water the person drinks. Furthermore, the rapid loss of fluid in the urine creates a constant thirst, which keeps the water flushing through the body.

The person with diabetes insipidus has a tendency to become dehydrated. However, this tendency is usually quite well offset by the increased thirst. Under conditions of circulatory stress or when water might not be adequately available, the fluid loss can become serious.

Diabetes insipidus occurs most frequently as a result of a tumor of the hypothalamus or hypophysis that destroys the portions of the hypothalamus that control ADH secretion. It can be treated easily by simply injecting vasopressin suspended in oil (for slow release) once every two days.

Excess Secretion of ADH — Syndrome of Inappropriate ADH Secretion. Occasionally, excess ADH is secreted by the hypophysial-posterior pituitary system or, much more usually by certain types of tumors in the body, particularly bronchogenic carcinomas of the lungs. When this occurs, a condition called *syndrome of inappropriate ADH secretion* develops, characterized by greatly decreased sodium ion concentration in the extracellular fluid but by only a few per cent increase in body water. The reasons for these effects were discussed in detail in Chapter 36; they especially point up the fact that ADH is much more a controller of sodium ion concentration than of body fluid volume.

Other Actions of ADH

Large doses of ADH can also cause contraction of almost any smooth muscle tissue in the body, including contractions of most of the intestinal musculature, the bile ducts, and the uterus. However, the concentrations required to cause these effects are far greater than that required to cause antidiuresis, and it is doubtful that these are significant physiological effects for normal function of the body.

Some psychologists have suggested that ADH secreted into the hypothalamus itself increases one's memory retention.

OXYTOCIC HORMONE

Effect on the Uterus. An oxytocic substance is one that causes contraction of the pregnant uterus. The hormone *oxytocin,* in accordance with its name, powerfully stimulates the pregnant uterus, especially so toward the end of gestation. Therefore, many obstetricians believe that this hormone is at least partially responsible for effecting birth of the baby. This is supported by the following facts: (1) In a hypophysectomized animal, the duration of labor is considerably prolonged, thus indicating a probable effect of oxytocin during delivery. (2) The amount of oxytocin in the plasma increases during labor, especially during the last stage. (3) Stimulation of the cervix in a pregnant animal elicits nervous signals that pass to the hypothalamus and cause increased secretion of oxytocin.

ADH also stimulates the pregnant uterus, though less than one-hundredth as strongly as oxytocin. Likewise, oxytocin excites contraction of a number of other smooth muscle structures in the body besides the uterus, though usually to a much less extent than does ADH, except for the myoepithelial cells of the breasts as discussed next. The fact that the functions of the two hormones partially overlap illustrates their physiological relation to each other; as will be seen, their chemical structures are also similar.

Effect of Oxytocin on Milk Ejection. Oxytocin has an especially important function in the process of lactation, for this hormone causes milk to be expressed from the alveoli into the ducts so that the baby can obtain it by suckling. This mechanism works as follows: The suckling stimuli on the nipple of the breast cause signals to be transmitted through the sensory nerves to the brain. The signals pass upward through the reticular areas of the brain stem and finally reach the oxytocin neurons in the paraventricular and supraoptic nuclei in the hypothalamus, to cause release of oxytocin. The oxytocin then is carried by the blood to the breasts where it causes contraction of *myoepithelial cells,* which lie outside of and form a latticework that surrounds the alveoli of the mammary glands. In less than a minute after the beginning of suckling, milk begins to flow. Therefore, this mechanism is frequently called *milk letdown* or *milk ejection.* This process is discussed further in Chapter 82 in relation to lactation.

Possible Effect of Oxytocin in Promoting Fertilization of the Ovum. Sexual stimulation of the female during intercourse increases the secretion of oxytocin, and the increased oxytocin has been postulated to be at least partially responsible for the uterine contractions that occur during the female orgasm. For these reasons, it has been proposed that oxytocin promotes fertilization of the ovum by causing uterine propulsion of the male semen upward through the fallopian tubes.

CHEMICAL NATURE OF ANTIDIURETIC HORMONE (VASOPRESSIN) AND OXYTOCIN

Both oxytocin and ADH (vasopressin) are polypeptides containing nine amino acids. The amino acid sequences of these are the following:

VASOPRESSIN:
Cys-Tyr-Phe-Gln-Asn-Cys-Pro-Arg-GlyNH$_2$
OXYTOCIN:
Cys-Tyr-Ile-Gln-Asn-Cys-Pro-Leu-GlyNH$_2$

Note that these two hormones are almost identical except that in vasopressin phenylalanine and arginine replace isoleucine and leucine of the oxytocin molecule. The similarity of the molecules explains their occasional functional similarities.

REFERENCES

Arimura, A.: Hypothalamic gonadotropin-releasing hormone and reproduction *Int. Rev. Physiol., 13*:1, 1977.
Assenmacher, I., and Farner, D. S. (eds.): Environmental Endocrinology. New York, Springer-Verlag, 1978.
Austin, C. R., and Short, R. V. (eds.): Mechanisms of Hormone Action. New York, Cambridge University Press, 1979.
Bargmann, W., *et al.* (eds.): Neurosection and Neuroendocrine Activity; Evolution, Structure and Function. New York, Springer-Verlag, 1978.

Barrington, E. J. W. (ed.): Hormones and Evolution. New York, Academic Press, 1979.

Baxter, J. D., and MacLeod, K. M.: Molecular basis for hormone action. *In* Bondy, P. K., and Rosenberg, L. E. (eds.): Metabolic Control and Disease, 8th Ed. Philadelphia, W. B. Saunders Co., 1980, p. 104.

Besser, G. M. (ed.): The Hypothalamus and Pituitary. Philadelphia, W. B. Saunders Co., 1977.

Bizollon, C. A. (ed.): Radioimmunology, 1979. New York, Elsevier/North-Holland, 1979.

Bowers, C. Y., *et al.*: Hypothalamic peptide hormones; Chemistry and physiology. *In* DeGroot, L. J., *et al.* (eds.): Endocrinology. Vol. 1. New York, Grune & Stratton, 1979, p. 65.

Brodish, A., and Lymangrover, J. R.: The hypothalamic-pituitary adrenocortical system. *Int. Rev. Physiol.*, 16:93, 1977.

Catt, K. J., and Dufau, M. L.: Peptide hormone receptors. *Annu. Rev. Physiol.*, 39:529, 1977.

Chiodini, P. G., and Liuzzi, A.: The Regulation of Growth Hormone Secretion. St. Albans, Vt., Eden Medical Research, 1979.

DaPrada, A. A. M., and Peckar, B. A. (eds.): Radioimmunoassay of Drugs and Hormones in Cardiovascular Medicine. New York, Elsevier/North-Holland, 1979.

Davis, J. C., and Hipkin. L. J.: Clinical Endocrine Pathology. Philadelphia, J. B. Lippincott Co., 1977.

DeGroot, L. J. (ed.): Endocrinology. New York, Grune & Stratton, 1979.

Dillon, R. S.: Handbook of Endocrinology: Diagnosis and Management of Endocrine and Metabolic Disorders, 2nd Ed. Philadelphia, Lea & Febiger, 1980.

Ezrin, C., *et al.* (eds.): Systematic Endocrinology, 2nd Ed. Hagerstown, Md., Harper & Row, 1979.

Ezrin, C., *et al.* (eds.): Pituitary Diseases. Boca Raton, Fla., CRC Press, 1980.

Fain, J. N., and Butcher, F. R.: Cyclic nucleotides in mode of hormone action. *Int. Rev. Physiol.*, 16:241, 1977.

Fink, G., and Geffen, L. B.: The hypothalamo-hypophysial system: Model for central peptidergic and monoaminergic transmission. *In* Porter, R. (ed.): International Review of Physiology: Neurophysiology III. Vol. 17. Baltimore, University Park Press, 1978, p. 1.

Friesen, S. R. (ed.): Surgical Endocrinology: Clinical Syndromes. Philadelphia, J. B. Lippincott Co., 1978.

Goss, R. J.: The Physiology of Growth. New York, Academic Press, 1977.

Gray, C. H., and James, V. H. T.: Hormones in Blood. New York, Academic Press, 1979.

Guillemin, R., and Burgus, R.: The hormones of the hypothalamus. *Sci. Am.*, 227:24, 1972.

Hall, K.: Human Somatomedin. Copenhagen, Periodica, 1972.

Hershman, J. M. (ed.): Management of Endocrine Disorders. Philadelphia, Lea & Febiger, 1980.

Jard, S., and Bockaert, J.: Stimulus-response coupling in neurohypophysial peptide target cells. *Physiol. Rev.*, 55:489, 1975.

Jeffcoate, S. L., and Hutchinson, J. S. M. (eds.): The Endocrine Hypothalamus. New York, Academic Press, 1978.

Johnston, F. E., *et al.* (eds.): Human Physical Growth and Maturation: Methodologies and Factors. New York, Plenum Press, 1980.

Joss, E. E.: Growth Hormone Deficiency in Childhood. New York, S. Karger, 1975.

Jubiz, W.: Endocrinology. New York, McGraw-Hill, 1979.

Kastrup, K. W., and Neilsen, J. H. (eds.): Growth Factors: Cellular Growth Processes, Growth Factors, Hormonal Control of Growth. New York, Pergamon Press, 1978.

Kleeman, C. R., and Beri, J.: The neurohypophysial hormones: Vasopressin. *In* DeGroot, L. J., *et al.* (eds.): Endocrinology. Vol. 1, New York, Grune & Stratton, 1979, p. 253.

Korenman, S. G., *et al.*: Endocrine Disease. Boston, Houghton Mifflin, 1978.

Kostyo, J. L., and Isaksson, O.: Growth hormone and the regulation of somatic growth. *Int. Rev. Physiol.*, 13:255, 1977.

Krulich, L.: Central neurotransmitters and the secretion of prolactin, GH, LH, and TSH. *Annu. Rev. Physiol.*, 41:603, 1979.

Krulich, L., and Fawcett, C. P.: The hypothalamic hypophysiotropic hormones. *Int. Rev. Physiol.*, 16:35, 1977.

Labrie, F., *et al.*: Mechanism of action of hypothalamic hormones in the adenohypophysis. *Annu. Rev. Physiol.*, 41:555, 1979.

Li, C. H. (ed.): Growth Hormone and Related Proteins. New York, Academic Press, 1977.

Li, C. H. (ed.): Hypothalamic Hormones. New York, Academic Press, 1979.

Li, C. H. (ed.): Lipotropin and Related Peptides. New York, Academic Press, 1978.

Marshall, W. A.: Human Growth and Its Disorders. New York, Academic Press, 1977.

McCann, S. M., and Ojeda, S. R.: The role of brain monoamines, acetylcholine and prostaglandins in the control of anterior pituitary function. *In* DeGroot, L. J., *et al.* (eds.): Endocrinology. Vol. 1. New York, Grune & Stratton, 1979, p. 55.

McCann, S. M., *et al.*: Hypothalamic hypophyseal releasing and inhibiting hormones. *In* MTP International Review of Science: Physiology. Vol. 5. Baltimore, University Park Press, 1974, p. 31.

McKerns, K. W., and Jutisz, M. (eds.): Synthesis and Release of Adenohypophyseal Hormones. New York, Plenum Press, 1979.

Merimee, T. J.: Growth hormone: Secretion and action. *In* DeGroot, L. J., *et al.* (eds.): Endocrinology. Vol. 1. New York, Grune & Stratton, 1979, p. 123.

Muller, E. E.: Growth hormone and the regulation of metabolism. *In* MTP International Review of Science: Physiology. Vol. 5. Baltimore, University Park Press, 1974, p. 141.

Nelson, D. H.: Regulatory mechanisms of the pituitary and pituitary-adrenal axis. *In* Frohlich, E. D. (ed.): Pathophysiology, 2nd Ed. Philadelphia, J. B. Lippincott Co., 1976, p. 327.

Ontjes, D. A., *et al.*: The anterior pituitary gland. *In* Bondy, P. K., and Rosenberg, L. E. (eds.): Metabolic Control and Disease, 8th Ed. Philadelphia, W. B. Saunders Co., 1980, p. 1165.

Polleri, A., and MacLeod, R. M. (eds.): Neuroendocrinology; Biological and Clinical Aspects. New York, Academic Press, 1979.

Reichlin, S., *et al.*: Hypothalamic Hormones. *Annu. Rev. Physiol.*, 38:389, 1976.

Rennels, E. G., and Herbert, D. C.: Functional correlates of anterior pituitary cytology. *In* Greep, R. O. (ed.): International Review of Physiology: Reproductive Physiology III. Vol. 22. Baltimore, University Park Press. 1980, p. 1.

Richenberg, H. V. (ed.): Biochemistry and Mode of Action of Hormones II. Baltimore, University Park Press, 1978.

Rosenfield, R. L.: Somatic growth and maturation. *In* DeGroot, L. J., *et al.* (eds.): Endocrinology. Vol. 3. New York, Grune & Stratton, 1979, p. 1805.

Ryan, W. G., *et al.* (eds.): Endocrine Disorders: A Pathophysiologic Approach, 2nd Ed. Chicago, Year Book Medical Publishers, 1980.

Schulster, D., and Levitski, A. (eds.): Cellular Receptors for Hormones and Neurotransmitters. New York, John Wiley & Sons, 1980.

Seif, S. M., and Robinson, A. G.: Localization and release of neurophysins. *Annu. Rev. Physiol.* 40:345, 1978.

Smith, M. J., Jr., *et al.*: Acute and chronic effects of vasopressin on blood pressure, electrolytes, and fluid volumes. *Am. J. Physiol.*, 237(3): F232, 1979.

Sowers, J. R. (eds.): Hypothalamic Hormones. Stroudsburg, Pa., Dowden, Hutchinson & Ross, 1980.

Tepperman, J.: Metabolic and Endocrine Physiology: An Introductory Text. Chicago, Year Book Medical Publishers, 1980.

Terry, L. C., and Martin, J. B.: Hypothalamic hormones: Subcellular distribution and mechanisms of release. *Annu. Rev. Pharmacol. Toxicol.*, 18:111, 1978.

Tindall, G. T., and Collins, W. F. (eds.): Clinical Management of Pituitary Disorders. New York, Raven Press, 1979.

Tolis, G., *et al.* (eds.): Clinical Neuroendocrinology. New York, Raven Press, 1979.

Vale, W., *et al.*: Regulatory peptides of the hypothalamus. *Annu. Rev. Physiol.*, 39:473, 1977.

Vorherr, H.: Oxytocin. *In* DeGroot, L. J., *et al.* (eds.): Endocrinology. Vol. 1. New York, Grune & Stratton, 1979, p. 277.

Weiner, R. I., and Ganong, W. F.: Role of brain monoamines and histamine in regulation of anterior pituitary secretion. *Physiol. Rev.*, 58:905, 1978.

Weitzman, R., and Kleeman, C. R.: Water metabolism and the neurohypophysial hormones. *In* Bondy, P. K., and Rosenberg, L. E. (eds.): Metabolic Control and Disease, 8th Ed. Philadelphia, W. B. Saunders Co., 1980, p. 1241.

Williams, R. H.: Textbook of Endocrinology, 5th Ed. Philadelphia, W. B. Saunders Co., 1975.

76

The Thyroid Hormones

The thyroid gland, which is located immediately below the larynx on either side of and anterior to the trachea, secretes two significant hormones, *thyroxine* and *triiodothyronine,* that have a profound effect on the metabolic rate of the body. It also secretes *calcitonin,* an important hormone for calcium metabolism that will be considered in detail in Chapter 79. Complete lack of thyroid secretion usually causes the basal metabolic rate to fall about 40 per cent below normal, and extreme excesses of thyroid secretion can cause the basal metabolic rate to rise as high as 60 to 100 per cent above normal. Thyroid secretion is controlled primarily by thyroid-stimulating hormone secreted by the anterior pituitary gland.

The purpose of this chapter is to discuss the formation and secretion of the thyroid hormones, their functions in the metabolic scheme of the body, and regulation of their secretion.

FORMATION AND SECRETION OF THE THYROID HORMONES

About 90 per cent of the hormone secreted by the thyroid gland is *thyroxine* and 10 per cent is *triiodothyronine.* However, a considerable portion of the thyroxine is converted to triiodothyronine in the peripheral tissues, so that both are very important functionally. The functions of these two hormones are qualitatively the same, but they differ in rapidity and intensity of action. Triiodothyronine is about 4 times as potent as thyroxine, but it is present in the blood in much smaller quantities and persists for a much shorter time than does thyroxine.

Physiologic Anatomy of the Thyroid Gland. The thyroid gland is composed, as shown in Figure 76–1, of large numbers of closed *follicles* (150 to 300 microns in diameter) filled with a secretory substance called *colloid* and lined with *cuboidal epithelioid cells* that secrete into the interior of the follicles. The major constituent of colloid is the large glycoprotein *thyroglobulin,* which contains the thyroid hormones. Once the secretion has entered the follicles, it must be absorbed back through the follicular epithelium into the blood before it can function in the body. The thyroid gland, having a blood flow about 5 times the weight of the gland each minute, has a blood supply as rich as that of any other area of the body with the probable exception of the adrenal cortex.

REQUIREMENTS OF IODINE FOR FORMATION OF THYROXINE

To form normal quantities of thyroxine, approximately 50 mg. of ingested iodine are required *each year,* or approximately *1 mg. per week.* To prevent iodine deficiency, common table salt is iodized with one part sodium iodide to every 100,000 parts sodium chloride.

Fate of Ingested Iodides. Iodides ingested orally are absorbed from the gastrointestinal tract into the blood in approximately the same manner as chlorides, but iodides do not remain in the circulatory system for a prolonged time because the kidneys have a very high plasma clearance for iodide ion — about 35 ml. per minute in comparison with only 1 ml. per minute for chloride ion. Within the first three days two thirds of the ingested iodides are normally lost into the urine, and almost all the remaining one third is selectively removed

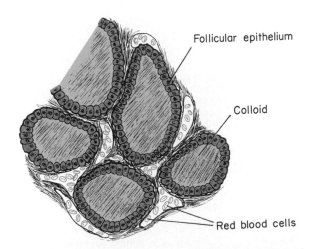

Figure 76–1. Microscopic appearance of the thyroid gland, showing the secretion of thyroglobulin into the follicles.

from the circulating blood by the cells of the thyroid gland and used for synthesis of thyroid hormones, which are stored in the form of *thyroglobulin* in the follicles and later secreted into the blood principally in the form of thyroxine, as is discussed below.

THE IODIDE PUMP (IODIDE TRAPPING)

The first stage in the formation of thyroid hormones, as shown in Figure 76–2, is transport of iodides from the extracellular fluid into the thyroid glandular cells and follicles. The basal membrane of the thyroid cell has a specific ability to transport the iodide actively to the interior of the cell, and it then diffuses through the apical border of the cell into the follicle as well. This is called *iodide trapping*. In a normal gland, the iodide pump can concentrate the iodide to about 40 times its concentration in the blood. However, when the thyroid gland becomes maximally active, the concentration ratio can rise to several times this value.

THYROGLOBULIN AND CHEMISTRY OF THYROXINE AND TRIIODOTHYRONINE FORMATION

Formation and Secretion of Thyroglobulin by the Thyroid Cells. The thyroid cells are typical protein-secreting glandular cells as illustrated in Figure 76–2. The endoplasmic reticulum and Golgi complex synthesize and secrete into the follicles a large glycoprotein molecule called *thyroglobulin* with a molecular weight of 660,000.

Each molecule of thyroglobulin contains 140 tyrosine amino acids, and these are the major substrates that combine with iodine to form the thyroid hormones. These hormones form *within* the thyroglobulin molecule. That is, the tyrosine amino acid residues, as well as the thyroxine and

triiodothyronine hormones formed from them, remain a part of the thyroglobulin molecule during the process of synthesis of the thyroid hormones.

In addition to secreting the thyroglobulin the glandular cells also provide the iodine, the enzymes, and other substances necessary for thyroid hormone synthesis.

Oxidation of the Iodide Ion. An essential step in the formation of the thyroid hormones is conversion of the iodide ions to an *oxidized form of iodine* that is then capable of combining directly with the amino acid tyrosine. The oxidized iodine is associated in some way with the enzyme that causes the iodine to bind to tyrosine.

This oxidation of iodine is promoted by the enzyme *peroxidase* and its accompanying *hydrogen peroxide,* which provide a potent system capable of oxidizing iodides. The peroxidase is located either in the apical membrane of the cell or in the cytoplasm immediately adjacent to this membrane, thus providing the oxidized iodine at exactly the point in the cell where the thyroglobulin molecule first issues forth from the Golgi complex. When the peroxidase system is blocked, or when it is hereditarily absent from the cells, the rate of formation of thyroid hormones falls to zero.

Iodination of Tyrosine and Formation of the Thyroid Hormones — "Organification" of Thyroglobulin. The binding of iodine with the thyroglobulin molecule is called *organification* of the thyroglobulin. Oxidized iodine even in the molecular form will bind directly but slowly with the amino acid tyrosine, but when the oxidized iodine is associated with an appropriate enzyme, this process can occur in seconds or minutes. Therefore, almost as rapidly as the thyroglobulin molecule is released from the Golgi apparatus, or as it is secreted through the apical cell membrane into the follicle, iodine binds with about one sixth of the tyrosine residues within the thyroglobulin molecule.

Figure 76–3 illustrates the successive stages of iodination of tyrosine and the final formation of the two important thyroid hormones, thyroxine and triiodothyronine. Tyrosine is first iodized to *monoiodotyrosine* and then to *diiodotyrosine.* Then during the next few minutes, hours, and even days, more and more of the diiodotyrosine residues become *coupled* with each other. The product of the coupling reaction is the molecule *thyroxine* that also remains part of the thyroglobulin molecule. Or one molecule of monoiodotyrosine couples with one molecule of diiodotyrosine to form *triiodothyronine.*

Storage of Thyroglobulin. After synthesis of the thyroid hormones has run its course, each thyroglobulin molecule contains from 5 to 6

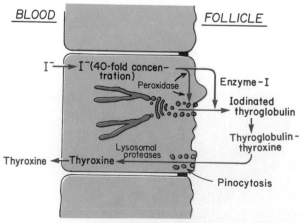

BLOOD   FOLLICLE

I⁻ → I⁻ (40-fold concentration)
Peroxidase
Enzyme-I
Iodinated thyroglobulin
Thyroglobulin-thyroxine
Lysosomal proteases
Thyroxine ← Thyroxine
Pinocytosis

Figure 76–2. Mechanisms of iodine transport, thyroxine formation, and thyroxine release into the blood. (Triiodothyronine formation and release parallels that of thyroxine.)

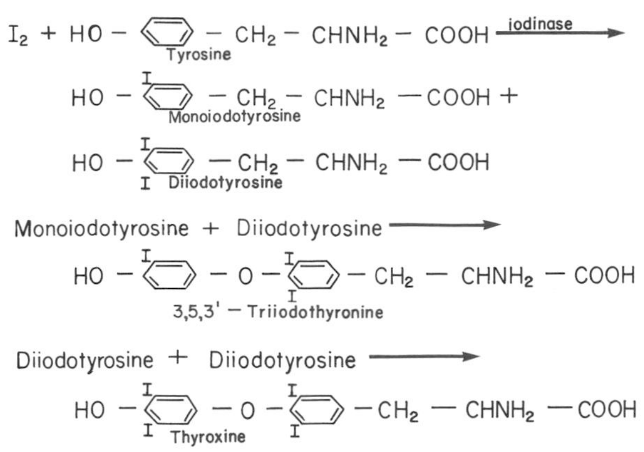

$$I_2 + HO - \langle \rangle - CH_2 - CHNH_2 - COOH \xrightarrow{\text{iodinase}}$$
Tyrosine

$$HO - \langle \rangle - CH_2 - CHNH_2 - COOH +$$
Monoiodotyrosine

$$HO - \langle \rangle - CH_2 - CHNH_2 - COOH$$
Diiodotyrosine

Monoiodotyrosine + Diiodotyrosine ⟶

$$HO - \langle \rangle - O - \langle \rangle - CH_2 - CHNH_2 - COOH$$
3,5,3' − Triiodothyronine

Diiodotyrosine + Diiodotyrosine ⟶

$$HO - \langle \rangle - O - \langle \rangle - CH_2 - CHNH_2 - COOH$$
Thyroxine

Figure 76–3. Chemistry of thyroxine and triiodothyronine formation.

thyroxine molecules, and there is an average of 1 triiodothyronine molecule for every 3 to 4 thyroglobulin molecules — about 18 molecules of thyroxine for every 1 molecule of triiodothyronine. In this form the thyroid hormones are often stored in the follicles for several months. In fact, the total amount stored is sufficient to supply the body with its normal requirements of thyroid hormones for one to three months. Therefore, even when synthesis of thyroid hormone ceases entirely, the effects of deficiency might not be observed for several months.

RELEASE OF THYROXINE AND TRIIODOTHYRONINE FROM THYROGLOBULIN

Thyroglobulin itself is not released into the circulating blood; instead, the thyroxine and triiodothyronine are first cleaved from the thyroglobulin molecule, and then these free hormones are released. This process occurs as follows: The apical surface of the thyroid cells normally sends out pseudopod extensions that close around small portions of the colloid to form pinocytic vesicles. Then lysosomes immediately fuse with these vesicles to form digestive vesicles containing the digestive enzymes from the lysosomes mixed with the colloid. The *proteinases* among these enzymes digest the thyroglobulin molecules and release the thyroxine and triiodothyronine, which then diffuse through the base of the thyroid cell, through the basement membrane, and finally into the surrounding capillaries. Thus, the thyroid hormones are released into the blood.

One-half or more of the iodinated tyrosine in the thyroglobulin never becomes thyroid hormones but instead remains monoiodotyrosine or diiodotyrosine. During the digestion of the thyroglobulin molecule to cause release of thyroxine and triiodothyronine, these iodinated tyrosines also are freed from the thyroid cells. However, they are not secreted into the blood. Instead, their iodine is cleaved from them by an *iodase enzyme* that makes most of this iodine available for recycling to the new thyroglobulin for formation of thyroid hormone. In the congenital absence of this iodase enzyme, persons frequently become iodine-deficient because of failure of this recycling process.

Daily Rate of Secretion of Thyroxine and Triiodothyronine. Over 90 per cent of the thyroid hormone released from the thyroid gland is thyroxine, and less than 10 per cent is triiodothyronine. During the ensuing few days while these hormones circulate in the blood, small portions of the thyroxine are slowly deiodinated to form additional triiodothyronine. Therefore, the quantities of the two hormones finally delivered to the tissues is approximately 90 micrograms of thyroxine per day and 60 micrograms of triiodothyronine per day.

Once the two hormones finally enter the peripheral tissue cells, triiodothyronine is about 4 times as potent in stimulating metabolism and causing other intracellular effects as thyroxine. On the other hand, the duration of action of thyroxine is about 4 times as long as the duration of action of triiodothyronine. Therefore, the integrated effect of each of the hormones over the period of its action per unit mass of hormone is probably about equal. This means, then, that roughly three fifths of the total hormone effect in the tissues is supplied by thyroxine and the remainder by triiodothyronine.

TRANSPORT OF THYROXINE AND TRIIODOTHYRONINE TO THE TISSUES

Binding of Thyroxine and Triiodothyronine with the Plasma Proteins. On entering the blood, all but minute portions of the thyroxine and triiodothyronine combine immediately with several of the plasma proteins. They combine approximately as follows: two thirds with *thyroxine-binding globulin*, which is a glycoprotein; about one fourth with *thyroxine-binding prealbumin;* and about one tenth with *albumin*. The quantity of thyroxine-binding globulin in the blood is only 1 to 1.5 milligrams per 100 milliliters of plasma, but its affinity for the thyroid hormones is so great that it still binds most of the hormones. Its affinity (and that of the other plasma proteins) is over 10 times as great for thyroxine as for triiodothyronine. This difference, plus the fact that the concentration of thyroxine in the plasma is considerably greater than that of triiodothyronine, causes the total amount of protein-bound thyroxine to be about 20 times as great as the protein-bound triiodothyronine.

Release of Thyroxine and Triiodothyronine to the Tissue Cells. Because of the very high affinity of the plasma-binding proteins for the thyroid hormones, these substances — in particular, thyroxine — are released to the tissue cells only very slowly. Half of the thyroxine in the blood is released to the tissue cells approximately every 6 days, whereas half of the triiodothyronine — because of its lower affinity — is released to the cells in approximately 1.3 days.

On entering the cells, both of these hormones again bind with intracellular proteins, the thyroxine once again binding more strongly than the triiodothyronine. Therefore, they are again stored, but this time in the functional cells themselves, and they are used slowly over a period of days or weeks.

Latency and Duration of Action of the Thyroid Hormones. After injection of a large quantity of thyroxine into a human being, essentially no effect on the metabolic rate can be discerned for two to three days, thereby illustrating that there is a *long latent* period before thyroxine activity begins. Once activity does begin, it increases progressively and reaches a maximum in 10 to 12 days, as shown in Figure 76–4. Thereafter, it decreases with a half-life of about 15 days. Some of the activity still persists as long as 6 weeks to 2 months later.

The actions of triiodothyronine occur about 4 times as rapidly as those of thyroxine, with the latent period as short as 6 to 12 hours and maximum cellular activity occurring within 2 to 3 days.

A large share of the latency and prolonged period of action of these hormones is caused by their binding with proteins both in the plasma and in the tissue cells, followed by their slow release. However, we shall see in subsequent discussions that part of the latent period also results from the manner in which these hormones perform their functions in the cells themselves.

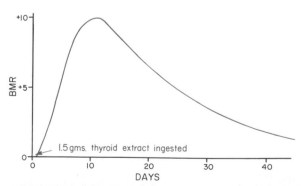

Figure 76–4. Prolonged effect on the basal metabolic rate of administering a large single dose of thyroid hormone.

FUNCTIONS OF THE THYROID HORMONES IN THE TISSUES

The thyroid hormones have two major effects on the body: (1) an increase in the overall metabolic rate, and (2) in children, stimulation of growth.

GENERAL INCREASE IN METABOLIC RATE

The thyroid hormones increase the metabolic activities of almost all tissues of the body (with a few notable exceptions such as the brain, retina, spleen, testes, and lungs). The basal metabolic rate can increase to as much as 60 to 100 per cent above normal when large quantities of the hormones are secreted. The rate of utilization of foods for energy is greatly accelerated. The rate of protein synthesis is at times increased, while at the same time the rate of protein catabolism is also increased. The growth rate of young persons is greatly accelerated. The mental processes are excited, and the activity of many other endocrine glands is often increased. Yet despite the fact that we know all these many changes in metabolism under the influence of the thyroid hormones, the basic mechanism (or mechanisms) by which they act is almost completely unknown. However, some of the possible mechanisms of action of the thyroid hormones are described in the following sections.

Effect of Thyroid Hormones on Causing Increased Protein Synthesis. When either thyroxine or triiodothyronine is given to an animal, protein synthesis increases in almost all tissues of the body. The first stage of the increased protein synthesis begins almost immediately and results from stimulation of the *translation* process — that is, increase in the rate of formation of proteins by the ribosomes. The second stage occurs hours to days later and is caused by increased RNA synthesis by the genes, the process of *transcription,* which leads to a generalized increase in synthesis of many types of proteins within the cells. It is believed that this stimulation of the genes occurs in the following way: (1) The thyroid hormone combines with a "receptor" protein in the cell nucleus. (2) This combination, or a product of it, then activates a large portion of the cellular genes to cause RNA formation and subsequent protein formation.

Effect of Thyroid Hormones on the Cellular Enzyme Systems. Within a week or so following administration of the thyroid hormones, at least 100 and probably many more intracellular enzymes are increased in quantity. As an example, one enzyme, α-glycerophosphate dehydrogenase, can be increased to an activity 6 times its normal level. Since this enzyme is particularly important in the degradation of carbohydrates, its increase could help to explain the rapid utilization of carbohydrates under the influence of thyroxine. Also, the oxidative enzymes and the elements of the electron transport system, both of which are normally found in mitochondria, are greatly increased.

Effect of Thyroid Hormones on Mitochondria. When thyroxine or triiodothyronine is given to an animal, the mitochondria in most cells of the body increase in size and also in number. Furthermore, the total membrane surface of the mitochondria increases almost directly in proportion to the increased metabolic rate of the whole animal. Therefore, it seems almost to be an obvious deduction that the principal function of thyroxine might be simply to increase the number and activity of mitochondria, and these in turn increase the rate of formation of ATP to energize cellular function. Unfortunately, though, the increase in number and activity of mitochondria could as well be the *result* of increased activity of the cells as be the cause of the increase.

When *extremely* high concentrations of thyroid hormone are administered, the mitochondria swell inordinately, and there is uncoupling of the oxidative phosphorylation process with production of large amounts of heat but little ATP. However, under natural conditions, it is questionable whether the concentration of thyroid hormones becomes high enough to cause this effect even in persons who have thyrotoxicosis.

Effect of Thyroid Hormone in Increasing Active Transport of Ions Through Cell Membranes. One of the enzymes that becomes increased in response to thyroid hormone is *Na-K ATPase*. This in turn increases the rate of transport of both sodium and potassium through the cell membranes of some tissues. Since this process utilizes energy and also increases the amount of heat produced in the body, it has also been suggested that this might be one of the mechanisms by which thyroid hormone increases the body's metabolic rate.

Summary. It is clear that we know many effects that occur in the cells throughout the body under the influence of thyroid hormone. Yet, a specific metabolic mechanism that leads to all of these effects has been elusive. At present, the most likely basic function of the thyroid hormones is their capability to activate the DNA transcription process in the cell nucleus with resulting formation of many new cellular proteins.

EFFECT OF THYROID HORMONE ON GROWTH

Thyroid hormone has both general and specific effects on growth. For instance, it has long been known that thyroid hormone is essential for the metamorphic change of the tadpole into the frog. In the human being, the effect of thyroid hormone on growth is manifest mainly in growing children. In those who are hypothyroid, the rate of growth is greatly retarded. In those who are hyperthyroid, excessive skeletal growth often occurs, causing the child to become considerably taller than otherwise. However, the epiphyses close at an early age so that the eventual height of the adult may be shortened.

The growth-promoting effect of thyroid hormone is presumably based on its ability to promote protein synthesis. On the other hand, a great excess of thyroid hormone can cause more rapid catabolism than synthesis of protein, so that the protein stores are then actually mobilized and amino acids released into the extracellular fluids.

EFFECTS OF THYROID HORMONE ON SPECIFIC BODILY MECHANISMS

Effect on Carbohydrate Metabolism. Thyroid hormone stimulates almost all aspects of carbohydrate metabolism, including rapid uptake of glucose by the cells, enhanced glycolysis, enhanced gluconeogenesis, increased rate of absorption from the gastrointestinal tract, and even increased insulin secretion with its resultant secondary effects on carbohydrate metabolism. All these effects probably result from the overall increase in enzymes caused by thyroid hormone.

Effect on Fat Metabolism. Essentially all aspects of fat metabolism are also enhanced under the influence of thyroid hormone. However, since fats are the major source of long-term energy supplies, the fat stores of the body are depleted to a greater extent than are most of the other tissue elements; in particular, lipids are mobilized from the fat tissue, which increases the free fatty acid concentration in the plasma, and thyroid hormone also greatly accelerates the oxidation of free fatty acids by the cells.

Effect on Blood and Liver Fats. Increased thyroid hormone *decreases* the quantity of cholesterol, phospholipids, and triglycerides in the blood, even though it *increases* the free fatty acids. On the other hand, decreased thyroid secretion greatly increases the concentrations of cholesterol, phospholipids, and triglycerides and almost always causes excessive deposition of fat in the liver. The large increase in circulating blood lipids in prolonged hypothyroidism is often associated with severe arteriosclerosis, which was discussed in Chapter 68.

Effect on Vitamin Metabolism. Because thyroid hormone increases the quantities of many of the different enzymes and because vitamins are essential parts of some of the enzymes or coenzymes, thyroid hormone causes increased need for vitamins. Therefore, a relative vitamin deficiency can occur when excess thyroid hormone is secreted, unless at the same time increased quantities of vitamins are available.

Effect on Basal Metabolic Rate. Because thyroid hormone increases metabolism in most cells of the body (with the exception of the brain, retina, spleen, testes,

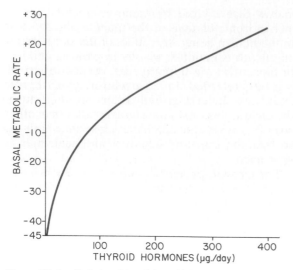

Figure 76–5. Relationship of thyroid hormone daily rate of secretion to the basal metabolic rate.

and lungs), excessive quantities of the hormone can occasionally increase the basal metabolic rate to as much as 60 to 100 per cent above normal. On the other hand, when no thyroid hormone is produced, the basal metabolic rate falls almost to half normal; that is, the basal metabolic rate becomes −30 to −45, as discussed in Chapter 71. Figure 76–5 shows the approximate relationship between the daily supply of thyroid hormones and the basal metabolic rate. Extreme amounts of the hormones are required to cause very high basal metabolic rates.

Effect on Body Weight. Greatly increased thyroid hormone production almost always decreases the body weight, and greatly decreased production almost always increases the body weight; but these effects do not always occur, because thyroid hormone increases the appetite, and this may overbalance the change in the metabolic rate.

Effect on the Cardiovascular System. *Blood Flow and Cardiac Output.* Increased metabolism in the tissues causes more rapid utilization of oxygen than normally and causes greater than normal quantities of metabolic end-products to be released from the tissues. These effects cause vasodilatation in most of the body tissues, thus increasing blood flow. Especially does the rate of blood flow in the skin increase because of the increased necessity for heat elimination.

As a consequence of the increased blood flow, the cardiac output also increases, sometimes rising to 50 per cent or more above normal when excessive thyroid hormone is present.

Heart Rate. The heart rate increases considerably more under the influence of thyroid hormone than would be expected simply because of the increased cardiac output. Therefore, thyroid hormone probably has a direct effect on the excitability of the heart, which in turn increases the heart rate. This effect is of particular importance because the heart rate is one of the most sensitive indices that the clinician has for determining whether a patient has excessive or diminished thyroid hormone production.

Strength of Heartbeat. The increased enzymatic activity caused by increased thyroid hormone production apparently increases the strength of the heart when only a slight excess of thyroid hormone is secreted. This is analogous to the increase in strength of heart beat that occurs in mild fevers and during exercise. However, when thyroid hormone is increased markedly, the heart muscle strength becomes depressed because of excessive protein catabolism. Indeed, some severely thyrotoxic patients die of cardiac decompensation secondary to myocardial failure and increased cardiac load imposed by the increased output.

Blood Volume. Thyroid hormone causes the blood volume to increase slightly. The effect probably results at least partly from the vasodilatation which allows increased quantities of blood to collect in the circulatory system.

Arterial Pressure. The increased cardiac output resulting from thyroid hormone tends to increase the arterial pressure. On the other hand, dilatation of the peripheral blood vessels due to the local effects of thyroid hormone and to excessive body heat tends to decrease the pressure. Therefore, the mean arterial pressure usually is unchanged. However, because of the increased rate of run-off of blood through the peripheral vessels, the pulse pressure is increased, with the systolic pressure elevated 10 to 20 mm. Hg, and the diastolic pressure correspondingly reduced.

Effect on Respiration. The increased rate of metabolism increases the utilization of oxygen and the formation of carbon dioxide; these effects activate all the mechanisms that increase the rate and depth of respiration.

Effect on the Gastrointestinal Tract. In addition to increased appetite and food intake, which has been discussed, thyroid hormone increases both the rate of secretion of the digestive juices and the motility of the gastrointestinal tract. Often, diarrhea results. Lack of thyroid hormone causes constipation.

Effect on the Central Nervous System. In general, thyroid hormone increases the rapidity of cerebration but also often dissociates this, while, on the other hand, lack of thyroid hormone decreases this function. The hyperthyroid individual is likely to develop extreme nervousness and is likely to have many psychoneurotic tendencies, such as anxiety complexes, extreme worry, or paranoias.

Effect on the Function of the Muscles. Slight increase in thyroid hormone usually makes the muscles react with vigor, but when the quantity of hormone becomes excessive, the muscles become weakened because of excess protein catabolism. On the other hand, lack of thyroid hormone causes the muscles to become extremely sluggish, and they relax slowly after a contraction.

Muscle Tremor. One of the most characteristic signs of hyperthyroidism is a fine muscle tremor. This is not the coarse tremor that occurs in Parkinson's disease or in shivering, for it occurs at the rapid frequency of 10 to 15 times per second. The tremor can be observed easily by placing a sheet of paper on the extended fingers and noting the degree of vibration of the paper. The cause of this tremor is not definitely known, but it is probably due to increased sensitivity of the neuronal synapses in the areas of the cord that control muscle tone. The tremor is an excellent means for assessing the degree of thyroid hormone effect on the central nervous system.

Effect on Sleep. Because of the exhausting effect of thyroid hormone on the musculature and on the central nervous system, the hyperthyroid subject often has a feeling of constant tiredness; but because of the excitable effects of thyroid hormone on the synapses, it is difficult to sleep. On the other hand, extreme somnolence is characteristic of hypothyroidism.

Effect on Other Endocrine Glands. Increased thyroid hormone increases the rates of secretion of most other endocrine glands, but it also increases the need of the tissues for the hormones. For instance, increased thyroxine secretion increases the rate of glucose metabolism everywhere in the body and therefore causes a corresponding need for increased insulin secretion by the pancreas. Also, thyroid hormone increases many metabolic activities related to bone formation and, as a consequence, increases the need for parathyroid hormone. And, finally, thyroid hormone increases the rate at which adrenal glucocorticoids are inactivated by the liver. This leads to feedback increase in ACTH production by the anterior pituitary and, therefore, increased rate of glucocorticoid secretion by the adrenal glands also.

Effect of Thyroid Hormone on Sexual Function. For normal sexual function to occur, thyroid secretion needs to be approximately normal — neither too great nor too little. In the male, lack of thyroid hormone is likely to cause complete loss of libido, while, on the other hand, great excesses of the hormone frequently cause impotence. In the female, lack of thyroid hormone often causes *menorrhagia* and *polymenorrhea*, which mean, respectively, excessive and frequent menstrual bleeding. In some women the lack may cause irregular periods and, occasionally, even total *amenorrhea*. A hypothyroid female, like the male, is also likely to have greatly decreased libido. Conversely, in the hyperthyroid female, *oligomenorrhea*, which means greatly reduced bleeding, is usual, and occasionally, amenorrhea results.

The action of thyroid hormone on the gonads cannot be pinpointed to a specific function but probably results from a combination of direct metabolic effects on the gonads and of excitatory and inhibitory effects operating through the anterior pituitary.

REGULATION OF THYROID HORMONE SECRETION

To maintain a normal basal metabolic rate, precisely the right amount of thyroid hormone must be secreted all the time, and, to provide this, a specific feedback mechanism operates through the hypothalamus and anterior pituitary gland to control the rate of thyroid secretion. This system can be explained as follows:

Effects of Thyroid-Stimulating Hormone on Thyroid Secretion. Thyroid-stimulating hormone (TSH), also known as *thyrotropin*, is an anterior pituitary hormone, a glycoprotein with a molecular weight of about 28,000, that was discussed in Chapter 75; it increases the secretion of thyroxine and triiodothyronine by the thyroid gland. Its specific effects on the thyroid gland are: (1) increased proteolysis of the thyroglobulin in the follicles, with resultant release of thyroid hormone into the circulating blood and diminishment of the follicular substance itself; (2) increased activity of the iodide pump, which increases the rate of "iodide trapping" in the glandular cells, sometimes increasing the ratio of intracellular to extracellular iodide concentration several-fold; (3) increased iodination of tyrosine and increased coupling to form the thyroid hormones; (4) increased size and increased secretory activity of the thyroid cells; and (5) increased number of thyroid cells, plus a change from cuboidal to columnar cells and much infolding of the thyroid epithelium into the follicles. In summary, thyroid-stimulating hormone *increases all the known activities of the thyroid glandular cells*.

The most important early effect following administration of thyroid-stimulating hormone is proteolysis of the thyroglobulin, which causes release of thyroxine and triiodothyronine into the blood within 30 minutes. The other effects require hours or even days and weeks to develop fully.

Role of Cyclic AMP in the Stimulatory Effect of TSH. In the past it was difficult to explain the many and varied effects of thyroid-stimulating hormone on the thyroid cell. However, it is now clear that at least most of these result from activation of the "second messenger" *cyclic AMP* system of the cell. The first event in this activation is binding of the thyroid-stimulating hormone with specific TSH receptors on the basal membrane surfaces of the cell. This then activates *adenyl-cyclase* in the membrane, which increases the formation of cyclic AMP in the cell. Finally, the cyclic AMP acts as a *second messenger* to activate essentially all systems of the thyroid cell. The result is both an immediate increase in secretion of thyroid hormones and prolonged growth of the thyroid glandular tissue itself. This method for control of thyroid cell activity is similar to the function of cyclic AMP in many other target tissues of the body.

Hypothalamic Regulation of TSH Secretion by the Anterior Pituitary — Thyrotropin-Releasing Hormone (TRH)

Electrical stimulation of several areas of the hypothalamus, but most particularly of the paraventricular and arcuate nuclei, increases the anterior pituitary secretion of TSH and correspondingly increases the activity of the thyroid gland. This control of anterior pituitary secretion is exerted by a hypothalamic hormone, *thyrotropin-releasing hormone* (TRH), which is secreted by nerve endings in the median eminence of the hypothalamus and then transported from there to the anterior pituitary in the hypothalamic-hypophysial portal blood, as was explained in Chapter 75. TRH has been obtained in pure form, and it has proved to be a very simple substance, a tripeptide amide — *pyroglutamyl-histidyl-proline-amide*. TRH directly affects the anterior pituitary gland cells to increase their output of thyroid-stimulating hormone. When the portal system from the hypothalamus to the anterior pituitary gland is completely blocked, the rate of secretion of TSH by the anterior pituitary is greatly decreased but not reduced to zero.

The hypothalamus can also inhibit anterior pituitary secretion of thyroid-stimulating hormone. It does this by secreting *somatostatin,* which inhibits the secretion of TSH at the same time that it also inhibits growth hormone secretion. But the role of somatostatin in the overall thyroid control system is unknown.

Effects of Cold and Other Neurogenic Stimuli on TRH and TSH Secretion.   One of the best-known stimuli for increasing the rate of TRH secretion by the hypothalamus, and therefore TSH secretion by the anterior pituitary, is exposure of an animal to cold. Exposure of rats for several weeks increases the output of thyroid hormones sometimes more than 100 per cent and can increase the basal metabolic rate as much as 50 per cent. Indeed, people moving to arctic regions have been known to develop basal metabolic rates 15 to 20 per cent above normal; however, the behavioral propensity of human beings to protect themselves from cold usually prevents a measurable effect.

Various emotional reactions can also affect the output of TRH and TSH and can, therefore, indirectly affect the secretion of thyroid hormones. On the other hand, excitement and anxiety — conditions that greatly stimulate the sympathetic nervous system — cause acute decrease in secretion of TSH, perhaps because these states increase the metabolic rate and the body heat.

Neither these emotional effects nor the effect of cold is observed when the hypophysial stalk is cut, illustrating that both these effects are mediated by way of the hypothalamus.

Inverse Feedback Effect of Thyroid Hormone on Anterior Pituitary Secretion of TSH — Feedback Regulation of Thyroid Secretion.   Increased thyroid hormone in the body fluids decreases the secretion of TSH by the anterior pituitary. When the rate of thyroid hormone secretion rises to about 1.75 times normal, the rate of TSH secretion

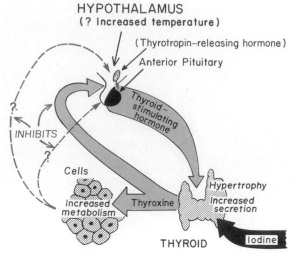

Figure 76–6.   Regulation of thyroid secretion.

falls essentially to zero. Most of this feedback depressant effect occurs even when the anterior pituitary has been completely separated from the hypothalamus, but the effect is somewhat greater if the hypothalamus and hypothalamic-hypophysial portal system are intact. Therefore, as illustrated in Figure 76–6, it is probable that increased thyroid hormone inhibits anterior pituitary secretion of TSH mainly by a direct feedback effect on the anterior pituitary itself but perhaps secondarily by much weaker effects acting through the hypothalamus.

One mechanism that has been suggested for the feedback effect on the anterior pituitary gland is that thyroid hormone reduces the number of TRH receptors on the cells that secrete thyroid-stimulating hormone. Therefore, the stimulating effect on these cells by thyrotropin-releasing hormone from the hypothalamus is greatly reduced.

Regardless of the mechanism of the feedback, its effect is to maintain an almost constant concentration of free thyroid hormone in the circulating body fluids. For instance, during periods of heavy exercise, thyroid hormone is consumed much more rapidly than under resting conditions; yet, because of appropriate feedback control, the rate of secretion of thyroid hormone rises to equal the rate of consumption, and the blood thyroid hormone concentration remains almost exactly constant.

If there is a feedback effect through the hypothalamus in addition to the direct feedback to the pituitary gland itself, this probably operates very slowly and might be caused at least partly by changes in the temperature of the hypothalamic thermostat, which are known to have significant effects in controlling the thyroid hormone system.

ANTITHYROID SUBSTANCES

Drugs that suppress thyroid secretion are called antithyroid substances. The three best known of these are thiocyanate, propylthiouracil, and high concentrations of inorganic iodides. The mechanism by which each of these blocks thyroid secretion is different from the others, and they can be explained as follows:

Decreased Iodide Trapping Caused by Thiocyanate Ions. Administration of thiocyanates (or perchlorates, nitrates, and many other similar compounds) decreases the rate at which iodide is pumped into the thyroid glandular cells, and, therefore, reduces the availability of iodides to the intracellular processes for the formation of thyroxine and triiodothyronine.

Lack of formation of thyroid hormone by the gland under the influence of thiocyanates causes the thyroid gland to enlarge — that is, to become a goiter. The mechanism of this is the following: Lack of thyroid hormone decreases the feedback inhibition of the anterior pituitary, which allows increased secretion of TSH by the pituitary. This then stimulates the glandular cells in the thyroid, making them secrete more and more thyroglobulin into the follicles, even though this does not contain significant quantities of thyroxine and triiodothyronine.

Depression of Thyroid Hormone Formation by Propylthiouracil. Propylthiouracil (and other similar compounds such as methimazole and carbimazole) prevents formation of thyroid hormone from iodides and tyrosine. The mechanism of this is partly to block iodination of tyrosine but to block many other aspects of thyroid hormone formation as well, especially the coupling of two iodinated tyrosines to form the thyroxine or triiodothyronine.

Propylthiouracil does not prevent formation of thyroglobulin, but the absence of thyroxine and triiodothyronine in the thyroglobulin leads to tremendous feedback enhancement of TSH secretion by the anterior pituitary gland. Therefore, in the same manner that thiocyanates cause the thyroid gland to enlarge, so also does propylthiouracil lead to enhanced growth of the glandular tissue, thus forming a goiter.

Decrease in Thyroid Activity Caused by Iodides. When iodides are present in the blood in high concentration (100 times the normal plasma level), most activities of the thyroid gland are decreased, but often they remain decreased only for a few weeks. The rate of iodide trapping is reduced, the rate of thyroid hormone formation is decreased, the secretory activity of the thyroid cells is decreased, and the rate of thyroid hormone release from the thyroglobulin is decreased. Since these are almost exactly opposite to the effects of TSH on the thyroid gland, it has been suggested that high concentrations of iodides in the blood directly inhibit the thyroid-stimulating effect of TSH.

Because iodides in high concentrations decrease all phases of thyroid activity, they decrease the size of the thyroid gland and especially decrease its blood supply, in contradistinction to the opposite effects caused by most of the other antithyroid agents. For this reason, iodides are frequently administered to patients for two or three weeks prior to surgical removal of the thyroid gland to decrease the necessary amount of surgery.

DISEASES OF THE THYROID

HYPERTHYROIDISM

Most effects of hyperthyroidism are obvious from the preceding discussion of the various physiological effects of thyroid hormone. However, some specific effects should be mentioned in connection especially with the development, diagnosis, and treatment of hyperthyroidism.

Causes of Hyperthyroidism (Toxic Goiter, Thyrotoxicosis, Graves' Disease). In the patient with hyperthyroidism the entire thyroid gland is usually markedly hyperplastic. It is increased to 2 to 3 times normal size, with tremendous folding of the follicular cell lining into the follicles so that the number of cells is increased several times as much as the size of the gland is increased. Also, each cell increases its rate of secretion several-fold; radioactive iodine uptake studies indicate that these hyperplastic glands secrete thyroid hormone at a rate as great as 5 to 15 times normal.

These changes in the thyroid gland are similar to those caused by excessive thyroid-stimulating hormone. However, radioimmunoassay studies have shown the plasma TSH concentrations to be less than normal rather than enhanced, and often to be essentially zero. On the other hand, some other substance or substances that have actions similar to that of TSH are found in the blood of almost all these patients. These substances are usually globulin antibodies that bind with the thyroid cell membranes. It is postulated that they bind with the same membrane receptors that bind TSH and that this induces continual activation of the cyclic AMP system of the cells, with the resultant development of hyperthyroidism. One of these antibodies, found in 50 to 80 per cent of thyrotoxic patients, is called *long-acting thyroid stimulator* (LATS). This has a prolonged stimulating effect on the thyroid gland, lasting as long as 12 hours in contrast to a little over 1 hour for TSH. The high level of thyroid hormone secretion caused by LATS, in turn, suppresses anterior pituitary formation of TSH.

The antibodies that cause hyperthyroidism almost certainly develop as the result of autoimmunity that has developed against thyroid tissue. Presumably, at some time in the history of the person an excess of thyroid cell antigens has been released from the thyroid cells, and this has resulted in the formation of antibodies against the thyroid gland itself.

Aside from the above variety of hyperthyroidism, the condition occasionally also results from a localized adenoma (a tumor) that develops in the thyroid tissue and secretes large quantities of thyroid hormone. This is different from the more usual type of hyperthyroidism in that it is not associated with any evidences of autoimmune disease. An interesting effect of the adenoma is that as long as it continues to secrete large quantities of thyroid hormone, function in the remainder of the thyroid gland is almost totally inhibited because the thyroid hormone from the adenoma depresses the production of TSH by the pituitary gland.

Symptoms of Hyperthyroidism

The symptoms of hyperthyroidism are obvious from the preceding discussion of the physiology of the thyroid hormones — intolerance to heat, increased sweating, mild to extreme weight loss (sometimes as much as 100 pounds), varying degrees of diarrhea, muscular weakness, nervousness or other psychic disorders, extreme fatigue but inability to sleep, and tremor of the hands.

Exophthalmos. Most, but not all, persons with hyperthyroidism develop some degree of protrusion of the eyeballs, as illustrated in Figure 76–7. This condition is called *exophthalmos*. A major degree of exophthalmos occurs in about one third of the hyperthyroid patients, and the condition on rare occasions becomes severe enough that the eyeball protrusion stretches the optic nerve. Much more often, the eyes are damaged because the eyelids do not close completely when the person blinks or is asleep. As a result, the dry epithelial surfaces of the eyes become irritated and often infected, resulting in ulceration of the cornea.

The cause of the protruding eyes is edematous swelling of the retro-orbital tissues and deposition of large quantities of mucopolysaccharides in the extracellular spaces; the factor or factors that initiate these changes are still in serious dispute. In most patients, antibodies can be found in the blood that react with the retro-orbital tissues. Therefore, there is much reason to believe that exophthalmos, like hyperthyroidism itself, is an autoimmune process. However, in many patients a hormonal substance called *exophthalmos-producing substance* can also be found in the plasma that will cause exophthalmos in animals into which it is injected. Some experiments have suggested that this substance is a split-off fragment of TSH and that it might be secreted by the anterior pituitary gland instead of TSH itself when hyperthyroidism develops. Thus, the riddle of

Figure 76–7. Patient with exophthalmic hyperthyroidism. Note protrusion of the eyes and retraction of the superior eyelids. The basal metabolic rate was +40. (Courtesy of Dr. Leonard Posey.)

exophthalmos still is not solved, though the most likely answer at present is that it is an autoimmune response. Usually, the exophthalmos disappears or at least greatly ameliorates with treatment of the hyperthyroidism.

Diagnostic Tests for Hyperthyroidism. For the usual case of hyperthyroidism, the most accurate diagnostic test is direct measurement of the concentration of "free" thyroxine in the plasma, using appropriate radioimmunoassay procedures.

Other tests that are frequently used are:

(1) The basal metabolic rate is usually increased to +30 or +60 in severe hyperthyroidism.

(2) The rate of uptake of a standard injected dose of radioactive iodine by the normal thyroid gland, when measured by a calibrated radioactive detector placed over the neck, is about 4 per cent per hour. In the hyperthyroid person, this can rise to as high as 20 to 25 per cent per hour.

(3) The amount of iodine bound to plasma proteins is usually, but not always, directly proportional to the amount of circulating thyroxine. Therefore, elevation of this is also often used in the diagnosis of hyperthyroidism.

Physiology of Treatment in Hyperthyroidism. The most direct treatment for hyperthyroidism is surgical removal of the thyroid gland. In general, it is desirable to prepare the patient for surgical removal of the gland prior to the operation. This is done by administering propylthiouracil, usually for several weeks, until the basal metabolic rate of the patient has returned to normal. Then administration of high concentrations of iodides for two weeks immediately prior to operation causes the gland itself to recede in size and its blood supply to diminish. By using these preoperative procedures, the operative mortality is less than 1 in 1000 in the better hospitals, whereas prior to development of these procedures the operative mortality was as great as 1 in 25.

Treatment of the Hyperplastic Thyroid Gland with Radioactive Iodine. As much as 80 to 90 per cent of an injected dose of iodide is absorbed by the hyperplastic, toxic thyroid gland within a day after injection. If this injected iodine is radioactive, it can destroy internally the secretory cells of the thyroid gland. Usually 5 millicuries of radioactive iodine is given to the patient, and then his condition is reassessed several weeks later. If he is still hyperthyroid, additional doses are repeated until he reaches a normal thyroid status. These quantities of radioiodine are about 1000 times as great as those used for diagnosis of hyperthyroidism as discussed above.

Thyroid Storm. Occasionally, patients with extremely severe thyrotoxicosis develop a state called *thyroid storm,* or *thyroid crisis* in which all the usual symptoms of thyrotoxicosis are excessively accentuated. The effects may be delirium, high fever, abnormal rhythm of the heart, extreme sweating, shock, vomiting, and dehydration. This condition is particularly likely to occur during the first day following surgical removal of a large hyperplastic thyroid gland, presumably because of excessive release of thyroid hormone into the circulatory system during the operative procedure. The extreme overactivity of the body's tissues can be so damaging that without treatment almost all persons entering this state die. However, by cooling the patient rapidly with ice or alcohol sponge baths and also administering large quantities of glucocorticoid hormones, it is now possible to save approximately half the patients. The glucocorticoids seem to be especially important, possibly because of their effect of reducing the breakdown of lysosomes in the damaged tissues and therefore of preventing autodigestion of the cells.

HYPOTHYROIDISM

The effects of hypothyroidism in general are opposite to those of hyperthyroidism, but here again, a few physiological mechanisms peculiar to hypothyroidism alone are involved.

Hypothyroidism, like hyperthyroidism, probably also results in most instances from autoimmunity against the thyroid gland, but immunity that destroys the gland rather than stimulating it. The thyroid glands of most of these patients first develop "thyroiditis," which means thyroid inflammation. This causes progressive deterioration and finally fibrosis of the gland, with resultant diminished or absent secretion of thyroid hormone. However, several other types of hypothyroidism also occur, often associated with development of enlarged thyroid glands, called thyroid goiter.

Endemic Colloid Goiter. The term goiter means a greatly enlarged thyroid gland. As pointed out in the discussion of iodine metabolism, about 50 mg. of iodine is necessary each year for the formation of adequate quantities of thyroid hormone. In certain areas of the world, notably in the Swiss Alps, in the Andes, and in the Great Lakes region of the United States, insufficient iodine is present in the soil for the foodstuffs to contain even this minute quantity of iodine. Therefore, in days prior to iodized table salt, many persons living in these areas developed extremely large thyroid glands called *endemic goiters*.

The mechanism for development of the large endemic goiters is the following: Lack of iodine prevents production of thyroid hormone by the thyroid gland, and, as a result, no hormone is available to inhibit production of TSH by the anterior pituitary; this allows the pituitary to secrete excessively large quantities of TSH. The TSH then causes the thyroid cells to secrete tremendous amounts of thyroglobulin (colloid) into the follicles, and the gland grows larger and larger. But unfortunately, due to lack of iodine, increased thyroxine and triiodothyronine production does not occur. The follicles become tremendous in size, and the thyroid gland may increase to as large as 300 to 500 grams or more.

Idiopathic Nontoxic Colloid Goiter. Enlarged thyroid glands similar to those of endemic colloid goiter frequently also occur in persons who do not have iodine deficiency. These goitrous glands may secrete normal quantities of thyroid hormones but more frequently the secretion of hormone is depressed, as in endemic colloid goiter.

The exact cause of the enlarged thyroid gland in patients with idiopathic colloid goiter is not known, but most of these patients show signs of mild thyroiditis; therefore, it has been suggested that the thyroiditis causes slight hypothyroidism, which then leads to increased TSH secretion and progressive growth of the noninflamed portions of the gland. This could explain why these glands usually are very nodular, with some portions of the gland growing while other portions are being destroyed by thyroiditis.

In some persons with colloid goiter, the thyroid glands have abnormal enzyme systems, which leads to diminished thyroid hormone formation and resultant excess stimulation of the thyroid gland by TSH. And, finally, some foods contain *goitrogenic substances* that have a propylthiouracil-type of anti-thyroid activity, thus also leading to TSH-stimulated enlargement of the thyroid gland. Such goitrogenic substances are found in some varieties of turnips and cabbages.

Characteristics of Hypothyroidism. Whether hypothyroidism is due to thyroiditis, endemic colloid goiter, idiopathic colloid goiter, destruction of the thyroid gland by irradiation, or surgical removal of the thyroid gland, the physiological effects are the same. These include fatigue and extreme somnolence with sleeping 14 to 16 hours a day, extreme muscular sluggishness, slowed heart rate, decreased cardiac output, decreased blood volume, sometimes increased weight, constipation, mental sluggishness, failure of many trophic functions in the body evidenced by depressed growth of hair and scaliness of the skin, development of a froglike husky voice, and, in severe cases, development of an edematous appearance throughout the body called myxedema.

Myxedema. The patient with almost total lack of thyroid function develops *myxedema*. Figure 76–8 shows such a patient, illustrating bagginess under the eyes and swelling of the face. In this condition, for reasons not yet explained, greatly increased quantities of proteoglycans, containing mainly hyaluronic acid, collect in the interstitial spaces, and this causes the total

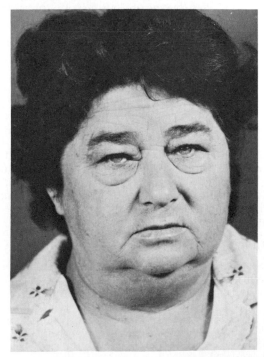

Figure 76–8. Patient with myxedema. (Courtesy of Dr. Herbert Langford.)

quantity of interstitial fluid also to increase. The fluid is adsorbed to the proteoglycans, thus greatly increasing the quantity of "ground substance" gel in the tissues. Because of the gel nature of the excess fluid, it is relatively immobile, and the edema is nonpitting in type.

Arteriosclerosis in Hypothyroidism. As pointed out in Chapter 68, lack of thyroid hormone increases the quantity of blood lipids, most importantly cholesterol, and the increase in blood cholesterol is usually associated with increased atherosclerosis and arteriosclerosis. Therefore, many hypothyroid patients, particularly those with myxedema, develop arteriosclerosis, which results in peripheral vascular disease, deafness, and often extreme coronary sclerosis with consequent early demise.

Diagnostic Tests in Hypothyroidism. The tests already described for diagnosis of hyperthyroidism give the opposite results in hypothyroidism. The free thyroxine in the blood is low. The basal metabolic rate in myxedema ranges between −30 and −40. The protein-bound iodine is as little as one-third normal. And the rate of radioactive iodine uptake by the thyroid gland (except in iodine deficiency hypothyroidism) measures less than 1 per cent per hour rather than the normal of approximately 4 per cent per hour. Probably more important for diagnosis than the various diagnostic tests, however, are the characteristic symptoms of hypothyroidism as just discussed.

Treatment of Hypothyroidism. Figure 76–4 shows the effect of thyroid hormone on the basal metabolic rate, illustrating that the hormone normally has a duration of action of more than one month. Consequently, it is easy to maintain a steady level of thyroid hormone activity in the body by daily oral ingestion of a tablet or so of desiccated thyroid gland or thyroid extract. Furthermore, proper treatment of the hypothyroid patient results in such complete normality that formerly myxedematous patients properly treated have lived into their 90's after treatment for over 50 years.

When a person with myxedema is first treated, immediate diuresis ensues, owing to removal of the myxedematous fluid from the interstitial spaces as the proteoglycan in this fluid is metabolized. Obviously, treatment increases the activity in the circulatory system, and, if the person has already developed severe coronary artery disease, the increased activity of the heart may lead to anginal pain or even to congestive heart failure. On the other hand, lack of treatment may lead to increased severity of the coronary disease. In this instance the clinician is in a dilemma and must proceed with progressive treatment slowly if at all.

Cretinism

Cretinism is the condition caused by extreme hypothyroidism during infancy and childhood, and it is characterized especially by failure of growth. Cretinism results from congenital lack of a thyroid gland (*congenital cretinism*), from failure of the thyroid gland to produce thyroid hormone because of a genetic deficiency of the gland, or from iodine lack in the diet (*endemic cretinism*). The severity of endemic cretinism varies greatly, depending on the amount of iodine in the diet,

and whole populaces of an endemic area have been known to have cretinoid tendencies.

A newborn baby without a thyroid gland may have absolutely normal appearance and function because he had been supplied with thyroid hormone by the mother while *in utero,* but a few weeks after birth his movements become sluggish, and both his physical and mental growth are greatly retarded. Treatment of the cretin at any time usually causes normal return of physical growth, but, unless the cretin is treated within a few months after birth, his mental growth will be permanently retarded. This is probably due to the fact that physical development of the neuronal cells of the central nervous system is rapid during the first year of life so that any retardation at this point is extremely detrimental.

Skeletal growth in the cretin is characteristically more inhibited than is soft tissue growth, though both are inhibited to a certain extent. However, as a result of this disproportionate rate of growth, the soft tissues are likely to enlarge excessively, giving the cretin the appearance of an obese and stocky, short child. Indeed, occasionally the tongue becomes so large in relation to the skeletal growth that it obstructs swallowing and breathing, inducing a characteristic guttural breathing that sometimes chokes the baby.

REFERENCES

Burrow, G. N.: The thyroid gland and reproduction. *In* Yen, S. S. C., and Jaffe, R. B. (eds.): Reproductive Endocrinology. Philadelphia, W. B. Saunders Co., 1978, p. 373.

Daniels, G. H., and Maloof, F.: Regulatory mechanisms of the pituitary-thyroid axis. *In* Frohlich, E. D. (ed.): Pathophysiology, 2nd Ed. Philadelphia, J. B. Lippincott Co., 1976, p. 341.

DeGroot, L. J.: Thyroid hormone action. *In* DeGroot, L. J., *et al.* (eds.): Endocrinology. Vol. 1. New York, Grune & Stratton, 1979, p. 357.

DeGroot, L. J., and Taurog, A.: Secretion of thyroid hormone. *In* DeGroot, L. J., *et al.* (eds.): Endocrinology. Vol. 1. New York, Grune & Stratton, 1979, p. 343.

Dumont, J. E., and Vassart, G.: Thyroid gland metabolism and the action of TSH. *In* DeGroot, L. J., *et al.* (eds.): Endocrinology. Vol. 1. New York, Grune & Stratton, 1979, p. 311.

Edelman, I. S.: Thyroid thermogenesis. *N. Engl. J. Med., 290:*1303, 1974.

Ekins, R., *et al.* (eds.): Free Thyroid Hormones. New York, Excerpta Medica, 1979.

Ermans, A. M.: Endemic goiter and enedmic cretinism. *In* DeGroot, L. J., *et al.* (eds.): Endocrinology. Vol. 1. New York, Grune & Stratton, 1979, p. 501.

Greenberg, A. H., *et al.:* Effects of thyroid hormone on growth, differentiation, and development. *In* Greep, R. O., and Astwood, E. B. (eds.): Handbook of Physiology. Sec. 7, Vol. 3. Baltimore, Williams & Wilkins, 1974, p. 377.

Greer, M. A., and Haibach, H.: Thyroid secretion. *In* Greep, R. O., and Astwood, E. B. (eds.): Handbook of Physiology. Sec. 7, Vol. 3. Baltimore, Williams & Wilkins, 1974, p. 135.

Hennessy, J. F.: Hypothyroidism: Discussions in Patient Management. Garden City, N.Y., Medical Examination Publishing Co., 1978.

Jensen, E. V., and DeSombre, E. R.: Steroid hormone receptors and action. *In* DeGroot, L. J., *et al.* (eds.): Endocrinology. Vol. 3. New York, Grune & Stratton, 1979, p. 2055.

Li, C. H. (ed.): Thyroid Hormones. New York, Academic Press, 1978.

Martin, J. B.: Regulation of the pituitary-thryoid axis. *In* MTP International Review of Science: Physiology. Vol. 5. Baltimore, University Park Press, 1974, p. 67.

McClung, M. R., and Greer, M. A.: Treatment of hyperthyroidism. *Annu. Rev. Med., 31:*385, 1980.

McKenzie, J. M., and Zakarija, M.: Hyperthyroidism. *In* DeGroot, L.

J., *et al.* (eds.): Endocrinology. Vol. 1. New York, Grune & Stratton, 1979, p. 429.

Middlesworth, L. V.: Metabolism and excretion of thyroid hormones. *In* Greep, R. O., and Astwood, E. B. (eds.): Handbook of Physiology. Sec. 7. Vol. 3. Baltimore, Williams &Wilkins, 1974, p. 215.

Ramsden, D. B.: Peripheral Metabolism and Action of Thyroid Hormones. Montreal, Eden Press, 1977.

Refetoff, S.: Thyroid function tests. *In* DeGroot, L. J., *et al.* (eds.): Endocrinology. Vol. 1. New York, Grune & Stratton, 1979, p. 387.

Robbins, J., *et al.*: The thyroid and iodine metabolism. *In* Bondy, P. K. and Rosenberg, L. E. (eds.): Metabolic Control and Disease, 8th Ed. Philadelphia, W. B. Saunders Co., 1980, p. 1325.

Roth, J.: Receptors for peptide hormones. *In* DeGroot, L. J., *et al.*

(eds.): Endocrinology. Vol. 3. New York, Grune & Stratton, 1979, p. 2037.

Stanbury, J. B. (ed.): Endemic Goiter and Endemic Cretinism. New York, John Wiley & Sons, 1980.

Stanbury, J. B., and Kroc, R. L. (eds.): Human Development and the Thyroid Gland. New York, Plenum Press, 1973.

Sterling, K., and Lazarus, J. H.: The thyroid and its control. *Annu. Rev. Physiol., 39*:349, 1977.

Utiger, R. D.: Hypothyroidism. *In* DeGroot, L. J., *et al.* (eds.): Endocrinology. Vol. 1. New York, Grune & Stratton, 1979, p. 471.

Wilber, J. F.: Human pituitary thyrotropin. *In* DeGroot, L. J., *et al.* (eds.): Endocrinology. Vol. 1. New York, Grune & Stratton, 1979, p. 141.

77

The Adrenocortical Hormones

The adrenal glands, which lie at the superior poles of the two kidneys, are each composed of two distinct parts, the *adrenal medulla* and the *adrenal cortex*. The adrenal medulla is functionally related to the sympathetic nervous system, and it secretes the hormones *epinephrine* and *norepinephrine* in response to sympathetic stimulation. In turn, these hormones cause almost the same effects as direct stimulation of the sympathetic nerves in all parts of the body. These hormones and their effects were discussed in detail in Chapter 57 in relation to the sympathetic nervous system.

The adrenal cortex secretes an entirely different group of hormones called *corticosteroids*. These hormones are all synthesized from the steroid cholesterol, and they all have similar chemical formulas. However, very slight differences in their molecular structures, which will be discussed later in the chapter, give them several very different but very important functions.

Mineralocorticoids and Glucocorticoids. The adrenocortical hormones do not all cause exactly the same effects in the body. Two major types of hormones, the *mineralocorticoids* and the *glucocorticoids*, are secreted by the adrenal cortex. In addition to these, small amounts of sex hormones are secreted, especially *androgenic hormones*, which exhibit the same effects in the body as the male sex hormone testosterone. These are normally unimportant, though in certain abnormalities of the adrenal cortices extreme quantities can be secreted (which is discussed later in the chapter) and can then result in masculinizing effects.

The *mineralocorticoids* have gained this name because they especially affect the electrolytes of the extracellular fluids — sodium and potassium, in particular. The *glucocorticoids* have gained this name because they exhibit an important effect in increasing blood glucose concentration. However, the glucocorticoids have additional effects on both protein and fat metabolism which may be equally

as important to body function as are their effects on carbohydrate metabolism.

Over 30 different steroids have been isolated from the adrenal cortex, but only two of these are of major importance to the endocrine functions of the body — *aldosterone,* which is the principal mineralocorticoid, and *cortisol,* which is the principal glucocorticoid.

FUNCTIONS OF THE MINERALOCORTICOIDS — ALDOSTERONE

Total loss of adrenocortical secretion usually causes death within three days to two weeks unless the person receives extensive salt therapy or mineralocorticoid therapy. Without mineralocorticoids, the potassium ion concentration of the extracellular fluid rises markedly, the sodium and chloride concentrations decrease, and the total extracellular fluid volume and blood volume also become greatly reduced. The person soon develops diminished cardiac output, which proceeds to a shocklike state followed by death. This entire sequence can be prevented by the administration of aldosterone or some other mineralocorticoid. Therefore, the mineralocorticoids are said to be the acute "life-saving" portion of the adrenocortical hormones, while the glucocorticoids are equally necessary to allow the person to resist the destructive effects of life's intermittent "stresses," as is discussed later in the chapter.

Aldosterone exerts at least 95 per cent of the mineralocorticoid activity of the adrenocortical secretion, but cortisol, the major glucocorticoid secreted by the adrenal cortex, also provides a small amount of mineralocorticoid activity. Other adrenal steroids secreted in small amounts that have mineralocorticoid effects are *corticosterone,* which also exerts glucocorticoid effects, and *des-*

oxycorticosterone, which has almost the same effects as aldosterone but with a potency one thirtieth that of aldosterone.

RENAL EFFECTS OF ALDOSTERONE

By far the most important function of aldosterone is to cause transport of sodium and potassium through the renal tubular walls and, to a lesser extent, transport of hydrogen ions. The mechanisms of these effects were discussed in detail in Chapters 34 through 37. However, let us summarize briefly the renal and body fluid effects of aldosterone.

Effect on Tubular Reabsorption of Sodium and Tubular Secretion of Potassium. It will be recalled from Chapter 34 that aldosterone causes an exchange transport of sodium and potassium — that is, absorption of sodium and simultaneous excretion of potassium by the tubular epithelial cells — in both the distal tubule and the collecting duct. Therefore, aldosterone causes sodium to be conserved in the extracellular fluid while potassium is excreted into the urine.

A high concentration of aldosterone in the plasma can decrease the sodium loss into the urine to as little as a few milligrams a day. At the same time, potassium loss into the urine increases many-fold. Conversely, total lack of aldosterone secretion can cause loss of as much as 20 grams of sodium in the urine a day, an amount equal to one fifth of all the sodium in the body. But, at the same time, potassium is conserved tenaciously in the extracellular fluid.

Therefore, the net effect of excess aldosterone in the plasma is to increase the total quantity of sodium in the extracellular fluid while decreasing the potassium. Yet, strangely enough, a large increase in aldosterone increases the sodium ion *concentration* in the extracellular fluid very little. Instead, it increases the extracellular fluid volume considerably. These differences can be explained as follows:

Mild Effect of Aldosterone on Sodium Ion Concentration. When aldosterone causes excess sodium reabsorption, this also causes reabsorption of almost an equivalent amount of water, an effect that was described in detail in Chapter 36. The main reason is that, when sodium is absorbed through the tubular epithelium, an osmotic gradient is created from the tubules toward the peritubular fluids, and water then "follows" the sodium. Secondly, if not enough water follows, the sodium concentration in the extracellular fluid rises slightly and elicits increased antidiuretic hormone secretion as well as increased thirst; these two together then enhance the amount of water in the extracellular fluid. Consequently, even vast amounts of aldos-

terone will rarely raise the sodium ion concentration more than 2 to 3 per cent, and total lack of aldosterone causes only a 5 to 8 per cent decrease in sodium concentration.

Effect on Extracellular Fluid Volume. Even though aldosterone increases the sodium ion concentration very little, the combined, almost equal absorption of sodium and water increases extracellular fluid volume much more, increasing the volume to as high as 20 per cent above normal with great excesses of aldosterone; or the volume may decrease to as low as 20 to 25 per cent below normal in the absence of aldosterone. Consequently, the extracellular fluid volume tends to change in proportion to the rate of aldosterone secretion.

Aldosterone Escape. When aldosterone is first administered, its effect on sodium retention and increase in extracellular fluid volume is maximal in about 3 days. However, after a few more days, diuresis and natriuresis occur, and most of the retained sodium and excess extracellular fluid volume is lost in the urine, leaving a final increase in extracellular fluid volume rarely exceeding 10 per cent. This secondary loss of sodium and extracellular fluid is called *aldosterone escape.* The precise mechanism of this escape is not known. Experiments in our laboratory have suggested that it results from a secondary rise in arterial blood pressure of 5 to 15 mm. Hg, which is enough to cause the pressure diuresis and natriuresis phenomena, as was explained in Chapters 22 and 36. However, other physiologists have suggested that some unknown hormonal changes are responsible or that some adaptation occurs in the kidney itself.

Hypokalemia and Muscle Paralysis; Hyperkalemia and Cardiac Toxicity. The excessive loss of potassium ions from the extracellular fluid into the urine under the influence of aldosterone causes serious decrease in the plasma potassium concentration, often decreasing it from the normal value of 4.5 mEq./l. to as low as 1 to 2 mEq./l. This condition is called *hypokalemia.* When the potassium ion concentration falls below approximately one-half normal, muscle paralysis or at least severe muscle weakness often develops. This is caused by hyperpolarization of the nerve and muscle fiber membranes (see Chapter 10), which prevents transmission of action potentials.

On the other hand, when aldosterone is deficient, the extracellular fluid potassium ion concentration can rise far above normal. When it rises to approximately double normal, serious cardiac toxicity, including weakness of contraction and arrhythmia, becomes evident; a slightly higher concentration of potassium leads inevitably to a cardiac death.

Effect of Aldosterone on Increasing Tubular Hydrogen Ion Secretion, with Resultant Mild Alkalosis. Though aldosterone mainly causes potassium to be secreted into the tubules in exchange for sodium reabsorption, to a much lesser extent it also causes tubular secretion of hydrogen ions in exchange for sodium. The obvious effect of this is to decrease the hydrogen ion concentration in the extracellular fluid. However, this effect is not a strong one, usually causing only a mild degree of alkalosis.

Effect of Aldosterone on Circulatory Function. The circulatory effects of aldosterone result almost entirely from the increase in extracellular fluid volume. In the absence of aldosterone secretion, with a decrease in extracellular fluid volume to 20 to 25 per cent below normal and a comparable decrease in plasma volume, circulatory shock develops rapidly. Indeed, in complete lack of aldosterone, a person not treated with extra intake of salt and/or administration of a mineralocorticoid drug is likely to die of circulatory shock within as few as four to eight days.

In the case of hypersecretion of aldosterone, not only is the extracellular fluid volume increased but the blood volume and cardiac output as well. Each of these can increase to as much as 20 to 30 per cent above normal in the first few days of excess aldosterone secretion, but after aldosterone escape occurs, the volumes and cardiac output usually return to no more than 5 to 10 per cent above normal. Nevertheless, over a prolonged period of time even these small increases are sufficient to cause moderate to severe hypertension, as we shall discuss later in the chapter in relation to primary aldosteronism.

EFFECTS OF ALDOSTERONE ON SWEAT GLANDS, SALIVARY GLANDS, AND INTESTINAL ABSORPTION

Aldosterone has almost the same effects on sweat glands and salivary glands that it has on the renal tubules. Both of these glands form a primary secretion that contains large quantities of sodium chloride, but much of the sodium chloride on passing through the excretory ducts is reabsorbed while potassium and bicarbonate ions are secreted. Aldosterone greatly increases the reabsorption of sodium chloride and the secretion of potassium. The effect on the sweat glands is important to conserve body salt in hot environments, and the effect on the salivary glands is necessary to conserve salt when excessive quantities of saliva are lost.

Aldosterone also greatly enhances sodium absorption by the intestines, which obviously prevents loss of sodium in the stools. On the other hand, in the absence of aldosterone, sodium absorption from the intestine can be very poor, leading to failure to absorb anions and water as well. The unabsorbed sodium chloride and water then lead to diarrhea, with further loss of salt from the body.

CELLULAR MECHANISM OF ALDOSTERONE ACTION

Although for many years we have known the overall effects of mineralocorticoids on the body, the basic action of aldosterone on the tubular cells to increase transport of sodium is still only partly understood. The sequence of events that leads to increased sodium reabsorption seems to be the following:

First, because of its lipid solubility in the cellular membranes aldosterone diffuses to the interior of the tubular epithelial cells.

Second, in the cytoplasm of the tubular cells aldosterone combines with a highly specific cytoplasmic *receptor protein,* a protein that has a stereomolecular configuration that will allow only aldosterone or extremely similar compounds to combine.

Third, the aldosterone-receptor complex diffuses into the nucleus where it may undergo further alterations, and then it induces specific portions of the DNA to form a type or types of messenger RNA related to the process of sodium and potassium transport.

Fourth, the messenger RNA diffuses back into the cytoplasm where, operating in conjunction with the ribosomes, it causes protein formation. The protein formed is one or more enzymes or carrier substances required for sodium transport, possibly a specific ATPase that catalyzes energy transfer from cytoplasmic ATP to the sodium transport mechanism of the cell membrane.

Thus, aldosterone does not have an immediate effect on sodium transport but must await the sequence of events that leads to the formation of the specific intracellular substance or substances required for sodium transport. Approximately 20 to 30 minutes are required before new RNA appears in the cells, and approximately 45 minutes are required before the rate of sodium transport begins to increase; the effect reaches maximum only after several hours.

REGULATION OF ALDOSTERONE SECRETION

The regulation of aldosterone secretion is so deeply intertwined with the regulation of extracellular fluid electrolyte concentrations, extracellular fluid volume, blood volume, arterial pressure, and many special aspects of renal function that it is not possible to discuss the regulation of aldosterone secretion independently of all these other factors. This subject has already been presented in great detail in Chapter 36, to which the reader is referred. However, it is important to list here as well the most important points of aldosterone secretion control.

Let us note first that aldosterone is secreted by the *zona glomerulosa*, a very thin zone of cells located on the surface of the adrenal cortex immediately beneath the capsule. These cells function almost entirely independently of the deeper cells in the zona reticularis and zona fasciculata, which secrete cortisol and the androgens. Furthermore, the regulation of aldosterone secretion is almost entirely independent of the regulation of these other hormones.

Four different factors are presently known to play essential roles in the regulation of aldosterone. In the probable order of their importance these are:

1. Potassium ion concentration of the extracellular fluid
2. Renin-angiotensin system
3. Quantity of body sodium
4. Adrenocorticotropic hormone (ACTH)

Effect of Potassium Ion Concentration on Aldosterone Secretion. An increase in potassium ion concentration of less than 1 mEq./liter will triple the rate of aldosterone secretion. Furthermore, this secretion will continue at an elevated level indefinitely under the stimulus of excess potassium ions.

This very potent effect of potassium ions is exceedingly important because it establishes a powerful feedback mechanism for control of extracellular fluid potassium ion concentration as follows: (1) An increase in potassium ion concentration causes increased secretion of aldosterone. (2) The aldosterone in turn has a potent effect on the kidneys, causing enhanced excretion of potassium. (3) Therefore, the potassium ion concentration returns to normal. Recent quantitative measurements of the feedback gain of this system show it to be by far the most potent of all the factors controlling aldosterone secretion, as has already been discussed in detail in Chapter 36. Serious interference with this control system can lead to a cardiac death if the potassium concentration rises too high or to a paralytic death if the potassium concentration falls too low.

This effect of potassium ions on aldosterone secretion is a direct effect of the potassium ions on the zona glomerulosa cells, though the cellular mechanism of the effect is unknown.

Effect of the Renin-Angiotensin System on Aldosterone Secretion. Infusion of large amounts of angiotensin into an animal can cause acute increases in aldosterone secretion of as much as 8-fold. However, if the angiotensin infusion is continued, the rate of aldosterone secretion falls in about 12 hours to only 50 to 100 per cent above normal. This effect is illustrated in Figure 77–1, which shows the effect of two different angiotensin infusion rates, one of which (the dashed curve) increased the extracellular fluid concentration of

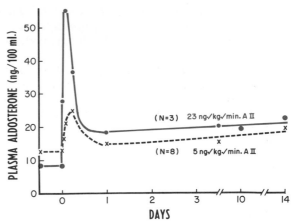

Figure 77–1. Effects on plasma aldosterone concentration caused by continuous infusion of angiotensin II at two different infusion rates for a period of two weeks. Note the very marked acute effect but the much weaker chronic effect. (Drawn from data in Cowley and McCaa, *Circ. Res.*, 39:788, 1976.)

angiotensin to about 3 times normal, and the other of which (the solid curve) increased the angiotensin concentration to about 15 times normal. From this data it can be calculated that a given per cent increase in angiotensin will cause about 60 times less increase in aldosterone secretion than will the same per cent increase in potassium concentration.

Nevertheless, elevated renin and angiotensin are found very frequently in the same clinical conditions in which aldosterone is also elevated. Therefore, the effect of angiotensin is often stated to be the most important of all the factors controlling aldosterone secretion.

Effect of Decreased Quantity of Total Body Sodium on Aldosterone Secretion. When an animal or human being is placed on a sodium-deficient diet, after several days the rate of aldosterone secretion increases markedly even though the sodium ion concentration of the body fluids does not fall significantly. A number of different suggestions for the cause of this effect have been the following:

1. The diminished sodium leads to diminished extracellular fluid volume, with resultant diminished cardiac output and renal blood flow. This causes enhanced formation of angiotensin, and the angiotensin stimulates aldosterone secretion.

2. Lack of sodium causes retention of potassium by the kidneys, as explained earlier. The elevated potassium could then cause the increased aldosterone secretion.

3. In a few experiments by McCaa and Young it has been shown that diminished sodium concentration possibly or probably causes the anterior pituitary gland to secrete some substance (not ACTH) that affects the adrenal glands to increase

aldosterone secretion. For the present, this substance is called the *unidentified pituitary factor*, though a recent study suggests that it may be beta-lipotropin.

4. Diminished sodium concentration directly affects the zona glomerulosal cells to enhance aldosterone secretion.

Unfortunately, the importance of each of these mechanisms is yet to be proved. Even so, prolonged sodium depletion is one of the most potent of all the stimulators of aldosterone secretion.

Effect of ACTH on Aldosterone Secretion. ACTH is the anterior pituitary hormone that controls the secretion of glucocorticoids. It also has a *permissive effect* on aldosterone secretion in the following way: All the above regulations of aldosterone secretion will occur if there is only a minimal amount of ACTH present. However, in the total absence of ACTH, the zona glomerulosa of the adrenal cortex partially atrophies and there is almost total atrophy of the remainder of the gland. Thus, total absence of ACTH can lead to a mild to moderate degree of aldosterone deficiency.

FUNCTIONS OF THE GLUCOCORTICOIDS

Even though mineralocorticoids can save the life of an acutely adrenalectomized animal, the animal still is far from normal. Instead, its metabolic systems for utilization of proteins, carbohydrates, and fats are considerably deranged. Furthermore, the animal cannot resist different types of physical or even mental stress, and minor illnesses such as respiratory tract infections can lead to death. Therefore, the glucocorticoids have functions just as important to long-continued life of the animal as do the mineralocorticoids. These are explained in the following sections.

At least 95 per cent of the glucocorticoid activity of the adrenocortical secretions results from the secretion of *cortisol,* known also as *hydrocortisone.* In addition to this, a small amount of glucocorticoid activity is provided by *corticosterone* and a minute amount by *cortisone.*

EFFECTS OF CORTISOL ON CARBOHYDRATE METABOLISM

Stimulation of Gluconeogenesis. By far the best-known metabolic effect of cortisol and other glucocorticoids on metabolism is their ability to stimulate gluconeogenesis by the liver, often increasing the rate of gluconeogenesis as much as 6- to 10-fold. This results from several different effects of cortisol:

First, cortisol increases the transport of amino acids from the extracellular fluids into the liver cells. This obviously increases the availability of amino acids to the cells for conversion into glucose.

Second, all the enzymes required to convert amino acids into glucose are increased in the liver cells. Also, the concentration of RNA is increased in the liver cells. Therefore, it is assumed that glucocorticoids activate DNA transcription in the liver cell nuclei, with formation of messenger RNAs that in turn lead to the array of enzymes required for gluconeogenesis.

Third, cortisol causes mobilization of amino acids from the extrahepatic tissues, mainly from muscle. As a result, more amino acids become available in the plasma to enter into the gluconeogenesis process of the liver and thereby to promote the formation of glucose.

One of the effects of increased gluconeogenesis is a marked increase in glycogen in the liver cells.

Decreased Glucose Utilization by the Cells. Cortisol also causes a moderate decrease in the rate of glucose utilization by the cells. Though the cause of this decrease is unknown, most physiologists believe that somewhere between the point of entry of glucose into the cells and its final degradation cortisol directly delays the rate of glucose utilization. A suggested mechanism for this effect is based on the observation that glucocorticoids depress the oxidation of NADH. Since NADH must be oxidized to allow rapid glycolysis, this effect could account for the diminished utilization of glucose by the cells.

Also, it is known that glucocorticoids slightly depress glucose transport into the cells, which could be an additional factor that depresses cellular glucose utilization.

Elevated Blood Glucose Concentration, and Adrenal Diabetes. Both the increased rate of gluconeogenesis and the moderate reduction in rate of glucose utilization by the cells cause the blood glucose concentration to rise. The increase in concentration is occasionally great enough — 50 per cent or more above normal — that the condition is called *adrenal diabetes,* and it has many similarities to pituitary diabetes, which was discussed in Chapter 75. Administration of insulin lowers the blood glucose concentration only a moderate amount in adrenal diabetes, not nearly so much as it does in pancreatic diabetes. On the other hand, insulin causes greater decrease in blood glucose concentration in adrenal diabetes than in pituitary diabetes. Therefore, *pituitary diabetes is said to be weakly insulin sensitive, adrenal diabetes moderately insulin sensitive, and pancreatic diabetes strongly insulin sensitive.*

EFFECTS OF CORTISOL ON PROTEIN METABOLISM

Reduction in Cellular Protein. One of the principal effects of cortisol on the metabolic systems of the body is reduction of the protein stores in essentially all body cells except those of the liver. This is caused both by decreased protein synthesis and increased catabolism of protein already in the cells. Both these effects may possibly result from decreased amino acid transport into extrahepatic tissues, as will be discussed later, but this probably is not the only cause since cortisol also depresses the formation of RNA in many extrahepatic tissues, especially muscle and lymphoid tissue.

In the presence of great excesses of cortisol the muscles can become so weak that the person cannot rise from the squatting position. And the immunity functions of the lymphoid tissue can be decreased to a small fraction of normal.

Increased Liver Protein and Plasma Proteins Caused by Cortisol. Coincidently with the reduced proteins elsewhere in the body, the liver proteins become enhanced. Furthermore, the plasma proteins (which are produced by the liver and then released into the blood) are also increased. Therefore, these are exceptions to the protein depletion that occurs elsewhere in the body. It is believed that this effect is caused both by the ability of cortisol to enhance amino acid transport into liver cells (but not into most other cells) and by enhancement of the liver enzymes required for protein synthesis.

Increased Blood Amino Acids, Diminished Transport of Amino Acids into Extrahepatic Cells, and Enhanced Transport into Hepatic Cells. Recent studies in isolated tissue have demonstrated that cortisol depresses amino acid transport into muscle cells and perhaps into other extrahepatic cells. But, in contrast to this, it enhances transport into liver cells.

Obviously, the decreased transport of amino acids into extrahepatic cells decreases their intracellular amino acid concentrations and as a consequence decreases the synthesis of protein. Yet catabolism of proteins in the cells continues to release amino acids from the already existing proteins, and these diffuse out of the cells to increase the plasma amino acid concentration. Therefore, it is said that *cortisol mobilizes amino acids from the tissues.*

The increased plasma concentration of amino acids, plus the fact that cortisol enhances transport of amino acids into the hepatic cells, could also account for enhanced utilization of amino acids by the liver in the presence of cortisol — such effects as: (1) increased rate of deamination of amino acids by the liver, (2) increased protein synthesis in the liver, (3) increased formation of plasma proteins by the liver, and (4) increased conversion of amino acids to glucose — that is, enhanced gluconeogenesis.

Thus, it is possible that many of the effects of cortisol on the metabolic systems of the body result mainly from this ability of cortisol to mobilize amino acids, though cortisol also increases the liver enzymes required for the hepatic effects, presumably by causing the formation of appropriate messenger RNAs.

EFFECTS OF CORTISOL ON FAT METABOLISM

Mobilization of Fatty Acids. In much the same manner that cortisol promotes amino acid mobilization from muscle, it also promotes mobilization of fatty acids from adipose tissue, but only weakly so. Yet, this does slightly increase the concentration of free fatty acids in the plasma, which also increases their utilization for energy. Cortisol moderately enhances the oxidation of fatty acids in the cells as well; this effect perhaps results secondarily to the reduced availability of glycolytic products for metabolism.

The mechanism by which cortisol promotes fatty acid mobilization is not yet understood. However, part of the effect probably results from diminished transport of glucose into the fat cells. It will be remembered that α-glycerophosphate, which is derived from glucose, is required both for deposition of and maintenance of triglycerides in these cells.

The increased mobilization of fats, combined with their increased oxidation in the cells, is one of the factors that helps to shift the metabolic systems of the cells in times of starvation or other stresses from utilization of glucose for energy to utilization of fatty acids instead. This cortisol mechanism, however, requires several hours to become fully developed — not nearly so rapid or nearly so powerful an effect as the similar shift elicited by a decrease in insulin. Nevertheless, it is probably an important factor for long-term conservation of body glucose and glycogen.

Ketogenic Effects of Cortisol. It is frequently said that cortisol has a *ketogenic effect* because ketosis usually will not develop unless cortisol is available to cause fat mobilization. However, this ketogenic effect occurs only under certain conditions, such as when insulin also is deficient.

Obesity Caused by Cortisol. Despite the fact that cortisol can cause a moderate degree of fatty acid mobilization from adipose tissue, persons with excess cortisol secretion frequently develop a peculiar type of obesity, with excess deposition of fat in the chest and head regions of the body, giving a buffalo-like torso. Careful experiments have shown that this obesity results from excess stimulation of food intake so that fat is generated in some tissues of the body at a rate even more rapidly than it is mobilized and oxidized.

OTHER EFFECTS OF CORTISOL

Function of Cortisol in Different Types of Stress

It is amazing that almost any type of stress, whether it be physical or neurogenic, will cause an immediate and marked increase in ACTH secretion, followed within minutes by greatly increased adrenocortical secretion of cortisol. Some of the different types of stress that increase cortisol release are the following:

1. Trauma of almost any type
2. Infection
3. Intense heat or cold
4. Injection of norepinephrine and other sympathomimetic drugs
5. Surgical operations
6. Injection of necrotizing substances beneath the skin
7. Restraining an animal so that it cannot move
8. Almost any debilitating disease

Thus, a wide variety of nonspecific stimuli can cause marked increase in the rate of cortisol secretion by the adrenal cortex.

Yet, even though we know that cortisol secretion often increases greatly in stressful situations, we still are not sure why this is of significant benefit to the animal. One guess, which is probably as good as any other, is that the glucocorticoids cause rapid mobilization of amino acids and fats from their cellular stores, making these available both for energy and for synthesis of other compounds, including glucose, needed by the different tissues of the body. Indeed, it is well known that when proteins are released from most of the tissue cells, the liver cells can use the mobilized amino acids to form new proteins. It has also been shown in a few instances that damaged tissues which are momentarily depleted of proteins can also utilize the newly available amino acids to form new proteins that are essential to the lives of the cells. Or perhaps the amino acids are also used to synthesize such essential intracellular substances as purines, pyrimidines, and creatine phosphate, which are necessary for maintenance of cellular life and reproduction of new cells.

But all this is mainly supposition. It is supported only by the fact that cortisol usually does not mobilize the basic functional proteins of the cells until almost all other proteins have been released. This preferential effect of cortisol in mobilizing labile proteins could make amino acids available to needy cells to synthesize substances essential to life.

Anti-Inflammatory Effects of Cortisol

When tissues are damaged by trauma, by infection with bacteria, or in almost any other way, they almost always will become inflamed. In some conditions the inflammation is more damaging than the trauma or disease itself. Administration of large amounts of cortisol can usually block this inflammation or even reverse many of its effects once it has begun. Before attempting to explain the way in which cortisol functions to block inflammation, let us first review the basic steps in the inflammation process, which was discussed in more detail in Chapter 5.

Basically, there are five main stages of inflammation: (1) release from the damaged tissue cells of chemical substances that activate the inflammation process — chemicals such as histamine, bradykinin, proteolytic enzymes, and so forth; (2) an increase in blood flow in the inflamed area caused by some of the released products from the tissues, a process called *erythema*; (3) leakage of large quantities of almost pure plasma out of the capillaries into the damaged areas, followed by clotting of the tissue fluid, thus causing a *nonpitting type of edema*; (4) infiltration of the area by leukocytes; and, finally, (5) tissue healing, which is often accomplished at least partially by ingrowth of fibrous tissue.

Yet, unfortunately, we still know very little about the basic mechanisms by which cortisol achieves its *anti-inflammatory effect*. However, very large amounts of cortisol are often required. When such large amounts of cortisol are secreted or are injected into a person, the cortisol has two basic effects: (1) it blocks the early stages of the inflammation process, and (2) if inflammation has already begun, it causes rapid resolution of the inflammation and increased rapidity of healing. These effects are explained as follows:

Prevention of the Development of Inflammation — Lysosome Stabilization and Other Effects. (1) One of the most important antiinflammation effects of cortisol is its ability to cause *stabilization of the lysosomal membranes*. That is, cortisol makes it much more difficult than is normal for the lysosomal membranes to rupture. Therefore, most of the substances released by damaged cells to cause inflammation, which are mainly formed in the lysosomes, are released in greatly decreased quantity. Other important effects are:

(2) Cortisol decreases the permeability of the capillaries, which prevents loss of plasma into the

tissues and also reduces the migration of white blood cells into the inflamed area.

(3) Cortisol depresses the ability of white blood cells to digest phagocytized substances, and this blocks further release of inflammatory materials.

(4) Cortisol suppresses the immune system. Therefore, reduced amounts of antibodies and sensitized leukocytes enter the inflamed area, which reduces tissue reactions that might be caused by these.

(5) Cortisol reduces fever, and this in turn reduces the degree of vasodilatation.

Effect of Cortisol in Causing Resolution of Inflammation. Even after inflammation has become well-established, administration of cortisol can often reduce inflammation within hours to several days. The immediate effect is to block most of the factors that are promoting the inflammation. Then, the rate of healing is also enhanced. This probably results from those same, mainly undefined factors that allow the body to resist many other types of physical stress when large quantities of cortisol are secreted: perhaps this results from the mobilization of amino acids and use of these to repair the damaged tissues; perhaps it results from increased amounts of glucose and fatty acids available for cellular energy; or perhaps it depends on some catalytic effect of cortisol to inactivate or remove inflammatory products.

Regardless of the precise mechanisms by which the anti-inflammatory effect occurs, this effect of cortisol can play a major role in combating certain types of diseases, such as rheumatoid arthritis, rheumatic fever, and acute glomerulonephritis. All of these are characterized by severe local inflammation, and the harmful effects to the body are caused mainly by the inflammation itself and not by other aspects of the disease. When cortisol or other glucocorticoids are administered to patients with these diseases, almost invariably the inflammation subsides within 24 to 48 hours. And even though the cortisol does not correct the basic disease condition but merely prevents the damaging effects of the inflammatory response, this alone can be a life-saving measure.

Effect on Blood Cells and on Immunity

Cortisol decreases the number of eosinophils and lymphocytes in the blood; this effect begins within a few minutes after the injection of cortisol and is marked within a few hours. Indeed, a finding of lymphocytopenia or eosinopenia is an important diagnostic criterion for overproduction of cortisol by the adrenal gland.

Likewise, administration of large doses of cortisol causes significant atrophy of all the lymphoid tissue throughout the body, which in turn decreases the output of both sensitized lymphocytes and antibodies from the lymphoid tissue. As a result, the level of immunity for almost all foreign invaders of the body is decreased. This can occasionally lead to fulminating infection and death from diseases that otherwise would not be lethal, such as fulminating tuberculosis in a person whose disease had previously been arrested. On the other hand, this ability of cortisol and other glucocorticoids to suppress immunity makes these among the most useful of all drugs to prevent immunological rejection of transplanted hearts, kidneys, and other tissues.

Cortisol increases the production of red blood cells, the cause of which is unknown. When excess cortisol is secreted by the adrenal glands, polycythemia often results, and, conversely, when the adrenal glands secrete no cortisol, anemia often results.

Effect on Allergy. Cortisol blocks the inflammatory response to allergic reactions in exactly the same way that it blocks other types of inflammatory responses. The basic allergic reaction between antigen and antibody is not affected by cortisol, and even some of the secondary effects of the allergic reaction, such as the release of histamine, still occur the same as ever. However, since the inflammatory response is responsible for many of the serious and sometimes lethal effects of allergic reactions, administration of cortisol can be lifesaving. For instance, cortisol effectively prevents shock or death in anaphylaxis which otherwise kills many persons, as explained in Chapter 7.

REGULATION OF CORTISOL SECRETION — ADRENOCORTICOTROPIC HORMONE (ACTH)

Control of Cortisol Secretion by ACTH. Unlike aldosterone secretion by the zona glomerulosa, which is controlled mainly by potassium and angiotensin acting directly on the adrenocortical cells, almost no stimuli have *direct* effects on the adrenal cells to control cortisol secretion. Instead, secretion of cortisol is controlled almost entirely by *adrenocorticotropic hormone* (ACTH) secreted by the anterior pituitary gland. This hormone, also called *corticotropin* or *adrenocorticotropin*, also enhances the production of adrenal androgens. As pointed out earlier in the chapter, small amounts of ACTH are also required for aldosterone secretion, providing a permissive role that allows the other, more important factors to exert their more powerful controls.

Control of ACTH Secretion by the Hypothalamus — Corticotropin Releasing Hormone (CRH). In the same way that other pituitary hormones are controlled by releasing hormones, also called releasing factors, from the hypothalamus, so also does an important releasing hormone control ACTH secretion. This is called *corticotropin releasing hormone* (CRH). It is secreted into the primary capillary plexus of the hypophysial portal system in the median eminence of the hypothalamus and then carried to the anterior pituitary gland where it induces ACTH secretion.

The anterior pituitary gland can secrete only small quantities of ACTH in the absence of CRH. Instead, most conditions that cause high ACTH secretory rates initiate this secretion by signals beginning in the hypothalamus and then transmitted by CRH to the anterior pituitary gland.

Effect of Physiological Stress on ACTH Secretion. It was pointed out earlier in the chapter that almost any type of physical or even mental stress can lead within minutes to greatly enhanced secretion of ACTH and consequently of cortisol as well, often increasing cortisol secretion as much as 20-fold. This effect is illustrated forcefully by the lowermost curve in Figure 77–2, which shows a many-fold increase in plasma corticosterone concentration in a rat within minutes after the tibia and fibula had been broken (corticosterone is the principal glucocorticoid secreted by the rat adrenal). It is believed that pain stimuli caused by the stress are first transmitted to the perifornical area of the hypothalamus. This area in turn transmits signals into other areas of the hypothalamus and eventually to the median eminence, as shown in Figure 77–3, where CRH is secreted into the hypophysial portal system. Within minutes the entire control sequence leads to large quantities of the glucocorticoids in the blood.

Inhibitory Effect of Cortisol on the Hypothalamus and on the Anterior Pituitary to Cause Decreased ACTH Secretion. Cortisol has direct negative feedback *effects* on (1) the hypothalamus to decrease the formation of CRH and (2) the an-

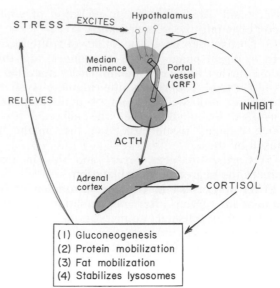

Figure 77–3. Mechanism for regulation of glucocorticoid secretion.

terior pituitary gland to decrease the formation of ACTH. These feedbacks help to regulate the plasma concentration of cortisol. That is, whenever the concentration becomes too great, the feedbacks automatically reduce the ACTH toward a normal control level. Or, if the level of cortisol falls too low, lack of the negative feedback will increase the cortisol again toward normal.

Summary of the Control System. Figure 77–3 illustrates the overall system for control of cortisol secretion. The central key to this control is the excitation of the hypothalamus by different types of stress. These activate the entire system to cause rapid release of cortisol, and the cortisol in turn initiates a series of metabolic effects directed toward relieving the damaging nature of the stressful state. In addition, there is also direct feedback of the cortisol to the hypothalamus and anterior pituitary gland to stabilize the concentration of cortisol in the plasma at times when the body is not experiencing stress. However, the stress stimuli are the prepotent ones; they can always break through this direct inhibitory feedback of cortisol.

Circadian Rhythm of Glucocorticoid Secretion. The secretory rates of CRH, ACTH, and cortisol are all high in the early morning but low in the late evening; the plasma cortisol level ranges between a high of about 20 μg/100 ml. and a low of about 5 μg./100 ml. This effect results from a 24-hour cyclic alteration in the signals from the hypothalamus that cause cortisol secretion. When a person changes daily sleeping habits, the cycle changes correspondingly. One of the reasons that the cycle is so important is that measurements of blood cortisol levels are meaningful only when expressed in terms of the time in the cycle at which the measurements are made.

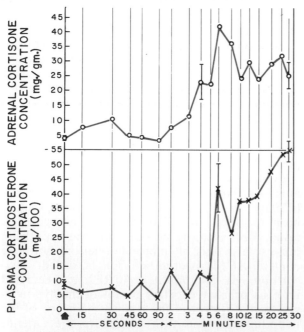

Figure 77–2. Rapid reaction of the adrenal cortex of a rat to stress caused by fracture of the tibia and fibula. (Courtesy of Drs. Guillemin, Dear, and Lipscomb.)

Secretion of Melanocyte-Stimulating Hormone, Lipotropin, and Endorphin in Association with ACTH

Usually when ACTH is secreted by the anterior pituitary gland, several other hormones that have similar chemical structures are secreted simultaneously, including especially *alpha-* and *beta-melanocyte stimulating hormone, beta-lipotropin,* and *beta-endorphin.* Under normal conditions, none of these is known to be secreted in enough quantity to have significant effect on the body, but this may not be true when the rate of secretion of ACTH is very high, as occurs in Addison's disease which will be discussed later.

Beta-endorphin is important because it has an opiate effect on the nervous system; in fact, the substance met-enkephalin, which is present in many areas of the brain, is itself a fragment of the beta-endorphin molecule. Some persons believe that the pituitary gland can secrete small amounts of this and other hormones into the hypothalamic tissues as well as into the circulating blood.

The functions of *beta-lipotropin* are yet unknown, but it has been suggested that it stimulates aldosterone secretion in the same way that ACTH stimulates cortisol secretion by the adrenal cortex.

Melanocyte Stimulating Hormone. Melanocyte stimulating hormone (MSH) occurs in two forms, an *alpha* and a *beta* form. The alpha form has exactly the same chemical structure as the first 13 amino acids of the 39 amino acid ACTH polypeptide chain.

MSH causes the *melanocytes,* which are located in abundance between the dermis and the epidermis of the skin, to form the pigment *melanin* and to disperse this in the cells of the epidermis. Injection of melanocyte stimulating hormone into a person over a period of eight to ten days can cause intense darkening of the skin. However, the effect is much greater in persons who have genetically dark skins than in light-skinned persons.

In some lower animals, an intermediate "lobe" of the pituitary gland, called the *pars intermedia,* is highly developed, lying between the anterior and the posterior pituitary lobes. This lobe secretes an especially large amount of melanocyte stimulating hormone. Furthermore, this secretion is independently controlled by the hypothalamus in response to the amount of light to which an animal is exposed or in response to other environmental factors. For instance, some arctic animals develop darkened fur in the summer and yet have entirely white fur in the winter.

ACTH, because of its similarity to MSH, has about one-thirtieth as much melanocyte-stimulating effect as MSH. Furthermore, because the quantities of MSH secreted in the human being are extremely small while those of ACTH are large, it is likely that ACTH is considerably more important than MSH in determining the amount of melanin in the skin.

CHEMISTRY OF ADRENOCORTICAL SECRETION

Locus of Hormone Formation in the Adrenal Cortex. Aldosterone is secreted by the *zona glomerulosa*

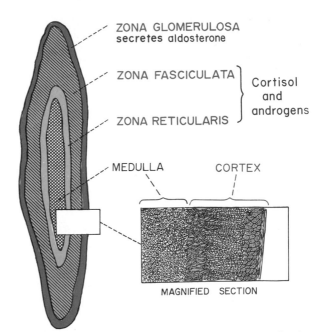

Figure 77–4. Secretion of adrenocortical hormones by the different zones of the adrenal cortex.

of the adrenal cortex, while the glucocorticoids and adrenal androgens are secreted by the *zona fasciculata* and *zona reticularis.* These zones are illustrated in Figure 77–4. During prolonged ACTH stimulation of the adrenal cortex, the zona fasciculata and zona reticularis both hypertrophy, while complete lack of ACTH causes these two zones to atrophy almost entirely while leaving the zona glomerulosa mainly intact. On the contrary, conditions that increase the output of aldosterone cause hypertrophy of the zona glomerulosa while not affecting the other two zones.

Chemistry of the Adrenocortical Hormones. All the adrenocortical hormones are steroid compounds, and they are mainly formed in the adrenal cortex from acetyl coenzyme A or to some extent from cholesterol preformed elsewhere in the body. Figure 77–5 illustrates a

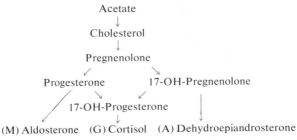

Figure 77–5. The major steps in the synthesis of the three principal adrenal steroids. The physiologic characteristics are designated (M) mineralocorticoid effect; (G) glucocorticoid effect; and (A) androgenic effect.

schema for formation of the three most important steroid products of the adrenal cortex. However, a change in even a single enzyme system somewhere in the scheme can cause vastly different types of hormones to be formed; occasionally, large quantities of masculinizing or, very rarely, feminizing sex hormones are secreted by adrenal tumors, as discussed later in the chapter.

Figure 77–6 illustrates the chemical formulas of aldosterone and cortisol. In addition, several steroids that do not naturally occur in the adrenal cortex but that have as potent or even more potent actions than the natural hormones have been synthesized. Thus, *dexamethasone,* which is a slight modification of cortisol, has 30 times as much glucocorticoid potency as cortisol but no significant increase in mineralocorticoid potency. Therefore, it is used clinically when one wishes maximum glucocorticoid activity and yet wishes to prevent electrolyte and water retention. *Fludrocortisone* has about one-third as much mineralocorticoid activity as aldosterone but is more readily available and can be used in quantities of only a few hundred micrograms a day to treat total deficiency of mineralocorticoid secretion.

On the other hand, other substances similar in chemical structure but having almost no mineralocorticoid activity — the *spirolactones,* for instance — become firmly fixed in the renal tubular epithelial cells to the intracellular receptors to which aldosterone normally binds. Therefore, these compounds block the effect of the body's own aldosterone. As a result, the kidneys then function as if no aldosterone were being secreted in the body.

Mechanism by Which ACTH Activates Adrenocortical Cells to Produce Steroids — Function of Cyclic AMP. The primary effect of ACTH on the adrenocortical cells is to activate *adenyl cyclase* in the cell membranes, and this then induces the formation of cyclic AMP in the cells, an effect that reaches maximum in three minutes. This substance, in turn, promotes the intracellular reactions for formation of adrenocortical hormones. Thus, this is another example of the function of cyclic AMP as a *second messenger* hormone.

Chemistry of ACTH. ACTH has been isolated in pure form from the anterior pituitary. It is a large polypeptide having a chain length of 39 amino acids. A digested product of ACTH having a chain length of 24 amino acids has all the trophic effects of the total molecule.

Chemistry of Corticotropin-Releasing Hormone. Several small peptides that have a chemical configuration similar to that of vasopressin and oxytocin have been isolated from the hypothalamus which, upon injection into an animal, cause rapid release of ACTH from the anterior pituitary. It is not known whether one of these is the specific corticotropin-releasing hormone.

Transport and Fate of the Adrenal Hormones. Cortisol combines with a globulin called *cortisol binding globulin* and, to a lesser extent, with albumin — about 98 per cent is normally transported in the bound form and about 2 per cent free. On the other hand, aldosterone combines only loosely with the plasma proteins so that about 50 per cent is in the free form. In both the combined and free forms the hormones are transported throughout the extracellular fluid compartment. In general, the hormones become fixed in the target tissues within an hour or two for cortisol and within about 30 minutes for aldosterone.

The adrenal steroids are degraded mainly in the liver and conjugated especially to form glucuronides and, to a lesser extent, sulfates. About 25 per cent of these are excreted in the bile and then in the feces and the remaining 75 per cent in the urine. The conjugated forms of these hormones are inactive.

The normal concentration of aldosterone in blood is about 8 nanograms (8 billionths of a gram) per 100 ml., and the secretory rate is about 150 μg. per day.

The concentration of cortisol in the blood averages 12 μg./100 ml., and the secretory rate averages 20 mg. per day.

THE ADRENAL ANDROGENS

Several moderately active male sex hormones called *adrenal androgens* (the most important of which, *dehydroepiandrosterone,* is illustrated in Fig. 77–5) are continually secreted by the adrenal cortex, especially so during fetal life as will be discussed in Chapter 82. Also, progesterone and estrogens, which are female sex hormones, have been extracted from the adrenal cortex, though these are secreted in only minute quantities.

In normal physiology of the human being, even the adrenal androgens have almost insignificant effects. However, it is possible that part of the early development of the male sex organs results from childhood secretion of adrenal androgens. The adrenal androgens also exert mild effects in the female, not only before

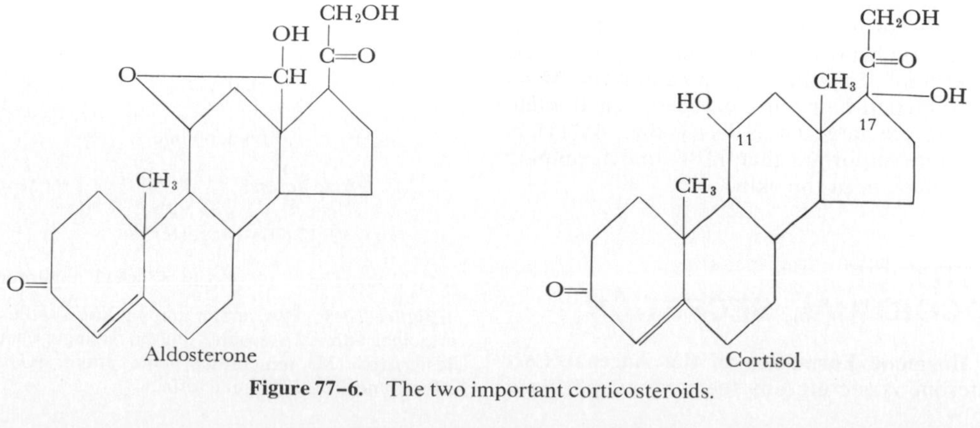

Figure 77–6. The two important corticosteroids.

puberty but also throughout life. Some of the adrenal androgens are converted to testosterone, the major male sex hormone, in the extra-adrenal tissues, which probably accounts for much of their androgenic activity. The physiological effects of androgens will be discussed in Chapter 80 in relation to male sexual function.

ABNORMALITIES OF ADRENOCORTICAL SECRETION

HYPOADRENALISM — ADDISON'S DISEASE

Addison's disease results from failure of the adrenal cortices to produce adrenocortical hormones, and this in turn is most frequently caused by *primary atrophy* of the adrenal cortices, probably resulting from autoimmunity against the cortices, but also frequently caused by tuberculous destruction of the adrenal glands or invasion of the adrenal cortices by cancer. Basically, the disturbances in Addison's disease are:

Mineralocorticoid Deficiency. Lack of aldosterone secretion greatly decreases sodium reabsorption and consequently allows sodium ions, chloride ions, and water to be lost into urine in great profusion. The net result is a greatly decreased extracellular fluid volume. Furthermore, the person develops hyperkalemia and mild acidosis because of failure of potassium and hydrogen ions to be secreted in exchange for sodium reabsorption.

As the extracellular fluid becomes depleted, the plasma volume falls, the red blood cell concentration rises markedly, the cardiac output decreases, and the patient dies in shock, death usually occurring in the untreated patient four days to two weeks after complete cessation of mineralocorticoid secretion.

Glucocorticoid Deficiency. Loss of cortisol secretion makes it impossible for the person with Addison's disease to maintain normal blood glucose concentration between meals because he cannot synthesize significant quantities of glucose by gluconeogenesis. Furthermore, lack of cortisol reduces the mobilization of both proteins and fats from the tissues, thereby depressing many other metabolic functions of the body. This sluggishness of energy mobilization when cortisol is not available is one of the major detrimental effects of glucocorticoid lack. However, even when excess quantities of glucose and other nutrients are available, the person's muscles are still weak, indicating that glucocorticoids are also needed to maintain other metabolic functions of the tissues besides simply energy metabolism.

Lack of adequate glucocorticoid secretion also makes the person with Addison's disease highly susceptible to the deteriorating effects of different types of stress, and even a mild respiratory infection can sometimes cause death.

Melanin Pigmentation. Another characteristic of most persons with Addison's disease is melanin pigmentation of the mucous membranes and skin. This melanin is not always deposited evenly but occasionally in blotches and especially in the thin skin areas, such as the mucous membranes of the lips and the thin skin of the nipples.

The cause of the melanin deposition is believed to be the following: When cortisol secretion is depressed, the normal negative feedback to the hypothalamus and anterior pituitary is also depressed, now allowing tremendous rates of ACTH secretion as well as simultaneous secretion of increased amounts of melanocyte stimulating hormone (MSH). Probably the tremendous amounts of ACTH cause most of the pigmenting effect because these can stimulate formation of melanin by the melanocytes in the same way that MSH does. However, it is possible that the MSH also formed in this condition plays a role. Even though this hormone has 30 times as much melanocyte-stimulating effect as ACTH, the amounts secreted by the human being are extremely small.

Diagnosis of Addison's Disease. A person with this disease has almost no urinary secretion of steroids derived from the adrenals. Also, immunoassay of the plasma aldosterone and competitive binding assay (see Chapter 75) of cortisol show either very low or unmeasurable values. In full-blown Addison's disease, less than one-tenth of the normal amount of 17-hydroxysteroids, which are derived almost entirely from cortisol secreted by the adrenals, is excreted in the urine in a 24-hour period. However, to be certain that the person has Addison's disease, ACTH is infused into the patient over a period of 8 hours or more, and the plasma cortisol level again measured. If the Addison's disease is caused by total destruction of the adrenal cortices, there will be no increase. If, on the other hand, the Addison's disease is caused by failure of the anterior pituitary to stimulate the adrenal cortices, successive administration of ACTH several days in a row will cause the adrenal glands to secrete normal quantities of corticosteroids.

Treatment of Persons with Addison's Disease. The untreated person with total adrenal destruction dies within a few days to a few weeks because of consuming weakness and eventual circulatory shock. Yet such a person can usually live for years if small quantities of mineralocorticoids and glucocorticoids are administered daily — about 0.2 milligram of fludrocortisone and 30 milligrams of cortisol.

The Addisonian Crisis. As noted earlier in the chapter, great quantities of glucocorticoids are occasionally secreted in response to different types of physical or mental stress. In the person with Addison's disease, the output of glucocorticoids does not increase during stress. Yet whenever different types of trauma, disease, or other stresses such as surgical operations supervene, a person is likely to develop an acute need for excessive amounts of glucocorticoids, and must be given as much as 10 or more times the normal quantities of glucocorticoids in order to prevent death.

This critical need for extra glucocorticoids and the associated severe debility in times of stress is called an Addisonian crisis.

HYPERADRENALISM — CUSHING'S SYNDROME

Hypersecretion of cortisol by the adrenal cortex causes a complex of hormonal effects called Cushing's disease, and this results from either a cortisol-secreting tumor of one adrenal cortex or general hyperplasia of both adrenal cortices. The hyperplasia in turn is caused by increased secretion of ACTH by the anterior pituitary or by "ectopic secretion" of ACTH by a tumor elsewhere in the body, such as an abdominal carcinoma. Most abnormalities of Cushing's syndrome are ascribable to abnormal amounts of cortisol, but secretion of androgens is also of significance.

A special characteristic of Cushing's disease is mobilization of fat from the lower part of the body, with concomitant extra deposition of fat in the thoracic and upper abdominal regions, giving rise to a so-called "buffalo" torso. The excess secretion of steroids also leads to an edematous appearance of the face, and the androgenic potency of some of the hormones sometimes causes acne and hirsutism (excess growth of facial hair). The total appearance of the face is frequently described as a "moon face," as illustrated to the left in Figure 77–7 in a patient with Cushing's syndrome prior to treatment. About 80 per cent of the patients have hypertension, presumably because of the slight mineralocorticoid effects of cortisol (because a pure glucocorticoid such as dexamethasone actually causes decreased arterial pressure).

Effects on Carbohydrate and Protein Metabolism. The abundance of cortisol secreted in Cushing's syndrome can cause increased blood glucose concentration, sometimes to values as high as 200 mg. per cent after meals. This effect results mainly from enhanced gluconeogenesis. If this "adrenal diabetes" lasts for many months, the beta cells in the islets of Langerhans occasionally "burn out" because the high blood glucose greatly overstimulates them to secrete insulin. The destruction of these cells then causes frank diabetes mellitus, which is permanent for the remainder of life.

The effects of glucocorticoids on protein catabolism are often profound in Cushing's syndrome, causing greatly decreased tissue proteins almost everywhere in the body with the exceptions of the liver and the plasma proteins. The loss of protein from the muscles in particular causes severe weakness. The loss of protein synthesis in the lymphoid tissues leads to a diminished immunity system, so that many of these patients die of infections. Even the collagen fibers in the subcutaneous tissue are diminished so that the subcutaneous tissues tear easily, resulting in development of large *purplish striae* where the subcutaneous tissues have torn apart. In addition, lack of protein deposition in the bones causes *osteoporosis* with consequent weakness of the bones.

Diagnosis and Treatment of Cushing's Disease. Diagnosis of Cushing's disease is most frequently made on the basis of such typical findings as the buffalo torso, puffiness of the face, mild masculinizing effects, elevated blood glucose concentration that is moderately insulin resistant, several times normal plasma levels of cortisol, and secretion of 3 to 5 times the normal quantities of 17-hydroxysteroids in the urine.

Treatment in Cushing's disease consists of removing an adrenal tumor if this is the cause, or of decreasing the secretion of ACTH, if this is possible. Hypertrophied pituitary glands or even small tumors in the pituitary that oversecrete ACTH can be surgically or microsurgically removed or destroyed by radiation. If ACTH secretion cannot easily be decreased, the only satisfactory treatment is usually bilateral total or partial adrenalectomy, followed by administration of adrenal steroids to make up for any insufficiency that develops.

PRIMARY ALDOSTERONISM

Occasionally a small tumor of the zona glomerulosa cells occurs and secretes large amounts of aldosterone.

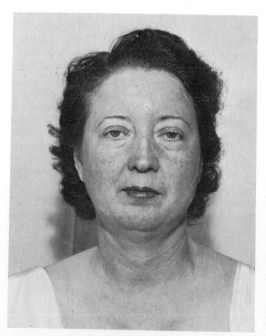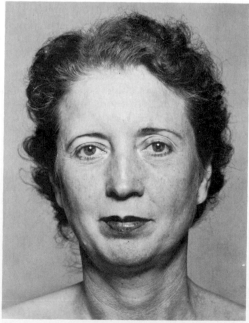

Figure 77–7. A person with Cushing's disease before subtotal adrenalectomy (left) and after subtotal adrenalectomy (right). (Courtesy of Dr. Leonard Posey.)

In a few instances, hyperplastic adrenal cortices secrete aldosterone rather than cortisol. The effects of the excess aldosterone are those discussed in detail earlier in the chapter. The most important effects are hypokalemia, slight increase in extracellular fluid volume and blood volume, very slight increase in plasma sodium concentration (usually not over a 2 to 3 per cent increase), and almost always hypertension. Especially interesting in primary aldosteronism are occasional periods of muscular paralysis caused by the hypokalemia. The paralysis is caused by hyperpolarization of the nerve fibers, as was explained in Chapter 10.

One of the diagnostic criteria of primary aldosteronism is a decreased plasma renin concentration. This results from feedback suppression of renin secretion caused by the excess aldosterone or by the excess extracellular fluid volume and arterial pressure resulting from the aldosteronism.

Treatment of primary aldosteronism is usually surgical removal of the tumor or of most of the adrenal tissue when hyperplasia is the cause. In about 30 per cent of the treated patients, the arterial pressure fails to return to normal even though plasma aldosterone concentrations do become normal; this presumably results from some long-term damage to the kidneys.

ADRENOGENITAL SYNDROME

An occasional adrenocortical tumor secretes excessive quantities of androgens that cause intense masculinizing effects throughout the body. If this occurs in a female, she develops virile characteristics, including growth of a beard, a much deeper voice, occasionally baldness if she also has the genetic inheritance for baldness, masculine distribution of hair on the body and on the pubis, growth of the clitoris to resemble a penis, and deposition of proteins in the skin and especially in the muscles to give typical masculine characteristics.

In the prepubertal male a virilizing adrenal tumor causes the same characteristics as in the female, plus rapid development of the male sexual organs and creation of male sexual desires. Typical development of the male sexual organs in a 4-year-old boy with the adrenogenital syndrome is shown in Figure 77–8.

In the adult male, the virilizing characteristics of the adrenogenital syndrome are usually completely obscured by the normal virilizing characteristics of the testosterone secreted by the testes. Therefore, it is often difficult to make a diagnosis of adrenogenital syndrome in the male adult.

In the adrenogenital syndrome, the excretion of 17-ketosteroids (which are derived from androgens) in the urine may be as much as 10 to 15 times normal.

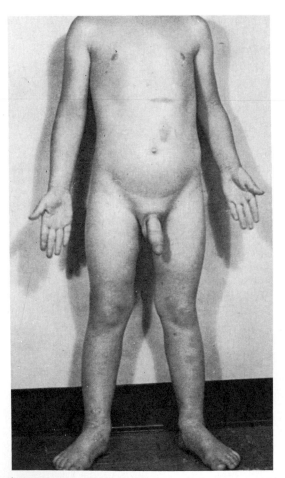

Figure 77–8. Adrenogenital syndrome in a 4-year-old boy. (Courtesy of Dr. Leonard Posey.)

REFERENCES

Azarnoff, D. L.: Steroid Therapy. Philadelphia, W. B. Saunders Co., 1975.

Baxter, J. D., and Rousseau, G. G.: Glucocorticoid Hormone Action. New York, Springer-Verlag, 1979.

Bondy, P. K.: The adrenal cortex. In Bondy, P. K., and Rosenberg, L. E. (eds.): Metabolic Control and Disease, 8th Ed. Philadelphia, W. B. Saunders Co., 1980, p. 1427.

Brodish, A., and Lymangrover, J. R.: The hypothalamic-pituitary adrenocortical system. Int. Rev. Physiol., 16:93, 1977.

Finkelstein, M., and Schaefer, J. M.: Inborn errors of steroid biosynthesis. Physiol. Rev., 59:353, 1979.

Genazzani, E., et al. (eds.): Pharmacological Modulation of Steroid Action. New York, Raven Press, 1980.

Gill, G. N., et al. (eds.): Pharmacology of Adrenal Cortical Hormones. New York, Pergamon Press, 1978.

Gill, J. R., Jr.: Bartter's syndrome. Annu. Rev. Med., 31:405, 1980.

Gorski, J., and Gannon, F.: Current models of steroid hormone action: A critique. Annu. Rev. Physiol., 38:425, 1976.

Hall, J. E., et al.: Control of arterial pressure and renal function during glucocorticoid excess in dogs. Hypertension, 2:139, 1980.

Harding, B. W.: Synthesis of adrenal cortical steroids and mechanism of ACTH effects. In DeGroot, L. J., et al. (eds.): Endocrinology. Vol. 2. New York, Grune & Stratton, 1979, p. 1131.

James, V. H. T. (ed.): The Endocrine Function of the Human Adrenal Cortex. New York, Academic Press, 1978.

James, V. H. (ed.): The Adrenal Gland. New York, Raven Press, 1979.

Kaye, A. M., and Kaye, M. (eds.): Development of Responsiveness to Steroid Hormones. New York, Pergamon Press, 1980.

Krieger, D. T.: Plasma ACTH and corticosteroids. In DeGroot, L. J., et al. (eds.): Endocrinology. Vol. 2. New York, Grune & Stratton, 1979, p. 1139.

Leavitt, W. W., and Clark, J. H. (eds.): Steroid Hormone Receptor Systems. New York, Plenum Press, 1979.

Lohmeier, T. E., et al.: Failure of chronic aldosterone infusion to increase arterial pressure in dogs with angiotensin-induced hypertension. Circ. Res., 43(3):381, 1978.

McCaa, R. E., et al.: Return of plasma aldosterone concentration to control levels in nephrectomized-decapitated dogs. Int. Cong. Endocrin., 1972.

McCaa, R. E., *et al.:* Evidence for a role of an unidentified pituitary factor in regulatory aldosterone secretion during altered sodium balance. *Circ. Res.,* (Suppl. 1)*34*:1, 1974; *35*:1, 1974.

McCaa, R. E., *et al.:* Role of aldosterone in experimental hypertension. *J. Endocrinol., 81*:69, 1979.

Melby, J. C.: Diagnosis and treatment of hyperaldosteronism and hypoaldosteronism. *In* DeGroot, L. J., *et al.* (eds.): Endocrinology. Vol. 2. New York, Grune & Stratton, 1979, p. 1225.

Nelson, D. H.: Regulatory mechanisms of the pituitary and pituitary-adrenal axis. *In* Frohlich, E. D. (ed.): Pathophysiology, 2nd Ed. Philadelphia, J. B. Lippincott Co., 1976, p. 327.

Nelson, D. H.: The Adrenal Cortex: Physiological Function and Disease. Philadelphia, W. B. Saunders Co., 1979.

Nelson, D. H.: Cushing's syndrome. *In* DeGroot, L. J., *et al.* (eds.): Endocrinology. Vol. 2. New York, Grune & Stratton, 1979, p. 1179.

Nelson, D. H.: Diagnosis and treatment of Addison's disease. *In* DeGroot, L. J., *et al.* (eds.): Endocinrology. Vol. 2. New York, Grune & Stratton, 1979, p. 1193.

O'Malley, B. W., and Schrader, W. T.: The receptors of steroid hormones. *Sci. Am., 234*(2):32, 1976.

Parrillo, J. E., and Fauci, A. S.: Mechanisms of glucocorticoid action on immune processes. *Annu. Rev. Pharmacol. Toxicol., 19*:179, 1979.

Raisz, L. G., *et al.:* Hormonal regulation of mineral metabolism. *Int. Rev. Physiol., 16*:199, 1977.

Roy, A. K., and Clark, J. H. (eds.): Gene Regulation by Steroid Hormones. New York, Springer-Verlag, 1979.

Ruhmann-Wennhold, A., and Nelson, D. H.: Pituitary adrenocorticotropin. *In* DeGroot, L. J., *et al.* (eds.): Endocrinology. Vol. 1. New York, Grune & Stratton, 1979, p. 133.

Tan, S. Y., and Mulrow, P. J.: Aldosterone in hypertension and edema. *In* Bondy, P. K., and Rosenberg, L. E. (eds.): Metabolic Control in Disease, 8th Ed. Philadelphia, W. B. Saunders Co., 1980, p. 1501.

Young, D. B., and Guyton, A. C.: Steady state aldosterone dose-response relationships. *Circ. Res., 40*(2):138, 1977.

78

Insulin, Glucagon, and Diabetes Mellitus

The pancreas, in addition to its digestive functions, secretes two important hormones, *insulin* and *glucagon*. The purpose of this chapter is to discuss the functions of these hormones in regulating glucose, lipid, and protein metabolism, as well as to discuss briefly the two diseases — *diabetes mellitus* and *hyperinsulinism* — caused, respectively, by hyposecretion of insulin and excess secretion of insulin.

Physiologic Anatomy of the Pancreas. The pancreas is composed of two major types of tissues, as shown in Figure 78–1: (1) the *acini*, which secrete digestive juices into the duodenum, and (2) the *islets of Langerhans,* which do not have any means for emptying their secretions externally but instead secrete insulin and glucagon directly into the blood. The digestive secretions of the pancreas were discussed in Chapter 64.

The islets of Langerhans of the human being contain three major types of cells, the *alpha, beta,* and *delta* cells, which are distinguished from one another by their morphology and staining characteristics. The beta cells secrete *insulin,* the alpha cells secrete *glucagon,* and the delta cells secrete a newly discovered hormone, *somatostatin,* which will be discussed later in the chapter.

Beta cells are usually still present in the pancreas of a person who has severe diabetes, but these cells then have a hyalinized appearance and contain no secretory granules; also, they do not exhibit staining reactions for insulin. Consequently, the hyalinized beta cells in diabetic patients are considered to be nonfunctional.

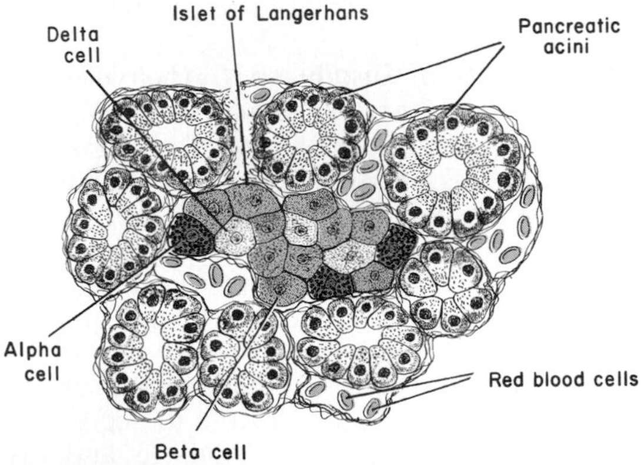

Chemistry of Insulin. Insulin is a small protein with a molecular weight of 5808 for human insulin. It is composed of two amino acid chains, illustrated in Figure 78–2, connected to each other by disulfide linkages. When the two amino acid chains are split apart, the functional activity of the insulin molecule is lost.

The chemical mechanism by which insulin exerts its function on cells is mainly unknown. However, insulin is known to bind with a large *receptor protein,* having a molecular weight of about 300,000, either in or on the surface of the cell membrane. There is some evidence that this binding might then activate the cyclic AMP system of the cell and that this in turn acts as a *second messenger* to promote the insulin effects. However, many of the effects of insulin have been demonstrated in the absence of increased cellular cyclic AMP, thus casting doubt on this as the principal mechanism of insulin action.

Once insulin has been secreted into the blood, it is almost as rapidly removed from the blood and then degraded. The liver is mainly responsible for this. Therefore, insulin circulates for less than 10 minutes. This rapid destruction is important because at times it is equally as important for the control functions of insulin to be turned off rapidly as for them to be turned on rapidly.

THE METABOLIC EFFECTS OF INSULIN

Insulin was first isolated from the pancreas in 1922 by Banting and Best, and almost overnight the outlook for the severely diabetic patient changed from one of rapid decline and death to that of a nearly normal person.

Historically, insulin has been associated with "blood sugar," and, true enough, insulin does have profound effects on carbohydrate metabolism. Yet, it is mainly abnormalities of fat me-

Figure 78–1. Physiologic anatomy of the pancreas.

959

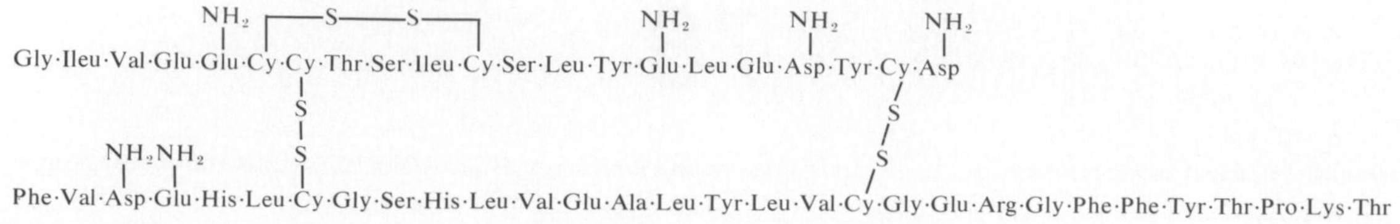

NH₂ ┌─── S ──── S ───┐ NH₂ NH₂ NH₂
│ │ │ │ │ │
Gly·Ileu·Val·Glu·Glu·Cy·Cy·Thr·Ser·Ileu·Cy·Ser·Leu·Tyr·Glu·Leu·Glu·Asp·Tyr·Cy·Asp
│ /
S S
│ /
NH₂NH₂ S S
│ │ │ /
Phe·Val·Asp·Glu·His·Leu·Cy·Gly·Ser·His·Leu·Val·Glu·Ala·Leu·Tyr·Leu·Val·Cy·Gly·Glu·Arg·Gly·Phe·Phe·Tyr·Thr·Pro·Lys·Thr

Figure 78–2. The human insulin molecule.

tabolism, which can cause acidosis and arteriosclerosis, that are the usual causes of death of a diabetic patient. And, in patients with prolonged diabetes, the inability to synthesize proteins leads to wasting of the tissues as well as many functional disorders. Therefore, it is clear that insulin affects fat and protein metabolism almost as much as it does carbohydrate metabolism.

EFFECT OF INSULIN ON CARBOHYDRATE METABOLISM

Immediately after a high carbohydrate meal, the glucose that is absorbed into the blood causes rapid secretion of insulin. The insulin in turn causes rapid uptake, storage, and use of glucose by almost all tissues of the body, but especially by the liver, muscles, and fat tissue. Therefore, let us discuss each of these.

Effect of Insulin in Promoting Liver Uptake, Storage, and Use of Glucose

One of the most important of all the effects of insulin is to cause most of the glucose absorbed after a meal to be stored almost immediately in the liver in the form of glycogen. Then, between meals, when insulin is not available and the blood glucose concentration begins to fall, the liver glycogen is split back into glucose, which is released back into the blood to keep the blood glucose concentration from falling too low.

The mechanism by which insulin causes glucose uptake and storage in the liver includes several almost simultaneous steps:

1. Insulin *inhibits phosphorylase*, the enzyme that causes liver glycogen to split into glucose. This obviously prevents breakdown of the glycogen that is already in the liver cells.

2. Insulin causes *enhanced uptake of glucose* from the blood by the liver cells. It does this by *increasing the activity of the enzyme glucokinase*, which is the enzyme that causes the initial phosphorylation of glucose after it diffuses into the liver cells. Once phosphorylated, the glucose is trapped inside the liver cells, because phosphorylated glucose cannot diffuse back through the cell membrane. But liver cells are highly permeable to free glucose so that still more glucose continues to diffuse inward.

3. Insulin also increases the activities of the enzymes that promote glycogen synthesis, including *phosphofructokinase* that causes the second stage in the phosphorylation of the glucose molecule and *glycogen synthetase* that is responsible for polymerization of the monosaccharide units to form the glycogen molecules.

The net effect of the above actions is to increase the amount of glycogen in the liver. The glycogen can increase to a total of about 5 to 6 per cent of the liver mass, which is equivalent to almost 100 grams of stored glycogen.

Release of Glycogen from the Liver Between Meals. After the meal is over and the blood glucose level begins to fall to a low level, several events now transpire that cause the liver to release glucose back into the circulating blood:

1. The decreasing blood glucose causes the pancreas to decrease its insulin secretion.

2. The lack of insulin then reverses all the effects listed above for glycogen storage, essentially stopping further synthesis of glycogen in the liver. This also prevents further uptake of glucose by the liver from the blood.

3. The lack of insulin also activates the enzyme *phosphorylase* which causes the splitting of glycogen into *glucose phosphate*.

4. The enzyme *glucose phosphatase* causes the phosphate radical to split away from the glucose, and this allows the free glucose to diffuse back into the blood.

Thus, the liver removes glucose from the blood when it is present in excess after a meal and returns it to the blood when it is needed between meals. Ordinarily, about 60 per cent of the glucose in the meal is stored in this way in the liver and then returned later.

Other Effects of Insulin on Carbohydrate Metabolism in the Liver. Insulin also *promotes the conversion of liver glucose into fatty acids,* and these fatty acids are subsequently transported to the adipose tissue and deposited as fat. This will be discussed in relation to insulin effects on fat metabolism. Insulin also *inhibits gluconeogenesis*. It does this mainly by decreasing the quantities and activities of the liver enzymes required for gluconeogenesis. However, part of the effect is also caused by an action of insulin to decrease the release of amino acids from muscle and other extrahepatic tissues, thus decreasing the availability of the necessary precursors required for gluconeogenesis. This will be discussed further in relation to the effect of insulin on protein metabolism.

Effect of Insulin in Promoting Glucose Metabolism in Muscle

During most of the day, muscle tissue depends not on glucose for its energy but instead on fatty acids. The principal reason for this is that the normal *resting muscle* membrane is almost impermeable to glucose except when the muscle fiber is stimulated by insulin. And, between meals, the amount of insulin that is secreted is too small to promote significant amounts of insulin entry into the muscle cells.

However, under two conditions the muscles do utilize large amounts of glucose for energy. One of these is during periods of heavy exercise. This usage of glucose does not require large amounts of insulin because exercising muscle fibers, for reasons not understood, become highly permeable to glucose even in the absence of insulin because of the contraction process itself.

The second condition for muscle usage of large amounts of glucose is during the few hours after a meal. At this time the blood glucose concentration is high; also, the pancreas is secreting large quantities of insulin, and the extra insulin causes rapid transport of glucose into the muscle cells. In addition, the insulin increases the activity of *phosphofructokinase* in the muscle, which catalyzes the complete phosphorylation of the glucose, making it available for use in the glycolytic energy system of the muscle cells. Therefore, the carbohydrate metabolic system of the muscle cell becomes temporarily energized, and this causes the muscle cell during this period of time to utilize carbohydrates preferentially over fatty acids.

Storage of Glycogen in Muscle. If the muscles are not exercising during the period following a meal and yet glucose is transported into the muscle cells in great abundance, then much of the glucose is stored in the form of muscle glycogen instead of being used for energy. However, the concentration of muscle glycogen rarely rises much above 1 per cent rather than the possible 5 to 6 per cent in liver cells. The glycogen can later be used for energy by the muscle. This glycogen is especially useful to provide anaerobic energy for a few minutes at a time by glycolytic breakdown of the glycogen to lactic acid, which can occur in the absence of oxygen.

Muscle glycogen is different from liver glycogen in that it cannot be reconverted into glucose and released into the body fluids. The reason for this is that there is no glucose phosphatase in muscle cells, in contrast to the liver cells.

Mechanism by Which Insulin Promotes Glucose Transport Through the Muscle Cell Membrane. Insulin promotes glucose transport into muscle cells quite differently from the way it promotes transport into liver cells. The transport into liver results mainly from a trapping mechanism caused by phosphorylation of the glucose under the influence of glucokinase. However, this is only a minor factor in the insulin effect on glucose transport into muscle cells. Of more importance, the insulin has a direct effect on the muscle cell membrane to facilitate glucose transport. This is illustrated by the experimental results depicted in Figure 78–3. This experiment was performed in muscle cells at a temperature of 4° C, a temperature at which glucose, upon entering the cell, cannot be phosphorylated. The lower curve labeled "control" shows the concentration of free glucose measured inside the cell, illustrating that the glucose concentration remained almost exactly zero despite increases in extracellular glucose concentration up to as high as 750 mg./100 ml. In contrast, the curve labeled "insulin" illustrates that the intracellular glucose concentration rose to as high as 400 mg./100 ml. when insulin was added. Thus, it is clear that insulin can increase the rate of transport of glucose into the resting muscle cell by at least 15- to 20-fold.

As pointed out in the discussion of glucose transport through the cell membrane in Chapter 4, glucose cannot pass through the membrane pores but instead must be transported through the lipid portion of the membrane. Figure 78–4 depicts the generally accepted method by which this is achieved, showing that glucose combines with a carrier substance in the cell membrane and then diffuses to the inside of the membrane where it is released to the interior of the cell. The carrier is then used again and again to transport additional

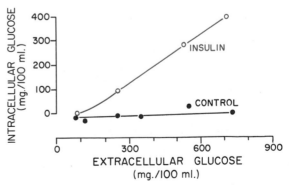

Figure 78–3. Effect of insulin in enhancing the concentration of glucose inside muscle cells. Note that in the absence of insulin (control) the intracellular glucose concentration remained near zero despite very high extracellular concentrations of glucose. (From Park, Morgan, Kaji, and Smith, in Eisenstein (ed.): The Biochemical Aspects of Hormone Action. Little, Brown and Co.)

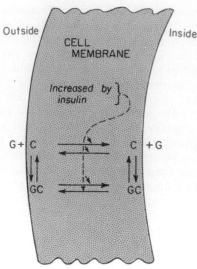

Figure 78–4. Effect of insulin in increasing glucose transport in either direction through the cell membrane.

quantities of glucose. This process can occur in *either direction,* as shown by the reversible arrows in the diagram.

Glucose transport through the cell membrane does not occur against a concentration gradient. That is, once the glucose concentration inside the cell rises to as high as the glucose concentration on the outside, additional glucose will not be transported to the interior. Therefore, the transport process is one of *facilitated diffusion,* which means simply that the carrier facilitates the diffusion of glucose through the membrane but cannot impart energy to the transport process to cause glucose movement against an energy gradient. The manner in which insulin enhances facilitated diffusion of glucose is still largely unknown. All that is known is that the insulin combines with a "receptor" protein in the cell membrane — a protein having a molecular weight of about 300,000. This may be the glucose carrier itself, or it may be merely the first step in a chain of events that leads to activation of the carrier system. The insulin increases the glucose transport within seconds to minutes, suggesting either a rapid direct action on the cell membrane itself or some other equally rapid mechanism.

Lack of Effect of Insulin on Glucose Uptake and Usage by the Brain

The brain is quite different from most other tissues of the body in that insulin has either little or no effect on uptake or use of glucose. Instead, the brain cells are permeable to glucose without the intermediation of insulin.

The brain cells normally use only glucose for energy. Therefore, it is essential that the blood

glucose level be maintained always above a critical level, which is one of the important functions of the blood glucose control system. When the blood glucose does fall too low, into the range of 20 to 50 mg./100 ml., symptoms of *hypoglycemic shock* develop, characterized by progressive irritability that leads to fainting, convulsions, and even coma.

Effect of Insulin on Carbohydrate Metabolism in Other Cells

Insulin increases glucose transport into and glucose usage by most other cells of the body (with the exception of the brain cells, as noted) in the same way that it affects glucose transport through the muscle cell membrane. The transport of glucose into adipose cells is essential for deposition of fat in these cells, as will be discussed next.

EFFECT OF INSULIN ON FAT METABOLISM

Though not quite as dramatic as the acute effects of insulin on carbohydrate metabolism, insulin also affects fat metabolism in ways that, in the long run, are perhaps even more important. Especially dramatic is the long-term effect of insulin lack in causing extreme atherosclerosis, often leading to heart attacks, cerebral strokes, and other vascular accidents. But, first, let us discuss the acute effects of insulin on fat metabolism.

Effect of Insulin Excess on Fat Synthesis and Storage

Insulin has several different effects that lead to fat storage in adipose tissue. One is the simple fact that insulin increases the rate of utilization of glucose by many of the body's tissues, and this functions as a "fat sparer." However, insulin also promotes fatty acid synthesis. Most of this synthesis occurs in the liver cells, and the fatty acids are then transported to the adipose cells to be stored. However, a small part of the synthesis occurs in the fat cells themselves. The different factors that lead to increased fatty acid synthesis in the liver include:

1. Insulin increases the transport of glucose into the liver cells. After the liver glycogen concentration increases to 5 to 6 per cent, this in itself inhibits further glycogen synthesis. Then, all the additional glucose entering the liver cells becomes available to form fat. The glucose is first split to pyruvate in the glycolytic pathway, and the pyruvate subsequently is converted to acetyl Co-A, the substrate from which fatty acids are synthesized.

2. An excess of *citrate* and *isocitrate ions* is formed by the citric acid cycle when excess amounts of glucose are being used for energy. These ions then have a direct effect in activating *acetyl Co-A carboxylase,* the enzyme required to initiate the first stage of fatty acid synthesis.

3. The fatty acids are then transported from the liver to the adipose cells where they are stored.

Storage of Fat in the Adipose Cells. Insulin has very much the same effect in adipose cells as in the liver in causing synthesis of fatty acids. However, only about one tenth as much glucose is transported into the fat cells of the human being as into the liver, so that the amount of fatty acids synthesized in adipose cells is rather small compared with the amount formed in the liver.

Yet, insulin has two other essential effects that are required for fat storage in adipose cells:

1. Insulin *inhibits the action of hormone-sensitive lipase.* Since this is the enzyme that causes hydrolysis of the triglycerides in fat cells, the release of fatty acids into the circulating blood is therefore inhibited.

2. Insulin *promotes glucose transport into the fat cells* in exactly the same way that it promotes glucose transport into muscle cells. The glucose is then utilized to synthesize fatty acids, as noted above, but, more important, it also forms another substance that is essential to the storage of fat. During the glycolytic breakdown of glucose, large quantities of the substance *α-glycerophosphate* is formed. This substance supplies the *glycerol* that binds with fatty acids to form triglycerides, the storage form of fat in adipose cells. Therefore, when insulin is not available to promote glucose entry into the fat cells, fat storage is either greatly inhibited or blocked.

Increased Metabolic Use of Fat Caused by Insulin Lack

All aspects of fat metabolism are greatly enhanced in the absence of insulin. This occurs even normally between meals when secretion of insulin is minimal, but it becomes extreme in diabetes when secretion of insulin is almost zero. The resulting effects are:

Lipolysis of Storage Fat and Release of Free Fatty Acids During Insulin Lack. In the absence of insulin, all the effects of insulin noted above causing storage of fat are reversed. The most important effect is that the enzyme *hormone-sensitive lipase* in the fat cells becomes strongly activated. This causes hydrolysis of the stored triglycerides, releasing large quantities of fatty acids and glycerol into the circulating blood. Consequently, the plasma concentration of free fatty acids rises within minutes to hours. This free fatty acid then becomes the main energy substrate used

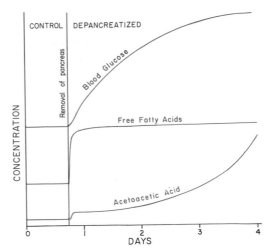

Figure 78–5. Effect of removing the pancreas on the concentrations of blood glucose, plasma free fatty acids, and acetoacetic acid.

by essentially all tissues of the body besides the brain. Figure 78–5 illustrates the effect of insulin lack on the plasma concentrations of free fatty acids, glucose, and acetoacetic acid. Note that immediately after removal of the pancreas the free fatty acid concentration in the plasma begins to rise, rising considerably more rapidly even than the concentration of glucose.

Effect of Insulin Lack on Causing a Fatty Liver. Strange as it may seem, though insulin lack causes dissolution of the fat stores in adipose tissue, it causes a great increase in the amount of stored triglycerides in the liver, leading to a very fatty liver. The reason is the following: The excess of free fatty acids in the blood causes rapid diffusion of fatty acids into the liver cells. Also, the glycerol released from the adipose cells is transported at the same time into the liver. The liver cells, unlike the adipose cells, have appropriate enzymes for converting glycerol into α-glycerophosphate, and this causes rapid binding of the fatty acids to form triglycerides. Thus, insulin lack causes triglycerides and glycerol to leave the adipose cells, but the liver captures a large share of the released fatty acids and glycerol and reconverts these to triglycerides that are mainly stored in the liver. In this manner, the liver can store 30 per cent or more of its weight in the form of fat.

Effect of Insulin Lack on Plasma Lipid Concentrations. The excess of fatty acids in the liver also promotes conversion of some of the fatty acids into phospholipids and cholesterol, two of the major products of fat metabolism. These two substances, along with some of the triglycerides formed in the liver, are then discharged into the blood in the lipoproteins. Occasionally, the plasma lipoproteins increase as much as 3-fold, giving a total concentration of plasma lipids of several per cent rather than the normal 0.6 per cent. This high lipid concentration — especially the high concentration of cholesterol — leads to rapid development of atherosclerosis in persons with serious diabetes.

Ketogenic and Acidotic Effect of Insulin Lack. Insulin lack also causes excessive amounts of *acetoacetic acid* to be formed in the liver cells. This results from the following effect: In the absence of insulin but in the presence of excess fatty acids in the liver cells, the carnitine transport mechanism for transporting fatty acids into the mitochondria becomes increasingly activated. In the mitochondria, beta oxidation of the fatty acids then proceeds exceedingly rapidly, releasing extreme amounts of acetyl Co-A. A part of this acetyl Co-A can be utilized in the liver itself for energy, but the excess is condensed to form acetoacetic acid, which, in turn, is released into the circulating blood. Most of this passes to the peripheral cells where it is again converted into acetyl Co-A and used for energy in the usual manner.

However, in the absence of insulin, so much acetoacetic acid is released from the liver that it cannot all be metabolized by the peripheral tissues. Therefore, as illustrated in Figure 78–5, its concentration rises during the days following cessation of insulin secretion, sometimes reaching a concentration as high as 10 or more milliequivalents per liter. As was explained in Chapter 68, some of the acetoacetic acid is also converted into β-hydroxybutyric acid and *acetone*. These two substances, along with the acetoacetic acid, are called *ketone bodies,* and their presence in large quantities in the body fluids is called *ketosis.* We shall see later that the acetoacetic acid and the β-hydroxybutyric acid can cause severe *acidosis* and *coma* in patients with severe diabetes. In the absence of heroic treatment, this leads almost inevitably to death.

Effect of Excess Plasma Fatty Acids on Inhibiting Glucose Utilization by Cells. It is already clear that lack of insulin decreases glucose uptake by most cells of the body and in this way greatly depresses glucose utilization by the cells when insulin is not available. However, still another effect occurs that further depresses glucose utilization. The increased fatty acids in the plasma caused by insulin lack have an additional direct effect on the cells themselves to depress glucose utilization. This results from the following mechanisms:

First, when the fatty acids enter the cells, they are immediately split to form acetyl Co-A. This then enters the citric acid cycle and provides the energy required by the cell.

Second, two of the important products of this energy system are *citrate ion* and *ATP,* both of which have a strong inhibitory effect on the enzyme *phosphofructokinase,* which is a rate-limiting enzyme for promoting glucose use in the cell. Therefore, use of glucose for energy almost ceases.

Third, since glucose cannot be utilized by the cell for energy, the mass law effect causes glucose uptake into the cell also to be greatly depressed.

This total process is called the *glucose–fatty acid cycle.* Its net effect is approximately to double the depressive action of insulin lack on cellular uptake and utilization of glucose. This obviously is important to prevent utilization of glucose by the cells of the body besides those of the brain when glucose is in short supply.

EFFECT OF INSULIN ON PROTEIN METABOLISM AND GROWTH

Effect of Insulin on Protein Synthesis and Storage. During the few hours following a meal when excess quantities of nutrients are available in the circulating blood, not only carbohydrates and fats but proteins as well are stored in the tissues; insulin is required for this to occur. The manner in which insulin causes protein storage is not as well understood as the mechanism for both glucose and fat storage. Some of the facts known are:

1. Insulin causes active transport of many of the amino acids into the cells. Among the amino acids most strongly transported are *valine, leucine, isoleucine, tyrosine,* and *phenylalanine.* Thus, insulin shares with growth hormone the capability of increasing the uptake of amino acids into cells. However, the amino acids affected are not necessarily the same ones.

2. Insulin has a direct effect on the ribosomes to *increase the translation of messenger RNA,* thus forming new proteins. In some unexplained way, insulin "turns on" the ribosomal machinery. In the absence of insulin the ribosomes simply stop working, almost as if insulin operates an "on-off" mechanism.

3. Over a longer period of time insulin also *increases the rate of transcription of DNA* in the cell nuclei, thus forming increased quantities of RNA. Eventually, it also increases the rate of formation of new DNA and even reproduction of cells. All these effects promote still more protein synthesis.

4. Insulin also *inhibits the catabolism of proteins,* thus decreasing the rate of amino acid release from the cells, especially from the muscle cells. Presumably this results from some ability of the insulin to diminish the normal degradation of proteins by the cellular lysosomes.

5. In the liver, large quantities of insulin *depress the rate of gluconeogenesis.* It does this by

decreasing the activity of the enzymes that promote gluconeogenesis. Since the substrates most used for synthesis of glucose by the process of gluconeogenesis are the plasma amino acids, this suppression of gluconeogenesis conserves the amino acids in the protein stores of the body.

In summary, insulin promotes entry of amino acids into cells, greatly enhances the rate of protein formation, and also prevents the degradation of proteins. However, we still know very little about the basic chemical mechanisms employed by insulin to achieve these ends. Part of the effects might result from increased availability of glucose-derived intermediary compounds that themselves play controlling roles in protein metabolism.

Protein Depletion and Increased Plasma Amino Acids Caused by Insulin Lack.   All protein storage comes to a complete halt when insulin is not available. The catabolism of proteins increases, protein synthesis stops, and large quantities of amino acid are dumped into the plasma. The plasma amino acid concentration rises considerably, and most of the excess amino acids are either used directly for energy or as substrates for gluconeogenesis. This degradation of the amino acids also leads to enhanced urea excretion in the urine. The resulting protein wasting is one of the most serious of all the effects of severe diabetes mellitus. It can lead to extreme weakness as well as to many deranged functions of the organs.

Effect of Insulin on Growth — Its Synergistic Effect with Growth Hormone.   Because insulin is required for the synthesis of proteins, it is equally as essential for growth of an animal as is growth hormone. This is illustrated in Figure 78–6, which shows that a depancreatized-hypophysectomized rat without therapy hardly grew at all. Furthermore, administration of neither growth hormone nor insulin one at a time caused significant growth. Yet a combination of both of these hormones did cause dramatic growth. Thus it appears that the two hormones function synergistically to promote growth, each performing its specific function that is separate from that of the other. Perhaps part of this necessity for both hormones results from the fact that each promotes cellular uptake of a different selection of amino acids, all of which are required if growth is to be achieved.

CONTROL OF INSULIN SECRETION

Formerly, it was believed that insulin secretion is controlled almost entirely by the blood glucose concentration. However, as more has been learned about the metabolic functions of insulin for protein and fat metabolism, it has been learned that blood amino acids and other factors also play important roles in controlling insulin secretion.

Stimulation of Insulin Secretion by Blood Glucose.   At the normal fasting level of blood glucose of 80 to 90 mg./100 ml., the rate of insulin secretion is minimal — in the order of 10 ng./min./kg. of body weight. If the blood glucose concentration is suddenly increased to a level 2 to 3 times normal and is kept at this high level thereafter, insulin secretion increases markedly in three separate stages:

1. Insulin secretion increases up to 10-fold about 5 minutes after acute elevation of the blood glucose; this results from immediate dumping of preformed insulin from the beta cells of the islets of Langerhans. However, this initial high rate of secretion is not maintained; instead, it decreases about halfway back toward normal in another 5 to 10 minutes.

2. After about 15 minutes, insulin secretion rises a second time, reaching a new plateau in 2 to 3 hours, this time usually at a rate of secretion even greater than that in the initial phase. This secretion results from activation of some enzyme system that synthesizes and releases insulin from the cells.

3. Over a period of a week or so, the rate of insulin secretion increases still more, often doubling the rate. This effect results from hypertrophy of the beta cells.

Relationship Between Blood Glucose Concentration and Insulin Secretion Rate.   As the concentration of blood glucose rises above 100 mg./100 ml. of blood, the rate of insulin secretion rises rapidly, reaching a peak some 10 to 20 times the basal level at blood glucose concentration between 300 and 400 mg./100 ml. Thus, the increase in insulin secretion under a glucose stimulus is dramatic both in its rapidity and in the tremendous level of secretion achieved. Furthermore, the turn-off of insulin secretion is almost equally as rapid, occurring within minutes after reduction in blood glucose concentration back to the fasting level.

This response of insulin secretion to an elevated blood glucose concentration provides an extremely

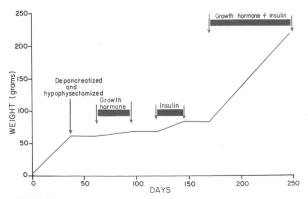

Figure 78–6.   Effect of (a) growth hormone, (b) insulin, and (c) growth hormone plus insulin on growth in a depancreatized and hypophysectomized rat.

important feedback mechanism for regulating blood glucose concentration. That is, the rise in blood glucose increases insulin secretion, and the insulin in turn causes transport of glucose into the liver, muscle, and other cells, thereby reducing the blood glucose concentration toward the normal value.

Other Factors That Stimulate Insulin Secretion

Amino Acids. In addition to excess glucose stimulating insulin secretion, many of the amino acids have a similar effect. The most potent of these are *arginine* and *lysine*. However, this effect differs from glucose stimulation of insulin secretion in the following way: Amino acids administered in the absence of a rise in blood glucose cause only a small increase in insulin secretion. However, when administered at the same time that the blood glucose concentration is elevated, the glucose-induced secretion of insulin may be as much as doubled in the presence of the excess amino acids. Thus, the amino acids very strongly potentiate the glucose stimulus for insulin secretion.

The stimulation of insulin secretion by amino acids seems to be a purposeful response because the insulin in turn promotes transport of the amino acids into the tissue cells and also promotes intracellular formation of protein. That is, the insulin is important for proper utilization of the excess amino acids.

Gastrointestinal Hormones. A mixture of several important gastrointestinal hormones — *gastrin, secretin, cholecystokinin,* and *gastric inhibitory peptide* — will cause a moderate increase in insulin secretion. These hormones are released in the gastrointestinal tract after a person eats a meal. They seem to cause an "anticipatory" increase in blood insulin in preparation for the glucose and amino acids to be absorbed from the meal. These gastrointestinal hormones almost double the rate of insulin secretion following an average meal.

Other Hormones. Other hormones that either directly increase insulin secretion or potentiate the glucose stimulus for insulin secretion include *glucagon, growth hormone, cortisol*, and to a lesser extent *progesterone* and *estrogen*. The importance of the stimulatory effects of these hormones is that prolonged secretion of any one of them in large quantities can occasionally lead to exhaustion of the beta cells of the islets of Langerhans and thereby cause diabetes. Indeed, diabetes often occurs in persons maintained on high pharmacological doses of some of these hormones. It is particularly common in giants or acromegalic persons with growth hormone–secreting tumors or in persons who already have a diabetic tendency to whom glucocorticoids are administered in high concentration.

ROLE OF INSULIN IN "SWITCHING" BETWEEN CARBOHYDRATE AND LIPID METABOLISM

From the above discussions it should already be clear that insulin promotes the utilization of carbohydrates for energy, while it depresses the utilization of fats. Conversely, lack of insulin causes fat utilization mainly to the exclusion of glucose utilization, except by brain tissue. Furthermore, the signal that controls this switching mechanism is principally the blood glucose concentration. When the glucose concentration is low, insulin secretion is suppressed and fat is utilized almost exclusively for energy everywhere except in the brain; when the glucose concentration is high, insulin secretion is stimulated, and carbohydrate is utilized instead of fat until the excess blood glucose is stored. Therefore, one of the most important functional roles of insulin in the body is to control which of these two foods from moment to moment will be utilized by the cells for energy.

At least two other hormones also play important roles in this switching mechanism: growth hormone from the anterior pituitary gland and cortisol from the adrenal cortex. Both of these are secreted in response to hypoglycemia, and both depress cellular utilization of glucose while promoting the utilization of fat.

Still another hormone that plays an important role in this overall mechanism is epinephrine. During periods of stress when the sympathetic nervous system is excited, epinephrine secreted by the adrenal medulla causes a large increase in both the blood sugar (resulting from glycogenolysis in the liver) and the blood concentration of fatty acids (caused by a direct lipolytic effect of epinephrine on the fat cells). Quantitatively, the enhancement of fatty acids is far greater than the enhancement of blood glucose. Therefore, epinephrine especially enhances the utilization of fat in such stressful states as exercise, circulatory shock, anxiety, and so forth.

GLUCAGON AND ITS FUNCTIONS

Glucagon, a hormone secreted by the alpha cells of the islets of Langerhans, has several functions that are diametrically opposed to those of insulin. Most important of these is to increase the blood glucose concentration, an effect which is exactly opposite to that of insulin.

Like insulin, glucagon is a small protein. It has a molecular weight of 3485 and is composed of a chain of 29 amino acids. On injection of purified glucagon into an animal, a profound *hyper*glycemic effect occurs. One microgram per kilogram of glucagon can elevate the blood glucose concentration approximately 20 mg./100 ml. of blood in about 20 minutes. For this reason, glucagon is frequently called *hyperglycemic factor*.

The two major effects of glucagon on glucose metabolism are (1) breakdown of liver glycogen (*glycogenolysis*) and (2) increased *gluconeogenesis*.

Glycogenolysis and Increased Blood Glucose Concentration Caused by Glucagon. The most dramatic effect of glucagon is its ability to cause glycogenolysis in the liver, which in turn increases the blood glucose concentration within minutes.

It does this by the following complex cascade of events:

1. Glucagon activates *adenylcyclase* in the hepatic cell membrane,
2. Which causes the formation of *cyclic AMP*,
3. Which activates *protein kinase regulator protein*,
4. Which activates *protein kinase*,
5. Which activates *phosphorylase b kinase*,
6. Which converts *phosphorylase b* into *phosphorylase a*,
7. Which promotes the degradation of glycogen into glucose-1-phosphate,
8. Which then is dephosphorylated and the glucose released from the liver cells.

This sequence of events is exceedingly important for several reasons. First, it is one of the most thoroughly studied of all the *second messenger* functions of cyclic AMP. Second, it illustrates a cascading system in which each succeeding product is produced in greater quantity than the preceding product. Therefore, it represents a potent *amplifying* mechanism. This explains how only 1 μg./kg. of glucagon can have the extreme effect of causing hyperglycemia.

Infusion of glucagon for about four hours can cause such intensive liver glycogenolysis that all of the liver stores of glycogen become totally depleted.

Gluconeogenesis Caused by Glucagon. Even after all the glycogen in the liver has been exhausted under the influence of glucagon, continued infusion of this hormone causes continued hyperglycemia. This results from an effect of glucagon in increasing the rate of gluconeogenesis in the liver cells. Unfortunately, the precise mechanism of this effect is unknown, but it is believed to result mainly from activation of the enzyme system for converting pyruvate to phosphoenolpyruvate, one of the rate-limiting steps in gluconeogenesis.

Other Effects of Glucagon. Most other effects of glucagon occur only when the concentration of this substance rises far above that normally found in the body fluids. Some of these include enhanced strength of the heart, enhanced bile secretion, enhanced secretion of calcitonin, and inhibition of gastric acid secretion. However, all these changes are probably unimportant in the normal function of the body.

REGULATION OF GLUCAGON SECRETION

Effect of Blood Glucose Concentration. Changes in blood glucose concentration have exactly the opposite effect on glucagon secretion as on insulin secretion. That is, a *decrease* in blood glucose increases glucagon secretion. When the blood glucose falls to as low as 70 mg./100 ml. of blood, the pancreas secretes large quantities of glucagon, and this secretion rapidly mobilizes glucose from the liver. Thus, glucagon helps to protect against hypoglycemia.

Effect of Exercise. In exercise the blood glucose concentration tends to decrease, and this increases glucagon secretion. The increased glucagon in turn plays an important role in mobilizing glucose from the liver for use by the muscles.

Effect of Amino Acids. Amino acids enhance the secretion of glucagon, an effect exactly opposite to that of glucose. The physiological importance of this is that it helps to prevent the hypoglycemia that would otherwise result when a meal of pure protein is ingested, because the amino acids from the protein enhance insulin secretion and thereby tend to decrease blood glucose. The increased glucagon secretion seems to nullify this effect.

Inhibition of Glucagon and Insulin Secretion by Somatostatin

The islets of Langerhans secrete still a third hormone, *somatostatin*, secreted by a third type of cell, the *delta cells*. However the role of this hormone as a controller of metabolic processes is not yet understood. It has the capability of inhibiting the secretion of both glucagon and insulin by the alpha and beta cells of the islets of Langerhans. Therefore, it has been suggested that somatostatin might help to control the secretion of one or both of these other hormones.

Somatostatin is the same as *growth hormone inhibitory hormone* that is secreted by the hypothalamus, a hormone that might help to control growth hormone secretion. It is also secreted by the mucosa of the upper gastrointestinal tract, but its function there also is unknown.

GLUCAGON-LIKE EFFECTS OF EPINEPHRINE

Epinephrine (and to a slight extent norepinephrine as well) is also a potent promoter of liver glycogenolysis, having an effect almost exactly the same as that of glucagon, though not quite as strong. However, some experiments have suggested that epinephrine can cause glycogenolysis without increasing the concentration of cyclic AMP in the liver cells, so that the mechanism of glycogenolysis activated by epinephrine might be different from that activated by glucagon.

SUMMARY OF BLOOD GLUCOSE REGULATION

In the normal person the blood glucose concentration is very narrowly controlled, usually in a range between 80 and 90 mg./100 ml. of blood in the fasting person each morning before breakfast. This concentration increases to 120 to 140 mg./100 ml. during the first hour or so following a meal, but the feedback systems for control of blood glucose return the glucose concentration very rapidly back to the control level, usually within two hours after the last absorption of carbohydrates. Conversely, in starvation the gluconeogenesis function of the liver provides the glucose that is required to maintain the fasting blood glucose level.

The mechanisms for achieving this high degree of control have been presented in this chapter. However, let us summarize these briefly:

1. The liver functions as a very important *blood glucose–buffer system*. That is, when the blood glucose rises to a very high concentration following a meal and the rate of insulin secretion also increases, as much as two thirds of the glucose absorbed from the gut is almost immediately stored in the liver in the form of glycogen. Then, during the succeeding hours, when both the blood glucose concentration and the rate of insulin secretion fall, the liver releases the glucose back into the blood. In this way, the liver decreases the variations in blood glucose concentration by about 3-fold. In fact, in patients with severe liver disease, it becomes almost impossible to maintain a narrow range of blood glucose concentration.

2. It is very clear that both insulin and glucagon function as important and separate feedback control systems for maintaining a normal blood glucose concentration. When the concentration rises to a level too high, insulin is secreted; the insulin in turn causes the blood glucose concentration to decrease toward normal. Conversely, a decrease in blood glucose stimulates glucagon secretion; the glucagon then functions in the opposite direction to increase the glucose up toward normal. Under most normal conditions, the insulin feedback mechanism probably is much more important than the glucagon mechanism, but in instances of diminished glucose intake or excessive utilization of glucose during exercise and other stressful situations, the glucagon mechanism is undoubtedly very valuable.

3. Also, in hypoglycemia, a direct effect of low blood glucose on the hypothalamus stimulates the sympathetic nervous system. In turn, the epinephrine secreted by the adrenal glands causes still further release of glucose from the liver. This, too, helps to protect against severe hypoglycemia.

4. And, finally, over a period of hours and days, both growth hormone and cortisol are secreted in response to prolonged hypoglycemia, and they both decrease the rate of glucose utilization by most cells of the body. This, too, helps to return the blood glucose concentration toward normal.

Importance of Blood Glucose Regulation. One might ask the question: Why is it important to maintain a constant blood glucose concentration, particularly since most tissues can shift to utilization of fats and proteins for energy in the absence of glucose? The answer is that glucose is the only nutrient that can be utilized by the *brain, retina,* and *germinal epithelium of the gonads* in sufficient quantities to supply them with their required energy. Therefore, it is important to maintain a blood glucose concentration at a sufficiently high level to provide this necessary nutrition.

Most of the glucose formed by gluconeogenesis during the interdigestive period is used for metabolism in the brain. Indeed, it is important that the pancreas not secrete any insulin during this time, for otherwise the scant supplies of glucose that are available would all go into the muscles and other peripheral tissues, leaving the brain without a nutritive source.

On the other hand, it is also important that the blood glucose concentration not rise too high for three reasons: First, glucose exerts a large amount of osmotic pressure in the extracellular fluid, and, if the glucose concentration rises to excessive values, this can cause considerable cellular dehydration. Second, an excessively high level of blood glucose concentration causes loss of glucose in the urine. And third, this causes osmotic diuresis by the kidneys, which can deplete the body of its fluids.

DIABETES MELLITUS

Etiology. Diabetes mellitus is caused in almost all instances by diminished rates of secretion of insulin by the beta cells of the islets of Langerhans. It is usually divided into two different types: *juvenile diabetes* that usually, but not always, begins in early life, and *maturity-onset diabetes* that usually, but not always, begins in later life and mainly in obese persons.

Heredity plays an important role in the development of both these types of diabetes. The juvenile type is usually rapid in onset and seems to result from hereditary predisposition to (a) development of antibodies against the beta cells, thus causing autoimmune destruction of these, (b) possible destruction of the beta cells by viral disease, or (c) possible simple degeneration of these cells. The maturity-onset type of diabetes seems to result from degeneration of the beta cells as a result of more rapid aging in susceptible persons than in others. Obesity predisposes to this type of diabetes because larger quantities of insulin are required for metabolic control in obese than in normal persons.

PATHOLOGICAL PHYSIOLOGY OF DIABETES

Most of the pathology of diabetes mellitus can be attributed to one of the following three major effects of insulin lack: (1) decreased utilization of glucose by the body cells, with a resultant increase in blood glucose concentration to as high as 300 to 1200 mg. per 100 ml.; (2) markedly increased mobilization of fats from the fat storage areas, causing abnormal fat metabolism as well as deposition of lipids in vascular walls to cause atherosclerosis; and (3) depletion of protein in the tissues of the body.

However, in addition, some special pathological, physiological problems occur in diabetes mellitus that are not so readily apparent. These are:

Loss of Glucose in the Urine of the Diabetic Person. Whenever the quantity of glucose entering the kidney tubules in the glomerular filtrate rises above approximately 225 mg. per minute, a significant proportion of the glucose begins to spill into the urine. If normal quantities of glomerular filtrate are formed per minute, this will occur when the blood glucose level rises over 180 mg. per cent. Consequently, it is frequently stated that the blood "threshold" for the appearance of glucose in the urine is approximately 180 mg. per cent. When the blood glucose level rises to 300 to 500 mg. per cent — common values in persons with severe untreated diabetes — a hundred or more grams of glucose can be lost into the urine each day.

Dehydrating Effect of Elevated Blood Glucose Levels in Diabetes. Blood glucose levels as high as 1200 mg. per cent, 12 times normal, can occur under certain conditions in extreme diabetes. Yet the only significant effect of the elevated glucose is dehydration of the tissue cells, for glucose does not diffuse easily through the pores of the cell membrane, and the increased osmotic pressure in the extracellular fluids causes osmotic transfer of water out of the cells.

In addition to the direct dehydrating effect of excessive glucose, the loss of glucose in the urine causes *diuresis* because of the osmotic effect of glucose in the tubules to prevent tubular reabsorption of fluid. The overall effect is dehydration of the extracellular fluid, which then causes compensatory dehydration of the intracellular fluid for reasons discussed in Chapter 33. Thus, one of the important features of diabetes is a tendency for extracellular and intracellular dehydration to develop and these are also often associated with collapse of the circulation.

Acidosis in Diabetes. The shift from carbohydrate to fat metabolism in diabetes has already been discussed. When the body depends almost entirely on fat for energy, the level of acetoacetic acid and β-hydroxybutyric acid in the body fluids may rise from 1 mEq./liter to as high as 10 mEq./liter. This, obviously, is likely to result in acidosis.

A second effect, which is usually even more important in causing acidosis than is the direct increase in keto acids, is a decrease in sodium concentration caused by the following effect: Keto acids have a low threshold for excretion by the kidneys; therefore, when the keto acid level rises in diabetes, as much as 100 to 200 grams of keto acids can be excreted in the urine each day. Because these are strong acids, having a pK averaging

4.0 or less, very little of them can be excreted in the acidic form but instead is excreted combined with sodium derived from the extracellular fluid. As a result, the sodium concentration in the extracellular fluid usually decreases, and the sodium is replaced by increased quantities of hydrogen ions, thus adding greatly to the acidosis.

Obviously, all the usual reactions that occur in metabolic acidosis take place in diabetic acidosis. These include *rapid and deep breathing* called Kussmaul respiration, which causes excessive expiration of carbon dioxide, and *marked decrease in bicarbonate content of the extracellular fluids*. Likewise, *large quantities of chloride ion are excreted by the kidneys* as an additional compensatory mechanism for correction of the acidosis. Though these extreme effects occur only in the most severe instances of uncontrolled diabetes, they can lead to acidotic coma and death within hours when they do occur. The overall changes in the electrolytes of the blood as a result of severe diabetic acidosis are illustrated in Figure 78–7.

Relationship of Other Diabetic Symptoms to the Pathological Physiology of Insulin Lack. *Polyuria* (excessive elimination of urine), *polydipsia* (excessive drinking of water), *polyphagia* (excessive eating), *loss of weight*, and *asthenia* (lack of energy) are the earliest symptoms of diabetes. As explained, the polyuria is due to the osmotic diuretic effect of glucose in the kidney tubules. In turn, the polydipsia is due to dehydration resulting from polyuria. The failure of glucose (and protein) utilization by the body causes loss of weight and a tendency toward polyphagia. The asthenia apparently also is caused mainly by loss of body protein.

PHYSIOLOGY OF DIAGNOSIS

The usual methods for diagnosing diabetes are based on various chemical tests of the urine and the blood.

Urinary Sugar. Simple office tests or more complicated quantitative laboratory tests may be used for

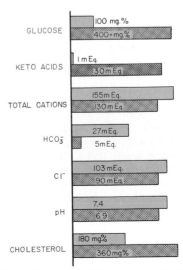

Figure 78–7. Changes in blood constituents in diabetic coma, showing normal values (light bars) and diabetic values (dark bars).

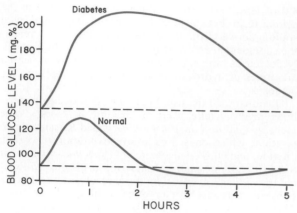

Figure 78–8. Glucose tolerance curve in the normal person and in a diabetic person.

determining the quantity of glucose lost in the urine. In general, the normal person loses undetectable amounts of glucose, whereas the diabetic loses glucose in small to large amounts, in proportion to the severity of disease and the intake of carbohydrates. (However, a condition known as *renal glycosuria* sometimes occurs even in persons without diabetes mellitus. This condition results from a low tubular maximum for glucose, as explained in Chapter 38, so that even though the blood glucose level is perfectly normal, a large quantity of glucose may still be lost in the urine.)

The Fasting Blood Glucose Level. The fasting blood sugar level in the early morning, at least eight hours after any previous meal, is normally 80 to 90 mg. per cent, and 110 mg. per cent is generally considered to be the upper limit of normal. A fasting blood sugar level above this value often indicates diabetes mellitus or, less commonly, either pituitary diabetes or adrenal diabetes.

The Glucose Tolerance Test. As illustrated by the bottom curve in Figure 78–8, when a normal, fasting person ingests 1 gram of glucose per kilogram of body weight, the blood glucose level rises from approximately 90 mg. per cent to 120 to 140 mg. per cent and falls back to below normal within about two hours.

Though an occasional diabetic person has a normal fasting blood glucose concentration, it is usually above 110 mg. per cent, and the glucose tolerance test is almost always abnormal. On ingestion of glucose, these persons exhibit a much greater than normal rise in blood glucose level, as illustrated by the upper curve in Figure 78–8, and the glucose level falls back to the control value only after some five to six hours; and it fails to fall below the control level. This slow fall of the curve and its failure to fall below the control level illustrates that the normal increase in insulin secretion following glucose ingestion does not occur in the diabetic person, and a diagnosis of diabetes mellitus can usually be definitely established on the basis of such a curve.

Insulin Sensitivity. To differentiate diabetes mellitus of pancreatic origin from high blood glucose levels resulting from excess secretion of adrenocortical or anterior pituitary hormones, an *insulin sensitivity test* can be performed. When little insulin is produced by the pancreas, a test dose of insulin causes the blood glucose

level to fall markedly, indicating greatly increased "insulin sensitivity." On the other hand, when the blood glucose level is high as a result of excessive adrenocortical or anterior pituitary secretion, the glucose level responds very poorly to the test dose of insulin because the pancreas is already secreting large quantities of insulin.

Acetone Breath. As pointed out in Chapter 68, small quantities of acetoacetic acid, which increase greatly in severe diabetes, can be converted to acetone, which is volatile and is vaporized into the expired air. Consequently, one frequently can make a diagnosis of diabetes mellitus simply by smelling acetone on the breath of a patient. Also, keto acids can be detected by chemical means in the urine, and their quantitation aids in determining the severity of the diabetes.

TREATMENT OF DIABETES

The theory of treatment of diabetes mellitus is to administer enough insulin so that the patient will have as nearly normal carbohydrate, fat, and protein metabolism as possible. Optimal therapy can prevent most acute effects of diabetes and greatly delay the chronic effects as well.

Figure 78–9 illustrates time-activity curves of different preparations of insulin following subcutaneous injection. Regular amorphous insulin and crystalline insulin have an activity lasting 3 to 6 hours, whereas insulin that has been precipitated slowly with zinc to form large crystalline particles (lente insulin) or insulin that has been precipitated with various protein derivatives (globin insulin and protamine zinc insulin) is relatively insoluble and is absorbed slowly. Such insulins may have activity durations of from 12 hours up to 48 to 72 hours. Ordinarily, the severely diabetic patient is given a single dose of one of the long-acting insulins each day; this increases overall carbohydrate metabolism throughout the day. Then additional quantities of regular insulin are given at those times of the day when the blood glucose level tends to rise too high, such as at meal times. Thus each patient is established on an individualized routine of treatment.

Diet of the Diabetic. The insulin requirements of a diabetic are established with the patient on a standard

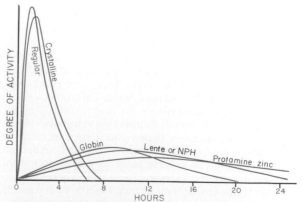

Figure 78–9. Time-action curves for different types of insulin.

diet containing normal, well-controlled amounts of carbohydrates, and any change in the quantity of carbohydrate intake changes the requirements for insulin. In the normal person, the pancreas has the ability to adjust the quantity of insulin produced to the intake of carbohydrate; but in the completely diabetic person, this control function is totally lost.

In the obesity maturity-onset type of diabetes, the disease can often be controlled by weight reduction alone. The decreased fat reduces the insulin requirements, and the pancreas can now supply the need.

Control of the Diabetic Patient in Fever and Exercise. Any abnormal state of metabolism in the diabetic patient alters the insulin requirement. Thus, fever, severe infection, and so forth frequently increase the requirement for insulin immensely, and failure to give the extra insulin can cause diabetic coma.

On the other hand, exercise frequently has exactly the opposite effect on insulin requirement, often reducing the requirement below that usually needed. The peculiar reason for this is that increased muscle activity increases the transport of glucose into the muscle cells even in the absence of insulin. Thus, *exercise actually has an insulin-like effect*. It is often remarked that children with severe diabetes require less insulin to control it if they live a very active life than if they lead an overprotected life.

Relationship of Treatment to Arteriosclerosis. Diabetic patients develop atherosclerosis, arteriosclerosis, severe coronary heart disease, and multiple microcirculatory lesions far more easily than do normal persons. Indeed, those who have relatively poorly controlled diabetes throughout childhood are likely to die of heart disease in their 20s.

In the early days of treating diabetes it was the tendency to reduce severely the carbohydrates in the diet so that the insulin requirements would be minimized. This procedure kept the blood sugar level down to normal values and prevented loss of glucose in the urine, but it did not prevent the abnormaliites of fat metabolism. Consequently, there is a tendency at present to allow the patient an almost normal carbohydrate diet and then to give simultaneously large quantities of insulin to metabolize the carbohydrates. This depresses the rate of fat metabolism and also depresses the high level of blood cholesterol that occurs in diabetes as a result of abnormal fat metabolism.

Because the complications of diabetes — such as atherosclerosis, greatly increased susceptibility to infection, diabetic retinopathy, cataracts, hypertension, and chronic renal disease — are more closely associated with the level of the blood lipids than with the level of blood glucose, it is the object of many clinics treating diabetes to administer sufficient glucose and insulin so that the quantity of blood lipids becomes normal.

Treatment with Drugs That Stimulate the Release of Insulin. Recently, several drugs that can be taken by mouth and that cause a hypoglycemic effect have been introduced for treatment of persons with mild diabetes (the maturity-onset type). The most useful drugs are *sulfonylurea compounds*, the most important of which is *tolbutamide* (Orinase).

These drugs act by stimulating insulin secretion by the islets of Langerhans. Obviously, they are of no value in the treatment of severe diabetes, for the islets have already lost all potential ability to secrete insulin.

DIABETIC COMA

If diabetes is not controlled satisfactorily, severe dehydration and acidosis may result; and sometimes, even when the person is receiving treatment, sporadic changes in metabolic rates of the cells, such as might occur during bouts of fever, can also precipitate dehydration and acidosis.

If the pH of the body fluids falls below approximately 7.0, the diabetic person develops coma. Also, in addition to the acidosis, dehydration is believed to exacerbate the coma. Once the diabetic person reaches this stage, the outcome is usually fatal unless immediate treatment is provided.

Physiological Basis of Treating Diabetic Coma. The patient with diabetic coma is extremely refractory to insulin because acidic plasma has an *insulin antagonist,* an alpha globulin, that opposes the action of the insulin. Also, the very high free fatty acid and acetoacetic acid levels in the blood inhibit cellular usage of glucose, as was discussed earlier. Therefore, instead of the usual 60 to 80 units of insulin per day, which is the dosage usually necessary for control of severe diabetes, several times this much insulin must often be given the first day of treatment of coma.

Administration of insulin alone is not likely to be sufficient to reverse the abnormal physiology and to effect a cure. In addition, it is usually necessary to correct both the dehydration and acidosis immediately. The dehydration is ordinarily corrected rapidly by administering large quantities of sodium chloride solution, and the acidosis is often corrected by administering sodium bicarbonate or sodium lactate solution.

HYPERINSULINISM

Though much rarer than diabetes, increased insulin production, which is known as *hyperinsulinism,* does occasionally occur. This usually results from an adenoma of an islet of Langerhans. About 10 to 15 per cent of these adenomas are malignant, and occasionally metastases from the islets of Langerhans spread throughout the body, causing tremendous production of insulin by both the primary and the metastatic cancers. Indeed, in order to prevent hypoglycemia, in some of these patients more than 1000 grams of glucose have had to be administered each 24 hours.

Diagnosis of hyperinsulinism is made with assurance by measuring very high levels of plasma insulin using the radioimmunoassay procedure — especially when the insulin remains high constantly throughout the day and without rising significantly with increased carbohydrate intake.

Insulin Shock and Hypoglycemia. As already emphasized, the central nervous system derives essentially all its energy from glucose metabolism, and insulin is not necessary for this use of glucose. However, if insulin causes the level of blood glucose to fall to low values, the metabolism of the central nervous system becomes depressed. Consequently, in patients with hyperinsulinism, or in diabetic patients who administer too much insulin to themselves, the syndrome called "insulin shock" may occur as follows:

As the blood sugar level falls into the range of 50 to 70 mg. per cent, the central nervous system usually becomes quite excitable, for this degree of hypoglycemia seems to facilitate neuronal activity. Sometimes various forms of hallucinations result, but more often the patient simply experiences extreme nervousness, trembles all over, and breaks out in a sweat. As the blood glucose level falls to 20 to 50 mg. per cent, clonic convulsions and loss of consciousness are likely to occur. As the glucose level falls still lower, the convulsions cease, and only a state of coma remains. Indeed, at times it is difficult to distinguish between diabetic coma as a result of insulin lack and coma due to hypoglycemia caused by excess insulin. However, the acetone breath and the rapid, deep breathing of diabetic coma are not present in hypoglycemic coma.

Obviously, proper treatment for a patient who has hypoglycemic shock or coma is immediate intravenous administration of large quantities of glucose. This usually brings the patient out of shock within a minute or more. Also, administration of glucagon (or, less effectively, epinephrine) can cause glycogenolysis in the liver and thereby increase the blood glucose level extremely rapidly.

If treatment is not effected immediately, permanent damage to the neuronal cells of the central nervous system occurs; this happens especially in prolonged hyperinsulinism due to pancreatic tumors. Hypoglycemic shock induced by insulin administration was in the past used for treatment of psychogenic disorders. This type of shock, like electric shock therapy, frequently benefitted especially the melancholic patient.

REFERENCES

Arky, R. A.: Hypoglycemia. *In* DeGroot, L. J., *et al.* (eds.): Endocrinology. Vol. 2. New York, Grune & Stratton, 1979, p. 1099.

Baba, S., *et al.* (eds.): Proinsulin, Insulin, C-Peptide. New York, Elsevier/North-Holland, 1979.

Bloom, A., and Ireland, J.: Color Atlas of Diabetes. Chicago, Year Book Medical Publishers, 1980.

Chick, W. L.: Microvascular pathology in diabetes. *In* Kaley, G., and Altura, B. M. (eds.): Microcirculation. Vol. III. Baltimore, University Park Press, 1977.

Clausen, T.: Cations, Glucose Metabolism, and Insulin Action. Aarhus, 1972.

Crepaldi, G., *et al.* (eds.): Diabetes, Obesity, and Hyperlipidemia. New York, Academic Press, 1978.

Diabetes Guidelines for Health Professionals. Silver Spring, Md., American Diabetes Association, 1979.

Ésmann, V. (ed.): Regulatory Mechanisms of Carbohydrate Metabolism. New York, Pergamon Press, 1978.

Fajans, S. S.: Diabetes mellitus: Description, etiology and pathogenesis, natural history and testing procedures. *In* DeGroot, L. J., *et al.*

(eds.): Encrinology. Vol. 2. New York, Grune & Stratton, 1979, p. 1007.

Fitzgerald, P. J., and Morrison, A. B. (eds.): The Pancreas. Baltimore, Williams & Wilkins, 1980.

Galbraith, R. M.: Immunological Aspects of Diabetes Mellitus. Boca Raton, Fla., CRC Press, 1979.

Gerich, J. E., *et al.*: Regulation of pancreatic insulin and glucagon secretion. *Annu. Rev. Physiol.*, 38:353, 1976.

Gorden, P.., *et al.*: Application of radioreceptor assay to circulating insulin, growth hormone, and to their tissue receptors in animals and man. *Pharmacol. Rev.*, 25:179, 1973.

Grave, G. D. (ed.): Early Detection of Potential Diabetes: The Problems and the Promise. New York, Raven Press, 1979.

Guyton, J. R., *et al.*: A model of glucose-insulin homeostasis in man that incorporates the heterogeneous fast pool theory of pancreatic insulin release. *Diabetes*, 27:1027, 1978.

Hedeskov, C. J.: Mechanism of glucose-induced insulin secretion. *Physiol. Rev.*, 60:442, 1980.

Katzen, H. M., and Mahler, R. J. (eds.): Diabetes, Obesity, and Vascular Disease: Metabolic and Molecular Interrelationships. New York, Halsted Press, 1977.

Klachko, D. M., *et al.* (eds.): The Endocrine Pancreas and Juvenile Diabetes. New York, Plenum Press, 1979.

Krall, L. P. (ed.): Joslin Diabetes Manual. Boston, G. K. Hall, 1979.

Liljenquist, J. E., *et al.*: Insulin and glucagon actions and consequences of derangements in secretion. *In* DeGroot, L. J., *et al.* (eds.): Endocrinology. Vol. 2. New York, Grune & Stratton, 1979, p. 981.

Lund-Anderson, H.: Transport of glucose from blood to brain. *Physiol. Rev.*, 59:305, 1979.

Matschinsky, F. M., *et al.*: Metabolism of pancreatic islets and regulation of insulin and glucagon secretion. *In* DeGroot, L. J., *et al.* (eds.): Endocrinology. Vol. 2. New York, Grune & Stratton, 1979, p. 935.

Notkins, A. L.: The causes of diabetes. *Sci. Am.*, 241(5):62, 1979.

Park, C. R., *et al.*: The action of insulin on the transport of glucose through the cell membrane. *Am. J. Med.*, 26:674, 1959.

Park, C. R., *et al.*: The regulation of glucose uptake in muscle as studied in the perfused rat heart. *Recent Prog. Horm. Res.*, 17:493, 1961.

Pfeifer, E. F.: Obesity, islet function, and diabetes mellitus. *Horm. Metab. Res., Suppl.* 4:143, 1974.

Pilkis, S. J., and Park, C. R.: Mechanism of action of insulin. *Annu. Rev. Pharmacol.*, 14:365, 1974.

Podolsky, S., and Viswanathan, M. (eds.): Secondary Diabetes: The Spectrum of the Diabetic Syndromes. New York, Raven Press, 1979.

Podolsky, S. (ed.): Clinical Diabetes: Modern Management. New York, Appleton-Century-Crofts, 1980.

Post, R. L., *et al.*: Regulation of glucose uptake in muscle. III. The interaction of membrane transport and phosphorylation in the control of glucose uptake. *J. Biol. Chem.*, 236:269, 1961.

Sherwin, R., and Felig, P.: Glucagon physiology in health and disease. *Int. Rev. Physiol.*, 16:151, 1977.

Sherwin, R., and Felig, P.: Treatment of diabetes mellitus. *In* DeGroot, L. J., *et al.* (eds.): Endocrinology. Vol. 2. New York, Grune & Stratton, 1979, p. 1061.

Unger, R. H.: The pancreas as a regulator of metabolism. *In* MTP International Review of Science: Physiology. Vol. 5. Baltimore, University Park Press, 1974, p. 179.

Unger, R. H., and Orci, L.: Physiology and pathophysiology of glucagon. *Physiol. Rev.*, 56:788, 1976.

Unger, R. H., and Orci, L.: Glucagon: secretion, transport, metabolism, physiologic regulation of secretion, and derangements in diabetes. *In* DeGroot, L. J., *et al.* (eds.): Endocrinology. Vol. 2. New York, Grune & Stratton, 1979, p. 959.

Unger, R. H., *et al.*: Insulin, glucagon, and somatostatin secretion in the regulation of metabolism. *Annu. Rev. Physiol.*, 40:307, 1978.

Winegrad, A. I., and Morrison, A. D.: Diabetic ketoacidosis, nonketotic hyperosmolar coma, and lactic acidosis. *In* DeGroot, L. J., *et al.* (eds.): Endocrinology. Vol. 2. New York, Grune & Stratton, 1979, p. 1025.

79

Parathyroid Hormone, Calcitonin, Calcium and Phosphate Metabolism, Vitamin D, Bone, and Teeth

The physiology of parathyroid hormone and of the hormone calcitonin is closely related to calcium and phosphate metabolism, the function of vitamin D, and the formation of bone and teeth. Therefore, these are discussed together in the present chapter.

CALCIUM AND PHOSPHATE IN THE EXTRACELLULAR FLUID AND PLASMA — FUNCTION OF VITAMIN D

ABSORPTION AND EXCRETION OF CALCIUM AND PHOSPHATE

Intestinal Absorption of Calcium and Phosphate. By far the major source of calcium in the diet is milk or milk products, which are also major sources of phosphate, but phosphate is also present in many other dietary foods, including especially meats.

Calcium is poorly absorbed from the intestinal tract because of the relative insolubility of many of its compounds and also because bivalent cations are poorly absorbed through the intestinal mucosa anyway. On the other hand, phosphate is absorbed exceedingly well most of the time except when excess calcium is in the diet; the calcium tends to form almost insoluble calcium phosphate compounds that fail to be absorbed but instead pass on through the bowels to be excreted in the feces.

Excretion of Calcium in Feces and Urine; Net Rate of Absorption. About five sixths of the daily intake of calcium is excreted in the feces, and the remaining sixth in the urine. The approximate daily turnover rates for calcium in the adult are the following:

Intake	800 mg.
Intestinal absorption	350 mg.
Secretion in gastrointestinal juices	190 mg.
Net absorption over secretion	170 mg.
Loss in the feces	630 mg.
Excretion in the urine	170 mg.

Excretion of calcium in the urine conforms to much the same principles as sodium excretion. All but a few per cent of the calcium in the glomerular filtrate is reabsorbed in the proximal tubules and ascending limbs of the loops of Henle. Then in the distal tubules and collecting ducts, further reabsorption of the remaining calcium is very selective, depending upon the calcium ion concentration in the blood. When low, this reabsorption is very great so that almost no calcium is lost in the urine. On the other hand, even a minute increase in calcium ion concentration above normal increases calcium excretion markedly. We shall see later in the chapter that one of the most important factors controlling reabsorption of calcium in the distal portions of the nephron, and therefore controlling the rate of calcium excretion, is parathyroid hormone.

Intestinal and Urinary Excretion of Phosphate. Except for the portion of phosphate that is excreted in the feces in combination with calcium, almost all the dietary phosphate is absorbed into the blood from the gut and later excreted in the urine.

Phosphate is a *threshold substance;* that is, when its concentration in the plasma is below the critical value of approximately 1 millimole/liter, no phosphate at all is lost into the urine; but, above this critical concentration, the rate of phosphate loss is directly proportional to the additional increase. Thus, the kidney regulates the phosphate concentration in the extracellular fluid by altering the rate of phosphate secretion in accordance with the plasma phosphate concentration.

As discussed later in the chapter, phosphate excretion

973

by the kidneys is greatly increased by parathyroid hormone, thereby playing an important role in the control of plasma phosphate concentration.

VITAMIN D AND ITS ROLE IN CALCIUM AND PHOSPHATE ABSORPTION

Vitamin D has a potent effect on increasing calcium absorption from the intestinal tract; it also has important effects on both bone deposition and bone reabsorption, as will be discussed later in the chapter. However, vitamin D itself is not the active substance that actually causes these effects. Instead, the vitamin D must first be converted through a succession of reactions in the liver and the kidney to the final active product, *1,25-dihydroxycholecalciferol*. Figure 79–1 illustrates the succession of steps that leads to the formation of this substance from vitamin D.

The Vitamin D Compounds. Several different compounds derived from sterols belong to the vitamin D family, and all these perform more or less the same functions. The most important of these, called vitamin D_3, is *cholecalciferol*. Most of this substance is formed in the skin as a result of irradiation of *7-dehydrocholesterol* by ultraviolet rays from the sun. Consequently, appropriate exposure to the sun prevents vitamin D deficiency.

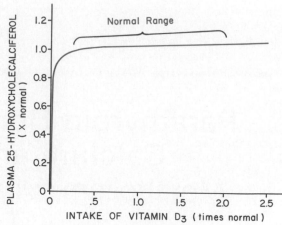

Figure 79–2. Effect of increasing vitamin D_3 intake on the plasma concentration of 25-hydroxycholecalciferol. This figure shows that tremendous changes in vitamin D intake have little effect on the final quantity of activated vitamin D that is formed.

Conversion of Cholecalciferol to 25-Hydroxycholecalciferol in the Liver and Its Feedback Control. The first step in the activation of cholecalciferol is to convert it to 25-hydroxycholecalciferol; this occurs in the liver. The process, however, is itself a limited one because the 25-hydroxycholecalciferol has a feedback inhibitory effect on the conversion reactions. This feedback effect is extremely important for two reasons:

First, the feedback mechanism regulates very precisely the concentration of 25-hydroxycholecalciferol in the plasma, an effect that is illustrated in Figure 79–2. Note that the intake of vitamin D_3 can change many-fold, and yet the concentration of 25-hydroxycholecalciferol still remains within a few per cent of its normal mean value. Indeed, as much as 1000 times normal quantities of vitamin D_3 can be administered to a person, and the concentration of 25-hydroxycholecalciferol will increase only three-fold. Obviously, this high degree of feedback control prevents excessive action of vitamin D when it is present in too great a quantity.

Second, this controlled conversion of vitamin D_3 to 25-hydroxycholecalciferol conserves the vitamin D for future use, because once it is converted, it persists in the body for only a short time thereafter, whereas in the vitamin D form it can be stored in the liver for as long as several months.

Formation of 1,25-Dihydroxycholecalciferol in the Kidneys and Its Control by Parathyroid Hormone. Figure 79–1 also illustrates the conversion in the kidneys of 25-hydroxycholecalciferol to 1,25-dihydroxycholecalciferol. This latter substance is the active form of vitamin D, for none of the previous products in the scheme of Figure 79–1 have very much vitamin D effect. Therefore,

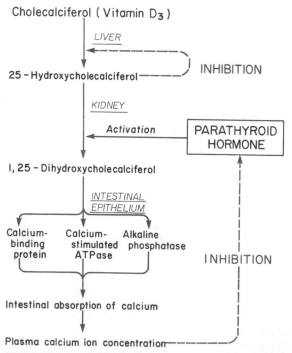

Figure 79–1. Activation of vitamin D_3 to form *1,25-dihydroxycholecalciferol;* and the role of vitamin D in controlling the plasma calcium concentration.

in the absence of the kidneys vitamin D is almost totally ineffective.

Note also in Figure 79–1 that the conversion of 25-hydroxycholecalciferol to 1,25-dihydroxycholecalciferol requires parathyroid hormone. In the absence of this hormone, either none or almost none of the 1,25-dihydroxycholecalciferol is formed. Therefore, parathyroid hormone exerts a potent effect in determining the functional effects of vitamin D in the body, specifically its effects on calcium absorption in the intestines and its effects on bone.

Hormonal Effect of 1,25-Dihydroxycholecalciferol on the Intestinal Epithelium in Promoting Calcium Absorption. 1,25-Dihydroxycholecalciferol has several effects on the intestinal epithelium, one or all of which may play important roles in promoting intestinal absorption of calcium. Probably the most important of these effects is that this "hormone" causes formation of a *calcium-binding protein* in the cytoplasm of the intestinal epithelial cells. The rate of calcium absorption seems to be directly proportional to the quantity of this calcium-binding protein. Furthermore, this protein remains in the cells for several weeks after the 1,25-dihydroxycholecalciferol has been removed from the body, thus causing a prolonged effect on calcium absorption.

Other effects of this "hormone," 1,25-dihydroxycholecalciferol, that might play a role in promoting calcium absorption are: (1) it causes the formation of a calcium-stimulated ATPase in the brush border of the epithelial cells; and (2) it causes the formation of an alkaline phosphatase in the epithelial cells. Unfortunately, the precise details of calcium absorption are still unknown.

Feedback Effect of Calcium Ion Concentration on 1,25-Dihydroxycholecalciferol. Later in the chapter we shall see that the rate of secretion of parathyroid hormone is controlled almost entirely and very potently by the plasma calcium ion concentration. When the calcium ion concentration rises, this change immediately inhibits parathyroid hormone secretion; in the absence of this secretion, 1,25-dihydroxycholecalciferol cannot be formed in the kidney. Thus, this rise in calcium ion concentration provides a *negative* feedback mechanism for control of both the plasma concentration of 1,25-dihydroxycholecalciferol and also of the plasma calcium ion concentration itself. That is, an increase in the calcium ion concentration decreases the vitamin D effect, decreases the absorption of calcium from the intestinal tract, and thus returns the calcium ion concentration back to its normal value. We shall see later in the chapter that this is one of the very important means by which the hormonal system of the body maintains a very constant calcium ion concentration.

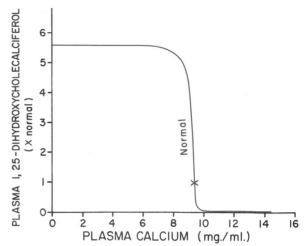

Figure 79–3. Effect of plasma calcium concentration on the plasma concentration of 1,25-dihydroxycholecalciferol. This figure shows that a very slight decrease in calcium concentration below normal causes marked formation of activated vitamin D, which in turn leads to greatly increased absorption of calcium from the intestine.

Figure 79–3 illustrates this feedback effect of plasma calcium concentration on the concentration of plasma 1,25-dihydroxycholecalciferol. Note that when the calcium ion concentration is only slightly greater than 10 mg. per cent, the concentration of 1,25-dihydroxycholecalciferol falls almost to zero, obviously causing a decrease in calcium absorption from the gut to a very low level. Conversely, the concentration of this "hormone" (the 1,25-dihydroxycholecalciferol) rises markedly when the calcium ion concentration falls even slightly below 10 mg. per cent, immediately turning on the mechanism for calcium absorption from the intestines.

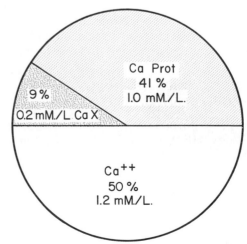

Figure 79–4. Distribution of ionic calcium (Ca^{++}), diffusible but un-ionized calcium ($Ca\ X$), and calcium proteinate ($Ca\ Prot$) in blood plasma.

Effect of Vitamin D on Phosphate Absorption. Much less is known about the effect of vitamin D on phosphate absorption than on calcium absorption. Also, this is much less important because phosphate is usually absorbed relatively easily anyway. However, phosphate flux through the gastrointestinal epithelium is enhanced by vitamin D. It is believed that this results from a direct effect of 1,25-dihydroxycholecalciferol, but it is possible that it results secondarily from this hormone's action on calcium absorption, the calcium in turn acting as a transport mediator for the phosphate.

THE CALCIUM IN THE PLASMA AND INTERSTITIAL FLUID

The concentration of calcium in the plasma is approximately 9.4 mg. per cent, normally varying between 9.0 and 10.0 mg. per cent. This is equivalent to approximately 2.4 millimoles per liter. It is apparent from these narrow limits of normality that the calcium level in the plasma is regulated exactly — and mainly by parathyroid hormone, as discussed later in the chapter.

The calcium in the plasma is present in three different forms, as shown in Figure 79–4. (1) Approximately 41 per cent (1.0 mM./liter) of the calcium is combined with the plasma proteins and consequently is nondiffusible through the capillary membrane. (2) Approximately 9 per cent of the calcium (0.2 mM./liter) is diffusible through the capillary membrane but is combined with other substances of the plasma and interstitial fluids (citrate and phosphate, for instance) in such a manner that it is not ionized. (3) The remaining 50 per cent of the calcium in the plasma is both diffusible through the capillary membrane and ionized. Thus, the plasma and interstitial fluids have a normal *calcium ion concentration of approximately 1.2 mM./liter.* This ionic calcium is important for most functions of calcium in the body, including the effect of calcium on the heart, on the nervous system, and on bone formation.

THE INORGANIC PHOSPHATE IN THE EXTRACELLULAR FLUIDS

Inorganic phosphate in the plasma is mainly in two forms: HPO_4^{--} and $H_2PO_4^-$. The concentration of HPO_4^{--} is approximately 1.05 mM./liter and the concentration of $H_2PO_4^-$ is approximately 0.26 mM./liter. When the total quantity of phosphate in the extracellular fluid rises, so does the quantity of each of these two types of phosphate ions. Furthermore, when the pH of the extracellular fluid becomes more acid, there is a relative increase in $H_2PO_4^-$ and decrease in the HPO_4^{--} while the opposite occurs when the extracellular fluid becomes alkaline. These relationships were presented in the discussion of acid-base balance in Chapter 37.

Because it is difficult to determine chemically the exact quantities of HPO_4^{--} and $H_2PO_4^-$ in the blood, ordinarily the total quantity of phosphate is often expressed in terms of milligrams of *phosphorus* per 100 ml. of blood. The average total quantity of inorganic phosphorus represented by both phosphate ions is about 4 mg./100 ml., varying between normal limits of 3.5 to 4 mg./100 ml. in adults and 4 to 5 mg./100 ml. in children.

EFFECTS OF ALTERED CALCIUM AND PHOSPHATE CONCENTRATIONS IN THE BODY FLUIDS

Changing the level of phosphate in the extracellular fluid from far below normal to as high as 3 to 4 times normal does not cause significant immediate effects on the body. There is a similar lack of effects for variations in most other anions, for even chloride ion can be substituted almost entirely by certain other anions without drastic effects.

On the other hand, elevation or depletion of calcium ion in the extracellular fluid causes extreme immediate effects. Both prolonged hypocalcemia and hypophosphatemia greatly decrease bone mineralization, as explained later in the chapter.

Tetany Resulting from Hypocalcemia. When the extracellular fluid concentration of calcium ion falls below normal, the nervous system becomes progressively more and more excitable because of increased neuronal membrane permeability. This increase in excitability occurs both in the central nervous system and in the peripheral nerves, though most symptoms are manifest peripherally. The nerve fibers become so excitable that they begin to discharge spontaneously, initiating nerve impulses that pass to the peripheral skeletal muscles where they elicit tetanic contraction. Consequently, hypocalcemia causes tetany. But it also occasionally causes convulsions because of its central action of increasing excitability.

Figure 79–5 illustrates tetany in the hand, which usually occurs before generalized tetany develops. This is called "carpopedal spasm."

Tetany ordinarily occurs when the blood concentration of calcium falls from its normal level of 9.4 mg. to approximately 6 mg. per cent, which is only 35 per cent below the normal calcium concentration, and it is usually lethal at about 4 mg. per cent.

When the calcium in the body fluids falls to a level not quite sufficient to cause tetany, "latent tetany" results; this can be diagnosed by weakly stimulating the nerves and noting the response. For instance, tapping on the 7th nerve where it passes over the angle of the jaw causes the facial

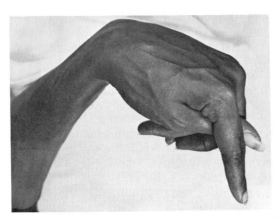

Figure 79–5. Hypocalcemic tetany in the hand, called "carpopedal spasm." (Courtesy of Dr. Herbert Langford.)

muscle to twitch. Second, placing a tourniquet on the upper arm causes ischemia of the peripheral nerves and also increases the excitability of the nerves, thus causing the muscles of the lower arm and hand to go into spasm. Finally, if the person with latent tetany hyperventilates, the resulting alkalinization of the body fluids increases the irritability of the nerves, causing overt signs of tetany.

In experimental animals, in which the level of calcium can be reduced beyond the normal lethal stage, extreme hypocalcemia can cause marked dilatation of the heart, changes in cellular enzyme activities, increased cell membrane permeability in other cells in addition to nerve cells, and impaired blood clotting.

Hypercalcemia. When the level of calcium in the body fluids rises above normal, the nervous system is depressed, and reflex activities of the central nervous system become sluggish. Also, increased calcium ion concentration decreases the QT interval of the heart, and it causes constipation and lack of appetite, probably because of depressed contractility of the muscle walls of the gastrointestinal tract.

The depressive effects of increased calcium level begin to appear when the blood level of calcium rises above approximately 12 mg. per cent, and they can become marked as the calcium level rises above 15 mg. per cent. When the level of calcium rises above approximately 17 mg. per cent in the body fluids, calcium phosphate is likely to precipitate throughout the body; this condition is discussed shortly in connection with parathyroid poisoning.

BONE AND ITS RELATIONSHIPS TO EXTRACELLULAR CALCIUM AND PHOSPHATES

Bone is composed of a tough *organic matrix* that is greatly strengthened by deposits of *calcium salts.* Average *compact bone* contains by weight approximately 30 per cent matrix and 70 per cent salts. However, *newly formed bone* may have a considerably higher percentage of matrix in relation to salts.

The Organic Matrix of Bone. The organic matrix of bone is 90 to 95 per cent *collagen fibers,* and the remainder is a homogeneous medium called *ground substance.* The collagen fibers extend primarily along the lines of tensional force. These fibers give bone its powerful tensile strength.

The ground substance is composed of extracellular fluid plus *proteoglycans,* especially *chondroitin sulfate* and *hyaluronic acid.* The precise function of these is not known, though perhaps they help to control the deposition of calcium salts.

The Bone Salts. The crystalline salts deposited in the organic matrix of bone are composed principally of *calcium* and *phosphate,* and the formula for the major crystalline salts, known as *hydroxyapatites,* is the following:

$$Ca^{++}_{10-x}(H_3O^+)_{2x} \cdot (PO_4)_6(OH^-)_2$$

Each crystal — about 400 A long, 10 to 30 A thick, and 100 A wide — is shaped like a long, flat plate. The relative ratio of calcium to phosphorus can vary markedly under different nutritional conditions, the Ca/P ratio on a weight basis varying between 1.3 and 2.0.

Magnesium, sodium, potassium, and *carbonate* ions are also present among the bone salts, though x-ray diffraction studies fail to show definite crystals formed by these. Therefore, they are believed to be conjugated to the hydroxyapatite crystals rather than organized into distinct crystals of their own. This ability of many different types of ions to conjugate to bone crystals extends to many ions normally foreign to bone, such as *strontium, uranium, plutonium, the other transuranic elements, lead, gold, other heavy metals,* and *at least 9 of 14 of the major radioactive products released by explosion of the hydrogen bomb.* Deposition of radioactive substances in the bone can cause prolonged irradiation of the bone tissues, and, if a sufficient amount is deposited, an osteogenic sarcoma (bone cancer) almost invariably eventually develops.

Tensile and Compressional Strength of Bone. Each collagen fiber of *compact* bone is composed of repeating periodic segments every 640 A along its length; hydroxyapatite crystals lie adjacent to each segment of the fiber, bound tightly to it. This intimate bonding prevents "shear" in the bone; that is, it prevents the crystals and collagen fibers from slipping out of place, which is essential in providing strength to the bone. In addition, the segments of adjacent collagen fibers overlap each other, also causing hydroxyapatite crystals to be overlapped like bricks keyed to each other in a brick wall.

The collagen fibers of bone, like those of tendons, have great tensile strength, while the calcium salts, which are similar in physical properties to marble, have great compressional strength. These combined properties, plus the degree of bondage between the collagen fibers and the crystals, provide a bony structure that has both extreme tensile and compressional strength. Thus, bones are constructed in exactly the same way that reinforced concrete is constructed. The steel of reinforced concrete provides the tensile strength, while the cement, sand, and rock provide the compressional strength. Indeed, the compressional strength of bone is greater than that of even the best reinforced concrete, and the tensile strength approaches that of reinforced concrete.

PRECIPITATION AND ABSORPTION OF CALCIUM AND PHOSPHATE IN BONE — EQUILIBRIUM WITH THE EXTRACELLULAR FLUIDS

Supersaturated State of Calcium and Phosphate Ions in Extracellular Fluids with Respect to Hydroxyapatite. The concentrations of calcium and phosphate ions in extracellular fluid are considerably greater than those required to cause precipitation of hydroxyapatite, if these two ions alone were in the fluid. However, inhibitors are present in most tissues of the body, as well as in plasma, to prevent such precipitation; one such inhibitor is pyrophosphate. Therefore, hydroxyapatite crystals fail to precipitate in normal tissues except for bone despite the state of supersaturation of the ions.

Mechanism of Bone Calcification. The initial stage in bone production is the secretion of collagen and ground substance by the osteoblasts. The collagen polymerizes rapidly to form collagen fibers, and the resultant tissue becomes *osteoid,* a cartilage-like material but differing from cartilage in that calcium salts precipitate in it. As the osteoid is formed, some of the osteoblasts become entrapped in the osteoid and then are called *osteocytes.* These may play an important role in the subsequent control of bone salts, though this is not certain.

Within a few days after the osteoid is formed, calcium salts begin to precipitate on the surfaces of the collagen fibers. The precipitates appear at periodic intervals along each collagen fiber, forming minute nidi that rapidly multiply and grow over a period of days and weeks into the finished product, *hydroxyapatite crystals.*

The initial calcium salts to be deposited probably are not hydroxyapatite crystals but, instead, are amorphous compounds (noncrystalline), a probable mixture of such salts as $CaHPO_4 \cdot 2H_2O$, $Ca_3(PO_4)_2 \cdot 3H_2O$, and others. Then by a process of substitution and addition of atoms, or reabsorption and reprecipitation, these salts are converted into the hydroxyapatite crystals. Yet, as much as 20 to 30 per cent may remain permanently in the amorphous form. This is important, because these salts can be absorbed rapidly when there is need for extra calcium in the extracellular fluid.

It is still not known what causes calcium salts to be deposited in osteoid. One theory suggests that the osteoblasts and the entrapped osteocytes in the osteoid play an important role in the following way: It is known that these cells concentrate large quantities of calcium and phosphate in their mitochondria and even precipitate calcium phosphate compounds in these. Electron micrographs indicate that calcium phosphate–containing vesicles break away from the mitochondria, migrate to the walls of the cell, and then extrude minute calcium phosphate crystals into the surrounding extracellular fluid. It may be these preformed calcium phosphate salts that attach to the collagen fibers to form the initial nidi for crystallization, and it is possible that subsequent vesicles help to supply the necessary calcium and phosphate ions for further growth of the crystals.

However, another theory holds that at the time of formation the collagen fibers are specially constituted in advance for causing precipitation of calcium salts. One variant of this theory suggests that the osteoblasts secrete a substance into the osteoid to neutralize an inhibitor (perhaps pyrophosphate) that normally prevents hydroxyapatite crystallization. Once the pyrophosphate has been neutralized, then the natural affinity of the collagen fibers for calcium salts supposedly causes the precipitation. In support of this theory is the fact that properly prepared collagen fibers from other tissues of the body besides bone will also cause precipitation of hydroxyapatite crystals from plasma.

The formation of initial crystals within the collagen fibers is called *crystal seeding* or *nucleation.*

The growth of hydroxyapatite crystals in newly forming bone reaches 75 per cent completion in a few days, but it usually takes months for the bone to achieve full calcification.

Precipitation of Calcium in Nonosseous Tissues Under Abnormal Conditions. Though calcium salts almost never precipitate in normal tissues besides bone, under abnormal conditions they do precipitate. For instance, they precipitate in arterial walls in the condition called arteriosclerosis and cause the arteries to become bonelike tubes. Likewise, calcium salts frequently deposit in degenerating tissues or in old blood clots. Presumably, in these instances, the inhibitor factors that normally prevent deposition of calcium salts disappear from the tissues, thereby allowing precipitation.

EXCHANGEABLE CALCIUM

If soluble calcium salts are injected intravenously, the calcium ion concentration can be made to increase immediately to very high levels. However, within minutes to an hour or more, the calcium ion concentration returns to normal. Likewise, if large quantities of calcium ions are removed from the circulating body fluids, the calcium ion concentration again returns to normal within minutes to hours. These effects result from the fact that the body contains a type of *exchangeable* calcium that is always in equilibrium with the calcium ions in the extracellular fluids. A small portion of this exchangeable calcium is that calcium found in all tissue cells, especially in highly permeable types of cells such as those of the liver and the gastrointestinal tract. However, most of the exchangeable calcium, as shown by studies using radioactively tagged calcium, is in the bone, and it normally amounts to about 0.4 to 1.0 per cent of the total bone calcium. Most of this calcium is probably deposited in the bones in the form of readily mobilizable salts such as $CaHPO_4$ and the other amorphous salts.

The importance of exchangeable calcium to the body is that it provides a rapid buffering mechanism to keep the calcium ion concentration in the extracellular fluids from rising to excessive levels or falling to very low levels under transient conditions of excess or hypoavailability of calcium.

DEPOSITION AND ABSORPTION OF BONE — REMODELING OF BONE

Deposition of Bone by the Osteoblasts. Bone is continually being deposited by *osteoblasts*, and it is continually being absorbed where *osteoclasts* are active. Osteoblasts are found on the outer surfaces of the bones and in the bone cavities. A small amount of osteoblastic activity occurs continually in all living bones (on about 4 per cent of all surfaces at any given time, as shown in Figure 79–6) so that at least some new bone is being formed constantly.

Absorption of Bone — Function of the Osteoclasts. Bone is also being continually absorbed in the presence of osteoclasts, which are normally active at any one time on less than 1 per cent of the outer surfaces and cavity surfaces. Later in the chapter we will see that parathyroid hormone controls the bone absorptive activity of osteoclasts.

Histologically, bone absorption occurs immediately adjacent to the osteoclasts, as illustrated in Figure 79–6. The mechanism of this absorption is

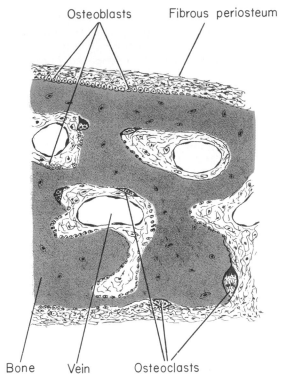

Figure 79–6. Osteoblastic and osteoclastic activity in the same bone.

believed to be the following: The osteoclasts send out villous-like projections toward the bone and from these "villi" secrete two types of substances: (1) proteolytic enzymes, released from the lysosomes of the osteoclasts, and (2) several acids, including citric acid and lactic acid. The enzymes presumably digest or dissolute the organic matrix of the bone, while the acids cause solution of the bone salts. Also, whole fragments of bone salts and collagen are literally gobbled up (phagocytosed) by the "villi" and then digested within the osteoclasts.

Equilibrium Between Bone Deposition and Absorption. Normally, except in growing bones, the rates of bone deposition and absorption are equal to each other so that the total mass of bone remains constant. Usually, osteoclasts exist in small but concentrated masses, and once a mass of osteoclasts begins to develop, it usually eats away at the bone for about three weeks, eating out a tunnel that may be as large as 1 mm. in diameter and several millimeters in length. At the end of this time the osteoclasts disappear and the tunnel is invaded by osteoblasts instead; then new bone begins to develop. Bone deposition then continues for several months, the new bone being laid down in successive layers on the inner surfaces of the cavity until the tunnel is filled. Deposition of new bone ceases when the bone begins to encroach on the blood vessels supplying the area. The canal through which these vessels run, called the *haversian canal,* therefore, is all that remains of the original cavity. Each new area of bone deposited in this way is called an *osteon,* as shown in Figure 79–7.

Value of Continual Remodeling of Bone. The continual deposition and absorption of bone has a number of physiologically important functions. First, bone ordinarily adjusts its strength in proportion to the degree of bone stress. Consequently, bones thicken when subjected to heavy loads. Second, even the shape of the bone can be rearranged for proper support of mechanical forces by deposition and absorption of bone in accordance with stress patterns. Third, since old bone becomes relatively weak and brittle, new organic matrix is needed as the old organic matrix degenerates. In this manner the normal toughness of bone is maintained. Indeed, the bones of children, in whom the rates of deposition and absorption are rapid, show little brittleness in comparison with the bones of old age, at which time the rates of deposition and absorption are slow.

Control of the Rate of Bone Deposition by Bone "Stress." Bone is deposited in proportion to the compressional load that the bone must carry. For instance, the bones of athletes become considerably heavier than those of nonathletes. Also, if a person has one leg in a cast but continues to walk on the opposite leg, the bone of the leg in the cast becomes thin and as much as 30 per cent decalcified within a few weeks, while the opposite bone remains thick and normally calcified. Therefore, continual physical stress stimulates calcification and osteoblastic deposition of bone.

Bone stress also determines the shape of bones under certain circumstances. For instance, if a long bone of the leg breaks in its center and then heals at an angle, the compression stress on the inside of the angle causes increased deposition of bone, while increased absorption occurs on the outer side of the angle where the bone is not compressed. After many years of increased deposition on the inner side of the angulated bone and absorption on the outer side, the bone becomes almost straight. This is especially true in children because of the rapid remodeling of bone at younger ages.

The deposition of bone at points of compressional stress has been suggested to be caused by a *piezoelectric* effect, as follows: Compression of bone causes a negative potential at the compressed site and a positive potential elsewhere in the bone. It has been shown that minute quantities of current flowing in bone cause osteoblastic activity at the negative end of the current flow, which could explain the increased bone deposition at compression sites. On the other hand, usual osteoclastic activity could account for reabsorption of bone at sites of tension.

Repair of a Fracture. A fracture of a bone in some way maximally activates all the periosteal and intraosseous osteoblasts involved in the break. Also, immense numbers of new osteoblasts are formed almost immediately from so-called *osteoprogenitor cells,* which are bone stem cells. Therefore, within a short time a large bulge of osteoblastic tissue and new organic bone matrix, followed shortly by the deposition of calcium salts, develops between the two broken ends of the bone. This is called a *callus.*

Many bone surgeons utilize the phenomenon of

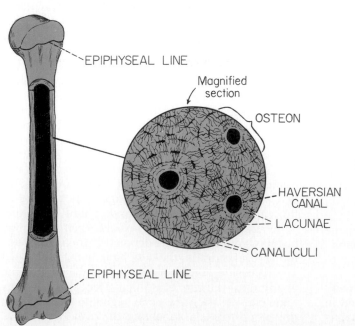

Figure 79–7. The structure of bone.

bone stress to accelerate the rate of fracture healing. This is done by use of special mechanical fixation apparatuses for holding the ends of the broken bone together so that the patient can use the bone immediately. This obviously causes stress on the opposed ends of the broken bones, which accelerates osteoblastic activity at the break and often shortens convalescence.

Blood Alkaline Phosphatase as an Indication of the Rate of Bone Deposition. The osteoblasts secrete large quantities of alkaline phosphatase when they are actively depositing bone matrix. This phosphatase is believed either to increase the local concentration of inorganic phosphate or to activate the collagen fibers in such a way that they cause the deposition of calcium salts. Since some alkaline phosphatase diffuses into the blood, the blood level of alkaline phosphatase is usually a good indicator of the rate of bone formation.

The alkaline phosphatase level is below normal in only a few diseases; this includes especially hypoparathyroidism. On the other hand, it is greatly elevated (1) during growth of children, (2) following major bone fractures, and (3) in almost any bone disease that causes bone destruction that must be repaired by osteoblastic activity, such as rickets, osteomalacia, and osteitis fibrosa cystica caused by excess parathyroid hormone.

PARATHYROID HORMONE

For many years it has been known that increased activity of the parathyroid gland causes rapid absorption of calcium salts from the bones with resultant hypercalcemia in the extracellular fluid; conversely, hypofunction of the parathyroid glands causes hypocalcemia, often with resultant tetany, as described earlier in the chapter. Also, parathyroid hormone is important in phosphate metabolism as well as in calcium metabolism.

Physiologic Anatomy of the Parathyroid Glands. Normally there are four parathyroid glands in the human being; these are located immediately behind the thyroid gland — one behind each of the upper and each of the lower poles of the thyroid. Each parathyroid gland is approximately 6 mm. long, 3 mm. wide, and 2 mm. thick, and has a macroscopic appearance of dark brown fat; therefore, the parathyroid glands are difficult to locate during thyroid operations. For this reason, before the importance of these glands was generally recognized, total or subtotal thyroidectomy frequently resulted in total removal of the parathyroid glands.

Removal of half the parathyroid glands usually causes little physiological abnormality. However, removal of three of four normal glands usually causes transient hypoparathyroidism. But even a small quantity of remaining parathyroid tissue is usually capable of hypertrophying satisfactorily to perform the function of all the glands.

The parathyroid gland of the adult human being, illustrated in Figure 79–8, contains mainly *chief cells* and *oxyphil cells,* but oxyphil cells are absent in many animals and in young human beings. The chief cells secrete most of the parathyroid hormone. The function

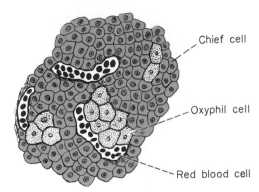

Figure 79–8. Histological structure of a parathyroid gland.

of the oxyphil cells is not certain; they are perhaps modified chief cells that still secrete some hormones.

Chemistry of Parathyroid Hormone. Parathyroid hormone has been isolated in a pure form. It is a small protein, having a molecular weight of approximately 9500, and is composed of 84 amino acids. Smaller compounds have also been isolated from the parathyroid glands that exhibit parathyroid hormone activity, but the activity is always slightly to much less than that of the larger protein molecule. Therefore, the smaller compounds are almost certainly breakdown products of the normal parathyroid hormone.

EFFECT OF PARATHYROID HORMONE ON CALCIUM AND PHOSPHATE CONCENTRATIONS IN THE EXTRACELLULAR FLUID

Figure 79–9 illustrates the effect on the blood calcium and phosphate concentrations caused by suddenly beginning to infuse parathyroid hormone into an animal and continuing this for an indefinite period of time. Note that at the onset of infusion the calcium ion concentration begins to rise and reaches a plateau level in about 4 hours. On the other hand, the phosphate concentration falls more rapidly and also reaches a depressed plateau level within a few hours. The rise in calcium concentration is caused principally by a direct

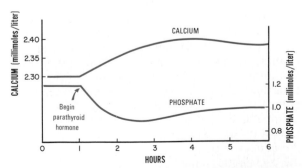

Figure 79–9. Approximate changes in calcium and phosphate concentrations during the first five hours of parathyroid hormone infusion at a moderate rate.

effect of parathyroid hormone in causing calcium and phosphate absorption from the bone. The decline in phosphate concentration, on the other hand, is caused by a very strong effect of parathyroid hormone on the kidney in causing excessive renal phosphate excretion, an effect that is usually great enough to override the increased phosphate absorption from the bone.

Calcium and Phosphate Absorption from the Bone Caused by Parathyroid Hormone. Parathyroid hormone seems to have two separate effects on bone in causing absorption of calcium and phosphate. One is a very rapid effect that takes place in minutes and probably results from activation of the already existing bone cells to promote the calcium and phosphate absorption. The second phase is a much slower one, requiring several days or even weeks to become fully developed, and it results from proliferation of the osteoclasts, followed by greatly increased osteoclastic reabsorption of the bone itself, not merely absorption of calcium phosphate salts from the bone.

The Rapid Phase of Calcium and Phosphate Absorption— The "Osteocytic Membrane" System. When large quantities of parathyroid hormone are injected, the calcium ion concentration in the blood begins to rise within minutes, long before any new bone cells can be developed. Histological studies have shown that the parathyroid hormone causes removal of bone salts from the bone matrix in the vicinity of the osteocytes lying within the bone itself and also in the vicinity of the osteoblasts. Yet, strangely enough, one does not usually think of either osteoblasts or osteocytes functioning to cause bone salt absorption, because both these types of cells are osteoblastic in nature and are normally associated with bone deposition and its calcification. However, recent studies have shown that the osteoblasts and osteocytes form a membrane system that spreads over all the bone surfaces except for small surface areas that are adjacent to the osteoclasts. Also, long filmy processes extend from osteocyte to osteocyte throughout the bone structure, and these processes also connect with the surface osteocytes and osteoblasts. There is much reason to believe that this extensive *osteocytic membrane system* provides a permeable membrane that separates the bone itself from the extracellular fluid. Between the osteocytic membrane and the bone, in turn, lies a small amount of fluid called simply *bone fluid*. Indirect experiments indicate that the osteocytic membrane pumps calcium ions from the bone fluid into the extracellular fluid, creating a calcium ion concentration in the bone fluid only one-third as great as the concentration in the extracellular fluid. When the osteocytic pump becomes excessively activated, the bone fluid calcium concentration falls even lower, and calcium phosphate salts are then ab-

sorbed from the bone. When the pump is inactivated, the bone fluid calcium concentration rises to a higher level, and calcium phosphate salts are then deposited.

But, where does parathyroid hormone fit into this picture? It seems that parathyroid hormone can activate the calcium pump strongly, thereby causing rapid removal of calcium phosphate salts from the amorphous bone crystals that lie near the bone surfaces adjacent to the bone fluid. The parathyroid hormone is believed to stimulate this pump by increasing the calcium permeability of the bone fluid side of the osteocytic membrane, thus allowing calcium ions to diffuse into the membrane cells from the bone fluid. Then the calcium pump on the other side of the cells transfers the calcium ions the rest of the way into the extracellular fluid.

The Slow Phase of Bone Absorption and Calcium Phosphate Release — Activation of the Osteoclasts. A much better known effect of parathyroid hormone, and one for which the evidence is also much clearer, is to activate the osteoclasts. These in turn set about their usual task of gobbling up the bone.

Activation of the osteoclastic system occurs in two stages: (1) immediate activation of the osteoclasts that are already formed, and (2) formation of new osteoclasts from *osteoprogenitor cells.* Usually several days of excess parathyroid hormone cause the osteoclastic system to become well developed, but it can continue to grow for literally months under the influence of very strong parathyroid hormone stimulation.

Parathyroid hormone also transiently depresses osteoblastic activity. However, after a few days to a few weeks the osteoclastic resorption of bone leads to weakened bones and secondary stimulation of the osteoblasts. Therefore, the late effect is actually to enhance both osteoblastic and osteoclastic activity. Still, even in the late stages, there is more bone absorption than bone deposition.

Bone contains such great amounts of calcium in comparison with the total amount in all the extracellular fluids (about 1000 times as much) that even when parathyroid hormone causes a great rise in calcium concentration in the fluids, it is impossible to discern any immediate effect at all on the bones. Yet prolonged administration or secretion of parathyroid hormone finally results in evident absorption in all the bones with development of large cavities filled with very large, multinucleated osteoclasts.

Effect of Parathyroid Hormone on Phosphate and Calcium Excretion by the Kidneys. Administration of parathyroid hormone causes immediate and rapid loss of phosphate in the urine. This effect is caused by diminished proximal tubular reabsorption of phosphate ions.

Parathyroid hormone also causes increased renal tubular reabsorption of calcium at the same time that it diminishes the rate of phosphate reabsorption. Moreover, it also increases the rate of reabsorption of magnesium ions and hydrogen ions, while it decreases the reabsorption of sodium, potassium, and amino acid ions in much the same way that it affects phosphate. However, the effects of increasing calcium absorption occurs in the distal tubules and collecting ducts instead of the proximal tubules.

Were it not for the *long-term* effect of parathyroid hormone on the kidneys to increase calcium reabsorption, the continual loss of calcium into the urine would occur even normally and would eventually deplete the bones of this mineral.

Effect of Parathyroid Hormone on Intestinal Absorption of Calcium and Phosphate. At this point we should be reminded again that parathyroid hormone greatly enhances both calcium and phosphate absorption from the intestines by increasing the formation of 1,25-dihydroxycholecalciferol from vitamin D, as was discussed earlier in the chapter.

Effect of Vitamin D on Bone and Its Relation to Parathyroid Activity. Vitamin D plays important roles in both bone absorption and bone deposition. Administration of large quantities of vitamin D causes absorption of bone in much the same way that administration of parathyroid hormone does. Also, in the absence of vitamin D, the effect of parathyroid hormone in causing bone absorption is greatly reduced or even prevented. Therefore, it is possible, if not likely, that parathyroid hormone functions in bone the same way that it functions in the kidneys and intestines — that is, by causing the conversion of vitamin D to 1,25-dihydroxycholecalciferol, this in turn acting to cause the bone absorption.

Vitamin D also promotes bone calcification. Obviously, one of the ways in which it does this is to increase calcium and phosphate absorption from the intestines. However, even in the absence of such increase, it still enhances the mineralization of bone. Here again, the mechanism of the effect is unknown, but it probably results from the ability of 1,25-dihydroxycholecalciferol to cause transport of calcium ions through cell membranes — perhaps through the osteoblastic or osteocytic cell membranes.

Role of Cyclic AMP as a Mediator of Parathyroid Stimulation. A large share of the effect of parathyroid hormone on its target organs is almost certainly mediated by the cyclic AMP *second messenger* mechanism. Within a few minutes after parathyroid hormone administration, the concentration of cyclic AMP increases in the osteoclasts and other target cells. This cyclic AMP, in turn, is probably responsible for such functions as osteoclastic secretion of enzymes and acids to cause

bone reabsorption, formation of 1,25-dihydroxycholecalciferol in the kidneys, and so forth.

CONTROL OF PARATHYROID SECRETION BY CALCIUM ION CONCENTRATION

Even the slightest decrease in calcium ion concentration in the extracellular fluid causes the parathyroid glands to increase their rate of secretion within minutes, and if the decreased calcium concentration persists, the glands will hypertrophy sometimes as much as five-fold or more. For instance, the parathyroid glands become greatly enlarged in *rickets*, in which the level of calcium is usually depressed only a few per cent; also they become greatly enlarged in pregnancy, even though the decrease in calcium ion concentration in the mother's extracellular fluid is hardly measurable; and, they are greatly enlarged during lactation because calcium is used for milk formation.

On the other hand, any condition that increases the calcium ion concentration causes decreased activity and reduced size of the parathyroid glands. Such conditions include (1) excess quantities of calcium in the diet, (2) increased vitamin D in the diet, and (3) bone absorption caused by factors other than parathyroid hormone (for example, bone absorption caused by disuse of the bones).

Figure 79–10 illustrates quantitatively the relationship between plasma calcium concentration and plasma parathyroid hormone concentration.

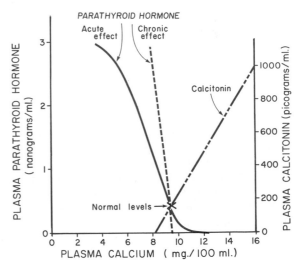

Figure 79–10. Approximate effect of plasma calcium concentration on the plasma concentrations of parathyroid hormone and calcitonin. Note especially that long-term, chronic changes in calcium concentration can cause as much as a 100 per cent change in parathyroid hormone concentration for only a 2 per cent change in calcium concentration.

The solid curve shows the acute relationship when the calcium concentration is changed over a period of a few hours. This shows that a decrease in calcium concentration from 9.4 mg. per cent to 8.4 mg. per cent approximately doubles the plasma parathyroid hormone. On the other hand, the chronic relationship that one finds when the calcium ion concentration changes over a period of many weeks is illustrated by the long-dashed line; this illustrates that about 0.1 mg. per cent decrease in plasma calcium concentration will double parathyroid hormone secretion. To state this still another way, the chronic relationship between plasma calcium and plasma parathyroid hormone shows that approximately a 1 per cent decrease in calcium can give as much as 100 per cent increase in parathyroid hormone. Obviously, this is the basis of the body's extremely potent feedback system for control of plasma calcium ion concentration.

CALCITONIN

About 20 years ago, a new hormone that has effects on blood calcium opposite to those of parathyroid hormone was discovered in several lower animals and at first was believed to be secreted by the parathyroid glands. This hormone was named *calcitonin* because it reduces the blood calcium ion concentration. Soon after its initial discovery, it was found to be secreted in the human being not by the parathyroid glands but instead by the thyroid gland, for which reason it has also been called *thyrocalcitonin*. Still more recently it was discovered that calcitonin is secreted by the *ultimobranchial glands* of fish, amphibia, reptiles, and birds, and that it plays an especially important role in helping to control the blood calcium ion concentration when these animals change their habitat from fresh water to sea water, where there are great excesses of calcium. Furthermore, its concentration in these ultimobranchial glands is extremely great. In the human being, ultimobranchial glands do not exist as such but have become incorporated into the thyroid gland. The so-called *parafollicular cells*, or "C" cells, in the interstitial tissue between the follicles of the human thyroid gland are remnants of the ultimobranchial glands of lower animals, and it is these cells that secrete the calcitonin.

Calcitonin is a large polypeptide with a molecular weight of approximately 3000 and having a chain of 32 amino acids.

Effect of Calcitonin in Decreasing Plasma Calcium Concentration. Calcitonin decreases blood calcium ion concentration very rapidly, beginning within minutes after injection of the calcitonin. Thus, the effect of calcitonin on blood calcium ion concentration is exactly opposite to that of para-

thyroid hormone, and it occurs several times as rapidly.

Calcitonin reduces plasma calcium concentration in three separate ways:

1. The immediate effect is to decrease the activity of the osteoclasts, an effect that is especially significant in growing children because of the rapid osteoclastic activity in children. A large dose of calcitonin can reduce osteoclastic activity as much as 70 per cent in 15 minutes.

2. The second effect, which can be seen within about an hour, is an increase in osteoblastic activity. However, this is a transient effect, lasting not more than a few days.

3. The third and most prolonged effect of calcitonin is to prevent formation of new osteoclasts from the osteoprogenitor cells. Also, since osteoclastic resorption of bone leads secondarily to osteoblastic activity, the depressed numbers of osteoclasts are followed by depressed numbers of osteoblasts as well. Therefore, over a long period of time the net result is simply greatly reduced osteoclastic and osteoblastic activity without a significant prolonged effect on plasma calcium ion concentration. That is, the effect on plasma calcium is mainly a transient one, lasting for a few days, at most. However, there is a prolonged effect of decreasing the rate of bone remodeling as well as of increasing the amount of calcium salts deposited in bone, under some conditions.

Importance of the Calcitonin Effect on Plasma Calcium Concentration. Calcitonin has only a very weak effect on plasma calcium concentration in the adult human being. The reason for this is twofold: First, the initial reduction of the calcium ion concentration leads within hours to a powerful stimulation of parathyroid hormone secretion, which almost completely overrides the calcitonin effect. Second, in the adult, osteoclastic absorption provides only 0.8 gram of calcium to the extracellular fluid each day, and the suppression of this amount of osteoclastic activity by calcitonin has very little effect on the plasma calcium. On the other hand, the effect in children is much more marked because bone remodeling occurs rapidly in children, with osteoclastic absorption of calcium as great as 5 or more grams per day — equal to 5 to 10 times the total calcium in all the extracellular fluid. Also, in certain bone diseases such as Paget's disease in which osteoclastic activity is greatly accelerated, calcitonin then has a potent effect of reducing the calcium absorption.

Effect of Plasma Calcium Concentration on the Secretion of Calcitonin

An increase in plasma calcium concentration of about 10 per cent causes an immediate two-fold increase in the rate of secretion of calcitonin, which is illustrated by the dot-dash line of Figure

79-10. This provides a second hormonal feedback mechanism for controlling the plasma calcium ion concentration, but one that works exactly oppositely to the parathyroid hormone system. That is, an increase in calcium concentration causes increased calcitonin secretion, and the increased calcitonin in turn reduces the plasma calcium concentration back toward normal.

However, there are two major differences between the calcitonin and the parathyroid feedback systems. First, the calcitonin mechanism operates more rapidly, reaching peak activity in less than an hour, in contrast to the several hours required for peak activity to be attained following parathyroid secretion.

The second difference is that the calcitonin mechanism acts mainly as a short-term regulator of calcium ion concentration because it is very rapidly overriden by the much more powerful parathyroid control mechanism. Therefore, over a prolonged period of time it is almost entirely the parathyroid system that sets the long-term level of calcium ions in the extracellular fluid. Yet, for short periods, such as for an hour or so after a high calcium meal, calcitonin does seem to play a significant role in decreasing the rise in the calcium ion concentration that otherwise would occur.

When the thyroid gland is removed and calcitonin is no longer secreted, the long-term blood calcium ion concentration is not measurably altered, but the amount of bone salts may become slightly decreased—again demonstrating the overriding effect of the parathyroid hormonal system, but possibly at the expense of increased bone absorption.

OVERALL CONTROL OF CALCIUM ION CONCENTRATION

At times the amount of calcium absorbed into or lost from the body fluids is as much as 0.3 gram in an hour. For instance, in cases of diarrhea, several grams of calcium can be secreted in the intestinal juices, passed into the intestinal tract, and lost into the feces each day. Conversely, after ingestion of large quantities of calcium, particularly when there is also an excess of vitamin D activity, a person may absorb as much as 0.3 gram in an hour. This figure compares with a *total quantity of calcium in all the extracellular fluid of about one gram.* The addition or subtraction of 0.3 gram from such a small amount of calcium in the extracellular fluid would obviously cause serious hyper- or hypocalcemia. However, there is a first line of defense to prevent this from occurring even before the parathyroid and calcitonin hormone feedback systems have a chance to act. This is the following mechanism:

Buffer Function of the Exchangeable Calcium in the Bones. The exchangeable calcium salts in the bones, which were discussed earlier in this chapter, are amorphous calcium phosphate compounds, probably mainly $CaHPO_4$ or some similar compound loosely bound in the bone and in reversible equilibrium with the calcium and phosphate ions in the extracellular fluid. The quantity of these salts that is immediately available for exchange is about 0.5 to 1 per cent of the total calcium salts of the bone, a total of 5 to 10 grams of calcium. Because of the ease of deposition of these exchangeable salts and their ease of resolubility, an increase in the concentrations of extracellular fluid calcium and phosphate ions above normal causes immediate deposition of exchangeable salt. Conversely, a decrease in these concentrations causes immediate absorption of exchangeable salt. This reaction is so rapid that a single passage through a bone of blood containing a high concentration of calcium will remove almost all the excess calcium. This rapid effect results from the fact that the amorphous bone crystals are extremely small, and their total surface area exposed to the fluids of the bone is perhaps an acre or more. Also, about 5 per cent of all the blood flows through the bones each minute — that is, about 1 per cent of all the extracellular fluid each minute. Therefore, about half of any excess calcium that appears in the extracellular fluid is removed by this buffer function of the bones in approximately 70 minutes.

In addition to the buffer function of the bones, the mitochondria of many of the tissues of the body, especially of the liver and intestine, also contain a reasonable amount of exchangeable calcium that provides an additional buffer system for maintaining constancy of the extracellular fluid calcium ion concentration.

Hormonal Control of Calcium Ion Exchange with Bone, the Second Line of Defense. Within 3 to 5 minutes after the calcium ion concentration is increased, both the calcitonin and parathyroid mechanisms for reducing calcium concentration begin to function, the calcitonin mechanism *increasing* the secretion of calcitonin and the parathyroid mechanism *decreasing* parathyroid hormone. The calcitonin acts within the first few minutes to reduce calcium absorption from the bones, which is probably an important factor in preventing excessive rise in the calcium ion concentration following a high calcium meal, as just noted. Then, within another few minutes the much more powerful parathyroid mechanism begins also to help return the calcium ion concentration to its normal level. Because of the combined effects of both these mechanisms, one often can hardly measure a change in the plasma calcium ion concentration after ingesting excessive amounts of calcium.

Conversely, when large quantities of calcium ions are lost from the body, parathyroid hormone begins to be secreted in great quantities within minutes, and this increased secretion not only continues as long as calcium continues to be lost, but it actually accelerates with time because of progressive hypertrophy of the parathyroid glands themselves. In fact, as was illustrated in Figure 79–10, the rate of parathyroid hormone secretion increases another 3- to 10-fold over a period of several weeks to several months. If the body's loss of calcium continues at the high rate for many months or years, the bones will literally become filled with osteoclasts stimulated to extreme activity by the excess parathyroid hormone. During all this time the blood calcium ion concentration will be maintained at almost exactly the normal level, until finally the bone is almost entirely depleted of calcium salts.

Intestinal and Renal Control of Plasma Calcium Concentration — Role of Parathyroid Hormone. Though it is frequently stated that the absorption and deposition of calcium in bone is *the* long-term controller of blood calcium ion concentration, this is true only as long as the bone does not become saturated with calcium or totally depleted. However, since the bone does have these limits, it is actually a large reservoir for long-term *buffering* of calcium ion concentration over a period of months or years. It is not, however, the eventual long-term controller of plasma calcium concentration. Instead, this is achieved by the control of absorption and excretion by the intestines and kidneys.

It has already been pointed out that an increase in parathyroid hormone causes an increase in net absorption of calcium from the intestines and also causes increased reabsorption of calcium from the renal tubules. When the bone has become saturated with calcium salts and can no longer function as a depository of additional calcium ions, the slight excess of extracellular calcium ions reduces parathyroid secretion, which then decreases calcium absorption in both the intestines and kidney tubules. Conversely, when the bone has even a slight deficit of calcium salts, parathyroid secretion increases; this increase can allow for maintenance of almost normal plasma calcium concentration by increasing calcium absorption from both the intestines and kidney tubules.

PHYSIOLOGY OF PARATHYROID AND BONE DISEASES

HYPOPARATHYROIDISM

When the parathyroid glands do not secrete sufficient parathyroid hormone, the osteoclasts of the bone be-come almost totally inactive. As a result, bone reabsorption is so depressed that the level of calcium in the body fluids decreases. Because calcium and phosphates are not being absorbed from the bone, the bone usually remains strong, and osteoblastic activity is concomitantly decreased.

When the parathyroid glands are suddenly removed, the calcium level in the blood falls from the normal of 9.4 mg. per cent to 6 to 7 mg. per cent within two to three days. When this level is reached, the usual signs of tetany develop. Among the muscles of the body especially sensitive to tetanic spasm are the laryngeal muscles. Spasm of these obstructs respiration, which is the usual cause of death in tetany unless appropriate treatment is applied.

If all four parathyroid glands are removed from an animal and the animal is prevented from dying of respiratory spasm by appropriate supportive measures, the total lack of parathyroid hormone secretion normally causes the blood level of calcium to fall to as low as approximately 4 to 5 mg. per cent while at the same time the level of phosphates increases from the normal of about 4 mg. per cent to approximately 12 mg. per cent because of decreased renal phosphate excretion.

Treatment of Hypoparathyroidism. *Parathyroid Hormone (Parathormone).* Parathyroid hormone is occasionally used for treating hypoparathyroidism. However, because of the expense of this hormone, because its effect lasts only a few hours, and because the tendency of the body to develop immune bodies against it makes it progressively less and less active in the body, treatment of hypoparathyroidism with parathyroid hormone is rare in present-day therapy.

Dihydrotachysterol and Vitamin D. In addition to its ability to cause increased absorption of calcium from the gastrointestinal tract, vitamin D also causes a moderate effect similar to that of parathyroid hormone in promoting calcium and phosphate absorption from bones. Therefore, a person with hypoparathyroidism can be treated satisfactorily by administration of *large quantities* of vitamin D. One of the vitamin D compounds, dihydrotachysterol (A.T. 10), has a more marked ability to cause bone absorption than do most of the other vitamin D compounds because it can be converted directly to 1,25-dihydroxycholecalciferol by the kidneys and is not limited by the normal liver feedback mechanism that controls the conversion of vitamin D_3 to the active form. Administration of calcium plus dihydrotachysterol three or more times a week can almost completely control the calcium level in the extracellular fluid of a hypoparathyroid person.

HYPERPARATHYROIDISM

The cause of hyperparathyroidism ordinarily is a tumor of one of the parathyroid glands; such tumors occur much more frequently in women than in men or children, probably because pregnancy, lactation, and perhaps other causes of prolonged low calcium levels, all of which stimulate the parathyroid gland, may predispose to the development of such a tumor.

In hyperparathyroidism extreme osteoclastic

activity occurs in the bones, and this elevates the calcium ion concentration in the extracellular fluid while usually (but not always) depressing slightly the concentration of phosphate ions because of increased renal excretion of phosphate.

Bone Disease in Hyperparathyroidism. Though in mild hyperparathyroidism new bone may be deposited rapidly enough to compensate for the increased osteoclastic reabsorption of bone, in severe hyperparathyroidism, the osteoclastic absorption soon far outstrips osteoblastic deposition, and the bone may be eaten away almost entirely. Indeed, the reason a hyperparathyroid person comes to the doctor is often a broken bone. X-ray film of the bone shows extensive decalcification and occasionally large punched-out cystic areas of the bone that are filled with osteoclasts in the form of so-called giant-cell "tumors." Obviously, multiple fractures of the weakened bones can result from only slight trauma, especially where cysts develop. The cystic bone disease of hyperparathyroidism is called *osteitis fibrosa cystica*.

As a result of increased osteoblastic activity that attempts to form new bone as rapidly as it is absorbed, the level of alkaline phosphatase in the body fluids rises markedly.

Effects of Hypercalcemia in Hyperparathyroidism. Hyperparathyroidism can at times cause the plasma calcium level to rise to as high as 12 to 15 mg. per cent and rarely to 15 to 20 mg. per cent. The effects of such elevated calcium levels, as detailed earlier in the chapter, are depression of the central and peripheral nervous systems, muscular weakness, constipation, abdominal pain, peptic ulcer, lack of appetite, and depressed relaxation of the heart during diastole.

Parathyroid Poisoning and Metastatic Calcification. When, on rare occasions, extreme quantities of parathyroid hormones are secreted, the level of calcium in the body fluids rises rapidly to very high values. Even the extracellular fluid phosphate concentration also often rises markedly instead of falling as is usually the case, probably because the kidneys cannot excrete rapidly enough all the phosphate being absorbed from the bone. Therefore, the calcium and phosphate in the body fluids become greatly supersaturated even for the deposition of calcium phosphate ($CaHPO_4$) crystals. Therefore, these crystals begin to deposit in the alveoli of the lungs, in the tubules of the kidneys, in the thyroid gland, in the acid-producing area of the stomach mucosa, and in the walls of the arteries throughout the body. This extensive *metastatic* deposition of calcium phosphate can develop within a few days.

Ordinarily, the level of calcium in the blood must rise above 17 mg. per cent before there is danger of parathyroid poisoning, but once such elevation develops along with some concurrent elevation of phosphate, death can occur in only a few days.

Formation of Kidney Stones in Hyperparathyroidism. Most patients with mild hyperparathyroidism show few signs of bone disease and few general abnormalities as a result of elevated calcium, but nevertheless do have an extreme tendency to form kidney stones. The reason for this is that all the excess calcium and phosphate absorbed from the intestines or mobilized from the bones in hyperparathyroidism is excreted by the kidneys, causing proportionate increase in the concentrations of these substances in the urine. As a result, crystals of calcium phosphate tend to precipitate in the kidney, forming calcium phosphate stones. Also, calcium oxalate stones develop as a result of the high level of calcium in the urine in association with normal levels of oxalate. Because the solubility of most renal stones is slight in alkaline media, the tendency for formation of renal calculi is considerably greater in alkaline urine than in acid urine. For this reason, acidotic diets and acidic drugs are frequently used for treating renal calculi.

Secondary Hyperparathyroidism. Because a low level of calcium ions in the body fluids directly increases the secretion of parathyroid hormone, any factor that causes a low level of calcium initiates the condition known as secondary hyperparathyroidism. This may result from low calcium diet, pregnancy, lactation, rickets, or osteomalacia. The hyperplasia of the parathyroid glands is a corrective measure for maintaining the level of calcium in the extracellular fluids at a nearly normal value.

RICKETS

Rickets occurs mainly in children as a result of calcium or phosphate deficiency in the extracellular fluid. Ordinarily, rickets is due to lack of vitamin D rather than to lack of calcium or phosphate in the diet. If the child is properly exposed to sunlight, the 7-dehydrocholesterol in the skin becomes activated by the ultraviolet rays and forms vitamin D_3, which prevents rickets by promoting calcium and phosphate absorption from the intestines, as discussed earlier in the chapter.

Children who remain indoors through the winter in general do not receive adequate quantities of vitamin D without some supplementary therapy in the diet. Rickets tends to occur especially in the spring months because vitamin D formed during the preceding summer is stored in the liver and is still available for use during the early winter months. Also, calcium and phosphate absorption from the bones must take place for several months before clinical signs of rickets become apparent.

Calcium and Phosphate in the Blood of Patients with Rickets. Ordinarily, the level of calcium in the blood in rickets is only slightly depressed, but the level of phosphate is greatly depressed. This is because the parathyroid glands prevent the calcium level from falling by promoting bone absorption every time the calcium level begins to fall. On the other hand, there is no good regulatory system for controlling a falling level of phosphate, and the increased parathyroid activity actually increases the excretion of phosphates in the urine.

Effect of Rickets on the Bone. During prolonged deficiency of calcium and phosphate in the body fluids, the resulting increase in parathyroid hormone secretion protects the body against hypocalcemia by causing osteoclastic absorption of the bone; this in turn causes the bone to become progressively weaker and imposes marked physical stress on the bone, resulting in rapid osteoblastic activity. The osteoblasts lay down large quantities of osteoid which does not become calcified because of insufficient calcium and phosphate ions. Consequently, the newly formed, uncalcified, and very weak osteoid gradually takes the place of other bone that is being reabsorbed.

Obviously, hyperplasia of the parathyroid glands is marked in rickets because of the decreased blood calcium level, and the alkaline phosphatase level in the blood is markedly increased as a result of the rapid osteoblastic activity.

Tetany in Rickets.   In the early stages of rickets, tetany almost never occurs because the parathyroid glands continually stimulate osteoclastic absorption of bone and therefore maintain an almost normal level of calcium in the body fluids. However, when the bones become exhausted of calcium, the level of calcium may fall rapidly. As the blood level of calcium falls below 7 mg. per cent, the usual signs of tetany develop, and the child may die of tetanic respiratory spasm unless intravenous calcium is administered, which relieves the tetany immediately.

Treatment of Rickets.   The treatment of rickets, obviously, depends on supplying adequate calcium and phosphate in the diet and also on administering adequate amounts of vitamin D. If vitamin D is not administered along with the calcium, little calcium is absorbed from the gut, and this calcium in turn carries large quantities of phosphate with it into the feces.

Tetany Resulting from Treatment.   Occasionally, when a child is treated for rickets, tetany occurs for the following reason: Administration of vitamin D without sufficient calcium often enhances the deposition of calcium in the newly formed osteoid and further depresses the blood calcium level. This results from the direct effect of vitamin D on bone calcification, an effect that has already been discussed.

Another treatment that can frequently lead to tetany is administration of calcium and phosphate in the diet without sufficient vitamin D. In this case the phosphate is often absorbed without absorption of calcium. The phosphate in turn causes deposition of the plasma calcium in the bone, which reduces the plasma calcium concentration.

These difficulties can be prevented by administering large quantities of calcium simultaneously with large quantities of vitamin D.

Osteomalacia.   Osteomalacia is rickets in adults and is frequently called "adult rickets."

Normal adults rarely have dietary lack of vitamin D or calcium because large quantities of calcium are not needed for bone growth as in children. However, lack of vitamin D and calcium occasionally occurs as a result of steatorrhea (failure to absorb fat), for vitamin D is fat soluble and calcium tends to form insoluble soaps with fat; consequently, in steatorrhea vitamin D and calcium tend to pass into the feces. Under these conditions an adult occasionally has such poor calcium and phosphate absorption that adult rickets can occur, though this almost never proceeds to the stage of tetany — but very often is a cause of severe bone disability.

Osteomalacia and Rickets Caused by Renal Disease.   "Renal rickets" is a type of osteomalacia resulting from prolonged kidney damage. The cause of this condition is mainly failure of the damaged kidneys to form 1,25-dihydroxycholecalciferol, the active form of vitamin D. In patients whose kidneys have been completely removed or destroyed and who are being treated by hemodialysis, the problem of renal rickets is often a very severe one.

Another type of renal disease that leads to rickets and osteomalacia is *congenital hypophosphatemia* resulting from congenitally reduced reabsorption of phosphates by the renal tubules. This type of rickets must be treated with phosphate compounds instead of calcium and vitamin D, and it is called *vitamin D–resistant* rickets.

OSTEOPOROSIS

Osteoporosis, the most common of all bone diseases in adults and especially in old age, is a different disease from osteomalacia and rickets, for it results from diminished organic matrix rather than abnormal bone calcification. Usually, in osteoporosis the osteoblastic activity in the bone is less than normal, and consequently the rate of bone deposition is depressed. But occasionally, as in hyperparathyroidism, the cause of the diminished bone is excess osteoclastic activity.

The many common causes of osteoporosis are (1) lack of use of the bones; (2) malnutrition to the extent that sufficient protein matrix cannot be formed; (3) lack of vitamin C, which is necessary for the secretion of intercellular substances by all cells, including the osteoblasts; (4) postmenopausal lack of estrogen secretion, for estrogens have an osteoblast-stimulating activity; (5) old age, in which many of the protein anabolic functions are poor anyway so that bone matrix cannot be deposited satisfactorily; (6) Cushing's disease, because massive quantities of glucocorticoids cause decreased deposition of protein throughout the body, cause increased catabolism of protein, and also have the specific effect of depressing osteoblastic activity; and (7) acromegaly, possibly because of lack of sex hormones, excess of adrenocortical hormones, and often lack of insulin because of the diabetogenic effect of growth hormone. Obviously, many diseases of protein metabolism can cause osteoporosis.

PHYSIOLOGY OF THE TEETH

The teeth cut, grind, and mix the food eaten. To perform these functions the jaws have powerful muscles capable of providing an occlusive force between the front teeth of as much as 50 to 100 pounds and as much as 150 to 200 pounds for the jaw teeth. Also, the upper and lower teeth are provided with projections and facets which interdigitate so that each set of teeth fits with the other. This fitting is called *occlusion*, and it allows even small particles of food to be caught and ground between the tooth surfaces.

FUNCTION OF THE DIFFERENT PARTS OF THE TEETH

Figure 79–11 illustrates a sagittal section of a tooth, showing its major functional parts: the *enamel, dentine,*

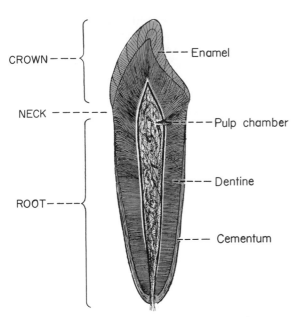

Figure 79–11. Functional parts of a tooth.

cementum, and pulp. The tooth can also be divided into the crown, which is the portion that protrudes out of the gum into the mouth, and the root, which is the portion that protrudes into the bony socket of the jaw. The collar between the crown and the root where the tooth is surrounded by the gum is called the neck.

Dentine. The main body of the tooth is composed of dentine, which has a strong, bony structure. Dentine is made up principally of hydroxyapatite crystals similar to those in the bone, but much more dense. These are embedded in a strong meshwork of collagen fibers. In other words, the principal constituents of dentine are very much the same as those of bone. The major difference is its histological organization, for dentine does not contain any osteoblasts, osteoclasts, or spaces for blood vessels or nerves. Instead, it is deposited and nourished by a layer of cells called odontoblasts, which line its inner surface along the wall of the pulp cavity.

The calcium salts in dentine make it extremely resistant to compressional forces, while the collagen fibers make it tough and resistant to tensional forces that might result when the teeth are struck by solid objects.

Enamel. The outer surface of the tooth is covered by a layer of enamel that is formed prior to eruption of the tooth by special epithelial cells called ameloblasts. Once the tooth has erupted, no more enamel is formed. Enamel is composed of large and very dense crystals of hydroxyapatite with adsorbed carbonate, magnesium, sodium, potassium, and other ions embedded in a fine meshwork of very strong and almost completely insoluble protein fibers that are similar in physical characteristics (but not chemically identical) to the keratin of hair. The crystalline structure of the salts makes the enamel extremely hard, much harder than the dentine. Also, the special protein fiber meshwork, though comprising only about one per cent of the enamel mass, nevertheless makes enamel very resistant to acids, enzymes, and other corrosive agents because this pro-

tein is one of the most insoluble and resistant proteins known.

Cementum. Cementum is a bony substance secreted by cells of the periodontal membrane, which lines the tooth socket. Many collagen fibers pass directly from the bone of the jaw, through the periodontal membrane, and then into the cementum. These collagen fibers and the cementum hold the tooth in place. When the teeth are exposed to excessive strain, the layer of cementum becomes thicker and stronger. Also, it increases in thickness and strength with age, causing the teeth to become progressively more firmly seated in the jaws as one reaches adulthood and older.

Pulp. The inside of each tooth is filled with pulp, which in turn is composed of connective tissue with an abundant supply of nerves, blood vessels, and lymphatics. The cells lining the surface of the pulp cavity are the odontoblasts, which, during the formative years of the tooth, lay down the dentine but at the same time encroach more and more on the pulp cavity, making it smaller. In later life the dentine stops growing and the pulp cavity remains essentially constant in size. However, the odontoblasts are still viable and send projections into small dentinal tubules that penetrate all the way through the dentine; these are of importance for providing nutrition.

DENTITION

Each human being and most other mammals develop two sets of teeth during a lifetime. The first teeth are called the deciduous teeth, or milk teeth, and they number 20 in the human being. These erupt between the seventh month and second year of life, and they last until the sixth to the thirteenth year. After each deciduous tooth is lost, a permanent tooth replaces it, and an additional 8 to 12 molars appear posteriorly in the jaw, making the total number of permanent teeth 28 to 32, depending on whether the four wisdom teeth finally appear, which does not occur in everyone.

Formation of the Teeth. Figure 79–12 illustrates the formation and eruption of teeth. Figure 79–12A shows invagination of the oral epithelium into the dental lamina; this is followed by the development of a tooth-producing organ. The epithelial cells above form ameloblasts, which form the enamel on the outside of the tooth. The epithelial cells below invaginate upward to

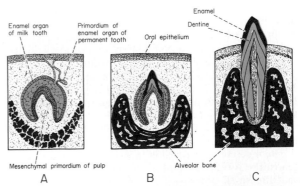

Figure 79–12. (A) Primordial tooth organ. (B) The developing tooth. (C) The erupting tooth.

form a pulp cavity and also to form the odontoblasts that secrete dentine. Thus, enamel is formed on the outside of the tooth, and dentine is formed on the inside, giving rise to an early tooth as illustrated in Figure 79–12B.

Eruption of Teeth. During early childhood, the teeth begin to protrude upward from the jaw bone through the oral epithelium into the mouth. The cause of "eruption" is unknown, though several theories have been offered in an attempt to explain this phenomenon. The most likely theory is that growth of the tooth root as well of the bone underneath the tooth progressively shoves the tooth forward.

Development of the Permanent Teeth. During embryonic life, a tooth-forming organ also develops in the dental lamina for each permanent tooth that will be needed after the deciduous teeth are gone. These tooth-producing organs slowly form the permanent teeth throughout the first 6 to 20 years of life. When each permanent tooth becomes fully formed, it, like the deciduous tooth, pushes upward through the bone of the jaw. In so doing it erodes the root of the deciduous tooth and eventually causes it to loosen and fall out. Soon thereafter, the permanent tooth erupts to take the place of the original one.

Metabolic Factors in Development of the Teeth. The rate of development and the speed of eruption of teeth can be accelerated by both thyroid and growth hormones. Also, the deposition of salts in the early forming teeth is affected considerably by various factors of metabolism, such as the availability of calcium and phosphate in the diet, the amount of vitamin D present, and the rate of parathyroid hormone secretion. When all these factors are normal, the dentine and enamel will be correspondingly healthy, but, when they are deficient, the calcification of the teeth also may be defective so that the teeth will be abnormal throughout life.

MINERAL EXCHANGE IN TEETH

The salts of teeth, like those of bone, are composed basically of hydroxyapatite with adsorbed carbonates and various cations bound together in a hard crystalline substance. Also, new salts are constantly being deposited while old salts are being reabsorbed from the teeth, as also occurs in bone. However, experiments indicate that deposition and reabsorption occur mainly in the dentine and cementum, while very little occurs in the enamel. And most of that which does occur in the enamel occurs by exchange of minerals with the saliva instead of with the fluids of the pulp cavity. The rate of absorption and deposition of minerals in the cementum is approximately equal to that in the surrounding bone of the jaw, while the rate of deposition and absorption of minerals in the dentine is only one-third that of bone. The cementum has characteristics almost identical with those of usual bone, including the presence of osteoblasts and osteoclasts, while dentine does not have these characteristics, as was explained above; this difference undoubtedly explains the different rates of mineral exchange.

The mechanism by which minerals are deposited and reabsorbed from the dentine is not clear. It is probable that the small processes of the odontoblasts that protrude into the tubules of the dentine are capable of absorbing salts and then of providing new salts to take the place of the old.

In summary, rapid mineral exchange occurs in the dentine and cementum of teeth, though the mechanism of this exchange in dentine is unclear. On the other hand, enamel exhibits extremely slow mineral exchange so that it maintains most of its original mineral complement throughout life.

DENTAL ABNORMALITIES

The two most common dental abnormalities are *caries* and *malocclusion*. Caries means erosions of the teeth, while malocclusion means failure of the projections of the upper and lower teeth to interdigitate properly.

Caries, and the Role of Fluorine. It is generally agreed by all research investigators of dental caries that caries results from the action of bacteria on the teeth, the most common of which is *Streptococcus mutans*. The first event in the development of caries is the deposit of *plaque*, a film of precipitated products of saliva and food, on the teeth. Large numbers of bacteria inhabit this plaque and are readily available to cause caries. However, these bacteria depend to a great extent on carbohydrates for their food. When carbohydrates are available, their metabolic systems are strongly activated and they also multiply. In addition, they form acids, particularly lactic acid, and proteolytic enzymes. The acids are the major culprit in the causation of caries, because the calcium salts of teeth are slowly dissolved in a highly acid medium. And once the salts have become absorbed, the remaining organic matrix is rapidly digested by the proteolytic enzymes.

Enamel is far more resistant to demineralization by acids than is dentine, primarily because the crystals of enamel are very dense and also are about 200 times as large as the dentine crystals. Therefore, the enamel of the tooth is the primary barrier to the development of caries. Once the carious process has penetrated through the enamel to the dentine, it then proceeds many times as rapidly because of the high degree of solubility of the dentine salts.

Because of the dependence of the caries bacteria on carbohydrates, it has frequently been taught that eating a diet high in carbohydrate content will lead to excessive development of caries. However, it is not the quantity of carbohydrate ingested but instead the frequency with which it is eaten that is important. If eaten in many small parcels throughout the day, such as in the form of candy, the bacteria are supplied with their preferential metabolic substrate for many hours of the day, and the development of caries is extreme. If eaten in large amounts only at mealtimes, the extensiveness of the caries is greatly reduced.

Some teeth are more resistant to caries than others. Studies show that teeth formed in children who drink water containing small amounts of fluorine develop enamel that is more resistant to caries than the enamel in children who drink water not containing fluorine. Fluorine does not make the enamel harder than usual, but instead it displaces hydroxyl ions in the hydroxyapatite crystals which, in turn, makes the enamel several times less soluble. It is also believed that the fluorine might be toxic to some of the bacteria as well. Regardless of the precise means by which fluorine protects the teeth, it is known that small amounts of fluorine deposited in enamel make teeth about 3 times as resistant to caries as are teeth without fluorine.

Malocclusion. Malocclusion is usually caused by a hereditary abnormality that causes the teeth of one jaw to grow to an abnormal position. In malocclusion, the teeth cannot perform their normal grinding or cutting action adequately. Occasionally malocclusion also results in abnormal displacement of the lower jaw in relation to the upper jaw, causing such undesirable effects as pain in the mandibular joint or deterioration of the teeth.

The orthodontist can often correct malocclusion by applying prolonged gentle pressure against the teeth with appropriate braces. The gentle pressure causes absorption of alveolar jaw bone on the compressed side of the tooth and deposition of new bone on the tensional side of the tooth. In this way the tooth gradually moves to a new position as directed by the applied pressure.

REFERENCES

Aegerter, E. E., and Kirkpatrick, J. A., Jr.: Orthopedic Diseases. Philadelphia, W. B. Saunders Co., 1975.

Avioli, L. V., and Raisz, L. G.: Bone metabolism and disease. *In* Bondy, P. K., and Rosenberg, L. E. (eds.): Metabolic Control and Disease, 8th Ed. Philadelphia, W. B. Saunders Co., 1980, p. 1709.

Barzel, U. S. (ed.): Osteoporosis II. New York, Grune & Stratton, 1979.

Bollet, A. J., and Eliel, L. P.: Mechanisms of bone and joint disease. *In* Frohlich, E. D. (ed.): Pathophysiology, 2nd Ed. Philadelphia, J. B. Lippincott Co., 1976, p. 425.

Borle, A. B.: Calcium and phosphate metabolism. *Annu. Rev. Physiol.*, 36:361, 1974.

Brighton, C. T., *et al.* (eds.): Electrical Properties of Bone and Cartilage: Experimental Effects and Clinical Applications. New York, Grune & Stratton, 1979.

Bringhurst, F. R., and Potts, J. T., Jr.: Calcium and phosphate distribution, turnover, and metabolic actions. *In* DeGroot, L. J., *et al.* (eds.): Endocrinology. Vol. 2. New York, Grune & Stratton, 1979, p. 551.

Burny, F., *et al.* (eds.): Electric Stimulation of Bone Growth and Repair. New York, Springer-Verlag, 1978.

Castleman, B., and Roth, S. I.: Tumors of the Parathyroid Glands. Washington, D.C., Armed Forces Institute of Pathology, 1978.

Coburn, J. W., *et al.:* Intestinal absorption of calcium and the effect of renal insufficiency. *Kidney Int.*, 4:96, 1973.

Copp, D. H.: Comparative endocrinology of calcitonin. *In* Greep, R. O., and Astwood, G. D. (eds.): Handbook of Physiology. Sec. 7, Vol. 7. Baltimore, Williams & Wilkins, 1976, p. 431.

Copp, D. H.: Calcitonin: Comparative endocrinology. *In* DeGroot, L. J., *et al.* (eds.): Endocrinology. Vol. 2. New York, Grune & Stratton, 1979, p. 637.

Dacke, C. G.: Calcium Regulation in Sub-Mammalian Vertebrates. New York, Academic Press, 1979.

DeLuca, H. F.: The kidney as an endocrine organ involved in the function of vitamin D. *Am. J. Med.*, 58:39, 1975.

Dennis, V. W., *et al.:* Renal handling of phosphate and calcium. *Annu. Rev. Physiol.*, 41:257, 1979.

Fink, G.: Feedback actions of target hormones on hypothalamus and pituitary, with special reference to gonadal steroids. *Annu. Rev. Physiol.*, 41:571, 1979.

Fraser, D. R.: Regulation of the metabolism of vitamin D. *Physiol. Rev.*, 60:551, 1980.

Glimcher, M. J.: Composition, structure, and organization of bone and other mineralized tissues and the mechanism of calcification. *In* Greep, R. O., and Astwood, G. D. (eds.): Handbook of Physiology. Sec. 7, Vol. 7. Baltimore, Williams & Wilkins, 1976, p. 25.

Goldberg, M., *et al.:* Renal handling of calcium and phosphate. *Int. Rev. Physiol.*, 11:211, 1976.

Gordan, G. S.: Drug treatment of the osteoporoses. *Annu. Rev. Pharmacol. Toxicol.*, 18:253, 1978.

Gray, T. K., *et al.:* Parathyroid hormone, thyrocalcitonin, and the control of mineral metabolism. *In* MTP International Review of Science: Physiology. Vol. 5. Baltimore, University Park Press, 1974, p. 239.

Habener, J. F., and Potts, J. T., Jr.: Diagnosis and differential diagnosis of hyperparathyroidism. *In* DeGroot, L. J., *et al.* (eds.): Endocrinology. Vol. 2. New York, Grune & Stratton, 1979, p. 703.

Habener, J. F., and Potts, J. T., Jr.: Chemistry, biosynthesis, secretion, and metabolism of parathyroid hormone. *In* Greep, R. O., and Astwood, G. D. (eds.): Handbook of Physiology. Sec. 7, Vol. 7. Baltimore, Williams & Wilkins, 1976, p. 313.

Ham, A. W.: Histophysiology of Cartilage, Bone, and Joints. Philadelphia, J. B. Lippincott Co., 1979.

Jaros, G. C., *et al.:* Model of short-term regulation of calcium ion concentration. Simulation, 32:193, 1979.

Johansen, E., *et al.* (eds.): Continuing Evaluation of the Use of Fluorides. Boulder, Colo., Westview Press, 1979.

Keutmann, H. T.: Chemistry of parathyroid hormone. *In* DeGroot, L. J., *et al.* (eds.): Endocrinology. Vol. 2. New York, Grune & Stratton, 1979, p. 593.

Lawson, D. E. M.: Vitamin D. New York, Academic Press, 1978.

Lechene, C. P., and Warner, R. R.: Ultramicroanalysis: X-ray spectrometry by electron probe excitation. *Annu. Rev. Biophys. Bioeng.*, 6:57, 1977.

MacIntyre, I.: Human Calcitonin and Paget's Disease. Berne, H. Huber, 1977.

Massry, S. G., and Fleisch, H. (eds.): Renal Handling of Phosphate. New York, Plenum Press, 1979.

Massry, S. G., *et al.* (eds.): Homeostasis of Phosphate and Other Minerals. New York, Plenum Press, 1978.

Mayer, G. P.: Parathyroid hormone secretion. *In* DeGroot, L. J., *et al.* (eds.): Endocrinology. Vol. 2. New York, Grune & Stratton, 1979, p. 607.

Munson, P. L.: Physiology and pharmacology of thyrocalcitonin. *In* Greep, R. O., and Astwood, G. D. (eds.): Handbook of Physiology. Sec. 7, Vol. 7. Baltimore, Williams & Wilkins, 1976, p. 443.

Myers, H. M.: Fluorides and Dental Fluorosis. New York, S. Karger, 1978.

Nellans, H. N., and Kimberg, D. V.: Intestinal calcium transport: Absorption, secretion, and vitamin D. *In* Crane, R. K. (ed.): International Review of Physiology: Gastrointestinal Physiology III. Vol. 19. Baltimore, University Park Press, 1979, p. 227.

Newman, H. N.: Dental Plaque; The Ecology of the Flora on Human Teeth. Springfield, Ill., Charles C Thomas, 1980.

Norman, A. W.: 1,25-Dihydroxyvitamin D_3: A kidney-produced steroid hormone essential to calcium homeostasis. *Am. J. Med.*, 57:21, 1974.

Norman, A. W.: Vitamin D: The Calcium Homeostatic Steroid Hormone. New York, Academic Press, 1979.

Parfitt, A. M.: Investigation of disorders of the parathyroid glands. *Clin. Endocrinol. Metab.*, 3:451, 1974.

Parsons, J. A.: Physiology of parathyroid hormone. *In* DeGroot, L. J., *et al.* (eds.): Endocrinology. Vol. 2. New York, Grune & Stratton, 1979, p. 621.

Phang, J. M., and Weiss, I. W.: Maintenance of calcium homeostasis in human beings. *In* Greep, R. O., and Astwood, G. D. (eds.): Handbook of Physiology. Sec. 7, Vol. 7. Baltimore, Williams & Wilkins, 1976, p. 157.

Prien, E. L., Jr., *et al.:* Secondary hyperparathyroidism. *In* Greep, R. O., and Astwood, G. D. (eds.): Handbook of Physiology. Sec. 7, Vol. 7. Baltimore, Williams & Wilkins, 1976, p. 383.

Raisz, L. G.: Mechanisms of bone resorption. *In* Greep, R. O., and Astwood, G. D. (eds.): Handbook of Physiology. Sec. 7, Vol. 7. Baltimore, Williams & Wilkins, 1976, p. 117.

Raisz, L. G. et al.: Hormonal regulation of mineral metabolism. *Int. Rev. Physiol.*, 16:199, 1977.

Rasmussen, H.: The Physiological and Cellular Basis of Metabolic Bone Disease. Baltimore, Williams & Wilkins, 1974.

Rasmussen, H., *et al.:* Hormonal control of skeletal and mineral homeostasis. *Am. J. Med.*, 56:751, 1974.

Schroeder, H. E., and Listgartern, M. A.: Fine Structure of the Developing Epithelial Attachment of Human Teeth. New York, S. Karger, 1977.

Seltzer, S., and Bender, I. B.: The Dental Pulp, 2nd Ed. Philadelphia, J. B. Lippincott Co., 1975.

Simmons, D. J., and Kunin, A. S. (eds.): Skeletal Research. New York, Academic Press, 1979.

Smith, R.: Biochemical Disorders of the Skeleton. Boston, Buttersworths, 1979.

Tada, M., *et al.:* Molecular mechanism of active calcium transport by sarcoplasmic reticulum. *Physiol. Rev.*, 58:1, 1978.

Talmage, R. V., and Cooper, C. W.: Physiology and mode of action of calcitonin. *In* DeGroot, L. J., *et al.* (eds.): Endocrinology. Vol. 2. New York, Grune & Stratton, 1979, p. 647.

Talmage, R. V., and Meyer, R. A., Jr.: Physiological role of parathyroid hormone. *In* Greep, R. O., and Astwood, G. D. (eds.): Handbook of Physiology. Sec. 7, Vol. 7. Baltimore, Williams & Wilkins, 1976, p. 343.

van Zwieten, P. A., and Schönbaum, E. (eds.): The Action of Drugs on Calcium Metabolism. New York, Fischer, 1978.

Wasserman, R. H., and Taylor, A. N.: Gastrointestinal absorption of calcium and phosphorus. *In* Greep, R. O., and Astwood, G. D. (eds.): Handbook of Physiology. Sec. 7, Vol. 7. Baltimore, Williams & Wilkins, 1976, p. 137.

Wheeler, R. C.: Dental Anatomy, Physiology and Occlusion, 5th Ed. Philadelphia, W. B. Saunders Co., 1974.

80

Reproductive and Hormonal Functions of the Male; and the Pineal Gland

The reproductive functions of the male can be divided into three major subdivisions: first, spermatogenesis, which means simply the formation of sperm; second, performance of the male sexual act; and third, regulation of male reproductive functions by the various hormones. Associated with these reproductive functions are the effects of the male sex hormones on the accessory sexual organs, on cellular metabolism, on growth, and on other functions of the body.

Physiologic Anatomy of the Male Sexual Organs. Figure 80–1 illustrates the various portions of the male reproductive system. Note that the testis is composed of a large number of coiled *seminiferous tubules* where the sperm are formed. (If all these tubules were placed end to end, they would be about 800 feet long.) The sperm then empty into the *epididymis,* another coiled tube approximately 20 feet long. The epididymis leads into the *vas deferens,* which enlarges into the *ampulla of the vas deferens* immediately before the vas enters the body of the *prostate gland.* A *seminal vesicle,* one located on each side of the prostate, empties into the prostatic end of the ampulla, and the contents from both the ampulla and the seminal vesicle pass into an *ejaculatory duct* leading through the body of the prostate gland to empty

into the *internal urethra. Prostate ducts* in turn empty into the ejaculatory duct. Finally, the *urethra* is the last connecting link from the testis to the exterior. The urethra is supplied with mucus derived from a large number of small *glands of Littré* located along its entire extent and also from large bilateral *bulbo-urethral glands* located near the origin of the urethra.

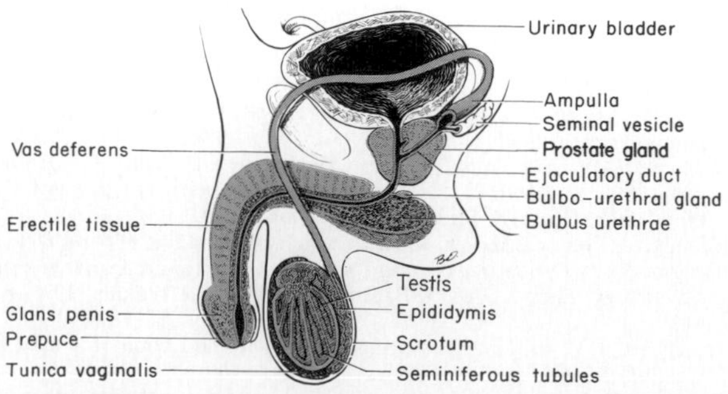

SPERMATOGENESIS

Spermatogenesis occurs in all the seminiferous tubules during active sexual life, beginning at an average age of 13 as the result of stimulation by adenohypophysial gonadotropic hormones and continuing throughout the remainder of life.

THE STEPS OF SPERMATOGENESIS

The seminiferous tubules, one of which is illustrated in Figure 80–2A, contain a large number of small to medium-sized germinal epithelial cells called *spermatogonia,* which are located in two to three layers along the outer border of the tubular epithelium. These continually proliferate to replenish themselves, and a portion of them differentiate through definite stages of development to form sperm, as shown in Figure 80–2B.

The first stage in spermatogenesis is growth of some spermatogonia to form considerably enlarged cells called *spermatocytes.* After the chromosomes of these cells reduplicate themselves to form chromatids (making a total of 92 chromatids in the cell), the spermatocyte then undergoes two rapid *meiotic divisions* without formation of any new chromatids. The net result of these two divisions is formation of four cells called spermatids, each containing only 23 unpaired chromosomes. The spermatids do not divide again but instead mature in an average of 74 days to become spermatozoa.

The net result of the meiotic divisions is that *only one chromosome* from each original pair of chromosomes is now in each spermatid. As noted in the following

Figure 80–1. The male reproductive system. (Modified from Bloom and Fawcett: Textbook of Histology, 10th Ed.)

Labels on figure: Urinary bladder, Ampulla, Seminal vesicle, Prostate gland, Ejaculatory duct, Bulbo-urethral gland, Bulbus urethrae, Testis, Epididymis, Scrotum, Seminiferous tubules, Vas deferens, Erectile tissue, Glans penis, Prepuce, Tunica vaginalis

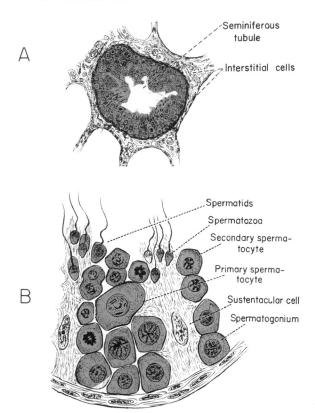

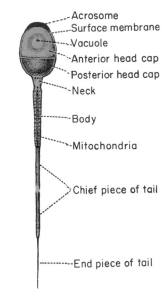

Figure 80–3. Structure of the human spermatozoon.

Figure 80–2. (A) Cross-section of a seminiferous tubule. (B) Spermatogenesis. (Modified from Arey: Developmental Anatomy, 7th Ed. Philadelphia, W. B. Saunders Company, 1974.)

chapter, a similar reduction in chromosomes occurs in the ovum during its maturation. Then, when the spermatozoon combines with the ovum during the fertilization process, the original complement of 46 chromosomes is again established.

The Sex Chromosomes. In each spermatogonium one of the 23 pairs of chromosomes carries the genetic information that determines the sex of the eventual offspring. This pair is composed of one "X" chromosome, which is called the *female chromosome*, and one "Y" chromosome, the *male chromosome*. During meiotic division the sex-determining chromosomes divide among the secondary spermatocytes so that half of the sperm become *male sperm* containing the Y chromosome and the other half *female sperm* containing the X chromosome. The sex of the offspring is determined by which of these two types of sperm fertilizes the ovum. This will be discussed further in Chapter 82.

Formation of Sperm. When the spermatids are first formed, they still have the usual characteristics of epithelioid cells, but soon each spermatid begins to elongate into a spermatozoon, illustrated in Figure 80–3, composed of a *head, neck, body,* and *tail.* To form the head, the nuclear material is condensed into a compact mass, and the cell membrane contracts around the nucleus. It is this nuclear material that fertilizes the ovum.

At the front of the sperm head is a small structure called the *acrosome,* which is formed from the Golgi apparatus and contains hyaluronidase and proteases that play important roles in the entry of the sperm into the ovum.

The *centrioles* are aggregated in the neck of the sperm and the *mitochondria* are arranged in a spiral in the body.

Extending beyond the body is a long tail, which is an outgrowth of one of the centrioles. This has almost the same structure as a cilium, which was described in detail in Chapter 2. The tail contains two paired microtubules down the center and nine double microtubules arranged around the border. It is covered by an extension of the cell membrane, and it contains large quantities of adenosine triphosphate (generated by the mitochondria in the body) which undoubtedly energize the movement of the tail. Upon release of sperm from the male genital tract into the female tract, the tail begins to wave back and forth and to move spirally near its top, providing snakelike propulsion that moves the sperm forward and rotating at a velocity of about 20 centimeters per hour.

Function of the Sertoli Cells. The Sertoli cells of the germinal epithelium, known also as the *sustentacular cells,* are illustrated in Figure 80–2B. These cells are large, extending from the base of the seminiferous epithelium all the way to the interior of the tubule. The spermatids attach themselves to the Sertoli cells, and a specific relationship exists between the spermatids and the Sertoli cells that causes the spermatids to change into spermatozoa. The Sertoli cells provide nutrient material, hormones, and possibly also enzymes that are necessary for causing appropriate changes in the spermatids. The Sertoli cells also remove the excess cytoplasm as the spermatids are converted to spermatozoa.

Maturation of Sperm in the Epididymis. Following formation in the seminiferous tubules, the sperm pass through the *vasa recta* into the *epididymis*. Sperm removed from the seminiferous tubules are completely nonmotile, and they cannot fertilize an ovum. However, after the sperm have been in the epididymis for some 18 hours to 10 days, they develop the capability of motility, even though some inhibitory factor still prevents motility until after ejaculation. The sperm also become capable of fertilizing the ovum, a process called *maturation*. It may not be any special function of the epididymis that changes the sperm from the nonmotile state into the motile and fertile state but, instead, simply an aging process. However, the epididymis secretes a copious quantity of fluid containing hormones, enzymes, and special nutrients that may be important or even essential for sperm maturation.

Storage of Sperm. A small quantity of sperm can be stored in the epididymis, but most sperm are stored in the vas deferens and ampulla of the vas deferens. They can remain stored, maintaining their fertility, in the genital ducts for several months, though it is doubtful that during normal sexual activity such prolonged storage ordinarily occurs. Indeed, with excessive sexual activity storage may be no longer than a few hours.

Physiology of the Mature Sperm. The usual motile and fertile sperm are capable of flagellated movement through the fluid media at a rate of approximately 1 to 4 mm. per minute. Furthermore, *normal* sperm tend to travel in a straight, rotating line rather than with a circuitous movement. The activity of sperm is greatly enhanced in neutral and slightly alkaline media as exists in the ejaculated semen, but it is greatly depressed in mildly acid media, and strong acid media can cause rapid death of sperm. The activity of sperm increases greatly with increasing temperature, but so does the rate of metabolism, causing the life of the sperm to be considerably shortened. Though sperm can live for many weeks in the genital ducts of the testes, the life of sperm in the female genital tract is only 1 to 3 days.

FUNCTION OF THE SEMINAL VESICLES

From early anatomical studies of the seminal vesicles it was erroneously believed that sperm were stored in these, whence came the name "seminal vesicles." However, these structures are only secretory glands instead of sperm storage areas.

The seminal vesicles are lined with a secretory epithelium that secretes a mucoid material containing an abundance of fructose and other nutrient substances, as well as large quantities of prostaglandins and fibrinogen. During the process of ejaculation each seminal vesicle empties its contents into the ejaculatory duct shortly after the vas deferens empties the sperm. This adds greatly to the bulk of the ejaculated semen, and the fructose and other substances in the seminal fluid are of considerable nutrient value for the ejaculated sperm until one of them fertilizes the ovum. The prostaglandins are believed to aid fertilization in two ways: (1) by reacting with the cervical mucus to make it more receptive to sperm, and (2) possibly causing reverse peristaltic contractions in the uterus and fallopian tubes to move the sperm toward the ovaries (a few sperm reach the upper end of the fallopian tubes within 5 minutes).

FUNCTION OF THE PROSTATE GLAND

The prostate gland secretes a thin, milky, alkaline fluid containing citric acid, calcium, acid phosphate, a clotting enzyme, and a profibrinolysin. During emission, the capsule of the prostate gland contracts simultaneously with the contractions of the vas deferens so that the thin, milky fluid of the prostate gland adds to the bulk of the semen. The alkaline characteristic of the prostatic fluid may be quite important for successful fertilization of the ovum, because the fluid of the vas deferens is relatively acidic owing to the presence of metabolic end-products of the sperm and, consequently, inhibits sperm fertility. Also, the vaginal secretions of the female are acidic (pH of 3.5 to 4.0). Sperm do not become optimally motile until the pH of the surrounding fluids rises to approximately 6 to 6.5. Consequently, it is probable that prostatic fluid neutralizes the acidity of these other fluids after ejaculation and greatly enhances the motility and fertility of the sperm.

SEMEN

Semen, which is ejaculated during the male sexual act, is composed of the fluids from the vas deferens, from the seminal vesicles, from the prostate gland, and from the mucous glands, especially the bulbo-urethral glands. The major bulk of the semen is seminal vesicle fluid (about 60 per cent), which is the last to be ejaculated and serves to wash the sperm out of the ejaculatory duct and urethra. The average pH of the combined semen is approximately 7.5, the alkaline prostatic fluid having neutralized the mild acidity of the other portions of the semen. The prostatic fluid gives the semen a milky appearance, while fluid from the seminal vesicles and from the mucous glands gives the semen a mucoid consistency. Indeed, the clotting enzyme of the prostatic fluid causes the fibrinogen of the seminal vesicle fluid to form a weak coagulum, which then dissolutes during the next 15 to 20 minutes because of lysis by fibrinolysin formed from the prostatic profibrinolysin. In the early minutes after ejaculation, the sperm remain relatively immobile, possibly because of the viscosity of the coagulum. However, after the coagulum dissolutes, the sperm simultaneously become highly motile.

Though sperm can live for many weeks in the male genital ducts, once they are ejaculated in the semen their maximal life span is only 24 to 72 hours at body temperature. At lowered temperatures, however, semen may be stored for several weeks; and when frozen at temperatures below $-100°$ C., sperm of some animals have been preserved for over a year.

Capacitation. In some lower animals, sperm are not capable of fertilizing the ovum immediately after being ejaculated but develop this capability during the next 4 to 6 hours. This phenomenon is called *capacitation*. It may be simply an aging process, or it may result from the action of components of the semen not derived from the vas deferens or of some of the female secretions. However, it is not known whether capacitation is required for human sperm to become fertile.

MALE FERTILITY

The seminiferous tubular epithelium can be destroyed by a number of different diseases. For instance, bilateral orchiditis resulting from mumps causes sterility in a large percentage of afflicted males. Also, many male infants are born with degenerate tubular epithelium as a result of strictures in the genital ducts or as a result of unknown causes. Finally, a cause of sterility, usually temporary, is excessive temperature of the testes, as follows:

Effect of Temperature on Spermatogenesis. Increasing the temperature of the testes can prevent spermatogenesis by causing degeneration of all the cells of the seminiferous tubules besides the spermatogonia.

It has often been stated that the reason the testicles are located in the dangling scrotum is to maintain the temperature of these glands below the temperature of the body. On cold days scrotal reflexes cause the musculature of the scrotum to contract, pulling the testicles close to the body, whereas on warm days the musculature of the scrotum becomes almost totally relaxed so that the testicles hang far from the body. Furthermore, the scrotum is well supplied with sweat glands which presumably aid in keeping the testicles cool. Thus the scrotum apparently is designed to act as a cooling mechanism for the testicles, without which spermatogenesis is said to be deficient during hot weather.

Cryptorchidism. Cryptorchidism means failure of a testis to descend from the abdomen into the scrotum. During the development of the male fetus, the testes are derived from the genital ridges in the abdomen. However, during the late stages of gestation, the testes descend through the inguinal canals into the scrotum. Occasionally this descent does not occur at all or occurs incompletely so that one or both testes remain in the abdomen, in the inguinal canal, or elsewhere along the route of descent.

A testicle that remains throughout life in the abdominal cavity is incapable of forming sperm. The tubular epithelium is completely degenerate, leaving only the interstitial structures of the testis. It is believed that even the few degrees higher temperature in the abdomen than in the scrotum is sufficient to cause degeneration of the tubular epithelium and consequently to cause sterility. For this reason operations to relocate the cryptorchid testes from the abdominal cavity into the scrotum prior to the beginning of adult sexual life are frequently performed on boys who have undescended testes.

Testosterone secretion by the fetal testes themselves is the stimulus that causes the testes to move into the scrotum from the abdomen. Therefore, many instances, if not most, of cryptorchidism are caused by abnormally formed testes that are unable to secrete enough testosterone.

Effect of Sperm Count on Fertility. The usual quantity of semen ejaculated at each coitus averages approximately 3.5 ml., and in each milliliter of semen is an average of approximately 120 million sperm, though even in "normal" persons this can vary from 35 million to 200 million. This means an average total of 400 million sperm are usually present in each ejaculate. When the number of sperm in each milliliter falls below approximately 20 million, the person is likely to be infertile. Thus, even though only a single sperm is necessary to fertilize the ovum, for reasons not yet completely understood the ejaculate must contain a tremendous number of sperm for at least one to fertilize the ovum.

Effect of Sperm Morphology and Motility on Fertility. Occasionally a man has a completely normal number of sperm but is still infertile. When this occurs, often as many as half of the sperm are found to be abnormal, such as having two heads, abnormally shaped heads, or abnormal tails, as illustrated in Figure 80–4; at other times, the sperm appear to be completely normal but for reasons not understood are either entirely nonmotile or relatively nonmotile. Whenever the majority of the sperm are abnormal in morphology or are found to be nonmotile the person is likely to be infertile even though the remainder of the sperm appear to be normal.

Function of Hyaluronidase and Proteinases Secreted by the Sperm for the Process of Fertilization. Stored in the acrosomes of the sperm are large quantities of hyaluronidase and proteinases. Hyaluronidase is an enzyme that depolymerizes the hyaluronic acid polymers that are

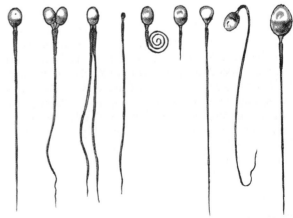

Figure 80–4. Abnormal sperm, compared with a normal sperm on the right.

present in the intercellular cement substance; proteinases can dissolute the proteins of tissues.

When the ovum is expelled from the follicle of the ovary into the abdominal cavity, it carries with it several layers of cells. Before a sperm can reach the ovum to fertilize it, these cells must be removed; it is believed that the hyaluronidase and proteinases released by the acrosomes play at least a small role (in addition to a much larger role played by sodium bicarbonate in the fallopian tube secretions) in causing these cells to break away from the ovum, thus allowing the sperm to reach the surface of the ovum. When sperm are insufficient in number — below 20 million per milliliter — the man is often sterile. This sterility has been postulated to result from insufficient enzymes to help remove the cell layers from the ovum.

Another possible function of the proteinases is to allow the sperm to penetrate the mucous plug that frequently forms in the cervix of the uterus. The proteinases act as mucolytic enzymes that presumably proceed in advance of the sperm and create channels through the mucous plug. It is believed that lack of appropriate mucolytic enzyme to perform this function is also occasionally responsible for male sterility.

THE MALE SEXUAL ACT

NEURONAL STIMULUS FOR PERFORMANCE OF THE MALE SEXUAL ACT

The most important source of impulses for initiating the male sexual act is the glans penis, for the glans contains a highly organized sensory end-organ system that transmits into the central nervous system a special modality of sensation called *sexual sensation*. The massaging action of intercourse on the glans stimulates the sensory end-organs, and the sexual sensations in turn pass through the pudendal nerve, thence through the sacral plexus into the sacral portion of the spinal cord, and finally up the cord to undefined areas of the cerebrum. Impulses may also enter the spinal cord from areas adjacent to the penis to aid in stimulating the sexual act. For instance, stimulation of the anal epithelium, the scrotum, and perineal structures in general can all send impulses into the cord which add to the sexual sensation. Sexual sensations can even originate in internal structures, such as irritated areas of the urethra, the bladder, the prostate, the seminal vesicles, the testes, and the vas deferens. Indeed, one of the causes of "sexual drive" is probably overfilling of the sexual organs with secretions. Infection and inflammation of these sexual organs sometimes cause almost continual sexual desire, and "aphrodisiac" drugs, such as cantharides, increase the sexual desire by irritating the bladder and urethral mucosa.

The Psychic Element of Male Sexual Stimula-

tion. Appropriate psychic stimuli can greatly enhance the ability of a person to perform the sexual act. Simply thinking sexual thoughts or even dreaming that the act of intercourse is being performed can cause the male sexual act to occur and to culminate in ejaculation. Indeed, *nocturnal emissions* during dreams occur in many males during some stages of sexual life, especially during the teens.

Integration of the Male Sexual Act in the Spinal Cord. Though psychic factors usually play an important part in the male sexual act and can actually initiate or inhibit it, the cerebrum is probably not absolutely necessary for its performance, for appropriate genital stimulation can cause ejaculation in some animals and occasionally in a human being after their spinal cords have been cut above the lumbar region. Therefore, the male sexual act results from inherent reflex mechanisms integrated in the sacral and lumbar spinal cord, and these mechanisms can be initiated by either psychic stimulation or actual sexual stimulation.

STAGES OF THE MALE SEXUAL ACT

Erection. Erection is the first effect of male sexual stimulation, and the degree of erection is proportional to the degree of stimulation, whether this be psychic or physical.

Erection is caused by parasympathetic impulses that pass from the sacral portion of the spinal cord through the nervi erigentes to the penis. These parasympathetic impulses dilate the arteries of the penis and simultaneously constrict the veins, thus allowing arterial blood to build up under high pressure in the *erectile tissue* of the penis, illustrated in Figure 80–5. This erectile tissue is nothing more than large, cavernous, venous sinusoids, which are normally relatively empty but which become dilated tremendously when arterial blood flows into them under pressure, since the venous outflow is partially occluded. Also, the erectile bodies are surrounded by strong fibrous coats; therefore, high pressure within the sinusoids causes ballooning of the erectile tissue to such an

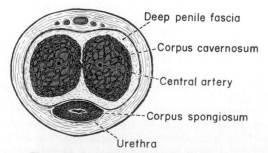

Figure 80–5.   Erectile tissue of the penis.

extent that the penis becomes hard and elongated.

Lubrication. During sexual stimulation, the parasympathetic impulses, in addition to promoting erection, cause the glands of Littré and the bulbo-urethral glands to secrete mucus. Thus mucus flows through the urethra during intercourse to aid in the lubrication of coitus. However, most of the lubrication of coitus is provided by the female sexual organs rather than by the male. Without satisfactory lubrication, the male sexual act is rarely successful because unlubricated intercourse causes pain impulses which inhibit rather than excite sexual sensations.

Emission and Ejaculation. Emission and ejaculation are the culmination of the male sexual act. When the sexual stimulus becomes extremely intense, the reflex centers of the spinal cord begin to emit sympathetic impulses that leave the cord at L-1 and L-2 and pass to the genital organs through the hypogastric plexus to initiate emission, which is the forerunner of ejaculation.

Emission is believed to begin with contraction of the epididymis, the vas deferens, and the ampulla to cause expulsion of sperm into the internal urethra. Then, contractions of the muscular coat of the prostate gland followed lastly by contraction of the seminal vesicles expel prostatic fluid and seminal fluid, forcing the sperm forward. All these fluids mix with the mucus already secreted by the bulbo-urethral glands to form the semen. The process to this point is *emission*.

The filling of the internal urethra then elicits signals that are transmitted through the pudendal nerves to the sacral regions of the cord. In turn, rhythmic nerve impulses are sent from the cord to skeletal muscles, mainly the bulbocavernosus muscles, that encase the base of the erectile tissue, causing rhythmic, wavelike increases in pressure in this tissue, which "ejaculates" the semen from the urethra to the exterior. This is the process of *ejaculation*.

TESTOSTERONE AND OTHER MALE SEX HORMONES

SECRETION, METABOLISM, AND CHEMISTRY OF THE MALE SEX HORMONE

Secretion of Testosterone by the Interstitial Cells of the Testes. The testes secrete several male sex hormones, which are collectively called *androgens*, including *testosterone, dihydrotestosterone,* and *androstenedione*. However, testosterone is so much more abundant and potent than the others that one can consider it to be the significant hormone responsible for the male hormonal effects.

Testosterone is formed by the *interstitial cells of Leydig*, which lie in the interstices between the seminiferous tubules and constitute about 20 per cent of the mass of the adult testes, as illustrated in Figure 80–6. Interstitial cells in the testes are not numerous in a child, but they *are* numerous in a newborn male infant and also in the adult male anytime after puberty; at both these times the testes secrete large quantities of testosterone. Furthermore, when tumors develop from the interstitial cells of Leydig, great quantities of testosterone are secreted. Finally, when the germinal epithelium of the testes is destroyed by x-ray treatment or by excessive heat, the interstitial cells, which are less easily destroyed, continue to produce testosterone.

Secretion of Androgens Elsewhere in the Body. The term *androgen* is used synonymously with the term male sex hormone, but it also includes male sex hormones produced elsewhere in the body besides the testes. For instance, the adrenal gland secretes at least five different androgens, though the total masculinizing activity of all these is normally so slight that they do not cause significant masculine characteristics even in women. But when an adrenal tumor of the androgen-producing cells occurs, the quantity of androgenic hormones may then become great enough to cause all the usual male secondary sexual characteristics. These effects were described in connection with the adrenogenital syndrome in Chapter 77.

Rarely, embryonic rest cells in the ovary can develop into a tumor which produces excessive quantities of androgens in the female; one such tumor is the *arrhenoblastoma*. The normal ovary also produces minute quantities of androgens, but these are not significant.

Chemistry of the Androgens. All androgens are steroid compounds, as illustrated by the formulas in

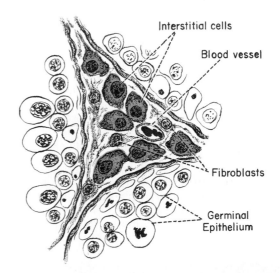

Figure 80–6. Interstitial cells located in the interstices between the seminiferous tubules. (Modified from Bloom and Fawcett: Textbook of Histology, 8th Ed.)

Interstitial cells

Blood vessel

Fibroblasts

Germinal Epithelium

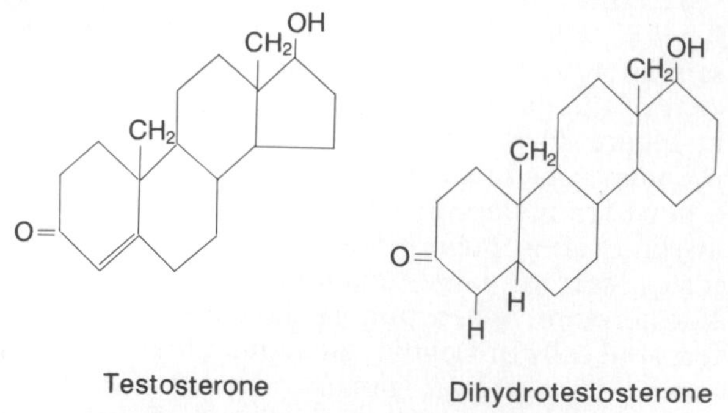

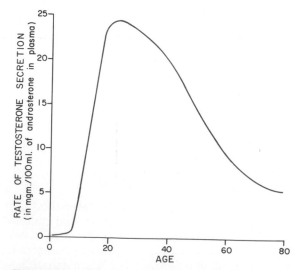

Testosterone   Dihydrotestosterone

Figure 80–7. Testosterone and dihydrotesterone.

Figure 80–7 for *testosterone* and *dihydrotestosterone*. Both in the testes and in the adrenals, the androgens can be synthesized either from cholesterol or directly from acetyl coenzyme A.

Metabolism of Testosterone. After secretion by the testes, testosterone, most of it loosely bound with plasma albumin or a β-globulin called gonadal steroid-binding globulin, circulates in the blood for about 15 to 30 minutes before it either becomes fixed to the tissues or is degraded into inactive products that are subsequently excreted.

Much of the testosterone that becomes fixed to the tissues is converted within the cells to *dihydrotestosterone,* especially in certain target organs such as the prostate gland in the adult and in the external genitalia of the fetal male. Some actions of testosterone are dependent on this conversion, while other actions are not. The intracellular functions will be discussed later in the chapter.

Degradation and Excretion of Testosterone. The testosterone that does not become fixed to the tissues is rapidly converted, mainly by the liver, into *androsterone* and *dehydroepiandrosterone* and simultaneously conjugated either as glucuronides or sulfates (glucuronides, particularly). These are excreted either into the gut in the bile or into the urine.

Production of Estrogen by the Testes. In addition to testosterone, small amounts of estrogens are formed in the male (about one-fifth the amount in the nonpregnant female), and a reasonable quantity of these can be recovered from a man's urine. The functions of estrogens in the male are unknown.

The exact source of the estrogens in the male is also still doubtful, but the following are known: (a) The quantity of estrogens decreases slightly when the germinal epithelium of the seminiferous tubules is destroyed. This indicates that the semiiferous tubules, probably the Sertoli cells, synthesize estrogens in men. (b) Estrogens are formed from testosterone and androstenedione in other tissues of the body, especially the liver, probably accounting for as much as 80 per cent of the total estrogen production.

FUNCTIONS OF TESTOSTERONE

In general, testosterone is responsible for the distinguishing characteristics of the masculine body. The testes are stimulated by chorionic gonadotropin from the placenta to produce moderate quantities of testosterone during fetal development and for a few weeks after birth; then essentially no testosterone is produced during childhood until approximately the age of 10 to 13. Then testosterone production increases rapidly at the onset of puberty and lasts throughout most of the remainder of life as illustrated in Figure 80–8, dwindling rapidly beyond the age of 40 to become perhaps one-fifth the peak value by the age of 80.

Functions of Testosterone During Fetal Development. Testosterone begins to be elaborated by the male at about the seventh week of embryonic life. Indeed, one of the major functional differences between the female and the male sex chromosome is that the male chromosome causes the newly developing genital ridge to secrete testosterone while the female chromosome causes this ridge to secrete estrogens. Injection of large quantities of male sex hormone into gravid animals causes development of male sexual organs even though the fetus is female. Also, removal of the testes in a male fetus causes development of female sexual organs. Therefore, testosterone secreted by the genital ridges and the subsequently developing testes is responsible for the development of the male sex characteristics, including the growth of a penis and a scrotum rather than the formation of a clitoris and a vagina. Also, it causes development of the prostate gland, the seminal vesicles, and the male genital ducts, while at the same time suppressing the formation of female genital organs.

Effect on the Descent of the Testes. The testes usually descend into the scrotum during the last

Figure 80–8. Rate of testosterone secretion at different ages, as judged from the concentrations of androsterone in the plasma.

two months of gestation when the testes are secreting reasonable quantities of testosterone. If a male child is born with undescended testes, administration of testosterone causes the testes to descend in the usual manner if the inguinal canals are large enough to allow the testes to pass. Or, administration of gonadotropic hormones, which stimulate the interstitial cells of the testes to produce testosterone, also causes the testes to descend. Thus, the stimulus for descent of the testes is testosterone, indicating again that testosterone is an important hormone for male sexual development during fetal life.

Effect of Testosterone on Development of Adult Primary and Secondary Sexual Characteristics. Testosterone secretion after puberty causes the penis, the scrotum, and the testes all to enlarge about 8-fold until the age of 20. In addition, testosterone causes the "secondary sexual characteristics" of the male to develop at the same time, beginning at puberty and ending at maturity. These secondary sexual characteristics, in addition to the sexual organs themselves, distinguish the male from the female as follows:

Effect on the Distribution of Body Hair. Testosterone causes growth of hair (1) over the pubis, (2) upward along the linea alba sometimes to the umbilicus and above, (3) on the face, (4) usually on the chest and (5) less often on other regions of the body, such as the back. It also causes the hair on most other portions of the body to become more prolific.

Baldness. Testosterone decreases the growth of hair on the top of the head; a man who does not have functional testes does not become bald. However, many virile men never become bald, for baldness is a result of two factors: first, a *genetic background* for the development of baldness and, second, superimposed on this genetic background, *large quantities of androgenic hormones.* A woman who has the appropriate genetic background and who develops a long-sustained androgenic tumor becomes bald in the same manner as a man.

Effect on the Voice. Testosterone secreted by the testes or injected into the body causes hypertrophy of the laryngeal mucosa and enlargement of the larynx. The effects cause at first a relatively discordant, "cracking" voice, but this gradually changes into the typical masculine bass voice.

Effect on the Skin, and Development of Acne. Testosterone increases the thickness of the skin over the entire body and increases the ruggedness of the subcutaneous tissues. Also, it causes increased quantities of melanin to be deposited by the skin, thereby deepening the hue of the skin.

Testosterone increases the rate of secretion by some or perhaps all the sebaceous glands. Especially important is the excessive secretion by the sebaceous glands of the face, for oversecretion of these can result in *acne.* Therefore, acne is one of the most common features of adolescence, when the male body is first becoming introduced to increased secretion of testosterone. After several years of testosterone secretion, the skin adapts itself to the testosterone in some way that allows it to overcome the acne.

Effect on Protein Formation and Muscular Development. One of the most important male characteristics is the development of increasing musculature following puberty. This is associated with increased protein in other parts of the body as well. Many of the changes in the skin are also due to deposition of proteins in the skin, and the changes in the voice could even result from this protein anabolic function of testosterone.

Testosterone has often been considered to be a "youth hormone" because of its effect on the musculature, and it is occasionally used for treatment of persons who have poorly developed muscles.

Effect on Bone Growth and Calcium Retention. Following puberty or following prolonged injection of testosterone, the bones grow considerably in thickness and also deposit considerable calcium salts. Thus, testosterone increases the total quantity of bone matrix, and it also causes calcium retention. The increase in bone matrix is believed to result from the general protein anabolic function of testosterone, and the deposition of calcium salts to result from increased bone matrix available to be calcified.

Because of the ability of testosterone to increase the size and strength of bones, testosterone is often used in old age to treat osteoporosis.

When great quantities of testosterone (or any other androgen) are secreted in the still-growing child, the rate of bone growth increases markedly, causing a spurt in total body growth as well. However, the testosterone also causes the epiphyses of the long bones to unite with the shafts of the bones at an early age in life. Therefore, despite the rapidity of growth, this early uniting of the epiphyses prevents the person from growing as tall as he would have grown had testosterone not been secreted at all. Even in normal men the final adult height is slightly less than that which would have been attained had the person been castrated prior to puberty.

Effect on Basal Metabolism. Injection of large quantities of testosterone can increase the basal metabolic rate by as much as 15 per cent, and it is believed that even the usual quantity of testosterone secreted by the testes during active sexual life increases the rate of metabolism some 5 to 10 per cent above the value that it would be were the testes not active. This increased rate of metabolism is possibly an indirect result of the effect of testosterone on protein anabolism, the increased quantity of proteins — the enzymes especially — increasing the activities of all cells.

Effect on the Red Blood Cells. When normal quantities of testosterone are injected into a castrated adult, the number of red blood cells per cubic millimeter of blood increases approximately 20 per cent. Also, the average man has about 700,000 more red blood cells per cubic millimeter than the average woman. However, this difference may be due partly to the increased metabolic rate following testosterone administration rather than to a direct effect of testosterone on red blood cell production.

Effect on Electrolyte and Water Balance. As pointed out in Chapter 77, many different steroid hormones can increase the reabsorption of sodium in the distal tubules of the kidneys. Testosterone has such an effect but only to a minor degree in comparison with the adrenal mineralocorticoids. Nevertheless, following puberty the blood and extracellular fluid volumes of the male in relation to his weight increase to a slight extent; this effect probably results at least partly from the sodium-retaining ability of testosterone.

BASIC INTRACELLULAR MECHANISM OF ACTION OF TESTOSTERONE

Though it is not known exactly how testosterone causes all the effects just listed, it is believed that they result mainly from increased rate of protein formation in cells. This has been studied extensively in the prostate gland, one of the organs that is most affected by testosterone. In this gland, testosterone enters the cells within a few minutes after secretion, is there converted to *dihydrotestosterone,* and binds with a cytoplasmic "receptor protein." This combination then migrates to the nucleus where it binds with a nuclear protein and induces the DNA-RNA transcription process. Within 30 minutes the concentration of RNA begins to increase in the cells, and this is followed by progressive increase in cellular protein. After several days the quantity of DNA in the gland has also increased, and there has been a simultaneous increase in the number of prostatic cells.

Therefore, it is assumed that testosterone greatly stimulates production of proteins in general, though increasing more specifically those proteins in "target" organs or tissues responsible for the development of secondary sexual characteristics.

CONTROL OF MALE SEXUAL FUNCTIONS BY THE GONADOTROPIC HORMONES — FSH AND LH

The anterior pituitary gland secretes two major gonadotropic hormones: (1) *follicle-stimulating*

hormone (FSH); and (2) *luteinizing hormone* (LH). These are glycoprotein hormones that play major roles in the control of both male and female sexual function.

Regulation of Testosterone Production by LH. Testosterone is produced by the interstitial cells of Leydig only when the testes are stimulated by LH from the pituitary gland, and the quantity of testosterone secreted varies approximately in proportion to the amount of LH available.

Injection of purified LH into a child causes fibroblasts in the interstitial areas of the testes to develop into interstitial cells of Leydig, though mature Leydig cells are not normally found in the child's testes until after the age of approximately 10. Also, simultaneous administration of *prolactin* (another pituitary hormone that is closely associated with the gonadotropic hormones) greatly potentiates the effect of LH in promoting testosterone production.

Effect of Human Chorionic Gonadotropin on the Fetal Testes. During gestation the placenta secretes large quantities of human chorionic gonadotropin, a hormone that has almost the same properties as LH. This hormone stimulates the formation of interstitial cells in the testes of the fetus and causes testosterone secretion. As pointed out earlier in the chapter, the secretion of testosterone during fetal life is important for promoting formation of male sexual organs.

Stimulation of Spermatogenesis by Follicle-Stimulating Hormone (FSH) and Testosterone. The conversion of primary spermatocytes into secondary spermatocytes in the seminiferous tubules is stimulated by FSH from the anterior pituitary gland. However, FSH cannot by itself cause complete formation of spermatozoa. For spermatogenesis to proceed to completion, testosterone must be secreted simultaneously by the interstitial cells. Thus, FSH seems to initiate the proliferative process of spermatogenesis, and testosterone diffusing from the interstitial cells into the seminiferous tubules apparently is necessary for final maturation of the spermatozoa. Indeed, once spermatogenesis has begun, testosterone alone can maintain this process for many weeks to months without additional FSH. Because testosterone is secreted by the interstitial cells under the influence of LH, both FSH and LH must be secreted by the anterior pituitary gland if spermatogenesis is to occur.

Regulation of Pituitary Secretion of LH and FSH by the Hypothalamus

The gonadotropins, like corticotropin and thyrotropin, are secreted by the anterior pituitary gland mainly in response to nervous activity in the hypothalamus. For instance, in the female rabbit, coitus with a male rabbit elicits nervous activity in the hypothalamus that in turn stimulates the anterior pituitary to secrete FSH and LH. These hormones then cause rapid ripening of follicles in the rabbit's ovaries, followed a few hours later by ovulation.

Many other types of nervous stimuli are also known to affect gonadotropin secretion. For instance, in sheep, goats, and deer, nervous stimuli in response to changes in weather and amount of light in the day increase the quantities of gonadotropins during one season of the year, the mating season, thus allowing birth of the young during an appropriate period for survival. Also, psychic stimuli can affect fertility of the male animal, as exemplified by the fact that transporting a bull under uncomfortable conditions can often cause almost complete temporary sterility. In the human being, it is known too that various psychogenic stimuli feeding into the hypothalamus can cause marked excitatory or inhibitory effects on gonadotropin secretion, in this way sometimes greatly altering the degree of fertility.

Luteinizing Hormone–Releasing Hormone (LHRH), the Hypothalamic Hormone That Stimulates Gonadotropin Secretion. In both the male and the female the hypothalamus controls gonadotropin secretion by way of the hypothalamic-hypophysial portal system, as was discussed in Chapter 75. Though there are two different gonadotropic hormones, luteinizing hormone and follicle-stimulating hormone, only one hypothalamic-releasing hormone has been discovered; this is *luteinizing hormone–releasing hormone* (LHRH). Though this hormone has an especially strong effect on inducing luteinizing hormone secretion by the anterior pituitary gland, it has a potent effect in causing follicle-stimulating hormone secretion as well. For this reason it is often also called *gonadotropin-releasing hormone*.

Luteinizing hormone–releasing hormone plays a similar role in controlling gonadotropin secretion in the female, where the interrelationships are far more complex. Therefore, its nature and its functions will be discussed in much more detail in the following chapter.

Reciprocal Inhibition of Hypothalamic–Anterior Pituitary Secretion of Gonadotropic Hormones by Testicular Hormones. *Feedback Control of Testosterone Secretion.* Injection of testosterone into either a male or a female animal strongly inhibits the secretion of luteinizing hormone but only slightly inhibits the secretion of the follicle-stimulating hormone. This inhibition depends on normal function of the hypothalamus; therefore, it is quite clear that the following negative feedback control system operates continuously to control very precisely the rate of testosterone secretion:

1. The hypothalamus secretes *luteinizing hormone–releasing hormone* which stimulates the anterior pituitary gland to secrete *luteinizing hormone*.

2. Luteinizing hormone in turn stimulates *hyperplasia of the Leydig cells* of the testes and also stimulates production of *testosterone* by these cells.

3. The testosterone in turn feeds back negatively to the hypothalamus, inhibiting production of luteinizing hormone–releasing hormone. This obviously limits the rate at which testosterone will be produced. On the other hand, when testosterone production is too low, lack of inhibition of the hypothalamus leads to subsequent return of testosterone secretion to the normal level.

Feedback Control of Spermatogenesis — Role of "Inhibin." It is known, too, that spermatogenesis by the testes inhibits the secretion of FSH. Conversely, failure of spermatogenesis causes markedly increased secretion of FSH; this is especially true when the seminiferous tubules are destroyed, including destruction of the Sertoli cells in addition to the germinal cells. Therefore, it is believed that the Sertoli cells secrete a hormone that has a direct inhibitory effect mainly on the anterior pituitary gland (but perhaps slightly on the hypothalamus as well) to inhibit the secretion of FSH. A hormone having a molecular weight between 25,000 and 100,000 and called *inhibin* has been discovered that probably is responsible for most of the feedback control of FSH secretion and of spermatogenesis. This feedback cycle is the following:

1. Follicle-stimulating hormone induces proliferation of the germinal epithelium of the seminiferous tubules and at the same time stimulates the Sertoli cells that provide nutrition for the developing spermatozoa.

2. The Sertoli cells (or less likely the germinal epithelial cells) release inhibin that in turn feeds back negatively to the anterior pituitary gland to inhibit the production of FSH. Thus, this feedback cycle maintains a rate of spermatogenesis that is required for male reproductive function.

Puberty and Regulation of Its Onset

Initiation of the onset of puberty has long been a mystery. In the earliest history of humanity, the belief was simply that the testicles "ripened" at this time. With the discovery of the gonadotropins ripening of the anterior pituitary gland was considered responsible. Now it is known from experiments in which both testicular and pituitary tissues have been transplanted from infant animals into adult animals that both the testes and the anterior pituitary of the infant are capable of performing adult functions if appropriately stimulated. Therefore, it is now almost certain that *during childhood the hypothalamus, not the two glands, is responsible: that is, the hypothalamus simply does not secrete gonadotropin-releasing hormones.* Also, it is known that even the minutest amount of testosterone inhibits the production of gonadotropin-releasing hormones by the childhood hypothalamus. This leads to the theory that the childhood hypothalamus is so extremely sensitive to inhibition that the minutest amount of testosterone secreted by the testicles inhibits the entire system. For reasons yet unknown, the hypothalamus loses this inhibitory sensitivity at the time of puberty, which allows the secretory mechanisms to develop full activity. Thus, puberty is now believed to result from a maturing process of the hypothalamic sexual control centers.

The Male Adult Sexual Life and the Male Climacteric. Following puberty, gonadotropic hormones are produced by the male pituitary gland for the remainder of life, and at least some spermatogenesis usually continues until death. Most men, however, begin to exhibit slowly decreasing sexual functions in their late 40s or 50s, and one study has shown that the average age for terminating intersexual relations was 68, though the variation was great. This decline is related to decrease in testosterone secretion, as depicted in Figure 80–8. The decrease in male sexual function is called the male climacteric. Occasionally, the male climacteric is associated with symptoms of hot flashes, suffocation, and psychic disorders similar to the menopausal symptoms of the female. These symptoms can be abrogated by administration of testosterone, synthetic methyl testosterone, or even estrogens that are used for treatment of menopausal symptoms in the female.

ABNORMALITIES OF MALE SEXUAL FUNCTION

THE PROSTATE GLAND AND ITS ABNORMALITIES

The prostate gland remains relatively small throughout childhood and begins to grow at puberty under the stimulus of testosterone. This gland reaches an almost stationary size by the age of about 20 and remains this size up to the age of approximately 40 to 50. At that time in some men it begins to degenerate along with the decreased production of testosterone by the testes. However, a benign prostatic fibroadenoma frequently develops in the prostate in older men and causes urinary obstruction. This hypertrophy is not caused by testosterone.

Cancer of the prostate gland is an extremely common cause of death, resulting in approximately 2 to 3 per cent of all male deaths.

Once cancer of the prostate gland does occur, the cancerous cells are usually stimulated to more rapid growth by testosterone and are inhibited by removal of the testes so that testosterone cannot be formed. Also, prostatic cancer can usually be inhibited by administration of estrogens. Some patients who have prostatic cancer that has already metastasized to almost all the bones of the body can be successfully treated for a few months to years by removal of the testes, by estrogen therapy, or by both; following this therapy the metastases degenerate and the bones heal. This treatment does not completely stop the cancer but does slow it and greatly diminishes the severe bone pain.

HYPOGONADISM IN THE MALE

Hypogonadism is caused by any of several abnormalities: First, the person may be born without functional testes. Second, he may have undeveloped testes, owing to failure of the anterior pituitary to secrete gonadotropic hormones. Third, cryptorchidism (undescended testes) may occur, associated with partial or total degeneration of the testes. Fourth, he may lose his testes, which is called *castration.*

When a boy loses his testes prior to puberty, a state of *eunuchism* ensues in which he continues to have infantile sexual characteristics throughout life. The height of the adult eunuch is slightly greater than that of the normal man, though the bones are quite thin, the muscles are considerably weaker than those of normal man, and, obviously, the sexual organs and secondary sexual characteristics are those of a child rather than those of an adult. The voice is childlike, there is no loss of hair on the head, and the normal masculine hair distribution on the face and elsewhere does not occur.

When a man is castrated following puberty, some male secondary sexual characteristics revert to those of a child, and others remain of masculine character. The sexual organs regress slightly in size but not to a childlike state, and the voice regresses from the bass quality only slightly. On the other hand, there is loss of masculine hair production, loss of the thick masculine bones, and loss of the musculature of the virile male.

In the castrated adult male, sexual desires are decreased but not totally lost, provided sexual activities have been practiced previously. Erection can still occur as before though with less ease, but it is rare that ejaculation can take place, primarily because the semen-forming organs become degenerate, and there has been a loss of the testosterone-driven psychic desire.

Adiposogenital Syndrome (Fröhlich's Syndrome). Damage to certain areas of the hypothalamus greatly decreases the secretion of gonadotropin-releasing hormones, and there is a corresponding decrease in the secretion of gonadotropic hormones by the anterior pituitary. If this occurs prior to puberty, it causes typical eunuchism. The hypothalamic lesion often causes simultaneous overeating because of its effect on the feeding center of the hypothalamus. Consequently, the person develops severe obesity along with the eunuchism. This condition is illustrated in Figure 80–9 and is called *adiposogenital syndrome, Fröhlich's syndrome,* or *hypothalamic eunuchism.*

TESTICULAR TUMORS AND HYPERGONADISM IN THE MALE

Interstitial cell tumors develop rarely in the testes, but when they do develop they sometimes produce as

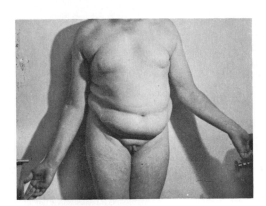

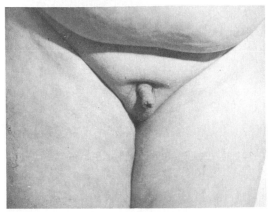

Figure 80–9. Adiposogenital syndrome in an adolescent male. Note the obesity and the childlike sexual organs. (Courtesy of Dr. Leonard Posey.)

much as 100 times the normal quantities of testosterone. When such tumors develop in young children, they cause rapid growth of the musculature and bones but also early uniting of the epiphyses so that the eventual adult height actually is less than that which would have been achieved otherwise. Obviously, such interstitial cell tumors cause excessive development of the sexual organs and of the secondary sexual characteristics. In the adult male, small interstitial cell tumors are difficult to diagnose because masculine features are already present. Diagnosis can be made, however, from urine tests that show greatly increased excretion of testosterone end-products.

Much more common than the interstitial cell tumors are tumors of the germinal epithelium. Because germinal cells are capable of differentiating into almost any type of cell, many of these tumors contain multiple tissues, such as placental tissue, hair, teeth, bone, skin, and so forth, all found together in the same tumorous mass called a *teratoma*. Often these tumors secrete no hormones, but if a significant quantity of placental tissue develops in the tumor, it may secrete large quantities of chorionic gonadotropin that has functions very similar to those of LH; these will be discussed in more detail in Chapter 82. Also, estrogenic hormones are frequently secreted by these tumors and cause the condition called *gynecomastia*, which means overgrowth of the breasts.

THE PINEAL GLAND — ITS FUNCTION IN CONTROLLING SEASONAL FERTILITY

For as long as the pineal gland has been known to exist, to it has been ascribed a myriad of different functions, even including being the seat of the soul. However, it is known from comparative anatomy studies that the pineal gland is a vestigial remnant of what was a third eye high in the back of the head in lower animals. Many physiologists have been content with the idea that this gland is in truth merely a vestigial remnant, but others have claimed for many years that it plays important roles in the control of sexual activities and reproduction, functions that still others said were nothing more than the zealous imaginings of physiologists preoccupied with sexual delusions.

But now, after years of turmoil and dispute, it looks as though the sex advocates have at last won, and that the pineal gland does indeed play an important regulatory role in sexual and reproduction function. For, in lower animals in which the pineal gland has been removed or in which the nervous circuits to the pineal gland have been sectioned, the normal annual periods of seasonal fertility are lost. To these animals such seasonal fertility is very important because it allows birth of the offspring in the spring and summer months when survival is most likely. Unfortunately, the mechanism of this effect is still not entirely clear but it seems to be the following:

First, the pineal gland is controlled by the amount of light seen by the eyes each day. For instance, in the hamster, greater than 13 hours of darkness each day activates the pineal gland, while less than that amount of darkness fails to activate it, with a very critical balance between activation and nonactivation. The nervous pathway involved is this: passage of light signals from the eyes to the suprachiasmal area of the brain stem, thence to the lateral hypothalamus, down to the sympathetic neurons in the spinal cord, upward through the sympathetic nerve fibers and superior cervical ganglion, and finally along the sympathetic fibers following the blood vessels to the pineal gland where the sympathetic nerve signals activate pineal secretion.

Second, the pineal gland secretes *melatonin* and several other similar substances. Either melatonin or one of the other substances than passes either by way of the blood or through the fluid of the third ventricle to the anterior pituitary gland to control gonadotropic hormone secretion.

In the presence of pineal gland secretion, the gonadotropic hormone secretion is suppressed, and the gonads become inhibited and even involuted. This is what occurs during the winter months. But after about four months of dysfunction, the gonadotropic hormone secretion breaks through the inhibitory effect of the pineal gland, and the gonads become functional once more, ready for a full springtime of activity.

But, does the pineal gland have a similar function for control of reproduction in man? The answer to this is still far from known. However, tumors often occur in the region of the pineal gland. Some of these are pineal tumors that secrete excessive quantities of pineal hormones, while others are tumors of surrounding tissue and press on the pineal gland to destroy it. Both types of tumors are often associated with serious hypo- or hypergonadal function. So, perhaps the pineal gland does play at least some role in controlling sexual drive and reproduction in men.

REFERENCES

Arimura, A.: Hypothalamic gonadotropin-releasing hormone and reproduction. *Int. Rev. Physiol. 13*:1, 1977.

Bardin, C. W.: Pituitary-testicular axis. *In* Yen, S. S. C., and Jaffe, R. B. (eds.): Reproductive Endocrinology. Philadelphia, W. B. Saunders Co., 1978, p. 110.

Bedford, J. M.: Maturation, transport, and fate of spermatozoa in the epididymis. *In* Greep, R. O., and Astwood, E. B. (eds.): Handbook of Physiology. Sec. 7, Vol. 5. Baltimore, Williams & Wilkins, 1975, p. 303.

Binkley, S.: A timekeeping enzyme in the pineal gland. *Sci. Am., 240*(4):66, 1979.

Buckner, W. P., Jr.: Medical Readings on Human Sexuality. Stanford, Cal., Medical Readings Inc., 1978.

Davidson, J. M.: Neurohormonal basis of male sexual behavior. *Int. Rev. Physiol., 13*:225, 1977.

Eik-Nes, K. B.: Biosynthesis and secretion of testicular steroids. *In* Greep, R. O., and Astwood, E. B. (eds.): Handbook of Physiology. Sec. 7, Vol, 5. Baltimore, Williams & Wilkins, 1975, p. 95.

Epel, D.: The program of fertilization. *Sci. Am., 237*(5):128, 1977.

Ewing, L. L., *et al.*: Regulation of testicular function: A spatial and temporal view. *In* Greep, R. O. (ed.): International Review of Physiology: Reproductive Physiology III. Vol. 22. Baltimore, University Park Press, 1980, p. 41.

Fawcett, D. W.: Ultrastructure and function of the Sertoli cell. *In* Greep, R. O., and Astwood, E. B. (eds.): Handbook of Physiology. Sec. 7, Vol. 5. Baltimore, Williams & Wilkins, 1975, p. 21.

Fink, G.: Feedback actions of target hormones on hypothalamus and pituitary, with special reference to gonadal steroids. *Annu. Rev. Physiol., 41*:571, 1979.

Fonkalsrud, E. W.: The Undescended Testis. Chicago, Year Book Medical Publishers, 1978.

Forleo, R., and Pasini, W. (eds.): Medical Sexology. Littleton, Mass., PSG Publishing Co., 1979.

Glover, T. D.: Recent progress in the study of male reproductive physiology: Testis stimulation, sperm formation, transport and maturation (epididymal physiology), semen analysis, storage, and artificial insemination. *In* MTP International Review of Science: Physiology. Baltimore, University Park Press, 1974, p. 221.

Griffin, J. E., and Wilson, J. D.: The testis. *In* Bondy, P. K., and Rosenberg, L. E. (eds.): Metabolic Control in Disease, 8th Ed. Philadelphia, W. B. Saunders Co., 1980, p. 1535.

Hafez, E. S. E. (ed.): Human Reproduction: Conception and Contraception. Hagerstown, Md., Harper & Row, 1979.

Hafez, E. S. E.: Human Reproductive Physiology. Ann Arbor Mich., Ann Arbor Science, 1978.

Hafez, E. S. E. (ed.): Descended and Cryptorchid Testis. Hingham, Mass., Kluwer Boston, 1980.

Hafez, E. S. E., and Spring-Mills, E. (eds.): Accessory Glands of the Male Reproductive Tract. Ann Arbor, Mich., Ann Arbor Science, 1979.

Hall, P. F.: Testicular hormones: Synthesis and control. *In* DeGroot, L. J., *et al.* (eds.): Endocrinology, Vol. 3. New York, Grune & Stratton, 1979, p. 1511.

Hawkins, D. F., and Elder, M. G.: Human Fertility Control: The Theory and Practice. Boston, Butterworths, 1979.

Holmes, R. L., and Fox, C. A.: Control of Human Reproduction. New York, Academic Press, 1977.

Huff, R. W., and Pauerstein, C. J.: Physiology and Pathophysiology of Human Reproduction. New York, John Wiley & Sons, 1978.

Jaffe, R. B.: Disorders of sexual development. *In* Yen, S. S. C., and Jaffe, R. B. (eds.): Reproductive Endocrinology. Philadelphia, W. B. Saunders Co., 1978, p. 271.

Kolodny, R. C., *et al.*: Textbook of Sexual Medicine. Boston, Little, Brown, 1979.

Lipsett, M. B.: Endocrine mechanisms of reproduction. *In* Frohlich, E. D. (ed.): Pathophysiology, 2nd Ed. Philadelphia, J. B. Lippincott Co., 1976, p. 405.

Lipsett, M. B.: Steroid hormones. *In* Yen, S. S. C., and Jaffe, R. B. (eds.): Reproductive Endocrinology. Philadelphia, W. B. Saunders Co., 1978, p. 80.

Means, A. R.: Biochemical effects of follicle-stimulating hormone on the testis. *In* Greep, R. O., and Astwood, E. B. (eds.): Handbook of Physiology. Sec. 7, Vol. 5. Baltimore, Williams & Wilkins, 1975, p. 203.

Mess, B., Trentini, G. P., and Tima, L.: Role of the Pineal Gland in the Regulation of Ovulation. New York, International Publications Service, 1978.

Money, J., and Higham, E.: Sexual behavior and endocrinology (normal and abnormal). *In* DeGroot, L. J., *et al.* (eds.): Endocrinology, Vol. 3. New York, Grune & Stratton, 1979, p. 1353.

Moore, P. Y.: The central nervous system and the neuroendocrine regulation of reproduction. *In* Yen, S. S. C., and Jaffe, R. B. (eds.): Reproductive Endocrinology. Philadelphia, W. B. Saunders Co., 1978, p. 3.

Odell, W. D.: The physiology of puberty: Disorders of the pubertal process. *In* DeGroot, L. J., *et al.* (eds.): Endocrinology. Vol. 3. New York, Grune & Stratton, 1979, p. 1363.

Phillips, D. M.: Spermiogenesis. New York, Academic Press, 1974.

Prasad, M. R. N., and Rajalakshmi, M.: Recent advances in the control of male reproductive functions. *Int. Rev. Physiol., 13*:153, 1977.

Reiter, R. J. (ed.): The Pineal and Reproduction. New York, S. Karger, 1978.

Relkin, R.: The Pineal. Montreal, Eden Press, 1976.

Ross, G. F., and Lipsett, M. B. (eds.): Reproductive Endocrinology. Clinics in Endocrinology and Metabolism. London, W. B. Saunders Co., Nov. 1978.

Roy, S., and Taneja, S. L.: Vasectomy, Vasocclusion, and Vasanastomosis: A Critical Appraisal. New Delhi, National Institute of Family Planning, 1974.

Segal, S. J.: The physiology of human reproduction. *Sci. Am., 231*(3):52, 1974.

Setchell, B. P.: The Mammalian Testis. Ithaca, N.Y., Cornell University Press, 1978.

Smith, K. D.: Testicular function in the aging male. *In* DeGroot, L. J., *et al.* (eds.): Endocrinology. Vol. 3. New York, Grune & Stratton, 1979, p. 1577.

Stangel, J. J.: Fertility and Conception: An Essential Guide for Childless Couples. New York, New American Library, 1980.

Steinberger, A., and Steinberger, E. (eds.): Testicular Development, Structure and Function. New York, Raven Press, 1980.

Steinberger, E.: Hormonal control of spermatogenesis. *In* DeGroot, L. J., *et al.* (eds.): Endocrinology. Vol. 3. New York, Grune & Stratton, 1979, p. 1535.

Steinberger, E.: Male infertility. *In* DeGroot, L. J., *et al.* (eds.): Endocrinology. Vol. 3. New York, Grune & Stratton, 1979, p. 1567.

Steinberger, E.: Structural consideration of the male reproductive system. *In* DeGroot, L. J., *et al.* (eds.): Endocrinology. Vol. 3. New York, Grune & Stratton, 1979, p. 1501.

Steinberger, E., and Steinberger, A.: Hormonal control of testicular function in mammals. *In* Greep, R. O., and Astwood, E. B. (eds.): Handbook of Physiology. Sec. 7, Vol. 4, Part 2. Baltimore, Williams & Wilkins, 1974, p. 325.

Steinberger, E., and Steinberger, A.: Spermatogenic function of the testis. *In* Greep, R. O., and Astwood, E. B. (eds.): Handbook of Physiology. Sec. 7, Vol. 5. Baltimore, Williams & Wilkins, 1975, p. 1.

Styne, D. M., and Grumbach, M. M.: Puberty in the male and female: Its physiology and disorders. *In* Yen, S. S. C., and Jaffe, R. B. (eds.): Reproductive Endocrinology, Philadelphia, W. B. Saunders Co., 1978, p. 189.

Talway, G. P. (ed.): Recent Advances in Reproduction and Regulation of Fertility. New York, Elsevier/North-Holland, 1979.

Thomas, J. A., and Singahl, R. L. (eds.): Sex Hormone Receptors in Endocrine Organs. Baltimore, Urban & Schwarzenberg, 1980.

Tollison, C. D., and Adams, H. E.: Sexual Disorders: Treatment, Theory and Research. New York, Gardner Press, 1979.

Tyson, J. E. (eds.): Symposium on Neuroendocrinology of Reproduction. Clinics in Obstetrics-Gynaecology. London, W. B. Saunders Co., Aug. 1978.

vanKeep, P. A., *et al.*: Female and Male Climacteric: Current Opinion, 1978. Baltimore, University Park Press, 1978.

Williams-Ashman, H. G.: Biochemical features of androgen physiology. *In* DeGroot, L. J., *et al.* (eds.): Endocrinology. Vol. 3. New York, Grune & Stratton, 1979, p. 1527.

Wilson, J. D.: Sexual differentiation. *Annu. Rev. Physiol., 40*:279, 1978.

81

Prepregnancy Reproductive Functions of the Female, and the Female Hormones

Female reproductive functions can be divided into two major phases: first, preparation of the body for conception and gestation, and second, the period of gestation itself. The present chapter is concerned with the preparation of the body for gestation, and the following chapter presents the physiology of pregnancy.

PHYSIOLOGIC ANATOMY OF THE FEMALE SEXUAL ORGANS

Figure 81–1 illustrates the principal organs of the human female reproductive tract, including the *ovaries,* the *fallopian tubes,* the *uterus,* and the *vagina.* Reproduction begins with the development of ova in the ovaries. A single ovum is expelled from an ovarian follicle into the abdominal cavity in the middle of each monthly sexual cycle. This ovum then passes through one of the fallopian tubes into the uterus, and, if it has been fertilized by a sperm, it implants in the uterus where it develops into a fetus, a placenta, and fetal membranes.

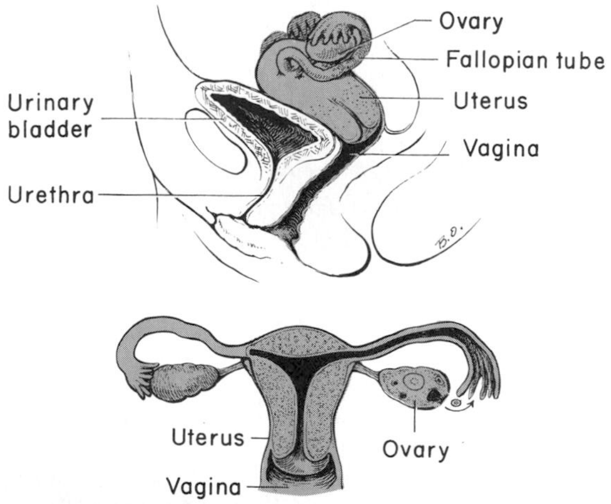

During fetal life the outer surface of the ovary is covered by a *germinal epithelium,* which embryologically is derived directly from the epithelium of the germinal ridges. As the fetus develops, *primordial ova* differentiate from the germinal epithelium and migrate into the substance of the ovarian cortex, carrying with them a layer of epithelioid *granulosa cells.* The ovum surrounded by a single layer of epithelioid granulosa cells is called a *primordial follicle.* At the 30th week of gestation the number of ova reaches about 7 million, but most of these soon degenerate so that less than 2 million are present in the two ovaries at birth, and only 300,000 at puberty. Then, during all the reproductive years of the female, only about 450 of these follicles develop enough to expel their ova; the remainder degenerate (become *atretic*). At the end of reproductive capability, at the *menopause,* only a few primordial follicles remain in the ovaries, and even these degenerate soon thereafter.

THE FEMALE HORMONAL SYSTEM

The female hormonal system, like that of the male, consists of three different hierarchies of hormones, as follows:

1. A hypothalamic-releasing hormone, *luteinizing hormone–releasing hormone* (LHRH).

2. The anterior pituitary hormones, *follicle-stimulating hormone* (FSH) and *luteinizing hormone* (LH), both of which are secreted in response to the releasing hormone from the hypothalamus.

3. The ovarian hormones, *estrogen* and *progesterone,* which are secreted by the ovaries in response to the two hormones from the anterior pituitary gland.

Figure 81–1. The female reproductive organs.

Labels: Ovary, Fallopian tube, Uterus, Vagina, Urinary bladder, Urethra, Uterus, Ovary, Vagina

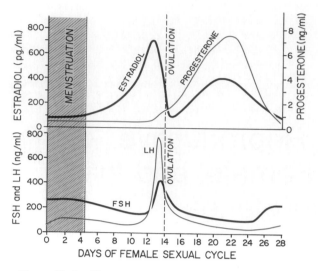

Figure 81–2. Plasma concentrations of the gonadotropins and ovarian hormones during the normal female sexual cycle.

The various hormones are not secreted in constant, steady amounts, but instead are secreted at drastically differing rates, during different parts of the female cycle. Figure 81–2 illustrates the changing concentrations of the anterior pituitary hormones, FSH and LH, and of the ovarian hormones, estradiol (estrogen) and progesterone. Although detailed measurements have not yet been made, it is reasonable to believe that the hypothalamic-releasing hormone LHRH also undergoes cyclic variations.

Before it is possible to discuss the interplay between these different hormones, it is first necessary to describe some of their specific functions and their relationships to function of the ovaries.

THE MONTHLY OVARIAN CYCLE AND FUNCTION OF THE GONADOTROPIC HORMONES

The normal reproductive years of the female are characterized by monthly rhythmic changes in the rates of secretion of the female hormones and corresponding changes in the ovaries and sexual organs as well. This rhythmic pattern is called the *female sexual cycle* (or less accurately, the *menstrual cycle*). The duration of the cycle averages 28 days. It may be as short as 20 days or as long as 45 days even in completely normal women, though abnormal cycle length is occasionally associated with decreased fertility.

The two significant results of the female sexual cycle are: First, only a *single* mature ovum is normally released from the ovaries each month so that only a single fetus can begin to grow at a time.

Second, the uterine endometrium is prepared for implantation of the fertilized ovum at the required time of the month.

The Gonadotropic Hormones

The ovarian changes during the sexual cycle depend completely on gonadotropic hormones secreted by the anterior pituitary gland. Ovaries that are not stimulated by gonadotropic hormones remain completely inactive, which is essentially the case throughout childhood when almost no gonadotropic hormones are secreted. However, at the age of about 8, the pituitary begins secreting progressively more gonadotropic hormones, which culminates in the initiation of monthly sexual cycles between the ages of 11 and 15; this culmination is called *puberty*.

(The ovaries also function during fetal life because of stimulation by gonadotropic hormones from the placenta. But within a few weeks to months after birth this stimulus is lost, and the ovaries become almost totally dormant until the prepubertal period.)

The anterior pituitary secretes two different hormones that are known to be essential for full function of the ovaries: (1) *follicle-stimulating hormone* (FSH), and (2) *luteinizing hormone* (LH). Both of these are small glycoproteins having molecular weights of about 30,000. The only significant effects of FSH and LH are on the ovaries in the female and the testes in the male.

During each month of the female sexual cycle, there is a cyclic increase and decrease of FSH and LH. These cyclic variations in turn cause the cyclic ovarian changes, which are explained in the following sections.

Both FSH and LH stimulate the ovarian cells by combining with highly specific cellular membrane FSH and LH receptors that in turn activate adenylcyclase. This then leads to an increase in cyclic AMP in the cells, causing growth and secretion by the specific ovarian cells. Thus, the mechanism of action of these hormones is typical of the usual cyclic AMP system for hormonal control, the details of which were explained in Chapter 75.

FOLLICULAR GROWTH — FUNCTION OF FOLLICLE-STIMULATING HORMONE (FSH)

Figure 81–3 depicts the various stages of follicular growth in the ovaries, illustrating, first, the primordial follicle (primary follicle). Throughout childhood the primordial follicles do not grow, but at puberty, when FSH and LH from the anterior pituitary gland begin to be secreted in large quantity, the entire ovaries and the follicles within them begin to grow. The first stage of follicular

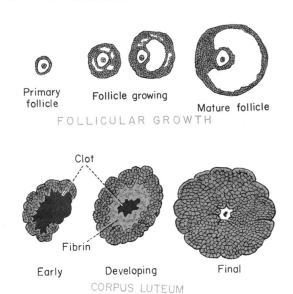

Figure 81–3. Stages of follicular growth in the ovary, showing also formation of the corpus luteum. (Modified from Arey: Developmental Anatomy, 7th Ed.)

growth is moderate enlargement of the ovum itself. This is followed by development of additional layers of granulosa cells around each ovum and development of several layers of theca cells around the granulosa cells. The theca cells originate from the stroma of the ovary and soon take on epithelioid characteristics; their total mass is called the *theca interna*. It is these cells that probably are destined to secrete most of the estrogens while the granulosa cells will be responsible mainly for secreting progesterone. Surrounding the theca interna, a connective tissue capsule known as the *theca externa* develops. This becomes the capsule of the developing follicle.

The Vesicular Follicles. At the beginning of each month of the female sexual cycle, at approximately the onset of menstruation, the concentrations of FSH and LH increase, the increase in FSH preceding that of LH by a few days. These increases cause accelerated growth mainly of the granulosa cells but to a lesser extent the theca cells as well in about 20 of the ovarian follicles each month. These cells also secrete a follicular fluid that contains a high concentration of estrogen, one of the important female sex hormones that will be discussed later. The accumulation of this fluid in the follicles causes an *antrum* to appear within the mass of granulosa cells, as illustrated in Figure 81–3.

After the antrum is formed, the granulosa and theca cells continue to proliferate, the rate of secretion accelerates, and each of the growing follicles becomes a *vesicular follicle*. This accelerating growth and increasing secretion is caused by two other factors in addition to the follicle-stimulating hormone. First, moderate amounts of luteinizing hormone are also secreted by the anterior pituitary gland; this hormone has a synergistic effect that supports the stimulatory effect of FSH. Second, the estrogen secreted into the follicle has a similar synergistic effect as well because it increases the number of FSH receptors on the granulosa cells.

As the vesicular follicle enlarges, the granulosa cells continue to develop and proliferate at one pole of the follicle. The ovum is located in this mass.

Atresia of All Follicles but One. After a week or more of growth — but before ovulation occurs — one of the follicles begins to outgrow all the others; the remainder begin to involute (a process called *atresia*), and these follicles are said to become *atretic*. The cause of the atresia is unknown, but it has been postulated to be the following: The one follicle that becomes more highly developed than the others also secretes more estrogen, and this positive feedback effect (a) enhances glomerulosa cell proliferation, and (b) enhances the number of FSH receptors on the glomerulosa cells, thus increasing the stimulatory effect also of FSH on this one follicle. However, the secreted estrogen acts on the hypothalamus to decrease the secretion of FSH and LH by the anterior pituitary gland, thus reducing the degree of stimulation of the less well-developed follicles. Therefore, the largest follicle continues to grow even without further increase in the gonadotropic hormones while all the other follicles stop growing and, indeed, involute.

This process of atresia obviously is important in that it allows only one of the follicles to grow large enough to ovulate. The single follicle reaches a size of approximately 1 to 1.5 centimeters at the time of ovulation.

Ovulation

Ovulation in a woman who has a normal 28-day female sexual cycle occurs 14 days after the onset of menstruation.

The process of ovulation has never been observed in the human being, but it has been observed and studied experimentally in rats and rabbits. Shortly before ovulation, the protruding outer wall of the follicle swells rapidly, and a small area in the center of the capsule, called the *stigma*, protrudes like a nipple. In another half hour or so, fluid begins to ooze from the follicle through the stigma. About two minutes later, as the follicle becomes smaller because of loss of fluid, the stigma ruptures widely, and a more viscous fluid that has occupied the central portion of the follicle is evaginated outward into the abdomen. This viscous fluid carries with it the ovum surrounded by several thousand small granulosa cells called the *corona radiata*.

Need for Luteinizing Hormone (LH) in Ovulation — Ovulatory Surge of LH. Luteinizing hormone is necessary for final follicular growth and ovulation. Without this hormone, even though large quantities of FSH are available the follicle will not progress to the stage of ovulation.

Approximately two days before ovulation, for reasons that are not completely known at present but which will be discussed in more detail later in the chapter, the rate of secretion of LH by the anterior pituitary gland increases markedly, rising six- to ten-fold and peaking about 18 hours before ovulation. FSH also increases about two-fold at the same time, and the two of these hormones act synergistically to cause rapid swelling of the follicle during the several days before ovulation. The LH also has the specific effect on the theca and granulosa cells of changing them into *lutein cells* that at first secrete less estrogen but progressively increasing amounts of progesterone. Therefore, the rate of secretion of estrogen begins to fall approximately one day prior to ovulation, while minute amounts of progesterone begin to be secreted.

It is in this environment of (a) very rapid growth of the follicle, (b) diminishing estrogen secretion after a prolonged phase of excessive estrogen secretion, and (c) beginning secretion of progesterone that ovulation occurs. Without the initial preovulatory surge of luteinizing hormone, ovulation will not take place.

Mechanism of Ovulation. Figure 81–4 illustrates the postulated mechanism of ovulation. It shows the initiating cause to be the large quantity of luteinizing hormone secreted by the anterior pituitary gland. The luteinizing hormone in turn causes rapid secretion of follicular steroid hormones containing a small amount of progesterone for the first time. Within a few hours two events occur, both of which are necessary for ovulation: (1) The theca externa (the capsule of the follicle) begins to form proteolytic enzymes that cause dissolution of the capsular wall and consequent weakening of the wall, resulting in further swelling of the entire follicle and degeneration at the stigma. (2) The luteinizing process of the theca and granulosa cells is accompanied by growth of new blood vessels into the follicle wall, and at the same time prostaglandins (local hormones that cause vasodilation) are secreted in the follicular tissues. These two effects in turn cause plasma transudation into the follicle, which also contributes to follicle swelling. Finally, the combined follicle swelling and simultaneous degeneration of the stigma causes follicle rupture with evagination of the ovum.

THE CORPUS LUTEUM— THE "LUTEAL" PHASE OF THE OVARIAN CYCLE

During the first few hours after expulsion of the ovum from the follicle, the remaining granulosa cells of the follicle undergo rapid physical and chemical change, a process called *luteinization,* and the mass of cells becomes a *corpus luteum.* Also incorporated in the corpus luteum are strands of theca cells. The *lutein cells* begin to secrete large amounts of progesterone and probably lesser amounts of estrogen, while the theca cells are believed to continue secreting estrogen. The lutein cells become greatly enlarged and develop lipid inclusions that give the cells a distinctive yellowish color, from which is derived the term *luteum.* A well-developed vascular supply also grows into the corpus luteum.

In the normal female, the corpus luteum grows to approximately 1.5 cm., reaching this stage of development approximately seven or eight days following ovulation. After this, it begins to involute and eventually loses its secretory function, as well as its lipid characteristics, approximately 12 days following ovulation, becoming then the so-called *corpus albicans;* during the ensuing few weeks this is replaced by connective tissue.

Luteinizing Function of Luteinizing Hormone (LH). The change of granulosa cells into lutein cells is mainly dependent on the LH secreted by the anterior pituitary gland. In fact, this function gave LH its name "luteinizing." However, lutein-

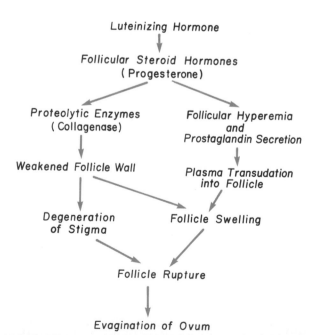

Luteinizing Hormone

Follicular Steroid Hormones
(Progesterone)

Proteolytic Enzymes
(Collagenase)

Follicular Hyperemia
and
Prostaglandin Secretion

Weakened Follicle Wall

Plasma Transudation
into Follicle

Degeneration
of Stigma

Follicle Swelling

Follicle Rupture

Evagination of Ovum

Figure 81–4. Postulated mechanism of ovulation. (Based primarily on the research studies of H. Lipner.)

ization of the granulosa cells also depends upon extrusion of the ovum from the follicle. A yet undiscovered local hormone that holds the luteinization process in check until after ovulation seems to be secreted by the ovum. Also for this reason, a corpus luteum does not develop in a follicle that does not ovulate.

Secretion by the Corpus Luteum: Function of LH. The corpus luteum is a highly secretory organ, secreting large amounts of both *progesterone* and *estrogen*. Once LH (mainly that which has been secreted during the ovulatory surge) has acted on the granulosa cells to cause luteinization, the newly formed lutein cells seem to be programmed to go through a preordained sequence of (a) proliferation, (b) enlargement, and (c) secretion, then to be followed by (d) degeneration. Even in the absence of further secretion of LH by the anterior pituitary gland, this process will still occur but it will last only 4 to 8 days. On the other hand, in the presence of LH the degree of growth of the corpus luteum is enhanced, its secretion is greater, and its life is extended. During this period of the ovarian cycle, the concentration of LH is very low; even so, this small amount of LH extends the life of the corpus luteum to a normal period of 12 days. We shall see in the discussion of pregnancy in the next chapter that another hormone that has almost exactly the same properties as luteinizing hormone, chorionic gonadotropin, which is secreted by the placenta, can also act on the corpus luteum to prolong its life — usually maintaining it for at least the first three to four months of pregnancy.

Termination of the Ovarian Cycle and Onset of the Next Cycle. During the luteal phase of the ovarian cycle, the large amounts of estrogen and progesterone secreted by the corpus luteum cause a feedback decrease in secretion of both FSH and LH by the anterior pituitary gland. Therefore, during this period no new follicles begin to grow in the ovary. However, when the corpus luteum degenerates completely at the end of 12 days of life (approximately also on the 26th day of the female sexual cycle), the lack of feedback suppression now allows the anterior pituitary gland to secrete several times as much FSH, followed a few days later by increasing quantities of LH as well. The FSH and LH initiate growth of new follicles to begin a new ovarian cycle. At the same time, the paucity of secretion of progesterone and estrogen leads to menstruation by the uterus, as will be explained later.

SUMMARY

Approximately each 28 days, gonadotropic hormones from the anterior pituitary gland cause new follicles to begin to grow in the ovaries. One of the follicles finally ovulates at the 14th day of the cycle. During growth of the follicles, estrogen is secreted.

Following ovulation, the secretory cells of the follicle develop into a corpus luteum that secretes large quantities of the female hormones progesterone and estrogen. After another two weeks the corpus luteum degenerates, whereupon the ovarian hormones estrogen and progesterone decrease greatly and menstruation begins. A new ovarian cycle then follows.

FUNCTIONS OF THE OVARIAN HORMONES — ESTROGENS AND PROGESTERONE

The two types of ovarian hormones are the *estrogens* and *progesterone*. The estrogens mainly promote proliferation and growth of specific cells in the body and are responsible for development of most secondary sexual characteristics of the female. On the other hand, progesterone is concerned almost entirely with final preparation of the uterus for pregnancy and the breasts for lactation.

CHEMISTRY OF THE SEX HORMONES

The Estrogens. In the normal, nonpregnant female, estrogens are secreted in major quantities only by the ovaries, though minute amounts are also secreted by the adrenal cortices. In pregnancy, tremendous quantities are also secreted by the placenta, indeed, up to 100 times the amount secreted by the ovaries during the normal monthly cycle.

At least six different natural estrogens have been isolated from the plasma of the human female, but only three are present in significant quantities, *β-estradiol, estrone,* and *estriol,* the formulas for which are illustrated in Figure 81–5. Both β-estradiol and estrone are present in large quantities in the venous blood from the ovaries, while estriol is an oxidative product derived from these first two, the conversion occurring mainly in the liver but also to some extent elsewhere in the body.

The estrogenic potency of β-estradiol is 12 times that of estrone and 80 times that of estriol. Considering these relative potencies, the total estrogenic effect of β-estradiol is usually many times that of the other two together. For this reason β-estradiol is considered to be the major estrogen, though the estrogenic effects of estrone are far from negligible.

Synthesis of the Estrogens. Note from the formulas of the estrogens in Figure 81–5 that all these are steroids. They are synthesized in the ovaries from cholesterol or acetyl coenzyme A, the acetate units of which can be conjugated to form the appropriate steroid nuclei. It is particularly interesting that progesterone as well as testosterone, the male sex hormone, are probably synthesized first, and then converted into the estrogens.

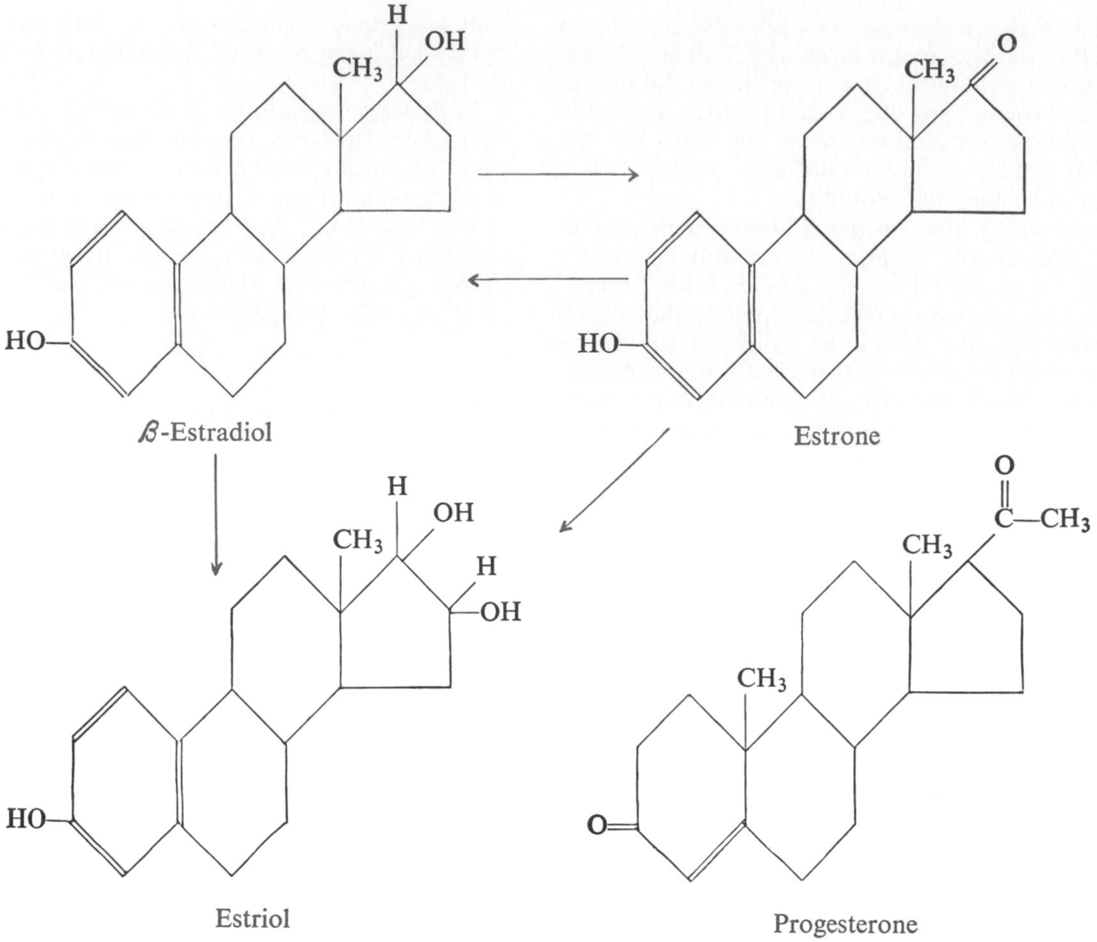

Figure 81–5. Chemical formulas of the principal female hormones.

Indeed, even normally about 1/15 as much testosterone is secreted by the ovaries as by the testes.

Fate of the Estrogens; Function of the Liver in Estrogen Degradation. The liver conjugates the estrogens to form glucuronides and sulfates, and about one-fifth of these conjugated products are excreted in the bile while most of the remainder are excreted in the urine. Also, the liver converts the potent estrogens, estradiol and estrone, into an almost totally impotent estrogen, estriol. Therefore, the liver plays an important role in removing those estrogens from the blood that do not enter other tissues to perform physiological functions. Indeed, diminished liver function actually *increases* the activity of estrogens in the body, sometimes causing *hyperestrinism.*

Progesterone. Almost all the progesterone in the nonpregnant female is secreted by the corpus luteum during the latter half of each ovarian cycle. However, the adrenal glands form a minute quantity of progesterone or compounds that have progesteronic activity, and during pregnancy progesterone is formed in extreme quantities by the placenta, about 10 times the normal monthly amount, especially after the fourth month of gestation.

Synthesis of Progesterone. Progesterone is a steroid having a molecular structure not far different from those of the other steroid hormones, the estrogens, testos-

terone, and the corticosteroids, as illustrated in Figure 81–5. Therefore, progesterone has some functions in common with all of these.

Progesterone is probably synthesized principally from acetyl coenzyme A. However, it can also be formed from cholesterol.

At least two other "progestin" hormones are secreted by the ovaries, but the quantities of these are so slight in comparison with that of progesterone that it is usually proper to consider progesterone as the single important progestin. However, at the time of ovulation, before full development of the corpus luteum, one of these, *17 α-hydroxyprogesterone,* is sometimes secreted in higher concentration than progesterone. The significance of this not yet known.

Fate of Progesterone. Progesterone is secreted in far greater quantities than the estrogens by the ovaries, but its potency per unit weight is much less than that of the estrogens. Within a few minutes after secretion, almost all the progesterone is degraded to other steroids that have no progesteronic effect. Here, as is also true with the estrogens, the liver is especially important for this metabolic degradation.

The major end-product of progesterone degradation is *pregnanediol.* Approximately 10 per cent of the original progesterone is excreted in the urine in this form. One can estimate the rate of progesterone formation in the body from the rate of this excretion, but because pregnanediol exerts no progesteronic effects it can be detected in the urine only by chemical means.

Transport of Estrogens and Progesterone in the Blood. Most estrogens and progesterone are transported in the blood loosely bound with the plasma albumin, but small amounts are also bound with specific binding globulins. Because of the loose bondage with the albumin, these hormones are readily released to the tissues.

FUNCTIONS OF THE ESTROGENS — EFFECTS ON THE PRIMARY AND SECONDARY SEXUAL CHARACTERISTICS

The principal function of the estrogens is to cause cellular proliferation and growth of the tissues of the sexual organs and of other tissues related to reproduction.

Effect on the Sexual Organs. During childhood, estrogens are secreted only in small quantities, but following puberty the quantity of estrogens secreted under the influence of the pituitary gonadotropic hormones increases some 20-fold or more. At this time the female sexual organs change from those of a child to those of an adult. The fallopian tubes, uterus, and vagina all increase in size. Also, the external genitalia enlarge, with deposition of fat in the mons pubis and labia majora and with enlargement of the labia minora.

In addition, estrogens change the vaginal epithelium from a cuboidal into a stratified type, which is considerably more resistant to trauma and infection than is the prepubertal epithelium. Infections in children, such as gonorrheal vaginitis, can actually be cured by administration of estrogens simply because of the resulting increased resistance of the vaginal epithelium.

During the few years following puberty, the size of the uterus increases two- to three-fold. More important, however, than increases in size are the changes that take place in the endometrium under the influence of estrogens, for estrogens cause marked proliferation of the endometrial stroma and also greatly increased development of the glands that will later be used to aid in nutrition of the implanting ovum. These effects are discussed later in the chapter in connection with the endometrial cycle.

Effect on the Fallopian Tubes. The estrogens have an effect on the mucosal lining of the fallopian tubes similar to that on the uterine endometrium: They cause the glandular tissues to proliferate; and, especially important, they cause the number of ciliated epithelial cells that line the fallopian tubes to increase. Also, the activity of the cilia is considerably enhanced, these always

beating toward the uterus. This helps to propel the fertilized ovum toward the uterus.

Effect on the Breasts. The primordial breasts of both female and male are exactly alike, and under the influence of appropriate hormones, the masculine breast, at least during the first two decades of life, can develop sufficiently to produce milk in the same manner as the female breast.

Estrogens cause fat deposition in the breasts, development of the stromal tissues of the breasts, and growth of an extensive ductile system. The lobules and alveoli of the breast develop to a slight extent, but it is progesterone and prolactin that cause the determinative growth and function of these structures. In summary, the estrogens initiate growth of the breasts and the breasts' milk-producing apparatus, and they are also responsible for the characteristic external appearance of the mature female breast, but they do not complete the job of converting the breasts into milk-producing organs.

Effect on the Skeleton. Estrogens cause increased osteoblastic activity. Therefore, at puberty, when the female enters her reproductive years, her growth rate becomes rapid for several years. However, estrogens have another potent effect on skeletal growth — that is, they cause early uniting of the epiphyses with the shafts of the long bones. This effect is much stronger in the female than is the similar effect of testosterone in the male. As a result, growth of the female usually ceases several years earlier than growth of the male. The female eunuch who is completely devoid of estrogen production usually grows several inches taller than the normal mature female because her epiphyses do not unite early.

Effect on Calcium and Phosphate Retention. As pointed out in the preceding chapter, testosterone causes greater than normal retention of calcium and phosphate. This effect is also true of the estrogens, but to a much less extent, because estrogens, like testosterone, promote growth of bones, entailing the deposition of increased amounts of bone matrix with subsequent retention of both calcium and phosphate.

Effect on Protein Deposition. Estrogens cause a slight increase in total body protein, which is evidenced by a slight positive nitrogen balance when estrogens are administered. This probably results from the growth-promoting effect of estrogen on the sexual organs and on a few other tissues of the body. The enhanced protein deposition caused by testosterone is much more general and many times as powerful as that caused by estrogens.

Effect on Metabolism and Fat Deposition. Estrogens increase the metabolic rate slightly but only about a third as much as the male sex hormone testosterone. They also cause deposition of increased quantities of fat in the subcutaneous tissues. As a result, the overall specific gravity of the female body, as judged by flotation in water, is considerably less than that of the male body which contains more protein and less fat. In addition to deposition of fat in the breasts and subcutaneous tissues, estrogens cause especially marked deposition of fat in the buttocks and thighs, causing the broadening of the hips that is characteristic of the feminine figure.

Effect on Hair Distribution.   Estrogens do not greatly affect hair distribution. However, hair develops in the pubic region and in the axillae after puberty. Probably androgens formed by the adrenal glands are mainly responsible for this.

Effect on the Skin.   Estrogens cause the skin to develop a texture that is soft and usually smooth but nevertheless thicker than that of the child or the female castrate. Also, estrogens cause the skin to become more vascular than normal; this effect is often associated with increased warmth of the skin, and it often results in greater bleeding of cut surfaces than is observed in men.

The adrenal androgens, which are secreted in increased quantities after puberty, cause increased secretion by the axillary sweat glands and also often cause acne.

Effect on Electrolyte Balance.   The similarity of estrogenic hormones to adrenocortical hormones has been pointed out; and estrogens, like adrenocortical hormones, cause sodium and water retention by the kidney tubules. However, this effect of estrogens is slight and rarely of significance except in pregnancy, as discussed in the following chapter.

Intracellular Functions of Estrogens.   Thus far we have discussed the gross effects of estrogens on the body. The cellular mechanism behind these effects are: Estrogens circulate in the blood for only a few minutes before they are delivered to the target cells. On entry into these cells, they combine within 10 to 15 seconds with a "receptor" protein in the cytoplasm and then, in combination with this protein, migrate to the nucleus. This in turn interacts with specific portions of the chromosomal DNA, and immediately initiates the process of transcription; therefore, RNA begins to be produced within a few minutes. In addition, over many hours new DNA also is produced, resulting eventually in division of the cell. The RNA diffuses to the cytoplasm, where it causes greatly increased protein formation and subsequently altered cellular function.

One of the principal differences between the protein anabolic effect of the estrogens and that of testosterone is that estrogen causes its effect almost exclusively in certain target organs, such as the uterus, the breasts, the skeleton, and certain

fatty areas of the body, while testosterone has a more generalized effect throughout the body.

FUNCTIONS OF PROGESTERONE

Effect on the Uterus.   By far the most important function of progesterone is *to promote secretory changes in the endometrium,* thus preparing the uterus for implantation of the fertilized ovum. This function is discussed later in connection with the endometrial cycle of the uterus.

In addition to this effect on the endometrium, progesterone decreases the frequency of uterine contractions, thereby helping to prevent expulsion of the implanted ovum, an effect discussed in the following chapter.

Effect on the Fallopian Tubes.   Progesterone also promotes secretory changes in the mucosal lining of the fallopian tubes. These secretions are necessary for nutrition of the fertilized, dividing ovum as it traverses the fallopian tube prior to implantation.

Effect on the Breasts.   Progesterone promotes development of the lobules and alveoli of the breasts, causing the alveolar cells to proliferate, to enlarge, and to become secretory in nature. However, progesterone does not cause the alveoli actually to secrete milk, for, as discussed in the following chapter, milk is secreted only after the prepared breast is further stimulated by prolactin from the anterior pituitary.

Progesterone also causes the breasts to swell. Part of this swelling is due to the secretory development in the lobules and alveoli, but part also seems to result from increased fluid in the subcutaneous tissue itself.

Effect on Electrolyte Balance.   Progesterone in very large quantity, like estrogens, testosterone, and adrenocortical hormones, can enhance sodium, chloride, and water reabsorption from the distal tubules of the kidney. Yet, strangely enough, progesterone more often causes increased sodium and water excretion. The cause of this is competition between progesterone and aldosterone, which probably occurs as follows: It is believed that these two substances combine with the same receptor proteins in the epithelial cells of the tubules. When progesterone combines with these, aldosterone cannot combine. Yet progesterone exerts many hundred times less sodium transport effect than does aldosterone. Therefore, despite the fact that under appropriate conditions progesterone can weakly promote sodium and water retention by the renal tubules, it blocks the far more potent effect of aldosterone, thus resulting in net loss of sodium and water from the body.

Protein Catabolic Effect.   Progesterone exerts a mild catabolic effect on the body's protein, similar to that of the glucocorticoids. Though in the normal sexual cycle this effect is probably not significant, it possibly is significant during pregnancy, when proteins must be mobilized for use by the fetus.

THE ENDOMETRIAL CYCLE AND MENSTRUATION

Associated with the cyclic production of estrogens and progesterone by the ovaries is an endometrial cycle operating through the following stages: first, proliferation of the uterine endometrium; second, secretory changes in the endometrium; and third, desquamation of the endometrium, which is known as *menstruation*. The various phases of the endometrial cycle are illustrated in Figure 81–6.

Proliferative Phase (Estrogen Phase) of the Endometrial Cycle. At the beginning of each monthly sexual cycle, most of the endometrium is desquamated by the process of menstruation. After menstruation, only a thin layer of endometrial stroma remains at the base of the original endometrium, and the only epithelial cells that are left are those located in the remaining deep portions of the glands and crypts of the endometrium. *Under the influence of estrogens,* secreted in increasing quantities by the ovary during the first part of the ovarian cycle, the stromal cells and the epithelial cells proliferate rapidly. The endometrial surface is re-epithelialized within three to seven days after the beginning of menstruation. For the first two weeks of the sexual cycle — that is, until ovulation — the endometrium increases greatly in thickness, owing to increasing numbers of stromal cells and to progressive growth of the endometrial glands and blood vessels into the endometrium, all of which effects are promoted by the estrogens. At the time of ovulation the endometrium is approximately 2 to 3 mm. thick.

Secretory Phase (Progestational Phase) of the Endometrial Cycle. During the latter half of the sexual cycle, progesterone as well as estrogen is secreted in large quantity by the corpus luteum. The estrogens cause only slight additional cellular proliferation in the endometrium during this phase of the endometrial cycle, but progesterone causes marked swelling and secretory development of the endometrium. The glands increase in tortuosity, secretory substances accumulate in the glandular epithelial cells, and the glands secrete small quantities of endometrial fluid. Also, the cytoplasm of the stromal cells increases, lipid and glycogen deposits increase greatly in the stromal cells, and the blood supply to the endometrium further increases in proportion to the developing secretory activity, the blood vessels becoming highly tortuous. The thickness of the endometrium approximately doubles during the secretory phase so that toward the end of the monthly cycle the endometrium has a thickness of 4 to 6 mm.

The whole purpose of all these endometrial changes is to produce a highly secretory endometrium containing large amounts of stored nutrients that can provide appropriate conditions for implantation of a fertilized ovum during the latter half of the monthly cycle. From the time fertilization first takes place until the ovum implants, the fallopian and uterine secretions, called "uterine milk," provide nutrition for the early dividing ovum. Then, once the ovum implants in the endometrium, the trophoblastic cells on the surface of the blastocyst begin to digest the endometrium and to absorb the substances digested, thus making still far greater quantities of nutrients available to the early embryo.

Menstruation. Approximately two days before the end of the monthly cycle, the ovarian hormones, estrogens and progesterone, decrease sharply to low levels of secretion, as was illustrated in Figure 81–2, and menstruation follows.

Menstruation is caused by the sudden reduction in both progesterone and estrogens at the end of the monthly ovarian cycle. The first effect is decreased stimulation of the endometrial cells by these two hormones, followed rapidly by involution of the endometrium itself to about 65 per cent of its previous thickness. During the 24 hours preceding the onset of menstruation, the tortuous blood vessels leading to the mucosal layers of the endometrium become vasospastic, presumably because of some effect of the involution, such as release of a vasoconstrictor material. The vasospasm and loss of hormonal stimulation cause beginning necrosis in the endometrium, especially of blood vessels in the stratum vasculare. As a result, blood seeps into the vascular layer of the endometrium, and the hemorrhagic areas grow over a period of 24 to 36 hours. Gradually, the necrotic outer layers of the endometrium separate from the uterus at the site of the hemorrhages, until, at approximately 48 hours following the onset of menstruation, all the superficial layers of the endometrium have desquamated. The desquamated tissue and blood in the uterine cavity initiate

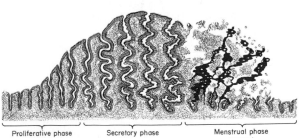

Proliferative phase (11 days)   Secretory phase (12 days)   Menstrual phase (5 days)

Figure 81–6. Phases of endometrial growth and menstruation during each monthly female sexual cycle.

uterine contractions that expel the uterine contents.

During normal menstruation, approximately 35 ml. of blood and an additional 35 ml. of serous fluid are lost. This menstrual fluid is normally nonclotting, because a *fibrinolysin* is released along with the necrotic endometrial material. However, if excessive bleeding occurs from the uterine surface, the quantity of fibrinolysin may not be sufficient to prevent clotting. The presence of clots during menstruation ordinarily is clinical evidence of uterine pathology.

Within three to seven days after menstruation starts, the loss of blood ceases, for by this time the endometrium has become completely re-epithelialized.

Leukorrhea During Menstruation. During menstruation tremendous numbers of leukocytes are released along with the necrotic material and blood. It is probable that some substance liberated by the endometrial necrosis causes this outflow of leukocytes. As a result of these many leukocytes and maybe still other factors, the uterus is resistant to infection during menstruation even though the endometrial surfaces are denuded. Obviously, this is of extreme protective value.

REGULATION OF THE FEMALE MONTHLY RHYTHM — INTERPLAY BETWEEN THE OVARIAN AND HYPOTHALAMIC-PITUITARY HORMONES

Now that we have presented the major cyclic changes that occur during the female sexual cycle, we can attempt to explain the basic rhythmic mechanism that causes these cyclic variations.

Function of the Hypothalamus in the Regulation of Gonadotropin Secretion — Luteinizing Hormone–Releasing Hormone (LHRH). As was pointed out in Chapter 75, secretion of most of the anterior pituitary hormones is controlled by releasing hormones formed in the hypothalamus and transported to the anterior pituitary gland by way of the hypothalamic-hypophysial portal system. In the case of the gonadotropins, at least one releasing factor, *luteinizing hormone–releasing hormone (LHRH)* is important. This has been purified and has been found to be a decapeptide having the following formula:

GLU-HIS-TRP-SER-TYR-GLY-LEU-ARG-
PRO-GLY-NH$_2$

Though some research workers believe that another substance similar to this is follicle-stimulating hormone–releasing hormone (FSHRH), it

has been found that the above purified LHRH causes release not only of luteinizing hormone but also of follicle-stimulating hormone. Therefore, since there is reason to believe that this decapeptide is in reality both LHRH and FSHRH combined in the same molecule, it is sometimes called simply *gonadotropin-releasing hormone* (GnRH).

Hypothalamic Centers for Stimulating Release of LHRH. Several different areas that profoundly influence the rate of secretion of the hypothalamic gonadotropin-releasing hormone LHRH have been found in the hypothalamus. In monkeys — and presumably in people — the midbasal region of the hypothalamus is the area most importantly involved. In lower animals, the area around the infundibulum causes a continuous tonic secretion of gonadotropin-releasing hormone, while two other areas modulate the rate of release. These areas are (1) a center in the preoptic area that causes cyclic variation in the secretory rate and (2) a center in the posterior hypothalamus that allows the psychic attitude of the animal to enhance or decrease the secretion of gonadotropin-releasing hormone.

Effect of Psychic Factors on the Female Sexual Cycle. It is well known that the young woman on first leaving home to go to college almost as often as not experiences disruption or irregularity of the female sexual cycle. Likewise, serious stresses of almost any type can interfere with the cycle. Finally, in many lower animals no ovulation occurs at all until after copulation; the sexual excitation attendant to the sexual act initiates a sequence of events that leads first to secretion in the hypothalamus of LHRH, then to secretion of the anterior pituitary gonadotropins, and finally to ovarian secretion of the female sex hormones and to ovulation. It is these effects that are believed to be mediated through the posterior hypothalamic center for modulating the output of LHRH.

Negative Feedback Effect of Estrogen and Progesterone on Secretion of Follicle-Stimulating Hormone and Luteinizing Hormone. Estrogen in small amounts and progesterone in large amounts inhibit the production of FSH and LH. Both of these feedback effects seem to operate mainly by the actions of these hormones on the hypothalamus, though it is possible that they have very slight direct feedback actions on the anterior pituitary gland as well.

Positive Feedback Effect of Estrogen Before Ovulation — the Preovulatory Luteinizing Hormone Surge. For reasons not completely understood, the anterior pituitary gland secretes greatly increased amounts of LH on the day immediately before ovulation. This effect is illustrated in Figure 81–2, and the figure shows a much smaller preovulatory surge of FSH as well.

Experiments have shown that infusion of estrogen into a female for a period of two to three days during the first half of the ovarian cycle will cause rapidly accelerating growth of the follicles and also rapidly accelerating secretion of ovarian estrogens. During this period the secretion of both the follicle-stimulating hormone and luteinizing hormone by the anterior pituitary gland is at first suppressed slightly. Then abruptly the secretion of luteinizing hormone increases about eight-fold, and the secretion of follicle-stimulating hormone increases about two-fold. This abrupt increase in secretion of the gonadotropins seems to result from a *positive feedback effect* of the estrogen in place of the normal negative feedback that occurs during the remainder of the female sexual cycle. The precise cause of the positive feedback is not known, but nevertheless it is an absolutely necessary and integral part of the control mechanism. Without the normal preovulatory surge of luteinizing hormone, ovulation will never occur.

FEEDBACK OSCILLATION OF THE HYPOTHALAMIC-PITUITARY-OVARIAN SYSTEM

Now, after discussing much of the known information about the interrelationships of the different components of the female hormonal system, we can digress from the area of proven fact into the realm of speculation and attempt to explain the feedback oscillation that controls the rhythm of the female sexual cycle. It seems to operate in approximately the following sequence of three successive events:

1. The Postovulatory Secretion of the Ovarian Hormones and Depression of Gonadotropins. The easiest part of the cycle to explain is the events that occur during the postovulatory phase — between ovulation and the beginning of menstruation. During this time the corpus luteum secretes very large quantities of both progesterone and estrogen. The combined effect of the estrogen and the progesterone on the hypothalamus is to inhibit the secretion of LHRH and therefore to cause strong negative feedback depression of secretion of the gonadotropins, both FSH and LH, during this period of time. These effects are illustrated in Figure 81–2.

2. The Follicular Growth Phase. A few days before menstruation, the corpus luteum involutes, and the secretion of both estrogen and progesterone decreases to a low ebb. This releases the hypothalamus from the feedback effect of the estrogen and progesterone so that LHRH increases again, followed in succession by a several hundred per cent increase also of FSH and LH. These hormones initiate new follicular growth and progressive increase in the secretion of estrogen, reaching a peak of estrogen secretion at about 12.5

to 13 days after the onset of menstruation. During the first 11 to 12 days of this follicular growth the rates of secretion of the gonadotropins FSH and LH decrease; then comes a sudden increase in secretion of both of these hormones, leading to the next stage of the cycle.

3. Preovulatory Surge of LH and FSH; Ovulation. At approximately 11.5 to 12 days after the onset of menstruation, the decline in secretion of FSH and LH comes to an abrupt halt. It is believed that the high level of estrogens at this time causes a positive feedback effect, as explained earlier, which leads to a terrific surge of secretion — especially of LH and to a lesser extent of FSH. This effect may be related to the fact that the follicular secretory cells are becoming exhausted so that their rate of secretion of estrogens had already begun to fall about one day prior to the LH surge. Whatever the cause of this preovulatory LH and FSH surge, the LH leads to both ovulation and formation of the corpus luteum. Thus, the hormonal system begins a new round of the female sexual cycle.

Anovulatory Cycles — the Sexual Cycle at Puberty

If the preovulatory surge of luteinizing hormone is not of sufficient magnitude, ovulation will not occur, and the cycle is then said to be "anovulatory." Most of the cyclic variations of the sexual cycle continue, but they are altered in the following ways: First, lack of ovulation causes failure of development of the corpus luteum and consequently there is only slight secretion of progesterone during the latter portion of the cycle. Second, the cycle is shortened by several days, but the rhythm continues. Therefore, it is likely that progesterone is not required for maintenance of the cycle itself though it can alter its rhythm.

Anovulatory cycles are usual during the first few cycles following puberty and for several years prior to menopause, presumably because the LH surge is not potent enough at these times to cause ovulation.

PUBERTY AND MENARCHE

Puberty means the onset of adult sexual life, and menarche means the onset of menstruation. As pointed out earlier in the chapter, the period of puberty is caused by a gradual increase in gonadotropic hormone secretion by the pituitary, beginning approximately the eighth year of life, as illustrated in Figure 81–7, and usually culminating in the onset of menstruation between the ages of 11 and 16.

In the female, as in the male, the infantile pituitary gland and ovaries are capable of full

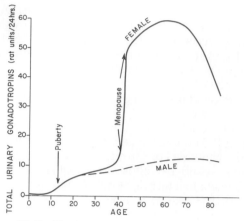

Figure 81–7. Total rates of secretion of gonadotropic hormones throughout the sexual lives of females and males, showing an especially abrupt increase in gonadotropic hormones at the menopause in the female.

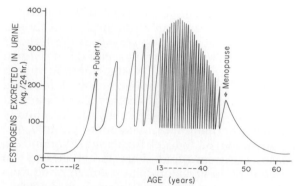

Figure 81–8. Estrogen secretion throughout sexual life.

function if appropriately stimulated. However, the hypothalamus is extremely sensitive to the inhibitory effects of estrogens, which keeps its stimulation of the pituitary almost completely suppressed throughout childhood. Then, at puberty, for reasons not understood, the hypothalamus matures; its excessive sensitivity to the negative feedback inhibition becomes greatly diminished, which allows enhanced production of gonadotropins and the onset of adult female sexual life.

Figure 81–8 illustrates (a) the increasing levels of estrogen secretion at puberty, (b) the cyclic variation during the monthly sexual cycles, (c) the further increase in estrogen secretion during the first few years of sexual life, (d) then progressive decrease in estrogen secretion toward the end of sexual life, and (e) finally almost no estrogen secretion beyond the menopause.

THE MENOPAUSE

At an average age of approximately 40 to 50 years the sexual cycles usually become irregular, and ovulation fails to occur during many of these cycles. After a few months to a few years, the cycles cease altogether, as illustrated in Figure 81–8. This period during which the cycles cease and the female sex hormones diminish rapidly to almost none at all is called the *menopause*.

The cause of the menopause is "burning out" of the ovaries. Throughout a woman's sexual life about 450 of the primordial follicles grow into vesicular follicles and ovulate, while literally thousands of the ova degenerate. At the age of about 45 only a few primordial follicles still remain to be stimulated by FSH and LH, and the production of estrogens by the ovary decreases as the number of primordial follicles approaches zero (also illustrated in Figure 81–8). When estrogen production falls

below a critical value, the estrogens can no longer inhibit the production of FSH and LH; nor can they cause an ovulatory surge of LH and FSH to cause oscillatory cycles. Instead, as illustrated in Figure 81–7, FSH and LH (mainly FSH) are produced thereafter in large and continuous quantities. Estrogens are produced in subcritical quantities for a short time after the menopause, but over a few years, as the final remaining primordial follicles become atretic, the production of estrogens by the ovaries falls almost to zero.

The Female Climacteric. The term "female climacteric" means the entire time, lasting from several months to several years, during which the sexual cycles become irregular and gradually stop. In this period the woman must readjust her life from one that has been physiologically stimulated by estrogen and progesterone production to one devoid of these feminizing hormones. The secretion of estrogens decreases rapidly, and essentially no progesterone is secreted after the last ovulatory cycle. The loss of the estrogens often causes marked physiological changes in the function of the body, including (1) "hot flashes" characterized by extreme flushing of the skin, (2) psychic sensations of dyspnea, (3) irritability, (4) fatigue, (5) anxiety, and (6) occasionally various psychotic states. These symptoms are of sufficient magnitude in approximately 15 per cent of women to warrant treatment. If psychotherapy fails, daily administration of an estrogen in small quantities will reverse the symptoms, and, by gradually decreasing the dose, the postmenopausal woman is likely to avoid severe symptoms; unfortunately, such treatment prolongs the symptoms.

INTERRELATIONSHIPS OF THE OVARIES WITH OTHER GLANDS

Relationship of the Ovaries to the Adrenal Glands. The adrenal glands normally secrete small quantities of both estrogen and progesterone, though these quantities are usually too small to exert major effects on the body. However, rare tumors of the adrenal gland secrete specifically increased quantities of female hormones and therefore cause feminizing characteristics.

Injection of estrogens causes the adrenal cortices to hypertrophy, an effect that is especially prominent during pregnancy because of large amounts of estrogens formed by the placenta. The effect is mediated through the pituitary, for estrogens, while inhibiting production of follicle-stimulating hormone, increase the secretion of ACTH.

Antagonistic Effects of Estrogen and Testosterone. Estrogens and testosterone exert opposite effects on the sexual organs and on many secondary sexual tissues such as the breasts and the prostate gland. These antagonistic effects are in part mediated through the hypothalamic-pituitary system, because both estrogens and testosterone are capable of decreasing the production of follicle-stimulating hormone and luteinizing hormone, which in turn abrogates secretion of the normal gonadal hormones. In general, in order to antagonize the effect of estrogens on the breast, approximately 50 times as much testosterone as β-estradiol must be administered.

ABNORMALITIES OF SECRETION BY THE OVARIES

Hypogonadism. Less than normal secretion by the ovaries can result from poorly formed ovaries or lack of ovaries. When ovaries are absent from birth or when they become nonfunctional before puberty, *female eunuchism* occurs. In this condition the usual secondary sexual characteristics do not appear, and the sexual organs remain infantile. Especially characteristic of this condition is excessive growth of the long bones because the epiphyses do not unite with the shafts of these bones at as early an age as in the normal adolescent woman. Consequently, the female eunuch is essentially as tall as, or perhaps even slightly taller than, her male counterpart of similar genetic background.

When the ovaries of a fully developed woman are removed, the sexual organs regress to some extent so that the uterus becomes almost infantile in size, the vagina becomes smaller, and the vaginal epithelium becomes thin and easily damaged. The breasts atrophy and become pendulous, and the pubic hair becomes considerably thinner. These same changes occur in women after the menopause.

Irregularity of Menses and Amenorrhea Due to Hypogonadism. As pointed out in the preceding discussion of the menopause, the quantity of estrogens produced by the ovaries must rise above a critical value if they are to be able to cause rhythmical sexual cycles. Consequently, in hypogonadism or when the gonads are secreting small quantities of estrogens as a result of other factors, such as *hypothyroidism,* the ovarian cycle likely will not occur normally. Instead, several months may elapse between menstrual periods, or menstruation may cease altogether (amenorrhea). Characteristically, prolonged ovarian cycles are frequently associated with failure of ovulation, presumably due to insufficient secretion of luteinizing hormone, which is necessary for ovulation.

Hypersecretion by the Ovaries. Extreme hypersecretion of ovarian hormones by the ovaries is a rare clinical entity, for excessive secretion of estrogens automatically decreases the production of gonadotropins by the pituitary, and this in turn limits the production of the ovarian hormones. Consequently, hypersecretion of feminizing hormones is recognized clinically only when a feminizing tumor develops.

A rare granulosa-theca cell tumor occasionally develops in an ovary, occurring more often after menopause than before. These tumors secrete large quantities of estrogens which exert the usual estrogenic effects, including hypertrophy of the uterine endometrium and irregular bleeding from this endometrium. In fact, bleeding is often the first indication that such a tumor exists.

Endometriosis. Endometriosis is the development and growth of endometrium in the peritoneal cavity, this growth usually occurring in the pelvis, closely associated with the sexual organs. There are several theories for explaining the origin of intra-abdominal endometrial tissue that causes endometriosis. Some believe that contraction of the uterus during menstruation occasionally expels viable endometrium backward through the fallopian tubes into the abdominal cavity, and that this endometrial tissue then implants on the peritoneum. In support of this is the fact that much of the endometrium sloughed during menstruation is still viable and will grow easily in tissue culture. Others believe that the endometrial tissue spreads by the lymphatics.

During each ovarian cycle the endometrium in the peritoneal cavity proliferates, secretes, and may even desquamate in the same manner that the intrauterine endometrium does. However, when desquamation occurs within the peritoneal cavity, the tissue and the hemorrhaging blood cannot be expelled to the exterior. Consequently, the quantity of endometrial tissue in the peritoneal cavity often increases progressively.

The presence of necrotic and hemorrhagic material in the abdominal cavity, and also the swelling of the endometrial tissue during each ovarian cycle, can cause considerable irritation of the peritoneum, sometimes producing severe abdominal pain. Also, fibrosis occurs in the areas of endometriosis, thereby promoting adhesions from one sexual organ to another or even causing intestinal adhesions and sometimes intestinal obstruction. Endometriosis is one of the prevalent causes of female infertility, owing especially to fibrotic immobilization of the fallopian tubes.

THE FEMALE SEXUAL ACT

Stimulation of the Female Sexual Act. As is true in the male sexual act, successful performance of the female sexual act depends on both psychic stimulation and local sexual stimulation.

The psychic factors that constitute "sex drive" in women are difficult to assess. The sex hormones, and the adrenocortical hormones as well, seem to exert a direct influence on the woman to create such a sex drive, but, on the other hand, the growing female child in modern society is often taught that sex is something to be hidden and that it is immoral. As a result of this training, much of the natural sex drive is inhibited, and whether the woman will have little or no sex drive ("frigidity") or will be more highly sexed probably depends partly on a balance between natural factors and previous training.

Local sexual stimulation in women occurs in more or less the same manner as in men, for massage, irritation, or other types of stimulation of the perineal region, sexual organs, and urinary tract create sexual sensations. The *clitoris* is especially sensitive for initiating sexual sensations. As in the male, the sexual sensory signals are mediated to the sacral segments of the spinal cord through the pudendal nerve and sacral plexus. Once these signals have entered the spinal cord, they are transmitted thence to the cerebrum. Also, local reflexes that are at least partly responsible for the female orgasm are integrated in the sacral and lumbar spinal cord.

Female Erection and Lubrication. Located around the introitus and extending into the clitoris is erectile tissue almost identical with the erectile tissue of the penis. This erectile tissue, like that of the penis, is controlled by the parasympathetic nerves that pass through the nervi erigentes from the sacral plexus to the external genitalia. In the early phases of sexual stimulation, the parasympathetics dilate the arteries and constrict the veins of the erectile tissues, and this allows rapid accumulation of blood in the erectile tissue so that the introitus tightens around the penis; this aids the male greatly in his attainment of sufficient sexual stimulation for ejaculation to occur.

Parasympathetic impulses also pass to the bilateral Bartholin's glands located beneath the labia minora to cause secretion of mucus immediately inside the introitus. This mucus is responsible for much of the lubrication during sexual intercourse, though much is also provided by other vaginal secretions as well and a small amount from the male urethral glands. The lubrication in turn is necessary for establishing during intercourse a satisfactory massaging sensation rather than an irritative sensation, which may be provoked by a dry vagina. A massaging sensation constitutes the optimal type of sensation for evoking the appropriate reflexes that culminate in both the male and female climaxes.

The Female Orgasm. When local sexual stimulation reaches maximum intensity, and especially when the local sensations are supported by appropriate psychic conditioning signals from the cerebrum, reflexes are initiated that cause the female orgasm, also called the *female climax*. The female orgasm is analogous to ejaculation in the male, and it perhaps helps promote fertilization of the ovum. Indeed, the human female is known to be somewhat more fertile when inseminated by normal sexual intercourse rather than by artificial methods, thus indicating an important function of the female orgasm. Possible effects that could result in this are:

First, during the orgasm the perineal muscles of the female contract rhythmically, which results from spinal cord reflexes similar to those that cause ejaculation in the male. It is possible, also, that these same reflexes increase uterine and fallopian tube motility during the orgasm, thus helping to transport the sperm toward the ovum, but the information on this subject is scanty.

Second, in many lower animals, copulation causes the posterior pituitary gland to secrete oxytocin; this effect is probably mediated through the amygdaloid nuclei and then through the hypothalamus to the pituitary. The oxytocin in turn causes increased contractility of the uterus, which also is believed to cause rapid transport of the sperm. Sperm have been shown to traverse the entire length of the fallopian tube in the cow in approximately five minutes, a rate at least 10 times as fast as that which the sperm themselves could achieve. Whether or not this occurs in the human female is unknown.

In addition to the possible effects of the orgasm on fertilization, the intense sexual sensations that develop during the orgasm also pass to the cerebrum and cause intense muscle tension throughout the body. But after culmination of the sexual act, this gives way during the succeeding minutes to a sense of satisfaction characterized by relaxed peacefulness, an effect called *resolution*.

FEMALE FERTILITY

The Fertile Period of Each Sexual Cycle. The ovum remains viable and capable of being fertilized after it is expelled from the ovary probably no longer than 24 hours. Therefore, sperm must be available soon after ovulation if fertilization is to take place. On the other hand, a few sperm can remain fertile in the female reproductive tract for up to 72 hours, though most of them for not more than 24 hours. Therefore, for fertilization to take place, intercourse usually must occur some time between one day prior to ovulation up to one day after ovulation. Thus, the period of female fertility during each sexual cycle is short.

The Rhythm Method of Contraception. One of the often practiced methods of contraception is to avoid intercourse near the time of ovulation. The difficulty with this method of contraception is the impossibility of predicting the exact time of ovulation. Yet the interval from ovulation until the next succeeding onset of menstruation is almost always between 13 and 15 days. Therefore, if the menstrual cycle is regular, with a periodicity of 28 days, ovulation usually occurs within one day of the 14th day of the cycle. If, on the other hand, the periodicity of the cycle is 40 days, ovulation usually occurs within one day of the 26th day of the cycle. Finally, if the periodicity of the cycle is 21 days, ovulation usually occurs within one day of the 7th day of the cycle. Therefore, it is usually stated that avoidance of intercourse for 4 days prior to the calculated day of ovulation and 3 days afterward prevents conception. Such a method of contraception can be used only when the periodicity of the menstrual cycle is regular, for otherwise it is impossible to determine the next onset of menstruation, and, therefore, it is impossible to predict the day of ovulation.

Hormonal Suppression of Fertility — "The Pill." It has long been known that administration of either estrogen or progesterone, if given in sufficient quantity, can inhibit ovulation. Though the exact mechanism of this effect is not clear, it is known that in the presence of enough of either or both of these hormones, the hypothalamus and pituitary gland fail to secrete the normal surge of LH-releasing hormone and its stimulatory product LH that usually occurs about 13 days after the onset of the monthly sexual cycle. From the discussion of this phenomenon earlier in the chapter, it will be recalled that this surge of LH is essential in causing ovulation.

The problem in devising methods for hormonal suppression of ovulation has been to develop appropriate combinations of estrogens and progestins that will suppress ovulation but that will not cause unwanted effects of these two hormones. For instance, too much of either of the hormones can cause abnormal menstrual bleeding patterns. However, use of a synthetic progestin in place of progesterone, especially the 19-norsteroids, along with small amounts of estrogens will usually prevent ovulation and yet, also, allow almost a normal pattern of menstruation. Therefore, almost all "pills" used for control of fertility consist of some combination of synthetic estrogens and synthetic progestins. The main reason for using synthetic estrogens and synthetic progestins is that the *natural* hormones are almost entirely destroyed by the liver within a short time after they are absorbed from the gastrointestinal tract into the portal circulation. However, many of the *synthetic* hormones can resist this destructive propensity of the liver, thus allowing oral administration.

Two of the most commonly used estrogens are *ethynyl estradiol* and *mestranol*. Among the most commonly used progestins are *norethindrone, norethynodrel, ethynodiol,* and *norgestrel.* The medication is usually begun in the early stages of the female sexual cycle and is continued beyond the time that ovulation normally would have occurred. Then the medication is stopped toward the end of the cycle, allowing menstruation to occur and a new cycle to begin.

Oral contraceptive regimens have also been devised in which very low dosage levels of estrogens and progestins are used. In these instances ovulation frequently does occur, but other effects prevent conception. These effects include (1) abnormal transport time through the fallopian tube (the usual time is almost exactly three days) so that implantation will not occur; (2) abnormal development of the endometrium so that it will not support a fertilized ovum; (3) abnormal characteristics of the cervical mucus, making this lethal to the sperm or in other ways blocking entry of the sperm to the uterus; and (4) abnormal contraction of the fallopian tubes and uterine musculature so that the ovum will be expelled rather than implanted.

Abnormal Conditions Causing Female Sterility. Approximately 1 of every 6 to 10 marriages is infertile; in about 60 per cent of these, the infertility is due to female sterility.

Occasionally, no abnormality whatsoever can be discovered in the female genital organs, in which case it must be assumed that the infertility is due either to abnormal physiological function of the genital system or to abnormal genetic development of the ova themselves.

However, probably by far the most common cause of female sterility is failure to ovulate. This can result from either hyposecretion of gonadotropic hormones, in which case the intensity of the hormonal stimuli simply is not sufficient to cause ovulation, or it can result from abnormal ovaries that will not allow ovulation. For instance, thick capsules occasionally exist on the outside of the ovaries that prevent ovulation.

Because of the high incidence of anovulation in sterile women, special methods are often utilized to determine whether or not ovulation occurs. These are all based on the effects of progesterone on the body, for the normal increase in progesterone secretion does not occur during the latter half of anovulatory cycles. In the absence of progesteronic effects, the cycle can be assumed to be anovulatory. One of these tests is simply to analyze the urine for a surge in pregnanediol, the end-product of progesterone metabolism, during the latter half of the sexual cycle, the lack of which indicates failure of ovulation. However, another common test is for the woman to chart her body temperature throughout the cycle. Secretion of progesterone during the latter half of the cycle raises the body temperature about one-half degree Fahrenheit, the temperature rise coming abruptly at the time of ovulation. Such a temperature chart, showing the point of ovulation, is illustrated in Figure 81–9.

Lack of ovulation caused by hyposecretion of the pituitary gonadotropic hormones can be treated by administration of *human chorionic gonadotropin,* a hormone that will be discussed in the following chapter and that is extracted from the human placenta. This hormone, though secreted by the placenta, has almost exactly the same effects as luteinizing hormone and, therefore, is a powerful stimulator of ovulation. However, excess use of this hormone can cause ovulation from many follicles simultaneously; and this results in multiple births, an effect that has caused as many as six children to be born to mothers treated for infertility with this hormone.

One of the most common causes of female sterility is also *endometriosis,* for, as described earlier, endometriosis causes fibrosis throughout the pelvis; and this fibrosis sometimes so enshrouds the ovaries that an

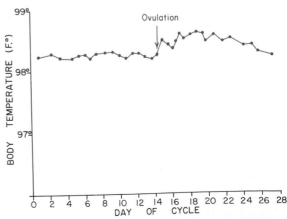

Figure 81–9. Elevation in body temperature shortly after ovulation.

ovum cannot be released into the abdominal cavity. Often, also, endometriosis occludes the fallopian tubes, either at the fimbriated ends or elsewhere along their extent. Another common cause of female infertility is salpingitis, that is, inflammation of the fallopian tubes; this causes fibrosis in the tubes, thereby occluding them. In past years, such inflammation was extremely common as a result of gonococcal infection, but with modern therapy this is becoming a less prevalent cause of female infertility.

Finally, still another cause of infertility that may be very important is abnormal secretion of mucus by the uterine cervix. Ordinarily, at the time of ovulation, the hormonal environment of estrogen causes secretion of a very thin mucus with special characteristics that will allow rapid mobility of sperm into the uterus and will actually guide the sperm up along mucus "threads." Abnormalities of the cervix itself, such as low-grade infection or inflammation, or abnormal hormonal stimulation of the cervix can lead to a viscous mucous plug that will prevent fertilization.

REFERENCES

Alexander, N. J.: Animal Models for Research on Contraception and Fertility. Proceedings of a Symposium. Hagerstown, Md., Harper & Row, 1979.

Arimura, A.: Hypothalamic gonadotropin-releasing hormone and reproduction. *Int. Rev. Physiol., 13*:1, 1977.

Aron, C.: Mechanisms of control of the reproductive function of olfactory stimuli in female mammals. *Physiol. Rev., 59*:229, 1979.

Beato, M. (ed.): Steroid-Induced Uterine Proteins. New York, Elsevier/North-Holland, 1980.

Behrman, H. R.: Prostaglandins in hypothalamo-pituitary and ovarian function. *Annu. Rev. Physiol., 41*:685, 1979.

Beller, F. K., and Schumacher, G. F. B. (eds.): The Biology of the Secretions of the Female Genital Tract. New York, Elsevier/North-Holland, 1979.

Benirschke, K.: The endometrium. *In* Yen, S. S. C., and Jaffe, R. B. (eds.): Reproductive Endocrinology. Philadelphia, W. B. Saunders Co., 1978, p. 241.

Briggs, M., and Briggs, M.: Oral Contraceptives. Montreal, Eden Press, 1977.

Brown-Grant, K.: Physiological aspects of the steroid hormone–gonadotropin interrelationship. *Int. Rev. Physiol., 13*:57, 1977.

Catt, K. J., and Pierce, J. G.: Gonadotropic hormones of the adenohypophysis (FSH, LH and Prolactin). *In* Yen, S. S. C., and Jaffe, R. B. (eds.): Reproductive endocrinology. Philadelphia, W. B. Saunders Co., 1978, p. 34.

Channing, C. P., and Marsh, J. M. (eds.): Ovarian Follicular and Corpus Luteum Function. New York, Plenum Press, 1979.

Channing, C. P., *et al.*: Ovarian follicular and luteal physiology. *In* Greep, R. O. (ed.): International Review of Physiology: Reproductive Physiology III. Vol. 22. Baltimore, University Park Press, 1980, p. 117.

Clark, J. H., and Peck, E. J., Jr.: Female Sex Steroids: Receptors and Function. New York, Springer-Verlag, 1979.

Crighton, D. B., *et al.*: Control of Ovulation. Boston, Butterworths, 1978.

Davajan, V., and Isreal, R.: Infertility: Causes, evaluation, and treatment. *In* DeGroot, L. J., *et al.* (eds.): Endocrinology. Vol. 3. New York, Grune & Stratton, 1979, p. 1459.

Diamond, M. C., and Korenbrot, C. C. (eds.): Hormonal Contraceptives and Human Welfare. New York, Academic Press, 1978.

Droegemueller, W., and Bressler, R.: Effectiveness and risks of contraception. *Annu. Rev. Med., 31*:329, 1980.

Epel, D.: The program of fertilization. *Sci. Am., 237*(5):128, 1977.

Fawcett, D. W., and Bedford, J. M. (eds.): The Spermatozoon. Baltimore, Urban & Schwarzenberg, 1979.

Glass, R. H.: Infertility. *In* Yen, S. S. C., and Jaffe, R. B. (eds.): Reproductive Endocrinology. Philadelphia, W. B. Saunders Co., 1978, p. 398.

Goldberg, V. J., and Ramwell, P. W.: Role of prostaglandins in reproduction. *Physiol. Rev., 55*:325, 1975.

Greenblatt, R. B. (ed.): Induction of Ovulation. Philadelphia, Lea & Febiger, 1979.

Hafez, E. S. E.: Human Reproductive Physiology. Ann Arbor, Mich., Ann Arbor Science, 1978.

Henzi, M. R.: Natural and synthetic female sex hormones. *In* Yen, S. S. C., and Jaffe, R. B. (eds.): Reproductive Endocrinology. Philadelphia, W. B. Saunders Co., 1978, p. 421.

Holmes, R. L., and Fox, C. A.: Control of Human Reproduction. New York, Academic Press, 1977.

Horton, E. W., and Poyser, N. L.: Uterine luteolytic hormone: A physiological role for prostaglandin F_{2a}. *Physiol. Rev., 56*:595, 1976.

Jaffe, R. B.: The menopause and perimenopausal period. *In* Yen, S. S. C., and Jaffe, R. B. (eds.): Reproductive Endocrinology. Philadelphia, W. B. Saunders Co., 1978, p. 261.

Kase, N. G., and Speroff, L.: The ovary. *In* Bondy, P. K., and Rosenberg, L. E. (eds.): Metabolic Control and Disease, 8th Ed. Philadelphia, W. B. Saunders Co., 1980, p. 1579.

Kumar, T. C. A. (ed.): Neuroendocrine Regulation of Fertility. New York, S. Karger, 1976.

Lande, I. J., and Scott, J. R.: Immunology and the Infertile Female. Chicago, Year Book Medical Publishers, 1977.

Lein, A.: The Cycling Female: Her Menstrual Rhythm. San Francisco, W. H. Freeman, 1979.

Lipner, H.: Mechanism of mammalian ovulation. *In*: Greep, R. O., and Astwood, E. B. (eds.): Handbook of Physiology. Sec. 7, Vol. 2, Part 1. Baltimore, Williams & Wilkins, 1973, p. 409.

Lipsett, M. B.: Endocrine mechanisms of reproduction. *In* Frohlich, E. D. (ed.): Pathophysiology, 2nd Ed. Philadelphia, J. B. Lippincott Co., 1976, p. 405.

Macvaugh, G. S.: Frigidity: Analysis and Treatment. New York, Pergamon Press, 1979.

Midgley, A. R., and Sadler, W. A. (eds.): Ovarian Follicular Development and Function. New York, Raven Press, 1979.

Mishell, D. R., Jr.: Contraception. *In* DeGroot, L. J., *et al.* (eds.): Endocrinology. Vol. 3. New York, Grune & Stratton, 1979, p. 1435.

Mishell, D. R., Jr., and Davajan, V. (eds.): Reproductive Endocrinology, Infertility, and Contraception. Philadelphia, F. A. Davis Co., 1979.

Odell, W. D.: FSH. *In* DeGroot, L. J., *et al.* (eds.): Endocrinology. Vol. 1. New York, Grune & Stratton, 1979, p. 149.

Odell, W. D.: LH. *In* DeGroot, L. J., *et al.* (eds.): Endocrinology. Vol. 1. New York, Grune & Stratton, 1979, p. 151.

Odell, W. D.: The menopause. *In* DeGroot, L. J., *et al.* (eds.): Endocrinology. Vol. 3. New York, Grune & Stratton, 1979, p. 1489.

Odell, W. D.: The reproductive system in women. *In* DeGroot, L. J., *et al.* (eds.): Endocrinology. Vol. 3. New York, Grune & Stratton, 1979, p. 1383.

Peters, H., and McNatty, K. P.: The Ovary: A Correlation of Structure and Function in Mammals. Berkeley, University of California Press, 1980.

Richards, J. S.: Maturation of ovarian follicles: Actions and interactions of pituitary and ovarian hormones on follicular cell differentiation. *Physiol. Rev., 60*:51, 1980.

Ross, G. T., and Schreiber, J. R.: The ovary. *In* Yen, S. S. C., and Jaffe, T. B. (eds.): Reproductive Endocrinology. Philadelphia, W. B. Saunders Co., 1978, p. 63.

Savoy-Moore, R. T., and Schwartz, N. B.: Differential control of FSH and LH secretion. *In* Greep, R. O. (ed.): International Review of Physiology: Reproductive Physiology III. Vol. 22. Baltimore, University Park Press, 1980, p. 203.

Segal, S. J.: The physiology of human reproduction. *Sci. Am., 231*(3):52, 1974.

Siiteri, P. K., and Febres, F.: Ovarian hormone synthesis, circulation, and mechanisms of action. *In* DeGroot, L. J., *et al.* (eds.): Endocrinology, Vol. 3. New York, Grune & Stratton, 1979, p. 1401.

Tollison, C. D., and Adams, H. E.: Sexual Disorders: Treatment, Theory and Research. New York, Gardner Press, 1979.

Tyson, J. E. (ed.): Symposium on Neuroendocrinology of Reproduction. Clinics in Obstetrics-Gynaecology. London, W. B. Saunders Co., Aug. 1978.

Wallach, E. E., and Kempers, R. D.: Modern Trends in Infertility and Conception Control. Baltimore, Williams & Wilkins, 1979.

Wilson, M. A.: The Ovulation Method of Birth Control: The Latest Advances for Achieving or Postponing Pregnancy — Naturally. New York, Van Nostrand Reinhold, 1979.

Yen, S. S. C.: Chronic anovulation due to inappropriate feedback system. *In* Yen, S. S. C., and Jaffe, R. B. (eds.): Reproductive Endocrinology. Philadelphia, W. B. Saunders Co., 1978, p. 297.

Yen, S. S. C.: The human menstrual cycle (integrative function of the hypothalamic-pituitary-ovarian-endometrial axis). *In* Yen, S. S. C., and Jaffe, R. B. (eds.): Reproductive Endocrinology. Philadelphia, W. B. Saunders Co., 1978, p. 126.

Yen, S. S. C., and Jaffe, R. B. (eds.): Reproductive Endocrinology: Physiology, Pathophysiology, and Clinical Management. Philadelphia, W. B. Saunders Co., 1978.

82

Pregnancy and Lactation

In the preceding two chapters the sexual functions of the male and female were described to the point of fertilization of the ovum. If the ovum becomes fertilized, a completely new sequence of events called *gestation*, or *pregnancy*, takes place, and the fertilized ovum eventually develops into a full-term fetus. The purpose of the present chapter is to discuss the early stages of ovum development after fertilization and then to discuss the physiology of pregnancy. In the following chapter some special problems of fetal and early childhood physiology are discussed.

MATURATION OF THE OVUM

Shortly before the ovum is released from the follicle, its nucleus divides by meiosis and a so-called *first polar body* is expelled from the nucleus of the ovum, which becomes the *secondary oocyte*. In this process, each of the 23 pairs of chromosomes loses one of the partners to the polar body so that 23 *unpaired* chromosomes remain in the secondary oocyte. It is at this point that fertilization occurs. Immediately after the sperm enters the ovum, the nucleus divides again, and a *second polar body* is expelled. This is also a meiotic division so that there still remain 23 unpaired chromosomes.

FERTILIZATION OF THE OVUM

After coitus, the first sperm are transported through the uterus to the ovarian end of the fallopian tubes within about five minutes. This is many times more rapid than the motility of the sperm themselves can account for, which indicates that propulsive movements of the uterus and fallopian tubes might be responsible for much of the sperm movement. It is known, for instance, that in some animals coitus causes the neurohypophysis to secrete oxytocin, which in turn enhances uterine contractions. Also, semen contains a *prostaglandin* that theoretically could add still further to the contractions. Even with these aids to their movement, of the half billion sperm deposited in the vagina only 1000 to 3000 succeed in traversing the fallopian tubes to reach the proximity of the ovum.

Sperm can remain fertile in the female genital tract for 24 to 72 hours but can remain highly fertile probably for only 12 to 24 hours. Furthermore, after ovulation a mature ovum also is fertilizable probably for up to 24 hours but maximally fertilizable for only 8 to 12 hours. Therefore, if coitus occurs when these two periods of fertility overlap, fertilization can take place.

Only one sperm is required for fertilization of the ovum, the process of which is illustrated in Figure 82–1. Furthermore, almost never does more than one sperm enter the ovum for the following reason: The zona pellucida of the ovum has a lattice-type structure, and once the ovum is punctured, some substance (perhaps one of the proteolytic enzymes of the sperm acrosome) seems to diffuse out of the ovum into the lattice to prevent penetration by additional sperm. Indeed, microscopic studies show that many sperm do attempt to penetrate the zona pellucida but become inactivated after traveling only part way through.

Once a sperm enters the ovum, its head swells rapidly to form a *male pronucleus*, which is also illustrated in Figure 82–1. Later, the 23 chromosomes of the male pronucleus and the 23 of the *female pronucleus* align themselves to reform a complete complement of 46 chromosomes (23 pairs) in the fertilized ovum.

Sex Determination. The sex of a child is determined by the type of sperm that fertilizes the ovum — that is, whether it is a male sperm or a female sperm. It will be recalled from Chapter 80 that a male sperm carries a *Y sex chromosome* and 22 *autosomal chromosomes,* while a female sperm carries the same 22 autosomal chromosomes but an *X sex chromosome*. On the other hand, the ovum always has an X sex chromosome and never a Y chromosome. After recombination of the male and female pronuclei during fertilization, the fertilized ovum then contains 44 autosomal chromosomes and either 2 X chromosomes, which causes a female child to develop, or an X and a Y chromosome, which causes a male child to develop.

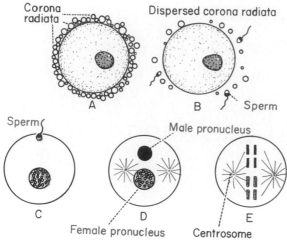

Figure 82–1. Fertilization of the ovum, showing (A) the mature ovum surrounded by the corona radiata, (B) dispersal of the corona radiata, (C) entry of the sperm, (D) formation of the male and female pronuclei, and (E) reorganization of a full complement of chromosomes and beginning division of the ovum. (Modified from Arey: Developmental Anatomy, 7th Ed.)

TRANSPORT AND IMPLANTATION OF THE DEVELOPING OVUM

Entry of the Ovum into the Fallopian Tube. When ovulation occurs, the ovum, along with its attached granulosa cells, the *cumulus oophorus*, is expelled directly into the peritoneal cavity and must then enter one of the fallopian tubes. The fimbriated ends of each fallopian tube fall naturally around the ovaries, and the inner surfaces of the fimbriated tentacles are lined with ciliated epithelium, the *cilia* of which continually beat toward the *abdominal ostium* of the fallopian tube. One can actually see a slow fluid current flowing toward the ostium. By this means the ovum enters one or the other fallopian tube.

It would seem likely that many ova might fail to enter the fallopian tubes. However, on the basis of conception studies it is probable that as many as 98 to 100 per cent succeed in this task. Indeed, cases are on record in which women with one ovary removed and the opposite fallopian tube removed have had as many as four children with relative ease of conception, thus illustrating that ova can even enter the opposite fallopian tube.

TRANSPORT OF THE OVUM THROUGH THE FALLOPIAN TUBE

Fertilization of the ovum normally takes place soon after the ovum enters the fallopian tube, but this process cannot occur until the granulosa cells attached to the outside of the ovum, the *cumulus oophorus*, are dispersed from the ovum. This dispersal probably results partly from the action of

hyaluronidase and proteolytic enzymes released from the acrosomes of the sperm, but most of the dispersion of these cells is caused by bicarbonate ions in the fallopian tube secretions. After fertilization has occurred, an additional three days are normally required for transport of the ovum through the tube into the cavity of the uterus. This transport is effected mainly by a feeble fluid current in the fallopian tube resulting from action of the ciliated epithelium that lines the tube, the cilia always beating toward the uterus. It is possible also that weak contractions of the fallopian tube aid in the passage of the ovum.

The fallopian tubes are lined with a rugged, cryptoid surface that actually impedes the passage of the ovum despite the fluid current. Also, the *isthmus* of the fallopian tube (the last two centimeters before the uterus is entered) remains spastically contracted for the first three days following ovulation. After this time, probably under the relaxing influence of the rapidly increasing progesterone, this region loses the spasm, thus allowing entry of the ovum to the uterus.

The delayed transport of the ovum through the fallopian tube allows several stages of division to occur before the ovum enters the uterus. During this time, large quantities of secretions are formed by secretory cells that alternate with ciliated cells lining the fallopian tube. These secretions are for nutrition of the developing ovum. Indeed, fertilized ova continue to divide in vitro as long as they are bathed in a solution of homogenized fallopian tube mucosa, but they will not divide in most other media.

IMPLANTATION OF THE OVUM IN THE UTERUS

After reaching the uterus, the developing ovum usually remains in the uterine cavity an additional four to five days before it implants in the endometrium, which means that implantation ordinarily occurs on the seventh or eighth day following ovulation. During this time the ovum obtains its nutrition from the endometrial secretions, called "uterine milk." Figure 82–2 shows a very early stage of implantation, illustrating that the developing ovum at this time is in the *blastocyst stage*.

Implantation results from the action of *trophoblast cells* that develop over the surface of the blastocyst. These cells secrete proteolytic enzymes that digest and liquefy the cells of the endometrium. Simultaneously, much of the fluid and nutrients thus released is actively absorbed into the blastocyst as a result of phagocytosis by the trophoblast cells; these absorbed substances provide the sustenance for further growth of the blastocyst. Also, at the same time, additional trophoblast cells form cords of cells that extend

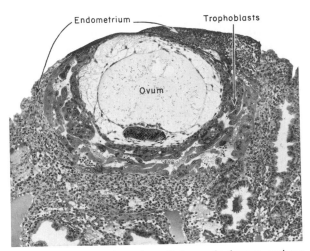

Figure 82–2. Implantation of the early human embryo, showing trophoblastic digestion and invasion of the endometrium. (Courtesy of Dr. Arthur Hertig.)

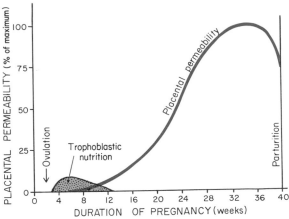

Figure 82–3. Nutrition of the fetus, illustrating that most of the early nutrition is due to trophoblastic digestion and absorption of nutrients from the endometrial decidua and that essentially all the later nutrition results from diffusion through the placental membrane.

into the deeper layers of the endometrium and attach to them. Thus, the blastocyst eats a hole in the endometrium and attaches to it at the same time.

Once implantation has taken place, the trophoblast and sub-lying cells proliferate rapidly; these form the placenta and the various membranes of pregnancy.

EARLY INTRAUTERINE NUTRITION OF THE EMBRYO

In the previous chapter it was pointed out that the progesterone secreted during the latter half of each sexual cycle has a special effect on the endometrium to convert the endometrial stromal cells into large swollen cells that contain extra quantities of glycogen, proteins, lipids, and even some necessary minerals for development of the ovum. If the fertilized ovum implants in the endometrium, the continued secretion of progesterone causes the stromal cells to swell still more and to store even more nutrients. These cells are now called *decidual cells*, and the total mass of cells is called the *decidua*.

As the trophoblast cells invade the decidua, digesting and imbibing it, the stored nutrients in the decidua are used by the embryo for appropriate growth and development. During the first week after implantation, this is the only means by which the embryo can obtain nutrients, and the embryo continues to obtain a large measure of its total nutrition in this way for 8 to 12 weeks, though the placenta also begins to provide slight amounts of nutrition after approximately the 16th day beyond fertilization (a little over a week after

implantation). Figure 82–3 depicts this trophoblastic period of nutrition, which gradually gives way to placental nutrition.

FUNCTION OF THE PLACENTA

DEVELOPMENTAL AND PHYSIOLOGIC ANATOMY OF THE PLACENTA

While the trophoblastic cords from the blastocyst are attaching to the uterus, blood capillaries grow into the cords from the vascular system of the embryo, and, by the 16th day after fertilization, blood begins to flow. Simultaneously, blood sinuses supplied with blood from the mother develop between the surface of the uterine endometrium and the trophoblastic cords. The trophoblast cells then gradually send out more and more projections, which become the *placental villi* into which fetal capillaries grow. Thus, the villi, carrying fetal blood, are surrounded by sinuses containing maternal blood.

The final structure of the placenta is illustrated in Figure 82–4. Note that the fetus' blood flows through two *umbilical arteries*, finally to the capillaries of the villi, and thence back through the *umbilical vein* into the fetus. On the other hand, the mother's blood flows from the *uterine arteries* into large *blood sinuses* surrounding the villi and then back into the *uterine veins* of the mother.

The lower part of Figure 82–4 illustrates the relationship between the fetal blood of the villus and the blood of the mother in the fully developed placenta. The capillaries of the villus are lined with an extremely thin endothelium and are surrounded by a layer of *mesenchymal tissue* that is covered on the outside of the villus by a layer of *syncytial trophoblast* cells. During the first 16 weeks of pregnancy, still an additional layer of cells is present immediately beneath the syncytial trophoblast layer. This layer is composed of distinct

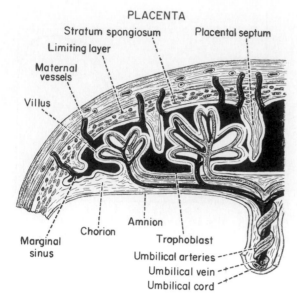

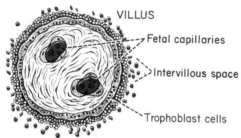

Figure 82–4. *Above:* Organization of the mature placenta. *Below:* Relationship of the fetal blood in the villus capillaries to the mother's blood in the intervillous spaces. (Modified from Gray and Goss: Anatomy of the Human Body. Lea & Febiger; and from Arey: Developmental Anatomy, 7th Ed.)

cuboidal cells called *cytotrophoblast cells*, or *Langerhans' cells*.

The total surface area of all the villi of the mature placenta is only a few square meters — many times less than the area of the pulmonary membrane. Also, remember that even at full maturity the placental membrane is still several cell layers thick, and the minimum distance between the maternal blood and the fetal blood is 3.5 microns, or almost 10 times the distance across the alveolar membranes of the lung. Nevertheless, many nutrients and other substances pass through the placental membrane by diffusion in very much the same manner as through the alveolar membranes of the lungs and the capillary membranes elsewhere in the body.

PERMEABILITY OF THE PLACENTAL MEMBRANE

Since the major function of the placenta is to allow *diffusion* of foodstuffs from the mother's blood into the fetus' blood and diffusion of excretory products from the fetus back into the mother, it is important to know the degree of permeability of the placental membrane. *Permeability is expressed as the total quantity of a given substance that crosses the entire placental membrane in a given time for a given concentration difference.*

In the early months of development, placental permeability is relatively slight, as illustrated in Figure 82–3, for two reasons: First, the total surface area of the placental membrane is still small at that time, and, second, the villar membranes have not yet been reduced to their minimum thickness. However, as the placenta becomes older, the permeability increases progressively until the last month or so of pregnancy when it begins to decrease again. This decrease results from deterioration of the placenta caused by its age and sometimes from destruction of whole segments due to infarction.

Occasionally, "breaks" occur in the placental membrane which allow fetal blood cells to pass into the mother or, more rarely, the mother's cells to pass into the fetus. Indeed, there are instances in which the fetus bleeds severely into the mother's circulation because of a ruptured placental membrane.

Diffusion of Oxygen Through the Placental Membrane. Almost exactly the same principles are applicable for the diffusion of oxygen through the placental membrane as through the pulmonary membrane; these principles were discussed in Chapter 40. The dissolved oxygen in the blood of the large placental sinuses simply passes into the fetal blood because of an oxygen pressure gradient from the mother's blood to the fetus' blood. The mean P_{O_2} in the mother's blood in the placental sinuses is approximately 50 mm. Hg toward the end of pregnancy, and the mean P_{O_2} in the fetal blood after it becomes "oxygenated" is about 30 mm. Hg. Therefore, the mean pressure gradient for diffusion of oxygen through the placental membrane is about 20 mm. Hg.

One might wonder how it is possible for a fetus to obtain sufficient oxygen when the fetal blood leaving the placenta has a P_{O_2} of only 30 mm. Hg. However, there are three different reasons why even this low P_{O_2} is capable of allowing the fetal blood to transmit almost as much oxygen to the fetal tissues as is transmitted by the mother's blood to her tissues:

First, the hemoglobin of the fetus is primarily *fetal hemoglobin*, a type of hemoglobin synthesized in the fetus prior to birth. Figure 82–5 illustrates the comparative oxygen dissociation curves of maternal hemoglobin and fetal hemoglobin, showing that the curve for fetal hemoglobin is shifted to the left of that for maternal hemoglobin. This means that at a given P_{O_2}, the fetal hemoglobin can carry as much as 20 to 30 per cent more oxygen than can maternal hemoglobin.

Second, the *hemoglobin concentration of the*

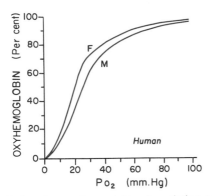

Figure 82–5. Oxygen-hemoglobin dissociation curves for maternal *(M)* and fetal *(F)* bloods, illustrating the ability of the fetal blood to carry a much greater quantity of oxygen than can maternal blood for a given blood P_{O_2}. (From Metcalfe, Moll, and Bartels: *Fed. Proc.,* 23:775, 1964.)

fetus is about 50 per cent greater than that of the mother; and this is an even more important factor than the first in enhancing the amount of oxygen transported to the fetal tissues.

Third, the *Bohr effect*, which was explained in relation to the exchange of carbon dioxide and oxygen in the lung in Chapter 41, provides another factor that enhances the transport of oxygen by the fetal blood. That is, hemoglobin can carry more oxygen at a low P_{CO_2} than it can at a high P_{CO_2}. The fetal blood entering the placenta carries large amounts of carbon dioxide, but much of this carbon dioxide diffuses from the fetal blood into the maternal blood. Loss of the carbon dioxide makes the fetal blood more alkaline while the increased carbon dioxide in the maternal blood makes this more acidic. These changes cause the combining capacity of fetal blood for oxygen to become increased and that of the maternal blood to become decreased. This forces more oxygen from the maternal blood while enhancing the oxygen in the fetal blood. Thus, the Bohr shift operates in one direction in the maternal blood and in the other in the fetal blood, these two effects adding to make the Bohr shift twice as important here as it is for oxygen exchange in the lungs.

By these three means, the fetus is capable of receiving more than adequate oxygen through the placenta despite the fact that the fetal blood leaving the placenta has a P_{O_2} of only 30 mm. Hg.

The total diffusing capacity of the placenta for oxygen at term is about 1.2 ml. of oxygen per minute per mm. Hg oxygen gradient. This compares favorably with that of the lungs of the newborn baby.

Diffusion of Carbon Dioxide Through the Placental Membrane. Carbon dioxide is continually formed in the tissues of the fetus in the same way that it is formed in maternal tissues. And the only means for excreting the carbon dioxide is through the placenta. The P_{CO_2} builds up in the fetal blood until it is about 48 mm. Hg, in contrast to about 40 to 45 mm. Hg in the maternal blood. Thus, a small pressure gradient for carbon dioxide develops across the placental membrane, and this is sufficient to allow adequate diffusion of carbon dioxide from the fetal blood into the maternal blood because the extreme solubility of carbon dioxide in the tissues of the placental membrane allows carbon dioxide to diffuse through this membrane rapidly, about 20 times as rapidly as oxygen.

Diffusion of Foodstuffs Through the Placental Membrane. Other metabolic substrates needed by the fetus diffuse into the fetal blood in the same manner as oxygen. For instance, the glucose level in the fetal blood ordinarily is 20 to 30 per cent lower than in the maternal blood, for glucose is being metabolized rapidly by the fetus. This in turn causes rapid diffusion of additional glucose from the maternal blood into the fetal blood.

Because of the high solubility of fatty acids in cell membranes, these also diffuse from the maternal blood into the fetal blood but more slowly than glucose so that glucose is preferentially used by the fetus for nutrition. Also, such substances as ketone bodies and potassium, sodium, and chloride ions diffuse from the maternal blood into the fetal blood.

Active Absorption by the Placental Membrane. As pointed out previously, early nutrition of the embryo depends on phagocytosis of fallopian tube and uterine secretions, and even on phagocytosis of the endometrial decidua. The trophoblast cells that line the outer surface of the villi can also actively absorb certain nutrients from the maternal blood in the placenta, at least during the first half of pregnancy and perhaps even throughout the entire pregnancy. For instance, the measured *amino acid* content of fetal blood is greater than that of maternal blood, and *calcium* and *inorganic phosphate* occur in greater concentration in fetal blood than in maternal blood, while *ascorbic acid* is as much as 3 times as concentrated in the fetal blood. These findings indicate that the placental membrane has the ability to absorb actively at least small amounts of certain substances even during the latter part of pregnancy.

Excretion Through the Placental Membrane. In the same manner that carbon dioxide diffuses from the fetal blood into the maternal blood, other excretory products formed in the fetus diffuse in the opposite direction into the maternal blood and then are excreted along with the excretory products of the mother. These include especially the *nonprotein nitrogens* such as *urea, uric acid,* and *creatinine*. The level of urea in the fetal blood is only slightly greater than that in maternal blood because urea diffuses through the placental membrane with considerable ease. On the other hand, creatinine, which does not diffuse as easily, has a considerably higher concentration gradient percentagewise. Therefore, insofar as is known, excretion from the fetus occurs entirely as a result of diffusion gradients across the placental membrane — that is, because of higher concentrations of the excretory products in the fetal blood than in the maternal blood.

STORAGE FUNCTION OF THE PLACENTA

During the first few months of pregnancy, the placenta grows tremendously in size while the fetus remains relatively diminutive. During this same time considerable quantities of metabolic substrates, including proteins, calcium, and iron, are stored in the placenta to be used in the latter months of pregnancy for growth by the fetus. Thus, in the early months of pregnancy, the placenta performs very much the same functions for the fetus that the liver performs for the adult human being, acting as a nutrient storehouse and helping to process some of the food substances that enter the fetus. For instance, in the early weeks of gestation the placenta is capable of storing glucose as glycogen and then of secreting glucose into the embryonic blood stream in much the same manner that the liver can secrete glucose into the adult blood stream. By this process, the placenta actually helps to control the fetal blood glucose concentration. Later in the growth of the fetus, these metabolic functions of the placenta become less and less important, while the fetal liver becomes progressively more important.

HORMONAL FACTORS IN PREGNANCY

In pregnancy, the placenta forms large quantities of *human chorionic gonadotropin, estrogens, progesterone,* and *human chorionic somatomammotropin,* the first three of which, and perhaps the fourth as well, are essential to the continuance of pregnancy. The functions of these hormones are discussed in the following sections.

HUMAN CHORIONIC GONADOTROPIN AND ITS EFFECT IN CAUSING PERSISTENCE OF THE CORPUS LUTEUM AND IN PREVENTING MENSTRUATION

Menstruation normally occurs approximately 14 days after ovulation, at which time most of the secretory endometrium of the uterus sloughs away from the uterine wall and is expelled to the exterior. If this should happen after an ovum has implanted, the pregnancy would terminate. However, this is prevented by the secretion of human chorionic gonadotropin in the following manner:

Coincidently with the development of the trophoblast cells from early fertilized ovum, the hormone *human chorionic gonadotropin* is secreted by the syncytial trophoblast cells into the fluids of the mother. As illustrated in Figure 82–6, the secretion of this hormone can first be measured in

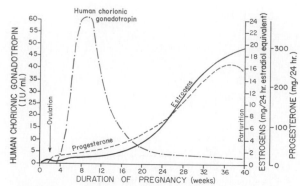

Figure 82–6. Rates of secretion of estrogens, progesterone, and chorionic gonadotropin at different stages of pregnancy.

the blood 6 to 8 days after ovulation, just as the ovum is first implanting in the endometrium. Then the rate of secretion rises rapidly to reach a maximum approximately 7 to 9 weeks after ovulation, and decreases to a relatively low value by 16 to 20 weeks after ovulation.

Function of Human Chorionic Gonadotropin. Human chorionic gonadotropin is a glycoprotein having a molecular weight of about 30,000 and very much the same molecular structure and function as luteinizing hormone secreted by the pituitary. By far its most important function is to prevent the normal involution of the corpus luteum at the end of the female sexual cycle. Instead, it causes the corpus luteum to secrete even larger quantities of its usual sex hormones, progesterone and estrogens. These sex hormones cause the endometrium to continue growing and to store large amounts of nutrients rather than to be passed in the menstruum. As a result, the *decidua-like cells* that develop in the endometrium during the normal female sexual cycle become actual, nutritious *decidual cells* soon after the blastocyst implants.

Under the influence of human chorionic gonadotropin, the corpus luteum grows to about 2 times its initial size by a month or so after pregnancy begins, and its continued secretion of estrogens and progesterone maintains the decidual nature of the uterine endometrium, which is necessary for the early development of the placenta and other fetal tissues. If the corpus luteum is removed before approximately the 7th week of pregnancy, spontaneous abortion almost always occurs and sometimes even up to the 12th week, though after this time the placenta itself secretes sufficient quantities of progesterone and estrogens to maintain pregnancy for the remainder of the gestation period. The corpus luteum then involutes slowly after the third to fourth month of gestation.

Effect of Human Chorionic Gonadotropin on the

Fetal Testes. Human chorionic gonadotropin also exerts an *interstitial cell–stimulating effect* on the testes, thus resulting in the production of testosterone in male fetuses. This small secretion of testosterone during gestation is the factor that causes the fetus to grow male sex organs. Near the end of pregnancy, the testosterone secreted by the fetal testes also causes the testicles to descend into the scrotum.

Clinically, human chorionic gonadotropin administered to the cryptorchid male (having undescended testes) often causes the testicles to descend into the scrotum by causing testosterone to be secreted, the testosterone in turn effecting the actual testicular descent.

SECRETION OF ESTROGENS BY THE PLACENTA

The placenta, like the corpus luteum, secretes both estrogens and progesterone. Both histochemical and physiological studies indicate that these two hormones, like most other placental hormones, are also secreted by the *syncytial trophoblast cells*.

Figure 82–6 shows that the daily production of placental estrogens increases markedly toward the end of pregnancy, to as much as several hundred times the daily production in the middle of a normal nonpregnant monthly cycle.

However, the secretion of estrogen by the placenta is quite different from the secretion by the ovaries in several ways, as follows: First, the estrogen that is secreted is about nine-tenths estriol, which, as will be recalled from Chapter 81, is formed in only small amounts in the nongravid female. Yet, because of the very low estrogenic potency of estriol, the total estrogenic activity rises only to about 30 times normal. Second, the estrogens secreted by the placenta are not synthesized *de novo* from basic substrates in the placenta. Instead, the steroid compound *dehydroepiandrosterone*, which is formed in the adrenal glands of the fetus and then transported by the fetal blood to the placenta, is converted into estradiol and estrone, and the steroid *16-hydroxydehydroepiandrosterone*, which is derived from dehydroepiandrosterone, is converted into estriol. (The cortices of the fetal adrenal glands are extremely large, composed almost entirely of the so-called *fetal zone*, whose primary function seems to be to secrete the dehydroepiandrosterone.)

Function of Estrogen in Pregnancy. In the discussions of estrogens in the preceding chapter it was pointed out that these hormones exert mainly a proliferative function on certain reproductive and associated organs. During pregnancy, the extreme quantities of estrogens cause (1) enlargement of the uterus, (2) enlargement of the breasts and growth of the breast ductal structure, and (3) enlargement of the female external genitalia.

The estrogens also relax the various pelvic ligaments so that the sacroiliac joints become relatively limber and the symphysis pubis becomes elastic. These changes obviously make for easier passage of the fetus through the birth canal.

There is much reason to believe that estrogens also affect the development of the fetus during pregnancy, for example, by affecting the rate of cell reproduction in the early embryo.

SECRETION OF PROGESTERONE BY THE PLACENTA

Progesterone is also a hormone essential for pregnancy. In addition to being secreted in moderate quantities by the corpus luteum at the beginning of pregnancy, it is secreted in tremendous quantities by the placenta, averaging about one-quarter gram per day toward the end of pregnancy. Indeed, the rate of progesterone secretion increases by as much as 10-fold during the course of pregnancy, as illustrated in Figure 82–6.

The special effects of progesterone that are essential for normal progression of pregnancy are the following:

1. As pointed out earlier, progesterone causes decidual cells to develop in the uterine endometrium, and these then play an important role in the nutrition of the early embryo.

2. Progesterone has a special effect on decreasing the contractility of the gravid uterus, thus preventing uterine contractions from causing spontaneous abortion.

3. Progesterone also contributes to the development of the ovum even prior to implantation, for it specifically increases the secretions of the fallopian tubes and uterus to provide appropriate nutritive matter for the developing *morula* and *blastocyst*. There are some reasons to believe, too, that progesterone even affects cell cleavage in the early developing embryo.

4. The progesterone secreted during pregnancy also helps to prepare the breasts for lactation, which is discussed later in the chapter.

HUMAN CHORIONIC SOMATOMAMMOTROPIN

Recently, a new placental hormone called *human chorionic somatomammotropin* has been discovered. This is a protein, having a molecular weight of about 38,000, that begins to be secreted about the fifth week of pregnancy and increases progressively throughout the remainder of pregnancy in direct proportion to the weight of the placenta.

Though the functions of chorionic somatomammotropin are still uncertain, this hormone has several important effects:

First, when administered to several different types of lower animals, human chorionic somatomammotropin causes at least partial development of the breasts and in some instances causes lactation. Since this was the first function of the hormone discovered, it was at first named *placental lactogen* and was believed to have functions similar to those of prolactin. However, attempts to promote lactation in the human being with its use have not been successful.

Second, this hormone has weak actions similar to those of growth hormone, causing deposition of protein tissues in the same way as growth hormone does. It also has a chemical structure similar to that of growth hormone, but 100 to 200 times as much human chorionic somatomammotropin is required to promote growth as is true for growth hormone.

Third, human chorionic somatomammotropin has recently been found to have important actions on both glucose metabolism and fat metabolism in the mother, effects that perhaps are very important for nutrition of the fetus. The hormone causes decreased insulin sensitivity and also decreased utilization of glucose by the mother, thereby making larger quantities of glucose available to the fetus. Since glucose is the major substrate used by the fetus for energy, the importance of this hormonal effect is obvious. Furthermore, the hormone promotes release of free fatty acids from the fat stores of the mother, thus providing this alternative source of energy for her metabolism.

Therefore, it is beginning to appear that human chorionic somatomammotropin is a general metabolic hormone that has specific nutritional implications for both the mother and the fetus.

OTHER HORMONAL FACTORS IN PREGNANCY

Almost all the nonsexual endocrine glands of the mother react markedly to pregnancy. This results mainly from the increased metabolic load on the mother but also to some extent from inverse effects of placental hormones on the pituitary and other glands. Some of the most notable effects are the following:

Pituitary Secretion. The anterior pituitary gland enlarges at least 50 per cent during pregnancy and increases its production of *corticotropin, thyrotropin*, and *prolactin*. On the other hand, production of follicle-stimulating hormone and luteinizing hormone is greatly suppressed as a result of the inhibitory effects of estrogens and progesterone from the placenta.

Corticosteroid Secretion. The rate of adrenocortical secretion of the *glucocorticoids* is moderately increased throughout pregnancy. It is possible that the glucocorticoids help to mobilize amino acids from the mother's tissues so that these can be used for synthesis of tissues in the fetus.

Pregnant women also usually have about a three-fold increase in secretion of *aldosterone,* reaching the peak at the end of gestation. This, along with the actions of the estrogens, causes a tendency for even the normal pregnant woman to reabsorb excess sodium from the renal tubules and therefore to retain fluid, often leading to hypertension.

Secretions by the Thyroid Gland. The thyroid gland ordinarily enlarges about 50 per cent during pregnancy and increases its production of thyroxine a corresponding amount. The increased thyroxine production is caused at least partly by human chorionic gonadotropin and by small quantities of a specific thyroid-stimulating hormone, *human chorionic thyrotropin*, secreted by the placenta.

Secretion by the Parathyroid Glands. The parathyroid glands also often enlarge during pregnancy; this is especially true if the mother is on a calcium-deficient diet. Enlargement of these glands causes calcium absorption from the mother's bones, thereby maintaining normal calcium ion concentration in the mother's extracellular fluids as the fetus removes calcium for ossifying its own bones. This secretion of parathyroid hormone is even more intensified during lactation following the birth of the baby, because the baby requires many times more calcium than the fetus.

Secretion of "Relaxin" by the Ovaries. An additional substance besides the estrogens and progesterone, a hormone called relaxin, can be isolated from the corpora lutea of the ovaries and from the placenta. Relaxin is a polypeptide having a molecular weight of about 9000. This hormone, when injected, causes relaxation of the ligaments of the symphysis pubis in the estrous rat and guinea pig. However, this effect is very poor in the pregnant woman. Instead, this role is probably played by both the estrogens and progesterone, which also cause relaxation of the pelvic ligaments.

It has also been claimed that relaxin has two other effects: (1) softening of the cervix of the pregnant woman at the time of delivery and (2) inhibition of uterine motility.

In summary, relaxin is a substance that can be isolated from the ovaries, but its functional importance is almost totally unknown.

RESPONSE OF THE MOTHER'S BODY TO PREGNANCY

Obviously, the presence of a growing fetus in the uterus adds an extra physiological load to the pregnant woman, and much of her response to pregnancy is due to this increased load. The hormones secreted during pregnancy either by the placenta or by the endocrine glands can also cause many reactions in the mother. Among the reactions are increased size of the various sexual organs. For instance, the uterus increases from about 30 grams to about 1100 grams, and the breasts approximately double in size. At the same time the vagina enlarges, and the introitus opens more widely. Also, the various hormones can cause marked changes in the appearance of the woman, sometimes resulting in the development of edema, acne, and masculine or acromegalic features.

CHANGES IN THE MATERNAL CIRCULATORY SYSTEM DURING PREGNANCY

Blood Flow Through the Placenta and Cardiac Output During Pregnancy. About 625 ml. of blood flows through the maternal circulation of the placenta each minute during the latter phases of gestation. This factor, plus a general increase in metabolism, increases the mother's cardiac output to 30 to 40 per cent above normal by the 27th week of pregnancy, but then, for reasons unexplained, the cardiac output falls to only a little above normal during the last eight weeks of pregnancy, despite the high uterine blood flow.

Blood Volume During Pregnancy. The maternal blood volume shortly before term is approximately 30 per cent above normal. This increase occurs mainly during the latter half of pregnancy, as illustrated by the curve of Figure 82–7. The cause of the increased volume is mainly hormonal, for both aldosterone and estrogens, which are greatly increased in pregnancy, cause increased fluid retention by the kidneys. Also, the bone marrow becomes increasingly active and produces an excess of red blood cells to go with the excess fluid volume. Therefore, at the time of birth of the baby, the mother has approximately 1 to 2 liters of extra blood in her circulatory system. Only about one-fourth of this amount is normally lost during delivery of the baby, thereby allowing a considerable safety factor for the mother.

WEIGHT GAIN IN THE PREGNANT WOMAN

During the first months of pregnancy, a woman ordinarily loses a few pounds of weight, possibly as a result of nausea, but during the entire pregnancy the average weight gain is about 24 pounds, most of this gain occurring during the last two trimesters. Of this increase in weight, approximately 7 pounds is fetus, and approximately 4 pounds is amniotic fluid, placenta, and fetal membranes. The uterus increases approximately 2 pounds, and the breasts approximately another 2 pounds, still leaving an average increase in weight of the woman's body of 9 pounds. About 6 pounds of this is fluid that is excreted in the urine during the first few days after birth, that is, after loss of the fluid-retaining hormones of the placenta.

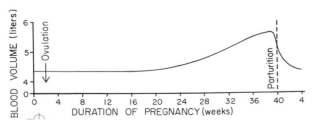

Figure 82–7. Effect of pregnancy on blood volume.

Often during pregnancy a woman has a greatly increased desire for food, partly as a result of fetal removal of food substrates from the mother's blood and partly because of hormonal factors. Without appropriate prenatal care some mothers eat tremendous quantities of food, and the weight gain, instead of averaging 24 pounds, may be as great as 75 pounds or more.

METABOLISM DURING PREGNANCY

As a consequence of the increased secretion of many different hormones during pregnancy, including thyroxine, adrenocortical hormones, and the sex hormones, the basal metabolic rate of the pregnant woman increases about 15 per cent during the latter half of the pregnancy. As a result, she frequently has sensations of becoming overheated. Also, owing to the extra load that she is carrying, greater amounts of energy than normal must be expended for muscular activity.

Nutrition During Pregnancy. The supplemental food needed by the woman during pregnancy to supply the needs of the fetus and fetal membranes includes extra dietary quantities of the various minerals, vitamins, and proteins. The growing fetus assumes priority in regard to many of the nutritional elements in the mother's body fluids, and many portions of the fetus continue to grow even though she does not eat a sufficient diet. For instance, in a woman lacking adequate nutrition, the length of the fetus increases almost normally. On the other hand, lack of adequate nutrition can decrease the fetus' weight considerably, can decrease ossification of the bones, and can cause anemia, hypoprothrombinemia, and decreased size of many bodily organs of the fetus.

By far the greatest growth of the fetus occurs during the last trimester of pregnancy; its weight almost doubles during the last two months of pregnancy. Ordinarily, the mother does not absorb sufficient protein, calcium, phosphates, and iron from the gastrointestinal tract during the last month of pregnancy to supply the fetus. However, from the beginning of pregnancy her body has been storing these substances to be used during the latter months of pregnancy. Some of this storage is in the placenta, but most is in the normal storage depots of the mother.

If appropriate nutritional elements are not present in the pregnant woman's diet, a number of maternal deficiencies can occur during pregnancy. Especially, deficiencies often occur for calcium, phosphates, iron, and the vitamins. For example, approximately 375 mg. of iron is needed by the fetus to form its blood and an additional 600 mg. is needed by the mother to form her own extra blood. The normal store of nonhemoglobin iron in the mother at the outset of pregnancy is often only 100 or so mg. and almost never over 700 mg. Therefore, without sufficient iron in her food, a pregnant woman usually develops anemia. Also, it is especially important that she receive vitamin D, for, even though the total quantity of calcium utilized by the fetus is small, calcium even normally is poorly absorbed by the gastrointestinal tract. Finally, shortly before birth of the baby vitamin K is often added to her diet so that the baby will have sufficient prothrombin to prevent hemorrhage, particularly brain hemorrhage, caused by the birth process.

RESPIRATION DURING PREGNANCY

Because of the increased basal metabolic rate of the pregnant woman and also because of her increase in size, the total amount of oxygen utilized by the mother shortly before birth of the baby is approximately 20 per cent above normal, and a commensurate amount of carbon dioxide is formed. These effects cause the minute ventilation to increase. It is also believed that the high levels of progesterone during pregnancy increase the minute ventilation still more, because progesterone increases the sensitivity of the respiratory center to carbon dioxide. The net result is an increase in minute ventilation of approximately 50 per cent and a decrease in arterial P_{CO_2} to slightly below that of the normal woman. Simultaneously, the growing uterus presses upward against the abdominal contents, and these in turn press upward against the diaphragm so that the total excursion of the diaphragm is decreased. Consequently, the respiratory rate is increased to maintain adequate ventilation.

FUNCTION OF THE MATERNAL URINARY SYSTEM DURING PREGNANCY

The rate of urine formation by the pregnant woman is usually slightly increased because of an increased load of excretory products. But, in addition, several special alterations of urinary functions occur:

First, reabsorption of sodium, chloride, and water by the renal tubules tends to be increased greatly as a consequence of increased production of steroid hormones by the placenta and adrenal cortex.

Second, the glomerular filtration rate often increases as much as 50 per cent during pregnancy, which tends to increase the rate of water and electrolyte loss in the urine. This factor normally almost balances the first so that the mother ordinarily has only moderate excess water and salt accumulation except when she develops *toxemia of pregnancy;* this disease is discussed later in the chapter.

THE AMNIOTIC FLUID AND ITS FORMATION

Normally, the volume of amniotic fluid is between 500 ml. and 1 liter, but it can be only a few milliliters or as much as several liters. Studies with isotopes on the rate of formation of amniotic fluid show that on the average the water in amniotic fluid is completely replaced once every three hours, and the electrolytes sodium and potassium are replaced once every 15 hours. A small portion of fluid is derived from renal excretion by the fetus. Likewise, a certain amount of absorption occurs by way of the gastrointestinal tract and lungs of the fetus. However, even after the death of a fetus, the rate of turnover of the amniotic fluid is still one-half as great as it is when the fetus is normal, which indicates that much of the fluid is formed and absorbed directly through the amniotic membranes. The total volume of amniotic fluid is probably regulated by the amniotic membranes, for as the volume increases the pressure

rises and presumably causes increased fluid absorption, thus returning the volume to normal.

ABNORMAL RESPONSES TO PREGNANCY

HYPEREMESIS GRAVIDARUM

In the earlier months of pregnancy, the pregnant woman frequently develops hyperemesis gravidarum, a condition characterized by nausea and vomiting and commonly known as "morning sickness." Occasionally, the vomiting becomes so severe that she becomes greatly dehydrated, and in rare instances the condition has been known to cause death.

The cause of the nausea and vomiting is unknown, but it occurs to its greatest extent during the same time that chorionic gonadotropin is secreted in large quantities by the placenta. Because of this coincidence, many clinicians have suggested that chorionic gonadotropin is in some way responsible for the nausea and vomiting; nevertheless, a causal relationship has never been proved.

On the other hand, during the first few months of pregnancy rapid trophoblastic invasion of the endometrium also takes place and, because the trophoblastic cells digest portions of the endometrium as they invade it, it is possible that degenerative products resulting from this invasion, instead of chorionic gonadotropin, are responsible for the nausea and vomiting. Indeed, degenerative processes in other parts of the body, such as following gamma ray irradiation and burns, can all cause similar nausea and vomiting.

Finally, another possible cause of the condition is the large quantity of estrogen secreted by the placenta. This theory is supported by the fact that estrogen injected daily into a person in large quantities for many weeks will often cause nausea and vomiting during the first few weeks of administration.

PREECLAMPSIA AND ECLAMPSIA

Approximately 4 per cent of all pregnant women develop a rapid rise in arterial blood pressure associated with loss of large amounts of protein in the urine at some time during the latter four months of pregnancy. This condition is called *preeclampsia*. It is often also characterized by salt and water retention by the kidneys, weight gain, and development of edema. In addition, arterial spasm occurs in many parts of the body, most significantly in the kidneys, brain, and liver. Both the renal blood flow and the glomerular filtration rate are decreased, which is exactly opposite to the changes that occur in the normal pregnant woman. The renal effects are caused by thickened glomerular tufts that contain a fibrinoid deposit in the basement membranes.

Various attempts have been made to prove that preeclampsia is caused by excessive secretion of placental or adrenal hormones, but proof of a hormonal basis is still lacking. Indeed, a more plausible theory is that preeclampsia results from some type of autoimmunity or allergy resulting from the presence of the fetus.

Indeed, the acute symptoms disappear within a few days after birth of the baby.

The severity of preeclampsia symptoms is closely associated with the retention of salt and water and the degree of increase in arterial pressure. In fact, an increasing pressure seems to set off a vicious circle that intensifies the arterial spasm and other pathological effects of preeclampsia. These effects can be greatly delayed by drastic limitation of salt intake and enforced bed rest during the latter months of pregnancy, which are the two cardinal features of therapy.

Eclampsia is a severe degree of preeclampsia characterized by extreme vascular spasticity throughout the body, clonic convulsions followed by coma, greatly decreased kidney output, malfunction of the liver, often extreme hypertension, and a generalized toxic condition of the body. Usually, it occurs shortly before parturition. Without treatment, a very high percentage of eclamptic patients die. However, with optimal and immediate use of rapidly acting vasodilating drugs to reduce the arterial pressure to normal, followed by immediate termination of pregnancy — by cesarean section if necessary — the mortality has been reduced to 1 per cent or less in some clinics.

PARTURITION

INCREASED UTERINE IRRITABILITY NEAR TERM

Parturition means simply the process by which the baby is born. At the termination of pregnancy the uterus becomes progressively more excitable until finally it begins strong rhythmic contractions with such force that the baby is expelled. The exact cause of the increased activity of the uterus is not known, but at least two major categories of effects lead up to the culminating contractions responsible for parturition: first, progressive hormonal changes that cause increased excitability of the uterine musculature, and, second, progressive mechanical changes.

Hormonal Factors That Cause Increased Uterine Contractility. *Ratio of Estrogens to Progesterone.* Progesterone inhibits uterine contractility during pregnancy, thereby helping to prevent expulsion of the fetus. On the other hand, estrogens have a definite tendency to increase the degree of uterine contractility. Both these hormones are secreted in progressively greater quantities throughout most of pregnancy, but from the seventh month onward estrogen secretion continues to increase while progesterone secretion remains constant or perhaps even decreases slightly. Therefore, it has been postulated that the *estrogen to progesterone ratio* increases sufficiently toward the end of pregnancy to be at least partly responsible for the increased contractility of the uterus.

Effect of Oxytocin on the Uterus. Oxytocin is a hormone secreted by the neurohypophysis that specifically causes uterine contraction (see Chapter 75). There are four reasons for believing that oxytocin might be particularly important in increasing the contractility of the uterus near term. (1) The uterus increases its responsiveness to a given dose of oxytocin during the latter few months of pregnancy. (2) The rate of oxytocin secretion by the neurohypophysis is considerably increased at the time of labor. (3) Though hypophysectomized animals and human beings can still deliver their young at term, labor is prolonged. (4) Recent experiments in animals indicate that irritation or stretching of the uterine cervix, as occurs during labor, can cause a neurogenic reflex to the neurohypophysis that increases the rate of oxytocin secretion.

Effect of Fetal Hormones on the Uterus. The fetus' pituitary gland also secretes increasing quantities of oxytocin that could possibly play a role in exciting the uterus. In addition, the fetal membranes release prostaglandins in high concentration at the time of labor. These, too, could increase the intensity of the uterine contractions.

Mechanical Factors That Increase the Contractility of the Uterus. *Stretch of the Uterine Musculature.* Simply stretching smooth muscle organs usually increases their contractility. Furthermore, intermittent stretch, as occurs repetitively in the uterus because of movements of the fetus, can also elicit smooth muscle contraction.

Note especially that twins are born on the average *19 days* earlier than a single child, which emphasizes the importance of mechanical stretch in eliciting uterine contractions.

Stretch or Irritation of the Cervix. There is much reason to believe that stretch or irritation of the uterine cervix is particularly important in eliciting uterine contractions. For instance, the obstetrician frequently induces labor by rupturing the membranes so that the head of the baby stretches the cervix more forcefully than usual or irritates it in some other way.

The mechanism by which cervical irritation excites the body of the uterus is not known. It has been supposed that stretch or irritation of neuronal cells in the cervix initiates reflexes to the body of the uterus, but the effect could also result simply from myogenic transmission of signals from the cervix to the body of the uterus.

ONSET OF LABOR — A POSITIVE FEEDBACK THEORY FOR ITS INITIATION

During most of the months of pregnancy the uterus undergoes periodic episodes of weak and slow rhythmic contractions called *Braxton-Hicks contractions*. These become progressively stronger toward the end of pregnancy; then they change rather suddenly, within hours, to become exceptionally strong contractions that start stretching the cervix and later force the baby through the birth canal, thereby causing parturition. This process is called *labor,* and the strong contractions that result in final parturition are called *labor contractions.*

Yet, strangely enough, we do not know what suddenly changes the slow and weak rhythmicity of the uterus into the strong labor contractions. However, on the basis of experience during the past few years with other types of physiological control systems, a theory has been proposed for explaining the onset of labor based on positive feedback. This theory suggests that stretch of the cervix by the fetus' head finally becomes great enough to elicit strong reflex increase in contractility of the uterine body. This pushes the baby forward, which stretches the cervix still more and initiates still more positive feedback to the uterine body. Thus, the process continues again and again until the baby is expelled. This theory is illustrated in Figure 82–8, and the data supporting it are the following:

First, labor contractions obey all the principles of positive feedback. That is, once the strength of uterine contraction becomes greater than a critical value, each contraction leads to subsequent contractions that become stronger and stronger until maximum effect is achieved. Referring to the discussion in Chapter 1 of positive feedback in control systems, we see that this is the precise nature of all positive feedback mechanisms with a feedback gain of more than unity.

Second, there are two known types of positive feedback that increase uterine contractions during labor: (1) Stretch of the cervix causes the entire body of the uterus to contract, and this stretches the cervix still more because of the downward thrust of the baby's head. (2) Cervical stretch also causes the pituitary gland to secrete oxytocin, which is still another means for increasing uterine contractility.

To summarize the theory, we can assume that multiple factors increase the contractility of the uterus toward the end of pregnancy. Eventually, a uterine contraction becomes strong enough to irritate the uterus, increase its contractility still more because of positive feedback, and result in a second contraction stronger than the first, and a third stronger than the second, and so forth. Once these contractions become strong enough to cause this type of feedback, with each succeeding contraction greater than the one preceding, the process proceeds to completion — all simply *because positive feedback initiates a vicious cycle when the gain of the feedback is greater than unity.*

One might immediately ask about the many instances of false labor in which the contractions become stronger and stronger and then fade away. Remember that for a vicious cycle to continue, *each* new cycle of the positive feedback must be stronger than the previous one. If, at any time after labor starts, some contractions fail to re-excite the uterus sufficiently, the positive feedback could go into a retrograde succession and the labor contractions would fade away.

ABDOMINAL MUSCLE CONTRACTION DURING LABOR

Once labor contractions become strong and painful, neurogenic reflexes, mainly from the birth canal to the spinal cord and thence back to the abdominal muscles, cause intense abdominal muscle contraction. This abdominal contraction adds greatly to the forces that cause expulsion of the baby.

MECHANICS OF PARTURITION

The uterine contractions during labor begin at the top of the uterine fundus and spread downward over the body of the uterus. Also, the intensity of contraction is great in the top and body of the uterus but weak in the lower segment of the uterus adjacent to the cervix. Therefore, each uterine contraction tends to force the baby downward toward the cervix.

In the early part of labor, the contractions might occur only once every 30 minutes. As labor progresses, the contractions finally appear as often as once every one to three minutes, and the intensity of contraction increases greatly, with only a short period of relaxation between contractions.

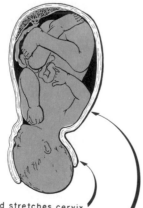

1. Baby's head stretches cervix...
2. Cervical stretch excites fundic contraction...
3. Fundic contraction pushes baby down and stretches cervix some more...
4. Cycle repeats over and over again...

Figure 82–8. Theory for the onset of intensely strong contractions during labor.

The combined contractions of the uterine and abdominal musculature during delivery of the baby cause downward force on the fetus of approximately 25 pounds during each strong contraction.

It is fortunate that the contractions of labor occur intermittently because strong contractions impede or sometimes even stop blood flow through the placenta and would cause death of the fetus were the contractions continuous. Indeed, in clinical use of various uterine stimulants, such as oxytocin, overuse of the drugs can cause uterine spasm rather than rhythmic contractions and can lead to death of the fetus.

In 19 of 20 births the head is the first part of the baby to be expelled, and in most of the remaining instances the buttocks are presented first. The head acts as a wedge to open the structures of the birth canal as the fetus is forced downward.

The first major obstruction to expulsion of the fetus is the uterine cervix. Toward the end of pregnancy the cervix becomes soft, which allows it to stretch when labor pains cause the body of the uterus to contract. The so-called *first stage of labor* is the period of progressive cervical dilatation, lasting until the opening is as large as the head of the fetus. This stage usually lasts 8 to 24 hours in the first pregnancy but often only a few minutes after many pregnancies.

Once the cervix has dilated fully, the fetus' head moves rapidly into the birth canal, and, with additional force from above, continues to wedge its way through the canal until delivery is effected. This is called the *second stage of labor,* and it may last from as little as a minute after many pregnancies up to half an hour or more in the first pregnancy.

SEPARATION AND DELIVERY OF THE PLACENTA

During the succeeding 10 to 45 minutes after birth of the baby, the uterus contracts to a very small size, which causes a *shearing* effect between the walls of the uterus and the placenta, thus separating the placenta from its implantation site. Obviously, separation of the placenta opens the placental sinuses and causes bleeding. However, the amount of bleeding is limited to an average of 350 ml. by the following mechanism: The smooth muscle fibers of the uterine musculature are arranged in figures of 8 around the blood vessels as they pass through the uterine wall. Therefore, contraction of the uterus following delivery of the baby constricts the vessels that had previously supplied blood to the placenta.

LABOR PAINS

With each uterine contraction the mother experiences considerable pain. The pain in early labor is probably caused mainly by hypoxia of the uterine muscle resulting from compression of the blood vessels to the uterus. This pain is not felt when the *hypogastric nerves,* which carry the sensory fibers leading from the uterus, have been sectioned. However, during the second stage of labor, when the fetus is being expelled through the birth canal, much more severe pain is caused by cervical stretch, perineal stretch, and stretch or tearing of structures in the vaginal canal itself. This pain is conducted by somatic nerves instead of by the hypogastric nerves.

INVOLUTION OF THE UTERUS

During the first four to five weeks following parturition, the uterus involutes. Its weight becomes less than one-half its immediate postpartum weight within a week, and in four weeks, if the mother lactates, the uterus may be as small as it had been prior to pregnancy. This effect of lactation results from the suppression of gonadotropin and ovarian hormone secretion during the first few months of lactation, as will be discussed later. During early involution of the uterus the placental site on the endometrial surface autolyzes, causing a vaginal discharge known as "lochia," which is first bloody and then serous in nature, continuing in all for approximately a week and a half. After this time, the endometrial surface will have become re-epithelialized and ready for normal, nongravid sex life again.

LACTATION

DEVELOPMENT OF THE BREASTS

The breasts begin to develop at puberty; this development is stimulated by the estrogens of the monthly sexual cycles that stimulate growth of the stroma and ductal system plus deposition of fat to give mass to the breasts. However, much additional growth occurs during pregnancy, and the glandular tissue only then becomes completely developed for actual production of milk.

Growth of the Ductal System — Role of the Estrogens. All through pregnancy, the tremendous quantities of estrogens secreted by the placenta cause the ductal system of the breasts to grow and to branch. Simultaneously, the stroma of the breasts also increases in quantity, and large quantities of fat are laid down in the stroma.

However, also important in growth of the ductal system are at least four other hormones: *growth hormone, prolactin,* the *adrenal glucocorticoids,* and *insulin.* Each of these is known to play at least some role in protein metabolism, which presumably explains their use for development of the breasts.

Development of the Lobule-Alveolar System — Role of Progesterone. Final development of the breasts into milk-secreting organs also requires the additional action of progesterone. Once the ductal system has developed, progesterone, acting synergistically with all the other hormones just mentioned, causes growth of the lobules, budding of alveoli, and development of secretory characteristics in the cells of the alveoli. These changes are analogous to the secretory effects of progesterone on the endometrium of the uterus during the latter half of the female sexual cycle.

INITIATION OF LACTATION — FUNCTION OF PROLACTIN

Though estrogen and progesterone are essential for the physical development of the breasts during pregnancy, both these hormones also have a specific effect on inhibiting the actual secretion of milk. On the other hand, the hormone *prolactin* has exactly the opposite effect, promotion of the secretion of milk. This hormone is secreted by the mother's pituitary gland, and its concentration in her blood rises steadily from the fifth week of pregnancy until birth of the baby, at which time it has risen to very high levels, usually about 10 times the normal nonpregnant level. This is illustrated in Figure 82–9. In addition, the placenta secretes large quantities of human chorionic somatomammotropin which also has mild lactogenic properties, thus supporting the prolactin from the mother's pituitary. Even so, only a few milliliters of fluid are secreted each day until after the baby is born. This fluid is called *colostrum;* it contains essentially the same concentrations of proteins and lactose as milk but almost no fat, and its maximum rate of production is about 1/100 the subsequent rate of milk production.

This absence of lactation during pregnancy is caused by the overriding suppressive effects of progesterone and estrogen, which are secreted in tremendous quantities as long as the placenta is still in the uterus and which completely subdue the lactogenic effects of both prolactin and human chorionic somatomammotropin. However, immediately after the baby is born, the sudden loss of both estrogen and progesterone secretion by the placenta now allows the lactogenic effect of the prolactin from the mother's pituitary gland to assume its natural role, and within two or three days the breasts begin to secrete copious quantities of milk instead of colostrum. This secretion of milk requires an adequate background secretion of most of the mother's other hormones as well, but most important of all are *growth hormone,* the *adrenal glucocorticoids,* and *parathyroid hormone.* These hormones are necessary to provide the amino acids, fatty acids, glucose, and calcium that are required for milk formation.

Following birth of the baby, the *basal level* of prolactin secretion returns during the next few weeks to the nonpregnant level, as shown in Figure 82–9. However, each time the mother nurses her baby, nervous signals from the nipples to the hypothalamus cause approximately a 10-fold surge in prolactin secretion lasting about one hour, which is also shown in the figure. This prolactin in turn acts on the breasts to provide the milk for the next nursing period. If this prolactin surge is absent, if it is blocked as a result of hypothalamic or pituitary damage, or if nursing does not continue, the breasts lose their ability to produce milk within a few days. However, milk production can continue for several years if the child continues to suckle, but the rate of milk formation normally decreases considerably within seven to nine months.

Hypothalamic Control of Prolactin Secretion. The hypothalamus plays an essential role in controlling prolactin secretion as it does for the control of secretion of almost all the other anterior pituitary hormones as well. However, this control is different in one aspect: the hypothalamus mainly *stimulates* the production of all the other hormones, but it mainly *inhibits* prolactin production. Consequently, damage to the hypothalamus or blockage of the hypothalamic-hypophysial portal system increases prolactin secretion while it depresses secretion of the other anterior pituitary hormones. Yet, under special conditions, such as when the baby suckles the breast, it seems that a different type of signal from the hypothalamus can then increase the secretion of prolactin.

Therefore, it is believed that two different hormones formed in the hypothalamus are transported to the anterior pituitary through the hypothalamic-hypophysial portal system to control prolactin release by the anterior pituitary gland. These are called *prolactin inhibitory hormone* (PIH), which is the dominant hormone under most normal conditions, and *prolactin-releasing hormone* (PRH) which can intermittently increase prolactin secretion. Yet, neither of these two hormones has been identified with certainty. PIH might be *dopamine,* which is known to be secreted in the hypothalamus and which also can decrease prolactin secretion as much as 10-fold.

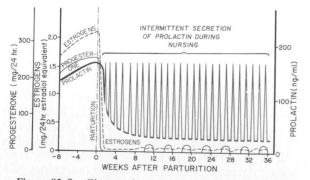

Figure 82–9. Changes in rates of secretion of estrogens, progesterone, and prolactin for 8 weeks prior to parturition and for 36 weeks thereafter. Note especially the decrease of prolactin secretion back to basal levels within a few weeks, but also the intermittent periods of marked prolactin secretion (for about one hour at a time) during and after periods of nursing.

Suppression of the Female Sexual Cycle During Nursing. In about half of nursing mothers, the ovarian cycle and ovulation do not resume until a few weeks after cessation of nursing the baby. Though the exact cause of this is not known, it is likely that the same nervous signals from the breasts to the hypothalamus that cause prolactin secretion during suckling simultaneously inhibit secretion of luteinizing hormone–releasing hormone by the hypothalamus, which in turn suppresses the formation of the gonadotropic hormones, luteinizing hormone and follicle-stimulating hormone. Yet, after several months of lactation, in about half of the mothers the pituitary begins again to secrete sufficient gonadotropic hormones to reinitiate the monthly sexual cycle. The rhythmic interplay between the ovarian and pituitary hormones during the sexual month does not require marked reduction in prolactin secretion.

THE EJECTION OR "LET-DOWN" PROCESS IN MILK SECRETION— FUNCTION OF OXYTOCIN

Milk is secreted continuously into the alveoli of the breasts, but milk does not flow easily from the alveoli into the ductal system and therefore does not continually leak from the breast nipples. Instead, the milk must be ejected or "let-down" from the alveoli to the ducts before the baby can obtain it. This process is caused by a combined neurogenic and hormonal reflex involving the hormone *oxytocin*.

When the baby suckles the breast, sensory impulses are transmitted through somatic nerves to the spinal cord and then to the hypothalamus, there causing *oxytocin* secretion at the same time that they cause prolactin secretion. The oxytocin is then carried in the blood to the breasts where it causes the *myoepithelial cells* that surround the outer walls of the alveoli to contract, thereby expressing the milk from the alveoli into the ducts. Thus, within 30 seconds to a minute after a baby begins to suckle the breast, milk begins to flow. This process is called milk ejection, or milk let-down.

Suckling on one breast causes milk flow not only in that breast but also in the opposite breast. It is especially interesting that fondling the baby or hearing the baby crying also is often enough of a signal to the hypothalamus to cause milk ejection.

Inhibition of Milk Ejection. A particular problem in nursing the baby comes from the fact that many psychogenic factors or generalized sympathetic stimulation throughout the body can inhibit oxytocin secretion and consequently depress milk

TABLE 82–1. PERCENTAGE COMPOSITION OF MILK

	Human Milk	Cow's Milk
Water	88.5	87
Fat	3.3	3.5
Lactose	6.8	4.8
Casein	0.9	2.7
Lactalbumin and other protein	0.4	0.7
Ash	0.2	0.7

ejection. For this reason, the mother must have an undisturbed puerperium if she is to be successful in nursing her baby.

MILK AND THE METABOLIC DRAIN ON THE MOTHER CAUSED BY LACTATION

Table 82–1 lists the contents of human milk and cow's milk. The concentration of lactose in human milk is approximately 50 per cent greater than that in cow's milk, but on the other hand the concentration of protein in cow's milk is ordinarily 2 or more times as great as that in human milk. Finally, the ash, which contains the minerals, is only one-third as much in human milk as in cow's milk.

At the height of lactation 1.5 liters of milk may be formed each day. With this degree of lactation great quantities of metabolic substrates are drained from the mother. For instance, approximately 50 grams of fat enter the milk each day; and approximately 100 grams of lactose, which must be derived from glucose, are lost from the mother each day. Also, 2 to 3 grams of calcium phosphate may be lost each day, and, unless the mother is drinking large quantities of milk and has an adequate intake of vitamin D, the output of calcium and phosphate by the lactating mammae will be much greater than the intake of these substances. To supply the needed calcium and phosphate, the parathyroid glands enlarge greatly, and the bones become progressively decalcified. The problem of decalcification is usually not very great during pregnancy, but it can be a distinct problem during lactation.

REFERENCES

Aragona, C., and Friesen, H. G.: Lactation and galactorrhea. *In* DeGroot, L. J., *et al.* (eds.): Endocrinology. Vol. 3. New York, Grune & Stratton, 1979, p. 1613.

Battaglia, F. C., and Meschia, G.: Principal substrates of fetal metabolism. *Physiol. Rev.*, 58:499, 1978.

Beaconfield, P., *et al.* (eds.): Placenta: A Neglected Experimental Animal. New York, Pergamon Press, 1979.

Biggers, J. D., and Borland, R. M.: Physiological aspects of growth and development of the preimplantation mammalian embryo. *Annu. Rev. Physiol.*, 38:95, 1976.

Brenner, R. M., and West, N. B.: Hormonal regulation of the reproductive tract in female mammals. *Annu. Rev. Physiol.*, 37:273, 1975.

Brinster, R. L.: Nutrition and metabolism of the ovum, zygote, and blastocyst. *In* Greep, R. O., and Astwood, E. B. (eds.): Handbook of Physiology. Sec. 7, Vol. 2, Part 2. Baltimore, Williams & Wilkins, 1973, p. 165.

Brudenell, M., *et al.* (ed.): Artificial Insemination. London, RCOG, 1976.

Buster, J. E., and Marshall, J. R.: Conception, gamete and ovum transport, implantation, fetal-placental hormones, hormonal preparation of parturition and parturition control. *In* DeGroot, L. J., *et al.* (eds.): Endocrinology. Vol. 3. New York, Grune & Stratton, 1979, p. 1595.

Challis, J. R. G.: Endocrinology of late pregnancy and parturition. *In* Greep, R. O. (ed.): International Review of Physiology: Reproductive Physiology III. Vol. 22. Baltimore, University Park Press, 1980, p. 277.

Chamberlain, G., and Wilkinson, A. (eds.): Placenta Transfer. Baltimore, University Park Press, 1979.

Coutinho, E. M., and Fuchs, F. (eds.): Physiology and Genetics of Reproduction. Basic Life Sciences Series. Vol. 4. New York, Plenum Press, 1974.

Cowie, A. T., *et al.:* Hormonal Control of Lactation. New York, Springer-Verlag, 1980.

Ensor, D. M.: Comparative Endocrinology of Prolactin. New York, Halsted Press, 1978.

Epel, D.: The program of fertilization. *Sci. Am., 237*(5):128, 1977.

Fairweather, D. V. I., and Eskes, T. K. A. B.: Amniotic Fluid: Research and Clinical Application. Amsterdam, Excerpta Medica, 1978.

Finn, C. A.: Recent research on implantation in animals. *Proc. R. Soc. Med., 67*:927, 1974.

Fisher, D. A.: Fetal endocrinology: Endocrine disease and pregnancy. *In* DeGroot, L. J., *et al.* (eds.): Endocrinology. Vol. 3. New York, Grune & Stratton, 1979, p. 1649.

Frantz, A. G.: Prolactin. *In* DeGroot, L. J., *et al.* (eds.): Endocrinology. Vol. 1. New York, Grune & Stratton, 1979, p. 153.

Gant, N. F. (ed.): Pregnancy-Induced Hypertension. New York, Grune & Stratton, 1978.

Goodlin, R. C.: Care of the Fetus. New York, Masson Publishing Co., 1979.

Grant, N. F., and Worley, R.: Hypertension in Pregnancy: Concepts and Management. New York, Appleton-Century-Crofts, 1980.

Grobstein, C.: External human fertilization. *Sci. Am., 240*(6):57, 1979.

Heap, R. B., *et al.:* The hormonal maintenance of pregnancy. *In* Greep, R. O., and Astwood, E. B. (eds.): Handbook of Physiology. Sec. 7, Vol. 2, Part 2. Baltimore, Williams & Wilkins, 1973, p. 217.

Hertig, A. T., and Barton, B. R.: Fine structure of mammalian oocytes and ova. *In* Greep, R. O., and Astwood, E. B. (eds.): Handbook of Physiology. Sec. 7, Vol. 2, Part 1. Baltimore, Williams & Wilkins, 1973, p. 317.

Hogarth, P. J.: Biology of Reproduction. New York, John Wiley & Sons, 1978.

Holmes, R. L., and Fox, C. A.: Control of Human Reproduction. New York, Academic Press, 1977.

Hytten, F. E., and Leitch, I.: The Physiology of Human Pregnancy. Philadelphia, J. B. Lippincott Co., 1974.

Inskeep, E. K., and Murdoch, W. J.: Relation of ovarian functions to uterine and ovarian secretion of prostaglandins during the estrous cycle and early pregnancy in the ewe and cow. *In* Greep, R. O. (ed.): International Review of Physiology: Reproductive Physiology III. Vol. 22. Baltimore, University Park Press, 1980, p. 325.

Jaffe, R. B.: The endocrinology of pregnancy. *In* Yen, S. S. C., and Jaffe, R. B. (eds.): Reproductive Endocrinology. Philadelphia, W. B. Saunders Co., 1978, p. 521.

Josimovich, J. B., *et al.* (eds.): Lactogenic Hormones, Fetal Nutrition, and Lactation. New York, John Wiley & Sons, 1974.

Klopper, A.: The hormones of the placenta and their role in the onset of labor. *In* MTP International Review of Science: Physiology. Vol. 8. Baltimore, University Park Press, 1974, p. 179.

Lawrence, R. A.: Breast-Feeding; A Guide for the Medical Profession. St. Louis, C. V. Mosby, 1979.

Li, C. H. (ed.): The Chemistry of Prolactin. New York, Academic Press, 1980.

Lipsett, M. B.: Endocrine mechanisms of reproduction. *In* Frohlich, E. D. (ed.): Pathophysiology, 2nd Ed. Philadelphia, J. B. Lippincott Co., 1976, p. 405.

Loke, Y, W.: Immunology and Immunopathology of the Human Foetal-Maternal Interaction. New York, Elsevier/North-Holland, 1978.

Marshall, J. M.: Effects of neurohypophyseal hormones on the myometrium. *In* Greep, R. O., and Astwood, E. B. (eds.): Handbook of Physiology. Sec. 7, Vol. 4, Part 1. Baltimore, Williams & Wilkins, 1974, p. 469.

Mittwoch, U.: Genetics of Sex Differentiation. New York, Academic Press, 1973.

Nathanielsz, P. W.: Endocrine mechanisms of parturition. *Annu. Rev. Physiol., 40*:411, 1978.

Pauerstein, C. J. (ed.): Seminar on Tubal Physiology and Biochemistry. New York, S. Karger, 1975.

Pitkin, R. M.: Calcium metabolism in pregnancy: A review. *Am. J. Obstet. Gynecol., 121*:724, 1975.

Psychoyos, A.: Endocrine control of egg implantation. *In* Greep, R. O., and Astwood, E. B. (eds.): Handbook of Physiology. Sec. 7, Vol. 2, Part 2. Baltimore, Williams & Wilkins, 1973, p. 187.

Ramsey, E. M.: Placental vasculature and circulation. *In* Greep, R. O., and Astwood, E. B. (eds.): Handbook of Physiology. Sec. 7, Vol. 2, Part 2. Baltimore, Williams & Wilkins, 1973, p. 323.

Smith, M. S.: Role of prolactin in mammalian reproduction. *In* Greep, R. O. (ed.): International Review of Physiology: Reproductive Physiology III. Vol. 22. Baltimore, University Park Press, 1980, p. 249.

Steinberger, E.: Genetics, anatomy, fetal endocrinology. *In* DeGroot, L. J., *et al.* (eds.): Endocrinology. Vol. 3. New York, Grune & Stratton, 1979, p. 1309.

Sutherland, H. W., and Stowers, J. M. (eds.): Carbohydrate Metabolism in Pregnancy and the Newborn 1979. New York, Springer-Verlag, 1979.

Symonds, E. M. (ed.): Hypertensive States in Pregnancy. Clinics in Obstetrics-Gynaecology. London, W. B. Saunders Co., Dec., 1977.

Thorburn, G. D., and Challis, J. R. G.: Endocrine control of parturition. *Physiol. Rev., 59*:863, 1979.

Tyson, J. E. (ed.): Symposium on Neuroendocrinology of Reproduction. Clinics in Obstetrics-Gynaecology. London, W. B. Saunders Company, Aug. 1978.

Vorherr, H. (ed.): Human Lactation. New York, Grune & Stratton, 1979.

Wilson, J. D.: Sexual differentiation. *Annu. Rev. Physiol., 40*:279, 1978.

Winick, M. (ed.): Nutrition, Pre- and Postnatal Development. New York, Plenum Press, 1979.

Yen, S. S. C.: Metabolic homeostasis during pregnancy. *In* Yen, S. S. C., and Jaffe, R. B. (eds.): Reproductive Endocrinology. Philadelphia, W. B. Saunders Co., 1978, p. 537.

Yen, S. S. C.: Physiology of human prolactin. *In* Yen, S. S. C., and Jaffe, R. B. (eds.): Reproductive Endocrinology. Philadelphia, W. B. Saunders Co., 1978, p. 152.

Yen, S. S. C., and Jaffe, R. B. (eds.): Reproductive Endocrinology: Physiology, Pathophysiology, and Clinical Management. Philadelphia, W. B. Saunders Co., 1978.

Yoshinaga, K.: Hormonal interplay in the establishment of pregnancy. *Int. Rev. Physiol., 13*:201, 1977.

83

Special Features of Fetal and Neonatal Physiology

A complete discussion of fetal development, functioning of the child immediately after birth, and growth and development through the early years of life lies within the province of formal courses in obstetrics and pediatrics. However, many aspects of these are strictly physiological problems, some of which relate to the physiological principles that we have discussed for the adult and some of which are peculiar to the infant itself. The present chapter delineates and discusses some of the most important of these special problems.

GROWTH AND FUNCTIONAL DEVELOPMENT OF THE FETUS

Early development of the placenta and of the fetal membranes occurs far more rapidly than development of the fetus itself. During the first two to three weeks the fetus remains almost microscopic in size, but thereafter, as illustrated in Figure 83–1, the dimensions of the fetus increase almost in proportion to age. At 12 weeks the length of the fetus is approximately 10 cm.; at 20 weeks, approximately 25 cm.; and at term (40 weeks), approximately 53 cm. (about 21 inches). Because the weight of the fetus is proportional to the cube of the length, the weight increases approximately in proportion to the cube of the age of the fetus. Note from Figure 83–1 that

the weight of the fetus remains almost nothing during the first months and reaches 1 pound only at five and a half months of gestation. Then, during the last trimester of pregnancy, the fetus gains tremendously so that two months prior to birth the weight averages 3 pounds, one month prior to birth 4.5 pounds, and at birth 7 pounds, this birth weight varying from as low as 4.5 pounds to as high as 11 pounds in completely normal infants with completely normal gestational periods.

DEVELOPMENT OF THE ORGAN SYSTEMS

Within one month after fertilization of the ovum all the different organs of the fetus have already been "blocked out," and during the next two to three months, most of the minute details of the different organs are established. Beyond the fourth month, the organs of the fetus are grossly the same as those of the newborn child. However, cellular development of these structures is usually far from complete at this time and requires the full remaining five months of pregnancy for complete development. Even at birth certain structures, particularly the nervous system, the kidneys, and the liver, still lack full development, which will be discussed in more detail later in the chapter.

The Circulatory System. The human heart begins beating during the fourth week following fertilization, contracting at the rate of about 65 beats per minute. This increases steadily as the fetus grows and reaches a rate of approximately 140 beats per minute immediately before birth.

Formation of Blood Cells. Nucleated red blood cells begin to be formed in the yolk sac and mesothelial layers of the placenta at about the third week of fetal development. This is followed a week later by the formation of non-nucleated red blood cells by the fetal mesenchyme and by the endothelium of the fetal blood vessels. Then at approximately six weeks, the liver begins to form blood cells, and in the third month the spleen and other lymphoid tissues of the body also begin forming blood cells. Finally, from approximately the third month on, the bone marrow also forms red and white blood cells. During the midportion of fetal life, the

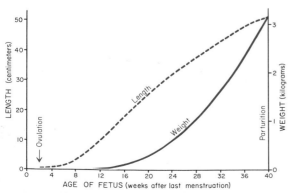

Figure 83–1. Growth of the fetus.

extramarrow areas are the major sources of the fetus' blood cells, but, during the latter three months of fetal life, the bone marrow gradually takes over while these other structures lose their ability completely to form blood cells.

The Respiratory System. Obviously, respiration cannot occur during fetal life. However, respiratory movements do take place beginning at the end of the first trimester of pregnancy. Tactile stimuli or fetal asphyxia especially cause respiratory movements.

However, during the latter three to four months of pregnancy, the respiratory movements of the fetus are mainly inhibited, for reasons unknown. This could possibly result from (1) special chemical conditions in the body fluids of the fetus, (2) the presence of fluid in the fetal lungs, or (3) other possible unexplored stimuli.

The inhibition of respiration during the latter months of fetal life prevents filling of the lungs with debris from the meconium excreted by the gastrointestinal tract into the amniotic fluid. Also, fluid is secreted into the lungs by the alveolar epithelium up until the moment of birth, thus filling the pulmonary spaces with this clean secretion.

FUNCTION OF THE NERVOUS SYSTEM

Most of the skin reflexes of the fetus are present by the third to fourth months of pregnancy. However, most higher functions of the central nervous system that involve the cerebral cortex are still undeveloped even at birth. Indeed, myelinization of some major tracts of the central nervous system becomes complete only after approximately a year of postnatal life.

FUNCTION OF THE GASTROINTESTINAL TRACT

Even in mid-pregnancy the fetus ingests and absorbs large quantities of amniotic fluid, and, during the last two to three months, gastrointestinal function approaches that of the normal newborn infant. Small quantities of *meconium* are continually formed in the gastrointestinal tract, and excreted from the bowels into the amniotic fluid. Meconium is composed partly of unabsorbed residue of amniotic fluid and partly of excretory products from the gastrointestinal mucosa and glands.

FUNCTION OF THE KIDNEYS

The fetal kidneys are capable of excreting urine during at least the latter half of pregnancy, and urination occurs normally *in utero*. However, the renal control systems for regulation of extracellular fluid electrolyte balances and especially acid-base balance are almost nonexistent until after mid-fetal life and do not reach full development until a few months after birth.

METABOLISM IN THE FETUS

The fetus utilizes mainly glucose for energy, and it has a high rate of storage of fat and protein, much if not

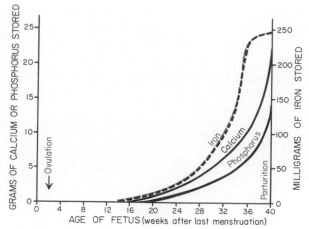

Figure 83–2. Calcium, phosphorus, and iron storage in the fetus at different stages of gestation.

most of the fat being synthesized from glucose rather than being absorbed from the mother's blood. Aside from these generalities there are some special problems of fetal metabolism in relation to calcium, phosphate, iron, and some vitamins, discussed next.

Metabolism of Calcium and Phosphate. Figure 83–2 illustrates the rate of calcium and phosphate accumulation in the fetus, showing that approximately 22.5 grams of calcium and 13.5 grams of phosphorus are accumulated in the average fetus during gestation. Approximately half of this accumulates during the last four weeks of gestation, which is also coincident with the period of rapid ossification of the fetal bones as well as with the period of rapid weight gain of the fetus.

During the earlier part of fetal life, the bones are relatively unossified, and have mainly a cartilaginous matrix. Indeed, x-ray pictures ordinarily will not show ossification until approximately the fourth month of pregnancy.

Note especially that the total amounts of calcium and phosphate needed by the fetus during gestation represent only about one-fiftieth the quantities of these substances in the mother's bones. Therefore, this is a minimal drain from the mother. However, a much greater drain occurs after birth during lactation.

Accumulation of Iron. Figure 83–2 shows that iron accumulates in the fetus somewhat more rapidly than calcium and phosphates. Most of the iron is in the form of hemoglobin, which begins to be formed as early as the third week following fertilization of the ovum.

Small amounts of iron are concentrated in the progestational endometrium even prior to implantation of the ovum; this iron is ingested into the embryo by the trophoblastic cells for early formation of the red blood cells.

Approximately one-third of the iron in a fully developed fetus is normally stored in the liver. This iron can then be used for several months after birth by the newborn infant for formation of additional hemoglobin.

Utilization and Storage of Vitamins. The fetus needs vitamins equally as much as the adult and in some instances to a far greater extent. In general, the vitamins function the same in the fetus as in the adult, as discussed in Chapter 74. Special functions of several vitamins should be mentioned, however.

The B vitamins, especially vitamin B_{12} and folic acid, are necessary for formation of red blood cells and also for overall growth of the fetus.

Vitamin C is necessary for appropriate formation of intercellular substances, especially the bone matrix and fibers of connective tissue.

Vitamin D probably is not necessary for fetal growth, although the mother needs it for adequate absorption of calcium from her gastrointestinal tract. If the mother has plenty of this vitamin in her body fluids, large quantities will be stored by the fetal liver to be used by the newborn child for several months after birth.

Vitamin E, though its precise function is unknown, is necessary for normal development of the early embryo. In its absence in experimental animals, spontaneous abortion usually occurs at an early age.

Vitamin K is used by the fetal liver for formation of factor VII, prothrombin, and several other blood coagulation factors. When vitamin K is insufficient in the mother, factor VII and prothrombin become deficient in the child as well as in the mother. Since most Vitamin K absorbed into the body is formed by bacterial action in the colon, the newborn child has no adequate source of vitamin K for the first week or so of life — that is, until a normal colonic bacterial flora becomes established. Therefore, prenatal storage of at least small amounts of vitamin K is helpful in preventing hemorrhage — particularly in the brain when the head is traumatized by squeezing through the birth canal.

ADJUSTMENTS OF THE INFANT TO EXTRAUTERINE LIFE

ONSET OF BREATHING

The most obvious effect of birth on the baby is loss of the placental connection with the mother, and therefore loss of this means for metabolic support. And by far the most important immediate adjustment required of the infant is the onset of breathing.

Cause of Breathing at Birth. Following completely normal delivery from a mother who has not been depressed by anesthetics, the child ordinarily begins to breathe immediately and has a completely normal respiratory rhythm from the onset. The promptness with which the fetus begins to breathe indicates that breathing is initiated by sudden exposure to the exterior world, probably resulting from a slightly asphyxiated state incident to the birth process but also from sensory impulses originating in the suddenly cooled skin. However, in an infant who does not breathe immediately, the body becomes progressively more hypoxic and hypercapnic, which provides additional stimulus to the respiratory center and usually causes breathing within a few seconds to a few minutes after birth.

Delayed and Abnormal Breathing at Birth — Danger of Hypoxia. If the mother has been depressed by an anesthetic during delivery, which at least partially also anesthetizes the child, respiration is likely to be delayed for several minutes, thus illustrating the importance of using as little obstetrical anesthesia as feasible. Also, many infants who have had head trauma during delivery or who undergo very prolonged delivery are slow to breathe or sometimes will not breathe at all. This can result from two possible effects: first, in a few infants, intracranial hemorrhage or brain contusion causes a concussion syndrome with a greatly depressed respiratory center. Second, and probably much more important, prolonged fetal hypoxia during delivery also causes serious depression of the respiratory center. Hypoxia frequently occurs during delivery because of (a) compression of the umbilical cord; (b) premature separation of the placenta; (c) excessive contraction of the uterus, which cuts off the blood flow to the placenta; or (d) excessive anesthesia of the mother, which depresses the oxygenation even of her blood.

Degree of Hypoxia That an Infant Can Tolerate. In the adult, failure to breathe for only four minutes often causes death, but a newborn infant often survives as long as 15 minutes of failure to breathe after birth. Unfortunately, though, permanent and very evident brain impairment often ensues if breathing is delayed more than 8 to 10 minutes. Indeed, actual lesions develop mainly in the thalamus, the inferior colliculi, and in other brain stem areas, thus affecting many of the stereotypical motor functions of the body.

Expansion of the Lungs at Birth. At birth, the walls of the alveoli are kept collapsed by the surface tension of the viscid fluid that fills them. More than 25 mm. Hg of negative pressure is required to oppose the effects of this surface tension and therefore to open the alveoli for the first time. But once the alveoli are open, further respiration can be effected with relatively weak respiratory movements. Fortunately, the first inspirations of the newborn infant are extremely powerful, usually capable of creating as much as 60 mm. Hg negative pressure in the intrapleural space.

Figure 83–3 illustrates the tremendous forces required to open the lungs at the onset of breathing. To the left is shown the pressure-volume curve (compliance curve) for the first breath after birth. Observe, first, the lowermost curve, which shows that the lungs hardly expand at all until the negative pressure has reached −40 cm. water (−30 mm. Hg). Then, as the negative pressure increases to −60 cm. water, about 40 ml. of air enters the lungs. To deflate the lungs, considerable positive pressure is required because of the viscous resistance offered by the fluid in the bronchioles.

Note that the second breath is much easier. However, breathing does not become completely normal until about 40 minutes after birth, as shown by the third

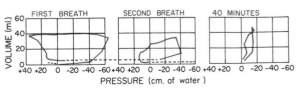

Figure 83–3. Pressure-volume curves of the lungs (compliance curves) of a newborn baby immediately after birth, showing (a) the extreme forces required for breathing during the first two breaths of life and (b) development of a nearly normal compliance within 40 minutes after birth. (From Smith: *Sci. Amer., 209*:32, 1963. © 1963 by Scientific American, Inc. All rights reserved.)

compliance curve, the shape of which compares favorably with that for the normal adult, as shown in Chapter 39.

Respiratory Distress Syndrome. A small number of infants, especially premature infants and infants born of diabetic mothers, develop severe respiratory distress during the few hours to several days following birth and frequently succumb within the next day or so. The alveoli of these infants at death contain large quantities of proteinaceous fluid, almost as if pure plasma had leaked out of the capillaries into the alveoli. The fluid also contains desquamated alveolar epithelial cells. This condition is also called *hyaline membrane disease* because microscopic slides of the lung show this alveolar material to look like a hyaline membrane.

Unfortunately, the cause of the respiratory distress syndrome is not certain. However, one of the most characteristic findings is failure to secrete adequate quantities of *surfactant,* a substance normally secreted into the alveoli, which decreases the surface tension of the alveolar fluid, therefore allowing the alveoli to open easily. Though this failure to secrete surfactant may be secondary to many other pathological changes in the pulmonary membrane, some research workers believe it to be the basic cause of the disease and that the pathology is the secondary result. The surfactant secreting cells (the type II alveolar epithelial cells) do not begin to secrete surfactant until the last one to three months of gestation. Therefore, many premature babies and some full-term babies are born without the capability of secreting surfactant, which therefore causes both a collapse tendency of the lungs and development of pulmonary edema. The role of surfactant in preventing these effects was discussed in Chapter 39.

CIRCULATORY READJUSTMENTS AT BIRTH

Almost as important as the onset of breathing at birth are the immediate circulatory adjustments that allow adequate blood flow through the lungs. Also, circulatory adjustments during the first few hours of life shunt more and more blood through the liver as well. To describe these readjustments we must first consider briefly the anatomical structure of the fetal circulation.

Specific Anatomic Structure of the Fetal Circulation. Because the lungs are mainly nonfunctional during fetal life and because the liver is only partially functional, it is not necessary for the fetal heart to pump much blood through either the lungs or the liver. On the other hand, the fetal heart must pump large quantities of blood through the placenta. Therefore, special anatomical arrangements cause the fetal circulatory system to operate considerably differently from that of the adult. First, as illustrated in Figure 83–4, blood returning from the placenta through the umbilical vein passes through the *ductus venosus,* mainly by-passing the liver. Then, most of the blood entering the right atrium from the inferior vena cava is directed in a straight pathway across the posterior aspect of the right atrium and thence through the *foramen ovale* directly into the left atrium. Thus, this well-oxygenated blood from the placenta enters the left side of the heart rather than the right side and is pumped by the left ventricle mainly into the vessels of the head and forelimbs.

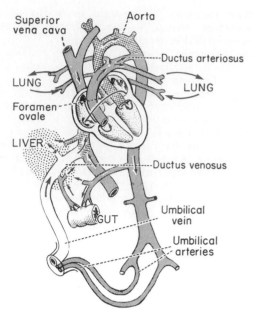

Figure 83–4. Organization of the fetal circulation. (Modified from Arey: Developmental Anatomy, 7th Ed.)

The blood entering the right atrium from the superior vena cava is directed downward through the tricuspid valve into the right ventricle. This blood is mainly deoxygenated blood from the head region of the fetus, and it is pumped by the right ventricle into the pulmonary artery, then mainly through the *ductus arteriosus* into the descending aorta and through the two umbilical arteries into the placenta. Thus, the deoxygenated blood becomes oxygenated.

Figure 83–5 illustrates the relative proportions of the total blood pumped by the heart that passes through the different vascular circuits of the fetus. This figure shows that 55 per cent of all the blood goes through the placenta, leaving only 45 per cent to pass through all the

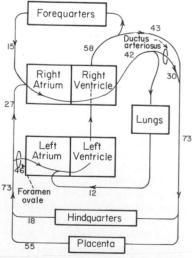

Figure 83–5. Diagram of the fetal circulatory system, showing relative distribution of blood flow to the different vascular areas. The numerals represent per cent of the total cardiac output flowing through the particular area.

tissues of the fetus. Furthermore, during fetal life, only 12 per cent of the blood flows through the lungs; immediately after birth 3 to 4 times this much must flow through the lungs, indicating a several-fold increase at birth.

Changes in Fetal Circulation at Birth. The basic changes in the fetal circulation at birth were discussed in Chapter 27 in relation to congenital anomalies of the ductus arteriosus and foramen ovale that persist throughout life. Briefly, these changes are the following:

Primary Changes in Pulmonary and Systemic Vascular Resistance at Birth. The primary changes in the circulation at birth are first, loss of the tremendous blood flow through the placenta, which *approximately doubles the systemic vascular resistance at birth.* This obviously *increases the aortic pressure* as well as the pressures in the left ventricle and left atrium.

Second, the *pulmonary vascular resistance greatly decreases* as a result of expansion of the lungs. In the unexpanded fetal lungs, the blood vessels are compressed because of the small volume of the lungs. Immediately upon expansion these vessels are no longer compressed, and the resistance to blood flow decreases several-fold. Also, in fetal life the hypoxia of the lungs causes considerable tonic vasoconstriction of the lung blood vessels, but vasodilation takes place when aeration of the lungs eliminates the hypoxia. These changes reduce the resistance to blood flow through the lungs as much as five-fold, which obviously *reduces the pulmonary arterial pressure,* the right ventricular pressure, and the right atrial pressure.

Closure of the Foramen Ovale. The *low right atrial pressure* and the *high left atrial pressure* that occur secondarily to the changes in pulmonary and systemic resistances at birth cause a tendency for blood to flow backward from the left atrium into the right atrium rather than in the other direction, as occurred during fetal life. Consequently, the small valve that lies over the foramen ovale on the left side of the atrial septum closes over this opening, thereby preventing further flow. In two-thirds of all persons the valve becomes adherent over the foramen ovale within a few months to a few years and forms a permanent closure. But, even if permanent closure does not occur, the left atrial pressure throughout life remains 2 to 4 mm. Hg greater than the right atrial pressure, and the back pressure keeps the valve closed.

Closure of the Ductus Arteriosus. The ductus arteriosus also closes, but for different reasons. First, the increased systemic resistance *elevates the aortic pressure* while the decreased pulmonary resistance *reduces the pulmonary arterial pressure.* As a consequence, within a few hours after birth, blood begins to flow backward from the aorta into the pulmonary artery rather than in the other direction as in fetal life. However, after only a few hours the muscular wall of the ductus arteriosus constricts markedly, and within one to eight days the constriction is sufficient to stop all blood flow. This is called *functional closure* of the ductus arteriosus. Then, during the next one to four months the ductus arteriosus ordinarily becomes anatomically *occluded* by growth of fibrous tissue into its lumen.

The causes of either functional closure or anatomical closure of the ductus are not completely known. How- ever, the most likely cause is increased oxygenation of the blood flowing through the ductus. In fetal life the P_{O_2} of the ductus blood is only 15 to 20 mm. Hg, but it increases to about 100 mm. Hg within a few hours after birth. Furthermore, many experiments have shown that the degree of contraction of the smooth muscle in the ductus wall is highly related to the availability of oxygen.

In one of several thousand infants the ductus fails to close, resulting in a *patent ductus arteriosus,* the consequences of which were discussed in Chapter 27.

Closure of the Ductus Venosus. In fetal life, the portal blood joins the blood from the umbilical vein and then passes through the ductus venosus directly into the vena cava, thus by-passing the liver. Immediately after birth, blood flow through the umbilical vein ceases, but most of the portal blood still flows through the ductus venosus, with only a small amount passing through the channels of the liver. However, within one to three hours the muscular wall of the ductus venosus contracts strongly and closes this avenue of flow. As a consequence, the portal venous pressure rises from about 0 mm. Hg to 6 to 10 mm. Hg, which is enough to force blood flow through the liver sinuses. Although the ductus venosus almost never fails to close, unfortunately we know almost nothing about what causes the closure.

NUTRITION OF THE NEWBORN INFANT

The fetus obtains almost all of its energy from glucose obtained from the mother's blood. Immediately after birth, the amount of glucose stored in the infant's body in the form of glycogen is sufficient to supply the infant's needs for only a few hours, and unfortunately the liver of the newborn infant is still far from functionally adequate at birth, which prevents significant gluconeogenesis. Therefore, the infant's blood glucose concentration frequently falls the first day to as low as 30 to 40 mg./100 ml. of plasma. Fortunately, the appropriate mechanisms are available so that the infant can utilize stored fats and proteins for metabolism until mother's milk can be provided two to three days later.

Special problems are also frequently associated with getting an adequate fluid supply to the newborn infant, because the infant's rate of body fluid turnover averages 7 times that of an adult, and the mother's milk supply requires several days to develop. Ordinarily, the infant's weight decreases 5 to 10 per cent and sometimes as much as 20 per cent within the first two to three days of life. Most of this weight loss is loss of fluid rather than of body solids.

SPECIAL FUNCTIONAL PROBLEMS IN THE NEONATAL INFANT

The most important characteristic of the newborn infant is instability of the various hormonal and neurogenic control systems. This results partly from the immature development of the different organs of the body and partly from the fact that the control systems simply have not become adjusted to the completely new way of life.

The Respiratory System.   The normal rate of respiration in the newborn is about 40 breaths per minute, and tidal air with each breath averages 16 ml. This gives a total minute respiratory volume of 640 ml. per minute, which is about 2 times as great in relation to the body weight as that of an adult. *The functional residual capacity of the infant is only half that of an adult in relation to body weight.* This allows cyclic changes in blood gas concentration when the respiration becomes slowed.

Blood Volume.   The blood volume of a newborn infant immediately after birth averages about 300 ml., but, if the infant is left attached to the placenta for a few minutes after birth or if the umbilical cord is stripped to force blood out of its vessels into the baby, an additional 75 ml. of blood enters the infant to make a total of 375 ml. Then, during the ensuing few hours, fluid is lost into the tissue spaces from this blood, which increases the hematocrit but returns the blood volume once again to the normal value of about 300 ml. Some pediatricians feel that this extra blood volume in some instances causes mild pulmonary edema with some degree of respiratory distress.

Cardiac Output.   The cardiac output of the newborn infant averages 550 ml./minute, which, like respiration and body metabolism, is about 2 times as much in relation to body weight as in the adult. Occasionally a child is born with an especially low cardiac output caused by hemorrhage through the placental membrane into the mother's body prior to birth.

Arterial Pressure.   The arterial pressure during the first day after birth averages about 70/50; this increases slowly during the next several months to approximately 90/60. Then there is a much slower rise during the subsequent years until the adult pressure of 120/80 is attained at adolescence.

Blood Characteristics.   The red blood cell count in the newborn infant averages about 4 million per cubic millimeter. If blood is stripped from the cord into the infant, the red blood cell count rises an additional 0.5 to 0.75 million during the first few hours of life, giving a red blood cell count of about 4.75 million per cubic mm., as illustrated in Figure 83–6. Subsequent to this, however, few new red blood cells are formed in the infant during the first few weeks of life, presumably because the hypoxic stimulus of fetal life is no longer present to stimulate red cell production. Thus, as shown in Figure 83–6, the average red blood cell count falls to 3.25 million per cubic mm. by about 8 to 10 weeks of age. From that time on, increasing activity by the fetus provides the appropriate stimulus for returning the red blood cell count to normal within another two to three months.

Immediately after birth, the white blood cell count of the infant is about 45,000 per cubic mm., which is about 5 times as great as that of the normal adult.

Neonatal Jaundice and Erythroblastosis Fetalis.   Bilirubin formed in the fetus can cross the placenta and be excreted through the liver of the mother, but immediately after birth the only means for ridding the infant of biliribin is through the infant's own liver, which, for the first week or so of life, still functions poorly and is incapable of conjugating significant quantities of biliru-

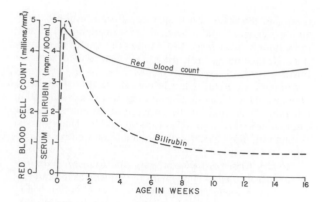

Figure 83–6.   Changes in the red blood cell count and in the serum bilirubin concentration during the first 16 weeks of life, showing "physiologic anemia" at 6 to 12 weeks of life and "physiologic hyperbilirubinemia" during the first two weeks of life.

bin with glucuronic acid for excretion into the bile. Consequently, the plasma bilirubin concentration rises from a normal value of less than 1 mg./100 ml. to an average of 5 mg./100 ml. during the first three days of life and then gradually falls back to normal as the liver becomes functional. This condition, called *physiological hyperbilirubinemia*, is illustrated in Figure 83–6, and it is associated with a mild *jaundice* (yellowness) of the infant's skin and especially of the sclerae of its eyes.

However, by far the most important cause of serious neonatal jaundice is *erythroblastosis fetalis*, which was discussed in detail in Chapter 5 in relation to Rh factor incompatibility between the infant and mother. Briefly, the baby inherits an Rh positive trait from the father while the mother is Rh negative. The mother then becomes immunized against the fetus' blood, and her antibodies in turn destroy the infant's red blood cells, releasing extreme quantities of bilirubin into the plasma. Before the advent of modern obstetrical methods, this condition occurred either mildly or seriously in 1 of every 50 to 100 newborn infants.

Fluid Balance, Acid-Base Balance, and Renal Function.   The rate of fluid intake and fluid excretion in the infant is 7 times as great in relation to weight as in the adult, which means that even a slight alteration of fluid balance can cause rapidly developing abnormalities. Second, the rate of metabolism in the infant is 2 times as great in relation to body mass as in the adult, which means that 2 times as much acid is normally formed, which leads to a tendency toward acidosis in the infant. Third, functional development of the kidneys is not complete until the end of approximately the first month of life. For instance, the kidneys of the newborn can concentrate urine to only 1.5 times the osmolality of the plasma instead of the normal 3 to 4 times in the adult.

Therefore, considering the immaturity of the kidney, together with the marked fluid turnover in the infant and rapid formation of acid, one can readily understand that among the most important problems of infancy are acidosis, dehydration, and rare instances of overhydration.

Liver Function.   During the first few days of life, liver function may be quite deficient, as evidenced by the following effects:

1. The liver of the newborn conjugates bilirubin with glucuronic acid poorly and therefore excretes bilirubin only slightly during the first few days of life.

2. The liver of the newborn is deficient in forming plasma proteins, so that the plasma protein concentration falls to 1 gram per cent less than that for older children. Occasionally, the protein concentration falls so low that the infant actually develops hypoproteinemic edema.

3. The gluconeogenetic function of the liver is particularly deficient. As a result, the blood glucose level of the unfed newborn infant falls to about 30 to 40 mg. per cent, and the infant must depend on its stored fats for energy until feeding can occur.

4. The liver of the newborn usually also forms too little of the factors needed for normal blood coagulation.

Digestion, Absorption, and Metabolism of Energy Foods. In general, the ability of the newborn infant to digest, absorb, and metabolize foods is not different from that of the older child, with the following three exceptions:

First, secretion of pancreatic amylase in the newborn infant is deficient so that the infant utilizes starches less adequately than do older children.

Second, absorption of fats from the gastrointestinal tract is somewhat less than that in the older child. Consequently, milk with a high fat content, such as cow's milk, is frequently inadequately utilized.

Third, because the liver functions are imperfect during at least the first week of life, the glucose concentration in the blood is unstable and also low.

The newborn is especially capable of synthesizing proteins and thereby storing nitrogen. Indeed, with a completely adequate diet, as much as 90 per cent of the ingested amino acids are utilized for formation of body proteins. This is a much higher percentage than in adults.

Metabolic Rate and Body Temperature. The normal metabolic rate of the newborn in relation to body weight is about 2 times that of the adult, which accounts also for the 2 times as great cardiac output and 2 times as great minute respiratory volume in the infant.

However, since the body surface area is very large in relation to the body mass, heat is readily lost from the body. As a result, the body temperature of the newborn infant, particularly of premature infants, falls easily. Figure 83–7 shows that the body temperature, of even the normal infant, falls several degrees during the first few hours after birth, but returns to normal in seven to eight hours. Still, the body temperature regulatory mechanisms remain poor during the early days of life, allowing marked deviations in temperature at first, which are also illustrated in Figure 83–7.

Nutritional Needs During the Early Weeks of Life. At birth, a newborn infant is usually in complete nutritional balance, provided the mother has had an adequate diet. Furthermore, function of the gastrointestinal system is usually more than adequate to digest and assimilate all the nutritional needs of the infant if these are provided in the diet. However, three specific problems do occur in the early nutrition of the infant:

Need for Calcium and Vitamin D. The newborn infant has only just begun rapid ossification of bones at birth so that a ready supply of calcium throughout infancy is

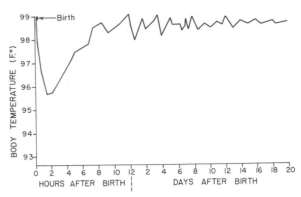

Figure 83–7. Fall in body temperature of the infant immediately after birth, and instability of body temperature during the first few days of life.

needed. This is ordinarily supplied adequately by its usual diet of milk. Yet absorption of calcium by the gastrointestinal tract is poor in the absence of vitamin D. Therefore, the vitamin D–deficient infant can develop severe rickets in only a few weeks. This is particularly true in premature babies since their gastrointestinal tracts absorb calcium even less effectively than those of normal infants.

Necessity for Iron in the Diet. If the mother has had adequate amounts of iron in her diet, the liver of the infant usually has stored enough iron to keep forming blood cells for four to six months after birth. But if the mother has had insufficient iron in her diet, severe anemia is likely to supervene in the infant after about three months of life. To prevent this possibility, early feeding of egg yolk, which contains reasonably large quantities of iron, or administration of iron in some other form is desirable by the second or third month of life.

Vitamin C Deficiency in Infants. Ascorbic acid (vitamin C) is not stored in significant quantities in the fetal tissues; yet it is required for proper formation of cartilage, bone, and other intercellular structures of the infant. Furthermore, milk has poor supplies of ascorbic acid, especially cow's milk, which has only one-fourth as much as mother's milk. For this reason, orange juice or other sources of ascorbic acid are usually prescribed by the third week of life.

Immunity. Fortunately, the neonate inherits much immunity from the mother because many antibodies diffuse from the mother's blood through the placenta into the fetus. However, the neonate does not form antibodies to a significant extent. By the end of the first month, the baby's gamma globulins, which contain the antibodies, have decreased to less than one-half the original level, with corresponding decrease in immunity. Thereafter, the baby's own immunization processes begin to form antibodies, and the gamma globulin concentration returns essentially to normal by the age of 12 to 20 months.

Despite the decrease in gamma globulins soon after birth, the antibodies inherited from the mother still protect the infant for about six months against most major childhood infectious diseases, including diphtheria, measles, smallpox, and polio. Therefore, immunization against these diseases before six months is usually unnecessary. On the other hand, the inherited antibodies against whooping cough are normally insufficient to protect the neonate; therefore, for full safety the infant requires immunization against this disease within the first month or so of life.

Allergy. Fortunately, the newborn infant is rarely subject to allergy. Several months later, however, when the neonate's own antibodies first begin to form, extreme allergic states can develop, often resulting in serious eczema, gastrointestinal abnormalities, or even anaphylaxis. As the child grows older and develops still higher degrees of immunity, these allergic manifestations usually disappear. This relationship of mild immunity to allergy was discussed in Chapter 7.

Endocrine Problems. Ordinarily the endocrine system of the infant is highly developed at birth, and the infant rarely exhibits any immediate endocrine abnormalities. However, there are special instances in which endocrinology of infancy is important.

1. If a pregnant mother bearing a female child is treated with an androgenic hormone or if she develops an androgenic tumor during pregnancy, the child will be born with a high degree of masculinization of her sexual organs, thus resulting in a type of *hermaphroditism.*

2. The sex hormones secreted by the placenta and by the mother's glands during pregnancy occasionally cause the neonate's breasts to form milk during the first few days of life. Sometimes the breasts then become inflamed or even develop infectious mastitis.

3. An infant born of a diabetic mother will have considerable hypertrophy and hyperfunction of the islets of Langerhans. As a consequence, the infant's blood glucose concentration may fall lower than 20 mg./100 ml. shortly after birth. Fortunately, the newborn infant, unlike the adult, only rarely develops insulin shock or coma from this low level of blood glucose concentration.

Because of metabolic deficits in the diabetic mother, the fetus is often stunted in growth, and growth and tissue maturation of the newborn infant are often impaired. Also, there is a high rate of intrauterine mortality, and of those fetuses that do come to term, there is still a high mortality rate. Two-thirds of the infants who die succumb to the respiratory distress syndrome, which was described earlier in the chapter.

4. Occasionally, a child is born with hypofunctional adrenal cortices, perhaps resulting from *agenesis* of the glands or *exhaustion atrophy,* which can occur when the adrenal glands have been overstimulated.

5. If a pregnant women has hyperthyroidism or is treated with excess thyroid hormone, the infant is likely to be born with a temporarily hyposecreting thyroid gland. On the other hand, if prior to pregnancy a woman had had her thyroid gland removed, her pituitary may secrete great quantities of thyrotropin during gestation, and the child might be born with temporary hyperthyroidism.

SPECIAL PROBLEMS OF PREMATURITY

All the problems just noted for neonatal life are especially exacerbated in prematurity. These can be categorized under the following two headings: (1) imma-

turity of certain organ systems and (2) instability of the different homeostatic control systems. Because of these effects, a premature baby rarely lives if it is born more than 2.5 to 3 months prior to term.

IMMATURE DEVELOPMENT OF THE PREMATURE INFANT

Amost all the organ systems of the body are immature in the premature infant, but some require particular attention if the life of the premature baby is to be saved.

Respiration. The respiratory system is especially likely to be underdeveloped in the premature infant. The vital capacity and the functional residual capacity of the lungs are especially small in relation to the size of the infant. Also, surfactant secretion is especially depressed. As a consequence, the respiratory distress syndrome is a common cause of death. Also, the low functional residual capacity in the premature infant is often associated with periodic breathing of the Cheyne-Stokes type.

Gastrointestinal Function. Another major problem of the premature infant is to ingest and absorb adequate food. If the infant is more than two months premature, the digestive and absorptive systems are almost always inadequate. The absorption of fat is so poor that the premature infant must have a low fat diet. Furthermore, the premature infant has unusual difficulty in absorbing calcium and therefore can develop severe rickets before one recognizes the difficulty. For this reason, special attention must be paid to adequate calcium and vitamin D intake.

Function of Other Organs. Immaturity of other organ systems that frequently causes serious difficulties in the premature infant includes: (1) immaturity of the liver, which results in poor intermediary metabolism and often also a bleeding tendency as a result of poor formation of coagulation factors; (b) immaturity of the kidneys, which are particularly deficient in their ability to rid the body of acids, thereby predisposing to acidosis as well as to serious fluid balance abnormalities; (c) immaturity of the blood-forming mechanism of the bone marrow, which allows rapid development of anemia; and (d) depressed formation of gamma globulin by the reticuloendothelial system, which is often associated with serious infection.

INSTABILITY OF THE CONTROL SYSTEMS IN THE PREMATURE INFANT

Immaturity of the different organ systems in the premature infant creates a high degree of instability in the homeostatic systems of the body. For instance, the acid-base balance can vary tremendously, particularly when the food intake varies from time to time. Likewise, the blood protein concentration is usually somewhat low because of immature liver development, often leading to *hypoproteinemic edema.* And inability of the infant to regulate its calcium ion concentration frequently brings on hypocalcemic tetany. Also, the blood glucose concentration can vary between the extremely wide limits of 20 mg./100 ml. to over 100 mg./100 ml., depending principally on the regularity of feeding. It is no wonder, then, with these extreme variations in the internal environment of the premature infant, that mortality is high.

Instability of Body Temperature. One of the particular problems of the premature infant is inability to maintain normal body temperature. Its temperature tends to approach that of its surroundings. At normal room temperature the temperature may stabilize in the low 90s or even in the 80s. Statistical studies show that a body temperature maintained below 96° F. is associated with a particularly high incidence of death, which explains the common use of the incubator in the treatment of prematurity.

DANGER OF OXYGEN THERAPY IN THE PREMATURE INFANT

Because the premature infant frequently develops respiratory distress, oxygen therapy has often been used in treating prematurity. However, it has been discovered that use of high oxygen concentrations in treating premature infants, especially in early prematurity, causes vascular ingrowth into the vitreous humor of the eyes when the infant is withdrawn from the oxygen. This is followed later by fibrosis. This condition, known as *retrolental fibroplasia*, causes permanent blindness. For this reason, it is particularly important to avoid treatment of premature infants with high concentrations of respiratory oxygen. Physiological studies indicate that the premature infant is probably safe up to 40 per cent oxygen, but some child physiologists believe that complete safety can be achieved only by normal oxygen concentration.

GROWTH AND DEVELOPMENT OF THE CHILD

The major physiological problems of the child beyond the neonatal period are related to special metabolic

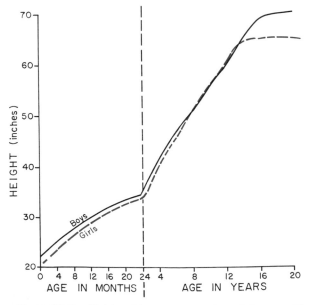

Figure 83–8. Height of boys and girls from infancy to 20 years of age.

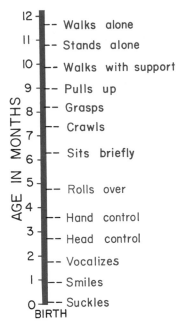

Figure 83–9. Behavioral development of the infant during the first year of life.

needs for growth, which have been fully covered in the sections on metabolism and endocrinology.

Figure 83–8 illustrates the changes in heights of boys and girls from the time of birth until the age of 20 years. Note especially that these parallel each other almost exactly until the end of the first decade of life. Between the ages of 11 and 13 the female estrogens cause rapid growth but early uniting of the epiphyses at about the 14th to 16th year of life, so that growth in height ceases. This contrasts with the effect of testosterone in the male, which causes growth at a slightly later age — mainly between ages 13 and 15. The male, however, undergoes much more prolonged growth so that his final height is considerably greater than that of the female.

BEHAVIORAL GROWTH

Behavioral growth is principally a problem of maturity of the nervous system. Here, it is extremely difficult to dissociate maturity of the anatomical structures of the nervous system from maturity caused by training. Anatomical studies show that certain major tracts in the central nervous system are not completely myelinated until the end of the first year of life. For this reason it is frequently stated that the nervous system is not fully functional at birth. The brain cortex and its associated mechanisms such as vision seem to require several months after birth for significant functional development.

At birth, the brain size is only 26 per cent of the adult size and 55 per cent at one year but reaches almost adult proportions by the end of the second year. This is also associated with closure of the fontanels and sutures of the skull, which allows only 20 per cent additional growth of the brain beyond the first two years of life.

Figure 83–9 illustrates a normal progress chart for the infant during the first year of life. Comparison of this with the baby's actual development is used for clinical assessment of mental and behavioral growth.

REFERENCES

Arey, L. B.: Developmental Anatomy, 7th Ed., Revised. Philadelphia, W. B. Saunders Co., 1974.

Babson, S. G., *et al.*: Diagnosis and Management of the Fetus and Neonate at Risk: A Guide for Team Care. St. Louis, C. V. Mosby, 1979.

Bachofen, M., *et al.*: Lung edema in the adult respiratory distress syndrome. *In* Fishman, A. P., and Renkin, E. M. (eds.): Pulmonary Edema. Baltimore, Waverly Press, 1979, p. 241.

Battaglia, F. C., and Meschia, G.: Principal substrates of fetal metabolism. *Physiol. Rev., 58*:499, 1978.

Biggers, J. D., and Borland, R. M.: Physiological aspects of growth and development of the preimplantation mammalian embryo. *Annu. Rev. Physiol., 38*:95, 1976.

Bradley, R. M., and Mistretta, C. M.: Fetal sensory receptors. *Physiol. Rev., 55*:352, 1975.

Cacciari, E., *et al.*: Obesity in Childhood. New York, Academic Press, 1978.

Cockburn, F., and Drillien, C. M.: Neonatal Medicine. Philadelphia, J. B. Lippincott Co., 1975.

Daughaday, W. H., *et al.*: The regulation of growth by endocrines. *Annu. Rev. Physiol., 37*:211, 1975.

Davis, J. A., and Dobbind, J.: Scientific Foundations of Pediatrics. Philadelphia, W. B. Saunders Co., 1975.

Falkner, F., and Tanner, J. M. (eds.): Human Growth. New York, Plenum Press, 1978.

Fenichel, G. M.: Neonatal Neurology. New York, Churchill Livingstone, 1980.

Fomon, S. J.: Infant Nutrition, 2nd Ed. Philadelphia, W. B. Saunders Co., 1975.

Gardner, L. I.: Endocrine and Genetic Diseases of Childhood and Adolescence, 2nd Ed. Philadelphia, W. B. Saunders Co., 1975.

Gluck, L. (ed.): Intrauterine Asphyxia and the Developing Fetal Brain. Chicago, Year Book Medical Publishers, 1977.

Godman, M. J., and Marquis, R. M.: Paedatric Cardiology: Heart Disease in the Neonate. New York, Churchill Livingstone, 1979.

Grundmann, E., and Kirsten, W. H. (eds.): Perinatal Pathology. New York, Springer-Verlag, 1979.

Hafez, E. S. E. (ed.): The Mammalian Fetus. Springfield, Ill., Charles C Thomas, 1975.

Haller, J. O., and Schneider, M.: Pediatric Ultrasound. Chicago, Year Book Medical Publishers, 1980.

Haymond, M. W., and Pagliara, A. S.: Endocrine and metabolic aspects of fuel homeostasis in the fetus and neonate. *In* DeGroot, L. J., *et al.*

(eds.): Endocrinology. Vol. 3. New York, Grune & Stratton, 1979, p. 1779.

Jellife, D. B., and Jellife, E. F. P. (eds.): Nutrition and Growth. New York, Plenum Press, 1979.

Klaus, M. H., and Fanaroff, A. A.: Care of the High-Risk Neonate. Philadelphia, W. B. Saunders Co., 1979.

Kumar, S., and Rathi, M. (eds.): Perinatal Medicine: Clinical and Biochemical Aspects of the Evaluation, Diagnosis and Management of the Fetus and Newborn. New York, Pergamon Press, 1978.

Lauer, R. M., and Shekelle, R. B. (eds.): Childhood Prevention of Atherosclerosis and Hypertension. New York, Raven Press, 1980.

Lough, M. D., *et al.* (eds.): Newborn Respiratory Care. Chicago, Year Book Medical Publishers, 1979.

Miller, D. R., *et al.* (eds.): Smith's Blood Diseases of Infancy and Childhood. St. Louis, C. V. Mosby, 1978.

Miller, H. C., and Merritt, T. A.: Fetal Growth in Humans. Chicago, Year Book Medical Publishers, 1979.

Murray, R. (ed.): Respiratory Distress Syndrome. Edison, N. J., Medical Research Book Publishers, 1979.

Nathan, D. G., and Oski, F. A. (eds.): Hematology of Infancy and Childhood, 2nd Ed. Philadelphia, W. B. Saunders Co., 1980.

Rudolph, A. M.: Fetal and neonatal pulmonary circulation. *Annu. Rev. Physiol., 41*:383, 1979.

Scarpelli, E. M., *et al.* (eds.): Pulmonary Disease of the Fetus and Child. Philadelphia, Lea & Febiger, 1978.

Schwartz, E. (ed.): Hemoglobinopathies in Children. Littleton, Mass., PSG Publishing Co., 1979.

Sinclair, D.: Human Growth after Birth. New York, Oxford University Press, 1978.

Sinclair, J. C. (ed.): Temperature Regulation and Energy Metabolism in the Newborn. New York, Grune & Stratton, 1978.

Smith, C. A., and Nelson, N. M.: The Physiology of the Newborn Infant. Springfield, Ill., Charles C Thomas, 1974.

Stern, L., *et al.* (eds.): Intensive Care in the Newborn, II. New York, Masson Publishing Co., 1978.

Strang, L. B.: Fetal and newborn lung. *In* MTP International Review of Science: Physiology. Vol. 2. Baltimore, University Park Press, 1974, p. 31.

Strang, L. B.: Neonatal Respiration: Physiological and Clinical Studies. Philadelphia, J. B. Lippincott Co., 1977.

Sutherland, H. W., and Stowers, J. M. (eds.): Carbohydrate Metabolism in Pregnancy and the Newborn 1979. New York, Springer-Verlag, 1979.

Van Leeuwen, G.: Van Leeuwen's Newborn Medicine. Chicago, Year Book Medical Publishers, 1979.

Warshaw, J. B. (ed.): Symposium on Fetal Disease. Philadelphia, W. B. Saunders Co., 1979.

Winick, M. (ed.): Nutrition, Pre- and Postnatal Development. New York, Plenum Press, 1979.

Wolff, P. H., and Ferber, P.: The development of behavior in human infants, premature and newborn. *In* Cowan, W. M., *et al.* (eds.): *Annu. Rev. Neurosci. 2*:291, 1979.

INDEX